The Oxford–Duden German
Desk Dictionary

The
Oxford-Duden
German
Desk
Dictionary

English–German
German–English

Edited by the Dudenredaktion
and the German Section of the
Oxford University Press
Dictionary Department

Chief Editors

M. CLARK
O. THYEN

OXFORD UNIVERSITY PRESS
1997

Oxford University Press, Great Clarendon Street, Oxford OX2 6DP

Oxford New York

Athens Auckland Bangkok Bogota Bombay
Buenos Aires Calcutta Cape Town Dar es Salaam
Delhi Florence Hong Kong Istanbul Karachi
Kuala Lumpur Madras Madrid Melbourne
Mexico City Nairobi Paris Singapore
Taipei Tokyo Toronto

and associated companies in
Berlin Ibadan

Oxford is a trade mark of Oxford University Press

Published in the United States by
Oxford University Press Inc., New York

ISBN 0–19–860147–6

Printed in Great Britain by
Mackays plc
Chatham, Kent

Foreword

The *Oxford–Duden German Desk Dictionary* has been designed to meet the needs of students, tourists, and all those who require quick and reliable answers to their translation questions. It provides clear guidance on selecting the most appropriate translation, numerous illustrative examples to help with problems of construction and usage, and precise information on grammar, style, and pronunciation.

Based on the much acclaimed *Oxford–Duden German Dictionary*, this easy-to-use compact dictionary carries the authority of two of the world's foremost dictionary publishers, Oxford University Press and the Dudenverlag, making use of the unparalleled databases maintained and continually expanded by the two publishers for their celebrated native-speaker dictionaries. Its reliability and clarity make it an invaluable aid to understanding, speaking, and writing everyday idiomatic German in the nineteen nineties.

MICHAEL CLARK
Oxford University Press

Editors and Contributors

in Oxford

Michael Clark
Bernadette Mohan
Maurice Waite
Ursula Lang
Trish Stableford
Tim Connell
Neil Morris
Ting Morris

in Mannheim

Olaf Thyen
Werner Scholze-Stubenrecht
Brigitte Alsleben
Ulrike Röhrenbeck
Magdalena Seubel
Eva Vennebusch

Key to entries

- pronunciation given in IPA
  Aussprache in internationaler Lautschrift

- indication of approximate equivalence
  Angabe ungefährer Entsprechungen

bailiff ['beɪlɪf] *n.* ≈ Gerichtsvollzieher, *der*

- cross-reference to a synonymous headword
  Verweis auf synonymes Stichwort

barrow ['bærəʊ] *n.* **a)** Karre, *die;* Karren, *der;* **b)** *see* **wheelbarrow**

- stress mark
  Betonungszeichen

- irregular plural
  unregelmäßige Pluralform

basis ['beɪsɪs] *n., pl.* **bases** ['beɪsiːz] Basis, *die;* Grundlage, *die*

- regional/national label
  räumliche Zuordnung

- gloss where no translation is possible
  Umschreibung, wenn eine Übersetzung nicht möglich ist

Belisha beacon [bəliːʃə 'biːkn] *n.* *(Brit.) gelbes Blinklicht an Zebrastreifen*

- compound block with a swung dash representing the first element of each compound
  Kompositablock mit Tilde für den ersten Teil jeder Zusammensetzung

black: **~berry** ['blækbəri] *n.* Brombeere, *die;* **~bird** *n.* Amsel, *die;* **~board** *n.* Wandtafel, *die;* **~currant** *n.* schwarze Johannisbeere

- idiomatic phrase
  feste Wendung

- swung dash representing the headword
  die Tilde vertritt das Stichwort

- style labels
  (also given for translations)
  Stilschichtangaben
  (auch für Übersetzungen)

boot [buːt] **1.** *n.* **a)** Stiefel, *der;* **give sb. the ~** *(fig. coll.)* jmdn. rausschmeißen *(ugs.);* **b)** *(Brit.: of car)* Kofferraum, *der.* **2.** *v.t. (coll.: kick)* kicken *(ugs.)*

- **irregular past tenses**
  unregelmäßige Verbformen

- **grammatical categories**
  Gliederung nach grammatischen Gesichtspunkten

- **phrasal verbs listed under main verb**
  Verben in festen Verbindungen mit Präpositionen oder Adverbien (Phrasal verbs) im Anschluß an das jeweilige einfache Verb

- **semantic categories**
  Gliederung nach Bedeutungsunterschieden

- **collocators**
  Kollokatoren

- **sense indicator**
  Indikator

- **information on syntax**
  syntaktische Angabe

- **usage example**
  Anwendungsbeispiel

- **subject labels**
  Bereichsangaben

- **cross-references for additional information**
  Verweise auf zusätzliche Informationen

---

break [breɪk] **1.** *v.t.,* **broke** [brəʊk], **broken** ['brəʊkn] **a)** brechen

break 'down 1. *v.i.* zusammenbrechen; ⟨*Verhandlungen:*⟩ scheitern; ⟨*Auto:*⟩ eine Panne haben. **2.** *v.t.* **a)** aufbrechen ⟨*Tür*⟩; brechen ⟨*Widerstand*⟩; niederreißen ⟨*Barriere, Schranke*⟩; **b)** *(analyse)* aufgliedern. **break 'in 1.** *v.i.* (*into building etc.*) einbrechen. **2.** *v.t.* **a)** zureiten ⟨*Pferd*⟩; **b)** einlaufen ⟨*Schuhe*⟩; **c)** ~ **the door in** die Tür aufbrechen. **break into** *see* ~ **2 b. break 'off 1.** *v.t.* abbrechen; abreißen ⟨*Faden*⟩; auflösen ⟨*Verlobung*⟩. **2.** *v.i.* **a)** abbrechen; **b)** *(cease)* aufhören. **break 'out** *v.i.* ausbrechen; ~ **out in spots/a rash** Pickel/einen Ausschlag bekommen

feature ['fiːtʃə(r)] **1.** *n.* **a)** *usu. in pl. (part of face)* Gesichtszug, *der;* **b)** *(characteristic)* [charakteristisches] Merkmal; **be a ~ of sth.** charakteristisch für etw. sein; **c)** *(Journ. etc.)* Feature, *das;* **d)** *(Cinemat.)* ~ [film] Hauptfilm, *der*

fifty ['fɪftɪ] **1.** *adj.* fünfzig. **2.** *n.* Fünfzig, *die.* See also **eight; eighty 2**

Erläuterungen zum Text

- grammatische Angaben
  grammatical information

- Bereichsangaben
  subject labels

- Verweise auf das Grundwort
  (bei einer Ableitung)
  references to root word
  (from a derivative)

- Betonung und Quantität
  stress and vowel length

- Aussprache in internationaler
  Lautschrift
  pronunciation given in IPA

- Kollokatoren
  collocators

- Indikatoren
  sense indicators

- die Tilde vertritt das Stichwort
  swung dash representing the
  headword

- Verweis auf ein Synonym
  cross-reference to a synonymous
  headword

- Anwendungsbeispiel
  usage example

- Redensart
  idiomatic phrase

- Sprichwort
  proverb

beschneiden *unr. tr. V.* **a)** cut ⟨*hedge*⟩; prune ⟨*bush*⟩; cut back ⟨*tree*⟩; **einem Vogel die Flügel ~:** clip a bird's wings; **b)** *(Med., Rel.)* circumcise. **Beschneidung die; ~, ~en a)** s. beschneiden a: cutting; pruning; cutting back; **b)** *(Med., Rel.)* circumcision

beschummeln *tr. V. (ugs.)* cheat; diddle *(Brit. coll.)*

Balkon [balˈkɔn, balˈkoːn] *der; ~s, ~s* [balˈkɔns] *od.* ~e [balˈkoːnə] **a)** balcony; **b)** *(im Theater, Kino)* circle

beschreiben *unr. tr. V.* **a)** write on; *(vollschreiben)* write ⟨*page, side, etc.*⟩; *(darstellen)* describe

bestanden *Adj.* **von** *od.* **mit etw. ~ sein** have sth. growing on it; **mit Tannen ~e Hügel** fir-covered hills

berappen *tr., itr. V. (ugs.)* s. **blechen**

Beruf *der; ~[e]s, ~e* occupation; *(akademischer)* profession; *(handwerklicher)* trade; **was sind Sie von ~?** what do you do for a living?

Besen *der; ~s, ~* broom; **ich fress' einen ~, wenn das stimmt** *(salopp)* I'll eat my hat if that's right *(coll.)*; **neue ~ kehren gut** *(Spr.)* a new broom sweeps clean *(prov.)*

- Kompositablock
  compound block

- Angaben zur Syntax
  information on syntax

- Gliederung nach grammatischen Gesichtspunkten
  grammatical categories

- Gliederung nach Bedeutungsunterschieden
  semantic categories

- Kompositionsfuge
  dot between elements of a compound

- räumliche Zuordnung des Stichworts/der Übersetzung
  regional labels for the headwords/translations

- zusätzliche Glosse zur Bedeutung des Stichworts
  additional gloss specifying the sense of the headword

- Angaben zur Stilschicht
  style labels

blut-, Blut-: **~unterlaufen** *Adj.* suffused with blood *postpos.*; bloodshot ⟨*eyes*⟩; **~vergießen** das; ~s bloodshed; **~vergiftung** die blood-poisoning *no indef. art., no pl.*; **~wurst** die black pudding

beschweren 1. *refl. V.* complain ⟨über + Akk.: wegen about⟩; **2.** *tr. V.* weight down

beschränkt 1. *Adj.* **a)** *(dumm)* dull-witted; **b)** *(engstirnig)* narrow-minded; **2.** *adv.* narrow-mindedly

Bei·name der epithet

Beis[e]l das; ~s, *od.* ~n *(österr.)* pub *(Brit. coll.)*; bar *(Amer.)*

Benzin das; ~s petrol *(Brit.)*; gasoline *(Amer.)*; gas *(Amer. coll.)*; *(Wasch~)* benzine

Bütten·papier das handmade paper *(with deckle-edge)*

bestatten *tr. V.* *(geh.)* inter *(formal)*; bury; **Bestattung die;** ~, ~en *(geh.)*

Contents

Phonetic symbols used in transcriptions /
Die bei den Ausspracheangaben verwendeten
Zeichen der Lautschrift   7

English abbreviations used in the Dictionary /
Im Wörterverzeichnis verwendete englische
Abkürzungen   9

German abbreviations used in the Dictionary /
Im Wörterverzeichnis verwendete deutsche
Abkürzungen   10

English–German Dictionary /
Englisch–deutsches Wörterverzeichnis   13

German–English Dictionary /
Deutsch–englisches Wörterverzeichnis   403

Englische unregelmäßige Verben   793

German irregular verbs   795

Weights and Measures /
Maße und Gewichte   799

Revisions to German spelling /
die neue Regelung der Rechtschreibung   801

Proprietary Names

Die für das Englische verwendeten Zeichen der Lautschrift

ɑ:	barb	bɑ:b		m	mat	mæt
ɑ̃	séance	'seɑ̃s		n	not	nɒt
æ	fat	fæt		ŋ	sing	sɪŋ
æ̃	lingerie	'læʒɪrɪ		ɒ	got	gɒt
aɪ	fine	faɪn		ɔ:	paw	pɔ:
aʊ	now	naʊ		ɔɪ	boil	bɔɪl
b	bat	bæt		p	pet	pet
d	dog	dɒg		r	rat	ræt
dʒ	jam	dʒæm		s	sip	sɪp
e	met	met		ʃ	ship	ʃɪp
eɪ	fate	feɪt		t	tip	tɪp
eə	fairy	'feərɪ		tʃ	chin	tʃɪn
əʊ	goat	gəʊt		θ	thin	θɪn
ə	ago	ə'gəʊ		ð	the	ðə
ɜ:	fur	fɜ:(r)		u:	boot	bu:t
f	fat	fæt		ʊ	book	bʊk
g	good	gʊd		ʊə	tourist	'tʊərɪst
h	hat	hæt		ʌ	dug	dʌg
ɪ	bit, lately	bɪt, 'leɪtlɪ		v	van	væn
ɪə	nearly	'nɪəlɪ		w	win	wɪn
i:	meet	mi:t		x	loch	lɒx
j	yet	jet		z	zip	zɪp
k	kit	kɪt		ʒ	vision	'vɪʒn
l	lot	lɒt				

:   Längezeichen, bezeichnet Länge des unmittelbar davor stehenden Vokals, z. B. boot [bu:t].

'   Betonung, steht unmittelbar vor einer betonten Silbe, z. B. ago [ə'gəʊ].

(r)   Ein „r" in runden Klammern wird nur gesprochen, wenn im Textzusammenhang ein Vokal unmittelbar folgt, z. B. pare [peə(r)]; pare away [peər ə'weɪ].

Phonetic information given in the German-English section

The pronunciation of German is largely regular, and phonetic transcriptions have only been given where additional help is needed. In all other cases only the position of the stressed syllable and the length of the vowel in that syllable are shown: a long vowel is indicated by an underline, e.g. Maß, a short vowel by a dot placed underneath, e.g. Masse.

8

Phonetic symbols used in transcriptions of German words

a	hat	hat	ŋ	lang	laŋ
aː	Bahn	baːn	o	Moral	moˈraːl
ɐ	Ober	ˈoːbɐ	oː	Boot	boːt
ɐ̯	Uhr	uːɐ̯	o̞	loyal	lo̞aˈjaːl
ã	Ensemble	ãˈsãːbl̩	õ	Fondue	fõˈdyː
ãː	Abonnement	abɔnəˈmãː	õː	Fond	fõː
ai̯	weit	vai̯t	ɔ	Post	pɔst
au̯	Haut	hau̯t	ø	Ökonom	økoˈnoːm
b	Ball	bal	øː	Öl	øːl
ç	ich	ɪç	œ	göttlich	ˈgœtlɪç
d	dann	dan	œː	Parfum	parˈfœː
dʒ	Gin	dʒɪn	ɔy̯	Heu	hɔy̯
e	egal	eˈgaːl	p	Pakt	pakt
eː	Beet	beːt	pf	Pfahl	pfaːl
ɛ	mästen	ˈmɛstn̩	r	Rast	rast
ɛː	wählen	ˈvɛːlən	s	Hast	hast
ɛ̃	Mannequin	ˈmanəkɛ̃	ʃ	schal	ʃaːl
ɛ̃ː	Cousin	kuˈzɛ̃ː	t	Tal	taːl
ə	Nase	ˈnaːzə	ts	Zahl	tsaːl
f	Faß	fas	tʃ	Matsch	matʃ
g	Gast	gast	u	kulant	kuˈlant
h	hat	hat	uː	Hut	huːt
i	vital	viˈtaːl	u̯	aktuell	akˈtu̯ɛl
iː	viel	fiːl	ʊ	Pult	pʊlt
i̯	Studie	ˈʃtuːdi̯ə	v	was	vas
ɪ	Birke	ˈbɪrkə	x	Bach	bax
j	ja	jaː	y	Physik	fyˈziːk
k	kalt	kalt	yː	Rübe	ˈryːbə
l	Last	last	ỹ	Nuance	ˈnỹãːsə
l̩	Nabel	ˈnaːbl̩	ʏ	Fülle	ˈfʏlə
m	Mast	mast	z	Hase	ˈhaːzə
n	Naht	naːt	ʒ	Genie	ʒeˈniː
n̩	baden	ˈbaːdn̩			

ˈ Glottal stop, e. g. beachten [bəˈʔaxtn̩].

ː Length sign, indicating that the preceding vowel is long, e. g. Chrom [kroːm].

˜ Indicates a nasal vowel, e. g. Fond [fõː].

ˈ Stress mark, immediately preceding a stressed syllable, e. g. Ballon [baˈlɔŋ].

English abbreviations used in the Dictionary/ Im Wörterverzeichnis verwendete englische Abkürzungen

abbr(s).	abbreviation(s)	Footb.	Football
abs.	absolute	Gastr.	Gastronomy
adj(s).	adjective(s)	Geog.	Geography
Admin.	Administration, Administrative	Geol.	Geology
adv.	adverb	Geom.	Geometry
Aeronaut.	Aeronautics	Her.	Heraldry
Agric.	Agriculture	Hist.	History, Historical
Amer.	American, America	Hort.	Horticulture
Anat.	Anatomy	imper.	imperative
arch.	archaic	impers.	impersonal
Archaeol.	Archaeology	incl.	including
Archit.	Architecture	indef.	indefinite
art.	article	Information Sci.	Information Science
Astrol.	Astrology	int.	interjection
Astron.	Astronomy	interrog.	interrogative
Astronaut.	Astronautics	Ir.	Irish, Ireland
attrib.	attributive	iron.	ironical
Austral.	Australian, Australia	joc.	jocular
Biol.	Biology	Journ.	Journalism
Bookk.	Bookkeeping	lang.	language
Bot.	Botany	Ling.	Linguistics
Brit.	British, Britain	Lit.	Literature
Chem.	Chemistry	lit.	literal
Cinemat.	Cinematography	masc.	masculine
coll.	colloquial	Math.	Mathematics
collect.	collective	Mech.	Mechanics
comb.	combination	Mech. Engin.	Mechanical Engineering
Commerc.	Commerce, Commercial	Med.	Medicine
compar.	comparative	Metalw.	Metalwork
condit.	conditional	Meteorol.	Meteorology
conj.	conjunction	Mil.	Military
def.	definite	Min.	Mineralogy
Dent.	Dentistry	Motor Veh.	Motor Vehicles
derog.	derogatory	Mus.	Music
dial.	dialect	Mythol.	Mythology
Diplom.	Diplomacy	n.	noun
Dressm.	Dressmaking	Naut.	Nautical
Eccl.	Ecclesiastical	neg.	negative
Ecol.	Ecology	N. Engl.	Northern English
Econ.	Economics	ns.	nouns
Educ.	Education	Nucl. Phys.	Nuclear Physics
Electr.	Electricity	obj.	object
ellipt.	elliptical	Ornith.	Ornithology
emphat.	emphatic	P	Proprietary name
esp.	especially	Parl.	Parliament
euphem.	euphemistic	pass.	passive
excl.	exclamation, exclamatory	Pharm.	Pharmacy
expr.	expressing	Philos.	Philosophy
fem.	feminine	Photog.	Photography
fig.	figurative	phr(s).	phrase(s)
		Phys.	Physics

Physiol.	Physiology	Scot.	Scottish, Scotland
pl.	plural	sing.	singular
poet.	poetical	sl.	slang
Polit.	Politics	Sociol.	Sociology
poss.	possessive	St. Exch.	Stock Exchange
postpos.	postpositive	sth.	something
p.p.	past participle	subord.	subordinate
pred.	predicative	suf.	suffix
pref.	prefix	superl.	superlative
prep.	preposition	Surv.	Surveying
pres.	present	symb.	symbol
pres. p.	present participle	tech.	technical
pr. n.	proper noun	Teleph.	Telephony
pron.	pronoun	Telev.	Television
prov.	proverbial	Theol.	Theology
Psych.	Psychology	Univ.	University
p.t.	past tense	usu.	usually
Railw.	Railways	v. aux.	auxiliary verb
RC Ch.	Roman Catholic Church	Vet. Med.	Veterinary Medicine
		v.i.	intransitive verb
refl.	reflexive	v. refl.	reflexive verb
rel.	relative	v.t.	transitive verb
Relig.	Religion	v.t. & i.	transitive and intransitive verb
rhet.	rhetorical		
sb.	somebody	Woodw.	Woodwork
Sch.	School	Zool.	Zoology
Sci.	Science		

German abbreviations used in the Dictionary/ Im Wörterverzeichnis verwendete deutsche Abkürzungen

a.	anderes; andere	berlin.	berlinisch
ä.	ähnliches; ähnliche	bes.	besonders
Abk.	Abkürzung	Bez.	Bezeichnung
adj.	adjektivisch	bibl.	biblisch
Adj.	Adjektiv	bild. Kunst	bildende Kunst
adv.	adverbial	Biol.	Biologie
Adv.	Adverb	Börsenw.	Börsenwesen
Akk.	Akkusativ	Bot.	Botanik
amerik.	amerikanisch	BRD	Bundesrepublik Deutschland
Amtsspr.	Amtssprache		
Anat.	Anatomie	brit.	britisch
Anthrop.	Anthropologie	Bruchz.	Bruchzahl
Archäol.	Archäologie	Buchf.	Buchführung
Archit.	Architektur	Buchw.	Buchwesen
Art.	Artikel	Bürow.	Bürowesen
Astrol.	Astrologie	chem.	chemisch
Astron.	Astronomie	christl.	christlich
A.T.	Altes Testament	Dat.	Dativ
attr.	attributiv	DDR	Deutsche Demokratische Republik
Bauw.	Bauwesen		
Bergmannsspr.	Bergmannssprache		

Dekl.	Deklination	jur.	juristisch
Demonstrativ-pron.	Demonstrativ-pronomen	Kardinalz.	Kardinalzahl
		kath.	katholisch
d. h.	das heißt	Kaufmannsspr.	Kaufmannssprache
dichter.	dichterisch	Kfz-W.	Kraftfahrzeugwesen
Druckerspr.	Druckersprache	Kinderspr.	Kindersprache
Druckw.	Druckwesen	Kochk.	Kochkunst
dt.	deutsch	Konj.	Konjunktion
DV	Datenverarbeitung	Kunstwiss.	Kunstwissenschaft
ehem.	ehemals, ehemalig	landsch.	landschaftlich
Eisenb.	Eisenbahn	Landw.	Landwirtschaft
elektr.	elektrisch	Literaturw.	Literaturwissenschaft
Elektrot.	Elektrotechnik	Luftf.	Luftfahrt
engl.	englisch	ma.	mittelalterlich
etw.	etwas	MA.	Mittelalter
ev.	evangelisch	marx.	marxistisch
fachspr.	fachsprachlich	Math.	Mathematik
fam.	familiär	Med.	Medizin
Ferns.	Fernsehen	Meeresk.	Meereskunde
Fernspr.	Fernsprechwesen	Met.	Meteorologie
fig.	figurativ	Metall.	Metallurgie
Finanzw.	Finanzwesen	Metallbearb.	Metallbearbeitung
Flugw.	Flugwesen	Milit.	Militär
Forstw.	Forstwesen	Mineral.	Mineralogie
Fot.	Fotografie	mod.	modifizierend
Frachtw.	Frachtwesen	Modalv.	Modalverb
Funkw.	Funkwesen	Münzk.	Münzkunde
Gastr.	Gastronomie	Mus.	Musik
Gattungsz.	Gattungszahl	Mythol.	Mythologie
Gaunerspr.	Gaunersprache	Naturw.	Naturwissenschaft
geh.	gehoben	Neutr.	Neutrum
Gen.	Genitiv	niederdt.	niederdeutsch
Geneal.	Genealogie	Nom.	Nominativ
Geogr.	Geographie	nordamerik.	nordamerikanisch
Geol.	Geologie	nordd.	norddeutsch
Geom.	Geometrie	nordostd.	nordostdeutsch
Handarb.	Handarbeit	nordwestd.	nordwestdeutsch
Handw.	Handwerk	ns.	nationalsozialistisch
Her.	Heraldik	N. T.	Neues Testament
hess.	hessisch	o.	ohne; oben
Hilfsv.	Hilfsverb	o. ä.	oder ähnliches;
hist.	historisch		oder ähnliche
Hochschulw.	Hochschulwesen	od.	oder
Holzverarb.	Holzverarbeitung	Ordinalz.	Ordinalzahl
Indefinitpron.	Indefinitpronomen	ostd.	ostdeutsch
indekl.	indeklinabel	österr.	österreichisch
Indik.	Indikativ	Päd.	Pädagogik
Inf.	Infinitiv	Papierdt.	Papierdeutsch
Informationst.	Informationstechnik	Parapsych.	Parapsychologie
Interj.	Interjektion	Parl.	Parlament
iron.	ironisch	Part.	Partizip
intr.	intransitiv	Perf.	Perfekt
Jagdw.	Jagdwesen	Pers.	Person
Jägerspr.	Jägersprache	pfälz.	pfälzisch
jmd.	jemand	Pharm.	Pharmazie
jmdm.	jemandem	Philos.	Philosophie
jmdn.	jemanden	Physiol.	Physiologie
jmds.	jemandes	Pl.	Plural
Jugendspr.	Jugendsprache	Plusq.	Plusquamperfekt

Postw.	Postwesen	Sup.	Superlativ
präd.	prädikativ	Textilw.	Textilwesen
Präp.	Präposition	Theol.	Theologie
Präs.	Präsens	thüring.	thüringisch
Prät.	Präteritum	Tiermed.	Tiermedizin
Pron.	Pronomen	tr.	transitiv
Psych.	Psychologie	Trenn.	Trennung
Raumf.	Raumfahrt	u.	und
Rechtsspr.	Rechtssprache	u. a.	und andere[s]
Rechtsw.	Rechtswesen	u. ä.	und ähnliches
refl.	reflexiv	ugs.	umgangssprachlich
regelm.	regelmäßig	unbest.	unbestimmt
Rel.	Religion	unpers.	unpersönlich
Relativpron.	Relativpronomen	unr.	unregelmäßig
rhein.	rheinisch	usw.	und so weiter
Rhet.	Rhetorik	v.	von
röm.	römisch	V.	Verb
röm.-kath.	römisch-katholisch	verächtl.	verächtlich
Rundf.	Rundfunk	veralt.	veraltet; veraltend
s.	siehe	Verhaltensf.	Verhaltensforschung
S.	Seite	verhüll.	verhüllend
scherzh.	scherzhaft	Verkehrsw.	Verkehrswesen
schles.	schlesisch	Versiche-	Versicherungswesen
schott.	schottisch	rungsw.	
Schülerspr.	Schülersprache	vgl.	vergleiche
Schulw.	Schulwesen	Vkl.	Verkleinerungsform
schwäb.	schwäbisch	Völkerk.	Völkerkunde
schweiz.	schweizerisch	Völkerr.	Völkerrecht
Seemannsspr.	Seemannssprache	Volksk.	Volkskunde
Seew.	Seewesen	volkst.	volkstümlich
Sexualk.	Sexualkunde	vulg.	vulgär
Sg.	Singular	Werbespr.	Werbesprache
s. o.	siehe oben	westd.	westdeutsch
Soldatenspr.	Soldatensprache	westfäl.	westfälisch
Sozialvers.	Sozialversicherung	Wieder-	Wiederholungs-
Soziol.	Soziologie	holungsz.	zahlwort
spött.	spöttisch	wiener.	wienerisch
Spr.	Sprichwort	Winzerspr.	Winzersprache
Sprachw.	Sprachwissenschaft	Wirtsch.	Wirtschaft
Steuerw.	Steuerwesen	Wissensch.	Wissenschaft
Stilk.	Stilkunde	Wz.	Warenzeichen
Studentenspr.	Studentensprache	Zahnmed.	Zahnmedizin
s. u.	siehe unten	z. B.	zum Beispiel
Subj.	Subjekt	Zeitungsw.	Zeitungswesen
subst.	substantivisch;	Zollw.	Zollwesen
	substantiviert	Zool.	Zoologie
Subst.	Substantiv	Zus.	Zusammensetzung
südd.	süddeutsch	Zusschr.	Zusammenschreibung
südwestd.	südwestdeutsch		

A

A, ¹a [eɪ] *n.* A, a, *das*
²a [ə, *stressed* eɪ] *indef. art.* ein/eine/
ein; **he is a gardener/a Frenchman** er
ist Gärtner/Franzose; **she did not say a
word** sie sagte kein Wort
AA *abbr. (Brit.)* **Automobile Association** *britischer Automobilklub*
aback [ə'bæk] *adv.* **be taken ~:** erstaunt sein
abandon [ə'bændən] *v. t.* verlassen
⟨*Ort, Person*⟩; aufgeben ⟨*Prinzip*⟩
abase [ə'beɪs] *v. t.* erniedrigen
abashed [ə'bæʃt] *adj.* beschämt
abate [ə'beɪt] *v. i.* nachlassen
abattoir ['æbətwɑː(r)] *n.* Schlachthof,
der
abbey ['æbɪ] *n.* Abtei, *die*
abbot ['æbət] *n.* Abt, *der*
abbreviate [ə'briːvɪeɪt] *v. t.* abkürzen.
abbreviation [əbriːvɪ'eɪʃn] *n.* Abkürzung, *die*
abdicate ['æbdɪkeɪt] *v. t.* abdanken.
abdication [æbdɪ'keɪʃn] *n.* Abdan-
kung, *die*
abdomen ['æbdəmɪn] *n.* Bauch, *der*
abdominal [æb'dɒmɪnl] *adj.* Bauch-
abduct [əb'dʌkt] *v. t.* entführen. **ab-
duction** [əb'dʌkʃn] *n.* Entführung,
die
aberration [æbə'reɪʃn] *n.* Abweichung, *die*
abet [ə'bet] *v. t.*, **-tt-** helfen (+ *Dat.*);
aid and ~: Beihilfe leisten (+ *Dat.*)
abhor [əb'hɔː(r)] *v. t.*, **-rr-** verabscheu-
en. **abhorrent** [əb'hɒrənt] *adj.* abscheulich
abide [ə'baɪd] **1.** *v. i.* **~ by** befolgen
⟨*Gesetz, Vorschrift*⟩; [ein]halten ⟨*Versprechen*⟩. **2.** *v. t.* ertragen; **I can't ~
dogs** ich kann Hunde nicht ausstehen
ability [ə'bɪlɪtɪ] *n.* **a)** *(capacity)* Fähigkeit, *die;* **have the ~ to do sth.** etw.
können; **b)** *(cleverness)* Intelligenz,
die; **c)** *(talent)* Begabung, *die*
abject ['æbdʒekt] *adj.* elend; bitter
⟨*Armut*⟩; demütig ⟨*Entschuldigung*⟩

ablaze [ə'bleɪz] *adj.* **be ~:** in Flammen
stehen
able ['eɪbl] *adj.* **a) be ~ to do sth.** etw.
können; **b)** *(competent)* fähig. **able-
bodied** ['eɪblbɒdɪd] *adj.* kräftig;
tauglich ⟨*Soldat, Matrose*⟩. **ably** ['eɪ-
blɪ] *adv.* geschickt; gekonnt
abnormal [æb'nɔːml] *adj.* abnorm;
a[b]normal ⟨*Interesse, Verhalten*⟩. **ab-
normality** [æbnɔː'mælɪtɪ] *n.* Abnor-
mität, *die*
aboard [ə'bɔːd] **1.** *adv.* an Bord. **2.**
prep. an Bord (+ *Gen.*); **~ the bus** im
Bus; **~ ship** an Bord
abode [ə'bəʊd] *n.* **of no fixed ~:** ohne
festen Wohnsitz
abolish [ə'bɒlɪʃ] *v. t.* abschaffen.
abolition [æbə'lɪʃn] *n.* Abschaffung,
die
abominable [ə'bɒmɪnəbl] *adj.* abscheulich; scheußlich
aborigine [æbə'rɪdʒɪnɪ] *n.* Ureinwoh-
ner, *der*
abort [ə'bɔːt] *v. t.* abtreiben ⟨*Baby*⟩.
abortion [ə'bɔːʃn] *n.* Abtreibung,
die. **abortive** [ə'bɔːtɪv] *adj.* mißlungen ⟨*Plan*⟩; fehlgeschlagen ⟨*Versuch*⟩
abound [ə'baʊnd] *v. i.* **~ in sth.** an etw.
(Dat.) reich sein
about [ə'baʊt] **1.** *adv.* **a)** *(all around)*
rings[her]um; *(here and there)* überall;
all ~: ringsumher; **b)** *(near)* **be ~:** dasein; hiersein; **c) be ~ to do sth.** gerade
etw. tun wollen; **d) be out and ~:** aktiv
sein; **e)** *(approximately)* ungefähr. **2.**
prep. **a)** *(all round)* um [... herum]; **b)**
(concerning) über (+ *Akk.*); **know ~
sth.** von etw. wissen; **a question ~ sth.**
eine Frage zu etw.; **what was it ~?**
worum ging es?
above [ə'bʌv] **1.** *adv.* **a)** *(position)* oben;
(higher up) darüber; **b)** *(direction)*
nach oben. **2.** *prep.* *(position)* über
(+ *Dat.*); *(direction, more than)* über
(+ *Akk.*); **~ all** vor allem. **a'bove-
mentioned** *adj.* oben genannt

abrasion [ə'breɪʒn] *n. (graze)* Hautabschürfung, *die*

abrasive [ə'breɪsɪv] **1.** *adj.* **a)** scheuernd; Scheuer-; **b)** *(fig.: harsh)* aggressiv. **2.** *n.* Scheuermittel, *das*

abreast [ə'brest] *adv.* **a)** nebeneinander; **b)** *(fig.)* keep ~ of sth. sich über etw. *(Akk.)* auf dem laufenden halten

abroad [ə'brɔ:d] *adv.* im Ausland; *(direction)* ins Ausland

abrupt [ə'brʌpt] *adj.,* **a'bruptly** *adv.* **a)** *(sudden[ly])* abrupt; plötzlich; **b)** *(brusque[ly])* schroff

abscess ['æbsɪs] *n.* Abszeß, *der*

abscond [əb'skɒnd] *v. i.* sich entfernen

absence ['æbsəns] *n.* Abwesenheit, *die;* the ~ of sth. der Mangel an etw. *(Dat.)*

absent ['æbsənt] *adj.* abwesend; be ~ from school/work in der Schule/am Arbeitsplatz fehlen. **absentee** [æbsən'ti:] *n.* Fehlende, *der/die;* Abwesende, *der/die.* **absent-minded** [æbsənt'maɪndɪd] *adj.* geistesabwesend; *(habitually)* zerstreut

absolute ['æbsəlu:t] *adj.* absolut; ausgemacht ⟨*Lüge, Skandal*⟩. **abso'lutely** *adv.* absolut; völlig ⟨*verrückt*⟩; you're ~ right! du hast völlig recht; ~ not! auf keinen Fall!

absolve [əb'zɒlv] *v. t.* ~ from entbinden von ⟨*Pflichten*⟩; lossprechen von ⟨*Schuld*⟩

absorb [əb'sɔ:b] *v. t.* **a)** aufsaugen ⟨*Flüssigkeit*⟩; **b)** abfangen ⟨*Schlag, Stoß*⟩; **c)** *(fig.: engross)* ausfüllen. **absorbent** [əb'sɔ:bənt] *adj.* saugfähig. **ab'sorbing** *adj.* faszinierend

abstain [əb'steɪn] *v. i.* ~ from sth. sich einer Sache *(Gen.)* enthalten; ~ |from voting| sich der Stimme enthalten

abstemious [əb'sti:mɪəs] *adj.* enthaltsam

abstention [əb'stenʃn] *n. (from voting)* Stimmenthaltung, *die*

abstinence ['æbstɪnəns] *n.* Abstinenz, *die*

abstract ['æbstrækt] **1.** *adj.* abstrakt. **2.** *n.* Zusammenfassung, *die*

absurd [əb'sɜ:d] *adj.* absurd; *(ridiculous)* lächerlich. **absurdity** [əb'sɜ:dɪtɪ] *n.* Absurdität, *die.* **ab'surdly** *adv.* lächerlich

abundance [ə'bʌndəns] *n.* |an| ~ of sth. eine Fülle von etw.

abundant [ə'bʌndənt] *adj.* reich (in an + *Dat.*)

abuse 1. [ə'bju:z] *v. t.* beschimpfen. **2.**

[ə'bju:s] *n.* Beschimpfungen *Pl.* **abusive** [ə'bju:sɪv] *adj.* beleidigend; become ~: ausfallend werden

abysmal [ə'bɪzml] *adj. (coll.: bad)* katastrophal *(ugs.)*

abyss [ə'bɪs] *n.* Abgrund, *der*

AC *abbr.* **alternating current** Ws

academic [ækə'demɪk] *adj.* akademisch

academy [ə'kædəmɪ] *n.* Akademie, *die*

accede [æk'si:d] *v. i.* **a)** zustimmen (to *Dat.*); **b)** ~ |to the throne| den Thron besteigen

accelerate [ək'seləreɪt] **1.** *v. t.* beschleunigen. **2.** *v. i.* sich beschleunigen; ⟨*Auto, Fahrer:*⟩ beschleunigen. **acceleration** [əkselə'reɪʃn] *n.* Beschleunigung, *die.* **accelerator** [ək'seləreɪtə(r)] *n.* ~ |pedal| Gas[pedal], *das*

accent ['æksənt] *n.* Akzent, *der.* **accentuate** [ək'sentjʊeɪt] *v. t.* betonen

accept [ək'sept] *v. t.* **a)** annehmen; entgegennehmen ⟨*Dank, Spende*⟩; übernehmen ⟨*Verantwortung*⟩; **b)** *(acknowledge)* akzeptieren. **acceptable** [ək'septəbl] *adj.* akzeptabel; annehmbar ⟨*Preis, Gehalt*⟩. **acceptance** [ək'septəns] *n.* **a)** Annahme, *die;* **b)** *(acknowledgement)* Anerkennung, *die*

access ['ækses] *n.* **a)** *(admission)* gain ~: Einlaß finden; **b)** *(opportunity to use or approach)* Zugang, der (to zu). **accessible** [ək'sesɪbl] *adj.* **a)** *(reachable)* erreichbar; **b)** *(available, understandable)* zugänglich (to für)

accession [ək'seʃn] *n.* Amtsantritt, *der;* ~ |to the throne| Thronbesteigung, *die*

accessory [ək'sesərɪ] *n.* **a)** accessories *pl.* Zubehör, *das;* **b)** *(dress article)* Accessoire, *das*

accident ['æksɪdənt] *n.* **a)** Unfall, *der;* **b)** *(chance)* Zufall, *der;* by ~: zufällig; **c)** *(mistake)* Versehen, *das;* by ~: versehentlich. **accidental** [æksɪ'dentl] *adj. (chance)* zufällig; *(unintended)* unbeabsichtigt. **acci'dentally** *adv. (by chance)* zufällig; *(by mistake)* versehentlich

acclaim [ə'kleɪm] *v. t.* feiern

acclimatize [ə'klaɪmətaɪz] *v. t.* get or become ~d sich akklimatisieren

accolade ['ækəleɪd] *n. (praise)* ~|s| Lob, *das*

accommodate [ə'kɒmədeɪt] *v. t.* **a)** unterbringen; *(hold)* Platz bieten (+ *Dat.*); **b)** *(oblige)* gefällig sein

(+ *Dat.*). **accommodating** [ə'kɒmədeɪtɪŋ] *adj.* zuvorkommend. **accommodation** [əkɒmə'deɪʃn] *n.* Unterkunft, *die*

accompaniment [ə'kʌmpənɪmənt] *n.* Begleitung, *die*

accompanist [ə'kʌmpənɪst] *n.* Begleiter, *der*/Begleiterin, *die*

accompany [ə'kʌmpənɪ] *v. t.* begleiten

accomplice [ə'kʌmplɪs] *n.* Komplize, *der*/Komplizin, *die*

accomplish [ə'kʌmplɪʃ] *v. t.* vollbringen ⟨*Tat*⟩; erfüllen ⟨*Aufgabe*⟩. **accomplished** [ə'kʌmplɪʃt] *adj.* fähig; he is an ~ speaker/dancer er ist ein erfahrener Redner/vollendeter Tänzer. **ac'complishment** *n.* a) *(completion)* Vollendung, *die;* b) *(achievement)* Leistung, *die; (skill)* Fähigkeit, *die*

accord [ə'kɔ:d] 1. *n.* Übereinstimmung, *die;* of one's own ~: aus eigenem Antrieb; with one ~: geschlossen. 2. *v. t.* ~ sb. sth. jmdm. etw. gewähren. **accordance** [ə'kɔ:dəns] *n.* in ~ with in Übereinstimmung mit. **ac'cording** *adv.* ~ to nach; ~ to him nach seiner Aussage. **ac'cordingly** *adv. (as appropriate)* entsprechend; *(therefore)* folglich

accordion [ə'kɔ:dɪən] *n.* Akkordeon, *das*

accost [ə'kɒst] *v. t.* ansprechen

account [ə'kaʊnt] *n.* a) *(Finance)* Rechnung, *die; (at bank, shop)* Konto, *das;* b) *(consideration)* take ~ of sth., take sth. into ~: etw. berücksichtigen; take no ~ of sth./sb. etw./jmdn. unberücksichtigt lassen; don't change your plans on my ~: ändert nicht meinetwegen eure Pläne; on ~ of wegen; on no ~: auf [gar] keinen Fall; c) *(report)* Bericht, *der;* d) call sb. to ~: jmdn. zur Rechenschaft ziehen. **ac'count for** *v.t.* Rechenschaft ablegen über (+*Akk.*); *(explain)* erklären

accountable [ə'kaʊntəbl] *adj.* verantwortlich

accountancy [ə'kaʊntənsɪ] *n.* Buchhaltung, *die*

accountant [ə'kaʊntənt] *n.* [Bilanz]buchhalter, *der*/-halterin, *die*

ac'count number *n.* Kontonummer, *die*

accrue [ə'kru:] *v. i.* ⟨*Zinsen:*⟩ auflaufen; ~ to sb. ⟨*Reichtümer, Einnahmen:*⟩ jmdm. zufließen

accumulate [ə'kju:mjʊleɪt] 1. *v. t.*

sammeln. 2. *v. i.* ⟨*Menge, Staub:*⟩ sich ansammeln; ⟨*Geld:*⟩ sich anhäufen. **accumulation** [əkju:mjʊ'leɪʃn] *n.* [An]sammeln, *das; (being accumulated)* Anhäufung, *die*

accuracy ['ækjʊrəsɪ] *n.* Genauigkeit, *die*

accurate ['ækjʊrət] *adj.,* '**accurately** *adv.* genau; *(correct[ly])* richtig

accusation [ækju:'zeɪʃn] *n.* Anschuldigung, *die; (Law)* Anklage, *die*

accusative [ə'kju:zətɪv] *adj. & n.* ~ [case] Akkusativ, *der*

accuse [ə'kju:z] *v. t.* beschuldigen; *(Law)* anklagen (of wegen + *Gen.*)

accustom [ə'kʌstəm] *v. t.* gewöhnen (to an + *Akk.*); grow/be ~ed to sth. sich an etw. *(Akk.)* gewöhnen/an etw. *(Akk.)* gewöhnt sein. **accustomed** [ə'kʌstəmd] *attrib. adj.* gewohnt; üblich

ace [eɪs] *n.* As, *das*

ache [eɪk] 1. *v. i.* schmerzen; weh tun. 2. *n.* Schmerz, *der*

achieve [ə'tʃi:v] *v. t.* zustande bringen; erreichen ⟨*Ziel, Standard*⟩. **a'chievement** *n.* a) *see* achieve: Zustandebringen, *das;* Erreichen, *das;* b) *(thing accomplished)* Leistung, *die*

acid ['æsɪd] 1. *adj.* sauer. 2. *n.* Säure, *die*. **acidic** [ə'sɪdɪk] *adj.* säuerlich. **acidity** [ə'sɪdɪtɪ] *n.* Säure, *die*

acid: ~ 'rain *n.* saurer Regen; ~ test *n. (fig.)* Feuerprobe, *die*

acknowledge [ək'nɒlɪdʒ] *v. t.* a) zugeben ⟨*Tatsache, Fehler, Schuld*⟩; b) sich erkenntlich zeigen für ⟨*Dienste, Bemühungen*⟩; erwidern ⟨*Gruß*⟩; c) bestätigen ⟨*Empfang, Bewerbung*⟩; ~ a letter den Empfang eines Briefes bestätigen. **acknowledg[e]ment** [ək'nɒlɪdʒmənt] *n.* a) *(admission)* Eingeständnis, *das;* b) *(thanks)* Dank, *der* (of für); c) *(of letter)* Bestätigung [des Empfangs]

acne ['æknɪ] *n.* Akne, *die*

acorn ['eɪkɔ:n] *n.* Eichel, *die*

acoustic [ə'ku:stɪk] *adj.* akustisch. **a'coustics** *n. pl.* Akustik, *die*

acquaint [ə'kweɪnt] *v. t.* be ~ed with sb. mit jmdm. bekannt sein. **acquaintance** [ə'kweɪntəns] *n.* a) ~ with sb. Bekanntschaft mit jmdm.; make sb.'s ~: jmds. Bekanntschaft machen; b) *(person)* Bekannte, *der/die*

acquiesce [ækwɪ'es] *v. i.* einwilligen (in in + *Akk.*)

acquire [ə'kwaɪə(r)] *v. t.* sich *(Dat.)* anschaffen ⟨*Gegenstände*⟩; erwerben

⟨*Besitz, Kenntnisse*⟩. **acquisition** [ækwɪ'zɪʃn] *n.* Erwerb, *der; (thing)* Anschaffung, *die.* **acquisitive** [ə'kwɪzɪtɪv] *adj.* raffsüchtig
acquit [ə'kwɪt] *v. t.,* -tt- freisprechen. **acquittal** [ə'kwɪtl] *n.* Freispruch, *der*
acre ['eɪkə(r)] *n.* Acre, *der*
acrid ['ækrɪd] *adj.* beißend ⟨*Geruch, Rauch*⟩; bitter ⟨*Geschmack*⟩
acrimonious [ækrɪ'məʊnɪəs] *adj.* bitter; erbittert ⟨*Streit*⟩
acrobat ['ækrəbæt] *n.* Akrobat, *der*/Akrobatin, *die.* **acrobatic** [ækrə-'bætɪk] *adj.* akrobatisch. **acrobatics** [ækrə'bætɪks] *n.* Akrobatik, *die*
acronym ['ækrənɪm] *n.* Akronym, *das*
across [ə'krɒs] 1. *adv. (from one side to the other)* darüber; *(from here to there)* hinüber; **be 9 miles** ~: 9 Meilen breit sein. 2. *prep.* über (+ *Akk.*); *(on the other side of)* auf der anderen Seite (+ *Gen.*)
act [ækt] 1. *n.* **a)** *(deed)* Tat, *die;* **b)** *(Theatre)* Akt, *der;* **c)** *(pretence)* Theater, *das;* **put on an** ~: Theater spielen; **d)** *(Law)* Gesetz, *das.* 2. *v. t.* spielen ⟨*Stück*⟩. 3. *v. i.* **a)** *(perform actions)* handeln; **b)** *(behave)* sich verhalten; ~ **as** fungieren als; **c)** *(perform play)* spielen; **d)** *(have effect)* ~ **on sth.** auf etw. *(Akk.)* wirken. '**acting** 1. *n.* *(Theatre etc.)* die Schauspielerei. 2. *adj. (temporary)* stellvertretend
action ['ækʃn] *n.* **a)** *(doing sth.)* Handeln, *das;* **take** ~: Schritte *od.* etwas unternehmen; **put a plan into** ~: einen Plan in die Tat umsetzen; **put sth. out of** ~: etw. außer Betrieb setzen; **b)** *(act)* Tat, *die;* **c)** *(legal process)* [Gerichts]verfahren, *das;* **d)** **die in** ~: im Kampf fallen. **action 'replay** *n.* Wiederholung [in Zeitlupe]
activate ['æktɪveɪt] *v. t.* **a)** in Gang setzen; **b)** *(Chem., Phys.)* aktivieren
active ['æktɪv] *adj.,* '**actively** *adv.* aktiv
activist ['æktɪvɪst] *n.* Aktivist, *der*/Aktivistin, *die*
activity [æk'tɪvɪtɪ] *n.* Aktivität, *die*
actor ['æktə(r)] *n.* Schauspieler, *der*
actress ['æktrɪs] *n.* Schauspielerin, *die*
actual ['æktʃʊəl] *adj.* eigentlich; wirklich ⟨*Name*⟩. '**actually** *adv. (in fact)* eigentlich; *(by the way)* übrigens; *(believe it or not)* sogar
acumen ['ækjʊmen] Scharfsinn, *der;* **business** ~: Geschäftssinn, *der*
acupuncture ['ækjʊpʌŋktʃə(r)] *n.* Akupunktur, *die*

acute [ə'kju:t] *adj.* **a)** spitz ⟨*Winkel*⟩; **b)** *(critical; Med.)* akut
AD *abbr.* **Anno Domini** n. Chr.
ad [æd] *n. (coll.)* Annonce, *die*
adamant ['ædəmənt] *adj.* unnachgiebig; **be** ~ **that** ...: darauf bestehen, daß ...
adapt [ə'dæpt] *v. t.* **a)** anpassen (to *Dat.*); ~ **oneself to sth.** sich an etw. *(Akk.)* gewöhnen; **b)** bearbeiten ⟨*Text, Theaterstück*⟩. **adaptable** [ə'dæptəbl] *adj.* anpassungsfähig. **adaptation** [ædəp'teɪʃn] *n.* **a)** Anpassung, *die;* **b)** *(version)* Adap[ta]tion, *die; (of story, text)* Bearbeitung, *die.* **adapter, adaptor** [ə'dæptə(r)] *n.* Adapter, *der*
add [æd] 1. *v. t.* hinzufügen (to *Dat.*); ~ **two and two** zwei und zwei zusammenzählen. 2. *v. i.* ~ **to** vergrößern ⟨*Schwierigkeiten, Einkommen*⟩. **add 'up** 1. *v. i.* ~ **up to sth.** *(fig.)* auf etw. *(Akk.)* hinauslaufen. 2. *v. t.* zusammenzählen
adder ['ædə(r)] *n.* Viper, *die*
addict 1. [ə'dɪkt] *v. t.* **be** ~**ed** süchtig sein (to nach). 2. ['ædɪkt] *n.* Süchtige, *der/die.* **addiction** [ə'dɪkʃn] *n.* Sucht, *die* (to nach). **addictive** [ə'dɪktɪv] *adj.* **be** ~: süchtig machen
addition [ə'dɪʃn] *n.* **a)** Hinzufügen, *das; (adding up)* Addieren, *das; (process)* Addition, *die;* **in** ~: außerdem; **in** ~ **to** zusätzlich zu; **b)** *(thing added)* Ergänzung, *die* (to zu). **additional** [ə'dɪʃənl] *adj.* zusätzlich
additive ['ædɪtɪv] *n.* Zusatz, *der*
address [ə'dres] 1. *v. t.* **a)** *(mark with* ~*)* adressieren (to an + *Akk.*); **b)** *(speak to)* anreden; sprechen zu ⟨*Zuhörern*⟩. 2. *n.* **a)** *(on letter)* Adresse, *die;* **b)** *(speech)* Ansprache, *die.* **addressee** [ædre'si:] *n.* Adressat, *der*/Adressatin, *die*
adept ['ædept] *adj.* geschickt (**in, at** in + *Dat.*)
adequate ['ædɪkwət] *adj.* **a)** angemessen (to *Dat.*); *(suitable)* passend; **b)** *(sufficient)* ausreichend. '**adequately** *adv.* **a)** *(sufficiently)* ausreichend; **b)** *(suitably)* angemessen ⟨*gekleidet, qualifiziert usw.*⟩
adhere [əd'hɪə(r)] *v. i.* haften, *(by glue)* kleben (**to an** + *Dat.*). **adhesion** [əd-'hi:ʒn] *n.* Haften, *das.* **adhesive** [əd-'hi:sɪv] 1. *adj.* gummiert ⟨*Briefmarke*⟩; Klebe⟨*band*⟩. 2. *n.* Klebstoff, *der*
adjacent [ə'dʒeɪsənt] *adj.* angrenzend; ~ **to** neben (*position:* + *Dat.; direction:* + *Akk.*)

adjective ['ædʒıktıv] *n.* Adjektiv, *das*

adjoin [ə'dʒɔın] *v.t.* grenzen an (+ *Akk.*)

adjourn [ə'dʒɜ:n] **1.** *v.t. (break off)* unterbrechen; *(put off)* aufschieben. **2.** *v.i.* sich vertagen; ~ **for lunch/half an hour** eine Mittagspause/halbstündige Pause einlegen. **a'djournment** *n. (of court)* Vertagung, *die; (of meeting)* Unterbrechung, *die*

adjudicate [ə'dʒu:dıkeıt] *v.i. (in court, tribunal)* das Urteil fällen; *(in contest)* entscheiden

adjust [ə'dʒʌst] **1.** *v.t.* einstellen; ~ **sth.** |**to sth.**| etw. |an etw. *(Akk.)*| anpassen. **2.** *v.i.* ⟨*Person:*⟩ sich anpassen **(to** an + *Akk.*). **adjustable** [ə'dʒʌstəbl] *adj.* einstellbar; verstellbar ⟨*Gerät*⟩. **a'djustment** *n.* Einstellung, *die; (to situation etc.)* Anpassung, *die*

ad-lib [æd'lıb] **1.** *adj.* improvisiert. **2.** *v.i.,* -bb- improvisieren

administer [æd'mınıstə(r)] **a)** *(manage)* verwalten; **b)** leisten ⟨*Hilfe*⟩; verabreichen ⟨*Medikamente*⟩. **administration** [ədmını'streıʃn] *n.* Verwaltung, *die.* **administrative** [əd'mınıstrətıv] *adj.* Verwaltungs-. **administrator** [əd'mınıstreıtə(r)] *n.* Administrator, *der;* Verwalter, *der*

admirable ['ædmərəbl] *adj.* bewundernswert

admiral ['ædmərəl] *n.* Admiral, *der*

admiration [ædmə'reıʃn] *n.* Bewunderung, *die* **(of, for** für)

admire [əd'maıə(r)] *v.t.* bewundern.

admirer [əd'maıərə(r)] *n.* Bewunderer, *der*/Bewunderin, *die*

admission [əd'mıʃn] *n.* **a)** *(entry)* Zutritt, *der;* **b)** *(charge)* Eintritt, *der;* **c)** *(confession)* Eingeständnis, *das*

admit [əd'mıt] *v.t.,* -tt-: **a)** *(let in)* hinein-/hereinlassen; **b)** *(acknowledge)* zugeben. **admittance** [əd'mıtəns] *n.* Zutritt, *der.* **admittedly** [əd'mıtıdlı] *adv.* zugeben[ermaßen]

admonish [əd'mɒnıʃ] *v.t.* ermahnen

ado [ə'du:] *n.* **without more** ~: ohne weiteres Aufhebens

adolescence [ædə'lesns] *n.* die Zeit des Erwachsenenwerdens. **adolescent** [ædə'lesnt] **1.** *n.* Heranwachsende, *der/die.* **2.** *adj.* heranwachsend

adopt [ə'dɒpt] *v.t.* **a)** adoptieren; **b)** *(take over)* annehmen ⟨*Glaube, Kultur*⟩; **c)** *(take up)* übernehmen ⟨*Methode*⟩; einnehmen ⟨*Standpunkt, Haltung*⟩. **adoption** [ə'dɒpʃn] *n.* **a)** Adoption, *die;* **b)** *(taking over)* Annahme,

die; **c)** *(taking up)* Übernahme, *die; (of point of view)* Einnahme, *die*

adorable [ə'dɔ:rəbl] *adj.* bezaubernd

adoration [ædə'reıʃn] *n.* Verehrung, *die*

adore [ə'dɔ:(r)] *v.t.* verehren

adorn [ə'dɔ:n] *v.t.* schmücken. **a'dornment** *n.* Verzierung, *die;* ~s Schmuck, *der*

adrenalin [ə'drenəlın] *n.* Adrenalin, *das*

Adriatic [eıdrı'ætık] *pr. n.* ~ |**Sea**| Adriatisches Meer

adrift [ə'drıft] *adj.* **be** ~: treiben

adroit [ə'drɔıt] *adj.* geschickt

adulation [ædjʊ'leıʃn] *n.* Vergötterung, *die*

adult ['ædʌlt, ə'dʌlt] **1.** *adj.* erwachsen. **2.** *n.* Erwachsene, *der/die*

adulterate [ə'dʌltəreıt] *v.t.* verunreinigen

adultery [ə'dʌltərı] *n.* Ehebruch, *der*

advance [əd'va:ns] **1.** *v.t.* **a)** *(also Mil.)* vorrücken lassen; **b)** *(put forward)* vorbringen ⟨*Plan, Meinung*⟩; **c)** *(further)* fördern; **d)** *(pay before due date)* vorschießen; ⟨*Bank:*⟩ leihen. **2.** *v.i.* **a)** *(also Mil.)* vorrücken; ⟨*Prozession:*⟩ sich vorwärts bewegen; **b)** *(fig.: make progress)* vorankommen. **3.** *n.* **a)** Vorrücken, *das; (fig.: progress)* Fortschritt, *der;* **b)** *usu. in pl. (personal approach)* Annäherungsversuch, *der;* **c)** *(on salary)* Vorschuß, *der;* **d) in** ~: im voraus. **advanced** [əd'va:nst] *adj.* fortgeschritten

advantage [əd'va:ntıdʒ] *n.* Vorteil, *der;* **take** ~ **of sb.** jmdn. ausnutzen; **be to one's** ~: für jmdn. von Vorteil sein; **turn sth. to** |**one's**| ~: etw. ausnutzen. **advantageous** [ædvən'teıdʒəs] *adj.* vorteilhaft

advent ['ædvent] *n.* Beginn, *der;* A~: Advent, *der*

adventure [əd'ventʃə(r)] *n.* Abenteuer, *das.* **adventurous** [əd'ventʃərəs] *adj.* abenteuerlustig

adverb ['ædvɜ:b] *n.* Adverb, *das*

adversary ['ædvəsərı] *n. (enemy)* Widersacher, *der*/Widersacherin, *die; (opponent)* Kontrahent, *der*/Kontrahentin, *die*

adverse ['ædvɜ:s] *adj.* **a)** *(unfavourable)* ungünstig; **b)** *(contrary)* widrig ⟨*Wind, Umstände*⟩. **adversity** [əd'vɜ:sıtı] *n.* **a)** *no pl.* Not, *die;* **b)** *usu. in pl.* Widrigkeit, *die*

advert ['ædvɜ:t] *(Brit. coll.) see* **advertisement**

advertise ['ædvətaɪz] **1.** *v. t.* werben für; *(by small ad)* inserieren; ausschreiben ⟨*Stelle*⟩. **2.** *v. i.* werben; *(in newspaper)* inserieren; annoncieren. **advertisement** [əd'vɜːtɪsmənt] *n.* Anzeige, *die;* TV ~: Fernsehspot, *der.* **advertiser** ['ædvətaɪzə(r)] *n. (in newspaper)* Inserent, *der*/Inserentin, *die.* **advertising** ['ædvətaɪzɪŋ] *n.* Werbung, *die; attrib.* Werbe-
advice [əd'vaɪs] *n.* Rat, *der;* take sb.'s ~: jmds. Rat *(Dat.)* folgen
advisable [əd'vaɪzəbl] *adj.* ratsam
advise [əd'vaɪz] *v. t.* beraten; ~ sth. zu etw. raten; *(inform)* unterrichten (of über + *Akk.*). **adviser, advisor** [əd'vaɪzə(r)] *n.* Berater, *der*/Beraterin, *die.* **advisory** [əd'vaɪzərɪ] *adj.* beratend
advocate 1. ['ædvəkət] *n. (of a cause)* Befürworter, *der*/Befürworterin, *die; (Law)* [Rechts]anwalt, *der*/-anwältin, *die.* **2.** ['ædvəkeɪt] *v. t.* befürworten
aerial ['eərɪəl] **1.** *adj.* Luft-. **2.** *n.* Antenne, *die*
aero- [eərəʊ] *in comb.* Aero-
aerody'namic *adj.* aerodynamisch
aeronautics [eərə'nɔːtɪks] *n.* Aeronautik, *die*
aeroplane ['eərəpleɪn] *n. (Brit.)* Flugzeug, *das*
aerosol ['eərəsɒl] *n. (spray)* Spray, *der od.* das; *(container)* ~ [spray] Spraydose, *die*
aesthetic [iːs'θetɪk] *adj.* ästhetisch
afar [ə'fɑː] *adv.* from ~: aus der Ferne
affable ['æfəbl] *adj.* freundlich
affair [ə'feə(r)] *n.* **a)** *(concern)* Angelegenheit, *die;* **b)** *in pl. (business)* Geschäfte *Pl.;* **c)** *(love ~)* Affäre, *die*
affect [ə'fekt] *v. t.* **a)** sich auswirken auf (+ *Akk.*); **b)** *(emotionally)* betroffen machen
affectation [æfek'teɪʃn] *n. (studied display)* Verstellung, *die; (artificiality)* Affektiertheit, *die*
affected [ə'fektɪd] *adj.* affektiert; gekünstelt ⟨*Sprache, Stil*⟩
affection [ə'fekʃn] *n.* Zuneigung, *die.* **affectionate** [ə'fekʃənət] *adj.* anhänglich; liebevoll ⟨*Umarmung*⟩. **af'fectionately** *adv.* liebevoll
affiliate [ə'fɪlɪeɪt] *v. t.* be ~d to sth. an etw. *(Akk.)* angegliedert sein
affinity [ə'fɪnɪtɪ] *n.* **a)** *(relationship)* Verwandtschaft, *die* (to mit); **b)** *(liking)* Neigung, *die* (for zu); feel an ~ to or for sb./sth. sich zu jmdm./etw. hingezogen fühlen

affirm [ə'fɜːm] *v. t. (assert)* bekräftigen ⟨*Absicht*⟩; beteuern ⟨*Unschuld*⟩; *(state as a fact)* bestätigen. **affirmation** [æfə'meɪʃn] *n. (of intention)* Bekräftigung, *die; (of fact)* Bestätigung, *die.* **affirmative** [ə'fɜːmətɪv] **1.** *adj.* affirmativ; bejahend ⟨*Antwort*⟩. **2.** *n.* answer in the ~: bejahend antworten
afflict [ə'flɪkt] *v. t. (physically)* plagen; *(mentally)* quälen; peinigen; be ~ed with sth. von etw. befallen sein. **affliction** [ə'flɪkʃn] *n.* Leiden, *das*
affluence ['æfluəns] *n.* Reichtum, *der.* **affluent** ['æfluənt] *adj.* reich
afford [ə'fɔːd] *v. t.* **a)** sich *(Dat.)* leisten; **b)** *(provide)* bieten; gewähren ⟨*Schutz*⟩
affray [ə'freɪ] *n.* Schlägerei, *die*
affront [ə'frʌnt] **1.** *v. t.* beleidigen. **2.** *n.* Beleidigung, *die*
afield [ə'fiːld] *adv.* far ~ *(direction)* weit hinaus; *(place)* weit draußen
afloat [ə'fləʊt] *pred. adj.* **a)** *(floating)* über Wasser; flott ⟨*Schiff*⟩; **b)** *(at sea)* auf See; be ~: auf dem Meer treiben
afoot [ə'fʊt] *pred. adj.* im Gange
aforementioned [ə'fɔːmenʃnd], **aforesaid** [ə'fɔːsed] *adjs.* obenerwähnt *od.* -genannt
afraid [ə'freɪd] *adj.* be ~ [of sb./sth.] [vor jmdm./etw.] Angst haben; be ~ to do sth. Angst davor haben, etw. zu tun; I'm ~ ~ not/not ich fürchte ja/nein
afresh [ə'freʃ] *adv.* von neuem
Africa ['æfrɪkə] *pr. n.* Afrika *(das).* **African** ['æfrɪkən] **1.** *adj.* afrikanisch. **2.** *n.* Afrikaner, *der*/Afrikanerin, *die*
after ['ɑːftə(r)] **1.** *adv.* **a)** *(later)* danach; **b)** *(behind)* hinterher. **2.** *prep.* **a)** *(in time)* nach; two days ~: zwei Tage danach; **b)** *(behind)* hinter (+ *Dat.*); **c)** ask ~ sb./sth. nach jmdm./etw. fragen; **d)** ~ all schließlich. **3.** *conj.* nachdem. 'after-care *n. (Med.)* Nachbehandlung, *die.* 'after-effect *n.* Nachwirkung, *die*
aftermath ['ɑːftəmæθ, 'ɑːftəmɑːθ] *n.* Nachwirkungen *Pl.*
after: ~noon *n.* Nachmittag, *der;* this/tomorrow ~: heute/morgen nachmittag; in the ~: am Nachmittag; *(regularly)* nachmittags; ~shave *n.* After-shave, *das;* ~thought *n.* nachträglicher Einfall
afterwards ['ɑːftəwədz] *adv.* danach
again [ə'gen, ə'geɪn] *adv.* wieder; *(one more time)* noch einmal; ~ and ~, time and [time] ~: immer wieder; back ~: wieder zurück·

against [ə'genst, ə'geɪnst] *prep.* gegen
age [eɪdʒ] **1.** *n.* **a)** Alter, *das;* **what ~
are you?** wie alt bist du?; **at the ~ of**
im Alter von; **come of ~:** volljährig
werden; **be under ~:** zu jung sein; **b)**
(great period) Zeitalter, *das;* **~s** *(coll.:
a long time)* eine Ewigkeit. **2.** *v. t.* altern lassen. **3.** *v. i.* altern. **aged** *adj.* **a)**
[eɪdʒd] **be ~ five** fünf Jahre alt sein; **a
boy ~ five** ein fünfjähriger Junge; **b)**
['eɪdʒɪd] *(elderly)* bejahrt
age: **~-group** *n.* Altersgruppe, *die;* **~
limit** *n.* Altersgrenze, *die*
agency ['eɪdʒənsɪ] *n. (business establishment)* Geschäftsstelle, *die; (news/
advertising ~)* Agentur, *die*
agenda [ə'dʒendə] *n.* Tagesordnung,
die
agent ['eɪdʒənt] *n.* Vertreter, *der/*Ver-
treterin, *die; (spy)* Agent, *der/*Agentin,
die
aggravate ['ægrəveɪt] *v. t.* **a)** *(make
worse)* verschlimmern; **b)** *(annoy)* aufregen; ärgern. **aggravating** ['ægrə-
veɪtɪŋ] *adj.* ärgerlich. **aggravation**
[ægrə'veɪʃn] *n.* **a)** Verschlimmerung,
die; **b)** *(annoyance)* Ärger, *der*
aggregate ['ægrɪgət] **1.** *n.* Gesamt-
menge, *die.* **2.** *adj.* gesamt
aggression [ə'greʃn] *n.* Aggression,
die
aggressive [ə'gresɪv] *adj.,* **ag'gressively** *adv.* aggressiv. **ag'gressive-
ness** *n.* Aggressivität, *die*
aggressor [ə'gresə(r)] *n.* Aggressor,
der
aggrieved [ə'griːvd] *v. t. (resentful)*
verärgert; *(offended)* gekränkt
aghast [ə'gɑːst] *pred. adj.* bestürzt
agile ['ædʒaɪl] *adj.* beweglich; flink
⟨*Bewegung*⟩. **agility** [ə'dʒɪlɪtɪ] *n.* Beweglichkeit, *die; (of movement)* Flink-
heit, *die*
agitate ['ædʒɪteɪt] **1.** *v. t.* **a)** *(shake)*
schütteln; **b)** *(disturb)* erregen. **2.** *v. i.*
agitieren. **agitation** [ædʒɪ'teɪʃn] *n.* **a)**
(shaking) Schütteln, *das;* **b)** *(emotional)* Erregung, *die.* **agitator**
['ædʒɪteɪtə(r)] *n.* Agitator, *der*
agnostic [æg'nɒstɪk] *n.* Agnostiker,
*der/*Agnostikerin, *die*
ago [ə'gəʊ] *adv.* **ten years ~:** vor zehn
Jahren; **[not] long ~:** vor [nicht] langer
Zeit
agog [ə'gɒg] *pred. adj.* gespannt
agonize ['ægənaɪz] *v. i.* **~ over sth.** sich
(Dat.) den Kopf über etw. *(Akk.)* zer-
martern
agony ['ægənɪ] *n.* Todesqualen *Pl.*

agree [ə'griː] **1.** *v. i.* **a)** *(consent)* einverstanden sein **(to, with** mit); **b)** *(hold
similar opinion)* einer Meinung sein;
they ~d [with me] sie waren derselben
Meinung [wie ich]; **c)** *(reach similar
opinion)* **~ on sth.** sich über etw.
(Akk.) einigen; **d)** *(harmonize)* übereinstimmen; **e)** **~ with sb.** *(suit)* jmdm.
bekommen. **2.** *v. t.* vereinbaren.
agreeable [ə'griːəbl] *adj.* **a)** *(pleasing)* angenehm; **b)** **be ~ [to sth.]** [mit
etw.] einverstanden sein. **agreeably**
[ə'griːəblɪ] *adv.* angenehm. **agreed**
[ə'griːd] *adj.* einig; vereinbart
⟨*Summe, Zeit*⟩. **a'greement** *n.* Übereinstimmung, *die;* **be in ~ [about sth.]**
sich *(Dat.)* [über etw. *(Akk.)*] einig
sein
agricultural [ægrɪ'kʌltʃərl] *adj.* land-
wirtschaftlich
agriculture ['ægrɪkʌltʃə(r)] *n.* Landwirtschaft, *die*
aground [ə'graʊnd] *adj.* **go** *or* **run ~:**
auf Grund laufen
ahead [ə'hed] *adv.* voraus; **~ of** vor
(+ *Dat.*); **be ~ of the others** *(fig.)* den
anderen voraus sein
aid [eɪd] **1.** *v. t.* **a)** **~ sb. [to do sth.]**
jmdm. helfen[, etw. zu tun]; **~ed by**
unterstützt von; **b)** *(promote)* fördern.
2. *n.* **a)** *(help)* Hilfe, *die;* **with the ~ of
sth./sb.** mit Hilfe einer Sache
*(Gen.)/*mit jmds. Hilfe; **in ~ of sb./sth.**
zugunsten von jmdm./etw.; **b)** *(source
of help)* Hilfsmittel, *das* **(to** für)
aide [eɪd] *n.* Berater, *der/*Beraterin, *die*
Aids [eɪdz] *n.* Aids *(das)*
ailment ['eɪlmənt] *n.* Gebrechen, *das*
aim [eɪm] **1.** *v. t.* ausrichten ⟨*Schußwaffe, Rakete*⟩; **~ sth. at sb./sth.** etw.
auf jmdn./etw. richten. **2.** *v. i.* zielen
(at auf + *Akk.*); **b)** **~ to do sth.** beabsichtigen, etw. zu tun; **~ at** *or* **for sth.**
(fig.) etwas anstreben. **3.** *n.* Ziel, *das;*
take ~ [at sth./sb.] [auf etw./jmdn.]
zielen. **'aimless** *adj.,* **'aimlessly**
adv. ziellos
air [eə(r)] **1.** *n.* **a)** Luft, *die;* **be/go on the
~:** senden; ⟨*Programm:*⟩ gesendet
werden; **by ~:** mit dem Flugzeug; *(by
~ mail)* mit Luftpost; **b)** *(facial expression)* Miene, *die;* **c)** **put on ~s** sich auf-
spielen. **2.** *v. t. (ventilate)* lüften;
(make public) [öffentlich] darlegen
air: **~-bed** *n.* Luftmatratze, *die;*
~borne *adj.* **be ~borne** sich in der
Luft befinden; **~-conditioned** *adj.*
klimatisiert; **~-conditioning** *n.* Kli-
maanlage, *die;* **~craft** *n., pl.* same

Flugzeug, *das;* ~**craft-carrier** *n.*
Flugzeugträger, *der;* ~ **fare** *n.* Flug-
preis, *der;* ~**field** *n.* Flugplatz, *der;* ~
force *n.* Luftwaffe, *die;* ~**gun** *n.*
Luftgewehr, *das;* ~ **hostess** *n.* Stewardeß, *die;* ~ **letter** *n.* Aerogramm,
das; ~**line** *n.* Fluggesellschaft, *die;* ~
mail *n.* Luftpost, *die;* by ~ **mail** mit
Luftpost; ~**man** ['eəmən] *n., pl.* ~**men**
[~mən] Flieger, *der;* ~**plane** *n.*
(Amer.) Flugzeug, *das;* ~**port** *n.*
Flughafen, *der;* ~ **raid** *n.* Luftangriff,
der; ~**-raid shelter** *n.* Luftschutzraum, *der;* ~**ship** *n.* Luftschiff, *das;*
~**sick** *adj.* luftkrank; ~**tight** *adj.*
luftdicht; ~**-traffic controller** *n.*
Fluglotse, *der*

'**airy** *adj.* luftig ⟨*Büro, Zimmer*⟩
aisle [aɪl] *n.* Gang, *der;* *(of church)* Sei-
tenschiff, *das*
ajar [ə'dʒɑː(r)] *adj.* be ~: einen Spalt-
breit offenstehen
akin [ə'kɪn] *adj.* be ~ to sth. einer Sache *(Dat.)* ähnlich sein
alarm [ə'lɑːm] **1.** *n.* **a)** Alarm, *der;* **give
or raise the** ~: Alarm schlagen; **b)**
(fear) Angst, *die.* **2.** *v. t.* aufschrecken.
a'larm clock *n.* Wecker, *der*
alas [ə'læs] *int.* ach
albatross ['ælbətrɒs] *n.* Albatros, *der*
album ['ælbəm] *n.* Album, *das*
alcohol ['ælkəhɒl] *n.* Alkohol, *der.* **alcoholic** [ælkə'hɒlɪk] **1.** *adj.* alkoholisch. **2.** *n.* Alkoholiker, *der/*Alkoho-
likerin, *die.* **alcoholism** ['ælkəhɒlɪzm]
n. Alkoholismus, *der*
alcove ['ælkəʊv] *n.* Alkoven, *der*
ale [eɪl] *n.* Ale, *das*
alert [ə'lɜːt] **1.** *adj.* wachsam. **2.** *n.*
Alarmbereitschaft, *die;* on the ~: auf
der Hut. **3.** *v. t.* alarmieren; ~ sb. |to
sth.| jmdn. [vor etw. *(Dat.)*] warnen
'**A level** *n. (Brit. Sch.)* ≈ Abitur, *das*
algebra ['ældʒɪbrə] *n.* Algebra, *die*
Algeria [æl'dʒɪərɪə] *pr. n.* Algerien
(das)
alias ['eɪlɪəs] **1.** *adv.* alias. **2.** *n.* ange-
nommener Name
alibi ['ælɪbaɪ] *n.* Alibi, *das*
alien ['eɪlɪən] **1.** *adj.* **a)** *(strange)*
fremd; **b)** *(foreign)* ausländisch. **2.** *n.*
a) *(from another world)* Außerirdi-
sche, *der/die;* **b)** *(Admin.: foreigner)*
Ausländer, *der/*Ausländerin, *die.*
alienate ['eɪlɪəneɪt] *v. t.* befremden.
alienation [eɪlɪə'neɪʃn] *n.* Entfremdung, *die*
'**alight** [ə'laɪt] *v. i.* **a)** aussteigen (from
aus); **b)** ⟨*Vogel:*⟩ sich niedersetzen

²**alight** *adj.* be/catch ~: brennen; set
sth. ~: etw. in Brand setzen
align [ə'laɪn] *v. t.* **a)** *(place in a line)*
ausrichten; **b)** *(bring into line)* in eine
Linie bringen. **a'lignment** *n.* Aus-
richtung, *die;* out of ~: nicht richtig
ausgerichtet
alike [ə'laɪk] *pred. adj.* ähnlich; *(indis-
tinguishable)* gleich
alimony ['ælɪmənɪ] *n.* Unterhaltszah-
lung, *die*
alive [ə'laɪv] *pred. adj.* **a)** lebendig; **b)**
(aware) be ~ to sth. sich *(Dat.)* einer
Sache *(Gen.)* bewußt sein; **c)** *(swarm-
ing)* be ~ with wimmeln von
alkali ['ælkəlaɪ] *n., pl.* ~s or ~es Alkali,
das
all [ɔːl] **1.** *attrib. adj.* **a)** *(entire extent or
quantity of)* ganz; ~ **day** den ganzen
Tag; ~ **my money** all mein Geld; mein
ganzes Geld; **b)** *(entire number of)* alle; ~ **the books** alle Bücher; ~ **my
books** all[e] meine Bücher; ~ **the
others** alle anderen; **c)** *(any whatever)*
jeglicher/jegliche/jegliches; **d)** *(greatest possible)* in ~ **innocence** in aller
Unschuld. **2.** *n.* **a)** *(~ persons)* alle; ~
of us wir alle; **the happiest of** ~: der/
die Glücklichste unter *od.* von allen;
b) *(every bit)* ~ **of it** alles; ~ **of the
money** das ganze Geld; **c)** ~ **of** *(coll.:
as much as)* be ~ **of seven feet tall** gut
sieben Fuß groß sein; **d)** *(~ things)* alles; ~ **I need is the money** ich brauche
nur das Geld; **that is** ~: das ist alles;
the most beautiful of ~: der/die/das
Schönste von allen; **most of** ~: am
meisten; **it was** ~ **but impossible** es war
fast unmöglich; **it's** ~ **the same to me**
es ist mir ganz egal; **can I help you at**
~? kann ich Ihnen irgendwie behilflich sein?; **she has no talent at** ~: sie
hat überhaupt kein Talent; **nothing at**
~: gar nichts; **not at** ~ **happy/well**
überhaupt nicht glücklich/gesund;
not at ~! überhaupt nicht!; *(acknowl-
edging thanks)* gern geschehen!; **if at**
~: wenn überhaupt; **in** ~: insgesamt;
e) *(Sport)* **two |goals|** ~: zwei zu zwei;
(Tennis) **thirty** ~: dreißig beide. **3.**
adv. ganz; ~ **but fast;** ~ **the better/
worse |for that|** um so besser/schlim-
mer; ~ **at once** *(suddenly)* plötzlich; **be**
~ '**in** *(exhausted)* total erledigt sein
(ugs.); **sth. is** ~ **right** etw. ist in Ordnung; *(tolerable)* etw. ist ganz gut; **I'm**
~ **right** mir geht es ganz gut; **yes,** ~
right ja, gut; **it's** ~ **right by me** das ist
mir recht

allay [ə'leı] *v. t.* zerstreuen ⟨*Besorgnis, Befürchtungen*⟩

all-'clear *n.* Entwarnung, *die*

allegation [ælı'geıʃn] *n.* Behauptung, *die*

allege [ə'ledʒ] *v. t.* behaupten. **alleged** [ə'ledʒd] *adj.*, **allegedly** [ə'ledʒɪdlı] *adv.* angeblich

allegiance [ə'li:dʒəns] *n.* Loyalität, *die* (to gegenüber)

allegory ['ælıgərı] *n.* Allegorie, *die*

allergic [ə'lɜːdʒɪk] *adj.* allergisch (to gegen)

allergy ['ælədʒı] *n.* Allergie, *die*

alleviate [ə'li:vıeıt] *v. t.* abschwächen

alley ['ælı] *n.* [schmale] Gasse

alliance [ə'laıəns] *n.* Bündnis, *das; (league)* Allianz, *die*

allied ['ælaıd] *adj.* be ~ to *or* with sb./ sth. mit jmdm./etw. verbündet sein

alligator ['ælıgeıtə(r)] *n.* Alligator, *der*

'all-in *adj.* Pauschal-

allocate ['æləkeıt] *v. t.* zuweisen, zuteilen (to *Dat.*). **allocation** [ælə-'keıʃn] *n.* Zuweisung, *die; (ration)* Zuteilung, *die*

allot [ə'lɒt] *v. t.*, -tt-: ~ sth. to sb. jmdm. etw. zuteilen. **al'lotment** *n. (Brit.: plot of land)* ≈ Schrebergarten, *der*

allow [ə'laʊ] 1. *v. t.* erlauben; zulassen; ~ sb. to do sth. jmdm. erlauben, etw. zu tun; be ~ed to do sth. etw. tun dürfen. 2. *v. i.* ~ for sth. etw. berücksichtigen. **allowance** [ə'laʊəns] *n.* a) Zuteilung, *die; (for special expenses)* Zuschuß, *der;* b) make ~s for sth./sb. etw./jmdn. berücksichtigen

alloy ['ælɔı] *n.* Legierung, *die*

all: ~-round *adj.* Allround-; ~-'rounder *n.* Allroundtalent, *das;* ~-time *adj.* ~-time record absoluter Rekord

allude [ə'lu:d] *v. i.* ~ to sich beziehen auf (+ *Akk.*); *(indirectly)* anspielen auf (+ *Akk.*). **allusion** [ə'lu:ʒn] *n.* Hinweis, *der; (indirect)* Anspielung, *die*

ally ['ælaı] *n.* Verbündete, *der/die;* the Allies die Alliierten

almighty [ɔ:l'maıtı] *adj.* allmächtig; the A~: der Allmächtige

almond ['ɑ:mənd] *n.* Mandel, *die*

almost ['ɔ:lməʊst] *adv.* fast; beinahe

alms [ɑ:mz] *n.* Almosen, *das*

alone [ə'ləʊn] 1. *pred. adj.* allein; alleine *(ugs.).* 2. *adv.* allein

along [ə'lɒŋ] 1. *prep.* entlang *(position:* + *Dat.; direction:* + *Akk.).* 2. *adv.*

weiter; I'll be ~ shortly ich komme gleich; all ~: die ganze Zeit [über].

along'side 1. *adv.* daneben. 2. *prep.* neben *(position:* + *Dat.; direction:* + *Akk.)*

aloof [ə'lu:f] 1. *adv.* abseits; hold ~ from sb. sich von jmdm. fernhalten. 2. *adj.* distanziert

aloud [ə'laʊd] *adv.* laut; read |sth.| ~: [etw.] vorlesen

alphabet ['ælfəbet] *n.* Alphabet, *das.* **alphabetical** [ælfə'betıkl] *adj.*, **alpha'betically** *adv.* alphabetisch

alpine ['ælpaın] *adj.* alpin

Alps [ælps] *pr. n. pl.* the ~: die Alpen

already [ɔ:l'redı] *adv.* schon

Alsation [æl'seıʃn] *n.* [deutscher] Schäferhund

also ['ɔ:lsəʊ] *adv.* auch; *(moreover)* außerdem

altar ['ɔ:ltə(r), 'ɒltə(r)] *n.* Altar, *der*

alter ['ɔ:ltə(r), 'ɒltə(r)] 1. *v. t.* ändern. 2. *v. i.* sich verändern. **alteration** [ɔ:ltə-'reıʃn, ɒltə'reıʃn] *n.* Änderung, *die*

alternate 1. [ɔ:l'tɜːnət] *adj.* sich abwechselnd. 2. ['ɔ:ltəneıt] *v. t.* abwechseln lassen. 3. ['ɔ:ltəneıt] *v. i.* sich abwechselnd. **al'ternately** *adv.* abwechselnd

alternative [ɔ:l'tɜːnətıv] 1. *adj.* alternativ; Alternativ-. 2. *n.* a) *(choice)* Alternative, *die;* b) *(possibility)* Möglichkeit, *die.* **al'ternatively** *adv.* oder aber; or ~: oder aber auch

although [ɔ:l'ðəʊ] *conj.* obwohl

altitude ['æltıtju:d] *n.* Höhe, *die*

altogether [ɔ:ltə'geðə(r)] *adv.* völlig; *(on the whole)* im großen und ganzen; *(in total)* insgesamt; not ~ |true/convincing| nicht ganz [wahr/überzeugend]

altruistic [æltrʊ'ıstık] *adj.* altruistisch

aluminium [æljʊ'mınıəm] *(Brit.)*, **aluminum** [ə'lu:mınəm] *(Amer.) ns.* Aluminium, *das*

always ['ɔ:lweız] *adv.* immer; *(repeatedly)* ständig

AM *abbr.* amplitude modulation AM

am *see* be

a.m. [eı'em] *adv.* vormittags; |at| one/ four ~: [um] ein/vier Uhr früh

amalgamate [ə'mælgəmeıt] 1. *v. t.* vereinigen. 2. *v. i.* sich vereinigen; ⟨*Firmen:*⟩ fusionieren. **amalgamation** [əmælgə'meıʃn] *n.* Vereinigung, *die; (of firms)* Fusion, *die*

amass [ə'mæs] *v. t.* anhäufen

amateur ['æmətə(r)] *n.* Amateur, *der; attrib.* Amateur-; Laien-. **amateur-**

ish ['æmətərıʃ] *adj.* laienhaft; amateurhaft

amaze [ə'meız] *v.t.* verblüffen; verwundern. **a'mazement** *n.* Verblüffung, *die;* Verwunderung, *die.* **amazing** [ə'meızıŋ] *adj. (remarkable)* erstaunlich; *(astonishing)* verblüffend

Amazon ['æməzən] *pr. n.* **the ~:** der Amazonas

ambassador [æm'bæsədə(r)] *n.* Botschafter, *der*/Botschafterin, *die*

amber ['æmbə(r)] 1. *n.* **a)** Bernstein, *der;* **b)** *(traffic light)* Gelb, *das.* 2. *adj.* Bernstein-; *(colour)* bernsteinfarben; gelb 〈*Verkehrslicht*〉

ambiguity [æmbı'gju:ıtı] *n.* Zweideutigkeit, *die*

ambiguous [æm'bıgjʊəs] *adj.* zweideutig

ambition [æm'bıʃn] *n.* Ehrgeiz, *der; (aspiration)* Ambition, *die.* **ambitious** [æm'bıʃəs] *adj.* ehrgeizig

ambivalent [æm'bıvələnt] *adj.* ambivalent

amble ['æmbl] *v.i.* schlendern

ambulance ['æmbjʊləns] *n.* Krankenwagen, *der;* Ambulanz, *die*

ambush ['æmbʊʃ] 1. *n.* Hinterhalt, *der;* **lie in ~:** im Hinterhalt liegen. 2. *v.t.* [aus dem Hinterhalt] überfallen

amen [ɑ:'men, eı'men] 1. *int.* amen. 2. *n.* Amen, *das*

amenable [ə'mi:nəbl] *adj.* zugänglich, aufgeschlossen (**to** *Dat.*)

amend [ə'mend] *v.t.* berichtigen; abändern 〈*Gesetzentwurf, Antrag*〉. **a'mendment** *n. (to motion)* Abänderungsantrag, *der; (to bill)* Änderungsantrag, *der*

amends [ə'mendz] *n. pl.* **make ~ |to sb.]** es [bei jmdm.] wiedergutmachen; **make ~ for sth.** etw. wiedergutmachen

amenity [ə'mi:nıtı] *n., usu. in pl.* **amenities** *(of town)* kulturelle und Freizeiteinrichtungen

America [ə'merıkə] *pr. n.* Amerika *(das).* **American** [ə'merıkən] 1. *adj.* amerikanisch; **sb. is ~:** jmd. ist Amerikaner/Amerikanerin. 2. *n. (person)* Amerikaner, *der*/Amerikanerin, *die.* **Americanize** [ə'merıkənaız] *v.t.* amerikanisieren

amiable ['eımıəbl] *adj.* umgänglich

amicable ['æmıkəbl] *adj.* freundschaftlich; gütlich 〈*Einigung*〉. **amicably** ['æmıkəblı] *adv.* in [aller] Freundschaft

amid[st] [ə'mıd(st)] *prep.* inmitten; *(fig.: during)* bei

amiss [ə'mıs] 1. *pred. adj.* verkehrt; **is anything ~?** stimmt irgend etwas nicht? 2. *adv.* **take sth. ~:** etw. übelnehmen

ammonia [ə'məʊnıə] *n.* Ammoniak, *das*

ammunition [æmjʊ'nıʃn] *n.* Munition, *die*

amnesia [æm'ni:zıə] Amnesie, *die*

amnesty ['æmnıstı] *n.* Amnestie, *die*

amok [ə'mɒk] *adv.* **run ~:** Amok laufen

among[st] [ə'mʌŋ(st)] *prep.* unter (+ *Dat.*); **~ other things** unter anderem; **they often quarrel ~ themselves** sie streiten oft miteinander

amoral [eı'mɒrl] *adj.* amoralisch

amorphous [ə'mɔ:fəs] *adj.* formlos; amorph 〈*Masse*〉

amount [ə'maʊnt] 1. *v.i.* **~ to sth.** *(fig.)* sich auf etw. *(Akk.)* belaufen; *(fig.)* etw. bedeuten. 2. *n.* **a)** *(total)* Betrag, *der;* Summe, *die;* **b)** *(quantity)* Menge, *die*

amp [æmp] *n.* Ampere, *das*

amphibian [æm'fıbıən] 1. *adj.* amphibisch. 2. *n.* Amphibie, *die.* **amphibious** [æm'fıbıəs] *adj.* amphibisch

amphitheatre ['æmfıθıətə(r)] *n.* Amphitheater, *das*

ample ['æmpl] *adj.* **a)** *(spacious)* weitläufig 〈*Garten, Räume*〉; reichhaltig 〈*Mahl*〉; **b)** *(enough)* **~ room/food** reichlich Platz/zu essen

amplifier ['æmplıfaıə(r)] *n.* Verstärker, *der*

amplify ['æmplıfaı] *v.t.* verstärken; *(enlarge on)* weiter ausführen

amputate ['æmpjʊteıt] *v.t.* amputieren. **amputation** [æmpjʊ'teıʃn] *n.* Amputation, *die*

amuse [ə'mju:z] *v.t.* **a)** *(interest)* unterhalten; **~ oneself by doing sth.** sich *(Dat.)* die Zeit damit vertreiben, etw. zu tun; **b)** *(make laugh or smile)* amüsieren. **a'musement** *n.* Belustigung, *die;* **~ arcade** Spielhalle, *die.* **amusing** [ə'mju:zıŋ] *adj.* amüsant

an [ən, *stressed* æn] *indef. art. see also* ²**a:** ein/eine/ein

anaemia [ə'ni:mıə] *n.* Blutarmut, *die;* Anämie, *die.* **anaemic** [ə'ni:mık] *adj.* blutarm; anämisch

anaesthetic [ænıs'θetık] *n.* Anästhetikum, *das; general* **~:** Narkosemittel, *das; local* **~:** Lokalanästhetikum, *das*

anagram ['ænəgræm] *n.* Anagramm, *das*

analogy [ə'nælədʒı] *n.* Analogie, *die*

analyse ['ænəlaız] *v.t.* analysieren.

23 **another**

analysis [ə'næləsıs] *n., pl.* **analyses** [ə'næləsi:z] Analyse, *die.* **analyst** ['ænəlıst] *n.* **a)** *(Psych.)* Analytiker, *der*/Analytikerin, *die;* **b)** *(Econ., Polit., etc.)* Experte, *der.* **analytic** [ænə'lıtık], **analytical** [ænə'lıtıkl] *adj.* analytisch. **analyze** *(Amer.) see* **analyse**

anarchist ['ænəkıst] *n.* Anarchist, *der*/Anarchistin, *die*

anarchy ['ænəkı] *n.* Anarchie, *die*

anatomical [ænə'tomıkl] *adj.* anatomisch

anatomy [ə'nætəmı] *n.* Anatomie, *die*

ancestor ['ænsestə(r)] *n.* Vorfahr, *der.*

ancestry ['ænsestrı] *n.* Abstammung, *die*

anchor ['æŋkə(r)] **1.** *n.* Anker, *der.* **2.** *v. t.* verankern. **3.** *v. i.* ankern. **anchorage** ['æŋkərıdʒ] *n.* Ankerplatz, *der*

anchovy ['æntʃəvı] *n.* Sardelle, *die*

ancient ['eınʃənt] *adj.* alt; historisch ⟨*Gebäude usw.*⟩; *(of antiquity)* antik

and [ənd, *stressed* ænd] *conj.* und; **for** weeks ~ weeks wochenlang; **better** ~ better immer besser

anecdote ['ænıkdəʊt] *n.* Anekdote, *die*

anemia, anemic *(Amer.) see* **anaemangel** ['eındʒl] *n.* Engel, *der.* **angelic** [æn'dʒelık] *adj.* engelhaft

anger ['æŋgə(r)] **1.** *n.* Zorn, *der* (at über + *Akk.*); *(fury)* Wut, *die* (at über + *Akk.*). **2.** *v. t.* verärgern; *(infuriate)* wütend machen

¹**angle** ['æŋgl] *n.* **a)** *(Geom.)* Winkel, *der;* **at an** ~ **of** 60° im Winkel von 60°; **at an** ~: schief; **b)** *(fig.)* Gesichtspunkt, *der*

²**angle** *v. i.* angeln; *(fig.)* ~ **for sth.** sich um etw. bemühen. **angler** ['æŋglə(r)] *n.* Angler, *der*/Anglerin, *die*

Anglican ['æŋglıkən] **1.** *adj.* anglikanisch. **2.** *n.* Anglikaner, *der*/Anglikanerin, *die*

Anglo- [æŋgləʊ] *in comb.* anglo-/Anglo-. **Anglo-Saxon** [~'sæksn] **1.** *n.* Angelsachse, *der*/Angelsächsin, *die; (language)* Angelsächsisch, *das.* **2.** *adj.* angelsächsisch

angrily ['æŋgrılı] *adv.* verärgert; *(stronger)* zornig

angry ['æŋgrı] *adj.* böse; verärgert ⟨*Person, Stimme, Geste*⟩; *(stronger)* zornig; wütend; **be** ~ **at** *or* **about sth.** wegen etw. böse sein; **be** ~ **with** *or* **at sb.** mit jmdm. *od.* auf jmdn. böse sein; **get** ~: böse werden

anguish ['æŋgwıʃ] *n.* Qualen *Pl.*

angular ['æŋgjʊlə(r)] *adj.* eckig ⟨*Gebäude, Struktur*⟩; kantig ⟨*Gesicht*⟩

animal ['ænıməl] **1.** *n.* Tier, *das.* **2.** *adj.* tierisch

animate 1. ['ænımeıt] *v. t.* beleben. **2.** ['ænımət] *adj.* beseelt ⟨*Leben, Körper*⟩; belebt ⟨*Objekt, Welt*⟩. **animated** ['ænımeıtıd] *adj.* lebhaft ⟨*Diskussion, Gebärde*⟩; ~ **cartoon** Zeichentrickfilm, *der.* **animation** [ænı'meıʃn] *n.* **a)** Lebhaftigkeit, *die;* **b)** *(Cinemat.)* Animation, *die*

animosity [ænı'mosıtı] *n.* Feindseligkeit, *die*

aniseed ['ænısi:d] *n.* Anis[samen], *der*

ankle ['æŋkl] *n.* Fußgelenk, *das*

annex 1. [ə'neks] *v. t.* annektieren ⟨*Land, Territorium*⟩. **2.** ['æneks] *n.* Anbau, *der.* **annexe** *see* **annex 2**

annihilate [ə'naıleıt] *v. t.* vernichten. **annihilation** [ənaıı'leıʃn] *n.* Vernichtung, *die*

anniversary [ænı'vɜ:sərı] *n.* Jahrestag, *der;* **wedding** ~: Hochzeitstag, *der*

annotate ['ænəteıt] *v. t.* kommentieren

announce [ə'naʊns] *v. t.* bekanntgeben; ansagen ⟨*Programm*⟩; *(over Tannoy etc.)* durchsagen; *(in newspaper)* anzeigen ⟨*Heirat usw.*⟩. **announcement** *n.* Bekanntgabe, *die; (over Tannoy etc.)* Durchsage, *die; (in newspaper)* Anzeige, *die.* **announcer** *n.* Ansager, *der*/Ansagerin, *die*

annoy [ə'nɔı] *v. t.* **a)** ärgern; **b)** *(harass)* schikanieren. **annoyance** [ə'nɔıəns] *n.* Verärgerung, *die; (nuisance)* Plage, *die.* **annoyed** [ə'nɔıd] *adj.* **be** ~ [at *or* with sb./sth.] ärgerlich [auf *od.* über jmdn./über etw.] sein; **he got very** ~: er hat sich darüber sehr geärgert. **annoying** *adj.* ärgerlich; lästig ⟨*Gewohnheit, Person*⟩

annual ['ænjʊəl] **1.** *adj.* **a)** *(reckoned by the year)* Jahres-; ~ **rainfall** jährliche Regenmenge; **b)** *(recurring yearly)* [all]jährlich ⟨*Ereignis, Feier*⟩; Jahres-⟨*bericht, -hauptversammlung*⟩. **2.** *n.* **a)** Jahrbuch, *das; (of comic etc.)* Jahresalbum, *das;* **b)** *(plant)* einjährige Pflanze. **'annually** *adv.* jährlich

annul [ə'nʌl] *v. t., -ll-* annullieren; auflösen ⟨*Vertrag*⟩

anonymous [ə'nonıməs] *adj.* anonym

anorak ['ænəræk] *n.* Anorak, *der*

anorexia [ænə'reksıə] *n.* Anorexie, *die (Med.);* Magersucht, *die (volkst.)*

another [ə'nʌðə(r)] **1.** *pron.* **a)** *(an*

additional one) noch einer/eine/eins; ein weiterer/eine weitere/ein weiteres; **b)** *(counterpart)* wieder einer/eine/eins; **c)** *(a different one)* ein anderer/eine andere/ein anderes. **2.** *adj.* **a)** *(additional)* noch ein/eine; ein weiterer/eine weitere/ein weiteres; **after ~ six weeks** nach weiteren sechs Wochen; **b)** *(different)* ein anderer/eine andere/ein anderes

answer ['ɑ:nsə(r)] **1.** *n.* **a)** *(reply)* Antwort, *die* (**to** auf + *Akk.*); **b)** *(to problem)* Lösung, *die* (**to** Gen.); *(to calculation)* Ergebnis, *das*. **2.** *v. t.* **a)** beantworten ⟨*Brief, Frage*⟩; antworten auf (+ *Akk.*) ⟨*Frage, Hilferuf, Einladung, Inserat*⟩; eingehen auf (+ *Akk.*) ⟨*Angebot, Vorschlag*⟩; sich stellen zu ⟨*Beschuldigung*⟩; erhören ⟨*Gebet*⟩; erfüllen ⟨*Bitte, Wunsch*⟩; ~ sb. jmdm. antworten; **b)** ~ **the door/bell** an die Tür gehen. **3.** *v. i.* **a)** *(reply)* antworten; ~ **to sth.** sich zu etw. äußern; **b)** *(be responsible)* ~ **for sth.** für etw. die Verantwortung übernehmen; **c)** ~ **to a description** einer Beschreibung *(Dat.)* entsprechen. **answerable** ['ɑ:nsərəbl] *adj.* verantwortlich (**for** für; **to** *Dat.*). '**answering machine** *n.* Anrufbeantworter, *der*

ant [ænt] *n.* Ameise, *die*

antagonism [æn'tægənɪzm] *n.* Feindseligkeit, *die* (**towards, against** gegenüber). **antagonist** [æn'tægənɪst] *n.* Gegner, *der*/Gegnerin, *die*. **antagonistic** [æntægə'nɪstɪk] *adj.* feindlich. **antagonize** [æn'tægənaɪz] *v. t.* ~ **sb.** sich *(Dat.)* jmdn. zum Feind machen

antarctic [ænt'ɑ:ktɪk] **1.** *adj.* antarktisch. **2.** *n.* **the A~:** die Antarktis

antelope ['æntɪləʊp] *n.* Antilope, *die*

antenna [æn'tenə] *n.* **a)** *pl.* ~**e** [æn'teni:] *(Zool.)* Fühler, *der*; **b)** *pl.* ~**s** *(Amer.: aerial)* Antenne, *die*

anthem ['ænθəm] *n.* Chorgesang, *der*

anthology [æn'θɒlədʒɪ] *n.* Anthologie, *die*

anthropology [ænθrə'pɒlədʒɪ] *n.* Anthropologie, *die*

anti- [æntɪ] *pref.* anti-/Anti-. **anti-'aircraft** *adj. (Mil.)* Flugabwehr-; ~ **gun** Flak, *die*

antibiotic [æntɪbaɪ'ɒtɪk] *n.* Antibiotikum, *das*

antic ['æntɪk] *n. (trick)* Mätzchen, *das (ugs.); (of clown)* Possen, *der*

anticipate [æn'tɪsɪpeɪt] *v. t.* **a)** *(expect)* erwarten; *(foresee)* voraussehen; ~

trouble mit Ärger rechnen; **b)** *(consider before due time)* vorwegnehmen. **anticipation** [æntɪsɪ'peɪʃn] *n.* Erwartung, *die*

anti'climax *n.* Abstieg, *der*

anti'clockwise *adv., adj.* gegen den Uhrzeigersinn

anti'cyclone *n.* Hochdruckgebiet, *das*

antidote ['æntɪdəʊt] *n.* Gegenmittel, *das* (**for, against, to** gegen)

'**antifreeze** *n.* Frostschutzmittel, *das*

antiquated ['æntɪkweɪtɪd] *adj.* antiquiert; veraltet

antique [æn'ti:k] **1.** *adj.* antik ⟨*Möbel, Schmuck usw.*⟩. **2.** *n.* Antiquität, *die*; ~ **shop** Antiquitätenladen, *der*

antiquity [æn'tɪkwɪtɪ] *n.* Altertum, *das;* Antike, *die*

anti'septic 1. *adj.* antiseptisch. **2.** *n.* Antiseptikum, *das*

anti'social *adj.* asozial

antithesis [æn'tɪθəsɪs] *n., pl.* **antitheses** [æn'tɪθəsi:z] Gegenstück, *das* (**of, to** zu)

antler ['æntlə(r)] *n.* Geweihsprosse, *die;* |**pair of**| ~**s** Geweih, *das*

anvil ['ænvɪl] *n.* Amboß, *der*

anxiety [æŋ'zaɪətɪ] *n.* Angst, *die; (concern about future)* Sorge, *die* (**about** wegen)

anxious ['æŋkʃəs] *adj.* **a)** *(troubled)* besorgt (**about** um); **b)** *(eager)* sehnlich; **be ~ for sth.** sich nach etw. sehnen. '**anxiously** *adv.* **a)** besorgt; **b)** *(eagerly)* sehnsüchtig

any ['enɪ] **1.** *adj.* **a)** *(some)* [irgend]ein/eine; **not ~:** kein/keine; **have you ~ wool/wine?** haben Sie Wolle/Wein?; **b)** *(one)* ein/eine; **c)** *(all, every)* jeder/jede/jedes; |**at**| ~ **time** jederzeit; **d)** *(whichever)* jeder/jede/jedes [beliebige]; **choose ~** |**one**| **book/~ books you like** suchen Sie sich *(Dat.)* irgendein Buch/irgendwelche Bücher aus. **2.** *pron.* **a)** *(some) in condit., interrog., or neg. sentence (replacing sing. n.)* einer/eine/ein[e]s; *(replacing collect. n.)* welcher/welche/welches; *(replacing pl. n.)* welche; **not ~:** keiner/keine/kein[e]s/ *Pl.* keine; **without ~:** ohne; **b)** *(no matter which)* irgendeiner/irgendeine/irgendein[e]s/irgendwelche *Pl.* **3.** *adv.* **do you feel ~ better today?** fühlen Sie sich heute [etwas] besser?; **if it gets ~ colder** wenn es noch kälter wird; **I can't wait ~ longer** ich kann nicht [mehr] länger warten

'**anybody** *n. & pron.* **a)** *(whoever)* je-

der; **b)** *(somebody)* [irgend]jemand; *after neg.* niemand
'anyhow *adv.* **a)** *see* **anyway; b)** *(haphazardly)* irgendwie
'anyone *see* **anybody**
'anything 1. *n. & pron.* **a)** *(whatever thing)* was [immer]; alles, was; **b)** *(something)* irgend etwas; *after neg.* nichts; **c)** *(a thing of any kind)* alles. **2.** *adv.* **not ~ like as ... as** keineswegs so ... wie
'anyway *adv.* **a)** *(in any case, besides)* sowieso; **b)** *(at any rate)* jedenfalls
'anywhere *adv.* **a)** *(in any place)* *(wherever)* überall, wo; wo [immer]; *(somewhere)* irgendwo; **not ~ near as ... as** *(coll.)* nicht annähernd so ... wie; **b)** *(to any place)* *(wherever)* wohin [auch immer]; *(somewhere)* irgendwohin
apart [ə'pɑːt] *adv.* **a)** *(separately)* getrennt; **~ from ...:** außer ...; **b)** *(into pieces)* auseinander
apartheid [ə'pɑːteit] *n.* Apartheid, *die*
apartment [ə'pɑːtmənt] *n.* **a)** *(room)* Apartment, *das;* **b)** *(Amer.: flat)* Wohnung, *die*
apathetic [æpə'θetɪk] *adj.* apathisch **(about** gegenüber**)**
apathy ['æpəθɪ] *n.* Apathie, *die* **(about** gegenüber**)**
ape [eɪp] **1.** *n.* [Menschen]affe, *der.* **2.** *v. t.* nachahmen
aperitif [əperɪ'tiːf] *n.* Aperitif, *der*
aperture ['æpətʃə(r)] *n.* Öffnung, *die*
apex ['eɪpeks] *n.* Spitze, *die*
aphrodisiac [æfrə'dɪzɪæk] *n.* Aphrodisiakum, *das*
apiece [ə'piːs] *adv.* je; **they cost a penny ~:** sie kosten einen Penny das Stück
apologetic [əpɒlə'dʒetɪk] *adj.* entschuldigend; **be ~:** sich entschuldigen
apologize [ə'pɒlədʒaɪz] *v. i.* sich entschuldigen **(to** bei**)**
apology [ə'pɒlədʒɪ] *n.* Entschuldigung, *die;* **make an ~:** sich entschuldigen **(to** bei**)**
apoplectic [æpə'plektɪk] *adj.* apoplektisch; **~ fit** Schlaganfall, *der*
apostle [ə'pɒsl] *n.* Apostel, *der*
apostrophe [ə'pɒstrəfɪ] *n.* Apostroph, *der;* Auslassungszeichen, *das*
appal *(Amer.:* **appall)** [ə'pɔːl] *v. t.,* -**ll**- entsetzen. **ap'palling** *adj.* entsetzlich
apparatus [æpə'reɪtəs] *n.* *(equipment)* Gerät, *das; (gymnastic ~)* Geräte *Pl.; (machinery, lit. or fig.)* Apparat, *der;* **a piece of ~:** ein Gerät

apparent [ə'pærənt] *adj.* **a)** *(clear)* offensichtlich; offenbar ⟨*Bedeutung, Wahrheit*⟩; **b)** *(seeming)* scheinbar. **ap'parently** *adv.* **a)** *(clearly)* offensichtlich; **b)** *(seemingly)* scheinbar
apparition [æpə'rɪʃn] *n.* [Geister]erscheinung, *die*
appeal [ə'piːl] **1.** *v. i.* **a)** *(Law etc.)* Einspruch einlegen; **b)** *(make earnest request)* **~ to sb. for sth./to do sth.** jmdn. um etw. ersuchen/jmdn. ersuchen, etw. zu tun; **c)** *(address oneself)* **~ to sb./sth.** an jmdn./etw. appellieren; **d)** *(be attractive)* **~ to sb.** jmdm. zusagen. **2.** *n.* **a)** *(Law etc.)* Einspruch, *der* **(to** bei**);** *(to higher court)* Berufung, *die* **(to** bei**); b)** *(request)* Appell, *der;* **an ~ to sb. for sth.** eine Bitte an jmdn. um etw.; **c)** *(attraction)* Reiz, *der.* **ap'pealing** *adj.* **a)** *(imploring)* flehend; **b)** *(attractive)* ansprechend; verlockend ⟨*Idee*⟩
appear [ə'pɪə(r)] *v. i.* **a)** *(become visible, arrive)* erscheinen; ⟨*Licht, Mond:*⟩ auftauchen; *(present oneself)* auftreten; **b)** *(occur)* vorkommen; **c)** *(seem)* **~ [to be] ...:** scheinen ... [zu sein]. **ap'pearance** [ə'pɪərəns] *n.* **a)** *(becoming visible)* Auftauchen, *das; (arrival)* Erscheinen, *das; (of performer etc.)* Auftritt, *der;* **b)** *(look)* Äußere, *das;* **to all ~s** allem Anschein nach; **c)** *(semblance)* Anschein, *der;* **d)** *(occurrence)* Vorkommen, *das*
appease [ə'piːz] *v. t.* besänftigen; *(Polit.)* beschwichtigen
append [ə'pend] *v. t.* anhängen **(to** an + *Akk.*); *(add)* anfügen **(+ *Dat.*). ap'pendage** [ə'pendɪdʒ] *n.* Anhängsel, *das; (addition)* Anhang, *der*
appendicitis [əpendɪ'saɪtɪs] *n.* Blinddarmentzündung, *die*
appendix [ə'pendɪks] *n., pl.* **appendices** [ə'pendɪsiːz] *or* -**es a)** Anhang, *der* **(to** zu**); b)** *(Anat.)* Blinddarm, *der*
appetite ['æpɪtaɪt] *n.* **a)** Appetit, *der* **(for** auf + *Akk.*); **b)** *(fig.)* Verlangen, *das* **(for** nach**). appetizer** ['æpɪtaɪzə(r)] *n.* Appetitanreger, *der.* **appetizing** ['æpɪtaɪzɪŋ] *adj.* appetitlich
applaud [ə'plɔːd] **1.** *v. i.* applaudieren; [Beifall] klatschen. **2.** *v. t.* applaudieren **(+ *Dat.*). applause** [ə'plɔːz] *n.* Beifall, *der;* Applaus, *der*
apple ['æpl] *n.* Apfel, *der*
appliance [ə'plaɪəns] *n.* Gerät, *das*
applicable [ə'plɪkəbl] *adj.* **a)** anwendbar **(to** auf + *Akk.*); **b)** *(appropriate)* geeignet; zutreffend ⟨*Fragebogenteil*⟩

applicant ['æplıkənt] *n.* Bewerber, *der*/Bewerberin, *die* (for um); *(claimant)* Antragsteller, *der*/-stellerin, *die*

application [æplı'keıʃn] *n.* **a)** *(request)* Bewerbung, *die* (for um); *(for passport, licence, etc.)* Antrag, *der* (for auf + *Akk.*); ~ **form** Antragsformular, *das;* **b)** *(putting)* Auftragen, *das* (to auf + *Akk.*); **c)** *(use)* Anwendung, *die*

apply [ə'plaı] **1.** *v.t.* **a)** auftragen ⟨*Creme, Farbe*⟩ (to auf + *Akk.*); **b)** *(make use of)* anwenden. **2.** *v.i.* **a)** *(have relevance)* zutreffen (to auf + *Akk.*); **b)** ~ |to sb.| for sth. [jmdn.] um etw. bitten; *(for passport etc.)* [bei jmdm.] etw. beantragen; *(for job)* sich [bei jmdm.] um etw. bewerben

appoint [ə'pɔɪnt] *v.t.* **a)** *(fix)* bestimmen; festlegen ⟨*Zeitpunkt, Ort*⟩; **b)** *(to job)* einstellen; *(to office)* ernennen. **ap'pointment** *n.* **a)** *(to job)* Einstellung, *die;* *(to office)* Ernennung, *die* (as zum/zur); **b)** *(job)* Stelle, *die;* *(arrangement)* Termin, *der;* **make an ~ with sb.** sich *(Dat.)* von jmdm. einen Termin geben lassen; **by ~:** nach Anmeldung

appreciable [ə'priːʃəbl] *adj.* **a)** *(perceptible)* nennenswert ⟨*Unterschied, Einfluß*⟩; spürbar ⟨*Veränderung, Wirkung*⟩; merklich ⟨*Verringerung, Anstieg*⟩; **b)** *(considerable)* beträchtlich. **appreciably** [ə'priːʃəblɪ] *adv.* **a)** *(perceptibly)* spürbar ⟨*verändern*⟩; merklich ⟨*sich unterscheiden*⟩; **b)** *(considerably)* beträchtlich

appreciate [ə'priːʃɪeɪt] **1.** *v.t.* **a)** *([correctly] estimate)* [richtig] einschätzen; *(understand)* verstehen; *(be aware of)* sich *(Dat.)* bewußt sein (+ *Gen.*); *(be grateful for)* schätzen; *(enjoy)* genießen. **2.** *v.i.* im Wert steigen. **appreciation** [əpriːʃɪ'eɪʃn] *n.* **a)** *([correct] estimation)* [richtige] Einschätzung; *(understanding)* Verständnis, *das* (of für); *(awareness)* Bewußtsein, *das;* **b)** *(gratefulness)* Dankbarkeit, *die; (enjoyment)* Gefallen, *das* (of an + *Dat.*). **appreciative** [ə'priːʃətɪv] *adj.* *(grateful)* dankbar (of für); *(approving)* anerkennend

apprehend [æprı'hend] *v.t.* **a)** *(arrest)* festnehmen; **b)** *(understand)* erfassen. **apprehension** [æprı'henʃn] *n.* Besorgnis, *die.* **apprehensive** [æprı'hensɪv] *adj.* besorgt

apprentice [ə'prentıs] *n.* Lehrling, *der* (to bei). **ap'prenticeship** *n.*

(training) Lehre, *die; (learning period)* Lehrzeit, *die*

approach [ə'prəʊtʃ] **1.** *v.i.* sich nähern; *(in time)* nahen. **2.** *v.t.* **a)** *(come near to)* sich nähern (+ *Dat.*); **b)** *(approximate to)* nahekommen (+ *Dat.*); **c)** *(appeal to)* sich wenden an (+ *Akk.*). **3.** *n.* **a)** [Heran]nahen, *das;* **b)** *(approximation)* Annäherung, *die* (to an + *Akk.*); **c)** *(appeal)* Herantreten, *das* (to an + *Akk.*); **d)** *(access)* Zugang, *der; (road)* Zufahrtsstraße, *die.* **approachable** [ə'prəʊtʃəbl] *adj.* **a)** *(friendly)* umgänglich; **b)** *(accessible)* zugänglich

appropriate **1.** [ə'prəʊprıət] *adj.* geeignet (to, for für). **2.** [ə'prəʊprıeɪt] *v.t.* sich *(Dat.)* aneignen. **appropriately** [ə'prəʊprıətlı] *adv.* gebührend; passend ⟨*gekleidet, genannt*⟩

approval [ə'pruːvl] *n.* **a)** *(sanctioning)* Genehmigung, *die; (of proposal)* Billigung, *die; (agreement)* Zustimmung, *die;* **b) on ~** *(Commerc.)* zur Probe

approve [ə'pruːv] **1.** *v.t.* **a)** *(sanction)* genehmigen ⟨*Plan, Projekt*⟩; billigen ⟨*Vorschlag*⟩; **b)** *(find good)* gutheißen. **2.** *v.i.* **of** billigen; zustimmen (+ *Dat.*) ⟨*Plan*⟩. **approving** [ə'pruːvɪŋ] *adj.* zustimmend ⟨*Worte*⟩; anerkennend ⟨*Blicke*⟩

approximate [ə'prɒksımət] *adj.* ungefähr *attr.* **ap'proximately** *adv.* ungefähr. **approximation** [əprɒksı'meɪʃn] *n.* **a)** Annäherung, *die* (to an + *Dat.*); **b)** *(estimate)* Annäherungswert, *der*

Apr. *abbr.* April Apr.

apricot ['eıprıkɒt] *n.* Aprikose, *die*

April ['eıprəl] *n.* April, *der;* ~ **fool** April[s]narr, *der; see also* **August**

apron ['eıprən] *n.* Schürze, *die*

apt [æpt] *adj.* **a)** *(suitable)* passend; treffend ⟨*Bemerkung*⟩; **b) be ~ to do sth.** dazu neigen, etw. zu tun

aptitude ['æptıtjuːd] *n.* Begabung, *die* **'aptly** *adv.* passend

aqualung ['ækwəlʌŋ] *n.* Tauchgerät, *das*

aquarium [ə'kweərıəm] *n., pl.* ~**s** *or* **aquaria** [ə'kweərıə] Aquarium, *das*

Aquarius [ə'kweərıəs] *n.* der Wassermann

aquatic [ə'kwætık] *adj.* aquatisch; Wasser-; ~ **plant** Wasserpflanze, *die*

aqueduct ['ækwıdʌkt] *n.* Aquädukt, *der od. das*

Arab ['ærəb] **1.** *adj.* arabisch. **2.** *n.* Araber, *der*/Araberin, *die*

Arabian [ə'reɪbɪən] **1.** adj. arabisch. **2.** n. Araber, der/Araberin, die
Arabic ['ærəbɪk] **1.** adj. arabisch. **2.** n. Arabisch, das; see also **English 2 a**
arbitrary ['ɑːbɪtrərɪ] adj. willkürlich
arbitrate ['ɑːbɪtreɪt] **1.** v. t. schlichten ⟨Streit⟩. **2.** v. i. ~ [upon sth.] [in einer Sache] vermitteln. **arbitration** [ɑːbɪ'treɪʃn] n. Vermittlung, die; (in industry) Schlichtung, die; **arbitrator** ['ɑːbɪtreɪtə(r)] n. Vermittler, der; (in industry) Schlichter, der
arc [ɑːk] n. [Kreis]bogen, der
arcade [ɑː'keɪd] n. Arkade, die
arch [ɑːtʃ] **1.** n. Bogen, der; (of foot) Wölbung, die. **2.** v. t. beugen ⟨Rücken⟩; ~ its **back** ⟨Katze:⟩ einen Buckel machen
arch- pref. Erz-
archaeological [ɑːkɪə'lɒdʒɪkl] adj. archäologisch
archaeologist [ɑːkɪ'ɒlədʒɪst] n. Archäologe, der/Archäologin, die
archaeology [ɑːkɪ'ɒlədʒɪ] n. Archäologie, die
archaic [ɑː'keɪɪk] adj. veraltet
arch'bishop n. Erzbischof, der
archeology etc. (Amer.) see **archaeology** etc.
archery ['ɑːtʃərɪ] n. Bogenschießen, das
archetype ['ɑːkɪtaɪp] n. (original) Urfassung, die; (typical specimen) Prototyp, der
architect ['ɑːkɪtekt] n. Architekt, der/Architektin, die
architectural [ɑːkɪ'tektʃərl] adj. architektonisch
architecture ['ɑːkɪtektʃə(r)] n. Architektur, die
archive ['ɑːkaɪv] **1.** n., usu. in pl. Archiv, das. **2.** v. t. archivieren
arctic ['ɑːktɪk] **1.** adj. arktisch; A~ Circle nördlicher Polarkreis; A~ Ocean Nordpolarmeer, das. **2.** n. the A~: die Arktis
ardent ['ɑːdənt] adj. leidenschaftlich; brennend ⟨Wunsch⟩; (eager) begeistert
ardor (Amer.), **ardour** (Brit.) ['ɑːdə(r)] n. Leidenschaft, die
arduous ['ɑːdjʊəs] adj. anstrengend
are see **be**
area ['eərɪə] n. a) (surface measure) Fläche, die; Flächeninhalt, der; b) (region) Gelände, das; (of wood, marsh, desert) Gebiet, das; (of city, country) Gegend, die; **parking/picnic** ~: Park-/Picknickplatz, der; c) (subject field) Gebiet, das

arena [ə'riːnə] n. Arena, die
aren't [ɑːnt] (coll.) = **are not;** see **be**
Argentina [ɑːdʒən'tiːnə] pr. n. Argentinien (das). **Argentinian** [ɑːdʒən'tɪnɪən] **1.** adj. argentinisch. **2.** n. Argentinier, der/Argentinierin, die
arguable ['ɑːgjʊəbl] adj. (questionable) fragwürdig. **arguably** ['ɑːgjʊəblɪ] adv. möglicherweise
argue ['ɑːgjuː] **1.** v. t. a) (maintain) ~ that ...: die Ansicht vertreten, daß ...; b) (with reasoning) darlegen ⟨Grund, Standpunkt⟩. **2.** v. i. ~ with sb. mit jmdm. streiten; ~ for/against sth. für/gegen etw. eintreten; ~ about sth. sich über/um etw. (Akk.) streiten. **argument** ['ɑːgjʊmənt] n. a) (reason) Begründung, die; ~s for/against sth. Argumente für/gegen etw.; b) (reasoning process) Argumentieren, das; c) (disagreement, quarrel) Auseinandersetzung, die. **argumentative** [ɑːgjʊ'mentətɪv] adj. widerspruchsfreudig
arid ['ærɪd] adj. trocken
Aries ['eəriːz] n. der Widder
arise [ə'raɪz] v. i., **arose** [ə'rəʊz], **arisen** [ə'rɪzn] a) (originate) entstehen; b) (present itself) auftreten; ⟨Gelegenheit:⟩ sich bieten; c) (result) ~ from or out of sth. von etw. herrühren
aristocracy [ærɪ'stɒkrəsɪ] n. Aristokratie, die
aristocrat ['ærɪstəkræt] n. Aristokrat, der/Aristokratin, die. **aristocratic** [ærɪstə'krætɪk] adj. aristokratisch
arithmetic [ə'rɪθmətɪk] n. Arithmetik, die
¹arm [ɑːm] n. Arm, der
²arm 1. n. a) usu. in pl. (weapon) Waffe, die; up in ~s (fig.) in Harnisch (about wegen); b) in pl. (heraldic device) Wappen, das. **2.** v. t. bewaffnen
arm: ~**band** n. Armbinde, die; ~**chair** n. Sessel, der
armed [ɑːmd] adj. bewaffnet; ~ forces Streitkräfte Pl.
armistice ['ɑːmɪstɪs] n. Waffenstillstand, der
armor (Amer.), **armour** (Brit.) ['ɑːmə(r)] n. a) (Hist.) Rüstung, die; b) (steel plates) Panzerung, die
'armpit n. Achselhöhle, die
army ['ɑːmɪ] n. Heer, das; join the ~: zum Militär gehen
aroma [ə'rəʊmə] n. Duft, der. **aromatic** [ærə'mætɪk] adj. aromatisch
arose see **arise**
around [ə'raʊnd] **1.** adv. a) (on every

side) |**all**| ~ : überall; **b)** *(round)* herum; **c)** *(in various places)* **ask/look** ~ : herumfragen/sich umsehen. **2.** *prep.* **a) an** [... herum]; **b)** *(approximately)* ~ **3 o'clock gegen 3 Uhr; sth.** |**costing**| ~ £2 etw. für ungefähr 2 Pfund

arouse [ə'raʊz] *v. t.* **a)** *(awake)* [auf]-wecken; **b)** *(excite)* erregen; erwecken ⟨*Interesse, Begeisterung*⟩; ~ **suspicion** Verdacht erregen

arrange [ə'reɪndʒ] **1.** *v. t.* **a)** *(order)* anordnen; **b)** *(settle, agree)* ausmachen, vereinbaren ⟨*Termin*⟩; planen ⟨*Urlaub*⟩; **they ~d to meet the following day** sie verabredeten sich für den nächsten Tag. **2.** *v. i.* *(plan)* sorgen (**for** für). **ar'rangement** *n.* **a)** *(ordering, order)* Anordnung, *die;* **b)** *(settling, agreement)* Vereinbarung, *die;* **c)** *in pl. (plans)* Vorkehrungen; **make ~s** Vorkehrungen treffen

arrears [ə'rɪəz] *n. pl.* Schulden *Pl.;* **be in ~ with sth.** mit etw. im Rückstand sein

arrest [ə'rest] **1.** *v. t.* **a)** verhaften, *(temporarily)* festnehmen ⟨*Person*⟩; **b)** *(stop)* aufhalten. **2.** *n.* Verhaftung, *die;* **under ~ :** festgenommen

arrival [ə'raɪvl] *n.* Ankunft, *die;* **new ~s** Neuankömmlinge

arrive [ə'raɪv] *v. i.* **a)** ankommen; ~ **at a conclusion/an agreement** zu einem Schluß/einer Einigung kommen; **b)** ⟨*Stunde, Tag, Augenblick:*⟩ kommen

arrogance ['ærəgəns] *n.* Arroganz, *die*

arrogant ['ærəgənt] *adj.* arrogant

arrow ['ærəʊ] *n.* Pfeil, *der*

arse [ɑːs] *n. (coarse)* Arsch, *der (derb)*

arsenal ['ɑːsənl] *n.* Waffenlager, *das*

arsenic ['ɑːsənɪk] *n.* **a)** Arsenik, *das;* **b)** *(element)* Arsen, *das*

arson ['ɑːsn] *n.* Brandstiftung, *die.* **arsonist** ['ɑːsənɪst] *n.* Brandstifter, *der/* Brandstifterin, *die*

art [ɑːt] *n.* **a)** Kunst, *die;* **works of** ~ : Kunstwerke *Pl.;* ~ **college** *or* **school** Kunsthochschule, *die;* ~ **s and crafts** Kunsthandwerk, *das;* **b)** *in pl. (branch of study)* Geisteswissenschaften

artery ['ɑːtərɪ] *n. (Anat.)* Schlagader, *die;* Arterie, *die; (bes. fachspr.)*

artful ['ɑːtfl] *adj.* schlau

'**art gallery** *n.* Kunstgalerie, *die*

arthritic [ɑː'θrɪtɪk] *adj.* arthritisch.

arthritis [ɑː'θraɪtɪs] *n.* Arthritis, *die (fachspr.);* Gelenkentzündung, *die*

artichoke ['ɑːtɪtʃəʊk] *n.* |**globe**| ~ : Artischocke, *die*

article ['ɑːtɪkl] *n.* **a)** *(in magazine,*

newspaper; Ling.) Artikel, *der;* **b) an** ~ **of furniture/clothing** ein Möbel-/Kleidungsstück; **an** ~ **of value** ein Wertgegenstand

articulate [ɑː'tɪkjʊlət] *adj.* redegewandt; **be ~/not very** ~ : sich gut/nicht sehr gut ausdrücken [können]

articulated [ɑː'tɪkjʊleɪtɪd] *adj.* ~ '**lorry** Sattelzug, *der*

artificial [ɑːtɪ'fɪʃl] *adj.* **a)** künstlich; Kunst-; *(not real)* unecht; ~ **limb** Prothese, *die;* **b)** *(affected)* gekünstelt

artificial: ~ **insemi'nation** *n.* künstliche Besamung; ~ **in'telligence** *n.* künstliche Intelligenz; ~ **respi'ration** *n.* künstliche Beatmung

artillery [ɑː'tɪlərɪ] *n.* Artillerie, *die*

artisan ['ɑːtɪzn, ɑːtɪ'zæn] *n.* [Kunst]-handwerker, *der*

artist ['ɑːtɪst] *n.* Künstler, *der/* Künstlerin, *die*

artiste [ɑː'tiːst] *n.* Artist, *der/* Artistin, *die*

artistic [ɑː'tɪstɪk] *adj.* **a)** *(of art)* Kunst-; künstlerisch; **b)** *(naturally skilled in art)* künstlerisch veranlagt

'**artless** *adj.* arglos

as [əz, *stressed* æz] **1.** *adv., conj.* **a) he is as tall as I am** er ist so groß wie ich; **as quickly as you can/as possible** so schnell du kannst/wie möglich; **b)** *(though)* **small as he was** obwohl er klein war; **c)** *(however much)* **try as he might/would, he could not concentrate** sosehr er sich auch bemühte, er konnte sich nicht konzentrieren; **d)** *expr. manner* wie; **as you may already have heard,** ... : wie Sie vielleicht schon gehört haben, ...; **as it were** sozusagen; **e)** *expr. time* als; während; **as we climbed the stairs** als wir die Treppe hinaufgingen; **as we were talking** während wir uns unterhielten; **f)** *expr. reason* da. **2.** *prep.* **a)** *(in the function of)* als; **as an artist** als Künstler; **speaking as a mother** ... : als Mutter ...; **b)** *(like)* wie; **c) the same as** ... : der-/die-/dasselbe wie ...; **such as** wie zum Beispiel. **3. as for** ... : was ... angeht *od.* betrifft; **as** |**it**| **is** wie die Dinge liegen; **the place is untidy enough as it is** es ist hier [so] schon unordentlich genug; **as of** ... *(Amer.)* von ... an; **as to** hinsichtlich (+ *Gen.*); **as yet** bis jetzt; noch

asbestos [æz'bestɒs] *n.* Asbest, *der*

ascend [ə'send] **1.** *v. i.* **a)** *(go up)* hinaufsteigen; *(climb up)* hinaufklettern; *(by vehicle)* hinauffahren; **b)** *(rise)* aufsteigen; ⟨*Hubschrauber:*⟩ höhersteigen; **c)** *(slope upwards)* ⟨*Hügel, Straße:*⟩ ansteigen. **2.** *v. t.* **a)** *(go up)* hinaufsteigen ⟨*Treppe, Leiter, Berg*⟩; **b)** ~ **the throne** den Thron besteigen

A'scension Day *n.* Himmelfahrtstag, *der*

ascent [ə'sent] *n.* Aufstieg, *der*

ascertain [æsə'teɪn] *v. t.* feststellen; ermitteln ⟨*Fakten, Daten*⟩

ascribe [ə'skraɪb] *v. t.* zuschreiben (**to** *Dat.*)

¹ash [æʃ] *n. (tree)* Esche, *die*

²ash *n. (from fire etc.)* Asche, *die*

ashamed [ə'ʃeɪmd] *adj.* beschämt; **be ~:** sich schämen (**of** wegen)

ashen ['æʃn] *adj.* aschfahl ⟨*Gesicht*⟩

ashore [ə'ʃɔ:(r)] *adv.* an Land

'ash-tray *n.* Aschenbecher, *der*

Ash 'Wednesday *n.* Aschermittwoch, *der*

Asia ['eɪʃə] *pr. n.* Asien *(das)*. **Asian** ['eɪʃən] **1.** *adj.* asiatisch. **2.** *n.* Asiat, *der/*Asiatin, *die*

aside [ə'saɪd] *adv.* beiseite; zur Seite

ask [ɑ:sk] **1.** *v. t.* **a)** fragen; ~ **sb.** |**sth.**| jmdn. [nach etw.] fragen; **b)** *(seek to obtain)* ~ **sth.** um etw. bitten; **how much are you ~ing for that car?** wieviel verlangen Sie für das Auto?; ~ **sb. to do sth.** jmdn. [darum] bitten, etw. zu tun; **c)** *(invite)* einladen. **2.** *v. i.* ~ **after sb./sth.** nach jmdm./etw. fragen; ~ **for sth./sb.** etw./jmdn. verlangen

askance [ə'skæns, ə'skɑ:ns] *adv.* **look ~ at sb.** jmdn. befremdet ansehen

askew [ə'skju:] *adv., pred. adj.* schief

asleep [ə'sli:p] *pred. adj.* schlafend; **be/lie ~:** schlafen; **fall ~:** einschlafen

asparagus [ə'spærəgəs] *n.* Spargel, *der*

aspect ['æspekt] *n.* Aspekt, *der*

aspersion [ə'spɜ:ʃn] *n.* **cast ~s on sb./ sth.** jmdn./etw in den Schmutz ziehen

asphalt ['æsfælt] *n.* Asphalt, *der*

asphyxiate [æs'fɪksɪeɪt] *v. t. & i.* ersticken

aspiration [æspə'reɪʃn] *n.* Streben, *das*

aspire [ə'spaɪə(r)] *v. i.* ~ **to** *or* **after sth.** nach etw. streben

aspirin ['æspərɪn] *n.* Aspirin ⓦ, *das;* Kopfschmerztablette, *die*

ass [æs] *n.* Esel, *der*

assailant [ə'seɪlənt] *n.* Angreifer, *der/* Angreiferin, *die*

assassin [ə'sæsɪn] *n.* Mörder, *der/* Mörderin, *die*. **assassinate** [ə'sæsɪneɪt] *v. t.* ermorden; **be ~d** einem Attentat zum Opfer fallen. **assassina-**

tion [əsæsɪ'neɪʃn] *n.* Mord, *der* (**of an** + *Dat.*); ~ **attempt** Attentat, *das* (**on** auf + *Akk.*)

assault [ə'sɔ:lt] **1.** *n.* Angriff, *der;* *(fig.)* Anschlag, *der.* **2.** *v. t.* angreifen

assemble [ə'sembl] **1.** *v. t.* **a)** zusammentragen; zusammenrufen ⟨*Menschen*⟩; **b)** *(fit together)* zusammenbauen. **2.** *v. i.* sich versammeln. **assembly** [ə'semblɪ] *n.* **a)** *(meeting)* Versammlung, *die;* *(in school)* Morgenandacht, *die;* **b)** *(fitting together)* Zusammenbau, *der.* **as'sembly line** *n.* Fließband, *das*

assent [ə'sent] **1.** *v. i.* zustimmen (**to** *Dat.*). **2.** *n.* Zustimmung, *die*

assert [ə'sɜ:t] *v. t.* **a)** geltend machen; ~ **oneself** sich durchsetzen; **b)** *(declare)* behaupten; beteuern ⟨*Unschuld*⟩. **assertion** [ə'sɜ:ʃn] *n.* **a)** Geltendmachen, *das;* **b)** *(declaration)* Behauptung, *die.* **assertive** [ə'sɜ:tɪv] *adj.* energisch ⟨*Person*⟩; bestimmt ⟨*Ton, Verhalten*⟩

assess [ə'ses] *v. t.* einschätzen; festsetzen ⟨*Steuer*⟩ (**at** auf + *Akk.*). **as'sessment** *n.* **a)** Einschätzung, *die;* **b)** *(tax to be paid)* Steuerbescheid, *der*

asset ['æset] *n.* **a)** Vermögenswert, *der;* **b)** *(useful quality)* Vorzug, *der* (**to** für); *(person)* Stütze, *die;* *(thing)* Hilfe, *die*

assiduous [ə'sɪdjʊəs] *adj.* **a)** *(diligent)* eifrig; **b)** *(conscientious)* gewissenhaft

assign [ə'saɪn] *v. t.* **a)** *(allot)* zuweisen (**to** *Dat.*); **b)** *(appoint)* zuteilen; ~ **sb. to do sth.** jmdn. damit betrauen, etw. zu tun. **as'signment** *n.* **a)** *(allotment)* Zuweisung, *die;* *(appointment)* Zuteilung, *die;* **b)** *(task)* Aufgabe, *die*

assimilate [ə'sɪmɪleɪt] *v. t.* angleichen (**to, with** an + *Akk.*). **assimilation** [əsɪmɪ'leɪʃn] *n.* Angleichung, *die* (**to, with** an + *Akk.*)

assist [ə'sɪst] **1.** *v. t.* helfen (+ *Dat.*). **2.** *v. i.* helfen; ~ **with sth./in doing sth.** bei etw. helfen/helfen, etw. zu tun. **assistance** [ə'sɪstəns] *n.* Hilfe, *die.* **assistant** [ə'sɪstənt] *n. (helper)* Helfer, *der/*Helferin, *die;* *(subordinate)* Mitarbeiter, *der/*Mitarbeiterin, *die;* *(of professor, artist)* Assistent, *der/*Assistentin, *die;* *(in shop)* Verkäufer, *der/*Verkäuferin, *die*

associate 1. [ə'səʊʃɪət, ə'səʊsɪət] *n.* *(partner)* Partner, *der/*Partnerin, *die;* *(colleague)* Kollege, *der/*Kollegin, *die.* **2.** [ə'səʊʃɪeɪt, ə'səʊsɪeɪt] *v. t.* in Verbindung bringen; **be ~d in** Verbin-

dung stehen. **3.** [ə'səʊʃɪeɪt, ə'səʊsɪeɪt]
v. i. ~ **with sb.** mit jmdm. Umgang haben. **association** [əsəʊsɪ'eɪʃn] *n.* **a)**
(organization) Vereinigung, *die;* **b)**
(mental connection) Assoziation, *die;*
c) *(connection)* Verbindung, *die*
assorted [ə'sɔːtɪd] *adj.* gemischt
assortment [ə'sɔːtmənt] *n.* Sortiment,
das; **a good ~ of hats** [to choose from]
eine gute Auswahl an Hüten
assume [ə'sjuːm] *v. t.* **a)** voraussetzen;
assuming that ...: vorausgesetzt,
daß ...; **b)** *(undertake)* übernehmen
⟨*Amt, Pflichten*⟩; **c)** *(take on)* annehmen ⟨*Namen, Rolle*⟩. **assumption**
[ə'sʌmpʃn] *n.* Annahme, *die;* **going on
the ~ that ...:** vorausgesetzt, daß ...
assurance [ə'ʃʊərəns] *n.* **a)** Zusicherung, *die;* **b)** *(self-confidence)* Selbstsicherheit, *die*
assure [ə'ʃʊə(r)] *v. t.* **a)** versichern
(+ *Dat.*); **b)** *(convince)* ~ **sb./oneself**
jmdn./sich überzeugen; **c)** *(make certain or safe)* gewährleisten. **assured**
[ə'ʃʊəd] *adj.* gewährleistet ⟨*Erfolg*⟩; **be
~ of sth.** sich *(Dat.)* einer Sache *(Gen.)*
sicher sein
asterisk [æstərɪsk] *n.* Sternchen, *das*
astern [ə'stɜːn] *adv.* *(Naut., Aeronaut.)*
achtern; *(towards the rear)* achteraus
asteroid ['æstərɔɪd] *n.* Asteroid, *der*
asthma ['æsmə] *n.* Asthma, *das.*
asthmatic [æs'mætɪk] **1.** *adj.*
asthmatisch. **2.** *n.* Asthmatiker,
*der/*Asthmatikerin, *die*
astonish [ə'stɒnɪʃ] *v. t.* erstaunen.
a'stonishing *adj.* erstaunlich.
a'stonishment *n.* Erstaunen, *das*
astound [ə'staʊnd] *v. t.* verblüffen.
a'stounding *adj.* erstaunlich
astray [ə'streɪ] *adv.* **sth. goes ~** *(is mislaid)* etw. wird verlegt; *(is lost)* etw.
geht verloren; **go/lead ~** *(fig.)* in die
Irre gehen/führen
astride [ə'straɪd] **1.** *adv.* rittlings ⟨*sitzen*⟩. **2.** *prep.* rittlings auf (+ *Dat.*)
astringent [ə'strɪndʒənt] *adj.* scharf
astrologer [ə'strɒlədʒə(r)] *n.* Astrologe, *der/*Astrologin, *die*
astrological [æstrə'lɒdʒɪkl] *adj.*
astrologisch
astrology [ə'strɒlədʒɪ] *n.* Astrologie,
die
astronaut ['æstrənɔːt] *n.* Astronaut,
*der/*Astronautin, *die*
astronomer [ə'strɒnəmə(r)] *n.* Astronom, *der/*Astronomin, *die*
astronomical [æstrə'nɒmɪkl] *adj.*
astronomisch

astronomy [ə'strɒnəmɪ] *n.* Astronomie, *die*
astute [ə'stjuːt] *adj.* scharfsinnig
asylum [ə'saɪləm] *n.* Asyl, *das*
at [ət, *stressed* æt] *prep.* **a)** *expr. place*
an (+ *Dat.*); **at the station** am Bahnhof; **at the baker's/butcher's/grocer's** at
the chemist's in der Apotheke/Drogerie; **at the supermarket** im Supermarkt; **at the party** auf der Party; **at
the office/hotel** im Büro/Hotel; **at
Dover** in Dover; **b)** *expr. time* **at
Christmas** [zu *od.* an] Weihnachten;
at six o'clock um sechs Uhr; **at midnight** um Mitternacht; **at midday** am
Mittag; **at** [the age of] **40** mit 40; im Alter von 40; **at this/the moment** in diesem/im Augenblick *od.* Moment; **c)**
expr. price **at £2.50** [each] zu *od.* für
[je] 2,50 Pfund; **d)** *expr. speed* **at 30
m. p. h.** *etc.* mit dreißig Meilen pro
Stunde *usw.;* **e)** **at that** *(at that point)*
dabei; *(at that provocation)* daraufhin; *(moreover)* noch dazu.
ate *see* **eat**
atheism ['eɪθɪɪzm] *n.* Atheismus, *der.*
atheist ['eɪθɪɪst] *n.* Atheist, *der/*Atheistin, *die*
Athens ['æθɪnz] *pr. n.* Athen *(das)*
athlete ['æθliːt] *n.* Athlet, *der/*Athletin, *die; (runner, jumper)* Leichtathlet,
*der/*Leichtathletin, *die.* **athletic** [æθ-
'letɪk] *adj.* sportlich. **ath'letics** *n.*
Leichtathletik, *die*
Atlantic [ət'læntɪk] **1.** *adj.* atlantisch;
~ **Ocean** Atlantischer Ozean. **2.** *pr. n.*
Atlantik, *der*
atlas ['ætləs] *n.* Atlas, *der*
atmosphere ['ætməsfɪə(r)] *n.* Atmosphäre, *die.* **atmospheric** [ætməs-
'ferɪk] *adj.* atmosphärisch
atom ['ætəm] *n.* Atom, *das.* '**atom
bomb** *n.* Atombombe, *die*
atomic [ə'tɒmɪk] *adj.* Atom-.
atomizer ['ætəmaɪzə(r)] *n.* Zerstäuber,
der
atone [ə'təʊn] *v. i.* es wiedergutmachen; ~ **for sth.** etw. wiedergutmachen. **a'tonement** *n.* Buße, *die*
atrocious [ə'trəʊʃəs] *adj.* grauenhaft;
scheußlich ⟨*Wetter, Benehmen*⟩.
a'trociously *adv.* grauenhaft;
scheußlich ⟨*sich benehmen*⟩. **atrocity**
[ə'trɒsɪtɪ] *n.* **a)** *(wickedness)* Grauenhaftigkeit, *die;* **b)** *(deed)* Greueltat, *die*
attach [ə'tætʃ] *v. t.* **a)** *(fasten)* befestigen (**to** an + *Dat.*); **please find ~ed a
copy of the letter** beigeheftet ist eine

Kopie des Briefes; **b)** *(fig.)* ~ **importance to sth.** einer Sache *(Dat.)* Gewicht beimessen
attaché [ə'tæʃeɪ] *n.* Attaché, *der.* **at-
'taché case** *n.* Diplomatenkoffer,
der
attached [ə'tætʃt] *adj. (emotionally)*
be ~ **to sb./sth.** an jmdm./etw. hängen
at'tachment *n.* **a)** *(act or means of
fastening)* Befestigung, *die;* **b)** *(affection)* Anhänglichkeit, *die* **(to an +
Akk.); c)** *(accessory)* Zusatzgerät, *das*
attack [ə'tæk] **1.** *v.t.* **a)** angreifen;
(ambush, raid) überfallen; *(fig.:
criticize)* attackieren; **b)** *(affect)*
⟨*Krankheit:*⟩ befallen. **2.** *v.i.* angreifen. **3.** *n.* Angriff, *der;* *(ambush)* Überfall, *der;* *(fig.: criticism)* Attacke, *die;*
(of illness) Anfall, *der.* **at'tacker** *n.*
Angreifer, *der*/Angreiferin, *die*
attain [ə'teɪn] *v.t.* erreichen. **at'tainment** *n.* Verwirklichung, *die*
attempt [ə'tempt] **1.** *v.t.* versuchen. **2.**
n. Versuch, *der*
attend [ə'tend] **1.** *v.i.* **a)** *(give care and
thought)* aufpassen; *(apply oneself)* ~
to sth. *(deal with sth.)* sich um etw.
kümmern; **b)** *(be present)* anwesend
sein **(at bei). 2.** *v.t.* **a)** *(be present at)*
teilnehmen an (+ *Dat.*); *(go regularly
to)* besuchen; **b)** *(wait on)* bedienen
(+ *Dat.*); **c)** ⟨*Arzt:*⟩ behandeln. **attendance** [ə'tendəns] *n.* Anwesenheit, *die;* *(number of people)* Teilneh-
merzahl, *die.* **attendant** [ə'tendənt]
n. **a)** |**lavatory|** ~: Toilettenmann,
der/-frau, *die;* |**cloakroom|** ~: Garde-
robenmann, *der*/-frau, *die;* **museum**
~: Museumswärter, *der*/-wärterin,
die; **b)** *(member of entourage)* Begleiter, *der*/Begleiterin, *die*
attention [ə'tenʃn] **1.** *n.* **a)** Aufmerk-
samkeit, *die;* **attract |sb.'s|** ~: |jmdn.|
auf sich *(Akk.)* aufmerksam machen;
pay ~ **to sb./sth.** jmdn./etw. beachten;
pay ~! gib acht!; paß auf!; **hold sb.'s**
~: jmds. Interesse wachhalten; ~
Miss Jones *(on letter)* zu Händen [von]
Miss Jones; **b)** *(Mil.)* **stand to** ~: stillstehen. **2.** *int.* **a)** Achtung; **b)** *(Mil.)*
stillgestanden
attentive [ə'tentɪv] *adj.* aufmerksam
attic ['ætɪk] *n. (room)* Dachboden, *der;*
(habitable) Dachkammer, *die*
attire [ə'taɪə(r)] *n.* Kleidung, *die*
attitude ['ætɪtjuːd] *n.* **a)** Haltung, *die;*
b) *(mental* ~*)* Einstellung, *die*
attorney [ə'tɜːnɪ] *n.* **a)** Bevollmächtigte, *der/die;* **power of** ~: Vollmacht,

die; **b)** *(Amer.: lawyer)* [Rechts]anwalt,
der/-anwältin, *die*
attract [ə'trækt] *v.t.* **a)** *(draw)* anziehen; auf sich *(Akk.)* ziehen ⟨*Interesse,
Blick, Kritik*⟩; **b)** *(arouse pleasure in)*
anziehend wirken auf (+ *Akk.*); **c)**
(arouse interest in) reizen **(about an +
Dat.).** **attraction** [ə'trækʃn] *n.* **a)** An-
ziehung, *die;* *(force, lit. or fig.)* Anzie-
hung[skraft], *die;* **b)** *(fig.: thing that at-
tracts)* Attraktion, *die;* *(charm)* Verlockung, *die;* Reiz, *der.* **attractive**
[ə'træktɪv] *adj.* **a)** anziehend; **b)** *(fig.)*
attraktiv; reizvoll ⟨*Vorschlag, Möglichkeit, Idee*⟩
attribute 1. ['ætrɪbjuːt] *n.* Eigenschaft, *die.* **2.** [ə'trɪbjuːt] *v.t.* zuschreiben **(to Dat.).** **attributive** [ə'trɪbjʊtɪv]
adj. (Ling.) attributiv
aubergine ['əʊbəʒiːn] *n.* Aubergine,
die
auburn ['ɔːbən] *adj.* rötlichbraun
auction ['ɔːkʃn] **1.** *n.* Versteigerung,
die. **2.** *v.t.* versteigern. **auctioneer**
[ɔːkʃə'nɪə(r)] *n.* Auktionator, *der*/
Auktionatorin, *die*
audacious [ɔː'deɪʃəs] *adj.* **a)** *(daring)*
kühn; verwegen; **b)** *(impudent)* dreist
audacity [ɔː'dæsɪtɪ] *n.* **a)** *(daringness)*
Kühnheit, *die;* Verwegenheit, *die;* **b)**
(impudence) Dreistigkeit, *die*
audible ['ɔːdɪbl] *adj.* hörbar
audience ['ɔːdɪəns] *n.* **a)** Publikum,
das; **b)** *(formal interview)* Audienz, *die*
(with bei)
audio ['ɔːdɪəʊ] *adj.* Ton-. **'audio typist** *n.* Phonotypist, *der*/-typistin, *die.*
audio'visual *adj.* audiovisuell
audit ['ɔːdɪt] **1.** *n.* ~ |**of the accounts|**
Rechnungsprüfung, *die.* **2.** *v.t.* prüfen
audition [ɔː'dɪʃn] **1.** *n. (singing)* Vorsingen, *das;* *(dancing)* Vortanzen, *das;*
(acting) Vorsprechen, *das.* **2.** *v.i.*
(sing) vorsingen; *(dance)* vortanzen;
(act) vorsprechen. **3.** *v.t.* vorsingen/
vortanzen/vorsprechen lassen
auditor ['ɔːdɪtə(r)] *n.* Buchprüfer,
der/-prüferin, *die*
auditorium [ɔːdɪ'tɔːrɪəm] *n.* Zuschau-
erraum, *der*
Aug. *abbr.* **August** Aug.
augment [ɔːg'ment] *v.t.* verbessern
⟨*Einkommen*⟩; aufstocken ⟨*Fonds*⟩
augur ['ɔːgə(r)] **1.** *v.t.* bedeuten; versprechen ⟨*Erfolg*⟩. **2.** *v.i.* ~ **well/ill for
sth./sb.** ein gutes/schlechtes Zeichen
für etw./jmdn. sein
August ['ɔːgəst] *n.* August, *der;* **in** ~:
im August; **last/next** ~: letzten/näch-

sten August; **the first of/on the first of**
~: der erste/am ersten August
aunt [ɑ:nt] *n.* Tante, *die*
auntie, aunty [ˈɑ:ntɪ] *n. (coll.)* Tänt-
chen, *das; (with name)* Tante, *die*
au pair [əʊ ˈpeə(r)] *n.* Au-pair-Mäd-
chen, *das*
aura [ˈɔ:rə] *n.* Aura, *die*
auspices [ˈɔ:spɪsɪs] *n. pl.* **under the ~
of sb./sth.** unter jmds./einer Sache
Schirmherrschaft
auspicious [ɔ:ˈspɪʃəs] *adj.* günstig;
vielversprechend ⟨*Anfang*⟩
austere [ɒˈstɪə(r)] *adj.* **a)** *(strict, stern)*
streng; **b)** *(severely simple)* karg. **aus-
terity** [ɒˈsterɪtɪ] *n.* **a)** *(strictness)*
Strenge, *die;* **b)** *(severe simplicity)*
Kargheit, *die;* **c)** *(lack of luxuries)*
wirtschaftliche Einschränkung
Australia [ɒˈstreɪlɪə] *pr. n.* Australien
(das). **Australian** [ɒˈstreɪlɪən] **1.** *adj.*
australisch. **2.** *n.* Australier, *der*/Au-
stralierin, *die*
Austria [ˈɒstrɪə] *pr. n.* Österreich
(das). **Austrian** [ˈɒstrɪən] **1.** *adj.*
österreichisch. **2.** *n.* Österreicher,
der/Österreicherin, *die*
authentic [ɔ:ˈθentɪk] *adj.* authentisch.
authenticity [ɔ:θenˈtɪsɪtɪ] *n.* Authen-
tizität, *die*
author [ˈɔ:θə(r)] *n.* Autor, *der*/Auto-
rin, *die; (profession)* Schriftsteller,
der/Schriftstellerin, *die*
authoritarian [ɔ:θɒrɪˈteərɪən] **1.** *adj.*
autoritär. **2.** *n.* autoritäre Person
authoritative [ɔ:ˈθɒrɪtətɪv] *adj.* maß-
gebend; zuverlässig ⟨*Bericht, Information*⟩
authority [ɔ:ˈθɒrɪtɪ] *n.* **a)** Autorität,
die; **in ~:** verantwortlich; **b) the authorities** die Behörde[n]
authorization [ɔ:θəraɪˈzeɪʃn] *n.* Genehmigung, *die*
authorize [ˈɔ:θəraɪz] *v. t.* **a)** ermächtigen; bevollmächtigen; **b)** *(sanction)*
genehmigen
auto [ˈɔ:təʊ] *n., pl.* ~**s** *(Amer. coll.)* Auto, *das*
auto- [ɔ:təʊ] *in comb.* auto-/Auto-
autobioˈgraphical *adj.* autobiogra-
phisch
autobiˈography *n.* Autobiographie,
die
autocratic [ɔ:təˈkrætɪk] *adj.* autokra-
tisch
autograph [ˈɔ:təgrɑ:f] **1.** *n.* Auto-
gramm, *das.* **2.** *v. t.* signieren
automate [ˈɔ:təmeɪt] *v. t.* automatisieren

automatic [ɔ:təˈmætɪk] **1.** *adj.* automatisch. **2.** *n. (weapon)* automatische
Waffe; *(vehicle)* Fahrzeug mit Auto-
matikgetriebe. **automatically** [ɔ:tə-
ˈmætɪkəlɪ] *adv.* automatisch
automation [ɔ:təˈmeɪʃn] *n.* Automation, *die*
automobile [ˈɔ:təməbi:l] *n. (Amer.)*
Auto, *das*
autonomous [ɔ:ˈtɒnəməs] *adj.* auto-
nom. **autonomy** [ɔ:ˈtɒnəmɪ] *n.* Auto-
nomie, *die*
autopsy [ˈɔ:tɒpsɪ] *n.* Autopsie, *die*
autumn [ˈɔ:təm] *n.* Herbst, *der;* **in |the|
~:** im Herbst. **autumnal** [ɔ:ˈtʌmnl]
adj. herbstlich
auxiliary [ɔ:gˈzɪljərɪ] **1.** *adj.* Hilfs-. **2.**
n. **a)** Hilfskraft, *die;* **b)** *(Ling.)* Hilfsverb, *das*
avail [əˈveɪl] **1.** *n.* **be of no ~:** nichts
nützen; **to no ~:** vergebens. **2.** *v. refl.*
~ oneself of sth. von etw. Gebrauch
machen
available [əˈveɪləbl] *adj.* **a)** *(at one's
disposal)* verfügbar; **b)** *(obtainable)*
erhältlich; lieferbar ⟨*Waren*⟩
avalanche [ˈævəlɑ:nʃ] *n.* Lawine, *die*
avarice [ˈævərɪs] *n.* Geldgier, *die;*
Habsucht, *die.* **avaricious** [ævəˈrɪ-
ʃəs] *adj.* geldgierig; habsüchtig
avenge [əˈvendʒ] *v. t.* rächen
avenue [ˈævənju:] *n.* Allee, *die; (fig.)*
Weg, *der* (to zu)
average [ˈævərɪdʒ] **1.** *n.* Durchschnitt,
der; **on ~:** im Durchschnitt; durchschnittlich. **2.** *adj.* durchschnittlich. **3.**
v. t. **a)** *(find the ~ of)* den Durchschnitt ermitteln von; **b)** *(amount on ~
to)* durchschnittlich betragen. **4.** *v. i.*
~ out at im Durchschnitt betragen
averse [əˈvɜ:s] *adj.* **be ~ to sth.** einer
Sache *(Dat.)* abgeneigt sein. **aversion** [əˈvɜ:ʃn] *n.* Abneigung, *die* (to
gegen)
avert [əˈvɜ:t] *v. t.* abwenden; verhüten
⟨*Unfall*⟩
aviary [ˈeɪvɪərɪ] *n.* Vogelhaus, *das*
aviation [eɪvɪˈeɪʃn] *n.* Luftfahrt, *die*
avid [ˈævɪd] *adj.* *(enthusiastic)* begeistert; **be ~ for sth.** *(eager, greedy)* begierig auf etw. *(Akk.)* sein
avocado [ævəˈkɑ:dəʊ] *n., pl.* ~**s:** ~
|pear| Avocado[birne], *die*
avoid [əˈvɔɪd] *v. t.* **a)** meiden ⟨*Ort*⟩; ~ **a
cyclist** einem Radfahrer ausweichen;
~ **the boss when he's in a temper** geh
dem Chef aus dem Weg, wenn er
schlechte Laune hat; **b)** *(refrain from,
escape)* vermeiden. **avoidable** [əˈvɔɪ-

backbone

dəbl] *adj.* vermeidbar. **avoidance** [ə'vɔɪdəns] *n.* Vermeidung, *die*

await [ə'weɪt] *v. t.* erwarten

awake [ə'weɪk] **1.** *v. i.,* awoke [ə'wəʊk], awoken [ə'wəʊkn] erwachen. **2.** *v. t.,* awoke, awoken wecken. **3.** *pred. adj.* wach; **wide ~:** hellwach

awaken [ə'weɪkn] *v. t. & i.* (*esp. fig.*) *see* **awake** 1, 2

award [ə'wɔːd] **1.** *v. t.* verleihen ⟨*Preis, Auszeichnung*⟩; zusprechen ⟨*Sorgerecht, Entschädigung*⟩; gewähren ⟨*Zahlung, Gehaltserhöhung*⟩. **2.** *n.* (*prize*) Auszeichnung, *die*

aware [ə'weə(r)] *adj.* **be ~ of sth.** sich (*Dat.*) einer Sache (*Gen.*) bewußt sein; **be ~ that ...:** sich (*Dat.*) [dessen] bewußt sein, daß ... **a'wareness** *n.* Bewußtsein, *das*

awash [ə'wɒʃ] *adj.* **be ~** (*flooded*) unter Wasser stehen

away [ə'weɪ] **1.** *adv.* **a)** (*at a distance*) entfernt; **play ~** (*Sport*) auswärts spielen; **b)** (*to a distance*) weg; fort; **c)** (*absent*) nicht da. **2.** *adj.* (*Sport*) auswärts präd.; Auswärts-

awe [ɔː] *n.* Ehrfurcht, *die* (**of** vor + *Dat.*)

awful ['ɔːfl] *adj.,* **'awfully** *adv.* furchtbar

awkward ['ɔːkwəd] *adj.* **a)** (*difficult to use*) ungünstig; **be ~ to use** unhandlich sein; **b)** (*clumsy*) unbeholfen; **c)** (*embarrassing*) peinlich; **d)** (*difficult*) schwierig; ungünstig ⟨*Zeitpunkt*⟩

awning ['ɔːnɪŋ] *n.* (*on house*) Markise, *die;* (*of tent*) Vordach, *das*

awoke, awoken *see* **awake**

awry [ə'raɪ] *adv.* schief; **go ~** (*fig.*) schiefgehen (*ugs.*); ⟨*Plan:*⟩ fehlschlagen

axe [æks] *n.* Axt, *die;* Beil, *das*

axis ['æksɪs] *n., pl.* axes ['æksiːz] Achse, *die*

axle ['æksl] *n.* Achse, *die*

B

B, b [biː] *n.* B, b, *das*

BA *abbr.* Bachelor of Arts

babble ['bæbl] *v. i.* **a)** (*talk incoherently*) stammeln; **b)** (*talk foolishly*)

[dumm] schwatzen; **c)** ⟨*Bach:*⟩ plätschern

baboon [bə'buːn] *n.* Pavian, *der*

baby ['beɪbɪ] *n.* **a)** Baby, *das;* **have a ~/be going to have a ~:** ein Kind bekommen; **b)** (*childish person*) **be a ~:** sich wie ein kleines Kind benehmen. **'baby-carriage** *n.* (*Amer.*) Kinderwagen, *der*

'babyish *adj.* kindlich ⟨*Aussehen*⟩; kindisch ⟨*Benehmen, Person*⟩

baby: ~-minder *n.* Tagesmutter, *die;* **~-sit** *v. i.,* forms as **sit** 1 babysitten (*ugs.*); auf das Kind/die Kinder aufpassen; **~-sitter** *n.* Babysitter, *der*

bachelor ['bætʃələ(r)] *n.* **a)** Junggeselle, *der;* **b)** (*Univ.*) **B~ of Arts/Science** Bakkalaureus der philosophischen Fakultät/der Naturwissenschaften

back [bæk] **1.** *n.* **a)** (*of person, animal*) Rücken, *der;* (*of house, cheque*) Rückseite, *die;* (*of vehicle*) Heck, *das;* (*inside car*) Rücksitz, *der;* **stand ~ to ~:** Rücken an Rücken stehen; **~ to front** verkehrt rum; **turn one's ~ on sb.** jmdm. den Rücken zuwenden; (*fig.*) jmdn. im Stich lassen; **turn one's ~ on sth.** (*fig.*) sich um etw. nicht kümmern; **get** *or* **put sb.'s ~ up** (*fig.*) jmdn. wütend machen; **be glad to see the ~ of sb./sth.** (*fig.*) froh sein, jmdn./etw. nicht mehr sehen zu müssen; **have one's ~ to the wall** (*fig.*) mit dem Rücken zur Wand stehen; **put one's ~ into sth.** (*fig.*) sich für etw. mit allen Kräften einsetzen; **with the ~ of one's hand** mit dem Handrücken; **at the ~** [**of the book**] hinten [im Buch]; **b)** (*Sport: player*) Verteidiger, *der.* **2.** *adj.* hinter-. **3.** *adv.* zurück; **two miles ~:** vor zwei Meilen; **~ and forth** hin und her; **there and ~:** hin und zurück; **a week/month ~:** vor einer Woche/vor einem Monat. **4.** *v. t.* **a)** (*assist*) unterstützen; **b)** (*bet on*) wetten auf; setzen auf (+ *Akk.*); **c)** zurücksetzen [mit] ⟨*Fahrzeug*⟩. **5.** *v. i.* zurücksetzen; **~ into/out of sth.** rückwärts in etw. (*Akk.*)/aus etw. fahren; **~ on to sth.** hinten an etw. (*Akk.*) grenzen. **back 'down** *v. i.* nachgeben. **back 'out** *v. i.* rückwärts herausfahren; **~ out of sth.** (*fig.*) von etw. zurücktreten. **back 'up** *v. t.* unterstützen; untermauern ⟨*Anspruch, These*⟩

back: ~ache *n.* Rückenschmerzen *Pl.;* **~-bencher** [bæk'bentʃə(r)] *n.* (*Brit. Parl.*) [einfacher] Abgeordneter/ [einfache] Abgeordnete; **~bone** *n.*

Rückgrat, *das;* ~**chat** *n. (coll.)* [freche] Widerrede; ~**date** *v. t.* zurückdatieren (to auf + *Akk.*); ~ '**door** *n.* Hintertür, *die*
'**backer** *n.* Geldgeber, *der*
back: ~-'**fire** *v. i.* knallen; *(fig.)* fehlschlagen; it ~**fired on me/him** *etc.* der Schuß ging nach hinten los *(ugs.);* ~**ground** *n.* Hintergrund, *der; (social status)* Herkunft, *die;* ~**hand** *(Tennis etc.)* 1. *adj.* Rückhand-; 2. *n.* Rückhand, *die;* ~-'**handed** *adj.* **a)** *(Tennis etc.)* Rückhand-; **b)** *(fig.)* indirekt; zweifelhaft ⟨*Kompliment*⟩; ~'**hander** *n. (sl.: bribe)* Schmiergeld, *das*
'**backing** *n. (support)* Unterstützung, *die*
back: ~**lash** *n. (fig.)* Gegenreaktion, *die;* ~**log** *n.* Rückstand, *der;* ~ **number** *n.* alte Nummer; ~-**pedal** *v. i.* die Pedale rückwärts treten; ~ '**seat** *n.* Rücksitz, *der;* ~**side** *n.* Hinterteil, *das (ugs.);* ~'**stage** *adv.* go ~**stage** hinter die Bühne gehen; ~ **street** *n.* kleine Seitenstraße; ~**stroke** *n.* Rückenschwimmen, *das*
backward ['bækwəd] 1. *adj.* **a)** rückwärts gerichtet; Rückwärts-; **b)** *(reluctant, shy)* zurückhaltend; **c)** *(underdeveloped)* rückständig ⟨*Land, Region*⟩. 2. *adv. see* **backwards**
backwards ['bækwədz] *adv.* **a)** nach hinten; **the child fell** |over| ~ **into the water** das Kind fiel rückwärts ins Wasser; **bend** *or* **lean over** ~ **to do sth.** *(fig. coll.)* sich zerreißen, um etw. zu tun *(ugs.);* **b)** *(oppositely to normal direction)* rückwärts; ~ **and forwards** hin und her
back: ~**water** *n. (fig.)* Kaff, *das (ugs.);* ~ '**yard** *n.* Hinterhof, *der*
bacon ['beɪkn] *n.* [Frühstücks]speck, *der*
bacterium [bæk'tɪərɪəm] *n., pl.* **bacteria** [bæk'tɪərɪə] Bakterie, *die*
bad [bæd] *adj.,* **worse** [wɜːs], **worst** [wɜːst] **a)** schlecht; *(rotten)* schlecht, verdorben ⟨*Fleisch, Fisch, Essen*⟩; faul ⟨*Ei, Apfel*⟩;**not** ~ *(coll.)* nicht schlecht; nicht übel; **b)** *(naughty)* ungezogen, böse ⟨*Kind, Hund*⟩; **c)** *(offensive)* |use| ~ **language** Kraftausdrücke [benutzen]; **d)** *(regretful)* feel ~ **about sth.** etw. bedauern; **I feel** ~ **about him** ich habe seinetwegen ein schlechtes Gewissen; **e)** *(serious)* schlimm ⟨*Sturz, Krise*⟩; schwer ⟨*Fehler, Krankheit, Unfall*⟩; **f)** *(Commerc.)* **a** ~ **debt** eine uneinbringliche Schuld

bade *see* **bid 1 b**
badge [bædʒ] *n.* Abzeichen, *das*
badger ['bædʒə(r)] *n.* Dachs, *der*
'**badly** *adv.,* **worse** [wɜːs], **worst** [wɜːst] **a)** schlecht; **b)** schwer ⟨*verletzt, beschädigt*⟩; **c)** *(urgently)* dringend
bad-mannered [bæd'mænəd] *adj.* **be** ~: schlechte Manieren haben
badminton ['bædmɪntən] *n.* Federball, *der; (als Sport)* Badminton, *das*
bad-tempered [bæd'tempəd] *adj.* griesgrämig
baffle ['bæfl] *v. t.* ~ **sb.** jmdm. unverständlich sein. **baffling** ['bæflɪŋ] *adj.* rätselhaft
bag [bæg] 1. *n.* Tasche, *die; (sack)* Sack, *der; (hand~)* Handtasche, *die; (plastic ~)* Beutel, *der; (small paper ~)* Tüte, *die;* ~**s of** *(sl.: large amount)* jede de Menge. 2. *v. t.,* -**gg**-: **a)** in Säcke/Beutel/Tüten füllen; **b)** *(claim possession of)* sich *(Dat.)* schnappen *(ugs.)*
baggage ['bægɪdʒ] *n.* Gepäck, *das.* '**baggage reclaim** *n.* Gepäckausgabe, *die*
baggy ['bægɪ] *adj.* weit [geschnitten] ⟨*Kleid, Hose*⟩; *(through long use)* ausgebeult ⟨*Hose*⟩
'**bagpipe[s]** *n.* Dudelsack, *der*
Bahamas [bə'hɑːməz] *pr. n. pl.* **the** ~: die Bahamas
[1]**bail** [beɪl] 1. *n.* Kaution, *die;* **be** |out| **on** ~: gegen Kaution auf freiem Fuß sein. 2. *v. t.* ~ **sb. out** jmdn. gegen Kaution freibekommen; *(fig.)* jmdm. aus der Klemme helfen *(ugs.)*
[2]**bail** *v. t. (scoop)* ~ |out| ausschöpfen. '**bail out** *v. i. (Pilot:)* abspringen
bailiff ['beɪlɪf] *n.* ≈ Gerichtsvollzieher, *der*
bait [beɪt] 1. *v. t.* mit einem Köder versehen. 2. *n.* Köder, *der*
bake [beɪk] *v. t. & i.* backen. '**baker** *n.* Bäcker, *der.* **bakery** ['beɪkərɪ] *n.* Bäckerei, *die*
baking: ~-**powder** *n.* Backpulver, *das;* ~-**tin** *n.* Backform, *die;* ~-**tray** *n.* Kuchenblech, *das*
balance ['bæləns] 1. *n.* **a)** *(instrument)* Waage, *die;* **b)** *(fig.)* **be** *or* **hang in the** ~: in der Schwebe sein; **c)** *(steady position)* Gleichgewicht, *das;* **keep/lose one's** ~: das Gleichgewicht halten/verlieren; *(fig.)* sein Gleichgewicht bewahren/verlieren; **strike a** ~ **between** *(fig.)* den Mittelweg finden zwischen (+ *Dat.*); **d)** *(Bookk.: difference)* Bilanz, *die; (state of bank account)* Kontostand, *der;* **on** ~ *(fig.)* al-

les in allem; ~ **sheet** Bilanz, *die;* e) *(Econ.)* ~ **of payments** Zahlungsbilanz, *die;* f) *(remainder)* Rest, *der.* 2. *v.t.* a) *(weigh up)* abwägen; b) *(bring into or keep in* ~*)* balancieren; auswuchten ⟨*Rad*⟩; c) *(equal, neutralize)* ausgleichen; ~ **each other, be** ~**d** sich *(Dat.)* die Waage halten. **'balanced** *adj.* ausgewogen; ausgeglichen ⟨*Person, Team, Gemüt*⟩

balcony ['bælkənɪ] *n.* Balkon, *der*
bald [bɔːld] *adj.* kahl ⟨*Kopf*⟩; kahlköpfig, glatzköpfig ⟨*Person*⟩
bale [beɪl] *n.* Ballen, *der*
balk [bɔːk] 1. *v.t.* **they were** ~**ed in their plan** ihr Plan wurde blockiert. 2. *v.i.* sich sträuben (**at** gegen)
Balkan ['bɔːlkn] 1. *adj.* Balkan-. 2. *n. pl.* **the** ~**s** der Balkan
¹ball [bɔːl] *n.* a) Ball, *der;* (*Billiards etc.,* *Croquet*) Kugel, *die;* **be on the** ~ *(coll.: be alert)* auf Zack sein *(ugs.);* b) *(of wool, string, fluff, etc.)* Knäuel, *das*
²ball *n. (dance)* Ball, *der*
ballad ['bæləd] *n.* Ballade, *die*
ballast ['bæləst] *n.* Ballast, *der*
ball-'bearing *n.* Kugellager, *das*
ballerina [bælə'riːnə] *n.* Ballerina, *die*
ballet ['bæleɪ] *n.* Ballett, *das;* ~ **dancer** Ballettänzer, *der*/Ballettänzerin, *die*
balloon [bə'luːn] *n.* a) Ballon, *der;* **hot-air** ~: Heißluftballon, *der;* b) *(toy)* Luftballon, *der*
ballot ['bælət] *n.* Abstimmung, *die;* [secret] ~: geheime Wahl
ball: ~**pen,** ~**point 'pen** *ns.* Kugelschreiber, *der;* ~**room** *n.* Tanzsaal, *der*
balm [bɑːm] *n.* Balsam, *der*
balmy ['bɑːmɪ] *adj. (mild)* mild
Baltic ['bɔːltɪk] 1. *pr. n.* Ostsee, *die.* 2. *adj.* ~ **Sea** Ostsee, *die*
balustrade [bælə'streɪd] *n.* Balustrade, *die*
bamboo [bæm'buː] *n.* Bambus, *der*
ban [bæn] 1. *v.t.,* -nn- verbieten; ~ **sb. from doing sth.** jmdm. verbieten, etw. zu tun. 2. *n.* Verbot, *das*
banal [bə'nɑːl] *adj.* banal
banana [bə'nɑːnə] *n.* Banane, *die*
band [bænd] 1. *n.* a) Band, *das;* **a** ~ **of light/colour** ein Streifen Licht/Farbe; b) *(range of values)* Bandbreite, *die;* c) *(organized group)* Gruppe, *die; (of robbers, outlaws, etc.)* Bande, *die;* d) *(Mus.)* [Musik]kapelle, *die; (pop group, jazz* ~) Band, *die.* 2. *v.i.* ~ **together [with sb.]** sich [mit jmdm.] zusammenschließen

bandage ['bændɪdʒ] 1. *n.* Verband, *der; (as support)* Bandage, *die.* 2. *v.t.* verbinden; bandagieren ⟨*[verstauchtes] Gelenk usw.*⟩
bandit ['bændɪt] *n.* Bandit, *der*
band: ~**stand** *n.* Musiktribüne, *die;* ~**wagon** *n.* **climb** *or* **jump on** [to] **the** ~**wagon** *(fig.)* auf den fahrenden Zug aufspringen *(fig.)*
¹bandy ['bændɪ] *v.t.* **they were** ~**ing words/insults** sie stritten sich/beschimpften sich gegenseitig
²bandy *adj.* krumm; **he is** ~-**legged** er hat O-Beine *(ugs.)*
bang [bæŋ] 1. *v.t.* knallen *(ugs.);* schlagen; zuknallen *(ugs.)* ⟨*Tür, Fenster, Deckel*⟩; ~ **one's head on sth.** mit dem Kopf an etw. *(Akk.)* knallen *(ugs.).* 2. *v.i. (strike)* ~ [**against sth.**] [gegen etw.] knallen *(ugs.);* ~ **shut** ⟨*Tür:*⟩ zuknallen *(ugs.).* 3. *n.* a) *(blow)* Schlag, *der;* b) *(noise)* Knall, *der.* 4. *adv.* **go** ~ ⟨*Gewehr, Feuerwerkskörper:*⟩ krachen
'banger *n. (sl.)* a) *(sausage)* Würstchen, *das;* b) *(firework)* Kracher, *der (ugs.);* c) *(car)* Klapperkiste, *die (ugs.)*
bangle ['bæŋgl] *n.* Armreif, *der*
banish ['bænɪʃ] *v.t.* verbannen (**from** aus)
banister ['bænɪstə(r)] *n.* [Treppen]geländer, *das*
banjo ['bændʒəʊ] *n., pl.* ~**s** *or* ~**es** Banjo, *das*
¹bank [bæŋk] *n.* a) *(slope)* Böschung, *die;* b) *(of river)* Ufer, *das*
²bank 1. *n. (Finance)* Bank, *die.* 2. *v.i.* ~ **at/with** ...: ein Konto haben bei ...; ~ **on sth.** *(fig.)* auf etw. *(Akk.)* zählen. 3. *v.t.* zur Bank bringen
bank: ~ **account** *n.* Bankkonto, *das;* ~ **card** *n.* Scheckkarte, *die;* ~ **clerk** *n.* Bankangestellte, *der/die*
'banker *n.* Bankier, *der*
bank 'holiday *n. (Brit.)* Feiertag, *der*
'banking *n.* Bankwesen, *das*
bank: ~ **manager** *n.* Zweigstellenleiter/-leiterin [einer/der Bank]; ~**note** *n.* Banknote, *die*
bankrupt ['bæŋkrʌpt] 1. *n.* Bankrotteur, *der.* 2. *adj.* **go** ~: Bankrott machen. 3. *v.t.* bankrott machen. **bankruptcy** ['bæŋkrʌptsɪ] *n.* Konkurs, *der;* Bankrott, *der*
banner ['bænə(r)] *n.* Banner, *das; (on two poles)* Spruchband, *das*
banns [bænz] *n. pl.* Aufgebot, *das*
banquet ['bæŋkwɪt] *n.* Bankett, *das*
baptism ['bæptɪzm] *n.* Taufe, *die*

Baptist ['bæptɪst] *n.* Baptist, *der*/Baptistin, *die*

baptize [bæp'taɪz] *v.t.* taufen

bar [bɑ:(r)] **1.** *n.* **a)** Stange, *die; (shorter, thinner also)* Stab, *der; (of cage, prison)* Gitterstab, *der;* **a ~ of soap** ein Stück Seife; **a ~ of chocolate** eine Tafel Schokolade; **b)** *(for refreshment)* Bar, *die; (counter)* Theke, *die.* **2.** *v.t.,* -rr-: **a)** *(fasten)* verriegeln; **b) ~ sb.'s way** jmdm. den Weg versperren; **c)** *(prohibit, hinder)* verbieten; **~ sb. from doing sth.** jmdn. daran hindern, etw. zu tun. **3.** *prep.* abgesehen von; **~ none** ohne Einschränkung

barb [bɑ:b] *n.* Widerhaken, *der*

barbarian [bɑ:'beərɪən] *n.* Barbar, *der*

barbaric [bɑ:'bærɪk] *adj.* barbarisch

barbarity [bɑ:'bærɪtɪ] *n.* Grausamkeit, *die*

barbecue ['bɑ:bɪkju:] **1.** *n.* **a)** *(party)* Grillparty, *die;* **b)** *(food)* Grillgericht, *das.* **2.** *v.t.* grillen

barbed wire [bɑ:bd 'waɪə(r)] *n.* Stacheldraht, *der*

barber ['bɑ:bə(r)] *n.* [Herren]friseur, *der*

'bar code *n.* Strichcode, *der*

bare [beə(r)] **1.** *adj.* nackt; *(leafless, unfurnished)* kahl; *(empty)* leer; äußerst ⟨*Notwendige*⟩; **do sth. with one's ~ hands** etw. mit den bloßen Händen tun. **2.** *v.t.* entblößen ⟨*Kopf, Arm, Bein*⟩; blecken ⟨*Zähne*⟩. **'barefaced** *adj. (fig.)* unverhüllt. **'barefoot 1.** *adj.* barfüßig. **2.** *adv.* barfuß

barely ['beəlɪ] *adv.* kaum; knapp ⟨*vermeiden, entkommen*⟩

bargain ['bɑ:gɪn] **1.** *n.* **a)** *(agreement)* Abmachung, *die;* **into the ~:** darüber hinaus; **b)** *(thing offered cheap)* günstiges Angebot; *(thing acquired cheaply)* guter Kauf. **2.** *v.i.* **a)** *(discuss)* handeln; **b) ~ for** *or* **on sth.** *(expect sth.)* mit etw. rechnen

barge [bɑ:dʒ] **1.** *n.* Kahn, *der.* **2.** *v.i.* **~ into sb.** jmdn. anrempeln; **~ in** *(intrude)* hineinplatzen/hereinplatzen *(ugs.)*

baritone ['bærɪtəʊn] **1.** *n.* Bariton, *der.* **2.** *adj.* Bariton-

¹bark [bɑ:k] *n. (of tree)* Rinde, *die*

²bark 1. *n. (of dog)* Bellen, *das.* **2.** *v.i.* bellen; **be ~ing up the wrong tree** auf dem Holzweg sein

barley ['bɑ:lɪ] *n.* Gerste, *die*

bar: ~maid *n. (Brit.)* Bardame, *die;* **~man** ['bɑ:mən] *n., pl.* **~men** ['bɑ:mən] Barmann, *der*

barmy ['bɑ:mɪ] *adj. (sl.: crazy)* bescheuert *(salopp)*

barn [bɑ:n] *n. (Brit.: for grain etc.)* Scheune, *die; (Amer.: for animals)* Stall, *der*

barnacle ['bɑ:nəkl] *n.* Rankenfüßer, *der*

barometer [bə'rɒmɪtə(r)] *n.* Barometer, *das*

baron ['bærn] *n.* Baron, *der;* Freiherr, *der;* **baroness** ['bærənɪs] *n.* Baronin, *die;* Freifrau, *die*

baroque [bə'rɒk, bə'rəʊk] **1.** *n.* Barock, *das.* **2.** *adj.* barock

barracks ['bærəks] *n. pl.* Kaserne, *die*

barrage ['bærɑ:ʒ] *n. (Mil.)* Sperrfeuer, *das;* **a ~ of questions** ein Bombardement von Fragen

barrel ['bærl] *n.* **a)** Faß, *das;* **b)** *(of gun)* Lauf, *der*

barren ['bærn] *adj.* unfruchtbar

barricade [bærɪ'keɪd] **1.** *n.* Barrikade, *die.* **2.** *v.t.* verbarrikadieren

barrier ['bærɪə(r)] *n.* Barriere, *die; (at level crossing etc.)* Schranke, *die*

barring ['bɑ:rɪŋ] *prep.* außer im Falle (+ *Gen.*)

barrister ['bærɪstə(r)] *n. (Brit.)* ~|-at-law] Barrister, *der;* ≈ [Rechts]anwalt/ -anwältin vor höheren Gerichten

barrow ['bærəʊ] *n.* **a)** Karre, *die;* Karren, *der;* **b)** *see* **wheelbarrow**

barter ['bɑ:tə(r)] **1.** *v.t.* [ein]tauschen; **~ sth. for sth.** |else| etw. für *od.* gegen etw. [anderes] [ein]tauschen. **2.** *v.i.* Tauschhandel treiben. **3.** *n.* Tauschhandel, *der*

base [beɪs] **1.** *n.* **a)** *(of lamp, mountain)* Fuß, *der; (of cupboard, statue)* Sockel, *der; (fig.: support)* Basis, *die;* **b)** *(Mil.)* Basis, *die;* Stützpunkt, *der.* **2.** *v.t.* **a)** **be ~d on sth.** sich auf etw. *(Akk.)* gründen; **~ sth. on sth.** etw. auf etw. *(Dat.)* aufbauen; **b)** *in pass.* **be ~d in Paris** *(permanently)* in Paris sitzen; *(temporarily)* in Paris sein

'baseball *n.* Baseball, *der*

basement ['beɪsmənt] *n.* Untergeschoß, *das;* **a ~ flat** eine Kellerwohnung

bash [bæʃ] *v.t.* [heftig] schlagen

bashful ['bæʃfl] *adj.* schüchtern

basic ['beɪsɪk] *adj.* grundlegend; Grund⟨*prinzip, -bestandteil, -lohn, -gehalt usw.*⟩; Haupt⟨*problem, -grund, -sache*⟩; **be ~ to sth.** wesentlich für etw. sein. **basically** ['beɪsɪkəlɪ] *adv.* im Grunde; grundsätzlich ⟨*übereinstimmen*⟩; *(mainly)* hauptsächlich

basil ['bæzɪl] *n.* Basilikum, *das*

basin ['beɪsn] *n.* **a)** Becken, *das;* *(wash-~)* Waschbecken, *das;* *(bowl)* Schüssel, *die;* **b)** *(of river)* Becken, *das*

basis ['beɪsɪs] *n., pl.* **bases** ['beɪsiːz] Basis, *die;* Grundlage, *die*

bask [bɑːsk] *v. i.* sich [wohlig] wärmen

basket ['bɑːskɪt] *n.* Korb, *der.* '**basketball** *n.* Basketball, *der*

Basle [bɑːl] *pr. n.* Basel *(das)*

bass [beɪs] **1.** *n.* **a)** Baß, *der;* **b)** *(coll.)* *(double-~)* [Kontra]baß, *der;* *(~ guitar)* Baß, *der.* **2.** *adj.* Baß-. **bass guitar** *n.* Baßgitarre, *die*

bassoon [bə'suːn] *n.* Fagott, *das*

bastard ['bɑːstəd] **1.** *adj.* unehelich. **2.** *n.* **a)** uneheliches Kind; **b)** *(coll. derog.: person)* Schweinehund, *der (derb)*

baste ['beɪst] *v. t.* [mit Fett] begießen

bastion ['bæstɪən] *n.* Bastei, *die*

¹**bat** [bæt] *n.* *(Zool.)* Fledermaus, *die*

²**bat 1.** *n.* *(Sport)* Schlagholz, *das;* *(for table-tennis)* Schläger, *der;* **do sth. off one's own ~** *(fig.)* etw. auf eigene Faust tun. **2.** *v. t.,* -**tt**- schlagen

³**bat** *v. t.* **not ~ an eyelid** nicht mit der Wimper zucken

batch [bætʃ] *n.* **a)** *(of loaves)* Schub, *der;* **b)** *(of people)* Gruppe, *die;* *(of books, papers)* Stapel, *der*

bated ['beɪtɪd] *v. t.* **with ~ breath** mit angehaltenem Atem

bath [bɑːθ] **1.** *n., pl.* ~**s** [bɑːðz] **a)** Bad, *das;* **have** *or* **take a ~:** ein Bad nehmen; **b)** *(tub)* Badewanne, *die;* **room with ~:** Zimmer mit Bad; **c)** *usu. in pl.* *(building)* Bad, *das.* **2.** *v. t. & i.* baden. '**bath cubes** *n. pl.* Badesalz, *das*

bathe [beɪð] *v. t. & i.* baden. **bather** ['beɪðə(r)] *n.* Badende, *der/die.* '**bathing** ['beɪðɪŋ] *n.* Baden, *das.* '**bathing-costume,** '**bathing-suit** *ns.* Badeanzug, *der*

bath: ~-**mat** *n.* Badematte, *die;* ~-**room** *n.* Badezimmer, *das;* ~-**salts** *n. pl.* Badesalz, *das;* ~-**towel** *n.* Badetuch, *das;* ~-**tub** *see* **bath 1 b**

baton ['bætn] *n.* **a)** *(truncheon)* Schlagstock, *der;* **b)** *(Mus.)* Taktstock, *der*

batsman ['bætsmən] *n., pl.* **batsmen** ['bætsmən] Schlagmann, *der*

battalion [bə'tæljən] *n.* Bataillon, *das*

¹**batter** ['bætə(r)] *v. t.* *(strike)* einschlagen auf (+ *Akk.*)

²**batter** *n.* *(Cookery)* [Back]teig, *der*

battery ['bætərɪ] *n.* Batterie, *die*

battery: ~ **charger** *n.* Batterieladegerät, *das;* ~ '**farming** *n.* Batteriehaltung, *die;* ~ '**hen** *n.* Batteriehuhn, *das*

battle ['bætl] **1.** *n.* Schlacht, *die;* *(fig.)* Kampf, *der.* **2.** *v. i.* kämpfen

battle: ~-**axe** *n.* *(coll.: woman)* Schreckschraube, *die (ugs.);* ~-**field,** ~-**ground** *ns.* Schlachtfeld, *das*

battlements ['bætlmənts] *n. pl.* Zinnen *Pl.*

'**battleship** *n.* Schlachtschiff, *das*

batty ['bætɪ] *adj.* *(sl.)* bekloppt *(salopp)*

bauble ['bɔːbl] *n.* Flitter, *der*

baulk *see* **balk**

Bavaria [bə'veərɪə] *pr. n.* Bayern *(das).* **Bavarian** [bə'veərɪən] **1.** *adj.* bay[e]risch. **2.** *n.* Bayer, *der*/Bayerin, *die*

bawdy ['bɔːdɪ] *adj.* zweideutig; *(stronger)* obszön

¹**bay** [beɪ] *n.* *(of sea)* Bucht, *die*

²**bay** *n.* **a)** *(space in room)* Erker, *der;* **b)** |parking-|~: Stellplatz, *der*

³**bay n.** **hold** *or* **keep sb./sth. at ~:** sich *(Dat.)* jmdn./etw. vom Leib halten

bayonet ['beɪənɪt] *n.* Bajonett, *das*

bay 'window *n.* Erkerfenster, *das*

bazaar [bə'zɑː(r)] *n.* Basar, *der*

BBC *abbr.* **British Broadcasting Corporation** BBC, *die*

BC *abbr.* **before Christ** v. Chr.

be [biː] *v., pres. t.* **I** am [əm, *stressed* æm], **he is** [ɪz], **we are** [ə(r), *stressed* ɑː(r)]; *p. t.* **I was** [wəz, *stressed* wɒz], **we were** [wə(r), *stressed* wɜː(r)]; *pres. p.* **being** ['biːɪŋ]; *p. p.* **been** [bɪn, *stressed* biːn] **1.** *copula* **a)** sein; **she is a mother/an Italian/a teacher** sie ist Mutter/Italienerin/Lehrerin; **be sensible!** sei vernünftig!; **be ill/unwell** krank sein/sich nicht wohl fühlen; **I am well** es geht mir gut; **I am hot** mir ist heiß; **I am freezing** mich friert es; **how are you/is she?** wie geht's *(ugs.)*/geht es ihr?; **it is the 5th today** heute haben wir den Fünften; **who's that?** wer ist das?; **if I were you** an deiner Stelle; **it's hers** es ist ihrs; **b)** *(cost)* kosten; **how much are the eggs?** was kosten die Eier?; **two times three is six, two threes are six** zweimal drei ist *od.* sind sechs; **c)** *(constitute)* bilden. **2.** *v. i.* **a)** *(exist)* [vorhanden] sein; **there is/are ...:** es gibt ...; **for the time being** vorläufig; **be that as it may** wie dem auch sei; **b)** *(remain)* bleiben; **I shan't be a moment** ich komme sofort; **let it be** laß es sein; **let him/her be** laß ihn/sie in Ruhe; **c)** *(happen)* stattfinden; sein; **d)** *(go, come)* **be off with you!** geh/geht!; **I'm off home** ich gehe jetzt nach Hause; **she's from Australia** sie stammt *od.* ist aus Australien; **e)** *(go or come* *on visit)* sein; **have you [ever] been to London?** bist du schon einmal in London gewesen?; **has anyone been?** ist jemand dagewesen? 3. *v. aux.* **a)** *forming passive* werden; **the child was found** das Kind wurde gefunden; **German is spoken here** hier wird Deutsch gesprochen; **b)** *forming continuous tenses, active* **he is reading** er liest [gerade]; **I am leaving tomorrow** ich reise morgen [ab]; **the train was departing when I got there** der Zug fuhr gerade ab, als ich ankam; **c)** *forming continuous tenses, passive* **the house is/was being built** das Haus wird/wurde [gerade] gebaut; **d)** *expr. arrangement, obligation* **be to** sollen; **I am to go/to inform you** ich soll gehen/Sie unterrichten; **e)** *expr. destiny* **they were never to meet again** sie sollten sich nie wieder treffen; **f)** *expr. condition* **if I were to tell you that ...**: wenn ich dir sagen würde, daß ... **4. bride-/husband-to-be** zukünftige Braut/zukünftiger Ehemann

beach [biːtʃ] *n.* Strand, *der;* **on the ~:** am Strand. '**beach wear** *n.* Strandkleidung, *die*

beacon ['biːkn] *n.* Leuchtfeuer, *das; (Naut.)* Leuchtbake, *die*

bead [biːd] *n.* Perle, *die;* **~s** Perlen *Pl.;* Perlenkette, *die;* **~s of dew/sweat** Tau-/Schweißtropfen

beak [biːk] *n.* Schnabel, *der*

beaker ['biːkə(r)] *n.* Becher, *der*

beam [biːm] **1.** *n.* **a)** *(timber etc.)* Balken, *der;* **b)** *(ray etc.)* [Licht]strahl, *der.* **2.** *v. i.* **a)** *(shine)* strahlen; glänzen; **b)** *(smile)* strahlen; **~ at sb.** jmdn. anstrahlen

bean [biːn] *n.* Bohne, *die;* **full of ~s** *(fig. coll.)* putzmunter *(ugs.)*

¹**bear** [beə(r)] *n.* Bär, *der*

²**bear** **1.** *v. t.,* **bore** [bɔː(r)], **borne** [bɔːn] **a)** tragen; aufweisen ⟨*Spuren, Ähnlichkeit*⟩; tragen, führen ⟨*Namen, Titel*⟩; **~ some/little relation to sth.** einen gewissen/wenig Bezug zu etw. haben; **b)** *(endure, tolerate)* ertragen ⟨*Schmerz, Kummer*⟩; *with neg.* ertragen, aushalten ⟨*Schmerz*⟩; ausstehen ⟨*Geruch, Lärm*⟩; **c)** *(be fit for)* vertragen; **it will not ~ scrutiny** es hält einer Überprüfung nicht stand; **it does not ~ thinking about** daran darf man gar nicht denken; **d)** *(give birth to)* gebären ⟨*Kind, Junges*⟩. **2.** *v. i.,* **bore, borne: ~ left** ⟨*Person:*⟩ sich links halten; **the path ~s to the left** der Weg führt nach

links. **bear 'out** *v. t. (fig.)* bestätigen ⟨*Bericht, Erklärung*⟩; **~ sb. out** jmdm. recht geben. '**bear with** *v. t.* Nachsicht haben mit

bearable ['beərəbl] *adj.* erträglich

beard [biəd] *n.* Bart, *der.* '**bearded** *adj.* bärtig. **be ~:** einen Bart haben

'**bearer** *n. (carrier)* Träger, *der/*Trägerin, *die; (of message, cheque)* Überbringer, *der/*Überbringerin, *die*

'**bearing** *n.* **a)** *(behaviour)* Verhalten, *das;* **b)** *(relation)* Bezug, *der;* **have some/no ~ on sth.** relevant/irrelevant für etw. sein; **c)** *(Mech. Engin.)* Lager, *das;* **d)** *(compass ~)* Position, *die;* **take a compass ~:** den Kompaßkurs feststellen; **get one's ~s** sich orientieren; *(fig.)* sich zurechtfinden

beast [biːst] *n.* Tier, *das; (fig.: brutal person)* Bestie, *die.* '**beastly** *adj., adv. (coll.)* scheußlich

beat [biːt] **1.** *v. t.,* **beat, beaten** ['biːtn] schlagen; klopfen ⟨*Teppich*⟩; *(surpass)* brechen ⟨*Rekord*⟩; **hard to ~:** schwer zu schlagen; **it ~s me how/why ...**: es ist mir ein Rätsel wie/warum ...; **~ time** den Takt schlagen; **~ it!** *(sl.)* hau ab! *(ugs.); see also* **beaten 2. 2.** *v. i.,* **beat, beaten** schlagen ⟨on auf + *Akk.*⟩; ⟨*Regen, Hagel:*⟩ prasseln ⟨**against** gegen⟩. **3.** *n.* **a)** *(stroke, throbbing)* Schlagen, *das; (Mus.) (rhythm)* Takt, *der; (single ~)* Schlag, *der;* **b)** *(of policeman)* Runde, *die.* **beat 'off** *v. t.* abwehren ⟨*Angriff*⟩. **beat 'up** *v. t.* zusammenschlagen ⟨*Person*⟩

beaten ['biːtn] **1.** *see* **beat 1, 2. 2.** *adj.* **a)** **off the ~ track** weit abgelegen; **b)** gehämmert ⟨*Silber, Gold*⟩

'**beating** *n.* **a)** *(punishment)* **a ~:** Schläge *Pl.;* Prügel *Pl.;* **b)** *(defeat)* Niederlage, *die;* **take some/a lot of ~:** nicht leicht zu übertreffen sein

'**beat-up** *adj. (sl.)* ramponiert *(ugs.)*

beautiful ['bjuːtɪfl] *adj.* schön; wunderschön ⟨*Augen, Aussicht, Morgen*⟩

beautify ['bjuːtɪfaɪ] *v. t.* verschönen

beauty ['bjuːtɪ] *n.* Schönheit, *die; (beautiful feature)* Schöne, *das;* **the ~ of it** das Schöne daran

beauty: **~ parlour** *see* **~ salon;** **~ queen** *n.* Schönheitskönigin, *die;* **~ salon** *n.* Kosmetiksalon, *der;* **~ spot** *n.* Schönheitsfleck, *der; (place)* schönes Fleckchen [Erde]

beaver ['biːvə(r)] *n.* Biber, *der*

became *see* **become**

because [bɪ'kɒz] **1.** *conj.* weil. **2.** *adv.* **~ of** wegen (+ *Gen.*)

beckon ['bekn] *v. t. & i.* winken (**to sb.** jmdm.); *(fig.)* locken

become [bɪ'kʌm] **1.** *copula,* **became** [bɪ'keɪm], **become** werden; ~ **a politician** Politiker werden; ~ **a nuisance/ rule** zu einer Plage/zur Regel werden. **2.** *v. i.,* **became, become,** **what has** ~ **of him?** was ist aus ihm geworden? **3.** *v. t.,* **became, become** *(suit)* **sb.** jmdm. stehen

becoming [bɪ'kʌmɪŋ] *adj.* **a)** *(fitting)* schicklich *(geh.);* **b)** *(flattering)* vorteilhaft ⟨*Hut, Kleid, Frisur*⟩

bed [bed] *n.* **a)** Bett, *das; (without bedstead)* Lager, *ein* ~: im Bett; ~ **and breakfast** Zimmer mit Frühstück; **get out of/into** ~: aufstehen/ins Bett gehen; **go to** ~: ins Bett gehen; **put sb. to** ~: jmdm. ins Bett bringen; **b)** *(flat base)* Unterlage, *die; (of machine)* Bett, *das;* **c)** *(in garden)* Beet, *das;* **d)** *(of sea, lake)* Grund, *der; (of river)* Bett, *das.* '**bedclothes** *n. pl.* Bettzeug, *das.* **bedding** ['bedɪŋ] *n.* Matratze und Bettzeug

bedlam ['bedləm] *n., no indef. art.* Tumult, *der*

'**bedpan** *n.* Bettpfanne, *die*

bedraggled [bɪ'drægld] *adj. (soaked)* durchnäßt; *(with mud)* verdreckt

bed: ~**ridden** *adj.* bettlägerig; ~**room** *n.* Schlafzimmer, *das;* ~**side** *n.* Seite des Bettes, *die;* ~**side table/ lamp** Nachttisch, *der/*Nachttischlampe, *die;* ~-**sit,** ~-'**sitter** *ns. (coll.)* Wohnschlafzimmer, *das;* ~**spread** *n.* Tagesdecke, *die;* ~**stead** *n.* Bettgestell, *das;* ~**time** *n.* Schlafenszeit, *die;* **at** ~**time** vor dem Zubettgehen; **a** ~**time story** eine Gutenachtgeschichte

bee [biː] *n.* Biene, *die*

beech [biːtʃ] *n.* Buche, *die*

beef [biːf] **1.** *n.* **a)** Rindfleisch, *das;* **b)** *(coll.: muscles)* Muskeln. **2.** *v. t.* ~ **up** stärken. **beefburger** ['biːfbɜːgə(r)] *n.* Beefburger, *der*

bee: ~**hive** *n.* Bienenstock, *der;* ~-**keeper** *n.* Imker, *der/*Imkerin, *die;* ~-**keeping** *n.* Imkerei, *die;* ~-**line** *n.* **make a** ~-**line for sth./sb.** schnurstracks auf etw./jmdn. zustürzen

been *see* **be**

beer [bɪə(r)] *n.* Bier, *das*

beet [biːt] *n.* Rübe, *die*

beetle ['biːtl] *n.* Käfer, *der*

'**beetroot** *n.* rote Beete *od.* Rübe

before [bɪ'fɔː(r)] **1.** *adv.* **a)** *(of time)* vorher; *(already)* schon; **the day** ~: am Tag zuvor; **never** ~: noch nie; **b)** *(ahead in position)* vor[aus]. **2.** *prep. (of time; position)* vor (+ *Dat.*); *(direction)* vor (+ *Akk.*); **the day** ~ **yesterday** vorgestern; ~ **now/then** früher/vorher; ~ **Christ** vor Christus; ~ **leaving,** **he phoned** bevor er wegging, rief er an. **3.** *conj.* bevor. **be'forehand** *adv.* vorher; *(in anticipation)* im voraus

beg [beg] **1.** *v. t.,* -**gg**-: **a)** betteln um; **b)** *(ask earnestly for)* ~ **sth.** um etw. bitten. **2.** *v. i.,* -**gg**- betteln **(for** um)

began *see* **begin**

beggar ['begə(r)] *n.* **a)** Bettler, *der/*Bettlerin, *die;* **b)** *(coll.)* **poor** ~: armer Teufel

begin [bɪ'gɪn] **1.** *v. t.,* -**nn**-, **began** [bɪ'gæn], **begun** [bɪ'gʌn] ~ **sth.** [mit] etw. beginnen; ~ **doing** *or* **to do sth.** anfangen *od.* beginnen, etw. zu tun. **2.** *v. i.,* -**nn**-, **began, begun** anfangen; ~ [up]on **sth.** etw. anfangen. **be'ginner** *n.* Anfänger, *der/*Anfängerin, *die.* **be'ginning** *n.* Anfang, *der;* **at** *or* **in the** ~: am Anfang; **at the** ~ **of February/the month** Anfang Februar/des Monats; **from the** ~: von Anfang an

begrudge [bɪ'grʌdʒ] *v. t.* ~ **sb. sth.** jmdm. etw. mißgönnen; ~ **doing sth.** etw. ungern tun

begun *see* **begin**

behalf [bɪ'hɑːf] *n.* **on** *or (Amer.)* **in** ~ **of sb./sth.** für jmdn./etw.; *(more formally)* im Namen von jmdm./etw.

behave [bɪ'heɪv] **1.** *v. i.* sich verhalten; sich benehmen; **well-/ill-** *or* **badly** ~**d** brav/ungezogen. **2.** *v. refl.* ~ **oneself** sich benehmen. **behaviour** [bɪ'heɪvjə(r)] *n.* Verhalten, *das*

behead [bɪ'hed] *v. t.* enthaupten

behind [bɪ'haɪnd] **1.** *adv.* hinten; *(further back)* **be miles** ~: kilometerweit zurückliegen; **stay** ~: dableiben; **leave sb./sth.** ~: jmdn./etw. zurücklassen; **fall** ~: zurückbleiben; *(fig.)* **be in** Rückstand geraten; **be/get** ~ **with one's payments/rent** mit seinen Zahlungen/der Miete im Rückstand sein/in Rückstand geraten. **2.** *prep.* **a)** hinter (+ *Dat.*); **one** ~ **the other** hintereinander; **b)** *(towards rear of)* hinter (+ *Akk.*)

being ['biːɪŋ] *n.* **a)** *(existence)* Dasein, *das;* **in** ~: bestehend; **come into** ~: entstehen; **b)** *(person etc.)* Wesen, *das*

belated [bɪ'leɪtɪd] *adj.,* **be'latedly** *adv.* verspätet

belch [beltʃ] **1.** *v. i.* heftig aufstoßen; rülpsen *(ugs.).* **2.** *n.* Rülpser, *der (ugs.)*

belfry ['belfrɪ] *n.* Glockenturm, *der*
Belgian ['beldʒən] **1.** *n.* Belgier, *der*/Belgierin, *die.* **2.** *adj.* belgisch
Belgium ['beldʒəm] *pr. n.* Belgien *(das)*
belie [bɪ'laɪ] *v. t.,* **belying** [bɪ'laɪɪŋ] hinwegtäuschen über ⟨*Tatsachen, wahren Zustand*⟩; nicht erfüllen ⟨*Versprechen*⟩; nicht entsprechen ⟨*Vorstellung (Dat.)*⟩
belief [bɪ'liːf] *n.* **a)** Glaube, *der* (in an + *Akk.*); **in the ~ that ...:** in der Überzeugung, daß ...; **b)** *(Relig.)* Glaube[n], *der*
believable [bɪ'liːvəbl] *adj.* glaubhaft
believe [bɪ'liːv] **1.** *v. i.* glauben (in an + *Dat.*); *(have faith)* glauben (in an + *Akk.*) ⟨*Gott, Himmel usw.*⟩; I ~ so/not ich glaube schon/nicht. **2.** *v. t.* glauben; ~ sb. jmdm. glauben; **I don't ~ you** das glaube ich dir nicht; **make ~ that ...:** so tun, als ob ...
Belisha beacon [bəliːʃə 'biːkn] *n. (Brit.)* gelbes Blinklicht an Zebrastreifen
belittle [bɪ'lɪtl] *v. t.* herabsetzen
bell [bel] *n.* Glocke, *die; (door~)* Klingel, *die*
belligerent [bɪ'lɪdʒərənt] *adj.* kriegführend ⟨*Nation*⟩; streitlustig ⟨*Person*⟩
bellow ['beləʊ] **1.** *v. i.* brüllen. **2.** *v. t.* ~ |out| brüllen ⟨*Befehl*⟩
bellows ['beləʊz] *n. pl.* Blasebalg, *der*
belly ['belɪ] *n.* Bauch, *der.* **'bellyache** *n.* Bauchschmerzen *Pl.*
belong [bɪ'lɒŋ] *v. i.* ~ to sb./sth. jmdm./zu etw. gehören; ~ to a club einem Verein angehören; **where does this ~?** wo gehört das hin? **be'longings** *n. pl.* Habe, *die;* Sachen *Pl.*
beloved [bɪ'lʌvɪd] **1.** *adj.* geliebt. **2.** *n.* Geliebte, *der/die*
below [bɪ'ləʊ] **1.** *adv.* **a)** *(position)* unten; *(lower down)* darunter; **from ~:** von unten [herauf]; **b)** *(direction)* nach unten; hinunter. **2.** *prep.* unter *(position: + Dat.; direction: + Akk.)*
belt [belt] *n.* Gürtel, *der; (for tools, weapons, ammunition)* Gurt, *der; (of trees)* Streifen, *der.* **belt 'up** *v. i. (Brit. sl.)* die Klappe halten *(salopp)*
bemused [bɪ'mjuːzd] *adj.* verwirrt
bench [bentʃ] *n.* Bank, *die; (worktable)* Werkbank, *die*
bend [bend] **1.** *n.* Beuge, *die; (in road)* Kurve, *die.* **2.** *v. t.,* **bent** [bent] biegen; beugen ⟨*Arm, Knie*⟩; anwinkeln ⟨*Bein*⟩. **3.** *v. i.,* **bent** sich biegen; *(bow)* sich bücken. **bend 'down** *v. i.* sich

bücken. **bend 'over** *v. i.* sich nach vorn beugen
beneath [bɪ'niːθ] *prep.* **a)** *(unworthy of)* ~ sb., ~ sb.'s dignity unter jmds. Würde *(Dat.);* **b)** *(arch./literary: under)* unter *(+ Dat.)*
benefactor ['benɪfæktə(r)] *n.* Wohltäter, *der; (patron)* Gönner, *der*
beneficial [benɪ'fɪʃl] *adj.* nützlich; vorteilhaft ⟨*Einfluß*⟩
benefit ['benɪfɪt] **1.** *n.* **a)** Vorteil, *der;* **be of ~ to sb./sth.** jmdm./einer Sache von Nutzen sein; **have the ~ of** den Vorteil (+ *Gen.*) haben; **with the ~ of** mit Hilfe (+ *Gen.*); **for sb.'s ~:** in jmds. Interesse *(Dat.);* **b)** *(allowance)* Beihilfe, *die;* **unemployment ~:** Arbeitslosenunterstützung, *die.* **2.** *v. t.* nützen (+ *Dat.*). **3.** *v. i.* ~ by/from sth. von etw. profitieren
benevolent [bɪ'nevələnt] *adj.* **a)** gütig; **b)** wohltätig ⟨*Institution, Verein*⟩
benign [bɪ'naɪn] *adj.* gütig; *(Med.)* gutartig
bent [bent] **1.** *see* bend 2, 3. **2.** *n. (liking)* Neigung, *die* (for zu). **3. a)** *adj.* krumm; **b)** *(Brit. sl.: corrupt)* link *(salopp)*
bequeath [bɪ'kwiːð] *v. t.* ~ sth. to sb. jmdm. etw. hinterlassen. **bequest** [bɪ'kwest] *n.* Legat, *das* (to an + *Akk.*)
bereaved [bɪ'riːvd] *n.* the ~: der/die Hinterbliebene/die Hinterbliebenen
beret ['bereɪ] *n.* Baskenmütze, *die*
Berlin [bɜː'lɪn] *pr. n.* Berlin *(das)*
Berne [bɜːn] *pr. n.* Bern *(das)*
berry ['berɪ] *n.* Beere, *die*
berserk [bə'sɜːk] *adj.* rasend; **go ~:** durchdrehen *(ugs.)*
berth [bɜːθ] *n. (for ship)* Liegeplatz, *der; (sleeping-place) (in ship)* Koje, *die; (in train)* Schlafwagenbett, *das*
beside [bɪ'saɪd] *prep.* **a)** neben (+ *Dat.*); ~ the sea/lake am Meer/See; **b)** be ~ the point nichts damit zu tun haben; **c)** ~ oneself außer sich
besides [bɪ'saɪdz] **1.** *adv.* außerdem. **2.** *prep.* außer
besiege [bɪ'siːdʒ] *v. t.* belagern
best [best] **1.** *adj.* best...; **the ~ part of an hour** fast eine ganze Stunde. **2.** *adv.* am besten. **3.** *n.* the ~: der/die/das Beste; **do one's ~:** sein bestes tun; **make the ~ of** aus dem Beste daraus machen; **at ~:** bestenfalls. **best 'man** *n.* Trauzeuge, *der (des Bräutigams).* **best 'seller** *n.* Bestseller, *der*
bet [bet] **1.** *v. t. & i.,* -tt-, ~ *or* ~ted wetten; I ~ him £10 ich habe mit ihm um

10 Pfund gewettet; ~ **on sth.** auf etw. *(Akk.)* setzen. **2.** *n.* Wette, *die; (fig. coll.)* Tip, *der*

betray [bɪ'treɪ] *v.t.* verraten **(to an** + *Akk.*). **betrayal** [bɪ'treɪəl] *n.* Verrat, *der*

better ['betə(r)] **1.** *adj.* besser; ~ **and** ~: immer besser; **be much** ~ *(recovered)* sich viel besser fühlen; **get** ~ *(recover)* gesund werden; **the** ~ **part of sth.** der größte Teil einer Sache *(Gen.)*. **2.** *adv.* besser; ~ **'off** *(financially)* besser gestellt; **be** ~ **off without sb./sth.** ohne jmdn./etw. besser dran sein; **I'd** ~ **be off now** ich gehe jetzt besser. **3.** *n.* **get the** ~ **of sb./sth.** jmdn./etw. unterkriegen *(ugs.);* **a change for the** ~: eine vorteilhafte Veränderung. **4.** *v.t.* übertreffen

'betting shop *n.* Wettbüro, *das*

between [bɪ'twiːn] **1.** *prep.* **a)** [in] ~: zwischen *(position:* + *Dat.; direction:* + *Akk.*); **b)** *(amongst)* unter (+ *Dat.*); ~ **ourselves,** ~ **you and me** unter uns *(Dat.)* gesagt; **c)** ~ **them/us** *(by joint action of)* gemeinsam; ~ **us we had 40p** wir hatten zusammen 40 Pence. **2.** *adv.* [in] ~: dazwischen; *(in time)* zwischendurch

beverage ['bevərɪdʒ] *n.* Getränk, *das*

beware [bɪ'weə(r)] *v.t. & i.; only in imper. and inf.* ~ [of] **sb./sth.** sich vor jmdm./etw. in acht nehmen; ~ **of doing sth.** sich davor hüten, etw. zu tun; '~ **of the dog'** „Vorsicht, bissiger Hund!"

bewilder [bɪ'wɪldə(r)] *v.t.* verwirren. **be'wilderment** *n.* Verwirrung, *die*

bewitch [bɪ'wɪtʃ] *v.t.* verzaubern; *(fig.)* bezaubern

beyond [bɪ'jɒnd] **1.** *adv.* **a)** *(in space)* jenseits; *(on other side of wall, mountain range, etc.)* dahinter; **b)** *(in time)* darüber hinaus; **c)** *(in addition)* außerdem. **2.** *prep.* **a)** *(at far side of)* jenseits (+ *Gen.*); **b)** *(later than)* nach; **c)** *(out of reach or comprehension or range)* über ... (+ *Akk.*) hinaus

bias ['baɪəs] **1.** *n.* Voreingenommenheit, *die.* **2.** *v.t.,* **-s-** *or* **-ss-** beeinflussen; **be** ~**ed in favour of/against sth./ sb.** für etw./jmdn. eingestellt sein/gegen etw./jmdn. voreingenommen sein

bib [bɪb] *n.* Lätzchen, *das*

Bible ['baɪbl] *n.* Bibel, *die.* **biblical** ['bɪblɪkl] *adj.* biblisch

bibliography [bɪblɪ'ɒgrəfɪ] *n.* Bibliographie, *die*

biceps ['baɪseps] *n.* Bizeps, *der*

bicker ['bɪkə(r)] *v.i.* sich zanken

bicycle ['baɪsɪkl] **1.** *n.* Fahrrad, *das; attrib.* Fahrrad-; ~ **clip** Hosenklammer, *die.* **2.** *v.i.* radfahren

bid [bɪd] **1.** *v.t.* **a)** -dd-, bid *(at auction)* bieten; **b)** -dd-, **bade** [bæd, beɪd] *or* **bid, bidden** ['bɪdn] *or* **bid:** ~ **sb. welcome/goodbye** jmdm. willkommen heißen/sich von jmdm. verabschieden. **2.** *v.i.,* -dd-, **bid a)** werben **(for** um); **b)** *(at auction)* bieten. **3.** *n.* **a)** *(at auction)* Gebot, *das;* **b)** *(attempt)* Versuch, *der*

bidden *see* **bid 1**

'bidder *n.* Bieter, *der*/Bieterin, *die*

bide ['baɪd] *v.t.* ~ **one's time** den richtigen Augenblick abwarten

bifocal [baɪ'fəʊkl] **1.** *adj.* Bifokal-. **2.** *n. in pl.* Bifokalgläser *Pl.*

big [bɪg] *adj.* groß

bigamy ['bɪgəmɪ] *n.* Bigamie, *die*

big-'headed *adj. (coll.)* eingebildet

bigoted ['bɪgətɪd] *adj.* eifernd

big: ~ **'toe** *n.* große Zehe; ~ **'top** *n.* Zirkuszelt, *das;* ~ **'wheel** *n. (at fair)* Riesenrad, *das*

bike [baɪk] *(coll.)* **1.** *n. (bicycle)* Rad, *das; (motor cycle)* Maschine, *die.* **2.** *v.i.* radfahren/[mit dem] Motorrad fahren

bikini [bɪ'kiːnɪ] *n.* Bikini, *der*

bilingual [baɪ'lɪŋgwəl] *adj.* zweisprachig

bilious ['bɪljəs] *adj. (Med.)* Gallen-; ~ **attack** Gallenanfall, *der*

¹bill [bɪl] *n. (of bird)* Schnabel, *der*

²bill *n.* **a)** *(Parl.)* Gesetzentwurf, *der;* **b)** *(note of charges)* Rechnung, *die;* **could we have the** ~ **please?** wir möchten zahlen; **c)** *(poster)* **'stick no** ~**s'** „[Plakate] ankleben verboten"

'billboard *n.* Reklametafel, *die*

billet ['bɪlɪt] **1.** *n.* Quartier, *das.* **2.** *v.t.* einquartieren **(with, on** bei)

'billfold *n. (Amer.)* Brieftasche, *die*

billiards ['bɪljədz] *n.* Billard[spiel], *das*

billion ['bɪljən] *n.* **a)** *(thousand million)* Milliarde, *die;* **b)** *(Brit.: million million)* Billion, *die*

billy-goat ['bɪlɪgəʊt] *n.* Ziegenbock, *der*

bin [bɪn] *n.* Behälter, *der; (for bread)* Brotkasten, *der; (for rubbish)* Mülleimer, *der*

binary ['baɪnərɪ] *adj.* binär

bind [baɪnd] *v.t.,* **bound** [baʊnd] **a)** fesseln ‹*Person, Tier*›; *(bandage)* wickeln ‹*Glied, Baum*›; verbinden ‹*Wunde*› **(with** mit); **b)** *(fasten together)* zusammenbinden; **c)** binden ‹*Buch*›; **d)** be

bound up with sth. *(fig.)* eng mit etw. verbunden sein; **e) be bound to do sth.** *(required)* verpflichtet sein, etw. zu tun; *(certain)* etw. ganz bestimmt tun; **it is bound to rain** es wird bestimmt regnen. **'binder** *n. (for papers)* Hefter, *der; (for magazines)* Mappe, *die.* **'binding 1.** *adj.* bindend ⟨*Vertrag, Abkommen*⟩ (on für). **2.** *n. (of book)* Einband, *der*

bingo ['bɪŋgəʊ] *n.* Bingo, *das*
binoculars [bɪ'nɒkjʊləz] *n. pl.* [a pair of] ~: Fernglas, *das*
biodegradable [baɪəʊdɪ'greɪdəbl] *adj.* biologisch abbaubar
biographer [baɪ'ɒgrəfə(r)] *n.* Biograph, *der*/Biographin, *die*
biographical [baɪə'græfɪkl] *adj.* biographisch
biography [baɪ'ɒgrəfɪ] *n.* Biographie, *die*
biological [baɪə'lɒdʒɪkl] *adj.* biologisch
biologist [baɪ'ɒlədʒɪst] *n.* Biologe, *der*/Biologin, *die*
biology [baɪ'ɒlədʒɪ] *n.* Biologie, *die*
biotechnology [baɪəʊtek'nɒlədʒɪ] *n.* Biotechnologie, *die*
birch [bɜːtʃ] *n.* Birke, *die*
bird [bɜːd] *n.* Vogel, *der*
bird: ~ cage *n.* Vogelkäfig, *der;* **~'s-eye 'view** *n.* Vogelperspektive, *die;* **~'s nest** *n.* Vogelnest, *das*
Biro, (P) ['baɪrəʊ] *n., pl.* ~s Kugelschreiber, *der;* Kuli, *der (ugs.)*
birth [bɜːθ] *n.* **a)** Geburt, *die;* **give ~** ⟨*Frau:*⟩ entbinden; ⟨*Tier:*⟩ jungen; werfen; **give ~ to a child** ein Kind zur Welt bringen; **b)** *(of movement, fashion, etc.)* Aufkommen, *das*
birth: ~ certificate *n.* Geburtsurkunde, *die;* **~ control** *n.* Geburtenkontrolle, *die;* **~day** *n.* Geburtstag, *der; attrib.* Geburtstags-; **~place** *n.* Geburtsort, *der*
biscuit ['bɪskɪt] *n. (Brit.)* Keks, *der*
bisect [baɪ'sekt] *v. t.* halbieren
bishop ['bɪʃəp] *n.* **a)** *(Eccl.)* Bischof, *der;* **b)** *(Chess)* Läufer, *der*
¹bit [bɪt] *n.* **a)** *(for horse)* Gebiß, *das;* **b)** *(of drill)* [Bohr]einsatz, *der*
²bit *n. (piece)* Stück, *das;* **not a or one ~** *(not at all)* überhaupt nicht; **a ~ tired/too early** ein bißchen müde/zu früh; **be a ~ of a coward/bully** ein ziemlicher Feigling sein/den starken Mann markieren *(ugs.)*
³bit *n. (Computing)* Bit, *das*
⁴bit *see* bite 1, 2

bitch [bɪtʃ] *n.* **a)** *(dog)* Hündin, *die;* **b)** *(sl. derog.: woman)* Miststück, *das (derb)*
bite [baɪt] **1.** *v. t.,* bit [bɪt], bitten ['bɪtn] beißen; ⟨*Moskito usw.:*⟩ stechen. **2.** *v. i.,* bit, bitten beißen/stechen; *(take bait)* anbeißen. **3.** *n.* Biß, *der; (piece)* Bissen, *der; (wound)* Bißwunde, *die; (by mosquito etc.)* Stich, *der.* **bite 'off** *v. t.* abbeißen
biting ['baɪtɪŋ] *adj.* beißend
bitten *see* bite 1, 2
bitter ['bɪtə(r)] *adj.* bitter. **'bitterly** *adv.* bitterlich ⟨*weinen, sich beschweren*⟩; **~ cold** bitterkalt. **'bitterness** *n.* Bitterkeit, *die*
bizarre [bɪ'zɑː(r)] *adj.* bizarr
black [blæk] **1.** *adj.* **a)** schwarz; **~ and blue** *(fig.)* grün und blau; **in ~ and white** *(fig.)* schwarz auf weiß; **in the ~** *(in credit)* in den schwarzen Zahlen; **b)** B~ *(dark-skinned)* schwarz. **2.** *n.* **a)** Schwarz, *das;* **b)** B~ *(person)* Schwarze, *der/die.* **3.** *v. t.* bestreiken ⟨*Betrieb*⟩; boykottieren ⟨*Arbeit*⟩. **black 'out 1.** *v. t.* verdunkeln. **2.** *v. i.* das Bewußtsein verlieren
black: ~berry ['blækbərɪ] *n.* Brombeere, *die;* **~bird** *n.* Amsel, *die;* **~board** *n.* [Wand]tafel, *die;* **'currant** *n.* schwarze Johannisbeere
blacken ['blækn] *v. t.* schwärzen; verfinstern ⟨*Himmel*⟩
black: ~ eye *n.* blaues Auge; B~ 'Forest *pr. n.* Schwarzwald, *der;* ~ 'ice *n.* Glatteis, *das;* **~leg** *n. (Brit.)* Streikbrecher, *der/*-brecherin, *die;* ~ list *n.* schwarze Liste; **~list** *v. t.* auf die schwarze Liste setzen; **~mail 1.** *v. t.* erpressen; **2.** *n.* Erpressung, *die;* **~ market** *n.* schwarzer Markt
'blackness *n.* Schwärze, *die; (darkness)* Finsternis, *die*
black: ~-out *n.* **a)** Verdunkelung, *die; (Theatre, Radio)* Blackout, *der;* **b)** *(Med.)* **have a ~-out** das Bewußtsein verlieren; B~ 'Sea *pr. n.* Schwarze Meer, *das;* **~smith** ['blæksmɪθ] *n.* Schmied, *der;* ~ spot *n.* Gefahrenstelle, *die*
bladder ['blædə(r)] *n.* Blase, *die*
blade [bleɪd] *n.* **a)** *(of sword, knife, razor, etc.)* Klinge, *die; (of saw, oar, propeller)* Blatt, *das;* **b)** *(of grass)* Spreite, *die*
blame [bleɪm] **1.** *v. t.* **~ sb. [for sth.]** jmdm. die Schuld [an etw. *(Dat.)*] geben; **be to ~ [for sth.]** an etw. *(Dat.)* schuld sein; **~ sth. [for sth.]** etw. [für

etw.] verantwortlich machen. 2. *n.* Schuld, *die.* '**blameless** *adj.* untadelig

blancmange [blə'mɒnʒ] *n.* Flammeri, *der*

bland [blænd] *adj.* mild; *(suave)* verbindlich

blank [blæŋk] 1. *adj.* a) leer; kahl ⟨*Wand, Fläche*⟩; b) *(empty)* frei. 2. *n.* a) *(space)* Lücke, *die;* b) *(cartridge)* Platzpatrone, *die;* c) **draw a** ~: kein Glück haben. **blank** '**cheque** *n.* Blankoscheck, *der; (fig.)* Blankovollmacht, *die*

blanket ['blæŋkɪt] *n.* Decke, *die;* **wet** ' ~ *(fig.)* Trauerkloß, *der (ugs.)*

blare ['bleə(r)] 1. *v. i.* ⟨*Lautsprecher:*⟩ plärren; ⟨*Trompete:*⟩ schmettern. 2. *v. t.* ~ **|out|** [hinaus]plärren ⟨*Worte*⟩; [hinaus]schmettern ⟨*Melodie*⟩

blasé ['blɑːzeɪ] *adj.* blasiert

blasphemous ['blæsfəməs] *adj.* lästerlich

blasphemy ['blæsfəmɪ] *n.* Blasphemie, *die*

blast [blɑːst] 1. *n.* a) **a** ~ **|of wind|** ein Windstoß; b) *(of horn)* Tuten, *das.* 2. *v. t. (blow up)* sprengen. **blast** '**off** *v. i.* abheben

'**blasted** *adj. (damned)* verdammt *(salopp)*

'**blast-off** *n.* Abheben, *das*

blatant ['bleɪtənt] *adj.* a) *(flagrant)* eklatant; b) *(unashamed)* unverhohlen; unverfroren ⟨*Lüge*⟩

blaze [bleɪz] 1. *n.* Feuer, *das.* 2. *v. i.* brennen; lodern *(geh.)*

blazer ['bleɪzə(r)] *n.* Blazer, *der*

bleach [bliːtʃ] 1. *v. t.* bleichen. 2. *n.* Bleichmittel, *das*

bleak ['bliːk] *adj.* a) öde ⟨*Landschaft usw.*⟩; b) *(unpromising)* düster

bleat [bliːt] *v. i.* ⟨*Schaf:*⟩ blöken; ⟨*Ziege:*⟩ meckern

bled *see* **bleed**

bleed [bliːd] *v. i.,* **bled** [bled] bluten

bleeper ['bliːpə(r)] *n.* Kleinempfänger, *der*

blemish ['blemɪʃ] *n.* Fleck, *der*

blend [blend] 1. *v. t.* mischen. 2. *v. i.* sich mischen lassen. 3. *n.* Mischung, *die.* '**blender** *n.* Mixer, *der*

bless [bles] *v. t.* segnen; ~ **you!** *(after sb. sneezes)* Gesundheit! **blessed** ['blesɪd] *adj.* a) *(revered)* heilig; b) *(cursed)* verdammt *(salopp).* '**blessing** *n.* Segen, *der*

blew *see* ¹**blow**

blight [blaɪt] *n. (fig.)* Fluch, *der*

blind [blaɪnd] 1. *adj.* blind; ~ **in one eye** auf einem Auge blind. 2. *adv.* blindlings. 3. *n.* Jalousie, *die; (made of cloth)* Rouleau, *das; (of shop)* Markise, *die.* 4. *v. t.* blenden. '**blindfold** 1. *v. t.* die Augen verbinden (+ *Dat.*). 2. *adj.* mit verbundenen Augen *nachgestellt.* '**blinding** *adj.* blendend. '**blindly** *adv.* [wie] blind; *(fig.)* blindlings. '**blindness** *n.* Blindheit, *die*

blink [blɪŋk] *v. i.* a) blinzeln; b) *(shine intermittently)* blinken

'**blinkers** *n. pl.* Scheuklappen *Pl.*

bliss [blɪs] *n.* [Glück]seligkeit, *die.*

blissful ['blɪsfl] *adj.* [glück]selig

blister ['blɪstə(r)] 1. *n.* Blase, *die.* 2. *v. i.* ⟨*Haut:*⟩ Blasen bekommen; ⟨*Anstrich:*⟩ Blasen werfen

blizzard ['blɪzəd] *n.* Schneesturm, *der*

blob [blɒb] *n. (drop)* Tropfen, *der; (small mass)* Klacks, *der (ugs.)*

block [blɒk] 1. *n.* a) Klotz, *der; (for chopping on)* Hackklotz, *der; (of concrete or stone, building-stone)* Block, *der;* b) *(building)* [Häuser]block, *der;* ~ **of flats/offices** Wohnblock, *der*/Bürohaus, *das.* 2. *v. t.* versperren ⟨*Tür, Straße, Durchgang, Sicht*⟩; verstopfen ⟨*Pfeife, Abfluß*⟩; verhindern ⟨*Fortschritt*⟩. **block** '**up** *v. t.* verstopfen; versperren ⟨*Eingang*⟩

blockade [blɒ'keɪd] 1. *n.* Blockade, *die.* 2. *v. t.* blockieren

blockage ['blɒkɪdʒ] *n.* Block, *der; (of pipe, gutter)* Verstopfung, *die*

block: ~ '**booking** *n.* Gruppenbuchung, *die;* ~ '**capital** *n.* Blockbuchstabe, *der;* ~ '**head** *n.* Dummkopf, *der;* ~ '**letters** *n. pl.* Blockschrift, *die*

bloke [bləʊk] *n. (Brit. coll.)* Typ, *der (ugs.)*

blonde [blɒnd] 1. *adj.* blond. 2. *n.* Blondine, *die*

blood [blʌd] *n.* Blut, *das*

blood: ~ **donor** *n.* Blutspender, *der*/-spenderin, *die;* ~ **group** *n.* Blutgruppe, *die;* ~**hound** *n.* Bluthund, *der;* ~ **pressure** *n.* Blutdruck, *der;* ~**shed** *n.* Blutvergießen, *das;* ~**shot** *adj.* blutunterlaufen; ~-**stained** *adj.* blutbefleckt; ~**stream** *n.* Blutstrom, *der;* ~ **test** *n.* Blutprobe, *die;* ~**thirsty** *adj.* blutrünstig; ~ **transfusion** *n.* Bluttransfusion, *die;* ~-**vessel** *n.* Blutgefäß, *das*

'**bloody** 1. *adj.* a) blutig; *(running with blood)* blutend; b) *(sl.: damned)* verdammt *(salopp).* 2. *adv. (sl.: damned)* verdammt *(salopp)*

bloom [blu:m] 1. *n.* Blüte, *die;* be in ~: in Blüte stehen. 2. *v. i.* blühen

blossom ['blɒsəm] 1. *n. (flower)* Blüte, *die; (mass)* Blütenmeer, *das (geh.).* 2. *v. i.* blühen; ⟨*Mensch:*⟩ aufblühen

blot [blɒt] 1. *n. (of ink)* Tintenklecks, *der; (stain)* Fleck, *der.* 2. *v. t.,* -tt- ablöschen ⟨*Tinte, Papier*⟩. **blot 'out** *v. t.. (fig.)* auslöschen

blotchy ['blɒtʃɪ] *adj.* fleckig

'blotting-paper *n.* Löschpapier, *das*

blouse [blaʊz] *n.* Bluse, *die*

¹blow [bləʊ] 1. *v. i.,* blew [blu:], blown [bləʊn] ⟨*Wind:*⟩ wehen; ⟨*Sturm:*⟩ blasen. 2. *v. t.,* blew, blown: a) blasen; ⟨*Wind:*⟩ wehen; machen ⟨*Seifenblase*⟩; ~ sb. a kiss jmdm. eine Kußhand zuwerfen; b) ~ one's nose sich *(Dat.)* die Nase putzen; c) ~ sth. to pieces etw. in die Luft sprengen. **blow 'out** 1. *v. t.* ausblasen. 2. *v. i.* ausgeblasen werden. **blow 'over** 1. *v. i.* umgeblasen werden; ⟨*Streit, Sturm:*⟩ sich legen. 2. *v. t.* umblasen. **blow 'up** 1. *v. t.* a) *(shatter)* [in die Luft] sprengen; b) aufblasen ⟨*Ballon*⟩; aufpumpen ⟨*Reifen*⟩; c) *(coll.: exaggerate)* hochspielen. 2. *v. i. (explode)* explodieren

²blow *n.* a) Schlag, *der; (with axe)* Hieb, *der;* come to ~s handgreiflich werden; b) *(disaster)* [schwerer] Schlag. **'blow-dry** *v. t.* fönen. **'blowlamp** *n.* Lötlampe, *die*

blown *see* ¹blow

blubber ['blʌbə(r)] *n.* Walspeck, *der*

blue [blu:] 1. *adj.* blau. 2. *n.* a) Blau, *das;* b) have the ~s deprimiert sein; c) *(Mus.)* the ~s der Blues; d) out of the ~: aus heiterem Himmel

blue: ~bell *n.* Glockenblume, *die;* ~bottle *n.* Schmeißfliege, *die;* ~-collar *adj.* ~-collar worker Arbeiter, *der/*Arbeiterin, *die;* ~ 'jeans *n. pl.* Blue jeans *Pl.;* ~print *n. (fig.)* Entwurf, *der*

bluff [blʌf] 1. *n.* Bluff, *der (ugs.);* call sb.'s ~: es darauf ankommen lassen *(ugs.).* 2. *v. i. & t.* bluffen *(ugs.)*

blunder ['blʌndə(r)] 1. *n.* [schwerer] Fehler. 2. *v. i.* a) *(make mistake)* einen [schweren] Fehler machen; b) *(move blindly)* tappen

blunt [blʌnt] 1. *adj.* a) stumpf; b) *(outspoken)* direkt; glatt *(ugs.)* ⟨*Ablehnung*⟩. 2. *v. t.* ~ [the edge of] stumpf machen. **'bluntly** *adv.* direkt; glatt ⟨*ablehnen*⟩

blur [blɜ:(r)] 1. *v. t.,* -rr-: a) verwischen;

b) *(become indistinct)* verschwimmen; his vision was ~red er sah alles verschwommen. 2. *n. (smear)* Fleck, *der; (dim image)* verschwommener Fleck

blurt [blɜ:t] *v. t.* ~ out herausplatzen mit *(ugs.)*

blush [blʌʃ] 1. *v. i.* rot werden. 2. *n.* Rotwerden, *das*

bluster ['blʌstə(r)] *v. i.* sich aufplustern *(ugs.)*

blustery ['blʌstərɪ] *adj.* stürmisch

boar [bɔ:(r)] *n.* |wild| ~: Keiler, *der*

board [bɔ:d] 1. *n.* a) Brett, *das; (black~)* Tafel, *die; (notice-~)* Schwarzes Brett; above ~ *(fig.)* korrekt; b) *(Commerc.)* ~ |of directors| Vorstand, *der; (supervisory* ~) Aufsichtsrat, *der;* c) *(Naut., Aeronaut.)* on ~: an Bord; d) ~ and lodging Unterkunft und Verpflegung; full ~: Vollpension, *die.* 2. *v. t.* ~ the ship/plane an Bord des Schiffes/Flugzeuges gehen; ~ the train/bus in den Zug/Bus einsteigen

'boarder *n. (Sch.)* Internatsschüler, *der/*-schülerin, *die*

'board game *n.* Brettspiel, *das*

boarding: ~-house *n.* Pension, *die;* ~ pass *n.* Bordkarte, *die;* ~-school *n.* Internat, *das*

board: ~ meeting *n.* Vorstandssitzung, *die;* ~room *n.* Sitzungssaal, *der*

boast [bəʊst] *v. i.* prahlen. **boastful** ['bəʊstfl] *adj.* prahlerisch

boat [bəʊt] *n.* Boot, *das*

¹bob [bɒb] *v. i.,* -bb-: ~ |up and down| sich auf und nieder bewegen

²bob *n. (~-sled)* Bob, *der*

bobbin ['bɒbɪn] *n.* Spule, *die*

bob: ~-sled, ~-sleigh *ns.* Bobschlitten, *der*

bodice ['bɒdɪs] *n.* Mieder, *das; (part of dress)* Oberteil, *das*

bodily ['bɒdɪlɪ] *adj.* körperlich; ~ needs leibliche Bedürfnisse

body ['bɒdɪ] *n.* a) Körper, *der;* b) *(corpse)* Leiche, *die;* c) *(group)* Gruppe, *die; (with particular function)* Organ, *das.* **bodyguard** *n. (single)* Leibwächter, *der; (group)* Leibwache, *die.* **bodywork** *n.* Karosserie, *die*

bog [bɒg] 1. *n.* Moor, *das; (marsh, swamp)* Sumpf, *der.* 2. *v. t.,* -gg-: be/get ~ged down *(fig.)* sich verzettelt haben/sich verzetteln

boggle ['bɒgl] *v. i. (coll.)* the mind ~s da kann man nur [noch] staunen

bogus ['bəʊgəs] *adj.* falsch

¹boil [bɔɪl] 1. *v. i. & t.* kochen. 2. *n.*

come to/go off the ~: zu kochen anfangen/aufhören; **bring to the** ~: zum Kochen bringen. **boil 'down** *v.i.* ~ **down to sth.** *(fig.)* auf etw. hinauslaufen. **boil 'over** *v.i.* überkochen
²**boil** *n. (Med.)* Furunkel, *der*
'**boiler** *n.* Kessel, *der*
'**boiling-point** *n.* Siedepunkt, *der*
boisterous ['bɔɪstərəs] *adj.* ausgelassen
bold [bəʊld] *adj.* **a)** *(courageous)* mutig; *(daring)* kühn; **b)** auffallend ⟨*Farbe, Muster*⟩. '**boldly** *adv. (courageously)* mutig; *(daringly)* kühn
Bolivia [bə'lɪvɪə] *pr. n.* Bolivien *(das)*
bollard ['bɒlɑːd] *n. (Brit.)* Poller, *der*
bolster ['bəʊlstə(r)] **1.** *n. (pillow)* Nackenrolle, *die*. **2.** *v.t. (fig.)* stärken
bolt [bəʊlt] **1.** *n.* **a)** *(on door or window)* Riegel, *der; (on gun)* Kammerverschluß, *der;* **b)** *(metal pin)* Schraube, *die; (without thread)* Bolzen, *der.* **2.** *v.i.* davonlaufen; ⟨*Pferd:*⟩ durchgehen; ⟨*Fuchs, Kaninchen:*⟩ flüchten. **3.** *v.t.* **a)** verriegeln ⟨*Tür, Fenster*⟩; **b)** *(fasten with* ~*s)* verschrauben/mit Bolzen verbinden; **c)** ~ **[down]** hinunterschlingen ⟨*Essen*⟩. **4.** *adv.* ~ **upright** kerzengerade
bomb [bɒm] **1.** *n.* Bombe, *die.* **2.** *v.t.* bombardieren
bombard [bɒm'bɑːd] *v.t.* beschießen.
bom'bardment *n.* Beschuß, *der*
bombastic [bɒm'bæstɪk] *adj.* bombastisch
bomber ['bɒmə(r)] *n. (Air Force)* Bomber, *der (ugs.)*
'**bomb-shell** *n.* Bombe, *die; (fig.)* Sensation, *die*
bond [bɒnd] *n.* **a)** Band, *das; in pl. (shackles)* Fesseln; **b)** *(adhesion)* Verbindung, *die;* **c)** *(Commerc.)* Anleihe, *die*
bone [bəʊn] **1.** *n.* Knochen, *der; (of fish)* Gräte, *die.* **2.** *v.t.* den/die Knochen herauslösen aus; entgräten ⟨*Fisch*⟩. **bone 'dry** *adj.* knochentrocken *(ugs.).* **bone 'idle** *adj.* stinkfaul *(salopp)*
bonfire ['bɒnfaɪə(r)] *n.* Freudenfeuer, *das; (for rubbish)* Feuer, *das*
bonnet ['bɒnɪt] *n.* **a)** *(woman's)* Haube, *die; (child's)* Häubchen, *das;* **b)** *(Brit. Motor Veh.)* Motorhaube, *die*
bonus ['bəʊnəs] *n.* zusätzliche Leistung; *(to shareholders)* Bonus, *der;* **Christmas** ~: Weihnachtsgratifikation, *die*
bony ['bəʊnɪ] *adj.* **a)** Knochen-; *(like*

bone) knochenartig; **b)** *(skinny)* knochendürr *(ugs.);* spindeldürr
boo [buː] **1.** *int. to surprise sb.* huh; *expr. disapproval, contempt* buh. **2.** *n.* Buh, *das (ugs.).* **3.** *v.t.* ausbuhen *(ugs.).* **4.** *v.i.* buhen *(ugs.)*
booby ['buːbɪ] *n.* Trottel, *der (ugs.).* '**booby prize** *n. Preis für den schlechtesten Teilnehmer an einem Wettbewerb.* '**booby trap** *n.* **a)** *Falle, mit der man jmdm. einen Streich spielen will;* **b)** *(Mil.)* versteckte Sprengladung
book [bʊk] **1.** *n.* Buch, *das; (for accounts)* Rechnungsbuch, *das; (for exercises)* [Schreib]heft, *das.* **2.** *v.t.* buchen ⟨*Reise, Flug, Platz [im Flugzeug]*⟩; [vor]bestellen ⟨*Eintrittskarte, Tisch, Zimmer, Platz [im Theater]*⟩. **3.** *v.i.* buchen. **book 'in 1.** *v.i.* sich eintragen. **2.** *v.t.* eintragen. **book 'up** *v.i. & t.* buchen; **be** ~**ed up** ⟨*Hotel usw.:*⟩ ausgebucht sein
book: ~case *n.* Bücherschrank, *der;* ~**ends** *n. pl.* Buchstützen
'**booking office** *n.* [Fahrkarten]schalter, *der*
book: ~keeper *n.* Buchhalter, *der/*-halterin, *die;* ~**keeping** *n.* Buchführung, *die;* Buchhaltung, *die*
booklet ['bʊklɪt] *n.* Broschüre, *die*
book: ~maker *n. (in betting)* Buchmacher, *der;* ~**mark** *n.* Lesezeichen, *das;* ~**seller** *n.* Buchhändler, *der/*-händlerin, *die;* ~**shelf** *n.* Bücherbord, *das;* ~**shop** *n.* Buchhandlung, *die;* ~**stall** *n.* Bücherstand, *der;* ~**store** *n. (Amer.)* Buchhandlung, *die;* ~ **token** *n.* Büchergutschein, *der;* ~**worm** *n.* Bücherwurm, *der*
¹**boom** [buːm] *n.* **a)** *(for camera or microphone)* Ausleger, *der;* **b)** *(Naut.)* Baum, *der*
²**boom 1.** *v.i.* **a)** dröhnen; **b)** ⟨*Geschäft, Verkauf, Gebiet:*⟩ sich sprunghaft entwickeln. **2.** *n.* **a)** Dröhnen, *das;* **b)** *(in business or economy)* Boom, *der*
boomerang ['buːməræŋ] *n.* Bumerang, *der*
boon [buːn] *n.* Segen, *der* (to für)
boorish ['bʊərɪʃ] *adj.* rüpelhaft
boost [buːst] **1.** *v.t.* in die Höhe treiben ⟨*Preis, Wert*⟩; stärken ⟨*Selbstvertrauen, Moral*⟩. **2.** *n.* Auftrieb, *der*
boot [buːt] **1.** *n.* **a)** Stiefel, *der;* **give sb. the** ~ *(fig. coll.)* jmdn. rausschmeißen *(ugs.);* **b)** *(Brit.: of car)* Kofferraum, *der.* **2.** *v.t. (coll.: kick)* kicken *(ugs.)*
booth [buːð] *n.* **a)** Bude, *die;* **b)** *(telephone* ~*)* Zelle, *die*

'**bootleg** *adj.* schwarz verkauft/gebrannt

booze [bu:z] *(coll.)* **1.** *v.i.* saufen *(derb).* **2.** *n.* Alkohol, *der*

border ['bɔ:də(r)] **1.** *n.* **a)** Rand, *der; (of table-cloth, handkerchief)* Bordüre, *die;* **b)** *(of country)* Grenze, *die;* **c)** *(flower-bed)* Rabatte, *die.* **2.** *attrib. adj.* Grenz⟨*stadt, -streit*⟩. **3.** *v.t.* **a)** *(adjoin)* [an]grenzen an (+ *Akk.*); **b)** *(put a ~ to, act as ~ to)* umranden; einfassen. **4.** *v.i.* ~ **on a)** *see* **3 a;** **b)** *(resemble)* grenzen an (+ *Akk.*). '**borderline 1.** *n.* Grenzlinie, *die.* **2.** *adj.* be ~: auf der Grenze liegen; **a ~ case/candidate** ein Grenzfall

'**bore** [bɔ:(r)] **1.** *v.t.* bohren. **2.** *n. (of firearm)* Kaliber, *das*

²**bore 1.** *n.* **a)** it's a real ~: es ist wirklich ärgerlich; **what a ~**! wie ärgerlich!; **b)** *(person)* Langweiler, *der (ugs.).* **2.** *v.t.* langweilen; **be ~d** sich langweilen

³**bore** *see* ²**bear**

boredom ['bɔ:dəm] *n.* Langeweile, *die*

'**borehole** *n.* Bohrloch, *das*

boring ['bɔ:rɪŋ] *adj.* langweilig

born [bɔ:n] **1.** be ~: geboren werden. **2.** *adj.* geboren; **be a ~ orator** der geborene Redner sein

borne *see* ²**bear**

borough ['bʌrə] *n. (town)* Stadt, *die; (village)* Gemeinde, *die*

borrow ['bɒrəʊ] *v.t.* leihen (**from** von, bei); *(from library)* entleihen. '**borrower** *n. (from bank)* Kreditnehmer, *der; (from library)* Entleiher, *der*

bosom ['bʊzəm] *n.* Brust, *die*

boss [bɒs] **1.** *n. (coll.)* Boß, *der (ugs.);* Chef, *der.* **2.** *v.t.* ~ [**about** *or* **around**] herumkommandieren *(ugs.).* '**bossy** *adj. (coll.)* herrisch

botanical [bə'tænɪkl] *adj.* botanisch

botanist ['bɒtənɪst] *n.* Botaniker, *der*/Botanikerin, *die*

botany ['bɒtənɪ] *n.* Botanik, *die*

botch [bɒtʃ] **1.** *v.t.* pfuschen bei *(ugs.).* **2.** *v.i.* pfuschen *(ugs.).* **botch 'up** *v.t. (bungle)* verpfuschen *(ugs.)*

both [bəʊθ] **1.** *adj.* beide; ~ [**the**] **brothers** beide Brüder. **2.** *pron.* beide; ~ [**of them**] **are dead** beide sind tot; ~ **of you/them are** ...: ihr seid/sie sind beide ... **3.** *adv.* ~ **A and B** sowohl A als [auch] B; **he and I were** ~ **there** er und ich waren beide da

bother ['bɒðə(r)] **1.** *v.t.* **a)** I **can't be ~ed** ich habe keine Lust; **b)** *(annoy)* lästig sein (+ *Dat.*); ⟨*Lärm, Licht:*⟩

stören; ⟨*Schmerz, Zahn:*⟩ zu schaffen machen (+ *Dat.*); **I'm sorry to ~ you, but** ...: es tut mir leid, wenn ich Sie störe, aber ...; **c)** *(worry)* Sorgen machen (+ *Dat.*); ⟨*Problem, Frage:*⟩ beschäftigen. **2.** *v.i.* **don't ~ to do it** Sie brauchen es nicht zu tun; **you needn't/shouldn't have ~ed** das wäre nicht nötig gewesen; **don't** ~! nicht nötig! **3.** *n.* **a)** *(trouble)* Ärger, *der;* **b)** *(effort)* Mühe, *die.* **4.** *int. (coll.)* wie ärgerlich!

bottle ['bɒtl] **1.** *n.* Flasche, *die;* **a ~ of beer** eine Flasche Bier. **2.** *v.t.* **a)** *(put into ~s)* in Flaschen [ab]füllen; **b)** *(preserve in jars)* einmachen. **bottle 'up** *v.t.* **a)** *(conceal)* in sich *(Dat.)* aufstauen; **b)** *(trap)* einschließen

bottle: ~ bank *n.* Altglasbehälter, *der;* **~-neck** *n. (fig.)* Flaschenhals, *der (ugs.);* **~-opener** *n.* Flaschenöffner, *der;* **~-top** *n.* Flaschenverschluß, *der*

bottom ['bɒtəm] **1.** *n.* **a)** unteres Ende; *(of cup, glass, box)* Boden, *der; (of valley, well, shaft)* Sohle, *die; (of hill, cliff, stairs)* Fuß, *der;* **b)** *(buttocks)* Hinterteil, *das (ugs.);* **c)** *(of sea, lake)* Grund, *der;* **d)** *(farthest point)* **at the ~ of the garden/street** hinten im Garten/am Ende der Straße; **e)** *(underside)* Unterseite, *die;* **f)** *(fig.)* **start at the ~:** ganz unten anfangen; **be ~ of the class** der/die Letzte in der Klasse sein. **2.** *adj.* **a)** *(lowest)* unterst...; *(lower)* unter...; **b)** *(fig.: last)* letzt... '**bottomless** *adj.* bodenlos; unendlich tief ⟨*Meer, Ozean*⟩

bough [baʊ] *n.* Ast, *der*

bought *see* **buy 1**

boulder ['bəʊldə(r)] *n.* Felsbrocken, *der*

boulevard ['bu:ləvɑ:d] *n.* Boulevard, *der*

bounce [baʊns] **1.** *v.i.* **a)** springen; **b)** *(coll.)* ⟨*Scheck:*⟩ platzen *(ugs.).* **2.** *v.t.* aufspringen lassen ⟨*Ball*⟩. **3.** *n.* Aufprall, *der.* '**bouncer** *n. (coll.)* Rausschmeißer, *der (ugs.).* **bouncing** ['baʊnsɪŋ] *adj.* stramm ⟨*Baby*⟩. **bouncy** ['baʊnsɪ] *adj.* gut springend ⟨*Ball*⟩; *(fig.: lively)* munter

'**bound** [baʊnd] **1.** *n., usu. in pl. (limit)* Grenze, *die;* **within the ~s of possibility** im Bereich des Möglichen; **sth. is out of ~s [to sb.]** der Zutritt zu etw. ist [für jmdn.] verboten. **2.** *v.t.* be ~**ed by sth.** durch etw. begrenzt werden

²**bound 1.** *v.i.* hüpfen. **2.** *n.* Satz, *der*

³**bound** *pred. adj.* be ~ for home/
Frankfurt auf dem Heimweg/nach
Frankfurt unterwegs sein; **homeward**
~: auf dem Weg nach Hause
⁴**bound** *see* **bind**
boundary ['baʊndərɪ] *n.* Grenze, *die*
'**boundless** *adj.* grenzenlos
bounty ['baʊntɪ] *n.* Kopfgeld, *das*
bouquet [bʊ'keɪ] *n.* [Blumen]strauß,
der
bourgeois ['bʊəʒwɑ:] 1. *n., pl. same*
Bürger, *der*/Bürgerin, *die.* 2. *adj.* bürgerlich
bout [baʊt] *n.* a) *(contest)* Wettkampf,
der; b) *(fit)* Anfall, *der*
boutique [bu:'ti:k] *n.* Boutique, *die*
¹**bow** [bəʊ] a) *(curve, weapon, Mus.)*
Bogen, *der;* b) *(knot, ribbon)* Schleife,
die
²**bow** [baʊ] 1. *v. i.* a) ~ |to sb.] sich [vor
jmdm.] verbeugen; b) *(submit)* sich
beugen (**to** *Dat.*). 2. *n.* Verbeugung,
die
³**bow** [baʊ] *n. (Naut.)* Bug, *der*
bowel ['baʊəl] *n. (Anat.)* ~s *pl.*, *(Med.)*
~: Darm, *der*
¹**bowl** [bəʊl] *n. (basin)* Schüssel, *die;*
(shallower) Schale, *die; (of spoon)*
Schöpfteil, *der; (of pipe)* Kopf, *der*
²**bowl** 1. *n.* a) *(ball)* Kugel, *die;* b) *in pl.*
(game) Bowls, *das.* 2. *v. i.* a) *(play ~s)*
Bowls spielen; b) *(Cricket)* werfen
bow-legged ['bəʊlegɪd] O-beinig
(ugs.)
¹**bowler** ['bəʊlə(r)] *n. (Cricket)* Werfer,
der
²**bowler** *n.* ~ |hat| Bowler, *der*
'**bowling** *n.* |ten-pin| ~: Bowling, *das;*
go ~: bowlen gehen. '**bowling-alley**
n. Bowlingbahn, *die.* '**bowling-
green** *n.* Rasenfläche für Bowls
bow [bəʊ]: ~-'**tie** *n.* Fliege, *die;*
~-**window** *n.* Erkerfenster, *das*
¹**box** [bɒks] *n.* Kasten, *der; (bigger)* Kiste, *die; (of cardboard)* Schachtel, *die*◄
²**box** 1. *n.* he gave him a ~ on the ear|s|
er gab ihm eine Ohrfeige. 2. *v. t.* a) he
~ed his ears *or* him round the ears er
ohrfeigte ihn; b) *(Sport)* ~ sb. gegen
jmdn. boxen. 3. *v. i.* boxen. '**boxer** *n.*
Boxer, *der.* '**boxing** *n.* Boxen, *das*
boxing: B~ **Day** *n.* zweiter Weihnachtsfeiertag; ~-**glove** *n.* Boxhand-
schuh, *der;* ~-**match** *n.* Boxkampf,
der; ~-**ring** *n.* Boxring, *der*
box: ~ **number** *n. (at newspaper office)* Chiffre, *die; (at post office)* Postfach, *das;* ~-**office** *n.* Kasse, *die;*
~-**room** *n. (Brit.)* Abstellraum, *der*

boy [bɔɪ] *n.* Junge, *der*
boycott ['bɔɪkɒt] 1. *v. t.* boykottieren.
2. *n.* Boykott, *der*
'**boy-friend** *n.* Freund, *der*
'**boyish** *adj.* jungenhaft
bra [brɑ:] *n.* BH, *der (ugs.)*
brace [breɪs] 1. *n.* a) *(connecting piece)*
Klammer, *die; (strut)* Strebe, *die;*
(Dent.) [Zahn]spange, *die;* b) *in pl.*
(trouser-straps) Hosenträger. 2. *v. refl.*
~ **oneself for sth.** sich auf etw. *(Akk.)*
vorbereiten
bracelet ['breɪslɪt] *n.* Armband, *das*
bracing ['breɪsɪŋ] *adj.* belebend
bracken ['brækn] *n.* [Adler]farn, *der*
bracket ['brækɪt] 1. *n.* a) *(support)*
Konsole, *die;* b) *(mark)* Klammer,
die. 2. *v. t.* einklammern
brag [bræg] *v. i. & t.*, -gg- prahlen
(**about mit**)
braid [breɪd] 1. *n.* a) *(plait)* Flechte, *die*
(geh.); Zopf, *der;* b) *(woven band)*
Borte, *die; (on uniform)* Litze, *die.* 2.
v. t. flechten
Braille [breɪl] *n.* Blindenschrift, *die*
brain [breɪn] *n.* Gehirn, *das*
brain: ~-**child** *n. (coll.)* Geistespro-
dukt, *das;* ~-**less** *adj.* hirnlos;
~**wash** *v. t.* einer Gehirnwäsche unterziehen; ~**wave** *n. (coll.: inspiration)* genialer Einfall
'**brainy** *adj.* intelligent
brake [breɪk] 1. *n.* Bremse, *die.* 2. *v. t.*
& *i.* bremsen; **braking distance** Brems-
weg, *der.* '**brake light** *n.* Bremslicht,
das
bramble ['bræmbl] *n.* Dornenstrauch,
der
bran [bræn] *n.* Kleie, *die*
branch [brɑ:ntʃ] 1. *n.* a) *(bough)* Ast,
der; (twig) Zweig, *der;* b) *(of artery,
antlers)* Ast, *der;* c) *(office)* Zweigstelle, *die; (shop)* Filiale, *die.* 2. *v. i.* sich
verzweigen. **branch 'off** *v. i.* abzweigen. **branch 'out** *v. i. (fig.)* ~ **out into**
sth. sich auch mit etw. befassen
'**branch line** *n.* Nebenstrecke, *die*
brand [brænd] *n.* a) *(trade mark)* Mar-
kenzeichen, *das; (goods of particular
make)* Marke, *die;* b) *(mark)* Brand-
mal, *das*
brandish ['brændɪʃ] *v. t.* schwenken;
schwingen ‹*Waffe*›
brand: ~-**name** *n.* Markenname, *der;*
~-'**new** *adj.* nagelneu *(ugs.)*
brandy ['brændɪ] *n.* Weinbrand, *der*
brash [bræʃ] *adj.* dreist
brass [brɑ:s] *n.* Messing, *das; attrib.*
Messing-; **the** ~ *(Mus.)* das Blech; ~

player *(Mus.)* Blechbläser, *der;* **get down to ~ tacks** zur Sache kommen.
brass 'band *n.* Blaskapelle, *die*
brassière ['bræzjə(r)] *n.* Büstenhalter, *der*
brat [bræt] *n.* Balg, *das od. der (ugs.)*
bravado [brə'vɑːdəʊ] *n.* **do sth. out of ~:** so waghalsig sein, etw. zu tun
brave [breɪv] **1.** *adj.* tapfer. **2.** *n.* [indianischer] Krieger. **3.** *v.t.* trotzen *(+ Dat.).* **'bravely** *adv.* tapfer.
bravery ['breɪvərɪ] *n.* Tapferkeit, *die*
bravo [brɑː'vəʊ] *int.* bravo
brawl [brɔːl] **1.** *v.i.* sich schlagen. **2.** *n.* Schlägerei, *die*
brawny ['brɔːnɪ] *adj.* muskulös
bray [breɪ] **1.** Iah, *das.* **2.** *v.i.* ⟨*Esel:*⟩ iahen
brazen ['breɪzn] **1.** *adj.* dreist; *(shameless)* schamlos. **2.** *v.t.* ~ [out] trotzen *(+ Dat.);* ~ **it out** *(deny guilt)* es abstreiten; *(not admit guilt)* es nicht zugeben
brazier ['breɪzɪə(r)] *n.* Kohlenbecken, *das*
Brazil [brə'zɪl] *pr. n.* Brasilien *(das).*
Bra'zil nut *n.* Paranuß, *die*
breach [briːtʃ] **1.** *n.* **a)** *(violation)* Verstoß, *der* (of gegen); ~ **of faith/duty** Vertrauensbruch, *der*/Pflichtverletzung, *die;* **b)** *(of relations)* Bruch, *der;* **c)** *(gap)* Bresche, *die; (fig.)* Riß, *der.* **2.** *v.t.* durchbrechen
bread [bred] *n.* Brot, *das;* **a piece of ~ and butter** ein Butterbrot
bread: **~-bin** *n.* Brotkasten, *der;* **~-board** *n.* [Brot]brett, *das;* **~crumb** *n.* Brotkrume, *die,* **~crumbs** *(coating)* Paniermehl, *das;* **~-knife** *n.* Brotmesser, *das;* **~line** *n.* **be or live on/below the ~line** gerade noch/nicht einmal mehr das Notwendigste zum Leben haben
breadth [bredθ] *n.* Breite, *die*
'bread-winner *n.* Ernährer, *der*/Ernährerin, *die*
break [breɪk] **1.** *v.t.,* **broke** [brəʊk], **broken** ['brəʊkn] **a)** brechen; *(so as to damage)* zerbrechen; kaputtmachen *(ugs.);* zerreißen ⟨*Seil*⟩; *(fig.: interrupt)* unterbrechen; brechen ⟨*Bann, Zauber, Schweigen*⟩; **the TV/my watch is broken** der Fernseher/meine Uhr ist kaputt *(ugs.);* ~ **the habit** es sich *(Dat.)* abgewöhnen; **b)** *(fracture)* sich *(Dat.)* brechen ⟨*Arm, Bein usw.*⟩; **c)** brechen ⟨*Vertrag, Versprechen*⟩; verstoßen gegen ⟨*Regel, Gesetz*⟩; **d)** *(surpass)* brechen ⟨*Rekord*⟩; **e)** *(cushion)*

auffangen ⟨*Schlag, jmds. Fall*⟩. **2.** *v.i.,* **broke, broken a)** kaputtgehen *(ugs.);* ⟨*Faden, Seil:*⟩ [zer]reißen; ⟨*Glas, Tasse, Teller:*⟩ zerbrechen; ⟨*Eis:*⟩ brechen; ~ **in two/in pieces** durchbrechen/zerbrechen; **b)** ~ **into** einbrechen in *(+ Akk.)* ⟨*Haus*⟩; aufbrechen ⟨*Auto, Safe*⟩; ~ **into laughter/tears** in Gelächter/Tränen ausbrechen; ~ **into a trot/run** zu traben/laufen anfangen; **c)** *(escape)* ~ **out of prison** aus dem Gefängnis ausbrechen; ~ **free or loose** sich losreißen; **d)** ⟨*Welle:*⟩ sich brechen **(on/against** an *(+ Dat.);* **e)** ⟨*Tag:*⟩ anbrechen; ⟨*Sturm:*⟩ losbrechen; **f)** sb's **voice is ~ing** jmd. kommt in den Stimmbruch. **3.** *n.* **a)** Bruch, *der; (of rope)* Reißen, *das;* **a ~ with sb./sth.** ein Bruch mit jmdm./etw.; **b)** *(gap)* Lücke, *die; (broken place)* Sprung, *der;* **c)** *(dash)* **they made a sudden ~:** sie stürmten plötzlich davon; **d)** *(interruption)* Unterbrechung, *die; (pause, holiday)* Pause, *die;* **take or have a ~:** Pause machen; **e)** *(coll. chance)* Chance, *die.* **break 'down 1.** *v.i.* zusammenbrechen; ⟨*Verhandlungen:*⟩ scheitern; ⟨*Auto:*⟩ eine Panne haben. **2.** *v.t.* **a)** aufbrechen ⟨*Tür*⟩; brechen ⟨*Widerstand*⟩; niederreißen ⟨*Barriere, Schranke*⟩; **b)** *(analyse)* aufgliedern. **break 'in 1.** *v.i. (into building etc.)* einbrechen. **2.** *v.t.* **a)** zureiten ⟨*Pferd*⟩; **b)** einlaufen ⟨*Schuhe*⟩; **c)** ~ **the door in** die Tür aufbrechen. **'break into** *see* ~ **2b. break 'off 1.** *v.t.* abbrechen; abreißen ⟨*Faden*⟩; auflösen ⟨*Verlobung*⟩. **2.** *v.i.* **a)** abbrechen; **b)** *(cease)* aufhören. **break 'out** *v.i.* ausbrechen; ~ **out in spots/a rash** Pickel/einen Ausschlag bekommen. **break 'up 1.** *v.t.* **a)** (~ **into pieces)** zerkleinern; ausschlachten ⟨*Auto*⟩; aufbrechen ⟨*Erde*⟩; **b)** *(disband)* auflösen. **2.** *v.i.* **a)** (~ **into pieces, lit. or fig.)** zerbrechen; **b)** *(disband)* sich auflösen; ⟨*Schule:*⟩ schließen; ⟨*Schüler, Lehrer:*⟩ in die Ferien gehen; **c)** ~ **up [with sb.]** sich [von jmdm.] trennen
breakable ['breɪkəbl] **1.** *adj.* zerbrechlich. **2.** *n.* **~s** zerbrechliche Dinge
breakage ['breɪkɪdʒ] *n.* Zerbrechen, *das;* **~s must be paid for** zerbrochene Ware muß bezahlt werden
'breakdown *n.* **a)** *(of vehicle)* Panne, *die; (in machine)* Störung, *die;* ~ **truck/van** Abschleppwagen, *der;* **b)** *(Med.)* Zusammenbruch, *der;* **c)** *(analysis)* Aufschlüsselung, *die*

49 **brim**

'breaker n. **a)** *(wave)* Brecher, der; **b)** ~'s |yard| Autoverwertung, die
breakfast ['brekfəst] **1.** n. Frühstück, das; **for** ~: zum Frühstück. **2.** v.i. frühstücken. **'breakfast cereal** n. ≈ Frühstücksflocken Pl. **breakfast 'television** n. Frühstücksfernsehen, das
'break-in n. Einbruch, der
'breaking n. ~ **and entering** *(Law)* Einbruch, der
break: ~**neck** adj. halsbrecherisch; ~**through** n. Durchbruch, der; ~**up** n. Auflösung, die; *(of relationship)* Bruch, der; ~**water** n. Wellenbrecher, der
breast [brest] n. Brust, die
breast: ~**bone** n. Brustbein, das; ~**-feed** v.t.&i. stillen; ~**-stroke** n. Brustschwimmen, das
breath [breθ] n. **a)** Atem, der; **get one's** ~ **back** wieder zu Atem kommen; **hold one's** ~: den Atem anhalten; **be out of** ~: außer Atem sein; **say sth. under one's** ~: etw. vor sich *(Akk.)* hin murmeln; **b)** *(one respiration)* Atemzug, der. **Breathalyser** *(Brit.),* **Breathalyzer (P)** ['breθəlaɪzə(r)] n. Alcotest-Röhrchen Ⓦ, das; ~ **test** Alcotest Ⓦ, der
breathe [bri:ð] **1.** v.i. atmen; ~ **in** einatmen; ~ **out** ausatmen. **2.** v.t. **a)** ~ |in/out| ein-/ausatmen; **b)** *(utter)* hauchen. **'breather** ['bri:ðə(r)] n. Verschnaufpause, die
'breathless adj. atemlos (with vor + Dat.)
'breath-taking adj. atemberaubend
bred see breed 1, 2
breeches ['brɪtʃɪz] n. pl. |pair of| ~: |Knie|bundhose, die; |riding-|~: Reithose, die
breed [bri:d] **1.** v.t., **bred** [bred] **a)** *(cause)* erzeugen; **b)** züchten ⟨Tiere, Pflanzen⟩. **2.** v.i., **bred** sich vermehren. **3.** n. *(of animals)* Rasse, die. **'breeding** n. |good| ~: gute Erziehung
breeze [bri:z] n. Brise, die. **breezy** ['bri:zɪ] adj. windig
brevity ['brevɪtɪ] n. Kürze, die
brew [bru:] **1.** v.t. brauen ⟨Bier⟩; ~ |up| kochen ⟨Kaffee, Tee usw.⟩. **2.** v.i. **a)** ⟨Bier:⟩ gären; ⟨Kaffee, Tee:⟩ ziehen; **b)** ⟨Unwetter:⟩ sich zusammenbrauen. **3.** n. *(brewed beer/tea)* Bier, das/Tee, der. **'brewer** n. Brauer, der; *(firm)* Brauerei, die. **brewery** ['bru:ərɪ] n. Brauerei, die

bribe [braɪb] **1.** n. Bestechung, die. **2.** v.t. bestechen; ~ **sb. to do/into doing sth.** jmdn. bestechen, damit er etw. tut. **bribery** ['braɪbərɪ] n. Bestechung, die
brick [brɪk] **1.** n. Ziegelstein, der; *(toy)* Bauklötzchen, das. **2.** adj. Ziegelstein-. **'bricklayer** n. Maurer, der. **'bricklaying** n. Mauern, das
bridal ['braɪdl] adj. Braut-
bride [braɪd] n. Braut, die. **'bridegroom** n. Bräutigam, der. **bridesmaid** ['braɪdzmeɪd] n. Brautjungfer, die
¹bridge [brɪdʒ] **1.** n. **a)** Brücke, die; **b)** *(Naut.)* |Kommando|brücke, die; **c)** *(of nose)* Nasenbein, das; **d)** *(of spectacles)* Steg, der. **2.** v.t. eine Brücke bauen über (+ Akk.)
²bridge n. *(Cards)* Bridge, das
bridle ['braɪdl] n. Zaum, der. **'bridle path** n. Reitweg, der
¹brief [bri:f] adj. **a)** kurz; gering ⟨Verspätung⟩; **b)** *(concise)* knapp; **in** ~, **to be** ~: kurz gesagt
²brief 1. n. *(instructions)* Instruktionen Pl.; *(Law: case)* Mandat, das. **2.** v.t. Instruktionen geben (+ Dat.); *(inform)* unterrichten. **'briefcase** n. Aktentasche, die. **'briefing** n. Briefing, das; *(of reporters)* Unterrichtung, die
'briefly adv. **a)** kurz; **b)** *(concisely)* knapp; kurz
briefs [bri:fs] n. pl. |pair of| ~: Slip, der
brigade [brɪ'geɪd] n. *(Mil.)* Brigade, die. **brigadier** [brɪgə'dɪə(r)] n. Brigadegeneral, der
bright [braɪt] adj. **a)** hell; grell ⟨Scheinwerfer|licht|, Sonnenlicht⟩; strahlend ⟨Sonnenschein, Augen, Tag⟩; leuchtend ⟨Farbe, Blume⟩; ~ **intervals/periods** Aufheiterungen; **b)** *(cheerful)* fröhlich; **c)** *(clever)* intelligent.
brighten ['braɪtn] **1.** v.t. ~ |up| aufhellen. **2.** v.i. **the weather** *or* **it is** ~**ing** |up| es klärt sich auf. **'brightly** adv. **a)** hell; **b)** *(cheerfully)* fröhlich. **'brightness** n. see bright: **a)** Helligkeit, die; Grelle, die; Strahlen, das; Leuchtkraft, die; **b)** Fröhlichkeit, die; **c)** Intelligenz, die
brilliance ['brɪlɪəns] n. see brilliant: **a)** Helligkeit, die; Leuchten, das; **b)** Genialität, die; Glanz, der
brilliant ['brɪljənt] adj. **a)** hell ⟨Licht⟩; leuchtend ⟨Farbe⟩; **b)** genial ⟨Mensch, Gedanke, Leistung⟩; glänzend ⟨Verstand, Aufführung, Idee⟩
brim [brɪm] **1.** n. Rand, der; *(of hat)* |Hut|krempe, die. **2.** v.i., **-mm-:** be

~**ming with sth.** randvoll mit etw. sein.
brim-'full *pred. adj.* randvoll (with mit)
brine [braɪn] *n.* Salzwasser, *das*
bring [brɪŋ] *v. t.,* brought [brɔːt] **a)** bringen; *(as a present or favour)* mitbringen; ~ sth. with one etw. mitbringen; **b)** ~ sb. to do sth. jmdn. dazu bringen, etw. zu tun; **I could not** ~ **myself to do it** ich konnte es nicht über mich bringen, es zu tun. **bring a'bout** *v. t.* verursachen. **bring 'back** *v. t.* **a)** *(return)* zurückbringen; *(from a journey)* mitbringen; **b)** *(recall)* in Erinnerung bringen; **c)** *(restore, reintroduce)* wieder einführen. **bring 'down** *v. t.* **a)** herunterbringen; **b)** *(kill, wound)* zur Strecke bringen; **c)** senken ⟨*Preise, Inflationsrate, Fieber*⟩. **bring 'forward** *v. t.* **a)** nach vorne bringen; **b)** vorbringen ⟨*Argument*⟩; zur Sprache bringen ⟨*Fall, Angelegenheit*⟩; **c)** vorverlegen ⟨*Termin*⟩ (to auf + *Akk.*). **bring 'in** *v. t.* hereinbringen; einbringen ⟨*Gesetzesvorlage, Verdienst, Summe*⟩. **bring 'off** *v. t. (conduct successfully)* zustande bringen. **bring 'on** *v. t.* **a)** *(cause)* verursachen; **b)** *(Sport)* einsetzen. **bring 'out** *v. t.* **a)** herausbringen; **b)** hervorheben ⟨*Unterschied*⟩; **c)** einführen ⟨*Produkt*⟩; herausbringen ⟨*Buch, Zeitschrift*⟩. **bring 'up** *v. t.* **a)** heraufbringen; **b)** *(educate)* erziehen; *(rear)* aufziehen; **c)** zur Sprache bringen ⟨*Angelegenheit, Thema, Problem*⟩
brink [brɪŋk] *n.* Rand, *der;* **be on the** ~ **of doing sth.** nahe daran sein, etw. zu tun
brisk [brɪsk] *adj.* flott ⟨*Gang*⟩; forsch ⟨*Person, Art*⟩; frisch ⟨*Wind*⟩; *(fig.)* rege ⟨*Handel, Nachfrage*⟩; lebhaft ⟨*Geschäft*⟩. **'briskly** *adv.* flott
bristle ['brɪsl] **1.** *n.* Borste, *die.* **2.** *v. i.* **a)** ~ [up] ⟨*Haare:*⟩ sich sträuben; **b)** ~ **with** *(fig.)* starren von (+ *Dat.*). **'bristly** ['brɪslɪ] *adj.* borstig
Britain ['brɪtn] *pr. n.* Großbritannien *(das)*
British ['brɪtɪʃ] **1.** *adj.* britisch; **he/she is** ~: er ist Brite/sie ist Britin. **2.** *n. pl.* **the** ~: die Briten. **British 'Isles** *pr. n. pl.* Britische Inseln
Briton ['brɪtn] *n.* Brite, *der*/Britin, *die*
Brittany ['brɪtənɪ] *pr. n.* Bretagne, *die*
brittle ['brɪtl] *adj.* spröde ⟨*Material*⟩
broach [brəʊtʃ] *v. t.* anschneiden ⟨*Thema*⟩
broad [brɔːd] *adj.* **a)** breit; *(extensive)* weit ⟨*Ebene, Land*⟩; ausgedehnt

⟨*Fläche*⟩; **b)** *(explicit)* klar ⟨*Hinweis*⟩; breit ⟨*Lächeln*⟩; **c)** *(main)* grob; *(generalized)* allgemein; **d)** stark ⟨*Akzent*⟩.
broad 'bean *n.* Saubohne, *die*
broadcast ['brɔːdkɑːst] **1.** *n.* Sendung, *die; (live)* Übertragung, *die.* **2.** *v. t.,* broadcast senden; übertragen ⟨*Livesendung*⟩. **3.** *v. i.,* broadcast senden. **'broadcasting** *n.* Senden, *das; (live)* Übertragen, *das;* **work in** ~: beim Funk arbeiten
broaden ['brɔːdn] **1.** *v. t.* **a)** verbreitern; **b)** ausweiten ⟨*Diskussion*⟩. **2.** *v. i.* sich verbreitern; *(fig.)* sich erweitern
'broadly *adv.* **a)** deutlich ⟨*hinweisen*⟩; breit ⟨*grinsen, lächeln*⟩; **b)** *(in general)* allgemein ⟨*beschreiben*⟩; ~ **speaking** allgemein gesagt
broad: ~-'minded *adj.* tolerant; ~side *n.* Breitseite, die
brocade [brə'keɪd] *n.* Brokat, *der*
broccoli ['brɒkəlɪ] *n.* Brokkoli, *der*
brochure ['brəʊʃə(r)] *n.* Broschüre, *die;* Prospekt, *der*
broil ['brɔɪl] *v. t. (esp. Amer.)* grillen
broke [brəʊk] **1.** *see* break 1, 2. **2.** *pred. adj. (coll.)* pleite *(ugs.)*
broken ['brəʊkn] **1.** *see* break 1, 2. **2.** *adj.* **a)** zerbrochen; gebrochen ⟨*Bein, Hals*⟩; verletzt ⟨*Haut*⟩; abgebrochen ⟨*Zahn*⟩; gerissen ⟨*Seil*⟩; kaputt *(ugs.)* ⟨*Uhr, Fernsehen, Fenster*⟩; ~ **glass** Glasscherben; **b)** *(imperfect)* gebrochen; **in** ~ **English** in gebrochenem Englisch; **c)** *(fig.)* ruiniert ⟨*Ehe*⟩; gebrochen ⟨*Mensch, Herz*⟩. **'broken-down** *adj.* baufällig ⟨*Gebäude*⟩; kaputt *(ugs.)* ⟨*Wagen*⟩. **broken-'hearted** *adj.* untröstlich
broker ['brəʊkə(r)] *n.* Makler, *der*
brolly ['brɒlɪ] *n. (Brit. coll.)* [Regen]schirm, *der*
bronchitis [brɒŋ'kaɪtɪs] *n.* Bronchitis, *die*
bronze [brɒnz] **1.** *n.* Bronze, *die.* **2.** *attrib. adj.* Bronze-; *(coloured like* ~*)* bronzefarben
brooch [brəʊtʃ] *n.* Brosche, *die*
brood [bruːd] **1.** *n.* Brut, *die.* **2.** *v. i.* [vor sich *(Akk.)* hin] brüten
brook [brʊk] *n.* Bach, *der*
broom [bruːm] *n.* **a)** Besen, *der;* **b)** *(Bot.)* Ginster, *der.* **'broomcupboard** *n.* Besenschrank, *der.* **'broomstick** *n.* Besenstiel, *der*
broth [brɒθ] *n.* Brühe, *die*
brothel ['brɒθl] *n.* Bordell, *das*
brother ['brʌðə(r)] *n.* Bruder, *der;* **my** ~**s and sisters** meine Geschwister.

'**brotherhood** *n. (organization)* Bruderschaft, *die.* '**brother-in-law** *n., pl.* brothers-in-law Schwager, *der*
brought *see* bring
brow [braʊ] *n.* a) *(eye~)* Braue, *die;* b) *(forehead)* Stirn, *die;* c) *(of hill)* Kuppe, *die*
'**browbeat** *v.t., forms as* beat 1 einschüchtern
brown [braʊn] 1. *adj.* braun. 2. *n.* Braun, *das.* **brown** '**bread** *n.* ≈ Mischbrot, *das*
Brownie ['braʊnı] *n.* Wichtel, *die*
brown '**paper** *n.* Packpapier, *das*
browse [braʊz] *v.i. (in shop)* sich umsehen; *(read)* blättern (**through** in + *Dat.*)
bruise [bru:z] 1. *n.* a) *(Med.)* blauer Fleck; b) *(on fruit)* Druckstelle, *die.* 2. *v.t.* quetschen ⟨*Obst, Pflanzen*⟩; ~ oneself/one's leg sich stoßen/sich am Bein stoßen
brunette [bru:'net] 1. *n.* Brünette, *die.* 2. *adj.* brünett
brunt [brʌnt] *n.* bear the ~ of the attack/financial cuts von dem Angriff/von den Einsparungen am meisten betroffen sein
brush [brʌʃ] 1. *n.* a) Bürste, *die; (for sweeping)* Besen, *der; (with short handle)* Handfeger, *der; (for painting or writing)* Pinsel, *der;* b) *(skirmish)* Zusammenstoß, *der;* c) *(light touch)* flüchtige Berührung. 2. *v.t.* a) kehren; fegen; abbürsten ⟨*Kleidung*⟩; ~ one's teeth/hair sich *(Dat.)* die Zähne putzen/die Haare bürsten; b) *(touch in passing)* streifen. 3. *v.i.* ~ past sb./sth. jmdn./etw. streifen. **brush** '**up** *v.t. & i.* ~ up |on| auffrischen ⟨*Kenntnisse usw.*⟩
brusque [brʌsk] *adj.,* '**brusquely** *adv.* schroff
Brussels ['brʌslz] *pr. n.* Brüssel *(das).* **Brussels** '**sprouts** *n. pl.* Rosenkohl, *der*
brutal ['bru:tl] *adj.* brutal. **brutality** [bru:'tælıtı] *n.* Brutalität, *die.* **brutally** ['bru:təlı] *adv.* brutal
brute [bru:t] 1. *n.* a) *(animal)* Bestie, *die;* b) *(person)* Rohling, *der.* 2. *attrib. adj.* by ~ force mit roher Gewalt
B.Sc. *abbr.* Bachelor of Science
BST *abbr.* British Summer Time Britische Sommerzeit
bubble ['bʌbl] 1. *n.* Blase, *die; (small)* Perle, *die.* 2. *v.i.* ⟨*Wasser, Schlamm, Lava:*⟩ Blasen bilden. '**bubble bath** *n.* Schaumbad, *das*

¹**buck** [bʌk] *n. (deer, chamois)* Bock, *der; (rabbit, hare)* Rammler, *der*
²**buck** *n.* pass the ~ to sb. jmdm. die Verantwortung aufhalsen
³**buck** *(coll.)* 1. *v.i.* ~ '**up** a) *(make haste)* sich ranhalten *(ugs.);* b) *(cheer up)* ~ up! Kopf hoch! 2. *v.t.* ~ one's ideas up *(coll.)* sich zusammenreißen
⁴**buck** *n. (Amer. sl.)* Dollar, *der*
bucket ['bʌkıt] *n.* Eimer, *der*
buckle ['bʌkl] 1. *n.* Schnalle, *die.* 2. *v.t.* a) zuschnallen; ~ sth. on/up etw. anschnallen/festschnallen; b) verbiegen ⟨*Stoßstange, Rad*⟩. 3. *v.i.* ⟨*Rad, Metallplatte:*⟩ [sich] verbiegen
bud [bʌd] 1. *n.* Knospe, *die;* come into ~/be in ~: Knospen treiben. 2. *v.i., -dd-* Knospen treiben
Buddhism ['bʊdızm] *n.* Buddhismus, *der.* **Buddhist** ['bʊdıst] 1. *n.* Buddhist, *der*/Buddhistin, *die.* 2. *adj.* buddhistisch
budge [bʌdʒ] 1. *v.i.* sich rühren; ⟨*Gegenstand:*⟩ sich bewegen. 2. *v.t.* bewegen
budgerigar ['bʌdʒərıgɑ:(r)] *n.* Wellensittich, *der*
budget ['bʌdʒıt] 1. *n.* Etat, *der;* Haushalt[splan], *der.* 2. *v.i.* ~ for sth. etw. [im Etat] einplanen
budgie ['bʌdʒı] *n. (coll.)* Wellensittich, *der*
buff [bʌf] 1. *adj.* gelbbraun. 2. *n. (coll.: enthusiast)* Fan, *der (ugs.)*
buffalo ['bʌfələʊ] *n., pl.* ~es or same Büffel, *der*
buffer ['bʌfə(r)] *n.* Prellbock, *der; (on vehicle; also fig.)* Puffer, *der*
buffet ['bʊfeı] *n.* Büfett, *das.* '**buffet car** *n.* Büfettwagen, *der*
bug [bʌg] *n. (also coll.: microphone)* Wanze, *die*
buggy ['bʌgı] *n. (pushchair)* Sportwagen, *der*
bugle ['bju:gl] *n.* Bügelhorn, *das*
build [bıld] 1. *v.t.,* built [bılt] bauen; *(fig.)* aufbauen ⟨*System, Gesellschaft, Zukunft*⟩. 2. *v.i.,* built bauen. 3. *n.* Körperbau, *der.* **build** '**in** *v.t.* einbauen. **build** '**on** aufbauen auf (+ *Dat.*); bebauen ⟨*Gebiet*⟩. **build** '**up** 1. *v.t.* aufhäufen ⟨*Reserven, Mittel*⟩; kräftigen ⟨*Personen, Körper*⟩; steigern ⟨*Produktion, Kapazität*⟩; stärken ⟨*[Selbst]vertrauen*⟩; aufbauen ⟨*Firma, Geschäft*⟩. 2. *v.i.* ⟨*Spannung, Druck:*⟩ zunehmen; ⟨*Schlange, Rückstau:*⟩ sich bilden; ⟨*Verkehr:*⟩ sich verdichten
'**builder** *n.* Bauunternehmer, *der*

'building *n.* **a)** Bau, *der;* **b)** *(structure)* Gebäude, *das.* 'building-site *n.* Baustelle, *die.* 'building society *n.* *(Brit.)* Bausparkasse, *die*

built *see* build 1, 2

built: ~-in *adj.* **a)** eingebaut; Einbau- ⟨schrank, küche usw.⟩; **b)** *(fig.: instinctive)* angeboren; ~-up *adj.* bebaut; ~-up area Wohngebiet, *das; (Motor Veh.)* geschlossene Ortschaft

bulb [bʌlb] *n.* **a)** *(Bot., Hort.)* Zwiebel, *die;* **b)** *(of lamp)* [Glüh]birne, *die*

Bulgaria [bʌl'geərɪə] *pr. n.* Bulgarien *(das).* Bulgarian [bʌl'geərɪən] **1.** *adj.* bulgarisch. **2.** *n.* **a)** *(person)* Bulgare, *der/*Bulgarin, *die;* **b)** *(language)* Bulgarisch, *das; see also* English 2 a

bulge [bʌldʒ] **1.** *n.* Ausbeulung, *die;* ausgebeulte Stelle. **2.** *v. i.* sich wölben

bulk [bʌlk] *n.* **a)** *(large quantity)* in ~: in großen Mengen; **b)** *(large shape)* massige Gestalt; **c)** *(size)* Größe, *die;* **d)** *(greater part)* der größte Teil; *(of population, votes)* Mehrheit, *die.* 'bulky *adj.* sperrig ⟨Gegenstand⟩; massig ⟨Gestalt, Körper⟩

bull [bʊl] *n.* Bulle, *der; (esp. for bullfight)* Stier, *der*

'bulldog *n.* Bulldogge, *die*

bulldozer ['bʊldəʊzə(r)] *n.* Planier- raupe, *die*

bullet ['bʊlɪt] *n.* Kugel, *die*

bulletin ['bʊlɪtɪn] *n.* Bulletin, *das*

'bulletproof *adj.* kugelsicher

'bullfight *n.* Stierkampf, *der*

bullion ['bʊljən] *n.* gold ~: Goldbarren *Pl.*

bullock ['bʊlək] *n.* Ochse, *der*

bull: ~ring *n.* Stierkampfarena, *die;* ~'s-eye *n.* *(of target)* Schwarze, *das*

bully ['bʊlɪ] **1.** *n.* *(schoolboy etc.)* ≈ Rabauke, *der; (boss)* Tyrann, *der.* **2.** *v. t.* schikanieren; *(frighten)* ein- schüchtern

¹bum [bʌm] *n.* *(Brit. sl.)* Hintern, *der (ugs.)*

²bum *n.* *(Amer. sl.: tramp)* Penner, *der (salopp)*

bumble-bee ['bʌmblbi:] *n.* Hummel, *die*

bump [bʌmp] **1.** *n.* **a)** *(sound)* Bums, *der; (impact)* Stoß, *der;* **b)** *(swelling)* Beule, *die;* **c)** *(hump)* Buckel, *der (ugs.).* **2.** *adv.* bums. **3.** *v. t.* anstoßen. 'bump into *v. t.* **a)** stoßen an (+ *Akk.*); **b)** *(meet by chance)* zufällig [wieder]treffen

'bumper **1.** *n.* Stoßstange, *die.* **2.** *attrib. adj.* Rekord⟨ernte, -jahr⟩

'bumpy *adj.* holp[e]rig ⟨Straße, Fahrt, Fahrzeug⟩; uneben ⟨Fläche⟩; unruhig ⟨Flug⟩

bun [bʌn] *n.* süßes Brötchen; *(currant ~)* Korinthenbrötchen, *das*

bunch [bʌntʃ] *n.* **a)** *(of flowers)* Strauß, *der; (of grapes, bananas)* Traube, *die; (of parsley, radishes)* Bund, *das; ~ of flowers/grapes* Blumenstrauß, *der/* Traube, *die;* **a ~ of keys** ein Schlüsselbund; **b)** *(lot)* Anzahl, *die;* the best or pick of the ~: der/die/das Beste [von allen]; **c)** *(of people)* Haufen, *der (ugs.)*

bundle ['bʌndl] *n.* Bündel, *das; (of papers)* Packen, *der*

bung [bʌŋ] **1.** *n.* Spund, *der.* **2.** *v. t. (sl.)* schmeißen *(ugs.).* bung 'up *v. i.* be/ get ~ed up verstopft sein/verstopfen

bungalow ['bʌŋgələʊ] *n.* Bungalow, *der*

bungle ['bʌŋgl] *v. t.* stümpern bei

bunk [bʌŋk] *n.* *(in ship, lorry)* Koje, *die; (in sleeping-car)* Bett, *das; (~-bed)* Etagenbett, *das*

bunker ['bʌŋkə(r)] *n.* Bunker, *der*

bunny ['bʌnɪ] *n.* Häschen, *das*

buoy [bɔɪ] *n.* Boje, *die*

buoyancy ['bɔɪənsɪ] *n.* Auftrieb, *der*

buoyant ['bɔɪənt] *adj.* schwimmend; be ~: schwimmen

burden ['bɜ:dn] **1.** *n.* Last, *die;* become a ~ *(fig.)* zur Last werden. **2.** *v. t.* belasten (with mit)

bureau ['bjʊərəʊ, bjʊə'rəʊ] *n.* **a)** *(Brit.: writing-desk)* Sekretär, *der;* **b)** *(office)* Büro, *das*

bureaucracy [bjʊə'rɒkrəsɪ] *n.* Bürokratie, *die.* bureaucrat ['bjʊərəkræt] *n.* Bürokrat, *der/*Bürokratin, *die.* bureaucratic [bjʊərə'krætɪk] *adj.* bürokratisch

burglar ['bɜ:glə(r)] *n.* Einbrecher, *der.* 'burglar alarm *n.* Alarmanlage, *die*

burglary ['bɜ:glərɪ] *n.* Einbruch, *der*

burgle ['bɜ:gl] *v. t.* einbrechen in (+ *Akk.*); the shop/he was ~d in dem Laden/bei ihm wurde eingebrochen

burial ['berɪəl] *n.* Begräbnis, *das*

burly ['bɜ:lɪ] *adj.* stämmig

Burma ['bɜ:mə] *pr. n.* Birma *(das)*

burn [bɜ:n] **1.** *n.* *(on the skin)* Verbrennung, *die; (on material)* Brandfleck, *der.* **2.** *v. t.*, ~t [bɜ:nt] or ~ed **a)** verbrennen; ~ oneself/one's hand sich verbrennen/sich *(Dat.)* die Hand verbrennen; ~ a hole in sth. ein Loch in etw. *(Akk.)* brennen; **b)** als Brennstoff verwenden ⟨Gas, Öl usw.⟩; heizen mit ⟨Kohle, Holz, Torf⟩; **c)** *(spoil)*

anbrennen lassen ⟨*Fleisch, Kuchen*⟩; **be ~t** angebrannt sein. **3.** *v.i.*, **~t** or **~ed** brennen; **~ to death** verbrennen; **she ~s easily** sie bekommt leicht einen Sonnenbrand. **burn 'down** *v.t. & i.* niederbrennen
'**burner** *n.* Brenner, *der*
'**burning** *adj.* glühend ⟨*Leidenschaft, Haß, Wunsch*⟩; brennend ⟨*Wunsch, Frage, Problem*⟩
burnt *see* burn 2, 3
burp [bɜːp] *(coll.)* **1.** *n.* Rülpser, *der (ugs.).* **2.** *v.i.* rülpsen *(ugs.)*
burrow ['bʌrəʊ] **1.** *n.* Bau, *der.* **2.** *v.i.* [sich *(Dat.)*] einen Gang graben
burst [bɜːst] **1.** *n.* **a)** *(split)* Bruch, *der;* **b)** *(of firing)* Salve, *die;* **c)** *(fig.)* **a ~ of applause/cheering** ein Beifallsausbruch/Beifallsrufe *Pl.* **2.** *v.t.*, **burst** zum Platzen bringen; platzen lassen ⟨*Luftballon*⟩; **~ pipe** Rohrbruch, *der.* **3.** *v.i.*, **burst a)** platzen; ⟨*Bombe:*⟩ explodieren; ⟨*Damm:*⟩ brechen; ⟨*Flußufer:*⟩ überschwemmt werden; ⟨*Furunkel, Geschwür:*⟩ aufgehen; **b) be ~ing** with sth. zum Bersten voll sein mit etw.; **be ~ing with pride/impatience/excitement** vor Stolz/Ungeduld platzen/vor Aufregung außer sich sein. **'burst into** *v.t.* **a)** eindringen in; **b) ~ into tears/laughter** in Tränen/ Gelächter ausbrechen; **~ into flames** in Brand geraten. **burst 'out** *v.i.* **a)** herausstürzen; **b)** *(exclaim)* losplatzen; **c) ~ out laughing/crying** in Lachen/Tränen ausbrechen
bury ['berɪ] *v.t.* **a)** begraben; **b)** *(hide)* vergraben; **~ one's face in one's hands** das Gesicht in den Händen vergraben; **c) ~ one's teeth in sth.** seine Zähne in etw. *(Akk.)* graben
bus [bʌs] *n.* Bus, *der*
bus: ~-conductor *n.* Busschaffner, *der;* **~-driver** *n.* Busfahrer, *der;* **~ fare** *n.* [Bus]fahrpreis, *der*
bush [bʊʃ] *n.* **a)** Busch, *der;* **b)** *(shrubs)* Gebüsch, *das.* '**bushy** *adj.* buschig
busily ['bɪzɪlɪ] *adj.* eifrig
business ['bɪznɪs] *n.* **a)** *(trading operation)* Geschäft, *das;* *(company, firm)* Betrieb, *der;* *(large)* Unternehmen, *das;* **b)** *(buying and selling)* Geschäfte *Pl.;* **c)** *(task, province)* Aufgabe, *die;* **mind your own ~!** kümmere dich um deine [eigenen] Angelegenheiten!; **d)** *(difficult matter)* Problem, *das*
business: ~ letter *n.* Geschäftsbrief, *der;* **~-like** *adj.* geschäftsmäßig ⟨*Art*⟩; geschäftstüchtig ⟨*Person*⟩;

~man *n.* Geschäftsmann, *der;* **~ school** *n.* kaufmännische Fachschule; **~woman** *n.* Geschäftsfrau, *die*
busker ['bʌskə(r)] *n.* Straßenmusikant, *der*
bus: ~-route *n.* Buslinie, *die;* **~ shelter** *n.* Wartehäuschen, *das;* **~-station** *n.* Omnibusbahnhof, *der;* **~-stop** *n.* Bushaltestelle, *die*
¹**bust** [bʌst] *n.* **a)** *(sculpture)* Büste, *die;* **b)** *(measurement)* Oberweite, *die*
²**bust** *(coll.)* **1.** *adj.* kaputt *(ugs.).* **2.** *v.t.*, **~ed** or **bust** *(break)* kaputtmachen *(ugs.);* **~ sth. open** etw. aufbrechen. **3.** *v.i.*, **~ed** or **bust** kaputtgehen *(ugs.)*
'**bus-ticket** *n.* Busfahrkarte, *die*
bustle ['bʌsl] **1.** *v.i.* **~ about** geschäftig hin und her eilen. **2.** *n.* Betrieb, *der.*
bustling ['bʌslɪŋ] *adj.* belebt ⟨*Straße, Stadt, Markt usw.*⟩; rege ⟨*Tätigkeit*⟩
busy ['bɪzɪ] **1.** *adj.* **a)** beschäftigt (at, with mit); arbeitsreich ⟨*Leben*⟩; ziemlich hektisch ⟨*Zeit*⟩; **I'm ~ now** ich habe jetzt zu tun; **he was ~ packing** er war mit Packen beschäftigt; **b)** *(Amer. Teleph.)* besetzt. **2.** *v.refl.* **~ oneself** sich beschäftigen (with mit). '**busybody** *n.* G[e]schaftlhuber, *der*
but 1. [bət, *stressed* bʌt] *conj.* aber; correcting after a negative sondern; **not that book ~ this one** nicht das Buch, sondern dieses. **2.** [bət] *prep.* außer (+ *Dat.*); **the next/last ~ one** der/die/ das übernächste/vorletzte
butcher ['bʊtʃə(r)] **1.** *n.* Fleischer, *der.* **2.** *v.t.* *(murder)* niedermetzeln
butler ['bʌtlə(r)] *n.* Butler, *der*
¹**butt** [bʌt] *n.* **a)** *(of rifle)* Kolben, *der;* **b)** *(of cigarette, cigar)* Stummel, *der*
²**butt** *n.* *(object of teasing or ridicule)* Zielscheibe, *die*
³**butt 1.** *n.* *(push)* *(by person)* [Kopf]stoß, *der;* *(by animal)* Stoß [mit den Hörnern]. **2.** *v.t. & i.* mit dem Kopf/den Hörnern stoßen. **butt 'in** *v.i.* dazwischenreden
butter ['bʌtə(r)] **1.** *n.* Butter, *die.* **2.** *v.t.* buttern
butter: ~-bean *n.* Mondbohne, *die;* **~cup** *n.* Butterblume, *die;* **~fly** *n.* **a)** Schmetterling, *der;* **b) ~ |stroke|** Delphinstil, *der*
buttock ['bʌtək] *n.* Hinterbacke, *die;* Gesäßhälfte, *die;* **~s** Gesäß, *das*
button ['bʌtn] **1.** *n.* Knopf, *der.* **2.** *v.t.* **~ |up|** zuknöpfen. '**buttonhole 1.** *n.* **a)** Knopfloch, *das;* **b)** *(flower)* Knopflochblume, *die.* **2.** *v.t.* zu fassen kriegen *(ugs.)*

buttress ['bʌtrɪs] *n. (Archit.)* Mauer-
stütze, *die*
buxom ['bʌksəm] *adj.* drall
buy [baɪ] **1.** *v.t.,* **bought** [bɔːt] kaufen;
~ **sb./oneself sth.** jmdm./sich etw.
kaufen. **2.** *n.* [Ein]kauf, *der;* **be a good**
~: preiswert sein. **buy 'up** *v.t.* auf-
kaufen
'**buyer** *n.* Käufer, *der*/Käuferin, *die*
buzz [bʌz] **1.** *n.* Summen, *das.* **2.** *v.i.*
summen. **buzz 'off** *v.i. (sl.)* abhauen
(salopp)
'**buzzer** *n.* Summer, *der*
by [baɪ] **1.** *prep.* **a)** *(near, beside)* an
(+ *Dat.*); bei; *(next to)* neben; ~ **the
window/river** am Fenster/Fluß; **b)** *(to
position beside)* zu; **c)** *(about, in the
possession of)* bei; **d)** [all] **by herself/
himself** *etc.* [ganz] allein[e]; **e)** *(along)*
entlang; *(via)* über (+ *Akk.*); **f)** *(passing)* vorbei an (+ *Dat.*); **g)** *(during)*
bei; **by day/night** bei Tag/Nacht; **h)**
(through the agency of) von; **written
by ...**: geschrieben von ...; **i)** *(through
the means of)* durch; **by bus/ship** *etc.*
mit dem Bus/Schiff *usw.;* **by air/sea**
mit dem Flugzeug/Schiff; **j)** *(not later
than)* bis; **by now/this time** inzwischen; **k)** *indicating unit* pro; **by the
minute/hour** pro Minute/Stunde; **day
by day/month by month** Tag für Tag/
Monat für Monat; **10 ft. by 20 ft.** 10
[Fuß] mal 20 Fuß; **l)** *indicating
amount* **one by one** einzeln; **two by
two/three by three** zu zweit/dritt; **m)**
indicating factor durch; **8 divided by 2
is 4** 8 geteilt durch 2 ist 4; **n)** *indicating
extent* um; **wider by a foot** um einen
Fuß breiter; **o)** *(according to)* nach. **2.**
adv. **a)** *(past)* vorbei; **b)** *(near)* **close/
near by** in der Nähe; **c) by and large** im
großen und ganzen
bye[-bye] ['baɪ(baɪ)] *int. (coll.)* tschüs
(ugs.)
bye-law *see* **by-law**
'**by-election** *n.* Nachwahl, *die*
bygone ['baɪgɒn] *adj.* vergangen
'**by-law** *n. (esp. Brit.)* Verordnung, *die*
'**bypass** **1.** *n.* Umgehungsstraße, *die.*
2. *v.t.* **a) the road** ~**es the town** die
Straße führt um die Stadt herum; **b)**
(fig.) übergehen
'**by-product** *n.* Nebenprodukt, *das*
'**by-road** *n.* Nebenstraße, *die*
bystander ['baɪstændə(r)] *n.* Zuschauer, *der*/Zuschauerin, *die*
byte [baɪt] *n. (Computing)* Byte, *das*
'**byway** *n.* Seitenweg, *der*
'**byword** *n.* Inbegriff, *der* **(for** *Gen.*)

C

C, c [siː] *n.* C, c, *das*
C. *abbr.* **a) Celsius** C; **b) Centigrade** C
cab [kæb] *n.* **a)** *(taxi)* Taxi, *das;* **b)** *(of
lorry, truck)* Fahrerhaus, *das; (of
train)* Führerstand, *der*
cabaret ['kæbəreɪ] *n.* Varieté, *das;
(satirical)* Kabarett, *das*
cabbage ['kæbɪdʒ] *n.* Kohl, *der;* **red/
white** ~: Rot-/Weißkohl, *der*
cabin ['kæbɪn] *n. (in ship) (for passengers)* Kabine, *die; (for crew)* Kajüte,
die; (in aircraft) Kabine, *die*
cabinet ['kæbɪnɪt] *n.* **a)** Schrank, *der;
(in bathroom, for medicines)* Schränk-
chen, *das;* [display] ~: Vitrine, *die;*
the C~ *(Polit.)* das Kabinett; **C~ Minister** Minister, *der*
cable ['keɪbl] **1.** *n.* **a)** *(rope)* Kabel, *das;
(of* ~*-car etc.)* Seil, *das;* **b)** *(Electr.,
Teleph.)* Kabel, *das;* **c)** *(message)* Kabel, *das.* **2.** *v.t.* kabeln ⟨*Mitteilung,
Nachricht*⟩. '**cable-car** *n.* Drahtseil-
bahn, *die.* **cable 'television** *n.* Ka-
belfernsehen, *das*
cache [kæʃ] *n.* geheimes [Waffen-/
Proviant-]lager
cackle ['kækl] **1.** *n.* **a)** *(of hen)*
Gackern, *das;* **b)** *(laughter)* [meckern-
des] Gelächter. **2.** *v.i.* **a)** ⟨*Henne:*⟩
gackern; **b)** *(laugh)* meckernd lachen
cactus ['kæktəs] *n., pl.* **cacti** ['kæktaɪ]
or ~**es** Kaktus, *der*
caddie ['kædɪ] *n. (Golf)* Caddie, *der*
caddy ['kædɪ] *n.* Dose, *die*
cadet [kə'det] *n.* Offiziersschüler, *der;
naval/police* ~: Marinekadett/Anwär-
ter für den Polizeidienst
cadge [kædʒ] *v.t.* [sich *(Dat.)*] erbetteln
café, cafe ['kæfeɪ] *n.* Lokal, *das; (tea-
room)* Café, *das*
cafeteria [kæfɪ'tɪərɪə] *n.* Cafeteria, *die*
caffeine ['kæfiːn] *n.* Koffein, *das*
cage [keɪdʒ] **1.** *n.* **a)** Käfig, *der;* **b)** *(of
lift)* Fahrkabine, *die.* **2.** *v.t.* einsperren
cagey ['keɪdʒɪ] *adj. (coll.)* zugeknöpft

(ugs.); **be ~ about sth.** mit etw. hinterm Berg halten *(ugs.).*

Cairo ['kaɪərəʊ] *pr. n.* Kairo *(das)*

cajole [kə'dʒəʊl] *v. t.* **~ sb. into sth./ doing sth.** jmdm. etw. einreden/jmdm. einreden, etw. zu tun

cake [keɪk] **1.** *n.* Kuchen, *der;* **a ~ of soap** ein Riegel *od.* Stück Seife. **2.** *v. t.* verkrusten; **~d with dirt/blood** schmutz-/blutverkrustet

calamity [kə'læmɪtɪ] *n.* Unheil, *das*

calcium ['kælsɪəm] *n.* Kalzium, *das*

calculate ['kælkjʊleɪt] **1.** *v. t.* **a)** berechnen; *(by estimating)* ausrechnen; **b) be ~d to do sth.** darauf abzielen, etw. zu tun. **2.** *v. i.* **~ on doing sth.** damit rechnen, etw. zu tun. **'calculated** *adj.* kalkuliert ⟨*Risiko*⟩; vorsätzlich ⟨*Handlung*⟩. **calculation** [kælkjʊ-'leɪʃn] *n.* **a)** *(result)* Rechnung, *die;* **he is out in his ~s** er hat sich verrechnet; **b)** *(calculating)* Berechnung, *die.* **calculator** ['kælkjʊleɪtə(r)] *n.* Rechner, *der*

calculus ['kælkjuːləs] *n.* **differential/ integral ~:** Differential-/Integralrechnung, *die*

calendar ['kælɪndə(r)] *n.* Kalender, *der; attrib.* Kalender-

¹calf [kɑːf] *n., pl.* **calves** Kalb, *das*

²calf *n., pl.* **calves** *(Anat.)* Wade, *die*

calibre *(Brit.; Amer.:* **caliber***)* ['kælɪbə(r)] *n.* Kaliber, *das*

calico ['kælɪkəʊ] *n.* Kattun, *der*

California [kælɪ'fɔːnɪə] *pr. n.* Kalifornien *(das)*

caliper *see* **calliper**

call [kɔːl] **1.** *v. i.* **a)** rufen; **~ to sb.** jmdm. etwas zurufen; **~ [out] for help** um Hilfe rufen; **b)** *(pay brief visit)* [kurz] besuchen *(at Akk.);* **~ on sb.** jmdn. besuchen; **~ round** vorbeikommen *(ugs.);* **~ at a port/station** einen Hafen anlaufen/an einem Bahnhof halten; **c)** *(Teleph.)* **who is ~ing, please?** wer spricht da, bitte?; **thank you for ~ing** vielen Dank für Ihren Anruf! **2.** *v. t.* **a)** rufen; aufrufen ⟨*Namen, Nummer*⟩; **b)** *(cry to, summon)* rufen; *(to a duty, to do sth.)* aufrufen; **c)** *(by radio/telephone)* rufen/ anrufen; *(initially)* Kontakt aufnehmen mit; **d)** *(rouse)* wecken; **e)** einberufen ⟨*Konferenz*⟩; ausrufen ⟨*Streik*⟩; **f)** *(name)* nennen; **he is ~ed Bob** er heißt Bob; **what is it ~ed in English?** wie heißt das auf englisch? **3.** *n.* **a)** Ruf, *der;* **a ~ for help** ein Hilferuf; **be on ~:** Bereitschaftsdienst haben; **b)**

(visit) Besuch, *der;* **make or pay a ~ on sb., make or pay sb. a ~:** jmdn. besuchen; **c)** *(telephone ~)* Anruf, *der;* **give sb. a ~:** jmdn. anrufen; **make a ~:** telefonieren; **d)** *(invitation, summons)* Aufruf, *der;* **e)** *(need, occasion)* Anlaß, *der.* **call 'back 1.** *v. t.* zurückrufen. **2.** *v. i.* zurückrufen; *(come back)* zurückkommen. **'call for** *v. t.* **a)** *(send for, order)* bestellen; **b)** *(collect)* abholen; **c)** *(require, demand)* erfordern; **this ~s for a celebration** das muß gefeiert werden. **call 'in 1.** *v. i.* vorbeikommen *(ugs.)* (on bei). **2.** *v. t.* zu Rate ziehen ⟨*Fachmann usw.*⟩. **call 'off** *v. t.* absagen ⟨*Treffen, Verabredung*⟩; rückgängig machen ⟨*Geschäft*⟩; lösen ⟨*Verlobung*⟩; *(end)* abbrechen ⟨*Streik*⟩. **'call on** *v. t.* **a)** *see* **~ 1b; b)** *see* **~ [up]on. call 'out 1.** *v. t.* alarmieren ⟨*Truppen*⟩; zum Streik aufrufen ⟨*Arbeitnehmer*⟩. **2.** *v. i. see* **~ 1 a. call 'up** *v. t.* **a)** *(by telephone)* anrufen; **b)** *(Mil.)* einberufen. **'call [up]on** *v. t.* **~ upon sb.'s generosity** an jmds. Großzügigkeit *(Akk.)* appellieren; **~ [up]on sb. to do sth.** jmdn. auffordern, etw. zu tun

'call-box *n.* Telefonzelle, *die*

'caller *n. (visitor)* Besucher, *der/*Besucherin, *die; (on telephone)* Anrufer, *der/*Anruferin, *die*

'call-girl *n.* Callgirl, *das*

'calling *n.* Beruf, *der*

calliper ['kælɪpə(r)] *n.* **a)** [a pair of] **~s** Tasterzirkel, *der;* **b)** *(Med.)* Beinschiene, *die*

callous ['kæləs] *adj.* gefühllos; herzlos ⟨*Handlung, Verhalten*⟩

'call-up *n. (Mil.)* Einberufung, *die*

calm [kɑːm] **1.** *n. (stillness)* Stille, *die; (serenity)* Ruhe, *die.* **2.** *adj.* ruhig. **3.** *v. t.* **~ sb. [down]** jmdn. beruhigen. **4.** *v. i.* **~ [down]** sich beruhigen. **'calmly** *adv.* ruhig; gelassen. **'calmness** *n.* Ruhe, *die; (of water)* Stille, *die*

Calor gas, (P) ['kælə gæs] *n.* Butangas, *das*

calorie ['kælərɪ] *n.* Kalorie, *die*

calves *pl. of* ¹,²**calf**

camber ['kæmbə(r)] *n.* Wölbung, *die*

came *see* **come**

camel ['kæml] *n.* Kamel, *das*

camera ['kæmərə] *n.* Kamera, *die*

'cameraman *n.* Kameramann, *der*

camouflage ['kæməflɑːʒ] **1.** *n.* Tarnung, *die.* **2.** *v. t.* tarnen

camp [kæmp] **1.** *n.* Lager, *das.* **2.** *v. i.* **~ [out]** campen; *(in tent)* zelten; **go ~ing** Campen/Zelten fahren/gehen

campaign [kæm'peɪn] **1.** *n.* **a)** *(Mil.)*
Feldzug, *der;* **b)** *(organized action)*
Kampagne, *die;* **publicity ~:** Werbekampagne, *die.* **2.** *v. i.* **~ for/against**
sth. sich für etw. einsetzen/gegen etw.
etwas unternehmen; **be ~ing** ⟨*Po-*
litiker:⟩ im Wahlkampf stehen
'**camp-bed** *n.* Campingliege, *die*
'**camper** *n.* *(person)* Camper, *der/*
Camperin, *die*
'**camping** *n.* Camping, *das;* *(in tent)*
Zelten, *das.* '**camping-ground**
(Amer.), '**camping site** *ns.* Campingplatz, *der*
'**campsite** *n.* Campingplatz, *der*
campus ['kæmpəs] *n.* Campus, *der*
¹**can** [kæn] **1.** *n.* **a)** *(milk ~, watering-~)*
Kanne, *die;* *(for oil, petrol)* Kanister,
der; *for refuse)* Eimer, *der;* **b)**
(for preserving) [Konserven]dose, *die;*
a ~ of tomatoes/beer eine Dose Toma-
ten/Bier. **2.** *v. t.,* **-nn-** konservieren
²**can** [kən, *stressed* kæn] *v. aux., only in*
pres. **can,** *neg.* **cannot** ['kænət], *(coll.)*
can't [kɑ:nt], *past* **could** [kʊd], *neg.*
(coll.) **couldn't** ['kʊdnt] können; *(have*
right, be permitted) dürfen; können; **I**
can't do that das kann ich nicht; *(it*
would be wrong) das kann ich nicht
tun; **you can't smoke here** hier dürfen
Sie nicht rauchen; **could you ring me**
tomorrow? könnten Sie mich morgen
anrufen?; **I could have killed him** ich
hätte ihn umbringen können; **[that]**
could be |so| das könnte *od.* kann sein
Canada ['kænədə] *pr. n.* Kanada
(das). **Canadian** [kə'neɪdɪən] **1.** *adj.*
kanadisch. **2.** *n.* Kanadier, *der/*Kana-
dierin, *die*
canal [kə'næl] *n.* Kanal, *der*
canary [kə'neərɪ] *n.* Kanarienvogel,
der
Ca'nary Islands *pr. n. pl.* Kanarische
Inseln *Pl.*
cancel ['kænsl] **1.** *v. t.,* *(Brit.)* **-ll-** absagen ⟨*Besuch, Urlaub, Reise, Sportver-*
anstaltung⟩; ausfallen lassen ⟨*Veran-*
staltung, Vorlesung, Zug, Bus⟩; fallenlassen ⟨*Pläne*⟩; rückgängig machen
⟨*Einladung, Vertrag*⟩; zurücknehmen
⟨*Befehl*⟩; stornieren ⟨*Bestellung, Auf-*
trag⟩; kündigen ⟨*Abonnement*⟩; abbe-
stellen ⟨*Zeitung*⟩. **2.** *v. i.,* *(Brit.)* **-ll-:** **~**
|out| sich [gegenseitig] aufheben. **can-**
cellation [kænsə'leɪʃn] *n. see* **cancel**
1: Absage, *die;* Ausfall, *der;* Fallenlassen, *das;* Rückgängigmachen, *das;*
Zurücknahme, *die;* Stornierung, *die;*
Kündigung, *die;* Abbestellung, *die*

cancer ['kænsə(r)] *n.* **a)** *(Med.)* Krebs,
der; **b)** **C~** *(Astrol., Astron.)* der Krebs
candelabra [kændɪ'lɑ:brə] *n.* Leuchter, *der*
candid ['kændɪd] *adj.* offen; ehrlich
⟨*Ansicht, Bericht*⟩
candidate ['kændɪdət, 'kændɪdeɪt] *n.*
Kandidat, *der/*Kandidatin, *die*
candle ['kændl] *n.* Kerze, *die*
candle: ~-light *n.* Kerzenlicht, *das;*
~stick *n.* Kerzenhalter, *der;* *(elabor-*
ate) Leuchter, *der;* **~wick** *n.* *(ma-*
terial) Frottierplüsch, *der*
candour *(Brit., Amer.:* **candor)**
['kændə(r)] *n. see* **candid:** Offenheit,
die; Ehrlichkeit, *die*
candy ['kændɪ] *n.* *(Amer.)* *(sweets)* Süßigkeiten *Pl.;* *(sweet)* Bonbon, *das*
od. der. '**candyfloss** ['kændɪflɒs] *n.*
Zuckerwatte, *die*
cane [keɪn] **1.** *n.* **a)** *(stem)* Rohr, *das;*
(of raspberry, blackberry) Sproß, *der;*
b) *(material)* Rohr, *das;* **c)** *(stick)*
[Rohr]stock, *der.* **2.** *v. t.* [mit dem
Stock] schlagen
canine ['keɪnaɪn] *adj.* **a)** *(of dog[s])*
Hunde-; **b)** **~ tooth** Eckzahn, *der*
canister ['kænɪstə(r)] *n.* Büchse, *die;*
(for petrol, oil, etc.) Kanister, *der*
cannabis ['kænəbɪs] *n.* *(hashish)* Ha-
schisch, *das;* *(marijuana)* Marihuana,
das
canned [kænd] *adj.* Dosen-; in Dosen
nachgestellt; **~ meat/fruit** Fleisch-/
Obstkonserven *Pl.;* **~ beer** Dosenbier;
~ food [Lebensmittel]konserven *Pl.;* **~**
music Musikkonserve, *die*
cannibal ['kænɪbl] *n.* Kannibale,
*der/*Kannibalin, *die.* **cannibalism**
['kænɪbəlɪzm] *n.* Kannibalismus, *der*
cannon ['kænən] **1.** *n.* Kanone, *die.* **2.**
v. i. *(Brit.)* **~ into sb./sth.** mit etw./
jmdm. zusammenprallen. '**cannon-**
ball *n.* Kanonenkugel, *die*
cannot *see* ²**can**
canny ['kænɪ] *adj.* *(shrewd)* schlau
canoe [kə'nu:] *n.* Paddelboot, *das;* *(In-*
dian ~, Sport) Kanu, *das.* **canoeist**
[kə'nu:ɪst] *n.* Paddelbootfahrer, *der/*
-fahrerin, *die*
canon ['kænən] *n.* **a)** *(general law,*
criterion) Grundregel, *die;* **b)** *(Eccl.:*
person) Kanoniker, *der*
canonize ['kænənaɪz] *v. t.* kanonisie-
ren ⟨*Heiligen*⟩; heiligsprechen ⟨*Mär-*
tyrer⟩
'**can-opener** *n.* Dosenöffner, *der*
canopy ['kænəpɪ] *n.* Baldachin, *der;*
(over entrance) Vordach, *das*

can't [kɑːnt] *(coll.)* = **cannot;** *see* ²**can**
cantankerous [kæn'tæŋkərəs] *adj.* streitsüchtig
canteen [kæn'tiːn] *n.* Kantine, *die*
canter ['kæntə(r)] **1.** *n.* Handgalopp, *der.* **2.** *v.i.* leicht galoppieren
canvas ['kænvəs] *n.* Leinwand, *die*
canvass ['kænvəs] **1.** *v.t.* Wahlwerbung treiben in ⟨*einem Wahlkreis, Gebiet*⟩; Wahlwerbung treiben bei ⟨*Wählern, Bürgern*⟩. **2.** *v.i.* werben (**on behalf of** für); ~ **for votes** um Stimmen werben. '**canvasser** *n.* *(for votes)* Wahlhelfer, *der/*-helferin, *die*
canyon ['kænjən] *n.* Cañon, *der*
cap [kæp] **1.** *n.* **a)** Mütze, *die; (nurse's, servant's)* Haube, *die; (with peak)* Schirmmütze, *die; (skull-~)* Kappe, *die;* **b)** *(of bottle, jar)* [Verschluß]kappe, *die; (petrol ~, radiator-~)* Deckel, *der.* **2.** *v.t.* **-pp-: a)** verschließen ⟨*Flasche*⟩; zudecken ⟨*Bohrloch*⟩; mit einer Schutzkappe versehen ⟨*Zahn*⟩; **b)** *(fig.)* überbieten; **to ~ it all** obendrein
capability [keɪpə'bɪlɪtɪ] *n.* Fähigkeit, *die*
capable ['keɪpəbl] *adj.* **a) be ~ of sth.** ⟨*Person:*⟩ zu etw. imstande sein; **b)** *(gifted, able)* fähig
capacity [kə'pæsɪtɪ] *n.* **a)** Fassungsvermögen, *das;* **the machine is working to** ~: die Maschine ist voll ausgelastet; **a seating ~ of 300** 300 Sitzplätze; **b)** *(measure)* Rauminhalt, *der;* Volumen, *das;* **measure of** ~: Hohlmaß, *das;* **c)** *(position)* Eigenschaft, *die;* **in his ~ as ...** : in seiner Eigenschaft als ...
¹**cape** [keɪp] *n.* *(garment)* Umhang, *der;* Cape, *das*
²**cape** *n.* *(Geog.)* Kap, *das;* **the C~ ⌊of Good Hope⌋** das Kap der guten Hoffnung; **C~ Town** Kapstadt *(das)*
caper ['keɪpə(r)] *v.i.* ~ [**about**] [herum]tollen
capital ['kæpɪtl] **1.** *attrib. adj.* **a)** Todes⟨*strafe, -urteil*⟩; Kapital⟨*verbrechen*⟩; **b)** groß, Groß⟨*buchstabe*⟩; **c)** *(principal)* Haupt⟨*stadt*⟩. **2.** *n.* **a)** *(letter)* Großbuchstabe, *der;* **b)** *(city, town)* Hauptstadt, *die;* **c)** *(stock, wealth)* Kapital, *das*
capitalism ['kæpɪtəlɪzm] *n.* Kapitalismus, *der.* **capitalist** ['kæpɪtəlɪst] **1.** *n.* Kapitalist, *der/*Kapitalistin, *die.* **2.** *adj.* kapitalistisch
capitalize ['kæpɪtəlaɪz] **1.** *v.t.* groß schreiben ⟨*Buchstaben, Wort*⟩. **2.** *v.i.* ~ **on sth.** aus etw. Kapital schlagen *(ugs.)*

capital 'punishment *n.* Todesstrafe, *die*
capitulate [kə'pɪtjʊleɪt] *v.i.* kapitulieren. **capitulation** [kəpɪtjʊ'leɪʃn] *n.* Kapitulation, *die*
capricious [kə'prɪʃəs] *adj.* launisch
Capricorn ['kæprɪkɔːn] *n.* der Steinbock
capsize [kæp'saɪz] **1.** *v.t.* zum Kentern bringen. **2.** *v.i.* kentern
capsule ['kæpsjuːl] *n.* Kapsel, *die*
captain ['kæptɪn] **1.** *n.* Kapitän, *der; (Army)* Hauptmann, *der.* **2.** *v.t.* ~ **a team** Kapitän einer Mannschaft sein
caption ['kæpʃn] *n. (heading)* Überschrift, *die; (under photograph, drawing)* Bildunterschrift, *die; (Cinemat., Telev.)* Untertitel, *der*
captivate ['kæptɪveɪt] *v.t.* fesseln. **captivating** ['kæptɪveɪtɪŋ] *adj.* bezaubernd; einnehmend ⟨*Lächeln*⟩
captive ['kæptɪv] **1.** *adj.* gefangen; **be taken ~** : gefangengenommen werden. **2.** *n.* Gefangener, *der/*Gefangene, *die.* **captivity** [kæp'tɪvɪtɪ] *n.* Gefangenschaft, *die;* **be held in ~** : gefangengehalten werden
captor ['kæptə(r)] *n.* **his ~** : der, der/ die, die ihn gefangennahm
capture ['kæptʃə(r)] **1.** *n.* **a)** *(of thief etc.)* Festnahme, *die; (of town)* Einnahme, *die;* **b)** *(thing, person)* Fang, *der.* **2.** *v.t.* festnehmen ⟨*Person*⟩; [ein]fangen ⟨*Tier*⟩; einnehmen ⟨*Stadt*⟩; gefangennehmen ⟨*Phantasie*⟩
car [kɑː(r)] *n.* Auto, *das;* Wagen, *der;* **by ~** : mit dem Auto
carafe [kə'ræf] *n.* Karaffe, *die*
caramel ['kærəmel] *n.* Karamel, *der; (toffee)* Karamelbonbon, *das*
carat ['kærət] *n.* Karat, *das;* **a 22-~ gold ring** ein 22karätiger Goldring
caravan ['kærəvæn] *n. (Brit.)* Wohnwagen, *der.* '**caravan site** *n.* Campingplatz für Wohnwagen
carbohydrate [kɑːbəʊ'haɪdreɪt] *n.* Kohlenhydrat, *das*
carbon ['kɑːbən] *n.* Kohlenstoff, *der* **carbon:** ~ '**copy** *n.* Durchschlag, *der;* ~ **dioxide** [~ daɪ'ɒksaɪd] *n.* Kohlendioxid, *das;* ~ **paper** *n.* Kohlepapier, *das*
carburettor *(Amer.:* **carburetor)** [kɑːbə'retə(r)] *n.* Vergaser, *der*
carcass *(Brit. also:* **carcase)** ['kɑːkəs] *n.* Kadaver, *der*
'**car crash** *n.* Autounfall, *der*
card [kɑːd] *n.* Karte, *die;* **play ~s** Karten spielen

card: ~**board** *n.* Pappe, *die;* ~**board box** *n.* [Papp]karton, *der; (smaller)* [Papp]schachtel, *die;* ~ **game** *n.* Kartenspiel, *das*

cardigan ['kɑːdɪgən] *n.* Strickjacke, *die*

cardinal ['kɑːdɪnl] **1.** *adj.* grundlegend ⟨*Frage, Doktrin, Pflicht*⟩; Kardinal- ⟨*fehler, -problem*⟩; Haupt⟨*punkt, -merkmal*⟩. **2.** *n. (Eccl.)* Kardinal, *der.* **cardinal 'number** *n.* Kardinalzahl, *die.* **cardinal 'sin** *n.* Todsünde, *die*

care [keə(r)] **1.** *n.* **a)** *(anxiety)* Sorge, *die;* **b)** *(pains)* Sorgfalt, *die;* **c)** *(caution)* Vorsicht, *die;* **take** ~: aufpassen; **d)** *medical* ~: ärztliche Betreuung; **e)** *(charge)* Obhut, *die (geh.);* **put sb. in** ~/**take sb. into** ~: jmdn. in Pflege geben/nehmen; ~ **of** *(on letter)* bei; **take** ~ **of sb./sth.** *(ensure safety of)* auf jmdn./etw. aufpassen; *(attend to)* sich um jmdn./etw. kümmern. **2.** *v. i.* ~ **for sb./sth.** *(look after)* sich um jmdn./ etw. kümmern; *(like)* jmdn./etw. mögen; ~ **to do sth.** etw. tun mögen; **I don't** ~ **[whether/how/what** *etc.*] es ist mir gleich[, ob/wie/was *usw.*]

career [kə'rɪə(r)] **1.** *n.* Beruf, *der.* **2.** *v. i.* rasen; ⟨*Pferd, Reiter:*⟩ galoppieren

carefree *adj.* sorgenfrei

careful ['keəfl] *adj. (thorough)* sorgfältig; *(cautious)* vorsichtig; **|be|** ~! Vorsicht!; **be** ~ **of sb./sth.** *(be cautious of)* sich vor jmdm./etw. in acht nehmen; **be** ~ **with sb./sth.** vorsichtig mit jmdm./etw. umgehen. '**carefully** *adv. (thoroughly)* sorgfältig; *(attentively)* aufmerksam; *(cautiously)* vorsichtig

careless ['keəlɪs] *adj.* **a)** *(inattentive)* unaufmerksam; *(thoughtless)* gedankenlos; leichtsinnig ⟨*Fahrer*⟩; nachlässig ⟨*Arbeiter, Arbeit*⟩; gedankenlos ⟨*Bemerkung, Handlung*⟩; unachtsam ⟨*Fahren*⟩; **b)** *(nonchalant)* ungezwungen. '**carelessly** *adv. (without care)* nachlässig; *(thoughtlessly)* gedankenlos. '**carelessness** *n. (lack of care)* Nachlässigkeit, *die; (thoughtlessness)* Gedankenlosigkeit, *die*

caress [kə'res] **1.** *n.* Liebkosung, *die.* **2.** *v. t.* liebkosen

'**caretaker** *n.* Hausmeister, *der*/-meisterin, *die*

'**car ferry** *n.* Autofähre, *die*

cargo ['kɑːgəʊ] *n.* Fracht, *die.* '**cargo boat,** '**cargo ship** *ns.* Frachter, *der*

Caribbean [kærɪ'biːən] **1.** *n.* the ~: die Karibik. **2.** *adj.* karibisch

caricature ['kærɪkətjʊə(r)] **1.** *n.* Karikatur, *die.* **2.** *v. t.* karikieren

carnage ['kɑːnɪdʒ] *n.* Gemetzel, *das*

carnal ['kɑːnl] *adj.* sinnlich

carnation [kɑː'neɪʃn] *n.* [Garten]nelke, *die*

carnet ['kɑːneɪ] *n. (of motorist)* Triptyk, *das;* |**camping|** ~: Ausweis für Camper

carnival ['kɑːnɪvl] *n.* Volksfest, *das*

carnivorous [kɑː'nɪvərəs] *adj.* fleischfressend

carol ['kærl] *n.* |**Christmas|** ~: Weihnachtslied, *das*

carp [kɑːp] *n., pl. same* Karpfen, *der*

'**car-park** *n.* Parkplatz, *der; (building)* Parkhaus, *das*

carpenter ['kɑːpɪntə(r)] *n.* Zimmermann, *der; (for furniture)* Tischler, *der*/Tischlerin, *die.* **carpentry** ['kɑːpɪntrɪ] *n.* Zimmerhandwerk, *das; (in furniture)* Tischlerhandwerk, *das*

carpet ['kɑːpɪt] *n.* Teppich, *der.* '**carpet-slipper** *n.* Hausschuh, *der.* '**carpet-sweeper** *n.* Teppichkehrer, *der*

'**car-port** *n.* Einstellplatz, *der*

carriage ['kærɪdʒ] *n.* **a)** *(horse-drawn)* Kutsche, *die;* **b)** *(Railw.)* Wagen, *der*

'**carriageway** *n.* Fahrbahn, *die*

carrier ['kærɪə(r)] *n.* **a)** *(bearer)* Träger, *der;* **b)** *(firm)* Transportunternehmen, *das.* '**carrier-bag** *n.* Tragetasche, *die.* '**carrier pigeon** *n.* Brieftaube, *die*

carrot ['kærət] *n.* Möhre, *die*

carry ['kærɪ] *v. t.* **a)** tragen; *(emphasizing destination)* bringen; **b)** *(possess)* besitzen ⟨*Autorität, Gewicht*⟩. **carry a'way** *v. t.* forttragen; **be** *or* **get carried away** sich hinreißen lassen. **carry 'on 1.** *v. t.* fortführen; ~ **on |doing sth.|** weiterhin etw. tun. **2.** *v. i.* weitermachen. **carry 'out** *v. t.* durchführen; ausführen ⟨*Anweisung, Auftrag*⟩; vornehmen ⟨*Verbesserungen*⟩

'**carry-cot** *n.* Babytragetasche, *die*

cart [kɑːt] **1.** *n.* Wagen, *der.* **2.** *v. t. (sl.)* schleppen

cartilage ['kɑːtɪlɪdʒ] *n.* Knorpel, *der*

carton ['kɑːtn] *n.* [Papp]karton, *der; (of drink)* Tüte, *die; (of cream, yoghurt)* Becher, *der*

cartoon [kɑː'tuːn] *n.* humoristische Zeichnung; *(satirical)* Karikatur, *die; (film)* Zeichentrickfilm, *der*

cartridge ['kɑːtrɪdʒ] *n.* **a)** *(for gun)* Patrone, *die;* **b)** *(of film; cassette)* Kassette, *die*

'**cart-wheel** n. (Gymnastics) Rad, das; **turn** or **do** ~**s** radschlagen

carve [kɑːv] **1.** v. t. **a)** tranchieren ⟨Fleisch, Braten, Hähnchen⟩; **b)** (from wood) schnitzen; (from stone) meißeln. **2.** v. i. ~ **in wood/stone** in Holz schnitzen/in Stein meißeln. **carving** ['kɑːvɪŋ] n. (in or from wood) Schnitzerei, die; (in or from stone) Skulptur, die. '**carving-knife** n. Tranchiermesser, das

'**car wash** n. Waschanlage, die

cascade [kæsˈkeɪd] n. Kaskade, die

¹**case** [keɪs] n. **a)** (instance, matter, set of arguments) Fall, der; **it is [not] the ~ that ...:** es trifft [nicht] zu, daß ...; **in ~ ...:** falls ...; **[just] in ~:** für alle Fälle; **in ~ of emergency** im Notfall; **in any ~:** jedenfalls; **in that ~:** in diesem Fall; **b)** (Med., Police, Soc. Serv., etc.) Fall, der; **c)** (Law) Fall, der; (action) Verfahren, das; **d)** (Ling.) Fall, der; Kasus, der (fachspr.)

²**case** n. **a)** Koffer, der; (brief-~) [Akten]tasche, die; **b)** (for spectacles, cigarettes) Etui, das; **c)** (crate) Kiste, die; **d)** [display-]~: Schaukasten, der

cash [kæʃ] **1.** n. Bargeld, das; **pay [in] ~, pay ~ down** bar zahlen. **2.** v. t. einlösen ⟨Scheck⟩

cash: ~ **and 'carry** n. cash and carry; (store) Cash-and-carry-Laden, der; ~**card** n. Geldautomatenkarte, die; ~ **desk** n. (Brit.) Kasse, die; ~ **dispenser** n. Geldautomat, der

cashier [kæˈʃɪə(r)] n. Kassierer, der/Kassiererin, die

cash: ~**point** n. Geldautomat, der; ~ **register** n. [Registrier]kasse, die

casino [kəˈsiːnəʊ] n. Kasino, das

cask [kɑːsk] n. Faß, das

casket ['kɑːskɪt] n. **a)** Kästchen, das; **b)** (Amer.: coffin) Sarg, der

casserole ['kæsərəʊl] n. Schmortopf, der

cassette [kəˈset, kæˈset] n. Kassette, die. **casˈsette-deck** n. Kassettendeck, das. **casˈsette recorder** n. Kassettenrecorder, der

cast [kɑːst] **1.** v. t., **cast a)** werfen; **b)** (shape, form) gießen; **c)** abgeben ⟨Stimme⟩. **2.** n. **a)** (Med.) Gipsverband, der; **b)** (actors) Besetzung, die. **cast a'side** v. t. beiseite schieben ⟨Vorschlag⟩; vergessen ⟨Sorgen⟩; fallenlassen ⟨Hemmungen⟩. **cast 'off** v. i. & t. (Naut.) losmachen

castanets [kæstəˈnets] n. pl. Kastagnetten Pl.

'**castaway** n. Schiffbrüchige, der/die

caste [kɑːst] n. Kaste, die

cast 'iron n. Gußeisen, das

castle ['kɑːsl] n. Burg, die; (mansion) Schloß, das

'**cast-offs** n. pl. abgelegte Sachen

castor ['kɑːstə(r)] n. (wheel) Rolle, die

castor: ~ **'oil** n. Rizinusöl, das; ~ **sugar** n. Raffinade, die

castrate [kæˈstreɪt] v. t. kastrieren. **castration** [kæˈstreɪʃn] n. Kastration, die

casual ['kæʒjʊəl] adj. ungezwungen; leger ⟨Kleidung⟩; beiläufig ⟨Bemerkung⟩; flüchtig ⟨Bekannter, Bekanntschaft, Blick⟩; unbekümmert ⟨Haltung, Einstellung⟩. '**casually** adv. ungezwungen; beiläufig ⟨bemerken⟩; flüchtig ⟨anschauen⟩; leger ⟨sich kleiden⟩

casualty ['kæʒjʊəltɪ] n. **a)** (injured person) Verletzte, der/die; (in battle) Verwundete, der/die; (dead person) Tote, der/die; **b)** (hospital department) Unfallstation, die

cat [kæt] n. Katze, die

catalogue (Amer.: **catalog**) ['kætəlɒg] **1.** n. Katalog, der. **2.** v. t. katalogisieren

catalyst ['kætəlɪst] n. Katalysator, der. **catalytic** [kætəˈlɪtɪk] adj. ~ **converter** Katalysator, der

catapult ['kætəpʌlt] **1.** n. Katapult, das. **2.** v. t. katapultieren

cataract ['kætərækt] n. **a)** Katarakt, der; **b)** (Med.) grauer Star

catarrh [kəˈtɑː(r)] n. Katarrh, der

catastrophe [kəˈtæstrəfɪ] n. Katastrophe, die. **catastrophic** [kætəˈstrɒfɪk] adj. katastrophal

catch [kætʃ] **1.** v. t., **caught** [kɔːt] **a)** fangen; ~ **hold of sb./sth.** jmdn./etw. festhalten; (to stop oneself falling) sich an jmdm./etw. festhalten; **get sth. caught** or ~ **sth. on/in sth.** mit etw. an/in etw. (Dat.) hängenbleiben; ~ **one's finger in the door** sich (Dat.) den Finger in der Tür einklemmen; **b)** (travel by) nehmen; (be in time for) [noch] erreichen; **c)** (surprise) ~ **sb. doing sth.** jmdn. [dabei] erwischen, wie er etw. tut (ugs.); **d)** (become infected with) sich (Dat.) zuziehen; ~ **sth. from sb.** sich bei jmdm. mit etw. anstecken; ~ **a cold** sich erkälten; ~ **it** (fig. coll.) etwas kriegen (ugs.); **e)** ~ **sb.'s attention/interest** jmds. Aufmerksamkeit erregen/jmds. Interesse wecken. **2.** v. i., **caught a)** (begin to burn) [anfangen zu] brennen; **b)** *(become hooked up)* hängenbleiben; ⟨*Haar, Faden:*⟩ sich verfangen. **3.** *n.* **a)** *(of ball)* **make a ~:** fangen; **b)** *(amount caught, lit. or fig.)* Fang, *der;* **c)** *(difficulty)* Haken, *der* **(in an +** *Dat.*)*;* **d)** *(of door)* Schnapper, *der.* **catch 'on** *v. i. (coll.)* **a)** *(become popular)* [gut] ankommen *(ugs.);* **b)** *(understand)* kapieren *(ugs.).* **catch 'up 1.** *v. t.* **~ sb. up** jmdn. einholen. **2.** *v. i.* **~ up** gleichziehen; **~ up on sth.** etw. nachholen

'**catching** *adj.* ansteckend

'**catchy** *adj.* eingängig

categorical [kætɪ'gɒrɪkl] *adj.* kategorisch

category ['kætɪgərɪ] *n.* Kategorie, *die*

cater ['keɪtə(r)] *v. i.* **~ for sb./sth.** für jmdn./etw. [die] Speisen und Getränke liefern; *(fig.)* auf jmdn./etw. eingestellt sein. '**caterer** *n.* Lieferant von Speisen und Getränken. '**catering** *n.* **a)** *(trade)* Gastronomie, *die;* **b)** *(service)* Lieferung von Speisen und Getränken

caterpillar ['kætəpɪlə(r)] *n.* Raupe, *die*

cathedral [kə'θi:drl] *n.* Dom, *der*

Catherine wheel ['kæθrɪn wi:l] *n.* Feuerrad, *das*

Catholic ['kæθəlɪk] **1.** *adj.* katholisch. **2.** *n.* Katholik, *der*/Katholikin, *die.* **Catholicism** [kə'θɒlɪsɪzm] *n.* Katholizismus, *der*

catkin ['kætkɪn] *n. (Bot.)* Kätzchen, *das*

'**Cat's-eye, (P)** *n. (Brit.: on road)* Bodenrückstrahler, *der*

cattle ['kætl] *n. pl.* Rinder *Pl.*

caught *see* **catch 1, 2**

cauldron ['kɔ:ldrən] *n.* Kessel, *der*

cauliflower ['kɒlɪflaʊə(r)] *n.* Blumenkohl, *der*

cause [kɔ:z] **1.** *n.* **a)** Ursache, *die* **(of** für *od. Gen.*); *(person)* Verursacher, *der*/Verursacherin, *die;* **be the ~ of sth.** etw. verursachen; **b)** *(reason)* Grund, *der;* **~ for sth.** Grund zu etw.; **c)** *(object of support)* Sache, *die;* **[in] a good ~:** [für] eine gute Sache. **2.** *v. t.* verursachen; erregen ⟨*Aufsehen, Ärgernis*⟩; hervorrufen ⟨*Unruhe, Verwirrung*⟩; **~ sb. worry/pain** jmdm. Sorge/Schmerzen bereiten; **~ sb. to do sth.** jmdn. veranlassen, etw. zu tun

causeway ['kɔ:zweɪ] *n.* Damm, *der*

caustic ['kɔ:stɪk] *adj.* ätzend; *(fig.)* bissig; beißend ⟨*Spott*⟩

caution ['kɔ:ʃn] **1.** *n.* **a)** Vorsicht, *die;* **b)** *(warning)* Warnung, *die.* **2.** *v. t.*

(warn) warnen; *(warn and reprove)* verwarnen **(for** wegen)

cautious ['kɔ:ʃəs] *adj.,* '**cautiously** *adv.* vorsichtig

cavalry ['kævəlrɪ] *n.* Kavallerie, *die*

cave [keɪv] *n.* Höhle, *die.* **cave 'in** *v. i.* einbrechen

'**caveman** *n.* Höhlenbewohner, *der*

cavern ['kævən] *n.* Höhle, *die.* **cavernous** ['kævənəs] *adj.* höhlenartig

caviar[e] ['kævɪɑ:(r)] *n.* Kaviar, *der*

cavity ['kævɪtɪ] *n.* Hohlraum, *der; (in tooth)* Loch, *das*

CB *abbr.* **citizen's band** CB

cc [si:'si:] *abbr.* **cubic centimetre(s)** cm³

CD *abbr.* **compact disc** CD

cease [si:s] **1.** *v. i.* aufhören. **2.** *v. t.* **a)** *(stop)* aufhören; **b)** *(end)* aufhören mit; einstellen ⟨*Bemühungen*⟩. '**cease-fire** *n.* Waffenruhe, *die*

cedar ['si:də(r)] *n.* Zeder, *die*

ceiling ['si:lɪŋ] *n.* **a)** Decke, *die;* **b)** *(upper limit)* Maximum, *das*

celebrate ['selɪbreɪt] *v. t. & i.* feiern. '**celebrated** *adj.* berühmt. **celebration** [selɪ'breɪʃn] *n.* Feier, *die.* **celebrity** [sɪ'lebrɪtɪ] *n.* Berühmtheit, *die*

celery ['selərɪ] *n.* Sellerie, *der od. die*

celibate ['selɪbət] *adj.* zölibatär *(Rel.);* ehelos

cell [sel] *n.* Zelle, *die*

cellar ['selə(r)] *n.* Keller, *der*

cellist ['tʃelɪst] *n.* Cellist, *der*/Cellistin, *die*

cello ['tʃeləʊ] *n., pl.* **~s** Cello, *das*

Cellophane, (P) ['seləfeɪn] *n.* Cellophan ⓦ, *das*

Celsius ['selsɪəs] *adj.* Celsius

cement [sɪ'ment] **1.** *n.* Zement, *der.* **2.** *v. t.* zementieren; *(stick together)* zusammenkleben. **ce'ment-mixer** *n.* Betonmischmaschine, *die*

cemetery ['semɪtərɪ] *n.* Friedhof, *der*

censor ['sensə(r)] **1.** *n.* Zensor, *der.* **2.** *v. t.* zensieren. '**censorship** *n.* Zensur, *die*

censure ['senʃə(r)] *v. t.* tadeln

census ['sensəs] *n.* Volkszählung, *die*

cent [sent] *n.* Cent, *der*

centenary [sen'ti:nərɪ] *adj. & n.* **~ [celebrations]** Hundertjahrfeier, *die*

center *(Amer.) see* **centre**

centigrade ['sentɪgreɪd] *see* **Celsius**

centimetre *(Brit.; Amer.:* **centimeter)** ['sentɪmi:tə(r)] *n.* Zentimeter, *der*

centipede ['sentɪpi:d] *n.* Tausendfüßler, *der*

central ['sentrl] *adj.* zentral

Central: ~ **A'merica** pr. n. Mittel-amerika (das); ~ **'Europe** pr. n. Mitteleuropa (das); ~ **Euro'pean** adj. mitteleuropäisch; **c~ 'heating** n. Zentralheizung, die
centralize ['sentrəlaɪz] v. t. zentralisieren
central reser'vation n. (Brit.) Mittelstreifen, der
centre ['sentə(r)] (Brit.) 1. n. a) Mitte, die; (of circle) Mittelpunkt, der; b) (of area, city) Zentrum, das. 2. adj. mittler... 3. v. i. ~ **on** sth. sich auf etw. (Akk.) konzentrieren; ~ |a|round sth. sich um etw. drehen. 4. v. t. a) in der Mitte anbringen; b) (concentrate) be ~d |a|round sth. etw. zum Mittelpunkt haben; ~ sth. on sth. etw. auf etw. (Akk.) konzentrieren. **centre-'forward** n. Mittelstürmer, der
centrifugal [sentrɪ'fju:gl] adj. ~ **force** Zentrifugalkraft, die; Fliehkraft, die
century ['sentʃərɪ] n. (hundred-year period from a year ...00) Jahrhundert, das; (hundred years) hundert Jahre
ceramic [sɪ'ræmɪk] adj. keramisch
cereal ['sɪərɪəl] n. Getreide, das; (breakfast dish) Getreideflocken Pl.
ceremonial [serɪ'məʊnɪəl] 1. adj. feierlich; (prescribed for ceremony) zeremoniell. 2. n. Zeremoniell, das
ceremony ['serɪmənɪ] n. Feier, die; (formal act) Zeremonie, die
certain ['sɜ:tn, 'sɜ:tɪn] adj. a) (settled, definite) bestimmt; b) be ~ to do sth. etw. bestimmt tun; c) (confident, sure to happen) sicher; d) (indisputable) unbestreitbar; e) a ~ **Mr Smith** ein gewisser Herr Smith; to a ~ **extent** in gewisser Weise. **'certainly** adv. a) (admittedly) sicher[lich]; (definitely) bestimmt; b) (in answer) [aber] sicher; |most| ~ **'not!** auf [gar] keinen Fall! **certainty** ['sɜ:tntɪ, 'sɜ:tɪntɪ] n. a) be a ~: sicher sein; b) (absolute conviction) Gewißheit, die
certificate [sə'tɪfɪkət] n. Urkunde, die; (of action performed) Schein, der
certify ['sɜ:tɪfaɪ] v. t. bescheinigen; this is to ~ that ...: hiermit wird bescheinigt od. bestätigt, daß ...
cf. abbr. compare vgl.
chafe [tʃeɪf] v. t. wund scheuern
chaff [tʃɑ:f] n. Spreu, die
chaffinch ['tʃæfɪntʃ] n. Buchfink, der
chagrin ['ʃægrɪn] n. Kummer, der
chain [tʃeɪn] 1. n. Kette, die; ~ of shops/hotels Laden-/Hotelkette, die. 2. v. t. [an]ketten (to an + Akk.)

chain: ~ **re'action** n. Kettenreaktion, die; **~-saw** n. Kettensäge, die; **~-smoker** n. Kettenraucher, der/-raucherin, die; ~ **store** Kettenladen, der
chair [tʃeə(r)] 1. n. a) Stuhl, der; (arm~, easy ~) Sessel, der; b) (professorship) Lehrstuhl, der; c) (at meeting) Vorsitz, der. 2. v. t. den Vorsitz haben bei
chair: **~-back** n. Rückenlehne, die; **~-lift** n. Sessellift, der; **~-man** ['tʃeəmən] n., pl. **~men** ['tʃeəmən] Vorsitzende, der/die
chalet ['ʃæleɪ] n. Chalet, das
chalk [tʃɔ:k] 1. n. Kreide, die. 2. v. t. mit Kreide schreiben/malen usw.
challenge ['tʃælɪndʒ] 1. n. Herausforderung, die. 2. v. t. a) (to contest etc.) herausfordern; b) (fig.) auffordern; (question) in Frage stellen. **'challenger** n. Herausforderer, der/Herausforderin, die. **challenging** ['tʃælɪndʒɪŋ] adj. herausfordernd; fesselnd ⟨Problem⟩; anspruchsvoll ⟨Arbeit⟩
chamber ['tʃeɪmbə(r)] n. Kammer, die
chamber: **~maid** n. Zimmermädchen, das; ~ **music** n. Kammermusik, die; **~-pot** n. Nachttopf, der
chameleon [kə'mi:ljən] n. Chamäleon
chamois ['ʃæmwɑ:] n. a) Gemse, die; b) ['ʃæmɪ] ~-[-leather] Chamois[leder], das
champagne [ʃæm'peɪn] n. Sekt, der; (from Champagne) Champagner, der
champion ['tʃæmpɪən] 1. n. a) (defender) Verfechter, der/Verfechterin, die; b) (Sport) Meister, der/Meisterin, die. 2. v. t. verfechten ⟨Sache⟩; sich einsetzen für ⟨Person⟩. **'championship** n. Meisterschaft, die
chance [tʃɑ:ns] 1. n. a) (fortune, trick of fate) Zufall, der; attrib. zufällig; ~ **encounter** Zufallsbegegnung, die; **game of** ~: Glücksspiel, das; **by** ~: zufällig; **take a** ~: es riskieren; **the ~s are that** ...: es ist wahrscheinlich, daß ...; **by |any|** ~, **by some ~** or other zufällig; b) (opportunity, possibility) Chance, die; **get a/the** ~ **to do** sth. eine/die Gelegenheit haben, etw. zu tun. 2. v. t. riskieren
chancellor ['tʃɑ:nsələ(r)] n. Kanzler, der; **C~ of the Exchequer** (Brit.) Schatzkanzler, der
chandelier [ʃændə'lɪə(r)] n. Kronleuchter, der
change [tʃeɪndʒ] 1. n. a) Verände-

rung, *die;* Änderung, *die; (of job, sur-roundings, government, etc.)* Wechsel, *der;* **b)** *(for the sake of variety)* Abwechslung, *die;* **for a ~:** zur Abwechslung; **c)** *(money)* Wechselgeld, *das;* |**loose** *or* **small**| ~: Kleingeld, *das;* |**here is**| **15 marks** ~: 15 Mark zurück; **keep the** ~: [es] stimmt so. **2.** *v. t.* **a)** *(switch)* wechseln; auswechseln ‹*Glühbirne, Batterie*›; ~ **one's clothes** sich umziehen; ~ **one's address/name** seine Anschrift/seinen Namen ändern; ~ **trains/buses** umsteigen; **b)** *(transform)* verwandeln (**into** in + *Akk.*); *(alter)* ändern; **c)** *(exchange)* eintauschen (**for** für); wechseln ‹*Geld*›. **3.** *v. i.* **a)** *(alter)* sich ändern; ‹*Person, Land:*› sich verändern; **b)** *(into something else)* sich verwandeln; **c)** *(put on other clothes)* sich umziehen. **change 'over** *v. i.* ~ **over from sth. to sth.** von etw. zu etw. übergehen
changeable ['tʃeɪndʒəbl] *adj.* veränderlich
'**changing room** *n. (Brit.)* Umkleideraum, *der*
channel ['tʃænl] **1.** *n. (also Telev.; Radio)* Kanal, *der;* **the C~** *(Brit.)* der [Ärmel]kanal. **2.** *v. t. (fig.)* lenken. '**Channel Islands** *pr. n. pl.* Kanalinseln *Pl.*
chant [tʃɑːnt] **1.** *v. t.* skandieren; *(Eccl.)* singen. **2.** *v. i.* Sprechchöre anstimmen; *(Eccl.)* singen. **3.** *n.* Sprechchor, *der; (Eccl.)* Gesang, *der*
chaos ['keɪɒs] *n.* Chaos, *das.* **chaotic** [keɪ'ɒtɪk] *adj.* chaotisch
¹**chap** [tʃæp] *n. (Brit. coll.)* Bursche, *der;* Kerl, *der*
²**chap** *v. t.,* **-pp-** aufplatzen lassen
chapel ['tʃæpl] *n.* Kapelle, *die*
chaperon ['ʃæpərəʊn] **1.** *n.* Anstandsdame, *die.* **2.** *v. t.* beaufsichtigen
chaplain ['tʃæplɪn] *n.* Kaplan, *der*
chapter ['tʃæptə(r)] *n.* Kapitel, *das*
char [tʃɑː(r)] *v. t. & i.,* **-rr-** verkohlen
character ['kærɪktə(r)] *n.* **a)** Charakter, *der;* **b)** *(in novel etc.)* Figur, *die;* **c)** *(coll.: extraordinary person)* Original, *das;* **d)** *(symbol)* Zeichen, *das.* **characteristic** [kærɪktə'rɪstɪk] **1.** *adj.* charakteristisch (**of** für). **2.** *n.* charakteristisches Merkmal. **characterize** ['kærɪktəraɪz] *v. t.* charakterisieren
charade [ʃə'rɑːd] *n.* Scharade, *die; (fig.)* Farce, *die*
charcoal ['tʃɑːkəʊl] *n.* Holzkohle, *die*
charge [tʃɑːdʒ] **1.** *n.* **a)** *(price)* Preis, *der; (for services)* Gebühr, *die;* **b) be in**

~ **of sth.** für etw. die Verantwortung haben; **take** ~: die Verantwortung übernehmen; **c)** *(Law: accusation)* Anklage, *die;* **d)** *(attack)* Angriff, *der;* **e)** *(of explosives, electricity)* Ladung, *die.* **2.** *v. t.* **a)** ~ **sb. sth.,** ~ **sth. to sb.** jmdm. etw. berechnen; **b)** *(Law: accuse)* anklagen (**with** wegen); **c)** *(Electr.)* [auf]laden ‹*Batterie*›; **d)** *(rush at)* angreifen. **3.** *v. i.* **a)** *(attack)* angreifen; **b)** *(coll.: hurry)* sausen
charitable ['tʃærɪtəbl] *adj.* **a)** wohltätig; **b)** *(lenient)* großzügig
charity ['tʃærɪtɪ] *n.* **a)** Wohltätigkeit, *die;* **b)** *(organization)* wohltätige Organisation
charlady ['tʃɑːleɪdɪ] *n. (Brit.)* Putzfrau, *die*
charlatan ['ʃɑːlətən] *n.* Scharlatan, *der*
charm [tʃɑːm] **1.** *n.* **a)** *(act)* Zauber, *der;* **b)** *(talisman)* Talisman, *der;* **c)** *(attractiveness)* Reiz, *der; (of person)* Charme, *der.* **2.** *v. t.* bezaubern. '**charming** *adj.* bezaubernd
chart [tʃɑːt] **1.** *n.* **a)** *(map)* Karte, *die;* **b)** *(graph etc.)* Schaubild, *das;* **c) the** ~**s** die Hitliste. **2.** *v. t. (fig.: describe)* schildern
charter ['tʃɑːtə(r)] **1.** *n.* **a)** Charta, *die;* **b) on** ~ gechartert. **2.** *v. t.* chartern ‹*Schiff, Flugzeug*›. **chartered accountant** *n. (Brit.)* Wirtschaftsprüfer, *der/*-prüferin, *die*
'**charter flight** *n.* Charterflug, *der*
charwoman ['tʃɑːwʊmən] *n.* Putzfrau, *die*
chase [tʃeɪs] **1.** *n.* Verfolgungsjagd, *die.* **2.** *v. t. (pursue)* jagen; ~ **sth.** *(fig.)* einer Sache *(Dat.)* nachjagen. **3.** *v. i.* ~ **after sb./sth.** hinter jmdm./etw. herjagen. **chase 'up** *v. t. (coll.)* ausfindig machen
chasm ['kæzm] *n.* Kluft, *die*
chassis ['ʃæsɪ] *n., pl. same* ['ʃæsɪz] Chassis, *das;* Fahrgestell, *das*
chaste [tʃeɪst] *adj.* keusch
chastening ['tʃeɪsənɪŋ] *adj.* ernüchternd
chastise [tʃæ'staɪz] *v. t.* züchtigen
chastity ['tʃæstɪtɪ] *n.* Keuschheit, *die*
chat [tʃæt] **1.** *n.* Schwätzchen, *das.* **2.** *v. i.,* **-tt-** plaudern; ~ **with** *or* **to sb. about sth.** mit jmdm. von etw. plaudern. **chat 'up** *v. t. (Brit. coll.)* anmachen *(ugs.)*
'**chat show** *n.* Talk-Show, *die*
chattels ['tʃætəlz] *n. pl.* bewegliche Habe *(geh.)*

chatter ['tʃætə(r)] 1. *v. i.* **a)** schwatzen; **b)** ⟨*Zähne:*⟩ klappern. 2. *n.* Schwatzen, *das.* '**chatterbox** *n.* Quasselstrippe, *die (ugs.)*

chatty ['tʃætɪ] *adj.* gesprächig

chauffeur ['ʃəʊfə(r)] 1. *n.* Fahrer, *der;* Chauffeur, *der.* 2. *v. t.* fahren

chauvinist ['ʃəʊvɪnɪst] *n.* Chauvinist, *der/*Chauvinistin, *die.* **chauvinistic** [ʃəʊvɪ'nɪstɪk] *adj.* chauvinistisch

cheap [tʃiːp] *adj., adv.* billig. **cheapen** ['tʃiːpn] *v. t. (fig.)* herabsetzen. '**cheaply** *adv.* billig

cheat [tʃiːt] 1. *n.* Schwindler, *der/* Schwindlerin, *die.* 2. *v. t. & i.* betrügen

¹**check** [tʃek] 1. *n.* **a)** Kontrolle, *die;* **make/keep a ~ on** kontrollieren; **b)** *(Amer.: bill)* Rechnung, *die.* 2. *v. t.* **a)** *(restrain)* unter Kontrolle halten; **b)** *(examine)* nachprüfen; kontrollieren ⟨*Fahrkarte*⟩; **c)** *(stop)* aufhalten. 3. *v. i.* **~ on sth.** etw. überprüfen; **~ with sb.** bei jmdm. nachfragen. **check 'in** *v. t. & i. (at airport)* einchecken. **check 'out** 1. *v. t.* überprüfen. 2. *v. i.* abreisen. **check 'up** *v. i.* **~ up |on|** überprüfen

²**check** *n. (pattern)* Karo, *das*

checkers ['tʃekəz] *(Amer.) see* **draughts**

check: ~-in *n.* Abfertigung, *die;* **~-list** *n.* Checkliste, *die;* **~mate** 1. *n.* [Schach]matt, *das;* 2. *int.* [schach]matt; **~-out [desk]** *n.* Kasse, *die;* **~-point** *n.* Kontrollpunkt, *der;* **~-up** *n. (Med.)* Untersuchung, *die*

cheek [tʃiːk] *n.* **a)** Backe, *die;* Wange, *die (geh.);* **b)** *(impertinence)* Frechheit, *die.* '**cheekily** *adv.*, '**cheeky** *adj.* frech

cheep [tʃiːp] 1. *v. i.* piep[s]en. 2. *n.* Piep[s]en, *das*

cheer [tʃɪə(r)] 1. *n.* **a)** *(applause)* Beifallsruf, *der;* **b)** *in pl. (Brit. coll.)* prost! 2. *v. t.* **a)** *(applaud)* ~ **sth./sb.** etw. bejubeln/jmdm. zujubeln; **b)** *(gladden)* aufmuntern. 3. *v. i.* jubeln. **cheer 'on** *v. t.* anfeuern ⟨*Sportler*⟩. **cheer 'up** 1. *v. t.* aufheitern. 2. *v. i.* bessere Laune bekommen; ~ **up!** Kopf hoch!

cheerful [['tʃɪəfl] *adj. (in good spirits)* fröhlich; *(bright, pleasant)* heiter. '**cheerfully** *adv.* vergnügt

'**cheering** 1. *adj.* fröhlich stimmend. 2. *n.* Jubeln, *das*

cheerio [tʃɪərɪ'əʊ] *int. (Brit. coll.)* tschüs *(ugs.)*

'**cheery** *adj.* fröhlich

cheese [tʃiːz] *n.* Käse, *der.* '**cheeseboard** *n.* Käseplatte, *die.* '**cheesecake** *n.* Käsetorte, *die*

cheetah ['tʃiːtə] *n.* Gepard, *der*

chef [ʃef] *n.* Küchenchef, *der; (as profession)* Koch, *der*

chemical ['kemɪkl] 1. *adj.* chemisch. 2. *n.* Chemikalie, *die*

chemist ['kemɪst] *n.* **a)** *(scientist)* Chemiker, *der/*Chemikerin, *die;* **b)** *(Brit.: pharmacist)* Drogist, *der/*Drogistin, *die;* ~ **'s |shop|**Drogerie, *die.* **chemistry** ['kemɪstrɪ] *n.* Chemie, *die*

cheque [tʃek] *n.* Scheck, *der;* **pay by** ~: mit [einem] Scheck bezahlen. '**cheque-book** *n.* Scheckbuch, *das.* '**cheque card** *n.* Scheckkarte, *die*

cherish ['tʃerɪʃ] *v. t.* hegen ⟨*Hoffnung, Gefühl*⟩; in Ehren halten ⟨*[Erinnerungs]gegenstand*⟩

cherry ['tʃerɪ] *n.* Kirsche, *die*

chess [tʃes] *n., no art.* das Schach[spiel]

chess: ~-board *n.* Schachbrett, *das;* **~-man** *n.* Schachfigur, *die;* **~-player** *n.* Schachspieler, *der/*-spielerin, *die*

chest [tʃest] *n.* **a)** Kiste, *die;* **b)** *(Anat.)* Brust, *die;* **get sth. off one's ~** *(fig. coll.)* sich *(Dat.)* etw. von der Seele reden; **c)** ~ **|measurement|** Brustumfang, *der*

chestnut ['tʃesnʌt] 1. *n.* **a)** Kastanie, *die;* **b)** *(colour)* Kastanienbraun, *das.* 2. *adj. (colour)* ~**|-brown|** kastanienbraun. '**chestnut-tree** *n.* Kastanie, *die*

chest of 'drawers *n.* Kommode, *die*

chew [tʃuː] *v. t. & i.* kauen. '**chewing-gum** *n.* Kaugummi, *der od. das*

chic [ʃiːk] *adj.* schick; elegant

chick [tʃɪk] *n.* **a)** Küken, *das;* **b)** *(sl.: young woman)* Biene, *die (ugs.)*

chicken ['tʃɪkɪn] 1. *n.* **a)** Huhn, *das; (grilled, roasted)* Hähnchen, *das;* **b)** *(coll.: coward)* Angsthase, *der.* 2. *adj. (coll.)* feig[e]. 3. *v. i.* ~ **out** *(sl.)* kneifen '**chicken-pox** [~pɒks] *n.* Windpocken *Pl.*

chicory ['tʃɪkərɪ] *n. (plant)* Chicorée, *der od. die; (for coffee)* Zichorie, *die*

chief [tʃiːf] 1. *n.* **a)** Oberhaupt, *das; (of tribe)* Häuptling, *der;* **b)** *(of department)* Leiter, *der;* ~ **of police** Polizeipräsident, *der.* 2. *adj., usu. attrib.* **a)** Haupt-; **b)** *(leading)* führend. '**chiefly** *adv.* hauptsächlich

chieftain ['tʃiːftən] *n.* Stammesführer, *der*

chilblain ['tʃɪlbleɪn] *n.* Frostbeule, *die*
child [tʃaɪld] *n., pl.* **~ren** ['tʃɪldrən] Kind, *das.* **'childbirth** *n.* Geburt, *die.* **'childhood** *n.* Kindheit, *die*
childish ['tʃaɪldɪʃ] *adj.,* **'childishly** *adv.* kindisch. **'childishness** *n. (behaviour)* kindisches Benehmen
child: **~less** *adj.* kinderlos; **~like** *adj.* kindlich; **~-minder** ['~maɪndə(r)] *n. (Brit.)* Tagesmutter, *die*
children *pl. of* **child**
'child's play *n. (fig.)* ein Kinderspiel
Chile ['tʃɪlɪ] *n.* Chile *(das)*
chill [tʃɪl] **1.** *n.* Kühle, *die; (illness)* Erkältung, *die.* **2.** *v. t.* kühlen
chilli ['tʃɪlɪ] *n., pl.* **~es** Chili, *der*
'chilly *adj.* kühl; **I am rather ~:** mir ist ziemlich kühl
chime [tʃaɪm] **1.** *n.* Geläute, *das.* **2.** *v. i.* läuten; ⟨*Turmuhr:*⟩ schlagen
chimney ['tʃɪmnɪ] *n.* Schornstein, *der.* **'chimney-sweep** *n.* Schornsteinfeger, *der*
chimpanzee [tʃɪmpən'ziː] *n.* Schimpanse, *der*
chin [tʃɪn] *n.* Kinn, *das*
China ['tʃaɪnə] *pr. n.* China *(das)*
china *n.* Porzellan, *das; (crockery)* Geschirr, *das*
Chinese [tʃaɪ'niːz] **1.** *adj.* chinesisch. **2.** *n.* **a)** *pl. same (person)* Chinese, *der*/Chinesin, *die;* **b)** *(language)* Chinesisch, *das; see also* **English 2 a**
chink *n. (gap)* Spalt, *der*
chip [tʃɪp] **1.** *n.* **a)** Splitter, *der;* **b)** *in pl. (Brit.: potato ~s.)* Pommes frites *Pl.s;* **c)** *(Gambling)* Chip, *der.* **2.** *v. t.,* **-pp-** anschlagen. **chip 'in** *(coll.)* **1.** *v. i.* **a)** *(interrupt)* sich einmischen; **b)** *(contribute money)* etwas beisteuern. **2.** *v. t. (contribute)* beisteuern
'chipboard *n.* Spanplatte, *die*
chipmunk ['tʃɪpmʌŋk] *n.* Chipmunk, *das*
chiropodist [kɪ'rɒpədɪst] *n.* Fußpfleger, *der*/-pflegerin, *die*
chiropody [kɪ'rɒpədɪ] *n.* Fußpflege, *die*
chirp [tʃɜːp] **1.** *v. i.* zwitschern; ⟨*Grille:*⟩ zirpen. **2.** *n.* Zwitschern, *das;* Zirpen, *das*
chisel ['tʃɪzl] **1.** *n.* Meißel, *der; (for wood)* Stemmeisen, *das.* **2.** *v. t., (Brit.)* **-ll-** meißeln; *(in wood)* hauen
chit [tʃɪt] *n.* Notiz, *die*
chit-chat ['tʃɪttʃæt] *n.* Plauderei, *die*
chivalrous ['ʃɪvlrəs] *adj.* ritterlich.
chivalry ['ʃɪvlrɪ] *n.* Ritterlichkeit, *die*
chives [tʃaɪvz] *n.* Schnittlauch, *der*

chloride ['klɔːraɪd] *n.* Chlorid, *das*
chlorine ['klɔːriːn] *n.* Chlor, *das*
chock [tʃɒk] *n.* Bremsklotz, *der.* **'chock-a-block** *pred adj.* vollgepfropft
chocolate ['tʃɒklət] *n.* Schokolade, *die*
choice [tʃɔɪs] **1.** *n.* **a)** Wahl, *die; from* ~: freiwillig; **b)** *(variety)* Auswahl, *die.* **2.** *adj.* ausgewählt
choir [kwaɪə(r)] *n.* Chor, *der.* **'choir-boy** *n.* Chorknabe, *der*
choke [tʃəʊk] **1.** *v. t.* **a)** ersticken; **b)** *(block up)* verstopfen. **2.** *v. i. (temporarily)* keine Luft [mehr] bekommen; *(permanently)* ersticken (**on an** + *Dat.*). **3.** *n. (Motor Veh.)* Choke, *der*
cholera ['kɒlərə] *n.* Cholera, *die*
cholesterol [kə'lestərɒl] *n.* Cholesterin, *das*
choose [tʃuːz] **1.** *v. t.,* **chose** [tʃəʊz], **chosen** ['tʃəʊzn] **a)** wählen; **b)** *(decide)* ~/~ **not to do sth.** sich dafür/dagegen entscheiden, etw. zu tun. **2.** *v. i.,* **chose**, **chosen** wählen (**between** zwischen); ~ **from sth.** aus etw./*(from several)* unter etw. *(Dat.)* [aus]wählen. **choos[e]y** ['tʃuːzɪ] *adj.* wählerisch
chop [tʃɒp] **1.** *n.* **a)** Hieb, *der;* **b)** *(of meat)* Kotelett, *das;* **c)** **get the** ~ *(coll.: be dismissed)* rausgeworfen werden *(ugs.).* **2.** *v. t.,* **-pp-** hacken ⟨*Holz*⟩; kleinschneiden ⟨*Fleisch, Gemüse*⟩. **'chopper** *n. (axe)* Beil, *das; (cleaver)* Hackbeil, *das*
'choppy *adj.* bewegt
choral ['kɔːrl] *adj.* Chor-
chord [kɔːd] *n. (Mus.)* Akkord, *der*
chore [tʃɔː(r)] *n.* [lästige] Routinearbeit
chortle ['tʃɔːtl] **1.** *v. i.* vor Lachen glucksen. **2.** *n.* Glucksen, *das*
chorus ['kɔːrəs] *n.* **a)** Chor, *der;* **b)** *(of song)* Chorus, *der*
chose, chosen *see* **choose**
chow [tʃaʊ] *n. (Amer. sl.: food)* Futter, *das (salopp)*
Christ [kraɪst] *n.* Christus *(der)*
christen ['krɪsn] *v. t.* taufen. **'christening** *n.* Taufe, *die*
Christian ['krɪstjən] **1.** *adj.* christlich. **2.** *n.* Christ, *der*/Christin, *die.* **Christianity** [krɪstɪ'ænɪtɪ] *n.* das Christentum
'Christian name *n.* Vorname, *der*
Christmas ['krɪsməs] *n.* Weihnachten, *das od. Pl.;* **merry** *or* **happy ~:** frohe *od.* fröhliche Weihnachten; **at ~:** [zu] Weihnachten

Christmas: ~ **'Day** *n*. erster Weihnachtsfeiertag; ~ **'Eve** *n*. Heiligabend, *der;* ~ **tree** *n*. Weihnachtsbaum, *der*
chrome [krəʊm], **chromium** ['krəʊmɪəm] *ns*. Chrom, *das*. **'chromiumplated** *adj*. verchromt
chronic ['krɒnɪk] *adj*. chronisch
chronicle ['krɒnɪkl] *n*. Chronik, *die*
chronological [krɒnə'lɒdʒɪkl] *adj*. chronologisch
chrysalis ['krɪsəlɪs] *n*., *pl*. ~**es** Puppe, *die*
chrysanthemum [krɪ'sænθɪməm] *n*. Chrysantheme, *die*
chubby ['tʃʌbɪ] *adj*. pummelig
chuck [tʃʌk] *v. t. (coll.)* schmeißen *(ugs.)*. **chuck 'away, chuck 'out** *v. t. (coll.)* wegschmeißen *(ugs.)*
chuckle ['tʃʌkl] **1.** *v. i*. leise [vor sich hin] lachen (**at** über + *Akk.*). **2.** *n*. leises, glucksendes Lachen
chug [tʃʌg] *v. i., -gg-* tuckern
chum [tʃʌm] *n. (coll.)* Kumpel, *der (salopp)*
chunk [tʃʌŋk] *n*. dickes Stück. **'chunky** *adj*. **a)** *(small and sturdy)* stämmig; **b)** dick ⟨*Pullover*⟩
church [tʃɜːtʃ] *n*. Kirche, *die;* **go to** ~: in die Kirche gehen; **the C~ of England** die Kirche von England. **'churchyard** *n*. Friedhof, *der (bei einer Kirche)*
churlish ['tʃɜːlɪʃ] *adj. (ill-bred)* ungehobelt; *(surly)* griesgrämig
churn [tʃɜːn] *n. (Brit.)* Butterfaß, *das*. **churn'out** *v. t*. massenweise produzieren *(ugs.)*
chute [ʃuːt] *n*. Schütte, *die; (for persons)* Rutsche, *die*
CIA *abbr. (Amer.)* **Central Intelligence Agency** CIA, *der od. die*
CID *abbr. (Brit.)* **Criminal Investigation Department** C.I.D.; **the** ~: die Kripo
cider ['saɪdə(r)] *n*. ≈ Apfelwein, *der*
cigar [sɪ'gɑː(r)] *n*. Zigarre, *die*
cigarette [sɪgə'ret] *n*. Zigarette, *die*
cigarette: ~**-end** *n*. Zigarettenstummel, *der;* ~**-lighter** *n*. Feuerzeug, *das;* ~**-packet** *n*. Zigarettenschachtel, *die*
cinders ['sɪndəz] *n. pl*. Asche, *die*
cine ['sɪnɪ]: ~ **camera** *n*. Filmkamera, *die;* ~ **film** *n*. Schmalfilm, *der*
cinema ['sɪnɪmə] *n*. Kino, *das;* **go to the** ~: ins Kino gehen
cinnamon ['sɪnəmən] *n*. Zimt, *der*
cipher ['saɪfə(r)] *n*. Geheimschrift, *die;* **in** ~: chiffriert

circle ['sɜːkl] **1.** *n*. Kreis, *der*. **2.** *v. i*. kreisen. **3.** *v. t*. umkreisen
circuit ['sɜːkɪt] *n*. **a)** *(Electr.)* Schaltung, *die;* **b)** *(Motor-racing)* Rundkurs, *der*
circular ['sɜːkjʊlə(r)] **1.** *adj. (round)* kreisförmig. **2.** *n*. Rundschreiben, *das*
circulate ['sɜːkjʊleɪt] **1.** *v. i*. zirkulieren; ⟨*Personen, Wein usw.*:⟩ herumgehen *(ugs.)*. **2.** *v. t*. in Umlauf setzen; herumgehen lassen ⟨*Buch, Bericht*⟩ (**around** in + *Dat.*). **circulation** [sɜːkjʊ'leɪʃn] *n*. **a)** *(Physiol.)* Kreislauf, *der;* **poor** ~: Kreislaufstörungen *Pl.;* **b)** *(copies sold)* verkaufte Auflage
circumcise ['sɜːkəmsaɪz] *v. t*. beschneiden
circumference [sə'kʌmfərəns] *n*. Umfang, *der*
circumstances ['sɜːkəmstənsɪz] *n. pl*. Umstände; **in** or **under the** ~: unter diesen Umständen; **under no** ~: unter keinen Umständen
circus ['sɜːkəs] *n*. Zirkus, *der*
CIS *abbr*. **Commonwealth of Independent States** GUS
cissy ['sɪsɪ] *see* sissy
cistern ['sɪstən] *n*. Wasserkasten, *der; (in roof)* Wasserbehälter, *der*
citation [saɪ'teɪʃn] *n*. Zitat, *das*
cite [saɪt] *v. t. (quote)* zitieren; anführen ⟨*Beispiel*⟩
citizen ['sɪtɪzən] *n*. **a)** *(of town, city)* Bürger, *der/*Bürgerin, *die;* **b)** *(of state)* [Staats]bürger, *der/*-bürgerin, *die*. **'citizenship** *n*. Staatsbürgerschaft, *die*
citrus ['sɪtrəs] *n*. ~ **[fruit]** Zitrusfrucht, *die*
city ['sɪtɪ] *n*. [Groß]stadt, *die*. **city 'centre** *n*. Stadtzentrum, *das*
civic ['sɪvɪk] *adj*. [staats]bürgerlich; ~ **centre** Verwaltungszentrum der Stadt
civil ['sɪvl] *adj*. **a)** *(not military)* zivil; **b)** *(polite, obliging)* höflich; **c)** *(Law)* Zivil-. **civil engi'neer** *n*. Bauingenieur, *der/*-ingenieurin, *die*. **civil engi'neering** *n*. Hoch- und Tiefbau, *der*
civilian [sɪ'vɪljən] **1.** *n*. Zivilist, *der*. **2.** *adj*. Zivil-
civility [sɪ'vɪlɪtɪ] *n*. Höflichkeit, *die*
civilization [sɪvɪlaɪ'zeɪʃn] *n*. Zivilisation, *die*
civilized ['sɪvɪlaɪzd] *adj*. zivilisiert
civil: ~ **'law** *n*. Zivilrecht, *das;* ~ **'rights** *n. pl*. Bürgerrechte; ~ **'servant** *n*. ≈ Staatsbeamte, *der/*-beamtin, *die;* **C~ 'Service** *n*. öffentlicher Dienst; ~ **'war** *n*. Bürgerkrieg, *der*

clad [klæd] *adj. (arch./literary)* gekleidet (**in** in + *Akk.*)

claim [kleɪm] **1.** *v. t.* **a)** beanspruchen ⟨*Thron, Gebiete*⟩; fordern ⟨*Lohnerhöhung, Schadenersatz*⟩; beantragen ⟨*Sozialhilfe usw.*⟩; **b)** *(assert)* behaupten. **2.** *v. i. (Insurance)* Ansprüche geltend machen. **3.** *n.* Anspruch, *der* (**to** auf + *Akk.*); **lay ~ to sth.** auf etw. *(Akk.)* Anspruch erheben. **claimant** ['kleɪmənt] *n.* Antragsteller, *der*/-stellerin, *die*

clairvoyant [kleə'vɔɪənt] **1.** *n.* Hellseher, *der*/Hellseherin, *die*. **2.** *adj.* hellseherisch

clam [klæm] **1.** *n.* Klaffmuschel, *die*. **2.** *v. i.,* **-mm-:** **~ up** *(coll.)* den Mund nicht [mehr] aufmachen

clamber ['klæmbə(r)] *v. i.* klettern

clammy ['klæmɪ] *adj.* feucht; kalt und schweißig ⟨*Haut*⟩; klamm ⟨*Kleidung*⟩

clamour *(Brit.; Amer.:* **clamor**) ['klæmə(r)] **1.** *n. (noise, shouting)* Lärm, *der*; lautes Geschrei. **2.** *v. i.* **~ for sth.** nach etw. schreien

clamp [klæmp] **1.** *n.* Klammer, *die*; *(Woodw.)* Schraubzwinge, *die*. **2.** *v. t.* klemmen; einspannen ⟨*Werkstück*⟩. **3.** *v. i. (fig.)* **~ down on sb./sth.** gegen jmdn./etw. rigoros vorgehen

clan [klæn] *n.* Sippe, *die*; *(of Scottish Highlanders)* Clan, *der*

clandestine [klæn'destɪn] *adj.* heimlich

clang [klæŋ] **1.** *n. (of bell)* Läuten, *das*; *(of hammer)* Klingen, *das*. **2.** *v. i.* ⟨*Glocke:*⟩ läuten; ⟨*Hammer:*⟩ klingen

clap [klæp] **1.** *n.* **a)** Klatschen, *das*; **b)** **~ of thunder** Donnerschlag, *der*. **2.** *v. i.,* **-pp-** klatschen. **3.** *v. t.,* **-pp-:** **~ one's hands** in die Hände klatschen; **~ sth.** etw. beklatschen; **~ sb.** jmdm. Beifall klatschen. **'clapping** *n.* Applaus, *der*

claret ['klærət] **1.** *n.* roter Bordeauxwein. **2.** *adj.* weinrot

clarification [klærɪfɪ'keɪʃn] *n.* Klarstellung, *die*

clarify ['klærɪfaɪ] *v. t.* klären ⟨*Situation usw.*⟩; *(by explanation)* klarstellen; erläutern ⟨*Bedeutung, Aussage*⟩

clarinet [klærɪ'net] *n.* Klarinette, *die*

clarity ['klærɪtɪ] *n.* Klarheit, *die*

clash [klæʃ] **1.** *v. i.* **a)** scheppern *(ugs.);* **b)** *(meet in conflict)* zusammenstoßen; **c)** *(disagree)* sich streiten; **d)** ⟨*Interesse, Ereignis:*⟩ kollidieren; ⟨*Farbe:*⟩ sich beißen *(ugs.)* (**with** mit). **2.** *v. t.* gegeneinanderschlagen. **3.** *n.* **a)** *(of cymbals)* Dröhnen, *das*; **b)** *(meeting in conflict)* Zusammenstoß, *der;* **c)** *(disagreement)* Auseinandersetzung, *die;* **d)** *(of personalities, colours)* Unverträglichkeit, *die; (of events)* Überschneiden, *das*

clasp [klɑːsp] **1.** *n.* Verschluß, *der*. **2.** *v. t.* umklammern

class [klɑːs] **1.** *n.* Klasse, *die; (in society)* Gesellschaftsschicht, *die; (Sch.: lesson)* Stunde, *die*. **2.** *v. t.* einstufen (**as** als). **'class-conscious** *adj.* klassenbewußt

classic ['klæsɪk] **1.** *adj.* klassisch. **2.** *n.* Klassiker, *der;* **~s** Altphilologie, *die*

classical ['klæsɪkl] *adj.* klassisch

classification [klæsɪfɪ'keɪʃn] *n.* Klassifikation, *die*

classified ['klæsɪfaɪd] *adj.* **a)** *(secret)* geheim; **b)** **~ advertisement** Kleinanzeige, *die*

classify ['klæsɪfaɪ] *v. t.* klassifizieren

class: **~-mate** *n.* Klassenkamerad, *der*/-kameradin, *die;* **~-room** *n.* Klassenzimmer, *das*

'classy *adj. (coll.)* klasse

clatter ['klætə(r)] **1.** *n.* Klappern, *das*. **2.** *v. i.* **a)** klappern; **b)** *(move or fall with a ~)* poltern

clause [klɔːz] *n.* **a)** Klausel, *die;* **b)** *(Ling.)* Teilsatz, *der;* |**subordinate**| **~ :** Nebensatz, *der*

claustrophobia [klɒstrə'fəʊbɪə] *n.* Klaustrophobie, *die*. **claustrophobic** [klɒstrə'fəʊbɪk] *adj.* beengend ⟨*Ort*⟩

claw [klɔː] **1.** *n.* Kralle, *die; (of crab etc.)* Schere, *die*. **2.** *v. t.* kratzen

clay [kleɪ] *n.* Lehm, *der; (for pottery)* Ton, *der*

clean [kliːn] **1.** *adj.* sauber; frisch ⟨*Wäsche, Hemd*⟩. **2.** *adv.* glatt. **3.** *v. t.* saubermachen; putzen ⟨*Zimmer, Schuh*⟩; reinigen ⟨*Teppich, Kleidung, Wunde*⟩; **~ one's teeth** sich *(Dat.)* die Zähne putzen. **4.** *n.* **give sth. a ~ :** etw. putzen. **clean 'out** *v. t.* **a)** saubermachen; **b)** *(sl.)* **~ sb. out** *(take all sb.'s money)* jmdn. [total] schröpfen *(ugs.).* **clean 'up 1.** *v. t.* **a)** aufräumen; **b)** *(fig.)* säubern. **2.** *v. i.* aufräumen

'cleaner *n.* **a)** Raumpfleger, *der*/-pflegerin, *die; (woman also)* Putzfrau, *die;* **b)** *usu. in pl.* (*dry-~*) Reinigung, *die;* **take sth. to the ~'s** etw. in die Reinigung bringen

cleanliness ['klenlɪnɪs] *n.* Reinlichkeit, *die*

cleanly ['kliːnlɪ] *adv.* sauber

cleanse [klenz] *v. t.* [gründlich] reinigen. **'cleanser** *n.* Reinigungsmittel, *das*

'clean-shaven *adj.* glattrasiert

clear [klıə(r)] **1.** *adj.* **a)** klar; scharf ⟨*Bild*⟩; **make oneself ~:** sich deutlich [genug] ausdrücken; **make it ~ [to sb.] that ...:** [jmdm.] klar und deutlich sagen, daß ...; **b)** *(complete)* **three ~ days** volle drei Tage; **c)** *(unobstructed)* frei; **keep sth. ~** *(not block)* etw. frei halten. **2.** *adv.* **keep ~ of sth./sb.** etw./jmdn. meiden; **please stand** *or* **keep ~ of the door** bitte von der Tür zurücktreten. **3.** *v. t.* **a)** räumen ⟨*Straße*⟩; abräumen ⟨*Schreibtisch*⟩; freimachen ⟨*Abfluß, Kanal*⟩; **~ a space for sb./sth.** für jmdn./etw. Platz machen; **b)** *(empty)* räumen, leeren ⟨*Briefkasten*⟩; **c)** *(remove)* wegräumen; beheben ⟨*Verstopfung*⟩; **d)** *(show to be innocent)* freisprechen; **e)** *(get permission for)* **~ sth. with sb.** etw. von jmdm. genehmigen lassen. **4.** *v. i.* **a)** ⟨*Wetter, Himmel:*⟩ sich aufheitern; **b)** *(disperse)* sich verziehen. **5.** *n.* **we're in the ~** *(free of suspicion)* auf uns fällt kein Verdacht; *(free of trouble)* wir haben es geschafft. **clear** '**off** *v. i.* abhauen *(salopp)*. **clear** '**out 1.** *v. t.* ausräumen. **2.** *v. i. (coll.)* verschwinden. **clear** '**up 1.** *v. t.* **a)** wegräumen ⟨*Abfall*⟩; aufräumen ⟨*Platz, Sachen*⟩; **b)** *(explain)* klären. **2.** *v. i.* **a)** aufräumen; **b)** ⟨*Wetter:*⟩ sich aufhellen

clearance ['klıərəns] *n.* **a)** *(of obstruction)* Beseitigung, *die;* **b)** *(clear space)* Spielraum, *der*

'clear cut *adj.* klar umrissen; klar ⟨*Abgrenzung, Ergebnis*⟩

'clearing *n.* Lichtung, *die*

'clearly *adv.* **a)** *(distinctly)* klar; deutlich ⟨*sprechen*⟩; **b)** *(manifestly, unambiguously)* eindeutig; klar ⟨*denken*⟩

'clearway *n. (Brit.)* Straße mit Halteverbot

cleaver ['kli:və(r)] *n.* Hackbeil, *das*

clef [klef] *n.* Notenschlüssel, *der*

cleft [kleft] *n.* Spalte, *die*

clench [klentʃ] *v. t.* zusammenpressen; **~ one's fist** *or* **fingers** die Faust ballen; **~ one's teeth** die Zähne zusammenbeißen

clergy ['klɜ:dʒı] *n. pl.* Geistlichkeit, *die;* Klerus, *der.* **clergyman** ['klɜ:dʒımən] *n., pl.* **~men** ['klɜ:dʒımən] Geistliche, *der*

clerical ['klerıkl] *adj.* Büro⟨*arbeit, -personal*⟩; **~ error** Schreibfehler, *der*

clerk [klɑ:k] *n. (in bank)* Bankangestellte, *der/die; (in office)* Büroangestellte, *der/die*

clever ['klevə(r)] *adj.* **a)** klug; **b)** *(skilful)* geschickt; **c)** *(ingenious)* geistreich ⟨*Idee, Argument*⟩; **d)** *(smart, cunning)* clever. **'cleverly** *adv.* **a)** klug; **b)** *(skilfully)* geschickt

cliché ['kli:ʃeı] *n.* Klischee, *das*

click [klık] **1.** *n.* Klicken, *das.* **2.** *v. i.* klicken

client ['klaıənt] *n.* **a)** Klient, *der/*Klientin, *die;* **b)** *(customer)* Kunde, *der/*Kundin, *die*

clientele [kli:ɒn'tel] *n. (of shop)* Kundschaft, *die*

cliff [klıf] *n.* Kliff, *das.* **'cliff-hanger** *n.* Thriller, *der*

climate ['klaımət] *n.* Klima, *das*

climax ['klaımæks] *n.* Höhepunkt, *der*

climb [klaım] **1.** *v. t.* hinaufsteigen; klettern auf ⟨*Baum*⟩; ⟨*Auto:*⟩ hinaufkommen ⟨*Hügel*⟩. **2.** *v. i.* **a)** klettern **(up** auf **+** *Akk.*); **b)** ⟨*Flugzeug, Sonne:*⟩ aufsteigen. **3.** *n.* Aufstieg, *der.* **climb** '**down** *v. i.* **a)** hinunterklettern; **b)** *(fig.)* nachgeben

'climb-down *n.* Rückzieher, *der (ugs.)*

climber ['klaımə(r)] *n.* Bergsteiger, *der*

clinch [klıntʃ] **1.** *v. t.* zum Abschluß bringen; perfekt machen *(ugs.)* ⟨*Geschäft*⟩. **2.** *n. (Boxing)* Clinch, *der*

cling [klıŋ] *v. i.,* **clung** [klʌŋ] sich klammern **(to** an **+** *Akk.*). **'cling film** *n.* Klarsichtfolie, *die*

clinic ['klınık] *n.* Klinik, *die.* **clinical** ['klınıkl] *adj.* **a)** *(Med.)* klinisch; **b)** *(dispassionate)* nüchtern

clink [klıŋk] **1.** *n. (of glasses)* Klirren, *das; (of coins)* Klimpern, *das.* **2.** *v. i.* ⟨*Flaschen:*⟩ klirren; ⟨*Münzen:*⟩ klimpern. **3.** *v. t.* klirren mit ⟨*Glas*⟩; klimpern mit ⟨*Kleingeld*⟩

¹clip [klıp] **1.** *n.* Klammer, *die; (for paper)* Büroklammer, *die.* **2.** *v. t.,* **-pp-** klammern **(Ion** to an **+** *Akk.*)

²clip *v. t.,* **-pp-** *(cut)* schneiden ⟨*Fingernägel, Haar, Hecke*⟩; stutzen ⟨*Flügel*⟩

clique [kli:k] *n.* Clique, *die*

cloak [kləʊk] **1.** *n.* Umhang, *der.* **2.** *v. t.* [ein]hüllen. **'cloakroom** *n.* Garderobe, *die; (Brit. euphem.: lavatory)* Toilette, *die*

clock [klɒk] **1.** *n.* **a)** Uhr, *die;* [work] **against the ~:** gegen die Zeit [arbeiten]; **round the ~:** rund um die Uhr; **b)** *(coll.) (speedometer)* Tacho, *der (ugs.); (milometer)* ≈ Kilometerzähler, *der.* **2.** *v. t.* **~ [up]** zu verzeichnen haben

3*

⟨*Erfolg*⟩; erreichen ⟨*Geschwindigkeit*⟩.
clock 'in, clock 'on *v. i.* [bei Arbeitsantritt] stechen. **clock 'off, clock 'out** *v. i.* [bei Arbeitsschluß] stechen

'**clockwise** *adv., adj.* im Uhrzeigersinn

'**clockwork** *n.* Uhrwerk, *das;* **a ~ car.** ein Aufziehauto; **as regular as ~** *(fig.)* absolut regelmäßig

clog [klɒg] **1.** *n.* Clog, *der; (traditional)* Holzschuh, *der.* **2.** *v. t.,* **-gg-:** ~ [up] verstopfen

cloister ['klɔɪstə(r)] *n.* Kreuzgang, *der*

clone [kləʊn] **1.** *n.* Klon, *der.* **2.** *v. t.* klonen

close 1. [kləʊs] *adj.* **a)** *(in space)* dicht; nahe; **be ~ to sth.** nahe bei *od.* an etw. *(Dat.)* sein; **at ~ quarters** aus der Nähe betrachtet; **b)** *(in time)* nahe **(to an + *Dat.*); c)** eng ⟨*Freund, Zusammenarbeit*⟩; nahe ⟨*Verwandte, Bekanntschaft*⟩; **d)** eingehend ⟨*Untersuchung, Prüfung usw.*⟩; **e)** hart ⟨*Wett[kampf], Spiel*⟩; knapp ⟨*Ergebnis*⟩; **that was a ~ call** *or* **shave** *(coll.)* das war knapp! **2.** [kləʊs] *adv.* nah[e]; **~ by** in der Nähe; **~ to sb./sth.** nahe bei jmdm./etw. **3.** [kləʊz] *v. t.* **a)** *(shut)* schließen; zuziehen ⟨*Vorhang*⟩; schließen ⟨*Laden, Fabrik*⟩; sperren ⟨*Straße*⟩; **b)** *(conclude)* schließen ⟨*Diskussion, Versammlung*⟩. **4.** [kləʊz] *v. i.* **a)** *(shut)* sich schließen; **b)** ⟨*Laden, Fabrik:*⟩ schließen, *(ugs.)* zumachen. **5.** [kləʊz] *n.* Ende, *das;* Schluß, *der;* **come or draw to a ~:** zu Ende gehen; **bring or draw sth. to a ~:** etw. zu Ende bringen. **close** [kləʊz] '**down 1.** *v. t.* schließen; stillegen ⟨*Werk*⟩. **2.** *v. i.* geschlossen werden; ⟨*Werk:*⟩ stillgelegt werden. **close 'in** *v. i.* ⟨*Nacht, Dunkelheit:*⟩ hereinbrechen; ⟨*Tage:*⟩ kürzer werden; **~ in on** umzingeln. **close 'off** *v. t.* [ab]sperren

closed [kləʊzd] *adj.* geschlossen; **we're ~:** wir haben geschlossen.

'**closed-circuit** *adj.* ~ **television** interne Fernsehanlage

close-down ['kləʊzdaʊn] *n. (Radio, Telev.)* Sendeschluß, *der*

closed 'shop *n.* Closed Shop, *der*

closely ['kləʊslɪ] *adv.* **a)** dicht; **b)** *(intimately)* eng; **c)** genau ⟨*befragen, prüfen*⟩; streng ⟨*bewachen*⟩

closet ['klɒzɪt] *n. (Amer.: cupboard)* Schrank, *der*

close-up ['kləʊsʌp] *n.* ~ [picture/shot] Nahaufnahme, *die*

closing ['kləʊzɪŋ]: ~ **date** *n. (for competition)* Einsendeschluß, *der; (to take part)* Meldefrist, *die;* ~-**time** *n. (of pub)* Polizeistunde, *die*

closure ['kləʊʒə(r)] *n.* Schließung, *die; (of road)* Sperrung, *die*

clot [klɒt] **1.** *n.* **a)** *(blood)* Gerinnsel, *das;* **b)** *(Brit. sl.: stupid person)* Trottel, *der.* **2.** *v. i.,* **-tt-** ⟨*Blut:*⟩ gerinnen

cloth [klɒθ] *n., pl.* ~**s** [klɒθs] **a)** Stoff, *der;* Tuch, *das;* **b)** *(dish-~)* Spültuch, *das; (table-~)* [Tisch]decke, *die*

clothe [kləʊð] *v. t.* kleiden

clothes [kləʊðz] *n. pl.* Kleider *Pl.;* **put one's ~ on** sich anziehen; **take one's ~ off** sich ausziehen

'**clothes:** ~-**brush** *n.* Kleiderbürste, *die;* ~-**line** *n.* Wäscheleine, *die;* ~-**peg** *(Brit.),* ~-**pin** *(Amer.) ns.* Wäscheklammer, *die*

clothing ['kləʊðɪŋ] *n.* Kleidung, *die*

clotted cream [klɒtɪd 'kri:m] *n. sehr fetter Rahm*

cloud [klaʊd] *n.* **a)** Wolke, *die;* **every ~ has a silver lining** *(prov.)* es hat alles sein Gutes; **b)** ~ **of dust/smoke** Staub-/Rauchwolke, *die.* **cloud 'over** *v. i.* sich bewölken

'**cloudburst** *n.* Wolkenbruch, *der*

'**cloudless** *adj.* wolkenlos

'**cloudy** *adj.* bewölkt ⟨*Himmel*⟩; trübe ⟨*Wetter, Flüssigkeit, Glas*⟩

clout [klaʊt] *(coll.)* **1.** *n.* Schlag, *der.* **2.** *v. t.* hauen *(ugs.)*

¹**clove** [kləʊv] *n.* ~ [of garlic] [Knoblauch]zehe, *die*

²**clove** *n. (spice)* [Gewürz]nelke, *die*

clover ['kləʊvə(r)] *n.* Klee, *der.*

'**cloverleaf** *n.* Kleeblatt, *das*

clown [klaʊn] **1.** *n.* Clown, *der.* **2.** *v. i.* ~ [about *or* around] den Clown spielen

cloying ['klɔɪŋ] *adj.* süßlich

club [klʌb] **1.** *n.* **a)** *(weapon)* Keule, *die; (golf-~)* Schläger, *der;* **b)** *(association)* Klub, *der;* Verein, *der;* **c)** *(Cards)* Kreuz, *das;* ~**s are trumps** Kreuz ist Trumpf; **the ace/seven of ~s** das Kreuzas/die Kreuzsieben. **2.** *v. t.,* **-bb-** *(beat)* prügeln; *(with ~)* knüppeln. **3.** *v. i.,* **-bb-:** ~ **together** *(to buy something)* zusammenlegen

cluck [klʌk] **1.** *n.* Gackern, *das.* **2.** *v. i.* gackern

clue [klu:] *n.* Anhaltspunkt, *der; (in criminal investigation)* Spur, *die;* **not have a ~:** keine Ahnung haben. '**clueless** *adj. (coll.)* unbedarft *(ugs.)*

clump [klʌmp] *n.* Gruppe, *die; (of grass)* Büschel, *das*

clumsy ['klʌmzɪ] *adj.* schwerfällig, unbeholfen 〈*Person, Bewegung*〉; plump 〈*Form, Figur, Nachahmung*〉

clung *see* **cling**

cluster ['klʌstə(r)] **1.** *n.* *(of grapes, berries)* Traube, *die;* *(of fruit, flowers)* Büschel, *das;* *(of stars, huts)* Haufen, *der.* **2.** *v. i.* ~ **[a]round** sb./sth. sich um jmdn./etw. scharen *od.* drängen

clutch [klʌtʃ] **1.** *v. t.* umklammern. **2.** *v. i.* ~ **at** sth. nach etw. greifen; *(fig.)* sich an etw. *(Akk.)* klammern. **3.** *n.* **a)** *in pl. (fig.: control)* Klauen; **b)** *(Motor Veh.)* Kupplung, *die*

clutter ['klʌtə(r)] **1.** *n.* Durcheinander, *das.* **2.** *v. t.* ~ **[up] the table/room** überall auf dem Tisch/im Zimmer herumliegen

cm. *abbr.* **centimetre[s]** cm

Co. *abbr.* **a) company** Co.; **b) county**

c/o *abbr.* **care of** bei; c/o

coach [kəʊtʃ] **1.** *n.* **a)** *(horse-drawn)* Kutsche, *die;* **b)** *(Railw.)* Wagen, *der;* **c)** *(bus)* [Reise]bus, *der;* **by** ~: mit dem Bus; **d)** *(Sport)* Trainer, *der*/Trainerin, *die.* **2.** *v. t.* trainieren. **'coach station** *n.* Busbahnhof, *der.* **'coach tour** *n.* Rundreise [im Omnibus]

coagulate [kəʊ'ægjʊleɪt] **1.** *v. t.* gerinnen lassen. **2.** *v. i.* gerinnen

coal [kəʊl] *n.* Kohle, *die.* **'coalfield** *n.* Kohlenrevier, *das*

coalition [kəʊə'lɪʃn] *n. (Polit.)* Koalition, *die*

coal: ~**-mine** *n.* [Kohlen]bergwerk, *das;* ~**-miner** *n.* [im Kohlenbergbau tätiger] Grubenarbeiter; ~**-mining** *n.* Kohlenbergbau, *der*

coarse [kɔːs] *adj.* **a)** *(in texture)* grob; **b)** *(unrefined, obscene)* derb

coast [kəʊst] **1.** *n.* Küste, *die.* **2.** *v. i.* im Freilauf fahren. **coastal** ['kəʊstl] *adj.* Küsten-. **'coaster** *n.* **a)** *(mat)* Untersetzer, *der;* **b)** *(ship)* Küstenmotorschiff, *das*

coast: ~**guard** *n.* Küstenwache, -wacht, *die;* ~**line** *n.* Küste, *die*

coat [kəʊt] **1.** *n.* **a)** Mantel, *der;* **b)** *(layer)* Schicht, *die;* *(of paint)* Anstrich, *der;* **c)** *(animal's hair, fur, etc.)* Fell, *das.* **2.** *v. t.* überziehen; *(with paint)* streichen

'coat-hanger *n.* Kleiderbügel, *der*

'coating *n.* Schicht, *die*

coat of 'arms *n.* Wappen, *das*

coax [kəʊks] *v. t.* überreden

cobble ['kɒbl] *n.* Pflasterstein, *der*

cobbler ['kɒblə(r)] *n.* Schuster, *der*

'cobble-stone *see* **cobble**

cobra ['kɒbrə] *n.* Kobra, *die*

cobweb ['kɒbweb] *n.* Spinnengewebe, *das;* Spinnennetz, *das*

cocaine [kə'keɪn] *n.* Kokain, *das*

cock [kɒk] **1.** *n.* Hahn, *der.* **2.** *v. t.* spitzen 〈*Ohren*〉; ~ **a/the gun** den Hahn spannen. **cock-a-hoop** [kɒkə'huːp] *adj.* überschwenglich

cockatoo [kɒkə'tuː] *n.* Kakadu, *der*

cockerel ['kɒkərəl] *n.* junger Hahn

cock-eyed ['kɒkaɪd] *adj.* **a)** *(crooked)* schief; **b)** *(absurd)* verrückt

cockle ['kɒkl] *n.* Herzmuschel, *die*

cockney ['kɒknɪ] **1.** *adj.* Cockney-. **2.** *n.* Cockney, *der*

'cockpit *n.* Cockpit, *das*

cockroach ['kɒkrəʊtʃ] *n.* [Küchen-, Haus-]schabe, *die*

cocktail ['kɒkteɪl] *n.* Cocktail, *der.* **'cocktail cabinet** *n.* Hausbar, *die.* **'cocktail party** *n.* Cocktailparty, *die*

cocoa ['kəʊkəʊ] *n.* Kakao, *der*

coconut ['kəʊkənʌt] *n.* Kokosnuß, *die*

cocoon [kə'kuːn] *(Zool.)* Kokon, *der*

cod [kɒd] *n., pl.* **same** Kabeljau, *der*

COD *abbr.* **cash on delivery,** *(Amer.)* **collect on delivery** p. Nachn.

code [kəʊd] *n.* **a)** *(statutes etc.)* Gesetzbuch, *das;* ~**s of behaviour** Verhaltensnormen; **b)** *(system of signals)* Code, *der;* **be in** ~: verschlüsselt sein. **2.** *v. t.* chiffrieren; verschlüsseln. **'code-name** *n.* Deckname, *der.* **'code-word** *n.* Kennwort, *das*

cod-liver 'oil *n.* Lebertran, *der*

co-driver ['kəʊdraɪvə(r)] *n.* Beifahrer, *der*/-fahrerin, *die*

coed ['kəʊed] *(esp. Amer. coll.)* **1.** *n.* Studentin, *die.* **2.** *adj.* ~ **school** gemischte Schule

coeducational [kəʊedjʊ'keɪʃnl] *adj.* koedukativ; Koedukations-

coerce [kəʊ'ɜːs] *v. t.* zwingen; ~ **sb. into sth.** jmdn. zu etw. zwingen. **coercion** [kəʊ'ɜːʃn] *n.* Zwang, *der*

coexist [kəʊɪg'zɪst] *v. i.* koexistieren. **coexistence** [kəʊɪg'zɪstəns] *n.* Koexistenz, *die*

C. of E. [siːəv'iː] *abbr.* **Church of England**

coffee ['kɒfɪ] *n.* Kaffee, *der;* **three black/white** ~**s** drei [Tassen] Kaffee ohne/mit Milch

coffee: ~ **bar** *n.* Café, *das;* ~**-bean** *n.* Kaffeebohne, *die;* ~**-break** *n.* Kaffeepause, *die;* ~**-cup** *n.* Kaffeetasse, *die;* ~**-pot** *n.* Kaffeekanne, *die;* ~ **shop** *n.* Kaffeestube, *die;* ~**-table** *n.* Couchtisch, *der*

coffin ['kɒfɪn] *n.* Sarg, *der*
cog [kɒg] *n. (Mech.)* Zahn, *der*
cogent ['kəʊdʒənt] *adj.* überzeugend ⟨*Argument*⟩; zwingend ⟨*Grund*⟩
cognac ['kɒnjæk] *n.* Cognac, *der* ⓌＺ
cog: ~-**railway** *n.* Zahnradbahn, *die;* ~-**wheel** *n.* Zahnrad, *das*
cohere [kəʊ'hɪə(r)] *v. i.* zusammenhalten. **coherent** [kəʊ'hɪərənt] *adj.* zusammenhängend
coil [kɔɪl] **1.** *v. t.* aufwickeln; *(twist)* aufdrehen. **2.** *v. i.* ~ **round sth.** etw. umschlingen. **3.** *n.* **a)** ~s of rope/wire aufgerollte Seile *Pl.*/aufgerollter Draht; **b)** *(single turn)* Windung, *die;* **c)** *(Electr.)* Spule, *die*
coin [kɔɪn] **1.** *n.* Münze, *die.* **2.** *v. t.* prägen ⟨*Wort, Redewendung*⟩
coincide [kəʊɪn'saɪd] *v. i.* **a)** *(in time)* zusammenfallen; **b)** *(agree)* übereinstimmen (**with** mit). **coincidence** [kəʊ'ɪnsɪdəns] *n.* Zufall, *der.* **coincidental** [kəʊɪnsɪ'dentl] *adj.* zufällig
coke [kəʊk] *n.* Koks, *der*
colander ['kʌləndə(r)] *n.* Sieb, *das*
cold [kəʊld] **1.** *adj.* **a)** kalt; **I feel** ~: mir ist kalt; **b)** *(fig.)* [betont] kühl ⟨*Person, Aufnahme, Begrüßung*⟩. **2.** *adv.* kalt. **3.** *n.* **a)** Kälte, *die;* **b)** *(illness)* Erkältung, *die;* ~ **[in the head]** Schnupfen, *der.* **cold-blooded** ['kəʊldblʌdɪd] *adj.* **a)** wechselwarm ⟨*Tier*⟩; **b)** kaltblütig ⟨*Person, Mord*⟩
coldly *adv.* [betont] kühl
coleslaw ['kəʊlslɔ:] *n.* Krautsalat, *der*
collaborate [kə'læbəreɪt] *v. i.* **a)** zusammenarbeiten; ~ **[with sb.] on sth.** zusammen [mit jmdm.] an etw. *(Dat.)* arbeiten; **b)** *(with enemy)* kollaborieren. **collaboration** [kəlæbə'reɪʃn] *n.* Zusammenarbeit, *die;* *(with enemy)* Kollaboration, *die.* **collaborator** [kə'læbəreɪtə(r)] *n.* Mitarbeiter, *der/*-arbeiterin, *die;* *(with enemy)* Kollaborateur, *der/*Kollaborateurin, *die*
collage ['kɒlɑ:ʒ] *n.* Collage, *die*
collapse [kə'læps] **1.** *n.* **a)** *(of person)* Zusammenbruch, *der;* **b)** *(of structure)* Einsturz, *der;* **c)** *(of negotiations)* Scheitern, *das;* *(of company)* Zusammenbruch, *der.* **2.** *v. i.* **a)** ⟨*Person:*⟩ zusammenbrechen; **b)** ⟨*Stuhl:*⟩ zusammenbrechen; ⟨*Gebäude:*⟩ einstürzen; **c)** ⟨*Verhandlungen:*⟩ scheitern; ⟨*Unternehmen:*⟩ zusammenbrechen; **d)** *(fold down)* ⟨*Regenschirm, Fahrrad, Tisch:*⟩ sich zusammenklappen lassen. **collapsible** [kə'læpsɪbl] *adj.* Klapp-⟨*stuhl, -tisch, -fahrrad*⟩

collar ['kɒlə(r)] **1.** *n.* **a)** Kragen, *der;* **b)** *(for dog)* [Hunde]halsband, *das.* **2.** *v. t.* schnappen *(ugs.).* **collar-bone** *n.* Schlüsselbein, *das*
colleague ['kɒli:g] *n.* Kollege, *der/*Kollegin, *die*
collect [kə'lekt] **1.** *v. i.* sich versammeln; ⟨*Staub, Müll usw.:*⟩ sich ansammeln. **2.** *v. t.* sammeln; aufsammeln ⟨*Müll, leere Flaschen usw.*⟩; *(coll.: fetch)* abholen ⟨*Menschen, Dinge*⟩; ~ **one's wits/thoughts** seine Gedanken sammeln. **collected** *adj.* **a)** *(gathered)* gesammelt; **b)** *(calm)* gesammelt; gelassen. **collection** [kə'lekʃn] *n.* **a)** *(collecting)* Sammeln, *das;* *(coll.: of goods, persons)* Abholen, *das;* **b)** *(amount of money collected)* Sammlung, *die;* *(in church)* Kollekte, *die;* **c)** *(from post-box)* Leerung, *die;* **d)** *(of stamps etc.)* Sammlung, *die.* **collective** [kə'lektɪv] *adj.* kollektiv *nicht präd.* **collective 'bargaining** *n.* Tarifverhandlungen *Pl.*
collector [kə'lektə(r)] *n.* **a)** *(of stamps etc.)* Sammler, *der/*Sammlerin, *die;* **b)** *(of taxes)* Einnehmer, *der/*Einnehmerin, *die.* **collector's item, collector's piece** *ns.* Sammlerstück, *das*
college ['kɒlɪdʒ] *n.* **a)** *(esp. Brit. Univ.)* College, *das;* **b)** *(place of further education)* Fach[hoch]schule, *die;* go to ~ *(esp. Amer.)* studieren
collide [kə'laɪd] *v. i.* zusammenstoßen (**with** mit)
collie ['kɒlɪ] *n.* Collie, *der*
colliery ['kɒljərɪ] *n.* Kohlengrube, *die*
collision [kə'lɪʒn] *n.* Zusammenstoß, *der;* **on a** ~ **course** *(lit. or fig.)* auf Kollisionskurs
colloquial [kə'ləʊkwɪəl] *adj.* umgangssprachlich
collusion [kə'lu:ʒn] *n.* geheime Absprache
Cologne [kə'ləʊn] **1.** *pr. n.* Köln *(das).* **2.** *attrib. adj.* Kölner
cologne *see* eau-de-Cologne
Colombia [kə'lɒmbɪə] *pr. n.* Kolumbien *(das)*
colon ['kəʊlən] *n.* Doppelpunkt, *der*
colonel ['kɜ:nl] *n.* Oberst, *der*
colonial [kə'ləʊnɪəl] *adj.* Kolonial-; kolonial
colonize ['kɒlənaɪz] *v. t.* kolonisieren
colony ['kɒlənɪ] *n.* Kolonie, *die*
color etc. *(Amer.) see* colour etc.
colossal [kə'lɒsl] *adj.* ungeheuer; gewaltig ⟨*Bauwerk*⟩
colour ['kʌlə(r)] *(Brit.)* **1.** *n.* Farbe,

die; **what** ~ **is it?** welche Farbe hat es?; **change** ~: die Farbe ändern; **he is off** ~: ihm ist nicht gut. **2.** *v. t.* **a)** *(give* ~ *to)* Farbe geben (+ *Dat.*); **b)** *(paint)* malen; **c)** *(stain, dye)* färben. **3.** *v. i.* ~ [up] erröten. **'colour-blind** *adj.* farbenblind **coloured** ['kʌləd] *(Brit.)* **1.** *adj.* **a)** farbig; **b)** *(of non-white descent)* farbig; ~ **people** Farbige *Pl.* **2.** *n.* Farbige, *der/ die* **'colour film** *n.* Farbfilm, *der* **colourful** ['kʌləfl] *adj. (Brit.)* bunt; anschaulich ⟨*Sprache, Stil, Bericht*⟩ **'colouring** *n. (Brit.)* **a)** *(colours)* Farben *Pl.;* **b)** ~ [matter] *(in food etc.)* Farbstoff, *der* **'colourless** *adj. (Brit.)* farblos **colour:** ~ **photograph** *n.* Farbaufnahme, *die;* ~ **scheme** *n.* Farb[en]zusammenstellung, *die;* ~ **supplement** *n.* Farbbeilage, *die;* ~ **television** *n.* Farbfernsehen, *das; (set)* Farbfernsehgerät, *das;* ~ **transparency** *n.* Farbdia, *das* **colt** [kəʊlt] *n.* [Hengst]fohlen, *das* **column** ['kɒləm] *n.* **a)** Säule, *die;* **b)** *(of page)* Spalte, *die; sports* ~: Sportteil, *der.* **columnist** ['kɒləmɪst] *n.* Kolumnist, *der/* Kolumnistin, *die* **coma** ['kəʊmə] *n.* Koma, *das;* **in a** ~: im Koma **comb** [kəʊm] **1.** *n.* Kamm, *der.* **2.** *v. t.* **a)** kämmen; ~ **sb.'s/one's hair** jmdm./ sich die Haare kämmen; **b)** *(search)* durchkämmen **combat** ['kɒmbæt] **1.** *n.* Kampf, *der.* **2.** *v. t.* bekämpfen. **combatant** ['kɒmbətənt] *n.* Kombattant, *der* **combination** [kɒmbɪ'neɪʃn] *n.* Kombination, *die.* **combi'nation lock** *n.* Kombinationsschloß, *das* **combine 1.** [kəm'baɪn] *v. t.* zusammenfügen *(into* zu); verbinden ⟨*Substanzen*⟩. **2.** *v. i. (join together)* ⟨*Stoffe:*⟩ sich verbinden. **3.** ['kɒmbaɪn] *n.* ~ [harvester] Mähdrescher, *der* **combustion** [kəm'bʌstʃn] *n.* Verbrennung, *die* **come** [kʌm] *v. i.,* **came** [keɪm], **come** [kʌm] kommen; ~ **here!** komm [mal] her!; [I'm] **coming!** [ich] komme schon!; **the train came into the station** der Zug fuhr in den Bahnhof ein; **Christmas is coming** bald ist Weihnachten; **the handle has** ~ **loose** der Griff ist lose; **nothing came of it** es ist nichts daraus geworden. **come a'bout** *v. i.* passieren. **come across**

1. [-'-'-] *v. i. (be understood)* verstanden werden. **2.** ['---] *v. t.* begegnen (+ *Dat.*). **come a'long** *v. i. (coll.)* **a)** *(hurry up)* ~ **along!** komm/kommt!; **b)** *(make progress)* ~ **along nicely** gute Fortschritte machen; **c)** *(to place)* mitkommen *(with* mit). **come 'back** *v. i.* zurückkommen. **come by 1.** ['--] *v. t. (obtain)* bekommen. **2.** [-'-] *v. i.* vorbeikommen. **come 'down** *v. i.* **a)** *(fall)* ⟨*Schnee, Regen, Preis:*⟩ fallen; **b)** *(~ lower)* herunterkommen; **c)** *(land)* [not]landen; *(crash)* abstürzen. **come 'in** *v. i. (enter)* hereinkommen; ~ **in!** herein! **'come into** *v. i.* **a)** *(enter)* hereinkommen in (+ *Akk.*); **b)** *(inherit)* erben. **come off 1.** [-'-] *v. i.* **a)** ⟨*Griff, Knopf:*⟩ abgehen; *(be removable)* sich abnehmen lassen; **b)** *(succeed)* ⟨*Pläne, Versuche:*⟩ Erfolg haben; **c)** *(take place)* stattfinden. **2.** ['--] *v. t.* ~ **off a horse/bike** vom Pferd/Fahrrad fallen; ~ **'off it!** *(coll.)* nun mach mal halblang! *(ugs.).* **come on 1.** [-'-] *v. i.* **a)** *(continue coming, follow)* kommen; ~ **on!** komm, komm/kommt, kommt!; *(encouraging)* na, komm; **b)** *(make progress)* ~ **on very well** gute Fortschritte machen. **2.** ['--] *v. t.* **see** ~ **upon. come 'out** *v. i.* **a)** herauskommen; **b)** *(fig.)* ⟨*Sonne, Wahrheit, Buch:*⟩ herauskommen; ~ **out with** herausrücken mit *(ugs.).* **come 'over 1.** *v. i.* herüberkommen. **2.** *v. t. (coll.)* kommen über (+ *Akk.*). **come 'round** *v. i.* **a)** *(visit)* vorbeischauen; **b)** *(recover)* wieder zu sich kommen. **come 'through 1.** *v. i.* durchkommen. **2.** *v. t. (survive)* überleben. **come to 1.** ['--] *v. t. (amount to)* ⟨*Rechnung, Kosten:*⟩ sich belaufen auf (+ *Akk.*). **2.** [-'-] *v. i.* wieder zu sich kommen. **'come under** *v. t.* **a)** *(be classed as or among)* kommen unter (+ *Akk.*); **b)** *(be subject to)* kommen unter (+ *Akk.*). **come 'up** *v. i.* **a)** *(~ higher)* hochkommen; **b)** ~ **up to sb.** *(approach for talk)* auf jmdn. zukommen; **c)** *(present itself)* sich ergeben; **d)** ~ **up to** *(reach)* reichen bis an (+ *Akk.*); entsprechen (+ *Dat.*) ⟨*Erwartungen*⟩; **e)** ~ **up against sth.** *(fig.)* auf etw. *(Akk.)* stoßen; **f)** ~ **up with** vorbringen ⟨*Vorschlag*⟩; wissen ⟨*Lösung, Antwort*⟩. **'come upon** *v. t. (meet by chance)* begegnen (+ *Dat.*) **'come-back** *n. (to profession etc.)* Comeback, *das* **comedian** [kə'miːdɪən] *n.* Komiker, *der.* **comedienne** [kəmi:dı'en] *n.* Komikerin, *die*

'**come-down** *n.* Abstieg, *der*

comedy ['kɒmıdı] **a)** *n.* Lustspiel, *das;* Komödie, *die;* **b)** *(humour)* Witz, *der;* Witzigkeit, *die*

comet ['kɒmıt] *n.* Komet, *der*

comeuppance [kʌm'ʌpəns] *n.* get one's ~: die Quittung kriegen *(fig.)*

comfort ['kʌmfət] **1.** *n.* **a)** *(consolation)* Trost, *der;* **b)** *(physical wellbeing)* Behaglichkeit, *die;* **c)** *in pl.* Komfort, *der.* **2:** *v.t.* trösten. **comfortable** ['kʌmfətəbl] *adj.* **a)** bequem ⟨*Bett, Schuhe*⟩; komfortabel ⟨*Haus, Zimmer*⟩; **a** ~ victory ein leichter Sieg; **b)** *(at ease)* **be/feel** ~: sich wohl fühlen. **comfortably** ['kʌmfətəblı] *adv.* bequem; leicht ⟨*gewinnen*⟩

'**comfort station** *n.* *(Amer.)* öffentliche Toilette

comfy ['kʌmfı] *adj. (coll.)* bequem; gemütlich ⟨*Haus, Zimmer*⟩

comic ['kɒmık] **1.** *adj.* komisch. **2.** *n.* **a)** *(comedian)* Komiker, *der*/Komikerin, *die;* **b)** *(periodical)* Comic-Heft, *das.* **comical** ['kɒmıkl] *adj.* komisch

coming ['kʌmıŋ] **1.** *adj.* in the ~ week kommende Woche. **2.** *n.* ~s and goings das Kommen und Gehen

comma ['kɒmə] *n.* Komma, *das*

command [kə'mɑːnd] **1.** *v.t.* **a)** *(order)* befehlen (sb. jmdm.); **b)** *(be in ~ of)* befehligen ⟨*Schiff, Armee*⟩; **c)** verfügen über (+ *Akk.*) ⟨*Gelder, Wortschatz*⟩. **2.** *n.* **a)** Kommando, *das; (in writing)* Befehl, *der;* **have/take** ~ of das Kommando über (+ *Akk.*) ... haben/übernehmen; **b)** *(mastery, possession)* Beherrschung, *die*

commandeer [kɒmən'dıə(r)] *v.t.* requirieren

com'mander *n.* Führer, *der*

com'manding *adj.* **a)** gebieterisch ⟨*Erscheinung, Stimme*⟩; imposant ⟨*Gestalt*⟩; **b)** beherrschend ⟨*Ausblick, Lage*⟩. **commanding 'officer** *n.* Befehlshaber, *der*/Befehlshaberin, *die*

com'mandment *n.* Gebot, *das*

commemorate [kə'meməreıt] *v.t.* gedenken (+ *Gen.*). **commemoration** [kəmemə'reıʃn] *n.* Gedenken, *das;* **in** ~ **of** zum Gedenken an (+ *Akk.*)

commence [kə'mens] *v.t. & i.* beginnen. **com'mencement** *n.* Beginn, *der*

commend [kə'mend] *v.t. (praise)* loben. **commendable** [kə'mendəbl] *adj.* lobenswert; löblich. **commen-**

dation [kɒmen'deıʃn] *n. (praise)* Lob, *das; (official)* Belobigung, *die; (award)* Auszeichnung, *die*

comment ['kɒment] **1.** *n.* Bemerkung, *die* (on über + *Akk.*); *(note)* Anmerkung, *die* (on über + *Akk.*); **no** ~! *(coll.)* kein Kommentar! **2.** *v.i.* ~ on sth. über etw. *(Akk.)* Bemerkungen machen; **he** ~**ed that ...**: er bemerkte, daß ... **commentary** ['kɒməntərı] *n.* **a)** Kommentar, *der* (on zu); **b)** *(Radio, Telev.)* [**live** or **running**] ~: Live-Reportage, *die.* **commentator** ['kɒmənteıtə(r)] *n.* Kommentator, *der*/Kommentatorin, *die; (Sport)* Reporter, *der*/Reporterin, *die*

commerce ['kɒmɜːs] *n.* Handel, *der*

commercial [kə'mɜːʃl] **1.** *adj.* Handels-; kaufmännisch ⟨*Ausbildung*⟩. **2.** *n.* Werbespot, *der.* **commercialism** [kə'mɜːʃəlızm] *n.* Kommerzialismus, *der.* **commercialize** [kə'mɜːʃəlaız] *v.t.* kommerzialisieren

commercial: ~ 'television *n.* Werbefernsehen, *das;* ~ 'vehicle *n.* Nutzfahrzeug, *das*

commiserate [kə'mızəreıt] *v.i.* ~ with sb. jmdm. sein Mitgefühl aussprechen (on zu)

commission [kə'mıʃn] **1.** *n.* **a)** *(official body)* Kommission, *die;* **b)** *(instruction, piece of work)* Auftrag, *der;* **c)** *(Mil.)* Ernennungsurkunde, *die;* **d)** *(pay of agent)* Provision, *die;* **e) in/out of** ~ ⟨*Auto, Maschine*⟩ in/außer Betrieb. **2.** *v.t.* beauftragen ⟨*Künstler*⟩; in Auftrag geben ⟨*Gemälde usw.*⟩

commissionaire [kəmıʃə'neə(r)] *n. (esp. Brit.)* Portier, *der*

commissioner [kə'mıʃənə(r)] *n. (of police)* Präsident, *der*

commit [kə'mıt] *v.t.,* **-tt-:** **a)** begehen ⟨*Verbrechen, Fehler, Ehebruch*⟩; **b)** *(pledge, bind)* ~ oneself/sb. to doing sth. sich/jmdn. verpflichten, etw. zu tun; **c)** *(entrust)* anvertrauen (to *Dat.*); **d)** ~ sb. for trial jmdm. dem Gericht überstellen. **com'mitment** *n.* Verpflichtung (**to** gegenüber). **com'mitted** *adj.* engagiert

committee [kə'mıtı] *n.* Ausschuß, *der*

commodity [kə'mɒdıtı] *n.* **a)** household ~: Haushaltsartikel, *der;* **b)** *(St. Exch.)* [vertretbare] Ware; *(raw material)* Rohstoff, *der*

common ['kɒmən] **1.** *adj.* **a)** *(belonging to all)* gemeinsam; **b)** *(public)* öffentlich; **c)** *(usual)* gewöhnlich; *(frequent)* häufig; allgemein verbreitet

⟨Sitte, Redensart⟩; **d)** *(vulgar)* ordinär.
2. *n.* **a)** *(land)* Gemeindeland, *das;* **b)**
have sth./nothing/a lot in ~ [**with sb.**]
etw./nichts/viel [mit jmdm.] ge-
mein[sam] haben. '**commoner** *n.*
Bürgerliche *der/die*

'**common-law** *adj.* she's his ~ wife sie
lebt mit ihm in eheähnlicher Gemeinschaft

'**commonly** *adv.* im allgemeinen
common: C~ '**Market** *n.* gemeinsa-
mer Markt; ~**place 1.** *n.* Gemeinplatz, *der;* **2.** *adj.* alltäglich

Commons ['kɒmənz] *n. pl.* **the** [House
of] ~: das Unterhaus

common: ~ '**sense** *n.* gesunder
Menschenverstand; ~**wealth** *n.* **the**
[British] **C~wealth** das Common-
wealth

commotion [kə'məʊʃn] *n.* Tumult,
der

communal ['kɒmjʊnl] *adj.* **a)** *(of or for
the community)* gemeindlich; **b)** *(for
common use)* gemeinsam

commune ['kɒmju:n] *n.* Kommune,
die

communicate [kə'mju:nɪkeɪt] **1.** *v. t.*
übertragen ⟨*Krankheit*⟩; übermitteln
⟨*Informationen*⟩; vermitteln ⟨*Gefühle,
Ideen*⟩. **2.** *v. i.* ~ **with sb.** mit jmdm.
kommunizieren. **communication**
[kəmju:nɪ'keɪʃn] *n.* **a)** *(of information)*
Übermittlung, *die;* **b)** *(message)* Mitteilung, *die* (**to an** + *Akk.*). **communi'cation-cord** *n.* Notbremse.
die. **communi'cations satellite** *n.*
Nachrichtensatellit, *der*

communicative [kə'mju:nɪkətɪv] *adj.*
gesprächig

Communion [kə'mju:nɪən] *n.* [Holy] ~
(Protestant Ch.) das [heilige] Abend-
mahl; *(RC Ch.)* die [heilige] Kommunion

communiqué [kə'mju:nɪkeɪ] *n.* Kom-
muniqué, *das*

communism ['kɒmjʊnɪzm] *n.* Kommunismus, *der;* **C~:** der Kommunismus. **Communist, communist**
['kɒmjʊnɪst] **1.** *n.* Kommunist,
der/Kommunistin, *die.* **2.** *adj.* kommunistisch

community [kə'mju:nɪtɪ] *n.* **a)** *(organized body)* Gemeinwesen, *das;* **the
Jewish** ~: die jüdische Gemeinde; **b)**
no pl. (public) Öffentlichkeit, *die.*
com'munity centre *n.* Gemeinde-
zentrum, *das*

commute [kə'mju:t] **1.** *v. t.* umwandeln ⟨*Strafe*⟩ (**to in** + *Akk.*). **2.** *v. i.*

pendeln. **com'muter** *n.* Pendler,
der/Pendlerin, *die*

¹**compact** [kəm'pækt] *adj.* kompakt
²**compact** ['kɒmpækt] *n.* Puderdose
[mit Puder(stein)]

compact 'disc *n.* Compact Disc, *die*
companion [kəm'pænjən] *n.* Begleiter, *der*/Begleiterin, *die.* **com'pan-
ionship** *n.* Gesellschaft, *die*

company ['kʌmpənɪ] *n.* **a)** *(persons as-
sembled, companioning)* Gesell-
schaft, *die;* **expect** ~: Besuch *od.* Gäste erwarten; **keep sb.** ~: jmdm. Gesellschaft leisten; **b)** *(firm)* Gesellschaft, *die;* ~ **car** Firmenwagen, *der;*
c) *(of actors)* Truppe, *die;* Ensemble,
das; **d)** *(Mil.)* Kompanie, *die*

comparable ['kɒmpərəbl] *adj.* ver-
gleichbar (**to, with mit**)

comparative [kəm'pærətɪv] **1.** *adj.* **a)**
(relative) relativ; **in** ~ **comfort** relativ
komfortabel; **b)** *(Ling.)* komparativ
(fachspr.); **a** ~ **adjective/adverb** ein
Adjektiv/Adverb im Komparativ. **2.**
n. (Ling.) Komparativ, *der.* **com-
'paratively** *adv.* verhältnismäßig

compare [kəm'peə(r)] **1.** *v. t.* vergleichen (**to, with mit**); ~**d with** *or* **to sb./
sth.** verglichen mit *od.* im Vergleich
zu jmdm./etw. **2.** *v. i.* sich vergleichen
lassen. **comparison** [kəm'pærɪsn] *n.*
Vergleich, *der;* **in** *or* **by** ~ [**with sb./
sth.**] im Vergleich [zu jmdm./etw.]

compartment [kəm'pɑ:tmənt] *n. (in
drawer, desk, etc.)* Fach, *das; (of railway carriage)* Abteil, *das*

compass ['kʌmpəs] *n.* **a)** *in pl.* [**a pair
of**] ~**es** ein Zirkel; **b)** *(for navigating)*
Kompaß, *der*

compassion [kəm'pæʃn] *n.* Mitgefühl, *das* (**for mit**). **compassionate**
[kəm'pæʃənət] *adj.* mitfühlend; **on** ~
grounds aus persönlichen Gründen;
(for family reasons) aus familiären
Gründen

compatible [kəm'pætɪbl] *adj.* vereinbar; zueinander passend ⟨*Personen*⟩;
(Computing) kompatibel

compel [kəm'pel] *v. t.,* **-ll-** zwingen
compendium [kəm'pendɪəm] *n.*
Kompendium, *das*

compensate ['kɒmpenseɪt] **1.** *v. i.* ~
for sth. etw. ersetzen. **2.** *v. t.* ~ **sb. for
sth.** jmdn. für etw. entschädigen.
compensation [kɒmpen'seɪʃn] *n.*
Ersatz, *der; (for damages, injuries,
etc.)* Schaden[s]ersatz, *der*

compère ['kɒmpeə(r)] *n. (Brit.)* Con-
férencier, *der*

compete [kəm'pi:t] *v. i.* konkurrieren (**for** um); *(Sport)* kämpfen

competence ['kɒmpɪtəns] *n.* Fähigkeiten *Pl.*

competent ['kɒmpɪtənt] *adj.* fähig; **not ~ to do sth.** nicht kompetent, etw. zu tun. '**competently** *adv.* kompetent

competition [kɒmpɪ'tɪʃn] *n.* **a)** *(contest)* Wettbewerb, *der;* (*in magazine etc.*) Preisausschreiben, *das;* **b)** *(those competing)* Konkurrenz, *die*

competitive [kəm'petɪtɪv] *adj.* wettbewerbsfähig ⟨*Preis, Unternehmen*⟩; ~ **sports** Leistungssport, *der*

competitor [kəm'petɪtə(r)] *n.* Konkurrent, *der*/Konkurrentin, *die;* (*in contest, race*) Teilnehmer, *der*/-nehmerin, *die*

compile [kəm'paɪl] *v. t.* zusammenstellen

complacency [kəm'pleɪsənsɪ] *n.* Selbstzufriedenheit, *die*

complacent [kəm'pleɪsənt] *adj.* selbstzufrieden

complain [kəm'pleɪn] *v. i.* sich beklagen (**about, at** über + *Akk.*); ~ **of sth.** über etw. *(Akk.)* klagen. **complaint** [kəm'pleɪnt] *n.* **a)** Beschwerde, *die;* **b)** *(ailment)* Leiden, *das*

complement 1. ['kɒmplɪmənt] *n.* **a)** *(what completes)* Vervollständigung, *die;* **b)** *(full number)* **a |full| ~:** die volle Zahl; *(of people)* die volle Stärke. **2.** ['kɒmplɪment] *v. t.* ergänzen. **complementary** [kɒmplɪ'mentərɪ] *adj.* **a)** *(completing)* ergänzend; **b)** *(completing each other)* einander ergänzend

complete [kəm'pli:t] **1.** *adj.* **a)** vollständig; *(in number)* vollzählig; **b)** *(finished)* fertig; **c)** *(absolute)* völlig ⟨*Idiot*⟩; absolut ⟨*Katastrophe*⟩; total, *(ugs.)* blutig ⟨*Anfänger*⟩. **2.** *v. t.* **a)** *(finish)* beenden; fertigstellen ⟨*Gebäude, Arbeit*⟩; **b)** ausfüllen ⟨*Formular*⟩. **com'pletely** *adv.* völlig; absolut ⟨*erfolgreich*⟩. **completion** [kəm'pli:ʃn] *n.* Beendigung, *die;* *(of building, work)* Fertigstellung, *die*

complex ['kɒmpleks] **1.** *adj.* kompliziert. **2.** *n.* Komplex, *der*

complexion [kəm'plekʃn] *n.* Gesichtsfarbe, *die;* *(fig.)* Gesicht, *das*

complexity [kəm'pleksɪtɪ] *n.* Kompliziertheit, *die*

complicate ['kɒmplɪkeɪt] *v. t.* komplizieren. '**complicated** *adj.* kompliziert. **complication** [kɒmplɪ'keɪʃn] *n.* Komplikation, *die*

complicity [kəm'plɪsɪtɪ] *n.* Mittäterschaft, *die* (**in** bei)

compliment 1. ['kɒmplɪmənt] *n.* Kompliment, *das; in pl. (formal greetings)* Grüße *Pl.;* **pay sb. a ~:** jmdn. ein Kompliment machen. **2.** ['kɒmplɪment] *v. t.* ~ **sb. on sth.** jmdm. Komplimente wegen etw. machen. **complimentary** [kɒmplɪ'mentərɪ] *adj.* **a)** schmeichelhaft; **b)** *(free)* Frei-

comply [kəm'plaɪ] *v. i.* ~ **with sth.** sich nach etw. richten; **he refused to ~:** er wollte sich nicht danach richten

component [kəm'pəʊnənt] **1.** *n.* Bestandteil, *der.* **2.** *adj.* **a ~ part** ein Bestandteil

compose [kəm'pəʊz] *v. t.* **a)** bilden; **be ~d of** sich zusammensetzen aus; **b)** verfassen ⟨*Rede, Gedicht*⟩; abfassen ⟨*Brief*⟩; **c)** *(Mus.)* komponieren. **com'poser** *n.* Komponist, *der*/Komponistin, *die.* **composition** [kɒmpə'zɪʃn] *n.* **a)** *(constitution)* (*of soil etc.*) Zusammensetzung, *die;* *(of picture)* Aufbau, *der;* **b)** *(essay)* Aufsatz, *der;* *(Mus.)* Komposition, *die*

compost ['kɒmpɒst] *n.* Kompost, *der.* '**compost heap** *n.* Komposthaufen, *der*

composure [kəm'pəʊʒə(r)] *n.* Gleichmut, *der*

¹**compound 1.** ['kɒmpaʊnd] *adj.* **a)** zusammengesetzt; **b)** *(Med.)* ~ **fracture** komplizierter Bruch. **2.** ['kɒmpaʊnd] *n.* **a)** *(mixture)* Mischung, *die;* **b)** *(Ling.)* Kompositum, *das;* **c)** *(Chem.)* Verbindung, *die.* **3.** ['kəm'paʊnd] *v. t.* verschlimmern ⟨*Schwierigkeiten, Verletzung usw.*⟩

²**compound** ['kɒmpaʊnd] *n.* umzäuntes Gelände

compound 'interest *n.* Zinseszinsen *Pl.*

comprehend [kɒmprɪ'hend] *v. t.* verstehen. **comprehensible** [kɒmprɪ'hensɪbl] *adj.* verständlich. **comprehension** [kɒmprɪ'henʃn] *n.* Verständnis, *das*

comprehensive [kɒmprɪ'hensɪv] **1.** *adj.* **a)** umfassend; **b)** ~ **school** Gesamtschule, *die;* **c)** *(insurance)* Vollkasko-. **2.** *n.* Gesamtschule, *die*

compress 1. [kəm'pres] *v. t.* **a)** *(squeeze)* zusammenpressen (**into** zu); **b)** komprimieren ⟨*Luft, Gas, Bericht*⟩. **2.** ['kɒmpres] *n.* Kompresse, *die.* **compression** [kəm'preʃn] *n.* Kompression, *die.* **compressor** [kəm'presə(r)] *n.* Kompressor, *der*

comprise [kəm'praız] *v. t. (include)* umfassen; *(consist of)* bestehen aus
compromise ['kɒmprəmaız] **1.** *n.* Kompromiß, *der.* **2.** *v. i.* Kompromisse/einen Kompromiß schließen. **3.** *v. t.* kompromittieren
compulsion [kəm'pʌlʃn] *n.* Zwang, *der;* **be under no ~ to do sth.** keineswegs etw. tun müssen. **compulsive** [kəm'pʌlsıv] *adj.* **a)** zwanghaft; **he is a ~ gambler** er ist dem Spiel verfallen; **b) this book is ~ reading** von diesem Buch kann man sich nicht losreißen. **compulsory** [kəm'pʌlsərı] *adj.* obligatorisch
compunction [kəm'pʌŋkʃn] *n.* Schuldgefühle
computer [kəm'pju:tə(r)] *n.* Computer, *der*
computer: ~-aided, ~-assisted *adjs.* computergestützt; **~ program** *n.* Programm, *das;* **~ programmer** *n.* Programmierer, *der*/Programmiererin, *die;* **~ programming** *n.* Programmieren, *das;* **~ terminal** *n.* Terminal, *das*
computing [kəm'pju:tıŋ] *n.* EDV, *die;* elektronische Datenverarbeitung
comrade ['kɒmreıd, 'kɒmrıd] *n.* Kamerad, *der*/Kameradin, *die.* '**comradeship** *n.* Kameradschaft, *die*
con [kɒn] *(coll.)* **1.** *n.* Schwindel, *der.* **2.** *v. t.,* **-nn-** reinlegen *(ugs.);* **~ sb. into sth.** jmdm. etw. aufschwatzen *(ugs.)*
concave ['kɒnkeıv] *adj.* konkav
conceal [kən'si:l] *v. t.* verbergen **(from** vor + *Dat.).* **con'cealment** *n.* Verbergen, *das*
concede [kən'si:d] *v. t.* zugeben
conceit [kən'si:t] *n.* Einbildung, *die.* **con'ceited** *adj.* eingebildet
conceivable [kən'si:vəbl] *adj.* vorstellbar; **it is scarcely ~ that ...:** man kann sich *(Dat.)* kaum vorstellen, daß ... **conceivably** [kən'sı:veblı] *adj.* möglicherweise; **he cannot ~ have done it** er kann es unmöglich getan haben
conceive [kən'si:v] **1.** *v. t.* **a)** empfangen ⟨*Kind*⟩; **b)** *(form in mind)* sich *(Dat.)* vorstellen; haben ⟨*Idee, Plan*⟩. **2.** *v. i.* **a)** *(become pregnant)* empfangen; **b) ~ of sth.** sich *(Dat.)* etw. vorstellen
concentrate ['kɒnsəntreıt] **1.** *v. t.* konzentrieren. **2.** *v. i.* sich konzentrieren **(on** auf + *Akk.).* '**concentrated** *adj.* konzentriert. **concentration** [kɒnsən'treıʃn] *n.* Konzentration, *die*

concentric [kən'sentrık] *adj.* konzentrisch
concept ['kɒnsept] *n.* Begriff, *der; (idea)* Vorstellung, *die.* **conception** [kən'sepʃn] **a)** Vorstellung, *die* **(of** von); **b)** *(of child)* Empfängnis, *die*
concern [kən'sɜ:n] **1.** *v. t.* **a)** *(affect)* betreffen; **so far as ... is ~ed** was ... betrifft; **'to whom it may ~'** ≈ „Bestätigung"; *(on certificate, testimonial)* ≈ „Zeugnis"; **b)** *(interest)* **~ oneself with** *or* **about sth.** sich mit etw. befassen; **c)** *(trouble)* beunruhigen. **2.** *n.* **a)** *(anxiety)* Besorgnis, *die; (interest)* Interesse, *das;* **b)** *(matter)* Angelegenheit, *die;* **d)** *(firm)* Unternehmen, *das.* **con'cerned** [kən'sɜ:nd] *adj.* **a)** *(involved)* betroffen; *(interested)* interessiert; **as** *or* **so far as I'm ~:** was mich betrifft; **b)** *(troubled)* besorgt. **con'cerning** *prep.* bezüglich
concert ['kɒnsət] *n.* Konzert, *das*
concerted [kən'sɜ:tıd] *adj.* vereint
concert: ~-goer *n.* Konzertbesucher, *der*/-besucherin, *die;* **~-hall** *n.* Konzertsaal, *der*
concertina [kɒnsə'ti:nə] *n.* Konzertina, *die*
concerto [kən'tʃeətəʊ] *n.* Konzert, *das*
concession [kən'seʃn] *n.* Konzession, *die.* **concessionary** [kən'seʃənərı] *adj.* Konzessions-; **~ rate/fare** ermäßigter Tarif
conciliatory [kən'sıljətərı] *adj.* versöhnlich
concise [kən'saıs] *adj.* kurz und prägnant; knapp, konzis ⟨*Stil*⟩
conclude [kən'klu:d] **1.** *v. t.* **a)** *(end)* beschließen; **b)** *(infer)* schließen **(from** aus); **c)** *(reach decision)* beschließen. **2.** *v. i. (end)* schließen. **concluding** [kən'klu:dıŋ] *adj.* abschließend. **conclusion** [kən'klu:ʒn] *n.* **a)** *(end)* Abschluß, *der;* **in ~:** zum Abschluß; **b)** *(result)* Ausgang, *der;* **c)** *(inference)* Schluß, *der;* **draw** *or* **reach a ~:** zu einem Schluß kommen. **conclusive** [kən'klu:sıv] *adj.,* **con'clusively** *adv.* schlüssig
concoct [kən'kɒkt] *v. t.* zubereiten; zusammenbrauen ⟨*Trank*⟩. **concoction** [kən'kɒkʃn] *n.* Gebräu, *das*
concourse ['kɒnkɔ:s] *n.* Halle, *die;* **station ~:** Bahnhofshalle, *die*
concrete ['kɒnkri:t] **1.** *adj.* konkret. **2.** *n.* Beton, *der.* Beton-; aus Beton *präd.* '**concrete-mixer** *n.* Betonmischer, *der*

concur [kən'kɜ:(r)] *v. i.,* -rr-: ~ [with sb.] [in sth.] [jmdm.] [in etw. *(Dat.)*] zustimmen. **concurrent** [kən'kʌrənt] *adj.,* **con'currently** *adv.* gleichzeitig
concussion [kən'kʌʃn] *n.* Gehirnerschütterung, *die*
condemn [kən'dem] *v. t.* **a)** *(censure)* verdammen; **b)** *(Law: sentence)* verurteilen (**to** zu); **c)** für unbewohnbar erklären ⟨*Gebäude*⟩. **condemnation** [kɒndem'neɪʃn] *n.* Verdammung, *die*
condensation [kɒnden'seɪʃn] *n.* **a)** *(condensing)* Kondensation, *die;* **b)** *(water)* Kondenswasser, *das*
condense [kən'dens] **1.** *v. t.* **a)** komprimieren; ~**d milk** Kondensmilch, *die;* **b)** *(Phys., Chem.)* kondensieren. **2.** *v. i.* kondensieren
condescend [kɒndɪ'send] *v. i.* ~ **to do sth.** sich dazu herablassen, etw. zu tun. **conde'scending** *adj.* herablassend
condition [kən'dɪʃn] *n.* **a)** *(stipulation)* [Vor]bedingung, *die;* **on** [**the**] ~ **that ...:** unter der Voraussetzung, daß ...; **b)** *in pl. (circumstances)* Umstände *Pl.;* **weather/living** ~**s** Witterungs-/Wohnverhältnisse; **working** ~**s** Arbeitsbedingungen; **c)** *(of athlete etc.)* Form, *die; (of thing)* Zustand, *der; (of patient)* Verfassung, *die;* **d)** *Med.)* Leiden, *das.* **conditional** [kən'dɪʃənl] *adj.* **a)** bedingt; **be** ~ [**up**]**on sth.** von etw. abhängen; **b)** *(Ling.)* Konditional-
con'ditioner *n.* Frisiermittel, *das*
condolence [kən'dəʊləns] *n.* Anteilnahme, die; **letter of** ~: Beileidsbrief, *der*
condom ['kɒndɒm] *n.* Kondom, *das od. der*
condominium ['kɒndə'mɪnɪəm] *n. (Amer.)* Appartementhaus [mit Eigentumswohnungen]
condone [kən'dəʊn] *v. t.* hinwegsehen über (+ *Akk.*); *(approve)* billigen
conducive [kən'dju:sɪv] *adj.* **be** ~ **to sth.** einer Sache *(Dat.)* förderlich sein
conduct 1. ['kɒndʌkt] *n.* **a)** *(behaviour)* Verhalten, *das;* **b)** *(way of ~ing)* Führung, *die.* **2.** [kən'dʌkt] *v. t.* **a)** führen; **b)** *(Mus.)* dirigieren; **c)** *(Phys.)* leiten; **d)** ~**ed tour** Führung, *die.* **conduction** [kən'dʌkʃn] *n. (Phys.)* Leitung, *die.* **conductor** [kən'dʌktə(r)] *n.* **a)** *(Mus.)* Dirigent, *der*/Dirigentin, *die;* **b)** *(of bus, tram)* Schaffner, *der.*
conductress [kən'dʌktrɪs] *n.* Schaffnerin, *die*

cone [kəʊn] *n.* **a)** Kegel, *der; (traffic* ~*)* Leitkegel, *der;* **b)** *(Bot.)* Zapfen, *der;* **c) ice-cream** ~: Eistüte, *die*
confectioner [kən'fekʃənə(r)] *n.* ~**'s** [**shop**] Süßwarengeschäft, *das.* **con'fectionery** *n.* Süßwaren *Pl.*
confederation [kən'fedə'reɪʃn] *n.* [Staaten]bund, *der*
confer [kən'fɜ:(r)] **1.** *v. t.,* -rr-: ~ **sth.** [**up**]**on sb.** jmdm. etw. verleihen. **2.** *v. i.,* -rr-: ~ **with sb.** sich mit jmdm. beraten
conference ['kɒnfərəns] *n.* **a)** Konferenz, *die;* **b) be in** ~: in einer Besprechung sein. **'conference-room** *n.* Konferenzraum, *der*
confess [kən'fes] **1.** *v. t.* **a)** gestehen; **b)** *(Eccl.)* beichten. **2.** *v. i.* **a)** ~ **to sth.** etw. gestehen; **b)** *(Eccl.)* beichten (**to sb.** jmdm.). **confession** [kən'feʃn] *n.* **a)** Geständnis, *das;* **b)** *(Eccl.: of sins etc.)* Beichte, *die*
confetti [kən'fetɪ] *n.* Konfetti, *das*
confide [kən'faɪd] **1.** *v. i.* ~ **in sb.** sich jmdm. anvertrauen. **2.** *v. t.* ~ **sth. to sb.** jmdm. etw. anvertrauen
confidence ['kɒnfɪdəns] *n.* **a)** *(firm trust)* Vertrauen, *das;* **have** ~ **in sb.**/**sth.** Vertrauen zu jmdm./etw. haben; **have** [**absolute**] ~ **that ...:** [absolut] sicher sein, daß ...; **b)** *(assured expectation)* Gewißheit, *die;* **c)** *(self-reliance)* Selbstvertrauen, *das;* **d) in** ~: im Vertrauen; **this is in** [**strict**] ~: das ist [streng] vertraulich. **'confidence trick** *n. (Brit.)* Trickbetrug, *der*
confident ['kɒnfɪdənt] *adj.* **a)** zuversichtlich (**about** in bezug auf + *Akk.*); **be** ~ **that ...:** sicher sein, daß ...; **b)** *(self-assured)* selbstbewußt
confidential [kɒnfɪ'denʃl] *adj.* vertraulich. **confidentiality** [kɒnfɪdenʃɪ'ælɪtɪ] *n.* Vertraulichkeit, *die.* **con fi'dentially** *adv.* vertraulich
'confidently *adv.* zuversichtlich
confine [kən'faɪn] *v. t.* **a)** einsperren; **be** ~**d to bed/the house** ans Bett/Haus gefesselt sein; **b)** *(fig.)* ~ **oneself to doing sth.** sich darauf beschränken, etw. zu tun. **con'fined** *adj.* begrenzt. **con'finement** *n. (imprisonment)* Einsperrung, *die.* **confines** ['kɒnfaɪnz] *n. pl.* Grenzen
confirm [kən'fɜ:m] *v. t.* bestätigen. **confirmation** [kɒnfə'meɪʃn] *n.* **a)** Bestätigung, *die;* **b)** *(Protestant Ch.)* Konfirmation, *die; (RC Ch.)* Firmung, *die.* **con'firmed** *adj.* eingefleischt ⟨*Junggeselle*⟩; überzeugt ⟨*Vegetarier*⟩
confiscate ['kɒnfɪskeɪt] *v. t.* beschlag-

nahmen. **confiscation** [kɒnfɪs'keɪʃn] *n.* Beschlagnahme, *die*

conflict 1. ['kɒnflɪkt] *n.* **a)** *(fight)* Kampf, *der;* **b)** *(clashing)* Konflikt, *der.* **2.** [kən'flɪkt] *v. i.* *(be incompatible)* sich *(Dat.)* widersprechen; **~ with sth.** einer Sache *(Dat.)* widersprechen. **con'flicting** *adj.* widersprüchlich

conform [kən'fɔːm] *v. i.* **a)** entsprechen (**to** *Dat.*); **b)** *(comply)* sich einfügen; **~ to** or **with sb./with sb.** sich nach etw./jmdm. richten. **conformist** [kən'fɔːmɪst] *n.* Konformist, *der*/Konformistin, *die.* **conformity** [kən'fɔːmɪtɪ] *n.* Übereinstimmung, *die* (**with, to** mit)

confound [kən'faʊnd] *v. t.* **a)** *(defeat)* vereiteln; **b)** *(confuse)* verwirren. **con'founded** *adj. (coll. derog.)* verdammt

confront [kən'frʌnt] *v. t.* **a)** gegenüberstellen; **~ sb. with sth./sb.** jmdn. mit etw./[mit] jmdm. konfrontieren; **b)** *(stand facing)* gegenüberstehen (+ *Dat.*). **confrontation** [kɒnfrən'teɪʃn] *n.* Konfrontation, *die*

confuse [kən'fjuːz] *v. t.* **a)** *(disorder)* durcheinanderbringen; **b)** *(mix up mentally)* verwechseln; **c)** *(perplex)* verwirren. **con'fused** *adj.* konfus; wirr ⟨*Gedanken, Gerüchte*⟩; verworren ⟨*Lage, Situation*⟩. **confusing** [kən'fjuːzɪŋ] *adj.* verwirrend. **confusion** [kən'fjuːʒn] *n.* **a)** Verwirrung, *die;* *(mixing up)* Verwechslung, *die;* **b)** *(embarrassment)* Verlegenheit, *die*

congeal [kən'dʒiːl] *v. i.* gerinnen

conger ['kɒŋɡə(r)] *n.* **~ [eel]** Seeaal, *der*

congested [kən'dʒestɪd] *adj.* verstopft ⟨*Straße, Nase*⟩. **congestion** [kən'dʒestʃn] *n. (of traffic)* Stauung, *die;* **nasal ~:** verstopfte Nase

conglomerate [kən'lɒmərət] *n. (Commerc.)* Großkonzern, *der.* **conglomeration** [kənɡlɒmə'reɪʃn] *n.* Anhäufung, *die*

congratulate [kən'ɡrætjʊleɪt] *v. t.* gratulieren (+ *Dat.*); **~ sb./oneself on sth.** jmdm./sich zu etw. gratulieren. **congratulations** [kənɡrætjʊ'leɪʃnz] **1.** *int.* **~!** herzlichen Glückwunsch! (**on** zu). **2.** *n. pl.* Glückwünsche *Pl.*

congregate ['kɒŋɡrɪɡeɪt] *v. i.* sich versammeln. **congregation** [kɒŋɡrɪ'ɡeɪʃn] *n. (Eccl.)* Gemeinde, *die*

congress ['kɒŋɡres] *n.* Kongreß, *der;* **C~** *(Amer.)* der Kongreß. **congressional** [kən'ɡreʃənl] *adj.* Kongreß-

conical ['kɒnɪkl] *adj.* kegelförmig

conifer ['kɒnɪfə(r)] *n.* Nadelbaum, *der*

conjecture [kən'dʒektʃə(r)] **1.** *n.* Vermutung, *die.* **2.** *v. t.* vermuten. **3.** *v. i.* Vermutungen anstellen

conjugate ['kɒndʒʊɡeɪt] *v. t. (Ling.)* konjugieren. **conjugation** [kɒndʒʊ'ɡeɪʃn] *n. (Ling.)* Konjugation, *die*

conjunction [kən'dʒʌŋkʃn] *n.* **a)** Verbindung, *die;* **in ~ with** in Verbindung mit; **b)** *(Ling.)* Konjunktion, *die*

conjure ['kʌndʒə(r)] *v. i.* zaubern; **conjuring trick** Zaubertrick, *der.* **conjure 'up** *v. t.* heraufbeschwören

conjurer, conjuror ['kʌndʒərə(r)] *n.* Zauberkünstler, *der*/-künstlerin, *die*

connect [kə'nekt] **1.** *v. t.* verbinden (**to, with** mit). **2.** *v. i.* **~ with sth.** mit etw. zusammenhängen. **con'nected** *adj.* zusammenhängend. **connection,** *(Brit.)* **connexion** [kə'nekʃn] *n.* **a)** *(act, state)* Verbindung, *die;* **b)** *(fig.: of ideas)* Zusammenhang, *der;* **in ~ with** im Zusammenhang mit; **c)** *(train, bus, etc.)* Anschluß, *der*

connoisseur [kɒnə'sɜː(r)] *n.* Kenner, *der*

connotation [kɒnə'teɪʃn] *n.* Assoziation, *die*

conquer ['kɒŋkə(r)] *v. t.* besiegen; erobern ⟨*Land*⟩. **conqueror** ['kɒŋkərə(r)] *n. (of a country)* Eroberer, *der*

conquest ['kɒŋkwest] *n.* Eroberung, *die*

conscience ['kɒnʃəns] *n.* Gewissen, *das;* **have a clear/guilty ~:** ein gutes/schlechtes Gewissen haben

conscientious [kɒnʃɪ'enʃəs] *adj.* pflichtbewußt; *(meticulous)* gewissenhaft; **~ objector** Wehrdienstverweigerer [aus Gewissensgründen]. **consci'entiously** *adv.* pflichtbewußt; *(meticulously)* gewissenhaft

conscious ['kɒnʃəs] *adj.* **a) he is not ~ of it** es ist ihm nicht bewußt; **b)** *pred. (awake)* bei Bewußtsein *präd.;* **c)** *(realized by doer)* bewußt ⟨*Versuch, Bemühung*⟩. **consciousness** ['kɒnʃəsnɪs] *n.* Bewußtsein, *das*

conscript 1. [kən'skrɪpt] *v. t.* einberufen. **2.** ['kɒnskrɪpt] *n.* Einberufene, *der/die.* **conscription** [kən'skrɪpʃn] *n.* Wehrpflicht, *die*

consecrate ['kɒnsɪkreɪt] *v. t.* weihen

consecutive [kən'sekjʊtɪv] *adj.* aufeinanderfolgend ⟨*Monate, Jahre*⟩; fortlaufend ⟨*Zahlen*⟩. **con'secutively** *adv.* hintereinander

consensus [kən'sensəs] *n.* Einigkeit, *die*

consent [kən'sent] **1.** *v. i.* zustimmen. **2.** *n. (agreement)* Zustimmung, *die* (to zu); **by common** *or* **general** ~: nach allgemeiner Auffassung

consequence ['kɒnsɪkwəns] *n.* **a)** *(result)* Folge, *die;* **in** ~: folglich; **as a** ~: infolgedessen; **b)** *(importance)* Bedeutung, *die.* **consequent** ['kɒnsɪkwənt] *adj.* daraus folgend. '**consequently** *adv.* infolgedessen

conservation [kɒnsə'veɪʃn] *n.* Erhaltung, *die;* **wildlife** ~: Schutz wildlebender Tierarten. **conservationist** [kɒnsə'veɪʃənɪst] *n.* Naturschützer, *der/*-schützerin, *die*

conservative [kən'sɜ:vətɪv] **1.** *adj.* **a)** konservativ; **b)** vorsichtig ⟨Schätzung⟩; **c)** C~ *(Brit. Polit.)* konservativ; **the C~ Party** die Konservative Partei. **2.** *n.* C~ *(Brit. Polit.)* Konservative, *der/die.* **con'servatively** *adv.* vorsichtig ⟨geschätzt⟩

conservatory [kən'sɜ:vətərɪ] *n.* Wintergarten, *der*

conserve [kən'sɜ:v] *v. t.* erhalten; schonen ⟨Kräfte⟩

consider [kən'sɪdə(r)] *v. t.* **a)** *(think about)* ~ **sth.** an etw. *(Akk.)* denken; **he's** ~**ing emigrating** er denkt daran, auszuwandern; **b)** *(reflect on)* sich *(Dat.)* überlegen; **c)** *(regard as)* halten für; **all things** ~**ed** alles in allem. **considerable** [kən'sɪdərəbl] *adj.,* **con'siderably** *adv.* erheblich. **considerate** [kən'sɪdərət] *adj.* rücksichtsvoll; *(thoughtfully kind)* entgegenkommend. **consideration** [kənsɪdə'reɪʃn] *n.* **a)** Überlegung, *die;* **take sth. into** ~: etw. berücksichtigen; **the matter is under** ~: die Angelegenheit wird geprüft; **b)** *(thoughtfulness)* Rücksichtnahme, *die.* **con'sidering** *prep.* ~ **sth.** wenn man etw. bedenkt; ~ **[that]** ...: wenn man bedenkt, daß ...

consign [kən'saɪn] *v. t.* anvertrauen (**to** *Dat.*). **con'signment** *n. (Commerc.)* Sendung, *die; (large)* Ladung, *die*

consist [kən'sɪst] *v. i.* ~ **of** bestehen aus. **consistency** [kən'sɪstənsɪ] *n.* **a)** *(density)* Konsistenz, *die;* **b)** *(being consistent)* Konsequenz, *die*

consistent [kən'sɪstənt] *adj.* **a)** *(compatible)* [miteinander] vereinbar; **b)** *(uniform)* gleichbleibend ⟨Qualität⟩; **c)** *(unchanging)* konsequent

consolation [kɒnsə'leɪʃn] *n.* Trost, *der.* **conso'lation prize** *n.* Trostpreis, *der*

console [kən'səʊl] *v. t.* trösten

consolidate [kən'sɒlɪdeɪt] *v. t.* festigen

consonant ['kɒnsənənt] *n.* Konsonant, *der*

consort [kən'sɔ:t] *v. i.* verkehren (**with** mit)

consortium [kən'sɔ:tɪəm] *n., pl.* **consortia** [kən'sɔ:tɪə] Konsortium, *das*

conspicuous [kən'spɪkjʊəs] *adj.* **a)** *(visible)* unübersehbar; **b)** *(obvious)* auffallend

conspiracy [kən'spɪrəsɪ] *n. (conspiring)* Verschwörung, *die; (plot)* Komplott, *das*

conspire [kən'spaɪə(r)] *v. i.* sich verschwören

constable ['kʌnstəbl, 'kɒnstəbl] *n. (Brit.)* Polizist, *der/*Polizistin, *die.* **constabulary** [kən'stæbjʊlərɪ] *n.* Polizei, *die*

constant ['kɒnstənt] *adj.* **a)** *(unceasing)* ständig; **b)** *(unchanging)* gleichbleibend. '**constantly** *adv.* **a)** *(unceasingly)* ständig; **b)** *(unchangingly)* konstant

constellation [kɒnstə'leɪʃn] *n.* Sternbild, *das*

consternation [kɒnstə'neɪʃn] *n.* Bestürzung, *die*

constipated ['kɒnstɪpeɪtɪd] *adj.* **be** ~: an Verstopfung leiden. **constipation** [kɒnstɪ'peɪʃn] *n.* Verstopfung, *die*

constituency [kən'stɪtjʊənsɪ] *n.* Wahlkreis, *der*

constituent [kən'stɪtjʊənt] *n.* **a)** *(part)* Bestandteil, *der;* **b)** *(Polit.)* Wähler, *der/*Wählerin, *die*

constitute [kən'stɪtju:t] *v. t.* **a)** *(form, be)* sein; ~ **a threat to** eine Gefahr sein für; **b)** *(make up)* bilden. **constitution** [kɒnstɪ'tju:ʃn] *n.* **a)** *(of person)* Konstitution, *die;* **b)** *(of state)* Verfassung, *die.* **constitutional** [kɒnstɪ'tju:ʃənl] *adj. (of constitution)* der Verfassung *nachgestellt; (in harmony with constitution)* verfassungsmäßig

constrain [kən'streɪn] *v. t.* zwingen. **constraint** [kən'streɪnt] *n. (limitation)* Einschränkung, *die*

constrict [kən'strɪkt] *v. t.* verengen. **constriction** [kən'strɪkʃn] *n.* Verengung, *die*

construct [kən'strʌkt] *v. t.* bauen; *(fig.)* erstellen ⟨Plan⟩. **construction** [kən'strʌkʃn] *n.* **a)** *(constructing)* Bau, *der;* **be under** ~: im Bau sein; **b)** *(thing constructed)* Bauwerk, *das.* **con-**

structive [kən'strʌktɪv] *adj.* konstruktiv

consul ['kɒnsl] *n.* Konsul, *der.* **consulate** ['kɒnsjʊlət] *n.* Konsulat, *das*

consult [kən'sʌlt] *v. t.* konsultieren ⟨*Arzt, Fachmann*⟩; ~ **a** book in einem Buch nachsehen. **consultant** [kən'sʌltənt] *n.* Berater, *der*/Beraterin, *die;* (*Med.*) Chefarzt, *der*/-ärztin, *die.* **consultation** [kɒnsəl'teɪʃn] *n.* Beratung, *die*

consume [kən'sju:m] *v. t.* verbrauchen; (*eat, drink*) konsumieren. **consumer** *n.* Verbraucher, *der*/Verbraucherin, *die.* **con'sumer goods** *n. pl.* Konsumgüter

consumption [kən'sʌmpʃn] *n.* Verbrauch, *der* (of an + *Dat.*); (*eating or drinking*) Verzehr, *der* (of von)

cont. *abbr.* continued Forts.

contact 1. ['kɒntækt] *n.* Berührung, *die;* (*fig.*) Kontakt, *der;* **be in** ~ **with sth.** etw. berühren; **be in** ~ **with sb.** (*fig.*) mit jmdm. Kontakt haben. **2.** ['kɒntækt, kən'tækt] *v. t.* sich in Verbindung setzen mit. **'contact lens** *n.* Kontaktlinse, *die*

contagious [kən'teɪdʒəs] *adj.* ansteckend

contain [kən'teɪn] *v. t.* **a)** (*hold, include*) enthalten; **b)** (*prevent from spreading*) aufhalten. **con'tainer** *n.* Behälter, *der;* (*cargo* ~) Container, *der;* **cardboard/wooden** ~: Pappkarton, *der*/Holzkiste, *die*

contaminate [kən'tæmɪneɪt] *v. t.* verunreinigen; (*with radioactivity*) verseuchen. **contamination** [kəntæmɪ'neɪʃn] *n.* Verunreinigung, *die;* (*with radioactivity*) Verseuchung, *die*

contemplate ['kɒntəmpleɪt] *v. t.* **a)** betrachten; (*mentally*) nachdenken über (+ *Akk.*); **b)** (*expect*) rechnen mit; (*consider*) ~ **sth./doing sth.** an etw. (*Akk.*) denken/daran denken, etw. zu tun. **contemplation** [kɒntəm'pleɪʃn] *n.* Betrachtung, *die;* (*mental*) Nachdenken, *das* (of über + *Akk.*)

contemporary [kən'tempərərɪ] **1.** *adj.* zeitgenössisch. **2.** *n.* Zeitgenosse, *der*/-genossin, *die*

contempt [kən'tempt] *n.* Verachtung, *die* (of, for für). **contemptible** [kən'temptɪbl] *adj.* verachtenswert. **contemptuous** [kən'temptjʊəs] *adj.* verächtlich

contend [kən'tend] *v. i.* **be able/have to** ~ **with** fertigwerden können/müssen mit. **con'tender** *n.* Bewerber, *der*/Bewerberin, *die*

¹content ['kɒntent] *n.* **a)** *in pl.* Inhalt, *der;* |table of| ~s Inhaltsverzeichnis, *das;* **b)** (*amount contained*) Gehalt, *der* (of an + *Dat.*)

²content [kən'tent] **1.** *pred. adj.* zufrieden. **2.** *v. t.* zufriedenstellen; ~ **oneself with sth./sb.** sich mit etw./jmdm. zufriedengeben. **con'tented** *adj.*, **con'tentedly** *adv.* zufrieden

contention [kən'tenʃn] *n.* **a)** Streit, *der;* **b)** (*point asserted*) Behauptung, *die.* **contentious** [kən'tenʃəs] *adj.* strittig ⟨*Punkt, Thema*⟩

con'tentment *n.* Zufriedenheit, *die*

contest 1. ['kɒntest] *n.* Wettbewerb, *der.* **2.** [kən'test] *v. t.* **a)** bestreiten; in Frage stellen ⟨*Behauptung*⟩; **b)** (*Brit.: compete for*) kandidieren für. **contestant** [kən'testənt] *n.* (*competitor*) Teilnehmer, *der*/Teilnehmerin, *die*

context ['kɒntekst] *n.* Kontext, *der;* **in/out of** ~: im/ohne Kontext, *der;* **in this** ~: in diesem Zusammenhang

continent ['kɒntɪnənt] *n.* Kontinent, *der;* **the C**~: das europäische Festland. **continental** [kɒntɪ'nentl] *adj.* **a)** kontinental; **b) C**~ (*mainland European*) kontinental[europäisch]. **continental 'breakfast** *n.* kontinentales Frühstück. **continental 'quilt** *n.* (*Brit.*) [Stepp]federbett, *das*

contingent [kən'tɪndʒənt] *n.* Kontingent, *das*

continual [kən'tɪnjʊəl] *adj.*, **con'tinually** *adv.* (*frequent[ly]*) ständig; (*without stopping*) unaufhörlich

continuation [kəntɪnjʊ'eɪʃn] *n.* Fortsetzung, *die*

continue [kən'tɪnju:] **1.** *v. t.* fortsetzen; '~**d on page 2**' „Fortsetzung auf S. 2"; ~ **doing** *or* **to do sth.** etw. weiter tun; **it** ~**d to rain** es regnete weiter. **2.** *v. i.* (*persist*) ⟨*Wetter, Zustand, Krise usw.*:⟩ andauern; (*persist in doing sth.*) nicht aufhören; ~ **with sth.** mit etw. fortfahren. **continuity** [kɒntɪ'nju:ɪtɪ] *n.* Kontinuität, *die.* **continuous** [kən'tɪnjʊəs] *adj.* **a)** ununterbrochen; anhaltend ⟨*Regen, Sonnenschein*⟩; ständig ⟨*Kritik, Streit*⟩; durchgezogen ⟨*Linie*⟩; **b)** (*Ling.*) ~ |form| Verlaufsform, *die.* **con'tinuously** *adv.* ununterbrochen; ständig ⟨*sich ändern*⟩

contort [kən'tɔ:t] *v. t.* verdrehen. **contortion** [kən'tɔ:ʃn] *n.* Verdrehung, *die*

contour ['kɒntʊə(r)] *n.* Kontur, *die;* ~ **map** Höhenlinienkarte, *die*

contraband ['kɒntrəbænd] *n.* Schmuggelware, *die*

contraception [kɒntrə'sepʃn] *n.* Empfängnisverhütung, *die*. **contraceptive** [kɒntrə'septɪv] **1.** *adj.* empfängnisverhütend. **2.** *n.* Verhütungsmittel, *das*

contract 1. ['kɒntrækt] *n.* Vertrag, *der;* ~ **of employment** Arbeitsvertrag, *der;* **be under** ~ **to do sth.** vertraglich verpflichtet sein, etw. zu tun. **2.** [kən'trækt] *v. t. (Med.)* sich *(Dat.)* zuziehen. **3.** *v. i.* **a)** ~ **to do sth.** sich vertraglich verpflichten, etw. zu tun; **b)** *(become smaller, be drawn together)* sich zusammenziehen. **contraction** [kən'trækʃn] *n.* Kontraktion, *die*. **contractor** [kən'træktə(r)] *n.* Auftragnehmer, *der/-nehmerin, die*

contradict [kɒntrə'dɪkt] *v. t.* widersprechen (+ *Dat.*). **contradiction** [kɒntrə'dɪkʃn] *n.* Widerspruch, *der;* **in** ~ **to sb./sth.** im Widerspruch zu jmdm./etw. **contradictory** [kɒntrə'dɪktərɪ] *adj.* widersprüchlich

contralto [kən'træltəʊ] *n., pl.* ~**s** Alt, *der*

contraption [kən'træpʃn] *n. (coll.)* [komisches] Gerät

contrary ['kɒntrərɪ] **1.** *adj.* **a)** entgegengesetzt; **be** ~ **to sth.** im Gegensatz zu etw. stehen; **b)** [kən'treərɪ] *(coll.: perverse)* widerspenstig. **2.** *n.* **the** ~: das Gegenteil; **on the** ~: im Gegenteil. **3.** *adv.* ~ **to sth.** entgegen einer Sache

contrast 1. [kən'trɑːst] *v. t.* gegenüberstellen. **2.** ['kɒntrɑːst] *n.* Kontrast, *der* (**with** zu); **in** ~, **...:** im Gegensatz dazu, ...; [**be**] **in** ~ **with sth.** im Gegensatz zu etw. [stehen]. **con'trasting** *adj.* gegensätzlich

contravene [kɒntrə'viːn] *v. t.* verstoßen gegen. **contravention** [kɒntrə'venʃn] *n.* Verstoß, *der* (**of** gegen)

contribute [kən'trɪbjuːt] **1.** *v. t.* ~ **sth.** [**to** or **towards sth.**] etw. [zu etw.] beitragen. **2.** *v. i.* ~ **to charity** für karitative Zwecke spenden; ~ **to the success of sth.** zum Erfolg einer Sache *(Gen.)* beitragen. **contribution** [kɒntrɪ'bjuːʃn] *n.* Beitrag, *der; (for charity)* Spende, *die* (**to** für); **make a** ~: einen Beitrag leisten; *(to charity)* etwas spenden. **contributor** [kən'trɪbjutə(r)] *n. (to encyclopaedia etc.)* Mitarbeiter, *der/*Mitarbeiterin, *die*

contrite ['kɒntraɪt] *adj.* zerknirscht

contrive [kən'traɪv] *v. t.* ~ **to do sth.** es fertigbringen, etw. zu tun

control [kən'trəʊl] **1.** *n.* **a)** Kontrolle, *die* (**of** über + *Akk.*); **keep** ~ **of sth.** etw. unter Kontrolle halten; **be in** ~ [**of sth.**] die Kontrolle [über etw. *(Akk.)*] haben; [**go** or **get**] **out of** ~: außer Kontrolle [geraten]; [**get sth.**] **under** ~: [etw.] unter Kontrolle [bringen]; **b)** *(device)* Regler, *der;* ~**s** Schalttafel, *die*. **2.** *v. t.,* -**ll**- kontrollieren; lenken ⟨*Auto*⟩; zügeln ⟨*Zorn*⟩; regeln ⟨*Verkehr*⟩. **con'trol centre** *n.* Kontrollzentrum, *das*. **con'trol desk** *n.* Schaltpult, *das*

con'troller *n. (director)* Leiter, *der/*Leiterin, *die*

control: ~ **panel** *n.* Schalttafel, *die;* ~ **room** *n.* Kontrollraum, *der;* ~ **tower** *n.* Kontrollturm, *der*

controversial [kɒntrə'vɜːʃl] *adj.* umstritten

controversy ['kɒntrəvɜːsɪ, kən'trɒvəsɪ] *n.* Auseinandersetzung, *die*

convalesce [kɒnvə'les] *v. i.* genesen. **convalescence** [kɒnvə'lesəns] *n.* Genesung, *die*

convection [kən'vekʃn] *n. (Phys., Meteorol.)* Konvektion, *die*

convector [kən'vektə(r)] *n.* Konvektor, *der*

convene [kən'viːn] **1.** *v. t.* einberufen. **2.** *v. i.* zusammenkommen

convenience [kən'viːnɪəns] *n.* **a) for sb.'s** ~ zu jmds. Bequemlichkeit; **at your** ~: wann es Ihnen paßt; **b)** *(toilet)* [**public**] ~: [öffentliche] Toilette. **con'venience food** *n.* Fertignahrung, *die*

convenient [kən'viːnɪənt] *adj.* günstig; *(useful)* praktisch; **would it be** ~ **to** or **for you?** würde es Ihnen passen? **con'veniently** *adv.* **a)** günstig ⟨gelegen, angebracht⟩; **b)** *(opportunely)* angenehmerweise

convent ['kɒnvənt] *n.* Kloster, *das*

convention [kən'venʃn] *n.* **a)** Brauch, *der;* **b)** *(assembly)* Konferenz, *die;* **c)** *(agreement)* Konvention, *die*. **conventional** [kən'venʃənl] *adj.* konventionell

converge [kən'vɜːdʒ] *v. i.* ~ [**on each other**] aufeinander zulaufen

conversant [kən'vɜːsənt] *pred. adj.* vertraut (**with** mit)

conversation [kɒnvə'seɪʃn] *n.* Unterhaltung, *die;* **have a** ~: ein Gespräch führen. **conversational** [kɒnvə'seɪʃənl] *adj.* ~ **English** gesprochenes Englisch

¹converse [kən'vɜːs] *v. i. (formal)* ~

81

cork

[with sb.] [about *or* on sth.] sich [mit jmdm.] [über etw. *(Akk.)*] unterhalten
²**converse** ['kɒnvɜːs] 1. *adj.* entgegengesetzt; umgekehrt ⟨*Fall, Situation*⟩. 2. *n.* Gegenteil, *das.* **conversely** [kən'vɜːslɪ] *adj.* umgekehrt
conversion [kən'vɜːʃn] *n.* **a)** Umwandlung, *die* (**into** in + *Akk.*); **b)** *(adaptation)* Umbau, *der;* **c)** *(of person)* Bekehrung, *die* (**to** zu)
convert [kən'vɜːt] 1. *v. t.* umwandeln (**into** in + *Akk.*); ~ **sb.** [**to sth.**] jmdn. [zu etw.] bekehren. 2. [kən'vɜːt] *v. i.* ~ **into sth.** sich in etw. *(Akk.)* umwandeln lassen. 3. ['kɒnvɜːt] *n.* Konvertit, *der*/Konvertitin, *die.* **convertible** [kən'vɜːtɪbl] 1. *adj.* **be** ~ **into sth.** sich in etw. *(Akk.)* umwandeln lassen. 2. *n.* Kabrio[lett], *das*
convex ['kɒnveks] *adj.* konvex
convey [kən'veɪ] *v. t.* befördern. **conveyance** [kən'veɪəns] *n.* **a)** *(transportation)* Beförderung, *die;* **b)** *(formal: vehicle)* Beförderungsmittel, *das.*
con'veyancing *n.* *(Law)* ~ [**of property**] [Eigentums]übertragung, *die.* **con'veyor** [kən'veɪə(r)] *n.* ~ [**belt**] Fließband, *das*
convict 1. ['kɒnvɪkt] *n.* Strafgefangene, *der/die.* 2. [kən'vɪkt] *v. t.* verurteilen. **conviction** [kən'vɪkʃn] *n.* **a)** *(Law)* Verurteilung, *die* (**for** wegen); **b)** *(belief)* Überzeugung, *die*
convince [kən'vɪns] *v. t.* überzeugen; ~ **sb. that** ...: jmdn. davon überzeugen, daß ...; **be** ~**d that** ...: davon überzeugt sein, daß ... **convincing** [kən'vɪnsɪŋ] *adj.,* **con'vincingly** *adv.* überzeugend
convivial [kən'vɪvɪəl] *adj.* fröhlich
convoluted ['kɒnvəluːtɪd] *adj. (complex)* kompliziert
convoy ['kɒnvɔɪ] *n.* Konvoi, *der;* **in** ~: im Konvoi
convulse [kən'vʌls] *v. t.* **be** ~**d with** sich krümmen vor (+ *Dat.*). **convulsions** [kən'vʌlʃnz] *n. pl.* Krämpfe
coo [kuː] *v. i.* gurren
cook [kʊk] 1. *n.* Koch, *der*/Köchin, *die.* 2. *v. t.* kochen ⟨*Mahlzeit*⟩; *(fry, roast)* braten; *(boil)* kochen. 3. *v. i.* kochen. **cook 'up** *v. t.* erfinden ⟨*Geschichte*⟩
'**cookbook** *n.* *(Amer.)* Kochbuch, *das*
'**cooker** *n.* *(Brit.)* Herd, *der*
'**cookery** [kʊkərɪ] *n.* Kochen, *das.* '**cookery book** *n.* *(Brit.)* Kochbuch, *das*
cookie ['kʊkɪ] *n.* *(Amer.)* Keks, *der*

'**cooking** *n.* Kochen, *das.* '**cooking apple** *n.* Kochapfel, *der.* '**cooking utensil** *n.* Küchengerät, *das*
cool [kuːl] 1. *adj.* **a)** kühl; **store in a** ~ **place** kühl aufbewahren; **b)** *(unemotional, unfriendly)* kühl; *(calm)* ruhig. 2. *n.* Kühle, *die.* 3. *v. i.* abkühlen. 4. *v. t.* kühlen; *(from high temperature)* abkühlen. **cool 'down, cool 'off** *v. i. & t.* abkühlen
coolly ['kuːllɪ] *adv. (calmly)* ruhig; *(unemotionally)* kühl
coop [kuːp] 1. *n.* *(for poultry)* Hühnerstall, *der.* 2. *v. t.* ~ **up** einpferchen
co-operate [kəʊ'ɒpəreɪt] *v. i.* mitarbeiten (**in** bei); *(with each other)* zusammenarbeiten (**in** bei). **co-operation** [kəʊɒpə'reɪʃn] *n.* Zusammenarbeit, *die.* **co-operative** [kəʊ'ɒpərətɪv] 1. *adj.* kooperativ; *(helpful)* hilfsbereit. 2. *n.* Genossenschaft, *die*
co-ordinate [kəʊ'ɔːdɪneɪt] *v. t.* koordinieren. **co-ordination** [kəʊɔː'dɪ'neɪʃn] *n.* Koordination, *die*
cop [kɒp] *n.* *(sl.: police officer)* Bulle, *der (salopp)*
cope [kəʊp] *v. i.* ~ **with sb./sth.** mit jmdm./etw. fertig werden
Copenhagen [kəʊpn'heɪgn] *pr. n.* Kopenhagen *(das)*
copier ['kɒpɪə(r)] *n. (machine)* Kopiergerät, *das*
co-pilot ['kəʊpaɪlət] *n.* Kopilot, *der*/Kopilotin, *die*
copious ['kəʊpɪəs] *adj.* reichhaltig
¹**copper** ['kɒpə(r)] *n.* Kupfer, *das*
²**copper** *(Brit. sl.)* see **cop**
coppice ['kɒpɪs], **copse** [kɒps] *ns.* Wäldchen, *das*
copulate ['kɒpjʊleɪt] *v. i.* kopulieren
copy ['kɒpɪ] 1. *n.* **a)** *(reproduction)* Kopie, *die;* **b)** *(specimen)* Exemplar, *das.* 2. *v. t. & i.* kopieren; *(transcribe)* abschreiben. '**copyright** *n.* Urheberrecht, *das*
coral ['kɒrl] *n.* Koralle, *die*
cord [kɔːd] *n.* **a)** Kordel, *die;* **b)** *(cloth)* Cord, *der;* **c)** *in pl. (trousers)* [**pair of**] ~**s** Cordhose, *die*
cordial ['kɔːdɪəl] 1. *adj.* herzlich. 2. *n.* *(drink)* Sirup, *der.* '**cordially** *adv.* herzlich
cordon ['kɔːdn] 1. *n.* Kordon, *der.* 2. *v. t.* ~ [**off**] absperren
corduroy ['kɔːdərɔɪ, 'kɔːdjʊrɔɪ] *n.* Cordsamt, *der*
core [kɔː(r)] 1. *n. (of fruit)* Kerngehäuse, *das.* 2. *v. t.* entkernen
cork [kɔːk] 1. *n.* **a)** *(bark)* Kork, *der;* **b)**

(bottle-stopper) Korken, *der.* **2.** *v. t.* zukorken. '**corkscrew** *n.* Korkenzieher, *der*

¹**corn** [kɔ:n] *n.* Getreide, *das*

²**corn** *n.* *(on foot)* Hühnerauge, *das*

corned beef [kɔ:nd 'bi:f] *n.* Corned beef, *das*

corner ['kɔ:nə(r)] **1.** *n.* **a)** Ecke, *die; (curve)* Kurve, *die;* **on the ~:** an der Ecke/in der Kurve; **b)** *(of mouth, eye)* Winkel, *der.* **2.** *v. t. (fig.)* in die Enge treiben. **3.** *v. i.* die Kurve nehmen. '**corner kick** *n. (Footb.)* Eckball, *der.* '**cornerstone** *n. (fig.)* Eckpfeiler, *der*

cornet ['kɔ:nɪt] *n.* **a)** *(Brit.: for ice-cream)* [Eis]tüte, *die;* **b)** *(Mus.)* Kornett, *das*

corn: ~flakes *n. pl.* Corn-flakes *Pl.;* **~flour** *(Brit.),* **~starch** *(Amer.)* ns. Maismehl, *das*

'**corny** *adj. (coll.: trite)* abgedroschen

coronation [kɒrə'neɪʃn] *n.* Krönung, *die*

coroner ['kɒrənə(r)] *n.* Coroner, *der; Beamter, der gewaltsame od. unnatürliche Todesfälle untersucht*

coronet ['kɒrənet] *n.* Krone, *die*

¹**corporal** ['kɔ:pərl] *adj.* körperlich

²**corporal** *n.* ≈ Hauptgefreite, *der*

corporation [kɔ:pə'reɪʃn] *n.* Stadtverwaltung, *die*

corps [kɔ:(r)] *n., pl. same* [kɔ:z] Korps, *das*

corpse [kɔ:ps] *n.* Leiche, *die*

corpulent ['kɔ:pjʊlənt] *adj.* korpulent

correct [kə'rekt] **1.** *v. t.* korrigieren. **2.** *adj.* korrekt; **that is ~:** das stimmt. **correction** [kə'rekʃn] *n.* Korrektur, *die.* **cor'rectly** *adv.* korrekt

correspond [kɒrɪ'spɒnd] *v. i.* **a) ~ |to each other|** einander entsprechen; **~ to sth.** einer Sache *(Dat.)* entsprechen; **b)** *(communicate)* **~ with sb.** mit jmdm. korrespondieren. **correspondence** [kɒrɪ'spɒndəns] *n.* **a)** Übereinstimmung, *die* (with, to mit); **b)** *(communication)* Briefwechsel, *der.* **correspondent** [kɒrɪ'spɒndənt] *n. (reporter)* Korrespondent, *der/*Korrespondentin, *die.* **corre'sponding** *adj.* entsprechend (to *Dat.*). **corre'spondingly** *adv.* entsprechend

corridor ['kɒrɪdɔ:(r)] *n.* **a)** Flur, *der;* **b)** *(Railw.)* [Seiten]gang, *der*

corroborate [kə'rɒbəreɪt] *v. t.* bestätigen

corrode [kə'rəʊd] **1.** *v. t.* zerfressen. **2.** *v. i.* zerfressen werden. **corrosion** [kə'rəʊʒn] *n.* Korrosion, *die*

corrugated ['kɒrəgeɪtɪd] *adj.* **~ cardboard** Wellpappe, *die;* **~ iron** Wellblech, *das*

corrupt [kə'rʌpt] **1.** *adj. (depraved)* verdorben *(geh.); (influenced by bribery)* korrupt. **2.** *v. t. (deprave)* korrumpieren; *(bribe)* bestechen. **corruption** [kə'rʌpʃn] *n. (moral deterioration)* Verdorbenheit, *die (geh.); (corrupt practices)* Korruption, *die*

corset ['kɔ:sɪt] *n.* Korsett, *das*

Corsica ['kɔ:sɪkə] *pr. n.* Korsika *(das)*

cortège [kɔ:'teɪʒ] *n.* Trauerzug, *der*

cosh [kɒʃ] *(Brit. coll.)* **1.** *n.* Totschläger, *der.* **2.** *v. t.* niederknüppeln

cosmetic [kɒz'metɪk] **1.** *adj.* kosmetisch. **2.** *n.* Kosmetikum, *das*

cosmic ['kɒzmɪk] *adj.* kosmisch

cosmonaut ['kɒzmənɔ:t] *n.* Kosmonaut, *der/*Kosmonautin, *die*

cosmopolitan [kɒzmə'pɒlɪtən] *adj.* kosmopolitisch

cosmos ['kɒzmɒs] *n.* Kosmos, *der*

cosset ['kɒsɪt] *v. t.* [ver]hätscheln

cost [kɒst] **1.** *n.* **a)** Kosten *Pl.;* **b)** *(fig.)* Preis, *der;* **at all ~s, at any ~:** um jeden Preis. **2.** *v. t.* **a)** *p.t., p.p.* cost *(lit. or fig.)* kosten; **how much does it ~?** was kostet es?; **b)** *p. t., p. p.* costed *(Commerc.: fix price of)* **~ sth.** den Preis für etw. kalkulieren. '**cost-effective** *adj.* rentabel

'**costly** *adj.* teuer

'**cost: ~ of 'living** *n.* Lebenshaltungskosten *Pl.;* **~ price** *n.* Selbstkostenpreis, *der*

costume ['kɒstju:m] *n.* Kleidermode, *die; (theatrical ~)* Kostüm, *das*

cosy ['kəʊzɪ] *adj.* gemütlich

cot [kɒt] *n.* Kinderbett, *das*

cottage ['kɒtɪdʒ] *n.* Cottage, *das*

cottage: ~ 'cheese *n.* Hüttenkäse, *der;* **~ industry** *n.* Heimarbeit, *die;* **~ 'pie** *n.* mit Kartoffelbrei überbackenes Hackfleisch

cotton ['kɒtn] **1.** *n.* Baumwolle, *die; (thread)* Baumwollgarn, *das.* **2.** *attrib. adj.* Baumwoll-. **3.** *v. i.* **~ 'on** *(coll.)* kapieren *(ugs.).* **cotton 'wool** *n.* Watte, *die*

couch [kaʊtʃ] *n.* Couch, *die*

couchette [ku:'ʃet] *n. (Railw.)* Liegesitz, *der*

cough [kɒf] **1.** *n.* Husten, *der.* **2.** *v. i.* husten. '**cough mixture** *n.* Hustensaft, *der*

could *see* ²**can**

couldn't ['kʊdnt] *(coll.)* **= could not;** *see* ²**can**

council ['kaʊnsl] n. Rat, der; local ~: Gemeinderat, der; city/town ~: Stadtrat, der. 'council flat n. Sozialwohnung, die. 'council house n. Haus des sozialen Wohnungsbaus

councillor ['kaʊnsələ(r)] n. Ratsmitglied, das

'council tax n. (Brit.) Gemeindesteuer, die

counsel ['kaʊnsl] 1. n. a) Rat[schlag], der; b) pl. same (Law) Rechtsanwalt, der/-anwältin, die. 2. v. t., (Brit.) -ll- beraten. counsellor, (Amer.) counselor ['kaʊnsələ(r)] n. Berater, der/Beraterin, die

¹count [kaʊnt] 1. n. Zählen, das; keep ~ |of sth.| [etw.] zählen; lose ~: sich verzählen. 2. v. t. a) zählen; b) (include) mitzählen; not ~ing abgesehen von; c) (consider) halten für; ~ oneself lucky sich glücklich schätzen können. 3. v. i. a) zählen; ~ |up| to ten bis zehn zählen; b) (be included) zählen. 'count on v. t. ~ on sb./sth. sich auf jmdn./etw. verlassen. count 'up v. t. zusammenzählen

²count n. (nobleman) Graf, der

'countdown n. Countdown, der od. das

countenance ['kaʊntɪnəns] 1. n. (literary: face) Antlitz, das. 2. v. t. (formal: approve) gutheißen

¹counter ['kaʊntə(r)] n. a) (in shop) Ladentisch, der; (in cafeteria) Büfett, das; (in bank) Schalter, der; b) (for games) Spielmarke, die

²counter 1. adj. Gegen-. 2. v. t. a) (oppose) begegnen (+ Dat.); b) (act against) kontern. 3. adv. go ~ to zuwiderlaufen (+ Dat.)

counter: ~'act v. t. entgegenwirken (+ Dat.); ~-attack n. Gegenangriff, der; ~balance v. t. (fig.) ausgleichen; ~-'espionage n. Spionageabwehr, die

counterfeit ['kaʊntəfɪt] 1. adj. gefälscht; ~ money Falschgeld, das. 2. v. t. fälschen. 'counterfeiter n. Fälscher, der/Fälscherin, die

counter: ~foil n. Kontrollabschnitt, der; ~part n. Gegenstück, das (of zu); ~-pro'ductive adj. sth. is ~-productive etw. bewirkt das Gegenteil des Gewünschten; ~sign v. t. gegenzeichnen

countess ['kaʊntɪs] n. Gräfin, die

'countless adj. zahllos

country ['kʌntrɪ] n. a) Land, das; sb's |home| ~: jmds. Heimat; b) (~side)

Landschaft, die; in the ~: auf dem Land. countryman ['kʌntrɪmən] n., pl. countrymen ['kʌntrɪmən] Landsmann, der. 'countryside n. a) (rural areas) Land, das; b) (rural scenery) Landschaft, die

county ['kaʊntɪ] n. (Brit.) Grafschaft, die

coup [ku:] n. a) Coup, der; b) see coup d'état. coup d'état [ku: deɪ'ta:] n. Staatsstreich, der

coupé ['ku:peɪ] n. Coupé, das

couple [kʌpl] n. 1. a) (pair) Paar, das; (married) [Ehe]paar, das; b) a ~ |of| (a few) ein paar; (two) zwei. 2. v. t. koppeln

coupon ['ku:pɒn] n. a) (for rations) Marke, die; b) (in advertisement) Coupon, der

courage ['kʌrɪdʒ] n. Mut, der. courageous [kə'reɪdʒəs] adj., cou'rageously adv. mutig

courgette [kʊə'ʒet] n. (Brit.) Zucchino, der

courier ['kʊrɪə(r)] n. a) (Tourism) Reiseleiter, der/-leiterin, die; b) (messenger) Kurier, der

course [kɔ:s] n. a) (of ship, plane) Kurs, der; ~ |of action| Vorgehensweise, die; b) of ~: natürlich; c) in due ~: zu gegebener Zeit; in the ~ of the day/ his life im Lauf[e] des Tages/seines Lebens; d) (of meal) Gang, der; e) (Sport) Kurs, der; |golf-|~: [Golf]platz, der; f) (Educ.) Kurs[us], der; g) (Med.) a ~ of treatment eine Kur

court [kɔ:t] 1. n. a) (of ship, plane) Hof, der; b) (Tennis, Squash) Platz, der; c) (Law) Gericht, das. 2. v. t. ~ sb. jmdn. umwerben

courteous ['kɜ:tɪəs] adj. höflich. courtesy ['kɜ:təsɪ] n. Höflichkeit, die

court: ~-house n. (Law) Gerichtsgebäude, das; ~ 'martial n., pl. ~s martial (Mil.) Kriegsgericht, das; ~yard n. Hof, der

cousin ['kʌzn] n. |first| ~: Cousin, der/Cousine, die

cove [kəʊv] n. (Geog.) [kleine] Bucht

covenant ['kʌvənənt] n. formelle Übereinkunft

cover ['kʌvə(r)] 1. n. a) (piece of cloth) Decke, die; (of cushion, bed) Bezug, der; (lid) Deckel, der; (of hole, engine, typewriter, etc.) Abdeckung, die; b) (of book) Einband, der; (of magazine) Umschlag, der; |send sth.| under separate ~: [etw.] mit getrennter Post [schicken]; d) take ~ |from sth.| Schutz [vor etw. *(Dat.)*] suchen; **under ~** *(from rain)* überdacht. **2.** *v. t.* **a)** bedecken; beziehen *(Sessel, Kisses);* zudecken *(Pfanne);* **the roses are ~ed with greenfly** die Rosen sind voller Blattläuse; **b)** *(include)* abdecken; **c)** *(Journ.)* berichten über (+ *Akk.*); **d)** decken *(Kosten).* **cover 'up** *v. t.* **1.** zudecken; *(fig.)* vertuschen. **2.** *v. i.* **~ up for sb.** jmdn. decken

coverage ['kʌvərɪdʒ] *n. (Journ.)* Berichterstattung, *die*

'**cover charge** *n.* [Preis für das] Gedeck

'**covering** *n.* Decke, *die; (of chair, bed)* Bezug, *der.* '**covering letter** *n.* Begleitbrief, *der*

covert ['kʌvət] *adj.* versteckt

'**cover-up** *n.* Verschleierung, *die*

covet ['kʌvɪt] *v. t.* begehren *(geh.).* **covetous** ['kʌvɪtəs] *adj.* begehrlich *(geh.)*

cow [kaʊ] *n.* Kuh, *die*

coward ['kaʊəd] *n.* Feigling, *der.* **cowardice** ['kaʊədɪs] *n.* Feigheit, *die.* '**cowardly** *adj.* feig[e]

'**cowboy** *n.* Cowboy, *der*

cower ['kaʊə(r)] *v. i.* sich ducken

cow: ~-shed *n.* Kuhstall, *der; ~slip* *n.* Schlüsselblume, *die*

coy [kɔɪ] *adj.* gespielt schüchtern

cozy *(Amer.) see* **cosy**

crab [kræb] *n.* Krabbe, *die.* '**crab-apple** *n.* Holzapfel, *der*

crack [kræk] **1.** *n.* **a)** *(noise)* Krachen, *das;* **b)** *(in china etc.)* Sprung, *der; (in rock)* Spalte, *die; (chink)* Spalt, *der;* **c)** *(coll.: try)* **have a ~ at sth./doing sth.** versuchen, etw. zu tun. **2.** *attrib. adj. (coll.)* erstklassig. **3.** *v. t.* **a)** knacken *(Nuß, Problem, Kode);* **b)** *(make a ~ in)* anschlagen *(Porzellan usw.);* **c)** **~ a joke** einen Witz machen; **d)** **~ a whip** mit einer Peitsche knallen. **4.** *v. i. (Porzellan usw.:)* einen Sprung/Sprünge bekommen. **crack 'down** *v. i. (coll.)* **~ down [on sb./sth.]** [gegen jmdn./etw.] [hart] vorgehen. **crack 'up** *v. i. (coll.) (Person:)* zusammenbrechen

cracked [krækt] *adj.* gesprungen *(Porzellan usw.);* rissig *(Verputz)*

cracker ['krækə(r)] *n.* **a)** [Christmas] ~ ≈ Knallbonbon, *der od. das;* **b)** *(biscuit)* Cracker, *der.* '**crackers** *pred. adj. (Brit. coll.)* übergeschnappt *(ugs.)*

crackle ['krækl] **1.** *v. i.* knistern; *(Feuer:)* prasseln. **2.** *n.* Knistern, *das*

cradle ['kreɪdl] **1.** *n.* Wiege, *die.* **2.** *v. t.* wiegen

craft [krɑːft] *n.* **a)** *(trade)* Handwerk, *das; (art)* Kunsthandwerk, *das;* **b)** *pl. same (boat)* Boot, *das.* **craftsman** ['krɑːftsmən] *n., pl.* **craftsmen** ['krɑːftsmən] Handwerker, *der.* '**crafty** *adj.* listig

crag [kræg] *n.* Felsspitze, *die.* '**craggy** *adj.* **a)** felsig; **b)** zerfurcht *(Gesicht)*

cram [kræm] **1.** *v. t.,* **-mm-** *(overfill)* vollstopfen *(ugs.); (force)* stopfen. **2.** *v. i.,* **-mm-** *(for exam)* büffeln *(ugs.)*

cramp [kræmp] **1.** *n. (Med.)* Krampf, *der.* **2.** *v. t.* einengen

cranberry ['krænbərɪ] *n.* Preiselbeere, *die*

crane [kreɪn] **1.** *n.* Kran, *der.* **2.** *v. t.* **~ one's neck** den Hals recken

¹**crank** [kræŋk] *n. (Mech. Engin.)* [Hand]kurbel, *die*

²**crank** *n.* Irre, *der/die (salopp)*

'**crankshaft** *n. (Mech. Engin.)* Kurbelwelle, *die*

'**cranky** *adj. (eccentric)* schrullig

cranny ['krænɪ] *n.* Ritze, *die*

crash [kræʃ] **1.** *n.* **a)** *(noise)* Krachen, *das;* **b)** *(collision)* Zusammenstoß, *der;* **have a ~:** einen Unfall haben. **2.** *v. i.* **a)** *(make a noise, go noisily)* krachen; **b)** *(have a collision)* einen Unfall haben; *(Flugzeug, Flieger:)* abstürzen; **~ into sth.** gegen etw. krachen. **3.** *v. t.* **a)** *(smash)* schmettern; **b)** *(cause to have collision)* einen Unfall haben mit

crash: ~ barrier *n.* Leitplanke, *die;* **~ course** *n.* Intensivkurs, *der;* **~-helmet** *n.* Sturzhelm, *der*

crass [kræs] *adj.* kraß

crate [kreɪt] *n.* Kiste, *die*

crater ['kreɪtə(r)] *n.* Krater, *der*

cravat [krə'væt] *n.* Krawatte, *die*

crave [kreɪv] *v. t.* **a)** *(beg)* erbitten; **b)** *(long for)* sich sehnen nach. '**craving** *n.* Verlangen, *das* (for nach)

crawl [krɔːl] **1.** *v. i.* **a)** kriechen; *(Baby, Insekt:)* krabbeln; **b)** *(coll.)* **~ to sb.** vor jmdm. kriechen. **2.** *n.* **a)** **go at a ~:** im Schneckentempo fahren; **b)** *(swimming-stroke)* Kraulen, *das.* '**crawler lane** *n.* Kriechspur, *die*

crayfish ['kreɪfɪʃ] *n., pl. same* Flußkrebs, *der*

crayon ['kreɪən] *n.* [coloured] **~:** Buntstift, *der; (wax)* Wachsmalstift, *der*

craze [kreɪz] *n.* Begeisterung, *die*

crazy ['kreɪzɪ] *adj.* verrückt; **be ~ about sb./sth.** *(coll.)* nach jmdm./etw. verrückt sein *(ugs.)*

creak [kriːk] **1.** *n.* Knarren, *das.* **2.** *v. i.* knarren

cream [kri:m] 1. *n.* **a)** Sahne, *die;* **b)** *(dessert, cosmetic)* Creme, *die.* 2. *adj.* ~-|-coloured] creme[farben]. **cream 'cheese** *n.* ≈ Frischkäse, *der*

'creamy *adj. (with cream)* sahnig; *(like cream)* cremig

crease [kri:s] 1. *n. (pressed)* Bügelfalte, *die; (accidental)* Falte, *die.* 2. *v. t. (press)* eine Falte bügeln in (+ *Akk.*); *(accidentally)* zerknittern. 3. *v. i.* Falten bekommen; knittern. **'crease-resistant** *adj.* knitterfrei

create [kri:'eɪt] *v. t.* schaffen; verursachen *⟨Verwirrung⟩* machen *⟨Eindruck⟩.* **creation** [kri:'eɪʃn] *n.* Schaffung, *die; (of the world)* Schöpfung, *die (geh.).* **creative** [kri:'eɪtɪv] *adj.* kreativ. **creator** [kri:'eɪtə(r)] *n.* Schöpfer, *der*/Schöpferin, *die*

creature ['kri:tʃə(r)] *n.* Geschöpf, *das*

crèche [kreʃ] *n.* [Kinder]krippe, *die*

credentials [krɪ'denʃlz] *n. pl.* Zeugnis, *das*

credibility [kredɪ'bɪlɪtɪ] *n.* Glaubwürdigkeit, *die*

credible ['kredɪbl] *adj.* glaubwürdig

credit ['kredɪt] 1. *n.* **a)** *(honour)* Ehre, *die;* **take the ~ for sth.** die Anerkennung für etw. einstecken; **b)** *(Commerc.)* Kredit, *der.* 2. *v. t.* **a)** glauben; **b)** *(Finance)* gutschreiben. **creditable** ['kredɪtəbl] *adj.* anerkennenswert

'credit card *n.* Kreditkarte, *die*

creditor ['kredɪtə(r)] *n.* Gläubiger, *der*/Gläubigerin, *die*

creed [kri:d] *n.* Glaubensbekenntnis, *das*

creek [kri:k] *n.* **a)** *(Brit.: of coast)* [kleine] Bucht; **b)** *(of river)* [kurzer] Flußarm

creep [kri:p] 1. *v. i.,* **crept** [krept] kriechen; *(move timidly, slowly, stealthily)* schleichen. 2. *n.* **a)** *(sl.: person)* Fiesling, *der (salopp);* **b)** *(coll.)* **give sb. the ~s** jmdn. nicht [ganz] geheuer sein. **'creeper** *n.* Kletterpflanze, *die.* **'creepy** *adj.* unheimlich

cremate [krɪ'meɪt] *v. t.* einäschern. **cremation** [krɪ'meɪʃn] *n.* Einäscherung, *die.* **crematorium** [kremə-'tɔ:rɪəm] *n.* Krematorium, *das*

creosote ['kri:əsəʊt] *n.* Kreosot, *das*

crept *see* **creep** 1

crescent ['kresənt] *n.* Mondsichel, *die*

cress [kres] *n.* Kresse, *die*

crest [krest] *n.* Kamm, *der.* **'crestfallen** *adj.* niedergeschlagen

Crete [kri:t] *pr. n.* Kreta *(das)*

cretin ['kretɪn] *n. (coll.)* Trottel, *der*

crevasse [krɪ'væs] *n.* Gletscherspalte, *die*

crevice ['krevɪs] *n.* Spalt, *der*

crew [kru:] *n.* Besatzung, *die.* **'crew-cut** *n.* Bürstenschnitt, *der*

crib [krɪb] 1. *n.* Krippe, *die.* 2. *v. t.,* -bb- *(coll.)* abkupfern *(salopp)*

crick [krɪk] *n.* **a ~ |in one's neck/back|** ein steifer Hals/Rücken

¹cricket ['krɪkɪt] *n.* Kricket, *das*

²cricket *n. (Zool.)* Grille, *die*

'cricket bat *n.* Schlagholz, *das*

'cricketer *n.* Kricketspieler, *der*/-spielerin, *die*

cried *see* **cry**

crime [kraɪm] *n.* **a)** Verbrechen, *das;* **b)** *collect.* **a wave of ~:** eine Welle von Straftaten; **~ doesn't pay** Verbrechen lohnen sich nicht

criminal ['krɪmɪnl] 1. *adj.* kriminell; strafbar; **~ act** *or* **deed/offence** Straftat, *die.* 2. *n.* Kriminelle, *der/die*

crimson ['krɪmzn] 1. *adj.* purpurrot. 2. *n.* Purpurrot, *das*

cringe [krɪndʒ] *v. i.* zusammenzucken

crinkle ['krɪŋkl] 1. *n.* Knitterfalte, *die.* 2. *v. t.* zerknittern. 3. *v. i.* knittern

cripple ['krɪpl] 1. *n.* Krüppel, *der.* 2. *v. t.* zum Krüppel machen; *(fig.)* lähmen. **crippled** ['krɪpld] *adj.* verkrüppelt

crisis ['kraɪsɪs] *n., pl.* **crises** ['kraɪsi:z] Krise, *die*

crisp [krɪsp] 1. *adj.* knusprig. 2. **a)** *n. usu. in pl. (Brit.: potato ~)* [Kartoffel]chip, *der;* **b)** **be burned to a ~:** verbrannt sein. **'crispbread** *n.* Knäckebrot, *das*

'crispy *adj.* knusprig

criss-cross ['krɪskrɒs] 1. *adj.* **~ pattern** Muster aus gekreuzten Linien. 2. *adv.* kreuz und quer. 3. *v. t.* wiederholt schneiden

criterion [kraɪ'tɪərɪən] *n., pl.* **criteria** [kraɪ'tɪərɪə] Kriterium, *das*

critic ['krɪtɪk] *n.* Kritiker, *der*/Kritikerin, *die.* **critical** ['krɪtɪkl] *adj.* kritisch; **be ~ of sb./sth.** jmdn./etw. kritisieren. **critically** ['krɪtɪkəlɪ] *adv.* kritisch; **~ ill** ernstlich krank

criticism ['krɪtɪsɪzm] *n.* Kritik, *die (of an + Dat.)*

criticize ['krɪtɪsaɪz] *v. t.* kritisieren (**for** wegen)

croak [krəʊk] 1. *n. (of frog)* Quaken, *das; (of person)* Krächzen, *das.* 2. *v. i.* *⟨Frosch:⟩* quaken; *⟨Person:⟩* krächzen. 3. *v. t.* krächzen

crochet ['krəʊʃeɪ] **1.** *n.* Häkelarbeit, *die;* ~ **hook** Häkelhaken, *der.* **2.** *v.t.* häkeln

crock [krɒk] *n. (coll.)* |old| ~ *(person)* altes Wrack, *das (fig.); (vehicle)* [alte] Klapperkiste *(ugs.)*

crockery ['krɒkərɪ] *n.* Geschirr, *das*

crocodile ['krɒkədaɪl] *n.* Krokodil, *das*

crocus ['krəʊkəs] *n.* Krokus, *der*

crony ['krəʊnɪ] *n.* Kumpel, *der (ugs.)*

crook [krʊk] *n.* **a)** *(coll.: rogue)* Gauner, *der;* **b)** *(shepherd's)* Hirtenstab, *der*

crooked ['krʊkɪd] *adj.* krumm; *(fig.: dishonest)* betrügerisch

crop [krɒp] *n.* **1.** [Feld]frucht, *die; (season's yield)* Ernte, *die.* **2.** *v.t.* stutzen ⟨*Haare usw.*⟩. **crop 'up** *v.i.* auftauchen

'**cropper** *n. (coll.)* **come a** ~: einen Sturz bauen *(ugs.)*

croquet ['krəʊkeɪ] *n.* Krocket[spiel], *das*

croquette [krə'ket] *n.* Krokette, *die*

cross [krɒs] **1.** *n.* **a)** Kreuz, *das;* **b)** *(mixture)* Mischung, *die* (**between** aus). **2.** *v.t.* **a)** [über]kreuzen; ~ **one's arms/legs** die Arme verschränken/die Beine übereinanderschlagen; **keep one's fingers ~ed** [for sb.] *(fig.)* |jmdm.| die *od.* den Daumen drücken; **b)** *(go across)* kreuzen; überqueren ⟨*Straße, Gebirge*⟩; durchqueren ⟨*Land, Zimmer*⟩; ~ **sb.'s mind** *(fig.)* jmdm. einfallen; **c)** *(Brit.)* **a ~ed cheque** ein Verrechnungsscheck; **d)** ~ **oneself** sich bekreuzigen. **3.** *v.i.* aneinander vorbeigehen; ~ |**in the post**| ⟨*Briefe:*⟩ sich kreuzen. **4.** *adj.* verärgert; **sb. will be** ~: jmd. wird ärgerlich *od.* böse werden; **be ~ with sb.** böse auf jmdn. sein. **cross 'out** *v.t.* ausstreichen. **cross 'over** *v.t.* überqueren; *abs.* hinübergehen

cross: ~**bar** *n.* **a)** [Fahrrad]stange, *die;* **b)** *(Sport)* Querlatte, *die;* ~**check 1.** *n.* Gegenprobe, *die;* **2.** *v.t.* [nochmals] nachprüfen; nachkontrollieren; ~**country 1.** *adj.* Querfeldein-; **2.** *adv.* querfeldein; ~**examination** *n.* Kreuzverhör, *das;* ~**examine** *v.t.* ins Kreuzverhör nehmen; ~**eyed** ['krɒsaɪd] *adj.* [nach innen] schielend; **be ~eyed** schielen; ~**fire** *n.* Kreuzfeuer, *das*

'**crossing** *n.* **a)** *(act)* Überquerung, *die;* **b)** *(pedestrian ~)* Überweg, *der*

'**crossly** *adv.* verärgert

cross: ~ '**purposes** *n. pl.* **talk at** ~

purposes aneinander vorbeireden; ~**reference** *n.* Querverweis, *der;* ~**roads** *n. sing.* Kreuzung, *die; (fig.)* Wendepunkt, *der;* ~**section** *n.* Querschnitt, *der;* ~**word** *n.* ~**word** |**puzzle**| Kreuzworträtsel, *das*

crotchet ['krɒtʃɪt] *n. (Brit. Mus.)* Viertelnote, *die*

crouch [kraʊtʃ] *v.i.* [sich zusammen]kauern

crow [krəʊ] *n.* Krähe, *die;* **as the** ~ **flies** Luftlinie

'**crowbar** *n.* Brechstange, *die*

crowd [kraʊd] **1.** *n.* [Menschen]menge, *die.* **2.** *vt.* füllen. **3.** *v.i.* sich sammeln. '**crowded** *adj.* überfüllt

crown [kraʊn] **1.** *n.* Krone, *die.* **2.** *v.t.* **a)** krönen; **b)** überkronen ⟨*Zahn*⟩

crucial ['kruːʃl] *adj.* entscheidend (**to** für)

crucifix ['kruːsɪfɪks] *n.* Kruzifix, *das.* **crucifixion** [kruːsɪ'fɪkʃn] *n.* Kreuzigung, *die*

crucify ['kruːsɪfaɪ] *v.t.* kreuzigen

crude [kruːd] *adj.* **a)** roh; ~ **oil** Rohöl, *das;* **b)** *(fig.)* grob ⟨*Entwurf, Worte*⟩

cruel ['kruːəl] *adj.* grausam. **cruelty** ['kruːəltɪ] *n.* Grausamkeit, *die*

cruise [kruːz] **1.** *v.i. (at random)* ⟨*Fahrzeug, Fahrer:*⟩ herumfahren. **2.** *n.* Kreuzfahrt, *die.* '**cruise missile** *n.* Marschflugkörper, *der.* '**cruiser** *n.* Kreuzer, *der*

crumb [krʌm] *n.* Krümel, *der*

crumble ['krʌmbl] **1.** *v.t.* zerkrümeln ⟨*Keks, Kuchen*⟩. **2.** *v.i.* ⟨*Mauer:*⟩ zusammenfallen. **crumbly** ['krʌmblɪ] *adj.* krümelig ⟨*Keks, Kuchen*⟩; bröckelig ⟨*Gestein*⟩

crumpet ['krʌmpɪt] *n.* weiches Hefeküchlein zum Toasten

crumple ['krʌmpl] **1.** *v.t.* **a)** *(crush)* zerdrücken; **b)** *(wrinkle)* zerknittern. **2.** *v.i.* knittern

crunch [krʌntʃ] **1.** *v.t.* [geräuschvoll] knabbern ⟨*Keks*⟩. **2.** *v.i.* ⟨*Schnee, Kies:*⟩ knirschen. **3.** *n.* Knirschen, *das;* **when it comes to the** ~: wenn es hart auf hart geht. '**crunchy** *adj.* knusprig

crusade [kruː'seɪd] **1.** *n. (Hist.; also fig.)* Kreuzzug, *der.* **2.** *v.i. (fig.)* zu Felde gehen. **cru'sader** *n. (Hist.)* Kreuzfahrer, *der*

crush [krʌʃ] **1.** *v.t.* **a)** quetschen; **b)** *(powder)* zerstampfen; **c)** *(fig.)* niederschlagen. **2.** *n. (crowd)* Gedränge, *das*

crust [krʌst] *n.* Kruste, *die.* '**crusty** *adj.* knusprig

87 **curriculum vitae**

crutch [krʌtʃ] n. Krücke, die; **go about on ~es** an Krücken gehen
crux [krʌks] n. **the ~ of the matter** der springende Punkt bei der Sache
cry [kraɪ] 1. n. (of grief) Schrei, der; (of words) Schreien, das; **a far ~ from ...** (fig.) etwas ganz anderes als ... 2. v.i. a) rufen; (loudly) schreien; b) (weep) weinen (over wegen). **cry off** v.i. absagen. **cry out** v.i. aufschreien
'**crying** adj. **it's a ~ shame** es ist eine wahre Schande
crypt [krɪpt] n. Krypta, die
cryptic ['krɪptɪk] adj. geheimnisvoll
crystal ['krɪstl] 1. n. a) Kristall, der; b) (glass) Bleikristall, das. 2. adj. (made of ~ glass) kristallen. **crystallize** ['krɪstəlaɪz] v.i. kristallisieren; (fig.) feste Form annehmen
cub [kʌb] n. a) Junge, das; (of wolf, fox, dog) Welpe, der; b) **Cub** see **Cub Scout**
Cuba ['kju:bə] n. Kuba (das)
cubby[-hole] ['kʌbɪ(həʊl)] n. Kämmerchen, das
cube [kju:b] n. Würfel, der. **cubic** ['kju:bɪk] adj. a) würfelförmig; b) Kubik(meter usw.)
cubicle ['kju:bɪkl] n. Kabine, die
'**Cub Scout** n. Wölfling, der
cuckoo ['kʊku:] n. Kuckuck, der. '**cuckoo clock** n. Kuckucksuhr, die
cucumber ['kju:kʌmbə(r)] n. [Salat]gurke, die
cuddle ['kʌdl] 1. n. enge Umarmung. 2. v.t. schmusen mit; hätscheln (kleines Kind). 3. v.i. schmusen
cuddly ['kʌdlɪ] adj. zum Schmusen nachgestellt. **cuddly toy** n. Plüschtier, das
cudgel ['kʌdʒl] n. Knüppel, der
¹**cue** [kju:] n. (Billiards etc.) Queue, das
²**cue** n. (Theatre) Stichwort, das
¹**cuff** [kʌf] n. a) Manschette, die; **off the ~** (fig.) aus dem Stegreif; b) (Amer.: trouser turn-up) [Hosen]aufschlag, der
²**cuff** 1. v.t. **~ sb.** jmdm. einen Klaps geben. 2. n. Klaps, der
'**cuff-link** n. Manschettenknopf, der
cul-de-sac ['kʌldəsæk] n. Sackgasse, die
culinary ['kʌlɪnərɪ] adj. kulinarisch
culminate ['kʌlmɪneɪt] v.i. gipfeln; **~ in sth.** in etw. (Dat.) seinen Höchststand erreichen. **culmination** [kʌlmɪ'neɪʃn] n. Höhepunkt, der
culottes [kju:'lɒts] n. pl. Hosenrock, der

culprit ['kʌlprɪt] n. Täter, der/Täterin, die
cult [kʌlt] n. Kult, der
cultivate ['kʌltɪveɪt] v.t. kultivieren (auch fig.); bestellen (Acker, Land); anbauen (Pflanzen). **cultivation** [kʌltɪ'veɪʃn] n. see **cultivate**: Kultivierung, die; Bestellen das; Anbau, der
culture ['kʌltʃə(r)] n. Kultur, die. '**cultured** adj. kultiviert
cumbersome ['kʌmbəsəm] adj. hinderlich (Kleider); sperrig (Pakete); schwerfällig (Arbeitsweise)
cunning ['kʌnɪŋ] 1. n. Schläue, die. 2. adj. schlau
cup [kʌp] n. a) Tasse, die; b) (prize, competition) Pokal, der; c) (~ful) Tasse, die; **a ~ of coffee/tea** eine Tasse Kaffee/Tee
cupboard ['kʌbəd] n. Schrank, der
'**Cup Final** n. Pokalendspiel, das
cupful ['kʌpfl] n. Tasse, die; **a ~ of water** eine Tasse Wasser
curable ['kjʊərəbl] adj. heilbar
curate ['kjʊərət] n. Kurat, der
curator [kjʊə'reɪtə(r)] n. (of museum) Direktor, der/Direktorin, die
curb [kɜ:b] v.t. zügeln
curdle ['kɜ:dl] v.i. gerinnen
cure [kjʊə(r)] 1. n. [Heil]mittel, das (for gegen); (fig.) Mittel, das. 2. v.t. a) heilen; b) [ein]pökeln (Fleisch)
curfew ['kɜ:fju:] n. Ausgangssperre, die
curiosity [kjʊərɪ'ɒsɪtɪ] n. a) Neugier[de], die; b) (object) Wunderding, das
curious ['kjʊərɪəs] adj. a) (inquisitive) neugierig; b) (strange, odd) seltsam
curl [kɜ:l] 1. n. Locke, die. 2. v.t. locken. 3. v.i. a) sich locken; b) (Straße, Fluß:) sich winden. '**curler** n. Lockenwickler, der. '**curly** adj. lockig
currant ['kʌrənt] n. Korinthe, die
currency ['kʌrənsɪ] n. (money) Währung, die; **foreign currencies** Devisen
current ['kʌrənt] 1. adj. a) verbreitet (Meinung); gebräuchlich (Wort); b) laufend (Jahr, Monat); c) (the present) aktuell (Ereignis, Mode); Tages(politik, -preis); **~ affairs** Tagespolitik, die. 2. n. a) (of water, air) Strömung, die; b) (Electr.) Strom, der. '**current account** n. Girokonto, das
'**currently** adv. zur Zeit
curriculum [kə'rɪkjʊləm] n. Lehrplan, der. **curriculum vitae** [~ 'vi:taɪ] n. Lebenslauf, der

¹**curry** ['kʌrɪ] *n.* Curry[gericht], *das*
²**curry** *v. t.* ~ **favour** [with sb.] sich [bei jmdm.] einschmeicheln
curse [kɜ:s] 1. *n.* Fluch, *der.* 2. *v. t.* verfluchen. 3. *v. i.* fluchen
cursory ['kɜ:sərɪ] *adj.* flüchtig
curt [kɜ:t] *adj.* kurz angebunden; kurz und schroff ⟨*Brief*⟩
curtain ['kɜ:tən] *n.* Vorhang, *der;* **draw** *or* **pull the** ~**s** *(open)* die Vorhänge aufziehen; *(close)* die Vorhänge zuziehen
curtsy ['kɜ:tsɪ] 1. *n.* Knicks, *der.* 2. *v. i.* einen Knicks machen (**to** vor + *Dat.)*
curve [kɜ:v] 1. *v. t.* krümmen. 2. *v. i.* ⟨*Straße, Fluß:*⟩ eine Biegung machen. 3. *n.* Kurve, *die*
cushion ['kʊʃn] 1. *n.* Kissen, *das.* 2. *v. t.* dämpfen ⟨*Aufprall, Stoß*⟩
cushy ['kʊʃɪ] *adj. (coll.)* bequem
custard ['kʌstəd] *n.* ≈ Vanillesoße, *die*
custodian [kʌs'təʊdɪən] *n. (of museum)* Wächter, *der*/Wächterin, *die; (of valuables)* Hüter, *der*/Hüterin, *die*
custody ['kʌstədɪ] *n.* **a)** *(care)* Obhut, *die;* **b)** *(imprisonment)* [be] **in** ~: in Haft [sein]
custom ['kʌstəm] *n.* **a)** Brauch, *der;* **b)** *in pl. (duty on imports)* Zoll, *der.* '**customs officer** *n.* Zollbeamter, *der*/-beamtin, *die.* **customary** ['kʌstəmərɪ] *adj.* üblich
customer ['kʌstəmə(r)] *n.* Kunde, *der*/Kundin, *die*
cut [kʌt] 1. *v. t.,* **-tt-,** **cut a)** schneiden; durchschneiden ⟨*Seil*⟩; ~ **one's leg** sich *(Dat. od. Akk.)* ins Bein schneiden; **b)** abschneiden ⟨*Scheibe*⟩; schneiden ⟨*Hecke*⟩; mähen ⟨*Getreide, Gras*⟩; ~ **one's nails** sich *(Dat.)* die Nägel schneiden; **c)** *(reduce)* senken ⟨*Preise*⟩; kürzen ⟨*Lohn*⟩; **d)** ~ **sth. short** *(interrupt)* etw. abbrechen. 2. *v. i.,* **-tt-,** **cut a)** ⟨*Messer:*⟩ schneiden; **b)** ~ **through** *or* **across the field/park** [quer] über das Feld/durch den Park gehen. 3. *n.* **a)** *(act of cutting)* Schnitt, *der;* **b)** *(stroke, blow) (with knife)* Schnitt, *der; (with sword, whip)* Hieb, *der;* **c)** *(reduction)* Kürzung, *die; (in prices)* Senkung, *die; (in services)* Verringerung, *die;* **d)** *(of meat)* Stück, *das.* **cut a'way** *v. t.* abschneiden. **cut 'back** *v. t.* **a)** *(reduce)* einschränken; **b)** *(prune)* stutzen. **cut 'down** 1. *v. t.* **a)** fällen ⟨*Baum*⟩; **b)** *(reduce)* einschränken. 2. *v. i.* ~ **down on sth.** etw. einschränken. **cut 'off** *v. t.* abschnei-

den; unterbrechen ⟨*Telefongespräch, Sprecher*⟩. **cut 'out 1.** *v. t.* **a)** ausschneiden *(of aus)*; **b)** **be** ~ **out for** geeignet sein zu. 2. *v. i.* ⟨*Motor:*⟩ aussetzen. **cut 'up** *v. t.* zerschneiden
cutlery ['kʌtlərɪ] *n.* Besteck, *das*
cutlet ['kʌtlɪt] *n.* Kotelett, *das*
'**cut-price** *adj.* herabgesetzt
'**cutting 1.** *adj.* beißend ⟨*Bemerkung, Antwort*⟩. 2. *n. (from newspaper)* Ausschnitt, *der*
c. v. *abbr.* **curriculum vitae**
cycle ['saɪkl] 1. *n.* **a)** *(recurrent period)* Zyklus, *der;* **b)** *(bicycle)* Rad, *das.* 2. *v. i.* radfahren. **cyclist** ['saɪklɪst] *n.* Radfahrer, *der*/-fahrerin, *die*
cylinder ['sɪlɪndə(r)] *n.* Zylinder, *der.* **cylindrical** [sɪ'lɪndrɪkl] *adj.* zylindrisch
cymbals ['sɪmblz] *n. pl.* Becken *Pl.*
cynic ['sɪnɪk] *n.* Zyniker, *der.* **cynical** ['sɪnɪkl] *adj.* zynisch; bissig ⟨*Bemerkung, Worte*⟩. **cynicism** ['sɪnɪsɪzm] *n.* Zynismus, *der*
Cyprus ['saɪprəs] *pr. n.* Zypern *(das)*
Czech [tʃek] 1. *adj.* tschechisch. 2. *n.* **a)** *(language)* Tschechisch, *das;* **b)** *(person)* Tscheche, *der*/Tschechin, *die*
Czechoslovakia [tʃekəʊslə'vækɪə] *pr. n. (Hist.)* die Tschechoslowakei. **Czechoslovakian** [tʃekəʊslə'vækɪən] *(Hist.)* 1. *adj.* tschechoslowakisch. 2. *n.* Tschechoslowake, *der*/Tschechoslowakin, *die*
Czech Republic *pr. n.* Tschechische Republik; Tschechien *(das)*

D

D, d [di:] *n.* D, d, *das*
dab [dæb] 1. *n.* Tupfer, *der.* 2. *v. t.,* **-bb-** abtupfen; ~ **sth. on** *or* **against sth.** etw. auf etw. *(Akk.)* tupfen
dabble ['dæbl] *v. i.* ~ **in sth.** sich in etw. *(Dat.)* versuchen
dachshund ['dækshʊnd] *n.* Dackel, *der*
dad [dæd] *n. (coll.)* Vater, *der*
daddy ['dædɪ] *n. (coll.)* Vati, *der (fam.).*
daddy-'long-legs *n.* Schnake, *die*
daffodil ['dæfədɪl] *n.* Osterglocke, *die*

daft [dɑːft] *adj.* doof *(ugs.)*

dagger ['dægə(r)] *n.* Dolch, *der*

daily ['deɪlɪ] **1.** *adj.* täglich; ~ |news|-**paper** Tageszeitung, *die.* **2.** *adv.* täglich. **3.** *n.* Tageszeitung, *die*

dainty ['deɪntɪ] *adj.* zierlich; anmutig ⟨*Bewegung, Person*⟩; zart ⟨*Gesichts-züge*⟩

dairy ['deərɪ] *n.* **a)** Molkerei, *die;* **b)** *(shop)* Milchladen, *der*

dais ['deɪɪs] *n.* Podium, *das*

daisy ['deɪzɪ] *n.* Gänseblümchen, *das*

dam [dæm] **1.** *n.* [Stau]damm, *der.* **2.** *v.t.,* **-mm-:** ~ |up| sth. etw. ab-blocken; **b)** aufstauen ⟨*Fluß*⟩

damage ['dæmɪdʒ] **1.** *n.* Schaden, *der.* **2.** *v.t.* beschädigen. **damaging** ['dæm-ɪdʒɪŋ] *adj.* schädlich (**to** für)

damn [dæm] **1.** *v.t.* verdammen. **2.** *adj., adv., int. (coll.)* verdammt *(ugs.).* **3.** *n.* he doesn't give *or* care a ~: ihm ist es völlig wurscht *(ugs.)*

damp [dæmp] **1.** *adj.* feucht. **2.** *v.t.* see **dampen. 3.** *n.* Feuchtigkeit, *die*

dampen ['dæmpn] *v.t.* befeuchten; *(fig.)* dämpfen ⟨*Begeisterung, Eifer*⟩

dampness *n.* Feuchtigkeit, *die*

dance [dɑːns] **1.** *v.i. & t.* tanzen. **2.** *n.* **a)** Tanz, *der;* **b)** *(party)* Tanzveranstal-tung, *die;* (private) Tanzparty, *die.* **'dance-hall** *n.* Tanzsaal, *der*

'dancer *n.* Tänzer, *der*/Tänzerin, *die*

dandelion ['dændɪlaɪən] *n.* Löwen-zahn, *der*

dandruff ['dændrʌf] *n.* [Kopf]schup-pen *Pl.*

Dane [deɪn] *n.* Däne, *der*/Dänin, *die*

danger ['deɪndʒə(r)] *n.* Gefahr, *die;* in/out of ~: in/außer Gefahr. **dan-gerous** ['deɪndʒərəs] *adj.,* **'danger-ously** *adv.* gefährlich

dangle ['dæŋgl] **1.** *v.i.* baumeln (**from** an + *Dat.*). **2.** *v.t.* baumeln lassen

Danish ['deɪnɪʃ] **1.** *adj.* dänisch; **sb. is** ~: jmd. ist Däne/Dänin. **2.** *n.* Dä-nisch, *das; see also* **English 2 a**

dank [dæŋk] *adj.* feucht

Danube ['dænjuːb] *pr. n.* Donau, *die*

dare [deə(r)] **1.** *v.t.* **a)** [es] wagen; ~ **to do sth.** [es] wagen, etw. zu tun; **b)** *(challenge)* ~ **sb. to do sth.** jmdn. auf-stacheln, etw. zu tun; **I ~ you!** trau dich! **2.** *n.* Mutprobe, *die.* **daring** ['deərɪŋ] *adj.* kühn

dark [dɑːk] **1.** *adj.* dunkel; *(dark-haired)* dunkelhaarig; **~-blue/-brown** dunkelblau/-braun; ~ **glasses** dunkle Brille. **2.** *n.* **a)** Dunkel, *das;* **in the** ~: im Dunkeln; **keep sb. in the** ~ *(fig.)*

jmdn. im dunkeln lassen; **b)** *no art. (nightfall)* Einbruch der Dunkelheit.

darken ['dɑːkn] *v.t.* verdunkeln.

'darkness *n.* Dunkelheit, *die*

'dark-room *n.* Dunkelkammer, *die*

darling ['dɑːlɪŋ] *n.* Liebling, *der*

darn [dɑːn] *v.t.* stopfen

dart [dɑːt] **1.** *n.* **a)** *(missile)* Pfeil, *der;* **b)** *(Sport)* Wurfpfeil, *der;* ~**s** *sing. (game)* Darts, *das.* **2.** *v.i.* sausen. **'dartboard** *n.* Dartscheibe, *die*

dash [dæʃ] **1.** *v.i.* sausen. **2.** *v.t. (fling)* schleudern. **3.** *n.* **a)** **make a** ~: rasen *(ugs.)* (**for** zu); **b)** *(horizontal stroke)* Gedankenstrich, *der;* **c)** *(small amount)* Schuß, *der*

'dashboard *n.* Armaturenbrett, *das*

data ['deɪtə, 'dɑːtə] *n.* Daten *Pl.* **data 'processing** *n.* Datenverarbeitung, *die*

¹date [deɪt] *n.* *(Bot.)* Dattel, *die*

²date 1. *n.* **a)** Datum, *das; (on coin etc.)* Jahreszahl, *die;* ~ **of birth** Geburtsda-tum, *das;* **be out of** ~: altmodisch sein; **to** ~: bis heute; **b)** *(coll.: appoint-ment)* Verabredung, *die;* **have/make a** ~ **with sb.** mit jmdm. verabredet sein/sich mit jmdm. verabreden. **2.** *v.t.* **a)** datieren; **b)** *(coll.: make seem old)* alt machen. **3.** *v.i.* ~ **back to**/~ **from** stammen aus. **dated** ['deɪtɪd] *adj.* alt-modisch. **'date-line** *n.* Datumsgren-ze, *die*

dative ['deɪtɪv] *adj. & n.* |case| Dativ, *der*

daub [dɔːb] *v.t. (smear)* beschmieren; *(put crudely)* schmieren

daughter ['dɔːtə(r)] *n.* Tochter, *die.* **'daughter-in-law** *n., pl.* **daughters-in-law** Schwiegertochter, *die*

daunt [dɔːnt] *v.t.* entmutigen

dawdle ['dɔːdl] *v.i.* bummeln *(ugs.)*

dawn [dɔːn] **1.** *v.i.* dämmern; **sth.** ~**s** |up|on **sb.** etw. dämmert jmdm. **2.** *n.* [Morgen]dämmerung, *die;* **at** ~: im Morgengrauen

day [deɪ] *n.* Tag, *der;* **all** ~ |long| den ganzen Tag [lang]; **for two** ~**s** zwei Ta-ge [lang]; **the** ~ **before yesterday/after tomorrow** vorgestern/übermorgen; ~ **after** ~: Tag für Tag; ~ **in** ~ **out** tag-aus, tagein; **in the** ~**s when** ...: zu der Zeit, als ...; **these** ~**s** heutzutage; **in those** ~**s** damals

day: ~**break** *n.* Tagesanbruch, *der;* ~**-dream 1.** *n.* Tagtraum, *der;* **2.** *v.i.* träumen; ~**-light** *n.* Tageslicht, *das;* **in broad** ~**light** am hellichten Tag[e]; ~**-re'turn** *n.* Tagesrückfahrkarte, *die;* ~**time** *n.* Tag, *der;* ~**-to-**~ *adj.* [tag]täglich; ~ **trip** *n.* Tagesausflug, *der*

daze ['deɪz] *v. t.* benommen machen.

dazed ['deɪzd] *adj.* benommen

dazzle ['dæzl] *v. t.* blenden

DC *abbr.* direct current GS

dead [ded] **1.** *adj.* **a)** tot; **b)** plötzlich ⟨*Halt*⟩; genau ⟨*Mitte*⟩; **c)** *(numb)* taub. **2.** *adv.* völlig; ~ **straight** schnurgerade; ~ **easy/slow** kinderleicht/ganz langsam; ~ **on time** auf die Minute; ~ **tired** todmüde

deaden ['dedn] *v. t.* dämpfen; betäuben ⟨*Schmerz*⟩

dead: ~ '**end** *n.* Sackgasse, *die;* ~ '**heat** *n.* totes Rennen; ~**line** *n.* [letzter] Termin; ~**lock** *n.* völliger Stillstand

deadly ['dedlɪ] *adj.* tödlich; *(fig. coll.: boring)* todlangweilig

Dead 'Sea *pr. n.* Tote Meer, *das*

deaf [def] *adj.* taub; ~ **and dumb** taubstumm. **deafen** ['defn] *v. t.* ~ **sb.** bei jmdm. zur Taubheit führen; **I was** ~**ed by the noise** *(fig.)* ich war von dem Lärm wie betäubt. '**deafening** *adj.* ohrenbetäubend. '**deafness** *n.* Taubheit, *die*

¹**deal** [diːl] **1.** *v. t.,* dealt [delt] **a)** *(Cards)* austeilen; **b)** ~ **sb. a blow** jmdm. einen Schlag versetzen. **2.** *v. i.,* dealt **a)** *(do business)* ~ **in sth.** mit etw. handeln; **b)** ~ **with sth.** *(occupy oneself)* sich mit etw. befassen; *(manage)* mit etw. fertig werden; *(be about)* von etw. handeln; ~ **with sb.** mit jmdm. fertig werden. **3.** *n. (coll.: arrangement)* Geschäft, *das.* **deal 'out** *v. t.* verteilen

²**deal** *n.* **a great** *or* **good** ~: viel; *(often)* ziemlich viel; **a great** *or* **good** ~ **of** viel

'**dealer** *n.* **a)** Händler, *der;* **b)** *(Cards)* Geber, *der;* **he's the** ~: er gibt

'**dealings** *n. pl.* **have** ~ **with sb.** mit jmdm. zu tun haben

dealt *see* ¹**deal 1, 2**

dean [diːn] *n. (Eccl.)* Dechant, *der*

dear [dɪə(r)] **1.** *adj.* **a)** lieb; **sb./sth. is** ~ **to sb.[/s heart]** jmd. liebt jmdn./etw.; *(beginning letter)* **D**~ **Sir/Madam** Sehr geehrter Herr/Sehr verehrte gnädige Frau; **D**~ **Mr Jones/Mrs Jones** Sehr geehrter Herr Jones/Sehr verehrte Frau Jones; **D**~ **Malcolm/Emily** Lieber Malcolm/Liebe Emily; **b)** *(expensive)* teuer. **2.** *int.* ~, ~!, ~ **me!**, **oh** ~! [ach] du liebe *od.* meine Güte!

'**dearly** *adv.* **a)** von ganzem Herzen; **b)** *(at high price)* teuer

dearth [dɜːθ] *n.* Mangel, *der* (of an + *Dat.*)

death [deθ] *n.* **a)** Tod, *der;* ... **to** ~: zu Tode ...; **bleed to** ~: verbluten; **b)** *(instance)* Todesfall, *der*

death: ~ **penalty** *n.* Todesstrafe, *die;* ~ **sentence** *n.* Todesurteil, *das;* ~**-trap** *n.* lebensgefährliche Sache

debatable [dɪ'beɪtəbl] *adj. (questionable)* fraglich

debate [dɪ'beɪt] *n.* Debatte, *die*

debit ['debɪt] **1.** *n.* Soll, *das.* **2.** *v. t.* belasten ⟨*Konto*⟩

debris ['debriː] *n.* Trümmer *Pl.*

debt [det] *n.* Schuld, *die;* **be in** ~: Schulden haben; **get into** ~: in Schulden geraten. **debtor** ['detə(r)] *n.* Schuldner, *der*/Schuldnerin, *die*

début *(Amer.:* **debut)** ['deɪbuː, 'deɪbjuː] *n.* Debüt, *das*

Dec. *abbr.* December Dez.

decade ['dekeɪd] *n.* Jahrzehnt, *das*

decadent ['dekədənt] *adj.* dekadent

decanter [dɪ'kæntə(r)] *n.* Karaffe, *die*

decay [dɪ'keɪ] **1.** *v. i.* verrotten; ⟨*Gebäude:*⟩ zerfallen; ⟨*Zahn:*⟩ faul werden. **2.** *n.* Verrotten, *das;* *(of building)* Zerfall, *der;* *(of tooth)* Fäule, *die*

deceased [dɪ'siːst] **1.** *adj.* verstorben. **2.** *n.* Verstorbene, *der/die*

deceit [dɪ'siːt] *n.* Täuschung, *die.* **deceitful** [dɪ'siːtfl] *adj.* falsch ⟨*Person, Art*⟩; hinterlistig ⟨*Trick*⟩

deceive [dɪ'siːv] *v. t.* täuschen; *(be unfaithful to)* betrügen

December [dɪ'sembə(r)] *n.* Dezember, *der; see also* **August**

decency ['diːsənsɪ] *n.* Anstand, *der*

decent ['diːsənt] *adj.* anständig

deception [dɪ'sepʃn] *n.* Betrug, *der;* *(being deceived)* Täuschung, *die.* **deceptive** [dɪ'septɪv] *adj.* trügerisch

decibel ['desɪbel] *n.* Dezibel, *das*

decide [dɪ'saɪd] **1.** *v. t.* **a)** *(settle, judge)* entscheiden über (+ *Akk.*); **b)** *(resolve)* ~ **that** ...: beschließen, daß ...; ~ **to do sth.** sich entschließen, etw. zu tun. **2.** *v. i.* sich entscheiden (**in favour of** zugunsten von, **against** gegen). **de'cided** *adj.,* **de'cidedly** *adv.* entschieden

deciduous [dɪ'sɪdjʊəs] *adj.* ~ **tree** ≈ Laubbaum, *der*

decimal ['desɪml] **1.** *n.* Dezimalbruch, *der.* **2.** *adj.* Dezimal-; ~ '**point** Komma, *das*

decimate ['desɪmeɪt] *v. t.* dezimieren

decipher [dɪ'saɪfə(r)] *v. t.* entziffern

decision [dɪ'sɪʒn] *n.* Entscheidung,

die. **decisive** [dɪ'saɪsɪv] *adj.* entscheidend

deck [dek] *n.* Deck, *das;* on ~: an Deck; **below ~|s|** unter Deck. **'deck-chair** *n.* Liegestuhl, *der*

declaration [deklə'reɪʃn] *n.* Erklärung, *die*

declare [dɪ'kleə(r)] *v. t.* erklären; kundtun *(geh.)* ⟨*Wunsch, Absicht*⟩; ~ **sth./sb. |to be|** sth. etw./jmdn. für etw. erklären

declension [dɪ'klenʃn] *n.* Deklination, *die*

decline [dɪ'klaɪn] 1. *v. i.* nachlassen; ⟨*Anzahl:*⟩ sinken. 2. *v. t.* **a)** ablehnen; **b)** *(Ling.)* deklinieren. 3. *n. see* 1: Nachlassen, *das*/Sinken, *das* (in *Gen.*); **be on the ~**: nachlassen/sinken

decode [diː'kəʊd] *v. t.* entziffern

decompose [diːkəm'pəʊz] *v. i.* sich zersetzen

décor ['deɪkɔː(r)] *n.* Ausstattung, *die*

decorate ['dekəreɪt] *v. t.* **a)** schmücken ⟨*Raum, Straße, Baum*⟩; verzieren ⟨*Kuchen, Kleid*⟩; *(paint)* streichen; *(wallpaper)* tapezieren; **b)** *(award medal etc. to)* auszeichnen.

decoration [dekə'reɪʃn] *n.* **a)** Schmücken, *das; (with paint)* Streichen, *das; (with wallpaper)* Tapezieren, *das; (of cake, dress)* Verzieren, *das;* **b)** *(adornment)* Schmuck, *der;* **c)** *(medal etc.)* Auszeichnung, *die.* **decorative** ['dekərətɪv] *adj.* dekorativ.

decorator ['dekəreɪtə(r)] *n.* Maler, *der; (paper-hanger)* Tapezierer, *der*

decorum [dɪ'kɔːrəm] *n.* Schicklichkeit, *die (geh.)*

decoy ['diːkɔɪ] *n.* Lockvogel, *der*

decrease 1. [dɪ'kriːs] *v. i.* abnehmen; ⟨*Stärke:*⟩ nachlassen. 2. [dɪ'kriːs] *v. t.* [ver]mindern ⟨*Wert, Lärm*⟩; schmälern ⟨*Popularität, Macht*⟩. 3. ['diːkriːs] *n.* Rückgang, *der; (in weight)* Abnahme, *die; (in strength)* Nachlassen, *das; (in value, noise)* Minderung, *die*

decree [dɪ'kriː] 1. *n.* Dekret, *das;* Erlaß, *der.* 2. *v. t.* verfügen

decrepit [dɪ'krepɪt] *adj.* altersschwach; *(dilapidated)* heruntergekommen

dedicate ['dedɪkeɪt] *v. t.* ~ **sth. to sb.** jmdm. etw. widmen. **'dedicated** *adj.* **a)** *(devoted)* **be ~ to sth./sb.** nur für etw./jmdn. leben; **b)** *(to vocation)* hingebungsvoll; **a ~ teacher** ein Lehrer mit Leib und Seele. **dedication** [dedɪ'keɪʃn] *n.* **a)** Widmung, *die* (to *Dat.*); **b)** *(devotion)* Hingabe, *die*

deduce [dɪ'djuːs] *v. t.* ~ **sth. |from sth.|** etw. [aus etw.] schließen

deduct [dɪ'dʌkt] *v. t.* ~ **sth. |from sth.|** etw. [von etw.] abziehen. **deduction** [dɪ'dʌkʃn] *n.* **a)** *(deducting)* Abzug, *der;* **b)** *(deducing, thing deduced)* Ableitung, *die;* **c)** *(amount)* Abzüge *Pl.*

deed [diːd] *n.* **a)** Tat, *die;* **b)** *(Law)* Urkunde, *die*

deem [diːm] *v. t.* erachten für

deep [diːp] 1. *adj. (lit. or fig.)* tief; tiefgründig ⟨*Bemerkung*⟩; **water ten feet ~**: drei Meter tiefes Wasser; **take a ~ breath** tief Atem holen; **be ~ in thought** in Gedanken versunken sein. 2. *adv.* tief. **'deepen** 1. *v. t.* vertiefen. 2. *v. i.* sich vertiefen. **deep-'freeze** *v. t.* tiefgefrieren. **'deeply** *adv. (lit. or fig.)* tief; äußerst ⟨*interessiert, dankbar*⟩

deer [dɪə(r)] *n., pl. same* Hirsch, *der; (roe~)* Reh, *das*

deface [dɪ'feɪs] *v. t.* verunstalten

defamation [defə'meɪʃn] *n.* Diffamierung, *die.* **defamatory** [dɪ'fæmətərɪ] *adj.* diffamierend

default [dɪ'fɔːlt, dɪ'fɒlt] 1. *n.* **lose/go by ~**: durch Abwesenheit verlieren/nicht zur Geltung kommen; **win by ~**: durch Nichterscheinen des Gegners gewinnen. 2. *v. i.* ~ **on one's payments/debts** seinen Zahlungsverpflichtungen nicht nachkommen

defeat [dɪ'fiːt] 1. *v. t.* besiegen. 2. *n. (being ~ed)* Niederlage, *die; (~ing)* Sieg, *der (of* über + *Akk.*). **de'featist** *adj.* defätistisch

defect 1. ['diːfekt] *n.* **a)** *(lack)* Mangel, *der;* **b)** *(shortcoming)* Fehler, *der.* 2. [dɪ'fekt] *v. i.* überlaufen (**to** zu). **defection** [dɪ'fekʃn] *n.* Flucht, *die.* **defective** [dɪ'fektɪv] *adj.* defekt ⟨*Maschine*⟩; fehlerhaft ⟨*Material, Arbeiten, Methode*⟩. **defector** [dɪ'fektə(r)] *n.* Überläufer, *der*/-läuferin, *die*

defence [dɪ'fens] *n. (Brit.)* Verteidigung, *die; (means of ~)* Schutz, *der.* **de'fenceless** *adj.* wehrlos

defend [dɪ'fend] *v. t.* verteidigen. **defendant** [dɪ'fendənt] *n. (Law) (accused)* Angeklagte, *der/die; (sued)* Beklagte, *der/die.* **de'fender** *n.* Verteidiger, *der*

defense etc. *(Amer.) see* **defence** etc.

defensive [dɪ'fensɪv] 1. *adj.* defensiv. 2. *n.* **be on the ~**: in der Defensive sein

¹defer [dɪ'fɜː(r)] *v. t.,* **-rr-** aufschieben

²defer *v. i.,* **-rr-**: ~ **|to sb.|** sich |jmdm.| beugen. **deference** ['defərəns] *n.* Respekt, *der;* in ~ to sb./sth. aus Achtung vor jmdm./etw. **deferential** [defə'renʃl] *adj.* respektvoll

defiance [dɪ'faɪəns] *n.* Trotz, *der;* in ~ of sb./sth. jmdm./einer Sache zum Trotz

defiant [dɪ'faɪənt] *adj.*, **de'fiantly** *adv.* trotzig

deficiency [dɪ'fɪʃənsɪ] *n.* Mangel, *der*

deficient [dɪ'fɪʃənt] *adj.* unzulänglich; sb./sth. is ~ in sth. jmdm./einer Sache mangelt es an etw. *(Dat.)*

deficit ['defɪsɪt] *n.* Defizit, *das* (of an + *Dat.*)

defile [dɪ'faɪl] *v. t.* verpesten ⟨*Luft*⟩; beflecken ⟨*Reinheit, Unschuld*⟩

define [dɪ'faɪn] *v. t.* definieren

definite ['defɪnɪt] *adj.* bestimmt; eindeutig ⟨*Antwort, Entscheidung, Beschluß, Verbesserung*⟩; klar umrissen ⟨*Ziel, Plan*⟩; klar ⟨*Vorstellung*⟩; genau ⟨*Zeitpunkt*⟩. **'definitely 1.** *adv.* bestimmt; eindeutig ⟨*festlegen, größer sein, verbessert*⟩; endgültig ⟨*entscheiden*⟩. **2.** *int. (coll.)* na, klar *(ugs.)*

definition [defɪ'nɪʃn] *n.* Definition, *die; (Telev., Phot.)* Schärfe, *die*

definitive [dɪ'fɪnɪtɪv] *adj.* endgültig ⟨*Beschluß, Antwort, Urteil*⟩; *(authoritative)* maßgeblich

deflate [dɪ'fleɪt] *v. t.* die Luft ablassen aus; *(fig.)* ernüchtern. **deflation** [dɪ-'fleɪʃn] *n. (Econ.)* Deflation, *die*

deflect [dɪ'flekt] *v. t.* brechen ⟨*Licht*⟩; ~ sb./sth. |from sb./sth.| jmdn./etw. |von jmdm./einer Sache| ablenken

deform [dɪ'fɔːm] *v. t.* deformieren. **deformed** [dɪ'fɔːmd] *adj.* entstellt ⟨*Gesicht*⟩; verunstaltet ⟨*Person, Körperteil*⟩. **deformity** [dɪ'fɔːmɪtɪ] *n. (malformation)* Verunstaltung, *die*

defraud [dɪ'frɔːd] *v. t.* ~ sb. |of sth.| jmdn. [um etw.] betrügen

defray [dɪ'freɪ] *v. t.* bestreiten

defrost [diː'frɒst] *v. t.* auftauen ⟨*Speisen*⟩; abtauen ⟨*Kühlschrank*⟩

deft [deft] *adj.*, **'deftly** *adv.* sicher und geschickt

defunct [dɪ'fʌŋkt] *adj.* defekt ⟨*Maschine*⟩; veraltet ⟨*Gesetz*⟩

defuse [diː'fjuːz] *v. t.* entschärfen

defy [dɪ'faɪ] *v. t.* **a)** *(resist openly)* ~ sb. jmdm. trotzen; **b)** *(refuse to obey)* ~ sb./sth. sich jmdm./einer Sache widersetzen

degenerate [dɪ'dʒenəreɪt] *v. i.* ~ |into sth.| [zu etw.] verkommen

degradation [degrə'deɪʃn] *n.* Erniedrigung, *die*

degrade [dɪ'greɪd] *v. t.* erniedrigen

degree [dɪ'griː] *n.* **a)** Grad, *der;* 20 ~s 20 Grad; **b)** *(academic rank)* [akademischer] Grad

de-ice [diː'aɪs] *v. t.* enteisen

deign [deɪn] *v. t.* ~ to do sth. sich [dazu] herablassen, etw. zu tun

deity ['diːɪtɪ] *n.* Gottheit, *die*

dejected [dɪ'dʒektɪd] *adj.* niedergeschlagen. **dejection** [dɪ'dʒekʃn] *n.* Niedergeschlagenheit, *die*

delay [dɪ'leɪ] **1.** *v. t. (make late)* aufhalten; verzögern ⟨*Ankunft, Abfahrt*⟩; the train has been ~ed der Zug hat Verspätung. **2.** *v. i.* warten. **3.** *n.* **a)** Verzögerung, *die* (to bei); **b)** *(Transport)* Verspätung, *die*

delectable [dɪ'lektəbl] *adj.* köstlich

delegate 1. ['delɪɡət] *n.* Delegierte, *der/die*. **2.** ['delɪɡeɪt] *v. t.* delegieren (to an + *Akk.*). **delegation** [delɪ'ɡeɪʃn] *n.* Delegation, *die*

delete [dɪ'liːt] *v. t.* streichen (from in + *Dat.*); *(Computing)* löschen. **deletion** [dɪ'liːʃn] *n.* Streichung, *die; (Computing)* Löschung, *die*

deliberate [dɪ'lɪbərət] *adj.* **a)** *(intentional)* absichtlich; bewußt ⟨*Lüge, Irreführung*⟩; **b)** *(fully considered)* wohlüberlegt. **de'liberately** *adv.* absichtlich. **deliberation** [dɪlɪbə'reɪʃn] *n.* Überlegung, *die; (discussion)* Beratung, *die*

delicacy ['delɪkəsɪ] *n.* **a)** *(tactfulness and care)* Feingefühl, *das;* **b)** *(food)* Delikatesse, *die*

delicate ['delɪkət] *adj.* zart; *(requiring careful handling)* empfindlich; delikat ⟨*Frage, Angelegenheit*⟩

delicatessen [delɪkə'tesən] *n.* Feinkostgeschäft, *das*

delicious [dɪ'lɪʃəs] *adj.* köstlich

delight [dɪ'laɪt] **1.** *v. t.* erfreuen. **2.** *v. i.* sb. ~s in doing sth. es macht jmdm. Freude, etw. zu tun. **3.** *n.* Freude, *die* (at über + *Akk.;* in an + *Dat.*). **de'lighted** *adj.* be ~ ⟨*Person:*⟩ hocherfreut sein; be ~ by or with sth. sich über etw. *(Akk.)* freuen. **delightful** [dɪ'laɪtfl] *adj.* wunderbar; köstlich ⟨*Geschmack*⟩; reizend ⟨*Person, Landschaft*⟩. **de'lightfully** *adv.* wunderbar

delinquent [dɪ'lɪŋkwənt] **1.** *n.* Randalierer, *der*. **2.** *adj.* kriminell

delirious [dɪ'lɪrɪəs] *adj.* be ~: im Delirium sein; be ~ [with sth.] *(fig.)* außer sich [vor etw. *(Dat.)*] sein

delirium [dɪ'lɪrɪəm] *n.* Delirium, *das*

93

deplorable

deliver [dɪ'lɪvə(r)] v. t. **a)** bringen; liefern ⟨Ware⟩; zustellen ⟨Post, Telegramm⟩; überbringen ⟨Botschaft⟩; **b)** halten ⟨Rede⟩. **delivery** [dɪ'lɪvərɪ] n. Lieferung, die; (of letters, parcels) Zustellung, die. **delivery van** n. Lieferwagen, der

delta ['deltə] n. Delta, das

delude [dɪ'lju:d] v. t. täuschen

deluge ['delju:dʒ] 1. n. sintflutartiger Regen. 2. v. t. überschwemmen

delusion [dɪ'lju:ʒn] n. Illusion, die

de luxe [də'lʌks] adj. Luxus-

demand [dɪ'mɑ:nd] 1. n. Forderung, die (for nach); (for commodity) Nachfrage, die; sth./sb. is in ~: etw. ist gefragt/jmd. ist begehrt. 2. v. t. verlangen (of, from von); fordern ⟨Recht⟩. **de'manding** adj. anspruchsvoll

demented [dɪ'mentɪd] adj. wahnsinnig

de'mobilize v. t. demobilisieren ⟨Armee, Kriegsschiff⟩; aus dem Kriegsdienst entlassen ⟨Soldat⟩

democracy [dɪ'mɒkrəsɪ] n. Demokratie, die. **Democrat** ['deməkræt] n. (Amer. Polit.) Demokrat, der/Demokratin, die. **democratic** [demə'krætɪk] adj., **democratically** [demə'krætɪkəlɪ] adv. demokratisch

demolish [dɪ'mɒlɪʃ] v. t. abreißen. **demolition** [demə'lɪʃn] n. Abriß, der; ~ work Abbruchsarbeit, die

demon ['di:mən] n. Dämon, der

demonstrate ['deménstreɪt] 1. v. t. zeigen; (be proof of) zeigen; beweisen. 2. v. i. demonstrieren. **demonstration** [demən'streɪʃn] n. (also Pol. etc.) Demonstration, die; (proof) Beweis, der. **demonstrative** [dɪ'mɒnstrətɪv] adj. **a)** offen ⟨Person⟩; **b)** (Ling.) Demonstrativ-. **demonstrator** ['deménstreɪtə(r)] n. (Pol. etc.) Demonstrant, der/Demonstrantin, die

demoralize [dɪ'mɒrəlaɪz] v. t. demoralisieren

demote [di:'məʊt] v. t. degradieren (to zu). **demotion** [di:'məʊʃn] n. Degradierung, die (to zu)

demur [dɪ'mɜ:(r)] v. i., -rr- Einwände erheben

demure [dɪ'mjʊə(r)] adj. betont zurückhaltend

den [den] n. Höhle, die

denial [dɪ'naɪəl] n. (refusal) Verweigerung, die; (of request) Ablehnung, die

denim ['denɪm] n. Denim ⓦ, der; Jeansstoff, der; ~ jacket Jeansjacke, die; ~s Bluejeans Pl.

Denmark ['denmɑ:k] pr. n. Dänemark (das)

denomination [dɪnɒmɪ'neɪʃn] n. (Relig.) Konfession, die

denote [dɪ'nəʊt] v. t. bezeichnen

denounce [dɪ'naʊns] v. t. denunzieren; (accuse publicly) beschuldigen

dense [dens] adj. **a)** dicht; massiv ⟨Körper⟩; **b)** (stupid) dumm. **'densely** adv. dicht; ~ packed dichtgedrängt. **density** ['densɪtɪ] n. Dichte, die

dent [dent] 1. n. Beule, die. 2. v. t. einbeulen

dental ['dentl] adj. Zahn-. **dental floss** ['dentl flɒs] n. Zahnseide, die. **dentist** ['dentɪst] n. Zahnarzt, der/-ärztin, die. **dentistry** ['dentɪstrɪ] n. Zahnheilkunde, die

denture ['dentʃə(r)] n. ~[s] Zahnprothese, die

denunciation [dɪnʌnsɪ'eɪʃn] n. Denunziation, die; (public accusation) Beschuldigung, die

deny [dɪ'naɪ] v. t. (declare untrue) bestreiten; (refuse) ~ sb. sth. jmdm. etw. verweigern; ~ sb.'s request jmdm. seine Bitte abschlagen

deodorant [di:'əʊdərənt] 1. adj. deodorierend. 2. n. Deodorant, das

depart [dɪ'pɑ:t] v. i. **a)** (go away) weggehen; **b)** (set out, leave) abfahren; (on one's journey) abreisen; **c)** (fig.: deviate) abweichen (from von)

department [dɪ'pɑ:tmənt] n. Abteilung, die; (government ~) Ministerium, das; (of university) Seminar, das. **de'partment store** n. Kaufhaus, das

departure [dɪ'pɑ:tʃə(r)] n. **a)** Abreise, die; (of train, bus, ship) Abfahrt, die; (of aircraft) Abflug, der; **b)** (deviation) ~ from sth. Abweichen von etw. **de'parture lounge** n. Abflughalle, die

depend [dɪ'pend] v. i. **a)** ~ [up]on abhängen von; **it/that ~s** es kommt drauf an; **b)** (rely, trust) ~ [up]on sich verlassen auf (+ Akk.); (have to rely on) angewiesen sein auf (+ Akk.). **dependable** [dɪ'pendəbl] adj. zuverlässig. **dependant** [dɪ'pendənt] n. Abhängige, der/die. **dependence** [dɪ'pendəns] n. Abhängigkeit, die. **dependent** [dɪ'pendənt] 1. n. see dependant. 2. adj. abhängig

depict [dɪ'pɪkt] v. t. darstellen

deplete [dɪ'pli:t] v. t. erheblich verringern

deplorable [dɪ'plɔ:rəbl] adj. beklagenswert

deplore [dɪ'plɔ:(r)] *v. t.* **a)** *(disapprove of)* verurteilen; **b)** *(regret)* beklagen

deploy [dɪ'plɔɪ] *v. t.* einsetzen

deport [dɪ'pɔ:t] *v. t.* ausweisen. **deportation** [di:pɔ:'teɪʃn] *n.* Ausweisung, *die*

depose [dɪ'pəʊz] *v. t.* absetzen

deposit [dɪ'pɒzɪt] **1.** *n.* **a)** *(in bank)* Depot, *das; (credit)* Guthaben, *das; (Brit.: at interest)* Sparguthaben, *das;* **b)** *(first instalment)* Anzahlung, *die;* put down a ~ on sth. eine Anzahlung für etw. leisten; **c)** *(on bottle)* Pfand, *das.* **2.** *v. t.* **a)** *(lay down)* ablegen; abstellen *(etw. Senkrechtes);* **b)** *(in bank)* deponieren. **de'posit account** *n. (Brit.)* Sparkonto, *das*

depot ['depəʊ] *n.* Depot, *das*

depraved [dɪ'preɪvd] *adj.* verdorben. **depravity** [dɪ'prævɪtɪ] *n.* Verdorbenheit, *die*

depreciate [dɪ'pri:ʃɪeɪt] *v. i.* an Wert verlieren. **depreciation** [dɪpri:ʃɪ'eɪʃn] *n.* Wertverlust, *der*

depress [dɪ'pres] *v. t.* **a)** *(deject)* deprimieren; **b)** *(push down)* herunterdrücken. **depressed** [dɪ'prest] *adj.* deprimiert. **de'pressing** *adj.,* **de'pressingly** *adv.* deprimierend. **depression** [dɪ'preʃn] *n.* **a)** Depression, *die;* **b)** *(sunk place)* Vertiefung, *die;* **c)** *(Meteorol.)* Tief[druckgebiet], *das;* **d)** *(Econ.)* Wirtschaftskrise, *die*

deprivation [deprɪ'veɪʃn] *n.* Entbehrung, *die*

deprive [dɪ'praɪv] *v. t.* ~ sb. of sth. jmdm. etw. nehmen; *(prevent from having)* jmdm. etw. vorenthalten. **deprived** [dɪ'praɪvd] *adj.* benachteiligt ⟨Kind, Familie usw.⟩

depth [depθ] *n.* **a)** Tiefe, *die;* in ~: gründlich; in the ~s of winter im tiefsten Winter. **'depth-charge** *n.* Wasserbombe, *die*

deputation [depjʊ'teɪʃn] *n.* Abordnung, *die*

deputize ['depjʊtaɪz] *v. i.* ~ for sb. jmdn. vertreten

deputy ['depjʊtɪ] *n.* [Stell]vertreter, *der/*-vertreterin, *die; attrib.* stellvertretend

derail [dɪ'reɪl] *v. t.* be ~ed entgleisen. **de'railment** *n.* Entgleisung, *die*

deranged [dɪ'reɪndʒd] *adj.* [mentally] ~: geistesgestört

derelict ['derɪlɪkt] **1.** *adj.* verlassen und verfallen. **2.** *n.* Ausgestoßene, *der/die*

deride [dɪ'raɪd] *v. t.* sich lustig machen

über (+ *Akk.*). **derision** [dɪ'rɪʒn] *n.* Spott, *der.* **derisive** [dɪ'raɪsɪv] *adj. (ironical)* spöttisch; *(scoffing)* verächtlich. **derisory** [dɪ'raɪzərɪ] *adj. (ridiculously inadequate)* lächerlich

derivation [derɪ'veɪʃn] *n.* Ableitung, *die*

derivative [dɪ'rɪvətɪv] **1.** *adj.* abgeleitet; *(lacking originality)* nachahmend. **2.** *n.* Ableitung, *die*

derive [dɪ'raɪv] **1.** *v. t.* ~ sth. from sth. etw. aus etw. gewinnen; ~ pleasure from sth. Freude an etw. *(Dat.)* haben. **2.** *v. i.* ~ from beruhen auf (+ *Dat.*)

derogatory [dɪ'rɒgətərɪ] *adj.* abfällig

derrick ['derɪk] *n.* [Derrick]kran, *der*

derv [dɜ:v] *n.* Diesel[kraftstoff], *der*

descend [dɪ'send] **1.** *v. i.* **a)** *(go down)* hinuntergehen / -steigen / -klettern / -fahren; *(come down)* herunterkommen; ⟨Fallschirm, Flugzeug:⟩ niedergehen; **b)** *(slope downwards)* abfallen; **c)** ~ on sb. jmdn. überfallen. **2.** *v. t. (go/come down)* hinunter- / heruntergehen / -steigen / -klettern / -fahren. **descendant** [dɪ'sendənt] *n.* Nachkomme, *der.* **de'scended** *adj.* be ~ from sb. von jmdm. abstammen. **descent** [dɪ'sent] *n.* **a)** Abstieg, *der; (of parachute, plane)* Niedergehen, *das;* **b)** *(lineage)* Herkunft, *die*

describe [dɪ'skraɪb] *v. t.* beschreiben. **description** [dɪ'skrɪpʃn] *n.* **a)** Beschreibung, *die;* **b)** *(sort, class)* Art, *die.* **descriptive** [dɪ'skrɪptɪv] *adj.* beschreibend; *(vivid)* anschaulich; a purely ~ report ein reiner Tatsachenbericht

desecrate ['desɪkreɪt] *v. t.* entweihen

¹desert ['dezət] *n.* Wüste, *die*

²desert [dɪ'zɜ:t] **1.** *v. t.* verlassen. **2.** *v. i.* ⟨Soldat:⟩ desertieren. **de'serted** *adj.* verlassen. **de'serter** *n.* Deserteur, *der.* **desertion** [dɪ'zɜ:ʃn] *n.* Desertion, *die*

desert 'island [dezət 'aɪlənd] *n.* einsame Insel

deserts [dɪ'zɜ:ts] *n. pl.* get one's [just] ~: das bekommen, was man verdient hat

deserve [dɪ'zɜ:v] *v. t.* verdienen. **deserving** [dɪ'zɜ:vɪŋ] *adj.* verdienstvoll; a ~ cause ein guter Zweck

design [dɪ'zaɪn] **1.** *n.* Entwurf, *der; (pattern)* Muster, *das; (established form of machine, engine, etc.)* Bauweise, *die; (general idea, construction)* Konstruktion, *die.* **2.** *v. t.* entwerfen; be ~ed to do sth. etw. tun sollen

designate ['dezɪgneɪt] *v. t.* **a)** bezeichnen; **b)** *(appoint)* designieren *(geh.).* **designation** [dezɪg'neɪʃn] *n.* Bezeichnung, *die*
'**designer** *n.* Designer, *der/* Designerin, *die; (of machines)* Konstrukteur, *der/*Konstrukteurin, *die; attrib.* Modell*⟨-kleidung, -jeans⟩*
desirability [dɪzaɪərə'bɪlɪtɪ] *n.* Wunschbarkeit, *die*
desirable [dɪ'zaɪərəbl] *adj.* wünschenswert
desire [dɪ'zaɪə(r)] **1.** *n.* Wunsch, *der* (for nach); *(longing)* Sehnsucht, *die* (for nach). **2.** *v. t.* sich *(Dat.)* wünschen; *(long for)* sich sehnen nach
desist [dɪ'zɪst] *v. i. (literary)* einhalten *(geh.);* ~ **from sth.** von etw. ablassen *(geh.)*
desk [desk] *n.* **a)** Schreibtisch, *der; (in school)* Tisch, *der;* **b)** *(cash ~)* Kasse, *die; (reception ~)* Rezeption, *die*
desolate ['desələt] *adj.* trostlos. **desolation** [desə'leɪʃn] *n.* Trostlosigkeit, *die*
despair [dɪ'speə(r)] **1.** *n.* Verzweiflung, *die;* **be the ~ of sb.** jmdn. zur Verzweiflung bringen. **2.** *v. i.* verzweifeln. **desperate** ['despərət] *adj.* verzweifelt; extrem *⟨Maßnahmen⟩;* **be ~ for sth.** etw. dringend brauchen. **desperation** [despə'reɪʃn] *n.* Verzweiflung, *die*
despicable [dɪ'spɪkəbl] *adj.* verabscheuungswürdig
despise [dɪ'spaɪz] *v. t.* verachten
despite [dɪ'spaɪt] *prep.* trotz
despondent [dɪ'spɒndənt] *adj.* bedrückt
despot ['despɒt] *n.* Despot, *der*
dessert [dɪ'zɜːt] *n.* Nachtisch, *der.* **des'sert spoon** *n.* Dessertlöffel, *der*
destination [destɪ'neɪʃn] *n.* Reiseziel, *das; (of goods)* Bestimmungsort, *der; (of train, bus)* Zielort, *der*
destine ['destɪn] *v. t.* bestimmen; **be ~d to do sth.** dazu bestimmt sein, etw. zu tun
destiny ['destɪnɪ] *n.* Schicksal, *das*
destitute ['destɪtjuːt] *adj.* mittellos
destroy [dɪ'strɔɪ] *v. t.* zerstören. **de'stroyer** *n. (also Naut.)* Zerstörer, *der.* **destruction** [dɪ'strʌkʃn] *n.* Zerstörung, *die.* **destructive** [dɪ'strʌktɪv] *adj.* zerstörerisch; verheerend *⟨Sturm, Feuer⟩*
detach [dɪ'tætʃ] *v. t.* entfernen; abnehmen *⟨wieder zu Befestigendes⟩;* herausnehmen *⟨innen Befindliches⟩.*

detachable [dɪ'tætʃəbl] *adj.* abnehmbar. **detached** [dɪ'tætʃt] *adj.* **a)** *(impartial)* unvoreingenommen; *(unemotional)* unbeteiligt; **b)** **a ~ house** ein Einzelhaus. **de'tachment** *n.* **a)** *see* **detach:** Entfernen, *das;* Abnehmen, *das;* Herausnehmen, *das;* **b)** *(Mil.)* Abteilung, *die*
detail ['diːteɪl] **1.** *n.* Einzelheit, *die;* Detail, *das;* **in ~:** Punkt für Punkt; **go into ~|s** ins Detail gehen. **2.** *v. t.* **a)** einzeln ausführen; **b)** *(Mil.)* abkommandieren. **detailed** ['diːteɪld] *adj.* detailliert; eingehend *⟨Studie⟩*
detain [dɪ'teɪn] *v. t.* **a)** festhalten; *(take into confinement)* verhaften; **b)** *(delay)* aufhalten. **detainee** [diːteɪ'niː] *n.* Verhaftete, *der/die*
detect [dɪ'tekt] *v. t.* entdecken; wahrnehmen *⟨Bewegung⟩;* aufdecken *⟨Irrtum, Verbrechen⟩.* **detection** [dɪ'tekʃn] *n.* Entdeckung, *die; (of error, crime)* Aufdeckung, *die.* **detective** [dɪ'tektɪv] *n.* Detektiv, *der;* **private ~:** Privatdetektiv, *der;* ~ **work** Ermittlungsarbeit, *die;* ~ **story** Detektivgeschichte, *die.* **detector** [dɪ'tektə(r)] *n.* Detektor, *der*
detention [dɪ'tenʃn] *n.* **a)** Festnahme, *die; (confinement)* Haft, *die;* **b)** *(Sch.)* Nachsitzen, *das*
deter [dɪ'tɜː(r)] *v. t.,* **-rr-** abschrecken
detergent [dɪ'tɜːdʒənt] *n.* Waschmittel, *das*
deteriorate [dɪ'tɪərɪəreɪt] *v. i.* sich verschlechtern; *⟨Haus:⟩* verfallen. **deterioration** [dɪtɪərɪə'reɪʃn] *n. see* **deteriorate:** Verschlechterung, *die;* Verfall, *der*
determination [dɪtɜːmɪ'neɪʃn] *n.* Entschlossenheit, *die*
determine [dɪ'tɜːmɪn] *v. t.* **a)** *(decide)* beschließen; **b)** *(be a decisive factor for)* bestimmen; **c)** *(ascertain)* feststellen. **determined** [dɪ'tɜːmɪnd] *adj.* **a)** **be ~ to do sth.** etw. unbedingt tun wollen; **b)** *(resolute)* entschlossen
deterrent [dɪ'terənt] *n.* Abschreckungsmittel, *das* (to für)
detest [dɪ'test] *v. t.* verabscheuen. **detestable** [dɪ'testəbl] *adj.* verabscheuenswert
detonate ['detəneɪt] **1.** *v. t.* zünden. **2.** *v. i.* detonieren. **detonation** [detə'neɪʃn] *n.* Detonation, *die.* **detonator** ['detəneɪtə(r)] *n.* Sprengkapsel, *die*
detour ['diːtʊə(r)] *n.* Umweg, *der; (diversion)* Umleitung, *die*

detract [dı'trækt] *v. i.* ~ **from sth.** etw. beeinträchtigen

detriment ['detrımənt] *n.* **to the** ~ **of sth.** zum Nachteil einer Sache *(Gen.)*. **detrimental** [detrı'mentl] *adj.* schädlich; **be** ~ **to sth.** einer Sache *(Dat.)* schaden

deuce [dju:s] *n. (Tennis)* Einstand, *der*

devaluation [di:vælju:'eıʃn] *n.* Abwertung, *die*

devalue [di:'vælju:] *v. t.* abwerten

devastate ['devəsteıt] *v. t.* verwüsten; *(fig.)* niederschmettern. **devastating** ['devəsteıtıŋ] *adj.* verheerend; *(fig.)* niederschmetternd. **devastation** [devə'steıʃn] *n.* Verwüstung, *die*

develop [dı'veləp] **1.** *v. t.* entwickeln; erschließen ⟨*natürliche Ressourcen*⟩; bekommen ⟨*Krankheit, Fieber, Lust*⟩; ~ **a taste for sth.** Geschmack an etw. *(Akk.)* finden. **2.** *v. i.* sich entwickeln **(from** aus; **into** zu). **de'veloper** *n.* **a)** *(Photog.)* Entwickler, *der;* **b)** *(of land)* Bauunternehmer, *der*

de'veloping country *n.* Entwicklungsland, *das*

de'velopment *n.* Entwicklung, *die* **(from** aus; **into** zu); *(of natural resources etc.)* Erschließung, *die*

deviant ['di:vıənt] *adj.* abweichend

deviate ['di:vıeıt] *v. i.* abweichen. **deviation** [di:vı'eıʃn] *n.* Abweichung, *die*

device [dı'vaıs] *n.* Gerät, *das; (as part of sth.)* Vorrichtung, *die;* **leave sb. to his own** ~s jmdn. sich *(Dat.)* selbst überlassen

devil ['devl] *n.* Teufel, *der;* **the D**~: der Teufel. **'devilish** *adj.* teuflisch

devious ['di:vıəs] *adj.* **a)** *(winding)* verschlungen; ~ **route** Umweg, *der;* **b)** *(unscrupulous, insincere)* hinterhältig

devise [dı'vaız] *v. t.* entwerfen; schmieden ⟨*Pläne*⟩

devoid [dı'vɔıd] *adj.* ~ **of sth.** *(lacking)* ohne etw.; *(free from)* frei von etw.

devolution [di:və'lu:ʃn] *n. (Polit.)* Dezentralisierung, *die*

devote [dı'vəʊt] *v. t.* widmen **(to** *Dat.*). **de'voted** *adj.* treu; aufrichtig ⟨*Freundschaft, Liebe, Verehrung*⟩; ~ **to sb.** jmdn. innig lieben. **devotion** [dı'vəʊʃn] *n.* ~ **to sb./sth.** Hingabe an jmdn./etw.

devour [dı'vaʊə(r)] *v. t.* verschlingen

devout [dı'vaʊt] *adj.* fromm

dew [dju:] *n.* Tau, *der*

dexterity [dek'sterıtı] *n.* Geschicklichkeit, *die*

dextrous ['dekstrəs] *adj.* geschickt

diabetes [daıə'bi:ti:z] *n.* Zuckerkrankheit, *die.* **diabetic** [daıə'betık] **1.** *adj.* zuckerkrank ⟨*Person*⟩. **2.** *n.* Diabetiker, *der*/Diabetikerin, *die*

diabolical [daıə'bɒlıkl] *adj.* teuflisch

diagnose [daıəg'nəʊz] *v. t.* diagnostizieren; feststellen ⟨*Fehler*⟩. **diagnosis** [daıəg'nəʊsıs] *n., pl.* **diagnoses** [daıəg'nəʊsi:z] Diagnose, *die;* **make a** ~: eine Diagnose stellen

diagonal [daı'ægənl] **1.** *adj.* diagonal. **2.** *n.* Diagonale, *die.* **di'agonally** *adv.* diagonal

diagram ['daıəgræm] *n.* Diagramm, *das*

dial ['daıəl] **1.** *n. (of clock or watch)* Zifferblatt, *das; (of gauge, meter, etc.)* Skala, *die; (Teleph.)* Wählscheibe, *die.* **2.** *v. t. & i., (Brit.)* **-ll-** *(Teleph.)* wählen; ~ **direct** selbst wählen; *(dial extension)* durchwählen

dialect ['daıəlekt] *n.* Dialekt, *der*

dialling *(Amer.:* **dialing):** ~ **code** *n.* Vorwahl, *die;* ~ **tone** Wählton, *der*

dialogue ['daıəlɒg] *n.* Dialog, *der*

'dial tone *n. (Amer.)* Wählton, *der*

diameter [daı'æmıtə(r)] *n.* Durchmesser, *der.* **diametrical** [daıə'metrıkl] *adj.,* **dia'metrically** *adv.* diametral

diamond ['daıəmənd] *n.* **a)** Diamant, *der;* **b)** *(figure)* Raute, *die;* **c)** *(Cards)* Karo, *das; see also* **club I c**

diaper ['daıəpə(r)] *n. (Amer.)* Windel, *die*

diaphragm ['daıəfræm] *n.* Diaphragma, *das*

diarrhoea *(Amer.:* **diarrhea)** [daıə'rı:ə] *n.* Durchfall, *der*

diary ['daıərı] *n.* **a)** Tagebuch, *das;* **b)** *(for appointments)* Terminkalender, *der*

dice [daıs] **1.** *n.* Würfel, *der.* **2.** *v. t. (Cooking)* würfeln

dicey ['daısı] *adj. (sl.)* riskant

dictate [dık'teıt] *v. t. & i.* diktieren; *(prescribe)* vorschreiben; ~ **to** Vorschriften machen (+ *Dat.).* **dictation** [dık'teıʃn] *n.* Diktat, *das.* **dictator** [dık'teıtə(r)] *n.* Diktator, *der.* **dictatorial** [dıktə'tɔ:rıəl] *adj.* diktatorisch. **dic'tatorship** *n.* Diktatur, *die*

dictionary ['dıkʃənərı] *n.* Wörterbuch, *das*

did *see* **do**

diddle ['dıdl] *v. t. (sl.)* übers Ohr hauen *(ugs.)*

didn't ['dıdnt] *(coll.)* = **did not**; *see* **do**

die [daı] *v. i., dying* ['daııŋ] sterben **(of,**

from an + *Dat.*); ⟨*Tier, Pflanze:*⟩ eingehen; **be dying to do sth.** darauf brennen, etw. zu tun; **be dying for sth.** etw. unbedingt brauchen. **die 'down** *v. i.* ⟨*Sturm, Wind, Protest:*⟩ sich legen; ⟨*Flammen:*⟩ kleiner werden; ⟨*Feuer:*⟩ herunterbrennen; ⟨*Lärm.:*⟩ leiser werden. **die 'out** *v. i.* aussterben

'**die-hard** *n.* Ewiggestrige, *der/die*

diesel ['di:zl] *n.* ~ **[engine]** Diesel[motor], *der;* ~ **[fuel]** Diesel[kraftstoff], *der*

diet ['daɪət] **1.** *n.* Diät, *die;* **be/go on a** ~: eine Schlankheitskur machen. **2.** *v. i.* eine Schlankheitskur machen

differ ['dɪfə(r)] *v. i.* *(be different)* sich unterscheiden

difference ['dɪfərəns] *n.* **a)** Unterschied, *der;* **make no** ~ **[to sb.]** [jmdm.] nichts ausmachen; **it makes a** ~: es ist ein *od. (ugs.)* macht einen Unterschied; **b)** *(disagreement)* Meinungsverschiedenheit, *die*

different ['dɪfərənt] *adj.* verschieden; *(pred. also)* anders; *(attrib. also)* ander...; **be** ~ **from** *or (esp. Brit.)* **to** *or (Amer.)* **than** ...: anders sein als ...

differentiate [dɪfə'renʃɪeɪt] *v. t. & i.* unterscheiden **(between** zwischen + *Dat.*)

'**differently** *adv.* anders **(from,** *esp. Brit.* **to** als)

difficult ['dɪfɪkəlt] *adj.* schwierig. '**difficulty** *n.* Schwierigkeit, *die;* **with [great]** ~: [sehr] mühsam; **get into difficulties** in Schwierigkeiten kommen

diffident ['dɪfɪdənt] *adj.* zaghaft; *(modest)* zurückhaltend

diffuse 1. [dɪ'fju:z] *v. t.* verbreiten. **2.** *v. i.* sich ausbreiten **(through** in + *Dat.*). **3.** [dɪ'fju:s] *adj.* diffus

dig [dɪg] **1.** *v. i.,* -gg-, **dug** [dʌg] graben **(for** nach). **2.** *v. t.,* -gg-, **dug** graben; umgraben ⟨*Erde, Garten*⟩. **dig 'out** *v. t.* ausgraben. **dig 'up** *v. t.* ausgraben; umgraben ⟨*Garten*⟩; aufreißen ⟨*Straße*⟩

digest [dɪ'dʒest, daɪ'dʒest] *v. t.* verdauen. **digestion** [dɪ'dʒestʃn, daɪ'dʒestʃn] *n.* Verdauung, *die*

'**digger** *n.* Bagger, *der*

digit ['dɪdʒɪt] *n.* Ziffer, *die*

digital ['dɪdʒɪtl] *adj.* Digital-

dignified ['dɪgnɪfaɪd] *adj.* würdig; *(stately)* würdevoll

dignify ['dɪgnɪfaɪ] *v. t.* Würde verleihen (+ *Dat.*)

dignitary ['dɪgnɪtərɪ] *n.* Würdenträger, *der;* **dignitaries** *(prominent people)* Honoratioren

dignity ['dɪgnɪtɪ] *n.* Würde, *die*

digress [daɪ'gres] *v. i.* abschweifen. **digression** [daɪ'greʃn] *n.* Abschweifung, *die*

dike [daɪk] *n.* Deich, *der*

dilapidated [dɪ'læpɪdeɪtɪd] *adj.* verfallen ⟨*Gebäude*⟩; verwahrlost ⟨*Erscheinung*⟩

dilate [daɪ'leɪt] **1.** *v. i.* sich weiten. **2.** *v. t.* ausdehnen

dilemma [dɪ'lemə, daɪ'lemə] *n.* Dilemma, *das*

diligence ['dɪlɪdʒəns] *n.* Fleiß, *der*

diligent ['dɪlɪdʒənt] *adj.,* '**diligently** *adv.* fleißig

dilute 1. [daɪ'lju:t, 'daɪlju:t] *adj.* verdünnt. **2.** [daɪ'lju:t] *v. t.* verdünnen

dim [dɪm] **1.** *adj.* **a)** schwach ⟨*Licht, Flackern*⟩; dunkel ⟨*Zimmer*⟩; verschwommen ⟨*Gestalt*⟩; **b)** *(vague)* verschwommen; **c)** *(coll.: stupid)* beschränkt. **2.** *v. i.* schwächer werden

dime [daɪm] *n. (Amer. coll.)* Zehncentstück, *das*

dimension [dɪ'menʃn, daɪ'menʃn] *n.* Dimension, *die,* ~**s** *(measurements)* Abmessungen; Maße

diminish [dɪ'mɪnɪʃ] **1.** *v. i.* nachlassen; ⟨*Vorräte, Einfluß:*⟩ abnehmen; ⟨*Wert, Ansehen:*⟩ geringer werden. **2.** *v. t.* verringern; schmälern ⟨*Ansehen, Ruf*⟩

dimple ['dɪmpl] *n.* Grübchen, *das*

dim: ~-**wit** *n. (coll.)* Dummkopf, *der (ugs.);* ~-**witted** ['dɪmwɪtɪd] *adj. (coll.)* dusselig *(salopp)*

din [dɪn] *n.* Lärm, *der*

dine [daɪn] *v. i.* [zu Mittag/zu Abend] essen. '**diner** *n.* Gast, *der*

dinghy ['dɪŋɪ, 'dɪŋɪ] *n.* Ding[h]i, *das; (inflatable)* Schlauchboot, *das*

dingy ['dɪndʒɪ] *adj.* schmuddelig

dining ['daɪnɪŋ]: ~-**car** *n.* Speisewagen, *der;* ~-**room** *n.* Eßzimmer, *das; (in hotel etc.)* Speisesaal, *der*

dinner ['dɪnə(r)] *n. (at midday)* Mittagessen, *das; (in the evening)* Abendessen, *das; (formal)* Diner, *das.* '**dinner-table** *n.* Eßtisch, *der.* '**dinner-time** *n.* Essenszeit, *die;* **at** ~-**time** zur Essenszeit; *(12–2 p. m.)* mittags

dinosaur ['daɪnəsɔ:(r)] *n.* Dinosaurier, *der*

dint [dɪnt] *n.* **by** ~ **of** durch; **by** ~ **of doing sth.** indem jmd. etw. tut

dip [dɪp] **1.** *v. t.,* -pp-: **a)** [ein]tauchen **(in** in + *Akk.*); **b)** ~ **one's headlights** abblenden. **2.** *v. i.* sinken; *(incline)* abfallen. **3.** *n.* **a)** *(in road)* Senke, *die;* **b)** *(coll.: bathe)* [kurzes] Bad

diphtheria [dɪfˈθɪərɪə] *n.* Diphtherie, *die*

diphthong [ˈdɪfθɒŋ] *n.* Diphthong, *der*

diploma [dɪˈpləʊmə] *n.* Diplom, *das*

diplomacy [dɪˈpləʊməsɪ] *n.* Diplomatie, *die*

diplomat [ˈdɪpləmæt] *n.* Diplomat, *der*/Diplomatin, *die*

diplomatic [dɪpləˈmætɪk] *adj.*, **diplomatically** *adv.* diplomatisch

dire [ˈdaɪə(r)] *adj.* furchtbar

direct [dɪˈrekt, daɪˈrekt] **1.** *v. t.* **a)** *(turn)* richten (to|wards| auf + *Akk.*); ~ **sb.** **to a place** jmdn. den Weg zu einem Ort weisen; **b)** *(control)* leiten; regeln ⟨*Verkehr*⟩; **c)** *(order)* anweisen; **d)** *(Theatre, Cinemat., etc.)* Regie führen bei. **2.** *adj.* direkt; durchgehend ⟨*Zug*⟩; unmittelbar ⟨*Ursache, Auswirkung, Erfahrung, Verantwortung*⟩; genau ⟨*Gegenteil*⟩; direkt ⟨*Widerspruch*⟩; diametral ⟨*Gegensatz*⟩; ~ **speech** direkte Rede. **3.** *adv.* direkt.

direct 'current *n.* Gleichstrom, *der*.

direct 'hit *n.* Volltreffer, *der*

direction [daɪˈrekʃn] *n.* **a)** Richtung, *die*; **in the** ~ **of London** in Richtung London; **b)** *(guidance)* Führung, *die*; **c)** *usu. in pl. (order)* Anordnung, *die*; **~s** |for use| Gebrauchsanweisung, *die*

di'rectly *adv.* **a)** direkt; unmittelbar ⟨*folgen, verantwortlich sein*⟩; **b)** *(exactly)* genau; **c)** *(at once)* umgehend; **d)** *(shortly)* gleich

di'rect object *n.* direktes Objekt

director [daɪˈrektə(r)] *n.* **a)** *(Commerc.)* Direktor, *der*/Direktorin, *die*; **board of** ~s Aufsichtsrat, *der*; **b)** *(Theatre, Cinemat., etc.)* Regisseur, *der*/Regisseurin, *die*

directory [daɪˈrektərɪ] *n. (telephone* ~*)* Telefonbuch, *das*; *(of tradesmen etc.)* Branchenverzeichnis, *das*; ~ **enquiries** *(Brit.)*, ~ **information** *(Amer.)* [Fernsprech]auskunft, *die*

dirt [dɜːt] *n.* Schmutz, *der*; ~ **cheap** spottbillig. **'dirty 1.** *adj.* schmutzig; **get sth.** ~: etw. schmutzig machen. **2.** *v. t.* schmutzig machen

disa'bility *n.* Behinderung, *die*

disabled [dɪsˈeɪbld] *adj.* behindert

disad'vantage *n.* Nachteil, *der*; **at a** ~: im Nachteil

disa'gree *v. i.* anderer Meinung sein; ~ **with sb./sth.** mit jmdm./etw. nicht übereinstimmen; ~ |with sb.| **about** *or* **over sth.** sich [mit jmdm.] über etw. *(Akk.)* nicht einig sein. **dis-**

a'greeable *adj.* unangenehm. **disa'greement** *n.* **a)** *(difference of opinion)* Uneinigkeit, *die*; **be in** ~ **with sb./sth.** mit jmdm./etw. nicht übereinstimmen; **b)** *(quarrel)* Meinungsverschiedenheit, *die*; **c)** *(discrepancy)* Diskrepanz, *die*

disal'low *v. t.* verbieten; *(Sport)* nicht geben ⟨*Tor*⟩

disap'pear *v. i.* verschwinden; ⟨*Brauch, Tierart:*⟩ aussterben. **disap'pearance** *n.* Verschwinden, *das*

disap'point *v. t.* enttäuschen. **disap'pointed** *adj.* enttäuscht. **disap'pointing** *adj.* enttäuschend. **disap'pointment** *n.* Enttäuschung, *die*

disap'proval *n.* Mißbilligung, *die*

disap'prove *v. i.* dagegen sein; ~ **of sb./sth.** jmdn. ablehnen/etw. mißbilligen

dis'arm *v. t.* entwaffnen. **disarmament** [dɪsˈɑːməmənt] *n.* Abrüstung, *die*

disarray [dɪsəˈreɪ] *n.* Unordnung, *die*; **in** ~: in Unordnung

disaster [dɪˈzɑːstə(r)] *n.* Katastrophe, *die*; ~ **area** Katastrophengebiet, *das*. **disastrous** [dɪˈzɑːstrəs] *adj.* katastrophal; verhängnisvoll ⟨*Irrtum, Entscheidung, Politik*⟩

dis'band 1. *v. t.* auflösen. **2.** *v. i.* sich auflösen

disbe'lief *n.* Unglaube, *der*; **in** ~: ungläubig

disbe'lieve *v. t.* ~ **sb./sth.** jmdm./etw. nicht glauben

disc [dɪsk] *n.* Scheibe, *die*; *(record)* Platte, *die*; **floppy** ~: Floppy disk, *die*; **hard** ~ *(fixed)* Festplatte, *die*

discard [dɪsˈkɑːd] *v. t.* wegwerfen; fallenlassen ⟨*Vorschlag, Idee*⟩

discern [dɪˈsɜːn] *v. t.* wahrnehmen. **discernible** [dɪˈsɜːnɪbl] *adj.* erkennbar. **dis'cerning** *adj.* kritisch

discharge 1. [dɪsˈtʃɑːdʒ] *v. t.* **a)** entlassen (**from** aus); freisprechen ⟨*Angeklagte*⟩; **b)** ablassen ⟨*Flüssigkeit, Gas*⟩. **2.** [ˈdɪstʃɑːdʒ] *n.* **a)** Entlassung, *die* (**from** aus); *(of defendant)* Freispruch, *der*; **b)** *(emission)* Ausfluß, *der*

disciple [dɪˈsaɪpl] *n.* **a)** *(Relig.)* Jünger, *der*; **b)** *(follower)* Anhänger, *der*/Anhängerin, *die*

disciplinary [dɪsɪˈplɪnərɪ] *adj.* disziplinarisch; ~ **action** Disziplinarmaßnahmen

discipline [ˈdɪsɪplɪn] **1.** *n.* Disziplin, *die*. **2.** *v. t.* disziplinieren; *(punish)* bestrafen

'disc jockey n. Diskjockey, der
dis'claim v. t. abstreiten
disclose [dıs'kləʊz] v. t. enthüllen; bekanntgeben ⟨Information, Nachricht⟩.
dis'closure n. Enthüllung, die; (of information, news) Bekanntgabe, die
disco ['dıskəʊ] n., pl. ~s (coll.) Disko, die
dis'colour (Brit.; Amer.: discolor) v. t. verfärben
dis'comfort n. a) no pl. (slight pain) Beschwerden Pl.; b) (hardship) Unannehmlichkeit, die
disconcert [dıskɒn'sɜːt] v. t. irritieren
discon'nect v. t. abtrennen; abstellen ⟨Telefon⟩
disconsolate [dıs'kɒnsələt] adj. a) (unhappy) unglücklich; b) (inconsolable) untröstlich
discon'tent n. Unzufriedenheit, die. discon'tented adj. unzufrieden
discon'tinue v. t. einstellen
discord ['dıskɔːd] n. a) Zwietracht, die; b) (Mus.) Dissonanz, die. discordant [dıs'kɔːdənt] adj. a) (conflicting) gegensätzlich; b) a ~ note ein Mißton
discothèque ['dıskətek] n. Diskothek, die
discount 1. ['dıskaʊnt] n. (Commerc.) Rabatt, der (on auf + Akk.). 2. [dı'skaʊnt] v. t. (disbelieve) unberücksichtigt lassen
discourage [dı'skʌrıdʒ] v. t. a) entmutigen; b) (advise against) abraten. di'scouragement n. a) Entmutigung, die; b) (depression) Mutlosigkeit, die. discouraging [dı'skʌrıdʒıŋ] adj. entmutigend
dis'courteous adj. unhöflich. dis'courtesy n. Unhöflichkeit, die
discover [dı'skʌvə(r)] v. t. a) entdecken; (by search) herausfinden. di'scovery n. Entdeckung, die
dis'credit 1. n. Mißkredit, der; bring ~ on sb./sth., bring sb./sth. into ~: jmdn./etw. in Mißkredit bringen. 2. v. t. in Mißkredit bringen
discreet [dı'skriːt] adj., di'screetly adv. diskret
discrepancy [dı'skrepənsı] n. Diskrepanz, die
discretion [dı'skreʃn] n. (prudence) Umsicht, die
discriminate [dı'skrımıneıt] v. i. a) unterscheiden; b) ~ against/in favour of sb. jmdn. diskriminieren/bevorzugen. discrimination [dıskrımı'neıʃn] n. a) Unterscheidung, die; b) Diskri-

minierung, die (against Gen.); ~ in favour of Bevorzugung (+ Gen.)
discus ['dıskəs] n. Diskus, der
discuss [dı'skʌs] v. t. besprechen; (debate) diskutieren über (+ Akk.). discussion [dı'skʌʃn] n. Gespräch, das; (debate) Diskussion, die
disdain [dıs'deın] 1. n. Verachtung, die. 2. v. t. verachten; ~ to do sth. zu stolz sein, etw. zu tun. disdainful [dıs'deınfl] adj. verächtlich
disease [dı'ziːz] n. Krankheit, die. diseased [dı'ziːzd] adj. krank
disem'bark v. i. von Bord gehen
disen'chant v. t. ernüchtern; he became ~ed with her/it sie/es hat ihn desillusioniert
disen'gage v. t. lösen (from aus, von); ~ the clutch auskuppeln
disen'tangle v. t. entwirren; (extricate) befreien (from aus)
dis'figure v. t. entstellen
disgrace [dıs'greıs] n. 1. Schande, die (to für). 2. v. t. Schande machen (+ Dat.); ~ oneself sich blamieren. di'sgraceful [dıs'greısfl] adj. skandalös; it's a ~: es ist ein Skandal
disgruntled [dıs'grʌntld] adj. verstimmt
disguise [dıs'gaız] 1. v. t. verkleiden ⟨Person⟩; verstellen ⟨Stimme⟩; tarnen ⟨Gegenstand⟩. 2. n. Verkleidung, die
disgust [dıs'gʌst] 1. n. (nausea) Ekel, der (at vor + Dat.); (revulsion) Abscheu, der (at vor + Dat.); (indignation) Empörung, die (at über + Akk.). 2. v. t. anwidern; (fill with nausea) ekeln; (fill with indignation) empören. dis'gusted adj. angewidert; (nauseated) angeekelt; (indignant) empört. dis'gusting adj. widerlich
dish [dıʃ] n. a) Schale, die; (deeper) Schüssel, die; ~es (crockery) Geschirr, das; wash or (coll.) do the ~es Geschirr spülen; b) (type of food) Gericht, das. dish 'out v. t. a) austeilen ⟨Essen⟩; b) (coll.: distribute) verteilen. dish 'up v. t. auftragen
'dishcloth n. Spültuch, das
dis'hearten v. t. entmutigen
dishevelled (Amer.: disheveled) [dı'ʃevld] adj. zerzaust ⟨Haar⟩; ungepflegt ⟨Erscheinung⟩
dis'honest adj., dis'honestly adv. unehrlich. dis'honesty n. Unehrlichkeit, die
dis'honour 1. n. Unehre, die. 2. v. t. beleidigen. dishonourable [dıs'ɒnərəbl] adj. unehrenhaft

'dishwasher n. Geschirrspülmaschi-
ne, die

disil'lusion 1. v.t. ernüchtern. 2. n.
Desillusion, die (with über + Akk.).
disil'lusionment n. Desillusionie-
rung, die

disin'fect v.t. desinfizieren. disinfectant [dısın'fektənt] 1. adj. desinfi-
zierend. 2. n. Desinfektionsmittel, das

dis'integrate v.i. zerfallen. disintegration [dısıntı'greıʃn] n. Zerfall,
der

dis'interested adj. a) (impartial) unvoreingenommen; b) (coll.: uninter-
ested) desinteressiert

disjointed [dıs'dʒɔıntıd] adj. unzusammenhängend

disk see disc

diskette [dı'sket] n. Diskette, die

dis'like 1. v.t. nicht mögen; ~ doing
sth. etw. ungern tun. 2. n. Abneigung,
die (of, for gegen); take a ~ to sb./sth.
eine Abneigung gegen jmdn./etw.
empfinden

dislocate ['dısləkeıt] v.t. ausrenken;
auskugeln ⟨Schulter, Hüfte⟩

dis'lodge v.t. entfernen (from aus)

dis'loyal adj. illoyal (to gegenüber).
dis'loyalty n. Illoyalität, die (to gegenüber)

dismal ['dızml] adj. trist

dismantle [dıs'mæntl] v.t. demontie-
ren; abbauen ⟨Schuppen, Gerüst⟩

dismay [dıs'meı] 1. v.t. bestürzen. 2.
n. Bestürzung, die (at über + Akk.)

dismiss [dıs'mıs] v.t. entlassen; (reject) ablehnen. dismissal [dıs'mısl]
n. Entlassung, die

dis'mount v.i. absteigen

diso'bedience n. Ungehorsam, der

diso'bedient adj. ungehorsam

diso'bey v.t. nicht gehorchen
(+ Dat.); nicht befolgen ⟨Befehl⟩

dis'order n. a) Durcheinander, das;
b) (Med.) Störung, die. dis'orderly
adj. (untidy) unordentlich; ~ conduct
ungebührliches Benehmen

dis'organized adj. chaotisch

dis'orientated, dis'oriented adj.
desorientiert

dis'own v.t. verleugnen

disparage [dı'spærıdʒ] v.t. herabsetzen. disparaging [dı'spærıdʒıŋ] adj.
abschätzig

disparity [dı'spærıtı] n. Ungleichheit,
die

dispatch [dı'spætʃ] 1. v.t. a)
schicken; b) (kill) töten. 2. n. Bericht,
der

dispel [dı'spel] v.t., -ll- vertreiben;
zerstreuen ⟨Besorgnis, Befürchtung⟩

dispensable [dı'spensəbl] adj. entbehrlich

dispensary [dı'spensərı] n. Apotheke,
die

dispense [dı'spens] v.i. ~ with verzichten auf (+ Akk.)

dispersal [dı'spɜːsl] n. Zerstreuung, die

disperse [dı'spɜːs] 1. v.t. zerstreuen.
2. v.i. sich zerstreuen

dispirited [dı'spırıtıd] adj. entmutigt

dis'place v.t. verschieben; (supplant)
ersetzen

display [dı'spleı] 1. v.t. zeigen; ausstellen ⟨Waren⟩. 2. n. Ausstellung, die;
(of goods) Auslage, die; (ostentatious
show) Zurschaustellung, die

dis'please v.t. ~ sb. jmds. Mißfallen
erregen. dis'pleasure n. Mißfallen,
das

disposable [dı'spəʊzəbl] adj. Weg-
werf-

disposal [dı'spəʊzl] n. Beseitigung,
die; have sth./sb. at one's ~: etw./
jmdn. zur Verfügung haben; be at
sb.'s ~: jmdm. zur Verfügung stehen

dispose [dı'spəʊz] v.t. ~ sb. to sth.
jmdn. zu etw. veranlassen; ~ sb. to do
sth. jmdn. dazu veranlassen, etw. zu
tun. di'spose of v.t. beseitigen;
(settle) erledigen

disposed [dı'spəʊzd] adj. be ~ to do
sth. dazu neigen, etw. zu tun; be well
~ towards sb./sth. jmdm. wohl gesinnt
sein/einer Sache (Dat.) positiv gegenüberstehen. disposition [dıspə'zıʃn]
n. Veranlagung, die; (nature) Art, die

dis'prove v.t. widerlegen

dispute [dı'spjuːt] 1. n. Streit, der
(over um). 2. v.t. a) (discuss) sich streiten über (+ Akk.); b) (oppose) bestreiten

disqualifi'cation n. Ausschluß, der;
(Sport) Disqualifikation, die

dis'qualify v.t. ausschließen (from
von); (Sport) disqualifizieren

disre'gard 1. v.t. ignorieren. 2. n.
Mißachtung, die (of, for Gen.); (of
wishes, feelings) Gleichgültigkeit, die
(for, of gegenüber)

dis'reputable adj. verrufen

disrepute [dısrı'pjuːt] n. Verruf, der;
bring sb./sth. into ~: jmdn./etw. in
Verruf bringen

disre'spect n. Mißachtung, die; show
~ for sb./sth. keine Achtung vor
jmdm./etw. haben. disre'spectful
adj. respektlos

disrupt [dɪs'rʌpt] *v. t.* stören. **disruption** [dɪs'rʌpʃn] *n.* Störung, *die.* **disruptive** [dɪs'rʌptɪv] *adj.* störend

dissatis'faction *n.* Unzufriedenheit, *die*

dis'satisfied *adj.* unzufrieden

dissect [dɪ'sekt] *v. t.* sezieren

dissent [dɪ'sent] **1.** *v. i.* **a)** *(refuse to assent)* nicht zustimmen; ~ **from sth.** mit etw. nicht übereinstimmen; **b)** *(disagree)* ~ **from sth.** von etw. abweichen. **2.** *n.* Ablehnung, *die; (from majority)* Abweichung, *die*

dissertation [dɪsə'teɪʃn] *n.* Dissertation, *die*

dis'service *n.* **do sb. a** ~: jmdm. einen schlechten Dienst erweisen

dissident ['dɪsɪdənt] *n.* Dissident, *der*/Dissidentin, *die*

dis'similar *adj.* unähnlich **(to** *Dat.*)

dissociate [dɪ'səʊʃɪeɪt] *v. t.* trennen; ~ **oneself** sich distanzieren **(from** von)

dissolve [dɪ'zɒlv] **1.** *v. t.* auflösen. **2.** *v. i.* sich auflösen

dissuade [dɪ'sweɪd] *v. t.* abbringen **(from** von)

distance ['dɪstəns] *n.* **a)** Entfernung, *die* **(from** zu); **b)** *(way to cover)* Strecke, *die;* **from a** ~: von weitem; **in/into the** ~: in der/die Ferne

distant ['dɪstənt] *adj.* **a)** fern; entfernt ⟨*Ähnlichkeit, Verwandtschaft, Verwandte*⟩; **b)** *(reserved)* distanziert

dis'taste *n.* Abneigung, *die* **(for** gegen). **dis'tasteful** *adj.* unangenehm

distend [dɪ'stend] *v. t.* erweitern

distil, *(Amer.)* **distill** [dɪ'stɪl] *v. t.* destillieren; brennen ⟨*Branntwein*⟩. **distillation** [dɪstɪ'leɪʃn] *n.* Destillation, *die.* **distillery** [dɪ'stɪlərɪ] *n.* Brennerei, *die*

distinct [dɪ'stɪŋkt] *adj.* deutlich; *(different)* verschieden. **distinction** [dɪ'stɪŋkʃn] *n.* Unterschied, *der.* **distinctive** [dɪ'stɪŋktɪv] *adj.* unverwechselbar. **dis'tinctly** *adv.* deutlich

distinguish [dɪ'stɪŋgwɪʃ] **1.** *v. t.* **a)** *(make out)* erkennen; **b)** *(differentiate)* unterscheiden; **c)** *(characterize)* kennzeichnen; **d)** ~ **oneself** [**by sth.**] sich [durch etw.] hervortun. **2.** *v. i.* unterscheiden; ~ **between** auseinanderhalten. **distinguished** [dɪ'stɪŋgwɪʃt] *adj.* angesehen; glänzend ⟨*Laufbahn*⟩; vornehm ⟨*Aussehen*⟩

distort [dɪ'stɔːt] *v. t.* verzerren; *(fig.)* verdrehen. **distortion** [dɪ'stɔːʃn] *n.* Verzerrung, *die; (fig.)* Verdrehung, *die*

distract [dɪ'strækt] *v. t.* ablenken; ~ sb.|'s attention from sth.| jmdn. [von etw.] ablenken. **di'stracted** *adj.* von Sinnen *nachgestellt; (mentally far away)* abwesend. **distraction** [dɪ'strækʃn] *n.* **a)** *(diversion)* Ablenkung, *die; (interruption)* Störung, *die;* **b) drive sb. to** ~: jmdn. zum Wahnsinn treiben

distraught [dɪ'strɔːt] *adj.* aufgelöst **(with** vor + *Dat.*); verstört ⟨*Blick*⟩

distress [dɪ'stres] **1.** *n.* **a)** Kummer, *der* **(at** über + *Akk.*); **b)** *(pain)* Qualen *Pl.;* **c) an aircraft/ship in** ~: ein Flugzeug in Not/ein Schiff in Seenot. **2.** *v. t.* nahegehen (+ *Dat.*). **di'stressing** *adj.* erschütternd. **di'stress signal** *n.* Notsignal, *das*

distribute [dɪ'strɪbjuːt] *v. t.* verteilen **(to an** + *Akk.;* **among** unter + *Akk.*); *(Commerc.)* vertreiben. **distribution** [dɪstrɪ'bjuːʃn] *n.* Verteilung, *die; (Commerc.)* Vertrieb, *der.* **distributor** [dɪ'strɪbjʊtə(r)] *n.* Verteiler, *der*/Verteilerin, *die; (Commerc.)* Vertreiber, *der*

district ['dɪstrɪkt] *n.* Gegend, *die; (Admin.)* Bezirk, *der.* **district 'nurse** *n. (Brit.)* Gemeindeschwester, *die*

dis'trust [dɪs'trʌst] **1.** *n.* Mißtrauen, *das* **(of** gegen). **2.** *v. t.* mißtrauen (+ *Dat.*).

disturb [dɪ'stɜːb] *v. t.* **a)** stören; '**do not** ~!' „bitte nicht stören!''; **b)** *(worry)* beunruhigen. **disturbance** [dɪ'stɜːbəns] *n.* Störung, *die;* **political** ~s politische Unruhen. **disturbed** [dɪ'stɜːbd] *adj.* besorgt; [**mentally**] ~: geistig gestört

disuse [dɪs'juːs] *n.* **fall into** ~: außer Gebrauch kommen

disused [dɪs'juːzd] *adj.* stillgelegt; leerstehend ⟨*Gebäude*⟩

ditch [dɪtʃ] **1.** *n.* Graben, *der.* **2.** *v. t. (sl.)* sausenlassen ⟨*Plan*⟩; sitzenlassen ⟨*Familie, Freund*⟩

dither ['dɪðə(r)] *v. i.* schwanken

ditto ['dɪtəʊ] *n., pl.* ~**s** ebenso; ditto; ~ **marks** Unterführungszeichen, *das*

divan [dɪ'væn] *n.* [Polster]liege, *die*

dive [daɪv] **1.** *v. i.,* **dived** *or (Amer.)* **dove** [dəʊv] **a)** einen Kopfsprung machen; *(when already in water)* tauchen; **b)** ⟨*Vogel, Flugzeug usw.:*⟩ einen Sturzflug machen. **2.** *n.* **a)** Kopfsprung, *der; (of bird, aircraft, etc.)* Sturzflug, *der;* **b)** *(coll.: place)* Spelunke, *die.* **'diver** *n.* **a)** *(Sport)* Kunstspringer, *der*/-springerin, *die;* **b)** *(as profession)* Taucher, *der*/Taucherin, *die*

diverge [daɪ'vɜ:dʒ] *v. i.* auseinandergehen. **divergent** [daɪ'vɜ:dʒənt] *adj.* auseinandergehend

diverse [daɪ'vɜ:s] *adj.* verschieden

diversion [daɪ'vɜ:ʃn] *n.* a) Ablenkung, *die;* create a ~: ein Ablenkungsmanöver durchführen; b) *(Brit.: alternative route)* Umleitung, *die*

diversity [daɪ'vɜ:sɪtɪ] *n.* Vielfalt, *die*

divert [daɪ'vɜ:t] *v. t.* umleiten ⟨Verkehr, Fluß⟩; ablenken ⟨Aufmerksamkeit⟩

divide [dɪ'vaɪd] **1.** *v. t.* a) teilen; ~ **sth. in two** etw. [in zwei Teile] zerteilen; *(distribute)* aufteilen (**among/between** unter + *Akk. od. Dat.*); c) *(Math.)* dividieren *(fachspr.)*, teilen (**by** durch). **2.** *v. i.* sich teilen; ~ |**from sth.**] von etw. abzweigen. **divide 'out** *v. t.* aufteilen (**among/between** unter + *Akk. od. Dat.*); *(distribute)* verteilen an (+ *Akk.*). **divide 'up** *v. t.* aufteilen

dividend ['dɪvɪdend] *n.* Dividende, *die*

dividers [dɪ'vaɪdəz] *n. pl.* Stechzirkel, *der*

divine [dɪ'vaɪn] *adj.* göttlich

diving ['daɪvɪŋ] *n.* Kunstspringen, *das.* '**diving-board** *n.* Sprungbrett, *das.* '**diving-suit** *n.* Taucheranzug, *der*

divinity [dɪ'vɪnɪtɪ] *n.* a) *(god)* Göttlichkeit, *die;* b) *(god)* Gottheit, *die*

divisible [dɪ'vɪzɪbl] *adj.* teilbar (**by** durch)

division [dɪ'vɪʒn] *n.* a) Teilung, *die;* b) *(Math.)* Dividieren, *das;* c) *(section, part)* Abteilung, *die;* d) *(group)* Gruppe, *die;* e) *(Mil. etc.)* Division, *die;* f) *(Footb. etc.)* Liga, *die;* Spielklasse, *die;* *(in British football)* Division, *die*

divorce [dɪ'vɔ:s] **1.** *n.* [Ehe]scheidung, *die.* **2.** *v. t.* ~ **one's husband/wife** sich von seinem Mann/seiner Frau scheiden lassen. **divorced** [dɪ'vɔ:st] *adj.* geschieden; **get** ~: sich scheiden lassen

divulge [daɪ'vʌldʒ] *v. t.* preisgeben

DIY *abbr.* **do-it-yourself**

dizzy ['dɪzɪ] *adj.* schwind[e]lig; **I feel** ~: mir ist schwindlig

do [də, *stressed* du:] **1.** *v. t., neg. coll.* **don't** [dəʊnt], *pres. t.* **he does** [dʌz], *neg. (coll.)* **doesn't** ['dʌznt], *p. t.* **did** [dɪd], *neg. (coll.)* **didn't** ['dɪdnt], *pres. p.* **doing** ['du:ɪŋ], *p. p.* **done** [dʌn] a) machen ⟨Hausaufgaben, Hausarbeit, Examen, Handstand⟩; erfüllen ⟨Pflicht⟩; verrichten ⟨Arbeit⟩; vorführen ⟨Trick, Nummer, Tanz⟩; durchführen ⟨Test⟩; machen ⟨Übersetzung, Kopie, Bett⟩; schaffen ⟨Pensum⟩; *(clean)* putzen;

(arrange) [zurecht]machen ⟨Haare⟩; schminken ⟨Lippen, Augen, Gesicht⟩; machen *(ugs.)* ⟨Nägel⟩; *(cut)* schneiden ⟨Nägel⟩; *(paint)* machen *(ugs.)* ⟨Zimmer⟩; streichen ⟨Haus, Möbel⟩; *(repair)* in Ordnung bringen; **do the shopping / washing-up / cleaning** einkaufen [gehen]/abwaschen/saubermachen; **what can I do for you?** *(in shop)* was darf's sein?; **do sth. about sth./sb.** etw. gegen etw./jmdn. unternehmen; b) *(cook)* braten; **well done** durch[gebraten]; c) *(solve)* lösen ⟨Problem, Rätsel⟩; machen ⟨Puzzle, Kreuzworträtsel⟩; d) *(sl.: swindle)* reinlegen *(ugs.);* **do sb. out of sth.** jmdn. um etw. bringen; e) *(satisfy)* zusagen (+ *Dat.*). **2.** *v. i., forms as* 1: a) *(act)* tun; **do as they do** mach es wie sie; b) *(fare)* **how are you doing?** wie geht's dir?; c) *(get on)* vorankommen; *(in exams)* abschneiden; **do well/badly at school** gut/schlecht in der Schule sein; d) **how do you do?** *(formal)* guten Tag/Morgen/Abend!; e) *(serve purpose)* es tun; *(suffice)* [aus]reichen; *(be suitable)* gehen; **that won't do** das geht nicht; **that will do!** jetzt aber genug! **3.** *v. substitute, forms as* 1: **you mustn't act as he does** du darfst nicht so wie er handeln; **You went to Paris, didn't you? – Yes, I did** Du warst doch in Paris, nicht wahr? – Ja[, stimmt]; **come in, do!** komm doch herein! **4.** *v. aux. forms as* 1: **I do love Greece** Griechenland gefällt mir wirklich gut; **little did he know that ...:** er hatte keine Ahnung, daß ...; **do you know him?** kennst du ihn?; **what does he want?** was will er?; **I don't** *or* **do not wish to take part** ich möchte nicht teilnehmen; **don't be so noisy!** seid [doch] nicht so laut! **5.** *n.* [du:], *pl.* **do's** *or* **dos** [du:z] *(Brit. coll.)* Feier, *die;* Fete, *die (ugs.).* **do a'way with** *v. t.* abschaffen. '**do for** *v. t. (coll.)* **do for sb.** jmdn. fertigmachen *(ugs.);* **be done for** erledigt sein. **do 'in** *v. t. (sl.)* kaltmachen *(salopp).* **do 'up** *v. t.* a) *(fasten)* zumachen; binden ⟨Schnürsenkel, Fliege⟩; b) *(wrap)* einpacken. '**do with** *v. t.* **I could do with ...:** ich brauche ... '**do without** *v. t.* **do without sth.** auf etw. *(Akk.)* verzichten

docile ['dəʊsaɪl] *adj.* sanft; *(submissive)* unterwürfig

¹**dock** [dɒk] **1.** *n.* a) Dock, *das;* b) *usu. in pl. (area)* Hafen, *der.* **2.** *v. t.* [ein]docken. **3.** *v. i.* anlegen

²dock n. *(in lawcourt)* Anklagebank, *die;* **stand/be in the ~:** ≈ auf der Anklagebank sitzen

'docker n. Hafenarbeiter, *der*

'dockyard n. Schiffswerft, *die*

doctor ['dɒktə(r)] **1.** n. **a)** Arzt, *der*/Ärztin, *die; as address* Herr/Frau Doktor; **b)** *(holder of degree)* Doktor, *der.* **2.** v. t. *(coll.)* verfälschen

doctrine ['dɒktrın] n. Lehre, *die*

document ['dɒkjʊmənt] n. Dokument, *das;* Urkunde, *die*

documentary [dɒkjʊ'mentərı] **1.** adj. dokumentarisch. **2.** n. *(film)* Dokumentarfilm, *der*

dodge [dɒdʒ] **1.** v. i. ausweichen. **2.** v. t. ausweichen (+ Dat.) ⟨Schlag, Hindernis usw.⟩; entkommen (+ Dat.) ⟨Polizei, Verfolger⟩. **3.** n. *(trick)* Trick, *der*

dodgems ['dɒdʒəmz] n. pl. [Auto]skooterbahn, *die;* **have a ride/go on the ~:** Autoskooter fahren

dodgy ['dɒdʒı] adj. *(Brit. coll.) (unreliable)* unsicher; *(risky)* gewagt

doe [dəʊ] n. *(deer)* Damtier, *das; (rabbit)* [Kaninchen]weibchen, *das*

does [dʌz] *see* do

doesn't ['dʌznt] *(coll.)* = does not; *see* do

dog [dɒg] **1.** n. Hund, *der.* **2.** v. t., **-gg-** verfolgen; *(fig.)* heimsuchen

dog: **~-biscuit** n. Hundekuchen, *der;* **~-collar** n. [Hunde]halsband, *das; (joc.: clerical collar)* Kollar, *das;* **~-eared** adj. **a ~-eared book** ein Buch mit Eselsohren

dogged ['dɒgıd] adj. hartnäckig ⟨Weigerung, Verurteilung⟩; zäh ⟨Durchhaltevermögen, Ausdauer⟩

dogma ['dɒgmə] n. Dogma, *das.* **dogmatic** [dɒg'mætık] adj. dogmatisch

doing ['du:ıŋ] n. Tun, *das*

do-it-yourself [du:ıtjə'self] **1.** adj. Do-it-yourself-. **2.** n. Heimwerken, *das*

doldrums ['dɒldrəmz] n. pl. **in the ~** *(in low spirits)* niedergeschlagen; *(Econ.)* in einer Flaute

dole [dəʊl] **1.** n. *(coll.)* **the ~:** Stempelgeld, *das (ugs.);* **be/go on the ~:** stempeln gehen *(ugs.).* **2.** v. t. **~ out** [in kleinen Mengen] verteilen

doll [dɒl] n. Puppe, *die*

dollar ['dɒlə(r)] n. Dollar, *der*

dollop ['dɒləp] n. *(coll.)* Klacks, *der (ugs.)*

'doll's house n. Puppenhaus, *das*

dolphin ['dɒlfın] n. Delphin, *der*

domain [də'meın] n. Gebiet, *das*

dome [dəʊm] n. Kuppel, *die*

domestic [də'mestık] adj. **a)** *(household)* häuslich; *(family)* familiär ⟨Angelegenheit, Reibereien⟩; **b)** *(Econ.)* inländisch; Binnen-; **c)** **~ animal/cat** Haustier, *das*/-katze, *die*

domesticated [də'mestıkeıtıd] adj. gezähmt ⟨Tier⟩; *(fig.)* häuslich

dominant ['dɒmınənt] adj. vorherrschend

dominate ['dɒmıneıt] v. t. beherrschen. **domination** [dɒmı'neıʃn] n. [Vor]herrschaft, *die* **(over** über + Akk.)

domineering [dɒmı'nıərıŋ] adj. herrisch

domino ['dɒmınəʊ] n. Domino[stein], *der;* **~es** sing. *(game)* Domino[spiel], *das;* **play ~es** Domino spielen

¹don [dɒn] v. t. *(Liter.)* anlegen *(geh.)*

²don n. *(Univ.)* Dozent, *der*

donate [dəʊ'neıt] v. t. spenden; *(on large scale)* stiften. **donation** [dəʊ'neıʃn] n. Spende, *die* **(to** für); *(large-scale)* Stiftung, *die*

done [dʌn] *see* do

donkey ['dɒŋkı] n. Esel, *der*

donor ['dəʊnə(r)] n. Spender, *der*/Spenderin, *die*

don't [dəʊnt] *(coll.)* = do not; *see* do

doodle ['du:dl] v. i. [herum]kritzeln

doom [du:m] **1.** n. Verhängnis, *das.* **2.** v. t. verurteilen; **be ~ed** verloren sein; **be ~ed to fail** *or* **failure** zum Scheitern verurteilt sein

door [dɔ:(r)] n. Tür, *die; (of castle, barn)* Tor, *das;* **out of ~s** im Freien; **go out of ~s** nach draußen gehen

door: **~bell** n. Türklingel, *die;* **~-handle** n. Türklinke, *die;* **~-mat** n. Fußmatte, *die;* **~-step** n. Türstufe, *die;* **on one's/the ~-step** *(fig.)* vor jmds. Tür; **~way** n. Eingang, *der*

dope [dəʊp] **1.** n. **a)** *(sl.: narcotic)* Stoff, *der (salopp); b) (coll.: fool)* Dussel, *der.* **2.** v. t. dopen ⟨Pferd, Athleten⟩

dormant ['dɔ:mənt] adj. ruhend ⟨Tier, Pflanze⟩; untätig ⟨Vulkan⟩

dormitory ['dɔ:mıtərı] n. Schlafsaal, *der*

dormouse ['dɔ:maʊs] n., pl. **dormice** ['dɔ:maıs] Haselmaus, *die*

dose [dəʊs] **1.** n. Dosis, *die.* **2.** v. t. **~ sb. with sth.** jmdm. etw. geben

dot [dɒt] n. Punkt, *der;* **on the ~:** auf den Punkt genau

dote [dəʊt] v. i. **~ on sb./sth.** jmdn./etw. abgöttisch lieben

dotted ['dɒtɪd] *adj.* gepunktet

dotty ['dɒtɪ] *adj. (coll.) (silly)* dümmlich; *(feeble-minded)* vertrottelt *(ugs.); (absurd)* blödsinnig *(ugs.)*

double ['dʌbl] 1. *adj.* doppelt; ~ **bed/room** Doppelbett, *das/*-zimmer, *das;* **be** ~ **the height/width/length** doppelt so hoch/breit/lang sein. 2. *adv.* doppelt. 3. *n.* a) Doppelte, *das;* b) *(twice as much)* doppelt soviel; *(twice as many)* doppelt so viele; c) *(person)* Doppelgänger, *der/*-gängerin, *die;* d) *in pl. (Tennis etc.)* Doppel, *das;* e) **at the** ~ *(Mil.)* im Laufschritt; *(fig.)* ganz schnell. 4. *v.t.* verdoppeln. 5. *v.i.* sich verdoppeln. **double 'back** *v.i.* kehrtmachen *(ugs.).* **double 'up** krümmen (with vor + *Dat.*)

double: ~-'**bass** *n.* Kontrabaß, *der;* ~-'**check** *v.t. (verify twice)* zweimal kontrollieren; *(verify in two ways)* zweifach überprüfen; ~ '**chin** *n.* Doppelkinn, *das;* ~-'**cross** *v.t.* ein Doppelspiel treiben mit; ~-**decker** *n.* [dʌbl'dekə(r)] *n.* Doppeldeckerbus, *der;* ~ '**glazing** *n.* Doppelverglasung, *die;* ~-'**jointed** *adj.* sehr gelenkig

doubly ['dʌblɪ] *adv.* doppelt

doubt [daʊt] 1. *n.* Zweifel, *der* (about, as to, of an + *Dat.*); ~|s| |about *or* as to sth./as to whether ...| *(as to future)* Ungewißheit, *(as to fact)* Unsicherheit [über etw. *(Akk.)*/darüber, ob ...|; there's no ~ that ...: es besteht kein Zweifel daran, daß ...; ~|s| *(hesitations)* Bedenken *Pl.* (about gegen); no ~ *(certainly)* gewiß; *(probably)* sicherlich. 2. *v.i.* zweifeln. 3. *v.t.* zweifeln an (+ *Dat.*); I don't ~ that *or* it ich bezweifle das nicht; I ~ whether *or* if *or* that ...: ich bezweifle, daß ... **doubtful** ['daʊtfl] *adj.* skeptisch ‹Wesen›; ungläubig ‹Blick›

dough [dəʊ] *n.* a) Teig, *der;* b) *(sl.: money)* Knete, *die (salopp).* '**doughnut** *n.* [Berliner] Pfannkuchen, *der*

douse [daʊs] *v.t.* übergießen; *(extinguish)* ausmachen

¹**dove** [dʌv] *n.* Taube, *die*

²**dove** [dəʊv] *see* dive 1

dowdy ['daʊdɪ] *adj.* unansehnlich; *(shabby)* schäbig

¹**down** [daʊn] *n. (feathers)* Daunen *Pl.*

²**down** 1. *adv.* a) *(to lower place)* herunter/hinunter; *(in lift)* abwärts; b) *(in lower place, downstairs)* unten; ~ **there/here** da/hier unten; **the next floor** ~: ein Stockwerk tiefer; **be** ~ **with an illness** eine Krankheit haben;

be three points/games ~: mit drei Punkten/Spielen zurückliegen. 2. *prep.* herunter/hinunter; **lower** ~ **the river** weiter unten am Fluß; **walk** ~ **the hill/road** den Berg/die Straße heruntergehen; **fall** ~ **the stairs/steps** die Treppe/Stufen herunterstürzen; **fall** ~ **a hole/ditch** in ein Loch/ einen Graben fallen; **go** ~ **the pub** in die Kneipe gehen; **live just** ~ **the road** ein Stück weiter unten in der Straße wohnen; **be** ~ **the pub/town** in der Kneipe/Stadt sein; **I've got coffee |all|** ~ **my skirt** mein ganzer Rock ist voll Kaffee. 3. *v.t. (coll.)* schlucken *(ugs.)* ‹Getränk›; ~ **tools** die Arbeit niederlegen

down: ~-**and-'out** *n.* Stadtstreicher, *der/*-streicherin, *die;* ~**cast** *adj.* niedergeschlagen; ~**fall** *n.* Untergang, *der;* ~-'**hearted** *adj.* niedergeschlagen; ~'**hill** *adv.* bergab; ~ **payment** *n.* Anzahlung, *die;* ~**pour** *n.* Regenguß, *der;* ~**right** *adj.* ausgemacht; glatt ‹Lüge›; ~**stairs** 1. [-'-] *adv.* die Treppe hinunter ‹gehen, fallen, kommen›; unten ‹wohnen, sein›; 2. ['--] *adj.* im Erdgeschoß *nachgestellt;* ~**stream** *adv.* flußabwärts; ~-**to-'earth** *adj.* sachlich; ~**town** *adv.* im/ *(direction)* ins Stadtzentrum; ~**trodden** *adj.* unterdrückt; ~ '**under** *adv. (coll.)* in/*(to)* nach Australien/Neuseeland

downward ['daʊnwəd] 1. *adj.* nach unten gerichtet. 2. *adv.* abwärts ‹sich bewegen›; nach unten ‹sehen, gehen›. **downwards** ['daʊnwədz] *see* downward 2

dowry ['daʊrɪ] *n.* Aussteuer, *die*

doze [dəʊz] 1. *v.i.* dösen *(ugs.).* 2. *n.* Nickerchen, *das (ugs.).* **doze 'off** *v.i.* eindösen *(ugs.)*

dozen ['dʌzn] *n.* a) Dutzend, *das;* **half a** ~: sechs; b) *in pl. (coll.: many)* Dutzende *Pl.*

Dr *abbr.* doctor Dr.

drab [dræb] *adj.* langweilig; trostlos ‹Landschaft›; eintönig ‹Leben›

draft [drɑːft] 1. *n.* a) *(of speech)* Konzept, *das;(of treaty, bill)* Entwurf, *der;* b) *(Amer.) see* draught. 2. *v.t.* entwerfen. **drafty** *(Amer.) see* draughty

drag [dræg] 1. *v.t.,* -gg- schleppen. 2. *v.i.,* -gg- schleifen; *(fig.: pass slowly)* sich [hin]schleppen. 3. *n. (sl.)* **in** ~: in Frauenkleidung. **drag 'on** *v.i.* sich [da]hin schleppen

dragon ['drægn] *n.* Drache, *der.* '**dragonfly** *n.* Libelle, *die*

drain [dreɪn] **1.** *n.* Abflußrohr, *das;* *(underground)* Kanalisationsrohr, *das;* *(grating at roadside)* Gully, *der;* **go down the ~** *(fig. coll.)* für die Katz sein *(ugs.).* **2.** *v.t.* **a)** trockenlegen ⟨*Teich*⟩; entwässern ⟨*Land*⟩; ableiten ⟨*Wasser*⟩; **b)** *(Cookery)* abgießen ⟨*Wasser, Gemüse*⟩; **c)** austrinken ⟨*Glas*⟩. **3.** *v.i.* ⟨*Flüssigkeit:*⟩ ablaufen; ⟨*Geschirr, Gemüse:*⟩ abtropfen. **drainage** ['dreɪnɪdʒ] *n.* Kanalisation, *die.* **'draining-board** (*Brit.; Amer.:* **'drainboard**) *n.* Abtropfbrett, *das.* **'drainpipe** *n.* Regen[abfall]rohr, *das*

drake [dreɪk] *n.* Enterich, *der*

drama ['drɑːmə] *n.* Drama, *das.* **dramatic** [drə'mætɪk] *adj.* dramatisch. **dramatist** ['dræmətɪst] *n.* Dramatiker, *der*/Dramatikerin, *die.* **dramatize** ['dræmətaɪz] *v.t.* dramatisieren

drank *see* **drink 2**

drape [dreɪp] **1.** *v.t.* drapieren. **2.** *n.* *(Amer.: curtain)* Vorhang, *der.* **'draper** *n.* *(Brit.)* Textilkaufmann, *der;* **~'s** [**shop**] Textilgeschäft, *das*

drastic ['dræstɪk] *adj.* drastisch

draught [drɑːft] *n.* [Luft]zug, *der;* **there's a ~:** es zieht. **'draughtboard** *n.* *(Brit.)* Damebrett, *das.* **'draughts** *n.* *(Brit.)* Damespiel, *das.* **'draughtsman** [~men] *n., pl.* **draughtsmen** [~mən] Zeichner, *der*/Zeichnerin, *die.* **'draughty** *adj.* zugig

draw [drɔː] **1.** *v.t.,* **drew** [druː], **drawn** [drɔːn] **a)** *(pull)* ziehen; **~ the curtains/blinds** *(close)* die Vorhänge zuziehen/die Jalousien herunterlassen; **~ sth. towards one** etw. zu sich heranziehen; **b)** *(attract)* anlocken; **be ~n to sb.** von jmdm. angezogen werden; **c)** *(take out)* herausziehen; schöpfen ⟨*Wasser*⟩; **~ money from the bank** Geld bei der Bank holen/abheben; **d)** beziehen ⟨*Gehalt, Rente, Arbeitslosenunterstützung*⟩; **e)** ziehen ⟨*Strich*⟩; zeichnen ⟨*geometrische Figur, Bild*⟩; **f)** ziehen ⟨*Parallele, Vergleich*⟩; herausstellen ⟨*Unterschied*⟩. **2.** *v.i.* **drew, drawn: ~ to an end** zu Ende gehen. **3.** *n.* **a)** *(raffle)* Tombola, *die;* **b)** *[result of] drawn game]* Unentschieden, *das.* **draw back 1.** *v.t.* zurückziehen. **2.** *v.i.* zurückweichen. **draw 'in** *v.i.* einfahren; ⟨*Tage:*⟩ kürzer werden. **draw 'out** *v.i.* abfahren; ⟨*Tage:*⟩ länger werden. **draw 'up 1.** *v.t.* **a)** aufsetzen

⟨*Vertrag*⟩; aufstellen ⟨*Liste*⟩; **b)** *(pull closer)* heranziehen. **2.** *v.i.* [an]halten

draw: ~-back *n.* Nachteil, *der;* **~-bridge** *n.* Zugbrücke, *die*

drawer [drɔː(r), 'drɔːə(r)] *n.* Schublade, *die*

'drawing *n.* *(sketch)* Zeichnung, *die*

drawing: ~-board *n.* Zeichenbrett, *das;* **~-pin** *n.* *(Brit.)* Reißzwecke, *die;* **~-room** *n.* Salon, *der*

drawl [drɔːl] **1.** *v.i.* gedehnt sprechen. **2.** *n.* gedehntes Sprechen

drawn *see* **draw 1, 2**

dread [dred] **1.** *v.t.* sich sehr fürchten vor (+ *Dat.*); **the ~ed day/moment** der gefürchtete Tag/Augenblick. **2.** *n.* Angst, *die.* **dreadful** ['dredfl] *adj.* schrecklich; *(coll.: very bad)* fürchterlich; **I feel ~** *(unwell)* ich fühle mich scheußlich *(ugs.).* **'dreadfully** *adv.* schrecklich; *(coll.: very badly)* fürchterlich

dream [driːm] **1.** *n.* Traum, *der;* **have a ~ about sb./sth.** von jmdm./etw. träumen. **2.** *v.i. & t.* **dreamt** [dremt] *or* **dreamed** träumen

dreary ['drɪərɪ] *adj.* trostlos

dredge [dredʒ] *v.t.* ausbaggern. **'dredger** *n.* Bagger, *der*

dregs [dregz] *n. pl.* [Boden]satz, *der*

drench [drentʃ] *v.t.* durchnässen

dress [dres] **1.** *n.* Kleid, *das; (clothing)* Kleidung, *die.* **2.** *v.t.* **a)** anziehen; **be well ~ed** gut gekleidet sein; **get ~ed** sich anziehen; **b)** verbinden ⟨*Wunde*⟩. **3.** *v.i.* sich anziehen. **dress 'up** *v.i.* sich feinmachen

'dresser *n.* **a)** Anrichte, *die;* **b)** *(Amer.)* *see* **dressing-table**

'dressing *n.* **a)** *no pl.* Anziehen, *das;* **b)** *(Cookery)* Dressing, *das;* **c)** *(Med.)* Verband, *der*

dressing: ~-gown *n.* Bademantel, *der;* **~-room** *n.* *(Sport)* Umkleideraum, *der; (for actor)* Garderobe, *die;* **~-table** *n.* Frisierkommode, *die*

dress: ~-maker *n.* Damenschneider, *der*/-schneiderin, *die;* **~-making** *n.* Damenschneiderei, *die;* **~ rehearsal** *n.* Generalprobe, *die*

drew *see* **draw 1, 2**

dribble ['drɪbl] *v.i.* **a)** *(slobber)* sabbern; **b)** *(Sport)* dribbeln

dried [draɪd] *adj.* getrocknet; **~ fruit[s]** Dörrobst, *das;* **~ milk** Trockenmilch, *die*

drier ['draɪə(r)] *n.* *(for hair)* Trockenhaube, *die; (hand-held)* Fön Ⓡ, *der; (for laundry)* [Wäsche]trockner, *der*

drift [drɪft] 1. *n.* a) *(of snow or sand)* Verwehung, *die;* b) *(gist)* **get** *or* **catch the ~ of** sth. etw im wesentlichen verstehen. 2. *v. i.* a) treiben; ⟨*Wolke:*⟩ ziehen; b) ⟨*Sand, Schnee:*⟩ zusammengeweht werden. '**driftwood** *n.* Treibholz, *das*

drill [drɪl] 1. *n.* a) *(tool)* Bohrer, *der;* b) *(Mil.: training)* Drill, *der.* 2. *v. t. & i.* bohren (for nach)

drink [drɪŋk] 1. *n.* Getränk, *das; (alcoholic)* Glas, *das; (not with food)* Drink, *der;* **have a ~:** [etwas] trinken; *(alcoholic)* ein Glas trinken. 2. *v. t. & i.* **drank** [dræŋk], **drunk** [drʌŋk] trinken. **drinkable** ['drɪŋkəbl] *adj.* trinkbar. '**drinking-water** *n.* Trinkwasser, *das*

drip [drɪp] 1. *n.* a) *(coll.: feeble person)* Schlappschwanz, *der (salopp).* 2. *v. i.,* -pp- tropfen; be **~ping with water/moisture** triefend naß sein. '**drip-dry** *adj.* bügelfrei

'**dripping** *n. (Cookery)* Schmalz, *das*

drive [draɪv] 1. *n.* a) Fahrt, *die;* b) *(private road)* Zufahrt, *die; (entrance) (to small building)* Einfahrt, *die; (to large building)* Auffahrt, *die;* c) *(energy)* Tatkraft, *die;* d) *(Psych.)* Trieb, *der;* e) *(Motor Veh.)* **left-hand/right-hand ~:** Links-/Rechtssteuerung, *die.* 2. *v. t.,* **drove** [drəʊv], **driven** ['drɪvn] a) fahren; b) treiben ⟨*Tier⟩;* c) *(compel to move)* vertreiben (out of, from aus); d) *(fig.)* **~ sb. to** sth. jmdn. zu etw. treiben; **~ sb. to do** sth. *or* **into doing** sth. jmdn. dazu treiben, etw. zu tun; c) *(power)* antreiben. 3. *v. i.,* **drove, driven** a) fahren; **can you ~?** kannst du Auto fahren?; b) *(go by car)* mit dem [eigenen] Auto fahren. '**drive at** *v. t. (fig.)* hinauswollen auf (+ *Akk.*); **what are you driving at?** worauf wollen Sie hinaus? **drive a'way** 1. *v. i.* wegfahren. 2. *v. t.* a) wegfahren; b) *(chase away)* vertreiben. **drive 'off** *see* drive away. **drive 'on** *v. i.* weiterfahren. **drive 'up** *v. i.* vorfahren (to vor + *Dat.*)

'**drive-in** *adj.* Drive-in-; **~ cinema** *or (Amer.)* **movie |theater|** Autokino, *das*

drivel ['drɪvl] *n.* Gefasel, *das (ugs.);* **talk ~:** faseln *(ugs.)*

driven *see* drive 2, 3

driver ['draɪvə(r)] *n.* Fahrer, *der/*Fahrerin, *die; (of locomotive)* Führer, *der/*Führerin, *die;* **~s license** *(Amer.)* Führerschein, *der*

driving ['draɪvɪŋ] 1. *n.* Fahren, *das.* 2. *adj.* peitschend ⟨*Regen⟩*

driving: ~-instructor *n.* Fahrlehrer, *der/*-lehrerin, *die;* **~-lesson** *n.* Fahrstunde, *die;* **~-licence** *n.* Führerschein, *der;* **~-school** *n.* Fahrschule, *die;* **~-test** *n.* Fahrprüfung, *die*

drizzle ['drɪzl] 1. *n.* Nieseln, *das.* 2. *v. i.* **it's drizzling** es nieselt

drone [drəʊn] 1. *v. i.* a) ⟨*Biene:*⟩ summen; ⟨*Maschine:*⟩ brummen; b) ⟨*Rezitator:*⟩ leiern. 2. *n. see* 1: Summen, *das;* Brummen, *das;* Geleier, *das*

drool [druːl] *v. i.* **~ over** eine kindische Freude haben an (+ *Dat.*)

droop [druːp] *v. i.* herunterhängen; ⟨*Blume:*⟩ den Kopf hängen lassen

drop [drɒp] 1. *n.* a) Tropfen, *der;* in **~s** tropfenweise; b) *(decrease)* Rückgang, *der.* 2. *v. i.,* -pp-: a) *(fall) (accidentally)* [herunter]fallen; *(deliberately)* sich [hinunter]fallen lassen; b) *(in amount etc.)* sinken; ⟨*Preis, Wert:*⟩ sinken, fallen; ⟨*Wind:*⟩ sich legen; ⟨*Stimme:*⟩ sich senken. 3. *v. t.,* -pp-: a) fallen lassen; abwerfen ⟨*Bomben, Nachschub⟩;* b) *(discontinue, abandon)* fallenlassen; c) *(omit)* auslassen. **drop 'by, drop 'in** *v. i.* vorbeikommen. **drop 'off** 1. *v. i.* a) *(fall off)* abfallen; b) *(fall asleep)* einnicken. 2. *v. t.* absetzen ⟨*Fahrgast⟩.* **drop 'out** *v. i.* a) herausfallen (of aus); b) *(withdraw)* aussteigen *(ugs.)* (of aus); *(beforehand)* seine Teilnahme absagen

'**drop-out** *n.* Aussteiger, *der/*Aussteigerin, *die*

drought [draʊt] *n.* Dürre, *die*

drove *see* drive 2, 3

drown [draʊn] 1. *v. i.* ertrinken. 2. *v. t.* ertränken; **be ~ed** ertrinken

drowsy ['draʊzɪ] *adj.* schläfrig; *(on just waking)* verschlafen

drudgery ['drʌdʒərɪ] *n.* Schufterei, *die*

drug [drʌg] 1. *n.* a) *(Med.)* [Arznei]mittel, *das;* b) *(narcotic)* Droge, *die;* **be on ~s** Rauschgift nehmen. 2. *v. t.,* -gg- betäuben ⟨*Person⟩;* **~ sb.'s food/drink** jmds. Essen/Getränk *(Dat.)* ein Betäubungsmittel beimischen

drug: ~ addict *n.* Drogensüchtige, *der/die;* **~ addiction** *n.* Drogensucht, *die;* **~-store** *n. (Amer.)* Drugstore, *der*

drum [drʌm] 1. *n.* a) Trommel, *die;* b) in *pl. (in jazz or pop)* Schlagzeug, *das;* c) *(container)* Faß, *das.* 2. *v. i.* trommeln. **drum 'up** *v. i.* auftreiben

'**drummer** *n.* Schlagzeuger, *der*

'**drumstick** *n.* a) Trommelschlegel, *der;* b) *(Cookery)* Keule, *die*

drunk [drʌŋk] **1.** *adj.* **be ~:** betrunken sein; **get ~:** betrunken werden (**on** von); *(intentionally)* sich betrinken (**on** mit). **2.** *n.* Betrunkene, *der/die*

drunkard ['drʌŋkəd] *n.* Trinker, *der/* Trinkerin, *die*

drunken ['drʌŋkn] *attrib. adj.* betrunken; *(habitually)* ständig betrunken; **~ driving** Trunkenheit am Steuer.

'**drunkenness** *n.* Betrunkenheit, *die; (habitual)* Trunksucht, *die*

dry [draɪ] **1.** *adj.* trocken; trocken, *(very ~)* herb ⟨*Wein*⟩; ausgetrocknet ⟨*Flußbett*⟩; **get** *or* **become ~:** trocknen. **2.** *v. t.* **a)** trocknen ⟨*Haare, Wäsche*⟩; abtrocknen ⟨*Geschirr, Baby*⟩; **~ one-self** sich abtrocknen; **~ one's eyes** *or* **tears/hands** sich ⟨*Dat.*⟩ die Tränen abwischen/die Hände abtrocknen; **b)** *(preserve)* trocknen; dörren ⟨*Obst, Fleisch*⟩. **3.** *v. i.* trocknen. **dry** '**out** *v. t. & i.* trocknen. **dry** '**up 1.** *v. t.* abtrocknen. **2.** *v. i.* **a)** *(~ the dishes)* abtrocknen; **b)** ⟨*Brunnen, Quelle:*⟩ versiegen; ⟨*Fluß, Teich:*⟩ austrocknen

dry: ~-'**clean** *v. t.* chemisch reinigen; ~-'**cleaner's** *n.* chemische Reinigung; ~-'**cleaning** *n.* chemische Reinigung

'**dryer** *see* drier

'**dryness** *n.* Trockenheit, *die*

dual ['dju:əl] *adj.* doppelt. **dual** '**carriageway** *n.* (*Brit.*) Straße mit Mittelstreifen. **dual-**'**purpose** *adj.* zweifach verwendbar

dubious ['dju:bɪəs] *adj.* *(doubting)* unschlüssig; *(suspicious)* zweifelhaft

duchess ['dʌtʃɪs] *n.* Herzogin, *die*

duck [dʌk] **1.** *n.* Ente, *die.* **2.** *v. i.* sich [schnell] ducken. **3.** *v. t.* **~ one's head** den Kopf einziehen

duckling ['dʌklɪŋ] *n.* Entenküken, *das*

duct [dʌkt] *n.* Rohr, *das; (for air)* Ventil, *das*

dud [dʌd] **1.** *n.* *(useless thing)* Niete, *die* (*ugs.*); *(counterfeit)* Fälschung, *die.* **2.** *adj.* mies (*ugs.*); schlecht; *(fake)* gefälscht; geplatzt ⟨*Scheck*⟩

due [dju:] **1.** *adj.* **a)** *(owed)* geschuldet; zustehend ⟨*Eigentum, Recht usw.*⟩; **there's sth. ~ to me, I've got sth. ~:** mir steht etw. zu; **b)** *(immediately payable)* fällig; **c)** *(that it is proper to give or use)* gebührend; angemessen ⟨*Belohnung*⟩; **be ~ to sb.** jmdm. gebühren; **with all ~ respect** bei allem gebotenen Respekt; **d)** *(attributable)* **the mistake was ~ to negligence** der Fehler war durch Nachlässigkeit verursacht; **it's ~ to**

her that we missed the train ihretwegen verpaßten wir den Zug; **e)** *(scheduled, expected)*; **be ~ to do sth.** etw. tun sollen; **be ~ |to arrive|** ankommen sollen; **f)** *(likely to get, deserving)* **be ~ for sth.** etw. verdienen. **2.** *adv.* **a)** **~ north** genau nach Norden; **b)** **~ to** auf Grund (+ *Gen.*); aufgrund (+ *Gen.*). **3.** *n.* **a)** **give sb. his ~:** jmdm. Gerechtigkeit widerfahren lassen; **b)** **~s** *(fees)* Gebühren *Pl.*

duel ['dju:əl] *n.* Duell, *das*

duet [dju:'et] *n.* *(for voices)* Duett, *das; (instrumental)* Duo, *das*

duffle ['dʌfl]: **~ bag** *n.* Matchbeutel, *der;* **~ coat** *n.* Dufflecoat, *der*

dug *see* dig

duke [dju:k] *n.* Herzog, *der*

dull [dʌl] **1.** *adj.* **a)** *(stupid)* beschränkt; *(slow to understand)* begriffsstutzig; **b)** *(boring)* langweilig; **c)** *(gloomy)* trübe ⟨*Wetter, Tag*⟩. **2.** *v. t.* abstumpfen ⟨*Geist, Sinne, Verstand*⟩

duly ['dju:lɪ] *adv.* ordnungsgemäß

dumb [dʌm] *adj.* **a)** stumm; **b)** *(coll.: stupid)* doof (*ugs.*)

dumbfounded [dʌm'faʊndɪd] *adj.* sprachlos

dummy ['dʌmɪ] *n.* **a)** *(of tailor)* Schneiderpuppe, *die; (in shop)* Schaufensterpuppe, *die; (of ventriloquist)* Puppe, *die; (stupid person)* Dummkopf, *der* (*ugs.*); **like a stuffed ~:** wie ein Ölgötze (*ugs.*); **b)** *(imitation)* Attrappe, *die;* **c)** *(esp. Brit.: for baby)* Schnuller, *der*

dump [dʌmp] **1.** *n.* **a)** *(place)* Müllkippe, *die; (heap)* Müllhaufen, *der; (permanent)* Mülhalde, *die;* **b)** *(Mil.)* Depot, *das;* **c)** *(coll.: town)* Kaff, *das* (*ugs.*). **2.** *v. t.* *(dispose of)* werfen; *(deposit)* abladen ⟨*Sand, Müll usw.*⟩; *(leave)* lassen; *(place)* abstellen

dumpling ['dʌmplɪŋ] *n.* Kloß, *der*

dumps *n. pl.* **be** *or* **feel down in the ~:** ganz down sein (*ugs.*)

dunce [dʌns] *n.* Null, *die* (*ugs.*)

dune [dju:n] *n.* Düne, *die*

dung [dʌŋ] *n.* Dung, *der*

dungarees [dʌŋgə'ri:z] *n. pl.* Latzhose, *die*

dungeon ['dʌndʒən] *n.* Kerker, *der*

dunk [dʌŋk] *v. t.* tunken

dupe [dju:p] **1.** *v. t.* übertölpeln. **2.** *n.* Dumme, *der/die*

duplex ['dju:pleks] *adj.* *(esp. Amer.)* *(two-storey)* zweistöckig ⟨*Wohnung*⟩; *(two-family)* Zweifamilien⟨*haus*⟩

duplicate 1. ['dju:plɪkət] *adj.* **a)** *(identical)* Zweit-; **b)** *(twofold)* doppelt. **2.** *n.*

Kopie, *die; (second copy of letter/document/key)* Duplikat, *das;* **in** ~: in doppelter Ausfertigung. **3.** ['dju:plɪkeɪt] *v. t.* **a)** *(make a copy of, make in* ~*)* ~ **sth.** eine zweite Anfertigung von etw. machen; **b)** *(on machine)* vervielfältigen; **c)** *(do twice)* noch einmal tun

durable ['djʊərəbl] *adj.* haltbar; dauerhaft ‹*Friede, Freundschaft usw.*›

duration [djʊə'reɪʃn] *n.* Dauer, *die*

duress [djʊə'res] *n.* Zwang, *der*

during ['djʊərɪŋ] *prep.* während; *(at a point in)* in (+ *Dat.*)

dusk [dʌsk] *n.* Einbruch der Dunkelheit

dust [dʌst] **1.** *n.* Staub, *der.* **2.** *v. t.* abstauben ‹*Möbel*›; ~ **a room/ house** in einem Zimmer/Haus Staub wischen. **3.** *v. i.* Staub wischen. **'dustbin** *n. (Brit.)* Mülltonne, *die.* **'dustcart** *(Brit.)* Müllwagen, *der*

'duster *n.* Staubtuch, *das*

dust: ~**-jacket** *n.* Schutzumschlag, *der;* ~**man** [~mən] *n., pl.* ~**men** [~mən] *(Brit.)* Müllmann, *der;* ~**pan** *n.* Kehrschaufel, *die*

'dusty *adj.* staubig; verstaubt ‹*Bücher, Möbel*›

Dutch [dʌtʃ] **1.** *adj.* holländisch; **sb. is** ~: jmd. ist Holländer/Holländerin. **2.** *n.* **a)** *(language)* Holländisch, *das; see also* **English 2 a**; **b) the** ~ *pl.* die Holländer

Dutch: ~ **'courage** *n.* angetrunkener Mut; ~**man** [~mən] *n., pl.* ~**men** [~mən] Holländer, *der;* ~**woman** *n.* Holländerin, *die*

dutiful ['dju:tɪfl] *adj.* pflichtbewußt

duty ['dju:tɪ] *n.* **a)** Pflicht, *die; (task)* Aufgabe, *die;* **be on** ~: Dienst haben; **off** ~: nicht im Dienst; **be off** ~: keinen Dienst haben; ‹*ab ... Uhr*› dienstfrei sein; **b)** *(tax)* Zoll, *der;* **pay** ~ **on sth.** Zoll für etw. bezahlen. **'duty-free** *adj.* zollfrei

duvet ['du:veɪ] *n.* Federbett, *das*

dwarf [dwɔ:f] *n., pl.* ~**s** *or* **dwarves** ['dwɔ:vz] Zwerg, *der*/Zwergin, *die*

dwell [dwel] *v. i.,* **dwelt** [dwelt] *(literary)* wohnen. **'dwell [up]on** *v. t. (in discussion)* sich ausführlich befassen mit; *(in thought)* in Gedanken verweilen bei

'dwelling *n.* Wohnung, *die*

dwelt *see* **dwell**

dwindle ['dwɪndl] *v. i.* ~ [**away**] abnehmen; ‹*Unterstützung, Interesse:*› nachlassen; ‹*Vorräte:*› schrumpfen

dye [daɪ] **1.** *n.* Färbemittel, *das.* **2.** *v. t.,* ~**ing** ['daɪɪŋ] färben

dying ['daɪɪŋ] *adj.* sterbend; absterbend ‹*Baum*›

dyke *see* **dike**

dynamic [daɪ'næmɪk] *adj.* dynamisch. **dynamism** ['daɪnəmɪzm] *n.* Dynamik, *die*

dynamite ['daɪnəmaɪt] *n.* Dynamit, *das*

dynamo ['daɪnəməʊ] *n.* Dynamo, *der; (in car)* Lichtmaschine, *die*

dynasty ['dɪnəstɪ] *n.* Dynastie, *die*

dysentry ['dɪsəntrɪ] *n.* Ruhr, *die*

E

E, e [i:] *n.* E, e, *das*

E. *abbr.* **a) east** O; **b) eastern** ö.

each [i:tʃ] **1.** *adj.* jeder/jede/jedes; **they cost** *or* **are a pound** ~: sie kosten ein Pfund pro Stück. **2.** *pron.* **a)** jeder/jede/jedes; **b)** ~ **other** sich

eager ['i:gə(r)] *adj.* eifrig; **be** ~ **to do sth.** etw. unbedingt tun wollen. **'eagerly** *adv.* eifrig; gespannt ‹*warten*›

eagle ['i:gl] *n.* Adler, *der*

¹ear [ɪə(r)] *n.* Ohr, *das*

²ear *n. (Bot.)* Ähre, *die*

ear: ~**ache** *n.* Ohrenschmerzen *Pl.;* ~**-drum** *n.* Trommelfell, *das*

earl [ɜ:l] *n.* Graf, *der*

'ear lobe *n.* Ohrläppchen, *das*

early ['ɜ:lɪ] **1.** *adj.* früh. **2.** *adv.* früh; **I am a bit** ~: ich bin etwas zu früh gekommen; ~ **next week** Anfang der nächsten Woche; ~ **in June** Anfang Juni; **from** ~ **in the morning till late at night** von früh [morgens] bis spät [nachts]; ~ **on** schon früh

ear: ~**mark** *v. t.* vorsehen; ~**-muffs** *n. pl.* Ohrenschützer, *Pl.*

earn [ɜ:n] *v. t.* verdienen; *(bring in as income or interest)* einbringen

earnest ['ɜ:nɪst] **1.** *adj.* ernsthaft. **2.** **in** ~: mit vollem Ernst

earnings ['ɜ:nɪŋz] *n. pl.* Verdienst, *der; (of business etc.)* Ertrag, *der*

ear: ~**phones** *n. pl.* Kopfhörer, *der;*

~-**plug** *n.* Ohropax, *das* Ⓦ; ~-**ring** *n.* Ohrring, *der;* ~-**shot** *n.* out of/within ~ shot außer/in Hörweite

earth [ɜːθ] 1. *n. (also Brit. Electr.)* Erde, *die;* how/what *etc.* on ~ ...? wie/was *usw.* in aller Welt ...? 2. *v.t. (Brit. Electr.)* erden

earthenware ['ɜːθnweə(r)] 1. *n.* Tonwaren *Pl.* 2. *adj.* Ton-

earth: ~**quake** *n.* Erdbeben, *das;* ~**worm** *n.* Regenwurm, *der*

'earthy *adj.* a) erdig; b) *(coarse)* derb

earwig ['ɪəwɪg] *n.* Ohrwurm, *der*

ease [iːz] 1. *n.* a) set sb. at ~: jmdn. beruhigen; at [one's] ~: entspannt; be *or* feel at [one's] ~: sich wohl fühlen; [stand] at ~! *(Mil.)* rührt euch!; b) with ~ *(without difficulty)* mit Leichtigkeit. 2. *v.t.* lindern ⟨Schmerz, Kummer⟩; entspannen ⟨Lage⟩; verringern ⟨Belastung, Druck, Spannung⟩. 3. *v.i.* nachlassen

easel ['iːzl] *n.* Staffelei, *die*

easily ['iːzɪlɪ] *adv.* leicht

easiness ['iːzɪnɪs] *n.* Leichtigkeit, *die*

east [iːst] 1. *n.* a) Osten, *der;* in/to[wards]/from the ~: im/nach/von Osten; to the ~ of östlich von; b) *usu.* E~ *(Geog., Polit.)* Osten, *der.* 2. *adj.* östlich; Ost⟨küste, -wind, -grenze⟩. 3. *adv.* nach Osten; ~ of östlich von.

'East Ber'lin *pr. n. (Hist.)* Ostberlin, *das.* **'eastbound** *adj.* ⟨Zug, Verkehr usw.⟩ in Richtung Osten

Easter ['iːstə(r)] *n.* Ostern, *das od. Pl.*

'Easter egg *n.* Osterei, *das*

easterly ['iːstəlɪ] *adj.* östlich; ⟨Wind⟩ aus östlichen Richtungen

eastern ['iːstən] *adj.* östlich; Ost⟨grenze, -hälfte, -seite⟩; ~ Germany Ostdeutschland, *das.* **Eastern 'Europe** *pr. n.* Osteuropa, *das*

Easter 'Sunday *n.* Ostersonntag, *der*

East: ~ **'German** *(Hist.)* 1. *adj.* ostdeutsch; 2. *n.* Ostdeutsche, *der/die;* ~ **'Germany** *pr. n. (Hist.)* Ostdeutschland *(das)*

eastward(s) [iːstwəd(z)] *adv.* ostwärts

easy ['iːzɪ] 1. *adj.* a) leicht; on ~ terms auf Raten ⟨kaufen⟩; b) sorglos ⟨Leben, Zeit⟩; c) *(free from constraint)* ungezwungen. 2. *adv.* leicht; **easier said than done** leichter gesagt als getan; **take it ~!** *(calm down!)* beruhige dich! **'easy chair** *n.* Sessel, *der.* **easy'going** *adj.* gelassen; *(lax)* nachlässig

eat [iːt] *v.t. & i.,* ate [et, eɪt], eaten ['iːtn] essen; ⟨Tier:⟩ fressen. **eat a'way** *v.t.*

⟨Rost, Säure:⟩ zerfressen. **eat 'out** *v.i.* essen gehen. **eat 'up** *v.t.* aufessen; ⟨Tier:⟩ auffressen

eaten *see* **eat**

eau-de-Cologne [əʊdəkə'ləʊn] *n.* Kölnisch Wasser, *das*

eaves [iːvz] *n. pl.* Dachgesims, *das.* **'eavesdrop** *v.i.* lauschen; ~ on belauschen. **'eavesdropper** *n.* Lauscher, *der*/Lauscherin, *die*

ebb [eb] 1. *n.* Ebbe, *die;* be at a low ~ *(fig.)* ⟨Person, Stimmung, Moral:⟩ auf dem Nullpunkt sein. 2. *v.i.* zurückgehen; ~ **away** *(fig.)* dahinschwinden. **'ebb-tide** *n.* Ebbe, *die*

ebony ['ebənɪ] *n.* Ebenholz, *das*

EC *abbr.* European Community EG

eccentric [ɪk'sentrɪk] 1. *adj.* exzentrisch. 2. *n.* Exzentriker, *der*/Exzentrikerin, *die.* **eccentricity** [eksen'trɪsɪtɪ] *n.* Exzentrizität, *die*

ecclesiastical [ɪkliːzɪ'æstɪkl] *adj.* kirchlich; geistlich ⟨Musik⟩

echo ['ekəʊ] 1. *n.* Echo, *das.* 2. *v.t.* zurückwerfen; *(fig.: repeat)* wiederholen

éclair [eɪ'kleə(r)] *n.* Eclair, *das*

eclipse [ɪ'klɪps] *n. (Astron.)* Finsternis, *die;* ~ of the sun Sonnenfinsternis, *die*

ecological [iːkə'lɒdʒɪkl] *adj.* ökologisch

ecology [ɪ'kɒlədʒɪ] *n.* Ökologie, *die*

economic [iːkə'nɒmɪk] *adj.* a) Wirtschafts⟨politik, -abkommen, -system⟩; wirtschaftlich ⟨Entwicklung, Zusammenbruch⟩; b) *(giving adequate return)* wirtschaftlich

economical [iːkə'nɒmɪkl] *adj.* wirtschaftlich; sparsam ⟨Person⟩; be ~ with sth. mit etw. haushalten. **eco'nomically** *adv.* wirtschaftlich; *(not wastefully)* sparsam

economics [iːkə'nɒmɪks] *n.* Wirtschaftswissenschaft, *die (meist Pl.)*

economist [ɪ'kɒnəmɪst] *n.* Wirtschaftswissenschaftler, *der*/-wissenschaftlerin, *die*

economize [ɪ'kɒnəmaɪz] *v.i.* sparen; ~ on sth. etw. sparen

economy [ɪ'kɒnəmɪ] *n.* a) *(frugality)* Sparsamkeit, *die;* b) *(instance)* Einsparung, *die;* make economies zu Sparmaßnahmen greifen; c) *(of country etc.)* Wirtschaft, *die.* **e'conomy size** *n.* Haushaltspackung, *die*

ecstasy ['ekstəsɪ] *n.* Ekstase, *die.* **ecstatic** [ɪk'stætɪk] *adj.* ekstatisch

ECU, ecu [eɪ'kjuː] *abbr.* European currency unit Ecu, *der od. die*

eddy ['edɪ] *n.* Strudel, *der*

edge [edʒ] 1. *n.* **a)** *(of knife, razor, weapon)* Schneide, *die;* on ~ *(fig.)* nervös *od.* gereizt (**about** wegen); **b)** *(of solid, bed, table)* Kante, *die; (of sheet of paper, road, forest, cliff)* Rand, *der.* **2.** *v. i.* sich schieben

edgy ['edʒɪ] *adj.* nervös

edible ['edɪbl] *adj.* eßbar

edict ['i:dɪkt] *n.* Erlaß, *der*

edit ['edɪt] *v. t.* herausgeben (*Zeitung*); redigieren (*Buch, Artikel, Manuskript*). **edition** [ɪ'dɪʃn] *n.* Ausgabe, *die.* **editor** ['edɪtə(r)] *n.* Redakteur, *der*/Redakteurin, *die; (of particular work)* Bearbeiter, *der*/Bearbeiterin, *die; (of newspaper)* Herausgeber, *der*/-geberin, *die.* **editorial** [edɪ-'tɔ:rɪəl] **1.** *n.* Leitartikel, *der.* **2.** *adj.* redaktionell

educate ['edjʊkeɪt] *v. t.* **a)** *(bring up)* erziehen; *(train mind and character of)* bilden; **b)** *(provide schooling for)* he was ~d at ...: er hat seine Ausbildung in ... erhalten. **educated** ['edjʊkeɪtɪd] *adj.* gebildet. **education** [edjʊ'keɪʃn] *n.* Erziehung, *die; (system)* Erziehungswesen, *das.* **educational** [edjʊ'keɪʃənl] *adj.* pädagogisch; Lehr(*film, -spiele, -anstalt*); Erziehungs(*methoden, -arbeit*)

EEC *abbr.* European Economic Community EWG

eerie ['ɪərɪ] *adj.* unheimlich

eel [i:l] *n.* Aal, *der*

effect [ɪ'fekt] *n.* **a)** Wirkung, *die* (on auf + *Akk.*); the ~s of sth. on sth. die Auswirkungen einer Sache *(Gen.)* auf etw. *(Akk.);* take ~: die erwünschte Wirkung erzielen; in ~: in Wirklichkeit; **b)** come into ~: gültig werden; (*Gesetz:*) in Kraft treten; put into ~: in Kraft setzen (*Gesetz*); verwirklichen (*Plan*); with ~ from 2 November/Monday mit Wirkung vom 2. November/von Montag

effective [ɪ'fektɪv] *adj.* **a)** wirksam (*Mittel, Maßnahmen*); be ~ (*Arzneimittel:*) wirken; **b)** *(in operation)* gültig; ~ from/as of mit Wirkung vom. **effectively** *adv.* *(in fact)* effektiv; *(with effect)* wirkungsvoll

effectual [ɪ'fektjʊəl] *adj.* wirksam

effeminate [ɪ'femɪnət] *adj.* unmännlich

effervescent [efə'vesənt] *adj.* sprudelnd; *(fig.)* übersprudelnd

efficiency [ɪ'fɪʃənsɪ] *n. (of person)* Fähigkeit, *die;* Tüchtigkeit, *die; (of machine, factory, engine)* Leistungsfähig-

keit, *die; (of organization, method)* gutes Funktionieren

efficient [ɪ'fɪʃənt] *adj.* fähig (*Person*); tüchtig (*Arbeiter, Sekretärin*); leistungsfähig (*Maschine, Motor, Fabrik*); gut funktionierend (*Methode, Organisation*). **efficiently** *adv.* gut

effigy ['efɪdʒɪ] *n.* Bildnis, *das*

effluent ['eflʊənt] Abwässer *Pl.*

effort ['efət] *n.* **a)** Anstrengung, *die;* Mühe, *die;* **make an/every** ~ *(physically)* sich anstrengen; *(mentally)* sich bemühen; **b)** *(attempt)* Versuch, *der.* **'effortless** *adj.* mühelos

effrontery [ɪ'frʌntərɪ] *n.* Dreistigkeit, *die;* **have the** ~ **to do sth.** die Stirn besitzen, etw. zu tun

effusive [ɪ'fju:sɪv] *adj.* überschwenglich; exaltiert *(geh.)* (*Person*)

e.g. [i:'dʒi:] *abbr.* for example z. B.

egg [eg] *n.* Ei, *das.* **egg 'on** *v. t.* anstacheln

egg: ~-**cup** *n.* Eierbecher, *der;* ~**shell** *n.* Eierschale, *die;* ~-**timer** *n.* Eieruhr, *die;* ~-**white** *n.* Eiweiß, *das;* ~ **yolk** *n.* Eigelb, *das*

ego ['egəʊ, 'i:gəʊ] *n., pl.* ~**s a)** *(Psych.)* Ego, *das;* **b)** *(self-esteem)* Selbstbewußtsein, *das*

Egypt ['i:dʒɪpt] *pr. n.* Ägypten *(das).*

Egyptian [ɪ'dʒɪpʃn] **1.** *adj.* ägyptisch. **2.** *n. (person)* Ägypter, *der*/Ägypterin, *die*

eiderdown ['aɪdədaʊn] *n.* Federbett, *das*

eight [eɪt] **1.** *adj.* acht; **at** ~: um acht; **half past** ~: halb neun; ~ **thirty** acht Uhr dreißig; ~ **ten/fifty** zehn nach acht/vor neun; *(esp. in timetable)* acht Uhr zehn/fünfzig; ~**year-old boy** achtjähriger Junge; **an** ~**year-old** ein Achtjähriger/eine Achtjährige; **at |the age of|** ~, **aged** ~: mit acht Jahren; ~ **times** achtmal. **2.** *n.* Acht, *die;* **the first/last** ~: die ersten/letzten acht; **there were** ~ **of us present** wir waren [zu] acht

eighteen [eɪ'ti:n] **1.** *adj.* achtzehn. **2.** *n.* Achtzehn, *die; See also* **eight.**

eighteenth [eɪ'ti:nθ] **1.** *adj.* achtzehnt... **2.** *n. (fraction)* Achtzehntel, *das. See also* **eighth**

eighth [eɪtθ] **1.** *adj.* acht...; **be/come** ~: achter sein/als achter ankommen; ~**largest** achtgrößt... **2.** *n. (in sequence)* achte, *der/die/das; (in rank)* Achte, *der/die/das; (fraction)* Achtel, *das;* **the** ~ **of May** der achte Mai

eightieth ['eɪtɪɪθ] *adj.* achtzigst...

eighty ['eɪtɪ] 1. *adj.* achtzig. 2. *n.* Achtzig, *die;* **the eighties** *(years)* die achtziger Jahre; **be in one's eighties** in den Achtzigern sein. *See also* **eight**

Eire ['eərə] *pr. n.* Irland, *das;* Eire, *das*

either ['aɪðə(r), 'iːðə(r)] 1. *adj.* **a)** *(each)* **at ~ end of the table** an beiden Enden des Tisches; **b)** *(one or other)* [irgend]ein ... [von beiden]; **take ~ one** nimm einen/eine/eins von [den] beiden. 2. *pron.* **a)** *(each)* beide *Pl.;* I **can't cope with ~:** ich kann mit keinem von beiden fertig werden; **b)** *(one or other)* einer/eine/ein[e]s [von beiden]. 3. *adv.* auch [nicht]; **'I don't like that ~:** ich mag es auch nicht. 4. *conj.* **~ ... or ...:** entweder ... oder ...; *(after negation)* weder ... noch ...

eject ['ɪdʒekt] 1. *v. t.* **a)** *(from hall, meeting)* hinauswerfen **(from** aus); **b)** ⟨*Gerät:*⟩ auswerfen; ⟨*Person:*⟩ herausholen ⟨*Kassette*⟩. 2. *v. i.* sich hinauskatapultieren. **ejector seat** *n.* ['ɪdʒektə siːt] *n.* Schleudersitz, *der*

eke out [iːk 'aʊt] *v. t.* strecken

elaborate 1. [ɪ'læbərət] *adj.* kompliziert; kunstvoll [gearbeitet] ⟨*Arrangement, Verzierung*⟩. 2. [ɪ'læbəreɪt] *v. i.* mehr ins Detail gehen; **~ on** näher ausführen

elapse [ɪ'læps] *v. i.* ⟨*Zeit:*⟩ vergehen

elastic [ɪ'læstɪk] 1. *adj.* elastisch. 2. *n.* **(~ band)** Gummiband, *das.* **elastic 'band** *n.* Gummiband, *das*

elated [ɪ'leɪtɪd] *adj.* freudig erregt; **be or feel ~:** in Hochstimmung sein.

elation [ɪ'leɪʃn] *n.* freudige Erregung

elbow ['elbəʊ] 1. *n.* Ell[en]bogen, *der.* 2. *v. t.* **~ sb. aside** jmdn. mit dem Ellenbogen zur Seite stoßen. **'elbow room** *n.* Ell[en]bogenfreiheit, *die*

¹**elder** ['eldə(r)] 1. *attrib. adj.* älter... 2. *n.* **a)** *(senior)* Ältere, *der/die;* **b)** *(village ~, church ~)* Älteste, *der/die*

²**elder** *n.* *(Bot.)* Holunder, *der.* **'elderberry** *n.* Holunderbeere, *die*

elderly ['eldəlɪ] 1. *adj.* älter. 2. *n. pl.* **the ~:** ältere Menschen

eldest ['eldɪst] *adj.* ältest...

elect [ɪ'lekt] 1. *adj. postpos.* gewählt; **the President ~:** der designierte Präsident. 2. *v. t.* wählen; **~ sb. chairman** jmdn. zum Vorsitzenden wählen. **election** [ɪ'lekʃn] *n.* Wahl, *die;* **general ~:** allgemeine Wahlen. **e'lection campaign** *n.* Wahlkampagne, *die*

electioneer [ɪlekʃə'nɪə(r)] *v. i.* **be/go ~ing** Wahlkampf machen

elector [ɪ'lektə(r)] *n.* Wähler, *der/* Wählerin, *die.* **electoral** [ɪ'lektərl] *adj.* Wahl-. **electorate** [ɪ'lektərət] *n.* Wähler *Pl.*

electric [ɪ'lektrɪk] *adj.* elektrisch; Elektro⟨*kabel, -motor, -herd, -kessel*⟩; Strom⟨*versorgung*⟩; *(fig.)* spannungsgeladen ⟨*Atmosphäre*⟩. **electrical** [ɪ'lektrɪkl] *adj.* elektrisch; Elektro⟨*abteilung, -handel, -geräte*⟩

electric: ~ 'blanket *n.* Heizdecke, *die;* **~ 'fire** *n.* [elektrischer] Heizofen

electrician [ɪlek'trɪʃn] *n.* Elektriker, *der/*Elektrikerin, *die*

electricity [ɪlek'trɪsɪtɪ] *n.* Elektrizität, *die*

electric 'shock *n.* Stromschlag, *der*

electrify [ɪ'lektrɪfaɪ] *v. t.* elektrifizieren; *(fig.)* elektrisieren

electrocute [ɪ'lektrəkjuːt] *v. t.* durch Stromschlag töten

electrode [ɪ'lektrəʊd] *n.* Elektrode, *die*

electron [ɪ'lektrɒn] *n.* Elektron, *das*

electronic [ɪlek'trɒnɪk] *adj.* elektronisch. **electronics** [ɪlek'trɒnɪks] *n.* Elektronik, *die*

elegance ['elɪgəns] *n.* Eleganz, *die*

elegant ['elɪgənt] *adj.* elegant

element ['elɪmənt] *n.* **a)** Element, *das;* **b)** *(Electr.)* Heizelement, *das;* **c)** **~s** *(rudiments)* Grundlagen *Pl.* **elementary** [elɪ'mentərɪ] *adj.* elementar; grundlegend ⟨*Fakten, Wissen*⟩; Grundschul⟨*bildung*⟩; Grund⟨*kurs, -ausbildung, -kenntnisse*⟩

elephant ['elɪfənt] *n.* Elefant, *der*

elevate ['elɪveɪt] *v. t.* [empor]heben. **elevation** [elɪ'veɪʃn] *n.* **a)** *(height)* Höhe, *die;* **b)** *(Archit.)* Aufriß, *der*

elevator ['elɪveɪtə(r)] *n.* *(Amer.)* Aufzug, *der;* Fahrstuhl, *der*

eleven [ɪ'levn] 1. *adj.* elf. 2. *n.* *(also Sport)* Elf, *die. See also* **eight**

elevenses [ɪ'levnzɪz] *n. sing. or pl.* *(Brit. coll.)* ≈ zweites Frühstück [gegen elf Uhr]

eleventh [ɪ'levnθ] 1. *adj.* elft...; **at the ~ hour** in letzter Minute. 2. *n.* *(fraction)* Elftel, *das. See also* **eighth**

elf [elf] *n., pl.* **elves** [elvz] Elf, *der/* Elfe, *die*

elicit [ɪ'lɪsɪt] *v. t.* entlocken **(from** *Dat.*); gewinnen ⟨*Unterstützung*⟩

eligible ['elɪdʒɪbl] *adj.* **be ~ for sth.** *(fit)* für etw. geeignet sein; *(entitled)* zu etw. berechtigt sein

eliminate [ɪ'lɪmɪneɪt] *v. t.* **a)** *(remove)* beseitigen; ausschließen ⟨*Möglich-*

keit〉; **b)** *(exclude)* ausschließen; **be ~d** *(Sport)* ausscheiden. **elimination** [ɪlɪmɪ'neɪʃn] *n.* **a)** *(removal)* Beseitigung, *die;* **process of ~:** Ausleseverfahren, *das;* **b)** *(exclusion)* Ausschluß, *der; (Sport)* Ausscheiden, *das*

élite [eɪ'li:t] *n.* Elite, *die*

ellipse [ɪ'lɪps] *n.* Ellipse, *die.* **elliptical** [ɪ'lɪptɪkl] *adj.* elliptisch

elm [elm] *n.* Ulme, *die*

elongated ['i:lɒŋɡeɪtɪd] *adj.* langgestreckt

elope [ɪ'ləʊp] *v. i.* durchbrennen *(ugs.)*

eloquence ['eləkwəns] *n.* Beredtheit, *die.* **eloquent** ['eləkwənt] *adj.* beredt 〈*Person*〉; gewandt 〈*Stil, Redner*〉

else [els] *adv.* **a)** *(besides)* sonst [noch]; **somebody/something ~:** [noch] jemand anders/noch etwas; **everybody/everything ~:** alle anderen/alles andere; **who/what/when/how ~?** wer/was/wann/wie sonst noch?; **why ~?** warum sonst?; **b)** *(instead)* ander...; **sb. ~'s hat** der Hut von jmd. anders; **anybody/anything ~?** [irgend] jemand anders/etwas anderes?; **somebody/something ~:** jemand anders/etwas anderes; **everybody/everything ~:** alle anderen/alles andere; **c)** *(otherwise)* sonst; **or ~:** oder aber; **do it or ~ ...!** tun Sie es, sonst ...! **elsewhere** *adv.* woanders

elude [ɪ'lu:d] *v.t.* *(avoid)* ausweichen (+ *Dat.*); *(escape from)* entkommen (+*Dat.*). **elusive** [ɪ'lu:sɪv] *adj.* schwer zu erreichen 〈*Person*〉; schwer zu fassen 〈*Straftäter*〉; schwer definierbar 〈*Begriff, Sinn*〉

elves *pl. of* elf

emaciated [ɪ'meɪsɪeɪtɪd] *adj.* abgezehrt

emancipated [ɪ'mænsɪpeɪtɪd] *adj.* emanzipiert: **become ~:** sich emanzipieren

emancipation [ɪmænsɪ'peɪʃn] *n.* Emanzipation, *die*

embalm [ɪm'bɑ:m] *v.t.* einbalsamieren

embankment [ɪm'bæŋkmənt] *n.* Damm, *der*

embargo [ɪm'bɑ:ɡəʊ] *n., pl.* **~es** Embargo, *das*

embark [ɪm'bɑ:k] *v. i.* **a)** sich einschiffen (**for** nach); **b)** **~ [up]on sth.** etw. in Angriff nehmen. **embarkation** [embɑ:'keɪʃn] *n.* Einschiffung, *die*

embarrass [ɪm'bærəs] *v.t.* in Verlegenheit bringen. **embarrassed** [ɪm'bærəst] *adj.* verlegen; **feel ~:** verlegen

sein. **embarrassing** *adj.* peinlich. **embarrassment** *n.* Verlegenheit, *die*

embassy ['embəsɪ] *n.* Botschaft, *die*

embellish [em'belɪʃ] *v. t.* beschönigen 〈*Wahrheit*〉; ausschmücken 〈*Geschichte, Bericht*〉

embers ['embəz] *n. pl.* Glut, *die*

embezzle [ɪm'bezl] *v. t.* unterschlagen

embitter [ɪm'bɪtə(r)] *v. t.* verbittern

emblem ['embləm] *n.* Emblem, *das*

embody [ɪm'bɒdɪ] *v. t.* verkörpern

embrace [ɪm'breɪs] **1.** *v. t.* umarmen; *(fig.: accept, adopt)* annehmen. **2.** *v. i.* sich umarmen. **3.** *n.* Umarmung, *die*

embroider [ɪm'brɔɪdə(r)] *v.t.* sticken 〈*Muster*〉; besticken 〈*Tuch, Kleid*〉; *(fig.)* ausschmücken. **embroidery** [ɪm'brɔɪdərɪ] *n.* Stickerei, *die*

embroil [ɪm'brɔɪl] *v.t.* **become/be ~ed in sth.** in etw. *(Akk.)* verwickelt werden/sein

embryo ['embrɪəʊ] *n.* Embryo, *der*

emerald ['emərəld] **1.** *n.* Smaragd, *der.* **2.** *adj.* smaragdgrün

emerge [ɪ'mɜ:dʒ] *v. i.* auftauchen (**from** aus, **from behind** hinter + *Dat.*); 〈*Wahrheit:*〉 an den Tag kommen; **it ~s that ...:** es stellt sich heraus, daß ...

emergency [ɪ'mɜ:dʒənsɪ] **1.** *n.* Notfall, *der;* **in an or in case of ~:** im Notfall. **2.** *adj.* Not-

emigrant ['emɪɡrənt] *n.* Auswanderer, *der*/Auswanderin, *die*

emigrate ['emɪɡreɪt] *v. i.* auswandern (**to** nach, **from** aus). **emigration** [emɪ'ɡreɪʃn] *n.* Auswanderung (**to** nach, **from** aus)

eminence ['emɪnəns] *n.* hohes Ansehen

eminent ['emɪnənt] *adj.* bedeutend; herausragend

emission [ɪ'mɪʃn] *n.* Emission, *die (fachspr.); (process also)* Abgabe, *die*

emit [ɪ'mɪt] *v.t.,* **-tt-** abgeben, emittieren *(fachspr.)* 〈*Wärme, Strahlung usw.*〉; ausstoßen 〈*Rauch*〉

emotion [ɪ'məʊʃn] *n.* Gefühl, *das.* **emotional** [ɪ'məʊʃənl] *adj.* emotional; Gemüts〈*zustand, -störung*〉; gefühlvoll 〈*Stimme*〉. **emotionally** *adv.* emotional; gefühlvoll 〈*sprechen*〉; **~ disturbed** seelisch gestört

emotive [ɪ'məʊtɪv] *adj.* emotional

emperor ['empərə(r)] *n.* Kaiser, *der*

emphasis ['emfəsɪs] *n., pl.* **emphases** ['emfəsi:z] Betonung, *die;* **lay** *or* **place** *or* **put ~ on sth.** etw. betonen

emphasize ['emfəsaɪz] *v. t.* betonen
emphatic [ɪm'fætɪk] *adj.* nachdrücklich; demonstrativ ⟨*Ablehnung*⟩; **be ~ that ...**: darauf bestehen, daß ... **em'phatically** *adv.* nachdrücklich
empire ['empaɪə(r)] *n.* Reich, *das*
employ [ɪm'plɔɪ] *v. t.* **a)** *(take on)* einstellen; *(have working for one)* beschäftigen; **be ~ed by a company** bei einer Firma arbeiten; **b)** *(use)* einsetzen (for, in, on für); anwenden ⟨*Methode, List*⟩ (for, in, on bei). **employee** (*Amer.*: **employe**) [emplɔɪ'i:, em'plɔɪi:] *n.* Angestellte, *der/die.* **employer** [ɪm'plɔɪə(r)] *n.* Arbeitgeber, *der/*-geberin, *die.* **employment** [ɪm'plɔɪmənt] *n.* **a)** *(work)* Arbeit, *die;* **b)** *(regular trade or profession)* Beschäftigung, *die.* **em'ployment agency** *n.* Stellenvermittlung, *die*
empower [ɪm'paʊə(r)] *v. t.* *(authorize)* ermächtigen; *(enable)* befähigen
empress ['emprɪs] *n.* Kaiserin, *die*
emptiness ['emptɪnɪs] *n.* Leere, *die*
empty ['emptɪ] **1.** *adj.* leer; frei ⟨*Sitz, Parkplatz*⟩. **2.** *v. t.* leeren; *(pour)* schütten (over über + *Akk.*). **3.** *v. i.* sich leeren. **'empty-handed** *adj.* mit leeren Händen
EMS *abbr.* **European Monetary System** EWS
emulate ['emjʊleɪt] *v. t.* nacheifern (+ *Dat.*)
emulsion [ɪ'mʌlʃn] *n.* Emulsion, *die*
enable [ɪ'neɪbl] *v. t.* **~ sb. to do sth.** es jmdm. ermöglichen, etw. zu tun
enamel [ɪ'næml] **1.** *n.* Email, *das.* **2.** *v. t., (Brit.)* -ll- emaillieren
enchant [ɪn'tʃɑ:nt] *v. t.* verzaubern; *(delight)* entzücken. **en'chanted** *adj.* verzaubert. **en'chanting** *adj.* entzückend. **en'chantment** *n.* Verzauberung, *die;* *(fig.)* Zauber, *der*
encircle [ɪn'sɜ:kl] *v. t.* umgeben
encl. *abbr.* **enclosed, enclosure[s]** Anl.
enclave ['enkleɪv] *n.* Enklave, *die*
enclose [ɪn'kləʊz] *v. t.* **a)** *(surround)* umgeben; *(shut up or in)* einschließen; **b)** *(with letter)* beilegen (with, in *Dat.*); **please find ~d** anbei erhalten Sie. **enclosure** [ɪn'kləʊʒə(r)] *n.* **a)** *(in zoo)* Gehege, *das;* **b)** *(with letter)* Anlage, *die*
encore ['ɒŋkɔ:(r)] **1.** *int.* Zugabe. **2.** *n.* Zugabe, *die*
encounter [ɪn'kaʊntə(r)] **1.** *v. t. (as adversary)* treffen auf (+ *Akk.*); *(by chance)* begegnen (+ *Dat.*); stoßen auf (+ *Akk.*) ⟨*Problem, Widerstand*

usw.⟩. **2.** *n. (chance meeting)* Begegnung, *die*
encourage [ɪn'kʌrɪdʒ] *v. t.* ermutigen; *(promote)* fördern. **encouragement** *n.* Ermutigung, *die* **(from durch)**
encroach [ɪn'krəʊtʃ] *v. i.* **~ on** eindringen in (+ *Akk.*); in Anspruch nehmen ⟨*Zeit*⟩
encumber [ɪn'kʌmbə(r)] *v. t.* belasten. **encumbrance** [ɪn'kʌmbrəns] *n.* Belastung, *die*
encyclopaedia [ɪnsaɪklə'pi:dɪə] *n.* Lexikon, *das;* Enzyklopädie, *die.* **encyclopaedic** [ɪnsaɪklə'pi:dɪk] *adj.* enzyklopädisch
end [end] **1.** *n.* **a)** Ende, *das;* *(of nose, hair, finger)* Spitze, *die;* **from ~ to ~**: von einem Ende zum anderen; **at the ~ of 1987/March** Ende 1987/März; **in the ~**: schließlich; **come to an ~**: ein Ende nehmen; **be at an ~**: zu Ende sein; **b)** *(of box, packet, etc.)* Schmalseite, *die;* *(top/bottom surface)* Ober-/ Unterseite, *die;* **on ~**: hochkant; **make ~s meet** *(fig.)* zurechtkommen; **no ~ of** *(coll.)* unendlich viel/viele; **c)** *(remnant)* Rest, *der;* *(of cigarette)* Stummel, *der;* **d)** *(purpose, object)* Ziel, *das;* **~ in itself** Selbstzweck, *der.* **2.** *v. t.* beenden. **3.** *v. i.* enden. **end 'up** *v. i.* enden; **~ up in** *(coll.)* landen in (+ *Dat.*); **~ up [as] a teacher** *(coll.)* schließlich Lehrer werden
endanger [ɪn'deɪndʒə(r)] *v. t.* gefährden
endear [ɪn'dɪə(r)] *v. t.* **~ sb./sth./oneself to sb.** jmdn./etw./sich bei jmdm. beliebt machen. **en'dearing** *adj.* reizend; gewinnend ⟨*Lächeln, Art*⟩
endeavour (*Brit.*; *Amer.*: **endeavor**) [ɪn'devə(r)] **1.** *v. i.* **~ to do sth.** sich bemühen, etw. zu tun. **2.** *n.* Bemühung, *die;* *(attempt)* Versuch, *der*
'ending *n.* Schluß, *der;* *(of word)* Endung, *die*
endive ['endaɪv] *n.* Endivie, *die*
'endless *adj.* endlos. **'endlessly** *adv.* unaufhörlich ⟨*streiten, schwatzen*⟩
endorse [ɪn'dɔ:s] *v. t.* **a)** indossieren ⟨*Scheck*⟩; **b)** beipflichten (+ *Dat.*) ⟨*Meinung*⟩; billigen ⟨*Entscheidung, Handlung*⟩; unterstützen ⟨*Vorschlag*⟩; **c)** *(Brit. Law)* einen Strafvermerk machen auf (+ *Akk. od. Dat.*). **en'dorsement** *n.* **a)** *(of cheque)* Indossament, *das;* **b)** *(support)* Billigung, *die;* *(of proposal)* Unterstützung, *die;* **c)** *(Brit. Law)* Strafvermerk, *der*
endow [ɪn'daʊ] *v. t.* [über Stiftungen/

eine Stiftung] finanzieren; stiften ⟨*Preis, Lehrstuhl*⟩; **be ~ed with charm/a talent for music** Charme/musikalisches Talent besitzen
endurable [ɪn'djʊərəbl] *adj.* erträglich
endurance [ɪn'djʊərəns] *n.* Ausdauer, *die*
endure [ɪn'djʊə(r)] *v. t.* ertragen
enema ['enəmə] *n.* Einlauf, *der*
enemy ['enəmɪ] **1.** *n.* Feind, *der* (**of, to** *Gen.*). **2.** *adj.* feindlich
energetic [enə'dʒetɪk] *adj.* energiegeladen; *(active)* tatkräftig
energy ['enədʒɪ] *n.* Energie, *die*
enforce [ɪn'fɔ:s] *v. t.* durchsetzen; sorgen für ⟨*Disziplin*⟩; **~d** erzwungen ⟨*Schweigen*⟩; unfreiwillig ⟨*Untätigkeit*⟩
engage [ɪn'geɪdʒ] **1.** *v. t.* **a)** *(hire)* einstellen ⟨*Arbeiter*⟩; engagieren ⟨*Sänger*⟩; **b)** wecken ⟨*Interesse*⟩; auf sich *(Akk.)* ziehen ⟨*Aufmerksamkeit*⟩; **c)** ~ **the clutch/first gear** einkuppeln/den ersten Gang einlegen. **2.** *v. i.* ~ **in** sth. sich an etw. *(Dat.)* beteiligen; ~ **in politics** sich politisch engagieren. **engaged** [ɪn'geɪdʒd] *adj.* **a) be ~ |to be married| |to sb.|** [mit jmdm.] verlobt sein; **get ~ |to be married| |to sb.|** sich [mit jmdm.] verloben; **b) be ~ in sth./in doing sth.** mit etw. beschäftigt sein/damit beschäftigt sein, etw. zu tun; **be otherwise ~:** etwas anderes vorhaben; **c)** besetzt ⟨*Toilette, [Telefon]anschluß, Nummer*⟩; ~ **signal** *or* **tone** *(Brit.)* Besetztzeichen, *das.* **en'gagement** *n.* **a)** *(to be married)* Verlobung, *die* (**to** mit); **b)** *(appointment)* Verabredung, *die.* **en'gagement ring** *n.* Verlobungsring, *der*
engaging [ɪn'geɪdʒɪŋ] *adj.* bezaubernd; einnehmend ⟨*Persönlichkeit, Art*⟩
engine ['endʒɪn] *n.* **a)** Motor, *der;* *(rocket/jet ~)* Triebwerk, *das;* **b)** *(locomotive)* Lok[omotive], *die.* **'engine driver** *n.* Lok[omotiv]führer, *der*
engineer [endʒɪ'nɪə(r)] **1.** *n.* **a)** Ingenieur, *der*/Ingenieurin, *die;* *(service ~, installation ~)* Techniker, *der*/Technikerin, *die;* **b)** *(Amer.: engine-driver)* Lok[omotiv]führer, *der.* **2.** *v. t.* arrangieren. **engi'neering** *n.* Technik, *die*
England ['ɪŋglənd] *pr. n.* England *(das)*
English ['ɪŋglɪʃ] **1.** *adj.* englisch; **he/ she is ~:** er ist Engländer/sie ist Engländerin. **2.** *n.* **a)** Englisch, *das;* **say sth. in ~:** etw. auf englisch sagen; **I**

cannot *or* **do not speak ~:** ich spreche kein Englisch; **translate into/from |the| ~:** ins Englische/aus dem Englischen übersetzen; **b)** *pl.* **the ~:** die Engländer
English: ~ **'Channel** *pr. n.* **the ~ Channel** der [Ärmel]kanal; **~man** [~mən] *n., pl.* **~men** [~mən] Engländer, *der;* **~woman** *n.* Engländerin, *die*
engrave [ɪn'greɪv] *v. t.* gravieren; eingravieren ⟨*Namen, Figur usw.*⟩. **engraving** [ɪn'greɪvɪŋ] *n.* Stich, *der;* *(from wood)* Holzschnitt, *der*
engross [ɪn'grəʊs] *v. t.* fesseln; **be ~ed in sth.** in etw. *(Akk.)* vertieft sein; **become** *or* **get ~ed in sth.** sich in etw. *(Akk.)* vertiefen
engulf [ɪn'gʌlf] *v. t.* verschlingen
enhance [ɪn'hɑ:ns] *v. t.* erhöhen ⟨*Wert, Aussichten, Schönheit*⟩; verstärken ⟨*Wirkung*⟩; heben ⟨*Aussehen*⟩
enigma [ɪ'nɪgmə] *n.* Rätsel, *das.* **enigmatic** [enɪg'mætɪk] *adj.* rätselhaft
enjoy [ɪn'dʒɔɪ] **1.** *v. t.* **a) I ~ed the book/work** das Buch/die Arbeit hat mir gefallen; **he ~s reading/travelling** er liest/reist gern; **b)** genießen ⟨*Rechte, Privilegien, Vorteile*⟩. **2.** *v. refl.* sich amüsieren. **enjoyable** [ɪn'dʒɔɪəbl] *adj.* schön; angenehm ⟨*Empfindung, Arbeit*⟩; unterhaltsam ⟨*Buch, Film, Stück*⟩. **en'joyment** *n.* Vergnügen, *das* (**of an** + *Dat.*)
enlarge [ɪn'lɑ:dʒ] **1.** *v. t.* vergrößern; verbreitern ⟨*Straße, Durchgang*⟩. **2.** *v. i.* ~ **upon** sth. etw. weiter ausführen. **en'largement** *n.* Vergrößerung, *die;* *(making wider)* Verbreiterung, *die*
enlighten [ɪn'laɪtn] *v. t.* aufklären (**on, as to** über + *Akk.*). **en'lightenment** *n.* Aufklärung, *die*
enlist [ɪn'lɪst] **1.** *v. t.* *(obtain)* gewinnen. **2.** *v. i.* ~ [**for the army/navy**] in die Armee/Marine eintreten; ~ [**as a soldier**] Soldat werden
enliven [ɪn'laɪvn] *v. t.* beleben
enmity ['enmɪtɪ] *n.* Feindschaft, *die*
enormous [ɪ'nɔ:məs] *adj.* enorm; riesig, gewaltig ⟨*Figur, Tier, Menge*⟩. **e'normously** *adv.* enorm
enough [ɪ'nʌf] **1.** *adj.* genug; **there's ~ room** es ist Platz genug. **2.** *n.* genug; **be ~ to do** sth. genügen, etw. zu tun; **have had ~ |of sb./sth.|** genug [von jmdm./etw.] haben; **I've had ~!** jetzt reicht's mir aber! **3.** *adv.* genug; **oddly/funnily ~:** merkwürdiger-/ *(ugs.)* komischerweise

enquire, enquiry *see* **inquir-**

enrage [ɪnˈreɪdʒ] *v. t.* wütend machen; **be ~d by sth.** über etw. *(Akk.)* wütend werden

enrich [ɪnˈrɪtʃ] *v. t.* reich machen; *(fig.)* bereichern

enrol *(Amer.:* **enroll)** [ɪnˈrəʊl] **1.** *v. i.,* **-ll-** sich einschreiben; **~ for a course** sich zu einem Kurs anmelden. **2.** *v. t.* einschreiben. **en'rolment** *(Amer.:* **en'rollment)** *n.* Einschreibung, *die*

en route [ɑ̃ ˈruːt] *adv.* unterwegs; **~ to Scotland/for Edinburgh** auf dem Weg nach Schottland/Edinburgh

ensign [ˈensaɪn, ˈensn] *n.* Hoheitszeichen, *das*

enslave [ɪnˈsleɪv] *v. t.* versklaven

ensue [ɪnˈsjuː] *v. i.* folgen **(from, on** aus); **the discussion which ~d** die anschließende Diskussion

ensure [ɪnˈʃʊə(r)] *v. t.* **~ that ...** *(see to it that)* gewährleisten, daß ...; **~ sth.** etw. gewährleisten

entail [ɪnˈteɪl] *v. t.* mit sich bringen; **sth. ~s doing sth.** etw. bedeutet, daß man etw. tun muß

entangle [ɪnˈtæŋgl] *v. t.* sich verfangen lassen; **get** *or* **become ~d in** *or* **with sth.** sich in etw. *(Dat.)* verfangen

enter [ˈentə(r)] **1.** *v. i.* **a)** hineingehen; ⟨*Fahrzeug:*⟩ hineinfahren; *(come in)* hereinkommen; *(into room)* eintreten; **b)** *(register as competitor)* sich zur Teilnahme anmelden **(for an** + *Dat.).* **2.** *v. t.* **a)** [hinein]gehen in (+ *Akk.*); ⟨*Fahrzeug:*⟩ [hinein]fahren in (+ *Akk.*); betreten ⟨*Gebäude, Zimmer*⟩; einlaufen in (+ *Akk.*) ⟨*Hafen*⟩; einreisen in (+ *Akk.*) ⟨*Land*⟩; *(come into)* [herein]kommen in (+ *Akk.*); **b)** teilnehmen an (+ *Dat.*) ⟨*Rennen, Wettbewerb*⟩; **c)** *(in book etc.)* eintragen **(in** in + *Akk.*). **'enter into** *v. t.* aufnehmen ⟨*Verhandlungen*⟩; eingehen ⟨*Verpflichtung*⟩; schließen ⟨*Vertrag*⟩. **'enter [up]on** *v. t.* beginnen

enterprise [ˈentəpraɪz] *n.* **a)** *(undertaking)* Unternehmen, *das;* **free/private ~:** freies/privates Unternehmertum; **b)** *(enterprising spirit)* Unternehmungsgeist, *der.* **enterprising** [ˈentəpraɪzɪŋ] *adj.* unternehmungslustig

entertain [entəˈteɪn] *v. t.* **a)** *(amuse)* unterhalten; **b)** *(receive as guest)* bewirten; **c)** haben ⟨*Vorstellung*⟩; hegen *(geh.)* ⟨*Gefühl, Verdacht, Zweifel*⟩; *(consider)* in Erwägung ziehen. **enter'tainer** *n.* Unterhalter, *der*/Unterhalterin, *die.* **enter'taining** *adj.* unterhaltsam. **enter'tainment** *n.* **a)** *(amusement)* Unterhaltung, *die;* **b)** *(performance, show)* Veranstaltung, *die*

enthral *(Amer.:* **enthrall)** [ɪnˈθrɔːl] *v. t.,* **-ll-** gefangennehmen *(fig.)*

enthuse [ɪnˈθjuːz] *(coll.)* **1.** *v. i.* in Begeisterung ausbrechen **(about** über + *Akk.*). **2.** *v. t.* begeistern

enthusiasm [ɪnˈθjuːzɪæzm] *n.* Begeisterung, *die.* **enthusiast** [ɪnˈθjuːzɪæst] *n.* Enthusiast, *der;* *(for sports)* Fan, *der;* **a DIY ~:** ein begeisterter Heimwerker. **enthusiastic** [ɪnθjuː-zɪˈæstɪk] *adj.* begeistert; **not be very ~ about doing sth.** keine große Lust haben, etw. zu tun

entice [ɪnˈtaɪs] *v. t.* locken **(into** in + *Akk.*); **~ sb. into doing** *or* **to do sth.** jmdn. dazu verleiten, etw. zu tun

entire [ɪnˈtaɪə(r)] *adj.* **a)** *(whole)* ganz; **b)** *(intact)* vollständig. **en'tirely** *adv.* **a)** *(wholly)* völlig; **b)** *(solely)* ganz ⟨*für sich behalten*⟩; voll ⟨*verantwortlich sein*⟩; **it's up to you ~:** es liegt ganz bei dir. **entirety** [ɪnˈtaɪərətɪ] *n.* **in its ~:** in seiner/ihrer Gesamtheit

entitle [ɪnˈtaɪtl] *v. t.* berechtigen **(to** zu); **~ sb. to do sth.** jmdm. das Recht geben, etw. zu tun; **be ~d to** [claim] **sth.** Anspruch auf etw. *(Akk.)* haben; **be ~d to do sth.** das Recht haben, etw. zu tun

entourage [ˈɒntʊrɑːʒ] *n.* Gefolge, *das*

entrails [ˈentreɪlz] *n. pl.* Eingeweide *Pl.*

¹entrance [ɪnˈtrɑːns] *v. t.* hinreißen

²entrance [ˈentrəns] *n.* *(way in)* Eingang, *der* **(to** *Gen. od.* zu); *(for vehicles)* Einfahrt, *die.* **'entrance fee** *n.* Eintrittsgeld, *das*

entrant [ˈentrənt] *n.* *(for competition, race, etc.)* Teilnehmer, *der*/Teilnehmerin, *die* **(for** *Gen.,* an + *Dat.*)

entreat [ɪnˈtriːt] *v. t.* anflehen. **en'treaty** *n.* flehentliche Bitte

entrepreneur [ɒntrəprəˈnɜː(r)] *n.* Unternehmer, *der*/Unternehmerin, *die*

entrust [ɪnˈtrʌst] *v. t.* **~ sb. with sth.** jmdm. etw. anvertrauen; **~ sb./sth. to sb./sth.** jmdn./etw. jmdm./einer Sache anvertrauen; **~ a task to sb.** jmdn. mit einer Aufgabe betrauen

entry [ˈentrɪ] *n.* **a)** Eintritt, *der* **(into** in + *Akk.*); *(into country)* Einreise, *die;* **'no ~'** *(for people)* „Zutritt verboten"; *(for vehicles)* „Einfahrt verboten"; **b)** *(way in)* Eingang, *der;* *(for vehicle)* Einfahrt, *die;* **c)** *(registration, item)*

Eintragung, *die* (**in, into** in + *Akk. od. Dat.*); *(in dictionary, encyclopaedia)* Eintrag, *der*

entry: ~ fee *n.* Eintrittsgeld, *das;* ~ **form** *n.* Anmeldeformular, *das;* ~ **visa** *n.* Einreisevisum, *das*

envelop [ɪn'veləp] *v. t.* [ein]hüllen (**in** in + *Akk.*); **be ~ed in flames** ganz von Flammen umgeben sein

envelope ['envələʊp, 'ɒnvələʊp] *n.* [Brief]umschlag, *der*

enviable ['envɪəbl] *adj.* beneidenswert

envious ['envɪəs] *adj.* neidisch (**of** auf + *Akk.*)

environment [ɪn'vaɪərənmənt] *n.* Umwelt, *die; (surrounding objects, region)* Umgebung, *die.* **environmental** [ɪnvaɪərən'mentl] *adj.* Umwelt-. **environ'mentalist** *n.* Umweltschützer, *der/*-schützerin, *die.* **environ'mentally** *adv.* ~ **friendly** umweltfreundlich

envisage [ɪn'vɪzɪdʒ] *v. t.* sich *(Dat.)* vorstellen

envoy ['envɔɪ] *n.* Gesandte, *der/*Gesandtin, *die*

envy ['envɪ] **1.** *n.* Neid, *der;* **you'll be the ~ of all your friends** alle deine Freunde werden dich beneiden. **2.** *v. t.* beneiden; ~ **sb. sth.** jmdn. um etw. beneiden

enzyme ['enzaɪm] *n.* Enzym, *das*

ephemeral [ɪ'femərl] *adj.* kurzlebig

epic ['epɪk] **1.** *adj.* episch. **2.** *n.* Epos, *das*

epidemic [epɪ'demɪk] **1.** *adj.* epidemisch. **2.** *n.* Epidemie, *die*

epilepsy ['epɪlepsɪ] *n.* Epilepsie, *die.* **epileptic** [epɪ'leptɪk] **1.** *adj.* epileptisch; epileptischer Anfall. **2.** *n.* Epileptiker, *der/*Epileptikerin, *die*

episode ['epɪsəʊd] *n.* **a)** Episode, *die;* **b)** *(of serial)* Folge, *die*

epitaph ['epɪtɑ:f] *n.* Grab[in]schrift, *die*

epitome [ɪ'pɪtəmɪ] *n.* Inbegriff, *der.* **epitomize** [ɪ'pɪtəmaɪz] *v. t.* ~ **sth.** der Inbegriff einer Sache *(Gen.)* sein

epoch ['i:pɒk] *n.* Epoche, *die.* **'epoch-making** *adj.* epochemachend

equal ['i:kwl] **1.** *adj.* **a)** gleich; ~ **in** or **of** ~ **height/size/importance** *etc.* gleich hoch/groß/wichtig *usw.;* **b)** **be ~ to sth./sb.** *(strong, clever, etc. enough)* einer Sache/jmdm. gewachsen sein. **2.** *n.* Gleichgestellte, *der/die;* **have no ~:** nicht seines-/ihresgleichen haben. **3.** *v. t., (Brit.)* **-ll-:** ~ **sb.** es

jmdm. gleich tun; **three times four ~s twelve** drei mal vier ist [gleich] zwölf. **equality** [ɪ'kwɒlɪtɪ] *n.* Gleichheit, *die; (equal rights)* Gleichberechtigung, *die.* **equalize** ['i:kwəlaɪz] *v. i. (Sport)* den Ausgleich[streffer] erzielen. **'equalizer** *n. (Sport)* Ausgleich[streffer], *der.* **equally** *adv.* gleich; *(just as)* ebenso; **in gleiche Teile** ⟨*aufteilen*⟩; gleichmäßig ⟨*verteilen*⟩.

equal oppor'tunity *n.* Chancengleichheit, *die.* **'equals sign** *n. (Math.)* Gleichheitszeichen, *das*

equanimity [ekwə'nɪmɪtɪ] *n.* Gelassenheit, *die*

equate [ɪ'kweɪt] *v. t.* gleichsetzen (**with** mit). **equation** [ɪ'kweɪʒn] *n. (Math.)* Gleichung, *die*

equator [ɪ'kweɪtə(r)] *n.* Äquator, *der*

equilibrium [i:kwɪ'lɪbrɪəm] *n., pl.* **equilibria** [i:kwɪ'lɪbrɪə] *or* ~**s** Gleichgewicht, *das*

equinox ['ekwɪnɒks] *n.* Tagundnachtgleiche, *die*

equip [ɪ'kwɪp] *v. t.,* **-pp-** ausrüsten ⟨*Fahrzeug, Armee*⟩; ausstatten ⟨*Küche*⟩; **fully ~ped** komplett ausgerüstet/ausgestattet; ~ **sb./oneself** [with sth.] jmdn./sich [mit etw.] ausrüsten. **e'quipment** *n.* Ausrüstung, *die; (of kitchen, laboratory)* Ausstattung, *die; (needed for activity)* Geräte

equivalent [ɪ'kwɪvələnt] **1.** *adj.* gleichwertig; **be ~ to sth.** einer Sache *(Dat.)* entsprechen. **2.** *n.* **a)** *(thing, person)* Pendant, *das;* Gegenstück, *das* (**of** zu); **b) be ~ of sth.** *(have same result)* einer Sache *(Dat.)* entsprechen

equivocal [ɪ'kwɪvəkl] *adj.* zweideutig

era ['ɪərə] *n.* Ära, *die*

eradicate [ɪ'rædɪkeɪt] *v. t.* ausrotten

erase [ɪ'reɪz] *v. t.* auslöschen; *(with rubber, knife)* ausradieren; *(from tape, also Computing)* löschen. **e'raser** *n.* |**pencil**| ~**:** Radiergummi, *der*

erect [ɪ'rekt] **1.** *adj.* aufrecht. **2.** *v. t.* errichten; aufstellen ⟨*Standbild, Mast, Verkehrsschild, Gerüst, Zelt*⟩. **erection** [ɪ'rekʃn] *n.* **a)** *see* erect 2: Errichtung, *die;* Aufstellen, *das;* **b)** *(Physiol.)* Erektion, *die*

ermine ['ɜ:mɪn] *n.* Hermelin, *der*

erode [ɪ'rəʊd] *v. t.* **a)** ⟨*Säure, Rost:*⟩ angreifen; ⟨*Wasser:*⟩ auswaschen; ⟨*Wind:*⟩ verwittern lassen; **b)** *(fig.)* unterminieren. **erosion** [ɪ'rəʊʒn] *n.* **a)** *see* erode **a:** Angreifen, *das;* Auswaschung, *die;* Verwitterung, *die;* **b)** *(fig.)* Unterminierung, *die*

erotic [ɪ'rɒtɪk] *adj.* erotisch
err [ɜ:(r)] *v. i.* sich irren
errand ['erənd] *n.* Botengang, *der;* *(shopping)* Besorgung, *die;* **go on** *or* **run an** ~: einen Botengang/eine Besorgung machen. **'errand boy** *n.* Laufbursche, *der*
erratic [ɪ'rætɪk] *adj.* unregelmäßig; sprunghaft ⟨*Wesen, Person, Art*⟩; launenhaft ⟨*Verhalten*⟩
erroneous [ɪ'rəʊnɪəs] *adj.* falsch; irrig ⟨*Schlußfolgerung, Annahme*⟩
error ['erə(r)] *n. (mistake)* Fehler, *der;* *(wrong opinion)* Irrtum, *der;* **in** ~: irrtümlich[erweise]
erudite ['eru:daɪt] *adj.* gelehrt
erupt [ɪ'rʌpt] *v. i.* ausbrechen. **eruption** [ɪ'rʌpʃn] *n.* Ausbruch, *der*
escalate ['eskəleɪt] *v. i.* sich ausweiten (**into** zu); ⟨*Preise, Kosten:*⟩ [ständig] steigen. **escalator** ['eskəleɪtə(r)] *n.* Rolltreppe, *die*
escapade [eskə'peɪd] *n.* Eskapade, *die (geh.)*
escape [ɪ'skeɪp] **1.** *n.* Flucht, *die* (**from** aus); **have a narrow** ~: gerade noch einmal davonkommen. **2.** *v. i.* **a)** fliehen (**from** aus); *(successfully)* entkommen (**from** *Dat.*); **b)** ⟨*Gas:*⟩ ausströmen; ⟨*Flüssigkeit:*⟩ auslaufen. **3.** *v. t.* **a)** entkommen (+ *Dat.*) ⟨*Verfolger, Feind*⟩; entgehen (+ *Dat.*) ⟨*Bestrafung, Gefangennahme, Tod*⟩; verschont bleiben von ⟨*Zerstörung, Auswirkungen*⟩; **b)** *(not be remembered by)* entfallen sein (+ *Dat.*). **e'scape route** *n.* Fluchtweg, *der*
escapism [ɪ'skeɪpɪzm] *n.* Realitätsflucht, *die*
escort 1. ['eskɔ:t] *n.* **a)** Begleitung, *die;* *(Mil.)* Eskorte, *die;* **b)** *(hired companion)* Begleiter, *der*/Begleiterin, *die.* **2.** [ɪ'skɔ:t] *v. t.* begleiten; *(lead)* führen; *(Mil.)* eskortieren
Eskimo ['eskɪməʊ] **1.** *adj.* Eskimo-. **2.** *n., pl.* ~s *or same* Eskimo, *der*/Eskimofrau, *die;* the ~[s] die Eskimos
esoteric [esəʊ'terɪk] *adj.* esoterisch
especial [ɪ'speʃl] *attrib. adj.* [ganz] besonder... **especially** [ɪ'speʃəlɪ] *adv.* besonders
espionage ['espɪənɑ:ʒ] *n.* Spionage, *die*
espresso [e'spresəʊ] *n., pl.* ~s *(coffee)* Espresso, *der.* **e'spresso bar** *n.* Espressobar, *die*
Esq. [ɪ'skwaɪə(r)] *abbr.* Esquire ≈ Hr.; *(on letter)* ≈ Hrn.; Jim Smith, ~ : Hr./Hrn. Jim Smith

essay ['eseɪ] *n.* Essay, *der;* Aufsatz, *der (bes. Schulw.)*
essence ['esəns] *n.* **a)** Wesen, *das;* *(gist)* Wesentliche, *das;* **in** ~ : im Wesentlichen; **b)** *(Cookery)* Essenz, *die*
essential [ɪ'senʃl] **1.** *adj.* **a)** *(fundamental)* wesentlich; **b)** *(indispensable)* unentbehrlich; lebensnotwendig ⟨*Versorgungseinrichtungen, Güter*⟩; unabdingbar ⟨*Qualifikation, Voraussetzung*⟩; **it is** ~ **that** ...: es ist unbedingt notwendig, daß ... **2.** *n. pl.* **the** ~**s** *(fundamentals)* das Wesentliche; *(items)* das Notwendigste. **es'sentially** *adv.* im Grunde
establish [ɪ'stæblɪʃ] *v. t.* **a)** schaffen ⟨*Einrichtung, Präzedenzfall*⟩; gründen ⟨*Organisation, Institut*⟩; errichten ⟨*Geschäft, System*⟩; **b)** *(secure acceptance for)* etablieren; **become** ~**ed** sich einbürgern; **c)** *(prove)* beweisen; **d)** *(discover)* feststellen. **established** [ɪ'stæblɪʃt] *adj.* bestehend ⟨*Ordnung*⟩; etabliert ⟨*Schriftsteller*⟩; *(accepted)* üblich; fest ⟨*Brauch*⟩; feststehend ⟨*Tatsache*⟩; **become** ~ : sich durchsetzen.
e'stablishment *n.* **a)** *(setting up, foundation)* Gründung, *die;* **b)** |**business**| ~ : Unternehmen, *das*
estate [ɪ'steɪt] *n.* **a)** *(landed property)* Gut, *das;* **b)** *(Brit.: housing* ~) [Wohn]siedlung, *die;* **c)** *(of deceased person)* Erbmasse, *die.* **e'state agent** *n. (Brit.)* Grundstücksmakler, *der;* **e'state car** *n. (Brit.)* Kombiwagen. *der*
esteem [ɪ'sti:m] **1.** *n.* Wertschätzung, *die (geh.)* (**for** *Gen.*, für). **2.** *v. t.* schätzen; **highly** ~**ed** hochgeschätzt
estimate 1. ['estɪmət] *n.* **a)** Schätzung, *die;* **at a rough** ~ : grob geschätzt; **b)** *(Commerc.)* Kostenvoranschlag, *der.* **2.** ['estɪmeɪt] *v. t.* schätzen (**at** auf + *Akk.*). **estimation** [estɪ'meɪʃn] *n.* Schätzung, *die;* **in sb.'s** ~ : nach jmds. Schätzung
estuary ['estjʊərɪ] *n.* [Trichter]mündung, *die*
etc. *abbr.* et cetera usw.
etch [etʃ] *v. t.* ätzen (**on** auf + *Akk.*); *(on metal also)* ⟨*bes. Künstler:*⟩ radieren; *(fig.)* einprägen (**in, on** *Dat.*). **'etching** *n. (Art)* Radierung, *die*
eternal [ɪ'tɜ:nl] *adj.,* **e'ternally** *adv.* ewig
eternity [ɪ'tɜ:nɪtɪ] *n.* Ewigkeit, *die*
ether ['i:θə(r)] *n.* Äther, *der.* **ethereal** [ɪ'θɪərɪəl] *adj.* ätherisch
ethical ['eθɪkl] *adj.* ethisch

ethics ['eθɪks] *n.* **a)** Moral, *die; (moral philosophy)* Ethik, *die;* **b)** *usu. constr. as pl. (moral code)* Ethik, *die (geh.)*

Ethiopia [iːθɪˈəʊpɪə] *pr. n.* Äthiopien *(das)*

ethnic ['eθnɪk] *adj.* ethnisch

etiquette ['etɪket] *n.* Etikette, *die*

etymology [etɪˈmɒlədʒɪ] *n.* Etymologie, *die*

eulogy ['juːlədʒɪ] *n.* Lobrede, *die*

euphemism ['juːfəmɪzm] *n.* Euphemismus, *der.* **euphemistic** [juːfəˈmɪstɪk] *adj.* verhüllend

euphoria [juːˈfɔːrɪə] *n.* Euphorie, *die (geh.)*

Euro- ['jʊərəʊ] *in comb.* euro-/Euro-. **Eurocheque** *n.* Euroscheck, *der*

Europe ['jʊərəp] *pr. n.* Europa *(das).* **European** [jʊərəˈpiːən] **1.** *adj.* europäisch; ~ |Economic| Community Europäische [Wirtschafts]gemeinschaft. **2.** *n.* Europäer, *der*/Europäerin, *die*

euthanasia [juːθəˈneɪzɪə] *n.* Euthanasie, *die*

evacuate [ɪˈvækjʊeɪt] *v. t.* evakuieren **(from** aus). **evacuation** [ɪvækjʊˈeɪʃn] *n.* Evakuierung, *die* **(from** aus)

evade [ɪˈveɪd] *v. t.* ausweichen *(+ Dat.)* ⟨*Angriff, Angreifer, Schlag, Problem, Frage*⟩; sich entziehen *(+ Dat.)* ⟨*Verhaftung, Verantwortung*⟩; entkommen *(+ Dat.)* ⟨*Verfolger, Verfolgung*⟩; hinterziehen ⟨*Steuern*⟩; ~ **doing sth.** vermeiden, etw. zu tun

evaluate [ɪˈvæljʊeɪt] *v. t.* einschätzen; bewerten ⟨*Daten*⟩

evangelical [iːvænˈdʒelɪkl] *adj.* missionarisch *(fig.); (Protestant)* evangelikal. **evangelist** [ɪˈvændʒəlɪst] *n.* Evangelist, *der*

evaporate [ɪˈvæpəreɪt] **1.** *v. i.* verdunsten. **2.** *v. t.* verdunsten lassen. **evaporated 'milk** *n.* Kondensmilch, *die* **evaporation** [ɪvæpəˈreɪʃn] *n.* Verdunstung, *die*

evasion [ɪˈveɪʒn] *n.* Umgehung, *die; (of responsibility, question)* Ausweichen, *das* (of vor + *Dat.*); **tax** ~: Steuerhinterziehung, *die.* **evasive** [ɪˈveɪsɪv] *adj.* **a)** be/become ~: ausweichen; **b)** ausweichend ⟨*Antwort*⟩

eve [iːv] *n.* Vorabend, *der* (of *Gen.*); *(day)* Vortag, *der* (of *Gen.*)

even ['iːvn] **1.** *adj.* **a)** eben ⟨*Boden, Fläche*⟩; gleich hoch ⟨*Stapel, Stuhl-, Tischbein*⟩; be ~ height/length gleich hoch/lang sein; **b)** gerade ⟨*Zahl, Seite, Hausnummer*⟩; **c)** be *or*

get ~ **with sb.** *(quits)* es jmdm. heimzahlen; **break** ~: die Kosten decken. **2.** *adv.* sogar; selbst; sogar noch ⟨*weniger, schlimmer usw.*⟩; ~ **if** selbst wenn; ~ **so** [aber] trotzdem; **not** *or* **never** ~ ...: [noch] nicht einmal ... **even 'up** *v. t.* ausgleichen

evening ['iːvnɪŋ] *n.* Abend, *der;* **this/tomorrow** ~: heute/morgen abend; **in the** ~: am Abend; *(regularly)* abends. **'evening class** *n.* Abendkurs, *der.* **'evening dress** *n.* Abendkleidung, *die*

'evenly *adv.* gleichmäßig

'even-numbered *adj.* gerade

event [ɪˈvent] *n.* **a)** **in the** ~ **of his dying** *or* **death im** Falle seines Todes; **in the** ~: letzten Endes; **in the** ~ **of rain** bei Regenwetter; **b)** *(occurrence)* Ereignis, *das.* **e'ventful** *adj.* ereignisreich

eventual [ɪˈventjʊəl] *adj.* **predict sb.'s** ~ **downfall** vorhersagen, daß jmd. schließlich zu Fall kommen wird; **the career of Napoleon and his** ~ **defeat** der Aufstieg Napoleons und schließlich seine Niederlage. **eventuality** [ɪventjʊˈælɪtɪ] *n.* Eventualität, *die.* **e'ventually** *adv.* schließlich

ever ['evə(r)] *adv.* **a)** *(always)* immer; **for** ~: für immer; ewig ⟨*lieben, dasein, leben*⟩; ~ **since** |then| seit [dieser Zeit]; **b)** *(at any time)* je[mals]; **hardly** ~: so gut wie nie; **c)** *in comb. with compar. adj. or adv.* noch; ~-**increasing** ständig zunehmend; **d)** *(coll.)* **what** ~ **does he want?** was will er nur?; **why** ~ **not?** warum denn nicht? **'evergreen 1.** *adj.* immergrün. **2.** *n.* immergrüne Pflanze. **ever'lasting** *adj.* **a)** *(eternal)* immerwährend; ewig ⟨*Leben*⟩; unvergänglich ⟨*Ruhm, Ehre*⟩; **b)** *(incessant)* endlos

every ['evrɪ] *adj.* **a)** jeder/jede/jedes; ~ **one** jeder/jede/jedes [einzelne]; **your** ~ **wish** all[e] deine Wünsche; **she comes** ~ **day** sie kommt jeden Tag; ~ **three/few days** alle drei/paar Tage; ~ **other** (~ *second, almost* ~) jeder/jede/jedes zweite; **b)** *(the greatest possible)* all ⟨*Respekt, Aussicht*⟩

every: ~**body** *n. & pron.* jeder; ~**body else** alle anderen; ~**day** *attrib. adj.* alltäglich; Alltags⟨*kleidung, -sprache*⟩; **in** ~**day life** im Alltag; ~**one** *see* ~**body;** ~**place** *(Amer.) see* ~**where;** ~**thing** *n. & pron.* alles; ~**where** *adv.* überall; ~**where you go/look** wohin man auch geht/sieht

evict [ɪˈvɪkt] *v. t.* ~ **sb.** |from his home|

jmdn. zur Räumung [seiner Wohnung] zwingen. **eviction** [ɪ'vɪkʃn] n. Zwangsräumung, die; **the ~ of the ten-ant** die zwangsweise Vertreibung des Mieters

evidence ['evɪdəns] n. a) Beweis, der; (indication) Anzeichen, das; **be ~ of sth.** etw. beweisen; b) (Law) Beweismaterial, das; **give ~:** aussagen

evident ['evɪdənt] adj. offensichtlich; **be ~ to sb.** jmdm. klar sein; **it soon became ~ that ...:** es stellte sich bald heraus, daß ... **evidently** adv. offensichtlich

evil ['iːvl, 'iːvɪl] 1. adj. böse; schlecht ⟨Charakter, Einfluß, System⟩. 2. n. a) Böse, das; b) (bad thing) Übel, das

evocative [ɪ'vɒkətɪv] adj. **be ~ of sth.** etw. heraufbeschwören

evoke [ɪ'vəʊk] v. t. heraufbeschwören; hervorrufen ⟨Bewunderung, Überraschung⟩; erregen ⟨Interesse⟩

evolution [iːvə'luːʃn] n. Entwicklung, die; (Biol.) Evolution, die

evolve [ɪ'vɒlv] 1. v. i. sich entwickeln (from aus, into zu). 2. v. t. entwickeln

ewe [juː] n. Mutterschaf, das

ex- pref. Ex-⟨Freundin, Präsident, Champion⟩; Alt⟨[bundes]kanzler⟩

exacerbate [ek'sæsəbeɪt] v. t. verschärfen ⟨Lage⟩; verschlechtern ⟨Zustand⟩

exact [ɪg'zækt] 1. adj. genau. 2. v. t. fordern; erheben ⟨Gebühr⟩. **exacting** [ɪg'zæktɪŋ] n. anspruchsvoll; hoch ⟨Anforderung⟩. **exactitude** [ɪg'zæktɪtjuːd] Genauigkeit, die. **exactly** [ɪg'zæktlɪ] adv. genau; **not ~** (coll. iron.) nicht gerade. **exactness** [ɪg'zæktnɪs] n. Genauigkeit, die

exaggerate [ɪg'zædʒəreɪt] v. t. übertreiben. **exaggeration** [ɪgzædʒə'reɪʃn] n. Übertreibung, die

exam [ɪg'zæm] (coll.) see examination b

examination [ɪgzæmɪ'neɪʃn] n. a) (inspection; Med.) Untersuchung, die; b) (Sch. etc.) Prüfung, die; (final ~ at university) Examen, das

examine [ɪg'zæmɪn] v. t. a) (inspect; Med.) untersuchen (for auf + Akk.); prüfen ⟨Dokument, Gewissen⟩; kontrollieren ⟨Ausweis, Gepäck⟩; b) (Sch. etc.) prüfen (in in + Dat.); c) (Law) verhören. **examiner** [ɪg'zæmɪnə(r)] n. Prüfer, der/Prüferin, die

example [ɪg'zɑːmpl] n. Beispiel, das; **for ~:** zum Beispiel; **make a ~ of sb.** ein Exempel an jmdm. statuieren

exasperate [ɪg'zæspəreɪt] v. t. (irrit-ate) verärgern; (infuriate) zur Verzweiflung bringen. **exasperation** [ɪg'zæspəreɪʃn] n. see exasperate: Ärger, der/Verzweiflung, die (with über + Akk.); **in ~:** verärgert/verzweifelt

excavate ['ekskəveɪt] v. t. a) ausschachten; (with machine) ausbaggern; b) (Archaeol.) ausgraben. **excavation** [ekskə'veɪʃn] n. a) Ausschachtung, die; (with machine) Ausbaggerung, die; b) (Archaeol.) Ausgrabung, die. **excavator** ['ekskəveɪtə(r)] n. Bagger, der

exceed [ɪk'siːd] v. t. a) (be greater than) übertreffen (in an + Dat.); ⟨Kosten, Summe, Anzahl:⟩ übersteigen (by um); b) (go beyond) überschreiten; hinausgehen über (+ Akk.) ⟨Auftrag, Befehl⟩. **exceedingly** adv. äußerst; ausgesprochen ⟨häßlich, dumm⟩

excel [ɪk'sel] 1. v. t., -ll- übertreffen; **~ oneself** (lit. or iron.) sich selbst übertreffen. 2. v. i., -ll- sich hervortun (at, in in + Dat.)

excellence ['eksələns] n. hervorragende Qualität. **excellent** ['eksələnt] adj. hervorragend

except [ɪk'sept] 1. prep. **~** [(coll.) for] außer (+ Dat.); **~ for** (in all respects other than) abgesehen von. 2. v. t. ausnehmen (from bei); **~ed** ausgenommen. **excepting** prep. außer (+ Dat.). **exception** [ɪk'sepʃn] n. Ausnahme, die; **take ~ to** Anstoß nehmen an (+ Dat.). **exceptional** [ɪk'sepʃənl] adj. außergewöhnlich. **exceptionally** adv. a) (as an exception) ausnahmsweise; b) (remarkably) ungewöhnlich

excerpt ['eksɜːpt] n. Auszug, der (from aus)

excess [ɪk'ses] n. a) Übermaß, das (of an + Dat.); **eat/drink to ~:** übermäßig essen/trinken; b) esp. in pl. (overindulgence) Exzeß, der; c) **be in ~ of** sth. etw. übersteigen; d) (surplus) Überschuß, der

excess ['ekses]: **~ baggage** n. Mehrgepäck, das; **~ fare** n. Mehrpreis, der; **pay the ~ fare** nachlösen

excessive [ɪk'sesɪv] adj. übermäßig; übertrieben ⟨Forderung, Lob, Ansprüche⟩; unmäßig ⟨Esser, Trinker⟩. **ex'cessively** adv. übertrieben; unmäßig ⟨essen, trinken⟩

exchange [ɪks'tʃeɪndʒ] 1. v. t. a) tauschen ⟨Plätze, Ringe, Küsse⟩; umtauschen ⟨Geld⟩; wechseln ⟨Blicke, Worte⟩; **~ insults** sich beleidigen; b) *(give in place of another)* eintauschen **(for** für, gegen); umtauschen *([gekaufte] Ware>* **(for** gegen). **2.** *n.* **a)** Tausch, *der;* in ~: dafür; in ~ **for sth.** für etw.; **b)** *(of money)* Umtausch, *der;* ~ **rate,** rate of ~: Wechselkurs, *der;* **c)** *(Teleph.)* Fernmeldeamt, *das*

exchequer [ɪks'tʃekə(r)] *n.* *(Brit.)* Schatzamt, *das*

excise ['eksaɪz] *n.* Verbrauchsteuer, *die;* **Customs and E~** *(Brit.)* Amt für Zölle und Verbrauchsteuer

excitable [ek'saɪtəbl] *adj.* leicht erregbar

excite [ɪk'saɪt] *v. t.* **a)** *(thrill)* begeistern; **b)** *(agitate)* aufregen. **ex'cited** *adj.* aufgeregt **(at** über + *Akk.);* **get** ~: sich aufregen. **ex'citement** *n.* Aufregung, *die; (enthusiasm)* Begeisterung, *die.* **exciting** [ɪk'saɪtɪŋ] *adj.* aufregend; *(full of suspense)* spannend

exclaim [ɪk'skleɪm] **1.** *v. t.* ausrufen. **2.** *v. i.* aufschreien. **exclamation** [eksklə'meɪʃn] *n.* Ausruf, *der.* **excla'mation mark,** *(Amer.)* **excla'mation point** *ns.* Ausrufezeichen, *das*

exclude [ɪk'sklu:d] *v. t.* ausschließen. **excluding** [ɪk'sklu:dɪŋ] *prep.* ~ **drinks/VAT** Getränke ausgenommen/ ohne Mehrwertsteuer. **exclusion** [ɪk-'sklu:ʒn] *n.* Ausschluß, *der.* **exclusive** [ɪk'sklu:sɪv] *adj.* **a)** alleinig *(Besitzer, Kontrolle>;* Allein<eigentum>; *(Journ.)* Exklusiv<bericht, -interview>; **b)** *(select)* exklusiv; **c)** ~ **of** ohne. **ex'clusively** *adv.* ausschließlich

excrement ['ekskrɪmənt] *n.* Kot, *der (geh.)*

excrete [ɪk'skri:t] *v. t.* ausscheiden

excruciating [ɪk'skru:'ʃieɪtɪŋ] *adj.* unerträglich

excursion [ɪk'skɜ:ʃn] *n.* Ausflug, *der*

excusable [ɪk'skju:zəbl] *adj.* entschuldbar; verzeihlich

excuse 1. [ɪk'skju:z] *v. t.* **a)** entschuldigen; ~ **oneself** sich entschuldigen; ~ **me** Entschuldigung; **b)** *(release, exempt)* befreien **(from** von). **2.** [ɪk'skju:s] *n.* Entschuldigung, *die*

ex-di'rectory *adj. (Brit. Teleph.)* Geheim<nummer, -anschluß>; **be** ~: nicht im Telefonbuch stehen

execute ['eksɪkju:t] *v. t.* **a)** hinrichten; **b)** *(put into effect)* ausführen. **execution** [eksɪ'kju:ʃn] *n.* **a)** Hinrichtung, *die;* **b)** *(putting into effect)* Ausführung, *die.* **exe'cutioner** *n.* Scharfrichter, *der*

executive [ɪg'zekjʊtɪv] **1.** *n.* leitender Angestellter/leitende Angestellte. **2.** *adj.* leitend <Stellung, Funktion>

executor [ɪg'zekjʊtə(r)] *n. (Law)* Testamentsvollstrecker, *der*

exemplary [ɪg'zemplərɪ] *adj.* **a)** *(model)* vorbildlich; **b)** *(deterrent)* exemplarisch

exemplify [ɪg'zemplɪfaɪ] *v. t.* veranschaulichen

exempt [ɪg'zempt] **1.** *adj.* [be] ~ [from sth.] [von etw.] befreit [sein]. **2.** *v. t.* befreien. **exemption** [ɪg'zempʃn] *n.* Befreiung, *die*

exercise ['eksəsaɪz] **1.** *n.* **a)** Übung, *die;* **b)** *no pl. (physical exertion)* Bewegung, *die;* **take** ~: sich *(Dat.)* Bewegung schaffen. **2.** *v. t.* ausüben *(Recht, Macht, Einfluß>;* walten lassen *(Vorsicht>.* **3.** *v. i.* sich *(Dat.)* Bewegung schaffen. **'exercise book** *n.* [Schul]heft, *das*

exert [ɪg'zɜ:t] **1.** *v. t.* aufbieten *(Kraft>;* ausüben *(Einfluß, Druck>.* **2.** *v. refl.* sich anstrengen. **exertion** [ɪg'zɜ:ʃn] *n.* **a)** *(of strength, force)* Aufwendung, *die; (of influence, pressure)* Ausübung, *die;* **b)** *(effort)* Anstrengung, *die*

exhale [eks'heɪl] *v. t. & i.* ausatmen

exhaust [ɪg'zɔ:st] **1.** *v. t.* erschöpfen; erschöpfend behandeln *(Thema>.* **2.** *n. (Motor Veh.)* Auspuff, *der; (gases)* Auspuffgase *Pl.* **ex'hausted** *adj.* erschöpft. **ex'hausting** *adj.* anstrengend. **exhaustion** [ɪg'zɔ:stʃn] *n.* Erschöpfung, *die.* **exhaustive** [ɪg'zɔ:stɪv] *adj.* umfassend. **ex'haustpipe** *n.* Auspuffrohr, *das*

exhibit [ɪg'zɪbɪt] **1.** *v. t.* ausstellen; zeigen *(Mut, Symptome, Angst usw.>.* **2.** *n.* Ausstellungsstück, *das.* **exhibition** [eksɪ'bɪʃn] *n.* Ausstellung, *die;* **make an** ~ **of oneself** sich unmöglich aufführen. **exhibitor** [ɪg'zɪbɪtə(r)] *n.* Aussteller, *der/*Ausstellerin, *die*

exhilarated [ɪg'zɪləreɪtɪd] *adj.* belebt. **exhilarating** [ɪg'zɪləreɪtɪŋ] *adj.* belebend. **exhilaration** [ɪgzɪlə'reɪʃn] *n.* [feeling of] ~: Hochgefühl, *das*

exhort [ɪg'zɔ:t] *v. t.* ermahnen

exile ['eksaɪl] **1.** *n.* **a)** Exil, *das;* **in/into** ~: im/ins Exil; **b)** *(person)* Verbannte, *der/die.* **2.** *v. t.* verbannen

exist [ɪg'zɪst] *v. i.* existieren; *(Zweifel, Gefahr, Problem, Einrichtung:>* bestehen; ~ **on sth.** von etw. leben. **existence** [ɪg'zɪstəns] *n.* Existenz, *die; (mode of living)* Dasein, *das;* **be in/ come into** ~: existieren/entstehen

exit ['eksɪt] *n. (way out)* Ausgang, *der* (from aus); *(for vehicle)* Ausfahrt, *die.* '**exit visa** *n.* Ausreisevisum, *das*
exonerate [ɪg'zɒnəreɪt] *v. t.* entlasten
exorbitant [ɪg'zɔ:bɪtənt] *adj.* [maßlos] überhöht
exorcize ['eksɔ:saɪz] *v. t.* austreiben
exotic [ɪg'zɒtɪk] *adj.* exotisch
expand [ɪk'spænd] **1.** *v. i.* **a)** sich ausdehnen; *(Commerc.)* expandieren; **b)** ~ **on** weiter ausführen. **2.** *v. t.* ausdehnen; *(Commerc.)* erweitern
expanse [ɪk'spæns] *n.* [weite] Fläche
expansion [ɪk'spænʃn] *n.* Ausdehnung, *die; (Commerc.)* Expansion, *die*
expect [ɪk'spekt] *v. t.* **a)** erwarten; ~ **to do sth.** damit rechnen, etw. zu tun; ~ **sb. to do sth.** damit rechnen, daß jmd. etw. tut; *(require)* von jmdm. erwarten, daß er etw. tut; **b)** *(coll.: think, suppose)* glauben; **I** ~ **so** ich glaube schon. **expectancy** [ɪk'spektənsɪ] *n.* Erwartung, *die.* **expectant** [ɪk'spektənt] *adj.* erwartungsvoll; ~ **mother** werdende Mutter. **ex'pectantly** *adv.* erwartungsvoll; gespannt ⟨*warten*⟩. **expectation** [ekspek'teɪʃn] *n.* Erwartung, *die*
expedient [ɪk'spi:dɪənt] **1.** *adj.* angebracht. **2.** *n.* Mittel, *das*
expedition [ekspɪ'dɪʃn] *n.* Expedition, *die*
expel [ɪk'spel] *v. t.,* **-ll-** ausweisen (**from** aus); ~ **sb. from school** jmdn. von der Schule verweisen
expend [ɪk'spend] *v. t.* **a)** aufwenden (**[up]on** für); **b)** *(use up)* aufbrauchen (**[up]on** für). **expendable** [ɪk'spendəbl] *adj.* entbehrlich; **be** ~: geopfert werden können
expenditure [ɪk'spendɪtʃə(r)] *n.* **a)** *(amount spent)* Ausgaben *Pl.* (**on** für); **b)** *(spending)* Ausgabe, *die*
expense [ɪk'spens] *n.* **a)** Kosten *Pl.;* **at sb.'s** ~: auf jmds. Kosten *(Akk.);* **at one's own** ~: auf eigene Kosten; **b)** *usu. in pl. (Commerc. etc.: amount spent [and repaid])* Spesen *Pl.;* **c)** *(fig.)* [**be**] **at the** ~ **of sth.** auf Kosten von etw. [gehen]. **ex'pense account** *n.* Spesenabrechnung, *die;* **put sth. on one's** ~: etw. als Spesen abrechnen. **expensive** [ɪk'spensɪv] *adj.,* **ex'pensively** *adv.* teuer
experience [ɪk'spɪərɪəns] **1.** *n.* Erfahrung, *die; (event)* Erlebnis, *das.* **2.** *v. t.* erleben; haben ⟨*Schwierigkeiten*⟩; verspüren ⟨*Kälte, Schmerz, Gefühl*⟩. **ex'perienced** *adj.* erfahren

experiment 1. [ɪk'sperɪmənt] *n.* **a)** Experiment, *das,* Versuch, *der* (**on an** + *Dat.*); **b)** *(fig.)* Experiment, *das.* **2.** [ɪk'sperɪment] *v. i.* Versuche anstellen (**on an** + *Dat.*). **experimental** [ɪksperɪ'mentl] *adj.* experimentell; Experimentier⟨*theater, -kino*⟩
expert ['ekspɜ:t] **1.** *adj.* ausgezeichnet; **be** ~ **in** *or* **at sth.** Fachmann *od.* Experte für etw. sein; **be** ~ **in** *or* **at doing sth.** etw. ausgezeichnet können. **2.** *n.* Fachmann, *der;* Experte, *der*/Expertin, *die;* **be an** ~ **in** *or* **at/on sth.** Fachmann *od.* Experte in etw. *(Dat.)*/für etw. sein. **expertise** [ekspɜ:'ti:z] *n.* Fachkenntnisse; *(skill)* Können, *das*
expire [ɪk'spaɪə(r)] *v. i.* ablaufen. **expiry** [ɪk'spaɪərɪ] *n.* Ablauf, *der*
explain [ɪk'spleɪn] **1.** *v. t., also abs.* erklären. **2.** *v. refl., often abs.* **please** ~ **[yourself]** bitte erklären Sie mir das. **explain a'way** *v. t.* eine [plausible] Erklärung finden für
explanation [eksplə'neɪʃn] *n.* Erklärung, *die;* **need** ~: einer Erklärung *(Gen.)* bedürfen
explanatory [ɪk'splænətərɪ] *adj.* erklärend; erläuternd ⟨*Bemerkung*⟩
explicable [ɪk'splɪkəbl] *adj.* erklärbar
explicit [ɪk'splɪsɪt] *adj.* klar; ausdrücklich ⟨*Zustimmung, Erwähnung*⟩. **ex'plicitly** *adv.* ausdrücklich; deutlich ⟨*beschreiben, ausdrücken*⟩
explode [ɪk'spləʊd] **1.** *v. i.* explodieren. **2.** *v. t.* zur Explosion bringen
exploit 1. ['eksplɔɪt] *n.* Heldentat, *die.* **2.** [ɪk'splɔɪt] *v. t.* ausbeuten ⟨*Arbeiter usw.*⟩; ausnutzen ⟨*Gutmütigkeit, Freund, Unwissenheit*⟩. **exploitation** [eksplɔɪ'teɪʃn] *n. see* **exploit 2:** Ausbeutung, *die;* Ausnutzung, *die*
exploration [eksplə'reɪʃn] *n.* Erforschung, *die; (fig.)* Untersuchung, *die*
exploratory [ɪk'splɒrətərɪ] *adj.* Forschungs-
explore [ɪk'splɔ:(r)] *v. t.* erforschen; *(fig.)* untersuchen. **ex'plorer** *n.* Entdeckungsreisende, *der/die*
explosion [ɪk'spləʊʒn] *n.* Explosion, *die.* **explosive** [ɪk'spləʊzɪv] **1.** *adj.* explosiv. **2.** *n.* Sprengstoff, *der*
export 1. [ɪk'spɔ:t, 'ekspɔ:t] *v. t.* exportieren; ausführen. **2.** ['ekspɔ:t] *n.* Export, *der.* **ex'porter** *n.* Exporteur, *der*
expose [ɪk'spəʊz] *v. t.* **a)** *(uncover)* freilegen; entblößen ⟨*Haut, Körper*⟩; **b)** offenbaren ⟨*Schwäche*⟩; aufdecken ⟨*Mißstände, Verbrechen*⟩; entlarven ⟨*Täter, Spion*⟩; **c)** *(subject)* ~ **to sth.** ei-

ner Sache *(Dat.)* aussetzen; **d)** *(Photog.)* belichten. **exposed** [ɪk-ˈspəʊzd] *adj. (unprotected)* ungeschützt; ~ **position** exponierte Stellung. **exposure** [ɪkˈspəʊʒə(r)] *n.* **a)** *(to cold etc.)* die of/suffer from ~: an Unterkühlung *(Dat.)* sterben/leiden; **b)** *(Photog.) (exposing time)* Belichtung, *die; (picture)* Aufnahme, *die.* **ex-ˈposure meter** *n.* Belichtungsmesser, *der*

expound [ɪkˈspaʊnd] *v.t.* darlegen

express [ɪkˈspres] **1.** *v.t.* ausdrücken; äußern ⟨*Meinung, Wunsch, Dank, Bedauern*⟩; ~ **oneself** sich ausdrücken. **2.** *attrib. adj.* **a)** Eil⟨*brief, -bote usw.*⟩; Schnell⟨*paket, -sendung*⟩; **b)** ausdrücklich ⟨*Wunsch, Absicht*⟩. **3.** *adv.* als Eilsache ⟨*senden*⟩. **4.** *n. (train)* Schnellzug, *der.* **expression** [ɪkˈspreʃn] *n.* Ausdruck, *der.* **ex-pressive** [ɪkˈspresɪv] *adj.* ausdrucksvoll

express: ~ ˈtrain *n.* D-Zug, *der;* ~**way** *n. (Amer.)* Schnellstraße, *die*
exˈpressly *adv.* ausdrücklich

expulsion [ɪkˈspʌlʃn] *n.* Ausweisung, *die* (from aus); *(from school)* Verweisung, *die* (from von)

exquisite [ˈekskwɪzɪt, ɪkˈskwɪzɪt] *adj.* erlesen. **exˈquisitely** *adv.* vorzüglich; kunstvoll ⟨*verziert, geschnitzt*⟩

extend [ɪkˈstend] **1.** *v.t.* verlängern; ausstrecken ⟨*Arm, Bein, Hand*⟩; ausziehen ⟨*Leiter, Teleskop*⟩; verlängern lassen ⟨*Leihbuch, Visum*⟩; ausdehnen ⟨*Einfluß, Macht*⟩; vergrößern ⟨*Haus, Geschäft, Fabrik*⟩; gewähren ⟨*[Gast]freundschaft, Hilfe, Kredit*⟩ (to *Dat.*); ~ **the time limit** den Termin hinausschieben. **2.** *v.i.* sich erstrecken; **the season ~s from November to March** die Saison geht von November bis März

extension [ɪkˈstenʃn] *n.* **a)** Verlängerung, *die;* **b)** *(part of house)* Anbau, *der;* **c)** *(telephone)* Nebenanschluß, *der; (number)* Apparat, *der.* **extensive** [ɪkˈstensɪv] *adj.* ausgedehnt; umfangreich ⟨*Reparatur, Wissen, Nachforschungen*⟩; beträchtlich ⟨*Schäden*⟩; weitreichend ⟨*Änderungen*⟩. **exˈtensively** *adv.* beträchtlich ⟨*ändern, beschädigen*⟩; ausführlich ⟨*berichten, schreiben*⟩

extent [ɪkˈstent] *n.* Ausdehnung, *die; (scope)* Umfang, *der; (of damage)* Ausmaß, *das;* **to what ~?** inwieweit?

exterior [ɪkˈstɪərɪə(r)] **1.** *adj.* äußer...;

Außen⟨*fläche, -wand*⟩. **2.** *n.* Äußere, *das; (of house)* Außenwände *Pl.*

exterminate [ɪkˈstɜːmɪneɪt] *v.t.* ausrotten; vertilgen ⟨*Ungeziefer*⟩. **extermination** [ɪkstɜːmɪˈneɪʃn] *n.* Ausrottung, *die; (of pests)* Vertilgung, *die*

external [ɪkˈstɜːnl] *adj.* äußer...; Außen⟨*fläche, -abmessungen*⟩; *purely* ~: rein äußerlich; **for** ~ **use only** nur äußerlich anzuwenden

extinct [ɪkˈstɪŋkt] *adj.* erloschen ⟨*Vulkan*⟩; ausgestorben ⟨*Art, Rasse, Gattung*⟩. **extinction** [ɪkˈstɪŋkʃn] *n.* Aussterben, *das*

extinguish [ɪkˈstɪŋgwɪʃ] *v.t.* löschen. **exˈtinguisher** *n.* Feuerlöscher, *der*

extol [ɪkˈstɒl] *v.t.,* **-ll-** rühmen; preisen

extort [ɪkˈstɔːt] *v.t.* erpressen (**out of** von). **extortion** [ɪkˈstɔːʃn] *n.* Erpressung, *die.* **extortionate** [ɪkˈstɔːʃənət] *adj.* Wucher⟨*preis, -zinsen usw.*⟩; maßlos überzogen ⟨*Forderung*⟩

extra [ˈekstrə] **1.** *adj.* zusätzlich; Mehr⟨*arbeit, -kosten, -ausgaben*⟩; Sonder⟨*bus, -zug*⟩. **2.** *adv.* **a)** *(more than usually)* besonders; extra ⟨*lang, stark, fein*⟩; **b)** *(additionally)* extra; **packing and postage** ~: zuzüglich Verpackung und Porto. **3.** *n.* **a)** *(added to services, salary, etc.)* zusätzliche Leistung; **b)** *(in play, film, etc.)* Statist, *der/*Statistin, *die*

extract 1. [ˈekstrækt] *n.* **a)** Extrakt, *der (fachspr. auch: das);* **b)** *(from book, music, etc.)* Auszug, *der.* **2.** [ɪkˈstrækt] *v.t.* ziehen ⟨*Zahn*⟩; herausziehen ⟨*Dorn, Splitter usw.*⟩. **extraction** [ɪkˈstrækʃn] *n. (of tooth)* Extraktion, *die; (of thorn, splinter, etc.)* Herausziehen, *das.* **exˈtractor fan** *n.* Entlüfter, *der*

extradite [ˈekstrədaɪt] *v.t.* ausliefern. **extradition** [ekstrəˈdɪʃn] *n.* Auslieferung, *die*

extraordinary [ɪkˈstrɔːdɪnərɪ] *adj.* außergewöhnlich; merkwürdig ⟨*Benehmen*⟩; **how** ~! wie seltsam!

extravagance [ɪkˈstrævəgəns] *n.* **a)** Extravaganz, *die;* **b)** *(extravagant thing)* Luxus, *der*

extravagant [ɪkˈstrævəgənt] *adj.* verschwenderisch; aufwendig ⟨*Lebensstil*⟩; teuer ⟨*Geschmack*⟩

extreme [ɪkˈstriːm] **1.** *adj.* **a)** äußerst... ⟨*Spitze, Rand, Ende*⟩; extrem ⟨*Gegensätze, Hitze, Kälte*⟩; höchst... ⟨*Gefahr*⟩; äußerst... ⟨*Notfall, Höflichkeit, Bescheidenheit*⟩; stärkst... ⟨*Schmerzen*⟩; größt... ⟨*Wichtigkeit*⟩; **at the** ~ **edge/left** ganz am Rand/ganz links; **b)**

(not moderate) extrem; drastisch ⟨*Maßnahme*⟩. **2.** *n.* Extrem, *das;* **go to ~s** vor nichts zurückschrecken; **go from one ~ to the other** von einem Extrem ins andere fallen. **ex'tremely** *adv.* äußerst. **extremist** [ɪk'striːmɪst] *n.* Extremist, *der*/Extremistin, *die; attrib.* extremistisch. **extremity** [ɪk'stremɪtɪ] *n.* äußerstes Ende

extricate ['ekstrɪkeɪt] *v.t.* **~ sth. from sth.** etw. aus etw. herausziehen; **~ oneself/sb. from sth.** sich/jmdn. aus etw. befreien

extrovert ['ekstrəvɜːt] **1.** *n.* extrovertierter Mensch; **be an ~:** extrovertiert sein. **2.** *adj.* extrovertiert

exuberant [ɪg'zjuːbərənt] *adj.* **be ~:** sich überschwenglich freuen

exude [ɪg'zjuːd] *v.t.* absondern; *(fig.)* ausstrahlen

exult [ɪg'zʌlt] *v.i.* jubeln **(in, at, over** über + *Akk.*)

eye [aɪ] **1.** *n.* **a)** Auge, *das;* **keep an ~ on sb./sth.** auf jmdn./etw. aufpassen; **see ~ to ~:** einer Meinung sein; **with one's ~s shut** *(fig.)* blind; *(easily)* im Schlaf; **be up to one's ~s in work/debt** bis über beide Ohren in Arbeit/Schulden stecken *(ugs.);* **b)** *(of needle)* Öhr, *das; (metal loop)* Öse, *die.* **2.** *v.t.,* beäugen; **~ sb. up and down** jmdn. von oben bis unten mustern

eye: ~ball *n.* Augapfel, *der;* **~brow** *n.* Augenbraue, *die;* **~lash** *n.* Augenwimper, *die;* **~level** *n.* Augenhöhe, *die; attrib.* in Augenhöhe *nachgestellt;* **at ~-level** in Augenhöhe; **~lid** *n.* Augenlid, *das;* **~-shadow** *n.* Lidschatten, *der;* **~sight** *n.* Sehkraft, *die;* **have good ~sight** gute Augen haben; **his ~sight is poor** er hat schlechte Augen; **~sore** *n.* Schandfleck, *der;* **~witness** *n.* Augenzeuge, *der/*-zeugin, *die*

F

F, f [ef] *n.* F, f, *das*

fable ['feɪbl] *n.* Fabel, *die; (myth, lie)* Märchen, *das*

fabric ['fæbrɪk] *n.* Gewebe, *das*

fabricate ['fæbrɪkeɪt] *v.t. (invent)* erfinden. **fabrication** [fæbrɪkeɪʃn] *n.* Erfindung, *die*

fabulous ['fæbjʊləs] *adj.* **a)** sagenhaft; **b)** *(coll.: marvellous)* fabelhaft *(ugs.)*

face [feɪs] **1.** *n.* **a)** Gesicht, *das;* **lie ~ down**|**ward**] ⟨*Person/Buch:*⟩ auf dem Bauch/Gesicht liegen; **make** *or* **pull a ~/~s** Grimassen schneiden; **on the ~ of it** dem Anschein nach; **in the ~ of sth.** trotz etw. *(Gen.);* **b)** *(of mountain, cliff)* Wand, *die; (of clock, watch)* Zifferblatt, *das; (of dice)* Seite, *die; (of coin, playing-card)* Vorderseite, *die.* **2.** *v.t.* **a)** sich wenden zu; |**stand**] **facing one another** sich *(Dat.)* gegenüber [stehen]; **b)** *(fig.)* ins Auge sehen (+ *Dat.)* ⟨*Tod, Vorstellung*⟩; stehen vor (+ *Dat.)* ⟨*Ruin, Entscheidung*⟩; **~ the facts** den Tatsachen ins Gesicht sehen; **be ~d with sth.** sich einer Sache *(Dat.)* gegenübersehen; **c)** *(coll.: bear)* verkraften. **3.** *v.i. (in train etc.)* **~ forwards/backwards** ⟨*Person:*⟩ in/entgegen Fahrtrichtung sitzen. **face 'up to** *v.t.* ins Auge sehen (+ *Dat.*); sich abfinden mit ⟨*Möglichkeit*⟩

face: ~-cream *n.* Gesichtscreme, *die;* **~-flannel** *n. (Brit.)* Waschlappen, *der;* **~-lift** *n.* **a)** Facelifting, *das;* **have** *or* **get a ~-lift** sich liften lassen; **b)** *(fig.)* Verschönerung, *die*

facet ['fæsɪt] *n.* Facette, *die; (fig.)* Aspekt, *der*

facetious [fə'siːʃəs] *adj.* [gewollt] witzig

face: ~-to-~: persönlich ⟨*Gespräch, Treffen*⟩; **~ value** *n.* Nennwert, *der;* **accept sth. at** |**its**] **~ value** *(fig.)* etw. für bare Münze nehmen

facial ['feɪʃl] *adj.* Gesichts-

facile ['fæsaɪl] *adj.* nichtssagend

facilities [fə'sɪlɪtɪz] *n. pl.* Einrichtungen; **cooking/washing ~:** Koch-/Waschgelegenheit, *die;* **sports ~:** Sportanlagen; **shopping ~:** Einkaufsmöglichkeiten

facsimile [fæk'sɪmɪlɪ] *n.* **a)** Faksimile, *das;* **b)** *see* **fax 1**

fact [fækt] *n.* Tatsache, *die;* **~s and figures** Fakten und Zahlen; **the ~ remains that ...:** Tatsache bleibt: ...; **the true ~s of the case** *or* **matter** der wahre Sachverhalt; **know for a ~ that ...:** genau wissen, daß ...; **in ~:** tatsächlich

faction ['fækʃn] *n.* Splittergruppe, *die*

factor ['fæktə(r)] *n.* Faktor, *der*

factory ['fæktərɪ] *n.* Fabrik, *die.* **'factory farm** *n.* Agrarfabrik, *die*

factual ['fæktjʊəl] *adj.* sachlich
faculty ['fækəltı] *n.* a) Fähigkeit, *die;*
mental ~: geistige Kraft; b) *(Univ.)*
Fakultät, *die*
fad [fæd] *n.* Marotte, *die*
fade [feɪd] *v. i.* a) ⟨*Blätter, Blumen:*⟩
[ver]welken; b) ~ [in colour] [ver]blei-
chen; **the light ~d** es dunkelte; c)
⟨*Laut:*⟩ verklingen; d) *(fig.)* verblassen; ⟨*Schönheit:*⟩ verblühen; ⟨*Hoffnung:*⟩ schwinden; e) *(blend)* übergehen (**into** in + *Akk.*). **fade a'way** *v. i.*
schwinden; ⟨*Laut:*⟩ verklingen (**into** in
+ *Dat.*)
faded ['feɪdɪd] *adj.* welk ⟨*Blume, Blatt,
Laub*⟩; verblichen ⟨*Stoff, Farbe*⟩
fag [fæg] *n.* a) *(Brit. coll.)* Schinderei,
die (ugs.); b) *(sl.: cigarette)* Stäbchen,
das (ugs.)
fail [feɪl] 1. *v. i.* a) scheitern; *(in examination)* nicht bestehen (**in** in + *Dat.*);
b) *(become weaker)* ⟨*Augenlicht, Gehör,
Stärke:*⟩ nachlassen; c) *(break down,
stop)* ⟨*Versorgung:*⟩ zusammenbrechen; ⟨*Motor:*⟩ aussetzen; ⟨*Batterie,
Pumpe:*⟩ ausfallen; ⟨*Bremse:*⟩ versagen. 2. *v. t.* a) ~ **to do sth.** *(not succeed
in doing)* etw. nicht tun (können); ~ **to
achieve one's purpose/aim** seine Ab-
sicht/sein Ziel verfehlen; b) *(be unsuccessful in)* nicht bestehen ⟨*Prüfung*⟩;
c) *(reject)* durchfallen lassen *(ugs.)*
⟨*Prüfling*⟩; d) ~ **to do sth.** *(not do)* etw.
nicht tun; *(neglect to do)* [es] versäumen, etw. zu tun; **not** ~ **to do sth.** etw.
tun; e) **words** ~ **me** mir fehlen die
Worte; **his courage** ~**ed him** ihn ver-
ließ der Mut. 3. *n.* **without** ~: auf jeden Fall. **'failing** 1. *n.* Schwäche, *die.*
2. *prep.* ~ **that** andernfalls. **failure**
['feɪljə(r)] *n.* a) *(omission, neglect)* Ver-
säumnis, *das;* b) *(lack of success)*
Scheitern, *das;* **end in** ~: scheitern; c)
(person or thing) Versager, *der;* **our
plan/attempt was a** ~: unser Plan/Ver-
such war fehlgeschlagen
faint [feɪnt] 1. *adj.* a) matt ⟨*Licht,
Farbe, Stimme, Lächeln*⟩; schwach
⟨*Geruch, Duft*⟩; leise ⟨*Flüstern, Geräusch, Stimme*⟩; entfernt ⟨*Ähnlichkeit*⟩; undeutlich ⟨*Umriß, Linie, Spur,
Fotokopie*⟩; b) *(giddy, weak)* matt; **she
felt** ~: ihr war schwindelig. 2. *v. i.*
ohnmächtig werden (**from** vor +
Dat.). 3. *n.* Ohnmacht, *die.* **'faintly**
adv. schwach; entfernt ⟨*sich ähneln*⟩
¹fair [feə(r)] *n. (fun-~)* Jahrmarkt, *der;*
(exhibition) Messe, *die;* **book/trade** ~:
Buch-/Handelsmesse, *die*

²fair 1. *adj.* a) *(just)* gerecht; begründet
⟨*Beschwerde, Annahme*⟩; fair ⟨*Spiel,
Kampf, Prozeß, Preis, Beurteilung,
Handel*⟩; ~ **play** Fairneß, *die;* b) *(not
bad, pretty good)* ganz gut ⟨*Bilanz, Anzahl, Chance*⟩; ziemlich ⟨*Maß, Geschwindigkeit*⟩; c) *(blond)* blond
⟨*Haar, Person*⟩; *(light)* hell ⟨*Haut*⟩;
(~-skinned) hellhäutig ⟨*Person*⟩; d)
schön ⟨*Wetter, Tag*⟩. **'fair-haired**
1. *adj.* blond. 2. *adv.* fair ⟨*kämpfen,
spielen*⟩. **'fairly** *adv.* a) fair ⟨*kämpfen,
spielen*⟩; gerecht ⟨*bestrafen, beurteilen,
behandeln*⟩; b) *(rather)* ziemlich. **'fairness** *n.* Gerechtigkeit, *die;* **in all** ~ [**to
sb.**] um fair [gegen jmdn.] zu sein
fairy ['feərı] *n.* Fee, *die*
fairy: ~ **'godmother** *n.* gute Fee; ~
story, ~**-tale** *ns.* Märchen, *das*
faith [feɪθ] *n.* a) *(reliance, trust)* Vertrauen, *das* (**in** zu); **have** ~ **in oneself**
Selbstvertrauen haben; **in good** ~: in
gutem Glauben; b) *(religious belief)*
Glaube, *der.* **faithful** ['feɪθfl] *adj.* a)
treu (**to** *Dat.*); b) *(conscientious)*
pflichtbewußt; [ge]treu ⟨*Diener*⟩; c)
(accurate) [wahrheits]getreu; original-
getreu ⟨*Wiedergabe, Kopie*⟩. **'faithfully** *adv.* a) treu ⟨*dienen*⟩; pflichtbewußt ⟨*überbringen, zustellen*⟩; hoch
und heilig ⟨*versprechen*⟩; b) *(accurately)* wahrheitsgetreu ⟨*erzählen*⟩; ori-
ginalgetreu ⟨*wiedergeben*⟩; genau ⟨*befolgen*⟩; c) **yours** ~: hochachtungsvoll
fake [feɪk] 1. *adj.* unecht; gefälscht
⟨*Dokument, Banknote, Münze*⟩. 2. *n.*
a) Imitation, *die;* *(painting)* Fälschung, *die;* b) *(person)* Schwindler,
der/Schwindlerin, *die.* 3. *v. t.* fälschen
⟨*Unterschrift*⟩; vortäuschen ⟨*Krankheit, Unfall*⟩
falcon ['fɔːlkn] *n.* Falke, *der*
fall [fɔːl] 1. *n.* a) Fallen, *das;* *(of person)*
Sturz, *der;* ~ **of snow/rain** Schnee-/
Regenfall, *der;* **have a** ~: stürzen; b)
(collapse, defeat) Fall, *der;* *(of dynasty,
empire)* Untergang, *der;* c) *(decrease)*
Rückgang, *der;* d) *(Amer.: autumn)*
Herbst, *der.* 2. *v. i.,* **fell** [fel], ~**en**
['fɔːln] a) fallen; ⟨*Baum:*⟩ umstürzen;
⟨*Pferd:*⟩ stürzen; ~ **off sth.,** ~ **down
from sth.** von etw. [herunter]fallen; ~
down [into] sth. in etw. *(Akk.)* [hin-
ein]fallen; ~ **to the ground** auf den Boden fallen; ~ **down the stairs** *or* **downstairs** die Treppe herunter-/hinunter-
fallen; b) ⟨*Nacht, Dunkelheit:*⟩ hereinbrechen; ⟨*Abend:*⟩ anbrechen; c)
⟨*Blätter:*⟩ [ab]fallen; d) *(sink)* sinken

far

⟨Barometer:⟩ fallen; ⟨Absatz, Verkauf:⟩ zurückgehen; ~ by 10 per cent/from 10[°C] to 0[°C] um 10%/von 10[°C] auf 0[°C] sinken; e) (be killed) ⟨Soldat:⟩ fallen; f) (collapse) einstürzen; ~ to pieces, ~ apart auseinanderfallen; g) (occur) fallen (on auf + Akk.). fall 'back v. i. zurückweichen. fall 'back on v. t. zurückgreifen auf (+ Akk.). fall 'down v. i. a) see fall 2 a; b) ⟨Brücke, Gebäude:⟩ einstürzen. 'fall for v. t. (coll.) ~ for sb. sich in jmdn. verknallen (ugs.); ~ for sth. auf etw. (Akk.) hereinfallen (ugs.). fall 'in v. i. a) hineinfallen; b) (Mil.) antreten; ~ in! angetreten!; c) ⟨Gebäude, Wand usw.:⟩ einstürzen. fall 'off v. i. a) herunterfallen; b) (diminish) nachlassen. fall 'out v. i. a) herausfallen; ⟨Haare, Federn⟩ ausfallen; b) (quarrel) ~ out [with sb.] sich [mit jmdm.] streiten. fall 'over v. i. umfallen; ⟨Person:⟩ [hin]fallen. fall 'through v. i. (fig.) ins Wasser fallen (ugs.)

fallacy ['fæləsɪ] n. Irrtum, der
fallen see fall 3
fallible ['fælɪbl] adj. nicht unfehlbar; fehlbar ⟨Person⟩
'fall-out n. radioaktiver Niederschlag
fallow ['fæləʊ] adj. brachliegend; ~ ground/land Brache, die/Brachland, das; lie ~: brachliegen
false [fɔːls, fɒls] adj. falsch; gefälscht ⟨Urkunde, Dokument⟩; künstlich ⟨Wimpern⟩; under a ~ name unter falschem Namen. 'falsely adv. falsch; fälschlich[erweise] ⟨annehmen, glauben, behaupten, beschuldigen⟩
false: ~ a'larm n. blinder Alarm; ~ 'start n. Fehlstart, der; ~ 'teeth n. pl. [künstliches] Gebiß
falsify ['fɔːlsɪfaɪ] v. t. (alter) fälschen; (misrepresent) verfälschen ⟨Tatsachen, Wahrheit⟩
falter ['fɔːltə(r)] v. i. stocken
fame [feɪm] n. Ruhm, der
familiar [fə'mɪljə(r)] adj. a) vertraut; bekannt ⟨Gesicht, Name, Lied⟩; he looks ~: er kommt mir bekannt vor; b) (informal) ungezwungen ⟨Sprache, Art⟩. familiarity [fəmɪlɪ'ærɪtɪ] n. Vertrautheit, die. familiarize [fə'mɪljəraɪz] v. t. vertraut machen (with mit)
family ['fæməlɪ] n. Familie, die
family: ~ name n. Familienname, der; ~ 'planning n. Familienplanung, die; ~ 'tree n. Stammbaum, der
famine ['fæmɪn] n. Hungersnot, die

famished ['fæmɪʃt] adj. ausgehungert; I'm absolutely ~ (coll.) ich sterbe vor Hunger (ugs.)
famous ['feɪməs] adj. berühmt
¹fan [fæn] 1. n. Fächer, der; (apparatus) Ventilator, der. 2. v. t., -nn- fächeln ⟨Gesicht⟩; anfachen ⟨Feuer⟩; ~ oneself/sb. sich/jmdm. Luft zufächeln. fan 'out v. i. fächern; ⟨Soldaten:⟩ ausfächern
²fan n. (devotee) Fan, der
fanatic [fə'nætɪk] n. Fanatiker, der/Fanatikerin, die. fanatical [fə'nætɪkl] adj. fanatisch. fanaticism [fə'nætɪsɪzm] n. Fanatismus, der
'fan belt n. Keilriemen, der
fanciful ['fænsɪfl] adj. überspannt ⟨Vorstellung, Gedanke⟩; phantastisch ⟨Gemälde, Design⟩
'fan club n. Fanklub, der
fancy ['fænsɪ] 1. n. a) (taste, inclination) he has taken a ~ to a new car/her ein neues Auto/sie hat es ihm angetan; take or catch sb.'s ~: jmdm. gefallen; b) (whim) Laune, die; tickle sb.'s ~: jmdn. reizen. 2. attrib. adj. kunstvoll ⟨Arbeit, Muster⟩; fein[st] ⟨Kuchen, Spitzen⟩. 3. v. t. a) (imagine) sich (Dat.) einbilden; ~ that! (coll.) sieh mal einer an!; b) (suppose) glauben; c) (wish to have) mögen; what do you ~ for dinner? was hättest du gern zum Abendessen? fancy 'dress n. [Masken]kostüm, das; in ~: kostümiert; fancy-dress party Kostümfest, das; fancy-dress ball Maskenball, der
fanfare ['fænfeə(r)] n. Fanfare, die
fang [fæŋ] n. Reißzahn, der; (of snake) Giftzahn, der
fan: ~ heater n. Heizlüfter, der; ~light n. Oberlicht, das; ~ mail n. Fanpost, die
fantastic [fæn'tæstɪk] adj. a) (grotesque, quaint) bizarr; b) (coll.: excellent) phantastisch (ugs.)
fantasy ['fæntəzɪ] n. Phantasie, die; (mental image) Phantasiegebilde, das
far [fɑː(r)] 1. adv. weit; ~ above/below hoch über/tief unter (+ Dat.); hoch oben/tief unten; as ~ as Munich/the church bis [nach] München/bis zur Kirche; ~ and wide weit und breit; from ~ and wide von fern und nah; ~ too viel zu; ~ longer/better weit[aus] länger/besser; as ~ as I remember/know soweit ich mich erinnere/weiß; go so ~ as to do sth. so weit gehen und etw. tun; so ~ (until now) bisher; so ~ so good so weit, so gut; by ~: bei wei-

tem; ~ **from easy/good** alles andere als leicht/gut. **2.** *adj.* **a)** *(remote)* weit entfernt; *(in time)* fern; **in the ~ distance** in weiter Ferne; **b)** *(more remote)* weiter entfernt; **the ~ bank of the river/side of the road** das andere Flußufer/die andere Straßenseite; **the ~ door/wall** *etc.* die hintere Tür/ Wand *usw.*

farce [fɑːs] *n.* Farce, *die.* **farcical** ['fɑːsɪkl] *adj. (absurd)* farcenhaft

fare [feə(r)] *n.* **a)** *(price)* Fahrpreis, *der; (money)* Fahrgeld, *das;* **what** *or* **how much is the ~?** was kostet die Fahrt?; **b)** *(food)* Kost, *die*

Far: ~ 'East *n.* **the ~ East** der Ferne Osten; ~ 'Eastern *adj.* fernöstlich; des Fernen Ostens *nachgestellt*

farewell [feə'wel] **1.** *int.* leb[e] wohl *(veralt.).* **2.** *n. attrib.* ~ **speech/gift** Abschiedsrede, *die/*-geschenk, *das*

far-'fetched *adj.* weit hergeholt

farm [fɑːm] **1.** *n.* [Bauern]hof, *der; (larger)* Gut, *das;* ~ **animals** Nutzvieh, *das.* **2.** *v.t.* bebauen ⟨*Land*⟩. **3.** *v.i.* Landwirtschaft treiben. **'farmer** *n.* Landwirt, *der/*-wirtin, *die*

'farmhouse *n.* Bauernhaus, *das; (larger)* Gutshaus, *das*

'farming *n.* Landwirtschaft, *die*

farm: ~**land** *n.* Acker- und Weideland, *das;* ~**yard** *n.* Hof, *der*

far: ~-'reaching *adj.* weitreichend; ~-'sighted *adj.* **a)** *(fig.)* weitblickend; **b)** *(Amer.: long-sighted)* weitsichtig

fart [fɑːt] *(coarse)* **1.** *v.i.* furzen *(derb).* **2.** *n.* Furz, *der (derb)*

farther ['fɑːðə(r)] *see* further 1a, 2

farthest ['fɑːðɪst] *see* furthest

fascinate ['fæsɪneɪt] *v.t.* fesseln; bezaubern. **fascination** [fæsɪ'neɪʃn] *n.* Zauber, *der;* **have a ~ for sb.** einen besonderen Reiz auf jmdn. ausüben

Fascism ['fæʃɪzm] *n.* Faschismus, *der.* **Fascist** ['fæʃɪst] **1.** *n.* Faschist, *der/*Faschistin, *die.* **2.** *adj.* faschistisch

fashion ['fæʃn] **1.** *n.* **a)** Mode, *die;* **b)** *(manner)* Art [und Weise]; **talk/behave in a peculiar ~:** merkwürdig sprechen/sich merkwürdig verhalten. **2.** *v.t.* formen (out of, from aus; [in]to zu). **fashionable** ['fæʃənəbl] *adj.* modisch; vornehm ⟨*Hotel, Restaurant*⟩; Mode⟨*farbe, -autor*⟩. **fashionably** ['fæʃənəblɪ] *adv.* modisch

¹fast [fɑːst] **1.** *v.i.* fasten. **2.** *n.* Fasten, *das*

²fast 1. *adj.* **a)** *(fixed, attached)* fest; **make [the boat] ~:** das Boot festmachen; **hard and ~:** fest; bindend ⟨*Regel*⟩; klar ⟨*Entscheidung*⟩; **b)** *(rapid)* schnell; ~ **train** Schnellzug, *der;* D-Zug, *der;* **c) be [ten minutes] ~** ⟨*Uhr:*⟩ [zehn Minuten] vorgehen. **2.** *adv.* **a) be ~ asleep** fest schlafen; *(when one should be awake)* fest eingeschlafen sein; **b)** *(quickly)* schnell

fasten ['fɑːsn] *v.t.* befestigen (on, to an + *Dat.*); zumachen ⟨*Kleid, Spange, Jacke*⟩; [ab]schließen ⟨*Tür*⟩; anstecken ⟨*Brosche*⟩ (to an + *Akk.*); ~ **one's seat-belt** sich anschnallen. **'fastener, 'fastening** *ns.* Verschluß, *der*

fastidious [fæ'stɪdɪəs] *adj.* wählerisch; *(hard to please)* heikel

'fast lane *n.* Überholspur, *die;* **life in the ~** *(fig.)* Leben auf vollen Touren *(ugs.)*

fat [fæt] **1.** *adj.* dick; rund ⟨*Wangen, Gesicht*⟩. **2.** *n.* Fett, *das*

fatal ['feɪtl] *adj.* **a)** *(disastrous)* verheerend (to für); **it would be ~:** das wäre das Ende; **b)** *(deadly)* tödlich ⟨*Unfall, Verletzung*⟩. **fatality** [fə'tælɪtɪ] *n.* Todesopfer, *das.* **'fatally** *adv.* tödlich; **be ~ ill** todkrank sein

fate [feɪt] *n.* Schicksal, *das*

'fat-head *n.* Dummkopf, *der (ugs.)*

father ['fɑːðə(r)] *n.* Vater, *der.* **Father 'Christmas** *n.* der Weihnachtsmann. **father-in-law** *n., pl.* ~**s-in-law** Schwiegervater, *der.* **'fatherly** *adj.* väterlich

fathom ['fæðəm] **1.** *n. (Naut.)* Faden, *der.* **2.** *v.t. (comprehend)* verstehen; ~ **sb./sth. out** jmdn./etw. ergründen

fatigue [fə'tiːg] **1.** *n.* Ermüdung, *die.* **2.** *v.t.* ermüden

'fatness *n.* Dicke, *die*

fatten ['fætn] *v.t.* herausfüttern ⟨*Person*⟩; mästen ⟨*Tier*⟩. **'fattening** *adj.* **be ~:** dick machen

fatty ['fætɪ] *adj.* fett ⟨*Fleisch, Soße*⟩; fetthaltig ⟨*Speise, Nahrungsmittel*⟩

faucet ['fɔːsɪt] *n. (Amer.)* Wasserhahn, *der*

fault [fɔːlt, fɒlt] *n.* **a)** Fehler, *der;* **b)** *(responsibility)* Schuld, *die;* **it's your ~:** du bist schuld; **it isn't my ~:** ich habe keine Schuld; **be at ~:** im Unrecht sein; **c)** *(in machinery; also Electr.)* Defekt, *der.* **'faultless** *adj.* einwandfrei. **'faulty** *adj.* fehlerhaft; defekt ⟨*Gerät, usw.*⟩

fauna ['fɔːnə] *n., pl.* ~**e** ['fɔːnɪ] *or* ~**s** Fauna, *die*

favor *etc. (Amer.) see* **favour** *etc.*

favour ['feɪvə(r)] **1.** *n.* **a)** Gunst, *die;* **b)** *(kindness)* Gefallen, *der;* **ask sb. a ~, ask a ~ of sb.** jmdn. um einen Gefallen bitten; **do sb. a ~, do a ~ for sb.** jmdm. einen Gefallen tun; **as a ~:** aus Gefälligkeit; **c) be in ~ of sth.** für etw. sein. **2.** *v. t.* bevorzugen

favourable ['feɪvərəbl] *adj. (Brit.)* **a)** günstig ‹*Eindruck, Licht*›; gewogen ‹*Haltung, Einstellung*›; freundlich ‹*Erwähnung*›; positiv ‹*Bericht[erstattung], Bemerkung*›; **b)** *(helpful)* günstig **(to** für) ‹*Wetter, Wind, Umstand*›.

favourably ['feɪvərəblɪ] *adv. (Brit.)* wohlwollend; **be ~ disposed towards sb./sth.** jmdm./einer Sache positiv gegenüberstehen

favourite ['feɪvərɪt] *(Brit.)* **1.** *adj.* Lieblings-. **2.** *n.* **a)** Liebling, *der; (food/country etc.)* Lieblingsessen, *das/-*land, *das usw.;* **this/he is my ~:** das/ihn mag ich am liebsten; **b)** *(Sport)* Favorit, *der/*Favoritin, *die.*

favouritism ['feɪvərɪtɪzm] *n. (Brit.)* Begünstigung, *die; (when selecting sb. for a post etc.)* Günstlingswirtschaft, *die*

fawn [fɔ:n] **1.** *n.* **a)** *(colour)* Rehbraun, *das;* **b)** *(young deer)* [Dam]kitz, *das.* **2.** *adj.* rehfarben

fax [fæks] **1.** *n.* [Tele]fax, *das.* **2.** *v. t.* faxen. **'fax machine** *n.* Faxgerät, *das*

FBI *abbr. (Amer.)* **Federal Bureau of Investigation** FBI, *das*

fear [fɪə(r)] **1.** *n.* Angst, *die* **(of** vor + *Dat.*); *(instance)* Befürchtung, *die;* **~ of death** *or* **dying/heights** Todes-/Höhenangst, *die;* **~ of doing sth.** Angst davor, etw. zu tun; **in ~:** angstvoll; **no ~!** *(coll.)* keine Bange! *(ugs.).* **2.** *v. t.* **a)** **~ sb./sth.** vor jmdn./etw. Angst haben; **~ to do** *or* **doing sth.** Angst haben, etw. zu tun; **b)** *(be worried about)* befürchten; **~ [that ...]** fürchten[, daß ...]. **fearful** ['fɪəfl] *adj.* **a)** *(terrible)* furchtbar; **b)** *(frightened)* ängstlich; **be ~ of sth./sb.** vor etw./jmdm. Angst haben. **'fearless** *adj.,* **'fearlessly** *adv.* furchtlos

feasibility [fi:zɪ'bɪlɪtɪ] *n.* Durchführbarkeit, *die*

feasible ['fi:zɪbl] *adj.* durchführbar

feast [fi:st] **1.** *n.* **a)** *(Relig.)* Fest, *das;* **b)** *(banquet)* Festessen, *das.* **2.** *v. i.* schlemmen; **~ on sth.** sich an etw. *(Dat.)* gütlich tun

feat [fi:t] *n.* Meisterleistung, *die*

feather ['feðə(r)] *n.* Feder, *die.*

'featherweight *n. (Boxing)* Federgewicht, *das*

feature ['fi:tʃə(r)] **1.** *n.* **a)** *usu. in pl. (of face)* Gesichtszug, *der;* **b)** *(characteristic)* [charakteristisches] Merkmal; **be a ~ of sth.** charakteristisch für etw. sein; **c)** *(Journ.)* Feature, *das;* **d)** *(Cinemat.)* **~ [film]** Hauptfilm, *der.* **2.** *v. t.* vorrangig vorstellen; *(in film)* in der Hauptrolle zeigen. **3.** *v. i.* vorkommen; **~ in** *(be important)* eine bedeutende Rolle haben bei

Feb. *abbr.* **February** Febr.

February ['febrʊərɪ] *n.* Februar, *der*

fed [fed] **1.** *see* **feed 1, 2. 2.** *pred. adj. (sl.)* **be/get ~ up with sb./sth.** jmdn./etw. satt haben/kriegen *(ugs.);* **I'm ~ up** Ich hab' die Nase voll *(ugs.)*

federal ['fedərl] *adj.* Bundes-; föderativ ‹*System*›. **federation** [fedə'reɪʃn] *n.* Föderation, *die*

fee [fi:] *n.* Gebühr, *die; (of doctor, lawyer, etc.)* Honorar, *das*

feeble ['fi:bl] *adj.* schwach; wenig überzeugend ‹*Entschuldigung*›; zaghaft ‹*Versuch*›; lahm *(ugs.)* ‹*Witz*›

feed [fi:d] **1.** *v. t.,* **fed** [fed] **a)** füttern; **~ sb./an animal with sth.** jmdm. etw. zu essen/einem Tier [etw.] zu fressen geben; **b)** *(provide food for)* ernähren **(on, with** mit). **2.** *v. i.,* **fed** ‹*Tier:*› fressen **(from** aus); ‹*Person:*› essen **(off** von); **~ on sth.** ‹*Tier:*› etw. fressen. **3.** *n.* **a)** *(for baby)* Mahlzeit, *die;* **b)** *(fodder)* Futter, *das.* **'feedback** *n.* Reaktion, *die*

feel [fi:l] **1.** *v. t.,* **felt** [felt] **a)** *(explore by touch)* befühlen; **b)** *(perceive by touch)* fühlen; *(become aware of)* bemerken; *(have sensation of)* spüren; **c)** *(experience)* empfinden; verspüren ‹*Drang*›; **~ the cold/heat** unter der Kälte/Hitze leiden; **~ [that] ...:** das Gefühl haben, daß ...; *(think)* glauben, daß ... **2.** *v. i.,* **felt** **a)** **~ [about] in sth. [for sth.]** in etw. *(Dat.)* [nach etw.] [herum]suchen; **b)** *(be conscious that one is)* sich ... fühlen; **~ angry/sure/disappointed** böse/sicher/enttäuscht sein; **~ like sth./doing sth.** auf etw. *(Akk.)* Lust haben/ Lust haben, etw. zu tun; **c)** *(be consciously perceived as)* sich ... anfühlen. **'feel for** *v. t.* **~ for sb.** mit jmdm. Mitleid haben

'feeler *n.* Fühler, *der.* **'feeling** *n.* **a)** Gefühl, *das; (sense of touch)* [sense of] **~:** Tastsinn, *der;* **hurt sb.'s ~s** jmdn. verletzen; **b)** *(opinion)* Ansicht, *die*

feet *pl. of* **foot**

feign [feɪn] *v. t.* vortäuschen; ~ **to do sth.** vorgeben, etw. zu tun

¹**fell** *see* **fall** 2

²**fell** [fel] *v. t.* fällen ‹*Baum*›

³**fell** *adj.* **in one ~ swoop** auf einen Schlag

fellow ['feləʊ] **1.** *n.* **a)** *(comrade)* Kamerad, *der;* **b)** *(Brit. Univ.)* Fellow, *der;* **c)** *(of academy or society)* Mitglied, *das;* **d)** *(coll.: man, boy)* Kerl, *der (ugs.).* **2.** *attrib. adj.* Mit-; ~ **man** *or* **human being** Mitmensch, *der*

¹**felt** [felt] *n.* Filz, *der*

²**felt** *see* **feel**

felt[-tipped] '**pen** *n.* Filzstift, *der*

female ['fiːmeɪl] **1.** *adj.* weiblich; Frauen‹stimme, -chor, -verein›. **2.** *n.* Frau, *die; (foetus, child)* Mädchen, *das; (animal)* Weibchen, *das*

feminine ['femɪnɪn] *adj.* weiblich; Frauen‹angelegenheit, -leiden›; *(womanly)* feminin. **feminist** ['femɪnɪst] **1.** *adj.* feministisch; Feministen‹bewegung, -gruppe›. **2.** *n.* Feministin, *die*/Feminist, *der*

fence [fens] **1.** *n.* Zaun, *der.* **2.** *v. i.* *(Sport)* fechten. **3.** *v. t.* ~ |**in**| einzäunen. '**fencer** *n.* Fechter, *der*/Fechterin, *die.* **fencing** ['fensɪŋ] *n.* *(Sport)* Fechten, *das*

fend [fend] *v. i.* ~ **for oneself** für sich selbst sorgen; *(in hostile surroundings)* sich allein durchschlagen. **fend** '**off** *v. t.* abwehren

fender ['fendə(r)] *n.* **a)** *(for fire)* Kaminschutz, *der;* **b)** *(Amer.) (car bumper)* Stoßstange, *die; (car mudguard)* Kotflügel, *der*

ferment [fə'ment] **1.** *v. i.* gären. **2.** *v. t.* zur Gärung bringen. **fermentation** [fɜːmen'teɪʃn] *n.* Gärung, *die*

fern [fɜːn] *n.* Farnkraut, *das*

ferocious [fə'rəʊʃəs] *adj.* wild. **ferocity** [fə'rɒsɪti] *n.* Wildheit, *die*

ferret ['ferɪt] *n.* Frettchen, *das*

ferry ['feri] **1.** *n.* Fähre, *die; (service)* Fährverbindung, *die.* **2.** *v. t. (in boat)* ~ |**across** *or* **over**| übersetzen

fertile ['fɜːtaɪl] *adj. (fruitful)* fruchtbar; *(capable of developing)* befruchtet. **fertility** [fɜː'tɪlɪti] *n.* Fruchtbarkeit, *die.* **fertilize** ['fɜːtɪlaɪz] *v. t.* befruchten. '**fertilizer** *n.* Dünger, *der*

fervent ['fɜːvənt] *adj.* leidenschaftlich; inbrünstig ‹*Gebet, Wunsch, Hoffnung*›. **fervour** *(Brit.; Amer.:* **fervor**) ['fɜːvə(r)] *n.* Leidenschaftlichkeit, *die*

fester ['festə(r)] *v. i.* eitern

festival ['festɪvl] *n.* **a)** *(feast day)* Fest, *das;* **b)** *(of music etc.)* Festival, *das*

festive ['festɪv] *adj.* festlich; fröhlich; **the ~ season** die Weihnachtszeit.

festivity [fe'stɪvɪti] *n.* **a)** *(gaiety)* Feststimmung, *die;* **b)** *(celebration)* Feier, *die;* **festivities** Feierlichkeiten *Pl.*

festoon [fe'stuːn] **1.** *n.* Girlande, *die.* **2.** *v. t.* schmücken **(with mit)**

fetch [fetʃ] *v. t.* **a)** holen; *(collect)* abholen **(from** von); ~ **sth.,** ~ **sth. for sb.** jmdm. etw. holen; **b)** *(be sold for)* erzielen ‹*Preis*›. '**fetching** *adj.* einnehmend

fête [feɪt] *n.* [Wohltätigkeits]basar, *der*

fetish ['fetɪʃ] *n.* Fetisch, *der.* **fetishism** ['fetɪʃɪzm] *n.* Fetischismus, *der.* **fetishist** ['fetɪʃɪst] *n.* Fetischist, *der*/Fetischistin, *die*

fetter ['fetə(r)] *v. t.* fesseln

feud [fjuːd] *n.* Fehde, *die*

feudal ['fjuːdl] *adj.* Feudal-; feudalistisch; ~ **system** Feudalsystem, *das*

fever ['fiːvə(r)] *n.* **a)** *(high temperature)* Fieber, *das;* **have a** |**high**| ~: [hohes] Fieber haben; **b)** *(disease)* Fieberkrankheit, *die.* '**feverish** *adj.* **a)** *(Med.)* fiebrig; **be** ~: Fieber haben; **b)** *(excited)* fiebrig

few [fjuː] **1.** *adj.* **a)** *(not many)* wenige; *abs.* nur wenige; **with very ~ exceptions** mit ganz wenigen Ausnahmen; **his ~ belongings** seine paar Habseligkeiten; **a ~ ...:** wenige ...; **b)** *(some)* wenige; **a ~ ...:** ein paar ...; **a ~ more ...:** noch ein paar ... **2.** *n.* **a)** *(not many)* wenige; **a ~:** wenige; **just a ~ of you/her friends** nur ein paar von euch/ihrer Freunde; **b)** *(some)* **with a ~ of our friends** mit einigen unserer Freunde; **quite a ~:** ziemlich viele

fiancé [fɪ'ɒseɪ] *n.* Verlobte, *der*

fiancée [fɪ'ɒseɪ] *n.* Verlobte, *die*

fiasco [fɪ'æskəʊ] *n., pl.* ~**s** Fiasko, *das*

fib [fɪb] **1.** *n.* Flunkerei, *die (ugs.);* **tell** ~**s** flunkern *(ugs.).* **2.** *v. i.,* **-bb-** flunkern *(ugs.)*

fibre *(Brit.; Amer.:* **fiber**) ['faɪbə(r)] *n.* **a)** Faser, *die;* **b)** *(material)* [Faser]gewebe, *das.* '**fibreglass** *n.* *(plastic)* glasfaserverstärkter Kunststoff

fiche [fiːʃ] *n., pl.* **same** *or* ~**s** Mikrofiche, *das. od.* **der**

fickle ['fɪkl] *adj.* unberechenbar

fiction ['fɪkʃn] *n.* erzählende Literatur; **a** ~/~**s** eine Erfindung. **fictional** ['fɪkʃənl] *adj.* erfunden ‹*Geschichte*›; fiktiv ‹*Figur*›

fictitious [fɪk'tɪʃəs] *adj.* fingiert; falsch ⟨*Name, Identität*⟩

fiddle ['fɪdl] **1.** *n.* **a)** *(Mus.) (coll./ derog.)* Fiedel, *die;* *(violin for traditional music)* Geige, *die;* |as| **fit as a ~**: kerngesund; **b)** *(sl.: swindle)* Gaunerei, *die.* **2.** *v.t. (sl.)* frisieren *(ugs.)* ⟨*Bücher, Rechnungen*⟩. **3.** *v.i.* herumspielen **(with** mit). **fiddler** ['fɪdlə(r)] *n.* Geiger, *der*/Geigerin, *die*

fiddly ['fɪdlɪ] *adj. (coll.)* knifflig

fidelity [fɪ'delɪtɪ] *n.* Treue, *die* **(to** zu)

fidget ['fɪdʒɪt] **1.** *v.i.* ~ |about| herumrutschen. **2.** *n. (person)* Zappelphilipp, *der (ugs.).* **fidgety** *adj.* unruhig; zappelig ⟨*Kind*⟩

field [fi:ld] *n.* **a)** Feld, *das;* **b)** *(for game)* Platz, *der;* [Spiel]feld, *das;* **c)** *(subject area)* [Fach]gebiet, *das;* in the ~ **of medicine** auf dem Gebiet der Medizin; **that is outside my ~**: das fällt nicht in mein Fach

field: **~-day** *n.* **have a ~day** seinen großen Tag haben; ~ **events** *n. pl.* technische Disziplinen; **~-glasses** *n. pl.* Feldstecher, *der;* **F-~** 'Marshal *n. (Brit. Mil.)* Feldmarschall, *der;* ~ **mouse** *n.* Brandmaus, *die*

fiend [fi:nd] *n.* **a)** *(wicked person)* Scheusal, *das;* **b)** *(evil spirit)* böser Geist. **fiendish** *adj.* **a)** teuflisch; **b)** *(very awkward)* höllisch

fierce ['fɪəs] *adj.* wild; erbittert ⟨*Widerstand, Kampf*⟩; scharf ⟨*Kritik*⟩. **fiercely** *adv.* heftig ⟨*angreifen, Widerstand leisten*⟩; wütend ⟨*brüllen*⟩; aufs heftigste ⟨*kritisieren, bekämpfen*⟩

fiery ['faɪərɪ] *adj.* glühend; *(looking like fire)* feurig; *(blazing red)* feuerrot

fifteen [fɪf'ti:n] **1.** *adj.* fünfzehn. **2.** *n.* Fünfzehn, *die. See also* **eight. fifteenth** [fɪf'ti:nθ] **1.** *adj.* fünfzehnt... **2.** *n. (fraction)* Fünfzehntel, *das. See also* **eighth**

fifth [fɪfθ] **1.** *adj.* fünft... **2.** *n. (in sequence)* fünfte, *der/die/das;* *(in rank)* Fünfte, *der/die/das; (fraction)* Fünftel, *das. See also* **eighth**

fiftieth ['fɪftɪθ] *adj.* fünfzigst...

fifty ['fɪftɪ] **1.** *adj.* fünfzig. **2.** *n.* Fünfzig, *die. See also* **eight; eighty 2**

fig [fɪg] *n.* Feige, *die*

fig. *abbr.* **figure** Abb.

fight [faɪt] **1.** *v.i.,* **fought** [fɔ:t] **a)** kämpfen; *(with fists)* sich schlagen; **b)** *(squabble)* [sich] streiten **(about** wegen). **2.** *v.t.,* **fought a)** ~ **sb./sth.** gegen jmdn./etw. kämpfen; *(using fists)* ~ **sb.** sich mit jmdm. schlagen; **b)** *(seek*

to overcome) bekämpfen; *(resist)* ~ **sb./sth.** gegen jmdn./etw. ankämpfen; **c)** ~ **a battle** einen Kampf austragen; **d)** kandidieren bei ⟨*Wahl*⟩. **3.** *n.* Kampf, *der* **(for** um). 'fight **against** *v.t.* kämpfen gegen; ankämpfen gegen ⟨*Wellen, Wind*⟩. **fight 'back 1.** *v.i.* zurückschlagen. **2.** *v.t. (suppress)* zurückhalten. **fight 'off** *v.t.* abwehren. 'fight **with** *v.t.* **a)** kämpfen mit; **b)** *(squabble with)* [sich] streiten mit

'fighter *n.* Kämpfer, *der*/Kämpferin, *die;* *(aircraft)* Kampfflugzeug, *das*

'fighting *n.* Kämpfe

figment ['fɪgmənt] *n.* **a** ~ **of one's or the imagination** pure Einbildung

'fig-tree *n.* Feigenbaum, *der*

figurative ['fɪgərətɪv] *adj.* übertragen

figure ['fɪgə(r)] **1.** *n.* **a)** *(shape)* Form, *die;* **b)** *(carving, sculpture, one's bodily shape)* Figur, *die;* **c)** *(illustration)* Abbildung, *die;* **d)** *(person as seen)* Gestalt, *die; (literary ~)* Figur, *die;* **e)** *(numerical symbol)* Ziffer, *die; (number)* Zahl, *die; (amount of money)* Betrag, *der;* **f)** ~ **of speech** Redewendung, *die.* **2.** *v.i.* **a)** vorkommen; **b) that ~s** *(coll.)* das kann gut sein. **figure 'out** *v.t.* **a)** *(by arithmetic)* ausrechnen; **b)** *(understand)* verstehen

filament ['fɪləmənt] *n.* **a)** Faden, *der;* **b)** *(Electr.)* Glühfaden, *der*

filch ['fɪltʃ] *v.t.* stibitzen *(ugs.)*

¹**file** [faɪl] **1.** *n.* Feile, *die.* **2.** *v.t.* feilen ⟨*Fingernägel*⟩; mit der Feile bearbeiten ⟨*Holz, Eisen*⟩

²**file 1.** *n.* **a)** *(holder)* Ordner, *der; (box)* Kassette, *die;* **b)** *(papers)* Ablage, *die; (cards)* Kartei, *die.* **2.** *v.t.* **a)** [in die Kartei] einordnen/[in die Akten] aufnehmen; **b)** einreichen ⟨*Antrag*⟩

³**file 1.** *n. (Mil. etc.)* Reihe, *die;* |in| **single or Indian ~**: [im] Gänsemarsch. **2.** *v.i.* ~ |in/out| in einer Reihe [hinein-/hinaus]gehen

filigree ['fɪlɪgri:] *n.* Filigran, *das*

'filing-cabinet *n.* Aktenschrank, *der*

'filings ['faɪlɪŋz] *n. pl.* Späne

fill [fɪl] **1.** *v.t.* **a)** füllen; besetzen ⟨*Sitzplätze*⟩; *(fig.)* ausfüllen ⟨*Gedanken, Zeit*⟩; *(pervade)* erfüllen; ~**ed with** voller ⟨*Reue, Bewunderung, Neid usw.*⟩; **b)** *(appoint sb. to)* besetzen ⟨*Posten*⟩. **2.** *v.i.* ~ |**with sth.**| sich |mit etw.| füllen. **3.** *n.* **eat/drink one's ~**: sich satt essen/trinken. **fill 'in 1.** *v.t.* **a)** füllen; zuschütten ⟨*Erdloch*⟩; **b)** *(complete)* ausfüllen; **c)** ~ **sb. in** |**on sth.**| *(coll.)* jmdn. |über etw. *(Akk.)*| ins

Bild setzen. **2.** *v. i.* ~ **in for sb.** für jmdn. einspringen. **fill 'out** *v. t.* ausfüllen. **fill 'up** *v. t.* **a)** füllen **(with** mit); **b)** *(put petrol into)* tanken

fillet ['fɪlɪt] **1.** *n.* Filet, *das.* **2.** *v. t.* entgräten ⟨*Fisch*⟩

'filling *n.* **a)** *(for teeth)* Füllung, *die;* **b)** *(for pancakes etc.)* Füllung, *die; (for sandwiches etc.)* Belag, *der; (for spreading)* Aufstrich, *der.* **'filling station** *n.* Tankstelle, *die*

filly ['fɪlɪ] *n.* junge Stute

film [fɪlm] **1.** *n.* **a)** Film, *der;* **b)** *(thin layer)* Schicht, *die.* **2.** *v. t.* filmen; drehen ⟨*Kinofilm, Szene*⟩. **'film script** *n.* Drehbuch, *das.* **'film star** *n.* Filmstar, *der*

Filofax, (P) ['faɪləʊfæks] *n.* ≈ Terminplaner, *der*

filter ['fɪltə(r)] **1.** *n.* Filter, *der.* **2.** *v. t.* filtern. **filter 'through** *v. t.* durchsickern

'filter-tip *n.* **a)** Filter, *der;* **b)** ~ |**cigarette**| Filterzigarette, *die*

filth [fɪlθ] *n.* Dreck, *der.* **'filthy** *adj.* schmutzig

fin [fɪn] *n.* Flosse, *die*

final ['faɪnl] **1.** *adj.* letzt...; End⟨*spiel, -stadium, -stufe, -ergebnis*⟩; endgültig ⟨*Entscheidung*⟩. **2.** *n.* **a)** *(Sport etc.)* Finale, *das;* **b)** ~s *pl. (university examination)* Examen, *das*

finale [fɪ'nɑːlɪ] *n.* Finale, *das*

finalist ['faɪnəlɪst] *n.* Teilnehmer/Teilnehmerin in der Endausscheidung; *(Sport)* Finalist, *der*/Finalistin, *die*

finalize ['faɪnəlaɪz] *v. t.* [endgültig] beschließen; *(complete)* zum Abschluß bringen

finally ['faɪnəlɪ] *adv.* **a)** *(in the end)* schließlich; **b)** *(expressing impatience etc.)* endlich; **c)** *(in conclusion)* abschließend; **c)** *(conclusively)* entschieden ⟨*sagen*⟩

finance [faɪ'næns, 'faɪnæns] **1.** *n.* **a)** *in pl. (resources)* Finanzen *Pl.;* **b)** *(management of money)* Geldwesen, *das;* **c)** *(support)* Geldmittel *Pl.* **2.** *v. t.* finanzieren. **financial** [faɪ'nænʃl] *adj.* finanziell; Finanz⟨*mittel, -experte, -lage*⟩; ~ **year** Geschäftsjahr, *das.* **financially** *adv.* finanziell. **financier** [faɪ'nænsɪə(r)] *n.* Finanzexperte, *der*/-expertin, *die*

finch [fɪntʃ] *n.* Fink[envogel], *der*

find [faɪnd] **1.** *v. t.,* **found** [faʊnd] finden; *(come across unexpectedly)* entdecken; auftreiben ⟨*Geld, Gegenstand*⟩; aufbringen ⟨*Kraft, Energie*⟩;

want to ~: suchen; ~ **that** ...: herausfinden, daß ...; ~ **sth. necessary** etw. für nötig erachten; ~ **sth./sb. to be** ...: herausfinden, daß etw./jmd. ... ist/ war; **you will** ~ |**that**| ...: Sie werden sehen, daß ... **2.** *n.* Fund, *der.* **find 'out** *v. t.* herausfinden

'finder *n.* Finder, *der*/Finderin, *die*

'findings *n. pl.* Ergebnisse

¹fine [faɪn] **1.** *n.* Geldstrafe, *die.* **2.** *v. t.* mit einer Geldstrafe belegen

²fine *adj.* **a)** hochwertig ⟨*Qualität, Lebensmittel*⟩; fein ⟨*Gewebe, Spitze*⟩; edel ⟨*Holz, Wein*⟩; **b)** *(delicate)* fein; zart ⟨*Porzellan*⟩; *(thin)* hauchdünn; **cut or run it** ~: knapp kalkulieren; **c)** *(in small particles)* [hauch]fein ⟨*Sand, Staub*⟩; ~ **rain** Nieselregen, *der;* **d)** *(sharp)* scharf ⟨*Spitze, Klinge*⟩; spitz ⟨*Nadel, Schreibfeder*⟩; **e)** *(excellent)* ausgezeichnet ⟨*Sänger, Schauspieler*⟩; **f)** *(satisfactory)* schön; **that's** ~ **by** *or* **with me** ja, ist mir recht; **g)** *(in good health or state)* gut; **feel** ~: sich wohl fühlen; **h)** schön ⟨*Wetter*⟩. **fine 'arts** *n. pl.* schöne Künste

finery ['faɪnərɪ] *n.* Pracht, *die; (garments etc.)* Staat, *der*

finger ['fɪŋgə(r)] **1.** *n.* Finger, *der.* **2.** *v. t.* berühren; *(meddle with)* befingern

finger: ~**mark** *n.* Fingerabdruck, *der;* ~**nail** *n.* Fingernagel, *der;* ~**print** *n.* Fingerabdruck, *der;* ~**tip** *n.* Fingerspitze, *die;* **have sth. at one's** ~**tips** *(fig.)* etw. im kleinen Finger haben *(ugs.)*

finish ['fɪnɪʃ] **1.** *v. t.* **a)** beenden ⟨*Unterhaltung*⟩; erledigen ⟨*Arbeit*⟩; abschließen ⟨*Kurs, Ausbildung*⟩; **have** ~**ed sth.** etw. fertig haben; ~ **writing/ reading sth.** etw. zu Ende schreiben/ lesen; **b)** aufessen ⟨*Mahlzeit*⟩; auslesen ⟨*Buch, Zeitung*⟩; austrinken ⟨*Flasche, Glas*⟩. **2.** *v. i.* **a)** aufhören; **have you** ~**ed?** sind Sie fertig?; **when does the concert** ~? wann ist das Konzert aus?; **b)** *(in race)* das Ziel erreichen. **3.** *n.* **a)** Ende, *das;* **b)** *(~ing line)* Ziel, *das.* **finish 'off** *v. t.* abschließen

'finishing post *n.* Zielpfosten, *der*

finite ['faɪnaɪt] *adj.* begrenzt

Finland ['fɪnlənd] *pr. n.* Finnland *(das)*

Finn [fɪn] *n.* Finne, *der*/Finnin, *die*

Finnish ['fɪnɪʃ] **1.** *adj.* finnisch. **2.** *n.* Finnisch, *das; see also* **English 2 a**

fiord [fɪ'ɔːd] *n.* Fjord, *der*

fir [fɜː(r)] *n.* Tanne, *die*

fire ['faɪə(r)] **1.** *n.* **a)** Feuer, *das;* **be on**

~: brennen; **catch** ~: Feuer fangen; ⟨*Wald, Gebäude:*⟩ in Brand geraten; **set** ~ **to sth.** etw. anzünden; **b)** *(in grate)* [offenes] Feuer; *(electric or gas* ~*)* Heizofen, *der;* **light the** ~: den Ofen anstecken; *(in grate)* das [Kamin]feuer anmachen; **c)** *(destructive burning)* Brand, *der;* **d)** *(of guns)* **come/be under** ~: unter Beschuß geraten/beschossen werden. **2.** *v. t.* **a)** abschießen ⟨*Gewehr*⟩; abfeuern ⟨*Kanone*⟩; abgeben ⟨*Schuß*⟩; ~ **one's gun/ pistol/rifle at sb.** auf jmdn. schießen; **two shots were** ~**d** es fielen zwei Schüsse; ~ **questions at sb.** jmdn. mit Fragen bombardieren; **b)** *(coll.: dismiss)* feuern *(ugs.).* **3.** *v. i.* feuern; ~ **at/on** schießen auf (+ *Akk.*); ~! Feuer!

fire: ~**-alarm** *n.* Feuermelder, *der;* ~**arm** *n.* Schußwaffe, *die;* ~ **brigade** *(Brit.),* ~ **department** *(Amer.)* *ns.* Feuerwehr, *die;* ~**-drill** *n.* Probe[feuer]alarm, *der;* ~**-engine** *n.* Löschfahrzeug, *das;* ~**-escape** *n. (staircase)* Feuertreppe, *die;* ~ **extinguisher** *n.* Feuerlöscher, *der;* ~ **hazard** *n.* Brandrisiko, *das;* ~**man** ['faɪəmən] *n., pl.* ~**men** [~mən] Feuerwehrmann, *der;* ~**place** *n.* Kamin, *der;* ~**side** *n.* **at** *or* **by the** ~**side** am Kamin; ~ **station** *n.* Feuerwache, *die;* ~**wood** *n.* Brennholz, *das;* ~**work** *n.* Feuerwerkskörper, *der;* ~**works** *(display)* Feuerwerk, *das*

¹**firm** [fɜ:m] *n.* Firma, *die*

²**firm** *adj.* **a)** fest; stabil ⟨*Konstruktion, Stuhl*⟩; **b)** *(resolute, strict)* bestimmt. **'firmly** *adv.* **a)** fest; **b)** *(resolutely, strictly)* bestimmt

first [fɜ:st] **1.** *adj.* erst...; **he was** ~ **to arrive** er kam als erster an. **2.** *adv.* **a)** *(before anyone else)* zuerst; als erster/erste ⟨*sprechen, ankommen*⟩; *(before anything else)* an erster Stelle ⟨*stehen, kommen*⟩; ~ **come** ~ **served** wer zuerst kommt, mahlt zuerst *(Spr.);* **b)** *(beforehand)* vorher; **c)** *(for the* ~ *time)* zum ersten Mal; **d)** ~ **of all** zuerst; *(in importance)* vor allem. **3.** *n.* **a)** **the** ~ *(in sequence)* der/die/das erste; *(in rank)* der/die/das Erste; **b)** **at** ~: zuerst; **from the** ~: von Anfang an

first: ~ **'aid** *n.* erste Hilfe; ~**-aid box/ kit** Verbandkasten, *der/*Erste-Hilfe-Ausrüstung, *die;* ~**-class 1.** ['--] *adj.* **a)** erster Klasse *nachgestellt;* Erste[r]-Klasse-⟨*Fahrkarte, Abteil, Post, Brief usw.*⟩; **b)** *(excellent)* erstklassig. **2.** [-'-] *adv.* erster Klasse ⟨*reisen*⟩

'firstly *adv.* zunächst [einmal]; *(followed by 'secondly')* erstens

first: ~ **name** *n.* Vorname, *der;* ~**-rate** *adj.* erstklassig; ~ **school** *n.* *(Brit.)* ≈ Grundschule, *die*

'fir tree *n.* Tanne, *die*

fish [fɪʃ] **1.** *n.* Fisch, *der.* **2.** *v. i.* fischen; *(with rod)* angeln; **go** ~**ing** fischen/angeln gehen. **fish 'out** *v. t. (coll.)* herausfischen *(ugs.)*

fisherman ['fɪʃəmən] *n., pl.* **fishermen** ['fɪʃəmən] Fischer, *der; (angler)* Angler, *der*

fish: ~ **'finger** *n.* Fischstäbchen, *das;* ~**-hook** *n.* Angelhaken, *der*

'fishing *n.* Fischen, *das; (with rod)* Angeln, *das*

fishing: ~ **boat** *n.* Fischerboot, *das;* ~**-net** *n.* Fischernetz, *das;* ~**-rod** *n.* Angelrute, *die*

fish: ~**monger** ['fɪʃmʌŋgə(r)] *n. (Brit.)* Fischhändler, *der/*-händlerin, *die;* ~ **shop** *n.* Fischgeschäft, *das*

'fishy *adj.* **a)** fischartig; Fisch⟨*geschmack, -geruch*⟩; **b)** *(coll.: suspicious)* verdächtig

fist [fɪst] *n.* Faust, *die*

¹**fit** [fɪt] *n.* Anfall, *der; (fig.)* [plötzliche] Anwandlung; **be in** ~**s of laughter** sich vor Lachen biegen; **in a** ~ **of ...**: in einem Anfall von ...

²**fit 1.** *adj.* **a)** *(suitable)* geeignet; ~ **to eat** eßbar; **b)** *(worthy)* würdig; wert; **c)** *(proper)* richtig; **see** *or* **think** ~ |**to do sth.|** es für richtig halten[, etw. zu tun]; **d)** *(healthy)* fit *(ugs.);* **keep** ~: sich fit halten. **2.** *n.* Paßform, *die;* **it is a good/ bad** ~: es sitzt *od.* paßt gut/nicht gut. **3.** *v. t.,* **-tt-: a)** ⟨*Kleider:*⟩ passen (+ *Dat.*); ⟨*Deckel, Bezug:*⟩ passen auf (+ *Akk.*); **b)** *(put into place)* anbringen (**to** an + *Dat. od. Akk.*); einbauen ⟨*Motor, Ersatzteil*⟩. **4.** *v. i.,* **-tt-** passen.

fit 'in 1. *v. t.* unterbringen. **2.** *v. i.* **a)** ⟨*Person:*⟩ sich anpassen (**with** an + *Akk.*); **b)** *(be in accordance with)* ~ **in with sth.** mit etw. übereinstimmen

fitful ['fɪtfl] *adj.* unbeständig; unruhig ⟨*Schlaf*⟩; launisch ⟨*Brise*⟩

'fitment *n.* Einrichtung, *die*

'fitness *n.* **a)** *(physical)* Fitneß, *die;* **b)** *(suitability)* Eignung, *die*

'fitted *adj.* **a)** *(suited)* geeignet (**for** für, zu); **b)** *(shaped)* tailliert ⟨*Kleider*⟩; Einbau⟨*küche, schrank*⟩; ~ **carpet** Teppichboden, *der*

'fitter *n.* Monteur, *der; (of pipes)* Installateur, *der; (of machines)* Maschinenschlosser, *der*

fitting 1. *adj.* *(appropriate)* passend; *(becoming)* schicklich *(geh.)* 〈*Benehmen*〉. 2. *n.* **a)** *usu. in pl. (fixture)* Anschluß, *der;* **~s** *(furniture)* Ausstattung, *die;* **b)** *(Brit.: size)* Größe, *die*

five [faɪv] 1. *adj.* fünf. 2. *n.* Fünf, *die. See also* **eight. fiver** ['faɪvə(r)] *n. (Brit. coll.)* Fünfpfundschein, *der*

fix [fɪks] 1. *v. t.* **a)** befestigen; **b)** festsetzen 〈*Termin, Preis, Grenze*〉; *(agree on)* ausmachen; **c)** *(repair)* reparieren; **d)** *(arrange)* arrangieren. 2. *n. (coll.: predicament)* Klemme, *die (ugs.);* **be in a ~:** in der Klemme sitzen. **fix 'up** *v. t.* **a)** *(arrange)* arrangieren; festsetzen 〈*Termin, Treffpunkt*〉; **b)** *(provide)* versorgen; **~ sb. up with sth.** jmdm. etw. verschaffen. **fixture** ['fɪkstʃə(r)] *n.* **a)** *(furnishing)* eingebautes Teil; **b)** *(Sport)* Veranstaltung, *die*

fizz [fɪz] *v. i.* [zischend] sprudeln

fizzle ['fɪzl] *v. i.* zischen. **fizzle 'out** *v. i.* 〈*Kampagne:*〉 im Sande verlaufen

fizzy ['fɪzɪ] *adj.* sprudelnd; **~ lemonade** Brause[limonade], *die*

flabbergast ['flæbəgɑːst] *v. t.* umhauen *(ugs.)*

flabby ['flæbɪ] *adj.* schlaff

¹flag [flæg] *n.* Fahne, *die; (national ~, on ship)* Flagge, *die*

²flag *v. i.,* **-gg-** 〈*Person:*〉 abbauen; 〈*Kraft, Begeisterung usw.:*〉 nachlassen

flagon ['flægn] *n.* Kanne, *die*

flag-pole *n.* Flaggenmast, *der*

flagrant ['fleɪgrənt] *adj.* eklatant; flagrant 〈*Verstoß*〉

flagstone *n.* Steinplatte, *die*

flair [fleə(r)] *n.* Gespür, *das; (special ability)* Talent, *das*

flake [fleɪk] 1. *n.* Flocke, *die; (of dry skin)* Schuppe, *die.* 2. *v. i.* abblättern. **flaky** ['fleɪkɪ] *adj.* blättrig 〈*Kruste*〉; **~ pastry** Blätterteig, *der*

flamboyant [flæm'bɔɪənt] *adj.* extravagant

flame [fleɪm] *n.* Flamme, *die;* **be in ~s** in Flammen stehen

flan [flæn] *n.* [fruit] **~:** [Obst]torte, *die*

flank [flæŋk] *n.* Seite, *die; (of animal; also Mil.)* Flanke, *die*

flannel ['flænl] *n.* **a)** *(fabric)* Flanell, *der;* **b)** *(Brit.: for washing)* Waschlappen, *der*

flap [flæp] 1. *v. t.,* **-pp-:** **~ its wings** mit den Flügeln schlagen. 2. *v. i.,* **-pp-** 〈*Flügel:*〉 schlagen; 〈*Segel, Fahne, Vorhang:*〉 flattern. 3. *n.* **a)** Klappe, *die; (envelope-seal, of shoe)* Lasche, *die;* **b)** *(fig. coll.)* **in a ~:** furchtbar aufgeregt

flare [fleə(r)] 1. *v. i.* flackern; *(fig.)* ausbrechen; **tempers ~d** die Gemüter erhitzten sich. 2. *n.* Leuchtsignal, *das.* **flare 'up** *v. i.* **a)** aufflackern; **b)** *(break out)* [wieder] ausbrechen

flash [flæʃ] 1. *n.* Aufleuchten, *das; (as signal)* Lichtsignal, *das;* **~ of lightning** Blitz, *der;* **in a ~:** *(quickly)* im Nu. 2. *v. t.* **a)** aufleuchten lassen; **~ one's headlights** die Lichthupe betätigen; **~ sb. a smile/glance** jmdm. ein Lächeln/einen Blick zuwerfen; **b)** *(display briefly)* kurz zeigen. 3. *v. i.* aufleuchten; **~ by** *or* **past** 〈*Zeit, Ferien:*〉 wie im Fluge vergehen

flash: **~-back** *n.* Rückblende, *die* **(to** auf + *Akk.*); **~ bulb** *n.* Blitzbirnchen, *das;* **~-cube** *n.* Blitzwürfel, *der;* **~-gun** *n.* Blitzgerät, *das;* **~light** *n.* **a)** *(for signals)* Blinklicht, *das;* **b)** *(Amer.: torch)* Taschenlampe, *die*

flashy *adj.* auffällig

flask [flɑːsk] *n.* **a)** *see* **Thermos; b)** *(for wine, oil)* [bauchige] Flasche; **c)** *(Chem.)* Kolben, *der*

¹flat [flæt] *n. (Brit.)* Wohnung, *die*

²flat 1. *adj.* **a)** flach; eben 〈*Fläche*〉; platt 〈*Nase, Reifen*〉; **b)** *(downright)* glatt *(ugs.)* 〈*Absage, Weigerung, Widerspruch*〉; **c)** -*(Mus.)* [um einen Halbton] erniedrigt 〈*Note*〉; **d)** schal, abgestanden 〈*Bier, Sekt*〉; **e)** leer 〈*Batterie*〉. 2. *adv. (Mus.)* zu tief

flat: **~ 'feet** *n. pl.* Plattfüße; **~-'fish** *n.* Plattfisch, *der;* **~-'footed** *adj.* plattfüßig

flatly *adv.* rundweg

flatten ['flætn] 1. *v. t.* flach drücken 〈*Schachtel*〉; dem Erdboden gleichmachen 〈*Stadt, Gebäude*〉. 2. *v. refl.* **~ oneself against sth.** sich flach gegen etw. drücken

flatter ['flætə(r)] *v. t.* schmeicheln (+ *Dat.*). **flattering** *adj.* schmeichelhaft. **flattery** *n.* Schmeichelei, *die*

flat 'tyre *n.* Reifenpanne, *die*

flaunt [flɔːnt] *v. t.* zur Schau stellen

flavor *etc. (Amer.) see* **flavour** *etc.*

flavour ['fleɪvə(r)] *(Brit.)* 1. *n.* **a)** Geschmack, *der;* **b)** *(fig.)* Anflug, *der.* 2. *v. t.* abschmecken. **flavouring** *n. (Brit.)* Aroma, *das*

flaw [flɔː] *n.* Fehler, *der; (imperfection)* Makel, *der; (in workmanship or goods)* Mangel, *der*

flax [flæks] *n.* Flachs, *der*

flea [fliː] *n.* Floh, *der.* **flea market** *n. (coll.)* Flohmarkt, *der*

fled *see* **flee**

flee [fli:] 1. *v. i.*, **fled** [fled] fliehen; ~ **from sth./sb.** aus etw./vor jmdm. flüchten. 2. *v. t.*, **fled** fliehen aus

fleece [fli:s] 1. *n.* [Schaf]fell, *das.* 2. *v. t. (fig.)* ausplündern. **fleecy** ['fli:sɪ] *adj.* flauschig

fleet [fli:t] *n.* Flotte, *die*

fleeting ['fli:tɪŋ] *adj.* flüchtig

flesh [fleʃ] *n.* Fleisch, *das; (of fruit, plant)* [Frucht]fleisch, *das.* **'fleshy** *adj.* fett; fleischig ⟨*Hände*⟩

flew *see* ²**fly** 1, 2

¹**flex** [fleks] *n. (Brit. Electr.)* Kabel, *das*

²**flex** *v. t.* beugen ⟨*Arm, Knie*⟩; ~ **one's muscles** seine Muskeln spielen lassen

flexible ['fleksɪbl] *adj.* **a)** biegsam; elastisch; **b)** *(fig.)* flexibel

flick [flɪk] *v. t.* schnippen; anknipsen ⟨*Schalter*⟩; verspritzen ⟨*Tinte*⟩. **'flick through** *v. t.* durchblättern

flicker ['flɪkə(r)] 1. *v. i.* flackern; ⟨*Fernsehapparat:*⟩ flimmern. 2. *n.* Flackern, *das; (of TV)* Flimmern, *das*

¹**flight** [flaɪt] *n.* **a)** Flug, *der;* **b)** ~ ⎮of stairs *or* steps⎮ Treppe, *die*

²**flight** *n. (fleeing)* Flucht, *die;* **take ~:** die Flucht ergreifen; **put to ~:** in die Flucht schlagen

'flight attendant *n.* Flugbegleiter, *der/*-begleiterin, *die*

flimsy ['flɪmzɪ] *adj.* **a)** dünn; nicht sehr haltbar ⟨*Verpackung*⟩; **b)** *(fig.)* fadenscheinig ⟨*Entschuldigung, Argument*⟩

flinch [flɪntʃ] *v. i.* zurückschrecken **(from** vor + *Dat.*); *(wince)* zusammenzucken

fling [flɪŋ] 1. *n.* **have a** *or* **one's ~:** sich ausleben. 2. *v. t.*, **flung** [flʌŋ] werfen; ~ **oneself into sth.** *(fig.)* sich in etw. *(Akk.)* stürzen

flint [flɪnt] *n.* Feuerstein, *der*

flip [flɪp] *v. t.*, **-pp-** schnipsen; ~ ⎮over⎮ *(turn over)* umdrehen. **'flip through** *v. t.* durchblättern

flippant ['flɪpənt] *adj.* leichtfertig

flipper ['flɪpə(r)] *n.* Flosse, *die*

flirt [flɜ:t] *v. i.* flirten. **flirtation** [flɜ:'teɪʃn] *n.* Flirt, *der*

flit [flɪt] *v. i.* huschen

float [fləʊt] 1. *v. i.* treiben; *(in air)* schweben. 2. *n. (for carnival)* Festwagen, *der.* 3. *v. t. (set afloat)* flott machen; *(fig.)* lancieren ⟨*Plan, Idee*⟩. **floating 'voter** *n.* Wechselwähler, *der/*-wählerin, *die*

flock [flɒk] 1. *n.* **a)** Herde, *die; (of birds)* Schwarm, *der;* **b)** *(of people)* Schar, *die.* 2. *v. i.* strömen; ~ **round sb.** sich um jmdn. scharen

flog [flɒg] *v. t.*, **-gg-: a)** auspeitschen; **b)** *(Brit. sl.: sell)* verscheuern *(salopp)*

flood [flʌd] 1. *n.* Überschwemmung, *die;* **the F~** *(Bibl.)* die Sintflut. 2. *v. i.* ⟨*Fluß:*⟩ über die Ufer treten; *(fig.)* strömen. 3. *v. t.* überschwemmen. **'floodlight** 1. *n.* Scheinwerfer, *der.* 2. *v. t.*, **floodlit** ['flʌdlɪt] anstrahlen

floor [flɔ:(r)] 1. *n.* **a)** Boden, *der;* **b)** *(storey)* Stockwerk, *das;* **first ~** *(Amer.)* Erdgeschoß, *das;* **first ~** *(Brit.),* **second ~** *(Amer.)* erster Stock; **ground ~:** Erdgeschoß, *das;* Parterre, *das.* 2. *v. t. (confound)* überfordern; **b)** *(knock down)* zu Boden schlagen

floor: ~**board** *n.* Dielenbrett, *das;* ~**-cloth** *n. (Brit.)* Scheuertuch, *das;* ~**-polish** *n.* Bohnerwachs, *das;* ~ **show** *n.* ≈ Unterhaltungsprogramm, *das*

flop [flɒp] 1. *v. i.*, **-pp-: a)** plumpsen; **b)** *(coll.: fail)* fehlschlagen; ⟨*Theaterstück, Show:*⟩ durchfallen. 2. *n. (coll.: failure)* Reinfall, *der (ugs.)*

floppy ['flɒpɪ] *adj.* weich und biegsam

flora ['flɔ:rə] *n.* Flora, *die*

floral ['flɔ:rl, 'flɒrl] *adj.* geblümt ⟨*Kleid, Stoff*⟩; Blumen⟨*muster*⟩

Florence ['flɒrəns] *pr. n.* Florenz *(das)*

florid ['flɒrɪd] *adj.* blumig ⟨*Stil, Redeweise*⟩; gerötet ⟨*Teint*⟩

florist ['flɒrɪst] *n.* Florist, *der/*Floristin, *die*

flotsam ['flɒtsəm] *n.* ~ ⎮and jetsam⎮ Treibgut, *das*

flounder ['flaʊndə(r)] *v. i.* taumeln

flour ['flaʊə(r)] *n.* Mehl, *das*

flourish ['flʌrɪʃ] 1. *v. i.* gedeihen; ⟨*Geschäft:*⟩ florieren, gutgehen. 2. *v. t.* schwingen. 3. *n.* **do sth. with a ~:** etw. schwungvoll tun

flout [flaʊt] *v. t.* mißachten

flow [fləʊ] 1. *v. i.* fließen; ⟨*Körner, Sand:*⟩ rinnen, rieseln; ⟨*Gas:*⟩ strömen. 2. *n.* **a)** Fließen, *das;* ~ **of water/ people** Wasser-/Menschenstrom, *der;* ~ **of information** Informationsfluß, *der;* **b)** *(of tide, river)* Flut, *die*

flower ['flaʊə(r)] 1. *n. (blossom)* Blüte, *die; (plant)* Blume, *die;* **come into ~:** zu blühen beginnen. 2. *v. i.* blühen. **'flower-bed** *n.* Blumenbeet, *das.* **'flower-pot** *n.* Blumentopf, *der*

'flowery *adj.* geblümt ⟨*Stoff, Muster*⟩; *(fig.)* blumig ⟨*Sprache*⟩

'flowing *adj.* fließend; wallend ⟨*Haar*⟩

flown *see* ²**fly** 1, 2

flu [flu:] *n. (coll.)* Grippe, *die*

fluctuate ['flʌktjʊeɪt] *v. i.* schwanken. **fluctuation** [flʌktjʊ'eɪʃn] *n.* Schwankung, *die*

fluency ['fluːənsɪ] *n.* Gewandtheit, *die; (spoken)* Redegewandtheit, *die*

fluent ['fluːənt] *adj.* gewandt ⟨*Stil, Redeweise, Redner, Schreiber*⟩; **be ~ in Russian, speak ~ Russian** fließend Russisch sprechen

fluff [flʌf] *n.* Flusen; Fusseln

fluffy ['flʌfɪ] *adj.* [flaum]weich ⟨*Kissen, Küken*⟩; flauschig ⟨*Spielzeug, Decke*⟩

fluid ['fluːɪd] **1.** *n.* Flüssigkeit, *die.* **2.** *adj.* flüssig

fluke [fluːk] *n. (piece of luck)* Glücksfall, *der*

flung *see* **fling** 2

fluorescent [fluːə'resnt] *adj.* fluoreszierend. **fluorescent 'light** *n.* Leuchtstofflampe, *die*

fluoride ['fluːəraɪd] *n.* Fluorid, *das;* **fluoride toothpaste** fluorhaltige Zahnpasta

flurry ['flʌrɪ] *n.* **a)** Aufregung, *die;* **b)** *(of rain/snow)* [Regen-/Schnee]schauer, *der*

¹flush [flʌʃ] **1.** *v. i.* rot werden. **2.** *v. t.* ausspülen ⟨*Becken*⟩; **~ the toilet** *or* **lavatory** spülen. **3.** *n.* Rotwerden, *das*

²flush *adj. (level)* bündig; **be ~ with sth.** mit etw. bündig abschließen

fluster ['flʌstə(r)] *v. t.* aus der Fassung bringen. **flustered** ['flʌstəd] *adj.* nervös

flute [fluːt] *n.* Flöte, *die*

flutter ['flʌtə(r)] **1.** *v. i.* flattern. **2.** *v. t.* flattern mit ⟨*Flügel*⟩

flux [flʌks] *n.* **in a state of ~:** im Fluß

¹fly [flaɪ] *n.* Fliege, *die*

²fly 1. *v. i.,* **flew** [fluː], **flown** [fləʊn] **a)** fliegen; **~ away** *or* **off** wegfliegen; **b)** *(fig.)* **~ [by** *or* **past]** wie im Fluge vergehen; **c)** ⟨*Fahne:*⟩ gehißt sein. **2.** *v. t.,* **flew, flown** fliegen ⟨*Flugzeug, Fracht, Einsatz usw.*⟩; fliegen über (+ *Akk.*) ⟨*Strecke*⟩. **3.** *n.* in *sing.* or *pl. (on trousers)* Hosenschlitz, *der.* **fly 'in** *v. i.* [mit dem Flugzeug] eintreffen (**from** aus). **fly 'out** *v. i.* abfliegen (**of** von)

flying ['flaɪɪŋ]: **~ 'saucer** *n.* fliegende Untertasse; **~ 'start** *n. (Sport)* fliegender Start; **~ 'visit** *n.* Stippvisite, *die (ugs.)*

fly: ~leaf *n.* Vorsatzblatt, *das;* **~over** *n. (Brit.)* [Straßen]überführung, *die*

foal [fəʊl] *n.* Fohlen, *das*

foam [fəʊm] **1.** *n.* Schaum, *der.* **2.** *v. i.* schäumen. **foam 'rubber** *n.* Schaumgummi, *der*

fob [fɒb] *v. t.,* **-bb-:** **~ sb. off with sth.** jmdn. mit etw. abspeisen *(ugs.)*

focus ['fəʊkəs] **1.** *n., pl.* **~es** *or* **foci** ['fəʊsaɪ] Brennpunkt, *der;* **out of/in ~:** unscharf/scharf eingestellt; unscharf/scharf ⟨*Foto, Film usw.*⟩; *(fig.)* **be the ~ of attention** im Brennpunkt des Interesses stehen. **2.** *v. t.,* **-s-** *or* **-ss-** einstellen (**on** auf + *Akk.*); bündeln ⟨*Licht, Strahlen*⟩. **3.** *v. i.,* **-s-** *or* **-ss-** *(fig.)* sich konzentrieren (**on** auf + *Akk.*)

fodder ['fɒdə(r)] *n.* [Vieh]futter, *das*

foe [fəʊ] *n. (poet./rhet.)* Feind, *der*

foetus ['fiːtəs] *n.* Fötus, *der*

fog [fɒg] *n.* Nebel, *der.* **'fog-light** *n. (Motor Veh.)* Nebelscheinwerfer, *der*

foggy ['fɒgɪ] *adj.* neblig

fogy ['fəʊgɪ] *n.* **|old| ~:** [alter] Opa *(salopp)*/[alte] Oma *(salopp)*

foible ['fɔɪbl] *n.* Eigenheit, *die*

¹foil [fɔɪl] *n.* Folie, *die*

²foil *v. t.* vereiteln

foist [fɔɪst] *v. t.* **~ |off| on to sb.** jmdm. andrehen *(ugs.);* auf jmdn. abwälzen ⟨*Probleme, Verantwortung*⟩

fold [fəʊld] **1.** *v. t.* [zusammen]falten; **~ one's arms** die Arme verschränken. **2.** *v. i.* **a)** *(become ~ed)* sich zusammenfalten; **b)** *(be able to be ~ed)* sich falten lassen. **3.** *n.* Falte, *die; (line made by ~ing)* Kniff, *der.* **fold 'up** *v. t.* zusammenfalten ⟨*Laken*⟩; zusammenklappen ⟨*Stuhl*⟩

'folder *n.* Mappe, *die*

foliage ['fəʊlɪɪdʒ] *n.* Blätter *Pl.; (of tree also)* Laub, *das*

folk [fəʊk] *n.* **a)** Volk, *das;* **b)** *in pl.* **~|s|** *(people)* Leute *Pl.*

folk: ~-dance *n.* Volkstanz, *der;* **~lore** [~lɔː(r)] *n.* Folklore, *die;* **~-music** *n.* Volksmusik, *die;* **~-song** Volkslied, *das; (modern)* Folksong, *der*

follow ['fɒləʊ] **1.** *v. t.* **a)** folgen (+ *Dat.*); **b)** entlanggehen/-fahren ⟨*Straße usw.*⟩; **c)** *(come after)* folgen auf (+ *Akk.*); **d)** *(result from)* die Folge sein von; **e)** *(treat or take as guide)* sich orientieren an (+ *Dat.*); **f)** folgen (+ *Dat.*) ⟨*Prinzip, Instinkt, Trend*⟩; verfolgen ⟨*Politik*⟩; befolgen ⟨*Regel, Vorschrift, Rat, Warnung*⟩; sich halten an (+ *Akk.*) ⟨*Konventionen, Diät*⟩; **g)** *(grasp meaning of)* folgen (+ *Dat.*); **do you ~ me?** verstehst du, was ich meine? **2.** *v. i.* **a)** *(go, come)* **~ after sb./sth.** jmdm./einer Sache folgen; **b)** *(come next in order or time)* folgen; **as ~s** wie folgt; **c)** **~ from sth.**

(result) die Folge von etw. sein; *(be deducible)* aus etw. folgen. **follow** 'on *v. i. (continue)* ~ on from sth. die Fortsetzung von etw. sein. **follow** 'up *v. t.* a) ausbauen ⟨*Erfolg, Sieg*⟩; b) nachgehen (+ *Dat.*) ⟨*Hinweis*⟩
'**follower** *n.* Anhänger, *der*/Anhängerin, *die*
'**following 1.** *adj.* folgend; **the** ~: folgendes. **2.** *prep.* nach. **3.** *n.* Anhängerschaft, *die*
folly ['fɒlɪ] *n.* Torheit, *die (geh.)*
fond [fɒnd] *adj.* liebevoll; lieb ⟨*Erinnerung*⟩; **be** ~ **of sb.** jmdn. mögen; **be** ~ **of doing sth.** etw. gern tun
fondle ['fɒndl] *v. t.* streicheln
'**fondness** *n.* Liebe, *die;* ~ **for sth.** Vorliebe für etw.
font [fɒnt] *n.* Taufstein, *der*
food [fu:d] *n.* a) Nahrung, *die; (for animals)* Futter, *das;* b) *(as commodity)* Lebensmittel *Pl.;* c) *(in solid form)* Essen, *das;* d) *(particular kind)* Nahrungsmittel, *das;* Kost, *die.* '**food poisoning** *n.* Lebensmittelvergiftung, *die.* '**food processor** *n.* Küchenmaschine, *die*
fool [fu:l] **1.** *n.* Dummkopf, *der (ugs.).* **2.** *v. t.* ~ **sb. into doing sth.** jmdn. [durch Tricks] dazu bringen, etw. zu tun. **fool a**'**bout, fool a**'**round** *v. i.* Unsinn machen
foolhardy ['fu:lhɑ:dɪ] *adj.* tollkühn
'**foolish** *adj.* töricht; verrückt *(ugs.)* ⟨*Idee, Vorschlag*⟩
'**foolproof** *adj. (infallible)* absolut sicher
foot [fʊt] **1.** *n., pl.* **feet** [fi:t] a) Fuß, *der;* **on** ~: zu Fuß; **put one's** ~ **in it** *(fig. coll.)* ins Fettnäpfchen treten *(ugs.);* b) *(far end)* unteres Ende; *(of bed)* Fußende, *das;* c) *(measure)* Fuß, *der (30,48 cm).* **2.** *v. t.* ~ **the bill** die Rechnung bezahlen
football ['fʊtbɔ:l] *n. (game, ball)* Fußball, *der.* '**football boot** *n.* Fußballschuh, *der.* '**footballer** *n.* Fußballspieler, *der*/-spielerin, *die.* '**football pools** *n. pl.* **the** ~: das Fußballtoto
foot: ~**brake** *n.* Fußbremse, *die;* ~**bridge** *n.* Fußgängerbrücke, *die;* ~**hold** *n.* Halt, *der*
'**footing** *n.* a) *(fig.)* **be on an equal** ~ |**with sb.**| [jmdm.] gleichgestellt sein; b) *(foothold)* Halt, *der*
foot: ~**note** *n.* Fußnote, *die;* ~**path** *n.* Fußweg, *der;* ~**print** *n.* Fußabdruck, *der;* ~**step** *n.* Schritt, *der;* follow in sb.'s ~steps *(fig.)* in jmds. Fuß-

stapfen *(Akk.)* treten; ~**wear** *n.* Schuhe *Pl.*
for [fə(r), *stressed* fɔ:(r)] **1.** *prep.* a) für; **what is it** ~? wofür ist das?; **reason** ~ **living** Grund zu leben; **a request** ~ **help** eine Bitte um Hilfe; **study** ~ **a university degree** auf einen Hochschulabschluß hin studieren; **take sb.** ~ **a walk** mit jmdm. einen Spaziergang machen; **be** '~ **doing sth.** *(in favour)* dafür sein, etw. zu tun; **cheque/bill** ~ **£5** Scheck/Rechnung über 5 Pfund; b) *(on account of, as penalty of)* wegen; **were it not** ~ **you/ your help** ohne dich/deine Hilfe; ~ **fear of** aus Angst vor (+ *Dat.*); c) *(in spite of)* ~ **all** ...: trotz ...; ~ **all that,** ...: trotzdem ...; d) ~ **all I know/care** ...: möglicherweise/ was mich betrifft, ...; ~ **one thing,** ...: zunächst einmal ...; e) *(during)* **stay** ~ **a week** eine Woche bleiben; **we've/we haven't been here** ~ **three years** wir sind seit drei Jahren hier/nicht mehr hier gewesen; f) **walk** ~ **20 miles** 20 Meilen gehen. **2.** *conj.* denn
forage ['fɒrɪdʒ] **1.** *n.* Futter, *das.* **2.** *v. i.* ~ **for sth.** auf der Suche nach etw. sein
forbad, forbade *see* **forbid**
forbid [fə'bɪd] *v. t.,* -dd-, **forbad** [fə'bæd] *or* **forbade** [fə'bæd, fə'beɪd], **forbidden** [fə'bɪdn] ~ **sb. to do sth.** jmdm. verbieten, etw. zu tun; ~ |**sb.**| **sth.** [jmdm.] etw. verbieten; **it is** ~**den** |**to do sth.**| es ist verboten[, etw. zu tun].
forbidden *see* **forbid.** for'**bidding** *adj.* furchteinflößend
force [fɔ:s] **1.** *n.* a) *(strength, power)* Stärke, *die; (of explosion, storm)* Wucht, *die; (Phys.; physical strength)* Kraft, *die;* **in** ~: mit einem großen Aufgebot; b) *(validity)* Kraft, *die;* **in** ~: in Kraft; **come into** ~ ⟨*Gesetz usw.*:⟩ in Kraft treten; c) *(violence)* Gewalt, *die;* **by** ~: gewaltsam; d) *(group) (of workers)* Kolonne, *die;* Trupp, *der; (of police)* Einheit, *die; (Mil.)* Armee, *die;* **the** ~**s** die Armee; **be in the** ~**s** beim Militär sein. **2.** *v. t.* a) zwingen; ~ **sth.** |**up**|**on sb.** jmdm. etw. aufzwingen; b) ~ |**open**| aufbrechen; ~ **one's way in** sich *(Dat.)* mit Gewalt Zutritt verschaffen. **forced** [fɔ:st] *adj.* a) *(contrived, unnatural)* gezwungen; b) *(compelled by force)* erzwungen; Zwangs⟨*arbeit*⟩. **forced** '**landing** *n.* Notlandung, *die.* '**force-feed** *v. t.* zwangsernähren. **forceful** ['fɔ:sfl] *adj.* stark ⟨*Persönlichkeit, Charakter*⟩;

energisch ⟨*Person, Art*⟩; eindrucksvoll ⟨*Sprache*⟩

forceps ['fɔ:seps] *n., pl. same* [pair of] ~: Zange, *die*

forcible ['fɔ:sɪbl] *adj.*, **forcibly** ['fɔ:sɪblɪ] *adv.* gewaltsam

ford [fɔ:d] 1. *n.* Furt, *die*. 2. *v. t.* durchqueren; *(wade through)* durchwaten

fore [fɔ:(r)] 1. *adj., esp. in comb.* vorder...; Vorder⟨*teil, -front usw.*⟩. 2. *n.* to the ~: im Vordergrund

'forearm *n.* Unterarm, *der*

foreboding [fɔ:'bəʊdɪŋ] *n.* Vorahnung, *die*

'forecast 1. *v. t.*, **forecast** *or* **forecasted** vorhersagen. 2. *n.* Voraussage, *die*

'forecourt *n.* Vorhof, *der*

'forefather *n., usu. in pl.* Vorfahr, *der*

'forefinger *n.* Zeigefinger, *der*

'forefront *n.* [be] in the ~ of in vorderster Linie (+ *Gen.*) [stehen]

'foregone *adj.* be a ~ conclusion von vornherein feststehen; *(be certain)* so gut wie sicher sein

'foreground *n.* Vordergrund, *der*

forehead ['fɒrɪd, 'fɔ:hed] *n.* Stirn, *die*

foreign ['fɒrɪn] *adj.* a) *(from abroad)* ausländisch; Fremd⟨*kapital, -sprache*⟩; he is ~: er ist Ausländer; b) *(abroad)* fremd; ~ country Ausland, *das*; Außen⟨*politik, -handel*⟩; c) *(from outside)* fremd; ~ body *or* substance Fremdkörper, *der*. **'foreigner** *n.* Ausländer, *der*/Ausländerin, *die*

foreign: ~ ex'change *n.* Devisen *Pl.*; F~ Office *n. (Brit. Hist./coll.)* Außenministerium, *das*; F~ 'Secretary *n. (Brit.)* Außenminister, *der*

foreman ['fɔ:mən] *n., pl.* **foremen** ['fɔ:mən] Vorarbeiter, *der*

foremost ['fɔ:məʊst, 'fɔ:məst] 1. *adj.* a) vorderst...; b) *(fig.)* führend. 2. *adv.* first and ~: zunächst einmal

'forename *n.* Vorname, *der*

'forerunner *n.* Vorläufer, *der*/Vorläuferin, *die*

foresaw *see* foresee

foresee [fɔ:'si:] *v. t.*, *forms as* see voraussehen. **foreseeable** [fɔ:'si:əbl] *adj.* vorhersehbar; in the ~ future in nächster Zukunft

foreseen *see* foresee

'foresight *n.* Weitblick, *der*

foreskin *n. (Anat.)* Vorhaut, *die*

forest ['fɒrɪst] *n.* Wald, *der*; *(commercially exploited)* Forst, *der*

fore'stall *v. t.* zuvorkommen (+ *Dat.*)

forestry ['fɒrɪstrɪ] *n.* Forstwirtschaft, *die*

'foretaste *n.* Vorgeschmack, *der*

fore'tell *v. t.*, **fore'told** voraussagen

forever [fə'revə(r)] *adv. (constantly)* ständig

fore'warn *v. t.* vorwarnen

'foreword *n.* Vorwort, *das*

forfeit ['fɔ:fɪt] 1. *v. t.* verlieren; verwirken *(geh.)* ⟨*Recht, jmds. Gunst*⟩. 2. *n.* Strafe, *die*; *(games)* Pfand, *das*

forgave *see* forgive

¹forge [fɔ:dʒ] 1. *n.* a) *(workshop)* Schmiede, *die*; b) *(blacksmith's hearth)* Esse, *die*. 2. *v. t.* a) schmieden (into zu); b) *(fig.)* schmieden ⟨*Plan*⟩; schließen ⟨*Vereinbarung, Freundschaft*⟩; c) *(counterfeit)* fälschen

²forge *v. i.* ~ ahead [das Tempo] beschleunigen; *(fig.)* Fortschritte machen

'forger *n.* Fälscher, *der*/Fälscherin, *die*

forgery ['fɔ:dʒərɪ] *n.* Fälschung, *die*

forget [fə'get] 1. *v. t.*, -tt-, **forgot** [fə'gɒt], **forgotten** [fə'gɒtn] vergessen; (~ learned ability) verlernen. 2. *v. i.*, -tt-, **forgot, forgotten** es vergessen; ~ about sth. etw. vergessen; ~ about it! *(coll.)* schon gut! **forgetful** [fə'getfl] *adj.* vergeßlich. **for'getfulness** *n.* Vergeßlichkeit, *die*. **for'get-me-not** *n. (Bot.)* Vergißmeinnicht, *das*

forgive [fə'gɪv] *v. t.*, **forgave** [fə'geɪv], **forgiven** [fə'gɪvn] verzeihen; vergeben ⟨*Sünden*⟩; ~ sb. [sth. or for sth.] jmdm. [etw.] verzeihen. **for'giveness** *n.* Verzeihung, *die*; *(of sins)* Vergebung, *die*

forgo [fɔ:'gəʊ] *v. t.*, *forms as* go verzichten auf (+ *Akk.*)

forgone *see* forgo

forgot, forgotten *see* forget

fork [fɔ:k] 1. *n.* a) Gabel, *die*; knives and ~s Besteck, *das*; b) *(in road)* Gabelung, *die*; *(one branch)* Abzweigung, *die*. 2. *v. i. (divide)* sich gabeln; *(turn)* abbiegen; ~ left links abbiegen. **fork 'out** *v. i. (sl.)* blechen *(ugs.)*

'fork-lift truck *n.* Gabelstapler, *der*

forlorn [fə'lɔ:n] *adj.* a) *(desperate)* verzweifelt; b) *(forsaken)* verlassen

form [fɔ:m] 1. *n.* a) *(shape, type, style)* Form, *die*; take ~: Gestalt annehmen; b) *(printed sheet)* Formular, *das*; c) *(Brit. Sch.)* Klasse, *die*; d) *(bench)* Bank, *die*; e) *(Sport: physical condition)* Form, *die*; *(fig.)* true to ~: wie üblich. 2. *v. t.* a) bilden; b) *(shape)* formen, gestalten (into zu); c) sich *(Dat.)* bilden ⟨*Meinung, Urteil*⟩; ge-

winnen ⟨*Eindruck*⟩; fassen ⟨*Plan*⟩;
entwickeln ⟨*Vorliebe, Gewohnheit*⟩;
schließen ⟨*Freundschaft*⟩; **d)** *(set up)*
bilden ⟨*Regierung*⟩; gründen ⟨*Bund,
Firma, Partei*⟩. **3.** *v. i.* sich bilden;
⟨*Idee:*⟩ Gestalt annehmen

formal ['fɔːml] *adj.* formell; förmlich
⟨*Person, Art, Einladung, Begrüßung*⟩;
(official) offiziell; **a** ~ 'yes'/'no' eine
bindende Zusage/endgültige Absage.

formality [fɔː'mælɪtɪ] *n.* **a)** *(requirement)* Formalität, *die;* **b)** *(being
formal)* Förmlichkeit, *die*

format ['fɔːmæt] *n.* Format, *das*

formation [fɔː'meɪʃn] *n.* **a)** *see* **form
2 a, d**: Bildung, *die;* Gründung, *die;* **b)**
(Mil., Aeronaut.) Formation, *die*

former ['fɔːmə(r)] *attrib. adj.* ehemalig; **in** ~ **times** früher; **the** ~: der/die/
das erstere; *pl.* die ersteren. **'formerly** *adv.* früher

formidable ['fɔːmɪdəbl] *adj.* gewaltig;
gefährlich ⟨*Herausforderung, Gegner*⟩

formula ['fɔːmjʊlə] *n.* Formel, *die.*

formulate ['fɔːmjʊleɪt] *v. t.* formulieren; *(devise)* entwickeln

forsake [fə'seɪk] *v. t.*, **forsook** [fə'sʊk],
~**n** [fə'seɪkn] **a)** *(give up)* verzichten
auf (+ *Akk.*); **b)** *(desert)* verlassen.

for'saken *adj.* verlassen

fort [fɔːt] *n. (Mil.)* Fort, *das*

forte ['fɔːteɪ] *n.* Stärke, *die*

forth [fɔːθ] *adv.* **and so** ~: und so weiter; *see also* **back 3**

forthcoming ['---, -'--] *adj.* **a)** *(ap-
proaching)* bevorstehend; in Kürze er-
scheinend ⟨*Buch usw.*⟩; **b)** *pred.* **be** ~
⟨*Geld, Antwort:*⟩ kommen; ⟨*Hilfe:*⟩ ge-
leistet werden; **not be** ~: ausbleiben;
c) *(responsive)* mitteilsam ⟨*Person*⟩

'forthright *adj.* direkt

forth'with *adv.* unverzüglich

fortieth ['fɔːtɪɪθ] *adj.* vierzigst ...

fortify ['fɔːtɪfaɪ] *v. t.* **a)** *(Mil.)* befestigen; **b)** *(strengthen)* stärken

fortitude ['fɔːtɪtjuːd] *n.* innere Stärke

fortnight ['fɔːtnaɪt] *n.* vierzehn Tage

fortress ['fɔːtrɪs] *n.* Festung, *die*

fortuitous [fɔː'tjuːɪtəs] *adj.,* **for'tuit-
ously** *adv.* zufällig

fortunate ['fɔːtʃənət] *adj.* glücklich.
'fortunately *adv.* glücklicherweise

fortune ['fɔːtʃən, 'fɔːtʃuːn] *n.* **a)**
(wealth) Vermögen, *das;* **b)** *(luck)*
Schicksal, *das;* **bad/good** ~: Pech/
Glück, *das.* **'fortune-teller** *n.* Wahrsager, *der*/Wahrsagerin, *die*

forty ['fɔːtɪ] *adj.* vierzig; **have** ~
'winks ein Nickerchen machen *(ugs.).*

2. *n.* Vierzig, *die. See also* **eight;
eighty 2**

forum ['fɔːrəm] *n.* Forum, *das*

forward ['fɔːwəd] **1.** *adv.* **a)** *(in direction faced)* vorwärts; **b)** *(to the front)*
nach vorn; vor⟨*laufen, -rücken,
-schieben*⟩; **c)** *(closer)* heran; **he came**
~ **to greet me** er kam auf mich zu, um
mich zu begrüßen; **d) come** ~ ⟨*Zeuge,
Helfer:*⟩ sich melden. **2.** *adj.* **a)** *(directed ahead)* vorwärts gerichtet; **b)** *(at
or to the front)* Vorder-; vorder... **3.** *n.*
(Sport) Stürmer, *der*/Stürmerin, *die.*
4. *v. t. (send on)* nachschicken ⟨*Post*⟩
(to an + *Akk.*)

forwards ['fɔːwədz] *see* **forward 1 a, b**

forwent *see* **forgo**

fossil ['fɒsɪl] *n.* Fossil, *das*

foster ['fɒstə(r)] **1.** *v. t.* **a)** fördern;
pflegen ⟨*Freundschaft*⟩; **b)** in Pflege
haben ⟨*Kind*⟩. **2.** *adj.* ~-: Pflege⟨*kind,
-eltern; -sohn usw.*⟩

fought *see* **fight 1, 2**

foul [faʊl] **1.** *adj.* **a)** abscheulich ⟨*Geruch, Geschmack*⟩; **b)** *(polluted)* ver-
schmutzt ⟨*Wasser, Luft*⟩; *(putrid)* faulig ⟨*Wasser*⟩; stickig ⟨*Luft*⟩; **c)** *(sl.:
awful)* scheußlich *(ugs.);* anstößig
⟨*Sprache*⟩. **2.** *n. (Sport)* Foul, *das.*
3. *v. t.* **a)** beschmutzen; verpesten
⟨*Luft*⟩; **b)** *(Sport)* foulen. **'foul-
smelling** *adj.* übelriechend

¹found [faʊnd] *v. t.* **a)** *(establish)* gründen; stiften ⟨*Krankenhaus, Kloster*⟩;
begründen ⟨*Wissenschaft, Religion*⟩;
b) *(fig.: base)* begründen; **be** ~**ed** |up|
on sth. [sich] auf etw. *(Akk.)* gründen

²found *see* **find 1**

foundation [faʊn'deɪʃn] *n.* **a)** Gründung, *die; (of hospital, monastery)*
Stiftung, *die;* **b)** *usu. in pl.* ~|s| *(lit. or
fig.)* Fundament, *das;* **be without** ~
(fig.) unbegründet sein. **foun'dation
stone** *n.* Grundstein, *der*

¹'founder *n.* Gründer, *der*/Gründerin,
die; (of hospital) Stifter, *der*/Stifterin,
die

²'founder *v. i.* **a)** ⟨*Schiff:*⟩ sinken; **b)**
(fig.: fail) sich zerschlagen

fountain ['faʊntɪn] *n.* Fontäne, *die;
(structure)* Springbrunnen, *der; (fig.)*
Quelle, *die.* **'fountain-pen** *n.* Füllfe-
derhalter, *der*

four [fɔː(r)] **1.** *adj.* vier. **2.** *n.* Vier, *die;*
on all ~**s** auf allen vieren *(ugs.). See
also* **eight. 'four-poster** *n.* ~ |bed|
Himmelbett, *das.* **foursome** ['fɔːsəm]
n. Quartett, *das;* **go in** *or* **as a** ~: zu
viert gehen

fourteen [fɔːˈtiːn] 1. *adj.* vierzehn.
2. *n.* Vierzehn, *die. See also* **eight.**
fourteenth [fɔːˈtiːnθ] 1. *adj.* vier-
zehnt... 2. *n. (fraction)* Vierzehntel,
das. See also **eighth**
fourth [fɔːθ] 1. *adj.* viert... 2. *n. (in sequence)* vierte, *der/die/das; (in rank)*
Vierte, *der/die/das; (fraction)* Viertel,
das. See also **eighth.** 'f**ourthly** *adv.*
viertens
fowl [faʊl] *n.* Haushuhn, *das; (collectively)* Geflügel, *das*
fox [fɒks] 1. *n.* Fuchs, *der.* 2. *v. t.* verwirren
foyer [ˈfɔɪeɪ] *n.* Foyer, *das*
fraction [ˈfrækʃn] *n.* **a)** *(Math.)* Bruch,
der; **b)** *(small part)* Bruchteil, *der*
fracture [ˈfræktʃə(r)] 1. *n.* Bruch, *der.*
2. *v. t.* brechen
fragile [ˈfrædʒaɪl] *adj.* zerbrechlich
fragment [ˈfrægmənt] *n.* Bruchstück,
das. **fragmentary** [ˈfrægməntərɪ]
adj. bruchstückhaft
fragrance [ˈfreɪɡrəns] *n.* Duft, *der.*
fragrant [ˈfreɪɡrənt] *adj.* duftend
frail [freɪl] *adj.* zerbrechlich; gebrechlich ‹Greis, Greisin›
frame [freɪm] 1. *n.* **a)** *(of vehicle)* Rahmen, *der; (of bed)* Gestell, *das;* **b)** *(bor-
der)* Rahmen, *der;* |**spectacle**|~s [Bril-
len]gestell, *das.* 2. *v. t.* **a)** rahmen; **b)**
formulieren ‹Frage, Antwort›; **c)** *(sl.:
incriminate)* ~ **sb.** jmdm. etwas anhängen *(ugs.).* 'f**rame-up** *n. (coll.)*
abgekartetes Spiel *(ugs.).* 'f**rame-
work** *n.* Gerüst, *das*
franc [fræŋk] *n.* Franc, *der; (Swiss)*
Franken, *der*
France [frɑːns] *pr. n.* Frankreich *(das)*
franchise [ˈfræntʃaɪz] *n.* **a)** Stimmrecht, *das;* **b)** *(Commerc.)* Lizenz, *die*
¹**frank** *adj.* offen; freimütig ‹Geständnis, Äußerung›; **be** ~ **with sb.** zu jmdm.
offen sein
²**frank** *v. t. (Post)* frankieren
frankfurter [ˈfræŋkfɜːtə(r)] *(Amer.:*
frankfurt [ˈfræŋkfɜːt]) *n.* Frankfurter
[Würstchen]
'f**rankly** *adv.* offen; *(honestly)* offen
gesagt
frantic [ˈfræntɪk] *adj.* **a)** verzweifelt
‹Hilferufe, Gestikulieren›; **be** ~ **with
fear/rage** *etc.* außer sich *(Dat.)* sein
vor Angst/Wut *usw.;* **b)** hektisch *(Aktivität, Suche).* **frantically** [ˈfræntɪ-
kəlɪ], 'f**ranticly** *adv.* verzweifelt
fraternize [ˈfrætənaɪz] *v. i.* ~ |**with sb.**|
sich verbrüdern [mit jmdm.]
fraud [frɔːd] *n.* **a)** *no pl.* Betrug, *der;* **b)**

(trick) Schwindel, *der;* **c)** *(person)* Betrüger, *der/*Betrügerin, *die.* **fraudu-
lent** [ˈfrɔːdjʊlənt] *adj.* betrügerisch
fraught [frɔːt] *adj.* **be** ~ **with danger**
voller Gefahren sein
¹**fray** [freɪ] *n.* [Kampf]getümmel, *das;*
enter *or* **join the** ~: sich in den Kampf
stürzen
²**fray** *v. i.* [sich] durchscheuern;
‹Hosenbein, Teppich, Seilende:› aus-
fransen; **our nerves/tempers began to**
~ *(fig.)* wir verloren langsam die Ner-
ven/unsere Gemüter erhitzten sich
freak [friːk] *n.* **a)** Mißgeburt, *die; attrib.* ungewöhnlich ‹Wetter, Ereignis›;
b) *(sl.: fanatic)* Freak, *der*
freckle [ˈfrekl] *n.* Sommersprosse, *die.*
'f**reckled** *adj.* sommersprossig
free [friː] 1. *adj.,* freer [ˈfriːə(r)], freest
[ˈfriːɪst] **a)** frei; **get** ~: freikommen; **set**
~: freilassen; ~ **of charge/cost** gebüh-
renfrei/kostenlos; **sb. is** ~ **to do sth.**
steht jmdm. frei, etw. zu tun; ~ **time**
Freizeit, *die;* **he's** ~ **in the mornings** er
hat morgens Zeit; **b)** *(without payment)* kostenlos; frei ‹Unterkunft,
Verpflegung›; Frei‹karte, -exemplar›;
Gratis‹probe›; '**admission** ~' „Eintritt
frei‟; **for** ~ *(coll.)* umsonst. 2. *adv.*
gratis; umsonst. 3. *v. t. (set at liberty)*
freilassen; *(disentangle)* befreien (of,
from von); ~ **sb./oneself from** jmdn./
sich befreien aus ‹Gefängnis, Sklaverei›. **freedom** [ˈfriːdəm] *n.* Freiheit,
die
free: ~ 'g**ift** *n.* Gratisgabe, *die; *~-**hold**
1. *n.* Besitzrecht, *das;* 2. *adj.* Eigen-
tums-; ~-**lance** 1. *n.* freier Mitarbei-
ter/freie Mitarbeiterin; 2. *adj.* freibe-
ruflich
'f**reely** *adv. (willingly)* großzügig; freimütig ‹eingestehen›; *(without restriction)* frei; *(frankly)* offen
free: F~**mason** *n.* Freimaurer, *der;*
~-**range** *adj.* freilaufend ‹Huhn›;
~-**range eggs** Eier von freilaufenden
Hühnern; ~ **speech** *n.* Redefreiheit,
die; ~-**way** *n. (Amer.)* Autobahn, *die;*
~-**wheel** *v. i.* im Freilauf fahren
freeze [friːz] 1. *n.,* froze [frəʊz],
frozen [ˈfrəʊzn] **a)** frieren; *(become
covered with ice)* zufrieren; ‹Straße:›
vereisen; ‹Flüssigkeit:› gefrieren;
‹Rohr, Schloß:› einfrieren; **b)** *(become
rigid)* steif frieren. 2. *v. t.,* froze,
frozen **a)** *(preserve)* tiefkühlen; **b)** einfrieren ‹Kredit, Löhne, Preise usw.›.
'f**reezer** *n.* Tiefkühltruhe, *die;* |**up-
right**| ~: Tiefkühlschrank, *der; *~ **com-**

partment Tiefkühlfach, *das.* **freezing** ['fri:zɪŋ] **1.** *adj. (lit. or fig.)* frostig; **it's** ~: es ist eiskalt. **2.** *n.* **above/below** ~: über/unter dem/den Gefrierpunkt

freight [freɪt] *n.* Fracht, *die.* **'freighter** *n.* Frachter, *der*

French [frentʃ] **1.** *adj.* französisch; **he/ she is** ~: er ist Franzose/sie ist Französin. **2.** *n.* **a)** *(language)* Französisch, *das; see also* **English 2 a;** **b)** **the** ~ *pl.* die Franzosen

French: ~ **'bean** *n. (Brit.)* Gartenbohne, *die;* ~ **'dressing** *n.* Vinaigrette, *die;* ~ **'fries** *n. pl.* Pommes frites *Pl.* ~**man** ['frentʃmən] *n., pl.* ~**men** ['frentʃmən] Franzose, *der;* ~ **'window** *n.* französisches Fenster; ~**woman** *n.* Französin, *die*

frenzied ['frenzɪd] *adj.* rasend

frenzy ['frenzɪ] *n.* **a)** Wahnsinn, *der; (fury)* Raserei, *die*

frequency ['fri:kwənsɪ] *n.* **a)** Häufigkeit, *die;* **b)** *(Phys.)* Frequenz, *die*

frequent 1. ['fri:kwənt] *adj.* **a)** häufig; **become less** ~: seltener werden; **b)** *(habitual)* eifrig. **2.** [fri:'kwent] *v. t.* häufig besuchen ⟨*Café, Klub, usw.*⟩. **frequently** ['fri:kwəntlɪ] *adv.* häufig

fresco *n. pl.* ~**es** *or* ~**s** Fresko, *das*

fresh [freʃ] *adj.* frisch; neu ⟨*Beweise, Anstrich, Mut, Energie*⟩; ~ **supplies** Nachschub, *der* (of an + *Dat.*); **make a** ~ **start** noch einmal von vorne anfangen; *(fig.)* neu beginnen

freshen ['freʃn] *v. i.* auffrischen. **freshen 'up** *v. i.* sich auffrischen. **'freshly** *adv.* frisch. **'freshness** *n.* Frische, *die.* **fresh 'water** *n.* Süßwasser, *das*

fret [fret] *v. i.,* **-tt-** sich *(Dat.)* Sorgen machen. **fretful** ['fretfl] *adj.* verdrießlich; quengelig *(ugs.)*. **'fretsaw** *n.* Laubsäge, *die*

Fri. *abbr.* Friday Fr.

friar ['fraɪə(r)] *n.* Ordensbruder, *der*

friction ['frɪkʃn] *n.* Reibung, *die*

Friday ['fraɪdeɪ, 'fraɪdɪ] *n.* Freitag, *der;* **on** ~: [am] Freitag; **on a** ~, **on** ~**s** freitags; ~ **13 August** Freitag, der 13. August; *(at top of letter etc.)* Freitag, den 13. August; **next/last** ~: [am] nächsten/letzten Freitag; **Good** ~: Karfreitag, *der*

fridge [frɪdʒ] *n. (Brit. coll.)* Kühlschrank, *der*

fried *see* **'fry**

friend [frend] *n.* **a)** Freund, *der/* Freundin, *die;* **be** ~**s with sb.** mit jmdm. befreundet sein; **make** ~**s [with**

sb.] [mit jmdm.] Freundschaft schließen. **friendliness** ['frendlɪnɪs] *n.* Freundlichkeit, *die.* **'friendly 1.** *adj.* freundlich **(to** zu); freundschaftlich ⟨*Rat, Beziehungen, Wettkampf*⟩. **2.** *n. (Sport)* Freundschaftsspiel, *das.* **'friendship** *n.* Freundschaft, *die*

frigate ['frɪgət] *n. (Naut.)* Fregatte, *die*

fright [fraɪt] *n.* Schreck, *der;* **take** ~: erschrecken. **frighten** ['fraɪtn] *v. t.* ⟨*Explosion, Schuß:*⟩ erschrecken; ⟨*Gedanke, Drohung:*⟩ angst machen (+ *Dat.*); **be** ~**ed at** *or* **by sth.** vor etw. *(Dat.)* erschrecken. **'frightful** *adj.,* **'frightfully** *adv.* furchtbar

frigid ['frɪdʒɪd] *adj.* frostig; *(sexually)* frigid[e]

frill [frɪl] *n.* **a)** Rüsche, *die;* **b)** *in pl. (embellishments)* Beiwerk, *das.* **'frilly** *adj.* mit Rüschen besetzt; Rüschen- ⟨*kleid, -bluse*⟩

fringe [frɪndʒ] *n.* **a)** Fransenkante, *die* **(on an** + *Dat.*); **b)** *(hair)* [Pony]fransen *(ugs.);* **c)** *(edge)* Rand, *der*

frisk [frɪsk] **1.** *v. i.* **[about]** [herum]springen. **2.** *v. t. (coll.)* filzen *(ugs.).* **'frisky** *adj.* munter

¹fritter ['frɪtə(r)] *n.* **apple** *etc.* ~**s** Apfelstücke *usw.* in Pfannkuchenteig

²fritter *v. t.* ~ **away** vergeuden

frivolity [frɪ'vɒlɪtɪ] *n.* Oberflächlichkeit, *die*

frivolous ['frɪvələs] *adj.* **a)** frivol; **b)** *(trifling)* belanglos

frizzy ['frɪzɪ] *adj.* kraus

fro [frəʊ] *adv. see* **to 2**

frock [frɒk] *n.* Kleid, *das*

frog [frɒg] *n.* Frosch, *der.* **frogman** ['frɒgmən] *n., pl.* ~**men** ['frɒgmən] Froschmann, *der.* **'frog-spawn** *n.* Froschlaich, *der*

frolic ['frɒlɪk] *v. i.,* **-ck-:** ~ **[about** *or* **around]** [herum]springen

from [frəm, *stressed* frɒm] *prep.* von; (~ *within; expr. origin)* aus; ~ **Paris** aus Paris; ~ **Paris to Munich** von Paris nach München; **be a mile** ~ **sth.** eine Meile von etw. entfernt sein; **where do you come** ~? **where are you** ~? woher kommen Sie?; **painted** ~ **life** nach dem Leben gemalt; **weak** ~ **hunger** schwach vor Hunger; ~ **the year 1972** seit 1972; ~ **[the age of] 18** ab 18 Jahre; ~ **4 to 6 eggs** 4 bis 6 Eier

front [frʌnt] **1.** *n.* **a)** Vorderseite, *die; (of house)* Vorderfront, *die;* **in** *or* **at the** ~ **[of sth.]** vorn [in etw. *position: Dat., movement: Akk.*]; **to the** ~: nach vorn; **in** ~: vorn[e]; **be in** ~ **of sth./sb.**

vor etw./jmdm. sein; **b)** *(Mil.)* Front, *die;* **c)** *(at seaside)* Strandpromenade, *die;* **d)** *(Metereol.)* Front, *die;* **e)** *(bluff)* Fassade, *die.* 2. *adj.* vorder...; Vorder- ⟨*rad, -zimmer, -zahn*⟩; ~ **garden** Vorgarten, *der;* ~ **row** erste Reihe.

frontal ['frʌntl] *adj.* Frontal-. **front 'door** *n. (of flat)* Wohnungstür, *die; (of house)* Haustür, *die*

frontier ['frʌntɪə(r)] *n.* Grenze, *die*

front 'page *n.* Titelseite, *die*

frost [frɒst] **1.** *n.* Frost, *der;* **ten degrees of ~** *(Brit.)* zehn Grad minus. **2.** *v. t.* **~ed glass** Mattglas, *das.* '**frostbite** *n.* Erfrierung, *die.* '**frosting** *n. (esp. Amer.)* Glasur, *die.* '**frosty** *adj.* frostig

froth [frɒθ] **1.** *n.* Schaum, *der.* **2.** *v. i.* schäumen. '**frothy** *adj.* schaumig

frown [fraʊn] **1.** *v. i.* die Stirn runzeln (**up|on** über + *Akk.*). **2.** *n.* Stirnrunzeln, *das*

froze *see* **freeze**

frozen ['frəʊzn] **1.** *see* **freeze. 2.** *adj.* **a)** zugefroren ⟨*Fluß, See*⟩; eingefroren ⟨*Wasserleitung*⟩; **I'm ~** *(fig.)* ich bin eiskalt; **b)** *(to preserve)* tiefgekühlt; **~ food** Tiefkühlkost, *die*

frugal ['fru:gl] *adj.* genügsam ⟨*Lebensweise, Mensch*⟩; frugal ⟨*Mahl*⟩

fruit [fru:t] *n.* Frucht, *die; collect.* Obst, *das*

fruitful ['fru:tfl] *adj.* fruchtbar

'**fruit juice** *n.* Obstsaft, *der*

'**fruitless** *adj.* nutzlos ⟨*Versuch, Gespräch*⟩; fruchtlos ⟨*Verhandlung, Suche*⟩

fruit: ~ machine *n. (Brit.)* Spielautomat, *der;* ~ '**salad** *n.* Obstsalat, *der*

'**fruity** *adj.* fruchtig ⟨*Geschmack, Wein*⟩

frustrate [frʌ'streɪt] *v. t.* vereiteln ⟨*Plan, Versuch*⟩; zunichte machen ⟨*Hoffnung, Bemühungen*⟩. '**frustrated** *adj.* frustriert. **frustration** [frʌ'streɪʃn] *n.* Frustration, *die*

¹**fry** [fraɪ] *v. t. & i.* braten; **fried egg** Spiegelei, *das*

²**fry** *n. (fishes)* Brut, *die;* **small ~** *(fig.)* unbedeutende Leute

'**frying-pan** *n.* Bratpfanne, *die*

ft. *abbr.* **feet, foot** ft.

fuck [fʌk] *(coarse)* **1.** *v. t. & i.* ficken *(vulg.).* **2.** *n.* Fick, *der (vulg.)*

fuddy-duddy ['fʌdɪdʌdɪ] *(sl.)* **1.** *adj.* verkalkt *(ugs.).* **2.** *n.* Fossil, *das (fig.)*

fudge [fʌdʒ] *n.* Karamelbonbon, *der od.* das

fuel ['fju:əl] *n.* Brennstoff, *der; (for*

vehicle) Kraftstoff, *der; (for ship, aircraft)* Treibstoff, *der*

fugitive ['fju:dʒɪtɪv] *n.* Flüchtige, *der/ die*

fugue [fju:g] *n. (Mus.)* Fuge, *die*

fulfil *(Amer.:* **fulfill)** [fʊl'fɪl] *v. t.,* **-ll-** erfüllen; halten ⟨*Versprechen*⟩. **ful'filment** *(Amer.:* **ful'fillment**) *n.* Erfüllung, *die*

full [fʊl] **1.** *adj.* **a)** voll; satt ⟨*Person*⟩; ~ **of** voller; **be ~ up** *(coll.)* voll [besetzt] sein; ⟨*Behälter:*⟩ randvoll sein; ⟨*Flug:*⟩ völlig ausgebucht sein; **I'm ~ |up|** *(coll.)* ich bin voll [bis obenhin] *(ugs.);* **be ~ of oneself** sehr von sich eingenommen sein; **b)** ausführlich ⟨*Bericht, Beschreibung*⟩; erfüllt ⟨*Leben*⟩; ganz ⟨*Stunde, Jahr, Monat, Seite*⟩; voll ⟨*Name, Bezahlung, Verständnis*⟩; ~ **details** alle Einzelheiten; **at ~ speed** mit Höchstgeschwindigkeit; **c)** voll ⟨*Gesicht*⟩; füllig ⟨*Figur*⟩; weit ⟨*Rock*⟩. **2.** *n.* **in ~:** vollständig. **3.** *adv. (exactly)* genau

full: ~ back *n.* Verteidiger, *der/*Verteidigerin, *die;* ~**-length** *adj.* lang ⟨*Kleid*⟩; ~ '**moon** *n.* Vollmond, *der;* ~**-scale** *adj.* **a)** in Originalgröße; **b)** großangelegt ⟨*Untersuchung, Suchaktion*⟩; ~ '**stop** *n.* Punkt, *der;* ~**-time** *adj.* ganztägig; ganztags⟨*arbeit*⟩

fully ['fʊlɪ] *adv.* voll [und ganz]; reich ⟨*belohnt*⟩; ausführlich ⟨*erklären*⟩

fulsome ['fʊlsəm] *adj.* übertrieben

fumble ['fʌmbl] *v. i.* ~ **at** *or* **with** [herum]fingern an (+ *Dat.*); ~ **in one's pockets** **for sth.** in seinen Taschen nach etw. kramen *(ugs.)*

fume [fju:m] **1.** *n. in pl.* **~s** Dämpfe. **2.** *v. i.* vor Wut schäumen

fumigate ['fju:mɪgeɪt] *v. t.* ausräuchern

fun [fʌn] *n.* Spaß, *der;* **have ~!** viel Spaß!; **make ~ of sb.** sich über jmdn. lustig machen; **for ~, for the ~ of it** zum Spaß

function ['fʌŋkʃn] **1.** *n.* Aufgabe, *die;* Funktion, *die; (formal event)* Veranstaltung, *die.* **2.** *v. i.* ⟨*Maschine, System:*⟩ funktionieren; ~ **as** fungieren als; *(serve as)* dienen als. **functional** ['fʌŋkʃənl] *adj.* **a)** *(useful)* funktionell; **b)** *(working)* funktionsfähig

fund [fʌnd] **1.** *n.* **a)** *(money)* Fonds, *der;* **b)** *(fig.: stock)* Fundus, *der (of* von, an + *Dat.*). **2.** *v. t.* finanzieren

fundamental [fʌndə'mentl] *adj.* grundlegend **(to** für); elementar ⟨*Bedürfnisse*⟩. **fundamentally** [fʌndə-

'məntəlı] *adv.* grundlegend; von Grund auf ⟨*verschieden, ehrlich*⟩

funeral ['fju:nərl] *n.* Beerdigung, *die.* ~ **director** Bestattungsunternehmer, *der;* ~ **service** Trauerfeier, *die*

'**fun-fair** *n. (Brit.)* Jahrmarkt, *der*

fungus ['fʌŋgəs] *n., pl.* **fungi** ['fʌŋgaɪ, 'fʌndʒaɪ] *or* ~**es** Pilz, *der*

funicular [fju:'nɪkjʊlə(r)] *adj. & n.* ~ |**railway**| [Stand]seilbahn, *die*

funnel ['fʌnl] *n.* Trichter, *der; (of ship etc.)* Schornstein, *der*

funnily ['fʌnɪlɪ] *adv.* komisch; ~ **enough** komischerweise

funny ['fʌnɪ] *adj.* **a)** komisch; lustig; witzig ⟨*Mensch, Einfall*⟩; **b)** *(strange)* komisch. '**funny-bone** *n.* Musikantenknochen, *der*

fur [fɜ:(r)] *n.* **a)** Fell, *das; (garment)* Pelz, *der;* ~ **coat** Pelzmantel, *der;* **b)** *(in kettle)* Kesselstein, *der*

furious ['fjʊərɪəs] *adj.* wütend; heftig ⟨*Streit*⟩; wild ⟨*Tanz, Tempo, Kampf*⟩; **be** ~ **with sb.** wütend auf jmdn. sein. '**furiously** *adv.* wütend; wild ⟨*kämpfen*⟩: wie wild *(ugs.)* arbeiten

furl [fɜ:l] *v. t.* einrollen ⟨*Segel, Flagge*⟩

furnace ['fɜ:nɪs] *n.* Ofen, *der*

furnish ['fɜ:nɪʃ] *v. t.* **a)** möblieren; **b)** *(supply)* liefern; ~ **sb. with sth.** jmdm. etw. liefern. '**furnishings** *n. pl.* Einrichtungsgegenstände

furniture ['fɜ:nɪtʃə(r)] *n.* Möbel *Pl.;* **piece of** ~: Möbel[stück], *das*

furrow ['fʌrəʊ] *n.* Furche, *die*

furry ['fɜ:rɪ] *adj.* haarig; ~ **animal** *(toy)* Plüschtier, *das*

further ['fɜ:ðə(r)] **1.** *adj.* **a)** *(in space)* weiter entfernt; **b)** *(additional)* weiter... **2.** *adv.* weiter. **3.** *v. t.* fördern. **further'more** *adv.* außerdem. '**furthermost** *adj.* äußerst ...

furthest ['fɜ:ðɪst] **1.** *adj.* am weitesten entfernt. **2.** *adv.* am weitesten ⟨*springen, laufen*⟩; am weitesten entfernt ⟨*sein, wohnen*⟩

furtive ['fɜ:tɪv] *adj.,* **furtively** *adv.* verstohlen

fury ['fjʊərɪ] *n.* Wut, *die; (of sea, battle)* Wüten, *das*

¹**fuse** [fju:z] **1.** *v. t. (blend)* verschmelzen (**into** zu). **2.** *v. i.* ~ **together** miteinander verschmelzen

²**fuse** *n.* |time-|~: [Zeit]zünder, *der; (cord)* Zündschnur, *die*

²**fuse** *(Electr.)* **1.** *n.* Sicherung, *die.* **2.** *v. i.* **the lights have** ~**d** die Sicherung ist durchgebrannt. '**fuse box** *n.* Sicherungskasten, *der*

fuselage ['fju:zəlɑ:ʒ] *n.* [Flugzeug]-rumpf, *der*

fusion ['fju:ʒn] *n.* **a)** Verschmelzung, *die;* **b)** *(Phys.)* Fusion, *die*

fuss [fʌs] **1.** *n.* Theater, *das (ugs.);* **make a** ~ |**about sth.**| einen Wirbel |um etw.| machen. **2.** *v. i.* Wirbel machen; *(get agitated)* sich [unnötig] aufregen. '**fussy** *adj. (fastidious)* eigen; penibel; **I'm not** ~ *(I don't mind)* ich bin nicht wählerisch

futile ['fju:taɪl] *adj.* vergeblich

future ['fju:tʃə(r)] **1.** *adj.* [zu]künftig; **at some** ~ **date** zu einem späteren Zeitpunkt. **2.** *n.* **a)** Zukunft, *die;* **in** ~: in Zukunft; künftig; **b)** *(Ling.)* Futur, *das;* Zukunft, *die.* **futuristic** [fju:tʃə-'rɪstɪk] *adj.* futuristisch

fuze [fju:z] *see* ²**fuse**

fuzzy ['fʌzɪ] *adj.* **a)** *(frizzy)* kraus; **b)** *(blurred)* verschwommen

G

G, g [dʒi:] *n.* G, g, *das*

gab [gæb] *n. (coll.)* **have the gift of the** ~: reden können

gabble ['gæbl] *v. i.* brabbeln *(ugs.)*

gable ['geɪbl] *n.* Giebel, *der*

gad [gæd] *v. i.,* -**dd-** *(coll.)* ~ **about** herumziehen

gadget ['gædʒɪt] *n.* Gerät, *das*

Gaelic ['geɪlɪk, 'gælɪk] **1.** *adj.* gälisch. **2.** *n.* Gälisch, *das*

gaffe [gæf] *n.* Fauxpas, *der*

gag [gæg] **1.** *n.* **a)** Knebel, *der;* **b)** *(joke)* Gag, *der.* **2.** *v. t.,* -**gg-** knebeln

gaiety ['geɪətɪ] *n.* Fröhlichkeit, *die*

gaily ['geɪlɪ] *adv.* fröhlich; in leuchtenden Farben ⟨*bemalt, geschmückt*⟩

gain [geɪn] **1.** *n.* **a)** Gewinn, *der;* **b)** *(increase)* Zunahme, *die* (**in** an + *Dat.*). **2.** *v. t.* **a)** gewinnen; finden ⟨*Zugang, Zutritt*⟩; erwerben ⟨*Wissen, Ruf*⟩; erlangen ⟨*Freiheit*⟩; erzielen ⟨*Vorteil, Punkte*⟩; verdienen ⟨*Lebensunterhalt, Geldsumme*⟩; ~ **weight/five pounds** |**in weight**| zunehmen/fünf Pfund zunehmen; ~ **speed** schneller werden; **b)** ⟨*Uhr:*⟩ vorgehen um. **3.** *v. i.* **a)** ~ **by**

sth. von etw. profitieren; **b)** ⟨*Uhr:*⟩ vorgehen

gait [geɪt] *n.* Gang, *der*

gala ['gɑːlə, 'geɪlə] *n.* Festveranstaltung, *die; attrib.* Gala⟨*abend, -vorstellung*⟩; **swimming ~:** Schwimmfest, *das*

galaxy ['gæləksɪ] *n.* Galaxie, *die*

gale [geɪl] *n.* Sturm, *der*

gall [gɔːl] *n. (sl.)* Unverschämtheit, *die*

gallant ['gælənt] *adj. (brave)* tapfer; *(chivalrous)* ritterlich. **gallantry** ['gæləntrɪ] *n. (bravery)* Tapferkeit, *die*

gall-bladder *n.* Gallenblase, *die*

gallery ['gælərɪ] *n.* **a)** Galerie, *die;* **b)** *(Theatre)* dritter Rang

galley ['gælɪ] *n.* **a)** *(ship's kitchen)* Kombüse, *die;* **b)** *(Hist.)* Galeere, *die*

gallivant ['gælɪvænt] *v. i. (coll.)* herumziehen *(ugs.)*

gallon ['gælən] *n.* Gallone, *die*

gallop ['gæləp] **1.** *n.* Galopp, *der.* **2.** *v. i.* ⟨*Pferd, Reiter:*⟩ galoppieren

gallows ['gæləʊz] *n. sing.* Galgen, *der*

galore [gə'lɔː(r)] *adv.* im Überfluß; in Hülle und Fülle

galvanize ['gælvənaɪz] *v. t.* wachrütteln; **~ sb. into action** jmdn. veranlassen, sofort aktiv zu werden

gambit ['gæmbɪt] *n.* Gambit, *das*

gamble ['gæmbl] *v. i.* **a)** [um Geld] spielen; **b)** *(fig.)* spekulieren; **~ on sth.** sich auf etw. *(Akk.)* verlassen. **gambler** ['gæmblə(r)] *n.* Glücksspieler, *der*

¹**game** [geɪm] *n.* **a)** Spiel, *das; (of [table-]tennis, chess, cards, cricket)* Partie, *die;* **b)** *(fig.: scheme)* Vorhaben, *das;* **c)** *in pl. (athletic contests)* Spiele; *(in school) (sports)* Sport, *der; (athletics)* Leichtathletik, *die;* **d)** *(Hunting, Cookery)* Wild, *das*

²**game** *adj.* mutig; **be ~ to do sth.** bereit sein, etw. zu tun

'**gamekeeper** *n.* Wildheger, *der*

gammon ['gæmən] *n.* Räucherschinken, *der*

gamut ['gæmət] *n.* Skala, *die*

gander ['gændə(r)] *n.* Gänserich, *der*

gang [gæŋ] **1.** *n.* Bande, *die; (of workmen, prisoners)* Trupp, *der.* **2.** *v. i.* **~ up against** *or* **on** *(coll.)* sich verbünden gegen

gangling ['gæŋglɪŋ] schlaksig *(ugs.)*

gangster ['gæŋstə(r)] *n.* Gangster, *der*

'**gangway** *n.* Gangway, *die; (Brit.: between seats)* Gang, *der*

gaol [dʒeɪl] *see* **jail**

gap [gæp] *n.* **a)** Lücke, *die;* **b)** *(in time)* Pause, *die;* **c)** *(divergence)* Kluft, *die*

gape [geɪp] *v. i.* **a)** den Mund aufsperren; ⟨*Loch, Abgrund, Wunde:*⟩ klaffen; **b)** *(stare)* Mund und Nase aufsperren *(ugs.);* **~ at sb./sth.** jmdn./ etw. mit offenem Mund anstarren

garage ['gærɪdʒ] *n.* Garage, *die; (selling petrol)* Tankstelle, *die; (for repairing cars)* [Kfz-]Werkstatt, *die*

garb [gɑːb] *n.* Tracht, *die*

garbage ['gɑːbɪdʒ] *n.* **a)** Abfall, *der;* Müll, *der;* **b)** *(coll.: nonsense)* Quatsch, *der (salopp).* '**garbage can** *n. (Amer.)* Mülltonne, *die*

garble ['gɑːbl] *v. t.* verstümmeln

garden ['gɑːdn] *n.* Garten, *der.* '**garden centre** *n.* Gartencenter, *das*

gardener ['gɑːdnə(r)] *n.* Gärtner, *der*/Gärtnerin, *die.* **gardening** ['gɑːdnɪŋ] *n.* Gartenarbeit, *die*

gargle ['gɑːgl] *v. i.* gurgeln

garish ['geərɪʃ] *adj.* grell ⟨*Farbe, Licht*⟩; knallbunt ⟨*Kleidung*⟩

garland ['gɑːlənd] *n.* Girlande, *die*

garlic ['gɑːlɪk] *n.* Knoblauch, *der*

garment ['gɑːmənt] *n.* Kleidungsstück, *das; ~s pl. (clothes)* Kleidung, *die;* Kleider

garnish ['gɑːnɪʃ] **1.** *v. t.* garnieren. **2.** *n.* Garnierung, *die*

garret ['gærɪt] *n.* Dachkammer, *die*

garrison ['gærɪsn] *n.* Garnison, *die*

garter ['gɑːtə(r)] *n.* Strumpfband, *das*

gas [gæs] **1.** *n.* **a)** *pl.* **~es** ['gæsɪz] Gas, *das;* **b)** *(Amer. coll.: petrol)* Benzin, *das.* **2.** *v. t.,* **-ss-** mit Gas vergiften. **gas 'cooker** *n. (Brit.)* Gasherd, *der.* **gas 'fire** *n.* Gasofen, *der*

gash [gæʃ] **1.** *n.* Schnittwunde, *die.* **2.** *v. t.* aufritzen ⟨*Haut*⟩; **~ one's finger** sich *(Dat. od. Akk.)* in den Finger schneiden

gas: ~ mask *n.* Gasmaske, *die; ~ meter** *n.* Gaszähler, *der*

gasoline (gasolene) ['gæsəliːn] *n. (Amer.)* Benzin, *das*

gasometer [gæ'sɒmɪtə(r)] *n.* Gasometer, *der*

gasp [gɑːsp] **1.** *v. i.* nach Luft schnappen (with vor); **he was ~ing for air** er rang nach Luft. **2.** *v. t.* **~ out** hervorstoßen. **3.** *n.* Keuchen, *das*

'**gas station** *n. (Amer.)* Tankstelle, *die*

gastronomy [gæ'strɒnəmɪ] *n.* Gastronomie, *die*

'**gasworks** *n. sing.* Gaswerk, *das*

'**gate** [geɪt] *n.* Tor, *das; (barrier)* Sperre, *die; (to field etc.)* Gatter, *das; (of level crossing)* [Bahn]schranke, *die; (in airport)* Flugsteig, *der*

gateau ['gætəʊ] *n., pl.* ~s *or* ~x ['gæ-təʊz] Torte, *die*

gate: ~**crasher** ['geɪtkræʃə(r)] *n.* ungeladener Gast; ~**way** *n.* Tor, *das*

gather ['gæðə(r)] **1.** *v. t.* **a)** sammeln; zusammentragen ⟨*Informationen*⟩; pflücken ⟨*Obst, Blumen*⟩; **b)** *(infer, deduce)* schließen (**from** aus); **c)** ~ **speed/force** schneller/stärker werden. **2.** *v. i.* sich versammeln; ⟨*Wolken:*⟩ sich zusammenziehen. '**gathering** *n.* Versammlung, *die*

gaudy ['gɔːdɪ] *adj.* protzig; grell ⟨*Farben*⟩

gauge [geɪdʒ] *n.* **1. a)** *(measure)* Maß, *das;* **b)** *(instrument)* Meßgerät, *das.* **2.** *v. t.* messen; *(fig.)* beurteilen

gaunt [gɔːnt] *adj.* hager

gauntlet ['gɔːntlɪt] *n.* Stulpenhandschuh, *der*

gauze [gɔːz] *n.* Gaze, *die*

gave *see* **give** 1, 2

gay [geɪ] **1.** *adj.* **a)** fröhlich; *(brightcoloured)* farbenfroh; **b)** *(coll.: homosexual)* schwul *(ugs.);* Schwulen⟨*lokal*⟩. **2.** *n. (coll.)* Schwule, *der (ugs.)*

gaze [geɪz] *v. i.* blicken; *(fixedly)* starren; ~ **at sb./sth.** jmdn./etw. anstarren

GB *abbr.* **Great Britain** GB

GCSE *abbr. (Brit.)* **General Certificate of Secondary Education**

gear [gɪə(r)] **1.** *n.* **a)** *(Motor Veh.)* Gang, *der;* **top/bottom** ~ *(Brit.)* der höchste/erste Gang; **change** *or* **shift** ~: schalten; **put the car into** ~: einen Gang einlegen; **out of** ~: im Leerlauf; **b)** *(coll.: clothes)* Aufmachung, *die;* **c)** *(equipment)* Gerät, *das;* Ausrüstung, *die.* **2.** *v. t.* ausrichten (**to** auf + *Akk.*). '**gearbox** *n.* Getriebekasten, *der.* '**gear-lever,** *(Amer.)* '**gear-shift** *ns.* Schalthebel, *der*

geese *pl. of* **goose**

geezer ['giːzə(r)] *(sl.: old man)* Opa, *der (ugs.)*

gel [dʒel] *n.* Gel, *das*

gelatin ['dʒelətɪn], *(Brit.)* **gelatine** ['dʒelətiːn] *n.* Gelatine, *die*

gelignite ['dʒelɪgnaɪt] *n.* Gelatinedynamit, *das*

gem [dʒem] *n.* Edelstein, *der*

Gemini ['dʒemɪnaɪ, 'dʒemɪnɪ] *n.* Zwillinge *Pl.*

gender ['dʒendə(r)] *n. (Ling.)* [grammatisches] Geschlecht

gene [dʒiːn] *n. (Biol.)* Gen, *das*

general ['dʒenrl] **1.** *adj.* allgemein; weitverbreitet ⟨*Ansicht*⟩; *(true of [nearly] all cases)* allgemeingültig; un-

gefähr ⟨*Vorstellung, Beschreibung usw.*⟩; **the** ~ **public** weite Kreise der Bevölkerung; **in** ~ **use** allgemein verbreitet; **as a** ~ **rule, in** ~: im allgemeinen. **2.** *n. (Mil.)* General, *der.*

general e'lection *see* **election**

generalization [dʒenrəlaɪ'zeɪʃn] *n.* Verallgemeinerung, *die*

generalize ['dʒenrəlaɪz] **1.** *v. t.* verallgemeinern. **2.** *v. i.* ~ **about sth.** [etw.] verallgemeinern

generally ['dʒenrəlɪ] *adv.* **a)** allgemein; ~ **available** überall erhältlich; ~ **speaking** im allgemeinen; **b)** *(usually)* im allgemeinen

general prac'titioner *n. (Med.)* Arzt/Ärztin für Allgemeinmedizin

generate ['dʒenəreɪt] *v. t.* erzeugen (**from** aus); *(result in)* führen zu.

generation [dʒenə'reɪʃn] *n.* **a)** Generation, *die;* **b)** *(production)* Erzeugung, *die.* **generator** ['dʒenəreɪtə(r)] *n.* Generator, *der*

generosity [dʒenə'rɒsɪtɪ] *n.* Großzügigkeit, *die*

generous ['dʒenərəs] *adj.* großzügig; reichlich ⟨*Vorrat, Portion*⟩. '**generously** *adv.* großzügig

genetic [dʒɪ'netɪk] *adj.* genetisch. **genetics** [dʒɪ'netɪks] *n.* Genetik, *die*

Geneva [dʒɪ'niːvə] **1.** *pr. n.* Genf *(das).* **2.** *attrib. adj.* Genfer

genial ['dʒiːnɪəl] *adj.* freundlich

genitals ['dʒenɪtlz] *n. pl.* Geschlechtsorgane

genitive ['dʒenɪtɪv] *adj. & n.* ~ |**case**| Genitiv, *der*

genius ['dʒiːnɪəs] *n.* **a)** *(person)* Genie, *das;* **b)** *(ability)* Talent, *das*

genre ['ʒɑ̃rə] *n.* Genre, *das*

gent [dʒent] *n.* **a)** *(coll./joc.)* Gent, *der (iron.);* **b) the G**~**s** *(Brit. coll.)* die Herrentoilette

genteel [dʒen'tiːl] *adj.* vornehm

gentle ['dʒentl] *adj.,* ~**r** ['dʒentlə(r)], ~**st** ['dʒentlɪst] sanft; liebenswürdig ⟨*Person, Verhalten*⟩; leicht, schwach ⟨*Brise*⟩; leise ⟨*Geräusch*⟩; gemächlich ⟨*Spaziergang, Tempo*⟩; mäßig ⟨*Hitze*⟩

gentleman ['dʒentlmən] *n., pl.* **gentlemen** ['dʒentlmən] Herr, *der;* *(well-mannered)* Gentleman, *der;* **Ladies and Gentlemen!** meine Damen und Herren!

'**gentleness** *n.* Sanftheit, *die;* *(of nature)* Sanftmütigkeit, *die*

gently ['dʒentlɪ] *adv,* *(tenderly)* zart; zärtlich; *(mildly)* sanft; *(carefully)* behutsam; *(quietly, softly)* leise

genuine ['dʒenjʊin] *adj.* **a)** *(real)* echt; **b)** *(true)* aufrichtig; wahr ⟨*Grund, Not*⟩. **'genuinely** *adv.* wirklich

genus ['dʒi:nəs, 'dʒenəs] *n., pl.* **genera** ['dʒenərə] *(Biol.)* Gattung, *die*

geographical [dʒi:ə'græfɪkl] *adj.* geographisch

geography [dʒɪ'ɒgrəfɪ] *n.* Geographie, *die;* Erdkunde, *die (Schulw.)*

geological [dʒi:ə'lɒdʒɪkl] *adj.* geologisch

geologist [dʒɪ'ɒlədʒɪst] *n.* Geologe, *der*/Geologin, *die*

geology [dʒɪ'ɒlədʒɪ] *n.* Geologie, *die*

geometric [dʒi:ə'metrɪk], **geometrical** [dʒi:ə'metrɪkl] *adj.* geometrisch

geometry [dʒɪ'ɒmɪtrɪ] *n.* Geometrie, *die*

geranium [dʒə'reɪnɪəm] *n.* Geranie, *die;* Pelargonie, *die*

geriatric [dʒerɪ'ætrɪk] *adj.* geriatrisch

germ [dʒɜ:m] *n.* Keim, *der*

German ['dʒɜ:mən] **1.** *adj.* deutsch; he/she is ~: er ist Deutscher/sie ist Deutsche. **2.** *n.* **a)** *(person)* Deutsche, *der/die;* **b)** *(language)* Deutsch, *das; see also* **English 2a**

German Democratic Re'public *pr. n. (Hist.)* Deutsche Demokratische Republik

Germanic [dʒɜ:'mænɪk] *adj.* germanisch

German 'measles *n.* Röteln *Pl.*

Germany ['dʒɜ:mənɪ] *pr. n.* Deutschland *(das);* Federal Republic of ~: Bundesrepublik Deutschland, *die*

germinate ['dʒɜ:mɪneɪt] *v. i.* keimen

gesticulate [dʒe'stɪkjʊleɪt] *v. i.* gestikulieren. **gesticulation** [dʒestɪkjʊ'leɪʃn] *n.* Gesten *Pl.*

gesture ['dʒestʃə(r)] *n.* Geste, *die*

get [get] **1.** *v. t.*, -tt-, got [gɒt], got *or (Amer.)* gotten ['gɒtn] **a)** *(obtain, receive)* bekommen; kriegen *(ugs.);* sich *(Dat.)* besorgen ⟨*Visum, Genehmigung*⟩; sich *(Dat.)* beschaffen ⟨*Geld*⟩; *(find)* finden ⟨*Zeit*⟩; *(fetch)* holen; *(buy)* kaufen; **where did you ~ that?** wo hast du das her? ~ **sb. a job/taxi,** ~ **a job/taxi for sb.** jmdm. einen Job verschaffen/ein Taxi besorgen; ~ **oneself sth.** sich *(Dat.)* etw. zulegen; **b)** ~ **the bus** etc. *(be in time for, catch)* den Bus *usw.* erreichen *od. (ugs.)* kriegen; *(travel by)* den Bus *usw.* nehmen; **c)** *(prepare)* machen *(ugs.)*, zubereiten ⟨*Essen*⟩; **d)** *(win)* bekommen; finden ⟨*Anerkennung*⟩; erzielen ⟨*Tor, Punkt, Treffer*⟩; gewinnen ⟨*Spiel, Preis, Be-*

lohnung⟩; ~ **permission** die Erlaubnis erhalten; **e)** finden ⟨*Schlaf, Ruhe*⟩; bekommen ⟨*Einfall, Vorstellung, Gefühl, Kopfschmerzen, Grippe*⟩; gewinnen ⟨*Eindruck*⟩; **f) have got** *(coll.: have)* haben; **have got a cold** eine Erkältung haben; **have got to do sth.** etw. tun müssen; **g)** *(succeed in placing, bringing, etc.)* bringen; kriegen *(ugs.);* ~ **a message to sb.** jmdm. eine Nachricht zukommen lassen; ~ **things going** *or* **started** die Dinge in Gang bringen; **h)** ~ **everything packed/prepared** alles [ein]packen/vorbereiten; ~ **sth. ready/done** etw. fertig machen; ~ **one's feet wet** nasse Füße kriegen; ~ **one's hands dirty** sich *(Dat.)* die Hände schmutzig machen; ~ **one's hair cut** sich die Haare schneiden lassen; ~ **sb. to do sth.** *(induce)* jmdn. dazu bringen, etw. zu tun; **i)** ~ **sb. [on the telephone]** jmdn. [telefonisch] erreichen; **j)** *(coll.) (understand)* kapieren *(ugs.);* *(hear)* mitkriegen *(ugs.).* **2.** *v. i.*, -tt-, got, got *or (Amer.)* gotten **a)** *(succeed in coming or going)* kommen; ~ **to London before dark** London vor Einbruch der Dunkelheit erreichen; **b)** *(come to be)* ~ **working** sich an die Arbeit machen; ~ **going** *or* **started** *(leave)* losgehen; *(become lively or operative)* in Schwung kommen; ~ **going on** *or* **with sth.** mit etw. anfangen; **c)** ~ **to know sb.** jmdn. kennenlernen; **d)** *(become)* werden; ~ **ready/washed** sich fertigmachen/waschen; ~ **frightened/hungry** Angst/Hunger kriegen. **get a'bout** *v. i.* **a)** *(travel)* herumkommen; **b)** ⟨*Gerücht:*⟩ sich verbreiten. **'get at** *v. t.* **a)** herankommen an (+ *Akk.*); **b)** *(find out)* [he]rausfinden ⟨*Wahrheit usw.*⟩; **what are you getting at?** worauf wollen Sie hinaus? **get a'way** *v. i.* **a)** *(leave)* wegkommen; **b)** *(escape)* entkommen. **get 'back 1.** *v. i.* zurückkommen; ~ **back home** nach Hause kommen. **2.** *v. t. (recover)* zurückbekommen; ~ **one's own back** *(sl.)* sich rächen. **get 'by** *v. i.* **a)** vorbeikommen; **b)** *(coll.: manage)* über die Runden kommen *(ugs.).* **get 'down 1.** *v. i.* hinunter-/heruntersteigen; ~ **down to sth.** *(start)* sich an etw. *(Akk.)* machen. **2.** *v. t.* **a)** ~ **sb./sth. down** jmdn./etw. hinunter-/herunterbringen; **b)** *(coll.: depress)* fertigmachen *(ugs.).* **get 'in 1.** *v. i. (into bus etc.)* einsteigen; *(arrive)* ankommen. **2.** *v. t. (fetch)* reinholen. **get 'off 1.** *v. i.* **a)** *(alight)* aussteigen; *(dismount)* absteigen; **b)** *(leave)* [weg]gehen; **c)** *(escape punishment)* davonkommen. **2.** *v. t.* **a)** *(remove)* ausziehen ‹*Kleidung usw.*›; entfernen ‹*Fleck usw.*›; abbekommen ‹*Deckel usw.*›; **b)** aussteigen aus; absteigen von ‹*Fahrrad*›; **c)** ~ **off the subject** vom Thema abkommen. **get 'on** *v. i.* **a)** *(mount)* aufsteigen; *(enter vehicle)* einsteigen; **b)** *(make progress)* vorankommen; **he's ~ting on well es** geht ihm gut; **c)** *(manage)* zurechtkommen. **get 'on with** *v. t.* **a)** weitermachen mit; **b)** ~ **on** [well] **with sb.** mit jmdm. [gut] auskommen. **get 'out 1.** *v. i.* **a)** rausgehen/rausfahren (of aus); **b)** *(alight)* aussteigen; **c)** *(escape)* ausbrechen (**of** aus); *(fig.)* herauskommen; ~ **out of** *(avoid)* herumkommen um *(ugs.).* **2.** *v. t.* **a)** *(cause to leave)* rausbringen; **b)** *(withdraw)* abheben ‹*Geld*› (**of** von). **get 'over** *v. t.* **a)** *(cross)* gehen über (+ *Akk.*); *(climb)* klettern über (+ *Akk.*); **b)** *(recover from)* überwinden; hinwegkommen über (+ *Akk.*). **get 'round** *v. i.* ~ **round to doing sth.** dazu kommen, etw. zu tun. **get 'through** *v. i.* durchkommen. **get 'up** *v. i.* aufstehen. **get 'up to** *v. t.* ~ **up to mischief** etwas anstellen

get: ~**away** *n.* Flucht, *die; attrib.* Flucht‹*plan, -wagen*› **make one's ~-away** entkommen; ~**-up** *n. (coll.)* Aufmachung, *die*

geyser ['giːzə(r)] *n.* **a)** *(spring)* Geysir, *der;* **b)** *(Brit.)* Durchlauferhitzer, *der*

ghastly ['gɑːstlɪ] *adj.* grauenvoll; entsetzlich ‹*Verletzungen*›; schrecklich ‹*Fehler*›

gherkin ['gɜːkɪn] *n.* Essiggurke, *die*

ghetto ['getəʊ] *n., pl.* ~s Getto, *das*

ghost [gəʊst] *n.* Geist, *der;* Gespenst, *das.* **'ghostly** *adj.* gespenstisch

giant ['dʒaɪənt] **1.** *n.* Riese, *der.* **2.** *attrib. adj.* riesig

gibberish ['dʒɪbərɪʃ] *n.* Kauderwelsch, *das*

gibe [dʒaɪb] *n.* Stichelei, *die*

giblets ['dʒɪblɪts] *n. pl.* [Geflügel]klein, *das*

giddiness ['gɪdɪnɪs] *n.* Schwindel, *der*

giddy ['gɪdɪ] *adj.* schwind[e]lig

gift [gɪft] *n.* **a)** Geschenk, *das;* **make sb. a ~ of sth., make a ~ of sth. to sb.** jmdm. etw. schenken; **a ~ box/pack** eine Geschenkpackung; **b)** *(talent)* Begabung, *die;* **have a ~ for languages/ mathematics** sprachbegabt/mathema-

tisch begabt sein. **'gifted** *adj.* begabt (**in, at** für). **'gift-wrap** *v. t.* als Geschenk einpacken

gigantic [dʒaɪˈgæntɪk] *adj.* gigantisch; riesig; enorm ‹*Verbesserung, Appetit*›

giggle ['gɪgl] **1.** *n.* Kichern, *das.* **2.** *v. i.* kichern

gild [gɪld] *v. t.* vergolden

gill [gɪl] *n.* Kieme, *die*

gilt [gɪlt] **1.** *n.* Goldauflage, *die; (paint)* Goldfarbe, *die.* **2.** *adj.* vergoldet

gimmick ['gɪmɪk] *n. (coll.)* Gag, *der*

gin [dʒɪn] *n.* Gin, *der*

ginger ['dʒɪndʒə(r)] *n.* **a)** Ingwer, *der;* **b)** *(colour)* Rötlichgelb, *das.* **ginger 'beer** *n.* Ingwerbier, *das.* **'gingerbread** *n.* Pfefferkuchen, *der*

gingerly ['dʒɪndʒəlɪ] *adv.* vorsichtig

gipsy *see* gypsy

giraffe [dʒɪˈrɑːf] *n.* Giraffe, *die*

girder ['gɜːdə(r)] *n.* Träger, *der*

girdle ['gɜːdl] *n.* Hüfthalter, *der*

girl [gɜːl] *n.* Mädchen, *das; (teenager)* junges Mädchen. **'girl-friend** *n.* Freundin, *die.* **'girlish** *adj.* mädchenhaft

giro ['dʒaɪərəʊ] *n.* **a)** Giro, *das; attrib.* Giro-; **bank** ~: Giroverkehr, *der;* **b)** *(coll.: cheque)* Scheck, *der*

girth [gɜːθ] *n.* **a)** Umfang, *der;* **b)** *(for horse)* Bauchgurt, *der*

gismo ['gɪzməʊ] *n. (sl.)* Ding, *das (ugs.)*

gist [dʒɪst] *n.* Wesentliche, *das; (of tale, question, etc.)* Kern, *der*

give [gɪv] **1.** *v. t.,* **gave** [geɪv], **given** ['gɪvn] **a)** geben (**to** *Dat.*); **b)** *(as gift)* schenken; ~ **sb. sth.,** ~ **sth. to sb.** jmdm. etw. schenken; ~ **and take** *(fig.)* Kompromisse eingehen; **c)** *(assign)* aufgeben ‹*Hausaufgaben usw.*›; *(grant, award, offer, allow to have)* geben; verleihen ‹*Preis, Titel usw.*›; lassen ‹*Wahl, Zeit*›; verleihen ‹*Gewicht, Nachdruck*›; bereiten, machen ‹*Freude, Mühe, Kummer*›; bieten ‹*Schutz*›; leisten ‹*Hilfe*›; gewähren ‹*Unterstützung*›; **be ~n sth.** etw. bekommen; ~**n that** *(because)* da; *(if)* wenn; ~ **sb. hope** jmdm. Hoffnung machen; **d)** *(tell)* angeben ‹*Namen, Anschrift, Alter, Grund*›; nennen ‹*Einzelheiten*›; geben ‹*Rat, Befehl, Anweisung, Antwort*›; fällen ‹*Urteil, Entscheidung*›; sagen ‹*Meinung*›; bekanntgeben ‹*Nachricht*›; ~ **him my best wishes** richte ihm meine besten Wünsche aus; **e)** *(perform, sing, etc.)* geben ‹*Vorstellung, Konzert*›; halten

⟨*Vortrag, Seminar*⟩; **f)** *(produce)* geben ⟨*Licht, Milch*⟩; ergeben ⟨*Zahlen, Resultat*⟩; **g)** *(make, show)* geben ⟨*Zeichen, Stoß, Tritt*⟩; machen ⟨*Satz, Ruck*⟩; ausstoßen ⟨*Schrei, Seufzer, Pfiff*⟩; ~ sb. **a [friendly] look** jmdm. einen [freundlichen] Blick zuwerfen; **h)** *(inflict)* versetzen ⟨*Schlag, Stoß*⟩; **sth. ~s me a headache** von etw. bekomme ich Kopfschmerzen; **i)** geben ⟨*Party, Essen usw.*⟩. **2.** *v. i.*, **gave, given** *(yield)* nachgeben; ⟨*Knie:*⟩ weich werden; ⟨*Bett:*⟩ federn. **3.** *n.* Nachgiebigkeit, *die; (elasticity)* Elastizität, *die.* **give a'way** *v. t.* **a)** verschenken; **b)** *(in marriage)* dem Bräutigam zuführen; **c)** *(betray)* verraten. **give 'back** *v. t.* zurückgeben. **give in 1.** ['--] *v. t.* abgeben. **2.** [-'-] *v. i.* nachgeben (**to** *Dat.*). **give 'off** *v. t.* ausströmen ⟨*Geruch*⟩; aussenden ⟨*Strahlen*⟩. **give 'up 1.** *v. i.* aufgeben. **2.** *v. t.* aufgeben; widmen ⟨*Zeit*⟩; ~ **sth. up** *(abandon habit)* sich *(Dat.)* etw. abgewöhnen; ~ **oneself up** sich stellen. **give 'way** *v. i.* **a)** *(yield)* nachgeben; **b)** *(in traffic)* ~ **way [to traffic from the right]** [dem Rechtsverkehr] die Vorfahrt lassen; **'G~ Way'** „Vorfahrt beachten"; **c)** *(collapse)* einstürzen

given *see* **give 1, 2**

gizmo *see* **gismo**

glacier ['glæsɪə(r)] *n.* Gletscher, *der*

glad [glæd] *adj.* froh; **be ~ of sth.** über etw. *(Akk.)* froh sein; für etw. dankbar sein. **gladden** ['glædn] *v. t.* erfreuen

glade [gleɪd] *n.* Lichtung, *die*

'gladly *adv.* gern

glamor *(Amer.) see* **glamour**

glamorous ['glæmərəs] *adj.* glanzvoll; glamourös ⟨*Filmstar*⟩

glamour ['glæmə(r)] *n.* Glanz, *der; (of person)* Ausstrahlung, *die*

glance [glɑːns] **1.** *n.* Blick, *der.* **2.** *v. i.* blicken; ~ **at sb./sth.** jmdn./etw. anblicken; ~ **at one's watch** auf seine Uhr blicken; ~ **at the newspaper** *etc.* einen Blick in die Zeitung *usw.* werfen; ~ **round [the room]** sich [im Zimmer] umsehen

gland [glænd] *n.* Drüse, *die.* **glandular** ['glændjʊlə(r)] *adj.* Drüsen-

glare [gleə(r)] **1.** *n.* **a)** grelles Licht; **b)** *(hostile look)* feindseliger Blick; **with a ~:** feindselig. **2.** *v. i. (glower)* [finster] starren; ~ **at sb./sth.** jmdn./etw. anstarren. **glaring** ['gleərɪŋ] *adj.* grell; *(fig.: conspicuous)* schreiend; grob ⟨*Fehler*⟩; kraß ⟨*Gegensatz*⟩

glass [glɑːs] *n.* **a)** *(substance)* Glas, *das;* **pieces of/broken ~:** Glasscherben *Pl.; (smaller)* Glassplitter *Pl.;* **b)** *(drinking ~)* Glas, *das;* **a ~ of milk** ein Glas Milch; **c)** *(pane)* [Glas]scheibe, *die;* **d)** *in pl. (spectacles)* [a pair of] ~es eine Brille. **'glassy** *adj.* gläsern

glaze [gleɪz] **1.** *n.* Glasur, *die.* **2.** *v. t.* **a)** glasieren; **b)** *(fit with glass)* verglasen. **glazier** ['gleɪzɪə(r)] *n.* Glaser, *der*

gleam [gliːm] **1.** *n.* Schein, *der; (fainter)* Schimmer, *der;* ~ **of hope** Hoffnungsschimmer, *der.* **2.** *v. i.* ⟨*Licht:*⟩ scheinen; ⟨*Fußboden, Stiefel:*⟩ glänzen; ⟨*Zähne:*⟩ blitzen; ⟨*Augen:*⟩ leuchten. **gleaming** *adj.* glänzend

glean [gliːn] *v. t.* zusammentragen ⟨*Informationen usw.*⟩; ~ **sth. from sth.** einer Sache *(Dat.)* etw. entnehmen

glee [gliː] *n.* Freude, *die; (gloating joy)* Schadenfreude, *die.* **gleeful** ['gliːfl] *adj.* freudig; *(gloating)* schadenfroh

glen [glen] *n.* [schmales] Tal

glib [glɪb] *adj.* aalglatt ⟨*Person*⟩; leicht dahingesagt ⟨*Antwort*⟩

glide [glaɪd] *v. i.* gleiten; *(through the air)* schweben. **'glider** *n.* Segelflugzeug, *das*

glimmer ['glɪmə(r)] **1.** *n.* Schimmer, *der* (**of** von); *(of fire)* Glimmen, *das.* **2.** *v. i.* glimmen

glimpse [glɪmps] **1.** *n.* [kurzer] Blick; **catch or have or get a ~ of sb./sth.** jmdn./etw. [kurz] zu sehen bekommen. **2.** *v. t.* flüchtig sehen

glint [glɪnt] **1.** *n.* Schimmer, *der.* **2.** *v. i.* blinken; glitzern

glisten ['glɪsn] *v. i.* glitzern

glitter ['glɪtə(r)] **1.** *v. i.* glitzern; ⟨*Juwelen, Sterne:*⟩ funkeln. **2.** *n.* Glitzern, *das; (of diamonds)* Funkeln, *das*

gloat [gləʊt] *v. i.* ~ **over sth.** sich hämisch über etw. *(Akk.)* freuen

global ['gləʊbl] *adj.* weltweit; ~ **warming** globaler Temperaturanstieg

globe [gləʊb] *n.* **a)** Kugel, *die;* **b)** Globus, *der;* **c)** *(world)* **the ~:** der Globus; der Erdball

gloom [gluːm] *n.* **a)** *(darkness)* Dunkel, *das; (geh.);* **b)** *(despondency)* düstere Stimmung. **'gloomy** *adj.* **a)** düster; finster; **b)** *(depressing)* düster; *(depressed)* trübsinnig ⟨*Person*⟩

glorify ['glɔːrɪfaɪ] *v. t.* verherrlichen; **a glorified messenger-boy** ein besserer Botenjunge

glorious ['glɔːrɪəs] *adj.* **a)** *(illustrious)* ruhmreich ⟨*Held, Sieg*⟩; **b)** *(delightful)* wunderschön; herrlich

glory ['glɔ:rɪ] **1.** *n.* **a)** *(splendour)* Schönheit, *die; (majesty)* Herrlichkeit, *die;* **b)** *(fame)* Ruhm, *der.* **2.** *v. i.* ~ **in sth.** *(be proud of)* sich einer Sache *(Gen.)* rühmen

gloss [glɒs] *n.* Glanz, *der;* ~ **paint** Lackfarbe, *die.* '**gloss over** *v. t.* bemänteln; beschönigen ‹*Fehler*›

glossary ['glɒsərɪ] *n.* Glossar, *das*

'**glossy** *adj.* glänzend

glove [glʌv] *n.* Handschuh, *der.* '**glove compartment** *n.* Handschuhfach, *das*

glow [gləʊ] *v. i.* **a)** glühen; ‹*Lampe, Leuchtfarbe:*› schimmern, leuchten; **b)** *(fig.) (with warmth or pride)* ‹*Gesicht, Wangen:*› glühen (**with** vor + *Dat.*); *(with health or vigour)* strotzen (**with** vor + *Dat.*)

glower ['glaʊə(r)] *v. i.* finster dreinblicken; ~ **at sb.** jmdn. finster anstarren

'**glowing** *adj.* glühend; begeistert ‹*Bericht*›

'**glow-worm** *n.* Glühwürmchen, *das*

glucose ['glu:kəʊz] *n.* Glucose, *die*

glue [glu:] **1.** *n.* Klebstoff, *der.* **2.** *v. t.* kleben; ~ **sth. to sth.** etw. an etw. *(Dat.)* an- *od.* festkleben

glum [glʌm] *adj.* verdrießlich

glut [glʌt] *n.* Überangebot, *das* (**of** an, von + *Dat.*)

glutton ['glʌtən] *n.* Vielfraß, *der (ugs.);* **a** ~ **for punishment** *(iron.)* ein Masochist *(fig.).* **gluttony** ['glʌtənɪ] *n.* Gefräßigkeit, *die*

glycerine ['glɪsəri:n] *(Amer.:* **glycerin** ['glɪsərɪn]) *n.* Glyzerin, *das*

GMT *abbr.* **Greenwich Mean Time** GMT; WEZ

gnarled [nɑ:ld] *adj.* knorrig; knotig ‹*Hand*›

gnash [næʃ] *v. t.* ~ **one's teeth** mit den Zähnen knirschen

gnat [næt] *n.* [Stech]mücke, *die*

gnaw [nɔ:] **1.** *v. i.* ~ [**away**] **at sth.** an etw. *(Dat.)* nagen. **2.** *v. t.* nagen an (+ *Dat.*); abnagen ‹*Knochen*›

gnome [nəʊm] *n.* Gnom, *der*

go [gəʊ] **1.** *v. i., pres.* **he goes** [gəʊz], *p. t.* **went** [went], *pres. p.* **going** ['gəʊɪŋ], *p. p.* **gone** [gɒn] **a)** *(Fahrzeug:)* fahren; ‹*Flugzeug:*› fliegen; ‹*Vierfüßer:*› laufen; *(on horseback etc.)* reiten; *(in lift)* fahren; *(on outward journey)* weg-, abfahren; *(travel regularly)* ‹*Verkehrsmittel:*› verkehren (**from** ... **to** zwischen + *Dat.* ... **und**); **go by bicycle/car/bus/train** *or* **rail/boat** *or* **sea** *or* **ship** mit dem [Fahr]rad/Auto/Bus/ Zug/Schiff fahren; **go by plane** *or* **air** fliegen; **go on foot** zu Fuß gehen; laufen *(ugs.);* **go on a journey** verreisen; **have far to go** es weit haben; **go to the toilet/cinema/a museum** auf die Toilette/ins Kino/ins Museum gehen; **go to the doctor[**'**s]** *etc.* zum Arzt *usw.* gehen; **go bathing** baden gehen; **go cycling** radfahren; **go to see sb.** jmdn. aufsuchen; **go and see whether** ...: nachsehen [gehen], ob ...; **I'll go!** ich geh schon!; *(answer phone)* ich geh ran *od.* nehme ab; *(answer door)* ich mache auf; **b)** *(start)* losgehen; *(in vehicle)* losfahren; **c)** *(pass, circulate)* gehen; **a shiver went up** *or* **down my spine** ein Schauer lief mir über den Rücken; **go to** *(be given to)* ‹*Preis, Gelder, Job:*› gehen an (+ *Akk.*); ‹*Titel, Besitz:*› übergehen auf (+ *Akk.*); **go towards** *(be of benefit to)* zugute kommen (+ *Dat.*); **d)** *(act, function effectively)* gehen; ‹*Mechanismus, Maschine:*› laufen; **keep going** *(in movement)* weitergehen/-fahren; *(in activity)* weitermachen; *(not fail)* sich aufrecht halten; **keep sth. going** etw. in Gang halten; **make sth. go, get/set sth. going** etw. in Gang bringen; **e)** **go to work** zur Arbeit gehen; **go to school** in die Schule gehen; **go to a comprehensive school** auf eine Gesamtschule gehen; **f)** *(depart)* gehen; ‹*Bus, Zug:*› [ab]fahren; ‹*Post:*› rausgehen *(ugs.);* **g)** *(cease to function)* kaputtgehen; ‹*Sicherung:*› durchbrennen; *(break)* brechen; ‹*Seil usw.:*› reißen; **h)** *(disappear)* weggehen; ‹*Mantel, Hut, Fleck:*› verschwinden; ‹*Geruch, Rauch:*› sich verziehen; ‹*Geld, Zeit:*› draufgehen *(ugs.)* (**in, on** für); **i) to go** *(still remaining)* **have sth.** [**still**] **to go** [noch] etw. übrig haben; **one week** *etc.* **to go to** ...: noch eine Woche *usw.* bis ...; **there's hours to go** es dauert noch Stunden; **j)** *(be sold)* weggehen *(ugs.);* verkauft werden; **going! going! gone!** zum ersten! zum zweiten! zum dritten!; **go to sb.** an jmdn. gehen; **k)** *(run)* ‹*Grenze, Straße usw.:*› verlaufen, gehen; *(lead)* gehen; führen; *(extend)* reichen; **as** *or* **so far as he/it goes** soweit; **l)** *(turn out, progress)* ‹*Projekt, Interview, Abend:*› verlaufen; **how did your holiday go?** wie war Ihr Urlaub?; **things have been going well/badly** in der letzten Zeit läuft alles gut/schief; **m)** *(be, have form or nature)* sein;

⟨*Sprichwort, Gedicht, Titel:*⟩ lauten; **that's the way it goes** so ist es nun mal; **go hungry** l.ungern; **go without food/ water** es ohne Essen/Wasser aushalten; **n)** *(become)* werden; **the tyre has gone flat** der Reifen ist platt; **o)** *(have usual place)* kommen; *(belong)* gehören; **where does the box go?** wo kommt *od.* gehört die Kiste hin?; **p)** *(fit)* passen; **go in[to] sth.** in etw. *(Akk.)* gehen *od.* [hinein]passen; **go through sth.** durch etw. [hindurch]gehen; **q)** *(match)* passen (**with** zu); **r)** ⟨*Turmuhr, Gong:*⟩ schlagen; ⟨*Glocke:*⟩ läuten; **s)** *(coll.: be acceptable or permitted)* erlaubt sein; **it/that goes without saying** es/das ist doch selbstverständlich. *See also* **going** 2. **2.** *n., pl.* **goes** [gəʊz] *(coll.)* **a)** *(attempt, try)* Versuch, *der;* *(chance)* Gelegenheit, *die;* **have a go** es versuchen; **let me have a go/can I have a go?** laß mich [auch ein]mal/kann ich [auch ein]mal? *(ugs.);* **it's 'my go** ich bin an der Reihe *od.* dran; **at one go** auf einmal; **at the first go** auf Anhieb; **b)** *(vigorous activity)* **it's all go** es ist alles eine einzige Hetzerei *(ugs.);* **be on the go** auf Trab sein *(ugs.);* **c)** *(success)* **make a go of sth.** mit etw. Erfolg haben. **go a'head** *v.i.* **a)** *(in advance)* vorausgehen (**of** *Dat.*); **b)** *(proceed)* weitermachen; *(make progress)* ⟨*Arbeit:*⟩ fortschreiten, vorangehen. **go a'way** *v.i.* weggehen; *(on holiday or business)* verreisen. **go 'back** *v.i.* zurückgehen/-fahren; *(restart)* ⟨*Schule, Fabrik:*⟩ wieder anfangen; *(fig.)* zurückgehen; **go back to the beginning** noch mal von vorne anfangen. **go by 1.** ['--] *v.t.* **go by sth.** sich nach etw. richten; *(adhere to)* sich an etw. *(Akk.)* halten. **2.** [-'-] *v.i.* ⟨*Zeit:*⟩ vergehen. **go 'down** *v.i.* hinuntergehen/-fahren; ⟨*Sonne:*⟩ untergehen; ⟨*Schiff:*⟩ untergehen; *(fall to ground)* ⟨*Flugzeug usw.:*⟩ abstürzen. **'go for** *v.t.* **go for sb./sth.** *(go to fetch)* jmdn./etw. holen; *(apply to)* für jmdn./etw. gelten; *(like)* jmdn./etw. gut finden. **go 'in** *v.i.* hineingehen; reingehen *(ugs.)*. **go 'off 1.** *v.i.* **a)** **go off with sb./sth.** sich mit jmdm./etw. auf- und davonmachen *(ugs.);* **b)** ⟨*Alarm, Schußwaffe:*⟩ losgehen; ⟨*Wecker:*⟩ klingeln; ⟨*Bombe:*⟩ hochgehen; **c)** *(turn bad)* schlecht werden; **d)** ⟨*Strom:*⟩ ausfallen. **2.** *v.t. (begin to dislike)* **go off sth.** von etw. abkommen. **go 'on** *v.i.* **a)** weitergehen/-fahren; **b)** *(continue)* weiterge-

chen; **c)** *(happen)* passieren. **go 'out** *v.i.* ausgehen; **go out to work/for a meal** arbeiten/essen gehen. **go over 1.** [-'--] *v.i.* hinübergehen. **2.** ['---, -'--] *v.t. (re-examine)* durchgehen. **go 'round** *v.i.* **a)** *(coll.)* **go round and or to see sb.** bei jmdm. vorbeigehen *(ugs.);* **b)** *(look round)* sich umschauen; **c)** *(suffice)* reichen; langen *(ugs.);* **d)** *(spin)* sich drehen. **go through 1.** [-'-] *v.i.* ⟨*Ernennung:*⟩ durchkommen; ⟨*Antrag:*⟩ durchgehen. **2.** ['--] **a)** *(rehearse)* durchgehen; **b)** *(examine)* durchsehen; **c)** *(endure)* durchmachen. **go 'under** *v.i.* untergehen; *(fig.: fail)* eingehen. **go 'up** *v.i.* **a)** hinaufgehen/-fahren; ⟨*Ballon:*⟩ aufsteigen; *(Theatre)* ⟨*Vorhang:*⟩ aufgehen; ⟨*Lichter:*⟩ angehen; **b)** *(increase)* ⟨*Zahl:*⟩ wachsen; ⟨*Preis, Wert, Niveau:*⟩ steigen; *(in price)* ⟨*Ware:*⟩ teurer werden. **go without 1.** ['---] *v.t.* verzichten auf (+ *Akk.*). **2.** [-'-] *v.i.* verzichten

goad [gəʊd] *v.t.* ~ **sb. into sth./doing sth.** jmdn. zu etw. anstacheln/dazu anstacheln, etw. zu tun

'go-ahead 1. *adj.* unternehmungslustig; *(progressive)* fortschrittlich. **2.** *n.* **give sb./sth. the ~:** jmdm./einer Sache grünes Licht geben

goal [gəʊl] *n.* **a)** *(aim)* Ziel, *das;* **b)** *(Footb., Hockey)* Tor, *das;* **score/kick a ~:** einen Treffer erzielen. **'goalkeeper** *n.* Torwart, *der*

goat [gəʊt] *n.* Ziege, *die*

gobble ['gɒbl] **1.** *v.t.* ~ **[down** *or* **up]** hinunterschlingen. **2.** *v.i.* schlingen

'go-between *n.* Vermittler, *der*/Vermittlerin, *die*

goblet ['gɒblɪt] *n.* Kelchglas, *das*

goblin ['gɒblɪn] *n.* Kobold, *der*

god [gɒd] *n.* **a)** Gott, *der;* **b)** God *(Theol.)* Gott. **'godchild** *n.* Patenkind, *das.* **'god-daughter** *n.* Patentochter, *die*

goddess ['gɒdɪs] *n.* Göttin, *die*

god: ~father *n.* Pate, *der;* **G~-forsaken** *adj.* gottverlassen; **~mother** *n.* Patentante, *die;* **~send** *n.* Gottesgabe, *die;* **be a ~send to sb.** für jmdn. ein Geschenk des Himmels sein; **~son** *n.* Patensohn, *der*

goggles ['gɒglz] *n. pl.* Schutzbrille, *die*

going ['gəʊɪŋ] **1.** *n.* *(progress)* Vorankommen, *das;* **while the ~ is good** solange es noch geht. **2.** *adj.* **a)** *(available)* erhältlich; **there is sth. ~:** es gibt etw.; **b)** **be ~ to do sth.** etw. tun [werden/wollen]; **I was ~ to say** ich wollte sagen; **it's ~ to snow** es wird schneien; **a ~ concern** eine gesunde Firma

goings-'on *n. pl.* Ereignisse

gold [gəʊld] **1.** *n.* Gold, *das.* **2.** *attrib. adj.* golden. Gold⟨*münze, -kette usw.*⟩

golden ['gəʊldn] *adj.* golden. **golden 'wedding** *n.* goldene Hochzeit

gold: ~**fish** *n.* Goldfisch, *der;* ~ **'medal** *n.* Goldmedaille, *die;* ~**-mine** *n.* Goldmine, *die; (fig.)* Goldgrube, *die;* ~-**'plated** *adj.* vergoldet; ~**smith** *n.* Goldschmied, *der/*-schmiedin, *die*

golf [gɒlf] *n.* Golf, *das*

golf: ~ **ball** *n.* Golfball, *der;* ~-**club** *n.* **a)** *(implement)* Golfschläger, *der;* **b)** *(association)* Golfclub, *der;* ~-**course** *n.* Golfplatz, *der*

'golfer *n.* Golfer, *der/*Golferin, *die*

gondola ['gɒndələ] *n.* Gondel, *die*

gone [gɒn] **1.** *see* **go 1. 2.** *pred. adj.* **a)** *(away)* weg; **it's time you were ~:** es ist *od.* wird Zeit, daß du gehst; **b)** *(of time: after)* nach; **it's ~ ten o'clock** es ist zehn Uhr vorbei

gong [gɒn] *n.* Gong, *der*

good [gʊd] **1.** *adj.,* **better** ['betə(r)], **best** [best] **a)** gut; günstig ⟨*Gelegenheit, Angebot*⟩; ausreichend ⟨*Vorrat*⟩; ausgiebig ⟨*Mahl*⟩; **as ~ as** so gut wie; **his ~ eye/leg** sein gesundes Auge/Bein; **in ~ time** frühzeitig; **all in ~ time** alles zu seiner Zeit; **be ~ at sth.** in etw. *(Dat.)* gut sein; **too ~ to be true** zu schön, um wahr zu sein; **apples are ~ for you** Äpfel sind gesund; **be too much of a ~ thing** zuviel des Guten sein; **~ times** eine schöne Zeit; **feel ~:** sich wohl fühlen; **take a ~ look round** sich gründlich umsehen; **give sb. a ~ beating/scolding** jmdn. tüchtig verprügeln/ausschimpfen; **~ afternoon/day** guten Tag!; **~ evening/morning** guten Abend/Morgen!; **~ night** gute Nacht!; **b)** *(enjoyable)* schön ⟨*Leben, Urlaub, Wochenende*⟩; **the ~ life** das angenehme[, sorglose] Leben; **have a ~ time!** viel Spaß!; **have a ~ journey!** gute Reise!; **c)** *(well-behaved)* gut; brav; **be ~!,** **be a ~ girl/boy!** sei brav *od.* lieb!; **[as] ~ as gold** ganz artig *od.* brav; **d)** *(virtuous)* rechtschaffen; *(kind)* nett; gut ⟨*Absicht, Wünsche, Benehmen, Tat*⟩; **be ~ to sb.** gut zu jmdm. sein; **would you be so ~ as to** *od.* **~ enough to do that?** wären Sie so freundlich *od.* nett, das zu tun'?; **that/it is ~ of you** das/es ist nett *od.* lieb

von dir; **e)** *(commendable)* gut; **~ for 'you** *etc. (coll.)* bravo!; **f)** *(attractive)* schön; gut ⟨*Figur*⟩; **look ~:** gut aussehen; **g)** *(considerable)* [recht] ansehnlich ⟨*Menschenmenge*⟩; ganz schön, ziemlich *(ugs.)* ⟨*Entfernung, Strecke*⟩; gut ⟨*Preis, Erlös*⟩; **h) make ~** *(succeed)* erfolgreich sein; *(compensate for)* wiedergutmachen; *(indemnify)* ersetzen. **2.** *n.* **a)** *(use)* Nutzen, *der;* **be some ~ to sb./sth.** jmdm./einer Sache nützen; **be no ~ to sb./sth.** für jmdn./etw. nicht zu gebrauchen sein; **it is no/not much ~ doing sth.** es hat keinen/kaum einen Sinn, etw. zu tun; **what's the ~ of ...?, what ~ is ...?** was nützt ...?; **b)** *(benefit)* **for your/his** *etc.* **own ~:** zu deinem/seinem *usw.* Besten; **do no/little ~:** nichts/wenig helfen *od.* nützen; **do sb./sth. ~:** jmdm./einer Sache nützen; ⟨*Ruhe, Erholung:*⟩ jmdm./einer Sache guttun; ⟨*Arznei:*⟩ jmdm./einer Sache helfen; **c)** *(goodness)* Gute, *das;* **be up to no ~:** nichts Gutes im Sinn haben; **d) for ~** *(finally)* ein für allemal; *(permanently)* für immer; **e)** *in pl. (wares etc.)* Waren; *(belongings)* Habe, *die;* *(Brit. Railw.)* Fracht, *die; attrib.* Güter⟨*wagen, -zug*⟩

good: ~-**'bye** *(Amer.:* ~-**'by)** *int.* auf Wiedersehen!; *(on telephone)* auf Wiederhören!; ~-**for-nothing 1.** *adj.* nichtsnutzig; **2.** *n.* Taugenichts, *der;* ~-**'looking** *adj.* gutaussehend

'goodness 1. *n.* Güte, *die.* **2.** *int.* [my] **~!** meine Güte! *(ugs.)*

good'will *n.* guter Wille; *attrib.* Goodwill⟨*botschaft, -reise usw.*⟩

'goody *n. (coll.: hero)* Gute, *der/die*

gooey ['gu:ɪ] *adj.,* **gooier** ['gu:ɪə(r)], **gooiest** ['gu:ɪɪst] *(coll.)* klebrig

goose [gu:s] *n., pl.* **geese** [gi:s] Gans, *die*

gooseberry ['gʊzbərɪ] *n.* Stachelbeere, *die*

'goose: ~-**pimples** *n. pl.* **have ~-pimples** eine Gänsehaut haben

¹gore [gɔ:(r)] *v. t.* [mit den Hörnern] aufspießen *od.* durchbohren

²gore *n.* Blut, *das*

gorge [gɔ:dʒ] **1.** *n.* Schlucht, *die.* **2.** *v. i. & refl.* ~ [oneself] sich vollstopfen *(ugs.)* (on mit)

gorgeous ['gɔ:dʒəs] *adj.* prächtig; hinreißend ⟨*Frau, Mann, Lächeln*⟩

gorilla [gə'rɪlə] *n.* Gorilla, *der*

gormless ['gɔ:mlɪs] *adj. (Brit. coll.)* dämlich *(ugs.)*

gorse [gɔ:s] *n.* Stechginster, *der*

gory ['gɔ:rɪ] *adj. (fig.)* blutrünstig
gosh [gɒʃ] *int. (coll.)* Gott!
'go-slow *n. (Brit.)* Bummelstreik, *der*
gospel ['gɒspl] *n.* Evangelium, *das*
gossamer ['gɒsəmə(r)] *n.* Altweibersommer, *der; attrib.* hauchdünn
gossip ['gɒsɪp] **1.** *n.* **a)** *(talk)* Klatsch, *der (ugs.);* **b)** *(person)* Klatschbase, *die (ugs.).* **2.** *v. i.* klatschen *(ugs.)*
got *see* get
Gothic ['gɒθɪk] *adj.* gotisch
gotten *see* get
gouge [gaʊdʒ] *v. t.* aushöhlen
goulash ['gu:læʃ] *n.* Gulasch, *das od. der*
gourmet ['gʊəmeɪ] *n.* Gourmet, *der*
gout [gaʊt] *n.* Gicht, *die*
govern ['gʌvn] **1.** *v. t.* **a)** regieren ⟨*Land, Volk*⟩; verwalten ⟨*Provinz*⟩; **b)** *(dictate)* bestimmen. **2.** *v. i.* regieren
governess ['gʌvənɪs] *n.* Gouvernante, *die (veraltet);* Hauslehrerin, *die*
government ['gʌvnmənt] *n.* Regierung, *die; attrib.*
governor ['gʌvənə(r)] *n.* **a)** *(of province etc.)* Gouverneur, *der;* **b)** *(of institution)* Direktor, *der*/Direktorin, *die;* |board of| ~s Vorstand, *der;* **c)** *(sl.: employer)* Boß, *der (ugs.)*
gown [gaʊn] *n.* **a)** [elegantes] Kleid; **b)** *(official or uniform robe)* Talar, *der*
GP *abbr.* general practitioner
grab [græb] **1.** *v. t.,* -bb- greifen nach; *(seize)* packen; ~ the chance die Gelegenheit ergreifen; ~ hold of sb./sth. sich *(Dat.)* jmdn./etw. schnappen *(ugs.).* **2.** *v. i.,* -bb-: ~ at sth. nach etw. greifen. **3.** *n.* make a ~ at *or* for sb./sth. nach jmdm./etw. greifen
grace [greɪs] *n.* **a)** *(charm)* Anmut, *die (geh.);* **b)** *(decency)* have the ~ to do sth. so anständig sein und etw. tun; **c)** *(delay)* Frist, *die;* give sb. a day's ~: jmdm. einen Tag Aufschub gewähren; **d)** *(prayers)* say ~: das Tischgebet sprechen. graceful ['greɪsfl] *adj.* elegant; graziös ⟨*Bewegung, Eleganz*⟩
gracious ['greɪʃəs] **1.** *adj.* **a)** liebenswürdig; **b)** *(merciful)* gnädig. **2.** *int.* good ~! [ach] du meine Güte!
grade [greɪd] **1.** *n.* **a)** Rang, *der; (Mil.)* Dienstgrad, *der;* **b)** *(position)* Stufe, *die;* **c)** *(Amer. Sch.: class)* Klasse, *die;* **d)** *(Sch., Univ.: mark)* Note, *die;* Zensur, *die.* **2.** *v. t.* **a)** einstufen ⟨*Schüler*⟩; [nach Größe/Qualität] sortieren ⟨*Eier, Kartoffeln*⟩; **b)** *(mark)* benoten
gradient ['greɪdɪənt] *n. (ascent)* Steigung, *die; (descent)* Gefälle, *das*

gradual ['grædʒʊəl] *adj.,* 'gradually *adv.* allmählich
graduate **1.** ['grædʒʊət] *n.* Graduierte, *der*/die; *(who has left university)* Akademiker, *der*/Akademikerin, *die;* university ~: Hochschulabsolvent, *der*/-absolventin, *die.* **2.** ['grædʒʊeɪt] *v. i.* einen akademischen Grad/Titel erwerben; *(Amer. Sch.)* die [Schul]abschlußprüfung bestehen (from an + *Dat.*)
graffiti [grəˈfiːtiː] *n. sing. or pl.* Graffiti *Pl.*
graft [grɑːft] **1.** *n.* **a)** *(Bot.)* Edelreis, *das;* **b)** *(Med.) (operation)* Transplantation, *die; (thing* – *ed)* Transplantat, *das;* **c)** *(Brit. sl.: work)* Plackerei, *die (ugs.).* **2.** *v. t.* **a)** *(Bot.)* pfropfen; **b)** *(Med.)* transplantieren. **3.** *v. i. (Brit. sl.)* schuften *(ugs.)*
grain [greɪn] *n.* **a)** Korn, *das; collect.* Getreide, *das;* **b)** *(particle)* Korn, *das;* **c)** *(in wood)* Maserung, *die; (in paper)* Faser, *die; (in leather)* Narbung, *die;* go against the ~ |for sb.| *(fig.)* jmdm. gegen den Strich gehen *(ugs.).* 'grainy *adj.* körnig; gemasert ⟨*Holz*⟩; genarbt ⟨*Leder*⟩
gram [græm] *n.* Gramm, *das*
grammar ['græmə(r)] *n.* Grammatik, *die.* 'grammar book *n.* Grammatik, *die.* 'grammar school *n. (Brit.)* ≈ Gymnasium, *das*
grammatical [grəˈmætɪkl] *adj.* **a)** grammat[ikal]isch richtig *od.* korrekt; **b)** *(of grammar)* grammatisch. grammatically [grəˈmætɪkəlɪ] *adv.* grammati[kal]isch ⟨*richtig, falsch*⟩
gramme *see* gram
gramophone ['græməfəʊn] *n.* Plattenspieler, *der*
granary ['grænərɪ] *n.* Getreidesilo, *der od. das;* Kornspeicher, *der*
grand [grænd] *adj.* **a)** *(most or very important)* groß; ~ finale großes Finale; **b)** *(splendid)* grandios; **c)** *(coll.: excellent)* großartig
grand: ~child *n.* Enkel, *der*/Enkelin, *die;* Enkelkind, *das;* ~dad[dy] ['grændæd(ɪ)] *n. (coll./child lang.);* Opa, *der (Kinderspr./ugs.);* ~daughter *n.* Enkelin, *die*
grandeur ['grændʒə(r), 'grændjə(r)] *n.* Erhabenheit, *die*
'grandfather *n.* Großvater, *der;* ~ clock Standuhr, *die*
grandiose ['grændɪəʊs] *adj.* grandios; *(pompous)* bombastisch
grand: ~ma *n. (coll./child lang.)* Oma,

die (Kinderspr./ugs.); ~**mother** *n.*
Großmutter, *die;* ~**pa** *n. (coll./child
lang.)* Opa, *der (Kinderspr./ugs.);*
~**parent** *n. (male)* Großvater, *der;
(female)* Großmutter, *die;* ~**parents**
Großeltern *Pl.;* ~ **pi'ano** *n.* [Kon-
zert]flügel, *der;* ~**son** *n.* Enkel, *der;*
~**stand** *n.* [Haupt]tribüne, *die*

granite ['grænɪt] *n.* Granit, *der*

granny ['grænɪ] *n. (coll./child lang.)*
Oma, *die (Kinderspr./ugs.)*

grant [grɑːnt] **1.** *v.t.* **a)** erfüllen
⟨*Wunsch*⟩; stattgeben (+ *Dat.*) ⟨*Gesuch*⟩; **b)** *(concede, give)* gewähren;
geben ⟨*Zeit*⟩; bewilligen ⟨*Geldmittel*⟩;
zugestehen ⟨*Recht*⟩; erteilen ⟨*Erlaubnis*⟩; **c)** *(in argument)* zugeben; **take
sb./sth. for** ~**ed** sich *(Dat.)* jmds. sicher sein/etw. für selbstverständlich
halten. **2.** *n.* Zuschuß, *der; (financial
aid [to student])* [Studien]beihilfe, *die;
(scholarship)* Stipendium, *das*

granulated sugar [grænjʊleɪtɪd 'ʃʊg-
ə(r)] *n.* Kristallzucker, *der*

granule ['grænjuːl] *n.* Körnchen, *das*

grape [greɪp] *n.* Weintraube, *die;* **a
bunch of** ~**s** eine Traube

grapefruit *n., pl. same* Grapefruit,
die

graph [grɑːf] *n.* graphische Darstellung; ~ **paper** Diagrammpapier, *das*

graphic ['græfɪk] *adj.* **a)** graphisch; **b)**
(vivid) plastisch; anschaulich. **graphically** ['græfɪkəlɪ] *adv.* **a)** *(vividly)* plastisch; **b)** *(using graphics)* graphisch.
graphics ['græfɪks] *n. (use of dia-
grams)* graphische Darstellung; **computer** ~: Computergraphik, *die*

grapple ['græpl] *v.i.* handgemein werden; ~ **with** *(fig.)* sich auseinandersetz-
zen mit

grasp [grɑːsp] **1.** *v.i.* ~ **at** ergreifen;
sich stürzen auf (+ *Akk.*) ⟨*Angebot*⟩.
2. *v.t.* **a)** *(seize)* ergreifen; **b)** *(hold
firmly)* festhalten; **c)** *(understand)* verstehen; erfassen ⟨*Bedeutung*⟩. **3.** *n.* **a)**
(firm hold) Griff, *der;* **b)** *(mental* ~*)*
have a good ~ **of sth.** etw. gut beherrschen. **grasping** *adj.* habgierig

grass [grɑːs] *n.* **a)** Gras, *das;* **b)** *(lawn)*
Rasen, *der;* **c)** *(Brit. sl.: police informer)* Spitzel, *der.* **grasshopper**
n. Grashüpfer, *der.* **grass-root[s]**
attrib. adj. (Polit.) Basis-

grate [greɪt] *n.* Rost, *der; (recess)* Kamin, *der*

grate *v.t.* **a)** reiben; *(less finely)* ras-
peln; **b)** *(grind)* ~ **one's teeth** mit den
Zähnen knirschen

grateful ['greɪtfl] *adj.* dankbar (**to**
Dat.). **gratefully** *adv.* dankbar

grater *n.* Reibe, *die;* Raspel, *die*

gratify ['grætɪfaɪ] *v.t.* freuen; **be grati-
fied by** *or* **with** *or* **at sth.** über etw.
(Akk.) erfreut sein. **gratifying** *adj.*
erfreulich

grating ['greɪtɪŋ] *n.* Gitter, *das*

gratitude ['grætɪtjuːd] *n.* Dankbarkeit, *die* (**to** gegenüber)

gratuitous [grə'tjuːɪtəs] *adj. (motive-
less)* grundlos

gratuity [grə'tjuːɪtɪ] *n.* Trinkgeld, *das*

grave [greɪv] *n.* Grab, *das*

grave *adj.* **a)** *(important, solemn)*
ernst; **b)** *(serious)* schwer ⟨*Fehler, Irrtum*⟩; ernst ⟨*Situation, Lage*⟩; groß
⟨*Gefahr*⟩; schlimm ⟨*Nachricht*⟩

grave-digger *n.* Totengräber, *der*

gravel ['grævl] *n.* Kies, *der*

grave: ~**stone** *n.* Grabstein, *der;*
~**yard** *n.* Friedhof, *der*

gravity ['grævɪtɪ] *n.* **a)** *(of mistake, offence)* Schwere, *die; (of situation)*
Ernst, *der;* **b)** *(Phys., Astron.)* Gravitation, *die;* Schwerkraft, *die*

gravy ['greɪvɪ] *n.* **a)** *(juices)* Bratensaft,
der; **b)** *(dressing)* [Braten]soße, *die*

gray *etc. (Amer.) see* **grey** *etc.*

graze [greɪz] *v.i.* grasen; weiden

graze *n.* **1.** *n.* Schürfwunde, *die.* **2.** *v.t.*
a) *(touch lightly)* streifen; **b)** *(scrape)*
abschürfen ⟨*Haut*⟩; zerkratzen ⟨*Oberfläche*⟩

grease [griːs] **1.** *n.* Fett, *das; (lubricant)* Schmierfett, *das.* **2.** *v.t.* einfet-
ten; *(lubricate)* schmieren. **grease-
proof** *adj.* fettdicht; ~ **paper** Perga-
ment- *od.* Butterbrotpapier, *das*

greasy ['griːsɪ] *adj.* fettig; fett ⟨*Essen*⟩;
(lubricated) geschmiert; *(dirty with
lubricant)* schmierig

great [greɪt] *adj.* **a)** groß; **a** ~ **many**
sehr viele; sehr gut ⟨*Freund*⟩; *(impressive; coll.: splendid)* großartig; **be
a** ~ **one for sth.** etw. sehr gern tun; **b)**
Groß⟨*onkel, -tante, -neffe, -nichte*⟩;
Ur⟨*großmutter, -großvater, -enkel, -en-
kelin*⟩. **Great 'Britain** *pr. n.* Großbritannien *(das).* **greatly** *adv.* sehr;
höchst ⟨*verärgert*⟩; stark ⟨*beeinflußt*⟩;
bedeutend ⟨*verbessert*⟩. **greatness**
n. Größe, *die*

Greece [griːs] *pr. n.* Griechenland
(das)

greed [griːd] *n.* Gier, *die* (**for** nach);
(gluttony) Gefräßigkeit, *die.* **greedy**
adj. gierig; *(gluttonous)* gefräßig

Greek [griːk] **1.** *adj.* griechisch; *sb.* **is**

~: jmd. ist Grieche/Griechin. **2.** *n.* **a)** *(person)* Grieche, *der*/Griechin, *die;* **b)** *(language)* Griechisch, *das; see also* **English 2 a**

green [gri:n] **1.** *adj.* **a)** grün; **b)** *(environmentally safe)* ökologisch; .**c)** *(gullible)* naiv; *(inexperienced)* grün; **d)** *(Polit.)* G~: grün; **the G~s** die Grünen. **2.** *n.* **a)** *(colour)* Grün, *das;* **b)** *(piece of land)* Grünfläche, *die;* **village** ~: Dorfanger, *der;* **c)** *in pl.* (~ *vegetables)* Grüngemüse, *das.* '**green belt** *n.* Grüngürtel, *der.* **green 'card** *n.* *(Motor Veh.)* grüne Karte

greenery ['gri:nərı] *n.* Grün, *das*

green: ~**fly** *n.* *(Brit.)* grüne Blattlaus; ~**gage** ['gri:ngeɪdʒ] *n.* Reineclaude, *die;* ~**grocer** *n.* *(Brit.)* Obst- und Gemüsehändler, *der*/-händlerin, *die;* ~**house** *n.* Gewächshaus, *das;* ~**house effect** Treibhauseffekt, *der*

Greenland ['gri:nlənd] *pr. n.* Grönland *(das)*

'**Green Party** *n.* *(Polit.)* die Grünen

greet [gri:t] *v. t.* begrüßen; *(in passing)* grüßen; *(receive)* empfangen. '**greeting** *n.* Begrüßung, *die;* *(in passing)* Gruß, *der;* *(words)* Grußformel, *die.* '**greetings card** *n.* Grußkarte, *die;* *(for birthday)* Glückwunschkarte, *die*

gregarious [grɪ'geərɪəs] *adj.* gesellig

grenade [grɪ'neɪd] *n.* Granate, *die*

grew *see* **grow**

grey [greɪ] **1.** *adj.* grau. **2.** *n.* Grau, *das.* '**greyhound** *n.* Windhund, *der*

grid [grɪd] *n.* **a)** *(grating)* Rost, *der;* **b)** *(of lines)* Gitter[netz], *das;* **c)** *(for supply)* Versorgungsnetz, *das*

grief [gri:f] *n.* Kummer, *der* **(over, at** über + *Akk.,* **um);** *(at loss of sb.)* Trauer, *die* **(for** um); **come to** ~ *(fail)* scheitern

grievance ['gri:vəns] *n.* *(complaint)* Beschwerde, *die;* *(grudge)* Groll, *der*

grieve [gri:v] **1.** *v. t.* betrüben; bekümmern. **2.** *v. i.* trauern **(for** um)

grievous ['gri:vəs] *adj.* schwer ‹*Verwundung, Krankheit*›

'**grill** [grɪl] **1.** *v. t.* *(cook)* grillen; *(fig.: question)* in die Mangel nehmen *(ugs.).* **2.** *n.* **a)** **mixed** ~: gemischte Grillplatte; **b)** *(on cooker)* Grill, *der*

grille (²**grill**) *n.* **a)** Gitter, *das;* **b)** *(Motor Veh.)* [Kühler]grill, *der*

grim [grɪm] *adj.* *(stern)* streng; grimmig ‹*Lächeln, Schweigen*›; *(unrelenting)* erbittert ‹*Widerstand, Kampf*›; *(ghastly)* grauenvoll ‹*Aufgabe, Nachricht*›; trostlos ‹*Aussichten*›

grimace [grɪ'meɪs] **1.** *n.* Grimasse, *die.* **2.** *v. i.* Grimassen schneiden; ~ **with pain** vor Schmerz das Gesicht verziehen

grime [graɪm] *n.* Schmutz, *der.* **grimy** ['graɪmɪ] *adj.* schmutzig

grin [grɪn] **1.** *n.* Grinsen, *das.* **2.** *v. i.,* **-nn-** grinsen; ~ **at sb.** jmdn. angrinsen

grind [graɪnd] **1.** *v. t.,* **ground** [graʊnd] **a)** ~ [up] zermahlen; mahlen ‹*Kaffee, Pfeffer, Getreide*›; **b)** *(sharpen)* schleifen ‹*Schere, Messer*›; schärfen ‹*Klinge*›; **c)** *(rub harshly)* zerquetschen; ~ **one's teeth** mit den Zähnen knirschen. **2.** *v. i.,* **ground:** ~ **to a halt** *(Fahrzeug:)* quietschend zum Stehen kommen; *(fig.)* ‹*Verkehr:*› zum Erliegen kommen. **3.** *n.* *(coll.)* Plackerei, *die* *(ugs.).* '**grinder** *n.* Schleifmaschine, *die;* *(coffee-~ etc.)* Mühle, *die.* '**grindstone** *n.* Schleifstein, *der*

grip [grɪp] **1.** *n.* **a)** *(firm hold)* Halt, *der;* *(fig.: power)* Umklammerung, *die;* **have a** ~ **on sth.** etw. festhalten; *(fig.)* etwas im Griff haben; **loosen one's** ~: loslassen; **lose one's** ~ *(fig.)* nachlassen; **b)** *(strength or way of* ~*ping)* Griff, *der.* **2.** *v. t.,* **-pp-** [fest] halten; ‹*Reifen:*› greifen; *(fig.)* fesseln ‹*Publikum, Aufmerksamkeit*›. **3.** *v. i.,* **-pp-** ‹*Räder, Bremsen usw.:*› greifen

gripe [graɪp] *v. i.* *(sl.)* meckern *(ugs.)* **(about** über + *Akk.*)

gripping ['grɪpɪŋ] *adj.* *(fig.)* packend

grisly ['grɪzlɪ] *adj.* grausig

gristle ['grɪsl] *n.* Knorpel, *der*

grit [grɪt] *n.* **a)** Sand, *der;* **b)** *(coll.: courage)* Schneid, *der* *(ugs.).* **2.** *v. t.,* **-tt-:** **a)** streuen ‹*Straßen*›; **b)** ~ **one's teeth** die Zähne zusammenbeißen *(ugs.)*

groan [grəʊn] **1.** *n.* Stöhnen, *das;* *(of thing)* Ächzen, *das.* **2.** *v. i.* [auf]stöhnen **(at** bei); ‹*Tisch, Planken:*› ächzen. **3.** *v. t.* stöhnen

grocer ['grəʊsə(r)] *n.* Lebensmittelhändler, *der*/-händlerin, *die.* **grocery** ['grəʊsərɪ] *n.* **a)** *in pl.* *(goods)* Lebensmittel *Pl.;* **b)** ~ **[store]** Lebensmittelgeschäft, *das*

groggy ['grɒgɪ] *adj.* groggy *präd.* *(ugs.)*

groin [grɔɪn] *n.* Leistengegend, *die*

groom [gru:m, grʊm] **1.** *n.* **a)** *(stableboy)* Stallbursche, *der;* **b)** *(bride~)* Bräutigam, *der.* **2.** *v. t.* striegeln ‹*Pferd*›; *(fig.)* vorbereiten **(for auf** + *Akk.*)

groove [gru:v] *n.* Rille, *die*

grope [grəʊp] *v. i.* tasten **(for** nach)

¹**gross** [grəʊs] *adj.* **a)** *(flagrant)* grob ⟨*Fahrlässigkeit, Fehler*⟩; **b)** *(obese)* fett; **c)** *(total)* Brutto-

²**gross** *n., pl. same* Gros, *das*

¹**grossly** *adj.* *(flagrantly)* äußerst; grob ⟨*übertreiben*⟩

grotesque [grəʊ'tesk] *adj.* grotesk

grotto ['grɒtəʊ] *n., pl.* ~es *or* ~s Grotte, *die*

grotty ['grɒtɪ] *adj. (Brit. sl.)* mies *(ugs.)*

¹**ground** [graʊnd] **1.** *n.* **a)** Boden, *der;* **get off the** ~ *(coll.)* konkrete Gestalt annehmen; **b)** |sports| ~: Sportplatz, *der;* **c)** *in pl. (attached to house)* Anlage, *die;* **d)** *(reason)* Grund, *der;* **on the** ~|s| **of** auf Grund (+ *Gen.*); **on the** ~|s| **that** ...: unter Berufung auf die Tatsache, daß ...; **e)** *in pl. (sediment)* Satz, *der.* **2.** *v. t. (Aeronaut.)* am Boden festhalten

²**ground** **1.** *see* **grind** 1, 2. **2.** *adj.* gemahlen ⟨*Kaffee, Getreide*⟩

ground 'floor *see* **floor** 1 b

'**grounding** *n.* Grundkenntnisse *Pl.*

'**groundless** *adj.* unbegründet

ground: ~**sheet** *n.* Bodenplane, *die;* ~**sman** ['graʊndzmən] *n., pl.* -**smen** ['graʊndzmən] *(Sport)* Platzwart, *der;* ~**work** *n.* Vorarbeiten *Pl.*

group [gru:p] **1.** *n.* Gruppe, *die.* **2.** *v. t.* gruppieren

¹**grouse** [graʊs] *n., pl. same* Rauhfußhuhn, *das;* |red| ~ *(Brit.)* Schottisches Moorschneehuhn

²**grouse** *v. i. (coll.)* meckern *(ugs.)*

grove [grəʊv] *n.* Wäldchen, *das*

grovel ['grɒvl] *v. i., (Brit.)* -ll- *(fig.)* katzbuckeln

grow [grəʊ] **1.** *v. i.,* grew [gru:], grown [grəʊn] **a)** wachsen; ~ **out of** *or* **from** sth. sich aus etw. entwickeln; *(from sth. abstract)* von etw. herrühren; ~ **in** gewinnen an (+ *Dat.*) ⟨*Größe, Bedeutung*⟩; **b)** *(become)* werden; ~ **apart** *(fig.)* sich auseinanderleben; ~ **to love/hate** sb./sth. jmdn./etw. liebenlernen/hassenlernen; ~ **to like** sb./sth. nach und nach Gefallen an jmdm./etw. finden. **2.** *v. t.,* grew, grown ziehen; *(on a large scale)* anpflanzen; züchten ⟨*Blumen*⟩. **grow 'up** *v. i.* **a)** aufwachsen; *(become adult)* erwachsen werden; **b)** ⟨*Legende:*⟩ entstehen

growl [graʊl] **1.** *n.* Knurren, *das; (of bear)* Brummen, *das.* **2.** *v. i.* knurren; ⟨*Bär:*⟩ [böse] brummen

grown [grəʊn] **1.** *see* **grow.** **2.** *adj.* erwachsen. '**grown-up** **1.** *n.* Erwachsene, *der/die.* **2.** *adj.* erwachsen

growth [grəʊθ] *n.* **a)** Wachstum, *das* (of, in *Gen.*); *(increase)* Zunahme, *die* (of, in *Gen.*); **b)** *(Med.)* Gewächs, *das*

grub [grʌb] *n.* **a)** Larve, *die; (maggot)* Made, *die;* **b)** *(sl.: food)* Fressen, *das (salopp)*

grubby ['grʌbɪ] *adj.* schmudd[e]lig *(ugs.)*

grudge [grʌdʒ] **1.** *v. t.* ~ sb. sth. jmdm. etw. mißgönnen; ~ **doing** sth. etw. ungern tun. **2.** *n.* Groll, *der;* **bear** sb. **a.** ~ *or* **a.** ~ **against** sb. jmdm. gegenüber nachtragend sein. **grudging** ['grʌdʒɪŋ] *adj.* widerwillig; widerwillig gewährt ⟨*Zuschuß*⟩. '**grudgingly** *adv.* wiederwillig

gruelling *(Amer.:* **grueling**) ['gru:əlɪŋ] *adj.* aufreibend; strapaziös ⟨*Reise*⟩

gruesome ['gru:səm] *adj.* grausig

gruff [grʌf] *adj.* barsch; rauh ⟨*Stimme*⟩

grumble ['grʌmbl] *v. i.* murren; ~ **about** *or* **over** sth. sich über etw. *(Akk.)* beklagen

grumpy ['grʌmpɪ] *adj.* unleidlich

grunt [grʌnt] **1.** *n.* Grunzen, *das.* **2.** *v. i.* grunzen

guarantee [gærən'ti:] **1.** *v. t.* **a)** garantieren für; [eine] Garantie geben auf (+ *Akk.*); **the clock is** ~**d for a year** die Uhr hat ein Jahr Garantie; **b)** *(promise)* garantieren *(ugs.); (ensure)* bürgen für ⟨*Qualität*⟩. **2.** *n.* **a)** *(Commerc. etc.)* Garantie, *die; (document)* Garantieschein, *der;* **b)** *(coll.: promise)* Garantie, *die (ugs.);* **give** sb. **a** ~ **that** ...: jmdm. garantieren, daß ...

guard [gɑ:d] **1.** *n.* **a)** *(guardsman)* Wachtposten, *der; (group of soldiers)* Wache, *die;* **be on** ~: Wache haben; **be on** |one's| ~ *(lit. or fig.)* sich hüten; **b)** *(Brit. Railw.)* [Zug]schaffner, *der/* -schaffnerin, *die;* **c)** *(Amer.: prison warder)* [Gefängnis]wärter, *der/*-wärterin, *die;* **d)** *(safety device)* Schutz, *der.* **2.** *v. t.* bewachen; hüten ⟨*Geheimnis*⟩; schützen ⟨*Leben*⟩; beschützen ⟨*Prominenten*⟩. '**guard against** *v. t.* sich hüten vor (+ *Dat.*); vorbeugen (+ *Dat.*) ⟨*Krankheit, Irrtum*⟩

'**guarded** *adj.* zurückhaltend

guardian ['gɑ:dɪən] *n.* **a)** Hüter, *der;* Wächter, *der;* **b)** *(Law)* Vormund, *der*

guerrilla [gə'rɪlə] *n.* Guerillakämpfer, *der/*-kämpferin, *die; attrib.* Guerilla-

guess [ges] **1.** *v. t.* **a)** *(estimate)* schätzen; *(surmise)* raten; *(surmise correctly)* erraten; raten ⟨*Rätsel*⟩; ~ **what!** *(coll.)* stell dir vor!; **b)** *(esp. Amer.:*

suppose) **I ~**: ich glaube. **2.** *v. i. (estimate)* schätzen; *(make assumption)* vermuten; *(surmise correctly)* es erraten; **~ at sth.** etw. schätzen; **keep sb. ~ing** *(coll.)* jmdn. im unklaren lassen. **3.** *n.* Schätzung, *die;* **make** *or* **have a ~**: schätzen. **'guesswork** *n.* **be ~**: eine Vermutung sein

guest [gest] *n.* Gast, *der.* **'guesthouse** *n.* Pension, *die*

guffaw [gʌ'fɔ:] **1.** *n.* brüllendes Gelächter. **2.** *v. i.* brüllend lachen

guidance ['gaɪdəns] *n.* **a)** *(leadership)* Führung, *die; (by teacher etc.)* [An]leitung, *die;* **b)** *(advice)* Rat, *der*

guide [gaɪd] **1.** *n.* **a)** Führer, *der*/Führerin, *die; (Tourism)* [Fremden]führer, *der/*-führerin, *die;* **b)** *(indicator)* **be a |good| ~ to sth.** ein [guter] Anhaltspunkt für etw. sein; **be no ~ to sth.** keine Rückschlüsse auf etw. *(Akk.)* zulassen; **c)** *(Brit.)* **|Girl| G~**: Pfadfinderin, *die;* **d)** *(handbook)* Handbuch, *das;* **e)** *(for tourists)* [Reise]führer, *der.* **2.** *v. t.* führen; *(fig.)* bestimmen ⟨*Handeln, Urteil*⟩; **be ~d by sth./sb.** sich von etw./jmdm. leiten lassen. **'guidebook** *n.* [Reise]führer, *der.* **guided 'missile** *n.* Lenkflugkörper, *der.* **'guide-dog** *n.* Blinden[führ]hund, *der.* **guided 'tour** *n.* Führung, *die* **(of** durch**).** **'guideline** *n.* Richtlinie, *die*

guild [gɪld] *n.* **a)** Verein, *der;* **b)** *(Hist.)* Gilde, *die;* Zunft, *die*

guile [gaɪl] *n.* Hinterlist, *die*

guillotine ['gɪləti:n] *n.* Guillotine, *die*

guilt [gɪlt] *n.* **a)** Schuld, *die* **(of, for an** + *Dat.*)**;** **b)** *(guilty feeling)* Schuldgefühle *Pl.* **'guilty** *adj.* **a)** schuldig; **be ~ of murder** des Mordes schuldig sein; **find sb. ~/not ~ |of sth.|** jmdn. [an etw. *(Dat.)*] schuldig sprechen/[von etw.] freisprechen; **feel ~** *(coll.)* ein schlechtes Gewissen haben; **b)** schuldbewußt ⟨*Miene, Blick, Verhalten*⟩; schlecht ⟨*Gewissen*⟩

guinea-pig ['gɪnɪpɪg] *n.* Meerschweinchen, *das; (fig.)* Versuchskaninchen, *das (ugs.)*

guise [gaɪz] *n.* Gestalt, *die;* **in the ~ of** in Gestalt (+ *Gen.*)

guitar [gɪ'tɑ:(r)] *n.* Gitarre, *die.* **guitarist** [gɪ'tɑ:rɪst] *n.* Gitarrist, *der*/Gitarristin, *die*

gulf [gʌlf] *n.* **a)** *(Geog.)* Golf, *der;* **b)** *(wide gap)* Kluft, *die*

gull [gʌl] *n.* Möwe, *die*

gullet ['gʌlɪt] *n.* **a)** Speiseröhre, *die;* **b)** *(throat)* Kehle, *die*

gullible ['gʌlɪbl] *adj.* leichtgläubig

gully ['gʌlɪ] *n. (artificial channel)* Abzugsrinne, *die; (drain)* Gully, *der*

gulp [gʌlp] **1.** *v. t.* hinunterschlingen; hinuntergießen ⟨*Getränk*⟩. **2.** *n.* **a)** Schlucken, *das;* **b)** *(large mouthful of drink)* kräftiger Schluck. **gulp 'down** *v. t.* hinunterschlingen; hinuntergießen ⟨*Getränk*⟩

¹**gum** [gʌm] *n. (Anat.)* **~|s|** Zahnfleisch, *das*

²**gum 1.** *n.* **a)** Gummi, *das; (glue)* Klebstoff, *der;* **b)** *(Amer.) see* chewing-gum. **2.** *v. t.,* **-mm-: a)** *(smear with ~)* mit Klebstoff bestreichen; gummieren ⟨*Briefmarken, Etiketten usw.*⟩; **b)** *(fasten with ~)* kleben, *(with delight)* **'gumboot** *n.* Gummistiefel, *der*

gumption ['gʌmpʃn] *n. (coll.)* Grips, *der*

gun [gʌn] *n.* Schußwaffe, *die; (rifle)* Gewehr, *das; (pistol)* Pistole, *die; (revolver)* Revolver, *der.* **gun 'down** *v. t.* niederschießen

gun: ~-fire *n.* Geschützfeuer, *das;* **~man** ['gʌnmən] *n., pl.* **~men** ['gʌnmən] bewaffneter Mann

gun: ~powder *n.* Schießpulver, *das;* **~shot** *n.* Schuß, *der*

gurgle ['gɜ:gl] **1.** *n.* Gluckern, *das; (of brook)* Plätschern, *das.* **2.** *v. i.* gluckern; ⟨*Bach:*⟩ plätschern; ⟨*Baby:*⟩ lallen; *(with delight)* glucksen

gush [gʌʃ] **1.** *n.* Schwall, *der.* **2.** *v. i.* **a)** strömen; **~ out** herausströmen; **b)** *(fig.: enthuse)* schwärmen

gust [gʌst] *n.* **~ |of wind|** Bö[e], *die*

gusto ['gʌstəʊ] *n.* Genuß, *der; (vitality)* Schwung, *der*

'gusty *adj.* böig

gut [gʌt] **1.** *n.* **a)** *(material)* Darm, *der;* **b)** *in pl. (bowels)* Eingeweide *Pl.;* Gedärme *Pl.;* **c)** *in pl. (coll.: courage)* Schneid, *der (ugs.).* **2.** *v. t.,* **-tt-: a)** *(remove ~s of)* ausnehmen; **b)** *(remove fittings from)* ausräumen; **the house was ~ted |by fire|** das Haus brannte aus

gutter ['gʌtə(r)] *n. (below edge of roof)* Dachrinne, *die; (at side of street)* Rinnstein, *der;* Gosse, *die*

guttural ['gʌtərl] *adj.* guttural; kehlig

guy [gaɪ] *n.* **a)** *(sl.: man)* Typ, *der (ugs.);* **b)** *in pl. (Amer.: everyone)* **|listen,| you ~s!** [hört mal,] Kinder! *(ugs.)*

guzzle ['gʌzl] **1.** *v. t. (eat)* hinunterschlingen; *(drink)* hinuntergießen. **2.** *v. i.* schlingen

gym [dʒɪm] *n. (coll.)* **a)** *(gymnasium)* Turnhalle, *die;* **b)** *(gymnastics)* Turnen, *das*

gymnasium *n.* [dʒɪm'neɪzɪəm] *n., pl.* ~s *or* **gymnasia** [dʒɪm'neɪzɪə] Turnhalle, *die*

gymnast ['dʒɪmnæst] *n.* Turner, *der/*Turnerin, *die*

gymnastic [dʒɪm'næstɪk] *adj.* turnerisch ‹*Können*›; ~ **equipment** Turngeräte. **gymnastics** [dʒɪm'næstɪks] *n.* Gymnastik, *die; (esp. with apparatus)* Tu.nen, *das*

'gym-slip *n.* Trägerrock, *der*

gynaecologist [gaɪnɪ'kɒlədʒɪst] *n.* Frauenarzt, *der/*Frauenärztin, *die*

gynaecology [gaɪnɪ'kɒlədʒɪ] *n.* Gynäkologie, *die*

gypsy, Gypsy ['dʒɪpsɪ] *n.* Zigeuner, *der/*Zigeunerin, *die*

gyrate [dʒaɪə'reɪt] *v. i.* sich drehen

H

¹H, h [eɪtʃ] *n.* H, h, *das*

haberdashery ['hæbədæʃərɪ] *n. (goods)* Kurzwaren *Pl.; (Amer.: menswear)* Herrenmoden *Pl.*

habit ['hæbɪt] *n.* **a)** Gewohnheit, *die;* **good/bad** ~: gute/schlechte [An]gewohnheit; **get** *or* **fall into a** *or* **the** ~ **of doing sth.** [es] sich *(Dat.)* angewöhnen, etw. zu tun; **b)** *(coll.: addiction)* Süchtigkeit, *die*

habitable ['hæbɪtəbl] *adj.* bewohnbar

habitat ['hæbɪtæt] *n.* Habitat, *das*

habitation [hæbɪ'teɪʃn] *n.* **fit/unfit for human** ~: bewohnbar/unbewohnbar

habitual [hə'bɪtjʊəl] *adj.* **a)** gewohnt; **b)** *(given to habit)* gewohnheitsmäßig; Gewohnheits‹*trinker*›. **ha'bitually** *adv. (regularly)* regelmäßig

¹hack [hæk] *v. t.* hacken ‹*Holz*›; ~ **sth. to bits** *or* **pieces** etw. in Stücke hacken. **hack 'off** *v. t.* abhacken. **hack 'out** *v. t.* heraushauen **(from** aus)

²hack *n. (derog.: writer)* Schreiberling, *der*

hackneyed ['hæknɪd] *adj.* abgegriffen; abgedroschen *(ugs.)*

'hack-saw *n.* [Metall]bügelsäge, *die* **had** *see* **have**

haddock ['hædək] *n., pl. same* Schellfisch, *der*

hadn't ['hædnt] *(coll.)* = **had not;** *see* **have**

haemorrhage ['hemərɪdʒ] *n.* Blutung, *die*

haemorrhoid ['hemərɔɪd] *n.* Hämorrhoide, *die*

hag [hæg] *n.* [alte] Hexe

haggard ['hægəd] *adj.* ausgezehrt; *(with worry)* abgehärmt

haggle ['hægl] *v. i.* sich zanken **(over, about** wegen); *(over price)* feilschen **(over, about** um)

Hague [heɪg] *pr. n.* **The** ~: Den Haag *(das)*

¹hail [heɪl] **1.** *n.* Hagel, *der.* **2.** *v. i.* **it** ~s *or* **is** ~**ing** es hagelt; ~ **down** *(fig.)* niederprasseln **(on** auf + *Akk.)*

²hail *v. t.* **a)** *(call out to)* anrufen; *(signal to)* anhalten ‹*Taxi*›; **b)** *(acclaim)* zujubeln (+ *Dat.)*; bejubeln **(as** als)

'hailstone *n.* Hagelkorn, *das*

hair [heə(r)] *n.* **a)** *(one strand)* Haar, *das;* **b)** *collect.* Haar, *das;* Haare *Pl.; attrib.* Haar-; **have** *or* **get one's** ~ **done** sich *(Dat.)* das Haar *od.* die Haare machen lassen *(ugs.)*

hair: ~**brush** *n.* Haarbürste, *die;* ~**-conditioner** *n.* Frisiermittel, *das;* ~**cut** *n.* **a)** *(act)* Haareschneiden, *das;* **go for/need a** ~**cut** zum Friseur gehen/müssen; **get/have a** ~**cut** sich *(Dat.)* die Haare schneiden lassen; **b)** *(style)* Haarschnitt, *der;* ~**-do** *n. (style)* Frisur, *die;* ~**dresser** *n.* Friseur, *der/*Friseuse, *die;* **go to the** ~**dresser's** zum Friseur gehen; ~**pin** *n.* Haarnadel, *die;* ~**pin 'bend** *n.* Haarnadelkurve, *die;* ~**-raising** ['heəreɪzɪŋ] *adj.* haarsträubend; ~**-style** *n.* Frisur, *die*

'hairy *adj.* **a)** behaart; flauschig ‹*Pullover, Teppich*›; **b)** *(sl.: difficult)* haarig

hale [heɪl] *adj.* ~ **and hearty** gesund und munter

half [hɑːf] **1.** *n., pl.* **halves** [hɑːvz] **a)** Hälfte, *die;* ~ |of sth.] die Hälfte [von etw.]; ~ **of Europe** halb Europa; **one and a** ~ **hours, one hour and a** ~: anderthalb *od.* eineinhalb Stunden; **divide sth. in** ~ *or* **into halves** etw. halbieren; **she is three and a** ~: sie ist dreieinhalb; **b)** *(Footb. etc.: period)* Halbzeit, *die.* **2.** *adj.* halb; ~ **the house/books/time** die Hälfte des Hauses/der Bücher/der Zeit; ~ **an hour** eine halbe Stunde. 3. *adv.* **a)** zur Hälfte; halb ⟨*schließen, aufessen, fertig, voll, geöffnet*⟩; *(almost)* fast ⟨*ersticken, tot sein*⟩; ~ **as much/many** halb so viel/viele; **only ~ hear what ...:** nur zum Teil hören, was ...; **b)** ~ *past or (coll.)* ~ **one/two/three** *etc.* halb zwei/drei/vier *usw.;* ~ **past twelve** halb eins

half: ~**-caste** *n.* Mischling, *der;* ~-'**hearted** *adj.* halbherzig; ~-'**hour** *n.* halbe Stunde; ~-'**mast** *n.* **be [flown] at** ~-**mast** auf Halbmast stehen; ~-**note** *n.* *(Amer. Mus.)* halbe Note; ~-'**price** **1.** *n.* halber Preis; **2.** *adj.* zum halben Preis *nachgestellt;* **3.** *adv.* zum halben Preis; ~-'**term** *n.* *(Brit.)* *(holiday)* ~**-term [holiday/break]** Ferien in der Mitte des Trimesters; ~-'**time** *n.* *(Sport)* Halbzeit, *die;* ~-'**way** **1.** *adj.* ~**-way point** Mitte, *die;* **2.** *adv.* die Hälfte des Weges ⟨*begleiten, fahren*⟩

hall [hɔːl] *n.* **a)** Saal, *der; (building)* Halle, *die;* **b)** *(entrance ~)* Flur, *der*

'**hallmark** *n.* [Feingehalts]stempel, *der; (fig.)* Kennzeichen, *das*

hallo [hə'ləʊ] *int.* **a)** *(to call attention)* hallo; **b)** *(Brit.) see* **hello**

Hallowe'en [hæləʊ'iːn] *n.* Halloween, *das; Abend vor Allerheiligen*

hallucination [həluːsɪ'neɪʃn] *n.* Halluzination, *die*

'**hallway** *n.* Flur, *der*

halo ['heɪləʊ] *n., pl.* ~**es** Heiligenschein, *der*

halt [hɒlt, hɔːlt] **1.** *n.* **a)** Pause, *die; (interruption)* Unterbrechung, *die;* **call a ~ to sth.** mit etw. Schluß machen; **b)** *(Brit. Railw.)* Haltepunkt, *der.* **2.** *v.i.* **a)** stehenbleiben; ⟨*Fahrer:*⟩ anhalten; *(for a rest)* eine Pause machen; *(esp. Mil.)* haltmachen; ~, **who goes there?** *(Mil.)* halt, wer da?; **b)** *(end)* eingestellt werden. **3.** *v.t.* anhalten; einstellen ⟨*Projekt*⟩. '**halting** *adj.* schleppend; zögernd ⟨*Antwort*⟩

halve [hɑːv] *v.t.* halbieren

halves *pl. of* **half**

ham [hæm] *n.* Schinken, *der*

hamburger ['hæmbɜːgə(r)] *n.* Hacksteak, *das; (in roll)* Hamburger, *der*

hamlet ['hæmlɪt] *n.* Weiler, *der*

hammer ['hæmə(r)] **1.** *n.* Hammer, *der.* **2.** *v.t.* hämmern. **3.** *v.i.* hämmern **(at an** + *Dat.*). **hammer 'out** *v.t.* ausklopfen ⟨*Delle, Beule*⟩; *(fig.:* devise) ausarbeiten

hammock ['hæmək] *n.* Hängematte, *die*

¹**hamper** ['hæmpə(r)] *n.* [Deckel]korb, *der*

²**hamper** *v.t.* behindern

hamster ['hæmstə(r)] *n.* Hamster, *der*

hand [hænd] **1.** *n.* **a)** Hand, *die;* **by** ~ *(manually)* mit der *od.* von Hand; **give** *or* **lend |sb.| a** ~ **|with** *or* **in sth.|** [jmdm.] [bei etw.] helfen; **b)** *(share)* **have a ~** in sth. bei etw. seine Hände im Spiel haben; **c)** *(worker)* Arbeiter, *der; (Naut.: seaman)* Matrose, *der;* **d)** *(of clock or watch)* Zeiger, *der;* **e) at** ~: in der Nähe; **on the one** ~ **...,** |**but**| **on the other** |~| **...:** einerseits ..., andererseits ...; **f)** *(Cards)* Karte, *die.* **2.** *v.t.* geben; ⟨*Überbringer:*⟩ übergeben ⟨*Sendung, Lieferung*⟩. **hand 'in** *v.t.* abgeben **(to, at** bei); einreichen ⟨*Petition*⟩. **hand 'out** *v.t.* austeilen. **hand 'over** *v.t.* übergeben **(to** *Dat.*)

hand: ~**bag** *n.* Handtasche, *die;* ~**-baggage** *n.* Handgepäck, *das;* ~**book** *n.* Handbuch, *das;* ~**-brake** *n.* Handbremse, *die;* ~**cuff** **1.** *n., usu.* **in** *pl.* Handschelle, *die;* **2.** *v.t.* ~**cuff sb.** jmdm. Handschellen anlegen

handful ['hændfʊl] *n.* Handvoll, *die;* **be a ~** *(fig. coll.)* einen ständig auf Trab halten *(ugs.)*

handicap ['hændɪkæp] **1.** *n.* **a)** *(Sport, also fig.)* Handikap, *das;* **b)** *(physical)* Behinderung, *die.* **2.** *v.t.,* -**pp**- benachteiligen. **handicapped** ['hændɪkæpt] *adj.* |**mentally/physically**| ~: [geistig/körperlich] behindert

handicraft ['hændɪkrɑːft] *n.* [Kunst]handwerk, *das; (needlework, knitting, etc.)* Handarbeit, *die*

handiwork ['hændɪwɜːk] *n.* handwerkliche Arbeit; **it's all his own ~:** das hat er selbst gemacht

handkerchief ['hæŋkətʃɪf] *n., pl.* ~**s** *or* **handkerchieves** ['hæŋkətʃiːvz] Taschentuch, *das*

handle ['hændl] **1.** *n.* Griff, *der; (of door)* Klinke, *die; (of axe, brush, comb, broom, saucepan)* Stiel, *der; (of cup, jug)* Henkel, *der.* **2.** *v.t.* **a)** *(touch, feel)* anfassen; **b)** *(control)* handhaben ⟨*Fahrzeug, Flugzeug*⟩; **c)** *(deal/cope with)* umgehen/fertigwerden mit. '**handlebars** *n. pl.* Lenkstange, *die*

hand: ~**-luggage** *n.* Handgepäck, *das;* ~**-made** *adj.* handgearbeitet; ~**shake** *n.* Händedruck, *der*

handsome ['hænsəm] *adj.* gutaussehend

hand: ~**stand** *n.* Handstand, *der;* ~**writing** *n.* [Hand]schrift, *die*

handy ['hændɪ] *adj.* greifbar; **keep/ have sth. ~:** etw. greifbar haben.

'**handyman** *n.* Handwerker, *der;* |home| ~: Heimwerker, *der*

hang [hæŋ] **1.** *v.t.* **a)** *p.t., p.p.* **hung** [hʌŋ] hängen; aufhängen ⟨*Bild, Gardinen*⟩; ankleben ⟨*Tapete*⟩; **b)** *p.t., p.p.* **hanged** *(execute)* hängen **(for we-** gen); ~ **oneself** sich erhängen. **2.** *v.i.,* **hung a)** hängen; ⟨*Kleid usw.:*⟩ fallen; **b)** *(be executed)* hängen. **3.** *n.* **get the ~ of sth.** *(coll.)* mit etw. klarkommen *(ugs.).* **hang a'bout, hang a'round** *v.i.* **a)** *(loiter)* herumlungern *(salopp);* **b)** *(coll.: wait)* warten. **hang 'on** *v.i.* **a)** sich festhalten **(to an** + *Dat.*); **b)** *(sl.: wait)* warten; **c)** ~ **on to** *(coll.: keep)* behalten. **hang 'out 1.** *v.t.* auf- hängen ⟨*Wäsche*⟩. **2.** *v.i.* **a)** heraus- hängen; **b)** *(sl.) (live)* wohnen; *(be often present)* sich herumtreiben *(ugs.).* **hang 'up** *v.t.* **1.** aufhängen. **2.** *v.i. (Teleph.)* auflegen

hangar ['hæŋə(r)] *n.* Hangar, *der*

'**hanger** *n.* Bügel, *der*

'**hang-glider** *n.* Drachen, *der*

'**hanging** *n. (execution)* Hinrichtung [durch den Strang]

hang: ~**man** [hæŋmən] *n., pl.* ~**men** [hæŋmən] Henker, *der;* ~**over** *n.* Ka- ter, *der (ugs.);* ~-**up** *n. (sl.)* Macke, *die (ugs.)*

hanker ['hæŋkə(r)] *v.i.* ~ **after** ein hef- tiges Verlangen haben nach

hanky ['hæŋkɪ] *n. (coll.)* Taschentuch, *das*

Hanover ['hænəʊvə(r)] *pr. n.* Hanno- ver *(das)*

haphazard [hæp'hæzəd] *adj.,* **hap-** '**hazardly** *adv.* willkürlich

happen ['hæpn] *v.i.* geschehen ⟨*Vor- hergesagtes:*⟩ eintreffen; ~ **to sb.** jmdm. passieren; ~ **to do sth./be sb.** zufällig etw. tun/jmd. sein; **as it ~s** *or* **it so ~s I have ...:** zufällig habe ich ...

'**happening** *n.* Ereignis, *das*

happily ['hæpɪlɪ] *adv.* **a)** glücklich ⟨*lä- cheln*⟩; vergnügt ⟨*spielen, lachen*⟩; **b)** *(gladly)* mit Vergnügen

happiness ['hæpɪnɪs] *n. see* **happy a:** Glück, *das;* Heiterkeit, *die;* Zufrie- denheit, *die*

happy ['hæpɪ] *adj.* **a)** *(joyful)* glück- lich; heiter ⟨*Bild, Veranlagung*⟩; er- freulich ⟨*Erinnerung, Szene*⟩; froh ⟨*Ereignis*⟩; *(contented)* zufrieden; **b)** **be ~ to do sth.** *(glad)* etw. gern tun.

happy-go-'lucky *adj.* sorglos

harass ['hærəs] *v.t.* schikanieren.

'**harassment** *n.* Schikanierung, *die;* **sexual ~:** [sexuelle] Belästigung

harbour *(Brit.; Amer.:* **harbor)** ['hɑːbə(r)] **1.** *n.* Hafen, *der;* **in ~:** im Hafen. **2.** *v.t.* Unterschlupf gewähren (+ *Dat.*) ⟨*Verbrecher, Flüchtling*⟩; he- gen *(geh.)* ⟨*Groll, Verdacht*⟩

hard [hɑːd] **1.** *adj.* **a)** hart; fest ⟨*Gelee*⟩; stark ⟨*Regen*⟩; streng ⟨*Frost, Winter*⟩; gesichert ⟨*Beweis, Daten*⟩; **b)** *(diffi- cult)* schwer; **this is ~ to believe** das ist kaum zu glauben; **do sth. the ~ way** es sich *(Dat.)* bei etw. unnötig schwer- machen; **c)** *(strenuous)* hart; **d)** *(vigor- ous)* kräftig ⟨*Schlag, Stoß, Tritt*⟩; **e)** *(harsh)* hart. **2.** *adv.* **a)** *(strenuously)* hart ⟨*arbeiten, trainieren*⟩; fleißig ⟨*stu- dieren, üben*⟩; genau ⟨*überlegen*⟩; gut ⟨*aufpassen, zuhören*⟩; **try ~:** sich sehr bemühen; **b)** *(vigorously)* heftig; fest ⟨*schlagen, drücken, klopfen*⟩; **c)** *(se- verely)* hart; **be ~ up** knapp bei Kasse sein *(ugs.);* **feel ~ done by** sich schlecht behandelt fühlen

hard: ~**back** *n.* gebundene Ausgabe; ~**board** *n.* Hartfaserplatte, *die;* ~-**boiled** *adj.* **a)** hartgekocht ⟨*Ei*⟩; **b)** *(tough)* hartgesotten

harden ['hɑːdn] **1.** *v.t.* härten; *(fig.)* abhärten **(to gegen). 2.** *v.i.* hart wer- den; *(become confirmed)* sich verhär- ten. **hardened** ['hɑːdnd] *adj.* abge- härtet **(to gegen);** hartgesotten ⟨*Ver- brecher*⟩

hard: ~-**headed** *adj.* nüchtern; ~-**hearted** *adj.* hartherzig **(towards** gegenüber)

hardly ['hɑːdlɪ] *adv.* kaum; ~ **anyone** *or* **anybody/anything** fast niemand/ nichts; ~ **ever** so gut wie nie; ~ **at all** fast überhaupt nicht

'**hardness** *n.* Härte, *die*

'**hardship** *n.* **a)** Not, *die;* Elend, *das;* **b)** *(instance)* Notlage, *die*

hard: ~ '**shoulder** *n. (Brit.)* Stand- spur, *die;* ~**ware** *n.* **a)** *(goods)* Eisen- waren *Pl.; attrib.* Eisenwaren⟨*ge- schäft*⟩; **b)** *(Computing)* Hardware, *die;* ~-**wearing** *adj.* strapazierfähig; ~-**working** *adj.* fleißig

hardy ['hɑːdɪ] *adj.* abgehärtet; zäh ⟨*Rasse*⟩; winterhart ⟨*Pflanze*⟩

hare [heə(r)] *n.* Hase, *der*

hark [hɑːk] *v.i.* |just| ~ **at him** hör ihn dir/hört ihn euch nur an!; ~ **back to** zurückkommen auf (+ *Akk.*)

harm [hɑːm] **1.** *n.* Schaden, *der;* **do sb. ~, do ~ to sb.** jmdm. schaden. **2.** *v.t.* etwas [zuleide] tun (+ *Dat.*); schaden

(+ *Dat.*) ⟨*Beziehungen, Land, Ruf*⟩.
harmful ['hɑːmfl] *adj.* schädlich (**to** für). '**harmless** *adj.* harmlos
harmonica [hɑːˈmɒnɪkə] *n.* Mundharmonika, *die*
harmonious [hɑːˈməʊnɪəs] *adj.* harmonisch
harmonize ['hɑːmənaɪz] **1.** *v. t.* aufeinander abstimmen. **2.** *v. i.* harmonieren (**with** mit)
harmony ['hɑːmənɪ] *n.* Harmonie, *die;* **be in** ~: harmonieren
harness ['hɑːnɪs] **1.** *n.* Geschirr, *das.* **2.** *v. t.* anschirren; *(fig.)* nutzen
harp [hɑːp] **1.** *n.* Harfe, *die.* **2.** *v. i.* ~ **on** [**about**] sth. immer wieder von etw. reden; *(critically)* auf etw. *(Dat.)* herumreiten *(salopp)*
harpoon [hɑːˈpuːn] *n.* Harpune, *die*
harrowing ['hærəʊɪŋ] *adj.* entsetzlich; grauenhaft ⟨*Anblick, Geschichte*⟩
harsh [hɑːʃ] *adj.* **a)** rauh ⟨*Gewebe, Klima*⟩; schrill ⟨*Ton, Stimme*⟩; grell ⟨*Licht*⟩; hart ⟨*Bedingungen, Leben*⟩; **b)** *(excessively severe)* [sehr] hart; (äußerst) streng ⟨*Disziplin*⟩; rücksichtslos ⟨*Tyrann, Herrscher, Politik*⟩. '**harshly** *adv.* [sehr] hart
harvest ['hɑːvɪst] **1.** *n.* Ernte, *die.* **2.** *v. t.* ernten
has *see* **have**
hash [hæʃ] *n.* **a)** *(Cookery)* Haschee, *das;* **b) make a** ~ **of** sth. *(coll.)* etw. verpfuschen *(ugs.)*
hasn't ['hæznt] = **has not;** *see* **have**
hassle ['hæsl] *(coll.)* **1.** *n.* Ärger, *der.* **2.** *v. t.* schikanieren
haste [heɪst] *n.* Eile, *die; (rush)* Hast, *die;* **make** ~: sich beeilen
hasten ['heɪsn] **1.** *v. t.* beschleunigen. **2.** *v. i.* eilen
hastily ['heɪstɪlɪ] *adv. (hurriedly)* eilig; *(rashly)* übereilt
hasty ['heɪstɪ] *adj.* eilig; flüchtig ⟨*Skizze, Blick*⟩; *(rash)* übereilt
hat [hæt] *n.* Hut, *der*
¹**hatch** [hætʃ] *n.* Luke, *die; (serving* ~*)* Durchreiche, *die*
²**hatch 1.** *v. t.* ausbrüten. **2.** *v. i.* [aus]schlüpfen. **hatch 'out 1.** *v. i.* ausschlüpfen. **2.** *v. t.* ausbrüten
'**hatchback** *n. (car)* Schräghecklimousine, *die*
hatchet ['hætʃɪt] *n.* Beil, *das;* **bury the** ~ *(fig.)* das Kriegsbeil begraben
hate [heɪt] **1.** *n.* Haß, *der.* **2.** *v. t.* hassen; **I** ~ **to say this** *(coll.)* ich sage das nicht gern. **hateful** ['heɪtfl] *adj.* abscheulich

hatred ['heɪtrɪd] *n.* Haß, *der*
haughty ['hɔːtɪ] *adj.* hochmütig
haul [hɔːl] **1.** *v. i. & t.* ziehen. **2.** *n.* **a)** Ziehen, *das;* **b)** *(catch)* Fang, *der; (fig.)* Beute, *die.* **haulage** ['hɔːlɪdʒ] *n.* Transport, *der*
haunch [hɔːntʃ] *n.* **sit on one's/its** ~**es** auf seinem Hinterteil sitzen
haunt [hɔːnt] *v. t.* ~ **a house/castle** in einem Haus/Schloß spuken; **a** ~**ed house** ein Haus, in dem es spukt. '**haunting** *adj.* sehnsüchtig
have 1. [hæv] *v. t., pres.* **he has** [hæz], *p. t. & p. p.* **had** [hæd] haben; *(obtain)* bekommen; *(take)* nehmen; bekommen ⟨*Kind*⟩; ~ **breakfast/dinner/lunch** frühstücken/zu Abend/zu Mittag essen; ~ **a cup of tea** eine Tasse Tee trinken; ~ **sb. to stay** jmdn. zu Besuch haben; **you've had it now** *(coll.)* jetzt ist es aus *(ugs.);* ~ **a game of football** Fußball spielen. **2.** [həv, əv, *stressed* hæv] *v. aux.,* **he has** [həz, əz, *stressed* hæz], **had** [həd, əd, *stressed* hæd] **I** ~/**I had read** ich habe/hatte gelesen; **I** ~/**I had gone** ich bin/war gegangen; **if I had known …:** wenn ich gewußt hätte …; ~ **sth. made** etw. machen lassen; ~ **to** müssen. **have 'on** *v. t.* **a)** *(wear)* tragen; **b)** *(Brit. coll.: deceive)* ~ **sb. on** jmdn. auf den Arm nehmen *(ugs.).* **have 'out** *v. t.* **a)** ~ **a tooth/one's tonsils out** sich *(Dat.)* einen Zahn ziehen lassen/sich *(Dat.)* die Mandeln herausnehmen lassen; **b)** ~ **it out with sb.** mit jmdm. offen sprechen
haven ['heɪvn] *n.* geschützte Anlegestelle, *die; (fig.)* Zufluchtsort, *der*
haven't ['hævnt] = **have not;** *see* **have**
haversack ['hævəsæk] *n.* Brotbeutel, *der*
havoc ['hævək] *n.* **a)** *(devastation)* Verwüstungen; **cause** *or* **wreak** ~: Verwüstungen anrichten; **b)** *(confusion)* Chaos; **play** ~ **with** sth. etw. völlig durcheinanderbringen
¹**hawk** [hɔːk] *n.* Falke, *der*
²**hawk** *v. t.* hausieren mit. '**hawker** *n.* Hausierer, *der*/Hausiererin, *die*
hay [heɪ] *n.* Heu, *das*
hay: ~ **fever** *n.* Heuschnupfen, *der;* ~**stack** *n.* Heuschober, *der (südd.);* Heudieme, *die (nordd.);* ~**wire** *adj. (coll.)* **go** ~**wire** ⟨*Instrument:*⟩ verrückt spielen *(ugs.)*
hazard ['hæzəd] **1.** *n.* Gefahr, *die.* **2.** *v. t.* ~ **a guess** mit Raten probieren. **hazardous** ['hæzədəs] *adj.* gefährlich

haze [heɪz] *n.* Dunst[schleier], *der*
hazelnut ['heɪzlnʌt] *n.* Haselnuß, *die*
hazy ['heɪzɪ] *adj.* dunstig; *(fig.)* vage
he [hɪ, *stressed* hiː] *pron.* er
head [hed] **1.** *n.* **a)** Kopf, *der;* ~ **first** mit dem Kopf voran; ~ **over heels** kopfüber; **keep/lose one's** ~: einen klaren Kopf behalten/den Kopf verlieren; **in one's** ~: im Kopf; **enter sb.'s** ~: jmdm. in den Sinn kommen; **use your** ~: gebrauch deinen Verstand; **a** *or* **per** ~: pro Kopf; **b)** *in pl. (on coin)* ~s Kopf; ~s **or tails?** Kopf oder Zahl?; **c)** *(leader)* Leiter, *der/*Leiterin, *die;* **d)** *(on beer)* Blume, *die.* **2.** *attrib. adj.* ~ **waiter** Oberkellner, *der;* ~ **office** Hauptverwaltung, *die.* **3.** *v. t.* **a)** *(stand at top of)* anführen ⟨ *Liste*⟩; *(lead)* leiten; führen ⟨ *Bewegung*⟩; **b)** *(Football)* köpfen. **4.** *v. i.* steuern; ~ **for London** ⟨*Flugzeug, Schiff:*⟩ Kurs auf London nehmen; ⟨*Auto:*⟩ in Richtung London fahren; **you're** ~**ing for trouble** du wirst Ärger bekommen.
'**headache** *n.* Kopfschmerzen *Pl.*
'**header** *n. (Footb.)* Kopfball, *der*
'**headgear** *n.* Kopfbedeckung, *die*
'**heading** *n.* Überschrift, *die*
head: ~**lamp** *n.* Scheinwerfer, *der;* ~**land** *n.* Landspitze, *die;* ~**light** *n.* Scheinwerfer, *der;* ~**line** *n.* Schlagzeile, *die;* ~**long** *adv.* kopfüber; ~'**master** *n.* Schulleiter, *der;* ~'**mistress** *n.* Schulleiterin, *die;* ~~**on 1.** ['~~] *adj.* frontal; Frontal⟨*zusammenstoß*⟩; **2.** [-'-] *adv.* frontal; ~~**phones** *n. pl.* Kopfhörer, *der;* ~**quarters** *n. sing. or pl.* Hauptquartier, *das;* ~~**rest** *n.* Kopfstütze, *die;* ~~**room** *n.* [lichte] Höhe, *die;* ~**strong** *adj.* eigensinnig; ~**way** *n.* **make** ~**way** Fortschritte machen; ~ **wind** *n.* Gegenwind, *der*
heady ['hedɪ] *adj.* berauschend
heal [hiːl] **1.** *v. t.* heilen. **2.** *v. i.* ~ [up] [ver]heilen
health [helθ] *n.* Gesundheit, *die;* **in good/very good** ~: bei guter/bester Gesundheit; **good** *or* **your** ~! auf deine Gesundheit!
health: ~ **centre** *n.* Poliklinik, *die;* ~ **food** *n.* Reformhauskost, *die;* ~~**food shop** Reformhaus, *das;* ~ **service** *n.* Gesundheitsdienst, *der*
healthy ['helθɪ] *adj.* gesund
heap [hiːp] **1.** *n.* Haufen, *der;* ~s **of** *(coll.)* jede Menge *(ugs.).* **2.** *v. t.* aufhäufen
hear [hɪə(r)] **1.** *v. t.,* **heard** [hɜːd] **a)** hö-

ren; **b)** *(understand)* verstehen. **2.** *v. i.,* **heard:** ~ **about sb./sth.** von jmdm./ etw. [etwas] hören; **he wouldn't** ~ **of it** er wollte nichts davon hören. **3.** *int.* H~! H~! bravo!; richtig! **hear 'out** *v. t.* ausreden lassen
heard *see* **hear 1, 2**
'**hearing** *n.* Gehör, *das;* **be hard of** ~: schwerhörig sein. '**hearing-aid** *n.* Hörgerät, *das*
hearsay ['hɪəseɪ] *n.* Gerücht, *das;* **it's only** ~: es ist nur ein Gerücht
hearse [hɜːs] *n.* Leichenwagen, *der*
heart [hɑːt] *n. (also Cards)* Herz, *das;* **by** ~: auswendig; **at** ~: im Grunde seines/ihres Herzens; **take/lose** ~: Mut schöpfen/verlieren; **my** ~ **sank** mein Mut sank; **the** ~ **of the matter** der wahre Kern der Sache; *see also* **club 1 c**
heart: ~ **attack** *n.* Herzanfall, *der;* *(fatal)* Herzschlag, *der;* ~**beat** *n.* Herzschlag, *der;* ~~**breaking** *adj.* herzzerreißend; ~~**broken** *adj.* **she was** ~~**broken** ihr Herz war gebrochen; ~**burn** *n.* Sodbrennen, *das*
hearten ['hɑːtn] *v. t.* ermutigen.
'**heartening** *adj.* ermutigend
heart: ~ **failure** *n.* Herzversagen, *das;* ~**felt** *adj.* tiefempfunden ⟨ *Beileid*⟩; aufrichtig ⟨ *Dankbarkeit*⟩
hearth [hɑːθ] *n.* Platz vor dem Kamin. '**hearth-rug** *n.* Kaminvorleger, *der*
heartily ['hɑːtɪlɪ] *adv.* von Herzen; **eat** ~: tüchtig essen
'**heartless** *adj.* herzlos
hearty ['hɑːtɪ] *adj.* herzlich; ungeteilt ⟨ *Zustimmung*⟩; herzhaft ⟨ *Mahlzeit*⟩
heat [hiːt] **1.** *n.* **a)** *(hotness)* Hitze, *die;* **b)** *(Phys.)* Wärme, *die;* **c)** *(Sport)* Vorlauf, *der.* **2.** *v. t.* heizen. **heat 'up** *v. t.* heiß machen
'**heated** *adj. (angry)* hitzig
'**heater** *n.* Ofen, *der;* (for water) Boiler, *der*
heath [hiːθ] *n.* Heide, *die*
heathen ['hiːðn] **1.** *adj.* heidnisch. **2.** *n.* Heide, *der/*Heidin, *die*
heather ['heðə(r)] *n.* Heidekraut, *das*
'**heating** *n.* Heizung, *die*
heat: ~~**stroke** *n.* Hitzschlag, *der;* ~**wave** *n.* Hitzewelle, *die*
heave [hiːv] **1.** *v. t.* **a)** heben; **b)** *(coll.: throw)* schmeißen *(ugs.);* **c)** ~ **a sigh** aufseufzen. **2.** *v. i. (pull)* ziehen. **3.** *n.* Zug, *der*
heaven ['hevn] *n.* Himmel, *der;* **in** ~: im Himmel; **for H**~'**s sake!** um Gottes willen! '**heavenly** *adj.* himmlisch

heavily ['hevɪlɪ] *adj.* schwer; *(to a great extent)* stark; schwer ⟨*bewaffnet*⟩; tief ⟨*schlafen*⟩; dicht ⟨*bevölkert*⟩; **smoke/drink ~:** ein starker Raucher/Trinker sein; **it rained/snowed ~:** es regnete/schneite stark

heavy ['hevɪ] *adj.* schwer; unmäßig ⟨*Trinken, Rauchen*⟩; **a ~ smoker/drinker** ein starker Raucher/Trinker; **be a ~ sleeper** sehr fest schlafen

heavy: **~-duty** *adj.* strapazierfähig ⟨*Kleidung, Material*⟩; schwer ⟨*Werkzeug, Maschine*⟩; **~ 'goods vehicle** *n.* *(Brit.)* Schwerlastwagen, *der;* **~weight** *n.* Schwergewicht, *das*

Hebrew ['hi:bru:] **1.** *adj.* hebräisch. **2.** *n.* *(language)* Hebräisch, *das*

heckle ['hekl] *v.t.* Zwischenrufe unterbrechen. **heckler** ['heklə(r)] *n.* Zwischenrufer, *der*

hectic ['hektɪk] *adj.* hektisch

he'd [hɪd, *stressed* hi:d] **a)** = **he had; b)** = **he would**

hedge [hedʒ] **1.** *n.* Hecke, *die.* **2.** *v.t.* **~ one's bets** *(fig.)* nicht alles auf eine Karte setzen. **3.** *v.i.* sich nicht festlegen

hedgehog ['hedʒhɒg] *n.* Igel, *der*

'hedgerow *n.* Hecke, *die* [als Feldbegrenzung]

heed [hi:d] **1.** *v.t.* beachten; beherzigen ⟨*Rat, Lektion*⟩; **~ the danger/risk** sich *(Dat.)* der Gefahr/des Risikos bewußt sein. **2.** *n.* **give** *or* **pay ~ to, take ~ of** Beachtung schenken (+ *Dat.*). **'heedless** *adj.* unachtsam; **be ~ of sth.** auf etw. *(Akk.)* nicht achten

heel [hi:l] *n.* Ferse, *die;* *(of shoe)* Absatz, *der;* **Achilles' ~** *(fig.)* Achillesferse, *die;* **down at ~** *(fig.)* heruntergekommen; **take to one's ~s** Fersengeld geben *(ugs.)*

hefty ['heftɪ] *adj.* kräftig; *(heavy)* schwer

height [haɪt] *n.* **a)** Höhe, *die;* *(of person, animal, building)* Größe, *die;* **b)** *(fig.: highest point)* Höhepunkt, *der.* **heighten** ['haɪtn] *v.t.* aufstocken; *(fig.)* verstärken

heir [eə(r)] *n.* Erbe, *der*/Erbin, *die.* **heiress** ['eərɪs] *n.* Erbin, *die*

heirloom ['eəlu:m] *n.* Erbstück, *das*

held *see* ²**hold** 1, 2

helicopter ['helɪkɒptə(r)] *n.* Hubschrauber, *der*

heliport ['helɪpɔ:t] *n.* Heliport, *der*

helium ['hi:lɪəm] *n.* Helium, *das*

hell [hel] *n.* **a)** Hölle, *die;* **b)** *(coll.)* |oh| **~!** verdammter Mist! *(ugs.);* **what the**

~! ach, zum Teufel! *(ugs.);* **run like ~:** wie der Teufel rennen *(ugs.)*

he'll [hɪl, *stressed* hi:l] = **he will**

hello [hə'ləʊ, he'ləʊ] *int.* *(greeting)* hallo; *(surprise)* holla

hell's 'angel *n.* Rocker, *der*

helm [helm] *n.* *(Naut.)* Ruder, *das*

helmet ['helmɪt] *n.* Helm, *der*

help [help] **1.** *v.t.* **a)** **~ sb.** |**to do sth.**| jmdm. helfen[, etw. zu tun]; **can I ~ you?** *(in shop)* was möchten Sie bitte?; **b)** *(serve)* **~ oneself** sich bedienen; **~ oneself to sth.** sich *(Dat.)* etw. nehmen; *(coll.: steal)* etw. mitgehen lassen *(ugs.);* **c)** *(avoid)* **if I/you can ~ it** wenn es irgend zu vermeiden ist; *(remedy)* **I can't ~ it** ich kann nichts dafür *(ugs.);* **it can't be ~ed** es läßt sich nicht ändern; **d)** *(refrain from)* **I can't ~ thinking** *or* **can't ~ but think that ...:** ich kann mir nicht helfen, ich glaube, ...; **I can't ~ laughing** ich muß einfach lachen. **2.** *n.* Hilfe, *die;* **with the ~ of ...:** mit Hilfe ... (+ *Gen.*); **be of** |**some**|/**no**/**much ~ to sb.** jmdm. eine gewisse/keine/eine große Hilfe sein.

help 'out 1. *v.i.* aushelfen. **2.** *v.t.* **~ sb. out** jmdm. helfen

'helper *n.* Helfer, *der*/Helferin, *die*

helpful ['helpfl] *adj.* *(willing)* hilfsbereit; *(useful)* hilfreich; nützlich

'helping 1. *adj.* lend |**sb.**| **a ~ hand** |**with sth.**| *(fig.)* |jmdm.| [bei etw.] helfen. **2.** *n.* Portion, *die*

'helpless *adj.,* **'helplessly** *adv.* hilflos

helter-skelter [heltə'skeltə(r)] *n.* [spiralförmige] Rutschbahn

hem [hem] **1.** *n.* Saum, *der.* **2.** *v.t.,* **-mm-** säumen. **hem 'in** *v.t.* einschließen; **feel ~med in** sich eingeengt fühlen

hemisphere ['hemɪsfɪə(r)] *n.* Halbkugel, *die*

'hem-line *n.* Saum, *der*

hemp [hemp] *n.* Hanf, *der*

hen [hen] *n.* Huhn, *das;* Henne, *die*

hence [hens] *adv.* *(therefore)* daher. **hence'forth** *adv.* von nun an

henchman ['hentʃmən] *n., pl.* **henchmen** ['hentʃmən] Handlanger, *der*

henpecked ['henpekt] *adj.* **a ~ husband** ein Pantoffelheld, *der (ugs.);* **be ~:** unter dem Pantoffel stehen *(ugs.)*

¹her [hə(r), *stressed* hɜ:(r)] *pron.* sie; *as indirect object* ihr; **it was ~:** sie war's

²her *poss. pron. attr.* ihr

herald ['herəld] **1.** *n.* Herold, *der.* **2.** *v.t.* ankündigen. **heraldic** [he'rældɪk] *adj.* heraldisch. **heraldry** ['herəldrı] *n.* Heraldik, *die*

herb [hɜ:b] *n.* Kraut, *das.* **herbaceous** [hɜ:'beɪʃəs] *adj.* krautartig; ~ **border** Staudenrabatte, *die.* **herbal** ['hɜ:bl] *attrib. adj.* Kräuter

herd [hɜ:d] **1.** *n.* Herde, *die;* (*of wild animals*) Rudel, *das.* **2.** *v. t.* **a)** treiben; ~ **people together** Menschen zusammenpferchen; **b)** (*tend*) hüten

here [hɪə(r)] **1.** *adv.* **a)** (*in or at this place*) hier; **down/in/up** ~: hier unten/drin/oben; ~ **you are** (*coll.: giving sth.*) hier; **b)** (*to this place*) hierher; **in|to|** ~: hierherein; **come/bring** ~: [hier]herkommen/-bringen. **2.** *int.* (*attracting attention*) he. **here'by** *adv.* (*formal*) hiermit

hereditary [hɪ'redɪtərɪ] *adj.* **a)** erblich ⟨*Titel, Amt*⟩; **b)** (*Biol.*) angeboren

heresy ['herɪsɪ] *n.* Ketzerei, *die*

heretic ['herɪtɪk] *n.* Ketzer, *der*/Ketzerin, *die*

here'with *adv.* in der Anlage

heritage ['herɪtɪdʒ] *n.* Erbe, *das*

hermetic [hɜ:'metɪk] *adj.* luftdicht. **hermetically** [hɜ:'metɪkəlɪ] *adv.* hermetisch

hermit ['hɜ:mɪt] *n.* Einsiedler, *der*/Einsiedlerin, *die*

hernia ['hɜ:nɪə] *n.* Bruch, *der*

hero ['hɪərəʊ] *n., pl.* ~**es** Held, *der.* **heroic** [hɪ'rəʊɪk] *adj.* heldenhaft

heroin ['herəʊɪn] *n.* Heroin, *das*

heroine ['herəʊɪn] *n.* Heldin, *die*

heroism ['herəʊɪzm] *n.* Heldentum, *das*

heron ['hern] *n.* Reiher, *der*

herring ['herɪŋ] *n.* Hering, *der*

hers [hɜ:z] *poss. pron. pred.* ihrer/ihre/ihres; **the book is** ~: das Buch gehört ihr

her'self *pron.* **a)** *emphat.* selbst; **[all] by** ~: [ganz] allein[e]; **b)** *refl.* sich; allein[e] ⟨*tun, wählen*⟩; **younger than/as heavy as** ~: jünger als/so schwer wie sie selbst

he's [hɪz, *stressed* hi:z] **a)** = **he is; b)** = **he has**

hesitant ['hezɪtənt] *adj.* zögernd ⟨*Reaktion*⟩; stockend ⟨*Rede*⟩

hesitate ['hezɪteɪt] *v. i.* zögern; (*falter*) ins Stocken geraten; ~ **to do sth.** Bedenken haben, etw. zu tun. **hesitation** [hezɪ'teɪʃn] *n.* **a)** (*indecision*) Unentschlossenheit, *die;* **without** ~: ohne zu zögern; **b)** (*instance of faltering*) Unsicherheit, *die;* (*reluctance*) Bedenken *Pl.*

heterosexual [hetərəʊ'seksjʊəl] **1.** *adj.* heterosexuell. **2.** *n.* Heterosexuelle, *der*/*die*

het up [het ʌp] *adj.* aufgeregt

hew [hju:] *v. t., p.p.* **hewn** [hju:n] *or* **hewed** [hju:d] hacken ⟨*Holz*⟩; losschlagen ⟨*Kohle, Gestein*⟩

hewn *see* **hew**

hexagon ['heksəgən] *n.* Sechseck, *das*

hey [heɪ] *int.* he; ~ **presto!** simsalabim!

heyday ['heɪdeɪ] *n.* Blütezeit, *die*

HGV *abbr.* (*Brit.*) **heavy goods vehicle**

hi [haɪ] *int.* hallo (*ugs.*)

hiatus [haɪ'eɪtəs] *n.* Unterbrechung, *die*

hibernate ['haɪbəneɪt] *v. i.* Winterschlaf halten. **hibernation** [haɪbə-'neɪʃn] *n.* Winterschlaf, *der*

hiccup ['hɪkʌp] **1.** *n.* **a)** Schluckauf, *der;* **have/get |the|** ~**s** den Schluckauf haben/bekommen; **b)** (*fig.: stoppage*) Störung, *die.* **2.** *v. i.* schlucksen. (*ugs.*)

hid *see* ¹**hide**

hidden *see* ¹**hide**

¹**hide** [haɪd] **1.** *v. t.,* **hid** [hɪd], **hidden** ['hɪdn] **a)** verstecken ⟨*Gegenstand, Person usw.*⟩ (**from** vor + *Dat.*); verbergen ⟨*Gefühle, Sinn usw.*⟩ (**from** vor + *Dat.*); verheimlichen ⟨*Tatsache, Absicht usw.*⟩ (**from** *Dat.*); **b)** (*obscure*) verdecken. **2.** *v. i.,* **hid, hidden** sich verstecken (**from** vor + *Dat.*)

²**hide** *n.* Haut, *die;* (*of furry animal*) Fell, *das;* (*dressed*) Leder, *das*

hide-and-'seek *n.* Versteckspiel, *das;* **play** ~: Verstecken spielen

hideous ['hɪdɪəs] *adj.* scheußlich

'hide-out *n.* Versteck, *das*

'hiding ['haɪdɪŋ] *n.* **go into** ~: sich verstecken; (*to avoid police, public attention*) untertauchen; **be in** ~: sich versteckt halten; (*to avoid police, public attention*) untergetaucht sein

²**hiding** *n.* (*coll.: beating*) Tracht Prügel; **give sb. a |good|** ~: jmdm. eine [ordentliche] Tracht Prügel verpassen (*ugs.*)

'hiding-place *n.* Versteck, *das*

hierarchy ['haɪərɑ:kɪ] *n.* Hierarchie, *die*

hi-fi ['haɪfaɪ] (*coll.*) **1.** *adj.* Hi-Fi-. **2.** *n.* Hi-Fi-Anlage, *die*

high [haɪ] **1.** *adj.* **a)** hoch; groß ⟨*Höhe*⟩; stark ⟨*Wind*⟩; **b)** (*coll.: on a drug*) high (*ugs.*); **c) it's** ~ **time you left** es ist höchste Zeit, daß du gehst. **2.** *adv.* hoch; **search** *or* **look** ~ **and low** überall suchen. **3.** *n.* **a)** (~**est level/figure**) Höchststand, *der;* **b)** (*Met-*

highbrow

162

eorol.) Hoch, *das.* **'highbrow** *(coll.)*
1. *n.* Intellektuelle, *der/die.* **2.** *adj.* intellektuell 〈*Person, Gerede usw.*〉; hochgestochen *(abwertend)* 〈*Person, Musik, Literatur usw.*〉. **'high chair** *n.* Hochstuhl, *der*

higher edu'cation *n.* Hochschulbildung, *die*

high: ~**-'handed** *adj.* selbstherrlich; ~**-heeled** [haɪ'hiːld] *adj.* 〈*Schuhe*〉 mit hohen Absätzen; ~ **jump** *n.* Hochsprung, *der;* ~**land** ['haɪlənd] *n.* Hochland, *das;* ~**light 1.** *n.* **a)** Höhepunkt, *der;* **b)** *(bright area)* Licht, *das;* **2.** *v. t.,* ~**lighted** ein Schlaglicht werfen auf (+ *Akk.*) 〈*Probleme usw.*〉

'highly *adv.* sehr; hoch〈*interessant, -angesehen, -bezahlt, -gebildet*〉; leicht 〈*entzündlich*〉; stark 〈*gewürzt*〉; **think** ~ **of sth./sb.** eine hohe Meinung von jmdm./etw. haben; **speak** ~ **of sb./sth.** jmdn./etw. sehr loben. **highly-strung** ['haɪlɪstrʌn] *adj.* übererregbar

Highness ['haɪnɪs] *n.* **His/her** *etc.* ~: Seine/Ihre *usw.* Hoheit

high: ~**-pitched** ['haɪpɪtʃt] *adj.* hoch 〈*Ton, Stimme*〉; ~ **'pressure** *n.* **a)** *(Meteorol.)* Hochdruck, *der;* **b)** *(Mech. Engin.)* Überdruck, *der;* ~**-rise** *adj.* ~**-rise building** Hochhaus, *das;* ~**-rise block of flats/office block** Wohn-/Bürohochhaus, *das;* ~ **school** *n.* ≈ Oberschule, *die;* ~ **season** *n.* Hochsaison, *die;* ~**way** *n.* öffentliche Straße

hijack ['haɪdʒæk] *v. t.* entführen. **'hijacker** *n.* Entführer, *der;* *(of aircraft)* Hijacker, *der*

hike [haɪk] *n.* Wanderung, *die.* **'hiker** *n.* Wanderer, *der/*Wanderin, *die*

hilarious [hɪ'leərɪəs] *adj.* urkomisch

hill [hɪl] *n.* Hügel, *der; (higher)* Berg, *der; (slope)* Hang, *der*

hill: ~**-billy** ['hɪlbɪlɪ] *n. (Amer.)* Hinterwäldler, *der/*Hinterwäldlerin, *die;* ~**side** *n.* Hang, *der;* ~**top** *n.* [Berg]gipfel, *der*

'hilly *adj.* hüg[e]lig

hilt [hɪlt] *n.* Griff, *der;* |**up**| **to the** ~ *(fig.)* voll und ganz

him [ɪm, *stressed* hɪm] *pron.* ihn; *as indirect object* ihm; **it was** ~: er war's

Himalayas [hɪmə'leɪəz] *pr. n. pl.* Himalaya, *der*

him'self *pron.* **a)** *emphat.* selbst; **b)** *refl.* sich. *See also* **herself**

hind [haɪnd] *adj.* hinter...; ~ **legs** Hinterbeine

hinder ['hɪndə(r)] *v. t. (impede)* behin-

dern; *(delay)* verzögern 〈*Vollendung einer Arbeit, Vorgang*〉; aufhalten 〈*Person*〉; ~ **sb. from doing sth.** jmdn. daran hindern, etw. zu tun

'hindquarters *n. pl.* Hinterteil, *das*
hindrance ['hɪndrəns] *n.* Hindernis, *das* (**to** für)

'hindsight *n.* **with** |**the benefit of**| ~: im nachhinein

Hindu ['hɪnduː, hɪn'duː] **1.** *n.* Hindu, *der.* **2.** *adj.* hinduistisch; Hindu〈*gott, -tempel*〉

hinge 〈hɪndʒ〉 **1.** *n.* Scharnier, *das.* **2.** *v. t.* mit Scharnieren versehen. **3.** *v. i. (depend)* abhängen (|**up**|**on** von)

hint [hɪnt] **1.** *n.* **a)** *(suggestion)* Wink, *der;* **b)** *(slight trace)* Spur, *die* (**of** von); **the** ~/**no** ~ **of a smile** der Anflug/nicht die Spur eines Lächelns; **c)** *(information)* Tip, *der* (**on** für). **2.** *v. i.* ~ **at** andeuten

hip [hɪp] *n.* Hüfte, *die*

hippie ['hɪpɪ] *n. (coll.)* Hippie, *der*

hippopotamus [hɪpə'pɒtəməs] *n.* Nilpferd, *das*

hippy *see* **hippie**

hire [haɪə(r)] **1.** *n.* Mieten, *das;* **be on** ~ |**to sb.**| |**an jmdn.**| vermietet sein; **for** ~: zu vermieten. **2.** *v. t.* **a)** *(employ)* anwerben; engagieren 〈*Anwalt, Berater usw.*〉; **b)** *(obtain use of)* mieten; ~ **sth. from sb.** etw. bei jmdm. mieten; **c)** *(grant use of)* ~ |**out**| vermieten; ~ **sth.** |**out**| **to sb.** etw. jmdm. *od.* an jmdn. vermieten. **'hire-car** *n.* Mietwagen, *der.* **hire-'purchase** *n. (Brit.)* Ratenkauf, *der; attrib.* Raten-; **pay for/buy sth. on** ~: etw. in Raten bezahlen/auf Raten kaufen

his [ɪz, *stressed* hɪz] *poss. pron.* **a)** *attrib.* sein; **b)** *pred.* seiner/seine/sein[e]s; *see also* **hers**

hiss [hɪs] **1.** *n.* Zischen, *das.* **2.** *v. i.* zischen

historian [hɪ'stɔːrɪən] *n.* Historiker, *der/*Historikerin, *die*

historic [hɪ'stɒrɪk] *adj.* historisch. **historical** [hɪ'stɒrɪkl] *adj.* historisch; geschichtlich 〈*Belege, Hintergrund*〉

history ['hɪstərɪ] *n.* Geschichte, *die*

hit [hɪt] **1.** *v. t.,* -tt-, **hit** schlagen; *(with missile)* treffen; 〈*Geschoß, Ball usw.*〉 treffen; 〈*Fahrzeug:*〉 prallen gegen; 〈*Schiff:*〉 laufen gegen; ~ **one's head on sth.** mit dem Kopf gegen etw. stoßen; ~ **it off with sb.** gut mit jmdm. auskommen. **2.** *v. i.,* -tt-, **hit** schlagen. **3.** *n.* **a)** *(blow)* Schlag, *der; (shot or bomb striking target)* Treffer, *der;* **b)** *(suc-*

cess) Erfolg, *der; (in entertainment)* Schlager, *der;* Hit, *der (ugs.).* **hit 'back** *v. t. & i.* zurückschlagen. '**hit [up]on** *v. t.* kommen auf (+ *Akk.*) ⟨*Idee*⟩; finden ⟨*richtige Antwort, Methode*⟩

hitch [hɪtʃ] **1.** *v. t.* **a)** binden ⟨*Seil*⟩ (**round** um + *Akk.*); [an]koppeln ⟨*Anhänger usw.*⟩ (**to** an + *Akk.*); spannen ⟨*Zugtier usw.*⟩ (**to** vor + *Akk.*); **b)** ~ **a lift** *or* **ride** *(coll.)* per Anhalter fahren. **2.** *n. (problem)* Problem, *das.* **hitch 'up** *v. t.* hochheben ⟨*Rock*⟩

'**hitch-hike** *v. i.* per Anhalter fahren. '**hitch-hiker** *n.* Anhalter, *der*/Anhalterin, *die*

'**hit parade** *n.* Hitparade, *die*

HIV *abbr.* **human immuno-deficiency virus** HIV

hive [haɪv] *n.* [Bienen]stock, *der*

HMS *abbr. (Brit.)* **Her/His Majesty's Ship** H.M.S.

hoard [hɔːd] **1.** *n.* Vorrat, *der.* **2.** *v. t.* [up] horten; hamstern ⟨*Lebensmittel*⟩

hoarding ['hɔːdɪŋ] *n. (fence)* Bauzaun, *der; (Brit.: for advertisements)* Reklamewand, *die*

hoar-frost ['hɔːfrɒst] *n.* [Rauh]reif, *der*

hoarse [hɔːs] *adj.* heiser

hoax [həʊks] **1.** *v. t.* anführen *(ugs.);* foppen. **2.** *n. (deception)* Schwindel, *der; (practical joke)* Streich, *der; (false alarm)* blinder Alarm

hob [hɒb] *n.* [Koch]platte, *die*

hobble ['hɒbl] *v. i.* ~ [about] [herum]humpeln

hobby ['hɒbɪ] *n.* Hobby, *das.* '**hobby-horse** *n.* Steckenpferd, *das*

hobnailed ['hɒbneɪld] *adj.* Nagel-⟨*schuh, -stiefel*⟩

hobo ['həʊbəʊ] *n., pl.* **-es** *(Amer.)* Landstreicher, *der*/-streicherin, *die*

hockey ['hɒkɪ] *n.* Hockey, *das.* '**hockey-stick** *n.* Hockeyschläger, *der*

hoe [həʊ] **1.** *n.* Hacke, *die.* **2.** *v. t. & i.* hacken

hog [hɒg] **1.** *n.* [Mast]schwein. **2.** *v. t.,* **-gg-** *(coll.)* mit Beschlag belegen

hoist [hɔɪst] **1.** *v. t.* hochziehen, hissen ⟨*Flagge usw.*⟩; hieven ⟨*Last*⟩; setzen ⟨*Segel*⟩. **2.** *n.* [Lasten]aufzug, *der*

¹**hold** [həʊld] *n. (of ship)* Laderaum, *der; (of aircraft)* Frachtraum, *der*

²**hold 1.** *v. t.,* **held** [held] **a)** halten; *(carry)* tragen; *(keep fast)* festhalten; ~ **the door open for sb.** jmdm. die Tür aufhalten; ~ **sth. in place** etw. halten;

b) *(contain)* enthalten; *(be able to contain)* fassen ⟨*Liter, Personen usw.*⟩; **c)** *(possess)* besitzen; haben; **d)** *(keep possession of)* halten ⟨*Stützpunkt, Stadt, Stellung*⟩; ~ **the line** *(Teleph.)* am Apparat bleiben; ~ **one's own** sich behaupten; **e)** *(cause to take place)* stattfinden lassen; abhalten ⟨*Veranstaltung, Konferenz, Gottesdienst, Sitzung*⟩; veranstalten ⟨*Festival, Auktion*⟩; austragen ⟨*Meisterschaften*⟩; führen ⟨*Unterhaltung, Gespräch*⟩; durchführen ⟨*Untersuchung*⟩; halten ⟨*Vortrag, Rede*⟩; **f)** *(think, believe)* ~ **a view** *or* **an opinion** eine Ansicht haben (**on** über + *Akk.*); ~ **that ...:** der Ansicht sein, daß ...; ~ **oneself responsible for sth.** sich für etw. verantwortlich fühlen; ~ **sth. against sb.** jmdm. etw. vorwerfen. **2.** *v. i.,* **held** halten; ⟨*Wetter:*⟩ sich halten. **3.** *n.* **a)** *(grasp)* Griff, *der;* **grab** *or* **seize** ~ **of sth.** etw. ergreifen; **get** *or* **lay** *or* **take** ~ **of sth.** etw. fassen *od.* packen; **keep** ~ **of sth.** etw. festhalten; **get** ~ **of sth.** *(fig.)* etw. auftreiben; **get** ~ **of sb.** *(fig.)* jmdn. erreichen; **b)** *(influence)* Einfluß, *der* (**on, over** auf + *Akk.*); **c)** *(Sport)* Griff, *der.* **hold 'back 1.** *v. t.* zurückhalten. **2.** *v. i.* zögern. **hold 'on 1.** *v. t.* [fest]halten. **2.** *v. i.* **a)** sich festhalten; ~ **on to sth.** sich festhalten an (+ *Dat.*); *(keep)* behalten; **b)** *(coll.: wait)* warten. **hold 'out 1.** *v. t.* ausstrecken ⟨*Hand, Arm usw.*⟩: hinhalten ⟨*Tasse, Teller*⟩. **2.** *v. i. (resist)* sich halten. **hold 'up** *v. t.* **a)** *(raise)* hochhalten; heben ⟨*Hand, Kopf*⟩; **b)** *(delay)* aufhalten; **c)** *(rob)* überfallen. '**hold with** *v. t.* **not** ~ **with sth.** etw. ablehnen

'**holdall** *n.* Reisetasche, *die*

'**holder** *n.* **a)** *(of post, title)* Inhaber, *der*/Inhaberin, *die;* **b)** ⟨*Zigaretten*⟩-spitze, *die;* ⟨*Papier-, Zahnputzglas*⟩-halter, *der*

'**hold-up** *n.* **a)** *(robbery)* [Raub]überfall, *der;* **b)** *(delay)* Verzögerung, *die*

hole [həʊl] *n.* Loch, *das; (of fox, badger, rabbit)* Bau, *der;* **pick ~s in** *(fig.)* zerpflücken *(ugs.)*

holiday ['hɒlɪdeɪ] *n.* **a)** [arbeits]freier Tag; *(public ~)* Feiertag, *der;* **b)** in *sing. or pl. (Brit.: vacation)* Urlaub, *der; (Sch.)* [Schul]ferien *Pl.* '**holiday-maker** *n.* Urlauber, *der*/Urlauberin, *die*

Holland ['hɒlənd] *pr. n.* Holland *(das)*

hollow ['hɒləʊ] **1.** *adj.* hohl; eingefallen ⟨*Wangen, Schläfen*⟩; *(fig.)* leer ⟨*Versprechen*⟩. **2.** *n.* [Boden]senke, *die.* **3.** *v. t.* ~ **out** aushöhlen

holly ['hɒlɪ] *n.* Stechpalme, *die*

hologram ['hɒləgræm] *n.* Hologramm, *der*

holster ['həʊlstə(r)] *n.* [Pistolen]halfter, *die od. das*

holy ['həʊlɪ] *adj.* heilig

Holy: ~ **'Ghost** *see* ~ **Spirit;** ~ **Land** *n.* the ~ **Land** das Heilige Land; ~ **'Spirit** *n.* Heiliger Geist

homage ['hɒmɪdʒ] *n.* Huldigung, *die* (**to an** + *Akk.*); **pay** *or* **do** ~ **to sb./sth.** jmdm./einer Sache huldigen

home [həʊm] **1.** *n.* **a)** Heim, *das; (flat)* Wohnung, *die; (house)* Haus, *das; (household)* [Eltern]haus, *das; (native country)* Heimat, *die;* **at** ~: zu Hause; **be/feel at** ~ *(fig.)* sich wohl fühlen; **make yourself at** ~: fühl dich wie zu Hause; **b)** *(institution)* Heim, *das.* **2.** *adj.* **a)** Haus-; **b)** *(Sport)* Heim-. **3.** *adv.* nach Hause

home: ~ **address** *n.* Privatanschrift, *die;* ~ **com'puter** *n.* Heimcomputer, *der;* ~**-grown** *adj.* selbstgezogen; ~**land** *n.* Heimat, *die*

'homeless 1. *adj.* obdachlos. **2.** *n.* the ~: die Obdachlosen. **'homelessness** *n.* Obdachlosigkeit, *die*

homely ['həʊmlɪ] *adj.* wohnlich ⟨*Zimmer usw.*⟩; behaglich ⟨*Atmosphäre*⟩

home: ~**-made** *adj.* selbstgemacht; selbstgebacken ⟨*Brot*⟩; hausgemacht ⟨*Lebensmittel*⟩; **H~ Office** *n. (Brit.)* Innenministerium, *das;* **H~ 'Secretary** *n. (Brit.)* Innenminister, *der;* ~**sick** *adj.* heimwehkrank; **become/ be** ~**sick** Heimweh bekommen/haben; ~ **'town** *n.* Heimatstadt, *die;* ~**work** *n. (Sch.)* Hausaufgaben *Pl.;* **piece of** ~**work** Hausaufgabe, *die*

homicide ['hɒmɪsaɪd] *n.* Tötung, *die; (manslaughter)* Totschlag, *der*

homosexual [həʊməʊ'seksjʊəl] **1.** *adj.* homosexuell. **2.** *n.* Homosexuelle, *der/die*

hone [həʊn] *v. t.* wetzen

honest ['ɒnɪst] *adj.* ehrlich. **'honestly** *adv.* ehrlich; redlich ⟨*handeln*⟩; ~! ehrlich!; *(annoyed)* also wirklich!

honesty ['ɒnɪstɪ] *n.* Ehrlichkeit, *die*

honey ['hʌnɪ] *n.* Honig, *der.* **'honeycomb** *n.* Honigwabe, *die.* **'honeymoon** *n.* Flitterwochen *Pl.; (journey)* Hochzeitsreise, *die*

honk [hɒŋk] **1.** *v. i.* ⟨*Fahrzeug, Fahrer:*⟩ hupen. **2.** *n.* Hupen, *das*

honor, honorable *(Amer.) see* honour, honourable

honorary ['ɒnərərɪ] *adj.* Ehren⟨*mitglied, -präsident, -doktor, -bürger*⟩

honour ['ɒnə(r)] *(Brit.)* **1.** *n.* **a)** Ehre, *die;* **b)** *(distinction)* Auszeichnung, *die.* **2.** *v. t.* ehren; *(Commerc.)* honorieren. **honourable** ['ɒnərəbl] *adj. (Brit.)* ehrenwert *(geh.)*

hood [hʊd] *n.* **a)** Kapuze, *die;* **b)** *(Amer. Motor Veh.)* Motorhaube, *die;* **c)** *(of pram)* Verdeck, *das*

hoodlum ['huːdləm] *n.* Rowdy, *der*

hoodwink ['hʊdwɪŋk] *v. t.* hinters Licht führen

hoof [huːf] *n., pl.* ~**s** *or* **hooves** [huːvz] Huf, *der*

hook [hʊk] **1.** *n.* Haken, *der;* **by** ~ **or by crook** mit allen Mitteln. **2.** *v. t.* **a)** *(grasp)* mit Haken/mit einem Haken greifen; **b)** *(fasten)* mit Haken/mit einem Haken befestigen (**to an** + *Dat.*); **c)** **be** ~**ed [on sth.]** *(addicted)* [von etw.] abhängig sein; *(harmlessly)* auf etw. stehen *(ugs.).* **hook 'up** *v. t.* festhaken (**to an** + *Akk.*)

hooligan ['huːlɪgən] *n.* Rowdy, *der.* **hooliganism** ['huːlɪgənɪzm] *n.* Rowdytum, *das*

hoop [huːp] *n.* Reifen, *der*

hooray [hʊ'reɪ] *int.* hurra

hoot [huːt] **1.** *v. i.* **a)** *(call out)* johlen; **b)** ⟨*Eule:*⟩ schreien; **c)** ⟨*Fahrzeug, Fahrer:*⟩ hupen. **2.** *n.* **a)** *(shout)* ~**s** of **derision** verächtliches Gejohle; **b)** *(of owl)* Schrei, *der;* **c)** *(of vehicle)* Hupen, *das.* **'hooter** *n. (Brit.: siren)* Sirene, *die*

hoover ['huːvə(r)] *(Brit.)* **1.** *n.* **a)** **H~ (P)** [Hoover]staubsauger, *der;* **b)** *(made by any company)* Staubsauger, *der.* **2.** *v. t.* staubsaugen

hooves *pl. of* **hoof**

¹hop [hɒp] *n.* **a)** *(plant)* Hopfen, *der;* **b)** in *pl. (Brewing)* Hopfen, *der*

²hop [hɒp] *v. i.,* -**pp**-: **a)** hüpfen; ⟨*Hase:*⟩ hoppeln; **b)** *(fig. coll.)* ~ **out of bed** aus dem Bett springen; ~ **into the car/on [to] the bus/train** sich ins Auto/in den Bus/Zug schwingen *(ugs.).* **2.** *v. t.,* -**pp**- *(Brit. sl.)* ~ **it** sich verziehen *(ugs.).* **3.** *n.* **a)** Hüpfer, *der;* **b)** *(Brit. coll.)* **catch sb. on the** ~: jmdn. überraschen

hope [həʊp] **1.** *n.* Hoffnung, *die;* **sb.'s** ~**[s] of sth.** jmds. Hoffnung auf etw. *(Akk.);* **raise sb.'s** ~**s** jmdm. Hoffnung machen. **2.** *v. i. & t.* hoffen (**for** auf + *Akk.*); **I** ~ **so/not** hoffentlich/hoffent-

lich nicht; ~ **for the best** das Beste
hoffen. **hopeful** ['həʊpfl] *adj.* **a)** zuversichtlich; **be ~ of sth./of doing sth.**
auf etw. *(Akk.)* hoffen/voller Hoffnung sein, etw. zu tun; **b)** *(promising)*
vielversprechend. **'hopefully** *adv.* **a)**
(expectantly) voller Hoffnung; **b)**
(coll.: it is hoped that) hoffentlich.
'hopeless *adj.* **a)** hoffnungslos; **b)**
(inadequate) miserabel. **'hopelessly**
adv. **a)** hoffnungslos; **b)** *(inadequately)* miserabel
hopscotch ['hɒpskɒtʃ] *n.* Himmel-
und-Hölle-Spiel, *das*
horde [hɔ:d] *n.* Horde, *die*
horizon [hə'raɪzn] *n.* Horizont, *der;* **on
the ~:** am Horizont
horizontal [hɒrɪ'zɒntl] *adj.* horizontal; waagerecht. **hori'zontally** *adv.*
horizontal; *(flat)* waagerecht
hormone ['hɔ:məʊn] *n.* Hormon, *das*
horn [hɔ:n] *n.* Horn, *das; (of vehicle)*
Hupe, *die*
hornet ['hɔ:nɪt] *n.* Hornisse, *die*
'horny *adj. (hard)* hornig
horoscope ['hɒrəskəʊp] *n.* Horoskop, *das*
horrible ['hɒrɪbl] *adj.* grauenhaft;
grausig ⟨*Monster*⟩; grauenvoll ⟨*Verbrechen, Alptraum*⟩
horrid ['hɒrɪd] *adj.* scheußlich
horrific [hə'rɪfɪk] *adj.* schrecklich
horrify ['hɒrɪfaɪ] *v. t.* mit Schrecken erfüllen; **be horrified** *(shocked, scan-
dalized)* entsetzt sein **(at,** by über +
Akk.). **'horrifying** *adj.* grauenhaft
horror ['hɒrə(r)] **1.** *n.* Entsetzen, *das*
(**at** über + *Akk.*); *(repugnance)* Grausen, *das; (horrifying thing)* Greuel,
der. **2.** *attrib. adj.* Horror-. **'horror-
stricken, 'horror-struck** *adjs.* von
Entsetzen gepackt
hors-d'œuvre [ɔ:'dɜ:vr] *n.* Hors-
d'œuvre, *das;* ≈ Vorspeise, *die*
horse [hɔ:s] *n.* Pferd, *das*
horse: ~back *n.* **on ~back** zu Pferd;
~man ['hɔ:smən] *n., pl.* **~men**
['hɔ:smən] *([skilled] rider)* [guter] Reiter; **~play** *n.* Balgerei, *die;* **~power**
n., pl. same (Mech.) Pferdestärke, *die;*
~racing *n.* Pferderennsport, *der;*
~radish *n.* Meerrettich, *der;* **~shoe**
n. Hufeisen, *das*
horticulture ['hɔ:tɪkʌltʃə(r)] *n.* Gartenbau, *der*
hose [həʊz], **'hose-pipe** *ns.*
Schlauch, *der*
hospice ['hɒspɪs] *n. (Brit.: for the ter-
minally ill)* Sterbeklinik, *die*

hospitable [hɒ'spɪtəbl] *adj.* gastfreundlich ⟨*Person, Wesensart*⟩; **be ~
to sb.** jmdn. gastfreundlich aufnehmen
hospital ['hɒspɪtl] *n.* Krankenhaus,
das; **in ~** *(Brit.),* **in the ~** *(Amer.)* im
Krankenhaus
hospitality [hɒspɪ'tælɪtɪ] *n.* Gast-
freundschaft, *die*
¹host [həʊst] *n. (large number)* Menge,
die; **a ~ of people/children** eine Menge Leute/eine Schar von Kindern
²host *n.* Gastgeber, *der/*-geberin, *die*
hostage ['hɒstɪdʒ] *n.* Geisel, *die*
hostel ['hɒstl] *n. (Brit.)* Wohnheim,
das
hostess ['həʊstɪs] *n.* Gastgeberin, *die;
(in night-club)* Animierdame, *die*
hostile ['hɒstaɪl] *adj.* **a)** feindlich; **b)**
(unfriendly) feindselig (**to**[**wards**] gegenüber); **be ~ to sth.** etw. ablehnen.
hostility [hɒ'stɪlɪtɪ] *n.* Feindseligkeit,
die
hot [hɒt] *adj.* **a)** heiß; warm ⟨*Mahlzeit,
Essen*⟩; **I am/feel ~:** mir ist heiß; **b)**
(pungent) scharf ⟨*Gewürz, Senf usw.*⟩;
scharf gewürzt ⟨*Essen*⟩; **c)** *(recent)*
noch warm ⟨*Nachrichten*⟩; **d)** *(sl.: illegally obtained)* heiß ⟨*Ware, Geld*⟩.
hot 'air *n. (sl.)* leeres Gerede *(ugs.).*
'hotbed *n. (Hort.)* Mistbeet, *das;
(fig.: of vice, corruption)* Brutstätte, *die*
(**of** für)
'hot dog *n. (coll.)* Hot dog, *der od. das*
hotel [həʊ'tel] *n.* Hotel, *das.* **ho'tel
room** *n.* Hotelzimmer, *das*
hot: ~house *n.* Treibhaus, *das;* **~
line** *n. (Polit.)* heißer Draht
'hotly *adv.* heftig
hot: ~plate *n.* Kochplatte, *die; (to
keep food ~)* Warmhalteplatte, *die;*
~-'water bottle *n.* Wärmflasche, *die*
hound [haʊnd] **1.** *n.* Jagdhund, *der.* **2.**
v. t. verfolgen
hour ['aʊə(r)] *n.* **a)** Stunde, *die;* **half an
~:** eine halbe Stunde; **an ~ and a half**
anderthalb Stunden; **be paid by the ~:**
stundenweise bezahlt werden; **the
24-~ clock** die Vierundzwanzigstun-
denuhr; **b)** *(time o'clock)* Zeit, *die;* **the
small ~s** [**of the morning**] die frühen
Morgenstunden; **0100/0200/0700/
1800 ~s** *(on 24-~ clock)* 1.00/2.00/
17.00/18.00 Uhr. **'hourly** *adj., adv.*
stündlich; **be paid ~:** stundenweise
bezahlt werden
house 1. [haʊs] *n., pl.* **~s** ['haʊzɪz]
Haus, *das;* **to/at my ~:** zu mir [nach
Hause]/bei mir [zu Hause]. **2.** [haʊz]

v. t. **a)** ein Heim geben (+ *Dat.*); **b)** *(keep, store)* unterbringen. **houseboat** ['haʊsbəʊt] *n.* Hausboot, *das* **household** ['haʊshəʊld] *n.* Haushalt, *der; attrib.* Haushalts-. **householder** *n.* Wohnungsinhaber, *der/* -inhaberin, *die*

house [haʊs]: **~keeper** *n.* Haushälterin, *die;* **~keeping** *n.* Hauswirtschaft, *die;* **~-plant** *n.* Zimmerpflanze, *die;* **~-trained** *adj. (Brit.)* stubenrein ⟨*Hund, Katze*⟩; **~-warming** *n.* **~-warming [party]** Einzugsfeier, *die;* **~wife** *n.* Hausfrau, *die;* **~work** *n.* Hausarbeit, *die*

housing ['haʊzɪŋ] *n. (dwellings)* Wohnungen; *(provision of dwellings)* Wohnungsbeschaffung, *die.* **'housing estate** *n. (Brit.)* Wohnsiedlung, *die* **hovel** ['hɒvl] *n.* [armselige] Hütte **hover** ['hɒvə(r)] *v. i.* **a)** schweben; **b)** *(linger)* sich herumdrücken *(ugs.).* **'hovercraft** *n., pl. same* Hovercraft, *das;* Luftkissenfahrzeug, *das.* **'hover mower** *n.* Luftkissenmäher, *der*

how [haʊ] *adv.* wie; **learn ~ to ride a bike/swim** radfahren/schwimmen lernen; **~ 'are you?** wie geht es dir?; *(greeting)* guten Morgen/Tag/ Abend!; **~ do you 'do?** *(formal)* guten Morgen/Tag/Abend!; **~ much?** wieviel?; **~ many?** wieviel?; wie viele?

however [haʊ'evə(r)] *adv.* **a)** wie ... auch; **b)** *(nevertheless)* jedoch; aber **howl** [haʊl] **1.** *n. (of animal)* Heulen, *das; (of distress)* Schrei, *der;* **~s of laughter** brüllendes Gelächter. **2.** *v. i.* ⟨*Tier, Wind:*⟩ heulen; *(with distress)* schreien. **3.** *v. t.* [hinaus]schreien **howler** ['haʊlə(r)] *n. (coll.: blunder)* Schnitzer, *der (ugs.)* **HP** *abbr. (Brit.)* **hire-purchase** **HQ** *abbr.* **headquarters** HQ **hub** [hʌb] *n.* [Rad]nabe, *die; (fig.)* Mittelpunkt, *der* **hubbub** ['hʌbʌb] *n.* Lärm, *der;* **a ~ of voices** ein Stimmengewirr **'hub-cap** *n.* Radkappe, *die* **huddle** ['hʌdl] *v. i.* sich drängen; **~ together** sich zusammendrängen. **huddle 'up** *v. i. (nestle up)* sich zusammenkauern; *(crowd together)* sich [zusammen]drängen **¹hue** [hju:] *n.* Farbton, *der* **²hue** *n.* **~ and cry** lautes Geschrei; *(protest)* Gezeter, *das* **huff** [hʌf] **1.** *v. i.* **~ and puff** schnaufen und keuchen. **2.** *n.* **in a ~:** beleidigt **hug** [hʌg] **1.** *n.* Umarmung, *die;* give

sb. a ~: jmdn. umarmen. **2.** *v. t.,* **-gg-** umarmen **huge** [hju:dʒ] *adj.* riesig; gewaltig ⟨*Unterschied, Verbesserung, Interesse*⟩ **hulking** ['hʌlkɪŋ] *adj. (coll.)* **~ great** klobig **hull** [hʌl] *n. (Naut.)* Schiffskörper, *der* **hum** [hʌm] **1.** *v. i.,* **-mm-:** **a)** summen; ⟨*Maschine:*⟩ brummen; **b)** **~ and haw** *(coll.)* herumdrucksen *(ugs.).* **2.** *v. t.,* **-mm-** summen. **3.** *n.* **a)** Summen, *das; (of machinery)* Brummen, *das;* **b)** *(of voices, conversation)* Gemurmel, *das; (of traffic)* Brausen, *das* **human** ['hju:mən] **1.** *adj.* menschlich; **the ~ race** die menschliche Rasse. **2.** *n.* Mensch, *der.* **human 'being** *n.* Mensch, *der* **humane** [hju:'meɪn] *adj.* human **humanitarian** [hju:mænɪ'teərɪən] *adj.* humanitär **humanity** [hju:'mænɪtɪ] *n.* **a)** *(mankind)* Menschheit, *die; (people collectively)* Menschen; **b)** *(being humane)* Humanität, *die* **humble** ['hʌmbl] **1.** *adj.* **a)** demütig; **b)** *(modest)* bescheiden; **c)** *(low-ranking)* einfach; niedrig ⟨*Status, Rang usw.*⟩. **2.** *v. t.* **a)** demütigen; **~ oneself** sich demütigen *od.* erniedrigen; **b)** *(defeat decisively)* [vernichtend] schlagen. **humbly** ['hʌmblɪ] *adv.* demütig **humdrum** ['hʌmdrʌm] *adj.* alltäglich; eintönig ⟨*Leben*⟩ **humid** ['hju:mɪd] *adj.* feucht. **humidity** [hju:'mɪdɪtɪ] *n.* Feuchtigkeit, *die* **humiliate** [hju:'mɪlɪeɪt] *v. t.* demütigen. **humiliation** [hju:mɪlɪ'eɪʃn] *n.* Demütigung, *die* **humility** [hju:'mɪlɪtɪ] *n.* Demut, *die* **humor** *(Amer.) see* **humour** **humorous** ['hju:mərəs] *adj.* lustig, komisch ⟨*Geschichte, Name, Situation*⟩; witzig ⟨*Bemerkung*⟩ **humour** ['hju:mə(r)] *(Brit.)* **1.** *n.* **a)** Humor, *der;* **sense of ~:** Sinn für Humor; **b)** *(mood)* Laune, *die.* **2.** *v. t.* **~ sb.** jmdm. seinen Willen lassen **hump** [hʌmp] *n.* **a)** *(of person)* Buckel, *der; (of animal)* Höcker, *der;* **b)** *(mound)* Hügel, *der.* **2.** *v. t. (Brit. sl.: carry)* schleppen. **humpback 'bridge** *n.* gewölbte Brücke **¹hunch** [hʌntʃ] *v. t.* **~ [up]** hochziehen **²hunch** *n. (feeling)* Gefühl, *das* **'hunchback** *n. (back)* Buckel, *der; (person)* Bucklige, *der/die;* **be a ~:** einen Buckel haben

hundred ['hʌndrəd] 1. *adj.* hundert; **a** *or* **one ~**: [ein]hundert; **two/several ~**: zweihundert/mehrere hundert; **a** *or* **one ~ and one** [ein]hundert[und]eins. 2. *n.* a) *(number)* hundert; **a** *or* **one/two ~**: [ein]hundert/zweihundert; **b)** *(written figure; group)* Hundert, *das;* **c)** *(indefinite amount)* **~s** Hunderte. *See also* **eight. hundredth** ['hʌndrədθ] 1. *adj.* hundertst...; **a ~ part** ein Hundertstel. 2. *n. (fraction)* Hundertstel, *das; (in sequence)* hundertste, *der/die/das; (in rank)* Hundertste, *der/die/das.* '**hundredweight** *n., pl. same (Brit.)* 50,8 kg; ≈ Zentner, *der*

hung *see* **hang 1, 2**

Hungarian [hʌŋ'geəriən] 1. *adj.* ungarisch; **sb. is ~**: jmd. ist Ungar/Ungarin. 2. *n.* a) *(person)* Ungar, *der/*Ungarin, *die;* b) *(language)* Ungarisch, *das; see also* **English 2 a**

Hungary ['hʌŋgəri] *pr. n.* Ungarn *(das)*

hunger ['hʌŋgə(r)] 1. *n.* Hunger, *der.* 2. *v. i.* **~ after** *or* **for sb./sth.** [heftiges] Verlangen nach jmdm./etw. haben. '**hunger-strike** *n.* Hungerstreik, *der;* **go on ~**: in den Hungerstreik treten

hungry ['hʌŋgrɪ] *adj.* hungrig; **be ~**: Hunger haben; **go ~**: hungern

hunk [hʌŋk] *n.* [großes] Stück

hunt [hʌnt] 1. *n.* Jagd, *die; (search)* Suche, *die.* 2. *v. t.* jagen; *(search for)* Jagd machen auf (+ *Akk.*) ⟨*Mörder usw.*⟩. 3. *v. i.* jagen; **go ~ing** auf die Jagd gehen; **~ after** *or* **for** Jagd machen auf (+ *Akk.*); *(seek)* suchen. '**hunter** *n.* Jäger, *der*

'**hunting** *n.* die Jagd (**of** auf + *Akk.*); *(searching)* Suche, *die* (**for** nach)

hurdle ['hɜːdl] *n.* Hürde, *die*

hurl [hɜːl] *v. t.* werfen; *(violently)* schleudern; **~ insults at sb.** jmdm. Beleidigungen ins Gesicht schleudern

hurrah [hʊ'rɑː], **hurray** [hʊ'reɪ] *int.* hurra

hurricane ['hʌrɪkən] *n.* Orkan, *der*

hurried ['hʌrɪd] *adj.* eilig; überstürzt ⟨*Abreise*⟩; in Eile ausgeführt ⟨*Arbeit*⟩

hurry ['hʌrɪ] 1. *n.* Eile, *die;* **in a ~**: eilig; **be in a ~**: es eilig haben; **there's no ~**: es eilt nicht. 2. *v. t.* antreiben ⟨*Person*⟩; hinunterschlingen ⟨*Essen*⟩; **~ one's work** seine Arbeit in zu großer Eile erledigen. 3. *v. i.* sich beeilen; *(to or from place)* eilen. **hurry 'up** 1. *v. i.* sich beeilen. 2. *v. t.* antreiben

hurt [hɜːt] 1. *v. t., hurt* **a)** weh tun (+ *Dat.*); *(injure)* verletzen; **~ oneself**

sich *(Dat.)* weh tun; *(injure oneself)* sich verletzen; **~ one's arm/back** sich *(Dat.)* am Arm/Rücken weh tun; *(injure)* sich *(Dat.)* den Arm/am Rücken verletzen; **b)** *(damage, be detrimental to)* schaden (+ *Dat.*); **c)** *(upset)* verletzen ⟨*Person, Stolz*⟩. 2. *v. i.,* **hurt a)** weh tun; **b)** *(cause damage, be detrimental)* schaden. 3. *adj.* gekränkt ⟨*Tonfall, Miene*⟩. 4. *n. (emotional pain)* Schmerz, *der.* **hurtful** ['hɜːtfl] *adj.* verletzend

hurtle ['hɜːtl] *v. i.* rasen *(ugs.)*

husband ['hʌzbənd] *n.* Ehemann, *der;* **my/your/her ~**: mein/dein/ihr Mann; **~ and wife** Mann und Frau

hush [hʌʃ] 1. *n. (silence)* Schweigen, *das; (stillness)* Stille, *die.* 2. *v. t. (silence)* zum Schweigen bringen; *(still)* beruhigen. 3. *v. i.* still sein; **~!** still! **hush 'up** *v. t.* vertuschen

husk [hʌsk] *n.* Spelze, *die*

¹**husky** ['hʌskɪ] *adj.* heiser

²**husky** *n.* Eskimohund, *der*

hustle ['hʌsl] 1. *v. t.* drängen (**into** zu). 2. *n.* **~ and bustle** geschäftiges Treiben

hut [hʌt] *n.* Hütte, *die*

hutch [hʌtʃ] *n.* Stall, *der*

hyacinth ['haɪəsɪnθ] *n.* Hyazinthe, *die*

hybrid ['haɪbrɪd] 1. *n.* Hybride, *die od. der* (**between** aus); *(fig.: mixture)* Mischung, *die.* 2. *adj.* hybrid ⟨*Züchtung*⟩

hydrangea [haɪ'dreɪndʒə] *n.* Hortensie, *die*

hydrant ['haɪdrənt] *n.* Hydrant, *der*

hydraulic [haɪ'drɔːlɪk] *adj.* hydraulisch

hydrochloric acid [haɪdrəklɔːrɪk 'æsɪd] *n.* Salzsäure, *die*

hydroelectric [haɪdrəʊɪ'lektrɪk] *adj.* hydroelektrisch; **~ power station** Wasserkraftwerk, *das*

hydrofoil ['haɪdrəfɔɪl] *n.* Tragflächenboot, *das*

hydrogen ['haɪdrədʒən] *n.* Wasserstoff, *der.* '**hydrogen bomb** *n.* Wasserstoffbombe, *die*

hyena [haɪ'iːnə] *n.* Hyäne, *die*

hygiene ['haɪdʒiːn] *n.* Hygiene, *die.* **hygienic** [haɪ'dʒiːnɪk] *adj.* hygienisch

hymn [hɪm] *n.* Hymne, *die; (sung in service)* Kirchenlied, *das.* '**hymnbook** *n.* Gesangbuch, *das*

hypermarket ['haɪpəmɑːkɪt] *n. (Brit.)* Verbrauchermarkt, *der*

hyphen ['haɪfn] 1. *n.* Bindestrich, *der.* 2. *v. t.* mit Bindestrich schreiben

hyphenate ['haɪfəneɪt] *see* **hyphen 2**
hypnosis [hɪp'nəʊsɪs] *n., pl.* **hypnoses**
[hɪp'nəʊsiːz] Hypnose, *die; (act, process)* Hypnotisierung, *die;* **under ~** : in
Hypnose *(Dat.)*. **hypnotic** [hɪp-
'nɒtɪk] *adj.* hypnotisch. **hypnotism**
['hɪpnətɪzm] *n.* Hypnotik, *die; (act)*
Hypnotisieren, *das.* **hypnotist** ['hɪp-
nətɪst] *n.* Hypnotiseur, *der*/Hypnoti-
seuse, *die.* **hypnotize** ['hɪpnətaɪz]
v. t. hypnotisieren
hypochondria [haɪpə'kɒndrɪə] *n.* Hy-
pochondrie, *die.* **hypochondriac**
[haɪpə'kɒndrɪæk] *n.* Hypochonder,
der
hypocrisy [hɪ'pɒkrɪsɪ] *n.* Heuchelei,
die. **hypocrite** ['hɪpəkrɪt] *n.* Heuchler, *der*/Heuchlerin, *die.* **hypocritical** [hɪpə'krɪtɪkl] *adj.* heuchlerisch
hypodermic [haɪpə'dɜːmɪk] *adj. & n.*
~ |syringe| Injektionsspritze, *die*
hypotenuse [haɪ'pɒtənjuːz] *n.* Hypotenuse, *die*
hypothesis [haɪ'pɒθɪsɪs] *n., pl.* **hypo-
theses** [haɪ'pɒθɪsiːz] Hypothese, *die.*
hypothetical [haɪpə'θetɪkl] *adj.* hypothetisch
hysteria [hɪ'stɪərɪə] *n.* Hysterie, *die.*
hysterical [hɪ'sterɪkl] *adj.* hysterisch. **hysterics** [hɪ'sterɪks] *n. pl.*
(laughter) hysterischer Lachanfall;
(crying) hysterischer Weinkrampf;
have ~ : hysterisch lachen/weinen

I

¹**I, i** [aɪ] *n.* I, i, *das*
²**I** *pron.* ich
ice [aɪs] **1.** *n.* **a)** Eis, *das;* **feel/be like ~**
(be very cold) eiskalt sein; **b)** *(~-*
cream) [Speise]eis, *das;* **an ~/two ~s**
ein/zwei Eis. **2.** *v. t.* glasieren ⟨*Ku-*
chen⟩. **ice 'over, ice 'up** *v. i.* ⟨*Gewäs-*
ser:⟩ zufrieren
'**ice age** *n.* Eiszeit, *die*
iceberg ['aɪsbɜːg] *n.* Eisberg, *der*
ice: ~box *n. (Amer.)* Kühlschrank,
der; **~-cold** *adj.* eiskalt; **~-'cream**
n. Eis, *das;* Eiscreme, *die;* **one**
~-cream/two/too many ~-creams ein/

zwei/zuviel Eis; **~-cube** *n.* Eiswür-
fel, *die;* **~ hockey** *n.* Eishockey, *das*
Iceland ['aɪslənd] *pr. n.* Island *(das).*
Icelandic [aɪs'lændɪk] **1.** *adj.* isländisch. **2.** *n.* Isländisch, *das; see also*
English 2 a
ice: ~ 'lolly *n.* Eis am Stiel; **~-rink** *n.*
Eisbahn, *die;* **~-skate 1.** *n.* Schlittschuh, *der;* **2.** *v. i.* Schlittschuh laufen; **~-skating** *n.* Schlittschuhlaufen, *das*
icicle ['aɪsɪkl] *n.* Eiszapfen, *der*
icing ['aɪsɪŋ] *n.* Zuckerguß, *der.* '**icing**
sugar *n. (Brit.)* Puderzucker, *der*
icon ['aɪkɒn] *n.* Ikone, *die*
icy ['aɪsɪ] *adj.* **a)** vereist ⟨*Berge, Land-*
schaft, Straße⟩; eisreich ⟨*Region,*
Land⟩; **in ~ conditions** bei Eis; **b)**
(very cold) eiskalt; eisig; *(fig.)* frostig
I'd [aɪd] **a)** = I had; **b)** = I would
idea [aɪ'dɪə] *n.* Idee, *die;* Gedanke,
der; (mental picture) Vorstellung, *die;*
(vague notion) Ahnung, *die;* **have you**
any ~ |of| how ...? weißt du ungefähr,
wie ...?; **have no ~ |of| where ...** : keine
Ahnung haben, wo ...; **not have the**
slightest or faintest ~ : nicht die leise-
ste Ahnung haben
ideal [aɪ'dɪəl] **1.** *adj.* ideal; vollendet
⟨*Ehemann, Gastgeber*⟩; vollkommen
⟨*Welt*⟩. **2.** *n.* Ideal, *das.* **idealism** [aɪ-
'dɪəlɪzm] *n.* Idealismus, *der.* **idealist**
[aɪ'dɪəlɪst] *n.* Idealist, *der*/Idealistin,
die. **idealistic** [aɪdɪə'lɪstɪk] *adj.* idealistisch. **idealize** [aɪ'dɪəlaɪz] *v. t.* idealisieren. **ideally** [aɪ'dɪəlɪ] *adv.* ideal;
~, ... : idealerweise *od.* im Idealfall ...
identical [aɪ'dentɪkl] *adj.* identisch;
be ~ : sich *(Dat.)* völlig gleichen; **~**
twins eineiige Zwillinge
identification [aɪdentɪfɪ'keɪʃn] *n.*
Identifizierung, *die; (of plants, an-*
imals⟩ Bestimmung, *die*
identify [aɪ'dentɪfaɪ] *v. t.* identifizieren; bestimmen ⟨*Pflanze, Tier*⟩
identity [aɪ'dentɪtɪ] *n.* Identität, *die;*
proof of ~ : Identitätsnachweis, *der;*
|case of| mistaken ~ : [Personen]ver-
wechslung, *die.* **i'dentity card** *n.*
[Personal]ausweis, *der*
idiocy ['ɪdɪəsɪ] *n.* Idiotie, *die*
idiom ['ɪdɪəm] *n.* [Rede]wendung, *die.*
idiomatic [ɪdɪə'mætɪk] *adj.* idioma-
tisch
idiosyncrasy [ɪdɪə'sɪŋkrəsɪ] *n.* Eigentümlichkeit, *die.* **idiosyncratic** [ɪdɪə-
sɪŋ'krætɪk] *adj.* eigenwillig
idiot ['ɪdɪət] *n.* Idiot, *der (ugs.).*
idiotic [ɪdɪ'ɒtɪk] *adj.* idiotisch *(ugs.).*

idle ['aɪdl] **1.** *adj.* **a)** *(lazy)* faul; **b)** *(not in use)* außer Betrieb *nachgestellt;* **be** ~ ⟨*Maschinen, Fabrik:*⟩ stillstehen; **c)** bloß ⟨*Neugier, Spekulation*⟩; leer ⟨*Geschwätz*⟩. **2.** *v. i.* ⟨*Motor:*⟩ leerlaufen. **idle a'way** *v. t.* vertun
'idleness *n.* Faulheit, *die*
idol ['aɪdl] *n.* Idol, *das.* **idolize** ['aɪdə-laɪz] *v. t.* vergöttern
idyllic [ɪ'dɪlɪk] *adj.* idyllisch
i.e. [aɪ'iː] *abbr.* that is d. h.; i. e.
if [ɪf] *conj.* **a)** wenn; **if anyone should ask ...:** falls jemand fragt, ...; **if I knew what to do ...:** wenn ich wüßte, was ich tun soll ...; **if I were you** an deiner Stelle; **if so/not** wenn ja/nein *od.* nicht; **if then/that/at all** wenn überhaupt; **as if** als ob; **if I only knew, if only I knew! wenn ich das nur wüßte!; if it isn't Ronnie!** das ist doch Ronnie!; **b)** *(whenever)* [immer] wenn; **c)** *(whether)* ob; **d)** *(though)* auch *od.* selbst wenn; **e)** *(despite being)* wenn auch
igloo ['ɪgluː] *n.* Iglu, *der od. das*
ignite [ɪg'naɪt] **1.** *v. t.* anzünden. **2.** *v. i.* sich entzünden. **ignition** [ɪg'nɪʃn] *n.* **a)** *(igniting)* Zünden, *das;* **b)** *(Motor Veh.)* Zündung, *die;* ~ **key** Zündschlüssel, *der*
ignorance ['ɪgnərəns] *n.* Unwissenheit, *die;* **keep sb. in** ~ **of sth.** jmdn. in Unkenntnis über etw. *(Akk.)* lassen
ignorant ['ɪgnərənt] *adj.* unwissend; **be** ~ **of sth.** *(uninformed)* über etw. *(Akk.)* nicht informiert sein
ignore [ɪg'nɔː(r)] *v. t.* ignorieren; nicht befolgen ⟨*Befehl, Rat*⟩; übergehen ⟨*Frage, Bemerkung*⟩
ill [ɪl] **1.** *adj.,* **worse** [wɜːs], **worst** [wɜːst] krank; **fall** ~: krank werden. **2.** *adv.* **be** ~ **at ease** sich unwohl fühlen. **3.** *n.* Übel, *das*
I'll [aɪl] **a)** = **I shall;** **b)** = **I will**
ill-advised *adj.* unklug
illegal [ɪ'liːgl] *adj.,* **il'legally** *adv.* illegal
illegible [ɪ'ledʒɪbl] *adj.* unleserlich
illegitimate [ɪlɪ'dʒɪtɪmət] *adj.* unehelich ⟨*Kind*⟩
ill 'health *n.* schwache Gesundheit
illicit [ɪ'lɪsɪt] *adj.* unerlaubt ⟨*Beziehung, [Geschlechts]verkehr*⟩; Schwarz-⟨*handel, -verkauf, -arbeit*⟩
'ill-informed *adj.* schlecht informiert; auf Unkenntnis beruhend ⟨*Bemerkung, Urteil*⟩
illiteracy [ɪ'lɪtərəsɪ] *n.* Analphabetentum, *das*
illiterate [ɪ'lɪtərət] *adj.* des Lesens und

Schreibens unkundig; analphabetisch ⟨*Bevölkerung*⟩
illness ['ɪlnɪs] *n.* Krankheit, *die*
illogical [ɪ'lɒdʒɪkl] *adj.* unlogisch
ill-'treat *v. t.* mißhandeln. **ill-'treatment** *n.* Mißhandlung, *die*
illuminate [ɪ'luːmɪneɪt] *v. t.* beleuchten. **illuminating** [ɪ'luːmɪneɪtɪŋ] *adj.* aufschlußreich. **illumination** [ɪluː-mɪ'neɪʃn] *n.* Beleuchtung, *die*
illusion [ɪ'luːʒn] *n.* Illusion, *die;* **be under the** ~ **that ...:** sich *(Dat.)* einbilden, daß ... **illusory** [ɪ'luːsərɪ] *adj.* illusorisch
illustrate ['ɪləstreɪt] *v. t.* **a)** *(serve as example of)* veranschaulichen; **b)** illustrieren ⟨*Buch, Erklärung*⟩. **illustration** [ɪlə'streɪʃn] *n.* **a)** *(example)* Beispiel, *das (of für)*; **b)** *(picture)* Abbildung, *die*
ill 'will *n.* Böswilligkeit, *die*
I'm [aɪm] = **I am**
image ['ɪmɪdʒ] *n.* **a)** Bildnis, *das (geh.);* **b)** *(Optics)* Bild, *das;* **c)** |public| ~: Image, *das*
imaginable [ɪ'mædʒɪnəbl] *adj.* **the best solution** ~: die denkbar beste Lösung
imaginary [ɪ'mædʒɪnərɪ] *adj.* imaginär *(geh.)*; eingebildet ⟨*Krankheit*⟩
imagination [ɪmædʒɪ'neɪʃn] *n.* **a)** Phantasie, *die;* **b)** *(fancy)* Einbildung, *die*
imaginative [ɪ'mædʒɪnətɪv] *adj.* phantasievoll; *(showing imagination)* einfallsreich
imagine [ɪ'mædʒɪn] *v. t.* **a)** sich *(Dat.)* vorstellen; **b)** *(coll.: suppose)* glauben; **c)** *(get the impression)* ~ **that ...:** sich *(Dat.)* einbilden[, daß ...]
imbalance [ɪm'bæləns] *n.* Unausgeglichenheit, *die*
imbecile ['ɪmbɪsiːl] *n.* Idiot, *der (ugs.)*
imitate ['ɪmɪteɪt] *v. t.* nachahmen. **imitation** [ɪmɪ'teɪʃn] *n.* **a)** Nachahmung, *die;* **b)** *(counterfeit)* Imitation, *die*
immaculate [ɪ'mækjʊlət] *adj.* *(spotless)* makellos; *(faultless)* tadellos
immaterial [ɪmə'tɪərɪəl] *adj.* unerheblich
immature [ɪmə'tjʊə(r)] *adj.* unreif; noch nicht voll entwickelt ⟨*Lebewesen*⟩. **immaturity** [ɪmə'tjʊərɪtɪ] *n.* Unreife, *die*
immediate [ɪ'miːdjət] *adj.* **a)** unmittelbar; *(nearest)* nächst... ⟨*Nachbar[schaft], Umgebung, Zukunft*⟩; engst... ⟨*Familie*⟩; **b)** *(occurring at once)* prompt; unverzüglich ⟨*Han-*

deln, Maßnahmen⟩; umgehend ⟨*Antwort*⟩. **im'mediately 1.** *adv.* **a)** unmittelbar; **b)** *(without delay)* sofort. **2.** *conj. (coll.)* sobald

immemorial [ɪmɪ'mɔ:rɪəl] *adj.* from time ~: seit undenklichen Zeiten

immense [ɪ'mens] *adj.* **a)** ungeheuer; **b)** *(coll.: great)* enorm. **im'mensely** *adv.* **a)** ungeheuer; **b)** *(coll.: very much)* unheimlich *(ugs.)*

immerse [ɪ'mɜ:s] *v.t.* [ein]tauchen; **be ~d in thought/one's work** in Gedanken versunken/in seine Arbeit vertieft sein. **immersion** [ɪ'mɜ:ʃn] *n.* Eintauchen, *das.* **im'mersion heater** *n.* Heißwasserbereiter, *der*

immigrant ['ɪmɪɡrənt] **1.** *n.* Einwanderer, *der*/Einwanderin, *die.* **2.** *adj.* Einwanderer-; ~ **workers** ausländische Arbeitnehmer

immigration [ɪmɪ'ɡreɪʃn] *n.* Einwanderung *die* (**into** nach, **from** aus); *attrib.* Einwanderungs⟨*kontrolle, -gesetz*⟩; ~ **officer** Beamter/Beamtin der Einwanderungsbehörde

imminent ['ɪmɪnənt] *adj.* unmittelbar bevorstehend; drohend ⟨*Gefahr*⟩; **be ~:** unmittelbar bevorstehen/drohen

immobile [ɪ'məʊbaɪl] *adj. (immovable)* unbeweglich. **immobilize** [ɪ'məʊbəlaɪz] *v.t.* verankern; *(fig.)* lähmen

immodest [ɪ'mɒdɪst] *adj.* unbescheiden; *(improper)* unanständig

immoral [ɪ'mɒrəl] *adj.* unmoralisch; *(in sexual matters)* sittenlos. **immorality** [ɪmə'rælɪtɪ] *n.* Unmoral, *die; (in sexual matters)* Sittenlosigkeit, *die*

immortal [ɪ'mɔ:tl] *adj.* unsterblich. **immortality** [ɪmɔ:'tælɪtɪ] *n.* Unsterblichkeit, *die.* **immortalize** [ɪ'mɔ:təlaɪz] *v.t.* unsterblich machen

immovable [ɪ'mu:vəbl] *adj.* unbeweglich; **be ~:** sich nicht bewegen lassen

immune [ɪ'mju:n] *adj.* **a)** *(exempt)* sicher (**from** vor + *Dat.*); **b)** *(not susceptible)* unempfindlich (**to** gegen); **c)** *(Med.)* immun (**to** gegen). **immunity** [ɪ'mju:nɪtɪ] *n.* **a) diplomatic ~:** diplomatische Immunität; **b)** *(Med.)* Immunität, *die.* **immunize** ['ɪmjʊnaɪz] *v.t.* immunisieren

imp [ɪmp] *n.* **a)** Kobold, *der;* **b)** *(coll.: child)* Racker, *der (fam.)*

impact ['ɪmpækt] *n.* **a)** *(on, against* auf + *Akk.*); *(collision)* Zusammenprall, *der;* **b)** *(fig.)* Wirkung, *die*

impair [ɪm'peə(r)] *v.t.* beeinträchtigen; schaden (+ *Dat.*)⟨*Gesundheit*⟩

impale [ɪm'peɪl] *v.t.* aufspießen

impart [ɪm'pɑ:t] *v.t.* **a)** *(give)* [ab]geben (**to an** + *Akk.*); **b)** *(communicate)* kundtun *(geh.)* (**to** *Dat.*); vermitteln ⟨*Kenntnisse*⟩ (**to** *Dat.*)

impartial [ɪm'pɑ:ʃl] *adj.* unparteiisch; gerecht ⟨*Entscheidung, Urteil*⟩

impassable [ɪm'pɑ:səbl] *adj.* unpassierbar (**to** für); *(to vehicles)* unbefahrbar (**to** für)

impasse ['æmpɑ:s] *n.* Sackgasse, *die*

impassive [ɪm'pæsɪv] *adj.* ausdruckslos

impatience [ɪm'peɪʃəns] *n.* Ungeduld, *die* (**at** über + *Akk.*)

impatient [ɪm'peɪʃənt] *adj.* ungeduldig; ~ **at sth./with sb.** ungeduldig über etw. *(Akk.)*/mit jmdm. **im'patiently** *adv.* ungeduldig

impeccable [ɪm'pekəbl] *adj.* makellos; tadellos ⟨*Manieren*⟩

impede [ɪm'pi:d] *v.t.* behindern. **impediment** [ɪm'pedɪmənt] *n.* **a)** Hindernis, *das* (**to** für); **b)** *(speech defect)* Sprachfehler, *der*

impel [ɪm'pel] *v.t.,* **-ll-** treiben; **feel ~led to do sth.** sich genötigt *od.* gezwungen fühlen, etw. zu tun

impending [ɪm'pendɪŋ] *adj.* bevorstehend

impenetrable [ɪm'penɪtrəbl] *adj.* undurchdringlich (**by, to** für)

imperative [ɪm'perətɪv] **1.** *adj.* dringend erforderlich. **2.** *n. (Ling.)* Imperativ, *der*

imperceptible [ɪmpə'septɪbl] *adj.* nicht wahrnehmbar; *(very slight or gradual)* unmerklich

imperfect [ɪm'pɜ:fɪkt] **1.** *adj.* **a)** *(incomplete)* unvollständig; **b)** *(faulty)* mangelhaft. **2.** *n. (Ling.)* Imperfekt, *das.* **imperfection** [ɪmpə'fekʃn] *n.* **a)** *(incompleteness)* Unvollständigkeit, *die;* **b)** *(fault)* Mangel, *der.* **im'perfectly** *adv.* **a)** *(incompletely)* unvollständig; **b)** *(faultily)* fehlerhaft

imperial [ɪm'pɪərɪəl] *adj.* kaiserlich. **imperialism** [ɪm'pɪərɪəlɪzm] *n.* Imperialismus, *der*

imperil [ɪm'perɪl] *v.t., (Brit.)* **-ll-** gefährden

imperious [ɪm'pɪərɪəs] *adj.* herrisch

impersonal [ɪm'pɜ:sənl] *adj.* unpersönlich

impersonate [ɪm'pɜ:səneɪt] *v.t.* sich ausgeben als; *(to entertain)* imitieren; nachahmen. **impersonator** [ɪm'pɜ:səneɪtə(r)] *n. (entertainer)* Imitator, *der*/Imitatorin, *die*

impertinence [ɪm'pɜːtɪnəns] *n.* Unverschämtheit, *die*

impertinent [ɪm'pɜːtɪnənt] *adj.* unverschämt

imperturbable [ɪmpə'tɜːbəbl] *adj.* gelassen; **be completely ~**: durch nichts zu erschüttern sein

impervious [ɪm'pɜːvɪəs] *adj.* undurchlässig; **be ~ to sth.** *(fig.)* unempfänglich für etw. sein

impetuous [ɪm'petjʊəs] *adj.* unüberlegt; impulsiv ⟨*Person*⟩

impetus ['ɪmpɪtəs] *n.* **a)** Kraft, *die*; **b)** *(fig.)* Motivation, *die*

impinge [ɪm'pɪndʒ] *v.i.* **~ on sth.** auf etw. *(Akk.)* Einfluß nehmen

impish *adj.* lausbübisch

implacable [ɪm'plækəbl] *adj.* unversöhnlich; erbittert ⟨*Gegner*⟩

implausible [ɪm'plɔːzɪbl] *adj.* unglaubwürdig

implement 1. ['ɪmplɪmənt] *n.* Gerät, *das*. **2.** ['ɪmplɪment] *v.t.* [in die Tat] umsetzen ⟨*Politik, Plan usw.*⟩

implicate ['ɪmplɪkeɪt] *v.t.* belasten; **be ~d in a scandal** in einen Skandal verwickelt sein. **implication** [ɪmplɪ'keɪʃn] *n.* Implikation, *die*; **by ~**: implizit

implicit [ɪm'plɪsɪt] *adj.* **a)** *(implied)* implizit *(geh.)*; unausgesprochen ⟨*Drohung, Zweifel*⟩; **b)** *(resting on authority)* unbedingt; blind ⟨*Vertrauen*⟩

implore [ɪm'plɔː(r)] *v.t.* anflehen **(for** um)

imply [ɪm'plaɪ] *v.t.* **a)** implizieren *(geh.)*; *(say indirectly)* hindeuten auf *(+ Akk.)*; **b)** *(insinuate)* unterstellen

impolite [ɪmpə'laɪt] *adj.* unhöflich

import 1. [ɪm'pɔːt] *v.t.* importieren, einführen ⟨*Waren*⟩ **(from** aus, **into** nach). **2.** ['ɪmpɔːt] *n.* **a)** *(process)* Import, *der*; **b)** *(article)* Importgut, *das*

importance [ɪm'pɔːtəns] *n.* Wichtigkeit, *die* **(to** für); *(significance)* Bedeutung, *die*; **be of ~**: wichtig sein; **full of one's own ~**: von seiner eigenen Wichtigkeit überzeugt

important [ɪm'pɔːtənt] *adj.* wichtig **(to** für); *(significant)* bedeutend

im'porter *n.* Importeur, *der*

impose [ɪm'pəʊz] *v.t.* auferlegen *(geh.)* ⟨*Bürde, Verpflichtung*⟩ **(up|on** *Dat.*); erheben ⟨*Steuer*⟩ **(on** auf + *Akk.*); verhängen ⟨*Kriegsrecht*⟩; anordnen ⟨*Rationierung*⟩. **im'pose on** *v.t.* ausnutzen ⟨*Gutmütigkeit, Toleranz usw.*⟩; **~ on sb.** sich jmdm. aufdrängen

imposing [ɪm'pəʊzɪŋ] *adj.* imposant

imposition [ɪmpə'zɪʃn] *n.* **a)** Auferlegung, *die*; *(of tax)* Erhebung, *die*; **b)** *(unreasonable demand)* Zumutung, *die*

impossibility [ɪmpɒsɪ'bɪlɪtɪ] *n.* Unmöglichkeit, *die*

impossible [ɪm'pɒsɪbl] *adj.*, **impossibly** [ɪm'pɒsɪblɪ] *adv.* unmöglich

impostor [ɪm'pɒstə(r)] *n.* Hochstapler, *der*/-staplerin, *die*; *(swindler)* Betrüger, *der*/Betrügerin, *die*

impound [ɪm'paʊnd] *v.t.* beschlagnahmen

impoverished [ɪm'pɒvərɪʃt] *adj.* **be/become ~**: verarmt sein/verarmen

impracticable [ɪm'præktɪkəbl] *adj.* undurchführbar

impractical [ɪm'præktɪkl] *adj.* **a)** *(unpractical)* unpraktisch; **b)** *see* **impracticable**

imprecise [ɪmprɪ'saɪs] *adj.* ungenau

impregnable [ɪm'pregnəbl] *adj.* uneinnehmbar ⟨*Festung, Bollwerk*⟩; *(fig.)* unanfechtbar ⟨*Ruf, Stellung*⟩

impregnate ['ɪmpregneɪt] *v.t.* imprägnieren

impress [ɪm'pres] *v.t.* beeindrucken; *abs.* Eindruck machen; **be ~ed by** *or* **with sth.** von etw. beeindruckt sein. **im'press [up]on** *v.t.* einschärfen **(+ *Dat.*)**; **~ sth. [up]on sb.'s memory** jmdm. etw. einschärfen. **impression** [ɪm'preʃn] *n.* **a)** Eindruck, *der*; **form an ~ of sb.** sich *(Dat.)* ein Bild von jmdm. machen; **b** *(impersonation)* **do an ~ of sb.** jmdn. imitieren; **do ~s** andere Leute imitieren. **impressionist** [ɪm'preʃənɪst] *n.* Impressionist, *der*/Impressionistin, *die*

impressive [ɪm'presɪv] *adj.* beeindruckend; imponierend

imprint 1. ['ɪmprɪnt] *n.* Abdruck, *der*; *(fig.)* Stempel, *der*. **2.** [ɪm'prɪnt] *v.t.* aufdrucken; *(fig.)* einprägen **(on** *Dat.*)

imprison [ɪm'prɪzn] *v.t.* in Haft nehmen; **be ~ed** sich in Haft befinden. **im'prisonment** *n.* Haft, *die*; **a long term of ~**: eine lange Haftstrafe

improbable [ɪm'prɒbəbl] *adj.* unwahrscheinlich

impromptu [ɪm'prɒmptjuː] **1.** *adj.* improvisiert; **an ~ speech** eine Stegreifrede. **2.** *adv.* aus dem Stegreif

improper [ɪm'prɒpə(r)] *adj.* **a)** *(wrong)* unrichtig; **b)** *(unseemly)* unpassend; *(indecent)* unanständig. **im'properly** *adv. see* **improper**: unrichtig; unpassend; unanständig

improvable [ɪm'pru:vəbl] *adj.* verbesserungsfähig

improve [ɪm'pru:v] **1.** *v. i.* besser werden; ⟨*Person, Wetter:*⟩ sich bessern. **2.** *v. t.* verbessern. **3.** *v. refl.* ~ **oneself** sich weiterbilden. **im'prove [up]on** *v. t.* überbieten ⟨*Rekord, Angebot*⟩; verbessern ⟨*Leistung*⟩. **improvement** [ɪm'pru:vmənt] *n.* Verbesserung, *die* (**on, over** gegenüber); **make** ~**s to sth.** Verbesserungen an etw. *(Dat.)* vornehmen

improvise ['ɪmprəvaɪz] *v. t.* improvisieren

impudence ['ɪmpjʊːdəns] *n.* Unverschämtheit, *die; (brazenness)* Dreistigkeit, *die*

impudent ['ɪmpjʊdənt] *adj.,* '**impudently** *adv.* unverschämt; *(brazen[ly])* dreist

impulse ['ɪmpʌls] *n.* Impuls, *der;* **on |an|** ~: impulsiv. **impulsive** [ɪm'pʌlsɪv] *adj.* impulsiv

impunity [ɪm'pju:nɪtɪ] *v. t.* **with** ~: ungestraft

impure [ɪm'pjʊə(r)] *adj.* unrein. **impurity** [ɪm'pjʊərɪtɪ] *n.* Unreinheit, *die; (foreign body)* Fremdstoff, *der*

impute [ɪm'pju:t] *v. t.* zuschreiben (**to** *Dat.*)

in [ɪn] **1.** *prep. (position; also fig.)* in (+ *Dat.*); *(into)* in (+ *Akk.*); **in this heat** bei dieser Hitze; **two feet in diameter** mit einem Durchmesser von zwei Fuß; **there are three feet in a yard** ein Yard hat drei Fuß; **draw in crayon/ink** mit Kreide/Tinte zeichnen; **pay in pounds/dollars** in Pfund/Dollars bezahlen; **in fog/rain** *etc.* bei Nebel/Regen *usw.;* **in the 20th century** im 20. Jahrhundert; **4 o'clock in the morning/afternoon** 4 Uhr morgens/abends; **in 1990** [im Jahre] 1990; **in three minutes/years** in drei Minuten/Jahren; **in doing this, he ...**; indem er das tut/tat, er ...; **in that ...**; insofern als. **2.** *adv.* **a)** *(inside)* hinein⟨*gehen usw.*⟩*;* herein⟨*kommen usw.*⟩*;* **b)** *(at home, work, etc.)* **be in** dasein; **c) have it in for sb.** es auf jmdn. abgesehen haben *(ugs.);* **sb. is in for sth.** *(about to undergo)* jmdm. steht etw. bevor. **3.** *adj. (coll.:* **in fashion)** in *(ugs.).* **4.** *n.* **know the ins and outs of sth.** sich in einer Sache genau auskennen

ina'bility *n.* Unfähigkeit, *die*

inaccessible [ɪnək'sesɪbl] *adj.* unzugänglich

in'accuracy *n.* **a)** *(incorrectness)* Unrichtigkeit, *die;* **b)** *(imprecision)* Ungenauigkeit, *die*

in'accurate *adj.* **a)** *(incorrect)* unrichtig; **b)** *(imprecise)* ungenau

in'active *adj.* untätig. **inac'tivity** *n.* Untätigkeit, *die*

in'adequate *adj.* unzulänglich; *(incompetent)* ungeeignet; **feel** ~: sich überfordert fühlen

inadvertent [ɪnəd'vɜ:tənt] *adj.,* **inad'vertently** *adv.* versehentlich

inad'visable *adj.* nicht ratsam

inane [ɪn'eɪn] *adj.* dümmlich

in'animate *adj.* unbelebt

inap'plicable *adj.* nicht zutreffend

inap'propriate *adj.* unpassend

in'apt *adj.* unpassend

inar'ticulate *adj.* **a) she's rather/very** ~: sie kann sich ziemlich/sehr schlecht ausdrücken; **b)** *(indistinct)* unverständlich

inat'tentive *adj.* unaufmerksam (**to** gegenüber)

in'audible *adj.* unhörbar

inau'spicious *adj. (ominous)* unheilvoll; *(unlucky)* unglücklich

'inborn *adj.* angeboren (**in** *Dat.*)

'in-built *adj.* jmdm./einer Sache eigen

incalculable [ɪn'kælkjʊləbl] *adj. (very great)* unermeßlich

in'capable *adj.* **a) be** ~ **of doing sth.** außerstande sein, etw. zu tun; **be** ~ **of sth.** zu etw. unfähig sein; **b) be** ~ **of** nicht zulassen ⟨*Beweis, Messung usw.*⟩

incapacitate [ɪnkə'pæsɪteɪt] *v. t.* unfähig machen

incarcerate [ɪn'kɑ:səreɪt] *v. t.* einkerkern *(geh.)*

incendiary [ɪn'sendɪərɪ] *adj. & n.* ~ **device** Brandsatz, *der;* ~ |**bomb**| Brandbombe, *die*

¹incense ['ɪnsens] *n.* Weihrauch, *der*

²incense [ɪn'sens] *v. t.* erzürnen

incentive [ɪn'sentɪv] *n.* Anreiz, *der*

incessant [ɪn'sesənt] *adj.,* **in'cessantly** *adv.* unablässig

incest ['ɪnsest] *n.* Inzest, *der.* **incestuous** [ɪn'sestjʊəs] *adj.* inzestuös

inch [ɪntʃ] **1.** *n.* Inch, *der;* Zoll, *der (veralt.).* **2.** *v. t. & i.* ~ |**one's way**| **forward** sich Zoll für Zoll vorwärtsbewegen

incident ['ɪnsɪdənt] *n.* **a)** *(notable event)* Vorfall, *der;* **b)** *(clash)* Zwischenfall, *der*

incidental [ɪnsɪ'dentl] *adj.* beiläufig ⟨*Bemerkung*⟩; Neben⟨*ausgaben, -einnahmen*⟩. **incidentally** [ɪnsɪ'dentəlɪ] *adv.* nebenbei [bemerkt]

incinerate [ɪn'sɪnəreɪt] *v. t.* verbrennen. **incinerator** [ɪn'sɪnəreɪtə(r)] *n.* Verbrennungsofen, *der*
incision [ɪn'sɪʒn] *n.* Einschnitt, *der*
incisive [ɪn'saɪsɪv] *adj.* schneidend ⟨*Ton*⟩; scharf ⟨*Verstand*⟩; scharfsinnig ⟨*Kritik, Frage, Bemerkung, Argument*⟩
incite [ɪn'saɪt] *v. t.* anstiften; aufstacheln ⟨*Massen, Volk*⟩. **in'citement** *n.* Anstiftung, *die*/Aufstachelung, *die*
inclination [ɪnklɪ'neɪʃn] *n.* Neigung, *die*
incline 1. [ɪn'klaɪn] *v. t.* **a)** *(bend)* neigen; **b)** *(dispose)* veranlassen. **2.** *v. i.* *(be disposed)* neigen (to[wards] zu). **3.** ['ɪnklaɪn] *n.* Steigung, *die*. **inclined** [ɪn'klaɪnd] *adj.* geneigt; **they are ~ to be slow** sie neigen zur Langsamkeit; **if you feel [so] ~:** wenn Sie Lust dazu haben
include [ɪn'klu:d] *v. t.* einschließen; *(contain)* enthalten; **~d in the price** im Preis inbegriffen. **including** [ɪn'klu:dɪŋ] *prep.* einschließlich (+ *Gen.*); **~ VAT** inklusive Mehrwertsteuer. **inclusion** [ɪn'klu:ʒn] *n.* Aufnahme, *die*. **inclusive** [ɪn'klu:sɪv] *adj.* einschließlich; **be ~ of sth.** etw. einschließen; **from 2 to 6 January ~:** vom 2. bis einschließlich 6. Januar; **cost £50 ~:** 50 Pfund kosten, alles inbegriffen
incognito [ɪnkɒg'ni:təʊ] *adj., adv.* inkognito
inco'herent *adj.* zusammenhanglos
income ['ɪnkəm] *n.* Einkommen, *das*. **'income tax** *n.* Einkommensteuer, *die*; *(on wages, salary)* Lohnsteuer, *die*
incoming *adj.* ankommend; landend ⟨*Flugzeug*⟩; einfahrend ⟨*Zug*⟩; eingehend ⟨*Telefongespräch, Auftrag*⟩
in'comparable *adj.* unvergleichlich
incom'patible *adj.* unvereinbar; **be ~** ⟨*Menschen:*⟩ nicht zueinander passen
in'competence [ɪn'kɒmpɪtəns] *n.* Unfähigkeit, *die*; Unvermögen, *das*
in'competent *adj.* unfähig
incom'plete *adj.* unvollständig
incompre'hensible *adj.* unbegreiflich; unverständlich ⟨*Rede, Argument*⟩
incon'ceivable *adj.* unvorstellbar
incon'clusive *adj.* ergebnislos; nicht schlüssig ⟨*Beweis, Argument*⟩
incongruous [ɪn'kɒŋgrʊəs] *adj.* unpassend
inconsequential [ɪnkɒnsɪ'kwenʃl] *adj.* belanglos
incon'siderate *adj.* rücksichtslos

incon'sistency *n. see* **inconsistent**: Widersprüchlichkeit, *die*; Inkonsequenz, *die*; Unbeständigkeit, *die*
incon'sistent *adj.* widersprüchlich; *(illogical)* inkonsequent; *(irregular)* unbeständig
inconsolable [ɪnkən'səʊləbl] *adj.* untröstlich
incon'spicuous *adj.* unauffällig
incontinence [ɪn'kɒntɪnəns] *n.* *(Med.)* Inkontinenz, *die*
incontinent [ɪn'kɒntɪnənt] *adj.* *(Med.)* inkontinent; **be ~:** an Inkontinenz leiden
incontrovertible [ɪnkɒntrə'vɜːtəbl] *adj.* unbestreitbar; unwiderlegbar ⟨*Beweis*⟩
incon'venience 1. *n.* Unannehmlichkeiten (to für); **put sb. to a lot of ~:** jmdm. große Unannehmlichkeiten bereiten. **2.** *v. t.* Unannehmlichkeiten bereiten (+ *Dat.*); *(disturb)* stören
incon'venient *adj.* unbequem; ungünstig ⟨*Lage, Standort*⟩; **come at an ~ time** zu ungelegener Zeit kommen; **be ~ for sb.** jmdm. nicht passen
incorporate [ɪn'kɔ:pəreɪt] *v. t.* aufnehmen (in[to], with in + *Akk.*)
incor'rect *adj.* **a)** unrichtig; **be ~:** nicht stimmen; **it is ~ to say that ...:** es stimmt nicht, daß ...; **b)** *(improper)* inkorrekt. **incor'rectly** *adv.* **a)** unrichtigerweise; falsch ⟨*beantworten, aussprechen*⟩; **b)** *(improperly)* inkorrekt
increase 1. [ɪn'kri:s] *v. i.* zunehmen; ⟨*Lärm:*⟩ größer werden; ⟨*Preise, Nachfrage:*⟩ steigen; **~ in weight/size/price** schwerer/größer/teurer werden. **2.** *v. t.* **a)** *(make greater)* erhöhen; **b)** *(intensify)* verstärken. **3.** ['ɪnkri:s] *n.* Zunahme, *die* (in *Gen.*); **be on the ~:** ständig zunehmen. **increasing** [ɪn'kri:sɪŋ] *adj.* steigend; **an ~ number of people** mehr und mehr Menschen. **in'creasingly** *adv.* in zunehmendem Maße; **become ~ apparent** immer deutlicher werden
in'credible *adj.* *(also coll.: remarkable)* unglaublich. **in'credibly** *adv.* *(also coll.: remarkably)* unglaublich
incredulous [ɪn'kredjʊləs] *adj.* ungläubig
incriminate [ɪn'krɪmɪneɪt] *v. t.* belasten
incubate ['ɪŋkjʊbeɪt] *v. t.* bebrüten; *(to hatching)* ausbrüten. **incubation** [ɪŋkjʊ'beɪʃn] *n.* Bebrütung, *die*. **incubator** [ɪŋkjʊ'beɪtə(r)] *n.* Inkubator, *der*; *(for babies also)* Brutkasten, *der*

incur [ɪn'kɜː(r)] *v. t., -rr-* sich *(Dat.)* zuziehen ⟨*Unwillen, Ärger*⟩; ~ **debts/expenses/risks** Schulden machen/Ausgaben haben/Risiken eingehen

in'curable *adj.* unheilbar

incursion [ɪn'kɜːʃn] *n.* Eindringen, *das; (by sudden attack)* Einfall, *der*

indebted [ɪn'detɪd] *pred. adj.* **be** |much| ~ **to sb. for sth.** jmdm. für etw. [sehr] zu Dank verpflichtet sein

in'decency *n.* Unanständigkeit, *die*

in'decent *adj.,* **in'decently** *adv.* unanständig

inde'cision *n.* Unentschlossenheit, *die*

inde'cisive *adj.* **a)** ergebnislos ⟨*Streit, Diskussion*⟩; nichtssagend ⟨*Ergebnis*⟩; **b)** *(hesitating)* unentschlossen

indeed [ɪn'diːd] *adv.* **a)** in der Tat; **thank you very much ~:** haben Sie vielen herzlichen Dank; ~ **it is** in der Tat; **b)** *(in fact)* ja sogar; ~, **he can ...:** ja, er kann sogar ...; **c)** *(admittedly)* zugegebenermaßen

in'definite *adj.* **a)** *(vague)* unbestimmt; **b)** *(unlimited)* unbegrenzt. **in'definitely** *adv.* **a)** *(vaguely)* unbestimmt; **b)** *(unlimitedly)* unbegrenzt; auf unbestimmte Zeit ⟨*verschieben*⟩

indelible [ɪn'delɪbl] *adj.* unauslöschlich; ~ **ink** Wäschetinte, *die*

indemnify [ɪn'demnɪfaɪ] *v. t.* absichern (**against** gegen); *(compensate)* entschädigen. **indemnity** [ɪn'demnɪtɪ] *n.* Absicherung, *die; (compensation)* Entschädigung, *die*

inde'pendence *n.* Unabhängigkeit, *die*

inde'pendent *adj.,* **inde'pendently** *adv.* unabhängig (**of** von)

indescribable [ɪndɪ'skraɪbəbl] *adj.* unbeschreiblich

indestructible [ɪndɪ'strʌktɪbl] *adj.* unzerstörbar

indeterminate [ɪndɪ'tɜːmɪnət] *adj.* unbestimmt; unklar ⟨*Konzept*⟩

index ['ɪndeks] **1.** *n.* Register, *das.* **2.** *v. t.* mit einem Register versehen. **'index finger** *n.* Zeigefinger, *der*

India ['ɪndɪə] *n.* Indien *(das).* **Indian** ['ɪndɪən] **1.** *adj.* **a)** indisch; **b)** |American| ~: indianisch. **2.** *n.* **a)** Inder, *der*/Inderin, *die;* **b)** |American| ~: Indianer, *der*/Indianerin, *die.* **Indian 'Ocean** *n.* Indischer Ozean

indicate ['ɪndɪkeɪt] **1.** *v. t.* **a)** *(be a sign of)* erkennen lassen; **b)** *(state briefly)* andeuten; **c)** *(mark, point out)* anzeigen; **d)** *(suggest, make evident)* zum Ausdruck bringen (**to** gegenüber). **2.**

v. i. (Motor Veh.) blinken. **indication** [ɪndɪ'keɪʃn] *n.* [An]zeichen, *das (of* Gen., für). **indicative** [ɪn'dɪkətɪv] **1.** *adj.* **a)** **be ~ of sth.** auf etw. *(Akk.)* schließen lassen; **b)** *(Ling.)* indikativisch. **2.** *n. (Ling.)* Indikativ, *der.* **indicator** ['ɪndɪkeɪtə(r)] *n. (on vehicle)* Blinker, *der*

indict [ɪn'daɪt] *v. t.* anklagen (**for, on a charge of** Gen.)

in'difference *n.* Gleichgültigkeit, *die* (**to|wards|** gegenüber)

in'different *adj.* **a)** gleichgültig; **b)** *(not good)* mittelmäßig

indi'gestion *n.* Magenverstimmung, *die; (chronic)* Verdauungsstörungen

indignant [ɪn'dɪgnənt] *adj.* entrüstet (**at, over, about** über + *Akk.*); indigniert ⟨*Blick, Geste*⟩. **in'dignantly** *adv.* entrüstet; indigniert. **indignation** [ɪndɪg'neɪʃn] *n.* Entrüstung, *die* (**about, at, against, over** über + *Akk.*)

in'dignity *n.* Demütigung, *die*

indigo ['ɪndɪgəʊ] **1.** *adj.* ~ |blue| indigoblau. **2.** *n.* ~ |blue| Indigoblau, *das*

indi'rect *adj.* indirekt; ~ **speech** indirekte Rede. **indi'rectly** *adv.* indirekt. **indirect 'object** *n.* indirektes Objekt; *(in German)* Dativobjekt, *das*

indi'screet *adj.* indiskret. **indi'scretion** *n.* Indiskretion, *die*

indiscriminate [ɪndɪ'skrɪmɪnət] *adj.* unkritisch

indi'spensable *adj.* unentbehrlich (**to** für); unabdingbar ⟨*Voraussetzung*⟩

indisputable [ɪndɪ'spjuːtəbl] *adj.,* **indisputably** [ɪndɪ'spjuːtəblɪ] *adv.* unbestreitbar

indi'stinct *adj.,* **indi'stinctly** *adv.* undeutlich

indi'stinguishable *adj.* nicht unterscheidbar

individual [ɪndɪ'vɪdjʊəl] **1.** *adj.* **a)** einzeln; **b)** *(distinctive, characteristic)* individuell. **2.** *n.* einzelne, *der/die.* **indi'vidually** *adv.* einzeln

indi'visible *adj.* unteilbar

indoctrinate [ɪn'dɒktrɪneɪt] *v. t.* indoktrinieren

indolence ['ɪndələns] *n.* Trägheit, *die*

indolent ['ɪndələnt] *adj.* träge

indomitable [ɪn'dɒmɪtəbl] *adj.* unbeugsam

Indonesia [ɪndə'niːʃə] *pr. n.* Indonesien *(das)*

'indoor *adj.* ~ **swimming-pool/sports** Hallenbad, *das/*-sport, *der;* ~ **plants** Zimmerpflanzen; ~ **games** Spiele im Haus; *(Sport)* Hallenspiele

indoors [ɪn'dɔːz] *adv.* drinnen; im Haus; **go/come ~:** nach drinnen gehen/kommen

induce [ɪn'djuːs] *v.t.* **~ sb. to do sth.** jmdn. dazu bringen, etw. zu tun. **in'ducement** *n. (incentive)* Anreiz, *der*

indulge [ɪn'dʌldʒ] **1.** *v.t.* **a)** nachgeben (+ *Dat.*) ⟨*Wunsch, Verlangen, Verlockung*⟩; frönen *(geh.)* (+ *Dat.*) ⟨*Leidenschaft*⟩; **b)** *(please)* verwöhnen. **2.** *v.i.* **~ in** frönen *(geh.)* (+ *Dat.*) ⟨*Leidenschaft*⟩. **indulgence** [ɪn'dʌldʒəns] *n.* **a)** Nachsicht, *die; (humouring)* Nachgiebigkeit, *die* (**with** gegenüber); *(thing indulged in)* Luxus, *der.* **indulgent** [ɪn'dʌldʒənt] *adj.* nachsichtig (**with, to|wards**] gegenüber)

industrial [ɪn'dʌstrɪəl] *adj.* industriell; Arbeits⟨*unfall, -medizin, -psychologie*⟩

industrial: ~ 'action *n.* Arbeitskampfmaßnahmen; **take ~ action:** in den Ausstand treten; **~ dispute** *n.* Arbeitskonflikt, *der;* **~ estate** *n.* Industriegebiet, *das*

industrialize [ɪn'dʌstrɪəlaɪz] *v.t.* industrialisieren

industrious [ɪn'dʌstrɪəs] *adj.* fleißig; *(busy)* emsig

industry ['ɪndəstrɪ] *n.* **a)** Industrie, *die;* **b)** *see* **industrious:** Fleiß, *der;* Emsigkeit, *die*

in'edible *adj.* ungenießbar

inef'fective *adj.* unwirksam; fruchtlos ⟨*Anstrengung, Versuch*⟩

ineffectual [ɪnɪ'fektjʊəl] *adj.* unwirksam; fruchtlos ⟨*Versuch, Bemühung*⟩; ineffizient ⟨*Methode, Person*⟩

inefficiency *n.* Leistungsschwäche, *die; (of organization, method)* schlechtes Funktionieren

inefficient *adj.* leistungsschwach; schlecht funktionierend ⟨*Organisation, Methode*⟩

in'elegant *adj.* unelegant

in'eligible *adj.* ungeeignet; **be ~ for** nicht in Frage kommen für ⟨*Beförderung, Position*⟩; nicht berechtigt sein zu ⟨*Leistungen des Staats usw.*⟩

inept [ɪ'nept] *adj.* unbeholfen

ine'quality *n.* Ungleichheit, *die*

inert [ɪ'nɜːt] *adj.* **a)** reglos; *(sluggish)* träge; **b)** *(Chem.)* inert; **~ gas** Edelgas, *das.* **inertia** [ɪ'nɜːʃə] *n.* Trägheit, *die*

inescapable [ɪnɪ'skeɪpəbl] *adj.* unausweichlich

ines'sential *adj.* unwesentlich; *(dispensable)* entbehrlich

inevitable [ɪn'evɪtəbl] *adj.* unvermeidlich; unabwendbar ⟨*Ereignis, Krieg,*

Schicksal⟩; zwangsläufig ⟨*Ergebnis, Folge*⟩. **inevitably** [ɪn'evɪtəblɪ] *adv.* zwangsläufig

ine'xact *adj.* ungenau

inex'cusable *adj.* unverzeihlich

inexhaustible [ɪnɪg'zɔːstɪbl] *adj.* unerschöpflich; unverwüstlich ⟨*Person*⟩

inexorable [ɪn'eksərəbl] *adj.* unerbittlich

inex'pensive *adj.* preisgünstig

inex'perience *n.* Unerfahrenheit, *die.* **inex'perienced** *adj.* unerfahren; **~ in sth.** wenig vertraut mit etw.

inex'plicable *adj.* unerklärlich

in'fallible *adj.* unfehlbar

infamous ['ɪnfəməs] *adj.* berüchtigt

infancy ['ɪnfənsɪ] *n.* frühe Kindheit; **be in its ~** *(fig.)* noch in den Anfängen stecken

infant ['ɪnfənt] *n.* kleines Kind. **infantile** ['ɪnfəntaɪl] *adj.* kindlich; *(childish)* kindisch

infantry ['ɪnfəntrɪ] *n.* Infanterie, *die*

'infant school *n. (Brit.)* ≈ Vorschule, *die*

infatuated [ɪn'fætjʊeɪtɪd] *adj.* **be ~ with sb.** in jmdn. vernarrt sein

infect [ɪn'fekt] *v.t.* anstecken; infizieren; **the wound became ~ed** die Wunde entzündete sich. **infection** [ɪn'fekʃn] *n.* Infektion, *die;* **throat/ear/eye ~:** Hals-/Ohren-/Augenentzündung, *die.* **infectious** [ɪn'fekʃəs] *adj.* ansteckend; **be ~** ⟨*Person:*⟩ eine ansteckende Krankheit haben

infer [ɪn'fɜː(r)] *v.t.,* **-rr-** schließen (**from** aus); ziehen ⟨*Schlußfolgerung*⟩. **inference** ['ɪnfərəns] *n.* [Schluß]folgerung, *die*

inferior [ɪn'fɪərɪə(r)] **1.** *adj. (of lower quality)* minderwertig ⟨*Ware*⟩; minder... ⟨*Qualität*⟩; unterlegen ⟨*Gegner*⟩; **~ to sth.** schlechter als etw.; **feel ~:** Minderwertigkeitsgefühle haben. **2.** *n.* Untergebene, *der/die.* **inferiority** [ɪnfɪərɪ'ɒrɪtɪ] *n.* Minderwertigkeit, *die/* Unterlegenheit, *die.* **inferi'ority complex** *n.* Minderwertigkeitskomplex, *der*

infernal [ɪn'fɜːnl] *adj.* **a)** *(of hell)* höllisch; **b)** *(coll.)* verdammt *(salopp)*

inferno [ɪn'fɜːnəʊ] *n.* Inferno, *das*

in'fertile *adj.* unfruchtbar. **infer'tility** *n.* Unfruchtbarkeit, *die*

infest [ɪn'fest] *v.t.* ⟨*Ungeziefer:*⟩ befallen; ⟨*Unkraut:*⟩ überwuchern; **~ed with** befallen/überwuchert von

infidelity [ɪnfɪ'delɪtɪ] *n.* Untreue, *die* (**to** gegenüber)

infiltrate ['ınfıltreıt] *v. t.* **a)** infiltrieren; unterwandern ⟨*Partei, Organisation*⟩; **b)** einschleusen ⟨*Agenten*⟩
infinite ['ınfınıt] *adj.* **a)** *(endless)* unendlich; **b)** *(very great)* ungeheuer
infinitive [ın'fınıtıv] *n.* *(Ling.)* Infinitiv, *der*
infinity [ın'fınıtı] *n.* Unendlichkeit, *die*
infirm [ın'fɜ:m] *adj.* gebrechlich. **infirmity** [ın'fɜ:mıtı] *n.* Gebrechlichkeit, *die*; *(malady)* Gebrechen, *das*
inflamed [ın'fleımd] *adj. (Med.)* **be/ become** ∼: entzündet sein/sich entzünden
inflammable [ın'flæməbl] *adj.* feuergefährlich
inflammation [ınflə'meıʃn] *n. (Med.)* Entzündung, *die*
inflammatory [ın'flæmətərı] *adj.* aufrührerisch; **an** ∼ **speech** eine Hetzrede
inflatable [ın'fleıtəbl] *adj.* aufblasbar; ∼ **dinghy** Schlauchboot, *das*
inflate [ın'fleıt] *v. t.* aufblasen; *(with pump)* aufpumpen
inflation [ın'fleıʃn] *n. (Econ.)* Inflation, *die*
in'flexible *adj.* **a)** *(stiff)* unbiegsam; **b)** *(obstinate)* [geistig] unbeweglich
inflict [ın'flıkt] *v. t.* zufügen ⟨*Leid, Schmerzen*⟩, beibringen ⟨*Wunde*⟩, versetzen ⟨*Schlag*⟩ (**on** *Dat.*)
influence ['ınfluəns] **1.** *n.* Einfluß, *der*; **be a good/bad** ∼ **[on sb.]** einen guten/schlechten Einfluß [auf jmdn.] ausüben. **2.** *v. t.* beeinflussen. **influential** [ınflu'enʃl] *adj.* einflußreich
influenza [ınflu'enzə] *n.* Grippe, *die*
influx ['ınflʌks] *n.* Zustrom, *der*
inform [ın'fɔ:m] **1.** *n.* informieren (**of, about** über + *Akk.*); **keep sb.** ∼**ed** jmdn. auf dem laufenden halten. **2.** *v. i.* ∼ **against** *or* **on sb.** jmdn. denunzieren (**to** bei)
in'formal *adj.* **a)** zwanglos; **b)** *(unofficial)* informell. **infor'mality** *n.* Zwanglosigkeit, *die*
informant [ın'fɔ:mənt] *n.* Informant, *der*/Informantin, *die*
information [ınfə'meıʃn] *n.* Informationen *Pl.*; **give** ∼ **on sth.** Auskunft über etw. *(Akk.)* erteilen; **piece** *or* **bit of** ∼: Information, *die*; ∼ **centre** Auskunftsbüro, *das*
informative [ın'fɔ:mətıv] *adj.* informativ; **not very** ∼: nicht sehr aufschlußreich ⟨*Dokument, Schriftstück*⟩
informed [ın'fɔ:md] *adj.* informiert
in'former *n.* Denunziant, *der*/Denunziantin, *die*

infra-red [ınfrə'red] *adj.* infrarot
in'frequent *adj.,* **in'frequently** *adv.* selten
infringe [ın'frındʒ] *v. t. & i.* ∼ **[on]** verstoßen gegen. **in'fringement** *n.* Verstoß, *der* (**of** gegen)
infuriate [ın'fjuərıeıt] *v. t.* wütend machen; **be** ∼**d** wütend sein (**by** über + *Akk.*). **infuriating** [ın'fjuərıeıtıŋ] *adj.* ärgerlich
ingenious [ın'dʒi:nıəs] *adj.* einfallsreich; genial ⟨*Methode, Idee*⟩; raffiniert ⟨*Spielzeug, Maschine*⟩. **ingenuity** [ındʒı'nju:ıtı] *n.* Genialität, *die*
ingot ['ıŋgət] *n.* Ingot, *der*
ingratiate [ın'greıʃıeıt] *v. refl.* ∼ **oneself with sb.** sich bei jmdm. einschmeicheln
in'gratitude *n.* Undankbarkeit, *die* (**to**|**wards**) gegenüber)
ingredient [ın'gri:dıənt] *n.* Zutat, *die*
ingrowing ['ıngrəuıŋ] *adj.* eingewachsen ⟨*Zehennagel usw.*⟩
inhabit [ın'hæbıt] *v. t.* bewohnen. **inhabitable** [ın'hæbıtəbl] *adj.* bewohnbar. **inhabitant** [ın'hæbıtənt] *n.* Bewohner, *der*/Bewohnerin, *die*
inhale [ın'heıl] *v. t. & i.* einatmen; inhalieren *(ugs.)* ⟨*Zigarettenrauch usw.*⟩
inherit [ın'herıt] *v. t.* erben. **inheritance** [ın'herıtəns] *n.* Erbe, *das*; *(inheriting)* Erbschaft, *die*
inhibit [ın'hıbıt] *v. t.* hemmen. **in'hibited** *adj.* gehemmt. **inhibition** [ınhı'bıʃn] *n.* Hemmung, *die*
inho'spitable *adj.* ungastlich ⟨*Person, Verhalten*⟩; unwirtlich ⟨*Gegend, Klima*⟩
in'human *adj.* unmenschlich
initial [ı'nıʃl] **1.** *adj.* anfänglich; Anfangs⟨*stadium, -schwierigkeiten*⟩. **2.** *n. esp. in pl.* Initiale, *die*. **3.** *v. t., (Brit.)* -ll- abzeichnen ⟨*Scheck, Quittung*⟩; paraphieren ⟨*Vertrag, Abkommen usw.*⟩. **i'nitially** *adv.* anfangs; am Anfang
initiate [ı'nıʃıeıt] *v. t.* **a)** *(introduce)* einführen (**into** in + *Akk.*); *(into knowledge, mystery)* einweihen (**into** in + *Akk.*); **b)** *(begin)* einleiten. **initiation** [ınıʃı'eıʃn] *n.* **a)** *(introduction)* Einführung, *die*; *(into knowledge, mystery)* Einweihung, *die*
initiative [ı'nıʃətıv] *n.* Initiative, *die*; **lack** ∼: keine Initiative haben
inject [ın'dʒekt] *v. t.* [ein]spritzen; injizieren *(Med.)*. **injection** [ın'dʒekʃn] *n.* Spritze, *die*; Injektion, *die*
injure ['ındʒə(r)] *v. t.* **a)** verletzen; **his**

leg was ~d er wurde/*(state)* war am Bein verletzt; **b)** *(impair)* schaden (+ *Dat.*). **injured** ['ɪndʒəd] *adj.* verletzt; verwundet ⟨*Soldat*⟩. **injury** ['ɪndʒərɪ] *n.* Verletzung, *die* (**to** *Gen.*)
in'justice *n.* Ungerechtigkeit, *die*
ink [ɪŋk] *n.* Tinte, *die*
inkling ['ɪŋklɪŋ] *n.* Ahnung, *die;* **have an ~ of sth.** etw. ahnen
inland ['ɪnlənd, 'ɪnlænd] *adj.* Binnen-; binnenländisch. **Inland 'Revenue** *n.* *(Brit.)* ≈ Finanzamt, *das*
'in-laws *n pl. (coll.)* Schwiegereltern
inlet ['ɪnlət] *n.* [schmale] Bucht
'inmate *n.* Insasse, *der*/Insassin, *die*
inn [ɪn] *n. (hotel)* Gasthof, *der; (pub)* Wirtshaus, *das*
innate [ɪ'neɪt] *adj.* angeboren
inner ['ɪnə(r)] *adj.* inner...; Innen⟨*hof, -tür, -fläche, -seite usw.*⟩; **~ tube** Schlauch, *der.* **innermost** ['ɪnəməʊst] *adj.* innerst...
innocence ['ɪnəsəns] *n.* **a)** Unschuld, *die;* **b)** *(naïvity)* Naivität, *die*
innocent ['ɪnəsənt] *adj.* **a)** unschuldig (**of** an + *Dat.*); **b)** *(naïve)* naiv
innocuous [ɪ'nɒkjʊəs] *adj.* harmlos
innovation [ɪnə'veɪʃn] *n.* Innovation, *die; (thing, change)* Neuerung, *die*
innumerable [ɪ'njuːmərəbl] *adj.* unzählig
inoculate [ɪ'nɒkjʊlət] *v.t.* impfen. **inoculation** [ɪnɒkjʊ'leɪʃn] *n.* Impfung, *die*
inof'fensive *adj.* harmlos
in'opportune *adj.* unpassend; unangebracht ⟨*Bemerkung*⟩
inordinate [ɪ'nɔːdɪnət] *adj.* unmäßig; ungeheuer ⟨*Menge*⟩
inor'ganic *adj.* anorganisch
'in-patient *n.* stationär behandelter Patient/behandelte Patientin
'input *n.* Input, *der od. das*
inquest ['ɪŋkwest] *n.* gerichtliche Untersuchung der Todesursache
inquire [ɪn'kwaɪə(r)] **1.** *v.i.* sich erkundigen (**about, after** nach, **of** bei); **~ into** untersuchen. **2.** *v.t.* sich erkundigen nach ⟨*Weg, Namen*⟩. **inquiry** [ɪn-'kwaɪərɪ] *n.* **a)** *(question)* Erkundigung, *die* (**into** über + *Akk.*); **make inquiries** Erkundigungen einziehen; **b)** *(investigation)* Untersuchung, *die*
inquisitive ~[ɪn'kwɪzɪtɪv] *adj.* neugierig
'inroad *n.* Eingriff, *der* (**on, into** in + *Akk.*); **make ~s into** sb.'s savings jmds. Ersparnisse angreifen
in'sane *adj.* geisteskrank
in'sanitary *adj.* unhygienisch

in'sanity *n.* Geisteskrankheit, *die*
insatiable [ɪn'seɪʃəbl] *adj.* unersättlich; unstillbar ⟨*Verlangen*⟩
inscribe [ɪn'skraɪb] *v.t.* schreiben; *(on stone, rock)* einmeißeln; mit einer Inschrift versehen ⟨*Denkmal, Grabstein*⟩. **inscription** [ɪn'skrɪpʃn] *n.* Inschrift, *die; (on coin)* Aufschrift, *die*
inscrutable [ɪn'skruːtəbl] *adj.* unergründlich; undurchdringlich ⟨*Miene*⟩
insect ['ɪnsekt] *n.* Insekt, *das.* **insecticide** [ɪn'sektɪsaɪd] *n.* Insektizid, *das.* **'insect repellent** *n.* Insektenschutzmittel, *das*
inse'cure *adj.* unsicher. **inse'curity** *n.* Unsicherheit, *die*
in'sensitive *adj.* **a)** gefühllos ⟨*Person, Art*⟩; *(unappreciative)* unempfänglich (**to** für); **b)** *(physically)* unempfindlich (**to** gegen)
in'separable *adj.* untrennbar; *(fig.)* unzertrennlich
insert [ɪn'sɜːt] *v.t.* einlegen ⟨*Film*⟩; einwerfen ⟨*Münze*⟩; hineinstecken ⟨*Schlüssel*⟩; einstechen ⟨*Nadel*⟩. **insertion** [ɪn'sɜːʃn] *n. see* **insert:** Einlegen, *das;* Einwerfen, *das;* Hineinstecken, *das;* Einstechen, *das*
inside 1. [-'-, '--] *n.* **a)** *(internal side)* Innenseite, *die;* **on the ~:** innen; **to/from the ~:** nach/von innen; **b)** *(inner part)* Innere, *das.* **2.** ['--] *adj.* inner...; Innen⟨*wand, -einrichtung, -ansicht*⟩; *(fig.)* intern. **3.** [-'-] *adv. (on or in the ~)* innen; *(to the ~)* nach innen hinein/ herein; *(indoors)* drinnen; **come ~:** hereinkommen; **take a look ~:** hineinsehen; **go ~:** [ins Haus] hineingehen; **turn a jacket ~ out** eine Jacke nach links wenden; **know sth. ~ out** in- und auswendig kennen. **4.** [-'-] *prep. (position)* in (+ *Dat.*); *(direction)* in (+ *Akk.*) hinein
insidious [ɪn'sɪdɪəs] *adj.* heimtückisch
'insight *n. (discernment)* Verständnis, *das;* **gain an ~ into sth.** Einblick in etw. *(Akk.)* gewinnen
insig'nificant *adj.* unbedeutend; geringfügig ⟨*Summe*⟩
insin'cere *adj.* unaufrichtig. **insin'cerity** *n.* Unaufrichtigkeit, *die*
insinuate [ɪn'sɪnjʊeɪt] *v.t.* andeuten (**to** sb. jmdm. gegenüber). **insinuation** [ɪnsɪnjʊ'eɪʃn] *n.* Anspielung, *die* (**about** auf + *Akk.*)
insipid [ɪn'sɪpɪd] *adj.* fade
insist [ɪn'sɪst] *v.i.* bestehen (**[up]on** auf + *Dat.*); **~ on doing sth./on sb.'s doing sth.** darauf bestehen, etw. zu

tun/daß jmd. etw. tut; **if you ~:** wenn du darauf bestehst. **insistence** [ɪn-'sɪstəns] *n.* Bestehen, *das* **(on auf + *Dat.*). insistent** [ɪn'sɪstənt] *adj.* **be ~ that ...:** darauf bestehen, daß ...
insolence ['ɪnsələns] *n.* Unverschämtheit, *die;* Frechheit, *die*
insolent ['ɪnsələnt] *adj.,* '**insolently** *adv.* unverschämt; frech
in·soluble *adj.* **a)** *(esp. Chem.)* unlöslich; **b)** *(not solvable)* unlösbar
in·solvent *adj.* zahlungsunfähig
insomnia [ɪn'sɒmnɪə] *n.* Schlaflosigkeit, *die.* **insomniac** [ɪn'sɒmnɪæk] *n.* **be an ~:** an Schlaflosigkeit leiden
inspect [ɪn'spekt] *v. t.* prüfend betrachten; *(examine officially)* überprüfen; kontrollieren ⟨*Räumlichkeiten*⟩. **inspection** [ɪn'spekʃn] *n.* Überprüfung, *die;* *(of premises)* Kontrolle, *die;* Inspektion, *die;* **on |closer| ~:** bei näherer Betrachtung. **inspector** [ɪn-'spektə(r)] *n.* **a)** *(on bus, train, etc.)* Kontrolleur, *der/* Kontrolleurin, *die;* **b)** *(Brit.)* ≈ Polizeiinspektor, *der*
inspiration [ɪnspə'reɪʃn] *n.* Inspiration, *die (geh.)*
inspire [ɪn'spaɪə(r)] *v. t.* **a)** inspirieren *(geh.)* ⟨*Person*⟩; **b)** *(instil)* einflößen (**in** *Dat.*). **inspiring** [ɪn'spaɪərɪŋ] *adj.* inspirierend *(geh.)*
insta·bility *n.* Instabilität, *die; (of person)* Labilität, *die*
install [ɪn'stɔːl] *v. t.* installieren; einbauen ⟨*Badezimmer*⟩; anschließen ⟨*Telefon, Herd*⟩; **~ oneself** sich installieren. **installation** [ɪnstə'leɪʃn] *n.* **a)** Installation, *die; (of bathroom)* Einbau, *der; (of telephone, cooker)* Anschluß, *der;* **b)** *(apparatus etc. installed)* Anlage, *die*
instalment *(Amer.:* **installment)** [ɪn'stɔːlmənt] *n.* **a)** *(part-payment)* Rate, *die;* **pay by** *or* **in ~s** in Raten zahlen; **b)** *(of serial, novel)* Fortsetzung, *die; (Radio, Telev.)* Folge, *die*
instance ['ɪnstəns] *n. (example)* Beispiel, *das* **(of für); for ~:** zum Beispiel; **in many ~s** *(cases)* in vielen Fällen; **in the first ~:** zunächst einmal
instant ['ɪnstənt] **1.** *adj.* unmittelbar; sofortig ⟨*Wirkung, Linderung, Ergebnis*⟩; **~ coffee/tea** Pulverkaffee/Instanttee, *der;* **~ potatoes** fertiger Kartoffelbrei. **2.** *n.* Augenblick, *der;* **at that very ~:** genau in dem Augenblick; **come here this ~:** komm sofort her; **in an ~:** augenblicklich. **instant·aneous** [ɪnstən'teɪnɪəs] *adj.* unmittel-

bar; **his reaction was ~:** er reagierte sofort. '**instantly** *adv.* sofort
instead [ɪn'sted] *adv.* statt dessen; **~ of doing sth.** [an]statt etw. zu tun; **~ of sth.** anstelle einer Sache *(Gen.);* **I will go ~ of you** ich gehe an deiner Stelle
'**instep** *n. (of foot)* Spann, *der;* Fußrücken, *der; (of shoe)* Blatt, *das*
instigate ['ɪnstɪgeɪt] *v. t.* anstiften (**to** zu); initiieren *(geh.)* ⟨*Reformen, Projekt usw.*⟩. **instigation** [ɪnstɪ'geɪʃn] *n.* Anstiftung, *die; (of reforms, project, etc.)* Initiierung, *die;* **at sb.'s ~:** auf jmds. Betreiben *(Akk.)*
instil *(Amer.:* **instill)** [ɪn'stɪl] *v. t.,* **-ll-** einflößen (**in** *Dat.*); beibringen ⟨*gutes Benehmen, Wissen*⟩ (**in** *Dat.*)
instinct ['ɪnstɪŋkt] *n.* Instinkt, *der.* **instinctive** [ɪn'stɪŋktɪv] *adj.,* **in·stinctively** *adv.* instinktiv
institute ['ɪnstɪtjuːt] **1.** *n.* Institut, *das.* **2.** *v. t.* einführen; einleiten ⟨*Suche, Verfahren*⟩; anstrengen ⟨*Prozeß*⟩
institution [ɪnstɪ'tjuːʃn] *n.* Institution, *die; (home)* Heim, *das;* Anstalt, *die*
instruct [ɪn'strʌkt] *v. t.* **a)** *(teach)* unterrichten ⟨*Klasse, Fach*⟩; **b)** *(direct, command)* anweisen. **instruction** [ɪn'strʌkʃn] *n.* **a)** *(teaching)* Unterricht, *der;* **b)** *esp. in pl. (direction, order)* Anweisung, *die;* **~ manual/~s for use** Gebrauchsanleitung, *die.* **in·structive** [ɪn'strʌktɪv] *adj.* aufschlußreich; lehrreich ⟨*Erfahrung, Buch*⟩. **instructor** [ɪn'strʌktə(r)] *n.* Lehrer, *der/* Lehrerin, *die; (Mil.)* Ausbilder, *der*
instrument ['ɪnstrʊmənt] *n.* Instrument, *das.* **instrumental** [ɪnstrə-'mentl] *adj.* **a)** *(Mus.)* Instrumental-; **b)** *(helpful)* dienlich (**to** *Dat.*); **he was ~ in finding me a job** er hat mir zu einer Stelle verholfen
insufferable [ɪn'sʌfərəbl] *adj. (unbearably arrogant)* unausstehlich
insuf·ficient *adj.* nicht genügend; unzulänglich ⟨*Beweise*⟩; unzureichend ⟨*Versorgung, Beleuchtung*⟩. **insuf·ficiently** *adv.* ungenügend
insulate ['ɪnsjʊleɪt] *v. t.* isolieren **(against, from gegen); insulating tape** Isolierband, *das.* **insulation** [ɪnsjʊ-'leɪʃn] *n.* Isolierung, *die*
insulin ['ɪnsjʊlɪn] *n.* Insulin, *das*
insult 1. ['ɪnsʌlt] *n.* Beleidigung, *die* **(to** *Gen.*). **2.** [ɪn'sʌlt] *v. t.* beleidigen. **insulting** [ɪn'sʌltɪŋ] *adj.* beleidigend
insuperable [ɪn'suːpərəbl] *adj.* unüberwindlich

insurance [ɪn'ʃʊərəns] *n.* Versicherung, *die; (fig.)* Sicherheit, *die;* **take out ~ against/on sth.** eine Versicherung gegen etw. abschließen/etw. versichern lassen; **travel ~:** Reisegepäck- und -unfallversicherung, *die.* **in'surance policy** *n.* Versicherungspolice, *die*

insure [ɪn'ʃʊə(r)] *v. t.* versichern ⟨*Person*⟩; versichern lassen ⟨*Gepäck, Gemälde usw.*⟩; **~ [oneself] against sth.** [sich] gegen etw. versichern

insurmountable [ɪnsə'maʊntəbl] *adj.* unüberwindlich

intact [ɪn'tækt] *adj.* **a)** *(entire)* unbeschädigt; intakt ⟨*Uhr, Maschine usw.*⟩; **b)** *(unimpaired)* unversehrt

'intake *n.* **a)** *(action)* Aufnahme, *die;* **b)** *(persons, things)* Neuzugänge; *(amount)* aufgenommene Menge

in'tangible *adj.* nicht greifbar; *(mentally)* unbestimmbar

integral ['ɪntɪgrl] *adj.* **a)** wesentlich ⟨*Bestandteil*⟩; **b)** *(whole)* vollständig

integrate ['ɪntɪgreɪt] *v. t.* integrieren (**into** in + *Akk.*). **integration** [ɪntɪ'greɪʃn] *n.* Integration, *die* (**into** in + *Akk.*)

integrity [ɪn'tegrɪtɪ] *n.* Redlichkeit, *die*

intellect ['ɪntəlekt] *n.* Verstand, *der;* Intellekt, *der.* **intellectual** [ɪntə'lektjʊəl] **1.** *adj.* intellektuell; geistig anspruchsvoll ⟨*Person, Publikum*⟩. **2.** *n.* Intellektuelle, *der/die*

intelligence [ɪn'telɪdʒəns] *n.* **a)** Intelligenz, *die;* **b)** *(information)* Informationen *Pl.;* **c) military ~** *(organization)* militärischer Geheimdienst. **intelligent** [ɪn'telɪdʒənt] *adj.* intelligent

intelligible [ɪn'telɪdʒɪbl] *adj.* verständlich (**to** für)

intend [ɪn'tend] *v. t.* beabsichtigen; **it was ~ed as a joke** das sollte ein Witz sein. **in'tended** *adj.* beabsichtigt ⟨*Wirkung*⟩; **be ~ for sb./sth.** für jmdn./etw. gedacht sein

intense [ɪn'tens] *adj.* **a)** intensiv; groß ⟨*Hitze, Belastung, Interesse*⟩; stark ⟨*Schmerzen*⟩; **b)** *(earnest)* ernst. **in'tensely** *adv.* äußerst; intensiv ⟨*studieren, fühlen*⟩

intensify [ɪn'tensɪfaɪ] **1.** *v. t.* intensivieren. **2.** *v. i.* zunehmen

intensity [ɪn'tensɪtɪ] *n. see* **intense a:** Intensität, *die;* Größe, *die;* Stärke, *die*

intensive [ɪn'tensɪv] *adj.* intensiv ⟨*kurs*⟩; **be in ~ care** auf der Intensivstation sein. **in'tensively** *adv.* intensiv

intent [ɪn'tent] **1.** *n.* Absicht, *die;* **to all ~s and purposes** im Grunde. **2.** *adj.* **be ~ on achieving sth.** etw. unbedingt erreichen wollen

intention [ɪn'tenʃn] *n.* Absicht, *die.* **intentional** [ɪn'tenʃənl] *adj.,* **in'tentionally** *adv.* absichtlich

in'tently *adv.* aufmerksam

interact [ɪntər'ækt] *v. i.* interagieren. **interaction** [ɪntər'ækʃn] *n.* Interaktion, *die*

intercede [ɪntə'siːd] *v. i.* sich einsetzen (**with** bei; **for, on behalf of** für)

intercept [ɪntə'sept] *v. t.* abfangen

interchange 1. ['ɪntətʃeɪndʒ] *n.* **a)** Austausch, *der;* **b)** *(road junction)* [Autobahn]kreuz, *das.* **2.** [ɪntə'tʃeɪndʒ] *v. t.* austauschen. **interchangeable** [ɪntə'tʃeɪndʒəbl] *adj.* austauschbar

inter-city [ɪntə'sɪtɪ] *adj.* Intercity-; **~ train** Intercity[-Zug], *der*

intercom ['ɪntəkɒm] *n. (coll.)* Gegensprechanlage, *die*

interconnect [ɪntəkə'nekt] **1.** *v. t.* miteinander verbinden. **2.** *v. i.* miteinander in Zusammenhang stehen

intercourse ['ɪntəkɔːs] *n. (sexual)* [Geschlechts]verkehr, *der*

interest ['ɪntrəst] **1.** *n.* **a)** Interesse, *das;* **take** *or* **have an ~ in sb./sth.** sich für jmdn./etw. interessieren; **[just] for** *or* **out of ~:** [nur] interessehalber; **with ~:** interessiert; **act in one's own/sb.'s ~[s]** im eigenen/in jmds. Interesse handeln; **be of ~:** interessant sein (**to** für); **b)** *(Finance)* Zinsen *Pl.* **2.** *v. t.* interessieren; **be ~ed** sich interessieren (**in** für). **'interesting** *adj.* interessant

interfere [ɪntə'fɪə(r)] *v. i.* sich einmischen (**in** in + *Akk.*); **~ with sth.** ⟨*Dat.*⟩ an etw. ⟨*Dat.*⟩ zu schaffen machen. **interference** [ɪntə'fɪərəns] *n.* **a)** *(interfering)* Einmischung, *die;* **b)** *(Radio, Telev.)* Störung, *die*

interim ['ɪntərɪm] **1.** *n.* **in the ~:** in der Zwischenzeit. **2.** *adj.* vorläufig

interior [ɪn'tɪərɪə(r)] **1.** *adj.* inner...; Innen⟨*fläche, -wand*⟩. **2.** *n.* Innere, *das*

interject [ɪntə'dʒekt] *v. t.* einwerfen. **interjection** [ɪntə'dʒekʃn] *n.* Ausruf, *der*

interloper ['ɪntələʊpə(r)] *n.* Eindringling, *der*

interlude ['ɪntəluːd] *n.* Pause, *die; (music)* Zwischenspiel, *das*

intermediate [ɪntə'miːdjət] *adj.* Zwischen-

interminable [ɪn'tɜːmɪnəbl] *adj.* endlos

intermission [ɪntə'mɪʃn] n. Pause, *die*
intermittent [ɪntə'mɪtənt] adj. in Abständen auftretend. **inter'mittently**
adv. in Abständen
intern [ɪn'tɜːn] v. t. gefangenhalten
internal [ɪn'tɜːnl] adj. inner...; Innen‹fläche, -abmessungen›. **internally**
[ɪn'tɜːnəlɪ] adv. innerlich
international [ɪntə'næʃənl] 1. adj. international. 2. n. a) *(Sport) (contest)*
Länderspiel, *das; (participant)* Nationalspieler, *der/-*spielerin, *die.* **inter'nationally** adv.: international
in'ternment n. Internierung, *die*
interplay ['ɪntəpleɪ] n. Zusammenspiel, *das*
interpret [ɪn'tɜːprɪt] 1. v. t. a) interpretieren; deuten ‹Traum, Zeichen›; b)
(between languages) dolmetschen. 2.
v. i. dolmetschen. **interpretation**
[ɪntɜːprɪ'teɪʃn] n. Interpretation, *die;
(of dream, symptoms)* Deutung, *die.*
in'terpreter n. Dolmetscher, *der/*
Dolmetscherin, *die*
interrogate [ɪn'terəgeɪt] v. t. verhören; ausfragen ‹Freund, Kind usw.›.
interrogation [ɪnterə'geɪʃn] n. Verhör, *das*
interrogative [ɪntə'rogətɪv] adj.
(Ling.) Interrogativ-
interrupt [ɪntə'rʌpt] 1. v. t. unterbrechen; **don't ~ me when I'm busy** stör
mich nicht, wenn ich zu tun habe. 2.
v. i. unterbrechen; stören. **interruption** [ɪntə'rʌpʃn] n. Unterbrechung,
die; Störung, *die*
intersect [ɪntə'sekt] v. i. a) ‹Straßen:›
sich kreuzen; b) *(Geom.)* sich schneiden. **intersection** [ɪntə'sekʃn] n. a)
(road junction) Kreuzung, *die;* b)
(Geom.) Schnitt₂unkt, *der*
intersperse [ɪntə'spɜːs] v. t. **be ~d
with** durchsetzt sein mit
interval ['ɪntəvl] n. a) [Zeit]abstand,
der; **at ~s** in Abständen; b) *(break;
also Brit. Theatre etc.)* Pause, *die;*
sunny ~s Aufheiterungen *Pl.*
intervene [ɪntə'viːn] v. i. a) [vermittelnd] eingreifen (**in** in + *Akk.*); b) **the
intervening years** die dazwischenliegenden Jahre. **intervention** [ɪntə-
'venʃn] n. Eingreifen, *das*
interview ['ɪntəvjuː] 1. n. a) *(for job)*
Vorstellungsgespräch, *das;* b) *(Journ.,
Radio, Telev.)* Interview, *das.* 2. v. t.
ein Vorstellungsgespräch führen mit;
interviewen ‹Politiker, Filmstar usw.›.
'interviewer n. Interviewer, *der/*Interviewerin, *die*

intestine [ɪn'testɪn] n. Darm, *der*
intimacy ['ɪntɪməsɪ] n. a) Vertrautheit,
die; b) *(sexual)* Intimität, *die*
intimate 1. ['ɪntɪmət] adj. a) ‹Freund, Verhältnis›; genau, *(geh.)* intim ‹Kenntnis›; b) *(sexually)* intim. 2.
['ɪntɪmeɪt] v. t. *(imply)* andeuten. **intimately** ['ɪntɪmətlɪ] adv. genau-
[estens] ‹kennen›; eng ‹verbinden›
intimidate [ɪn'tɪmɪdeɪt] v. t. einschüchtern. **intimidation** [ɪntɪmɪ-
'deɪʃn] n. Einschüchterung, *die*
into *[before vowel* 'ɪntʊ, *before consonant* 'ɪntə] prep. in (+ *Akk.*); *(against)*
gegen; **I went out ~ the street** ich ging
auf die Straße hinaus; **translate sth. ~
English** etw. ins Englische übersetzen
in'tolerable adj. unerträglich
in'tolerance n. Intoleranz, *die*
in'tolerant adj. intolerant (**of** gegenüber)
intonation [ɪntə'neɪʃn] n. Intonation,
die
intoxicate [ɪn'tɒksɪkeɪt] v. t. betrunken machen. **intoxication** [ɪntɒksɪ-
'keɪʃn] n. Rausch, *der*
intractable [ɪn'træktəbl] adj. hartnäckig ‹Problem›
intransigent [ɪn'trænsɪdʒənt] adj. unnachgiebig
in'transitive adj. *(Ling.)* intransitiv
'in-tray n. Eingangskorb, *der*
intrepid [ɪn'trepɪd] adj. unerschrocken
intricacy ['ɪntrɪkəsɪ] n. Kompliziertheit, *die*
intricate ['ɪntrɪkət] adj. kompliziert
intrigue [ɪn'triːg] v. t. faszinieren. **intriguing** [ɪn'triːgɪŋ] adj. faszinierend
intrinsic [ɪn'trɪnsɪk] adj. innewohnend; inner...; **~ value** innerer Wert
introduce [ɪntrə'djuːs] v. t. einführen;
~ oneself/sb. |to sb.| sich/jmdn.
[jmdm.] vorstellen. **introduction** [ɪn-
trə'dʌkʃn] n. Einführen, *das;* Einführung, *die; (to person)* Vorstellung, *die;
(to book)* Einleitung, *die.* **introductory** [ɪntrə'dʌktərɪ] adj. einleitend;
Einführungs‹kurs, -vortrag›
introspective [ɪntrə'spektɪv] adj. in
sich *(Akk.)* gerichtet
introvert ['ɪntrəvɜːt] 1. n. Introvertierte, *der/die;* **be an ~:** introvertiert sein.
2. adj. introvertiert
intrude [ɪn'truːd] v. i. stören. **in'truder** n. Eindringling, *der.* **intrusion** [ɪn'truːʒn] n. Störung, *die.* **intrusive** [ɪn'truːsɪv] adj. aufdringlich
intuition [ɪntjuː'ɪʃn] n. Intuition, *die*

intuitive [ɪn'tjuːɪtɪv] *adj.*, **in'tuitively** *adv.* intuitiv

inundate ['ɪnəndeɪt] *v. t.* überschwemmen

inure [ɪ'njʊə(r)] *v. t.* gewöhnen (**to** an + *Akk.*)

invade [ɪn'veɪd] *v. t.* einfallen in (+ *Akk.*). **in'vader** *n.* Angreifer, *der*

¹**invalid** ['ɪnvəlɪd] *(Brit.)* 1. *n.* Kranke, *der/die; (disabled)* Körperbehinderte, *der/die.* 2. *adj.* körperbehindert

²**invalid** [ɪn'vælɪd] *adj.* nicht schlüssig ⟨*Argument, Theorie*⟩; ungültig ⟨*Fahrkarte, Garantie, Vertrag*⟩. **invalidate** [ɪn'vælɪdeɪt] *v. t.* aufheben; widerlegen ⟨*Theorie, These*⟩

in'valuable *adj.* unersetzlich ⟨*Person*⟩; unschätzbar ⟨*Dienst, Hilfe*⟩; außerordentlich wichtig ⟨*Rolle*⟩

in'variable *adj.* unveränderlich. **invariably** [ɪn'veərɪəblɪ] *adv.* immer; ausnahmslos ⟨*falsch, richtig*⟩

invasion [ɪn'veɪʒn] *n.* Invasion, *die*

invective [ɪn'vektɪv] *n.* Beschimpfungen *Pl.*

invent [ɪn'vent] *v. t.* erfinden. **invention** [ɪn'venʃn] *n.* Erfindung, *die.* **inventive** [ɪn'ventɪv] *adj.* **a)** schöpferisch ⟨*Person, Begabung*⟩; **b)** *(original)* originell. **inventor** [ɪn'ventə(r)] *n.* Erfinder, *der/*Erfinderin, *die*

inventory ['ɪnvəntərɪ] *n.* Bestandsliste, *die;* **make** *or* **take an ~ of sth.** von etw. ein Inventar aufstellen

inverse ['ɪnvɜːs] *adj.* umgekehrt

invert [ɪn'vɜːt] *v. t.* umstülpen

in'vertebrate *n.* wirbelloses Tier

inverted 'commas *n. pl. (Brit.)* Anführungszeichen *Pl.*

invest [ɪn'vest] *v. t.* **a)** *(Finance)* anlegen (**in** in + *Dat.*); investieren (**in** in + *Dat. od. Akk.*); **b)** *(fig.)* investieren; **~ sb. with sth.** jmdm. etw. übertragen; **~ sth. with sth.** einer Sache *(Dat.)* etw. verleihen

investigate [ɪn'vestɪgeɪt] *v. t.* untersuchen. **investigation** [ɪnvestɪ'geɪʃn] *n.* Untersuchung, *die.* **investigator** [ɪn'vestɪgeɪtə(r)] *n.* [**private**] **~:** [Privat]detektiv, *der/*-detektivin, *die*

in'vestment *n.* Investition, *die; (money invested)* angelegtes Geld; **be a good ~** *(fig.)* sich bezahlt machen. **investor** [ɪn'vestə(r)] *n.* [Kapital]anleger, *der/*-anlegerin, *die*

inveterate [ɪn'vetərət] *adj.* eingefleischt ⟨*Trinker, Raucher*⟩; unverbesserlich ⟨*Lügner*⟩

invigorate [ɪn'vɪgəreɪt] *v. t.* stärken;

(physically) kräftigen. **invigorating** [ɪn'vɪgəreɪtɪŋ] *adj.* kräftigend ⟨*Getränk, Klima*⟩

invincible [ɪn'vɪnsɪbl] *adj.* unbesiegbar

in'visible *adj.* unsichtbar

invitation [ɪnvɪ'teɪʃn] *n.* Einladung, *die;* **at sb.'s ~:** auf jmds. Einladung *(Akk.)*

invite [ɪn'vaɪt] *v. t.* **a)** *(request to come)* einladen; **b)** *(request to do sth.)* auffordern; **c)** *(bring on)* herausfordern ⟨*Kritik, Verhängnis*⟩. **inviting** [ɪn'vaɪtɪŋ] *adj.* einladend; verlockend ⟨*Gedanke, Vorstellung*⟩

invoice ['ɪnvɔɪs] 1. *n. (bill)* Rechnung, *die.* 2. *v. t.* **~ sb.** jmdm. eine Rechnung schicken; **~ sb. for sth.** jmdm. etw. in Rechnung stellen

invoke [ɪn'vəʊk] *v. t.* anrufen

in'voluntarily *adv.*, **in'voluntary** *adj.* unwillkürlich

involve [ɪn'vɒlv] *v. t.* **a)** *(implicate)* verwickeln; **b)** **become** *or* **get ~d in a fight** in eine Schlägerei verwickelt werden; **get ~d with sb.** sich mit jmdm. einlassen; **c)** *(entail)* mit sich bringen. **involved** [ɪn'vɒlvd] *adj.* verwickelt; *(complicated)* kompliziert

invulnerable [ɪn'vʌlnərəbl] *adj.* unverwundbar; *(fig.)* unantastbar

inward ['ɪnwəd] 1. *adj.* inner... 2. *adv.* einwärts ⟨*gerichtet, gebogen*⟩; **open ~:** nach innen öffnen. **'inwardly** *adv.* im Inneren; innerlich. **inwards** ['ɪnwədz] *see* **inward** 2

iodine ['aɪədiːn] *n.* Jod, *das*

ion ['aɪən] *n.* Ion, *das*

iota [aɪ'əʊtə] *n.* **not one** *or* **an ~:** nicht ein Jota *(geh.)*

IOU [aɪəʊ'juː] *n.* Schuldschein, *der*

Iran [ɪ'rɑːn] *pr. n.* Iran, *der od. (das)*

Iraq [ɪ'rɑːk] *pr. n.* Irak, *der od. (das)*

irate [aɪ'reɪt] *adj.* wütend

Ireland ['aɪələnd] *pr. n.* Irland *(das)*

iris ['aɪərɪs] *n. (Bot., Anat.)* Iris, *die*

Irish ['aɪərɪʃ] 1. *adj.* irisch; **sb. is ~:** jmd. ist Ire/Irin. 2. *n.* **a)** *(language)* Irisch, *das; see also* **English** 2 a; **b)** *constr. as pl.* **the ~:** die Iren

Irish: ~man ['aɪərɪʃmən] *n., pl.* **~men** ['aɪərɪʃmən] Ire, *der;* **~ Re'public** *pr. n.* Irische Republik; **~ 'Sea** *pr. n.* Irische See; **~woman** *n.* Irin, *die*

irk [ɜːk] *v. t.* ärgern. **irksome** ['ɜːksəm] *adj.* lästig

iron ['aɪən] 1. *n.* **a)** *(metal)* Eisen, *das;* **b)** *(for smoothing)* Bügeleisen, *das.* 2. *attrib. adj.* eisern; Eisen⟨platte *usw.*⟩.

3. *v. t. & i.* bügeln. **iron 'out** *v. t.* herausbügeln; *(fig.)* aus dem Weg räumen

Iron 'Curtain *n. (Hist.)* Eiserner Vorhang

ironic [aɪ'rɒnɪk], **ironical** [aɪ'rɒnɪkl] *adj.* ironisch

ironing ['aɪənɪŋ] *n.* Bügeln, *das; (items)* Bügelwäsche, *die;* **do the ~:** bügeln. **'ironing-board** *n.* Bügelbrett, *das*

ironmonger ['aɪənmʌŋɡə(r)] *n. (Brit.)* Eisenwarenhändler, *der/*-händlerin, *die*

irony ['aɪərənɪ] *n.* Ironie, *die;* **the ~ was that ...:** die Ironie lag darin, daß ...

irradiate [ɪ'reɪdɪeɪt] *v. t.* bestrahlen

irrational [ɪ'ræʃənl] *adj.* irrational

irreconcilable [ɪ'rekənsaɪləbl] *adj. (incompatible)* unvereinbar

irrefutable [ɪrɪ'fju:təbl] *adj.* unwiderlegbar

irregular [ɪ'reɡjʊlə(r)] *adj.* unregelmäßig; unkorrekt ⟨*Verhalten, Handlung usw.*⟩. **irregularity** [ɪreɡjʊ'lærɪtɪ] *n. see* **irregular:** Unregelmäßigkeit, *die;* Unkorrektheit, *die*

irrelevant [ɪ'relɪvənt] *adj.* belanglos; irrelevant *(geh.)*

irreparable [ɪ'repərəbl] *adj.* nicht wiedergutzumachen *nicht präd.;* irreparabel *(geh., Med.)*

irreplaceable [ɪrɪ'pleɪsəbl] *adj.* unersetzlich

irrepressible [ɪrɪ'presɪbl] *adj.* nicht zu unterdrücken *nicht präd.;* **she is ~:** sie ist nicht unterzukriegen *(ugs.)*

irreproachable [ɪrɪ'prəʊtʃəbl] *adj.* untadelig

irresistible [ɪrɪ'zɪstɪbl] *adj.* unwiderstehlich; bestechend ⟨*Argument*⟩

irresolute [ɪ'rezəlu:t] *adj.* unentschlossen

irrespective [ɪrɪ'spektɪv] *adj.* **~ of** ungeachtet (+ *Gen.*)

irresponsible [ɪrɪ'spɒnsɪbl] *adj.* verantwortungslos ⟨*Person*⟩; unverantwortlich ⟨*Benehmen*⟩

irretrievable [ɪrɪ'tri:vəbl] *adj.* nicht mehr wiederzubekommen *nicht attr.*

irreverent [ɪ'revərənt] *adj.* respektlos

irreversible [ɪrɪ'vɜ:sɪbl], **irrevocable** [ɪ'revəkəbl] *adjs.* unwiderruflich

irrigate ['ɪrɪɡeɪt] *v. t.* bewässern. **irrigation** [ɪrɪ'ɡeɪʃn] *n.* Bewässerung, *die*

irritable ['ɪrɪtəbl] *adj. (quick to anger)* reizbar; *(temporarily)* gereizt

irritant ['ɪrɪtənt] *n.* Reizstoff, *der*

irritate ['ɪrɪteɪt] *v. t.* **a)** ärgern; **get ~d**

ärgerlich werden; **be ~d by sth.** sich über etw. *(Akk.)* ärgern; **b)** *(Med.)* reizen. **irritating** ['ɪrɪteɪtɪŋ] *adj.* lästig. **irritation** [ɪrɪ'teɪʃn] *n.* **a)** Ärger, *der;* **b)** *(Med.)* Reizung, *die*

is *see* **be**

Islam ['ɪzlɑ:m] *n.* Islam, *der*

island ['aɪlənd] *n.* Insel, *die.* **'islander** *n.* Inselbewohner, *der/*-bewohnerin, *die*

isle [aɪl] *n.* Insel, *die*

isn't ['ɪznt] *(coll.)* = **is not;** *see* **be**

isolate ['aɪsəleɪt] *v. t.* isolieren. **isolated** ['aɪsəleɪtɪd] *adj.* **a)** *(single)* einzeln; **~ cases/instances** Einzelfälle; **b)** *(remote)* abgelegen. **isolation** [aɪsə'leɪʃn] *n.* **a)** *(act)* Isolierung, *die;* **b)** *(state)* Isolation, *die*

Israel ['ɪzreɪl] *pr. n.* Israel *(das).* **Israeli** [ɪz'reɪlɪ] **1.** *adj.* israelisch. **2.** *n.* Israeli, *der/die*

issue ['ɪʃu:, 'ɪsju:] **1.** *n.* **a)** *(point in question)* Frage, *die;* **make an ~ of sth.** etw. aufbauschen; **evade** *or* **dodge the ~:** ausweichen; **b)** *(of magazine etc.)* Ausgabe, *die;* **c)** *(result, outcome)* Ergebnis, *das.* **2.** *v. t.* **a)** *(give out)* ausgeben; ausstellen ⟨*Paß*⟩; erteilen ⟨*Lizenz, Befehl*⟩; **~ sb. with sth.** etw. an jmdn. austeilen; **b)** *(publish)* herausgeben ⟨*Publikation*⟩

it [ɪt] *pron.* **a)** es; **I can't cope with it any more** ich halte das nicht mehr länger aus; **what is it?** was ist los?; **b)** *(the thing, animal, young child previously mentioned)* er/sie/es; *as direct obj.* ihn/sie/es; *as indirect obj.* ihm/ihr/ihm; **c)** *(the person in question)* **who is it?** wer ist da?; **it was the children** es waren die Kinder; **is it you, Dad?** bist du es, Vater?

Italian [ɪ'tæljən] **1.** *adj.* italienisch; **sb. is ~:** jmd. ist Italiener/Italienerin. **2.** *n.* **a)** *(person)* Italiener, *der/*Italienerin, *die;* **b)** *(language)* Italienisch, *das; see also* **English 2 a**

italic [ɪ'tælɪk] **1.** *adj.* kursiv. **2.** *n. in pl.* Kursivschrift, *die;* **in ~s** kursiv

Italy ['ɪtəlɪ] *pr. n.* Italien *(das)*

itch [ɪtʃ] **1.** *n.* Juckreiz, *der;* **I have an ~:** es juckt mich. **2.** *v. i.* **a)** einen Juckreiz haben; **it ~es** es juckt; **b)** **~** *or* **be ~ing to do sth.** darauf brennen, etw. zu tun. **'itchy** *adj.* kratzig; **be ~:** ⟨*Körperteil:*⟩ jucken

it'd ['ɪtəd] *(coll.)* **a)** = **it had; b)** = **it would**

item ['aɪtəm] *n.* **a)** Ding, *das;* Sache, *die; (in shop, catalogue)* Artikel, *der;*

(on radio, TV) Nummer, *die;* ~ **of clothing** Kleidungsstück, *das;* **b)** ~ |**of news**| Nachricht, *die.* **itemize** ['aɪtə-maɪz] *v. t.* einzeln aufführen

itinerary [aɪ'tɪnərərɪ] *n.* Reiseroute, *die*

it'll [ɪtl] *(coll.)* = it will

its [ɪts] *poss. pron. attrib.* sein/ihr/sein

it's [ɪts] **a)** = it is; **b)** = it has

itself [ɪt'self] *pron.* **a)** *emphat.* selbst; **b)** *refl.* sich

I've [aɪv] = I have

ivory ['aɪvərɪ] *n.* Elfenbein, *das; attrib.* elfenbeinern; Elfenbein-

ivy ['aɪvɪ] *n.* Efeu, *der*

J

J, j [dʒeɪ] *n.* J, j, *das*

jab [dʒæb] **1.** *v. t.,* -**bb**- stoßen. **2.** *n.* **a)** Stoß, *der; (with needle)* Stich, *der;* **b)** *(Brit. coll.: injection)* Spritze, *die*

jabber ['dʒæbə(r)] *v. i.* plappern *(ugs.)*

jack [dʒæk] *n.* **a)** *(for car)* Wagenheber, *der;* **b)** *(Cards)* Bube, *der*

jackal ['dʒækl] *n.* Schakal, *der*

jackdaw ['dʒækdɔ:] *n.* Dohle, *die*

jacket ['dʒækɪt] *n.* **a)** Jacke, *die; (of suit)* Jackett, *das;* **sports** ~: Sakko, *der;* **b)** *(of book)* Schutzumschlag, *der;* **c)** ~ **potatoes** in der Schale gebackene Kartoffeln

'**jackpot** *n.* Jackpot, *der;* **hit the ~** *(fig.)* das große Los ziehen

jaded ['dʒeɪdɪd] *adj.* abgespannt

jagged ['dʒægɪd] *adj.* gezackt

jaguar ['dʒægjʊə(r)] *n.* Jaguar, *der*

jail [dʒeɪl] **1.** *n.* Gefängnis, *das.* **2.** *v. t.* ins Gefängnis bringen. '**jailbreak** *n.* Gefängnisausbruch, *der.* **jailer, jailor** ['dʒeɪlə(r)] *n.* Gefängniswärter, *der/-wärterin, die*

'**jam** [dʒæm] **1.** *v. t.,* -**mm**-: **a)** *(between two surfaces)* einklemmen; **b)** *(make immovable)* blockieren; *(fig.)* lähmen. **2.** *v. i.,* -**mm**-: **a)** *(become wedged)* sich verklemmen; **b)** *⟨Maschine:⟩* klemmen. **3.** *n.* **a)** *(crush, stoppage)* Blockierung, *die;* **b)** *(coll.: dilemma)* **be in a ~:** in der Klemme stecken

(ugs.). **jam 'on** *v. t.* ~ **the brakes |full| on** [voll] auf die Bremse steigen *(ugs.)*

²**jam** *n.* Marmelade, *die*

Jamaica [dʒə'meɪkə] *pr. n.* Jamaika *(das)*

Jan. *abbr.* **January** Jan.

jangle ['dʒæŋgl] **1.** *v. i.* klimpern; *⟨Klingel:⟩* bimmeln. **2.** *v. t.* rasseln mit

janitor ['dʒænɪtə(r)] *n.* Hausmeister, *der*

January ['dʒænjʊərɪ] *n.* Januar, *der; see also* **August**

Japan [dʒə'pæn] *n.* Japan *(das).* **Japanese** [dʒæpə'ni:z] **1.** *adj.* japanisch. **2.** *n., pl. same* **a)** *(person)* Japaner, *der/*Japanerin, *die;* **b)** *(language)* Japanisch, *das; see also* **English 2 a**

¹**jar** [dʒɑ:(r)] **1.** *v. i.,* -**rr**- quietschen; *(fig.)* ~ **on sb./sb.'s nerves** jmdm. auf die Nerven gehen. **2.** *v. t.,* -**rr**- erschüttern

²**jar** *n.* Topf, *der; (glass ~)* Glas, *das*

jargon ['dʒɑ:gən] *n.* Jargon, *der*

jasmin[e] ['dʒæsmɪn] *n.* Jasmin, *der*

jaundice ['dʒɔ:ndɪs] *n. (Med.)* Gelbsucht, *die.* **jaundiced** ['dʒɔ:ndɪst] *adj. (fig.)* verbittert

jaunt [dʒɔ:nt] *n.* Ausflug, *der*

javelin ['dʒævlɪn] *n.* **a)** Speer, *der;* **b)** *(Sport: event)* Speerwerfen, *das*

jaw [dʒɔ:] *n.* Kiefer, *der.* '**jawbone** *n.* Kieferknochen, *der*

jay [dʒeɪ] *n.* Eichelhäher, *der*

jazz [dʒæz] **1.** *n.* Jazz, *der; attrib.* Jazz-. **2.** *v. t.* ~ **up** aufpeppen *(ugs.)*

jealous ['dʒeləs] *adj.* eifersüchtig (of auf + *Akk.*). '**jealousy** *n.* Eifersucht, *die*

jeans [dʒi:nz] *n. pl.* Jeans *Pl.*

jeer [dʒɪə(r)] *v. i.* höhnen *(geh.);* ~ **at sb.** jmdn. verhöhnen

jelly ['dʒelɪ] *n.* Gelee, *das; (dessert)* Götterspeise, *die.* '**jellyfish** *n.* Qualle, *die*

jeopardize ['dʒepədaɪz] *v. t.* gefährden

jeopardy ['dʒepədɪ] *n.* **in ~:** in Gefahr; gefährdet

jerk [dʒɜ:k] **1.** *n.* Ruck, *der.* **2.** *v. t.* reißen an (+ *Dat.*). **3.** *v. i.* zucken

jersey ['dʒɜ:zɪ] *n.* Pullover, *der; (Sport)* Trikot, *das*

jest [dʒest] **1.** *n.* Scherz, *der;* **in ~:** im Scherz. **2.** *v. i.* scherzen

Jesus ['dʒi:zəs] *pr. n.* Jesus *(der)*

jet [dʒet] *n.* **a)** *(stream)* Strahl, *der;* **b)** *(nozzle)* Düse, *die;* **c)** *(aircraft)* Düsenflugzeug, *das;* Jet, *der*

jet: ~-**black** *adj.* pechschwarz; ~ **en-**

gine *n.* Düsentriebwerk, *das;* ~ **lag** *n.* Jet-travel-Syndrom, *das;* ~-**propelled** *adj.* düsengetrieben

jetsam ['dʒetsəm] *n. see* flotsam

'**jet-set** *n.* Jet-set, *der*

jettison ['dʒetɪsən] *v. t.* über Bord werfen; *(discard)* wegwerfen

jetty ['dʒetɪ] *n.* Landungsbrücke, *die*

Jew [dʒuː] *n.* Jude, *der*/Jüdin, *die*

jewel ['dʒuːəl] *n.* Juwel, *das od. der.* **jeweller** (*Amer.:* **jeweler**) ['dʒuːə-lə(r)] *n.* Juwelier, *der.* **jewellery** (*Brit.*), **jewelry** ['dʒuːəlrɪ] *n.* Schmuck, *der*

Jewish ['dʒuːɪʃ] *adj.* jüdisch

jib [dʒɪb] *v. i.,* -bb- sich sträuben (**at** gegen)

jibe *see* gibe

jiffy ['dʒɪfɪ] *n. (coll.)* **in a** ~: sofort

jig [dʒɪg] *n.* Jig, *die*

'**jigsaw** *n.* ~ [**puzzle**] Puzzle, *das*

jilt [dʒɪlt] *v. t.* sitzenlassen *(ugs.)*

jingle ['dʒɪŋgl] 1. *n. (Commerc.)* Werbespruch, *der.* 2. *v. i.* klimpern; ⟨*Glöckchen:*⟩ bimmeln. 3. *v. t.* klimpern mit ⟨*Münzen, Schlüsseln*⟩

jinx [dʒɪŋks] *(coll.)* 1. *n.* Fluch, *der.* 2. *v. t.* verhexen

jitters ['dʒɪtəz] *n. pl. (coll.)* großes Zittern. **jittery** ['dʒɪtərɪ] *adj. (coll.) (nervous)* nervös; *(frightened)* verängstigt

job [dʒɒb] *n.* **a)** *(piece of work)* Arbeit, *die;* **I have a** ~ **for you** ich habe eine Aufgabe für dich; **b)** *(employment)* Stelle, *die;* Job, *der (ugs.).* '**job-centre** *n. (Brit.)* Arbeitsvermittlungsstelle, *die.* '**jobless** *adj.* arbeitslos

jockey ['dʒɒkɪ] *n.* Jockei, *der*

jocular ['dʒɒkjʊlə(r)] *adj.* lustig

jodhpurs ['dʒɒdpəz] *n. pl.* Reithose, *die*

jog [dʒɒg] 1. *v. t.,* -gg-: **a)** *(shake)* rütteln; **b)** *(nudge)* [an]stoßen; **c)** ~ **sb.'s memory** jmds. Gedächtnis *(Dat.)* auf die Sprünge helfen. 2. *v. i.,* -gg-: **a)** *(up and down)* auf und ab hüpfen; **b)** *(trot)* ⟨*Pferd:*⟩ [dahin]trotten; **c)** *(Sport)* joggen. 3. *n.* **go for a** ~: joggen gehen. '**jogging** *n.* Jogging, *das*

join [dʒɔɪn] 1. *v. t.* **a)** *(connect)* verbinden (**to** mit); **b)** *(come into company of)* sich gesellen zu; **c)** eintreten in (+ *Akk.*) ⟨*Armee, Firma, Verein, Partei*⟩. 2. *v. i.* ⟨*Straßen:*⟩ zusammenlaufen. **join in** 1. [ˈ-ˈ] *v. i.* mitmachen (**with** bei). 2. [ˈ--] *v. t.* mitmachen bei. **join 'up** 1. *v. i. (Mil.)* einrücken. 2. *v. t.* miteinander verbinden

'**joiner** *n.* Tischler, *der*/Tischlerin, *die*

joint [dʒɔɪnt] 1. *n.* **a)** *(Building)* Fuge, *die;* **b)** *(Anat.)* Gelenk, *das;* **c) a** ~ [**of meat**] ein Stück Fleisch; *(for roasting)* ein Braten; **d)** *(sl.: place)* Laden, *der.* 2. *adj.* **a)** *(of two or more)* gemeinsam; **b)** Mit⟨*autor, -erbe, -besitzer*⟩. '**jointly** *adv.* gemeinsam

joist [dʒɔɪst] *n. (Building)* Deckenbalken, *der; (steel)* [Decken]träger, *der*

joke [dʒəʊk] 1. *n.* Witz, *der;* Scherz, *der.* 2. *v. i.* scherzen, Witze machen (**about** über + *Akk.*); **joking apart** Scherz beiseite! '**joker** *n.* **a)** Spaßvogel, *der;* **b)** *(Cards)* Joker, *der*

jollity ['dʒɒlɪtɪ] *n.* Fröhlichkeit, *die; (merry-making)* Festlichkeit, *die*

jolly ['dʒɒlɪ] 1. *adj.* fröhlich. 2. *adv. (Brit. coll.)* ganz schön *(ugs.);* ~ **good!** ausgezeichnet!

jolt [dʒəʊlt] 1. *v. t.* ⟨*Fahrzeug:*⟩ durchrütteln. 2. *v. i.* ⟨*Fahrzeug:*⟩ holpern. 3. *n.* **a)** *(jerk)* Stoß, *der;* Ruck, *der;* **b)** *(fig.: shock)* Schock, *der*

Jordan ['dʒɔːdn] *pr. n.* Jordanien *(das)*

jostle ['dʒɒsl] 1. *v. i.* ~ [**against each other**] aneinanderstoßen. 2. *v. t.* stoßen

jot [dʒɒt] *n.* [**not**] **a** ~: [k]ein bißchen. **jot 'down** *v. t.* [rasch] aufschreiben

jotter ['dʒɒtə(r)] *n.* Notizblock, *der*

journal ['dʒɜːnl] *n.* Zeitschrift, *die*

journalism ['dʒɜːnəlɪzm] *n.* Journalismus, *der.* **journalist** ['dʒɜːnəlɪst] *n.* Journalist, *der*/Journalistin, *die*

journey ['dʒɜːnɪ] *n.* **a)** Reise, *die;* **b)** *(of vehicle)* Fahrt, *die*

jovial ['dʒəʊvɪəl] *adj.* herzlich ⟨*Gruß*⟩; fröhlich ⟨*Person*⟩

joy [dʒɔɪ] *n.* Freude, *die.* **joyful** ['dʒɔɪfl] *adj.* froh[gestimmt] ⟨*Person*⟩; freudig ⟨*Blick, Ereignis, Gesang*⟩. **joyride** *n. (coll.)* Spritztour, *die*

JP *abbr.* **Justice of the Peace**

jubilant ['dʒuːbɪlənt] *adj.* jubelnd; **be** ~ ⟨*Person:*⟩ frohlocken. **jubilation** [dʒuːbɪ'leɪʃn] *n.* Jubel, *der*

jubilee ['dʒuːbɪliː] *n.* Jubiläum, *das*

judge [dʒʌdʒ] 1. *n.* **a)** Richter, *der*/Richterin, *die;* **b)** *(in contest)* Preisrichter, *der*/-richterin, *die;* **c)** *(fig.: critic)* Kenner, *der*/Kennerin, *die.* 2. *v. t.* **a)** *(sentence)* richten *(geh.);* **b)** *(form opinion about)* [be]urteilen. '**judg[e]ment** *n.* **a)** Urteil, *das;* **b)** *(critical faculty)* Urteilsvermögen, *das*

judicial [dʒuː'dɪʃl] *adj.* gerichtlich

judicious [dʒuː'dɪʃəs] *adj.* klarblickend

judo ['dʒuːdəʊ] *n.* Judo, *das*

jug [dʒʌg] *n.* Krug, *der; (with lid, water-~)* Kanne, *die*

juggernaut ['dʒʌgənɔːt] *n. (Brit.: lorry)* schwerer Brummer *(ugs.)*

juggle ['dʒʌgl] *v. i.* jonglieren. **juggler** ['dʒʌglə(r)] *n.* Jongleur, *der*/Jongleuse, *die*

juice [dʒuːs] *n.* Saft, *der.* **juicy** ['dʒuː-sɪ] *adj.* saftig

juke-box ['dʒuːkbɒks] *n.* Jukebox, *die;* Musikbox, *die*

Jul. *abbr.* July Jul.

July [dʒʊ'laɪ] *n.* Juli, *der; see also* **August**

jumble ['dʒʌmbl] **1.** *v. t.* ~ up durcheinanderbringen. **2.** *n.* Durcheinander, *das.* '**jumble sale** *n. (Brit.)* Trödelmarkt, *der*

jumbo jet [dʒʌmbəʊ 'dʒet] *n.* Jumbo-Jet, *der*

jump [dʒʌmp] **1.** *n.* **a)** Sprung, *der;* **b)** *(in prices)* sprunghafter Anstieg. **2.** *v. i.* **a)** springen; ~ **for joy** einen Freudensprung machen; **b)** ~ **to conclusions** voreilige Schlüsse ziehen. **3.** *v. t.* **a)** überspringen; **b)** ~ **the queue** *(Brit.)* sich vordrängeln. **jump a'bout, jump a'round** *v. i.* herumspringen *(ugs.).* '**jump at** *v. t. (fig.)* sofort zugreifen bei ⟨*Angebot, Gelegenheit*⟩

'**jumper** *n.* Pullover, *der*

jumpy ['dʒʌmpɪ] *adj.* nervös

Jun. *abbr.* June Jun.

junction ['dʒʌŋkʃn] *n.* **a)** *(of railway lines, roads)* ≈ Einmündung, *die;* **b)** *(crossroads)* Kreuzung, *die*

juncture ['dʒʌŋktʃə(r)] *n.* **at this ~:** zu diesem Zeitpunkt

June [dʒuːn] *n.* Juni, *der; see also* **August**

jungle ['dʒʌŋgl] *n.* Dschungel, *der*

junior ['dʒuːnɪə(r)] *adj.* **a)** *(in age)* jünger; ~ **team** *(Sport)* Juniorenmannschaft, *die;* **b)** *(in rank)* rangniedriger ⟨*Person*⟩; niedriger ⟨*Rang*⟩. '**junior school** *n. (Brit.)* Grundschule, *die*

junk [dʒʌŋk] *n.* Trödel, *der (ugs.); (trash)* Ramsch, *der (ugs.).* '**junk food** *n.* minderwertige Kost. '**junk shop** *n.* Trödelladen, *der (ugs.)*

Jupiter ['dʒuːpɪtə(r)] *pr. n. (Astron.)* Jupiter, *der*

jurisdiction [dʒʊərɪs'dɪkʃn] *n.* Gerichtsbarkeit, *die*

juror ['dʒʊərə(r)] *n.* Geschworene, *der/die*

jury ['dʒʊərɪ] *n.* **a)** *(in court)* **the ~:** die Geschworenen; **b)** *(in competition)* Jury, *die*

just [dʒʌst] **1.** *adj. (morally right)* gerecht. **2.** *adv.* **a)** *(exactly)* genau; ~ **then/enough** gerade da/genug; ~ **as** *(exactly as)* genauso wie; *(when)* gerade, als; ~ **as you like** *or* **please** ganz wie Sie wünschen/du magst; ~ **as good** *etc.* genauso gut *usw.;* **b)** *(barely)* gerade [eben]; *(with little time to spare)* gerade noch; *(no more than)* nur; ~ **under £10** nicht ganz zehn Pfund; **c)** *(at this moment)* gerade; **not ~ now** im Moment nicht; **d)** *(coll.) (simply)* einfach; *(only)* nur; *esp. with imper.* mal [eben]; ~ **look at that!** guck dir das mal an!; ~ **a moment** einen Moment mal; ~ **in case** für alle Fälle

justice ['dʒʌstɪs] *n.* **a)** Gerechtigkeit, *die;* **b)** *(magistrate)* Schiedsrichter, *der*/-richterin, *die;* **J~** **of the Peace** Friedensrichter, *der*/-richterin, *die*

justifiable [dʒʌstɪ'faɪəbl] *adj.* berechtigt. **justifiably** [dʒʌstɪ'faɪəblɪ] *adv.* zu Recht

justification [dʒʌstɪfɪ'keɪʃn] *n.* Rechtfertigung, *die*

justify ['dʒʌstɪfaɪ] *v. t.* rechtfertigen; **be justified in doing sth.** etw. zu Recht tun

jut [dʒʌt] *v. i.*, **-tt-:** ~ [out] [her]vorragen; herausragen

juvenile ['dʒuːvənaɪl] **1.** *adj.* **a)** jugendlich; **b)** *(immature)* kindisch. **2.** *n.* Jugendliche, *der/die.* **juvenile delinquency** [~ dɪ'lɪŋkwənsɪ] *n.* Jugendkriminalität, *die.* **juvenile delinquent** [~ dɪ'lɪŋkwənt] *n.* jugendlicher Straftäter/jugendliche Straftäterin

juxtapose [dʒʌkstə'pəʊz] *v. t.* nebeneinanderstellen (**with,** to und). **juxtaposition** [dʒʌkstəpə'zɪʃn] *n.* Nebeneinanderstellung, *die*

K

K, k [keɪ] *n.* K, k, *das*

kaleidoscope [kə'laɪdəskəʊp] *n.* Kaleidoskop, *das*

kangaroo [kæŋgə'ruː] *n.* Känguruh, *das*

karate [kəˈrɑːtɪ] *n.* Karate, *das*
keel [kiːl] *n. (Naut.)* Kiel, *der*
keen [kiːn] *adj.* **a)** *(sharp)* scharf; **b)** *(cold)* schneidend ⟨*Wind, Kälte*⟩; **c)** *(eager)* begeistert ⟨*Fußballfan, Sportler*⟩; lebhaft ⟨*Interesse*⟩; **be ~ to do sth.** darauf erpicht sein, etw. zu tun; **d)** *(sensitive)* scharf ⟨*Augen*⟩; fein ⟨*Sinne*⟩. **'keenly** *adv.* **a)** *(sharply)* scharf; **b)** *(eagerly)* eifrig; brennend ⟨*interessiert sein*⟩; **c)** *(acutely)* **be ~ aware of sth.** sich *(Dat.)* einer Sache *(Gen.)* voll bewußt sein
keep [kiːp] **1.** *v. t.,* **kept** [kept] **a)** halten ⟨*Versprechen, Schwur, Sabbat, Fasten*⟩; einhalten ⟨*Verabredung, Vereinbarung*⟩; begehen, feiern ⟨*Fest*⟩; **b)** *(have charge of)* aufbewahren; **c)** *(retain)* behalten; *(not lose or destroy)* aufheben ⟨*Quittung, Rechnung*⟩; **d)** halten ⟨*Bienen, Hund usw.*⟩; **e)** führen ⟨*Tagebuch, Geschäft, Ware*⟩; **f)** *(support)* versorgen ⟨*Familie*⟩; **g)** *(detain)* festhalten; **~ sb. waiting** jmdn. warten lassen; **what kept you?** wo bleibst du denn?; **h)** *(reserve)* aufheben. **2.** *v. i.,* **kept a)** *(remain)* bleiben; **are you ~ing well?** geht's dir gut?; **b)** **~ [to the] left/right** sich links/rechts halten; **~ doing sth.** *(repeatedly)* etw. immer wieder tun; **~ talking/working** *etc.* **until ...:** weiterreden/-arbeiten *usw.,* bis ...; **c)** *(remain good)* ⟨*Lebensmittel:*⟩ sich halten. **3.** *n.* **a)** *(maintenance)* Unterhalt, *der;* **b)** **for ~s** *(coll.)* auf Dauer; **c)** *(Hist.: tower)* Bergfried, *der.* **keep 'back 1.** *v. i.* zurückbleiben. **2.** *v. t.* **a)** *(restrain)* zurückhalten ⟨*Menschenmenge, Tränen*⟩; **b)** *(withhold)* verschweigen ⟨*Informationen, Tatsachen*⟩ *(from Dat.).* **keep 'down 1.** *v. i.* unten bleiben. **2.** *v. t.* **a)** niedrig halten ⟨*Steuern, Preise usw.*⟩; **keep one's weight down** nicht zunehmen; **b) keep your voice down!** rede nicht so laut! **keep 'off 1.** *v. i.* ⟨*Person:*⟩ wegbleiben. **2.** *v. t.* fernhalten; **'keep off the grass'** „Betreten des Rasens verboten". **keep 'out 1.** *v. i.* **'keep out'** „Zutritt verboten". **2.** *v. t.* nicht hereinlassen. **keep 'up 1.** *v. i.* **keep up with sb./sth.** mit jmdm./etw. Schritt halten. **2.** *v. t.* aufrechterhalten ⟨*Freundschaft, jmds. Moral*⟩; **keep one's strength up** sich bei Kräften halten; **keep it up!** weiter so!
keep-'fit *n.* Fitneßtraining, *das*
'keeping *n.* **be in ~ with sth.** einer Sache *(Dat.)* entsprechen

'keepsake *n.* Andenken, *das*
keg [keg] *n.* [kleines] Faß
kennel [ˈkenl] *n.* Hundehütte, *die*
Kenya [ˈkenjə] *pr. n.* Kenia *(das)*
kept *see* **keep 1, 2**
kerb [kɜːb], **'kerbstone** *ns. (Brit.)* Bordstein, *der*
kernel [ˈkɜːnl] *n.* Kern, *der*
ketchup [ˈketʃʌp] *n.* Ketchup, *der od. das*
kettle [ˈketl] *n.* [Wasser]kessel, *der*
key [kiː] *n.* **a)** Schlüssel, *der;* **b)** *(on piano, typewriter, etc.)* Taste, *die;* **c)** *(Mus.)* Tonart, *die*
key: **~board** *n. (of piano etc.)* Klaviatur, *die; (of typewriter etc.)* Tastatur, *die;* **~hole** *n.* Schlüsselloch, *das;* **~ring** *n.* Schlüsselring, *der*
kg. *abbr.* **kilogram[s]** kg
khaki [ˈkɑːkɪ] **1.** *adj.* khakifarben. **2.** *n. (cloth)* Khaki, *der*
kick [kɪk] **1.** *n.* **a)** [Fuß]tritt, *der; (Footb.)* Schuß, *der;* **give sb. a ~:** jmdm. einen Tritt geben; **b)** *(coll.: thrill)* **do sth. for ~s** etw. zum Spaß tun; **he gets a ~ out of it** er hat Spaß daran. **2.** *v. i.* treten; ⟨*Pferd:*⟩ ausschlagen. **3.** *v. t.* einen Tritt geben (+ *Dat.*) ⟨*Person, Hund*⟩; treten gegen ⟨*Gegenstand*⟩; kicken *(ugs.),* schießen ⟨*Ball*⟩. **kick a'bout, kick a'round** *v. t.* [in der Gegend] herumkicken *(ugs.).* **kick 'off** *v. i. (Footb.)* anstoßen. **kick 'up** *v. t. (coll.)* **~ up a fuss/row** Krach schlagen/anfangen *(ugs.)*
kid [kɪd] **1.** *n.* **a)** *(young goat)* Kitz, *der;* **b)** *(coll.: child)* Kind, *das.* **2.** *v. t.,* **-dd-** *(coll.)* auf den Arm nehmen *(ugs.);* **~ oneself** sich *(Dat.)* was vormachen
kidnap [ˈkɪdnæp] *v. t., (Brit.)* **-pp-** entführen. **'kidnapper** *n.* Entführer, *der*/Entführerin, *die*
kidney [ˈkɪdnɪ] *n.* Niere, *die.* **'kidney machine** *n.* künstliche Niere
kill [kɪl] *v. t.* **a)** töten; *(deliberately)* umbringen; **be ~ed in action** im Kampf fallen; **be ~ed in a car crash** bei einem Autounfall ums Leben kommen; **b)** **~ time** die Zeit totschlagen. **'killer** *n.* Mörder, *der*/Mörderin, *die.* **'killing** *n.* **a)** Töten, *das;* **b)** **make a ~** *(coll.: great profit)* einen [Mords]reibach machen *(ugs.).* **'killjoy** *n.* Spielverderber, *der*/-verderberin, *die*
kiln [kɪln] *n.* Brennofen, *der*
kilo [ˈkiːləʊ] *n., pl.* **~s** Kilo, *das*
kilogram, kilogramme [ˈkɪləgræm] *n.* Kilogramm, *das*

kilometre (*Brit.; Amer.:* **kilometer**) ['kɪləmiːtə(r) *(Brit.),* kɪ'lɒmɪtə(r)] *n.* Kilometer, *der*

kilowatt *n.* ['kɪləwɒt] Kilowatt, *das*

kilt [kɪlt] *n.* Kilt, *der*

kin [kɪn] *n.* Verwandte

¹kind [kaɪnd] *n.* **a)** *(class, sort)* Art, *die;* **several ~s of apples** mehrere Sorten Äpfel; **all ~s of things/excuses** alles mögliche/alle möglichen Ausreden; **no ... of any ~:** keinerlei ...; **what ~ is it?** was für einer/eine/eins ist es?; **what ~ of [a] tree is this?** was für ein Baum ist das?; **b)** *(implying vagueness)* **a ~ of ...:** [so] eine Art ...; **~ of cute** *(coll.)* irgendwie niedlich *(ugs.)*

²kind *adj.* liebenswürdig; *(showing friendliness)* freundlich; **be ~ to animals** gut zu Tieren sein; **how ~!** wie nett [von ihm/Ihnen *usw.*]!

kindergarten ['kɪndəgɑːtn] *n.* Kindergarten, *der*

kindle ['kɪndl] *(fig.)* wecken

kindly ['kaɪndlɪ] **1.** *adv.* **a)** freundlich; nett; **b)** *in polite request etc.* freundlicherweise; **thank you ~:** herzlichen Dank. **2.** *adj.* freundlich; nett; *(kindhearted)* gütig

'kindness *n.* **a)** *no pl. (kind nature)* Freundlichkeit, *die;* **b) do sb. a ~** *(kind act)* jmdm. eine Gefälligkeit erweisen

kindred ['kɪndrɪd] *adj.* verwandt; **~ 'spirit** Gleichgesinnte, *der/die*

king [kɪŋ] *n.* König, *der.* **kingdom** ['kɪŋdəm] *n.* Königreich, *das*

'kingfisher *n.* Eisvogel, *der*

'king-size[d] *adj.* extragroß; King-size-⟨Zigaretten⟩

kink [kɪŋk] *n. (in pipe, wire, etc.)* Knick, *der; (in hair, wool)* Welle, *die*

'kinky *adj. (coll.)* spleenig; *(sexually)* abartig

kiosk ['kiːɒsk] *n.* **a)** Kiosk, *der;* **b)** *(telephone booth)* [Telefon]zelle, *die*

kip [kɪp] *n. (Brit. sl.: sleep)* **have a/get some ~:** eine Runde pennen *(salopp)*

kipper ['kɪpə(r)] *n.* Kipper, *der*

kiss [kɪs] **1.** *n.* Kuß, *der.* **2.** *v. t.* küssen; **~ sb. good night/goodbye** jmdm. einen Gutenacht-/Abschiedskuß geben. **3.** *v. i.* **they ~ed** sie küssten sich

kit [kɪt] *n.* **a)** *(Brit.: set of items)* Set, *das;* **b)** *(Brit.: clothing etc.)* **sports ~:** Sportzeug, *das;* **riding-/skiing-~:** Reit-/Skiausrüstung, *die.* **'kitbag** *n.* Tornister, *der*

kitchen ['kɪtʃɪn] *n.* Küche, *die; attrib.* Küchen-. **kitchen 'sink** *n.* [Küchen]ausguß, *der*

kite [kaɪt] *n.* Drachen, *der*

kith [kɪθ] *n.* **~ and kin** Freunde und Verwandte

kitten ['kɪtn] *n.* Kätzchen, *das*

kitty ['kɪtɪ] *n. (money)* Kasse, *die*

kleptomania [kleptə'meɪnɪə] *n.* Kleptomanie, *die.* **kleptomaniac** [kleptə'meɪnɪæk] *n.* Kleptomane, *der/*Kleptomanin, *die*

km. *abbr.* **kilometre[s]** km

knack [næk] *n.* Talent, *das;* **get the ~ [of doing sth.]** den Bogen rauskriegen [, wie man etw. macht] *(ugs.);* **have lost the ~:** es nicht mehr zustande bringen

knapsack ['næpsæk] *n.* Rucksack, *der; (Mil.)* Tornister, *der*

knead [niːd] *v. t.* kneten

knee [niː] *n.* Knie, *das*

knee: ~cap *n.* Kniescheibe, *die;* **~-deep** *adj.* knietief; **~-high** *adj.* kniehoch; **~-jerk reaction** *n. (fig.)* automatische Reaktion; **~-joint** *n.* Kniegelenk, *das*

kneel [niːl] *v. i.,* **knelt** [nelt] *or (esp. Amer.)* **kneeled** knien; **~ down** niederknien

knelt *see* **kneel**

knew *see* **know**

knickers ['nɪkəz] *n. pl. (Brit.)* [Damen]schlüpfer, *der*

knife [naɪf] **1.** *n., pl.* **knives** [naɪvz] Messer, *das.* **2.** *v. t. (stab)* einstechen auf (+ *Akk.*); *(kill)* erstechen

knight [naɪt] *n.* **a)** *(Hist.)* Ritter, *der;* **b)** *(Chess)* Springer, *der.* **'knighthood** *n.* Ritterwürde, *die*

knit [nɪt] *v. t.,* **-tt-** stricken; **~ one's brow** die Stirn runzeln. **'knitting** *n.* Stricken, *das; (work being knitted)* Strickarbeit, *die.* **'knitting needle** *n.* Stricknadel, *die.* **'knitwear** *n.* Strickwaren *Pl.*

knives *pl. of* **knife 1**

knob [nɒb] *n.* **a)** *(on door, walkingstick, etc.)* Knauf, *der;* **b)** *(control on radio etc.)* Knopf, *der;* **c)** *(of butter)* Klümpchen, *das*

knock [nɒk] **1.** *v. t.* **a)** *(strike) (lightly)* klopfen an (+ *Akk.*); *(forcefully)* schlagen gegen *od.* an (+ *Akk.*); **~ a hole in sth.** ein Loch in etw. (+ *Akk.*) schlagen; **b)** *(sl.: criticize)* herziehen über (+ *Akk.*) *(ugs.).* **2.** *v. i.* klopfen (**at** *od.* **on** + *Akk.*). **3.** *n.* Klopfen, *das.* **knock 'down** *v. t.* **a)** *(in car)* umfahren; **b)** *(demolish)* abreißen. **knock 'off 1.** *v. t.* **a)** **~ off work** *(coll.: leave)* Feierabend machen; **b)** *(deduct)* **~ five pounds off the price** es fünf Pfund bil-

liger machen; c) *(coll.: do quickly)* aus
dem Ärmel schütteln *(ugs.);* d) *(sl.:
steal)* klauen *(salopp).* 2. *v. i. (coll.)*
Feierabend machen. **knock 'out** *v. t.*
a) *(make unconscious)* bewußtlos umfallen lassen; b) *(Boxing)* k. o. schlagen; c) *(sl.: exhaust)* kaputtmachen
(ugs.). **knock 'over** *v. t.* umstoßen;
⟨*Fahrer, Fahrzeug:*⟩ umfahren ⟨*Person*⟩
'knock-down *adj.* ~ **prices** Schleuderpreise
'knocker *n.* [Tür]klopfer, *der*
knock: ~**-kneed** ['nɒkniːd] *adj.*
X-beinig ⟨*Person*⟩; ~**-out** *n. (Boxing)*
K.-o.-Schlag, *der*
knot [nɒt] 1. *n.* Knoten, *der.* 2. *v. t.,*
-tt- knoten ⟨*Seil, Faden usw.*⟩
'knotty *adj. (fig.: puzzling)* verwickelt
know [nəʊ] *v. t.,* knew [njuː], known
[nəʊn] a) *(recognize)* erkennen (by an
+ *Dat.,* for als + *Akk.*); b) *(be able to
distinguish)* ~ **sth. from sth.** etw. von
etw. unterscheiden können; c) *(be
aware of)* wissen; d) *(have understanding of)* können ⟨*ABC, Einmaleins,
Deutsch usw.*⟩; ~ **how to mend fuses**
wissen, wie man Sicherungen repa-
riert; ~ **how to drive a car** Auto fahren
können; e) kennen ⟨*Person*⟩. **'know-
all** *n.* Neunmalkluge, *der/die.*
'know-how *n.* praktisches Wissen
'knowing *adj.* a) wissend ⟨*Blick, Lächeln*⟩; b) *(cunning)* verschlagen.
'knowingly *adv.* a) *(intentionally)*
wissentlich; b) vielsagend ⟨*lächeln,
anblicken*⟩
knowledge ['nɒlɪdʒ] *n.* a) *(familiarity)* Kenntnisse (of in + *Dat.*); b)
(awareness) Wissen, *das;* **have no ~ of
sth.** nichts von etw. wissen; keine
Kenntnis von etw. haben *(geh.);* c) |a|
~ **of languages/French** Sprach-/Fran-
zösischkenntnisse *Pl.* **knowledge-
able** ['nɒlɪdʒəbl] *adj.* be ~ **about** *or* **on
sth.** viel über etw. *(Akk.)* wissen
known [nəʊn] 1. *see* **know.** 2. *adj.* bekannt
knuckle ['nʌkl] *n.* [Finger]knöchel,
der
Korea [kə'rɪə] *pr. n.* Korea *(das)*
kosher ['kəʊʃə(r)] *adj.* koscher
kudos ['kjuːdɒs] *n. (coll.)* Prestige, *das*
kW *abbr.* **kilowatt|s|** kW

L

L, l [el] *n.* L, l, *das*
£ *abbr.* **pound|s|** £; **cost £5** 5£ *od.*
Pfund kosten
l. *abbr.* **litre|s|** l
lab [læb] *n. (coll.)* Labor, *das*
label ['leɪbl] 1. *n.* Schildchen, *das; (on
bottles, in clothes)* Etikett, *das; (tied/
stuck to an object)* Anhänger/Aufkle-
ber, *der.* 2. *v. t., (Brit.)* **-ll-:** a) etikettieren; auszeichnen ⟨*Waren*⟩; *(write on)*
beschriften; b) *(fig.)* ~ **sb./sth. |as| sth.**
jmdn./etw. als etw. etikettieren
labor *(Amer.) see* **labour**
laboratory [lə'bɒrətərɪ] *n.* Labor[ato-
rium], *das*
labored, laborer *(Amer.) see* **labour-**
laborious [lə'bɔːrɪəs] *adj.* mühsam.
la'boriously *adv.* mühevoll
labour ['leɪbə(r)] *(Brit.)* 1. *n.* a) Arbeit,
die; b) *(workers)* Arbeiterschaft, *die;*
immigrant ~: ausländische Arbeitskräfte; c) **L~,** **the ~ Party** *(Polit.)* die
Labour Party; d) *(childbirth)* Wehen
Pl.; **be in ~:** in den Wehen liegen. 2.
v. i. hart arbeiten (**at, on** an + *Dat.*).
3. *v. t.* ~ **the point** sich lange darüber
verbreiten
laboured ['leɪbəd] *adj. (Brit.)* mühsam; schwerfällig ⟨*Stil*⟩; **his breathing
was ~:** er atmete schwer
'labourer *n. (Brit.)* Arbeiter, *der/*Arbeiterin, *die*
'labour-saving *adj.* arbeit[s]sparend
labyrinth ['læbərɪnθ] *n.* Labyrinth,
das
lace [leɪs] 1. *n.* a) *(for shoe)* Schnürsenkel, *der;* b) *(fabric)* Spitze, *die; attrib.*
Spitzen-. 2. *v. t.* ~ **|up|** [zu]schnüren
lacerate ['læsəreɪt] *v. t.* aufreißen
'lace-up 1. *attrib. adj.* Schnür-. 2. *n.*
Schnürschuh/-stiefel, *der*
lack [læk] 1. *n.* Mangel, *der* (of an +
Dat.). 2. *v. t.* **sb./sth. ~s sth.** jmdm./
einer Sache fehlt es an etw. *(Dat.)*
lackey ['lækɪ] *n.* Lakai, *der*
'lacking *adj.* be ~: fehlen
laconic lə'kɒnɪk] *adj.* lakonisch

lacquer ['lækə(r)] *n.* Lack, *der*
lacrosse [lə'krɒs] *n.* Lacrosse, *das*
lacy ['leɪsɪ] *adj.* Spitzen-
lad [læd] *n.* Junge, *der*
ladder ['lædə(r)] **1.** *n.* **a)** Leiter, *die;* **b)** *(Brit.: in tights etc.)* Laufmasche, *die.* **2.** *v.i. (Brit.)* Laufmaschen/eine Laufmasche bekommen. **3.** *v.t. (Brit.)* Laufmaschen/eine Laufmasche machen in (+ *Akk.*)
laden ['leɪdn] beladen (**with** mit)
ladle ['leɪdl] *n.* Schöpfkelle, *die*
lady ['leɪdɪ] *n.* **a)** Dame, *die;* **~-in-waiting** *(Brit.)* Hofdame, *die;* **b)** 'Ladies' *(WC)* „Damen"; **c)** *as form of address* **Ladies** meine Damen; **d)** *(Brit.) as title* L~: Lady
lady: **~bird,** *(Amer.)* **~bug** *ns.* Marienkäfer, *der;* **~like** *adj.* damenhaft
¹**lag** [læg] *v.i.,* -gg-: **~** [**behind**] zurückbleiben; *(fig.)* im Rückstand sein
²**lag** *v.t.,* -gg- *(insulate)* isolieren
lager ['lɑːgə(r)] *n.* Lagerbier, *das*
lagging *n.* Isolierung, *die*
lagoon [lə'guːn] *n.* Lagune, *die*
laid *see* ²**lay**
'**laid-back** *adj. (coll.)* gelassen
lain *see* ²**lie**
lair [leə(r)] *n. (of wild animal)* Unterschlupf, *der;* *(of pirates, bandits)* Schlupfwinkel, *der*
lake [leɪk] *n.* See, *der*
lamb [læm] *n.* **a)** Lamm, *das;* **b)** *(meat)* Lamm[fleisch], *das.* **lamb 'chop** *n.* Lammkotelett, *das.* **lamb's-wool** *n.* Lambswool, *die*
lame [leɪm] *adj.,* **lamely** *adv.* lahm
lament [lə'ment] **1.** *n.* Klage, *die* (for um). **2.** *v.t.* **~ that** ...: beklagen, daß ... **3.** *v.i.* klagen *(geh.);* **~ over sth.** etw. beklagen *(geh.).* **lamentable** ['læməntəbl] *adj.* beklagenswert
laminated ['læmɪneɪtɪd] *adj.* lamelliert; **~ glass** Verbundglas, *das*
lamp [læmp] *n.* Lampe, *die;* *(in street)* [Straßen]laterne, *die.* '**lamppost** *n.* Laternenpfahl, *der.* '**lampshade** *n.* Lampenschirm, *der*
lance [lɑːns] **1.** *n.* Lanze, *die.* **2.** *v.t. (Med.)* mit der Lanzette öffnen
lance-'corporal *n.* Obergefreite, *der*
land [lænd] **1.** *n.* Land, *das;* **have or own ~:** Grundbesitz haben. **2.** *v.t.* **a)** *(set ashore)* [an]landen; **b)** *(Aeronaut.)* landen; **c)** **~ oneself in trouble** sich in Schwierigkeiten bringen; **~ sb. with sth.,** **~ sth. on sb.** jmdm. etw. aufhalsen *(ugs.).* **3.** *v.i.* **a)** ⟨*Boot usw.:*⟩ anlegen, landen; ⟨*Passagier:*⟩ aussteigen

(from aus); **we ~ed at Dieppe** wir gingen in Dieppe an Land; **b)** *(Aeronaut.)* landen; **c)** **~ on one's feet** *(fig.)* [wieder] auf die Füße fallen. '**landed** *adj.* **~ gentry/aristrocracy** Landadel, *der.*
'**landing** *n.* **a)** *(of ship, aircraft)* Landung, *die;* **b)** *(on stairs)* Treppenabsatz, *der;* *(passage)* Treppenflur, *der.*
'**landing-card** *n.* Landekarte, *die.*
'**landing-stage** *n.* Landesteg, *der*
land: **~lady** *n.* **a)** *(of rented property)* Vermieterin, *die;* **b)** *(of public house)* [Gast]wirtin, *die;* **~-locked** *adj.* vom Land eingeschlossen ⟨*Bucht, Hafen*⟩; ⟨*Staat*⟩ ohne Zugang zum Meer; **~lord** *n.* **a)** *(of rented property)* Vermieter, *der;* **b)** *(of public house)* [Gast]wirt, *der;* **~mark** *n.* **a)** Orientierungspunkt, *der;* **b)** *(fig.)* Markstein, *der;* **~owner** *n.* Grundbesitzer, *der/* -besitzerin, *die;* **~scape** ['lændskeɪp] *n.* Landschaft, *die;* **~slide** *n.* Erdrutsch, *der*
lane [leɪn] *n.* **a)** *(in the country)* Landsträßchen, *das;* Weg, *der;* **b)** *(in town)* Gasse, *die;* **c)** *(part of road)* [Fahr]spur, *die;* '**get in ~**' „bitte einordnen"; **d)** *(Sport)* Bahn, *die*
language ['læŋgwɪdʒ] *n.* Sprache, *die;* *(style)* Ausdrucksweise, *die*
languid ['læŋgwɪd] *adj.* träge
languish ['læŋgwɪʃ] *v.i.* **a)** *(lose vitality)* ermatten *(geh.);* **b)** **~ under sth.** unter etw. *(Dat.)* schmachten *(geh.)*
lank [læŋk] *adj.* **a)** hager; **b)** glatt herabhängend ⟨*Haar*⟩
lanky ['læŋkɪ] *adj.* schlaksig *(ugs.)*
lantern ['læntən] *n.* Laterne, *die*
¹**lap** [læp] *n. (part of body)* Schoß, *der*
²**lap** *n. (Sport)* Runde, *die*
³**lap** **1.** *v.i.,* -pp- schlecken. **2.** *v.t.,* -pp-: **~** [**up**] [auf]schlecken. **lap 'up** *v.t. (fig.)* schlucken
lapel [lə'pel] *n.* Revers, *das*
Lapland ['læplænd] *pr. n.* Lappland *(das)*
lapse [læps] **1.** *n.* **a)** *(interval)* **a/the ~ of** ...: eine/die Zeitspanne von ...; **b)** *(mistake)* Fehler, *der;* **~ of memory** Gedächtnislücke, *die.* **2.** *v.i.* **a)** ⟨*Vertrag, usw.:*⟩ ungültig werden; **b)** **~ into** verfallen in (+ *Akk.*)
larceny ['lɑːsənɪ] *n.* Diebstahl, *der*
lard [lɑːd] *n.* Schweineschmalz, *das*
larder ['lɑːdə(r)] *n.* Speisekammer, *die*
large [lɑːdʒ] **1.** *adj.* groß. **2.** *n.* **at ~** *(not in prison etc.)* auf freiem Fuß. **3.** *adv.* see **by** 2 d. '**largely** *adv.* weitgehend
'**large-size[d]** *adj.* groß

¹**lark** [lɑːk] *n. (Ornith.)* Lerche, *die*

²**lark** *(coll.)* 1. *n.* Jux, *der (ugs.).* 2. *v. i.* ~ [about *or* around] herumalbern *(ugs.)*

larva [ˈlɑːvə] *n., pl.* ~e [ˈlɑːviː] Larve, *die*

laryngitis [lærɪnˈdʒaɪtɪs] *n.* Kehlkopfentzündung, *die*

larynx [ˈlærɪŋks] *n.* Kehlkopf, *der*

lascivious [ləˈsɪvɪəs] *adj.* lüstern *(geh.)*

laser [ˈleɪzə(r)] *n.* Laser, *der.* '**laser beam** *n.* Laserstrahl, *der*

lash [læʃ] 1. *n.* **a)** *(stroke)* [Peitschen]hieb, *der;* **b)** *(on eyelid)* Wimper, *die.* 2. *v. i.* ⟨*Welle, Regen:*⟩ peitschen (**against** gegen, **on** auf + *Akk.*). 3. *v. t.* **a)** *(fasten)* festbinden (**to an** + *Dat.*); **b)** *(as punishment)* auspeitschen. **lash** '**down** 1. *v. t.* festbinden. 2. *v. i.* ⟨*Regen:*⟩ niederprasseln. **lash** '**out** *v. i.* **a)** *(hit out)* um sich schlagen; ~ **out at sb.** nach jmdm. schlagen; **b)** ~ **out on sth.** *(coll.: spend freely)* sich *(Dat.)* etw. leisten

lashings [ˈlæʃɪŋz] *n. pl.* ~ **of sth.** Unmengen von etw.

lass [læs] *n.* Mädchen, *das*

lasso [ləˈsuː] Lasso, *das*

¹**last** [lɑːst] 1. *adj.* letzt...; **be** ~ **to arrive** als letzter/letzte ankommen; ~ **night** gestern nacht. 2. *adv.* **a)** [ganz] zuletzt; als letzter/letzte ⟨*sprechen, ankommen*⟩; **b)** *(on* ~ *previous occasion)* das letzte Mal; zuletzt. 3. *n.* **a)** *(person or thing)* letzter...; **b)** **at** [long] ~: endlich

²**last** *v. i.* **a)** *(continue)* dauern; ⟨*Wetter, Ärger:*⟩ anhalten; **b)** *(suffice)* reichen

'**last-ditch** *adj.* ~ **attempt** letzter verzweifelter Versuch

'**lasting** *adj.* bleibend; dauerhaft ⟨*Beziehung*⟩; nachhaltig ⟨*Eindruck, Wirkung*⟩

'**lastly** *adv.* schließlich

latch [lætʃ] *n.* Riegel, *der;* **on the** ~: nur eingeklinkt. **latch** '**on to** *v. t. (coll.: understand)* kapieren *(ugs.)*

late [leɪt] 1. *adj.* **a)** spät; **am I** ~? komme ich zu spät?; **be** ~ **for the train** den Zug verpassen; **the train is** [an hour] ~: der Zug hat [eine Stunde] Verspätung; ~ **shift** Spätschicht, *die;* ~ **summer** Spätsommer, *der;* **b)** *(dead)* verstorben; **c)** *(former)* ehemalig. *See also* **later** 1; **latest.** 2. *adv.* **a)** *(after proper time)* verspätet; **b)** *(at/till a* ~ *hour)* spät; **be up** ~: bis spät in die Nacht aufbleiben; **work** ~ **at the office** [abends] lange im Büro arbeiten; [**a bit**] ~ **in the day** *(fig. coll.)* reichlich

spät. 3. *n.* **of** ~: in letzter Zeit. **latecomer** [ˈleɪtkʌmə(r)] *n.* Zuspätkommende, *der/die.* '**lately** *adv.* in letzter Zeit. '**lateness** *n.* **a)** *(delay)* Verspätung, *die;* **b)** **the** ~ **of the performance** der späte Beginn der Vorstellung

latent [ˈleɪtənt] *adj.* latent

later [ˈleɪtə(r)] 1. *adv.* ~ [on] später. 2. *adj.* später; *(more recent)* neuer

lateral [ˈlætərl] *adj.* seitlich (**to** von); ~ **thinking** Querdenken, *das*

latest [ˈleɪtɪst] *adj.* **a)** *(modern)* neu[e]st...; **b)** *(most recent)* letzt...; **at** [the] ~/**the very** ~: spätestens/allerspätestens

lathe [leɪð] *n.* Drehbank, *die*

lather [ˈlɑːðə(r)] 1. *n.* [Seifen]schaum, *der.* 2. *v. t.* einschäumen

Latin [ˈlætɪn] 1. *adj.* lateinisch. 2. *n.* Latein, *das; see also* **English** 1 2 a. **Latin A'merica** *pr. n.* Lateinamerika *(das).* **Latin-A'merican** *adj.* lateinamerikanisch

latitude [ˈlætɪtjuːd] *n.* **a)** *(freedom)* Freiheit, *die;* **b)** *(Geog.)* Breite, *die*

latrine [ləˈtriːn] *n.* Latrine, *die*

latter [ˈlætə(r)] *attrib. adj.* letzter...; **the** ~: der/die/das letzere; *pl.* die letzteren. '**latterly** *adv.* in letzter Zeit

lattice [ˈlætɪs] *n.* Gitter, *das*

laudable [ˈlɔːdəbl] *adj.* lobenswert

laugh [lɑːf] 1. *n.* Lachen, *das; (continuous)* Gelächter, *das.* 2. *v. i.* lachen; ~ **out loud** laut auflachen; ~ **at sb./sth.** über jmdn./etw. lachen; *(jeer)* jmdn. auslachen/etw. verlachen. **laugh** '**off** *v. t.* mit einem Lachen abtun

laughable [ˈlɑːfəbl] *adj.* lachhaft; lächerlich

'**laughing** *n.* **be no** ~ **matter** nicht zum Lachen sein. '**laughing-gas** *n.* Lachgas, *das.* '**laughing-stock** *n.* **make sb. a** ~, **make a** ~ **of sb.** jmdn. zum Gespött machen

laughter [ˈlɑːftə(r)] *n.* Lachen, *das; (continuous)* Gelächter, *das*

launch [lɔːntʃ] *v. t.* **a)** zu Wasser lassen ⟨*Boot*⟩; vom Stapel lassen ⟨*neues Schiff*⟩; abschießen ⟨*Harpune, Torpedo*⟩; schleudern ⟨*Speer*⟩; **b)** *(fig.)* auf den Markt bringen ⟨*Produkt*⟩; vorstellen ⟨*Buch, Schallplatte, Sänger*⟩; ~ **an attack** einen Angriff durchführen. '**launching pad,** **launch pad** *ns.* [Raketen]abschußrampe, *die*

launder [ˈlɔːndə(r)] *v. t.* waschen und bügeln. **launderette** [lɔːndəˈret],

laundrette [lɔ:n'dret],   (Amer.)
laundromat ['lɔ:ndrəmæt] ns.
Waschsalon, der. **laundry** ['lɔ:ndrɪ]
n. **a)** (place) Wäscherei, die; **b)** (clothes
etc.) Wäsche, die

lava ['lɑ:və] n. Lava, die

lavatory ['lævətərɪ] n. Toilette, die

lavender ['lævɪndə(r)] n. Lavendel,
der

lavish ['lævɪʃ] **1.** adj. großzügig. **2.** v. t.
~ sth. on sb. jmdn. mit·etw. überhäu-
fen

law [lɔ:] n. **a)** Gesetz, das; **break the** ~ :
gegen das Gesetz verstoßen; **take the**
~ **into one's own hands** sich (Dat.)
selbst Recht verschaffen; ~ **and order**
Ruhe und Ordnung; **b)** (of game) Regel, die; **c)** (as subject) Jura o. Art.

law: ~**-abiding** ['lɔ:əbaɪdɪŋ] adj. ge-
setzestreu; ~**court** n. Gerichtsgebäude, das; (room) Gerichtssaal, der;
~**ful** ['lɔ:fl] adj. rechtmäßig ⟨Besitzer,
Erbe⟩; legal, gesetzmäßig ⟨Vorgehen,
Maßnahme⟩; ~**less** adj. gesetzlos

lawn [lɔ:n] n. Rasen, der. '**lawnmower** n. Rasenmäher, der

'**law suit** n. Prozeß, der

lawyer ['lɔ:jə(r)] n. Rechtsanwalt,
der/Rechtsanwältin, die

lax [læks] adj. lax

laxative ['læksətɪv] n. Abführmittel,
das

laxity ['læksɪtɪ], '**laxness** ns. Laxheit,
die

'**lay** [leɪ] adj. Laien-

²**lay** v. t., **laid** [leɪd] **a)** legen ⟨Teppich-
boden, Rohr, Kabel⟩; **b)** (impose) auferlegen ⟨Verantwortung, Verpflich-
tung⟩ (on Dat.); verhängen ⟨Strafe⟩
(on über + Akk.); **c)** ~ **the table** den
Tisch decken; **d)** (Biol.) legen ⟨Ei⟩.
lay a'side v. t. beiseite legen. **lay 'by**
v. t. beiseite legen. **lay 'down** v. t. **a)**
hinlegen; **b)** festlegen ⟨Regeln, Bedin-
gungen⟩. **lay 'off 1.** v. t. (from work)
vorübergehend entlassen. **2.** v. i. (coll.:
stop) aufhören. **lay 'out** v. t. **a)**
(spread out) ausbreiten; **b)** anlegen
⟨Garten⟩. **lay 'up** v. t. **a)** (store) lagern;
b) I was laid up in bed for a week ich
mußte eine Woche mein Bett hüten

³**lay** see ²**lie**

lay: ~**about** n. (Brit.) Gammler, der
(ugs.); ~**-by** n., pl. ~**-bys** (Brit.) Park-
bucht, die; Haltebucht, die

layer ['leɪə(r)] n. Schicht, die

layette [leɪ'et] n. [baby's] ~ : Babyaus-
stattung, die

lay: ~**man** ['leɪmən] n., pl. ~**men** ['leɪ-

mən] Laie, der; ~**out** n. (of garden,
park) Anlage, die; (of book, advertisement, etc.) Layout, das

laze [leɪz] v. i. faulenzen; ~ **around** or
about herumfaulenzen (ugs.)

lazily ['leɪzɪlɪ] adv. faul

laziness ['leɪzɪnɪs] n. Faulheit, die

lazy ['leɪzɪ] adj. faul. '**lazy-bones** n.
sing. Faulpelz, der

lb. abbr. pound[s] ≈ Pfd.

'**lead** [led] **1.** n. **a)** (metal) Blei, das; **b)**
(in pencil) [Bleistift]mine, die. **2.** attrib.
adj. Blei-

²**lead** [li:d] **1.** v. t., **led** [led] **a)** führen; ~
sb. to do sth. (fig.) jmdn. dazu bringen,
etw. zu tun; **b)** (fig.: influence) ~ sb. to
do sth. jmdn. veranlassen, etw. zu tun;
be easily led sich leicht beeinflussen
lassen; **he led me to believe that ...**: er
machte mich glauben, daß ...; **c)** (be
first in) anführen; **d)** (direct) anführen
⟨Bewegung, Abordnung⟩; leiten ⟨Dis-
kussion, Orchester⟩. **2.** v. i., **led a)**
⟨Straße usw., Tür:⟩ führen; **b)** (be first)
führen; (go in front) vorangehen. **3.** n.
a) (precedent) Beispiel, das; (clue) Anhaltspunkt, der; **follow sb.'s** ~ : jmds.
Beispiel (Dat.) folgen; **b)** (first place)
Führung, die; **be in the** ~ : in Führung
liegen; **c)** (distance ahead) Vorsprung,
der; **d)** (leash) Leine, die; **on a** ~ : an
der Leine; **e)** (Electr.) Kabel, das; **f)**
(Theatre) Hauptrolle, die. **lead
a'way** v. t. abführen ⟨Gefangenen,
Verbrecher⟩. **lead 'off 1.** v. t. abführ-
ren. **2.** v. i. beginnen. **lead 'on 1.** v. t.
~ **sb. on** (entice) jmdn. reizen; (deceive) jmdn. auf den Leim führen. **2.**
v. i. ~ **on to the next topic** etc. zum
nächsten Thema usw. führen. **lead
'up to** v. t. schließlich führen zu

'**leader** n. **a)** Führer, der/Führerin,
die; (of political party) Vorsitzende,
der/die; (of expedition) Leiter,
der/Leiterin, die; **b)** (Brit. Journ.)
Leitartikel, der. '**leadership** n. Führung, die

lead-free ['ledfri:] adj. bleifrei

leading ['li:dɪŋ] adj. führend

leading: ~ '**lady** n. Hauptdarstelle-
rin, die; ~ '**man** n. Hauptdarsteller,
der; ~ '**question** n. Suggestivfrage,
die; ~ **role** n. Hauptrolle, die; (fig.)
führende Rolle

lead [led]: ~**-'pencil** n. Bleistift, der;
~**-poisoning** n. Bleivergiftung, die

leaf [li:f] n., pl. **leaves** [li:vz] Blatt, das;
(of table) Platte die. **leaf 'through**
v. t. durchblättern

leaflet ['li:flıt] *n.* [Hand]zettel, *der;* *(advertising)* Reklamezettel, *der;* *(political)* Flugblatt, *das*

'**leafy** *adj.* belaubt

league [li:g] *n.* a) *(agreement)* Bündnis, *das;* be in ~ with sb. mit jmdm. im Bunde sein; b) *(Sport)* Liga, *die*

leak [li:k] 1. *n.* a) *(hole)* Leck, *das; (in roof, tent; also fig.)* undichte Stelle; b) *(escaping gas)* durch ein Leck austretendes Gas. 2. *v.i.* a) *(escape)* austreten (**from** aus); b) 〈*Faß, Tank, Schiff:*〉 lecken; 〈*Rohr, Leitung, Dach:*〉 undicht sein; 〈*Gefäß, Füller:*〉 auslaufen; c) *(fig.)* ~ |out| durchsickern. 3. *v.t.* ~ **sth. to sb.** jmdm. etw. zuspielen. **leakage** ['li:kıdʒ] *n.* Auslaufen, *das; (of fluid, gas)* Ausströmen, *das; (fig.: of information)* Durchsickern, *das.* '**leaky** *adj.* undicht; leck 〈*Boot*〉

¹**lean** [li:n] 1. *adj.* mager. 2. *n. (meat)* Magere, *das*

²**lean** 1. *v.i.,* leaned [li:nd, lent] *or (Brit.)* **leant** [lent] a) sich beugen; ~ **against the door** sich gegen die Tür lehnen; ~ **down/forward** sich herab-/vorbeugen; ~ **back** sich zurücklehnen; b) *(support oneself)* ~ **against/on sth.** sich gegen/an etw. *(Akk.)* lehnen; c) *(be supported)* lehnen (**against** an + *Dat.*); d) *(fig.)* ~ |up|on sb. *(rely)* auf jmdn. bauen; ~ to|wards| **sth.** *(tend)* zu etw. neigen. 2. *v.t.,* leaned *or (Brit.)* leant lehnen (**against** gegen *od.* an + *Akk.*). **lean** '**over** *v.i.* sich hinüberbeugen

'**leaning** *n.* Neigung, *die*

leant *see* ²**lean**

leap [li:p] 1. *v.i.,* leaped [li:pt, lept] *or* leapt [lept] a) springen; 〈*Herz:*〉 hüpfen; b) *(fig.)* ~ **at the chance** die Gelegenheit beim Schopf packen. 2. *v.t.,* leaped *or* leapt überspringen. 3. *n.* Sprung, *der;* with *or* in one ~: mit einem Satz; by ~s and bounds *(fig.)* mit Riesenschritten. '**leap-frog** 1. *n.* Bockspringen, *das.* 2. *v.i.,* -gg- Bockspringen machen

leapt *see* leap 1, 2

'**leap year** *n.* Schaltjahr, *das*

learn [lɜ:n] 1. *v.t.,* learned [lɜ:nd, lɜ:nt] *or* learnt [lɜ:nt] a) lernen; ~ **to swim** schwimmen lernen; b) *(find out)* erfahren. 2. *v.i.,* learned *or* learnt a) lernen; ~ **about sth.** etwas über etw. *(Akk.)* lernen; b) *(get to know)* erfahren (**of** von). **learned** ['lɜ:nıd] *adj.* gelehrt. '**learner** *n. (beginner)* Anfänger, *der/*Anfängerin, *die;* ~ |driver|

Fahrschüler, *der/*-schülerin, *die.* '**learning** *n. (of person)* Gelehrsamkeit, *die*

learnt *see* learn

lease [li:s] 1. *n. (of land, business premises)* Pachtvertrag, *der; (of house, flat, office)* Mietvertrag, *der.* 2. *v.t.* a) *(grant ~ on)* verpachten 〈*Grundstück, Geschäft, Rechte*〉; vermieten 〈*Haus, Wohnung, Büro*〉; b) *(take ~ on)* pachten 〈*Grundstück, Geschäft*〉; mieten 〈*Haus, Wohnung, Büro*〉. '**leasehold** *n. see* lease 2: have the ~ of *or* on sth. etw. gepachtet/gemietet haben

leash [lı:ʃ] *n.* Leine, *die*

least [li:st] 1. *adj. (smallest)* kleinst...; *(in quantity)* wenigst...; *(in status)* geringst... 2. *n.* Geringste, *das;* the ~ I can do das mindeste, was ich tun kann; at ~: mindestens; *(anyway)* wenigstens; at the |very| ~: [aller]mindestens; not |in| the ~: nicht im geringsten. 3. *adv.* am wenigsten

leather ['leðə(r)] 1. *n.* Leder, *das.* 2. *adj.* ledern; Leder〈*jacke, -mantel*〉. '**leather goods** *n.* Lederwaren *Pl.* '**leathery** *adj.* ledern

¹**leave** [li:v] *n.* a) *(permission)* Erlaubnis, *die;* b) *(from duty or work)* Urlaub, *der;* ~ |of absence| Urlaub, *der;* c) take one's ~ sich verabschieden

²**leave** *v.t.,* left [left] a) *(make or let remain)* hinterlassen; ~ **sb. to do sth.** es jmdm. überlassen, etw. zu tun; *(in will)* ~ **sb. sth.,** ~ **sth. to sb.** jmdm. etw. hinterlassen; b) *(refrain from doing, using, etc.)* stehenlassen 〈*Abwasch, Essen*〉; c) *(in given state)* lassen; ~ **sb. alone** *(allow to be alone)* jmdn. allein lassen; *(stop bothering)* jmdn. in Ruhe lassen; d) *(refer, entrust)* ~ **sth. to sb./ sth.** etw. jmdm./einer Sache überlassen; e) *(go away from, quit, desert)* verlassen; ~ **home at 6 a.m.** um 6 Uhr früh von zu Hause weggehen/-fahren; ~ **Bonn at 6 p.m.** *(by car, in train)* um 18 Uhr von Bonn abfahren; *(by plane)* um 18 Uhr in Bonn abfliegen; *abs.* the train ~s at 8.30 a.m. der Zug fährt *od.* geht um 8.30 Uhr; ~ **on the 8 a.m. train/flight** mit dem Acht-Uhr-Zug fahren/der Acht-Uhr-Maschine fliegen. **leave a'side** *v.t.* beiseite lassen. **leave be'hind** *v.t.* zurücklassen; *(by mistake)* vergessen; liegenlassen. **leave** '**off** *v.t. (stop)* aufhören mit; *abs.* aufhören. **leave** '**out** *v.t.* auslassen. **leave** '**over** *v.t.* be left over übrig [geblieben] sein

leaves *pl. of* **leaf**
Lebanon ['lebənən] *pr. n.* |the| ~: [der] Libanon
lecherous ['letʃərəs] *adj.* lüstern *(geh.)*
lecture ['lektʃə(r)] **1. a)** *n.* Vortrag, *der; (Univ.)* Vorlesung, *die;* **b)** *(reprimand)* Strafpredigt, *die (ugs.).* **2.** *v. i.* ~ |to sb.| |on sth.| [vor jmdm.] einen Vortrag/*(Univ.)* eine Vorlesung [über etw. *(Akk.)*] halten. **3.** *v. t. (scold)* ~ **sb.** jmdm. eine Strafpredigt halten. **'lecturer** *n.* Vortragende, *der/die;* **senior** ~: Dozent, *der/*Dozentin, *die*
led *see* ²**lead** 1, 2
ledge [ledʒ] *n.* Sims, *der od. das; (of rock)* Vorsprung, *der*
ledger ['ledʒə(r)] *n. (Commerc.)* Hauptbuch, *das*
lee [li:] *n.* **a)** *(shelter)* Schutz, *der;* **b)** ~ |side| *(Naut.)* Leeseite, *die*
leech [li:tʃ] *n.* [Blut]egel, *der*
leek [li:k] *n.* Stange Porree *od.* Lauch; ~**s** Porree, *der;* Lauch, *der*
leer [lɪə(r)] **1.** *n.* anzüglicher/spöttischer Blick. **2.** *v. i.* ~ **at sb.** jmdm. einen anzüglichen/spöttischen [Seiten]blick zuwerfen
leeward ['li:wəd] **1.** *adj.* to/on the ~ side of the ship nach/in Lee. **2.** *n.* Leeseite, *die;* **to** ~: leewärts
'leeway *n.* **a)** *(Naut.)* Leeweg, *der;* Abdrift, *die;* **b)** *(fig.)* Spielraum, *der*
¹**left** *see* ²**leave**
²**left** [left] **1.** *adj.* **a)** link...; **on the** ~ **side** auf der linken Seite; links; **b)** L~ *(Polit.)* link... **2.** *adv.* nach links. **3.** *n.* **a)** *(~-hand side)* linke Seite; **on** *or* **to the** ~ |of sb./sth.| links [von jmdm./ etw.]; **b)** *(Polit.)* **the L~**: die Linke
left: ~**-hand** *adj.* link...; ~**-handed 1.** *adj.* linkshändig; *(Werkzeug)* für Linkshänder; **be** ~**-handed** Linkshänder/Linkshänderin sein; **2.** *adv.* linkshändig; ~**-luggage [office]** *n. (Brit. Railw.)* Gepäckaufbewahrung, *die;* ~**-overs** *n. pl.* Reste; ~ **'wing** *n.* linker Flügel; ~**-wing** *adj. (Polit.)* linksgerichtet; Links⟨*extremist, -intellektueller*⟩; ~**-'winger** *n.* **a)** *(Sport)* Linksaußen, *der;* **b)** *(Polit.)* Angehöriger/Angehörige des linken Flügels
leg [leg] *n.* **a)** Bein, *das;* **pull sb.'s** ~ *(fig.)* jmdn. auf den Arm nehmen *(ugs.);* **stretch one's** ~**s** sich *(Dat.)* die Beine vertreten; **b)** ~ **of lamb** Lammkeule, *die;* **c)** *(of journey)* Etappe, *die*
legacy ['legəsɪ] *n.* Vermächtnis, *das (Rechtsspr.);* Erbschaft, *die*

legal ['li:gl] *adj.* **a)** *(concerning the law)* juristisch; Rechts⟨*beratung, -streit, -experte, -schutz*⟩; gesetzlich ⟨*Vertreter*⟩; rechtlich ⟨*Gründe, Stellung*⟩; Gerichts⟨*kosten*⟩; **b)** *(required by law)* gesetzlich ⟨*Verpflichtung*⟩; gesetzlich verankert ⟨*Recht*⟩; **c)** *(lawful)* legal; rechtsgültig ⟨*Vertrag, Testament*⟩.
legality [lɪ'gælɪtɪ] *n.* Legalität, *die.*
legalize ['li:gəlaɪz] *v. t.* legalisieren
legend ['ledʒənd] *n.* Sage, *die; (unfounded belief)* Legende, *die.* **legendary** ['ledʒəndərɪ] *adj.* legendär
legibility [ledʒɪ'bɪlɪtɪ] *n.* Leserlichkeit, *die*
legible ['ledʒɪbl] *adj.* leserlich; **easily/ scarcely** ~: leicht/kaum lesbar
legion ['li:dʒn] *n.* Legion, *die*
legislate ['ledʒɪsleɪt] *v. i.* Gesetze verabschieden. **legislation** [ledʒɪs'leɪʃn] *n.* **a)** *(laws)* Gesetze; **b)** *(legislating)* Gesetzgebung, *die.* **legislative** ['ledʒɪslətɪv] *adj.* gesetzgebend. **legislator** ['ledʒɪsleɪtə(r)] *n.* Gesetzgeber, *der.* **legislature** ['ledʒɪsleɪtʃə(r)] *n.* Legislative, *die*
legitimate [lɪ'dʒɪtɪmət] *adj.* **a)** *(lawful)* legitim; rechtmäßig ⟨*Besitzer, Regierung*⟩; **b)** *(valid)* berechtigt; **c)** ehelich ⟨*Kind*⟩
leisure ['leʒə(r)] *n.* Freizeit, *die; attrib.* Freizeit-. **'leisurely** *adj.* gemächlich
lemon ['lemən] *n.* Zitrone, *die.* **lemonade** [lemə'neɪd] *n.* [Zitronen]limonade, *die*
lend [lend] *v. t., lent* [lent] leihen; ~ **sth. to sb.** jmdm. etw. leihen. **'lender** *n.* Verleiher, *der/*Verleiherin, *die*
length [leŋθ, leŋkθ] *n.* **a)** *(also of time)* Länge, *die;* **be six feet in** ~: sechs Fuß lang sein; **a short** ~ **of time** kurze Zeit; **b)** **at** ~ *(for a long time)* lange; *(eventually)* schließlich; **at |great|** ~ *(in great detail)* lang und breit; **at some** ~: ziemlich ausführlich; **c)** **go to any/ great** ~**s** alles nur/alles Erdenkliche tun; **d)** *(piece of material)* Länge, *die;* Stück, *das.* **lengthen** ['leŋθən] **1.** *v. i.* länger werden. **2.** *v. t.* verlängern; länger machen ⟨*Kleid*⟩. **lengthways** ['leŋθweɪz] *adv.* der Länge nach; längs. **'lengthy** *adj.* überlang
lenient ['li:nɪənt] *adj.* nachsichtig
lens [lenz] *n.* Linse, *die*
Lent [lent] *n.* Fastenzeit, *die*
lent *see* **lend**
lentil ['lentl] *n.* Linse, *die*
Leo ['li:əʊ] *n., pl.* ~**s** der Löwe
leopard ['lepəd] *n.* Leopard, *der*

leotard ['li:ɑtɑ:d] *n.* Turnanzug, *der*
leper ['lepə(r)] *n.* Leprakranke, *der/die*
leprosy ['leprəsɪ] *n.* Lepra, *die*
lesbian ['lezbɪən] **1.** *n.* Lesbierin, *die.*
2. *adj.* lesbisch
less [les] **1.** *adj.* weniger; **of ~ value/**
importance weniger wertvoll/wichtig.
2. *adv.* weniger; **~ and ~:** immer we-
niger; **~ and ~ |often|** immer seltener.
3. *n.* weniger. **4.** *prep. (deducting)* ten
~ three zehn weniger drei. **lessen**
['lesn] **1.** *v. t.* verringern. **2.** *v. i.* sich
verringern. **lesser** ['lesə(r)] *attrib.
adj.* geringer...
lesson ['lesn] *n.* **a)** *(class)* [Unter-
richts]stunde, *die;* **b)** *(example, warning)* Lehre, *die;* **c)** *(Eccl.)* Lesung, *die*
let [let] **1.** *v. t.,* -tt-, let **a)** *(allow to)* lassen; **~ sb. do sth.** jmdn. etw. tun las-
sen; **~ alone** *(far less)* geschweige
denn; **b)** *(cause to)* **~ sb. know** jmdn.
wissen lassen; **c)** *(Brit.: rent out)* vermieten. **2.** *v. aux.,* -tt-, let lassen; **Let's**
go to the cinema. – Yes, ~'s/No, ~'s
not Komm/Kommt, wir gehen ins Kino. – Ja, gut/Nein, lieber nicht; **~**
them come in sie sollen hereinkommen. **let 'down** *v. t.* **a)** *(lower)* herun-
ter-/hinunterlassen; **b)** *(Dressm.)* auslassen; **c)** *(disappoint, fail)* im Stich
lassen. **let 'in** *v. t.* **a)** *(admit)* herein-/
hineinlassen; **b)** **~ oneself in for sth.**
sich auf etw. *(Akk.)* einlassen; **c)** **~ sb.**
in on a secret/plan *etc.* jmdn. in ein
Geheimnis/einen Plan *usw.* einwei-
hen. **let into** *v. t.* **a)** *(admit into)* lassen in (+ *Akk.*); **b)** *(fig.: acquaint
with)* **~ sb. into a secret** jmdn. in ein
Geheimnis einweihen. **let 'off** *v. t.* **a)**
(excuse) laufenlassen *(ugs.);* **~ sb. off**
sth. jmdm. etw. erlassen; **b)** *(allow to
alight)* aussteigen lassen; **c)** abbren-
nen ⟨*Feuerwerk⟩*. **let 'on** *(sl.)* **1.** *v. i.*
don't ~ on! nichts verraten! **2.** *v. t.* sb.
~ on to me that ...: man hat mir ge-
steckt, daß ... *(ugs.).* **let 'out** *v. t.* **a)** **~**
sb./an animal out jmdn./ein Tier her-
aus-/hinauslassen; **b)** ausstoßen
⟨*Schrei⟩;* **~ out a groan** aufstöhnen; **c)**
verraten ⟨*Geheimnis⟩;* **d)** *(Dressm.)*
auslassen; **e)** *(Brit.: rent out)* vermieten. **let 'through** *v. t.* durchlassen.
let 'up *v. i. (coll.)* nachlassen
'let-down *n.* Enttäuschung, *die*
lethal ['li:θl] *adj.* tödlich
lethargic [lɪ'θɑ:dʒɪk] *adj.* träge; *(apathetic)* lethargisch
lethargy ['leθədʒɪ] *n.* Trägheit, *die;
(apathy)* Lethargie, *die*

letter ['letə(r)] **a)** Brief, *der* (**to an**
+ *Akk.*); **b)** *(of alphabet)* Buchstabe,
der. **'letter bomb** *n.* Briefbombe,
die. **'letter-box** *n.* Briefkasten, *der*
'lettering *n.* Typographie, *die*
lettuce ['letɪs] *n.* [Kopf]salat, *der*
leukaemia, *(Amer.)* **leukemia**
[lu:'ki:mɪə] *n.* Leukämie, *die*
level ['levl] **1.** *n.* **a)** Höhe, *die; (storey)*
Etage, *die;* **b)** *(fig.: steady state)* Ni-
veau, *das;* **be on a ~ |with sb./sth.|** auf
dem gleichen Niveau sein [wie jmd./
etw.]. **2.** *adj.* **a)** waagerecht; eben
⟨*Boden, Land⟩;* **b)** *(on a ~)* be **~ |with**
sth./sb.| auf gleicher Höhe [mit etw./
jmdm.] sein; **c)** *(fig.)* **keep a ~ head** einen kühlen Kopf bewahren; **do one's**
~ best *(coll.)* sein möglichstes tun. **3.**
v. t., *(Brit.)* -ll-: **a)** *(make ~)* ebnen; **b)**
(aim) richten ⟨*Blick, Gewehr⟩* (**at auf**
+ *Akk.*); *(fig.)* richten ⟨*Kritik usw.⟩*
(**at** gegen). **level 'crossing** *n. (Brit.
Railw.)* [schienengleicher] Bahnüber-
gang. **level-'headed** *adj.* besonnen
lever ['li:və(r)] **1.** *n.* Hebel, *der.* **2.** *v. t.*
~ sth. open etw. aufhebeln. **leverage**
['li:vərɪdʒ] *n.* Hebelwirkung, *die*
levity ['levɪtɪ] *n. (frivolity)* Unernst, *der*
levy ['levɪ] **1.** *n. (tax)* Steuer, *die.* **2.** *v. t.*
erheben
lewd [lju:d] geil; anzüglich ⟨*Geste⟩;*
schlüpfrig ⟨*Witz⟩*
liability [laɪə'bɪlɪtɪ] *n.* **a)** Haftung, *die;*
b) *(handicap)* Belastung, *die* (**to** für)
liable ['laɪəbl] *pred. adj.* **a)** *(legally
bound)* **be ~ for sth.** für etw. haftbar
sein *od.* haften; **b)** *(prone)* **be ~ to sth.**
⟨*Person:⟩* zu etw. neigen; **be ~ to do**
sth. ⟨*Sache:⟩* leicht etw. tun; ⟨*Person:⟩*
dazu neigen, etw. zu tun
liaise [lɪ'eɪz] *v. i. (coll.)* eine Verbindung herstellen; **~ on a project** bei einem Projekt zusammenarbeiten. **li-**
aison [lɪ'eɪzɒn] *n. (co-operation)* Zusammenarbeit, *die*
liar ['laɪə(r)] *n.* Lügner, *der/*Lügnerin,
die
libel ['laɪbl] **1.** *n.* Verleumdung, *die.* **2.**
v. t., *(Brit.)* -ll- verleumden. **libellous**
(Amer.: **libelous**) ['laɪbələs] *adj.* verleumderisch
liberal ['lɪbərl] **1.** *adj.* **a)** großzügig; **b)**
(Polit.) liberal; **the L~ Democrats**
(Brit.) die Liberaldemokraten. **2.** *n.*
L~ *(Polit.)* Liberale, *der/die*
liberate ['lɪbəreɪt] *v. t.* befreien (**from**
aus). **liberation** [lɪbə'reɪʃn] *n.* Befreiung, *die.* **liberator** ['lɪbəreɪtə(r)] *n.*
Befreier, *der/*Befreierin, *die*

liberty ['lıbǝtı] *n.* Freiheit, *die;* **take the ~ of doing sth.** sich *(Dat.)* die Freiheit nehmen, etw. zu tun; **take liberties with sb.** sich *(Dat.)* Freiheiten gegen jmdn. herausnehmen *(ugs.)*

Libra ['li:brǝ] *n.* Waage, *die*

librarian [laı'breǝrıǝn] *n.* Bibliothekar, *der/*Bibliothekarin, *die*

library ['laıbrǝrı] *n.* Bibliothek, *die;* **public ~:** öffentliche Bücherei. 'library **book** *n.* Buch aus der Bibliothek

Libya ['lıbıǝ] *pr. n.* Libyen *(das)*

lice *pl. of* **louse**

licence ['laısǝns] **1.** *n.* [behördliche] Genehmigung; Lizenz, *die;* |driving-| ~: Führerschein, *der.* **2.** *v. t. see* **license 1**

license ['laısǝns] **1.** *v. t.* ermächtigen; **get a car ~d ≈** die Kfz-Steuer für ein Auto bezahlen. **2.** *n. (Amer.) see* **licence 1**

licentious [laı'senfǝs] *adj.* zügellos ⟨*Person*⟩; unzüchtig ⟨*Benehmen*⟩

lichen ['laıkn, 'lıtʃn] *n.* Flechte, *die*

lick [lık] **1.** *v. t.* **a)** lecken; **b)** *(sl.: beat)* verdreschen *(ugs.).* **2.** *n.* Lecken, *das.* **lick 'off** *v. t.* ablecken

lid [lıd] *n.* **a)** Deckel, *der;* **b)** *(eyelid)* Lid, *das*

lido ['li:dǝʊ] *n., pl.* ~s Freibad, *das*

¹lie [laı] **1.** *n.* Lüge, *die;* **tell ~s/a ~:** lügen. **2.** *v. i.,* **lying** ['laıɪŋ] lügen; **~ to sb.** jmdn. be- *od.* anlügen

²lie *v. i.,* **lying** ['laıɪŋ], **lay** [leı], **lain** [leın] **a)** liegen; *(assume horizontal position)* sich legen; **b)** ~ **idle** ⟨*Maschine, Fabrik:*⟩ stillstehen. **lie a'bout, lie a'round** *v. i.* herumliegen *(ugs.).* **lie 'back** *v. i.* sich zurücklegen; *(sitting)* sich zurücklehnen. **lie 'down** *v. i.* sich hinlegen

lie-detector ['laıdı'tektǝ(r)] *n.* Lügendetektor, *der*

'lie-in *n. (coll.)* **have a ~:** [sich] ausschlafen

lieu [lju:] *n.* **in ~ of sth.** anstelle einer Sache *(Gen.);* **get holiday in ~:** statt dessen Urlaub bekommen

lieutenant [lef'tenǝnt] *n. (Army)* Oberleutnant, *der*

life [laıf] *n., pl.* **lives** [laıvz] Leben, *das;* **for ~:** lebenslänglich ⟨*inhaftiert*⟩; **true to ~:** wahrheitsgetreu

life: ~**belt** *n.* Rettungsring, *der;* ~**boat** *n.* Rettungsboot, *das;* ~**buoy** *n.* Rettungsring, *der;* ~ **cycle** *n.* Lebenszyklus, *der;* ~-**guard** *n.* Rettungsschwimmer, *der/*-schwimmerin,

die; ~-**insurance** *n.* Lebensversicherung, *die;* ~-**jacket** *n.* Schwimmweste, *die;* ~-**less** *adj.* leblos; *(fig.)* farblos; ~-**like** *adj.* lebensecht; ~-**line** *n.* Rettungsleine, *die; (fig.)* Rettungsanker, *der;* ~-**long** *adj.* lebenslang; ~-**saving** *n.* Rettungsschwimmen, *das; attrib.* Rettungs-; ~ **sentence** *n.* lebenslängliche Freiheitsstrafe; ~-**size,** ~-**sized** *adj.* lebensgroß; **in Lebensgröße** *nachgestellt;* ~-**style** *n.* Lebensstil, *der;* ~-**time** *n.* Lebenszeit, *die;* **during my ~time** während meines Lebens; **the chance of a ~time** eine einmalige Gelegenheit

lift [lıft] **1.** *v. t.* heben; *(fig.)* erheben ⟨*Gemüt, Geist*⟩. **2.** *n.* **a)** *(in vehicle)* **get a ~:** mitgenommen werden; **give sb. a ~:** jmdn. mitnehmen; **b)** *(Brit.: elevator)* Aufzug, *der.* **3.** *v. i.* ⟨*Nebel:*⟩ sich auflösen. **'lift off** *v. t. & i.* abheben. **lift 'up** *v. t.* hochheben; heben ⟨*Kopf*⟩ **'lift-off** *n.* Abheben, *das*

ligament ['lıgǝmǝnt] *n.* Band, *das*

¹light [laıt] **1.** *n.* **a)** Licht, *das;* ~ **of day** Tageslicht, *das;* **b)** *(lamp)* Licht, *das; (fitting)* Lampe, *die;* **c)** *(signal to traffic)* Ampel, *die;* **d)** *(to ignite)* **have you got a ~?** haben Sie Feuer? **set ~ to sth.** etw. anzünden; **e)** **bring sth. to ~:** etw. ans [Tages]licht bringen; **throw** *or* **shed ~ |up|on sth.** Licht in etw. *(Akk.)* bringen; **f)** *(aspect)* **in that ~:** aus dieser Sicht; **seen in this ~:** so gesehen; **in the ~ of** angesichts (+ *Gen.*); **show sb. in a bad ~:** ein schlechtes Licht auf jmdn. werfen. **2.** *adj.* hell; ~-**blue/-brown** *etc.* hellblau/-braun *usw.* **3.** *v. t.,* **lit** [lıt] *or* **lighted a)** *(ignite)* anzünden; **b)** *(illuminate)* erhellen. **light 'up 1.** *v. i.* **a)** *(become lit)* erleuchtet werden; **b)** *(become bright)* aufleuchten (**with** vor). **2.** *v. t.* **a)** *(illuminate)* erleuchten; **b)** anzünden ⟨*Zigarette*⟩

²light **1.** *adj.* leicht; *(mild)* mild ⟨*Strafe*⟩. **2.** *adv.* **travel ~:** mit wenig *od.* leichtem Gepäck reisen

'light-bulb *n.* Glühbirne, *die*

'lighted *adj.* brennend ⟨*Kerze, Zigarette*⟩; angezündet ⟨*Streichholz*⟩

¹lighten ['laıtn] *v. t. (make less heavy, difficult)* leichter machen

²lighten 1. *v. t. (make brighter)* aufhellen; heller machen ⟨*Raum*⟩. **2.** *v. i.* sich aufhellen

'lighter *n.* Feuerzeug, *das*

light: ~-'**headed** *adj.* leicht benommen; ~-'**hearted** *adj.* **a)** *(humorous)* unbeschwert; **b)** *(optimistic)* unbekümmert; ~**house** n. Leuchtturm, der

'**lighting** n. Beleuchtung, die

'**lightly** adv. **a)** leicht; **b)** (without serious consideration) leichtfertig; **c)** (cheerfully) leichthin; **not treat sth. ~:** etw. nicht auf die leichte Schulter nehmen; **d) get off ~:** glimpflich davonkommen

¹**lightness** n. (of weight; also fig.) Leichtigkeit, die

²**lightness** n. (of colour) Helligkeit, die

lightning ['laɪtnɪŋ] n. Blitz, der; **flash of ~:** Blitz, der. '**lightning-conductor** n. Blitzableiter, der

'**lightweight 1.** adj. leicht. **2.** n. Leichtgewicht, das

¹**like** [laɪk] **1.** adj. **a)** (resembling) wie; **your dress is ~ mine** dein Kleid ist so ähnlich wie meins; **in a case ~ that** in so einem Fall; **what is sb./sth. ~?** wie ist jmd./etw.?; **b)** (characteristic of) typisch für ⟨dich, ihn usw.⟩; **c)** (similar) ähnlich. **2.** prep. (in the manner of) wie; [just] ~ **that** [einfach] so. **3.** n. **a)** (equal) his/her ~: seines-/ihresgleichen; **b)** (similar things) the ~: so etwas; **and the ~:** und dergleichen

²**like 1.** v. t. (be fond of, wish for) mögen; ~ **vegetables** Gemüse mögen; gern Gemüse essen; ~ **doing sth.** etw. gern tun; **would you ~ a drink?** möchtest du etwas trinken?; **would you ~ me to do it?** möchtest du, daß ich es tue?; **how do you ~ it?** wie gefällt es dir?; **if you ~,** expr. assent wenn du willst. **2.** n., in pl. ~**s and dislikes** Vorlieben und Abneigungen. **likeable** ['laɪkəbl] adj. nett; sympathisch

likelihood ['laɪklɪhʊd] n. Wahrscheinlichkeit, die

likely ['laɪklɪ] **1.** adj. wahrscheinlich; **there are ~ to be |traffic| hold-ups** man muß mit [Verkehrs]staus rechnen; **they are |not| ~ to come** sie werden wahrscheinlich [nicht] kommen; **is it ~ to rain tomorrow?** wird es morgen wohl regnen?; **this is not ~ to happen** es ist unwahrscheinlich, daß das geschieht. **2.** adv. wahrscheinlich; **as ~ as not** höchstwahrscheinlich; **not ~!** (coll.) auf keinen Fall!

'**like-minded** adj. gleichgesinnt

liken ['laɪkn] v. t. ~ **sth./sb. to sth./sb.** etw./jmdn. mit etw./jmdm. vergleichen

'**likeness** n. Ähnlichkeit, die (**to** mit)

likewise ['laɪkwaɪz] adv. ebenso

liking ['laɪkɪŋ] n. Vorliebe, die; **take a**

~ **to sb./sth.** an jmdm./etw. Gefallen finden; **sth. is |not| to sb.'s ~:** etw. ist [nicht] nach jmds. Geschmack

lilac ['laɪlək] n. **a)** (Bot.) Flieder, der; **b)** (colour) Zartlila, das

lily ['lɪlɪ] n. Lilie, die

limb [lɪm] n. **a)** (Anat.) Glied, das; **b) be out on a ~** (fig.) exponiert sein

limber up ['lɪmbər 'ʌp] v. i. (loosen up) die Muskeln lockern

¹**lime** [laɪm] n. |quick|~: [ungelöschter] Kalk

²**lime** n. (fruit) Limone, die

³**lime** see lime-tree

'**limelight** n. **be in the ~:** im Rampenlicht [der Öffentlichkeit] stehen

limerick ['lɪmərɪk] n. Limerick, der

'**lime-tree** n. Linde, die

limit ['lɪmɪt] **1.** n. **a)** Grenze, die; **set or put a ~ on sth.** etw. begrenzen; **be over the ~** ⟨Autofahrer:⟩ zu viele Promille haben; **lower/upper ~:** Untergrenze/ Höchstgrenze, die; **without ~:** unbegrenzt; **within ~s** inerhalb gewisser Grenzen; **b)** (coll.) **this is the ~!** das ist [doch] die Höhe!; **he/she is the |very| ~:** er/sie ist [einfach] unmöglich. **2.** v. t. begrenzen (**to** auf + Akk.); einschränken ⟨Freiheit⟩. **limitation** [lɪmɪ'teɪʃn] n. Beschränkung, die. '**limited** adj. **a)** (restricted) begrenzt; **b)** (intellectually narrow) beschränkt. '**limitless** adj. grenzenlos

limousine ['lɪməʊziːn] n. Limousine, die

'**limp** [lɪmp] **1.** v. i. hinken. **2.** n. Hinken, das

²**limp** adj. schlaff. '**limply** adv. schlaff; (weakly) schwach

limpet ['lɪmpɪt] n. (Zool.) Napfschnecke, die

limpid ['lɪmpɪd] adj. klar

linctus ['lɪŋktəs] n. Hustensaft, der

'**line** [laɪn] **1.** n. **a)** (string, cord, rope, etc.) Leine, die; **b)** (telephone cable) Leitung, die; **c)** (long mark; also Math., Phys.) Linie, die; **d)** (row, series) Reihe, die; (Amer.: queue) Schlange, die; **bring sb. into ~:** dafür sorgen, daß jmd. nicht aus der Reihe tanzt (ugs.); **e)** (row of words on a page) Zeile, die; **f)** (wrinkle) Falte, die; **g)** (direction, course) Richtung, die; **on the ~s of** nach Art (+ Gen.); **be on the right/wrong ~s** in die richtige/falsche Richtung gehen; **along or on the same ~s** in der gleichen Richtung; **h)** (Railw.) Bahnlinie, die; (track) Gleis, das; **i)** (field of activity)

Branche, *die;* **j)** *(Commerc.: product)*
Artikel, *der;* Linie, *die (fachspr.).* **2.**
v. t. **a)** linieren ⟨*Papier*⟩; **a ~d face** ein
faltiges Gesicht; **b)** säumen *(geh.)*
⟨*Straße, Strecke*⟩. **line 'up 1.** *v. t.* antreten lassen ⟨*Gefangene, Soldaten
usw.*⟩; [in einer Reihe] aufstellen
⟨*Gegenstände*⟩. **2.** *v. i.* ⟨*Gefangene,
Soldaten:*⟩ antreten; *(queue up)* sich
anstellen

²**line** *v. t.* füttern ⟨*Kleidungsstück*⟩;
ausschlagen ⟨*Schublade usw.*⟩

lineage ['lɪnɪɪdʒ] *n.* Abstammung, *die*

linear ['lɪnɪə(r)] *adj.* linear

linen ['lɪnɪn] **1.** *n.* **a)** Leinen, *das;* **b)**
(shirts, sheets, etc.) Wäsche, *die.* **2.**
adj. Leinen⟨*faden, -bluse*⟩; Lein⟨*tuch*⟩

liner ['laɪnə(r)] *n.* Linienschiff, *das*

'line-up *n.* Aufstellung, *die*

linger ['lɪŋgə(r)] *v. i.* verweilen *(geh.);*
bleiben

lingerie ['læʒərɪ] *n.* |**women's**| ~: Da-
menunterwäsche, *die*

lingo ['lɪŋgəʊ] *n. (coll.)* Sprache, *die*

linguist ['lɪŋgwɪst] *n.* Sprachkundige,
der/die

linguistic [lɪŋ'gwɪstɪk] *adj. (of ~s)* lin-
guistisch; *(of language)* sprachlich.
linguistics [lɪŋ'gwɪstɪks] *n.* Lingui-
stik, *die*

lining ['laɪnɪŋ] *n. (of clothes)* Futter,
das; (of objects, machines, etc.) Aus-
kleidung, *die*

link [lɪŋk] **1.** *n.* **a)** *(of chain)* Glied, *das;*
b) *(connection)* Verbindung, *die.* **2.**
v. t. verbinden; ~ **arms** sich unterha-
ken. **link 'up** *v. t.* miteinander verbinden

links [lɪŋks] *n.* |**golf**| ~: Golfplatz, *der*

lino ['laɪnəʊ] *n., pl.* ~s Linoleum, *das*

linseed ['lɪnsiːd] *n.* Leinsamen, *der.*
linseed 'oil *n.* Leinöl, *das*

lint [lɪnt] *n.* Mull, *der*

lintel ['lɪntl] *n. (Archit.)* Sturz, *der*

lion ['laɪən] *n.* Löwe, *der.* **lioness**
['laɪənɪs] *n.* Löwin, *die*

lip [lɪp] *n.* **a)** Lippe, *die;* **lower/upper** ~:
Unter-/Oberlippe, *die;* **b)** *(of cup)*
[Gieß]rand, *der; (of jug)* Schnabel, *der*
lip: ~-**read** *v. i.* von den Lippen lesen;
~-**reading** *n.* Lippenlesen, *das;*
~-**service** *n.* **pay** ~-**service to sth.** ein
Lippenbekenntnis zu etw. ablegen;
~**stick** *n.* Lippenstift, *der*

liquefy ['lɪkwɪfaɪ] **1.** *v. t.* verflüssigen.
2. *v. i.* sich verflüssigen

liqueur [lɪ'kjʊə(r)] *n.* Likör, *der*

liquid ['lɪkwɪd] **1.** *adj.* flüssig. **2.** *n.*
Flüssigkeit, *die*

liquidate ['lɪkwɪdeɪt] *v. t. (Commerc.)*
liquidieren. **liquidation** [lɪkwɪ'deɪʃn]
n. (Commerc.) Liquidation, *die*

liquidize ['lɪkwɪdaɪz] *v. t.* auflösen;
(Cookery) [im Mixer] pürieren.
'liquidizer *n.* Mixer, *der*

liquor ['lɪkə(r)] *n. (drink)* Alkohol, *der*

liquorice ['lɪkərɪs] *n.* Lakritze, *die*

Lisbon ['lɪzbən] *pr. n.* Lissabon *(das)*

lisp [lɪsp] **1.** *v. i. & t.* lispeln. **2.** *n.* Lispeln, *das*

¹**list** [lɪst] **1.** *n.* Liste, *die.* **2.** *v. t.* aufführen; auflisten; *(verbally)* aufzählen

²**list** *v. i. (Naut.)* Schlagseite haben

listen ['lɪsn] *v. i.* zuhören; ~ **to music/
the radio** Musik/Radio hören; **they
~ed to his words** sie hörten ihm zu.

listener ['lɪsnə(r)] *n.* Zuhörer,
*der/*Zuhörerin, *die; (to radio)* Hörer,
*der/*Hörerin, *die*

listless ['lɪstlɪs] *adj.* lustlos

lit *see* ¹**light 3**

litany ['lɪtənɪ] *n.* Litanei, *die*

liter *(Amer.) see* **litre**

literacy ['lɪtərəsɪ] *n.* Lese- und
Schreibfertigkeit, *die*

literal ['lɪtərl] *adj.* **a)** wörtlich; **b)** *(not
exaggerated)* buchstäblich. **literally**
['lɪtərəlɪ] *adv.* **a)** wörtlich; **b)** *(actually)*
buchstäblich; **c)** *(coll.: with some exaggeration)* geradezu

literary ['lɪtərərɪ] *adj.* literarisch

literate ['lɪtərət] *adj.* des Lesens und
Schreibens kundig; *(educated)* gebildet

literature ['lɪtrətʃə(r)] *n.* Literatur, *die*

lithe [laɪð] *adj.* geschmeidig

litigation [lɪtɪ'geɪʃn] *n.* Rechtsstreit,
der

litre ['liːtə(r)] *n. (Brit.)* Liter, *der od. das*

litter ['lɪtə(r)] **1.** *n.* **a)** *(rubbish)* Abfall,
der; **b)** *(of animals)* Wurf, *der.* **2.** *v. t.*
verstreuen. **'litter-basket** *n.* Abfall-
korb, *der.* **'litter-bin** Abfalleimer, *der*

little ['lɪtl] **1.** *adj.,* ~**r** ['lɪtlə(r)], ~**st** ['lɪt-
lɪst] (*Note: it is more common to use the
compar. and superl. forms* **smaller,
smallest**) **a)** klein; **a ~ way** ein kurzes
Stück; **after a ~ while** nach kurzer
Zeit; **b)** *(not much)* wenig; **there is very
~ tea left** es ist kaum noch Tee da; **a
~ ...** *(a small quantity of)* etwas ...; **ein
bißchen ... 2.** *n.* wenig; **a ~** *(a small
quantity)* etwas; *(somewhat)* ein wenig; **~ by ~:** nach und nach

liturgy ['lɪtədʒɪ] *n.* Liturgie, *die*

¹**live** [laɪv] **1.** *adj.* **a)** *attrib. (alive)* lebend; **b)** *(Radio, Telev.)* ~ **perform-
ance** Live-Aufführung, *die;* ~ **broad-**

cast Live-Sendung, *die;* c) *(Electr.)* stromführend. 2. *adv. (Radio, Telev.)* live *⟨übertragen usw.⟩*

²**live** [lɪv] 1. *v. i.* a) leben; b) *(make permanent home)* wohnen; leben. 2. *v. t.* leben. **live 'down** *v. t.* Gras wachsen lassen über (+ *Akk.*); **he will never be able to ~ it down** das wird ihm ewig anhängen. **live on** 1. ['--] *v. t.* leben von. 2. [-'-] *v. i.* weiterleben. **live 'up to** *v. t.* gerecht werden (+ *Dat.*)

livelihood ['laɪvlɪhʊd] *n.* Lebensunterhalt, *der*

liveliness ['laɪvlɪnɪs] *n.* Lebhaftigkeit, *die*

lively ['laɪvlɪ] *adj.* lebhaft; lebendig ⟨*Schilderung*⟩; rege ⟨*Handel*⟩

liven up [laɪvn 'ʌp] 1. *v. t.* Leben bringen in (+ *Akk.*). 2. *v. i.* ⟨*Person:*⟩ aufleben

liver ['lɪvə(r)] *n.* Leber, *die*

livery ['lɪvərɪ] *n.* Livree, *die*

lives *pl. of* **life**

livestock ['laɪvstɒk] *n. pl.* Vieh, *das*

livid ['lɪvɪd] *adj. (Brit. coll.)* fuchtig *(ugs.)*

living ['lɪvɪŋ] 1. *n.* a) Leben, *das;* b) **make a ~:** seinen Lebensunterhalt verdienen; c) **the ~:** die Lebenden. 2. *adj.* lebend; **within ~ memory** seit Menschengedenken. **'living-room** *n.* Wohnzimmer, *das*

lizard ['lɪzəd] *n.* Eidechse, *die*

llama ['lɑːmə] *n.* Lama, *das*

load [ləʊd] 1. *n. (burden, weight; also fig.)* Last, *die; (amount carried)* Ladung, *die.* 2. *v. t.* a) *(put ~ on)* beladen; *(put as load)* ~ **sb. with work** *(fig.)* jmdm. Arbeit auftragen; b) laden ⟨*Gewehr*⟩; ~ **a camera** einen Film [in einen Fotoapparat] einlegen. **load 'up** *v. i.* laden (with *Akk.*)

'loaded *adj.* **a ~ question** eine suggestive Frage; **be ~** *(sl.: rich)* [schwer] Kohle haben *(salopp)*

¹**loaf** [ləʊf] *n., pl.* **loaves** [ləʊvz] Brot, *das;* [Brot]laib, *der;* **a ~ of bread** ein Laib Brot

²**loaf** *v. i.* ~ **round town/the house** in der Stadt/zu Hause herumlungern *(ugs.)*

loan [ləʊn] 1. *n.* a) *(thing lent)* Leihgabe, *die;* **be out on ~:** ausgeliehen sein; **have sth. on ~** [from sb.] etw. [von jmdm.] geliehen haben; b) *(money lent)* Darlehen, *das.* 2. *v. t.* ~ **sth. to sb.** jmdm. etw. leihen

loath [ləʊθ] *pred. adj.* **be ~ to do sth.** etw. ungern tun

loathe [ləʊð] *v. t.* verabscheuen.

loathing ['ləʊðɪŋ] *n.* Abscheu, *der* (of, for vor + *Dat.*). **loathsome** ['ləʊðsəm] *adj.* abscheulich; widerlich

loaves *pl. of* ¹**loaf**

lobby ['lɒbɪ] *n.* a) *(pressure group)* Lobby, *die;* b) *(of hotel)* Eingangshalle, *die; (of theatre)* Foyer, *das*

lobe [ləʊb] *n. (ear~)* Ohrläppchen, *das*

lobster ['lɒbstə(r)] *n.* Hummer, *der*

local ['ləʊkl] 1. *adj.* lokal *(bes. Zeitungsw.);* Kommunal⟨*wahl, -abgaben*⟩; *(of this area)* hiesig; *(of that area)* dortig; ortsansässig ⟨*Firma, Familie*⟩; ⟨*Wein, Produkt, Spezialität*⟩ [aus] der Gegend; **she's a ~ girl** sie ist von hier/dort. 2. *n.* a) *(person)* Einheimische, *der/die;* b) *(Brit. coll.: pub)* [Stamm]kneipe, *die*

local: ~ **anaes'thetic** *n.* Lokalanästhetikum, *das;* ~ **au'thority** *n. (Brit.)* Kommunalverwaltung, *die;* ~ **call** *n. (Teleph.)* Ortsgespräch, *das;* ~ **'government** *n.* Kommunalverwaltung, *die*

locality [ləʊ'kælɪtɪ] *n.* Ort, *der*

'locally *adv.* im/am Ort

locate [ləʊ'keɪt] *v. t.* a) **be ~d** liegen; b) *(determine position of)* ausfindig machen. **location** [ləʊ'keɪʃn] *n.* a) Lage, *die;* b) *(Cinemat.)* **be on ~:** bei Außenaufnahmen sein

loch [lɒx, lɒk] *n. (Scot.)* See, *der*

¹**lock** [lɒk] *n. (of hair)* [Haar]strähne, *die*

²**lock** 1. *n.* a) *(of door etc.)* Schloß, *das;* b) *(on canal etc.)* Schleuse, *die.* 2. *v. t.* zuschließen. 3. *v. i.* ⟨*Tür, Kasten usw.:*⟩ sich zuschließen lassen. **lock a'way** *v. t.* einschließen; einsperren ⟨*Person*⟩. **lock 'in** *v. t.* einschließen; *(deliberately)* einsperren. **lock 'out** *v. t.* aussperren (of aus); ~ **oneself out** sich aussperren. **lock 'up** 1. *v. i.* abschließen. 2. *v. t.* a) abschließen ⟨*Haus, Tür*⟩; b) *(imprison)* einsperren

locker ['lɒkə(r)] *n.* Schließfach, *das*

locket ['lɒkɪt] *n.* Medaillon, *das*

lock: ~**jaw** *n. (Med.)* Kieferklemme, *die;* ~**out** *n.* Aussperrung, *die;* ~**smith** *n.* Schlosser, *der*

locomotive [ləʊkə'məʊtɪv] *n.* Lokomotive, *die*

locust ['ləʊkəst] *n.* Heuschrecke, *die*

lodge [lɒdʒ] 1. *n.* a) *(cottage)* Pförtner-/Gärtnerhaus, *das;* b) *(porter's room)* [Pförtner]loge, *die.* 2. *v. t.* a) einlegen ⟨*Beschwerde, Protest usw.*⟩; b) einreichen ⟨*Klage*⟩. 3. *v. i.* [zur Miete] wohnen. **'lodger** *n.* Untermieter,

der/Untermieterin, die. **lodging** ['lɒdʒɪŋ] n. [möbliertes] Zimmer
loft [lɒft] n. (attic) [Dach]boden, der
lofty ['lɒftɪ] adj. a) (exalted) hoch; b) (haughty) hochmütig
log [lɒg] n. a) (timber) [geschlagener] Baumstamm; (as firewood) [Holz]scheit, das; b) ~|-book| (Naut.) Logbuch, das. **log 'cabin** n. Blockhütte, die. **log-'fire** n. Holzfeuer, das
loggerheads ['lɒgəhedz] n. pl. be at ~ with sb. mit jmdm. im Clinch liegen
logic ['lɒdʒɪk] n. Logik, die. **logical** ['lɒdʒɪkl] adj. logisch; **she has a ~ mind** sie denkt logisch. **logically** ['lɒdʒɪkəlɪ] adv. logisch
logo ['ləʊgəʊ] n., pl. ~s Signet, das
loin [lɔɪn] n. Lende, die. **'loincloth** n. Lendenschurz, der
loiter ['lɔɪtə(r)] v.i. trödeln; (linger suspiciously) herumlungern
loll [lɒl] v.i. sich lümmeln (ugs.)
lollipop ['lɒlɪpɒp] n. Lutscher, der
London ['lʌndən] 1. pr. n. London (das). 2. attrib. adj. Londoner. **'Londoner** pr. n. Londoner, der/Londonerin, die
lone [ləʊn] attrib. adj. einsam. **loneliness** ['ləʊnlɪnɪs] n. Einsamkeit, die
lonely ['ləʊnlɪ] adj. einsam
loner ['ləʊnə(r)] n. Einzelgänger, der/-gängerin, die
lonesome ['ləʊnsəm] adj. einsam
¹long [lɒŋ] 1. adj., ~er ['lɒŋgə(r)], ~est ['lɒŋgɪst] a) lang; weit ⟨Reise, Weg⟩; b) (elongated) länglich; schmal; c) **in the ~ run** auf die Dauer. 2. n. (~ interval) take ~: lange dauern; for ~: lange; (since ~ ago) seit langem; before ~: bald. 3. adv., ~er, ~est a) lang[e]; as or so ~ as solange; **you should have finished ~ before now** du hättest schon längst fertig sein sollen; **much ~er** viel länger; b) as or so ~ as (provided that) solange; wenn
²long v.i. ~ for sb./sth. sich nach jmdm./etw. sehnen; ~ to do sth. sich danach sehnen, etw. zu tun
long-distance 1. ['---] adj. Fern⟨gespräch, -verkehr usw.⟩; Langstrecken-⟨läufer, -flug usw.⟩. 2. [-'--] adv. phone ~: ein Ferngespräch führen
longevity [lɒn'dʒevɪtɪ] n. Langlebigkeit, die
'longhand n. Langschrift, die
longing 1. n. Sehnsucht, die. 2. adj. sehnsüchtig. **'longingly** adv. sehnsüchtig
longitude ['lɒŋgɪtjuːd] n. Länge, die

long: ~ **jump** n. (Brit. Sport) Weitsprung, der; ~**-lived** ['lɒŋlɪvd] adj. langlebig; ~**-playing 'record** n. Langspielplatte, die; ~**-range** adj. a) Langstrecken⟨flugzeug, -rakete usw.⟩; b) (relating to time) langfristig; ~**-sighted** [lɒŋ'saɪtɪd] adj. weitsichtig; (fig.) weitblickend; ~**-sleeved** ['lɒŋsliːvd] adj. langärmelig; ~**-standing** attrib. adj. seit langem bestehend; alt ⟨Schulden, Streit⟩; ~**-suffering** adj. schwer geprüft; ~**-term** adj. langfristig; ~ **wave** n. (Radio) Langwelle, die; ~**-winded** [lɒŋ'wɪndɪd] adj. langatmig
loo [luː] n. (Brit. coll.) Klo, das (ugs.)
look [lʊk] 1. v.i. a) sehen; gucken (ugs.); b) (search) nachsehen; c) (face) zugewandt sein (to|wards| Dat.); d) (appear) aussehen; ~ well/ill gut/schlecht aussehen. 2. n. a) Blick, der; **have or take a ~ at sb./sth.** sich (Dat.) jmdn./etw. ansehen; b) (appearance) Aussehen, das. **look 'after** v.t. (care for) sorgen für. **look a'head** v.i. (fig.) an die Zukunft denken. **'look at** v.t. a) (regard) ansehen; b) (consider) betrachten. **look 'back** v.i. a) sich umsehen; b) ~ **back on** or **to sth.** an etw. (Akk.) zurückdenken. **look 'down [up]on** v.t. a) herunter-/hinuntersehen auf (+ Akk.); b) (fig.: despise) herabsehen auf (+ Akk.). **'look for** v.t. a) (seek) suchen nach; b) (expect) erwarten. **look 'out** v.i. a) hinaus-/heraussehen (of aus); b) (take care) aufpassen; c) ~ **out on sth.** ⟨Zimmer, Wohnung usw.:⟩ zu etw. hin liegen. **look 'out for** v.t. (be prepared for) achten auf (+ Akk.); (keep watching for) Ausschau halten nach ⟨Arbeit, Gelegenheit, Sammelobjekt usw.⟩. **look 'over** v.t. a) sehen über (+ Akk.); b) (survey) sich (Dat.) ansehen ⟨Haus⟩. **look 'round** v.i. sich umsehen. **'look through** v.t. a) ~ **through sth.** durch etw. [hindurch] sehen; b) (inspect) durchsehen ⟨Papiere⟩. **look 'to** v.t. (rely on) ~ **to sb./sth. for sth.** etw. von jmdm./etw. erwarten. **look 'up** 1. v.i. a) aufblicken; b) (improve) besser werden. 2. v.t. nachschlagen ⟨Wort⟩; heraussuchen ⟨Telefonnummer, Zugverbindung usw.⟩. **look 'up to** v.t. ~ **up to sb.** zu jmdm. aufsehen
'look-alike n. Doppelgänger, der/-gängerin, die
looker-'on n. Zuschauer, der/Zuschauerin, die

'**looking-glass** *n.* Spiegel, *der*

'**look-out** *n., pl.* ~s a) *(observation post)* Ausguck, *der;* b) *(person)* Wache, *die;* c) *(Brit. fig.)* that's a bad ~: das sind schlechte Aussichten; that's his ~: das ist sein Problem; d) keep a ~ [for sb./sth.] [nach etw./jmdm.] Ausschau halten

¹**loom** [lu:m] *n. (Weaving)* Webstuhl, *der*

²**loom** *v. i.* auftauchen

loop [lu:p] 1. *n.* a) *(not firm)* Schleife, *die;* b) *(cord)* Schlaufe, *die.* 2. *v. t.* zu einer Schlaufe formen. '**loophole** *n. (fig.)* Lücke, *die*

loose [lu:s] *adj.* a) *(not firm)* locker ⟨Zahn, Schraube, Knopf⟩; b) *(not fixed)* lose ⟨Knopf, Buchseite, Brett, Stein⟩; offen ⟨Haar⟩; c) be at a ~ end *(fig.)* nichts zu tun haben; d) *(inexact)* ungenau. '**loose-fitting** *adj.* bequem geschnitten. '**loose-leaf** *adj.* Loseblatt-; ~ file Ringbuch, *das*

'**loosely** *adv.* locker; lose ⟨zusammenhängen⟩; frei ⟨übersetzen⟩

loosen ['lu:sn] *v. t.* lockern. **loosen up** *v. i.* sich auflockern; *(relax)* auftauen

'**looseness** *n.* Lockerheit, *die*

loot [lu:t] 1. *v. t.* plündern. 2. *n.* Beute, *die.* '**looter** *n.* Plünderer, *der*

lop [lɒp] *v. t.* ~ sth. [off or away] etw. abbauen *od.* abhacken

lopsided [lɒp'saɪdɪd] *adj.* schief

lord [lɔ:d] 1. *n.* a) *(master)* Herr, *der;* b) L~ *(Relig.)* Herr, *der;* c) *(Brit.: as title)* Lord, *der;* the House of L~s *(Brit.)* das Oberhaus. 2. *int. (coll.)* Gott; oh/good L~! du lieber Himmel! '**lordship** *n.* Lordschaft, *die*

lore [lɔ:(r)] *n.* Überlieferung, *die*

lorry ['lɒrɪ] *n. (Brit.)* Lastwagen, *der;* Lkw, *der.* '**lorry-driver** *n. (Brit.)* Lastwagenfahrer, *der;* Lkw-Fahrer, *der*

lose [lu:z] 1. *v. t.,* lost [lɒst] a) verlieren; ~ one's way sich verlaufen/verfahren; b) ⟨Uhr:⟩ nachgehen; c) *(waste)* vertun ⟨Zeit⟩; *(miss)* versäumen ⟨Gelegenheit⟩; d) ~ weight abnehmen. 2. *v. i.,* lost a) *(in match, contest)* verlieren; b) ⟨Uhr:⟩ nachgehen. '**loser** *n.* Verlierer, *der/*Verliererin, *die*

loss [lɒs] *n.* a) Verlust, *der* (of *Gen.*); sell at a ~: mit Verlust verkaufen; b) be at a ~: nicht [mehr] weiterwissen; be at a ~ for words um Worte verlegen sein; be at a ~ what to do nicht wissen, was zu tun ist

lost [lɒst] 1. *see* lose. 2. *adj.* a) verloren; get ~ ⟨Person:⟩ sich verlaufen/verfahren; get ~! *(sl.)* verdufte! *(salopp);* ~ cause aussichtslose Sache, *die;* b) *(wasted)* vertan ⟨Zeit⟩; *(missed)* versäumt ⟨Gelegenheit⟩

lot [lɒt] *n.* a) *(destiny)* Los, *das;* b) *(set of persons)* Haufen, *der;* the ~: [sie] alle; c) *(set of things)* Menge, *die;* the ~: alle/alles; d) *(coll.: large quantity)* ~s or a ~ of money *etc.* viel *od.* eine Menge Geld *usw.;* sing *etc.* a ~: viel singen *usw.;* like sth. a ~: etw. sehr mögen; have ~s to do viel zu tun haben; e) *(for choosing)* Los, *das;* draw/cast/throw ~s [for sth.] um etw. losen

lotion ['ləʊʃn] *n.* Lotion, *die*

lottery ['lɒtərɪ] *n.* Lotterie, *die*

loud [laʊd] 1. *adj.* a) laut; lautstark ⟨Protest, Kritik⟩; b) *(flashy, conspicuous)* aufdringlich; grell ⟨Farbe⟩. 2. *adv.* laut; laugh out ~: laut auflachen; say sth. out ~: etw. aussprechen. **loud'hailer** *n.* Megaphon, *das*

'**loudly** *adv.* laut

'**loudness** *n.* Lautstärke, *die*

loud'speaker *n.* Lautsprecher, *der*

lounge [laʊndʒ] 1. *v. i.* ~ [about or around] [faul] herumliegen/-sitzen/-stehen. 2. *n.* a) *(in hotel)* [Hotel]halle, *die;* (at airport) Wartehalle, *die;* b) *(sitting-room)* Wohnzimmer, *das*

louse [laʊs] *n., pl.* lice [laɪs] Laus, *die*

lousy ['laʊzɪ] *adj. (sl.)* a) *(disgusting)* ekelhaft; b) *(very poor)* lausig *(ugs.);* feel ~: sich mies *(ugs.)* fühlen

lout [laʊt] *n.* Rüpel, *der;* Flegel, *der*

louver, louvre ['lu:və(r)] *n.* ~ window Jalousiefenster, *das;* ~ door Jalousietür, *die*

lovable ['lʌvəbl] *adj.* liebenswert

love [lʌv] 1. *n.* a) Liebe, *die* (of, for zu); in ~ [with] verliebt [in (+ *Akk.*)]; fall in ~ [with] sich verlieben [in (+ *Akk.*)]; for ~: aus Liebe; ~ from Beth *(in letter)* herzliche Grüße von Beth; send one's ~ to sb. jmdn. grüßen lassen; b) *(sweetheart)* Geliebte, *der/die;* [my] ~ *(coll.: form of address)* [mein] Liebling *od.* Schatz; c) *(Tennis)* fifteen/thirty ~: fünfzehn/dreißig null. 2. *v. t.* a) lieben; our/their ~d ones unsere/ihre Lieben; b) *(like)* I'd ~ a cigarette ich hätte sehr gerne eine Zigarette; ~ to do *or* ~ doing sth. etw. gern tun

love: ~ affair *n.* Liebesverhältnis, *das;* ~-letter *n.* Liebesbrief, *der;* ~-life *n.* Liebesleben, *das*

loveliness ['lʌvlɪnɪs] n. Schönheit, die
lovely ['lʌvlɪ] adj. [wunder]schön; herrlich ⟨Tag, Essen⟩
lover ['lʌvə(r)] n. **a)** Liebhaber, der; Geliebte, der; (woman) Geliebte, die; be ~s ein Liebespaar sein; **b)** (person who likes sth.) Freund, der/Freundin, die
love: ~**sick** adj. an Liebeskummer leidend; liebeskrank (geh.); ~**song** n. Liebeslied, das; ~**story** n. Liebesgeschichte, die
loving ['lʌvɪŋ] adj. **a)** (affectionate) liebend; **b)** (expressing love) liebevoll. **'lovingly** adv. liebevoll
low [ləʊ] **1.** adj. **a)** niedrig; tief ausgeschnitten ⟨Kleid⟩; tief ⟨Ausschnitt⟩; tiefliegend ⟨Grund⟩; **b)** (of humble rank) nieder...; niedrig; **c)** (inferior) niedrig; gering ⟨Intelligenz, Bildung⟩; **d)** (in pitch) tief; (in loudness) leise. **2.** adv. **a)** (to a ~ position) tief; **b)** (not loudly) leise; **c)** lie ~ (hide) untertauchen. **'lowbrow** adj. (coll.) schlicht ⟨Person⟩; [geistig] anspruchslos ⟨Buch, Programm⟩. **'low-cut** adj. [tief] ausgeschnitten ⟨Kleid⟩
¹lower ['ləʊə(r)] v.t. **a)** herab-/hinablassen; **b)** senken ⟨Blick⟩; auslassen ⟨Saum⟩; senken ⟨Preis, Miete, Zins usw.⟩; ~ one's voice leiser sprechen
²lower 1. compar. adj. unter...; Unter⟨grenze-, arm, -lippe usw.⟩. **2.** compar. adv. tiefer
low: ~**fat** adj. fettarm; ~**grade** adj. minderwertig; ~**land** ['ləʊlənd] n. Tiefland, das
lowly ['ləʊlɪ] adj. (modest) bescheiden
low: ~**lying** adj. tiefliegend; ~ **point** n. Tiefpunkt, der; ~ **pressure** n. (Meteorol.) Tiefdruck, der
loyal ['lɔɪəl] adj. treu. **loyalty** ['lɔɪəltɪ] n. Treue, die
lozenge ['lɒzɪndʒ] n. Pastille, die
LP abbr. **long-playing record** LP, die
Ltd. abbr. **Limited** GmbH
lubricant ['luːbrɪkənt] n. Schmiermittel, das
lubricate ['luːbrɪkeɪt] v.t. schmieren. **lubrication** [luːbrɪ'keɪʃn] n. Schmierung, die; attrib. Schmier⟨system, -vorrichtung⟩
lucid ['luːsɪd] adj. klar. **lucidity** [luː'sɪdɪtɪ] n. Klarheit, die
luck [lʌk] n. Glück, das; good ~: Glück, das; bad or hard ~: Pech, das; good ~! viel Glück!; be in/out of ~: Glück/kein Glück haben; no such ~: schön wär's. **luckily** ['lʌkɪlɪ] adv.

glücklicherweise. **lucky** ['lʌkɪ] adj. **a)** glücklich; be ~: Glück haben; **b)** (bringing good luck) Glücks⟨zahl, -tag usw.⟩; ~ **charm** Glücksbringer, der
lucrative ['luːkrətɪv] adj. einträglich; lukrativ
ludicrous ['luːdɪkrəs] adj. lächerlich; lachhaft ⟨Angebot, Ausrede⟩
lug [lʌg] v.t., -gg- (drag) schleppen
luggage ['lʌgɪdʒ] n. Gepäck, das. **'luggage-locker** n. [Gepäck]schließfach, das. **'luggage-rack** n. Gepäckablage, die
lugubrious [luː'guːbrɪəs] adj. (mournful) kummervoll; (dismal) düster
lukewarm ['luːkwɔːm] adj. lauwarm
lull [lʌl] **1.** v.t. **a)** (soothe) lullen; **b)** (fig.) einlullen; ~ sb. into a false sense of security jmdn. in einer trügerischen Sicherheit wiegen. **2.** n. Pause, die
lullaby ['lʌləbaɪ] n. Schlaflied, das
lumbago [lʌm'beɪgəʊ] n., pl. ~s (Med.) Hexenschuß, der
lumber ['lʌmbə(r)] **1.** n. **a)** (furniture) Gerümpel, das; **b)** (useless material) Kram, der (ugs.); **c)** (Amer.: timber) [Bau]holz, das. **2.** v.t. ~ sb. with sth./ sb. jmdm. etw./jmdn. aufhalsen (ugs.)
'lumbering adj. schwerfällig
lumberjack ['lʌmbədʒæk] n. (Amer.) Holzfäller, der
luminous ['luːmɪnəs] adj. [hell] leuchtend; Leucht⟨anzeige, -zeiger usw.⟩
lump [lʌmp] **1.** n. **a)** Klumpen, der; (of sugar, butter, etc.) Stück, das; (of wood) Klotz, der; (of dough) Kloß, der; (of bread) Brocken, der; **b)** (swelling) Beule, die. **2.** v.t. ~ sth. with sth. etw. und etw. zusammentun. **lump 'sum** n. Pauschalsumme, die
'lumpy adj. klumpig ⟨Brei⟩; ⟨Kissen, Matratze⟩ mit klumpiger Füllung
lunacy ['luːnəsɪ] n. Wahnsinn, der
lunar ['luːnə(r)] adj. Mond-
lunatic ['luːnətɪk] **1.** adj. wahnsinnig. **2.** n. Wahnsinnige, der/die; Irre, der/ die. **'lunatic asylum** n. (Hist.) Irrenanstalt, die (veralt., ugs.)
lunch [lʌntʃ] **1.** n. Mittagessen, das; have or eat [one's] ~: zu Mittag essen. **2.** v.i. zu Mittag essen
luncheon voucher ['lʌntʃn vaʊtʃə(r)] n. (Brit.) Essenmarke, die
lunch: ~**hour** n. Mittagspause, die; ~**time** n. Mittagszeit, die; at ~**time** mittags
lung [lʌŋ] n. (right or left) Lungenflügel, der; ~s Lunge, die. **'lung cancer** n. Lungenkrebs, der

lunge [lʌndʒ] **1.** *n.* Sprung nach vorn. **2.** *v. i.* ~ **at sb. with a knife** jmdn. mit einem Messer angreifen

¹lurch [lɜːtʃ] *n.* **leave sb. in the** ~: jmdn. im Stich lassen

²lurch 1. *n.* Rucken, *das.* **2.** *v. i.* rucken; ⟨*Betrunkener:*⟩ torkeln

lure [ljʊə(r), lʊə(r)] **1.** *v. t.* locken. **2.** *n.* Lockmittel, *das*

lurid ['ljʊərɪd, 'lʊərɪd] *adj.* **a)** *(in colour)* grell; **b)** *(sensational)* reißerisch

lurk [lɜːk] *v. i.* lauern

luscious ['lʌʃəs] *adj.* köstlich [süß]; saftig [süß] ⟨*Obst*⟩

lush [lʌʃ] *adj.* saftig ⟨*Wiese*⟩; grün ⟨*Tal*⟩; üppig ⟨*Vegetation*⟩

lust [lʌst] **1.** *n.* **a)** *(sexual)* Sinnenlust, *die;* **b)** *(strong desire)* Gier, *die* (for nach). **2.** *v. i.* ~ **after** [lustvoll] begehren *(geh.).* **lustful** ['lʌstfl] *adj.* lüstern *(geh.)*

lustily ['lʌstɪlɪ] *adv.* kräftig; aus voller Kehle ⟨*rufen, singen*⟩

lustre ['lʌstə(r)] *n. (Brit.)* **a)** Schimmer, *der;* **b)** *(fig.: splendour)* Glanz, *der*

lusty ['lʌstɪ] *adj.* kräftig

Luxembourg, Luxemburg ['lʌksəmbɜːg] *pr. n.* Luxemburg *(das)*

luxuriant [lʌg'zjʊərɪənt] *adj.* üppig

luxuriate [lʌg'zjʊərɪeɪt] *v. i.* ~ **in** sich aalen in (+ *Dat.*)

luxurious [lʌg'zjʊərɪəs] *adj.* luxuriös

luxury ['lʌkʃərɪ] *n.* **a)** Luxus, *der;* **b)** *(article)* Luxusgegenstand, *der;* **luxuries** Luxus, *der*

LW *abbr. (Radio)* long wave LW

lying ['laɪŋ] *adj.* verlogen ⟨*Person*⟩. *See also* **¹lie 2**

lynch [lɪntʃ] *v. t.* lynchen

lyric ['lɪrɪk] **1.** *adj.* lyrisch; ~ **poetry** Lyrik, *die.* **2.** *n. in pl. (of song)* Text, *der.* **lyrical** ['lɪrɪkl] *adj.* **a)** lyrisch; **b)** *(coll.: enthusiastic)* gefühlvoll

M

M, m [em] *n.* M, m, *das*

m. *abbr.* **a)** masculine m.; **b)** metre|s| m; **c)** million|s| Mill.; **d)** minute|s| Min.

MA *abbr.* Master of Arts M. A.

mac [mæk] *n. (Brit. coll.)* Regenmantel, *der*

macaroni [mækə'rəʊnɪ] *n.* Makkaroni *Pl.*

machine [mə'ʃiːn] *n.* Maschine, *die.* **ma'chine-gun** *n.* Maschinengewehr, *das*

machinery [mə'ʃiːnərɪ] *n.* **a)** *(machines)* Maschinen *Pl.;* **b)** *(mechanism)* Mechanismus, *der*

machine: ~ **tool** *n.* Werkzeugmaschine, *die;* ~**-washable** *adj.* waschmaschinenfest

machinist [mə'ʃiːnɪst] *n.* Maschinist, *der*/Maschinistin, *die;* |sewing-|~: |Maschinen|näherin, *die*/-näher, *der*

macho ['mætʃəʊ] *adj.* Macho-; **he is** ~: er ist ein Macho

mackerel ['mækərl] *n., pl. same or* ~s Makrele, *die*

mackintosh ['mækɪntɒʃ] *n.* Regenmantel, *der*

mad [mæd] *adj.* **a)** *(insane)* geisteskrank; **b)** *(frenzied)* wahnsinnig; **drive sb. mad** jmdn. um den Verstand bringen; **c)** *(foolish)* verrückt *(ugs.);* **d)** *(very enthusiastic)* **be** ~ **about** *or* **on sb./sth.** auf jmdn./etw. wild sein *(ugs.);* **e)** *(coll.: annoyed)* ~ |with *or* at sb.| sauer [auf jmdn.] *(ugs.);* **f)** *(with rabies)* toll|wütig|; |run *etc.*| like ~: wie wild [laufen *usw.*]

madam ['mædəm] *n.* gnädige Frau; **Dear M**~ *(in letter)* Sehr verehrte gnädige Frau

madden ['mædn] *v. t. (irritate)* [ver]ärgern. **maddening** ['mædənɪŋ] *adj. (irritating)* [äußerst] ärgerlich

made *see* **make 1**

'madly *adv. (coll.)* wahnsinnig *(ugs.)*

madman ['mædmən] *n., pl.* **madmen** ['mædmən] *n.* Wahnsinnige, *der*

'madness *n.* Wahnsinn, *der*

magazine [mægə'ziːn] *n.* **a)** Zeitschrift, *die;* **b)** *(of firearm)* Magazin, *das*

maggot ['mægət] *n.* Made, *die*

magic ['mædʒɪk] **1.** *n.* **a)** Magie, *die;* **work like** ~: wie ein Wunder wirken; **b)** *(conjuring)* Zauberei, *die.* **2.** *adj.* **a)** magisch; Zauber⟨*trank, -baum*⟩; **b)** *(fig.)* wunderbar. **magical** ['mædʒɪkl] *adj.* zauberhaft. **magician** [mə'dʒɪʃn] *n.* Magier, *der*/Magierin, *die; (conjurer)* Zauberer, *der*/Zauberin, *die*

magistrate ['mædʒɪstreɪt] *n.* Friedensrichter, *der*/-richterin, *die*

magnanimity [mægnə'nɪmɪtɪ] *n.* Großmut, *die*

magnanimous [mæg'nænɪməs] *adj.* großmütig (**towards gegen**)

magnate ['mægneɪt] *n.* Magnat, *der*/Magnatin, *die*

magnesium [mæg'ni:zɪəm] *n.* Magnesium, *das*

magnet ['mægnɪt] *n.* Magnet, *der.* **magnetic** [mæg'netɪk] *adj.* magnetisch. **magnetic 'tape** *n.* Magnetband, *das*

magnetism ['mægnɪtɪzm] *n.* **a)** *(force, lit. or fig.)* Magnetismus, *der;* **b)** *(fig.: charm)* Anziehungskraft, *die*

magnetize ['mægnɪtaɪz] *v. t.* magnetisieren

magnification [mægnɪfɪ'keɪʃn] *n.* Vergrößerung, *die*

magnificence [mæg'nɪfɪsəns] *n.* Pracht, *die; (beauty)* Herrlichkeit, *die; (lavishness)* Üppigkeit, *die*

magnificent [mæg'nɪfɪsənt] *adj.* **a)** prächtig; herrlich ⟨*Garten, Kunstwerk, Wetter*⟩; *(lavish)* üppig ⟨*Mahl*⟩; **b)** *(coll.: excellent)* fabelhaft *(ugs.)*

magnify ['mægnɪfaɪ] *v. t.* **a)** vergrößern; **b)** *(exaggerate)* aufbauschen. **'magnifying glass** *n.* Lupe, *die*

magnitude ['mægnɪtju:d] *n.* **a)** *(size)* Größe, *die;* **b)** *(importance)* Wichtigkeit, *die*

magpie ['mægpaɪ] *n.* Elster, *die*

mahogany [mə'hɒgənɪ] *n.* Mahagoni[holz], *das; attrib.* Mahagoni-

maid [meɪd] *n.* Dienstmädchen, *das*

maiden ['meɪdn] **1.** *n.* Jungfrau, *die.* **2.** *adj.* **a)** *(unmarried)* unverheiratet; **b)** *(first)* ~ **voyage/speech** Jungfernfahrt/-rede, *die.* **'maiden name** *n.* Mädchenname, *der*

mail [meɪl] **1.** *n. see* ²**post** 1. **2.** *v. t.* abschicken

mail: ~bag *n.* Postsack, *der;* **~box** *n.* *(Amer.)* Briefkasten, *der;* **~ing list** *n.* Adressenliste, *die;* **~man** *n. (Amer.)* Briefträger, *der;* ~ **order** *n.* Bestellung per Post

maim [meɪm] *v. t.* verstümmeln

main [meɪn] **1.** *n.* **a)** *(channel, pipe)* Hauptleitung, *die;* **~s** *(Electr.)* Stromnetz, *das;* **b) in the** ~**:** im großen und ganzen. **2.** *attrib. adj.* Haupt-; **the ~ thing is that …:** die Hauptsache ist, daß … **mainland** ['meɪnlənd] *n.* Festland, *das*

'mainly *adv.* hauptsächlich

main: ~stay *n.* [wichtigste] Stütze; ~ **street** [*Brit.* -'-, *Amer.* '--] *n.* Hauptstraße, *die*

maintain [meɪn'teɪn] *v. t.* **a)** *(keep up)* aufrechterhalten; **b)** *(provide for)* ~ **sb.** für jmds. Unterhalt aufkommen; **c)** *(preserve)* instand halten; warten ⟨*Maschine*⟩; **d)** ~ **that …:** behaupten, daß … **maintenance** ['meɪntənəns] *n.* **a)** *(keeping up)* Aufrechterhaltung, *die;* **b)** *(preservation)* Instandhaltung, *die; (of machinery)* Wartung, *die;* **c)** *(Law: money paid to support sb.)* Unterhalt, *der*

maison[n]ette [meɪzə'net] *n.* [zweistöckige] Wohnung

maize [meɪz] *n.* Mais, *der*

majestic [mə'dʒestɪk] *adj.,* **majestically** [mə'dʒestɪkəlɪ] *adv.* majestätisch

majesty ['mædʒɪstɪ] *n.* Majestät, *die (geh.);* **Your/Her** *etc.* **M~:** Eure/Seine *usw.* Majestät

major ['meɪdʒə(r)] **1.** *adj.* **a)** *attrib. (greater)* größer…; **b)** *attrib. (important)* bedeutend…; *(serious)* schwer; ~ **road** Hauptverkehrsstraße, *die;* **c)** *(Mus.)* Dur-; **C** ~: C-Dur. **2.** *n. (Mil.)* Major, *der.* **3.** *v. i. (Amer. Univ.)* ~ **in sth.** etw. als Hauptfach haben

Majorca [mə'jɔ:kə] *pr. n.* Mallorca *(das)*

majority [mə'dʒɒrɪtɪ] *n.* Mehrheit, *die;* **be in the** ~: in der Mehrzahl sein

make [meɪk] **1.** *v. t.,* **made** [meɪd] **a)** machen *(of aus);* bauen ⟨*Straße, Flugzeug*⟩; anlegen ⟨*Teich, Weg usw.*⟩; zimmern ⟨*Tisch, Regal*⟩; basteln ⟨*Spielzeug, Vogelhäuschen usw.*⟩; nähen ⟨*Kleider*⟩; *(manufacture)* herstellen; *(prepare)* zubereiten ⟨*Mahlzeit*⟩; machen, kochen ⟨*Kaffee, Tee*⟩; backen ⟨*Brot, Kuchen*⟩; **b)** *(establish, enact)* treffen ⟨*Unterscheidung, Übereinkommen*⟩; ziehen ⟨*Vergleich*⟩; erlassen ⟨*Gesetz*⟩; aufstellen ⟨*Regeln, Behauptung*⟩; stellen ⟨*Forderung*⟩; geben ⟨*Bericht*⟩; vornehmen ⟨*Zahlung*⟩; erheben ⟨*Protest, Beschwerde*⟩; **c)** *(cause to be or become)* ~ **happy/known** *etc.* glücklich/bekannt *usw.* machen; ~ **sb. captain** jmdn. zum Kapitän machen; **d)** ~ **sb. do sth.** *(cause)* jmdn. dazu bringen, etw. zu tun; *(compel)* jmdn. zwingen, etw. zu tun; **be made to do sth.** etw. tun müssen; **e)** *(earn)* machen ⟨*Profit, Verlust*⟩; verdienen ⟨*Lebensunterhalt*⟩; **f) what do you** ~ **of him?** was hältst du von ihm?; **g)** *(arrive at)* erreichen; **make it** *(succeed in arriving)* es schaffen; **h)** ~ **'do with/without sth.** mit/ohne etw. auskommen. **2.** *n.*

(brand) Marke, *die.* '**make for** *v. t.* zusteuern auf (+ *Akk.*). **make 'off** *v. i.* sich davonmachen. **make 'off with** *v. t.* ~ **off with sb./sth.** sich mit jmdm./etw. [auf und] davonmachen. **make 'out 1.** *v. t.* **a)** *(write)* ausstellen; **b)** *(claim)* behaupten; **c)** *(manage to see or hear)* ausmachen; *(manage to read)* entziffern; **d)** *(pretend)* vorgeben. **2.** *v. i. (coll.)* zurechtkommen (**at** bei). **make 'over** *v. t.* überschreiben. **make 'up 1.** *v. t.* **a)** *(assemble)* zusammenstellen; **b)** *(invent)* erfinden; **c)** *(constitute)* bilden; **be made up of** ...: bestehen aus ...; **d)** *(apply cosmetics to)* schminken; ~ **up one's face** sich schminken. **2.** *v. i. (be reconciled)* sich wieder vertragen. **make 'up for** *v. t.* wiedergutmachen; ~ **up for lost time** Versäumtes nachholen

'**make-believe 1.** *n.* **it's only** ~: das ist bloß Phantasie. **2.** *adj.* nicht echt

'**maker** *n. (manufacturer)* Hersteller, *der*

make: ~**shift** *adj.* behelfsmäßig; ~-**up** *n. (Cosmetics)* Make-up, *das*

making ['meɪkɪŋ] *n.* **in the** ~: im Entstehen; **have the** ~**s of a leader** das Zeug zum Führer haben *(ugs.)*

maladjusted [mælə'dʒʌstɪd] *adj.* verhaltensgestört

malady ['mælədɪ] *n.* Leiden, *das*

malaise [mə'leɪz] *n.* Unbehagen, *das*

malaria [mə'leərɪə] *n.* Malaria, *die*

Malaysia [mə'leɪzɪə] *pr. n.* Malaysia *(das)*

male [meɪl] **1.** *adj.* männlich; Männer- ⟨*stimme, -chor, -verein*⟩; ~ **doctor/ nurse** Arzt, *der*/Krankenpfleger, *der.* **2.** *n. (person)* Mann, *der; (animal)* Männchen, *das*

malevolence [mə'levələns] *n.* Boshaftigkeit, *die*

malevolent [mə'levələnt] *adj.* boshaft

malfunction [mæl'fʌŋkʃn] **1.** *n.* Störung, *die; (Med.)* Funktionsstörung, *die.* **2.** *v. i.* nicht richtig funktionieren

malice ['mælɪs] *n.* Bosheit, *die.* **malicious** [mə'lɪʃəs] *adj.* böse

malign [mə'laɪn] *v. t.* verleumden

malignant [mə'lɪgnənt] *adj.* bösartig

malinger [mə'lɪŋgə(r)] *v. i.* simulieren. **ma'lingerer** *n.* Simulant, *der*/Simulantin, *die*

malleable ['mælɪəbl] *adj.* formbar

mallet ['mælɪt] *n.* Holzhammer, *der*

malnutrition [mælnju:'trɪʃn] *n.* Unterernährung, *die*

malt [mɔːlt] *n.* Malz, *das*

Malta ['mɔːltə] *pr. n.* Malta *(das)*

maltreat [mæl'tri:t] *v. t.* mißhandeln. **mal'treatment** *n.* Mißhandlung, *die*

mammal ['mæml] *n.* Säugetier, *das*

mammoth ['mæməθ] **1.** *n.* Mammut, *das.* **2.** *adj.* Mammut-; gigantisch ⟨*Vorhaben*⟩

man [mæn] **1.** *n.* **a)** *pl.* **men** [men] Mann, *der;* **b)** *(human race)* der Mensch. **2.** *v. t.,* -**nn**- bemannen ⟨*Schiff*⟩; besetzen ⟨*Büro, Stelle usw.*⟩; bedienen ⟨*Telefon, Geschütz*⟩

manacle ['mænəkl] **1.** *n., usu. in pl.* [Hand]fessel, *die.* **2.** *v. t.* Handfesseln anlegen (+ *Dat.*)

manage ['mænɪdʒ] **1.** *v. t.* **a)** leiten ⟨*Geschäft*⟩; **b)** *(Sport)* betreuen ⟨*Mannschaft*⟩; **c)** *(cope with)* schaffen; **d)** ~ **to do sth.** es fertigbringen, etw. zu tun; **he** ~**d to do it** es gelang ihm, es zu tun. **2.** *v. i.* zurechtkommen; ~ **without sth.** ohne etw. auskommen; **I can** ~: es geht. **manageable** ['mænɪdʒəbl] *adj.* leicht frisierbar ⟨*Haar*⟩; fügsam ⟨*Person, Tier*⟩; überschaubar ⟨*Größe, Menge*⟩. '**management** *n.* **a)** *(of a business)* Leitung, *die;* **b)** *(managers)* **the** ~: die Geschäftsleitung. '**manager** *n. (of shop or bank)* Filialleiter, *der*/-leiterin, *die; (of football team)* [Chef]trainer, *der*/-trainerin, *die; (of restaurant, shop, hotel)* Geschäftsführer, *der*/-führerin, *die.* **manageress** [mænɪdʒə'res] *n.* Geschäftsführerin, *die.* **managing** ['mænɪdʒɪŋ] *adj.* ~ **director** Geschäftsführer, *der*/-führerin, *die*

¹**mandarin** ['mændərɪn] *n.* ~ [orange] Mandarine, *die*

²**mandarin** *n. (bureaucrat)* Bürokrat, *der*/Bürokratin, *die*

mandarine ['mændəri:n] *see* ¹**mandarin**

mandate ['mændeɪt] *n.* Mandat, *das*

mandatory ['mændətərɪ] *adj.* obligatorisch

mandolin[e] [mændə'lɪn] *n.* Mandoline, *die*

mane [meɪn] *n.* Mähne, *die*

maneuver[able] *(Amer.) see* **manœuvr-**

manful ['mænfl] *adj.,* **manfully** ['mænfəlɪ] *adv.* mannhaft

manger ['meɪndʒə(r)] *n.* Krippe, *die*

mangle ['mæŋgl] *v. t.* verstümmeln ⟨*Person*⟩; demolieren ⟨*Sache*⟩

mangy ['meɪndʒɪ] *adj.* **a)** *(Vet. Med.)* räudig; **b)** *(shabby)* schäbig

man: ~**handle** *v. t.* **a)** von Hand bewegen *(Gegenstand)*; **b)** grob behandeln *(Person)*; **~hole** n. Mannloch, *das*

'manhood n. Mannesalter, *das*

man: **~-hour** n. Arbeitsstunde, *die;* **~-hunt** n. Verbrecherjagd, *die*

mania ['meɪnɪə] n. Manie, *die*

manicure ['mænɪkjʊə(r)] **1.** n. Maniküre, *die.* **2.** v. t. maniküren

manifest ['mænɪfest] **1.** adj. offenkundig. **2.** v. t. *(reveal)* offenbaren. **'manifestly** adv. offenkundig

manifesto [mænɪ'festəʊ] n., pl. **~s** Manifest, *das*

manifold ['mænɪfəʊld] adj. *(literary)* mannigfaltig *(geh.)*

manipulate [mə'nɪpjʊleɪt] v. t. **a)** manipulieren; **b)** *(handle)* handhaben. **manipulation** [mənɪpjʊ'leɪʃn] n. **a)** Manipulation, *die;* **b)** *(handling)* Handhabung, *die*

mankind [mæn'kaɪnd] n. Menschheit, *die*

manly ['mænlɪ] adj. männlich

'man-made adj. künstlich; *(synthetic)* Kunst*(faser, -stoff)*

manned [mænd] adj. bemannt

manner ['mænə(r)] n. **a)** Art, *die;* Weise, *die;* **in this ~:** auf diese Art und Weise; **b)** *(general behaviour)* Art, *die;* **c)** in pl. Manieren *Pl.* **mannerism** ['mænərɪzm] n. Eigenart, *die*

manœuvrable [mə'nu:vrəbl] adj. *(Brit.)* manövrierfähig

manœuvre [mə'nu:və(r)] *(Brit.)* **1.** n. Manöver, *das.* **2.** v. t. & i. manövrieren

manor ['mænə(r)] n. **a)** *(land)* [Land]gut, *das;* **b)** see **manor-house.** **'manor-house** n. Herrenhaus, *das*

'manpower n. Arbeitskräfte *Pl.*

mansion ['mænʃn] n. Herrenhaus, *das*

manslaughter ['mænslɔːtə(r)] n. Totschlag, *der*

mantelpiece ['mæntlpiːs] n. Kaminsims, *der od. das*

mantle ['mæntl] n. Umhang, *der*

manual ['mænjʊəl] **1.** adj. **a)** manuell; **~ work** Handarbeit; **b)** *(not automatic)* handbetrieben; *(Bedienung, Schaltung)* von Hand. **2.** n. Handbuch, *das*

manufacture [mænjʊ'fæktʃə(r)] **1.** n. Herstellung, *die.* **2.** v. t. herstellen. **manu'facturer** n. Hersteller, *der*

manure [mə'njʊə(r)] **1.** n. Dung, *der.* **2.** v. t. düngen

manuscript ['mænjʊskrɪpt] n. Manuskript, *das*

many ['menɪ] **1.** adj. viele; **how ~**

people/books? wie viele *od.* wieviel Leute/Bücher? **2.** n. viele [Leute]; **~ of us** viele von uns; **a good/great ~:** eine Menge

map [mæp] **1.** n. [Land]karte, *die; (street plan)* Stadtplan, *der.* **2.** v. t., **-pp-** kartographieren. **~ 'out** v. t. im einzelnen festlegen

maple ['meɪpl] n. Ahorn, *der*

mar [mɑː(r)] v. t. verderben

marathon ['mærəθən] n. **a)** Marathon[lauf], *der;* **b)** *(fig.)* Marathon, *das*

marauder [mə'rɔːdə(r)] n. Plünderer, *der*

marble ['mɑːbl] n. **a)** *(stone)* Marmor, *der;* **b)** *(toy)* Murmel, *die;* **[game of] ~s** Murmelspiel, *das*

March [mɑːtʃ] n. März, *der; see also* **August**

march **1.** n. Marsch, *der;* **[protest] ~:** Protestmarsch, *der.* **2.** v. i. marschieren. **march 'off** **1.** v. i. losmarschieren. **2.** v. t. abführen. **march 'past** v. i. vorbeimarschieren

'marcher n. **[protest] ~:** Demonstrant, *der/*Demonstrantin, *die*

mare [meə(r)] n. Stute, *die*

margarine [mɑːdʒə'riːn], *(coll.)* **marge** [mɑːdʒ] ns. Margarine, *die*

margin ['mɑːdʒɪn] n. **a)** *(of page)* Rand, *der;* **b)** *(extra amount)* Spielraum, *der;* **[profit] ~:** [Gewinn]spanne, *die;* **by a narrow ~:** knapp. **marginal** ['mɑːdʒɪnl] adj.; **'marginally** adv. unwesentlich

marigold ['mærɪɡəʊld] n. Ringelblume, *die*

marijuana [mærɪjʊ'ɑːnə] n. Marihuana, *das*

marina [mə'riːnə] n. Jachthafen, *der*

marinade [mærɪ'neɪd] **1.** n. Marinade, *die.* **2.** v. t. marinieren

marine [mə'riːn] **1.** adj. Meeres-; See*(versicherung, -recht usw.)*; Schiffs*(ausrüstung, -turbine usw.)*. **2.** n. *(person)* Marineinfanterist, *der.* **mariner** ['mærɪnə(r)] n. Seemann, *der*

marionette [mærɪə'net] n. Marionette, *die*

marital ['mærɪtl] adj. ehelich; **~ status** Familienstand, *der*

maritime ['mærɪtaɪm] adj. See-

'mark [mɑːk] **1.** n. **a)** *(trace)* Spur, *die; (stain etc.)* Fleck, *der; (scratch)* Kratzer, *der;* **b)** *(sign)* Zeichen, *das;* **c)** *(Sch.)* Note, *die;* **d)** *(target)* Ziel, *das.* **2.** v. t. **a)** *(dirty)* schmutzig machen; *(scratch)* zerkratzen; **b)** *(put distinguishing ~ on)* kennzeichnen, markieren (**with** mit); **c)** *(Sch.) (correct)* korrigieren; *(grade)* benoten; **d)** ~ **time** auf der Stelle treten. **mark 'off** *v. t.* abgrenzen (**from** von, gegen). **mark 'out** *v. t.* markieren

²**mark** *n. (monetary unit)* Mark, *die*

marked [mɑːkt] *adj.,* **markedly** ['mɑːkɪdlɪ] *adv.* deutlich

'**marker** *n.* Markierung, *die.* '**marker pen** *n.* Markierstift, *der*

market ['mɑːkɪt] **1.** *n.* Markt, *der.* **2.** *v. t.* vermarkten. '**marketing** *n.* Marketing, *das.* '**market-place** *n.* Marktplatz, *der; (fig.)* Markt, *der*

'**marking** *n.* **a)** Markierung, *die;* **b)** *(on animal)* Zeichnung, *die*

marksman ['mɑːksmən] *n., pl.* **marksmen** ['mɑːksmən] Scharfschütze, *der*

marmalade ['mɑːməleɪd] *n.* |orange| ~: Orangenmarmelade, *die*

¹**maroon** [məˈruːn] **1.** *adj.* kastanienbraun. **2.** *n.* Kastanienbraun, *das*

²**maroon** *v. t.* **a)** *(Naut.: put ashore)* aussetzen; **b)** ⟨*Flut, Hochwasser:*⟩ von der Außenwelt abschneiden

marquee [mɑːˈkiː] *n.* Festzelt, *das*

marquess, marquis ['mɑːkwɪs] *n.* Marquis, *der*

marriage ['mærɪdʒ] *n.* **a)** Ehe, *die* (**to** mit); **b)** *(wedding)* Hochzeit, *die*

married ['mærɪd] *adj.* **a)** verheiratet; ~ **couple** Ehepaar, *das;* **b)** *(marital)* Ehe⟨*leben, -name*⟩

marrow ['mærəʊ] *n.* **a)** |vegetable| ~: Speisekürbis, *der;* **b)** *(Anat.)* |Knochen|mark, *das*

marry ['mærɪ] **1.** *v. t.* **a)** heiraten; **b)** *(join)* trauen; **they were** *or* **got/have got married** sie haben geheiratet. **2.** *v. i.* heiraten

Mars [mɑːz] *pr. n. (Astron.)* Mars, *der*

marsh [mɑːʃ] *n.* Sumpf, *der*

marshal ['mɑːʃl] **1.** *n.* **a)** *(officer in army)* Marschall, *der;* **b)** *(Sport)* Ordner, *der.* **2.** *v. t., (Brit.)* **-ll-** aufstellen ⟨*Truppen*⟩; ordnen ⟨*Fakten*⟩. '**marshalling yard** *n.* Rangierbahnhof, *der*

marshmallow [mɑːʃˈmæləʊ] *n. (sweet)* ≈ Mohrenkopf, *der*

'**marshy** *adj.* sumpfig

marsupial [mɑːˈsjuːpɪəl] *n.* Beuteltier, *das*

martial ['mɑːʃl] *adj.* kriegerisch. **martial 'law** *n.* Kriegsrecht, *das*

martyr ['mɑːtə(r)] **1.** *n.* Märtyrer, *der*/Märtyrerin, *die.* **2.** *v. t.* **be ~ed** den Märtyrertod sterben

marvel ['mɑːvl] **1.** *n.* Wunder, *das.* **2.**

v. i., (Brit.) **-ll-** *(literary)* ~ **at** sth. über etw. *(Akk.)* staunen. **marvellous** ['mɑːvələs] *adj.,* '**marvellously** *adv.* wunderbar

marvelous[ly] *(Amer.) see* **marvellous[ly]**

Marxism ['mɑːksɪzm] *n.* Marxismus, *der.* **Marxist** ['mɑːksɪst] **1.** *n.* Marxist, *der*/Marxistin, *die.* **2.** *adj.* marxistisch

marzipan ['mɑːzɪpæn] *n.* Marzipan, *das*

mascara [mæˈskɑːrə] *n.* Mascara, *das*

mascot ['mæskɒt] *n.* Maskottchen, *das*

masculine ['mæskjʊlɪn] *adj.* männlich. **masculinity** [mæskjuˈlɪnɪtɪ] *n.* Männlichkeit, *die*

mash [mæʃ] **1.** *n.* **a)** Brei, *der;* **b)** *(Brit. coll.:* ~*ed potatoes)* Kartoffelbrei, *der.* **2.** *v. t.* zerdrücken; **~ed potatoes** Kartoffelbrei, *der*

mask [mɑːsk] **1.** *n.* Maske, *die.* **2.** *v. t.* maskieren

masochism ['mæsəkɪzm] *n.* Masochismus, *der.* **masochist** ['mæsəkɪst] *n.* Masochist, *der*/Masochistin, *die.* **masochistic** [mæsəˈkɪstɪk] *adj.* masochistisch

mason ['meɪsn] *n.* **a)** Steinmetz, *der;* **b)** **M~** *(Free~)* |Frei|maurer, *der.* **Masonic** [məˈsɒnɪk] *adj.* |frei|maurerisch; ~ **lodge** |Frei|maurerloge, *die.* **masonry** ['meɪsnrɪ] *n.* Mauerwerk, *das*

masquerade [mæskəˈreɪd, mɑːskəˈreɪd] **1.** *n.* Maskerade, *die.* **2.** *v. i.* ~ **as** sb./sth. sich als jmd./etw. ausgeben

¹**mass** [mæs] *n. (Eccl.)* Messe, *die*

²**mass** [mæs] **1.** *n.* **a)** Masse, *die;* **b)** **a** ~ **of** ...: eine Unmenge von ... **2.** *v. t.* anhäufen. **3.** *v. i.* sich ansammeln; ⟨*Truppen:*⟩ sich massieren

massacre ['mæsəkə(r)] **1.** *n.* Massaker, *das.* **2.** *v. t.* massakrieren

massage ['mæsɑːʒ] **1.** *n.* Massage, *die.* **2.** *v. t.* massieren

massive ['mæsɪv] *adj.* massiv; gewaltig ⟨*Aufgabe*⟩; enorm ⟨*Schulden*⟩

mass: ~ '**media** *n. pl.* Massenmedien *Pl.;* ~-**pro'duced** *adj.* serienmäßig produziert; ~ **pro'duction** *n.* Massenproduktion, *die*

mast [mɑːst] *n.* Mast, *der*

master ['mɑːstə(r)] **1.** *n.* **a)** Herr, *der;* **b)** *(of dog)* Herrchen, *das; (of ship)* Kapitän, *der;* **c)** *(Sch.: teacher)* Lehrer, *der;* **d)** *(expert, great artist)* Meister, *der* (**at in** + *Dat.*); **e)** **M~ of**

Arts/Science Magister Artium/rerum naturalium. **2.** *adj.* Haupt-. **3.** *v. t. (learn)* erlernen; **have ~ed a language** eine Sprache beherrschen. **masterful** ['mɑːstəfl] *adj. (masterly)* meisterhaft

'**master-key** *n.* Hauptschlüssel, *der*
masterly ['mɑːstəlı] *adj.* meisterhaft
master: ~mind 1. *n.* führender Kopf; **2.** *v. t.* **~mind the plot** der Kopf des Komplotts sein; **~piece** *n. (work of art)* Meisterwerk, *das;* **~-stroke** *n.* Meisterstück, *das; ~* **switch** *n.* Hauptschalter, *der*
mastery ['mɑːstərı] *n.* **a)** *(skill)* Meisterschaft, *die;* **b)** *(knowledge)* Beherrschung, *die* (**of** *Gen.*)
masturbate ['mæstəbeıt] *v. i. & t.* masturbieren. **masturbation** [mæstə'beıʃn] *n.* Masturbation, *die*
mat [mæt] *n.* **a)** Matte, *die;* **b)** *(to protect table etc.)* Untersetzer, *der*
¹**match** [mætʃ] **1.** *n.* **a) be no ~ for sb.** sich mit jmdm. nicht messen können; **meet one's ~:** seinen Meister finden; **b) be a [good** *etc.*] **~ for sth.** [gut *usw.*] zu etw. passen; **c)** *(Sport)* Spiel, *das; (Boxing)* Kampf, *der.* **2.** *v. t.* **a)** *(equal)* **~ sb. at chess** es mit jmdm. im Schach aufnehmen [können]; **b)** *(harmonize with)* passen zu; **a handbag and ~ing shoes** eine Handtasche und [dazu] passende Schuhe; **~ each other** zueinander passen. **3.** *v. i.* zusammenpassen
²**match** *n. (~stick)* Streichholz, *das*
'**matchless** *adj.* unvergleichlich
'**matchstick** *n.* Streichholz, *das*
¹**mate** [meıt] *n.* **a)** Kumpel, *der (ugs.);* **b)** *(Naut.)* ≈ Kapitänleutnant, *der;* **c)** *(workman's assistant)* Gehilfe, *der;* **d)** *(Zool.) (male)* Männchen, *das; (female)* Weibchen, *das.* **2.** *v. i.* sich paaren. **3.** *v. t.* paaren ⟨*Tiere*⟩
²**mate** *(Chess) see* **checkmate**
material [mə'tıərıəl] **1.** *adj.* **a)** materiell; **b)** *(relevant)* wesentlich. **2.** *n.* **a)** **~[s]** Material, *das;* **building/writing ~s** Bau-/Schreibmaterial, *das;* **b)** *(cloth)* Stoff, *der.* **materialism** [mə'tıərıəlızm] *n.* Materialismus, *der.* **materialistic** [mətıərıə'lıstık] *adj.* materialistisch. **materialize** [mə'tıərıəlaız] *v. i.* ⟨*Plan, Idee:*⟩ sich verwirklichen; ⟨*Treffen:*⟩ zustande kommen
maternal [mə'tɜːnl] *adj.* mütterlich; Mutter⟨*instinkt*⟩
maternity [mə'tɜːnıtı] *n.* Mutterschaft, *die.* **ma'ternity dress** *n.*

Umstandskleid, *das.* **ma'ternity hospital** *n.* Entbindungsheim, *das*
matey ['meıtı] *adj.,* **matier** ['meıtıə(r)], **matiest** ['meıtııst] *(Brit. coll.)* kameradschaftlich
math [mæθ] *(Amer. coll.) see* **maths**
mathematical [mæθı'mætıkl] *adj.,* **mathematically** [mæθı'mætıkəlı] *adv.* mathematisch
mathematician [mæθımə'tıʃn] *n.* Mathematiker, *der*/Mathematikerin, *die*
mathematics [mæθı'mætıks] *n.* Mathematik, *die*
maths [mæθs] *n. (Brit. coll.)* Mathe, *die (Schülerspr.)*
matinée ['mætıneı] *n.* Nachmittagsvorstellung, *die*
matrices *pl. of* **matrix**
matriculate [mə'trıkjʊleıt] **1.** *v. t.* immatrikulieren (**in an** + *Dat.*). **2.** *v. i.* sich immatrikulieren. **matriculation** [mətrıkjʊ'leıʃn] *n.* Immatrikulation, *die*
matrimonial [mætrı'məʊnıəl] *adj.* Ehe-
matrimony ['mætrımənı] *n.* Ehestand, *der*
matrix ['meıtrıks] *n., pl.* **matrices** ['meıtrısiːz] *or* **~es** Matrix, *die*
matron ['meıtrən] *n. (in school)* ≈ Hausmutter, *die; (in hospital)* Oberschwester, *die*
matt [mæt] *adj.* matt
'**matted** *adj.* verfilzt
matter ['mætə(r)] **1.** *n.* **a)** *(affair)* Angelegenheit, *die;* **~s** die Dinge; **money ~s** Geldangelegenheiten; **b) it's a ~ of taste** das ist Geschmackssache; [**only**] **a ~ of time** [nur noch] eine Frage der Zeit; **c) what's the ~?** was ist [los]?; **d)** *(physical material)* Materie, *die.* **2.** *v. i.* etwas ausmachen; **what does it ~?** was macht das schon?; [**it**] **doesn't ~:** [das] macht nichts *(ugs.).* '**matter -of-fact** *adj.* sachlich
mattress ['mætrıs] *n.* Matratze, *die*
mature [mə'tjʊə(r)] **1.** *adj.* reif; ausgereift ⟨*Stil, Käse, Portwein, Sherry*⟩. **2.** *v. t.* reifen lassen. **3.** *v. i.* reifen. **maturity** [mə'tjʊərıtı] *n.* Reife, *die*
Maundy Thursday [mɔːndı 'θɜːzdı] *n.* Gründonnerstag, *der*
mausoleum [mɔːsə'liːəm] *n.* Mausoleum, *das*
mauve [məʊv] *adj.* mauve
mawkish ['mɔːkıʃ] *adj.* rührselig
max. *abbr.* **maximum** *(adj.)* max., *(n.)* Max.

maxim ['mæksɪm] *n.* Maxime, *die*
maximum ['mæksɪməm] **1.** *n., pl.*
maxima ['mæksɪmə] Maximum, *das.*
2. *adj.* maximal; Maximal-; ~ **speed/
temperature** Höchstgeschwindigkeit,
die/-temperatur, *die*
May [meɪ] *n.* Mai, *der; see also* **August**
may *v. aux., only in pres.* **may,** *neg.
(coll.)* **mayn't** [meɪnt], *past* **might**
[maɪt], *neg. (coll.)* **mightn't** ['maɪtnt] **a)**
expr. possibility können; **it ~ be true**
das kann stimmen; **I ~ be wrong** vielleicht irre ich mich; **it ~ not be
possible** das wird vielleicht nicht möglich sein; **he ~ have missed his train**
vielleicht hat er seinen Zug verpaßt; **it
~** *or* **might rain** es könnte regnen; **we
~** *or* **might as well go** wir könnten eigentlich ebensogut [auch] gehen; **b)**
expr. permission dürfen; **c)** *expr. wish*
mögen; **~ the best man win!** auf daß
der Beste gewinnt!
maybe ['meɪbi:, 'meɪbɪ] *adv.* vielleicht
mayn't [meɪnt] *(coll.)* **= may not;** *see*
may
mayonnaise [meɪə'neɪz] *n.* Mayonnaise, *die*
mayor [meə(r)] *n.* Bürgermeister, *der*
mayoress ['meərɪs] *n. (woman mayor)*
Bürgermeisterin, *die; (mayor's wife)*
[Ehe]frau des Bürgermeisters
maze [meɪz] *n.* Labyrinth, *das*
me [mɪ, *stressed* mi:] *pron.* mich; *as indirect object* mir; **who, me?** wer, ich?;
not me ich nicht; **it's me** ich bin's
meadow ['medəʊ] *n.* Wiese, *die*
meagre ['miːgə(r)] *adj.* dürftig
meal [miːl] *n.* Mahlzeit, *die; go out for
a ~:** essen gehen. **'mealtime** *n.* Essenszeit, *die*
¹mean [miːn] *n.* **a)** Mittelweg, *der;* **b)**
(Math.) Mittelwert, *der*
²mean *adj.* **a)** *(miserly)* geizig; **b)** *(unkind)* gemein; **c)** *(shabby)* schäbig
³mean *v. t.,* **meant** [ment] **a)** *(intend)*
beabsichtigen; **~ to do sth.** etw. tun
wollen; **b)** *(design, destine)* **be ~t to do
sth.** etw. tun sollen; **c)** *(intend to convey)* meinen; **I [really] ~** it ich meine
das ernst; **what do you ~ by that?** was
hast du damit gemeint?; **d)** *(signify)*
bedeuten
meander [mɪ'ændə(r)] *v. i.* **a)** ⟨*Fluß:*⟩
sich winden; **b)** ⟨*Person:*⟩ schlendern
'meaning *n.* Bedeutung, *die; (of text
etc., life)* Sinn, *der.* **meaningful**
['miːnɪŋfl] *adj.* bedeutungsvoll ⟨*Blick,
Ergebnis*⟩; sinnvoll ⟨*Aufgabe, Gespräch*⟩. **'meaningless** *adj.* ⟨*Wort,*

Gespräch:⟩ ohne Sinn; sinnlos ⟨*Aktivität*⟩
means [miːnz] *n.* **a)** Möglichkeit, *die;*
[Art und] Weise; **by this ~:** hierdurch;
~ of transport Transportmittel, *das;* **b)**
pl. (resources) Mittel *Pl.;* **live within/
beyond one's ~:** seinen Verhältnissen
entsprechend/über seine Verhältnisse
leben; **c) by all ~!** selbstverständlich!;
by no [manner of] ~: ganz und gar
nicht; **by ~ of** durch; mit [Hilfe von]
'means test *n.* Überprüfung der Be-
dürftigkeit
meant *see* **³mean**
mean: ~time *n.* **in the ~time** inzwischen; **~time, ~while** *advs.* inzwischen
measles ['miːzlz] *n.* Masern *Pl.*
measly ['miːzlɪ] *adj. (coll.)* pop[e]lig
(ugs.)
measurable ['meʒərəbl] *adj.* meßbar
measure ['meʒə(r)] **1.** *n.* **a)** Maß, *das;*
for good ~: sicherheitshalber; *(as an
extra)* zusätzlich; **made to ~:** maßgeschneidert; **b)** *(degree)* **in some/large
~:** in gewisser Hinsicht/ in hohem
Maße; **c)** *(for measuring)* Maß, *das;* **d)**
(step) Maßnahme, *die;* **take ~s** Maßnahmen treffen. **2.** *v. t.* messen
⟨*Größe, Menge usw.*⟩; ausmessen
⟨*Raum*⟩. **3.** *v. i.* messen. **measure 'up
to** *v. t.* entsprechen (+ *Dat.*)
'measured ['meʒəd] *adj.* gemessen
⟨*Schritt, Worte*⟩
'measurement *n.* **a)** Messung, *die;* **b)**
(dimension) Maß, *das*
meat [miːt] *n.* Fleisch, *das.* **'meaty**
adj. **a)** fleischig; **b)** *(fig.)* gehaltvoll
mechanic [mɪ'kænɪk] *n.* Mechaniker,
*der/*Mechanikerin, *die*
mechanical [mɪ'kænɪkl] *adj.,*
me'chanically *adv.* mechanisch.
mechanical 'pencil *n.* *(Amer.)*
Drehbleistift, *der*
me'chanics *n.* **a)** Mechanik, *die;* **b)**
pl. (mechanism) Mechanismus, *der*
mechanism ['mekənɪzm] *n.* Mechanismus, *der*
mechanization [mekənaɪ'zeɪʃn] *n.*
Mechanisierung, *die*
mechanize ['mekənaɪz] *v. t.* mechani-
sieren
medal ['medl] *n.* Medaille, *die; (decoration)* Orden, *der*
medallion [mɪ'dæljən] *n.* [große] Medaille
medallist ['medəlɪst] *n.* Medaillenge-
winner, *der/*-gewinnerin, *die*
meddle ['medl] *v. i.* **~ with sth.** sich

(Dat.) an etw. *(Dat.)* zu schaffen machen; ~ **in** sth. sich in etw. *(Akk.)* einmischen

media ['mi:dɪə] *see* **mass media; medium** 1

mediaeval *see* **medieval**

mediate ['mi:dɪeɪt] *v. i.* vermitteln. **mediator** ['mi:dɪeɪtə(r)] *n.* Vermittler, *der*/Vermittlerin, *die*

medical ['medɪkl] *adj.* medizinisch; ärztlich ⟨*Behandlung, Untersuchung*⟩

medical: ~ **certificate** *n.* Attest, *das;* ~ **school** *n.* medizinische Hochschule; ~ **student** *n.* Medizinstudent, *der*/-studentin *die*

medicated ['medɪkeɪtɪd] *adj.* medizinisch

medication [medɪ'keɪʃn] *n. (medicine)* Medikament, *das*

medicinal [mɪ'dɪsɪnl] *adj.* medizinisch

medicine ['medsən, 'medɪsɪn] *n.* **a)** *(science)* Medizin, *die;* **b)** *(preparation)* Medikament, *das*

medieval [medɪ'i:vl] *adj.* mittelalterlich

mediocre [mi:dɪ'əʊkə(r)] *adj.* mittelmäßig. **mediocrity** [mi:dɪ'ɒkrɪtɪ] *n.* Mittelmäßigkeit, *die*

meditate ['medɪteɪt] *v. i.* nachdenken, *(esp. Relig.)* meditieren (**lup|on** über + *Akk.*). **meditation** [medɪ'teɪʃn] *n.* **a)** *(act)* Nachdenken, *das;* **b)** *(Relig.)* Meditation, *die*

Mediterranean [medɪtə'reɪnɪən] *pr. n.* **the** ~: das Mittelmeer

medium ['mi:dɪəm] **1.** *n., pl.* **media** ['mi:dɪə] *or* ~**s a)** *(substance)* Medium, *das;* **b)** *(means)* Mittel, *das;* **by** or **through the** ~ **of** durch; **c)** *pl.* ~**s** *(Spiritualism)* Medium, *das;* **d)** *in pl.* **media** *(mass media)* Medien *Pl.* **2.** *adj.* mittler ...; medium *nur präd.* ⟨*Steak*⟩.

'**medium-size[d]** *adj.* mittelgroß

medley ['medlɪ] *n.* **a)** buntes Gemisch; **b)** *(Mus.)* Potpourri, *das*

meek [mi:k] *adj.* **a)** *(humble)* sanftmütig; **b)** *(submissive)* zu nachgiebig

meet [mi:t] **1.** *v. t.,* met [met] **a)** treffen; *(collect)* abholen; **b)** *(make the acquaintance of)* kennenlernen; **pleased to** ~ **you** [sehr] angenehm; **c)** *(experience)* stoßen auf (+ *Akk.*) ⟨*Widerstand, Problem*⟩; **d)** *(satisfy)* entsprechen (+ *Dat.*) ⟨*Wunsch, Bedürfnis, Kritik*⟩; **e)** *(pay)* decken ⟨*Kosten*⟩; bezahlen ⟨*Rechnung*⟩. **2.** *v. i.,* **met a)** *(by chance)* sich *(Dat.)* begegnen; *(by arrangement)* sich treffen; **we've met before** wir kennen uns bereits; **b)** ⟨*Komi-*

tee, Ausschuß usw.:⟩ tagen. **meet 'up** *v. i.* sich treffen; ~ **up with sb.** *(coll.)* sich treffen. '**meet with** *v. t.* **a)** begegnen (+ *Dat.*); **b)** *(experience)* haben ⟨*Erfolg, Unfall*⟩; stoßen auf (+ *Akk.*) ⟨*Widerstand*⟩

'**meeting** *n.* **a)** Begegnung, *die; (by arrangement)* Treffen, *das;* **b)** *(assembly)* Versammlung, *die; (of committee etc.)* Sitzung, *die*

megalomania [megələ'meɪnɪə] *n.* Größenwahn, *der*

megaphone ['megəfəʊn] *n.* Megaphon, *das*

melancholic [melən'kɒlɪk] *adj.* melancholisch

melancholy ['melənkəlɪ] **1.** *n.* Melancholie, *die.* **2.** *adj.* melancholisch

mellow ['meləʊ] **1.** *adj.* **a)** *(softened by age or experience)* abgeklärt; **b)** *(ripe, well-matured)* reif. **2.** *v. i.* reifen

melodic [mɪ'lɒdɪk], **melodious** [mɪ'ləʊdɪəs] *adjs.,* **me'lodiously** *adv.* melodisch

melodrama ['melədrɑ:mə] *n.* Melodrama, *das.* **melodramatic** [melədrə'mætɪk] *adj.* melodramatisch

melody ['melədɪ] *n.* Melodie, *die*

melon ['melən] *n.* Melone, *die*

melt [melt] **1.** *v. i.* schmelzen. **2.** *v. t.* schmelzen; zerlassen ⟨*Butter*⟩. **melt a'way** *v. i.* [weg]schmelzen. **melt 'down 1.** *v. i.* schmelzen. **2.** *v. t.* einschmelzen

melting: ~-**point** *n.* Schmelzpunkt, *der;* ~-**pot** *n. (fig.)* Schmelztiegel, *der*

member ['membə(r)] *n.* **a)** Mitglied, *das;* **be a** ~: Mitglied sein; ~ **of a/the family** Familienangehörige, *der*/*die;* **b)** **M~ lof Parliament|** *(Brit.)* Abgeordnete [des Unterhauses], *der/die.* '**membership** *n.* **a)** Mitgliedschaft, *die* (of in + *Dat.*); **b)** *(number of members)* Mitgliederzahl, *die;* **c)** *(members)* Mitglieder *Pl.*

membrane ['membreɪn] *n. (Biol.)* Membran, *die*

memento [mɪ'mentəʊ] *n., pl.* ~**es** *or* ~**s** Andenken, *das* (of an + *Akk.*)

memo ['meməʊ] *n., pl.* ~**s** *(coll.) see* **memorandum**

memoirs ['memwɑ:z] *n. pl.* Memoiren *Pl.*

memorable ['memərəbl] *adj.* denkwürdig ⟨*Ereignis, Tag*⟩; unvergeßlich ⟨*Film, Buch, Aufführung*⟩

memorandum [memə'rændəm] *n., pl.* **memoranda** [memə'rændə] *or* ~**s** Mitteilung, *die*

memorial [mɪˈmɔːrɪəl] **1.** *adj.* Gedenk-. **2.** *n.* Denkmal, *das* (**to** für)
memorize [ˈmeməraɪz] *v. t.* sich *(Dat.)* merken *od.* einprägen; *(learn by heart)* auswendig lernen
memory [ˈmemərɪ] *n.* **a)** Gedächtnis, *das;* **b)** *(thing remembered, act of remembering)* Erinnerung, *die* (**of** an + *Akk.*); **from** ~: aus dem Gedächtnis; **in** ~ **of** zur Erinnerung an (+ *Akk.*); **c)** *(Computing)* Speicher, *der*
men *pl. of* **man**
menace [ˈmenɪs] **1.** *v. t.* bedrohen. **2.** *n.* Plage, *die.* **'menacing** [ˈmenəsɪŋ] *adj.* drohend
mend [mend] **1.** *v. t.* reparieren; ausbessern ⟨*Kleidung*⟩; kleben ⟨*Glas, Porzellan*⟩. **2.** *v. i.* ⟨*Knochen, Bein usw.*:⟩ heilen. **3.** *n.* **be on the** ~: auf dem Wege der Besserung sein
'menfolk *n. pl.* Männer
menial [ˈmiːnɪəl] *adj.* niedrig; untergeordnet ⟨*Aufgabe*⟩
meningitis [menɪnˈdʒaɪtɪs] *n.* Hirnhautentzündung, *die*
menopause [ˈmenəpɔːz] *n.* Wechseljahre *Pl.*
menstruate [ˈmenstrʊeɪt] *v. i.* menstruieren. **menstruation** [menstrʊˈeɪʃn] *n.* Menstruation, *die*
menswear [ˈmenzweə(r)] *n.* Herrenbekleidung, *die*
mental [ˈmentl] *adj.* **a)** *(of the mind)* geistig; Geistes⟨*zustand, -störung*⟩; **b)** *(Brit. coll.: mad)* verrückt *(salopp)*
mental: ~ **a'rithmetic** *n.* Kopfrechnen, *das;* ~ **hospital** *n.* Nervenklinik, *die (ugs.);* ~ **'illness** *n.* Geisteskrankheit, *die*
mentality [menˈtælɪtɪ] *n.* Mentalität, *die*
'mentally *adv.* geistig
mention [ˈmenʃn] **1.** *n.* Erwähnung, *die.* **2.** *v. t.* erwähnen (**to** gegenüber); **don't** ~ **it** keine Ursache
menu [ˈmenjuː] *n.* [Speise]karte, *die*
mercenary [ˈmɜːsɪnərɪ] **1.** *adj.* gewinnsüchtig. **2.** *n.* Söldner, *der*
merchandise [ˈmɜːtʃəndaɪz] *n.* [Handels]ware, *die*
merchant [ˈmɜːtʃənt] *n.* Kaufmann, *der.* **merchant 'bank** *n.* Handelsbank, *die.* **merchant 'navy** *n. (Brit.)* Handelsmarine, *die*
merciful [ˈmɜːsɪfl] *adj.* gnädig. **mercifully** [ˈmɜːsɪfəlɪ] *adv. (fortunately)* glücklicherweise
merciless [ˈmɜːsɪlɪs] *adj.,* **'mercilessly** *adv.* gnadenlos

mercury [ˈmɜːkjʊrɪ] **1.** *n.* Quecksilber, *das.* **2.** *pr. n.* **M~** *(Astron.)* Merkur, *der*
mercy [ˈmɜːsɪ] *n.* Erbarmen, *das* (**on** mit); **show sb. [no]** ~: mit jmdm. [kein] Erbarmen haben; **be at the** ~ **of sb./sth.** jmdm./einer Sache [auf Gedeih und Verderb] ausgeliefert sein
mere [mɪə(r)] *adj.,* **'merely** *adv.* bloß
merge [mɜːdʒ] **1.** *v. t.* **a)** *(combine)* zusammenschließen; **b)** *(blend gradually)* verschmelzen (**with** mit). **2.** *v. i.* **a)** *(combine)* fusionieren (**with** mit); **b)** ⟨*Straße:*⟩ zusammenlaufen (**with** mit).
merger [ˈmɜːdʒə(r)] *n.* Fusion, *die*
meringue [məˈræŋ] *n.* Meringe, *die;* Baiser, *das*
merit [ˈmerɪt] **1.** *n.* **a)** *(worth)* Verdienst, *das;* **b)** *(good feature)* Vorzug, *der.* **2.** *v. t.* verdienen
mermaid [ˈmɜːmeɪd] *n.* Nixe, *die*
merrily [ˈmerɪlɪ] *adv.* munter
merriment [ˈmerɪmənt] *n.* Fröhlichkeit, *die*
merry [ˈmerɪ] *adj.* fröhlich; ~ **Christmas!** frohe *od.* fröhliche Weihnachten! **'merry-go-round** *n.* Karussell, *das.* **'merry-making** *n.* Feiern, *das*
mesh [meʃ] *n.* **a)** Masche, *die;* **b)** *(netting; also fig.: network)* Geflecht, *das;* **wire** ~: Maschendraht, *der*
mesmerize [ˈmezməraɪz] *v. t.* faszinieren
mess [mes] *n.* **a)** *(dirty/untidy state)* **[be] a** ~ *or* **in a** ~: schmutzig/unaufgeräumt [sein]; **what a** ~! was für ein Dreck *(ugs.)*/Durcheinander!; **b)** *(bad state)* **be [in] a** ~: sich in einem schlimmen Zustand befinden; ⟨*Person:*⟩ schlimm dran sein; **get into a** ~: in Schwierigkeiten geraten; **make a** ~ **of** verpfuschen *(ugs.)* ⟨*Arbeit, Leben*⟩; **c)** *(Mil.)* Kasino, *das.* **mess a'bout, mess a'round 1.** *v. i. (potter)* herumwerken; *(fool about)* herumalbern. **2.** *v. t.* ~ **sb. about** *or* **around** mit jmdm. nach Belieben umspringen. **mess 'up** *v. t.* **a)** *(make dirty)* schmutzig machen; *(make untidy)* in Unordnung bringen; **b)** *(bungle)* ~ **it/things up** Mist bauen *(ugs.)*
message [ˈmesɪdʒ] *n.* Nachricht, *die;* **give sb. a** ~: jmdm. etwas ausrichten
messenger [ˈmesɪndʒə(r)] *n.* Bote, *der*/Botin, *die*
Messiah [mɪˈsaɪə] *n.* Messias, *der*
Messrs [ˈmesəz] *n. pl.* **a)** *(in name of firm)* ≈ Fa.; **b)** *pl. of* **Mr;** *(in list of names)* ~ **A and B** die Herren A und B

'**messy** *adj. (dirty)* schmutzig; *(untidy)*
unordentlich

met *see* **meet**

metabolism [mɪ'tæbəlɪzm] *n.* Stoff-
wechsel, *der*

metal ['metl] **1.** *n.* Metall, *das.* **2.** *adj.*
Metall-. **metallic** [mɪ'tælɪk] *adj.* metallisch; **have a ~ taste** nach Metall
schmecken. **metallurgy** [mɪ'tælədʒɪ]
n. Metallurgie, *die*

metamorphosis [metə'mɔːfəsɪs] *n.,*
pl. **metamorphoses** [metə'mɔːfəsiːz]
Metamorphose, *die*

metaphor ['metəfə(r)] *n.* Metapher,
die. **metaphorical** [metə'fɒrɪkl] *adj.,*
metaphorically [metə'fɒrɪkəlɪ] *adv.*
metaphorisch

meteor ['miːtɪə(r)] *n.* Meteor, *der.*
meteoric [miːtɪ'ɒrɪk] *adj. (fig.)* ko-
metenhaft

meteorological [miːtɪərə'lɒdʒɪkl]
adj. meteorologisch ⟨*Instrument*⟩;
Wetter⟨*ballon, -bericht*⟩

meteorologist [miːtɪə'rɒlədʒɪst] *n.*
Meteorologe, *der*/Meteorologin, *die*

meteorology [miːtɪə'rɒlədʒɪ] *n.* Meteorologie, *die*

¹**meter** ['miːtə(r)] *n.* a) Zähler, *der; (for*
coins) Münzzähler, *der;* b) *(parking-*
~) Parkuhr, *die*

²**meter** *(Amer.) see* ¹, ²**metre**

method ['meθəd] *n.* Methode, *die.*
methodical [mɪ'θɒdɪkl] *adj.,* **me-**
'**thodically** *adv.* systematisch

Methodist ['meθədɪst] *n.* Methodist,
der/Methodistin, *die*

meths [meθs] *n. (Brit. coll.)* [Brenn]-
spiritus, *der*

methylated spirit[s] [meθɪleɪtɪd
'spɪrɪt(s)] *n. [pl.]* Brennspiritus, *der*

meticulous [mɪ'tɪkjʊləs] *adj.,* **me-**
'**ticulously** *adv. (scrupulous[ly])* sorgfältig; *(over-scrupulous[ly])* übergenau

¹**metre** ['miːtə] *n. (Brit.: poetic rhythm)*
Metrum, *das*

²**metre** *n. (Brit.: unit)* Meter, *der od.*
das. **metric** ['metrɪk] *adj.* metrisch; ~
system metrisches System. **metrica-**
tion [metrɪ'keɪʃn] *n.* Umstellung auf
das metrische System

metronome ['metrənəʊm] *n.* Metro-
nom, *das*

metropolis [mɪ'trɒpəlɪs] *n.* Metropole, *die.* **metropolitan** [metrə'pɒlɪtən]
adj. ~ **New York** der Großraum New
York; ~ **London** Großlondon *(das)*

Mexican ['meksɪkən] **1.** *adj.* mexikanisch. **2.** *n.* Mexikaner, *der*/Mexika-
nerin, *die*

Mexico ['meksɪkəʊ] *pr. n.* Mexiko
(das)

miaow [mɪ'aʊ] **1.** *v. i.* miauen. **2.** *n.*
Miauen, *das*

mice *pl. of* **mouse**

microbe ['maɪkrəʊb] *n.* Mikrobe, *die*

micro ['maɪkrəʊ]: **~chip** *n.* Mikro-
chip, *der;* **~computer** *n.* Mikrocom-
puter, *der;* **~fiche** *n.* Mikrofiche, *das*
od. der; **~film** **1.** *n.* Mikrofilm, *der;* **2.**
v. t. auf Mikrofilm aufnehmen

microphone ['maɪkrəfəʊn] *n.* Mikro-
phon, *das*

microprocessor [maɪkrəʊ'prəʊse-
sə(r)] *n.* Mikroprozessor, *der*

microscope ['maɪkrəskəʊp] *n.* Mi-
kroskop, *das.* **microscopic** [maɪkrə-
'skɒpɪk] *adj.* mikroskopisch; *(fig.:*
very small) winzig

microwave *n.* Mikrowelle, *die;* ~
[oven] Mikrowellenherd, *der*

mid- [mɪd] *in comb.* **in ~-air** in der
Luft; **in ~-sentence** mitten im Satz;
~July Mitte Juli; **the ~-60s** die Mitte
der sechziger Jahre; **a man in his ~-**
fifties ein Mittfünfziger; **be in one's**
~-thirties Mitte Dreißig sein

midday ['mɪddeɪ, mɪd'deɪ] *n.* a) *(noon)*
zwölf Uhr; b) *(middle of day)* Mittag,
der; attrib. Mittags-

middle ['mɪdl] **1.** *attrib. adj.* mittler...
2. *n.* a) Mitte, *die;* **in the ~ of the**
forest/night mitten im Wald/in der
Nacht; b) *(waist)* Taille, *die*

middle: ~ '**age** *n.* mittleres [Lebens]alter; **~-aged** [mɪdleɪdʒd] *adj.*
mittleren Alters *nachgestellt;* **M~**
'**Ages** *n. pl.* **the M~ Ages** das Mittel-
alter; ~ '**class** *n.* Mittelstand, *der;*
~-class *adj.* bürgerlich; **M~ 'East**
pr. n. **the M~ East** der Nahe [und
Mittlere] Osten; **M~ 'Eastern** *adj.*
nahöstlich

middling ['mɪdlɪŋ] *adj.* mittelmäßig

midge [mɪdʒ] *n.* Stechmücke, *die*

midget ['mɪdʒɪt] **1.** *n.* Liliputaner,
der/Liliputanerin, *die.* **2.** *adj.* winzig

Midlands ['mɪdləndz] *n. pl.* **the ~**
(Brit.) Mittelengland

'**midnight** *n.* Mitternacht, *die*

'**midpoint** *n.* Mitte, *die*

midriff ['mɪdrɪf] *n.* Bauch, *der*

midst [mɪdst] *n.* **in the ~ of sth.** mitten
in einer Sache; **in our/their/your ~:** in
unserer/ihrer/eurer Mitte

midsummer ['---, -'--] *n.* die [Zeit der]
Sommersonnenwende

midway ['--, -'-] *adv.* auf halbem
Weg[e] ⟨*sich treffen, sich befinden*⟩

'**midwife** *n., pl.* '**midwives** Hebamme, *die*

mid'winter *n.* die [Zeit der] Wintersonnenwende

'**might** [maɪt] *see* may

²**might** *n.* a) *(force)* Gewalt, *die;* b) *(power)* Macht, *die*

mightn't ['maɪtnt] *(coll.)* = might not; *see* may

mighty ['maɪtɪ] 1. *adj.* mächtig. 2. *adv. (coll.)* verdammt *(ugs.)*

migraine ['mi:greɪn] *n.* Migräne, *die*

migrant ['maɪgrənt] *n.* a) Auswanderer, *der*/Auswanderin, *die;* b) *(bird)* Zugvogel, *der*

migrate [maɪ'greɪt] *v. i.* a) *(to a town)* abwandern; *(to another country)* auswandern; b) ⟨*Vogel:*⟩ fortziehen. **migration** [maɪ'greɪʃn] *n.* a) *(to a town)* Abwandern, *das; (to another country)* Auswandern, *das;* b) *(of birds)* Zug, *der*

mike [maɪk] *n. (coll.)* Mikro, *das*

Milan [mɪ'læn] *pr. n.* Mailand *(das)*

mild [maɪld] *adj.* mild; sanft ⟨*Person*⟩

mildew ['mɪldju:] *n.* a) Schimmel, *der;* b) *(on plant)* Mehltau, *der*

'**mildly** *adv.* a) *(gently)* mild[e]; b) *(slightly)* ein bißchen; c) to put it ~: gelinde gesagt

mile [maɪl] *n.* a) Meile, *die;* b) *(fig. coll.)* ~s better/too big tausendmal besser/viel zu groß; be ~s ahead of sb. jmdm. weit voraus sein. **mileage** ['maɪlɪdʒ] *n.* [Anzahl der] Meilen; a low ~: ein niedriger Meilenstand.

'**milestone** *n.* Meilenstein, *der*

militant ['mɪlɪtənt] 1. *adj.* militant. 2. *n.* Militante, *der/die*

military ['mɪlɪtərɪ] 1. *adj.* militärisch; Militär⟨*regierung, -akademie, -uniform, -parade*⟩; ~ service Militärdienst, *der.* 2. *n.* the ~: das Militär

militate ['mɪlɪteɪt] *v. i.* ~ against/in favour of sth. [deutlich] gegen/für etw. sprechen

militia [mɪ'lɪʃə] *n.* Miliz, *die*

milk [mɪlk] 1. *n.* Milch, *die.* 2. *v. t.* melken

milk: ~ '**chocolate** *n.* Milchschokolade, *die;* ~ **jug** *n.* Milchkännchen, *das;* ~**man** ['mɪlkmən] *n., pl.* ~**men** ['mɪlkmən] Milchmann, *der;* ~ **shake** *n.* Milchshake, *der;* ~-**tooth** *n.* Milchzahn, *der*

'**milky** *adj.* milchig. **Milky 'Way** *n.* Milchstraße, *die*

mill [mɪl] 1. *n.* a) Mühle, *die;* b) *(factory)* Fabrik, *die.* 2. *v. t.* a) mahlen ⟨Getreide⟩; b) fräsen ⟨*Metallgegenstand*⟩. **mill a'bout** *(Brit.),* **mill a'round** *v. i.* durcheinanderlaufen

'**miller** *n.* Müller, *der*

millet ['mɪlɪt] *n.* Hirse, *die*

milligram ['mɪlɪgræm] *n.* Milligramm, *das*

millilitre *(Brit.; Amer.:* **milliliter)** ['mɪlɪli:tə(r)] *n.* Milliliter, *der od. das*

millimetre *(Brit.; Amer.:* **millimeter)** ['mɪlɪmi:tə(r)] *n.* Millimeter, *der*

milliner ['mɪlɪnə(r)] *n.* Modist, *der*/Modistin, *die.* '**millinery** *n.* Hutmacherei, *die*

million ['mɪljən] 1. *adj.* a or one/two ~: eine Million/zwei Millionen; half a ~: eine halbe Million. 2. *n.* a) Million, *die;* b) *(indefinite amount)* ~s of people eine Unmenge Leute. **millionaire** [mɪljə'neə(r)] *n.* Millionär, *der*/Millionärin, *die.* **millionth** ['mɪljənθ] 1. *adj.* millionst... 2. *n. (fraction)* Millionstel, *das*

'**millstone** *n.* Mühlstein, *der*

mime [maɪm] 1. *n.* a) *(performance)* Pantomime, *die;* b) *(art)* Pantomimik, *die.* 2. *v. i.* pantomimisch agieren. 3. *v. t.* pantomimisch darstellen

mimic ['mɪmɪk] 1. *n.* Imitator, *der.* 2. *v. t.,* -**ck**- nachahmen

min. *abbr.* a) **minute[s]** Min.; b) **minimum** *(adj.)* mind., *(n.)* Min.

mince [mɪns] 1. *n.* Hackfleisch, *das.* 2. *v. t.* durch den [Fleisch]wolf drehen ⟨*Fleisch*⟩. '**mincemeat** *n.* a) Hackfleisch, *das;* b) *(sweet)* süße Pastetenfüllung aus Obst, Rosinen, Gewürzen, Nierenfett usw. **mince 'pie** *n.* mit „mincemeat b" gefüllte Pastete

'**mincer** *n.* Fleischwolf, *der*

mind [maɪnd] 1. *n.* a) Geist, *der;* b) *(remembrance)* bear or keep sth. in ~: an etw. *(Akk.)* denken; **have [got] sb./sth.** in ~: an jmdn./etw. denken; c) *(opinion)* give sb. a piece of one's ~: jmdm. gründlich die Meinung sagen; to my ~: meiner Meinung *od.* Ansicht nach; change one's ~: seine Meinung ändern; I have a good ~ to do that ich hätte große Lust, das zu tun; make up one's ~, make one's ~ up sich entscheiden; d) *([normal] mental powers)* Verstand, *der;* be out of one's ~: den Verstand verloren haben; e) frame of ~: [seelische] Verfassung. 2. *v. t.* a) I can't afford a bicycle, never ~ a car ich kann mir kein Fahrrad leisten, geschweige denn ein Auto; b) *usu. neg.*

or interrog. (object to) **would you ~ opening the door?** würdest du bitte die Tür öffnen?; **I wouldn't ~ a walk** ich hätte nichts gegen einen Spaziergang; **c)** *(take care)* **~ you don't go too near the cliff-edge!** paß auf, daß du nicht zu nah an den Klippenrand gehst!; **~ how you go!** paß auf! **d)** *(have charge of)* aufpassen auf (+ *Akk.*). **3.** *v.i.* **a)** **~!** Vorsicht!; Achtung!; **b)** *(care, object)* **do you ~ if I smoke?** stört es Sie, wenn ich rauche?; **c)** **never ~** *(it's not important)* macht nichts. **mind 'out** *v.i.* aufpassen **(for** auf + *Akk.*); **~ out!** Vorsicht!

'**minded** *adj.* **mechanically ~:** technisch veranlagt; **not politically ~:** unpolitisch

mindful ['maɪndfl] *adj.* **be ~ of sth.** etw. berücksichtigen

'**mindless** *adj.* geistlos ⟨*Person*⟩; sinnlos ⟨*Gewalt*⟩

¹**mine** [maɪn] *n.* **a)** Bergwerk, *das;* **b)** *(explosive)* Mine, *die*

²**mine** *poss. pron. pred.* meiner/meine/mein[e]s; *see also* **hers**

'**minefield** *n.* Minenfeld, *das*

'**miner** *n.* Bergmann, *der*

mineral ['mɪnərl] **1.** *adj.* mineralisch; Mineral⟨*salz, -quelle*⟩. **2.** *n.* **a)** Mineral, *das;* **b)** *(Brit.: soft drink)* Erfrischungsgetränk, *das.* '**mineral water** *n.* Mineralwasser, *das*

minesweeper ['maɪnswiːpə(r)] *n.* Minensuchboot, *das*

mingle ['mɪŋgl] **1.** *v.t.* [ver]mischen. **2.** *v.i.* sich [ver]mischen **(with** mit)

mini ['mɪnɪ] *n. (coll.)* **a)** *(car)* **M~, (P)** Mini, *der;* **b)** *(skirt)* Mini, *der (ugs.)*

mini- ['mɪnɪ] *in comb.* Mini-; Klein-⟨*bus, -wagen, -taxi*⟩

miniature ['mɪnɪtʃə(r)] **1.** *n. (picture)* Miniatur, *die.* **2.** *adj.* Miniatur-

mini-: **~bus** *n.* Kleinbus, *der;* **~cab** *n.* Minicar, *das*

minim ['mɪnɪm] *n. (Brit. Mus.)* halbe Note

minimal ['mɪnɪml] *adj.* minimal

minimize ['mɪnɪmaɪz] *v.t.* **a)** *(reduce)* auf ein Mindestmaß reduzieren; **b)** *(understate)* bagatellisieren

minimum ['mɪnɪməm] **1.** *n., pl.* **minima** ['mɪnɪmə] Minimum, *das* **(of** an + *Dat.*). **2.** *attrib. adj.* Mindest-

mining ['maɪnɪŋ] *n.* Bergbau, *der; attrib.* Bergbau-. '**mining industry** *n.* Bergbau, *der.* '**mining town** *n.* Bergbaustadt, *die*

minion ['mɪnjən] *n.* Lakai, *der*

minister ['mɪnɪstə(r)] **1.** *n.* **a)** *(Polit.)* Minister, *der*/Ministerin, *die;* **b)** *(Eccl.)* Geistliche, *der/die;* Pfarrer, *der*/Pfarrerin, *die.* **2.** *v.i.* **~ to sb.** sich um jmdn. kümmern. **ministerial** [mɪnɪ'stɪərɪəl] *adj. (Polit.)* Minister-; ministeriell. **ministry** ['mɪnɪstrɪ] *n.* **a)** *(Polit.)* Ministerium, *das;* **b)** *(Eccl.)* geistliches Amt

mink [mɪŋk] *n.* Nerz, *der*

minnow ['mɪnəʊ] *n.* Elritze, *die*

minor ['maɪnə(r)] **1.** *adj.* **a)** *(lesser)* kleiner...; **b)** *(unimportant)* weniger bedeutend; *(not serious)* leicht; **~ road** kleine Straße; *(Mus.)* Moll-; **A ~:** a-Moll. **2.** *n.* Minderjährige, *der/die.* **minority** [maɪ'nɒrɪtɪ, mɪ'nɒrɪtɪ] *n.* Minderheit, *die;* **in the ~:** in der Minderheit

minstrel ['mɪnstrl] *n.* fahrender Sänger

¹**mint** [mɪnt] **1.** *n. (place)* Münzanstalt, *die.* **2.** *adj.* funkelnagelneu *(ugs.);* **in ~ condition** in tadellosem Zustand. **3.** *v.t.* prägen

²**mint** *n.* **a)** *(plant)* Minze, *die;* **b)** *(peppermint)* Pfefferminz, *das; attrib.* Pfefferminz-

minuet [mɪnjʊ'et] *n.* Menuett, *das*

minus ['maɪnəs] *prep.* minus; weniger; *(without)* abzüglich (+ *Gen.*)

minuscule ['mɪnəskjuːl] *adj.* winzig

¹**minute** ['mɪnɪt] *n.* **a)** Minute, *die; (moment)* Moment, *der;* **b)** **~s** *(of meeting)* Protokoll, *das;* **take the ~s of a meeting** bei einer Sitzung [das] Protokoll führen

²**minute** [maɪ'njuːt] *adj. (tiny)* winzig

miracle ['mɪrəkl] *n.* Wunder, *das.* **miraculous** [mɪ'rækjʊləs] *adj.* wunderbar

mirage ['mɪrɑːʒ] *n.* Fata Morgana, *die*

mirror ['mɪrə(r)] **1.** *n.* Spiegel, *der.* **2.** *v.t.* [wider]spiegeln

misadventure [mɪsəd'ventʃə(r)] *n.* Mißgeschick, *das*

misapprehension [mɪsæprɪ'henʃn] *n.* Mißverständnis, *das;* **be under a ~:** einem Irrtum unterliegen

misbehave [mɪsbɪ'heɪv] *v.i. & refl.* sich schlecht benehmen. **misbehaviour** (*Amer.:* **misbehavior**) [mɪsbɪ'heɪvjə(r)] *n.* schlechtes Benehmen

miscalculate [mɪs'kælkjʊleɪt] **1.** *v.t.* falsch berechnen; *(misjudge)* falsch einschätzen. **2.** *v.i.* sich verrechnen. **miscalculation** [mɪskælkjʊ'leɪʃn] *n.* Rechenfehler, *der; (misjudgement)* Fehleinschätzung, *die*

miscarriage [mɪsˈkærɪdʒ] *n.* **a)** Fehlgeburt, *die;* **b)** ~ **of justice** Justizirrtum, *das*

miscellaneous [mɪsəˈleɪnɪəs] *adj.* **a)** [kunter]bunt; **b)** *with pl. n.* verschieden. **miscellany** [mɪˈseləni] *n.* [bunte] Sammlung; [buntes] Gemisch

mischief [ˈmɪstʃɪf] *n.* **a)** Unfug, *der;* **get up to** ~: etwas anstellen; **b)** *(harm)* Schaden, *der.* **mischievous** [ˈmɪstʃɪvəs] *adj.* spitzbübisch; schelmisch

misconception [mɪskənˈsepʃn] *n.* falsche Vorstellung (about von); **be [labouring] under a** ~ **about sth.** sich *(Dat.)* eine falsche Vorstellung von etw. machen

misconduct [mɪsˈkɒndʌkt] *n.* unkorrektes Verhalten

misconstrue [mɪskənˈstruː] *v. t.* mißverstehen

miscount [mɪsˈkaʊnt] **1.** *v. i.* sich verzählen. **2.** *v. t.* falsch zählen

misdeed [mɪsˈdiːd] *n.* Missetat, *die (veralt., scherzh.)*

misdemeanour *(Amer.:* **misdemeanor)** [mɪsdɪˈmiːnə(r)] *n.* Missetat, *die (veralt., scherzh.)*

misdirect [mɪsdɪˈrekt, mɪsdaɪˈrekt] *v. t.* falsch adressieren 〈*Brief*〉; in die falsche Richtung schicken 〈*Person*〉

miser [ˈmaɪzə(r)] *n.* Geizhals, *der*

miserable [ˈmɪzərəbl] *adj.* **a)** unglücklich; **feel** ~: sich elend fühlen; **b)** trist 〈*Wetter, Urlaub*〉. **miserably** [ˈmɪzərəblɪ] *adv.* unglücklich; jämmerlich 〈*versagen*〉; ~ **poor** bettelarm

miserly [ˈmaɪzəlɪ] *adj.* geizig

misery [ˈmɪzərɪ] *n.* **a)** Elend, *das;* **b)** *(coll.: discontented person)* ~[**-guts**] Miesepeter, *der (ugs.)*

misfire [mɪsˈfaɪə(r)] *v. i.* **a)** 〈*Motor:*〉 Fehlzündungen haben; **b)** 〈*Plan, Versuch:*〉 fehlschlagen; 〈*Streich, Witz:*〉 danebengehen

misfit [ˈmɪsfɪt] *n.* Außenseiter, *der/* Außenseiterin, *die*

misfortune [mɪsˈfɔːtʃuːn] *n.* Mißgeschick, *das*

misgiving [mɪsˈgɪvɪŋ] *n.* ~[**s**] Bedenken *Pl.*

misguided [mɪsˈgaɪdɪd] *adj.* töricht

mishandle [mɪsˈhændl] *v. t.* falsch behandeln

mishap [ˈmɪshæp] *n.* Mißgeschick, *das*

mishear [mɪsˈhɪə(r)] **1.** *v. i.,* **misheard** [mɪsˈhɜːd] sich verhören. **2.** *v. t.,* **misheard** falsch verstehen

mishit 1. [ˈmɪshɪt] *n.* Fehlschlag, *der.*

2. [mɪsˈhɪt] *v. t.,* -**tt-,** **mishit** verschlagen

mishmash [ˈmɪʃmæʃ] *n.* Mischmasch, *der (ugs.)* (of aus)

misinform [mɪsɪnˈfɔːm] *v. t.* falsch informieren

misinterpret [mɪsɪnˈtɜːprɪt] *v. t. (make wrong inference from)* falsch deuten; mißdeuten. **misinterpretation** [mɪsɪntɜːprɪˈteɪʃn] *n.* **be open to** ~: leicht mißdeutet werden können

misjudge [mɪsˈdʒʌdʒ] *v. t.* falsch einschätzen; falsch beurteilen 〈*Person*〉. **misjudgement,** **misjudgment** [mɪsˈdʒʌdʒmənt] *n.* Fehleinschätzung, *die;* *(of person)* falsche Beurteilung

mislay [mɪsˈleɪ] *v. t.,* **mislaid** [mɪsˈleɪd] verlegen

mislead [mɪsˈliːd] *v. t.,* **misled** [mɪsˈled] irreführen. **misˈleading** *adj.* irreführend

mismanage [mɪsˈmænɪdʒ] *v. t.* schlecht abwickeln 〈*Geschäft, Projekt*〉. **mismanagement** [mɪsˈmænɪdʒmənt] *n.* schlechte Abwicklung

misnomer [mɪsˈnəʊmə(r)] *n.* unzutreffende Bezeichnung

misplace [mɪsˈpleɪs] *v. t.* an den falschen Platz stellen/legen/setzen *usw.*

misprint 1. [ˈmɪsprɪnt] *n.* Druckfehler, *der.* **2.** [mɪsˈprɪnt] *v. t.* verdrucken

mispronounce [mɪsprəˈnaʊns] *v. t.* falsch aussprechen

misread [mɪsˈriːd] *v. t.,* **misread** [mɪsˈred] falsch lesen

misrepresent [mɪsreprɪˈzent] *v. t.* falsch darstellen. **misrepresentation** [mɪsreprɪzenˈteɪʃn] *n.* falsche Darstellung

Miss [mɪs] *n. (unmarried woman)* Frau; Fräulein *(veralt.);* *(girl)* Fräulein

miss 1. *n.* Fehlschlag, *der; (shot)* Fehlschuß, *der; (throw)* Fehlwurf, *der.* **2.** *v. t.* **a)** *(fail to hit)* verfehlen; **b)** *(let slip)* verpassen; ~ **an opportunity** sich *(Dat.)* eine Gelegenheit entgehen lassen; **c)** *(fail to catch)* verpassen 〈*Zug*〉; **d)** *(fail to take part in)* versäumen; ~ **school** in der Schule fehlen; **e)** *(fail to see)* übersehen; *(fail to hear)* nicht mitbekommen; **f)** *(feel the absence of)* vermissen; **she** ~**es him** er fehlt ihr. **3.** *v. i. (not hit sth.)* nicht treffen. **miss 'out 1.** *v. t.* weglassen. **2.** *v. i.* ~ **out on sth.** *(coll.)* sich *(Dat.)* etw. entgehen lassen

misshapen [mɪsˈʃeɪpn] *adj.* mißgebildet

missile ['mɪsaɪl] *n.* **a)** *(thrown)* [Wurf]geschoß, *das;* **b)** *(Mil.)* Rakete, *die.* '**missile base,** 'missile site *ns.* Raketenbasis, *die*

'**missing** *adj.* fehlend; **be ~:** fehlen; ⟨*Person:*⟩ *(Mil. etc.)* vermißt werden; *(not present)* fehlen; **~ person** Vermißte, *der/die*

mission ['mɪʃn] *n.* **a)** Mission, *die;* **b)** *(planned operation)* Einsatz, *der.* **missionary** ['mɪʃənərɪ] *n.* Missionar, *der/*Missionarin, *die*

misspell [mɪs'spel] *v. t., forms as* ¹**spell** falsch schreiben

mist [mɪst] *n.* *(fog)* Nebel, *der;* *(haze)* Dunst, *der;* *(on windscreen etc.)* Beschlag, *der.* **mist 'up** *v. i.* [sich] beschlagen

mistake [mɪ'steɪk] **1.** *n.* Fehler, *der;* **by ~:** versehentlich. **2.** *v. t., forms as* **take 1: a)** falsch verstehen; **b) ~ x for y** x mit y verwechseln. **mistaken** [mɪ'steɪkn] *adj.* **be ~:** sich täuschen; **a case of ~ identity** eine Verwechslung. **mi'stakenly** *adv.* irrtümlicherweise

mistletoe ['mɪsltəʊ] *n.* Mistel, *die*

mistook *see* **mistake 2**

mistress ['mɪstrɪs] *n.* **a)** *(Brit. Sch.: teacher)* Lehrerin, *die;* **b)** *(lover)* Geliebte, *die*

mistrust [mɪs'trʌst] **1.** *v. t.* mißtrauen (+ *Dat.*). **2.** *n.* Mißtrauen, *das* (of gegenüber + *Dat.*). **mistrustful** [mɪs'trʌstfl] *adj.* mißtrauisch (of gegenüber)

'**misty** *adj.* dunstig

misunderstand [mɪsʌndə'stænd] *v. t., forms as* **understand** mißverstehen. **misunder'standing** *n.* Mißverständnis, *das*

misuse 1. [mɪs'juːz] *v. t.* mißbrauchen. **2.** [mɪs'juːs] *n.* Mißbrauch, *der*

mite [maɪt] *n.* **a)** *(Zool.)* Milbe, *die;* **b)** *(small child)* Würmchen, *das (fam.);* **poor little ~:** armes Kleines

miter *(Amer.) see* **mitre**

mitigate ['mɪtɪgeɪt] *v. t.* **a)** *(reduce)* lindern; **b)** *(make less severe)* mildern; **mitigating circumstances** mildernde Umstände

mitre ['maɪtə(r)] *n.* *(Brit. Eccl.)* Mitra, *die*

mitten ['mɪtn] *n.* Fausthandschuh, *der*

mix [mɪks] **1.** *v. t.* [ver]mischen; verrühren ⟨*Zutaten*⟩. **2.** *v. i.* **a)** *(become ~ed)* sich vermischen; **b)** *(be sociable, participate)* Umgang mit anderen [Menschen] haben; **~ with** Umgang haben mit; **~ well** kontaktfreudig sein. **3.** *n.*

(coll.) Mischung, *die;* |**cake-**|**~:** Backmischung, *die.* **mix 'up** *v. t.* **a)** vermischen; **b)** *(muddle)* durcheinanderbringen; *(confuse)* verwechseln; **c) be/get ~ed up in sth.** in etw. *(Akk.)* verwickelt sein/werden

mixed [mɪkst] *adj.* **a)** gemischt; **b)** *(diverse)* unterschiedlich. **mixed 'grill** *n.* Mixed grill, *der.* **mixed 'up** *adj.* *(coll.)* verwirrt ⟨*Person*⟩; **be/feel very ~:** völlig durcheinander sein

'**mixer** *n.* *(for food)* Mixer, *der*

mixture ['mɪkstʃə(r)] *n.* **a)** Mischung, *die* (of aus); **b)** *(Med.)* Mixtur, *die*

'**mix-up** *n.* Durcheinander, *das;* *(misunderstanding)* Mißverständnis, *das*

mm. *abbr.* **millimetre[s]** mm

moan [məʊn] **1.** *n.* **a)** Stöhnen, *das;* **b)** **have a ~** *(complain)* jammern. **2.** *v. i.* **a)** stöhnen (with vor + *Dat.*); **b)** *(complain)* jammern (about über + *Akk.*). **3.** *v. t.* stöhnen

moat [məʊt] *n.* |**castle**| **~:** Burggraben, *der*

mob [mɒb] **1.** *n.* **a)** *(rabble)* Mob, *der;* **b)** *(sl.: group)* Peter **and his ~:** Peter und seine ganze Blase *(salopp).* **2.** *v. t., -bb-** belagern *(ugs.)* ⟨*Star*⟩

mobile ['məʊbaɪl] **1.** *adj.* beweglich; *(on wheels)* fahrbar. **2.** *n.* Mobile, *die.* **mobile 'home** *n.* transportable Wohneinheit. **mobile 'phone** *n.* Mobiltelefon, *das*

mobility [mə'bɪlɪtɪ] *n.* Beweglichkeit, *die*

mobilization [məʊbɪlaɪ'zeɪʃn] *n.* Mobilisierung, *die*

mobilize ['məʊbɪlaɪz] *v. t.* mobilisieren

moccasin ['mɒkəsɪn] *n.* Mokassin, *der*

mocha ['mɒkə] *n.* Mokka, *der*

mock [mɒk] **1.** *v. t.* sich lustig machen über (+ *Akk.*). **2.** *v. i.* sich lustig machen (at über + *Akk.*). **3.** *adj.* Schein-⟨*kampf, -angriff, -ehe*⟩. **mockery** ['mɒkərɪ] *n.* Spott, *der;* **make a ~ of sth.** etw. zur Farce machen

'**mock-up** *n.* Modell [in Originalgröße]

mode [məʊd] *n.* **a)** Art [und Weise], *die;* **b)** *(fashion)* Mode, *die*

model ['mɒdl] **1.** *n.* **a)** Modell, *das;* **b)** *(example to be imitated)* Vorbild, *das;* **c)** *(Art)* Modell, *das;* *(Fashion)* Mannequin, *das;* *(male)* Dressman, *der.* **2.** *adj.* **a)** *(exemplary)* Muster-; **b)** *(miniature)* Modell-. **3.** *v. t., (Brit.) -ll-:* **a)** modellieren; **~ sth. after** *or* |**up**|**on sth.** etw. einer Sache *(Dat.)* nachbilden; **b)**

(Fashion) vorführen. **4.** *v. i. (Fashion)* als Mannequin/Dressman arbeiten; *(Art)* Modell stehen/sitzen

modem ['məʊdem *n.* Modem, *der*

moderate 1. ['mɒdərət] *adj.* **a)** gemäßigt ⟨*Ansichten*⟩; maßvoll ⟨*Trinker, Forderungen*⟩; **b)** mittler... ⟨*Größe, Menge, Wert*⟩; *(reasonable)* angemessen ⟨*Preis, Summe*⟩. **2.** ['mɒdərət] *n.* Gemäßigte, *der/die.* **3.** ['mɒdəreɪt] *v. t.* mäßigen. **4.** ['mɒdəreɪt] *v.i.* nachlassen. **moderately** ['mɒdərətlɪ] *adv.* einigermaßen; mäßig ⟨*begeistert, groß, begabt*⟩. **moderation** [mɒdə'reɪʃn] *n.* Mäßigkeit, *die;* **in** ~: mit Maßen

modern ['mɒdn] *adj.* modern; heutig ⟨*Zeit[alter], Welt, Mensch*⟩; ~ **languages** neuere Sprachen. **modernize** ['mɒdənaɪz] *v. t.* modernisieren

modest ['mɒdɪst] *adj.* bescheiden; einfach ⟨*Haus, Kleidung*⟩. **modestly** *adv.* bescheiden. **modesty** *n.* Bescheidenheit, *die*

modification [mɒdɪfɪ'keɪʃn] *n.* [Ab]änderung, *die*

modify ['mɒdɪfaɪ] *v. t.* [ab]ändern

modulate ['mɒdjʊleɪt] *v. t. & i.* modulieren. **modulation** [mɒdjʊ'leɪʃn] *n.* Modulation, *die*

module ['mɒdju:l] *n.* **a)** Bauelement, *das;* **b)** *(Astronaut.)* **command** ~: Kommandoeinheit, *die*

mohair ['məʊheə(r)] *n.* Mohair, *der*

moist [mɔɪst] *adj.* feucht (**with** von). **moisten** ['mɔɪsn] *v.t.* anfeuchten. **moisture** ['mɔɪstʃə(r)] *n.* Feuchtigkeit, *die.* **moisturizer** ['mɔɪstʃəraɪzə(r)], **moisturizing cream** ['mɔɪstʃəraɪzɪŋ kri:m] *ns.* Feuchtigkeitscreme, *die*

molar ['məʊlə(r)] *n.* Backenzahn, *der*
molasses [mə'læsɪz] *n.* Melasse, *die*
mold *(Amer.)* see [1,2]**mould**
molder, molding, moldy *(Amer.)* see **mould-**
[1]**mole** [məʊl] *n. (on skin)* Leberfleck, *der*
[2]**mole** *n. (animal)* Maulwurf, *der*
molecular [mə'lekjʊlə(r)] *adj.* molekular
molecule ['mɒlɪkju:l] *n.* Molekül, *das*
molehill *n.* Maulwurfshügel, *der*
molest [mə'lest] *v. t.* belästigen
mollify ['mɒlɪfaɪ] *v.t.* besänftigen
mollusc, *(Amer.)* **mollusk** ['mɒləsk] *n.* Weichtier, *das*
mollycoddle ['mɒlɪkɒdl] *v. t.* [ver]hätscheln
molt *(Amer.)* see **moult**

molten ['məʊltn] *adj.* geschmolzen
mom [mɒm] *(Amer. coll.)* see [2]**mum**
moment ['məʊmənt] *n.* Augenblick, *der;* **at any** ~, *(coll.)* **any** ~: jeden Augenblick; **one** *or* **just a** *or* **wait a** ~! einen Augenblick!; **in a** ~ *(very soon)* sofort; **at the** ~: im Augenblick; **the** ~ **of truth** die Stunde der Wahrheit. **momentarily** ['məʊməntərɪlɪ] *adv.* einen Augenblick lang. **momentary** ['məʊməntərɪ] *adj.* kurz
momentous [mə'mentəs] *adj. (important)* bedeutsam; *(of consequence)* folgenschwer
momentum [mə'mentəm] *n.* Schwung, *der*
Mon. *abbr.* Monday Mo.
monarch ['mɒnək] *n.* Monarch, *der*/Monarchin, *die.* **monarchy** *n.* Monarchie, *die*
monastery ['mɒnəstrɪ] *n.* Kloster, *das.* **monastic** [mə'næstɪk] *adj.* mönchisch
Monday ['mʌndeɪ, 'mʌndɪ] *n.* Montag, *der; see also* **Friday**
monetary ['mʌnɪtərɪ] *adj.* **a)** *(of currency)* monetär; Währungs⟨*politik, -system*⟩; **b)** *(of money)* finanziell
money ['mʌnɪ] *n.* Geld, *das;* **make** ~ ⟨*Person:*⟩ [viel] Geld verdienen; ⟨*Geschäft:*⟩ etwas einbringen; **for 'my** ~: wenn man mich fragt
money: ~**bag** *n.* Geldsack, *der;* ~**box** *n.* Sparbüchse, *die;* ~**making** *adj.* gewinnbringend; ~ **order** *n.* Postanweisung, *die*
Mongolia [mɒŋ'gəʊlɪə] *pr. n.* Mongolei, *die.* **Mongolian** [mɒŋ'gəʊlɪən] **1.** *adj.* mongolisch. **2.** *n. (person)* Mongole, *der*/Mongolin, *die*
mongrel ['mʌŋgrəl] *n.* ~ [**dog**] Promenadenmischung, *die*
monitor ['mɒnɪtə(r)] **1.** *n.* **a)** *(Sch.)* Aufsichtsschüler, *der*/-schülerin, *die;* **b)** *(Med., Telev., etc.)* Monitor, *der.* **2.** *v. t.* beobachten ⟨*Wetter, Flugzeug*⟩; abhören ⟨*Sendung, Telefongespräch*⟩
monk [mʌŋk] *n.* Mönch, *der*
monkey ['mʌŋkɪ] *n.* Affe, *der*
monkey: ~ **business** *n. (coll.: mischief)* Schabernack, *der;* ~**nut** *n.* Erdnuß, *die;* ~**wrench** *n.* Universalschraubenschlüssel, *der*
mono ['mɒnəʊ] *adj.* Mono⟨*platte[nspieler], -wiedergabe*⟩
monocle ['mɒnəkl] *n.* Monokel, *das*
monologue *(Amer.:* **monolog)** ['mɒnəlɒg] *n.* Monolog, *der*
monopolize [mə'nɒpəlaɪz] *v. t.*

(Econ.) monopolisieren; *(fig.)* mit Beschlag belegen; ~ **the conversation** den/die anderen nicht zu Wort kommen lassen

monopoly [mə'nɒpəlɪ] *n.* **a)** *(Econ.)* Monopol, *das (of auf + Dat.);* **b)** *(exclusive possession)* alleiniger Besitz

monotone ['mɒnətəʊn] *n.* gleichbleibender Ton. **monotonous** [mə'nɒtənəs] *adj.,* **mo'notonously** *adv.* eintönig. **monotony** [mə'nɒtənɪ] *n.* Eintönigkeit, *die*

monsoon [mɒn'suːn] *n.* Monsun, *der*

monster ['mɒnstə(r)] *n.* **a)** *(creature)* Ungeheuer, *das;* **b)** *(huge thing)* Ungetüm, *das;* **b)** *(inhuman person)* Unmensch, *der.* **monstrosity** [mɒn'strɒsɪtɪ] *n.* **a)** *(outrageous thing)* Ungeheuerlichkeit, *die;* **b)** *(hideous building etc.)* Ungetüm, *das.* **monstrous** ['mɒnstrəs] *adj.* **a)** *(huge)* riesig; **b)** *(outrageous)* ungeheuerlich; **c)** *(atrocious)* scheußlich

month [mʌnθ] *n.* Monat, *der;* **for a** ~/~s einen Monat [lang]/monatelang. **'monthly 1.** *adj.* monatlich; Monats⟨einkommen, -gehalt⟩. **2.** *adv.* einmal im Monat. **3.** *n.* Monatsschrift, *die*

monument ['mɒnjʊmənt] *n.* Denkmal, *das.* **monumental** [mɒnjʊ'mentl] *adj.* **a)** *(massive)* monumental; **b)** gewaltig ⟨Mißerfolg, Irrtum⟩

moo [muː] **1.** *n.* Muhen, *das.* **2.** *v. i.* muhen

mooch [muːtʃ] *v. i. (sl.)* ~ **about** *or* **around/along** herumschleichen *(ugs.)/* zockeln *(ugs.)*

mood [muːd] *n.* **a)** Stimmung, *die;* **be in a good/bad** ~; **[bei]** guter/schlechter Laune sein; **I'm not in the** ~: ich hab' keine Lust dazu; **b)** *(bad* ~*)* Verstimmung, *die.* **'moody** *adj.* **a)** *(sullen)* mißmutig; **b)** *(subject to moods)* launenhaft

moon [muːn] *n.* Mond, *der*

moon-: ~**beam** *n.* Mondstrahl, *der;* ~**light 1.** *n.* Mondlicht, *das;* Mondschein, *der;* **2.** *v. i. (coll.)* nebenberuflich abends arbeiten; ~**lit** *adj.* mondbeschienen *(geh.)*

¹**moor** [mʊə(r), mɔː(r)] *n. (Geog.)* [Hoch]moor, *das*

²**moor** *v. t. & i.* festmachen; vertäuen. **'mooring** *n.* ~**[s]** Anlegestelle, *die*

moose [muːs] *n., pl. same* Amerikanischer Elch

moot [muːt] **1.** *adj.* umstritten; offen ⟨Frage⟩; strittig ⟨Punkt⟩. **2.** *v. t.* erörtern ⟨Frage, Punkt⟩

mop [mɒp] **1.** *n.* **a)** Mop, *der;* **b)** ~ **[of hair]** Wuschelkopf, *der.* **2.** *v. t.,* -pp- moppen ⟨Fußboden⟩; *(wipe)* abwischen ⟨Träne, Schweiß, Stirn⟩. **mop 'up** *v. t.* aufwischen

mope [məʊp] *v. i.* Trübsal blasen

moped ['məʊped] *n.* Moped, *das*

moral ['mɒrl] **1.** *adj.* **a)** moralisch; sittlich ⟨Wert⟩; Moral⟨begriff, -prinzip⟩; **b)** *(virtuous)* moralisch ⟨Leben, Person⟩. **2.** *n.* **a)** Moral, *die;* **b)** *in pl. (habits)* Moral, *die*

morale [mə'rɑːl] Moral, *die;* **low/high** ~: schlechte/gute Moral

morality [mə'rælɪtɪ] *n.* Moral, *die*

morbid ['mɔːbɪd] *adj.* krankhaft; morbid *(geh.)* ⟨Faszination, Neigung⟩

more [mɔː(r)] **1.** *adj.* mehr; **any** *or* **some** ~ ⟨apples, books, etc.⟩ noch welche; **any** *or* **some** ~ *(tea, paper, etc.)* noch etwas; **any** *or* **some** ~ **apples/tea** noch Äpfel/Tee; **I haven't any** ~ **[apples/tea]** ich habe keine [Apfel]/keinen [Tee] mehr; ~ **and** ~: immer mehr. **2.** *n.* mehr; ~ **and** ~: immer mehr; **six or** ~: mindestens sechs. **3.** *adv.* **a)** mehr; ~ **interesting** interessanter; **b)** *(nearer, rather)* eher; **c)** *(again)* wieder; **no** ~, **not any** ~: nicht mehr; **once** ~: noch einmal; **d)** ~ **and** ~: immer mehr; ~ **and** ~ **absurd** immer absurder; **e)** ~ **or less** *(fairly)* mehr oder weniger; *(approximately)* annähernd. **more'over** *adv.* und außerdem

morgue [mɔːg] *see* **mortuary**

morning ['mɔːnɪŋ] *n.* Morgen, *der;* *(not afternoon)* Vormittag, *der; attrib.* morgendlich; Morgen-; **this** ~: heute morgen; **tomorrow** ~, *(coll.)* **in the** ~: morgen früh; **[early] in the** ~: am [frühen] Morgen; *(regularly)* [früh] morgens

Moroccan [mə'rɒkən] **1.** *adj.* marokkanisch. **2.** *n.* Marokkaner, *der/*Marokkanerin, *die*

Morocco [mə'rɒkəʊ] *pr. n.* Marokko *(das)*

moron ['mɔːrɒn] *n. (coll.)* Schwachkopf, *der (ugs.)*

Morse [code] [mɔːs ('kəʊd)] *n.* Morsealphabet, *das*

mortal ['mɔːtl] **1.** *adj.* **a)** sterblich; **b)** *(fatal)* tödlich (to für). **2.** *n.* Sterbliche, *der/die.* **mortality** [mɔː'tælɪtɪ] *n.* **a)** Sterblichkeit, *die;* **b)** ~ **[rate]** Sterblichkeitsrate, *die.* **'mortally** *adv.* tödlich

mortar ['mɔːtə(r)] *n.* Mörtel, *der*

mortgage ['mɔːgɪdʒ] **1.** *n.* Hypothek,

die. 2. *v. t.* mit einer Hypothek belasten

mortuary ['mɔ:tjʊərɪ] *n.* Leichenschauhaus, *das*

mosaic [məʊ'zeɪɪk] *n.* Mosaik, *das*

Moscow ['mɒskəʊ] *pr. n.* Moskau *(das)*

Moselle [məʊ'zel] *pr. n.* Mosel, *die*

Moslem ['mɒzləm] *see* **Muslim**

mosque [mɒsk] *n.* Moschee, *die*

mosquito [mɒs'ki:təʊ] *n., pl.* ~**es** Stechmücke, *die; (in tropics)* Moskito, *der*

moss [mɒs] *n.* Moos, *das.* '**mossy** *adj.* moosig

most [məʊst] 1. *adj. (in number, majority of)* die meisten; *(in amount)* meist...; **make the ~ mistakes/the ~ noise** die meisten Fehler/den größten Lärm machen; **for the ~ part** größtenteils. 2. *n.* **a)** *(greatest amount)* **the ~ it will cost is £10** es wird höchstens zehn Pfund kosten; **pay the ~:** am meisten bezahlen; **b)** *(greater part)* **~ of the girls** die meisten Mädchen; **~ of his friends** die meisten seiner Freunde; **~ of the poem** der größte Teil des Gedichts; **~ of the time** die meiste Zeit; **c)** *(on ~ occasions)* meistens. 3. *adv.* **a)** am meisten; **the ~ interesting book** das interessanteste Buch; **~ often** am häufigsten; **b)** *(exceedingly)* äußerst. '**mostly** *adv. (most of the time)* meistens; *(mainly)* größtenteils

MOT *see* **MOT test**

motel [məʊ'tel] *n.* Motel, *das*

moth [mɒθ] *n.* Nachtfalter, *der; (in clothes)* Motte, *die.* '**mothball** *n.* Mottenkugel, *die.* '**moth-eaten** *adj.* von Motten zerfressen

mother ['mʌðə(r)] 1. *n.* Mutter, *die.* 2. *v. t. (over-protect)* bemuttern. '**motherhood** *n.* Mutterschaft, *die*

mother: ~-in-law *n., pl.* ~**s-in-law** Schwiegermutter, *die;* ~**land** *n.* Vaterland, *das*

motherly ['mʌðəlɪ] *adj.* mütterlich; ~ **love** Mutterliebe, *die*

mother: ~-of-'pearl *n.* Perlmutt, *das;* **M~'s Day** *n.* Muttertag, *der;* ~**'tongue** *n.* Muttersprache, *die*

'**moth-proof** *adj.* mottenfest

motif [məʊ'ti:f] *n.* Motiv, *das*

motion ['məʊʃn] 1. *n.* **a)** Bewegung, *die;* **b)** *(proposal)* Antrag, *der.* 2. *v. t. & i.* ~ **[to]** sb. to do sth. jmdm. bedeuten *(geh.),* etw. zu tun. **motionless** *adj.* bewegungslos

motivate ['məʊtɪveɪt] *v. t.* motivieren.

motivation [məʊtɪ'veɪʃn] *n.* Motivation, *die*

motive ['məʊtɪv] *n.* Beweggrund, *der;* **the ~ for the crime** das Tatmotiv

motley ['mɒtlɪ] *adj.* buntgemischt

motor ['məʊtə(r)] 1. *n.* **a)** Motor, *der;* **b)** *(Brit.:~ car)* Auto, *das.* 2. *adj.* Motor*(mäher, -jacht usw.).* 3. *v. i. (Brit.)* [mit dem Auto] fahren

motor: ~bike *n. (coll.)* Motorrad, *das;* ~ **boat** *n.* Motorboot, *das;* ~ **car** *n. (Brit.)* Kraftfahrzeug, *das;* ~**cycle** *n.* Motorrad, *das*

'**motoring** *n. (Brit.)* Autofahren, *das*

'**motorist** *n.* Autofahrer, *der/*-fahrerin, *die*

motorize ['məʊtəraɪz] *v. t.* motorisieren

motor: ~-racing *n.* Autorennsport, *der;* ~ **vehicle** *n.* Kraftfahrzeug, *das;* ~**way** *n. (Brit.)* Autobahn, *die*

MOT test *n. (Brit.)* ≈ TÜV, *der*

mottled ['mɒtld] *adj.* gesprenkelt

motto ['mɒtəʊ] *n., pl.* ~**es** Motto, *das*

¹**mould** [məʊld] 1. *n. (hollow container)* Form, *die.* 2. *v. t.* formen *(out of, from* aus)

²**mould** *n. (Bot.)* Schimmel, *der*

moulder ['məʊldə(r)] *v. i.* ~ **[away]** [ver]modern

'**moulding** *n.* **a)** Formteil, *das* (of, in aus); *(Archit.)* Zierleiste, *die;* **b)** *(wooden)* Leiste, *die*

'**mouldy** *adj.* schimmlig; **go ~:** schimmeln

moult [məʊlt] *v. i.* ⟨*Vogel:*⟩ sich mausern; ⟨*Hund, Katze:*⟩ sich haaren

mound [maʊnd] *n.* **a)** *(of earth)* Hügel, *der;* **b)** *(heap)* Haufen, *der*

mount [maʊnt] 1. *n.* **a)** M~ **Vesuvius/ Everest** der Vesuv/der Mount Everest; **b)** *(animal)* Reittier, *das; (horse)* Pferd, *das;* **c)** *(of picture, photograph)* Passepartout, *das;* **d)** *(for gem)* Fassung, *die.* 2. *v. t.* **a)** hinaufsteigen ⟨*Treppe*⟩; steigen auf (+ *Akk.*) ⟨*Plattform, Reittier, Fahrzeug*⟩; **b)** aufziehen ⟨*Bild*⟩; einfassen ⟨*Edelstein usw.*⟩; **c)** inszenieren ⟨*Stück, Oper*⟩; organisieren ⟨*Ausstellung*⟩; durchführen ⟨*Angriff, Operation*⟩. 3. *v. i.* ~ **[up]** *(increase)* steigen (**to** auf + *Akk.*)

mountain ['maʊntɪn] *n.* Berg, *der;* **in the ~s** im Gebirge. **mountaineer** [maʊntɪ'nɪə(r)] *n.* Bergsteiger, *der/* Bergsteigerin, *die.* **mountai'neering** *n.* Bergsteigen, *das.* **mountainous** ['maʊntɪnəs] *adj.* **a)** gebirgig; **b)** *(huge)* riesig

mourn [mɔ:n] **1.** *v. i.* trauern; ~ **for** *or* **over** trauern um ⟨*Toten*⟩. **2.** *v. t.* betrauern. '**mourner** *n.* Trauernde, *der/die.* **mournful** ['mɔ:nfl] *adj.* klagend ⟨*Stimme, Ton, Schrei*⟩; trauervoll *(geh.)* ⟨*Person*⟩. '**mourning** *n.* Trauer, *die;* **be in/go into** ~: Trauer tragen/anlegen

mouse [maʊs] *n., pl.* **mice** [maɪs] Maus, *die.* '**mouse trap** *n.* Mausefalle, *die*

mousse [mu:s] *n.* Mousse, *die*

moustache [mə'stɑ:ʃ] *n.* Schnurrbart, *der*

mousy ['maʊsɪ] *adj.* **a)** mattbraun ⟨*Haar*⟩; **b)** *(timid)* scheu

mouth 1. [maʊθ] *n.* **a)** *(of person)* Mund, *der; (of animal)* Maul, *das;* **with one's** ~ **open/full** mit offenem/vollem Mund; **b)** *(harbour entrance)* [Hafen]einfahrt, *die; (of tunnel, cave)* Eingang, *der; (of river)* Mündung, *die.* **2.** [maʊð] *v. t.* mit Lippenbewegungen sagen. **mouthful** ['maʊθfʊl] *n.* Mundvoll, *der*

mouth: ~**organ** *n.* Mundharmonika, *die;* ~**piece** *n.* **a)** Mundstück, *das;* **b)** *(fig.)* Sprachrohr, *das*

movable ['mu:vəbl] *adj.* beweglich

move [mu:v] **1.** *n.* **a)** *(change of home)* Umzug, *der;* **b)** *(action taken)* Schritt, *der; (Footb. etc.)* Spielzug, *der;* **c)** *(turn in game)* Zug, *der;* **make a** ~: ziehen; **it's your** ~: du bist am Zug; **d)** **be on the** ~ ⟨*Person:*⟩ unterwegs sein; **make a** ~ *(do sth.)* etwas tun; *(coll.: leave)* losziehen *(ugs.);* **f) get a** ~ **on** *(coll.)* einen Zahn zulegen *(ugs.);* **get a** ~ **on!** *(coll.)* [mach] Tempo! *(ugs.).* **2.** *v. t.* **a)** *(change position of)* bewegen; wegräumen ⟨*Hindernis, Schutt*⟩; *(transport)* befördern; ~ **sth. to a new position** etw. an einen neuen Platz bringen; ~ **house** umziehen; **b)** *(in game)* ziehen; **c)** *(affect)* bewegen; ~ **sb. to tears** jmdn. zu Tränen rühren; **be** ~**d by sth.** über etw. *(Akk.)* gerührt sein; *(prompt)* ~ **sb. to do sth.** jmdn. dazu bewegen, etw. zu tun; **e)** *(propose)* beantragen. **3.** *v. i.* **a)** sich bewegen; *(in vehicle)* fahren; **b)** *(in games)* ziehen; **c)** *(do sth.)* handeln; **d)** *(change home)* umziehen **(to** nach); ~ **into a flat** in eine Wohnung einziehen; ~ **out of a flat** aus einer Wohnung ausziehen; ~ **to London** nach London ziehen; **e)** *(change posture or state)* sich bewegen; **don't** ~! keine Bewegung! **move a'bout 1.** *v. i.* zu-

gange sein; *(travel)* unterwegs sein. **2.** *v. t.* herumräumen. **move a'long 1.** *v. i.* **a)** gehen/fahren; **b)** ~ **along, please!** gehen/fahren Sie bitte weiter! **2.** *v. t.* zum Weitergehen/-fahren auffordern. **move 'in 1.** *v. i.* **a)** *(to home etc.)* einziehen; **b)** ~ **in on** ⟨*Truppen, Polizeikräfte:*⟩ vorrücken gegen. **2.** *v. t.* hineinbringen. **move 'off** *v. i.* sich in Bewegung setzen. **move 'on 1.** *v. i.* weitergehen/-fahren; ~ **on to another question** *(fig.)* zu einer anderen Frage übergehen. **2.** *v. t.* zum Weitergehen/-fahren auffordern. **move 'out** *v. t.* ausziehen (**of** aus). **move 'over** *v. i.* rücken. **move 'up** *v. i.* **a)** rücken; **b)** *(in queue, hierarchy)* aufrücken

'**movement** *n.* **a)** Bewegung, *die; (trend, tendency)* Tendenz, *die* **(towards** zu); **b)** *in pl.* Aktivitäten *Pl.;* **c)** *(Mus.)* Satz, *der*

movie ['mu:vɪ] *n. (Amer. coll.)* Film, *der;* **the** ~**s** der Film; **go to the** ~**s** ins Kino gehen

moving ['mu:vɪŋ] *adj.* **a)** beweglich; **b)** *(affecting)* ergreifend

mow [məʊ] *v. t., p.p.* **mown** [məʊn] *or* **mowed** [məʊd] mähen. **mow 'down** *v. t. (shoot)* niedermähen ⟨*Menschen*⟩

'**mower** *n.* Rasenmäher, *der*

mown *see* **mow**

MP *abbr.* **Member of Parliament**

m.p.g. *abbr.* **miles per gallon**

m.p.h. *abbr.* **miles per hour**

Mr ['mɪstə(r)] *n.* Herr; *(in an address)* Herrn

Mrs ['mɪsɪz] *n.* Frau

Ms [mɪz] *n.* Frau

Mt. *abbr.* **Mount**

much [mʌtʃ] **1.** *adj., more* [mɔ:(r)], *most* [məʊst] viel; **too** ~: zuviel *indekl.* **2.** *n.* vieles; ~ **of the day** der Großteil des Tages; **not be** ~ **to look at** nicht sehr ansehnlich sein. **3.** *adv., more, most* **a)** viel ⟨*besser, schöner usw.*⟩; ~ **more lively/attractive** viel lebhafter/attraktiver; **b)** mit Abstand ⟨*der/die/das beste, klügste usw.*⟩; **c)** *(greatly)* sehr ⟨*lieben, genießen usw.*⟩; *(for* ~ *of the time)* viel ⟨*lesen, spielen usw.*⟩; *(often)* oft ⟨*sehen, besuchen usw.*⟩; **d)** |**pretty** *or* **very**| ~ **the same** fast [genau] der-/die-/dasselbe

muck [mʌk] *n.* **a)** *(coll.: something disgusting)* Dreck, *der (ugs.);* **b)** *(coll.: nonsense)* Mist, *der (ugs.).* **muck a'bout, muck a'round** *(Brit. sl.) v. i.* **a)** herumalbern *(ugs.);* **b)** *(tinker)* her-

umfummeln (**with** an + *Dat.*). **muck
'in** *v. i. (coll.)* mit anpacken (**with** bei).
muck 'up *v. t.* **a)** *(Brit. coll.: bungle)*
vermurksen *(ugs.);* **b)** *(make dirty)*
dreckig machen *(ugs.);* **c)** *(coll.: spoil)*
vermasseln *(salopp)*
'**mucky** *adj.* dreckig *(ugs.)*
mucus ['mju:kəs] *n.* Schleim, *der*
mud [mʌd] *n.* Schlamm, *der*
muddle ['mʌdl] **1.** *n.* Durcheinander,
das. **2.** *v. t.* ~ |**up**| durcheinanderbrin-
gen; ~ **up** *(mix up)* verwechseln (**with**
mit). **muddle a'long, muddle 'on**
v. i. vor sich *(Akk.)* hin wursteln
(ugs.). **muddle 'through** *v. i.* sich
durchwursteln *(ugs.)*
muddy ['mʌdɪ] *adj.* schlammig; **get** *or*
become ~: verschlammen
'**mudguard** *n.* Schutzblech, *das; (of
car)* Kotflügel, *der*
¹**muff** [mʌf] *n.* Muff, *der*
²**muff** *v. t.* verpatzen *(ugs.)*
muffle ['mʌfl] *v. t.* **a)** *(envelop)* ~ |**up**|
einhüllen; **b)** dämpfen *(Geräusch).*
'**muffler** *n.* **a)** *(wrap, scarf)* Schal,
der; **b)** *(Amer. Motor Veh.)* Schalldämpfer, *der*
mug [mʌg] **1.** *n.* **a)** Becher, *der (meist
mit Henkel); (for beer etc.)* Krug, *der;*
b) *(sl.: face, mouth)* Visage, *die (salopp).* **c)** *(Brit. sl.: gullible person)* Trottel, *der (ugs.).* **2.** *v. t.,* -**gg**- *(rob)* überfallen und berauben. '**mugger** *n.*
Straßenräuber, *der*/-räuberin, *die.*
'**mugging** *n.* Straßenraub, *der*
muggy ['mʌgɪ] *adj.* schwül
mule [mju:l] *n.* Maultier, *das*
multicoloured *(Brit., Amer.:* **multi-
colored**) ['mʌltɪkʌləd] *adj.* mehrfar-
big; bunt *⟨Stoff, Kleid⟩*
multinational [mʌltɪ'næʃənl] **1.** *adj.*
multinational. **2.** *n.* multinationaler
Konzern, *der;* Multi, *der (ugs.)*
multiple ['mʌltɪpl] *adj.* mehrfach.
multiple-'choice *adj.* Multiple-
choice-*⟨Test, Frage⟩.* **multiple 'store**
n. (Brit.: shop) Kettenladen, *der*
multiplication [mʌltɪplɪ'keɪʃn] *n.*
Multiplikation, *die*
multiply ['mʌltɪplaɪ] **1.** *v. t.* multiplizieren, malnehmen (**by** mit). **2.** *v. i.*
sich vermehren
multi-storey ['mʌltɪstɔːrɪ] *adj.* mehr-
stöckig; ~ **car park/block of flats**
Parkhaus/Wohnhochhaus, *das*
multitude ['mʌltɪtjuːd] *n. (crowd)*
Menge, *die; (great number)* Vielzahl,
die
¹**mum** [mʌm] *(coll.)* **1.** *int.* ~**'s the word**

nicht weitersagen! **2.** *adj.* **keep** ~: den
Mund halten *(ugs.)*
²**mum** *n. (Brit. coll.: mother)* Mama, *die
(fam.)*
mumble ['mʌmbl] *v. i. & t.* nuscheln
(ugs.)
mumps [mʌmps] *n.* Mumps, *der*
munch [mʌntʃ] *v. t. & i.* ~ |**one's food**|
mampfen *(salopp)*
mundane [mʌn'deɪn] *adj.* **a)** *(dull)* banal; **b)** *(worldly)* weltlich
Munich ['mju:nɪk] *pr. n.* München
(das)
municipal [mju:'nɪsɪpl] *adj.* kommunal; Kommunal*⟨politik, -verwaltung⟩*
mural ['mjʊərl] *n.* Wandbild, *das*
murder ['mɜ:də(r)] **1.** *n.* Mord, *der (of*
an + *Dat.).* **2.** *v. t.* ermorden. '**murderer** *n.* Mörder, *der*/Mörderin, *die.*
murderess ['mɜ:dərɪs] *n.* Mörderin,
die. **murderous** ['mɜ:dərəs] *adj.* tödlich; Mord*⟨absicht, -drohung⟩;* mörderisch *(ugs.) ⟨Kampf⟩*
murk [mɜ:k] *n.* Dunkelheit, *die.*
'**murky** *adj.* **a)** *(dark)* düster; **b)**
(dirty) schmutzig-trüb *⟨Wasser⟩*
murmur ['mɜ:mə(r)] **1.** *n.* **a)** *(subdued
sound)* Rauschen, *das;* **b)** *(expression
of discontent)* Murren, *das;* **c)** *(soft
speech)* Murmeln, *das.* **2.** *v. t.* murmeln. **3.** *v. i. ⟨Person:⟩* murmeln; *(complain)* murren
muscle ['mʌsl] *n.* Muskel, *der.* **muscular** ['mʌskjʊlə(r)] *adj.* **a)** *(Anat.)*
Muskel-; **b)** *(strong)* muskulös
muse [mju:z] *(literary) v. i.* [nach]sin-
nen *(geh.)* (**on,** over über + *Akk.)*
museum [mju:'zi:əm] *n.* Museum, *das*
mush [mʌʃ] *n.* Brei, *der*
mushroom ['mʌʃrʊm, 'mʌʃru:m] **1.** *n.*
Pilz, *der; (cultivated)* Champignon,
der. **2.** *v. i.* wie Pilze aus dem Boden
schießen
'**mushy** *adj.* breiig
music ['mju:zɪk] *n.* **a)** Musik, *die;*
piece of ~: Musikstück, *das;* **set sth. to**
~: etw. vertonen; **b)** *(score)* Noten *Pl.*
musical ['mju:zɪkl] **1.** *adj.* musikalisch; Musik*⟨instrument, -verständnis,
-notation, -abend⟩.* **2.** *n.* Musical, *das*
Muslim ['mʊslɪm, 'mʌzlɪm] **1.** *adj.*
moslemisch. **2.** *n.* Moslem, *der*/Mos-
lime, *die*
muslin ['mʌzlɪn] *n.* Musselin, *der*
mussel ['mʌsl] *n.* Muschel, *die*
must [məst, *stressed* mʌst] **1.** *v. aux.,
only in pres., neg. (coll.)* **mustn't**
['mʌsnt] müssen; *with neg.* dürfen. **2.**
n. (col!.) Muß, *das*

mustache *see* **moustache**
mustard ['mʌstəd] *n.* Senf, *der*
muster ['mʌstə(r)] 1. *n.* **pass ~:** akzeptabel sein. 2. *v. t.* versammeln; *(Mil., Naut.)* [zum Appell] antreten lassen; *(fig.)* zusammennehmen ⟨*Kraft, Mut, Verstand*⟩. 3. *v. i.* sich [ver]sammeln.
muster 'up *v. t.* aufbringen
mustn't ['mʌsnt] *(coll.)* = **must not;** *see* **must 1**
musty ['mʌstɪ] *adj.* muffig
mutant ['mjuːtənt] 1. *adj.* mutiert. 2. *n.* Mutante, *die*
mutation [mjuː'teɪʃn] *n.* Mutation, *die*
mute [mjuːt] 1. *adj.* stumm. 2. *n.* Stumme, *der/die.* **'muted** *adj.* gedämpft
mutilate ['mjuːtɪleɪt] *v. t.* verstümmeln. **mutilation** [mjuːtɪ'leɪʃn] *n.* Verstümmelung, *die*
mutinous ['mjuːtɪnəs] *adj.* meuternd
mutiny ['mjuːtɪnɪ] 1. *n.* Meuterei, *die.* 2. *v. i.* meutern
mutter ['mʌtə(r)] *v. i. & t.* murmeln. **'muttering** *n.* Gemurmel, *das*
mutton ['mʌtn] *n.* Hammelfleisch, *das*
mutual ['mjuːtjʊəl] *adj.* **a)** gegenseitig; **b)** *(coll.: shared)* gemeinsam. **'mutually** *adv.* **a)** gegenseitig; **be ~ exclusive** sich [gegenseitig] ausschließen; **b)** *(in common)* gemeinsam
muzzle ['mʌzl] 1. *n.* **a)** *(of dog)* Schnauze, *die; (of horse, cattle)* Maul, *das;* **b)** *(of gun)* Mündung, *die;* **c)** *(put over animal's mouth)* Maulkorb, *der.* 2. *v. t.* **a)** einen Maulkorb anlegen (+ *Dat.*) ⟨*Hund*⟩; **b)** *(fig.)* mundtot machen *(ugs.)* (+ *Dat.*)
MW *abbr. (Radio)* **medium wave** MW
my [maɪ] *poss. pron. attrib.* mein; **my|, my!|, |my|** oh my! [ach du] meine Güte! *(ugs.)*
myself [maɪ'self] *pron.* **a)** *emphat.* selbst; **I thought so ~:** das habe ich auch gedacht; **b)** *refl.* mich/mir. *See also* **herself**
mysterious [mɪ'stɪərɪəs] *adj.* rätselhaft; geheimnisvoll ⟨*Fremder, Orient*⟩. **my'steriously** *adv.* auf rätselhafte Weise; geheimnisvoll ⟨*lächeln usw.*⟩
mystery ['mɪstərɪ] *n.* **a)** Rätsel, *das;* **b)** *(secrecy)* Geheimnis, *das.* **'mystery tour** *n.* Fahrt ins Blaue *(ugs.)*
mystic ['mɪstɪk] 1. *adj.* mystisch. 2. *n.* Mystiker, *der*/Mystikerin, *die.* **mystical** ['mɪstɪkl] *adj.* mystisch
mystify ['mɪstɪfaɪ] *v. t.* verwirren
myth [mɪθ] *n.* Mythos, *der.* **mythical** ['mɪθɪkl] *adj.* **a)** *(based on myth)* my-

thisch; **b)** *(invented)* fiktiv. **mythological** [mɪθə'lɒdʒɪkl] *adj.* mythologisch. **mythology** [mɪ'θɒlədʒɪ] *n.* Mythologie, *die*

N

N, n [en] *n.* N, n, *das*
N. *abbr.* **a) north** N; **b) northern** n.
NAAFI ['næfɪ] *abbr. (Brit.)* **Navy, Army and Air Force Institutes** Kaufhaus für Angehörige der britischen Truppen
nab [næb] *v. t.,* **-bb-** *(sl.)* **a)** *(arrest)* schnappen *(ugs.);* **b)** *(seize)* sich *(Dat.)* schnappen
nag [næg] *v. i. & t.* **-gg-: ~ |at|** sb. an jmdm. herumnörgeln; **~ |at|** sb. **to do** sth. jmdm. zusetzen *(ugs.),* daß er etw. tut. **'nagging** 1. *adj. (persistent)* quälend; bohrend ⟨*Schmerz*⟩. 2. *n.* Genörgel, *das*
nail [neɪl] 1. *n.* Nagel, *der;* **hit the ~ on the head** *(fig.)* den Nagel auf den Kopf treffen *(ugs.).* 2. *v. t.* nageln (**to** an + *Akk.*). **nail 'down** *v. t.* festnageln; zunageln ⟨*Kiste*⟩
nail: ~-brush *n.* Nagelbürste, *die;* **~-clippers** *n. pl.* |pair of| **~-clippers** Nagelknipser, *der;* **~-file** *n.* Nagelfeile, *die; ~* **polish** *n.* Nagellack, *der;* **~-polish remover** Nagellackentferner, *der;* **~-scissors** *n. pl.* |pair of| **~-scissors** Nagelschere, *die; ~* **varnish** *(Brit.) see ~* **polish**
naïve, naive [naɪ'iːv] *adj.,* **na'ïvely, na'ively** *adv.* naiv
naked ['neɪkɪd] *adj.* nackt; **visible to** *or* **with the ~ eye** mit bloßem Auge zu erkennen. **'nakedness** *n.* Nacktheit, *die*
name [neɪm] 1. *n.* **a)** Name, *der;* **what's your ~/the ~ of this place?** wie heißt du/dieser Ort? **my ~ is Jack** ich heiße Jack; **last ~:** Nachname, *der;* **by ~:** namentlich ⟨*erwähnen, aufrufen usw.*⟩; **know sb. by ~:** jmdn. mit Namen kennen; **b)** *(reputation)* Ruf, *der;* **make a ~ for oneself** sich *(Dat.)* einen Namen machen; **c) call sb. ~s** jmdn. beschimpfen. 2. *v. t.* **a)** *(give ~ to)* ei-

nen Namen geben (+ *Dat.*); ~ **sb.**
John jmdn. John nennen; ~ **sb./sth.**
after *or* (*Amer.*) **for sb.** jmdn./etw.
nach jmdm. benennen; **be ~d John**
John heißen; **a man ~d Smith** ein
Mann namens Smith; **b)** (*call by right*
~) benennen; **c)** (*nominate*) ~ **sb.** [as]
sth. jmdn. zu etw. ernennen. **'name-**
less *adj.* namenlos. **'namely** *adv.*
nämlich. **'namesake** *n.* Namensvet-
ter, *der*/-schwester, *die*
nanny ['nænı] *n.* (*Brit.*) Kindermäd-
chen, *das.* **'nanny-goat** *n.* Ziege, *die*
nap [næp] **1.** *n.* Nickerchen, *das*
(*fam.*); **have a ~:** ein Nickerchen halten. **2.** *v. i.,* -pp- dösen (*ugs.*); **catch sb.**
~**ping** (*fig.*) jmdn. überrumpeln
nape [neıp] *n.* ~ [of the neck] Nacken,
der; Genick, *das*
napkin ['næpkın] *n.* Serviette, *die*
Naples ['neıplz] *pr. n.* Neapel (*das*)
nappy ['næpı] *n.* (*Brit.*) Windel, *die*
narcissus [nɑː'sısəs] *n., pl.* **narcissi**
[nɑː'sısaı] *or* ~**es** Narzisse, *die*
narcotic [nɑː'kɒtık] **1.** *n.* **a)** (*drug*)
Rauschgift, *das;* **b)** (*active ingredient*)
Betäubungsmittel, *das.* **2.** *adj.* **a)** nar-
kotisch; ~ **drug** Rauschgift, *das;* **b)**
(*causing drowsiness*) einschläfernd
narrate [nə'reıt] *v. t.* erzählen; kommentieren (*Film*). **narration** [nə-
'reıʃn] *n.* Erzählung, *die.* **narrative**
['nærətıv] **1.** *n.* Erzählung, *die.* **2.** *adj.*
erzählend. **narrator** [nə'reıtə(r)] *n.*
Erzähler, *der*/Erzählerin, *die*
narrow ['nærəʊ] **1.** *adj.* **a)** schmal;
schmal geschnitten (*Rock, Hose,*
Ärmel usw.); eng (*Tal, Gasse*); **b)**
(*limited*) eng; begrenzt (*Auswahl*); **c)**
knapp (*Sieg, Mehrheit*); **have a ~ es-**
cape mit knapper Not entkommen
(**from** *Dat.*); **d)** (*not tolerant*) engstirnig. **2.** *v. i.* sich verschmälern; (*Tal:*)
sich verengen. **3.** *v. t.* verschmälern;
(*fig.*) einengen. **narrow 'down** *v. t.*
einengen (**to** auf + *Akk.*)
narrow-'minded *adj.* engstirnig
nasal ['neızl] *adj.* **a)** (*Anat.*) Nasen-; **b)**
näselnd; **speak in a ~ voice** näseln
nastily ['nɑːstılı] *adv.* **a)** (*unpleasantly*)
scheußlich; **b)** (*ill-naturedly*) gemein;
behave ~: häßlich sein
nasty ['nɑːstı] *adj.* **a)** (*unpleasant*)
scheußlich (*Geruch, Geschmack*); gemein (*Trick, Person*); häßlich (*Ange-*
wohnheit); **that was a ~ thing to say/**
do das war gemein; **b)** (*ill-natured*) böse; **be ~ to sb.** häßlich zu jmdm. sein;
c) (*serious*) übel; schlimm (*Krankheit,*

Husten, Verletzung); **she had a ~ fall**
sie ist übel gefallen
nation ['neıʃn] *n.* Nation, *die;* (*people*)
Volk, *das.* **national** ['næʃənl] **1.** *adj.*
national; National(*flagge, -held,*
-theater, -gericht, -charakter); Staats-
(*sicherheit, -religion*); überregional
(*Rundfunkstation, Zeitung*); landes-
weit (*Streik*). **2.** *n.* (*citizen*) Staatsbürger, *der*/-bürgerin, *die;* **foreign ~:**
Ausländer, *der*/Ausländerin, *die*
national: ~ 'anthem *n.* National-
hymne, *die;* ~ **'costume** *n.* National-
tracht, *die;* **N~ 'Health [Service]** *n.*
(*Brit.*) staatlicher Gesundheitsdienst;
N~ Health doctor/patient/spectacles
≈ Kassenarzt, *der*/-patient, *der*/-bril-
le, *die;* **N~ In'surance** *n.* (*Brit.*) Sozi-
alversicherung, *die*
nationalism ['næʃənəlızm] *n.* Nationalismus, *der.* **nationalist** ['næʃənə-
lıst] **1.** *n.* Nationalist, *der*/Nationali-
stin, *die.* **2.** *adj.* nationalistisch
nationality [næʃə'nælıtı] *n.* Staatsangehörigkeit, *die;* **what's his ~?** welche
Staatsangehörigkeit hat er?
nationalization [næʃənəlaı'zeıʃn] *n.*
Verstaatlichung, *die*
nationalize ['næʃənəlaız] *v. t.* ver-
staatlichen
'nationally *adv.* landesweit
native ['neıtıv] **1.** *n.* **a)** (*of specified*
place) **a ~ of Britain** ein gebürtiger
Brite/eine gebürtige Britin; **b)** (*person*
born in a place) Eingeborene, *der/die;*
c) (*local inhabitant*) Einheimische,
der/die. **2.** *adj.* eingeboren; einheimisch (*Pflanze, Tier*); ~ **inhabitant**
Eingeborene/Einheimische, *der/die;*
~ **land** Geburts- *od.* Heimatland, *das;*
~ **language** Muttersprache, *die*
nativity [nə'tıvıtı] *n.* **the N~** [of Christ]
die Geburt Christi. **na'tivity play** *n.*
Krippenspiel, *das*
NATO, Nato ['neıtəʊ] *abbr.* North Atlantic Treaty Organization NATO, *die*
natter ['nætə(r)] (*Brit. coll.*) **1.** *v. i.*
quatschen (*ugs.*). **2.** *n.* **have a ~:** quatschen (*ugs.*)
natural ['nætʃrəl] *adj.* natürlich; Na-
tur(*zustand, -seide, -gewalt*). **natural**
'gas *n.* Erdgas, *das.* **natural 'his-**
tory *n.* Naturkunde, *die*
naturalism ['nætʃrəlızm] *n.* Naturalismus, *der*
naturalist ['nætʃrəlıst] *n.* Naturfor-
scher, *der*/-forscherin, *die*
naturalization [nætʃrəlaı'zeıʃn] *n.*
Einbürgerung, *die*

223

necktie

naturalize ['nætʃrəlaɪz] *v.t.* einbürgern

'**naturally** *adv.* **a)** *(by nature)* von Natur aus ⟨blaß, fleißig usw.⟩; *(in a true-to-life way)* naturgetreu; **b)** *(of course)* natürlich

'**naturalness** *n.* Natürlichkeit, *die*

nature ['neɪtʃə(r)] *n.* **a)** Natur, *die;* **b)** *(essential qualities)* Beschaffenheit, *die;* in the ~ of things naturgemäß; **c)** *(kind)* Art, *die;* things of this ~: derartiges; **d)** *(character)* Wesen, *das;* be proud/friendly *etc.* by ~: ein stolzes/freundliches *usw.* Wesen haben. '**nature reserve** *n.* Naturschutzgebiet, *das.* '**nature study** *n.* Naturkunde, *die.* '**nature trail** *n.* Naturlehrpfad, *der*

naught [nɔːt] *n.* *(arch./dial.)* **come to** ~: zunichte werden

naughtily ['nɔːtɪlɪ] *adv.* ungezogen

naughtiness ['nɔːtɪnɪs] *n.* Ungezogenheit, *die*

naughty ['nɔːtɪ] *adj.* ungezogen; **you ~ boy/dog** du böser Junge/Hund

nausea ['nɔːzɪə] *n.* Übelkeit, *die.* **nauseate** ['nɔːzɪeɪt] *v.t.* *(disgust)* anwidern. '**nauseating** *adj.* *(disgusting)* widerlich. **nauseous** ['nɔːzɪəs] *adj.* **sb. is** *or* **feels ~:** jmdm. ist übel

nautical ['nɔːtɪkl] *adj.* nautisch. **nautical 'mile** *n.* Seemeile, *die*

naval ['neɪvl] *adj.* Marine-; See-⟨schlacht, -macht, -streitkräfte⟩; ~ **ship** Kriegsschiff, *das*

nave [neɪv] *n.* [Mittel]schiff, *das*

navel ['neɪvl] *n.* Nabel, *der*

navigate ['nævɪgeɪt] *v.t.* **a)** navigieren ⟨Schiff, Flugzeug⟩; **b)** befahren ⟨Fluß usw.⟩. **navigation** [nævɪ'geɪʃn] *n.* Navigation, *die.* **navigator** ['nævɪgeɪtə(r)] *n.* Navigator, *der*/Navigatorin, *die*

navy ['neɪvɪ] *n.* **a)** [Kriegs]marine, *die;* **b)** *see* **navy blue. navy 'blue** *n.* Marineblau, *das.* '**navy-blue** *adj.* marineblau

Nazi ['nɑːtsɪ] **1.** *n.* Nazi, *der.* **2.** *adj.* nazistisch; Nazi-

NB *abbr.* **nota bene** NB

NCO *abbr.* **non-commissioned officer** Uffz.

NE *abbr.* **north-east** NO

near [nɪə(r)] **1.** *adv.* nah[e]; **stand/live |quite| ~:** [ganz] in der Nähe stehen/wohnen; **come** *or* **draw ~/~er** ⟨Tag, Zeitpunkt:⟩ nahen/näherrücken; **get ~er together** näher zusammenrücken; **~ at hand** in Reichweite *(Dat.);* ⟨Ort⟩

ganz in der Nähe; ~ **to = 2. 2.** *prep.* **a)** *(position)* nahe an/bei (+ *Dat.*); *(fig.)* in der Nähe (+ *Gen.*); **keep ~ me** halte dich in meiner Nähe; **it's ~ here** es ist hier in der Nähe; **b)** *(motion)* nahe an (+ *Akk.*); *(fig.)* in der Nähe (+ *Gen.*); **don't come ~ me** komm mir nicht zu nahe. **3.** *adj.* **a)** *(in space or time)* nahe; **in the ~ future** in nächster Zukunft; **the ~est man** der am nächsten stehende Mann; **b)** *(in nature)* £30 or ~/~est offer 30 Pfund oder nächstbestes Angebot; ~ **escape** Entkommen mit knapper Not; **that was a ~ miss/thing!** das war knapp! **4.** *v.t.* sich nähern (+ *Dat.*); **the building is ~ing completion** das Gebäude steht kurz vor seiner Vollendung. **5.** *v.i.* ⟨Zeitpunkt:⟩ näherrücken. '**nearby** *adj.* nahe gelegen

'**nearly** *adv.* fast; **be ~ in tears** den Tränen nahe sein; **it is ~ six o'clock** es ist kurz vor sechs Uhr; **are you ~ ready?** bist du bald fertig?

'**nearness** *n.* Nähe, *die*

'**near-sighted** *adj.* *(Amer.)* kurzsichtig

neat [niːt] *adj.* **a)** *(tidy)* ordentlich; **b)** *(undiluted)* pur; **c)** *(smart)* gepflegt ⟨Erscheinung, Kleidung⟩; **d)** *(deft)* geschickt. '**neatly** *adv. see* **neat a, c, d:** ordentlich; gepflegt; geschickt. '**neatness** *n. see* **neat a, c, d:** Ordentlichkeit, *die;* Gepflegtheit, *die;* Geschicktheit, *die*

necessarily [nesɪ'serɪlɪ] *adv.* zwangsläufig; **it is not ~ true** es muß nicht [unbedingt] stimmen

necessary ['nesɪsərɪ] **1.** *adj.* nötig; notwendig; **do everything ~:** das Nötige *od.* Notwendige tun. **2.** *n.* **the necessaries of life** das Lebensnotwendige

necessitate [nɪ'sesɪteɪt] *v.t.* erforderlich machen

necessity [nɪ'sesɪtɪ] *n.* **a)** *(need, necessary thing)* Notwendigkeit, *die;* **do sth. out of** *or* **from ~:** etw. notgedrungen tun; **of ~:** notwendigerweise; **b)** *(want)* Not, *der*

neck [nek] *n.* **a)** Hals, *der;* **be a pain in the ~** *(coll.)* jmdm. auf die Nerven gehen *(ugs.);* **break one's ~** *(fig. coll.)* sich den Hals brechen; ~ **and ~:** Kopf an Kopf; **b)** *(of garment)* Kragen, *der*

neck: ~**lace** ['neklɪs] *n.* [Hals]kette, *die;* *(with jewels)* Kollier, *das;* ~**line** *n.* [Hals]ausschnitt, *der;* ~**tie** *n.* Krawatte, *die*

nectar ['nektə(r)] *n.* Nektar, *der*
née (*Amer.:* **nee**) [neɪ] *adj.* geborene
need [niːd] **1.** *n.* **a)** Notwendigkeit, *die* (for, of *Gen.*); (demand) Bedarf, *der* (for, of an + *Dat.*); **as the ~ arises** nach Bedarf; **if ~ be** nötigenfalls; **there's no ~ for that** [das ist] nicht nötig; **there's no ~ to do sth.** es ist nicht nötig, etw. zu tun; **be in ~ of sth.** etw. brauchen; **there's no ~ for you to come** du brauchst nicht zu kommen; **b)** *no pl.* (*emergency*) Not, *die;* **in case of ~:** im Notfall; **c)** (*thing*) Bedürfnis, *das.* **2.** *v. t.* **a)** (*require*) brauchen; **sth. that urgently ~s doing** etw., was dringend gemacht werden muß; **it ~s a coat of paint** es muß gestrichen werden; **b)** *expr. necessity* müssen; **I ~ to do it** ich muß es tun; **it ~s/doesn't ~ to be done** es muß getan werden/es braucht nicht getan zu werden; **c)** *pres.* **he ~**, *neg.* **~ not** *or* (*coll.*) **~n't** ['niːdnt] *expr. desirability* müssen; *with neg.* brauchen zu
needle ['niːdl] **1.** *n.* Nadel, *die.* **2.** *v. t.* (*coll.*) nerven (*ugs.*)
needless ['niːdlɪs] *adj.* unnötig; **~ to add** *or* **say, ...:** überflüssig zu sagen, daß ... '**needlessly** *adv.* unnötig
'**needlework** *n.* Handarbeit, *die;* **do ~:** handarbeiten
needn't ['niːdnt] (*coll.*) = need not; *see* need 2 c
'**needy** *adj.* notleidend; bedürftig
negation [nɪ'geɪʃn] *n.* Verneinung, *die*
negative ['negətɪv] **1.** *adj.* negativ. **2.** *n.* **a)** (*Photog.*) Negativ, *das;* **b)** (~ statement) negative Aussage; (*answer*) Nein, *das.* '**negatively** *adv.* negativ
neglect [nɪ'glekt] **1.** *v. t.* vernachlässigen; **she ~ed to write** sie hat es versäumt zu schreiben. **2.** *n.* Vernachlässigung, *die;* **be in a state of ~** (*Gebäude:*) verwahrlost sein. **neglectful** [nɪ'glektfl] *adj.* gleichgültig (of gegenüber); **be ~ of** sich nicht kümmern um
negligence ['neglɪdʒəns] *n.* Nachlässigkeit, *die;* (*Law, Insurance, etc.*) Fahrlässigkeit, *die*
negligent ['neglɪdʒənt] *adj.* nachlässig; **be ~ about sth.** sich um etw. nicht kümmern
negligible ['neglɪdʒɪbl] *adj.* unerheblich
negotiable [nɪ'gəʊʃəbl] *adj.* **a)** verhandlungsfähig (*Forderung, Bedingungen*); **b)** passierbar (*Straße, Fluß*)
negotiate [nɪ'gəʊʃɪeɪt] **1.** *v. i.* verhandeln (for, on, about über + *Akk.*). **2.**

v. t. **a)** (*arrange*) aushandeln; **b)** überwinden (*Hindernis*); passieren (*Straße, Fluß*); nehmen (*Kurve*).
negotiation [nɪgəʊʃɪ'eɪʃn] *n.* Verhandlung, *die.* **negotiator** [nɪ'gəʊʃɪeɪtə(r)] *n.* Unterhändler, *der/*-händlerin, *die*
Negress ['niːgrɪs] *n.* Negerin, *die*
Negro ['niːgrəʊ] **1.** *n., pl.* **~es** Neger, *der.* **2.** *adj.* Neger-
neigh [neɪ] **1.** *v. i.* wiehern. **2.** *n.* Wiehern, *das*
neighbor *etc.* (*Amer.*) *see* **neighbour** *etc.*
neighbour ['neɪbə(r)] **1.** *n.* Nachbar, *der/* Nachbarin, *die;* **my next-door ~s** meine Nachbarn von nebenan. **2.** *v. t. & i.* **~ [upon]** grenzen an (+ *Akk.*). '**neighbourhood** *n.* (*district*) Gegend, *die;* (*neighbours*) Nachbarschaft, *die;* [**somewhere**] **in the ~ of £100** [so] um [die] 100 Pfund. '**neighbouring** *adj.* Nachbar-; angrenzend (*Felder*)
neither ['naɪðə(r), 'niːðə(r)] **1.** *adj.* keiner/keine/keins der beiden. **2.** *pron.* keiner/keine/keins von *od.* der beiden. **3.** *adv.* (*also not*) auch nicht; **~ am I,** (*sl.*) **me ~:** ich auch nicht. **4.** *conj.* (*not either*) weder; **~ ... nor ...:** weder ... noch ...
neon ['niːɒn] *n.* Neon, *das*
neon: **~ 'light** *n.* Neonlampe, *die;* '**sign** *n.* Neonreklame, *die*
nephew ['nevjuː, 'nefjuː] *n.* Neffe, *der*
nepotism ['nepətɪzm] *n.* Vetternwirtschaft, *die*
Neptune ['neptjuːn] *pr. n.* (*Astron.*) Neptun, *der*
nerve [nɜːv] *n.* Nerv, *der;* **get on sb.'s ~s** jmdm. auf die Nerven gehen (*ugs.*); **lose one's ~:** die Nerven verlieren; **what** [**a**] **~!** [so eine] Frechheit! '**nerve gas** *n.* Nervengas, *das.* '**nerve-racking** *adj.* nervenaufreibend
nervous ['nɜːvəs] *adj.* **a)** (*Anat., Med.*) Nerven-; **~ breakdown** Nervenzusammenbruch, *der;* **b)** (*having delicate nerves*) nervös; **be a ~ wreck** mit den Nerven völlig am Ende sein; **c)** (*Brit.: timid*) **be ~ of** *or* **about** Angst haben vor (+ *Dat.*); **be a ~ person** ängstlich sein. '**nervously** *adv.* nervös. '**nervousness** *n.* Ängstlichkeit, *die*
nervy ['nɜːvɪ] *adj.* **a)** nervös; **b)** (*Amer. coll.: impudent*) unverschämt
nest **1.** *n.* Nest, *das.* **2.** *v. i.* nisten. '**nest-egg** *n.* (*fig.*) Notgroschen, *der*

nestle ['nesl] *v. i.* **a)** sich schmiegen (**to, up against** an + *Akk.*); **b)** *(lie half hidden)* eingebettet sein

¹net [net] **1.** *n.* Netz, *das.* **2.** *v. t.*, -tt- [mit einem Netz] fangen

²net *adj.* **a)** netto; Netto⟨einkommen, -[verkaufs]preis usw.⟩; ~ **weight** Nettogewicht, *das;* **b)** *(ultimate)* End⟨ergebnis, -effekt⟩

net: ~**ball** *n.* Netzball, *der.* ~ 'cur-tain *n.* Store, *der*

Netherlands ['neðələndz] *pr. n. sing. or pl.* Niederlande *Pl.*

nett *see* ²**net a**

'**netting** *n.* ([piece of] net) Netz, *das;* wire ~: Maschendraht, *der*

nettle ['netl] *n.* Nessel, *die*

'**network** *n.* Netz, *das*

neuralgia [njʊə'rældʒə] *n.* Neuralgie, *die*

neurosis [njʊə'rəʊsɪs] *n., pl.* **neuroses** [njʊə'rəʊsiːz] Neurose, *die.* **neurotic** [njʊə'rɒtɪk] *adj.* **a)** nervenkrank; **b)** *(coll.)* neurotisch

neuter ['njuːtə(r)] *adj.* sächlich

neutral ['njuːtrl] **1.** *adj.* neutral. **2.** *n.* (~ **gear**) Leerlauf, *der.* **neutrality** [njuː'trælɪti] *n.* Neutralität, *die*

neutralize ['njuːtrəlaɪz] *v. t.* neutralisieren

neutron ['njuːtrɒn] *n.* Neutron, *das*

never ['nevə(r)] *adv.* **a)** nie; ~-**ending** endlos; **b)** *(coll.)* you ~ **believed that, did you?** du hast das doch wohl nicht geglaubt?; **well, I ~ [did]!** [na] so was!

neverthe'less *adv.* trotzdem

new [njuː] *adj.* neu

new: ~-**born** *adj.* neugeboren; ~**comer** ['njuːkʌmə(r)] *n.* Neuankömmling, *der;* ~**fangled** ['njuːfæŋgld] *adj.* neumodisch; ~-**found** *adj.* neu; ~-**laid** *adj.* frisch [gelegt]

'**newly** *adv.* *(recently)* neu; ~ **married** seit kurzem verheiratet. '**newly-wed** *n.* Jungverheiratete, *der/die*

new 'moon *n.* Neumond, *der*

'**newness** *n.* Neuheit, *die*

news [njuːz] *n., no pl.* **a)** Nachricht, *die;* be in the ~: Schlagzeilen machen; **good/bad** ~: schlechte/gute Nachrichten *Pl.;* **b)** *(Radio, Telev.)* Nachrichten *Pl.*

news: ~**agent** *n.* Zeitungshändler, *der/*-händlerin, *die;* ~ **bulletin** *n.* Nachrichten *Pl.* ~**caster** *n.* Nachrichtensprecher, *der/*-sprecherin, *die;* ~**flash** *n.* Kurzmeldung, *die;* ~ '**headline** *n.* Schlagzeile, *die;* ~**letter** *n.* Rundschreiben, *das;*

~**paper** ['njuːspeɪpə(r)] *n.* **a)** Zeitung, *die;* **b)** *(material)* Zeitungspapier, *das;* ~**reader** *n.* Nachrichtensprecher, *der/*-sprecherin, *die;* ~**reel** *n.* Wochenschau, *die;* ~-**sheet** *n.* Informationsblatt, *das;* ~ **summary** *n.* Kurznachrichten *Pl.;* ~**worthy** *adj.* [für die Medien] interessant

newt [njuːt] *n.* [Wasser]molch, *der*

New: **new 'year** *n.* Neujahr, *das;* **over the new year** über Neujahr; **a Happy ~ Year** ein glückliches *od.* gutes neues Jahr. ~ '**Year's** *(Amer.),* ~ **Year's 'Day** *ns.* Neujahrstag, *der;* ~ **Year's 'Eve** *n.* Silvester, *der od. das;* ~ **Zealand** [~ 'ziːlənd] *pr. n.* Neuseeland *(das);* ~ '**Zealander** *n.* Neuseeländer, *der/*-länderin, *die*

next [nekst] **1.** *adj.* nächst...; **the ~ but one** der/die/das übernächste; ~ **to** *(fig.: almost)* fast; nahezu; |**the**| ~ **time** das nächste Mal; **the ~ best** der/die/das nächstbeste; **am I ~?** komme ich jetzt dran? **2.** *adv. (in the ~ place)* als nächstes; *(on the ~ occasion)* das nächste Mal; **it's my turn ~:** ich komme als nächster dran; **sit/stand ~ to sb.** neben jmdm. stehen/sitzen; **place sth. ~ to sb./sth.** etw. neben jmdn./etw. stellen. **3.** *n.* **a)** **the week after ~:** [die] übernächste Woche; **b)** *(person)* ~ **of kin** nächster/nächste Angehörige; **~, please!** der nächste, bitte! '**next-door** *adj.* gleich nebenan *nachgestellt*

NHS *abbr. (Brit.)* **National Health Service**

nib [nɪb] *n.* Feder, *die*

nibble ['nɪbl] *v. t. & i.* knabbern (**at, on** an + *Dat.*)

nice [naɪs] *adj.* nett; angenehm ⟨*Stimme*⟩; schön ⟨*Wetter*⟩; *(iron.: disgraceful, difficult)* schön; ~ |**and**| **warm/fast** schön warm/schnell; ~-**looking** hübsch. '**nicely** *adv. (coll.)* **a)** *(well)* nett; gut ⟨*arbeiten, sich benehmen, plaziert sein*⟩; **b)** *(all right)* gut; **that will do ~:** das reicht völlig. **niceties** ['naɪsɪtɪz] *n. pl.* Feinheiten

niche [nɪtʃ, niːʃ] *n.* **a)** *(in wall)* Nische, *die;* **b)** *(fig.: suitable place)* Platz, *der*

nick *n.* **a)** *(notch)* Kerbe, *die;* **b)** *(sl. prison)* Knast, *der (salopp);* **c)** *(Brit.: police station)* Wache, *die;* **d)** **in good/ poor ~** *(coll.)* gut/nicht gut im Schuß *(ugs.);* **e)** **in the ~ of time** gerade noch rechtzeitig. **2.** *v. t.* **a)** einkerben; **b)** *(Brit. sl.: arrest)* einlochen *(salopp);* **c)** *(Brit. sl.: steal)* klauen *(salopp)*

nickel ['nɪkl] *n.* **a)** Nickel, *das;* **b)** *(Amer. coll.: coin)* Fünfcentstück, *das*

nickname ['nɪkneɪm] *n.* Spitzname, *der; (affectionate)* Koseform, *die*

nicotine ['nɪkəti:n] *n.* Nikotin, *das*

niece [ni:s] *n.* Nichte, *die*

Nigeria [naɪ'dʒɪərɪə] *pr. n.* Nigeria *(das)*

niggardly ['nɪgədlɪ] *adj.* knaus[e]rig *(ugs.)*

niggling ['nɪglɪŋ] *adj.* **a)** *(petty)* belanglos; **b)** *(trivial)* nichtssagend; **c)** *(nagging)* nagend

night [naɪt] *n.* Nacht, *die; (evening)* Abend, *der;* **the following** ~: die Nacht/der Abend darauf; **the previous** ~: die vorausgegangene Nacht/der vorausgegangene Abend; **on Sunday** ~: Sonntag nacht/[am] Sonntag abend; **for the** ~: über Nacht; **at** ~: nachts/abends; **late at** ~: spätabends

night: ~**cap** *n. (drink)* Schlaftrunk, *der;* ~**club** *n.* Nachtklub, *der;* ~**dress** *n.* Nachthemd, *das;* ~**fall** *n.* Einbruch der Dunkelheit

nightie ['naɪtɪ] *n. (coll.)* Nachthemd, *das*

nightingale ['naɪtɪŋgeɪl] *n.* Nachtigall, *die*

'night-life *n.* Nachtleben, *das*

nightly ['naɪtlɪ] **1.** *adj. (happening every night/evening)* allnächtlich/allabendlich. **2.** *adv. (every night)* jede Nacht; *(every evening)* jeden Abend

night: ~**mare** *n.* Alptraum, *der;* ~**school** *n.* Abendschule, *die;* ~**shift** *n.* Nachtschicht, *die;* ~**time** *n.* Nacht, *die;* **in the** *or* **at** ~**time** nachts; ~**'watchman** *n.* Nachtwächter, *der*

nil [nɪl] *n.* null

Nile [naɪl] *pr. n.* Nil, *der*

nimble ['nɪmbl] *adj.,* **nimbly** ['nɪmblɪ] *adv.* flink

nine [naɪn] **1.** *adj.* neun. **2.** *n.* Neun, *die. See also* **eight**

nineteen [naɪn'ti:n] **1.** *adj.* neunzehn. **2.** *n.* Neunzehn, *die. See also* **eight**.

nineteenth [naɪn'ti:nθ] **1.** *adj.* neunzehnt... **2.** *n. (fraction)* Neunzehntel, *das. See also* **eighth**

ninetieth ['naɪntɪɪθ] *adj.* neunzigst...

ninety ['naɪntɪ] **1.** *adj.* neunzig. **2.** *n.* Neunzig, *die. See also* **eight; eighty 2**

ninth [naɪnθ] **1.** *adj.* neunt... **2.** *n. (in sequence)* neunte, *der/die/das; (in rank)* Neunte, *der/die/das; (fraction)* Neuntel, *das. See also* **eighth**

nip 1. *v. t.,* -pp- zwicken. **2.** *v. i.,* -pp- *(Brit. sl.)* ~ **in** hinein-/hereinflitzen

(ugs.); ~ **out** hinaus-/herausflitzen *(ugs.)*. **3.** *n. (pinch, squeeze)* Kniff, *der; (bite)* Biß, *der.* **'nipper** *n. (Brit. coll.: child)* Balg, *das (ugs.)*

nipple ['nɪpl] *n.* **a)** Brustwarze, *die;* **b)** *(of feeding-bottle)* Sauger, *der*

nitric acid ['naɪtrɪk æsɪd] *n.* Salpetersäure, *die*

nitrogen ['naɪtrədʒən] *n.* Stickstoff, *der*

nitwit ['nɪtwɪt] *n. (coll.)* Trottel, *der (ugs.)*

no [nəʊ] **1.** *adj.* kein. **2.** *adv.* **a)** *(by no amount)* nicht; **no less |than|** nicht weniger [als]; **no more wine?** keinen Wein mehr?; **b)** *(as answer)* nein. **3.** *n., pl.* **noes** [nəʊz] Nein, *das*

No. *abbr.* **number** Nr.

Noah's ark [nəʊəz 'ɑ:k] *n.* die Arche Noah

nobility [nə'bɪlɪtɪ] *n.* Adel, *der;* **many of the** ~: viele Adlige

noble ['nəʊbl] **1.** *adj.* ad[e]lig; edel ⟨Gedanken, Gefühle⟩. **2.** *n.* Adlige, *der/die.* **nobleman** ['nəʊblmən] *n., pl.* **noblemen** ['nəʊblmən] Adlige, *der*

nobly ['nəʊblɪ] *adv.* **a)** edel[gesinnt]; **b)** *(generously)* edelmütig *(geh.)*

nobody ['nəʊbədɪ] *n. & pron.* niemand; keiner; *(person of no importance)* Niemand, *der*

nocturnal [nɒk'tɜ:nl] *adj.* nächtlich; ~ **animal/bird** Nachttier, *das/*-vogel, *der*

nod [nɒd] **1.** *v. i.,* -dd- nicken. **2.** *v. t.,* -dd-: ~ **one's head |in greeting|** [zum Gruß] mit dem Kopf nicken. **3.** *n.* [Kopf]nicken, *das.* **nod 'off** *v. i.* einnicken *(ugs.)*

noise [nɔɪz] *n.* Geräusch, *das; (loud, harsh, unwanted)* Lärm, *der.* **'noiseless** *adj.,* **'noiselessly** *adv.* lautlos. **noisily** ['nɔɪzɪlɪ] *adv.,* **noisy** ['nɔɪzɪ] *adj.* laut

nomad ['nəʊmæd] *n.* Nomade, *der.* **nomadic** [nəʊ'mædɪk] *adj.* nomadisch; ~ **tribe** Nomadenstamm, *der*

'no man's land *n.* Niemandsland, *das*

nominal ['nɒmɪnl] *adj.* nominell; äußerst niedrig ⟨Preis, Miete⟩

nominate ['nɒmɪneɪt] *v. t.* **a)** *(propose)* nominieren; **b)** *(appoint)* ernennen. **nomination** [nɒmɪ'neɪʃn] *n. see* **nominate:** Nominierung, *die;* Ernennung, *die*

nominative ['nɒmɪnətɪv] *adj. & n.* ~ |**case**| Nominativ, *der*

nominee [nɒmɪ'ni:] *n. (candidate)* Kandidat, *der/*Kandidatin, *die*

non- [nɒn] *pref.* nicht-
nonchalant ['nɒnʃələnt] *adj.* unbekümmert
non-commissioned 'officer *n.* Unteroffizier, *der*
non-committal [nɒnkə'mɪtl] *adj.* unverbindlich; **he was ~:** er hat sich nicht klar geäußert
nondescript ['nɒndɪskrɪpt] *adj.* unscheinbar; undefinierbar ⟨Farbe⟩
none [nʌn] **1.** *pron.* kein...; **~ of them** keiner/keine/keines von ihnen; **~ of this** nichts davon. **2.** *adv.* keineswegs; **I'm ~ the wiser now** jetzt bin ich um nichts klüger; **~ the less** nichtsdestoweniger
nonentity [nɒ'nɛntɪtɪ] *n.* Nichts, *das*
non-existent [nɒnɪg'zɪstənt] *adj.* nicht vorhanden
non-'fiction *n.* Sachliteratur, *die*
non-'iron *adj.* bügelfrei
non-'member *n.* Nichtmitglied, *das*
nonplus [nɒn'plʌs] *v.t.,* **-ss-** verblüffen
nonsense ['nɒnsəns] **1.** *n.* Unsinn, *der.* **2.** *int.* Unsinn. **nonsensical** [nɒn'sensɪkl] *adj.* unsinnig
non-'smoker *n.* **a)** *(person)* Nichtraucher, *der*/-raucherin, *die;* **b)** *(train compartment)* Nichtraucherabteil, *das*
non-'stick *adj.* **~ frying-pan** *etc.* Bratpfanne *usw.* mit Antihaftbeschichtung
non-stop 1. ['--] *adj.* durchgehend ⟨Zug, Busverbindung⟩; Nonstop⟨flug, -revue⟩. **2.** [-'-] *adv.* ohne Unterbrechung ⟨tanzen, reden, reisen, senden⟩; nonstop ⟨fliegen, tanzen, fahren⟩
noodle ['nu:dl] *n., usu. pl.* Nudel, *die*
nook [nʊk] *n.* Winkel, *der;* Ecke, *die*
noon [nu:n] *n.* Mittag, *der;* zwölf Uhr [mittags]; **at/before ~:** um/vor zwölf [Uhr mittags]
'no one *pron. see* **nobody**
noose [nu:s] *n.* Schlinge, *die*
nor [nə(r), *stressed* nɔ:(r)] *conj.* noch; **neither/not ... ~ ...:** weder ... noch ...
norm [nɔ:m] *n.* Norm, *die*
normal ['nɔ:ml] **1.** *adj.* normal. **2.** *n.* **a)** *(~ value)* Normalwert, *der;* **b)** *(usual state)* normaler Stand; **everything is back to** *or* **has returned to ~:** es hat sich wieder alles normalisiert. **normality** [nɔ:'mælɪtɪ] Normalität, *die.* **'normally** *adv.* **a)** *(in normal way)* normal; **b)** *(ordinarily)* normalerweise
north [nɔ:θ] **1.** *n.* **a)** Norden, *der;* **in/to[wards]/from the ~:** im/nach/von Norden; **to the ~ of** nördlich von; **b)**

usu. **N~** *(Geog., Polit.)* Norden, *der.* **2.** *adj.* nördlich; Nord⟨wind, -küste, -grenze⟩. **3.** *adv.* nach Norden; **~ of** nördlich von
north: ~ 'Africa *pr. n.* Nordafrika *(das);* **N~ A'merica** *pr. n.* Nordamerika *(das);* **N~ A'merican 1.** *adj.* nordamerikanisch; **2.** *n.* Nordamerikaner, *der*/-amerikanerin, *die;* **~bound** *adj.* ⟨Zug, Verkehr usw.⟩ in Richtung Norden; **~-'east 1.** *n.* Nordosten, *der;* **2.** *adj.* nordöstlich; Nordost⟨wind, -küste⟩; **3.** *adv.* nordostwärts; nach Nordosten; **~-'eastern** *adj.* nordöstlich
northerly ['nɔ:ðəlɪ] *adj.* nördlich; ⟨Wind⟩ aus nördlichen Richtungen
northern ['nɔ:ðən] *adj.* nördlich; Nord⟨grenze, -hälfte, -seite⟩. **Northern 'Ireland** *pr. n.* Nordirland *(das)*
North: ~ 'Germany *pr. n.* Norddeutschland *(das);* **~ 'Pole** *pr. n.* Nordpol, *der;* **~ 'Sea** *pr. n.* Nordsee, *die*
northward[s] ['nɔ:θwəd(z)] *adv.* nordwärts
north: ~-'west 1. *n.* Nordwesten, *der;* **2.** *adj.* nordwestlich; Nordwest⟨wind, -küste⟩; **3.** *adv.* nordwestwärts; nach Nordwesten; **~-'western** *adj.* nordwestlich
Norway ['nɔ:weɪ] *pr. n.* Norwegen *(das).* **Norwegian** [nɔ:'wi:dʒn] **1.** *adj.* norwegisch; **sb. is ~:** jmd. ist Norweger/Norwegerin. **2.** *n.* **a)** *(person)* Norweger, *der*/Norwegerin, *die;* **b)** *(language)* Norwegisch, *das; see also* **English 2 a**
Nos. *abbr.* numbers Nrn.
nose [nəʊz] **1.** *n.* Nase, *die.* **2.** *v.t.* **~ one's way** sich *(Dat.)* vorsichtig seinen Weg bahnen. **3.** *v.i.* sich vorsichtig bewegen. **nose a'bout, nose a'round** *v.i. (coll.)* herumschnüffeln *(ugs.)*
nose: ~bleed *n.* Nasenbluten, *das;* **~dive 1.** *n.* Sturzflug, *der;* **2.** *v.i.* im Sturzflug hinuntergehen
nosey *see* **nosy**
nostalgia [nɒ'stældʒə] *n.* Nostalgie, *die;* **~ for sth.** Sehnsucht nach etw.
nostalgic [nɒ'stældʒɪk] *adj.* nostalgisch
nostril ['nɒstrɪl] *n.* Nasenloch, *das; (of horse)* Nüster, *die*
nosy ['nəʊzɪ] *adj. (sl.)* neugierig
not [nɒt] *adv.* nicht; **he is ~ a doctor** er ist kein Arzt; **~ at all** überhaupt nicht; **~ ... but ...:** nicht ..., sondern ...; **~ a thing** gar nichts

notable ['nəʊtəbl] *adj.* bemerkenswert; **be ~ for** sth. für etw. bekannt sein. **notably** ['nəʊtəblɪ] *adv.* besonders

notation [nəʊ'teɪʃn] *n.* Notierung, *die*

notch [nɒtʃ] **1.** *n.* Kerbe, *die.* **2.** *v. t.* kerben. **notch 'up** *v. t.* erreichen

note [nəʊt] **1.** *n.* **a)** *(Mus.) (sign)* Note, *die; (key of piano)* Taste, *die; (sound)* Ton, *der;* **b)** *(jotting)* Notiz, *die;* **take** *or* **make ~s** sich *(Dat.)* Notizen machen; **take** *or* **make a ~ of** sth. sich *(Dat.)* etw. notieren; **c)** *(comment, footnote)* Anmerkung, *die;* **d)** *(short letter)* [kurzer] Brief; **e)** *(importance)* **a person/something of ~:** eine bedeutende Persönlichkeit/etwas Bedeutendes; **be of ~:** bedeutend sein. **2.** *v. t.* **a)** *(pay attention to)* beachten; **b)** *(notice)* bemerken; **c)** *(write)* ~ |**down**| [sich *(Dat.)*] notieren. **'notebook** *n.* Notizbuch, *das*

'noted *adj.* bekannt (**for** für, wegen). **note:** **~pad** *n.* Notizblock, *der;* **~paper** *n.* Briefpapier, *das;* **~worthy** *adj.* bemerkenswert

nothing ['nʌθɪŋ] *n.* nichts; **~ interesting** nichts Interessantes; **~ much** nichts Besonderes; **~ more than** nur; **~ more, ~ less** nicht mehr, nicht weniger; **next to ~:** so gut wie nichts; **have** |got| *or* **be ~ to do with** sb./sth. *(not concern)* nichts zu tun haben mit jmdm./etw.; **have ~ to do with** sb. *(avoid)* jmdm. aus dem Weg gehen

notice ['nəʊtɪs] **1.** *n.* **a)** Anschlag, *der; (in newspaper)* Anzeige, *die;* **b)** *(warning)* **at short/a moment's ~:** kurzfristig/von einem Augenblick zum andern; **c)** *(formal notification)* Ankündigung, *die;* **until further ~:** bis auf weiteres; **d)** *(ending an agreement)* Kündigung, *die;* **give** sb. **a month's ~:** jmdm. mit einer Frist von einem Monat kündigen; **hand in one's ~, give ~** *(Brit.),* **give one's ~** *(Amer.)* kündigen; **e)** *(attention)* **bring** sb./sth. **to** sb.'s **~:** jmdn. auf jmdn./etw. aufmerksam machen; **take no ~ of** sb./sth. *(disregard)* keine Notiz von jmdm./etw. nehmen; **take no ~:** sich nicht darum kümmern. **2.** *v. t.* bemerken. **noticeable** ['nəʊtɪsəbl] *adj.* wahrnehmbar ‹*Fleck, Schaden, Geruch*›; merklich ‹*Verbesserung*›; spürbar ‹*Mangel*›. **'notice-board** *n. (Brit.)* Anschlagbrett, *das;* Schwarzes Brett

notification [nəʊtɪfɪ'keɪʃn] *n.* Mitteilung, *die* (**of** sth. über etw. *[Akk.]*)

notify ['nəʊtɪfaɪ] *v. t.* **a)** *(make known)* ankündigen; **b)** *(inform)* benachrichtigen (**of** über + *Akk.*)

notion ['nəʊʃn] *n.* Vorstellung, *die;* **not have the faintest/least ~ of how/ what** *etc.* nicht die blasseste/geringste Ahnung haben, wie/was *usw.*

notoriety [nəʊtə'raɪətɪ] *n.* traurige Berühmtheit

notorious [nə'tɔːrɪəs] *adj.* berüchtigt (**for** wegen); notorisch ‹*Lügner*›

nougat ['nuːgɑː] *n.* Nougat, *das od. der*

nought [nɔːt] *n.* Null, *die*

noun [naʊn] *n. (Ling.)* Substantiv, *das*

nourish ['nʌrɪʃ] *v. t.* ernähren (**on** mit). **'nourishing** *adj.* nahrhaft. **'nourishment** *n.* Nahrung, *die*

Nov. *abbr.* November Nov.

novel ['nɒvl] **1.** *n.* Roman, *der.* **2.** *adj.* neuartig. **novelist** ['nɒvəlɪst] *n.* Romanautor, *der/*-autorin, *die*

novelty ['nɒvltɪ] *n.* **a)** **be a/no ~:** etwas/nichts Neues sein; **b)** *(newness)* Neuheit, *die;* **c)** *(gadget)* Überraschung, *die*

November [nə'vembə(r)] *n.* November, *der; see also* **August**

novice ['nɒvɪs] *n.* Anfänger, *der/*Anfängerin, *die*

now [naʊ] **1.** *adv.* jetzt; *(nowadays)* heutzutage; *(immediately)* [jetzt] sofort; **just ~** *(very recently)* gerade eben; |**every**| **~ and then** *or* **again** hin und wieder; **well ~:** also; **~, ~:** na, na; **~ then** na *(ugs.).* **2.** *conj.* **~** |**that**| ...: jetzt, wo ... **3.** *n.* **before ~:** früher; **by ~:** inzwischen; **a week from ~:** [heute] in einer Woche. **nowadays** ['naʊədeɪz] *adv.* heutzutage

nowhere ['nəʊweə(r)] *adv.* nirgends; nirgendwo; *(to no place)* nirgendwohin

nozzle ['nɒzl] *n.* Düse, *die*

nuance ['njuːɑ̃s] *n.* Nuance, *die*

nuclear ['njuːklɪə(r)] *adj.* Atom-; Kern‹*explosion*›; atomar ‹*Antrieb, Gefechtskopf, Wettrüsten, Abrüstung*›; nuklear ‹*Abschreckung, Sprengkörper*›; atomgetrieben ‹*Unterseeboot*›

nucleus ['njuːklɪəs] *n., pl.* **nuclei** ['njuːklɪaɪ] Kern, *der*

nude [njuːd] **1.** *adj.* nackt. **2.** *n.* **a)** *(figure)* Akt, *der;* **b)** **in the ~:** nackt

nudge [nʌdʒ] **1.** *v. t.* anstoßen. **2.** *n.* Stoß, *der*

nudism ['njuːdɪzm] *n.* Nudismus, *der;* Freikörperkultur, *die.* **nudist** ['njuːdɪst] *n.* Nudist, *der/*Nudistin, *die;* at-

trib. Nudisten-. **nudity** ['nju:dɪtɪ] *n.*
Nacktheit, *die*

nugget ['nʌgɪt] *n.* Klumpen, *der; (of
gold)* Goldklumpen, *der; (fig.)* **~s of
wisdom** goldene Weisheiten

nuisance ['nju:səns] *n.* Ärgernis, *das;*
what a ~! so etwas Dummes!

null [nʌl] *adj.* **~ and void** null und nichtig

numb [nʌm] **1.** *adj.* gefühllos, taub
(with vor + *Dat.*); *(without emotion)*
benommen. **2.** *v. t.* betäuben

number ['nʌmbə(r)] **1.** *n.* **a)** *(in series)*
Nummer, *die;* **you've got the wrong ~**
(Teleph.) Sie sind falsch verbunden;
dial a wrong ~: sich verwählen *(ugs.);*
b) *(esp. Math.: numeral)* Zahl, *die;* **c)**
(sum, total, quantity) [An]zahl, *die;* **a
~ of people/things** einige Leute/Din-
ge; **a ~ of times** mehrmals. **2.** *v. t.* **a)**
(assign ~ to) numerieren; **b)** *(amount
to, comprise)* zählen; **c)** *(include)* zählen **(among, with** zu**); d) sb.'s days are
~ed** jmds. Tage sind gezählt. '**number-plate** *n.* Nummernschild, *das*

numeral ['nju:mərl] *n.* Ziffer, *die*

numerate [nju:mərət] *adj.* **be ~:** rechnen können

numerical [nju:'merɪkl] *adj.* nume-
risch; Zahlen⟨*wert, -folge*⟩; zahlenmäßig ⟨*Stärke, Überlegenheit*⟩

numerous ['nju:mərəs] *adj.* zahlreich

nun [nʌn] *n.* Nonne, *die*

nurse [nɜːs] **1.** *n.* Krankenschwester,
die; |**male**| **~:** Krankenpfleger, *der.* **2.**
v. t. **a)** pflegen ⟨*Kranke*⟩; **b)** *(fig.)* hegen *(geh.)* ⟨*Gefühl, Groll*⟩

nursery ['nɜːsərɪ] *n.* **a)** *(room)* Kinder-
zimmer, *das;* **b)** *(crèche)* Kindertages-
stätte, *die;* **c)** *see* **nursery school; d)**
(for plants) Gärtnerei, *die.* '**nursery
rhyme** *n.* Kinderreim, *der.* '**nursery
school** *n.* Kindergarten, *der*

nursing ['nɜːsɪŋ] *n.* Krankenpflege,
die; attrib. Pflege⟨*personal, -beruf*⟩.
'**nursing home** *n.* Pflegeheim, *das*

nurture ['nɜːtʃə(r)] *v. t. (rear)* aufziehen; *(fig.)* nähren

nut [nʌt] *n.* **a)** Nuß, *die;* **b)** *(Mech.
Engin.)* [Schrauben]mutter, *die;* **c)**
(crazy person) Verrückte, *der/die
(ugs.).* '**nut-case** *n. (sl.)* Verrückte,
der/die (ugs.). '**nutcrackers** *n. pl.*
Nußknacker, *der*

nutmeg ['nʌtmeg] *n.* Muskat, *der*

nutrient ['nju:trɪənt] *n.* Nährstoff, *der*

nutrition [nju:'trɪʃn] *n.* Ernährung,
die; (food) Nahrung, *die.* **nutritious**
[nju:'trɪʃəs] *adj.* nahrhaft

'**nutshell** *n.* Nußschale, *die;* **in a ~**
(fig.) in aller Kürze

nutty ['nʌtɪ] *adj.* **a)** *(in taste)* nussig; **b)**
(sl.: crazy) verrückt *(ugs.)*

nuzzle ['nʌzl] *v. i.* sich kuscheln **(up to,
against** an + *Akk.*)

NW *abbr.* **north-west** NW

nylon ['naɪlɒn] *n.* **a)** Nylon, *das; attrib.* Nylon-; **b)** *in pl. (stockings)* Ny-
lonstrümpfe

nymph [nɪmf] *n.* Nymphe, *die*

O

O, o [əʊ] *n.* O, o, *das*

oaf [əʊf] *n.* Stoffel, *der (ugs.)*

oak [əʊk] *n.* Eiche, *die*

OAP *abbr. (Brit.)* **old-age pensioner**
Rentner, *der*/Rentnerin, *die*

oar [ɔː(r)] *n.* Ruder, *das*

oasis [əʊ'eɪsɪs] *n., pl.* **oases** [əʊ'eɪsiːz]
Oase, *die*

oat [əʊt] *n.* **~s** Hafer, *der*

oath [əʊθ] *n.* **a)** Eid, *der;* Schwur, *der;*
b) *(swear-word)* Fluch, *der*

obedience [ə'biːdɪəns] *n.* Gehorsam,
der

obedient [ə'biːdɪənt] *adj.* gehorsam;
be ~ to sb./sth. jmdm./einer Sache gehorchen. **o'bediently** *adv.* gehorsam

obelisk ['ɒbəlɪsk] *n.* Obelisk, *der*

obese [əʊ'biːs] *adj.* fettleibig.
obesity [əʊ'biːsɪtɪ] *n.* Fettleibigkeit,
die

obey [əʊ'beɪ] **1.** *v. t.* gehorchen
(+ *Dat.*); sich halten an (+ *Akk.*)
⟨*Vorschrift, Regel*⟩; befolgen ⟨*Befehl*⟩.
2. *v. i.* gehorchen

obituary [ə'bɪtjʊərɪ] *n.* Nachruf, *der*
(to, of auf + *Akk.*)

object 1. ['ɒbdʒɪkt] *n.* **a)** *(thing)* Gegenstand, *der;* **b)** *(purpose)* Ziel, *das;*
c) *(obstacle)* **money/time** etc. **is no ~:**
Geld/Zeit *usw.* spielt keine Rolle; **d)**
(Ling.) Objekt, *das.* **2.** [əb'dʒekt] *v. i.*
a) Einwände/einen Einwand erheben
(to gegen); **b)** *(have objection or dislike)* etwas dagegen haben); **~ to sth./
sth.** etwas gegen jmdn./etw. haben. **3.**
v. t. einwenden. **objection** [əb-

'dʒekʃn] *n.* **a)** Einwand, *der;* **raise** *or* **make an ~** [to sth.] einen Einwand [gegen etw.] erheben; **b)** *(dislike)* Abneigung, *die;* **have an/no ~ to sb./sth.** etw./nichts gegen jmdn./etw. haben; **have no ~s** nichts dagegen haben. **objectionable** [əb'dʒekʃənəbl] *adj.* unangenehm ⟨Anblick, Geruch⟩; anstößig ⟨Bemerkung, Wort, Benehmen⟩

objective [əb'dʒektɪv] **1.** *adj.* objektiv. **2.** *n. (goal)* Ziel, *das.* **ob'jectively** *adv.* objektiv. **objectivity** [ɒbdʒek-'tɪvɪtɪ] *n.* Objektivität, *die*

obligation [ɒblɪ'geɪʃn] *n.* Verpflichtung, *die;* **be under an ~ to sb.** jmdm. verpflichtet sein; **without ~:** unverbindlich

obligatory [ə'blɪgətərɪ] *adj.* obligatorisch; **it has become ~ to ...:** es ist jetzt Pflicht, zu ...

oblige [ə'blaɪdʒ] *v. t.* **a)** *(be binding on)* **~ sb. to do sth.** jmdm. vorschreiben, etw. zu tun; **b)** *(compel)* zwingen; **be ~d to do sth.** gezwungen sein, etw. zu tun; **feel ~d to do sth.** sich verpflichtet fühlen, etw. zu tun; **c)** *(be kind to)* **~ sb. by doing sth.** jmdm. den Gefallen tun und etw. tun; **d)** *(grateful)* **be much/greatly ~d to sb.** [for sth.] jmdm. [für etw.] sehr verbunden sein; **much ~d!** besten Dank! **obliging** [ə'blaɪdʒɪŋ] *adj.* entgegenkommend

oblique [ə'bli:k] *adj.* schief ⟨Gerade, Winkel⟩; *(fig.)* indirekt

obliterate [ə'blɪtəreɪt] *v. t.* auslöschen

oblivion [ə'blɪvɪən] *n.* Vergessenheit, *die;* **sink** *or* **fall into ~:** in Vergessenheit geraten

oblivious [ə'blɪvɪəs] *adj.* **be ~ to** *or* **of sth.** sich *(Dat.)* einer Sache *(Gen.)* nicht bewußt sein

oblong ['ɒblɒŋ] **1.** *adj.* rechteckig. **2.** *n.* Rechteck, *das*

obnoxious [əb'nɒkʃəs] *adj.* widerlich

oboe ['əʊbəʊ] *n.* Oboe, *die*

obscene [əb'si:n] *adj.* obszön. **obscenity** [əb'senɪtɪ] *n.* Obszönität, *die*

obscure [əb'skjʊə(r)] **1.** *adj.* **a)** *(unexplained)* dunkel; **b)** *(hard to understand)* schwer verständlich ⟨Argument, Dichtung, Autor, Stil⟩; **c)** *(unknown)* unbekannt. **2.** *v. t.* **a)** *(make indistinct)* verdunkeln; versperren ⟨Aussicht⟩; **b)** *(make unintelligible)* unverständlich machen

obsequious [əb'si:kwɪəs] *adj.* unterwürfig

observance [əb'zɜ:vəns] *n.* Einhaltung, *die*

observant [əb'zɜ:vənt] *adj.* aufmerksam

observation [ɒbzə'veɪʃn] *n.* **a)** Beobachtung, *die;* **be [kept] under ~:** beobachtet werden; *(by police)* überwacht werden; **b)** *(remark)* Bemerkung, *die* (**on** über + *Akk.*)

observatory [əb'zɜ:vətərɪ] *n. (Astron.)* Sternwarte, *die*

observe [əb'zɜ:v] *v. t.* **a)** *(watch)* beobachten; *(perceive)* bemerken; **b)** *(abide by, keep)* einhalten; **c)** *(say)* bemerken. **ob'server** *n.* Beobachter, *der*/Beobachterin, *die*

obsess [əb'ses] *v. t.* **~ sb.** von jmdm. Besitz ergreifen *(fig.);* **be/become ~ed with** *or* **by sb./sth.** von jmdm./etw. besessen sein/werden. **obsession** [əb-'seʃn] *n.* Zwangsvorstellung, *die.* **obsessive** [əb'sesɪv] *adj.* zwanghaft; **be ~ about sth.** von etw. besessen sein

obsolete ['ɒbsəli:t] *adj.* veraltet

obstacle ['ɒbstəkl] *n.* Hindernis, *das* (**to** für)

obstinacy ['ɒbstɪnəsɪ] *n. see* **obstinate:** Starrsinn, *der;* Hartnäckigkeit, *die*

obstinate ['ɒbstɪnət] *adj.* starrsinnig; *(adhering to particular course of action)* hartnäckig

obstruct [əb'strʌkt] *v. t.* **a)** *(block)* blockieren; behindern ⟨Verkehr⟩; **~ sb.'s view** jmdm. die Sicht versperren; **b)** *(fig.: impede; also Sport)* behindern. **obstruction** [əb'strʌkʃn] *n.* Blockierung, *die; (of progress; also Sport)* Behinderung, *die.* **obstructive** [əb'strʌktɪv] *adj.* hinderlich; obstruktiv ⟨Politik, Taktik⟩; **be ~** ⟨Person:⟩ sich querlegen *(ugs.)*

obtain [əb'teɪn] *v. t.* bekommen; erzielen ⟨Resultat, Wirkung⟩. **obtainable** [əb'teɪnəbl] *adj.* erhältlich

obtrusive [əb'tru:sɪv] *adj.* aufdringlich; *(conspicuous)* auffällig

obtuse [əb'tju:s] *adj.* **a)** stumpf ⟨Winkel⟩; **b)** *(stupid)* begriffsstutzig

obvious ['ɒbvɪəs] *adj.* offenkundig; *(easily seen)* augenfällig; **be ~** [to sb.] **that ...:** [jmdm.] klar sein, daß ... **'obviously** *adv.* offenkundig; sichtlich ⟨enttäuschen, überraschen usw.⟩

occasion [ə'keɪʒn] **1.** *n.* **a)** Gelegenheit, *die;* **rise to the ~:** sich der Situation gewachsen zeigen; **on several ~s** bei mehreren Gelegenheiten; **on ~[s]** gelegentlich; **b)** *(special occurrence)* Anlaß, *der;* **it was quite an ~:** es war ein Ereignis; **c)** *(reason)* Grund, *der*

(for zu**). 2.** *v. t.* verursachen. **occasional** [ə'keɪʒənl] *adj.* gelegentlich; vereinzelt ⟨*Regenschauer*⟩. **oc'casionally** *adv.* gelegentlich; |only| very ~: gelegentlich einmal

occult [ɒ'kʌlt, 'ɒkʌlt] *adj.* okkult; **the** ~: das Okkulte

occupant ['ɒkjʊpənt] *n.* Bewohner, *der*/Bewohnerin, *die*; *(of car, bus, etc.)* Insasse, *der*/Insassin, *die*

occupation [ɒkjʊ'peɪʃn] *n.* **a)** *(Mil.)* Besetzung, *die*; *(period)* Besatzungszeit, *die*; **b)** *(activity)* Beschäftigung, *die*; **c)** *(profession)* Beruf, *der*. **occupational** [ɒkjʊ'peɪʃənl] *adj.* Berufs-⟨*beratung, -risiko*⟩; ~ **therapy** Beschäftigungstherapie, *die*

occupier ['ɒkjʊpaɪə(r)] *n.* *(Brit.)* Besitzer, *der*/Besitzerin, *die*; *(tenant)* Bewohner, *der*/Bewohnerin, *die*

occupy ['ɒkjʊpaɪ] *v. t.* **a)** *(Mil.; as demonstration)* besetzen; **b)** *(live in)* bewohnen; **c)** *(take up, fill)* einnehmen; belegen ⟨*Zimmer*⟩; in Anspruch nehmen ⟨*Zeit, Aufmerksamkeit*⟩; **d)** *(busy, employ)* beschäftigen

occur [ə'kɜː(r)] *v. i.,* **-rr-: a)** *(be met with)* vorkommen; ⟨*Gelegenheit:*⟩ sich bieten; ⟨*Problem:*⟩ auftreten; **b)** *(happen)* ⟨*Veränderung:*⟩ eintreten; ⟨*Unfall, Vorfall:*⟩ sich ereignen; **c)** ~ **to sb.** *(be thought of)* jmdm. in den Sinn kommen; ⟨*Idee:*⟩ jmdm. kommen.

occurrence [ə'kʌrəns] *n.* **a)** *(incident)* Ereignis, *das*; Begebenheit, *die*; **b)** *(occurring)* Vorkommen, *das*

ocean ['əʊʃn] *n.* Ozean, *der*; Meer, *das*

o'clock [ə'klɒk] *adv.* **it is two/six** ~: es ist zwei/sechs Uhr; **at two/six** ~: um zwei/sechs Uhr; **six** ~ *attrib.* Sechs-Uhr-⟨*Zug, Maschine, Nachrichten*⟩

Oct. *abbr.* **October** Okt.

octagon ['ɒktəgən] *n.* Achteck, *das*

octane ['ɒkteɪn] *n.* Oktan, *das*

octave ['ɒktɪv] *n.* Oktave, *die*

October [ɒk'təʊbə(r)] *n.* Oktober, *der*; *see also* **August**

octopus ['ɒktəpəs] *n.* Tintenfisch, *der*

odd [ɒd] *adj.* **a)** *(surplus, spare)* übrig ⟨*Stück, Silbergeld*⟩; **£25 and a few** ~ **pence** 25 Pfund und ein paar Pence; **b)** *(occasional)* gelegentlich; ~ **job/~-job man** Gelegenheitsarbeit, *die*/-arbeiter, *der*; **c)** *(one of pair or group)* einzeln; ~ **socks** nicht zusammengehörende Socken; **be the ~ man out** ⟨*Gegenstand:*⟩ nicht dazu passen; **d)** *(uneven)* ungerade ⟨*Zahl, Seite, Hausnummer*⟩; **e)** *(plus something)*

forty ~: über vierzig; **twelve pounds** ~: etwas mehr als zwölf Pfund; **f)** *(strange, eccentric)* seltsam. **oddity** ['ɒdɪtɪ] *n.* *(object, event)* Kuriosität, *die*. **'oddly** *adv.* seltsam; ~ **enough** seltsamerweise. **'odd-numbered** *adj.* ungerade

odds [ɒdz] *n. pl.* **a)** *(Betting)* Odds *Pl.*; **b)** |the| ~ **are that she did it** wahrscheinlich hat sie es getan; **the** ~ **are against/in favour of sb./sth.** jmds. Aussichten/die Aussichten für etw. sind gering/gut; **c)** ~ **and ends** Kleinigkeiten; *(of food)* Reste; **d)** **be at** ~ **with sb. over sth.** mit jmdm. in etw. *(Dat.)* uneinig sein; **e)** **it makes no/little** ~ |whether ...| es ist völlig/ziemlich gleichgültig[, ob ...]

odious ['əʊdɪəs] *adj.* widerwärtig

odor *etc.* *(Amer.)* see **odour** *etc.*

odour ['əʊdə(r)] *n.* Geruch, *der*. **'odourless** *adj.* geruchlos

of [əv, *stressed* ɒv] *prep.* von; *indicating material, substance* aus; **articles of clothing** Kleidungsstücke; **a friend of mine** ein Freund von mir; **where's that pencil of mine?** wo ist mein Bleistift?; **it was clever of you to do that** es war klug von dir, das zu tun; **the approval of sb.** jmds. Zustimmung; **the works of Shakespeare** Shakespeares Werke; **be made of ...**: aus ... [hergestellt] sein; **the fifth of January** der fünfte Januar; **his love of his father** seine Liebe zu seinem Vater; **person of extreme views** Mensch mit extremen Ansichten; **a boy of 14 years** ein vierzehnjähriger Junge; **the five of us** wir fünf

off [ɒf] **1.** *adv.* **a)** *(away)* **be a few miles** ~: wenige Meilen entfernt sein; **the lake is not far** ~: der See ist nicht weit [weg]; **I'm** ~ **now** ich gehe jetzt; ~ **we go!** los geht's!; **b)** *(not on or attached or supported)* ab; **get the lid** ~: den Deckel abbekommen; **c)** **be** ~ *(switched or turned* ~*)* ⟨*Wasser, Gas, Strom:*⟩ abgestellt sein; **the light/radio** *etc.* **is** ~: das Licht/Radio *usw.* ist aus; **d)** **the meat** *etc.* **is** ~: das Fleisch *usw.* ist schlecht [geworden]; **e)** **be** ~ *(cancelled)* abgesagt sein; ⟨*Verlobung:*⟩ [auf]gelöst sein; ~ **and on** immer und wieder *(ugs.)*; **f)** *(not at work)* frei; **on my day** ~: an meinem freien Tag; **have a week** ~: eine Woche Urlaub bekommen; **g)** *(no longer available)* |the| **soup** *etc.* **is** ~: es gibt keine Suppe *usw.* mehr; **h)** *(situated as regards money etc.)* **he is badly** *etc.* ~: er ist schlecht *usw.* gestellt. **2.** *prep.* von; **be ~ school/work** in der Schule/am Arbeitsplatz fehlen; **be ~ one's food** keinen Appetit haben; **just ~ the square** ganz in der Nähe des Platzes

offal ['ɒfl] *n.* Innereien *Pl.*

offence [ə'fens] *n.* *(Brit.)* **a)** *(hurting of sb.'s feelings)* Kränkung, *die;* **I meant no ~:** ich wollte Sie/ihn *usw.* nicht kränken; **b)** *(annoyance)* **give ~:** Mißfallen erregen; **take ~:** verärgert sein; **c)** *(crime)* Straftat, *die;* **criminal ~:** strafbare Handlung

offend [ə'fend] **1.** *v. i.* verstoßen **(against** gegen). **2.** *v. t.* **~ sb.** bei jmdm. Anstoß erregen; *(hurt feelings of)* jmdn. kränken. **offender** *n.* Straffällige, *der/die*

offense *(Amer.)* see **offence**

offensive [ə'fensɪv] **1.** *adj.* **a)** *(aggressive)* offensiv; Angriffs⟨waffe⟩; **b)** *(giving offence)* ungehörig; *(indecent)* anstößig. **2.** *n.* Offensive, *die;* **take the** *or* **go on the ~:** in die *od.* zur Offensive übergehen

offer ['ɒfə(r)] **1.** *v. t.* anbieten; vorbringen ⟨*Entschuldigung*⟩; bieten ⟨*Chance*⟩; aussprechen ⟨*Beileid*⟩; **~ to help** seine Hilfe anbieten; **~ resistance** Widerstand leisten. **2.** *n.* Angebot, *das;* [**have/be**] **on ~:** im Angebot [haben/sein]

offhand 1. *adv.* **a)** *(without preparation)* auf Anhieb ⟨*sagen, wissen*⟩; spontan ⟨*beschließen, entscheiden*⟩; **b)** *(casually)* leichthin. **2.** *adj.* **a)** *(without preparation)* spontan; **b)** *(casual)* beiläufig; **be ~ with sb.** zu jmdm. kurz angebunden sein

office ['ɒfɪs] *n.* **a)** Büro, *das;* **b)** *(branch)* Zweigstelle, *die;* **c)** *(position)* Amt, *das;* **hold ~:** amtieren. **'office hours** *n. pl.* Dienststunden *Pl.*

officer ['ɒfɪsə(r)] *n.* **a)** *(Army etc.)* Offizier, *der;* **b)** *(official)* Beamte, *der*/Beamtin, *die;* **c)** *(constable)* Polizeibeamte, *der*/-beamtin, *die*

official [ə'fɪʃl] **1.** *adj.* offiziell; amtlich ⟨*Verlautbarung*⟩; regulär ⟨*Streik*⟩. **2.** *n.* Beamte, *der*/Beamtin, *die; (party, union, or sports ~)* Funktionär, *der*/Funktionärin, *die.* **officially** *adv.* offiziell

officious [ə'fɪʃəs] *adj.* übereifrig

offing ['ɒfɪŋ] *n.* **be in the ~:** bevorstehen; ⟨*Gewitter:*⟩ aufziehen

off: **~-licence** *n.* *(Brit.)* ≈ Wein- und Spirituosenladen, *der;* **~-load** *v. t.* abladen; **~-putting** ['ɒfpʊtɪŋ] *adj.*

(Brit. coll.) abstoßend; **~set** ['--, -'-] *v. t., forms as* **set:** ausgleichen; **~shore** *adj.* küstennah; **~'side** *adj.* Abseits-; **be ~-side** abseits sein; **~spring** *n., pl. same* Nachkommenschaft, *die; (of animal)* Junge *Pl.*

often ['ɒfn, 'ɒftn] *adv.* oft; **every so ~:** gelegentlich

oh [əʊ] *int.* oh; *expr. pain* au

oil [ɔɪl] **1.** *n.* Öl, *das.* **2.** *v. t.* ölen

oil: **~field** *n.* Ölfeld, *das;* **~painting** *n.* Ölgemälde, *das;* **~refinery** *n.* [Erd]ölraffinerie, *die;* **~ rig** see ¹**rig** 1; **~skins** *n. pl.* Ölzeug, *das;* **~-slick** *n.* Ölteppich, *der;* **~-tanker** *n.* Öltanker, *der;* **~ well** *n.* Ölquelle, *die*

oily ['ɔɪlɪ] *adj.* ölig; ölverschmiert ⟨*Gesicht, Hände*⟩

ointment ['ɔɪntmənt] *n.* Salbe, *die*

OK [əʊ'keɪ] *(coll.)* **1.** *adj.* in Ordnung; okay *(ugs.).* **2.** *adv.* gut. **3.** *int.* okay *(ugs.).* **4.** *v. t. (approve)* zustimmen (+ *Dat.*); **be ~'d by sb.** von jmdm. das Okay bekommen *(ugs.)*

okay [əʊ'keɪ] see **OK**

old [əʊld] *adj.* alt; **be [more than] 30 years ~:** [über] 30 Jahre alt sein

old: **~ 'age** *n.* [fortgeschrittenes] Alter; **~-age** *attrib. adj.* Alters⟨rente, -ruhegeld⟩; **~-age pensioner** Rentner, *der*/Rentnerin, *die;* **~-fashioned** [əʊld'fæʃnd] *adj.* altmodisch

olive ['ɒlɪv] *n.* Olive, *die.* **olive 'oil** *n.* Olivenöl, *das*

Olympic [ə'lɪmpɪk] *adj.* olympisch; **~ Games** Olympische Spiele

omelette (omelet) ['ɒmlɪt] *n.* Omelett, *das*

omen ['əʊmən] *n.* Vorzeichen, *das*

ominous ['ɒmɪnəs] *adj. (of evil omen)* ominös; *(worrying)* beunruhigend

omission [ə'mɪʃn] *n.* Auslassung, *die; (failure to act)* Unterlassung, *die*

omit [ə'mɪt] *v. t.,* **-tt-** weglassen; **~ to do sth.** es versäumen, etw. zu tun

on [ɒn] **1.** *prep.* auf *(position:* + *Dat.; direction:* + *Akk.); (attached to)* an (+ *Dat./Akk.); (concerning, about)* über (+ *Akk.); in expressions of time* an ⟨*einem Abend, Tag usw.*⟩; **write sth. on the wall** etw. an die Wand schreiben; **be hanging on the wall** an der Wand hängen; **have sth. on one** etw. bei sich haben; **on the bus/train** im Bus/Zug; *(by bus/train)* mit dem Bus/Zug; **on Oxford 56767** unter der Nummer Oxford 56767; **on Sundays** sonntags; **on [his] arrival** bei seiner

Ankunft; **on entering the room ...:**
beim Betreten des Zimmers ...; **it's just
on 9** es ist fast 9 Uhr; **the drinks are on
me** *(coll.)* die Getränke gehen auf
mich. **2.** *adv.* **with/without a hat/coat
on** mit/ohne Hut/Mantel; **have a hat
on** einen Hut aufhaben; **on and on immer** weiter; **speak/wait/work** *etc.* **on**
weiterreden/-warten/-arbeiten *usw.;*
from now on von jetzt an; **the light/
radio** *etc.* **is on** das Licht/Radio *usw.*
ist an; **is Sunday's picnic on?** findet
das Picknick am Sonntag statt?;
what's on at the cinema? was läuft im
Kino?; **on and off** immer mal wieder
(ugs.); **on to, onto** auf (+ *Akk.*)
once [wʌns] **1.** *adv.* **a)** einmal; ~ **a
week/month/year** einmal die Woche/
im Monat/im Jahr; ~ **again** *or* **more**
noch einmal; ~ **[and] for all** ein für allemal; **never/not** ~: nicht ein einziges
Mal; **b)** *(multiplied by one)* einmal; **c)**
(formerly) früher einmal; ~ **upon a
time there lived a king** es war einmal
ein König; **d) at** ~ *(immediately)* sofort; *(at the same time)* gleichzeitig;
all at ~ *(suddenly)* plötzlich; *(simultaneously)* alle[s] zugleich. **2.** *conj.*
wenn; *(with past tense)* als. **3.** *n.* [just
or only] **this** ~: [nur] dieses eine Mal
'**oncoming** *adj.* entgegenkommend
⟨*Fahrzeug, Verkehr*⟩
one [wʌn] **1.** *adj.* ein; *see also* **eight 1;**
(single, only) einzig; **no/not** ~: kein;
the ~ **thing** das einzige; **at** ~ **time** einmal; ~ **morning/night** eines Morgens/
Nachts. **2.** *n.* **a)** eins; **b)** *(number, symbol)* Eins, *die;* **c)** *(unit)* **in** ~**s** einzeln.
3. *pron.* **a)** ein... (of + *Gen.*); **big** ~**s
and little** ~**s** große und kleine; **the
older/younger** ~: der/die/das ältere/
jüngere; **this** ~: dieser/diese/dieses
[da]; **that** ~: der/die/das [da]; **which
~?** welcher/welche/welches?; **which
~s?** welche?; ~ **by** ~: einzeln; **love/
hate** ~ **another** sich lieben/hassen; **be
kind to** ~ **another** nett zueinander
sein; **b)** *(people in general; coll.: I, we)*
man; *as indirect object* einem; *as
direct object* einen; ~**'s** sein
one- : ~'**self** *pron.* **a)** *emphat.* selbst; **be
~self** man selbst sein; **b)** *refl.* sich; *see
also* **herself;** ~**-sided** *adj.* einseitig;
~**-way** *adj.* **a)** in einer Richtung
nachgestellt ⟨*straße, -ver-
kehr*⟩; **b)** einfach ⟨*Fahrpreis, Flug*⟩
onion ['ʌnjən] *n.* Zwiebel, *die*
'**onlooker** *n.* Zuschauer, *der/*Zu-
schauerin, *die*

only ['əʊnlɪ] **1.** *attrib. adj.* einzig...; **the
~ person** der/die einzige; **an ~ child**
ein Einzelkind. **2.** *adv.* nur; **we had
been waiting ~ 5 minutes when ...:** wir
hatten erst 5 Minuten gewartet, als ...;
it's ~/~ just 6 o'clock es ist erst 6 Uhr/
gerade erst 6 Uhr vorbei; **he ~ just
made it** er hat es gerade noch geschafft; ~ **if** nur [dann] ..., wenn; ~ **the
other day/week** erst neulich
'**onset** *n.* *(of winter)* Einbruch, *der; (of
disease)* Ausbruch, *der*
'**onslaught** ['ɒnslɔ:t] *n.* [heftige] Attacke *(fig.)*
onus ['əʊnəs] *n.* **the ~ is on him to do it**
es ist seine Sache, es zu tun
onward[s] ['ɒnwədz] *adv. (in space)*
vorwärts; **from X ~:** von X an; **from
that day ~:** von diesem Tag an
ooze [u:z] **1.** *v. i.* sickern **(from** aus). **2.**
v. t. triefen von *od.* vor (+ *Dat.*);
(fig.) ausstrahlen
opaque [əʊ'peɪk] *adj.* lichtundurch-
lässig; opak *(fachspr.)*
open ['əʊpn] **1.** *adj.* **a)** offen; *(not
blocked or obstructed)* frei; *(available)*
frei ⟨*Stelle*⟩; **in the ~ air** im Freien; **be
~** ⟨*Laden, Museum, Bank usw.*⟩: geöffnet sein; **have an ~ mind about** *or*
on sth. einer Sache gegenüber aufgeschlossen sein; **b)** unverhohlen ⟨*Bewunderung, Haß, Verachtung*⟩; **c)**
(frank, communicative) offen ⟨*Wesen,
Streit, Abstimmung, Regierungsstil*⟩;
(not secret) öffentlich ⟨*Wahl*⟩; **d)** geöffnet ⟨*Regenschirm*⟩; aufgeblüht
⟨*Blume, Knospe*⟩; aufgeschlagen ⟨*Zeitung, Landkarte*⟩. **2.** *n.* **in the ~** *(out-
doors)* unter freiem Himmel; **[out] in
the ~** *(fig.)* öffentlich bekannt. **3.** *v. t.*
a) öffnen; **b)** eröffnen ⟨*Konferenz,
Diskussion, Laden*⟩; beginnen ⟨*Verhandlungen, Spiel*⟩; **c)** *(unfold, spread
out)* aufschlagen ⟨*Zeitung, Land-
karte*⟩; öffnen ⟨*Schirm*⟩. **4.** *v. i.* **a)** sich
öffnen; ~ **into/on to sth.** zu etw. führen; **b)** *(become ~ to customers)* öffnen; *(start trading etc.)* eröffnet werden; **c)** *(start)* beginnen; *(Ausstellung:)* eröffnet werden; ⟨*Theaterstück:*⟩ Premiere haben. **open 'up 1.**
v. t. öffnen; *(establish)* eröffnen. **2.**
v. i. sich öffnen; ⟨*Filiale:*⟩ eröffnet
werden; ⟨*Firma:*⟩ sich niederlassen
'**open-air** *attrib. adj.* Openair⟨*konzert*⟩; ~ **[swimming-]pool** Freibad, *das*
'**opener** *n.* Öffner, *der*
'**opening 1.** *n.* **a)** Öffnen, *das; (becoming open)* Sichöffnen, *das; (of exhibi-*

tion, new centre) Eröffnen, *das;* **b)** *(establishment, ceremony)* Eröffnung, die; **c)** *(initial part)* Anfang, *der;* **d)** *(gap, aperture)* Öffnung, *die;* **e)** *(opportunity)* Möglichkeit, *die; (vacancy)* freie Stelle. **2.** *adj.* einleitend. **'opening hours** *n. pl.* Öffnungszeiten *Pl.* **'openly** *adv.* **a)** *(publicly)* in der Öffentlichkeit; öffentlich ⟨*zugeben, verurteilen*⟩; **b)** *(frankly)* offen

open: ~-'minded *adj.* aufgeschlossen; ~-'plan *adj.* ~-plan office Großraumbüro, *das;* ~ 'sandwich *n.* belegtes Brot

opera ['ɒpərə] *n.* Oper, *die*

opera: ~-glasses *n. pl.* Opernglas, *das;* ~-house *n.* Opernhaus, *das;* ~-singer *n.* Opernsänger, *der/*-sängerin, *die*

operate ['ɒpəreɪt] **1.** *v. i.* **a)** *(be in action)* in Betrieb sein; ⟨*Bus, Zug usw.:*⟩ verkehren; **b)** *(function)* arbeiten; **the torch ~s on batteries** das Taschenlampe arbeitet mit Batterien; ~ [on sb.] *(Med.)* [jmdn.] operieren. **2.** *v. t.* bedienen ⟨*Maschine*⟩; unterhalten ⟨*Busverbindung, Telefondienst*⟩; betätigen ⟨*Hebel, Bremse*⟩. **'operating-theatre** *n. (Brit. Med.)* Operationssaal, *der*

operation [ɒpə'reɪʃn] *n.* **a)** *(causing to work) (of machine)* Bedienung, *die; (of bus service, telephone service, etc.)* Unterhaltung, *die; (of lever, brake)* Betätigung, *die;* **b) come into** ~ ⟨*Gesetz, Gebühr usw.:*⟩ in Kraft treten; **be in/out of** ~ ⟨*Maschine, Gerät usw.:*⟩ in/außer Betrieb sein; **c)** *(Med.)* Operation, *die;* **have an** ~: operiert werden

operational [ɒpə'reɪʃənl] *adj. (esp. Mil.: ready to function)* einsatzbereit

operative ['ɒpərətɪv] *adj.* **become** ~ ⟨*Gesetz:*⟩ in Kraft treten; **the scheme is** ~: das Programm läuft

operator ['ɒpəreɪtə(r)] *n.* [Maschinen]bediener, *der/*-bedienerin, *die; (Teleph.) (at exchange)* Vermittlung, *die; (at switchboard)* Telefonist, *der/* Telefonistin, *die*

opinion [ə'pɪnjən] *n.* Meinung, *die* (**on** über + *Akk.,* zu); **have a high/low ~ of sb.** eine/keine hohe Meinung von jmdm. haben; **in my ~:** meiner Meinung nach. **opinionated** [ə'pɪnjəneɪtɪd] *adj.* rechthaberisch

opium ['əʊpɪəm] *n.* Opium, *das*

opponent [ə'pəʊnənt] *n.* Gegner, *der/* Gegnerin, *die*

opportune ['ɒpətjuːn] *adj.* **a)** *(favour-*

able) günstig; **b)** *(well-timed)* zur rechten Zeit nachgestellt. **opportunism** [ɒpə'tjuːnɪzm] *n.* Opportunismus, *der.* **opportunist** [ɒpə'tjuːnɪst] *n.* Opportunist, *der/*Opportunistin, *die*

opportunity [ɒpə'tjuːnɪtɪ] *n.* Gelegenheit, *die*

oppose [ə'pəʊz] **1.** *v. t.* sich wenden gegen. **2.** *v. i.* **the opposing team** die gegnerische Mannschaft. **opposed** [ə'pəʊzd] *adj.* **as** ~ **to** im Gegensatz zu; **be** ~ **to sth.** ⟨*Personen:*⟩ gegen etw. sein

opposite ['ɒpəzɪt] **1.** *adj.* gegenüberliegend ⟨*Straßenseite, Ufer*⟩; entgegengesetzt ⟨*Ende, Weg, Richtung*⟩; **the** ~ **sex** das andere Geschlecht. **2.** *n.* Gegenteil, *das* (**of** von). **3.** *adv.* gegenüber. **4.** *prep.* gegenüber

opposition [ɒpə'zɪʃn] *n.* **a)** Opposition, *die; (resistance)* Widerstand, *der* (**to** gegen); **in** ~ **to** entgegen; **b)** *(Brit. Polit.)* **the O~:** die Opposition

oppress [ə'pres] *v. t.* unterdrücken; *(fig.)* ⟨*Gefühl:*⟩ bedrücken. **oppression** [ə'preʃn] *n.* Unterdrückung, *die.* **oppressive** [ə'presɪv] *adj.* repressiv; *(fig.)* bedrückend ⟨*Ängste, Atmosphäre*⟩; *(hot and close)* drückend ⟨*Wetter, Klima, Tag*⟩

opt [ɒpt] *v. i.* sich entscheiden (**for** für); ~ **to do sth.** sich dafür entscheiden, etw. zu tun; ~ **out** nicht mitmachen/*(stop taking part)* nicht länger mitmachen (**of** bei)

optical ['ɒptɪkl] *adj.* optisch

optician [ɒp'tɪʃn] *n.* Optiker, *der/*Optikerin, *die*

optima *pl. of* optimum

optimism ['ɒptɪmɪzm] *n.* Optimismus, *der.* **optimist** ['ɒptɪmɪst] *n.* Optimist, *der/*Optimistin, *die.* **optimistic** [ɒptɪ'mɪstɪk] *adj.* optimistisch

optimum ['ɒptɪməm] **1.** *n., pl.* **optima** ['ɒptɪmə] Optimum, *das.* **2.** *adj.* optimal

option ['ɒpʃn] *n. (choice)* Wahl, *die; (thing)* Wahlmöglichkeit, *die.* **optional** ['ɒpʃənl] *adj.* nicht zwingend; ~ **subject** Wahlfach, *das*

opulence ['ɒpjʊləns] *n.* Wohlstand, *der*

opulent ['ɒpjʊlənt] *adj.* wohlhabend; feudal ⟨*Auto, Haus usw.*⟩

or [ə(r), *stressed* ɔː(r)] *conj.* **a)** oder; **he cannot read or write** er kann weder lesen noch schreiben; **without food or water** ohne Essen und Wasser; **15 or 20 minutes** 15 bis 20 Minuten; **in a day**

or two in ein, zwei Tagen; **b)** *introdu-*
cing explanation das heißt; **or rather**
beziehungsweise
oracle ['ɒrəkl] *n.* Orakel, *das*
oral ['ɔ:rl] *adj.* mündlich; *(Med.)* oral
orange ['ɒrɪndʒ] **1.** *n.* **a)** *(fruit)* Orange, *die;* Apfelsine, *die;* **b)** *(colour)*
Orange, *das.* **2.** *adj.* orange[farben]
orator ['ɒrətə(r)] *n.* Redner, *der/*Rednerin, *die*
oratory ['ɒrətərɪ] *n.* Redekunst, *die*
orbit ['ɔ:bɪt] **1.** *n. (Astron.)* [Um-
lauf]bahn, *die.* **2.** *v. i.* kreisen. **3.** *v. t.*
umkreisen. **orbital** ['ɔ:bɪtl] *adj.* ~
road Ringstraße, *die*
orchard ['ɔ:tʃəd] *n.* Obstgarten, *der;*
(commercial) Obstplantage, *die*
orchestra ['ɔ:kɪstrə] *n.* Orchester,
das. **orchestral** [ɔ:'kestrl] *adj.* Or-
chester-
orchid ['ɔ:kɪd] *n.* Orchidee, *die*
ordain [ɔ:'deɪn] *v. t.* **a)** *(Eccl.)* ordinieren; **b)** *(decree)* verfügen
ordeal [ɔ:'di:l] *n.* Qual, *die*
order ['ɔ:də(r)] **1.** *n.* **a)** *(sequence)* Reihenfolge, *die;* **out of ~**: durcheinan-
der; **b)** *(regular arrangement, normal*
state) Ordnung, *die;* **be/not be in ~**: in
Ordnung/nicht in Ordnung sein
(ugs.); **be out of/in ~** *(not in/in work-*
ing condition) nicht funktionieren/
funktionieren; **'out of ~'** „außer Betrieb"; **in good/bad ~**: in gutem/
schlechtem Zustand; **c)** *(command)*
Anweisung, *die;* (Mil.) Befehl, *der;* **d)**
in ~ to do sth. um etw. zu tun; **e)**
(Commerc.) Auftrag, *der* (**for** über +
Akk.); *(to waiter, ~ed goods)* Bestellung, *die;* **f) keep ~**: Ordnung
[be]wahren; *see also* **law b; g)** *(reli-*
gious ~) Orden, *der.* **2.** *v. t.* **a)** *(com-*
mand) ⟨*Richter:*⟩ verfügen; **~**
sb. to do sth. jmdn. anweisen/
(Milit.) jmdm. befehlen, etw. zu tun;
b) *(Commerc.)* bestellen (**from** bei); **c)**
(arrange) ordnen
orderly ['ɔ:dəlɪ] **1.** *adj.* friedlich; diszipliniert ⟨*Menge*⟩; *(methodical)* methodisch; *(tidy)* ordentlich. **2.** *n.* **a)** *(Mil.)*
[Offiziers]bursche, *der;* **b) medical ~**:
≈ Krankenpflegehelfer, *der*
ordinal ['ɔ:dɪnl] *adj. & n.* **~ [number]**
Ordinalzahl, *die*
ordinary ['ɔ:dɪnərɪ] *adj. (normal)* normal ⟨*Gebrauch*⟩; üblich ⟨*Verfahren*⟩;
(not exceptional) gewöhnlich
ordination [ɔ:dɪ'neɪʃn] *n. (Eccl.)* Ordination, *die;* Ordinierung, *die*
ore [ɔ:(r)] *n.* Erz, *das*

organ ['ɔ:gən] *n.* **a)** *(Mus.)* Orgel, *die;*
b) *(Biol.)* Organ, *das*
organic [ɔ:'gænɪk] *adj.* organisch;
biologisch-dynamisch ⟨*Ackerbau*⟩;
biodynamisch ⟨*Nahrungsmittel*⟩
organism ['ɔ:gənɪzm] *n.* Organismus,
der
organist ['ɔ:gənɪst] *n.* Organist,
*der/*Organistin, *die*
organization [ɔ:gənaɪ'zeɪʃn] *n.* Organisation, *die;* **~ of time/work** Zeit-/Ar-
beitseinteilung, *die*
organize ['ɔ:gənaɪz] *v. t.* organisieren;
einteilen ⟨*Arbeit, Zeit*⟩; veranstalten
⟨*Konferenz, Festival*⟩; **~ into groups** in
Gruppen einteilen. **'organizer** *n.* Organisator, *der/*Organisatorin, *die;* (of
event, festival) Veranstalter, *der/*Veranstalterin, *die*
orgasm ['ɔ:gæzm] *n.* Orgasmus, *der*
orgy ['ɔ:dʒɪ] *n.* Orgie, *die*
orient 1. ['ɔ:rɪənt] *n.* **the O~**: der
Orient. **2.** ['ɒrɪent] *v. t.* ausrichten (**to-**
wards nach); **~ oneself** sich orientieren
oriental [ɒrɪ'entl] **1.** *adj.* orientalisch.
2. *n.* Asiat, *der/*Asiatin, *die*
orientate ['ɒrɪənteɪt] *see* **orient 2.**
orientation [ɒrɪən'teɪʃn] *n.* Orientierung, *die*
orienteering [ɒrɪən'tɪərɪŋ] *n. (Brit.)*
Orientierungsrennen, *das*
orifice ['ɒrɪfɪs] *n.* Öffnung, *die*
origin ['ɒrɪdʒɪn] *n. (derivation)* Herkunft, *die;* *(beginnings)* Anfänge *Pl.;*
(source) Ursprung, *der;* **country of ~**:
Herkunftsland, *das;* **have its ~ in sth.**
seinen Ursprung in etw. *(Dat.)* haben.
original [ə'rɪdʒɪnl] **1.** *adj.* ursprünglich; Ur⟨*text, -fassung*⟩; eigenständig
⟨*Forschung*⟩; *(inventive)* originell; **an**
~ painting ein Original. **2.** *n.* Original,
das. **originality** [ərɪdʒɪ'nælɪtɪ] *n.* Originalität, *die.* **originally** [ə'rɪdʒɪnəlɪ]
adv. **a)** ursprünglich; **b)** originell
⟨*schreiben usw.*⟩. **originate** [ə'rɪdʒɪ-
neɪt] *v. i.* **~ from** entstehen aus; **~ in**
seinen Ursprung haben in (+ *Dat.*)
ornament ['ɔ:nəmənt] *n.* Ziergegen-
stand, *der.* **ornamental** [ɔ:nə'mentl]
adj. dekorativ; Zier⟨*pflanze, -naht*
usw.⟩
ornate [ɔ:'neɪt] *adj.* reich verziert;
prunkvoll ⟨*Dekoration*⟩
ornithology [ɔ:nɪ'θɒlədʒɪ] *n.* Ornitho-
logie, *die*
orphan ['ɔ:fn] **1.** *n.* Waise, *die.* **2.** *v. t.*
be ~ed [zur] Waise werden. **orphan-**
age ['ɔ:fənɪdʒ] *n.* Waisenhaus, *das*

orthodox ['ɔːθədɒks] *adj.* orthodox
oscillate ['ɒsɪleɪt] *v. i.* schwingen. **os-cillation** [ɒsɪ'leɪʃn] *n.* Schwingen, *das; (single ~)* Schwingung, *die*
ostensible [ɒ'stensɪbl] *adj.* vorgeschoben. **ostensibly** [ɒ'stensɪblɪ] *adv.* vorgeblich
ostentatious [ɒsten'teɪʃəs] *adj.* prunkhaft ⟨*Kleidung, Schmuck*⟩; prahlerisch ⟨*Art*⟩
osteopath ['ɒstɪəpæθ] *n.* Spezialist für Knochenleiden
ostrich ['ɒstrɪtʃ] *n.* Strauß, *der*
other ['ʌðə(r)] **1.** *adj.* **a)** *(not the same)* ander...; **the ~ two/three** *etc. (the remaining)* die beiden/drei *usw.* anderen; **the ~ one** der/die/das andere; **some ~ time** ein andermal; **b)** *(further)* **one ~ thing** noch eins; **some/six ~ people** noch ein paar/noch sechs [andere *od.* weitere] Leute; **no ~ questions** keine weiteren Fragen; **c) ~ than** *(different from)* anders als; *(except)* außer; **d) the ~ day/evening** neulich/neulich abends. **2.** *n.* anderer/andere/anderes; **there are six ~s** es sind noch sechs andere da; **any ~:** irgendein anderer/-eine andere/-ein anderes; **not any ~:** kein anderer/keine andere/kein anderes; **one after the ~:** einer/eine/eins nach dem/der/dem anderen. **3.** *adv.* anders; **~ than that, ...:** abgesehen davon, ...
otherwise ['ʌðəwaɪz] **1.** *adv.* **a)** *(in a different way)* anders; **b)** *(or else)* anderenfalls; **c)** *(in other respects)* im übrigen. **2.** *pred. adj.* anders
otter ['ɒtə(r)] *n.* [Fisch]otter, *der*
ouch [aʊtʃ] *int.* autsch
ought [ɔːt] *v. aux. only in pres. and past* ought, *neg. (coll.)* oughtn't ['ɔːtnt] **I ~ to do/have done it** *expr. moral duty* ich müßte es tun/hätte es tun müssen; *expr. desirability* ich sollte es tun/hätte es tun sollen; **~ not** *or* **~n't you to have left by now?** müßtest du nicht schon weg sein?; **one ~ not to do it** man sollte es nicht tun; **he ~ to be hanged/in hospital** er gehört an den Galgen/ins Krankenhaus; **that ~ to be enough** das dürfte reichen; **he ~ to win** er müßte [eigentlich] gewinnen
oughtn't ['ɔːtnt] *(coll.)* = **ought not**
ounce [aʊns] *n. (measure)* Unze, *die*
our [aʊə(r)] *poss. pron. attrib.* unser
ours ['aʊəz] *poss. pron. pred.* unserer/unsere/unseres; *see also* **hers**
ourselves [aʊə'selvz] *pron.* **a)** *emphat.* selbst; **b)** *refl.* uns. *See also* **herself**

oust [aʊst] *v. t.* verdrängen; **~ sb. from his job/from power** jmdn. von seinem Arbeitsplatz vertreiben/jmdn. entmachten
out [aʊt] *adv.* **a)** *(away from place)* **~ here/there** hier/da draußen; **be ~ in the garden** draußen im Garten sein; **what's it like ~?** wie ist es draußen?; **go ~ shopping** *etc.* einkaufen *usw.* gehen; **be ~** *(not at home, not in one's office, etc.)* nicht dasein; **she was ~ all night** sie war eine/die ganze Nacht weg; **have a day ~ in London** einen Tag in London verbringen; **row ~ to ...:** hinaus-/herausrudern zu ...; **~ at sea** auf See sein; **b)** *(Sport, Games)* **be ~** ⟨*Ball:*⟩ aus *od.* im Aus sein; ⟨*Mitspieler:*⟩ ausscheiden; ⟨*Schlagmann:*⟩ aus[geschlagen] sein; **not ~:** nicht aus; **c) be ~** *(asleep)* weg sein *(ugs.); (unconscious)* bewußtlos sein; **d)** *(no longer burning)* aus[gegangen]; **e)** *(in error)* **be 3% ~ in one's calculations** sich um 3% verrechnet haben; **this is £5 ~:** das stimmt um 5 Pfund nicht; **f)** *(not in fashion)* passé *(ugs.);* out *(ugs.);* **g) say it ~ loud** es laut sagen; **~ with it!** heraus mit der Sprache; **their secret is ~:** ihr Geheimnis ist bekannt geworden; |**the**| **truth will ~:** die Wahrheit wird an den Tag kommen; **the sun/moon is ~:** die Sonne/der Mond scheint; **the third volume is just ~:** der dritte Band ist soeben erschienen; **the roses are ~:** die Rosen blühen; **h) be ~ for sth./to do sth.** auf etw. *(Akk.)* aussein/darauf aussein, etw. zu tun; **be ~ for trouble** Streit suchen; **i)** *(to or at an end)* **before the day/month was ~:** am selben Tag/vor Ende des Monats. *See also* **out of**
out: **~'bid** *v. t.,* **~bid** überbieten; **~board** *adj.* **~board motor** Außenbordmotor, *der;* **~break** *n.* Ausbruch, *der;* **at the ~break of war** bei Kriegsausbruch; **an ~break of flu** eine Grippeepidemie; **~building** *n.* Nebengebäude, *das;* **~burst** *n.* Ausbruch, *der;* **an ~burst of weeping/laughter** ein Weinkrampf/Lachanfall; **an ~burst of temper** ein Wutanfall; **~cast** *n.* Ausgestoßene, *der/die;* **a social ~cast** ein Geächteter/eine Geächtete; **~come** *n.* Ergebnis, *das;* Resultat, *das;* **~cry** *n.* [Aufschrei der] Empörung; **~'dated** *adj.* überholt; **~'do** *v. t.* überbieten (**in** an + *Dat.*); **~door** *adj.* **~door shoes/things** Stra-

ßenschuhe/-kleidung, *die;* ~**door games/pursuits** Spiele/Beschäftigungen im Freien; ~**door swimming-pool** Freibad, *das;* ~'**doors** 1. *adv.* draußen; **go** ~**doors** nach draußen gehen; 2. *n.* **the |great| ~doors** die freie Natur

outer ['aʊtə(r)] *adj.* äußer...; Außen⟨*fläche, -seite, -wand, -tür*⟩. **outer** '**space** *n.* Weltraum, *der*

out: ~**fit** *n.* **a)** *(clothes)* Kleider *Pl.;* **b)** *(equipment)* Ausrüstung, *die;* **c)** *(coll.: organization)* Laden, *der (ugs.);* ~**going** 1. *adj.* **a)** [aus dem Amt] scheidend ⟨*Regierung, Präsident*⟩; **b)** *(friendly)* kontaktfreudig ⟨*Person*⟩; 2. *n., in pl.* ~**s** *(expenditure)* Ausgaben *Pl.;* ~'**grow** *v. t., forms as* **grow** herauswachsen aus ⟨*Kleider*⟩; *(leave behind)* entwachsen (+ *Dat.*); ~**house** *n.* Nebengebäude, *das*

'**outing** *n.* Ausflug, *der*

out: ~**landish** [aʊt'lændɪʃ] *adj.* ausgefallen; ~**law** 1. *n.* Bandit, *der*/Banditin, *die;* 2. *v. t.* verbieten; ~**lay** *n.* Ausgaben *Pl.* (**on** für); ~**let** ['aʊtlet, 'aʊtlɪt] *n.* **a)** Ablauf, -fluß, *der;* **b)** *(fig.)* Ventil, *das;* ~**line** 1. *n.* **a)** *in sing. or pl.* Umriß, *der;* **b)** *(short account)* Grundriß, *der;* *(of topic)* Übersicht, *die* (**of** über + *Akk.*); 2. *v. t. (describe)* umreißen ⟨*Kleider*⟩; ~**live** [aʊt'lɪv] *v. t.* überleben; ~**look** *n.* **a)** *(view)* Aussicht, *die* (**over** über + *Akk.,* **on to** auf + *Akk.*); *(fig.; Meteorol.)* Aussichten *Pl.;* **b)** *(mental attitude)* Einstellung, *die* (**on** zu); ~**lying** *adj.* entlegen; ~**moded** [aʊt'məʊdɪd] *adj.* antiquiert; ~'**number** *v. t.* zahlenmäßig überlegen sein (+ *Dat.*)

out of *prep.* **a)** *(from within)* aus; **go** ~ **the door** zur Tür hinausgehen; **b)** *(not within)* **be** ~ **the country** im Ausland sein; **be** ~ **town/the room** nicht in der Stadt/im Zimmer sein; **feel** ~ **it** *or* **things** sich ausgeschlossen fühlen; **c)** *(from among)* **one** ~ **every three smokers** jeder dritte Raucher; **58** ~ **every 100** 58 von hundert; **d)** *(beyond range of)* außer ⟨*Reich-/Hörweite, Sicht, Kontrolle*⟩; **e)** *(from)* aus; **get money** ~ **sb.** Geld aus jmdm. herausholen; **do well** ~ **sb./sth.** von jmdm./ etw. profitieren; **f)** aus ⟨*Mitleid, Furcht, Neugier usw.*⟩; **g)** *(without)* ~ **money** ohne Geld; **we're** ~ **tea** wir haben keinen Tee mehr; **h)** *(away from)* **von ... entfernt; ten miles** ~ **London** 10 Meilen außerhalb von London

out: ~-**of-**'**date** *attrib. adj.* veraltet;

(expired) ungültig ⟨*Karte*⟩; ~-**patient** *n.* ambulanter Patient/ambulante Patientin; ~-**patients|' department|** Poliklinik, *die;* ~'**play** *v. t. (Sport)* besser spielen als; ~**post** *n.* Außenposten, *der; (of civilization etc.; also Mil.)* Vorposten, *der;* ~**put** *n.* Produktion, *die; (of liquid, electricity, etc.)* Leistung, *die; (Computing)* Ausgabe, *die*

outrage 1. ['aʊtreɪdʒ] *n.* **a)** *(deed)* Verbrechen, *das; (during war)* Greueltat, *die; (against decency)* grober Verstoß; **b)** *(strong resentment)* Empörung, *die* (**at** gegen). 2. [aʊt'reɪdʒ] *v. t.* empören.

outrageous [aʊt'reɪdʒəs] *adj.* unverschämt; unverschämt hoch ⟨*Preis*⟩; unerhört ⟨*Frechheit, Skandal*⟩

out: ~**right** 1. [-'-] *adv.* **a)** ganz, komplett ⟨*kaufen, verkaufen*⟩; **b)** *(openly)* freiheraus ⟨*erzählen, sagen, lachen*⟩; 2. ['--] *adj.* ausgemacht ⟨*Unehrlichkeit*⟩; glatt *(ugs.)* ⟨*Ablehnung, Absage, Lüge*⟩; klar ⟨*Sieg, Niederlage, Sieger*⟩; ~**set** *n.* Anfang, *der;* **at the** ~**set** zu Anfang; **from the** ~**set** von Anfang an

outside 1. [-'-, '--] *n.* **a)** Außenseite, *die;* **on the** ~: außen; **to/from the** ~: nach/von außen; **b)** *(external appearance)* Äußere, *das;* **c)** **at the |very|** ~ *(coll.)* äußerstenfalls; höchstens. 2. ['--] *adj.* **a)** äußer...; Außen⟨*wand, -antenne, -kajüte, -toilette, -durchmesser*⟩; ~ **lane** Überholspur, *die;* **b)** **have only an** ~ **chance** nur eine sehr geringe Chance haben. 3. [-'-] *adv.* **(on the** ~*)* draußen; *(to the* ~*)* nach draußen. 4. [-'-] *prep.* **a)** *(position)* außerhalb (+ *Gen.*); ~ **the door** vor der Tür; **b)** *(to the* ~ *of)* aus ... hinaus; **go** ~ **the house** nach draußen gehen. **out-**'**sider** *n. (Sport; also fig.)* Außenseiter, *der*

out: ~**size** *adj.* überdimensional; ~**size clothes** Kleidung in Übergröße; ~**skirts** *n. pl.* Stadtrand, *der;* **the** ~**skirts of the town** die Außenbezirke der Stadt; ~'**spoken** *adj.* freimütig; **be** ~ **about sth.** sich freimütig über etw. äußern; ~-'**standing** *adj.* **a)** *(exceptional)* hervorragend; überragend ⟨*Bedeutung*⟩; außergewöhnlich ⟨*Person, Mut, Fähigkeit*⟩; **b)** *(not yet settled)* ausstehend ⟨*Schuld, Geldsumme*⟩; unbezahlt ⟨*Rechnung*⟩; ungelöst ⟨*Problem*⟩; ~**standingly** *adv.* außergewöhnlich; ~**stretched** *adj.* ausgestreckt; *(spread out)* ausgebreitet; ~'**strip** *v. t. (pass in running)* überholen; *(in competition)* überflügeln;

~-**tray** *n.* Ablage für Ausgänge; ~'**vote** *v. t.* überstimmen

outward ['aʊtwəd] **1.** *adj.* **a)** *(external, apparent)* [rein] äußerlich; äußere ⟨*Erscheinung, Bedingung*⟩; **b)** Hin⟨*reise, -fracht*⟩. **2.** *adv.* nach außen ⟨*aufgehen, richten*⟩. '**outwardly** *adv.* nach außen hin ⟨*Gefühle zeigen*⟩; öffentlich ⟨*Loyalität erklären*⟩. '**outwards** *see* **outward 2**

out: ~'**weigh** *v. t.* schwerer wiegen als; überwiegen ⟨*Nachteile*⟩; ~'**wit** *v. t.,* -tt- überlisten

oval ['əʊvl] **1.** *adj.* oval. **2.** *n.* Oval, *das*

ovation [əʊ'veɪʃn] *n.* Ovation, *die;* **a standing** ~: stehende Ovationen

oven ['ʌvn] *n.* [Back]ofen, *der*

oven: ~-**glove** *n.* Topfhandschuh, *der;* ~-**proof** *adj.* feuerfest; ~-**ready** *adj.* backfertig ⟨*Pommes frites, Pastete*⟩; bratfertig ⟨*Geflügel*⟩

over ['əʊvə(r)] **1.** *adv.* **a)** *(outward and downward)* hinüber; **climb/look/jump** ~: hinüber- *od.* *(ugs.)* rüberklettern/-sehen/-springen; **b)** *(so as to cover surface)* **board/cover** ~: zunageln/-decken; **c)** *(across a space)* hinüber; *(towards speaker)* herüber; **he swam** ~ **to us/the other side** er schwamm zu uns herüber/hinüber zur anderen Seite; ~ **here/there** *(direction)* hier herüber/dort hinüber; *(location)* hier/dort; |**come in, please,**| ~ *(Radio)* übernehmen Sie bitte; ~ **and out** *(Radio)* Ende; **d)** *(in excess etc.)* **children of 12 and** ~: Kinder im Alter von zwölf Jahren und darüber; **be** |**left**| ~: übrig[geblieben] sein; **e)** *(from beginning to end)* von Anfang bis Ende; **say sth. twice** ~: etw. zweimal sagen; |**all**| ~ **again,** *(Amer.)* ~: noch einmal [ganz von vorn]; ~ **and** ~ |**again**| immer wieder; **f)** *(at an end)* vorbei, vorüber; **be** ~: vorbei sein; ⟨*Aufführung:*⟩ zu Ende sein; **get sth.** ~ **with** etw. hinter sich *(Akk.)* bringen; **be** ~ **and done with** erledigt sein; **g) all** ~ *(completely finished)* aus [und vorbei]; **I ache all** ~: mir tut alles weh; **be shaking all** ~: am ganzen Körper zittern. **2.** *prep.* **a)** *(above, on, round about)* über *(position:* + *Dat.; direction:* + *Akk.); (across)* über (+ *Akk.*); **look** ~ **a wall** über eine Mauer sehen; **fall** ~ **a cliff** von einem Felsen stürzen; **the pub** ~ **the road** die Wirtschaft gegenüber; **hit sb.** ~ **the head** jmdm. auf den Kopf schlagen; ~ **the page** auf der nächsten Seite; **b)** *(in or across every part of)*

[überall] in (+ *Dat.*); *(to and fro upon)* über (+ *Akk.*); *(all through)* durch; **all** ~ *(in or on all parts of)* überall in (+ *Dat.*); **travel all** ~ **the country** das ganze Land bereisen; **all** ~ **Spain** in ganz Spanien; **all** ~ **the world** in der ganzen Welt; **c)** *(on account of)* wegen; **d)** *(engaged with)* bei; **take trouble** ~ **sth.** sich *(Dat.)* mit etw. Mühe geben; **be a long time** ~ **sth.** lange für etw. brauchen; ~ **work/dinner** bei der Arbeit/beim Essen; **e)** *(superior to, in charge of)* über (+ *Akk.*); **have command/authority** ~ **sb.** Befehlsgewalt über jmdn./Weisungsbefugnis gegenüber jmdm. haben; **be** ~ **sb.** *(in rank)* über jmdm. stehen; **f)** *(beyond, more than)* über (+ *Akk.*); ~ **and above** zusätzlich zu; **g)** *(throughout, during)* über (+ *Akk.*); ~ **the weekend/summer** übers Wochenende/den Sommer über; ~ **the past years** in den letzten Jahren

over: ~**all 1.** *n. (Brit.: garment)* Arbeitskittel, *der;* **2.** *adj.* **a)** Gesamt-⟨*breite, -einsparung, -abmessung*⟩; **have an** ~**all majority** die absolute Mehrheit haben; **b)** *(general)* allgemein; **3.** ['---, --'-] *adv.* **a)** *(in all parts)* insgesamt; **b)** *(taken as a whole)* im großen und ganzen; ~'**awe** *v. t.* Ehrfurcht einflößen (+ *Dat.*); ~'**balance** *v. i.* das Gleichgewicht verlieren; ~'**bearing** *adj.* herrisch; ~**board** *adv.* über Bord; **fall** ~**board** über Bord gehen; ~**cast** *adj.* trübe; bewölkt ⟨*Himmel*⟩; ~-'**charge** *v. t.* **a)** *(beyond reasonable price)* zuviel abverlangen (+ *Dat.*); **b)** *(beyond right price)* zuviel berechnen (+ *Dat.*); ~**coat** *n.* Mantel, *der;* ~'**come** *v. t.,* forms as **come: a)** überwinden; bezwingen ⟨*Feind*⟩; ⟨*Dämpfe:*⟩ betäuben; **b) he was** ~**come by grief/with emotion** Kummer/Rührung überwältigte ihn; ~-'**cooked** *adj.* verkocht; ~'**crowded** *adj.* überfüllt; ~'**do** *v. t. (carry to excess)* übertreiben; ~**do it** *or* **things** *(work too hard)* sich übernehmen; ~'**done** *adj.* **a)** *(exaggerated)* übertrieben; **b)** *(~-cooked)* verkocht; verbraten ⟨*Fleisch*⟩; ~**dose** *n.* Überdosis, *die;* ~**draft** *n.* Kontoüberziehung, *die;* **have an** ~**draft of £50** sein Konto um 50 Pfund überzogen haben; ~'**draw** *v. t.,* forms as **draw 1** überziehen ⟨*Konto*⟩; ~'**drawn** *adj.* überzogen ⟨*Konto*⟩; **I am** ~**drawn** |**at the bank**| mein Konto ist überzogen;

~**drive** *n.* Schongang, *der;* ~'**due**
adj. überfällig; **the train is 15 minutes**
~**due** der Zug hat schon 15 Minuten
Verspätung; ~'**eat** *v. i., forms as* **eat**
zuviel essen; ~**estimate** 1. [~'estɪ-
meɪt] *v. t.* überschätzen; 2. [~'estɪmət]
n. zu hohe Schätzung; ~'**fill** *v. t.* zu
voll machen; ~**flow** 1. [--'-] *v. t.* laufen über (+ *Akk.*) *(Rand); (flow over
brim of)* überlaufen aus; ~**flow its
banks** ⟨*Fluß:*⟩ über die Ufer treten; 2.
[--'-] *v. i.* überlaufen; 3. ['---] *n.* ~**flow
|pipe|** Überlauf, *der;* ~'**full** *adj.* zu
voll; übervoll; ~**grown** *adj.* über-
wachsen (with von); ~**hang** 1. [--'-]
v. t., ~**hung** [əʊvə'hʌŋ] ⟨*Felsen, Stockwerk:*⟩ hinausragen über (+ *Akk.*); 2.
['---] *n.* Überhang, *der;* ~'**hanging**
adj. überhängend; ~**haul** 1. [--'-] *v. t.*
überholen; überprüfen ⟨*System*⟩; 2.
['---] *n.* Überholung, *die;* ~**head** 1.
[--'-] *adv.* über mir/ihm/uns *usw.;* 2.
['---] *adj.* ~**head wires** Hochleitung,
die; 3. ['---] *n.* ~**heads,** *(Amer.)* ~**head**
(Commerc.) Gemeinkosten *Pl.;*
~'**hear** *v. t., forms as* **hear** 1 *(accidentally)* zufällig [mit]hören; *(intentionally)* belauschen; ~'**heat** *v. i.* zu heiß
werden; ⟨*Maschine, Lager:*⟩ heißlau-
fen

overjoyed [əʊvə'dʒɔɪd] *adj.* über-
glücklich (**at** über + *Akk.*)

over: ~**lap** 1. [--'-] *v. t.* überlappen; 2.
[--'-] *v. i.* ⟨*Flächen, Dachziegel:*⟩ sich
überlappen; ⟨*Aufgaben:*⟩ sich überschneiden; 3. *n.* Überlappung, *die;*
~'**leaf** *adv.* auf der Rückseite; ~'**load**
v. t. überladen; ~'**look** *v. t.* **a)** ⟨*Hotel,
Zimmer, Haus:*⟩ Aussicht bieten auf
(+ *Akk.*); **b)** *(ignore, not see)* übersehen; *(allow to go unpunished)* hinwegsehen über (+ *Akk.*)

'**overly** *adv.* allzu

over: ~**night** 1. [--'-] *adv. (also fig.:
suddenly)* über Nacht; **stay** ~ **night**
übernachten; 2. ['---] *adj.* ~**night stay**
Übernachtung, *die;* **be an** ~**night success** *(fig.)* über Nacht Erfolg haben;
~'**pay** *v. t., forms as* **pay** 2 überbezah-
len; ~'**power** *v. t.* überwältigen;
~'**powering** *adj.* überwältigend;
durchdringend ⟨*Geruch*⟩; ~-'**priced**
adj. zu teuer; ~'**rate** *v. t.* überschätzen; ~-**re'act** *v. i.* unangemessen heftig reagieren (**to** auf + *Akk.*); ~-
re'action *n.* Überreaktion, *die* (**to**
auf + *Akk.*); ~'**ride** *v. t. forms as* **ride**
3 sich hinwegsetzen über (+ *Akk.*);
~**ripe** *adj.* überreif; ~'**rule** *v. t.* auf-

heben ⟨*Entscheidung*⟩; zurückweisen
⟨*Einwand, Argument*⟩; ~**rule sb.** jmds.
Vorschlag ablehnen; ~'**run** *v. t., forms
as* **run** 3: **be** ~**run with** überlaufen sein
von ⟨*Touristen*⟩; überwuchert sein
von ⟨*Unkraut*⟩; ~**seas** 1. [--'-] *adv.* in
Übersee ⟨*leben, sein*⟩; nach Übersee
⟨*gehen*⟩; 2. ['---] *adj.* Übersee-; ~'**see**
v. t., forms as **see** 1 überwachen; *(manage)* leiten ⟨*Abteilung*⟩; ~'**shadow**
v. t. überschatten; ~'**shoot** *v. t., forms
as* **shoot** 1 hinausschießen über
(+ *Akk.*); ~**shoot |the runway|** ⟨*Pilot,
Flugzeug:*⟩ zu weit kommen; ~'**sight**
n. Versehen, *das;* ~'**sleep** *v. i., forms
as* **sleep** 2 verschlafen; ~'**spend** *v. i.,
forms as* **spend** zuviel [Geld] ausgeben; ~**statement** *n.* Übertreibung,
die; ~'**step** *v. t.* überschreiten

overt [əʊ'vɜːt] *adj.* unverhohlen

over: ~'**take** *v. t.* überholen; '**no**
~**taking'** *(Brit.)* „Überholen verboten"; ~'**throw** 1. [--'-] *v. t., forms as*
throw 1 stürzen; 2. ['---] *n.* Sturz, *der;*
~**time** 1. *n.* Überstunden; 2. *adv.*
work ~**time** Überstunden machen;
~**tone** *n. (fig.)* Unterton, *der*

overture ['əʊvətjʊə(r)] *n. (Mus.)* Ouvertüre, *die*

over: ~'**turn** 1. *v. t.* umstoßen; 2. *v. i.*
⟨*Auto, Boot:*⟩ umkippen; ⟨*Boot:*⟩ kentern; ~-**use** [əʊvə'juːz] *v. t.* zu oft verwenden; ~**weight** *adj.* übergewich-
tig ⟨*Person*⟩; **be** ~**weight** Übergewicht
haben

overwhelm [əʊvə'welm] *v. t.* überwältigen. **over'whelming** *adj.* überwältigend

over: ~'**work** 1. *v. t.* mit Arbeit über-
lasten; 2. *v. i.* sich überarbeiten;
~'**wrought** *adj.* überreizt

owe [əʊ] *v. t., owing* ['əʊɪŋ] schulden; ~
sb. sth., ~ **sth. to sb.** jmdm. etw. schulden; *(fig.)* jmdm. etw. verdanken.
owing ['əʊɪŋ] *pred. adj.* ausstehend;
be ~: ausstehen. '**owing to** *prep.* wegen

owl [aʊl] *n.* Eule, *die*

own [əʊn] 1. *adj.* eigen; **be sb.'s** ~
|**property|** jmdm. selbst gehören; **a
house/ideas of one's** ~: ein eigenes
Haus/eigene Ideen; **on one's/its** ~: allein. 2. *v. t.* besitzen; **be** ~**ed by sb.**
jmdm. gehören. **own 'up** *v. i.* gestehen; ~ **up to sth.** etw. zugeben

'**owner** *n.* Besitzer, *der*/Besitzerin,
die; (of shop, hotel, firm, etc.) Inhaber,
der/Inhaberin, *die.* '**ownership** *n.*
Besitz, *der*

ox [ɒks] *n., pl.* **oxen** ['ɒksn] Ochse, *der*
oxygen ['ɒksɪdʒən] *n.* Sauerstoff, *der*
oyster ['ɔɪstə(r)] *n.* Auster, *die*
oz. *abbr.* **ounce[s]**
ozone ['əʊzəʊn] *n.* Ozon, *das.*
'**ozone-friendly** *adj.* ozonsicher;
(not using (CFCs) FCKW-frei. '**ozone layer** *n.* Ozonschicht, *die*

P

P, p [piː] *n.* P, p, *das*
p. *abbr.* **a)** page S.; **b)** *(Brit.)* penny/
pence p
pace [peɪs] **1.** *n.* **a)** *(step)* Schritt, *der;*
b) *(speed)* Tempo, *das;* **keep ~ with**
Schritt halten mit. **2.** *v. i.* **~ up and
down** auf und ab gehen. **3.** *v. t.* auf-
und abgehen in (+ *Dat.*)
'**pacemaker** *n. (Sport, Med.)* Schritt-
macher, *der*
Pacific [pə'sɪfɪk] **1.** *adj. (Geog.)* ~
Ocean Pazifischer *od.* Stiller Ozean.
2. *n.* **the ~:** der Pazifik
pacifier ['pæsɪfaɪə(r)] *n. (Amer.:
dummy)* Schnuller, *der*
pacifism ['pæsɪfɪzm] *n.* Pazifismus,
der. **pacifist** ['pæsɪfɪst] **1.** *n.* Pazifist,
*der/*Pazifistin, *die.* **2.** *adj.* pazifistisch
pacify ['pæsɪfaɪ] *v. t.* besänftigen
pack [pæk] **1.** *n.* **a)** *(bundle)* Bündel,
das; (Mil.) Tornister, *der; (rucksack)*
Rucksack, *der;* **b)** *(derog.: lot) (people)*
Bande, *die;* **a ~ of lies/nonsense** ein
Sack voll Lügen/eine Menge Unsinn;
c) *(Brit.)* ~ [of cards] [Karten]spiel,
das; **d)** *(wolves, wild dogs)* Rudel, *das;
(hounds)* Meute, *die;* **e)** *(packet)*
Packung, *die.* **2.** *v. t.* **a)** einpacken;
(fill) packen; **~ one's bags** seine Koffer packen; **b)** *(cram)* vollstopfen
(ugs.); **c)** *(wrap)* verpacken (**in** in +
Dat. od. Akk.). **3.** *v. i* packen; **send sb.
~ing** *(fig.)* jmdn. rausschmeißen
(ugs.). **pack 'up 1.** *v. t.* zusammen-
packen ⟨*Sachen, Werkzeug*⟩; packen
⟨*Paket*⟩. **2.** *v. i. (coll.: stop)* aufhören
package ['pækɪdʒ] **1.** *n.* Paket, *das.* **2.**
v. t. verpacken

package: ~ deal *n.* Paket, *das;* ~

holiday, ~ tour *ns.* Pauschalreise,
die
packed [pækt] *adj.* **a)** gepackt; ~
lunch Lunchpaket, *das;* **b)** *(crowded)*
[über]voll; **~ out** gerammelt voll *(ugs.)*
packet ['pækɪt] *n.* Päckchen, *das;
(box)* Schachtel, *die;* **a ~ of cigarettes**
ein Päckchen/eine Schachtel Zigaretten
'**packing** *n. (material)* Verpackungs-
material, *das;* **postage and ~:** Porto
und Verpackung. '**packing-case** *n.*
[Pack]kiste, *die*
pact [pækt] *n.* Pakt, *der*
'**pad** [pæd] **1.** *n.* Polster, *das; (block of
paper)* Block, *der.* **2.** *v. t.,* **-dd-** polstern ⟨*Jacke, Schulter*⟩. **pad 'out** *v. t.
(fig.)* auswalzen
'**pad** *v. i.,* **-dd-** tappen
padding ['pædɪŋ] *n.* Polsterung, *die;
(fig.)* Füllsel, *das*
'**paddle** ['pædl] **1.** *n.* [Stech]paddel,
das. **2.** *v. t. & i.* paddeln
'**paddle 1.** *v. i. (with feet)* planschen.
2. *n.* **have a/go for a ~:** ein bißchen planschen/planschen gehen.
paddling-pool ['pædlɪŋpuːl] *n.*
Planschbecken, *das*
paddock ['pædək] *n.* Koppel, *die*
'**padlock 1.** *n.* Vorhängeschloß, *das.*
2. *v. t.* [mit einem Vorhängeschloß]
verschließen
pagan ['peɪgən] **1.** *n.* Heide, *der/*Heidin, *die.* **2.** *adj.* heidnisch
'**page** [peɪdʒ] *n. (boy)* Page, *der*
'**page** *n. (of book etc.)* Seite, *die*
pageant ['pædʒənt] *n. (spectacle)*
Schauspiel, *das.* **pageantry** ['pæ-
dʒəntrɪ] *n.* Prunk, *der*
paid [peɪd] **1.** *see* **pay** 2, 3. **2.** *adj.* **a)** bezahlt ⟨*Urlaub, Arbeit*⟩; **b)** **put ~ to**
(Brit. coll.) zunichte machen; kurzen
Prozeß machen mit *(ugs.)* ⟨*Person*⟩
pail [peɪl] *n.* Eimer, *der*
pain [peɪn] *n.* **a)** *(suffering)* Schmerzen; *(mental ~)* Qualen; **be in ~:**
Schmerzen haben; **b)** *(instance)*
Schmerz, *der;* **I have a ~ in my knee/
stomach** mein Knie/Magen tut weh; **c)**
in pl. (trouble taken) Mühe, *die;* **take
~s** sich *(Dat.)* Mühe geben (**over** mit,
bei). **painful** ['peɪnfl] *adj.* **a)** schmerzhaft; **be ~** ⟨*Körperteil:*⟩ weh tun; **b)**
(distressing) schmerzlich ⟨*Gedanke,
Erinnerung*⟩; traurig ⟨*Pflicht*⟩. '**pain-
killer** *n.* schmerzstillendes Mittel.
'**painless** *adj.* schmerzlos; *(fig.)* un-
problematisch. **painstaking** ['peɪnz-
teɪkɪŋ] *adj.* gewissenhaft

paint [peɪnt] **1.** *n.* Farbe, *die;* (on car) Lack, *der.* **2.** *v. t.* (cover, colour) [an]streichen; (make picture of, make by ~ing) malen; bemalen ⟨ *Wand, Vase, Decke*⟩. **'paintbox** *n.* Malkasten, *der;* ~**brush** *n.* Pinsel, *der*

'painter *n.* Maler, *der/*Malerin, *die*

'painting *n.* (art) Malerei, *die;* (picture) Gemälde, *das;* Bild, *das*

pair [peə(r)] **1.** *n.* Paar, *das;* **a** ~ **of** gloves/socks/shoes *etc.* ein Paar Handschuhe/Socken/Schuhe *usw.;* in ~s paarweise; **a** ~ **of trousers/jeans** eine Hose/Jeans. **2.** *v. t.* paaren. **pair 'off** *v. i.* Zweiergruppen bilden

pajamas [pə'dʒɑːməz] *(Amer.) see* **pyjamas**

Pakistan [pɑːkɪ'stɑːn] *pr. n.* Pakistan *(das).* **Pakistani** [pɑːkɪ'stɑːnɪ] **1.** *adj.* pakistanisch. **2.** *n.* Pakistani, *der/die*

pal [pæl] *n.* (coll.) Kumpel, *der* (ugs.)

palace ['pælɪs] *n.* Palast, *der*

palate ['pælət] *n.* Gaumen, *der*

palatial [pə'leɪʃl] *adj.* palastartig

'pale [peɪl] *adj.* blaß, (nearly white) bleich ⟨ *Gesichtsfarbe, Haut, Gesicht*⟩; blaß ⟨ *Farbe*⟩; fahl ⟨ *Licht*⟩; **go** ~: blaß/bleich werden; (fig.) ~ **imitation** schlechte Nachahmung

²pale *n.* **beyond the** ~: unmöglich

Palestine ['pælɪstaɪn] *pr. n.* Palästina *(das).* **Palestinian** [pælɪ'stɪnɪən] **1.** *adj.* palästinensisch. **2.** *n.* Palästinenser, *der/*Palästinenserin, *die*

palette ['pælɪt] *n.* Palette, *die*

'pall [pɔːl] *n.* **a)** (over coffin) Sargtuch, *das;* **b)** (fig.) Schleier, *der*

²pall *v. i.* ~ [on sb.] [jmdm.] langweilig werden

pallor ['pælə(r)] *n.* Blässe, *die*

'palm [pɑːm] *n.* (tree) Palme, *die*

²palm *n.* Handteller, *der.* **palm 'off** *v. t.* ~ sth. off on sb., ~ sb. off with sth. jmdm. etw. andrehen (ugs.)

palmistry ['pɑːmɪstrɪ] *n.* Handlesekunst, *die*

palm: P~ 'Sunday *n.* Palmsonntag, *der;* ~-**tree** *n.* Palme, *die*

paltry ['pɔːltrɪ, 'pɒltrɪ] *adj.* schäbig

pamper ['pæmpə(r)] *v. t.* verhätscheln; ~ **oneself** sich verwöhnen

pamphlet ['pæmflɪt] *n.* (leaflet) Prospekt, *der;* (booklet) Broschüre, *die*

pan [pæn] *n.* [Koch]topf, *der;* (for frying) Pfanne, *die*

panacea [pænə'sɪə] *n.* Allheilmittel, *das*

Panama [pænə'mɑː] *pr. n.* Panama *(das);* ~ **Ca'nal** Panamakanal, *der*

'pancake *n.* Pfannkuchen, *der*

panda ['pændə] *n.* Panda, *der*

pandemonium [pændɪ'məʊnɪəm] *n.* Chaos, *das;* (uproar) Tumult, *der*

pander ['pændə(r)] *v. i.* ~ **to** allzu sehr entgegenkommen (+ *Dat.*)

pane [peɪn] *n.* Scheibe, *die*

panel ['pænl] *n.* **a)** Paneel, *das;* **b)** (esp. Telev., Radio, etc.) (quiz team) Rateteam, *das;* (in public discussion) Podium, *das*

pang [pæŋ] *n.* (of pain) Stich, *der;* **feel** ~**s of conscience/guilt** Gewissensbisse haben; ~[s] **of hunger** quälender Hunger

panic ['pænɪk] **1.** *n.* Panik, *die;* **hit the** ~ **button** (fig. coll.) Alarm schlagen; (~) durchdrehen (ugs.). **2.** *v. i.,* -ck- in Panik (Akk.) geraten; **don't** ~! nur keine Panik! **'panic-stricken, 'panic-struck** *adjs.* von Panik erfaßt

panorama [pænə'rɑːmə] *n.* Panorama, *das*

pansy ['pænzɪ] *n.* Stiefmütterchen, *das*

pant [pænt] *v. i.* keuchen; ⟨ *Hund:*⟩ hecheln

panther ['pænθə(r)] *n.* Panther, *der*

panties ['pæntɪz] *n. pl.* (coll.) [pair of] ~: Schlüpfer, *der*

pantomime ['pæntəmaɪm] *n.* (Brit.) Märchenspiel im Varietéstil, *das um Weihnachten aufgeführt wird*

pantry ['pæntrɪ] *n.* Speisekammer, *die*

pants [pænts] *n. pl.* **a)** (esp. Amer. coll.: trousers) [pair of] ~: Hose, *die;* **b)** (Brit. coll.: underpants) Unterhose, *die*

paper ['peɪpə(r)] **1.** *n.* **a)** (material) Papier, *das;* **b)** in *pl.* (documents) Unterlagen *Pl.;* (to prove identity etc.) Papiere *Pl.;* **c)** (in examination) (Univ.) Klausur, *die;* (Sch.) Arbeit, *die;* **d)** (newspaper) Zeitung, *die;* **e)** (learned article) Referat, *das.* **2.** *adj.* aus Papier nachgestellt; Papier⟨ *mütze, -taschentuch*⟩. **3.** *v. t.* tapezieren

paper: ~**back 1.** *n.* Paperback, *das;* **2.** *adj.* ~**back book** Paperback, *das;* ~**'bag** *n.* Papiertüte, *die;* ~-**clip** *n.* Büroklammer, *die;* (larger) Aktenklammer, *die;* ~**weight** *n.* Briefbeschwerer, *der;* ~-**work** *n.* Schreibarbeit, *die*

par [pɑː(r)] *n.* **feel below** ~: nicht ganz auf dem Posten sein (ugs.); **be on a** ~ **with** sb./sth. jmdm./einer Sache gleichkommen

parable ['pærəbl] *n.* Gleichnis, *das*

parachute ['pærəʃuːt] **1.** *n.* Fallschirm, *der.* **2.** *v. i.* ⟨ *Truppen:*⟩ abspringen (into über + *Dat.*)

parade [pə'reɪd] 1. *n.* **a)** *(display)* Zurschaustellung, *die;* **b)** *(Mil.)* Appell, *der;* **c)** *(procession)* Umzug, *der; (of troops)* Parade, *die.* 2. *v. t.* zur Schau stellen. 3. *v. i.* paradieren
paradise ['pærədaɪs] *n.* Paradies, *das*
paradox ['pærədɒks] *n.* Paradox[on], *das.* **paradoxical** [pærə'dɒksɪkl] *adj.* paradox
paraffin ['pærəfɪn] *n.* Paraffin, *das; (Brit.: fuel)* Petroleum, *das*
paragon ['pærəgən] Muster, *das (of an + Dat.);* ~ **of virtue** Tugendheld, *der*
paragraph ['pærəgrɑːf] *n.* Absatz, *der*
parallel ['pærəlel] 1. *adj.* parallel; *(fig.: similar)* vergleichbar; ~ **bars** Barren, *der.* 2. *n.* Parallele, *die;* ~ **|of latitude|** Breitenkreis, *der*
paralyse ['pærəlaɪz] *v. t.* lähmen; *(fig.)* lahmlegen ⟨*Verkehr, Industrie*⟩. **paralysis** [pə'rælɪsɪs] *n.* Lähmung, *die*
paralyze *(Amer.) see* **paralyse**
paramount ['pærəmaʊnt] *adj.* größt... ⟨*Wichtigkeit*⟩; Haupt⟨*überlegung*⟩; **be** ~: Vorrang haben
paranoia [pærə'nɔɪə] *n.* Paranoia, *die (Med.); (tendency)* Verfolgungswahn, *der.* **paranoid** ['pærənɔɪd] *adj.* **be** ~ ⟨*Person:*⟩ an Verfolgungswahn leiden
parapet ['pærəpɪt] *n.* Brüstung, *die*
paraphernalia [pærəfə'neɪlɪə] *n. sing.* Apparat, *der*
paraphrase ['pærəfreɪz] 1. *n.* Umschreibung, *die.* 2. *v. t.* umschreiben
parasite ['pærəsaɪt] *n.* Schmarotzer, *der.* **parasitic** [pærə'sɪtɪk] *adj.* **a)** *(Biol.)* parasitisch; **b)** *(fig.)* schmarotzerhaft
parasol ['pærəsɒl] *n.* Sonnenschirm, *der*
paratroops ['pærətruːps] *n. pl.* Fallschirmjäger *Pl.*
parcel ['pɑːsl] *n.* Paket, *das*
parched [pɑːtʃt] *adj.* ausgedörrt; trocken ⟨*Lippen*⟩
parchment ['pɑːtʃmənt] *n.* Pergament, *das*
pardon ['pɑːdn] 1. *n.* Verzeihung, *die;* **beg sb.'s** ~: jmdn. um Entschuldigung bitten; **I beg your** ~: entschuldigen Sie bitte. 2. *v. t.* **a)** ~ **sb. |for| sth.** jmdm. etw. verzeihen; **b)** *(excuse)* entschuldigen. **pardonable** ['pɑːdənəbl] *adj.* verzeihlich
pare [peə(r)] *v. t. (trim)* schneiden; *(peel)* schälen
parent ['peərənt] *n.* Elternteil, *der;* ~**s** Eltern *Pl.*

parenthesis [pə'renθɪsɪs] *n., pl.* **parentheses** [pə'renθɪsiːz] *(bracket)* runde Klammer
Paris ['pærɪs] *pr. n.* Paris *(das)*
parish ['pærɪʃ] *n.* Gemeinde, *die.* **parishioner** [pə'rɪʃənə(r)] *n.* Gemeinde[mit]glied, *das*
park [pɑːk] 1. *n.* Park, *der.* 2. *v. i.* parken. 3. *v. t.* abstellen; parken ⟨*Kfz*⟩; **a** ~**ed car** ein parkendes Auto. **parking** *n.* Parken, *das;* 'No ~' „Parken verboten"
parking: ~-**light** *n.* Parkleuchte, *die;* ~-**lot** *n. (Amer.)* Parkplatz, *der;* ~-**meter** *n.* Parkuhr, *die;* ~-**space** *n.* **a)** *no pl.* Parkraum, *der;* **b)** *(single space)* Parkplatz, *der;* ~-**ticket** *n.* Strafzettel [für falsches Parken]
parliament ['pɑːləmənt] *n.* Parlament, *das;* [**Houses of**] **P**~ *(Brit.)* Parlament, *das.* **parliamentary** [pɑːlə'mentərɪ] *adj.* parlamentarisch; Parlaments⟨*geschäfte, -wahlen, -reform*⟩
parlour *(Brit.; Amer.:* **parlor**) ['pɑːlə(r)] *n. (dated)* Wohnzimmer, *das*
parochial [pə'rəʊkɪəl] *adj.* krähwinklig
parody ['pærədɪ] 1. *n.* Parodie, *die (of auf + Akk.).* 2. *v. t.* parodieren
parole [pə'rəʊl] *n.* bedingter Straferlaß *(Rechtsw.);* **on** ~: auf Bewährung
parquet ['pɑːkɪ, 'pɑːkeɪ] *n.* ~ |**floor/flooring**| Parkett, *das*
parrot ['pærət] *n.* Papagei, *der*
parry ['pærɪ] *v. t.* abwehren ⟨*Faustschlag*⟩; *(Fencing; also fig.)* parieren
parsley ['pɑːslɪ] *n.* Petersilie, *die*
parsnip ['pɑːsnɪp] *n.* Gemeiner Pastinak, *der*
parson ['pɑːsn] *n.* Pfarrer, *der*
part [pɑːt] 1. *n.* **a)** Teil, *der;* **the greater** ~: der größte Teil; der Großteil; **for the most** ~: größtenteils; **in** ~: teilweise; **in large** ~: groß[en]teils; **in** ~**s** zum Teil; **b)** *(of machine)* [Einzel]teil, *das;* **c)** *(share)* Anteil, *der;* **d)** *(Theatre)* Rolle, *die;* **e)** *(Mus.)* Part, *der;* Stimme, *die;* **f)** *usu. in pl. (region)* Gegend, *die; (of continent, world)* Teil, *der;* **g)** *(side)* Partei, *die;* **take sb.'s** ~: jmds. od. für jmdn. Partei ergreifen; **h)** **take** |**no**| ~ |**in sth.**| sich [an etw. *(Dat.)*] [nicht] beteiligen; **i)** **take sth. in good** ~: etw. nicht übelnehmen. 2. *adv.* teils. 3. *v. i.* **a)** *(divide into* ~*s)* teilen; scheiteln ⟨*Haar*⟩; **b)** *(separate)* trennen. 4. *v. i.* ⟨*Seil, Tau, Kette:*⟩ reißen; ⟨*Wege, Personen:*⟩ sich trennen; ~ **with** sich trennen von ⟨*Besitz, Geld*⟩

partial ['pɑ:ʃl] *adj.* **a)** *(biased)* voreingenommen; **b) be/not be ~ to sth.** eine Schwäche/keine besondere Vorliebe für etw. haben; **c)** partiell ⟨*Lähmung, Sonnenfinsternis*⟩; **a ~ success** ein Teilerfolg. **'partially** *adv.* teilweise
participant [pɑ:'tısıpənt] *n.* Beteiligte, *der/die* (**in an** + *Dat.*)
participate [pɑ:'tısıpeıt] *v.i.* sich beteiligen (**in an** + *Dat.*); *(in arranged event)* teilnehmen (**in an** + *Dat.*).
participation [pɑ:tısı'peıʃn] *n.* Beteiligung, *die* (**in an** + *Dat.*); *(in arranged event)* Teilnahme, *die* (**in bei,** an + *Dat.*)
participle ['pɑ:tısıpl] *n.* Partizip, *das*
particle ['pɑ:tıkl] *n.* Teilchen, *das*
particular [pə'tıkjʊlə(r)] **1.** *adj.* **a)** besonder...; **here in ~:** besonders hier; **nothing/anything** [**in**] ~: nichts/irgend etwas Besonderes; **b)** *(fastidious)* genau; **I am not ~:** es ist mir gleich; **be ~ about sth.** es mit etw. genau nehmen. **2.** *n., in pl.* Einzelheiten; Details; *(of person)* Personalien *Pl.* **par'ticularly** *adv.* besonders
'parting 1. *n.* **a)** [**final**] ~: Abschied, *der;* **b)** *(Brit.: in hair)* Scheitel, *der.* **2.** *attrib. adj.* Abschieds-
partisan ['pɑ:tızæn] *n.* Partisan, *der/*Partisanin,
partition [pɑ:'tıʃn] **1.** *n.* **a)** *(Polit.)* Teilung, *die;* **b)** *(room-divider)* Trennwand, *die.* **2.** *v.t.* **a)** *(divide)* aufteilen ⟨*Land, Zimmer*⟩; **b)** *(Polit.)* teilen ⟨*Land*⟩. **partition 'off** *v.t.* abteilen
'partly *adv.* zum Teil; teilweise
partner ['pɑ:tnə(r)] *n.* Partner, *der/*Partnerin, *die;* **'partnership** *n.* Partnerschaft, *die;* **business ~:** [Personen]gesellschaft, *die*
partridge ['pɑ:trıdʒ] *n., pl. same or* ~s Rebhuhn, *das*
part-time 1. ['--] *adj.* Teilzeit⟨*arbeit, -arbeiter*⟩. **2.** [-'-] *adv.* stundenweise, halbtags ⟨*arbeiten, studieren*⟩
party ['pɑ:tı] *n.* **a)** *(Polit., Law)* Partei, *die; attrib.* Partei-; **b)** *(group)* Gruppe, *die;* **c)** *(social gathering)* Party, *die*
pass [pɑ:s] **1.** *n.* **a)** *(passing of an examination)* bestandene Prüfung; **'~'** *(mark)* Ausreichend, *das;* **get a ~ in maths** die Mathematikprüfung bestehen; **b)** *(written permission)* Ausweis, *der;* **c)** *(Footb.)* Paß, *der (fachspr.);* Ballabgabe, *die;* **d)** *(in mountains)* Paß, *der.* **2.** *v.i.* **a)** *(go by)* ⟨*Fußgänger:*⟩ vorbeigehen; ⟨*Fahrer, Fahrzeug:*⟩ vorbeifahren; ⟨*Zeit, Se-*

kunde:⟩ vergehen; *(by chance)* ⟨*Person, Fahrzeug:*⟩ vorbeikommen; **b)** *(come to an end)* vorbeigehen; ⟨*Gewitter, Unwetter:*⟩ vorüberziehen; **c)** *(be accepted)* durchgehen (**as** als, **for** für); **d)** *(in exam)* bestehen. **3.** *v.t.* **a)** ⟨*Fußgänger:*⟩ vorbeigehen an (+ *Dat.*); ⟨*Fahrer, Fahrzeug:*⟩ vorbeifahren an (+ *Dat.*); *(by chance)* ⟨*Person, Fahrzeug:*⟩ vorbeikommen an (+ *Dat.*); **b)** *(overtake)* vorbeifahren an (+ *Dat.*); **c)** bestehen ⟨*Prüfung*⟩; **d)** *(approve)* verabschieden ⟨*Gesetzentwurf*⟩; annehmen ⟨*Vorschlag*⟩; bestehen lassen ⟨*Prüfungskandidaten*⟩; **e)** *(Footb. etc.)* abgeben (**to an** + *Akk.*); **f)** *(spend)* verbringen ⟨*Leben, Zeit, Tag*⟩; **g)** *(hand)* ~ **sb. sth.** jmdm. etw. reichen *od.* geben; **h)** fällen ⟨*Urteil*⟩; machen ⟨*Bemerkung*⟩; **i)** ~ **water** Wasser lassen. **pass a'way** *v.i.* *(euphem.)* die Augen schließen *(verhüll.).* **pass 'off** *v.t.* ~ **sth. off as sth.** etw. als etw. ausgeben. **pass 'on** *v.t.* weitergeben (**to an** + *Akk.*). **pass 'out** *v.i.* ohnmächtig werden. **pass 'up** *v.t.* entgehen lassen ⟨*Gelegenheit*⟩; ablehnen ⟨*Angebot*⟩
passable ['pɑ:səbl] *adj.* **a)** *(acceptable)* passabel; **b)** befahrbar ⟨*Straße*⟩
passage ['pæsıdʒ] *n.* **a)** *(voyage)* Überfahrt, *die;* **b)** *(way)* Durchgang, *der; (corridor)* Korridor, *der;* **c)** *(part of book etc.)* Textstelle, *die; (Mus.)* Stelle, *die*
passenger ['pæsındʒə(r)] *n.* Passagier, *der; (on train)* Reisende, *der/die; (on bus, in taxi)* Fahrgast, *der; (in car, on motor cycle)* Mitfahrer, *der/*Mitfahrerin, *die; (in front seat of car)* Beifahrer, *der/*Beifahrerin, *die.* **'passenger seat** *n.* Beifahrersitz, *der*
passer-by [pɑ:sə'baı] *n.* Passant, *der/*Passantin, *die*
'passing 1. *n.* *(of time, years)* Lauf, *der;* **in ~:** beiläufig ⟨*bemerken usw.*⟩. **2.** *adj.* **a)** vorbeifahrend ⟨*Zug, Auto*⟩; vorbeikommend ⟨*Person*⟩; **b)** flüchtig ⟨*Blick*⟩; vorübergehend ⟨*Mode, Interesse*⟩; flüchtig ⟨*Bekanntschaft*⟩
passion ['pæʃn] *n.* Leidenschaft, *die; (enthusiasm)* leidenschaftliche Begeisterung; **he has a ~ for steam engines** Dampfloks sind seine Leidenschaft.
passionate ['pæʃənət] *adj.* leidenschaftlich; heftig ⟨*Verlangen*⟩
passive ['pæsıv] **1.** *adj.* **a)** passiv; **b)** *(Ling.)* Passiv-. **2.** *n.* *(Ling.)* Passiv, *das*

pass: ~**port** *n*. **a)** [Reise]paß, *der; attrib*. Paß-; **b)** *(fig.)* Schlüssel, *der* (to zu); ~**word** *n*. **a)** Parole, *die;* Losung, *die;* **b)** *(Computing)* Paßwort, *das*

past [pɑːst] **1.** *adj*. **a)** *pred. (over)* vorbei; **b)** *attrib. (previous)* früher; vergangen; ehemalig ⟨*Präsident, Vorsitzende usw.*⟩; **c)** *attrib. (just gone by)* letzt...; vergangen; **in the** ~ **few days** während der letzten Tage; **d)** *(Ling.)* ~ **tense** Vergangenheit, *die*. **2.** *n*. Vergangenheit, *die;* **in the** ~: früher; in der Vergangenheit ⟨*leben*⟩; **be a thing of the** ~: der Vergangenheit angehören. **3.** *prep. (in time)* nach; *(in place)* hinter (+ *Dat.*); **five |minutes|** ~ **two** fünf [Minuten] nach zwei; **gaze/walk** ~ **sb./sth.** an jmdm./etw. vorbeiblicken/vorbeigehen; ~ **repair** nicht mehr zu reparieren. **4.** *adv*. vorbei; **hurry** ~: vorübereilen

pasta ['pæstə] *n*. Nudeln *Pl*.

paste [peɪst] **1.** *n*. **a)** Brei, *der;* **b)** *(glue)* Kleister, *der;* **c)** *(of meat, fish, etc.)* Paste, *die*. **2.** *v. t.* kleben; ~ **sth. into sth.** etw. in etw. *(Akk.)* einkleben

pastel ['pæstl] **1.** *n. (crayon)* Pastellstift, *der*. **2.** *adj*. pastellfarben; Pastell⟨*farben, -töne, -zeichnung*⟩

pasteurize ['pɑːstʃəraɪz] *v. t.* pasteurisieren

pastille ['pæstɪl] *n*. Pastille, *die*

pastime ['pɑːstaɪm] *n*. Zeitvertreib, *der; (person's specific* ~*)* Hobby, *das*

pastor ['pɑːstə(r)] *n*. Pfarrer, *der*/Pfarrerin, *die;* Pastor, *der*/Pastorin, *die*

pastoral ['pɑːstərl] *adj*. Weide-; ländlich ⟨*Reiz, Idylle, Umgebung*⟩

pastry ['peɪstrɪ] *n*. Teig, *der; (article of food)* Gebäckstück, *das;* **pastries** *collect.* [Fein]gebäck, *das*

pasture ['pɑːstʃə(r)] *n*. Weide, *die*

pasty ['pæstɪ] *n*. Pastete, *die*

¹pat [pæt] **1.** *n*. **a)** *(tap)* Klaps, *der;* **b)** *(of butter)* Stückchen, *das*. **2.** *v. t.,* -**tt**- leicht klopfen auf (+ *Akk.*); tätscheln; *(once)* einen Klaps geben (+ *Dat.*) ⟨*Person, Hund, Pferd*⟩; ~ **sb. on the arm/head** jmdm. den Arm/Kopf tätscheln

²pat *adv*. **have sth. off** ~: etw. parat haben

patch [pætʃ] **1.** *n*. **a)** Stelle, *die;* **fog** ~**es** Nebelfelder; **b)** *(on worn garment)* Flicken, *der;* **be not a** ~ **on sth.** *(fig. coll.)* nichts gegen etw. sein. **2.** *v. t.* flicken. **patch 'up** *v. t.* reparieren; *(fig.)* beilegen ⟨*Streit*⟩

patchy ['pætʃɪ] *adj*. uneinheitlich ⟨*Qualität*⟩; ungleichmäßig ⟨*Arbeit*⟩; sehr lückenhaft ⟨*Wissen*⟩

pâté ['pæteɪ] *n*. Pastete, *die*

patent ['peɪtənt, 'pætənt] **1.** *adj. (obvious)* offenkundig. **2.** *n*. Patent, *das*. **3.** *v. t.* patentieren lassen. **patent 'leather** *n*. Lackleder, *das;* ~ **shoes** Lackschuhe. '**patently** *adv*. offenkundig; ~ **obvious** ganz offenkundig

paternal [pə'tɜːnl] *adj*. väterlich

path [pɑːθ] *n*. Weg, *der; (line of motion)* Bahn, *die*

pathetic [pə'θetɪk] *adj*. **a)** *(pitiful)* mitleiderregend; **b)** *(contemptible)* armselig ⟨*Entschuldigung*⟩; erbärmlich ⟨*Person, Leistung*⟩

'**pathway** *n*. Weg, *der*

patience ['peɪʃəns] *n*. Geduld, *die*

patient ['peɪʃənt] **1.** *adj*. geduldig. **2.** *n*. Patient, *der*/Patientin, *die*. '**patiently** *adv*. geduldig

patio ['pætɪəʊ] *n., pl*. ~**s** Veranda, *die;* Terrasse, *die*

patriot ['peɪtrɪət] *n*. Patriot, *der*/Patriotin, *die*. **patriotic** [peɪtrɪ'ɒtɪk] *adj*. patriotisch. **patriotism** ['peɪtrɪətɪzm] *n*. Patriotismus, *der*

patrol [pə'trəʊl] **1.** *n. (Police)* Streife, *die; (Mil.)* Patrouille, *die;* **be on** ~: patrouillieren. **2.** *v. i.,* -**ll**- patrouillieren; ⟨*Polizei:*⟩ Streife laufen/fahren. **3.** *v. t.,* -**ll**- patrouillieren durch (+ *Akk.*); abpatrouillieren ⟨*Straßen, Gegend, Lager*⟩; patrouillieren vor (+ *Dat.*) ⟨*Küste, Grenze*⟩; ⟨*Polizei:*⟩ Streife laufen/fahren in (+ *Dat.*) ⟨*Straßen, Stadtteil*⟩. **pa'trol boat** *n*. Patrouillenboot, *das*. **pa'trol car** *n*. Streifenwagen, *der*

patron ['peɪtrən] *n*. **a)** Gönner, *der*/Gönnerin, *die; (of institution, campaign)* Schirmherr, *der*/Schirmherrin, *die;* **b)** *(customer) (of shop)* Kunde, *der*/Kundin, *die; (of restaurant, hotel)* Gast, *der; (of theatre, cinema)* Besucher, *der*/Besucherin, *die;* **c)** ~ **|saint|** Schutzheilige, *der/die*. **patronage** ['pætrənɪdʒ] *n*. Gönnerschaft, *die; (for campaign, institution)* Schirmherrschaft, *die*

patronize ['pætrənaɪz] *v. t.* **a)** *(frequent)* besuchen; **b)** *(condescend to)* ~ **sb.** jmdn. herablassend behandeln. **patronizing** ['pætrənaɪzɪŋ] *adj*. gönnerhaft; herablassend

patter ['pætə(r)] **1.** *n. (of rain)* Prasseln, *das; (of feet)* Trappeln, *das*. **2.** *v. i.* ⟨*Regen:*⟩ prasseln

pattern ['pætən] *n.* Muster, *das;* *(model)* Vorlage, *die;* *(for sewing)* Schnittmuster, *das;* *(for knitting)* Strickmuster, *das*

paunch [pɔ:ntʃ] *n.* Bauch, *der*

pauper ['pɔ:pə(r)] *n.* Arme, *der/die*

pause [pɔ:z] **1.** *n.* Pause, *die.* **2.** *v. i.* eine Pause machen; ⟨*Redner:*⟩ innehalten; *(hesitate)* zögern

pave [peɪv] *v. t.* befestigen; *(with stones)* pflastern; ~ **the way for sth.** *(fig.)* einer Sache *(Dat.)* den Weg ebnen. '**pavement** *n.* **a)** *(Brit.: footway)* Bürgersteig, *der;* **b)** *(Amer.: roadway)* Fahrbahn, *die*

pavilion [pə'vɪljən] *n.* Pavillon, *der;* *(Brit. Sport)* Klubhaus, *das*

paw [pɔ:] *n.* Pfote, *die;* *(of bear, lion, tiger)* Pranke, *die*

¹**pawn** [pɔ:n] *n.* *(Chess)* Bauer, *der;* *(fig.)* Schachfigur, *die*

²**pawn** **1.** *n.* Pfand, *das;* **in ~:** verpfändet. **2.** *v. t.* verpfänden. '**pawnbroker** *n.* Pfandleiher, *der/*-leiherin, *die.* '**pawnshop** *n.* Leihhaus, *das*

pay [peɪ] **1.** *n.* *(wages)* Lohn, *der;* *(salary)* Gehalt, *das;* **be in the ~ of sb./sth.** für jmdn./etw. arbeiten. **2.** *v. t.,* **paid** [peɪd] bezahlen; zahlen ⟨*Geld*⟩; ~ **sb. to do sth.** jmdn. dafür bezahlen, daß er etw. tut; ~ **sb. £10** jmdm. 10 Pfund zahlen. **3.** *v. i.,* **paid a)** zahlen; ~ **for sth./sb./** für jmdn. bezahlen; **sth. ~s for itself** etw. macht sich bezahlt; **b)** *(be profitable)* sich lohnen; ⟨*Geschäft:*⟩ rentabel sein; **it ~s to be careful** es lohnt sich, vorsichtig zu sein. *See also* **paid. pay 'back** *v. t.* zurückzahlen; **I'll ~ you back later** ich gebe dir das Geld später zurück. **pay 'in** *v. t.* einzahlen. **pay 'off** *v. t.* auszahlen ⟨*Arbeiter*⟩; abbezahlen ⟨*Schulden*⟩; ablösen ⟨*Hypothek*⟩; befriedigen ⟨*Gläubiger*⟩. **pay 'out** *v. t.* auszahlen; *(spend)* ausgeben. **pay 'up** *v. i.* zahlen

payable ['peɪəbl] *adj.* zahlbar; **be ~ to sb.** an jmdn. zu zahlen sein; **make a cheque ~ to the Post Office/to sb.** einen Scheck auf die Post/auf jmds. Namen ausstellen

payee [peɪ'i:] *n.* Zahlungsempfänger, *der/*-empfängerin, *die*

'**payment** *n.* **a)** *(of sum, bill, debt, fine)* Bezahlung, *die;* *(of interest, instalment, tax, fee)* Zahlung, *die;* **in ~ [for sth.]** als Bezahlung [für etw.]; **b)** *(amount)* Zahlung, *die*

pay: **~-packet** *n.* *(Brit.)* Lohntüte, *die;* ~ **phone** *n.* Münzfernsprecher, *der;* **~-rise** *n.* Lohn-/Gehaltserhöhung, *die;* **~-roll** *n.* Lohnliste, *die;* **be on sb.'s ~-roll** für jmdn. arbeiten; **~-slip** *n.* Lohnstreifen, *der/*Gehaltszettel, *der;* ~ **station** *n.* *(Amer.) see* ~ **phone**

PC *abbr.* **a)** *(Brit.)* **police constable** Wachtm.; **b) personal computer** PC

PE *abbr.* **physical education**

pea [pi:] *n.* Erbse, *die*

peace [pi:s] *n.* Frieden, *der;* *(tranquillity)* Ruhe, *die;* ~ **of mind** Seelenfrieden, *der.* **peaceable** ['pi:səbl] *adj.* friedfertig; *(calm)* friedlich. **peaceful** ['pi:sfl] *adj.* friedlich; friedfertig ⟨*Person, Volk*⟩. '**peacefully** *adv.* friedlich; **die ~:** sanft entschlafen

'**peacetime** *n.* Friedenszeiten *Pl.*

peach [pi:tʃ] *n.* Pfirsich, *der*

'**peacock** *n.* Pfau, *der*

peak [pi:k] **1.** *n.* **a)** *(of cap)* Schirm, *der;* **b)** *(of mountain)* Gipfel, *der;* *(fig.)* Höhepunkt, *der.* **2.** *attrib. adj.* Höchst-, Spitzen⟨*preise, -werte*⟩; **~-hour traffic** Stoßverkehr, *der.* **peaked** [pi:kt] *adj.* ~ **cap** Schirmmütze, *die*

peal [pi:l] *n.* Läuten, *das;* ~ **of bells** Glockenläuten, *das;* **a ~/~s of laughter** schallendes Gelächter

peanut ['pi:nʌt] *n.* Erdnuß, *die;* ~ **butter** Erdnußbutter, *die;* **~s** *(coll.: little money)* ein paar Kröten *(salopp)*

pear [peə(r)] *n.* Birne, *die*

pearl [pɜ:l] *n.* Perle, *die*

'**pear-tree** *n.* Birnbaum, *der*

peasant ['pezənt] *n.* [armer] Bauer, *der;* Landarbeiter, *der*

peat [pi:t] *n.* Torf, *der*

pebble ['pebl] *n.* Kiesel[stein], *der*

peck [pek] **1.** *v. t.* hacken; picken ⟨*Körner*⟩. **2.** *v. i.* picken (**at** nach); ~ **at one's food** in seinem Essen herumstochern. **3.** *n.* *(kiss)* flüchtiger Kuß. '**pecking order** *n.* Hackordnung, *die*

peckish ['pekɪʃ] *adj.* *(coll.)* **feel/get ~:** Hunger haben/bekommen

peculiar [pɪ'kju:lɪə(r)] *adj.* **a)** *(strange)* seltsam; **I feel [slightly] ~:** mir ist [etwas] komisch; **b)** *(especial)* besonder...; **c)** *(belonging exclusively)* eigentümlich (**to** *Dat.*). **peculiarity** [pɪkju:lɪ'ærɪtɪ] *n.* **a)** *(odd trait)* Eigentümlichkeit, *die;* **b)** *(distinguishing characteristic)* [charakteristisches] Merkmal. **pe'culiarly** *adv.* **a)** *(strangely)* seltsam; **b)** *(especially)* besonders

pedal ['pedl] **1.** *n.* Pedal, *das.* **2.** *v. i.,*

(Brit.) **-ll-** in die Pedale treten.
'**pedal-bin** n. Treteimer, *der*
pedant ['pedənt] n. Pedant, *der*/Pedantin, *die.* **pedantic** [pɪ'dæntɪk] *adj.* pedantisch
peddle ['pedl] *v. t.* auf der Straße verkaufen; *(door to door)* hausieren mit
pedestal ['pedɪstl] n. Sockel, *der*
pedestrian [pɪ'destrɪən] 1. *adj. (uninspired)* trocken; langweilig. 2. n. Fußgänger, *der*/-gängerin, *die.* **pedestrian 'crossing** n. Fußgängerüberweg, *der*
pedigree ['pedɪgriː] 1. n. Stammbaum, *der.* 2. *adj.* mit Stammbaum *nachgestellt*
pedlar ['pedlə(r)] n. Straßenhändler, *der*/-händlerin, *die; (door to door)* Hausierer, *der*/Hausiererin, *die*
pee [piː] *(coll.)* 1. *v. i.* pinkeln *(salopp);* Pipi machen *(Kinderspr.).* 2. n. **a) have a ~:** pinkeln *(salopp);* **b)** *(urine)* Pipi, *das (Kinderspr.)*
peek [piːk] *see* ²peep
peel [piːl] 1. *v. t.* schälen. 2. *v. i.* ⟨*Person, Haut:*⟩ sich schälen; ⟨*Farbe:*⟩ abblättern. 3. n. Schale, *die.* '**peelings** n. pl. Schalen
¹**peep** [piːp] 1. *v. i.* ⟨*Maus, Vogel:*⟩ piep[s]en. 2. n. Piepsen, *das; (coll.: remark etc.)* Piep[s], *der*
²**peep** 1. *v. i.* gucken *(ugs.); (furtively)* verstohlen gucken *(ugs.).* 2. n. kurzer/verstohlener Blick. '**peep-hole** n. Guckloch, *das.* **peeping 'Tom** n. Spanner, *der (ugs.)*
¹**peer** [pɪə(r)] n. Peer, *der; (equal)* Gleichgestellte, *der/die*
²**peer** *v. i.* forschend schauen; *(with difficulty)* angestrengt schauen; **~ at sth./sb.** [sich *(Dat.)*] etw. genau ansehen/jmdn. forschend ansehen; *(with difficulty)* [sich *(Dat.)*] etw./jmdn. angestrengt ansehen
peerage ['pɪərɪdʒ] n. Peerswürde, *die*
peevish ['piːvɪʃ] *adj.* nörgelig
peg [peg] n. *(for holding together)* Stift, *der; (for tying things to)* Pflock, *der; (for hanging things on)* Haken, *der; (clothes-~)* Wäscheklammer, *die; (tent-~)* Hering, *der;* **off the ~** *(Brit.: ready-made)* von der Stange *(ugs.)*
pejorative [pɪ'dʒɒrətɪv] *adj.,* **pe'joratively** *adv.* abwertend
pelican ['pelɪkən] n. Pelikan, *der.* '**pelican crossing** n. *(Brit.)* Ampelübergang, *der*
pellet ['pelɪt] n. Kügelchen, *das*
pelmet ['pelmɪt] n. Blende, *die*

¹**pelt** [pelt] n. Fell, *das*
²**pelt** 1. *v. t.* **~ sb. with sth.** jmdn. mit etw. bewerfen. 2. *v. i.* **a) it was ~ing down [with rain]** es goß wie aus Kübeln *(ugs.);* **b)** *(run fast)* rasen *(ugs.)*
pelvis ['pelvɪs] n., pl. **pelves** ['pelviːz] *or* **-es** *(Anat.)* Becken, *das*
¹**pen** [pen] 1. n. *(enclosure)* Pferch, *der.* 2. *v. t.,* **-nn-: ~ sb. in a corner** jmdn. in eine Ecke drängen. **pen 'in** *v. t.* einpferchen
²**pen** 1. n. Federhalter, *der; (fountain-~)* Füller, *der; (ball-~)* Kugelschreiber, *der; (felt-tip ~)* Filzstift, *der.* 2. *v. t.,* **-nn-** schreiben
penal ['piːnl] *adj.* Straf-
penalize ['piːnəlaɪz] *v. t.* bestrafen; *(Sport)* eine Strafe verhängen gegen
penalty ['penltɪ] n. **a)** Strafe, *die;* **pay the ~/the ~ for** *or* **of sth.** dafür/für etw. büßen [müssen]; **b)** *(Footb.)* Elfmeter, *der*
penance ['penəns] n. Buße, *die;* **act of ~:** Bußwerk, *das;* **do ~:** Buße tun
pence *see* penny
pencil ['pensl] 1. n. Bleistift, *der;* **red/coloured ~:** Rot-/Buntstift, *der.* 2. *v. t., (Brit.)* **-ll-** mit einem Bleistift/Farbstift schreiben. '**pencil-case** n. Griffelkasten, *der; (of soft material)* Federmäppchen, *das.* '**pencil-sharpener** n. Bleistiftspitzer, *der*
pendant ['pendənt] n. Anhänger, *der*
pending ['pendɪŋ] 1. *adj.* unentschieden ⟨*Angelegenheit, Sache*⟩; schwebend ⟨*Verfahren*⟩. 2. *prep.* **~ his return** bis zu seiner Rückkehr
pendulum ['pendjʊləm] n. Pendel, *das*
penetrate ['penɪtreɪt] *v. t.* eindringen in (+ *Akk.); (pass through)* durchdringen. **penetrating** ['penɪtreɪtɪŋ] *adj.* durchdringend. **penetration** [penɪ'treɪʃn] n. Eindringen, *das* (of in + *Akk.); (passing through)* Durchdringen, *das*
'**pen-friend** n. Brieffreund, *der*/-freundin, *die*
penguin ['peŋgwɪn] n. Pinguin, *der*
penicillin [penɪ'sɪlɪn] n. Penizillin, *das*
peninsula [pɪ'nɪnsjʊlə] n. Halbinsel, *die*
penis ['piːnɪs] n. Penis, *der*
penitence ['penɪtəns] n. Reue, *die*
penitent ['penɪtənt] *adj.* reuevoll *(geh.);* reuig *(geh.)* ⟨*Sünder*⟩
penitentiary [penɪ'tenʃərɪ] n. *(Amer.)* Straf[vollzugs]anstalt, *die*
'**penknife** n. Taschenmesser, *das*

pennant ['penənt] *n.* Wimpel, *der;* *(on official car etc.)* Ständer, *der*

penniless ['penɪlɪs] *adj.* mittellos

penny ['penɪ] *n., pl. usu.* **pennies** ['penɪz] *(for separate coins),* **pence** [pens] *(for sum of money)* Penny, *der;* **fifty pence** fünfzig Pence; **two/fifty pence |piece|** Zwei-/Fünfzigpencestück, *das*

pension ['penʃn] *n.* Rente, *die; (payment to retired civil servant also)* Pension, *die;* **be on a ~:** eine Rente beziehen; **widow's ~:** Witwenrente, *die.* **pension 'off** *v. t.* berenten *(Amts-spr.);* auf Rente setzen *(ugs.);* pensionieren ⟨*Lehrer, Beamten*⟩

'**pensioner** *n.* Rentner, *der*/Rentnerin, *die; (retired civil servant)* Pensionär, *der*/Pensionärin, *die*

pensive ['pensɪv] *adj.* nachdenklich

pentagon ['pentəgən] *n.* Fünfeck, *das*

pent: **~house** *n.* Penthaus, *das;* **~-up** *adj.* angestaut ⟨*Ärger, Wut*⟩; unterdrückt ⟨*Sehnsucht, Gefühle*⟩

penultimate [pe'nʌltɪmət] *adj.* vorletzt...

people ['pi:pl] *n.* **a)** *constr. as pl.* Leute *Pl.;* Menschen; *(as opposed to animals)* Menschen *Pl.;* **city/country ~** *(inhabitants)* Stadt-/Landbewohner; **local ~:** Einheimische *Pl.;* **working ~:** arbeitende Menschen; **coloured/white ~:** Farbige/Weiße; **~ say ...:** man sagt ...; **a crowd of ~:** eine Menschenmenge; **b)** *(nation)* Volk, *das*

pepper ['pepə(r)] **1.** *n.* **a)** Pfeffer, *der;* **b)** *(vegetable)* Paprikaschote, *die;* **red/green ~:** roter/grüner Paprika. **2.** *v. t.* **a)** pfeffern; **b)** *(pelt)* bombardieren *(ugs.)*

pepper: **~corn** *n.* Pfefferkorn, *das;* **~mint** *n. (sweet)* Pfefferminz, *das;* **~-pot** *n.* Pfefferstreuer, *der*

per [pə(r), *stressed* pɜ:(r)] *prep.* pro

perceive [pə'si:v] *v. t.* wahrnehmen; *(with the mind)* spüren; **~d** vermeintlich ⟨*Bedrohung, Gefahr, Wert*⟩

per cent (*Brit.; Amer.:* **percent**) [pə'sent] **1.** *adv.* **ninety ~ effective** zu 90 Prozent wirksam. **2.** *adj.* **a 5 ~ increase** ein Zuwachs von 5 Prozent. **3.** *n.* **a)** Prozent, *das;* **b)** *see* percentage

percentage [pə'sentɪdʒ] *n.* Prozentsatz, *der*

perceptible [pə'septɪbl] *adj.* wahrnehmbar

perception [pə'sepʃn] *n. (act)* Wahrnehmung, *die; (result)* Erkenntnis, *die; (faculty)* Wahrnehmungsvermögen, *das*

perceptive [pə'septɪv] *adj.* einfühlsam ⟨*Person, Bemerkung*⟩

perch [pɜ:tʃ] **1.** *n.* Sitzstange, *die.* **2.** *v. i.* **a)** sich niederlassen; **b)** *(be supported)* sitzen. **3.** *v. t.* setzen/stellen/legen

percolate ['pɜ:kəleɪt] *v. i.* [durch]sickern. **percolator** ['pɜ:kəleɪtə(r)] *n.* Kaffeemaschine, *die*

percussion [pə'kʌʃn] *n. (Mus.)* Schlagzeug, *das;* **~ instrument** Schlaginstrument, *das*

perennial [pə'renjəl] **1.** *adj.* **a)** *(Bot.)* ausdauernd; **b)** immer wieder auftretend ⟨*Problem*⟩. **2.** *n. (Bot.)* ausdauernde Pflanze

perfect 1. ['pɜ:fɪkt] *adj.* vollkommen; perfekt ⟨*Englisch, Timing*⟩; tadellos ⟨*Zustand*⟩; *(coll.: unmitigated)* absolut; **a ~ stranger** ein völlig Fremder. **2.** [pə'fekt] *v. t.* vervollkommnen. **perfection** [pə'fekʃn] *n.* Perfektion, *die;* **to ~:** perfekt. **perfectionism** [pə'fekʃənɪzm] *n.* Perfektionismus, *der.* **perfectionist** [pə'fekʃənɪst] *n.* Perfektionist, *der*/Perfektionistin, *die.* '**perfectly** *adv.* **a)** *(completely)* vollkommen; **be ~ entitled to do sth.** durchaus berechtigt sein, etw. zu tun; **b)** *(faultlessly)* perfekt; tadellos ⟨*sich verhalten*⟩

perforate ['pɜ:fəreɪt] *v. t.* perforieren; *(make opening into)* durchlöchern. **perforation** [pɜ:fə'reɪʃn] *n.* **a)** *(hole)* Loch, *das;* **b)** *in pl.* **~s** Perforation, *die; (in sheets of stamps)* Zähnung, *die*

perform [pə'fɔ:m] **1.** *v. t.* ausführen ⟨*Arbeit, Operation*⟩; erfüllen ⟨*Pflicht, Aufgabe*⟩; vollbringen ⟨[*Helden]tat, Leistung*⟩; ausfüllen ⟨*Funktion*⟩; vollbringen ⟨*Wunder*⟩; anstellen ⟨*Berechnungen*⟩; durchführen ⟨*Experiment, Sektion*⟩; vorführen ⟨*Trick*⟩; aufführen ⟨*Theaterstück, Scharade*⟩; vortragen ⟨*Lied, Sonate usw.*⟩. **2.** *v. i.* eine Vorführung geben; *(sing)* singen; *(play)* spielen. **performance** [pə'fɔ:məns] *n.* **a)** *(of duty, task)* Erfüllung, *die;* **b)** *([notable] achievement; Motor Veh.)* Leistung, *die;* **c)** *(at theatre, cinema, etc.)* Vorstellung, *die;* **her ~ as Desdemona** ihre Darstellung der Desdemona; **the ~ of a play/opera** die Aufführung eines Theaterstücks/einer Oper. **per'former** *n.* Künstler, *der*/Künstlerin, *die.* **per'forming** *attrib. adj.* dressiert ⟨*Tier*⟩

perfume ['pɜ:fju:m] *n.* Duft, *der; (fluid)* Parfüm, *das*

perfunctory [pə'fʌŋktərɪ] *adj.* oberflächlich ‹*Arbeit, Überprüfung*›; flüchtig ‹*Erkundigung, Bemerkung*›

perhaps [pə'hæps] *adv.* vielleicht

peril ['perɪl] *n.* Gefahr, *die*. **perilous** ['perələs] *adj.* gefahrvoll; **be** ~: gefährlich sein

perimeter [pə'rɪmɪtə(r)] *n.* [äußere] Begrenzung; Grenze, *die*

period ['pɪərɪəd] **1.** *n.* **a)** *(of history or life)* Periode, *die*; Zeit, *die*; *(any portion of time)* Zeitraum, *der*; **the Classical/Romantic** ~: die Klassik/Romantik; **b)** *(Sch.)* Stunde, *die*; **chemistry/English** ~: Chemie-/Englischstunde, *die*; **c)** *(menstruation)* Periode, *die*; **d)** *(punctuation mark)* Punkt, *der*. **2.** *adj.* zeitgenössisch ‹*Tracht, Kostüm*›; antik ‹*Möbel*›. **periodic** [pɪərɪ'ɒdɪk] *adj.* regelmäßig; *(intermittent)* gelegentlich. **periodical** [pɪərɪ'ɒdɪkl] **1.** *adj. see* **periodic. 2.** *n.* Zeitschrift, *die*; **weekly/monthly** ~: Wochenzeitschrift/Monatsschrift, *die*. **peri'odically** *adv.* regelmäßig; *(intermittently)* gelegentlich

peripheral [pə'rɪfərl] *adj.* peripher *(geh.)*; Rand‹*problem, -erscheinung*›

periphery [pə'rɪfərɪ] *n.* Peripherie, *die*

periscope ['perɪskəʊp] *n.* Periskop, *das*

perish ['perɪʃ] *v.i.* **a)** *(die)* umkommen; **b)** *(rot)* verderben; ‹*Gummi:*› altern. **perishable** ['perɪʃəbl] *adj.* [leicht] verderblich

'**perishing** *(coll.)* **1.** *adj.* mörderisch ‹*Kälte*›; **it's/I'm** ~: es ist bitterkalt/ ich komme um vor Kälte *(ugs.)*. **2.** *adv.* mörderisch ‹*kalt*›

perjury ['pɜːdʒərɪ] *n.* Meineid, *der*; **commit** ~: einen Meineid leisten

¹**perk** [pɜːk] *(coll.)* **1.** *v.i.* ~ **up** munter werden. **2.** *v.t.* ~ **up** aufmuntern

²**perk** *n.* *(Brit. coll.)* [Sonder]vergünstigung, *die*

perky ['pɜːkɪ] *adj.* lebhaft; munter

perm [pɜːm] **1.** *n.* Dauerwelle, *die*. **2.** *v.t.* **have one's hair** ~**ed** sich *(Dat.)* eine Dauerwelle machen lassen

permanence ['pɜːmənəns] *n.* Dauerhaftigkeit, *die*

permanent ['pɜːmənənt] *adj.* fest ‹*Sitz, Bestandteil, Mitglied*›; ständig ‹*Wohnsitz, Adresse, Kampf*›; Dauer‹*stellung, -visum*›; bleibend ‹*Schaden*›. '**permanently** *adv.* dauernd; auf Dauer ‹*verhindern, bleiben*›

permeable ['pɜːmɪəbl] *adj.* durchlässig; **be** ~ **to sth.** etw. durchlassen

permeate ['pɜːmɪeɪt] **1.** *v.t.* dringen durch; **be** ~**d with** *or* **by sth.** *(fig.)* von etw. durchdrungen sein. **2.** *v.i.* ~ **through sth.** etw. durchdringen

permissible [pə'mɪsɪbl] *adj.* zulässig; **be** ~ **to** *or* **for sb.** jmdm. erlaubt sein

permission [pə'mɪʃn] *n.* Erlaubnis, *die*; *(given by official body)* Genehmigung, *die*; **give sb.** ~ **to do sth.** jmdm. erlauben, etw. zu tun

permissive [pə'mɪsɪv] *adj.* **the** ~ **society** die permissive Gesellschaft

permit 1. [pə'mɪt] *v.t.,* -tt- zulassen ‹*Berufung, Einspruch usw.*›; ~ **sb. sth.** jmdm. etw. erlauben; **sb. is** ~**ted to do sth.** es ist jmdm. erlaubt, etw. zu tun. **2.** *v.i.,* -tt- es zulassen. **3.** ['pɜːmɪt] *n.* Genehmigung, *die*

pernicious [pə'nɪʃəs] *adj.* verderblich; bösartig ‹*Krankheit*›

perpendicular [pɜːpən'dɪkjʊlə(r)] *adj.* senkrecht

perpetrate ['pɜːpɪtreɪt] *v.t.* begehen; verüben ‹*Greuel*›

perpetual [pə'petjʊəl] *adj.* **a)** *(eternal)* ewig; **b)** *(continuous; coll.: repeated)* ständig. **per'petually** *adv.* **a)** *(eternally)* ewig; **b)** *(continuously; coll.: repeatedly)* ständig

perpetuate [pə'petjʊeɪt] *v.t.* aufrechterhalten

perplex [pə'pleks] *v.t.* verwirren. **perplexed** [pə'plekst] *adj.* verwirrt; *(puzzled)* ratlos. **perplexity** [pə'pleksɪtɪ] *n.* Verwirrung, *die*; *(puzzlement)* Ratlosigkeit, *die*

persecute ['pɜːsɪkjuːt] *v.t.* verfolgen. **persecution** [pɜːsɪ'kjuːʃn] *n.* Verfolgung, *die*

perseverance [pɜːsɪ'vɪərəns] *n.* Beharrlichkeit, *die*; Ausdauer, *die*

persevere [pɜːsɪ'vɪə(r)] *v.i.* ausharren; ~ **with** *or* **at** *or* **in sth.** bei etw. dabeibleiben

Persian ['pɜːʃn] *adj.* persisch; Perser‹*katze, -teppich*›

persist [pə'sɪst] *v.i.* **a)** nicht nachgeben; ~ **in doing sth.** etw. weiterhin [beharrlich] tun; **b)** *(continue to exist)* anhalten. **persistence** [pə'sɪstəns] Hartnäckigkeit, *die*. **persistent** [pə'sɪstənt] *adj.* **a)** hartnäckig; **b)** *(constantly repeated)* dauernd; hartnäckig ‹*Gerüchte*›. **per'sistently** *adv.* hartnäckig

person ['pɜːsn] *n.* Mensch, *der*; **in** ~: persönlich; selbst

personal ['pɜːsənl] *adj.* persönlich; Privat‹*angelegenheit, -leben*›; ~ **com-**

puter Personalcomputer, *der;* ~ **stereo** Walkman, *der;* ~ **hygiene** Körperpflege, *die.* **personal as'sistant** *n.* persönlicher Referent/persönliche Referentin

personality [pɜ:sə'nælɪtɪ] *n.* Persönlichkeit, *die*

'personally *adv.* persönlich

personification [pəsɒnɪfɪ'keɪʃn] *n.* Verkörperung, *die*

personify [pə'sɒnɪfaɪ] *v. t.* verkörpern; **be kindness personified** die Freundlichkeit in Person sein

personnel [pɜ:sə'nel] *n.* Belegschaft, *die; (of shop, restaurant, etc.)* Personal, *das; attrib.* Personal-

perspective [pə'spektɪv] *n.* Perspektive, *die; (fig.)* Blickwinkel, *der*

perspiration [pɜ:spɪ'reɪʃn] *n.* Schweiß, *der*

perspire [pə'spaɪə(r)] *v. i.* schwitzen

persuade [pə'sweɪd] *v. t.* **a)** *(convince)* überzeugen **(of** von); ~ **oneself [that]** ...: sich *(Dat.)* einreden, daß ...; **b)** *(induce)* überreden. **persuasion** [pə'sweɪʒn] *n.* Überzeugung, *die;* **it didn't take much** ~ : es brauchte nicht viel Überredungskunst. **persuasive** [pə'sweɪsɪv] *adj.,* **per'suasively** *adv.* überzeugend

pert [pɜ:t] *adj.* keck

pertinent ['pɜ:tɪnənt] *adj.* relevant **(to** für)

perturb [pə'tɜ:b] *v. t.* beunruhigen

Peru [pə'ru:] *pr. n.* Peru *(das).* **Peruvian** [pə'ru:vɪən] **1.** *adj.* peruanisch. **2.** *n.* Peruaner, *der*/Peruanerin, *die*

pervade [pə'veɪd] *v. t.* durchdringen. **pervasive** [pə'veɪsɪv] *adj.* durchdringend *(Geruch, Kälte);* weit verbreitet *(Ansicht);* sich ausbreitend *(Gefühl)*

perverse [pə'vɜ:s] *adj.* starrköpfig

perversion [pə'vɜ:ʃn] *n.* **a)** *(sexual)* Perversion, *die;* **b)** ~ **of justice** Rechtsbeugung, *die.* **pervert 1.** [pə'vɜ:t] *v. t. (morally)* verderben. **2.** ['pɜ:vɜ:t] *n.* perverser Mensch. **perverted** [pə'vɜ:tɪd] *adj. (sexually)* pervers

pessimism ['pesɪmɪzm] *n.* Pessimismus, *der.* **pessimist** ['pesɪmɪst] *n.* Pessimist, *der*/Pessimistin, *die.* **pessimistic** [pesɪ'mɪstɪk] *adj.* pessimistisch

pest [pest] *n. (thing)* Ärgernis, *das; (person)* Nervensäge, *die (ugs.); (animal)* Schädling, *der*

pester ['pestə(r)] *v. t.* belästigen; nerven *(ugs.);* ~ **sb. for sth.** jmdm. wegen etw. in den Ohren liegen

pesticide ['pestɪsaɪd] *n.* Pestizid, *das*

pet [pet] **1.** *n.* **a)** *(animal)* Haustier, *das;* **b)** *(as term of endearment)* Schatz, *der.* **2.** *adj. (favourite)* Lieblings-. **3.** *v. i.,* -tt- knutschen *(ugs.)*

petal ['petl] *n.* Blütenblatt, *das*

peter ['pi:tə(r)] *v. i.* ~ **out** [allmählich] zu Ende gehen; *(Weg.:)* sich verlieren

petite [pə'ti:t] *adj.* zierlich

petition [pə'tɪʃn] **1.** *n.* Petition, *die;* Eingabe, *die.* **2.** *v. t.* eine Eingabe richten an *(+ Akk.)*

petrify ['petrɪfaɪ] *v. t.* **be petrified with fear/shock** starr vor Angst/Schrecken sein

petrol ['petrl] *n. (Brit.)* Benzin, *das*

petroleum [pɪ'trəʊlɪəm] *n.* Erdöl, *das*

petrol: ~-**pump** *n. (Brit.)* Zapfsäule, *die;* ~-**station** *n. (Brit.)* Tankstelle, *die;* ~-**tank** *n. (Brit.)* Benzintank, *der;* ~-**tanker** *n. (Brit.)* Benzintankwagen, *der*

'pet shop *n.* Tierhandlung, *die*

petticoat ['petɪkəʊt] *n.* Unterrock, *der*

petty ['petɪ] *adj.* kleinlich *(Vorschrift, Einwand);* belanglos *(Detail, Sorgen)*

petulant ['petjʊlənt] *adj.* bockig

pew [pju:] *n. (Eccl.)* Kirchenbank, *die*

pewter ['pju:tə(r)] *n.* Pewter, *der*

phantom ['fæntəm] *n.* Phantom, *das*

pharmacist ['fɑ:məsɪst] *n.* Apotheker, *der*/Apothekerin, *die*

pharmacy ['fɑ:məsɪ] *n. (dispensary)* Apotheke, *die*

phase [feɪz] *n.* Phase, *die.* **phase 'in** *v. t.* stufenweise einführen. **phase 'out** *v. t.* allmählich abschaffen *(Verfahrensweise, Methode);* *(stop producing)* [langsam] auslaufen lassen

Ph.D. [pi:eɪtʃ'di:] *abbr.* **Doctor of Philosophy** Dr. phil.

pheasant ['fezənt] *n.* Fasan, *der*

phenomenal [fɪ'nɒmɪnl] *adj.* phänomenal

phenomenon [fɪ'nɒmɪnən] *n., pl.* **phenomena** [fɪ'nɒmɪnə] Phänomen, *das*

phew [fju:] *int.* puh

Philippines ['fɪlɪpi:nz] *pr. n. pl.* Philippinen *Pl.*

philistine ['fɪlɪstaɪn] *n.* Banause, *der*/Banausin, *die*

philosopher [fɪ'lɒsəfə(r)] *n.* Philosoph, *der*/Philosophin, *die*

philosophical [fɪlə'sɒfɪkl] *adj.* **a)** philosophisch; **b)** *(resigned)* abgeklärt

philosophy [fɪ'lɒsəfɪ] *n.* Philosophie, *die*

phlegm [flem] *n.* Schleim, *der*

phobia ['fəʊbɪə] *n.* Phobie, *die*

phone [fəʊn] *(coll.)* **1.** *n.* Telefon, *das;* by ~: telefonisch; be on the ~: Telefon haben; *(be phoning)* telefonieren. **2.** *v. t. & i.* anrufen. **phone 'back** *v. t. & i.* zurückrufen; *(make further call)* wieder anrufen. **phone 'up** *v. t. & i.* anrufen

phone: ~ **book** *n.* Telefonbuch, *das;* ~ **box** *n.* Telefonzelle, *die;* ~ **call** *n.* Anruf, *der;* ~ **card** *n.* Telefonkarte, *die;* ~ **number** *n.* Telefonnummer, *die*

phonetic [fə'netɪk] *adj.* phonetisch. **phonetics** [fə'netɪks] *n.* Phonetik, *die*

phoney ['fəʊnɪ] *adj. (coll.) (sham)* falsch; gefälscht ⟨Brief, Dokument⟩

phonograph ['fəʊnəɡrɑːf] *n. (Amer.)* Plattenspieler, *der*

phony see **phoney**

phosphorus ['fɒsfərəs] *n.* Phosphor, *der*

photo ['fəʊtəʊ] *n., pl.* ~s Foto, *das*

photo: ~**copier** *n.* Fotokopiergerät, *das;* ~**copy 1.** *n.* Fotokopie, *die;* **2.** *v. t.* fotokopieren

photogenic [fəʊtə'dʒiːnɪk] *adj.* fotogen

photograph ['fəʊtəɡrɑːf] **1.** *n.* Fotografie, *die;* Foto, *das;* take a ~ [of sb./ sth.] [jmdn./etw.] fotografieren. **2.** *v. t. & i.* fotografieren. **photographer** [fə'tɒɡrəfə(r)] *n.* Fotograf, *der*/Fotografin, *die.* **photographic** [fəʊtə'ɡræfɪk] *adj.* fotografisch; Foto⟨ausrüstung, -apparat, -ausstellung⟩. **photography** [fə'tɒɡrəfɪ] *n.* Fotografie, *die*

phrase [freɪz] **1.** *n.* [Rede]wendung, *die.* **2.** *v. t.* formulieren. **'phrasebook** *n.* Sprachführer, *der*

physical ['fɪzɪkl] *adj.* **a)** physisch ⟨Gewalt⟩; dinglich ⟨Welt, Universum⟩; **b)** *(of physics)* physikalisch; **c)** *(bodily)* körperlich. **physical edu'cation** *n. (Sch.)* Sport, *der.* **'physically** *adv. (relating to the body)* körperlich

physician [fɪ'zɪʃn] *n.* Arzt, *der*/Ärztin, *die*

physicist ['fɪzɪsɪst] *n.* Physiker, *der*/ Physikerin, *die*

physics ['fɪzɪks] *n.* Physik, *die*

physiology [fɪzɪ'ɒlədʒɪ] *n.* Physiologie, *die*

physiotherapy [fɪzɪəʊ'θerəpɪ] *n.* Physiotherapie, *die*

physique [fɪ'ziːk] *n.* Körperbau, *der*

pianist ['pɪːənɪst] *n.* Pianist, *der*/Pianistin, *die*

piano [pɪ'ænəʊ] *n., pl.* ~s *(upright)* Klavier, *das; (grand)* Flügel, *der.* **piano-ac'cordion** *n.* Akkordeon, *das*

¹pick [pɪk] *n. (tool)* Spitzhacke, *die*

²pick 1. *n.* **a)** *(choice)* Wahl, *die;* take your ~: du hast die Wahl; **b)** *(best part)* Elite, *die;* the ~ of the fruit die besten Früchte. **2.** *v. t.* **a)** pflücken ⟨Blumen, Äpfel usw.⟩; lesen ⟨Trauben⟩; **b)** *(select)* auswählen; ~ one's way sich *(Dat.)* vorsichtig [s]einen Weg suchen; **c)** ~ one's nose in der Nase bohren; **d)** ~ sb.'s pocket jmdn. bestehlen; he had his pocket ~ed er wurde von einem Taschendieb bestohlen; **e)** ~ a lock ein Schloß knacken *(salopp).* **3.** *v. i.* ~ and choose wählerisch sein. **'pick at** *v. t.* herumstochern in (+ *Dat.*) ⟨Essen⟩. **'pick on** *v. t. (victimize)* es abgesehen haben auf (+ *Akk.*). **pick 'out** *v. t.* **a)** *(choose)* auswählen; *(for oneself)* sich *(Dat.)* aussuchen; **b)** *(distinguish)* entdecken ⟨Detail, jmds. Gesicht in der Menge⟩. **pick up 1.** [--] *v. t.* **a)** [in die Hand] nehmen; hochnehmen ⟨Baby⟩; *(after dropping)* aufheben; aufnehmen ⟨Masche⟩; ~ up the telephone den [Telefon]hörer abnehmen; **b)** *(collect)* mitnehmen; *(by arrangement)* abholen (at, from von); *(obtain)* holen; **c)** *(become infected by)* sich *(Dat.)* holen *(ugs.)* ⟨Virus, Grippe⟩; **d)** ⟨Bus, Autofahrer:⟩ mitnehmen; **e)** *(rescue from the sea)* [aus Seenot] bergen; **f)** empfangen ⟨Signal, Funkspruch usw.⟩; **g)** *(coll.: make acquaintance of)* aufreißen *(ugs.).* **2.** [-'-] *v. i.* **a)** sich bessern; **b)** ⟨Wind:⟩ auffrischen

'pickaxe *(Amer.:* **'pickax)** *see* **¹pick**

picket ['pɪkɪt] **1.** *n.* Streikposten, *der.* **2.** *v. i.* Streikposten stehen. **3.** *v. t.* Streikposten stellen vor (+ *Dat.*). **'picket-line** *n.* Streikpostenkette, *die*

pickle ['pɪkl] **1.** *n., usu. in pl. (food)* [Mixed] Pickles *Pl.* **2.** *v. t.* einlegen ⟨Gurken, Zwiebeln, Eier⟩; marinieren ⟨Hering⟩

pick: ~**-me-up** *n.* Stärkungsmittel, *das;* ~**pocket** *n.* Taschendieb, *der*/-diebin, *die;* ~**-up** *n.* **a)** ~ [truck] Kleinlastwagen, *der;* **b)** *(of record-player, guitar)* Tonabnehmer, *der*

picnic ['pɪknɪk] **1.** *n.* Picknick, *das;* go for *or* on/have a ~: ein Picknick machen. **2.** *v. i.,* -ck- picknicken; Picknick machen. **'picnic site** *n.* Picknickplatz, *der*

pictorial [pɪk'tɔːrɪəl] *adj.* illustriert ⟨*Bericht, Zeitschrift*⟩; bildlich ⟨*Darstellung*⟩

picture ['pɪktʃə(r)] **1.** *n.* **a)** Bild, *das;* **get the ~** *(coll.)* verstehen[, worum es geht]; **put sb. in the ~:** jmdn. ins Bild setzen; **b)** *(film)* Film, *der;* **c)** *in pl. (Brit.: cinema)* Kino, *das;* **go to the ~s** ins Kino gehen; **what's on at the ~s?** was läuft im Kino? **2.** *v. t.* ~ [**to oneself**] sich *(Dat.)* vorstellen. '**picturebook** *n.* Bilderbuch, *das.* **picture 'postcard** *n.* Ansichtskarte, *die*

picturesque [pɪktʃə'resk] *adj.* malerisch

pidgin ['pɪdʒɪn] *n.* Pidgin, *das.* **pidgin 'English** *n.* Pidgin-Englisch, *das*

pie [paɪ] *n. (of meat, fish, etc.)* Pastete, *die; (of fruit etc.)* ≈ Obstkuchen, *der*

piece [piːs] **1.** *n.* **a)** Stück, *das; (of broken glass or pottery)* Scherbe, *die; (of jigsaw puzzle, crashed aircraft, etc.)* Teil, *der; (Amer.: distance)* [kleines] Stück; **a ~ of meat/cake** ein Stück Fleisch/Kuchen; ~ **of furniture/luggage** Möbel-/Gepäckstück, *das;* **a three-~ suite** eine dreiteilige Sitzgarnitur; ~ **of luck** Glücksfall, *der;* ~ **of news/gossip/information** Nachricht, *die*/Klatsch, *der*/Information, *die;* **b)** *(Chess)* Figur, *die;* **c)** *(coin)* gold ~: Goldstück, *das;* **a 10p ~:** ein 10-Pence-Stück; **d)** *(literary or musical composition)* Stück, *das;* ~ **of music** Musikstück, *das.* **2.** *v. t.* ~ **to'gether** zusammenfügen (**from** aus)

piece: ~**meal** *adv., adj.* stückweise; ~**work** *n.* Akkordarbeit, *die*

pier [pɪə(r)] *n. (at seaside)* Pier, *der*

pierce [pɪəs] *v. t. (prick)* durchstechen; *(penetrate)* [ein]dringen in (+ *Akk.*) ⟨*Körper, Fleisch, Herz*⟩; ~ **a hole in sth.** ein Loch in etw. *(Akk.)* stechen.

piercing ['pɪəsɪŋ] *adj.* durchdringend ⟨*Stimme, Schrei, Blick*⟩

piety ['paɪətɪ] *n.* Frömmigkeit, *die*

pig [pɪg] *n.* **a)** Schwein, *das;* ~**s might fly** *(iron.)* da müßte schon ein Wunder geschehen; **b)** *(coll.: greedy person)* Vielfraß, *der (ugs.)*

pigeon ['pɪdʒɪn] *n.* Taube, *die.* '**pigeon-hole** *n.* [Ablage]fach, *das; (for letters)* Postfach, *das*

piggy ['pɪgɪ]: ~**back** *n.* **give sb. a** ~**back** jmdn. huckepack nehmen; ~**bank** *n.* Sparschwein[chen], *das*

pig'headed *adj.* dickschädelig *(ugs.)*

pigment ['pɪgmənt] *n.* Pigment, *das*

pig: ~**sty** *n. (lit. or fig.)* Schweinestall,

der; ~**tail** *n. (plaited)* Zopf, *der;* ~**tails** *(at either side of head)* Rattenschwänzchen *Pl. (ugs.)*

pike [paɪk] *n., pl. same* Hecht, *der*

pilchard ['pɪltʃəd] *n.* Sardine, *die*

¹**pile** [paɪl] **1.** *n.* **a)** *(of dishes, plates)* Stapel, *der; (of paper, books, letters)* Stoß, *der; (of clothes)* Haufen, *der;* **b)** *(coll.: large quantity)* Haufen, *der (ugs.).* **2.** *v. t.* **a)** *(load)* [voll] beladen; **b)** *(heap up)* aufstapeln ⟨*Holz, Steine*⟩; aufhäufen ⟨*Abfall, Schnee*⟩. **pile 'in** *v. i. (seen from outside)* hineindrängen; *(seen from inside)* hereindrängen. '**pile into** *v. t.* sich zwängen in (+ *Akk.*) ⟨*Auto, Zimmer, Zugabteil*⟩. **pile 'on 1.** *v. i. see* **pile in. 2.** *v. t. (fig.)* ~ **on the pressure** Druck machen. '**pile on to** *v. t.* drängen in (+ *Akk.*) ⟨*Bus usw.*⟩. **pile 'out** *v. i.* nach draußen drängen. **pile 'up 1.** *v. i.* **a)** ⟨*Waren, Post, Arbeit, Schnee:*⟩ sich auftürmen; ⟨*Verkehr:*⟩ sich stauen; **b)** *(crash)* aufeinander auffahren. **2.** *v. t.* aufstapeln ⟨*Steine, Bücher usw.*⟩; aufhäufen ⟨*Abfall, Schnee*⟩

²**pile** *n. (of fabric etc.)* Flor, *der*

³**pile** *n. (stake)* Pfahl, *der.* '**pile-driver** *n.* [Pfahl]ramme, *die*

piles [paɪlz] *n. pl. (Med.)* Hämorrhoiden *Pl.*

'**pile-up** *n.* Massenkarambolage, *die*

pilfer ['pɪlfə(r)] *v. t.* stehlen

pilgrim ['pɪlgrɪm] *n.* Pilger, *der*/Pilgerin, *die.* **pilgrimage** ['pɪlgrɪmɪdʒ] *n.* Pilgerfahrt, *die*

pill [pɪl] *n.* **a)** Tablette, *die;* Pille, *die (ugs.);* **b)** *(coll.: contraceptive)* **the ~** or **P~:** die Pille *(ugs.);* **be on the ~:** die Pille nehmen *(ugs.)*

pillage ['pɪlɪdʒ] *v. t.* [aus]plündern

pillar ['pɪlə(r)] *n.* Säule, *die.* '**pillarbox** *n. (Brit.)* Briefkasten, *der*

pillion ['pɪljən] *n.* Beifahrersitz, *der;* **ride ~:** als Beifahrer/Beifahrerin mitfahren

pillow ['pɪləʊ] *n.* [Kopf]kissen, *das.* '**pillowcase,** '**pillowslip** *ns.* [Kopf]kissenbezug, *der*

pilot ['paɪlət] **1.** *n.* **a)** *(Aeronaut.)* Pilot, *der*/Pilotin, *die;* **b)** *(Naut.)* Lotse, *der.* **2.** *adj.* Pilot⟨*programm, -studie, -projekt usw.*⟩. **3.** *v. t.* **a)** *(Aeronaut.)* fliegen; **b)** *(Naut.; fig.)* lotsen

'**pilot-light** *n.* Zündflamme, *die*

pimp [pɪmp] *n.* Zuhälter, *der*

pimple ['pɪmpl] *n.* Pickel, *der*

pin 1. *n.* **a)** Stecknadel, *die;* ~**s and needles** *(fig.)* Kribbeln, *das;* **b)** *(peg)*

Stift, *der;* c) *(Electr.)* **a two-/three-~ plug** ein zwei-/dreipoliger Stecker. **2.** *v.t.,* **-nn-:** a) nageln 〈*Knochen, Bein*〉; **~ a badge to one's lapel** sich *(Dat.)* ein Abzeichen ans· Revers stecken; b) *(fig.)* **~ one's hopes on sb./sth.** seine [ganze] Hoffnung auf jmdn./etw. setzen; **~ the blame for sth. on sb.** jmdm. die Schuld an etw. *(Dat.)* zuschieben; c) **~ sb. against the wall** jmdn. an die Wand drängen. **pin 'down** *v.t.* **a)** *(fig.)* festnageln **(to** *or* **on auf +** *Akk.);* b) *(trap)·* festhalten. **pin 'up** *v.t.* aufhängen 〈*Bild, Foto*〉; anschlagen 〈*Bekanntmachung, Liste*〉; aufstecken 〈*Haar*〉; heften 〈*Saum*〉

pinafore ['pɪnəfɔː(r)] *n.* Schürze, *die (mit Oberteil)*

pincers ['pɪnsəz] *n. pl.* a) |pair of| **~:** Beißzange, *die;* b) *(of crab etc.)* Schere, *die*

pinch [pɪntʃ] **1.** *n.* a) *(squeezing)* Kniff, *der;* **give sb. a ~ on the arm/cheek** jmdn. in den Arm/die Backe kneifen; b) *(fig.)* **feel the ~:** knapp bei Kasse sein *(ugs.);* **at a ~:** zur Not; c) *(small amount)* Prise, *die.* **2.** *v.t.* a) kneifen; **~ sb.'s cheek/bottom** jmdn. in die Wange/den Hintern *(ugs.)* kneifen; b) *(coll.: steal)* klauen *(salopp)*

'pincushion *n.* Nadelkissen, *das*

¹pine [paɪn] *n. (tree)* Kiefer, *die*

²pine *v.i.* sich [vor Kummer] verzehren *(geh.).* **pine a'way** *v.i.* dahinkümmern

pineapple ['paɪnæpl] *n.* Ananas, *die*

'pine-tree *n.* Kiefer, *die*

ping-pong *(Amer.:* **Ping-Pong,** P) ['pɪŋpɒŋ] *n.* Tischtennis, *das*

pink [pɪŋk] **1.** *n.* Pink, *das;* Rosa, *das.* **2.** *adj.* pinkfarben; rosa

pinkie ['pɪŋkɪ] *n. (Amer., Scot.)* kleiner Finger

'pin-money *n.* Taschengeld, *das*

pinnacle ['pɪnəkl] *n.* Gipfel, *der; (fig.)* Höhepunkt, *der*

'pin-point *v.t.* genau festlegen

pint [paɪnt] *n.* Pint, *das;* ≈ halber Liter

'pin-up *(coll.) n.* Pin-up-Girl, *das; (picture) (of beautiful girl)* Pin-up[-Foto], *das; (of sports, film or pop star)* Starfoto, *das*

pioneer [paɪə'nɪə(r)] **1.** *n.* Pionier, *der.* **2.** *v.t.* Pionierarbeit leisten für

pious ['paɪəs] *adj.* fromm

pip [pɪp] *n. (seed)* Kern, *der*

pipe [paɪp] **1.** *n.* a) *(tube)* Rohr, *das;* b)

(Mus.) Pfeife, *die;* c) |tobacco-|**~:** [Tabaks]pfeife, *die.* **2.** *v.t.* [durch ein Rohr/durch Rohre] leiten. **pipe 'down** *v.i. (coll.)* ruhig sein. **pipe 'up** *v.i. (coll.)* etwas sagen

'pipeline *n.* Pipeline, *die;* **in the ~** *(fig.)* in Vorbereitung

piper ['paɪpə(r)] *n.* Pfeifer, *der/*Pfeiferin, *die; (bagpiper)* Dudelsackspieler, *der/*-spielerin, *die*

piping hot ['paɪpɪŋ hɒt] *adj.* kochendheiß

piquant ['piːkənt] *adj.* pikant

pique [piːk] *n.* **in a |fit of| ~:** verstimmt

piracy ['paɪrəsɪ] *n.* Seeräuberei, *die*

pirate ['paɪrət] *n.* a) Pirat, *der;* Seeräuber, *der;* b) *(Radio)* **~ radio station** Piratensender, *der*

Pisces ['paɪsiːz] *n.* Fische *Pl.*

piss [pɪs] *(coarse)* **1.** *n.* a) *(urine)* Pisse, *die (derb);* b) **have a/go for a ~:** pissen/pissen gehen *(derb).* **2.** *v.i.* pissen *(derb)*

pistol ['pɪstl] *n.* Pistole, *die*

piston ['pɪstən] *n.* Kolben, *der*

pit [pɪt] **1.** *n. (hole, mine)* Grube, *die; (natural)* Vertiefung, *die.* **2.** *v.t.,* **-tt-:** **~ one's wits/skill** *etc.* **against sth.** seinen Verstand/sein Können *usw.* an etw. *(Dat.)* messen

¹pitch [pɪtʃ] **1.** *n.* a) *(Brit.: usual place)* [Stand]platz, *der; (Sport: playing-area)* Feld, *das;* Platz, *der;* b) *(Mus.)* Tonhöhe, *die;* c) *(slope)* Neigung, *die.* **2.** *v.t.* a) *(erect)* aufschlagen; **~ camp** ein/das Lager aufschlagen; b) *(throw)* werfen. **3.** *v.i.* stürzen; 〈*Schiff:*〉 stampfen; **~ forward** vornüberstürzen

²pitch *n. (substance)* Pech, *das.* **pitch-'black** *adj.* pechschwarz; stockdunkel *(ugs.)* 〈*Nacht*〉. **pitch-'dark** *adj.* stockdunkel *(ugs.)*

pitcher ['pɪtʃə(r)] *n.* [Henkel]krug, *der*

'pitchfork *n.* Heugabel, *die*

'pitfall *n.* Fallstrick, *der*

pith [pɪθ] *n.* a) *(of orange etc.)* weiße Haut; b) *(fig.)* Kern, *der.* **'pithy** *adj. (fig.)* prägnant

pitiable ['pɪtɪəbl], **pitiful** ['pɪtɪfl] *adjs.* a) mitleiderregend; b) *(contemptible)* jämmerlich

pitiless *adj.* unbarmherzig

pittance ['pɪtəns] *n.* Hungerlohn, *der*

pity ['pɪtɪ] **1.** *n.* Mitleid, *das;* **feel ~ for sb.** Mitgefühl für jmdn. empfinden; **have/take ~ on sb.** Erbarmen mit jmdm. haben; **[what a] ~!** [wie] schade! **2.** *v.t.* bemitleiden; **I ~ you** du tust mir leid

pivot ['pɪvət] **1.** *n.* [Dreh]zapfen, *der.*
2. *v. i.* sich drehen
pixie ['pɪksɪ] *n.* Kobold, *der*
pizza ['pi:tsə] *n.* Pizza, *die*
placard ['plækɑ:d] *n.* Plakat, *das*
placate [plə'keɪt] *v. t.* beschwichtigen
place [pleɪs] **1.** *n.* **a)** Ort, *der; (spot)*
Stelle, *die;* **a [good] ~ to park/to stop**
ein [guter] Platz zum Parken/eine [gu-
te] Stelle zum Halten; **do you know a
good/cheap ~ to eat?** weißt du, wo
man gut/billig essen kann?; **~ of worship** Andachtsort, *der;* **all over the ~:**
überall; *(coll.: in a mess)* ganz durcheinander *(ugs.);* **b)** *(rank, position)*
Stellung, *die;* **put sb. in his ~:** jmdn. in
seine Schranken weisen; **c)** *(country,
town)* Ort, *der;* **~ of birth** Geburtsort,
der; '**go ~s** *(coll.: fig.)* es [im Leben] zu
was bringen *(ugs.);* **d)** *(coll.: premises)*
Bude, *die (ugs.);* **she is at his ~:** sie ist
bei ihm; **e)** *(seat etc.)* [Sitz]platz, *der;*
change ~s |with sb.| [mit jmdm.] die
Plätze tauschen; *(fig.)* [mit jmdm.]
tauschen; **f)** *(step, stage)* **in the first ~:**
zuerst; **why didn't you say so in the first
~?** warum hast du das nicht gleich gesagt?; **g)** *(proper ~)* Platz, *der;* **everything fell into ~** *(fig.)* alles wurde klar;
out of ~: nicht am richtigen Platz;
(several things) in Unordnung; **h)** *(position ir. competition)* Platz, *der.* **2.** *v. t.*
a) *(vertically)* stellen; *(horizontally)* legen; **b)** *in p.p. (situated)* gelegen; **c)**
(find situation or home for) unterbringen *(with bei);* **d)** *(class)* einordnen;
einstufen; **be ~d second in the race** im
Rennen den zweiten Platz belegen
placid ['plæsɪd] *adj.* ruhig
plagiarism ['pleɪdʒərɪzm] *n.* Plagiat,
das. **plagiarize** ['pleɪdʒəraɪz] *v. t.*
plagiieren
plague [pleɪg] **1.** *n.* **a)** *(esp. Hist.:
epidemic)* Seuche, *die;* **the ~** *(bubonic)*
die Pest; **b)** *(infestation)* **~ of rats** Rat-
tenplage, *die.* **2.** *v. t.* plagen; **~d with**
or **by sth.** von etw. geplagt
plaice [pleɪs] *n., pl. same* Scholle, *die*
plain [pleɪn] **1.** *adj.* **a)** *(clear)* klar; *(obvious)* offensichtlich; **b)** *(frank)* offen;
schlicht ‹*Wahrheit*›; **be ~ sailing** *(fig.)*
[ganz] einfach sein; **c)** *(unsophisticated)* einfach; schlicht ‹*Kleidung*›;
unliniert ‹*Papier*›; ‹*Stoff*› ohne Muster; **d)** wenig attraktiv ‹*Mädchen*›. **2.**
adv. **a)** *(clearly)* deutlich; **b)** *(simply)*
einfach. **3.** *n.* Ebene, *die.* **plain
'chocolate** *n.* halbbittere Schokolade. **plain 'clothes** *n. pl.* **in ~:** in Zivil

plainly *adv.* **a)** *(clearly)* deutlich;
b) *(obviously)* offensichtlich; *(undoubtedly)* eindeutig; **c)** *(frankly)* offen; **d)** *(simply)* schlicht
plaintiff ['pleɪntɪf] *n.* Kläger, *der/*Klägerin, *die*
plaintive ['pleɪntɪv] *adj.* klagend
plait [plæt] **1.** *n.* Zopf, *der.* **2.** *v. t.*
flechten
plan [plæn] **1.** *n.* Plan, *der;* **[go] according to ~:** nach Plan [gehen]; planmä-
ßig [verlaufen]. **2.** *v. t.,* **-nn-** planen;
(design) entwerfen. **3.** *v. i.,* **-nn-** planen
¹**plane** [pleɪn] *n.* **~|-tree|** Platane, *die*
²**plane 1.** *n. (tool)* Hobel, *der.* **2.** *v. t.*
hobeln
³**plane** *n.* **a)** *(Geom.: fig.)* Ebene, *die;*
b) *(aircraft)* Flugzeug, *das;* Maschine,
die (ugs.)
planet ['plænɪt] *n.* Planet, *der*
plank [plæŋk] *n.* Brett, *das; (thicker)*
Bohle, *die; (on ship)* Planke, *die*
plankton ['plæŋktən] *n.* Plankton, *das*
'**planner** *n.* Planer, *der/*Planerin, *die*
'**planning** *n.* Planen, *das;* Planung, *die*
plant [plɑ:nt] **1.** *n.* **a)** *(Bot.)* Pflanze,
die; **b)** *no indef. art. (machinery)* Maschinen, *die;* **c)** *(factory)* Fabrik, *die;*
Werk, *das.* **2.** *v. t.* **a)** pflanzen; **b)** *(sl.:
conceal)* anbringen ‹*Wanze*›; legen
‹*Bombe*›; **~ sth. on sb.** jmdm. etw. unterschieben. **plantation** [plɑ:n'teɪʃn]
n. Plantage, *die*
plaque [plɑ:k, plæk] *n.* **a)** Platte, *die;*
(commemorating sb.) [Gedenk]tafel,
die; **b)** *(Dent.)* Plaque, *die*
plaster ['plɑ:stə(r)] **1.** *n.* **a)** *(for walls
etc.)* [Ver]putz, *der;* **b)** **~ |of Paris|**
Gips, *der;* **c)** *see* **sticking-plaster. 2.**
v. t. **a)** verputzen ‹*Wand*›; **b)** *(daub)* **~
sth. on sth.** etw. dick auf etw. *(Akk.)*
auftragen. **plastered** ['plɑ:stəd] *adj.*
(sl.: drunk) voll *(salopp).* '**plasterer**
n. Gipser, *der*
plastic ['plæstɪk] **1.** *n.* Plastik, *das;*
Kunststoff, *der.* **2.** *adj.* aus Plastik *od.*
Kunststoff *nachgestellt;* **~ bag** Pla-
stiktüte, *die;* **~ surgery** plastische
Chirurgie
Plasticine, (P) ['plæstɪsi:n] *n.* Plasti-
lin, *das*
plate [pleɪt] **1.** *n.* **a)** Teller, *der; (serving
~)* Platte, *die;* **b)** *(metal – with name
etc.)* Schild, *das;* **c)** *(for printing)* Platte, *die; (illustration)* [Bild]tafel, *die.* **2.**
v. t. **~ sth. |with gold/silver|** etw. ver-
golden/versilbern
plateau ['plætəʊ] *n., pl.* **~x** ['plætəʊz]
or **~s** Hochebene, *die;* Plateau, *das*

plate 'glass n. Flachglas, das
platform ['plætfɔ:m] n. **a)** (Brit. Railw.) Bahnsteig, der; ~ 4 Gleis 4; **b)** (stage) Podium, das
platinum ['plætɪnəm] n. Platin, das
platitude ['plætɪtjuːd] n. Platitüde, die (geh.); Gemeinplatz, der
platoon [plə'tuːn] n. (Mil.) Zug, der
plausible ['plɔːzɪbl] adj. plausibel; einleuchtend
play [pleɪ] **1.** n. **a)** (Theatre) [Theater]stück, das; television ~: Fernsehspiel, das; **b)** (recreation) Spielen, das; Spiel, das; ~ **on words** Wortspiel, das; **c)** (Sport) Spiel, das; **d) come into ~, be brought** or **called into ~:** ins Spiel kommen. **2.** v. i. **a)** spielen; ~ **safe** sichergehen; ~ **for time** Zeit gewinnen wollen; **b)** (Mus.) spielen (on auf + Dat.). **3.** v. t. (also Sport, Theatre, Cards, Mus.) spielen; abspielen ⟨Schallplatte, Tonband⟩; schlagen ⟨Ball⟩; spielen gegen ⟨Mannschaft, Gegner⟩; ~ **the violin** etc. Geige usw. spielen; ~ **a trick/joke on sb.** jmdn. hereinlegen (ugs.)/jmdm. einen Streich spielen; ~ **one's cards right** (fig.) es richtig anfassen (fig.). **play a'bout, play a'round** v. i. spielen; **stop ~ing about** or **around** hör doch auf mit dem Unsinn! **play a'long** v. i. mitspielen. **play 'back** v. t. abspielen ⟨Tonband⟩. **play 'down** v. t. herunterspielen. **play 'up 1.** v. i. (coll.) ⟨Kinder:⟩ nichts als Ärger machen. **2.** v. t. (coll.: annoy) ärgern
'playboy n. Playboy, der
'player n. Spieler, der/Spielerin, die
playful ['pleɪfl] adj. spielerisch; (frolicsome) verspielt
play: ~**ground** n. Spielplatz, der; (Sch.) Schulhof, der; ~ **group** n. Spielgruppe, die
playing: ~~**card** n. Spielkarte, die; ~~**field** n. Sportplatz, der
play: ~**mate** n. Spielkamerad, der/Spielkameradin, die; ~~**off** n. Entscheidungsspiel, das; ~~**pen** Laufgitter, das; ~**thing** n. Spielzeug, das; ~**wright** ['pleɪraɪt] n. Dramatiker, der/Dramatikerin, die
PLC, plc abbr. (Brit.) **public limited company** ≈ GmbH
plea [pliː] n. Appell, der (for zu)
plead [pliːd] **1.** v. i. **a)** inständig bitten (for um); (imploringly) flehen (for um); ~ **with sb. for sth.** jmdn. inständig um etw. bitten; **b)** (Law; also fig.) plädieren. **2.** v. t. **a)** inständig bitten;

(imploringly) flehen; **b)** (Law) ~ **guilty/not guilty** sich schuldig/nicht schuldig bekennen. **'pleading** adj. flehend
pleasant ['plezənt] adj. angenehm
please [pliːz] **1.** v. t. gefallen (+ Dat.); ~ **oneself** tun, was man will; ~ **yourself** ganz wie du willst. **2.** v. i. **I come and go as I ~:** ich komme und gehe, wie es mir gefällt; **if you ~:** bitte schön. **3.** int. bitte; ~ **do!** aber bitte od. gern! **pleased** [pliːzd] adj. (satisfied) zufrieden (by mit); (happy) erfreut (by über + Akk.); **be ~ at** or **about sth.** sich über etw. (Akk.) freuen. **pleasing** ['pliːzɪŋ] adj. gefällig
pleasure ['pleʒə(r)] n. (joy) Freude, die; (enjoyment) Vergnügen, das; **have the ~ of doing sth.** das Vergnügen haben, etw. zu tun; **with ~:** mit Vergnügen
pleat [pliːt] n. Falte, die. **'pleated** adj. gefältelt; Falten⟨rock⟩
pledge [pledʒ] **1.** n. Versprechen, das. **2.** v. t. versprechen; geloben ⟨Treue⟩
plentiful ['plentɪfl] adj. reichlich; **be ~:** reichlich vorhanden sein
plenty ['plentɪ] n. ~ **of** viel; eine Menge; (coll.: enough) genug
pleurisy ['plʊərɪsɪ] n. Pleuritis, die; Brustfellentzündung, die
pliable ['plaɪəbl] adj. biegsam
plied see **ply**
pliers ['plaɪəz] n. pl. |**pair of**| ~: Zange, die
plight [plaɪt] n. Notlage, die
plimsoll ['plɪmsl] n. (Brit.) Turnschuh, der
plinth [plɪnθ] n. Sockel, der
plod [plɒd] v. i., **-dd-** trotten. **plod 'on** v. i. (fig.) sich weiterkämpfen
plonk [plɒŋk] n. (sl.) [billiger] Wein
plot [plɒt] **1.** n. **a)** (conspiracy) Verschwörung, die; **b)** (of play, novel) Handlung, die; **c)** (of ground) Stück Land. **2.** v. t., **-tt-: a)** [heimlich] planen; **b)** (mark on map) einzeichnen. **3.** v. i., **-tt-:** ~ **against sb.** sich gegen jmdn. verschwören. **'plotter** n. Verschwörer, der/Verschwörerin, die
plough [plaʊ] **1.** n. Pflug, der. **2.** v. t. pflügen. **plough 'back** v. t. (Finance) reinvestieren
plow (Amer./arch.) see **plough**
ploy [plɔɪ] n. Trick, der
pluck [plʌk] **1.** v. t. **a)** pflücken ⟨Obst⟩; ~ |**out**| auszupfen ⟨Federn, Haare⟩; **b)** (pull at) zupfen an (+ Dat.); **c)** (strip of feathers) rupfen. **2.** v. i. ~ **at sth.** an

etw. *(Dat.)* zupfen. **3.** *n.* Mut, *der.*

pluck 'up *v. t.* ~ **up |one's| courage** all seinen Mut zusammennehmen

pluckily ['plʌkılı] *adv.,* **'plucky** *adj.* tapfer

plug [plʌg] **1.** *n.* **a)** *(filling hole)* Pfropfen, *der;* *(in cask)* Spund, *der;* *(for basin etc.)* Stöpsel, *der;* **b)** *(Electr.)* Stecker, *der.* **2.** *v. t.,* **-gg-: a)** ~ |up| zustopfen ⟨*Loch usw.*⟩*;* **b)** *(coll.: advertise)* Schleichwerbung machen für.

plug 'in *v. t.* anschließen

'plug-hole *n.* Abfluß, *der*

plum [plʌm] *n.* **a)** Pflaume, *die;* **b)** *(fig.)* Leckerbissen, *der;* **a** ~ **job** ein Traumjob *(ugs.)*

plumage ['plu:mıdʒ] *n.* Gefieder, *das*

¹plumb [plʌm] **1.** *v. t.* loten. **2.** *adv.* **a)** lotrecht; **b)** *(fig.)* genau

²plumb *v. t.* ~ **in** fest anschließen

plumber ['plʌmə(r)] *n.* Klempner, *der.* **plumbing** ['plʌmıŋ] *n.* **a)** Klempnerarbeiten *Pl.;* **b)** *(waterpipes)* Wasserleitungen *Pl.*

'plumb-line *n.* Lot, *das*

plume [plu:m] *n.* Feder, *die;* *(ornamental bunch)* Federbusch, *der*

plummet ['plʌmıt] *v. i.* stürzen

plump [plʌmp] *adj.* mollig; rundlich.

'plump for *v. t.* sich entscheiden für

plunder ['plʌndə(r)] **1.** *v. t.* |aus|plündern ⟨*Gebäude, Gebiet*⟩. **2.** *n.* Plünderung, *die;* *(booty)* Beute, *die*

plunge [plʌndʒ] **1.** *v. t.* stecken; *(into liquid)* tauchen. **2.** *v. i.* **a)** ~ **into sth.** in etw. *(Akk.)* stürzen; **b)** ⟨*Straße usw.*⟩ steil abfallen. **3.** *n.* Sprung, *der;* **take the** ~ *(fig. coll.)* den Sprung wagen

plural ['plʊərl] **1.** *adj.* pluralisch; Plural-; ~ **noun** Substantiv im Plural. **2.** *n.* Mehrzahl, *die;* Plural, *der*

plus [plʌs] **1.** *prep.* plus (+ *Dat.*). **2.** *n.* *(advantage)* Pluspunkt, *der*

plush [plʌʃ] **1.** *n.* Plüsch, *der.* **2.** *adj.* *(coll.)* feudal *(ugs.)*

Pluto ['plu:təʊ] *pr. n.* *(Astron.)* Pluto, *der*

ply [plaı] **1.** *v. t.* **a)** *(use)* gebrauchen; **b)** nachgehen (+ *Dat.*) ⟨*Handwerk, Arbeit*⟩*;* **c)** *(supply)* ~ **sb. with sth.** jmdn. mit etw. versorgen; **d)** *(assail)* überhäufen. **2.** *v. i.* ~ **between** zwischen ⟨*Orten*⟩ [hin- und her]pendeln

'plywood *n.* Sperrholz, *das*

p.m. [pi:'em] *adv.* nachmittags; **one** ~: ein Uhr mittags

pneumatic [nju:'mætık] *adj.* pneumatisch. **pneumatic 'drill** *n.* Preßluftbohrer, *der*

pneumonia [nju:'məʊnıə] *n.* Lungenentzündung, *die*

PO *abbr.* **a) postal order** PA; **b) Post Office** PA

¹poach [pəʊtʃ] *v. t.* **a)** *(catch illegally)* wildern; illegal fangen ⟨*Fische*⟩*;* **b)** stehlen, *(ugs.)* klauen ⟨*Idee*⟩

²poach *v. t.* *(Cookery)* pochieren ⟨*Ei*⟩*;* dünsten ⟨*Fisch, Fleisch, Gemüse*⟩

'poacher *n.* Wilderer, *der*

pocket ['pɒkıt] **1.** *n.* **a)** Tasche, *die;* **b)** *(fig.)* **be in** ~: Geld verdient haben; **be out of** ~: draufgelegt haben. **2.** *adj.* Taschen⟨*rechner, -uhr, -ausgabe*⟩. **3.** *v. t.* **a)** einstecken; **b)** *(steal)* in die eigene Tasche stecken *(ugs.)*. **'pocketbook** *n.* *(wallet)* Brieftasche, *die;* *(notebook)* Notizbuch, *das.* **'pocketmoney** *n.* Taschengeld, *das*

'pock-marked *adj.* **a)** pockennarbig ⟨*Gesicht, Haut*⟩*;* **b) a wall** ~ **with bullets** eine mit Einschüssen übersäte Wand

pod [pɒd] *n.* Hülse, *die;* *(of pea)* Schote, *die*

podgy ['pɒdʒı] *adj.* dicklich

poem ['pəʊım] *n.* Gedicht, *das*

poet ['pəʊıt] *n.* Dichter, *der.* **poetic** [pəʊ'etık] *adj.* dichterisch

poetry ['pəʊıtrı] *n.* [Vers]dichtung, *die;* Lyrik, *die*

poignant ['pɔınjənt] *adj.* tief ⟨*Bedauern, Trauer*⟩*;* ergreifend ⟨*Anblick*⟩

point [pɔınt] **1.** *n.* **a)** *(tiny mark, dot)* Punkt, *der;* **b)** *(of tool, pencil, etc.)* Spitze, *die;* **c)** *(single item; unit of scoring)* Punkt, *der;* **d)** *(stage, degree)* **up to a** ~: bis zu einem gewissen Grad; **he gave up at this** ~: an diesem Punkt gab er auf; **e)** *(moment)* Zeitpunkt, *der;* **be on the** ~ **of doing sth.** etw. gerade tun wollen; **f)** *(distinctive trait)* Seite, *die;* **best/strong** ~: starke Seite; Stärke, *die;* **g)** *(thing to be discussed)* **come to** *or* **get to the** ~: zum Thema kommen; **be beside the** ~: keine Rolle spielen; **make a** ~ **of doing sth.** [großen] Wert darauf legen, etw. zu tun; **h)** *(of story, joke, remark)* Pointe, *die;* **i)** *(purpose)* Zweck, *der;* Sinn, *der;* **j)** *(precise place, spot)* Punkt, *der;* Stelle, *die;* ~ **of view** *(fig.)* Standpunkt, *der;* **k)** *(Brit.)* |power *or* electric| ~: Steckdose, *die;* **l)** *usu in pl.* *(Brit. Railw.)* Weiche, *die.* **2.** *v. i.* **a)** zeigen, weisen **(to, at** auf + *Akk.*)*;* **b)** ~ **towards** *or* **to** *(fig.)* [hin]deuten auf (+ *Akk.*). **3.** *v. t.* richten ⟨*Waffe, Kamera*⟩ **(at** auf + *Akk.*)*;* ~ **one's finger at sth./sb.** mit

dem Finger auf etw./jmdn. zeigen.
point 'out *v. t.* hinweisen auf
(+ *Akk.*); ~ **sth./sb. out to sb.** jmdn.
auf etw./jmdn. hinweisen
point-'blank 1. *adj. (lit. or fig.)* direkt;
glatt ⟨*Weigerung*⟩; ~ **range** kürzeste
Entfernung. **2.** *adv. (at very close
range)* aus kürzester Entfernung
'pointed *adj.* **a)** spitz; **b)** *(fig.)* unmißverständlich
'pointer *n.* **a)** Zeiger, *der;* (rod) Zeigestock, *der;* **b)** *(coll.: indication)* Hinweis, *der* (**to** auf + *Akk.*)
'pointless *adj.* sinnlos; belanglos
⟨*Bemerkung, Geschichte*⟩
poise [pɔɪz] *n. (composure)* Haltung,
die; (self-confidence) Selbstvertrauen,
das. **poised** [pɔɪzd] *adj.* selbstsicher
poison ['pɔɪzn] **1.** *n.* Gift, *das.* **2.** *v. t.*
vergiften. **'poisoning** *n.* Vergiftung,
die. **poisonous** ['pɔɪzənəs] *adj.* giftig
poke 1. *v. t.* **a)** ~ **sth.** |**with sth.**| [mit
etw.] gegen etw. stoßen; ~ **sth. into
sth.** etw. in etw. *(Akk.)* stoßen; ~ **the
fire** das Feuer schüren; **b)** stecken
⟨*Kopf*⟩. **2.** *v. i.* **a)** [herum]stochern (**at,
in, among** in + *Dat.*); **b)** *(pry)* schnüffeln *(ugs.).* **3.** *n.* **a)** *(thrust)* Stoß, *der;*
give sb. a ~ |**in the ribs**| jmdm. einen
[Rippen]stoß versetzen; **give the fire a**
~: das Feuer [an]schüren. **poke
a'bout, poke a'round** *v. i.* herumschnüffeln *(ugs.)*
¹'poker *n.* Schüreisen, *das*
²'poker *n. (Cards)* Poker, *das od. der*
'poker-faced *adj.* mit unbewegter
Miene *nachgestellt*
poky ['pəʊkɪ] *adj.* winzig
Poland ['pəʊlənd] *pr. n.* Polen *(das)*
polar ['pəʊlə(r)] *adj.* polar ⟨*Kaltluft,
Gewässer*⟩; Polar⟨*eis, -gebiet, -fuchs*⟩.
polar 'bear *n.* Eisbär, *der*
Pole [pəʊl] *n.* Pole, *der/*Polin, *die*
¹pole *n. (support)* Stange, *die;* **drive sb.
up the** ~ *(Brit. sl.)* jmdn. zum Wahnsinn treiben *(ugs.)*
²pole *n. (Astron., Geog., Magn.,
Electr., fig.)* Pol, *der.* **'pole-star** *n.*
Polarstern, *der*
'pole-vault *n.* Stabhochsprung, *der*
police [pə'li:s] **1.** *n. pl.* Polizei, *die;
(members)* Polizisten *Pl.;* attrib. Polizei-. **2.** *v. t.* [polizeilich] überwachen
⟨*Fußballspiel*⟩; kontrollieren ⟨*Gebiet*⟩.
police: ~ **force** *n.* **the** ~ **force** die Polizei; ~**man** [pə'li:smən] *n., pl.* -**men**
[pə'li:smən] Polizist, *der;* ~ **station**
n. Polizeirevier, *das;* ~**woman** *n.*
Polizistin, *die*

¹policy ['pɒlɪsɪ] *n.* Politik, *die*
²policy *n. (Insurance)* Police, *die*
polio ['pəʊlɪəʊ] *n., no art.* Polio, *die;*
[spinale] Kinderlähmung
Polish ['pəʊlɪʃ] **1.** *adj.* polnisch; **sb. i**
~: jmd. ist Pole/Polin. **2.** *n.* Polnisch,
das; see also **English 2 a**
polish ['pɒlɪʃ] **1.** *v. t.* **a)** polieren; bohnern ⟨*Fußboden*⟩; putzen ⟨*Schuhe*⟩; **b)**
(fig.) ausfeilen ⟨*Text, Theorie, Stil*⟩. **2.**
n. **a)** *(smoothness)* Glanz, *der;* **b)** *(substance)* Politur, *die;* **c)** *(fig.)* Schliff,
der. **polish 'off** *v. t. (coll.)* **a)** *(consume)* verdrücken *(ugs.);* **b)** *(complete
quickly)* durchziehen *(ugs.).* **polish
'up** *v. t.* **a)** polieren; **b)** ausfeilen
⟨*Stil*⟩; aufpolieren ⟨*Kenntnisse*⟩
polite [pə'laɪt] *adj.,* ~**r** [pə'laɪtə(r)], ~**st**
[pə'laɪtɪst] höflich. **po'liteness** *n.*
Höflichkeit, *die*
political [pə'lɪtɪkl] *adj.* politisch
politician [pɒlɪ'tɪʃn] *n.* Politiker,
*der/*Politikerin, *die*
politics ['pɒlɪtɪks] *n.* Politik, *die; (of
individual)* politische Einstellung
polka ['pɒlkə, 'pəʊlkə] *n.* Polka, *die.*
'polka dot *n.* [großer] Tupfen
poll [pəʊl] **1.** *n. (voting)* Abstimmung, *die; (to elect sb.)* Wahl, *die;* **go
to the** ~**s** zur Wahl gehen; **b)** *(opinion
~)* Umfrage, *die.* **2.** *v. t.* **a)** *(take vote[s]
of)* abstimmen/wählen lassen; **b)**
(take opinion of) befragen
pollen ['pɒlən] *n.* Pollen, *der;* Blütenstaub, *der.* **'pollen count** *n.* Pollenmenge, *die*
'polling-booth *n.* Wahlkabine, *die*
'poll-tax *n.* Kopfsteuer, *die*
pollutant [pə'lu:tənt] *n.* [Umwelt]schadstoff, *der*
pollute [pə'lu:t] *v. t.* verschmutzen
⟨*Luft, Boden, Wasser*⟩. **pollution** [pə
'lu:ʃn] *n.* [Umwelt]verschmutzung, *die*
polo ['pəʊləʊ] *n.* Polo, *das.* **'polo-
neck** *n.* Rollkragen, *der*
polyester [pɒlɪ'estə(r)] *n.* Polyester,
der
polystyrene [pɒlɪ'staɪri:n] *n.* Polystyrol, *das;* ~ **foam** Styropor ⓦ, *das*
polytechnic [pɒlɪ'teknɪk] *n. (Brit.)* ≈
technische Hochschule
polythene ['pɒlɪθi:n] *n.* Polyäthylen,
das; ~ **bag** Plastikbeutel, *der*
pomegranate ['pɒmɪɡrænɪt] *n.* Granatapfel, *der*
'pommel-horse *n.* Seitpferd, *das*
pomp [pɒmp] *n.* Pomp, *der*
pom-pom ['pɒmpɒm] *n.* Pompon,
der; ~ **hat** Pudelmütze, *die*

pompous ['pɒmpəs] *adj.* großspurig; gespreizt ⟨*Sprache*⟩

pond [pɒnd] *n.* Teich, *der*

ponder ['pɒndə(r)] 1. *v. t.* nachdenken über (+ *Akk.*) ⟨*Frage, Ereignis*⟩; abwägen ⟨*Vorteile, Worte*⟩. 2. *v. i.* nachdenken (**over, on** über + *Akk.*)

ponderous ['pɒndərəs] *adj.* schwer

pong [pɒŋ] *(Brit. coll.)* 1. *n.* Mief, *der (ugs.).* 2. *v. i.* miefen *(ugs.)*

pony ['pəʊnɪ] *n.* Pony, *das.* 'pony-tail *n.* Pferdeschwanz, *der.* **pony-trekking** ['pəʊnɪtrekɪŋ] *n. (Brit.)* Ponyreiten, *das*

poodle ['pu:dl] *n.* Pudel, *der*

[^1]**pool** [pu:l] *n.* **a)** Tümpel, *der;* **b)** *(temporary)* Lache, *die;* ~ **of blood** Blutlache, *die;* **c)** *(swimming-~)* Schwimmbecken, *das; (public)* Schwimmbad, *das; (in house or garden)* Pool, *der*

[^2]**pool** 1. *n.* **a)** *(Gambling)* [gemeinsame Spiel]kasse; **the ~s** *(Brit.)* das Toto; **b)** *(common supply)* Topf, *der;* **a ~ of experience** ein Erfahrungsschatz; **c)** *(game)* Pool[billard], *das.* 2. *v. t.* zusammenlegen ⟨*Geld, Ersparnisse*⟩; bündeln ⟨*Anstrengungen*⟩

poor [pʊə(r)] 1. *adj.* **a)** arm; **b)** *(inadequate)* schlecht; schwach ⟨*Spiel, Gesundheit, Leistung, Rede*⟩; dürftig ⟨*Kleidung, Essen, Unterkunft*⟩; **~ of quality** minderer Qualität; **c)** *(paltry)* schwach ⟨*Trost*⟩; schlecht ⟨*Aussichten*⟩; **d)** *(unfortunate)* arm *(auch iron.);* **e)** karg ⟨*Boden*⟩; **f)** *(deficient)* arm (**in an** + *Dat.*); **~ in vitamins** vitaminarm. 2. *n. pl.* **the ~:** die Armen. **poorly** ['pʊəlɪ] *adv., pred. adj.* schlecht

[^1]**pop** [pɒp] 1. *v. i.,* **-pp-:** **a)** *(make sound)* knallen; **b)** *(coll.: go quickly)* **let's ~ round to Fred's** komm, wir gehen kurz bei Fred vorbei *(ugs.).* 2. *v. t.,* **-pp-:** **a)** *(coll.: put)* ~ **the meat in the fridge** das Fleisch in den Kühlschrank tun; **b)** platzen ⟨*Luftballon*⟩. 3. *n.* **a)** Knall, *der;* Knallen, *das;* **b)** *(coll.: drink)* Brause, *die (ugs.).* 4. *adv.* **go ~:** knallen. **pop 'out** *v. i.* hervorschießen; **~ out to the shops** schnell einkaufen gehen

[^2]**pop** *(coll.)* 1. *n.* Popmusik, *die;* Pop, *der.* 2. *adj.* Pop⟨*star, -musik usw.*⟩

'**popcorn** *n.* Popcorn, *das*

pope [pəʊp] *n.* Papst, *der/*Päpstin, *die*

poplar ['pɒplə(r)] *n.* Pappel, *die*

popper ['pɒpə(r)] *n. (Brit. coll.)* Druckknopf, *der*

poppy ['pɒpɪ] *n.* Mohn, *der*

popular ['pɒpjʊlə(r)] *adj.* **a)** *(well*

liked) beliebt; populär ⟨*Entscheidung, Maßnahme*⟩; **b)** verbreitet ⟨*Aberglaube, Irrtum, Meinung*⟩; allgemein ⟨*Wahl, Unterstützung*⟩. **popularity** [pɒpjʊ'lærɪtɪ] *n.* Beliebtheit, *die; (of decision, measure)* Popularität, *die.*

popularize ['pɒpjʊləraɪz] *v. t.* **a)** *(make popular)* populär machen; **b)** *(make understandable)* breiteren Kreisen zugänglich machen. '**popularly** *adv.* allgemein

populate ['pɒpjʊleɪt] *v. t.* bevölkern; bewohnen ⟨*Insel*⟩. **population** [pɒpjʊ'leɪʃn] *n.* Bevölkerung, *die;* **Britain has a ~ of 56 million** Großbritannien hat 56 Millionen Einwohner

porcelain ['pɔ:slɪn] *n.* Porzellan, *das*

porch [pɔ:tʃ] *n.* Vordach, *das; (with side walls)* Vorbau, *der; (enclosed)* Windfang, *der*

porcupine ['pɔ:kjʊpaɪn] *n.* Stachelschwein, *das*

[^1]**pore** [pɔ:(r)] *n.* Pore, *die*

[^2]**pore** *v. i.* ~ **over sth.** etw. [genau] studieren

pork [pɔ:k] *n.* Schweinefleisch, *das; attrib.* Schweine-. **pork 'chop** *n.* Schweinekotelett, *das.* **pork 'pie** *n.* Schweinepastete, *die*

porn [pɔ:n] *n. (coll.)* Pornographie, *die;* Pornos *(ugs.)*

pornographic [pɔ:nə'græfɪk] *adj.* pornographisch; Porno- *(ugs.)*

pornography [pɔ:'nɒgrəfɪ] *n.* Pornographie, *die*

porous ['pɔ:rəs] *adj.* porös

porridge ['pɒrɪdʒ] *n.* [Hafer]brei, *der*

[^1]**port** [pɔ:t] 1. *n.* **a)** Hafen, *der;* **b)** *(Naut., Aeronaut.: left side)* Backbord, *das.* 2. *adj. (Naut., Aeronaut.: left)* Backbord-; backbordseitig

[^2]**port** *n. (wine)* Portwein, *der*

portable ['pɔ:təbl] *adj.* tragbar

[^1]**porter** ['pɔ:tə(r)] *n. (Brit.: doorman)* Pförtner, *der; (of hotel)* Portier, *der*

[^2]**porter** *n.* [Gepäck]träger, *der/*-trägerin, *die; (in hotel)* Hausdiener, *der*

portfolio [pɔ:t'fəʊlɪəʊ] *n. pl.* **~s a)** *(Polit.)* Geschäftsbereich, *der;* **b)** *(case, contents)* Mappe, *die*

porthole ['pɔ:thəʊl] *n. (Naut.)* Seitenfenster, *das; (round)* Bullauge, *das*

portion ['pɔ:ʃn] *n.* **a)** *(part)* Teil, *der; (of ticket)* Abschnitt, *der;* **b)** *(of food)* Portion, *die*

portly ['pɔ:tlɪ] *adj.* beleibt

portrait ['pɔ:trɪt] *n.* Porträt, *das*

portray [pɔ:'treɪ] *v. t.* darstellen; *(make likeness of)* porträtieren

Portugal ['pɔːtjʊgl] *pr. n.* Portugal *(das).* **Portuguese** [pɔːtjʊ'giːz] **1.** *adj.* portugiesisch; **sb. is ~:** jmd. ist Portugiese/Portugiesin. **2.** *n., pl. same* **a)** *(person)* Portugiese, *der/*Portugiesin, *die·* **b)** *(language)* Portugiesisch, *das; see also* **English 2 a**

pose [pəʊz] **1.** *v. t.* aufwerfen ⟨*Frage, Problem*⟩; darstellen ⟨*Bedrohung*⟩; mit sich bringen ⟨*Schwierigkeiten*⟩. **2.** *v. i.* **a)** *(assume attitude)* posieren; *(fig.)* sich geziert benehmen; **b) ~ as** sich geben als. **3.** *n.* Pose, *die;* **strike a ~:** eine Pose einnehmen. **poser** ['pəʊzə(r)] *n. (question)* knifflige Frage

posh [pɒʃ] *adj. (coll.)* vornehm; nobel *(spött.);* stinkvornehm *(salopp)*

position [pə'zɪʃn] **1.** *n.* **a)** *(place occupied)* Platz, *der; (of player in team, of plane, ship, etc.)* Position, *die; (of hands of clock, words, stars)* Stellung, *die; (of building)* Lage, *die;* **be in/out of ~:** an seinem Platz/nicht an seinem Platz sein; **b)** *(Mil.)* Stellung, *die;* **c)** *(fig.: mental attitude)* Standpunkt, *der;* **d)** *(fig.: situation)* **be in a good ~ [financially]** [finanziell] gut gestellt sein; **be in a ~ of strength** eine starke Position haben; **e)** *(rank)* Stellung, *die;* **f)** *(job)* Stelle, *die;* **g)** *(posture)* Haltung, *die.* **2.** *v. t.* plazieren; postieren ⟨*Polizisten, Wachen*⟩; **~ oneself** sich stellen/*(sit)* setzen

positive ['pɒzɪtɪv] *adj.* **a)** *(also Math.)* positiv; konstruktiv ⟨*Vorschlag*⟩; *(definite)* eindeutig; *(convinced)* sicher; **I'm ~ of it** ich bin [mir] [dessen] ganz sicher; **b)** *(Electr.)* positiv ⟨*Elektrode, Ladung*⟩; Plus⟨*platte, -leiter*⟩; **c)** *as intensifier (coll.)* echt

possess [pə'zes] *v. t.* besitzen; *(as faculty or quality)* haben; ⟨*Furcht usw.*⟩ ergreifen; **what ~ed you?** *(coll.)* was ist in dich gefahren? **possessed** [pə'zest] *adj.* besessen. **possession** [pə'zeʃn] *n.* **a)** *(thing possessed)* Besitz, *der;* **some of my ~s** einige meiner Sachen; **b)** *in pl. (property)* Besitz, *der;* **c)** *(possessing)* Besitz, *der;* **be in ~ of sth.** im Besitz einer Sache *(Gen.)* sein; **take ~ of** in Besitz nehmen; beziehen ⟨*Haus, Wohnung*⟩. **possessive** [pə'zesɪv] *adj.* **a)** besitzergreifend; **be ~ about sth./sb.** etw. eifersüchtig hüten/ an jmdn. Besitzansprüche stellen; **b)** *(Ling.)* possessiv. **possessor** [pə'zesə(r)] *n.* Besitzer, *der/*Besitzerin, *die* **possibility** [pɒsɪ'bɪlɪtɪ] *n.* Möglichkeit, *die*

possible ['pɒsɪbl] *adj.* möglich; *(likely)* [gut] möglich; **if ~:** wenn möglich; **as ... as ~:** so ... wie möglich; möglichst ... **possibly** ['pɒsɪblɪ] *adv.* **a) as often as I ~ can** so oft ich irgend kann; **I cannot ~ commit myself** ich kann mich unmöglich festlegen; **b)** *(perhaps)* möglicherweise

¹post [pəʊst] *n.* **a)** *(as support)* Pfosten, *der;* **b)** *(stake)* Pfahl, *der;* **c)** *(starting/finishing ~)* Start-/Zielpfosten, *der*

²post 1. *n.* **a)** *(Brit.: one dispatch/delivery of letters)* Postausgang, *der/* Post[zustellung], *die;* **by return of ~:** postwendend; **b)** *no indef. art. (Brit.: official conveying)* Post, *die;* **by ~:** mit der Post; per Post; **c)** *(~ office)* Post, *die.* **2.** *v. t.* **a)** abschicken; **b)** *(fig. coll.)* **keep sb. ~ed** jmdn. auf dem laufenden halten

³post 1. *n.* **a)** *(job)* Stelle, *die;* Posten, *der;* **b)** *(Mil.; also fig.)* Posten, *der.* **2.** *v. t.* postieren; aufstellen

postage ['pəʊstɪdʒ] *n.* Porto, *das*

postal ['pəʊstl] *adj.* Post-; postalisch ⟨*Aufgabe, Einrichtung*⟩; *(by post)* per Post *nachgestellt.* **'postal order** *n.* ≈ Postanweisung, *die*

post: ~box *n. (Brit.)* Briefkasten, *der;* **~card** *n.* Postkarte, *die;* **~code** *n. (Brit.)* Postleitzahl, *die;* **~-'date** *v. t. (give later date to)* vordatieren

poster ['pəʊstə(r)] *n.* Plakat, *das*

posterior [pɒ'stɪərɪə(r)] *n. (joc.)* Hinterteil, *das (ugs.)*

posterity [pɒ'sterɪtɪ] *n., no art.* Nachwelt, *die*

posthumous ['pɒstjʊməs] *adj.* postum

post: ~man ['pəʊstmən], *pl.* **~men** ['pəʊstmən] *n.* Briefträger, *der;* **~mark 1.** *n.* Poststempel, *der;* **2.** *v. t.* abstempeln

post-mortem [pəʊst'mɔːtəm] *n.* Obduktion, *die*

post office *n.* **a)** *(organization)* **the P~ Office** die Post; **b)** *(place)* Postamt, *das;* Post, *die*

postpone [pə'spəʊn] *v. t.* verschieben; *(for an indefinite period)* aufschieben. **post'ponement** *n.* Verschiebung, *die/*Aufschub, *der*

postscript ['pəʊskrɪpt] *n.* Nachschrift, *die; (fig.)* Nachtrag, *der*

posture ['pɒstʃə(r)] *n.* [Körper]haltung, *die*

'post-war *adj.* Nachkriegs-; der Nachkriegszeit *nachgestellt*

posy ['pəʊzɪ] *n.* Sträußchen, *das*

pot [pɒt] 1. *n.* a) [Koch]topf, *der;* go to ~ *(coll.)* den Bach runtergehen *(ugs.);* b) *(container, contents)* Topf, *der; (tea-pot, coffee-pot)* Kanne, *die;* c) *(coll.: large sum)* a ~ of/~s of massenweise. 2. *v.t.* ~ |up| eintopfen ‹*Pflanze*›

potassium [pə'tæsɪəm] *n.* Kalium, *das*

potato [pə'teɪtəʊ] *n., pl.* ~es Kartoffel, *die*

potent ['pəʊtənt] *adj.* [hoch]wirksam ‹*Droge*›; stark ‹*Schnaps usw.*›; schlag-kräftig ‹*Waffe*›

potential [pə'tenʃl] 1. *adj.* potentiell *(geh.);* möglich. 2. *n.* Potential, *das (geh.);* Möglichkeiten

'**pot-hole** *n.* a) Schlagloch, *das;* b) *(cave)* [tiefe] Höhle. '**pot-holer** *n.* Höhlenforscher, *der/*-forscherin, *die*

'**pot-shot** *n.* take a ~ |at sb./sth.| aufs Geratewohl [auf jmdn./etw.] schießen

'**potted** *adj.* a) *(planted)* Topf-; b) *(abridged)* kurzgefaßt

¹'**potter** *n.* Töpfer, *der/*Töpferin, *die*

²'**potter** *v.i.* ~ |about| [he]rumwerkeln *(ugs.)*

pottery ['pɒtərɪ] *n.* a) Töpferware, *die;* b) *(workshop, craft)* Töpferei, *die*

¹'**potty** ['pɒtɪ] *adj. (Brit. sl.)* verrückt *(ugs.)* **(about, on** nach)

²'**potty** *n. (Brit. coll.)* Töpfchen, *das*

pouch [paʊtʃ] *n.* Beutel, *der*

pouffe [pu:f] *n.* Sitzpolster, *das*

poultry ['pəʊltrɪ] *n.* Geflügel, *das*

pounce [paʊns] *v.i.* a) sich auf sein Opfer stürzen; ‹*Raubvogel:*› herabsto-ßen auf (+ *Akk.*); b) *(fig.)* ~ |up|on/at sich stürzen auf (+ *Akk.*)

¹**pound** [paʊnd] *n.* a) *(unit of weight)* [britisches] Pfund (453,6 *Gramm);* two ~|s| of apples 2 Pfund Äpfel; b) *(unit of currency)* Pfund, *das*

²**pound** *n. (enclosure)* Pferch, *der; (for stray dogs)* Zwinger; *(for cars)* Ab-stellplatz, *der*

³**pound** 1. *v.t. (crush)* zerstoßen. 2. *v.i.* a) *(make one's way heavily)* stampfen; b) ‹*Herz:*› heftig schlagen

pour [pɔ:(r)] 1. *v.t.* gießen; *(into cup, glass)* einschenken. 2. *v.i.* a) *(flow)* strömen; ‹*Rauch:*› hervorquellen *(from* aus); ~ |with rain| in Strömen regnen; b) *(fig.)* strömen; ~ in her-ein-/hineinströmen; ~ out heraus-/hinausströmen. **pour 'down** *v.i.* it's ~ing down es gießt [in Strömen] *(ugs.)*

pout [paʊt] 1. *v.i.* einen Schmollmund machen. 2. *v.t.* aufwerfen ‹*Lippen*›

poverty ['pɒvətɪ] *n.* Armut, *die*

powder ['paʊdə(r)] 1. *n.* a) Pulver, *das;* b) *(cosmetic)* Puder, *der.* 2. *v.t.* a) pudern; b) *(reduce to* ~) pulverisie-ren; ~ed milk Milchpulver, *das.* '**powdery** *adj.* pulv[e]rig

power ['paʊə(r)] 1. *n.* a) *(ability)* Kraft, *die;* do all in one's ~ to help sb. alles in seiner Macht Stehende tun, um jmdm. zu helfen; b) *(faculty)* Fä-higkeit, *die;* c) *(strength, intensity)* Kraft, *die; (of blow)* Wucht, *die;* d) *(authority, political* ~) Macht, *die* (over über + *Akk.*); come into ~: an die Macht kommen; e) *(authorization)* Vollmacht, *die;* f) *(State)* Macht, *die;* g) *(Math.)* Potenz, *die;* h) *(Mech., Electr.)* Kraft, *die; (electric current)* Strom, *der.* 2. *v.t.* ‹*Treibstoff, Strom:*› antreiben; ‹*Batterie:*› mit Energie ver-sorgen. **powerful** ['paʊəfl] *adj.* a) *(strong)* stark; kräftig ‹*Tritt, Schlag, Tier*›; heftig ‹*Gefühl, Empfindung*›; hell, strahlend ‹*Licht*›; b) *(mighty)* mächtig ‹*Clique, Person, Herrscher*›. '**power-less** *adj.* machtlos. '**power station** *n.* Kraftwerk, *das*

p.p. [pi:'pi:] *abbr.* by proxy pp[a].

pp. *abbr.* pages

practicable ['præktɪkəbl] *adj.* durch-führbar ‹*Projekt, Plan*›

practical ['præktɪkl] *adj.* a) praktisch; praktisch veranlagt ‹*Person*›; b) *(vir-tual)* tatsächlich; c) *(feasible)* mög-lich. **practical 'joke** *n.* Streich, *der.* '**practically** *adv.* praktisch; *(almost)* so gut wie; praktisch *(ugs.)*

¹'**practice** ['præktɪs] *n.* a) *(repeated exercise)* Übung, *die;* be out of ~: au-ßer Übung sein; b) *(session)* Übungen *Pl.;* piano ~: Klavierüben, *das;* c) *(of doctor, lawyer, etc.)* Praxis, *die;* d) *(ac-tion)* put sth. into ~: etw. in die Praxis umsetzen; e) *(custom)* Gewohnheit, *die;* regular ~: Brauch, *der*

²'**practice, practiced, practicing** *(Amer.) see* practis-

practise ['præktɪs] 1. *v.t.* a) *(apply)* anwenden; praktizieren; b) ausüben ‹*Beruf, Religion*›; c) trainieren in (+ *Dat.*) ‹*Sportart*›; ~ the piano/flute Klavier/Flöte üben. 2. *v.i.* üben.

practised ['præktɪst] *adj.* geübt.

practising ['præktɪsɪŋ] *adj.* prakti-zierend ‹*Arzt, Katholik usw.*›

pragmatic [præg'mætɪk] *adj.* pragma-tisch

Prague [prɑ:g] *pr. n.* Prag *(das)*

prairie ['preərɪ] *n.* Grassteppe, *die; (in North America)* Prärie, *die*

praise [preız] 1. *v. t.* loben; *(more strongly)* rühmen. 2. *n.* Lob, *das.* '**praiseworthy** *adj.* lobenswert

pram [præm] *n. (Brit.)* Kinderwagen, *der*

prance [prɑːns] *v. i.* a) ⟨*Pferd:*⟩ tänzeln; b) *(fig.)* stolzieren; ~ **about** *or* **around** herumhüpfen

prank [præŋk] *n.* Streich, *der*

prattle ['prætl] 1. *v. i.* plappern *(ugs.).* 2. *n.* Geplapper, *das (ugs.)*

prawn [prɔːn] *n.* Garnele, *die*

pray [preı] *v. i.* beten (for um). **prayer** [preə(r)] *n.* a) Gebet, *das;* b) *no art. (praying)* Beten, *das*

preach [priːtʃ] 1. *v. i.* predigen (to zu, vor + *Dat.;* on über + *Akk.*). 2. *v. t.* halten ⟨*Predigt*⟩; predigen ⟨*Evangelium, Botschaft*⟩. '**preacher** *n.* Prediger, *der*/Predigerin, *die*

precarious [prı'keərıəs] *adj.* a) *(uncertain)* labil; prekär; **make a ~ living** eine unsichere Existenz haben; b) *(insecure, dangerous)* gefährlich

precaution [prı'kɔːʃn] *n.* Vorsichts-, Schutzmaßnahme, *die;* **as a ~:** vorsichtshalber

precede [prı'siːd] *v. t. (in order or time)* vorangehen (+ *Dat.*). **precedence** ['presıdəns] *n.* Priorität, *die (geh.),* Vorrang, *der* (**over** vor + *Dat.*). **precedent** ['presıdənt] *n.* Präzedenzfall, *der*

precinct ['priːsıŋkt] *n.* a) |pedestrian| ~: Fußgängerzone, *die;* b) *(Amer.: district)* Bezirk, *der*

precious ['preʃəs] 1. *adj.* a) kostbar ⟨*Schmuckstück, Zeit*⟩; b) *(beloved)* lieb; c) *(affected)* affektiert. 2. *adv. (coll.)* herzlich ⟨*wenig, wenige*⟩

precipice ['presıpıs] *n.* Abgrund, *der*

precipitate 1. [prı'sıpıtət] *adj.* eilig ⟨*Flucht*⟩; übereilt ⟨*Entschluß*⟩. 2. [prı'sıpıteıt] *v. t. (hasten)* beschleunigen; *(trigger)* auslösen

precipitation [prısıpı'teıʃn] *n. (Meteorol.)* Niederschlag, *der*

precipitous [prı'sıpıtəs] *adj.* a) *(steep)* sehr steil; b) *see* **precipitate 1**

précis ['preısiː] *n., pl. same* [preısiːz] Zusammenfassung, *die*

precise [prı'saıs] *adj.* genau; präzise; fein ⟨*Instrument*⟩; förmlich ⟨*Art*⟩; **be |more| ~:** sich präzise[r] ausdrücken. **pre'cisely** *adv.* genau. **precision** [prı'sıʒn] *n.* Genauigkeit, *die*

preclude [prı'kluːd] *v. t.* ausschließen

precocious [prı'kəʊʃəs] *adj.* frühreif ⟨*Kind*⟩; altklug ⟨*Äußerung*⟩

preconceived [priːkən'siːvd] *adj.* vorgefaßt ⟨*Ansicht, Vorstellung*⟩. **preconception** [priːkən'sepʃn] *n.* vorgefaßte Meinung (**of** über + *Akk.*)

precondition [priːkən'dıʃn] *n.* Vorbedingung, *die* (**of** für)

precursor [priː'kɜːsə(r)] *n.* Wegbereiter, *der*/-bereiterin, *die*

predator ['predətə(r)] *n.* Raubtier, *das; (fish)* Raubfisch, *der.* '**predatory** *adj.* räuberisch; ~ **animal** Raubtier, *das*

predecessor ['priːdısesə(r)] *n.* Vorgänger, *der*/-gängerin, *die*

predestine [priː'destın] *v. t.* von vornherein bestimmen (**to** zu)

predicament [prı'dıkəmənt] *n.* Dilemma, *das*

predicate ['predıkət] *n. (Ling.)* Prädikat, *das.* **predicative** [prı'dıkətıv] *adj. (Ling.)* prädikativ

predict [prı'dıkt] *v. t.* voraus-, vorhersagen; vorhersehen ⟨*Folgen*⟩. **predictable** [prı'dıktəbl] *adj.* voraussagbar; vorhersehbar ⟨*Ereignis, Reaktion*⟩; berechenbar ⟨*Person*⟩. **prediction** [prı'dıkʃn] *n.* Vorhersage, *die*

predominance [prı'dɒmınəns] *n.* a) *(control)* Vorherrschaft, *die* (**over** über + *Akk.*); b) *(majority)* Überzahl, *die* (**of** von)

predominant [prı'dɒmınənt] *adj. (having more power)* dominierend; *(prevailing)* vorherrschend

predominate [prı'dɒmıneıt] *v. i. (be more powerful)* dominierend sein; *(be more important)* vorherrschen

pre-eminent [priː'emınənt] *adj.* herausragend

pre-empt [priː'empt] *v. t.* zuvorkommen (+ *Dat.*)

preen [priːn] *v. t.* putzen ⟨*Federn*⟩

prefab ['priːfæb] *n. (coll.)* Fertighaus, *das.* **prefabricated** [priː'fæbrıkeıtıd] *adj.* vorgefertigt

preface ['prefəs] 1. *n.* Vorwort, *das* (**to** Gen.). 2. *v. t. (introduce)* einleiten

prefect ['priːfekt] *n. (Sch.)* die Aufsicht führender älterer Schüler/führende ältere Schülerin

prefer [prı'fɜː(r)] *v. t.,* **-rr-** vorziehen; ~ **to do sth.** etw. lieber tun; ~ **sth. to sth.** etw. einer Sache (*Dat.*) vorziehen. **preferable** ['prefərəbl] *adj.* vorzuziehen präd.; vorzuziehend attr.; besser (**to** als). **preferably** ['prefərəblı] *adv.* am besten; *(as best liked)* am liebsten; **Wine or beer? – Wine, ~!** Wein oder Bier? – Lieber Wein! **preference**

['prefərəns] *n.* **a)** *(greater liking)* Vorliebe, *die;* for ~ *see* preferably; have a
~ for sth. |over sth.| etw. [einer Sache
(Dat.)] vorziehen; do sth. in ~ to sth.
else etw. lieber als etw. anderes tun; **b)**
(thing preferred) **what are your ~s?** was
wäre dir am liebsten?; **c)** give ~ to sb.
jmdn. bevorzugen. **preferential**
[prefə'renʃl] *adj.* bevorzugt 〈*Behandlung*〉

prefix ['pri:fɪks] *n.* Präfix, *das*

pregnancy ['pregnənsɪ] *n. (of woman)*
Schwangerschaft, *die; (of animal)*
Trächtigkeit, *die*

pregnant ['pregnənt] *adj.* schwanger
〈*Frau*〉; trächtig 〈*Tier*〉

prehistoric [pri:hɪ'stɒrɪk] *adj.* prähi-
storisch. **prehistory** [pri:'hɪstərɪ]
Vorgeschichte, *die*

prejudge [pri:'dʒʌdʒ] *v. t.* vorschnell
urteilen über (+ *Akk.*)

prejudice ['predʒʊdɪs] **1.** *n.* Vorurteil,
das. **2.** *v. t.* beeinflussen. **prejudiced**
['predʒʊdɪst] *adj.* voreingenommen
(**about** gegenüber, **against** gegen)

preliminary [prɪ'lɪmɪnərɪ] **1.** *adj.* Vor-;
vorbereitend 〈*Forschung, Maßnah-
me*〉. **2.** *n., usu. in pl.* **preliminaries**
Präliminarien *Pl.;* **as a ~ to sth.** als
Vorbereitung auf etw. *(Akk.)*

prelude ['prelju:d] *n.* **a)** *(introduction)*
Anfang, *der* (to Gen.); **b)** *(Theatre,
Mus.)* Vorspiel, *das*

premature ['premətjʊə(r)] *adj.* **a)**
(hasty) übereilt; **b)** *(early)* vorzeitig
〈*Altern, Ankunft*〉; verfrüht 〈*Bericht,
Eile*〉; ~ **baby** Frühgeburt, *die.*
prematurely *adv. (early)* vorzeitig;
zu früh 〈*geboren werden*〉; *(hastily)*
übereilt

premeditated [pri:'medɪteɪtɪd] *adj.*
vorsätzlich

premier ['premɪə(r)] *n.* Premier[mini-
ster], *der*/Premierministerin, *die*

première ['premjeə(r)] *n.* Premiere,
die; Erstaufführung, *die*

premise ['premɪs] *n.* **a)** ~**s** *pl. (building)* Gebäude, *das; (buildings and
land)* Gelände, *das; (rooms)* Räum-
lichkeiten *Pl.;* **b)** *see* **premiss**

premiss ['premɪs] *n.* Prämisse, *die*

premium ['pri:mɪəm] *n.* Prämie, *die;*
be at a ~ *(fig.)* sehr gefragt sein. **Premium Bond** *n. (Brit.)* Prämienanlei-
he, *die;* Losanleihe, *die*

premonition [premə'nɪʃn] *n.* Vorahnung, *die*

preoccupation [prɪɒkjʊ'peɪʃn] *n.*
Sorge, *die* (**with** um)

preoccupied [prɪ'ɒkjʊpaɪd] *adj. (lost
in thought)* gedankenverloren; *(concerned)* besorgt (**with** um)

pre-'packed *adj.* abgepackt

preparation [prepə'reɪʃn] *n.* Vorbereitung, *die;* ~**s** *pl.* Vorbereitungen *Pl.*
(**for** für). **preparatory** [prɪ'pærətərɪ]
1. *adj.* vorbereitend 〈*Maßnahme,
Schritt*〉; ~ **work** Vorarbeiten *Pl.* **2.**
adv. ~ **to sth.** vor etw. *(Dat.)*

prepare [prɪ'peə(r)] **1.** *v. t.* **a)** vorbereiten; ausarbeiten 〈*Plan, Rede*〉; vorbe-
reiten 〈*Person*〉 (**for** auf + *Akk.*); **be
~d to do sth.** *(be willing)* bereit sein,
etw. zu tun; **b)** herstellen 〈*Chemikalie
usw.*〉; zubereiten 〈*Essen*〉. **2.** *v. i.* sich
vorbereiten (**for** auf + *Akk.*)

prepaid [pri:'peɪd] *adj.* ~ **envelope**
frankierter Umschlag

preponderance [prɪ'pɒndərəns] *n.*
Überlegenheit, *die* (**over** über + *Akk.*)

preposition [prepə'zɪʃn] *n. (Ling.)*
Präposition, *die*

prepossessing [pri:pə'zesɪŋ] *adj.*
einnehmend

preposterous [prɪ'pɒstərəs] *adj.* absurd; grotesk 〈*Äußeres, Kleidung*〉

prerequisite [pri:'rekwɪzɪt] **1.** *n.*
[Grund]voraussetzung, *die.* **2.** *adj.* unbedingt erforderlich

prerogative [prɪ'rɒgətɪv] *n.* Privileg,
das; Vorrecht, *das*

Presbyterian [prezbɪ'tɪərɪən] **1.** *adj.*
presbyterianisch. **2.** *n.* Presbyterianer,
der/Presbyterianerin, *die*

prescribe [prɪ'skraɪb] *v. t.* **a)** *(impose)*
vorschreiben; **b)** *(Med.; also fig.)* verschreiben. **prescription** [prɪ'skrɪpʃn]
n. **a)** Vorschreiben, *das;* **b)** *(Med.)* Rezept, *das*

presence ['prezəns] *n.* **a)** *(of person)*
Anwesenheit, *die; (of things)* Vorhan-
densein, *das;* **in the ~ of** in Anwesenheit (+ *Gen.*); **b)** ~ **of mind** Geistesge-
genwart, *die*

¹**present** ['prezənt] **1.** *adj.* **a)** anwesend (**at** bei); **all those ~:** alle Anwesenden; **b)** *(existing now)* gegenwärtig; jetzig 〈*Bischof, Chef usw.*〉; **c)**
(Ling.) ~ **tense** Präsens, *das;* Gegenwart, *die.* **2.** *n.* **a)** **the ~:** die Gegenwart; **at ~:** zur Zeit; **for the ~:** vorläufig; **b)** *(Ling.)* Präsens, *das;* Gegenwart, *die*

²**present 1.** ['prezənt] *n. (gift)* Geschenk, *das.* **2.** [prɪ'zent] *v. t.* **a)** schenken; überreichen 〈*Preis, Medaille, Geschenk*〉; ~ **sth. to sb.** *or* **sb. with sth.**
jmdm. etw. schenken/überreichen; ~

sb. with difficulties/a problem jmdn. vor Schwierigkeiten/ein Problem stellen; **b)** überreichen ⟨*Gesuch*⟩ **(to** bei); vorlegen ⟨*Scheck, Bericht, Rechnung*⟩ **(to** *Dat.*); **~ one's case** seinen Fall darlegen; **c)** *(exhibit)* zeigen; bereiten ⟨*Schwierigkeit*⟩; **d)** *(introduce)* vorstellen **(to** *Dat.*); vorlegen ⟨*Abhandlung*⟩; moderieren ⟨*Sendung*⟩. **3.** *v. refl.* ⟨*Problem:*⟩ auftreten; ⟨*Möglichkeit:*⟩ sich ergeben; **~ oneself for an interview** zu einem Gespräch erscheinen.

presentable [prɪ'zentəbl] *adj.* ansehnlich; **I'm not ~:** ich kann mich nicht so zeigen. **presentation** [prezən'teɪʃn] *n.* **a)** *(giving)* Schenkung, *die;* *(of prize, medal)* Überreichung, *die;* **b)** *(ceremony)* Verleihung, *die;* **c)** *(of petition)* Überreichung, *die; (of cheque, report, account)* Vorlage, *die; (of case)* Darlegung, *die*

present-'day *adj.* heutig

presenter [prɪ'zentə(r)] *n.* *(Radio, Telev.)* Moderator, *der*/Moderatorin, *die*

presentiment [prɪ'zentɪmənt] *n.* Vorahnung, *die*

presently ['prezəntlɪ] *adv.* bald; *(Amer., Scot.: now)* zur Zeit

preservation [prezə'veɪʃn] *n.* Erhaltung, *die; (of leather, wood, etc.)* Konservierung, *die.* **preservative** [prɪ'zɜːvətɪv] *n.* Konservierungsmittel, *das*

preserve [prɪ'zɜːv] **1.** *n.* **a)** *in sing. or pl. (fruit)* Eingemachte, *das;* **b)** *(fig.: special sphere)* Domäne, *die (geh.);* **c)** **wildlife/game ~:** Tierschutzgebiet, *das*/Wildpark, *der.* **2.** *v. t.* **a)** *(keep safe)* schützen **(from** vor **+** *Dat.*); **b)** bewahren ⟨*Brauch*⟩; wahren ⟨*Anschein, Reputation*⟩; **c)** *(keep from decay)* konservieren; einmachen ⟨*Obst, Gemüse*⟩; **d)** *(protect)* hegen ⟨*Tierart, Wald*⟩

preside [prɪ'zaɪd] *v. i.* präsidieren, vorsitzen **(over** *Dat.*); *(at meeting etc.)* den Vorsitz haben **(at** bei)

presidency ['prezɪdənsɪ] *n.* **a)** Präsidentschaft, *die;* **b)** *(of society)* Vorsitz, *der*

president ['prezɪdənt] *n.* **a)** Präsident, *der*/Präsidentin, *die;* **b)** *(of society)* Vorsitzende, *der/die.* **presidential** [prezɪ'denʃl] *adj.* Präsidenten-

¹press [pres] **1.** *n.* **a)** *(newspapers etc.)* Presse, *die; attrib.* Presse-; **b)** *see* **printing-press; c)** *(for flattening, compressing, etc.)* Presse, *die.* **2.** *v. t.* **a)** drücken; drücken auf **(+** *Akk.*) ⟨*Klin-*

gel, Knopf⟩; treten auf **(+** *Akk.*) ⟨*Gas-, Brems-, Kupplungspedal usw.*⟩; **b)** *(urge)* drängen ⟨*Person*⟩; *(force)* aufdrängen (**|up|on** *Dat.*); nachdrücklich vorbringen ⟨*Forderung, Argument*⟩; **he did not ~ the point** er ließ die Sache auf sich beruhen; **c)** *(compress)* pressen; auspressen ⟨*Orangen, Saft*⟩; keltern ⟨*Trauben, Äpfel*⟩; **d)** *(iron)* bügeln; **e) be ~ed for time/ money** zu wenig Zeit/Geld haben. **3.** *v. i.* **a)** *(exert pressure)* drücken; **b)** *(be urgent)* drängen; **c)** *(make demand)* **~ for sth.** auf etw. *(Akk.)* drängen.

press a'head, press 'on *v. i. (continue)* [zügig] weitermachen; *(continue travelling)* [zügig] weitergehen/-fahren; **~ on with one's work** sich mit der Arbeit ranhalten *(ugs.)*

²press *v. t.* **~ into service/use** in Dienst nehmen; einsetzen

'press conference *n.* Pressekonferenz, *die*

'pressing *adj. (urgent)* dringend

press: ~ release *n.* Presseinformation, *die;* **~-up** *n.* Liegestütz, *der*

pressure ['preʃə(r)] **1.** *n.* Druck, *der;* **put ~ on sb.** jmdn. unter Druck setzen; **atmospheric ~:** Luftdruck, *der.* **2.** *v. t.* unter Druck setzen ⟨*Person*⟩; **~ sb. into doing sth.** jmdn. [dazu] drängen, etw. zu tun. **'pressure-cooker** *n.* Schnellkochtopf, *der.* **'pressure group** *n.* Pressure-group, *die*

pressurize ['preʃəraɪz] *v. t.* **a)** *see* **pressure 2; b) ~d cabin** Druckkabine, *die*

prestige [pre'stiːʒ] *n.* Prestige, *das.* **prestigious** [pre'stɪdʒəs] *adj.* angesehen

presumably [prɪ'zjuːməblɪ] *adv.* vermutlich

presume [prɪ'zjuːm] **1.** *v. t.* **a) ~ to do sth.** sich *(Dat.)* anmaßen, etw. zu tun; *(take the liberty)* sich *(Dat.)* erlauben, etw. zu tun; **b)** *(suppose)* annehmen. **2.** *v. i.* **|up|on sth.** etw. ausnützen. **presumption** [prɪ'zʌmpʃn] *n.* **a)** *(arrogance)* Anmaßung, *die;* **b)** *(assumption)* Annahme, *die.* **presumptuous** [prɪ'zʌmptjʊəs] *adj.* anmaßend

presuppose [priːsə'pəʊz] *v. t.* voraussetzen

pretence [prɪ'tens] *n. (Brit.)* **a)** *(pretext)* Vorwand, *der;* **b) no art.** *(make believe, insincere behaviour)* Verstellung, *die;* **it is all** *or* **just a ~:** das ist alles nicht echt

pretend [prɪ'tend] **1.** *v. t.* **a)** vorgeben

she ~ed to be asleep sie tat, als ob sie
schlief[e]; b) *(imagine in play)* ~ to be
sth. so tun, als ob man etw. sei. 2. *v.i.*
sich verstellen; she's only ~ing sie tut
nur so
pretense *(Amer.) see* pretence
pretension [prɪ'tenʃn] *n.* a) Anspruch, *der* (to auf + *Akk.*); b)
(pretentiousness) Überheblichkeit,
die. **pretentious** [prɪ'tenʃəs] *adj.*
hochgestochen; wichtigtuerisch ⟨*Person*⟩; *(ostentatious)* großspurig
pretext ['priːtekst] *n.* Vorwand, *der;*
|up|on *or* under the ~ of doing sth. un-
ter dem Vorwand, etw. tun zu wollen
prettily ['prɪtɪlɪ] *adv.* hübsch; sehr
schön ⟨*singen, tanzen*⟩
pretty ['prɪtɪ] 1. *adj. (also iron.)*
hübsch. 2. *adv.* ziemlich; I am ~ well
es geht mir ganz gut
prevail [prɪ'veɪl] *v.i.* a) die Oberhand
gewinnen (**against,** over über +
Akk.); ~ |up|on sb. to do sth. jmdn. da-
zu bewegen, etw. zu tun; b) *(predominate)* ⟨*Zustand, Bedingung:*⟩ vorherrschen; c) *(be current)* herrschen
prevalence ['prevələns] *n.* Vorherrschen, *das*
prevalent ['prevələnt] *adj.* a) *(existing)* herrschend; weit verbreitet
⟨*Krankheit*⟩; b) *(predominant)* vorherrschend
prevent [prɪ'vent] *v.t. (hinder)* verhindern; *(forestall)* vorbeugen; ~ sb.
from doing sth., ~ sb.'s doing sth.,
(coll.) ~ sb. doing sth. jmdn. daran
hindern, etw. zu tun. **prevention**
[prɪ'venʃn] *n.* Verhinderung, *die;*
(forestalling) Vorbeugung, *die.*
preventive [prɪ'ventɪv] *adj.* vorbeu-
gend; Präventiv⟨*maßnahme*⟩
preview ['priːvjuː] *n. (of film, play)*
Voraufführung, *die; (of exhibition)*
Vernissage, *die (geh.)*
previous ['priːvɪəs] 1. *adj.* a) früher
⟨*Anstellung, Gelegenheit*⟩; vorherig
⟨*Abend*⟩; vorig ⟨*Besitzer, Wohnsitz*⟩;
the ~ page die Seite davor; b) *(prior)* ~
to vor (+ *Dat.*). 2. *adv.* ~ to vor
(+ *Dat.*). **'previously** *adv.* vorher
pre-war ['priːwɔː(r)] *adj.* Vorkriegs-
prey [preɪ] 1. *n., pl. same* a) *(animal[s])*
Beute, *die;* **beast/bird of** ~: Raubtier,
das/-vogel, *der;* b) *(victim)* Opfer,
das. 2. *v.i.* ~ |up|on ⟨*Raubtier, Raub-
vogel:*⟩ schlagen; *(plunder)* ausplündern ⟨*Person*⟩; Jagd machen auf
(+ *Akk.*); ~ |up|on sb.'s mind jmdm.
keine Ruhe lassen

price [praɪs] *n. (lit. or fig.)* Preis, *der;* **at
a** ~ of zum Preis von; **what is the** ~ **of
this?** was kostet das?; **at/not at any** ~:
um jeden/keinen Preis. **'priceless**
adj. a) *(invaluable)* unbezahlbar; b)
(coll.: amusing) köstlich
price: ~-**list** *n.* Preisliste, *die;* ~-**rise**
n. Preisanstieg, *der;* ~-**tag** *n.* Preisschild, *das*
prick [prɪk] 1. *v.t.* stechen; stechen in
⟨*Ballon*⟩; aufstechen ⟨*Blase*⟩. 2. *v.i.*
stechen. 3. *n.* Stich, *der.* **'prick up**
v.t. aufrichten ⟨*Ohren*⟩; ~ up one's/its
ears die Ohren spitzen
prickle ['prɪkl] 1. *n.* a) Dorn, *der;* b)
(Zool., Bot.) Stachel, *der.* 2. *v.i.* kratzen. **prickly** ['prɪklɪ] *adj.* dornig; stachelig; *(fig.)* empfindlich
pride [praɪd] 1. *n.* a) Stolz, *der; (arrog-
ance)* Hochmut, *der;* **take** |a| ~ **in** sb./
sth. auf jmdn./etw. stolz sein; sb's ~
and joy jmds. ganzer Stolz; b) *(of
lions)* Rudel, *das.* 2. *v. refl.* ~ oneself
|up|on sth. auf etw. *(Akk.)* stolz sein
pried *see* pry
priest [priːst] *n.* Priester, *der.* **'priest-
hood** *n.* geistliches Amt
prim [prɪm] *adj.* spröde; *(prudish)* zimperlich
primarily ['praɪmərɪlɪ] *adv.* in erster
Linie
primary ['praɪmərɪ] 1. *adj.* a) *(first)*
primär *(geh.);* grundlegend; b) *(chief)*
Haupt⟨rolle, -ziel, -zweck⟩. 2. *n.*
(Amer.: election) Vorwahl, *die.* **'primary school** *n.* Grundschule, *die*
primate ['praɪmeɪt] *n.* a) *(Eccl.)* Pri-
mas, *der;* b) *(Zool.)* Primat, *der*
'prime [praɪm] 1. *n.* Höhepunkt, *der;*
be in one's ~: in den besten Jahren
sein. 2. *adj.* a) Haupt-; hauptsächlich;
b) *(excellent)* erstklassig; vortrefflich
⟨*Beispiel*⟩
²prime *v.t.* a) *(equip)* vorbereiten; ~
sb. with information/advice jmdn. in-
struieren/jmdm. Ratschläge erteilen;
b) grundieren ⟨*Wand, Decke*⟩; c)
schärfen ⟨*Sprengkörper*⟩
prime: ~ 'minister *n.* Premiermini-
ster, *der/*-ministerin, *die;* ~ 'number
n. (Math.) Primzahl, *die*
'primer *n.* a) *(explosive)* Zündvorrich-
tung, *die;* b) *(paint)* Grundierlack, *der*
primeval [praɪ'miːvl] *adj.* urzeitlich;
Ur⟨zeiten, -wälder⟩
primitive ['prɪmɪtɪv] *adj.* primitiv;
(prehistoric) urzeitlich ⟨*Mensch*⟩
primrose ['prɪmrəʊz] *n.* gelbe Schlüsselblume

Primus, (P) ['praɪməs] *n.* ~ |stove| Primuskocher, *der*

prince [prɪns] *n.* Prinz, *der.* '**princely** *adj.* fürstlich

princess [prɪn'ses] *n.* Prinzessin, *die; (wife of prince)* Fürstin, *die*

principal ['prɪnsɪpl] **1.** *adj.* Haupt-; *(most important)* wichtigst... **2.** *n. (of college)* Rektor, *der*/Rektorin, *die*

principality [prɪnsɪ'pælɪtɪ] *n.* Fürstentum, *das*

'**principally** *adv.* in erster Linie

principle ['prɪnsɪpl] *n.* Prinzip, *das;* **on the ~ that ...**: nach dem Grundsatz, daß ...; **in ~**: im Prinzip; **do sth. on ~** *or* **as a matter of ~**: etw. prinzipiell *od.* aus Prinzip tun

print [prɪnt] **1.** *n.* **a)** *(impression)* Abdruck, *der; (finger~)* Fingerabdruck, *der;* **b)** *(~ed lettering)* Gedruckte, *das; (type-face)* Druck, *der;* **c) be in/out of ~** ⟨*Buch:*⟩ erhältlich/vergriffen sein; **d)** *(~ed picture or design)* Druck, *der;* **e)** *(Photog.)* Abzug, *der.* **2.** *v. t.* **a)** drucken ⟨*Buch, Zeitschrift usw.*⟩; **b)** *(write)* in Druckschrift schreiben. **print 'out** *v. t. (Computing)* ausdrucken

'**printed** *adj.* **a)** gedruckt; **b)** *(published)* veröffentlicht. '**printed matter** *n. (Post)* Drucksachen *Pl.*

'**printer** *n.* **a)** *(worker)* Drucker, *der*/Druckerin, *die; (firm)* Druckerei, *die;* **b)** *(Computing)* Drucker, *der*

'**printing** *n.* **a)** Drucken, *das;* **b)** *(writing like print)* Druckschrift, *die;* **c)** *(edition)* Auflage, *die.* '**printing-press** *n.* Druckerpresse, *die*

'**printout** *n. (Computing)* Ausdruck, *der*

prior ['praɪə(r)] **1.** *adj.* vorherig ⟨*Warnung, Zustimmung usw.*⟩; früher ⟨*Verabredung*⟩; Vor⟨*geschichte, -kenntnis*⟩. **2.** *adv.* **~ to** vor (+ *Dat.*); **~ to doing sth.** bevor man etw. tut/tat; **~ to that** vorher. **priority** [praɪ'ɒrɪtɪ] *n.* **a)** *(precedence)* Vorrang, *der; attrib.* vorrangig; **have** *or* **take ~**: Vorrang haben (**over** vor + *Dat.*); **have ~** *(on road)* Vorfahrt haben; **give ~ to sb./ sth.** jmdm./einer Sache den Vorrang geben; **give top ~ to sth.** einer Sache *(Dat.)* höchste Priorität einräumen; **b)** *(matter)* vordringliche Angelegenheit

prism ['prɪzm] *n.* Prisma, *das*

prison ['prɪzn] *n.* **a)** Gefängnis, *das; attrib.* Gefängnis-; **b)** *(custody)* Haft, *die;* **in ~**: im Gefängnis; **go to ~**: ins Gefängnis gehen. '**prisoner** *n.* Gefangene, *der/die;* **take sb. ~**: jmdn. gefangennehmen

pristine ['prɪstiːn] *adj.* unberührt; **in ~ condition** in tadellosem Zustand

privacy ['prɪvəsɪ] *n.* Privatsphäre, *die; (being undisturbed)* Ungestörtheit, *die;* **invasion of ~**: Eindringen in die Privatsphäre; **in the strictest ~**: unter strengster Geheimhaltung

private ['praɪvət] **1.** *adj.* **a)** *(outside State system)* privat; Privat⟨*schule, -industrie, -klinik usw.*⟩; **b)** persönlich ⟨*Dinge, Meinung, Interesse*⟩; nichtöffentlich ⟨*Versammlung, Sitzung*⟩; privat ⟨*Telefongespräch, Vereinbarung*⟩; Privat⟨*strand, -parkplatz, -leben*⟩; geheim ⟨*Verhandlung, Geschäft*⟩; persönlich ⟨*Gründe*⟩; *(confidential)* vertraulich. **2.** *n.* **a)** *(Brit. Mil.)* einfacher Soldat; **b) in ~**: privat; **in kleinem Kreis** ⟨*feiern*⟩; *(confidentially)* ganz im Vertrauen. '**privately** *adv.* privat ⟨*erziehen, zugeben*⟩; vertraulich ⟨*jmdn. sprechen*⟩; insgeheim ⟨*denken, glauben*⟩; **~ owned** in Privatbesitz

privation [praɪ'veɪʃn] *n.* Not, *die;* **suffer many ~s** viele Entbehrungen erleiden

privatize ['praɪvətaɪz] *v. t.* privatisieren

privet ['prɪvɪt] *n.* Liguster, *der*

privilege ['prɪvɪlɪdʒ] *n. (right, immunity)* Privileg, *das; (special benefit)* Sonderrecht, *das; (honour)* Ehre, *die.* '**privileged** *adj.* privilegiert

privy ['prɪvɪ] *adj.* **be ~ to sth.** in etw. *(Akk.)* eingeweiht sein

¹**prize** [praɪz] **1.** *n.* **a)** *(reward, money)* Preis, *der;* **win** *or* **take first ~**: den ersten Preis gewinnen; **b)** *(in lottery)* Gewinn, *der.* **2.** *v. t.* **~ sth. |highly|** etw. hoch schätzen

²**prize** *v. t.* **~ |open|** aufstemmen

prize: **~-giving** *n.* Preisverleihung, *die;* **~-money** *n.* Geldpreis, *der; (Sport)* Preisgeld, *das;* **~-winner** *n.* Preisträger, *der*/-trägerin, *die; (in lottery)* Gewinner, *der*/Gewinnerin, *die*

pro [prəʊ] *n. in pl.* **the ~s and cons** das Pro und Kontra

probability [prɒbə'bɪlɪtɪ] *n.* Wahrscheinlichkeit, *die;* **in all ~**: aller Wahrscheinlichkeit nach

probable ['prɒbəbl] *adj.* wahrscheinlich; **highly ~**: höchstwahrscheinlich

probably ['prɒbəblɪ] *adv.* wahrscheinlich

probation [prə'beɪʃn] *n.* **a)** Probezeit, *die;* **b)** *(Law)* Bewährung, *die;* **on ~**

auf Bewährung. **probationary** [prə-'beɪʃənərɪ] *adj.* Probe-; ~ **period** Probezeit, *die*

probe [prəʊb] **1.** *n.* **a)** Untersuchung, *die* (**into** *Gen.*); **b)** (*Med., Astron.*) Sonde, *die*. **2.** *v. t.* untersuchen

problem ['prɒbləm] *n.* Problem, *das;* (*puzzle*) Rätsel, *das;* **what's the ~?** (*coll.*) wo fehlt's denn?; **the ~ about** *or* **with sb./sth.** das Problem mit jmdm./ bei etw. **problematic** [prɒblə'mætɪk], **problematical** [prɒblə'mætɪkl] *adj.* problematisch

procedure [prə'siːdjə(r)] *n.* Verfahren, *das*

proceed [prə'siːd] *v. i.* (*formal*) **a)** (*on foot*) gehen; (*as or by vehicle*) fahren; (*after interruption*) weitergehen/-fahren; **b)** (*begin and carry on*) beginnen; (*after interruption*) fortfahren; ~ **in** *or* **with sth.** (*begin*) [mit] etw. beginnen; (*continue*) etw. fortsetzen; **c)** (*be under way*) ⟨Verfahren:⟩ laufen; (*be continued after interruption*) fortgesetzt werden. **pro'ceedings** *n. pl.* **a)** (*events*) Vorgänge; **b)** (*Law*) Verfahren, *das;* **legal ~:** Gerichtsverfahren, *das;* **start/take [legal] ~:** gerichtlich vorgehen (**against** gegen)

proceeds ['prəʊsiːdz] *n. pl.* Erlös, *der* (**from** aus)

¹**process** ['prəʊses] **1.** *n.* **a)** (*of time or history*) Lauf, *der;* **he learnt a lot in the ~:** er lernte eine Menge dabei; **be in the ~ of doing sth.** gerade etw. tun; **b)** (*proceeding, natural operation*) Vorgang, *der;* **c)** (*method*) Verfahren, *das.* **2.** *v. t.* verarbeiten ⟨Rohstoff, Signal⟩; bearbeiten ⟨Antrag, Akte⟩; (*Photog.*) entwickeln ⟨Film⟩

²**process** [prə'ses] *v. i.* ziehen. **procession** [prə'seʃn] *n.* Zug, *der;* (*religious*) Prozession, *die;* (*festive*) Umzug, *der;* **go/march in ~:** ziehen

proclaim [prə'kleɪm] *v. t.* erklären ⟨Absicht⟩; geltend machen ⟨Recht, Anspruch⟩; verkünden ⟨Amnestie⟩; ausrufen ⟨Republik⟩. **proclamation** [prɒklə'meɪʃn] *n.* **a)** (*proclaiming*) Verkündung, *die;* **b)** (*notice*) Bekanntmachung, *die;* (*decree*) Erlaß, *der*

procure [prə'kjʊə(r)] *v. t.* beschaffen

prod **1.** *v. t.,* **-dd-** (*poke*) stupsen (*ugs.*); stoßen mit ⟨Stock, Finger usw.⟩; ~ **sb.** gently jmdn. anstupsen. **2.** *n.* Stupser, *der;* **give sb. a ~:** jmdm. einen Stupser geben

prodigal ['prɒdɪgl] *adj.* verschwenderisch; ~ **son** verlorener Sohn

prodigious [prə'dɪdʒəs] *adj.* ungeheuer

prodigy ['prɒdɪdʒɪ] *n.* [außergewöhnliches] Talent; **child ~:** Wunderkind, *das*

produce **1.** ['prɒdjuːs] *n.* Produkte *Pl.;* Erzeugnisse *Pl.* **2.** [prə'djuːs] *v. t.* **a)** vorzeigen ⟨Paß, Fahrkarte⟩; **b)** produzieren ⟨Show, Film⟩; inszenieren ⟨Theaterstück, Hörspiel⟩; herausgeben ⟨Schallplatte, Buch⟩; **c)** (*manufacture*) herstellen; (*in nature; Agric.*) produzieren; **d)** (*cause*) hervorrufen; bewirken ⟨Änderung⟩; **e)** (*bring into being*) erzeugen; führen zu ⟨Situation⟩; **f)** (*yield*) geben ⟨Milch⟩; legen ⟨Eier⟩; **g)** ⟨Baum, Blume:⟩ tragen ⟨Früchte, Blüten⟩; entwickeln ⟨Triebe⟩; bilden ⟨Keime⟩. **producer** [prə'djuːsə(r)] *n.* **a)** (*Cinemat., Theatre, Radio, Telev.*) Produzent, *der*/Produzentin, *die;* **b)** (*Brit. Theatre/Radio/Telev.*) Regisseur, *der*/Regisseurin, *die*

product ['prɒdʌkt] *n.* **a)** Produkt, *das;* (*of industrial process*) Erzeugnis, *das;* (*of art or intellect*) Werk, *das;* **b)** (*result*) Folge, *die;* **c)** (*Math.*) Produkt, *das* (**of** aus)

production [prə'dʌkʃn] *n.* **a)** (*Cinemat.*) Produktion, *die;* (*Theatre*) Inszenierung, *die;* (*of record, book*) Herausgabe, *die;* **b)** (*making*) Produktion, *die;* (*manufacturing*) Herstellung, *die;* (*thing produced*) Produkt, *das;* (*thing created*) Werk, *das;* **c)** (*yielding*) Produktion, *die;* (*yield*) Ertrag, *der.* **pro'duction line** *n.* Fertigungsstraße, *die*

productive [prə'dʌktɪv] *adj.* leistungsfähig ⟨Betrieb, Bauernhof⟩; fruchtbar ⟨Gespräch, Verhandlungen⟩. **productivity** [prɒdʌk'tɪvɪtɪ] *n.* Produktivität, *die*

Prof. [prɒf] *abbr.* Professor Prof.

profane [prə'feɪn] *adj.* **a)** (*irreligious*) gotteslästerlich; **b)** (*secular*) weltlich; **c)** (*irreverent*) respektlos ⟨Bemerkung⟩; profan ⟨Sprache⟩

profess [prə'fes] *v. t.* **a)** (*declare openly*) bekunden ⟨Vorliebe, Abneigung⟩; ~ **to be/do sth.** erklären, etw. zu sein/tun; **b)** (*claim*) vorgeben; ~ **to be/do sth.** behaupten, etw. zu sein/tun

profession [prə'feʃn] *n.* **a)** Beruf, *der;* **be a pilot by ~:** von Beruf Pilot sein; **b)** (*body of people*) Berufsstand, *der*

professional [prə'feʃənl] **1.** *adj.* **a)** Berufs⟨ausbildung, -leben⟩; beruflich ⟨Qualifikation⟩; **b)** (*worthy of profes-*

sion) (in technical expertise) fachmännisch; *(in attitude)* professionell; *(in experience)* routiniert; **c)** ~ **people** Angehörige hochqualifizierter Berufe; **d)** *(by profession)* gelernt; *(not amateur)* Berufs⟨*musiker, -sportler*⟩; Profi⟨*sportler*⟩; **e)** *(paid)* Profi⟨*sport, -boxen*⟩. **2.** *n. (trained person)* Fachmann, *der*/Fachfrau, *die; (non-amateur; also Sport)* Profi, *der*

professor [prə'fesə(r)] *n.* **a)** *(Univ.)* Professor, *der*/Professorin, *die (of* für); **b)** *(Amer.: teacher at university)* Dozent, *der*/Dozentin, *die*

proficiency [prə'fɪʃənsɪ] *n.* Können, *das*

proficient [prə'fɪʃənt] *adj.* fähig; gut ⟨*Pianist, Reiter usw.*⟩; geschickt ⟨*Radfahrer, Handwerker*⟩; **be** ~ **at** *or* **in maths** viel von Mathematik verstehen

profile ['prəʊfaɪl] *n.* **a)** *(side aspect)* Profil, *das;* **b)** *(biographical sketch)* Porträt, *das;* **c)** *(fig.)* **keep a low** ~: sich zurückhalten

profit ['prɒfɪt] *n.* Gewinn, *der;* Profit, *der;* **make a** ~ **from** *or* **out of sth.** mit etw. Geld verdienen; **make |a few pence|** ~ **on sth.** [ein paar Pfennige] an etw. *(Dat.)* verdienen. **'profit by** *v. t.* profitieren von; Nutzen ziehen aus ⟨*Fehler, Erfahrung*⟩. **'profit from** *v. t.* profitieren von

profitable ['prɒfɪtəbl] *adj.* rentabel; einträglich; *(fruitful)* nützlich

profiteer [prɒfɪ'tɪə(r)] **1.** *n.* Profitmacher, *der*/-macherin, *die.* **2.** *v. i.* sich bereichern. **profi'teering** *n.* Wucher, *der*

profligate ['prɒflɪgət] *adj.* verschwenderisch; **be** ~ **of** *or* **with sth.** verschwenderisch umgehen mit etw.

profound [prə'faʊnd] *adj.* tief; nachhaltig ⟨*Wirkung, Einfluß*⟩; tiefgreifend ⟨*Wandel, Veränderung*⟩; tiefempfunden ⟨*Beileid, Mitgefühl*⟩; tiefsitzend ⟨*Mißtrauen*⟩

program ['prəʊgræm] **1.** *n.* **a)** *(Amer.)* see **programme 1;** **b)** *(Computing)* Programm, *das.* **2.** *v. t.,* **-mm-** *(Computing)* programmieren

programme ['prəʊgræm] *n.* **a)** *([notice of] events)* Programm, *das;* **b)** *(Radio, Telev.)* Sendung, *die;* **c)** *(plan, instructions for machine)* Programm, *das*

progress **1.** ['prəʊgres] *n.* **a)** no pl., no indef. art. *(onward movement)* [Vorwärts]bewegung, *die; (advance)* Fortschritt, *der;* **make** ~: vorankommen; ⟨*Student, Patient:*⟩ Fortschritte ma-

chen; **in** ~ : im Gange. **2.** [prə'gres] *v. i.* **a)** *(move forward)* vorankommen; **b)** *(be carried on, develop)* Fortschritte machen. **progression** [prə'greʃn] *n.* **a)** *(development)* Fortschritt, *der;* **b)** *(succession)* Folge, *die.* **progressive** [prə'gresɪv] *adj.* **a)** fortschreitend ⟨*Verbesserung, Verschlechterung*⟩; schrittweise ⟨*Reform*⟩; allmählich ⟨*Veränderung*⟩; **b)** *(favouring reform; in culture)* fortschrittlich; progressiv. **pro'gressively** *adv.* immer ⟨*schlechter, weiter*⟩

prohibit [prə'hɪbɪt] *v. t. (forbid)* verbieten; ~ **sb.'s doing sth.,** ~ **sb. from doing sth.** jmdm. verbieten, etw. zu tun. **prohibition** [prəʊhɪ'bɪʃn, prəʊɪ'bɪʃn] *n.* Verbot, *das.* **prohibitive** [prə'hɪbɪtɪv] *adj.* unerschwinglich ⟨*Preis, Miete*⟩; untragbar ⟨*Kosten*⟩

project 1. [prə'dʒekt] *v. t.* werfen ⟨*Schein*⟩; senden ⟨*Strahl*⟩; *(Cinemat.)* projizieren. **2.** [prə'dʒekt] *v. i. (jut out)* ⟨*Felsen:*⟩ vorspringen; ⟨*Zähne, Brauen:*⟩ vorstehen. **3.** ['prɒdʒekt] *n.* Projekt, *das*

projectile [prə'dʒektaɪl] *n.* Geschoß, *das*

projection [prə'dʒekʃn] *n.* **a)** *(protruding thing)* Vorsprung, *der;* **b)** *(estimate)* Hochrechnung, *die; (forecast)* Voraussage, *die*

projector [prə'dʒektə(r)] *n.* Projektor, *der*

proliferate [prə'lɪfəreɪt] *v. i. (increase)* sich ausbreiten. **proliferation** [prəlɪfə'reɪʃn] *n.* starke Zunahme

prolific [prə'lɪfɪk] *adj.* **a)** *(fertile)* fruchtbar; **b)** *(productive)* produktiv

prologue *(Amer.:* **prolog)** ['prəʊlɒg] *n.* Prolog, *der* **(to** zu)

prolong [prə'lɒŋ] *v. t.* verlängern. **prolonged** [prə'lɒŋd] *adj.* lang; langanhaltend ⟨*Beifall*⟩

promenade [prɒmə'nɑːd] *n.* Promenade, *die*

prominence ['prɒmɪnəns] *n.* **a)** *(conspicuousness)* Auffälligkeit, *die;* **b)** *(distinction)* Bekanntheit, *die*

prominent ['prɒmɪnənt] *adj.* **a)** *(conspicuous)* auffallend; **b)** *(foremost)* herausragend; **he was** ~ **in politics** er war ein prominenter Politiker; **c)** *(projecting)* vorspringend; vorstehen ⟨*Backenknochen, Brauen*⟩

promiscuity [prɒmɪ'skjuːɪtɪ] *n.* Promiskuität, *die (geh.)*

promiscuous [prə'mɪskjʊəs] *adj.* promiskuitiv *(geh.);* **a** ~ **man** ein Mann, der häufig die Partnerin wechselt

promise ['prɒmɪs] 1. *n.* **a)** Versprechen, *das; sb.'s ~s* ~s jmds. Versprechungen; **give** *or* **make a ~ [to sb.]** [jmdm.] ein Versprechen geben; **give** *or* **make a ~ [to sb.] to do sth.** [jmdm.] versprechen, etw. zu tun; **b)** *(fig.: reason for expectation)* Hoffnung, *die;* **a painter of** *or* **with** ~: ein vielversprechender Maler. 2. *v.t.* **a)** versprechen; **~ sth. to sb., ~ sb. sth.** jmdm. etw. versprechen; **b)** *(fig.: give reason for expectation of)* verheißen *(geh.);* **~ sb. sth.** jmdm. etw. in Aussicht stellen. 3. *v.i.* **~ well** *or* **favourably** vielversprechend sein; **I can't ~:** ich kann es nicht versprechen. **promising** ['prɒmɪsɪŋ] *adj.* vielversprechend

promote [prə'məʊt] *v.t.* **a)** *(to more senior job)* befördern; **b)** *(encourage)* fördern; **c)** *(publicize)* Werbung machen für; **d)** *(Footb.)* **be ~d** aufsteigen.

pro'moter *n.* Veranstalter, *der/*Veranstalterin, *die.* **promotion** [prə'məʊʃn] *n.* **a)** Beförderung, *die;* **win** *or* **gain ~:** befördert werden; **b)** *(furtherance)* Förderung, *die;* **c)** *(publicization)* Werbung, *die; (instance)* Werbekampagne, *die;* **d)** *(Footb.)* Aufstieg, *der.* **promotional** [prə'məʊʃənl] *adj.* Werbe⟨*kampagne, -broschüre usw.*⟩

prompt [prɒmpt] 1. *adj.* **a)** *(ready to act)* bereitwillig; **be ~ in doing sth.** *or* **to do sth.** etw. unverzüglich tun; **b)** *(done readily)* sofortig; **her ~ answer** ihre prompte Antwort; **take ~ action** sofort handeln; **c)** *(punctual)* pünktlich. 2. *adv.* pünktlich; **at 6 o'clock ~:** Punkt 6 Uhr. 3. *v.t.* **a)** *(incite)* veranlassen; **b)** *(supply with words)* soufflieren (+ *Dat.*); *(give suggestion to)* weiterhelfen (+ *Dat.*); **c)** hervorrufen ⟨*Kritik*⟩; provozieren ⟨*Antwort*⟩. **'promptly** *adv.* **a)** *(quickly)* prompt; **b)** *(punctually)* pünktlich

prone [prəʊn] *adj. (liable)* **be ~ to** anfällig sein für ⟨*Krankheiten*⟩; **be ~ to do sth.** dazu neigen, etw. zu tun

prong [prɒŋ] *n. (of fork)* Zinke, *die*

pronoun ['prəʊnaʊn] *n. (Ling.)* Pronomen, *das;* Fürwort, *das*

pronounce [prə'naʊns] 1. *v.t.* **a)** *(declare)* verkünden; **~ sb./sth. [to be] sth.** jmdn./etw. für etw. erklären; **~ sb. fit for work** jmdn. für arbeitsfähig erklären; **b)** aussprechen ⟨*Wort, Buchstaben usw.*⟩. 2. *v.i.* **~ on sth.** zu etw. Stellung nehmen; **~ for** *or* **in favour of/against sth.** sich für/gegen etw. aussprechen. **pronounced** [prə-'naʊnst] *adj. (marked)* ausgeprägt. **pro'nouncement** *n.* Erklärung, *die;* **make a ~ [about sth.]** eine Erklärung [zu etw.] abgeben

pronunciation [prənʌnsɪ'eɪʃn] *n.* Aussprache, *die;* **what is the ~ of this word?** wie wird dieses Wort ausgesprochen?

proof [pruːf] 1. *n.* **a)** *(fact, evidence)* Beweis, *der;* **b)** *no indef. art. (Law)* Beweismaterial, *das;* **c)** *(proving)* **in ~ of** zum Beweis (+ *Gen.*); **d)** *no art. (standard of strength)* Proof *o. Art.;* **100° ~** *(Brit.),* **128° ~** *(Amer.)* 64 Vol.-% Alkohol; **e)** *(Printing)* Abzug, *der.* 2. *adj.* **a)** **be ~ against sth.** unempfindlich gegen etw. sein; *(fig.)* gegen etw. immun sein; **b)** *in comb.* ⟨*kugel-, einbruch-, idioten*⟩sicher; ⟨*schall-, wasser*⟩dicht; **flame-~:** nicht brennbar

'proof-read *v.t.* Korrektur lesen. **'proof-reader** *n.* Korrektor, *der/*Korrektorin, *die*

prop [prɒp] 1. *n.* Stütze, *die; (Mining)* Strebe, *die.* 2. *v.t.,* **-pp-** stützen; **the ladder was ~ped against the house** die Leiter war gegen das Haus gelehnt. **prop 'up** *v.t.* stützen; *(fig.)* vor dem Konkurs bewahren ⟨*Firma*⟩; stützen ⟨*Regierung*⟩

propaganda [prɒpə'gændə] *n.* Propaganda, *die*

propagate ['prɒpəgeɪt] 1. *v.t.* **a)** *(Hort., Agric.)* vermehren (**from, by** durch); **b)** *(spread)* verbreiten. 2. *v.i.* **a)** *(Bot.)* sich vermehren; **b)** *(spread)* sich ausbreiten. **propagation** [prɒpə'geɪʃn] *n.* **a)** *(Hort., Agric.)* Züchtung, *die;* **b)** *(Bot.)* Vermehrung, *die;* **c)** *(spreading)* Verbreitung, *die*

propel [prə'pel] *v.t.,* **-ll-** antreiben. **pro'peller** *n.* Propeller, *der.* **propelling 'pencil** *n. (Brit.)* Drehbleistift, *der*

propensity [prə'pensɪtɪ] *n.* **have a ~ to do sth.** *or* **for doing sth.** dazu neigen, etw. zu tun

proper ['prɒpə(r)] *adj.* **a)** *(accurate)* richtig; zutreffend ⟨*Beschreibung*⟩; eigentlich ⟨*Wortbedeutung*⟩; **b)** *postpos. (strictly so called)* im engeren Sinn nachgestellt; **in London ~:** in London selbst; **c)** *(genuine)* echt; richtig ⟨*Wirbelsturm, Schauspieler*⟩; **d)** *(satisfactory)* richtig; zufriedenstellend ⟨*Antwort*⟩; **e)** *(suitable)* angemessen; *(morally fitting)* gebührend; **do sth. the ~ way** etw. richtig machen; **f)** *attrib.*

(coll.: thorough) richtig. **'properly**
adv. richtig; *(rightly)* zu Recht; ~
speaking genaugenommen
proper: ~ **'name,** ~ **'noun** *ns. (Ling.)*
Eigenname, *der*
property ['prɒpətɪ] *n.* **a)** *(posses-sion[s])* Eigentum, *das;* **b)** *(estate)* Be-sitz, *der;* Immobilie, *die (fachspr.);* **c)**
(attribute) Eigenschaft, *die; (effect,
special power)* Wirkung, *die*
prophecy ['prɒfɪsɪ] *n. (prediction)* Vor-hersage, *die; (prophetic utterance)* Pro-phezeiung, *die*
prophesy ['prɒfɪsaɪ] *v. t. (predict)* vor-hersagen; *(fig.)* prophezeien 〈*Un-glück*〉; *(as fortune-teller)* weissagen
prophet ['prɒfɪt] *n.* Prophet, *der.* **pro-phetic** [prə'fetɪk] *adj.* prophetisch
proportion [prə'pɔ:ʃn] **1.** *n.* **a)** *(por-tion)* Teil, *der;* **b)** *(ratio)* Verhältnis,
das; **the** ~ **of sth. to sth.** das Verhältnis
von etw. zu etw.; **c)** *(correct relation;
Math.)* Proportion, *die;* **be in** ~ |**to** *or*
with sth.| im richtigen Verhältnis [zu
od. mit etw.] stehen; **keep things in** ~
(fig.) die Dinge im richtigen Licht se-hen; **be out of** ~/**all** *or* **any** ~ |**to** *or*
with sth.| in keinem/keinerlei Verhält-nis zu etw. stehen; **d)** *in pl. (size)* Di-mension, *die.* **2.** *v. t.* proportionieren.
proportional [prə'pɔ:ʃənl] *adj.* **a)** *(in
proportion)* entsprechend; **be** ~ **to sth.**
einer Sache *(Dat.)* entsprechen; **b)**
(Math.) **be directly/indirectly** ~ **to sth.**
einer Sache *(Dat.)* direkt/umgekehrt
proportional sein. **proportionate**
[prə'pɔ:ʃənət] *see* **proportional a**
proposal [prə'pəʊzl] *n.* Vorschlag,
der; (offer) Angebot, *das;* ~ |**of mar-riage**| [Heirats]antrag, *der*
propose [prə'pəʊz] **1.** *v. t.* **a)** vorschla-gen; ~ **sth. to sb.** jmdm. etw. vorschla-gen; ~ **marriage** |**to sb.**| [jmdm.] einen
Heiratsantrag machen; **b)** *(nominate)*
~ **sb. as/for sth.** jmdn. als/für etw.
vorschlagen; **c)** *(intend)* ~ **doing** *or* **to
do sth.** beabsichtigen, etw. zu tun. **2.**
v. i. (offer marriage) ~ |**to sb.**| jmdm.
einen Heiratsantrag machen. **pro-position** [prɒpə'zɪʃn] *n.* **a)** *(proposal)*
Vorschlag, *der;* **make** *or* **put a** ~ **to sb.**
jmdm. einen Vorschlag machen; **b)**
(statement; Logic) Aussage, *die*
propound [prə'paʊnd] *v. t.* darlegen
proprietary [prə'praɪətərɪ] *adj.* ~
name *or* **term** Markenname, *der*
proprietor [prə'praɪətə(r)] *n.* Inhaber,
*der/*Inhaberin, *die*
propriety [prə'praɪətɪ] *n.* Anstand,

der; **breach of** ~: Verstoß gegen die
guten Sitten
propulsion [prə'pʌlʃn] *n.* Antrieb, *der*
prosaic [prə'zeɪɪk] *adj.* prosaisch
(geh.); nüchtern
proscribe [prə'skraɪb] *v. t.* verbieten
prose [prəʊz] *n.* Prosa, *die; attrib.* Pro-sa〈*werk, -stil*〉
prosecute ['prɒsɪkju:t] **1.** *v. t.* straf-rechtlich verfolgen; ~ **sb. for sth./
doing sth.** jmdn. wegen etw. straf-rechtlich verfolgen/jmdn. strafrecht-lich verfolgen, weil er etw. tut/getan
hat. **2.** *v. i.* Anzeige erstatten. **pro-secution** [prɒsɪ'kju:ʃn] *n. (bringing to
trial)* [strafrechtliche] Verfolgung;
(court procedure) Anklage, *die; (pro-secuting party)* Anklage[vertretung],
die; **the** ~: die Anklage. **prosecutor**
['prɒsɪkju:tə(r)] *n.* Ankläger, *der/*An-klägerin, *die;* **public** ~ ≈ General-staatsanwalt, *der/*-anwältin, *die*
prospect 1. ['prɒspekt] *n.* **a)** *(expecta-tion)* Erwartung, *die (of hinsichtlich);*
|**at the**| ~ **of sth./doing sth.** [bei der]
Aussicht auf etw.*(Akk.)/*[darauf], etw.
zu tun; **b)** *in pl. (hope of success)* Zu-kunftsaussichten; **a man with** |**good**| ~**s**
ein Mann mit Zukunft; **sb.'s** ~**s of
sth./doing sth.** jmds. Chancen auf etw.
*(Akk.)/*darauf, etw. zu tun; **the** ~**s for
sb./sth.** die Aussichten für jmdn./etw.
2. [prə'spekt] *v. i.* nach Bodenschätzen
suchen. **prospective** [prə'spektɪv]
adj. voraussichtlich; zukünftig 〈*Erbe,
Braut*〉; potentiell 〈*Käufer, Kandi-dat*〉. **prospector** [prə'spektə(r)] *n.*
Prospektor, *der; (for gold)* Goldsu-cher, *der*
prospectus [prə'spektəs] *n.* Prospekt,
der; (Brit. Univ.) Studienführer, *der*
prosper [prɒspə(r)] *v. i.* gedeihen;
〈*Geschäft:*〉 florieren; 〈*Berufstätiger:*〉
Erfolg haben. **prosperity** [prɒ'sperɪ-tɪ] *n.* Wohlstand, *der.* **prosperous**
['prɒspərəs] *adj.* wohlhabend; florie-rend 〈*Unternehmen*〉
prostitute ['prɒstɪtju:t] *n.* Prostituier-te, *die.* **prostitution** [prɒstɪ'tju:ʃn] *n.*
Prostitution, *die*
prostrate 1. ['prɒstreɪt] *adj.* [auf dem
Bauch] ausgestreckt. **2.** [prə'streɪt] *v.
refl.* ~ **oneself** |**at sth./before sb.**| sich
[vor etw./jmdm.] niederwerfen
protagonist [prəʊ'tægənɪst] *n. (Lit.)*
Protagonist, *der/*Protagonistin, *die*
protect [prə'tekt] *v. t.* **a)** schützen
(from vor + *Dat.,* **against** gegen); **b)**
(preserve) unter [Natur]schutz stellen

⟨*Pflanze, Tier*⟩. **protection** [prə-'tekʃn] *n*. Schutz, *der* (**from** vor + *Dat.*, **against** gegen). **protective** [prə'tektɪv] *adj*. schützend; Schutz-⟨*hülle, -anstrich, -vorrichtung, -maske*⟩; **be ~ towards** sb. fürsorglich gegenüber jmdm. sein

protein ['prəʊtiːn] *n*. Protein, *das* (*fachspr.*); Eiweiß, *das*

protest 1. ['prəʊtest] *n*. **a)** Beschwerde, *die*; **make** *or* **lodge a ~ |against** sb./ sth.] eine Beschwerde [gegen jmdn./ etw.] einreichen; **b)** (*gesture of disapproval*) ~|s| Protest, *der*; **under ~**: unter Protest; **in ~ |against** sth.] aus Protest [gegen etw.]; **c)** *no art.* (*dissent*) Protest, *der.* **2.** [prə'test] *v. t.* (*affirm*) beteuern. **3.** [prə'test] *v. i.* protestieren (**about** gegen); (*make written or formal* ~) Protest einlegen (**to** bei)

Protestant ['prɒtɪstənt] **1.** *n*. Protestant, *der*/Protestantin, *die*. **2.** *adj*. protestantisch; evangelisch

pro'tester *n*. Protestierende, *der/die*; (*at demonstration*) Demonstrant, *der*/ Demonstrantin, *die*

protocol ['prəʊtəkɒl] *n*. Protokoll, *das*

proton ['prəʊtən] *n*. Proton, *das*

prototype ['prəʊtətaɪp] *n*. Prototyp, *der*

protract [prə'trækt] *v. t.* verlängern. **protractor** [prə'træktə(r)] *n*. (*Geom.*) Winkelmesser, *der*

protrude [prə'truːd] *v. i.* herausragen (**from** aus); ⟨*Zähne*⟩ vorstehen

proud [praʊd] **1.** *adj*. **a)** stolz; **~ to do** sth. *or* **to be doing** sth. stolz darauf, etw. zu tun; **~ of** sb./sth./doing sth. stolz auf jmdn./etw./darauf, etw. zu tun; **b)** (*arrogant*) hochmütig. **2.** *adv.* (*Brit. coll.*) **do** sb. **~**: jmdn. verwöhnen. **'proudly** *adv.* **a)** stolz; **b)** (*arrogantly*) hochmütig

prove [pruːv] **1.** *v. t., p.p.* **~d** *or* **proven** ['pruːvn] beweisen; nachweisen ⟨*Identität*⟩; **~ one's ability** sein Können unter Beweis stellen; **~** sb. **right/ wrong** ⟨*Ereignis:*⟩ jmdm. recht/unrecht geben; **be ~d wrong** *or* **to be false** ⟨*Theorie:*⟩ widerlegt werden; **~ one's/sb.'s case** *or* **point** beweisen, daß man recht hat/jmdm. recht geben. **2.** *v. refl., p. p.* **proved** *or* **proven**: **~ oneself** sich bewähren. **3.** *v. i., p. p.* **proved** *or* **proven**: **~ |to be|** sich erweisen als

proven *see* **prove**

proverb ['prɒvɜːb] *n*. Sprichwort, *das*. **proverbial** [prə'vɜːbɪəl] *adj*. sprichwörtlich

provide [prə'vaɪd] *v. t.* **a)** besorgen; liefern ⟨*Beweis*⟩; bereitstellen ⟨*Dienst, Geld*⟩; **~ a home/a car for** sb. jmdm. Unterkunft/ein Auto [zur Verfügung] stellen; **b)** ⟨*Vertrag, Gesetz:*⟩ vorsehen. **pro'vide for** *v. t.* **a)** (*make provision for*) vorsorgen für; ⟨*Plan, Gesetz:*⟩ vorsehen; **b)** (*maintain*) sorgen für, versorgen ⟨*Familie, Kind*⟩. **pro'vided** *conj.* **~ |that|...**: vorausgesetzt, [daß]...

providence ['prɒvɪdəns] *n*. **a)** |divine| **~**: die [göttliche] Vorsehung; **b)** P~ (*God*) der Himmel

province ['prɒvɪns] *n*. **a)** Provinz, *die*; **b) the ~s** (*regions outside capital*) die Provinz; **c)** (*sphere of action*) [Tätigkeits]bereich, *der;* (*area of responsibility*) Zuständigkeitsbereich, *der*. **provincial** [prə'vɪnʃl] *adj*. Provinz-

provision [prə'vɪʒn] *n*. **a)** (*providing*) Bereitstellung, *die;* **make ~ for** vorsorgen *od*. Vorsorge treffen für ⟨*Notfall*⟩; **b) ~s** *pl.* (*food*) Lebensmittel

provisional [prə'vɪʒənl] *adj..* **provisionally** [prə'vɪʒənəlɪ] *adv.* vorläufig; provisorisch

proviso [prə'vaɪzəʊ] *n., pl.* **~s** Vorbehalt, *der*

provocation [prɒvə'keɪʃn] *n*. Provokation, *die*

provocative [prə'vɒkətɪv] *adj*. provozierend; (*sexually*) aufreizend

provoke [prə'vəʊk] *v. t.* **a)** provozieren ⟨*Person*⟩; reizen ⟨*Person, Tier*⟩; **~** sb. **into doing** sth. jmdn. so sehr provozieren, daß er etw. tut; **b)** (*give rise to*) hervorrufen; erregen

prow [praʊ] *n*. (*Naut.*) Bug, *der*

prowl [praʊl] **1.** *v. i.* streifen. **2.** *v. t.* durchstreifen. **3.** *n*. **be on the ~**: auf einem Streifzug sein

proximity [prɒk'sɪmɪtɪ] *n*. Nähe, *die*

proxy ['prɒksɪ] *n*. **by ~**: durch einen Bevollmächtigten/eine Bevollmächtigte

prude [pruːd] *n*. prüder Mensch

prudence ['pruːdəns] *n*. Besonnenheit, *die*

prudent ['pruːdənt] *adj*. **a)** (*careful*) besonnen; **b)** (*circumspect*) vorsichtig

prudish ['pruːdɪʃ] *adj*. prüde

¹prune [pruːn] *n*. Backpflaume, *die*

²prune *v. t.* **a)** (*trim*) [be]schneiden; **b)** (*fig.*) reduzieren

pry [praɪ] *v. i.* neugierig sein. **'pry into** *v. t.* seine Nase stecken in (+ *Akk.*) (*ugs.*) ⟨*Angelegenheit*⟩

PS *abbr.* **postscript** PS

psalm [sɑːm] *n*. Psalm, *der*

pseudonym ['sju:dənɪm] n. Pseudonym, das

psychiatric [saɪkɪ'ætrɪk] adj. psychiatrisch

psychiatrist [saɪ'kaɪətrɪst] n. Psychiater, der/Psychiaterin, die

psychiatry [saɪ'kaɪətrɪ] n. Psychiatrie, die

psychic ['saɪkɪk] adj. be ~: übernatürliche Fähigkeiten haben

psychoanalyse [saɪkəʊænəlaɪz] v.t. psychoanalysieren. **psychoa'nalysis** n. Psychoanalyse, die. **psycho'analyst** n. Psychoanalytiker, der/-analytikerin, die

psychological [saɪkə'lɒdʒɪkl] adj. psychologisch; psychisch ⟨Problem⟩

psychologist [saɪ'kɒlədʒɪst] n. Psychologe, der/Psychologin, die

psychology [saɪ'kɒlədʒɪ] n. Psychologie, die

psychopath ['saɪkəpæθ] n. Psychopath, der/Psychopathin, die

PTO abbr. **please turn over** b. w.

pub [pʌb] n. (Brit. coll.) Kneipe, die (ugs.)

puberty ['pju:bətɪ] n., no art. Pubertät, die

public ['pʌblɪk] **1.** adj. öffentlich; **make sth. ~:** etw. bekannt machen. **2.** n., sing. or pl. **a)** (the people) Öffentlichkeit, die; **b)** (section of community) Publikum, das; **c) in ~:** öffentlich

publican ['pʌblɪkən] n. (Brit.) [Gast]wirt, der/-wirtin, die

publication [pʌblɪ'keɪʃn] n. Veröffentlichung, die

public: ~ con'venience n. öffentliche Toilette; **~ 'holiday** n. gesetzlicher Feiertag; **~ 'house** n. (Brit.) Gastwirtschaft, die; Gaststätte, die

publicity [pʌb'lɪsɪtɪ] n. Publicity, die; (advertising) Werbung, die; **~ campaign** Werbekampagne, die

publicize ['pʌblɪsaɪz] v.t. publik machen ⟨Ungerechtigkeit⟩; werben für, Reklame machen für ⟨Produkt⟩

public 'library n. öffentliche Bücherei

'publicly adv. öffentlich; **~ owned** staatseigen

public: ~ re'lations n., sing. or pl. Public Relations Pl.; **~ school** n. **a)** (Brit.) Privatschule, die; **b)** (Scot., Amer.) staatliche od. öffentliche Schule; **~ 'transport** n. öffentlicher Personenverkehr

publish ['pʌblɪʃ] v.t. ⟨Verlag:⟩ verlegen ⟨Buch, Zeitschrift, Musik usw.⟩;

⟨Autor:⟩ veröffentlichen ⟨Text⟩. **'publisher** n. Verleger, der/Verlegerin, die; **~[s]** (company) Verlag, der. **'publishing** n., no art. Verlagswesen, das

puck [pʌk] n. (Ice Hockey) Puck, der

pucker ['pʌkə(r)] **1.** v.t. **~ |up|** runzeln ⟨Brauen, Stirn⟩; kräuseln ⟨Lippen⟩. **2.** v.i. **~ |up|** ⟨Stoff:⟩ sich kräuseln

pudding ['pʊdɪŋ] n. **a)** Pudding, der; **b)** (dessert) süße Nachspeise

puddle ['pʌdl] n. Pfütze, die

puerile ['pjʊəraɪl] adj. kindisch

puff [pʌf] **1.** n. **a)** Stoß, der; **~ of breath/wind** Atem-/Windstoß, der; **b) ~ of smoke** Rauchstoß, der; **c)** (pastry) Blätterteigteilchen, das. **2.** v.i. **a) ~ |and blow|** schnaufen [und keuchen]; **b)** (~ cigarette smoke etc.) paffen (ugs.) (at an + Dat.); **c)** ⟨Person:⟩ keuchen; ⟨Zug, Lokomotive⟩ schnaufend fahren. **3.** v.t. blasen ⟨Rauch⟩; stäuben ⟨Puder⟩. **puff 'out** v.t. **a)** bauschen ⟨Segel⟩; **b)** (put out of breath) außer Atem bringen ⟨Person⟩; **be ~ed |out|** außer Atem sein

puff 'pastry n. Blätterteig, der

puffy ['pʌfɪ] adj. verschwollen

pugnacious [pʌg'neɪʃəs] adj. kampflustig

pull [pʊl] **1.** v.t. **a)** (draw, tug) ziehen an (+ Dat.); ziehen ⟨Hebel⟩; **~ sb.'s or sb. by the hair/ears/sleeve** jmdn. an den Haaren/Ohren/am Ärmel ziehen; **~ sth. over one's ears/head** sich (Dat.) etw. über die Ohren/den Kopf ziehen; **~ to pieces** in Stücke reißen; (fig.) zerpflücken ⟨Argument usw.⟩; **b)** (extract) [her]ausziehen; [heraus]ziehen ⟨Zahn⟩; **c)** (strain) sich (Dat.) zerren ⟨Muskel⟩. **2.** v.i. **a)** ziehen; **'P~'** „Ziehen"; **b) ~ |to the left/right|** ⟨Auto, Boot:⟩ [nach links/rechts] ziehen; **c)** (pluck) **~ at** ziehen an (+ Dat.); **~ at sb.'s sleeve** jmdn. am Ärmel ziehen. **3.** n. **a)** Zug, der; **b)** (influence) Einfluß, der (with auf + Akk., bei). **pull a'part** v.t. **a)** (take to pieces) auseinandernehmen; **b)** (fig.: criticize) zerpflücken; verreißen ⟨Buch, [literarisches] Werk⟩. **pull 'down** v.t. **a)** herunterziehen; **b)** (demolish) abreißen. **pull 'in 1.** v.t. hereinziehen. **2.** v.i. **a)** ⟨Zug:⟩ einfahren; **b)** (move to side of road) an die Seite fahren; (stop) anhalten. **pull 'off** v.t. **a)** (remove) abziehen; (violently) abreißen; **b)** (accomplish) an Land ziehen (ugs.). **pull 'out 1.** v.t. herausziehen. **2.** v.i. **a)** (depart) abfahren; **b)**

(away from roadside) ausscheren. **pull 'through** *v. i.* ⟨*Patient:*⟩ durchkommen. **pull to'gether** *v. refl.* sich zusammennehmen. **pull 'up 1.** *v. t.* **a)** hochziehen; **b)** [he]rausziehen ⟨*Unkraut, Pflanze*⟩; **c)** *(reprimand)* zurechtweisen. **2.** *v. i. (stop)* anhalten.
pulley ['pʊlɪ] *n.* Rolle, *die*
pullover ['pʊləʊvə(r)] *n.* Pullover, *der*
pulp [pʌlp] **1.** *n.* Brei, *der.* **2.** *v. t.* zerdrücken ⟨*Rübe*⟩; einstampfen ⟨*Druckerzeugnis*⟩
pulpit ['pʊlpɪt] *n.* Kanzel, *die*
pulsate [pʌl'seɪt] *v. i.* pulsieren
¹pulse [pʌls] *n.* Puls, *der; (single beat)* Pulsschlag, *der*
²pulse *n. (Cookery)* Hülsenfrucht, *die*
pulverize ['pʌlvəraɪz] *v. t.* pulverisieren
puma ['pjuːmə] *n.* Puma, *der*
pumice ['pʌmɪs] *n.* ~[-stone] Bimsstein, *der*
pummel ['pʌml] *v. t., (Brit)* -ll- einschlagen auf (+ *Akk.*)
pump [pʌmp] **1.** *n.* Pumpe, *die.* **2.** *v. i.* pumpen. **3.** *v. t.* pumpen; ~ **sth. dry** etw. leerpumpen; ~ **sb. for information** Auskünfte aus jmdm. herausholen; ~ **up** aufpumpen
pumpkin ['pʌmpkɪn] *n.* Kürbis, *der*
pun [pʌn] *n.* Wortspiel, *das*
¹punch 1. *v. t.* **a)** *(with fist)* boxen; **b)** *(pierce)* lochen; ~ **a hole** ein Loch stanzen; ~ **a hole/holes in sth.** lochen. **2.** *n.* **a)** *(blow)* Faustschlag, *der;* **b)** *(for making holes) (in leather, tickets)* Lochzange, *die; (in paper)* Locher, *der*
²punch *n. (drink)* Punsch, *der*
punch: ~ **line** *n.* Pointe, *die;* ~**-up** *n. (Brit. coll.)* Prügelei, *die*
punctual ['pʌŋktjʊəl] *adj.* pünktlich. **punctuality** [pʌŋktjʊ'ælɪtɪ] *n.* Pünktlichkeit, *die.* '**punctually** *adv.* pünktlich
punctuate ['pʌŋktjʊət] *v. t.* mit Satzzeichen versehen. **punctuation** [pʌŋktjʊ'eɪʃn] *n.* Zeichensetzung, *die.* **punctu'ation mark** *n.* Satzzeichen, *das*
puncture ['pʌŋktʃə(r)] **1.** *n.* **a)** *(flat tyre)* Reifenpanne, *die;* **b)** *(hole)* Loch, *das.* **2.** *v. t.* durchstechen; **be ~d** ⟨*Reifen:*⟩ platt sein
pundit ['pʌndɪt] *n.* Experte, *der*/Expertin, *die*
pungent ['pʌndʒənt] *adj.* beißend, ätzend ⟨*Rauch*⟩; scharf ⟨*Soße*⟩; stechend riechend ⟨*Gas*⟩

punish ['pʌnɪʃ] *v. t.* bestrafen. **punishable** ['pʌnɪʃəbl] *adj.* strafbar. '**punishment** *n.* **a)** *(punishing)* Bestrafung, *die;* **b)** *(penalty)* Strafe, *die*
punitive ['pjuːnɪtɪv] *adj.* **a)** *(penal)* Straf-; **b)** *(severe)* [allzu] rigoros
punk [pʌŋk] *n.* **a)** *(Amer. sl.: worthless person)* Dreckskerl, *der (salopp);* **b)** *(admirer of* ~ *rock)* Punk, *der; (performer)* Punk[rock]er, *der*/-[rock]erin, *die;* **c)** *(music)* Punkrock, *der*
punt [pʌnt] *n.* Stechkahn, *der*
puny ['pjuːnɪ] *adj.* **a)** *(undersized)* zu klein ⟨*Baby, Junge*⟩; **b)** *(feeble)* gering ⟨*Kraft*⟩; schwach ⟨*Waffe, Person*⟩
pup [pʌp] *n.* Welpe, *der*
pupa ['pjuːpə] *n., pl.* ~**e** ['pjuːpiː] Puppe, *die.* **pupate** [pjuː'peɪt] *v. i.* sich verpuppen
pupil ['pjuːpɪl] *n.* **a)** Schüler, *der*/Schülerin, *die;* **b)** *(Anat.)* Pupille, *die*
puppet ['pʌpɪt] *n.* Puppe, *die;(marionette; also fig.)* Marionette, *die*
puppy ['pʌpɪ] *n.* Hundejunge, *das;* Welpe, *der*
purchase ['pɜːtʃəs] **1.** *n.* **a)** Kauf, *der;* **make a** ~: etwas kaufen; **b)** *(hold)* Halt, *der; (leverage)* Hebelwirkung, *die.* **2.** *v. t.* kaufen. '**purchaser** *n.* Käufer, *der*/Käuferin, *die*
pure [pjʊə(r)] *adj.* rein
purée ['pjʊəreɪ] *n.* Püree, *das*
'**purely** *adv.* **a)** *(solely)* rein; **b)** *(merely)* lediglich
purgatory ['pɜːgətərɪ] *n.* it was ~ *(fig.)* es war eine Strafe
purge [pɜːdʒ] **1.** *v. t.* **a)** *(cleanse)* reinigen (of von); **b)** *(remove)* entfernen; **c)** *(rid)* säubern ⟨*Partei*⟩ (of von). **2.** *n.* Säuberung[saktion], *die*
purification [pjʊərɪfɪ'keɪʃn] *n.* Reinigung, *die*
purify ['pjʊərɪfaɪ] *v. t.* reinigen
purist ['pjʊərɪst] *n.* Purist, *der*/Puristin, *die*
puritan, *(Hist.)* **Puritan** ['pjʊərɪtn] *n.* Puritaner, *der*/Puritanerin, *die.* **puritanical** [pjʊərɪ'tænɪkl] *adj.* puritanisch
purity ['pjʊərɪtɪ] *n.* Reinheit, *die*
purl [pɜːl] **1.** *n.* linke Masche. **2.** *v. t.* ~ **three [stitches]** drei linke Maschen stricken
purple ['pɜːpl] **1.** *adj.* lila; violett. **2.** *n.* Lila, *das;* Violett, *das*
purport [pə'pɔːt] *v. t.* ~ **to do sth.** *(profess)* [von sich] behaupten, etw. zu tun; *(be intended to seem)* den Anschein erwecken sollen, etw. zu tun

purpose ['pɜːpəs] *n.* a) *(object)* Zweck, *der*; *(intention)* Absicht, *die*; **what is the ~ of doing that?** was hat es für einen Zweck, das zu tun?; **on ~:** mit Absicht; absichtlich; b) *(effect)* **to no ~:** ohne Erfolg; **to some/good ~:** mit einigem/gutem Erfolg; c) *(determination)* Entschlossenheit, *die*. **purposeful** ['pɜːpəsfl] *adj.* zielstrebig; *(with specific aim)* entschlossen. **'purposely** *adv.* absichtlich

purr [pɜː(r)] 1. *v.i.* schnurren. 2. *n.* Schnurren, *das*

purse [pɜːs] 1. *n.* Portemonnaie, *das*. 2. *v.t.* kräuseln ⟨*Lippen*⟩

purser ['pɜːsə(r)] *n.* Zahlmeister, *der*/ -meisterin, *die*

pursue [pə'sjuː] *v.t.* a) *(chase)* verfolgen; b) *(look into)* nachgehen (+ *Dat.*); c) *(engage in)* betreiben. **pursuer** [pə'sjuːə(r)] *n.* Verfolger, *der*/Verfolgerin, *die*. **pursuit** [pə-'sjuːt] *n.* a) Verfolgung, *die*; *(of knowledge, truth, etc.)* Streben, *das* (of nach); **in ~ of** auf der Jagd nach ⟨*Wild, Dieb usw.*⟩; in Ausführung (+ *Gen.*) ⟨*Beschäftigung*⟩; **with the police in |full| ~:** mit der Polizei [dicht] auf den Fersen; b) *(pastime)* Beschäftigung, *die*

pus [pʌs] *n.* Eiter, *der*

push [puʃ] 1. *v.t.* a) schieben; *(make fall)* stoßen; drücken gegen ⟨*Tür*⟩; **~ one's way through/into/on to** *etc.* **sth.** sich *(Dat.)* einen Weg durch/in/auf usw. etw. *(Akk.)* bahnen; b) *(fig.: impel)* drängen; c) *(tax)* **~ sb. |hard|** jmdn. [stark] fordern; **be ~ed for sth.** *(coll.: find it difficult to provide sth.)* mit etw. knapp sein; **be ~ed for money or cash** knapp bei Kasse sein *(ugs.)*; d) *(sell illegally, esp. drugs)* pushen *(Drogenjargon)*. 2. *v.i.* a) schieben; *(in queue)* drängeln; *(at door)* drücken; **~ and shove** schubsen und drängeln; b) *(make demands)* **~ for sth.** etw. fordern; c) *(make one's way)* **he ~ed between us** er drängte sich zwischen uns; **~ through the crowd** sich durch die Menge drängeln. 3. *n.* a) Stoß, *der*; **give sth. a ~:** etw. schieben; b) *(effort)* Anstrengungen *Pl.*; *(Mil.: attack)* Vorstoß, *der*; c) *(crisis)* **when it comes to the ~,** *(Amer. coll.)* **when ~ comes to shove** wenn es ernst wird; d) *(Brit. sl.: dismissal)* **get the ~:** rausfliegen *(ugs.)*. **push a'head** *v.i.* **~ ahead with sth.** etw. vorantreiben. **push 'in** *v.i.* sich hineindrängen. **push 'off** *v.i.* a)

(Boating) abstoßen; b) *(sl.: leave)* abhauen *(salopp)*. **push 'on** 1. *v.i. (with plans etc.)* weitermachen. 2. *v.t.* draufdrücken ⟨*Deckel usw.*⟩. **push 'up** *v.t.* hochschieben; *(fig.)* hochtreiben

push: **~-button** *n.* [Druck]knopf, *der*; Drucktaste, *die*; **~-chair** *n. (Brit.)* Sportwagen, *der*; **~-over** *n. (coll.)* Kinderspiel, *das*

pushy ['puʃɪ] *adj. (coll.)* [übermäßig] ehrgeizig ⟨*Person*⟩

pussy ['pusɪ] *n. (child lang.: cat)* Miezekatze, *die (fam.)*

put [put] 1. *v.t.,* -tt-, put a) *(place)* tun; *(vertically)* stellen; *(horizontally)* legen; **~ plates on the table** Teller auf den Tisch stellen; **~ a stamp on the letter** eine Briefmarke auf den Brief kleben; **~ the letter in an envelope/the letter-box** den Brief in einen Umschlag/in den Briefkasten stecken; **~ sth. in one's pocket** etw. in die Tasche stecken; **~ petrol in the tank** Benzin in den Tank füllen; **~ the car in|to| the garage** das Auto in die Garage stellen; **~ the plug in the socket** den Stecker in die Steckdose stecken; **~ one's hands over one's eyes** sich *(Dat.)* die Hände auf die Augen legen; **where shall I ~ it?** wo soll ich es hintun *(ugs.)*/-stellen/-legen usw.?; *(fig.)* **be ~ in a difficult position** in eine schwierige Lage geraten; **~ sb. on to sth.** jmdn. auf etw. *(Akk.)* hinweisen; **~ sb. to work** jmdn. arbeiten lassen; **~ sb. on antibiotics** jmdn. auf Antibiotika setzen; **~ oneself in sb.'s place** *or* **situation** sich in jmds. Lage *(Akk.)* versetzen; b) *(submit)* unterbreiten ⟨*Vorschlag, Plan*⟩ (to *Dat.*); c) *(express)* ausdrücken; **let's ~ it like this: ...:** sagen wir so: ...; **~ sth. into English** *etc.* etw. ins Englische usw. übertragen; **~ sth. into words** etw. in Worte fassen; d) *(write)* schreiben; **~ one's name on the list** seinen Namen auf die Liste setzen; **~ sth. on the bill** etw. auf die Rechnung setzen; e) *(stake)* setzen (on auf + *Akk.*); f) *(estimate)* **~ sb./sth. at** jmdn./etw. schätzen auf (+ *Akk.*). 2. *v.i.* -tt-, put *(Naut.)* **~ |out| to sea** in See stechen. **put a'cross** *v.t.* a) *(communicate)* vermitteln (to *Dat.*); b) *(make acceptable)* ankommen mit. **put a'way** *v.t.* a) wegräumen; reinstellen ⟨*Auto*⟩; *(in file)* abheften; b) *(save)* beiseite legen; c) *(coll.)* *(eat)* verdrücken *(ugs.)*; *(drink)* runterkippen *(ugs.)*; d) *(coll.:*

confine) einsperren *(ugs.).* **put 'back**
v. t. **a)** ~ **the book back** das Buch zu-
rücktun; **b)** ~ **the clock back** die Uhr
zurückstellen; **c)** *(postpone)* verschieben. **put 'down** *v. t.* **a)** *(set down) (vertically)* hinstellen; *(horizontally)* hinlegen; auflegen ⟨*Hörer*⟩; **b)** *(suppress)*
niederwerfen; **c)** *(humiliate)* herabsetzen; **d)** *(kill)* töten; **e)** *(write)* notieren;
f) *(attribute)* ~ **sth. down to sth.** etw.
auf etw. *(Akk.)* zurückführen. **put
'forward** *v. t.* **a)** *(propose)* aufwarten
mit; **b)** *(nominate)* vorschlagen; **c)** ~
the clock forward die Uhr vorstellen.
put 'in 1. *v. t.* **a)** *(install)* einbauen; **b)**
(submit) stellen ⟨*Forderung*⟩; einrei-
chen ⟨*Bewerbung*⟩; **c)** *(devote)* aufwenden ⟨*Mühe*⟩; *(perform)* einlegen
⟨*Sonderschicht, Überstunden*⟩. **2.** *v. i.* ~
in for sich bewerben um ⟨*Stellung*⟩;
beantragen ⟨*Urlaub*⟩. **put 'off** *v. t.* **a)**
(postpone) verschieben **(until** auf +
Akk.); *(postpone engagement with)*
vertrösten **(until** auf + *Akk.*); **b)**
(switch off) ausmachen; **c)** *(repel)* abstoßen; ~ **sb. off sth.** jmdm. etw. ver-
leiden; **d)** *(distract)* stören; **e)** *(dissuade)* ~ **sb. off doing sth.** jmdn. davon abbringen, etw. zu tun. **put 'on**
v. t. **a)** anziehen ⟨*Kleidung, Hose
usw.*⟩; aufsetzen ⟨*Hut, Brille*⟩; drauf-
setzen ⟨*Deckel*⟩; ~ **it on** *(coll.)* [nur]
Schau machen *(ugs.)*; **b)** anmachen
⟨*Radio, Licht*⟩; aufsetzen ⟨*Wasser,
Kessel*⟩; **c)** *(gain)* ~ **on weight** zunehmen; **d)** *(stage)* spielen ⟨*Stück*⟩; zeigen ⟨*Film*⟩. **put 'out** *v. t.* **a)** rausbringen; **b)** ausmachen ⟨*Licht*⟩; löschen
⟨*Feuer*⟩; **c)** *(inconvenience)* in Verlegenheit bringen. **put 'through** *v. t.* **a)**
(carry out) durchführen ⟨*Plan, Programm*⟩; **b)** *(Teleph.)* verbinden (**to**
mit). **put 'up 1.** *v. t.* **a)** heben ⟨*Hand*⟩;
errichten ⟨*Gebäude, Denkmal*⟩; aufstellen ⟨*Gerüst*⟩; **b)** *(display)* aushän-
gen; **c)** hochnehmen ⟨*Fäuste*⟩; leisten
⟨*Widerstand, Gegenwehr*⟩; **d)** *(propose)*
vorschlagen; *(nominate)* aufstellen; **e)**
(incite) ~ **sb. up to sth.** jmdn. zu etw.
anstiften; **f)** *(accommodate)* unterbringen; **g)** *(increase)* [he]raufsetzen
⟨*Preis, Miete*⟩. **2.** *v. i.* *(lodge)* übernachten. **put 'up with** *v. t.* sich *(Dat.)*
bieten lassen ⟨*Beleidigung, Beneh-
men*⟩; sich abfinden mit ⟨*Lärm,
Elend*⟩; sich abgeben mit ⟨*Person*⟩
putrefy ['pju:trɪfaɪ] *v. i.* sich zersetzen
putrid ['pju:trɪd] *adj. (rotten)* faul; ~
smell Fäulnisgeruch, *der*

putt [pʌt] *(Golf)* **1.** *v. i. & t.* putten. **2.** *n.*
Putt, *der*. **'putter** *n.* Putter, *der*
putty ['pʌtɪ] *n.* Kitt, *der*
'put-up *adj.* **a** ~ **job** ein abgekartetes
Spiel *(ugs.)*
puzzle ['pʌzl] **1.** *n. (problem, enigma)*
Rätsel, *das; (toy)* Geduldsspiel, *das.*
2. *v. t.* rätselhaft *od.* ein Rätsel sein
(+ *Dat.*). **3.** *v. i.* ~ **over** *or* **about sth.**
sich *(Dat.)* über etw. den Kopf zerbre-
chen. **puzzled** ['pʌzld] *adj.* ratlos.
puzzling ['pʌzlɪŋ] *adj.* rätselhaft
PVC *abbr.* **polyvinyl chloride** PVC, *das*
pygmy ['pɪgmɪ] *n.* Pygmäe, *der*
pyjamas [pɪ'dʒɑːməz] *n. pl.* |**pair of**| ~ :
Schlafanzug, *der*
pylon ['paɪlən] *n.* Mast, *der*
pyramid ['pɪrəmɪd] *n.* Pyramide, *die*
Pyrenees [pɪrə'niːz] *pr. n. pl.* **the** ~ :
die Pyrenäen
python ['paɪθən] *n.* Python, *die*

Q

Q, q [kjuː] *n.* Q, q, *das*
quack [kwæk] **1.** *v. i.* ⟨*Ente:*⟩ quaken.
2. *n.* Quaken, *das*
quadrangle ['kwɒdræŋgl] *n.* [vier-
eckiger] Innenhof
quadruped ['kwɒdrʊped] *n.* Vierfüß-
ler, *der*
quadruple ['kwɒdrʊpl] **1.** *adj.* vier-
fach. **2.** *v. t.* vervierfachen. **3.** *v. i.* sich
vervierfachen
quagmire ['kwægmaɪə(r)] *n.* Sumpf,
der; Morast, *der*
¹quail [kweɪl] *n. (Ornith.)* Wachtel, *die*
²quail *v. i.* ⟨*Person:*⟩ [ver]zagen
quaint [kweɪnt] *adj.* drollig ⟨*Häuschen, Einrichtung*⟩; malerisch ⟨*Ort*⟩;
(odd) kurios ⟨*Bräuche, Anblick*⟩
quake [kweɪk] **1.** *n. (coll.)* [Erd]beben,
das. **2.** *v. i.* beben; ~ **with fear** vor
Angst zittern
Quaker ['kweɪkə(r)] *n.* Quäker, *der/*
Quäkerin, *die*
qualification [kwɒlɪfɪ'keɪʃn] *n.* **a)**
Qualifikation, *die; (condition)* Voraussetzung, *die;* **b)** *(limitation)* Vorbehalt,
der; **without** ~ : vorbehaltlos

qualified ['kwɒlɪfaɪd] *adj.* **a)** qualifiziert; *(by training)* ausgebildet; **b)** *(restricted)* nicht uneingeschränkt; **a ~ success** kein voller Erfolg; **~ acceptance** bedingte Annahme

qualify ['kwɒlɪfaɪ] **1.** *v. t.* **a)** *(make competent)* berechtigen **(for** zu); **b)** *(modify)* einschränken. **2.** *v. i.* **a) ~ in law/medicine** seinen [Studien]abschluß in Jura/Medizin machen; **~ as a doctor/lawyer** sein Examen als Arzt/ Anwalt machen; **b)** *(fulfil a condition)* in Frage kommen **(for** für); **c)** *(Sport)* sich qualifizieren

quality ['kwɒlɪtɪ] **1.** *n.* **a)** Qualität, *die;* **b)** *(characteristic)* Eigenschaft, *die.* **2.** *adj.* Qualitäts-

qualm [kwɑːm] *n.* Bedenken, *das* **(over, about** gegen)

quantity ['kwɒntɪtɪ] *n.* **a)** Quantität, *die;* **b)** *(amount, sum)* Menge, *die*

quarantine ['kwɒrəntiːn] *n.* Quarantäne, *die;* **be in ~:** unter Quarantäne stehen

quarrel ['kwɒrl] **1.** *n.* **a)** Streit, *der;* **have/pick a ~ with sb. [about/over sth.]** sich mit jmdm. [über etw. *(Akk.)*] streiten/Streit anfangen; **b)** *(cause of complaint)* Einwand, *der* **(with** gegen). **2.** *v. i., (Brit.)* **-ll-** [sich] streiten **(over** um, **about** über + *Akk.*); **~ with each other** [sich] [miteinander] streiten; *(fall out)* sich [zer]streiten **(over** um, **about** über + *Akk.*). **quarrelsome** ['kwɒrlsəm] *adj.* streitsüchtig

¹quarry ['kwɒrɪ] *n.* Steinbruch, *der*

²quarry *n. (prey)* Beute, *die*

quart [kwɔːt] *n.* Quart, *das*

quarter ['kwɔːtə(r)] **1.** *n.* **a)** Viertel, *das;* **a** *or* **one ~ of** ein Viertel (+ *Gen.*); **a ~ of a mile/an hour** eine Viertelmeile/-stunde; **b)** *(of year)* Quartal, *das;* Vierteljahr, *das;* **c)** **[a] ~ to/past six** Viertel vor/nach sechs; **d)** *(direction)* Richtung, *die;* **e)** *(area of town)* [Stadt]viertel, *das;* **f)** **~s** *pl. (lodgings)* Quartier, *das (bes. Milit.);* Unterkunft, *die;* **g)** *(Amer. coin)* Vierteldollar, *der.* **2.** *v. t.* **a)** *(divide)* vierteln; **b)** *(lodge)* einquartieren ⟨*Soldaten*⟩. **quarter-'final** *n.* Viertelfinale, *das.* **'quarterly 1.** *adj.* vierteljährlich. **2.** *n.* Vierteljahr[e]sschrift, *die*

quartet [kwɔː'tet] *n.* Quartett, *das*

quartz [kwɔːts] *n.* Quarz, *der*

quash [kwɒʃ] *v. t.* **a)** *(annul)* aufheben; **b)** *(suppress)* niederschlagen

quaver ['kweɪvə(r)] **1.** *n. (Brit. Mus.)* Achtelnote, *die.* **2.** *v. i. (vibrate)* zittern

quay [kiː] *n.* **'quayside** *ns.* Kai, *der*

queasy ['kwiːzɪ] *adj.* unwohl

queen [kwiːn] *n.* **a)** Königin, *die;* **b)** *(Chess, Cards)* Dame, *die.* **queen 'mother** *n.* Königinmutter, *die*

queer [kwɪə(r)] **1.** *adj.* **a)** *(strange)* sonderbar; *(eccentric)* verschroben; **b)** *(shady)* merkwürdig; **c)** *(out of sorts)* unwohl; **d)** *(sl. derog.: homosexual)* schwul *(ugs.).* **2.** *n. (sl. derog.: homosexual)* Schwule, *der (ugs.)*

quell [kwel] *v. t. (literary)* niederschlagen ⟨*Aufstand*⟩; zügeln ⟨*Furcht*⟩

quench [kwentʃ] *v. t.* löschen

query ['kwɪərɪ] **1.** *n.* Frage, *die.* **2.** *v. t.* in Frage stellen ⟨*Anweisung, Glaubwürdigkeit*⟩; beanstanden ⟨*Rechnung*⟩

quest [kwest] *n.* Suche, *die* **(for** nach)

question ['kwestʃn] **1.** *n.* **a)** Frage, *die;* **ask sb. a ~:** jmdm. eine Frage stellen; **b)** *(doubt, objection)* Zweifel, *der* **(about an** + *Dat.);* **there is no ~ about sth.** es besteht kein Zweifel an etw. *(Dat.);* **beyond all** *or* **without ~:** ohne Frage; **c)** *(problem, concern)* Frage, *die;* **sth./it is only a ~ of time** etw./ es ist [nur] eine Frage der Zeit; **it is [only] a ~ of doing sth.** es geht [nur] darum, etw. zu tun; **the person/thing in ~:** die fragliche Person/Sache; **sth./it is out of the ~:** etw./es ist ausgeschlossen. **2.** *v. t.* **a)** befragen; ⟨*Polizei, Gericht usw.:*⟩ vernehmen; **b)** *(throw doubt upon, raise objections to)* bezweifeln. **questionable** ['kwestʃənəbl] *adj.* fragwürdig. **'question mark** *n.* Fragezeichen, *das*

questionnaire [kwestʃə'neə(r)] *n.* Fragebogen, *der*

queue [kjuː] **1.** *n.* Schlange, *die.* **join the ~:** sich anstellen. **2.** *v. i.* **~ [up]** Schlange stehen

quibble ['kwɪbl] **1.** *n.* Spitzfindigkeit, *die.* **2.** *v. i.* streiten

quiche [kiːʃ] *n.* Quiche, *die*

quick [kwɪk] **1.** *adj.* schnell; kurz ⟨*Rede, Pause*⟩; flüchtig ⟨*Kuß, Blick*⟩; **be ~!** mach schnell! *(ugs.);* **be ~ to do sth.** etw. schnell tun; **a ~ temper** ein aufbrausendes Wesen. **2.** *adv.* schnell. **3.** *n.* empfindliches Fleisch; **be cut to the ~** *(fig.)* tief getroffen sein.

quicken ['kwɪkn] **1.** *v. t.* beschleunigen. **2.** *v. i.* sich beschleunigen. **'quickly** *adv.* schnell. **'quickness** *n.* **a)** *(speed)* Schnelligkeit, *die;* **b)** *(~ of perception)* Schärfe, *die*

quick: ~**sand** *n.* Treibsand, *der;*
~**-tempered** [~'tempəd] *adj.* hitzig;
be ~**-tempered** leicht aufbrausen;
~**-witted** *adj.* geistesgegenwärtig

quid [kwɪd] *n., pl. same (Brit. sl.)*
Pfund, *das*

quiet ['kwaɪət] **1.** *adj.,* ~**er** ['kwaɪə-
tə(r)], ~**est** ['kwaɪətɪst] **a)** *(silent)* still;
(not loud) leise; **keep** ~ **about** stn. *(fig.)*
etw. geheimhalten; **b)** *(peaceful, not
busy)* ruhig; **c)** *(not overt)* versteckt;
on the ~: still und heimlich. **2.** *n.* Ruhe, *die; (silence, stillness)* Stille, *die.*
quieten ['kwaɪətn] *v. t.* beruhigen.
quieten 'down *v. i.* sich beruhigen
'**quietly** *adv.* **a)** *(silently)* still; *(not
loudly)* leise; **b)** *(peacefully)* ruhig
'**quietness** *n. (absence of noise)* Stille,
die; (peacefulness) Ruhe, *die*

quill [kwɪl] *n. (feather)* Kielfeder, *die;
(of porcupine)* Stachel, *der*

quilt [kwɪlt] **1.** *n.* Schlafdecke, *die.* **2.**
v. t. wattieren

quince [kwɪns] *n.* Quitte, *die*

quintet [kwɪn'tet] *n.* Quintett, *das*

quip [kwɪp] **1.** *n.* Witzelei, *die.* **2.** *v. i.,*
-**pp**- witzeln (**at** über + *Akk.*)

quirk [kwɜːk] *n.* Marotte, *die;* **a** ~ **of
fate** eine Laune des Schicksals

quit [kwɪt] *v. t.,* -**tt**-, *(Amer.)* quit *(give
up)* aufgeben; *(stop)* aufhören mit; ~
doing stn. aufhören, etw. zu tun; **they
were given notice to** ~ |**the flat**| ihnen
wurde |die Wohnung| gekündigt

quite [kwaɪt] *adv.* **a)** *(entirely)* ganz;
völlig; fest ‹entschlossen›; ~ |**so!**| [ja.]
genau!; **b)** *(to some extent)* ziemlich;
ganz ‹gern›; ~ **a few** ziemlich viele

quits [kwɪts] *pred. adj.* be ~ |**with sb.**|
[mit jmdm.] quitt sein *(ugs.)*

¹**quiver** ['kwɪvə(r)] *v. i.* zittern (**with** vor
+ *Dat.*); ‹Stimme, Lippen:› beben
(geh.); ‹Lid:› zucken

²**quiver** *n. (for arrows)* Köcher, *der*

quiz [kwɪz] **1.** *n., pl.* ~**zes** Quiz, *das.* **2.**
v. t., -**zz**- ausfragen (**about** stn. nach
etw., **about** sb. über jmdn.). **quizzical**
['kwɪzɪkl] *adj.* fragend

quoit [kɔɪt] *n.* [Gummi]ring, *der*

quorum ['kwɔːrəm] *n.* Quorum, *das*

quota ['kwəʊtə] *n.* **a)** *(share)* Anteil,
der; **b)** *(goods to be produced)* Produk-
tionsmindestquote, *die;* **c)** *(maximum
number)* Höchstquote, *die*

quotation [kwəʊ'teɪʃn] *n.* **a)** Zitieren,
das; (passage) Zitat, *das;* **b)** *(estimate)*
Kosten[vor]anschlag, *der.* **quo'ta-
tion-marks** *n. pl.* Anführungszeichen *Pl.*

quote [kwəʊt] **1.** *v. t. also abs.* zitieren
(**from** aus); zitieren aus ‹*Buch, Text*›;
(mention) anführen; nennen ‹*Preis*›. **2.**
n. (coll.) **a)** *(passage)* Zitat, *das;* **b)** *(estimate)* Kosten[vor]anschlag, *der;* **c)**
usu. in pl. (quotation-mark) Anführungszeichen, *das*

R

R, r [ɑː(r)] *n.* R, r, *das*

R. *abbr.* River Fl.

rabbi ['ræbaɪ] *n.* Rabbi[ner], *der; (as
title)* Rabbi, *der*

rabbit ['ræbɪt] *n.* Kaninchen, *das*

rabbit: ~**-burrow** *n.* Kaninchenbau,
der; ~**-hutch** *n. (also fig.)* Kanin-
chenstall, *der;* ~**-warren** *n.* Kanin-
chengehege, *das; (fig.)* Labyrinth, *das*

rabble ['ræbl] *n.* Mob, *der*

rabid ['ræbɪd] *adj.* **a)** tollwütig; **b)** *(extreme)* fanatisch

rabies ['reɪbiːz] *n.* Tollwut, *die*

¹**race** [reɪs] **1.** *n.* Rennen, *das; (fig.)* **a** ~
against time ein Wettlauf mit der Zeit.
2. *v. i.* **a)** *(in swimming, running, etc.)*
um die Wette schwimmen/laufen
usw. (**with, against** mit); **b)** ‹*Motor:*›
durchdrehen; ‹*Puls:*› jagen; **c)** *(rush)*
sich sehr beeilen; ~ **after** sb. jmdm.
hinterherhetzen. **3.** *v. t.* um die Wette
schwimmen/laufen *usw.* mit

²**race** *n. (Anthrop., Biol.)* Rasse, *die;*
the human ~: die Menschheit

race: ~**-course** *n.* Rennbahn, *die;*
~**horse** *n.* Rennpferd, *das;* ~**-track**
n. Rennbahn, *die*

racial ['reɪʃl] *adj.* Rassen‹diskriminierung, -konflikt, -gleichheit›; rassisch
‹Gruppe, Minderheit›. **racialism** ['reɪ-
ʃəlɪzm] *n.* Rassismus, *der.* **racialist**
['reɪʃəlɪst] **1.** *n.* Rassist, *der*/Rassistin,
die. **2.** *adj.* rassistisch

racing ['reɪsɪŋ] *n.* Rennsport, *der; (with
horses)* Pferdesport, *der.* '**racing-car**
n. Rennwagen, *der.* '**racing driver**
n. Rennfahrer, *der*/-fahrerin, *die*

racism ['reɪsɪzm] *n.* Rassismus, *der.*

racist ['reɪsɪst] **1.** *n.* Rassist, *der*/Ras-
sistin, *die.* **2.** *adj.* rassistisch

rack [ræk] 1. *n.* *(for luggage)* Ablage, *die; (for toast, plates)* Ständer, *der; (on bicycle, motor cycle)* Gepäckträger, *der.* 2. *v. t.* ~ **one's brain|s|** *(fig.)* sich *(Dat.)* den Kopf zerbrechen *(ugs.)*

¹racket ['rækɪt] *n.* Schläger, *der*

²racket *n.* a) *(disturbance)* Lärm, *der;* Krach, *der;* b) *(scheme)* Schwindelgeschäft, *das (ugs.).* **racketeer** [rækɪ'tɪə(r)] *n.* Ganove, *der; (profiteer)* Wucherer, *der*

racoon [rə'ku:n] *n.* Waschbär, *der*

racy ['reɪsɪ] *adj.* flott *(ugs.)* ⟨*Stil*⟩

radar ['reɪda:(r)] *n.* Radar, *das od. der*

radiant ['reɪdɪənt] *adj.* strahlend; fröhlich ⟨*Stimmung*⟩; **be** ~: strahlen (**with** vor + *Dat.*)

radiate ['reɪdɪeɪt] 1. *v. i.* a) ⟨*Hitze, Wärme:*⟩ ausstrahlen; ⟨*Schein, Wellen:*⟩ ausgehen (**from** von); b) *(from central point)* strahlenförmig ausgehen (**from** von). 2. *v. t.* ausstrahlen ⟨*Licht, Wärme; Glück, Liebe*⟩; aussenden ⟨*Strahlen, Wellen*⟩. **radiation** [reɪdɪ'eɪʃn] *n. (of energy)* Emission, *die; (of signals)* Ausstrahlung, *die; (energy transmitted)* Strahlung, *die.* **radiator** ['reɪdɪeɪtə(r)] *n.* a) *(for heating)* Heizkörper, *der;* b) *(Motor Veh.)* Kühler, *der*

radical ['rædɪkl] *adj.* 1. a) *(thorough; also Polit.)* radikal; drastisch ⟨*Maßnahme*⟩; b) *(progressive)* radikal; c) *(fundamental)* grundlegend. 2. *n. (Polit.)* Radikale, *der*

radio ['reɪdɪəʊ] 1. *n., pl.* ~**s** a) *no indef. art.* Funk, *der; (for private communication)* Sprechfunk, *der;* b) *no indef. art. (Broadcasting)* Rundfunk, *der;* **on the** ~: im Radio; c) *(apparatus)* Radio, *das.* 2. *attrib. adj. (Broadcasting)* Rundfunk-; Radio⟨*welle, -teleskop*⟩; Funk⟨*mast, -turm, -taxi*⟩. 3. *v. t.* funken

radio: ~**'active** *adj.* radioaktiv; ~**ac'tivity** *n.* Radioaktivität, *die;* ~**con'trolled** *adj.* funkgesteuert

radish ['rædɪʃ] *n.* Rettich, *der; (small, red)* Radieschen, *das*

radius ['reɪdɪəs] *n., pl.* **radii** ['reɪdɪaɪ] *or* ~**es** *(Math.)* Radius, *der; (fig.)* Umkreis, *der*

RAF [a:reɪ'ef, *(coll.)* ræf] *abbr.* **Royal Air Force**

raffle ['ræfl] 1. *n.* Tombola, *die;* ~ **ticket** Los, *das.* 2. *v. t.* ~ |**off**| verlosen

raft [ra:ft] *n.* Floß, *das*

rafter ['ra:ftə(r)] *n.* Sparren, *der*

¹rag [ræg] *n.* a) [Stoff]fetzen, *der;* b) *(old and torn clothes)* Lumpen *Pl.;* c) *(derog.: newspaper)* Käseblatt, *das (salopp)*

²rag *v. t.,* -**gg**- *(tease)* aufziehen

rag: ~**bag** *n. (fig.)* Sammelsurium, *das;* ~ **doll** *n.* Stoffpuppe, *die*

rage [reɪdʒ] 1. *n.* a) *(violent anger)* Wut, *die; (fit of anger)* Wutausbruch, *der;* b) **sth. is |all| the** ~: etw. ist [ganz] groß in Mode. 2. *v. i.* a) *(rave)* toben; ~ **at** *or* **against sth./sb.** gegen etw./jmdn. wüten; b) *(be violent, unchecked)* toben; ⟨*Krankheit:*⟩ wüten

ragged ['rægɪd] *adj.* zerrissen

raid [reɪd] 1. *n.* Einfall, *der;* Überfall, *der; (Mil.)* Überraschungsangriff, *der; (by police)* Razzia, *die* (**on** in + *Dat.*). 2. *v. t. (Polizei:)* eine Razzia machen auf (+ *Akk.*); ⟨*Räuber, Soldaten:*⟩ überfallen. **raider** *n.* Räuber, *der/* Räuberin, *die*

rail [reɪl] *n.* a) Stange, *die; (on ship)* Reling, *die; (as protection against contact)* Barriere, *die;* b) *(Railw.: of track)* Schiene, *die;* c) *(~way)* [Eisen]bahn, *die; attrib.* Bahn-; **by** ~: mit der Bahn

railing ['reɪlɪŋ] *n. (round park)* Zaun, *der; (on staircase)* Geländer, *das*

'railroad *(Amer.),* **'railway** *ns.* a) *(track)* Bahnlinie, *die;* Bahnstrecke, *die;* b) *(system)* [Eisen]bahn, *die*

railway: ~ **carriage** *n.* Eisenbahnwagen, *der;* ~ **engine** *n.* Lokomotive, *die;* ~ **line** *n.* [Eisen]bahnlinie, *die;* ~ **station** *n.* Bahnhof, *der*

rain [reɪn] 1. *n.* a) Regen, *der;* b) *(fig.: of arrows, blows, etc.)* Hagel, *der.* 2. *v. i. impers.* **it is** ~**ing** es regnet. 3. *v. t.* hageln lassen ⟨*Schläge, Hiebe*⟩

rain: ~**bow** ['reɪnbəʊ] *n.* Regenbogen, *der;* ~**check** *n. (Amer. fig.)* **take a** ~**check on sth.** auf etw. *(Akk.)* später wieder zurückkommen; ~**coat** *n.* Regenmantel, *der;* ~**fall** *n.* Niederschlag, *der;* ~**proof** *adj.* regendicht; ~**water** *n.* Regenwasser, *das*

'rainy *adj.* regnerisch ⟨*Tag, Wetter*⟩; regenreich ⟨*Gebiet, Sommer*⟩; ~ **season** Regenzeit, *die;* **keep sth. for a** ~ **day** *(fig.)* sich *(Dat.)* etw. für schlechte Zeiten aufheben

raise [reɪz] *v. t.* a) *(lift up)* heben; erhöhen ⟨*Temperatur, Miete, Gehalt*⟩; hochziehen ⟨*Fahne*⟩; aufziehen ⟨*Vorhang*⟩; hochheben ⟨*Arm*⟩; ~ **one's glass to sb.** das Glas auf jmdn. erheben; b) *(set upright)* aufrichten; erheben ⟨*Banner*⟩; ~ **sb.'s spirits** jmds. Stimmung heben; c) erheben ⟨*Forde-*

rungen, Einwände⟩; aufwerfen ⟨*Frage*⟩; zur Sprache bringen ⟨*Thema, Problem*⟩; **d)** aufziehen ⟨*Vieh, [Haus]tiere*⟩; großziehen ⟨*Familie, Kinder*⟩; **e)** aufbringen ⟨*Geld, Betrag*⟩; **f)** aufheben ⟨*Belagerung, Blockade, Embargo, Verbot*⟩

raisin ['reɪzn] *n.* Rosine, *die*

rake [reɪk] **1.** *n.* Rechen, *der;* Harke, *die.* **2.** *v.t.* **a)** harken; **b)** ~ **the fire** die Asche entfernen; **c)** *(with eyes, shots)* bestreichen. **rake in** *v.t. (coll.)* scheffeln *(ugs.).* **rake up** *v.t.* zusammenharken; *(fig.)* wieder ausgraben

'rake-off *n. (coll.)* [Gewinn]anteil, *der*

rakish ['reɪkɪʃ] *adj.* flott; keß

rally ['rælɪ] **1.** *v.i. (regain health)* sich wieder [ein wenig] erholen. **2.** *v.t.* **a)** *(reassemble)* wieder zusammenrufen; **b)** einigen ⟨*Partei, Kräfte*⟩; sammeln ⟨*Anhänger*⟩. **3.** *n.* **a)** *(mass meeting)* Versammlung, *die;* **b)** |**motor**| ~: Rallye, *die;* **c)** *(Tennis)* Ballwechsel, *der*

ram [ræm] **1.** *n. (Zool.)* Schafbock, *der;* Widder, *der.* **2.** *v.t.,* **-mm-:** **a)** *(force)* stopfen; ~ **a post into the ground** einen Pfosten in die Erde rammen; ~ **sth. home to sb.** jmdm. etw. deutlich vor Augen führen; **b)** *(collide with)* rammen

ramble ['ræmbl] **1.** *n.* |**nature**| ~: Wanderung, *die.* **2.** *v.i.* **a)** *(walk)* umherstreifen **(through, in** in + *Dat.*); **b)** *(in talk)* zusammenhangloses Zeug reden; **keep rambling on about sth.** sich endlos über etw. *(Akk.)* auslassen.

rambler ['ræmblə(r)] *n.* Wanderer, *der*/Wanderin, *die.* **rambling** ['ræmblɪŋ] **1.** *n.* Wandern, *das.* **2.** *adj.* **a)** *(irregularly arranged)* verschachtelt; verwinkelt ⟨*Straßen*⟩; **b)** *(incoherent)* unzusammenhängend ⟨*Erklärung*⟩; **c)** ~ **rose** Kletterrose, *die*

ramp [ræmp] *n.* Rampe, *die*

rampage 1. ['ræmpeɪdʒ] *n.* Randale, *die (ugs.);* **be/go on the** ~ *(coll.)* randalieren. **2.** [ræm'peɪdʒ] *v.i.* randalieren

rampant ['ræmpənt] *adj.* zügellos ⟨*Gewalt, Rassismus*⟩; steil ansteigend ⟨*Inflation*⟩; üppig ⟨*Wachstum*⟩

rampart ['ræmpɑːt] *n.* Wehrgang, *der*

'ramshackle *adj.* klapprig ⟨*Auto*⟩; verkommen ⟨*Gebäude*⟩

ran *see* **run** 2, 3

ranch [rɑːntʃ] *n.* Ranch, *die*

rancid ['rænsɪd] *adj.* ranzig

rancour *(Brit.; Amer.:* **rancor)** ['ræŋkə(r)] *n.* (tiefe) Verbitterung

random ['rændəm] **1.** *n.* **at** ~: wahllos;

willkürlich; *(aimlessly)* ziellos; **choose at** ~: aufs Geratewohl wählen. **2.** *adj.* willkürlich

randy ['rændɪ] *adj.* geil; scharf *(ugs.)*

rang *see* ²**ring** 2, 3

range [reɪndʒ] **1.** *n.* **a)** ~ **of mountains** Bergkette, *die;* **b)** *(of subjects)* Palette, *die; (of knowledge, voice)* Umfang, *der;* **c)** *(of missile etc.)* Reichweite, *die;* **at a** ~ **of 200 metres** auf eine Entfernung von 200 Metern; **d)** *(series, selection)* Kollektion, *die;* **e)** *(stove)* Herd, *der.* **2.** *v.i.* ⟨*Preise, Temperaturen:*⟩ schwanken, sich bewegen **(from ... to** zwischen [+ *Dat.*] ... und)

'ranger *n.* Förster, *der*/Försterin, *die*

¹**rank** [ræŋk] **1.** *n.* **a)** *(position in hierarchy)* Rang, *der; (Mil. also)* Dienstgrad, *der;* **b)** *(social position)* [soziale] Stellung; **c)** *(row)* Reihe, *die;* **the** ~ **and file** *(fig.)* die breite Masse; **the** ~**s** *(enlisted men)* die Mannschaften und Unteroffiziere. **2.** *v.t.* ~ **among** zählen zu. **3.** *v.i.* ~ **among** zählen zu

²**rank** *adj.* **a)** kraß ⟨*Außenseiter*⟩; **b)** ~ **weeds** [wild]wucherndes Unkraut

ransack ['rænsæk] *v.t.* **a)** *(search)* durchsuchen **(for** nach); **b)** *(pillage)* plündern

ransom ['rænsəm] *n.* ~ |**money**| Lösegeld, *das;* **hold to** ~: als Geisel festhalten

rant [rænt] *v.i.* ~ |**and rave**| wettern *(ugs.)* **(about** über + *Akk.*)

rap [ræp] **1.** *n.* [energisches] Klopfen. **2.** *v.t.,* **-pp-** klopfen. **3.** *v.i.,* **-pp-** klopfen **(on** an + *Akk.*)

¹**rape** [reɪp] **1.** *n.* Vergewaltigung, *die.* **2.** *v.t.* vergewaltigen

²**rape** *n. (Bot., Agric.)* Raps, *der*

rapid ['ræpɪd] **1.** *adj.* schnell ⟨*Bewegung, Wachstum, Puls*⟩; rasch ⟨*Fortschritt, Ausbreitung*⟩. **2.** *n.* **in pl.** Stromschnellen. **rapidity** [rə'pɪdɪtɪ] *n.* Schnelligkeit, *die.* **'rapidly** *adv.* schnell

rapist ['reɪpɪst] *n.* Vergewaltiger, *der*

rapport [ra'pɔː(r)] *n.* [harmonisches] Verhältnis

rapt [ræpt] *adj.* gespannt ⟨*Miene*⟩

rapture ['ræptʃə(r)] *n.* |**state of**| ~: Verzückung, *die.* **rapturous** ['ræptʃərəs] *adj.* begeistert

¹**rare** [reə(r)] *adj.,* **'rarely** *adv.* selten

²**rare** *adj. (Cookery)* englisch gebraten

rarity ['reərɪtɪ] *n.* Seltenheit, *die*

¹**rash** [ræʃ] *n.* [Haut]ausschlag, *der*

²**rash** *adj.* voreilig ⟨*Urteil, Entscheidung*⟩; überstürzt ⟨*Versprechung*⟩

rasher ['ræʃə(r)] n. Speckscheibe, die
'**rashly** adv. voreilig
rasp [rɑ:sp] n. (tool) Raspel, die
raspberry ['rɑ:zbərɪ] n. Himbeere, die
rat [ræt] n. a) Ratte, die; smell a ~ (fig.) Lunte riechen (ugs.); b) (coll. derog.. person) Ratte, die (derb)
rate [reɪt] 1. n. a) (proportion) Rate, die; b) (tariff) Satz, der; ~ |of pay| Lohnsatz, der; c) (speed) Geschwindigkeit, die; Tempo, das; d) (Brit.: levy) |local or council| ~s Gemeindeabgaben; e) (coll.) at any ~ (at least) zumindest; wenigstens; (whatever happens) auf jeden Fall; at this ~ we won't get any work done so kriegen wir gar nichts fertig (ugs.). 2. v. t. a) einschätzen (Intelligenz, Leistung); b) (consider) betrachten; rechnen (among zu). 3. v. i. ~ as gelten als
rather ['rɑ:ðə(r)] adv. a) (by preference) lieber; b) (somewhat) ziemlich; I ~ think that ...: ich bin ziemlich sicher, daß ...; c) (more truly) vielmehr; or ~: beziehungsweise
ratification [rætɪfɪ'keɪʃn] n. Ratifizierung, die
ratify ['rætɪfaɪ] v. t. ratifizieren
rating ['reɪtɪŋ] n. a) (estimated standing) Einschätzung, die; b) (Radio, Telev.) |popularity| ~: Einschaltquote, die; c) (Brit. Navy) Matrose, der
ratio ['reɪʃɪəʊ] n., pl. ~s Verhältnis, das
ration ['ræʃn] 1. n. ~|s| Ration, die (of an + Dat.). 2. v. t. rationieren (Benzin, Zucker usw.)
rational ['ræʃənl] adj. (having reason) rational (Wesen); (sensible) vernünftig (Person, Art usw.)
rationalize ['ræʃənəlaɪz] v. t. rationalisieren
'**rat race** n. erbarmungsloser Konkurrenzkampf
rattle ['rætl] 1. v. i. a) (Fenster:) klappern; (Flaschen:) klirren; (Kette:) rasseln; b) (Zug, Bus:) rattern. 2. v. t. a) klappern mit (Würfel, Geschirr); klirren lassen (Fenster[scheiben]); rasseln mit (Kette); b) (sl.: disconcert) ~ sb., get sb. ~d jmdn. durcheinanderbringen. 3. n. a) (of baby) Rassel, die; b) (sound) Klappern, das. **rattle 'off** v. t. (coll.) herunterrasseln (ugs.)
'**rattlesnake** n. Klapperschlange, die
raucous ['rɔ:kəs] adj. rauh
ravage ['rævɪdʒ] 1. v. t. heimsuchen (Gebiet, Stadt). 2. n. in pl. verheerende Wirkung
rave [reɪv] 1. v. i. a) (talk wildly) irrere-

den; b) (speak admiringly) schwärmen (about von). 2. attrib. adj. (coll.) begeistert (Kritik)
raven ['reɪvn] n. Rabe, der
ravenous ['rævənəs] adj. I'm ~: ich habe einen Bärenhunger (ugs.)
ravine [rə'vi:n] n. Schlucht, die
raving ['reɪvɪŋ] 1. adj. irreredend (Idiot). 2. adv. be ~ mad völlig verrückt sein (ugs.)
ravish ['rævɪʃ] v. t. (charm) entzücken.
'**ravishing** adj. bildschön (Anblick, Person); hinreißend (Schönheit)
raw [rɔ:] adj. a) (uncooked) roh; b) (inexperienced) unerfahren; c) (stripped of skin) blutig (Fleisch); offen (Wunde); d) (chilly) naßkalt. **raw ma'terial** n. Rohstoff, der
ray [reɪ] n. Strahl, der; ~ of sunshine/light Sonnen-/Lichtstrahl, der
raze [reɪz] v. t. ~ to the ground dem Erdboden gleichmachen
razor ['reɪzə(r)] n. Rasiermesser, das; |electric| ~: |elektrischer| Rasierapparat. '**razor-blade** n. Rasierklinge, die
RC abbr. **Roman Catholic** r.-k.; röm.-kath.
Rd. abbr. **road** Str.
RE abbr. (Brit.) **Religious Education** Religionslehre, die
re [ri:] prep. (Commerc.) betreffs
reach [ri:tʃ] 1. v. t. a) (arrive at) erreichen; ankommen in (+ Dat.) (Stadt, Land); erzielen (Übereinstimmung); kommen zu (Entscheidung; Ausgang, Eingang); you can ~ her at this number du kannst sie unter dieser Nummer erreichen; b) (extend to) (Straße:) führen bis zu; (Leiter, Haar:) reichen bis zu. 2. v. i. a) (stretch out hand) ~ for sth. nach etw. greifen; ~ across the table über den Tisch langen; b) (be long/tall enough) sth. will/won't ~: etw. ist/ist nicht lang genug; I can't ~: ich komme nicht daran; c) (go as far as) (Wasser, Gebäude, Besitz:) reichen (|up| to bis [hinauf] zu). 3. n. Reichweite, die; be within easy ~: leicht erreichbar sein; be out of ~: nicht erreichbar sein. **reach 'out** v. i. Hand ausstrecken (for nach)
react [rɪ'ækt] v. i. reagieren (to auf + Akk.). **reaction** [rɪ'ækʃn] n. Reaktion, die (to auf + Akk.)
reactionary [rɪ'ækʃənərɪ] (Polit.) 1. adj. reaktionär. 2. n. Reaktionär, der/Reaktionärin, die
reactor [rɪ'æktə(r)] n. |nuclear| ~: Kernreaktor, der

read [ri:d] **1.** *v.t.*, read [red] **a)** lesen; ~
sb. sth., ~ sth. to sb. jmdm. etwas vor-
lesen; ~ **the gas meter** das Gas able-
sen; **b)** *(interpret)* deuten; ~ **between
the lines** zwischen den Zeilen lesen; **c)**
(study) studieren. **2.** *v.i.*, **read a)** lesen; ~ **to sb.** jmdm. vorlesen; **b)** *(convey meaning)* lauten; **the contract** ~s
as follows der Vertrag hat folgenden
Wortlaut. **read 'out** *v.t.* laut vorlesen. **read 'over, read 'through** *v.t.*
durchlesen. **read 'up** *v.t.* sich informieren (**on** über + *Akk.*)
readable ['ri:dəbl] *adj.* **a)** *(pleasant to
read)* lesenswert; **b)** *(legible)* leserlich
'**reader** *n.* **a)** Leser, *der*/Leserin, *die;*
b) *(book)* Lesebuch, *das*
'**readership** *n.* Leserschaft, *die*
readily ['redɪlɪ] *adv.* **a)** *(willingly)* bereitwillig; **b)** *(easily)* ohne weiteres
readiness ['redɪnɪs] *n.* Bereitschaft,
die; **be in** ~: bereit sein (**for** für)
'**reading** *n.* **a)** Lesen, *das;* **b)** *(figure
shown)* Anzeige, *die;* **c)** *(recital)* Lesung, *die* (**from** aus); **d)** *(Parl.)* Lesung, *die.* '**reading-lamp, 'reading-
light** *ns.* Leselampe, *die.* '**reading-
matter** *n.* Lesestoff, *der;* Lektüre, *die*
ready ['redɪ] **1.** *adj.* **a)** *(prepared)*
fertig; **be** ~ **to do sth.** bereit sein, etw.
zu tun; **get** ~: sich fertigmachen. **b)**
(willing) bereit; **c)** *(within reach)* griff-
bereit. **2.** *adv.* fertig. **3.** *n.* **at the** ~
⟨*Schußwaffe*⟩ im Anschlag
ready: ~ '**cash** *see* ~ **money;**
~-'**made** *adj.* **a)** Konfektions⟨anzug,
-kleidung⟩; **b)** *(fig.)* vorgefertigt; ~
'**money** *n.* Bargeld, *das*
real [rɪəl] *adj.* **a)** *(actually existing)*
real ⟨*Ereignis, Lebewesen*⟩; wirklich
⟨*Macht*⟩; **b)** *(genuine)* echt ⟨*Interesse,
Gold, Seide*⟩; **c)** *(complete)* total *(ugs.)*
⟨*Desaster, Enttäuschung*⟩; **d)** *(true)*
wahr ⟨*Grund, Name, Glück*⟩; echt
⟨*Mitleid, Sieg*⟩; **the** ~ **thing** der/die/
das Echte; **e) be for** ~ *(sl.)* echt sein.
'**real estate** *n.* Immobilien *Pl.*
realism ['rɪəlɪzm] *n.* Realismus, *der*
'**realist** *n.* Realist, *der*/Realistin, *die*
realistic [rɪə'lɪstɪk] *adj.* realistisch
reality [rɪ'ælɪtɪ] *n.* Realität, *die*
realization [rɪəlaɪ'zeɪʃn] *n.* Erkenntnis, *die*
realize ['rɪəlaɪz] *v.t.* **a)** *(be aware of)*
bemerken; erkennen ⟨*Fehler*⟩; **I didn't**
~ *(abs.)* ich habe es nicht gewußt; ~
[**that**]...: merken, daß...; **b)** *(make happen)* verwirklichen; **c)** erbringen
⟨*Summe, Preis*⟩

really ['rɪəlɪ] *adv.* wirklich; **not** ~: ei-
gentlich nicht; |**well,**| ~! [also] so was!
realm [relm] *n.* Königreich, *das*
realtor ['ri:əltə(r)] *(Amer.)* Grundstücksmakler, *der*
reap [ri:p] *v.t.* *(cut)* schneiden ⟨*Getreide*⟩; *(gather in)* einfahren ⟨*Getreide, Ernte*⟩
reappear [ri:ə'pɪə(r)] *v.i.* wieder auftauchen; *(come back)* [wieder] zurückkommen
¹**rear** [rɪə(r)] **1.** *n.* **a)** *(back part)* hinterer
Teil; **b)** *(back)* Rückseite, *die;* **c)** *(Mil.)*
Rücken, *der.* **2.** *adj.* hinter...; ~ **axle**
Hinterachse, *die*
²**rear 1.** *v.t.* großziehen ⟨*Kind,
Familie*⟩; halten ⟨*Vieh*⟩. **2.** *v.i.* ⟨*Pferd:*⟩
sich aufbäumen
rear: ~**guard** *n.* *(Mil.)* Nachhut, *die;*
~-**light** *n.* Rücklicht, *das*
rearm [ri:'ɑ:m] *v.i. & t.* wiederaufrü-
sten
rearrange [ri:ə'reɪndʒ] *v.t.* umräumen
⟨*Möbel*⟩; verlegen ⟨*Spiel*⟩ (**for** auf +
Akk.); ändern ⟨*Programm*⟩
rear-view 'mirror *n.* Rückspiegel,
der
reason ['ri:zn] **1.** *n.* **a)** *(cause)* Grund,
der; **have no** ~ **to complain** sich nicht
beklagen können; **for that** |**very**| ~:
aus [eben] diesem Grund; **b)** *(power to
understand; sense)* Vernunft, *die;*
(power to think) Verstand, *der;* **in** *or*
within ~: innerhalb eines vernünfti-
gen Rahmens. **2.** *v.i.* **a)** schlußfolgern
(**from** aus); **b)** ~ **with** diskutieren mit
(**about, on** über + *Akk.*); **you can't** ~
with her mit ihr kann man nicht vernünftig reden. **3.** *v.t.* schlußfolgern
reasonable ['ri:zənəbl] *adj.* **a)** vernünftig; **b)** *(inexpensive)* günstig.
reasonably ['ri:zənəblɪ] *adv.* **a)**
(within reason) vernünftig; **b)** *(fairly)*
ganz ⟨*gut*⟩; ziemlich ⟨*gesund*⟩
reassurance [ri:ə'ʃʊərəns] *n.* **a)**
(calming) **give sb.** ~: jmdn. beruhigen;
b) *(confirmation)* Bestätigung, *die*
reassure [ri:ə'ʃʊə(r)] *v.t.* beruhigen;
~ **sb. about his health.** jmdm. versichern, daß er gesund ist. **reassuring**
[ri:ə'ʃʊərɪŋ] *adj.* beruhigend
rebate ['ri:beɪt] *n.* **a)** *(refund)* Rückzahlung, *die;* **b)** *(discount)* Preisnach-
laß, *der* (**on** auf + *Akk.*)
rebel 1. ['rebl] *n.* Rebell, *der*/Rebellin,
die. **2.** *attrib. adj.* Rebellen-. **3.** [rɪ'bel]
v.i., -**ll**- rebellieren. **rebellion** [rɪ'bel-
jən] *n.* Rebellion, *die.* **rebellious** [rɪ-
'beljəs] *adj.* rebellisch

rebound 1. [rɪ'baʊnd] *v. i.* **a)** *(spring back)* abprallen (**from** von); **b)** *(fig.)* zurückfallen (**upon** auf + *Akk.*). **2.** ['ri:baʊnd] *n.* Abprall, *der*

rebuff [rɪ'bʌf] **1.** *n.* [schroffe] Abweisung. **2.** *v. t.* [schroff] zurückweisen

rebuild [ri:'bɪld] *v. t.,* **rebuilt** [ri:'bɪlt] wieder aufbauen

rebuke [rɪ'bju:k] **1.** *v. t.* tadeln, rügen (**for** wegen). **2.** *n.* Rüge, *die*

recall 1. [rɪ'kɔ:l] *v. t.* **a)** *(remember)* sich erinnern an (+ *Akk.*); **b)** *(serve as reminder of)* erinnern an (+ *Akk.*); **c)** abberufen ⟨*Botschafter*⟩. **2.** [rɪ'kɔ:l, 'ri:kɔ:l] *n.* **a)** |**powers of**| ~: Gedächtnis, *das;* **b) beyond** ~: unwiderruflich

recant [rɪ'kænt] *v. i.* [öffentlich] widerrufen

recap ['ri:kæp] *v. t. & i.,* **-pp-** *(coll.)* rekapitulieren

recapitulate [ri:kə'pɪtjʊleɪt] *v. t. & i.* rekapitulieren

recapture [ri:'kæptʃə(r)] *v. t.* wieder ergreifen ⟨*Gefangenen*⟩; wieder einfangen ⟨*Tier*⟩

recede [ri:'si:d] *v. i.* ⟨*Hochwasser, Flut:*⟩ zurückgehen; ~ |**into the distance**| in der Ferne verschwinden. **receding** [rɪ'si:dɪŋ] *adj.* fliehend ⟨*Kinn, Stirn*⟩

receipt [rɪ'si:t] *n.* **a)** *(receiving)* Empfang, *der;* **b)** *(written acknowledgement)* Quittung, *die;* **c)** *in pl. (amount received)* Einnahmen (**from** aus)

receive [rɪ'si:v] *v. t.* **a)** *(get)* erhalten; beziehen ⟨*Gehalt, Rente*⟩; **b)** *(accept)* entgegennehmen ⟨*Bukett, Lieferung*⟩; **c)** *(entertain)* empfangen ⟨*Gast*⟩. **re'ceiver** *n.* **a)** Empfänger, *der*/Empfängerin, *die;* **b)** *(Teleph.)* [Telefon]hörer, *der;* **c)** *(of stolen goods)* Hehler, *der*/Hehlerin, *die*

recent ['ri:sənt] *adj.* jüngst ⟨*Ereignisse, Vergangenheit usw.*⟩; **the** ~ **closure of the factory** die kürzlich erfolgte Schließung der Fabrik. **'recently** *adv. (a short time ago)* vor kurzem; *(in the recent past)* in der letzten Zeit

receptacle [rɪ'septəkl] *n.* Behälter, *der;* Gefäß, *das*

reception [rɪ'sepʃn] *n.* **a)** *(welcome)* Aufnahme, *die;* **b)** *(party)* Empfang, *der;* **c)** *(Brit.: foyer)* die Rezeption. **re'ceptionist** *n. (in hotel)* Empfangschef, *der*/-dame, *die; (at doctor's)* Sprechstundenhilfe, *die.* **re'ception desk** *n.* Rezeption, *die*

receptive [rɪ'septɪv] *adj.* aufgeschlossen, empfänglich (**to** für)

recess [rɪ'ses, 'ri:ses] *n.* **a)** *(alcove)* Ni-

sche, *die;* **b)** *(Brit. Parl.; Amer.: short vacation)* Ferien *Pl.; (Amer. Sch.; between classes)* Pause, *die*

recharge [ri:'tʃɑ:dʒ] *v. t.* aufladen ⟨*Batterie*⟩

recipe ['resɪpɪ] *n.* Rezept, *das*

recipient [rɪ'sɪpɪənt] *n.* Empfänger, *der*/Empfängerin, *die*

reciprocal [rɪ'sɪprəkl] *adj.* gegenseitig ⟨*Abkommen, Zuneigung*⟩

reciprocate [rɪ'sɪprəkeɪt] *v. t.* erwidern

recital [rɪ'saɪtl] *n. (performance)* [Solisten]konzert, *das; (of literature also)* Rezitation, *die*

recitation [resɪ'teɪʃn] *n.* Rezitation, *die*

recite [rɪ'saɪt] *v. t.* **a)** rezitieren ⟨*Gedicht*⟩; **b)** *(list)* aufzählen

reckless ['reklɪs] *adj.* unbesonnen; rücksichtslos ⟨*Fahrweise*⟩; ~ **of the dangers/consequences** ungeachtet der Gefahren/Folgen

reckon ['rekn] *v. t.* **a)** *(work out)* ausrechnen ⟨*Kosten*⟩; bestimmen ⟨*Position*⟩; **b)** *(consider)* halten (**as** für); *(estimate)* schätzen. '**reckon on** *v. t.* **a)** *(rely on)* zählen auf (+ *Akk.*); **b)** *(expect)* rechnen mit. '**reckon with** *v. i.* rechnen mit

reckoning *n.* Berechnung, *die;* **by my** ~: nach meiner Rechnung

reclaim [rɪ'kleɪm] *v. t.* **a)** zurückbekommen ⟨*Steuern*⟩; **b)** urbar machen ⟨*Land*⟩

recline [rɪ'klaɪn] *v. i.* liegen; **reclining seat** Liegesitz, *der*

recluse [rɪ'klu:s] *n.* Einsiedler, *der*/Einsiedlerin, *die*

recognition [rekəg'nɪʃn] *n.* **a)** Wiedererkennen, *das;* **be beyond all** ~: nicht wiederzuerkennen sein; **b)** *(acknowledgement)* Anerkennung, *die;* **in** ~ **of** als Anerkennung für

recognize ['rekəgnaɪz] *v. t.* **a)** *(know again)* wiedererkennen (**by** an + *Dat.,* **from** durch); **b)** *(acknowledge)* erkennen; anerkennen ⟨*Gültigkeit, Land*⟩; **be** ~**d as** gelten als

recoil 1. [rɪ'kɔɪl] *v. i.* zurückfahren. **2.** ['ri:kɔɪl, rɪ'kɔɪl] *n.* Rückstoß, *der*

recollect [rekə'lekt] **1.** *v. t.* sich erinnern an (+ *Akk.*). **2.** *v. i.* sich erinnern. **recollection** [rekə'lekʃn] *n.* Erinnerung, *die*

recommend [rekə'mend] *v. t.* empfehlen. **recommendation** [rekəmen'deɪʃn] *n.* Empfehlung, *die;* **on sb.'s** ~: auf jmds. Empfehlung *(Akk.)*

recompense ['rekəmpens] 1. *v. t.* entschädigen. 2. *n.* Entschädigung, *die*
reconcile ['rekənsaıl] *v. t.* a) *(restore to friendship)* versöhnen; b) ~ oneself to sth. sich mit etw. versöhnen
reconnaissance [rı'kɒnısəns] *n. (Mil.)* Aufklärung, *die*
reconnoitre *(Brit.; Amer.:* **reconnoiter)** [rekə'nɔıtə(r)] *v. i.* auf Erkundung [aus]gehen
reconsider [ri:kən'sıdə(r)] *v. t.* [noch einmal] überdenken
reconstruct [ri:kən'strʌkt] *v. t.* wieder aufbauen; *(fig.)* rekonstruieren. **reconstruction** [ri:kən'strʌkʃn] *n.* Wiederaufbau, *der; (thing reconstructed)* Rekonstruktion, *die*
record 1. [rı'kɔ:d] *v. t.* a) aufzeichnen; ~ a new LP eine neue LP aufnehmen; b) *(register officially)* dokumentieren; protokollieren ⟨Verhandlung⟩. 2. ['rekɔ:d] *n.* a) be on ~ ⟨Prozeß, Verhandlung, Besprechung:⟩ protokolliert sein; **have sth. on** ~: etw. dokumentiert haben; b) *(report)* Protokoll, *das;* c) *(document)* Dokument, *das;* |strictly| off the ~: [ganz] inoffiziell; d) *(for ~-player)* [Schall]platte, *die;* e) **have a |criminal/police|** ~: vorbestraft sein; f) *(Sport)* Rekord, *der*
recorded [rı'kɔ:dıd] *adj.* aufgezeichnet ⟨Konzert, Rede⟩; ~ music Musikaufnahmen. **recorded de'livery** *n. (Brit. Post.)* eingeschriebene Sendung *(ohne Versicherung)*
recorder [rı'kɔ:də(r)] *n. (Mus.)* Blockflöte, *die*
recording [rı'kɔ:dıŋ] *n.* a) *(process)* Aufzeichnung, *die;* b) *(what is recorded)* Aufnahme, *die.* **re'cording studio** *n.* Tonstudio, *das*
record ['rekɔ:d]: ~-player *n.* Plattenspieler, *der;* ~ token *n.* [Schall]plattengutschein, *der*
re-count 1. [ri:'kaunt] *v. t.* [noch einmal] nachzählen. 2. ['ri:kaunt] *n.* Nachzählung, *die*
recoup [rı'ku:p] *v. t.* [wieder] hereinbekommen ⟨[Geld]einsatz⟩
recourse [rı'kɔ:s] *n.* have ~ to sb./sth. bei jmdm./zu etw. Zuflucht nehmen
recover [rı'kʌvə(r)] 1. *v. t.* zurückbekommen. 2. *v. i.* ~ from sth. sich von etw. [wieder] erholen; be |fully| ~ed *(völlig)* wiederhergestellt sein. **recovery** [rı'kʌvərı] *n.* Erholung, *die;* **make a quick/good** ~: sich schnell/gut erholen
recreation [rekrı'eıʃn] *n.* Freizeitbe-

schäftigung, *die;* Hobby, *das.* **recreational** [rekrı'eıʃənl] *adj.* Freizeit-
recruit [rı'kru:t] 1. *n.* a) *(Mil.)* Rekrut, *der;* b) *(new member)* neues Mitglied. 2. *v. t. (Mil.: enlist)* anwerben; *(into party etc.)* werben ⟨Mitglied⟩; einstellen ⟨neuen Mitarbeiter⟩. **re'cruitment** *n. (Mil.)* Anwerbung, *die; (of new staff)* Neueinstellung, *die;* ~ of members Mitgliederwerbung, *die*
rectangle ['rektæŋgl] *n.* Rechteck, *das.* **rectangular** [rek'tæŋgjulə(r)] *adj.* rechteckig
rector ['rektə(r)] *n.* a) Pfarrer, *der;* b) *(Univ.)* Rektor, *der*/Rektorin, *die.* **rectory** ['rektərı] *n.* Pfarrhaus, *das*
recuperate [rı'kju:pəreıt] *v. i.* sich erholen. **recuperation** [rıkju:pə'reıʃn] *n.* Erholung, *die*
recur [rı'kɜ:(r)] *v. i.,* -rr- sich wiederholen; ⟨Krankheit:⟩ wiederkehren; ⟨Symptom:⟩ wieder auftreten. **recurrence** [rı'kʌrəns] *n.* Wiederholung, *die; (of illness, thought, feeling)* Wiederkehr, *die; (of symptom)* Wiederauftreten, *das.* **recurrent** [rı'kʌrənt] *adj.* immer wiederkehrend
recycle [ri:'saıkl] *v. t.* wiederverwerten. **recycling** [ri:'saıklıŋ] *n.* Recycling, *das*
red [red] 1. *adj.* rot. 2. *n.* Rot, *das.* **Red 'Cross** *n.* Rotes Kreuz. **red'currant** *n.* [rote] Johannisbeere
redden ['redn] *v. i.* ⟨Gesicht, Himmel:⟩ sich röten; ⟨Person:⟩ rot werden
reddish ['redıʃ] *adj.* rötlich
redecorate [ri:'dekəreıt] *v. t.* renovieren; *(with wallpaper)* neu tapezieren; *(with paint)* neu streichen
redeem [rı'di:m] *v. t.* a) [wieder] einlösen ⟨Pfand⟩; einlösen ⟨Gutschein, Coupon⟩; b) *(save)* retten. **redemption** [rı'dempʃn] *n. (from sin)* Erlösung, *die*
redeploy [ri:dı'plɔı] *v. t.* woanders einsetzen ⟨Arbeitskräfte⟩
red: ~-'handed *adj.* catch sb. ~-handed jmdn. auf frischer Tat ertappen; ~ 'herring *n. (fig.)* Ablenkungsmanöver, *das;* ~-hot *adj.* [rot]glühend; **Red 'Indian** *(Brit.)* 1. *n.* Indianer, *der*/Indianerin, *die;* 2. *adj.* Indianer-
redirect [ri:daı'rekt] *v. t.* nachsenden ⟨Post, Brief usw.⟩; umleiten ⟨Verkehr⟩
rediscover [ri:dı'skʌvə(r)] *v. t.* wiederentdecken
red: ~-'letter day *n.* großer Tag; ~ 'light *n.* rotes Warnlicht; *(traffic-*

light) rote Ampel; **drive through a ~ light** bei rot über die Ampel fahren; **~-'light district** *n.* Strich, *der (salopp)*

redo [ri:'du:] *v. t. forms as* do noch einmal machen ⟨*Bett, Hausaufgabe*⟩; neu frisieren ⟨*Haare*⟩

redouble [ri:'dʌbl] *v. t.* verdoppeln

redress [rɪ'dres] **1.** *n.* Entschädigung, *die.* **2.** *v. t.* wiedergutmachen; **~ the balance** das Gleichgewicht wiederherstellen

red 'tape *n. (fig.)* [unnötige] Bürokratie

reduce [rɪ'dju:s] *v. t.* **a)** senken ⟨*Preis, Gebühr, Fieber, Aufwendungen, Blutdruck usw.*⟩; reduzieren ⟨*Geschwindigkeit, Gewicht*⟩; **at ~d prices** zu herabgesetzten Preisen; **b)** ~ **to silence/tears** verstummen lassen/zum Weinen bringen. **reduction** [rɪ'dʌkʃn] *n. (in price, costs, speed, etc.)* Senkung, *die* (**in** *Gen.*); ~ **in wages/weight** Lohnsenkung, *die*/Gewichtsabnahme, *die*

redundancy [rɪ'dʌndənsɪ] *n. (Brit.)* Arbeitslosigkeit, *die;* **redundancies** Entlassungen

redundant [rɪ'dʌndənt] *adj. (Brit.)* arbeitslos; **be made ~:** den Arbeitsplatz verlieren; **make ~:** entlassen

red 'wine *n.* Rotwein, *der*

reed [ri:d] *n.* Schilf[rohr], *das*

reef [ri:f] *n.* Riff, *das*

'reef-knot *n.* Kreuzknoten, *der*

reek [ri:k] *v. i.* stinken (of nach)

reel [ri:l] **1.** *n.* ⟨*Garn-, Angel*⟩rolle, *die;* ⟨*Film-, Tonband*⟩spule, *die.* **2.** *v. i.* **a)** *(be in a whirl)* sich drehen; **b)** *(sway)* torkeln

refectory [rɪ'fektərɪ] *n.* Mensa, *die*

refer [rɪ'fɜ:(r)] **1.** *v. i.,* **-rr-:** **a)** ~ **to** *(allude to)* sich beziehen auf (+ *Akk.*) ⟨*Buch, Person usw.*⟩; *(speak of)* sprechen von ⟨*Person, Problem usw.*⟩; **b)** ~ **to** *(apply to, relate to)* betreffen; **c)** ~ **to** *(consult, cite as proof)* nachsehen in (+ *Dat.*). **2.** *v. t.,* **-rr-:** ~ **sb./sth. to sb./ sth.** jmdn./etw. an jmdn./auf etw. *(Akk.)* verweisen

referee [refə'ri:] *(Sport)* **1.** *n. (umpire)* Schiedsrichter, *der*/-richterin, *die; (Boxing)* Ringrichter, *der.* **2.** *v. t.* als Schiedsrichter/-richterin leiten

reference ['refrəns] *n.* **a)** *(allusion)* Hinweis, *der* (**to** auf + *Akk.*); **make no ~ to sth.** etw. nicht ansprechen; **b)** *(testimonial)* Zeugnis, *das*

referendum [refə'rendəm] *n.* Volksentscheid, *der*

refill 1. [ri:'fɪl] *v. t.* nachfüllen; ~ **the glasses** nachschenken. **2.** ['ri:fɪl] *n. (for ball-pen)* Ersatzmine, *die*

refine [rɪ'faɪn] *v. t.* **a)** *(purify)* raffinieren; **b)** *(make cultured)* kultivieren; **c)** *(improve)* verbessern; verfeinern ⟨*Stil, Technik*⟩. **refined** [rɪ'faɪnd] *adj.* kultiviert. **re'finement** *n.* Kultiviertheit, *die; (improvement)* Verbesserung, *die.*

refinery [rɪ'faɪnərɪ] *n.* Raffinerie, *die*

reflect [rɪ'flekt] *v. t.* **a)** reflektieren; **b)** *(fig.)* widerspiegeln ⟨*Ansichten*⟩; **c)** *(contemplate)* nachdenken über (+ *Akk.*); ~ **what/how ...:** überlegen, was/wie ... **re'flect [up]on** *(consider)* nachdenken über (+ *Akk.*); **b)** ~ **badly [up]on sb./sth.** auf jmdn./ etw. ein schlechtes Licht werfen. **re'flection** [rɪ'flekʃn] *n.* **a)** Reflexion, *die; (by surface of water)* Spiegelung, *die;* **b)** *(image)* Spiegelbild, *das;* **c)** *(consideration)* Nachdenken, *das* (**upon** über + *Akk.*); **on ~:** bei weiterem Nachdenken. **reflective** [rɪ'flektɪv] *adj.* **a)** reflektierend; **b)** *(thoughtful)* nachdenklich. **reflector** [rɪ'flektə(r)] *n.* Rückstrahler, *der*

reflex ['ri:fleks] **1.** *n.* Reflex, *der.* **2.** *adj.* ~ **action** Reflexhandlung, *die*

reflexive [rɪ'fleksɪv] *adj. (Ling.)* reflexiv

reform [rɪ'fɔ:m] **1.** *v. t. (make better)* bessern ⟨*Person*⟩; reformieren ⟨*Institution*⟩. **2.** *n.* Reform, *die* (**in** *Gen.*). **reformation** [refə'meɪʃn] *n. (of character)* Wandlung, *die;* **the R~** *(Hist.)* die Reformation. **re'former** *n.* **[political]** ~**:** Reformpolitiker, *der*/Reformpolitikerin, *die*

refract [rɪ'frækt] *v. t. (Phys.)* brechen

¹refrain [rɪ'freɪn] *n.* Refrain, *der*

²refrain *v. i.* ~ **from doing sth.** es unterlassen, etw. zu tun

refresh [rɪ'freʃ] *v. t.* erfrischen. **re'freshing** *adj.* erfrischend; wohltuend ⟨*Abwechslung*⟩. **re'freshment** *n.* Erfrischung, *die*

refrigerate [rɪ'frɪdʒəreɪt] *v. t.* **a)** kühl lagern ⟨*Lebensmittel*⟩; **b)** *(chill)* kühlen; *(freeze)* einfrieren. **refrigeration** [rɪfrɪdʒə'reɪʃn] *n.* kühle Lagerung; *(chilling)* Kühlung, *die; (freezing)* Einfrieren, *das.* **refrigerator** [rɪ'frɪdʒəreɪtə(r)] *n.* Kühlschrank, *der*

refuel [ri:'fju:əl], *(Brit.)* **-ll-: 1.** *v. t.* auftanken. **2.** *v. i.* [auf]tanken

refuge ['refju:dʒ] *n.* Zuflucht, *die;* **take ~ in** Schutz *od.* Zuflucht suchen in (+ *Dat.*) (**from** vor + *Dat.*)

refugee [refjʊ'dʒiː] *n.* Flüchtling, *der*
refund 1. [riː'fʌnd] *v.t. (pay back)* zurückzahlen ⟨*Geld*⟩; erstatten ⟨*Kosten*⟩. **2.** ['riːfʌnd] *n.* Rückzahlung, *die; (of expenses)* [Rück]erstattung, *die*
refusal [rɪ'fjuːzl] *n.* Ablehnung, *die; (after a period of time)* Absage, *die;* ~ **to do sth.** Weigerung, etw. zu tun
¹refuse [rɪ'fjuːz] **1.** *v.t.* ablehnen; verweigern ⟨*Zutritt, Einreise, Erlaubnis*⟩; ~ **sb. admittance/entry/permission** jmdm. den Zutritt/die Einreise/die Erlaubnis verweigern; ~ **to do sth.** sich weigern, etw. zu tun. **2.** *v.i.* ablehnen; *(after request)* sich weigern
²refuse ['refjuːs] *n.* Abfall, *der*
refuse ['refjuːs]: ~ **collection** *n.* Müllabfuhr, *die;* ~ **collector** *n.* Müllwerker, *der;* ~ **disposal** *n.* Abfallbeseitigung, *die*
refute [rɪ'fjuːt] *v.t.* widerlegen
regain [rɪ'geɪn] *v.t.* zurückgewinnen ⟨*Zuversicht, Vertrauen, Augenlicht*⟩; ~ **one's strength** wieder zu Kräften kommen
regal ['riːgl] *adj.* majestätisch
regalia [rɪ'geɪlɪə] *n. pl. (of royalty)* Krönungsinsignien
regard [rɪ'gɑːd] **1.** *v.t.* **a)** *(look at)* betrachten; **b)** *(give heed to)* beachten; **c)** *(fig.: look upon, contemplate)* betrachten; ~ **sb. as a friend/fool/genius** jmdn. als Freund betrachten/für einen Dummkopf/ein Genie halten; **be** ~**ed as** gelten als; **d)** *(concern, have relation to)* betreffen; **as** ~**s sb./sth.,** ~**ing sb./sth.** was jmdn./etw. angeht *od.* betrifft. **2.** *n.* **a)** *(attention)* **pay** *or* **have** ~ **to sb./sth.** jmdm./etw. Beachtung schenken; **without** ~ **to** ohne Rücksicht auf (+ *Akk.*); **b)** *(esteem)* Achtung, *die;* **hold sb./sth. in high** ~: jmdn./etw. sehr schätzen; **c)** *in pl.* Grüße; **give her my** ~**s** grüße sie von mir; **with kind[est]** ~**s** mit herzlich[st]en Grüßen. **re'gardless** *adj.* ohne Rücksicht (**of** auf + *Akk.*)
regatta [rɪ'gætə] *n.* Regatta, *die*
regenerate [rɪ'dʒenəreɪt] *v.t.* erneuern
regime, régime [reɪ'ʒiːm] *n.* [Regierungs]system, *das*
regiment ['redʒɪmənt, 'redʒmənt] *n.* Regiment, *das.* **regimental** [redʒɪ'mentl] *adj.* Regiments-
region ['riːdʒn] *n.* **a)** *(area)* Gebiet, *das;* **b)** *(administrative division)* Bezirk, *der;* **in the** ~ **of** *(fig.)* ungefähr.
regional ['riːdʒnl] *adj.* regional

register ['redʒɪstə(r)] **1.** *n.* Register, *das; (at school)* Klassenbuch, *das.* **2.** *v.t.* **a)** *(enter)* registrieren; *(cause to be entered)* registrieren lassen; anmelden ⟨*Auto, Patent*⟩; *(at airport)* einchecken ⟨*Gepäck*⟩; *abs. (at hotel)* sich ins Fremdenbuch eintragen; ~ **with the police** sich polizeilich anmelden; **b)** *(enrol)* anmelden; *(Univ.)* sich einschreiben; **c)** zum Ausdruck bringen ⟨*Überraschung*⟩; ~ **a protest** Protest anmelden. **registered** ['redʒɪstəd] *adj.* eingetragen ⟨*Firma*⟩; eingeschrieben ⟨*Student, Brief*⟩; ~ **trade mark** eingetragenes Warenzeichen; **by** ~ **post** per Einschreiben
registrar ['redʒɪstrɑː(r), redʒɪ'strɑː(r)] *n.* Standesbeamte, *der/*-beamtin, *die*
registration [redʒɪ'streɪʃn] *n.* Registrierung, *die; (enrolment)* Anmeldung, *die; (of students)* Einschreibung, *die.* **regi'stration document** *n.* *(Brit.)* Kraftfahrzeugbrief, *der.* **regi'stration number** *n.* amtliches Kennzeichen
registry ['redʒɪstrɪ] *n.* ~ **[office]** Standesamt, *das*
regret [rɪ'gret] **1.** *v.t.,* -tt- bedauern; **I** ~ **to say that** ...: ich muß leider sagen, daß ... **2.** *n.* Bedauern, *das;* **have no** ~**s** nichts bereuen. **regretfully** [rɪ'gretfəlɪ] *adv.* mit Bedauern. **regrettable** [rɪ'gretəbl] *adj.* bedauerlich. **regrettably** [rɪ'gretəblɪ] *adv.* bedauerlicherweise
regroup [riː'gruːp] **1.** *v.t.* umgruppieren. **2.** *v.i.* **a)** *(form new group)* sich neu gruppieren; **b)** *(Mil.)* sich neu formieren
regular ['regjʊlə(r)] **1.** *adj.* regelmäßig; geregelt ⟨*Arbeit*⟩; fest ⟨*Anstellung*⟩; ~ **customer** Stammkunde, *der/*-kundin, *die;* ~ **army** reguläre Armee. **2.** *n. (coll.:* ~ *customer)* Stammkunde, *der/*-kundin, *die; (in pub)* Stammgast, *der.* **regularity** [regjʊ'lærɪtɪ] *n.* Regelmäßigkeit, *die.* **'regularly** *adv.* regelmäßig
regulate ['regjʊleɪt] *v.t. (control)* regeln; *(restrict)* begrenzen; *(adjust)* regulieren. **regulation** [regjʊ'leɪʃn] *n.* **a)** *see* **regulate:** Regelung, *die;* Begrenzung, *die;* Regulierung, *die;* **b)** *(rule)* Vorschrift, *die*
rehabilitate [riːhə'bɪlɪteɪt] *v.t.* rehabilitieren; ~ **[back into society]** wieder [in die Gesellschaft] eingliedern
rehash 1. [riː'hæʃ] *v.t.* aufwärmen. **2.** ['riːhæʃ] *n.* Aufguß, *der*

rehearsal [rɪ'hɜːsl] *n.* Probe, *die.* **rehearse** [rɪ'hɜːs] *v. t.* proben

reign [reɪn] 1. *n.* Herrschaft, *die.* 2. *v. i.* herrschen (over über + *Akk.*)

rein [reɪn] *n.* Zügel, *der*

reincarnation [riːɪŋkɑːˈneɪʃn] *n.* *(Relig.)* Reinkarnation, *die*

reindeer ['reɪndɪə(r)] *n., pl. same* Ren[tier], *das*

reinforce [riːɪnˈfɔːs] *v. t.* verstärken; ~d concrete Stahlbeton, *der.* rein'forcement *n.* Verstärkung, *die;* ~[s] *(additional men etc.)* Verstärkung, *die*

reinstate [riːɪn'steɪt] *v. t. (in job)* wieder einstellen

reinvigorate [riːɪnˈvɪɡəreɪt] *v. t.* neu beleben; feel ~d sich gestärkt fühlen

reiterate [riːˈɪtəreɪt] *v. t.* wiederholen

reject 1. [rɪˈdʒekt] *v. t.* ablehnen; zurückweisen ⟨*Bitte, Annäherungsversuch*⟩. 2. ['riːdʒekt] *(thing)* Ausschuß, *der.* rejection [rɪˈdʒekʃn] *n.* Ablehnung, *die*/Zurückweisung, *die*

rejoice [rɪˈdʒɔɪs] *v. i.* sich freuen (over, at über + *Akk.*)

¹rejoin [rɪˈdʒɔɪn] *v. t. (reply)* erwidern (to auf + *Akk.*)

²rejoin [rɪ:ˈdʒɔɪn] *v. t.* wieder eintreten in (+ *Akk.*) ⟨*Partei, Verein*⟩

rejoinder [rɪˈdʒɔɪndə(r)] *n.* Erwiderung, *die* (to auf + *Akk.*)

rejuvenate [rɪˈdʒuːvəneɪt] *v. t.* verjüngen

rekindle [riːˈkɪndl] *v. t.* wieder anfachen; wieder aufleben lassen ⟨*Verlangen, Hoffnungen*⟩

relapse [rɪ'læps] 1. *v. i.* ⟨*Kranker:*⟩ einen Rückfall bekommen. 2. *n.* Rückfall, *der*

relate [rɪ'leɪt] 1. *v. t.* a) erzählen ⟨*Geschichte*⟩; erzählen von ⟨*Abenteuer*⟩; b) *(bring into relation)* in Zusammenhang bringen (to, with mit). 2. *v. i.* a) ~ to *(have reference)* in Zusammenhang stehen mit; betreffen ⟨*Person*⟩; b) ~ to *(feel involved with)* eine Beziehung haben zu. re'lated *adj.* verwandt (to mit). relation [rɪ'leɪʃn] *n.* a) *(connection)* Beziehung, *die,* Zusammenhang, *der* (of ... and zwischen ... und); in *or* with ~ to in bezug auf (+ *Akk.*); b) *in pl. (dealings)* Verhältnis, *das* (with zu); c) *(relative)* Verwandte, *der/die.* re'lationship *n.* a) *(mutual tie)* Beziehung, *die* (with zu); b) *(kinship)* Verwandtschaftsverhältnis, *das;* c) *(connection)* Beziehung, *die; (between cause and effect)* Zusammenhang, *der;* d) *(sexual)* Verhältnis, *das*

relative ['relətɪv] 1. *n.* Verwandte, *der/die.* 2. *adj.* relativ. 'relatively *adv.* relativ; verhältnismäßig. relative 'pronoun *n. (Ling.)* Relativpronomen, *das*

relax [rɪ'læks] 1. *v. t.* a) entspannen ⟨*Muskel, Körper[teil]*⟩; lockern ⟨*Griff*⟩; b) *(make less strict)* lockern ⟨*Gesetz, Disziplin*⟩. 2. *v. i.* sich entspannen. relaxation [riːlækˈseɪʃn] *n.* Entspannung, *die;* for ~: zur Entspannung. relaxed [rɪˈlækst] *adj.* entspannt, gelöst ⟨*Atmosphäre, Person*⟩. re'laxing *adj.* entspannend

relay 1. ['riːleɪ] *n.* a) *(race)* Staffel, *die;* b) *(gang)* Schicht, *die;* work in ~s schichtweise arbeiten; c) *(Electr.)* Relais, *das.* 2. [riːˈleɪ] *v. t.* a) weiterleiten; b) *(Radio, Telev.)* übertragen. 'relay race *n.* Staffellauf, *der; (Swimming)* Staffelschwimmen, *das*

release [rɪ'liːs] 1. *v. t.* a) *(free)* freilassen ⟨*Tier, Häftling, Sklaven*⟩; *(from jail)* entlassen (from aus); b) *(let go)* loslassen; lösen ⟨*Handbremse*⟩; c) *(make known)* veröffentlichen ⟨*Erklärung, Nachricht*⟩; *(issue)* herausbringen ⟨*Film, Schallplatte*⟩. 2. *n.* a) *see* 1 a: Freilassung, *die;* Entlassung, *die;* b) *(of published item)* Veröffentlichung, *die;* c) *(handle, lever, button)* Auslöser, *der*

relegate ['relɪɡeɪt] *v. t.* a) ~ sb. to the position of ...: jmdn. zu ... degradieren; b) *(Sport)* absteigen lassen; be ~d absteigen (to in + *Akk.*). relegation [relɪ'ɡeɪʃn] *n. (Sport)* Abstieg, *der*

relent [rɪ'lent] *v. i.* nachgeben. re'lentless *adj.,* re'lentlessly *adv.* unerbittlich

relevance ['relɪvəns] *n.* Relevanz, *die* (to für)

relevant ['relɪvənt] *adj.* relevant (to für); wichtig ⟨*Information*⟩

reliability [rɪlaɪə'bɪlɪtɪ] *n.* Zuverlässigkeit, *die*

reliable [rɪ'laɪəbl] *adj.,* reliably [rɪ'laɪəblɪ] *adv.* zuverlässig

reliance [rɪ'laɪəns] *n.* Abhängigkeit, *die* (on von)

reliant [rɪ'laɪənt] *adj.* be ~ on sb./sth. auf jmdn./etw. angewiesen sein

¹relief [rɪ'liːf] *n.* a) Erleichterung, *die;* give [sb.] ~ [from pain] [jmdm.] [Schmerz]linderung verschaffen; what a ~!, that's a ~! da bin ich aber erleichtert!; b) *(assistance)* Hilfe, *die*

²relief *n. (Art)* Relief, *das*

relief: ~ bus *n.* Entlastungsbus, *der;*

(as replacement) Ersatzbus, *der;* ~
map *n.* Reliefkarte, *die;* ~ **road** *n.*
Entlastungsstraße, *die*

relieve [rɪ'liːv] *v. t.* **a)** erleichtern; unterbrechen ⟨*Eintönigkeit*⟩; abbauen
⟨*Anspannung*⟩; stillen ⟨*Schmerzen*⟩; **I
am** *or* **feel** ~**d to hear that** ...: es erleichtert mich zu hören, daß ...; **b)** ablösen ⟨*Wache, Truppen*⟩

religion [rɪ'lɪdʒn] *n.* Religion, *die*
religious [rɪ'lɪdʒəs] *adj.* religiös; Re-
ligions⟨*freiheit, -unterricht*⟩. **re'li-
giously** *adv. (conscientiously)* gewissenhaft

relinquish [rɪ'lɪŋkwɪʃ] *v. t.* **a)** *(give up)*
aufgeben; **b)** ~ **one's hold** *or* **grip on**
sb./sth. jmdn./etw. loslassen

relish ['relɪʃ] **1.** *n.* **a)** *(liking)* Vorliebe,
die; **do sth. with ⟨great⟩** ~: etw. mit
[großem] Genuß tun; **b)** *(condiment)*
Relish, *das.* **2.** *v. t.* genießen

relive [riː'lɪv] *n.* noch einmal durchle-
ben

reload [riː'ləʊd]   *v. t.*   nachladen
⟨*Schußwaffe*⟩

reluctance [rɪ'lʌktəns] *n.* Widerwille,
der; **have a ⟨great⟩** ~ **to do sth.** etw. nur
mit Widerwillen tun

reluctant [rɪ'lʌktənt] *adj.* unwillig; **be**
~ **to do sth.** etw. nur ungern tun. **re-
'luctantly** *adv.* nur ungern

rely [rɪ'laɪ] *v. i. (have trust)* sich verlas-
sen/*(be dependent)* angewiesen sein
(up⟨on + *Akk.*⟩

remain [rɪ'meɪn] *v. i.* **a)** *(be left over)*
übrigbleiben; **b)** *(stay)* bleiben; ~ **behind** noch dableiben; **c)** *(continue to
be)* bleiben; **it** ~**s to be seen** es wird
sich zeigen. **remainder** [rɪ'meɪndə(r)]
n. Rest, *der.* **re'maining** *adj.* restlich.
re'mains *n. pl.* **a)** Reste; **b)** *(human)*
sterbliche [Über]reste *(verhüll.)*

remand [rɪ'mɑːnd] **1.** *v. t.* ~ **sb. [in cus-
tody]** jmdn. in Untersuchungshaft behalten. **2.** *n.* **on** ~: in Untersuchungshaft

remark [rɪ'mɑːk] **1.** *v. t.* bemerken **(to**
gegenüber). **2.** *v. i.* eine Bemerkung
machen **(up⟨on** zu, über + *Akk.*⟩. **3.** *n.*
Bemerkung, *die* **(on** über + *Akk.*⟩

remarkable [rɪ'mɑːkəbl] *adj.* **a)** *(notable)* bemerkenswert; **b)** *(extraordinary)* außergewöhnlich. **remarkably**
[rɪ'mɑːkəblɪ] *adv.* **a)** *(notably)* bemerkenswert; **b)** *(exceptionally)* außergewöhnlich

remarry [riː'mærɪ] *v. i. & t.* wieder heiraten

remedy ['remɪdɪ] **1.** *n.* [Heil]mittel,

das **(for** gegen). **2.** *v. t.* beheben ⟨*Problem*⟩; retten ⟨*Situation*⟩

remember [rɪ'membə(r)] *v. t.* **a)** sich
erinnern an (+ *Akk.*); **I** ~**ed to bring
the book** ich habe daran gedacht, das
Buch mitzubringen; **an evening to** ~:
ein unvergeßlicher Abend; **b)** *(convey
greetings)* ~ **me to them** grüße sie von
mir. **remembrance** [rɪ'membrəns] *n.*
Gedenken, *das;* **in** ~ **of sb.** zu jmds.
Gedächtnis

remind [rɪ'maɪnd] *v. t.* erinnern **(of an**
+ *Akk.*); ~ **sb. to do sth.** jmdn. daran
erinnern, etw. zu tun; **that** ~**s me, ...:**
dabei fällt mir ein, ... **re'minder** *n.*
Erinnerung, *die* **(of an** + *Akk.*); *(letter)* Mahnung, *die;* Mahnbrief, *der*

reminisce [remɪ'nɪs] *v. i.* sich in Erinnerungen *(Dat.)* ergehen **(about an** +
Akk.). **reminiscences** [remɪ'nɪsən-
sɪz] *n. pl.* Erinnerungen; *(memoirs)*
[Lebens]erinnerungen *Pl.* **reminis-
cent** [remɪ'nɪsənt] *adj.* **be** ~ **of sth.** an
etw. *(Akk.)* erinnern

remiss [rɪ'mɪs] *adj.* nachlässig **(of** von)
remission [rɪ'mɪʃn] *n.* **a)** *(of debt, punishment)* Erlaß, *der;* **b)** *(of prison sentence)* Straferlaß, *der*

remit [rɪ'mɪt] *v. t.,* -tt- *(send)* überweisen ⟨*Geld*⟩. **remittance** [rɪ'mɪtəns] *n.*
Überweisung, *die*

remnant ['remnənt] *n.* Rest, *der*
remonstrate ['remənstreɪt] *v. i.* protestieren **(against** gegen); ~ **with sb.**
jmdm.   Vorenthaltungen   machen
(about, on wegen)

remorse [rɪ'mɔːs] *n.* Reue, *die* **(for,
about** über + *Akk.*). **re'morseful** [rɪ-
'mɔːsfl] *adj.* reumütig. **re'morseless**
adj. unerbittlich

remote [rɪ'məʊt] *adj.,* ~**r** [rɪ'məʊtə(r)],
~**st** [rɪ'məʊtɪst] **a)** fern ⟨*Vergangenheit,
Zukunft, Zeit*⟩; abgelegen ⟨*Ort, Gebiet*⟩; ~ **from** weit entfernt von; **b)**
(slight) gering ⟨*Chance*⟩. **remote
con'trol** *n. (of vehicle)* Fernlenkung,
die; (for TV set) Fernbedienung, *die.*
remote-con'trol[led] *adj.* fernge-
lenkt; fernbedient ⟨*Anlage*⟩
re'motely *adv.* entfernt ⟨*verwandt*⟩;
they are not ~ **alike** sie haben nicht die
entfernteste Ähnlichkeit

removable [rɪ'muːvəbl] *adj.* abnehmbar; entfernbar ⟨*Trennwand*⟩; heraus-
nehmbar ⟨*Futter*⟩

removal [rɪ'muːvl] *n.* **a)** Entfernung,
die; (of obstacle, problem) Beseitigung,
die; **b)** *(transfer of furniture)* Umzug,
der

removal: ~ **firm** n. Spedition, die; ~ **man** n. Möbelpacker, der; ~ **van** n. Möbelwagen, der

remove [rɪ'muːv] v. t. entfernen; beseitigen ⟨Spur, Hindernis⟩; (take off) abnehmen; ausziehen ⟨Kleidungsstück⟩; ~ **a book from the shelf** ein Buch vom Regal nehmen. **re'mover** n. **a)** (of paint/varnish/hair/rust) Farb- / Lack- / Haar- / Rostentferner, der; **b)** (man) Möbelpacker, der; [firm of] ~**s** Spedition[sfirma], die

remunerate [rɪ'mjuːnəreit] v. t. bezahlen. **remuneration** [rɪmjuːnə'reɪʃn] n. Bezahlung, die

Renaissance [rə'neɪsəns, rɪ'neɪsəns] n. (Hist.) Renaissance, die

rename [riː'neɪm] v. t. umbenennen

render ['rendə(r)] v. t. **a)** (make) machen; **b)** erweisen ⟨Dienst⟩; **c)** (translate) übersetzen (by mit). **'rendering** n. (translation) Übersetzung, die

rendezvous ['rɒndeɪvuː] n., pl. same ['rɒndeɪvuːz] **a)** (meeting-place) Treffpunkt, der; **b)** (meeting) Verabredung, die

renegade ['renɪgeɪd] **1.** n. Abtrünnige, der/die. **2.** adj. abtrünnig

renew [rɪ'njuː] v. t. erneuern; fortsetzen ⟨Angriff, Bemühungen⟩; (extend) erneuern ⟨Vertrag, Ausweis usw.⟩; ~ **a library book** ⟨Bibliothekar/Benutzer:⟩ ein Buch [aus der Bücherei] verlängern/verlängern lassen. **renewal** [rɪ'njuːəl] n. Erneuerung, die

renounce [rɪ'naʊns] v. t. verzichten auf (+ Akk.); verstoßen ⟨Person⟩; ~ **the devil/one's faith** dem Teufel/seinem Glauben abschwören

renovate ['renəveɪt] v. t. renovieren ⟨Gebäude⟩; restaurieren ⟨Möbel usw.⟩. **renovation** [renə'veɪʃn] n. Renovierung, die/Restaurierung, die

renown [rɪ'naʊn] n. Renommee, das. **renowned** [rɪ'naʊnd] adj. berühmt (for wegen, für)

rent [rent] **1.** n. (for house etc.) Miete, die; (for land) Pacht, die. **2.** v. t. **a)** (use) mieten ⟨Haus, Wohnung usw.⟩; pachten ⟨Land⟩; mieten ⟨Auto⟩; **b)** (let) vermieten ⟨Haus, Auto usw.⟩ (to Dat., an + Akk.); verpachten ⟨Land⟩ (to Dat., an + Akk.). **rent 'out** v. t. see **rent 2 b**

rental ['rentl] n. Miete, die

renunciation [rɪnʌnsɪ'eɪʃn] n. see **renounce:** Verzicht, der/Verstoßung, die

reopen [riː'əʊpn] **1.** v. t. wieder öffnen; wieder aufmachen; wiedereröffnen ⟨Geschäft, Lokal usw.⟩; wiederaufnehmen ⟨Diskussion, Verhandlung⟩. **2.** v. i. ⟨Geschäft, Lokal usw.:⟩ wieder öffnen

reorder [riː'ɔːdə(r)] v. t. **a)** (Commerc.) nachbestellen ⟨Ware⟩; **b)** (rearrange) umordnen

reorganization [riːɔːgənaɪ'zeɪʃn] n. Umorganisation, die; (of time, work) Neueinteilung, die

reorganize [riː'ɔːgənaɪz] v. t. umorganisieren; neu einteilen ⟨Zeit, Arbeit⟩

rep [rep] n. (coll.: representative) Vertreter, der/Vertreterin, die

repaid see **repay**

repair [rɪ'peə(r)] **1.** v. t. (mend) reparieren; ausbessern ⟨Kleidung, Straße⟩. **2.** n. Reparatur, die; **be in good/bad** ~: in gutem/schlechtem Zustand sein. **re'pair man** n. Mechaniker, der; (in house) Handwerker, der. **re'pair shop** n. Reparaturwerkstatt, die

repatriate [riː'pætrɪeɪt] v. t. repatriieren. **repatriation** [riːpætrɪ'eɪʃn] n. Repatriierung, die

repay [riː'peɪ] v. t., **repaid** [riː'peɪd] zurückzahlen ⟨Schulden usw.⟩; erwidern ⟨Besuch, Gruß, Freundlichkeit⟩; ~ **sb. for sth.** jmdm. etw. vergelten. **re'payment** n. Rückzahlung, die

repeal [rɪ'piːl] **1.** v. t. aufheben ⟨Gesetz, Erlaß usw.⟩. **2.** n. Aufhebung, die

repeat [rɪ'piːt] **1.** n. Wiederholung, die. **2.** v. t. wiederholen; **please** ~ **after me:** ...: sprich/sprecht/sprechen Sie mir bitte nach: ... **re'peated** adj. wiederholt; (several) mehrere; **make** ~ **efforts to** ...: wiederholt od. mehrfach versuchen, ...zu... **re'peatedly** adv. mehrmals

repel [rɪ'pel] v. t., **-ll-: a)** (drive back) abwehren; **b)** (be repulsive to) abstoßen. **repellent** [rɪ'pelənt] adj. abstoßend

repent [rɪ'pent] v. i. bereuen (of Akk.). **repentance** [rɪ'pentəns] n. Reue, die. **repentant** [rɪ'pentənt] adj. reuig

repercussion [riːpə'kʌʃn] n., usu. in pl. Auswirkung, die (|up|on auf + Akk.)

repertoire ['repətwɑː(r)] n. Repertoire, das

repertory ['repətərɪ] n. (Theatre) Repertoiretheater, das. **'repertory company** n. Repertoiretheater, das

repetition [repɪ'tɪʃn] n. Wiederholung, die

repetitious [repɪ'tɪʃəs] adj. sich immer wiederholend attr.

repetitive [rɪ'petɪtɪv] *adj.* eintönig

rephrase [riː'freɪz] *v. t.* umformulieren; I'll ~ **that** ich will es anders ausdrücken

replace [rɪ'pleɪs] *v. t.* **a)** *(vertically)* zurückstellen; *(horizontally)* zurücklegen; **b)** *(take place of)* ersetzen; ~ A **with** *or* **by** B A durch B ersetzen; **c)** *(exchange)* austauschen, auswechseln ⟨Maschinen[teile] *usw.*⟩. **re'placement** *n.* **a)** *see* **replace a:** Zurückstellen, *das;* Zurücklegen, *das;* **b)** *(provision of substitute for)* Ersatz, *der; attrib.* Ersatz-; **c)** *(substitute)* Ersatz, *der;* ~ |part| Ersatzteil, *das*

replay 1. [riː'pleɪ] *v. t.* wiederholen ⟨Spiel⟩; nochmals abspielen ⟨Tonband *usw.*⟩. **2.** ['riː'pleɪ] *n.* Wiederholung, *die;* *(match)* Wiederholungsspiel, *das*

replenish [rɪ'plenɪʃ] *v. t.* auffüllen

replica ['replɪkə] *n.* Nachbildung, *die*

reply [rɪ'plaɪ] **1.** *v. i.* ~ |to **sb./sth.**| [jmdm./auf etw. *(Akk.)*] antworten. **2.** *v. t.* ~ **that** ...: antworten, daß ... **3.** *n.* Antwort, *die* (to auf + *Akk.*)

report [rɪ'pɔːt] **1.** *v. t.* **a)** *(relate)* berichten/*(in writing)* einen Bericht schreiben über (+ *Akk.*); *(state formally also)* melden; **b)** *(name to authorities)* melden (to *Dat.*); *(for prosecution)* anzeigen (to bei). **2.** *v. i.* **a)** Bericht erstatten (on über + *Akk.*); berichten (on über + *Akk.*); **b)** *(present oneself)* sich melden (to bei). **3.** *n.* **a)** *(account)* Bericht, *der* (on, about über + *Akk.*); **b)** *(Sch.)* Zeugnis, *das;* **c)** *(of gun)* Knall, *der.* **reportedly** [rɪ'pɔːtɪdlɪ] *adv.* wie verlautet. **reported speech** *n.* indirekte Rede. **re'porter** *n.* Reporter, *der*/Reporterin, *die*

repossess [riːpə'zes] *v. t.* wieder in Besitz nehmen

reprehensible [reprɪ'hensɪbl] *adj.* tadelnswert

represent [reprɪ'zent] *v. t.* **a)** darstellen (as als); **b)** *(act for)* vertreten. **representation** [reprɪzen'teɪʃn] *n.* **a)** *(depicting, image)* Darstellung, *die;* **b)** *(acting for sb.)* Vertretung, *die;* **c)** **make ~s to sb.** bei jmdm. Protest einlegen. **representative** [reprɪ'zentətɪv] **1.** *n.* **a)** *(Commerc.)* Vertreter, *der*/Vertreterin, *die;* **b)** R~ *(Amer. Polit.)* Abgeordneter/Abgeordnete. **2.** *adj.* *(typical)* repräsentativ (of für)

repress [rɪ'pres] *v. t.* unterdrücken. **repression** [rɪ'preʃn] *n.* Unterdrückung, *die.* **repressive** [rɪ'presɪv] *adj.* repressiv

reprieve [rɪ'priːv] **1.** *v. t.* ~ **sb.** *(postpone execution)* jmdm. Strafaufschub gewähren; *(remit execution)* jmdn. begnadigen. **2.** *n.* Strafaufschub, *der* (of für)/Begnadigung, *die;* *(fig.)* Gnadenfrist, *die*

reprimand ['reprɪmɑːnd] **1.** *n.* Tadel, *der.* **2.** *v. t.* tadeln

reprint 1. [riː'prɪnt] *v. t.* wieder abdrucken. **2.** ['riː'prɪnt] *n.* Nachdruck, *der*

reprisal [rɪ'praɪzl] *n.* Vergeltungsakt, *der* (for gegen)

reproach [rɪ'prəʊtʃ] **1.** *v. t.* ~ **sb.** jmdm. Vorwürfe machen. **2.** *n.* Vorwurf, *der.* **reproachful** [rɪ'prəʊtʃfl] *adv.* vorwurfsvoll

reproduce [riːprə'djuːs] **1.** *v. t.* wiedergeben. **2.** *v. i.* *(multiply)* sich fortpflanzen. **reproduction** [riːprə'dʌkʃn] *n.* **a)** Wiedergabe, *die;* **b)** *(producing offspring)* Fortpflanzung, *die;* **c)** *(copy)* Reproduktion, *die*

reprove [rɪ'pruːv] *v. t.* tadeln

reptile ['reptaɪl] *n.* Reptil, *das*

republic [rɪ'pʌblɪk] *n.* Republik, *die.* **republican** [rɪ'pʌblɪkən] **1.** *adj.* republikanisch. **2.** *n.* R~ *(Amer. Polit.)* Republikaner, *der*/Republikanerin, *die*

repudiate [rɪ'pjuːdɪeɪt] *v. t.* zurückweisen

repugnance [rɪ'pʌgnəns] *n.* Abscheu, *der* (to|wards| vor + *Dat.*)

repugnant [rɪ'pʌgnənt] *adj.* widerlich (to *Dat.*)

repulse [rɪ'pʌls] *v. t.* abwehren

repulsion [rɪ'pʌlʃn] *n.* *(disgust)* Widerwille, *der* (towards gegen)

repulsive [rɪ'pʌlsɪv] *adj.* abstoßend

reputable ['repjʊtəbl] *adj.* angesehen ⟨Person, Beruf, Zeitung *usw.*⟩; anständig ⟨Verhalten⟩; seriös ⟨Firma⟩

reputation [repjʊ'teɪʃn] *n.* **a)** Ruf, *der;* **have a ~ for** *or* **of doing/being sth.** in dem Ruf stehen, etw. zu tun/sein; **b)** *(good name)* Name, *der*

repute [rɪ'pjuːt] **1.** *v. t.* in pass. **be ~d** |to be| **sth.** als etw. gelten; **she is ~d to have/make ...:** man sagt, daß sie ... hat/macht. **2.** *n.* Ruf, *der.* **reputed** [rɪ'pjuːtɪd] *adj.,* **re'putedly** *adv.* angeblich

request [rɪ'kwest] **1.** *v. t.* bitten; ~ **sth. of** *or* **from sb.** jmdn. um etw. bitten. **2.** *n.* Bitte, *die* (for um); **at sb.'s ~:** auf jmds. Bitte *(Akk.)* [hin]. **re'quest stop** *n.* *(Brit.)* Bedarfshaltestelle, *die*

require [rɪ'kwaɪə(r)] *v. t.* **a)** *(need)* brauchen; **b)** *(order, demand)* verlangen (of von); be ~d to do sth. etw. tun müssen. re'quirement *n.* a) *(need)* Bedarf, *der;* b) *(condition)* Erfordernis, *das*

requisite ['rekwɪzɪt] 1. *adj.* notwendig (to, for für). 2. *n. in pl.* toilet/travel ~s Toiletten-/Reiseartikel *Pl.*

requisition [rekwɪ'zɪʃn] 1. *n. (order for sth.)* Anforderung, *die* (for *Gen.*). 2. *v. t.* anfordern

rescind [rɪ'sɪnd] *v. t.* für ungültig erklären

rescue ['reskjuː] 1. *v. t.* retten (from aus). 2. *n.* Rettung, *die; attrib.* Rettungs⟨dienst, -mannschaft⟩; go/come to the/sb.'s ~: jmdm. zu Hilfe kommen. rescuer ['reskjuːə(r)] *n.* Retter, *der*/Retterin, *die*

research [rɪ'sɜːtʃ, 'riːsɜːtʃ] 1. *n.* Forschung, *die* (into, on über + *Akk.*); ~ work Recherchen *Pl.* 2. *v. i.* forschen; ~ into sth. etw. erforschen. researcher [-'--, '---] *n.* Forscher, *der*/Forscherin, *die*

resell [riː'sel] *v. t.*, resold [riː'səʊld] weiterverkaufen (to an + *Akk.*)

resemblance [rɪ'zembləns] *n.* Ähnlichkeit, *die* (to mit)

resemble [rɪ'zembl] *v. t.* ähneln, gleichen (+ *Dat.*)

resent [rɪ'zent] *v. t.* übelnehmen. resentful [rɪ'zentfl] *adj.* übelnehmerisch, nachtragend ⟨Person, Art⟩; be ~ of *or* feel ~ about sth. etw. übelnehmen. re'sentment *n.* Groll, *der (geh.);* feel ~ towards *or* against sb. einen Groll auf jmdn. haben

reservation [rezə'veɪʃn] *n.* a) Reservierung, *die;* have a ~ [for a room] ein Zimmer reserviert haben; b) *(doubt)* Vorbehalt, *der* (about gegen); Bedenken (about bezüglich + *Gen.*); without ~: ohne Vorbehalt

reserve [rɪ'zɜːv] 1. *v. t.* reservieren lassen ⟨Zimmer, Tisch, Platz⟩; (set aside) reservieren; ~ the right to do sth. sich *(Dat.)* [das Recht] vorbehalten, etw. zu tun. 2. *n.* a) *(extra amount)* Reserve, *die* (of an + *Dat.*) have/hold *or* keep sth. in ~: etw. in Reserve haben/halten; b) *(place set apart)* Reservat, *das;* c) *(Sport)* Reservespieler, *der*/-spielerin, *die;* the R~s die Reserve; d) *(reticence)* Zurückhaltung, *die.* reserved [rɪ'zɜːvd] *adj. (reticent)* reserviert

reservoir ['rezəvwɑː(r)] *n. ([artificial] lake)* Reservoir, *das*

reshape [riː'ʃeɪp] *v. t.* umgestalten

reshuffle [riː'ʃʌfl] 1. *v. t.* a) umbilden ⟨Kabinett⟩; b) *(Cards)* neu mischen. 2. *n.* Umbildung, *die*

reside [rɪ'zaɪd] *v. i. (formal)* wohnen; wohnhaft sein *(Amtsspr.).* residence ['rezɪdəns] *n.* a) *(abode)* Wohnsitz, *der; (of ambassador etc.)* Residenz, *die;* b) *(stay)* Aufenthalt, *der.* 'residence permit *n.* Aufenthaltsgenehmigung, *die.* resident ['rezɪdənt] 1. *adj.* wohnhaft; be ~ in England sein Wohnsitz in England haben. 2. *n. (inhabitant)* Bewohner, *der*/Bewohnerin, *die; (at hotel)* Hotelgast, *der.* residential [rezɪ'denʃl] *adj.* Wohn⟨gebiet, -siedlung, -straße⟩; ~ hotel Hotel für Dauergäste

residue ['rezɪdjuː] *n.* a) Rest, *der;* b) *(Chem.)* Rückstand, *der*

resign [rɪ'zaɪn] 1. *v. t.* zurücktreten von ⟨Amt⟩. 2. *v. refl.* ~ oneself to sth./to doing sth. sich mit etw. abfinden/sich damit abfinden, etw. zu tun. 3. *v. i.* ⟨Arbeitnehmer:⟩ kündigen; ⟨Regierungsbeamter:⟩ zurücktreten (from von). resignation [rezɪg'neɪʃn] *n.* a) *see* resign 3: Kündigung, *die;* Rücktritt, *der;* tender one's ~: seine Kündigung/seinen Rücktritt einreichen; b) *(being resigned)* Resignation, *die;* with ~: resigniert. resigned [rɪ'zaɪnd] *adj.* resigniert; be ~ to sth. sich mit etw. abgefunden haben

resilience [rɪ'zɪlɪəns] *n.* a) Elastizität, *die;* b) *(fig.)* Unverwüstlichkeit, *die.* resilient [rɪ'zɪlɪənt] *adj.* elastisch; *(fig.)* unverwüstlich

resin ['rezɪn] *n.* Harz, *das*

resist [rɪ'zɪst] 1. *v. t.* a) standhalten (+ *Dat.*) ⟨Frost, Hitze, Feuchtigkeit usw.⟩; b) *(oppose)* sich widersetzen (+ *Dat.*); widerstehen (+ *Dat.*) ⟨Versuchung⟩. 2. *v. i. see* 1 b: sich widersetzen; widerstehen. resistance [rɪ'zɪstəns] *n.* Widerstand, *der* (to gegen). resistant [rɪ'zɪstənt] *adj.* a) *(opposed)* be ~ to sich widersetzen (+ *Dat.*); b) *(having power to resist)* widerstandsfähig (to gegen)

resold *see* resell

resolute ['rezəluːt] *adj.* resolut, energisch ⟨Person⟩; entschlossen ⟨Tat⟩

resolution [rezə'luːʃn] *n.* a) *(firmness)* Entschlossenheit, *die;* b) *(decision)* Entschließung, *die; (Polit. also)* Resolution, *die;* c) *(resolve)* Vorsatz, *der;* make a ~: einen Vorsatz fassen

resolve [rɪ'zɒlv] 1. *v. t.* a) lösen ⟨Problem, Rätsel⟩; ausräumen ⟨Schwierig-

keit⟩; **b)** *(decide)* beschließen; **c)** *(settle)* beilegen ⟨*Streit*⟩; regeln ⟨*Angelegenheit*⟩. **2.** *n.* **a)** Vorsatz, *der;* **b)** *(resoluteness)* Entschlossenheit, *die.* **resolved** [rɪ'zɒlvd] *adj.* ~ [to do sth.] entschlossen[, etw. zu tun]

resonant ['rezənənt] *adj.* hallend ⟨*Ton, Klang*⟩

resort [rɪ'zɔ:t] **1.** *n.* **a)** *(place)* Aufenthalt[sort], *der;* [holiday] ~: Ferienort, *der;* **ski** ~: Skiurlaubsort, *der;* **seaside** ~: Seebad, *das;* **b)** *(recourse)* **as a last** ~: als letzter Ausweg. **2.** *v. i.* ~ **to sth./ sb.** zu etw. greifen/sich an jmdn. wenden **(for** um**)**

resound [rɪ'zaʊnd] *v.i.* widerhallen. **re'sounding** *adj.* hallend ⟨*Lärm*⟩; überwältigend ⟨*Sieg, Erfolg*⟩

resource [rɪ'sɔ:s, rɪ'zɔ:s] *n. usu. in pl. (stock)* Mittel *Pl.;* Ressource, *die.* **resourceful** [rɪ'sɔ:sfl, rɪ'zɔ:sfl] *adj.* findig ⟨*Person*⟩

respect [rɪ'spekt] **1.** *n.* **a)** *(esteem)* Respekt, *der,* Achtung, *die* **(for** vor + *Dat.*⟩; **show** ~ **for sb./sth.** Respekt vor jmdm./etw. zeigen; **b)** *(aspect)* Hinsicht, *die;* **in some** ~s in mancher Hinsicht; **c) with** ~ **to** ...: in bezug auf ... *(Akk.)*; was ... [an]betrifft. **2.** *v. t.* respektieren; achten. **respectable** [rɪ'spektəbl] *adj.* angesehen ⟨*Bürger usw.*⟩; ehrenwert ⟨*Motive*⟩; *(decent)* ehrbar *(geh.)* ⟨*Leute, Kaufmann*⟩; anständig, respektabel ⟨*Beschäftigung usw.*⟩. **respectful** [rɪ'spektfl] *adj.* respektvoll **(to[wards]** gegenüber**)**. **re'spectfully** *adv.* respektvoll

respective [rɪ'spektɪv] *adj.* jeweilig. **re'spectively** *adv.* beziehungsweise

respiration [respɪ'reɪʃn] *n.* Atmung, *die*

respite ['respaɪt] *n.* Ruhepause, *die; (delay)* Aufschub, *der;* **without** ~: ohne Pause

resplendent [rɪ'splendənt] *adj.* prächtig

respond [rɪ'spɒnd] **1.** *v. i.* **a)** *(answer)* antworten **(to** auf + *Akk.*); **b)** *(react)* reagieren **(to** auf + *Akk.*); ⟨*Patient, Bremsen:*⟩ ansprechen **(to** auf + *Akk.*). **2.** *v. t.* antworten; erwidern **response** [rɪ'spɒns] *n.* **a)** *(answer)* Antwort, *die* **(to** auf + *Akk.*); **in** ~ [to] als Antwort [auf (+ *Akk.*)]; **b)** *(reaction)* Reaktion, *die*

responsibility [rɪspɒnsɪ'bɪlɪtɪ] *n.* **a)** *(being responsible)* Verantwortung, *die;* **b)** *(duty)* Verpflichtung, *die*

responsible [rɪ'spɒnsɪbl] *adj.* **a)** ver-

antwortlich; **be** ~ **to sb.** jmdm. gegenüber verantwortlich sein **(for** für**)**; **b)** *(trustworthy)* verantwortungsvoll. **re'sponsibly** [rɪ'spɒnsɪblɪ] *adv.* verantwortungsbewußt

responsive [rɪ'spɒnsɪv] *adj.* aufgeschlossen ⟨*Person*⟩; **be** ~ **to sth.** auf etw. *(Akk.)* reagieren

¹rest [rest] **1.** *v. i.* ruhen; ~ **on** ruhen auf (+ *Dat.*); ~ **from sth.** sich von etw. ausruhen; ~ **assured that ...**: seien Sie versichert, daß ...; ~ **with sb.** ⟨*Verantwortung:*⟩ bei jmdm. liegen. **2.** *v. i.* ~ **sth. against sth.** etw. an etw. *(Akk.)* lehnen; **b)** ausruhen ⟨*Augen*⟩. **3.** *n.* **a)** *(repose)* Ruhe, *die;* **b)** *(break, relaxation)* Ruhe[pause], *die;* Erholung, *die* **(from** von**)**; **take a** ~: sich ausruhen **(from** von**)**; **c)** *(pause)* **have a** ~: [eine] Pause machen; ~ **period** [Ruhe]pause, *die*

²rest *n.* **the** ~: der Rest; **we'll do the** ~: alles Übrige erledigen wir

restaurant ['restərɔ̃, 'restərɒnt] *n.* Restaurant, *das*

rested ['restɪd] *adj.* ausgeruht

restful ['restfl] *adj.* ruhig ⟨*Tag, Woche*⟩

restive ['restɪv] *adj.* unruhig

'restless *adj.* unruhig ⟨*Nacht, Schlaf, Bewegung*⟩; ruhelos ⟨*Person*⟩

restoration [restə'reɪʃn] *n.* **a)** *(of peace, health)* Wiederherstellung, *die; (of work of art, building)* Restaurierung, *die;* **b) the R~** *(Brit. Hist.)* die Restauration

restore [rɪ'stɔ:(r)] *v. t.* **a)** *(give back)* zurückgeben; **b)** restaurieren ⟨*Bauwerk, Kunstwerk usw.*⟩; ~ **sb. to health** jmdn. wiederherstellen; **c)** wiederherstellen ⟨*Ordnung, Ruhe*⟩

restrain [rɪ'streɪn] *v. t.* zurückhalten ⟨*Gefühl, Lachen, Person*⟩; bändigen ⟨*unartiges Kind, Tier*⟩; ~ **sb./oneself from doing sth.** jmdn. davon abhalten/ sich zurückhalten, etw. zu tun. **restrained** [rɪ'streɪnd] *adj.* zurückhaltend ⟨*Wesen, Kritik*⟩; beherrscht ⟨*Reaktion, Worte*⟩. **restraint** [rɪ'streɪnt] *n.* **a)** *(restriction)* Einschränkung, *die;* **b)** *(reserve)* Zurückhaltung, *die;* **c)** *(self-control)* Selbstbeherrschung, *die*

restrict [rɪ'strɪkt] *v. t.* beschränken **(to** auf + *Akk.*). **re'stricted** *adj.* beschränkt. **restriction** [rɪ'strɪkʃn] *n.* Beschränkung, *die* **(on** *Gen.*). **restrictive** [rɪ'strɪktɪv] *adj.* restriktiv

'rest room *n. (esp. Amer.)* Toilette, *die*

result [rɪ'zʌlt] **1.** *v. i.* **a)** *(follow)* ~ **from sth.** die Folge einer Sache *(Gen.)* sein;

b) *(end)* ~ **in sth.** in etw. *(Dat.)* resultieren. **2.** *n.* Ergebnis, *das;* **be the ~ of sth.** die Folge einer Sache *(Gen.)* sein; **as a ~** |of this| infolgedessen. **re'sultant** [rɪ'zʌltənt] *attrib. adj.* daraus resultierend

resume [rɪ'zju:m] *v. t.* wiederaufnehmen; fortsetzen ⟨*Reise*⟩

résumé ['rezʊmeɪ] *n.* Zusammenfassung, *die*

resumption [rɪ'zʌmpʃn] *n.* Wiederaufnahme, *die*

resurrection [rezə'rekʃn] *n.* *(Relig.)* Auferstehung, *die*

resuscitate [rɪ'sʌsɪteɪt] *v. t.* wiederbeleben

retail ['ri:teɪl] **1.** *adj.* Einzel⟨*handel*⟩; Einzelhandels⟨*geschäft, -preis*⟩. **2.** *adv.* **buy/sell** ~: en détail kaufen/verkaufen. **'retailer** *n.* Einzelhändler, *der/-*händlerin, *die.* **retail 'price index** *n.* *(Brit.)* Preisindex des Einzelhandels

retain [rɪ'teɪn] *v. t.* behalten; ein-, zurückbehalten ⟨*Gelder*⟩

retaliate [rɪ'tælɪeɪt] *v. i.* Vergeltung üben **(against** an + *Dat.).* **retaliation** [rɪtælɪ'eɪʃn] *n.* Vergeltung, *die;* **in ~ for** als Vergeltung für

retarded [rɪ'tɑːdɪd] *adj.* |mentally| ~: [geistig] zurückgeblieben

retch [retʃ] *v. i.* würgen

retentive [rɪ'tentɪv] *adj.* gut ⟨*Gedächtnis*⟩

rethink [ri:'θɪŋk] *v. t.,* **rethought** [ri:'θɔːt] noch einmal überdenken

reticence ['retɪsəns] *n.* Zurückhaltung, *die*

reticent ['retɪsənt] *adj.* zurückhaltend **(on, about** in bezug auf + *Akk.*)

retina ['retɪnə] *n.* Netzhaut, *die*

retinue ['retɪnjuː] *n.* Gefolge, *das*

retire [rɪ'taɪə(r)] *v. i.* **a)** ⟨*Angestellter, Arbeiter:*⟩ in Rente *(Akk.)* gehen; ⟨*Beamter, Militär:*⟩ in Pension *od.* den Ruhestand gehen; **b)** *(withdraw)* sich zurückziehen **(to** in + *Akk.*). **retired** [rɪ'taɪəd] *adj.* aus dem Berufsleben ausgeschieden; ⟨*Beamter, Soldat*⟩ im Ruhestand, pensioniert. **re'tirement** *n.* Ruhestand, *der*

retiring [rɪ'taɪərɪŋ] *adj. (shy)* zurückhaltend

retort [rɪ'tɔːt] **1.** *n.* Entgegnung, *die* **(to** auf + *Akk.*). **2.** *v. t.* entgegnen

retrace [rɪ'treɪs] *v. t.* zurückverfolgen; ~ **one's steps** denselben Weg noch einmal zurückgehen

retract [rɪ'trækt] *v. t.* zurücknehmen

retrain [ri:'treɪn] **1.** *v. i.* [sich] umschulen [lassen]. **2.** *v. t.* umschulen

retreat [rɪ'tri:t] **1.** *n.* **a)** *(withdrawal)* Rückzug, *der;* **beat a ~** *(fig.)* das Feld räumen; **b)** *(place)* Zufluchtsort, *der.* **2.** *v. i.* sich zurückziehen

retribution [retrɪ'bju:ʃn] *n.* Vergeltung, *die*

retrieval [rɪ'tri:vl] *n.* **a)** *(of situation)* Rettung, *die;* **beyond** *or* **past ~:** hoffnungslos; **b)** *(rescue)* Rettung, *die; (from wreckage)* Bergung, *die*

retrieve [rɪ'tri:v] *v. t.* **a)** *(rescue)* retten **(from** aus); *(from wreckage)* bergen **(from** aus); **b)** *(recover)* zurückholen ⟨*Brief*⟩; wiederholen ⟨*Ball*⟩; wiederbekommen ⟨*Geld*⟩; **c)** *(Computing)* wiederauffinden ⟨*Informationen*⟩; **d)** ⟨*Hund:*⟩ apportieren; **e)** retten ⟨*Situation*⟩. **re'triever** *n.* Apportierhund, *der; (breed)* Retriever, *der*

return [rɪ'tɜːn] **1.** *v. i. (come back)* zurückkommen; *(go back)* zurückgehen; *(by vehicle)* zurückfahren. **2.** *v. t.* **a)** *(bring back)* zurückbringen; zurückgeben ⟨geliehenen/gestohlenen Gegenstand⟩; ~**ed with thanks** mit Dank zurück; **b)** erwidern ⟨*Besuch, Gruß, Liebe*⟩; sich revanchieren für *(ugs.)* ⟨*Freundlichkeit, Gefallen*⟩; **c)** *(elect)* wählen ⟨*Kandidaten*⟩; **d)** ~ **a verdict of guilty/not guilty** ⟨*Geschworene:*⟩ auf „schuldig"/„nicht schuldig" erkennen. **3.** *n.* **a)** Rückkehr, *die;* **many happy ~s** |of the day|! herzlichen Glückwunsch [zum Geburtstag]!; **b) by ~** |of post| postwendend; **c)** *(ticket)* Rückfahrkarte, *die; (for flight)* Rückflugschein, *der;* **b)** ~|s| *(proceeds)* Gewinn, *der* **(on, from** aus); **e)** *(bringing back)* Zurückbringen, *das; (of property, goods, book)* Rückgabe, *die* **(to** an + *Akk.*); **receive/get sth. in ~** |for sth.| etw. [für etw.] bekommen

return: ~ **'fare** *n.* Preis für eine Rückfahrkarte/*(for flight)* einen Rückflugschein; ~ **'flight** *n.* Rückflug, *der;* ~ **'journey** *n.* Rückreise, *die;* Rückfahrt, *die;* ~ **'match** *n.* Rückspiel, *das;* ~ **'ticket** *n.* *(Brit.)* Rückfahrkarte, *die; (for flight)* Rückflugschein, *der*

retype [ri:'taɪp] *v. t.* neu tippen

reunion [ri:'ju:njən] *n.* *(gathering)* Treffen, *das*

reunite [ri:jʊ'naɪt] *v. t.* wieder zusammenführen

reuse 1. [ri:'ju:z] *v. t.* wiederverwenden. **2.** [ri:'ju:s] *n.* Wiederverwendung, *die*

rev [rev] *(coll.)* **1.** *n., usu. in pl.* Umdrehung, *die.* **2.** *v. i.,* -vv- hochtourig laufen. **3.** *v. t.,* -vv- aufheulen lassen. **rev 'up** *v. t.* aufheulen lassen

Rev. ['revərənd, *(coll.)* rev] *abbr.* Reverend Rev.

reveal [rɪ'viːl] *v. t.* enthüllen *(geh.);* be ~ed *(Wahrheit:)* ans Licht kommen. **re'vealing** *adj.* aufschlußreich

revel ['revl] *v. i.,* -ll- genießen (**in** *Akk.*); ~ **in doing sth.** es [richtig] genießen, etw. zu tun

revelation [revə'leɪʃn] *n.* **a)** Enthüllung, *die (geh.);* be **a ~:** einem die Augen öffnen; **b)** *(Relig.)* Offenbarung, *die*

revelry ['revəlrɪ] *n.* Feiern, *das*

revenge [rɪ'vendʒ] **1.** *v. t.* rächen *(Person, Tat).* **2.** *n. (action)* Rache, *die;* **take ~** *or* **have one's ~** [**on** sb.] [**for** sth.] Rache [an jmdm.] [für etw.] nehmen; **in ~ for sth.** als Rache für etw.

revenue ['revənjuː] *n.* ~[s] Einnahmen

revere [rɪ'vɪə(r)] *v. t.* verehren. **reverence** ['revərəns] *n.* Ehrfurcht, *die*

Reverend ['revərənd] *adj.* **the ~ John Wilson** Hochwürden John Wilson

reverent ['revərənt] *adj.* ehrfürchtig

reverie ['revərɪ] *n.* Träumerei, *die*

reversal [rɪ'vɜːsl] *n.* Umkehrung, *die*

reverse [rɪ'vɜːs] **1.** *adj.* entgegengesetzt *(Richtung);* Rück*(seite);* umgekehrt *(Reihenfolge).* **2.** *n.* **a)** *(contrary)* Gegenteil, *das;* **b)** *(Motor Veh.)* Rückwärtsgang, *der;* **put the car into ~, go into ~:** den Rückwärtsgang einlegen. **3.** *v. t.* **a)** umkehren *(Reihenfolge);* ~ **the charge[s]** *(Brit.)* ein R-Gespräch anmelden; **b)** zurücksetzen *(Fahrzeug).* **4.** *v. i.* zurücksetzen; rückwärts fahren. **reverse 'gear** *n. (Motor Veh.)* Rückwärtsgang, *der;* see also **gear 1 a**

reversible [rɪ'vɜːsɪbl] *adj.* beidseitig tragbar *(Kleidungsstück);* Wende*(mantel, -jacke)*

re'versing light *n.* Rückfahrscheinwerfer, *der*

revert [rɪ'vɜːt] *v. i.* ~ **to** zurückkommen auf (+ *Akk.*) *(Thema, Frage);* ~ **to savagery** in den Zustand der Wildheit zurückfallen

review [rɪ'vjuː] **1.** *n.* **a)** *(survey)* Überblick, *der* (**of** über + *Akk.*); **b)** *(reexamination)* [nochmalige] Überprüfung; **c)** *(of book, play, etc.)* Kritik, *die;* Rezension, *die.* **2.** *v. t.* **a)** *(survey)* untersuchen; prüfen; **b)** *(re-examine)* überprüfen; **c)** *(Mil.)* inspizieren; **d)**

(write a criticism of) rezensieren. **re-'viewer** *n.* Rezensent, *der*/Rezensentin, *die*

revile [rɪ'vaɪl] *v. t.* schmähen *(geh.)*

revise [rɪ'vaɪz] *v. t.* **a)** *(check over)* durchsehen *(Manuskript);* **b)** *(for exam)* wiederholen; *abs.* lernen. **revision** [rɪ'vɪʒn] *n.* **a)** *(checking over)* Durchsicht, *die;* **b)** *(amended version)* revidierte Fassung; **c)** *(for exam)* Wiederholung, *die*

revisit [riː'vɪzɪt] *v. t.* wieder besuchen

revitalize [riː'vaɪtəlaɪz] *v. t.* neu beleben

revival [rɪ'vaɪvl] *n.* Neubelebung, *die*

revive [rɪ'vaɪv] **1.** *v. i. (come back to consciousness)* wieder zu sich kommen; *(be reinvigorated)* zu neuem Leben erwachen. **2.** *v. t.* **a)** *(restore to consciousness)* wiederbeleben; *(reinvigorate)* wieder zu Kräften kommen lassen; **b)** wieder wecken *(Lebensgeister, Interesse)*

revoke [rɪ'vəʊk] *v. t.* aufheben *(Entscheidung);* widerrufen *(Befehl);* widerrufen *(Erlaubnis, Genehmigung)*

revolt [rɪ'vəʊlt] **1.** *v. i.* revoltieren (**against** gegen). **2.** *v. t.* mit Abscheu erfüllen. **3.** *n.* Revolte, *die (auch fig.);* Aufstand, *der.* **re'volting** *adj.* abscheulich; *(coll.: unpleasant)* widerlich

revolution [revə'luːʃn] *n.* Revolution, *die.* **revolutionary** [revə'luːʃənərɪ] **1.** *adj.* revolutionär. **2.** *n.* Revolutionär, *der*/Revolutionärin, *die*

revolve [rɪ'vɒlv] **1.** *v. t.* drehen. **2.** *v. i.* sich drehen (**round, about, on** um)

revolver [rɪ'vɒlvə(r)] *n.* [Trommel]revolver, *der*

revolving [rɪ'vɒlvɪŋ] *attrib. adj.* Dreh*(bühne, -tür)*

revue [rɪ'vjuː] *n.* Kabarett, *das; (musical show)* Revue, *die*

revulsion [rɪ'vʌlʃn] *n.* Abscheu, *der* (**at** vor + *Dat.,* gegen)

reward [rɪ'wɔːd] **1.** *n.* Belohnung, *die.* **2.** *v. t.* belohnen. **re'warding** *adj.* lohnend; **be ~/financially ~:** sich lohnen/einträglich sein

rewind [riː'waɪnd] *v. t.,* **rewound** [riː'waʊnd] **a)** wieder aufziehen *(Uhr);* **b)** zurückspulen *(Film, Band)*

reword [riː'wɜːd] *v. t.* umformulieren

rewrite [riː'raɪt] *v. t.,* **rewrote** [riː'rəʊt], **rewritten** [riː'rɪtn] noch einmal [neu] schreiben; *(write differently)* umschreiben

rhetoric ['retərɪk] *n.* [**art of**] ~: Rede-

kunst, *die;* Rhetorik, *die.* **rhetorical**
[rɪ'tɒrɪkl] *adj.* rhetorisch

rheumatic [ru:'mætɪk] *adj.* rheumatisch

rheumatism ['ru:mətɪzm] *n.* Rheumatismus, *der;* Rheuma, *das (ugs.)*

Rhine [raɪn] *pr. n.* Rhein, *der*

rhino ['raɪnəʊ] *n., pl.* **same** *or* ~**s** *(coll.),*
rhinoceros [raɪ'nɒsərəs] *n., pl. same*
or ~**es** Nashorn, *das;* Rhinozeros, *das*

rhododendron [rəʊdə'dendrən] *n.*
Rhododendron, *der*

rhubarb ['ru:bɑ:b] *n.* Rhabarber, *der*

rhyme [raɪm] **1.** *n.* Reim, *der;* **without**
~ **or reason** ohne Sinn und Verstand.
2. *v. i.* sich reimen **(with** auf + *Akk.)*

rhythm ['rɪðm] *n.* Rhythmus, *der.*
rhythmic ['rɪðmɪk], **rhythmical**
['rɪðmɪkl] *adj.* rhythmisch

rib [rɪb] **1.** *n.* Rippe, *die.* **2.** *v. t.,* -**bb**-
(coll.) aufziehen *(ugs.)*

ribald ['rɪbəld] *adj.* zotig

ribbon ['rɪbn] *n.* Band, *das; (on typewriter)* [Farb]band, *das*

rice [raɪs] *n.* Reis, *der.* **rice 'pudding**
n. Milchreis, *der.* **'rice wine** *n.* Reis-
wein, *der*

rich [rɪtʃ] **1.** *adj.* **a)** reich **(in** an +
Dat.); *(fertile)* fruchtbar ⟨*Land,
Boden*⟩; **b)** *(splendid)* prachtvoll; **c)**
(containing much fat, oil, eggs, etc.) gehaltvoll; **d)** *(deep, full)* voll[tönend]
⟨*Stimme*⟩; voll ⟨*Ton*⟩; satt ⟨*Farbe,
Farbton*⟩. **2.** *n. pl.* **the** ~: die Reichen;
~ **and poor** Arm und Reich. **riches**
['rɪtʃɪz] *n. pl.* Reichtum, *der.* **'richly**
adv. **a)** *(splendidly)* reich; üppig ⟨*aus-
gestattet*⟩; prächtig ⟨*gekleidet*⟩; **b)**
(fully) voll und ganz; ~ **deserved** wohl-
verdient. **'richness** *n.* **a)** *(of food)*
Reichhaltigkeit, *die;* **b)** *(of voice)* voller Klang; *(of colour)* Sattheit, *die*

rickets ['rɪkɪts] *n.* Rachitis, *die*

rickety ['rɪkɪtɪ] *adj.* wack[e]lig

ricochet ['rɪkəʃeɪ] **1.** *n.* **a)** Abprallen,
das; **b)** *(hit)* Abpraller, *der.* **2.** *v. i.,*
~**ed** ['rɪkəʃeɪd] abprallen **(off** von)

rid [rɪd] *v. t.,* -**dd**-, **rid:** ~ **sth. of sth.** etw.
von etw. befreien; ~ **oneself of sb./sth.**
sich von jmdm./etw. befreien; **be** ~ **of**
sb./sth. jmdn./etw. los sein *(ugs.);* **get**
~ **of sb./sth.** jmdn./etw. loswerden

riddance ['rɪdəns] *n.* **good** ~! Gott sei
Dank ist er/es *usw.* weg!

ridden *see* ride 2, 3

¹riddle ['rɪdl] *n.* Rätsel, *das*

²riddle *v. t.* durchlöchern; ~**d with bullets** von Kugeln durchsiebt

ride [raɪd] **1.** *n. (on horseback)* [Aus]ritt,

der; (in vehicle, at fair) Fahrt, die; ~ **in**
a train/coach Zug-/Busfahrt, *die;* **go**
for a ~: ausreiten; **go for a** [bi]cycle ~:
radfahren; **go for a** ~ [in the car] [mit
dem Auto] wegfahren; **take sb. for a** ~
(fig. sl.: deceive) jmdn. reinlegen
(ugs.). **2.** *v. i.,* **rode** [rəʊd], **ridden** ['rɪdn]
(on horse) reiten; *(on bicycle, in
vehicle)* fahren; ~ **to town on one's
bike/in one's car/on the train** mit dem
Rad/Auto/Zug in die Stadt fahren. **3.**
v. t., **rode, ridden** reiten ⟨*Pferd usw.*⟩;
fahren mit ⟨*Fahrrad*⟩. **ride a'way,
ride 'off** *v. i.* wegreiten/-fahren

'rider *n.* **a)** Reiter, *der*/Reiterin, *die; (of
cycle)* Fahrer, *der*/Fahrerin, *die;* **b)**
(addition) Zusatz, *der*

ridge [rɪdʒ] *n.* **a)** *(of roof)* First, *der;* **b)**
(long hilltop) Grat, *der;* Kamm, *der;* **c)**
(Meteorol.) ~ [of high pressure] langge-
strecktes Hoch

ridicule ['rɪdɪkju:l] **1.** *n.* Spott, *der.* **2.**
v. t. verspotten

ridiculous [rɪ'dɪkjʊləs] *adj.* lächerlich

riding ['raɪdɪŋ] *n.* Reiten, *das.* **'riding
lesson** *n.* Reitstunde, *die.* **'riding-
school** *n.* Reitschule, *die*

rife [raɪf] *pred. adj.* weit verbreitet

riff-raff ['rɪfræf] *n.* Gesindel, *das*

rifle ['raɪfl] **1.** *n.* Gewehr, *das.* **2.** *v. t.*
durchwühlen. **3.** *v. i.* ~ **through sth.**
etw. durchwühlen

rift [rɪft] *n.* Unstimmigkeit, *die*

¹rig [rɪg] *n. (for oil-well)* [Öl]förderturm,
der; (off shore) Förderinsel, *die.* **rig
'out** *v. t.* ausstaffieren. **rig 'up** *v. t.*
aufbauen

²rig *v. t.,* -**gg**- manipulieren ⟨*[Wahl]-
ergebnis*⟩; fälschen ⟨*Wahl*⟩

rigging ['rɪgɪŋ] *n.* Takelung, *die*

right [raɪt] **1.** *adj.* **a)** *(just, morally
good, sound)* richtig; **b)** *(correct, true)*
richtig; **you're** [quite] ~: du hast [völ-
lig] recht; **be** ~ **in sth.** recht mit etw.
haben; **is that clock** ~? geht die Uhr
da richtig?; **put** *or* **set** ~: richtigstellen
⟨*Irrtum, Behauptung*⟩; wiedergutma-
chen ⟨*Unrecht*⟩; berichtigen ⟨*Fehler*⟩;
richtig stellen ⟨*Uhr*⟩; **put** *or* **set sb.** ~:
jmdn. berichtigen; **that's** ~: ja[wohl];
so ist es; **is that** ~? stimmt das?; *(indeed?)* aha!; [**am I**] ~? nicht [wahr]?;
c) *(preferable, most suitable)* richtig;
recht; **do sth. the** ~ **way** etw. richtig
machen; **d)** *(opposite of left)* recht...;
on the ~ **side** rechts; **e)** R~ *(Polit.)*
recht... **2.** *v. t.* aus der Welt schaffen
⟨*Unrecht*⟩. **3.** *n.* **a)** *(fair claim, authority)* Recht, *das;* **have a/no** ~ **to sth.**

ein/kein Anrecht *od.* Recht auf etw. *(Akk.)* haben; **in one's own ~:** aus eigenem Recht; **~ of way** Vorfahrtsrecht, *das;* **have ~ of way** Vorfahrt haben; **b)** *(what is just)* Recht, *das;* **by ~[s]** von Rechts wegen; **in the ~:** im Recht; **c)** *(~-hand side)* rechte Seite; **on** *or* **to the ~** [of sb./sth.] rechts [von jmdm./etw.]; **d)** *(Polit.)* **the R~:** die Rechte. **4.** *adv.* **a)** *(correctly)* richtig; **b)** *(to the ~-hand side)* nach rechts; **c)** *(completely)* ganz; **d)** *(exactly)* genau; **~ 'now** im Moment; jetzt sofort ⟨*handeln*⟩; **e)** *(straight)* direkt

'right angle *n.* rechter Winkel; **at ~s to** sth. rechtwinklig zu etw.

righteous ['raɪtʃəs] *adj.* rechtschaffen

rightful ['raɪtfl] *adj.* rechtmäßig ⟨*Besitzer, Herrscher*⟩

right: **~-hand** *adj.* recht...; **~-'handed 1.** *adj.* rechtshändig; ⟨*Werkzeug*⟩ für Rechtshänder; **be ~-handed** ⟨*Person:*⟩ Rechtshänder/Rechtshänderin sein; **2.** *adv.* rechtshändig; **~-hand 'man** *n.* rechte Hand

'rightly *adv.* zu Recht

right: **~-'minded** *adj.* gerecht denkend; **~ 'wing** *n.* rechter Flügel; **~-wing** *adj.* *(Polit.)* rechtsgerichtet; Rechts⟨*extremist, -intellektueller*⟩; **~-winger** *n.* **a)** *(Sport)* Rechtsaußen, *der;* **b)** *(Polit.)* Rechte, *der/die*

rigid ['rɪdʒɪd] *adj.* **a)** starr; *(stiff)* steif; **b)** *(strict)* streng; unbeugsam ⟨*System*⟩. **rigidity** [rɪ'dʒɪtɪ] *n. see* rigid: Starrheit, *die;* Steifheit, *die;* Strenge, *die*

rigmarole ['rɪgmərəʊl] *n.* **a)** *(talk)* langatmiges Geschwafel *(ugs.);* **b)** *(procedure)* Zirkus, *der*

rigor ['rɪgə(r)] *(Amer.) see* rigour

rigor mortis [rɪgə 'mɔːtɪs] *n.* Totenstarre, *die*

rigorous ['rɪgərəs] *adj.* streng

rigour ['rɪgə(r)] *n. (Brit.)* Strenge, *die*

rile [raɪl] *v.t. (coll.)* ärgern

rim [rɪm] *n.* Rand, *der;* *(of wheel)* Felge, *die*

rind [raɪnd] *n. (of fruit)* Schale, *die; (of cheese)* Rinde, *die; (of bacon)* Schwarte, *die*

¹ring [rɪŋ] **1.** *n.* **a)** Ring, *der;* **b)** *(Boxing)* Ring, *der; (in circus)* Manege, *die.* **2.** *v.t. (surround)* umringen; einkreisen ⟨*Wort usw.*⟩.

²ring **1.** *n.* **a)** *(act of sounding bell)* Läuten, *das;* Klingeln, *das;* **b)** *(Brit. coll.: telephone call)* Anruf, *der;* **give sb. a**

~: jmdn. anrufen; **c)** *(fig.: impression)* **have the ~ of truth** [about it] glaubhaft klingen. **2.** *v.i.,* **rang** [ræŋ], **rung** [rʌŋ] **a)** *(sound clearly)* [er]schallen; ⟨*Hammer:*⟩ [er]dröhnen; **b)** *(be sounded)* ⟨*Glocke, Klingel, Telefon:*⟩ läuten; ⟨*Wecker, Telefon, Kasse:*⟩ klingeln; **the doorbell rang** es klingelte; **c)** *(~ bell)* läuten (for nach); **d)** *(Brit.: make telephone call)* anrufen. **3.** *v.t.,* **rang, rung a)** läuten ⟨*Glocke*⟩; **~ the** [door]bell läuten; klingeln; **it ~s a bell** *(fig. coll.)* es kommt mir [irgendwie] bekannt vor; **b)** *(Brit.: telephone)* anrufen. **ring 'back** *(Brit.) v.t. & i.* **a)** *(again)* wieder anrufen; **b)** *(in return)* zurückrufen. **ring 'off** *v.i. (Brit.)* auflegen. **ring 'out** *v.i.* ertönen

ring: **~ binder** *n.* Ringbuch, *das;* **~-finger** *n.* Ringfinger, *der*

ringing ['rɪŋɪŋ] *n.* Läuten, *das; (Brit. Teleph:)* **~ tone** Freiton, *der*

'ringleader *n.* Anführer, *der/*Anführerin, *die*

ringlet ['rɪŋlɪt] *n.* [Ringel]löckchen, *das*

'ring road *n.* Ringstraße, *die*

rink [rɪŋk] *n. (for ice-skating)* Eisbahn, *die; (for roller-skating)* Rollschuhbahn, *die*

rinse [rɪns] **1.** *v.t.* **a)** *(wash out)* ausspülen ⟨*Mund, Gefäß usw.*⟩; **b)** [aus]spülen ⟨*Wäsche usw.*⟩; abspülen ⟨*Hände, Geschirr*⟩. **2.** *n.* Spülen, *das;* **give sth. a** [good/quick] **~:** etw. [gut/schnell] ausspülen/abspülen/spülen. **rinse 'out** *v.t.* ausspülen

riot ['raɪət] **1.** *n.* Aufruhr, *der;* **~s** Unruhen *Pl.;* **run ~:** randalieren. **2.** *v.i.* randalieren. **'rioter** *n.* Randalierer, *der.* **riotous** ['raɪətəs] *adj.* **a)** gewalttätig; **b)** *(unrestrained)* wild

rip [rɪp] **1.** *n.* Riß, *der.* **2.** *v.t.,* **-pp-** zerreißen; **~ open** aufreißen. **rip 'off** *v.t.* **a)** *(remove from)* reißen von; *(remove)* abreißen; **b)** *(sl.: defraud)* übers Ohr hauen *(ugs.).* **rip 'out** *v.t.* herausreißen (of aus)

RIP *abbr.* rest in peace R.I.P.

'rip-cord *n.* Reißleine, *die*

ripe [raɪp] *adj.* reif (for zu). **ripen** ['raɪpn] **1.** *v.t.* zur Reife bringen. **2.** *v.i.* reifen. **'ripeness** *n.* Reife, *die*

'rip-off *n. (sl.)* Nepp, *der (ugs.)*

riposte [rɪ'pɒst] **1.** *n. (retort)* [rasche] Entgegnung. **2.** *v.i.* [rasch] antworten

ripple ['rɪpl] **1.** *n.* kleine Welle. **2.** *v.i.* ⟨*See:*⟩ sich kräuseln; ⟨*Welle:*⟩ plätschern. **3.** *v.t.* kräuseln

rise [raɪz] **1.** *n.* **a)** *(advancement)* Aufstieg, *der;* **b)** *(in value, price, cost)* Steigerung, *die; (in population, temperature)* Zunahme, *die;* **c)** *(Brit.)* |pay| ~ *(in wages)* Lohnerhöhung, *die; (in salary)* Gehaltserhöhung, *die;* **d)** *(hill)* Anhöhe, *die;* **e)** give ~ to führen zu; Anlaß geben zu ⟨*Spekulation*⟩. **2.** *v.i.,* rose [rəʊz], risen ['rɪzn] **a)** *(go up)* aufsteigen; **b)** ⟨*Sonne, Mond:*⟩ aufgehen; **c)** *(increase, reach higher level)* steigen; **d)** *(advance)* ⟨*Person:*⟩ aufsteigen; **e)** ⟨*Teig, Kuchen:*⟩ aufgehen; **f)** *(Theatre)* ⟨*Vorhang:*⟩ aufgehen; **g)** ⟨*Fluß:*⟩ entspringen. **rise 'up** *v.i.* **a)** ~ up |in revolt| aufbegehren *(geh.);* **b)** ⟨*Berg:*⟩ aufragen
risen *see* rise 2
'riser *n.* early ~: Frühaufsteher, *der*/Frühaufsteherin, *die*
rising ['raɪzɪŋ] **1.** *n. (of sun, moon, etc.)* Aufgang, *der.* **2.** *adj.* **a)** aufgehend ⟨*Sonne, Mond usw.*⟩; **b)** steigend ⟨*Kosten, Temperatur, Wasser, Flut*⟩; **c)** *(sloping upwards)* ansteigend
risk [rɪsk] **1.** *n.* Gefahr, *die; (chance taken)* Risiko, *das;* at one's own ~: auf eigene Gefahr *od.* eigenes Risiko; take the ~ of doing sth. es riskieren, etw. zu tun; be at ~ ⟨*Zukunft, Plan:*⟩ gefährdet sein. **2.** *v.t.* riskieren; I'll ~ it ich lasse es darauf ankommen. **'risky** *adj.* gefährlich; gewagt ⟨*Experiment, Projekt*⟩
risqué ['rɪskeɪ] *adj.* gewagt
rissole ['rɪsəʊl] *n.* Rissole, *die*
rite [raɪt] *n.* Ritus, *der*
ritual ['rɪtʃʊəl] **1.** *adj.* rituell; Ritual-⟨*mord, -tötung*⟩. **2.** *n.* Ritual, *das*
rival ['raɪvl] **1.** *n. (competitor)* Rivale, *der*/Rivalin, *die;* business ~s Konkurrenten. **2.** *v.t., (Brit.)* -ll- nicht nachstehen (+ *Dat.*). **rivalry** ['raɪvlrɪ] *n.* Rivalität, *die (geh.)*
river ['rɪvə(r)] *n.* Fluß, *der.* **'river-bed** *n.* Flußbett, *das.* **'riverside 1.** *n.* Flußufer, *das.* **2.** *attrib. adj.* am Fluß gelegen; am Fluß *nachgestellt*
rivet ['rɪvɪt] **1.** *n.* Niete, *die.* **2.** *v.t.* **a)** [ver]nieten; **b)** *(fig.)* fesseln. **'riveting** *adj.* fesselnd
RN *abbr. (Brit.)* **Royal Navy** Königl. Mar.
road [rəʊd] *n.* Straße, *die;* across *or* over the ~ |from us| [bei uns] gegenüber; by ~ *(by car/bus/lorry)* per Auto/Bus/Lkw; be on the ~: auf Reisen *od.* unterwegs sein; ⟨*Theaterensemble usw.:*⟩ auf Tournee *od.* Tour sein

road: ~ **accident** *n.* Verkehrsunfall, *der;* ~**-block** *n.* Straßensperre, *die;* ~**-hog** *n.* Verkehrsrowdy, *der;* ~**-map** *n.* Straßenkarte, *die;* ~ **safety** *n.* Verkehrssicherheit, *die;* ~ **sense** *n.* Gespür für Verkehrssituationen; ~**side** *n.* Straßenrand, *der;* at *or* by/along the ~**side** am Straßenrand; ~ **sign** *n.* Verkehrszeichen, *das;* Straßenschild, *das (ugs.);* ~**-sweeper** *n.* Straßenkehrer, *der*/-kehrerin, *die;* ~**-user** *n.* Verkehrsteilnehmer, *der*/-teilnehmerin, *die;* ~**way** *n.* Fahrbahn, *die;* ~**works** *n. pl.* Straßenbauarbeiten *Pl.;* ~**worthy** *adj.* fahrtüchtig
roam [rəʊm] *v.i.* umherstreifen. *v.t.* streifen durch
roar [rɔː(r)] **1.** *n. (of wild beast)* Gebrüll, *das; (of applause)* Tosen, *das; (of engine, traffic)* Dröhnen, *das;* ~s/a ~ |of laughter| dröhnendes Gelächter. **2.** *v.i.* brüllen (with vor + *Dat.*); ⟨*Motor:*⟩ dröhnen. **'roaring** *adj.* **a)** bullernd *(ugs.)* ⟨*Feuer*⟩; **b)** a ~ success ein Bombenerfolg; do a ~ trade ein Bombengeschäft machen
roast [rəʊst] **1.** *v.t.* braten; rösten ⟨*Kaffeebohnen, Kastanien*⟩. **2.** *attrib. adj.* gebraten ⟨*Fleisch, Ente usw.*⟩; Brat⟨*hähnchen, -kartoffeln*⟩; Röst⟨*kastanien*⟩; ~ beef *(sirloin)* Roastbeef, *das.* **3.** *n.* Braten, *der*
rob [rɒb] *v.t.,* -bb- ausrauben ⟨*Bank, Safe, Kasse*⟩; berauben ⟨*Person*⟩. **robber** ['rɒbə(r)] *n.* Räuber, *der*/Räuberin, *die.* **robbery** ['rɒbərɪ] *n.* Raub, *der;* robberies Raubüberfälle
robe [rəʊb] *n.* Gewand, *das (geh.); (of judge, vicar)* Talar, *der*
robin ['rɒbɪn] *n.* ~ |redbreast| Rotkehlchen, *das*
robot ['rəʊbɒt] *n.* Roboter, *der*
robust [rəʊ'bʌst] *adj.* robust
¹rock [rɒk] *n.* **a)** *(piece of ~)* Fels, *der;* **b)** *(large ~, hill)* Felsen, *der;* **c)** *(substance)* Fels, *der; (esp. Geol.)* Gestein, *das;* **d)** *(boulder)* Felsbrocken, *der; (Amer.: stone)* Stein, *der;* **e)** stick of ~: Zuckerstange, *die;* **f)** be on the ~s *(fig. coll.)* ⟨*Ehe, Firma:*⟩ kaputt sein *(ugs.)*
²rock 1. *v.t.* wiegen; *(in cradle)* schaukeln. **2.** *v.i.* **a)** schaukeln; **b)** *(sway)* schwanken. **3.** *n. (Mus.)* Rock, *der; attrib.* Rock-; ~ and *or* 'n' roll |music| Rock and Roll, *der*
rock: ~**-'bottom** *(coll.)* **1.** *adj.* ~-bottom prices Schleuderpreise *(ugs.);* **2.** *n.* reach *or* touch ~-bottom ⟨*Handel,*

Preis:) in den Keller fallen *(ugs.)*; **her spirits reached ~-bottom** ihre Stimmung war auf dem Tiefpunkt; **~-climbing** n. [Fels]klettern, *das*

rockery ['rɒkərɪ] n. Steingarten, *der*

rocket ['rɒkɪt] 1. *n.* Rakete, *die.* 2. *v.i.* ⟨*Preise:*⟩ in die Höhe schnellen

rocking: ~-chair n. Schaukelstuhl, *der;* **~-horse** n. Schaukelpferd, *das*

rocky adj. **a)** felsig; **b)** *(coll.: unsteady)* wackelig *(ugs.)*

rod [rɒd] n. Stange, *die; (for punishing)* Rute, *die; (for fishing)* [Angel]rute, *die*

rode see **ride** 2, 3

rodent ['rəʊdənt] n. Nagetier, *das*

¹roe [rəʊ] n. *(of fish)* |hard] ~: Rogen, *der;* |soft] ~: Milch, *die*

²roe n. ~ |deer] Reh, *das*

rogue [rəʊg] n. Gauner, *der*

role, rôle [rəʊl] n. Rolle, *die*

¹roll [rəʊl] n. **a)** Rolle, *die; (of cloth etc.)* Ballen, *der;* ~ **of film** Rolle Film; **b)** |bread] ~: Brötchen, *das*

²roll 1. *n. (of drum)* Wirbel, *der.* 2. *v.t.* **a)** rollen; *(between surfaces)* drehen; **b)** *(shape by ~ing)* rollen; drehen ⟨*Zigarette*⟩; **c)** walzen ⟨*Rasen, Metall usw.*⟩; ausrollen ⟨*Teig*⟩. 3. *v.i.* **a)** rollen; **b)** ⟨*Maschine:*⟩ laufen; **get sth. ~ing** *(fig.)* etw. ins Rollen bringen; **c)** **be ~ing in money** *or* **in it** *(coll.)* im Geld schwimmen *(ugs.)*. **roll a'bout** *v.i.* herumrollen; ⟨*Schiff:*⟩ schlingern; ⟨*Kind, Hund:*⟩ sich wälzen. **roll 'back** *v.t.* zurückrollen. **roll 'by** *v.i.* ⟨*Zeit:*⟩ vergehen. **roll 'in** *v.i. (coll.)* ⟨*Briefe, Geldbeträge:*⟩ eingehen. **roll 'out** *v.t.* ausrollen ⟨*Teig, Teppich*⟩. **roll 'over** *v.i.* ⟨*Person:*⟩ sich umdrehen, *(to make room)* sich zur Seite rollen. **roll 'up** 1. *v.t.* aufrollen ⟨*Teppich*⟩; zusammenrollen ⟨*Landkarte, Dokument usw.*⟩; hochkrempeln ⟨*Ärmel*⟩. 2. *v.i. (coll.: arrive)* aufkreuzen *(salopp)*

'roll-call n. Ausrufen aller Namen; *(Mil.)* Zählappell, *der*

'roller n. **a)** Rolle, *die; (for lawn, road, etc.)* Walze, *die;* **b)** *(for hair)* Lockenwickler, *der*

roller: ~ blind n. Rouleau, *das;* **~-coaster** n. Achterbahn, *die;* **~-skate** 1. n. Rollschuh, *der;* 2. *v.i.* Rollschuh laufen; **~-skating** n. Rollschuhlaufen, *das*

'rolling adj. wellig ⟨*Gelände*⟩; ~ **hills** sanfte Hügel

rolling: ~-pin n. Teigrolle, *die;* **~-stock** n. *(Brit. Railw.)* Fahrzeugbestand, *der*

ROM [rɒm] abbr. *(Computing)* **read only memory** ROM

Roman ['rəʊmən] 1. *n.* Römer, *der/* Römerin, *die.* 2. *adj.* römisch.

Roman 'Catholic 1. *adj.* römischkatholisch. 2. *n.* Katholik, *der/*Katholikin, *die;* **is a ~:** jmd. ist römischkatholisch

romance [rə'mæns] n. **a)** *(love affair)* Romanze, *die;* **b)** *(love-story)* [romantische] Liebesgeschichte

Romania [rəʊ'meɪnɪə] pr. n. Rumänien *(das).* **Romanian** [rəʊ'meɪnɪən] 1. *adj.* rumänisch. 2. *n.* **a)** *(person)* Rumäne, *der/*Rumänin, *die;* **b)** *(language)* Rumänisch, *das; see also* **English** 2 a

Roman 'numeral n. römische Ziffer

romantic [rəʊ'mæntɪk] adj. romantisch

romanticism [rəʊ'mæntɪsɪzm] n. *(Lit., Art., Mus.)* Romantik, *die*

Romany ['rəʊmənɪ] 1. **a)** *(person)* Rom, *der;* **b)** *(language)* Romani, *das.* 2. *adj.* Roma-; *(Ling.)* Romani-

Rome [rəʊm] pr. n. Rom *(das)*

romp [rɒmp] 1. *v.i.* **a)** [herum]tollen; **b)** ~ **home** *or* **in** *(coll.: win easily)* spielend gewinnen. 2. *n.* Tollerei, *die*

rompers ['rɒmpəz] n. pl. Spielhöschen, *das*

roof [ruːf] 1. *n.* **a)** Dach, *das;* **b)** ~ **of the mouth** Gaumen, *der.* 2. *v.t.* bedachen. **'roofing** n. *(material)* Deckung, *die*

roof: ~-rack n. Dachgepäckträger, *der;* **~-top** n. Dach, *das*

¹rook [rʊk] n. *(Ornith.)* Saatkrähe, *die*

²rook n. *(Chess)* Turm, *der*

room [ruːm, rʊm] n. **a)** *(in building)* Zimmer, *das; (for function)* Saal, *der;* **b)** *(space)* Platz, *der;* **make ~** |for sb./ sth.] [jmdm./einer Sache] Platz machen; **there is still ~ for improvement in his work** seine Arbeit ist noch verbesserungsfähig

room: ~-mate n. Zimmergenosse, *der/*-genossin, *die;* ~ **service** n. Zimmerservice, *der;* ~ **temperature** n. Zimmertemperatur, *die*

roomy ['ruːmɪ] adj. geräumig

roost [ruːst] 1. *n.* [Sitz]stange, *die.* 2. *v.i.* ⟨*Vogel:*⟩ sich [zum Schlafen] niederlassen

¹root [ruːt] 1. *n.* Wurzel, *die;* **put down ~s/take** ~: Wurzeln schlagen. 2. *v.i.* ⟨*Pflanze:*⟩ wurzeln. 3. *v.t.* **stand ~ed to the spot** wie angewurzelt dastehen. **root 'out** *v.t.* ausrotten

²**root** *v. i* **a)** *(turn up ground)* wühlen (for nach); **b)** *(coll.)* ~ **for** *(cheer)* anfeuern

rope [rəʊp] **1.** *n.* **a)** *(cord)* Seil, *das;* **b)** **know the** ~**s** sich auskennen. **2.** *v. t.* festbinden. **rope 'in** *v. t. (fig.)* einspannen *(ugs.)*

rope-'ladder *n.* Strickleiter, *die*

rosary ['rəʊzərɪ] *n.* Rosenkranz, *der*

¹**rose** [rəʊz] *n.* **a)** *(plant, flower)* Rose, *die;* **b)** *(colour)* Rosa, *das*

²**rose** *see* rise 2

rosé [rəʊ'zeɪ, 'rəʊzeɪ] *n.* Rosé, *der*

rose: ~**-bed** *n.* Rosenbeet, *das;* ~**-bud** *n.* Rosenknospe, *die;* ~**-bush** *n.* Rosenstrauch, *der*

rosemary ['rəʊzmərɪ] *n.* Rosmarin, *der*

'**rose petal** *n.* Rosen[blüten]blatt, *das*

rosette [rəʊ'zet] *n.* Rosette, *die*

roster ['rɒstə(r)] *n.* Dienstplan, *der*

rostrum ['rɒstrəm] *n., pl.* **rostra** ['rɒstrə] *or* ~**s** Podium, *das*

rosy ['rəʊzɪ] *adj.* rosig

rot [rɒt] **1.** *n.* **a)** *see* 2: Verrottung, *die;* Fäulnis, *die; (fig.: deterioration)* Verfall, *der;* **stop the** ~ *(fig.)* dem Verfall Einhalt gebieten; **b)** *(sl.: nonsense)* Quark, *der (salopp).* **2.** *v. i.,* **-tt-** verrotten; ⟨*Fleisch, Gemüse, Obst:*⟩ verfaulen. **3.** *v. t.,* **-tt-** verrotten lassen; verfaulen lassen ⟨*Fleisch, Gemüse, Obst*⟩; zerstören ⟨*Zähne*⟩

rota ['rəʊtə] *n. (Brit.) (order of rotation)* Turnus, *der; (list)* Arbeitsplan, *der*

rotary ['rəʊtərɪ] *adj.* rotierend

rotate [rəʊ'teɪt] **1.** *v. i. (revolve)* rotieren; sich drehen. **2.** *v. t.* in Rotation versetzen. **rotation** [rəʊ'teɪʃn] *n.* **a)** Rotation, *die,* Drehung, *die* (**about** um); **b)** *(succession)* turnusmäßiger Wechsel; **in** *or* **by** ~: im Turnus

rote [rəʊt] *n.* **by** ~: auswendig

rotten ['rɒtn] *adj.,* ~**er** ['rɒtnə(r)], ~**est** ['rɒtnɪst] **a)** *(decayed)* verrottet; verfault ⟨*Obst, Gemüse*⟩; faul ⟨*Ei, Holz, Zähne*⟩; ~ **to the core** *(fig.)* verdorben bis ins Mark; **b)** *(corrupt)* verdorben; **c)** *(sl.: bad)* mies *(ugs.)*

rotund [rəʊ'tʌnd] *adj.* **a)** *(round)* rund; **b)** *(plump)* rundlich

rouble ['ruːbl] *n.* Rubel, *der*

rouge [ruːʒ] *n.* Rouge, *das*

rough [rʌf] **1.** *adj.* **a)** *(coarse, uneven)* rauh; holp[e]rig ⟨*Straße usw.*⟩; uneben ⟨*Gelände*⟩; unruhig ⟨*Überfahrt*⟩; **b)** *(violent)* grob ⟨*Person, Worte, Behandlung*⟩; **c)** *(trying)* hart; **this is** ~ **on him** das ist hart für ihn; **sth. is** ~ **going**

etw. ist nicht einfach; **d)** *(approximate)* grob ⟨*Skizze, Schätzung*⟩; vag ⟨*Vorstellung*⟩; ~ **paper/notebook** Konzeptpapier, *das*/Kladde, *die;* **e)** *(coll.: ill)* angeschlagen *(ugs.).* **2.** *n.* **[be] in** ~: [sich] im Rohzustand [befinden]. **3.** *adv.* rauh ⟨*spielen*⟩; **sleep** ~: im Freien schlafen. **4.** *v. t.* ~ **it** primitiv leben. **rough 'out** *v. t.* grob entwerfen. **rough 'up** *v. t. (sl.)* anrempeln *(ugs.)*

roughage ['rʌfɪdʒ] *n.* Ballaststoffe *Pl.*

rough: ~**-and-ready** *adj.* provisorisch; ~**-and-'tumble** *n.* [milde] Rauferei; ~ **copy,** ~ **draft** *ns.* grobe Skizze; grober Entwurf

roughen ['rʌfn] *v. t.* aufrauhen

'**roughly** *adv.* **a)** *(violently)* roh; grob; **b)** *(crudely)* leidlich; grob ⟨*skizzieren, bearbeiten, bauen*⟩; **c)** *(approximately)* ungefähr; grob ⟨*geschätzt*⟩

'**roughness** *n.* **a)** Rauheit, *die; (unevenness)* Unebenheit, *die;* **b)** *(violence)* Roheit, *die*

'**roughshod** *adj.* **ride** ~ **over sb./sth.** jmdn./etw. mit Füßen treten

roulette [ruː'let] *n.* Roulette, *das*

round [raʊnd] **1.** *adj.* rund; **in** ~ **figures** rund gerechnet. **2.** *n.* **a)** *(recurring series)* Serie, *die;* ~ **of talks/negotiations** Gesprächs-/Verhandlungsrunde, *die;* **the daily** ~: der Alltag; **b)** *(of ammunition)* Ladung, *die;* **50** ~**s |of ammunition|** 50 Schuß Munition; **c)** *(of game or contest)* Runde, *die;* **d)** *(burst)* ~ **of applause** Beifallssturm, *der;* **e)** ~ **|of drinks|** Runde, *die;* **f)** *(regular calls)* Runde, *die;* Tour, *die;* **go |on|** *or* **make one's** ~**s** seine Runden machen; **g)** **a** ~ **of toast/sandwiches** eine Scheibe Toast/eine Portion Sandwiches. **3.** *adv.* **a)** **all the year** ~: das ganze Jahr hindurch; **the third time** ~: beim dritten Mal; **have a look** ~: sich umsehen; **ask sb.** ~ **|for a drink|** jmdn. [zu einem Gläschen zu sich] einladen; **b)** *(by indirect way)* herum; **walk** ~: außen herum gehen; **c)** *(here)* hier; *(there)* dort; **I'll go** ~ **tomorrow** ich gehe morgen hin. **4.** *prep.* **a)** um [... herum]; **travel** ~ **England** durch England reisen; **run** ~ **the streets** durch die Straßen rennen; **walk** ~ **and** ~ **sth.** immer wieder um etw. herumgehen; **b)** *(in various directions from)* um [... herum]; rund um ⟨*einen Ort*⟩. **5.** *v. i.* ~ **a bend** um eine Kurve fahren/gehen/ kommen *usw.* **round 'off** *v. t.* abrunden. **round 'up** *v. t.* verhaften ⟨*Verdächtige*⟩; zusammentreiben ⟨*Vieh*⟩

round: ~ **a'bout** *adv. (on all sides)* ringsum; ~**about** 1. *n.* **a)** *(Brit.: merry-go-round)* Karussell, *das;* **b)** *(Brit.: road junction)* Kreisverkehr, *der.* 2. *adj.* umständlich

rounders ['raʊndəz] *n. sing. (Brit.)* Rounders, *das*

round: ~ **'number** *n.* runde Zahl; ~-**shouldered** [raʊnd'ʃəʊldəd] *adj.* ⟨Person⟩ mit einem Rundrücken; ~ **'trip** *n.* Rundreise, *die*

rouse [raʊz] *v. t.* wecken **(from** aus)

rousing ['raʊzɪŋ] *adj.* mitreißend ⟨Lied⟩; leidenschaftlich ⟨Rede⟩

rout [raʊt] 1. *n.* [wilde] Flucht; *(defeat)* verheerende Niederlage. 2. *v. t.* aufreiben ⟨Feind, Truppen⟩; vernichtend schlagen ⟨Gegner⟩

route [ruːt] *n.* Route, *die;* Weg, *der*

routine [ruːˈtiːn] 1. *n.* **a)** Routine, *die;* **b)** *(coll.: set speech)* Platte, *die (ugs.);* **c)** *(Theatre)* Nummer, *die;* *(Dancing, Skating)* Figur, *die.* 2. *adj.* routinemäßig; Routine⟨arbeit⟩

roux [ruː] *n.* Mehlschwitze, *die*

¹row [raʊ] 1. *(coll.) n.* **a)** *(noise)* Krach, *der;* **make a** ~: Krach machen; **b)** *(quarrel)* Krach, *der (ugs.);* **have/start a** ~: Krach haben/anfangen *(ugs.).* 2. *v. i* sich streiten

²row [rəʊ] *n.* Reihe, *die;* **in a** ~: in einer Reihe

³row [rəʊ] *v. i. & t. (with oars)* rudern

rowan ['rəʊən] *n.* ~[-**tree**] Eberesche, *die*

row-boat ['rəʊbəʊt] *n. (Amer.)* Ruderboot, *das*

rowdy ['raʊdɪ] 1. *adj.* rowdyhaft; **the party was** ~: auf der Party ging es laut zu. 2. *n.* Krawallmacher, *der*

rowing-boat ['rəʊɪŋbəʊt] *n. (Brit.)* Ruderboot, *das*

royal ['rɔɪəl] *adj.* königlich

royal: R~ **'Air Force** *n. (Brit.)* Königliche Luftwaffe; ~ **'blue** *n. (Brit.)* Königsblau, *das;* ~ **'family** *n.* königliche Familie; R~ **'Navy** *n. (Brit.)* Königliche Kriegsmarine

royalty ['rɔɪəltɪ] *n.* **a)** *(payment)* Tantieme, *die* **(on** für); **b)** *collect. (royal persons)* Mitglieder des Königshauses

RSPCA *abbr. (Brit.)* **Royal Society for the Prevention of Cruelty to Animals** *britischer Tierschutzverein*

rub [rʌb] 1. *v. t.,* -**bb**- reiben **(on, against** an + *Dat.*); *(to remove dirt etc.)* abreiben; *(to dry)* trockenreiben; ~ **sth. off sth.** etw. von etw. reiben. 2. *v. i.,* -**bb**- reiben **(**[up]**on, against** an +

Dat.). 3. *n.* **give it a** ~: reib es ab; **there's the** ~ *(fig.)* da liegt der Haken [dabei] *(ugs.).* **rub 'down** *v. t.* abreiben. **rub 'in** *v. t.* einreiben; **there's no need to** *or* **don't** ~ **it in** *(fig.)* reib es mir nicht [dauernd] unter die Nase. **rub 'off** *v. t.* wegreiben; wegwischen. **rub 'out** 1. *v. t.* ausreiben; *(using eraser)* ausradieren. 2. *v. i.* sich ausreiben/ sich ausradieren lassen

rubber ['rʌbə(r)] *n.* **a)** Gummi, *das od. der;* **b)** *(eraser)* Radiergummi, *der*

rubber: ~ **'band** *n.* Gummiband, *das;* ~ **plant** *n.* Gummibaum, *der;* ~ **'stamp** *n.* Gummistempel, *der;* ~-**stamp** *v. t. (fig.)* absegnen *(ugs.)*

rubbish ['rʌbɪʃ] 1. *n.* **a)** *(refuse)* Abfall, *der;* *(to be collected and dumped)* Müll, *der;* **b)** *(worthless material)* Plunder, *der (ugs.);* **be** ~: nichts taugen; **c)** *(nonsense)* Quatsch, *der (ugs.).* 2. *int.* Quatsch *(ugs.).* **'rubbish-bin** *n.* Abfall-/Mülleimer, *der.* **'rubbish dump** *n.* Müllkippe, *die*

rubble ['rʌbl] *n.* Trümmer *Pl.*

ruby ['ruːbɪ] *n.* Rubin, *der*

rucksack ['rʌksæk, 'rʊksæk] *n.* Rucksack, *der*

rudder ['rʌdə(r)] *n.* Ruder, *das*

ruddy ['rʌdɪ] *adj.* **a)** *(reddish)* rötlich; **b)** *(Brit. sl.: bloody)* verdammt *(salopp)*

rude [ruːd] *adj.* **a)** unhöflich; *(stronger)* rüde; **be** ~ **to sb.** zu jmdm. grob unhöflich sein/jmdn. rüde behandeln; **b)** *(abrupt)* unsanft; ~ **awakening** böses Erwachen. **'rudely** *adv.* **a)** *(impolitely)* unhöflich; rüde; **b)** *(abruptly)* jäh *(geh.).* **'rudeness** *n.* *(bad manners)* ungehöriges Benehmen

rudimentary [ruːdɪˈmentərɪ] elementar; primitiv ⟨Gebäude⟩

rudiments ['ruːdɪmənts] *n. pl.* Grundlagen *Pl.*

rueful ['ruːfl] *adj.* reumütig

ruffian ['rʌfɪən] *n.* Rohling, *der*

ruffle ['rʌfl] *v. t.* **a)** kräuseln; ~ **sb.'s hair** jmdm. durch die Haare fahren; **b)** *(upset)* aus der Fassung bringen

rug [rʌg] *n.* [kleiner, dicker] Teppich

Rugby ['rʌgbɪ] *n.* Rugby, *das*

rugged ['rʌgɪd] *adj.* **a)** *(uneven)* zerklüftet; unwegsam ⟨Land⟩; zerfurcht ⟨Gesicht⟩; **b)** *(sturdy)* robust

ruin ['ruːɪn] 1. *n.* **a)** *in sing. or pl. (remains)* Ruine, *die;* **in** ~**s** in Trümmern; **b)** *(downfall)* Ruin, *der.* 2. *v. t.* ruinieren; verderben ⟨Urlaub, Abend⟩; ~**ed** *(reduced to ruins)* verfal-

len; **a ~ed castle/church** eine Burg-/
Kirchenruine. **ruinous** ['ru:ɪnəs] *adj.*
ruinös

rule [ru:l] **1.** *n.* **a)** Regel, *die;* **the ~s of
the game** die Spielregeln; **be against
the ~s** regelwidrig sein; *(fig.)* gegen
die Spielregeln verstoßen; **as a ~:** in
der Regel; **~ of thumb** Faustregel, *die;*
b) *no pl. (government)* Herrschaft, *die*
(**over** über + *Akk.*). **2.** *v. t.* **a)** *(control)*
beherrschen; **b)** *(be the ruler of)* regieren; ⟨*Monarch, Diktator usw.:*⟩ herrschen über (+ *Akk.*). **3.** *v. i.* **a)** *(govern)* herrschen; **b)** *(decide)* entscheiden (**against** gegen; **in favour of** für).
rule 'out *v. t.* ausschließen; *(prevent)*
unmöglich machen

ruled [ru:ld] *adj.* liniert ⟨*Papier*⟩
ruler ['ru:lə(r)] *n.* **a)** *(person)* Herrscher, *der*/Herrscherin, *die;* **b)** *(for
measuring)* Lineal, *das*
ruling ['ru:lɪŋ] **1.** *adj.* herrschend
⟨*Klasse*⟩; regierend ⟨*Partei*⟩. **2.** *n.* Entscheidung, *die*
rum [rʌm] *n.* Rum, *der*
Rumania *etc.* [ru:'meɪnɪə] *see* **Romania** *etc.*
rumble ['rʌmbl] **1.** *n.* Grollen, *das.* **2.**
v. i. **a)** grollen; ⟨*Magen:*⟩ knurren; **b)**
⟨*Fahrzeug:*⟩ rumpeln *(ugs.)*
ruminate ['ru:mɪneɪt] *v. i.* **~ on** *or* **over
sth.** über etw. *(Akk.)* grübeln
rummage ['rʌmɪdʒ] *v. i.* wühlen; **~
through sth.** etw. durchwühlen *(ugs.)*
rummy ['rʌmɪ] *n.* Rommé, *das*
rumour *(Brit.; Amer.:* **rumor)**
['ru:mə(r)] **1.** *n.* Gerücht, *das;* **there is
a ~ that ...:** es geht das Gerücht, daß ...
2. *v. t.* **it is ~ed that ...:** es geht das Gerücht, daß ...
rump [rʌmp] *n.* **a)** *(buttocks)* Hinterteil, *das (ugs.);* **b)** *(remnant)* Rest, *der*
rumple ['rʌmpl] *v. t.* **a)** *(crease)* zerknittern; **b)** *(tousle)* zerzausen
'rump steak *n.* Rumpsteak, *das*
rumpus ['rʌmpəs] *n. (coll.)* Krach, *der
(ugs.);* **kick up** *or* **make a ~:** einen
Spektakel veranstalten *(ugs.)*
run [rʌn] **1.** *n.* **a)** Lauf, *der;* **on the ~:**
auf der Flucht; **b)** *(trip in vehicle)*
Fahrt, *die; (for pleasure)* Ausflug, *der;*
c) *(continuous stretch)* Länge, *die;* **d)**
(spell) **she has had a long ~ of success**
sie war lange [Zeit] erfolgreich; **have a
long ~** ⟨*Stück, Show:*⟩ viele Auffüh-
rungen erleben; **e)** *(succession)* Serie,
die; (Cards) Sequenz, *die;* **a ~ of vic-
tories** eine Siegesserie; **f)** *(use)* **have
the ~ of sth.** etw. zu seiner freien Ver-

fügung haben; **g)** *(enclosure)* Auslauf,
der; **h)** *(in stocking etc.)* Laufmasche,
die. **2.** *v. i.,* **-nn-, ran** [ræn], **run a)** laufen; **~ for the bus** laufen, um den Bus
zu kriegen *(ugs.);* **~ to help sb.** jmdm.
zu Hilfe eilen; **b)** *(roll, slide)* laufen;
⟨*Ball, Kugel:*⟩ rollen, laufen; ⟨*Schlitten, [Schiebe]tür:*⟩ gleiten; **c)** ⟨*Rad,
Maschine:*⟩ laufen; **d)** *(operate on a
schedule)* fahren; **~ between two places**
⟨*Zug, Bus:*⟩ zwischen zwei Orten verkehren; **e)** *(flow)* laufen; ⟨*Fluß:*⟩ fließen; ⟨*Augen:*⟩ tränen; **his nose was
~ning** ihm lief die Nase; **f)** ⟨*Vertrag,
Theaterstück:*⟩ laufen; **g)** *(have wording)* lauten; ⟨*Geschichte:*⟩ gehen *(fig.);*
h) ⟨*Butter, Eis:*⟩ zerlaufen; ⟨*Farben:*⟩
auslaufen; **i)** *(in election)* kandidieren.
3. *v. t.,* **-nn-, ran, run a)** laufen lassen;
(drive) fahren; **~ one's hand/fingers
through/along** *or* **over sth.** mit der
Hand/den Fingern durch etw. fahren/
über etw. *(Akk.)* streichen; **~ an** *or*
one's eye along *or* **down** *or* **over sth.**
(fig.) etw. überfliegen; **b)** *(cause to
flow)* [ein]laufen lassen; **~ a bath** ein
Bad einlaufen lassen; **c)** *(organize,
manage)* führen, leiten ⟨*Geschäft
usw.*⟩; veranstalten ⟨*Wettbewerb*⟩; **d)**
(operate) bedienen ⟨*Maschine*⟩; ein-
kehren lassen ⟨*Verkehrsmittel*⟩; einsetzen ⟨*Sonderbus, -zug*⟩; laufen lassen ⟨*Motor*⟩; **e)** *(own and use)* sich
(Dat.) halten ⟨*Auto*⟩; **f)** **~ sb. into town**
etc. jmdn. in die Stadt usw. fahren.
run a'cross *v. t.* **~ across sb./sth.**
jmdn. treffen/auf etw. *(Akk.)* stoßen.
run a'way *v. i.* **a)** *(flee)* weglaufen;
fortlaufen; **b)** *(abscond)* **~ away [from
home]** [von zu Hause] weglaufen. **run
'down 1.** *v. t.* **a)** *(collide with)* überfahren; **b)** *(criticize)* heruntermachen
(ugs.); **c)** *(reduce)* abbauen. **2.** *v. i.* **a)**
hin-/herunterlaufen; **b)** *(decline)* sich
verringern; **c)** ⟨*Uhr, Spielzeug:*⟩ ablaufen; ⟨*Batterie*⟩ leer werden. **'run into**
v. t. **a)** **~ into a tree** gegen einen Baum
fahren; **b)** *(meet)* **~ into sb.** jmdm. in
die Arme laufen *(ugs.);* **c)** stoßen auf
(+ *Akk.*) ⟨*Schwierigkeiten, Widerstand
usw.*⟩; **d)** *(amount to)* **~ into thousands**
in die Tausende gehen. **run 'off 1.** *v. i.*
weglaufen. **2.** *v. t.* abziehen ⟨*Kopien*⟩.
run 'out *v. i.* **a)** hin-/herauslaufen; **b)**
⟨*Vorräte, Bestände:*⟩ zu Ende gehen.
run 'out of *v. t.* **sb. ~s out of sth.**
jmdm. geht etw. aus; **I'm ~ning out of
patience** meine Geduld geht zu Ende.
run 'over 1. ['---] *v. t. (knock down)*

überfahren. 2. [-'--] *v. i.* überlaufen.
'**run through** *v. t.* durchspielen
⟨*Theaterstück*⟩. '**run to** *v. t.* **a)** *(amount to)* sich belaufen auf *(Akk.)*; **b)** *(be sufficient for)* sth. will ~ to sth. etw. reicht für etw. **run 'up** 1. *v. i.* hinlaufen; **come ~ning up** hingelaufen kommen. 2. *v. t.* **a)** rasch nähen ⟨*Kleidungsstück*⟩; **b)** zusammenkommen lassen ⟨*Schulden, Rechnung*⟩. **run 'up against** *v. t.* stoßen auf (+ *Akk.*) ⟨*Probleme, Widerstand usw.*⟩

run: **~away** 1. *n.* Ausreißer, *der*/Ausreißerin, *die* *(ugs.)*; 2. *attrib. adj.* durchgegangen ⟨*Pferd*⟩; außer Kontrolle geraten ⟨*Fahrzeug, Preise*⟩; galoppierend ⟨*Inflation*⟩; **~-down** 1. ['--] *n. (coll.: briefing)* Übersicht, *die* (on über + *Akk.*); 2. [-'-] *adj. (tired)* mitgenommen
¹**rung** [rʌŋ] *n.* Sprosse, *die*
²**rung** *see* ²**ring** 2, 3
'**runner** *n.* **a)** Läufer, *der*/Läuferin, *die;* **b)** *(Bot.)* Ausläufer, *der;* **c)** *(on sledge)* Kufe, *die.* '**runner bean** *n.* *(Brit.)* Stangenbohne, *die.* **runner-'up** *n.* Zweite, *der/die;* **the runners-up** die Plazierten
'**running** 1. *n.* **a)** *(management)* Leitung, *die;* **b)** *(action)* Laufen, *das;* **in/ out of the ~:** im/aus dem Rennen. 2. *adj. (in succession)* hintereinander; **win for the third year ~:** schon drei Jahre hintereinander gewinnen. **running 'commentary** *n. (Broadcasting; also fig.)* Live-Kommentar, *der*
runny ['rʌnɪ] *adj.* **a)** laufend ⟨*Nase*⟩; **b)** zu dünn ⟨*Farbe, Marmelade*⟩
run: -of-the-'mill *adj.* ganz gewöhnlich; **~-up** *n.* **a)** *during or* in the **~-up to an event** im Vorfeld eines Ereignisses; **b)** *(Sport)* Anlauf, *der;* **~way** *n. (for take-off)* Startbahn, *die; (for landing)* Landebahn, *die*
rupture ['rʌptʃə(r)] 1. *n.* Bruch, *der.* 2. *v. t.* **~ oneself** sich *(Dat.)* einen Bruch zuziehen
rural ['rʊərl] *adj.* ländlich
ruse [ruːz] *n.* List, *die*
¹**rush** [rʌʃ] *n. (Bot.)* Binse, *die*
²**rush** 1. *n.* **a)** *(hurry)* Eile, *die;* **what's all the ~?** wozu diese Hast?; **be in a |great| ~:** in [großer] Eile sein; **b)** *(period of great activity)* Hochbetrieb, *der; (~-hour)* Stoßzeit, *die;* **c) make a ~ for sth.** sich auf etw. *(Akk.)* stürzen. 2. *v. t.* **a)** **~ sb./sth. somewhere** jmdn./ etw. auf schnellstem Wege irgendwohin bringen; **be ~ed** *(have to hurry)* in

Eile sein; **~ sb. into doing sth.** jmdn. dazu drängen, etw. zu tun; **b)** *(perform quickly)* auf die Schnelle erledigen; **~ it zu schnell machen. 3. *v. i.* **a)** *(move quickly)* eilen; ⟨*Hund, Pferd:*⟩ laufen; **~ to help sb.** jmdm. zu Hilfe eilen; **b)** *(hurry unduly)* sich zu sehr beeilen; **don't ~!** nur keine Eile! **rush a'bout, rush a'round** *v. i.* herumhetzen
'**rush-hour** *n.* Stoßzeit, *die*
rusk [rʌsk] *n.* Zwieback, *der*
Russia ['rʌʃə] *pr. n.* Rußland *(das).* **Russian** ['rʌʃn] 1. *adj.* russisch; **sb. is ~:** jmd. ist Russe/Russin. 2. *n.* **a)** *(person)* Russe, *der*/Russin, *die;* **b)** *(language)* Russisch, *das; see also* **English** 2 a
rust [rʌst] 1. *n.* Rost, *der.* 2. *v. i.* rosten
rustic ['rʌstɪk] *adj.* **a)** ländlich; **b)** rustikal ⟨*Mobiliar*⟩
rustle ['rʌsl] 1. *n.* Rascheln, *das.* 2. *v. i.* rascheln. 3. *v. t.* **a)** rascheln lassen; **b)** *(Amer.: steal)* stehlen. **rustle 'up** *v. t.* zusammenzaubern ⟨*Mahlzeit*⟩
'**rust-proof** *adj.* rostfrei
'**rusty** *adj.* rostig
rut [rʌt] *n.* Spurrille, *die;* **be in a ~** *(fig.)* aus dem [Alltags]trott nicht mehr herauskommen
ruthless ['ruːθlɪs] *adj.* rücksichtslos
rye [raɪ] *n.* Roggen, *der*

S

S, s [es] *n.* S, s, *das*
S. *abbr.* **a)** south S; **b)** southern s.
sabbath ['sæbəθ] *n.* Sabbath, *der*
sabbatical [sə'bætɪkl] 1. *adj.* **~ term/ year** Forschungssemester/-jahr, *das.* 2. *n.* Forschungsurlaub, *der*
sabotage ['sæbətɑːʒ] 1. *n.* Sabotage, *die.* 2. *v. t.* einen Sabotageakt verüben auf (+ *Akk.*); *(fig.)* sabotieren
saccharin ['sækərɪn] *n.* Saccharin, *das*
sachet ['sæʃeɪ] *n.* Beutel, *der; (cushion-shaped)* Kissen, *das*
sack [sæk] 1. *n.* **a)** Sack, *der;* **b)** *(coll.: dismissal)* Rausschmiß, *der (ugs.);* **get the ~:** rausgeschmissen werden *(ugs.);* **give sb. the ~:** jmdn. raus-

schmeißen *(ugs.)*. **2.** *v. t. (coll.)* rausschmeißen *(ugs.)* **(for wegen)**
sacrament ['sækrəmənt] *n.* Sakrament, *das*
sacred ['seɪkrɪd] *adj.* heilig
sacrifice ['sækrɪfaɪs] **1.** *n.* Opfer, *das*.
2. *v. t.* opfern
sacrilege [sækrɪlɪdʒ] *n.* |act of| ~: Sa-
krileg, *das*
sad [sæd] *adj.* traurig **(at, about** über
+ *Akk.*); schmerzlich ⟨*Tod, Verlust*⟩;
feel ~: traurig sein. **sadden** ['sædn]
v. t. traurig stimmen
saddle ['sædl] **1.** *n.* Sattel, *der*. **2.** *v. t.*
a) satteln ⟨*Pferd usw.*⟩; **b)** *(fig.)* ~ **sb.**
with sth. jmdm. etw. aufbürden *(geh.)*.
'**saddle-bag** *n.* Satteltasche, *die*
sadism ['seɪdɪzm] *n.* Sadismus, *der*.
sadist ['seɪdɪst] *n.* Sadist, *der*/Sadi-
stin, *die*. **sadistic** [sə'dɪstɪk] *adj.*, **sa-**
'**distically** *adv.* sadistisch
'**sadly** *adv.* **a)** *(with sorrow)* traurig; **b)**
(unfortunately) leider
'**sadness** *n.* Traurigkeit, *die*
safari [sə'fɑːrɪ] *n.* Safari, *die;* **on** ~: auf
Safari
safe [seɪf] **1.** *n.* Safe, *der;* Geld-
schrank, *der.* **2.** *adj.* **a)** *(out of danger)*
sicher **(from** vor + *Dat.*); **he's** ~: er ist
in Sicherheit; ~ **and sound** sicher und
wohlbehalten; **b)** *(free from danger)*
ungefährlich; sicher ⟨*Ort, Hafen*⟩;
wish sb. a ~ **journey** jmdm. eine gute
Reise wünschen; **to be on the** ~ **side**
zur Sicherheit; **c)** *(reliable)* sicher
⟨*Methode, Investition*⟩. '**safeguard 1.**
n. Schutz, *der.* **2.** *v. t.* schützen.
'**safely** *adv.* sicher; **did the parcel ar-**
rive ~**?** ist das Paket heil angekom-
men? **safety** ['seɪftɪ] *n.* Sicherheit, *die*
safety: ~**-belt** *n.* Sicherheitsgurt,
der; ~ **helmet** *n.* Schutzhelm, *der;* ~
margin *n.* Spielraum, *der;* ~**-pin** *n.*
Sicherheitsnadel, *die;* ~**-valve** *n.* Si-
cherheitsventil, *das; (fig.)* Ventil, *das*
sag [sæg] *v. i.,* **-gg-** durchhängen;
(sink) sich senken
saga ['sɑːgə] *n.* **a)** *(story of adventure)*
Heldenepos, *das; (medieval narrative)*
Saga, *die;* **b)** *(coll.: long involved story)*
[ganzer] Roman *(fig.)*
¹**sage** [seɪdʒ] *n. (Bot.)* Salbei, *der od.*
die
²**sage 1.** *adj.* weis. **2.** *n.* Weise, *der*
Sagittarius [sædʒɪ'teərɪəs] *n.* der
Schütze
Sahara [sə'hɑːrə] *pr. n.* **the** ~ **|Desert|**
die [Wüste] Sahara
said *see* **say 1**

sail [seɪl] **1.** *n.* **a)** Segelfahrt, *die;* **b)**
(piece of canvas) Segel, *das.* **2.** *v. i.* **a)**
(travel on water) fahren; *(in sailing*
boat) segeln; **b)** *(start voyage)* auslaufen **(for** nach). **3.** *v. t.* **a)** steuern ⟨*Boot,*
Schiff⟩; segeln mit ⟨*Segeljacht,*
-schiff⟩; **b)** durchfahren/⟨*Segelschiff:*⟩
durchsegeln ⟨*Meer*⟩
sail: ~**board** *n.* Surfbrett, *das (zum*
Windsurfen); ~**-boarding** *n.* Windsurfen, *das;* ~**boat** *n. (Amer.)* Segelboot, *das*
'**sailing** *n.* Segeln, *das.* '**sailing boat**
n. Segelboot, *das.* '**sailing ship** *n.*
Segelschiff, *das*
sailor ['seɪlə(r)] *n.* Seemann, *der; (in*
navy) Matrose, *der*
saint 1. [sənt] *adj.* **S**~ **Michael** der heilige Michael; Sankt Michael. **2.** [seɪnt]
n. Heilige, *der/die.* '**saintly** ['seɪntlɪ]
adj. heilig
sake [seɪk] *n.* **for the** ~ **of** um ... *(Gen.)*
willen; **for my** *etc.* ~: um meinetwil-
len *usw.;* mir *usw.* zuliebe
salad ['sæləd] *n.* Salat, *der.* '**salad**
cream *n.* ≈ Mayonnaise, *die.* '**salad**
dressing *n.* Salatsoße, *die*
salary ['sælərɪ] *n.* Gehalt, *das*
sale [seɪl] *n.* **a)** Verkauf, *der; (at re-*
duced prices) Ausverkauf, *der;* |up| **for**
~: zu verkaufen; **b)** ~**s** *(amount sold)*
Verkaufszahlen *Pl.* (of für); Absatz,
der; **c)** |jumble *or* rummage| ~: [Wohl-
tätigkeits]basar, *der*
salesman ['seɪlzmən] *n., pl.* ~**men**
['seɪlzmən] Verkäufer, *der.* '**sales-**
manship *n.* Kunst des Verkaufens
'**saleswoman** *n.* Verkäuferin, *die*
salient ['seɪlɪənt] *adj.* auffallend
saliva [sə'laɪvə] *n.* Speichel, *der*
sallow ['sæləʊ] *adj.* blaßgelb
salmon ['sæmən] *n.* Lachs, *der*
saloon [sə'luːn] *n.* **a)** *(Brit.)* ~ |bar| se-
parater Teil eines Pubs mit mehr Kom-
fort; **b)** *(Brit.)* ~ |car| Limousine, *die*
salt [sɔːlt, sɒlt] **1.** *n.* |common| ~:
[Koch]salz, *das.* **2.** *adj. (containing or*
tasting of ~) salzig; *(preserved with* ~)
gepökelt ⟨*Fleisch*⟩; gesalzen ⟨*Butter*⟩.
3. *v. t.* **a)** salzen; **b)** *(cure)* [ein]pökeln;
c) ~ **the roads** Salz auf die Straßen
streuen. '**salt-cellar** *n.* Salzstreuer,
der. **salt 'water** *n.* Salzwasser, *das*
'**salty** *adj.* salzig
salute [sə'luːt] **1.** *v. t.* grüßen. **2.** *v. i.*
(Mil., Navy) [militärisch] grüßen. **3.** *n.*
Salut, *der;* militärischer Gruß
salvage ['sælvɪdʒ] **1.** *n.* Bergung, *die.*
2. *v. t.* bergen

salvation [sæl'veɪʃn] *n*. Erlösung, *die*.
Salvation 'Army *n*. Heilsarmee, *die*
salvo ['sælvəʊ] *n*. Salve, *die*
Samaritan [sə'mærɪtən] *n*. **good ~**:
[barmherziger] Samariter; **the ~s** (*organization*) ≈ die Telefonseelsorge
same [seɪm] **1.** *adj*. **the ~**: der/die/das
gleiche; **the ~** |**thing**| (*identical*) der-/
die-/dasselbe. **2.** *adv*. **all** *or* **just the ~**:
trotzdem
sample ['sɑːmpl] **1.** *n*. (*example*) [Muster]beispiel, *das*; (*specimen*) Probe,
die; |**commercial**| **~**: Muster, *das*. **2.**
v. t. probieren
sanctify ['sæŋktɪfaɪ] *v. t.* heiligen
sanctimonious [sæŋktɪ'məʊnɪəs] *adj*.
scheinheilig
sanction ['sæŋkʃn] **1.** *n*. Sanktion,
die. **2.** *v. t.* sanktionieren
sanctity ['sæŋktɪtɪ] *n*. Heiligkeit, *die*
sanctuary ['sæŋktʃʊərɪ] *n*. **a)** (*holy
place*) Heiligtum, *das*; **b)** (*refuge*) Zufluchtsort, *der*; **c)** (*for animals*) Naturschutzgebiet, *das*
sand [sænd] **1.** *n*. Sand, *der*. **2.** *v. t.* **~
sth.** |**down**| etw. [ab]schmirgeln
sandal ['sændl] *n*. Sandale, *die*
sand: **~bag** *n*. Sandsack, *der*; **2.**
v. t. mit Sandsäcken schützen;
~bank *n*. Sandbank, *die*; **~castle**
n. Sandburg, *die*; **~paper 1.** *n*. Sandpapier, *das*; **2.** *v. t.* [mit Sandpapier]
[ab]schmirgeln; **~-pit** *n*. Sandkasten,
der; **~stone** *n*. Sandstein, *der*
sandwich ['sænwɪdʒ] **1.** *n*. Sandwich,
der od. das; ≈ [zusammengeklapptes]
belegtes Brot; **cheese ~**: Käsebrot,
das. **2.** *v. t.* einschieben (**between** zwischen + *Akk.*; **into** in + *Akk.*)
'sandy *adj*. **a)** sandig; Sand⟨*boden,
-strand*⟩; **b)** rotblond ⟨*Haar*⟩
sane [seɪn] *adj*. **a)** geistig gesund; **b)**
(*sensible*) vernünftig
sang *see* **sing**
sanitary ['sænɪtərɪ] *adj*. sanitär ⟨*Verhältnisse, Anlagen*⟩. **'sanitary napkin** (*Amer.*), **'sanitary towel** (*Brit.*)
ns. Damenbinde, *die*
sanitation [sænɪ'teɪʃn] *n*. Kanalisation und Abfallbeseitigung
sanity ['sænɪtɪ] *n*. geistige Gesundheit; **lose one's ~**: den Verstand verlieren
sank *see* **sink 2, 3**
Santa Claus ['sæntə klɔːz] *n*. der
Weihnachtsmann
sap [sæp] **1.** *n*. Saft, *der*. **2.** *v. t.*, **-pp-**
zehren an (+ *Dat.*)
sapling ['sæplɪŋ] *n*. junger Baum

sarcasm ['sɑːkæzm] *n*. Sarkasmus,
der. **sarcastic** [sɑː'kæstɪk] *adj*. sarkastisch
sardine [sɑː'diːn] *n*. Sardine, *die*
Sardinia [sɑː'dɪnɪə] *pr. n*. Sardinien
(*das*)
sardonic [sɑː'dɒnɪk] *adj*. höhnisch;
sardonisch ⟨*Lächeln*⟩
sash [sæʃ] *n*. Schärpe, *die*
sat *see* **sit**
Sat. *abbr*. Saturday Sa.
Satan ['seɪtən] *pr. n*. Satan, *der*. **satanic** [sə'tænɪk] *adj*. satanisch
satchel ['sætʃl] *n*. [Schul]ranzen, *der*
satellite ['sætəlaɪt] *n*. Satellit, *der*.
satellite: **~ 'broadcasting** *n*. Satellitenfunk, *der*; **~-dish** *n*. Satellitenschüssel, *die*; **~ 'television** *n*. Satellitenfernsehen, *das*
satin ['sætɪn] *n*. Satin, *der*
satire ['sætaɪə(r)] *n*. Satire, *die* (**on** auf
+ *Akk.*). **satirical** [sə'tɪrɪkl] *adj*. satirisch
satisfaction [sætɪs'fækʃn] *n*. Befriedigung, *die* (**at, with** über + *Akk.*);
meet with sb.'s |**complete**| **~**: jmdn. [in
jeder Weise] zufriedenstellen
satisfactory [sætɪs'fæktərɪ] *adj*. zufriedenstellend
satisfy ['sætɪsfaɪ] *v. t.* **a)** befriedigen;
zufriedenstellen ⟨*Kunden*⟩; stillen
⟨*Hunger, Durst*⟩; **b)** (*convince*) **~ sb.**
|**of sth.**| jmdn. [von etw.] überzeugen.
'satisfying *adj*. befriedigend; sättigend ⟨*Gericht, Speise*⟩
saturate ['sætʃəreɪt] *v. t.* durchnässen;
[mit Feuchtigkeit durch]tränken
⟨*Boden, Erde*⟩. **saturated** ['sætʃəreɪtɪd] *adj*. durchnäßt. **saturation** [sætʃə'reɪʃn] *n*. Durchnässung, *die*
Saturday ['sætədeɪ, 'sætədɪ] *n*. Sonnabend, *der*; Samstag, *der*; *see also* **Friday**
Saturn ['sætən] *pr. n*. (*Astron.*) Saturn,
der
sauce [sɔːs] *n*. **a)** Soße, *die*; **b)** (*impudence*) Frechheit, *die*. **saucepan**
['sɔːspən] *n*. Kochtopf, *der*; (*with
straight handle*) Kasserolle, *die*
saucer ['sɔːsə(r)] *n*. Untertasse, *die*
saucy ['sɔːsɪ] *adj*. **a)** (*rude*) frech; **b)**
(*pert, jaunty*) keck
Saudi Arabia [saʊdɪ ə'reɪbɪə] *pr. n*.
Saudi-Arabien (*das*)
sauna ['sɔːnə, 'saʊnə] *n*. Sauna, *die*
saunter ['sɔːntə(r)] *v. i.* schlendern
sausage ['sɒsɪdʒ] *n*. Wurst, *die*. **sausage 'roll** *n*. Blätterteig mit Wurstfüllung

savage ['sævɪdʒ] **1.** *adj.* **a)** *(uncivilized)* primitiv; wild ⟨*Volksstamm*⟩; unzivilisiert ⟨*Land*⟩; **b)** *(fierce)* brutal; wild ⟨*Tier*⟩. **2.** *n.* Wilde, *der/die (veralt.).*
savagery ['sævɪdʒrɪ] *n.* Brutalität, *die*
save [seɪv] **1.** *v.t.* **a)** *(rescue)* retten (**from** vor + *Dat.*); ~ **oneself from falling** sich [beim Hinfallen] fangen; **b)** *(put aside)* aufheben; sparen ⟨*Geld*⟩; sammeln ⟨*Briefmarken usw.*⟩; *(conserve)* sparsam umgehen mit; **c)** *(make unnecessary)* sparen ⟨*Geld, Zeit, Energie*⟩; ~ **sb./oneself sth.** jmdm./sich etw. ersparen; **d)** *(Sport)* abwehren ⟨*Schuß, Ball*⟩. **2.** *v.i.* sparen (**on** *Akk.*). **3.** *n.* *(Sport)* Abwehr, *die.*
save 'up 1. *v.t.* sparen. **2.** *v.i.* sparen (**for** für, auf + *Akk.*)
'**saver** *n.* Sparer, *der/*Sparerin, *die*
'**saving** ['seɪvɪŋ] **1.** *n. in pl.* Ersparnisse *Pl.* **2.** *adj.* ⟨*kosten-, benzin*⟩sparend
savings: ~ **account** *n.* Sparkonto, *das;* ~ **bank** *n.* Sparkasse, *die*
saviour ['seɪvjə(r)] *n.* **a)** Retter, *der/*Retterin, *die;* **b)** *(Relig.)* **the S~:** der Heiland
savor *etc. (Amer.) see* **savour** *etc.*
savour ['seɪvə(r)] *(Brit.)* **1.** *n. (flavour)* Geschmack, *der.* **2.** *v.t.* genießen
savoury ['seɪvərɪ] *(Brit.)* **1.** *adj.* **a)** pikant; salzig; **b)** *(appetizing)* appetitanregend. **2.** *n.* [pikantes] Häppchen
¹**saw** [sɔː] **1.** *n.* Säge, *die. v.t., p.p.* **sawn** [sɔːn] *or* **sawed** [zer]sägen; ~ **in half** in der Mitte durchsägen. **3.** *v.i., p.p.* **sawn** *or* **sawed** sägen; ~ **through sth.** etw. durchsägen
²**saw** *see* **see**
'**sawdust** *n.* Sägemehl, *das*
sawn *see* ¹**saw** 2, 3
saxophone ['sæksəfəʊn] *n.* Saxophon, *das*
say [seɪ] **1.** *v.t. pres. t.* **he says** [sez], *p.t. & p.p.* **said** [sed] **a)** sagen; **that is to** ~: das heißt; **do as** *or* **what I** ~: tun Sie, was ich sage; **when all is said and done** letztes Endes; **go without** ~**ing** sich von selbst verstehen; **she is said to be clever/to have done it** man sagt, sie sei klug/habe es getan; **b)** *(recite)* sprechen ⟨*Gebet, Text*⟩; **c)** *(have specified wording or reading)* sagen; ⟨*Zeitung:*⟩ schreiben; ⟨*Uhr:*⟩ zeigen ⟨*Uhrzeit*⟩; **what does it** ~ **here?** was steht hier? **2.** *n.* **have a** *or* **some** ~: ein Mitspracherecht haben (**in** bei); **have no** ~: nichts zu sagen haben; **have one's** ~: seine Meinung sagen. '**saying** *n.* Redensart, *die*

scab [skæb] *n.* [Wund]schorf, *der*
scaffold ['skæfəld] *n.* Schafott, *das*
'**scaffolding** *n.* Gerüst, *das*
scald [skɔːld, skɒld] **1.** *n.* Verbrühung, *die.* **2.** *v.t.* verbrühen
¹**scale** [skeɪl] *n.* **a)** *(of fish, reptile, etc.)* Schuppe, *die;* **b)** *(in kettle etc.)* Kesselstein, *der; (on teeth)* Zahnstein, *der*
²**scale** *n.* **a)** *in sing. or pl. (weighing-instrument)* ~**|s|** Waage, *die;* **b)** *(dish of balance)* Waagschale, *die*
³**scale 1.** *n.* **a)** *(series of degrees)* Skala, *die;* **b)** *(Mus.)* Tonleiter, *die;* **c)** *(dimensions)* Ausmaß, *das;* **be on a small** ~: bescheidenen Umfang haben; **d)** *(ratio of reduction)* Maßstab, *der;* **what is the** ~ **of the map?** welchen Maßstab hat diese Karte?; **e)** *(indication) (on map)* Maßstab, *der; (on thermometer)* [Anzeige]skala, *die.* **2.** *v.t.* ersteigen ⟨*Mauer, Leiter, Gipfel*⟩.
scale 'down *v.t.* [entsprechend] drosseln ⟨*Produktion*⟩; Abstriche machen bei ⟨*Planungen*⟩
scalp [skælp] *n.* Kopfhaut, *die*
scalpel ['skælpl] *n.* Skalpell, *das*
scam [skæm] *n. (Amer. sl.)* Masche, *die (ugs.).*
scamper ['skæmpə(r)] *v.i.* ⟨*Person:*⟩ flitzen; ⟨*Tier:*⟩ huschen
scampi ['skæmpɪ] *n. pl.* Scampi *Pl.*
scan [skæn] **1.** *v.t., -nn-:* **a)** *(search thoroughly)* absuchen (**for** nach); **b)** *(look over cursorily)* flüchtig ansehen; überfliegen ⟨*Zeitung, Liste usw.*⟩ (**for** auf der Suche nach); **c)** *(Med.)* szintigraphisch untersuchen. **2.** *v.i., -nn-* ⟨*Vers[zeile]:*⟩ das richtige Versmaß haben. **3.** *n. (Med.)* szintigraphische Untersuchung.
scandal ['skændl] *n.* **a)** Skandal, *der* (**about/of** um); *(story)* Skandalgeschichte, *die;* **b)** *(outrage)* Empörung, *die;* **c)** *(gossip)* Klatsch, *der (ugs.).*
scandalize ['skændəlaɪz] *v.t.* schockieren. **scandalous** ['skændələs] *adj.* skandalös; schockierend ⟨*Bemerkung*⟩
Scandinavia [skændɪ'neɪvɪə] *pr. n.* Skandinavien *(das)*
scant [skænt] *adj.* wenig. **scanty** ['skæntɪ] *adj.* spärlich; knapp ⟨*Bikini*⟩
scapegoat ['skeɪpgəʊt] *n.* Sündenbock, *der;* **make sb. a** ~: jmdn. zum Sündenbock machen
scar [skɑː(r)] **1.** *n.* Narbe, *die.* **2.** *v.t., -rr-:* ~ **sb./sb.'s face** bei jmdm./in jmds. Gesicht *(Dat.)* Narben hinterlassen

scarce [skeəs] *adj.* **a)** *(insufficient)* knapp; **b)** *(rare)* selten; **make oneself ~** *(coll.)* sich aus dem Staub machen *(ugs.).* **'scarcely** *adv.* kaum. **scarcity** ['skeəsɪtɪ] *n.* Knappheit, *die (of* an + *Dat.)*

scare [skeə(r)] **1.** *n.* **a)** *(sensation of fear)* Schreck[en], *der;* **give sb. a ~:** jmdm. einen Schreck[en] einjagen; **b)** *(general alarm)* [allgemeine] Hysterie; **bomb ~:** Bombendrohung, *die.* **2.** *v.t.* *(frighten)* Angst machen (+ *Dat.*); *(startle)* erschrecken. **scare a'way, scare 'off** *v.t.* verscheuchen **'scarecrow** *n.* Vogelscheuche, *die*

scared [skeəd] *adj.* **be ~ of sb./sth.** vor jmdm./etw. Angst haben; **be ~ of doing/to do sth.** sich nicht [ge]trauen, etw. zu tun

scarf [skɑːf] *n., pl.* **~s** *or* **scarves** [skɑːvz] Schal, *der; (square)* Halstuch, *das; (worn over hair)* Kopftuch, *das*

scarlet ['skɑːlɪt] **1.** *n.* Scharlach, *der.* **2.** *adj.* scharlachrot. **scarlet 'fever** *n.* Scharlach, *der*

scarves *see* **scarf**

scary ['skeərɪ] *adj.* furchterregend ⟨*Anblick*⟩; schaurig ⟨*Film, Geschichte*⟩

scatter ['skætə(r)] **1.** *v.t.* **a)** vertreiben; auseinandertreiben ⟨*Menge*⟩; **b)** *(distribute irregularly)* verstreuen. **2.** *v.i.* sich auflösen; ⟨*Menge:*⟩ sich zerstreuen; *(in fear)* auseinanderstieben. **scattered** ['skætəd] *adj.* verstreut; vereinzelt ⟨*Regenschauer*⟩

scavenge ['skævɪndʒ] *v.i.* **~ for sth.** nach etw. suchen. **'scavenger** *n.* *(animal)* Aasfresser, *der; (fig. derog.: person)* Aasgeier, *der (ugs.)*

scene [siːn] *n.* **a)** *(place of event)* Schauplatz, *der; ~* **of the crime** Tatort, *der;* **b)** *(division of act)* Auftritt, *der;* **c)** *(view)* Anblick, *der;* **d) behind the ~s** hinter den Kulissen. **scenery** ['siːnərɪ] *n.* **a)** Landschaft, *die;* **b)** *(Theatre)* Bühnenbild, *das.* **scenic** ['siːnɪk] *adj.* landschaftlich schön

scent [sent] **1.** *n.* **a)** *(smell)* Duft, *der;* **b)** *(Hunting; also fig.: trail)* Fährte, *die;* **be on the ~ of sb./sth.** *(fig.)* jmdm./einer Sache auf der Spur sein; **c)** *(Brit.: perfume)* Parfüm, *das.* **2.** *v.t.* wittern

sceptic ['skeptɪk] *n.* Skeptiker, *der/*Skeptikerin, *die.* **sceptical** ['skeptɪkl] *adj.* skeptisch; **be ~ about** *or* **of sb./sth.** jmdm./einer Sache skeptisch gegenüberstehen. **scepticism** ['skeptɪsɪzm] *n.* Skepsis, *die*

schedule ['ʃedjuːl] **1.** *n.* **a)** *(list)* Tabelle, *die; (for event)* Programm, *das;* **b)** *(of work)* Zeitplan, *der;* **c) on ~:** plangemäß. **2.** *v.t.* zeitlich planen. **'scheduled flight** *n.* Linienflug, *der*

scheme [skiːm] *n.* **a)** *(arrangement)* Anordnung, *die;* **b)** *(plan)* Programm, *das; (project)* Projekt, *das;* **c)** *(dishonest plan)* Intrige, *die*

schizophrenia [ˌskɪtsəˈfriːnɪə] *n.* Schizophrenie, *die.* **schizophrenic** [ˌskɪtsəˈfrenɪk, ˌskɪtsəˈfriːnɪk] *adj.* schizophren

scholar ['skɒlə(r)] *n.* Gelehrte, *der/ die.* **'scholarly** *adj.* wissenschaftlich; gelehrt ⟨*Person*⟩. **'scholarship** *n.* **a)** *(award)* Stipendium, *das;* **b)** *(scholarly work)* Gelehrsamkeit, *die*

school [skuːl] *n.* Schule, *die; (Amer.: college)* Hochschule, *die;* **be at** *or* **in ~:** in der Schule sein; *(attend ~)* zur Schule gehen; **go to ~:** zur Schule gehen; **~ holidays/exchange** Schulferien *Pl./*Schüleraustausch, *der*

school: ~boy *n.* Schüler, *der;* **~girl** *n.* Schülerin, *die;* **~master** *n.* Lehrer, *der;* **~mistress** *n.* Lehrerin, *die;* **~teacher** *n.* Lehrer, *der/*Lehrerin, *die*

sciatica [saɪˈætɪkə] *n.* Ischias, *die*

science ['saɪəns] *n.* Wissenschaft, *die.* **science 'fiction** *n.* Science-fiction, *die.* **scientific** [ˌsaɪənˈtɪfɪk] *adj.* wissenschaftlich. **scientist** ['saɪəntɪst] *n.* Wissenschaftler, *der/*Wissenschaftlerin, *die*

scintillating ['sɪntɪleɪtɪŋ] *adj.* *(fig.)* geistsprühend

scissors ['sɪzəz] *n. pl.* **[pair of] ~:** Schere, *die*

¹scoff [skɒf] *v.i.* *(mock)* spotten; **~ at** sich lustig machen über (+ *Akk.*)

²scoff *v.t.* *(sl.: eat greedily)* verschlingen

scold [skəʊld] *v.t.* ausschimpfen **(for** wegen**)**; **she ~ed him for being late** sie schimpfte ihn aus, weil er zu spät kam

scone [skɒn, skəʊn] *n.* weicher, oft zum Tee gegessener kleiner Kuchen

scoop [skuːp] **1.** *n.* **a)** Schaufel, *die; (for ice-cream etc.)* Portionierer, *der;* **b)** *(Journ.)* Knüller, *der (ugs.).* **2.** *v.t.* schaufeln ⟨*Kohlen, Zucker*⟩; schöpfen ⟨*Flüssigkeit*⟩. **scoop 'out** *v.t.* **a)** *(hollow out)* aushöhlen; schaufeln ⟨*Loch, Graben*⟩; **b)** [her]ausschöpfen ⟨*Flüssigkeit*⟩; auslöffeln ⟨*Fruchtfleisch*⟩; *(with a knife)* herausschneiden ⟨*Gehäuse, Fruchtfleisch*⟩. **scoop 'up** *v.t.*

schöpfen ⟨Flüssigkeit, Suppe⟩; schaufeln ⟨Erde⟩

scooter ['sku:tə(r)] *n.* **a)** *(toy)* Roller, *der;* **b)** **|motor|** ~: [Motor]roller, *der*

scope [skəʊp] *n.* **a)** Bereich, *der; (of discussion etc.)* Rahmen, *der;* **b)** *(opportunity)* Entfaltungsmöglichkeiten *Pl.*

scorch [skɔ:tʃ] *v.t.* versengen. '**scorching** *adj.* glühend heiß

score [skɔ:(r)] **1.** *n.* **a)** *(points)* [Spiel]stand, *der; (made by one player)* Punktzahl, *die;* **keep** [the] ~: zählen; **b)** *(Mus.)* Partitur, *die; (Cinemat.)* [Film]musik, *die;* **c)** *pl. same or* ~s *(group of 20)* zwanzig; **d)** *in pl. (great numbers)* ~s **|and** ~s**| of** zig *(ugs.)*; Dutzende [von]; **e) on that** ~: was das betrifft; **f) pay off** *or* **settle an old** ~ *(fig.)* eine alte Rechnung begleichen. **2.** *v.t.* erzielen ⟨Erfolg, Punkt usw.⟩; **a goal** ein Tor schießen. **3.** *v.i.* **a)** *(make* ~*)* Punkte/einen Punkt erzielen; *(*~ *goal/goals)* ein Tor/Tore schießen/werfen; **b)** *(keep* ~*)* aufschreiben. '**score-board** *n.* Anzeigetafel, *die.* '**scorer** *n.* **a)** *(recorder)* Anschreiber, *der/*Anschreiberin, *die;* **b)** *(Footb.)* Torschütze, *der/*-schützin, *die*

scorn [skɔ:n] **1.** *n.* Verachtung, *die.* **2.** *v.t.* verachten; in den Wind schlagen ⟨Rat⟩; ausschlagen ⟨Angebot⟩. **scornful** ['skɔ:nfl] *adj.* verächtlich ⟨Lächeln, Blick⟩; **be** ~ **of sth.** für etw. nur Verachtung haben

Scorpio ['skɔ:pɪəʊ] *n.* der Skorpion

scorpion ['skɔ:pɪən] *n.* Skorpion, *der*

Scot [skɒt] *n.* Schotte, *der/*Schottin, *die*

Scotch [skɒtʃ] **1.** *adj. see* **Scottish. 2.** *n.* Scotch, *der;* schottischer Whisky

scotch *v.t.* den Boden entziehen (+ *Dat.*) ⟨Gerücht⟩; zunichte machen ⟨Plan⟩

Scotch: ~ '**egg** *n. hartgekochtes Ei in Wurstbrät;* ~ '**whisky** *n.* schottischer Whisky

scot-'free *adj.* **|get off/go|** ~: ungeschoren [davonkommen *od.* bleiben]

Scotland ['skɒtlənd] *pr. n.* Schottland *(das)*

Scots [skɒts] **1.** *adj. (esp. Scot.)* schottisch; **sb. is** ~: jmd. ist Schotte/Schottin. **2.** *n. (dialect)* Schottisch, *das.* **Scotsman** ['skɒtsmən] *n., pl.* **Scotsmen** ['skɒtsmən] Schotte, *der.* '**Scotswoman** *n.* Schottin, *die*

Scottish ['skɒtɪʃ] *adj.* schottisch; **sb. is** ~: jmd. ist Schotte/Schottin

scoundrel ['skaʊndrl] *n.* Schuft, *der*

[1]**scour** [skaʊə(r)] *v.t. (search)* durchkämmen **(for** nach)

[2]**scour** *v.t.* scheuern ⟨Topf, Metall⟩. '**scourer** *n.* Topfreiniger, *der*

scourge [skɜ:dʒ] *n.* Geißel, *die*

scout [skaʊt] **1.** *n.* **a)** **|Boy|** S~: Pfadfinder, *der;* **b)** *(Mil.)* Späher, *der.* **2.** *v.i.* ~ **for** Ausschau halten nach

scowl [skaʊl] **1.** *v.i.* ein mürrisches Gesicht machen. **2.** *n.* mürrischer [Gesichts]ausdruck

scram [skræm] *v.i.,* **-mm-** *(sl.)* abhauen *(salopp)*

scramble ['skræmbl] **1.** *v.i.* **a)** *(clamber)* klettern; ~ **through a hedge** sich durch eine Hecke zwängen; **b)** *(move hastily)* rennen *(ugs.);* ~ **for sth.** um etw. rangeln. **2.** *v.t. (Teleph., Radio)* verschlüsseln. **scrambled 'egg** *n.* Rührei, *das*

[1]**scrap** [skræp] **1.** *n.* **a)** *(of paper)* Fetzen, *der; (of food)* Bissen, *der;* **b)** *in pl. (odds and ends) (of food)* Reste *Pl.;* **c)** *(smallest amount)* **not a** ~ **of** kein bißchen; *(of sympathy, truth also)* nicht ein Fünkchen; **not a** ~ **of evidence** nicht die Spur eines Beweises; **d)** ~ **|metal|** Schrott, *der;* ~ **iron** Alteisen, *das.* **2.** *v.t.,* **-pp-** wegwerfen; *(send for* ~*)* verschrotten; *(fig.)* aufgeben

[2]**scrap** *(coll.)* **1.** *n. (fight)* Rauferei, *die.* **2.** *v.i.,* **-pp-** sich raufen

'**scrap-book** *n.* [Sammel]album, *das*

scrape [skreɪp] **1.** *v.t.* **a)** *(make smooth)* schaben ⟨Häute, Möhren, Kartoffeln usw.⟩; abziehen ⟨Holz⟩; *(damage)* verschrammen ⟨Fußboden, Auto⟩; **b)** *(remove)* [ab]kratzen ⟨Farbe, Schmutz, Rost⟩ **(off, from** von**); c)** *(draw along)* schleifen; **d)** ~ **together** *(raise)* zusammenkratzen *(ugs.); (save up)* zusammensparen. **2.** *v.i.* **a)** *(move with sound)* schleifen; **b)** *(emit scraping noise)* ein schabendes Geräusch machen; **c)** *(rub)* streifen **(against, over** Akk.**). 3.** *n.* **a)** *(act, sound)* Kratzen, *das* **(against an** + *Dat.***); b)** *(predicament)* Schwulitäten *Pl. (ugs.).* **scrape 'by** *v.i. (fig.)* sich über Wasser halten **(on** mit**). scrape 'out** *v.t.* **a)** *(excavate)* buddeln *(ugs.);* scharren; **b)** *(clean)* auskratzen. **scrape through** [**|**--**|**] *v.t.* sich zwängen durch; *(fig.)* mit Hängen und Würgen kommen durch ⟨Prüfung⟩. **2.** [**|**-'-**|**] *v.i.* sich durchzwängen; *(fig.: in examination)* mit Hängen und Würgen durchkommen '**scraper** n. *(for shoes)* Kratzeisen, *das;* *(grid)* Abtreter, *der;* *(tool, kitchen utensil)* Schaber, *der;* *(for removing ice from car windows)* [Eis]kratzer, *der*

scrap: ~-**heap** n. Schutthaufen, *der;* ~ '**paper** n. Schmierpapier, *das*

scrappy ['skræpɪ] adj. lückenhaft

'**scrap-yard** n. Schrottplatz, *der*

scratch [skrætʃ] 1. *v. t.* a) *(score surface of)* zerkratzen; *(score skin of)* kratzen; b) *(get scratch[es] on)* ~ **oneself/one's hands** etc. sich schrammen/ sich *(Dat.)* die Hände *usw.* zerkratzen; c) *(scrape without marking)* kratzen; kratzen an (+ *Dat.*) *(Insektenstich usw.)*; ~ **oneself/one's arm** sich kratzen/sich *(Dat.)* den Arm *od.* am Arm kratzen. 2. *v. i.* kratzen; *(~ oneself)* sich kratzen. 3. n. a) *(mark, wound)* Kratzer, *der (ugs.);* Schramme, *die;* b) *(sound)* Kratzen, *das;* c) **have a [good]** ~: sich [ordentlich] kratzen; d) **start from** ~: bei Null anfangen *(ugs.);* **be up to** ~ *(Arbeit, Leistung:)* nichts zu wünschen übriglassen; *(Person:)* den Anforderungen genügen. **scratch a'bout, scratch a'round** *v. i.* scharren; *(fig.: search)* suchen **(for** nach). **scratch 'out** *v. t.* auskratzen *(Auge)*

scrawl [skrɔ:l] 1. *v. t.* hinkritzeln. 2. *v. i.* kritzeln. 3. n. Gekritzel, *das; (handwriting)* Klaue, *die (salopp)*

scrawny ['skrɔ:nɪ] adj. hager; dürr

scream [skri:m] 1. *v. i.* schreien **(with** vor + *Dat.).* 2. *v. t.* schreien. 3. n. Schrei, *der; (of jet engine)* Heulen, *das;* ~s **of pain** Schmerzensschreie

screech [skri:tʃ] 1. *v. i. & t.* kreischen. 2. n. Kreischen, *das*

screen [skri:n] 1. n. a) *(partition)* Trennwand, *die; (piece of furniture)* Wandschirm, *der;* b) *(of trees, persons, fog)* Wand, *die;* c) *(Cinemat.)* Leinwand, *die;* [**TV**] ~: Bildschirm, *der.* 2. *v. t.* a) *(shelter)* schützen **(from** vor + *Dat.);* *(conceal)* verdecken; b) vorführen *(Film);* *(for disease)* untersuchen. '**screenplay** n. Drehbuch, *das*

screw [skru:] 1. n. Schraube, *die.* 2. *v. t.* schrauben **(to** an + *Akk.);* ~ **together** zusammenschrauben; ~ **down** festschrauben. **screw 'up** *v. t.* a) *(crumple up)* zusammenknüllen *(Blatt Papier);* b) verziehen *(Gesicht);* zusammenkneifen *(Augen, Mund);* c) *(sl.: bungle)* vermurksen *(salopp);* ~ **it/ things up** Mist bauen *(salopp)*

screw: ~-**cap** n. Schraubverschluß,

der; ~-**driver** n. Schraubenzieher, *der;* ~-**top** *see* ~-**cap**

screwy ['skru:ɪ] adj. *(sl.)* spinnig *(ugs.)*

scribble ['skrɪbl] 1. *v. t.* hinkritzeln. 2. *v. i.* kritzeln. 3. n. Gekritzel, *das*

script [skrɪpt] n. a) *(handwriting)* Handschrift, *die;* b) *(of play)* Regiebuch, *das; (of film)* [Dreh]buch, *das;* c) *(for broadcaster)* Manuskript, *das*

scripture ['skrɪptʃə(r)] n. a) [Holy] S~, **the** [Holy] S~s **die** [Heilige] Schrift; b) *(Sch.)* Religion, *die*

'**script-writer** n. *(of film)* Drehbuchautor, *der/*-autorin, *die*

scroll [skrəʊl] n. *(roll)* Rolle, *die*

scrounge ['skraʊndʒ] *(coll.)* 1. *v. t.* schnorren *(ugs.)* **(off, from** von). 2. *v. i.* schnorren *(ugs.)* **(from** bei). '**scrounger** n. *(coll.)* Schnorrer, *der/* Schnorrerin, *die*

'**scrub** [skrʌb] 1. *v. t.,* -**bb**-: a) schrubben *(ugs.);* scheuern; b) *(coll.: cancel)* zurücknehmen *(Befehl);* sausenlassen *(ugs.)* *(Plan).* 2. *v. i.,* -**bb**- schrubben *(ugs.);* scheuern. 3. n. **give sth. a** ~: etw. schrubben *(ugs.) od.* scheuern

²**scrub** n. *(brushwood)* Buschwerk, *das; (area)* Buschland, *das*

'**scruff** [skrʌf] n. **by the** ~ **of the neck** beim Genick

²**scruff** n. *(Brit. coll.)* *(man)* vergammelter Typ *(ugs.);* *(woman, girl)* Schlampe, *die.* '**scruffy** adj. vergammelt *(ugs.)*

scrum [skrʌm] n. Gedränge, *das*

scruple ['skru:pl] n. Skrupel, *der;* **have no** ~s **about doing sth.** keine Skrupel haben, etw. zu tun

scrupulous ['skru:pjʊləs] adj. gewissenhaft *(Person);* unbedingt *(Ehrlichkeit);* peinlich *(Sorgfalt)*

scrutinize ['skru:tɪnaɪz] *v. t.* [genau] untersuchen *([Forschungs]gegenstand);* [über]prüfen *(Rechnung, Paß, Fahrkarte);* mustern *(Person)*

scrutiny ['skru:tɪnɪ] n. a) *(critical gaze)* musternder Blick; b) *(examination) (of recruit)* Musterung, *die; (of bill, passport, ticket)* [Über]prüfung, *die*

scuff [skʌf] 1. *v. t.* streifen; verschrammen *(Schuhe, Fußboden).* 2. n. Schramme, *die*

scuffle ['skʌfl] 1. n. Handgreiflichkeiten *Pl.* 2. *v. i.* handgreiflich werden **(with** gegen)

scullery ['skʌlərɪ] n. Spülküche, *die*

sculptor ['skʌlptə(r)] n. Bildhauer, *der/*-hauerin, *die*

sculpture ['skʌlptʃə(r)] n. a) *(art)*

Bildhauerei, *die;* **b)** *(piece of work)* Skulptur, *die;* Plastik, *die; (pieces collectively)* Skulpturen

scum [skʌm] *n.* **a)** Schmutzschicht, *die; (film)* Schmutzfilm, *der;* **b)** *(fig. derog.)* Abschaum, *der*

scurry ['skʌrɪ] *v. i.* huschen

¹scuttle ['skʌtl] *n.* Kohlenfüller, *der*

²scuttle *(Naut.) v. t.* versenken

³scuttle *v. i.* rennen; flitzen *(ugs.);* ⟨*Maus, Krabbe:*⟩ huschen

scythe [saɪð] *n.* Sense, *die*

SE *abbr.* south-east SO

sea [si:] *n.* **a)** Meer, *das;* the ~: das Meer; die See; by ~: mit dem Schiff; by the ~: am Meer; at ~: auf See *(Dat.);* be all at ~ *(fig.)* nicht mehr weiter wissen; put |out| to ~: in See *(Akk.)* gehen; **b)** *(specific tract of water)* Meer, *das*

sea: ~ 'air *n.* Seeluft, *die;* ~-'bed *n.* Meeresboden, *der;* ~-gull *n.* [See]möwe, *die*

¹seal [si:l] *n. (Zool.)* Robbe, *die;* |common| ~: [Gemeiner] Seehund

²seal **1.** *n. (wax etc., stamp, impression)* Siegel, *das.* **2.** *v. t.* **a)** *(stamp, affix ~ to)* siegeln ⟨*Dokument*⟩; *(fasten with ~)* verplomben ⟨*Tür, Stromzähler*⟩; **b)** *(close securely)* abdichten ⟨*Behälter, Rohr usw.*⟩; zukleben ⟨*Umschlag, Paket*⟩; **c)** *(stop up)* verschließen; abdichten ⟨*Leck*⟩; verschmieren ⟨*Riß*⟩. **seal** 'off *v. t.* abriegeln

sea: ~-legs *n. pl.* Seebeine *Pl. (Seemannsspr.);* get *or* find one's ~-legs sich *(Dat.)* Seebeine wachsen lassen; ~-level *n.* Meeresspiegel, *der*

'**sealing-wax** *n.* Siegellack, *der*

'**sea-lion** *n.* Seelöwe, *der*

seam [si:m] *n.* **a)** Naht, *die;* **b)** *(of coal)* Flöz, *das*

seaman ['si:mən] *n., pl.* **seamen** ['si:mən] Matrose, *der*

'**sea mist** *n.* Küstennebel, *der*

'**seamless** *adj.* nahtlos

'**seamy** *adj.* the ~ side |of life *etc.*| *(fig.)* die Schattenseite[n] [des Lebens *usw.*]

seance ['seɪəns], **séance** ['seɪɑ̃s] *n.* Séance, *die*

sea: ~plane *n.* Wasserflugzeug, *das;* ~port *n.* Seehafen, *der*

sear ['sɪə(r)] *v. t.* versengen

search [sɜ:tʃ] **1.** *v. t.* durchsuchen (for nach); absuchen ⟨*Gebiet, Fläche*⟩ (for nach); *(fig.: probe)* erforschen ⟨*Herz, Gewissen*⟩; suchen in *(+ Dat.)* ⟨*Gedächtnis*⟩ (for nach). **2.** *v. i.* suchen (for nach). **3.** *n.* Suche, *die* (for nach);

(of building, room, etc.) Durchsuchung, *die;* **in ~ of sb./sth.** auf der Suche nach jmdm./etw. '**searching** *adj.* prüfend, forschend ⟨*Blick*⟩; bohrend ⟨*Frage*⟩

search: ~light *n.* Suchscheinwerfer, *der;* ~-party *n.* Suchtrupp, *der;* ~-warrant *n.* Durchsuchungsbefehl, *der*

sea: ~-shore *n.* [Meeres]küste, *die; (beach)* Strand, *der;* ~-sick *adj.* seekrank; ~-sickness *n.* Seekrankheit, *die;* ~-side *n.* [Meeres]küste, *die;* by/to/at the ~-side am/ans/am Meer; ~side town Seestadt, *die*

season ['si:zn] **1.** *n.* **a)** Jahreszeit, *die;* nesting ~: Nistzeit, *die;* **b)** *(period of social activity)* |opera/football| ~: [Opern-/Fußball]saison, *die;* holiday *or (Amer.)* vacation ~: Urlaubszeit, *die;* tourist ~: Reisezeit, *die;* **c)** raspberries are in/out of *or* not in ~: jetzt ist die/nicht die Saison *od.* Zeit für Himbeeren; be in ~ *(on heat)* brünstig sein; **d)** *see* season-ticket. **2.** *v. t.* würzen ⟨*Fleisch, Rede*⟩. **seasonable** ['si:zənəbl] *adj.* der Jahreszeit gemäß. '**seasoned** *adj. (fig.)* erfahren. '**seasoning** *n.* Gewürze *Pl.;* Würze, *die.* '**season-ticket** *n.* Dauerkarte, *die*

seat [si:t] **1.** *n.* **a)** Sitzgelegenheit, *die; (in vehicle, cinema, etc.)* Sitz, *der; (of toilet)* [Klosett]brille, *die (ugs.);* **b)** *(place)* Platz, *der; (in vehicle)* [Sitz]platz, *der;* have *or* take a ~: sich [hin]setzen; **c)** *(part of chair)* Sitzfläche, *die;* **d)** *(buttocks)* Gesäß, *das; (part of clothing)* Gesäßpartie, *die; (of trousers)* Sitz, *der.* **2.** *v. t.* **a)** *(cause to sit)* setzen; ⟨*Platzanweiser:*⟩ einen Platz anweisen *(+ Dat.);* ~ oneself sich setzen; **b)** *(have ~s for)* Sitzplätze bieten *(+ Dat.);* ~ 500 people 500 Sitzplätze haben. '**seat-belt** *n.* Sicherheitsgurt, *der.* '**seated** *adj.* sitzend; remain ~: sitzen bleiben. '**seating** *n.* Sitzplätze; *attrib.* Sitz⟨*ordnung, -plan*⟩

sea: ~-urchin *n.* Seeigel, *der;* ~-wall *n.* Strandmauer, *die;* ~-water *n.* Meerwasser, *das;* ~weed *n.* [See]tang, *der;* ~worthy *adj.* seetüchtig

secluded [sɪ'klu:dɪd] *adj. (hidden)* versteckt; *(isolated)* abgelegen; zurückgezogen ⟨*Leben*⟩. **seclusion** [sɪ'klu:ʒn] *n. (remoteness)* Abgelegenheit, *die; (privacy)* Zurückgezogenheit, *die*

'**second** ['sekənd] **1.** *adj.* zweit...; ~

largest/highest *etc.* zweitgrößt.../-höchst... *usw.;* **come/be ~:** zweiter/ zweite werden/sein. **2.** *n.* **a)** *(unit of time or angle)* Sekunde, *die;* **b)** *(coll.: moment)* Sekunde, *die (ugs.);* **wait a few ~s** einen Moment warten; **in a ~** *(immediately)* sofort *(ugs.); (very quickly)* im Nu *(ugs.);* **just a ~!** *(coll.)* einen Moment!; **c) the ~** *(in sequence)* der/die/das Zweite; **d)** *in pl. (helping of food)* zweite Portion. **3.** *v. t. (support)* unterstützen

²**second** [sɪˈkɒnd] *v. t. (transfer)* vorübergehend versetzen

secondary [ˈsekəndərɪ] *adj. (of less importance)* zweitrangig; Neben-*(sache);* **be ~ to sth.** einer Sache *(Dat.)* untergeordnet sein. **secondary school** *n.* höhere Schule

second: ~-best 1. [ˈ---] *adj.* zweitbest...; **2.** [--ˈ-] *n.* Zweitbeste, *der/die/das;* **~-class 1.** [ˈ---] *adj. (of lower class)* zweiter Klasse nachgestellt; Zweite[r]-Klasse-*(Fahrkarte, Abteil, Post, Brief usw.);* **~-class stamp** Briefmarke für einen Zweiter-Klasse-Brief; **2.** [--ˈ-] *adv.* zweiter Klasse *(fahren);* **~ floor** *see* **floor 1 b; ~ hand** *n.* Sekundenzeiger, *der;* **~-hand 1.** [ˈ---] *adj.* **a)** gebraucht *(Kleidung, Auto usw.);* antiquarisch *(Buch);* **b)** *(selling used goods)* Gebrauchtwaren-; Secondhand*(laden);* **c)** *(Nachrichten, Bericht)* aus zweiter Hand; **2.** [--ˈ-] *adv.* aus zweiter Hand

secondly *adv.* zweitens

second: ~ name *n.* Nachname, *der;* **~-'rate** *adj.* zweitklassig; **~ 'thoughts** *n. pl.* **have ~ thoughts** es sich *(Dat.)* anders überlegen *(about* mit); **we've had ~ thoughts about buying it** wir wollen es nun doch nicht kaufen; **but on ~ thoughts ...:** wenn ich's mir [noch mal] überlege, ...

secrecy [ˈsiːkrɪsɪ] *n.* **a)** *(keeping of secret)* Geheimhaltung, *die;* **b)** *(secretiveness)* Heimlichtuerei, *die;* **c) in ~:** im geheimen

secret [ˈsiːkrɪt] **1.** *adj.* geheim; Geheim*(fach, -tür, -abkommen, -kode);* heimlich *(Trinker, Liebhaber);* **keep sth. ~:** etw. geheimhalten *(from* vor + *Dat.).* **2.** *n.* **a)** Geheimnis, *das;* **make no ~ of sth.** kein Geheimnis aus etw. machen; *(fig.)* keinen Hehl aus etw. machen; **keep ~s/ a ~:** schweigen *(fig.);* **b) in ~:** im geheimen. **secret 'agent** *n.* Geheimagent, *der/* -agentin, *die*

secretarial [sekrəˈteərɪəl] *adj.* Sekretärinnen*(kursus, -tätigkeit); (Arbeit)* als Sekretärin

secretary [ˈsekrətərɪ] *n.* Sekretär, *der/*Sekretärin, *die*

secretive [ˈsiːkrɪtɪv] *adj.* verschlossen *(Person);* **be ~:** geheimnisvoll tun **(about** mit)

'secretly *adv.* heimlich; insgeheim *(etw. glauben)*

sect [sekt] *n.* Sekte, *die*

section [ˈsekʃn] *n.* **a)** *(part cut off)* Abschnitt, *der;* Stück, *das; (part of divided whole)* Teil, *der;* **b)** *(of firm)* Abteilung, *die; (of organization)* Sektion, *die;* **c)** *(of chapter, book)* Abschnitt, *der; (of statute etc.)* Paragraph, *der*

sector [ˈsektə(r)] *n.* Sektor, *der*

secular [ˈsekjʊlə(r)] *adj.* weltlich

secure [sɪˈkjʊə(r)] **1.** *adj.* sicher; *(firmly fastened)* fest; **~ against** burglars gegen Einbruch geschützt. **2.** *v. t.* **a)** sichern *(for* Dat.); beschaffen *(Auftrag) (for* Dat.); *(for oneself)* sich *(Dat.)* sichern; **b)** *(fasten)* sichern. **se-'curely** *adv. (firmly)* fest *(verriegeln, zumachen);* sicher *(befestigen, untergebracht sein).* **security** [sɪˈkjʊərɪtɪ] *n.* **a)** Sicherheit, *die;* **~ [measures]** Sicherheitsmaßnahmen; **b)** *(Finance)* **securities** *pl.* Wertpapiere

security: ~ forces *n. pl.* Sicherheitskräfte *Pl.;* **~ guard** *n.* Wächter, *der/* Wächterin, *die;* **~ risk** *n.* Sicherheitsrisiko, *das*

sedan [sɪˈdæn] *n. (Amer. Motor Veh.)* Limousine, *die*

sedate [sɪˈdeɪt] **1.** *adj.* bedächtig; gesetzt *(alte Dame);* gemächlich *(Tempo, Leben).* **2.** *v. t.* sedieren. **sedation** [sɪˈdeɪʃn] *n.* Sedation, *die;* **be under ~:** sediert sein. **sedative** [ˈsedətɪv] **1.** *n.* Beruhigungsmittel, *das.* **2.** *adj.* sedativ

sedentary [ˈsedəntərɪ] *adj.* sitzend

sediment [ˈsedɪmənt] *n.* Ablagerung, *die; (of tea, coffee, etc.)* Bodensatz, *der*

seduce [sɪˈdjuːs] *v. t.* verführen. **seduction** [sɪˈdʌkʃn] *n.* Verführung, *die.* **seductive** [sɪˈdʌktɪv] *adj.* verführerisch; verlockend *(Angebot)*

see [siː] **1.** *v. t.,* **saw** [sɔː], **seen** [siːn] **a)** sehen; **I can ~ it's hard for you** ich verstehe, daß es nicht leicht für dich ist; **I ~ what you mean** ich verstehe[, was du meinst]; **b)** *(meet [with])* sehen; treffen; *(meet socially)* sich treffen mit; **I'll ~ you there/at five** wir sehen uns

dort/um fünf; ~ you!, |I'll| be ~ing you! *(coll.)* bis bald! *(ugs.);* c) *(speak to)* sprechen ⟨*Person*⟩ **(about** wegen); *(visit)* gehen zu ⟨*Arzt, Anwalt usw.*⟩; *(receive)* empfangen; **d)** *(find out)* feststellen; *(by looking)* nachsehen; **e)** *(make sure)* ~ |**that**| ...: darauf achten, daß ...; **f)** *(imagine)* sich *(Dat.)* vorstellen; **g)** *(escort)* begleiten. **2.** *v. i.,* **saw, seen a)** sehen; **b)** *(make sure)* nachsehen; **c)** I ~: ich verstehe; **you** ~: weißt du/wißt ihr/wissen Sie. **'see about** *v. t.* sich kümmern um. **see 'off** *v. t.* **a)** *(say goodbye to)* verabschieden; **b)** *(chase away)* vertreiben. **see 'out** *v. t.* *(escort)* hinausbegleiten **(of** aus); ~ **oneself out** allein hinausfinden. **see through** *v. t.* **a)** ['--] hindurchsehen durch; *(fig.)* durchschauen; **b)** [-'-] *(not abandon)* zu Ende bringen. **'see to** *v. t.* sich kümmern um
seed [si:d] **1.** *n.* **a)** Samen, *der; (of grape etc.)* Kern, *der;* **b)** *no pl., no indef. art.* *(~s collectively)* Samen[körner] *Pl.; (as collected for sowing)* Saatgut, *das; (for birds)* Körner *Pl.;* **go or run to** ~: Samen bilden; *(fig.)* herunterkommen *(ugs.);* **c)** *(Sport)* gesetzter Spieler/gesetzte Spielerin. **2.** *v. t.* **a)** *(place ~s in)* besäen; **b)** *(Sport)* setzen ⟨*Spieler*⟩; **be ~ed number one** als Nummer eins gesetzt werden/sein. **'seedless** *adj.* kernlos
seedling ['si:dlɪŋ] *n.* Sämling, *der*
'seedy *adj.* **a)** *(coll.: unwell)* **feel** ~: sich [leicht] angeschlagen fühlen; **b)** *(shabby)* schäbig, *(ugs.)* vergammelt ⟨*Aussehen*⟩; heruntergekommen ⟨*Stadtteil*⟩; **c)** *(disreputable)* zweifelhaft
'seeing *conj.* ~ |**that**| ...: da ...; wo ... *(ugs.)*
seek [si:k] *v. t.,* **sought** [sɔ:t] suchen; anstreben ⟨*Posten, Amt*⟩; sich bemühen um ⟨*Anerkennung, Interview, Einstellung*⟩; *(try to reach)* aufsuchen
seem [si:m] *v. i.* scheinen; **you** ~ **tired** du wirkst müde; **she** ~**s nice** sie scheint nett zu sein. **'seeming** *adj.* scheinbar. **'seemingly** *adv.* **a)** *(evidently)* offensichtlich; **b)** *(to outward appearance)* scheinbar
seemly ['si:mlɪ] *adj.* schicklich
seen *see* see
seep [si:p] *v. i.* ~ |**away**| [ab]sickern
'see-saw *n.* Wippe, *die*
seethe [si:ð] *v. i.* **a)** ~ |**with anger/inwardly**| vor Wut/innerlich schäumen; **b)** ⟨*Straßen usw.*⟩: wimmeln **(with** von)

'see-through *adj.* durchsichtig
segment ['segmənt] *n.* *(of orange, pineapple, etc.)* Scheibe, *die*
segregate ['segrɪgeɪt] *v. t.* trennen; *(racially)* absondern. **segregation** [segrɪ'geɪʃn] *n.* Trennung, *die;* |**racial**| ~: Rassentrennung, *die*
seismic ['saɪzmɪk] *adj.* seismisch
seize [si:z] **1.** *v. t.* **a)** ergreifen; ~ **power** die Macht ergreifen; ~ **sb. by the arm/ collar** jmdn. am Arm/Kragen packen; ~ **the opportunity** |**to do sth.**| die Gelegenheit ergreifen [und etw. tun]; ~ **any/a** or **the chance** |**to do sth.**| jede/ die Gelegenheit nutzen[, um etw. zu tun]; **be ~d with remorse/panic** von Gewissensbissen geplagt/von Panik ergriffen werden; **b)** *(capture)* gefangennehmen ⟨*Person*⟩; kapern ⟨*Schiff*⟩; mit Gewalt übernehmen ⟨*Flugzeug, Gebäude*⟩; einnehmen ⟨*Festung, Brücke*⟩; **c)** *(confiscate)* beschlagnahmen. **2.** *v. i.* *see* ~ **up.** **'seize on** *v. t.* sich *(Dat.)* vornehmen ⟨*Einzelheit, Aspekt, Schwachpunkt*⟩; aufgreifen ⟨*Idee, Vorschlag*⟩. **seize 'up** *v. i.* sich festfressen
seizure ['si:ʒə(r)] *n.* **a)** *see* seize 1 b, c: Gefangennahme, *die;* Kapern, *das;* Übernahme, *die;* Einnahme, *die;* Beschlagnahme, *die;* **b)** *(Med.)* Anfall, *der*
seldom ['seldəm] *adv.* selten
select [sɪ'lekt] **1.** *adj.* ausgewählt. **2.** *v. t.* auswählen. **selection** [sɪ'lekʃn] *n.* **a)** *(what is selected [from])* Auswahl, *die* **(of** an + *Dat.,* **from** aus); **b)** *(act of choosing)* [Aus]wahl, *die.* **selective** [sɪ'lektɪv] *adj.* *(using selection)* selektiv; *(careful in one's choice)* wählerisch
self [self] *n., pl.* **selves** [selvz] Selbst, *das (geh.);* Ich, *das*
self- *in comb.* selbst-/Selbst-
self: ~-**ad'dressed** *adj.* ~-**addressed envelope** adressierter Rückumschlag; ~-**ad'hesive** *adj.* selbstklebend; ~-**ap'pointed** *adj.* selbsternannt; ~-**as'surance** *n.* Selbstsicherheit, *die;* ~-**as'sured** *adj.* selbstsicher; ~-**'catering 1.** *adj.* mit Selbstversorgung nachgestellt; **2.** *n.* Selbstversorgung, *die;* ~-**'centred** *adj.* egozentrisch; ~-**'confidence** *n.* Selbstbewußtsein, *das;* ~-**'confident** *adj.* selbstbewußt; ~-**'conscious** *adj.* unsicher; ~-**'consciousness** *n.* Unsicherheit, *die;* ~-**con'tained** *adj.* abgeschlossen ⟨*Wohnung*⟩; ~-**con'trol** *n.* Selbstbeherrschung, *die;* ~-

con'trolled adj. voller Selbstbeherrschung nachgestellt; ~-**de'ception** n.
Selbsttäuschung, die; ~-**de'fence** n.
Notwehr, die; **in ~-defence** aus Not-
wehr; ~-**drive** adj. ~-**drive hire** |com-
pany| Autovermietung, die; ~-**drive
vehicle** Mietwagen, der; ~-**em-
'ployed** adj. selbständig; ~-'**evident** adj. offenkundig; ~-**ex'planat-
ory** adj. ohne weiteres verständlich;
be ~-explanatory für sich selbst sprechen; ~-'**help** n. Selbsthilfe, die;
~-**im'portant** adj. eingebildet;
~-**in'dulgent** adj. maßlos; ~-'**interest** n. Eigeninteresse, das
'**selfish** adj., '**selfishly** adv. selbst-
süchtig. '**selfishness** n. Selbstsucht,
die
self: ~-'**pity** n. Selbstmitleid, das;
~-'**portrait** n. Selbstporträt, das;
~-**pos'sessed** adj. selbstbeherrscht;
~-'**raising flour** n. (Brit.) mit Backpulver versetztes Mehl; ~-**re'spect**
n. Selbstachtung, die; ~-**re'specting**
adj. **no ~-respecting person** ...: niemand, der etwas auf sich hält, ...;
~-'**righteous** adj. selbstgerecht;
~-'**sacrifice** n. Selbstaufopferung,
die; ~-'**satisfied** adj. selbstzufrieden; (smug) selbstgefällig; ~-'**service** n. Selbstbedienung, die; attrib.
Selbstbedienungs-; ~-**sufficient**
adj. unabhängig; selbständig (Person); ~-'**willed** adj. eigenwillig
sell [sel] 1. v. t., **sold** [səʊld] ~ **sth. to
sb.,** ~ **sb. sth.** jmdm. etw. verkaufen;
be sold out ausverkauft sein. 2. v. i.,
sold sich verkaufen; (Person:) verkaufen. **sell 'off** v. t. verkaufen. **sell 'out**
1. v. t. a) ausverkaufen; b) (coll.: betray) verpfeifen (ugs.). 2. v. i. **we have
or are sold out** wir sind ausverkauft
'**sell-by date** n. ≈ Mindesthaltbar-
keitsdatum, das
'**seller** n. a) Verkäufer, der/Verkäufe-
rin, die; b) (product) **be a good/slow ~:**
sich gut/nur langsam verkaufen
Sellotape, (P) ['seləteɪp] n. ≈ Tesafilm, der(W)
'**sellotape** v. t. mit Tesafilm kleben
'**sell-out** n. **be a ~:** ausverkauft sein;
(coll.: betrayal) Verrat sein
selves pl. of **self**
semaphore ['seməfɔ:(r)] 1. n. (system)
Winken, das. 2. v. i. ~ **to sb.** jmdm. ein
Winksignal übermitteln
semblance ['sembləns] n. Anschein,
der
semen ['si:mən] n. Samen, der

semi- [semɪ] pref. halb-/Halb-
semi: ~-**breve** n. (Brit. Mus.) ganze
Note; ~-**circle** n. Halbkreis, der;
~-'**circular** adj. halbkreisförmig;
~-'**colon** n. Semikolon, das; ~-
de'tached adj. & n. ~-**detached**
|house| Doppelhaushälfte, die;
~-**'final** n. Halbfinale, das
seminar ['semɪnɑ:(r)] n. Seminar, das
'**semitone** n. (Mus.) Halbton, der
semolina [semə'li:nə] n. Grieß, der
senate ['senət] n. Senat, der. **senator**
['senətə(r)] n. Senator, der
send [send] v. t., **sent** [sent] schicken;
senden (geh.). **send a'way** 1. v. t.
˜wegschicken. 2. v. i. ~ **away** |to sb.| for
sth. etw. [bei jmdm.] anfordern. **send
'back** v. t. zurückschicken. '**send for**
v. t. a) (tell to come) holen lassen; rufen (Polizei, Arzt usw.); b) (order from
elsewhere) anfordern. **send 'off** v. t.
1. a) (dispatch) abschicken (Sache); b)
(Sport) vom Platz stellen. 2. v. i. see
send away 2. **send 'up** v. t. (Brit. coll.:
parody) parodieren
'**sender** n. Absender, der
'**send-off** n. Verabschiedung, die
senile ['si:naɪl] adj. senil. **senility**
[sɪ'nɪlɪtɪ] n. Senilität, die
senior ['si:nɪə(r)] 1. adj. a) (older) älter; b) höher (Rang, Beamter, Stellung); leitend (Angestellter, Stellung).
2. n. (older) Ältere, der/die; (of higher
rank) Vorgesetzte, der/die. **senior
'citizen** n. Senior, der/Seniorin, die.
seniority [si:nɪ'ɒrɪtɪ] n. (greater
length of service) höheres Dienstalter;
(higher rank) höherer Rang
sensation [sen'seɪʃn] n. a) (feeling)
Gefühl, das; b) (person, event, etc.)
Sensation, die. **sensational** [sen'sei-
ʃənl] adj. sensationell
sense [sens] 1. n. a) (faculty) Sinn,
der; ~ **of smell/touch/taste** Geruchs-/
Tast-/Geschmackssinn, der; **come to
one's ~s** das Bewußtsein wiedererlan-
gen; b) **in** pl. (normal state of mind)
Verstand, der; **have taken leave of
one's ~s** den Verstand verloren haben; c) (consciousness) Gefühl, das; ~
of responsibility/guilt Verantwor-
tungs-/Schuldgefühl, das; d) (practical wisdom) Verstand, der; **sound or
good ~:** [gesunder Menschen]ver-
stand; **not have the ~ to do sth.** nicht
so schlau sein, etw. zu tun; **there is no
~ in doing that** es hat keinen Sinn, das
zu tun; e) (meaning) Sinn, der; (of
word) Bedeutung, die; **make ~:** einen

Sinn ergeben; **in a** *or* **one** ~: in gewisser Hinsicht; **make** ~ **of sth.** etw. verstehen. **2.** *v. t.* spüren. **'senseless** *adj.* **a)** *(unconscious)* bewußtlos; **b)** *(purposeless)* sinnlos

sensible ['sensɪbl] *adj.* **a)** *(reasonable)* vernünftig; **b)** *(practical)* zweckmäßig. **sensibly** ['sensɪblɪ] *adv.* **a)** *(reasonably)* vernünftig; **b)** *(practically)* zweckmäßig

sensitive ['sensɪtɪv] *adj.* empfindlich; **be** ~ **to sth.** empfindlich auf etw. *(Akk.)* reagieren. **sensitivity** [sensɪ-'tɪvɪtɪ] *n.* Empfindlichkeit, *die*

sensory ['sensərɪ] *adj.* Sinnes-

sensual ['sensjʊəl] *adj.* sinnlich

sensuous ['sensjʊəs] *adj.* sinnlich

sent *see* **send**

sentence ['sentəns] **1.** *n.* **a)** *(Law)* [Straf]urteil, *das;* **b)** *(Ling.)* Satz, *der.* **2.** *v. t.* verurteilen (**to** zu)

sentiment ['sentɪmənt] *n.* **a)** Gefühl, *das;* **b)** *(sentimentality)* Sentimentalität, *die;* **c)** *(opinion)* Gedanke, *der.* **sentimental** [sentɪ'mentl] *adj.* sentimental. **sentimentality** [sentɪmen-'tælɪtɪ] *n.* Sentimentalität, *die*

sentry ['sentrɪ] *n.* Wache, *die*

separable ['sepərəbl] *adj.* trennbar

separate 1. ['sepərət] *adj.* verschieden ⟨*Fragen, Probleme, Gelegenheiten*⟩; gesondert ⟨*Teil*⟩; separat ⟨*Eingang, Toilette, Blatt Papier, Abteil*⟩; *(one's own, individual)* eigen ⟨*Zimmer, Identität, Organisation*⟩; **keep two things** ~: zwei Dinge auseinanderhalten. **2.** ['sepəreɪt] *v. t.* trennen; **they are** ~**d** *(no longer live together)* sie leben getrennt. **3.** *v. i.* **a)** *(disperse)* sich trennen; **b)** ⟨*Ehepaar:*⟩ sich trennen. **separately** ['sepərətlɪ] *adv.* getrennt. **separation** [sepə'reɪʃn] *n.* Trennung, *die*

Sept. *abbr.* **September** Sept.

September [sep'tembə(r)] *n.* September, *der; see also* **August**

septic ['septɪk] *adj.* septisch; **go** ~: eitrig werden

sequel ['si:kwl] *n.* **a)** *(consequence, result)* Folge, *die* (**to** von); **b)** *(continuation)* Fortsetzung, *die*

sequence ['si:kwəns] *n.* **a)** Reihenfolge, *die;* **b)** *(part of film)* Sequenz, *die*

sequin ['si:kwɪn] *n.* Paillette, *die*

serenade [serə'neɪd] **1.** *n.* Ständchen, *das.* **2.** *v. t.* ~ **sb.** jmdm. ein Ständchen bringen

serene [sɪ'ri:n] *adj.* gelassen. **serenity** [sɪ'renɪtɪ] *n.* Gelassenheit, *die*

sergeant ['sɑ:dʒənt] *n.* *(Mil.)* Unterof-

fizier, *der; (police officer)* ≈ Polizeimeister, *der.* **sergeant-'major** *n.* ≈ [Ober]stabsfeldwebel, *der*

serial ['sɪərɪəl] *n.* Fortsetzungsgeschichte, *die; (Radio, Telev.)* Serie, *die.* **serialize** ['sɪərɪəlaɪz] *v. t.* in Fortsetzungen veröffentlichen; *(Radio, Telev.)* in Fortsetzungen senden

series ['sɪərɪːz, 'sɪərɪz] *n., pl. same* **a)** *(sequence)* Reihe, *die; (of events, misfortunes)* Folge, *die;* **b)** *(set of successive issues)* Serie, *die;* **radio/TV** ~: Hörfunkreihe/Fernsehserie, *die;* **c)** *(set of books)* Reihe, *die*

serious ['sɪərɪəs] *adj.* **a)** *(earnest)* ernst; **b)** *(important, grave)* ernst ⟨*Angelegenheit, Lage, Problem, Zustand*⟩; ernsthaft ⟨*Frage, Einwand, Kandidat*⟩; schwer ⟨*Krankheit, Unfall, Fehler, Niederlage*⟩; ernstzunehmend ⟨*Rivale*⟩; ernstlich ⟨*Gefahr, Bedrohung*⟩; bedenklich ⟨*Mangel*⟩. **'seriously** *adv.* **a)** *(earnestly)* ernst; **take sth./sb.** ~: etw./jmdn. ernst nehmen; **b)** *(severely)* ernstlich; schwer ⟨*verletzt*⟩. **'seriousness** *n.* Ernst, *der;* **in all** ~: ganz im Ernst

sermon ['sɜ:mən] *n.* Predigt, *die*

serrated [se'reɪtɪd] *adj.* gezackt; ~ **knife** Sägemesser, *das*

serum ['sɪərəm] *n.* Serum, *das*

servant ['sɜ:vənt] *n.* Diener, *der*/Dienerin, *die*

serve [sɜ:v] **1.** *v. t.* **a)** *(work for)* dienen (+ *Dat.*); **b)** *(be useful to)* dienlich sein (+ *Dat.*); **c)** *(meet needs of)* nutzen (+ *Dat.*); ~ **a/no purpose** einen Zweck erfüllen/keinen Zweck haben; **d)** durchlaufen ⟨*Lehre*⟩; verbüßen ⟨*Haftstrafe*⟩; **e)** *(dish up)* servieren; *(pour out)* einschenken (**to** *Dat.*); **f)** ~|**s**| *or* **it** ~**s him right!** *(coll.)* [das] geschieht ihm recht! **2.** *v. i.* **a)** dienen; ~ **as chairman** das Amt des Vorsitzenden innehaben; ~ **as a Member of Parliament** Mitglied des Parlaments sein; ~ **on a jury** Geschworener/Geschworene sein; **b)** *(be of use)* ~ **to do sth.** dazu dienen, etw. zu tun; ~ **to show sth.** etw. zeigen; ~ **for** *or* **as** dienen als; **c)** *(Sport)* aufschlagen. **3.** *n. see* **service 1 g. serve 'up** *v. t.* **a)** servieren; **b)** *(offer for consideration)* auftischen *(ugs.)*

service ['sɜ:vɪs] **1.** *n.* **a)** Dienst, *der;* **do sb. a** ~: jmdm. einen guten Dienst erweisen; **b)** *(Eccl.)* Gottesdienst, *der;* **c)** *(attending to customer)* Service, *der; (in shop, garage, etc.)* Bedienung, *die,*

d) *(system of transport)* Verbindung, *die;* **there is no |bus| ~ on Sundays** Sonntags verkehren keine Busse; **e)** *(provision of maintenance)* |after-sale| ~: Kundendienst, *der;* **take one's car in for a ~:** sein Auto zur Inspektion bringen; **f)** *(operation)* Betrieb, *der;* **out of ~:** außer Betrieb; **g)** *(Sport)* Aufschlag, *der;* **whose ~ is it?** wer hat Aufschlag?; **h)** *(crockery set)* Service, *das;* **i)** *(assistance)* **can I be of ~ |to you|?** kann ich Ihnen behilflich sein?; **I'm at your ~:** ich stehe zu Ihren Diensten; **j)** *(Mil.)* **the |armed or fighting| ~s** die Streitkräfte; **in the ~s** beim Militär; **k)** |motorway| **~s** [Autobahn]- raststätte, *die.* **2.** *v. t.* warten ⟨Wagen, Waschmaschine, Heizung⟩. **service- able** ['sɜːvɪsəbl] *adj.* **a)** *(useful)* nütz- lich; **b)** *(durable)* haltbar

service: **~ area** *n.* Raststätte, *die;* **~ charge** *n.* Bedienungsgeld, *das;* **~ industry** *n.* Dienstleistungsbetrieb, *der;* **~man** ['sɜːvɪsmən] *n., pl.* **~men** ['sɜːvɪsmən] Militärangehörige, *der;* **~ station** *n.* Tankstelle, *die*

serviette [sɜːvɪ'et] *n.* *(Brit.)* Serviette, *die*

servile ['sɜːvaɪl] *adj.* unterwürfig

serving ['sɜːvɪŋ] *n.* Portion, *die.* **'serv- ing spoon** *n.* Vorlegelöffel, *der*

session ['seʃn] *n.* *(meeting)* Sitzung, *die;* **be in ~:** tagen

set [set] **1.** *v. t.,* -tt-, **set a)** *(put)* *(hori- zontally)* legen; *(vertically)* stellen; **~ sb. ashore** jmdn. an Land setzen; **~ sth./things right** *or* **in order** etw./die Dinge in Ordnung bringen; **b)** *(apply)* setzen; **~ a match to sth.** ein Streich- holz an etw. *(Akk.)* halten; *see also* **fire 1 a;** **¹light 1 d;** **c)** *(adjust)* einstel- len (**at** auf + *Akk.*); aufstellen ⟨Falle⟩; stellen ⟨Uhr⟩; **~ the alarm for 5.30 a.m.** den Wecker auf 5.30 Uhr stellen; **d)** be ~ ⟨Buch, Film:⟩ spielen; **e)** *(specify)* festlegen ⟨Bedingungen⟩; festsetzen ⟨Termin, Ort usw.⟩ (**for** auf + *Akk.*); **~ limits** Grenzen setzen; **f)** **~ sb. thinking that ...:** jmdn. auf den Gedanken bringen, daß ...; **g)** *(put for- ward)* stellen ⟨Frage, Aufgabe⟩; aufge- ben ⟨Hausaufgabe⟩; aufstellen ⟨Re- kord⟩; *(compose)* zusammenstellen ⟨Rätsel, Fragen⟩; **~ sb. an example, ~ an example to sb.** jmdm. ein Beispiel geben; **~ sb. a task/problem** jmdm. ei- ne Aufgabe stellen/jmdn. vor ein Pro- blem stellen; **~ |sb./oneself| a target** |jmdm./sich| ein Ziel setzen; **h)** *(Med.:*

put into place) [ein]richten; einrenken ⟨verrenktes Gelenk⟩; **i)** legen ⟨Haare⟩; **j)** decken ⟨Tisch⟩; auflegen ⟨Gedeck⟩; **k)** fassen ⟨Edelstein⟩. **2.** *v. i.,* -tt-, **set a)** *(solidify)* fest werden; **b)** *(go down)* ⟨Sonne, Mond:⟩ untergehen. **3.** *n.* **a)** *(group)* Satz, *der;* **~ |of two|** Paar, *das;* **a ~ of chairs** eine Sitzgruppe; **b)** *(radio, TV)* Gerät, *das;* **c)** *(Tennis)* Satz, *der;* **d)** *(of hair)* Legen, *das;* **e)** *(Theatre: scenery)* Bühnenbild, *das; (area of performance) (of film)* Dre- hort, *der; (of play)* Bühne, *die;* **f)** *(of people)* Kreis, *der;* **g)** *(Math.)* Menge, *die.* **4.** *adj.* **a)** *(fixed)* ⟨Absichten, Ziel- vorstellungen, Zeitpunkt⟩; **be ~ in one's ways** *or* **habits** in seinen Ge- wohnheiten festgefahren sein; **~ meal** *or* **menu** Menü, *das;* **b)** vorgeschrie- ben ⟨Buch, Lektüre⟩; **c)** *(ready)* be |all| **~ for sth.** zu etw. bereit sein; **be |all| ~ to do sth.** bereit sein, etw. zu tun; **d)** *(determined)* **be ~ on sth./doing sth.** zu etw. entschlossen sein/entschlossen sein, etw. zu tun. **'set about** *v. t.* **~ about sth.** sich an etw. *(Akk.)* machen; **~ about doing sth.** sich daranmachen, etw. zu tun. **set a'side** *v. t.* **a)** beiseite legen; **b)** aufheben ⟨Urteil, Entschei- dung⟩. **set 'back** *v. t.* **a)** aufhalten ⟨Entwicklung⟩; zurückwerfen ⟨Projekt, Programm⟩; **b)** *(coll.: cost)* kosten ⟨Person⟩; **c)** *(place at a distance)* zu- rücksetzen. **set 'down** *v. t.* **a)** abset- zen ⟨Fahrgast⟩; **b)** *(record)* nieder- schreiben. **set 'off 1.** *v. i.* *(begin jour- ney)* aufbrechen; *(start to move)* los- laufen; ⟨Fahrzeug:⟩ losfahren. **2.** *v. t.* **a)** *(cause to explode)* explodieren las- sen; abbrennen ⟨Feuerwerk⟩; **b)** auslö- sen ⟨Reaktion, Alarmanlage⟩. **set 'out 1.** *v. i.* **a)** *(begin journey)* aufbrechen (**for** nach/zu); **b)** **~ out to do sth.** sich *(Dat.)* vornehmen, etw. zu tun. **2.** *v. t.* darlegen. **set 'up 1.** *v. t.* **a)** errichten ⟨Straßensperre, Denkmal⟩; aufbauen ⟨Zelt, Klapptisch⟩; **b)** *(establish)* grün- den ⟨Firma, Organisation⟩; einrichten ⟨Büro⟩. **2.** *v. i.* **~ up in business** ein Ge- schäft aufmachen.

'set-back *n.* Rückschlag, *der*

settee [se'tiː] *n.* Sofa, *das*

'setting *n.* **a)** *(Mus.)* Vertonung, *die;* **b)** *(surroundings)* Rahmen, *der; (of novel etc.)* Schauplatz, *der*

settle ['setl] **1.** *v. t.* **a)** *(horizontally)* [sorgfältig] legen; *(vertically)* [sorgfäl- tig] stellen; *(at an angle)* [sorgfältig] lehnen; **b)** *(determine, resolve)* sich einigen auf ⟨*Preis*⟩; beilegen ⟨*Streit, Konflikt, Meinungsverschiedenheit*⟩; ausräumen ⟨*Zweifel*⟩; entscheiden ⟨*Frage, Spiel*⟩; **c)** bezahlen ⟨*Rechnung, Betrag*⟩; erfüllen ⟨*Forderung, Anspruch*⟩; ausgleichen ⟨*Konto*⟩. **2.** *v. i.* **a)** *(become established)* sich niederlassen; *(as colonist)* sich ansiedeln; **b)** *(pay)* abrechnen; **c)** *(in chair, in front of fire, etc.)* sich niederlassen; *(to work etc.)* sich konzentrieren **(to auf +** *Akk.*)**;** *(into way of life, retirement, etc.)* sich gewöhnen **(into an +** *Akk.*)**; d)** *(subside)* ⟨*Haus, Fundament, Boden:*⟩ sich senken; **e)** ⟨*Schnee:*⟩ liegenbleiben. **settle 'down 1.** *v. i.* **a)** *(make oneself comfortable)* sich niederlassen **(in in +** *Dat.*)**; b)** *(in town or house)* heimisch werden. **2.** *v. t.* **a)** ~ oneself **down** sich [gemütlich] hinsetzen; **b)** *(calm down)* beruhigen. **'settle for** *v. t. (agree to)* sich zufriedengeben mit. **settle 'in** *v. i. (in new home)* sich einleben. **'settle on** *v. t. (decide on)* sich entscheiden für. **settle 'up** *v. i.* abrechnen; ~ **up with the waiter** beim Kellner bezahlen

'settlement *n.* **a)** *(of argument, conflict, dispute, differences)* Beilegung, *die; (of question)* Klärung, *die; (of bill, account)* Bezahlung, *die; (of court case)* Vergleich, *der;* **b)** *(colony)* Siedlung, *die*

settler ['setlə(r)] *n.* Siedler, *der*/Siedlerin, *die*

set: ~**-to** *n., pl.* ~**-tos: have a** ~**-to** Streit haben; *(with fists)* sich prügeln; ~**-up** *n.* System, *das*

seven ['sevn] **1.** *adj.* sieben. **2.** *n.* Sieben, *die. See also* **eight**

seventeen [sevn'ti:n] **1.** *adj.* siebzehn. **2.** *n.* Siebzehn, *die. See also* **eight.**

seventeenth [sevn'ti:nθ] **1.** *adj.* siebzehnt... **2.** *n. (fraction)* Siebzehntel, *das. See also* **eighth**

seventh ['sevnθ] **1.** *adj.* sieb[en]t... **2.** *n. (in sequence)* sieb[en]te, *der/die/das; (in rank)* Sieb[en]te, *der/die/das; (fraction)* Sieb[en]tel, *das. See also* **eighth**

seventieth ['sevntɪɪθ] *adj.* siebzigst...

seventy ['sevntɪ] **1.** *adj.* siebzig. **2.** *n.* Siebzig, *die. See also* **eight; eighty 2**

sever ['sevə(r)] *v. t.* **a)** *(cut)* durchtrennen; *(fig.)* abbrechen ⟨*Beziehungen*⟩; **b)** *(separate)* abtrennen; *(with axe etc.)* abhacken

several ['sevrl] **1.** *adv.* mehrere; einige; ~ **times** mehrmals. **2.** *pron.* einige;

~ **of us** einige von uns; ~ **of the buildings** einige *od.* mehrere [der] Gebäude

severe [sɪ'vɪə(r)] *adj.*, ~**r** [sɪ'vɪərə(r)], ~**st** [sɪ'vɪərɪst] hart ⟨*Urteil, Strafe, Kritik, Test, Prüfung*⟩; streng ⟨*Frost, Stil, Schönheit*⟩; schwer ⟨*Dürre, Verlust, Behinderung, Verletzung, Krankheit*⟩; rauh ⟨*Wetter*⟩; heftig ⟨*Anfall, Schmerz*⟩; bedrohlich ⟨*Mangel, Knappheit*⟩; stark ⟨*Blutung*⟩. **se-'verely** *adv.* hart; schwer ⟨*verletzt, behindert*⟩. **severity** [sɪ'verɪtɪ] *n.* Strenge, *die; (of drought, shortage)* großes Ausmaß; *(of criticism)* Schärfe, *die*

sew [səʊ] *v. t. & i., p.p.* **sewn** [səʊn] *or* **sewed** [səʊd] nähen. **sew 'on** *v. t.* annähen ⟨*Knopf*⟩; aufnähen ⟨*Abzeichen, Band*⟩. **sew 'up** *v. t.* nähen ⟨*Saum, Naht, Wunde*⟩

sewer ['sju:ə(r), 'su:ə(r)] *n. (tunnel)* Abwasserkanal, *der; (pipe)* Abwasserleitung, *die*

'sewing *n.* Näharbeit, *die.* **'sewing-machine** *n.* Nähmaschine, *die*

sewn *see* **sew**

sex [seks] *n.* **a)** Geschlecht, *das;* **b)** *(sexuality; coll.: intercourse)* Sex, *der (ugs.);* **have** ~ **with sb.** *(coll.)* mit jmdm. schlafen

sexism ['seksɪzm] *n.* Sexismus, *der*

sexist ['seksɪst] *adj.* sexistisch

'sex maniac *n.* Triebverbrecher, *der*

sexual ['sekʃʊəl] *adj.* sexuell. **sexual 'intercourse** *n.* Geschlechtsverkehr, *der.* **sexuality** [sekʃʊ'ælɪtɪ] *n.* Sexualität, *die*

'sexy *adj.* sexy *(ugs.)*

shabbily ['ʃæbɪlɪ] *adv.,* **shabby** ['ʃæbɪ] *adj.* schäbig

shack [ʃæk] *n.* [armselige] Hütte

shackle ['ʃækl] **1.** *n., usu. in pl.* Fessel, *die.* **2.** *v. t.* anketten **(to an +** *Akk.*)

shade [ʃeɪd] **1.** *n.* **a)** Schatten, *der;* **b)** *(colour)* Ton, *der; (fig.)* Schattierung, *die;* **c)** *(lamp~)* [Lampen]schirm, *der.* **2.** *v. t.* **a)** *(screen)* beschatten; **b)** *(darken with lines)* ~ **[in]** [ab]schattieren. **3.** *v. i.* übergehen **(into in +** *Akk.*)

shadow ['ʃædəʊ] **1.** *n.* Schatten, *der.* **2.** *v. t. (follow)* beschatten. **'shadowy** *adj. (indistinct)* schattenhaft

shady ['ʃeɪdɪ] *adj.* **a)** schattig; **b)** *(disreputable)* zwielichtig

shaft [ʃɑ:ft] *n.* **a)** *(of tool, golf club)* Schaft, *der;* **b)** *(Mech. Engin.)* Welle, *die;* **c)** *(of mine, lift)* Schacht, *der;* **d)** *(of light, lightning)* Strahl, *der*

shaggy ['ʃægɪ] *adj.* zottelig

shake [ʃeɪk] **1.** *n.* Schütteln, *das;* **give**

sb./sth. a ~: jmdn./etw. schütteln. 2.
v. t., shook [ʃʊk], shaken ['ʃeɪkn] a)
(move violently) schütteln; ~ one's
fist/a stick at sb. jmdm. mit der Faust/
einem Stock drohen; ~ hands sich
(Dat.) die Hand geben; b) *(cause to
tremble)* erschüttern ⟨Gebäude usw.⟩;
~ one's head den Kopf schütteln; c)
(shock) erschüttern. 3. *v. i.*, shook,
shaken wackeln; ⟨Boden, Stimme:⟩ beben; ⟨Hand:⟩ zittern. shake 'off *v. t.*
abschütteln. shake 'up *v. t.* a) *(upset,
shock)* einen Schrecken einjagen
(+ *Dat.*); b) *(coll.: reorganize)* um-
krempeln *(ugs.)*

shaken *see* shake 2, 3

shaky ['ʃeɪkɪ] *adj.* wack[e]lig ⟨Möbel-
stück, Leiter⟩; zittrig ⟨Hand, Stimme,
Greis⟩; feel ~: sich zittrig fühlen

shall [ʃl, *stressed* ʃæl] *v. aux. only in
pres.* shall, *neg. (coll.)* shan't [ʃɑːnt],
past should [ʃəd, *stressed* ʃʊd], *neg.
(coll.)* shouldn't ['ʃʊdnt] a) *expr. simple
future* werden; b) should *expr. conditional* würde/würdest/würden/wür-
det; I should have been killed if I had
let go ich wäre getötet worden, wenn
ich losgelassen hätte; if we should be
defeated falls wir unterliegen [sollten];
c) *expr. will or intention* what ~ we do?
was sollen wir tun?; let's go in, ~ we?
gehen wir doch hinein, oder?; we
should be safe by now jetzt dürften wir
in Sicherheit sein; he shouldn't do
things like that! er sollte so etwas
nicht tun!

shallot [ʃə'lɒt] *n.* Schalotte, *die*

shallow ['ʃæləʊ] *adj.* seicht ⟨Wasser,
Fluß⟩; flach ⟨Schüssel, Teller, Was-
ser⟩; *(fig.)* flach ⟨Person⟩

sham [ʃæm] 1. *adj.* unecht; imitiert
⟨Leder, Holz, Pelz⟩. 2. *n. (pretence)*
Heuchelei, *die; (person)* Heuchler,
der/Heuchlerin, *die.* 3. *v. t.*, -mm- vortäuschen. 4. *v. i.*, -mm- simulieren

shambles ['ʃæmblz] *n. (coll.)* Chaos,
das; the room was a ~: das Zimmer
glich einem Schlachtfeld

shame [ʃeɪm] *n.* a) Scham, *die;* b)
(state of disgrace) Schande, *die;* put
sb./sth. to ~: jmdn. beschämen/etw.
in den Schatten stellen; c) what a ~!
wie schade! 'shamefaced *adj.* betreten. shameful ['ʃeɪmfl] *adj.* beschämend. 'shameless *adj.* schamlos

shampoo [ʃæm'puː] 1. *v. t.* schampo-
nieren. 2. *n.* Shampoo[n], *das*

shamrock ['ʃæmrɒk] *n.* Klee, *der*

shandy ['ʃændɪ] *n.* Bier mit Limonade

shan't [ʃɑːnt] *(coll.)* = shall not

'shanty ['ʃæntɪ] *n. (hut)* [armselige]
Hütte

²shanty *n. (song)* Shanty, *das*

'shanty town *n.* Elendsviertel, *das*

shape [ʃeɪp] 1. *v. t.* formen; bearbeiten
⟨Holz, Stein⟩ (into zu). 2. *n.* Form, *die;*
take ~: Gestalt annehmen. shape
'up *v. i.* sich entwickeln

'shapeless *adj.* formlos; unförmig
⟨Kleid, Person⟩

shapely ['ʃeɪplɪ] *adj.* wohlgeformt
⟨Beine, Busen⟩; gut ⟨Figur⟩

share [ʃeə(r)] 1. *n.* a) *(portion)* Teil, *der
od. das;* |fair| ~: Anteil, *der;* fair ~s
gerechte Teile; do more than one's
|fair| ~ of the work mehr als seinen
Teil zur Arbeit beitragen. b) *(Commerc.)* Aktie, *die.* 2. *v. t.* teilen; gemeinsam tragen ⟨Verantwortung⟩. 3.
v. i. ~ in teilnehmen an (+ *Dat.*); beteiligt sein an (+ *Dat.*) ⟨Gewinn⟩; teilen ⟨Freude, Erfahrung⟩. share 'out
v. t. aufteilen (among unter + *Akk.*)

share: ~holder *n.* Aktionär, *der*/
Aktionärin, *die;* ~-out *n.* Aufteilung,
die

shark [ʃɑːk] *n.* Hai[fisch], *der*

sharp [ʃɑːp] 1. *adj.* a) scharf; spitz
⟨Nadel, Bleistift, Gipfel, Winkel⟩;
deutlich ⟨Unterscheidung⟩; sauer
⟨Apfel⟩; herb ⟨Wein⟩; *(shrill, piercing)*
schrill ⟨Schrei, Pfiff⟩; heftig ⟨Schmerz,
Krampf, Kampf⟩; begabt ⟨Schüler,
Student⟩; b) *(derog.: dishonest)* gerissen; c) *(Mus.)* [um einen Halbton] er-
höht ⟨Note⟩. 2. *adv.* a) *(punctually)* at
six o'clock ~: Punkt sechs Uhr; b)
turn ~ right/left scharf nach rechts/
links abbiegen; c) look ~! halt dich
ran! *(ugs.)*; d) *(Mus.)* zu hoch ⟨singen,
spielen⟩. sharpen ['ʃɑːpn] *v. t.* schärfen; [an]spitzen ⟨Bleistift⟩. 'sharp-
ener *n. (for pencils)* Spitzer, *der
(ugs.)*. 'sharp-eyed *adj.* scharfäu-
gig; be ~: scharfe Augen haben.
'sharply *adv.* scharf; in scharfem
Ton ⟨antworten⟩. 'sharpness *n.*
Schärfe, *die; (fineness of point)* Spitz-
heit, *die*

shatter ['ʃætə(r)] 1. *v. t.* zertrümmern;
zerbrechen ⟨Glas, Fenster⟩; zerschlagen ⟨Hoffnungen⟩. 2. *v. i.* zerbrechen

shattered ['ʃætəd] *adj.* a) zerbrochen ⟨Glas, Fenster⟩; *(fig.)* zerstört
⟨Hoffnungen⟩; zerrüttet ⟨Nerven⟩; b)
(coll.: greatly upset) she was ~ by the
news die Nachricht hat sie schwer mitgenommen; I'm ~! ich bin ganz er-

schüttert!; *(Brit. coll.: exhausted)* ich bin kaputt! *(ugs.)*. **'shattering** *adj.* verheerend ⟨*Wirkung*⟩; vernichtend ⟨*Schlag, Niederlage*⟩

shave [ʃeɪv] **1.** *v. t.* rasieren; abrasieren ⟨*Haare*⟩. **2.** *v. i.* sich rasieren. **3.** *n.* Rasur, *die;* **have a** ~: sich rasieren. **shave 'off** *v. t.* abrasieren

'shaven ['ʃeɪvn] *adj.* rasiert; [kahl]geschoren ⟨*Kopf*⟩

'shaver *n.* Rasierapparat, *der.* **'shaver point** *n.* Anschluß für den Rasierapparat

shaving ['ʃeɪvɪŋ] *n.* **a)** Rasieren, *das;* **b)** *in pl. (of wood, metal, etc.)* Späne

shaving: **~-brush** *n.* Rasierpinsel, *der;* **~-cream** *n.* Rasiercreme, *die;* **~-foam** *n.* Rasierschaum, *der*

shawl [ʃɔːl] *n.* Schultertuch, *das*

she [ʃɪ, *stressed* ʃiː] *pron.* sie

sheaf [ʃiːf] *n., pl.* **sheaves** [ʃiːvz] *(of corn etc.)* Garbe, *die; (of paper, arrows, etc.)* Bündel, *das*

shear [ʃɪə(r)] *v. t., p. p.* **shorn** [ʃɔːn] *or* **sheared** *(clip)* scheren. **shears** [ʃɪəz] *n. pl.* |pair of| ~: Schere, *die;* **garden** ~: Gartenschere, *die*

sheath [ʃiːθ] *n., pl.* ~**s** [ʃiːðz, ʃiːθs] **a)** *(for knife, sword, etc.)* Scheide, *die;* **b)** *(condom)* Gummischutz, *der*

sheaves *pl. of* **sheaf**

¹shed [ʃed] *v. t.,* -dd-, **shed a)** verlieren; abwerfen ⟨*Laub, Geweih*⟩; **b)** vergießen ⟨*Blut, Tränen*⟩; **c)** verbreiten ⟨*Licht*⟩

²shed *n.* Schuppen, *der*

she'd [ʃɪd, *stressed* ʃiːd] **a)** = **she had;** **b)** = **she would**

sheen [ʃiːn] *n.* Glanz, *der*

sheep [ʃiːp] *n., pl. same* Schaf, *das.* **'sheep-dog** *n.* Hütehund, *der*

sheepish ['ʃiːpɪʃ] *adj.* verlegen

'sheepskin *n.* Schaffell, *das*

sheer [ʃɪə(r)] *adj.* **a)** rein; blank ⟨*Unsinn, Gewalt*⟩; **by** ~ **chance** rein zufällig; **b)** schroff ⟨*Felsen, Abfall*⟩

sheet [ʃiːt] *n.* **a)** Laken, *das;* **b)** *(of thin metal or plastic)* Folie, *die; (of iron, tin)* Blech, *das; (of glass)* Platte, *die; (of paper)* Bogen, *der;* Blatt, *das;* **c)** ⟨*Eis-, Nebel*⟩decke, *die*

sheik[h] [ʃeɪk, ʃiːk] *n.* Scheich, *der*

shelf [ʃelf] *n., pl.* **shelves** [ʃelvz] Brett, *das;* **shelves** *(set)* Regal, *das.* **'shelf-life** *n.* Lagerfähigkeit, *die*

shell [ʃel] **1.** *n.* **a)** Schale, *die; (of snail)* Haus, *das; (of turtle, tortoise)* Panzer, *der; (on beach)* Muschel, *die;* **b)** *(Mil.) (bomb)* Granate, *die.* **2.** *v. t.* **a)** *(take*

out of ~*)* schälen; **b)** *(Mil.)* |mit Artillerie| beschießen. **shell 'out** *v. t. & i. (sl.)* blechen *(ugs.)* **(on** für)

she'll [ʃɪl, *stressed* ʃiːl] = **she will**

'shellfish *n., pl. same* **a)** Schal[en]tier, *das; (oyster, clam)* Muschel, *die; (crustacean)* Krebstier, *das;* **b)** *in pl. (Gastr.)* Meeresfrüchte *Pl.*

shelter ['ʃeltə(r)] **1.** *n.* **a)** *(shield)* Schutz, *der* **(against** vor + *Dat.,* gegen**)**; **bomb** *or* **air-raid** ~: Luftschutzraum, *der;* **get under** ~: sich unterstellen; **b)** *no pl. (place of safety)* Zuflucht, *die.* **2.** *v. t.* schützen **(from** vor + *Dat.*)**; ** Unterschlupf gewähren (+ *Dat.*) ⟨*Flüchtling*⟩. **3.** *v. i.* Schutz suchen **(from** vor + *Dat.*)**. 'sheltered** *adj.* geschützt; behütet ⟨*Leben*⟩

shelve [ʃelv] **1.** *v. t. (defer)* auf Eis legen *(ugs.)*. **2.** *v. i. (slope)* abfallen

shelves *pl. of* **shelf**

'shelving *n.* Regale *Pl.*

shepherd ['ʃepəd] **1.** *n.* Schäfer, *der.* **2.** *v. t.* führen. **'shepherdess** *n.* Schäferin, *die*

shepherd: ~**'s 'crook** *n.* Schäferstock, *der;* ~**'s 'pie** *n. Auflauf aus Hackfleisch mit einer Schicht Kartoffelbrei darüber*

sherry ['ʃerɪ] *n.* Sherry, *der*

she's [ʃɪz, *stressed* ʃiːz] **a)** = **she is;** **b)** = **she has**

shield [ʃiːld] **1.** *n.* Schild, *der.* **2.** *v. t.* schützen **(from** vor + *Dat.*)

shift [ʃɪft] **1.** *v. t.* **a)** *(move)* umstellen ⟨*Möbel*⟩; wegnehmen ⟨*Arm, Hand, Fuß*⟩; wegräumen ⟨*Schutt*⟩; entfernen ⟨*Schmutz, Fleck*⟩; ~ **the responsibility/ blame on to sb.** die Verantwortung/ Schuld auf jmdn. schieben; **b)** *(Amer. Motor Veh.)* ~ **gears** schalten. **2.** *v. i.* **a)** ⟨*Wind:*⟩ drehen **(to** nach); ⟨*Ladung:*⟩ verrutschen; **b)** *(sl.: move quickly)* rasen. **3.** *n.* **a)** ~ **in emphasis** eine Verlagerung des Akzents; **a** ~ **in public opinion** ein Umschwung der öffentlichen Meinung; **b)** *(for work)* Schicht, *die;* **eight-hour/late** ~: Achtstunden-/ Spätschicht, *die;* **do** *or* **work the late** ~: Spätschicht haben. **'shift work** *n.* Schichtarbeit, *die*

shifty ['ʃɪftɪ] *adj.* verschlagen

shilling ['ʃɪlɪŋ] *n. (Hist.)* Shilling, *der*

shilly-shally ['ʃɪlɪʃælɪ] *v. i.* zaudern; **stop** ~**ing!** entschließ dich endlich!

shimmer ['ʃɪmə(r)] **1.** *v. i.* schimmern. **2.** *n.* Schimmer, *der*

shin [ʃɪn] **1.** *n.* Schienbein, *das.* **2.** *v. i.,* -nn-: ~ **up/down a tree** *etc.* einen

Baum *usw.* hinauf-/hinunterklettern.
'**shin-bone** *n.* Schienbein, *das*
shine [ʃaɪn] **1.** *v.i.,* **shone** [ʃɒn]
⟨*Lampe, Licht, Stern:*⟩ leuchten;
⟨*Sonne, Mond:*⟩ scheinen; *(reflect light)* glänzen. **2.** *v.t.,* **shone:** ~ **a light on sth./in sb.'s eyes** etw. anleuchten/ jmdm. in die Augen leuchten. **3.** *n.* Glanz, *der*
shingle ['ʃɪŋgl] *n.* *(pebbles)* Kies, *der*
'**shingles** *n.* *(Med.)* Gürtelrose, *die*
shin: ~-**guard,** ~-**pad** *ns.* Schienbeinschutz, *der*
shiny ['ʃaɪnɪ] *adj.* glänzend
ship [ʃɪp] **1.** *n.* Schiff, *das.* **2.** *v.t.,* -**pp**- *(transport by sea)* verschiffen; *(send by road, train, or air)* verschicken ⟨*Waren*⟩. '**shipbuilding** *n.* Schiffbau, *der*
'**shipment** *n.* **a)** Versand, *der;* *(by sea)* Verschiffung, *die;* **b)** *(amount)* Sendung, *die*
'**shipowner** *n.* Schiffseigentümer, *der*/-eigentümerin, *die;* *(of several ships)* Reeder, *der*/Reederin, *die*
'**shipper** *n.* Spediteur, *der*/Spediteurin, *die;* *(company)* Spedition, *die*
'**shipping** *n.* **a)** *(ships)* Schiffe; *(traffic)* Schiffahrt, *die;* **b)** *(transporting)* Versand, *der*
ship: ~-**shape** *adj.* in bester Ordnung; ~-**wreck 1.** *n.* Schiffbruch, *der.* **2.** *v.t.* **be** ~-**wrecked** Schiffbruch erleiden; ~-**yard** *n.* [Schiffs]werft, *die*
shirk [ʃɜːk] *v.t.* sich drücken vor (+ *Dat.*). '**shirker** *n.* Drückeberger, *der (ugs.)*
shirt [ʃɜːt] *n.* [**man's**] ~: [Herren- od. Ober]hemd, *das;* [**woman's**] ~: Hemdbluse, *die.* '**shirt-sleeve** *n.* Hemdsärmel, *der;* **in** ~**s** in Hemdsärmeln
shit [ʃɪt] *(coarse)* **1.** *v.i.,* -**tt**-, **shitted** or **shit** scheißen *(derb).* **2.** *n.* **a)** Scheiße, *die (derb);* **have** *(Brit.)* or *(Amer.)* **take a** ~: scheißen *(derb);* **b)** *(person)* Scheißkerl, *der (derb);* **c)** *(nonsense)* Scheiß, *der (salopp)*
shiver ['ʃɪvə(r)] **1.** *v.i.* zittern (**with** vor + *Dat.*). **2.** *n.* Schau[d]er, *der (geh.)*
shoal [ʃəʊl] *n.* *(of fish)* Schwarm, *der*
shock [ʃɒk] **1.** *n.* **a)** Schock, *der;* **give sb. a** ~: jmdm. einen Schock versetzen; **b)** *(violent impact)* Erschütterung, *die (of durch);* **c)** *(Electr.)* Schlag, *der;* **d)** *(Med.)* Schock, *der.* **2.** *v.t.* ~ **sb.** [**deeply**] ein [schwerer] Schock für jmdn. sein; *(scandalize)* jmdn. schockieren. '**shock absorber** *n.* Stoßdämpfer, *der*

'**shocking** *adj.* **a)** schockierend; **b)** *(coll.: very bad)* fürchterlich *(ugs.)*
'**shock-proof** *adj.* stoßfest
shod *see* **shoe 2**
shoddy ['ʃɒdɪ] *adj.* schäbig; minderwertig ⟨*Arbeit, Stoff, Artikel*⟩
shoe [ʃuː] **1.** *n.* Schuh, *der;* *(of horse)* [Huf]eisen, *das;* **put oneself into sb.'s** ~**s** *(fig.)* sich in jmds. Lage *(Akk.)* versetzen. **2.** *v.t.,* ~**ing, shod** [ʃɒd] beschlagen ⟨*Pferd*⟩
shoe: ~-**horn** *n.* Schuhlöffel, *der;* ~-**lace** *n.* Schnürsenkel, *der;* ~-**maker** *n.* Schuhmacher, *der;* ~-**polish** *n.* Schuhcreme, *die;* ~-**shop** *n.* Schuhgeschäft, *das;* ~-**string** *n.* **on a** ~-**string** *(coll.)* mit ganz wenig Geld
shone *see* **shine 1, 2**
shoo [ʃuː] **1.** *int.* sch. **2.** *v.t.* scheuchen; ~ **away** fortscheuchen
shook *see* **shake 2, 3**
shoot [ʃuːt] **1.** *v.i.,* **shot** [ʃɒt] **a)** schießen (**at** auf + *Akk.*); **b)** *(move rapidly)* schießen *(ugs.).* **2.** *v.t.,* **shot a)** *(wound)* anschießen; *(kill)* erschießen; *(hunt)* schießen; ~ **sb. dead** jmdn. erschießen; **b)** schießen mit ⟨*Bogen, Munition, Pistole*⟩; abschießen ⟨*Pfeil, Kugel*⟩ (**at** auf + *Akk.*); **c)** *(Cinemat.)* drehen ⟨*Film, Szene*⟩. **3.** *n.* *(Bot.)* Trieb, *der.* **shoot 'down** *v.t.* niederschießen ⟨*Person*⟩; abschießen ⟨*Flugzeug*⟩. **shoot 'out** *v.i.* hervorschießen. **shoot 'up** *v.i.* in die Höhe schießen; ⟨*Preise, Kosten, Temperatur:*⟩ in die Höhe schnellen
shooting: ~-**range** *n.* Schießstand, *der;* ~ '**star** *n.* Sternschnuppe, *die*
'**shoot-out** *n.* Schießerei, *die*
shop [ʃɒp] **1.** *n.* Laden, *der;* Geschäft, *das;* **go to the** ~**s** einkaufen gehen; **talk** ~: fachsimpeln *(ugs.).* **2.** *v.i.,* -**pp**- einkaufen; **go** ~**ping** einkaufen gehen. **shop a'round** *v.i.* sich umsehen (**for** nach)
shop: ~ **assistant** *n.* *(Brit.)* Verkäufer, *der*/Verkäuferin, *die;* ~-**keeper** *n.* Ladenbesitzer, *der*/-besitzerin, *die;* ~-**lifter** *n.* Ladendieb, *der*/-diebin, *die;* ~-**lifting** *n.* Ladendiebstahl, *der;* ~-**owner** *see* ~-**keeper**
'**shopper** *n.* Käufer, *der*/Käuferin, *die*
'**shopping** *n.* **a)** Einkaufen, *das;* **do the/one's** ~: einkaufen/[seine] Einkäufe machen; **b)** *(items bought)* Einkäufe *Pl.*
shopping: ~-**bag** *n.* Einkaufstasche, *die;* ~-**basket** *n.* Einkaufskorb, *der;*

~ **centre** *n*. Einkaufszentrum, *das;* ~**list** *n*. Einkaufszettel, *der;* ~ **mall** [~ mæl] *n*. Einkaufszentrum, *das;* ~ **street** *n*. Geschäftsstraße, *die;* ~ **trolley** *n*. Einkaufswagen, *der*

shop: ~**-soiled** *adj. (Brit.) (slightly damaged)* leicht beschädigt; *(slightly dirty)* angeschmutzt; ~ '**window** *n*. Schaufenster, *das*

shore [ʃɔ:(r)] *n*. Ufer, *das; (beach)* Strand, *der.* **shore 'up** *v. t.* abstützen ⟨*Mauer, Haus*⟩; *(fig.)* stützen

shorn *see* **shear**

short [ʃɔ:t] **1.** *adj.* **a)** kurz; **in a ~ time** *or* **while** *(soon)* bald; in Kürze; **a ~ time** *or* **while ago/later** vor kurzem/ kurze Zeit später; **in ~,** ...: kurz, ...; **b)** klein ⟨*Person, Wuchs*⟩; **c)** *(deficient, scanty)* knapp; **go ~** ⟨*of sth.*⟩ [an etw. *(Dat.)*] Mangel leiden; **sb. is ~ of sth.** jmdm. fehlt es an etw. *(Dat.);* **time is getting/is ~:** die Zeit wird/ist knapp; **be in ~ supply** knapp sein; **be ~** ⟨*of cash*⟩ knapp [bei Kasse] sein *(ugs.).* **2.** *adv.* **a)** *(abruptly)* plötzlich; **stop ~:** plötzlich abbrechen; **stop sb. ~:** jmdm. ins Wort fallen; **b)** **stop ~ of doing sth.** davor zurückschrecken, etw. zu tun. **shortage** [ʃɔ:tɪdʒ] *n*. Mangel, *der* ⟨**of an** + *Dat.*⟩; ~ **of fruit/teachers** Obstknappheit, *die/* Lehrermangel, *der*

short: ~**bread** *n*. Shortbread, *das;* Keks aus Butterteig; ~ '**circuit** *n*. *(Electr.)* Kurzschluß, *der;* ~**coming** *n.,* usu. *in pl.* Unzulänglichkeit, *die;* ~ '**cut** *n*. Abkürzung, *die;* **take a ~ cut** den Weg abkürzen

shorten [ʃɔ:tn̩] **1.** *v. i.* kürzer werden. **2.** *v. t.* kürzen; verkürzen ⟨*Besuch, Wartezeit*⟩

short: ~**hand** *n*. Stenographie, *die;* ~**hand typist** Stenotypist, *der/*-typi-stin, *die;* ~ **list** *n*. *(Brit.)* engere Aus-wahl; **be on/put sb. on the ~ list** in der engeren Auswahl sein/jmdn. in die engere Auswahl nehmen; ~**list** *v. t.* in die engere Auswahl nehmen; ~**lived** *adj.* kurzlebig

'**shortly** *adv.* in Kürze; gleich *(ugs.);* ~ **before/after sth.** kurz vor/nach etw.

'**short-range** *adj.* **a)** Kurzstrecken-⟨*flugzeug, -rakete usw.*⟩; **b)** *(relating to time)* kurzfristig

shorts [ʃɔ:ts] *n. pl.* **a)** *(trousers)* kurze Hose[n]; Shorts *Pl.;* **b)** *(Amer.: under-pants)* Unterhose, *die*

short: ~'**sighted** *adj.* kurzsichtig; ~**sleeved** [~'sli:vd] *adj.* kurzärm[e]-

lig; ~**-staffed** [~'sta:ft] *adj.* **be** [very] ~**-staffed** [viel] zu wenig Personal ha-ben; ~ '**story** *n*. Kurzgeschichte, *die;* ~**-term** *adj.* kurzfristig; *(provisional)* vorläufig ⟨*Lösung*⟩; ~ **wave** *n*. *(Radio)* Kurzwelle, *die*

shot [ʃɒt] **1.** *n*. **a)** Schuß, *der;* **fire a ~:** einen Schuß abgeben **(at auf** + *Akk.*); **like a ~** *(fig.)* wie der Blitz *(ugs.);* **I'd do it like a ~:** ich würde es auf der Stelle tun; **b)** *(Athletics)* **put the ~:** die Kugel stoßen; **[putting] the ~:** Kugel-stoßen, *das;* **c)** *(Sport: stroke, kick, throw)* Schuß, *der;* **d)** *(Photog.)* Auf-nahme, *die; (Cinemat.)* Einstellung, *die.* **2.** *see* **shoot** 1, 2. **3.** *adj.* **be/get ~ of** *(sl.)* los sein/loswerden. '**shotgun** *n*. Schrotflinte, *die*

should *see* **shall**

shoulder [ʃəʊldə(r)] **1.** *n*. Schulter, *die.* **2.** *v. t.* schultern; *(fig.)* überneh-men

shoulder: ~**-bag** *n*. Umhängetasche, *die;* ~**-blade** *n*. Schulterblatt, *das;* ~**-strap** *n*. *(on garment)* Schulter-klappe, *die; (on bag)* Tragriemen, *der*

shouldn't [ʃʊdnt] *(coll.)* = **should not;** *see* **shall**

shout [ʃaʊt] **1.** *n*. Ruf, *der; (inarticu-late)* Schrei, *der.* **2.** *v. i. & t.* schreien. **shout 'down** *v. t.* niederschreien. **shout 'out 1.** *v. i.* aufschreien. **2.** *v. t.* [laut] rufen

'**shouting** *n*. Geschrei, *das*

shove [ʃʌv] **1.** *n*. Stoß, *der.* **2.** *v. t.* sto-ßen; schubsen *(ugs.); (coll.: put)* tun. **shove a'way** *v. t. (coll.)* wegschub-sen *(ugs.).* **shove 'off** *v. i. (sl.: leave)* abschieben *(ugs.)*

shovel [ʃʌvl] **1.** *n*. Schaufel, *die.* **2.** *v. t., (Brit.)* -ll- schaufeln

show [ʃəʊ] **1.** *n*. **a)** *(entertainment, per-formance)* Show, *die; (Theatre)* Vor-stellung, *die; (Radio, Telev.)* [Unter-haltungs]sendung, *die;* **b)** *(exhibition)* Ausstellung, *die;* Schau, *die;* **put sth. on ~:** etw. ausstellen; **be on ~:** ausge-stellt sein; **c)** *(appearance)* Anschein, *der;* **be for ~:** reine Angeberei sein *(ugs.).* **2.** *v. t., p.p.* **shown** [ʃəʊn] **a)** zeigen; vorzeigen ⟨*Paß, Fahrschein usw.*⟩; ~ **sb. sth.,** ~ **sth. to sb.** jmdm. etw. zeigen; **b)** beweisen ⟨*Mut, Urteils-vermögen usw.*⟩; ~ **sb. that ...:** jmdm. beweisen, daß ...; ~ [**sb.**] **kindness/ mercy** freundlich [zu jmdm.] sein/Er-barmen [mit jmdm.] haben; **c)** ⟨*Ther-mometer, Uhr usw.*⟩ anzeigen; **d)** *(ex-hibit in a show)* ausstellen; zeigen

⟨*Film*⟩. 3. *v. i.*, *p. p.* **shown** a) *(be visible)* sichtbar *od.* zu sehen sein; *(come into sight)* sich zeigen; b) *(be ~n)* ⟨*Film:*⟩ laufen. **show 'in** *v. t.* hinein-/hereinführen. **show 'off** *v. i.* angeben *(ugs.)*; prahlen. **show 'out** *v. t.* hinausführen. **show 'round** *v. t.* herumführen. **show 'through** *v. i.* durchscheinen. **show 'up** 1. *v. t.* a) *(make visible)* [deutlich] sichtbar machen; b) *(coll.: embarrass)* blamieren. 2. *v. i.* a) *(be visible)* [deutlich] zu sehen sein; b) *(coll.: arrive)* sich blicken lassen *(ugs.)*

'**show-down** *n. (fig.)* Kraftprobe, *die;* **have a ~ [with sb.]** sich [mit jmdm.] auseinandersetzen

shower ['ʃaʊə(r)] 1. *n.* a) Schauer, *der;* ~ **of rain/hail** Regen-/Hagelschauer, *der;* b) *(for washing)* Dusche, *die;* **have** *or* **take a [cold/quick] ~:** [kalt/schnell] duschen. 2. *v. t. (lavish)* ~ **sth. [up]on sb.**, ~ **sb. with sth.** jmdn. mit etw. überhäufen. 3. *v. i. (have a ~)* duschen

shower: ~-curtain *n.* Duschvorhang, *der;* ~ **gel** *n.* Duschgel, *das;* ~**-proof** *adj.* [bedingt] regendicht

'**showery** *adj.* **it is ~:** es gibt immer wieder kurze Schauer; **a ~ day** ein Tag mit Schauerwetter

'**show-jumping** *n.* Springreiten, *das*

shown *see* show 2, 3

show: ~-off *n. (coll.)* Angeber, *der/* Angeberin, *die;* ~**-piece** *n. (of exhibition, collection)* Schaustück, *das; (highlight)* Paradestück, *das;* ~**room** *n.* Ausstellungsraum, *der*

'**showy** *adj.* protzig *(ugs.)*

shrank *see* shrink

shred [ʃred] 1. *n.* Fetzen, *der; (fig.)* Spur, *die;* **tear sth. to ~s** etw. zerfetzen; *(fig.)* etw. zerpflücken. 2. *v. t.*, -dd- [im Reißwolf] zerkleinern

shrew [ʃru:] *n. (Zool.)* Spitzmaus, *die*

shrewd [ʃru:d] *adj.* klug; genau ⟨*[Ein]schätzung*⟩

shriek [ʃri:k] 1. *n.* [Auf]schrei, *der.* 2. *v. i.* [auf]schreien. 3. *v. t.* schreien

shrift [ʃrɪft] *n.* **give sb. short ~:** jmdn. kurz abfertigen *(ugs.)*; **get short ~** kurz abgefertigt werden *(ugs.)*

shrill [ʃrɪl] *adj.* schrill

shrimp [ʃrɪmp] *n.* Garnele, *die*

shrine [ʃraɪn] *n. (tomb)* Grab, *das*

shrink [ʃrɪŋk] 1. *v. i.*, **shrank** [ʃræŋk], **shrunk** [ʃrʌŋk] a) schrumpfen; ⟨*Kleidung, Stoff:*⟩ einlaufen; ⟨*Metall, Holz:*⟩ sich zusammenziehen; b) *(recoil)* ~ **from sb./sth.** vor jmdm. zu

rückweichen/vor etw. *(Dat.)* zurückschrecken; ~ **from doing sth.** sich scheuen, etw. zu tun. 2. *v. t.*, **shrank, shrunk** einlaufen lassen ⟨*Textilien*⟩

shrivel ['ʃrɪvl] *v. i.*, *(Brit.)* -ll-: ~ [up] verschrumpeln; ⟨*Pflanze, Blume:*⟩ welk werden

shroud [ʃraʊd] 1. *n.* Leichentuch, *das.* 2. *v. t.* ~ **sth. in sth.** etw. in etw. *(Akk.)* hüllen

Shrove [ʃrəʊv] '**Tuesday** *n.* Fastnachtsdienstag, *der*

shrub [ʃrʌb] *n.* Strauch, *der*

shrug [ʃrʌg] 1. *v. t. & i.*, -gg-: ~ [one's shoulders] die Achseln zucken. 2. *n.* ~ [of one's *or* the shoulders] Achselzucken, *das.* **shrug 'off** *v. t.* in den Wind schlagen

shrunk *see* shrink

shrunken ['ʃrʌŋkn] *adj.* verhutzelt *(ugs.)* ⟨*Person*⟩; schrump[e]lig ⟨*Apfel*⟩

shudder ['ʃʌdə(r)] 1. *v. i.* zittern (with vor + *Dat.*). 2. *n.* Zittern, *das*

shuffle ['ʃʌfl] 1. *n.* a) Schlurfen, *das;* **walk with a ~:** schlurfen; b) *(Cards)* Mischen, *das;* **give the cards a [good] ~:** die Karten [gut] mischen. 2. *v. t.* a) *(Cards)* mischen; b) ~ **one's feet** von einem Fuß auf den anderen treten

shun [ʃʌn] *v. t.*, -nn- meiden

shunt [ʃʌnt] *v. t. (Railw.)* rangieren

shush [ʃʊʃ] *int.* still

shut [ʃʌt] 1. *v. t.*, -tt-, **shut** zumachen; schließen; zusammenklappen ⟨*Klappmesser, Fächer*⟩; ~ **one's finger in the door** sich *(Dat.)* den Finger in der Tür einklemmen. 2. *v. i.*, -tt-, **shut** schließen; ⟨*Blüte:*⟩ sich schließen. **shut 'down** 1. *v. t.* a) schließen, zumachen ⟨*Deckel*⟩; b) stillegen ⟨*Fabrik*⟩; abschalten ⟨*Kernreaktor*⟩. 2. *v. i.* ⟨*Laden, Fabrik:*⟩ geschlossen werden. **shut 'out** *v. t.* aussperren. **shut 'up** 1. *v. t.* abschließen; einsperren ⟨*Tier, Person*⟩. 2. *v. i. (coll.: be quiet)* den Mund halten

shutter ['ʃʌtə(r)] *n.* a) [Fenster]laden, *der;* b) *(Photog.)* Verschluß, *der;* ~ **release** Auslöser, *der;* ~ **speed** Verschlußzeit, *die*

shuttle ['ʃʌtl] 1. *n. (in loom)* Schiffchen, *das.* 2. *v. i.* pendeln. '**shuttlecock** *n.* Federball, *der.* '**shuttle service** *n.* Pendelverkehr, *der*

shy [ʃaɪ] *adj.*, ~**er** *or* **shier** ['ʃaɪə(r)], ~**est** *or* **shiest** ['ʃaɪɪst] scheu; *(diffident)* schüchtern. **shy a'way** *v. i.* ~ **away from sth./doing sth.** etw. scheuen/sich scheuen, etw. zu tun

'**shyness** *n.* Scheuheit, *die; (diffidence)* Schüchternheit, *die*

Siamese [saɪə'mi:z]: ~ '**cat** *n.* Siamkatze, *die;* ~ '**twins** *n. pl.* siamesische Zwillinge

Siberia [saɪ'bɪərɪə] *pr. n.* Sibirien *(das)*

Sicily ['sɪsɪlɪ] *pr. n.* Sizilien *(das)*

sick [sɪk] *adj.* **a)** *(ill)* krank; **be off** ~: krank [gemeldet] sein; **b)** *(Brit.: vomiting or about to vomit)* **be** ~: sich erbrechen; **I'm going to be** ~: ich muß mich erbrechen; **sb. gets/feels** ~: jmdm. wird/ist [es] übel *od.* schlecht; **be/get** ~ **of sb./sth.** *(fig.)* jmdn./etw. satt haben/allmählich satt haben; **make sb.** ~ *(disgust)* jmdn. anekeln. '**sicken** ['sɪkn] **1.** *v. i.* **be** ~**ing for sth.** *(Brit.)* krank werden; etw. ausbrüten *(ugs.).* **2.** *v. t. (disgust)* anwidern. '**sickening** *adj.* ekelerregend, widerlich ⟨Anblick, Geruch⟩

sickle ['sɪkl] *n.* Sichel, *die*

'**sick-leave** *n.* Urlaub wegen Krankheit; **be on** ~ ≈ krank geschrieben sein

sickly ['sɪklɪ] *adj.* kränklich

'**sickness** *n.* Krankheit, *die; (nausea)* Übelkeit, *die*

sick: ~**-pay** *n.* Entgeltfortzahlung im Krankheitsfalle; *(paid by insurance)* Krankengeld, *das;* ~**-room** *n.* Krankenzimmer, *das*

side [saɪd] **1.** *n.* **a)** Seite, *die;* ~ **of beef** Rinderhälfte, *die;* ~ **of bacon** Speckseite, *die;* **walk/stand** ~ **by** ~: nebeneinander gehen/stehen; **work/fight** ~ **by** ~ [**with sb.**] Seite an Seite [mit jmdm.] arbeiten/kämpfen; **live** ~ **by** ~ [**with sb.**] in [jmds.] unmittelbarer Nachbarschaft leben; **to one** ~: zur Seite; **on one** ~: an der Seite; **on the** ~ *(as ~line)* nebenbei; **take** ~**s** [**with/against sb.**] [für/gegen] jmdn. Partei ergreifen; **b)** *(Sport: team)* Mannschaft, *die.* **2.** *v. i.* ~ **with sb.** sich auf jmds. Seite *(Akk.)* stellen. **3.** *adj.* Seiten-

side: ~**board** *n.* Anrichte, *die;* ~**-car** *n.* Beiwagen, *der;* ~**-dish** *n.* Beilage, *die;* ~**-door** *n.* Seitentür, *die;* ~**-effect** *n.* Nebenwirkung, *die;* ~**-entrance** *n.* Seiteneingang, *der;* ~**-exit** *n.* Seitenausgang, *der;* ~**light** *n.* Begrenzungsleuchte, *die;* **drive on** ~**lights** mit Standlicht fahren; ~**line** *n. (occupation)* Nebenbeschäftigung, *die;* ~**-road** *n.* Seitenstraße, *die;* ~**-show** *n.* Nebenattraktion, *die;* ~**-step 1.** *n.* Schritt zur Seite; **2.** *v. t.*

ausweichen (+ *Dat.*); ~**-street** *n.* Seitenstraße, *die;* ~**track** *v. t.* **get** ~**tracked** abgelenkt werden; ~**walk** *n. (Amer.)* Bürgersteig, *der;* ~**ways** ['saɪdweɪz] **1.** *adv.* **look at sb./sth.** ~**ways** jmdn./etw. von der Seite ansehen; **2.** *adj.* seitlich

siding ['saɪdɪŋ] *n.* Abstellgleis, *das*

sidle ['saɪdl] *v. i.* schleichen [**up to zu**]

siege [si:dʒ] *n.* Belagerung, *die; (by police)* Umstellung, *die;* **lay** ~ **to sth.** etw. belagern

sieve [sɪv] **1.** *n.* Sieb, *das.* **2.** *v. t.* sieben

sift [sɪft] *v. t.* sieben; ~ **sth. from sth.** etw. von etw. trennen. **sift** '**out** *v. t.* aussieben

sigh [saɪ] **1.** *n.* Seufzer, *der;* **breathe** *or* **give** *or* **heave a** ~: einen Seufzer ausstoßen; ~ **of relief/contentment** Seufzer der Erleichterung/Zufriedenheit. **2.** *v. i.* seufzen; ~ **with relief/despair** erleichtert/verzweifelt seufzen

sight [saɪt] **1.** *n.* **a)** *(faculty)* Sehvermögen, *das;* **know sb. by** ~: jmdn. vom Sehen kennen; **b)** *(act of seeing; spectacle)* Anblick, *der;* **catch/lose** ~ **of sb./sth.** jmdn./etw. erblicken/aus dem Auge verlieren; **at first** ~: auf den ersten Blick; ~**s** *(places of interest)* Sehenswürdigkeiten; **see the** ~**s** die Sehenswürdigkeiten ansehen; **d)** *(range)* Sichtweite, *die;* **in** ~: in Sicht; **within** *or* **in** ~ **of sb./sth.** *(able to see)* in jmds. Sichtweite *(Dat.)*/in Sichtweite einer Sache; **out of** ~: außer Sicht; **e)** *(of gun)* Visier, *das;* **set/have** [**set**] **one's** ~**s on sth.** *(fig.)* etw. anpeilen. **2.** *v. t.* sichten ⟨Land, Schiff, Flugzeug⟩; sehen ⟨Entflohenen, Vermißten⟩. '**sightseeing** *n.* **go** ~: Besichtigungen machen. **sightseer** ['saɪtsi:ə(r)] *n.* Tourist *(der die Sehenswürdigkeiten besichtigt)*

sign [saɪn] **1.** *n.* **a)** *(symbol, signal, indication)* Zeichen, *das; (of future event)* Anzeichen, *das;* **as a** ~ **of** als Zeichen (+ *Gen.*); **b)** *(Astrol.)* ~ [**of the zodiac**] Sternzeichen, *das;* **c)** *(notice; on shop etc.)* Schild, *das.* **2.** *v. t. & i.* unterschreiben; ~ **one's name** [**mit seinem Namen**] unterschreiben. **sign** '**on** *v. i. (as unemployed)* sich arbeitslos melden. **sign** '**up** *v. i.* sich [vertraglich] verpflichten (**with** bei); *(for course)* sich einschreiben

signal ['sɪgnl] **1.** *n.* Signal, *das;* **a** ~ **for sth./to sb.** ein Zeichen zu etw./für jmdn. **2.** *v. i., (Brit.)* **-ll-** signalisieren; Signale geben; ⟨Kraftfahrer:⟩ blinken *(with hand)* anzeigen; ~ **to sb.** |**to do sth.**| jmdm. ein Zeichen geben[, etw. zu tun]. '**signal-box** n. Stellwerk, *das*

signature ['sɪgnətʃə(r)] n. Unterschrift, *die; (on painting)* Signatur, *die.* '**signature tune** n. Erkennungsmelodie, *die*

'**signboard** n. Schild, *das*

signet-ring ['sɪgnɪt rɪŋ] n. Siegelring, *der*

significance [sɪg'nɪfɪkəns] n. Bedeutung, *die;* **be of** |**no**| ~: [nicht] von Bedeutung sein

significant [sɪg'nɪfɪkənt] adj. **a)** *(noteworthy, important)* bedeutend; **b)** *(full of meaning)* bedeutsam. **sig'nificantly** adv. **a)** *(meaningfully)* bedeutungsvoll; ~ |**enough**| bedeutsamerweise; **b)** *(notably)* bedeutend

signify ['sɪgnɪfaɪ] v. t. bedeuten

'**signpost** n. Wegweiser, *der*

silence ['saɪləns] **1.** n. Schweigen, *das; (keeping a secret)* Verschwiegenheit, *die; (stillness)* Stille, *die;* **there was** ~: es herrschte Schweigen/Stille; **in** ~: schweigend. **2.** v. t. zum Schweigen bringen; *(fig.)* ersticken ⟨*Proteste*⟩; mundtot machen ⟨*Gegner*⟩. '**silencer** n. *(Arms; Brit. Motor Veh.)* Schalldämpfer, *der*

silent ['saɪlənt] adj. stumm; *(noiseless)* unhörbar; *(still)* still; **be** ~ *(say nothing)* schweigen; ~ **film** Stummfilm, *der.* '**silently** adv. schweigend; stumm ⟨*weinen, beten*⟩; *(noiselessly)* lautlos

silhouette [sɪlʊ'et] **1.** n. **a)** *(picture)* Schattenriß, *der;* **b)** *(appearance against the light)* Silhouette, *die.* **2.** v. t. **be** ~**d against** sth. sich als Silhouette gegen etw. abheben

silicon ['sɪlɪkən] n. Silicium, *das;* ~ **chip** Siliciumchip, *der*

silk [sɪlk] **1.** n. Seide, *die.* **2.** attrib. adj. seiden; Seiden-. '**silkworm** n. Seidenraupe, *die.* '**silky** adj. seidig

sill [sɪl] n. *(of door)* [Tür]schwelle, *die; (of window)* Fensterbank, *die*

silly ['sɪlɪ] adj. dumm; *(imprudent, unwise)* töricht; *(childish)* albern

silo ['saɪləʊ] n., pl. ~**s** Silo, *der*

silt [sɪlt] n. Schlamm, *der;* Schlick, *der*

silver ['sɪlvə(r)] **1.** n. Silber, *das.* **2.** attrib. adj. silbern; Silber⟨pokal, -münze⟩

silver: ~ '**medal** n. Silbermedaille, *die;* ~ '**paper** n. Silberpapier, *das;* ~**-plated** adj. versilbert; ~ '**wedding** n. Silberhochzeit, *die*

similar ['sɪmɪlə(r)] adj. ähnlich **(to** *Dat.*). **similarity** [sɪmɪ'lærɪtɪ] n. Ähnlichkeit, *die* **(to** mit). '**similarly** adv. ähnlich; *(in exactly the same way)* ebenso

simile ['sɪmɪlɪ] n. Vergleich, *der*

simmer ['sɪmə(r)] **1.** v. i. ⟨*Flüssigkeit:*⟩ sieden. **2.** v. t. köcheln lassen. **simmer 'down** v. i. sich abregen *(ugs.)*

simple ['sɪmpl] adj. einfach; *(unsophisticated, not elaborate)* schlicht ⟨*Mobiliar, Schönheit, Kunstwerk, Kleidung*⟩; **it was a** ~ **misunderstanding** es war [ganz] einfach ein Mißverständnis. '**simple-minded** adj. **a)** *(unsophisticated)* schlicht; **b)** *(unintelligent)* beschränkt. **simpleton** ['sɪmpltən] n. Einfaltspinsel, *der (ugs.).* **simplicity** [sɪm'plɪsɪtɪ] n. Einfachheit, *die; (unpretentiousness, lack of sophistication)* Schlichtheit, *die.* **simplification** [sɪmplɪfɪ'keɪʃn] n. Vereinfachung, *die.* **simplify** ['sɪmplɪfaɪ] v. t. vereinfachen. **simplistic** [sɪm'plɪstɪk] adj. [all]zu simpel. **simply** ['sɪmplɪ] adv. einfach; *(in an unsophisticated manner)* schlicht; *(merely)* nur; **it** ~ **isn't true** es ist einfach nicht wahr; **I was** ~ **trying to help** ich wollte nur helfen

simulate ['sɪmjʊleɪt] v. t. **a)** *(feign)* vortäuschen; **b)** simulieren ⟨*Bedingungen, Wetter usw.*⟩

simultaneous [sɪml'teɪnɪəs] adj., **simul'taneously** adv. gleichzeitig

sin [sɪn] **1.** n. Sünde, *die.* **2.** v. i., -**nn**- sündigen

since [sɪns] **1.** adv. seitdem. **2.** prep. seit; ~ **seeing you ...:** seit ich dich gesehen habe; ~ **then/that time** inzwischen. **3.** conj. **a)** seit; **it is a long time/so long/not so long** ~ **...:** es ist lange/so lange/gar nicht lange her, daß ...; **b)** *(seeing that, as)* da

sincere [sɪn'sɪə(r)] adj., ~**r** [sɪn'sɪərə(r)], ~**st** [sɪn'sɪərɪst] aufrichtig; herzlich ⟨*Grüße, Glückwünsche usw.*⟩. **sin'cerely** adv. aufrichtig; **yours** ~: mit freundlichen Grüßen. **sincerity** [sɪn'serɪtɪ] n. Aufrichtigkeit, *die*

sinew ['sɪnju:] n. Sehne, *die*

sinful ['sɪnfl] adj. sündig; *(reprehensible)* sündhaft; **it is** ~ **to ...:** es ist eine Sünde, ... zu ...

sing [sɪŋ] v. i. & t., **sang** [sæŋ], **sung** [sʌŋ] singen. **sing 'up** v. i. lauter singen

singe [sɪndʒ] v. t. & i., ~**ing** versengen

singer ['sɪŋə(r)] n. Sänger, *der*/Sängerin, *die*

single ['sɪŋgl] 1. *adj.* **a)** einfach; *(sole)* einzig; *(separate, individual, isolated)* einzeln; **not a ~ one** kein einziger/keine einzige/kein einziges; **every ~ one** jeder/jede/jedes einzelne; **every ~ day** jeden Tag; **~ ticket** *(Brit.)* einfache Fahrkarte; **b)** *(for one person)* Einzel-⟨bett, -zimmer⟩; **c)** *(unmarried)* ledig; **a ~ man/woman** ein Lediger/eine Ledige; **~ people** Ledige. 2. *n.* **a)** *(Brit.: ticket)* einfache Fahrkarte; |**a**| **~/two ~s to Manchester, please** einfach/zweimal einfach nach Manchester, bitte; **b)** *(record)* Single, *die;* **c)** *in pl. (Tennis etc.)* Einzel, *das.* **single 'out** *v. t.* **~ sb./sth. out as/for sth.** jmdn./etw. als/für etw. auswählen

single: **~-decker** 1. *n.* **be a ~-decker** ⟨Bus, Straßenbahn:⟩ nur ein Deck haben; 2. *adj.* **~-decker bus/tram** Bus/Straßenbahn mit [nur] einem Deck; **~ [European] market** *n.* [europäischer] Binnenmarkt; **~-'handed** *adv.* allein; **~-minded** *adj.* zielstrebig

singlet ['sɪŋglɪt] *n.* *(Brit.)* *(vest)* Unterhemd, *das;* *(Sport)* Trikot, *das*

singly ['sɪŋglɪ] *adv.* einzeln

singular ['sɪŋgjʊlə(r)] 1. *adj.* **a)** *(Ling.)* singularisch; Singular-; **~ noun** Substantiv im Singular; **b)** *(extraordinary)* einmalig. 2. *n.* *(Ling.)* Einzahl, *die;* Singular, *der.* **'singularly** *adv.* *(extraordinarily)* außerordentlich

sinister ['sɪnɪstə(r)] *adj.* finster; *(of evil omen)* unheilverkündend

sink [sɪŋk] 1. *n.* Spülbecken, *das.* 2. *v. i.,* **sank** [sæŋk] *or* **sunk** [sʌŋk], **sunk** sinken. 3. *v. t.,* **sank** *or* **sunk, sunk** **a)** versenken ⟨Schiff⟩; **b)** niederbringen ⟨Schacht⟩. **sink 'in** *v. i.* *(fig.)* jmdm. ins Bewußtsein dringen; ⟨Warnung, Lektion:⟩ verstanden werden

'sinner *n.* Sünder, *der*/Sünderin, *die*

sinus ['saɪnəs] *n.* Nebenhöhle, *die*

sip [sɪp] 1. *v. t.,* **-pp-:** **~** |**up**| schlürfen. 2. *v. i.,* **-pp-:** **~ at/from sth.** an etw. *(Dat.)* nippen. 3. *n.* Schlückchen, *das*

siphon ['saɪfn] 1. *n.* Siphon, *der.* 2. *v. t.* [durch einen Saugheber] laufen lassen

sir [sɜː(r)] *n.* **a)** *(formal address)* der Herr; *(to teacher)* Herr Meier/Schmidt *usw.;* **b)** *(in letter)* **Dear Sir** Sehr geehrter Herr; **Dear Sirs** Sehr geehrte [Damen und] Herren; **Dear Sir or Madam** Sehr geehrte Dame/Sehr geehrter Herr; **c)** **Sir** [sə(r)] *(title of knight etc.)* Sir

siren ['saɪrən] *n.* Sirene, *die*

sirloin ['sɜːlɔɪn] *n.* **a)** *(Brit.)* Roastbeef, *das;* **~ steak** Rumpsteak, *das;* **b)** *(Amer.)* Rumpsteak, *das*

sissy ['sɪsɪ] 1. *n.* Waschlappen, *der.* 2. *adj.* feige

sister ['sɪstə(r)] *n.* **a)** Schwester, *die;* **b)** *(Brit.: nurse)* Oberschwester, *die.* **'sister-in-law** *n., pl.* **sisters-in-law** Schwägerin, *die*

sit [sɪt] 1. *v. i.,* **-tt-, sat** [sæt] **a)** *(become seated)* sich setzen; **~ on** *or* **in a chair/in an armchair** sich auf einen Stuhl/in einen Sessel setzen; **b)** *(be seated)* sitzen. 2. *v. t.,* **-tt-, sat a)** setzen; **b)** *(Brit.)* machen ⟨Prüfung⟩. **sit 'back** *v. i.* sich zurücklehnen; *(fig.)* sich im Sessel zurücklehnen. **sit 'down** *v. i.* **a)** *(become seated)* sich setzen (**on/in** auf/in + *Akk.*); **b)** *(be seated)* sitzen. **sit 'up** 1. *v. i.* **a)** *(rise)* sich aufsetzen; **b)** *(be sitting erect)* [aufrecht] sitzen; **c)** *(stay up)* aufbleiben. 2. *v. t.* aufsetzen

site [saɪt] 1. *n.* **a)** *(land)* Grundstück, *das;* **b)** *(location)* Sitz, *der; (of new factory etc.)* Standort, *der.* 2. *v. t.* stationieren ⟨Raketen⟩; **~ a factory in London** London als Standort einer Fabrik wählen; **be ~d** gelegen sein

'sitting *n.* Sitzung, *die;* **the first ~ |for lunch|** der erste Schub [zum Mittagessen]

situate ['sɪtjʊeɪt] *v. t.* legen. **'situated** *adj.* gelegen; **be ~:** liegen. **situation** [sɪtjʊ'eɪʃn] *n.* **a)** *(location)* Lage, *die;* **b)** *(circumstances)* Situation, *die;* **c)** *(job)* Stelle, *die*

six [sɪks] 1. *adj.* sechs. 2. *n.* Sechs, *die. See also* **eight**

sixteen [sɪks'tiːn] 1. *adj.* sechzehn. 2. *n.* Sechzehn, *die. See also* **eight**. **sixteenth** [sɪks'tiːnθ] 1. *adj.* sechzehnt-. 2. *n.* *(fraction)* Sechzehntel, *das. See also* **eighth**

sixth [sɪksθ] 1. *adj.* sechst... 2. *n.* *(in sequence)* sechste, *der/die/das; (in rank)* Sechste, *der/die/das; (fraction)* Sechstel, *das. See also* **eighth**

sixtieth ['sɪkstɪɪθ] *adj.* sechzigst...

sixty ['sɪkstɪ] 1. *adj.* sechzig. 2. *n.* Sechzig, *die. See also* **eight; eighty 2**

size [saɪz] *n.* Größe, *die; (of paper)* Format, *das;* **be twice the ~ of sth.** zweimal so groß wie etw. sein; **a ~ 8 dress** ein Kleid [in] Größe 8; **be ~ 8** ⟨*Person:*⟩ Größe 8 haben. **size 'up** *v. t.* taxieren ⟨Lage⟩

sizeable ['saɪzəbl] *adj.* ziemlich groß; beträchtlich ⟨Summe, Einfluß⟩

sizzle ['sɪzl] *v. i.* zischen

skate 1. *n. (ice-~)* Schlittschuh, *der; (roller-~)* Rollschuh, *der.* **2.** *v. i. (ice-~)* Schlittschuh laufen; *(roller-~)* Rollschuh laufen. '**skateboard 1.** *n.* Skateboard, *das;* Rollerbrett, *das.* **2.** *v. i.* Skateboard fahren. '**skater** *n. (ice-~)* Eisläufer, *der*/Eisläuferin, *die; (roller-~)* Rollschuhläufer, *der*/-läuferin, *die.* **skating** ['skeɪtɪŋ] *n. (ice-~)* Schlittschuhlaufen, *das; (roller-~)* Rollschuhlaufen, *das.* '**skating rink** *n. (ice)* Eisbahn, *die; (for roller-skating)* Rollschuhbahn, *die*

skeleton ['skelɪtn] *n.* Skelett, *das.* '**skeleton key** *n.* Dietrich, *der.* '**skeleton staff** *n.* Minimalbesetzung, *die*

skeptic *etc. (Amer.) see* **sceptic** *etc.*

sketch [sketʃ] **1.** *n.* **a)** *(drawing)* Skizze, *die.* **b)** *(play)* Sketch, *der.* **2.** *v. t.* skizzieren. '**sketch-book** *n.* Skizzenbuch, *das.* '**sketch map** *n.* Faustskizze, *die*

'**sketchy** *adj.* skizzenhaft; lückenhaft ⟨*Informationen, Bericht*⟩

skew [skju:] **1.** *adj.* schräg. **2.** *n.* **on the ~:** schief

skewer ['skju:ə(r)] **1.** *n.* Bratspieß, *der.* **2.** *v. t.* aufspießen

ski [ski:] **1.** *n.* **a)** Ski, *der;* **b)** *(on vehicle)* Kufe, *die.* **2.** *v. i.* Ski laufen *od.* fahren. '**ski boot** *n.* Skistiefel, *der.* '**ski-lift** *n.* Skilift, *der*

skid [skɪd] **1.** *v. i.,* -dd- schlittern; *(from one side to the other; spinning round)* schleudern. **2.** *n.* Schlittern/Schleudern, *das.* '**skid marks** *n. pl.* Schleuderspur, *die*

skier ['ski:ə(r)] *n.* Skiläufer, *der*/-läuferin, *die*

skiing ['ski:ɪŋ] *n.* Skilaufen, *das; (Sport)* Skisport, *der*

skilful ['skɪlfl] *adj.* geschickt; gewandt ⟨*Redner*⟩; gut ⟨*Beobachter, Lehrer*⟩

skill [skɪl] *n.* **a)** *(expertness)* Geschick, *das; (of artist)* Können, *das;* **b)** *(technique)* Fertigkeit, *die; (of weaving, bricklaying)* Technik, *die.* **skilled** ['skɪld] *adj.* **a)** *see* **skilful;** **b)** qualifiziert ⟨*Arbeit, Tätigkeit*⟩; ~ **trade** Ausbildungsberuf, *der;* **c)** *(trained)* ausgebildet. '**skillful** *(Amer.) see* **skilful**

skim [skɪm] *v. t.,* -mm-: **a)** *(remove)* abschöpfen; **b)** abrahmen ⟨*Milch*⟩; **c)** *see* ~ **through. skim 'off** *v. t.* abschöpfen. '**skim through** *v. t.* überfliegen ⟨*Buch, Zeitung*⟩

skimmed 'milk *n.* entrahmte Milch

skimp [skɪmp] **1.** *v. t.* sparen an

(+ *Dat.*). **2.** *v. i.* sparen (**with, on** an + *Dat.*). '**skimpy** *adj.* winzig ⟨*Badeanzug*⟩; spärlich ⟨*Wissen*⟩

skin [skɪn] **1.** *n.* **a)** Haut, *die;* **b)** *(fur)* Fell, *das;* **c)** *(peel)* Schale, *die.* **2.** *v. t.,* -nn- häuten; schälen ⟨*Frucht*⟩

skin: ~ cream *n.* Hautcreme, *die;* ~-'**deep** *adj. (fig.)* oberflächlich; ~-'**diver** *n.* Taucher, *der*/Taucherin, *die;* ~**flint** *n.* Geizhals, *der;* ~**head** *n. (Brit.)* Skinhead, *der*

skinny ['skɪnɪ] *adj.* mager

'**skin-tight** *adj.* hauteng

¹**skip** [skɪp] **1.** *v. i.,* **a)** hüpfen; **b)** *(with skipping-rope)* seilspringen. **2.** *v. t.,* -pp- *(omit)* überspringen; ~ **breakfast/lunch** das Frühstück/Mittagessen auslassen. **3.** *n.* Hüpfer, *der*

²**skip** *n. (Building)* Container, *der*

ski: ~ pass *n.* Skipaß, *der;* ~ **pole** *n.* Skistock, *der*

skipper ['skɪpə(r)] *n.* Kapitän, *der*

'**skipping-rope** *(Brit.),* '**skip-rope** *(Amer.) ns.* Sprungseil, *das*

'**ski-resort** *n.* Skiurlaubsort, *der*

skirmish ['skɜ:mɪʃ] *n. (Mil.)* **a)** Gefecht, *das;* **b)** *(argument)* Auseinandersetzung, *die*

skirt [skɜ:t] **1.** *n.* Rock, *der.* **2.** *v. t.* herumgehen um. **skirt 'round** *v. t.* herumgehen um; *(fig.)* umgehen

'**skirting. ~-[board]** *(Brit.)* Fußleiste, *die*

ski: ~-run *n.* Skihang, *der; (prepared)* [Ski]piste, *die;* ~-**stick** *n.* Skistock, *der*

skittle ['skɪtl] *n.* **a)** Kegel, *der;* **b)** ~**s** *sing. (game)* Kegeln, *das*

skive [skaɪv] *v. i. (Brit. sl.)* sich drücken *(ugs.).* **skive 'off** *(Brit. sl.)* **1.** *v. i.* sich verdrücken *(ugs.).* **2.** *v. t.* schwänzen *(ugs.)*

skulk [skʌlk] *v. i.* lauern

skull [skʌl] *n.* Schädel, *der*

skunk [skʌŋk] *n.* Stinktier, *das*

sky [skaɪ] *n.* Himmel, *der;* **in the ~:** am Himmel

sky: ~-high 1. *adj.* himmelhoch; astronomisch *(ugs.)* ⟨*Preise usw.*⟩; **2.** *adv.* **go ~-high** ⟨*Preise usw.*⟩ in astronomische Höhen klettern *(ugs.);* ~**light** *n.* Dachfenster, *das;* ~**scraper** *n.* Wolkenkratzer, *der*

slab [slæb] *n.* **a)** *(flat stone etc.)* Platte, *die;* **b)** *(thick slice)* [dicke] Scheibe; *(of cake)* [dickes] Stück; *(of chocolate, toffee)* Tafel, *die*

slack [slæk] **1.** *adj.* **a)** *(lax)* nachlässig; schlampig *(ugs.);* **b)** *(loose)* schlaff;

locker ⟨*Verband*⟩. **2.** *n.* **take in** *or* **up the ~:** das Seil/die Schnur *usw.* straffen. **3.** *v. i. (coll.)* bummeln *(ugs.)*

slacken ['slækn] **1.** *v. i.* **a)** *(loosen)* sich lockern; **b)** *(diminish)* nachlassen; ⟨*Geschwindigkeit:*⟩ sich verringern. **2.** *v. t.* **a)** *(loosen)* lockern; **b)** *(diminish)* verringern

slacks [slæks] *n. pl.* **|pair of|** **~:** lange Hose; Slacks *Pl. (Mode)*

slag [slæg] *n.* Schlacke, *die*

slain *see* **slay**

slake [sleɪk] *v. t.* stillen

slam [slæm] **1.** *v. t.,* -mm-: **a)** *(shut)* zuschlagen; **b)** *(put violently)* knallen *(ugs.).* **2.** *v. i.,* -mm- zuschlagen

slander ['slɑːndə(r)] **1.** *n.* Verleumdung, *die* (**on** *Gen.*). **2.** *v. t.* verleumden. **slanderous** ['slɑːndərəs] *adj.* verleumderisch

slang [slæŋ] *n.* Slang, *der;* ⟨*Theater-, Soldaten-, Juristen*⟩jargon, *der; attrib.* Slang⟨*wort,* -*ausdruck*⟩

slant [slɑːnt] **1.** *v. i.* ⟨*Fläche:*⟩ sich neigen; ⟨*Linie:*⟩ schräg verlaufen. **2.** *v. t.* **a)** abschrägen; **b)** *(fig.:* bias) [so] hinbiegen *(ugs.)* ⟨*Meldung, Bemerkung*⟩. **3.** *n.* Schräge, *die;* **on the** *or* **a ~:** schräg

slap [slæp] **1.** *v. t.,* -pp-: **a)** schlagen; **b)** *(put)* knallen *(ugs.).* **2.** *v. i.,* -pp- schlagen; klatschen. **3.** *n.* Schlag, *der.* **4.** *adv.* voll; **~ in the middle** genau in der Mitte. **'slapdash** *adj.* schludrig *(ugs.).* **'slap-up** *attrib. adj. (sl.)* ⟨*Essen*⟩ mit allen Schikanen *(ugs.)*

slash [slæʃ] **1.** *v. t.* **a)** aufschlitzen; **b)** *(fig.)* [drastisch] reduzieren; [drastisch] kürzen ⟨*Gehalt, Umfang*⟩. **2.** *n.* **a)** *(slit)* Schlitz, *der;* **b)** *(~ing stroke)* Hieb, *der*

slat [slæt] *n.* Latte, *die*

slate [sleɪt] **1.** *n.* **a)** *(Geol.)* Schiefer, *der;* **b)** *(Building)* Schieferplatte, *die.* **2.** *v. t. (Brit. coll.: criticize)* in der Luft zerreißen *(ugs.)*

slaughter ['slɔːtə(r)] **1.** *n.* Schlachten, *das; (massacre)* Gemetzel, *das.* **2.** *v. t.* schlachten; *(massacre)* abschlachten

slave [sleɪv] **1.** *n.* Sklave, *der/*Sklavin, *die.* **2.** *v. i.* **~** [**away**] schuften *(ugs.);* sich abplagen (**at** mit). **'slave-driver** *n. (fig.)* Sklaventreiber, *der/*-treiberin, *die.* **slavery** ['sleɪvərɪ] *n.* Sklaverei, *die.* **slavish** ['sleɪvɪʃ] *adj.* sklavisch

slay [sleɪ] *v. t.,* **slew** [sluː], **slain** [sleɪn] *(literary)* ermorden

sleazy ['sliːzɪ] *adj.* schäbig; *(disreputable)* anrüchig

sled [sled], **sledge** [sledʒ] *ns.* Schlitten, *der.* '**sledge-hammer** *n.* Vorschlaghammer, *der*

sleek [sliːk] *adj. (glossy)* seidig

sleep [sliːp] **1.** *n.* Schlaf, *der;* **get/go to ~:** einschlafen; **put to ~:** einschläfern ⟨*Tier*⟩. **2.** *v. i.,* **slept** [slept] schlafen. **3.** *v. t.* **slept: the hotel ~s 80** das Hotel hat 80 Betten. '**sleeper** *n.* **a)** **be a heavy/light ~:** einen tiefen/leichten Schlaf haben; **b)** *(Brit. Railw.: support)* Schwelle, *die;* **c)** *(Railw.: coach)* Schlafwagen, *der; (train)* **|night| ~:** Nachtzug mit Schlafwagen

sleeping: ~-bag *n.* Schlafsack, *der;* **~-car** *n.* Schlafwagen, *der;* **~-pill,** **~-tablet** *ns.* Schlaftablette, *die*

sleep: ~less *adj.* schlaflos; **~-walk** *v. i.* schlafwandeln; **~-walker** *n.* Schlafwandler, *der/*-wandlerin, *die*

'**sleepy** *adj.* schläfrig

sleet [sliːt] **1.** *n.* Schneeregen, *der.* **2.** *v. i. impers.* **it is ~ing** es gibt Schneeregen

sleeve [sliːv] *n.* **a)** Ärmel, *der; (fig.)* **have sth. up one's ~:** etw. in petto haben *(ugs.);* **roll up one's ~s** die Ärmel hochkrempeln *(ugs.);* **b)** *(for record)* Hülle, *die.* '**sleeveless** *adj.* ärmellos

sleigh [sleɪ] *n.* Schlitten, *der*

sleight [slaɪt] **of 'hand** *n.* Fingerfertigkeit, *die*

slender ['slendə(r)] *adj.* **a)** *(slim)* schlank; schmal ⟨*Buch, Band*⟩; **b)** gering ⟨*Chance, Mittel, Hoffnung*⟩

slept *see* **sleep 2, 3**

sleuth [sluːθ] *n.* Detektiv, *der*

'**slew** [sluː] *v. i. & t.* schwenken

²**slew** *see* **slay**

slice [slaɪs] **1.** *n.* Scheibe, *die; (of apple, melon, peach, cake, pie)* Stück, *das;* **a ~ of cake** ein Stück Kuchen. **2.** *v. t.* in Scheiben schneiden; in Stücke schneiden ⟨*Bohnen, Apfel, Kuchen usw.*⟩; **~d bread** Schnittbrot, *das*

slick [slɪk] **1.** *adj. (coll.)* **a)** *(dexterous)* professionell; **b)** *(pretentiously dexterous)* clever *(ugs.).* **2.** *n.* **|oil-|~:** Ölteppich, *der*

slid *see* **slide 1, 2**

slide [slaɪd] **1.** *v. i.,* **slid** [slɪd] rutschen; ⟨*Kolben, Schublade, Feder:*⟩ gleiten. **2.** *v. t.,* **slid** schieben. **3.** *n.* **a)** *(children's ~)* Rutschbahn, *die;* **b)** *(Photog.)* Dia[positiv], *das.* **sliding** ['slaɪdɪŋ] **door** *n.* Schiebetür, *die*

slight [slaɪt] **1.** *adj.* leicht; schwach ⟨*Hoffnung, Aussichten, Wirkung*⟩; **not in the ~est** nicht im geringsten. **2.** *n.*

Verunglimpfung, die (on Gen.); (lack of courtesy) Affront, der (on gegen).

'**slightly** adv. ein bißchen; leicht ⟨verletzen, riechen nach, gewürzt sein, ansteigen⟩; flüchtig ⟨jmdn. kennen⟩; oberflächlich ⟨etw. kennen⟩

slim [slɪm] 1. adj. schlank; schmal ⟨Band, Buch⟩; schwach ⟨Aussicht, Hoffnung⟩; gering ⟨Gewinn, Chancen⟩. 2. v. i., -mm- abnehmen

slime [slaɪm] n. Schleim, der. **slimy** ['slaɪmɪ] adj. schleimig

sling [slɪŋ] 1. n. (Med.) Schlinge, die. 2. v. t., slung [slʌŋ] (coll.: throw) schmeißen (ugs.). **sling 'out** v. t. (coll.) wegschmeißen (ugs.); ~ sb. out jmdn. rausschmeißen (ugs.)

slink [slɪŋk] v. i., slunk [slʌŋk] schleichen. **slink a'way, slink 'off** v. i. davonschleichen

slip [slɪp] 1. v. i., -pp-: a) (slide) rutschen; ⟨Messer:⟩ abrutschen; (and fall) ausrutschen; b) (escape) schlüpfen; c) (go) ~ **to the butcher's** etc. [rasch] zum Fleischer usw. rüberspringen (ugs.). 2. v. t., -pp-: a) stecken; ~ **the dress over one's head** das Kleid über den Kopf streifen; b) ~ **sb.'s mind** or **memory** jmdm. entfallen. 3. n. a) (fall) after his ~: nachdem er ausgerutscht [und gestürzt] war; b) (mistake) Versehen, das; ~ **of the tongue** Versprecher, der; c) (underwear) Unterrock, der; d) (piece of paper) Zettel, der; e) give sb. the ~: jmdm. entwischen (ugs.). **slip a'way** v. i. a) ⟨Person:⟩ sich fortschleichen; b) ⟨Zeit:⟩ verfliegen. **slip 'down** v. i. runterrutschen (ugs.). **slip 'in** v. i. ⟨Person:⟩ sich hineinschleichen. '**slip into** v. t. schlüpfen in (+ Akk.) ⟨Kleidungsstück⟩. **slip 'off** 1. v. i. a) runterrutschen (ugs.); b) see **slip away** a. 2. v. t. abstreifen ⟨Schmuck, Handschuh⟩; schlüpfen aus ⟨Kleid, Schuh⟩. **slip 'on** v. t. überstreifen ⟨Handschuh, Ring⟩; schlüpfen in (+ Akk.) ⟨Kleid, Schuh⟩. **slip 'out** v. i. ⟨Person:⟩ sich hinausschleichen. **slip 'over** v. i. (fall) ausrutschen. **slip 'up** v. i. (coll.) einen Schnitzer machen (ugs.)

slipped [slɪpt] '**disc** n. Bandscheibenvorfall, der

'**slipper** n. Hausschuh, der

slippery ['slɪpərɪ] adj. schlüpfrig

slip: ~-**road** n. (Brit.) (to motorway) Auffahrt, die; (from motorway) Ausfahrt, die; ~**shod** adj. schludrig (ugs.); ~-**up** n. (coll.) Schnitzer, der

slit [slɪt] 1. n. Schlitz, der. 2. v. t., -tt-, slit aufschlitzen; ~ **sb.'s throat** jmdm. die Kehle durchschneiden

slither ['slɪðə(r)] v. i. rutschen

sliver ['slɪvə(r)] n. Splitter, der

slobber ['slɒbə(r)] v. i. sabbern (ugs.)

slog [slɒg] 1. v. t., -gg- (in boxing, fight) voll treffen. 2. v. i., -gg- (work) schuften (ugs.). 3. n. a) (hit) wuchtiger Schlag; b) (work) Plackerei, die (ugs.)

slogan ['sləʊgən] n. Slogan, der; (advertising ~) Werbeslogan, der

slop [slɒp] 1. v. i. schwappen (out of, from aus). 2. v. t. schwappen; (intentionally) kippen. **slop 'over** v. i. überschwappen

slope [sləʊp] 1. n. a) (slant) Neigung, die; b) (slanting ground) Hang, der. 2. v. i. (slant) sich neigen; ⟨Boden, Garten:⟩ abschüssig sein; ~ **downwards/upwards** ⟨Straße:⟩ abfallen/ansteigen. **slope a'way** v. i. abfallen. **slope 'off** v. i. (sl.) sich verdrücken (ugs.)

sloppy ['slɒpɪ] adj. schludrig (ugs.)

slosh [slɒʃ] adj. 1. v. i. platschen (ugs.); ⟨Flüssigkeit:⟩ schwappen. 2. v. t. (coll.: pour clumsily) schwappen

slot [slɒt] 1. n. a) (hole) Schlitz, der; b) (groove) Nut, die. 2. v. t., -tt-: ~ **sth. into place/sth.** etw. einfügen/in etw. (Akk.) einfügen. **slot 'in** 1. v. t. einfügen. 2. v. i. sich einfügen

sloth [sləʊθ] n. a) (lethargy) Trägheit, die; b) (Zool.) Faultier, das

slot-machine n. Automat, der; (for gambling) Spielautomat, der

slouch [slaʊtʃ] v. i. sich schlecht halten

slovenly ['slʌvnlɪ] adj. schlampig (ugs.)

slow [sləʊ] 1. adj. langsam; langwierig ⟨Arbeit⟩; **be |ten minutes|** ~ ⟨Uhr:⟩ [zehn Minuten] nachgehen. 2. adv. langsam. 3. v. i. langsamer werden; ~ **to a halt** anhalten. **slow 'down, slow 'up** v. i. langsamer werden

'**slowcoach** n. Trödler, der/Trödlerin, die (ugs.)

'**slowly** adv. langsam

slow 'motion n. **in** ~: in Zeitlupe

slowness n. Langsamkeit, die

sludge [slʌdʒ] n. Schlamm, der

slug [slʌg] n. Nacktschnecke, die

sluggish ['slʌgɪʃ] adj. träge; schleppend ⟨Nachfrage⟩

sluice [slu:s] 1. n. Schütz, das. 2. v. t. ~ |**down**| abspritzen

slum [slʌm] n. Slum, der; (single house or apartment) Elendsquartier, das

slumber ['slʌmbə(r)] *(poet./rhet.)* **1.** *n.*
~|s| Schlummer, *der (geh.).* **2.** *v.i.*
schlummern *(geh.)*

slump [slʌmp] **1.** *n.* Sturz, *der (fig.); (in
demand, investment, sales)* starker
Rückgang (in *Gen.*); *(economic depression)* Depression, *die.* **2.** *v.i.* **a)**
(Commerc.) stark zurückgehen; ⟨*Preise, Kurse:*⟩ stürzen; **b)** *(collapse)* ⟨*Person:*⟩ fallen; ~ed in a chair in einem
Sessel zusammengesunken

slung *see* sling 2

slunk *see* slink

slur [slɜː(r)] **1.** *v.t.,* -rr-: ~ one's words/
speech undeutlich sprechen. **2.** *n.* Beleidigung, *die* (on für)

slurp [slɜːp] *(coll.)* **1.** *v.t.* ~ |up| schlür-
fen. **2.** *n.* Schlürfen, *das*

slush [slʌʃ] *n.* Schneematsch, *der.*
'slushy *adj.* **a)** matschig; **b)** *(sloppy)*
sentimental

slut [slʌt] *n.* Schlampe, *die (ugs.)*

sly [slaɪ] **1.** *adj.* schlau; gerissen *(ugs.)*
⟨*Geschäftsmann, Trick*⟩; verschlagen
⟨*Blick*⟩. **2.** *n.* on the ~: heimlich

¹smack [smæk] **1.** *n.* **a)** *(sound)*
Klatsch, *der;* **b)** *(blow)* Schlag, *der; (on
child's bottom)* Klaps, *der (ugs.).* **2.**
v.t. **a)** [mit der flachen Hand] schlagen; **b)** ~ one's lips [mit den Lippen]
schmatzen. **3.** *adv. (coll.)* direkt

²smack *v.i.* ~ of schmecken nach;
(fig.) riechen nach *(ugs.)*

small [smɔːl] **1.** *adj.* klein; gering ⟨*Wirkung, Appetit, Fähigkeit*⟩; schmal
⟨*Taille*⟩; dünn ⟨*Stimme*⟩; make sb. feel
~: jmdn. beschämen. **2.** *n.* ~ of the
back Kreuz, *das.* **3.** *adv.* klein

small: ~ ad *n. (coll.)* Kleinanzeige,
die; ~ 'change *n.* Kleingeld, *das;*
~holding *n.* landwirtschaftlicher
Kleinbetrieb; ~'minded *adj.* kleinlich; ~pox *n.* Pocken *Pl.;* ~ talk *n.*
leichte Unterhaltung; *(at parties)*
Smalltalk, *der;* make ~ talk |with sb.|
[mit jmdm.] Konversation machen

smarmy ['smɑːmɪ] *adj. (coll.)* krieche-
risch

smart [smɑːt] **1.** *adj.* **a)** *(clever)* clever;
(ingenious) raffiniert; **b)** *(neat)* schick;
schön ⟨*Haus, Garten, Auto*⟩; **c)** *attrib.
(fashionable)* elegant; smart. **2.** *v.i.*
schmerzen. **smart alec[k]** ['smɑːt
ælɪk] *n. (coll.)* Besserwisser, *der.*
smarten ['smɑːtn] *v.t.* herrichten; ~
oneself |up| auf sein Äußeres achten.
'smartly *adv.* **a)** *(cleverly)* clever; **b)**
(neatly) schmuck ⟨*[an]gestrichen*⟩;
smart, flott ⟨*gekleidet, geschnitten*⟩

smash [smæʃ] **1.** *v.t.* **a)** zerschlagen;
b) ~ sb. in the face/mouth jmdm. [hart]
ins Gesicht/auf den Mund schlagen;
c) *(Tennis etc.)* schmettern. **2.** *v.i.* **a)**
zerbrechen; **b)** *(crash)* krachen (into
gegen). **3.** *n.* **a)** *(sound)* Krachen, *das;*
b) *see* smash-up; **c)** *(Tennis)* Schmet-
terball, *der.* **smash 'in** *v.t.* zerschmettern; einschlagen ⟨*Tür, Schädel*⟩. **smash 'up** *v.t.* zertrümmern

smash-and-'grab [raid] *n. (coll.)*
Schaufenstereinbruch, *der*

'smashing *adj. (coll.)* toll *(ugs.)*

'smash-up *n.* schwerer Zusammenstoß

smattering ['smætərɪŋ] *n.* |have| a ~ of
German *etc.* ein paar Brocken
Deutsch *usw.* [können]

smear [smɪə(r)] **1.** *v.t.* **a)** *(daub)* beschmieren; *(put on or over)* schmie-
ren; **b)** *(smudge)* verwischen; **c)** *(fig.)*
in den Schmutz ziehen. **2.** *n.* **a)**
(blotch) [Schmutz]fleck, *der;* **b)** *(fig.)*
Beschmutzung, *die* (on *Gen.*)

smell [smel] **1.** *n.* **a)** have a good/bad
sense of ~: einen guten/schlechten
Geruchssinn haben; **b)** *(odour)* Geruch, *der* (of nach); *(pleasant also)*
Duft, *der* (of nach); a ~ of burning/
gas ein Brand-/Gasgeruch; **c)** *(stink)*
Gestank, *der.* **2.** *v.t.,* smelt |smelt| *or*
smelled |smeld| **a)** *(perceive)* riechen;
b) *(inhale ~ of)* riechen an (+ *Dat.*). **3.**
v.i., smelt *or* smelled **a)** *(emit ~)* riechen; *(pleasantly also)* duften; **b)** ~ of
sth. *(lit. or fig.)* nach etw. riechen; **c)**
(stink) riechen. **'smelly** *adj.* stin-
kend; be ~: stinken

smelt *see* smell 2, 3

smile [smaɪl] **1.** *n.* Lächeln, *das;* give
sb. a ~: jmdn. anlächeln. **2.** *v.i.* lächeln; ~ at sb./sth. jmdn. anlächeln/
über etw. *(Akk.)* lächeln

smirk [smɜːk] **1.** *v.t.* grinsen. **2.** *n.*
Grinsen, *das*

smith [smɪθ] *n.* Schmied, *der*

smithereens [smɪðə'riːnz] *n. pl.* blow/
smash sth. to ~: etw. in tausend
Stücke sprengen/schlagen

smock [smɒk] *n.* Kittel, *der*

smog [smɒg] *n.* Smog, *der*

smoke [sməʊk] **1.** *n.* Rauch, *der.* **2.**
v.i. & t. rauchen. **smoked** [sməʊkt]
adj. (Cookery) geräuchert

smoke: ~ detector *n.* Rauchmelder,
der; ~less *adj.* rauchlos; rauchfrei
⟨*Zone*⟩

'smoker *n.* **a)** Raucher, *der*/Rauche-
rin, *die;* **b)** *(Railw.)* Raucherabteil, *das*

'**smoke-screen** n. [künstliche] Nebelwand; *(fig.)* Vernebelung *die (for Gen.)*

smoking ['sməʊkɪŋ] n. **a)** Rauchen, *das;* '**no ~**' „Rauchen verboten"; **b)** *(seating area)* **|do you want to sit in|** ~ or non-~? möchten Sie für Raucher oder Nichtraucher?

smoky ['sməʊkɪ] *adj. (emitting smoke)* rauchend; *(smoke-filled)* verräuchert

smooth [smu:ð] **1.** *adj.* **a)** *(even)* glatt; eben ‹*Straße, Weg*›; **b)** *(mild)* weich; **c)** *(not jerky)* geschmeidig ‹*Bewegung*›; ruhig ‹*Fahrt, Flug*›; weich ‹*Landung*›; **d)** *(without problems)* reibungslos. **2.** *v.t.* glätten. '**smoothly** *adv.* **a)** *(evenly)* glatt; **b)** *(not jerkily)* geschmeidig ‹*sich bewegen*›; weich ‹*landen*›; reibungslos ‹*funktionieren*›

smother ['smʌðə(r)] *v.t.* ersticken; *(fig.)* unterdrücken ‹*Gähnen*›; ersticken ‹*Gelächter, Schreie*›

smoulder ['sməʊldə(r)] *v.i.* schwelen; she was ~ing with rage Zorn schwelte in ihr

smudge [smʌdʒ] **1.** *v.t.* verwischen. **2.** *v.i.* schmieren. **3.** *n.* Fleck, *der*

smug [smʌg] *adj.* selbstgefällig

smuggle ['smʌgl] *v.t.* schmuggeln. **smuggle 'in** *v.t.* einschmuggeln; hinein-/hereinschmuggeln ‹*Person*›. **smuggle 'out** *v.t.* hinaus-/herausschmuggeln

smuggler ['smʌglə(r)] n. Schmuggler, *der*/Schmugglerin, *die*

smuggling ['smʌglɪŋ] n. Schmuggel, *der*

smutty ['smʌtɪ] *adj. (lewd)* schmutzig

snack [snæk] n. Imbiß, *der.* '**snackbar** n. Schnellimbiß, *der*

snag [snæg] n. *(problem)* Haken, *der;* what's the ~? wo klemmt es? *(ugs.)*

snail [sneɪl] n. Schnecke, *die;* at [a] ~'s pace im Schneckentempo *(ugs.)*

snake [sneɪk] n. Schlange, *die*

snap [snæp] **1.** *v.t.,* **-pp-:** **a)** *(break)* zerbrechen; ~ **sth. in two** *or* **in half** etw. in zwei Stücke brechen; ~ one's fingers mit den Fingern schnalzen; **c)** ~ sth. home *or* into place etw. einschnappen lassen; ~ shut zuschnappen ‹*Portemonnaie, Schloß*›; zuklappen ‹*Buch, Etui*›; ~ sth. open etw. aufschnappen lassen; **d)** *(take photograph of)* knipsen; **e)** *(say sharply)* fauchen; *(speak crisply or curtly)* bellen. **2.** *v.i.,* **-pp-:** **a)** *(break)* brechen; **b)** *(fig.: give way under strain)* ausrasten *(ugs.);* my patience

has finally ~ped nun ist mir der Geduldsfaden aber gerissen. **3.** *n. (Photog.)* Schnappschuß, *der.* **snap at** *v.t. (speak sharply to)* anfauchen *(ugs.).* **snap 'off** *v.t. & i.* abbrechen. **snap 'up** *v.t. (fig. coll.)* [sich *(Dat.)*] schnappen *(ugs.).*

'**snapshot** n. Schnappschuß, *der*

snare [sneə(r)] **1.** *n.* Schlinge, *die.* **2.** *v.t.* [mit einer Schlinge] fangen

¹**snarl** [snɑ:l] **1.** *v.i.* knurren. **2.** *n.* Knurren, *das*

²**snarl** n. *(tangle)* Knoten, *der.* **snarl 'up** *v.t. (bring to a halt)* zum Erliegen bringen; get ~ed up in the traffic im Verkehr steckenbleiben

'**snarl-up** n. Stau, *der*

snatch [snætʃ] **1.** *v.t.* **a)** *(grab)* schnappen; ~ **sth. from sb.** jmdm. etw. wegreißen; ~ some sleep ein bißchen schlafen; **b)** *(steal)* klauen *(ugs.).* **2.** *v.i.* einfach zugreifen. **3.** *n.* ~es of talk/conversation Gesprächsfetzen *Pl.*

sneak [sni:k] **1.** *v.t.* schmuggeln; ~ a look at schielen nach. **2.** *v.i.* **a)** schleichen; **b)** *(Brit. Sch. sl.: tell tales)* petzen *(Schülerspr.).* **3.** *n. (Brit. Sch. sl.)* Petzer, *der (Schülerspr.)*

sneer [snɪə(r)] *v.i.* höhnisch lächeln/ grinsen. '**sneer at** *v.t.* höhnisch anlächeln/angrinsen; *(scorn)* verhöhnen

sneeze [sni:z] **1.** *v.i.* niesen. **2.** *n.* Niesen, *das*

sniff [snɪf] **1.** *n.* Schnuppern, *das; (with running nose, while crying)* Schniefen, *das.* **2.** *v.i.* schniefen; *(to detect a smell)* schnuppern. **3.** *v.t.* riechen *od.* schnuppern an (+ *Dat.*). '**sniff at** *v.t.* **a)** *see* sniff 3; **b)** *(show contempt for)* die Nase rümpfen über

snigger ['snɪgə(r)] **1.** *v.i.* [boshaft] kichern. **2.** *n.* [boshaftes] Kichern

snip [snɪp] **1.** *v.t.,* **-pp-** schnippeln *(ugs.),* schneiden ‹*Loch*›; schnippeln *(ugs.) od.* schneiden an (+ *Dat.*) ‹*Tuch, Haaren, Hecke*›; *(cut off)* abschnippeln *(ugs.);* abschneiden. **2.** *n. (cut)* Schnitt, *der;* Schnipser, *der (ugs.)*

snipe [snaɪp] *v.i.* ~ at aus dem Hinterhalt beschießen. '**sniper** n. Heckenschütze, *der*

snippet ['snɪpɪt] n. *(of information in newspaper)* Notiz, *die; (of conversation)* Gesprächsfetzen, *der;* useful ~s of information nützliche Hinweise

snivel ['snɪvl] *v.i., (Brit.)* **-ll-** schniefen

snob [snɒb] n. Snob, *der.* **snobbery** ['snɒbərɪ] n. Snobismus, *der.* **snobbish** ['snɒbɪʃ] *adj.* snobistisch

snooker ['snu:kə(r)] *n.* Snooker, *das*
snoop [snu:p] *v. i.* schnüffeln *(ugs.)*
snooty ['snu:tɪ] *adj. (coll.)* hochnäsig *(ugs.)*
snooze [snu:z] *(coll.)* 1. *v. i.* dösen *(ugs.).* 2. *n.* Nickerchen, *das (fam.)*
snore [snɔ:(r)] 1. *v. i.* schnarchen. 2. *n.* Schnarcher, *der (ugs.);* ~s Schnarchen, *das*
snorkel ['snɔ:kl] *n.* Schnorchel, *der*
snort [snɔ:t] *v. i.* schnauben (**with, in** vor + *Dat.*)
snot [snɒt] *n. (sl.)* Rotz, *der (derb).* **snotty** *adj.* rotznäsig *(salopp);* ~ **child/nose** Rotznase, *die (salopp)*
snout [snaʊt] *n.* Schnauze, *die; (of pig)* Rüssel, *der*
snow [snəʊ] 1. *n.* Schnee, *der.* 2. *v. i. impers.* it ~s/is ~ing es schneit. **snow 'in** *v. t.* they are ~ed in sie sind eingeschneit. **snow 'under** *v. t.* be ~ed **under** *(with work)* erdrückt werden; *(with gifts, mail)* überschüttet werden
snow: ~**ball** 1. *n.* Schneeball, *der;* 2. *v. i. (fig.)* lawinenartig zunehmen; ~**bound** *adj.* eingeschneit; ~**-drift** *n.* Schneewehe, *die;* ~**drop** *n.* Schneeglöckchen, *das;* ~**fall** *n.* Schneefall, *der;* ~**flake** *n.* Schneeflocke, *die;* ~**man** *n.* Schneemann, *der;* ~**-plough** *n.* Schneepflug, *der;* ~**storm** *n.* Schneesturm, *der*
'snowy *adj.* schneereich ⟨*Gegend*⟩; schneebedeckt ⟨*Berge*⟩
snub [snʌb] 1. *v. t.,* -bb-: a) *(rebuff)* brüskieren; b) *(reject)* ablehnen. 2. *n.* Abfuhr, *die*
snub-'nosed *adj.* stupsnasig
¹snuff [snʌf] *n.* Schnupftabak, *der;* **take a pinch of** ~: eine Prise schnupfen
²snuff *v. t.* ~ |out| löschen ⟨*Kerze*⟩
snuffle ['snʌfl] *v. i.* schnüffeln
snug [snʌg] *adj.* gemütlich; behaglich; **be a** ~ **fit** genau passen
snuggle ['snʌgl] *v. i.* ~ **up to sb.** sich an jmdn. kuscheln; ~ **together** sich aneinanderkuscheln; ~ **up** *or* **down in bed** sich ins Bett kuscheln
so [səʊ] 1. *adv.* so; **as winter draws near, so it gets darker** je näher der Winter rückt, desto dunkler wird es; **so ... as so ... wie; so far** bis hierher; *(until now)* bisher; *(to such a distance)* **so weit; so much the better** um so besser; **so long!** bis dann! *(ugs.);* **and so on** |and so forth| und so weiter |und so fort|; **so as to um ... zu; so |that|** damit; **I'm so glad/tired!** ich bin ja so froh/

müde!; **It's a rainbow! – So it is!** Es ist ein Regenbogen! – Ja, wirklich!; **'You suggested it. – So I did** Du hast es vorgeschlagen. – Das stimmt; **is that so?** so? *(ugs.);* wirklich?; **so am/have/would/could/will/do** I ich auch. 2. *pron.* **he suggested that I take the train, and if I had done so, ...:** er riet mir, den Zug zu nehmen, und wenn ich es getan hätte, ...; **I'm afraid so;** ich told you so ich habe es dir [doch] gesagt; **a week or so** etwa eine Woche; **very much so** in der Tat. 3. *conj. (therefore)* daher; **so there you 'are!** ich habe also recht!; **so 'there!** [und] fertig!; **so?** na und?; **so you see ...:** du siehst also ...; **so where have you been?** wo warst du denn?

soak [səʊk] 1. *v. t.* a) einweichen ⟨*Wäsche in Lauge*⟩; eintauchen ⟨*Brot in Milch*⟩; b) *(wet)* naß machen. 2. *v. i.* a) *(steep)* put sth. in sth. to ~: etw. in etw. *(Dat.)* einweichen; b) *(drain)* ⟨*Feuchtigkeit, Nässe:*⟩ sickern. **'soaking** *adj. & adv.* ~ |wet| völlig durchnäßt
'so-and-so *n., pl.* ~'s a) *(person not named)* [Herr/Frau] Soundso; b) *(coll.: disliked person)* Biest, *das (ugs.)*
soap [səʊp] *n.* Seife, *die;* **with** ~ **and water** mit Wasser und Seife
soap: ~ **opera** *n.* Seifenoper, *die (ugs.);* ~ **powder** *n.* Seifenpulver, *das;* ~**suds** *n. pl.* Seifenschaum, *der*
'soapy *adj.* seifig; ~ **water** Seifenlauge, *die*
soar [sɔ:(r)] *v. i.* aufsteigen; *(fig.)* ⟨*Preise, Kosten usw.:*⟩ in die Höhe schießen *(ugs.)*
sob [sɒb] 1. *v. i.,* -bb- schluchzen (**with** vor + *Dat.*). 2. *n.* Schluchzer, *der*
sober ['səʊbə(r)] *adj.* a) *(not drunk)* nüchtern; b) *(serious)* ernst. **sober 'up** 1. *v. i.* nüchtern werden. 2. *v. t.* ausnüchtern
'sobering *adj.* ernüchternd
so-called ['səʊkɔ:ld] *adj.* sogenannt; *(alleged)* angeblich
soccer ['sɒkə(r)] *n.* Fußball, *der*
sociable ['səʊʃəbl] *adj.* gesellig
social ['səʊʃl] *adj.* a) sozial; gesellschaftlich; b) *(of ~ life)* gesellschaftlich; gesellig ⟨*Abend, Beisammensein*⟩
socialism ['səʊʃəlɪzm] *n.* Sozialismus, *der.* **socialist** ['səʊʃəlɪst] 1. *n.* Sozialist, *der*/Sozialistin, *die.* 2. *adj.* sozialistisch
socialize ['səʊʃəlaɪz] *v. i.* geselligen Umgang pflegen; ~ **with sb.** *(chat)* sich mit jmdm. unterhalten

'socially *adv.* **meet** ~: sich privat treffen; ~ **deprived** sozial benachteiligt
social: ~ **se'curity** *n.* **a)** *(Brit.: benefit)* Sozialhilfe, *die;* **b)** *(system)* soziale Sicherheit; ~ **'service** *n.* staatliche Sozialleistung; ~ **work** *n.* Sozialarbeit, *die;* ~ **worker** *n.* Sozialarbeiter, *der/-arbeiterin, die*
society [sə'saɪətɪ] *n.* **a)** Gesellschaft, *die;* **high** ~: High-Society, *die;* **b)** *(club, association)* Verein, *der*
sociologist [səʊsɪ'ɒlədʒɪst] *n.* Soziologe, *der/Soziologin, die*
sociology [səʊsɪ'ɒlədʒɪ] *n.* Soziologie, *die*
¹sock [sɒk] *n.* Socke, *die*
²sock *v.t. (coll.: hit)* hauen *(ugs.)*
socket ['sɒkɪt] *n.* **a)** *(Anat.) (of eye)* Höhle, *die; (of joint)* Pfanne, *die;* **b)** *(Electr.)* Steckdose, *die*
soda ['səʊdə] *n.* Soda, *das.* **'soda water** *n.* Soda[wasser], *das*
sodden ['sɒdn] *adj.* durchnäßt **(with** von**)**
sodium ['səʊdɪəm] *n.* Natrium, *das*
sofa ['səʊfə] *n.* Sofa, *das*
soft [sɒft] *adj.* weich; *(quiet)* leise; *(gentle)* sanft; **have a ~ spot for sb.** eine Vorliebe für jmdn. haben. **'soft-boiled** *adj.* weichgekocht ‹*Ei*›. **'soft drink** *n.* alkoholfreies Getränk
soften ['sɒfn] **1.** *v.i.* weicher werden. **2.** *v.t.* aufweichen ‹*Boden*›; enthärten ‹*Wasser*›; mildern ‹*Farbe*›
'softly *adv. (quietly)* leise; *(gently)* sanft
soft: ~ **toy** *n.* Stofftier, *das;* ~**ware** *n. (Computing)* Software, *die*
soggy ['sɒgɪ] *adj.* aufgeweicht
¹soil [sɔɪl] *n.* Erde, *die;* Boden, *der*
²soil *v.t.* beschmutzen
solace ['sɒləs] *n.* Trost, *der;* **take** *or* **find** ~ **in sth.** Trost in etw. *(Dat.)* finden
solar ['səʊlə(r)] *adj.* Sonnen-
sold *see* **sell**
solder ['səʊldə(r)] **1.** *n.* Lot, *das.* **2.** *v.t.* löten
soldier ['səʊldʒə(r)] *n.* Soldat, *der*
¹sole [səʊl] *n. (of foot/shoe)* Sohle, *die*
²sole *adj.* einzig; alleinig ‹*Verantwortung, Recht*›; Allein‹*erbe, -eigentümer*›. **'solely** *adv.* einzig und allein
solemn ['sɒləm] *adj.* feierlich; ernst ‹*Anlaß, Gespräch*›
solicitor [sə'lɪsɪtə(r)] *n. (Brit.: lawyer)* Rechtsanwalt, *der/-anwältin, die*
solid ['sɒlɪd] **1.** *adj.* **a)** *(rigid)* fest; **b)** *(of the same substance all through)*

massiv; **c)** *(well-built)* stabil; solide gebaut ‹*Haus, Mauer usw.*›; **d)** *(complete)* ganz; **a good** ~ **meal** eine kräftige Mahlzeit. **2.** *n.* fester Körper
solidarity [sɒlɪ'dærɪtɪ] *n.* Solidarität, *die*
solidify [sə'lɪdɪfaɪ] *v.i* fest werden
solitary ['sɒlɪtərɪ] *adj.* **a)** einsam; ~ **confinement** Einzelhaft, *die;* **b)** *(sole)* einzig
solitude ['sɒlɪtjuːd] *n.* Einsamkeit, *die*
solo ['səʊləʊ] **1.** *n., pl.* ~**s** *(Mus.)* Solo, *das.* **2.** *adj.* **a)** *(Mus.)* Solo-; **b)** ~ **flight** Alleinflug, *der.* **3.** *adv.* **a)** *(Mus.)* solo; **b) go/fly** ~ *(Aeronaut.)* einen Alleinflug machen. **soloist** ['səʊləʊɪst] *n. (Mus.)* Solist, *der/Solistin, die*
solstice ['sɒlstɪs] *n.* Sonnenwende, *die*
soluble ['sɒljʊbl] *adj.* **a)** *(esp. Chem.)* löslich; **b)** *(solvable)* lösbar
solution [sə'luːʃn] *n.* **a)** *(esp. Chem.)* Lösung, *die;* **b)** *[(result of) solving]* Lösung, *die* (**to** Gen.); **find a** ~ **to sth.** eine Lösung für etw. finden; etw. lösen
solvable ['sɒlvəbl] *adj.* lösbar
solve [sɒlv] *v.t.* lösen
solvent ['sɒlvənt] **1.** *adj.* **a)** *(esp. Chem.)* lösend; **b)** *(Finance)* solvent. **2.** *n. (esp. Chem.)* Lösungsmittel, *das*
sombre *(Amer.:* **somber**) ['sɒmbə(r)] *adj.* dunkel; düster ‹*Stimmung, Atmosphäre*›
some [səm, *stressed* sʌm] **1.** *adj.* **a)** *(one or other)* [irgend]ein; ~ **day** eines Tages; **b)** *(a considerable quantity of)* einig...; **c)** *(a small quantity of)* ein bißchen; **would you like** ~ **wine/cherries?** möchten Sie [etwas] Wein/[ein paar] Kirschen?; **do** ~ **shopping/reading** einkaufen/lesen; **d)** *(to a certain extent)* ~ **guide** eine gewisse Orientierungshilfe. **2.** *pron.* einig...; **would you like** ~? möchtest du etwas/*(plural)* welche?; ~ **...**, **others ...:** manche ..., andere ...
somebody ['sʌmbədɪ] *n. & pron.* jemand; ~ **or other** irgend jemand
'somehow *adv.* ~ **[or other]** irgendwie
someone ['sʌmwʌn] *see* **somebody**
somersault ['sʌməsɔːlt] *n.* Purzelbaum, *der (ugs.);* Salto, *der (Sport);* **turn a** ~: einen Purzelbaum schlagen *(ugs.)*/einen Salto springen
'something *n. & pron.* etwas; ~ **new** etwas Neues; ~ **or other** irgend etwas; **see** ~ **of sb.** jmdn. sehen
'sometime 1. *adj.* ehemalig. **2.** *adv.* irgendwann
'sometimes *adv.* manchmal

'**somewhat** *adv.* ziemlich

'**somewhere** 1. *adv.* a) *(in a place)* irgendwo; b) *(to a place)* irgendwohin. 2. *n.* look for ~ to stay sich nach einer Unterkunft umsehen

son [sʌn] *n.* Sohn, *der*

sonata [sə'nɑːtə] *n.* Sonate, *die*

song [sɒŋ] *n.* a) Lied, *das;* b) *(bird cry)* Gesang, *der*

'**son-in-law** *n., pl.* **sons-in-law** Schwiegersohn, *der*

soon [suːn] *adv.* a) bald; *(quickly)* schnell; b) *(early)* früh; **none too ~:** keinen Augenblick zu früh; **~er or later** früher oder später; c) **we'll set off as ~ as he arrives** sobald er ankommt, machen wir uns auf den Weg; **as ~ as possible** so bald wie möglich; d) *(willingly)* **just as ~ [as ...]** genauso gern [wie ...]; **she would ~er die than ...:** sie würde lieber sterben, als ...

soot [sʊt] *n.* Ruß, *der*

soothe [suːð] *v. t.* a) *(calm)* beruhigen; b) lindern ⟨*Schmerz*⟩

'**sooty** *adj.* verrußt; rußig

sophisticated [sə'fɪstɪkeɪtɪd] *adj.* a) *(cultured)* kultiviert; b) *(elaborate, complex)* hochentwickelt; subtil ⟨*Argument, System*⟩

soporific [sɒpə'rɪfɪk] *adj.* einschläfernd

sopping ['sɒpɪŋ] *adj. & adv.* ~ [wet] völlig durchnäßt

soppy ['sɒpɪ] *adj. (Brit. coll.)* rührselig; sentimental ⟨*Person*⟩

soprano [sə'prɑːnəʊ] *n.* Sopran, *der; (female also)* Sopranistin, *die*

sordid ['sɔːdɪd] *adj.* dreckig; unerfreulich ⟨*Detail, Geschichte*⟩

sore [sɔː(r)] 1. *adj.* weh; *(inflamed or injured)* wund; **a ~ throat** Halsschmerzen *Pl.;* **sb. has a ~ back/foot** *etc.* jmdm. tut der Rücken/Fuß *usw.* weh. 2. *n.* wunde Stelle. '**sorely** *adv.* sehr; dringend ⟨*nötig*⟩; ~ **tempted** stark versucht

sorrow ['sɒrəʊ] *n.* Kummer, *der*

sorry ['sɒrɪ] *adj.* a) **sb. is ~ that ...:** es tut jmdm. leid, daß ...; **sb. is ~ about sth.** jmdm. tut etwas leid; **I am or feel ~ for him** er tut mir leid; **sb. is or feels ~ for sth.** jmd. bedauert etw.; **~!** Entschuldigung!; **~?** wie bitte?; **I'm ~ to say** leider; **you'll be ~!** das wird dir noch leid tun; b) *(wretched)* traurig

sort [sɔːt] 1. *n.* a) Art, *die; (type)* Sorte, *die;* **a new ~ of bicycle** ein neuartiges Fahrrad; **all ~s of ...:** alle möglichen ...; **there are all ~s of things to do** es gibt alles mögliche *od.* allerlei zu tun; ~ **of** *(coll.: more or less)* mehr oder weniger; **nothing of the ~:** nichts dergleichen; b) **be out of ~s** nicht in Form sein. 2. *v. t.* sortieren. **sort 'out** *v. t.* a) *(settle)* klären; schlichten ⟨*Streit*⟩; beenden ⟨*Verwirrung*⟩; b) *(select)* aussuchen

'**sort code** *n.* Bankleitzahl, *die*

sortie ['sɔːtiː] *n.* Ausfall, *der; (flight)* Einsatz, *der*

SOS *n.* SOS, *das*

'**so so, 'so-so** *adj., adv.* so lala *(ugs.)*

soufflé ['suːfleɪ] *n.* Soufflé, *das*

sought *see* seek

soul [səʊl] *n.* Seele, *die;* **not a ~:** keine Menschenseele

'**soul-destroying** *adj.* a) *(boring)* nervtötend; b) *(depressing)* deprimierend

soulful ['səʊlfl] *adj.* gefühlvoll; *(sad)* schwermütig

soul: ~ **mate** *n.* Seelenverwandte, *der/die;* ~-**searching** *n.* Gewissenskampf, *der*

¹**sound** [saʊnd] 1. *adj.* a) *(healthy)* gesund; intakt ⟨*Gebäude, Mauerwerk*⟩; **of ~ mind** im Vollbesitz der geistigen Kräfte; b) *(well-founded)* vernünftig ⟨*Argument, Rat*⟩; klug ⟨*Wahl*⟩; **it makes ~ sense** es ist sehr vernünftig; c) *(Finance: secure)* gesund, solide ⟨*Basis*⟩; klug ⟨*Investition*⟩. 2. *adv.* fest, tief ⟨*schlafen*⟩

²**sound** 1. *n.* a) *(Phys.)* Schall, *der;* b) *(noise)* Laut, *der; (of wind, sea, car, footsteps, breaking glass or twigs)* Geräusch, *das; (of voices, laughter, bell)* Klang, *der;* **do sth. without a ~:** etw. lautlos tun; c) *(Radio, Telev., Cinemat.)* Ton, *der;* d) *(fig.: impression)* **I like the ~ of your plan** ich finde, Ihr Plan hört sich gut an; **I don't like the ~ of this** das hört sich nicht gut an. 2. *v. i.* klingen; **it ~s as if .../like ...:** es klingt, als .../wie ...; **that ~s a good idea to me** ich finde, die Idee hört sich gut an; **that ~s odd to me** das hört sich seltsam an, finde ich; ~**s good to me!** klingt gut! *(ugs.).* 3. *v. t.* a) ertönen lassen; b) *(utter)* ~ **a note of caution** zur Vorsicht mahnen. **sound 'off** *v. i.* tönen *(ugs.),* schwadronieren **(on, about,** von). **sound 'out** *v. i.* ausfragen ⟨*Person*⟩; ~ **sb. out on sth.** bei jmdm. wegen etw. vorfühlen

sound: ~ **barrier** *n.* Schallmauer, *die;* ~ **effect** *n.* Geräuscheffekt, *der*

'**sounding-board** *n.* a) *(Mus.)* Decke,

die; b) *(fig.: trial audience)* ≈ Testgruppe, *die*

'**soundless** *adj.* lautlos

'**soundly** *adv.* a) *(solidly)* stabil, solide ⟨*bauen*⟩; b) *(deeply)* tief, fest ⟨*schlafen*⟩; c) *(thoroughly)* ordentlich *(ugs.)* ⟨*verhauen*⟩; vernichtend ⟨*schlagen, besiegen*⟩

sound: ~-**proof** 1. *adj.* schalldicht; 2. *v. t.* schalldicht machen; ~-**track** *n.* Soundtrack, *der;* ~-**wave** *n.* Schallwelle, *die*

soup [su:p] *n.* Suppe, *die;* **be/land in the** ~ *(fig. sl.)* in der Patsche sitzen/landen *(ugs.)*

soup: ~-**plate** *n.* Suppenteller, *der;* ~-**spoon** *n.* Suppenlöffel, *der*

sour ['saʊə(r)] *adj.* a) sauer; b) *(morose)* griesgrämig; säuerlich ⟨*Blick*⟩; c) *(unpleasant)* bitter

source [sɔ:s] *n.* Quelle, *die;* ~ **of income/infection** Einkommensquelle, *die/*Infektionsherd, *der;* **at** ~: an der Quelle

south [saʊθ] 1. *n.* a) Süden, *der;* **in/to|wards|/from the** ~: im/nach/von Süden; **to the** ~ **of** südlich von; b) *usu.* S~ *(Geog., Polit.)* Süden, *der.* 2. *adj.* südlich; Süd⟨*küste, -wind, -grenze*⟩. 3. *adv.* nach Süden; ~ **of** südlich von

South: ~ '**Africa** *pr. n.* Südafrika *(das);* ~ '**African** *adj.* südafrikanisch; ~ **A'merica** *pr. n.* Südamerika *(das);* ~ **A'merican** *adj.* südamerikanisch; **s~-bound** *adj.* ⟨*Zug, Verkehr usw.*⟩ in Richtung Süden; **s~-'east** 1. *n.* Südosten, *der;* 2. *adj.* südöstlich; Südost⟨*wind, -küste*⟩; 3. *adv.* südostwärts; nach Südosten; **s~-'eastern** *adj.* südöstlich

southerly ['sʌðəlɪ] *adj.* südlich; ⟨*Wind*⟩ aus südlichen Richtungen

southern ['sʌðən] *adj.* südlich; Süd⟨*grenze, -hälfte, -seite*⟩

South: ~ '**Germany** *pr. n.* Süddeutschland *(das);* ~ '**Pole** *pr. n.* Südpol, *der*

southward[s] ['saʊθwəd(s)] *adv.* südwärts

south: ~-'**west** 1. *n.* Südwesten, *der;* 2. *adj.* südwestlich; Südwest⟨*wind, -küste*⟩; 3. *adv.* südwestwärts; nach Südwesten; ~-'**western** *adj.* südwestlich

souvenir [su:vəˈnɪə(r)] *n.* Souvenir, *das* (of aus)

sovereign ['sɒvrɪn] *n.* *(ruler)* Souverän, *der.* **sovereignty** ['sɒvrɪntɪ] *n.* Souveränität, *die*

Soviet ['səʊvɪət, 'sɒvɪət] *adj. (Hist.)* sowjetisch; Sowjet⟨*bürger, -literatur*⟩

Soviet 'Union *pr. n. (Hist.)* Sowjetunion, *die*

¹**sow** [səʊ] *v. t., p. p.* **sown** [səʊn] *or* **sowed** [səʊd] a) *(plant)* [aus]säen; b) einsäen ⟨*Feld, Boden*⟩

²**sow** [saʊ] *n. (female pig)* Sau, *die*

sown *see* ¹**sow**

soya [bean] ['sɔɪə (bi:n)] *n.* Sojabohne, *die*

spa [spɑ:] *n.* a) *(place)* Bad, *das;* Badeort, *der;* b) *(spring)* Mineralquelle, *die*

space [speɪs] *n.* a) Raum, *der;* b) *(interval between points)* Platz, *der;* **clear a** ~: Platz schaffen; c) **the wide open** ~**s** das weite, flache Land; d) *(Astron.)* Weltraum, *der;* e) *(blank between words)* Zwischenraum, *der;* f) *(interval of time)* Zeitraum, *der;* **in the** ~ **of a minute/an hour** innerhalb einer Minute/Stunde; **in a short** ~ **of time he was back** nach kurzer Zeit war er zurück. **space 'out** *v. t.* verteilen

space: ~ **age** *n.* [Welt]raumzeitalter, *das;* ~-**bar** *n.* Leertaste, *die;* ~**craft** *n.* Raumfahrzeug, *das;* ~-**saving** *adj.* platzsparend; ~**ship** *n.* Raumschiff, *das;* ~**suit** *n.* Raumanzug, *der;* ~ **travel** *n.* Raumfahrt, *die*

spacious ['speɪʃəs] *adj.* geräumig

spade [speɪd] *n.* a) Spaten, *der;* b) *(Cards)* Pik, *das; see also* **club I c**

spaghetti [spəˈgetɪ] *n.* Spaghetti *Pl.*

Spain [speɪn] *pr. n.* Spanien *(das)*

span [spæn] 1. *n.* a) Spanne, *die;* Zeitspanne, *die;* b) *(of bridge)* Spannweite, *die.* 2. *v. t.,* -**nn**- überspannen ⟨*Fluß*⟩; umfassen ⟨*Zeitraum*⟩

Spaniard ['spænjəd] *n.* Spanier, *der/*Spanierin, *die*

Spanish ['spænɪʃ] 1. *adj.* spanisch; **sb. is** ~: jmd. ist Spanier/Spanierin. 2. *n.* a) *(language)* Spanisch, *das; see also* **English 2a;** b) **the** ~ *pl.* die Spanier

spank [spæŋk] 1. *n.* ≈ Klaps, *der (ugs.).* 2. *v. t.* ~ **sb.** jmdm. den Hintern versohlen *(ugs.)*

spanner ['spænə(r)] *n. (Brit.)* Schraubenschlüssel, *der*

spar [spɑ:(r)] *v. i.,* -**rr**-: a) *(Boxing)* sparren; b) *(fig.: argue)* [sich] zanken

spare [speə(r)] 1. *adj.* a) *(not in use)* übrig; ~ **time/moment** Freizeit, *die/*freier Augenblick; **there is one** ~ **seat** ein Platz ist noch frei; b) *(for use when needed)* zusätzlich, Extra⟨*bett, -tasse*⟩; ~ **room** Gästezimmer, *das.* 2. *n.* Ersatzteil, *das/*-reifen, *der usw.* 3.

v. t. **a)** entbehren; **we arrived with ten minutes to ~:** wir kamen zehn Minuten früher an; **b)** *(not inflict on)* **~ sb. sth.** jmdm. etw. ersparen; **c)** *(not hurt)* [ver]schonen; **d)** *(fail to use)* **not ~ any expense/pains** *or* **efforts** keine Kosten/Mühe scheuen; **no expense ~d** an nichts gespart

spare: **~ 'part** *n.* Ersatzteil, *das;* **~ 'tyre** *n.* Reserve-, Ersatzreifen, *der;* **~ 'wheel** *n.* Ersatzrad, *das*

sparing ['speərɪŋ] *adj.* sparsam

spark [spɑːk] **1.** *n.* **a)** Funke, *der;* (fig.) **a ~ of generosity/decency** ein Funke[n] Großzügigkeit/Anstand; **b) a bright ~** *(person, also iron.)* ein schlauer Kopf. **2.** *v. t.* **~** [off] zünden; *(fig.)* auslösen

sparkle ['spɑːkl] **1.** *v. i.* **a)** ⟨*Diamant:*⟩ glitzern; ⟨*Augen:*⟩ funkeln; **b)** *(be lively)* sprühen **(with** *vor* + *Dat.).* **2.** *n.* Funkeln, *das.* **sparkling** ['spɑːklɪŋ] *adj.* glitzernd ⟨*Diamant*⟩; funkelnd ⟨*Augen*⟩. **sparkling 'wine** *n.* Schaumwein, *der*

'spark-plug *n.* Zündkerze, *die*

sparrow ['spærəʊ] *n.* Spatz, *der*

sparse [spɑːs] *adj.* spärlich; dünn ⟨*Besiedlung*⟩

spasm ['spæzm] *n.* Krampf, *der*

spasmodic [spæz'mɒdɪk] *adj.* **a)** *(marked by spasms)* krampfartig; **b)** *(intermittent)* sporadisch

spastic ['spæstɪk] **1.** *n.* Spastiker, *der/* Spastikerin, *die.* **2.** *adj.* spastisch

spat *see* spit 1, 2

spate [speɪt] *n.* **a) the river is in** [full] **~:** der Fluß führt Hochwasser; **b)** *(fig.)* **a ~ of sth.** eine Flut von etw.; **a ~ of burglaries** eine Einbruchsserie

spatial ['speɪʃl] *adj.* räumlich

spatter ['spætə(r)] *v. t.* spritzen; **~ sb./sth. with sth.** jmdn./etw. mit etw. bespritzen

spatula ['spætjʊlə] *n.* Spachtel, *die*

spawn [spɔːn] **1.** *v. t.* (fig.) hervorbringen. **2.** *v. i.* *(Zool.)* laichen. **3.** *n.* *(Zool.)* Laich, *der*

speak [spiːk] **1.** *v. i.,* spoke [spəʊk], spoken ['spəʊkn] **a)** sprechen; **~** [with sb.] on *or* about sth. [mit jmdm.] über etw. *(Akk.)* sprechen; **~ for/against sth.** sich für/gegen etw. aussprechen; **b)** *(on telephone)* **Is Mr Grant there? – S~ing!** Ist Mister Grant da? – Am Apparat!; **who is ~ing, please?** wer ist am Apparat, bitte? **2.** *v. t.,* spoke, spoken sprechen ⟨*Satz, Wort, Sprache*⟩; sagen ⟨*Wahrheit*⟩; **~ one's mind** sagen, was man denkt. **'speak**

for *v. t.* sprechen für; **sth. is spoken for** *(reserved)* etw. ist schon vergeben. **'speak of** *v. t.* sprechen von; **~ing of Mary** da wir gerade von Mary sprechen; **nothing to ~ of** nichts Besonderes. **'speak to** *v. t.* sprechen *od.* reden mit. **speak 'up** *v. i.* lauter sprechen

'speaker *n.* **a)** *(in public)* Redner, *der/*Rednerin, *die;* **b)** *(of a language)* Sprecher *der/*Sprecherin, *die;* **be a 'French ~:** Französisch sprechen; **c)** *(loudspeaker)* Lautsprecher, *der*

'speaking 1. *n.* Sprechen, *das;* **~ clock** *(Brit.)* telefonische Zeitansage. **2.** *adv.* **strictly/generally ~:** genaugenommen/im allgemeinen

spear [spɪə(r)] *n.* Speer, *der.* **'spearhead 1.** *n.* (fig.) Speerspitze, *die.* **2.** *v. t.* (fig.) anführen. **'spearmint** *n.* Grüne Minze; **~ chewing-gum** Pfefferminzkaugummi, *der od. das*

special ['speʃl] *adj.* speziell; besonder...; **nobody ~:** niemand Besonderer. **special de'livery** *n.* *(Post)* Eilzustellung, *die*

specialist ['speʃəlɪst] *n.* **a)** Spezialist, *der/*Spezialistin, *die* **(in** für**);** **b)** *(Med.)* Facharzt, *der/*-ärztin, *die*

speciality [speʃɪ'ælɪtɪ] *n.* Spezialität, *die*

specialize ['speʃəlaɪz] *v. i.* sich spezialisieren **(in** auf + *Akk.*)

'specially *adv.* **a)** speziell; **make sth. ~:** etw. speziell *od.* extra anfertigen; **b)** *(especially)* besonders

special 'offer *n.* Sonderangebot, *das;* **on ~:** im Sonderangebot

specialty ['speʃltɪ] *(esp. Amer.) see* speciality

species ['spiːʃiːz] *n., pl.* same Art, *die*

specific [spɪ'sɪfɪk] *adj.* bestimmt; **could you be more ~?** kannst du dich genauer ausdrücken? **specifically** [spɪ'sɪfɪkəlɪ] *adv.* ausdrücklich; eigens; extra *(ugs.)*

specification [spesɪfɪ'keɪʃn] *n., often pl.* *(details)* technische Daten; *(for building)* Baubeschreibung, *die*

specify ['spesɪfaɪ] *v. t.* ausdrücklich sagen; **unless otherwise specified** wenn nicht anders angegeben

specimen ['spesɪmən] *n.* **a)** *(example)* Exemplar, *das;* **b)** *(sample)* Probe, *die*

speck [spek] *n.* **a)** *(spot)* Fleck, *der;* **b)** *(particle)* Teilchen, *das;* **~ of soot/dust** Rußflocke, *die/*Staubkörnchen, *das*

specs [speks] *n.pl.* *(coll.: spectacles)* Brille, *die*

spectacle ['spektəkl] *n.* **a)** *in pl.* |pair of| ~s Brille, *die;* **b)** *(public show)* Spektakel, *das;* **c)** *(object of attention)* Anblick, *der.* '**spectacle case** *n.* Brillenetui, *das*

spectacular [spek'tækjʊlə(r)] *adj.* spektakulär

spectator [spek'teɪtə(r)] *n.* Zuschauer, *der*/Zuschauerin, *die*

specter *(Amer.) see* **spectre**

spectra *pl. of* **spectrum**

spectre ['spektə(r)] *n. (Brit.)* **a)** *(ghost)* Gespenst, *das;* **b)** *(fig.)* Schreckgespenst, *das*

spectrum ['spektrəm] *n., pl.* **spectra** ['spektrə] Spektrum, *das*

speculate ['spekjʊleɪt] *v. i.* spekulieren (**about, on** über + *Akk.*). **speculation** [spekjʊ'leɪʃn] *n.* Spekulation, *die* (**over** über + *Akk.*). **speculative** ['spekjʊlətɪv] *adj.* spekulativ. **speculator** ['spekjʊleɪtə(r)] *n.* Spekulant, *der*/Spekulantin, *die*

sped *see* **speed** 2

speech [spi:tʃ] *n.* **a)** *(public address)* Rede, *die;* **make** *or* **deliver** *or* **give a** ~: eine Rede halten; **b)** *(faculty or manner of speaking)* Sprache, *die.* '**speechless** *adj.* sprachlos (**with** vor + *Dat.*)

speed [spi:d] **1.** *n.* Geschwindigkeit, *die;* Schnelligkeit, *die;* **at a** ~ **of …:** mit einer Geschwindigkeit von … **2.** *v. i.* **a)** *p. t. & p. p.* **sped** [sped] *or* **speeded** schnell fahren; rasen *(ugs.);* **b)** *p. t. & p. p.* **speeded** *(go too fast)* zu schnell fahren; rasen *(ugs.).* '**speedboat** *n.* Rennboot, *das*

'**speeding** *n.* Geschwindigkeitsüberschreitung, *die*

'**speed limit** *n.* Geschwindigkeitsbeschränkung, *die*

speedometer [spi:'dɒmɪtə(r)] *n.* Tachometer, *der od. das*

'**speedy** *adj.* schnell; umgehend, prompt *(Antwort)*

¹**spell** [spel] **1.** *v. t.,* **spelt** [spelt] *(Brit.)* *or* **spelled a)** schreiben; *(aloud)* buchstabieren; **b)** *(fig.: mean)* bedeuten. **2.** *v. i.,* **spelt** *(Brit.)* *or* **spelled** *(say)* buchstabieren; *(write)* richtig schreiben

²**spell** *n. (period)* Weile, *die;* **a cold** ~: eine Kälteperiode

³**spell** *n.* **a)** *(magic charm)* Zauberspruch, *der;* **cast a** ~ **on sb.** jmdn. verzaubern; **b)** *(fascination)* Zauber, *der;* **break the** ~: den Bann brechen. '**spellbound** *adj.* verzaubert

'**spelling** *n.* Rechtschreibung, *die*

spelt *see* ¹**spell**

spend [spend] *v. t.,* **spent** [spent] **a)** *(pay out)* ausgeben; ~ **a penny** *(fig. coll.)* mal verschwinden *(ugs.);* **b)** verbringen ⟨*Zeit*⟩. '**spendthrift** *n.* Verschwender, *der*/Verschwenderin, *die*

spent 1. *see* **spend. 2.** *adj.* **a)** *(used up)* verbraucht; **b)** *(drained of energy)* erschöpft

sperm ['spɜ:m] *n. pl* ~s *or same (Biol.)* Sperma, *der*

spew [spju:] *v. t.* spucken

sphere [sfɪə(r)] *n.* **a)** *(field of action)* Bereich, *der;* Sphäre, *die (geh.);* **b)** *(Geom.)* Kugel, *die.* **spherical** ['sferɪkl] *adj.* kugelförmig

spice [spaɪs] **1.** *n.* Gewürz, *das; (fig.)* Würze, *die.* **2.** *v. t.* würzen. **spicy** ['spaɪsɪ] *adj.* pikant; würzig

spider ['spaɪdə(r)] *n.* Spinne, *die*

spike [spaɪk] *n.* Stachel, *der.* **spiky** ['spaɪkɪ] *adj.* stachelig

spill [spɪl] **1.** *v. t.,* **spilt** [spɪlt] *or* **spilled** verschütten ⟨*Flüssigkeit*⟩; ~ **sth. on** ɔth. etw. auf etw. *(Akk.)* schütten; ~ **the beans** aus der Schule plaudern. **2.** *v. i.,* **spilt** *or* **spilled** überlaufen

spilt *see* **spill**

spin [spɪn] **1.** *v. t.,* -**nn**-, **spun** [spʌn] **a)** spinnen; ~ **yarn** Garn spinnen; **b)** *(in washing-machine etc.)* schleudern. **2.** *v. i.,* -**nn**-, **spun** sich drehen; **my head is** ~**ning** *(fig.)* mir schwirrt der Kopf. **spin 'out** *v. t. (prolong)* in die Länge ziehen

spinach ['spɪnɪdʒ] *n.* Spinat, *der*

spinal ['spaɪnl] *adj.* Wirbelsäulen-; Rückgrat[s]-. **spinal 'column** *n.* Wirbelsäule, *die.* **spinal 'cord** *n.* Rückenmark, *das*

spindle ['spɪndl] *n.* Spindel, *die.* **spindly** ['spɪndlɪ] *adj.* spindeldürr

spin-'drier *n.* Wäscheschleuder, *die*

spin-'dry *v. t.* schleudern

spine [spaɪn] *n.* **a)** *(backbone)* Wirbelsäule, die; **b)** *(Bot., Zool.)* Stachel, *der.* '**spineless** *adj. (fig.)* rückgratlos

'**spin-off** *n.* Nebenprodukt, *das*

spinster ['spɪnstə(r)] *n.* ledige Frau

spiny ['spaɪnɪ] *adj.* stachelig

spiral ['spaɪrl] **1.** *adj.* spiralförmig. **2.** *n.* Spirale, *die.* **3.** *v. i., (Brit.)* -**ll**- ⟨*Weg:*⟩ sich hochwinden; ⟨*Kosten:*⟩ in die Höhe klettern; ⟨*Rauch:*⟩ in einer Spirale aufsteigen. **spiral 'staircase** *n.* Wendeltreppe, *die*

spire ['spaɪə(r)] *n.* Turmspitze, *die*

spirit ['spɪrɪt] *n.* **a)** *in pl. (distilled liquor)* Spirituosen *Pl.;* **b)** *(mental atti-*

tude) Geisteshaltung, *die;* **in the right/ wrong** ~: mit der richtigen/falschen Einstellung; **take sth. in the wrong** ~: etw. falsch auffassen; **c)** *(courage)* Mut, *der;* **d)** *(mental tendency)* Geist, *der;* **high** ~s gehobene Stimmung; **in poor** *or* **low** ~s niedergedrückt.

'**spirited** *adj.* beherzt

'**spirit-level** *n.* Wasserwaage, *die*

spiritual ['spɪrɪtʃʊəl] *adj.* spirituell *(geh.)*

spit [spɪt] **1.** *v. i.,* -tt-, **spat** [spæt] *or* **spit** spucken. **2.** *v. t.,* -tt-, **spat** *or* **spit** spucken. **3.** *n.* Spucke, *die.* **spit 'out** *v. t.* ausspucken

spite [spaɪt] **1.** *n.* **a)** Boshaftigkeit, *die;* **b) in** ~ **of** trotz; **in** ~ **of oneself** obwohl man es eigentlich nicht will. **2.** *v. t.* ärgern. **spiteful** ['spaɪtfl] *adj.* gehässig

spittle ['spɪtl] *n.* Spucke, *die*

splash [splæʃ] **1.** *v. t.* spritzen; ~ **sth. on |to|** *or* **over sb./sth.** jmdn./etw. mit etw. bespritzen. **2.** *v. i.* **a)** spritzen; **b)** *(in water)* platschen *(ugs.).* **3.** *n.* **a)** *(liquid)* Spritzer, *der;* **b)** *(noise)* Plätschern, *das*

splendid ['splendɪd] *adj.* *(excellent)* großartig; *(magnificent)* prächtig

splendour *(Brit.; Amer.:* **splendor)** ['splendə(r)] *n.* Pracht, *die*

splint [splɪnt] *n.* Schiene, *die*

splinter ['splɪntə(r)] *n.* Splitter, *der*

split [splɪt] **1.** *n.* **a)** *(tear)* Riß, *der;* **b)** *(division into parts)* [Auf]teilung, *die;* *(fig.)* Spaltung, *die.* **2.** *adj.* gespalten; **be** ~ **on a question** *(Dat.)* in einer Frage uneins sein. **3.** *v. t.,* -tt-, **split a)** *(tear)* zerreißen; **b)** *(divide)* teilen. **4.** *v. i.,* -tt-, **split a)** ⟨*Holz:*⟩ splittern; ⟨*Stoff, Seil:*⟩ reißen; ~ **apart** zersplittern; **b)** *(divide into parts)* sich teilen. **split 'up 1.** *v. t.* aufteilen. **2.** *v. i.* *(coll.)* sich trennen; ~ **up with sb.** sich von jmdm. trennen

splutter ['splʌtə(r)] *v. i.* ⟨*Person:*⟩ prusten; ⟨*Motor:*⟩ stottern

spoil [spɔɪl] **1.** *v. t.,* **spoilt** [spɔɪlt] *or* **spoiled a)** *(impair)* verderben; **b)** *(pamper)* verwöhnen; **be** ~**t for choice** die Qual der Wahl haben. **2.** *v. i.,* **spoilt** *or* **spoiled a)** verderben; **b) be** ~**ing for a fight** Streit suchen. **3.** *n.* ~|s *pl.*| Beute, *die.* '**spoilsport** *n.* Spielverderber, *der/*-verderberin, *die*

spoilt *see* spoil 1, 2

¹**spoke** [spəʊk] *n.* Speiche, *die*

²**spoke, spoken** *see* speak

spokesman ['spəʊksmən] *n., pl.* **spokesmen** ['spəʊksmən] Sprecher, *der*

sponge [spʌndʒ] **1.** *n.* Schwamm, *der.* **2.** *v. t.* mit einem Schwamm waschen. '**sponge on** *v. t.* ~ **on sb.** bei *od.* von jmdm. schnorren *(ugs.)*

sponge: ~**bag** *n.* *(Brit.)* Kulturbeutel, *der;* ~**cake** *n.* Biskuitkuchen, *der*

sponger ['spʌndʒə(r)] *n.* Schmarotzer, *der/*Schmarotzerin, *die*

spongy ['spʌndʒɪ] *adj.* schwammig

sponsor ['spɒnsə(r)] **1.** *n.* Sponsor, *der.* **2.** *v. t.* **a)** sponsern; **b)** *(Polit.)* ~ **sb.** jmds. Kandidatur unterstützen

spontaneous [spɒn'teɪnɪəs] *adj.* spontan

spooky ['spuːkɪ] *adj.* gespenstisch

spool [spuːl] *n.* Spule, *die*

spoon [spuːn] *n.* **a)** Löffel, *der;* **b)** *(amount)* *see* spoonful. **spoonful** ['spuːnfʊl] *n.* **a** ~ **of sugar** ein Löffel [voll] Zucker

sporadic [spə'rædɪk] *adj.* sporadisch

spore [spɔː(r)] *n.* Spore, *die*

sport [spɔːt] **1.** *n.* **a)** Sport, *der;* ~s Sportarten; **water/indoor** ~: Wasser-/ Hallensport, *der;* **b)** *(fun)* Spaß, *der;* **c) be a |real|** ~ *(coll.)* ein prima Kerl sein *(ugs.);* **be a** ~! sei kein Spielverderber! **2.** *v. t.* stolz tragen. '**sporting** *adj.* **a)** sportlich; **b) give sb. a** ~ **chance** jmdm. eine [faire] Chance geben

sports: ~**car** *n.* Sportwagen, *der;* ~**jacket** *n.* sportlicher Sakko; ~**man** ['spɔːtsmən] *n., pl.* ~**men** ['spɔːtsmən] Sportler, *der;* ~**manship** ['spɔːtsmənʃɪp] *n.* *(fairness)* [sportliche] Fairneß; ~**wear** *n.* Sport[be]kleidung, *die;* ~**woman** *n.* Sportlerin, *die*

'**sporty** *adj.* sportlich

spot [spɒt] **1.** *n.* **a)** *(precise place)* Stelle, *die;* **on this** ~: an dieser Stelle; **be in a tight** ~ *(fig. coll.)* in der Klemme sitzen *(ugs.);* **put sb. on the** ~ *(fig. coll.)* jmdn. in Verlegenheit bringen; **b)** *(suitable area)* Platz, *der;* **c)** *(dot)* Tupfen, *der;* **d)** *(stain)* ~ |of blood/grease/ ink| [Blut-/Fett-/Tinten]fleck, *der;* **e)** *(Brit. coll.: small amount)* **do a** ~ **of work/sewing** ein bißchen arbeiten/nähen; **f)** *(drop)* **a** ~ *or* **a few** ~**s of rain** ein paar Regentropfen; **g)** *(Med.)* Pickel, *der.* **2.** *v. t.,* -tt- *(detect)* entdecken; erkennen ⟨*Gefahr*⟩

spot: ~ '**check** *n.* Stichprobe, *die;* ~**less** *adj.* fleckenlos; **her house is absolutely** ~ *(fig.)* ihr Haus ist makellos sauber; ~**light** *n.* Scheinwerfer, *der;* **be in the** ~**light** *(fig.)* im Rampenlicht stehen

spotted ['spɒtɪd] *adj.* gepunktet

'**spotty** *adj. (pimply)* picklig

spouse [spaʊs] *n.* [Ehe]gatte, *der/*-gattin, *die*

spout [spaʊt] 1. *n.* Schnabel, *der; (of tap)* Ausflußrohr, *das.* 2. *v. i. (gush)* schießen (from aus)

sprain [spreɪn] 1. *v. t.* verstauchen. 2. *n.* Verstauchung, *die*

sprang *see* spring 2, 3

sprawl [sprɔ:l] *v. i.* **a)** sich ausstrecken; *(fall)* der Länge nach hinfallen; **b)** *(straggle)* sich ausbreiten

'**spray** [spreɪ] *(bouquet)* Strauß, *der*

²**spray** 1. *v. t.* spritzen; sprühen *⟨Parfüm⟩;* besprühen ⟨*Haar, Pflanze⟩.* 2. *n.* **a)** *(drops)* Sprühnebel, *der;* **b)** *(liquid)* Spray, *der od. das*

spread [spred] 1. *v. t.,* **spread a)** ausbreiten ⟨*Tuch, Landkarte⟩* (on auf + *Dat.*); streichen ⟨*Butter, Farbe, Marmelade⟩;* **b)** *(extend range of)* verbreiten; **c)** *(distribute)* verteilen. 2. *v. i.,* **spread** sich ausbreiten. 3. *n.* **a)** Verbreitung, *die; (of city, poverty)* Ausbreitung, *die;* **b)** *(coll.: meal)* Festessen, *das; (paste)* Brotaufstrich, *der.*

spread 'out 1. *v. t.* ausbreiten. 2. *v. i.* sich verteilen

spree [spri:] *n.* go on a shopping ~: ganz groß einkaufen gehen

sprig [sprɪg] *n.* Zweig, *der*

sprightly ['spraɪtlɪ] *adj.* munter

spring [sprɪŋ] 1. *n.* **a)** *(season)* Frühling, *der;* in |the| ~: im Frühling *od.* Frühjahr; **b)** *(water)* Quelle, *die;* **c)** *(Mech.)* Feder, *die;* **d)** *(jump)* Sprung, *der.* 2. *v. i.,* **sprang** [spræŋ] *or (Amer.)* **sprung** [sprʌŋ], **sprung a)** *(jump)* springen; ~ to life *(fig.)* [plötzlich] zum Leben erwachen; **b)** *(arise)* entspringen (from *Dat.*). 3. *v. t.,* **sprang** *or (Amer.)* **sprung, sprung:** ~ sth. on sb. jmdm. mit etw. überfallen

spring: ~board *n.* Sprungbrett, *das;* ~-'clean 1. *n.* Frühjahrsputz, *der;* 2. *v. t.* Frühjahrsputz machen; ~ 'onion *n.* Frühlingszwiebel, *die;* ~time *n.* Frühling, *der*

sprinkle ['sprɪŋkl] *v. t.* streuen; sprengen ⟨*Flüssigkeit⟩.* **sprinkler** ['sprɪŋklə(r)] *n. (Hort.)* Sprinkler, *der*

sprint [sprɪnt] 1. *v. t. & i.* rennen; sprinten *(bes. Sport).* 2. *n.* Sprint, *der*

sprout [spraʊt] 1. *n.* **a)** Brussels ~s Rosenkohl, *der;* **b)** *(Bot.)* Trieb, *der.* 2. *v. i.* sprießen *(geh.)*

spruce [spru:s] 1. *adj.* gepflegt. 2. *n.* Fichte, *die*

sprung [sprʌŋ] 1. *see* spring 2, 3. 2. *attrib. adj.* gefedert

spud [spʌd] *n. (sl.)* Kartoffel, *die*

spun *see* spin

spur [spɜ:(r)] 1. *n.* Sporn, *der; (fig.)* Ansporn, *der;* on the ~ of the moment ganz spontan. 2. *v. t.,* -rr- *(fig.)* anspornen

spurious ['spjʊərɪəs] *adj.* gespielt ⟨*Interesse⟩;* zweifelhaft ⟨*Anspruch⟩*

spurn [spɜ:n] *v. t.* zurückweisen

'**spurt** [spɜ:t] *n.* Spurt, *der;* put on a ~: einen Spurt einlegen

²**spurt** 1. *v. i.* ~ out |from *or* of| herausspritzen [aus]. 2. *n.* Strahl, *der*

spy [spaɪ] 1. *n.* Spion, *der/*Spionin, *die.* 2. *v. i.* spionieren; ~ on sb. jmdm. nachspionieren

squabble ['skwɒbl] 1. *n.* Streit, *der.* 2. *v. i.* sich zanken (over, about wegen)

squad [skwɒd] *n.* **a)** *(Mil.)* Gruppe, *die;* **b)** *(group)* Mannschaft, *die*

squadron ['skwɒdrən] *n.* **a)** *(Navy)* Geschwader, *das;* **b)** *(Air Force)* Staffel, *die*

squalid ['skwɒlɪd] *adj.* **a)** *(dirty)* schmutzig; **b)** *(poor)* schäbig

squall [skwɔ:l] *n. (gust)* Bö, *die*

squalor ['skwɒlə(r)] *n.* Schmutz, *der*

squander ['skwɒndə(r)] *v. t.* vergeuden

square [skweə(r)] 1. *n.* **a)** *(Geom.)* Quadrat, *das;* **b)** *(open area)* Platz, *der.* 2. *adj.* **a)** quadratisch; **b)** a ~ foot/mile ein Quadratfuß/eine Quadratmeile. 3. *v. t.* **a)** *(Math.)* quadrieren; **b)** ~ it with sb. es mit jmdm. klären. 4. *v. i. (agree)* übereinstimmen.

square 'up *v. i. (settle up)* abrechnen

square 'root *n.* Quadratwurzel, *die*

squash [skwɒʃ] 1. *v. t. (crush)* zerquetschen; ~ sth. flat etw. platt drücken. 2. *n.* **a)** Fruchtsaftgetränk, *das;* **b)** *(Sport)* Squash, *das*

squat [skwɒt] *v. i.,* -tt-: **a)** *(crouch)* hocken; **b)** ~ in a house ein Haus besetzen. '**squatter** *n.* Hausbesetzer, *der/*-besetzerin, *die*

squawk [skwɔ:k] *v. i.* ⟨*Krähe:⟩* krähen; ⟨*Huhn:⟩* kreischen

squeak [skwi:k] 1. *n.* **a)** *(of animal)* Quieken, *das;* **b)** *(of brakes, hinge, etc.)* Quietschen, *das.* 2. *v. i.* **a)** ⟨*Tier:⟩* quieken; **b)** ⟨*Scharnier, Tür, Bremse, Schuh usw.:⟩* quietschen

squeal [skwi:l] 1. *v. i.* **a)** ~ with pain/in fear ⟨*Person:⟩* vor Schmerz/Angst aufschreien; ⟨*Tier:⟩* vor Schmerz/Angst laut quieken; **b)** ⟨*Bremsen, Räder:⟩*

kreischen; ⟨*Reifen:*⟩ quietschen. **2.** *n.*
Kreischen, *das; (of tyres)* Quietschen,
das; (of animal) Quieken, *das*

squeamish ['skwi:mɪʃ] *adj.* be ~:
zartbesaitet sein

squeeze [skwi:z] **1.** *n.* Druck, *der; give
sth. a small* ~: etw. [leicht] drücken. **2.**
v. t. **a)** *(press)* drücken; drücken auf
(+ Akk.)⟨*Tube, Plastikflasche*⟩; *(to get
juice)* auspressen; **b)** *(extract)* drücken
(*out of* aus); ~ **out** sth. etw. heraus-
drücken; **c)** *(force)* zwängen

squelch [skweltʃ] *v. i.* quatschen *(ugs.)*

squid [skwɪd] *n.* Kalmar, *der*

squiggle ['skwɪɡl] *n.* Schnörkel, *der*

squint [skwɪnt] **1.** *n.* Schielen, *das.* **2.**
v. i. **a)** *(Med.)* schielen; **b)** *(with half-
closed eyes)* blinzeln

squire ['skwaɪə(r)] *n.* ≈ Gutsherr, *der*

squirm [skwɜ:m] *v. i.* sich winden
(*with* vor + *Dat.*)

squirrel ['skwɪrl] *n.* Eichhörnchen, *das*

squirt [skwɜ:t] **1.** *v. t.* spritzen; sprühen ⟨*Spray, Puder*⟩; ~ **sth. at sb.** jmdn.
mit etw. bespritzen/besprühen. **2.** *v. i.*
spritzen. **3.** *n.* Spritzer, *der*

St *abbr.* Saint St.

St. *abbr.* Street Str.

st. *abbr. (Brit.: unit of weight)* **stone**

stab [stæb] **1.** *v. t.,* -bb- stechen; ~ **sb.
in the chest** jmdm. in die Brust ste-
chen. **2.** *v. i.,* -bb- stechen. **3.** *n.* **a)**
Stich, *der;* **b)** *(coll.: attempt)* **make** *or*
have a ~ [at it] [es] probieren

stability [stə'bɪlɪtɪ] *n.* Stabilität, *die*

stabilize ['steɪbɪlaɪz] **1.** *v. t.* stabilisieren. **2.** *v. i.* sich stabilisieren

¹**stable** ['steɪbl] *adj.* stabil; gefestigt
⟨*Person*⟩

²**stable** *n.* Stall, *der*

stack [stæk] **1.** *n.* **a)** *(pile)* Stoß, *der;*
Stapel, *der;* **b)** *(coll.: large amount)*
Haufen, *der (ugs.);* **c)** |chimney-|~:
Schornstein, *der.* **2.** *v. t.* ~ |up|
[auf]stapeln

stadium ['steɪdɪəm] *n.* Stadion, *das*

staff [stɑ:f] **1.** *n.* **a)** *(stick)* Stock, *der;*
b) *(personnel)* Personal, *das; (of
school)* Lehrkollegium, *das.* **2.** *v. t.* mit
Personal ausstatten. '**staff-room** *n.*
(Sch.) Lehrerzimmer, *das*

stag [stæɡ] *n.* Hirsch, *der*

stage [steɪdʒ] **1.** *n.* **a)** *(Theatre)* Bühne,
die; **b)** *(part of process)* Stadium, *das;*
at this ~: in diesem Stadium; **do sth.
by** ~**s** etw. abschnittsweise tun; **in the
final** ~**s** in der Schlußphase; **c)** *(distance)* Etappe, *die.* **2.** *v. t.* **a)** *(present)*
inszenieren; **b)** *(arrange)* veranstalten

stage: ~-**coach** *n.* Postkutsche, *die;*
~ **door** *n.* Bühneneingang, *der;* ~
fright *n.* Lampenfieber, *das;*
~-**manage** *v. t. (fig.)* veranstalten

stagger ['stæɡə(r)] **1.** *v. i.* schwanken.
2. *v. t. (astonish)* die Sprache verschlagen (+ *Dat.*).

stagnant ['stæɡnənt] *adj.* **a)** stehend
⟨*Gewässer*⟩; **b)** *(Econ.)* stagnierend

stagnate [stæɡ'neɪt] *v. i.* **a)** ⟨*Wasser:*⟩
abstehen; **b)** ⟨*Wirtschaft, Geschäft:*⟩
stagnieren; ⟨*Person:*⟩ abstumpfen.

stagnation [stæɡ'neɪʃn] *n.* **a)** *(of
water)* Stehen, *das;* **b)** *(Econ.)* Stagnation, *die*

staid [steɪd] *adj.* gesetzt

stain [steɪn] **1.** *v. t.* **a)** verfärben; *(make
~s on)* Flecken hinterlassen auf
(+ *Dat.*); **b)** *(colour)* beizen ⟨*Holz*⟩. **2.**
n. Fleck, *der.* **stained 'glass** *n.* far-
biges Glas; ~ '**window** Fenster mit
Glasmalerei

'**stainless** *adj.* fleckenlos. **stainless
'steel** *n.* Edelstahl, *der*

stair [steə(r)] *n. (step)* [Treppen]stufe,
die; ~**s** Treppe, *die.* '**staircase** *n.*
Treppenhaus, *das*

stake [steɪk] *n.* **a)** *(pointed stick)* Pfahl,
der; **b)** *(wager)* Einsatz, *der;* **be at** ~:
auf dem Spiel stehen

stale [steɪl] *adj.* alt; muffig; abgestanden ⟨*Luft*⟩; alt[backen] ⟨*Brot*⟩; schal
⟨*Bier, Wein usw.*⟩

'**stalemate** *n.* Patt, *das*

¹**stalk** [stɔ:k] *v. t.* sich heranpirschen
an (+ *Akk.*)

²**stalk** *n. (Bot.) (main stem)* Stengel,
der; (of leaf, flower, fruit) Stiel, *der*

stall [stɔ:l] **1.** *n.* **a)** Stand, *der;* **b)** *(Brit.
Theatre)* ~**s** Parkett, *das.* **2.** *v. t.* ab-
würgen *(ugs.)* ⟨*Motor*⟩. **3.** *v. i.* ⟨*Motor:*⟩
stehenbleiben

stallion ['stæljən] *n.* Hengst, *der*

stalwart ['stɔ:lwət] *adj. (determined)*
entschieden; *(loyal)* treu

stamina ['stæmɪnə] *n.* Ausdauer, *die*

stammer ['stæmə(r)] **1.** *v. i.* stottern. **2.**
v. t. stammeln. **3.** *n.* Stottern, *das*

stamp [stæmp] **1.** *v. t.* **a)** *(impress, imprint sth. on)* [ab]stempeln; **b)** ~ **one's
foot** mit dem Fuß stampfen; **c)** *(put
postage* ~ *on)* frankieren; ~**ed addressed envelope** frankierter Rückumschlag; **d) become** *or* **be** ~**ed on sb.'s
memory** *or* **mind** sich jmdm. fest einprägen. **2.** *v. i.* aufstampfen. **3.** *n.* Marke, *die; (postage* ~) Briefmarke, *die;
(instrument for* ~*ing)* Stempel, *der.*
'**stamp on** *v. t.* **a)** zertreten ⟨*Insekt*⟩;

~ **on sb's foot** jmdm. auf den Fuß treten; b) *(suppress)* durchgreifen gegen. **stamp 'out** *v. t.* [aus]stanzen; *(fig.)* ausmerzen

stamp: ~ **album** *n.* Briefmarkenalbum, *das;* ~**-collecting** *n.* Briefmarkensammeln, *das*

stampede [stæm'pi:d] *n.* Stampede, *die*

stand [stænd] **1.** *v. i.,* stood [stʊd] **a)** stehen; b) **my offer/promise still ~s** mein Angebot/Versprechen gilt nach wie vor; **as it ~s, as things ~:** wie die Dinge [jetzt] liegen; **I'd like to know where I ~** *(fig.)* ich möchte wissen, wo ich dran bin; c) *(be candidate)* kandidieren; d) |not| ~ **in sb.'s way** *(fig.)* jmdm. [keine] Steine in den Weg legen; e) *(be likely)* ~ **to win** *or* **gain/lose sth.** etw. gewinnen/verlieren können. **2.** *v. t.,* stood **a)** *(set in position)* stellen; b) *(endure)* ertragen; **I cannot ~ |the sight of| him/her** ich kann ihn/sie nicht ausstehen; **he can't ~ the pressure/strain** er ist dem Druck/den Strapazen nicht gewachsen; **I can't ~ it any longer!** ich halte es nicht mehr aus!; c) *(buy)* ~ **sb. sth.** jmdm. etw. ausgeben. **3.** *n.* **a)** *(support)* Ständer, *der;* b) *(stall; at exhibition)* Stand, *der;* c) *(raised structure)* Tribüne, *die.* **stand a'bout, stand a'round** *v. i.* herumstehen. **stand a'side** *v. i.* zur Seite treten. **'stand between** *v. t.* **sth.** ~**s between sb. and sth.** *(fig.)* etw. steht jmdm. bei etw. im Wege. **stand by 1.** [-'-] *v. i.* **a)** *(be near)* daneben stehen; b) *(be ready)* sich zur Verfügung halten. **2.** ['--] *v. t.* **a)** *(support)* ~ **by sb./ one another** jmdm. /sich [gegenseitig] beistehen; b) *(adhere to)* ~ **by sth.** zu etw. stehen. **'stand for** *v. t.* **a)** *(signify)* bedeuten; b) *(coll.: tolerate)* sich *(Dat.)* bieten lassen. **stand 'in** *v. i.* aushelfen; ~ **in for sb.** für jmdn. einspringen. **stand 'out** *v. i. (be prominent)* herausragen; ~ **out a mile** *(fig.)* nicht zu übersehen sein. **'stand over** *v. t.* beaufsichtigen. **stand 'up** *v. i.* **a)** aufstehen; ~ **up straight** sich aufrecht hinstellen; b) ~ **up well |in comparison with sb./sth.|** [im Vergleich zu jmdm./ etw.] gut abschneiden; ~ **up for sb./ sth.** für jmdn./etw. Partei ergreifen; ~ **up to sb.** sich jmdm. entgegenstellen

standard ['stændəd] **1.** *n.* **a)** Maßstab, *der;* **safety** ~**s** Sicherheitsnormen; **above/below/up to** ~**:** überdurchschnittlich [gut]/unter dem Durchschnitt/der Norm entsprechend; b) *(degree)* Niveau, *das;* ~ **of living** Lebensstandard, *der;* c) ~**s** *(morals)* Prinzipien; d) *(flag)* Standarte, *die.* **2.** *adj.* Standard-; **be ~ practice** allgemein üblich sein. **standardize** ['stændədaɪz] *v. t.* standardisieren

'standard lamp *n.* Stehlampe, *die*

stand: ~**-by 1.** *n.* **be on ~-by** einsatzbereit sein; **2.** *adj.* Ersatz-; ~**-in 1.** *n.* Ersatz, *der;* **2.** *adj.* Ersatz-

'standing 1. *n.* **a)** *(repute)* Ansehen, *das;* b) *(duration)* **of long/short ~:** von langer/kurzer Dauer. **2.** *adj.* **a)** *(erect)* stehend; b) fest ⟨*Regel, Brauch*⟩. **standing:** ~ **'order** *n.* Dauerauftrag, *der;* ~ **o'vation** *n.* stürmischer Beifall; ~**-room** *n.* Stehplätze

stand: ~**-pipe** *n.* Standrohr, *das;* ~**point** *n. (fig.)* Standpunkt, *der;* ~**still** *n.* Stillstand, *der;* **be at a ~still** stillstehen; **come to a ~still** zum Stehen kommen

stank *see* **stink 1**

staple ['steɪpl] **1.** *n.* [Heft]klammer, *die.* **2.** *v. t.* heften **(on to** an + *Akk.*). **stapler** ['steɪplə(r)] *n.* [Draht]hefter, *der*

star [stɑ:(r)] **1.** *n.* **a)** Stern, *der;* b) *(prominent person)* Star, *der.* **2.** *v. i.* ~ **in a film** in einem Film die Hauptrolle spielen

starboard ['stɑ:bəd] *n.* Steuerbord, *das*

starch [stɑ:tʃ] *n.* Stärke, *die*

stardom ['stɑ:dəm] *n.* Starruhm, *der*

stare [steə(r)] *v. i.* starren; ~ **at sb./sth.** jmdn./etw. anstarren

'starfish *n.* Seestern, *der*

stark [stɑ:k] **1.** *adj.* scharf ⟨*Kontrast, Umriß*⟩. **2.** *adv.* völlig; ~ **naked** splitternackt *(ugs.)*

starling ['stɑ:lɪŋ] *n.* Star, *der*

starry ['stɑ:rɪ] *adj.* sternklar

start [stɑ:t] **1.** *v. i.* **a)** *(begin)* anfangen; ~ **on sth.** etw. beginnen; b) *(set out)* aufbrechen; c) *(begin to function)* anlaufen; ⟨*Auto, Motor usw.*⟩: anspringen. **2.** *v. t.* **a)** *(begin)* beginnen [mit]; ~ **doing** *or* **to do sth.** [damit] anfangen, etw. zu tun; b) *(cause)* auslösen; anfangen ⟨*Streit, Schlägerei*⟩; legen/*(accidentally)* verursachen ⟨*Brand*⟩; c) *(set up)* ins Leben rufen ⟨*Organisation, Projekt*⟩; d) *(switch on)* einschalten; anlassen ⟨*Motor, Auto*⟩. **3.** *n.* **a)** Anfang, *der;* Beginn, *der; (of race)* Start, *der;* **from the ~:** von Anfang an; **from ~ to finish** von Anfang bis Ende;

make a ~: anfangen (on mit); *(on journey)* aufbrechen; **b)** *(Sport: ~ing-place)* Start, *der*. '**starter** *n*. **a)** *(food)* Vorspeise, *die;* **b)** *(Sport)* Starter, *der*

startle ['sta:tl] *v.t.* erschrecken; **be ~d by sth.** über etw. *(Akk.)* erschrecken. **startling** ['sta:tlɪŋ] *adj.* erstaunlich

starvation [sta:'veɪʃn] *n.* Verhungern, *das*

starve [sta:v] *v.i.* **~ [to death]** verhungern

state [steɪt] **1.** *n.* **a)** *(condition)* Zustand, *der;* **b)** *(nation)* Staat, *der;* **c) be in a ~**: aufgeregt sein; **d) lie in ~**: aufgebahrt sein. **2.** *v.t. (express)* erklären; angeben ⟨*Alter usw.*⟩

stately ['steɪtlɪ] *adj.* majestätisch; stattlich ⟨*Körperbau, Gebäude*⟩. **stately 'home** *n.* Herrensitz, *der*

'**statement** *n.* **a)** *(stating, account)* Aussage, *die; (declaration)* Erklärung, *die;* **b)** |bank| **~**: Kontoauszug, *der*

statesman ['steɪtsmən] *n., pl.* **statesmen** ['steɪtsmən] Staatsmann, *der*

static ['stætɪk] *adj.* statisch

station ['steɪʃn] **1.** *n.* **a)** *see* railway-station; **b)** *(status)* Rang, *der*. **2.** *v.t.* aufstellen ⟨*Wache*⟩

stationary ['steɪʃənərɪ] *adj.* stehend; **be ~**: stehen

stationer ['steɪʃənə(r)] *n.* **~'s** |shop| Schreibwarengeschäft, *das*. **stationery** ['steɪʃənərɪ] *n.* **a)** *(writing materials)* Schreibwaren *Pl.;* **b)** *(writing-paper)* Briefpapier, *das*

'**station-wagon** *n. (Amer.)* Kombiwagen, *der*

statistical [stə'tɪstɪkl] *attrib. adj.* **statistically** [stə'tɪstɪkəlɪ] *adv.* statistisch

statistics [stə'tɪstɪks] *n.* Statistik, *die*

statue ['stætʃu:, 'stætjuː] *n.* Statue, *die*

stature ['stætʃə(r)] *n.* Statur, *die; (fig.)* Format, *das*

status ['steɪtəs] *n.* Rang, *der;* **social ~**: [gesellschaftlicher] Status. '**status symbol** *n.* Statussymbol, *das*

statute ['stætjuːt] *n.* Gesetz, *das.* **statutory** ['stætjʊtərɪ] *adj.* gesetzlich

staunch [stɔːntʃ] *adj.* treu ⟨*Freund*⟩; überzeugt ⟨*Katholik usw.*⟩

stave [steɪv] *v.t.* **~ 'off** abwenden; stillen ⟨*Hunger*⟩

stay [steɪ] **1.** *n.* Aufenthalt, *der; (visit)* Besuch, *der;* **come/go for a short ~ with sb.** jmdn. kurz besuchen. **2.** *v.i.* bleiben; **~ put** *(coll.)* ⟨*Person:*⟩ bleiben[, wo man ist]; **~ the night in a hotel** die Nacht in einem Hotel ver-

bringen. **3.** *v.t.* **~ the course** *(fig.)* durchhalten. **stay a'way** *v.i.* wegbleiben. **stay be'hind** *v.i.* zurückbleiben. **stay 'in** zu Hause bleiben. **stay 'out** *v.i.* **a)** *(not go home)* wegbleiben *(ugs.);* **b)** *(remain outside)* draußen bleiben. **stay 'up** *v.i.* aufbleiben

stead [sted] *n.* **a) in sb.'s ~**: an jmds. Stelle *(Dat.);* **b) stand sb. in good ~**: jmdm. zustatten kommen

steadfast ['stedfɑːst] *adj.* standhaft; zuverlässig ⟨*Freund*⟩

steadily ['stedɪlɪ] *adv.* **a)** *(stably)* fest; **b)** *(continuously)* stetig

steady ['stedɪ] **1.** *adj.* **a)** *(stable)* stabil; *(not wobbling)* standfest; **b)** *(still)* ruhig; **c)** *(regular, constant)* stetig; gleichmäßig ⟨*Arbeit, Tempo*⟩; stabil ⟨*Preis, Lohn*⟩; gleichbleibend ⟨*Temperatur*⟩; **we had ~ rain/drizzle** wir hatten Dauerregen/es nieselte [bei uns] ständig; **d) a ~ job** eine feste Stelle; **a ~ boy-friend** ein fester Freund. **2.** *v.t.* festhalten ⟨*Leiter*⟩; beruhigen ⟨*Nerven*⟩

steak [steɪk] *n.* Steak, *das*

steal [stiːl] **1.** *v.t.,* **stole** [stəʊl], **stolen** ['stəʊln] stehlen **(from** *Dat.*). **2.** *v.i.,* **stole, stolen a)** stehlen; **~ from sb.** jmdn. bestehlen; **b) ~ in/out** sich hinein-/hinausstehlen

stealth [stelθ] *n.* Heimlichkeit, *die;* **by ~**: heimlich. **stealthy** ['stelθɪ] *adj.* heimlich

steam [stiːm] **1.** *n.* Dampf, *der;* **let off ~** *(fig.)* Dampf ablassen *(ugs.);* **run out of ~** *(fig.)* den Schwung verlieren; **under one's own ~** *(fig.)* aus eigener Kraft. **2.** *v.t. (Cookery)* dämpfen; dünsten. **3.** *v.i.* dämpfen; **~ing hot** dampfend heiß. **steam 'up** *v.i.* beschlagen

'**steam engine** *n.* Dampflok[omotive], *die*

'**steamer** *n.* Dämpfer, *der*

'**steamroller** *n.* Dampfwalze, *die*

'**steam train** *n.* Dampfzug, *der*

'**steamy** *adj.* dunstig; beschlagen ⟨*Glas*⟩

steel [stiːl] **1.** *n.* Stahl, *der.* **2.** *attrib. adj.* stählern; Stahl⟨*helm, -block, -platte*⟩. **3.** *v.t.* **~ oneself for/against sth.** sich für/gegen etw. wappnen *(geh.);* **~ oneself to do sth.** allen Mut zusammennehmen, um etw. zu tun. '**steelworks** *n. sing. or pl.* Stahlwerk, *das*

'**steep** [stiːp] *adj.* **a)** steil; **b)** *(coll.: ex-*

cessive) happig *(ugs.);* **the bill is | a bit|** ~: die Rechnung ist [ziemlich] gesalzen *(ugs.)*

²**steep** *v. t. (soak)* einweichen

steeple ['sti:pl] *n.* Kirchturm, *der*

steer [stɪə(r)] **1.** *v. t.* steuern; lenken. **2.** *v. i.* steuern; ~ **clear of sb./sth.** *(fig. coll.)* jmdm./einer Sache aus dem Weg[e] gehen. '**steering** *n. (Motor Veh.)* Lenkung, *die.* '**steering-wheel** *n.* Lenkrad, *das*

¹**stem** [stem] **1.** *n.* **a)** *(Bot.)* Stiel, *der;* **b)** *(Ling.)* Stamm, *der.* **2.** *v. i.,* -mm-: ~ **from sth.** auf etw. *(Akk.)* zurückzuführen sein

²**stem** *v. t.,* -mm- *(check, dam up)* aufhalten; eindämmen ⟨*Flut*⟩; stillen ⟨*Blutung*⟩

stench [stentʃ] *n.* Gestank, *der*

stencil ['stensl] *n.* Schablone, *die; (for duplicating)* Matritze, *die*

step [step] **1.** *n.* **a)** Schritt, *der;* **take a** ~ **back/forwards** einen Schritt zurücktreten/nach vorn treten; **b)** *(stair)* Stufe, *die;* **a flight of** ~s eine Treppe; |**pair of|** ~s *(ladder)* Stehleiter, *die;* **c)** **be in** ~: im Schritt sein; *(with music)* im Takt sein; **d) take** ~s **to do sth.** Schritte unternehmen, um etw. zu tun; **e)** *(stage)* ~ **by** ~: Schritt für Schritt; **what is the next** ~? wie geht es weiter?; **f)** *(grade)* Stufe, *die.* **2.** *v. i.,* -pp- treten; ~ **inside** eintreten; ~ **into sb's shoes** *(fig.)* an jmds. Stelle *(Akk.)* treten; ~ **over sb./sth.** über jmdn./etw. steigen. **step 'back** *v. i.* zurücktreten. **step 'in** *v. i.* **a)** eintreten; **b)** *(fig.) (take sb.'s place)* einspringen; *(intervene)* eingreifen. **step 'up 1.** *v. i. (ascend)* hinaufsteigen. **2.** *v. t.* erhöhen; verstärken ⟨*Anstrengungen*⟩

step: ~child *n.* Stiefkind, *das;* **~daughter** *n.* Stieftochter, *die;* **~father** *n.* Stiefvater, *der;* **~ladder** *n.* Stehleiter, *die;* **~mother** *n.* Stiefmutter, *die*

'**stepping-stone** *n.* Trittstein, *der; (fig.)* Sprungbrett, das **(to** für)

stereo ['sterɪəʊ] **1.** *n.* Stereo, *das; (equipment)* Stereoanlage, *die.* **2.** *adj.* stereo; Stereo⟨*aufnahme, -platte*⟩

stereophonic [sterɪə'fɒnɪk] *adj.* stereophon

stereotype ['sterɪətaɪp] **1.** *n.* Stereotyp, *das.* **2.** *v. t.* in ein Klischee zwängen; ~d stereotyp

sterile ['steraɪl] *adj.* steril

sterilize ['sterɪlaɪz] *v. t.* sterilisieren

sterling ['stɜ:lɪŋ] **1.** *n.* Sterling, *der;* in

~: in Pfund [Sterling]. **2.** *attrib. adj.* **a)** ~ **silver** Sterlingsilber, *das;* **b)** *(fig.)* gediegen

¹**stern** [stɜ:n] *adj.* streng; ernst ⟨*Warnung*⟩

²**stern** *n. (Naut.)* Heck, *das*

'**sternly** *adv.* streng

steroid ['sterɔɪd] *n.* Steroid, *das*

stethoscope ['steθəskəʊp] *n.* Stethoskop, *das*

stew [stju:] **1.** *n.* Eintopf, *der.* **2.** *v. t.* schmoren [lassen]

steward ['stju:əd] *n.* **a)** *(on ship, plane)* Steward, *der;* **b)** *(at public meeting etc.)* Ordner, *der.* '**stewardess** *n.* Stewardeß, *die*

stick [stɪk] **1.** *v. t.,* **stuck** [stʌk] **a)** *(thrust point of)* stecken; ~ **sth. in|to| sth.** mit etw. in etw. *(Akk.)* stechen; **b)** *(coll.: put)* stecken; ~ **a picture on the wall/a vase on the shelf** ein Bild an die Wand hängen/eine Vase aufs Regal stellen; ~ **sth. in the kitchen** etw. in die Küche tun *(ugs.);* **c)** *(with glue etc.)* kleben; **d) the car is stuck in the mud** das Auto ist im Schlamm steckengeblieben; **the door is stuck** die Tür klemmt [fest]. **2.** *v. i.,* **stuck a)** *(be fixed by point)* stecken; **b)** *(adhere)* kleben; ~ **to sth.** an etw. *(Dat.)* kleben; **c)** *(become immobile)* ⟨*Auto, Räder:*⟩ steckenbleiben; ⟨*Schublade, Tür, Griff, Bremse:*⟩ klemmen; ⟨*Schlüssel:*⟩ feststecken. **3.** *n.* Stock, *der;* **a** ~ **of chalk** ein Stück Kreide; **a** ~ **of celery/rhubarb** eine Stange Sellerie/Rhabarber. **stick a'bout, stick a'round** *v. i. (coll.)* dableiben; *(wait)* warten. '**stick by** *v. t. (fig.)* stehen zu. **stick 'on** *v. t. (glue on)* aufkleben. **stick 'out 1.** *v. t.* **a)** herausstrecken ⟨*Zunge*⟩; **b)** ~ **it out** *(coll.)* durchhalten. **2.** *v. i.* **a)** ⟨*Bauch:*⟩ vorstehen; **his ears** ~ **out** er hat abstehende Ohren; **b)** *(fig.: be obvious)* sich abheben; ~ **out a mile** *(sl.)* [klar] auf der Hand liegen; ~ **out like a sore thumb** *(coll.)* ins Auge springen. '**stick to** *v. t.* **a)** *(be faithful to)* halten ⟨*Versprechen*⟩; bleiben bei ⟨*Entscheidung*⟩; **b)** ~ **to the point** beim Thema bleiben. **stick 'up 1.** *v. t.* **a)** *(coll.)* anschlagen ⟨*Poster*⟩; ~ **up one's hand** die Hand heben; **b)** *(seal)* zukleben. **2.** *v. i.* ~ **up for sb./sth.** für jmdn./etw. eintreten; ~ **up for yourself!** setz dich zur Wehr!

'**sticker** *n.* Aufkleber, *der*

'**sticking-plaster** *n.* Heftpflaster, *das*

'**sticky** *adj.* **a)** klebrig; ~ **label** Aufkleber, *der;* b) *(humid)* schwül ⟨*Klima, Luft*⟩

stiff [stıf] *adj.* a) *(rigid)* steif; hart ⟨*Bürste, Stock*⟩; **be frozen** ~: steif vor Kälte sein; b) *(intense, severe)* hartnäckig; c) *(formal)* steif; d) *(difficult)* hart ⟨*Test*⟩; schwer ⟨*Frage, Prüfung*⟩; e) *(coll.)* **be bored/scared** ~: sich zu Tode langweilen/eine wahnsinnige Angst haben *(ugs.).* **stiffen** ['stıfn] 1. *v. t.* steif machen. 2. *v. i.* steifer werden; ⟨*Person:*⟩ erstarren. **'stiffness** *n.* Steifheit, *die*

stifle ['staıfl] 1. *v. t.* ersticken; *(fig.)* unterdrücken. 2. *v. i.* ersticken. **stifling** ['staıflıŋ] *adj.* stickig; drückend ⟨*Hitze*⟩

stigma ['stıgmə] *n.* Stigma, *das (geh.)*

stile [staıl] *n.* Zauntritt, *der*

stiletto [stı'letəu] *n.* ~ [**heel**] Stöckelabsatz, *der*

¹still [stıl] 1. *pred. adj.* still; **be** ~: [still] stehen; **hold sth.** ~: etw. ruhig halten; **keep** *or* **stay** ~: stillhalten; **stand** ~: stillstehen. 2. *adv.* a) *(without change)* noch; *expr. surprise or annoyance* immer noch; b) *(nevertheless)* trotzdem; c) *with comparative (even)* noch

²still *n.* Destillierapparat, *der*

still: ~**born** *adj.* totgeboren; ~ '**life** *n.* *(Art)* Stilleben, *das*

stilt [stılt] *n.* Stelze, *die.* **'stilted** *adj.* gestelzt

stimulant ['stımjulənt] *n.* Stimulans, *das*

stimulate ['stımjuleıt] *v. t.* anregen. **stimulation** [stımju'leıʃn] *n.* Anregung, *die*

stimulus ['stımjuləs] *n., pl.* **stimuli** ['stımjulaı] Ansporn, *der*

sting [stıŋ] 1. *n.* a) *(wounding)* Stich, *der; (by jellyfish, nettles)* Verbrennung, *die;* b) *(from ointment, wind)* Brennen, *das.* 2. *v. t.,* **stung** [stʌŋ] stechen. 3. *v. i.,* **stung** brennen. **'stinging-nettle** *n.* Brennessel, *die*

stingy ['stındʒı] *adj.* geizig; knaus[e]rig *(ugs.)*

stink [stıŋk] 1. *v. i.,* **stank** [stæŋk] *or* **stunk** [stʌŋk], **stunk** stinken (**of** nach). 2. *n.* Gestank, *der*

stint [stınt] 1. *v. i.* ~ **on sth.** an etw. *(Dat.)* sparen. 2. *n.* [Arbeits]pensum, *das*

stipulate ['stıpjuleıt] *v. t.* *(demand)* fordern; *(lay down)* festlegen. **stipulation** [stıpju'leıʃn] *n.* *(condition)* Bedingung, *die*

stir [stɜ:(r)] 1. *v. t.,* **-rr-:** a) *(mix)* rühren; umrühren ⟨*Tee, Kaffee*⟩; b) *(move)* bewegen. 2. *v. i.,* **-rr-** *(move)* sich rühren. 3. *n.* Aufregung, *die.* **stir 'in** *v. t.* einrühren. **stir 'up** *v. t.* a) *(disturb)* aufrühren; b) *(fig.: arouse)* wecken ⟨*Interesse, Leidenschaft*⟩

stirring ['stɜ:rıŋ] *adj.* bewegend ⟨*Musik, Poesie*⟩; mitreißend ⟨*Rede*⟩

stirrup ['stırəp] *n.* Steigbügel, *der*

stitch [stıtʃ] 1. *n.* a) *(Sewing)* Stich, *der; (Knitting)* Masche, *die;* b) *(pain)* **have a** ~: Seitenstechen haben. 2. *v. t.* nähen

stoat [stəut] *n.* Hermelin, *das*

stock [stɒk] 1. *n.* a) *(origin, family, breed)* Abstammung, *die;* b) *(supply, store)* Vorrat, *der; (in shop etc.)* Warenbestand, *der;* **be in/out of** ~ ⟨*Ware:*⟩ vorrätig/nicht vorrätig sein; **have sth. in** ~: etw. auf Lager haben; **take** ~ **of sth.** *(fig.)* über etw. *(Akk.)* Bilanz ziehen; c) *(Cookery)* Brühe, *die.* 2. *v. t.* a) *(supply with* ~) beliefern; b) *(Commerc.: keep in* ~) auf Lager haben. 3. *attrib. adj.* Standard-

stock: ~**broker** Effektenmakler, *der/*-maklerin, *die;* ~ **cube** *n.* Brühwürfel, *der;* ~ **exchange** *n.* Börse, *die*

stocking ['stɒkıŋ] *n.* Strumpf, *der*

'stockist *n.* Fachhändler, *der/*-händlerin, *die*

stock: ~-**market** *n.* a) Börse, *die;* b) *(trading)* Börsengeschäft, *das;* ~**pile** 1. *n.* Vorrat, *der; (weapons)* Arsenal, *das;* 2. *v. t.* horten; anhäufen ⟨*Waffen*⟩; ~-'**still** *pred. adj.* bewegungslos; ~-**taking** *n.* Inventur, *die*

stocky ['stɒkı] *adj.* stämmig

stodgy ['stɒdʒı] *adj.* pappig

stoical ['stəuıkl] *adj.* stoisch

stoke [stəuk] *v. t.* heizen ⟨*Ofen, Kessel*⟩; unterhalten ⟨*Feuer*⟩

stole *see* **steal**

stolen ['stəuln] 1. *see* **steal.** 2. *attrib. adj.* heimlich ⟨*Vergnügen, Kuß*⟩

stolid ['stɒlıd] *adj.* stur *(ugs.)*

stomach ['stʌmək] 1. *n.* a) Magen, *der;* b) *(abdomen)* Bauch, *der.* 2. *v. t.* *(fig.: tolerate)* ausstehen. **'stomachache** *n.* Magenschmerzen *Pl.* **'stomach upset** *n.* Magenverstimmung, *die*

stone [stəun] 1. *n.* a) Stein, *der;* **a** ~**'s throw [away]** *(fig.)* nur einen Steinwurf weit entfernt; b) *(Brit.: weight unit)* Gewicht von 6,35 kg. 2. *adj.* steinern; Stein⟨*mauer, -brücke*⟩. 3. *v. t.* mit Steinen bewerfen

stone: S~ Age n. Steinzeit, die; **~-cold** adj. eiskalt; **~-'deaf** adj. stocktaub (ugs.)

stony ['stəʊnɪ] adj. steinig

stood see stand 1, 2

stool [stu:l] n. Hocker, der

stoop [stu:p] 1. v.i. ~ |down| sich bücken. 2. n. **walk with a ~**: gebeugt gehen

stop [stɒp] 1. v.t., -pp-: a) anhalten ⟨Person, Fahrzeug⟩; aufhalten ⟨Fortschritt, Verkehr, Feind⟩; b) (not let continue) unterbrechen ⟨Redner, Spiel, Gespräch⟩; beenden ⟨Krieg, Arbeit⟩; stoppen ⟨Produktion, Uhr⟩; einstellen ⟨Zahlung, Lieferung⟩; **~ that!** hör damit auf!; **~ smoking/crying** aufhören zu rauchen/weinen; c) (not let happen) verhindern ⟨Verbrechen, Unfall⟩; **~ sth. |from| happening** verhindern, daß etw. geschieht; d) (switch off) abstellen ⟨Maschine⟩; e) (block up) zustopfen ⟨Loch⟩; verschließen ⟨Wasserhahn, Flasche⟩; f) **~ a cheque** einen Scheck sperren lassen. 2. v.i., -pp-: a) (not extend further) aufhören ⟨Zahlungen, Lieferungen:⟩ eingestellt werden; b) (not move further) ⟨Fahrzeug, Fahrer:⟩ halten; ⟨Maschine, Motor:⟩ stillstehen; ⟨Uhr, Fußgänger, Herz:⟩ stehenbleiben. 3. n. a) (halt) Halt, der; **bring to a ~**: zum Stehen bringen ⟨Fahrzeug⟩; zum Erliegen bringen ⟨Verkehr⟩; unterbrechen ⟨Arbeit⟩; **come to a ~**: stehenbleiben; ⟨Fahrzeug:⟩ zum Stehen kommen; ⟨Arbeit, Verkehr:⟩ zum Erliegen kommen; **put a ~ to** abstellen ⟨Mißstände, Unsinn⟩; b) (place) Haltestelle, die. **stop 'by** v.i. (Amer.) vorbeischauen (ugs.). **stop 'out** v.i. (coll.) draußen bleiben. **stop 'over** v.i. übernachten (at bei). **stop 'up** 1. v.t. zustopfen ⟨Loch, Öffnung⟩. 2. v.i. (coll.) see **stay up**

stop: ~gap n. Notlösung, die; **~-light** n. (traffic-light) rotes Licht; **~over** n. Stopover, der

stoppage ['stɒpɪdʒ] n. a) (halt) Stillstand, der; (strike) Streik, der; b) (deduction) Abzug, der

stopper ['stɒpə(r)] n. Stöpsel, der

stop: ~-press n. letzte Meldung/ Meldungen; **~ sign** n. Stoppschild, das; **~watch** n. Stoppuhr, die

storage ['stɔ:rɪdʒ] n. Lagerung, die; (of films, books, documents) Aufbewahrung, die; (of data, water, electricity) Speicherung, die. **'storage**

heater n. [Nacht]speicherofen, der. **'storage tank** n. Sammelbehälter, der

store [stɔ:(r)] 1. n. a) (Amer.: shop) Laden, der; b) (Brit.: large general shop) Kaufhaus, das; c) (warehouse) Lager, das; **put sth. in ~**: etw. einlagern; d) (stock) Vorrat, der (of an + Dat.); **be or lie in ~ for sb.** jmdn. erwarten; e) **set |great| ~ by or on sth.** [großen] Wert auf etw. (Akk.) legen. 2. v.t. einlagern; speichern ⟨Getreide, Energie, Wissen, Daten⟩. **store 'up** v.t. speichern; **~ up provisions** sich (Dat.) Vorräte anlegen

store: ~house n. Lager[haus], das; **~-room** n. Lagerraum, der

storey ['stɔ:rɪ] n. Stockwerk, das

stork [stɔ:k] n. Storch, der

storm [stɔ:m] 1. n. Unwetter, das; (thunder~) Gewitter, das. 2. v.t. & i. stürmen. **'stormy** adj. stürmisch

¹story ['stɔ:rɪ] n. a) Geschichte, die; b) (news item) Bericht, der; c) (coll.: lie) Märchen, das

²story (Amer.) see **storey**

stout [staʊt] adj. a) (strong) fest; b) (fat) beleibt

stove [stəʊv] n. Ofen, der; (for cooking) Herd, der

stow [stəʊ] v.t. verstauen (into in + Dat.). **stow a'way** 1. v.t. verwahren. 2. v.i. als blinder Passagier reisen

straddle ['strædl] v.t. **~ a fence/chair** rittlings auf einem Zaun/Stuhl sitzen

straggle ['strægl] v.i. **~ |along| behind the others** den anderen hinterherzockeln (ugs.). **straggler** ['stræglə(r)] n. Nachzügler, der

straight [streɪt] 1. adj. a) gerade; glatt ⟨Haar⟩; **in a ~ line** in gerader Linie; b) (undiluted) **drink whisky ~**: Whisky pur trinken; c) (direct) direkt ⟨Blick, Schuß, Weg⟩; **be ~ with sb.** zu jmdm. offen sein; **get sth. ~** (fig.) etw. genau verstehen; **put or set the record ~**: die Sache richtigstellen. 2. adv. a) gerade; b) (directly) geradewegs; **~ after** sofort nach; **come ~ to the point** direkt zur Sache kommen; **look sb. ~ in the eye** jmdm. direkt in die Augen blicken; **~ ahead** or on immer geradeaus; c) (frankly) aufrichtig; d) (clearly) klar ⟨sehen, denken⟩. **straight a'way** adv. (coll.) sofort

straighten ['streɪtn] 1. v.t. a) geradeziehen ⟨Teppich⟩; glätten ⟨Kleidung, Haare⟩; b) (put in order) aufräumen. 2. v.i. gerade werden. **straighten**

'**out 1.** *v.t.* **a)** geradebiegen; glätten ⟨*Decke, Teppich*⟩; **b)** *(clear up)* klären. **2.** *v.i.* gerade werden. **straighten 'up 1.** *v.t. see* **tidy up. 2.** *v.i.* sich aufrichten

straight'forward *adj.* **a)** *(frank)* freimütig; schlicht ⟨*Stil, Sprache, Bericht*⟩; klar ⟨*Anweisung, Vorstellungen*⟩; **b)** *(simple)* einfach

strain [streɪn] **1.** *n.* **a)** *(pull)* Belastung, *die; (on rope)* Spannung, *die;* **b)** *(tension)* Streß, *der;* **be under [a great deal of]** ~: unter großem Streß stehen; **c)** *(person, thing)* **be a** ~ **on sb./sth.** jmdn./etw. belasten; **d)** *(muscular injury)* Zerrung, *die.* **2.** *v.t.* **a)** *(overexert)* überanstrengen; zerren ⟨*Muskel*⟩; **b)** *(stretch tightly)* [fest] spannen; **c)** *(filter)* durchseihen; seihen **(through** durch**). 3.** *v.i. (strive intensely)* sich anstrengen. **strained** [streɪnd] *adj.* gezwungen ⟨*Lächeln*⟩; ~ **relations** gespannte Beziehungen

'**strainer** *n.* Sieb, *das*

strait [streɪt] *n.* **a)** *in sing. or pl. (Geog.)* Meerenge, *die;* **b)** *usu. in pl. (distress, difficulty)* Schwierigkeiten. '**straitjacket** *n.* Zwangsjacke, *die.* **strait-laced** [streɪt'leɪst] *adj.* puritanisch

'**strand** [strænd] *n. (thread)* Faden, *der; (of beads)* Kette, *die; (of hair)* Strähne, *die; (of rope)* Strang, *der*

²**strand** *v.t. (leave behind)* trocken setzen; **be [left]** ~**ed** *(fig.)* seinem Schicksal überlassen sein

strange [streɪndʒ] *adj. (peculiar)* seltsam; sonderbar; ~ **to say** seltsamerweise; **feel** ~: sich komisch fühlen. **stranger** ['streɪndʒə(r)] *n.* Fremde, *der/die;* **he is a** ~ **here/to the town** er ist hier/in der Stadt fremd

strangle ['stræŋgl] *v.t.* erwürgen. '**stranglehold** *n.* Würgegriff, *der.* **strangulation** [stræŋgjʊ'leɪʃn] *n.* Erwürgen, *das*

strap [stræp] **1.** *n.* **a)** *(leather)* Riemen, *der; (textile)* Band, *das; (shoulder-~)* Träger, *der; (for watch)* Armband, *das;* **b)** *(to grasp in vehicle)* Halteriemen, *der.* **2.** *v.t.,* **-pp-:** ~ **[into position]/down** festschnallen; ~ **oneself in** sich anschnallen. '**strapless** *adj.* trägerlos

strapping ['stræpɪŋ] *adj.* stramm

strata *pl. of* **stratum**

strategic [strə'tiːdʒɪk] *adj.* strategisch

strategist ['strætɪdʒɪst] *n.* Stratege, *der/*Strategin, *die*

strategy ['strætɪdʒɪ] *n.* Strategie, *die*

stratosphere ['strætəsfɪə(r)] *n.* Stratosphäre, *die*

stratum ['strɑːtəm] *n., pl.* **strata** ['strɑːtə] Schicht, *die*

straw [strɔː] *n.* **a)** *no pl.* Stroh, *das;* **b)** *(single stalk)* Strohhalm, *der;* **that's the last** *or* **final** ~: jetzt reicht's aber; **c)** |drinking-|~: Strohhalm, *der*

strawberry ['strɔːbərɪ] *n.* Erdbeere, *die*

stray [streɪ] **1.** *v.i.* **a)** *(wander)* streunen; **b)** *(deviate)* abweichen **(from** von**). 2.** *n. (animal)* streunendes Tier. **3.** *adj.* **a)** streunend; **b)** *(occasional)* vereinzelt

streak [striːk] *n.* Streifen, *der; (in hair)* Strähne, *die;* **have a jealous/cruel** ~: zur Eifersucht/Grausamkeit neigen. '**streaky** *adj.* streifig; ~ **bacon** durchwachsener Speck

stream [striːm] **1.** *n. (of water)* Wasserlauf, *der; (brook)* Bach, *der.* **2.** *v.i.* strömen; ⟨*Sonnenlicht:*⟩ fluten. '**streamline** *v.t.* [eine] Stromlinienform geben **(+** *Dat.***);** **be** ~**lined** eine Stromlinienform haben

street [striːt] *n.* Straße, *die;* **in the** ~: auf der Straße; **in** *(Brit.) or* **on ... Street** in der ...straße

street: ~**car** *n. (Amer.)* Straßenbahn, *die;* ~**lamp** *n.* Straßenlaterne, *die;* ~**lighting** *n.* Straßenbeleuchtung, *die;* ~**map** *n.* Stadtplan, *der;* ~**market** *n.* Markt, *der;* ~**plan** *n.* Stadtplan, *der;* ~**wise** *adj. (coll.)* ~**wise** wissen, wo es langgeht

strength [streŋθ] *n. (power)* Kraft, *die; (strong point, force, intensity, amount of ingredient)* Stärke, *die; (of poison, medicine)* Wirksamkeit, *die;* **not know one's own** ~: nicht wissen, wie stark man ist; **give sb.** ~: jmdn. stärken; **go from** ~ **to** ~: immer erfolgreicher werden; **on the** ~ **of sth./that** auf Grund einer Sache *(Gen.)*/dessen; **in full** ~: in voller Stärke; **the police were there in** ~: ein starkes Polizeiaufgebot war da. **strengthen** ['streŋθn] *v.t.* stärken; *(reinforce, intensify)* verstärken

strenuous ['strenjʊəs] *adj.* **a)** *(energetic)* energisch; gewaltig ⟨*Anstrengung*⟩; **b)** *(requiring exertion)* anstrengend

stress [stres] **1.** *n.* **a)** *(strain)* Streß, *der;* **be under** ~: unter Streß *(Dat.)* stehen; **b)** *(emphasis)* Betonung, *die.* **2.** *v.t. (emphasize)* betonen

stretch [stretʃ] **1.** *v.t.* **a)** *(lengthen)* strecken ⟨*Arm, Hand*⟩; recken ⟨*Hals*⟩;

dehnen *(Gummiband)*; *(tighten)* spannen; b) *(widen)* dehnen. **2. a)** v. i. *(extend in length)* sich dehnen; b) ~ to sth. *(be sufficient for)* für etw. reichen. **3.** v. refl. sich strecken. **4.** n. **a) have a** ~: sich strecken; b) at a ~ *(fig.)* wenn es sein muß; c) *(expanse)* Abschnitt, *der;* a ~ of road ein Stück Straße; d) *(period)* a four-hour ~: eine [Zeit]spanne von vier Stunden; at a ~: ohne Unterbrechung. **5.** adj. Stretch- *(hose, -gewebe)*

stretcher ['stretʃə(r)] n. [Trag]bahre, *die*

strew [stru:] v. t., p. p. strewed [stru:d] *or* strewn [stru:n] streuen

stricken ['strɪkn] adj. *(afflicted)* heimgesucht; havariert *(Schiff)*; be ~ with fear/grief angsterfüllt/gramgebeugt

strict [strɪkt] adj. **a)** *(firm)* streng; in ~ confidence streng vertraulich; **b)** *(precise)* streng. **'strictly** adv. streng; ~ |speaking] strenggenommen

stride [straɪd] **1.** n. Schritt, *der;* put sb. off his ~ *(fig.)* jmdn. aus dem Konzept bringen; take sth. in one's ~ *(fig.)* mit etw. gut fertig werden. **2.** v. i., strode [strəʊd], stridden ['strɪdn] [mit großen Schritten] gehen

strident ['straɪdənt] adj. schrill

strife [straɪf] n. Streit, *der*

strike [straɪk] **1.** n. *(Industry)* Streik, *der;* Ausstand, *der;* be on/go |out] *or* come out on ~: in den Streik getreten sein/in den Streik treten. **2.** v. t., struck [strʌk] **a)** *(hit)* schlagen; *(Schlag, Geschoß:)* treffen; *(Blitz:)* [ein]schlagen in (+ Akk.); **b)** *(delete)* streichen (from, off aus); **c)** *(ignite)* anzünden *(Streichholz)*; **d)** *(chime)* schlagen; **e)** *(impress)* beeindrucken; ~ sb. as [being] silly jmdm. dumm erscheinen; it ~s sb. that ...: es scheint jmdm., daß ...; **f)** *(occur to)* einfallen (+ Dat.). **3.** v. i., struck **a)** *(deliver a blow)* zuschlagen; *(Blitz:)* einschlagen; *(Unheil, Katastrophe:)* hereinbrechen *(geh.)*; *(hit)* schlagen (against gegen, |up]on auf + Akk.); **b)** *(ignite)* zünden; **c)** *(chime)* schlagen; **d)** *(Industry)* streiken. **strike 'back** v. i. zurückschlagen. **strike 'off** v. t. (~ off list) streichen *(Namen)*; *(from professional body)* die Zulassung entziehen (+ Dat.). **'strike up** v. t. beginnen *(Unterhaltung)*; schließen *(Freundschaft)*

'strike pay n. Streikgeld, *das*

'striker n. Streikende, *der/die*

striking ['straɪkɪŋ] adj. auffallend; erstaunlich *(Ähnlichkeit)*; schlagend *(Beispiel)*

string [strɪŋ] **1.** n. **a)** *(thin cord)* Schnur, *die;* *(to tie up parcels etc. also)* Bindfaden, *der;* pull |a few *or* some] ~s *(fig.)* seine Beziehungen spielen lassen; with no ~s attached ohne Bedingung[en]; **b)** *(of bow)* Sehne, *die;* *(of racket, musical instrument)* Saite, *die.* **2.** v. t., strung [strʌŋ] *(thread)* auffädeln. **string a'long** v. t. *(deceive)* an der Nase herumführen *(ugs.)*. **string to'gether** v. t. auffädeln; miteinander verknüpfen *(Wörter)*. **string 'up** v. t. aufhängen

string 'bag n. [Einkaufs]netz, *das*

stringent ['strɪndʒənt] adj. streng

¹strip [strɪp] **1.** v. t., -pp- ausziehen *(Person)*. **2.** v. i., -pp- sich ausziehen

²strip n. *(narrow piece)* Streifen, *der*

stripe · [straɪp] n. Streifen, *der.* **striped** [straɪpt] adj. gestreift

strip: ~ light n. Neonröhre, *die;* ~ lighting n. Neonbeleuchtung, *die*

stripper ['strɪpə(r)] n. Stripper, *der/* Stripperin, *die (ugs.)*

stripy ['straɪpɪ] adj. gestreift; Streifen- *(muster)*

strive [straɪv] v. i., strove [strəʊv], striven ['strɪvn] sich bemühen; ~ after *or* for sth. nach etw. streben

strode see stride 2

¹stroke [strəʊk] n. **a)** *(act of striking)* Schlag, *der;* **b)** *(Med.)* Schlaganfall, *der;* **c)** *(sudden impact)* ~ of lightning Blitzschlag, *der;* ~ of |good] luck Glücksfall, *der;* **d)** at a *or* one ~: auf einen Schlag; not do a ~ |of work] keinen [Hand]schlag tun; ~ of genius genialer Einfall; **e)** *(in swimming)* Zug, *der;* **f)** *(of clock)* Schlag, *der;* on the ~ of nine Punkt neun [Uhr]

²stroke 1. v. t. streicheln. **2.** n. give sb./sth. a ~: jmdn./etw. streicheln

stroll [strəʊl] **1.** v. i. spazierengehen. **2.** n. go for a ~: einen Spaziergang machen

strong [strɒŋ] adj., ~er ['strɒŋgə(r)], ~est ['strɒŋgɪst] stark; fest *(Fundament, Schuhe)*; robust *(Konstitution, Magen)*; kräftig *(Arme, Muskeln, Tritt, Zähne)*; leistungsfähig *(Wirtschaft)*; gut, handfest *(Grund, Beispiel, Argument)*; glühend *(Anhänger)*; kräftig *(Geruch, Geschmack, Stimme)*; there is a ~ possibility that ...: es ist sehr wahrscheinlich, daß ...; take ~ measures/action energisch vorgehen.

'**stronghold** *n.* Festung, *die;* *(fig.)* Hochburg, *die.* **strong 'language** *n.* derbe Ausdrucksweise

'**strongly** *adv.* stark; solide 〈*gearbeitet*〉; energisch 〈*protestieren, bestreiten*〉; nachdrücklich 〈*unterstützen*〉; dringend 〈*raten*〉; fest 〈*glauben*〉

strong: ~**man** *n.* Muskelmann, *der (ugs.);* ~-**'minded** *adj.* willensstark; ~-**room** *n.* Tresorraum, *der*

strove *see* strive

struck *see* strike 2, 3

structural ['strʌktʃərl] *adj.* baulich

structure ['strʌktʃə(r)] *n.* a) Struktur, *die;* b) *(something constructed)* Konstruktion, *die; (building)* Bauwerk, *das*

struggle ['strʌgl] 1. *v. i.* kämpfen; ~ to do sth. sich abmühen, etw. zu tun; ~ against *or* with sb./sth. mit jmdm./ etw. od. gegen jmdn./etw. kämpfen; ~ with sth. *(try to cope)* mit etw. kämpfen. 2. *n.* Kampf, *der*

strum [strʌm] 1. *v. i.,* -mm- klimpern *(ugs.)* (on auf + *Dat.*). 2. *v. t.,* -mm-klimpern *(ugs.)* auf (+ *Dat.*)

strung *see* string 2

¹**strut** [strʌt] 1. *v. i.,* -tt- stolzieren. 2. *n.* stolzierender Gang

²**strut** *n. (support)* Strebe, *die*

stub [stʌb] 1. *n.* a) *(remaining portion)* Stummel, *der; (of cigarette)* Kippe, *die;* b) *(counterfoil)* Abschnitt, *der.* 2. *v. t.,* -bb-: a) ~ one's toe [against *or* on sth.] sich *(Dat.)* den Zeh [an etw. *(Dat.)*] stoßen; b) ausdrücken 〈*Zigarette*〉. **stub 'out** *v. t.* ausdrücken

stubble ['stʌbl] *n.* Stoppeln *Pl.*

stubborn ['stʌbən] *adj.* a) *(obstinate)* starrköpfig; störrisch 〈*Tier, Gesicht, Haltung*〉; b) *(resolute)* hartnäckig. '**stubbornness** *n. see* stubborn: Starrköpfigkeit, *die;* Hartnäckigkeit, *die*

stuck *see* stick 1, 2

'**stuck up** *adj. (conceited)* eingebildet

student ['stju:dənt] *n.* Student, *der/* Studentin, *die; (in school or training establishment)* Schüler, *der/*Schülerin, *die;* be a ~ of sth. etw. studieren

studio ['stju:dɪəʊ] *n., pl.* ~s a) *(workroom)* Atelier, *das;* b) *(Cinemat., Radio, Telev.)* Studio, *das*

studious ['stju:dɪəs] *adj.* lerneifrig

study ['stʌdɪ] 1. *n.* a) Studium, *das;* b) *(room)* Arbeitszimmer, *das.* 2. *v. t.* studieren; sich *(Dat.)* [sorgfältig] durchlesen 〈*Prüfungsfragen, Bericht*〉

stuff [stʌf] 1. *n. (material[s])* Zeug, *das (ugs.).* 2. *v. t.* a) stopfen; zustopfen

〈*Loch, Ohren*〉; *(Cookery)* füllen; ~ sth. with *or* full of sth. etw. mit etw. vollstopfen *(ugs.);* b) *(sl.)* ~ him! zum Teufel mit ihm! '**stuffing** *n.* a) *(material)* Füllmaterial, *das;* b) *(Cookery)* Füllung, *die*

'**stuffy** ['stʌfɪ] *adj.* stickig

stumble ['stʌmbl] *v. i.* stolpern (over über + *Akk.*). **stumbling-block** ['stʌmblɪŋblɒk] *n.* Stolperstein, *der*

stump [stʌmp] 1. *n. (of tree, branch, tooth)* Stumpf, *der; (of cigar, pencil)* Stummel, *der.* 2. *v. t.* verwirren; be ~ed ratlos sein. '**stumpy** *adj.* gedrungen; ~ tail Stummelschwanz, *der*

stun [stʌn] *v. t.,* -nn- *(knock senseless)* betäuben; be ~ned at *or* by sth. *(fig.)* von etw. wie betäubt sein

stung *see* sting 2, 3

stunk *see* stink 1

¹**stunt** [stʌnt] *v. t.* hemmen

²**stunt** *n.* halsbrecherisches Kunststück; *(Cinemat.)* Stunt, *der*

stupendous [stju:'pendəs] *adj.* gewaltig

stupid ['stju:pɪd] *adj.* dumm; *(ridiculous)* lächerlich; it would be ~ to do sth. es wäre töricht, etw. zu tun. **stupidity** [stu:'pɪdɪtɪ] *n.* Dummheit, *die.* '**stupidly** *adv.* dumm

stupor ['stju:pə(r)] *n.* Benommenheit, *die;* in a drunken ~: sinnlos betrunken

sturdy ['stɜ:dɪ] *adj.* kräftig; stämmig 〈*Beine, Arme*〉

stutter ['stʌtə(r)] 1. *v. i.* stottern. 2. *n.* Stottern, *das*

¹**sty** [staɪ] *see* pigsty

²**sty, stye** [staɪ] *n. (Med.)* Gerstenkorn, *das*

style [staɪl] *n.* Stil, *der;* dress in the latest ~: sich nach der neuesten Mode kleiden; |hair-|~: Frisur, *die*

styli *pl. of* stylus

stylish ['staɪlɪʃ] *adj.* stilvoll; elegant 〈*Kleidung, Auto, Person*〉

stylist ['staɪlɪst] *n. (hair-~)* Haarstilist, *der/*-stilistin, *die*

stylus ['staɪləs] *n., pl.* **styli** ['staɪlaɪ] *or* ~es [Abtast]nadel, *die*

suave [swɑ:v] *adj.* gewandt

sub- [sʌb] *pref.* unter-; sub-

sub'conscious *(Psych.)* 1. *adj.* unterbewußt. 2. *n.* Unterbewußtsein, *das*

'**subcontract** *v. t.* an einen Subunternehmer vergeben

subdivide ['---, --'-] *v. t.* unterteilen

subdue [səb'dju:] *v. t.* bändigen 〈*Kind, Tier*〉; dämpfen 〈*Zorn, Lärm, Licht*〉

subdued [səb'dju:d] *adj.* gedämpft

subject 1. ['sʌbdʒɪkt] *n.* **a)** Staatsbürger, *der/*-bürgerin, *die; (to monarch)* Untertan, *der/*Untertanin, *die;* **b)** *(topic)* Thema, *das; (of study)* Fach, *das;* **change the ~:** das Thema wechseln. **2.** ['sʌbdʒɪkt] *adj.* **be ~ to sth.** von etw. abhängen. **3.** [səb'dʒekt] *v. t.* unterwerfen **(to** *Dat.***);** *(expose)* **~ sb./ sth. to sth.** jmdn./etw. einer Sache *(Dat.)* aussetzen

subjective [səb'dʒektɪv] *adj.* subjektiv

subjugate ['sʌbdʒʊgeɪt] *v. t.* unterjochen **(to unter** + *Akk.*)

subjunctive [səb'dʒʌŋktɪv] *(Ling.) n.* Konjunktiv, *der*

sub'let *v. t.,* **-tt-, sublet** untervermieten

sublime [sə'blaɪm] *adj.* erhaben

submarine [sʌbmə'riːn] *n.* Unterseeboot, *das;* U-Boot, *das*

submerge [səb'mɜːdʒ] *v. t.* **a) ~ sth. [in the water]** etw. eintauchen; **b)** *(flood)* ⟨*Wasser:*⟩ überschwemmen; **be ~d in water** unter Wasser stehen

submission [səb'mɪʃn] *n.* **a)** *(surrender, meekness)* Unterwerfung, *die;* **b)** *(presentation)* Einreichung, *die* **(to** bei)

submissive [səb'mɪsɪv] *adj.* gehorsam

submit [səb'mɪt] *v. t.,* **-tt-** *(present)* einreichen; vorbringen ⟨*Vorschlag*⟩; **~ sth. to sb.** jmdm. etw. vorlegen

subordinate 1. [sə'bɔːdɪnət] *adj.* untergeordnet. **2.** [sə'bɔːdɪnət] *n.* Untergebene, *der/die.* **3.** [sə'bɔːdɪneɪt] *v. t.* unterordnen **(to** *Dat.*)

subscribe [səb'skraɪb] *v. i.* **a)** *(support)* **~ to sth.** sich einer Sache *(Dat.)* anschließen; **b)** *(make contribution)* **~ to sth.** eine Spende für etw. zusichern.

sub'scriber *n.* **(to** *newspaper etc.)* Abonnent, *der/*Abonnentin, *die* **(to** *Gen.).* **subscription** [səb'skrɪpʃn] *n. (membership fee)* Mitgliedsbeitrag, *der* **(to** für); **(to** *newspaper etc.)* Abonnement, *das*

subsequent ['sʌbsɪkwənt] *adj.* folgend; später ⟨*Gelegenheit*⟩

subservient [səb'sɜːvɪənt] *adj.* untergeordnet **(to** *Dat.*); *(servile)* unterwürfig

subside [səb'saɪd] *v. i.* **a)** *(sink lower)* ⟨*Flut, Fluß:*⟩ sinken; ⟨*Boden, Haus:*⟩ sich senken; **b)** *(abate)* nachlassen

subsidiary [səb'sɪdɪərɪ] **1.** *adj.* untergeordnet ⟨*Funktion, Stellung*⟩; Neben⟨*fach, -aspekt*⟩. **2.** *n. (Commerc.)* Tochtergesellschaft, *die*

subsidize ['sʌbsɪdaɪz] *v. t.* subventio-

nieren. **subsidy** ['sʌbsɪdɪ] *n.* Subvention, *die*

subsist [səb'sɪst] *v. i.* **~ on sth.** von etw. leben. **subsistence** [səb'sɪstəns] *n.* [Über]leben, *das*

substance ['sʌbstəns] *n.* **a)** Stoff, *der;* **b)** *(solidity)* Substanz, *die;* **c)** *(content)* Inhalt, *der*

sub'standard *adj.* unzulänglich

substantial [səb'stænʃl] *adj.* **a)** *(considerable)* beträchtlich; **b)** gehaltvoll ⟨*Essen*⟩; **c)** *(solid)* solide ⟨*Möbel, Haus*⟩; wesentlich ⟨*Unterschied*⟩. **sub'stantially** *adv.* **a)** *(considerably)* wesentlich; **b)** *(solidly)* solide; **c)** *(essentially)* im wesentlichen

substitute ['sʌbstɪtjuːt] **1.** *n.* **~[s** *pl.*] Ersatz, *der.* **2.** *v. t.* **~ A for B** B durch A ersetzen. **substitution** [sʌbstɪ'tjuːʃn] *n.* Ersetzung, *die*

'subtitle *n.* Untertitel, *der*

subtle ['sʌtl] *adj.* subtil *(geh.);* zart ⟨*Duft, Parfüm, Hinweis*⟩; fein ⟨*Geschmack, Unterschied, Humor*⟩

subtract [səb'trækt] *v. t.* abziehen. **subtraction** [səb'trækʃn] *n.* Subtraktion, *die*

suburb ['sʌbɜːb] *n.* Vorort, *der.* **suburban** [sə'bɜːbən] *adj.* Vorort-; ⟨*Leben, Haus*⟩ am Stadtrand

subversive [səb'vɜːsɪv] *adj.* subversiv

'subway *n.* **a)** *(passage)* Unterführung, *die;* **b)** *(Amer.: railway)* Untergrundbahn, *die;* U-Bahn, *die*

succeed [sək'siːd] *v. i.* **a)** Erfolg haben; **sb. ~s in sth.** jmdm. gelingt etw.; jmd. schafft etw.; **sb. ~s in doing sth.** es gelingt jmdm., etw. zu tun; jmd. schafft es, etw. zu tun; **~ in business/ college** geschäftlich/im Studium erfolgreich sein; **I did not ~ in doing it** ich habe es nicht geschafft; **b)** *(come next)* die Nachfolge antreten

success [sək'ses] *n.* Erfolg, *der;* **make a ~ of sth.** bei etw. Erfolg haben. **successful** [sək'sesfl] *adj.* erfolgreich; **be ~ in sth./doing sth.** Erfolg bei etw. haben/dabei haben, etw. zu tun. **successfully** *adv.* erfolgreich

succession [sək'seʃn] *n.* **a)** Folge, *die;* **in ~:** hintereinander; **b)** *(series)* Serie, *die;* **c)** *(to throne)* Erbfolge, *die*

successive [sək'sesɪv] *adj.* aufeinanderfolgend. **successively** *adv.* hintereinander

successor [sək'sesə(r)] *n.* Nachfolger, *der/*Nachfolgerin, *die*

succinct [sək'sɪŋkt] *adj.* **a)** *(terse)* knapp; **b)** *(clear)* prägnant

succulent ['sʌkjʊlənt] *adj.* saftig

succumb [sə'kʌm] *v. i.* unterliegen; ~ to sth. einer Sache *(Dat.)* erliegen

such [sʌtʃ] 1. *adj.* a) *(of that kind)* solch...; ~ a person ein solcher Mensch; ~ a book ein solches Buch; ~ people solche Leute; ~ things so etwas; I said no ~ thing ich habe nichts dergleichen gesagt; there is no ~ bird einen solchen Vogel gibt es nicht; or some ~ thing oder so etwas; you'll do no ~ thing das wirst du nicht tun; experiences ~ as these solche Erfahrungen; b) *(so great)* solch...; derartig; I got ~ a fright that ...: ich bekam einen derartigen *od. (ugs.)* so einen Schrecken, daß ...; ~ was the force of the explosion that ...: die Explosion war so stark, daß ...; to ~ an extent dermaßen; c) *with adj.* so; ~ a big house ein so großes Haus; ~ a long time so lange. 2. *pron.* as ~: als solcher/solche/solches; *(strictly speaking)* im Grunde genommen; an sich; ~ as wie [zum Beispiel]; ~ is life so ist das Leben. **such-and-such** ['sʌtʃənsʌtʃ] *adj.* at ~ a time um die und die Zeit. **'suchlike** *pron.* derlei

suck [sʌk] *v. t.* saugen (out of aus); lutschen ⟨*Bonbon*⟩. **suck 'up** 1. *v. t.* aufsaugen. 2. *v. i.* ~ up to sb. *(sl.)* jmdm. in den Hintern kriechen *(salopp)*

'sucker *n.* a) *(suction pad)* Saugfuß, *der; (Zool.)* Saugnapf, *der;* b) *(sl.: dupe)* Dumme, *der/die*

suckle ['sʌkl] *v. t.* säugen

suction ['sʌkʃn] *n.* Saugwirkung, *die*

Sudan [su:'dɑ:n] *pr. n.* [the] ~: [der] Sudan

sudden ['sʌdn] 1. *adj.* a) *(unexpected)* plötzlich; b) *(abrupt)* jäh ⟨*Abgrund, Übergang, Ruck*⟩; there was a ~ bend in the road plötzlich machte die Straße eine Biegung. 2. *n.* all of a ~: plötzlich. **'suddenly** *adv.* plötzlich. **'suddenness** *n.* Plötzlichkeit, *die*

suds [sʌdz] *n.* [soap-]~: [Seifen]lauge, *die; (froth)* Schaum, *der*

sue [su:] 1. *v. t.* verklagen (for auf + *Akk.*). 2. *v. i.* klagen (for auf + *Akk.*)

suede [sweɪd] *n.* Wildleder, *das*

suet ['su:ɪt] *n.* Talg, *der*

Suez ['su:ɪz, 'sju:ɪz] *pr. n.* Suez *(das);* ~ Canal Suez-Kanal, *der*

suffer ['sʌfə(r)] 1. *v. t.* erleiden; durchmachen ⟨*Schweres, Kummer*⟩; dulden ⟨*Unverschämtheit*⟩. 2. *v. i.* leiden. **'suffer from** *v. t.* leiden unter (+ *Dat.*); leiden an (+ *Dat.*) ⟨*Krankheit*⟩

sufferance ['sʌfərəns] *n.* Duldung, *die;* he remains here on ~ only er ist hier bloß geduldet

'suffering *n.* Leiden, *das*

suffice [sə'faɪs] 1. *v. i.* genügen; ~ it to say ...: nur soviel sei gesagt: ... 2. *v. t.* genügen (+ *Dat.*)

sufficiency [sə'fɪʃənsɪ] *n.* Zulänglichkeit, *die*

sufficient [sə'fɪʃənt] *adj.* genug; ~ money/food genug Geld/genug zu essen; be ~: genügen; ~ reason Grund genug; have you had ~? *(food, drink)* haben Sie schon genug? **suf'ficiently** *adv.* genug; *(adequately)* ausreichend; ~ large groß genug; a ~ large number eine genügend große Zahl

suffix ['sʌfɪks] *n.* Nachsilbe, *die*

suffocate ['sʌfəkeɪt] 1. *v. t.* ersticken; he was ~d by the smoke der Rauch erstickte ihn. 2. *v. i.* ersticken. **suffocation** [sʌfə'keɪʃn] *n.* Erstickung, *die;* a feeling of ~: das Gefühl, zu ersticken

sugar ['ʃʊgə(r)] 1. *n.* Zucker, *der;* two ~s, please *(lumps)* zwei Stück Zucker, bitte; *(spoonfuls)* zwei Löffel Zucker, bitte. 2. *v. t.* zuckern

sugar: ~ basin *see* ~-bowl; ~-beet *n.* Zuckerrübe, *die;* ~-bowl *n.* Zuckerschale, *die; (covered)* Zuckerdose, *die;* ~-cane *n.* Zuckerrohr, *das;* ~-coated *adj.* gezuckert; mit Zucker überzogen ⟨*Dragee usw.*⟩; ~-lump *n.* Zuckerstück, *das; (when counted)* Stück Zucker

'sugary *adj.* süß; *(fig.)* süßlich

suggest [sə'dʒest] *v. t.* a) *(propose)* vorschlagen; ~ sth. to sb. jmdm. etw. vorschlagen; he ~ed going to the cinema er schlug vor, ins Kino zu gehen; b) *(assert)* are you trying to ~ that he is lying wollen Sie damit sagen, daß er lügt?; c) *(make one think of)* suggerieren; ⟨*Symptome, Tatsachen:*⟩ schließen lassen auf (+ *Akk.*). **suggestion** [sə'dʒestʃn] *n.* a) Vorschlag, *der;* at or on sb.'s ~: auf jmds. Vorschlag *(Akk.);* b) *(insinuation)* Andeutungen *Pl.;* c) *(fig.: trace)* Spur, *die.* **suggestive** [sə'dʒestɪv] *adj.* a) be ~ of sth. auf etw. *(Akk.)* schließen lassen; b) *(indecent)* anzüglich

suicidal [su:ɪ'saɪdl] *adj.* selbstmörderisch; I felt *or* was quite ~: ich hätte mich am liebsten gleich umgebracht

suicide ['su:ɪsaɪd] *n.* Selbstmord, *der.* **'suicide attempt** *n.* Selbstmordversuch, *der*

suit [su:t] **1.** *n.* **a)** *(for men)* Anzug, *der;* *(for women))* Kostüm, *das;* **b)** *(Law)* ~ |at law| Prozeß, *der;* **c)** *(Cards)* Farbe, *die;* **follow** ~ *(fig.)* das Gleiche tun. **2.** *v. t.* **a)** anpassen (to *Dat.*); **b) be** ~ed |to sth./one another| [zu etw./zueinander] passen; **c)** *(satisfy needs of)* passen (+ *Dat.*); **will Monday** ~ **you?** paßt Ihnen Montag?; **does the climate** ~ **you?** bekommt Ihnen das Klima?; **d)** *(go well with)* passen zu; **does this hat** ~ **me?** steht mir dieser Hut?; **black** ~s **her** Schwarz steht ihr gut. **3.** *v. refl.* ~ **oneself** tun, was man will; ~ **yourself!** [ganz] wie du willst!

suitability [su:tə'bɪlɪtɪ] *n.* Eignung, *die* (for für)

suitable ['su:təbl] *adj.* geeignet; angemessen ⟨*Kleidung*⟩; *(convenient)* passend; **Monday is the most** ~ **day** |for me| Montag paßt |mir| am besten. **suitably** ['su:təblɪ] *adv.* angemessen; entsprechend ⟨*gekleidet*⟩

'**suitcase** *n.* Koffer, *der*

suite [swi:t] *n.* **a)** *(of furniture)* Garnitur, *die;* **three-piece** ~: Polstergarnitur, *die;* **b)** *(of rooms)* Suite, *die*

suitor ['su:tə(r)] *n.* Freier, *der*

sulfur, sulfuric *(Amer.)* see **sulph**-

sulk [sʌlk] *v. i.* schmollen. '**sulky** *adj.* schmollend; eingeschnappt *(ugs.)*

sullen ['sʌlən] *adj.* mürrisch

sulphur ['sʌlfə(r)] *n.* Schwefel, *der.* **sulphuric** [sʌl'fjʊərɪk] *adj.* ~ **acid** Schwefelsäure, *die*

sultan ['sʌltən] *n.* Sultan, *der*

sultana [sʌl'tɑ:nə] *n.* Sultanine, *die*

sultry ['sʌltrɪ] *adj.* schwül

sum [sʌm] *n.* **a)** Summe, *die* (of aus); ~ |total| Ergebnis, *das;* **b)** *(Arithmetic)* Rechenaufgabe, *die;* **do** ~s *(coll.)* rechnen; **she is good at** ~s *(coll.)* sie kann gut rechnen. **sum 'up 1.** *v. t.* **a)** zusammenfassen; **b)** *(Brit.: assess)* einschätzen. **2.** *v. i.* ein Fazit ziehen

summarily ['sʌmərɪlɪ] *adv.* summarisch; ~ **dismissed** fristlos entlassen

summarize ['sʌməraɪz] *v. t.* zusammenfassen

summary ['sʌmərɪ] **1.** *adj.* summarisch; fristlos ⟨*Entlassung*⟩. **2.** *n.* Zusammenfassung, *die*

summer ['sʌmə(r)] *n.* Sommer, *der;* **in** |the| ~: im Sommer. '**summer-house** *n.* |Garten|laube, *die.* '**summertime** *n.* Sommer, *der*

'**summery** *adj.* sommerlich

summing 'up *n.* Zusammenfassung, *die*

summit ['sʌmɪt] *n.* Gipfel, *der*

summon ['sʌmən] *v. t.* **a)** rufen (to zu); holen ⟨*Hilfe*⟩; **b)** *(Law)* vorladen.

summon 'up *v. t.* aufbringen

summons ['sʌmənz] *n.* Vorladung, *die*

sump [sʌmp] *n.* Ölwanne, *die*

sumptuous ['sʌmptjʊəs] *adj.* üppig; luxuriös ⟨*Möbel, Kleidung*⟩

sun [sʌn] **1.** *n.* Sonne, *die;* **catch the** ~ *(be in sunny position)* viel Sonne abbekommen; *(get* ~*burnt)* einen Sonnenbrand bekommen. **2.** *v. refl.,* **-nn-** sich sonnen

Sun. *abbr.* **Sunday** So.

sun: ~**bathe** *v. i.* sonnenbaden; ~**bathing** *n.* Sonnenbaden, *das;* ~**beam** *n.* Sonnenstrahl, *der;* ~**bed** *n.* *(with UV lamp)* Sonnenbank, *die;* *(in garden)* Gartenliege, *die;* ~**burn** *n.* Sonnenbrand, *der;* ~**burnt** *adj.* **be/get** ~**burnt** einen Sonnenbrand haben/bekommen

sundae ['sʌndeɪ] *n.* |ice-cream| ~: Eisbecher, *der*

Sunday ['sʌndeɪ, 'sʌndɪ] *n.* Sonntag, *der; see also* **Friday**

'**sundial** *n.* Sonnenuhr, *die*

sundry ['sʌndrɪ] **1.** *adj.* verschieden. **2.** *n. in pl.* Verschiedenes

'**sunflower** *n.* Sonnenblume, *die*

sung *see* **sing**

sun: ~**glasses** *n. pl.* Sonnenbrille, *die;* ~**hat** *n.* Sonnenhut, *der*

sunk *see* **sink** 2, 3

sun: ~**lamp** *n.* Höhensonne, *die;* ~**lit** *adj.* sonnenbeschienen; ~**light** *n.* Sonnenlicht, *das*

sunny ['sʌnɪ] *adj.* sonnig; ~ **intervals** Aufheiterungen

sun: ~**rise** *n.* Sonnenaufgang, *der;* ~**roof** *n.* *(Motor Veh.)* Schiebedach, *das;* ~**set** *n.* Sonnenuntergang, *der;* ~**shade** *n.* Sonnenschirm, *der;* ~**shine** *n.* Sonnenschein, *der;* ~**stroke** *n.* Sonnenstich, *der;* ~**tan** *n.* [Sonnen]bräune, *die;* **get a** ~**tan** braun werden; ~**tan lotion** *n.* Sonnencreme, *die;* ~**tanned** *adj.* braun[gebrannt]; ~**tan oil** *n.* Sonnenöl, *das*

super ['su:pə(r)] *adj.* *(coll.)* super *(ugs.)*

superb [su:'pɜ:b] *adj.* einzigartig; erstklassig ⟨*Essen*⟩

supercilious [su:pə'sɪlɪəs] *adj.* hochnäsig

superficial [su:pə'fɪʃl] *adj.* oberflächlich

superfluous [sʊ'pɜ:flʊəs] *adj.* überflüssig

super: ~**glue** *n.* Sekundenkleber, *der;* ~**human** *adj.* übermenschlich
superintendent [su:pərɪnˈtendənt] *n. (Brit. Police)* Kommissar, *der/*Kommissarin, *die*
superior [su:ˈpɪərɪə(r)] 1. *adj.* **a)** *(of higher quality)* besonders gut ⟨*Restaurant, Qualität, Stoff*⟩; überlegen ⟨*Technik, Intelligenz*⟩; **he thinks he is ~ to us** er hält sich für besser als wir; **b)** *(having higher rank)* höher...; **be ~ to sb.** einen höheren Rang als jmd. haben. 2. *n.* Vorgesetzte, *der/die.* **superiority** [su:pɪərɪˈɒrɪtɪ] *n.* Überlegenheit, *die* (**to** über + *Akk.*)
superlative [su:ˈpɜ:lətɪv] 1. *adj.* **a)** unübertrefflich; **b)** *(Ling.)* **a ~ adjective/adverb** ein Adjektiv/Adverb im Superlativ. 2. *n. (Ling.)* Superlativ, *der*
super: ~**market** *n.* Supermarkt, *der;* ~**natural** *adj.* übernatürlich; ~**power** *n. (Polit.)* Supermacht, *die*
supersede [su:pəˈsi:d] *v.t.* ablösen (**by durch**)
supersonic [su:pəˈsɒnɪk] *adj.* Überschall-
superstition [su:pəˈstɪʃn] *n.* Aberglaube, *der.* **superstitious** [su:pəˈstɪʃəs] *adj.* abergläubisch
supervise [ˈsu:pəvaɪz] *v.t.* beaufsichtigen. **supervision** [su:pəˈvɪʒn] *n.* Aufsicht, *die.* **supervisor** [ˈsu:pəvaɪzə(r)] *n.* Aufseher, *der/*Aufseherin, *die*
supper [ˈsʌpə(r)] *n.* Abendessen, *das;* **have |one's| ~:** zu Abend essen. **'supper-time** *n.* Abendbrotzeit, *die;* **it's ~:** es ist Zeit zum Abendessen
supplant [səˈplɑ:nt] *v.t.* ablösen, ersetzen (**by durch**)
supple [ˈsʌpl] *adj.* geschmeidig
supplement [ˈsʌplɪmənt] 1. *n.* **a)** Ergänzung, *die* (**to** + *Gen.*); *(addition)* Zusatz, *der;* **b)** *(of book)* Nachtrag, *der;* **c)** *(to fare)* Zuschlag, *der.* 2. *v.t.* ergänzen. **supplementary** [sʌplɪˈmentərɪ] *adj.* zusätzlich; ~ **fare/charge** Zuschlag, *der*
supplier [səˈplaɪə(r)] *n. (Commerc.)* Lieferant, *der/*Lieferantin, *die*
supply [səˈplaɪ] 1. *v.t.* liefern ⟨*Waren usw.*⟩; beliefern ⟨*Kunden, Geschäft*⟩; ~ **sth. to sb.,** ~ **sb. with sth.** jmdn. mit etw. versorgen/*(Commerc.)* beliefern. 2. *n.* Vorräte *Pl.;* **military/medical supplies** militärischer/medizinischer Nachschub; ~ **and demand** *(Econ.)* Angebot und Nachfrage
support [səˈpɔ:t] 1. *v.t.* **a)** *(hold up)* stützen ⟨*Mauer, Verletzten*⟩; *(bear*

weight of) tragen; **b)** unterstützen ⟨*Politik, Verein*⟩; *(Footb.)* ~ **Spurs** Spurs-Fan sein; **c)** *(provide for)* ernähren ⟨*Familie, sich selbst*⟩; **d)** *(speak in favour of)* befürworten. 2. *n.* **a)** Unterstützung, *die;* **in** ~: zur Unterstützung; **speak in ~ of sb./sth.** jmdn. unterstützen/etw. befürworten; **b)** *(money)* Unterhalt, *der;* **c)** *(sb./sth. that ~s)* Stütze, *die.* **sup'porter** *n.* Anhänger, *der/*Anhängerin, *die;* **football ~:** Fußballfan, *der.* **sup'porting** *adj. (Cinemat., Theatre)* ~ **role** Nebenrolle, *die;* ~ **actor/actress** Schauspieler/-spielerin in einer Nebenrolle; ~ **film** Vorfilm, *der.* **supportive** [səˈpɔ:tɪv] *adj.* hilfreich; **be very ~ |to sb.|** |jmdm.| eine große Hilfe *od.* Stütze sein
suppose [səˈpəʊz] *v.t.* **a)** *(assume)* annehmen; ~ *or* **supposing |that| he ...:** angenommen, |daß| er ...; **b)** *(presume)* vermuten; **I ~ so** *(doubtfully)* ja, vermutlich; *(more confidently)* ich glaube schon; **c) be ~d to do/be sth.** *(be generally believed to do/be sth.)* etw. tun/sein sollen; **d)** *(allow)* **you are not ~d to do that** das darfst du nicht; **I'm not ~d to be here** ich dürfte eigentlich gar nicht hier sein. **supposedly** [səˈpəʊzɪdlɪ] *adv.* angeblich
supposition [sʌpəˈzɪʃn] *n.* Annahme, *die;* Vermutung, *die*
suppress [səˈpres] *v.t.* unterdrücken. **suppression** [səˈpreʃn] *n.* Unterdrückung, *die*
supremacy [su:ˈpreməsɪ] *n.* **a)** *(supreme authority)* Souveränität, *die;* **b)** *(superiority)* Überlegenheit, *die*
supreme [su:ˈpri:m] *adj.* höchst...
surcharge [ˈsɜ:tʃɑ:dʒ] *n.* Zuschlag, *der*
sure [ʃʊə(r)] 1. *adj.* sicher; **be ~ of sth.** sich *(Dat.)* einer Sache *(Gen.)* sicher sein; ~ **of oneself** selbstsicher; **don't be too ~:** da wäre ich mir nicht so sicher; **there is ~ to be a garage** es gibt bestimmt eine Tankstelle; **don't worry, it's ~ to turn out well** keine Sorge, es wird schon alles gutgehen; **for ~** *(coll.)* auf jeden Fall; **make ~ |of sth.|** sich |einer Sache| vergewissern; **make** *or* **be ~ you do it, be ~ to do it** *(do not fail to do it)* sieh zu, daß du es tust; *(do not forget)* vergiß nicht, es zu tun; **a ~ winner** ein todsicherer Tip *(ugs.).* 2. *adv.* ~ **enough** tatsächlich. 3. *int.* ~!, ~ **thing!** *(Amer.)* na klar! *(ugs.).*
sure-footed [ˈʃʊəfʊtɪd] *adj.* trittsi-

cher. **'surely 1.** *adv.* **a)** *as sentence-modifier* doch; ~ **we've met before?** wir kennen uns doch, oder?; **b)** *(steadily)* sicher; **slowly but ~:** langsam, aber sicher; **c)** *(certainly)* sicherlich. **2.** *int. (Amer.)* natürlich

surf [sɜːf] *n.* Brandung, *die*

surface ['sɜːfɪs] **1.** *n.* Oberfläche, *die;* **outer ~:** Außenfläche, *die;* **the earth's ~:** die Erdoberfläche; **on the ~:** an der Oberfläche; *(fig.)* oberflächlich betrachtet. **2.** *v. i.* auftauchen; *(fig.)* hochkommen. **'surface area** *n.* Oberfläche, *die.* **'surface mail** *n.* gewöhnliche Post

'surfboard *n.* Surfbrett, *das*

surfeit ['sɜːfɪt] *n.* Übermaß, *das*

'surfer *n.* Surfer, *der*/Surferin, *die*

'surfing *n.* Surfen, *das*

surge [sɜːdʒ] *v. i.* ⟨*Wellen:*⟩ branden; **the crowd ~d forward** die Menschenmenge drängte sich nach vorn

surgeon ['sɜːdʒən] *n.* Chirurg, *der*/Chirurgin, *die*

surgery ['sɜːdʒərɪ] *n.* **a)** Chirurgie, *die;* **undergo ~:** sich einer Operation *(Dat.)* unterziehen; **b)** *(Brit.: place)* Praxis, *die;* **doctor's/dental ~:** Arzt-/Zahnarztpraxis, *die;* **c)** *(Brit.: time)* Sprechstunde, *die*

surgical ['sɜːdʒɪkl] *adj.* chirurgisch; ~ **treatment** Operation, *die*/Operationen

surly ['sɜːlɪ] *adj.* mürrisch

surmise [sə'maɪz] **1.** *n.* Vermutung, *die.* **2.** *v. t.* mutmaßen

surmount [sə'maʊnt] *v. t.* überwinden

surname ['sɜːneɪm] *n.* Nachname, *der;* Zuname, *der*

surpass [sə'pɑːs] *v. t.* übertreffen; ~ **oneself** sich selbst übertreffen

surplus ['sɜːpləs] **1.** *n.* Überschuß, *der* (of an + *Dat.*). **2.** *adj.* überschüssig; **be ~ to sb.'s requirements** von jmdm. nicht benötigt werden

surprise [sə'praɪz] **1.** *n.* **a)** Überraschung, *die;* **take sb. by ~:** jmdn. überrumpeln; **to my great ~, much to my ~:** zu meiner großen Überraschung; **it came as a ~ to us** es war für uns eine Überraschung; **b)** *attrib.* überraschend, unerwartet ⟨*Besuch*⟩; **a ~ attack** ein Überraschungsangriff. **2.** *v. t.* überraschen; überrumpeln ⟨*Feind*⟩; **I shouldn't be ~d if ...:** es würde mich nicht wundern, wenn ...; **be ~d at sb./sth.** sich über jmdn./etw. wundern. **surprising** [sə'praɪzɪŋ] *adj.* überraschend

surreal [sə'rɪːəl] *adj.* surrealistisch

surrealism [sə'rɪːəlɪzm] *n.* Surrealismus, *der*

surrender [sə'rendə(r)] **1.** *n. (to enemy)* Kapitulation, *die; (of possession)* Aufgabe, *die.* **2.** *v. i.* kapitulieren. **3.** *v. t.* aufgeben

surreptitious [sʌrəp'tɪʃəs] *adj.* heimlich; verstohlen ⟨*Blick*⟩

surrogate ['sʌrəgət] *n.* Ersatz, *der*

surround [sə'raʊnd] *v. t.* **a)** *(come or be all round)* umringen; ⟨*Truppen, Heer:*⟩ umzingeln ⟨*Stadt, Feind*⟩; **b)** *(encircle)* umgeben; **be ~ed by** *or* **with sth.** von etw. umgeben sein. **sur'rounding** *adj.* umliegend; **the ~ countryside** die [Landschaft in der] Umgebung. **sur'roundings** *n. pl.* Umgebung, *die*

surveillance [sə'veɪləns] *n.* Überwachung, *die;* **be under ~:** überwacht werden

survey 1. [sə'veɪ] *v. t.* betrachten; überblicken ⟨*Landschaft*⟩; inspizieren ⟨*Gebäude*⟩; bewerten ⟨*Situation*⟩. **2.** ['sɜːveɪ] *n.* Überblick, *der* (of über + *Akk.*); *(poll)* Umfrage, *die; (Surv.)* Vermessung, *die.* **surveyor** [sə'veɪə(r)] *n. (of building)* Gutachter, *der*/Gutachterin, *die; (of land)* Landvermesser, *der*/-vermesserin, *die*

survival [sə'vaɪvl] *n.* Überleben, *das;* **fight for ~:** Existenzkampf, *der*

survive [sə'vaɪv] **1.** *v. t.* überleben. **2.** *v. i.* ⟨*Person:*⟩ überleben; ⟨*Schriften, Traditionen:*⟩ erhalten bleiben. **survivor** [sə'vaɪvə(r)] *n.* Überlebende, *der/die*

susceptible [sə'septɪbl] *adj.* empfänglich **(to für);** *(to illness)* anfällig **(to für)**

suspect 1. [sə'spekt] *v. t.* **a)** *(imagine to be likely)* vermuten; ~ **the worst** das Schlimmste befürchten; ~ **sb. to be sth.,** ~ **that sb. is sth.** glauben *od.* vermuten, daß jmd. etw. ist; **b)** *(mentally accuse)* verdächtigen; ~ **sb. of sth./of doing sth.** jmdn. einer Sache verdächtigen/jmdn. verdächtigen, etw. zu tun. **2.** ['sʌspekt] *adj.* fragwürdig; verdächtig ⟨*Stoff, Paket*⟩. **3.** ['sʌspekt] *n.* Verdächtige, *der/die*

suspend [sə'spend] *v. t.* **a)** *(hang up)* [auf]hängen; **b)** *(stop)* suspendieren; **c)** *(from work)* ausschließen **(from** von); sperren ⟨*Sportler*⟩. **suspended 'sentence** *n. (Law)* Strafe mit Bewährung

suspender belt [sə'spendə belt] *n. (Brit.)* Strumpfbandgürtel, *der*

suspenders [sə'spendəz] *n. pl.* **a)**
(Brit.: for stockings) Strumpfbänder;
b) *(Amer.: for trousers)* Hosenträger
suspense [sə'spens] *n.* Spannung,
die; **keep sb. in ~:** jmdn. auf die Folter
spannen. **suspension** [sə'spenʃn] *n.*
(Motor Veh.) Federung, *die.* **su'spen-
sion bridge** *n.* Hängebrücke, *die*
suspicion [sə'spɪʃn] *n.* **a)** *(uneasy feeling)* Mißtrauen, *das* **(of** gegenüber);
(unconfirmed belief) Verdacht, *der;*
have a ~ that ...: den Verdacht haben,
daß ...; **b)** *(suspecting)* Verdacht, *der*
(of auf + *Akk.*); **on ~ of murder** wegen Mordverdachts; **be under ~:** ver-
dächtigt werden
suspicious [sə'spɪʃəs] *adj.* **a)** *(tending
to suspect)* mißtrauisch **(of** gegenüber); **be ~ of sb./sth.** jmdm./einer
Sache mißtrauen; **b)** *(arousing suspicion)* verdächtig
sustain [sə'steɪn] *v. t.* **a)** *(support)* tragen *‹Gewicht›; (fig.)* aufrechterhalten;
b) erleiden *‹Verlust, Verletzung›*
sustenance ['sʌstɪnəns] *n.* Nahrung,
die
SW *abbr.* **a)** south-west SW; **b)** *(Radio)*
short wave KW
swab [swɒb] *n. (Med.: pad)* Tupfer,
der
swagger ['swægə(r)] *v. i.* großspurig
stolzieren
¹swallow ['swɒləʊ] **1.** *v. t.* schlucken;
(by mistake) verschlucken. **2.** *v. i.*
schlucken. **3.** *n.* Schluck, *der.* **swallow 'up** *v. t.* verschlucken
²swallow *n.* Schwalbe, *die*
swam *see* swim 1
swamp [swɒmp] **1.** *n.* Sumpf, *der.* **2.**
v. t. überschwemmen. **'swampy** *adj.*
sumpfig
swan [swɒn] *n.* Schwan, *der*
swap [swɒp] **1.** *v. t.,* -pp- tauschen **(for**
gegen). **2.** *v. i.,* -pp- tauschen. **3.** *n.*
Tausch, *der*
swarm [swɔ:m] **1.** *n.* Schwarm, *der.*
2. *v. i.* schwärmen; *(teem)* wimmeln
(with von)
swarthy ['swɔ:ðɪ] *adj.* dunkel
swastika ['swɒstɪkə] *n.* Hakenkreuz,
das
swat [swɒt] *v. t.,* -tt- totschlagen
sway [sweɪ] **1.** *v. i.* [hin und her]
schwanken; *(gently)* sich wiegen. **2.**
v. t. **a)** wiegen; **b)** *(influence)* beeinflussen. **3.** *n. (fig.)* Herrschaft, *die;*
hold ~ over sb. über jmdn. herrschen
swear [sweə(r)] **1.** *v. t.,* swore [swɔ:(r)],
sworn [swɔ:n] schwören *‹Eid usw.›*. **2.**

v. i., swore, sworn **a)** fluchen; **b)** ~ **to**
sth. etw. beschwören. **'swear at** *v. t.*
beschimpfen. **'swear by** *v. t. (coll.)*
schwören auf *(+ Akk.)*
'swear-word *n.* Kraftausdruck, *der*
sweat [swet] **1.** *n.* Schweiß, *der.* **2.** *v. i.*
schwitzen
sweater ['swetə(r)] *n.* Pullover, *der*
'sweaty *adj.* schweißig
Swede [swi:d] *n.* Schwede, *der/*
Schwedin, *die*
swede *n.* Kohlrübe, *die*
Sweden ['swi:dn] *pr. n.* Schweden
(das)
Swedish ['swi:dɪʃ] **1.** *adj.* schwedisch; **sb. is ~:** jmd. ist Schwede/
Schwedin. **2.** *n.* Schwedisch, *das; see
also* **English 2 a**
sweep [swi:p] **1.** *v. t.,* swept [swept] **a)**
fegen; kehren; **b)** ~ **the country**
‹Epidemie, Mode:› das Land überrol-
len. **2.** *v. i.,* swept **a)** fegen; kehren; **b)**
(go fast) *‹Person, Auto:›* rauschen;
‹Wind usw.:› fegen. **3.** *n.* **a) give sth. a**
~: etw. fegen *od.* kehren; **b)** *(curve)*
Bogen, *der.* **sweep 'up** *v. t.* zusam-
menfegen; zusammenkehren
'sweeping *adj.* pauschal; weitreichend *‹Einsparung›;* umwälzend *‹Veränderung›*
sweet [swi:t] **1.** *adj.* süß; reizend
‹Wesen, Gesicht, Mädchen›; **have a ~**
tooth gern Süßes mögen; **how ~ of**
you! wie nett *od.* lieb von dir! **2.** *n.*
(Brit.) **a)** *(candy)* Bonbon, *das od. der;*
b) *(dessert)* Nachtisch, *der.* **sweet-
and-'sour** *attrib. adj.* süßsauer.
'sweet corn *n.* Zuckermais, *der*
sweeten ['swi:tn] *v. t.* süßen. **'sweetener** *n.* Süßstoff, *der*
'sweetheart *n.* Schatz, *der*
'sweetness *n.* Süße, *die*
sweet: ~ **'pea** *n.* Wicke, *die;* ~**-shop**
n. (Brit.) Süßwarengeschäft, *das*
swell [swel] **1.** *v. t.,* swelled, swollen
['swəʊlən] *or* swelled anschwellen lassen. **2.** *v. i.,* swelled, swollen *or* swelled
a) *(expand)* *‹Körperteil:›* anschwellen;
‹Segel:› sich blähen; *‹Material:›* auf-
quellen; **b)** *‹Anzahl:›* zunehmen.
'swelling *n.* Schwellung, *die*
swelter ['sweltə(r)] *v. i.* ~**ing** glühend
heiß *‹Tag, Wetter›;* ~**ing heat** Bruthit-
ze, *die*
swept *see* sweep 1, 2
swerve [swɜ:v] **1.** *v. i.* einen Bogen
machen; ~ **to the right/left** nach
rechts/links [aus]schwenken. **2.** *n.* Bogen, *der*

swift [swɪft] 1. *adj.* schnell. 2. *n.* Mauersegler, *der.* '**swiftly** *adv.* schnell
swig [swɪg] *(coll.)* Schluck, *der*
swill [swɪl] *v. t.* ~ |out| [aus]spülen
swim [swɪm] 1. *v. i.,* -mm-, swam [swæm], swum [swʌm] schwimmen; **my head was** ~ming mir war schwindelig. 2. *n.* **have a/go for a** ~: schwimmen/schwimmen gehen. '**swimmer** *n.* Schwimmer, *der*/Schwimmerin, *die;* **be a good/poor** ~: gut/schlecht schwimmen können. '**swimming** *n.* Schwimmen, *das*
swimming: ~-baths *n. pl.* Schwimmbad, *das;* ~-costume *n.* Badeanzug, *der;* ~-pool *n.* Schwimmbecken, *das; (building)* Schwimmbad, *das;* ~-trunks *n. pl.* Badehose, *die*
'**swim-suit** *n.* Badeanzug, *der*
swindle ['swɪndl] 1. *v. t.* betrügen; ~ **sb. out of sth.** jmdn. um etw. betrügen. 2. *n.* Schwindel, *der;* Betrug, *der.* **swindler** ['swɪndlə(r)] *n.* Schwindler, *der*/Schwindlerin, *die*
swine [swaɪn] *n.* Schwein, *das*
swing [swɪŋ] 1. *n.* a) Schaukel, *die;* b) *(~ing)* Schaukeln, *das;* **in full** ~ *(fig.)* in vollem Gang[e]. 2. *v. i.,* swung [swʌŋ] a) schwingen; *(in wind)* schaukeln; b) *(go in sweeping curve)* schwenken. 3. *v. t.,* swung schwingen. **swing-'door** *n.* Pendeltür, *die*
swipe [swaɪp] *(coll.) v. t.* a) *(hit)* knallen *(ugs.);* b) *(sl.: steal)* klauen *(ugs.)*
swirl [swɜːl] 1. *v. i.* wirbeln. 2. *v. t.* umherwirbeln. 3. *n.* Spirale, *die*
swish [swɪʃ] 1. *v. i.* zischen. 2. *n.* Zischen, *das.* 3. *adj. (coll.)* schick *(ugs.)*
Swiss [swɪs] 1. *adj.* Schweizer; schweizerisch; **sb. is** ~: jmd. ist Schweizer/Schweizerin. 2. *n.* Schweizer, *der*/Schweizerin, *die;* **the** ~ *pl.* die Schweizer. **Swiss 'roll** *n.* Biskuitrolle, *die*
switch [swɪtʃ] 1. *n.* a) *(esp. Electr.)* Schalter, *der;* b) *(change)* Wechsel, *der.* 2. *v. t.* a) *(change)* ~ **sth. |over| to sth.** etw. auf etw. *(Akk.)* umstellen *od. (Electr.)* umschalten; b) *(exchange)* tauschen. 3. *v. i.* wechseln; ~ |over| to **sth.** auf etw. *(Akk.)* umstellen *od. (Electr.)* umschalten. **switch 'off** *v. t. & i.* ausschalten; *(also fig. coll.)* abschalten. **switch 'on** 1. *v. t.* einschalten; anschalten. 2. *v. i.* sich anschalten
switch: ~back *n.* Achterbahn, *die;* ~board *n.* [Telefon]zentrale, *die*

Switzerland ['swɪtsələnd] *pr. n.* die Schweiz
swivel ['swɪvl] 1. *v. i.,* -ll- sich drehen. 2. *v. t.,* -ll- drehen. '**swivel chair** *n.* Drehstuhl, *der*
swollen ['swəʊlən] 1. *see* swell. 2. *adj.* geschwollen; angeschwollen ‹Fluß›
swoon [swuːn] *(literary) v. i.* ohnmächtig werden
swoop [swuːp] 1. *n.* a) Sturzflug, *der;* b) *(coll.: raid)* Razzia, *die.* 2. *v. i.* herabstoßen; ~ **on sb.** sich auf jmdn. stürzen
sword [sɔːd] *n.* Schwert, *das.* '**swordfish** *n.* Schwertfisch, *der*
swore, sworn *see* swear
swot [swɒt] *(Brit. coll.)* 1. *n.* Streber, *der*/Streberin, *die.* 2. *v. i.,* -tt- büffeln *(ugs.)*
swum *see* swim 1
swung *see* swing 2, 3
sycamore ['sɪkəmɔː(r)] *n.* Bergahorn, *der*
sycophant ['sɪkəfænt] *n.* Kriecher, *der*
syllable ['sɪləbl] *n.* Silbe, *die*
syllabus ['sɪləbəs] *n.* Lehrplan, *der; (for exam)* Studienplan, *der*
symbol ['sɪmbl] *n.* Symbol, *das* (of für)
symbolic [sɪm'bɒlɪk], **symbolical** [sɪm'bɒlɪkl] *adj.* symbolisch. **symbolism** ['sɪmbəlɪzm] *n.* Symbolik, *die.* **symbolize** ['sɪmbəlaɪz] *v. t.* symbolisieren
symmetrical [sɪ'metrɪkl] *adj.* symmetrisch
symmetry ['sɪmɪtrɪ] *n.* Symmetrie, *die*
sympathetic [sɪmpə'θetɪk] *adj.* mitfühlend
sympathize ['sɪmpəθaɪz] *v. i.* a) ~ **with sb.** mit jmdm. [mit]fühlen; b) ~ **with** *(understand)* Verständnis haben für
sympathy ['sɪmpəθɪ] *n.* Mitgefühl, *das;* **in deepest** ~: mit aufrichtigem Beileid
symphonic [sɪm'fɒnɪk] *adj.* sinfonisch
symphony ['sɪmfənɪ] *n.* Sinfonie, *die*
symptom ['sɪmptəm] *n.* Symptom, *das.* **symptomatic** [sɪmptə'mætɪk] *adj.* symptomatisch (of für)
synagogue *(Amer.:* **synagog**) ['sɪnəgɒg] *n.* Synagoge, *die*
synchromesh ['sɪŋkrəmeʃ] *n. (Motor Veh.)* Synchrongetriebe, *das*
synchronize ['sɪŋkrənaɪz] *v. t.* synchronisieren; gleichstellen ‹Uhren›
syndicate ['sɪndɪkət] *n.* Syndikat, *das*

syndrome ['sındrəʊm] *n.* Syndrom, *das*

synonym ['sınənım] *n.* Synonym, *das.* **synonymous** [sı'nɒnıməs] *adj.* **a)** *(Ling.)* synonym (with mit); **b)** ~ with *(fig.)* gleichbedeutend mit

synopsis [sı'nɒpsıs] *n., pl.* **synopses** [sı'nɒpsiːz] Inhaltsangabe, *die*

syntactic [sın'tæktık] *adj.* syntaktisch

syntax ['sıntæks] *n.* Syntax, *die*

synthesis ['sınθısıs] *n., pl.* **syntheses** ['sınθısiːz] Synthese, *die*

synthesize ['sınθısaız] *v.t.* zur Synthese bringen; *(Chem.)* synthetisieren. **synthesizer** ['sınθısaızə(r)] *n. (Mus.)* Synthesizer, *der*

synthetic [sın'θetık] *adj.* synthetisch

syphilis ['sıfılıs] *n.* Syphilis, *die*

syphon *see* **siphon**

Syria ['sırıə] *pr. n.* Syrien *(das)*

syringe [sı'rındʒ] **1.** *n.* Spritze, *die.* **2.** *v.t.* spritzen; ausspritzen ⟨*Ohr*⟩

syrup ['sırəp] *n.* Sirup, *der*

system ['sıstəm] *n.* System, *das.* **systematic** [sıstə'mætık] *adj.,* **systematically** [sıstə'mætıkəlı] *adv.* systematisch. **systematize** ['sıstəmətaız] *v.t.* systematisieren. **'systems analyst** *n.* Systemanalytiker, *der/* -analytikerin, *die*

T

T, t [tiː] *n.* T, t, *das;* **to a T** ganz genau; **T-junction** Einmündung, *die;* **T-bone steak** T-bone-Steak, *das;* **T-shirt** T-shirt, *das*

ta [tɑː] *int. (Brit. coll.)* danke

tab [tæb] *n.* **a)** *(projecting flap)* Zunge, *die; (on clothing)* Etikett, *das; (with name)* Namensschild, *das;* **b) pick up the ~** *(Amer. coll.)* die Zeche bezahlen; **c) keep ~s** *or* **a ~ on** *(watch)* [genau] beobachten

tabby ['tæbı] *n.* **~ [cat]** Tigerkatze, *die*

table ['teıbl] **1.** *n.* **a)** *(list)* Tisch, *der;* **b)** *(list)* Tabelle, *die;* **~ of contents** Inhaltsverzeichnis, *das.* **2.** *v.t.* einbringen

tableau ['tæbləʊ] *n., pl.* **~x** ['tæbləʊz] Tableau, *das*

table: ~-cloth *n.* Tischdecke, *die;* **~ manners** *n. pl.* Tischmanieren *Pl.;* **~-mat** *n.* Set, *das;* **~ salt** *n.* Tafelsalz, *das;* **~spoon** *n.* Servierlöffel, *der;* **~spoonful** *n.* ≈ Eßlöffel[voll], *der*

tablet ['tæblıt] *n.* **a)** Tablette, *die;* **b)** *(of soap)* Stück, *das*

table: ~ tennis *n.* Tischtennis, *das;* **~ tennis bat** Tischtennisschläger, *der;* **~ wine** *n.* Tischwein, *der*

tabloid ['tæblɔıd] *n.* Boulevardzeitung, *die*

taboo, tabu [tə'buː] **1.** *n.* Tabu, *das.* **2.** *adj.* Tabu⟨*wort*⟩; **be ~:** tabu sein

tabulate ['tæbjʊleıt] *v.t.* tabellarisch darstellen. **tabulator** ['tæbjʊleıtə(r)] *n.* Tabulator, *der*

tacit ['tæsıt] *adj.,* **'tacitly** *adv.* stillschweigend

taciturn ['tæsıtɜːn] *adj.* schweigsam; wortkarg

tack [tæk] **1.** *n.* **a)** *(nail)* kleiner Nagel; **b)** *(stitch)* Heftstich, *der;* **c)** *(Naut., also fig.)* Kurs, *der.* **2.** *v.t.* **a)** *(nail)* festnageln; **b)** *(stitch)* heften. **3.** *v.i. (Naut.)* kreuzen

tackle ['tækl] **1.** *v.t.* **a)** angehen ⟨*Problem usw.*⟩; **~ sb. about/on/over sth.** jmdn. auf etw. *(Akk.)* ansprechen; *(ask for sth.)* jmdn. um etw. angehen; **b)** *(Sport)* angreifen ⟨*Spieler*⟩; *(Amer. Footb.; Rugby)* fassen. **2.** *n.* **a)** *(equipment)* Ausrüstung, *die;* **b)** *(Sport)* Angriff, *der; (sliding ~)* Tackling, *das; (Amer. Footb.; Rugby)* Fassen und Halten

tacky ['tækı] *adj.* klebrig

tact [tækt] *n.* Takt, *der;* **he has no ~:** er hat kein Taktgefühl. **tactful** ['tæktfl] *adj.,* **'tactfully** *adv.* taktvoll

tactical ['tæktıkl] *adj.* taktisch

tactics ['tæktıks] *n. pl.* Taktik, *die*

'tactless *adj.,* **'tactlessly** *adv.* taktlos

tadpole ['tædpəʊl] *n.* Kaulquappe, *die*

¹tag [tæg] *n.* Schild, *das.* **tag a'long** *v.i.* mitkommen

²tag *n. (game)* Fangen, *das*

tail [teıl] **1.** *n.* **a)** Schwanz, *der;* **b)** *in pl. (on coin)* **~s** [it is] Zahl. **2.** *v.t. (sl.: follow)* beschatten. **tail 'back** *v.i.* sich stauen. **tail 'off** *v.i.* **a)** zurückgehen; **b)** *(into silence)* verstummen

tail: ~back *n. (Brit.)* Rückstau, *der;* **~-end** *n.* Ende, *das;* **~-gate** *n. (Motor Veh.)* Heckklappe, *die;* **~-light** *n.* Rücklicht, *das*

tailor ['teılə(r)] *n.* Schneider, *der/*

Schneiderin, *die.* **'tailor-made** *adj.* maßgeschneidert

'tail wind *n.* Rückenwind, *der*

taint [teɪnt] *v. t.* verderben; **be ~ed with sth.** mit etw. behaftet sein *(geh.)*

Taiwan [taɪˈwɑːn] *pr. n.* Taiwan *(das)*

take [teɪk] **1.** *v. t.,* took [tʊk], taken ['teɪkn] **a)** *(get hold of, grasp, seize)* nehmen; **b)** *(capture)* einnehmen ⟨*Stadt, Festung*⟩; machen ⟨*Gefangenen*⟩; **c)** *(gain, earn)* ⟨*Laden*:⟩ einbringen; ⟨*Person*:⟩ einnehmen; ⟨*Film, Stück*:⟩ einspielen; *(win)* gewinnen ⟨*Satz, Spiel, Preis, Titel*⟩; **d)** *(~ away with one)* mitnehmen; *(steal)* mitnehmen *(verhüll.)*; ~ **place** stattfinden; *(spontaneously)* sich ereignen; ⟨*Wandlung*:⟩ sich vollziehen; **e)** *(avail oneself of, use)* nehmen; machen ⟨*Pause, Ferien, Nickerchen*⟩; ~ **the opportunity to do/of doing sth.** die Gelegenheit dazu benutzen, etw. zu tun; **f)** *(carry, guide, convey)* bringen; ~ **sb. to visit sb.** jmdn. zu Besuch bei jmdm. mitnehmen; ~ **home** mit nach Hause nehmen; *(earn)* nach Hause bringen ⟨*Geld*⟩; *(accompany)* nach Hause bringen; **g)** *(remove)* nehmen; *(deduct)* abziehen; ~ **sth./sb. from sb.** jmdm. etw./jmdm. wegnehmen; **h)** *(make)* machen ⟨*Foto, Kopie*⟩; *(photograph)* aufnehmen; aufnehmen ⟨*Brief, Diktat*⟩; machen ⟨*Prüfung, Sprung, Spaziergang, Reise*⟩; ablegen ⟨*Gelübde, Eid*⟩; treffen ⟨*Entscheidung*⟩; **i)** *(conduct)* halten ⟨*Gottesdienst, Unterricht*⟩; **Ms X ~s us for maths** in Mathe haben wir Frau X; **j)** *(eat, drink)* nehmen ⟨*Zucker, Milch, Tabletten, Überdosis*⟩; trinken ⟨*Tee, Kaffee, Kognak usw.*⟩; **k)** *(need, require)* brauchen ⟨*Platz, Zeit*⟩; haben ⟨*Objekt, Plural-s*⟩; gebraucht werden mit ⟨*Kasus*⟩; **sth. ~s an hour/a year/all day** etw. dauert eine Stunde/ein Jahr/einen ganzen Tag; **l)** *(ascertain and record)* notieren ⟨*Namen, Adresse, Autonummer usw.*⟩; fühlen ⟨*Puls*⟩; messen ⟨*Temperatur, Größe usw.*⟩; **m)** *(assume)* ~ **it [that]** ...: annehmen, daß ...; ~ **sb./sth. for/to be sth.** jmdn./etw. für etw. halten; **n)** *(react to)* aufnehmen; ~ **sth. well/badly** etw. gut/nur schwer verkraften; ~ **sth. calmly** *or* **coolly** etw. gelassen [auf]nehmen; **o)** *(accept)* annehmen; **p)** *(adopt, choose)* ergreifen ⟨*Maßnahmen*⟩; unternehmen ⟨*Schritte*⟩; ~ **the wrong road** die falsche Straße nehmen; **q)** **be ~n ill** krank werden; **r)** ~

sth. to bits *or* **pieces** etw. auseinandernehmen. **2.** *v. i.,* **took, taken a)** ⟨*Transplantat*:⟩ vom Körper angenommen werden; ⟨*Sämling, Pflanze*:⟩ angehen; **b)** *(detract)* ~ **from sth.** etw. schmälern. **'take after** *v. t. (resemble)* jmdm. ähnlich sein; ⟨~ **as one's example**⟩ es jmdm. gleichtun. **take a'way** *v. t.* **a)** *(remove)* wegnehmen; *(to a distance)* mitnehmen; ~ **sth. away from sb.** jmdm. etw. abnehmen; **to ~ away** ⟨*Pizza, Snack usw.*⟩ zum Mitnehmen; **b)** *(Math.: deduct)* abziehen. **take a'way from** *v. t.* schmälern. **take 'back** *v. t.* zurücknehmen; *(return)* zurückbringen. **take 'down** *v. t.* **a)** *(carry or lead down)* hinunterbringen; **b)** abnehmen ⟨*Bild, Ankündigung, Weihnachtsschmuck*⟩; herunterziehen ⟨*Hose*⟩; ~ **sth. down from a shelf** etw. von einem Regal herunternehmen; **c)** *(write down)* aufnehmen. **take 'in** *v. t.* **a)** hineinbringen; *(bring indoors)* hereinholen; **b)** enger machen ⟨*Kleidungsstück*⟩; **c)** *(understand)* begreifen; **d)** *(cheat)* hereinlegen *(ugs.)*; *(deceive)* täuschen. **take 'off 1.** *v. t.* **a)** abnehmen ⟨*Deckel, Hut, Tischtuch, Verband*⟩; abziehen ⟨*Kissenbezug*⟩; ausziehen ⟨*Schuhe, Handschuhe*⟩; ablegen ⟨*Mantel, Schmuck*⟩; **b)** *(deduct)* abziehen; ~ **sth. off etw.** von etw. abziehen; **c)** ~ **a day** *etc.* **off** sich *(Dat.)* einen Tag *usw.* frei nehmen *(ugs.)*; **d)** *(mimic)* nachahmen. **2.** *v. i. (Aeronaut.)* starten. **take 'on** *v. t.* **a)** *(undertake)* übernehmen; auf sich *(Akk.)* nehmen ⟨*Bürde*⟩; **b)** *(employ)* einstellen; **c)** *(as opponent)* es aufnehmen mit; *(Sport: meet)* antreten gegen. **take 'out** *v. t.* **a)** *(remove)* herausnehmen; ziehen ⟨*Zahn*⟩; ~ **sth. out of sth.** etw. aus etw. [heraus]nehmen; **b)** *(withdraw)* abheben ⟨*Geld*⟩; **c)** *(go out with)* ~ **sb. out** mit jmdm. ausgehen; ~ **sb. out to** *or* **for lunch** jmdn. zum Mittagessen einladen; **d)** *(get issued)* abschließen ⟨*Versicherung*⟩; ausleihen ⟨*Bücher*⟩; ~ **out a subscription to sth.** etw. abonnieren; **e)** ~ **it out on sb.** seine Wut an jmdm. auslassen. **take 'over 1.** *v. t.* übernehmen. **2.** *v. i.* übernehmen; ⟨*Manager, Firmenleiter*:⟩ die Geschäfte übernehmen; ⟨*Regierung, Präsident*:⟩ die Amtsgeschäfte übernehmen; ~ **over from sb.** jmdn. ersetzen; *(temporarily)* jmdn. vertreten. **'take to** *v. t.* **a)** *(get into habit of)* ~ **to doing sth.** es sich *(Dat.)* angewöhnen, etw. zu tun; **b)** *(like)* sich hingezogen fühlen zu ‹*Person*›; sich erwärmen für ‹*Sache*›. **take 'up 1.** *v.t.* **a)** *(lift up)* hochheben; *(pick up)* aufheben; herausreißen ‹*Dielen*›; aufreißen ‹*Straße*›; **b)** *(carry or lead up)* hinaufbringen; **c)** in Anspruch nehmen ‹*Zeit*›; brauchen/*(undesirably)* wegnehmen ‹*Platz*›; **d)** *(start)* ergreifen ‹*Beruf*›; anfangen ‹*Tennis, Schach, Gitarre usw.*›; aufnehmen ‹*Arbeit, Kampf*›; antreten ‹*Stelle*›; ~ **up a hobby** sich *(Dat.)* ein Hobby zulegen; **e)** *(pursue further)* ~ **sth. up with sb.** sich in einer Sache an jmdn. wenden. **2.** *v.i.* ~ **up with sb.** *(coll.)* sich mit jmdm. einlassen

'take-away *n.* *(meal)* Essen zum Mitnehmen; *(restaurant)* Restaurant mit Straßenverkauf

taken *see* **take**

take: **~-off** *n.* **a)** *(Aeronaut.)* Start, *der;* **b)** *(coll.: caricature)* Parodie, *die;* **~-over** *n.* Übernahme, *die*

takings ['teɪkɪŋz] *n. pl.* Einnahmen

talcum ['tælkəm] *n.* ~ **[powder]** Körperpuder, *der*

tale [teɪl] *n.* Erzählung, *die;* Geschichte, *die* (of von, about über + *Akk.*)

talent ['tælənt] *n.* Talent, *das;* **have [great/no** *etc.*] ~ **[for sth.]** [viel/kein *usw.*] Talent [zu *od.* für etw.] haben.

'talented *adj.* talentiert

talk [tɔːk] **1.** *n.* **a)** *(discussion)* Gespräch, *das;* **have a** ~ **[with sb.] [about sth.]** [mit jmdm.] [über etw. *(Akk.)*] sprechen; **have** *or* **hold** ~**s [with sb.]** [mit jmdm.] Gespräche führen; **b)** *(speech, lecture)* Vortrag, *der.* **2.** *v.i.* sprechen (**with, to** mit); *(lecture)* sprechen; *(converse)* sich unterhalten; *(have* ~*s)* Gespräche führen; *(gossip)* reden; ~ **on the phone** telefonieren. **3.** *v.t.* reden; ~ **sb. into/out of sth.** jmdn. zu etw. überreden/jmdm. etw. ausreden. **talk 'over** *v.t.* besprechen. **talk 'round** *v.t.* ~ **sb. round** jmdn. überreden

talkative ['tɔːkətɪv] *adj.* gesprächig

talking: ~ **point** *n.* Gesprächsthema, *das;* **~-to** *n.* *(coll.)* Standpauke, *die (ugs.)*

tall [tɔːl] *adj.* hoch; groß ‹*Person, Tier*›; **that's a** ~ **order** das ist ziemlich viel verlangt; ~ **story** unglaubliche Geschichte

tally ['tælɪ] **1.** *n.* **keep a** ~ **of sth.** über etw. *(Akk.)* Buch führen. **2.** *v.i.* übereinstimmen

talon ['tælən] *n.* Klaue, *die*

tambourine [tæmbəˈriːn] *n.* Tamburin, *das*

tame [teɪm] **1.** *adj.* zahm; *(fig.: spiritless)* lahm *(ugs.).* **2.** *v.t.* zähmen

tamper ['tæmpə(r)] *v.i.* ~ **with** sich *(Dat.)* zu schaffen machen an (+ *Dat.*)

tampon ['tæmpɒn] *n.* Tampon, *der*

tan [tæn] **1.** *v.t.,* **-nn-** gerben ‹*Tierhaut, Fell*›. **2.** *v.i.,* **-nn-** braun werden. **3.** *n.* **a)** *(colour)* Gelbbraun, *das;* **b)** *(sun-~)* Bräune, *die;* **have/get a** ~: braun sein/werden. **4.** *adj.* gelbbraun

tandem ['tændəm] *n.* ~ **[bicycle]** Tandem, *das*

tang [tæŋ] *n.* *(taste)* Geschmack, *der;* *(smell)* Geruch, *der*

tangent ['tændʒənt] *n.* Tangente, *die;* **go off at a** ~ *(fig.)* plötzlich vom Thema abschweifen

tangible ['tændʒɪbl] *adj.* greifbar; spürbar ‹*Unterschied, Verbesserung*›; handfest ‹*Beweis*›

tangle ['tæŋgl] **1.** *n.* Gewirr, *das;* *(in hair)* Verfilzung, *die.* **2.** *v.t.* verheddern *(ugs.);* verfilzen ‹*Haar*›. **tangle 'up** *v.t.* verheddern *(ugs.)*

tango ['tæŋgəʊ] *n., pl.* ~**s** Tango, *der*

tank [tæŋk] *n.* **a)** Tank, *der;* **b)** *(Mil.)* Panzer, *der*

tankard ['tæŋkəd] *n.* Krug, *der*

tanker ['tæŋkə(r)] *n.* *(ship)* Tanker, *der;* *(vehicle)* Tank[last]wagen, *der*

tanned [tænd] *adj.* braungebrannt

tantalize ['tæntəlaɪz] *v.t.* reizen. **tantalizing** ['tæntəlaɪzɪŋ] *adj.* verlockend

tantamount ['tæntəmaʊnt] *adj.* **be ~ to sth.** gleichbedeutend mit etw. sein

tantrum ['tæntrəm] *n.* Wutanfall, *der;* *(of child)* Trotzanfall, *der;* **throw a** ~: einen Wutanfall/Trotzanfall bekommen

'tap [tæp] **1.** *n.* Hahn, *der;* **hot/cold[-water]** ~: Warm-/Kaltwasserhahn, *der;* **be on** ~ *(fig.)* zur Verfügung stehen. **2.** *v.t.,* **-pp-: a)** erschließen ‹*Reserven, Markt*›; **b)** *(Teleph.)* abhören; anzapfen *(ugs.)*

²tap 1. *v.t.,* **-pp-** klopfen an (+ *Akk.*); *(on upper surface)* klopfen auf (+ *Akk.*). **2.** *v.i.,* **-pp-:** ~ **at/on sth.** an etw. *(Akk.)* klopfen; *(on upper surface)* auf etw. *(Akk.)* klopfen. **3.** *n.* Klopfen, *das.* **'tap-dance 1.** *n.* Step[tanz], *der.* **2.** *v.i.* steptanzen; steppen

tape [teɪp] **1.** *n.* **a)** Band, *das;* **adhesive** *or* *(coll.)* **sticky** ~: Klebeband, *das;* **b)**

(for recording) [Ton]band, *das (of mit)*; **make a ~ of sth.** etw. auf Band aufnehmen. **2.** *v. t.* **a)** *(record on ~)* [auf Band] aufnehmen; **b)** *(bind with ~)* [mit Klebeband] zukleben; **c)** **have got sb./sth.~-d** *(sl.)* jmdn. durchschaut haben/etw. im Griff haben

tape: ~ cassette *n.* Tonbandkassette, *die;* **~ deck** *n.* Tapedeck, *das;* **~-measure** *n.* Bandmaß, *das*

taper ['teɪpə(r)] **1.** *v. i.* sich verjüngen; **~ |to a point|** spitz zulaufen. **2.** *n.* |wax| ~: Wachsstock, *der*

tape: ~ recorder *n.* Tonbandgerät, *das;* **~ recording** *n.* Tonbandaufnahme, *die*

tapestry ['tæpɪstrɪ] *n.* Gobelingewebe, *das; (wall-hanging)* Bildteppich, *der*

'tapeworm *n.* Bandwurm, *der*

'tap-water *n.* Leitungswasser, *das*

tar [tɑː(r)] **1.** *n.* Teer, *der.* **2.** *v. t.,* **-rr-** teeren

target ['tɑːgɪt] *n.* **a)** Ziel, *das;* **hit/miss the/its ~:** [das Ziel] treffen/das Ziel verfehlen; **b)** *(Sport)* Zielscheibe, *die*

tariff ['tærɪf] *n.* **a)** *(tax)* Zoll, *der;* **b)** *(list of charges)* Tarif, *der*

tarnish ['tɑːnɪʃ] **1.** *v. t.* stumpf werden lassen ⟨Metall⟩; *(fig.)* beflecken ⟨Ruf⟩. **2.** *v. i.* stumpf werden

tarpaulin [tɑːˈpɔːlɪn] *n.* Persenning, *die*

¹tart [tɑːt] *adj.* herb; sauer ⟨Obst⟩; *(fig.)* scharfzüngig

²tart *n.* **a)** *(Brit.) (filled pie)* ≈ Obstkuchen, *der; (small pastry)* Obsttörtchen, *das;* **b)** *(sl.: prostitute)* Nutte, *die (salopp).* **tart 'up** *v. t. (Brit. coll.)* **~ oneself up, get ~ed up** sich auftakeln *(ugs.)*

tartan ['tɑːtən] **1.** *n.* Schotten[stoff], *der.* **2.** *adj.* Schotten⟨rock, -jacke⟩

tartar ['tɑːtə(r)] *n.* Zahnstein, *der*

tartar sauce ['tɑːtə ˈsɔːs] *n.* Remoulade[nsoße], *die*

task [tɑːsk] *n.* Aufgabe, *die;* **take sb. to ~:** jmdm. eine Lektion erteilen. **'task force** *n.* Sonderkommando, *das*

tassel ['tæsl] *n.* Quaste, *die*

taste [teɪst] **1.** *v. t.* **a)** schmecken; *(try a little)* probieren; **b)** *(recognize flavour of)* [heraus]schmecken. **2.** *v. i.* schmecken **(of** nach); **not ~ of anything** nach nichts schmecken. **3.** *n.* **a)** *(flavour)* Geschmack, *der;* |sense of| Geschmack[ssinn], *der;* **b)** *(discernment)* Geschmack, *der;* **c)** *(sample)* Kostprobe, *die.* **tasteful** ['teɪstfl]

adj., **'tastefully** *adv.* geschmackvoll. **'tasteless** *adj.* geschmacklos. **tasty** ['teɪstɪ] *adj.* lecker

tat [tæt] *n. see* ²**tit**

tattered ['tætəd] *adj.* zerlumpt ⟨Kleidung⟩; zerfleddert ⟨Buch⟩. **tatters** ['tætəz] *n. pl.* Fetzen; **be in ~:** in Fetzen sein; *(fig.)* ruiniert sein

tattoo [təˈtuː] **1.** *v. t.* tätowieren. **2.** *n.* Tätowierung, *die*

tatty ['tætɪ] *adj. (coll.)* schäbig

taught *see* **teach**

taunt [tɔːnt] **1.** *v. t.* verspotten **(about** wegen). **2.** *n.* spöttische Bemerkung

Taurus ['tɔːrəs] *n.* der Stier

taut [tɔːt] *adj.* straff ⟨Seil, Kabel⟩; gespannt ⟨Muskel⟩

tavern ['tævən] *n.* Schenke, *die*

tawny ['tɔːnɪ] *adj.* gelbbraun

tax [tæks] **1.** *n.* Steuer, *die.* **2.** *v. t.* **a)** besteuern; versteuern ⟨Einkommen⟩; **b)** *(fig.)* strapazieren ⟨Kräfte, Geduld⟩. **taxable** ['tæksəbl] *adj.* steuerpflichtig. **taxation** [tækˈseɪʃn] *n.* Besteuerung, *die; (taxes payable)* Steuern. **'tax-free** *adj.* steuerfrei

taxi ['tæksɪ] **1.** *n.* Taxi, *das.* **2.** *v. i.,* **~ing** *or* **taxying** ⟨Flugzeug:⟩ rollen. **'taxi-driver** *n.* Taxifahrer, *der/*-fahrerin, *die*

'tax inspector *n.* Steuerinspektor, *der/*-inspektorin, *die*

taxi: ~-rank *(Brit.),* **~ stand** *(Amer.) ns.* Taxistand, *der*

tax: ~-payer *n.* Steuerzahler, *der/*-zahlerin, *die;* **~ return** *n.* Steuererklärung, *die*

tea [tiː] *n.* **a)** Tee, *der;* **b)** *(meal)* |high| ~: Abendessen, *das.* **'tea-bag** *n.* Teebeutel, *der.* **'tea-break** *n. (Brit.)* Teepause, *die*

teach [tiːtʃ] **1.** *v. t.,* **taught** [tɔːt] unterrichten; *(at university)* lehren; **~ sb./oneself/an animal sth.** jmdm./sich/einem Tier etw. beibringen; **~ sb. to ride** jmdm. das Reiten beibringen. **2.** *v. i.,* **taught** unterrichten. **'teacher** *n.* Lehrer, *der/*Lehrerin, *die*

tea: ~-cloth *n.* Geschirrtuch, *das;* **~-cup** *n.* Teetasse, *die*

teak [tiːk] *n.* Teak[holz], *das*

'tea-leaf *n.* Teeblatt, *das*

team [tiːm] *n.* Team, *das; (Sport also)* Mannschaft, *die.* **team 'up** *v. i.* sich zusammentun *(ugs.)*

'team-work *n.* Teamarbeit, *die*

'teapot *n.* Teekanne, *die*

¹tear [teə(r)] **1.** *n.* Riß, *der.* **2.** *v. t.,* **tore** [tɔː(r)]**, torn** [tɔːn] **a)** *(rip)* zerreißen;

(pull apart) auseinanderreißen; *(damage)* aufreißen; ~ **open** aufreißen ⟨*Brief, Paket*⟩; **b)** ~ **sth. out of sb.'s hands** jmdm. etw. aus der Hand reißen. **3.** *v. i.,* **tore, torn a)** *(rip)* [zer]reißen; **b)** *(move hurriedly)* rasen *(ugs.)*. **tear a'way** *v. t.* wegreißen; ~ **oneself away** *(fig.)* sich losreißen. **tear 'up** *v. t.* zerreißen

²**tear** [tɪə(r)] *n.* Träne, *die.* **tearful** ['tɪəfl] *adj.* weinend

tear [tɪə(r)]: ~**-drop** *n.* Träne, *die;* ~**-gas** *n.* Tränengas, *das*

tease [tiːz] **1.** *v. t.* necken **(about wegen)**; aufziehen *(ugs.)* **(about mit)**. **2.** *v. i.* seine Späße machen

tea: ~**-shop** *n.* *(Brit.)* ≈ Café, *das;* ~**spoon** *n.* Teelöffel, *der;* ~**-strainer** *n.* Teesieb, *das*

teat [tiːt] *n.* **a)** Zitze, *die;* **b)** *(of rubber or plastic)* Sauger, *der*

tea: ~**-time** *n.* Teezeit, *die;* ~**-towel** *n.* Geschirrtuch, *das*

technical ['teknɪkl] *adj.* technisch ⟨*Problem, Daten, Fortschritt*⟩; Fach-⟨*kenntnis, -sprache, -begriff, -wörterbuch*⟩; ~ **term** Fachbegriff, *der;* Fachausdruck, *der.* **technicality** [teknɪ'kælɪtɪ] *n.* technisches Detail

technician [tek'nɪʃn] *n.* Techniker, *der/*Technikerin, *die*

technique [tek'niːk] *n.* Technik, *die; (procedure)* Methode, *die*

technological [teknə'lɒdʒɪkl] *adj.* technisch; technologisch

technology [tek'nɒlədʒɪ] *n.* Technik, *die; (application of science)* Technologie, *die*

teddy ['tedɪ] *n.* ~ **[bear]** Teddy[bär], *der*

tedious ['tiːdɪəs] *adj.* langwierig ⟨*Reise, Arbeit*⟩; *(uninteresting)* langweilig

tee [tiː] *(Golf)* Tee, *das*

teem [tiːm] *v. i.* wimmeln **(with von)**

teenage[d] ['tiːneɪdʒ(d)] *attrib.* im Teenageralter *nachgestellt.* **teenager** ['tiːneɪdʒə(r)] *n.* Teenager, *der; (loosely)* Jugendliche, *der/die*

teens [tiːnz] *n. pl.* Teenagerjahre

teeter ['tiːtə(r)] *v. i.* wanken; ~ **on the edge of sth.** schwankend am Rande einer Sache *(Gen.)* stehen

teeth *pl. of* **tooth**

teething troubles ['tiːðɪŋ trʌblz] *n. pl.* **have** ~ *(fig.)* Anfangsschwierigkeiten haben

teetotal [tiː'təʊtl] *adj.* abstinent lebend. **teetotaller** [tiː'təʊtələ(r)] *n.* Abstinenzler, *der/*Abstinenzlerin, *die*

telecommunications [telɪkəmjuːnɪ'keɪʃnz] *n. pl.* Fernmelde- *od.* Nachrichtentechnik, *die*

telegram ['telɪgræm] *n.* Telegramm, *das*

telegraph ['telɪgrɑːf] *n.* Telegraf, *der;* ~ **pole** Telegrafenmast, *der*

telepathy [tɪ'lepəθɪ] *n.* Telepathie, *die*

telephone ['telɪfəʊn] **1.** *n.* Telefon, *das; attrib.* Telefon-; **answer the ~:** Anrufe entgegennehmen; *(on one occasion)* ans Telefon gehen; *(speak)* sich melden; **be on the ~:** Telefon haben; *(be speaking)* telefonieren **(to mit)**. **2.** *v. t.* anrufen. **3.** *v. i.* anrufen; ~ **for a taxi** nach einem Taxi telefonieren

telephone: ~ **book** *n.* Telefonbuch, *das;* ~ **booth,** *(Brit.)* ~**-box** *ns.* Telefonzelle, *die;* ~ **call** *n.* Telefongespräch, *das;* ~ **directory** *n.* Telefonverzeichnis, *das;* ~ **exchange** *n.* Fernmeldeamt, *das;* ~ **number** *n.* Telefonnummer, *die;* ~ **operator** *n.* Telegrafist, *der/*Telegrafistin, *die*

telephoto [telɪ'fəʊtəʊ] *adj. (Photog.)* ~ **lens** Teleobjektiv, *das*

teleprinter ['telɪprɪntə(r)] *n.* Fernschreiber, *der*

telescope ['telɪskəʊp] *n.* Teleskop, *das;* Fernrohr, *das.* **telescopic** [telɪ'skɒpɪk] *adj. (collapsible)* ausziehbar; Teleskop⟨*antenne*⟩

televise ['telɪvaɪz] *v. t.* im Fernsehen senden *od.* übertragen

television ['telɪvɪʒn, telɪ'vɪʒn] *n.* **a)** *no art.* das Fernsehen; **on ~:** im Fernsehen; **watch ~:** fernsehen; **b)** *(~ set)* Fernsehapparat, *der;* Fernseher, *der (ugs.)*

television: ~ **channel** *n.* [Fernseh]kanal, *der;* ~ **programme** *n.* Fernsehsendung, *die;* ~ **set** *n.* Fernsehgerät, *das*

Telex, telex ['teleks] **1.** *n.* Telex, *das.* **2.** *v. t.* ein Telex schicken (+ *Dat.*); telexen ⟨*Nachricht*⟩

tell [tel] **1.** *v. t.,* **told** [təʊld] **a)** *(relate)* erzählen; *(make known)* sagen ⟨*Name, Adresse*⟩; anvertrauen ⟨*Geheimnis*⟩; ~ **sb. sth.** *or* **sth. to sb.** jmdm. etw. erzählen/sagen/anvertrauen; ~ **sb. the way to the station** jmdm. den Weg zum Bahnhof beschreiben; ~ **sb. the time** jmdm. die Uhrzeit sagen; ~ **tales** *(lie)* Lügengeschichten erzählen; *(gossip)* tratschen *(ugs.);* **b)** *(instruct)* sagen; ~ **sb. [not] to do sth.** jmdm. sagen, er soll[e] etw. [nicht] tun; **c)** *(determine)* feststellen; *(see, recognize)* erkennen **(by an +** *Dat.);* *(with reference to the future)* [vorher]sagen; **d)** *(distinguish)* unterscheiden; **e) all told** insgesamt. **2.** *v.i.,* **told a)** *(determine)* **how can you ~?** wie kann man das feststellen *od.* wissen?; **you never can ~:** man kann nie wissen; **b)** *(give information)* erzählen **(of, about** von); **c)** *(reveal secret)* es verraten; **time will ~:** das wird sich zeigen; **d)** *(produce an effect)* sich auswirken. **tell a'part** *v.t.* auseinanderhalten. **tell 'off** *v.t. (coll.)* **~ sb. off [for sth.]** jmdn. [für *od.* wegen etw.] ausschimpfen

teller ['telə(r)] *n.* **a)** *(in bank) see* **cashier; b)** *(counting votes)* Stimmenzähler, *der/*-zählerin, *die*

telly ['telɪ] *n. (Brit. coll.)* Fernseher, *der (ugs.)*

temp [temp] *n. (Brit. coll.)* Zeitarbeitskraft, *die*

temper ['tempə(r)] **1.** *n.* **a)** Naturell, *das;* **be in a good/bad ~:** gute/schlechte Laune haben; **keep/lose one's ~:** sich beherrschen/die Beherrschung verlieren; **b)** *(anger)* fit of **~:** Wutanfall, *der;* **have a ~:** jähzornig sein. **2.** *v.t.* mäßigen; mildern ‹*Kritik*›

temperament ['temprəmənt] *n. (nature)* Veranlagung, *die;* Natur, *die; (disposition)* Temperament, *das.* **temperamental** [temprə'mentl] *adj.* launenhaft

temperate ['tempərət] *adj.* gemäßigt

temperature ['temprɪtʃə(r)] *n.* Temperatur, *die;* **have** *od.* **run a ~** *(coll.)* Temperatur *od.* Fieber haben

template ['templɪt] *n.* Schablone, *die*

¹**temple** ['templ] *n.* Tempel, *der*

²**temple** *n. (Anat.)* Schläfe, *die*

tempo ['tempəʊ] *n., pl.* **~s** *or* **tempi** ['tempiː] Tempo, *das*

temporary ['tempərərɪ] *adj.* vorübergehend; provisorisch ‹*Gebäude, Büro*›

tempt [tempt] *v.t.* **a) ~ sb. to do sth.** jmdn. geneigt machen, etw. zu tun; **be ~ed to do sth.** versucht sein, etw. zu tun; **~ sb. out** jmdn. hinauslocken; **b)** *(provoke)* herausfordern; **~ fate** das Schicksal herausfordern. **temptation** [temp'teɪʃn] *n.* **a)** *no pl. (attracting)* Verlockung, *die; (being attracted)* Versuchung, *die;* **b)** *(thing)* Verlockung, *die.* '**tempting** *adj.* verlockend

ten [ten] **1.** *adj.* zehn. **2.** *n.* Zehn, *die. See also* **eight**

tenable ['tenəbl] *adj.* haltbar ‹*Theorie*›; vertretbar ‹*Standpunkt*›

tenacious [tɪ'neɪʃəs] *adj.* hartnäckig. **tenacity** [tɪ'næsɪtɪ] *n.* Hartnäckigkeit, *die*

tenant ['tenənt] *n. (of flat, residential building)* Mieter, *der/*Mieterin, *die; (of farm, shop)* Pächter, *der/*Pächterin, *die*

¹**tend** [tend] *v.i.* **~ to do sth.** dazu neigen *od.* tendieren, etw. zu tun; **~ to sth.** zu etw. neigen; **he ~s to get upset if ...:** er regt sich leicht auf, wenn ...

²**tend** *v.t.* sich kümmern um; hüten ‹*Schafe*›; bedienen ‹*Maschine*›

tendency ['tendənsɪ] *n. (inclination)* Tendenz, *die;* **have a ~ to do sth.** dazu neigen, etw. zu tun

¹**tender** ['tendə(r)] *adj.* **a)** *(not tough)* zart; **b)** *(loving)* zärtlich; **c)** *(sensitive)* empfindlich

²**tender 1.** *v.t.* **a)** *(present)* einreichen ‹*Rücktritt*›; vorbringen ‹*Entschuldigung*›; **b)** *(offer as payment)* anbieten. **2.** *n.* Angebot, *das*

'**tenderly** *adv. (gently)* behutsam; *(lovingly)* zärtlich

'**tenderness** *n. see* ¹**tender:** Zartheit, *die;* Zärtlichkeit, *die;* Empfindlichkeit, *die*

tendon ['tendən] *n. (Anat.)* Sehne, *die*

tenement ['tenɪmənt] *n.* Mietshaus, *das*

tenet ['tenɪt] *n.* Grundsatz, *der*

tenner ['tenə(r)] *n. (Brit. coll.)* Zehnpfundschein, *der*

tennis ['tenɪs] *n.* Tennis, *das*

tennis: ~-ball *n.* Tennisball, *der;* **~-court** *n. (for lawn ~)* Tennisplatz, *der; (indoor)* Tennishalle, *die;* **~-racket** *n.* Tennisschläger, *der*

tenor ['tenə(r)] *n. (Mus.)* Tenor, *der*

¹**tense** [tens] *n. (Ling.)* Zeit, *die*

²**tense 1.** *adj.* gespannt. **2.** *v.i.* **sb. ~s** jmds. Muskeln spannen sich an. **3.** *v.t.* anspannen. **tension** ['tenʃn] *n.* **a)** Spannung, *die;* **b)** *(mental strain)* Anspannung, *die*

tent [tent] *n.* Zelt, *das*

tentacle ['tentəkl] *n.* Tentakel, *der od. das*

tentative ['tentətɪv] *adj.* **a)** *(not definite)* vorläufig; **b)** *(hesitant)* zaghaft

tenterhooks ['tentəhʊks] *n. pl.* **be on ~:** [wie] auf glühenden Kohlen sitzen

tenth [tenθ] **1.** *adj.* zehnt... **2.** *n. (in sequence)* zehnte, *der/die/das; (in rank)* Zehnte, *der/die/das; (fraction)* Zehntel, *das. See also* **eighth**

'**tent-peg** *n.* Zeltpflock, *der*

tenuous ['tenjʊəs] *adj.* dünn ‹*Atmo-*

sphäre); dürftig ⟨*Argument*⟩; unbegründet ⟨*Anspruch*⟩

tepid ['tepɪd] *adj.* lauwarm

term [tɜ:m] **1.** *n.* **a)** [Fach]begriff, *der;* **b)** *in pl. (conditions)* Bedingungen; **come to ~s with sth.** mit etw. zurechtkommen; *(resign oneself to sth.)* sich mit etw. abfinden; **c)** *in pl. (charges)* Konditionen; **d) in the short/long/medium ~:** kurz-/lang-/mittelfristig; **e)** *(Sch.)* Halbjahr, *das; (Univ.: one of two/three divisions per year)* Semester, *das/*Trimester, *das;* **f)** *(limited period)* Zeitraum, *der; ~ |of office|* Amtszeit, *die;* **g)** *in pl. (mode of expression)* Worte; **h)** *in pl. (relations)* **be on good/bad ~s with sb.** jit jmdm. auf gutem/ schlechtem Fuß stehen. **2.** *v. t.* nennen

terminal ['tɜ:mɪnl] **1.** *n.* **a)** *(for train or bus)* Bahnhof, *der; (for airline passengers)* Terminal, *der od. das;* **b)** *(Teleph., Computing)* Terminal, *das.* **2.** *adj. (Med.)* unheilbar

terminate ['tɜ:mɪneɪt] *v. t.* **a)** beenden; lösen ⟨*Vertrag*⟩; **b)** *(Med.)* unterbrechen ⟨*Schwangerschaft*⟩. **termination** [tɜ:mɪ'neɪʃn] *n.* **a)** *no pl.* Beendigung, *die; (of lease)* Ablauf, *der;* **b)** *(Med.)* Schwangerschaftsabbruch, *der*

termini *pl. of* **terminus**

terminology [tɜ:mɪ'nɒlədʒɪ] *n.* Terminologie, *die*

terminus ['tɜ:mɪnəs] *n., pl.* **~es** *or* **termini** ['tɜ:mɪnaɪ] Endstation, *die*

terrace ['terəs, 'terɪs] *n.* Häuserreihe, *die.* **terraced house** ['terəst haʊs], 'terist haʊs] *n.* Reihenhaus, *das*

terrain [te'reɪn] *n.* Gelände, *das*

terrible ['terɪbl] *adj.* **a)** *(coll.: very great or bad)* schrecklich *(ugs.);* **b)** *(coll.: incompetent)* schlecht; **c)** *(causing terror)* furchtbar. **terribly** ['terɪblɪ] *adv.* **a)** *(coll.: very)* unheimlich *(ugs.);* **b)** *(coll.: appallingly)* furchtbar *(ugs.);* **c)** *(coll.: incompetently)* schlecht; **d)** *(fearfully)* auf erschreckende Weise

terrier ['terɪə(r)] *n.* Terrier, *der*

terrific [tə'rɪfɪk] *adj. (coll.)* **a)** *(great, intense)* irrsinnig *(ugs.);* **b)** *(magnificent)* sagenhaft *(ugs.);* **c)** *(highly expert)* klasse *(ugs.)*

terrify ['terɪfaɪ] *v. t.* **a)** angst machen (+ *Dat.*); **be terrified that ...:** Angst haben, daß ...; **b)** *(scare)* Angst einjagen (+ *Dat.*). **'terrifying** *adj.* entsetzlich ⟨*Erlebnis, Buch*⟩; furchterregend ⟨*Anblick*⟩; beängstigend ⟨*Geschwindigkeit*⟩

territorial [terɪ'tɔ:rɪəl] *adj.* territorial;

Gebiets⟨*anspruch usw.*⟩. **territory** ['terɪtrɪ] *n.* Gebiet, *das*

terror ['terə(r)] *n.* [panische] Angst; Schrecken, *der.* **terrorism** ['terərɪzm] *n.* Terrorismus, *der; (terrorist acts)* Terror, *der.* **'terrorist** *n.* Terrorist, *der/*Terroristin, *die.* **terrorize** ['terəraɪz] *v. t.* **a)** *(frighten)* in [Angst und] Schrecken versetzen; **b)** *(coerce)* terrorisieren

terse [tɜ:s] *adj.* **a)** *(concise)* kurz und bündig; **b)** *(curt)* knapp

test [test] **1.** *n.* **a)** *(Sch.)* Klassenarbeit, *die; (Univ.)* Klausur, *die;* **put sb./sth. to the ~:** jmdn./etw. erproben; **b)** *(analysis)* Test, *der.* **2.** *v. t.* untersuchen ⟨*Wasser, Augen*⟩; testen ⟨*Gehör, Augen*⟩; prüfen ⟨*Schüler*⟩; **~ sb. for Aids** jmdn. auf Aids untersuchen. **'test out** *v. t.* ausprobieren ⟨*Produkte*⟩ **(on** an + *Dat.*); erproben ⟨*Theorie, Idee*⟩

Testament ['testəmənt] *n.* **Old/New ~** *(Bibl.)* Altes/Neues Testament

testicle ['testɪkl] *n.* Testikel, *der (fachspr.);* Hoden, *der*

testify ['testɪfaɪ] **1.** *v. i.* **a)** **~ to sth.** etw. bezeugen; **b)** *(Law)* **~ against sb.** gegen jmdn. aussagen. **2.** *v. t.* bestätigen

testimonial [testɪ'məʊnɪəl] Zeugnis, *das;* Referenz, *die*

testimony ['testɪmənɪ] *n.* Aussage, *die*

'test-tube *n.* Reagenzglas, *das*

testy ['testɪ] *adj.* leicht reizbar ⟨*Person*⟩; gereizt ⟨*Antwort*⟩

tetanus ['tetənəs] *n.* Tetanus, *der*

tetchy ['tetʃɪ] *adj.* leicht reizbar; gereizt

tether ['teðə(r)] **1.** *n.* **be at the end of one's ~:** am Ende [seiner Kraft] sein. **2.** *v. t.* anbinden **(to** an + *Dat. od. Akk.*)

text [tekst] *n.* Text, *der.* **'textbook** *n.* Lehrbuch, *das*

textile ['tekstaɪl] *n.* Stoff, *der; ~s* Textilien *Pl.*

texture ['tekstʃə(r)] *n.* Beschaffenheit, *die; (of fabric)* Struktur, *die*

Thai [taɪ] **1.** *adj.* thailändisch. **2.** *n.* **a)** *pl. same or* **~s** Thai, *der/die;* **b)** *(language)* Thai, *das.* **Thailand** ['taɪlænd] *pr. n.* Thailand *(das)*

Thames [temz] *pr. n.* Themse, *die*

than [ðən, *stressed* ðæn] *conj.* als; **I know you better ~ |I do| him** ich kenne dich besser als ihn

thank [θæŋk] *v. t.* **~ sb. |for sth.|** jmdm. [für etw.] danken; **~ God** *or* **goodness** *or* **heaven|s|** Gott sei Dank; **|I| ~ you**

danke; **no**, ~ **you** nein, danke; **yes**, ~ **you** ja, bitte; ~ **you very much** vielen herzlichen Dank. **thankful** ['θæŋkfl] *adj.* dankbar. '**thankless** *adj.* undankbar. **thanks** [θæŋks] *n.pl.* **a)** *(gratitude)* Dank, *der;* ~ **to** *(with the help of)* dank; *(on account of the bad influence of)* wegen; **b)** *(formula expr. gratitude)* danke; **no**, ~: nein, danke; **yes**, ~: ja, bitte; **many** ~ *(coll.)* vielen Dank. '**thank-you** *n.* *(coll.)* Dankeschön, *das*

that 1. [ðæt] *adj., pl.* **those** [ðəʊz] **a)** dieser/diese/dieses; **b)** *(coupled or contrasted with 'this')* der/die/das. **2.** [ðæt] *pron., pl.* **those a)** der/die/das; **what bird is** ~? was für ein Vogel ist das?; **like** ~: so; **[just] like** ~ *(without effort, thought)* einfach so; ~'**s right!** gut *od.* recht so; *(iron.)* nur so weiter!; ~ **will do** das reicht; **b)** *(Brit.)* **who is** ~? wer ist da?; *(on telephone)* wer ist am Apparat? **3.** [ðət] *rel. pron., pl.* **same** der/die/das; **everyone** ~ **I know** jeder, den ich kenne; **this is all [the money]** ~ **I have** das ist alles [Geld], was ich habe. **4.** [ðæt] *adv. (coll.)* so. **5.** [ðət] *rel. adv.* der/die/das; **the day** ~ **I first met her** der Tag, an dem ich sie zum ersten Mal sah. **6.** [ðət, *stressed* ðæt] *conj.* daß; **[in order]** ~: damit

thatch [θætʃ] *n. (of straw)* Strohdach, *das; (of reeds)* Schilfdach, *das; (roofing)* Dachbedeckung, *die.* **thatched** [θætʃt] *adj.* stroh-/schilfgedeckt

thaw [θɔː] **1.** *n.* Tauwetter, *das.* **2.** *v.i.* **a)** tauen; **b)** *(melt)* auftauen. **3.** *v.t.* auftauen. **thaw 'out** *see* **thaw 2, 3**

the [*before vowel* ðı, *before consonant* ðə, *when stressed* ðiː] **1.** *def. art.* der/die/das. **2.** *adv.* ~ **more I practise** ~ **better I play** je mehr ich übe, desto *od.* um so besser spiele ich; **so much** ~ **worse for sb./sth.** um so schlimmer für jmdn./etw.

theatre *(Amer.:* **theater)** ['θıətə(r)] *n.* **a)** Theater, *das;* **b)** *(lecture* ~) Hörsaal, *der;* **c)** *(Brit. Med.) see* **operating theatre. theatrical** [θı'ætrıkl] *adj.* **a)** schauspielerisch; **b)** *(showy)* theatralisch

theft [θeft] *n.* Diebstahl, *der*

their [ðeə(r)] *poss. pron. attrib.* ihr

theirs [ðeəz] *poss. pron. pred.* ihrer/ihre/ihres

them [ðəm, *stressed* ðem] *pron.* sie; *(as indirect object)* ihnen; *see also* ¹**her**

theme [θiːm] *n.* Thema, *das*

themselves [ðəm'selvz] *pron.* **a)** em-

phat. selbst; **b)** *refl.* sich ⟨*waschen usw.*⟩; sich selbst ⟨*die Schuld geben, regieren*⟩. *See also* **herself**

then [ðen] **1.** *adv.* **a)** *(at that time)* damals; ~ **and there** auf der Stelle; **b)** *(after that)* dann; ~ **[again]** *(and also)* außerdem; **but** ~ *(after all)* aber schließlich; **c)** *(in that case)* dann; **but** ~ **again** aber andererseits. **2.** *n.* **before** ~: vorher; davor; **since** ~: seitdem. **3.** *adj.* damalig

theological [θiə'lɒdʒıkl] *adj.* theologisch; Theologie⟨*student*⟩

theology [θı'ɒlədʒı] *n.* Theologie, *die*

theoretical [θıə'retıkl] *adj.* theoretisch

theory ['θıərı] *n.* Theorie, *die;* **in** ~: theoretisch

therapeutic [θerə'pjuːtık] *adj.* therapeutisch

therapist ['θerəpıst] *n.* Therapeut, *der*/Therapeutin, *die*

therapy ['θerəpı] *n.* Therapie, *die*

there [ðeə(r)] **1.** *adv.* **a)** *(in/at that place)* da; dort; *(fairly close)* da; **be down/in/up** ~: da unten/drin/oben sein; **b)** *(calling attention)* **hello** *or* **hi** ~! hallo!; **you** ~! Sie da!; **c)** *(in that respect)* da; **so** ~: und damit basta *(ugs.);* **d)** *(to that place)* dahin, dorthin ⟨*gehen, fahren, rücken*⟩; **down/up** ~: dort hinunter/hinauf; **e)** [ðə(r), *stressed* ðeə(r)] **was** ~ **anything in it?** war da irgendwas drin?; ~ **was once** es war einmal; ~ **is enough food** es gibt genug zu essen. **2.** *int.* ~, ~ na, na *(ugs.);* ~ **[you are]!** da, siehst du! **3.** *n.* da; dort; **near** ~: da *od.* dort in der Nähe. **thereabouts** ['ðeərəbaʊts] *adv.* **a)** da [in der Nähe]; **b)** *(near that number)* ungefähr. **therefore** ['ðeəfɔː(r)] *adv.* deshalb; also

thermal ['θɜːml] *adj.* thermisch; ~ **underwear** kälteisolierende Unterwäsche

thermometer [θə'mɒmıtə(r)] *n.* Thermometer, *das*

Thermos, thermos, (P) ['θɜːmɒs] *n.* ~ **[flask/jug/bottle]** Thermosflasche, *die* ⓦ

thermostat ['θɜːməstæt] *n.* Thermostat, *der*

these *pl. of* **this**

thesis ['θiːsıs] *n., pl.* **theses** ['θiːsiːz] **a)** *(proposition)* These, *die;* **b)** *(dissertation)* Dissertation, *die* (**on** über + *Akk.*)

they [ðeı] *pron.* **a)** sie; **b)** *(people in general)* man

they'd [ðeɪd] a) = they would; b) = they had

they'll [ðeɪl] = they will

they're [ðeə(r)] = they are

they've [ðeɪv] = they have

thick [θɪk] **1.** *adj.* **a)** dick; **a rope two inches ~, a two-inch ~ rope** ein zwei Zoll starkes *od.* dickes Seil; **b)** *(dense)* dicht ⟨*Haar, Nebel, Wolken usw.*⟩; **c)** *(filled)* **~ with** voll von; **d)** dickflüssig ⟨*Sahne*⟩; dick ⟨*Suppe, Schlamm, Kleister*⟩; **e)** *(stupid)* dumm. **2.** *n.* **in the ~ of** mitten in (+ *Dat.*). **thick 'ear** *n.* **give sb. a ~** *(Brit. sl.)* jmdm. ein paar hinter die Ohren geben *(ugs.)*

thicken ['θɪkn] **1.** *v. t.* dicker machen; eindicken ⟨*Sauce*⟩. **2.** *v. i.* **a)** dicker werden; **b)** ⟨*Nebel:*⟩ dichter werden; **c)** **the plot ~s** die Sache wird kompliziert

'thickly *adv.* **a)** *(in a thick layer)* dick; **b)** *(densely)* dicht

'thickness *n.* **a)** Dicke, *die;* **be two metres in ~:** zwei Meter dick sein; **b)** *(denseness)* Dichte, *die*

thick: **~-set** *adj.* gedrungen; **~-skinned** *adj.* *(fig.)* dickfellig *(ugs.)*

thief [θi:f] *n., pl.* **thieves** [θi:vz] Dieb, *der*/Diebin, *die*

thieve [θi:v] *v. i.* stehlen

thieves *pl. of* **thief**

thigh [θaɪ] *n.* Oberschenkel, *der*

thimble ['θɪmbl] *n.* Fingerhut, *der*

thin [θɪn] **1.** *adj.* **a)** dünn; **a tall, ~ man** ein großer, hagerer Mann; **b)** *(sparse)* dünn, schütter ⟨*Haar*⟩. **2.** *adv.* dünn. **3.** *v. t.,* **-nn-:** **a)** dünner machen; **b)** *(dilute)* verdünnen. **thin 'out** *v. i.* ⟨*Menschenmenge:*⟩ sich verlaufen; ⟨*Verkehr:*⟩ abnehmen

thing [θɪŋ] *n.* **a)** Sache, *die;* Ding, *das;* **what's that ~ in your hand?** was hast du da in der Hand?; **be a rare ~:** etwas Seltenes sein; **b)** *(action)* **it was the right ~ to do** es war das einzig Richtige; **that was a foolish/friendly ~ to do** das war eine große Dummheit/ das war sehr freundlich; **c)** *(fact)* [Tat]sache, *die;* **it's a strange ~ that ...:** es ist seltsam, daß ...; **the best/worst ~ about her** das Beste/Schlimmste an ihr; **d)** *(idea)* **say the first ~ that comes into one's head** das sagen, was einem gerade so einfällt; **what a ~ to say!** wie kann man nur so etwas sagen!; **e)** *(task)* **she has a reputation for getting ~s done** sie ist für ihre Tatkraft bekannt; **a big ~ to undertake** ein großes Unterfangen; **f)** *(affair)* Sache, *die;* Angelegenheit, *die;* **g)** *(circumstance)*

take ~s too seriously alles zu ernst nehmen; **how are ~s?** wie geht's [dir]?; **h)** *(individual, creature)* Ding, *das;* **i)** *in pl. (personal belongings, clothes)* Sachen; **j)** *(product of work)* Sache, *die;* **the latest ~:** der letzte Schrei; **k)** *(what is important or proper)* das Richtige; **the ~ is ...** *(question)* die Frage ist ...

think [θɪŋk] **1.** *v. t.,* **thought** [θɔ:t] **a)** *(consider)* meinen; **we ~ [that] he will come** wir denken *od.* glauben, daß er kommt; **what do you ~?** was meinst du? **do you really ~ so?** findest du wirklich?; **what do you ~ of him/it?** was hältst du von ihm/davon?; **...**, **don't you ~?** ... , findest *od.* meinst du nicht auch?; **I ~ so/not** ich glaube schon/nicht; **I ~ I'll try** ich glaube, ich werde es versuchen; **b)** *(imagine)* sich *(Dat.)* vorstellen. **2.** *v. i.,* **thought** [nach]denken; **I need time to ~:** ich muß es mir erst überlegen; **I've been ~ing** ich habe nachgedacht; **~ twice** es sich *(Dat.)* zweimal überlegen. **'think of** *v. t.* **a)** denken an (+ *Akk.*); **he ~s of everything** er denkt einfach an alles; **b)** *(have as idea)* **we'll ~ of something** wir werden uns etwas einfallen lassen; **can you ~ of anyone who ...?** fällt dir jemand ein, der ...?; **c)** *(remember)* sich erinnern an (+ *Akk.*); **I just can't ~ of her name** ich komme einfach nicht auf ihren Namen; **d)** **~ little/nothing of sb./sth.** *(consider contemptible)* wenig/nichts von jmdm./ etw. halten. **think 'over** *v. t.* sich *(Dat.)* überlegen. **think 'through** *v. t.* [gründlich] durchdenken. **think 'up** *v. t. (coll.)* sich *(Dat.)* ausdenken

'thinker *n.* Denker, *der*/Denkerin, *die*

third [θɜ:d] **1.** *adj.* dritt... **2.** *n. (in sequence)* dritte, *der/die/das;* *(in rank)* Dritte, *der/die/das;* *(fraction)* Drittel, *das.* See also **eighth.** **'thirdly** *adv.* drittens

'third-rate *adj.* drittklassig

Third 'World *n.* dritte Welt

thirst [θɜ:st] **1.** *n.* Durst, *der;* **die of ~:** verdursten. **2.** *v. i.* **~ for revenge/ knowledge** nach Rache/Wissen dürsten *(geh.).* **'thirsty** *adj.* durstig; **be ~:** Durst haben

thirteen [θɜ:'ti:n] **1.** *adj.* dreizehn. See also **eight. 2.** *n.* Dreizehn, *die.* See also **eight. thirteenth** [θɜ:'ti:nθ] *adj.* dreizehnt... See also **eighth**

thirtieth ['θɜ:tɪɪθ] **1.** *adj.* dreißigst... **2.** *n. (fraction)* Dreißigstel, *das.* See also **eighth**

thrilling

thirty ['θɜ:tɪ] **1.** *adj.* dreißig. **2.** *n.* Dreißig, *die. See also* **eight; eighty 2**
this [ðɪs] **1.** *adj., pl.* **these** [ði:z] dieser/diese/dieses; *(with less emphasis)* der/die/das; **at ~ time** zu dieser Zeit; **by ~ time** inzwischen; mittlerweile; **these days** heut[zutag]e; **before ~ time** vorher; zuvor; **all ~ week** die[se] ganze Woche; **~ morning/evening** *etc.* heute morgen/abend *usw.;* **these last three weeks** die letzten drei Wochen; **~ Monday** *(to come)* nächsten Montag. **2.** *pron., pl.* **these a)** **what's ~?** was ist [denn] das?; **fold it like ~!** falte es so!; **b)** *(the present)* **before ~:** bis jetzt; **c)** *(Brit. Teleph.: person speaking)* **~ is Andy** hier [spricht *od.* ist] Andy; *(Amer. Teleph.)* **who did you say ~ was?** wer ist am apparat?; **d)** **~ and that** dies und das
thistle ['θɪsl] *n.* Distel, *die*
thorn [θɔ:n] *n.* **a)** *(part of plant)* Dorn, *der;* **b)** *(plant)* Dornenstrauch, *der.* **'thorny** *adj.* **a)** dornig; **b)** *(fig.)* heikel
thorough ['θʌrə] *adj.* gründlich
thorough: ~bred *n.* reinrassiges Tier; *(horse)* Rassepferd, *das;* **~fare** *n.* Durchfahrtsstraße, *die;* **'no ~fare'** „Durchfahrt verboten"; *(on foot)* „kein Durchgang"
'thoroughly *adv.* gründlich ⟨*untersuchen*⟩; gehörig ⟨*erschöpft*⟩; so richtig ⟨*genießen*⟩; zutiefst ⟨*beschämt*⟩; total ⟨*verdorben, verwöhnt*⟩; **be ~ fed up with sth.** *(sl.)* von etw. die Nase gestrichen voll haben *(ugs.).* **'thoroughness** *n.* Gründlichkeit, *die*
those *see* **that 1, 2**
though [ðəʊ] **1.** *(conj.)* **a)** *(despite the fact that)* obwohl; **late ~ it was** obwohl es so spät war; **the car, ~ powerful, is also economical** der Wagen ist zwar stark, aber [zugleich] auch wirtschaftlich; **b)** *(but nevertheless)* aber; **a slow ~ certain method** eine langsame, aber *od.* wenn auch sichere Methode; **c)** *(even if)* **[even] ~:** auch wenn; **d)** *(and yet)* **~ you never know** obwohl man nie weiß. **2.** *adv. (coll.)* trotzdem
thought [θɔ:t] **1.** *see* **think. 2.** *n.* **a)** *no pl.* Denken, *das;* **b)** *no pl., no art. (reflection)* Überlegung, *die;* Nachdenken, *das;* **c)** *(consideration)* Rücksicht, *die* **(for** auf + *Akk.);* **d)** *(idea, conception)* Gedanke, *der;* **it's the ~ that counts** der gute Wille zählt; **give up all ~[s] of sth.** sich *(Dat.)* etw. aus dem Kopf schlagen. **thoughtful** ['θɔ:tfl] *adj.* **a)** nachdenklich; **b)** *(considerate)*

rücksichtsvoll; *(helpful)* aufmerksam. **'thoughtfully** *adv.* **a)** nachdenklich; **b)** *(considerately)* rücksichtsvollerweise. **'thoughtless** *adj.* **a)** gedankenlos; **b)** *(inconsiderate)* rücksichtslos. **'thoughtlessly** *adv.* **a)** gedankenlos; **b)** *(inconsiderately)* aus Rücksichtslosigkeit
thousand ['θaʊznd] **1.** *adj.* **a)** tausend; **a** *or* **one ~:** eintausend; **two/several ~:** zweitausend/mehrere tausend; **a** *or* **one ~ and one** [ein]tausend[und]eins; **b)** **a ~** [and one] *(fig.: innumerable)* tausend *(ugs.).* **2.** *n.* **a)** *(number)* tausend; **a** *or* **one/two ~:** ein-/zweitausend; **b)** *(written figure; group)* Tausend, *das;* **c)** *(indefinite amount)* **~s** Tausende. **thousandth** ['θaʊznθ] **1.** *adj.* tausendst... **2.** *n.* *(fraction)* Tausendstel, *das; (in sequence)* Tausendste, *der/die/das*
thrash [θræʃ] *v.t.* **a)** verprügeln; **b)** *(defeat)* vernichtend schlagen. **thrash 'out** *v.t.* ausdiskutieren
thread [θred] **1.** *n.* **a)** Faden, *der;* **b)** *(of screw)* Gewinde, *das.* **2.** *v.t.* **a)** einfädeln; auffädeln ⟨*Perlen*⟩; **b)** **~ one's way through sth.** sich durch etw. schlängeln. **'threadbare** *adj.* abgenutzt; abgetragen ⟨*Kleidung*⟩; *(fig.)* abgedroschen ⟨*Argument*⟩
threat [θret] *n.* Drohung, *die.* **threaten** ['θretn] *v.t.* **a)** bedrohen; **~ sb. with sth.** jmdm. etw. androhen; **b)** **~ to do sth.** damit drohen, etw. zu tun; **c)** drohen mit ⟨*Gewalt, Rache usw.*⟩. **threatening** ['θretnɪŋ] *adj.* drohend
three [θri:] **1.** *adj.* drei. **2.** *n.* Drei, *die. See also* **eight**
three: ~-dimensional [θri:dɪ'menʃnl] *adj.* dreidimensional; **~fold** *adj., adv.* dreifach; **a ~fold increase** ein Anstieg auf das Dreifache; **~-quarters 1.** *n.* **a)** drei Viertel *pl.* **(of** + *Gen.);* **~-quarters of an hour** eine Dreiviertelstunde; **2.** *adv.* dreiviertel ⟨*voll*⟩; **~some** ['θri:səm] *n.* Dreigespann, *das;* Trio, *das*
thresh [θreʃ] *v.t.* dreschen
threshold ['θreʃəʊld] *n.* Schwelle, *die*
threw *see* **throw 1**
thrift [θrɪft] *n.* Sparsamkeit, *die.* **'thrifty** *adj.* sparsam
thrill [θrɪl] **1.** *v.t.* **a)** *(excite)* faszinieren; **b)** *(delight)* begeistern. **2.** *n.* **a)** Erregung, *die;* **b)** *(exciting experience)* aufregendes Erlebnis. **'thriller** *n.* Thriller, *der.* **'thrilling** *adj.* aufregend; spannend ⟨*Buch, Film*⟩

thrive [θraɪv] *v. i.,* **thrived** *or* **throve** [θrəʊv], **thrived** *or* **thriven** ['θrɪvn] **a)** ⟨*Pflanze:*⟩ wachsen und gedeihen; **b)** *(prosper)* aufblühen (**on** bei)

throat [θrəʊt] *n.* Hals, *der; (esp. inside)* Kehle, *die;* **a [sore]** ~: Halsschmerzen

throb [θrɒb] **1.** *v. i.,* **-bb-** pochen; ⟨*Motor:*⟩ dröhnen. **2.** *n.* Pochen, *das; (of engine)* Dröhnen, *das*

throes [θrəʊz] *n. pl.* Qual, *die;* **be in the ~ of sth.** *(fig.)* mitten in etw. *(Dat.)* stecken *(ugs.)*

thrombosis [θrɒm'bəʊsɪs] *n., pl.* **thromboses** [θrɒm'bəʊsi:z] Thrombose, *die*

throne [θrəʊn] *n.* Thron, *der*

throng [θrɒŋ] *n.* [Menschen]menge, *die*

throttle ['θrɒtl] *v. t.* erdrosseln

through [θru:] **1.** *prep.* **a)** durch; **b)** *(Amer.: up to and including)* bis [einschließlich]; **c)** *(by reason of)* durch; infolge von ⟨*Vernachlässigung, Einflüssen*⟩. **2.** *adv.* **a)** let sb. ~: jmdn. durchlassen; **b)** *(Teleph.)* **be** ~: durch sein *(ugs.);* **be** ~ **to sb.** mit jmdm. verbunden sein. **3.** *attrib. adj.* durchgehend ⟨*Zug*⟩. **through'out 1.** *prep.* ~ **the war/period** den ganzen Krieg/die ganze Zeit hindurch; ~ **the country** im ganzen Land. **2.** *adv. (entirely)* ganz; *(always)* stets; die ganze Zeit [hindurch]

throve *see* **thrive**

throw [θrəʊ] **1.** *v. t.,* **threw** [θru:], **thrown** [θrəʊn] **a)** werfen; ~ **sth. to sb.** jmdm. etw. zuwerfen; ~ **sth. at sb.** etw. nach jmdm. werfen; **b)** *(bring to the ground)* zu Boden werfen; abwerfen ⟨*Reiter*⟩; **c)** *(coll.: disconcert)* ⟨*Frage:*⟩ aus der Fassung bringen. **2.** *n.* Wurf, *der.* **throw a'way** *v. t.* **a)** wegwerfen; **b)** *(lose by neglect)* verschenken ⟨*Vorteil, Spiel usw.*⟩. **throw 'up 1.** *v. t.* **a)** hochwerfen ⟨*Arme, Hände*⟩; **b)** *(produce)* hervorbringen ⟨*Ideen usw.*⟩. **2.** *v. i. (coll.)* brechen *(ugs.)*

'throw-away *adj.* **a)** Wegwerf-; Einweg-; **b)** beiläufig ⟨*Bemerkung*⟩

thrown *see* **throw** 1

thrush [θrʌʃ] *n. (Ornith.)* Drossel, *die*

thrust [θrʌst] **1.** *v. t.,* **thrust** stoßen; ~ **aside** *(fig.)* beiseite schieben. **2.** *n.* Stoß, *der*

thud [θʌd] *n.* dumpfer Schlag

thug [θʌg] *n.* Schläger, *der*

thumb [θʌm] **1.** *n.* Daumen, *der;* **get the ~s up** ⟨*Person, Projekt:*⟩ akzeptiert

werden; **be under sb.'s** ~: unter jmds. Fuchtel stehen. **2.** *v. t.* ~ **a lift** per Anhalter fahren. **'thumb through** *v. t.* durchblättern

thumb: ~ **index** *n.* Daumenregister, *das;* ~**tack** *n. (Amer.)* Reißzwecke, *die*

thump [θʌmp] **1.** *v. t.* [mit Wucht] schlagen. **2.** *v. i.* **a)** hämmern (**at, on** gegen); **b)** ⟨*Herz:*⟩ heftig pochen. **3.** *n. (blow)* Schlag, *der; (sound)* Bums, *der (ugs.);* dumpfer Schlag

thunder ['θʌndə(r)] **1.** *n.* Donner, *der.* **2.** *v. i.* donnern. **'thunderclap** *n.* Donnerschlag, *der.* **'thunderstorm** *n.* Gewitter, *das.* **'thundery** *adj.* gewittrig

Thurs. *abbr.* **Thursday** Do.

Thursday ['θɜ:zdeɪ, 'θɜ:zdɪ] *n.* Donnerstag, *der; see also* **Friday**

thus [ðʌs] *adv.* so

thwart [θwɔ:t] *v. t.* durchkreuzen ⟨*Pläne*⟩; vereiteln ⟨*Versuch*⟩; ~ **sb.** jmdm. einen Strich durch die Rechnung machen

thyme [taɪm] *n.* Thymian, *der*

thyroid ['θaɪrɔɪd] *n.* Schilddrüse, *die*

tiara [tɪ'ɑ:rə] *n.* Diadem, *das*

tick [tɪk] **1.** *v. i.* ticken. **2.** *v. t.* **a)** mit einem Häkchen versehen; **b)** *see* ~ **off a. 3.** *n.* **a)** *(of clock etc.)* Ticken, *das;* **b)** *(mark)* Häkchen, *das.* **tick 'off** *v. t.* **a)** *(cross off)* abhaken; **b)** *(coll.: reprimand)* rüffeln *(ugs.)*

ticket ['tɪkɪt] *n.* Karte, *die; (for bus, train)* Fahrschein, *der; (for aeroplane)* Flugschein, *der; (for lottery, raffle)* Los, *das; (for library)* Ausweis, *der;* **price** ~: Preisschild, *das.* **'ticket-collector** *n. (on train)* Schaffner, *der*/Schaffnerin, *die; (on station)* Fahrkartenkontrolleur, *der*/-kontrolleurin, *die.* **'ticket-office** *n.* Fahrkartenschalter, *der; (for advance booking)* Kartenvorverkaufsstelle, *die*

tickle ['tɪkl] **1.** *v. t.* kitzeln. **2.** *v. i.* kitzeln. **ticklish** ['tɪklɪʃ] *adj.* kitzlig

tidal ['taɪdl] *adj.* Gezeiten-. **tidal wave** *n.* Flutwelle, *die*

tiddly-winks ['tɪdlɪwɪŋks] *n. sing. (game)* Flohhüpfen, *das*

tide [taɪd] **1.** *n.* Tide, *die (nordd.);* **high** ~: Flut, *die;* **low** ~: Ebbe, *die;* **the ~s** die Gezeiten; **the** ~ **is in/out** es ist Flut/Ebbe. **2.** *v. t.* ~ **sb. over** jmdm. über die Runden helfen *(ugs.)*

tidiness ['taɪdɪnɪs] *n.* Ordentlichkeit, *die*

tidy ['taɪdɪ] **1.** *adj.* ordentlich; aufge-

räumt ⟨Zimmer, Schreibtisch⟩. 2. v. t.
aufräumen; ~ oneself sich zurechtma-
chen. **tidy 'up** v. i. aufräumen

tie [taɪ] 1. v. t., **tying** ['taɪɪŋ] binden (**to**
an + Akk., **into** zu); ~ **a knot** einen
Knoten machen; (Sport) ~ **the match**
unentschieden spielen. 2. v. i., **tying a)**
(be fastened) it ~**s at the back** es wird
hinten gebunden; **b)** (have equal
scores) ~ **for second place** mit gleicher
Punktzahl den zweiten Platz errei-
chen. 3. n. **a)** Krawatte, die; **b)** (bond)
Band, das; (restriction) Bindung, die;
c) (equality of scores) Punktgleichheit,
die; **d)** (Sport: match) Begegnung, die.
tie 'in v. i. ~ **in with sth.** zu etw. passen. **tie 'up** v. t. **a)** festbinden; ~ **up a**
parcel ein Paket verschnüren; **b)** (keep
busy) beschäftigen

tier [tɪə(r)] n. **a)** Rang, der; **b)** (unit)
Stufe, die

tiger ['taɪɡə(r)] n. Tiger, der

tight [taɪt] 1. adj. **a)** (firm) fest; fest angezogen ⟨Schraube, Mutter⟩; festsit-
zend ⟨Deckel⟩; **b)** (close-fitting) eng
⟨Kleid, Schuh usw.⟩; **c)** (impermeable)
~ **seal/joint** dichter Verschluß/dichte
Fuge; **d)** (taut) straff; **e)** (difficult to
negotiate) **a ~ corner** eine enge Kurve;
be in a ~ corner (fig.) in der Klemme
sein (ugs.); **f)** (strict) streng ⟨Kontrolle,
Disziplin⟩; **g)** (coll.: stingy) knauserig
(ugs.); **h)** (coll.: drunk) voll (salopp). 2.
adv. fest; **hold ~!** halt dich fest! 3. n.
in pl. **a)** (Brit.) |pair of| ~s Strumpfho-
se, die; **b)** (of dancer etc.) Trikothose,
die. **tighten** ['taɪtn] 1. v. t. **a)** (fest| an-
ziehen ⟨Knoten, Schraube⟩; straffzie-
hen ⟨Seil⟩; **b)** verschärfen ⟨Kontrolle⟩.
2. v. i. sich spannen. **tight-fisted**
[taɪt'fɪstɪd] adj. geizig. **'tightrope** n.
Drahtseil, das

tile [taɪl] 1. n. (on roof) Ziegel, der; (on
floor, wall) Fliese, die; Kachel, die. 2.
v. t. [mit Ziegeln] decken ⟨Dach⟩; flie-
sen ⟨Wand, Fußboden⟩; kacheln
⟨Wand⟩

¹**till** [tɪl] 1. prep. bis; (followed by article
+ noun) bis zu; **not |...| ~:** erst. 2.
conj. bis

²**till** n. Kasse, die

tilt [tɪlt] 1. v. i. kippen. 2. v. t. kippen;
neigen ⟨Kopf⟩. 3. n. **a)** Schräglage,
die; **a 45° ~:** eine Neigung von 45°; **b)**
|at| **full ~:** mit voller Wucht

timber ['tɪmbə(r)] n. [Bau]holz, das

time [taɪm] 1. n. **a)** Zeit, die; **in |the**
course of| ~, as ~ goes on/went on mit
der Zeit; im Laufe der Zeit; **in ~, with**

~ (sooner or later) mit der Zeit; **in**
|good| ~ (not late) rechtzeitig; **all the**
or this ~: die ganze Zeit; (without
ceasing) ständig; **a short ~ ago** vor
kurzem; ~ **off** or **out** freie Zeit; **in 'no**
~: im Handumdrehen; **in a**
week's/year's ~: in einer Woche/in einem Jahr; **harvest/Christmas** ~: Ern-
te-/Weihnachtszeit, die; **on** ~ (punctually) pünktlich; **ahead of** ~: zu früh
⟨ankommen⟩; vorzeitig ⟨fertig wer-
den⟩; **have a good** ~: sich amüsieren;
Spaß haben (ugs.); **b)** (occasion) Mal,
das; **for the first** ~: zum ersten Mal;
at ~**s** gelegentlich; ~ **and again,** ~
after ~: immer [und immer] wieder; **at**
one ~, **at |one and| the same** ~ (simulta-
neously) gleichzeitig; **one at a** ~: einzeln; **two at a** ~: jeweils zwei; **c)** (point
in day etc.) [Uhr]zeit, die; **tell the** ~:
die Uhr lesen; **what ~ is it?, what is the**
~? wie spät ist es?; **by this/that** ~: in-
zwischen; **by the** ~ |that| **we arrived** bis
wir hinkamen; **d)** (multiplication) mal;
three ~**s four** drei mal vier; **e)** (Mus.)
Takt, der; **in** ~: im Takt. 2. v. t. **a)** zeitlich abstimmen; **be well** ~**d** zur richtigen Zeit kommen; **b)** (set to operate at
correct ~) einstellen; **c)** (measure ~
taken by) stoppen

time: ~ **bomb** n. Zeitbombe, die;
~-**lag** n. zeitliche Verzögerung;
~-**limit** n. Frist, die

timely ['taɪmlɪ] adj. rechtzeitig

time: ~-**scale** n. Zeitskala, die;
~-**switch** n. Zeitschalter, der;
~-**table** n. **a)** (scheme of work) Zeitplan, der; (Educ.) Stundenplan, der;
b) (Transport) Fahrplan, der; ~-**zone**
n. Zeitzone, die

timid ['tɪmɪd] adj. **a)** scheu ⟨Tier⟩; **b)**
zaghaft ⟨Mensch⟩; (shy) schüchtern

timing ['taɪmɪŋ] n. **a)** **that was perfect**
~! du kommst gerade im richtigen
Augenblick!; **b)** (Theatre, Sport)
Timing, das

tin [tɪn] 1. n. **a)** (metal) Zinn, das;
~-|**plate**| Weißblech, das; **b)** (Brit.: for
preserving) [Konserven]dose, die. 2.
v. t., -nn- (Brit.) zu Konserven verarbeiten. **tin 'foil** n. Stanniol, das; Alu-
folie, die

tinge [tɪndʒ] 1. v. t., ~**ing** ['tɪndʒɪŋ] tönen. 2. n. [leichte] Färbung; (fig.)
Hauch, der

tingle ['tɪŋɡl] v. i. kribbeln

tinker ['tɪŋkə(r)] 1. n. Kesselflicker,
der. 2. v. i. ~ **with sth.** an etw. (Dat.)
herumbasteln (ugs.)

tinkle ['tɪŋkl] 1. *n.* Klingeln, *das.* 2. *v.i.* klingeln

tinned [tɪnd] *adj. (Brit.)* Dosen-

tin: ~-opener *n. (Brit.)* Dosenöffner, *der.* **~pot** *attrib. adj. (derog.)* schäbig

tinsel ['tɪnsl] *n.* Lametta, *das*

tint [tɪnt] 1. *n.* Farbton, *der.* 2. *v.t.* tönen; kolorieren ⟨*Zeichnung*⟩

tiny ['taɪnɪ] *adj.* winzig

¹tip [tɪp] *n. (end, point)* Spitze, *die*

²tip 1. *v.i.,* -pp- *(lean, fall)* kippen; ~ **over** umkippen. 2. *v.t.,* -pp-: a) *(make tilt)* kippen; b) *(make overturn)* umkippen; *(Brit.: discharge)* kippen; c) voraussagen ⟨*Sieger*⟩; ~ sb. **to win** auf jmds. Sieg tippen; d) *(reward)* ~ **sb.** jmdm. Trinkgeld geben. 3. *n.* a) *(money)* Trinkgeld, *das;* b) *(special information)* Hinweis, *der;* Tip, *der (ugs.);* c) *(Brit.)* Müllkippe, *die.* **tip 'off** *v.t.* ~ **sb. off** jmdm. einen Hinweis *od. (ugs.)* Tip geben

'tip-off *n.* Hinweis, *der*

tipsy ['tɪpsɪ] *adj. (coll.)* angeheitert; beschwipst *(ugs.)*

tip: ~toe 1. *v.i.* auf Zehenspitzen gehen; 2. *n.* **on ~toe|s|** auf Zehenspitzen; **~top** *adj.* tipptopp *(ugs.)*

¹tire ['taɪə(r)] *(Amer.) see* tyre

²tire 1. *v.t.* ermüden. 2. *v.i.* müde werden; ermüden; ~ **of sth./doing sth.** einer Sache *(Gen.)* überdrüssig werden. **tire 'out** *v.t.* erschöpfen; ~ **oneself out doing sth.** etw. bis zur Erschöpfung tun

tired ['taɪəd] *adj.* a) *(weary)* müde; b) *(fed up)* **be ~ of sth./doing sth.** etw. satt haben/es satt haben etw. zu tun. **'tireless** *adj.* unermüdlich. **tiresome** ['taɪəsəm] *adj.* a) *(wearisome)* mühsam; b) *(annoying)* lästig. **tiring** ['taɪərɪŋ] *adj.* ermüdend

tissue ['tɪʃuː, 'tɪsjuː] *n.* a) Gewebe, *das;* b) |paper| ~: Papiertuch, *das; (handkerchief)* Papiertaschentuch, *das;* c) ~ |paper| Seidenpapier, *das*

¹tit [tɪt] *n. (Ornith.)* Meise, *die*

²tit *n.* **it's ~ for tat** wie du mir, so ich dir

'titb... *n.* a) *(food)* Häppchen, *das (ugs.);* b) *(piece of news)* Neuigkeit, *die*

title ['taɪtl] *n.* Titel, *der.* **'title-role** *n.* Titelrolle, *die*

tittle-tattle ['tɪtltætl] *n.* Klatsch, *der (ugs.)*

to 1. *[before vowel* tʊ, *before consonant* tə, *stressed* tuː*] prep.* a) *(in the direction of and reaching)* zu; *(with name of place)* nach; **go to work/to the theatre** zur Arbeit/ins Theater gehen; **to France** nach Frankreich; b) *(as far as)* bis zu; **from London to Edinburgh** von London [bis] nach Edinburgh; **increase from 10% to 20%** von 10% auf 20% steigen; c) *introducing relationship or indirect object to* sb./sth. jmdm./einer Sache *(Dat.);* **lend/explain** *etc.* **sth. to sb.** jmdm. etw. leihen/erklären *usw.;* **speak to sb.** mit jmdm. sprechen; **that's all there is to it** mehr ist nicht dazu zu sagen; **what's that to you?** was geht das dich an?; **to me** *(my opinion)* meiner Meinung nach; **14 miles to the gallon** 14 Meilen auf eine Gallone; d) *(until)* bis; **to the end** bis zum Ende; **to this day** bis heute; **five |minutes| to eight** fünf [Minuten] vor acht; e) *with infinitive of a verb* zu; *expr. purpose, or after* **too** um [...] zu; **want to know** wissen wollen; **do sth. to annoy sb.** etw. tun, um jmdn. zu ärgern; **too hot to drink** zu heiß zum Trinken; **he would have phoned but forgot to** er hätte angerufen, aber er vergaß es. 2. *[tuː] adv.* **to and fro** hin und her

toad [təʊd] *n. (also fig. derog.)* Kröte, *die*

'toadstool *n.* Giftpilz, *der*

toast [təʊst] 1. *n.* a) *no pl.* Toast, *der;* **a piece of ~:** eine Scheibe Toast; b) *(call to drink)* Toast, *der;* **drink a ~ to sb./sth.** auf jmdn./etw. trinken. 2. *v.t.* a) rösten; toasten ⟨*Brot*⟩; b) *(drink to)* trinken auf (+ *Akk.*). **'toaster** *n.* Toaster, *der*

tobacco [tə'bækəʊ] *n., pl.* ~s Tabak, *der.* **tobacconist** [tə'bækənɪst] *n.* Tabak[waren]händler, *der/*-händlerin, *die*

toboggan [tə'bɒɡən] 1. *n.* Schlitten, *der.* 2. *v.i.* Schlitten fahren

today [tə'deɪ] 1. *n.* heute; **~'s newspaper** die Zeitung von heute. 2. *adv.* heute

toddler ['tɒdlə(r)] *n.* ≈ Kleinkind, *das*

to-do [tə'duː] *n.* Getue, *das (ugs.)*

toe [təʊ] 1. *n.* Zeh, *der;* Zehe, *die; (of footwear)* Spitze, *die.* 2. *v.t.,* **~ing** *(fig.)* ~ **the line** *or (Amer.)* **mark** sich einordnen. **'toe-nail** *n.* Zeh[en]nagel, *der*

toffee ['tɒfɪ] *n.* Karamel, *der; (Brit.: piece)* Toffee, *das;* Sahnebonbon, *das*

together [tə'ɡeðə(r)] *adv.* a) *(in or into company)* zusammen; b) *(simultaneously)* gleichzeitig; c) *(one with another)* miteinander

toil [tɔil] **1.** *v. i.* schwer arbeiten. **2.** *n.* [harte] Arbeit

toilet ['tɔilit] *n.* Toilette, *die.* **'toilet-bag** *n.* Kulturbeutel, *der.* **'toilet-paper** *n.* Toilettenpapier, *das*

toiletries ['tɔilitriz] *n. pl.* Körperpflegemittel; Toilettenartikel

toilet: ~-**roll** *n.* Rolle Toilettenpapier; ~ **water** *n.* Toilettenwasser, *das;* Eau de Toilette, *das*

token ['təʊkn] **1.** *n.* **a)** *(voucher)* Gutschein, *der;* **b)** *(counter, disc)* Marke, *die;* **c)** *(sign)* Zeichen, *das.* **2.** *attrib. adj.* symbolisch ⟨*Preis*⟩

Tokyo ['təʊkjəʊ] *pr. n.* Tokio *(das)*

told *see* **tell**

tolerable ['tɒlərəbl] *adj.* **a)** *(endurable)* erträglich (**to, for** für); **b)** *(fairly good)* leidlich; annehmbar. **tolerance** ['tɒlərəns] *n.* Toleranz, *die.* **tolerant** ['tɒlərənt] *adj.* tolerant (**of, towards** gegen[über]). **tolerate** ['tɒləreit] *v. t.* dulden; *(bear)* ertragen ⟨*Schmerzen*⟩. **toleration** [tɒlə'reiʃn] *n.* Tolerierung, *die (geh.)*

¹**toll** [təʊl] *n.* Gebühr, *die;* **b)** *(damage etc.)* Aufwand, *der;* **take its ~ of sth.** einen Tribut an etw. *(Dat.)* fordern *(fig.)*

²**toll** *v. i.* ⟨*Glocke:*⟩ läuten

'toll-bridge *n.* gebührenpflichtige Brücke

tom [tɒm] *n. (cat)* Kater, *der*

tomato [tə'mɑːtəʊ] *n., pl.* ~**es** Tomate, *die.* **to'mato juice** *n.* Tomatensaft, *der.* **tomato 'purée** *n.* Tomatenmark, *das*

tomb [tuːm] *n.* Grab, *das;* (*monument)* Grabmal, *das*

'tomboy *n.* Wildfang, *der*

'tombstone *n.* Grabstein, *der*

'tom-cat *n.* Kater, *der*

tome [təʊm] *n.* dicker Band; Wälzer, *der (ugs.)*

tomfoolery [tɒm'fuːləri] *n.* Blödsinn, *der (ugs.)*

tomorrow [tə'mɒrəʊ] **1.** *n.* morgen; ~ **morning/afternoon/evening/night** morgen früh *od.* vormittag/nachmittag/abend/nacht; ~'**s newspaper** die morgige Zeitung. **2.** *adv.* morgen; **see you ~!** *(coll.)* bis morgen!; **the day after ~:** übermorgen

ton [tʌn] *n.* Tonne, *die*

tone [təʊn] **1.** *n.* **a)** *(sound)* Klang, *der;* *(Teleph.)* Ton, *der;* **b)** *(style of speaking)* Ton, *der;* **c)** *(tint, shade)* [Farb]ton, *der;* **d)** *(fig.: character)* **lower/raise the ~ of sth.** das Niveau

einer Sache *(Gen.)* senken/erhöhen; **set the ~:** den Ton angeben. **2.** *v. t.* tönen; abtönen ⟨*Farbe*⟩. **tone 'down** *v. t.* [ab]dämpfen ⟨*Farbe*⟩; *(fig.)* mäßigen ⟨*Sprache*⟩

tongs [tɒŋz] *n. pl.* |**pair of**| ~: Zange, *die*

tongue [tʌŋ] *n.* Zunge, *die;* **bite one's ~** *(lit. or fig.)* sich auf die Zunge beißen; **find one's ~:** seine Sprache wiederfinden; **hold one's ~:** stillschweigen; **he made the remark ~ in cheek** *(fig.)* er meinte die Bemerkung nicht ernst. **'tongue-twister** *n.* Zungenbrecher, *der (ugs.)*

tonic ['tɒnik] **1.** *n.* **a)** *(Med.)* Tonikum, *das;* **b)** *(fig.: invigorating influence)* Wohltat, *die (geh.);* **c)** *(~ water)* Tonic, *das.* **2.** *attrib. adj.* kräftigend; *(fig.)* wohltuend ⟨*Wirkung*⟩. **'tonic water** *n.* Tonic[wasser], *das*

tonight [tə'nait] **1.** *n.* **a)** *(this evening)* heute abend; ~'**s performance** die heutige [Abend]vorstellung; **b)** *(this or the coming night)* heute nacht. **2.** *adv.* **a)** *(this evening)* heute abend; **b)** *(during this or the coming night)* heute nacht; |**I'll**| **see you ~!** bis heute abend!

tonne [tʌn] *n.* [metrische] Tonne

tonsil ['tɒnsl] *n.* [Gaumen]mandel, *die;* **have one's ~s out** sich *(Dat.)* die Mandeln herausnehmen lassen. **tonsillitis** [tɒnsə'laitis] *n.* Mandelentzündung, *die*

too [tuː] *adv.* **a)** *(excessively)* zu; ~ **difficult a task** eine zu schwierige Aufgabe; **b)** *(also)* auch; **c)** *(coll.: very)* besonders; **not ~ pleased** nicht gerade erfreut

took *see* **take**

tool [tuːl] *n.* Werkzeug, *das;* *(garden ~)* Gerät, *das;* |**set of**| ~**s** Werkzeug, *das.* **'tool box** *n.* Werkzeugkasten, *der.* **'tool kit** *n.* Werkzeug, *das*

toot [tuːt] **1.** *v. i. (on car etc. horn)* hupen. **2.** *n.* Tuten, *das*

tooth [tuːθ] *n., pl.* **teeth** [tiːθ] **a)** Zahn, *der;* **b)** *(of rake, fork, comb)* Zinke, *die;* *(of cog-wheel, saw)* Zahn, *der*

tooth: ~-**ache** *n.* Zahnschmerzen *Pl.;* ~-**brush** *n.* Zahnbürste, *die;* ~-**paste** *n.* Zahnpasta, *die;* ~-**pick** *n.* Zahnstocher, *der*

¹**top** [tɒp] **1.** *n.* **a)** *(highest part)* Spitze, *die;* *(of table)* Platte, *die;* *(~ end)* oberes Ende; *(of tree)* Wipfel, *der;* *(~ floor)* oberstes Stockwerk; *(rim of glass)* Rand, *der;* **on ~ of one another**

top

aufeinander; **on ~ of sth.** *(fig.: in addition)* zusätzlich zu etw.; **from ~ to bottom** von oben bis unten; **at the ~:** oben; **at the ~ of the building/hill/ pile/stairs** oben im Gebäude/[oben] auf dem Hügel/[oben] auf dem Stapel/oben an der Treppe; **b)** *(highest rank)* Spitze, *die;* **~ of the table** *(Sport)* Tabellenspitze, *die;* **be |at the] ~ of the class** der/die Klassenbeste sein; **c)** *(upper surface)* Oberfläche, *die;* *(of cupboard, chest)* Oberseite, *die;* **on ~ of sth.** [oben] auf etw. *(position: Dat.; direction: Akk.);* **d)** *(folding roof)* Verdeck, *das;* **e)** *(upper deck of bus)* Oberdeck, *das;* **f)** *(cap of pen)* [Verschluß]kappe, *die;* **g)** *(upper garment)* Oberteil, *das;* **h)** *(lid)* Deckel, *der;* *(of bottle)* Stöpsel, *der.* **2.** *adj.* oberst...; höchst... ⟨*Ton, Preis*⟩; **~ end** oberes Ende; **the ~ pupil** der beste Schüler; **~ speed** Spitzen- od. Höchstgeschwindigkeit, *die.* **3.** *v.t.* **a)** *(be taller than)* überragen; **b)** *(surpass)* übertreffen. **top 'up** *(Brit. coll.)* *v.t.* auffüllen ⟨*Tank, Flasche, Glas*⟩

²top *n. (toy)* Kreisel, *der*

top: **~ 'hat** *n.* Zylinder[hut], *der;* **~-heavy** *adj.* oberlastig

topic ['tɒpɪk] *n.* Thema, *das.* **topical** ['tɒpɪkl] *adj.* aktuell

topless *adj.* **a ~ dress/swimsuit** ein busenfreies Kleid/ein Oben-ohne-Badeanzug

topmost ['tɒpməʊst, 'tɒpməst] *adj.* oberst...; höchst... ⟨*Gipfel, Note*⟩

topple ['tɒpl] **1.** *v.i.* fallen. **2.** *v.t.* stürzen. **topple 'down** *v.i.* hinab-/herabfallen. **topple 'over** *v.i.* umfallen

top 'secret *adj.* streng geheim

topsy-turvy [tɒpsɪ'tɜ:vɪ] *adv.* verkehrtrum *(ugs.);* **turn sth. ~:** etw. auf den Kopf stellen *(ugs.)*

torch [tɔ:tʃ] *n. (Brit.)* Taschenlampe, *die*

tore, torn *see* ¹**tear** 2, 3

tornado [tɔ:'neɪdəʊ] *n., pl.* **~es** Wirbelsturm, *der;* *(in North America)* Tornado, *der*

torpedo [tɔ:'pi:dəʊ] **1.** *n., pl.* **~es** Torpedo, *der.* **2.** *v.t.* torpedieren

torrent ['tɒrənt] *n.* reißender Bach; *(fig.)* Flut, *die.* **torrential** [tə'renʃl] *adj.* wolkenbruchartig ⟨*Regen*⟩

torso ['tɔ:səʊ] *n., pl.* **~s** Rumpf, *der;* **bare ~:** nackter Oberkörper

tortoise ['tɔ:təs] *n.* Schildkröte, *die.* **tortoiseshell** ['tɔ:təsʃel] *n.* Schildpatt, *das*

tortuous ['tɔ:tjʊəs] *adj.* verschlungen; *(fig.)* umständlich

torture ['tɔ:tʃə(r)] **1.** *n.* Folter, *die.* **2.** *v.t.* foltern; *(fig.)* quälen

toss [tɒs] **1.** *v.t.* **a)** *(throw upwards)* hochwerfen; **~ a pancake** einen Pfannkuchen [durch Hochwerfen] wenden; **b)** *(throw casually)* werfen; schmeißen *(ugs.);* **c)** **~ a coin** eine Münze werfen; **d)** *(Cookery: mix)* wenden; mischen ⟨*Salat*⟩. **2.** *v.i.* **a)** **~ and turn** sich [schlaflos] im Bett wälzen; **b)** ⟨*Schiff:*⟩ hin und her geworfen werden; **c)** *(~ coin)* eine Münze werfen; **~ for sth.** mit einer Münze um etw. losen. **3.** *n.* **a)** **~ of a coin** Hochwerfen einer Münze; **b)** *(throw)* Wurf, *der.* **toss 'up** *v.i.* eine Münze werfen; **~ up for sth.** mit einer Münze um etw. losen

¹tot [tɒt] *n. (coll.)* **a)** kleines Kind; **b)** *(of liquor)* Gläschen, *das*

²tot *(coll.) v.t.,* **-tt-:** **~ 'up** zusammenziehen *(ugs.)*

total ['təʊtl] **1.** *adj.* **a)** gesamt; Gesamt-⟨*gewicht, -wert, usw.*⟩; **b)** *(absolute)* völlig *nicht präd.;* **a ~ beginner** ein absoluter Anfänger. **2.** *n. (number)* Gesamtzahl, *die;* *(amount)* Gesamtbetrag, *der;* *(result of addition)* Summe, *die;* **a ~ of 200** insgesamt 200; **in ~:** insgesamt. **3.** *v.t., (Brit.)* **-ll-: a)** addieren, zusammenzählen ⟨*Zahlen*⟩; **b)** *(amount to)* [insgesamt] betragen

totalitarian [təʊtælɪ'teərɪən] *adj.* totalitär

'totally *adv.* völlig

totter ['tɒtə(r)] *v.i.* wanken; taumeln

touch [tʌtʃ] **1.** *v.t.* **a)** berühren; **b)** *(harm)* anrühren; **c)** *(fig.: rival)* **~ sth.** an etw. *(Akk.)* heranreichen; **d)** *(affect emotionally)* rühren. **2.** *v.i.* sich berühren; **don't ~!** nicht anfassen! **3.** *n.* **a)** Berührung, *die;* **b)** *no art. (faculty)* **|sense of] ~:** Tastsinn, *der;* **c)** *(small amount)* **a ~ of salt/pepper** *etc.* eine Spur Salz/Pfeffer *usw.;* **a ~ of irony** *etc.* ein Anflug von Ironie *usw.;* **d)** *(fig.)* Detail, *das;* **e)** *(communication)* **be in/out of ~ |with sb.]** [mit jmdm.] Kontakt/keinen Kontakt haben; **get in ~:** mit jmdm. Kontakt aufnehmen. **touch 'down** *v.i.* ⟨*Flugzeug:*⟩ landen. **'touch on** *v.t. (mention)* ansprechen. **touch 'up** *v.t. (improve)* ausbessern **'touch: ~-and-go** *adj.* **it is ~-and-go |whether...]** es steht auf des Messers Schneide [, ob...]; **~down** *n. (Aeronaut.)* Landung, *die* 'touching *adj.* rührend. touchy ['tʌtʃi] *adj.* empfindlich; heikel ⟨*Thema*⟩

tough [tʌf] *adj.* a) fest ⟨*Material, Stoff*⟩; zäh ⟨*Fleisch; fachspr.: Werkstoff, Metall*⟩; widerstandsfähig ⟨*Belag, Glas, Haut*⟩; b) *(hardy)* zäh ⟨*Person*⟩; c) *(difficult)* schwierig; d) *(severe, harsh)* hart; e) *(coll.)* ~ luck Pech, *das.*

toughen ['tʌfn] *v.t.* ~ |up| abhärten ⟨*Person*⟩; verschärfen ⟨*Gesetz*⟩

tour [tʊə(r)] 1. *n.* a) [Rund]reise, *die*; Tour, *die (ugs.)*; b) *(Theatre, Sport)* Tournee, *die*; c) *(of house etc.)* Besichtigung, *die*; d) ~ |of duty| Dienstzeit, *die.* 2. *v.i.* a) ~/go ~ing in *or* through a country eine Reise *od. (ugs.)* Tour durch ein Land machen; b) *(Theatre, Sport)* eine Tournee machen. 3. *v.t.* a) besichtigen ⟨*Stadt, Gebäude*⟩; ~ a country/region eine Reise *od. (ugs.)* Tour durch ein Land/Gebiet machen; b) *(Theatre, Sport)* ~ a country/the provinces eine Tournee durch das Land/die Provinz machen

tourism ['tʊərɪzm] *n.* a) Tourismus, *der*; b) *(operation of tours)* Touristik, *die.* tourist ['tʊərɪst] 1. *n.* Tourist, *der*/Touristin, *die.* 2. *attrib. adj.* Touristen-. **tourist infor'mation centre, 'tourist office** *ns.* Fremdenverkehrsbüro, *das*

tournament ['tʊənəmənt] *n. (Hist.; Sport)* Turnier, *das*

'tour operator *n.* Reiseveranstalter, *der*/-veranstalterin, *die*

tousle ['taʊzl] *v.t.* zerzausen

tout [taʊt] 1. *v.i.* ~ for customers Kunden anreißen *(ugs.) od.* werben. 2. *n.* Anreißer, *der*/Anreißerin, *die (ugs.)*; ticket ~: Kartenschwarzhändler, *der*/-händlerin, *die*

tow [təʊ] 1. *v.t.* schleppen; ziehen ⟨*Anhänger, Wasserskiläufer*⟩. 2. *n.* Schleppen, *das*; give a car a ~: einen Wagen schleppen; on ~: im Schlepp[tau]. tow a'way *v.t.* abschleppen

toward [tə'wɔːd], towards [tə'wɔːdz] *prep.* a) *(in direction of)* ~ sb./sth. auf jmdn./etw. zu; turn ~ sb. sich zu jmdm. umdrehen; b) *(in relation to)* gegenüber; feel sth. ~ sb. jmdm. gegenüber etw. empfinden; c) *(for)* a contribution ~ sth. ein Beitrag zu etw.; proposals ~ solving a problem Vorschläge zur Lösung eines Problems; d) *(near)* gegen; ~ the end of May [gegen] Ende Mai

towel ['taʊəl] *n.* Handtuch, *das*

tower ['taʊə(r)] 1. *n.* Turm, *der.* 2. *v.i.* in die Höhe ragen. 'tower above *v.t.* ~ above sb./sth. jmdn./etw. überragen

'tower block *n.* Hochhaus, *das*

'towering *attrib. adj.* hoch aufragend; *(fig.)* herausragend ⟨*Leistung*⟩

town [taʊn] *n.* Stadt, *die*; the ~ of Cambridge die Stadt Cambridge; in |the| ~: in der Stadt; the ~ *(people)* die Stadt; be in/out of ~: in der Stadt/ nicht in der Stadt sein

town: ~ 'centre *n.* Stadtmitte, *die*; Stadtzentrum, *das*; ~ 'hall *n.* Rathaus, *das*; ~ 'planning *n.* Stadtplanung, *die*

tow: ~-path *n.* Leinpfad, *der*; ~-rope *n.* Abschleppseil, *das*

toxic ['tɒksɪk] *adj.* giftig

toy [tɔɪ] 1. *n.* Spielzeug, *das*; ~s Spielzeug, *das.* 2. *adj.* Spielzeug-. 3. *v.i.* ~ with the idea of doing sth. mit dem Gedanken spielen, etw. zu tun. 'toyshop *n.* Spielwarengeschäft, *das*

trace [treɪs] 1. *v.t.* a) *(copy)* durchpausen; abpausen; b) zeichnen ⟨*Linie*⟩; c) *(follow track of)* folgen (+ *Dat.*); verfolgen; d) *(find)* finden. 2. *n.* Spur, *die.* 'tracing paper ['treɪsɪŋ peɪpə(r)] *n.* Pauspapier, *das*

track [træk] 1. *n.* a) Spur, *die*; *(of wild animal)* Fährte, *die*; ~s *(footprints)* [Fuß]spuren; *(of animal also)* Fährte, *die*; keep ~ of sb./sth. jmdn./etw. im Auge behalten; b) *(path)* Weg, *der*; *(footpath)* Pfad, *der*; c) *(Sport)* Bahn, *die*; cycling/greyhound ~: Radrennbahn, *die*/Windhundrennbahn, *die*; d) *(Railw.)* Gleis, *das*; e) *(course taken)* Route, *die*; *(of rocket, satellite)* Bahn, *die.* 2. *v.t.* ~ an animal die Spur/Fährte eines Tieres verfolgen; the police ~ed him |to Paris| die Polizei folgte seiner Spur [bis nach Paris]. track 'down *v.t.* aufspüren

track: ~ events *n. pl.* Laufwettbewerbe; ~ suit *n.* Trainingsanzug, *der*

¹tract [trækt] *n. (area)* Gebiet, *das*

²tract *n. (pamphlet)* [Flug]schrift, *die*

tractor ['træktə(r)] *n.* Traktor, *der*

trade [treɪd] *n.* a) *(line of business)* Gewerbe, *das*; he's a butcher/lawyer etc. by ~: er ist von Beruf Metzger/ Rechtsanwalt *usw.*; b) *no indef. art (commerce)* Handel, *der*; c) *(craft)* Handwerk, *das.* 2. *v.i. (buy and sell)* Handel treiben. 3. *v.t.* tauschen; austauschen ⟨*Waren, Grüße*⟩; sich *(Dat.)*

sagen ⟨Beleidigungen⟩; ~ sth. for sth.
etw. gegen etw. tauschen. **trade** 'in
v. t. in Zahlung geben
'**trade mark** n. Warenzeichen, das;
leave one's ~ on sth. (fig.) einer Sache
(Dat.) seinen Stempel aufdrücken
'**trader** n. Händler, der/Händlerin, die
trade: ~ '**union** n. Gewerkschaft, die;
attrib. Gewerkschafts-; ~·'**unionist**
n. Gewerkschaft[l]er, der/Gewerk-
schaft[l]erin, die
trading ['treidɪŋ] n. Handel, der.
'**trading estate** n. (Brit.) Gewerbe-
gebiet, das. '**trading stamp** n. Ra-
battmarke, die
tradition [trə'dɪʃn] n. Tradition, die.
traditional [trə'dɪʃənl] adj. traditionell; herkömmlich ⟨Erziehung, Metho-
de⟩. **tra**'**ditionally** adv. traditionell
traffic ['træfɪk] 1. n. a) no indef. art.
Verkehr, der; b) (trade) Handel, der.
2. v. i., -ck-: ~ in sth. mit etw. handeln
traffic: ~ **circle** n. (Amer.) Kreisverkehr, der; ~ **jam** n. [Verkehrs]stau,
der; ~ **lights** n. pl. [Verkehrs]ampel,
die; ~ **sign** n. Verkehrszeichen, das;
~ **signals** see ~ **lights**; ~ **warden** n.
(Brit.) Hilfspolizist, der; (woman)
Hilfspolizistin, die; Politesse, die
tragedy ['trædʒɪdɪ] n. Tragödie, die.
tragic ['trædʒɪk] adj. tragisch
trail [treil] 1. n. a) Spur, die; ~ of
smoke/dust Rauch-/Staubfahne, die;
b) (Hunting) Spur, die; Fährte, die; c)
(path) Pfad, der; Weg, der. 2. v. t. a)
(pursue) verfolgen; b) (drag) ~ sth.
[after or behind one] etw. hinter sich
(Dat.) herziehen. 3. v. i. a) (be
dragged) schleifen; b) (lag) hinterher-
trotten; c) ⟨Pflanze:⟩ kriechen
trailer ['treilə(r)] n. a) Anhänger, der;
(Amer.: caravan) Wohnanhänger, der;
b) (Cinemat., Telev.) Trailer, der
train [trein] 1. v. t. a) ausbilden (in in
+ Dat.); erziehen ⟨Kind⟩; abrichten
⟨Hund⟩; dressieren ⟨Tier⟩; b) (Sport)
trainieren; c) (Hort.) ziehen. 2. v. i. a)
eine Ausbildung machen; he is ~ing
as or to be a doctor/engineer er macht
eine Arzt-/Ingenieursausbildung; b)
(Sport) trainieren. 3. n. a) (Railw.)
Zug, der; on the ~: im Zug; b) (of skirt
etc.) Schleppe, die; c) ~ of thought
Gedankengang, der. '**train-driver** n.
Lokomotivführer, der/-führerin, die
trained [treind] adj. ausgebildet ⟨Ar-
beiter, Lehrer, Arzt, Stimme⟩; abge-
richtet ⟨Hund⟩; dressiert ⟨Tier⟩; ge-
schult ⟨Geist, Auge, Ohr⟩

trainee [trei'ni:] n. Auszubildende,
der/die
'**trainer** n. [Konditions]trainer, der/
-trainerin, die
'**train fare** n. Fahrpreis, der
'**training** n. a) Ausbildung, die; b)
(Sport) Training, das
train: ~ **journey** n. Bahnfahrt, die;
(long) Bahnreise, die; ~ **set** n. [Mo-
dell]eisenbahn, die; ~ **station** n.
(Amer.) Bahnhof, der
trait [treit] n. Eigenschaft, die
traitor ['treitə(r)] n. Verräter, der/Ver-
räterin, die
tram [træm] n. (Brit.) Straßenbahn,
die; ~**lines** Straßenbahnschienen
tramp [træmp] 1. n. Landstreicher,
der/-streicherin, die; (in city) Stadtstreicher, der/-streicherin, die. 2. v. i.
a) (tread heavily) trampeln; b) (walk)
marschieren
trample ['træmpl] 1. v. t. zertrampeln.
2. v. i. trampeln. '**trample on** v. t.
herumtrampeln auf (+ Dat.)
trampoline ['træmpəli:n] n. Trampo-
lin, das
trance [trɑ:ns] n. Trance, die; be in a
~: in Trance sein
tranquil ['træŋkwɪl] adj. ruhig. **tran**-
quillity [træŋ'kwɪlɪtɪ] Ruhe, die.
tranquillizer ['træŋkwɪlaɪzə(r)] n.
Beruhigungsmittel, das
transact [træn'zækt] v. t. ~ **business**
Geschäfte tätigen. **transaction**
[træn'zækʃn] n. Geschäft, das; (financial) Transaktion, die
transcend [træn'send] v. t. übersteigen
transcript ['trænskrɪpt] n. Abschrift,
die; (of trial) Protokoll, das
transfer 1. [træns'fɜ:(r)] v. t., -rr-: a)
(move) verlegen (to nach); überweisen
⟨Geld⟩ (to auf + Akk.); übertragen
⟨Befugnis, Macht⟩ (to Dat.); b) übereignen ⟨Gegenstand, Grundbesitz⟩ (to
Dat.); c) versetzen ⟨Arbeiter, Ange-
stellte⟩; (Footb.) transferieren. 2.
[træns'fɜ:(r)] v. i., -rr-: a) (when travelling) umsteigen; b) (change job etc.)
wechseln. 3. ['trænsfə:(r)] n. a) (moving) Verlegung, die; (of powers) Übertragung, die (to an + Akk.); (of
money) Überweisung, die; b) (of employee etc.) Versetzung, die; (Footb.)
Transfer, der; c) (picture) Abziehbild,
das. **transferable** [træns'fɜ:rəbl]
adj. übertragbar
transform [træns'fɔ:m] v. t. verwandeln. **transformation** [trænsfə-

'meiʃn] *n.* Verwandlung, *die.* **trans-former** *n. (Electr.)* Transformator, *der*

transfusion [træns'fju:ʒn] *n. (Med.)* Transfusion, *die*

transient ['trænzɪənt] *adj.* kurzlebig; vergänglich

transistor [træn'zɪstə(r)] *n.* Transistor, *der*

transit ['trænsɪt] *n.* **in** ~: auf der Durchreise; ⟨*Waren*⟩ auf dem Transport; **passengers in** ~: Transitreisende

transition [træn'sɪʒn, træn'zɪʃn] *n.* Übergang, *der;* Wechsel, *der*

transitive ['trænsɪtɪv] *adj. (Ling.)* transitiv

transitory ['trænsɪtərɪ] *adj.* vergänglich; *(fleeting)* flüchtig

translate [træns'leɪt] *v. t.* übersetzen. **translation** [træns'leɪʃn] *n.* Übersetzung, *die.* **translator** [træns'leɪtə(r)] *n.* Übersetzer, *der/*Übersetzerin, *die*

translucent [træns'lu:sənt] *adj.* durchscheinend

transmission [træns'mɪʃn] *n.* **a)** Übertragung, *die;* **b)** *(Motor Veh.)* Antrieb, *der; (gearbox)* Getriebe, *das*

transmit [træns'mɪt] *v. t.,* -tt-: **a)** *(pass on)* übersenden; übertragen; **b)** durchlassen ⟨*Licht*⟩; leiten ⟨*Wärme*⟩. **trans'mitter** *n.* Sender, *der*

transparency [træns'pærənsɪ] *n.* **a)** Durchsichtigkeit, *die;* **b)** *(Photog.)* Transparent, *das; (slide)* Dia, *das*

transparent [træns'pærənt] *adj.* durchsichtig

transpire [træn'spaɪə(r)] *v. i.* sich herausstellen; *(coll.: happen)* passieren

transplant **1.** [træns'plɑ:nt] *v. t.* **a)** verpflanzen ⟨*Organ*⟩; **b)** *(plant in another place)* umpflanzen. **2.** ['trænsplɑ:nt] *n. (Med.)* Transplantation, *die;* Verpflanzung, *die*

transport **1.** [træns'pɔ:t] *v. t.* transportieren; befördern. **2.** ['trænspɔ:t] *n.* **a)** Transport, *der;* Beförderung, *die; attrib.* Beförderungs-; **b)** *(means of conveyance)* Verkehrsmittel, *das;* **be without** ~: kein [eigenes] Fahrzeug haben

transpose [træns'pəʊz] *v. t.* vertauschen; umstellen

transvestite [træns'vestaɪt] *n.* Transvestit, *der*

trap [træp] **1.** *n.* **a)** Falle, *die;* **set** *or* **lay a** ~ **for an animal** eine Falle für ein Tier legen *od.* aufstellen; **set** *or* **lay a** ~ **for sb.** *(fig.)* jmdm. eine Falle stellen; **fall into a/sb.'s** ~ *(fig.)* in die/ jmdm. in die Falle gehen; **b)** *(sl.: mouth)* Klappe, *die (salopp).* **2.** *v. t.,*

-pp-: **a)** [in *od.* mit einer Falle] fangen ⟨*Tier*⟩; *(fig.)* in eine Falle locken ⟨*Person*⟩; **be** ~**ped** *(fig.)* in eine Falle gehen/in der Falle sitzen; **be** ~**ped in a cave/by the tide** in einer Höhle festsitzen/von der Flut abgeschnitten sein; **b)** *(confine)* einschließen; einklemmen ⟨*Körperteil*⟩. **trap'door** *n.* Falltür, *die*

trapeze [trə'pi:z] *n.* Trapez, *das*

trash [træʃ] *n., no indef. art.* **a)** *(rubbish)* Abfall, *der;* **b)** *(badly made thing)* Mist, *der (ugs.); (bad literature)* Schund, *der (ugs.)*

trauma ['trɔ:mə] *n., pl.* ~**ta** ['trɔ:mətə] *or* ~**s** Trauma, *das.* **traumatic** [trɔ:'mætɪk] *adj.* traumatisch

travel ['trævl] **1.** *n.* Reisen, *das; attrib.* Reise-. **2.** *v. i., (Brit.)* -ll- reisen; *(go in vehicle)* fahren. **3.** *v. t., (Brit.)* -ll- zurücklegen ⟨*Strecke, Entfernung*⟩; benutzen ⟨*Weg, Straße*⟩; **we had** ~**led 10 miles** wir waren 10 Meilen gefahren. **'travel agency** *n.* Reisebüro, *das.* **'travel agent** *n.* Reisebürokaufmann, *der/*-kauffrau, *die*

traveler, traveling *(Amer.) see* **travell-**

traveller ['trævlə(r)] *n. (Brit.)* **a)** Reisende, *der/die;* **b)** *in pl. (gypsies etc.)* fahrendes Volk. **'traveller's cheque** *n.* Reisescheck, *der*

travelling ['trævlɪŋ] *attrib. adj. (Brit.)* Wander⟨*zirkus, -ausstellung*⟩

trawler ['trɔ:lə(r)] *n.* [Fisch]trawler, *der*

tray [treɪ] *n.* Tablett, *das; (for correspondence)* Abschlagekorb, *der*

treacherous ['tretʃərəs] *adj.* **a)** treulos ⟨*Person*⟩; **b)** *(deceptive)* tückisch. **treachery** ['tretʃərɪ] *n.* Verrat, *der*

treacle ['tri:kl] *n. (Brit.)* Sirup, *der*

tread [tred] **1.** *n.* **a)** *(of tyre, boot, etc.)* Lauffläche, *die;* **2 millimetres of** ~ **on a tyre** 2 Millimeter Profil auf einem Reifen; **b)** *(sound of walking)* Schritt, *der.* **2.** *v. i.,* trod [trɒd], **trodden** ['trɒdn] *or* trod treten (in/on *od.* in/auf + *Akk.*); *(walk)* gehen. **3.** *v. t.* trod, **trodden** *or* trod treten auf (+ *Akk.*); stampfen ⟨*Weintrauben*⟩

treason ['tri:zn] *n.* **|high|** ~: Hochverrat, *der*

treasure ['treʒə(r)] **1.** *n.* Schatz, *der;* Kostbarkeit, *die;* **art** ~**s** Kunstschätze. **2.** *v. t.* in Ehren halten. **'treasure-hunt** *n.* Schatzsuche, *die*

treasurer ['treʒərə(r)] *n.* Kassenwart, *der/*-wartin, *die.* **treasury** ['treʒərɪ] *n.* **the T**~: das Finanzministerium

treat [tri:t] **1.** *n.* **a)** [besonderes] Vergnügen; **b)** *(entertainment)* Vergnügen, für dessen Kosten *jmd. anderes aufkommt;* **lay on a special ~ for sb.** jmdm. etwas Besonderes bieten. **2.** *v. t.* **a)** behandeln; **~ sth. as a joke** etw. als Witz nehmen; **~ sth. with contempt** für etw. nur Verachtung haben; **b)** *(Med.)* behandeln; **~ sb. for sth.** jmdn. wegen etw. behandeln; *(before confirmation of diagnosis)* jmdn. auf etw. *(Akk.)* behandeln; **c)** klären ‹*Abwässer*›; **d)** *(provide with at own expense)* einladen; **~ sb. to sth.** jmdm. etw. spendieren; **~ oneself to a new hat** sich *(Dat.)* einen neuen Hut leisten

treatise ['tri:tɪs, 'tri:tɪz] *n.* Abhandlung, *die*

treatment *n.* Behandlung, *die*

treaty ['tri:tɪ] *n.* [Staats]vertrag, *der*

treble ['trebl] **1.** *adj.* **a)** dreifach; **b)** *(Brit. Mus.)* **~ voice** Sopranstimme, *die.* **2.** *n.* **a)** *(~ quantity)* Dreifache, *das;* **b)** *(Mus.)* **he is a ~:** er singt Sopran. **3.** *v. t.* verdreifachen. **4.** *v. i.* sich verdreifachen. **'treble clef** *n. (Mus.)* Violinschlüssel, *der*

tree [tri:] *n.* Baum, *der*

trek [trek] **1.** *v. i.,* **-kk-** ziehen **(across** durch**). 2.** *n.* [schwierige] Reise

trellis ['trelɪs] *n.* Gitter, *das; (for plants)* Spalier, *das*

tremble ['trembl] *v. i.* zittern **(with** vor + *Dat.*)

tremendous [trɪ'mendəs] *adj.* gewaltig; *(coll.: wonderful)* großartig

tremor ['tremə(r)] *n.* **a)** Zittern, *das;* **b)** |earth| ~: leichtes Erdbeben

trench [trentʃ] *n.* Graben, *der; (Mil.)* Schützengraben, *der*

trend [trend] *n.* **a)** Trend, *der;* **upward ~:** steigende Tendenz; **b)** *(fashion)* Mode, *die;* [Mode]trend, *der.* **'trendy** *adj. (Brit. coll.)* modisch; Schickimicki‹*kneipe*› *(ugs.)*

trepidation [trepɪ'deɪʃn] *n.* Beklommenheit, *die*

trespass ['trespəs] *v. i.* **~ on** unerlaubt betreten ‹*Grundstück*›. **'trespasser** *n.* Unbefugte, *der/die*

trial ['traɪəl] *n.* **a)** *(Law)* [Gerichts]verfahren, *das;* **be on ~** |for murder| [wegen Mordes] angeklagt sein; **b)** *(testing)* Test, *der;* **employ sb. on ~:** jmdn. probeweise einstellen; [by| ~ **and error** [durch] Ausprobieren; **c)** *(trouble)* Problem, *das;* **d)** *(Sport) (competition)* Prüfung, *die; (for selection)* Testspiel, *das*

triangle ['traɪæŋgl] *n.* **a)** Dreieck, *das;* **b)** *(Mus.)* Triangel, *das od. der.* **triangular** [traɪ'æŋgjʊlə(r)] *adj.* dreieckig

tribe [traɪb] *n.* Stamm, *der*

tribulation [trɪbjʊ'leɪʃn] *n.* Kummer, *der*

tribunal [traɪ'bju:nl] *n.* Schiedsgericht, *das*

tributary ['trɪbjʊtərɪ] *n.* Nebenfluß, *der*

tribute ['trɪbju:t] *n.* Tribut, *der* (**to** an + *Akk.*); **pay ~ to sb./sth.** jmdm./einer Sache den schuldigen Tribut zollen *(geh.)*

trice [traɪs] *n.* **in a ~:** im Handumdrehen

trick [trɪk] **1.** *n.* **a)** Trick, *der;* **it was all a ~:** das war [alles] nur Bluff; **b)** *(feat of skill etc.)* Kunststück, *das;* **that should do the ~** *(coll.)* damit dürfte es klappen *(ugs.);* **c)** *(knack)* **get** *or* **find the ~** |of doing sth.| den Dreh finden[, wie man etw. tut]; **d)** *(prank)* Streich, *der;* **play a ~ on sb.** jmdm. einen Streich spielen; **e)** *(Cards)* Stich, *der.* **2.** *v. t.* täuschen; hereinlegen; **~ sb. out of/into sth.** jmdm. etw. ablisten **3.** *adj.* **~ photograph** Trickaufnahme, *die;* **~ question** Fangfrage, *die.* **trickery** ['trɪkərɪ] *n.* [Hinter]list, *die*

trickle ['trɪkl] *v. i.* rinnen; *(in drops)* tröpfeln

trickster ['trɪkstə(r)] *n.* Schwindler, *der*/Schwindlerin, *die*

'tricky *adj.* verzwickt *(ugs.)*

tricycle ['traɪsɪkl] *n.* Dreirad, *das*

tried *see* **try 2, 3**

trifle ['traɪfl] *n.* **a)** *(Brit. Gastron.)* Trifle, *das;* **b)** *(thing of slight value)* Kleinigkeit, *die.* **trifling** ['traɪflɪŋ] *adj.* unbedeutend ‹*Angelegenheit*›; gering ‹*Wert*›

trigger ['trɪgə(r)] **1.** *n.* **a)** *(of gun)* Abzug, *der; (of machine)* Drücker, *der;* **b)** *(fig.)* Auslöser, *der.* **2.** *v. t.* **~** |off| auslösen

trigonometry [trɪgə'nɒmɪtrɪ] *n.* Trigonometrie, *die*

trim [trɪm] **1.** *v. t.,* **-mm-: a)** schneiden ‹*Hecke*›; [nach]schneiden ‹*Haar*›; beschneiden ‹*Papier, Hecke, Budget*›; **b)** *(ornament)* besetzen **(with** mit). **2.** *adj.* proper; gepflegt ‹*Garten*›. **3.** *n.* **a)** **be in ~** *(healthy)* in Form *od.* fit sein; **b)** *(cut)* Nachschneiden, *das.* **'trimming** *n.* **a)** *(decorations)* Verzierung, *die;* **b)** *in pl. (coll.: accompaniments)* Beilagen; **with all the ~s** mit allem Drum und Dran *(ugs.)*

Trinity ['trɪnɪtɪ] *n. (Theol.)* **the [Holy]** ~: die Heilige Dreieinigkeit

trinket ['trɪŋkɪt] *n.* kleines, billiges Schmuckstück

trio ['tri:əʊ] *n., pl.* ~s Trio, *das*

trip [trɪp] **1.** *n.* **a)** Reise, *die; (shorter)* Ausflug, *der;* **b)** *(coll.: drug-induced hallucinations)* Trip, *der.* **2.** *v.i.,* -pp- stolpern (on über + *Akk.*). **trip 'up 1.** *v.i.* **a)** stolpern; **b)** *(fig.)* einen Fehler machen. **2.** *v.t.* **a)** stolpern lassen; **b)** *(fig.)* aufs Glatteis führen *(fig.)*

tripe [traɪp] *n.* **a)** Kaldaunen *Pl.;* **b)** *(sl.: rubbish)* Quatsch, *der (ugs.)*

triple ['trɪpl] **1.** *adj.* **a)** *(threefold)* drei- fach; **b)** *(three times greater than)* ~ **the ...:** der/die/das dreifache ... **2.** *n.* Dreifache, *das.* **3.** *v.i.* sich verdreifa- chen. **4.** *v.t.* verdreifachen

triplet ['trɪplɪt] *n.* Drilling, *der*

triplicate ['trɪplɪkət] *n.* **in** ~: in dreifa- cher Ausfertigung

tripod ['traɪpɒd] *n.* Dreibein, *das*

'tripper *n. (Brit.)* Ausflügler, *der*/Aus- flüglerin, *die*

trite [traɪt] *adj.* banal

triumph ['traɪəmf, 'traɪʌmf] **1.** *n.* Tri- umph, *der* **(over** über + *Akk.*). **2.** *v.i.* triumphieren **(over** über + *Akk.*).

triumphant [traɪ'ʌmfənt] *adj.* **a)** siegreich; **b)** triumphierend *(Blick)*

trivial ['trɪvɪəl] *adj.* belanglos. **tri- viality** [trɪvɪ'ælɪtɪ] *n.* Belanglosigkeit, *die*

trod, trodden *see* tread 2, 3

trolley ['trɒlɪ] *n.* **a)** *(for serving food)* Servierwagen, *der;* **b)** |supermarket| ~: Einkaufswagen, *der*

trombone [trɒm'bəʊn] *n.* Posaune, *die*

troop [tru:p] **1.** *n.* **a)** *in pl.* Truppen; **b)** *(fig.)* Schar, *die.* **2.** *v.i.* ~ **in/out** hin- ein-/hinausströmen

trophy ['trəʊfɪ] *n.* Trophäe, *die*

tropic ['trɒpɪk] *n.* **the T**~s *(Geog.)* die Tropen; **the** ~ **of Cancer/Capricorn** *(Astron., Geog.)* der Wendekreis des Krebses./Steinbocks. **tropical** ['trɒ- pɪkl] *adj.* tropisch; Tropen*(krankheit, -kleidung)*

trot [trɒt] **1.** *n. (coll.)* **on the** ~: hinter- einander; **be on the** ~: auf Trab sein *(ugs.)*. **2.** *v.i.,* -tt- traben

trouble ['trʌbl] **1.** *n.* **a)** Ärger, *der;* Schwierigkeiten *Pl.;* **there'll be** ~ |if ...| es wird Ärger geben, [wenn ...]; **what's the** ~? was ist denn?; **b)** engine/brake ~: Probleme mit dem Motor/der Bremse; **suffer from heart/liver** ~: Probleme mit dem Herz/der Leber haben; **c)** *(inconvenience)* Mühe, *die;* **take a lot of** ~: sich *(Dat.)* sehr viel Mühe geben; **it's more** ~ **than it's worth** es lohnt sich nicht; **d)** *in sing. or pl. (unrest)* Unruhen. **2.** *v.t.* **a)** *(agit- ate)* beunruhigen; **don't let it** ~ **you** mach dir deswegen keine Sorgen; **b)** *(inconvenience)* stören. **3.** *v.i. (make an effort)* sich bemühen. **troubled** ['trʌbld] *adj.* **a)** *(worried)* besorgt; **b)** *(restless)* unruhig. **'trouble-maker** *n.* Unruhestifter *der*/-stifterin, *die* **troublesome** ['trʌblsəm] *adj.* schwierig; lästig *(Krankheit)*

trough [trɒf] *n.* Trog, *der*

troupe [tru:p] *n.* Truppe, *die*

trousers ['traʊzəz] *n.pl.* |pair of| ~: Hose, *die*

'trouser suit *n. (Brit.)* Hosenanzug, *der*

trousseau ['tru:səʊ] *n., pl.* ~s or ~x ['tru:səʊz] Aussteuer, *die*

trout [traʊt] *n., pl.* same Forelle, *die*

trowel ['traʊəl] *n.* Kelle, *die; (Hort.)* Pflanzkelle, *die*

truant ['tru:ənt] *n.* **play** ~: [die Schule] schwänzen *(ugs.)*

truce [tru:s] *n.* Waffenstillstand, *der*

truck [trʌk] *n.* **a)** Last[kraft]wagen, *der;* Lkw, *der;* **b)** *(Brit. Railw.)* offener Güterwagen

truculent ['trʌkjʊlənt] *adj.* aufsässig

trudge [trʌdʒ] *v.i.* trotten; *(through snow etc.)* stapfen

true [tru:] *adj.,* ~**r** ['tru:ə(r)], ~**st** ['tru:ɪst] **a)** wahr; wahrheitsgetreu *(Bericht);* richtig *(Vorteil);* *(rightly so called)* eigentlich; echt, wahr *(Freund);* **is it** ~ **that ...?** stimmt es, daß ...?; ~ **to life** lebensecht; **b)** *(loyal)* treu

truffle ['trʌfl] *n.* Trüffel, *die od. (ugs.) der*

truism ['tru:ɪzm] *n.* Binsenweisheit, *die*

truly ['tru:lɪ] *adv.* **a)** wirklich; **b)** *(ac- curately)* zutreffend; **yours** ~: mit freundlichen Grüßen

trump [trʌmp] *(Cards)* **1.** *n.* Trumpf, *der.* **2.** *v.t.* übertrumpfen. **'trump up** *v.t. (coll.)* konstruieren

trumpet ['trʌmpɪt] *n.* Trompete, *die.* **'trumpeter** *n.* Trompeter, *der*/Trom- peterin, *die*

truncheon ['trʌntʃn] *n.* Schlagstock, *der*

trundle ['trʌndl] *v.t. & i.* rollen

trunk [trʌŋk] *n.* **a)** *(of elephant etc.)*

Rüssel, *der;* b) *(large box)* Schrank-koffer, *der;* c) *(of tree)* Stamm, *der;* d) *(of body)* Rumpf, *der;* e) *(Amer.: of car)* Kofferraum, *der;* f) *in pl. (Brit.)* |swimming| ~s Badehose, *die*

truss [trʌs] *n. (Med.)* Bruchband, *das*

trust [trʌst] 1. *n.* a) Vertrauen, *das;* **place** *or* **put one's ~ in sb./sth.** sein Vertrauen auf *od.* in jmdn./etw. setzen; **take sth. on ~:** etw. einfach glauben; b) *(organization managed by trustees)* Treuhandgesellschaft, *die;* |charitable| ~: Stiftung, *die;* *(association of companies)* Trust, *der;* c) *(Law)* **hold in ~:** treuhänderisch verwalten. 2. *v. t. (rely on)* trauen (+ *Dat.*); vertrauen (+ *Dat.*) ⟨*Person*⟩; ~ **sb. with sth.** jmdm. etw. anvertrauen. 3. *v. i.* a) ~ **to** sich verlassen auf (+ *Akk.*); b) *(believe)* ~ **in sb./sth.** auf jmdn./etw. vertrauen. **trustee** [trʌˈstiː] *n.* Treuhänder, *der*/Treuhän-derin, *die*. **trustful** [ˈtrʌstfl], **'trust-ing** *adjs.* vertrauensvoll. **'trust-worthy** *adj.* vertrauenswürdig

truth [truːθ] *n., pl.* ~s [truːðz, truːθs] Wahrheit, *die;* **tell the |whole| ~:** die |ganze| Wahrheit sagen. **truthful** [ˈtruːθfl] *adj.* ehrlich

try [traɪ] 1. *n.* Versuch, *der;* **have a ~ at sth./doing sth.** etw. versuchen/versuchen, etw. zu tun; **give it a ~, have a ~:** es versuchen. 2. *v. t.* a) *(attempt)* versuchen; b) *(test usefulness of)* probieren; c) *(test)* auf die Probe stellen ⟨*Fähigkeit, Kraft, Geduld*⟩; d) *(Law.: take to trial)* ~ **a case** einen Fall verhandeln; ~ **sb. |for sth.|** jmdn. [wegen einer Sache] vor Gericht stellen. 3. *v. i.* es versuchen; ~ **hard/harder** sich *(Dat.)* viel/mehr Mühe geben. **try 'on** *v. t.* anprobieren ⟨*Kleidungsstück*⟩. **try 'out** *v. t.* ausprobieren

'trying *adj.* a) *(testing)* schwierig; b) *(difficult to endure)* anstrengend

T-shirt *n.* T-Shirt, *das*

tub [tʌb] *n.* Kübel, *der; (for ice-cream etc.)* Becher, *der*

tuba [ˈtjuːbə] *n. (Mus.)* Tuba, *die*

tubby [ˈtʌbɪ] *adj.* rundlich

tube [tjuːb] *n.* a) *(for conveying liquids etc.)* Rohr, *das;* b) *(small cylinder)* Tu-be, *die; (for sweets, tablets)* Röhrchen, *das;* c) *(Anat., Zool.)* Röhre, *die;* d) *(of TV etc.)* Röhre, *die;* e) *(Brit. coll.: underground railway)* U-Bahn, *die*

tuber [ˈtjuːbə(r)] *n. (Bot.)* Knolle, *die*

tuberculosis [tjuːbɜːkjʊˈləʊsɪs] *n.* Tu-berkulose, *die*

tube: ~ **station** *n. (Brit. coll.)* U-Bahnhof, *der;* ~ **train** *n. (Brit. coll.)* U-bahn-Zug, *der*

tubing [ˈtjuːbɪŋ] *n.* Rohre *Pl.*

tubular [ˈtjuːbjʊlə(r)] *adj.* rohrförmig

tuck [tʌk] 1. *v. t.* stecken. 2. *n. (in fabric) (for decoration)* Biese, *die; (to tighten)* Abnäher, *der.* **tuck 'in** 1. *v. t.* hineinstecken. 2. *v. i. (coll.)* zulangen *(ugs.).* **tuck 'up** *v. t.* a) hochkrempeln ⟨*Ärmel, Hose*⟩; hochnehmen ⟨*Rock*⟩; b) *(cover snugly)* zudecken

Tue., Tues. *abbrs.* **Tuesday** Di.

Tuesday [ˈtjuːzdeɪ, ˈtjuːzdɪ] *n.* Diens-tag, *der; see also* **Friday**

tuft [tʌft] *n.* Büschel, *das*

tug [tʌg] 1. *n.* a) Ruck, *der;* ~ **of war** Tauziehen, *das;* b) ~ |boat| Schlepper, *der.* 2. *v. t.,* -gg- ziehen. 3. *v. i.,* -gg-zerren (**at an** + *Dat.*)

tuition [tjuːˈɪʃn] *n.* Unterricht, *der*

tulip [ˈtjuːlɪp] *n.* Tulpe, *die*

tumble [ˈtʌmbl] 1. *v. i.* stürzen; fallen. 2. *n.* Sturz, *der.* **'tumble-drier** *n.* Wäschetrockner, *der.* **'tumble-dry** *v. t.* im Automaten trocknen

tumbler [ˈtʌmblə(r)] *n. (short)* Whisky-glas, *das; (long)* Wasserglas, *das*

tummy [ˈtʌmɪ] *n. (child lang./coll.)* Bäuchlein, *das.* **'tummy-ache** *n. (child lang./coll.)* Bauchweh, *das*

tumour *(Brit.; Amer.:* **tumor**) [ˈtjuːmə(r)] *n.* Tumor, *der*

tumult [ˈtjuːmʌlt] *n.* Tumult, *der*

tuna [ˈtjuːnə] *n., pl. same or* ~s Thun-fisch, *der*

tune [tjuːn] 1. *n.* a) *(melody)* Melodie, *die;* **change one's ~** *(fig.)* sein Verhalten ändern; **call the ~:** den Ton angeben; b) *(correct pitch)* **sing in/out of ~:** richtig/falsch singen; **be in/out of ~** ⟨*Instrument:*⟩ richtig .gestimmt/ver-stimmt sein. 2. *v. t.* a) *(Mus.: put in ~)* stimmen; b) *(Radio, Telev.)* einstellen (**to** auf + *Akk.*); c) einstellen ⟨*Motor, Vergaser*⟩. **tune 'in** *(Radio, Telev.)* ~ **to a station** einen Sender einstellen

tuneful [ˈtjuːnfl] *adj.* melodisch

tunic [ˈtjuːnɪk] *n. (of soldier)* Uniform-jacke, *die; (of schoolgirl)* Kittel, *der*

'tuning-fork [ˈtjuːnɪŋfɔːk] *n.* Stimm-gabel, *die*

Tunisia [tjuːˈnɪzɪə] *pr. n.* Tunesien *(das)*

tunnel [ˈtʌnl] 1. *n.* Tunnel, *der; (dug by animal)* Gang, *der.* 2. *v. i., (Brit.)* -ll-einen Tunnel graben

turban [ˈtɜːbən] *n.* Turban, *der*

turbine [ˈtɜːbaɪn] *n.* Turbine, *die*

turbulence ['tɜːbjʊlǝns] *n.* **a)** Aufge-
wühltheit, *die;* *(fig.)* Aufruhr, *der;* **b)**
(Phys.) Turbulenz, *die*
turbulent ['tɜːbjʊlǝnt] *adj.* **a)** aufge-
wühlt; **b)** *(Phys.)* turbulent
tureen [tjʊǝˈriːn] *n.* Terrine, *die*
turf [tɜːf] *n., pl.* ~s *or* **turves** [tɜːvz] **a)** *no
pl.* Rasen, *der;* **b)** *(segment)* Rasen-
stück, *das.* **turf 'out** *v. t. (sl.)* rausschmeißen *(ugs.)*
Turk [tɜːk] *n.* Türke, *der*/Türkin, *die*
Turkey ['tɜːkɪ] *pr. n.* die Türkei
turkey *n.* Truthahn, *der*/Truthenne,
die; (esp. as food) Puter, *der*/Pute, *die*
Turkish ['tɜːkɪʃ] **1.** *adj.* türkisch; *sb. is*
~: jmd. ist Türke/Türkin. **2.** *n.* Türkisch, *das; see also* **English 2 a**
turmoil ['tɜːmɔɪl] *n.* Aufruhr, *der*
turn [tɜːn] **1.** *n.* **a)** it is sb.'s ~ to do sth.
jmd. ist an der Reihe, etw. zu tun; **it's
your ~ |next|** du bist als nächster/
nächste dran *(ugs.) od.* an der Reihe;
out of ~: außer der Reihe; *(fig.)* an
der falschen Stelle ⟨lachen⟩; **take |it in|**
~s sich abwechseln; **b)** *(rotary motion)*
Drehung, *die;* **c)** *(change of direction)*
Wende, *die;* **take a** ~ **to the right/left,
do** *or* **take a right/left** ~: nach rechts/
links abbiegen; *(fig.)* **take a favourable** ~ sich zum Guten wenden; **the** ~
of the year/century die Jahres-/Jahr-
hundertwende; **d)** *(bend)* Kurve, *die;*
(corner) Ecke, *die;* **e)** *(short performance)* Nummer, *die;* **f)** *(service)* **do sb.
a good** ~: jmdm. einen guten Dienst
erweisen; **g)** *(coll.: fright)* **give sb. quite
a** ~: jmdm. einen gehörigen Schrek-
ken einjagen *(ugs.).* **2.** *v. t.* **a)** *(make revolve)* drehen; **b)** *(reverse)* umdrehen;
wenden ⟨Pfannkuchen, Auto, Heu⟩; ~
sth. upside down *or* **on its head** *(lit. or
fig.)* etw. auf den Kopf stellen; ~ **the
page** umblättern; **c)** *(give new direction
to)* drehen, wenden ⟨Kopf⟩; ~ **a hose/
gun on sb./sth.** einen Schlauch/ein
Gewehr auf jmdn./etw. richten; ~
one's attention/mind to sth. sich/seine
Gedanken einer Sache *(Dat.)* zuwenden; **d)** ~ **sb. loose on sb./sth.** jmdn.
auf jmdn./etw. loslassen; **e)** *(cause to
become)* verwandeln; ~ **the lights
|down|** low das Licht dämpfen; ~ **a
play/book into a film** ein Theater-
stück/Buch verfilmen; **f)** *(shape in
lathe)* drechseln ⟨Holz⟩; drehen ⟨Me-
tall⟩; **g)** drehen ⟨Pirouette⟩; schlagen
⟨Purzelbaum⟩. **3.** *v. i.* **a)** *(revolve)* sich
drehen; **b)** *(reverse direction)* ⟨Person:⟩
sich herumdrehen; ⟨Auto:⟩ wenden;

c) *(take new direction)* sich wenden; *(~
round)* sich umdrehen; ~ **to the left/
right** nach links/rechts abbiegen; **d)**
(become) werden; ~ **|in|to** sth. zu etw.
werden; *(be transformed)* sich in etw.
(Akk.) verwandeln; **e)** *(become sour)*
⟨Milch:⟩ sauer werden. **turn a'way 1.**
v. i. sich abwenden. **2.** *v. t.* **a)** *(avert)*
abwenden; **b)** *(send away)* wegschik-
ken. **turn 'down** *v. t.* **a)** herunter-
schlagen ⟨Kragen⟩; **b)** niedriger stellen ⟨Heizung⟩; herunterdrehen ⟨Gas⟩;
leiser stellen ⟨Ton, Radio, Fernseher⟩;
c) *(reject)* ablehnen; abweisen ⟨Kandi-
daten usw.⟩. **turn 'in 1.** *v. t.* **a)** nach innen drehen; **b)** *(hand in)* abgeben. **2.**
v. i. **a)** *(enter)* einbiegen; **b)** *(coll.: go to
bed)* in die Falle gehen *(salopp).* **turn
'off 1.** *v. t.* abschalten; abstellen
⟨Wasser, Gas⟩; zudrehen ⟨Wasser-
hahn⟩. **2.** *v. i.* abbiegen. **turn on** *v. t.*
a) [-'-] anschalten; aufdrehen ⟨Wasser-
hahn, Gas⟩; **b)** ['--] *(attack)* angreifen.
turn 'out 1. *v. t.* **a)** *(expel)* hinauswerfen *(ugs.);* **b)** *(switch off)* ausschalten;
abdrehen ⟨Gas⟩; **c)** *(produce)* produzieren; **d)** *(Brit.)* *(empty)* ausräumen;
leeren; *(get rid of)* wegwerfen. **2.** *v. i.*
a) *(prove to be)* sb./sth. ~s out to be
sth. jmd./etw. stellt sich als jmd./etw.
heraus; **everything** ~ed **out well/all
right in the end** alles endete gut; **b)**
(appear) ⟨Fans usw.:⟩ erscheinen. **turn
'over 1.** *v. t.* umdrehen. **2.** *v. i.* **a)** *(tip
over)* umkippen; ⟨Boot:⟩ kentern;
⟨Auto, Flugzeug:⟩ sich überschlagen;
b) *(from one side to the other)* sich umdrehen; **c)** *(~ a page)* weiterblättern.
turn 'round *v. i.* sich umdrehen.
'turn to *v. t. (fig.)* ~ **to sb.** sich an
jmdn. wenden; ~ **to sb. for help/ad-
vice** bei jmdm. Hilfe/Rat suchen; ~ **to
drink** sich in den Alkohol flüchten.
turn 'up 1. *v. i.* **a)** ⟨Person:⟩ erscheinen; **b)** *(present itself)* auftauchen;
⟨Gelegenheit:⟩ sich bieten. **2.** *v. t.* **a)**
hochschlagen ⟨Kragen⟩; **b)** lauter stellen ⟨Ton, Radio, Fernseher⟩; aufdrehen ⟨Heizung, Gas⟩; heller machen
⟨Licht⟩
'turning *n.* Abzweigung, *die.*
'turning-point *n.* Wendepunkt, *der*
turnip ['tɜːnɪp] *n.* Kohlrübe, *die*
turn: ~-**out** *n. (of people)* Beteiligung,
die (for an + *Dat.*); ~**over** *n.* **a)**
(Commerc.) Umsatz, *der; (of stock)*
Umschlag, *der;* **b)** *(of staff)* Fluktuati-
on, *die;* ~**pike** *n. (Amer.)* gebührenpflichtige Autobahn; ~**stile** *n.* Dreh-

kreuz, *das;* ~**table** *n.* Plattenteller, *der;* ~-**up** *n.* *(Brit. Fashion)* Aufschlag, *der*

turpentine ['tɜːpntaɪn] *n.* Terpentin, *das*

turquoise ['tɜːkwɔɪz] 1. *n.* a) Türkis, *der;* b) *(colour)* Türkis, *das.* 2. *adj.* türkis[farben]

turret ['tʌrɪt] *n.* Türmchen, *das*

turtle ['tɜːtl] *n.* a) Meeresschildkröte, *die;* b) *(Amer.: freshwater reptile)* Wasserschildkröte, *die*

turves *see* turf b·

tusk [tʌsk] *n.* Stoßzahn, *der*

tussle ['tʌsl] 1. *n.* Gerangel, *das (ugs.).* 2. *v. i.* sich balgen

tutor ['tjuːtə(r)] *n.* |private| ~: [Privat]lehrer, *der/*-lehrerin, *die*

tuxedo [tʌkˈsiːdəʊ] *n., pl.* ~s *or* ~es *(Amer.)* Smoking, *der*

TV [tiːˈviː] *n.* a) Fernsehen, *das;* b) *(television set)* Fernseher, *der (ugs.)*

twaddle ['twɒdl] *n.* Gewäsch, *das (ugs.)*

twang [twæŋ] 1. *v. t.* zupfen ⟨Saite⟩. 2. *n.* |nasal| ~: Näseln, *das*

tweed [twiːd] *n.* Tweed, *der*

tweezers ['twiːzəz] *n. pl.* |pair of| ~: Pinzette, *die*

twelfth [twelfθ] 1. *adj.* zwölft... 2. *n.* *(fraction)* Zwölftel, *das.* See also **eighth**

twelve [twelv] 1. *adj.* zwölf. 2. *n.* Zwölf, *die.* See also **eight**

twentieth ['twentɪɪθ] 1. *adj.* zwanzigst... 2. *n.* *(fraction)* Zwanzigstel, *das.* See also **eighth**

twenty ['twentɪ] 1. *adj.* zwanzig. 2. *n.* Zwanzig, *die.* See also **eight**; **eighty 2**

twice [twaɪs] *adv.* a) zweimal; b) *(doubly)* doppelt

twiddle ['twɪdl] *v. t.* herumdrehen an (+ *Dat.*) *(ugs.)*

¹**twig** [twɪg] *n.* Zweig, *der*

²**twig** *(coll.)* 1. *v. t.,* -gg- kapieren *(ugs.).* 2. *v. i.,* -gg- es kapieren *(ugs.)*

twilight ['twaɪlaɪt] *n.* a) *(evening light)* Dämmerlicht, *das;* b) *(period of half-light)* Dämmerung, *die*

twin [twɪn] 1. *attrib. adj.* a) Zwillings-; b) *(forming a pair)* Doppel-. 2. *n.* Zwilling, *der.* **twin 'beds** *n. pl.* zwei Einzelbetten

twine [twaɪn] 1. *n.* Bindfaden, *der.* 2. *v. i.* sich winden **(about, around** um)

twinge [twɪndʒ] *n.* Stechen, *das;* ~|s| **of conscience** *(fig.)* Gewissensbisse

twinkle ['twɪŋkl] 1. *v. i.* funkeln **(with** vor + *Dat.*). 2. *n.* Funkeln, *das*

twinkling ['twɪŋklɪŋ] *n.* **in a** ~, **in the** ~ **of an eye** im Handumdrehen

'**twin town** *n.* *(Brit.)* Partnerstadt, *die*

twirl [twɜːl] 1. *v. t.* [schnell] drehen. 2. *v. i.* wirbeln **(around** über + *Akk.*)

twist [twɪst] 1. *v. t.* a) verdrehen ⟨Worte, Bedeutung⟩; ~ **one's ankle** sich *(Dat.)* den Knöchel verrenken; ~ **sb.'s arm** jmdm. den Arm umdrehen; *(fig.)* jmdm. [die] Daumenschrauben anlegen; b) *(rotate)* drehen. 2. *v. i.* sich winden. 3. *n.* a) *(motion)* Drehung, *die;* b) *(unexpected occurrence)* überraschende Wendung

twit [twɪt] *n.* *(Brit. sl.)* Trottel, *der (ugs.)*

twitch [twɪtʃ] 1. *v. i.* ⟨Mund, Lippe:⟩ zucken. 2. *n.* Zucken, *das*

twitter ['twɪtə(r)] 1. *n.* Zwitschern, *das.* 2. *v. i.* zwitschern

two [tuː] 1. *adj.* zwei. 2. *n.* Zwei, *die.* See also **eight**

two: ~-**faced** ['tuːfeɪst] *adj.* *(fig.)* falsch; ~-**fold** *adj.; adv.* zweifach; a ~**fold increase** ein Anstieg auf das Doppelte; ~-**piece** 1. *n.* Zweiteiler, *der;* 2. *adj.* zweiteilig; ~**some** ['tuːsəm] *n.* Paar, *das;* ~-**way** *adj.* a) zweibahnig *(Verkehrsw.);* '~-**way traffic ahead'** „Achtung Gegenverkehr"; b) ~-**way mirror** Einwegspiegel, *der*

tycoon [taɪˈkuːn] *n.* Magnat, *der*

tying *see* tie 1, 2

type [taɪp] 1. *n.* a) Art, *die;* *(person)* Typ, *der;* **what** ~ **of car ...?** was für ein Auto ...?; b) *(Printing)* Drucktype, *die.* 2. *v. t.* [mit der Maschine] schreiben; tippen *(ugs.).* 3. *v. i.* maschineschreiben. **type 'out** *v. t.* [mit der Schreibmaschine] abschreiben; abtippen *(ugs.)*

'**typewriter** *n.* Schreibmaschine, *die.* '**typewritten** *adj.* maschine[n]geschrieben

typhoid ['taɪfɔɪd] *n.* ~ Typhus, *der*

typhoon [taɪˈfuːn] *n.* Taifun, *der*

typical ['tɪpɪkl] *adj.* typisch **(of** für)

typify ['tɪpɪfaɪ] *v. t.* ~ **sth.** als typisches Beispiel für etw. dienen

typing ['taɪpɪŋ] *n.* Maschineschreiben, *das*

typist ['taɪpɪst] *n.* Schreibkraft, *die*

tyrannical [tɪˈrænɪkl] *adj.* tyrannisch

tyranny ['tɪrənɪ] *n.* Tyrannei, *die*

tyrant ['taɪərənt] *n.* Tyrann, *der*

tyre ['taɪə(r)] *n.* Reifen, *der*

U

U, u [ju:] *n.* U, u, *das*
ubiquitous [ju:'bɪkwɪtəs] *adj.* allgegenwärtig
udder ['ʌdə(r)] *n.* Euter, *das*
ugliness ['ʌglɪnɪs] *n.* Häßlichkeit, *die*
ugly ['ʌglɪ] *adj.* **a)** häßlich; **b)** *(nasty)* übel ⟨*Wunde, Laune usw.*⟩
UK *abbr.* United Kingdom
Ukraine [ju:'kreɪn] *pr. n.* Ukraine, *die*
ulcer ['ʌlsə(r)] *n.* Geschwür, *das*
ulterior [ʌl'tɪərɪə(r)] *adj.* hintergründig; ~ **motive** Hintergedanke, *der*
ultimate ['ʌltɪmət] **1.** *attrib. adj.* **a)** *(final)* letzt...; *(eventual)* endgültig ⟨*Sieg*⟩; **b)** *(fundamental)* tiefst... **2.** *n.* the ~ **in comfort/luxury** der Gipfel an Bequemlichkeit/Luxus. '**ultimately** *adv.* **a)** *(in the end)* schließlich; **b)** *(in the last analysis)* letzten Endes
ultimatum [ʌltɪ'meɪtəm] *n., pl.* ~s *or* **ultimata** [ʌltɪ'meɪtə] Ultimatum, *das*
ultra'violet *adj. (Phys.)* ultraviolett; UV-⟨*Lampe, Filter*⟩
umbilical cord [ʌm'bɪlɪkl kɔ:d] *n.* Nabelschnur, *die*
umbrage ['ʌmbrɪdʒ] *n.* take ~ [at sth.] [an etw. (+ *Dat.*)] Anstoß nehmen
umbrella [ʌm'brelə] *n.* [Regen]schirm, *der*
umpire ['ʌmpaɪə(r)] *n.* Schiedsrichter, *der*/-richterin, *die*
umpteen [ʌmp'ti:n] *adj. (coll.)* zig *(ugs.)*; x *(ugs.)*
unabashed [ʌnə'bæʃt] *adj.* ungeniert
unable [ʌn'eɪbl] *pred. adj.* be ~ to do sth. etw. nicht tun können
unabridged [ʌnə'brɪdʒd] *adj.* ungekürzt
unac'ceptable *adj.* unannehmbar
unac'countable *adj.* unerklärlich. **unaccountably** [ʌnə'kaʊntəblɪ] *adv.* unerklärlicherweise
unac'customed *adj.* ungewohnt; be ~ to sth. etw. *(Akk.)* nicht gewöhnt sein
unadulterated [ʌnə'dʌltəreɪtɪd] *adj.* **a)** *(pure)* unverfälscht; **b)** *(utter)* völlig

unaided [ʌn'eɪdɪd] *adj.* ohne fremde Hilfe
unanimity [ju:nə'nɪmɪtɪ] *n.* Einmütigkeit, *die*
unanimous [ju:'nænɪməs] *adj.* einstimmig; be ~ in doing sth. etw. einmütig tun
unarmed [ʌn'ɑ:md] *adj.* unbewaffnet; ~ combat Kampf ohne Waffen
unassuming [ʌnə'sju:mɪŋ] *adj.* bescheiden
unattached [ʌnə'tætʃt] *adj.* **a)** nicht befestigt; **b)** *(without a partner)* ungebunden
unat'tended *adj.* **a)** ~ to *(not dealt with)* unerledigt; nicht bedient ⟨*Kunde*⟩; nicht behandelt ⟨*Patient*⟩; **b)** *(not supervised)* unbewacht ⟨*Parkplatz, Gepäck*⟩
unat'tractive *adj.* unattraktiv
unauthorized [ʌn'ɔ:θəraɪzd] *adj.* unbefugt; no entry for ~ persons Zutritt für Unbefugte verboten
una'vailable *adj.* nicht erhältlich ⟨*Ware*⟩; be ~ for comment zur Stellungnahme nicht zur Verfügung stehen
una'voidable *adj.* unvermeidlich
unaware [ʌnə'weə(r)] *adj.* be ~ of sth. sich *(Dat.)* einer Sache *(Gen.)* nicht bewußt sein. **unawares** [ʌnə'weəz] *adv.* catch sb. ~: jmdn. überraschen
unbalanced [ʌn'bælənst] *adj.* **a)** unausgewogen; **b)** *(mentally ~)* unausgeglichen
un'bearable *adj.*, **unbearably** [ʌn'beərəblɪ] *adv.* unerträglich
unbeatable [ʌn'bi:təbl] *adj.* unschlagbar *(ugs.)*
un'beaten *adj.* **a)** ungeschlagen; **b)** *(not surpassed)* unerreicht; ungebrochen ⟨*Rekord*⟩
unbe'lievable *adj.* **a)** unglaublich; **b)** *(tremendous)* unwahrscheinlich
unbiased, unbiassed [ʌn'baɪəst] *adj.* unvoreingenommen
unblemished [ʌn'blemɪʃt] *adj.* makellos ⟨*Haut, Ruf*⟩
un'block *v. t.* frei machen
un'bolt *v. t.* aufriegeln ⟨*Tür*⟩
unborn [ʌn'bɔ:n, *attrib.* 'ʌnbɔ:n] *adj.* ungeboren
un'breakable *adj.* unzerbrechlich
unburden [ʌn'bɜ:dn] *v. t.* ~ oneself sein Herz ausschütten
un'button *v. t.* aufknöpfen
uncalled-for [ʌn'kɔ:ldfɔ:(r)] *adj.* unangebracht
uncanny [ʌn'kænɪ] *adj.* unheimlich

uncaring [ʌnˈkeərɪŋ] *adj.* gleichgültig

unceasing [ʌnˈsiːsɪŋ] *adj.* unaufhörlich

unceremonious [ˌʌnserɪˈməʊnɪəs] *adj.* **a)** *(informal)* formlos; **b)** *(abrupt)* brüsk. **unceˈremoniously** *adv.* ohne Umschweife

unˈcertain *adj.* **a)** *(not sure)* be ~ |whether ...| sich *(Dat.)* nicht sicher sein[, ob ...]; **b)** *(not clear)* ungewiß ⟨*Ergebnis, Zukunft*⟩; **of ~ age/origin** unbestimmten Alters/unbestimmter Herkunft; **c)** *(ambiguous)* vage; **in no ~ terms** ganz eindeutig. **uncertainty** [ʌnˈsɜːtntɪ] *n.* **a)** Ungewißheit, *die;* **b)** *(hesitation)* Unsicherheit, *die*

unchanged [ʌnˈtʃeɪndʒd] *adj.* unverändert

unˈcharitable *adj.* **uncharitably** [ʌnˈtʃærɪtəblɪ] *adv.* lieblos

unˈcivil *adj.* unhöflich

uncle [ˈʌŋkl] *n.* Onkel, *der*

unˈcomfortable *adj.* **a)** unbequem; **b)** *(feeling discomfort)* be ~ : sich unbehaglich fühlen; **c)** *(uneasy, disconcerting)* unangenehm; peinlich ⟨*Stille*⟩. **unˈcomfortably** *adv.* unbequem; be ~ **aware of sth.** sich *(Dat.)* einer Sache peinlich bewußt sein

unˈcommon *adj.* ungewöhnlich

uncompliˈmentary *adj.* wenig schmeichelhaft

uncompromising [ʌnˈkɒmprəmaɪzɪŋ] *adj.* kompromißlos

unconˈditional *adj.* bedingungslos ⟨*Kapitulation*⟩; kategorisch ⟨*Ablehnung*⟩; ⟨*Versprechen*⟩ ohne Vorbehalte

unˈconscious **1.** *adj.* **a)** *(Med.)* bewußtlos; **b)** *(unaware)* be ~ **of sth.** sich einer Sache *(Gen.)* nicht bewußt sein; **c)** *(not intended; Psych.)* unbewußt. **2.** *n.* Unbewußte, *das.* **unˈconsciously** *adv.* unbewußt

unconˈventional *adj.,* **unconˈventionally** *adv.* unkonventionell

uncoˈoperative *adj.* unkooperativ; *(unhelpful)* wenig hilfsbereit

unˈcork *v. t.* entkorken

uncouth [ʌnˈkuːθ] *adj.* ungehobelt ⟨*Person, Benehmen*⟩; grob ⟨*Bemerkung*⟩

unˈcover *v. t.* aufdecken

undaunted [ʌnˈdɔːntɪd] *adj.* unverzagt

undecided [ʌndɪˈsaɪdɪd] *adj.* **a)** *(not settled)* nicht entschieden; **b)** *(hesitant)* unentschlossen

undeniably [ʌndɪˈnaɪəblɪ] *adv.* unbestreitbar

under [ˈʌndə(r)] **1.** *prep.* **a)** *(underneath, below)* unter *(position: + Dat.; motion: + Akk.);* **from ~ the table/bed** unter dem Tisch/Bett hervor; **b)** *(undergoing)* ~ **treatment** in Behandlung; ~ **repair** in Reparatur; ~ **construction** im Bau; **c)** *(in conditions of)* bei ⟨*Streß, hohen Temperaturen usw.*⟩; **d)** *(subject to)* unter (+ *Dat.*); ~ **the terms of the contract** nach den Bestimmungen des Vertrags; **e)** *(with the use of)* unter (+ *Dat.*); ~ **an assumed name** unter falschem Namen; **f)** *(less than)* unter (+ *Dat.*). **2.** *adv.* **a)** *(in or to a lower or subordinate position)* darunter; **b)** *(in/into a state of unconsciousness)* be ~/put sb. ~ : in Narkose liegen/jmdn. in Narkose versetzen

under: ~**carriage** *n.* Fahrwerk, *das;* ~**clothes** *n. pl.,* ~**clothing** *see* underwear; ~**cover** *adj. (disguised)* getarnt; *(secret)* verdeckt; ~**cover agent** Geheimagent, *der;* ~**current** *n.* Unterströmung, *die; (fig.)* Unterton, *der;* ~ˈ**cut** *v. t.,* ~**cut** unterbieten; ~**dog** *n.* **a)** *(in fight)* Unterlegene, *der/die;* **b)** *(fig.)* Benachteiligte, *der/die;* ~ˈ**done** *adj.* halbgar; ~**estimate** [ʌndərˈestɪmeɪt] **1.** *v. t.* unterschätzen; **2.** [ʌndərˈestɪmət] *n.* Unterschätzung, *die;* ~ˈ**fed** *adj.* unterernährt; ~ˈ**foot** *adv.* am Boden; **be trampled** ~**foot** mit Füßen zertrampelt werden; ~ˈ**go** *v. t., forms as* go 1 durchmachen; ~**go treatment** sich einer Behandlung unterziehen; ~**go a change** sich verändern; ~ˈ**graduate** *n.* ~**graduate |student|** Student/Studentin vor der ersten Prüfung; ~**ground** 1. |--ˈ-| *adv.* **a)** unter der Erde; *(Mining)* unter Tage; **b)** *(fig.) (in hiding)* im Untergrund; *(into hiding)* in den Untergrund; **2.** |ˈ---| *adj.* unterirdisch ⟨*Höhle, See*⟩; ~**ground railway** Untergrundbahn, *die;* U-Bahn, *die;* ~**ground car-park** Tiefgarage, *die;* **3.** *n. (railway)* U-Bahn, *die;* ~ **station/train** U-Bahnhof, *der/*U-Bahn-Zug, *der;* ~**growth** *n.* Unterholz, *das;* ~**hand,** ~**handed** *adj.* **a)** *(secret)* heimlich; **b)** *(crafty)* hinterhältig; ~**lay** *n.* Unterlage, *die;* ~ˈ**lie** *v. t., forms as* ²lie: ~**lie sth.** *(fig.)* einer Sache *(Dat.)* zugrundeliegen; ~**lying cause** eigentliche Ursache; ~ˈ**line** *v. t.* unterstreichen

underling [ˈʌndəlɪŋ] *n.* Untergebene, *der/die*

under: ~ˈ**lying** *see* underlie; ~ˈ**mine**

v.t. **a)** unterhöhlen; **b)** *(fig.)* untergraben; unterminieren ⟨*Autorität*⟩

underneath [ʌndəˈniːθ] **1.** *prep.* unter *(position:* + *Dat.; motion:* + *Akk.).* **2.** *adv.* darunter

under: ~'**paid** *adj.* unterbezahlt; ~**pants** *n.pl.* Unterhose, *die;* ~**pass** *n.* Unterführung, *die;* ~'**play** *v.t.* herunterspielen; ~'**privileged** *adj.* unterprivilegiert; ~'**rate** *v.t.* unterschätzen; ~**seal** *n.* Unterbodenschutz, *der*

understand [ʌndəˈstænd] **1.** *v.t.,* **understood** [ʌndəˈstʊd] **a)** verstehen; **make oneself understood** sich verständlich machen; **b)** *(have heard)* gehört haben; **c)** *(take as implied)* **it was understood that ...:** es wurde allgemein angenommen, daß ... **2.** *v.i.,* **understood a)** verstehen; **b)** *(gather, hear)* **if I ~ correctly** wenn ich mich nicht irre; **he is, I ~, no longer here** er ist, wie ich höre, nicht mehr hier. **understandably** [ʌndəˈstændəblɪ] *adv.* verständlicherweise. **under'standing 1.** *adj.* verständnisvoll. **2.** *n.* **a)** *(agreement)* Verständigung, *die;* **reach an ~ with sb.** sich mit jmdm. verständigen; **on the ~ that ...:** unter der Voraussetzung, daß ...; **b)** *(intelligence)* Verstand, *der;* **c)** *(insight)* Verständnis, *das* (**of, for** für)

under: ~**statement** *n.* Untertreibung, *die;* ~**study** *n.* Ersatzspieler, *der/*-spielerin, *die;* ~'**take** *v.t., forms as take 1* unternehmen; ~**take a task** eine Aufgabe übernehmen; ~**take to do sth.** sich verpflichten, etw. zu tun; ~**taker** *n.* Leichenbestatter, *der/*-bestatterin, *die;* ~'**taking** *n.* **a)** *(task)* Aufgabe, *die;* **b)** *(pledge)* Versprechen, *das;* ~**tone** *n.* **in** ~**tones** *or* **an** ~**tone** mit gedämpfter Stimme; ~**tone of criticism** kritischer Unterton; ~**tow** *n.* Unterströmung, *die;* ~'**value** *v.t.* unterbewerten; ~**water 1.** [ˈ----] *attrib. adj.* Unterwasser-; **2.** [--ˈ--] *adv.* unter Wasser; ~**wear** *n.* Unterwäsche, *die;* ~'**weight** *adj.* untergewichtig; ~**world** *n.* Unterwelt, *die*

unde'sirable *adj.* unerwünscht; **it is ~ that ...:** es ist nicht wünschenswert, daß ...

undeveloped [ʌndɪˈveləpt] *adj.* **a)** *(immature)* nicht voll ausgebildet; **b)** *(not built on)* nicht bebaut

undies [ˈʌndɪz] *n. pl. (coll.)* Unterwäsche, *die*

un'dignified *adj.* blamabel

undo [ʌnˈduː] *v.t.,* **undoes** [ʌnˈdʌz], **un-** doing [ʌnˈduːɪŋ], **undid** [ʌnˈdɪd], **undone** [ʌnˈdʌn] *(unfasten)* aufmachen

un'done *adj.* **a)** *(not accomplished)* unerledigt; **b)** *(not fastened)* offen

undoubted [ʌnˈdaʊtɪd] *adj.* unzweifelhaft. **un'doubtedly** *adv.* zweifellos

un'dress 1. *v.t.* ausziehen; **get ~ed** sich ausziehen **2.** *v.i.* sich ausziehen

un'due *attrib. adj.* übertrieben; übermäßig

undulating [ˈʌndjʊleɪtɪŋ] *adj.* Wellen- ⟨*linie*⟩; **~ country** sanfte Hügellandschaft

unduly [ʌnˈdjuːlɪ] *adv.* übermäßig

undying [ʌnˈdaɪɪŋ] *adj.* ewig; unsterblich ⟨*Ruhm*⟩

unearth [ʌnˈɜːθ] *v.t.* **a)** ausgraben; **b)** *(fig.: discover)* aufdecken

unearthly [ʌnˈɜːθlɪ] *adj.* unheimlich; **at an ~ hour** in aller Herrgottsfrühe

un'easy *adj.* **a)** *(anxious)* besorgt; **he felt ~:** ihm war unbehaglich zumute; **b)** *(restless)* unruhig

uneatable [ʌnˈiːtəbl] *adj.* ungenießbar

uneco'nomic *adj.* unrentabel. **uneco'nomical** *adj.* **~ [to run]** unwirtschaftlich

unemployed [ʌnɪmˈplɔɪd] **1.** *adj.* arbeitslos. **2.** *n.pl.* **the ~:** die Arbeitslosen. **unem'ployment** *n.* Arbeitslosigkeit, *die.* **unem'ployment benefit** *n.* Arbeitslosengeld, *das*

un'ending *adj.* endlos

un'equal *adj.* unterschiedlich; ungleich ⟨*Kampf*⟩; **be ~ to sth.** einer Sache *(Dat.)* nicht gewachsen sein. **un'equalled** (*Amer.:* **unequaled**) [ʌnˈiːkwld] *adj.* unerreicht

une'quivocal *adj.* eindeutig

unerring [ʌnˈɜːrɪŋ] *adj.* unfehlbar

un'ethical *adj.* unmoralisch

un'even *adj.* **a)** *(not smooth)* uneben; **b)** *(not uniform)* ungleichmäßig; **c)** *(odd)* ungerade ⟨*Zahl*⟩. **un'evenly** *adv.* ungleichmäßig

unex'pected *adj.* unerwartet

un'fair *adj.* unfair; ungerecht. **un'fairly** *adv.* **a)** *(unjustly)* ungerecht; unfair ⟨*spielen*⟩; **b)** *(unreasonably)* zu Unrecht. **un'fairness** *n.* Ungerechtigkeit, *die*

un'faithful *adj.* untreu

unfa'miliar *adj.* **a)** *(strange)* unbekannt; ungewohnt ⟨*Arbeit*⟩; **b)** **be ~ with sth.** sich mit etw. nicht auskennen

un'fasten *v.t.* **a)** öffnen; **b)** *(detach)* lösen

un'favourable *adj.* ungünstig. **un'favourably** *adv.* ungünstig; **be ~ disposed towards sb./sth.** jmdm./etw. gegenüber ablehnend eingestellt sein

un'feeling *adj.* gefühllos

unfinished [ʌnˈfɪnɪʃt] *adj.* unvollendet ⟨*Werk*⟩; unerledigt ⟨*Arbeit*⟩

un'fit *adj.* **a)** ungeeignet; **b)** *(not physically fit)* nicht fit *(ugs.);* **~ for military service** [wehrdienst]untauglich

un'flattering *adj.* wenig schmeichelhaft

un'flinching *adj.* unerschrocken

un'fold 1. *v. t.* entfalten; ausbreiten ⟨*Zeitung, Landkarte*⟩. **2.** *v. i.* sich entfalten; *(develop)* sich entwickeln

unfore'seen *adj.* unvorhergesehen

unforgettable [ʌnfəˈgetəbl] *adj.* unvergeßlich

un'fortunate *adj.* unglücklich. **un'fortunately** *adv.* leider

un'founded *adj. (fig.)* unbegründet

un'freeze *v. t. & i.,* **unfroze** [ʌnˈfrəʊz], **unfrozen** [ʌnˈfrəʊzn] auftauen

un'friendly *adj.* unfreundlich; feindlich ⟨*Staat*⟩

un'furl 1. *v. t.* aufrollen; losmachen ⟨*Segel*⟩. **2.** *v. i.* sich aufrollen

un'furnished *adj.* unmöbliert

ungainly [ʌnˈgeɪnlɪ] *adj.* unbeholfen

ungram'matical *adj.* ungrammatisch

un'grateful *adj.* undankbar

un'happily *adv.* **a)** unglücklich; **b)** *(unfortunately)* leider

un'happiness *n.* Bekümmertheit, *die*

un'happy *adj.* unglücklich; *(not content)* unzufrieden (**about** with); **be** *or* **feel ~ about doing sth.** Bedenken haben, etw. zu tun

un'harmed *adj.* unbeschädigt; *(uninjured)* unverletzt

un'healthy *adj.* ungesund

un'helpful *adj.* wenig hilfsbereit ⟨*Person*⟩; ⟨*Bemerkung, Kritik*⟩ die einem nicht weiterhilft

un'hook *v. t.* vom Haken nehmen; aufhaken ⟨*Kleid*⟩

un'hurt *adj.* unverletzt

unhy'gienic *adj.* unhygienisch

unicorn [ˈjuːnɪkɔːn] *n.* Einhorn, *das*

uni'dentified *adj.* nicht identifiziert; **~ flying object** unbekanntes Flugobjekt

unification [juːnɪfɪˈkeɪʃn] *n.* Einigung, *die*

uniform [ˈjuːnɪfɔːm] **1.** *adj.* einheitlich; **be ~ in shape/size** die gleiche Form/Größe haben. **2.** *n.* Uniform, *die;* **in/out of ~:** in/ohne Uniform.

uniformity [juːnɪˈfɔːmɪtɪ] *n.* Einheitlichkeit, *die.* **'uniformly** *adv.* einheitlich

unify [ˈjuːnɪfaɪ] *v. t.* einigen

unilateral [juːnɪˈlætərl] *adj.* einseitig

uni'maginable *adj.* unvorstellbar

uni'maginative *adj.* phantasielos

unim'portant *adj.* unwichtig; bedeutungslos

unin'habitable *adj.* unbewohnbar

unin'habited *adj.* unbewohnt

un'injured *adj.* unverletzt

uninspiring [ʌnɪnˈspaɪərɪŋ] *adj.* langweilig

unin'telligent *adj.* nicht intelligent

unin'telligible *adj.* unverständlich

unin'tended *adj.* unbeabsichtigt

unin'tentional *adj.,* **unin'tentionally** *adv.* unabsichtlich

un'interested *adj.* desinteressiert (**in** an + *Dat.*)

union [ˈjuːnɪən] *n.* **a)** *(trade ~)* Gewerkschaft, *die;* **b)** *(Polit.)* Union, *die.* **Union 'Jack** *n. (Brit.)* Union Jack, *der*

unique [juːˈniːk] *adj.* einzigartig

unison [ˈjuːnɪsən] *n.* Unisono, *das;* **in ~:** einstimmig; **act in ~** *(fig.)* vereint handeln

unit [ˈjuːnɪt] *n.* **a)** *(also Mil., Math.)* Einheit, *die;* **~ of length/monetary ~:** Längen-/Währungseinheit, *die;* **b)** *(piece of furniture)* Element, *das;* **kitchen ~:** Küchenelement, *das*

unite [juːˈnaɪt] **1.** *v. t.* vereinigen; einen, einigen ⟨*Partei, Mitglieder*⟩. **2.** *v. i.* sich vereinigen. **u'nited** *adj.* **a)** *(harmonious)* einig; **b)** *(combined)* gemeinsam

United: ~ 'Kingdom *pr. n.* Vereinigtes Königreich [Großbritannien und Nordirland]; **~ 'Nations** *pr. n. sing.* Vereinte Nationen *Pl.;* **~ States [of A'merica]** *pr. n. sing.* Vereinigte Staaten [von Amerika]

unity [ˈjuːnɪtɪ] *n.* Einheit, *die*

universal [juːnɪˈvɜːsl] *adj.,* **uni'versally** *adv.* allgemein

universe [ˈjuːnɪvɜːs] *n.* Universum, *das*

university [juːnɪˈvɜːsɪtɪ] *n.* Universität, *die; attrib.* Universitäts-

un'just *adj.* ungerecht

unkempt [ʌnˈkempt] *adj.* ungepflegt

un'kind *adj.,* **un'kindly** *adv.* unfreundlich. **un'kindness** *n.* Unfreundlichkeit, *die*

un'known 1. *adj.* unbekannt. **2.** *adv.* **~ to sb.** ohne daß jmd. davon weiß/ wußte

un'lawful *adj.* ungesetzlich

unless [ən'les] *conj.* es sei denn; wenn ... nicht

un'like 1. *adj.* nicht ähnlich. **2.** *prep.* be ~ sb./sth. jmdm./einer Sache nicht ähnlich sein; ~ **him**, ...: im Gegensatz zu ihm ...

un'likely *adj.* unwahrscheinlich; **be ~ to do sth.** etw. wahrscheinlich nicht tun

un'limited *adj.* unbegrenzt

un'load *v. t.* entladen ⟨*Lastwagen, Waggon*⟩; löschen ⟨*Schiff, Schiffsladung*⟩; ausladen ⟨*Gepäck*⟩

un'lock *v. t.* aufschließen

un'lucky *adj.* **a)** unglücklich; *(not successful)* glücklos; **be |very| ~:** [großes] Pech haben; **b)** *(bringing bad luck)* **an ~ number** eine Unglückszahl; **be ~:** Unglück bringen

un'manned *adj.* unbemannt

un'married *adj.* unverheiratet; ledig

un'mask *v. t.* *(fig.)* entlarven

unmi'stakable *adj.* deutlich; unverwechselbar ⟨*Handschrift, Stimme*⟩. **unmistakably** [ʌnmɪ'stəɪkəblɪ] *adv.* unverkennbar

un'mitigated *adj.* vollkommen; **be an ~ disaster** *(coll.)* eine einzige Katastrophe sein

un'natural *adj.,* **un'naturally** *adv.* unnatürlich; *(abnormal)* nicht normal

un'necessarily *adv.,* **un'necessary** *adj.* unnötig

unofficial *adj.,* **unofficially** *adv.* inoffiziell

un'pack *v. t. & i.* auspacken

un'paid *adj.* unbezahlt; nicht bezahlt; ~ **for** nicht bezahlt

unpalatable [ʌn'pælətəbl] *adj.* ungenießbar

un'paralleled *adj.* beispiellos

un'pardonable *adj.* unverzeihlich

un'pleasant *adj.,* **un'pleasantly** *adv.* unangenehm. **un'pleasantness** *n.* *(bad feeling)* Verstimmung, die

un'plug *v. t.,* -gg-: ~ **a lamp** den Stecker einer Lampe herausziehen

un'popular *adj.* unbeliebt ⟨*Lehrer, Regierung usw.*⟩, unpopulär ⟨*Maßnahme, Politik*⟩ **(with** bei)

un'precedented *adj.* beispiellos

unpre'dictable *adj.* unberechenbar

unpre'pared *adj.* unvorbereitet

unprepos'sessing *adj.* wenig attraktiv

unpre'tentious *adj.* einfach ⟨*Wein, Stil, Haus*⟩; bescheiden ⟨*Person*⟩

unprincipled [ʌn'prɪnsɪpld] *adj.* skrupellos

unprintable [ʌn'prɪntəbl] *adj.* nicht druckreif

unpro'ductive *adj.* fruchtlos ⟨*Diskussion, Nachforschung*⟩; unproduktiv ⟨*Zeit, Arbeit*⟩

unpro'fessional *adj.* *(contrary to standards)* standeswidrig

un'profitable *adj.* unrentabel

un'promising *adj.* nicht sehr vielversprechend

un'qualified *adj.* **a)** unqualifiziert; **b)** *(absolute)* uneingeschränkt; voll ⟨*Erfolg*⟩

un'questionable *adj.* unbezweifelbar ⟨*Tatsache*⟩; unbestreitbar ⟨*Recht, Ehrlichkeit*⟩. **unquestionably** [ʌn'kwestʃənəblɪ] *adv.* ohne Frage

unravel [ʌn'rævl] **1.** *v. t.,* *(Brit.)* -ll- entwirren; *(undo)* aufziehen; *(fig.)* ~ **a mystery/the truth** ein Geheimnis enträtseln/die Wahrheit aufdecken. **2.** *v. i.,* *(Brit.)* -ll- sich aufziehen

un'real *adj.* unwirklich

unrea'listic *adj.* unrealistisch

un'reasonable *adj.* unvernünftig; übertrieben ⟨*Ansprüche, Forderung, Preis, Kosten*⟩

unrecognizable [ʌn'rekəgnaɪzəbl] *adj.* **be |absolutely** *or* **quite| ~:** [überhaupt] nicht wiederzuerkennen sein

unre'lated *adj.* **be ~:** nicht miteinander zusammenhängen; *(by family)* nicht verwandt sein

unre'liable *adj.* unzuverlässig

unrequited [ʌnrɪ'kwaɪtɪd] *adj.* unerwidert

unreservedly [ʌnrɪ'zɜːvɪdlɪ] *adv.* uneingeschränkt

un'rest *n.* Unruhen *Pl.*

un'ripe *adj.* unreif

un'rivalled *(Amer.:* **un'rivaled)** *adj.* unübertroffen

un'roll 1. *v. t.* aufrollen. **2.** *v. i.* sich aufrollen

unruly [ʌn'ruːlɪ] *adj.* ungebärdig

un'safe *adj.* nicht sicher; **feel ~:** sich unsicher fühlen

un'said *adj.* ungesagt

un'salted *adj.* ungesalzen

unsatis'factory *adj.* unbefriedigend

un'savoury *(Amer.:* **un'savory)** *adj.* unangenehm; zweifelhaft ⟨*Angelegenheit*⟩; unerfreulich ⟨*Einzelheiten*⟩

unscathed [ʌn'skeɪðd] *adj.* unversehrt

un'screw 1. *v. t.* abschrauben. **2.** *v. i.* sich abschrauben lassen

un'scrupulous *adj.* skrupellos
un'seemly *adj.* unschicklich
unself'conscious *adj.* unbefangen
un'selfish *adj.* selbstlos. un'selfish-
ress *n.* Selbstlosigkeit, *die*
un'settled *adj. (changeable)* wechselhaft; *(fig.)* ruhelos ⟨*Leben*⟩; unruhig
⟨*Zeit, Land*⟩
un'settling *adj.* störend
unshak[e]able [ʌnˈʃeɪkəbl] *adj.* un-
erschütterlich
un'shaven *adj.* unrasiert
un'sightly *adj.* unschön
un'skilled *adj.* ungelernt ⟨*Arbeiter*⟩
un'sociable *adj.* ungesellig
unso'phisticated *adj.* einfach
un'sound *adj.* **a)** *(diseased)* nicht gesund; krank; **b)** baufällig ⟨*Gebäude*⟩;
c) *(ill-founded)* wenig stichhaltig;
nicht vertretbar ⟨*Ansicht, Methode*⟩;
d) of ~ mind unzurechnungsfähig
unspeakable [ʌnˈspiːkəbl] *adj.* unbeschreiblich; *(very bad)* unsäglich
un'stable *adj.* nicht stabil; |mentally/
emotionally| ~: [psychisch] labil
un'steadily *adv.* unsicher
un'steady *adj.* unsicher; wackelig
⟨*Leiter, Tisch*⟩
un'stuck *adj.* come ~: sich lösen; *(fig.
coll.: fail)* ⟨*Person:*⟩ baden gehen *(ugs.)*
(over mit)
unsuc'cessful *adj.* erfolglos; be ~:
keinen Erfolg haben. unsuc'cess-
fully *adv.* erfolglos
un'suitable *adj.* ungeeignet
unsu'specting *adj.* nichtsahnend
un'sweetened *adj.* ungesüßt
unsympa'thetic *adj.* wenig mitfühlend; be ~: kein Mitgefühl zeigen
unthinkable [ʌnˈθɪŋkəbl] *adj.* unvorstellbar
un'tidily *adv.* unordentlich
un'tidiness *n. see* untidy: Ungepflegt-
heit, *die;* Unaufgeräumtheit, *die*
un'tidy *adj.* ungepflegt ⟨*Äußeres, Person, Garten*⟩; unaufgeräumt ⟨*Zimmer*⟩
un'tie *v.t.*, un'tying aufknüpfen ⟨*Seil,
Paket*⟩; aufbinden ⟨*Knoten*⟩; losbin-
den ⟨*Pferd, Boot*⟩
until [ənˈtɪl] **1.** *prep.* bis; ~ |the| evening
bis zum Abend; ~ then bis dahin; not
~ |Christmas/the summer| erst [Weih-
nachten/im Sommer]. **2.** *conj.* bis
un'timely *adj.* **a)** ungelegen; **b)** *(premature)* vorzeitig
un'tiring *adj.* unermüdlich
un'told *adj.* unbeschreiblich; unermeßlich ⟨*Reichtümer, Anzahl*⟩
untoward [ʌntəˈwɔːd, ʌnˈtəʊəd] *adj.*

ungünstig; nothing ~ happened es gab
keine Schwierigkeiten
untranslatable [ʌntrænsˈleɪtəbl] *adj.*
unübersetzbar
un'true *adj.* unwahr; that's ~: das ist
nicht wahr
un'trustworthy *adj.* unzuverlässig
un'truth *n.* Unwahrheit, *die*
¹unused [ʌnˈjuːzd] *adj. (new, fresh)* un-
benutzt; *(not utilized)* ungenutzt
²unused [ʌnˈjuːst] *adj. (unaccustomed)*
be ~ to sth./doing sth. etw. *(Akk.)*
nicht gewohnt sein/nicht gewohnt
sein, etw. zu tun
un'usual *adj.*, un'usually *adv.* ungewöhnlich
un'veil *v.t.* enthüllen; *(fig.)* vorstellen
⟨*Produkt*⟩; enthüllen ⟨*Plan*⟩
un'versed *adj.* nicht bewandert (in in
+ *Dat.*)
un'wanted *adj.* unerwünscht
un'warranted *adj.* ungerechtfertigt
un'welcome *adj.* unwillkommen
un'well *adj.* unwohl; look ~: nicht
wohl *od.* gut aussehen; he feels ~
(poorly) er fühlt sich nicht wohl
unwieldy [ʌnˈwiːldɪ] *adj.* sperrig
un'willing *adj.* widerwillig; be ~ to do
sth. etw. nicht tun wollen. un'willingly *adv.* widerwillig
unwind [ʌnˈwaɪnd] **1.** *v.t.*, unwound
[ʌnˈwaʊnd] abwickeln. **2.** *v.i.*, unwound **a)** sich abwickeln; **b)** *(coll.:
relax)* sich entspannen
un'wise *adj.* unklug
unwitting [ʌnˈwɪtɪŋ] *adj.*, un'wittingly *adv.* unwissentlich
un'workable *adj.* undurchführbar
⟨*Plan*⟩
un'worthy *adj.* unwürdig; be ~ of sth.
einer Sache *(Gen.)* nicht würdig sein;
be ~ of sb./sth. ⟨*Verhalten:*⟩ einer Per-
son/Sache *(Gen.)* unwürdig sein
un'wrap *v.t.*, **-pp-** auswickeln
un'written *adj.* ungeschrieben
un'zip *v.t.*, **-pp-**: ~ a dress/bag *etc.* den
Reißverschluß eines Kleides/einer
Tasche *usw.* öffnen
up [ʌp] **1.** *adv.* **a)** *(to higher place)* nach
oben; *(in lift)* aufwärts; the bird flew
up to the roof der Vogel flog aufs
Dach [hinauf]; up into the air in die
Luft [hinauf]; up here/there hier her-
auf/dort hinauf; higher/a little way up
höher/ein kurzes Stück hinauf; come
on up! komm [hier/weiter] herauf!;
b) *(to upstairs)* herauf/hinauf; nach
oben; **c)** *(in higher place, upstairs)*
oben; up here/there hier/da oben; the

next floor up ein Stockwerk höher; **d)** *(out of bed)* **be up** aufsein; **e)** *(in price, value, amount)* **prices have gone up/are up** die Preise sind gestiegen; **butter is up [by ...]** Butter ist [...] teurer; **f)** *(as far as)* **up to sth.** bis zu etw.; **up to here/there** bis hier[hin]/bis dorthin; **up to sth.** *(capable of sth.)* einer Sache *(Dat.)* [nicht] gewachsen sein/ sich einer Sache *(Dat.)* [nicht] ge- wachsen fühlen; **[not] be/feel up to doing sth.** [nicht] in der Lage sein/sich [nicht] in der Lage fühlen, etw. zu tun; **h) be up to sth.** *(doing)* etw. anstellen *(ugs.)*; **it is [not] up to sb. to do sth.** *(sb.'s duty)* es ist [nicht] jmds. Sache, etw. zu tun; **i) be three points/games up** mit drei Punkten/Spielen vorn lie- gen; **j) walk up and down** auf und ab gehen; **k) time is up** die Zeit ist abge- laufen. **2.** *prep.* herauf/hinauf; **walk up the hill/road** den Berg/die Straße hinaufgehen; **walk up and down the platform** auf dem Bahnsteig auf und ab gehen; **further up the ladder/coast** weiter oben auf der Leiter/an der Kü- ste; **live just up the road** ein Stück wei- ter oben in der Straße wohnen. **3.** *adj.* *(coll.: amiss)* **what's up?** was ist los? *(ugs.)*; **something is up** irgendwas ist los *(ugs.)*. **4.** *v. t.* **-pp-** *(coll.: increase)* erhöhen

'**upbringing** *n.* Erziehung, *die*

up'date *v. t.* auf den aktuellen Stand bringen

up'grade *v. t.* **a)** aufwerten ⟨*Stellung*⟩; **b)** *(improve)* verbessern

upheaval [ʌp'hiːvl] *n.* Aufruhr, *der;* *(disturbance)* Durcheinander, *das*

up'hill 1. *adj.* *(fig.)* **an ~ task/struggle** eine mühselige Aufgabe/ein harter Kampf. **2.** *adv.* bergauf

uphold *v. t.*, **upheld** unterstützen; wahren ⟨*Tradition*⟩

upholster [ʌp'həʊlstə(r)] *v. t.* polstern. **up'holsterer** *n.* Polsterer, *der*/Polste- rin, *die.* **up'holstery** *n.* **a)** *(craft)* Polster[er]handwerk, *das;* **b)** *(pad- ding)* Polsterung, *die*

'**upkeep** *n.* Unterhalt, *der*

'**up-market** *adj.* exklusiv

upon [ə'pɒn] *prep.* auf *(direction:* + *Akk.; position:* + *Dat.)*

upper ['ʌpə(r)] **1.** *compar. adj.* ober...; Ober⟨*grenze, -lippe, -arm usw.*⟩; **~ circle** oberer Rang; **~ class[es]** Ober- schicht, *die;* **have/get/gain the ~ hand** die Oberhand haben/gewinnen/erhal- ten. **2.** *n.* Oberteil, *das.* **upper 'deck**

n. Oberdeck, *das.* '**uppermost 1.** *adj.* oberst... **2.** *adv.* ganz oben

'**upright 1.** *adj.* aufrecht. **2.** *n.* seitliche Leiste

'**uprising** *n.* Aufstand, *der*

'**uproar** *n.* Aufruhr, *der*

up'root *v. t.* [her]ausreißen; ⟨*Sturm:*⟩ entwurzeln

upset 1. [ʌp'set] *v. t.*, **-tt-, upset a)** *(overturn)* umkippen; *(accidentally)* umstoßen ⟨*Tasse, Milch usw.*⟩; **b)** *(dis- tress)* erschüttern; *(make angry)* auf- regen; **don't let it ~ you** nimm es nicht so schwer; **c)** *(make ill)* **sth. ~s sb.** etw. bekommt jmdm. nicht; **d)** durchein- anderbringen ⟨*Plan*⟩. **2.** *v. i.*, **-tt-, upset** umkippen. **3.** *adj. (distressed)* be- stürzt; *(agitated)* aufgeregt; **get ~ [about/over sth.]** sich [über etw. *(Akk.)*] aufregen. **4.** ['ʌpset] *n.* **a)** *(agitation)* Aufregung, *die;* *(annoyance)* Verärge- rung, *die;* **b) stomach ~:** Magenver- stimmung, *die;* **c)** *(upheaval)* Aufruhr, *der.* **up'setting** *adj.* erschütternd; ⟨*sad*⟩ traurig; *(annoying)* ärgerlich

'**upshot** *n.* Ergebnis, *das*

upside 'down 1. *adv.* verkehrt her- um; **turn sth. ~:** etw. auf den Kopf stellen. **2.** *adj.* auf dem Kopf stehend ⟨*Bild*⟩; **be ~:** auf dem Kopf stehen

upstairs 1. [-'-] *adv.* nach oben ⟨*gehen, kommen*⟩; oben ⟨*sein, wohnen*⟩. **2.** ['--] *adj.* im Obergeschoß *nachgestellt*

'**upstart** *n.* Emporkömmling, *der*

up'stream *adv.* flußaufwärts

'**uptake** *n.* **be quick/slow on the ~** *(coll.)* schnell begreifen/schwer von Begriff sein *(ugs.)*

uptight [-'-, '--] *adj. (coll.: tense)* ner- vös *(about wegen)*

up to 'date *adj.* **be/keep ~:** auf dem neusten Stand sein/bleiben; **bring sth. ~:** etw. auf den neusten Stand brin- gen. **up-to-'date** *attrib. adj. (current)* aktuell; *(modern)* modern

'**upturn** *n.* Aufschwung, *der* *(in Gen.)*

upward ['ʌpwəd] **1.** *adj.* nach oben ge- richtet. **2.** *adv.* aufwärts ⟨*sich bewe- gen*⟩; nach oben ⟨*sehen, gehen*⟩. **up- wards** ['ʌpwədz] *adv.* **a)** *see* upward 2; **b) ~ of** über (+ *Akk.*)

uranium [jʊə'reɪnɪəm] *n.* Uran, *das*

Uranus ['jʊərənəs, jʊə'reɪnəs] *pr. n.* *(Astron.)* Uranus, *der*

urban ['ɜːbn] *adj.* städtisch; Stadt⟨*ge- biet, -bevölkerung, -planung*⟩

urchin ['ɜːtʃɪn] *n.* Strolch, *der*

urge [ɜːdʒ] **1.** *v. t.* **~ sb. to do sth.** jmdn. drängen, etw. zu tun. **2.** *n.* Trieb, *der.*

urge 'on *v. t.* antreiben; *(encourage)* anfeuern

urgency ['ɜːdʒənsɪ] *n.* Dringlichkeit, *die*

urgent ['ɜːdʒənt] *adj.* dringend; *(to be dealt with immediately)* eilig; **be in ~ need of sth.** etw. dringend brauchen.
'urgently *adv.* dringend; *(immediately)* eilig

urinate ['juərɪneɪt] *v. i.* urinieren

urine ['juərɪn] *n.* Urin, *der;* Harn, *der*

urn [ɜːn] *n.* **a) tea/coffee ~:** Tee-/Kaffeemaschine, *die;* **b)** *(vessel)* Urne, *die*

Uruguay ['juərəgwaɪ] *pr. n.* Uruguay *(das)*

US *abbr.* United States USA

us [əs, *stressed* ʌs] *pron.* uns; **it's us** wir sind's *(ugs.)*

USA *abbr.* United States of America USA

usage ['juːzɪdʒ, 'juːsɪdʒ] *n.* **a)** Brauch, *der;* **b)** *(Ling.)* Sprachgebrauch, *der*

use 1. [juːs] *n.* **a)** Gebrauch, *der; (of dictionary, calculator, room)* Benutzung, *die; (of word, pesticide, spice)* Verwendung, *die;* |not| **be in ~:** [nicht] in Gebrauch sein; **be no longer in ~:** nicht mehr verwendet werden; **make ~ of sb./sth.** jmdn./etw. gebrauchen/ *(exploit)* ausnutzen; **make good ~ of, turn** *or* **put to good ~:** gut nutzen ⟨*Zeit, Talent, Geld*⟩; **put sth. to ~:** etw. verwenden; **b)** *(usefulness)* Nutzen, *der;* **is it of |any| ~?** ist das [irgendwie] von Nutzen?; **be |of| no ~ |to sb.|** [jmdm.] nicht nützen; **it's no ~ |doing that|** es hat keinen Sinn[, das zu tun]; **c)** *(purpose)* Verwendung, *die;* **have/find a ~ for sth./sb.** für etw./jmdn. Verwendung haben/finden; **have no/not much ~ for sth./sb.** etw./jmdn. nicht/kaum brauchen. **2.** [juːz] *v. t.* **a)** benutzen; nutzen ⟨*Gelegenheit*⟩; anwenden ⟨*Gewalt*⟩; in Anspruch nehmen ⟨*Firma, Dienstleistung*⟩; nutzen ⟨*Zeit, Gelegenheit*⟩; verwenden ⟨*Kraftstoff, Butter, Wort*⟩; **b) ~d to** ['juːst tə]: **I ~d to live in London** früher habe ich in London gelebt. **use 'up** *v. t.* aufbrauchen; verbrauchen ⟨*Geld, Energie*⟩

used 1. *adj.* **a)** [juːzd] gebraucht; gestempelt ⟨*Briefmarke*⟩; **~ car** Gebrauchtwagen, *der;* **b)** [juːst] **~ to sth.** [an] etw. *(Akk.)* gewöhnt. **2.** [juːst] *see* use 2 b

useful ['juːsfl] *adj.* nützlich; praktisch ⟨*Werkzeug*⟩; hilfreich ⟨*Rat, Idee*⟩.
'usefulness *n.* Nützlichkeit, *die*

'useless *adj.* unbrauchbar ⟨*Werkzeug,*

Rat, Idee⟩; nutzlos ⟨*Wissen, Information, Protest, Anstrengung, Kampf*⟩; zwecklos ⟨*Widerstand, Protest*⟩

'user *n.* Benutzer, *der*/Benutzerin, *die.*
'user-friendly *adj.* benutzerfreundlich

usher ['ʌʃə(r)] **1.** *n. (in court)* Gerichtsdiener, *der; (at cinema, church)* Platzanweiser, *der.* **2.** *v. t.* führen. **usher 'in** *v. t.* hineinführen; *(fig.)* einläuten

usherette [ʌʃə'ret] *n.* Platzanweiserin, *die*

USSR *abbr. (Hist.)* Union of Soviet Socialist Republics UdSSR, *die*

usual ['juːʒʊəl] *adj.* üblich. **usually** ['juːʒʊəlɪ] *adv.* gewöhnlich

usurp [juː'zɜːp] *v. t.* sich *(Dat.)* widerrechtlich aneignen

utensil [juː'tensɪl] *n.* Utensil, *das;* **writing ~s** Schreibutensilien; **kitchen ~s** Küchengeräte

uterus ['juːtərəs] *n.* Gebärmutter, *die*

utility [juː'tɪlɪtɪ] *n.* **a)** Nutzen, *der;* **b)** |public| **~:** öffentlicher Versorgungsbetrieb. **u'tility room** *n. Raum, in den [größere] Haushaltsgeräte (z. B. Waschmaschine) installiert sind*

utilize ['juːtɪlaɪz] *v. t.* nutzen

utmost ['ʌtməʊst] **1.** *adj.* äußerst...; größt... ⟨*Höflichkeit, Eleganz, Einfachheit, Geschwindigkeit*⟩. **2.** *n.* Äußerste, *das;* **do** *or* **try one's ~ to do sth.** mit allen Mitteln versuchen, etw. zu tun

¹utter ['ʌtə(r)] *adj.* völlig; vollkommen; **~ fool** Vollidiot, *der (ugs.)*

²utter *v. t.* **a)** von sich geben ⟨*Schrei, Seufzer*⟩; **b)** *(say)* sagen. **utterance** ['ʌtərəns] *n.* Worte *Pl.*

'utterly *adv.* völlig; vollkommen; äußerst ⟨*dumm, lächerlich*⟩

'U-turn *n.* Wende [um 180°]; *(fig.)* Kehrtwendung, *die;* **make a ~:** wenden; 'No ~s' „Wenden verboten"

V

¹V, v [viː] *n.* V, v, *das*
²V *abbr.* volt|s| V
v. *abbr.* versus gg.
vacancy ['veɪkənsɪ] *n.* **a)** *(job)* freie

Stelle; b) *(room)* freies Zimmer; '**vacancies**' „Zimmer frei"; '**no vacancies**'
„belegt"

vacant ['veɪkənt] *adj.* a) frei; '**situations ~**' „Stellenangebote"; b) *(mentally)* leer

vacate [və'keɪt] *v. t.* räumen

vacation [və'keɪʃn] *n.* a) *(Brit. Univ.)*
Ferien *Pl.;* b) *(Amer.) see* holiday b

vaccinate ['væksɪneɪt] *v. t.* impfen.
vaccination [væksɪ'neɪʃn] *n.* Impfung, *die;* **have a ~**: geimpft werden

vaccine ['væksi:n] *n.* Impfstoff, *der*

vacuum ['vækjʊəm] 1. *n.* a) Vakuum,
das; **live in a ~**: im luftleeren Raum
leben; b) *(coll.: ~ cleaner)* Sauger, *der
(ugs.).* 2. *v. t. & i.* [staub]saugen

vacuum: ~ cleaner *n.* Staubsauger,
der; ~ **flask** *n. (Brit.)* Thermosflasche, *die; ~-***packed** *adj.* vakuum-
verpackt

vagaries ['veɪgərɪz] *n. pl.* Launen *Pl.*

vagina [və'dʒaɪnə] *n.* Scheide, *die*

vagrant ['veɪgrənt] *n.* a) Landstreicher,
*der/*Landstreicherin, *die; (in cities)*
Stadtstreicher, *der/*Stadtstreicherin,
die

vague [veɪg] *adj.* vage; verschwommen ⟨*Form, Umriß⟩; (absent-minded)*
geistesabwesend; **not have the ~st idea**
or **notion** nicht die blasseste *od.* leise-
ste Ahnung haben. '**vaguely** *adv.* vage; entfernt ⟨*bekannt sein, erinnern
an⟩;* schwach ⟨*sich erinnern⟩*

vain [veɪn] *adj.* a) *(conceited)* eitel; b)
(useless) leer; vergeblich ⟨*Hoffnung,
Versuch⟩;* **in ~**: vergeblich. '**vainly**
adv. vergebens

vale [veɪl] *n. (arch./poet.)* Tal, *das*

valentine ['væləntaɪn] *n. ~* [card]
Grußkarte zum Valentinstag

valet ['væleɪ] *n.* Kammerdiener, *der*

valiant ['væliənt] *adj.,* '**valiantly** *adv.*
tapfer

valid ['vælɪd] *adj.* a) *(legally acceptable)* gültig; berechtigt ⟨*Anspruch⟩;* b)
(justifiable) stichhaltig ⟨*Argument⟩;*
triftig ⟨*Grund⟩;* begründet ⟨*Einwand,
Entschuldigung⟩.* **validate** ['vælɪdeɪt]
v. t. rechtskräftig machen. **validity**
[və'lɪdɪtɪ] *n.* Gültigkeit, *die*

valley ['vælɪ] *n.* Tal, *das*

valour *(Amer.:* **valor**) ['vælə(r)] *n.*
Tapferkeit, *die*

valuable ['væljʊəbl] 1. *adj.* wertvoll;
be ~ to sb. für jmdn. wertvoll sein. 2.
*n. ~***s** Wertsachen

valuation [væljʊ'eɪʃn] *n.* Schätzung,
die

value ['vælju:] 1. *n.* Wert, *der;* **be of
great/little/some/no ~** [to sb.] [für
jmdn.] von großem/geringem/eini-
gem/keinerlei Nutzen sein; **know the
~ of sth.** wissen, was etw. wert ist;
something/nothing of ~: etwas/nichts
Wertvolles. 2. *v. t.* schätzen. **value
added 'tax** *n.* Mehrwertsteuer, *die*

valve [vælv] *n.* a) Ventil, *das;* b)
(Anat.) Klappe, *die*

vampire ['væmpaɪə(r)] *n.* Vampir, *der*

van [væn] *n.* [delivery] ~: Lieferwagen,
der

vandal ['vændl] *n.* Rowdy, *der.* **van-
dalism** ['vændəlɪzm] *n.* Wandalis-
mus, *der.* **vandalize** ['vændəlaɪz] *v. t.*
[mutwillig] beschädigen

vanilla [və'nɪlə] 1. *n.* Vanille, *die.* 2.
adj. Vanille-

vanish ['vænɪʃ] *v. i.* verschwinden

vanity ['vænɪtɪ] *n.* Eitelkeit, *die.* '**vanity bag** *n.* Kosmetiktäschchen, *das*

vantage-point ['vɑ:ntɪdʒ pɔɪnt] *n.*
Aussichtspunkt, *der*

vapour *(Brit.; Amer.:* **vapor**) ['veɪ-
pə(r)] *n.* Dampf, *der*

variable ['veərɪəbl] *adj.* a) *(alterable)*
veränderbar; **be ~:** verändert werden
können; b) *(inconsistent)* unbeständig
⟨*Wetter, Wind, Leistung⟩;* wechselhaft
⟨*Wetter, Launen, Qualität⟩*

variance ['veərɪəns] *n.* **be at ~** [with
sth.] [mit etw.] nicht übereinstimmen

variant ['veərɪənt] *n.* Variante, *die*

variation [veərɪ'eɪʃn] *n.* a) *(varying)*
Veränderung, *die; (difference)* Unterschied, *der;* b) *(variant)* Variante, *die
(of, on Gen.)*

varicose vein [værɪkəʊs 'veɪn] *n.*
Krampfader, *die*

varied ['veərɪd] *adj.* unterschiedlich;
abwechslungsreich ⟨*Diät, Leben⟩*

variety [və'raɪətɪ] *n.* a) *(diversity)* Viel-
fältigkeit, *die; (in diet, routine)* Abwechslung, *die;* **add** *or* **give ~ to sth.**
etw. abwechslungsreicher gestalten;
b) *(assortment)* Auswahl, *die* (**of** an +
Dat., von); **for a ~ of reasons** aus ver-
schiedenen Gründen; c) *(Theatre)* Varieté, *das;* d) *(form)* Art, *die; (of fruit,
vegetable)* Sorte, *die; (cultivated)*
Züchtung, *die*

various ['veərɪəs] *adj.* a) *pred. (different)* verschieden; unterschiedlich; b)
attrib. (several) verschiedene; **at ~
times** mehrere Male. '**variously** *adv.*
unterschiedlich

varnish ['vɑ:nɪʃ] 1. *n.* Lasur, *die.* 2.
v. t. lasieren

vary ['veərɪ] **1.** *v. t.* verändern; ändern ⟨*Bestimmungen, Programm, Methode, Route*⟩; *(add variety to)* abwechslungsreicher gestalten. **2.** *v. i. (become different)* sich ändern; ⟨*Preis, Qualität:*⟩ schwanken; *(be different)* unterschiedlich sein. '**varying** *adj.* wechselnd; *(different)* unterschiedlich

vase [vɑːz] *n.* Vase, *die*

vast [vɑːst] *adj.* **a)** *(huge)* riesig; weit ⟨*Fläche, Meer*⟩; **b)** *(coll.: great)* enorm; Riesen⟨*menge, -summe*⟩. '**vastly** *adv. (coll.)* enorm; weitaus ⟨*besser*⟩; weit ⟨*überlegen, unterlegen*⟩

VAT [viːeɪ'tiː, væt] *abbr.* **value added tax** MwSt.

vat [væt] *n.* Bottich, *der*

Vatican ['vætɪkən] *pr. n.* Vatikan, *der*

¹**vault** [vɔːlt, vɒlt] *n.* **a)** *(Archit.)* Gewölbe, *das;* **b)** *(in bank)* Tresorraum, *der;* **c)** *(tomb)* Gruft, *die*

²**vault 1.** *v. i.* sich schwingen. **2.** *v. t.* sich schwingen über (+ *Akk.*). **3.** *n.* Sprung, *der*

VD *n.* Geschlechtskrankheit, *die*

VDU *abbr.* **visual display unit**

veal [viːl] *n.* Kalb[fleisch], *das; attrib.* Kalbs-

veer [vɪə(r)] *v. i.* ⟨*Auto:*⟩ ausscheren. **veer a'way, veer 'off** *v. i.* ⟨*Auto:*⟩ ausscheren; ⟨*Fahrer, Straße:*⟩ abbiegen

veg [vedʒ] *n., pl. same (coll.)* Gemüse, *das*

vegetable ['vedʒɪtəbl] *n.* Gemüse, *das;* **fresh ~s** frisches Gemüse; *attrib.* Gemüse⟨*suppe, -extrakt, -garten*⟩. '**vegetable oil** *n.* Pflanzenöl, *das*

vegetarian [vedʒɪ'teərɪən] **1.** *n.* Vegetarier, *der/*Vegetarierin, *die.* **2.** *adj.* vegetarisch

vegetate ['vedʒɪteɪt] *v. i.* nur noch [dahin]vegetieren. **vegetation** [vedʒɪ-'teɪʃn] *n.* Vegetation, *die*

vehement ['viːəmənt] *adj.,* '**vehemently** *adv.* heftig

vehicle ['viːɪkl] *n.* **a)** Fahrzeug, *das;* **b)** *(fig.: medium)* Vehikel, *das*

veil [veɪl] **1.** *n.* Schleier, *der.* **2.** *v. t.* verschleiern

vein [veɪn] *n.* **a)** Vene, *die; (any blood-vessel)* Ader, *die;* **b)** *(fig.: mood)* Stimmung, *die;* **in a similar ~:** vergleichbarer Art

Velcro, (P) ['velkrəʊ] *n.* Klettverschluß, *der* Ⓦ

velocity [vɪ'lɒsɪtɪ] *n.* Geschwindigkeit, *die*

velvet ['velvɪt] **1.** *n.* Samt, *der.* **2.** *adj.* aus Samt *nachgestellt;* Samt-. '**velvety** *adj.* samtig

vendetta [ven'detə] *n.* Hetzkampagne, *die; (feud)* Fehde, *die*

vending-machine ['vendɪŋ məʃiːn] *n.* [Verkaufs]automat, *der*

vendor ['vendə(r)] *n.* Verkäufer, *der/* Verkäuferin, *die*

veneer [vɪ'nɪə(r)] *n.* Furnier, *das*

venerable ['venərəbl] *adj.* ehrwürdig

ve'nereal disease *n. (Med.)* Geschlechtskrankheit, *die*

venetian blind [vɪ'niːʃn blaɪnd] *n.* Jalousie, *die*

Venezuela [venɪ'zweɪlə] *pr. n.* Venezuela *(das)*

vengeance ['vendʒəns] *n.* **a)** Rache, *die;* **take ~ [up]on** sb. [**for** sth.] sich an jmdm. [für etw.] rächen; **b)** **with a ~** *(coll.)* gewaltig *(ugs.)*

Venice ['venɪs] *pr. n.* Venedig *(das)*

venison ['venɪsn, 'venɪzn] *n.* Hirsch, *der;* Hirschfleisch, *das; (roe deer)* Reh[fleisch], *das*

venom ['venəm] *n.* Gift, *das.* **venomous** ['venəməs] *adj.* giftig

¹**vent** [vent] **1.** *n.* **a)** Öffnung, *die;* **b)** *(fig.)* Ventil, *das (fig.);* **give ~ to** Luft machen (+ *Dat.*). **2.** *v. t. (fig.)* Luft machen (+ *Dat.*)

²**vent** *n. (in garment)* Schlitz, *der*

ventilate ['ventɪleɪt] *v. t.* belüften. **ventilation** [ventɪ'leɪʃn] *n.* Belüftung, *die.* **ventilator** ['ventɪleɪtə(r)] *n.* **a)** Ventilator, *der;* **b)** *(Med.)* Beatmungsgerät, *das*

ventriloquist [ven'trɪləkwɪst] *n.* Bauchredner, *der/*-rednerin, *die*

venture ['ventʃə(r)] **1.** *n.* Unternehmung, *die.* **2.** *v. i.* **a)** *(dare)* wagen; **b)** *(dare to go)* sich wagen. **3.** *v. t.* wagen. **venture 'out** *v. i.* sich hinauswagen

venue ['venjuː] *n. (Sport)* [Austragungs]ort, *der; (Mus., Theatre)* [Veranstaltungs]ort, *der; (meeting-place)* Treffpunkt, *der*

Venus ['viːnəs] *pr. n. (Astron.)* Venus, *die*

veranda[h] [və'rændə] *n.* Veranda, *die*

verb [vɜːb] *n.* Verb, *das.* **verbal** ['vɜːbl] *adj.,* **verbally** ['vɜːbəlɪ] *adv.* **a)** *(relating to words)* sprachlich; **b)** *(oral[ly])* mündlich

verbatim [və'beɪtɪm] *adj., adv.* [wort]wörtlich

verbose [və'bəʊs] *adj.* weitschweifig ⟨*Roman, Autor*⟩; langatmig ⟨*Rede, Redner*⟩

verdict ['vɜ:dɪkt] *n.* Urteil, *das;* ~ **of guilty/not guilty** Schuld-/Freispruch, *der;* **reach a** ~: zu einem Urteil kommen

verge [vɜ:dʒ] *n.* **a)** Rasensaum, *der; (on road)* Bankette, *die;* **b)** *(fig.)* **be on the** ~ **of war/tears** am Rande des Krieges stehen/den Tränen nahe sein; **be on the** ~ **of doing sth.** kurz davor stehen, etw. zu tun. **'verge on** *v. t.* [an]grenzen an (+ *Akk.*)

verger ['vɜ:dʒə(r)] *n.* Küster, *der*

verification [verɪfɪ'keɪʃn] *n.* **a)** *(check)* Überprüfung, *die;* **b)** *(confirmation)* Bestätigung, *die*

verify ['verɪfaɪ] *v. t.* **a)** *(check)* überprüfen; **b)** *(confirm)* bestätigen

vermin ['vɜ:mɪn] *n.* Ungeziefer, *das*

vernacular [və'nækjʊlə(r)] *n.* Landessprache, *die*

versatile ['vɜ:sətaɪl] *adj.* vielseitig; *(having many uses)* vielseitig verwendbar. **versatility** [vɜ:sə'tɪlɪtɪ] *n.* Vielseitigkeit, *die*

verse [vɜ:s] *n.* **a)** *(stanza)* Strophe, *die;* **b)** *(poetry)* Lyrik, *die;* **write some** ~: einige Verse schreiben; **piece of** ~: Gedicht, *das;* **written in** ~: in Versform; **c)** *(in Bible)* Vers, *der.* **versed** [vɜ:st] *adj.* **be [well]** ~ **in sth.** sich in etw. *(Dat.)* [gut] auskennen

version ['vɜ:ʃn] *n.* Version, *die; (in another language)* Übersetzung, *die; (of vehicle, machine, tool)* Modell, *das*

versus ['vɜ:səs] *prep.* gegen

vertebra ['vɜ:tɪbrə] *n., pl.* ~e ['vɜ:tɪbri:] Wirbel, *der.* **vertebrate** ['vɜ:tɪbrət] *n.* Wirbeltier, *das*

vertical ['vɜ:tɪkl] *adj.* senkrecht; **be** ~: senkrecht stehen. **vertically** ['vɜ:tɪkəlɪ] *adv.* senkrecht

vertigo ['vɜ:tɪgəʊ] *n.* Schwindel, *der*

verve [vɜ:v] *n.* Schwung, *der*

very ['verɪ] **1.** *attrib. adj.* **a)** *(precise, exact)* genau; **you're the** ~ **person I wanted to see** genau dich wollte ich sehen; **at the** ~ **moment when** ...: im selben Augenblick, als ...; **at the** ~ **centre** genau in der Mitte; **the** ~ **thing** genau das Richtige; **b)** *(extreme)* **at the** ~ **back/front** ganz hinten/vorn; **at the** ~ **end/beginning** ganz am Ende/Anfang; **from the** ~ **beginning** von Anfang an; **only a** ~ **little** nur ein ganz kleines bißchen; **c)** *(mere)* bloß ⟨*Gedanke*⟩; **d)** *(absolute)* absolut ⟨*Minimum, Maximum*⟩; **the** ~ **most I can offer is** ...: ich kann allerhöchstens ... anbieten; **for the** ~ **last time** zum allerletzten Mal;

e) *emphat.* **before their** ~ **eyes** vor ihren Augen. **2.** *adv.* **a)** *(extremely)* sehr; **it's** ~ **near** es ist ganz in der Nähe; ~ **probably** höchstwahrscheinlich; **not** ~ **much** nicht sehr; ~ **little** [nur] sehr wenig ⟨*verstehen, essen*⟩; **thank you** [~,] ~ **much** [vielen,] vielen Dank; **b)** *(absolutely)* aller⟨*best..., -letzt..., -leichtest...*⟩; **at the** ~ **latest** allerspätestens; **c)** *(precisely)* **the** ~ **same one** genau der-/die-/dasselbe

vessel ['vesl] *n.* **a)** *(receptacle)* Gefäß, *das;* |drinking-|~: Trinkgefäß, *das;* **b)** *(Naut.)* Schiff, *das*

vest [vest] **1.** *n.* **a)** *(Brit.)* Unterhemd, *das;* **b)** *(Amer.: waistcoat)* Weste, *die.* **2.** *v. t.* ~ **sb. with sth.,** ~ **sth. in sb.** jmdm. etw. verleihen. **'vested** *adj.* **have a** ~ **interest in sth.** ein persönliches Interesse an etw. *(Dat.)* haben

vestige ['vestɪdʒ] *n.* Spur, *die;* **not a** ~ **of truth** kein Fünkchen Wahrheit

vestment ['vestmənt] *n.* [Priester]gewand, *das*

vestry ['vestrɪ] *n.* Sakristei, *die*

vet [vet] **1.** *n.* Tierarzt, *der/*-ärztin, *die.* **2.** *v. t.,* **-tt-** überprüfen

veteran ['vetərən] *n.* Veteran, *der/*Veteranin, *die.* **veteran 'car** *n.* *(Brit.)* Veteran, *der*

veterinarian [vetərɪ'neərɪən] *n.* *(Amer.)* Tierarzt, *der/*-ärztin, *die*

veterinary ['vetərɪnərɪ] *adj.* tiermedizinisch. **veterinary 'surgeon** *n.* *(Brit.)* Tierarzt, *der/*-ärztin, *die*

veto ['vi:təʊ] **1.** *n., pl.* ~**es** Veto, *das.* **2.** *v. t.* sein Veto einlegen gegen

vex [veks] *v. t.* [ver]ärgern; *(cause to worry)* beunruhigen; **be** ~**ed with sb.** sich über jmdn. ärgern. **vexation** [vek'seɪʃn] *n.* Verärgerung, *die.* **vexed** [vekst] *adj.* **a)** verärgert; **b)** ~ **question** vieldiskutierte Frage

VHF *abbr.* **Very High Frequency** UKW

via ['vaɪə] *prep.* über (+ *Akk.*) ⟨*Ort, Sender, Telefon*⟩; durch ⟨*Eingang, Schornstein, Person*⟩; per ⟨*Post*⟩

viability [vaɪə'bɪlɪtɪ] *n.* *(feasibility)* Realisierbarkeit, *die*

viable ['vaɪəbl] *adj. (feasible)* realisierbar

viaduct ['vaɪədʌkt] *n.* Viadukt, *das od. der*

vibrant ['vaɪbrənt] *adj.* lebenssprühend ⟨*Atmosphäre*⟩; lebhaft ⟨*Farbe*⟩

vibrate [vaɪ'breɪt] **1.** *v. i.* vibrieren; *(under strong impact)* beben. **2.** *v. t.* vibrieren lassen. **vibration** [vaɪ'breɪʃn] *n.* Vibrieren/Beben, *das*

vicar ['vɪkə(r)] *n.* Pfarrer, *der.* **vicarage** ['vɪkərɪdʒ] *n.* Pfarrhaus, *das*
vicarious [vɪ'keərɪəs] *adj.* nachempfunden
¹**vice** [vaɪs] *n.* Laster, *das*
²**vice** *n. (Brit.: tool)* Schraubstock, *der*
vice: ~-'**chairman** *n.* stellvertretender Vorsitzender; ~-'**president** *n.* Vizepräsident, *der*/-präsidentin, *die*
vice versa [vaɪsɪ 'vɜːsə] *adv.* umgekehrt
vicinity [vɪ'sɪnɪtɪ] *n.* Umgebung, *die;* **in the** ~ [**of a place**] in der Nähe [eines Ortes]
vicious ['vɪʃəs] *adj.* a) *(malicious)* böse; bösartig ⟨*Tier*⟩; b) *(violent)* brutal. **vicious** '**circle** *n.* Teufelskreis, *der* '**viciously** *adv.* a) *(maliciously)* boshaft; b) *(violently)* brutal
victim ['vɪktɪm] *n.* Opfer, *das; (of sarcasm, abuse)* Zielscheibe, *die (fig.).* **victimization** [vɪktɪmaɪ'zeɪʃn] *n.* Schikanierung, *die.* **victimize** ['vɪktɪmaɪz] *v.t.* schikanieren
victor ['vɪktə(r)] *n.* Sieger, *der*/Siegerin, *die*
victorious [vɪk'tɔːrɪəs] *adj.* siegreich
victory ['vɪktərɪ] *n.* Sieg, *der* (**over** über + *Akk.*); *attrib.* Sieges-
video ['vɪdɪəʊ] 1. *adj.* Video-. 2. *n., pl.* ~**s** (~ *recorder*) Videorecorder, *der; (~tape,* ~ *recording)* Video, *das (ugs.).* 3. *v.t. see* **videotape** 2
video: ~ **camera** *n.* Videokamera, *die;* ~ **cas'sette** *n.* Videokassette, *die;* ~ **game** *n.* Videospiel, *das;* ~ '**nasty** *n.* Horrorvideo, *das;* ~ **recorder** *n.* Videorecorder, *der;* ~ **recording** *n.* Videoaufnahme, *die;* ~**tape** 1. *n.* Videoband, *das;* 2. *v.t.* [auf Videoband *(Akk.)*] aufnehmen
vie [vaɪ] *v.i.,* **vying** ['vaɪɪŋ] ~ [**with sb.**] **for sth.** [mit jmdm.] um etw. wetteifern
Vienna [vɪ'enə] 1. *pr. n.* Wien *(das).* 2. *attrib. adj.* Wiener. **Viennese** [vɪə'niːz] 1. *adj.* Wiener. 2. *n., pl. same* Wiener, *der*/Wienerin, *die*
Vietnam [vɪet'næm] *pr. n.* Vietnam *(das).* **Vietnamese** [vɪetnə'miːz] 1. *adj.* vietnamesisch. 2. *n., pl. same* a) *(person)* Vietnamese, *der*/Vietnamesin, *die;* b) *(language)* Vietnamesisch, *das*
view [vjuː] 1. *n.* a) *(range of vision)* Sicht, *die;* **be out of/in** ~: nicht zu sehen/zu sehen sein; b) *(what is seen)* Aussicht, *die;* c) *(picture)* Ansicht, *die;* d) *(opinion)* Ansicht, *die;* **what is your** ~ **or are your** ~**s on this?** was meinst

du dazu?; **hold** *or* **take the** ~ **that ...:** der Ansicht sein, daß ...; **in my** ~: meiner Ansicht nach; **e) be on** ~: besichtigt werden können; **in** ~ **of sth.** *(fig.)* angesichts einer Sache; **with a** ~ **to doing sth.** in der Absicht, etw. zu tun. 2. *v.t.* **a)** *(look at)* sich *(Dat.)* ansehen; **b)** *(consider)* betrachten; **c)** *(inspect)* besichtigen. 3. *v.i. (Telev.)* fernsehen. '**viewer** *n.* **a)** *(Telev.)* [Fernseh]zuschauer, *der*/-zuschauerin, *die;* **b)** *(for slides)* Diabetrachter, *der*
view: ~**finder** *n.* Sucher, *der;* ~**point** *n.* Standpunkt, *der*
vigil ['vɪdʒɪl] *n.* Wachen, *das;* **keep** ~: wachen
vigilance ['vɪdʒɪləns] *n.* Wachsamkeit, *die*
vigilant ['vɪdʒɪlənt] *adj.* wachsam
vigor *(Amer.) see* **vigour**
vigorous ['vɪgərəs] *adj.* kräftig; heftig ⟨*Attacke, Protest*⟩; energisch ⟨*Versuch, Anstrengung, Leugnen, Maßnahme*⟩. '**vigorously** *adv.* heftig; kräftig ⟨*schrubben, drücken*⟩
vigour ['vɪgə(r)] *n. (Brit.) (of person)* Vitalität, *die; (of body)* Kraft, *die; (of protest, attack)* Heftigkeit, *die*
vile [vaɪl] *adj.* gemein ⟨*Verleumdung*⟩; vulgär ⟨*Sprache*⟩; *(repulsive)* widerwärtig; *(coll.: very unpleasant)* scheußlich *(ugs.)*
villa ['vɪlə] *n.* **a)** [**holiday**] ~: Ferienhaus, *das;* **b)** [**country**] ~: Landhaus, *das*
village ['vɪlɪdʒ] *n.* Dorf, *das; attrib.* Dorf-. '**villager** *n.* Dorfbewohner, *der*/-bewohnerin, *die*
villain ['vɪlən] *n.* **a)** Verbrecher, *der;* **b)** *(Theatre)* Bösewicht, *der.* **villainous** ['vɪlənəs] *adj.* gemein
vindicate ['vɪndɪkeɪt] *v.t.* **a)** *(justify)* rechtfertigen; **b)** *(clear)* rehabilitieren. **vindication** [vɪndɪ'keɪʃn] *n.* **a)** *(justification)* Rechtfertigung, *die;* **b)** *(clearing)* Rehabilitierung, *die*
vindictive [vɪn'dɪktɪv] *adj.* nachtragend
vine [vaɪn] *n.* Weinrebe, *die*
vinegar ['vɪnɪgə(r)] *n.* Essig, *der*
vineyard ['vɪnjɑːd, 'vɪnjəd] *n.* Weinberg, *der*
vintage ['vɪntɪdʒ] 1. *n.* Jahrgang, *der.* 2. *adj.* erlesen ⟨*Wein*⟩. **vintage** '**car** *n. (Brit.)* Oldtimer, *der*
vinyl ['vaɪnɪl] *n.* Vinyl, *das*
viola [vɪ'əʊlə] *n.* Bratsche, *die*
violate ['vaɪəleɪt] *v.t.* **a)** verletzen; brechen ⟨*Vertrag, Versprechen, Gesetz*⟩;

b) *(profane, rape)* schänden. **violation** [vaɪə'leɪʃn] *n. see* violate: Verletzung, *die;* Bruch, *der;* Schändung, *die*
violence ['vaɪələns] *n.* **a)** *(force)* Heftigkeit, *die;* *(of blow)* Wucht, *die;* **b)** *(brutality)* Gewalt, *die;* *(at public event)* Gewalttätigkeiten; **resort to** *or* **use** ~: Gewalt anwenden
violent ['vaɪələnt] *adj.* gewalttätig; *(fig.)* heftig; wuchtig ⟨*Schlag, Stoß*⟩; Gewalt⟨*verbrecher, -tat*⟩. '**violently** *adv.* brutal; *(fig.)* heftig
violet ['vaɪələt] **1.** *n.* **a)** Veilchen, *das;* **b)** *(colour)* Violett, *das.* **2.** *adj.* violett
violin [vaɪə'lɪn] *n.* Violine, *die;* Geige, *die.* **vio'linist** *n.* Geiger, *der/*Geigerin, *die*
VIP [vi:aɪ'pi:] *n.* Prominente, *der/die;* **the** ~**s** die Prominenz
viper ['vaɪpə(r)] *n.* Viper, *die*
virgin ['vɜ:dʒɪn] **1.** *n.* **a)** Jungfrau, *die;* **b) the [Blessed] V~ [Mary]** die [Heilige] Jungfrau [Maria]. **2.** *adj. (unspoiled)* unberührt. **virginity** [və'dʒɪnɪtɪ] *n.* Unschuld, *die*
Virgo ['vɜ:gəʊ] *n., pl.* ~**s** die Jungfrau
virile ['vɪraɪl] *adj.* männlich. **virility** [vɪ'rɪlɪtɪ] *n.* Männlichkeit, *die*
virtual ['vɜ:tjʊəl] *adj.* **a** ~ ...: so gut wie ein/eine ...; **the traffic came to a** ~ **standstill** der Verkehr kam praktisch zum Stillstand *(ugs.).* '**virtually** *adv.* so gut wie; praktisch *(ugs.)*
virtue ['vɜ:tju:] *n.* **a)** *(moral excellence)* Tugend, *die;* **b)** *(advantage)* Vorteil, *der;* **c) by** ~ **of** auf Grund (+ *Gen.*)
virtuoso [vɜ:tjʊ'əʊzəʊ] *n., pl.* **virtuosi** [vɜ:tjʊ'əʊzi:] *or* ~**s** Virtuose, *der/*Virtuosin, *die*
virtuous ['vɜ:tjʊəs] *adj.* rechtschaffen ⟨*Person*⟩; tugendhaft ⟨*Leben*⟩
virulent ['vɪrʊlənt] *adj.* **a)** *(Med.)* virulent; starkwirkend ⟨*Gift*⟩; **b)** *(fig.)* heftig; scharf ⟨*Angriff*⟩
virus ['vaɪərəs] *n.* Virus, *das*
visa ['vi:zə] *n.* Visum, *das*
vis-à-vis [vi:zɑː'vi:] *prep. (in relation to)* bezüglich (+ *Gen.*)
viscount ['vaɪkaʊnt] *n.* Viscount, *der*
viscous ['vɪskəs] *adj.* dickflüssig
visibility [vɪzɪ'bɪlɪtɪ] *n.* **a)** Sichtbarkeit, *die;* **b)** *(range of vision)* Sicht, *die;* *(Meteorol.)* Sichtweite, *die*
visible ['vɪzɪbl] *adj.* sichtbar. '**visibly** *adv.* sichtlich
vision ['vɪʒn] *n.* **a)** *(sight)* Sehkraft, *die;* **b)** *(dream)* Vision, *die;* **c)** *usu. pl. (imaginings)* Phantasien; **d)** *(insight, foresight)* Weitblick, *der*

visit ['vɪzɪt] **1.** *v. t.* besuchen; aufsuchen ⟨*Arzt*⟩. **2.** *v. i.* einen Besuch/Besuche machen. **3.** *n.* Besuch, *der;* **pay** *or* **make a** ~ **to sb.**, **pay sb. a** ~: jmdm. einen Besuch abstatten *(geh.)*
'**visiting:** ~ **card** *n.* Visitenkarte, *die;* ~ **hours** *n. pl.* Besuchszeiten
visitor ['vɪzɪtə(r)] *n.* Besucher, *der/*Besucherin, *die;* *(to hotel)* Gast, *der;* **have** ~**s/a** ~: Besuch haben
visual ['vɪzjʊəl, 'vɪʒʊəl] *adj.* visuell; optisch ⟨*Eindruck, Darstellung*⟩. **visual 'aids** *n. pl.* Anschauungsmaterial, *das.* **visual dis'play unit** *n.* Bildschirmgerät, *das*
visualize ['vɪzjʊəlaɪz, 'vɪʒʊəlaɪz] *v. t.* **a)** *(imagine)* sich *(Dat.)* vorstellen; **b)** *(envisage)* voraussehen
'**visually** *adv.* bildlich
vital ['vaɪtl] *adj.* **a)** *(essential to life)* lebenswichtig; **b)** *(essential)* unbedingt notwendig; **c)** *(crucial)* entscheidend (**to** für); **it is** ~ **that you** ...: es ist von entscheidender Bedeutung, daß Sie ...
vitality [vaɪ'tælɪtɪ] *n.* Vitalität, *die.* '**vitally** *adv.* ~ **important** von allergrößter Wichtigkeit; *(crucial)* von entscheidender Bedeutung
vitamin ['vɪtəmɪn, 'vaɪtəmɪn] *n.* Vitamin, *das*
vivacious [vɪ'veɪʃəs] *adj.* lebhaft. **vivacity** [vɪ'væsɪtɪ] *n.* Lebhaftigkeit, *die*
vivid ['vɪvɪd] *adj.* lebhaft ⟨*Farbe, Erinnerung*⟩; lebendig ⟨*Schilderung*⟩. '**vividly** *adv.* lebendig ⟨*beschreiben*⟩; **remember sth.** ~: sich lebhaft an etw. *(Akk.)* erinnern
vixen ['vɪksn] *n.* Füchsin, *die*
vocabulary [və'kæbjʊlərɪ] *n.* **a)** *(list)* Vokabelverzeichnis, *das;* **learn** ~: Vokabeln lernen; **b)** *(range of language)* Wortschatz, *der*
vocal ['vəʊkl] *adj.* **a)** *(concerned with voice)* stimmlich; **b)** lautstark ⟨*Minderheit, Protest*⟩. '**vocal cords** *n. pl.* Stimmbänder
vocalist ['vəʊkəlɪst] *n.* Sänger, *der/*Sängerin, *die*
vocation [və'keɪʃn] *n.* Berufung, *die*
vocational [və'keɪʃənl] *adj.* berufsbezogen. **vocational 'guidance** *n.* Berufsberatung, *die.* **vocational 'training** *n.* berufliche Bildung
vociferous [və'sɪfərəs] *adj.* laut; lautstark ⟨*Forderung, Protest*⟩
vodka ['vɒdkə] *n.* Wodka, *der*
vogue [vəʊg] *n.* Mode, *die;* **be in/come into** ~: in Mode sein/kommen
voice [vɔɪs] **1.** *n.* Stimme, *die;* **in a**

firm/loud/soft ~: mit fester/lauter/sanfter Stimme. **2.** *v. t.* zum Ausdruck bringen

void [vɔɪd] **1.** *adj.* **a)** *(empty)* leer; **b)** *(invalid)* ungültig; **c)** ~ **of** ohne [jeden/jedes/jede]. **2.** *n.* Nichts, *das*

volatile ['vɒlətaɪl] *adj.* **a)** *(Chem.)* flüchtig; **b)** *(fig.)* impulsiv; brisant 〈*Lage*〉

volcanic [vɒl'kænɪk] *adj.* vulkanisch

volcano [vɒl'keɪnəʊ] *n., pl.* **~es** Vulkan, *der*

volition [və'lɪʃn] *n.* Wille, *der;* **of one's own** ~: aus eigenem Willen

volley ['vɒlɪ] *n.* **a)** *(of missiles)* Salve, *die;* **a** ~ **of arrows** ein Hagel von Pfeilen; **b)** *(Tennis)* Volley, *der.* '**volleyball** *n.* Volleyball, *der*

volt [vəʊlt] *n.* Volt, *das.* **voltage** ['vəʊltɪdʒ] *n.* Spannung, *die*

voluble ['vɒljʊbl] *adj.* redselig

volume ['vɒljuːm] *n.* **a)** *(book)* Band, *der;* **b)** *(loudness)* Lautstärke, *die; (of voice)* Volumen, *das;* **c)** *(space)* Rauminhalt, *der; (amount of substance)* Teil, *der.* '**volume control** *n.* Lautstärkeregler, *der*

voluntarily ['vɒləntərɪlɪ] *adv.,* **voluntary** ['vɒləntərɪ] *adj.* freiwillig

volunteer [vɒlən'tɪə(r)] **1.** *n.* Freiwillige, *der/die.* **2.** *v. t.* anbieten 〈*Hilfe, Dienste*〉; herausrücken mit 〈*Informationen*〉. **3.** *v. i.* sich [freiwillig] melden; ~ **to do** *or* ~ **for the shopping** sich zum Einkaufen bereiterklären

voluptuous [və'lʌptjʊəs] *adj.* üppig

vomit ['vɒmɪt] **1.** *v. t.* erbrechen. **2.** *v. i.* sich übergeben. **3.** *n.* Erbrochene, *das*

voracious [və'reɪʃəs] *adj.* gefräßig 〈*Person*〉; unbändig 〈*Appetit*〉

vote [vəʊt] **1.** *n.* **a)** *(individual* ~*)* Stimme, *die;* **b)** *(act of voting)* Abstimmung, *die;* **take a** ~ **on sth.** über etw. *(Akk.)* abstimmen; **c)** *(right to* ~*)* Stimmrecht, *das.* **2.** *v. i.* abstimmen; *(in election)* wählen; ~ **for/against** stimmen für/gegen; ~ **to do sth.** beschließen, etw. zu tun; ~ **Labour/Conservative** *etc.* Labour/die Konservativen *usw.* wählen. **3.** *v. t.* wählen; ~ **sb. Chairman/President** *etc.* jmdn. zum Vorsitzenden/Präsidenten *usw.* wählen. '**voter** *n.* Wähler, *der*/Wählerin, *die*

vouch [vaʊtʃ] *v. i.* ~ **that ...**: sich dafür verbürgen, daß ... **2.** *v. i.* ~ **for sb./sth.** sich für jmdn./etw. verbürgen

'**voucher** *n.* Gutschein, *der*

vow [vaʊ] **1.** *n.* Gelöbnis, *das; (Relig.)* Gelübde, *das.* **2.** *v. t.* geloben

vowel ['vaʊəl] *n.* Vokal, *der*

voyage ['vɔɪɪdʒ] **1.** *n.* Reise, *die; (sea* ~*)* Seereise, *die;* **outward/homeward** ~, ~ **out/home** Hin-/Rückreise, *die;* **a** ~ **to the moon** ein Mondflug. **2.** *v. i. (literary)* reisen

vulgar ['vʌlgə(r)] *adj.* vulgär; ordinär 〈*Person, Benehmen, Witz*〉. **vulgarity** [vʌl'gærɪtɪ] *n.* Vulgarität, *die*

vulnerable ['vʌlnərəbl] *adj.* **a)** *(exposed to danger)* angreifbar; **be** ~ **to sth.** für etw. anfällig sein; **be** ~ **to attack/in a** ~ **position** leicht angreifbar sein; **b)** *(without protection)* schutzlos

vulture ['vʌltʃə(r)] *n.* Geier, *der*

vying *see* **vie**

W

¹**W, w** ['dʌbljuː] *n.* W, w, *das*

²**W** *abbr.* watt[s] W

W. *abbr.* **a)** west W.; **b)** western w.

wad [wɒd] *n.* **a)** Knäuel, *das; (smaller)* Pfropfen, *der;* **b)** *(of papers)* Bündel, *das.* '**wadding** *n.* Futter, *das*

waddle ['wɒdl] **1.** *v. i.* watscheln. **2.** *n.* watschelnder Gang

wade [weɪd] *v. i.* waten. '**wade through** *v. t. (fig. coll.)* durchackern *(ugs.)* 〈*Buch*〉

wafer ['weɪfə(r)] *n.* Waffel, *die.* '**wafer-thin** *adj.* hauchdünn

¹**waffle** ['wɒfl] *n. (Gastr.)* Waffel, *die*

²**waffle** *(Brit. coll.: talk)* **1.** *v. i.* schwafeln *(ugs.).* **2.** *n.* Geschwafel, *das (ugs.)*

waft [wɒft, wɑːft] **1.** *v. t.* weben. **2.** *v. i.* ziehen

wag [wæg] **1.** *v. t.,* **-gg-** 〈*Hund:*〉 wedeln mit 〈*Schwanz*〉; ~ **one's finger at sb.** jmdm. mit dem Finger drohen. **2.** *v. i.,* **-gg-** 〈*Schwanz:*〉 wedeln

wage [weɪdʒ] **1.** *n. in sing. or pl.* Lohn, *der.* **2.** *v. t.* führen. '**wage increase** *n.* Lohnerhöhung, *die.* '**wage packet** *n.* Lohntüte, *die*

wager ['weɪdʒə(r)] *(dated/formal)* **1.** *n.* Wette, *die;* **lay a** ~ **on sth.** auf etw. *(Akk.)* wetten. **2.** *v. t. & i.* wetten

waggle ['wægl] *(coll.)* **1.** *v. t.* ~ **its tail**

⟨*Hund:*⟩ mit dem Schwanz wedeln. 2.
v. i. hin und her schlagen

waggon *(Brit.),* **wagon** ['wægən] *n.*
Wagen, *der*

wail [weɪl] **1.** *v. i.* **a)** klagen *(geh.)* (for
um); ⟨*Kind:*⟩ heulen. **2.** *n.* klagender
Schrei; ~s Geheul, *das*

waist [weɪst] *n.* Taille, *die;* **tight round
the ~:** eng in der Taille. **waistcoat**
['weɪskəʊt] *n. (Brit.)* Weste, *die.*
'**waistline** *n.* Taille, *die;* **be bad for
the ~:** schlecht für die schlanke Linie
sein

wait [weɪt] **1.** *v. i.* **a)** warten; ~ [for] an
hour eine Stunde warten; ~ **a moment**
Moment mal; **keep sb. ~ing, make sb.
~:** jmdn. warten lassen; **b)** ~ **at table**
servieren. **2.** *v. t. (await)* warten auf
(+ *Akk.*); ~ **one's turn** warten, bis
man drankommt. **3.** *n.* **a)** **after a long/
short ~:** nach langer/kurzer Warte-
zeit; **b)** **lie in** ~ **for sb./sth.** jmdn./ei-
ner Sache auflauern. **wait be'hind**
v. i. noch hier-/dableiben. '**wait for**
v. t. warten auf (+ *Akk.*); ~ **for sb. to
do sth.** darauf warten, daß jmd. etw.
tut; ~ **for the rain to stop** warten, bis
der Regen aufhört. '**wait on** *v. t.*
(serve) bedienen. **wait 'up** *v. i.* aufbleiben (for wegen)

'**waiter** *n.* Kellner, *der;* ~! Herr Ober!
'**waiting:** ~**-list** *n.* Warteliste, *die;*
~**-room** *n.* Wartezimmer, *das;*
(Railw.) Warteraum, *der*

waitress ['weɪtrɪs] *n.* Serviererin, *die;*
~! Fräulein! *(veralt.)*

waive [weɪv] *v. t.* verzichten auf
(+ *Akk.*)

¹**wake** [weɪk] **1.** *v. i.,* **woke** [wəʊk],
woken ['wəʊkn] aufwachen. **2.** *v. t.,*
woke, woken wecken. **3.** *n. (by corpse)*
Totenwache, *die.* **wake 'up 1.** *v. i.*
aufwachen; ~ **up to sth.** *(fig.: realize)*
etw. erkennen. **2.** *v. t.* **a)** wecken; **b)**
(fig.: enliven) wachrütteln

²**wake** *n.* Kielwasser, *das;* **in the ~ of
sth.** *(fig.)* im Gefolge von etw.

waken ['weɪkn] **1.** *v. t.* wecken. **2.** *v. i.*
aufwachen

Wales [weɪlz] *pr. n.* Wales *(das)*

walk [wɔːk] **1.** *v. i.* **a)** laufen; *(not run)*
gehen; *(not drive)* zu Fuß gehen; **learn
to ~:** laufen lernen; **b)** *(exercise)* gehen. **2.** *v. t.* **a)** *(lead)* führen; ausführen ⟨*Hund*⟩; **b)** *(accompany)* bringen.
3. *n.* **a)** Spaziergang, *der;* **go [out] for
or take or have a ~:** einen Spazier-
gang machen; **ten minutes' ~ from
here** zehn Minuten zu Fuß von hier;

b) *(gait)* Gang, *der;* **c)** *(path)* [Spa-
zier]weg, *der.* **walk a'way with** *v. t.*
(coll.: win easily) spielend leicht gewinnen. '**walk into** *v. t. (hit by accident)* laufen gegen ⟨*Pfosten, Laternenpfahl*⟩; ~ **into sb.** mit jmdm. zusammenstoßen; ~ **into a trap** in eine Falle
gehen. **walk 'off with** *v. t.* sich davonmachen mit. **walk 'out** *v. i.* **a)**
(leave in protest) aus Protest den Saal
verlassen; **b)** *(go on strike)* in den
Streik treten. **walk 'out of** *v. t. (leave
in protest)* aus Protest verlassen. **walk
'out on** *v. t. (coll.)* verlassen

'**walker** *n.* Spaziergänger, *der/*-gänge-
rin, *die;* *(rambler)* Wanderer, *der/*
Wanderin, *die*

walkie-talkie [wɔːkɪ'tɔːkɪ] *n.* Walkie-
talkie, *das*

'**walking** *n.* [Spazieren]gehen, *das;* **at
~ pace** im Schritttempo; **be within ~
distance** zu Fuß zu erreichen sein

walking: ~ **holiday** *n.* Wanderur-
laub, *der;* ~ **shoe** *n.* Wanderschuh,
der; ~**-stick** *n.* Spazierstock, *der*

walk: ~**-out** *n.* Arbeitsniederlegung,
die; ~**-over** *n. (fig.: easy victory)* Spaziergang, *der (ugs.)*

wall [wɔːl] *n.* Wand, *die;* *(free-
standing)* Mauer, *die;* **drive sb. up the
~** *(fig. coll.)* jmdn. auf die Palme bringen *(ugs.);* **go to the ~** *(fig.)* an die
Wand gedrückt werden. **wall 'up** *v. t.*
zumauern

wallet ['wɒlɪt] *n.* Brieftasche, *die*

'**wallflower** *n.* Goldlack, *der*

wallop ['wɒləp] *(coll.)* **1.** *v. t.* schlagen.
2. *n.* Schlag, *der*

wallow ['wɒləʊ] *v. i.* **a)** sich wälzen; **b)**
(fig.) schwelgen (in in + *Dat.*)

wall: ~**-painting** *n.* Wandgemälde,
das; ~**paper 1.** *n.* Tapete, *die;* **2.** *v. t.*
tapezieren; ~**-to-~** *adj.* ~**-to-~ car-
peting** Teppichboden, *der*

walnut ['wɔːlnʌt] *n.* Walnuß, *die*

walrus ['wɔːlrəs] *n.* Walroß, *das*

waltz [wɔːlts, wɒls] **1.** *n.* Walzer, *der.*
2. *v. i.* Walzer tanzen

wan [wɒn] *adj.* bleich

wand [wɒnd] *n.* Stab, *der*

wander ['wɒndə(r)] **1.** *v. i. (go aimlessly)* umherirren; *(walk slowly)* bummeln. **2.** *v. t.* wandern durch. **3.** *n.
(coll.)* Spaziergang, *der.* **wander
a'bout** *v. i.* sich herumtreiben. **wander 'off** *v. i. (stray)* weggehen

wane [weɪn] *v. i.* abnehmen

wangle ['wæŋgl] *v. t. (coll.)* organisieren *(ugs.)*

want [wɒnt] **1.** *v. t.* **a)** *(desire)* wollen; ~ **to do sth.** etw. tun wollen; **I ~ it done by tonight** ich will, daß es bis heute abend fertig wird; **b)** *(require, need)* brauchen; **'W~ed – cook'** „Koch/Köchin gesucht"; **you're ~ed on the phone** du wirst am Telefon verlangt; **the windows ~ painting** die Fenster müßten gestrichen werden; **you ~ to be [more] careful** du solltest vorsichtig[er] sein; **c)** ~**ed [by the police]** [polizeilich] gesucht. **2.** *n.* **a)** *(lack)* Mangel, *der* (**of** an + *Dat.*); **for ~ of sth.** aus Mangel an etw. *(Dat.)*; **b)** *(need)* Not, *der;* **c)** *(desire)* Bedürfnis, *das.* **'want for** *v. t.* **sb.** ~**s for nothing** *or* **doesn't ~ for anything** jmdm. fehlt es an nichts

'wanting *adj.* **be ~:** fehlen; **sb./sth. is ~ in sth.** jmdm./einer Sache fehlt es an etw. *(Dat.);* **be found ~:** für unzureichend befunden werden

wanton ['wɒntən] *adj.,* **'wantonly** *adv.* mutwillig

war [wɔ:(r)] *n.* Krieg, *der;* **between the ~s** zwischen den Weltkriegen; **declare ~:** den Krieg erklären (**on** *Dat.*); **be at ~:** sich im Krieg befinden; **make ~:** Krieg führen (**on** gegen)

warble ['wɔ:bl] *v. t. & i.* trällern

ward [wɔ:d] *n.* **a)** *(in hospital)* Station, *die;* **she's in W~ 3** sie liegt auf Station 3; **b)** *(child)* Mündel, *das. die;* **c)** *(electoral division)* Wahlbezirk, *der.* **ward 'off** *v. t.* abwehren

warden ['wɔ:dn] *n.* **a)** *(of hostel)* Heimleiter, *der/*-leiterin, *die;/(of youth hostel)* Herbergsvater, *der/*-mutter, *die;* **b)** *(supervisor)* Aufseher, *der/*Aufseherin, *die*

'warder *n. (Brit.)* Wärter, *der*

wardrobe ['wɔ:drəʊb] *n.* **a)** Kleiderschrank, *der;* **b)** *(clothes)* Garderobe, *die*

warehouse ['weəhaʊs] *n.* Lagerhaus, *das; (part of building)* Lager, *das*

wares [weəz] *n. pl.* Ware, *die*

warfare ['wɔ:feə(r)] *n.* Krieg, *der*

'warhead *n.* Sprengkopf, *der*

warily ['weərɪlɪ] *adv.* vorsichtig; *(suspiciously)* mißtrauisch

'warlike *adj.* kriegerisch

warm [wɔ:m] **1.** *adj.* **a)** warm; **I am [very] ~:** mir ist [sehr] warm; **b)** *(enthusiastic)* herzlich ⟨*Grüße, Dank*⟩. **2.** *v. t.* wärmen; warm machen ⟨*Flüssigkeit*⟩; ~ **one's hands** sich *(Dat.)* die Hände wärmen. **3.** *v. i.* ~ **to sb./sth.** *(come to like)* sich für jmdn./etw. er-

wärmen. **warm 'up 1.** *v. i.* warm werden; ⟨*Sportler:*⟩ sich aufwärmen. **2.** *v. t.* aufwärmen ⟨*Speisen*⟩; erwärmen ⟨*Raum, Zimmer*⟩

warm: ~**-blooded** ['wɔ:mblʌdɪd] *adj.* warmblütig; ~**-hearted** ['wɔ:mhɑ:tɪd] *adj.* warmherzig ⟨*Person*⟩

'warmly *adv.* **a)** warm; **b)** *(fig.)* herzlich ⟨*willkommen heißen, gratulieren, begrüßen, grüßen, danken*⟩

warmonger ['wɔ:mʌŋgə(r)] *n.* Kriegshetzer, *der/*-hetzerin, *die*

warmth [wɔ:mθ] *n.* **a)** Wärme, *die;* **b)** *(fig.)* Herzlichkeit, *die*

warn [wɔ:n] *v. t.* **a)** *(inform, give notice)* warnen (**against, of, about** vor + *Dat.*); ~ **sb. that ...:** jmdn. darauf hinweisen, daß ...; ~ **sb. not to do sth.** jmdn. davor warnen, etw. zu tun; **b)** *(admonish)* ermahnen; *(officially)* abmahnen. **'warning 1.** *n.* **a)** *(advance notice)* Vorwarnung, *die;* **b)** *(lesson)* **let that be a ~ to you** laß dir das eine Warnung sein; **c)** *(caution)* Verwarnung, *die; (less official)* Warnung, *die.* **2.** *adj.* Warn⟨*schild, -signal usw.*⟩

warp [wɔ:p] **1.** *v. i.* sich verbiegen; ⟨*Holz, Schallplatte:*⟩ sich verziehen. **2.** *v. t.* **a)** verbiegen; **b)** *(fig.)* **a ~ed sense of humour** ein abartiger Humor

war: ~**-path** *n.* **be on the ~-path** *(fig.)* in Rage sein; ~**plane** *n.* Kampfflugzeug, *das*

warrant ['wɒrənt] **1.** *n. (for sb.'s arrest)* Haftbefehl, *der; [search]* ~**:** Durchsuchungsbefehl, *der.* **2.** *v. t.* **a)** *(justify)* rechtfertigen; **b)** *(guarantee)* garantieren. **'warranty** *n.* Garantie, *die*

warrior ['wɒrɪə(r)] *n. (esp. literary)* Krieger, *der (geh.)*

Warsaw ['wɔ:sɔ:] **1.** *pr. n.* Warschau *(das).* **2.** *attrib. adj.* Warschauer

'warship *n.* Kriegsschiff, *das*

wart [wɔ:t] *n.* Warze, *die*

'wartime *n.* **a)** Kriegszeit, *die;* **in** *or* **during ~:** im Krieg; **b)** *attrib.* Kriegs⟨*rationierung, -evakuierung usw.*⟩

wary ['weərɪ] *adj.* vorsichtig; *(suspicious)* mißtrauisch (**of** gegenüber); **be ~ of sb./sth.** sich vor jmdm./etw. in acht nehmen

was *see* **be**

wash [wɒʃ] **1.** *v. t.* **a)** waschen; ~ **oneself** sich waschen; ~ **one's hands/face/hair** sich *(Dat.)* die Hände/das Gesicht/die Haare waschen; ~ **the clothes** Wäsche waschen; ~ **the dishes** [Geschirr] spülen; ~ **the floor** den Fußboden aufwischen; **b)** *(remove)* waschen ⟨*Fleck*⟩ **(out of** aus); abwaschen ⟨*Schmutz*⟩ **(off** von); **c)** *(carry along)* spülen. **2.** *v. i.* **a)** sich waschen; **b)** ⟨*Stoff, Kleidungsstück:*⟩ sich waschen lassen. **3.** *n.* **a) give sb./sth. a [good]** ~: jmdn./etw. [gründlich] waschen; **b)** *(laundering)* Wäsche, *die;* **c)** *(of ship)* Sog, *der.* **wash 'down** *v. t.* abspritzen ⟨*Auto, Deck, Hof*⟩. **wash 'off 1.** *v. t.* ~ **sth. off** etw. abwaschen. **2.** *v. i.* abgehen; *(from fabric etc.)* herausgehen. **wash 'out** *v. t.* ausscheuern ⟨*Topf*⟩; ausspülen ⟨*Mund*⟩; ~ **dirt/ marks out of clothes** Schmutz/Flecken aus Kleidern [her]auswaschen. **wash 'up 1.** *v. t. (Brit.)* ~ **the dishes up** das Geschirr spülen. **2.** *v. i.* spülen

washable ['wɒʃəbl] *adj.* waschbar

'wash-basin *n.* Waschbecken, *das*

'washing *n.* Wäsche, *die;* **do the** ~: waschen

washing: ~**-machine** *n.* Waschmaschine, *die;* ~**-powder** *n.* Waschpulver, *das;* ~**-'up** *n. (Brit.)* Abwasch, *der;* **do the** ~**-up** spülen; ~**'up liquid** *n.* Spülmittel, *das*

wasn't ['wɒznt] *(coll.)* = **was not;** *see* **be**

wasp *n.* Wespe, *die*

waste [weɪst] **1.** *n.* **a)** *(useless remains)* Abfall, *der;* **kitchen** ~: Küchenabfälle *Pl.;* **b)** *(extravagant use)* Verschwendung, *die;* **it's a** ~ **of time/money/energy** das ist Zeit-/Geld-/Energieverschwendung. **2.** *v. t. (squander)* verschwenden; **all his efforts were** ~**d** all seine Mühe war umsonst; **don't** ~ **my time!** stehlen Sie mir nicht die Zeit! **3.** *adj.* **a)** ~ **material** Abfall, *der;* **b) lay sth.** ~: etw. verwüsten. **waste a'way** *v. i.* immer mehr abmagern

waste: ~ **disposal** *n.* Abfallbeseitigung, *die;* ~**-disposal unit** *n.* Müllzerkleinerer, *der*

wasteful ['weɪstfl] *adj.* **a)** *(extravagant)* verschwenderisch; **b)** *(causing waste)* unwirtschaftlich

waste: ~**-land** *n.* Ödland, *das;* ~ **'paper** *n.* Papierabfall, *der;* ~**-'paper basket** *n.* Papierkorb, *der*

watch [wɒtʃ] **1.** *n.* **a)** |wrist-/pocket-|~: [Armband-/Taschen]uhr, *die;* **keep** ~: Wache halten; **keep |a|** ~ **for sb./ sth.** auf jmdn./etw. achten; **c)** *(Naut.)* Wache, *die.* **2.** *v. i.* ~ **for sb./sth.** auf jmdn./etw. warten. **3.** *v. t.* **a)** *(observe)* sich *(Dat.)* ansehen ⟨*Sportveranstaltung, Fernsehsendung*⟩; ~ **|the| television** *or* **TV** fernsehen; ~ **sb. do** *or*

doing **sth.** zusehen, wie jmd. etw. tut; **we are being** ~**ed** wir werden beobachtet; **b)** *(be careful of, look after)* achten auf (+ *Akk.*). **watch 'out** *v. i.* **a)** *(be careful)* aufpassen; ~ **out!** Vorsicht!; **b)** *(look out)* ~ **out for sb./sth.** auf jmdn./etw. achten

'watch-dog *n.* Wachhund, *der;* *(fig.)* |**public**| ~: *[Leiter/Leiterin einer] Aufsichtsbehörde*

watchful ['wɒtʃfl] *adj.* wachsam

watch: ~**maker** *n.* Uhrmacher, *der/* Uhrmacherin, *die;* ~**man** ['wɒtʃmən] *n., pl.* ~**men** ['wɒtʃmən] Wachmann, *der;* ~**-strap** *n.* [Uhr]armband, *das;* ~**-tower** *n.* Wachturm, *der*

water ['wɔːtə(r)] **1.** *n.* **a)** Wasser, *das;* **b)** *in pl. (part of the sea etc.)* Gewässer *Pl.* **2.** *v. t.* **a)** bewässern ⟨*Land*⟩; wässern ⟨*Pflanzen*⟩; ~ **the flowers** die Blumen [be]gießen; **b)** verwässern ⟨*Bier usw.*⟩; **c)** tränken ⟨*Tier*⟩. **3.** *v. i.* ⟨*Augen:*⟩ tränen; **my mouth was** ~**ing** mir lief das Wasser im Munde zusammen. **water 'down** *v. t.* verwässern

water: ~**-butt** *n.* Regentonne, *die;* ~**-colour** *n.* **a)** *(paint)* Wasserfarbe, *die;* **b)** *(picture)* Aquarell, *das;* ~**cress** *n.* Brunnenkresse, *die;* ~**fall** *n.* Wasserfall, *der*

'watering-can *n.* Gießkanne, *die*

water: ~**-lily** *n.* Seerose, *die;* ~**-line** *n. (Naut.)* Wasserlinie, *die;* ~**logged** ['wɔːtəlɒgd] *adj.* naß ⟨*Boden*⟩; aufgeweicht ⟨*Sportplatz*⟩; ~**-main** *n.* Hauptwasserleitung, *die;* ~**mark** *n.* Wasserzeichen, *das;* ~**-melon** *n.* Wassermelone, *die;* ~ **meter** *n.* Wasseruhr, *die;* ~ **polo** *n.* Wasserball, *der;* ~**proof 1.** *adj.* wasserdicht; **2.** *v. t.* wasserdicht machen; imprägnieren ⟨*Stoff*⟩; ~**shed** *n. (fig.)* Wendepunkt, *der;* ~**-ski** **1.** *n.* Wasserski, *der;* **2.** *v. i.* Wasserski laufen; ~**-skiing** *n.* Wasserskilaufen, *das;* ~**tight** *adj.* wasserdicht; ~**-tower** *n.* Wasserturm, *der;* ~**way** *n.* Wasserstraße, *die*

watery *adj.* wäßrig

watt [wɒt] *n.* Watt, *das*

wave [weɪv] **1.** *n.* **a)** Welle, *die;* **b)** *(gesture)* **give sb. a** ~: jmdm. zuwinken; **with a** ~ **of one's hand** mit einem Winken. **2.** *v. i.* **a)** ⟨*Fahne, Flagge, Wimpel:*⟩ wehen; ⟨*Baum, Gras, Korn:*⟩ sich wiegen; **b)** *(with hand)* winken; ~ **at** *or* **to sb.** jmdm. winken. **3.** *v. t.* schwenken; schwingen ⟨*Schwert*⟩; ~ **one's hand at** *or* **to sb.** jmdm. winken; ~

~ **goodbye to sb.** jmdm. zum Abschied zuwinken. **wave a'side** *v. t.* **a)** abtun ⟨*Zweifel, Einwand*⟩; **b)** *(signal to move)* ~ **sb. aside** [jmdm.] abwinken

wave: ~**band** *n.* Wellenbereich, *der;* ~**length** *n.* Wellenlänge, *die;* **be on the same** ~**length [as sb.]** *(fig.)* die gleiche Wellenlänge [wie jmd.] haben

waver ['weɪvə(r)] *v. i.* schwanken

wavy ['weɪvɪ] *adj.* wellig; ~ **line** Schlangenlinie, *die*

¹**wax** [wæks] **1.** *n.* **a)** Wachs, *das;* **b)** *(in ear)* Schmalz, *das.* **2.** *v. t.* wachsen

²**wax** *v. i.* **a)** ⟨*Mond:*⟩ zunehmen; **b)** *(become)* werden

wax: ~**work** *n.* Wachsfigur, *die;* ~**works** *n. sing., pl. same* Wachsfigurenkabinett, *das*

'**waxy** *adj.* wachsweich

way [weɪ] **1.** *n.* **a)** Weg, *der;* **ask the** *or* **one's** ~: nach dem Weg fragen; '**W-In/Out'** „Ein-/Ausgang"; **by** ~ **of Switzerland** über die Schweiz; **lead the** ~: vorausgehen; **go out of one's** ~: einen Umweg machen; *(fig.)* keine Mühe scheuen; **b)** *(method)* Art und Weise, *die;* **do it this** ~: mach es so; **c)** *(distance)* Stück, *das;* **it's a long** ~ **off** *or* **a long** ~ **from here** es ist weit weg von hier; **all the** ~: den ganzen Weg; **d)** *(direction)* Richtung, *die;* **she went this/that/the other** ~: sie ist in diese/die/die andere Richtung gegangen; **stand sth. the right/wrong** ~ **up** etw. richtig/falsch herum stellen; **e)** *(respect)* **in [exactly] the same** ~: [ganz] genauso; **in some** ~s in gewisser Hinsicht; **in one** ~: auf eine Art; **in every** ~: in jeder Hinsicht; **in a** ~: auf eine Art; **f)** *(custom)* Art, *die;* **g) get** *or* **have one's [own]** ~, **have it one's [own]** ~: seinen Willen kriegen; **be in sb.'s** *or* **the** ~: [jmdm.] im Weg sein; **make** ~ **for sth.** für etw. Platz machen; *(fig.)* einer Sache *(Dat.)* Platz machen; **in a bad** ~: schlecht; **either** ~: so oder so; **by the** ~: übrigens. **2.** *adv.* weit; ~ **back** *(coll.)* vor langer Zeit. **way'lay** *v. t., forms as* ²**lay 1: a)** *(ambush)* überfallen; **b)** *(stop for conversation)* abfangen. **way-'out** *adj. (coll.)* verrückt

WC *abbr.* **water-closet** WC, *das*

we [wɪ, *stressed* wiː] *pl. pron.* wir

weak [wiːk] *adj.* **a)** schwach; *(easily led)* labil ⟨*Charakter, Person*⟩; **b)** dünn ⟨*Getränk*⟩

weaken ['wiːkn] **1.** *v. t.* schwächen; beeinträchtigen ⟨*Augen*⟩. **2.** *v. i.* ⟨*Entschlossenheit, Kraft:*⟩ nachlassen

weakling ['wiːklɪŋ] *n.* Schwächling, *der*

'**weakly** *adv.* schwach

'**weakness** *n.* Schwäche, *die*

wealth [welθ] *n.* **a)** *(abundance)* Fülle, *die;* **b)** *(riches, being rich)* Reichtum, *der.* '**wealthy 1.** *adj.* reich. **2.** *n. pl.* **the** ~: die Reichen

wean [wiːn] *v. t.* abstillen; ~ **sb. [away] from sth.** *(fig.)* jmdm. etw. abgewöhnen

weapon ['wepən] *n.* Waffe, *die*

wear [weə(r)] **1.** *n.* **a)** ~ **[and tear]** Abnutzung, *die;* **b)** *(clothes)* Kleidung, *die.* **2.** *v. t.,* **wore** [wɔː(r)], **worn** [wɔːn] **a)** *(have on)* tragen ⟨*Schmuck, Brille, Kleidung, Perücke*⟩; **I haven't a thing to** ~: ich habe überhaupt nichts anzuziehen; **b)** *(rub)* abtragen ⟨*Kleidungsstück*⟩; abnutzen ⟨*Teppich*⟩; **a [badly] worn tyre** ein [stark] abgefahrener Reifen. **3.** *v. i.,* **wore, worn a)** ⟨*Kleider:*⟩ sich durchscheuern; ⟨*Absätze:*⟩ sich ablaufen; ⟨*Teppich:*⟩ sich abnutzen; **b)** *(endure rubbing)* halten; ~ **well/badly** sich gut/schlecht tragen. **wear a'way 1.** *v. t.* abschleifen. **2.** *v. i.* sich abnutzen. **wear 'down** *v. t. (fig.)* zermürben. **wear 'off** *v. i.* ⟨*Schicht:*⟩ abgehen; ⟨*Wirkung, Schmerz:*⟩ nachlassen. **wear 'out 1.** *v. t.* **a)** aufbrauchen; auftragen ⟨*Kleidungsstück*⟩; **b)** *(fig.: exhaust)* kaputtmachen *(ugs.);* **be worn out** kaputt sein *(ugs.).* **2.** *v. i.* kaputtgehen *(ugs.)*

wearable ['weərəbl] *adj.* **sth. is [not]** ~: man kann etw. [nicht] anziehen

wearily ['wɪərɪlɪ] *adv.* müde

weary ['wɪərɪ] **1.** *adj.* **a)** *(tired)* müde; **b) be** ~ **of sth.** einer Sache *(Gen.)* überdrüssig sein. **2.** *v. t.* **be wearied by sth.** durch etw. erschöpft sein. **3.** *v. i.* ~ **of sth./sb.** einer Sache/jmds. überdrüssig werden

weasel ['wiːzl] *n.* Wiesel, *das*

weather ['weðə(r)] **1.** *n.* Wetter, *das;* **what's the** ~ **like?** wie ist das Wetter?; **in all** ~s bei jedem Wetter; **he is feeling under the** ~ *(fig.)* er ist [zur Zeit] nicht ganz auf dem Posten. **2.** *v. t.* abwettern ⟨*Sturm*⟩; *(fig.)* durchstehen ⟨*schwere Zeit*⟩

weather: ~**beaten** *adj.* wettergegerbt ⟨*Gesicht*⟩; verwittert ⟨*Felsen, Gebäude*⟩; ~**cock** *n.* Wetterhahn, *der;* ~ **forecast** *n.* Wettervorhersage, *die;* ~**man** *n.* Meteorologe, *der;* ~ **report** *n.* Wetterbericht, *der;* ~**vane** *n.* Wetterfahne, *die*

¹**weave** [wi:v] **1.** *n.* Bindung, *die.* **2.**
v. t., wove [wəʊv], woven ['wəʊvn] **a)**
weben; flechten ⟨*Korb, Kranz*⟩; **b)**
(fig.) einflechten ⟨*Thema usw.*⟩ (**into**
in + *Akk.*)

²**weave** *v. i. (take intricate course)* sich
schlängeln

'**weaver** *n.* Weber, *der*/Weberin, *die*

web [web] *n.* Netz, *das;* **spider's ~:**
Spinnennetz, *das;* **webbed feet**
[webd 'fi:t] *n. pl.* Schwimmfüße

we'd [wɪd, *stressed* wi:d] **a)** = **we had;**
b) = **we would**

Wed. *abbr.* **Wednesday** Mi.

wedding ['wedɪŋ] *n.* Hochzeit, *die*

wedding: **~ anniversary** *n.* Hochzeitstag, *der;* **~-cake** *n.* Hochzeits-
kuchen, *der;* **~ day** *n.* Hochzeitstag,
der; **~ dress** *n.* Brautkleid, *das;* **~**
present *n.* Hochzeitsgeschenk, *das;*
~-ring *n.* Ehering, *der*

wedge [wedʒ] **1.** *n.* Keil, *der.* **2.** *v. t.*
verkeilen; **~ a door/window open** eine
Tür/ein Fenster festklemmen, damit
sie/es offen bleibt. '**wedge-shaped**
adj. keilförmig

wedlock ['wedlɒk] *n.* **born in/out of**
~: ehelich/unehelich geboren

Wednesday ['wenzdeɪ, 'wenzdɪ] *n.*
Mittwoch, *der; see also* **Friday**

¹**wee** [wi:] *adj. (child lang./Scot.)* klein

²**wee** *see* **wee-wee**

weed [wi:d] **1.** *n.* **~[s]** Unkraut, *das.* **2.**
v. t. jäten. **weed 'out** *v. t. (fig.)* aussieben

'**weed-killer** *n.* Unkrautvertilgungs-
mittel, *das*

'**weedy** *adj.* spillerig *(ugs.)* ⟨*Person*⟩

week [wi:k] *n.* Woche, *die;* **for several**
~s mehrere Wochen lang; **once a ~,**
every ~: einmal in der Woche; **three**
times a ~: dreimal in der Woche; **a**
two-~ visit ein zweiwöchiger Besuch;
a ~ today/tomorrow heute/morgen in
einer Woche; **a ~ on Monday, Mon-**
day ~: Montag in einer Woche.
'**weekday** *n.* Wochentag, *der.* **week-**
end [-'-, '--] *n.* Wochenende, *das;* **at**
the ~: am Wochenende; **go away for**
the ~: übers Wochenende wegfahren

weekly ['wi:klɪ] **1.** *adj.* wöchentlich;
Wochen⟨zeitung, -zeitschrift, -lohn⟩. **2.**
adv. wöchentlich. **3.** *n. (newspaper)*
Wochenzeitung, *die; (magazine)* Wo-
chenzeitschrift, *die*

weep [wi:p] *v. i. & t.,* **wept** [wept] weinen. **weeping 'willow** *n.* Trauer-
weide, *die*

'**wee-wee** *(coll.)* **1.** *n.* Pipi, *das (ugs.);*

do a ~: Pipi machen *(ugs.).* **2.** *v. i.* Pipi
machen *(ugs.)*

weigh [weɪ] *v. t. & i.* wiegen. **weigh**
'**down** *v. t. (fig.: depress)* niederdrücken. **weigh 'up** *v. t.* abwägen

weight [weɪt] *n.* Gewicht, *das;* **what is**
your ~? wieviel wiegen Sie?; **be**
under/over ~: zuwenig/zuviel wiegen.
'**weighting** *n.* Zulage, *die.* '**weight-**
lessness *n.* Schwerelosigkeit, *die*

weight: **~-lifter** *n.* Gewichtheber,
der/-heberin, *die;* **~-lifting** *n.* Ge-
wichtheben, *das*

'**weighty** *adj.* **a)** *(heavy)* schwer; **b)**
(important) gewichtig

weir [wɪə(r)] *n.* Wehr, *das*

weird [wɪəd] *adj. (coll.: odd)* bizarr

welcome ['welkəm] **1.** *int.* willkommen; **~ home/to England!** willkommen zu Hause/in England! **2.** *n.* **a)**
Willkommen, *das;* **b)** *(reception)*
Empfang, *der.* **3.** *v. t.* begrüßen. **4.**
adj. **a)** willkommen; gefällig ⟨*An-*
blick⟩; **b)** *pred.* **you are ~ to take it** du
kannst es gern nehmen; **you're ~:**
gern geschehen!

weld [weld] *v. t. (join)* verschweißen;
(repair, make, attach) schweißen (**[on]**
to an + *Akk.*). '**welder** *n.* Schweißer,
*der/*Schweißerin, *die*

welfare ['welfeə(r)] *n.* Wohl, *das.*
Welfare 'State *n.* Wohlfahrtsstaat,
der. '**welfare work** *n.* Sozialarbeit,
die

¹**well** [wel] *n.* **a)** Brunnen, *der;* **b)** *see*
oil well; c) *(stair~)* Treppenloch, *das*

²**well 1.** *int.* **~!** meine Güte!; **~, let's**
forget that na ja, lassen wir das; **~,**
who was it? nun *od.* und, wer war's?;
oh ~[, never mind] na ja[, lassen wir's]
nichts]; **~?** na? **2.** *adv.,* **better** ['bet-
ə(r)], **best** [best] gut; gründlich ⟨*trock-*
nen, schütteln⟩; **the business/patient is**
doing ~: das Geschäft geht gut/dem
Patienten geht es gut; **~ done!** großartig!; **he is ~ over forty** er ist weit über
vierzig; **as ~** *(in addition)* auch; **A as ~**
as B: B und auch [noch] A. **3.** *adj. (in*
good health) **How are you feeling**
now? – Quite ~, thank you Wie fühlen
Sie sich jetzt? – Ganz gut, danke; **look**
~: gut aussehen; **feel ~:** sich wohl
fühlen; **he isn't [very] ~:** es geht ihm
nicht [sehr] gut; **get ~ soon!** gute Besserung!; **make sb. ~:** jmdn. gesund
machen

we'll [wɪl, *stressed* wi:l] = **we will**

well: **~-behaved** *see* **behave 1;**
~-being *n.* Wohl, *das;* **~-bred** *adj.*

anständig; ~-**built** adj. ⟨Person:⟩ mit
guter Figur; be ~-**built** eine gute Figur
haben; ~ **done** adj. (Cookery) durch-
gebraten; ~-**dressed** adj. gutgeklei-
det; ~-**educated** adj. gebildet; ~-
heeled adj. (coll.) gutbetucht (ugs.)
wellington ['welɪŋtən] n. ~ [**boot**]
Gummistiefel, der
well: ~-**known** adj. bekannt; ~
made adj. gut [gearbeitet]; ~-**mannered** adj. ⟨Person⟩ mit guten Manieren; be ~-**mannered** gute Manieren
haben; ~-**meaning** adj. wohlmei-
nend; be ~-**meaning** es gut meinen;
~-**meant** adj. gutgemeint; ~ **off** adj.
wohlhabend; sb. is ~ **off** jmdm. geht
es [finanziell] gut; ~-**read** ['welred]
adj. belesen; ~-**timed** adj. zeitlich
gut gewählt; ~-**to-do** adj. wohlhabend; ~-**wisher** n. Sympathisant,
der/Sympathisantin, die
Welsh [welʃ] 1. adj. walisisch; sb. is
~: jmd. ist Waliser/Waliserin. 2. n. a)
(language) Walisisch, das; see also
English 2 a; b) pl. the ~: die Waliser.
Welshman ['welʃmən] n., pl. ~**men**
['welʃmən] Waliser, der. ~ '**rabbit,** ~
rarebit ['reəbɪt] ns. Käsetoast, der
went see go 1
wept see weep
were see be
we're [wɪə(r)] = we are
weren't (coll.) = were not; see be
west [west] 1. n. a) Westen, der; in/
to|wards|/from the ~: im/nach/von
Westen; to the ~ of westlich von; b)
usu. W~ (Geog., Polit.) Westen, der.
2. adj. westlich; West⟨küste, -wind,
-grenze, -tor⟩. 3. adv. nach Westen; ~
of westlich von
West: ~ **Ber'lin** pr. n. (Hist.) West-
Berlin (das); **w~bound** adj. ⟨Zug,
Verkehr usw.⟩ in Richtung Westen; ~
Country n. (Brit.) Westengland, das;
~ '**End** n. (Brit.) Westend, das
westerly ['westəlɪ] adj. westlich;
⟨Wind⟩ aus westlichen Richtungen
western ['westən] 1. adj. westlich;
West⟨grenze, -hälfte, -seite⟩; ~ **Germany** Westdeutschland, das. 2. n.
Western, der. **Western 'Europe** pr.
n. Westeuropa (das)
West: ~ '**German** (Hist.) 1. adj. west-
deutsch; 2. n. Westdeutsche, der/die;
~ '**Germany** pr. n. (Hist.) Westdeutschland (das); ~ '**Indian** 1. adj.
westindisch; 2. n. Westinder, der/-in-
derin, die; ~ '**Indies** pr. n. pl. westin-
dische Inseln

westward[s] ['westwəd(z)] adv.
westwärts
wet [wet] 1. adj. a) naß; b) (rainy) regnerisch; feucht ⟨Klima⟩; c) frisch
⟨Farbe⟩; '~ **paint**' „frisch gestrichen";
d) (sl.: feeble) schlapp (ugs.). 2. v. t.,
wet or wetted befeuchten. 3. n. a)
(moisture) Feuchtigkeit, die; b) in the
~: im Regen. '**wetness** n. Nässe, die
'**wet suit** n. Tauchanzug, der
we've [wɪv, stressed wiːv] = we have
whack [wæk] (coll.) 1. v. t. hauen
(ugs.). 2. n. Schlag, der
whale [weɪl] n. a) Wal, der; b) (coll.)
we had a ~ of a [good] time wir haben
uns bombig (ugs.) amüsiert
wharf [wɔːf] n., pl. **wharves** [wɔːvz] or
~s Kai, der
what [wɒt] 1. adj. welch...; ~ **book?**
welches Buch?; ~ **time does it start?**
um wieviel Uhr fängt es an?; ~ **kind
of man is he?** was für ein Mensch ist
er?; ~ **a fool you are!** was für ein
Dummkopf du doch bist!; ~ **cheek/
luck!** was für eine Frechheit/ein
Glück!; **I will give you ~ help I can** ich
werde dir helfen, so gut ich kann. 2.
adv. ~ **do I care?** was kümmert's
mich?; ~ **does it matter?** was macht's?
3. pron. was; ~? wie?; was? (ugs.); ~
is your name? wie heißt du/heißen
Sie?; ~ **about ...?** (~ will become of ...?)
was ist mit ...?; ~ **about a game of
chess?** wie wär's mit einer Partie
Schach?; ~'s-**his**/-**her**/-**its-name** wie
heißt er/sie/es noch; ~ **for?** wozu?; ~
is it like? wie ist es?; **so ~?** na und?;
do ~ I tell you tu, was ich dir sage
whatever [wɒt'evə(r)] 1. adj. ~ **problems you have** was für Probleme Sie
auch haben; **nothing** ~: absolut
nichts. 2. pron. **do ~ you like** mach,
was du willst; ~ **happens, ...:** was auch
geschieht, ...; **or** ~: oder was auch immer; ~ **does he want?** (coll.) was will er
nur?
wheat [wiːt] n. Weizen, der
wheedle ['wiːdl] v. t. ~ **sb. into doing
sth.** jmdm. so lange gut zureden, bis er
etw. tut; ~ **sth. out of sb.** jmdm. etw.
abschwatzen (ugs.)
wheel [wiːl] 1. n. a) Rad, das; |**potter's**|
~: Töpferscheibe, die; b) (steering ~)
~: Lenkrad, das; (ship's ~) Steuerrad,
das; **at** or **behind the ~** (of car) am
Steuer. 2. v. t. (push) schieben. 3. v. i.
a) (turn round) kehrtmachen; b)
(circle) kreisen
wheel: ~**barrow** n. Schubkarre, die;

~chair *n.* Rollstuhl, *der;* **~clamp** *n.* Parkkralle, *die*

wheeze [wi:z] *v.t.* schnaufen

when [wen] **1.** *adv.* wann; **the time ~** ...: die Zeit, zu der/*(with past tense)* als ...; **the day ~** ...: der Tag, an dem/ *(with past tense)* als ... **2.** *conj.* **a)** *(at the time that)* als; *(with present or future tense)* wenn; **~ reading |a newspaper|** beim Lesen [einer Zeitung]; **b)** *(whereas)* **why do you go abroad ~ it's cheaper here?** warum fährst du ins Ausland, wo es doch hier billiger ist? **3.** *pron.* **by/till ~** ...?; **bis wann ...?; since ~** ...? seit wann ...?

whence [wens] *adv., conj. (arch./ literary)* woher

whenever [wen'evə(r)] **1.** *adv.* wann immer; **or ~:** oder wann immer; **~ did he do it?** *(coll.)* wann hat er es nur getan? **2.** *conj.* jedesmal wenn

where [weə(r)] **1.** *adv.* **a)** *(position)* wo; **~ shall we sit?** wohin wollen wir uns setzen?; **b)** *(to ~)* wohin. **2.** *conj.* wo. **3.** *pron.* **near/not far from ~ it happened** nahe der Stelle/nicht weit von der Stelle, wo es passiert ist

whereabouts 1. [weərə'baʊts] *adv. (where)* wo; *(to where)* wohin. **2.** ['weə- rəbaʊts] *n., sing. or pl. (of thing)* Verbleib, *der; (of person)* Aufenthalt[sort], *der*

where: ~'as *conj.* während; **he is very quiet, ~as she is an extrovert** er ist sehr ruhig, sie dagegen ist eher extravertiert; **~'by** *adv.* mit dem/der/denen; **~upon** [weərə'pɒn] *adv.* worauf

wherever [weər'evə(r)] **1.** *adv.* **a)** *(position)* wo immer; **sit ~ you like** setz dich, wohin du magst; **or ~:** oder wo immer; **b)** *(direction)* wohin immer; **or ~:** oder wohin immer; **c) ~ have you been?** *(coll.)* wo hast du bloß gesteckt? **2.** *conj.* **a)** *(position)* überall [da], wo; **~ possible** wo *od.* wenn [irgend] möglich; **b)** *(direction)* wohin auch; **~ he went** wohin er auch ging

whet [wet] *v.t., -tt-:* **a)** *(sharpen)* wetzen; **b)** *(fig.)* anregen ⟨*Appetit*⟩

whether ['weðə(r)] *conj.* ob; **I don't know ~ to go |or not|** ich weiß nicht, ob ich gehen soll [oder nicht]

which [wɪtʃ] **1.** *adj.* welch...; **~ one** welcher/welche/welches; **~ ones** welche; **~ way** *(how)* wie; *(in ~ direction)* wohin. **2.** *pron.* **a)** *interrog.* welcher/ welche/welches; **~ of you?** wer von euch?; **b)** *rel.* der/die/das; **of ~:** dessen/deren; **after ~:** worauf[hin]

whichever [wɪtʃ'evə(r)] **1.** *adj.* welcher/welche/welches ... auch. **2.** *pron.* **a)** welcher/welche/welches ... auch; **b)** *(coll.)* **~ could it be?** welcher/welche/ welches könnte das nur sein?

whiff [wɪf] *n. (puff; fig.: trace)* Hauch, *der; (smell)* leichter Geruch

while [waɪl] **1.** *n.* Weile, *die;* |**for**| **a ~:** eine Weile; **a long ~:** lange; **for a little or short ~:** eine kleine Weile; **be worth sb.'s ~:** sich [für jmdn.] lohnen. **2.** *conj.* **a)** während; *(as long as)* solange; **b)** *(although)* obgleich; **c)** *(whereas)* während. **while a'way** *v.t.* **~ away the time** sich *(Dat.)* die Zeit vertreiben (**by, with** mit)

whilst [waɪlst] *(Brit.) see* **while 2**

whim [wɪm] *n.* Laune, *die*

whimper ['wɪmpə(r)] **1.** *n.* ~|s| Wimmern, *das; (of dog etc.)* Winseln, *das.* **2.** *v.i.* wimmern; ⟨*Hund:*⟩ winseln

whimsical ['wɪmzɪkl] *adj.* launenhaft; *(odd, fanciful)* spleenig

whine [waɪn] **1.** *v.i.* **a)** heulen; ⟨*Hund:*⟩ jaulen; **b)** *(complain)* jammern. **2.** *n.* **a)** Heulen, *das; (of dog)* Jaulen, *das;* **b)** *(complaint)* ~|s| Gejammer, *das*

whip [wɪp] **1.** *n.* **a)** Peitsche, *die;* **b)** *(Brit. Parl.)* Fraktionsgeschäftsführer, *der/*-führerin, *die.* **2.** *v.t., -pp-:* **a)** peitschen; **b)** *(Cookery)* schlagen; **c)** *(move quickly)* reißen; **d)** *(sl.: steal)* klauen *(ugs.).* **whip 'out** *v.t.* [blitzschnell] herausziehen. **whip 'up** *v.t.* **a)** *(arouse)* anheizen *(ugs.);* **b)** *(coll.: make quickly)* schnell hinzaubern ⟨*Gericht, Essen*⟩

whipped 'cream *n.* Schlagsahne, *die*

whirl [wɜ:l] **1.** *v.t.* [im Kreis] herumwirbeln. **2.** *v.i.* wirbeln. **3.** *n.* **a)** Wirbeln, *das;* **she was** *or* **her thoughts were in a ~** *(fig.)* ihr schwirrte der Kopf; **b)** *(bustle)* Trubel, *der.* **whirl 'round 1.** *v.t.* [im Kreis] herumwirbeln. **2.** *v.i.* [im Kreis] herumwirbeln; ⟨*Rad, Rotor:*⟩ wirbeln

whirl: ~pool *n.* Strudel, *der; (bathing pool)* Whirlpool, *der;* **~wind** *n.* Wirbelwind, *der*

whirr [wɜ:(r)] **1.** *v.i.* surren. **2.** *n.* Surren, *das*

whisk [wɪsk] **1.** *n. (Cookery)* Schneebesen, *der; (part of mixer)* Rührbesen, *der.* **2.** *v.t.* **a)** *(Cookery)* [mit dem Schnee-/Rührbesen] schlagen; **b)** *(convey rapidly)* in Windeseile bringen. **whisk a'way** *v.t.* **a)** *(remove suddenly)* **~ sth. away |from sb.|**

[jmdm.] etw. [plötzlich] wegreißen; **b)** *(convey rapidly)* in Windeseile wegbringen

whisker ['wɪskə(r)] *n.* **a)** ~s *(on man's cheek)* Backenbart, *der;* **b)** *(of cat, mouse, rat)* Schnurrhaar, *das*

whiskey *(Amer., Ir.),* **whisky** ['wɪskɪ] *n.* Whisky, *der; (American or Irish)* Whiskey, *der*

whisper ['wɪspə(r)] **1.** *v. i.* flüstern; ~ **to sb.** jmdm. etwas zuflüstern. **2.** *v. t.* flüstern; ~ **sth. to sb.** jmdm. etw. zuflüstern. **3.** *n.* **a)** Flüstern, *das;* **in a ~, in** ~s im Flüsterton; **b)** *(rumour)* Gerücht, *das*

whistle ['wɪsl] **1.** *v. i.* pfeifen; ~ **at sb.** *(in disapproval)* jmdn. auspfeifen. **2.** *v. t.* pfeifen. **3.** *n.* **a)** *(sound)* Pfiff, *der; (whistling)* Pfeifen, *das;* **b)** *(instrument)* Pfeife, *die;* **blow a/one's ~:** pfeifen

white [waɪt] **1.** *adj.* weiß. **2.** *n.* **a)** *(colour)* Weiß, *das;* **b)** *(of egg)* Eiweiß, *das;* **c) W~** *(person)* Weiße, *der/die*

white: ~ **bread** *n.* Weißbrot, *das;* ~ **'coffee** *n. (Brit.)* Kaffee mit Milch; ~-**'collar worker** *n.* Angestellte, *der/die;* **W~ House** *pr. n. (Amer. Polit.)* **the W~ House** das Weiße Haus

whiten ['waɪtn] **1.** *v. t.* weiß machen; weißen ‹*Wand, Schuhe*›. **2.** *v. i.* weiß werden

whiteness *n.* Weiß, *das*

white: W~ 'Paper *n. (Brit.)* öffentliches Diskussionspapier über Vorhaben der Regierung; ~**wash 1.** *n.* [weiße] Tünche; *(fig.)* Schönfärberei, *die;* **2.** *v. t.* [weiß] tünchen; ~ **'wine** *n.* Weißwein, *der*

Whit [wɪt] **'Monday** *n.* Pfingstmontag, *der*

Whitsun ['wɪtsn] *n.* Pfingsten, *das od. Pl.;* **at ~:** zu *od.* an Pfingsten

whittle ['wɪtl]: ~ **a'way** *v. t.* ~ **away sb.'s rights/power** jmdm. nach und nach alle Rechte/Macht nehmen; ~ **'down** *v. t.* allmählich reduzieren ‹*Anzahl, Gewinn*›; verkürzen ‹*Liste*›

whiz, whizz [wɪz] **1.** *v. i.,* -**zz**- zischen. **2.** *n.* Zischen, *das.* **'whiz[z]-kid** *n. (coll.)* Senkrechtstarter, *der*

who [hʊ, *stressed* hu:] *pron.* **a)** *interrog.* wer; *(coll.: whom)* wen; *(coll.: to whom)* wem; **b)** *rel.* der/die/das; *pl.* die; *(coll.: whom)* den/die/das; *(coll.: to whom)* dem/der/denen; **anyone/ those ~ ...:** wer ...; **everybody ~ ...:** jeder, der ...

whoa [wəʊ] *int.* brr

who'd [hʊd, *stressed* hu:d] **a)** = **who had; b)** = **who would**

whoever [hu:'evə(r)] *pron.* **a)** wer [immer]; **b)** *(no matter who)* wer ... auch; **c)** *(coll.)* ~ **could it be?** wer könnte das nur sein?

whole [həʊl] **1.** *adj.* ganz; **the ~ lot [of them]** [sie] alle. **2.** *n.* Ganze, *das;* **the ~:** das Ganze; **the ~ of my money/the village/London** mein ganzes Geld/das ganze Dorf/ganz London; **as a ~:** als Ganzes; **on the ~:** im großen und ganzen

whole: ~-**hearted** [həʊl'hɑːtɪd] *adj.* herzlich ‹*Dank[barkeit]*›; rückhaltlos ‹*Unterstützung*›; ~**meal** *adj.* Vollkorn-; ~ *note n. (Amer. Mus.)* ganze Note; ~ **'number** *n.* ganze Zahl; ~**sale 1.** *adj.* **a)** Großhandels-; **b)** *(fig.: on a large scale)* massenhaft; Massen-; **2.** *adv.* **a)** en gros; **b)** *(fig.: on a large scale)* massenweise; ~**saler** ['həʊlseɪlə(r)] *n.* Großhändler, *der/* -händlerin, *die*

wholesome ['həʊlsəm] *adj.* gesund

who'll [hʊl, *stressed* hu:l] = **who will**

wholly ['həʊllɪ] *adv.* völlig

whom [hu:m] *pron.* **a)** *interrog.* wen; *as indirect object* wem; **b)** *rel.* den/die/ das; *pl.* die; *as indirect object* dem/ der/dem; *pl.* denen

whooping cough ['hu:pɪŋ kɒf] *n.* Keuchhusten, *der*

whopper ['wɒpə(r)] *n. (coll.)* **a)** Riese, *der;* **b)** *(lie)* faustdicke Lüge

whopping ['wɒpɪŋ] *adj. (coll.)* riesig; Riesen- *(ugs.);* faustdick ‹*Lüge*›

whore [hɔː(r)] *n.* Hure, *die*

who's [hu:z] **a)** = **who is; b)** = **who has**

whose [hu:z] *pron.* **a)** *interrog.* wessen; ~ **[book] is that?** wem gehört das [Buch]?; **b)** *rel.* dessen/deren/dessen; *pl.* deren

who've [hʊv, *stressed* hu:v] = **who have**

why [waɪ] **1.** *adv.* **a)** *(for what reason)* warum; *(for what purpose)* wozu; ~ **is that?** warum das?; **b)** *(on account of which)* **the reason ~ he did it** der Grund, warum er es tat. **2.** *int.* ~**, certainly/of course!** aber sicher!

wick [wɪk] *n.* Docht, *der*

wicked ['wɪkɪd] *adj.* böse. **'wickedness** *n.* Bosheit, *die*

wicker ['wɪkə(r)] *n.* Korbgeflecht, *das; attrib.* Korb‹*waren,* -*stuhl*›. **'wickerwork** *n.* **a)** *(material)* Korbgeflecht, *das;* **b)** *(articles)* Korbwaren

wicket ['wɪkɪt] *n. (Cricket)* Tor, *das*

wide [waɪd] **1.** *adj.* **a)** *(broad)* breit; groß ⟨*Abstand, Winkel*⟩; **three feet** ~: drei Fuß breit; **b)** *(extensive)* weit; umfassend ⟨*Lektüre, Wissen, Kenntnisse*⟩; reichhaltig ⟨*Auswahl, Sortiment*⟩; **c)** *(off target)* **be** ~ **of sth.** etw. verfehlen. **2.** *adv.* **a)** ~ **awake** hellwach; **b)** *(off target)* **shoot** ~: danebenschießen; **go** ~: das Ziel verfehlen. **wide-angle 'lens** *n.* Weitwinkelobjektiv, *das*

'widely *adv.* **a)** *(over a wide area)* weit ⟨*verbreitet, gestreut*⟩; **b)** *(by many people)* weithin ⟨*bekannt, akzeptiert*⟩; **a** ~ **held view** eine weitverbreitete Ansicht; **c)** *(greatly)* erheblich ⟨*sich unterscheiden*⟩

widen ['waɪdn] **1.** *v. t.* verbreitern. **2.** *v. i.* sich verbreitern

wide: ~**-open** *attrib. adj.*, ~ **'open** *pred. adj.* weit geöffnet ⟨*Fenster, Tür*⟩; weit aufgerissen ⟨*Mund, Augen*⟩; **be** ~ **open** ⟨*Fenster, Tür:*⟩ weit offenstehen; ~**spread** *adj.* weitverbreitet *präd.* getrennt geschr.

widow ['wɪdəʊ] *n.* Witwe, *die.* **widowed** ['wɪdəʊd] *adj.* verwitwet. **widower** ['wɪdəʊə(r)] *n.* Witwer, *der*

width [wɪdθ] *n.* Breite, *die; (of garment)* Weite, *die*

wield [wiːld] *v. t.* schwingen; *(fig.)* ausüben ⟨*Macht, Einfluß*⟩

wife [waɪf] *n., pl.* **wives** [waɪvz] Frau, *die*

wig [wɪg] *n.* Perücke, *die*

wiggle ['wɪgl] *(coll.)* **1.** *v. t.* hin und her bewegen. **2.** *v. i.* wackeln

wild [waɪld] **1.** *adj.* **a)** wildlebend ⟨*Tier*⟩; wildwachsend ⟨*Pflanze*⟩; **b)** wild ⟨*Landschaft*⟩; **c)** *(unrestrained)* wild ⟨*Erregung*⟩; **run** ~ ⟨*Pferd, Hund:*⟩ frei herumlaufen; ⟨*Kind:*⟩ herumtoben; **send** *or* **drive sb.** ~: jmdn. rasend vor Erregung machen; **d)** *(coll.: very keen)* **be** ~ **about sb./sth.** wild auf jmdn./etw. sein. **2.** *n.* **the** ~**[s]** die Wildnis; **see an animal in the** ~: ein Tier in freier Wildbahn sehen

wilderness ['wɪldənɪs] *n.* Wildnis, *die; (desert)* Wüste, *die*

wild: ~**-'goose chase** *n. (fig.)* aussichtslose Suche; ~**life** *n.* die Tier- und Pflanzenwelt; ~**life park/reserve/ sanctuary** Naturpark, *der/*-reservat, *das/*-schutzgebiet, *das*

'wildly *adv.* wild; **be** ~ **excited about sth.** über etw. *(Akk.)* ganz aus dem Häuschen sein *(ugs.);* ~ **inaccurate** völlig ungenau

wilful ['wɪlfl] *adj.*, **wilfully** ['wɪlfəlɪ] *adv.* **a)** *(deliberate[ly])* vorsätzlich; **b)** *(obstinate[ly])* starrsinnig

¹will [wɪl] *v. aux., only in: pres.* **will,** *neg. (coll.)* **won't** [wəʊnt], *past* **would** [wʊd], *neg. (coll.)* **wouldn't** ['wʊdnt] He won't help me. W~/Would you? Er will mir nicht helfen. Bist du bereit?; **the car won't start** das Auto springt nicht an; ~/would you pass the salt, please? gibst du bitte mal das Salz rüber?/würdest du bitte mal das Salz rübergeben?; ~ you be quiet! willst du wohl ruhig sein!; **he** ~ **sit there hour after hour** er pflegt dort stundenlang zu sitzen; **he** '~ **insist on doing it** er besteht unbedingt darauf, es zu tun; ~ **you have some more cake?** möchtest *od.* willst du noch etwas Kuchen?; **the box** ~ **hold 5 lb. of tea** in die Kiste gehen 5 Pfund Tee; **tomorrow he** ~ **be in Oxford** morgen ist er in Oxford; **I promise I won't do it again** ich verspreche, ich mach's nicht noch mal; **if he tried, he would succeed** wenn er es versuchen würde, würde er es erreichen; ~ **you please tidy up** würdest du bitte aufräumen?

²will *n.* **a)** *(faculty)* Wille, *der;* **b)** *(Law: testament)* Testament, *das;* **c)** *(desire)* **at** ~: nach Belieben; ~ **to live** Lebenswille, *der;* **against one's/sb.'s** ~: gegen seinen/jmds. Willen

'willing *adj.* willig; **ready and** ~: bereit; **be** ~ **to do sth.** bereit sein, etw. zu tun. **'willingly** *adv.* **a)** *(with pleasure)* gern[e]; **b)** *(voluntarily)* freiwillig. **'willingness** *n.* Bereitschaft, *die*

willow ['wɪləʊ] *n.* Weide, *die*

'will-power *n.* Willenskraft, *die*

willy-nilly [wɪlɪ'nɪlɪ] *adv.* wohl oder übel ⟨*etw. tun müssen*⟩

wilt [wɪlt] *v. i.* ⟨*Pflanze, Blumen:*⟩ welk werden, welken

wily ['waɪlɪ] *adj.* listig; gewieft ⟨*Person*⟩

wimp [wɪmp] *n. (coll.)* Schlappschwanz, *der (ugs.)*

win [wɪn] **1.** *v. t.,* **-nn-,** **won** [wʌn] gewinnen; bekommen ⟨*Stipendium, Vertrag, Recht*⟩; ~ **sb. sth.** jmdm. etw. einbringen. **2.** *v. i.,* **-nn-,** **won** gewinnen. **3.** *n.* Sieg, *der;* **have a** ~: gewinnen. **win 'over, win 'round** *v. t.* bekehren; *(to one's side)* auf seine Seite bringen; *(convince)* überzeugen. **win 'through** *v. i.* Erfolg haben

wince [wɪns] *v. i.* zusammenzucken **(at** bei)

winch [wɪntʃ] 1. *n.* Winde, *die.* 2. *v.t.*
winden; ~ **up** hochwinden
¹wind [wɪnd] 1. *n.* Wind, *der; (Med.)*
Blähungen; get ~ of sth. *(fig.)* Wind
von etw. bekommen; **be in the** ~ *(fig.)*
in der Luft liegen; **get/have the** ~ **up**
(sl.) Manschetten *(ugs.)* kriegen/ha-
ben. 2. *v.t.* **the blow** ~**ed him** der
Schlag nahm ihm den Atem
²wind [waɪnd] 1. *v.i., wound* [waʊnd] **a)**
(curve) sich winden; *(move)* sich
schlängeln; **b)** *(coil)* sich wickeln. 2.
v.t., wound **a)** *(coil)* wickeln; ~ **sth. on**
|to| sth. etw. auf etw. *(Akk.)* [auf]-
wickeln; **b)** aufziehen ⟨*Uhr*⟩. **wind**
'**down** *v.t.* **a)** herunterdrehen ⟨*Autofenster*⟩; **b)** *(fig.: reduce gradually)*
einschränken. **wind** '**up** 1. *v.t.* **a)**
hochdrehen ⟨*Autofenster*⟩; **b)** *(coil)*
aufwickeln; **c)** aufziehen ⟨*Uhr*⟩; **d)**
(coll.: annoy deliberately) auf die Palme bringen *(ugs.);* **e)** beschließen ⟨*Debatte*⟩; **f)** *(Finance, Law)* auflösen. 2.
v.i. **a)** *(conclude)* schließen; **b)** *(coll.:
end up)* ~ **up in prison/hospital** [zum
Schluß] im Gefängnis/Krankenhaus
landen *(ugs.)*
wind [wɪnd]: ~**break** *n.* Windschutz,
der; ~-**chill factor** *n.* Wind-chill-In-
dex, *der (Meteor.)*
winder ['waɪndə(r)] *n. (of watch)* Krone, *die; (of clock, toy)* Aufziehschrau-
be, *die*
wind [wɪnd]: ~**fall** *n.* **a)** *(fruit)* ~**falls**
Fallobst, *das;* **b)** *(fig.)* warmer Regen
(ugs.); ~ **farm** *n.* Windpark, *der;*
Windfarm, *die;* ~ **instrument** *n.*
(Mus.) Blasinstrument, *das;* ~**mill** *n.*
Windmühle, *die*
window ['wɪndəʊ] *n.* Fenster, *das;*
(shop-~) [Schau]fenster, *das;* **break a**
~: eine Fensterscheibe zerbrechen
window: ~-**box** *n.* Blumenkasten,
der; ~-**cleaner** *n.* Fensterputzer,
der/-putzerin, *die;* ~-**dressing** *n.*
(fig.) Schönfärberei, *die;* ~-**pane** *n.*
Fensterscheibe, *die;* ~-**shopping** *n.*
Schaufensterbummeln, *das;* **go**
~-**shopping** einen Schaufensterbum-
mel machen; ~-**sill** *n. (inside)* Fensterbank, *die; (outside)* Fenstersims,
der od. das
wind [wɪnd]: ~**pipe** *n. (Anat.)* Luft-
röhre, *die;* ~**screen,** *(Amer.)*
~**shield** *ns. (Motor Veh.)* Wind-
schutzscheibe, *die;* ~**screen/**~**shield**
wiper Scheibenwischer, *der;* ~**screen/**
~**shield washer** Scheibenwaschanlage,
die; ~**surfer** *n.* Windsurfer, *der;*

~**surfing** *n.* Windsurfen, *das;*
~**swept** *adj.* windgepeitscht; vom
Wind zerzaust ⟨*Person, Haare*⟩;
~-**tunnel** *n.* Windkanal, *der*
windward ['wɪndwəd] *adj.* ~ **side**
Windseite, *die*
'**windy** *adj.* windig
wine [waɪn] *n.* Wein, *der*
wine: ~-**bar** *n.* Weinstube, *die;*
~-**cellar** *n.* [Wein]keller, *das;*
~**glass** *n.* Weinglas, *das;* ~-**list** *n.*
Weinkarte, *die;* ~-**tasting** ['waɪn-
teɪstɪŋ] *n.* Weinprobe, *die*
wing [wɪŋ] *n.* **a)** *(Ornith., Archit.,
Sport)* Flügel, *der;* **b)** *(Aeronaut.)*
Tragfläche, *die;* **c)** *(Brit. Motor. Veh.)*
Kotflügel, *der*
wink [wɪŋk] 1. *v.i.* **a)** blinzeln; *(as signal)* zwinkern; ~ **at sb.** jmdm. zuzwinkern; **b)** *(flash)* blinken. 2. *n.* **a)** Blinzeln, *das; (signal)* Zwinkern, *das;* **give
sb. a** ~: jmdm. zuzwinkern; **b) not
sleep a** ~: kein Auge zutun
'**winner** *n.* Sieger, *der/*Siegerin, *die;
(of competition or prize)* Gewinner,
*der/*Gewinnerin, *die*
'**winning** *adj.* **a)** *attrib.* siegreich; ~
number Gewinnzahl, *die;* **b)** *(charming)* einnehmend; gewinnend ⟨*Lächeln*⟩. '**winning-post** *n.* Zielpfosten, *der.* '**winnings** *n. pl.* Gewinn,
der
winter ['wɪntə(r)] *n.* Winter, *der;* **in**
|the| ~: im Winter. **winter** '**sports** *n.
pl.* Wintersport, *der*
wintry ['wɪntrɪ] *adj.* winterlich; ~
shower Schneegestöber, *das*
wipe [waɪp] 1. *v.t.* **a)** abwischen;
[auf]wischen ⟨*Fußboden*⟩; *(dry)* abtrocknen; ~ **one's mouth/eyes/nose**
sich *(Dat.)* den Mund/die Tränen/die
Nase abwischen; ~ **one's feet/shoes**
[sich *(Dat.)*] die Füße/Schuhe abtre-
ten; **b)** *(get rid of)* [ab]wischen; ~
one's/sb.'s tears sich/jmdm. die Tränen abwischen. 2. *n.* **give sth. a** ~: etw.
abwischen. **wipe** '**down** *v.t.* abwischen; *(dry)* abtrocknen. **wipe** '**off**
v.t. **a)** *(remove)* wegwischen; löschen
⟨*Bandaufnahme*⟩; **b)** *(pay off)* zurückzahlen ⟨*Schulden*⟩. **wipe** '**out** *v.t.* **a)**
(remove) wegwischen; *(erase)* auslöschen; **b)** *(cancel)* tilgen; zunichte
machen ⟨*Vorteil, Gewinn usw.*⟩; **c)**
(destroy) ausrotten ⟨*Rasse, Tierart,
Feinde*⟩; ausmerzen ⟨*Seuche, Korrup-
tion*⟩. **wipe** '**up** *v.t.* **a)** aufwischen; **b)**
(dry) abtrocknen
'**wiper** *n. (Motor Veh.)* Wischer, *der*

wire ['waɪə(r)] 1. *n.* a) Draht, *der;* b) *(Electr., Teleph.)* Leitung, *die; (coll.: telegram)* Telegramm, *das.* 2. *v.t.* a) *(fasten)* ~ sth. **together** etw. mit Draht verbinden; b) *(Electr.)* ~ **sth. to sth.** etw. an etw. *(Akk.)* anschließen; ~ **a house** in einem Haus die Stromleitungen legen; c) *(coll.: telegraph)* ~ **sb.** jmdm. *od.* an jmdm. telegrafieren. **'wireless** *n. (Brit.)* Radio, *das.* **wire 'netting** *n.* Maschendraht, *der*

wiring ['waɪərɪŋ] *n.* [elektrische] Leitungen

wisdom ['wɪzdəm] *n.* a) Weisheit, *die;* b) *(prudence)* Klugheit, *die.* **'wisdom tooth** *n.* Weisheitszahn, *der*

wise [waɪz] *adj.* a) weise; vernünftig ⟨*Meinung*⟩; *(prudent)* klug; c) **be none the ~r** kein bißchen klüger als vorher sein. **'wisely** *adv.* weise; *(prudently)* klug

wish [wɪʃ] 1. *v.t.* wünschen; **I ~ I was** *or* **were rich** ich wollte, ich wäre reich; **I ~ to go** ich möchte gehen; ~ **sb. luck/success** *etc.* jmdm. Glück/Erfolg *usw.* wünschen; ~ **sb. well** jmdm. alles Gute wünschen. 2. *v.i.* wünschen; ~ **for sth.** sich *(Dat.)* etw. wünschen. 3. *n.* Wunsch, *der;* **make a ~:** sich *(Dat.)* etwas wünschen; **get** *or* **have one's ~:** seinen Wunsch erfüllt bekommen. **wishful thinking** [wɪʃfl 'θɪŋkɪŋ] *n.* Wunschdenken, *das*

wishy-washy ['wɪʃɪwɒʃɪ] *adj.* labberig *(ugs.); (fig.)* laff

wisp [wɪsp] *n. (of straw)* Büschel, *das;* ~ **of hair** Haarsträhne, *die;* ~ **of cloud/smoke** Wolkenfetzen, *der/* Rauchfahne, *die*

wistful ['wɪstfl] *adj.,* **'wistfully** *adv.* wehmütig

wit [wɪt] *n.* a) *(humour)* Witz, *der;* b) *(intelligence)* Geist, *der;* **be at one's ~'s** *or* **~s' end** sich *(Dat.)* keinen Rat mehr wissen; **be frightened** *or* **scared out of one's ~s** Todesangst haben; **have/keep one's ~s about one** auf Draht *(ugs.)* sein/nicht den Kopf verlieren; c) *(person)* geistreicher Mensch

witch [wɪtʃ] *n.* Hexe, *die*

witch: ~**craft** *n.* Hexerei, *die;* ~**-doctor** *n.* Medizinmann, *der;* ~**-hunt** *n.* Hexenjagd, *die* (for auf + *Akk.*)

with [wɪð] *prep.* mit; **put sth. ~ sth.** etw. zu etw. stellen/legen; **have nothing to write ~:** nichts zum Schreiben haben; **I'm not '~ you** *(coll.)* ich kom-

me nicht mit; **tremble ~ fear** vor Angst zittern; **I have no money ~ me** ich habe kein Geld dabei *od.* bei mir; **sleep ~ the window open** bei offenem Fenster schlafen

with'draw 1. *v.t., forms as* **draw 1** zurückziehen; abziehen ⟨*Truppen*⟩; ~ **sth. from an account** etw. von einem Konto abheben. 2. *v.i., forms as* **draw 1** sich zurückziehen. **withdrawal** [wɪð'drɔːəl] *n.* a) Zurücknahme, *die; (of troops)* Abzug, *der; (of money)* Abhebung, *die;* b) *(from drugs)* Entzug, *der;* ~ **symptoms** Entzugserscheinungen. **with'drawn** *adj. (unsociable)* verschlossen

wither ['wɪðə(r)] 1. *v.t.* verdorren lassen. 2. *v.i.* [ver]welken. **wither a'way** *v.i.* dahinwelken *(geh.)*

with'hold *v.t., forms as* ²**hold:** ~ **sth. from sb.** jmdm. etw. vorenthalten

within [wɪ'ðɪn] *prep.* innerhalb; **stay/ be ~ the law** den Boden des Gesetzes nicht verlassen; ~ **eight miles of sth.** acht Meilen im Umkreis von etw.

without [wɪ'ðaʊt] *prep.* ohne; ~ **doing sth.** ohne etw. zu tun; ~ **his knowing** ohne daß er davon weiß/wußte

with'stand *v.t.,* **withstood** [wɪθ'stʊd] standhalten (+ *Dat.*); aushalten ⟨*Beanspruchung, hohe Temperaturen*⟩

witness ['wɪtnɪs] 1. *n.* Zeuge, *der/* Zeugin, *die* (of, to Gen.). 2. *v.t.* a) *(see)* ~ **sth.** Zeuge/Zeugin einer Sache *(Gen.)* sein; b) bestätigen ⟨*Unterschrift*⟩. **'witness-box** *(Brit.),* **'witness-stand** *(Amer.)* ns. Zeugenstand, *der*

witticism ['wɪtɪsɪzm] *n.* Witzelei, *die*

wittingly ['wɪtɪŋlɪ] *adv.* wissentlich

witty ['wɪtɪ] *adj.* witzig; geistreich ⟨*Person*⟩

wives *pl. of* **wife**

wizard ['wɪzəd] *n.* Zauberer, *der.* **wizardry** ['wɪzədrɪ] *n.* Zauberei, *die*

wizened ['wɪzənd] *adj.* runz[e]lig

wobble ['wɒbl] *v.i.* wackeln. **wobbly** ['wɒblɪ] *adj.* wack[e]lig

woe [wəʊ] *n. (arch./literary/joc.)* ~[s] Jammer, *der;* ~ **betide you!** wehe dir!

woke, woken *see* ¹**wake** 1, 2

wolf [wʊlf] 1. *n., pl.* **wolves** [wʊlvz] Wolf, *der.* 2. *v.t.* ~ [**down**] verschlingen

woman ['wʊmən] *n., pl.* **women** ['wɪmɪn] Frau, *die;* ~ **doctor** Ärztin, *die;* ~ **friend** Freundin, *die.* **womanizer** ['wʊmənaɪzə(r)] *n.* Schürzenjäger, *der.* **'womanly** *adj.* fraulich

womb [wu:m] *n.* Gebärmutter, *die*

women *pl. of* **woman**

women: ~**folk** *n. pl.* Frauen; **W~'s
'Lib** *(coll.),* **W~'s Libe'ration** *ns.* die
Frauenbewegung; ~**'s 'rights** *n. pl.*
die Rechte der Frau

won *see* **win 1, 2**

wonder ['wʌndə(r)] **1.** *n.* **a)** *(thing)*
Wunder, *das;* **b)** *(feeling)* Staunen,
das. **2.** *adj.* Wunder-. **3.** *v. i.* sich wundern; staunen (**at** über + *Akk.*). **4.** *v. t.*
sich fragen; **I ~ what the time is** wieviel Uhr mag es wohl sein?; **I ~
whether I might open the window** dürfte
ich vielleicht das Fenster öffnen?

wonderful ['wʌndəfl] *adj.,* **wonderfully** ['wʌndəfəlɪ] *adv.* wunderbar

won't [wəʊnt] *(coll.)* = **will not;** *see*
¹**will**

woo [wu:] *v. t.* **a)** *(literary: court)* ~ **sb.**
um jmdn. werben *(geh.);* **b)** umwer-
ben ⟨*Kunden, Wähler*⟩

wood [wʊd] *n.* **a)** Holz, *das;* **touch ~**
(Brit.), **knock on ~** *(Amer.)* unberufen!; **b)** *(trees)* Wald, *der.* **'woodcut**
n. Holzschnitt, *der.* **'woodcutter** *n.*
Holzfäller, *der*

'wooded *adj.* bewaldet

wooden ['wʊdn] *adj.* **a)** hölzern;
Holz-; **b)** *(fig.: stiff)* hölzern

wood: ~**land** ['wʊdlənd] *n.* Wald-
land, *das;* ~**pecker** *n.* Specht, *der;*
~**-wind** *n.* **the ~-wind |section|** die
Holzbläser; ~**-wind instrument** Holz-
blasinstrument, *das;* ~**work** *n.* **a)**
(craft) Arbeiten mit Holz; **b)** *(things)*
Holzarbeit[en]; ~**worm** *n.* Holz-
wurm; **it's got ~worm** da ist der Holz-
wurm drin *(ugs.)*

'woody *adj.* **a)** *(wooded)* waldreich; **b)**
(consisting of wood) holzig

wool [wʊl] *n.* Wolle, *die; attrib.* Woll-.
woollen *(Amer.:* **woolen)** ['wʊlən]
1. *adj.* wollen. **2.** *n.* ~**s** Wollsachen *Pl.*

'woolly *adj.* **a)** wollig; Woll⟨*pullover,
-mütze*⟩; **b)** *(confused)* verschwommen

word [wɜ:d] **1.** *n.* Wort, *das;* ~**s** *(of
song or actor)* Text, *der;* **in other ~s**
mit anderen Worten; ~ **for** ~: Wort
für Wort; **too funny** *etc.* **for ~s** unsag-
bar komisch *usw.;* **have ~s** einen
Wortwechsel haben; **have a ~ |with
sb.| about sth.** [mit jmdm.] über etw.
(Akk.) sprechen; **could I have a ~ |with
you|?** kann ich dich mal sprechen?;
say a few ~s ein paar Worte sprechen;
keep/break one's ~: sein Wort halten/
brechen; **by ~ of mouth** durch mündli-
che Mitteilung; **send ~ that ...:** Nach-

richt geben, daß ... **2.** *v. t.* formulieren.
'wording *n.* Formulierung, *die*

word: ~ **order** *n.* Wortstellung, *die;*
~ **processing** *n.* Textverarbeitung,
die; ~ **processor** *n.* Textverarbei-
tungssystem, *das*

wore *see* **wear 2, 3**

work [wɜ:k] **1.** *n.* **a)** Arbeit, *die;* **at ~**
(engaged in ~ing) bei der Arbeit; *(fig.:
operating)* am Werk; *(at job)* auf der
Arbeit; **out of ~:** arbeitslos; **be in ~:**
eine Stelle haben; **b)** ~**s** *sing. or pl.*
(factory) Werk, *das;* **c)** ~**s** *pl. (~ing
parts)* Werk, *das; (operations)* Arbeiten; **d)** *(thing made or achieved)* Werk,
das; **a ~ of art/literature** ein Kunst-
werk/literarisches Werk. **2.** *v. i.* **a)** arbeiten; **b)** *(function effectively)* funktionieren; **make the television ~:** den
Fernsehapparat in Ordnung bringen;
c) *(have an effect)* wirken (**on** auf
+ *Akk.*); **d)** ~ **loose** sich lockern. **3.**
v. t. **a)** bedienen ⟨*Maschine*⟩; betätigen ⟨*Bremse*⟩; **b)** *(get labour from)* arbeiten lassen; **c)** ausbeuten ⟨*Steinbruch, Grube*⟩; **d)** *(cause to go gradually)* führen; ~ **one's way up/into sth.**
sich hocharbeiten/in etw. hineinar-
beiten. **work 'off** *v. t.* **a)** *(get rid of)*
loswerden; abreagieren ⟨*Wut*⟩; **b)** ab-
arbeiten ⟨*Schuld*⟩. **'work on** *v. t.* **a)** ~
on sth. an etw. *(Dat.)* arbeiten; **b)** *(try
to persuade)* ~ **on sb.** jmdn. bearbeiten
(ugs.). **work 'out 1.** *v. t.* **a)** *(calculate)*
ausrechnen; **b)** *(solve)* lösen; **c)** *(devise)* ausarbeiten. **2.** *v. i.* **a)** **sth. ~s out
at £2** etw. ergibt 2 Pfund; **b)** *(have
result)* laufen; **things ~ed out |well| in
the end** es ist schließlich doch alles
gutgegangen. **work 'up 1.** *v. t. (excite)*
aufpeitschen ⟨*Menge*⟩; **get ~ed up** sich
aufregen. **2.** *v. i.* ~ **up to sth.** ⟨*Musik:*⟩
sich zu etw. steigern; ⟨*Geschichte,
Film:*⟩ auf etw. *(Akk.)* zusteuern

workable ['wɜ:kəbl] *adj. (feasible)*
durchführbar

workaholic [wɜ:kə'hɒlɪk] *n. (coll.)* ar-
beitswütiger Mensch

'worker *n.* Arbeiter, *der*/Arbeiterin,
die

'workforce *n.* Belegschaft, *die*

'working *adj.* **a)** *(in work)* werktätig;
b) ~ **model** funktionsfähiges Modell

working: ~ **'class** *n.* Arbeiterklasse,
die; ~**-class** *adj.* der Arbeiterklasse
nachgestellt; **sb. is ~-class** jmd. gehört
zur Arbeiterklasse; ~ **clothes** *n. pl.*
Arbeitskleidung, *die;* ~ **'day** *n.* **a)**
(portion of day) Arbeitstag, *der;* **b)**

(day when work is done) Werktag, *der;* ~ '**order** *n.* be in [good] ~ **order** betriebsbereit sein; ⟨*Auto:*⟩ fahrbereit sein

workman ['wɜːkmən] *n., pl.* ~**men** ['wɜːkmən] Arbeiter, *der.* '**workmanship** *n. (quality)* Kunstfertigkeit, *die*

work: ~-**out** *n.* [Fitneß]training, *das;* ~**shop** *n.* a) *(room)* Werkstatt; *die,* b) *(building)* Werk, *das*

world [wɜːld] *n.* a) Welt, *die;* **in the** ~: auf der Welt; **the tallest building in the** ~: das höchste Gebäude der Welt; **all over the** ~: in *od.* auf der ganzen Welt; b) *(vast amount)* **it will do him a** *or* **the** ~ **of good** es wird ihm unendlich guttun; **a** ~ **of difference** ein weltweiter Unterschied. **world** 'cham**pion** *n.* Weltmeister, *der/*-meisterin, *die.* **world-'famous** *adj.* weltberühmt

'**worldly** *adj.* weltlich; weltlich eingestellt ⟨*Person*⟩

world-wide 1. ['--] *adj.* weltweit *nicht präd.* **2.** [-'-] *adv.* weltweit

worm [wɜːm] **1.** *n.* Wurm, *der.* **2.** *v. t.* a) ~ **oneself into sb.'s favour** sich in jmds. Gunst *(Akk.)* schleichen; b) ~ **sth. out of sb.** etw. aus jmdm. herausbringen *(ugs.).* '**worm-eaten** *adj.* wurmstichig

worn *see* wear 2, 3

'**worn-out** *adj.* a) abgetragen ⟨*Kleidungsstück*⟩; abgenutzt ⟨*Teppich*⟩; b) erschöpft ⟨*Person*⟩

worried ['wʌrɪd] *adj.* besorgt

worry ['wʌrɪ] **1.** *v. t.* a) beunruhigen; b) *(bother)* stören. **2.** *v. i.* sich *(Dat.)* Sorgen machen. '**worrying** *adj.* a) *(causing worry)* beunruhigend; b) *(full of worry)* sorgenvoll ⟨*Zeit, Woche*⟩

worse [wɜːs] **1.** *adj.* schlechter; schlimmer ⟨*Schmerz, Krankheit, Benehmen*⟩. **2.** *adv.* schlechter/ schlimmer. **3.** *n.* Schlimmeres. **worsen** ['wɜːsn] **1.** *v. t.* verschlechtern. **2.** *v. i.* sich verschlechtern

worship ['wɜːʃɪp] **1.** *v. t., (Brit.)* -pp-: a) anbeten; b) *(idolize)* abgöttisch verehren. **2.** *v. i., (Brit.)* -pp- am Gottesdienst teilnehmen. **3.** *n.* a) Anbetung, *die; (service)* Gottesdienst, *der;* b) **Your/His W**~: ≈ Euer/seine Ehren. '**worshipper** *(Amer.:* **worshiper)** *n.* Gottesdienstbesucher, *der/*-besucherin, *die*

worst [wɜːst] **1.** *adj.* schlechtest...; schlimmst... ⟨*Schmerz, Krankheit, Benehmen*⟩. **2.** *adv.* am schlechtesten/

schlimmsten. **3.** *n.* a) **the** ~: der/die/ das Schlimmste; **get** *or* **have the** ~ **of it** *(suffer the most)* am meisten zu leiden haben; **if the** ~ **comes to the** ~: wenn es zum Schlimmsten kommt; b) *(poorest in quality)* Schlechteste, *der/ die/das*

worsted ['wʊstɪd] *n.* Kammgarn, *das*

worth [wɜːθ] **1.** *adj.* wert; it's ~ £80 es ist 80 Pfund wert; **is it** ~ **hearing/the effort?** ist es hörenswert/der Mühe wert?; **is it** ~ **doing?** lohnt es sich?; it **isn't** ~ **it** es lohnt sich nicht. **2.** *n.* Wert, *der;* **ten pounds'** ~ **of petrol** Benzin für zehn Pfund. '**worthless** *adj.* a) *(valueless)* wertlos; b) *(having bad qualities)* nichtswürdig. '**worthwhile** *adj.* lohnend

worthy ['wɜːðɪ] *adj.* würdig

wouldn't ['wʊdnt] *(coll.)* = would not; *see* ¹will

¹**wound** [wuːnd] **1.** *n.* Wunde, *die.* **2.** *v. t.* verwunden; *(fig.)* verletzen

²**wound** *see* ²wind

wove, woven *see* ¹weave 2

wrangle ['ræŋgl] **1.** *v. i.* [sich] streiten. **2.** *n.* Streit, *der*

wrap [ræp] **1.** *v. t.,* -pp- einwickeln; *(fig.)* hüllen; ~**ped** abgepackt ⟨*Brot usw.*⟩; ~ **sth. [a]round sth.** etw. um etw. wickeln. **2.** *n.* Umschlag[e]tuch, *das.* **wrap 'up** *v. t.* a) *see* wrap 1; b) *(conclude)* abschließen; c) **be** ~**ped up in one's work** in seine Arbeit völlig versunken sein

'**wrapper** *n.* a) **sweet-/toffee-**~[s] Bonbonpapier, *das;* b) *(of book)* Schutzumschlag, *der*

'**wrapping** *n.* Verpackung, *die.* '**wrapping-paper** *n. (strong)* Packpapier, *das; (decorative)* Geschenkpapier, *das*

wrath [rɒθ] *n.* Zorn, *der*

wreak [riːk] *v. t.* a) *(cause)* anrichten; b) ~ **vengeance on sb.** an jmdm. Rache nehmen

wreath [riːθ] *n., pl.* **wreaths** [riːðz, riːθs] Kranz, *der*

wreck [rek] **1.** *n.* a) Wrack, *das; (destruction of ship)* Schiffbruch, *der.* **2.** *v. t.* a) *(destroy)* ruinieren; zu Schrott fahren ⟨*Auto*⟩; **be** ~**ed** *(shipwrecked)* Schiffbruch erleiden; b) *(fig.: ruin)* zerstören; ruinieren ⟨*Gesundheit, Urlaub*⟩. **wreckage** ['rekɪdʒ] *n.* Wrackteile *Pl.; (fig.)* Trümmer *Pl.*

wren *n.* Zaunkönig, *der*

wrench [rentʃ] **1.** *n.* a) *(tool)* verstellbarer Schraubenschlüssel; b) *(violent* *twist)* Verrenkung, *die;* c) *(fig.)* **be a great ~ [for sb.]** sehr schmerzhaft für jmdn. sein. **2.** *v. t.* **a)** reißen; **~ sth. from sb.** jmdm. etw. entreißen; **b) ~ one's ankle** sich *(Dat.)* den Knöchel verrenken

wrest [rest] *v. t.* **~ sth. from sb.** jmdm. etw. entreißen

wrestle ['resl] *v. i.* ringen. **wrestler** ['reslə(r)] *n.* Ringer, *der/*Ringerin, *die.* **wrestling** ['reslɪŋ] *n.* Ringen, *das*

wretch [retʃ] *n.* Kreatur, *die*

wretched ['retʃɪd] *adj.* **a)** *(miserable)* unglücklich; **b)** *(coll.: damned)* elend; **c)** *(very bad)* erbärmlich

wriggle ['rɪgl] **1.** *v. i.* **a)** sich winden; ⟨*Fisch:*⟩ zappeln; **b)** *(move)* sich schlängeln. **2.** *v. t.* **~ one's way** sich schlängeln. **3.** *n.* Windung, *die*

wring [rɪŋ] *v. t.,* **wrung** [rʌŋ] **a)** wringen; **~ out** auswringen; **b) ~ sb.'s hand** jmdm. fest die Hand drücken; **~ the neck of an animal** einem Tier den Hals umdrehen; **c) ~ sth. from** *or* **out of sb.** *(fig.)* jmdm. etw. abpressen. **wringing 'wet** *adj.* tropfnaß

wrinkle ['rɪŋkl] **1.** *n.* Falte, *die; (in paper)* Knick, *der.* **2.** *v. t.* falten. **3.** *v. i.* sich in Falten legen. **wrinkled** ['rɪŋkld], **wrinkly** ['rɪŋklɪ] *adjs.* runz[e]lig

wrist [rɪst] *n.* Handgelenk, *das.* **'wrist-watch** *n.* Armbanduhr, *die*

writ [rɪt] *n. (Law)* Verfügung, *die*

write [raɪt] **1.** *v. i.,* **wrote** [rəʊt], **written** ['rɪtn] schreiben; **~ to sb./a firm** jmdm./an eine Firma schreiben. **2.** *v. t.,* **wrote, written** schreiben; ausschreiben ⟨*Scheck*⟩; **the written language** die Schriftsprache; **written applications** schriftliche Anträge. **write 'back** *v. i.* zurückschreiben. **write 'down** *v. t.* aufschreiben. **write 'off 1.** *v. t.* **a)** abschreiben ⟨*Schulden, Verlust*⟩; **b)** *(destroy)* zu Schrott fahren. **2.** *v. i.* **~ off for sth.** etw. [schriftlich] anfordern

'write-off *n.* Totalschaden, *der*

writer ['raɪtə(r)] *n.* Schriftsteller, *der/*Schriftstellerin, *die; (of letter, article)* Verfasser, *der/*Verfasserin, *die*

'write-up *n. (by critic)* Kritik, *die*

writhe [raɪð] *v. i.* sich winden

writing ['raɪtɪŋ] *n.* **a)** Schreiben, *das;* **put sth. in ~:** etw. schriftlich machen *(ugs.);* **b)** *(handwriting, something written)* Schrift, *die.* **'writing paper** *n.* Schreibpapier, *das*

written *see* write

wrong [rɒŋ] **1.** *adj.* **a)** *(morally bad)* unrecht *(geh.); (unfair)* ungerecht; **b)** *(mistaken)* falsch; **be ~** ⟨*Person:*⟩ sich irren; **the clock is ~:** die Uhr geht falsch; **c)** *(not suitable)* falsch; **give the ~ answer** eine falsche Antwort geben; **[the] ~ way round** verkehrt herum; **d)** *(out of order)* nicht in Ordnung; **what's ~?** ist etwas nicht in Ordnung. **2.** *adv.* falsch. **3.** *n.* Unrecht, *das;* **do ~:** unrecht tun. **4.** *v. t.* **~ sb.** jmdn. ungerecht behandeln. **wrongful** ['rɒŋfl] *adj.* **a)** *(unfair)* unrecht *(geh.);* **b)** *(unlawful)* rechtswidrig. **'wrongfully** *adv.* **a)** *(unfairly)* unrecht *(geh.)* ⟨handeln⟩; zu Unrecht ⟨*beschuldigen*⟩; **b)** *(unlawfully)* rechtswidrig. **'wrongly** *adv.* **a)** falsch; **b)** *(mistakenly)* zu Unrecht; **c)** *see* wrongfully a

wrote *see* write

wrought iron [rɔːt 'aɪən] *n.* Schmiedeeisen, *das; attrib.* schmiedeeisern

wrung *see* wring

wry [raɪ] *adj.,* **~er** *or* **wrier** ['raɪə(r)], **~est** *or* **wriest** ['raɪɪst] ironisch ⟨*Blick*⟩; fein ⟨*Humor, Witz*⟩

X

X, x [eks] *n.* X, x, *das*

Xerox, (P), **xerox** ['zɪərɒks] **1.** *n.* *(copy)* Xerokopie, *die.* **2. xerox** *v. t.* xerokopieren

Xmas ['krɪsməs, 'eksməs] *n. (coll.)* Weihnachten, *das*

'X-ray 1. *n. (picture)* Röntgenaufnahme, *die.* **2.** *v. t.* röntgen; durchleuchten ⟨*Gepäck*⟩

Y

Y, y [waɪ] *n.* Y, y, *das*

yacht [jɒt] *n.* **a)** *(for racing)* Segeljacht, *die;* **b)** *(for pleasure)* Jacht, *die.* **'yachting** *n.* Segeln, *das*

Yank [jæŋk] *n. (Brit. coll.: American)* Ami, *der (ugs.)*

yank *(coll.)* **1.** *v. t.* reißen an (+ *Dat.*). **2.** *n.* Reißen, *das*

yap [jæp] *v. i.,* -pp- kläffen

'yard [jɑ:d] *n. (measure)* Yard, *das*

'yard *n.* **a)** *(attached to building)* Hof, *der;* **in the** ~: auf dem Hof; **b)** *(for storage)* Lager, *das*

'yardstick *n. (fig.)* Maßstab, *der*

yarn [jɑ:n] *n.* **a)** *(thread)* Garn, *das;* **b)** *(coll.: story)* Geschichte, *die*

yawn [jɔ:n] **1.** *n.* Gähnen, *das.* **2.** *v. i.* gähnen. **'yawning** *adj.* gähnend

year [jɪə(r)] *n.* **a)** Jahr, *das;* **for [many]** ~s jahrelang; **once a** ~, **every** ~: einmal im Jahr; **a ten-~-old** ein Zehnjähriger/eine Zehnjährige; **b)** *(group of students, vintage of wine)* Jahrgang, *der.* **'yearbook** *n.* Jahrbuch, *das.* **'yearly 1.** *adj.* jährlich; Einjahres‹abonnement›. **2.** *adv.* jährlich

yearn [jɜ:n] *v. i.* ~ **for** *or* **after sth./for sb.** sich nach etw./jmdm. sehnen; ~ **to do sth.** sich danach sehnen, etw. zu tun. **'yearning** *n.* Sehnsucht, *die*

yeast [ji:st] *n.* Hefe, *die*

yell [jel] **1.** *n.* gellender Schrei. **2.** *v. t. & i.* [gellend] schreien

yellow ['jeləʊ] **1.** *adj.* gelb. **2.** *n.* Gelb, *das.* **'yellowish** *adj.* gelblich

yelp [jelp] **1.** *v. i.* jaulen. **2.** *n.* Jaulen, *das*

yen [jen] *n. (coll.: longing)* **sb. has a** ~ **to do sth.** es drängt jmdn. danach, etw. zu tun

yes [jes] **1.** *adv.* ja; *(in contradiction)* doch. **2.** *n., pl.* ~es Ja, *das*

yesterday ['jestədeɪ, 'jestədɪ] **1.** *n.* gestern; **the day before** ~: vorgestern; ~**'s paper** die gestrige Zeitung. **2.** *adv.* gestern; **the day before** ~: vorgestern

yet [jet] **1.** *adv.* **a)** *(still)* noch; ~ **again** noch einmal; **b)** *(hitherto)* bisher; **his best** ~: sein bisher bestes; **c)** *neg.* **not [just]** ~: [jetzt] noch nicht; **d)** *(before all is over)* doch noch; **he could win** ~: er könnte noch gewinnen; **e)** *with compar. (even)* noch; **f)** *(nevertheless)* doch. **2.** *conj.* doch

yew [ju:] *n.* ~[-**tree**] Eibe, *die*

Yiddish ['jɪdɪʃ] **1.** *adj.* jiddisch. **2.** *n.* Jiddisch, *das; see also* **English 2 a**

yield [ji:ld] **1.** *v. t. (give)* bringen; hervorbringen ‹Ernte›; abwerfen ‹Gewinn›. **2.** *v. i.* **a)** sich unterwerfen; **b)** *(give right of way)* Vorfahrt gewähren. **3.** *n.* Ertrag, *der*

yodel ['jəʊdl] *v. i. & t., (Brit.)* -ll- jodeln

yoga ['jəʊgə] *n.* Joga, *der od. das*

yoghurt, yogurt ['jɒgət] *n.* Joghurt, *der od. das*

yoke [jəʊk] *n.* Joch, *das*

yokel ['jəʊkl] *n.* [Bauern]tölpel, *der*

yolk [jəʊk] *n.* Dotter, *der;* Eigelb, *das*

yonder ['jɒndə(r)] *(literary)* **1.** *adj.* ~ **tree** jener Baum dort *(geh.).* **2.** *adv.* dort drüben

you [jʊ, *stressed* ju:] *pron.* **a)** *sing./pl.* du/ihr; *(polite) sing. or pl.* Sie; *as direct object* dich/euch/Sie; *as indirect object* dir/euch/Ihnen; *refl.* dich/dir/ euch; *(polite)* sich; **it was** ~: du warst/ ihr wart/Sie waren es; **b)** *(one)* man

you'd [jʊd, *stressed* ju:d] **a)** = **you had; b)** = **you would**

you'll [jʊl, *stressed* ju:l] **a)** = **you will; b)** = **you shall**

young [jʌŋ] **1.** *adj.,* ~**er** ['jʌŋgə(r)], ~**est** ['jʌŋgɪst] jung. **2.** *n. pl. (of animals)* Junge: **the** ~ *(~ people)* die jungen Leute. **youngster** ['jʌŋstə(r)] *n.* **a)** *(child)* Kleine, *der/die/das;* **b)** *(young person)* Jugendliche, *der/die*

your [jə(r), *stressed* jʊə(r), jɔ:(r)] *poss. pron. attrib.: sing.* dein; *pl.* euer; *(polite) sing. or pl.* Ihr

you're [jə(r), *stressed* jʊə(r), jɔ:(r)] = **you are**

yours [jʊəz, jɔ:z] *poss. pron. pred.: sing.* deiner/deine/dein[e]s; *pl.* eurer/ eure/eures; *(polite) sing. or pl.* Ihrer/ Ihre/Ihr[e]s; *see also* **hers**

yourself [jə'self, jʊə'self, jɔ:'self] *pron.* **a)** *emphat.* selbst; **b)** *refl.* dich/dir/ *(polite)* sich. *See also* **herself**

yourselves [jə'selvz, jʊə'selvz, jɔ:'selvz] *pron.* **a)** *emphat.* selbst; **b)** *refl.* euch/*(polite)* sich. *See also* **herself**

youth [ju:θ] *n.* **a)** Jugend, *die;* **b)** *pl.* ~s [ju:ðz] *(young man)* Jugendliche, *der.* **'youth club** *n.* Jugendklub, *der*

youthful ['ju:θfl] *adj.* jugendlich

'youth hostel *n.* Jugendherberge, *die*

you've [jʊv, *stressed* ju:v] = **you have**

Yugoslav ['ju:gəslɑ:v] *see* **Yugoslavian**

Yugoslavia [ju:gə'slɑ:vɪə] *(Hist.)* Jugoslawien *(das).* **Yugoslavian** [ju:gə'slɑ:vɪən] *(Hist.)* **1.** *adj.* jugoslawisch. **2.** *n.* Jugoslawe, *der/* Jugoslawin, *die*

Z

Z, z [zed, *(Amer.)* zi:] *n.* Z, z, *das*
Zaire [zɑːˈɪə(r)] *pr. n.* Zaire *(das)*
Zambia [ˈzæmbɪə] *pr. n.* Sambia *(das)*
zany [ˈzeɪnɪ] *adj.* irre komisch *(ugs.);* Wahnsinns⟨*humor, -komiker*⟩
zeal [ziːl] *n.* Eifer, *der.* **zealous** [ˈzeləs] *adj.* eifrig
zebra [ˈzebrə, ˈziːbrə] *n.* Zebra, *das.* **zebra ˈcrossing** *n. (Brit.)* Zebrastreifen, *der*
zenith [ˈzenɪθ] *n.* Zenit, *der*
zero [ˈzɪərəʊ] *n., pl.* ~s Null, *die*
zest [zest] *n. (enthusiasm)* Begeisterung, *die;* ~ **for living** Lebenslust, *die*
zigzag [ˈzɪgzæg] **1.** *adj.* zickzackför-

mig; Zickzack⟨*muster, -anordnung*⟩. **2.** *n.* Zickzacklinie, *die*
Zimbabwe [zɪmˈbɑːbwɪ] *pr. n.* Simbabwe *(das)*
zinc [zɪŋk] *n.* Zink, *das*
zip [zɪp] **1.** *n.* Reißverschluß, *der.* **2.** *v. t.,* **-pp-:** ~ |up| sth. den Reißverschluß an etw. *(Dat.)* zuziehen
ˈZip code *n. (Amer.)* Postleitzahl, *die*
zip-fastener *see* zip 1
zipper [ˈzɪpə(r)] *see* zip 1
zither [ˈzɪðə(r)] *n.* Zither, *die*
zodiac [ˈzəʊdɪæk] *n.* Tierkreis, *der;* **sign of the** ~: Tierkreiszeichen, *das*
zombie *(Amer.:* **zombi**) [ˈzɒmbɪ] *n.* Zombie, *der*
zone [zəʊn] *n.* Zone, *die*
zoo [zuː] *n.* Zoo, *der*
zoological [zəʊəˈlɒdʒɪkl] *adj.* zoologisch
zoologist [zəʊˈɒlədʒɪst] *n.* Zoologe, *der*/Zoologin, *die*
zoology [zəʊˈɒlədʒɪ] *n.* Zoologie, *die*
zoom [zuːm] *v. i.* rauschen. **zoom ˈin on** *v. t. (Cinemat., Telev.)* zoomen auf (+ *Akk.*)
ˈzoom lens *n.* Zoomobjektiv, *das*

A

a, A [a:] das; ~, ~ **a)** (Buchstabe) a/A; das A und O (fig.) the essential thing/ things (Gen. for); **von A bis Z** (fig. ugs.) from beginning to end; **b)** (Musik) [key of] A

a Abk. Ar, Are

à [a] Präp. mit Nom., Akk. (Kaufmannsspr.) **zehn Marken à 50 Pfennig** ten stamps at 50 pfennigs each

A Abk. Autobahn ≈ M

Aal der; ~|e|s, ~e eel; ~ **grün** (Kochk.) green eels; stewed eels; **aalen** refl. V. (ugs.) stretch out; **aal·glatt** (abwertend) **1.** Adj. slippery; ~ **sein** be as slippery as an eel; **2.** adv. smoothly

Aas das; ~es, ~e od. Äser **a)** o. Pl. carrion no art.; **b)** Pl. ~e [rotting] carcass; **c)** Pl. Äser (salopp) (abwertend) swine; (anerkennend) devil

ab 1. Präp. mit Dat. **a)** from; **ab 1980** as from 1980; **ab Werk** (Kaufmannsspr.) ex works; **ab Frankfurt fliegen** fly from Frankfurt; **b)** ([Rang]folge) from ... on[wards]; **ab 20 DM** from 20 DM [upwards]; **2.** Adv. **a)** (weg) off; away; **b)** (ugs.: Aufforderung) off; away; **ab nach Hause** get off home; **c)** **Gewehr ab!** (milit. Kommando) order arms!; **d)** **ab und zu** od. **an** now and then

ab|ändern tr. V. alter; amend ⟨text⟩; **Ab·änderung** die alteration; (eines Textes) amendment

ab|arbeiten tr. V. work for ⟨meal⟩; work off ⟨debt, amount⟩

ab·artig Adj. deviant; abnormal

Abb. Abk. Abbildung Fig.

Ab·bau der **a)** dismantling; (von Zelten, Lagern) striking; **b)** s. **abbauen c:** cutback (Gen. in); pruning; reduction; **c)** (Bergbau) mining; (von Stein) quarrying

ab|bauen tr. V. **a)** dismantle; strike ⟨tent, camp⟩; **b)** (beseitigen) gradually remove; break down ⟨prejudices, inhibitions⟩; **c)** (verringern) cut back ⟨staff⟩; prune ⟨jobs⟩; reduce ⟨wages⟩; **d)** (Bergbau) mine; quarry ⟨stone⟩

ab|beißen 1) unr. tr. V. bite off; **2.** unr. itr. V. have a bite

ab|bekommen unr. tr. V. **a)** get; **b)** einen Schlag/ein paar Kratzer ~: get hit/get a few scratches; etwas ~ (getroffen werden) be hit; (verletzt werden) be hurt; **c)** (los-, herunterbekommen) get ⟨paint, lid, chain⟩ off

ab|berufen unr. tr. V. recall ⟨ambassador, envoy⟩ (aus, von from)

ab|bestellen tr. V. cancel

ab|bezahlen tr. V. pay off

ab|biegen unr. itr. V.; mit sein turn off; **links/rechts ~:** turn [off] left/right

Ab·bild das (eines Menschen) likeness; (eines Gegenstandes) copy; (fig.) portrayal; **ab|bilden** tr. V. copy; reproduce ⟨object, picture⟩; depict ⟨person, landscape⟩; (fig.) portray; **Abbildung die** illustration

ab|binden unr. tr. V. **a)** (losbinden) untie; undo; **b)** (abschnüren) put a tourniquet on ⟨artery, arm, leg, etc.⟩; tie ⟨umbilical cord⟩

ab|blasen unr. tr. V. (ugs.) call off

ab|blättern itr. V.; mit sein flake off

ab|blenden itr. V. dip (Brit.) or (Amer.) dim one's headlight; **Ab·blend·licht** das dipped (Brit.) or (Amer.) dimmed beam

ab|blitzen itr. V.; mit sein (ugs.) **sie ließ alle Verehrer ~:** she gave all her admirers the brush-off

ab|brausen tr. V.; s. **abduschen**

ab|brechen 1. unr. tr. V. **a)** break off; break ⟨needle, pencil⟩; **b)** (zerlegen) strike ⟨tent, camp⟩; **c)** (abreißen) demolish, pull down ⟨building⟩; **d)** (beenden) break off ⟨negotiations, [diplomatic] relations, discussion, activity⟩; (vorzeitig) cut short ⟨conversation, holiday, activity⟩; **2.** unr. itr. V. **a)** mit sein break [off]; **b)** (aufhören) break off

ạb|bremsen 1. *tr. V.* **a)** brake; **b)** retard ⟨*motion*⟩; **2.** *itr. V.* brake

ạb|brennen 1. *unr. itr. V.; mit sein* **a)** be burned down; **das Haus ist abgebrannt** the house has burned down; **b)** ⟨*fuse*⟩ burn away; ⟨*candle*⟩ burn down; **2.** *unr. tr. V.* **a)** let off ⟨*firework*⟩; **b)** burn down ⟨*building*⟩

ạb|bringen *unr. tr. V.* **jmdn. davon ~, etw. zu tun** dissuade sb. from doing sth. **jmdn. vom Kurs ~:** make sb. change course

ạb|bröckeln *itr. V.; mit sein (auch fig.)* crumble away

Ạb·bruch der a) *o. Pl.* demolition; pulling down; **b)** *(Beendigung)* breaking-off; **c)** **einer Sache** *(Dat.)* **[keinen] ~ tun** do [no] harm to sth.

ạb|buchen *tr. V.* ⟨*bank*⟩ debit (von to); ⟨*creditor*⟩ claim by direct debit (von to)

ạb|bürsten *tr. V.* **a)** brush off; **b)** *(säubern)* brush ⟨*garment*⟩

ạb|büßen *tr. V.* serve [out] ⟨*prison sentence*⟩

Abc [a(:)be(:)'tse:] **das;** **~** *(auch fig.)* ABC; **Abc-Schütze der** child just starting school

ạb|dampfen *itr. V.; mit sein (ugs.: abfahren)* set off

ạb|danken *itr. V.* ⟨*ruler*⟩ abdicate; ⟨*government, minister*⟩ resign; **Ạb·dankung die; ~, ~en** *s.* **abdanken:** abdication; resignation

ạb|decken 1. *tr. V.* **a)** open up; ⟨*gale*⟩ take the roof/roofs off ⟨*house*⟩, take the tiles off ⟨*roof*⟩; **b)** *(herunternehmen, -reißen)* take off; **c)** *(abräumen)* clear ⟨*table*⟩; clear away ⟨*dishes*⟩; **d)** *(schützen)* cover ⟨*person*⟩

ạb|dichten *tr. V.* seal

ạb|drängen *tr. V.* push away

ạb|drehen 1. *tr. V.* **a)** *(ausschalten)* turn off; **den Hahn ~** *(fig.)* turn off the supply; **b)** *(abtrennen)* twist off; **2.** *itr. V.; meist mit sein* turn off

Ạb·druck der; *Pl.* **Ạbdrücke** mark; *(Fuß~)* footprint; *(Wachs~)* impression; *(Gips~)* cast; **ạb|drücken 1.** *itr. V.* pull the trigger; shoot; **2.** *tr. V.* *(zudrücken)* constrict

ạb|dunkeln *tr. V.* darken ⟨*room*⟩; dim ⟨*light*⟩

ạb|duschen *tr. V.* **sich/jmdn. [warm] ~:** take/give sb. a [hot] shower

ạbend *Adv.* **heute/morgen/gestern ~:** this/tomorrow/yesterday evening; **Ạbend der; ~s, ~e** evening; **guten ~!** good evening; **am [frühen/späten] ~:** [early/late] in the evening; **zu ~ essen**

have dinner; *(allgemeiner)* have one's evening meal; **ein bunter ~:** a social [evening]

Abend-: ~an·zug der evening suit; **~brot das** supper; **~essen das** dinner; **~kasse die** box-office *(open on the evening of the performance);* **~kleid das** evening dress; **~kurs[us] der** evening class; **~land das;** *o. Pl.* West

abendlich *Adj.* evening; ⟨*quiet, coolness*⟩ of the evening

Abend-: ~mahl das; *o. Pl. (Rel.)* Communion; *(N.T.)* Last Supper; **~programm das** evening programmes *pl.;* **~rot das** red glow of the sunset sky

abends *Adv.* in the evenings; **um sechs Uhr ~:** at six o'clock in the evening

Abend-: ~schule die night school; **~stern der** evening star; **~stunde die** evening hour; **~vorstellung die** evening performance

Ạbenteuer das; ~s, ~ a) *(auch fig.)* adventure; **b)** *(Unternehmen)* adventure; **c)** *(Liebesaffäre)* affair; **ạbenteuerlich** *Adj.* **a)** *(riskant)* risky; **b)** *(bizarr)* bizarre; **Ạbenteuer·roman der** adventure novel; **Ạbenteurer der; ~s, ~:** adventurer

ạber 1. *Konj.* but; **2.** *Partikel* **~ ja/nein!** why, yes/no! **~ natürlich!** but of course!; **du bist ~ groß!** aren't you tall!

Ạber·glaube[n] der superstition; **ạber·gläubisch** *Adj.* superstitious

ạbermals *Adv.* once again; once more

Ạbf. *Abk.* **Abfahrt** dep.

ạb|fahren 1. *unr. itr. V.; mit sein* **a)** *(wegfahren)* leave; **wo fährt der Zug nach Paris ab?** where does the Paris train leave from?; **b)** *(hinunterfahren)* drive down; *(Skisport)* ski down; **2.** *unr. tr. V.* **a)** *(abtransportieren)* take away; **b)** *(abnutzen)* wear out; **abgefahrene Reifen** worn tyres; **Ạb·fahrt die a)** departure; **b)** *(Skisport)* descent; *(Strecke)* run; **Ạbfahrtslauf der** *(Skisport)* downhill [racing]; **Ạbfahrt[s]·zeit die** time of departure; departure time

Ạb·fall der *(Küchen~ o.ä.)* rubbish, *(Amer.)* trash *no indef. art., no pl.;* *(Fleisch~)* offal *no indef. art., no pl.;* *(Industrie~)* waste *no indef. art.; (auf der Straße)* litter *no indef. art., no pl.;* **Ạbfall·eimer der** rubbish bin; trash can *(Amer.); (auf der Straße)* litter bin;

trash can *(Amer.);* **ab|fallen** *unr. itr.
V.; mit sein* **a)** fall off; **b)** *(abschüssig
sein)* ⟨*land, road, etc.*⟩ drop away,
slope; **c)** *(übrigbleiben)* be left [over];
für dich wird |dabei| auch etwas ~:
you'll get something out of it too; **d)
von jmdm. ~:** leave sb.; **vom Glauben
~:** desert the faith; **ab · fällig 1.** *Adj.*
disparaging; **2.** *adv.* **sich ~ über jmdn.
äußern** make disparaging remarks
about sb.

ab|fangen *unr. tr. V.* catch; intercept
⟨*agent, message, aircraft*⟩; **b)** repel
⟨*charge, assault*⟩; ward off ⟨*blow, attack*⟩

ab|färben *itr. V.* **a)** ⟨*colour, garment,
etc.*⟩ run; **b) auf jmdn./etw. ~** *(fig.)* rub
off on sb./sth.

ab|fassen *tr. V.* write ⟨*report, letter,
etc.*⟩; draw up ⟨*will*⟩

ab|fegen *tr. V.* **a)** brush off; **etw. von
etw. ~:** brush sth. off sth.; **b)** *(säubern)*
etw. ~: brush sth. clean

ab|fertigen *tr. V.* dispatch ⟨*mail*⟩;
deal with ⟨*applicant*⟩; handle ⟨*passengers*⟩; serve ⟨*customer*⟩; clear ⟨*ship*⟩
for sailing; clear ⟨*aircraft*⟩ for take-
off; clear ⟨*lorry*⟩ for departure

ab|feuern *tr. V.* fire

ab|finden 1. *unr. tr. V.* **a)** jmdn. mit
etw. ~: compensate sb. with sth.; **seine
Gläubiger ~:** settle with one's cred-
itors; **2.** *unr. refl. V.* **sich ~:** resign
oneself; **sich ~ mit** come to terms
with; learn to live with ⟨*noise, heat*⟩;
Abfindung *die* ~, **~en** settlement;
eine ~ in Höhe von ... zahlen make a
settlement of ...

ab|flauen *itr. V.; mit sein* die down;
subside; ⟨*interest, conversation*⟩ flag;
⟨*business*⟩ become slack; ⟨*noise*⟩ abate

ab|fliegen *unr. itr. V.; mit sein* leave

ab|fließen *unr. itr. V.; mit sein* flow
off

Ab · flug *der* departure

Ab · fluß *der* drain; *(Rohr)* drain-pipe;
(für Abwasser) waste-pipe

ab|fragen *tr. V.* test; **jmdn.** *od.* **jmdm.
die Vokabeln ~:** test sb. on his/her vocabulary

Abfuhr *die* ~, **~en a)** removal; **b)**
jmdm. eine ~ erteilen *(fig. ugs.)* rebuff
sb.; **ab|führen 1.** *tr. V.* **a)** *(nach Fest-
nahme)* take away; **b)** *(zahlen)* pay
out; **c)** *(abbringen)* take away; **2.** *itr. V.*
(für Stuhlgang sorgen) be a laxative;
Abführ · mittel *das* laxative

ab|füllen *tr. V. (in Flaschen)* bottle; *(in
Dosen)* can

Ab · gabe *die* **a)** handing in; *(eines
Briefes, Pakets, Telegramms)* delivery;
(eines Gesuchs, Antrags) submission;
b) *(Steuer, Gebühr)* tax; *(auf Produkte)*
duty; **c)** *(Ausstrahlung)* release;
emission; **d)** *(Sport: Abspiel)* pass

Ab · gang *der* **a)** leaving; departure;
(Abfahrt) departure; *(Theater)* exit; **b)**
(jmd., der ausscheidet) departure;
(Schule) leaver; **c)** *(bes. Amtsspr.: Todesfall)* death; **d)** *(Turnen)* dismount

Ab · gas *das* exhaust

abgearbeitet *Adj.* work-worn
⟨*hands*⟩

ab|geben 1. *unr. tr. V.* **a)** *(aushändi-
gen)* hand over; deliver ⟨*letter, parcel,
telegram*⟩; hand in, submit ⟨*application*⟩; hand in ⟨*school work*⟩; **den
Mantel in der Garderobe ~:** leave
one's coat in the cloakroom; **b) auch**
itr. **jmdm. |etwas| von etw. ~:** let sb.
have some of sth.; **c)** *(abfeuern)* fire;
2. *unr. refl. V.* **sich mit jmdm./etw. ~:**
spend time on sb./sth.; *(geringschätzig)* waste one's time on sb./sth

ab · gebrannt *Adj. (ugs.)* broke *(coll.)*

abgebrüht *Adj. (ugs.)* hardened

ab · gedroschen *Adj. (ugs.)* hackneyed

ab · gegriffen *Adj.* battered

ab|gehen *unr. itr. V.; mit sein* **a)** *(sich
entfernen)* leave; *(Theater)* exit; **b)**
(ausscheiden) leave; **c)** *(abfahren)*
⟨*train, ship, bus*⟩ leave, depart; **d)** *(ab-
geschickt werden)* ⟨*message, letter*⟩ be
sent [off]; **e)** *(abzweigen)* branch off; **f)**
(sich lösen) come off

abgehetzt *Adj.* exhausted

ab · gelegen *Adj.* remote; *(einsam)*
isolated; out-of-the-way ⟨*district*⟩

ab · geneigt *Adj.* averse *(Dat.* to);
|nicht| ~ sein, etw. zu tun [not] be averse
to doing sth.

Abgeordnete *der/die;* *adj. Dekl.*
member [of parliament]; *(z. B. in
Frankreich)* deputy

ab · gerissen *Adj.* ragged

ab · geschlagen *Adj. (Sport)* [well]
beaten

ab · geschlossen *Adj.* secluded

ab · gesehen *Adv.* **~ von** apart from; **~
davon, daß ...** apart from the fact that ...

ab · gespannt *Adj.* weary; exhausted

ab · gestanden *Adj.* flat

ab · gestorben *Adj.* dead ⟨*branch,
tree*⟩; numb ⟨*fingers, legs, etc.*⟩

ab · getreten *Adj.* worn down

abgewetzt *Adj.* well-worn; battered
⟨*case etc.*⟩

ab|gewöhnen *tr. V.:* jmdm. etw. ~:
make sb. give up sth.; **sich** *(Dat.)* etw.
~: give up sth.

ab|gießen *unr. tr. V.* pour away ⟨*li-
quid*⟩; drain ⟨*potatoes*⟩

abgöttisch *Adj.* idolatrous

ab|grenzen *tr. V.* **a)** bound; etw. gegen
od. von etw. ~: separate sth. from sth.;
b) *(unterscheiden)* distinguish

Ab·grund der abyss; chasm; *(Abhang)*
precipice

ab|hacken *tr. V.* chop off; jmdm. die
Hand *usw.* ~: chop sb.'s hand *etc.*

ab|haken *tr. V.* tick off; check off
(Amer.)

ab|halten *unr. tr. V.* **a)** jmdn./etw. |von
jmdm./etw.| ~: keep sb./sth. off
[sb./sth.]; **b)** jmdn. davon ~, etw. zu tun
stop sb. doing sth.; **c)** *(durchführen)*
hold ⟨*elections, meeting, referendum*⟩

ab|handeln *tr. V.* **a)** jmdm. etw. ~: do a
deal with sb. for sth.; **b)** *(darstellen)*
deal with

abhanden *Adv.* ~ kommen get lost; go
astray; etw. kommt jmdm. ~: sb. loses
sth.

Ab·handlung die treatise (über +
Akk. on)

Ab·hang der slope; incline; **¹ab|hän-
gen** *unr. itr. V.* von jmdm./etw. ~: de-
pend on sb./sth.; **²ab|hängen 1.** *tr. V.*
a) take down; **b)** *(abkuppeln)* un-
couple; **c)** *(ugs.)* shake off *(coll.)* ⟨*pur-
suer, competitor*⟩. **2.** *itr. V. (den Hörer
auflegen)* hang up; **abhängig** *Adj.*
dependent **(von on)**; *(süchtig)* addicted
(von to); von jmdm./etw. ~ sein depend
on sb./sth.; **Abhängigkeit die;** ~,
~en dependence; *(Sucht)* addiction

ab|härten *tr. V.* harden

ab|hauen 1. *unr. tr. V.* **a)** *Prät.* **haute ab**
knock off; **b)** *Prät.* **hieb** *(geh.) od.*
haute ab *(mit Schwert, Axt usw.)* chop
off; **2.** *unr. itr. V.; mit sein; Prät.* **haute
ab** *(salopp)* beat it *(sl.)*

ab|heben 1. *unr. tr. V.* **a)** lift off ⟨*lid,
cover, etc.*⟩; |den Hörer| ~: answer [the
telephone]; **b)** *(von einem Konto)* withdraw ⟨*money*⟩; **2.** *unr. itr. V.* ⟨*balloon*⟩
rise; ⟨*aircraft, bird*⟩ take off; ⟨*rocket*⟩
lift off; **3.** *unr. refl. V.* stand out **(von**
against)

ab|heften *tr. V.* file

ab|hetzen *refl. V.* rush [around]; *s.
auch* abgehetzt

Ab·hilfe die; *o. Pl.* action to improve
matters; ~ **schaffen** put things right

ab|holen *tr. V.* collect, pick up ⟨*parcel,
book, tickets, etc.*⟩; pick up ⟨*person*⟩

ab|hören *tr. V.* **a)** jmdm. *od.* jmdn. Vo-
kabeln ~: test sb.'s vocabulary [orally];
das Einmaleins ~: ask questions on the
multiplication table; **b)** tap ⟨*telephone
conversation, telephone*⟩; bug *(coll.)*
⟨*conversation, premises*⟩; jmdn. ~: tap
sb.'s telephone

Abi das; ~s, ~s *(Schülerspr.),* **Abitur
das;** ~s, ~e Abitur *(school-leaving examination at grammar school needed
for entry to higher education);* ≈ A
levels *(Brit.);* **Abiturient der;** ~en,
~en *sb. who is taking/has passed the
'Abitur'*

ab|jagen *tr. V.* jmdm. etw. ~: finally
get sth. away from sb.

Abk. *Abk.* Abkürzung abbr.

ab|kaufen *tr. V.* jmdm. etw. ~: buy sth.
from sb.

ab|klopfen *tr. V.* **a)** knock off; **b)** *(säu-
bern)* knock the dirt/snow/crumbs *etc.*
off; **c)** *(untersuchen)* tap

ab|knicken 1. *tr. V.* snap off; **2.** *itr. V.;
mit sein* snap

ab|kochen *tr. V.* boil

ab|kommen *unr. itr. V.; mit sein* **a)**
vom Weg ~: lose one's way; vom Kurs
~: go off course; von der Fahrbahn ~:
leave the road; vom Thema ~: stray
from the topic; **b)** von einem Plan ~:
abandon a plan; **Ab·kommen das;**
~s, ~: agreement; **abkömmlich** *Adj.*
free; available

ab|kratzen 1. *tr. V.* **a)** *(mit den Fin-
gern)* scratch off; *(mit einem Werkzeug)* scrape off; **b)** *(säubern)* scrape
[clean]; **2.** *itr. V.; mit sein (derb)* snuff it
(sl.)

ab|kriegen *tr. V. (ugs.) s.* abbekommen

ab|kühlen 1. *tr. V.* cool down; **2.** *itr.,
refl. V.; itr. meist mit sein* cool down;
Ab·kühlung die cooling

ab|kürzen *tr., itr. V.* **a)** *(räumlich)*
shorten; den Weg ~: take a shorter
route; **b)** *(zeitlich)* cut short; **c)** *(kürzer
schreiben)* abbreviate **(mit** to); **Abkürzung die a)** *(Weg)* short cut; **b)** *(Wort)*
abbreviation

ab|laden *unr. tr., itr. V.* unload

ab|lagern *tr. V.* deposit

ab|lassen 1. *unr. tr. V.* let out **(aus** of);
let off ⟨*steam*⟩; **2.** *unr. itr. V.* **a)** von
jmdm./etw. ~: leave sb./sth. alone; **b)**
von etw. ~: *(etw. aufgeben)* give sth. up

Ab·lauf der a) *(Verlauf)* course; *(einer
Veranstaltung)* passing off; **b)** *o. Pl.
(Ende)* nach ~ eines Jahres after a year;
nach ~ einer Frist at the end of a
period of time; **ab|laufen** *unr. itr. V.;*

mit sein **a)** flow away; *(aus einem Behälter)* run out; **b)** *(verlaufen)* pass off; **c)** ⟨*alarm clock*⟩ run down; ⟨*parking meter*⟩ expire; **d)** ⟨*period, contract, passport*⟩ expire

ab|lecken *tr. V.* **a)** lick off; **b)** *(säubern)* lick clean

ab|legen 1. *tr. V.* **a)** lay *or* put down; **b)** *(Bürow.)* file; **c)** stop wearing ⟨*clothes*⟩; **d)** give up ⟨*habit*⟩; lose ⟨*shyness*⟩; **e)** swear ⟨*oath*⟩; sit ⟨*examination*⟩; make ⟨*confession*⟩; 2. *tr., itr. V.* take off; **möchten Sie ~?** would you like to take your coat off?; 3. *itr. V.* |vom Kai| ~: cast off; **Ableger** der; ~s, ~: layer; *(Steckling)* cutting

ab|lehnen *tr. V.* **a)** decline; decline, turn down ⟨*money, invitation, position*⟩; reject ⟨*suggestion, applicant*⟩; **b)** *(mißbilligen)* diapprove of; **Ablehnung die**; ~, ~en **a)** rejection; **b)** *(Mißbilligung)* disapproval

ab|leiten *tr. V.* **a)** divert; **b)** *(herleiten)* etw. aus/von etw. ~: derive sth. from sth.; **Ab·leitung die** derivation

ab|lenken *tr. V.* **a)** deflect; **b)** jmdn. von etw. ~: distract sb. from sth.; **c)** *(zerstreuen)* divert; **sich** ~: amuse oneself; **Ab·lenkung die** *s.* **ablenken:** deflection; distraction; diversion

ab|lesen *unr. tr. V.* **a)** read ⟨*speech, lecture*⟩; **werden Sie frei sprechen oder ~?** will you be talking from notes or reading your speech?; **b)** read ⟨*gas meter, thermometer, etc.*⟩; check ⟨*time, speed, temperature*⟩; **c)** *(erkennen)* see

ab|liefern *tr., itr. V.* hand in; deliver ⟨*goods*⟩

ab|lösen 1. *tr. V.* **a)** etw. |von etw.| ~: get sth. off [sth.]; **b)** jmdn. ~: relieve sb.; **sich** *od.* **einander** ~: take turns; 2. *refl. V.* sich |von etw.| ~: come off [sth.]

Ab·lösung die *(eines Postens)* changing; **ich schicke Ihnen jemanden zur** ~: I'll send someone to relieve you

ab|machen *tr. V.* **a)** *(ugs.)* take off; take down ⟨*sign, rope*⟩; **b)** *(vereinbaren)* agree; **Abmachung die**; ~, ~en agreement

ab|magern *itr. V.*; *mit sein* become thin; *(absichtlich)* slim; **Abmagerungs·kur die** reducing diet

ab|marschieren *itr. V.*; *mit sein* depart; *(Milit.)* march off

ab|melden *tr. V.* **a)** sich/jmdn. ~: report that one/sb. is leaving; **b)** *(Umzug melden)* notify the authorities that one is moving from an address; **c)** ein Auto ~: cancel a car's registration; **Ab·mel-**

dung die a) *(beim Weggehen)* report that one is leaving; **b)** *(beim Umzug)* registration of a move with the authorities at one's old address; **c)** ~ eines Autos cancellation of a car's registration

Ab·messung die *meist Pl.* *(Dimension)* dimension; measurement

ab|montieren *tr. V.* take off ⟨*part*⟩; dismantle ⟨*machine, equipment*⟩

ab|mühen *refl. V.* toil; **sie mühte sich mit dem schweren Koffer ab** she struggled with the heavy suitcase

Abnahme die; ~, ~n **a)** *o. Pl.* *(das Entfernen)* removal; **b)** *(Verminderung)* decrease; **ab|nehmen** 1. *unr. tr. V.* **a)** *(entfernen)* take off; take down ⟨*picture, curtain, lamp*⟩; **b)** jmdm. den Koffer ~: take sb.'s suitcase; jmdm. eine Arbeit ~: save sb. a job; **c)** jmdm. ein Versprechen/einen Eid ~: make sb. give a promise/swear an oath; **d)** *(prüfen)* inspect and approve; test and pass ⟨*vehicle*⟩; **e)** jmdm. etw. ~ *(wegnehmen)* take sth. off sb.; **f)** *(beim Telefon)* answer ⟨*telephone*⟩; pick up ⟨*receiver*⟩; **g)** *(Handarb.)* decrease; **h)** das nehme ich dir/ihm *usw.* nicht ab I won't buy that *(coll.)*; 2. *unr. itr. V.* **a)** *(Gewicht verlieren)* lose weight; **b)** *(sich verringern)* decrease; drop; ⟨*attention, interest*⟩ flag; ⟨*brightness*⟩ diminish; **wir haben ~den Mond** there is a waning moon; **c)** *(beim Telefon)* answer the telephone

Ab·neigung die dislike (gegen for)

ab|nutzen, *(landsch.:)* **ab|nützen** *tr., refl. V.* wear out; **abgenutzt** worn

Abonnement [abɔnə'mãː] das; ~s, ~s subscription (*Gen.* to); **abonnieren** *tr. V.* subscribe to

Ab·ordnung die delegation

ab|passen *tr. V.* **a)** *(abwarten)* wait for; **b)** *(aufhalten)* catch

ab|pausen *tr. V.* trace

ab|pfeifen *(Sport)* 1. *itr. V.* blow the whistle; 2. *tr. V.* [blow the whistle to] stop; **Ab·pfiff der** *(Sport)* final whistle; *(Halbzeit~)* half-time whistle

ab|pflücken *tr. V.* pick

ab|plagen *refl. V.* slave away

ab|prallen *itr. V.*; *mit sein* rebound; ⟨*bullet, missile*⟩ ricochet

ab|putzen *tr. V. (ugs.)*; **a)** wipe off; **b)** *(säubern)* wipe; jmdm./sich das Gesicht ~: clean sb.'s/one's face

ab|quälen *refl. V.* sich |mit etw.| ~: struggle [with sth.]

ab|rackern *refl. V. (ugs.)* flog oneself to death *(coll.)*

ab|rasieren *tr. V.* shave off

ab|raten *unr. itr. V.* **jmdm. von etw. ~:** advise sb. against sth.

ab|räumen *tr. V.* **a)** clear away; **b)** *(leer machen)* clear ⟨*table*⟩

ab|rechnen 1. *itr. V.* cash up; **mit jmdm. ~** *(fig.)* call sb. to account; **2.** *tr. V.* **die Kasse ~:** reckon up the till; **seine Spesen ~:** claim one's expenses; **Ab·rechnung die a)** cashing up *no art.; (Aufstellung)* statement; **b)** *(fig.: Vergeltung)* reckoning

ab|reiben *unr. tr. V.* **a)** rub off; **b)** *(säubern)* rub

Ab·reise die departure **(nach** for); **bei meiner ~:** when I left/leave; **ab|reisen** *itr. V.; mit sein* leave **(nach** for)

ab|reißen 1. *unr. tr. V.* **a)** tear off; tear down ⟨*poster, notice*⟩; pull off ⟨*button*⟩; **b)** *(niederreißen)* demolish, pull down ⟨*building*⟩; **2.** *unr. itr. V.; mit sein* **a)** fly off; ⟨*shoe-lace*⟩ break off; **b)** *(aufhören)* come to an end; ⟨*connection, contact*⟩ be broken off

ab|richten *tr. V.* train

Ab·riß der a) *o. Pl.: s.* **abreißen 1 b:** demolition; pulling down; **b)** *(knappe Darstellung)* outline

ab|rollen 1. *tr. V.* unwind; **2.** *itr. V.; mit sein* unwind [itself]

ab|rücken 1. *tr. V. (wegschieben)* move away; **2.** *itr. V.; mit sein* move away

Ab·ruf der: auf ~: on call; *(DV)* in retrievable form; **ab|rufen** *unr. tr. V.* summon; call

ab|runden *tr. V.* **a)** *(auch fig.)* round off; **b)** round ⟨*figure*⟩ up/down **(auf +** Akk. to); **etw. nach oben/unten ~:** round sth. up/down

abrupt [ap'rʊpt] **1.** *Adj.* abrupt; **2.** *adv.* abruptly

ab|rüsten *itr., tr. V.* disarm; **Ab·rüstung die; ~:** disarmament

ab|rutschen *itr. V.; mit sein* **a)** slip; **b)** *(nach unten rutschen)* slide down

Abs. *Abk.* **a)** Absender; **b)** Absatz

Ab·sage die *(auf eine Einladung)* refusal; *(auf eine Bewerbung)* rejection; **ab|sagen 1.** *tr. V.* cancel; withdraw ⟨*participation*⟩; **2.** *itr. V.* **jmdm. ~:** tell sb. one cannot come

ab|sägen *tr. V.* saw off

Ab·satz der a) *(am Schuh)* heel; **b)** *(Textunterbrechung)* break; **c)** *(Textabschnitt)* paragraph; **d)** *(Kaufmannsspr.)* sales *pl.*

ab|saufen *unr. itr. V.; mit sein (ugs.)* ⟨*engine, car*⟩ flood

ab|saugen *tr. V.* **a)** suck away; **b)** *(säubern)* hoover *(Brit. coll.)*

ab|schaben *tr. V.* **a)** scrape off; **b)** *(säubern)* scrape [clean]

ab|schaffen *tr. V.* **a)** *(beseitigen)* abolish ⟨*capital punishment, regulation, customs duty, institution*⟩; repeal ⟨*law*⟩; put an end to ⟨*injustice, abuse*⟩; **b)** *(weggeben)* get rid of; **Ab·schaffung die** abolition; *(von Gesetzen)* repeal; *(von Unrecht, Mißstand)* ending

ab|schalten *tr., itr. V.* switch off; shut down ⟨*power-station*⟩

abschätzig 1. *Adj.* derogatory; **2.** *adv.* derogatorily

Ab·scheu der; ~s detestation; abhorrence; **abscheulich 1.** *Adj.* **a)** disgusting ⟨*smell, taste*⟩; repulsive ⟨*sight*⟩; **b)** *(verwerflich)* disgraceful ⟨*behaviour*⟩; abominable ⟨*crime*⟩; **2.** *adv.* disgracefully

ab|schicken *tr. V.* send [off]

ab|schieben *unr. tr. V.* **a)** push away; **b)** *(abwälzen)* shift ⟨*responsibility, blame*⟩; **c)** *(außer Landes bringen)*; deport

Abschied der; ~[e]s, ~e parting **(von** from); farewell **(von** to); **~ nehmen** take one's leave **(von** of)

Abschieds-: ~brief der farewell letter; **~geschenk das** parting gift; **~gruß der** goodbye; farewell

ab|schießen *unr. tr. V.* **a)** shoot down ⟨*aeroplane*⟩; fire ⟨*arrow*⟩; launch ⟨*spacecraft*⟩; **c)** *(töten)* take

ab|schirmen *tr. V.* **a)** *(schützen)* shield; **b)** *(fernhalten)* screen off ⟨*light, radiation*⟩

ab|schlachten *tr. V.* slaughter

Ab·schlag der a) *(Kaufmannsspr.)* discount; **b)** *(Teilzahlung)* interim payment; *(Vorschuß)* advance; **c)** *(Fußball)* goalkeeper's kick out; **ab|schlagen 1.** *unr. tr. V.* **a)** knock off; *(mit dem Beil, Schwert usw.)* chop off; **b)** *(ablehnen)* refuse; **c)** *(abwehren)* beat off; **2.** *unr. itr. V. (Fußball)* kick the ball out

ab|schleifen *unr. tr. V. (von Holz)* sand off; *(von Metall, Glas usw.)* grind off

Abschlepp·dienst der breakdown recovery service; tow[ing] service *(Amer.)*; **ab|schleppen** tow away; take ⟨*ship*⟩ in tow; **ein Auto zur Werkstatt ~:** tow a car to the garage; **Abschlepp·seil das** tow-rope; *(aus Draht)* towing cable

ab|schließen 1. *unr. tr. V.* **a)** *auch itr.*

(zuschließen) lock ⟨*door, gate, cupboard*⟩; lock [up] ⟨*house, flat, room, park*⟩; **b)** *(verschließen)* seal; **etw. luftdicht ~:** seal sth. hermetically; **c)** *(begrenzen)* border; **d)** *(zum Abschluß bringen)* conclude; **e)** *(vereinbaren)* strike ⟨*bargain, deal*⟩; make ⟨*purchase*⟩; enter into ⟨*agreement*⟩; **2.** *unr. itr. V. (aufhören, enden)* end; **~d sagte er ~:** in conclusion he said ...; **Abschluß der** *(Beendigung)* conclusion; end

ab|schmecken *tr. V.* **a)** *(kosten)* taste; try; **b)** *(würzen)* season

ab|schmieren *tr. V. (Technik)* grease

ab|schminken *tr. V.* **jmdn./sich ~:** remove sb.'s/one's make-up

ab|schmirgeln *tr. V.* rub off with emery; *(mit Sandpapier)* sand off

ab|schnallen *tr. V.* unfasten

ab|schneiden 1. *unr. tr. V.* **a)** *(auch fig.: isolieren)* cut off; cut down ⟨*sth. hanging*⟩; **etw. von etw. ~:** cut sth. off sth.; **sich** *(Dat.)* **eine Scheibe Brot ~:** cut oneself a slice of bread; **b)** *(kürzer schneiden)* cut; **c) jmdm. den Weg ~:** take a short cut to get ahead of sb.; **2.** *unr. itr. V.* **bei etw. gut/schlecht ~:** do well/badly in sth.; **Ab·schnitt der a)** *(Kapitel)* section; **b)** *(Zeitspanne)* phase; **d)** *(Teil eines Formulars)* [detachable] portion

ab|schrauben *tr. V.* unscrew [and remove]

ab|schrecken *tr. V.* **a)** deter; **b)** *(fernhalten)* scare off; **c)** *(Kochk.)* pour cold water over; **Abschreckung die; ~, ~en** deterrence

ab|schreiben 1. *unr. tr. V.* **a)** copy out; **etw. bei** *od.* **von jmdm. ~** *– (in der Schule)* copy sth. off sb.; *(als Plagiator)* plagiarize sth. from sb.; **b)** *(Wirtsch.)* amortize; **2.** *unr. itr. V.* **bei** *od.* **von jmdm. ~** *– (in der Schule)* copy off sb.; *(als Plagiator)* copy from sb.; **Ab·schreibung die** *(Wirtsch.)* amortization; **Ab·schrift die** copy

ab|schürfen *tr. V.* graze

Ab·schuß der a) *(eines Flugzeugs)* shooting down; **b)** *(von Geschossen)* firing; *(eines Raumschiffs)* launching

abschüssig *Adj.* downward sloping ⟨*land*⟩

ab|schütteln *tr. V.* shake off; *(herunterschütteln)* shake down

ab|schwächen 1. *tr. V.* **a)** *(mildern)* tone down ⟨*statement, criticism*⟩; **b)** *(verringern)* lessen ⟨*effect, impression*⟩; cushion ⟨*blow, impact*⟩; **2.** *refl.*

V. ⟨*interest, demand*⟩ wane; **Abschwächung die; ~, ~en a)** *(Milderung)* toning down; **b)** *(eines Aufpralls, Stoßes usw.)* cushioning

ab|schweifen *itr. V.; mit sein* digress; **Abschweifung die; ~, ~en** digression

ab|schwören *unr. itr. V.* **dem Teufel/ seinem Glauben ~:** renounce the Devil/one's faith; **dem Alkohol/Laster ~:** forswear alcohol/vice

absehbar *Adj.* foreseeable; **in ~er Zeit** within the foreseeable future;

ab|sehen 1. *unr. tr. V.* **a)** *(voraussehen)* predict; foresee ⟨*event*⟩; **b) es auf etw.** *(Akk.)* **abgesehen haben** be after sth.; **er hat es darauf abgesehen, uns zu ärgern** he's out to annoy us; **der Chef hat es auf ihn abgesehen** the boss has got it in for him; **2.** *unr. itr. V.* **a) von etw. ~** *(etw. nicht beachten)* leave aside sth.; *s. auch* **abgesehen; b) von etw. ~** *(auf etw. verzichten)* refrain from sth.

ab|seilen 1. *tr. V.* lower [with a rope]. **2.** *refl. V. (Bergsteigen)* abseil

ab|sein *unr. itr. V.; mit sein (Zusschr. nur im Inf. u. Part.) (abgegangen sein)* have come off

abseits 1. *Präp. mit Gen.* away from; **2.** *Adv.* **a)** far away; **b)** *(Ballspiele)* **~ sein** *od.* **stehen** be offside; **Abseits das; ~, ~: das war ein klares ~:** that was clearly offside

ab|senden *unr. od. regelm. tr. V.* dispatch; **Ab·sender der** sender; *(Anschrift)* sender's address

ab|setzen 1. *tr. V.* **a)** take off ⟨*hat, glasses, etc.*⟩; **b)** *(hinstellen)* put down ⟨*bag, suitcase*⟩; **c)** *(aussteigen lassen)* **jmdn. ~** *(im öffentlichen Verkehr)* put sb. down; let sb. out *(Amer.)*; *(im privaten Verkehr)* drop sb. [off]; **d)** remove ⟨*chancellor, judge*⟩ from office; depose ⟨*king, emperor*⟩; **2.** *refl. V.* **a)** *(sich ablagern)* be deposited; **b)** *(flüchten)* get away

Absetzung die; ~, ~en *s.* **absetzen 1 d:** removal; deposition

ab|sichern 1. *tr. V.* make safe; **2.** *refl. V.* safeguard oneself

Ab·sicht die; ~, ~en intention; **etw. mit ~ tun** do sth. intentionally; **etw. ohne** *od.* **nicht mit ~ tun** do sth. unintentionally; **ab·sichtlich 1.** *Adj.* intentional; deliberate; **2.** *adv.* intentionally; deliberately

ab|sinken *unr. itr. V.; mit sein* sink

absolut *Adj.* absolute; **Absolutismus der; ~** *(hist.)* absolutism *no art.*

Absolvent der; ~en, ~en *(einer Schule)* one who has taken the leaving *or* *(Amer.)* final examination; *(einer Akademie)* graduate; **absolvieren** *tr. V.* complete; **Absolvierung die**; ~: completion

ab·sonderlich *Adj.* strange; odd; **ab|sondern 1.** *tr. V.* exude; *(Physiol.)* secrete; **2.** *refl. V.* isolate oneself

absorbieren *tr. V.* absorb

ab|speisen *tr. V.* jmdn. mit etw. ~: fob sb. off with sth.

abspenstig *Adj.* jmdm. etw. ~ **machen** get sb. to part with sth.

ab|sperren *tr. V.* seal off; close off

Ab·spiel das *(Ballspiele)* passing; **ab|spielen 1.** *tr. V.* **a)** play ⟨record, tape⟩; **b)** vom Blatt ~: play ⟨piece of music⟩ at sight; **c)** *(Ballspiele)* pass; **2.** *refl. V.* take place

Ab·sprache die arrangement; **eine ~ treffen** make an arrangement; **ab|sprechen** *unr. tr. V.* **a)** jmdm. etw. ~: deny that sb. has sth.; **b)** *(vereinbaren)* arrange

ab|springen *unr. itr. V.; mit sein* jump off; *(herunterspringen)* jump down; **vom Fahrrad** ~: jump off one's bicycle; **Ab·sprung der** take-off; *(das Herunterspringen)* jump

ab|spülen 1. *tr. V.* **a)** wash off; **b)** *(reinigen)* rinse off; **sich** *(Dat.)* **die Hände usw.** ~: rinse one's hands *etc.*; **das Geschirr** ~ *(bes. südd.)* wash the dishes; **2.** *itr. V. (bes. südd.)* wash up

ab|stammen *itr. V.* be descended (von from); **Abstammung die**; ~, ~en descent

Ab·stand der a) distance; **in 20 Meter** ~: at a distance of 20 metres; **b)** *(Unterschied)* gap

ab|stauben *tr., itr. V.* dust

Abstecher der; ~s, ~: side-trip

ab|stehen *unr. itr. V.* ⟨hair⟩ stand up; ⟨pigtail[s]⟩ stick out; ~**de Ohren** protruding ears

Ab·steige die; ~, ~n *(ugs. abwertend)* cheap and crummy hotel *(sl.)*; **ab|steigen** *unr. itr. V.; mit sein* **a)** |vom Pferd/Fahrrad| ~: get off [one's horse/bicycle]; **b)** *(abwärts gehen)* go down

ab|stellen *tr. V.* **a)** put down; **b)** *(unterbringen)* put; *(parken)* park; **c)** *(ausschalten, abdrehen)* turn off; **d)** *(unterbinden)* put a stop to

Abstell-: ~**kammer die,** ~**raum der** lumber-room

ab|stempeln *tr. V.* **a)** frank ⟨letter⟩;

cancel ⟨stamp⟩; **b)** *(fig.)* label, brand **(zu, als** as)

ab|sterben *unr. itr. V.; mit sein* **a)** [gradually] die; **b)** *(gefühllos werden)* go numb

Abstieg der; ~|e|s, ~e **a)** descent; **b)** *(Niedergang)* decline

ab|stimmen 1. *itr. V.* vote (über + Akk. on); **2.** *tr. V.* etw. mit jmdm. ~: discuss and agree on sth. with sb.; **Ab·stimmung die a)** vote; **während der** ~; during the voting; **b)** *(Absprache)* agreement

abstinent [apsti'nɛnt] *Adj.* teetotal; ~ **sein** be a teetotaller; **Abstinenz die**; ~: teetotalism; **Abstinenzler der**; ~s, ~: teetotaller

ab|stoppen 1. *tr. V.* halt; stop; check ⟨advance⟩; **2.** *itr. V.* come to a halt; ⟨person⟩ stop

Ab·stoß der *(Fußball)* goal-kick; **ab|stoßen** *unr. tr. V.* **a)** push off; **b)** *(beschädigen)* chip ⟨crockery, paintwork, plaster⟩; **c)** *(verkaufen)* sell off; **d)** *(anwidern)* repel; put off; **abstoßend** *Adj.* repulsive

abstrakt [ap'strakt] *Adj.* abstract

ab|streifen *tr. V.* pull off; strip off ⟨berries⟩; **die Asche |von der Zigarette/Zigarre|** ~: remove the ash [from one's cigarette/cigar]

ab|streiten *unr. tr. V.* deny

Ab·strich der a) *(Med.)* swab; **einen** ~ **machen** take a swab; **b)** *(Streichung, Kürzung)* cut; ~**e machen** make cuts **(an** + *Dat.* in)

ab|stumpfen *itr. V.; mit sein* jmd. **stumpft ab** *(wird unsensibel)* sb.'s mind becomes deadened

Ab·sturz der fall; *(eines Flugzeugs)* crash; **ab|stürzen** *itr. V.; mit sein* fall; ⟨aircraft, pilot, passenger⟩ crash

ab|stützen 1. *refl. V.* support oneself **(mit** on, **an** + *Dat.* against); **2.** *tr. V.* support

ab|suchen *tr. V.* search **(nach** for)

absurd *Adj.* absurd

Abszeß der; **Abszesses, Abszesse a)** *(Med.)* abscess; **b)** *(Geschwür)* ulcer

Abszisse die; ~, ~en *(Math.)* abscissa

Abt der; ~|e|s, **Äbte** abbot

Abt. *Abk.* Abteilung

ab|tasten *tr. V.* etw. ~: feel sth. all over

ab|tauen 1. *itr. V.; mit sein (eis-/schneefrei werden)* become clear of ice/snow; ⟨refrigerator⟩ defrost; **2.** *tr. V.* melt; thaw; defrost ⟨refrigerator⟩

Abtei die; ~, ~en abbey

Abteil das; ~[e]s, ~e compartment; **Ab·teilung** die department; **Abteilungs·leiter** der head of department

ab|tippen tr. V. (ugs.) type out

Äbtissin die; ~, ~nen abbess

ab|tönen tr. V. tint

ab|töten tr. V. destroy ⟨parasites, germs⟩; deaden ⟨nerve, feeling⟩

ab|tragen unr. tr. V. (abnutzen) wear out; abgetragen well worn

abträglich Adj. (geh.) einer Sache (Dat.) ~ sein be detrimental to sth.

Ab·transport der s. abtransportieren: taking away; removal; **ab|transportieren** tr. V. take away; remove ⟨dead, injured⟩

ab|treiben 1. unr. tr. V. a) carry away; jmdn./ein Schiff vom Kurs ~: drive sb./a ship off course; b) abort ⟨foetus⟩; ein Kind ~ lassen have an abortion; 2. unr. itr. V.; mit sein be carried away; ⟨ship⟩ be drives off course; **Abtreibung** die; ~, ~en abortion

ab|trennen tr. V. detach

ab|treten 1. unr. tr. V. a) sich (Dat.) die Füße/Schuhe ~: wipe one's feet; b) jmdm. etw. ~: let sb. have sth.; 2. unr. itr. V.; mit sein a) (Theater) exit; (fig.) make one's exit; b) (zurücktreten) step down; ⟨monarch⟩ abdicate; **Abtreter** der; ~s, ~: doormat

ab|trocknen tr. V. dry; sich (Dat.) die Hände/die Tränen ~: dry one's hands/tears

ab|tropfen itr. V.; mit sein drip off

abtrünnig Adj. (einer Partei) renegade; (einer Religion, Sekte) apostate; der Kirche/dem Glauben ~ werden desert the Church/the faith

ab|tun unr. tr. V. dismiss

ab|wägen unr. od. regelm. tr., itr. V. weigh up; abgewogen carefully weighted; balanced ⟨judgement⟩

ab|wählen tr. V. vote out; drop ⟨school subject⟩

ab|wandeln tr. V. adapt

ab|wandern itr. V.; mit sein migrate; (in ein anderes Land) emigrate; **Abwanderung** die migration; (in ein anderes Land) emigration

Ab·wandlung die adaptation

ab|warten 1. itr. V. wait; sie warteten ab they awaited events; warte ab! wait and see; (als Drohung) just you wait!; 2. tr. V. wait for

abwärts Adv. downwards; (bergab) downhill; den Fluß ~: downstream

Abwasch der; ~[e]s washing-up (Brit.); washing dishes (Amer.); den ~ machen do the washing-up/wash the dishes; **abwaschbar** Adj. washable; **ab|waschen** 1. unr. tr. V. a) wash off; b) (reinigen) wash down; wash [up] ⟨dishes⟩; 2. unr. itr. V. wash up (Brit.); wash the dishes (Amer.)

Ab·wasser das; Pl. -wässer sewage

ab|wechseln refl., itr. V. alternate; wir wechselten uns ab we took turns; **abwechselnd** Adv. alternately; **Abwechslung** die; ~, ~en variety; (Wechsel) change; zur ~: for a change

Ab·weg der; auf ~e kommen od. geraten go astray; **abwegig** erroneous; false ⟨suspicion⟩

Ab·wehr die; ~ a) repulsion; (von Schlägen) fending off; (Sport) clearance; clearing (Amer.); b) (Sport: Hintermannschaft) defence; **ab|wehren** tr. V. a) repulse; fend off ⟨blow⟩; (Sport) clear ⟨ball, shot⟩; b) avert ⟨danger, consequences⟩

ab|weichen unr. itr. V.; mit sein a) deviate; b) (sich unterscheiden) differ; **Abweichung** die; ~, ~en a) deviation; b) (Unterschied) difference

ab|weisen unr. tr. V. turn away; turn down ⟨applicant, suitor⟩; **abweisend** Adj. cold ⟨look, tone of voice⟩; in ~em Ton coldly; **Ab·weisung** die; ~, ~en s. abweisen: turning away; turning down

ab|wenden 1. unr. od. regelm. tr. V. a) turn away; b) nur regelm. (verhindern) avert. 2. unr. od. regelm. refl. V. turn away

ab|werben unr. tr. V. lure away

ab|werfen 1. unr. tr. V. a) drop; throw off ⟨clothing⟩; jettison ⟨ballast⟩; throw ⟨rider⟩; b) (ins Spielfeld werfen) throw out ⟨ball⟩; c) (einbringen) bring in; 2. unr. itr. V. (Sport) throw the ball out

ab|werten tr., itr. V. devalue; **abwertend** Adj. derogatory ⟨term⟩; **Ab·wertung** die devaluation

abwesend Adj. absent; **Abwesenheit** die; ~ absence

ab|wickeln tr. V. a) unwind; b) (erledigen) deal with ⟨case⟩; do ⟨business⟩; **Abwicklung** die; ~, ~en s. abwickeln 1 b: dealing (Gen. with); doing

ab|wiegen unr. tr. V. weigh out; weigh ⟨single item⟩

ab|wimmeln tr. V. (ugs.) get rid of

ab|wischen tr. V. a) wipe away; b) (säubern) wipe

Ab·wurf der **a)** dropping; *(von Ballast)* jettisoning; **b)** beim ~ **stolperte der Torwart** the goalkeeper stumbled as he threw the ball out

ab|zahlen *tr. V.* pay off ⟨*debt, loan*⟩

ab|zählen *tr. V.* count

Ab·zahlung die paying off; **etw. auf** ~ **kaufen/verkaufen** buy/sell sth. on easy terms

Ab·zeichen das emblem; *(Ansteckna-del, Plakette)* badge

ab|zeichnen 1. *tr. V.* **a)** *(kopieren)* copy; **b)** *(signieren)* initial; **2.** *refl. V.* stand out; *(fig.)* begin to emerge

Abzieh·bild das transfer; **ab|ziehen** 1. *unr. tr. V.* **a)** pull off; peel off ⟨*skin*⟩; strip ⟨*bed*⟩; **b)** *(Fot.)* make a print/prints of; **c)** *(Milit., auch fig.)* withdraw; **d)** *(subtrahieren)* subtract; take away; *(abrechnen)* deduct; **2.** *unr.. itr. V.; mit sein (sich verflüchtigen)* escape; **b)** *(Milit.)* withdraw; **Ab·zug** der **a)** *(an einer Schußwaffe)* trigger; **b)** *(Fot.)* print; **c)** *(Verminderung)* deduction; **abzüglich** *Präp. mit Gen. (Kaufmannsspr.)* less

ab|zweigen 1. *itr. V.; mit sein* branch off; **2.** *tr. V.* put aside; **Abzweigung** die; ~, ~en turn-off; *(Gabelung)* fork

ach *Interj.* **a)** *(betroffen, mitleidig)* oh [dear]; **b)** *(bedauernd, unwirsch)* oh; **c)** *(klagend)* ah; **d)** *(erstaunt)* oh; ~, **wirklich?** no, really?; ~, **der!** oh, him!; **e)** ~ **so!** oh, I see; ~ **was** *od.* **wo!** of course not

Achat der; ~[e]s, ~e *(Min.)* agate

Achse die; ~, ~n **a)** *(Rad~)* axle; **b)** *(Dreh~, Math., Astron.)* axis

Achsel die; ~, ~n *(Schulter)* shoulder; *(~höhle)* armpit

Achsel-: ~**haare** *Pl.* armpit hair *sing.*; ~**höhle** die armpit

¹**acht** *Kardinalz.* eight; **um** ~ **[Uhr]** at eight [o'clock]; **um halb** ~: at half past seven; **dreiviertel** ~, **Viertel vor** ~: [a] quarter to eight; **es steht** ~ **zu** ~/~ **zu 2** *(Sport)* the score is eight all/eight to two; ²**acht: sie waren zu** ~: there were eight of them

³**acht:** **etw. außer** ~ **lassen** disregard sth.; **sich in** ~ **nehmen** be careful; **sich vor jmdm./etw. in** ~ **nehmen** be wary of sb./sth.

acht... *Ordinalz.* eighth; **der** ~**e September** the eighth of September; **München, [den] 8. Mai 1984** Munich, 8 May 1984; **Acht** die; ~, ~**en a)** eight; **b)** *(Figur)* figure eight; **c)** *(Verbiegung)* buckle; **mein Rad hat eine** ~: my

wheel is buckled; **Achte** der/die; *adj. Dekl.* eighth

acht-, Acht-: ~**eck** das; ~**s**, ~**e** octagon; ~**eckig** *Adj.* octagonal; ~**einhalb** *Bruchz.* eight and a half

achtel *Bruchz.* eighth; **Achtel** das *(schweiz. meist* der*)*; ~**s**, ~ eighth; **Achtel·note** die *(Musik)* quaver

achten 1. *tr. V.* respect; **2.** *itr. V.* **auf etw.** *(Akk.)* ~: pay heed to sth.

achtens *Adv.* eighthly; **Achterbahn** die roller-coaster; **acht·fach** *Vervielfältigungsz.* eightfold; **die** ~**fache Menge** eight times the quantity; ~**fach vergrößert/verkleinert** magnified/reduced eight times; **das Achtfache kosten** cost eight times as much

acht|geben *unr. itr. V.* **a) auf jmdn./ etw.** ~: take care of sb./sth.; **b)** *(vorsichtig sein)* be careful

acht-: ~**hundert** *Kardinalz.* eight hundred; ~**jährig** *Adj. (8 Jahre alt)* eight-year-old *attrib.;* eight years old *pred.; (8 Jahre dauernd)* eight-year *attrib.;* ~**köpfig** *Adj.* ⟨*family, committee*⟩ of eight

acht·los 1. *Adj.* heedless; **2.** *adv.* heedlessly

acht-: ~**mal** *Adv.* eight times; ~**spurig** *Adj.* eight-lane ⟨*road*⟩; eight-track ⟨*cassette*⟩; ~**stellig** *Adj.* eight-figure *attrib.;* ~**stellig sein** have eight figures; ~**stimmig** 1. *Adj.* eight-part *attrib.;* **2.** *adv.* in eight parts; ~**stöckig** *Adj.* eight-storey *attrib.;* ~**tägig** *Adj. (8 Tage alt)* eight-day-old *attrib.; (8 Tage dauernd)* eight-day[-long] *attrib.;* ~**tausend** *Kardinalz.* eight thousand; ~**teilig** *Adj.* eight-piece ⟨*tea-service, tool-set, etc.*⟩; eight-part ⟨*series, serial*⟩

Achtung die; ~ **a)** respect (vor + *Dat., Gen.* for); **b)** ~**!** watch out!; ~, **fertig, los!** on your marks, get set, go!

acht·zehn *Kardinalz.* eighteen; **18 Uhr 33** 6.33 p.m.; *(auf der 24-Stunden-Uhr)* 1833; **achtzehn·jährig** *Adj. (18 Jahre alt)* eighteen-year-old *attrib.;* eighteen years old *pred.; (18 Jahre dauernd)* eighteen-year *attrib.*

achtzig *Kardinalz.* eighty; **mit** ~ **[km/ h] fahren** drive at *or (coll.)* do eighty [k.p.h.]; **über/etwa** ~ **[Jahre alt] sein** be over/about eighty [years old]; **mit** ~ **[Jahren]** *od.* **Achtzig** at eighty [years of age]; **achtzig·jährig** *Adj. (80 Jahre alt)* eighty-year-old *attrib.;* eighty years old *pred.; (80 Jahre dauernd)* eighty-year *attrib.*

ächzen *itr. V.* groan

Acker der; ~s, **Äcker** field; **Acker-**
bau der; *o. Pl.* arable farming *no in-*
def. art.

A. D. *Abk.* **Anno Domini** AD

ADAC [a:de:a:'tse:] der; ~ *Abk.* **Allge-**
meiner Deutscher Automobilclub

Adams·apfel der *(ugs.)* Adam's
apple

adäquat [at|ε'kva:t] *Adj.* ap- propriate
(Dat. to); suitable *(Dat.* for)

addieren 1. *tr. V.* add [up]; 2. *itr. V.*
add; **Addition** die; ~, ~en addition

ade *Interj. (veralt., landsch.)* farewell
(dated); bye *(coll.)*

Adel der; ~s nobility; **der niedere/hohe**
~: the lesser nobility/the aristocracy;
adelig *s.* adlig; **Adelige** *s.* Adlige;
adeln *tr. V.* jmdn. ~: give sb. a title;
(in den hohen Adel erheben) raise sb. to
the peerage

Adels-: ~**geschlecht das** noble family; ~**stand** der nobility; *(hoher Adel)*
nobility; ~**titel** der title of nobility

Ader die; ~, ~n a) blood-vessel; b) *o.*
Pl. (Anlage, Begabung) streak; c)
(Bot., Geol.) vein; d) *(Elektrot.)* core

adieu [a'diø:] *Interj. (veralt.)* adieu

Adjektiv das; ~s, ~e *(Sprachw.)* adjective

Adjutant der; ~en, ~en adjutant

Adler der; ~s, ~: eagle

adlig *Adj.* noble; ~ **sein** be a noble
[man/woman]; **Adlige** der/die *adj.*
Dekl. noble [man/woman]

Admiral der; ~s, ~e *od.* **Admiräle** admiral

adoptieren *tr. V.* adopt; **Adoption**
die; ~, ~en adoption

Adoptiv-: ~**eltern** *Pl.* adoptive parents; ~**kind** das adopted child

Adressat der; ~en, ~en, **Adressatin**
die; ~, ~nen addressee; **Adreß-**
buch das directory; **Adresse** die; ~,
~n address; **bei jmdm. an die falsche** ~
kommen *od.* **geraten** *(fig. ugs.)* come
to the wrong address *(fig.);* **adres-**
sieren *tr. V.* address

adrett 1. *Adj.* smart. 2. *adv.* smartly

Advent [at'vεnt] der; ~s a) Advent; b)
(Adventssonntag) Sunday in Advent

Advents-: ~**kalender** der Advent
calendar; ~**kranz** der garland of ever-
greens with four candles for the Sundays in Advent

Adverb [at'vεrp] das; ~s, ~ien
(Sprachw.) adverb; **adverbial**
(Sprachw.) 1. *Adj.* adverbial; 2. *adv.*
adverbially

Advokat [atvo'ka:t] der; ~en, ~en
(österr., schweiz., sonst veralt.) lawyer;
advocate *(arch.)*

Aero- [aero- *od.* ε:ro-]: ~**gramm** das
air[-mail] letter; ~**sol** das; ~s, ~e
aerosol

Affäre die; ~, ~n affair; **sich aus der** ~
ziehen *(ugs.)* get out of it

Affe der; ~n, ~n a) monkey; *(Men-*
schen~) ape; b) *(salopp) (dummer*
Kerl) oaf; clot *(Brit. sl.); (Geck)* dandy

Affekt der; ~[e]s, ~e emotion; **im** ~:
in the heat of the moment; **affek-**
tiert *(abwertend)* 1. *Adj.* affected; 2.
adv. affectedly

Affen·theater das *(salopp)* farce

Afghane [af'ga:nə] der; ~n, ~n a) Afghan; b) *(Hund)* Afghan hound;
afghanisch *Adj.* Afghan; **Afghani-**
stan [af'ga:nιsta:n] **(das);** ~s Afghanistan

Afrika·(das); ~s Africa; **Afrikaner**
der; ~s, ~African; **afrikanisch** *Adj.*
African

After der; ~s, ~: anus

AG [a:'ge:] *Abk.* **die;** ~, ~s **Aktienge-**
sellschaft PLC *(Brit.);* Ltd. *(private*
company) (Brit.); Inc. *(Amer.)*

Agent der; ~en, ~en; **Agentin** die;
~, ~nen agent; **Agentur** die; ~, ~en
agency

Aggregat das; ~[e]s, ~e *(Technik)*
unit; *(Elektrot.)* set; **Aggregat·zu-**
stand der *(Chemie)* state

Aggression die; ~, ~en aggression;
aggressiv 1. *Adj.* aggressive; 2. *adv.*
aggressively; **Aggressivität** die; ~:
aggressiveness; **Aggressor** der; ~s,
~en aggressor

Agitation die; ~: agitation; **agi-**
tieren *itr. V.* agitate

Agrar·land das; *Pl.* ~länder agrarian
country

Ägypten (das); ~s Egypt; **Ägypter**
der; ~s, ~Egyptian; **ägyptisch** *Adj.*
Egyptian

ah *Interj. (verwundert)* oh; *(freudig, ge-*
nießerisch) ah; *(verstehend)* oh; ah

äh [ε(:)] *Interj.* **a)** *(angeekelt)* ugh; **b)**
(stotternd) er; hum

aha [a'ha(:)] *Interj. (verstehend)* oh[, I
see]; *(triumphierend)* aha

Ahn der; ~[e]s, *od.* ~en, ~en *(geh.),*
Ahne der; ~n, ~n forebear; ancestor

ähneln *itr. V.* jmdm. ~: resemble *or* be
like sb.; **jmdm. sehr/wenig** ~: strongly
resemble *or* be very like sb./bear little
resemblance to sb.; **einer Sache** *(Dat.)*
~: be similar to sth.; be like sth.; **sich**

(Dat.) ~: resemble one another; be alike

ahnen *tr. V.* **a)** *(im voraus fühlen)* have a premonition of; **b)** *(vermuten)* suspect; **das konnte ich doch nicht ~!** I had no way of knowing that

ähnlich 1. *Adj.* similar; **jmdm. ~ sein** be like sb.; **~ wie** like; **2.** *adv.* similarly; ⟨*answer, react*⟩ in a similar way; **3.** *Präp. mit Dat.* like; **Ähnlichkeit die;** ~, ~en similarity; **mit jmdm. ~ haben** be like sb.

Ahnung die; ~, ~en **a)** *(Vorgefühl)* premonition; **b)** *(ugs.: Kenntnisse)* knowledge; **von etw. |viel| ~ haben** know [a lot] about sth.; **keine ~!** [I've] no idea; **ahnungs·los** *Adj. (nichts ahnend)* unsuspecting; *(naiv, unwissend)* naïve

ahoi *Interj. (Seemannsspr.)* ahoy

Ahorn ['a:hɔrn] der; ~s, ~e maple

Ähre die; ~, ~n ear

Aids [e:ts] das; ~: Aids

Aids-: ~**kranke** der/die person suffering from Aids; ~**test der** Aids test

Akademie die; ~, ~n academy; *(Bergbau, Forst~, Bau~)* school; college; **Akademiker der;** ~s, ~, **Akademikerin die;** ~, ~nen [university/college] graduate; **akademisch 1.** *Adj.* academic; **2.** *adv.* academically

Akazie [a'ka:tsiə] die; ~, ~n acacia

akklimatisieren *refl. V.* become *or* get acclimatized

Akkord der; ~|e|s, ~e **a)** *(Musik)* chord; **b)** *(Wirtsch.) (~arbeit)* piecework; *(~lohn)* piece-work pay *no indef. art., no pl.; (~satz)* piece-rate

Akkordeon das; ~s, ~s accordion

Akku der; ~s, ~s *(ugs.)*, **Akkumulator der;** ~s, ~en accumulator *(Brit.);* storage battery

akkurat 1. *Adj.* meticulous; **2.** *adv.* meticulously

Akkusativ der; ~s, ~e *(Sprachw.)* accusative [case]; **Akkusativ·objekt das** *(Sprachw.)* accusative *or* direct object

Akne die; ~, ~n *(Med.)* acne

Akrobat der; ~en, ~en acrobat; **Akrobatik die;** ~ acrobatics *pl.;* **akrobatisch** *Adj.* acrobatic

Akt der; ~|e|s, ~e **a)** *(auch Theater, Zirkus~, Varieté~)* act; **b)** *(Zeremonie)* ceremony; **c)** *(Geschlechts~)* sexual act; **d)** *(bild. Kunst)* nude; **Akt·bild das** nude [picture]

Akte die; ~, ~n file

Akten-: ~**deckel der** folder; ~**kof-**

fer der attaché case; ~**mappe die** brief-case; ~**ordner der** file; ~**tasche die** brief-case; ~**zeichen das** reference

Akteur [ak'tø:ɐ̯] der; ~s, ~e person involved

Akt·foto das nude photo

Aktie ['aktsiə] die; ~, ~n *(Wirtsch.)* share; ~**n** shares *(Brit.);* stock *(Amer.);* **die ~n fallen/steigen** share *or* stock prices are falling/rising; **Aktien·gesellschaft die** joint-stock company

Aktion die; ~, ~en **a)** action *no indef. art.; (militärisch)* operation; **b)** *(Kampagne)* campaign

Aktionär der; ~s, ~e shareholder

aktiv 1. *Adj.* **a)** active; **b)** *(Milit.)* serving *attrib.* ⟨*officer, soldier*⟩; **2.** *adv.* actively; **Aktiv das;** ~s, ~e *(Sprachw.)* active; **Aktive der/die;** *adj. Dekl. (Sport)* participant; **aktivieren** *tr. V.* mobilize ⟨*party members, group, class, etc.*⟩; **den Kreislauf ~:** stimulate the circulation; **Aktivität die;** ~, ~en activity

Akt·modell das nude model

Aktualität die; ~, ~en **a)** *(Gegenwartsbezug)* relevance [to the present]; **b)** *(von Nachrichten usw.)* topicality; **aktuell** *Adj.* topical; *(gegenwärtig)* current; *(neu)* up-to-the-minute; **eine ~e Sendung** *(Ferns., Rundf.)* a [news and] current affairs programme

Akupunktur die; ~, ~en *(Med.)* acupuncture

Akustik die; ~ **a)** *(Lehre vom Schall)* acoustics *sing., no art.;* **b)** *(Schallverhältnisse)* acoustics *pl.;* **akustisch 1.** *Adj.* acoustic. **2.** *adv.* acoustically

akut *Adj. (auch Med.)* acute; pressing, urgent ⟨*question, issue*⟩

Akzent der; ~|e|s, ~e **a)** *(Sprachw.) (Betonung)* stress; *(Betonungszeichen)* accent; **b)** *(Sprachmelodie, Aussprache)* accent

akzeptabel 1. *Adj.* acceptable; **2.** *adv.* acceptably; **akzeptieren** *tr. V.* accept

à la [a la] *(Gastr., ugs.)* à la

Alabaster der; ~s, ~: alabaster

à la carte [ala'kart] *(Gastr.)* à la carte

Alarm der; ~|e|s, ~e alarm; *(Flieger~)* air-raid warning; ~ **geben/***(fig. ugs.)* **schlagen** raise the alarm; **blinder ~:** false alarm

alarm-, Alarm-: ~**anlage die** alarm system; ~**bereit** *Adj.* on alert *postpos.;* ~**bereitschaft die** alert

alarmieren *tr. V.* **a)** alarm; **b)** *(zu Hilfe rufen)* call [out] ⟨*doctor, police, fire brigade, etc.*⟩

Alarm-: **~sirene** die warning siren; **~stufe** die alert stage

Albaner der; ~s, ~Albanian; **Albanien** [al'ba:niən] (das); ~s Albania; **albanisch** *Adj.* Albanian

Albatros der; ~, ~se *(Zool.)* albatross

Alben *s.* Album

albern *Adj.* **a)** silly; **sich ~ benehmen** act silly; **b)** *(ugs.: nebensächlich)* silly; stupid; **Albernheit** die; ~, ~en silliness

Albino der; ~s, ~s albino

Album das; ~s, **Alben** album

Alge die; ~, ~n alga

Algebra [*österr.*: al'ge:bra] die; ~: algebra

Algerien [al'ge:riən] (das); ~s Algeria; **Algerier** der; ~s, ~: Algerian; **algerisch** *Adj.* Algerian

alias *Adv.* alias

Alibi das; ~s, ~s alibi

Alkohol der; ~s, ~e alcohol; **alkoholfrei** *Adj.* non-alcoholic; **Alkoholiker** der; ~s, ~: alcoholic; **alkoholisch** *Adj.* alcoholic; **Alkoholismus** der; ~: alcoholism *no art.*

all *Indefinitpron. u. unbest. Zahlw.* **1.** *attr. (ganz, gesamt...)* all; **~es andere/Weitere/übrige** everything else; **~es Schöne** everything *or* all that is beautiful; **~es Gute!** all the best!; **wir/ihr/sie ~e** all of us/you/them; **~e Anwesenden** all those present; **~e Bewohner der Stadt** all the inhabitants of the town; **~e Jahre wieder** every year; **~e fünf Minuten/Meter** every five minutes/metres; **Bücher ~er Art** all kinds of books; **in ~er Ruhe** in peace and quiet; **2.** *alleinstehend* **a)** ~e all; ~e, die...: all those who ...; **b)** ~es *(auf Sachen bezogen)* everything; *(auf Personen bezogen)* everybody; **das ~es** all that; **trotz ~em** in spite of everything; **~es in ~em** all in all; **vor ~em** above all; **das ist ~es** that's all *or (coll.)* it; **ist das ~es?** is that all *or (coll.)* it?; ~es **mal herhören!** *(ugs.)* listen everybody!; **~es aussteigen!** *(ugs.)* everyone out!; *(vom Schaffner gesagt)* all change!

All das; ~s *s.* **Weltall**

alle *Adj.; nicht attr.*: **~ sein** be all gone; **~ werden** run out

alle·dem *Pron.* **trotz ~**: in spite of *or* despite all that

Allee die; ~, ~n avenue

allein [a'lain] **1.** *Adj.; nicht attr.* **a)** *(für sich)* alone; on one's/its own; by oneself/itself; **ganz ~**: all on one's/its own; **b)** *(einsam)* alone; **2.** *adv. (ohne Hilfe)* by oneself/itself; on one's/its own; **etw. ~ tun** do sth. oneself; **von ~** *(ugs.)* by oneself/itself; **3.** *Adv.* **a)** *(geh.: ausschließlich)* alone; **b)** |schon| **~ der Gedanke/**|schon| **der Gedanke ~**: the mere thought [of it]; **alleine** *(ugs.) s.* **allein 1a, 2, 3b; alleinig** *Adj.; nicht präd.* sole

allein-, Allein-: **~gang** der *(fig.)* independent initiative; **im ~gang** off one's own bat; **~stehend** *Adj.* *(person)* living alone; *(ledig)* single ⟨*person*⟩; **~stehende der/die**; *adj. Dekl.* person living alone; *(Ledige[r])* single person

alle·mal *Adv.* **a)** *(ugs.)* any time *(coll.)*; **was der kann, das kann ich doch ~**: anything he can do, I can do too; **b)** **ein für ~**: once and for all; **allenfalls** *Adv.* **a)** *(höchstens)* at [the] most; **b)** *(bestenfalls)* at best

aller-: **~dings** *Adv.* **a)** *(einschränkend)* though; **es stimmt ~dings, daß ...**: it's true though that ...; **b)** *(zustimmend)* [yes,] certainly; **das war ~dings Pech** that was bad luck, to be sure; **~erst...** *Adj.; nicht präd.* very first; **der/die/das ~erste** the very first; **b)** *(best...)* very best

Allergie die; ~, ~n *(Med.)* allergy; **allergisch 1.** *Adj. (Med.)* allergic; **2.** *adv.* **auf etw. (Akk.) ~ reagieren** have an allergic reaction to sth.

aller-, Aller-: **~größt...** *Adj.* utmost ⟨*trouble, care, etc.*⟩; biggest ⟨*car, house, town, etc.*⟩; tallest ⟨*person*⟩ of all; **am ~größten sein** be [the] biggest/tallest of all; **~hand** *indekl. unbest. Gattungsz. (ugs.)* **a)** *attr.* all kinds *or* sorts of; **b)** *alleinstehend* all kinds *or* sorts of things; **das ist ~hand** *(viel)* that's a lot; **das ist ja ~hand!** that's just not on! *(Brit. coll.)*; **~heiligen** das; ~; *(bes. kath. Kirche)* All Saints' Day; **~herzlichst 1.** *Adj.* warmest ⟨*thanks, greetings, congratulations*⟩; most cordial ⟨*reception, welcome, invitation*⟩; **2.** most warmly; **~höchst... 1.** *Adj.* highest ⟨*building, tree, etc.*⟩ of all; **2.** *adv.* **am ~höchsten** ⟨*fly, jump, etc.*⟩ the highest of all; **~höchstens** *Adv.* at the very most

allerlei *indekl. unbest. Gattungsz.*: *attr.* all kinds *or* sorts of; *alleinstehend* all kinds *or* sorts of things; **Al-**

lerlei das; ~s, ~s *(Gemisch)* potpourri; *(Durcheinander)* jumble

aller-, Aller-: ~**letzt...** *Adj.; nicht präd.* **a)** very last; **b)** *(ugs. abwertend)* most dreadful *(coll.);* **das ist das Allerletzte** that is the absolute limit; ~**liebst...** 1) *Adj.* most favourite; **es wäre mir am** ~**liebsten** *od.* **das** ~**liebste, wenn ...:** I should like it best of all if ...; 2. *adv.* **etw. am** ~**liebsten tun** like doing sth. best of all; ~**meist...** 1. *Indefinitpron. u. unbest. Zahlw.* by far the most *attrib.;* **das** ~**meiste/am** ~**meisten** most of all/by far the most; 2. *Adv.* **am** ~**meisten** most of all; ~**mindest...** *Adj.*slightest; least; **das** ~**mindeste** the very least; ~**nächst...** 1. *Adj.* very nearest *attrib.; (Reihenfolge ausdrückend)* very next *attrib.;* 2. *adv.* **am** ~**nächsten** nearest of all; ~**neu[e]st...** *Adj.* very latest *attrib.;* **das Allerneue|e|ste** the very latest; ~**schlimmst...** *Adj.* very worst *attrib.:* ~**schönst...** 1. *Adj.* most beautiful *attrib.;* loveliest *attrib.; (angenehmst...)* very nicest *attrib.;* 2. *adv.* **er singt am** ~**schönsten** his singing is the most beautiful of all; ~**seits** *Adv.* **guten Morgen** ~**seits!** good morning everyone

Allerwelts-: ~**gesicht das** nondescript face; ~**wort das** hackneyed word

allerwenigst... 1. *Adj.* lest ... of all; *Pl.* fewest ... of all; 2. *adv.* **am** ~**wenigsten** least of all

alle·samt *Indefinitpron. u. unbest. Zahlw. (ugs.)* all [of you/us/them]; **wir** ~**:** we all

Alles·kleber der all-purpose adhesive

all·gemein 1. *Adj.* general; universal ⟨*conscription, suffrage*⟩; **im** ~**en Interesse** in the common interest; **im** ~**en** in general; 2. *adv.* **a)** generally; *(ausnahmslos)* universally; **es ist** ~ **bekannt, daß ...:** it is common knowledge that ...; **b)** *(unverbindlich)* ⟨*write, talk, discuss*⟩ in general terms

Allgemein-: ~**befinden das** *(Med.)* general state of health; ~**bildung die;** *o. Pl.* general education

Allgemeinheit die; ~ **a)** generality; **b) die** ~**:** the general public

Allgemein-: ~**medizin die;** *o. Pl.* general medicine; ~**wohl das** public good

All·heilmittel das *(auch fig.)* cureall; panacea

Alligator der; ~s, ~en alligator

Alliierte der; *adj. Dekl.* ally; **die** ~**n** the Allies

all-: ~**jährlich** 1. *Adj.* annual; yearly; 2. *adv.* annually; every year; ~**mächtig** *Adj.* all-powerful

all·mählich 1. *Adj.* gradual; 2. *adv.* gradually; 3. *Adv.* **wir sollten** ~ **gehen** it's time we got going

all-, All-: ~**morgendlich** 1. *Adj.* regular morning; 2. *adv.* every morning; ~**seitig** 1. *Adj.* general; all-round, *(Amer.)* all-around *attrib.;* 2. *adv.* generally; ~**seits** *Adv.* on all sides; ~**tag der a)** *(Werktag)* weekday; **b)** *o. Pl. (Einerlei)* daily routine; **der graue** ~**:** the dull routine of everyday life; ~**täglich** *Adj.* ordinary ⟨*face, person, appearance, etc.*⟩; everyday ⟨*topic, event, sight*⟩; commonplace ⟨*remark*⟩; **ein nicht** ~**täglicher Anblick** a sight one doesn't see every day; ~**tags** *Adv.* [on] weekdays; ~**zu** *Adv.* all too; **nicht** ~**zu viele** not too many

allzu-: ~**früh** *Adv.* all too early; *(*~*bald)* all too soon; ~**lang[e]** *Adv.* too long; ~**oft** *Adv.* too often; ~**sehr** *Adv.* too much; **nicht** ~**sehr** not too much; ~**viel** *Adv.* too much

Alm die; ~, ~**en** mountain pasture; Alpine pasture; **Alm·hütte die** Alpine hut

Almosen das; ~s, ~ alms *pl.*

Alp die; ~, ~**en** *(bes. schweiz.) s.* Alm

Alpaka das; ~s, ~s alpaca

Alpen *Pl. die* ~**:** the Alps

Alpen-: ~**rose die** rhododendron; ~**veilchen das** cyclamen

Alpha das; ~|s|, ~|s| alpha; **Alphabet das;** ~|e|s, ~e alphabet; **alphabetisch** 1. *Adj.* alphabetical; 2. *adv.* alphabetically

Alp·horn das alpenhorn; **alpin** *Adj.* Alpine; **Alpinist der;** ~en, ~en Alpinist

als *Konj.* **a)** *(zeitlich)* when; **damals,** ~**:** [in the days] when; **gerade** ~**:** just as; **b)** *(kausal)* **um so mehr,** ~**:** all the more since *or* in that; **c)** *Vergleichspartikel* **größer/älter/mehr/weniger** ~**:** bigger/older/more/less than; **anders** ~ **wir sein/leben** be different/live differently from us; **soviel/soweit** ~ **möglich** as much/as far as possible; **so bald/schnell** ~ **möglich** as soon/as quickly as possible; ~ **|wenn od. ob|** (+ *Konjunktiv II)* as if; as though; ~ **ob ich das nicht wüßte!** as if I didn't know; **d)**

~ **Rentner/Arzt** as a pensioner/a doctor sich ~ **wahr/Lüge erweisen** prove to be true/a lie

also 1. *Adv.* so; therefore; **2.** *Partikel* a) *(das heißt)* that is; b) *(nach Unterbrechung)* well [then]; c) *(verstärkend)* **na** ~! there you are[, you see]; ~ **schön** well all right then

alt, älter, ältest... *Adj.* a) old; ~ **und jung** old and young; **seine ~en Eltern** his aged parents; **wie ~ bist du?** how old are you?; **mein älterer/ältester Bruder** my elder/eldest brother; b) *(nicht mehr frisch)* old; ~**es Brot** stale bread; c) *(vom letzten Jahr)* old; ~**e Äpfel/Kartoffeln** last year's apples/potatoes; d) *(langjährig)* long-standing *⟨acquaintance⟩*; e) *(antik, klassisch)* ancient; f) *(vertraut)* old familiar *⟨streets, sights, etc.⟩*; **ganz der/die ~e sein** be just the same

¹**Alt** der; ~s, ~e *(Musik)* alto; *(Frauenstimme)* contralto; *(im Chor)* contraltos *pl.*

²**Alt** das; ~[s], ~: *top fermented, dark beer*

Altar der; ~[e]s, **Altäre** altar

alt-, Alt-: ~**bau·wohnung** die flat *(Brit.)* or *(Amer.)* apartment in an old building; ~**bekannt** *Adj.* well-known; ~**bier** das s. ²**Alt**

Alte der/die; *adj. Dekl.* a) *(alter Mensch)* old man/woman; *Pl.* old people; b) *(salopp) (Vater, Ehemann)* old man *(coll.); (Mutter, Ehefrau)* old woman *(coll.); (Chef)* governor *(sl.); (Chefin)* boss *(coll.);* **die ~n** *(Eltern)* my/his *etc.* old man and old woman 4(coll.); c) *Pl. (Tiereltern)* parents

alt·ehrwürdig *Adj. (geh.)* venerable; time-honoured *⟨customs⟩*; **Alt·englisch** das Old English

Alter das; ~s, ~: age; *(hohes ~)* old age; **im ~:** in one's old age; **im ~ von** at the age of; **älter 1.** *s.* **alt; 2.** *Adj. (nicht mehr jung)* elderly; **altern** *itr. V.; mit sein* age

alters-, Alters-: ~**genosse** der, ~**genossin** die contemporary; ~**gruppe** die age-group; ~**schwach** *Adj.* old and infirm *⟨person⟩;* old and weak *⟨animal⟩;* ~**schwäche** die; o. *Pl. (bei Menschen)* [old] age and infirmity; *(bei Tieren)* [old] age and weakness; ~**stufe** die age; ~**unterschied** der age difference; ~**versorgung** die provision for one's old age; *(System)* pension scheme

Altertum das; ~s antiquity *no art.*

Älteste der/die; *adj. Dekl.* a) *(Dorf~, Vereins~, Kirchen~ usw.)* elder; b) *(Sohn, Tochter)* eldest

alt-, Alt-: ~**griechisch** das classical or ancient Greek; ~**hochdeutsch** das Old High German; ~**klug**; ~**kluger..., klugst...** 1. *Adj.* precocious; **2.** *adv.* precociously; ~**last** die *(Ökologie)* old, improperly disposed of harmful waste; *(fig.)* inherited problem

ältlich *Adj.* rather elderly

alt-, Alt-: ~**modisch** 1. *Adj.* old-fashioned; **2.** *adv.* in an old-fashioned way; ~**rosa** old rose; ~**stadt** die old [part of the] town; ~**waren·händler** der second-hand dealer

Alu das; ~s *(ugs.)* aluminium; **Alu·folie** die aluminium foil; **Aluminium** das; ~s aluminium

am *Präp. + Art.* a) = **an dem;** b) **Frankfurt am Main** Frankfurt on [the] Main; **am Marktplatz** on the market square; **am Meer/Fluß** by the sea/on *or* by the river; **am Anfang/Ende** at the beginning/end; **am 19. November** on 19 November; **am schnellsten laufen** run [the] fastest; **am Verwelken sein** be wilting

Amateur [ama'tøːɐ̯] der; ~s, ~e amateur

Amazonas der; ~: Amazon

Amboß der; **Ambosses, Ambosse** anvil

ambulant *(Med.)* 1. *Adj.* out-patient *attrib.;* **2.** *adv.* **jmdn.** ~ **behandeln** give sb. out-patient treatment; **Ambulanz** die; ~, ~**en** a) *(in Kliniken)* out-patient[s'] department; b) *(Krankenwagen)* ambulance

Ameise die; ~, ~**n** ant

Ameisen-: ~**bär** der ant-eater; ~**haufen** der anthill

amen *Adv.* amen; **Amen** das; ~s, ~: Amen

Amerika (das); ~s America; **Amerikaner** der; ~s, ~ a) American; b) *(Gebäck) small, flat iced cake;* **Amerikanerin** die; ~, ~**nen** American; **amerikanisch** *Adj.* American

Amino·säure die *(Chemie)* amino acid

Ammann der; ~[e]s, **Ammänner** *(schweiz.) (Gemeinde~, Bezirks~)* ≈ mayor; *(Land~)* cantonal president

Amme die; ~, ~**n** wet-nurse

Amnestie [amnɛsˈtiː] die; ~, ~**n** amnesty; **amnestieren** *tr. V.* grant an amnesty to

Amöbe die; ~, ~n *(Biol.)* amoeba
Amok der: ~ **laufen** run amok; **Amok·läufer** der madman
Ampel die; ~, ~n a) *(Verkehrs~)* traffic lights *pl.;* b) *(für Pflanzen)* hanging flowerpot
Amphibie [am'fi:biə] die; ~, ~n *(Zool.)* amphibian; **Amphibien·fahrzeug** das amphibious vehicle
Amphi·theater das amphitheatre
Ampulle die; ~, ~n *(Med.)* ampoule
Amputation die; ~, ~en *(Med.)* amputation; **amputieren** *tr. V.* amputate
Amsel die; ~, ~n blackbird
Amt das; ~|e|s, **Ämter** a) *(Stellung)* post; position; *(hohes politisches od. kirchliches ~)* office; **im** ~ **sein** be in office; b) *(Aufgabe)* task; job; c) *(Behörde)* office; d) *(Fernsprechvermittlung)* exchange; **amtieren** *itr. V.* a) hold office; b) *(vorübergehend)* act (**als** as); **amtlich** 1. *Adj.* a) official; *(ugs.: sicher)* definite; 2. *adv.* officially; **Amt·mann** der; *Pl.* ...**männer** *od.* ...**leute**, **Amt·männin** die; ~, ~**nen** *senior civil servant*
Amts-: ~**arzt** der medical officer; ~**gericht** das local *or* district court; ~**geschäfte** *Pl.* official duties; ~**leitung** die *(Fernspr.)* exchange line
Amulett das; ~|e|s, ~e amulet; charm
amüsant 1. *Adj.* entertaining; 2. *adv.* in an entertaining way; **amüsieren** 1. *refl. V.* a) *(sich vergnügen)* enjoy oneself; **sich mit jmdm.** ~: have fun *or* a good time with sb.; b) *(belustigt sein)* be amused; **sich über jmdn./etw.** ~: find sb./sth. funny; 2. *tr. V.* amuse
an 1. *Präp. mit Dat.* a) *(räumlich)* at; *(auf)* on; **Frankfurt an der Oder** Frankfurt on [the] Oder; **Tür an Tür** next door to one another; **an ... vorbei** past; b) *(zeitlich)* on; **an jedem Sonntag** every Sunday; **an Ostern** *(bes. südd.)* at Easter; c) **arm/reich an Vitaminen** low/rich in vitamins; **jmdn. an etw. erkennen** recognize sb. by sth.; **an etw. leiden** suffer from sth.; **an einer Krankheit sterben** die of a disease; d) **an |und für| sich** actually; 2. *Präp. mit Akk.* a) to; *(auf, gegen)* on; b) **an etw./ jmdn. glauben** believe in sth./sb.; **an etw. denken** think of sth.; **sich an etw. erinnern** remember sth.; 3. *Adv.* a) *(Verkehrsw.)* **Köln an: 9.15** arriving Cologne 09.15; b) *(ugs.: in Betrieb)* on; **die Waschmaschine/der Fernseher**

ist an the washing-machine/television is on; *s. auch* **ansein**; c) *(ugs.: ungefähr)* around; about; **an |die| 20 000 DM** around *or* about 20,000 DM
Analyse die; ~, ~n analysis; **analysieren** *tr. V.* analyse; **analytisch** 1. *Adj.* analytical; 2. *adv.* analytically
Ananas die; ~, ~ *od.* ~**se** pineapple
Anarchie die; ~, ~n anarchy; **Anarchist** der; ~**en**, ~**en** anarchist
Anatomie die; ~, ~n anatomy; **anatomisch** *Adj.* anatomical
an|bahnen 1. *tr. V.* initiate ⟨*negotiations, talks, process, etc.*⟩; develop ⟨*relationship, connection*⟩; 2. *refl. V.* ⟨*development*⟩ be in the offing; ⟨*friendship, relationship*⟩ start to develop
an|bändeln *itr. V.* **mit jmdm.** ~ *(ugs.)* get off with sb. *(Brit. coll.);* pick sb. up
An·bau der; *Pl.* **Anbauten** a) *o. Pl.* building; b) *(Gebäude)* extension; c) *o. Pl. (das Anpflanzen)* growing
an|bauen 1. *tr. V.* a) build on; b) *(anpflanzen)* grow; 2. *itr. V. (das Haus vergrößern)* build an extension
an·bei *Adv. (Amtsspr.)* herewith; **Rückporto** ~: return postage enclosed
an|beißen 1. *unr. tr. V.* bite into; take a bite of; 2. *unr. itr. V. (auch fig. ugs.)* bite
an|belangen *tr. V.* **was mich/dies** *usw.* **anbelangt** as far as I am/this matter is *etc.* concerned
an|beten *tr. V. (auch fig.)* worship
An·betracht der: **in** ~ **einer Sache** *(Gen.)* in view of sth.
an|betreffen *unr. tr. V. s.* **anbelangen**
an|betteln *tr. V.* **jmdn.** ~: beg from sb.; **jmdn. um etw.** ~: beg sb. for sth.
Anbetung die; ~, ~en *(auch fig.)* worship
an|biedern *refl. V.* **sich |bei jmdm.|** ~: curry favour [with sb.]
an|bieten 1. *unr. tr. V.* offer; **jmdm. etw.** ~: offer sb. sth.; 2. *unr. refl. V.* a) offer one's services; **sich** ~, **etw. zu tun** offer to do sth.; b) *(fig.)* ⟨*possibility, solution*⟩ suggest itself
an|binden *unr. tr. V.* tie [up] (**an** + *Dat. od. Akk.* to); tie up, moor ⟨*boat*⟩ (**an** + *Dat. od. Akk.* to); tether ⟨*animal*⟩ (**an** + *Dat. od. Akk.* to)
an|blasen *unr. tr. V.* a) blow at; b) *(anfachen)* blow on
An·blick der the sight; **an|blicken** *tr. V.* look at
an|blinzeln *tr. V.* a) blink at; b) *(zuzwinkern)* wink at

an|brechen 1. *unr. tr. V.* **a)** crack; **b)** *(öffnen)* open; **c)** *(zu verbrauchen beginnen)* break into ⟨*supplies, reserves*⟩; 2. *unr. itr. V.; mit sein (geh.: beginnen)* ⟨*dawn, day*⟩ break; ⟨*age, epoch*⟩ dawn

an|brennen 1. *unr. tr. V. (anzünden)* light; 2. *unr. itr. V.; mit sein* burn

an|bringen *unr. tr. V.* **a)** *(befestigen)* put up ⟨*sign, aerial, curtain, plaque*⟩ (**an** + *Dat.* on); **b)** *(äußern)* make ⟨*request, complaint, comment*⟩; **c)** *(zeigen)* demonstrate ⟨*knowledge, experience*⟩; **d)** *(ugs.: herbeibringen)* bring

An·bruch der *o. Pl. (geh.: Beginn)* dawn[ing]; **der ~ des Tages** daybreak

an|brüllen *tr. V. (ugs.)* bellow at

Andacht die; ~, ~en **a)** *o. Pl. (Sammlung)* rapt attention; *(im Gebet)* silent worship; **b)** *(Gottesdienst)* prayers *pl.;*

andächtig 1. *Adj.* rapt; *(ins Gebet versunken)* devout; 2. *adv.* with rapt attention; *(ins Gebet versunken)* devoutly

an|dauern *itr. V.* ⟨*negotiations*⟩ continue, go on; ⟨*weather, rain*⟩ last

andauernd 1. *Adj.* continual; constant; 2. *adv.* continually; constantly

Anden *Pl.* **die ~**: the Andes

An·denken das; ~s, ~ **a)** *o. Pl.* memory; **zum ~ an** jmdn./etw. to remind you/us *etc.* of sb./sth.; **b)** *(Erinnerungsstück)* memento; *(Reise~)* souvenir

ander... *Indefinitpron.* 1. *attr.* **a)** other; **ein ~er/eine ~e/ein ~es** another; **das Kleid gefällt mir nicht, haben Sie noch ~e/ein ~es?** I don't like that dress, do you have any others/another?; **jemand ~er** *od.* **~es** someone else; *(in Fragen)* anyone else; **niemand ~er** *od.* **~es** nobody else; **etwas ~es** something else; *(in Fragen)* anything else; **nichts ~es** nothing else; not anything else; **b)** *(verschieden)* different; 2. *alleinstehend* **ein ~r/eine ~e**: another [one]; **nicht drängeln, einer nach dem ~n** don't push, one after the other; **ein ~er/eine ~e/ein ~es** another [one]; **ein[e]s nach dem ~[e]n** first things first; **ich will weder das eine noch das ~e** I don't want either; **anderen·falls** *Adv.* otherwise; **anderer·seits** *Adv.* on the other hand; **ander·mal** *Adv.* **ein ~**: another time; **andern·falls** *Adv.* otherwise

ändern 1. *tr. V.* change; alter ⟨*garment*⟩; change ⟨*person*⟩; 2. *refl. V.* change

anders *Adv.* **a)** *(verschieden)* ⟨*think,*

act, feel, do⟩ differently (**als** from *or esp. Brit.* to); ⟨*be, look, sound, taste*⟩ different (**als** from *or esp. Brit.* to); **es war alles ganz ~**: it was all quite different; **b)** *(sonst)* else; **niemand ~**: nobody else; **jemand ~**: someone else; *(in Fragen)* anyone else

anders-, Anders-: **~artig** *Adj.* different; **~farbig** *Adj.* different-coloured *attrib.;* of a different colour *postpos.;* **~gläubige** der/die person of a different religion; **~herum** *Adv.* the other way round *or (Amer.)* around; **~herum gehen/fahren** go/drive round *or (Amer.)* around the other way; **~wo** *Adv. (ugs.)* elsewhere; **~woher** *Adv. (ugs.)* from somewhere else; **~wohin** *Adv. (ugs.)* somewhere else

andert·halb *Bruchz.* one and a half; **~ Stunden** an hour and a half

Änderung die; ~, ~en change (*Gen.* in); alteration (*Gen.* to)

anderweitig 1. *Adj.* other; 2. *adv.* in another way

an|deuten 1. *tr. V.* **a)** *(zu verstehen geben)* hint; **b)** *(nicht vollständig ausführen)* outline; *(kurz erwähnen)* indicate; 2. *refl. V.* be indicated;

An·deutung die hint

An·drang der; *o. Pl.* crowd; *(Gedränge)* crush

andre... *s.* **ander...**

an|drehen *tr. V.* **a)** *(einschalten)* turn on; **b)** jmdm. etw. **~** *(ugs.)* palm sb. off with sth.

andrer·seits *Adv.* on the other hand

an|drohen *tr. V.* jmdm. etw. **~**: threaten sb. with sth.; **An·drohung** die threat

an|drücken *tr. V.* press down

an|ecken *itr. V.; mit sein* **bei** jmdm. **~** *(fig. ugs.)* rub sb. [up *(Brit.)*] the wrong way

an|eignen *refl. V.* **a)** appropriate; **b)** *(lernen)* acquire; learn

an·einander *Adv.* **~ denken** think of each other *or* one another; **~ vorbeigehen** pass each other *or* one another

aneinander: **~|binden** *unr. tr. V.* tie together; **~|legen** *tr. V.* put *or* place next to each other *or* one another; **~|liegen** *unr. itr. V.* lie next to each other

Anekdote die; ~, ~n anecdote

an|ekeln *tr. V.* disgust

Anemone die; ~, ~n anemone

an|erkennen *unr. tr. V.* **a)** recognize ⟨*country, record, verdict, qualification,* *document*⟩; acknowledge ⟨*debt*⟩; accept ⟨*demand, bill, conditions, rules*⟩; allow ⟨*claim, goal*⟩; **b)** *(nicht leugnen)* acknowledge; **c)** *(würdigen)* appreciate; respect ⟨*viewpoint, opinion*⟩; **ein ~der Blick** an appreciative look; **ạnerkennens·wert** *Adj.* commendable; **Ạnerkennung die; ~, ~en** *s.* **anerkennen: a)** recognition; acknowledgement; acceptance; allowance; **b)** acknowledgement; **c)** appreciation; respect *(Gen.* for)

ạn|fachen *tr. V.* fan; *(fig.)* arouse ⟨*anger, curiosity, enthusiasm*⟩; inflame ⟨*passion*⟩; stir up ⟨*hatred*⟩; inspire ⟨*hope*⟩; ferment ⟨*discord, war*⟩

ạn|fahren 1. *unr. tr. V.* **a)** run into; hit; **b)** *(herbeifahren)* deliver; **c)** *(ansteuern)* stop at ⟨*village etc.*⟩; ⟨*ship*⟩ put in at ⟨*port*⟩; **d)** *(zurechtweisen)* shout at; **2.** *unr. itr. V.; mit sein* **a)** *(starten)* start off; **b) angefahren kommen** come driving/riding up; **Ạn·fahrt die a)** *(das Anfahren)* journey; **b)** *(Weg)* approach

Ạn·fall der attack; *(epileptischer ~, fig.)* fit; **einen ~ bekommen** *od. (ugs.)* **kriegen** have an attack/a fit; **ạn·fallen 1.** *unr. tr. V.* attack; **2.** *unr. itr V.; mit sein* ⟨*costs*⟩ be incurred; ⟨*interest*⟩ accrue; ⟨*work*⟩ come up; **ạn·fällig** *Adj.* ⟨*person*⟩ with a delicate constitution; ⟨*machine*⟩ susceptible to faults; **gegen** *od.* **für etw. ~ sein** be susceptible to sth.

Ạn·fang der beginning; start; *(erster Abschnitt)* beginning; **am** *od.* **zu ~:** at first; **von ~ an** from the outset; **~ 1984/der Woche** *usw.* at the beginning of 1984/of the week *etc.*; **ạn·fangen 1.** *unr. itr. V.* **a)** begin; start; **mit etw. ~:** start [on] sth.; **~, etw. zu tun** start to do sth.; **b)** *(zu sprechen ~)* begin; **von etw. ~:** start on about sth.; **c)** *(eine Stelle antreten)* start; **2.** *unr. tr. V.* **a)** begin; start; *(anbrechen)* start; **b)** *(machen)* do; **Ạn·fänger der; ~s, ~:** beginner; **ạnfänglich** *Adj.* initial; **ạnfangs** *Adv.* at first; initially

Ạnfangs-: **~buchstabe der** initial [letter]; **~stadium das** initial stage

ạn|fassen 1. *tr. V.* **a)** *(fassen, halten)* take hold of; **b)** *(berühren)* touch; **c) jmdn. ~** *(an der Hand nehmen)* take sb.'s hand; **d)** *(angehen)* tackle ⟨*problem, task, etc.*⟩; **e)** *(behandeln)* treat ⟨*person*⟩; **2.** *itr. V.* |**mit**| **~:** lend a hand

ạnfechtbar *Adj.: s.* **anfechten a:** disputable; contestable; challengeable;

ạn|fechten *unr. tr. V.* **a)** dispute ⟨*statement, contract*⟩; contest ⟨*will*⟩; challenge ⟨*decision, law, opinion*⟩; **b)** *(beunruhigen)* trouble

ạn|fertigen *tr. V.* make

ạn|feuchten *tr. V.* moisten ⟨*lips, stamp*⟩; dampen ⟨*ironing, cloth, etc.*⟩

ạn|feuern *tr. V.* spur on

ạn|flehen *tr. V.* beseech; implore

ạn|fliegen 1. *unr. itr. V.; mit sein* fly in; **angeflogen kommen** come flying in; **gegen den Wind ~:** fly into the wind; **2.** *unr. tr. V.* fly to ⟨*city, country, airport*⟩; **Ạn·flug der a)** approach; **b)** *(Hauch)* hint; **c)** *(Anwandlung)* fit; **in einem ~ von Großzügigkeit** in a fit of generosity

ạn|fordern *tr. V.* ask for; order ⟨*goods, materials*⟩; send for ⟨*ambulance*⟩; **Ạn·forderung die a)** *o. Pl. (das Anfordern)* request *(Gen.* for); **b)** *(Anspruch)* demand

Ạn·frage die inquiry; *(Parl.)* question; **ạn|fragen** *itr. V.* inquire; ask

ạn|freunden *refl. V.* become friends

ạn|fügen *tr. V.* add

ạn|fühlen *refl. V.* feel

ạn|führen *tr. V.* **a)** lead; **b)** *(zitieren)* quote; **c)** *(nennen)* give ⟨*example, reason, details, proof*⟩; **d)** *(ugs.: hereinlegen)* have on *(Brit. coll.);* dupe; **Ạn·führer der** leader; *(Rädelsführer)* ringleader; **Ạn·führung die a)** *(das Zitieren, Zitat)* quotation; **b)** *(Nennung)* giving

Ạnführungs-: **~strich der,** **~zeichen das** quotation-mark

Ạn·gabe die a) *(das Mitteilen)* giving; **b)** *(Information)* piece of information; **~n** information *sing.;* **c)** *(Ballspiele)* service; serve; **ạn|geben 1.** *unr. tr. V.* **a)** give ⟨*reason*⟩; declare ⟨*income, dutiable goods*⟩; name ⟨*witness*⟩; **b)** *(bestimmen)* set ⟨*course, direction*⟩; **den Takt ~:** keep time; **2.** *unr. itr. V.* **a)** *(prahlen)* boast; brag; *(sich angeberisch benehmen)* show off; **b)** *(Ballspiele)* serve; **Ạngeber der; ~s, ~:** braggart; **Angeberẹi die; ~:** showing-off; **ạngeblich 1.** *Adj.* alleged; **2.** *adv.* supposedly; allegedly

ạn·geboren *Adj.* innate ⟨*characteristic*⟩; congenital ⟨*disease*⟩

Ạn·gebot das a) offer; **b)** *(Wirtsch.)* supply; *(Sortiment)* range

ạn·gebracht *Adj.* appropriate

ạn·gegriffen *Adj.* weakened ⟨*health, stomach*⟩; strained ⟨*nerves, voice*⟩

angeheitert *Adj.* tipsy

an|gehen 1. *unr. itr. V.; mit sein* **a)** ⟨*radio, light, heating*⟩ come on; ⟨*fire*⟩ catch; **b)** *(anwachsen, wachsen)*⟨*plant*⟩ take root; **e) es mag noch ~:** it's [just about] acceptable; **f) gegen etw./jmdn. ~:** fight sth./sb.; **2.** *unr. tr. V.* **a)** *(angreifen)* attack; **b)** *(in Angriff nehmen)* tackle ⟨*problem, difficulty*⟩; take ⟨*fence, bend*⟩; **c)** *(bitten)* ask (**um** for); **d)** *(betreffen)* concern; **das geht dich nichts an** it's none of your business; **angehend** *Adj.* budding; *(zukünftig)* prospective

an|gehören *itr. V.* **jmdm./einer Sache ~:** belong to sb./sth.; **der Regierung/ einer Familie ~:** be a member of the government/a family; **an·gehörig** *Adj.* belonging (*Dat.* to); **Angehörige der/die;** *adj. Dekl.* **a)** *(Verwandte)* relative; relation; **b)** *(Mitglied)* member

Angeklagte der/die; *adj. Dekl.* accused; defendant

Angel die; ~, ~n a) fishing-rod; **b)** *(Tür~, Fenster~ usw.)* hinge; **etw. aus den ~n heben** *(fig.)* turn sth. upside down

An·gelegenheit die matter; *(Aufgabe, Problem)* affair

Angel·haken der fish-hook; **angeln 1.** *tr. V. (zu fangen suchen)* fish for; *(fangen)* catch. **2.** *itr. V.* angle; fish; **Angel·rute die** fishing-rod

Angel·sachse der Anglo-Saxon

Angel·schnur die fishing-line

an·gemessen *Adj.* appropriate; reasonable, fair ⟨*price, fee*⟩

an·genehm 1. *Adj.* pleasant; **~e Reise/Ruhe!** [have a] pleasant journey/ have a good rest; **|sehr| ~!** delighted to meet you; **2.** *adv.* pleasantly

an·gesehen *Adj.* respected

angesichts *Präp. mit Gen. (geh.)* **a)** in the face of; **b)** *(fig.: in Anbetracht)* in view of

angespannt *Adj.* **a)** close ⟨*attention*⟩; taut ⟨*nerves*⟩; **b)** tense ⟨*situation*⟩; tight ⟨*market, economic situation*⟩

angestellt *Adj.* **bei jmdm. ~ sein** be employed by sb.; work for sb.; **Angestellte der/die;** *adj. Dekl.* [salaried] employee

an·getan *Adj.* **von jmdm./etw. ~ sein** be taken with sb./sth.

an·getrunken *Adj.* [slightly] drunk

an·gewiesen *Adj.* **auf jmdn./etw. ~ sein** have to rely on sb./sth.

an|gewöhnen *tr. V.* **jmdm. etw. ~:** get sb. used to sth.; **jmdm. ~, etw. zu tun** get sb. used to doing sth.; **sich** *(Dat.)* **etw. ~:** get into the habit of sth.; **|es| sich** *(Dat.)* **~, etw. zu tun** get into the habit of doing sth.; **An·gewohnheit die** habit

an|gleichen 1. *unr. tr. V.* **etw. einer Sache** *(Dat.)* **od. an etw.** *(Akk.)* **~:** bring sth. into line with sth.; **2.** *unr. refl. V.* **sich jmdm./einer Sache od. an jmdn./etw. ~:** become like sb./sth.; **An·gleichung die: die ~ der Löhne an die Preise** bringing wages into line with prices

Angler der; ~s, ~ angler

Anglistik die; ~: English studies *pl.*, *no art.*

Angola (das); ~s Angola

Angora-: ~katze die angora cat; **~wolle die** angora [wool]

an|greifen 1. *unr. tr. V.* **a)** *(auch fig.)* attack; **b)** *(schwächen)* affect ⟨*health, heart, stomach, intestine, voice*⟩; weaken ⟨*person*⟩; **2.** *unr. itr. V. (auch fig.)* attack; **An·greifer der** *(auch fig.)* attacker; **An·griff der a)** attack; **zum ~ blasen** *(auch fig.)* sound the attack; **b) etw. in ~ nehmen** tackle sth.

angst *Adj.* **jmdm. ist/wird |es| ~ |und bange|** sb. is/becomes frightened

Angst die; ~, Ängste a) *(Furcht)* fear; **~ bekommen** *od. (ugs.)* **kriegen** become frightened; **~ haben** be frightened (**vor** + *Dat.* of); **b)** *(Sorge)* anxiety; **~ haben** be anxious (**um** about); **keine ~, ich vergesse es schon nicht!** don't worry, I won't forget [it]!; **ängstigen 1.** *tr. V.* frighten; *(beunruhigen)* worry; **2.** *refl. V.* be frightened; *(sich sorgen)* worry; **ängstlich 1.** *Adj.* anxious; **2.** *adv.* anxiously; **Ängstlichkeit die; ~:** timidity

an|gucken *tr. V. (ugs.)* look at; **sich** *(Dat.)* **etw./jmdn. ~:** have a look at sth./sb.

an|gurten *tr. V.* strap in; **sich ~:** put on one's seat-belt

an|haben *unr. tr. V.* **a)** *(ugs.: am Körper tragen)* have on; **b) jmdm./einer Sache etwas ~ können** be able to harm sb./sth.

an|halten 1. *unr. tr. V.* **a)** stop; **b)** *(auffordern)* urge; **2.** *unr. itr. V.* **a)** stop; **b)** *(andauern)* go on; last; **anhaltend 1.** *Adj.* constant; continuous; **2.** *adv.* constantly; continuously; **An·halter der** hitch-hiker; **per ~ fahren** hitch[-hike]; **An·halterin die** hitch-hiker; **Anhalts·punkt der** clue (**für** to); *(für eine Vermutung)* grounds *pl.*

an·hand 1. *Präp. mit Gen.* with the help of; **2.** *Adv.* ~ **von** with the help of
An·hang der a) *(Buchw.)* appendix; **b)** *(Anhängerschaft)* following; **c)** *(Verwandtschaft)* family; **an|hängen 1.** *tr. V.* **a)** hang up **(an + Akk.** on); **b)** *(ankuppeln)* couple on **(an + Akk.** to); hitch up ‹*trailer*› **(an + Akk.** to); **c)** *(anfügen)* add **(an + Akk.** to); **2.** *refl. V.* **a)** hang on **(an + Akk.** to); **b)** *(ugs.: sich anschließen)* sich [an jmdn. *od.* bei jmdm.] ~: tag along [with sb.] *(coll.)*; **An·hänger der a)** *(Mensch)* supporter; **b)** *(Wagen)* trailer; **c)** *(Schmuckstück)* pendant; **d)** *(Schildchen)* tag; **Anhängerschaft die;** ~, ~en supporters *pl.*; **anhänglich** *Adj.* devoted ‹*dog, friend*›; **Anhänglichkeit die;** ~: devotion
an|hauchen *tr. V.* breathe on ‹*mirror, glasses*›; blow on ‹*fingers, hands*›
an|häufen *tr. V.* accumulate; **Anhäufung die** accumulation
an|heben *unr. tr. V.* **a)** lift [up]; **b)** *(erhöhen)* raise ‹*prices, wages, etc.*›
an|heften *tr. V.* attach ‹*label, list*›; put up ‹*sign, notice*›
anheim|stellen *(geh.) tr. V.* |es] jmdm. ~, etw. zu tun leave it to sb. to do sth.
An·hieb der: auf ~ *(ugs.)* straight off
an|himmeln *tr. V.* worship
An·höhe die rise
an|hören 1. *tr. V.* listen to; sich *(Dat.)* jmdn./etw. ~: listen to sb./sth.; **2.** *refl. V.* sound
animieren *tr. V.* encourage
Anis der; ~|es] aniseed
Ank. *Abk.* **Ankunft** arr.
An·kauf der purchase; **an|kaufen** *tr. V.* purchase; buy
Anker der; ~s, ~ anchor; **vor** ~ **gehen/liegen** drop anchor/lie at anchor; ~ **werfen** drop anchor; **ankern** *itr. V.* **a)** anchor; **b)** *(vor Anker liegen)* be anchored; **Anker·platz der** anchorage
An·klage die a) charge; **unter** ~ **stehen** have been charged **(wegen** with); **b)** *(~vertretung)* prosecution; **Anklage·bank die** dock; **auf der** ~ **sitzen** *(auch fig.)* be in the dock; **an|klagen** *tr. V.* **a)** *(Rechtsw.)* charge **(Gen.,** wegen with); accuse; **b)** *(geh.: beschuldigen)* accuse; **An·kläger der** prosecutor
an|klammern 1. *tr. V.* peg *(Brit.)*, pin *(Amer.)* ‹*clothes, washing*› up; clip ‹*sheet etc.*›; *(mit Heftklammern)* staple ‹*sheet etc.*›; **2.** *refl. V.* sich an jmdn./etw. ~: cling to sb./sth.

An·klang der: |bei jmdm.| ~ **finden** meet with [sb.'s] approval
an|kleben 1. *tr. V.* stick up ‹*poster, etc.*›; **2.** *itr. V.; mit sein* stick
an|kleiden *tr. V. (geh.)* dress; sich ~: dress
an|klopfen *itr. V.* knock
an|knüpfen 1. *tr. V.* **a)** tie on **(an +** Akk. to); **b)** *(beginnen)* start up ‹*conversation*›; establish ‹*relations, business links*›; form ‹*relationship*›; **2.** *itr. V.* an etw. *(Akk.)* ~: take sth. up; ich knüpfe dort an, wo ... I'll pick up where ...
an|kommen *unr. itr. V.; mit sein* **a)** *(eintreffen)* arrive; seid ihr gut angekommen? did you arrive safely?; **b)** |bei jmdm.| |gut| ~ *(fig. ugs.)* go down [very] well [with sb.]; **c)** gegen jmdn./etw. ~: be able to deal with sb./fight sth.; **d)** *unpers.* es kommt auf jmdn./etw. an *(jmd./etw. ist ausschlaggebend)* it depends on sb./sth.; **es kommt auf etw.** *(Akk.)* **an** *(etw. ist wichtig)* sth. matters *(Dat.* to); **es kommt |ganz| darauf** *od.* **drauf an** *(ugs.)* it [all] depends; **e)** *unpers.* **es darauf** *od.* **drauf** ~ **lassen** *(ugs.)* chance it; **es auf etw.** *(Akk.)* ~ **lassen** [be prepared to] risk sth.
an|koppeln 1. *tr. V.* couple ‹*carriage*› up; hitch ‹*trailer*› up; dock ‹*spacecraft*›; **2.** *itr. V.* ‹*spacecraft*› dock
an|kreuzen *tr. V.* mark with a cross
an|kündigen 1. *tr. V.* announce; **2.** *refl. V.* announce itself; **An·kündigung die** announcement
Ankunft die; ~, **Ankünfte** arrival; „~" 'arrivals'
Ankunfts-: ~**halle die** arrival[s] hall; ~**tafel die** arrivals board
an|kuppeln *tr. V. s.* ankoppeln 1
an|kurbeln *tr. V.* **a)** crank [up]; **b)** *(fig.)* boost ‹*economy, production, etc.*›
Anl. *Abk.* **Anlage** encl.
an|lächeln *tr. V.* smile at; **an|lachen 1.** *tr. V.* smile at. **2.** *refl. V.* sich *(Dat.)* jmdn. ~ *(ugs.)* get off with sb. *(Brit. coll.)*; pick sb. up
An·lage die a) *o. Pl. (das Anlegen)* *(einer Kartei)* establishment; *(eines Parks, Gartens usw.)* laying out; *(eines Parkplatzes, Stausees)* construction; **b)** *(Grün~)* park; *(um ein Schloß usw. herum)* grounds *pl.*; **c)** *(Einrichtung)* facilities *pl.*; **militärische** ~**n** military installations; **d)** *(Werk)* plant; **e)** *(Musik~, Lautsprecher~ usw.)* system; **f)** *(Geld~)* investment; **g)** *(Konzeption)* conception; *(Struktur)* structure; **h)**

(Veranlagung) aptitude; *(Neigung)* tendency; **i)** *(Beilage zu einem Brief)* enclosure

Anlaß der; Anlasses, Anlässe a) cause (zu for); **etw. zum ~ nehmen, etw. zu tun** take sth. as an opportunity to do sth.; **aus aktuellem ~**: because of current events; **b)** *(Gelegenheit)* occasion

an|lassen 1. *unr. tr. V.* **a)** leave ⟨light, radio, heating, etc.⟩ on; leave ⟨engine⟩ running; leave ⟨candle⟩ burning; **b)** keep ⟨coat, gloves, etc.⟩ on; **c)** *(in Gang setzen)* start [up]; **2.** *unr. refl. V.* **sich gut/schlecht ~**: get off to a good/bad start; **Anlasser der; ~s, ~** starter

an·läßlich *Präp. mit Gen.* on the occasion of

An·lauf der a) run-up; **[mehr] ~ nehmen** take [more of] a run-up; **b)** *(Versuch)* attempt; **beim** *od.* **im ersten/ dritten ~**: at the first/third attempt; **an|laufen 1.** *unr. itr. V.; mit sein* **a) angelaufen kommen** come running along; *(auf einen zu)* come running up; **b) gegen jmdn./etw. ~**: run at sb./ sth.; **c)** *(Anlauf nehmen)* take a run-up; **d)** *(zu laufen beginnen)* ⟨engine⟩ start [up]; *(fig.)* ⟨film⟩ open; ⟨production, campaign, search⟩ start; **e) rot/ dunkel** *usw.* **~**: go *or* turn red/dark *etc.*; **f)** *(beschlagen)* mist up; **2.** *unr. tr. V.* put in at ⟨port⟩

an|legen 1. *tr. V.* **a)** put *or* lay ⟨domino, card⟩ [down] **(an** + *Akk.* next to); place, position ⟨ruler, protractor⟩ **(an** + *Akk.* on); put ⟨ladder⟩ up **(an** + *Akk.* against); **b) die Flügel/Ohren ~**: close its wings/lay its ears back; **die Arme ~**: put one's arms to one's sides; **c)** *(geh.: anziehen, umlegen)* don; **d)** *(schaffen, erstellen)* lay out ⟨town, garden, plantation, street⟩; start ⟨file, album⟩; compile ⟨statistics, index⟩; **e)** *(investieren)* invest; **f)** *(ausgeben)* spend **(für** on); **g) es darauf ~, etw. zu tun** be determined to do sth.; **2.** *itr. V.* **a)** *(landen)* moor; **b)** *(Kartenspiele)* lay a card/cards; **c)** *(Domino)* play [a domino/dominoes]; **d)** *(das Gewehr~)* aim **(auf** + *Akk.* at); **3.** *refl. V.* **sich mit jmdm. ~**: pick an argument with sb.

Anlege-: ~platz der berth; **~steg der** jetty

an|lehnen 1. *tr. V.* **a)** lean **(an** + *Akk. od. Dat.* against); **b)** leave ⟨door, window⟩ slightly open; **2.** *refl. V.* **sich [an jmdn.** *od.* **jmdm./etw.] ~**: lean [on sb./ against sth.]; **Anlehnung die; ~,**

~en: in ~an (+ *Akk.)* in imitation of; following

Anleihe die; ~, ~n loan

an|leiten *tr. V.* instruct; **An·leitung die** instructions *pl.*

an|lernen *tr. V.* train

an|liegen *unr. itr. V.* **a)** ⟨pullover etc.⟩ fit tightly; **b)** *(ugs.: vorliegen)* be on; **An·liegen das; ~s, ~** *(Bitte)* request; *(Angelegenheit)* matter; **anliegend** *Adj.* **a)** *(angrenzend)* adjacent; **b)** *(beiliegend)* enclosed; **Anlieger der; ~s, ~**: resident; **„~ frei"** 'except for access'

an|locken *tr. V.* attract ⟨customers, tourists, etc.⟩; lure ⟨bird, animal⟩

an|lügen *tr. V.* lie to

an|machen *tr. V.* **a)** put ⟨light, radio, heating⟩ on; light ⟨fire⟩; **b)** mix ⟨cement, plaster, paint, etc.⟩; dress ⟨salad⟩

an|malen *tr. V.* paint

an|maßen *refl. V.* **sich** *(Dat.)* **etw. ~**: claim sth. [for oneself]; **an·maßend 1.** *Adj.* presumptuous; *(arrogant)* arrogant; **2.** *adv.* presumptuously; *(arrogant)* arrogantly; **Anmaßung die; ~, ~en** presumption; *(Arroganz)* arrogance

an|melden *tr. V.* **a)** *(als Teilnehmer)* enrol **(zu** for); **sich ~**: enrol **(zu** for); **b)** *(melden, anzeigen)* license ⟨radio, television⟩; apply for ⟨patent⟩; register ⟨domicile, car, trade mark⟩; **sich ~**: register one's new address; **c)** *(ankündigen)* announce; **sind Sie angemeldet?** do you have an appointment?; **sich beim Arzt ~**: make an appointment to see the doctor; **d)** *(geltend machen)* express ⟨reservation, doubt, wish⟩; put forward ⟨demand⟩; **An·meldung die a)** *(zur Teilnahme)* enrolment; **b)** *s.* **anmelden b:** licensing; application *(Gen.* for); registration; **c)** *(Ankündigung)* announcement; *(beim Arzt, Rechtsanwalt usw.)* making an appointment

an|merken *tr. V.* **a) jmdm. seinen Ärger/seine Verlegenheit** *usw.* **~**: notice that sb. is annoyed/embarrassed *etc.;* **man merkt ihm [nicht] an, daß er krank ist** you can[not] tell that he is ill; **sich nichts ~ lassen** not let it show; **b)** *(geh.: bemerken)* note; **Anmerkung die; ~, ~en a)** *(Fußnote)* note; **b)** *(geh.: Bemerkung)* comment

An·mut die; ~ *(geh.)* grace; **an·mutig** *(geh.)* **1.** *Adj.* graceful ⟨girl, movement, dance⟩; charming, delightful

⟨girl, smile, picture, landscape⟩; 2. adv. ⟨move, dance⟩ gracefully; ⟨smile, greet⟩ charmingly

an|nähen tr. V. sew on

an|nähern 1. refl. V. get closer (Dat. to sth); 2. tr. V. bring closer (Dat. to); **annähernd** 1. Adv. almost; (ungefähr) approximately; 2. adj. approximate

Annahme die; ~, ~n a) (das Annehmen) acceptance; b) (Vermutung) assumption; **in der** ~, **daß** ...: on the assumption that ...; **annehmbar** 1. Adj. a) acceptable; b) (recht gut) reasonable; 2. adv. reasonably [well]; **an|nehmen** 1. unr. tr. V. a) accept; take; accept ⟨alms, invitation, condition, help⟩; take ⟨food, telephone call⟩; accept, take up ⟨offer, challenge⟩; b) (Sport) take ⟨ball, pass, etc.⟩; c) (billigen) approve; d) (aufnehmen) take on ⟨worker, patient, pupil⟩; e) (hinnehmen) accept ⟨fate, verdict, punishment⟩; f) (adoptieren) adopt; g) (haften lassen) take ⟨dye, ink⟩; h) (sich aneignen) adopt ⟨habit, mannerism, name, attitude⟩; i) (bekommen) take on ⟨look, appearance, form, dimension⟩; j) (vermuten, voraussetzen) assume; **angenommen, [daß]** ...: assuming [that] ...; 2. unr. refl. V. (geh.) **sich jmds./einer Sache** ~: look after sb./sth.; **Annehmlichkeit** die; ~, ~en comfort; (Vorteil) advantage

annektieren tr. V. annex

Annonce [a'nõ:sə] die; ~, ~n advertisement; advert (Brit. coll.); **annoncieren** itr. V. advertise

annullieren tr. V. annul

anonym 1. Adj. anonymous; 2. adv. anonymously; **Anonymität** die; ~: anonymity

Anorak der; ~s, ~s anorak

an|ordnen tr. V. a) (arrangieren) arrange; b) (befehlen) order; **An·ordnung** die s. **anordnen: a)** arrangement; b) order

an·organisch Adj. inorganic

an|packen 1. tr. V. a) (ugs.: anfassen) grab hold of; b) (angehen) tackle; 2. itr. V. |mit| ~ (ugs.: mithelfen) lend a hand

an|passen 1. tr. V. a) (passend machen) fit; b) (abstimmen) suit (Dat. to); 2. refl. V. adapt [oneself] (Dat. to); ⟨animal⟩ adapt; **Anpassung** die; ~, ~en adaptation (**an** + Akk. to); **anpassungs·fähig** Adj. adaptable

an|pfeifen 1. unr. tr. V. **das Spiel/die zweite Halbzeit** ~: blow the whistle to start the game/the second half; 2. unr. itr. V. blow the whistle; **An·pfiff der a)** (Sport) whistle for the start of play; b) (salopp: Zurechtweisung) bawling-out (coll.)

an|pflanzen tr. V. a) plant; b) (anbauen) grow

an|pöbeln tr. V. (ugs.) abuse

an|prangern tr. V. denounce (**als** as)

an|preisen unr. tr. V. extol

An·probe die fitting; **an|probieren** tr. V. try on

an|rechnen tr. V. a) count; b) **jmdm. etw.** ~ (in Rechnung stellen) charge sb. for sth.

An·recht das right; **ein** ~ **auf etw.** (Akk.) **haben** be entitled to sth.

An·rede die form of address; **an|reden** tr. V. address

an|regen tr. V. a) stimulate ⟨imagination, digestion⟩; whet ⟨appetite⟩; b) (ermuntern) prompt; (vorschlagen) propose; **anregend** Adj. stimulating; **An·regung die a)** s. **anregen a:** stimulation; whetting; b) (Denkanstoß) stimulus; c) (Vorschlag) proposal

an|reichern 1. tr. V. enrich; 2. refl. V. accumulate

An·reise die journey [there/here]; **an|reisen** itr. V.; mit sein travel there/here; **mit der Bahn** ~: go/come by train

An·reiz der incentive

an|rempeln tr. V. barge into; (absichtlich) jostle

Anrichte die; ~, ~n sideboard; **an|richten** tr. V. a) arrange ⟨food⟩; (servieren) serve; b) cause ⟨disaster, confusion, devastation, etc.⟩

anrüchig Adj. a) disreputable; b) (unanständig) indecent

an|rücken itr. V.; mit sein ⟨troops⟩ advance; ⟨firemen, police⟩ move in

An·ruf der call; **Anruf·beantworter der;** ~s, ~: [telephone-]answering machine; **an|rufen** unr. tr. V. a) call or shout to ⟨friend, passer-by⟩; call ⟨sleeping person⟩; b) (geh.: angehen, bitten) appeal to ⟨person, court⟩ (**um** for); call upon ⟨God⟩; c) auch itr. (telefonisch ~) call; **Anrufer der;** ~s, ~: caller

an|rühren tr. V. a) touch; b) (bereiten) mix

ans Präp. + Art. a) = **an das;** b) **sich** ~ **Arbeiten machen** set to work

An·sage die announcement; **an|sagen** *tr. V.* **a)** announce; **b)** *(Kartenspiele)* bid

an|sammeln 1. *tr. V.* accumulate; amass *(riches, treasure)*; **2.** *refl. V.* accumulate; *(fig.) (anger, excitement)* build up; **An·sammlung die a)** collection; **b)** *(Auflauf)* crowd

ansässig *Adj.* resident

An·satz der *(erstes Zeichen, Beginn)* beginnings *pl.*

an|schaffen *tr. V.* [sich *(Dat.)*] etw. ~ get [oneself] sth.; **An·schaffung die** purchase

an|schalten *tr. V.* switch on

an|schauen *tr. V. (bes. südd., österr., schweiz.) s.* ansehen; **anschaulich 1.** vivid; **2.** *adv.* vividly; **Anschauung die;** ~, ~en **a)** *(Wahrnehmung)* experience; **b)** *(Auffassung)* view

An·schein der appearance; **allem** *od.* **dem** ~ **nach** to all appearances; **an·scheinend** *Adv.* apparently

an|schieben *unr. tr. V.* push *(vehicle)*

an|schießen *unr. tr. V.* shoot and wound

An·schlag der a) *(Bekanntmachung)* notice; *(Plakat)* poster; **b)** *(Attentat)* assassination attempt; *(auf ein Gebäude, einen Zug o. ä.)* attack; **c)** *(Texterfassung)* keystroke; **mit dem Gewehr im** ~: with rifle/rifles levelled; **an|schlagen** *unr. tr. V.* put up, *(notice, announcement, message)* (an + *Akk.* on); **b)** *(beschädigen)* chip

an|schließen 1. *unr. tr. V.* **a)** connect (an + *Akk. od. Dat.* to); connect up *(electrical device)*; **b)** *(festschließen)* lock, secure (an + *Dat. od. Akk.* to); **2.** *unr. refl. V.* sich jmdm./einer Sache ~: join sb./sth.; **An·schluß der** connection; **Anschluß·zug der** connecting train

an|schnallen *tr. V.* put on *(skis, skates)*; sich ~ *(im Auto)* put on one's seatbelt; *(im Flugzeug)* fasten one's seatbelt

an|schrauben *tr. V.* screw on (an + *Akk.* to)

an|schreien *unr. tr. V.* shout at

An·schrift die address

Anschuldigung die; ~, ~en accusation

an|schwellen *unr. itr. V.; mit sein* **a)** swell [up]; *(fig.)* swell *(water, river)* rise; **b)** *(lauter werden)* grow louder; *(noise)* rise

an|schwemmen *tr. V.* wash ashore

an|sehen *unr. tr. V.* **a)** look at; watch *(television programme)*; see *(play, film)*; jmdn. **groß/böse** ~: stare at sb./give sb. an angry look; **hübsch** *usw.* **anzusehen sein** be pretty *etc.* to look at; **sieh [mal] [einer] an!** *(ugs.)* well, I never! *(coll.)*; **b)** *(erkennen)* **man sieht ihm sein Alter nicht an** he does not look his age; **man sieht ihr die Strapazen an** she's showing the strain; **c)** *(zusehen bei)* etw. [mit] ~: watch sth.; **das kann man doch nicht [mit]** ~: I/you can't just stand by and watch that; **Ansehen das;** ~s [high] standing; **an·sehnlich** *Adj.* **a)** *(beträchtlich)* considerable; **b)** *(gut aussehend, stattlich)* handsome

an|sein *unr. itr. V.; mit sein (ugs.)* *(light, gas, etc.)* be on

an|setzen *tr. V.* **a)** *(in die richtige Stellung bringen)* position *(ladder, jack, drill, saw)*; **b)** *(anfügen)* put on (an + *Akk. od. Dat.* to); **c)** *(festlegen)* fix *(meeting etc.)* (für, auf + *Akk.* for); fix, set *(deadline, date, price)*; **d)** *(veranschlagen)* estimate; **e)** *(anrühren)* mix

An·sicht die a) *(Meinung)* opinion; view; **meiner** ~ **nach** in my opinion *or* view; **b)** *(Bild)* view; **Ansichts·karte die** picture postcard

an|spannen 1. *tr. V.* **a)** harness *(horse etc.)* (an + *Akk.* to); yoke up *(oxen)* (an + *Akk.* to); hitch up *(carriage, cart, etc.)* (an + *Akk.* to); **b)** *(anstrengen)* strain; **An·spannung die** strain

an|spielen *itr. V.* auf jmdn./etw. ~: allude to sb./sth.; **Anspielung die;** ~, ~en allusion (auf + *Akk.* to); *(verächtlich, böse)* insinuation (auf + *Akk.* about)

An·sporn der incentive; **an|spornen** *tr. V.* spur on

An·sprache die speech; address; **an|sprechen 1.** *unr. tr. V.* **a)** speak to; **b)** *(gefallen)* appeal to; **2.** *unr. itr. V. (reagieren)* respond (auf + *Akk.* to)

an|springen 1. *unr. itr. V.; mit sein* *(car, engine)* start; **2.** *unr. tr. V.* jump up at

An·spruch der a) claim; *(Forderung)* demand; **[keine] Ansprüche stellen** make [no] demands; **in** ~ **nehmen** take advantage of *(offer)*; exercise *(right)*; take up *(time)*; **b)** *(Anrecht)* right

an·spruchs-: ~**los 1.** *Adj.* **a)** *(genügsam)* undemanding; **b)** *(schlicht)* unpretentious; **2.** *adv.* **a)** *(genügsam)* undemandingly; *(live)* modestly; **b)** *(schlicht)* unpretentiously; ~**voll 1.** *Adj.* discriminating *(reader, audience,* *gourmet*⟩; *(schwierig)* demanding; ambitious ⟨*subject*⟩

an|spucken *tr. V.* spit at

Anstalt die; ~, ~en institution

An·stand der *o. Pl.* decency; **anständig 1.** *Adj.* **a)** decent; *(ehrbar)* respectable; **2.** *adv.* decently; *(ordentlich)* properly

an|starren *tr. V.* stare at

an·statt *Konj.* ~ zu arbeiten/~, daß er arbeitet instead of working

an|stecken 1. *tr. V.* **a)** pin on ⟨*badge, brooch*⟩; put on ⟨*ring*⟩; **b)** *(infizieren, auch fig.)* infect; **2.** *itr. V.* be infectious; **ansteckend** *Adj.* infectious; *(durch Berührung)* contagious; **Ansteckung** die; ~, ~en infection; *(durch Berührung)* contagion

an|stehen *unr. itr. V. (Schlange stehen)* queue [up], *(Amer.)* stand in line **(nach** for)

an·stelle 1. *Präp. mit Gen.* instead of; **2.** *Adv.* ~ **von** instead of

an|stellen 1. *refl. V.* queue [up], *(Amer.)* stand in line **(nach** for); **2.** *tr. V.* **a)** *(aufdrehen)* turn on; **b)** *(einschalten)* switch on; **c)** *(einstellen)* employ; **An·stellung** die **a)** *o. Pl.* employment; **b)** *(Stellung)* job

Anstieg der; ~|e|s rise, increase (+ *Gen.* in)

an|stiften *tr. V.* incite; **An·stifter** der, **An·stifterin** die instigator; **An·stiftung** die incitement

an|stimmen *tr. V.* start singing ⟨*song*⟩; start playing ⟨*piece of music*⟩; **ein Geschrei** ~: start shouting

An·stoß der **a)** stimulus **(zu** for); **den |ersten|** ~ **zu etw. geben** initiate sth.; **b)** ~ **erregen** cause offence **(bei** to); **|keinen|** ~ **an etw.** *(Dat.)* **nehmen** [not] object to sth.; **an|stoßen 1.** *unr. itr. V.* **a)** *mit sein* **an etw.** *(Akk.)* ~: bump into sth.; **b)** |mit den Gläsern| ~: clink glasses; **auf jmdn./etw.** ~: drink to sb./sth.; **2.** *unr. tr. V.* **jmdn./etw.** ~: give sb./sth. a push; **jmdn. aus Versehen** ~: knock into sb. inadvertently; **anstößig 1.** *Adj.* offensive; **2.** *adv.* offensively

an|strahlen *tr. V.* **a)** illuminate; *(mit Scheinwerfer)* floodlight; **b)** *(anblicken)* beam at

an|streben *tr. V. (geh.)* aspire to; *(mit großer Anstrengung)* strive for

an|streichen *unr. tr. V.* **a)** paint; **b)** *(markieren)* mark

an|strengen 1. *refl. V.* make an effort; **sich mehr/sehr** ~: make more of an ef-

fort/a great effort; **2.** *tr. V.* strain ⟨*eyes, ears, voice*⟩; be a strain on ⟨*person*⟩; **seine Phantasie** ~: exercise one's imagination; **anstrengend** *Adj. (körperlich)* strenuous; *(geistig)* demanding; **Anstrengung** die; ~, ~en **a)** effort; **große ~en machen, etw. zu tun** make every effort to do sth.; **b)** *(Strapaze)* strain

An·strich der paint

An·sturm der rush **(auf** + *Akk.* to); *(auf Banken, Waren)* run **(auf** + *Akk.* on)

Antarktika (das); ~s Antarctica; **Antarktis** die; ~: **die** ~: the Antarctic; **antarktisch** *Adj.* Antarctic

An·teil der share **(an** + *Dat.* of); ~ **an etw.** *(Dat.)* **nehmen** take an interest in sth.; **An·teilnahme** die **a)** interest **(an** + *Dat.* in); **b)** *(Mitgefühl)* sympathy **(an** + *Dat.* with)

Antenne die; ~, ~n aerial; antenna *(Amer.)*

anthrazit [antra'tsi:t] *Adj.; nicht attr.* anthracite[-grey]; **anthrazit·grau** *Adj.* anthracite-grey

anti-, Anti- anti-; **Anti·alkoholiker** der teetotaller; **Antibiotikum** das; ~s, **Antibiotika** *(Med.)* antibiotic

antik *Adj.* **a)** classical; **b)** *(aus vergangenen Zeiten)* antique ⟨*furniture, fittings, etc.*⟩; **Antike** die; ~ classical antiquity *no art.*

Antilope die; ~, ~n antelope

Antipathie die; ~, ~n antipathy

Antiquariat antiquarian bookshop/ department; *(mit neueren gebrauchten Büchern)* second-hand bookshop/department; **Antiquität** die; ~, ~en antique

an|treffen *unr. tr. V.* find; *(zufällig)* come across

an|treiben *unr. tr. V.* **a)** drive ⟨*animals, column of prisoners*⟩ on or along; *(fig.)* urge; **b)** *(in Bewegung setzen)* drive; power ⟨*ship, aircraft*⟩

an|treten 1. *unr. itr. V.; mit sein* **a)** form up; *(in Linie)* line up; *(Milit.)* fall in; **b)** *(sich stellen)* meet one's opponent; *(als Mannschaft)* line up; **gegen jmdn.** ~: meet sb./line up against sb.; **2.** *unr. tr. V.* **a)** start ⟨*job, apprenticeship*⟩; take up ⟨*position, appoint-*

ment); set out on ⟨*journey*⟩; begin ⟨*prison sentence*⟩; come into ⟨*inheritance*⟩

An·trieb der drive

An·tritt der: vor ~ Ihres Urlaubs before you go on holiday *(Brit.)* or *(Amer.)* vacation; vor ~ der Reise before setting out on the journey

an|tun *unr. tr. V.* a) jmdm. ein Leid ~: hurt sb.; jmdm. etwas Böses/ein Unrecht ~: do sb. harm/an injustice; b) jmd./etw. hat es jmdm. angetan sb. was taken with sb./sth.; *s. auch* angetan

Antwort die; ~, ~en a) answer; reply; er gab mir keine ~: he didn't answer [me] *or* reply; b) *(Reaktion)* response; **antworten** *itr. V.* a) answer; reply; auf etw. *(Akk.)* ~: answer sth.; reply to sth.; jmdm. ~: answer sb.; reply to sb.; b) *(reagieren)* respond **(auf** + *Akk.* to)

an|vertrauen 1. *tr. V.* jmdm. etw. ~: entrust sb. with sth.; *(fig.: mitteilen)* confide sth. to sb.; 2. *refl. V.* sich jmdm./einer Sache ~: put one's trust in sb./sth.; sich jmdm. ~ *(fig.: sich jmdm. mitteilen)* confide in sb.

an|wachsen *unr. itr. V.; mit sein* a) grow on; b) *(Wurzeln schlagen)* take root; c) *(zunehmen)* grow

Anwalt der; ~[e]s, Anwälte, **Anwältin** die; ~, ~en a) *(Rechts~)* lawyer; solicitor *(Brit.);* attorney *(Amer.); (vor Gericht)* barrister *(Brit.);* attorney[-at-law] *(Amer.);* advocate *(Scot.);* b) *(Fürsprecher)* advocate

An·wärter der candidate **(auf** + *Akk.* for); *(Sport)* contender **(auf** + *Akk.* for)

an|weisen *unr. tr. V.* instruct; *s. auch* angewiesen; **An·weisung** die instruction

an|wenden *unr. (auch regelm.) tr. V.* use, employ ⟨*process, trick, method, violence, force*⟩; use ⟨*medicine, money, time*⟩; apply ⟨*rule, paragraph, proverb, etc.*⟩ **(auf** + *Akk.* to); **An·wendung** die *s.* anwenden: use; employment; application

An·wesen das property

anwesend *Adj.* present (bei at); die Anwesenden those present; **Anwesenheit** die; ~: presence

an|widern *tr. V.* nauseate

An·zahl die; ~: number; eine ganze ~: a whole lot

an|zahlen *tr. V.* put down ⟨*sum*⟩ as a deposit **(auf** + *Akk.* on); *(bei Ratenzahlung)* make a down payment of

⟨*sum*⟩ **(auf** + *Akk.* on); **An·zahlung** die deposit; *(bei Ratenzahlung)* down payment

An·zeichen das sign; indication

Anzeige die; ~, ~n a) *(Straf~)* report; b) *(Inserat)* advertisement; c) *(eines Instruments)* display; **an|zeigen** *tr. V.* a) *(Strafanzeige erstatten)* jmdn./ etw. ~: report sb./sth. to the police/ the authorities; b) *(zeigen)* show; indicate; show ⟨*time, date*⟩; **Anzeigen·teil** der advertisement section *or* pages *pl.*

an|ziehen *unr. tr. V.* a) *(auch fig.)* attract; b) draw up ⟨*knees, feet, etc.*⟩; c) tighten ⟨*rope, wire, screw, knot, belt, etc.*⟩; put on ⟨*handbrake*⟩; d) *(ankleiden)* dress; sich ~: get dressed; e) *(anlegen)* put on ⟨*clothes*⟩; **anziehend** *Adj.* attractive; **An·zug** der a) suit; b) im ~ sein ⟨*storm*⟩ be approaching; ⟨*fever, illness*⟩ be coming on; ⟨*enemy*⟩ be advancing; **anzüglich** 1. *Adj.* insinuating ⟨*remark, question*⟩; 2. *adv.* in an insinuating way

an|zünden *tr. V.* light; set fire to ⟨*building etc.*⟩

an|zweifeln *tr. V.* doubt; question

apart 1. *Adj.* individual *attrib.;* 2. *adv.* in an individual style

Apartment das; ~s, ~s studio flat *(Brit.);* studio apartment *(Amer.)*

Apathie die; ~, ~n apathy; **apathisch** 1. *Adj.* apathetic; 2. *adv.* apathetically

Aperitif [aperi'ti:f] der; ~s, ~s aperitif

Apfel der; ~s, Äpfel apple

Apfel-: ~baum der apple-tree; ~kuchen der apple-cake; *(mit Äpfeln belegt)* apple flan; ~mus das apple purée; ~saft der apple-juice

Apfelsine die; ~, ~n orange

Apfel-: ~strudel der apfelstrudel; ~wein der cider

Apostel der; ~s, ~: apostle

Apotheke die; ~, ~n a) chemist's [shop] *(Brit.);* drugstore *(Amer.);* b) *(Haus~)* medicine cabinet; *(Reise~, Bord~)* first-aid kit; **Apotheker** der; ~s, ~, **Apothekerin** die; ~, ~nen [dispensing] chemist *(Brit.);* druggist

App. *Abk.* Apparat ext.

Apparat der; ~[e]s, ~e a) apparatus *no pl.; (Haushaltsgerät)* appliance; *(kleiner)* gadget; b) *(Radio~)* radio; *(Fernseh~)* television; *(Foto~)* camera; c) *(Telefon)* telephone; *(Nebenstelle)* extension; d) *(Personen und Hilfsmittel)* organization; *(Verwaltungs~)* system

Appartement [apartə'mã:, *schweiz. auch:* -'mɛnt] **das;** ~s, ~s *(schweiz. auch:* ~e) **a)** *s.* **Apartment; b)** *(Hotelsuite)* suite

Appell der; ~s, ~e **a)** appeal **(zu** for, **an** + *Akk.* to); **b)** *(Milit.)* muster; *(Anwesenheits~)* roll-call; **appellieren** *itr. V.* appeal **(an** + *Akk.* to)

Appetit der; ~[e]s, ~e appetite **(auf** + *Akk.* for); **guten** ~! enjoy your meal!; **appetitlich** *Adj.* **a)** appetizing; **b)** *(sauber, ansprechend)* attractive and hygienic; **Appetit·losigkeit die;** ~: lack of appetite

applaudieren *itr. V.* applaud; **Applaus der;** ~es, ~e applause

Aprikose die; ~, ~n apricot

April der; ~[s], ~e April; **der** ~: April

Aquädukt der *od.* **das;** ~[e]s, ~e aqueduct

Aquaplaning das; ~[s] aquaplaning

Aquarell das; ~s, ~e water-colour [painting]

Aquarium das; ~s, **Aquarien** aquarium

Äquator der; ~s equator

Ar das *od.* **der;** ~s, ~e are

Ära die; ~, **Ären** era

Araber der; ~s, ~ Arab; **Arabien** [a'ra:biən] **(das);** ~s Arabia; **arabisch** *Adj.* Arabian; Arabic *(language, numeral, literature, etc.)*

Arbeit die; ~, ~en **a)** work *no indef. art.;* **vor/nach der** ~ *(ugs.)* before/after work; **b)** *(Produkt, Werk)* work; **c)** *(Aufgabe)* job; **d)** *(Klassen~)* test; **arbeiten 1.** *itr. V.* work; **2.** *tr. V. (herstellen)* make; **Arbeiter der;** ~s, ~: worker; *(Bau~, Land~)* labourer; **Arbeiter·klasse die** working class[es *pl.*]; **Arbeiterschaft die;** ~: workers *pl.;* **Arbeit·geber der;** ~s, ~: employer; **Arbeitnehmer der;** ~s, ~: employee

arbeits-, Arbeits-: ~**amt das** job centre *(Brit.);* ~**bedingungen** *Pl.* working conditions; ~**fähig** *Adj.* fit for work *postpos.; (grundsätzlich)* able to work *postpos.;* ~**gang der** operation; ~**kraft die a)** capacity for work; **b)** *(Mensch)* worker; ~**los** *Adj.* unemployed; ~**lose der/die;** *adj. Dekl.* unemployed person/man/woman *etc.;* **die** ~**n** the unemployed; ~**losigkeit die;** ~: unemployment *no indef. art.;* ~**markt der** labour market; ~**platz der a)** work-place; **b)** *(~stätte)* place of work; **c)** *(~verhältnis)* job; ~**scheu** *Adj.* work-shy; ~**tag der**

working day; ~**teilung die** division of labour; ~**unfähig** *Adj.* unfit for work *postpos.; (grundsätzlich)* unable to work *postpos.;* ~**zeit die** working hours *pl.;* **die tägliche** ~**zeit** the working day

Archäologe der; ~n, ~n archaeologist; **Archäologie die;** ~: archaeology *no art.;* **archäologisch** *Adj.* archaeological

Arche die; ~, ~n ark; **die** ~ **Noah** Noah's Ark

Architekt der; ~en, ~en architect; **Architektur die;** ~ architecture

Archiv das; ~s, ~e archives *pl.;* archive

Ären *s.* **Ära**

Arena die; ~, **Arenen** arena; *(Stierkampf~, Manege)* ring

arg, ärger, ärgst... *(geh., landsch.)* **1.** *Adj.* **a)** *(schlimm)* bad; **im** ~ **sein** be in a sorry state; **b)** *(unangenehm groß, stark)* severe ⟨pain, hunger, shock, disappointment⟩; serious ⟨error, dilemma⟩; extreme ⟨embarrassment⟩; gross ⟨exaggeration, injustice⟩; **2.** *adv. (äußerst, sehr)* extremely

Ärger der; ~s **a)** annoyance; **b)** *(Unannehmlichkeiten)* trouble; ~ **bekommen** get into trouble; **ärgerlich 1.** *Adj.* **a)** annoyed; **b)** *(Ärger erregend)* annoying; **2.** *adv.* **a)** with annoyance; **b)** *(Ärger erregend)* annoyingly; **ärgern 1.** *tr. V.* **a)** annoy; **b)** *(reizen)* tease; **2.** *refl. V.* **sich** [über jmdn./etw.] ~: be/get annoyed [at sb./about sth.]; **Ärgernis das;** ~ses, ~se annoyance; *(etw. Anstößiges)* nuisance

arg-, Arg-: ~**listig** *Adj.* deceitful; *(heimtückisch)* malicious; ~**los 1.** *Adj.* unsuspecting; **2.** *adv.* unsuspectingly; ~**losigkeit die;** ~: unsuspecting nature

ärgst... *s.* **arg**

Argument das; ~[e]s, ~e argument; **Argumentation die;** ~, ~en argumentation; **argumentieren** *itr. V.* argue

Argwohn der; ~[e]s suspicion; **argwöhnisch** *(geh.)* **1.** *Adj.* suspicious; **2.** *adv.* suspiciously

Arie ['a:riə] **die;** ~, ~n aria

Aristokrat der; ~en, ~en aristocrat; **Aristokratin die;** ~, ~nen aristocrat; **Aristokratie die;** ~, ~n aristocracy; **aristokratisch 1.** *Adj.* aristocratic; **2.** *adv.* aristocratically

arithmetisch 1. *Adj.* arithmetical; **2.** *adv.* arithmetically

Arkade die; ~, ~n arcade

Arktis die; ~: Arctic; **arktisch** *Adj.* Arctic; *(fig.)* arctic

arm, **ärmer**, **ärmst...** *Adj.* poor; ~ **und reich** *(veralt.)* rich and poor [alike]; ~ **an Nährstoffen** poor in nutrients; **der/ die Ärmste** *od.* **Arme** the poor man/ boy/woman/girl

Arm der; ~[e]s, ~e arm; **jmdm. [mit etw.] unter die ~e greifen** help sb. out [with sth.]; **ein Hemd mit halbem ~:** a short-sleeved shirt

Armaturen·brett das instrument panel; *(im Kfz)* dashboard

Arm-: ~**band** das bracelet; *(Uhr~)* strap; ~**band·uhr** die wrist-watch

Armee die; ~, ~n *(auch fig.)* army

Ärmel der; ~s, ~: sleeve; **[sich** *(Dat.)]* **etw. aus dem ~ schütteln** *(ugs.)* produce sth. just like that

Ärmel·kanal der [English] Channel

ärmer *s.* arm; **ärmlich 1.** *Adj.* cheap ⟨clothing⟩; shabby ⟨flat, office⟩; meagre ⟨meal⟩. **2.** *adv.* cheaply ⟨furnished, dressed⟩

Arm·reif der armlet

arm·selig *Adj.* **a)** miserable; pathetic ⟨result, figure⟩; meagre ⟨meal, food⟩; paltry ⟨return, salary, sum, fee⟩; **b)** *(abwertend: erbärmlich)* miserable; **ärmst...** *s.* arm; **Armut** die; ~ poverty

Aroma das; ~s, **Aromen** *(Duft)* aroma; *(Geschmack)* flavour; **aromatisch** *Adj.* aromatic; distinctive ⟨taste⟩; ~ **duften** give off an aromatic fragrance

arrangieren [arã'zi:rən] **1.** *tr. V. (geh., Musik)* arrange; **2.** *refl. V.* **sich** ~: adapt; **sich mit jmdm.** ~: come to an accommodation with sb.

Arrest der; ~[e]s, ~e detention

arrogant 1. *Adj.* arrogant; **2.** *adv.* arrogantly; **Arroganz** die; ~ arrogance

Arsch der; ~[e]s, **Ärsche** *(derb)* **a)** arse *(Brit. coarse);* ass *(Amer. sl.);* **leck mich am ~!** *(fig.)* piss off *(coarse);* **im ~ sein** *(fig.)* be buggered *(coarse);* **b)** *(widerlicher Mensch)* arse-hole *(Brit. coarse);* ass-hole *(Amer. sl.);* **Arsch·loch** das *(derb) s.* Arsch b

Art die; ~, ~en **a)** kind; sort; **Bücher aller** ~: all kinds *or* sorts of books; **[so] eine** ~ ...: a sort of ...; **aus der ~ schlagen** not be true to type; *(in einer Familie)* be different from all the rest of the family; **b)** *(Biol.)* species; **c)** *o. Pl. (Wesen)* nature; *(Verhaltensweise)* way; *(gutes Benehmen)* behaviour; **die feine englische ~** *(ugs.)* the proper way

to behave; **d)** *(Weise)* way; **auf diese** ~: in this way; ~ **und Weise** way; *(Kochk.)* **nach ~ des Hauses** à la maison; **nach Schweizer ~:** Swiss style

Arterie [ar'te:riə] die; ~, ~n artery

artig *Adj.* well-behaved; **sei** ~: be a good boy/girl/dog *etc.*

Artikel der; ~s, ~ **a)** article; **b)** *(Ware)* item

Artillerie die; ~, ~n artillery

Artischocke die; ~, ~n artichoke

Artist der; ~en, ~en [variety/circus] performer

Arznei die; ~, ~en *(veralt.),* **Arznei·mittel** das medicine

Arzt der; ~es, **Ärzte**, **Ärztin** die; ~, ~nen doctor; **ärztlich 1.** *Adj.* medical; **auf** ~**e Verordnung** on doctor's orders; **2.** *adv.* **sich** ~ **behandeln lassen** have medical treatment

As das; ~ses, ~se ace

Asbest der; ~[e]s, ~e asbestos

Asche die; ~, ~n ash[es *pl.*]; *(sterbliche Reste)* ashes *pl.*

Aschen-: ~**becher** der ashtray; ~**brödel** das; ~s, ~ *(auch fig.)* Cinderella

Ascher·mittwoch der Ash Wednesday

Äser *s.* Aas

Asiat der; ~en, ~en, **Asiatin** die; ~, ~nen Asian; **asiatisch** *Adj.* Asian; **Asien** ['a:ziən] *(das)* ~s Asia

Askese die; ~: asceticism; **Asket** der; ~en, ~en ascetic; **asketisch 1.** *Adj.* ascetic; **2.** *adv.* ascetically

asozial 1. *Adj.* asocial; **2.** *adv.* asocially

Aspekt der; ~[e]s, ~e aspect

Asphalt der; ~[e]s, ~e asphalt

Aspik der *(österr. auch* das); ~s, ~e aspic

aß *1. u. 3. Pers. Sg. Prät. v.* essen

Assistent der; ~en, ~en, **Assistentin** die; ~, ~nen assistant

Ast der; ~[e]s, **Äste** branch; **sich** *(Dat.)* **einen** ~ **lachen** *(ugs.)* split one's sides [with laughter]

Aster die; ~, ~n aster; *(Herbst~)* Michaelmas daisy

ästhetisch 1. *Adj.* aesthetic; **2.** *adv.* aesthetically

Asthma das; ~s asthma

Astrologe der; ~n, ~n astrologer; **Astrologie** die; ~: astrology *no art.;* **Astrologin** die; ~, ~nen astrologer

Astronaut der; ~en, ~en, **Astronautin** die; ~, ~nen astronaut

Astronom der; ~en, ~en astronomer;

Astronomie die; ~: astronomy *no art.;* **astronomisch** *Adj.* astronomical

Asyl das; ~s, ~e a) asylum; b) *(Obdachlosen~)* hostel; **Asylant** der; ~en, ~en, **Asylantin** die; ~, ~nen person granted [political] asylum; **Asyl·bewerber** der person seeking [political] asylum

Atelier [ata'lie:] das; ~s, ~s studio

Atem der; ~s breath; **außer** ~ **sein/geraten** be/get out of breath

atem-, Atem-: ~**beraubend 1.** *Adj.* breath-taking; **2.** *adv.* breathtakingly; ~**los 1.** *Adj.* breathless; **2.** *adv.* breathlessly; ~**pause** die breathing space; ~**zug** der breath

Atheismus der; ~: atheism *no art.;* **Atheist** der; ~en, ~en atheist; **atheistisch 1.** *Adj.* atheistic; **2.** *adv.* atheistically

Athen (das); ~s Athens

Äther der; ~s, ~ ether

Äthiopien [ɛ'tio:piən] (das); ~s Ethiopia

Athlet der; ~en, ~en a) *(Sportler)* athlete; b) *(ugs.: kräftiger Mann)* muscleman; **athletisch** *Adj.* athletic

Atlanten s. ¹**Atlas**

Atlantik der; -s Atlantic; **atlantisch** *Adj.* Atlantic; **der Atlantische Ozean** the Atlantic Ocean

Atlas der; ~ *od.* ~ses, **Atlanten** *od.* ~se atlas

atmen *itr., tr. V.* breathe

Atmosphäre [atmo'sfɛ:rə] die; ~, ~n *(auch fig.)* atmosphere

Atmung die; ~: breathing

Atom das; ~s, ~e atom; **atomar** *Adj.* atomic; *(Atomwaffen betreffend)* nuclear

Atom-: ~**bombe** die atom bomb; ~**energie** die; *o. Pl.* nuclear energy *no indef. art.;* ~**kern** der atomic nucleus; ~**kraft** die; *o. Pl.* nuclear power *no indef. art.;* ~**kraftwerk** das nuclear power-station; ~**krieg** der nuclear war; ~**müll** der nuclear waste; ~**physik** die nuclear physics *sing., no art.;* ~**pilz** der mushroom cloud; ~**reaktor** der nuclear reactor; ~**waffe** die nuclear weapon; ~**waffen·frei** *Adj.* nuclear-free; ~**zeitalter** das; *o. Pl.* nuclear age

Attacke die; ~, ~n *(auch Med.)* attack **(auf** + *Akk.* on)

Attentat das; ~[e]s, ~e assassination attempt; *(erfolgreich)* assassination; **Attentäter** der; ~s, ~, **Attentäte-**

rin die; ~, ~nen would-be assassin; *(erfolgreich)* assassin

Attest das; ~[e]s, ~e medical certificate

Attraktion die; ~, ~en attraction; **attraktiv 1.** *Adj.* attractive; **2.** *adv.* attractively; **Attraktivität** die; ~: attractiveness

Attrappe die; ~, ~n dummy

Attribut das; ~[e]s, ~e attribute

ätzen 1. *tr. V.* etch; **2.** *itr. V.* corrode; **ätzend 1.** *Adj.* corrosive; *(fig.)* caustic ⟨*wit, remark, criticism*⟩; pungent ⟨*smell*⟩; **2.** *adv.* caustically ⟨*ironic, critical*⟩

au *Interj.* **a)** *(bei Schmerz)* ouch; **b)** *(bei Überraschung, Begeisterung)* oh

Aubergine [obɛr'ʒi:nə] die; ~, ~n aubergine *(Brit.);* egg-plant

auch 1. *Adv.* **a)** as well; too; also; **Klaus war** ~ **dabei** Klaus was there as well *or* too; Klaus was also there; **Ich gehe jetzt. – Ich** ~: I'm going now – So am I; **Mir ist warm. – Mir** ~: I feel warm – So do I; **das weiß ich** ~ **nicht** I don't know either; **b)** *(sogar, selbst)* even; ~ **wenn, wenn** ~: even if; **2.** *Partikel* **a)** **etwas anderes habe ich** ~ **nicht erwartet** I never expected anything else; **nun hör aber** ~ **zu!** now listen!; **b) bist du dir** ~ **im klaren, was das bedeutet?** are you sure you unterstand what that means?; **bist du** ~ **glücklich?** are you truly happy?; **lügst du** ~ **nicht?** you're not lying, are you?; **c) wo .../wer .../was ...** *usw.* ~: wherever/whoever/whatever *etc. ...;* **wie dem** ~ **sei** however that may be; **d) mag er** ~ **noch so klug sein** no matter how clever he is

Audienz die; ~, ~en audience

auf 1. *Präp. mit Dat.* **a)** on; ~ **See** at sea; ~ **dem Baum** in the tree; ~ **der Erde** on earth; ~ **der Welt** in the world; ~ **der Straße** in the street; **b)** at ⟨*post office, town, hall, police station*⟩; ~ **seinem Zimmer** *(ugs.)* in his room; **Geld** ~ **der Bank haben** have money in the bank; ~ **der Schule/Uni** at school/university; **c)** at ⟨*party, wedding*⟩; on ⟨*course, trip, walk, holiday, tour*⟩; **2.** *Präp. mit Akk.* **a)** on; ~ **einen Berg steigen** climb up a mountain; ~ **die Straße gehen** go [out] into the street; **b)** ~ **die Schule/Uni gehen** go to school/university; ~ **einen Lehrgang gehen** go on a course; **c)** ~ **10 km** [**Entfernung**] for [a distance of] 10 km; **wir näherten uns der Hütte** [**bis**] ~ **30 m** we

approached to within 30 m of the hut;
d) ~ Jahre |hinaus| for years [to come];
etw. ~ nächsten Mittwoch verschieben
postpone sth. until next Wednesday;
die Nacht von Sonntag ~ Montag Sunday night; **das fällt ~ einen Montag** it
falls on a Monday; **e) ~ diese Art und
Weise** in this way; **~ deutsch** in German; **~ das sorgfältigste** *(geh.)* most
carefully; **f) ~ Wunsch** on request; **~
meine Bitte** at my request; **~ Befehl** on
command; **g) ein Teelöffel ~ einen Liter Wasser** one teaspoon to one litre
of water; **~ die Sekunde/den Millimeter |genau|** [precise] to the second/
millimetre; **~ deine Gesundheit!** your
health; **~ bald/morgen!** *(bes. südd.)*
see you soon/tomorrow; **3.** *Adv.* **a) ~!**
(steh/steht auf!) up you get!; **b) sie waren längst ~ und davon** they had made
off long before; **c) ~!** *(bes. südd.: los)*
come on; **~ geht's** off we go; **~ ins
Schwimmbad!** come on, off to the
swimming-pool!; **d) ~ und ab** *(hin und
her)* up and down; to and fro; **e)
Helm/Hut/Brille ~!** helmet/hat/
glasses on!; **f) Fenster/Mund ~!** open
the window/your mouth!

auf|atmen *itr. V.* breathe a sigh of relief

auf|bahren *tr. V.* lay out; **aufgebahrt
sein** lie in state

Auf·bau der; **~|e|s, ~ten a)** *o. Pl.*
building; **b)** *o. Pl. (Struktur)* structure;
c) *Pl. (Schiffbau)* superstructure *sing.*

auf|bauen *tr. V.* **a)** erect ⟨*hut, kiosk,
podium*⟩; set up ⟨*equipment, train set*⟩;
build ⟨*house, bridge*⟩; put up ⟨*tent*⟩; **b)**
(hinstellen, arrangieren) lay *or* set out
⟨*food, presents, etc.*⟩; **c)** *(fig.: schaffen)*
build ⟨*state, economy, etc.*⟩; build up
⟨*business, organization, army, spy network*⟩; **d)** *(fig.: strukturieren)* structure

auf|bäumen *refl. V.* rear up; **sich gegen jmdn./etw. ~** *(fig.)* rise up against
sb./sth.

auf|bessern *tr. V.* improve; increase
⟨*pension, wages, etc.*⟩

auf|bewahren *tr. V.* keep; **etw. kühl
~:** store sth. in a cool place; **Auf·bewahrung** die keeping

auf|bieten *unr. tr. V.* exert ⟨*strength,
energy, will-power, influence, authority*⟩; call on ⟨*skill, wit, powers of persuasion or eloquence*⟩

auf|blasen *unr. tr. V.* blow up; inflate

auf|bleiben *unr. itr. V.; mit sein* **a)** *(geöffnet bleiben)* stay open; **b)** *(nicht zu
Bett gehen)* stay up

auf|blenden *itr. V.* switch to full
beam

auf|blicken *itr. V.* **a)** look up; *(kurz)*
glance up; **b) zu jmdm. ~** *(fig.)* look up
to sb.

auf|blühen *itr. V.; mit sein* **a)** come
into bloom; ⟨*bud*⟩ open; **b)** *(fig.: aufleben)* blossom [out]

auf|brauchen *tr. V.* use up

auf|brechen **1.** *unr. tr. V.* break open
⟨*lock, safe, box, crate, etc.*⟩; break into
⟨*car*⟩; force [open] ⟨*door*⟩; **2.** *unr. itr.
V.; mit sein* **a)** ⟨*bud*⟩ open; ⟨*ice [sheet],
surface, ground*⟩ break up; ⟨*wound*⟩
open; **b)** *(losgehen, -fahren)* set off

auf|bringen *unr. tr. V.* **a)** find; raise
⟨*money*⟩; *(fig.)* summon [up] ⟨*strength,
energy, courage*⟩; find ⟨*patience*⟩; **b)**
(kreieren) start ⟨*fashion, custom, rumour*⟩; introduce ⟨*slogan, theory*⟩; **c)**
jmdn. ~: make sb. angry; **d) jmdn. gegen jmdn./etw. ~:** set sb. against sb./
sth.

Auf·bruch der departure

auf|brühen *tr. V.* brew [up]

auf|decken *tr. V.* **a)** uncover; **b)** *(Kartenspiele)* show; **c)** *(fig.)* reveal; uncover; *(enthüllen)* expose

auf|drängen **1.** *tr. V.* **jmdm. etw. ~:**
force sth. on sb.; **2.** *refl. V.* **sich jmdm.
~:** force oneself on sb.

auf|drehen *tr. V.* **a)** unscrew ⟨*bottlecap, nut*⟩; undo ⟨*screw*⟩; turn on ⟨*tap,
gas, water*⟩; open ⟨*valve, bottle, vice*⟩;
b) *(ugs.)* turn up ⟨*radio, record-player,
etc.*⟩

auf·dringlich **1.** *Adj.* pushy *(coll.)*
⟨*person*⟩; *(fig.)* insistent ⟨*music, advertisement*⟩; pungent ⟨*perfume, smell*⟩;
loud ⟨*colour, wallpaper*⟩; **2.** *adv.* ⟨*behave*⟩ pushily, *(coll.)*; **Aufdringlichkeit** die; **~s. aufdringlich:** pushiness
(coll.); insistent manner; pungency

auf·einander *Adv.* on top of one another

aufeinander-: **~|folgen** *itr. V.; mit
sein* follow one another; **~folgend**
successive; **~|legen** **1.** *tr. V.* lay
⟨*planks etc.*⟩ one on top of the other; **2.** *refl. V.* lie on top of one another;
~|liegen *unr. itr. V.* lie on top of each
other *or* one another; **~|prallen** *itr.
V.; mit sein* crash into one another;
collide; *(fig.)* ⟨*opinions*⟩ clash; **~|treffen** *unr. itr. V.; mit sein (fig.)* meet

Aufenthalt der; **~|e|s, ~e a)** stay; **b)**
(Fahrtunterbrechung) stop

Aufenthalts-: **~erlaubnis** die
residence permit; **~raum** der *(in einer*

Schule o. ä.) common-room *(Brit.); (in einer Jugendherberge)* day-room; *(in einem Betrieb o. ä.)* recreation-room

auf|essen *unr. tr. (auch itr.) V.* eat up

auf|fahren 1. *unr. itr. V.; mit sein* **a)** **auf ein anderes Fahrzeug ~** *(aufprallen)* drive into the back of another vehicle; **b) auf den Vordermann zu dicht ~:** drive too close to the car in front; **c)** *(vorfahren)* drive up; **d)** *(in Stellung gehen)* move up [into position]; **2.** *unr. tr. V.* **a)** *(in Stellung bringen)* move up; **b)** *(ugs.: auftischen)* serve up; **Auf·fahrt die a)** drive up; **b)** *(Weg)* drive; **c)** *(Autobahn~)* slip-road *(Brit.);* access road *(Amer.);* **d)** *(schweiz.) s.* **Himmelfahrt**

auf|fallen *unr. itr. V.; mit sein* stand out; **jmdm. fällt etw. auf** sb. notices sth.; **auffallend 1.** *Adj.* conspicuous; *(eindrucksvoll, bemerkenswert)* striking; **2.** *adv.* conspicuously; *(eindrucksvoll, bemerkenswert)* strikingly; **auf·fällig 1.** *Adj.* conspicuous; garish *(colour);* **2.** *adv.* conspicuously

auf|fangen *unr. tr. V.* **a)** catch; **b)** *(aufnehmen, sammeln)* collect

auf|fassen *tr. V.* grasp; **a) etw. als etw. ~:** regard sth. as sth.; **etw. persönlich/falsch ~:** take sth. personally/misunderstand sth.; **Auf·fassung die** *(Ansicht)* view; *(Begriff)* conception; **der ~ sein, daß ...:** take the view that ...

auffindbar *Adj.* findable; **auf|finden** *unr. tr. V.* find

auf|fordern *tr. V.* **jmdn. ~, etw. zu tun** call upon sb. to do sth.; *(einladen, ermuntern)* ask sb. to do sth.; **jmdn. |zum Tanz| ~:** ask sb. to dance; **Auf·forderung die** request; *(nachdrücklicher)* demand; *(Einladung, Ermunterung)* invitation

auf|fressen *unr. tr. V. (auch fig.)* eat up

auf|führen 1. *tr. V.* **a)** put on *(film);* stage *(play, ballet, opera);* perform *(piece of music);* **b)** *(auflisten)* list; **2.** *refl. V.* behave; **Auf·führung die** performance

Auf·gabe die a) task; **b)** *(fig.: Zweck, Funktion)* function; **c)** *(Schulw.) (Übung)* exercise; *(Prüfungs~)* question; *(Haus~) s.* **Haus~; d)** *(Rechen~, Mathematik~)* problem; **e)** *(Kapitulation)* retirement; *(im Schach)* resignation; **jmdn. zur ~ zwingen** force sb. to retire/resign; **f)** *(das Aufgeben a)* giving up; **g)** *(einer Postsendung)* posting *(Brit.);* mailing *(Amer.);* *(eines Tele-*

gramms) handing in; *(einer Bestellung, einer Annonce)* placing; **h)** *(von Gepäck)* checking in

Auf·gang der a) *(eines Gestirns)* rising; **b)** *(Treppe)* stairs *pl.;* staircase; stairway; *(in einem Bahnhof, zu einer Galerie, einer Tribüne)* steps *pl.*

auf|geben 1. *unr. tr. V.* **a)** give up; *(Sport)* retire from *(race, competition);* **b)** *(übergeben, übermitteln)* post *(Brit.),* mail *(letter, parcel);* hand in, *(telefonisch)* phone in *(telegram);* place *(advertisement, order);* check *(luggage, baggage)* in; **c)** *(Schulw.: als Hausaufgabe)* set *(Brit.);* assign *(Amer.);* **d)** **jmdm. ein Rätsel ~:** set *(Brit.) or (Amer.)* assign sb. a puzzle; **2.** *unr. itr. V.* **a)** give up; *(im Sport)* retire; *(im Schach)* resign

Auf·gebot das a) contingent; **ein gewaltiges ~ an Polizisten/Fahrzeugen/Material** a huge force of police/array of vehicles/materials; **b)** *(zur Heirat)* notice of an/the intended marriage; *(kirchlich)* banns *pl.*

auf|gehen *unr. itr. V.; mit sein* **a)** rise; **b)** *(sich öffnen [lassen]) (door, parachute, wound)* open; *(stage curtain)* go up; *(knot, button, zip, bandage, shoelace, stitching)* come undone; *(boil, pimple, blister)* burst; *(flower, bud)* open [up]; **c)** *(keimen)* come up; **d)** *(aufgetrieben werden) (dough, cake)* rise; **e)** *(Math.) (calculation)* work out; *(equation)* come out; **f)** **etw. geht jmdm. auf** sb. realizes sth.

aufgeklärt *Adj.* enlightened; **~ sein** *(sexualkundlich)* know the facts of life

auf·gelegt *Adj.* **gut/schlecht usw. ~ sein** be in a good/bad *etc.* mood; **zu etw. ~ sein** be in the mood for sth.

auf·gelöst *Adj.* distraught *(person)*

aufgeregt 1. *Adj.* excited; *(nervös, beunruhigt)* agitated; **2.** *adv.* exitedly; *(nervös, beunruhigt)* agitatedly

auf·geschlossen *Adj.* open-minded *(gegenüber* as regards, about); *(interessiert, empfänglich)* receptive *(Dat., für* to); *(zugänglich)* approachable; **Auf·geschlossenheit die** *s.* **aufgeschlossen:** open-mindedness; receptiveness; approachableness

aufgeweckt *Adj.* bright; **Aufgewecktheit die; ~:** brightness

auf|gießen *unr. tr. V.* make *(coffee, tea)*

auf|gliedern *tr. V.* subdivide, break down **(in +** *Akk.* into); **Auf·gliederung die** subdivision; breakdown

auf|greifen *unr. tr. V.* pick up
auf Grund, aufgrund *s.* **Grund** c
auf|haben *(ugs.)* **1.** *unr. tr. V.* **a)** *(aufgesetzt haben)* have on; **b)** *(geöffnet haben)* have ⟨zip⟩ undone; have ⟨door, window, jacket, blouse⟩ open; **2.** *unr. itr. V.* ⟨shop, office⟩ be open
auf|halten 1. *unr. tr. V.* **a)** halt; **b)** *(stören)* hold up; **c)** *(ugs.: geöffnet halten)* hold ⟨sack, door, etc.⟩ open; **die Augen [und Ohren]** ~: keep one's eyes [and ears] open; **2.** *unr. refl. V.* **a)** stay; **b)** **sich mit jmdm./etw.** ~: spend [a long] time on sb./sth.
auf|hängen 1. *tr. V.* **a)** hang up; hang ⟨picture, curtains⟩; **b)** *(erhängen)* hang; **2.** *refl. V.* hang oneself; **Aufhänger** der; ~s, ~ loop
auf|heben *unr. tr. V.* **a)** pick up; **b)** *(aufbewahren)* keep; **c)** *(abschaffen)* abolish; repeal ⟨law⟩; rescind ⟨order, instruction⟩; cancel ⟨contract⟩; lift ⟨ban, prohibition⟩; **d)** *(ausgleichen)* cancel out; neutralize ⟨effect⟩; **Aufheben das;** ~s: **viel** ~**[s]/kein** ~ **von jmdm./etw. machen** make a great fuss/not make any fuss about sb./sth.
auf|heitern 1. *tr. V.* cheer up; **2.** *refl. V.* ⟨weather⟩ brighten up
auf|hetzen *tr. V.* incite
auf|holen 1. *tr. V.* make up ⟨time, delay⟩; pull back ⟨lead⟩; **2.** *itr. V.* catch up; ⟨athlete, competitor⟩ make up ground
auf|horchen *itr. V.* prick up one's ears
auf|hören *itr. V.* stop; **[damit]** ~, **etw. zu tun** stop doing sth.
auf|kaufen *tr. V.* buy up
auf|klappen *tr. V.* open, fold open ⟨chair, table⟩; open [up] ⟨suitcase, trunk⟩; open ⟨book, knife⟩
auf|klären 1. *tr. V.* **a)** clear up ⟨matter, mystery, question, misunderstanding, error, confusion⟩; solve ⟨crime, problem⟩; explain ⟨event, incident, cause⟩; resolve ⟨contradiction, disagreement⟩; **b)** *(unterrichten)* enlighten; **ein Kind** ~ *(sexualkundlich)* tell a child the facts of life; **2.** *refl. V.* **a)** ⟨misunderstanding, mystery⟩ be cleared up; **b)** ⟨weather⟩ brighten [up]; ⟨sky⟩ brighten; **Auf·klärung die** *s.* **aufklären 1: a)** clearing up; solution; explanation; resolution; **b)** enlightenment; **die** ~ **der Kinder** *(über Sexualität)* telling the children the facts of life
auf|kleben *tr. V.* stick on; *(mit Kle-*

ster) paste on; **Auf·kleber** der sticker
auf|knöpfen *tr. V.* unbutton; undo
auf|kochen 1. *tr. V.* bring to the boil; **2.** *itr. V. mit sein* come to the boil
auf|kommen *unr. itr. V.; mit sein* **a)** ⟨wind⟩ spring up; ⟨storm, gale⟩ blow up; ⟨fog⟩ come down; ⟨rumour⟩ start; ⟨suspicion, doubt, feeling⟩ arise; ⟨fashion, style, invention⟩ come in; ⟨boredom⟩ set in; ⟨mood, atmosphere⟩ develop; **b)** ~ **für** *(bezahlen)* bear ⟨costs⟩; pay for ⟨damage⟩; pay ⟨expenses⟩; be liable for ⟨debts⟩; stand ⟨loss⟩; **c)** ~ **für** *(Verantwortung tragen für)* be responsible for
auf|krempeln *tr. V.* roll up
auf|laden 1. *unr. tr. V.* **a)** load **(auf +** *Akk.* **on [to]);** **b)** **jmdm. etw.** ~ *(ugs.)* load sb. with sth.; *(fig.)* saddle sb. with sth.; **c)** charge [up] ⟨battery⟩; **2.** *unr. refl. V.* ⟨battery⟩ charge
Auf·lage die a) *(Buchw.)* edition; **b)** *(Verpflichtung)* condition
auf|lassen *unr. tr. V.* *(ugs.)* **a)** leave ⟨door, window, jacket, etc.⟩ open; **b)** keep on ⟨hat, glasses, etc.⟩
auf|lauern *itr. V.* **jmdm.** ~: lie in wait for sb.
Auf·lauf der a) *(Menschen~)* crowd; **b)** *(Speise)* soufflé
auf|leben *itr. V.; mit sein* revive; *(fig.: wieder munter werden)* come to life
auf|legen 1. *tr. V.* **a)** put on; **den Hörer** ~: put down the receiver; **b)** *(Buchw.)* publish; **2.** *itr. V. (den Hörer* ~**)** hang up
auf|lehnen *refl. V.* rebel; **Auflehnung der;** ~, ~**en** rebellion
auf|leuchten *itr. V.; auch mit sein* light up; *(für kurze Zeit)* flash
auf|lockern *tr. V.* **a)** loosen; break up ⟨soil⟩; **b)** *(fig.)* introduce some variety into ⟨landscape, lesson, lecture⟩; relieve ⟨pattern, façade⟩; make ⟨mood, atmosphere, evening⟩ more relaxed; **Auf·lockerung die a)** *s.* **auflockern a:** loosening; breaking up; **b) zur** ~ **der Stimmung/des Abends** to make the mood/evening more relaxed
auf|lösen 1. *tr. V.* dissolve; resolve ⟨difficulty, contradiction⟩; solve ⟨puzzle, equation⟩; break off ⟨engagement⟩; cancel ⟨arrangement, contract, agreement⟩; dissolve ⟨organization⟩; **2.** *refl. V.* dissolve **(in +** *Akk.* into); ⟨parliament⟩ dissolve itself; ⟨crowd, demonstration⟩ break up; ⟨fog, mist⟩ lift; *(fig.)* ⟨empire, social order⟩ dis-

integrate; **Auf·lösung die a)** *s.* **auflösen 1:** dissolving; resolution; solution; breaking off; cancellation; dissolution; **b)** *s.* **auflösen 2:** dissolving; breaking up lifting; disintegration

auf|machen 1. *tr. V.* **a)** open; undo ⟨*button, knot*⟩; **b)** *(ugs.: eröffnen)* open [up] ⟨*shop, business, etc.*⟩; **2.** *itr. V.* **a)** ⟨*shop, office, etc.*⟩ open; **b)** *(ugs.: die Tür öffnen)* open the door; **jmdm. ~:** open the door to sb.; **c)** *(ugs.: eröffnet werden)* ⟨*shop, business*⟩ open [up]; **Aufmachung die; ~, ~en** presentation; *(Kleidung)* get-up

auf|marschieren *itr. V.; mit sein* assemble; *(heranmarschieren)* march up; **Truppen sind an der Grenze aufmarschiert** troops were deployed along the border

aufmerksam 1. *Adj.* **a)** attentive; sharp ⟨*eyes*⟩; **jmdn. auf jmdn./etw. ~ machen** draw sb.'s attention to sth.; **auf jmdn./etw. ~ werden** become aware of sb./sth.; **~ werden** notice; **b)** *(höflich)* attentive; **2.** *adv.* attentively; **Aufmerksamkeit die; ~, ~en a)** *o. Pl.* attention; **b)** *(Höflichkeit)* attentiveness; **c)** *(Geschenk)* small gift

auf|muntern *tr. V.* **a)** cheer up; **b)** *(beleben)* liven up; **c)** *(ermutigen)* encourage; **Aufmunterung die; ~** *s.* **aufmuntern:** cheering up; livening up; encouragement

Aufnahme die; ~, ~n a) *s.* **aufnehmen b:** opening; establishment; taking up; **b)** *(Empfang)* reception; **c)** *s.* **aufnehmen d:** admission (**in** + *Akk.* into); **d)** *(Einschließung)* inclusion; **e)** *(Finanzw.)* raising; **f)** *(Aufzeichnung)* taking down; *(von Personalien, eines Diktats)* taking [down]; **g)** *s.* **aufnehmen k:** taking: photographing; filming; **h)** *(Bild)* shot; **i)** *(das Aufnehmen auf Tonträger, das Aufgenommene)* recording; **j)** *(Anklang)* reception; response *(Gen.* to); **k)** *(Einverleibung, Absorption)* absorption

auf|nehmen *unr. tr. V.* **a)** *(aufheben)* pick up; *(fig.)* take up ⟨*idea, theme, etc.*⟩; **es mit jmdm./etw. ~/nicht ~ können** *(fig.)* be a/no match for sb./ sth.; **b)** *(beginnen mit)* open ⟨*negotiations, talks*⟩; establish ⟨*relations, contacts*⟩; take up ⟨*studies, activity, occupation*⟩; start ⟨*production, investigation*⟩; **c)** *(empfangen)* receive; *(beherbergen)* take in; **d)** *(beitreten lassen)* admit (**in** + *Akk.* to); **e)** *(einschließen, verzeichnen)* include; **f)** *(erfassen)*

take in ⟨*impressions, information, etc.*⟩; **g)** *(absorbieren)* absorb; **h)** *(Finanzw.)* raise ⟨*mortgage, money, loan*⟩; **i)** *(reagieren auf)* receive; **j)** *(aufschreiben)* take down; take [down] ⟨*dictation, particulars*⟩; **k)** *(fotografieren)* take ⟨*picture*⟩; photograph, take a photograph of ⟨*scene, subject*⟩; *(filmen)* film; **l)** *(auf Tonträger)* record

auf|opfern *refl. V.* devote oneself sacrificingly (**für** to); **aufopfernd 1.** *Adj.* self-sacrificing; **2.** *adv.* self-sacrificingly

auf|passen *itr. V.* **a)** watch out; *(konzentriert sein)* pay attention; **paß mal auf!** *(ugs.: hör mal zu!)* now listen; **b)** **auf jmdn./etw. ~:** keep an eye on sb./ sth.

auf|platzen *itr. V.; mit sein* burst open; ⟨*seam, cushion*⟩ split open; ⟨*wound*⟩ open up

Auf·prall der; ~[e]s, ~e impact; **auf|prallen** *itr. V.; mit sein* **auf etw.** *(Akk.)* **~:** hit sth.

Auf·preis der additional charge

auf|pumpen *tr. V.* pump up

auf|putschen *tr. V.* stimulate; arouse ⟨*passions, urge*⟩; **Aufputsch·mittel das** stimulant

auf|räumen *tr., itr. V.* clear up

auf·recht 1. *Adj. (auch fig.)* upright; **2.** *adv.* ⟨*walk, sit, hold oneself*⟩ straight; **aufrecht|erhalten** *unr. tr. V.* maintain; keep up ⟨*deception, fiction, contact, custom*⟩

auf|regen 1. *tr. V.* excite; *(ärgern)* annoy; irritate; *(beunruhigen)* agitate; **2.** *refl. V.* get worked up (**über** + *Akk.* about); **Auf·regung die** excitement *no pl.; (Beunruhigung)* agitation *no pl.;* **jmdn. in ~ versetzen** make sb. excited/agitated

auf|reißen 1. *unr. tr. V.* **a)** *(öffnen)* tear open; wrench open ⟨*drawer*⟩; fling open ⟨*door, window*⟩; **die Augen/ den Mund ~:** open one's eyes/mouth wide; **b)** *(beschädigen)* tear open; tear ⟨*clothes*⟩; break up ⟨*road, soil*⟩; **2.** *itr. V.; mit sein* ⟨*clothes*⟩ tear; ⟨*seam*⟩ split; ⟨*wound*⟩ open; ⟨*cloud*⟩ break up

auf|reizen *tr. V.* excite; **auf·reizend 1.** *Adj.* provocative; **2.** *adv.* provocatively

auf|richten 1. *tr. V.* erect; put up; **den Oberkörper ~:** raise one's upper body; **jmdn. [wieder] ~** *(fig.)* give fresh heart to sb.; **2.** *refl. V.* stand up [straight]; **sich an jmdm./etw. [wieder] ~** *(fig.)* take heart from sb./sth.

auf·richtig 1. *Adj.* sincere; 2. *adv.* sincerely; **Auf·richtigkeit die** sincerity

auf|rücken *itr. V.; mit sein* move up

Auf·ruf der a) call; **b)** *(Appell)* appeal **(an** + *Akk.* to); **auf|rufen** *unr. tr. V.* **a)** call; **b)** jmdn. ~, etw. zu tun call upon sb. to do sth.; **c)** *(Rechtsw.)* appeal for ⟨witnesses⟩

Aufruhr der; ~s, ~e **a)** *(Widerstand)* rebellion; **b)** *o. Pl. (Erregung)* turmoil; **aufrührerisch** *Adj.* inflammatory

Auf·rüstung die armament

aufs *Präp.* + *Art.* = **auf das**

auf|sagen *tr. V.* recite

auf|sammeln *tr. V.* gather up

aufsässig 1. *Adj.* recalcitrant; 2. *adv.* recalcitrantly

Auf·satz der *(Text)* essay

auf|saugen *unr. (auch regelm.) tr. V.* soak up; *(fig.)* absorb

auf|schieben *unr. tr. V.* postpone

Auf·schlag der a) *(Aufprall)* impact; **b)** *(Preis~)* surcharge; **c)** *(Ärmel~)* cuff; *(Hosen~)* turn-up; *(Revers)* lapel; **d)** *(Tennis usw.)* serve

auf|schlagen 1. *unr. itr. V.* **a)** *mit sein* **auf etw.** *(Dat. od. Akk.)* ~: hit sth.; **b)** *(teurer werden)* ⟨price, rent, costs⟩ go up; **c)** *(Tennis usw.)* serve.; 2. *unr. tr. V.* **a)** *(öffnen)* crack ⟨nut, egg⟩ [open]; knock a hole in ⟨ice⟩; **sich** *(Dat.)* **das Knie/den Kopf** ~: cut one's knee/head; **b)** open ⟨book, newspaper, one's eyes⟩; **schlagt S. 15 auf!** turn to page 15; **c)** turn up ⟨collar, sleeve, trouserleg⟩; **d)** *(aufbauen)* set up ⟨camp⟩; pitch ⟨tent⟩; put up ⟨bed, hut, scaffolding⟩; **e)** **5% auf etw.** *(Akk.)* ~: put 5% on sth.

auf|schließen 1. *unr. tr. V.* unlock; 2. *unr. itr. V.* |jmdm.| ~: unlock the door/gate *etc.* [for sb.]; **Auf·schluß der** information *no pl.*

auf|schneiden 1. *unr. tr. V.* **a)** cut open; **b)** *(zerteilen)* cut; 2. *unr. itr. V.* *(ugs.: prahlen)* boast **(mit** about); **Auf·schnitt der;** *o. Pl.* [assorted] cold meats *pl.*/cheeses *pl.*

auf|schnüren *tr. V.* undo

auf|schrauben *tr. V.* unscrew; unscrew the top of ⟨bottle, jar, etc.⟩

auf|schreiben *unr. tr. V.* write down; [sich *(Dat.)*] etw. ~: make a note of sth.; **Auf·schrift die** inscription

Auf·schub der postponement; **die Sache duldet keinen** ~: the matter brooks no delay

Auf·schwung der upturn *(Gen.* in)

Aufsehen das; ~s stir; |großes| ~ erregen cause a [great] stir; **Auf·seher der** *(im Gefängnis)* warder *(Brit.)*; [prison] guard *(Amer.)*; *(im Park)* park-keeper; *(im Museum, auf dem Parkplatz)* attendant; *(auf einem Gut, Sklaven~)* overseer

auf|sein *unr. itr. V.; mit sein; nur im Inf. und Part. zusammengeschrieben (ugs.)* **a)** be open; **b)** *(nicht im Bett sein)* be up

auf|setzen 1. *tr. V.* **a)** put on; **b)** *(verfassen)* draw up ⟨text⟩; 2. *refl. V.* sit up

Auf·sicht die supervision; *(bei Prüfungen)* invigilation *(Brit.)*; proctoring *(Amer.)*

auf|springen *unr. itr. V.; mit sein* **a)** jump up; **b)** *(hinaufspringen)* jump on **(auf** + *Akk.* to); **c)** *(rissig werden)* crack

Auf·stand der rebellion; **auf·ständisch** *Adj.* rebellious

auf|stehen *unr. itr. V. mit sein* stand up; *(aus dem Liegen)* get up

auf|steigen *unr. itr. V.; mit sein* **a)** *(auf ein Fahrzeug)* get on; **auf etw.** *(Akk.)* ~: get on [to] sth.; **b)** *(bergan steigen)* climb; **c)** *(hochsteigen)* ⟨sap, smoke, mist⟩ rise; **d)** *(beruflich, gesellschaftlich)* rise **(zu** to); **zum Direktor** ~: rise to be manager

auf|stellen 1. *tr. V.* **a)** put up **(auf** + *Akk.* on); set up ⟨skittles⟩; *(postieren)* post; **b)** *(aufrecht hinstellen)* stand up; **c)** *(Sport)* select, pick ⟨team, player⟩; **d)** *(bilden)* put together ⟨team of experts⟩; raise ⟨army⟩; **e)** *(nominieren)* nominate; put up; 2. *refl. V.* position oneself; **Auf·stellung die a)** *s.* **aufstellen 1a:** putting up; setting up; posting; **b)** *s.* **aufstellen b:** standing up; **c)** *s.* **aufstellen c:** selection; picking; **c)** *s.* **aufstellen d:** putting together; raising; **d)** *(Nominierung)* nomination

Aufstieg der; ~|e|s, ~e **a)** climb; **b)** *s.* **aufsteigen d:** rise

auf|stoßen 1. *unr. tr. V.* push open; 2. *unr. itr. V.* belch; ⟨baby⟩ bring up wind

Auf·strich der spread

auf|stützen 1. *tr. V.* rest ⟨one's arms etc.⟩; 2. *refl. V.* support oneself; **die Arme auf etw.** *(Akk. od. Dat.)* ~: rest one's arms on sth.

auf|suchen *tr. V.* call on; go to ⟨doctor⟩

Auf·takt der *(fig.)* start

auf|tauchen *itr. V.; mit sein* a) surface; b) *(sichtbar werden)* appear

auf|tauen 1. *tr. V.* thaw; 2. *itr. V.; mit sein (auch fig.)* thaw

auf|teilen *tr. V.* a) divide [up]; b) *(verteilen)* share out

Auftrag der; ~|e|s, Aufträge a) instructions *pl.;* in jmds. ~ *(Dat.)* on sb.'s instructions; *(für jmdn.)* on behalf of sb; b) *(Bestellung)* order; *(bei Künstlern, Architekten usw.)* commission; c) *(Mission)* task; *(Aufgabe)* job; **auf|tragen** *unr. tr. V.* a) jmdm. ~, etw. zu tun instruct sb. to do sth.; b) *(aufstreichen)* put on ⟨paint, make-up, etc.⟩; **Auftrag·geber** der client

auf|treten *unr. itr. V.; mit sein* a) tread; b) *(sich benehmen)* behave; c) *(eine Vorstellung geben)* appear; **als Zeuge/Kläger ~:** appear as a witness/a plaintiff; d) *(auftauchen)* ⟨problem, difficulty, difference of opinion⟩ arise; ⟨symptom, danger, pest⟩ appear; **Auftreten** das; ~s *(Benehmen)* manner

Auf·trieb der a) *(Physik) (statischer ~)* buoyancy; *(dynamischer ~)* lift; b) *(fig.)* impetus; **das hat ihm ~/neuen ~ gegeben** that has given him a lift/given him new impetus

Auf·tritt der a) *(Vorstellung)* appearance; b) *(Theater: das Auftreten)* entrance; *(Szene)* scene

auf|tun *unr. refl. V. (geh.)* open; *(fig.)* open up

auf|wachen *itr. V.; mit sein* wake up, awaken **(aus** from); *(aus Ohnmacht, Narkose)* come round **(aus** from)

auf|wachsen *unr. itr. V.; mit sein* grow up

Auf·wand der; ~|e|s cost; expense

auf|wärmen 1. *tr. V.* heat *or* warm up ⟨food⟩; 2. *refl. V.* warm oneself up

aufwärts *Adv.* upwards

auf|wecken *tr. V.* wake [up]; waken

auf|weichen 1. *tr. V.* soften; 2. *itr. V.; mit sein* become soft; soften up

auf·wendig 1. *Adj.* lavish; *(kostspielig)* costly; expensive; 2. *adv.* lavishly; *(kostspielig)* expensively

auf|wiegeln *tr. V.* incite; stir up

auf|wirbeln *tr. V.* swirl up

auf|wischen *tr. V.* a) wipe *or* mop up; b) *(säubern)* wipe ⟨floor⟩; *(mit Wasser)* wash ⟨floor⟩

auf|zählen *tr. V.* list; **Auf·zählung** die a) listing; b) *(Liste)* list

auf|zeichnen *tr. V.* a) record; b) *(zeichnen)* draw; **Auf·zeichnung**

die record; *(Film~, Ton~)* recording; ~en *(Notizen)* notes

auf|ziehen 1. *unr. tr. V.* a) pull open ⟨drawer⟩; open, draw [back] ⟨curtains⟩; undo ⟨zip⟩; b) wind up ⟨clock, toy, etc.⟩. 2. *unr. itr. V.; mit sein* come up; ⟨clouds, storm⟩ gather

Auf·zucht die raising; rearing

Auf·zug der a) *(Lift)* lift *(Brit.);* elevator *(Amer.);* b) *(abwertend: Aufmachung)* get-up; c) *(Theater: Akt)* act

Aug·apfel der eyeball; **Auge** das; ~s, ~n eye; **gute/schlechte ~n haben** have good/poor eyesight; **auf einem ~ blind** blind in one eye; **da wird er ~n machen** *(fig. ugs.)* his eyes will pop out of his head; **ich traute meinen ~n nicht** *(ugs.)* I couldn't believe my eyes; **ein ~ od. beide ~n zudrücken** *(fig.)* turn a blind eye; **jmdn./etw. nicht aus den ~n lassen** not take one's eyes off sb./sth.; **ins ~ gehen** *(fig. ugs.)* end in disaster; **unter vier ~n** *(fig.)* in private

Augen-: ~**arzt** der eye specialist; ~**blick** [*auch*: --'-] der *s.* ¹Moment; ~**blicklich** [*auch*: --'--] 1. *Adj.* a) *(sofortig)* immediate; b) *(gegenwärtig)* present; 2. *adv.* a) *(sofort)* at once; b) *(zur Zeit)* at the moment; ~**braue die** eyebrow; ~**lid das** eyelid; ~**zeuge** der eyewitness

August der; ~|e|s *od.* ~, ~e August

Auktion die; ~, ~en auction

Aula die; ~, Aulen *od.* ~s hall

aus 1. *Präp. mit Dat.* a) *(aus dem Inneren von)* out of; b) *(Herkunft, Quelle, Ausgangspunkt angebend, auch zeitlich)* from; ~ **Spanien/Köln** *usw.* from Spain/Cologne *etc.;* c) ~ **der Mode/Übung sein** be out of fashion/training; d) *(Grund, Ursache angebend)* out of; **etw. ~ Erfahrung wissen** know sth. from experience; ~ **Versehen** by mistake; e) *(bestehend ~)* of; *(hergestellt ~)* made of; ~ **etw. bestehen** consist of sth.; f) ~ **ihm ist ein guter Arzt geworden** he made a good doctor; 2. *Adv.* a) *(ugs.: vorbei)* ~ **jetzt!** that's enough; b) „~" *(an Lichtschaltern)* 'out'; *(an Geräten)* 'off'; c) **vom Fenster/obersten Stockwerk ~:** from the window/top storey; **von mir ~** *(ugs.)* if you like; **von sich** *(Dat.)* ~: of one's own accord

aus|atmen *itr., tr. V.* breathe out

Aus·bau der; ~|e|s a) *(Erweiterung)* extension; b) *(Ausgestaltung)* conversion **(zu** into); **aus|bauen** *tr. V.* a) *(demontieren)* remove **(aus** from); b) *(erweitern)* extend

Aus·beute die yield; **aus|beuten** tr. V. exploit

aus|bilden tr. V. **a)** train; **b)** (entwickeln) develop; **Aus·bildung** die **a)** training; **b)** (Entwicklung) development

Aus·blick der view (**auf** + Akk. of)

aus|brechen unr. itr. V.; mit sein **a)** break out (**aus** of); (fig.) break free (**aus** from); **b)** jmdm. bricht der Schweiß aus sb. breaks into a sweat; (fig.) ⟨volcano⟩ erupt; **d)** (beginnen) break out; ⟨crisis⟩ break; **e)** in Gelächter/Weinen ~: burst out laughing/crying; in Beifall/Tränen ~: burst into applause/tears

aus|breiten 1. tr. V. spread; spread [out] ⟨map, cloth, sheet, etc.⟩; open out ⟨fan, newspaper⟩; (nebeneinanderlegen) spread out; 2. refl. V. spread

Aus·bruch der **a)** (Flucht) escape (**aus** from); **b)** (Beginn) outbreak; **c)** (Gefühls~) outburst; **d)** (eines Vulkans) eruption

aus|brüten tr. V. hatch out; (im Brutkasten) incubate

Aus·dauer die stamina; **aus·dauernd** Adj. with stamina postpos.

aus|dehnen 1. tr. V. **a)** stretch; (fig.) extend (**auf** + Akk. to); (zeitlich) prolong; 2. refl. V. expand; (zeitlich) go on; **Aus·dehnung** die expansion; (fig.) extension; (zeitlich) prolongation

aus|denken unr. refl. V. sich (Dat.) etw. ~: think sth. up

Aus·druck der; ~[e]s, Ausdrücke expression; (Terminus) term; etw. zum ~ bringen express sth.; **aus|drücken** 1. tr. V. **a)** (auspressen) squeeze [juice] out; squeeze [out] ⟨lemon, grape, orange, etc.⟩; squeeze out ⟨sponge⟩; squeeze ⟨boil, pimple⟩; **b)** stub out ⟨cigarette⟩; **c)** (mitteilen) express; 2. refl. V. **a)** express oneself; **b)** (offenbar werden) be expressed; **ausdrücklich** [od. -'--] 1. Adj. express attrib. ⟨command, wish, etc.⟩; explicit ⟨reservation⟩; 2. adv. expressly; ⟨mention⟩ explicitly; **ausdrucks·los** 1. Adj. expressionless; 2. adv. expressionlessly; **ausdrucks·voll** 1. Adj. expressive; 2. adv. expressively

aus einander Adv. apart; etw. ~ schreiben write sth. as separate words **auseinander-, Auseinander-:** ~|brechen 1. unr. itr. V.; mit sein (auch fig.) break up; 2. unr. tr. V. break ⟨sth.⟩ up; ~|gehen unr. itr. V.;

mit sein **a)** part; ⟨crowd⟩ disperse; **b)** (fig.) ⟨opinions, views⟩ differ; ~|halten unr. tr. V. tell ⟨things, people⟩ apart; ~|nehmen unr. tr. V. take ⟨sth.⟩ apart; ~|setzen 1. tr. V. jmdm. etw. ~setzen explain sth. to sb.; 2. refl. V. sich mit jmdm. ~setzen have it out with sb.; sich mit etw. ~setzen concern oneself with sth.; ~setzung die ~, ~en **a)** (Streit) argument; **b)** (Kampfhandlungen) clash

Aus·fahrt die exit

Aus·fall der **a)** (das Nichtstattfinden) cancellation; **b)** (Einbuße, Verlust) loss; **c)** (eines Motors) failure; (einer Maschine, eines Autos) breakdown; **aus|fallen** unr. itr. V.; mit sein **a)** fall out; **b)** (nicht stattfinden) be cancelled; etw. ~ lassen cancel sth.; **c)** (ausscheiden) drop out; **d)** (nicht mehr funktionieren) ⟨engine, brakes, signal⟩ fail; ⟨machine, car⟩ break down; **e)** (ein bestimmtes Ergebnis zeigen) turn out; **ausfallend** Adj. [gegen jmdn.] ~ sein/werden be/become abusive [towards sb.]; **Ausfall·straße** die main road out of the/a town/city

aus·findig Adv. jmdn./etw. ~ machen find sb./sth.

Aus·flug der outing; **Ausflügler** der; ~s, ~: day-tripper; excursionist (Amer.)

Ausflugs-: ~dampfer der pleasure steamer; ~lokal das restaurant/café catering for [day-]trippers

aus|fragen tr. V. jmdn. ~: question sb., ask sb. questions (nach, über + Akk. about)

aus|fransen itr. V.; mit sein fray

Aus·fuhr die; ~, ~en s. Export; **aus|führen** tr. V. **a)** (ausgehen mit) take ⟨person⟩ out; **b)** (spazierenführen) take ⟨person, animal⟩ for a walk; **c)** (exportieren) export; **d)** (durchführen) carry out; ⟨Sport⟩ take ⟨penalty, free kick, corner⟩; **ausführlich** [auch: -'--] 1. Adj. detailed; full; 2. adv. in detail; **Aus·führung** die (Durchführung) carrying out; (Sport) taking

aus|füllen tr. V. **a)** fill; fill in ⟨form, crossword puzzle⟩; **b)** (beanspruchen, einnehmen) take up ⟨space⟩

Aus·gabe die **a)** o. Pl. giving out; (von Essen) serving; **b)** (Geld~) item of expenditure; ~n expenditure sing. (für on); **c)** (Edition) edition

Aus·gang der **a)** o. Pl. (Erlaubnis zum Ausgehen) time off; (von Soldaten) leave; **b)** (Tür ins Freie) exit (Gen.

from); **c)** *(Anat.)* outlet; **d)** *(Ende)* end; *(eines Romans, Films usw.)* ending; **e)** *o. Pl. (Ergebnis)* outcome; *(eines Wettbewerbs)* result; **ein Unfall mit tödlichem ~:** an accident with fatal consequences; **Ausgangspunkt der** starting-point

aus|geben *unr. tr. V.* **a)** give out; serve *(food, drinks)*; **b)** *(verbrauchen)* spend *(money)* (für on)

ausgebucht *Adj.* booked up

aus·gefallen *Adj.* unusual

ausgeglichen *Adj.* balanced; well-balanced *(person)*; equable *(climate)*

aus|gehen *unr. itr. V.; mit sein* **a)** go out; **b)** *(fast aufgebraucht sein)* run out; **c)** *(enden)* end; **gut/schlecht ~:** turn out well/badly; *(story, film)* end happily/unhappily; **d) von jmdm./etw. ~:** come from sb./sth.; **e) von etw. ~** *(etw. zugrunde legen)* take sth. as one's starting-point; *(etw. annehmen)* assume sth.

aus·gelassen 1. *Adj.* exuberant *(mood, person)*; lively *(party, celebration)*; *(wild)* boisterous; 2. *adv.* exuberantly; *(wild)* boisterously

aus·genommen *Konj.* except

ausgeprägt *Adj.* marked

ausgerechnet *Adv.* *(ugs.)* **~ heute/ morgen** today/tomorrow of all days; **~ hier** here of all places; **~ Sie** you of all people

aus·geschlossen *Adj.* **das ist ~:** that is out of the question

aus·geschnitten *Adj.* low-cut *(dress, blouse, etc.)*

aus·gestorben *Adj.* [wie] **~:** deserted

ausgezeichnet [*od.* '--'--] 1. *Adj.* excellent; outstanding *(expert)*; 2. *adv.* excellently

ausgiebig 1. *Adj.* substantial *(meal)*; 2. *adv.* *(profit)* handsomely; *(read)* extensively; **von etw. ~ Gebrauch machen** make full use of sth.

aus|gießen *unr. tr. V.* **a)** pour out (aus of); **b)** *(leeren)* empty

Ausgleich der; ~[e]s, ~e a) *s.* ausgleichen a: evening out; reconciliation; **b)** *(Schadensersatz)* compensation; **als** *od.* **zum ~ für etw.** to make up sth.; **aus|gleichen** *unr. tr. V.* **a)** even out; reconcile *(differences of opinions, contradictions)*; **b)** compensate for *(damage)*; make up for *(misfortune, lack)*; **etw. durch etw. ~:** make up for sth. with sth.; **sich ~:** balance out; *(sich gegenseitig aufheben)* cancel each other out

aus|graben *unr. tr. V.* dig up; *(Archäol.)* excavate; **Aus·grabung die** *(Archäol.)* excavation

aus|halten *unr. tr. V.* stand; bear; endure; withstand *(attack, load, pressure, test, wear and tear)*; **er konnte es zu Hause nicht mehr ~:** he couldn't stand it at home any more; **es ist nicht zum Aushalten** it is unbearable

aus|handeln *tr. V.* negotiate

aus|händigen *tr. V.* hand over

Aus·hang der notice

aus|heben *unr. tr. V.* dig out *(earth etc.)*; dig *(trench, grave, etc.)*

aus|helfen *unr. itr. V.* help out; **jmdm. ~:** help sb. out (mit, bei with); **Aus·hilfe die a)** *o. Pl. (das Aushelfen)* help; **b)** *s.* Aushilfskraft; **Aushilfs·kraft die** temporary worker; *(in Läden, Gaststätten)* temporary assistant; *(Sekretärin)* temporary secretary; temp *(coll.)*

aus|holen *itr. V.* **[mit dem Arm] ~:** draw back one's arm; *(zum Schlag)* raise one's arm

aus|kennen *unr. refl. V.* *(an einem Ort usw.)* know one's way around; *(in einem Fach, einer Angelegenheit usw.)* know what's what; **sie kennt sich in dieser Stadt aus** she knows her way around the town; **sich [gut] mit/in etw.** *(Dat.)* **~:** know [a lot] about sth.

Aus·klang der *(geh.)* end; **zum ~ des Festes** to end *or* close the festival

aus|kleiden *tr. V.* *(geh.)* undress; **sich ~:** undress

aus|klingen *unr. itr. V. mit sein* end

aus|klopfen *tr. V.* **a)** beat out (aus + *Dat.* of); **b)** *(säubern)* beat *(carpet)*; knock *(pipe)* out

aus|kochen *tr. V.* boil; *(keimfrei machen)* sterilize *(instruments etc.)* [in boiling water]

aus|kommen *unr. itr. V.; mit sein* **a)** manage (mit on); **b) mit jmdm. [gut] ~:** get on [well] with sb.

Auskommen das; ~s livelihood

Auskunft die; ~, Auskünfte **a)** piece of information; **Auskünfte** information *sing.; [jmdm. über etw. (Akk.)] ~* geben give [sb.] information [about sth.]; **b)** *o. Pl. (Stelle)* information desk/counter/office/centre *etc.*; *(Fernspr.)* directory enquiries *no art.* *(Brit.)*; directory information *no art.* *(Amer.)*

aus|lachen *tr. V.* laugh at

aus|laden *unr. tr. V.* unload *(goods etc.)*

Aus·lage die a) *Pl. (Unkosten)* expenses; **b)** *(ausgestellte Ware)* item on display; ~n goods on display

Aus·land das; *o. Pl.* foreign countries *pl.*; **im/ins ~:** abroad; **aus dem ~:** from abroad; **Ausländer der;** ~s, ~, **Ausländerin die;** ~, ~nen foreigner; **ausländisch** *Adj.* foreign

Auslands-: ~**aufenthalt der** stay abroad; ~**gespräch das** *(Fernspr.)* international call; ~**korrespondent der** foreign correspondent; ~**reise die** trip abroad

aus|lassen *unr. tr. V.* **a)** *(weglassen)* leave out; **b)** *(versäumen)* miss ⟨*opportunity, chance, etc.*⟩

Auslauf der a) *o. Pl.* **keinen/zuwenig~ haben** have no/too little chance to run around outside; **b)** *(Raum)* space to run around in; **aus|laufen** *unr. itr. V.; mit sein* **a)** run out **(aus** of); **b)** *(leer laufen)* empty; ⟨*egg*⟩ run out; **c)** *(in See stechen)* sail **(nach** for); **d)** *(erlöschen)* ⟨*contract, agreement, etc.*⟩ run out; **Aus·läufer der a)** *(Geogr.)* foothill *usu. in pl.*; **b)** *(Met.)* *(eines Hochs)* ridge; *(eines Tiefs)* trough

aus|legen *tr. V.* **a)** *(hinlegen)* lay out; display ⟨*goods, exhibits*⟩; **b)** etw. mit Fliesen/Teppichboden ~: tile/carpet sth.; **c)** *(leihen)* lend; **d)** *(interpretieren)* interpret; **etw. falsch ~:** misinterpret sth.; **Auslegung die;** ~, ~en interpretation

aus|leihen *unr. tr. V. s.* **leihen**

aus|liefern *tr. V.* jmdm. etw. *od.* etw. an jmdn. ~: hand sth. over to sb.

aus|löschen *tr. V.* **a)** extinguish; **b)** *(beseitigen)* erase ⟨*drawing, writing*⟩

aus|losen *tr. V.* etw. ~: draw lots for sth.

aus|lösen *tr. V.* **a)** trigger ⟨*mechanism, device, alarm, etc.*⟩; release ⟨*camera shutter*⟩; **b)** provoke ⟨*discussion, anger, laughter, reaction, outrage, heart attack*⟩; cause ⟨*sorrow, horror, surprise, disappointment, panic, war*⟩; excite, arouse ⟨*interest, enthusiasm*⟩; **Auslöser der;** ~s, ~ *(Fot.)* shutter release

aus|machen *tr. V.* **a)** *(ugs.)* put out ⟨*light, fire, cigarette, candle*⟩; switch off ⟨*television, radio, hi-fi*⟩; turn off ⟨*gas*⟩; **b)** *(vereinbaren)* agree [on]; ~, daß ...: agree that ...; **c)** *(auszeichnen, kennzeichnen)* make up; **d)** **wenig/ nichts/viel ~:** make little/no/a great difference; **e) das macht mir nichts aus** I don't mind

Aus·maß das a) *(Größe)* size; **b)** *(Grad)* extent

aus|messen *unr. tr. V.* measure up

Aus·nahme die; ~, ~n exception; **mit ~ von** with the exception of; **bei** jmdm. eine ~ machen make an exception in sb.'s case; **Ausnahme·zustand der** state of emergency; **ausnahms·weise** *Adv.* by way of an exception; **Dürfen wir mitkommen? – Ausnahmsweise [ja]** May we come too? – Yes, just this once

aus|nehmen *unr. tr. V.* **a)** ⟨*fish, rabbit, chicken*⟩; **b)** *(ausschließen von)* exclude; *(gesondert behandeln)* make an exception of

aus|nutzen, *(bes. südd., österr.)* **aus|nützen** *tr. V.* **a)** take advantage of; **b)** *(ausbeuten)* exploit

aus|packen *tr., itr. V.* unpack; *(auswickeln)* unwrap

aus|pressen *tr. V.* squeeze out ⟨*juice*⟩; squeeze ⟨*orange, lemon*⟩; *(keltern)* press ⟨*grapes etc.*⟩

aus|probieren *tr. V.* try out

Aus·puff der exhaust

aus|radieren *tr. V.* rub out; erase

aus|rauben *tr. V.* rob

aus|räumen 1. *tr. V.* **a)** clear out **(aus** of); **b)** *(fig.)* clear up; dispel ⟨*prejudice, suspicion, misgivings*⟩; **2.** *itr. V.* clear everything out

aus|rechnen *tr. V.* work out; **das kannst du dir leicht ~** *(ugs.)* you can easily work that out [for yourself]

Aus·rede die excuse; **aus|reden 1.** *itr. V.* finish [speaking]; **2.** *tr. V.* jmdm. etw. ~: talk sb. out of sth.

aus|reichen *itr. V.* be enough *or* sufficient **(zu** for); **ausreichend 1.** *Adj.* sufficient; enough; *(als Note)* fair. **2.** *adv.* sufficiently

aus|reißen 1. *unr. tr. V.* tear out; pull out ⟨*plants, weeds*⟩; **2.** *unr. itr. V.; mit sein* **a)** *(sich lösen)* come off; **b)** *(ugs.: weglaufen)* run away *(Dat.* from)

aus|renken *tr. V.* dislocate

aus|richten *tr. V.* **a)** jmdm. etw. ~: tell sb. sth.; **b)** *(einheitlich anordnen)* line up; **c)** *(erreichen)* achieve

aus|rollen *tr. V.* roll out

aus|rotten *tr. V.* eradicate

Aus·ruf der cry; **aus|rufen** *unr. tr. V.* **a)** call out; **"Schön!" rief er aus** 'Lovely', he exclaimed; **b)** *(offiziell verkünden)* proclaim; declare ⟨*state of emergency*⟩; **c)** *(zum Kauf anbieten)* cry; **Ausrufe·zeichen das** exclamation mark

aus|ruhen *refl., itr. V.* have a rest; |sich| **ein wenig/richtig** ~: rest a little/ have a good rest; **ausgeruht sein** be rested

aus|rüsten *tr. V.* equip; **Aus·rü· stung die** a) *o. Pl.* equipping; b) *(~sgegenstände)* equipment *no pl.*

aus|rutschen *itr. V.; mit sein* slip

Aus·sage die statement; **aus|sagen** 1. *tr. V.* a) say; c) *(vor Gericht, vor der Polizei)* state; *(unter Eid)* testify; 2. *itr. V.* make a statement; *(unter Eid)* testify

aus|schalten *tr. V.* a) switch *or* turn off; b) *(fig.)* eliminate; exclude ⟨*emotion, influence*⟩; dismiss ⟨*doubt, objection*⟩; shut out ⟨*feeling, thought*⟩

Aus·schank der; ~|e|s, serving

Aus·schau die: nach jmdm./etw. ~ **halten** keep a look-out for sb./sth.; **aus|schauen** *itr. V.* **nach jmdm./etw.** ~ look out for sb./sth.

aus|scheiden 1. *unr. itr. V.; mit sein* a) **aus etw.** ~: leave sth.; **aus dem Amt** ~: leave office; b) *(Sport)* be eliminated; c) **diese Möglichkeit/dieser Kandidat scheidet aus** this possibility/candidate has to be ruled out; 2. *unr. tr. V. (Physiol.)* excrete ⟨*waste*⟩; eliminate, expel ⟨*poison*⟩; exude ⟨*sweat*⟩; **Aus·scheidung die** a) *(Physiol.) s.* **ausscheiden** 2: excretion; elimination; expulsion; exudation; **~en** *(Ausgeschiedenes)* excreta; b) *(Sport)* qualifier

aus|schenken *tr. V.* serve

aus|schimpfen *tr. V.* jmdn. ~: tell sb. off

aus|schlafen 1. *unr. itr., refl. V.* have a good sleep; 2. *unr. tr. V.* **seinen Rausch** ~: sleep off the effects of alcohol

Aus·schlag der a) *(Haut~)* rash; b) *(eines Zeigers, einer Waage)* deflection; *(eines Pendels)* swing; **den** ~ **geben** *(fig.)* tip the scales *(fig.);* **aus|schlagen** 1. *unr. tr. V.* a) knock out; b) *(ablehnen)* turn down; 2. *unr. itr. V.* a) ⟨*horse*⟩ kick; b) ⟨*needle, pointer*⟩ be deflected, swing; c) *(sprießen)* come out [in bud]; **ausschlag·gebend** *Adj.* decisive

aus|schließen *unr. tr. V.* a) *(ausstoßen)* expel (aus from); b) *(nicht teilnehmen lassen)* exclude (aus from); c) *(fig.)* rule out ⟨*possibility*⟩; **jeden Irrtum** ~: rule out all possibility of error; e) *(aussperren)* lock out; **ausschließlich** [*od.* '-'--, -'--] 1. *Adj.* ex-

clusive; 2. *Adv.* exclusively; 3. *Präp. mit Gen.* excluding; **Aus·schluß der** exclusion (von from); *(aus einer Gemeinschaft)* expulsion (aus from); **unter** ~ **der Öffentlichkeit** with the public excluded; *(Rechtsw.)* in camera

aus|schmücken *tr. V.* deck out

aus|schneiden *unr. tr. V.* cut out; **Aus·schnitt der** a) *(Zeitungs~)* cutting; clipping; b) *(Hals~)* neck; **ein tiefer** ~: a plunging neck-line; c) *(Teil)* part; *(eines Textes)* excerpt; *(eines Films)* clip; *(Bild~)* detail

aus|schreiben *unr. tr. V.* a) *(nicht abgekürzt schreiben)* **etw.** ~: write sth. out in full; b) *(ausstellen)* make out ⟨*cheque, invoice, receipt*⟩; c) *(bekanntgeben)* call ⟨*election, meeting*⟩; advertise ⟨*flat, job*⟩; put ⟨*supply order etc.*⟩ out to tender; **Aus·schreibung die** *s.* **ausschreiben** c: calling; advertisement; invitation to tender

Ausschreitungen *Pl.* acts of violence

Aus·schuß der committee

aus|schütten *tr. V.* tip out ⟨*water, sand, coal, etc.*⟩; *(ausleeren)* empty ⟨*bucket, bowl, container*⟩

ausschweifend 1. *Adj.* wild ⟨*imagination, emotion, hope, desire, orgy*⟩; extravagant ⟨*idea*⟩; riotous, wild ⟨*enjoyment*⟩; dissolute ⟨*life, person*⟩; 2. *adv.* ~ **leben** lead a dissolute life; **Ausschweifung die;** ~, **~en** *(im Genießen)* dissolution

aus|sehen *unr. itr. V.* look (wie like); **so siehst du aus!** *(ugs.)* that's what you think!; **Aussehen das;** ~s appearance

aus|sein *unr. itr. V.; mit sein; nur im Inf. und Part. zusammengeschrieben* a) ⟨*play, film, war*⟩ be over; **wann ist die Vorstellung aus?** what time does the performance end?; **die Schule ist aus** school is out; b) ⟨*fire, candle, etc.*⟩ be out; c) ⟨*radio, light, etc.*⟩ be off

außen *Adv.* outside; **die Vase ist** ~ **bemalt** the vase is painted on the outside; **das Fenster geht nach** ~ **auf** the window opens outwards; **von** ~: from the outside

außen-, Außen-: ~**handel der;** *o. Pl.* foreign trade *no art.;* ~**minister der** Foreign Minister; ~**ministerium das** Foreign Ministry; ~**politik die** foreign politics *sing.;* ~**politisch** 1. *Adj.* ⟨*question*⟩ relating to foreign policy; 2. *adv.* as regards foreign policy; ~**seite die** outside

Au̱ßenseiter der; ~s, ~, **Au̱ßensei-terin** die; ~, ~nen outsider

au̱ßer 1. *Präp. mit Dat.* **a)** *(abgesehen von)* apart from; aside from *(Amer.)*; **b)** *(außerhalb von)* out of; ~ sich sein be beside oneself **(vor** + *Dat.* with); **c)** *(zusätzlich zu)* in addition to; **2.** *Präp. mit Akk.* ~ sich geraten become beside oneself **(vor** + *Dat.* with); **3.** *Konj.* except; **ä̱ußer...** *Adj.* outer; outside *(pocket)*; outlying *(district, area)*; external *(injury, form, circumstances, cause, force)*; outward *(appearance, similarity, effect, etc.)*; foreign *(affairs)*

au̱ßer·dem [*auch:* --'-] *Adv.* as well; *(überdies)* besides

Ä̱ußere das; ~n [outward] appearance

au̱ßer-: ~**ehelich 1.** *Adj.* extramarital; illegitimate *(child, birth)*; **2.** *adv.* outside marriage; ~**gewöhnlich 1.** *Adj.* **a)** unusual; **b)** *(das Gewohnte übertreffend)* exceptional; **2.** *adv.* **a)** unusually; **b)** *(sehr)* exceptionally; ~**halb** *Präp. mit Gen.* outside

ä̱ußerlich 1. *Adj.* external *(use, injury)*; outward *(appearance, calm, similarity, etc.)*; **2.** *adv.:* s. *Adj.:* externally; outwardly

ä̱ußern 1. *tr. V.* express, voice *(opinion, view, criticism, reservations, disapproval, doubt)*; voice *(wish)*; voice *(suspicion)*; **2.** *refl. V.* **a)** sich über etw. *(Akk.)* ~: give one's view on sth.; **b)** *(illness)* manifest itself **(in** + *Dat.*, durch in)

au̱ßer·ordentlich 1. *Adj.* **a)** extraordinary; **b)** *(das Gewohnte übertreffend)* exceptional; **2.** *adv.* *(sehr)* exceptionally; extremely *(pleased, relieved)*

ä̱ußerst *Adv.* extremely; **ä̱ußerst...** *Adj.* **a)** extreme; **b)** *(letztmöglich)* latest possible *(date, deadline)*; *(höchst...)* highest *(price)*; *(niedrigst...)* lowest *(price)*; **c)** *(schlimmst...)* worst

au̱ßersta̱nde *Adv.* ~ sein, etw. zu tun *(nicht befähigt)* be unable to do sth.; *(nicht in der Lage)* not be in a position to do sth.

Ä̱ußerung die; ~, ~en comment

au̱s|setzen 1. *tr. V.* **a)** expose *(Dat.* to); Belastungen ausgesetzt sein be subject to strains; **b)** *(sich selbst überlassen)* abandon *(baby, animal)*; *(auf einer einsamen Insel)* maroon; **c)** an jmdm./etw. etwas auszusetzen haben find fault with sb./sth.; **2.** *itr. V.* **a)** *(aufhören)* stop; *(engine, machine)* cut

out; **b)** *(pausieren)* *(player)* miss a turn; **mit der Arbeit/dem Training |ein paar Wochen|** ~: stop work/training [for a few weeks]

Au̱s·sicht die **a)** view **(auf** + *Akk.* of); **b)** *(fig.)* prospect; ~ auf etw. *(Akk.)* haben, etw. in ~ haben have the prospect of sth.

au̱ssichts-, Au̱ssichts-: ~los **1.** *Adj.* hopeless. **2.** *adv.* hopelessly; ~**reich** *Adj.* promising; ~**turm** der look-out tower

au̱s|sortieren *tr. V.* sort out

au̱s|spannen *itr. V.* take *or* have a break

au̱s|sperren 1. *tr. V.* lock out; shut *(animal)* out; **2.** *itr. V.* lock the workforce out; **Au̱s·sperrung** die lockout

au̱s|spielen *tr. V.* **a)** auch *itr.* *(Kartenspiel)* lead; **b)** jmdn./etw. gegen jmdn./etw. ~: play sb./sth. off against sb./sth.

Au̱s·sprache die **a)** pronunciation; **b)** *(Gespräch)* discussion

au̱s|sprechen 1. *unr. tr. V.* **a)** pronounce; **b)** *(ausdrücken)* express; voice *(suspicion, request)*; **2.** *unr. refl. V.* **a) b)** sich lobend/mißbilligend usw. über jmdn./etw. ~: speak highly/disapprovingly of *etc.* sb./sth.; **c)** *(offen sprechen)* say what's on one's mind; sich bei jmdm. ~: have a heart-to-heart talk with sb.; **d)** *(Strittiges klären)* talk things out **(mit** with); **3.** *unr. itr. V.* *(zu Ende sprechen)* finish [speaking]; **Au̱s·spruch** der remark

au̱s|spucken 1. *itr. V.* spit.; **2.** *tr. V.* spit out

au̱s|spülen *tr. V.* rinse out

Au̱s·stand der strike

au̱s|statten ['au̱sʃtatn̩] *tr. V.* provide **(mit** with); *(mit Gerät)* equip; *(mit Möbeln, Teppichen, Gardinen usw.)* furnish; **Au̱sstattung** die; ~, ~en **a)** s. ausstatten provision; equipping; furnishing; **b)** *(Ausrüstung)* equipment; *(Innen~ eines Autos)* trim; **c)** *(Einrichtung)* furnishings *pl.*

au̱s|stehen 1. *unr. itr. V.* noch ~ *(debt)* be outstanding; *(decision)* be still to be taken; *(solution)* be still to be found; **2.** *unr. tr. V.* ich kann ihn/das nicht ~: I can't stand him/it

au̱s|steigen *unr. itr. V.;* mit sein get out; *(aus einem Zug, Bus)* get off

au̱s|stellen *tr. V.* **a)** put on display; display; *(im Museum, auf einer Messe)* exhibit; **b)** *(ausfertigen)* make out

⟨cheque, prescription, receipt, bill⟩; issue ⟨visa, passport, certificate⟩; c) *(ugs.: ausschalten)* switch off ⟨cooker, radio, heating, engine⟩; **Aus·stellung die a)** exhibition; **b)** *s.* **ausstellen b:** making out; issuing

aus|sterben *unr. itr. V.; mit sein* die out; ⟨species⟩ become extinct

Aus·steuer die trousseau *(consisting mainly of household linen)*

Ausstieg der: ~|e|s, ~e *(Tür)* exit

aus|stopfen *tr. V.* stuff

aus|stoßen *unr. tr. V.* **a)** expel; give off, emit ⟨gas, fumes, smoke⟩; **b)** give ⟨cry, whistle, laugh, sigh, etc.⟩; let out ⟨cry, scream, yell⟩; utter ⟨curse, threat, etc.⟩

aus|strahlen 1. *tr. V.* **a)** *(auch fig.)* radiate; ⟨lamp⟩ give out ⟨light⟩; **b)** *(Rundf., Ferns.)* broadcast; **2.** *itr. V.* **a)** radiate; ⟨light⟩ be given out; *(fig.)* ⟨pain⟩ spread; **b)** **auf jmdn./etw. ~** *(fig.)* communicate itself to sb./influence sth.; **Aus·strahlung die** *(fig.)* charisma

aus|strecken 1. *tr. V.* stretch out; put out ⟨feelers⟩; **2.** *refl. V.* stretch out

aus|streichen *unr. tr. V.* cross out

aus|strömen *itr. V.; mit sein* pour out; ⟨gas, steam⟩ escape

aus|suchen *tr. V.* choose; pick

Aus·tausch der a) exchange; **im ~ für** *od.* **gegen** in exchange for; **b)** *(das Ersetzen)* replacement **(gegen** with); **aus|tauschen** *tr. V.* **a)** exchange **(gegen** for); **b)** *(ersetzen)* replace **(gegen** with); **Austausch·motor der** replacement engine

aus|teilen *tr. V.* distribute **an** + *Akk.* to); *(aushändigen)* hand out ⟨books, post, etc.⟩ **(an** + *Akk.* to); give ⟨orders⟩; deal [out] ⟨cards⟩; give out ⟨marks, grades⟩; serve ⟨food etc.⟩

Auster die: ~, ~n oyster

aus|tragen *unr. tr. V.* **a)** deliver ⟨newspapers, post⟩; **b)** ⟨pregnant woman⟩ carry ⟨child⟩ to full term; *(nicht abtreiben)* have ⟨child⟩; **c)** *(ausfechten)* settle ⟨conflict, differences⟩; fight out ⟨battle⟩

Australien [aʊsˈtraːliən] **(das);** ~s Australia; **Australier der;** ~s, ~: Australian; **australisch** *Adj.* Australian

aus|treiben *unr. tr. V.* **a)** exorcize, cast out ⟨evil spirit, demon⟩; **b)** **jmdm. etw. ~:** cure sb. of sth.

aus|treten 1. *unr. tr. V.* **a)** tread out ⟨spark, cigarette-end⟩; trample out

⟨fire⟩; **b)** *(bahnen)* tread out ⟨path⟩; **c)** wear out ⟨shoes⟩; **2.** *unr. itr. V.; mit sein* **a)** *(ugs.: zur Toilette gehen)* pay a call *(coll.)*; **b)** **aus etw. ~** *(ausscheiden)* leave sth.

aus|trinken *tr. V.* drink up ⟨drink⟩; finish ⟨glass, cup, etc.⟩

Aus·tritt der leaving

aus|trocknen 1. *tr. V.* dry out; dry up ⟨river bed, marsh⟩; **2.** *itr. V.; mit sein* dry out; ⟨river bed, pond, etc.⟩ dry up; ⟨skin, hair⟩ become dry

aus|üben *tr. V.* practise ⟨art, craft⟩; follow ⟨profession⟩; carry on ⟨trade⟩; do ⟨job⟩; hold ⟨office⟩; wield ⟨power, right, control⟩

Aus·verkauf der sale; **ausverkauft** *Adj.* sold out

Aus·wahl die a) choice; **b)** *(Sortiment)* range; **viel/wenig ~ haben** have a wide/limited selection **(an** + *Dat.*, **von** of); **aus|wählen** *tr. V.* choose **(aus** from)

Aus·wanderer der emigrant; **aus|-wandern** *itr. V.; mit sein* emigrate; **Aus·wanderung die** emigration

auswärtig *Adj.* **a)** non-local; **b)** *(das Ausland betreffend)* foreign; **auswärts** *Adv.* **a)** *(nach außen)* outwards; **b)** *(nicht zu Hause)* ⟨sleep⟩ away from home; **~essen** eat out; **c)** *(nicht am Ort)* in another town; *(Sport)* away; **Auswärts·spiel das** *(Sport)* away match

aus|waschen *unr. tr. V.* wash out

aus|wechseln *tr. V.* **a)** change **(gegen** + *Akk.* for); **b)** *(ersetzen)* replace **(gegen** with); *(Sport)* substitute ⟨player⟩

Aus·weg der way out **(aus** of); **ausweg·los 1.** *Adj.* hopeless. **2.** *adv.* hopelessly

aus|weichen *unr. itr. V.; mit sein* get out of the way **(**Dat. of); *(Platz machen)* make way **(**Dat. for); **einem Schlag/ Angriff ~:** dodge a blow/ evade an attack; **dem Feind ~:** avoid [contact with] the enemy; **einer Frage ~:** evade a question; **eine ~de Antwort** an evasive answer

Ausweis der; ~es, ~e card; *(Personal~)* identity card; **aus|weisen 1.** *unr. tr. V.* **a)** expel **(aus** from); **b)** **jmdn. als etw. ~:** show that sb. is/was sth.; **2.** *unr. refl. V.* prove *or* establish one's identity [by showing one's papers]; **können Sie sich ~?** do you have any means of identification?; **Aus·weisung die** expulsion **(aus** from)

aus|weiten tr. V. stretch

aus·wendig Adv. etw. ~ können/lernen know/learn sth. [off] by heart

aus|werten tr. V. analyse and evaluate; **Aus·wertung die** analysis and evaluation

aus|wirken refl. V. have an effect (auf + Akk. on); sich günstig ~: have a favourable effect; **Aus·wirkung die** effect (auf + Akk. on)

Aus·wuchs der a) (Wucherung) growth; excrescence (Med., Bot.); b) (fig.) unhealthy product; (Exzeß) excess

aus|zahlen 1. tr. V. a) pay out ⟨money⟩; b) pay off ⟨employee, worker⟩; buy out ⟨business partner⟩; 2. refl. V. pay

aus|zählen tr. V. a) count [up] ⟨votes etc.⟩; b) (Boxen) count out

aus|zeichnen tr. V. a) (mit einem Preisschild) mark; b) (ehren) honour; **Aus·zeichnung die** a) o. Pl. (von Waren) marking; b) (Ehrung) honouring; (Orden) decoration; (Preis) award

aus|ziehen 1. unr. tr. V. a) pull out ⟨couch⟩; extend ⟨table, tripod, etc.⟩; b) (ablegen) take off ⟨clothes⟩; c) (entkleiden) undress; sich ~: get undressed; 2. unr. itr. V.; mit sein move out (aus of); **Aus·zug der** a) (das Ausziehen) move; b) (Bankw.) statement; c) (Textpassage) extract

Auto das; ~s, ~s car; automobile (Amer.); ~ fahren drive; (mitfahren) go in the car

Auto-: ~**bahn die** motorway (Brit.); expressway (Amer.); ~**biographie** [-----'-] die autobiography; ~**bus der** s. Bus; ~**fähre die** car ferry; ~**fahren das** driving; motoring; ~**fahrer der** [car-]driver; ~**fahrt die** drive; ~**gramm** [--'-] das; ~s, ~e autograph; ~**kino das** drive-in cinema

Automat der; ~en, ~en a) (Verkaufs~) [vending-] machine; (Spiel~) slot-machine; b) (in der Produktion) robot; **Automatik die**; ~, ~en automatic control mechanism; (Getriebe~) automatic transmission; **automatisch** (auch fig.) 1. Adj. automatic; 2. adv. automatically; **automatisieren** tr. V. automate; **Automatisierung die**; ~, ~en automation

auto-, Auto-: ~**mobil** [---'-] das; ~s, ~e (geh.) motor car; automobile (Amer.); ~**nom** [--'-] 1. Adj. autonomous; 2. adv. autonomously; ~**nomie**

[---'-] die; ~, ~n autonomy; ~**nummer die** [car] registration number

Autopsie [autɔ'psi:] die; ~, ~n post-mortem [examination]

Autor der; ~s, ~en author

Auto-: ~**radio das** car radio; ~**reifen der** car tyre; ~**reisezug der** Motorail train (Brit.); auto train (Amer.)

Autorin die; ~, ~nen authoress; author

autoritär Adj. authoritarian; **Autorität die**; ~, ~en authority

Auto-: ~**schlange die** queue of cars; ~**schlüssel der** car key; ~**stopp der** hitch-hiking; per ~**stopp fahren**, ~**stopp machen** hitch-hike; ~**unfall der** car accident; ~**vermietung die** car rental firm; ~**werkstatt die** garage

Avocado [avo'ka:do] die; ~, ~s avocado [pear]

Axt die; ~, Äxte axe

Azalee [atsa'le:ə] die; ~, ~n azalea

B

b, B [be:] das; ~, ~ a) (Buchstabe) b/B; b) (Musik) [key of] B flat

B Abk. Bundesstraße ≈ A (Brit.)

Baby ['be:bi] das; ~s, ~s baby; **Baby·sitter** [-sɪtɐ] der; ~s, ~: babysitter

Bach der; ~[e]s, Bäche a) stream; brook; b) (Rinnsal) stream [of water]

Back·blech das baking-sheet

Back·bord das (Seew., Luftf.) port [side]; **Backe die**; ~, ~n cheek

backen 1. unr. itr. V. bake; 2. unr. tr. V. a) bake; b) (bes. südd.) s. braten

Backen·zahn der molar

Bäcker der; ~s, ~: baker; er ist ~: he is a baker; zum/beim ~: to the/at the baker's; **Bäckerei die**; ~, ~en baker's [shop]

Back-: ~**fisch der** fried fish (in breadcrumbs); ~**form die** baking-tin (Brit.); baking-pan (Amer.); ~**hähnchen das**, ~**hendl das** (österr.), ~**huhn das** fried chicken (in breadcrumbs); ~**ofen der** oven; ~**pulver**

das baking-powder; **~stein** der brick; **~waren** *Pl.* bread, cakes, and pastries

Bad das; ~|e|s, Bäder **a)** bath; *(das Schwimmen)* swim; *(im Meer o. ä.)* bathe; **ein ~ nehmen** *(geh.)* take a bath; *(schwimmen)* go for a swim; *(im Meer o. ä.)* bathe; **b)** *(Badezimmer)* bathroom; **ein Zimmer mit ~:** a room with [private] bath; **c)** *(Schwimm~)* [swimming-]pool; **d)** *(Heil~)* spa; *(See~)* [seaside] resort

Bade-: ~**an·zug** der bathing costume; ~**hose die** bathing trunks *pl*; ~**mantel** der dressing-gown; bathrobe; ~**meister** der swimming-pool attendant; ~**mütze die** bathing cap

baden 1. *itr. V.* **a)** have a bath; **b)** *(schwimmen)* bathe; ~ **gehen** go for a bathe; 2. *tr. V.* bath ⟨*child, patient, etc.*⟩; bathe ⟨*wound, eye, etc.*⟩

Bäder *s.* Bad

Bade-: ~**strand** der bathing-beach; ~**tuch** das bath towel; ~**wanne die** bath[-tub]; ~**wasser** das bath water; ~**zimmer** das bathroom

Bagatelle die; ~, ~n trifle

Bagger der; ~s, ~: excavator; *(Schwimm~)* dredger; **Bagger·see** der flooded gravel-pit

Bahn die; ~, ~en **a)** *(Weg)* path; **b)** *(Route)* path; *(eines Geschosses)* trajectory; **c)** *(Sport)* track; *(für Pferderennen)* course *(Brit.);* track *(Amer.);* *(für einzelne Teilnehmer)* lane; *(Kegel~)* alley; *(Bowling~)* lane; **d)** *(Eisen~)* railways *pl.;* railroad *(Amer.);* *(Zug)* train; **jmdn. zur ~ bringen** take sb. to the station; **|mit der| ~ fahren** go by train; **f)** *(Straßen~)* tram; streetcar *(Amer.)*

bahn-, Bahn-: ~**brechend** *Adj.* pioneering; ~**bus** der railway bus; ~**damm** der railway embankment

bahnen *tr. V.* clear ⟨*way, path*⟩; **jmdm./einer Sache einen Weg ~** *(fig.)* pave the way for sb./sth.

Bahn-: ~**fahrt** die train journey; ~**hof** der [railway *or (Amer.)* railroad] station; ~**reise** die train journey; ~**schranke** die level-crossing *(Brit.)* *or (Amer.)* grade crossing barrier/gate; ~**steig** der; ~|e|s, ~e [station] platform; ~**übergang** der level-crossing *(Brit.);* grade crossing *(Amer.);* ~**verbindung** die train connection

Bahre die; ~, ~n **a)** *(Kranken~)* stretcher; **b)** *(Toten~)* bier

Baiser [bɛ'ze:] das; ~s, ~s meringue

Bajonett das; ~|e|s, ~e bayonet

Bakterie [bak'te:riə] die; ~, ~n bacterium

Balance [ba'lãsə] die; ~, ~n balance; **balancieren** *itr., tr. V.; itr. mit sein* balance

bald *Adv.* **a)** soon; *(leicht, rasch)* quickly; easily; **wird's ~?** get a move on, will you; **bis ~!** see you soon; **b)** *(ugs.: fast)* almost

Baldrian ['baldria:n] der; ~s, ~e valerian

Balkan ['balka:n] der; ~s: **der ~:** the Balkans *pl.;* *(Gebirge)* the Balkan Mountains *pl.* **auf dem ~:** in the Balkans

Balken der; ~s, ~: beam

Balkon [bal'kɔŋ, bal'ko:n] der; ~s, ~s [bal'kɔŋs] *od.* ~e [bal'ko:nə] **a)** balcony; **b)** *(im Theater, Kino)* circle

Ball der; ~|e|s, Bälle **a)** ball; ~ **spielen** play ball; **b)** *(Fest)* ball

Ballade die; ~, ~n ballad

Ballast der; ~|e|s, ~e ballast

ballen 1. *tr. V.* clench ⟨*fist*⟩; 2. *refl. V.* ⟨*fist*⟩ clench; **Ballen der;** ~s, ~ **a)** *(Packen)* bale; **b)** *(Hand~, Fuß~)* ball

Ballett das; ~|e|s, ~e ballet

Ball-: ~**junge** der ballboy; ~**kleid** das ball gown

Ballon [ba'lɔŋ] der; ~s, ~s balloon

Ball-: ~**saal** der ballroom; ~**spiel** das ball game; ~**spielen** das; ~s playing ball *no art.*

Ballungs·gebiet das conurbation

Balsam der; ~s, ~e balsam; *(fig.)* balm

Balte der; ~n, ~n, **Baltin** die; ~, ~nen Balt; **Baltikum** das; ~s Baltic States *pl.;* **baltisch** *Adj.* Baltic

Bambus der; ~ *od.* ~ses, ~se bamboo

banal *Adj.* **a)** banal; **b)** *(gewöhnlich)* commonplace

Banane die; ~, ~n banana

Banause der; ~n, ~n *(abwertend)* philistine

band *1. u. 3. Pers. Sg. Prät. v.* binden

¹Band das; ~|e|s, Bänder **a)** ribbon; *(Haar~, Hut~)* band; *(Schürzen~)* string; **b)** *(Klebe~, Isolier~, Ton~ usw.)* tape; **etw. auf ~** *(Akk.)* **aufnehmen** tape[-record] sth.; **c)** *s.* Förderband; **d)** *s.* Fließband; **e)** **am laufenden ~** *(ugs.)* nonstop; **f)** *(Anat.)* ligament

²Band der; ~|e|s, Bände ['bɛndə] volume

³Band [bɛnt] die; ~, ~s band; *(Beat~, Rock~ usw.)* group

¹**Bạnde** die; ~, ~n a) gang; b) *(ugs.: Gruppe)* mob *(sl.)*

²**Bạnde** die; ~, ~n *(Sport)* [perimeter] barrier; *(mit Reklame)* billboards *pl.; (Billard)* cushion

Bänder *s.* ¹**Band**

bändigen *tr. V.* tame ⟨animal⟩; control ⟨person, anger, urge⟩

Bandịt der; ~en, ~en bandit

Bạnd·scheibe die [intervertebral] disc

bang, bạnge; bạnger, bạngst... *od.* **bänger, bängst...:** 1. *Adj.* afraid; scared; *(besorgt)* anxious; **mir ist/wurde ~ [zumute]** I am/became scared; 2. *adv.* anxiously; **bạngen** *itr. V.* be anxious

¹**Bạnk** die; ~, **Bänke** bench; *(mit Lehne)* bench seat; *(Kirchen~)* pew; **etw. auf die lange ~ schieben** *(ugs.)* put sth. off

²**Bạnk** die; ~, ~en bank

¹**Bankẹtt** das; ~[e]s, ~e banquet

²**Bankẹtt** das; ~[e]s, ~e *(an Straßen)* shoulder; *(unbefestigt)* verge

Bankier [baŋ'kie:] der; ~s, ~s banker

Bạnk-: ~**konto** das bank account; ~**leitzahl** die bank sorting code number; ~**note** die banknote; bill *(Amer.);* ~**raub** der bank robbery; ~**räuber** der bank robber

bankrọtt *Adj.* bankrupt; ~ **gehen** go bankrupt; **Bankrọtt** der; ~[e]s, ~e bankruptcy; ~ **machen** go bankrupt

Bạnn der; ~[e]s *(fig. geh.)* spell

bạr 1. *Adj.* cash; 2. *adv.* in cash

Bạr die; ~, ~s bar

Bär der; ~en, ~en bear

Barạcke die; ~, ~n hut

Barbạr der; ~en, ~en barbarian; **Barbarei** die; ~, ~en a) *(Roheit)* barbarity; b) *(Kulturlosigkeit)* barbarism *no indef. art.;* **barbạrisch** 1. *Adj.* a) *(roh)* barbarous; b) *(unzivilisiert)* barbaric; 2. *adv.* a) *(roh)* barbarously; b) *(unzivilisiert)* barbarically

Bạr·dame die barmaid

Barẹtt das; ~[e]s, ~e *(eines Geistlichen)* biretta; *(eines Richters, Professors)* cap; *(Baskenmütze)* beret

bar·fuß *indekl. Adj.; nicht attr.* barefooted; ~ **herumlaufen/gehen** run about/go barefoot

bạrg *1. u. 3. Pers. Sg. Prät. v.* bergen

Bạr-: ~**geld** das cash; ~**hocker** der bar stool

Bariton ['ba(:)ritɔn] der; ~s, ~e baritone

Barkạsse die; ~, ~n launch

barmhẹrzig *(geh.)* 1. *Adj.* merciful; 2. *adv.* mercifully; **Barmhẹrzigkeit** die; ~ *(geh.)* mercy

Barọck das *od.* der; ~[s] a) baroque; b) *(Zeit)* baroque age

Baro·mẹter das barometer

Barọn der; ~s, ~e baron; *(als Anrede)* **[Herr]** ~: ≈ my lord; **Barọnin** die; ~, ~nen baroness; *(als Anrede)* **[Frau]** ~: ≈ my Lady

Bạrren der; ~s, ~ a) *(Gold~, Silber~ usw.)* bar; b) *(Turngerät)* parallel bars *pl.*

Barriere [ba'rie:rə] die; ~, ~n *(auch fig.)* barrier

Barrikạde die; ~, ~n barricade

bạrsch 1. *Adj.* curt; 2. *adv.* curtly

Bạrsch der; ~[e]s, ~e perch

bạrst *1. u. 3. Pers. Sg. Prät. v.* bersten

Bạrt der; ~[e]s, **Bärte** a) beard; *(Oberlippen~, Schnurr~)* moustache; b) *(von Katzen, Mäusen, Robben)* whiskers *pl.;* c) *(am Schlüssel)* bit; **bärtig** *Adj.* bearded; **Bạrt·wuchs** der growth of beard

Bạr·zahlung die cash payment

Basạlt der; ~[e]s, ~e basalt

Basạr der; ~s, ~e bazaar

Bạsis die; ~, **Basen** a) *(Grundlage)* basis; b) *(Math., Archit., Milit.)* base

Bạske der; ~n, ~n, **Bạskin** die; ~, ~nen Basque

Bạsken-: ~**land** das Basque region; ~**mütze** die beret

Basket·ball ['ba(:)skət-] der basketball

Bạß der; **Bạsses, Bässe** *(Musik)* a) bass; b) *(Instrument)* double-bass

Bassin [ba'sẽ:] das; ~s, ~s *(Schwimm~)* pool; *(im Garten)* pond

Bassịst der; ~en, ~en *(Musik)* a) *(Sänger)* bass; b) *(Instrumentalist)* double-bass player; bassist; *(in einer Rockband)* bass guitarist

Bạst der; ~[e]s, ~e bast; *(Raffia~)* raffia

bạsta *Interj. (ugs.)* that's enough; **und damit ~!** and that's that!

Bastelei die; ~, ~en; a) *(Gegenstand)* piece of handicraft work; b) *(ugs.: das Basteln)* handicraft work; **bạsteln** 1. *tr. V.* make; 2. *itr. V.* make things [with one's hands]

Bastịon die; ~, ~en bastion

bạt *1. u. 3. Pers. Sg. Prät. v.* bitten

Bataillon [batal'jo:n] das; ~s, ~e *(Milit.)* battalion

Batik der; ~s, ~en *od.* die; ~, ~en batik

Batist der; ~|e|s, ~e batiste
Batterie die; ~, ~n battery
Batzen der; ~s, ~ *(ugs.)* a) *(Klumpen)* lump; b) *(Menge)* pile *(coll.)*
¹**Bau** der; ~|e|s, ~ten a) o. Pl. *(Errichtung)* building; im ~ sein be under construction; b) *(Gebäude)* building; c) auf dem ~ arbeiten *(Bauarbeiter sein)* be in the building trade; d) o. Pl. *(Struktur)* structure
²**Bau** der; ~|e|s, ~e *(Kaninchen~)* burrow; hole; *(Fuchs~)* earth
Bau·arbeiten Pl. building work *sing.*
Bauch der; ~|e|s, Bäuche *(auch fig.: von Schiffen, Flugzeugen)* belly; **bauchig** Adj. bulbous
Bauch-: ~**laden** der vendor's tray; ~**landung** die belly-landing; ~**nabel** der *(ugs.)* belly-button *(coll.)*; ~**redner** der ventriloquist; ~**schmerzen** Pl. stomach-ache *sing.*; ~**speichel·drüse** die pancreas; ~**tanz** der belly-dance; ~**tänzerin** die belly-dancer; ~**weh** das *(ugs.)* tummy-ache *(coll.)*; stomach-ache
bauen 1. *tr. V.* build; 2. *itr. V.* a) build; wir wollen ~: we want to build a house; *(bauen lassen)* we want to have a house built; b) auf jmdn./etw. ~ *(fig.)* rely on sb./sth.
¹**Bauer** der; ~n, ~n a) farmer; *(mit niedrigem sozialem Status)* peasant; b) *(Schachfigur)* pawn; c) *(Kartenspiele)* s. **Bube**
²**Bauer** das od. der; ~s, ~: [bird-]cage
Bäuerin die; ~, ~nen a) s. ¹ **Bauer** a: [lady] farmer; peasant [woman]; b) *(Frau eines Bauern)* farmer's wife; **bäuerlich** Adj. farming attrib.; *(ländlich)* rural
Bauern-: ~**haus** das farmhouse; ~**hof** der farm
bau-, Bau-: ~**fällig** Adj. ramshackle; unsafe *(roof, ceiling)*; ~**jahr** das year of construction; *(bei Autos)* year of manufacture; ~**kasten** der construction set; *(mit Holzklötzen)* box of bricks; ~**klotz** der building-brick
baulich Adj.; nicht präd. structural
Baum der; ~|e|s, Bäume tree; **Bäumchen** das; ~s, ~ small tree
Bau·meister der *(hist.)* [architect and] master builder
baumeln itr. V. *(ugs.)* dangle (an + Dat. from)
Baum-: ~**schule** die tree nursery; ~**stamm** der tree-trunk; ~**stumpf** der tree-stump; ~**wolle** die cotton

Bau·platz der site for building
bäurisch *(abwertend)* 1. Adj. boorish; 2. adv. boorishly
Bau·satz der kit
Bausch der; ~|e|s, ~e od. Bäusche a) *(Watte~)* a wad; b) etw. in ~ und Bogen verwerfen/verdammen reject/condemn sth. wholesale; **bauschen** 1. tr. V. billow *(sail, curtains, etc.)*; 2. refl. V. *(dress, sleeve)* puff out; *(ungewollt)* bunch up; *(im Wind)* *(curtain, flag, etc.)* billow [out]; **bauschig** Adj. puffed *(dress)*; baggy *(trousers)*
bau-, Bau-: ~**sparen** itr. V.; nur Inf. gebr. save with a building society; ~**spar·kasse** die ≈ building society; ~**stein** der a) building stone; b) *(Bestandteil)* element; *(Elektronik, DV)* module; c) *(~klotz)* building-brick; ~**stelle** die building site; *(beim Straßenbau)* road-works pl.
Bauten Pl.: s. **Bau**
Bau-: ~**unternehmer** der building contractor; ~**weise** die method of construction; ~**werk** das building; *(Brücke, Staudamm)* structure
Bayer der; ~n, ~n Bavarian; **bay[e]risch** Adj. Bavarian; **Bayern** (das); ~s Bavaria
Bazille die; ~, ~n *(ugs.)* s. **Bazillus** a; **Bazillus** der; ~, Bazillen a) bacillus; b) *(fig.)* cancer
Bd. Abk. Band Vol.
beabsichtigen tr. V. intend
beachten tr. V. a) follow *(rule, regulations, instruction)*; heed, follow *(advice)*; obey *(traffic signs)*; observe *(formalities)*; b) *(berücksichtigen)* take account of; *(achten auf)* pay attention to; **beachtlich** 1. Adj. considerable; 2. adv. considerably; **Beachtung** die a) s. beachten a: following; heeding; obeying; b) *(Berücksichtigung)* consideration; c) *(Aufmerksamkeit)* attention
Beamte der; adj. Dekl. official; *(Staats~)* [permanent] civil servant; *(Kommunal~)* [established] local government officer; *(Polizei~)* [police] officer; **Beamtin** die; ~, ~nen s. Beamte
beängstigend Adj. worrying
beanspruchen tr. V. a) claim; etw. ~ können be entitled to expect sth.; b) *(ausnutzen)* make use of *(person, equipment)*; take advantage of *(hospitality, services)*; c) *(erfordern)* demand *(energy, attention, stamina)*; take up *(time, space, etc.)*; **Bean-**

spruchung die; ~, ~en demands *pl.* (*Gen.* on); **die ~ durch den Beruf** the demands of his/her job

beanstanden *tr. V.* take exception to; *(sich beklagen über)* complain about; **Beanstandung** die; ~, ~en complaint

beantragen *tr. V.* apply for

beantworten *tr. V.* answer; reply to ⟨*letter*⟩; return ⟨*greeting*⟩

bearbeiten *tr. V.* a) deal with; handle ⟨*case*⟩; b) *(adaptieren)* adapt (**für** for); **Bearbeitung** die; ~, ~en a) **die ~ eines Antrags/eines Falles** *usw.* dealing with an application/handling a case *etc.;* b) *(Adaption)* adaptation

beaufsichtigen *tr. V.* supervise; look after ⟨*child*⟩

beauftragen *tr. V.* entrust

bebauen *tr. V.* build on; develop; **Bebauung** die; ~, ~en a) development; b) *(Gebäude)* buildings *pl.*

beben *itr. V.* shake; **Beben** das; ~s, ~ *(Erd~)* earthquake

bebildern *tr. V.* illustrate

Becher der; ~s, ~ *(Glas~, Porzellan~)* glass; tumbler; *(Plastik~)* beaker; cup; *(Eis~) (aus Glas, Metall)* sundae dish; *(aus Pappe)* tub; *(Joghurt~)* carton

Becken das; ~s, ~ a) *(Wasch~)* basin; *(Abwasch~)* sink; *(Toiletten~)* pan; b) *(Anat.)* pelvis; c) *Pl. (Musik)* cymbals

bedacht *Adj.* **auf etw.** *(Akk.)* ~ **sein** be intent on sth.; **bedächtig** 1. *Adj.* a) deliberate; measured ⟨*steps, stride, speech*⟩; b) *(besonnen)* thoughtful; well-considered ⟨*words*⟩; 2. *adv.* a) deliberately; b) *(besonnen)* thoughtfully

bedanken *refl. V.* say thank you; **sich bei jmdm. [für etw.] ~:** thank sb. [for sth.]

Bedarf der; ~[e]s need (**an** + *Dat.* of); requirement (**an** + *Dat.* for); *(Bedarfsmenge)* needs *pl.;* requirements *pl.;* **bei ~:** if required

bedauerlich *Adj.* regrettable; **bedauerlicher·weise** *Adv.* regrettably; **bedauern** *tr., itr. V.* a) feel sorry for; **sie läßt sich gerne ~:** she likes being pitied; b) *(schade finden)* regret; **ich bedauere sehr, daß ...:** I am very sorry that ...; **Bedauern** das; ~s regret; **zu meinem ~:** to my regret; **bedauerns·wert** *Adj. (geh.)* unfortunate ⟨*person*⟩

bedecken *tr. V.* cover; **bedeckt** *Adj.* overcast ⟨*sky*⟩

bedenken *unr. tr. V.* a) consider; b) *(beachten)* take into consideration; **Bedenken** das; ~s, ~ reservation (**gegen** about); **ohne ~:** without hesitation; **bedenken·los** 1. *Adj.* unhesitating; *(skrupellos)* unscrupulous; 2. *adv.* without hesitation; *(skrupellos)* unscrupulously; **bedenklich** 1. *Adj.* a) dubious ⟨*methods, transactions, etc.*⟩; b) *(bedrohlich)* alarming; 2. *adv.* alarmingly; **Bedenk·zeit** die; *o. Pl.* time for reflection

bedeuten *tr. V.* a) mean; **was soll das ~?** what does that mean?; b) *(sein)* represent; **das bedeutet ein Wagnis** that is being really daring; **bedeutend** 1. *Adj.* a) important; b) *(groß)* substantial; considerable ⟨*success*⟩; 2. *adv.* considerably; **Bedeutung** die; ~, ~en a) meaning; b) *o. Pl. (Wichtigkeit)* importance

bedeutungs-: **~los** *Adj.* insignificant; **~voll** 1. *Adj.* a) significant; b) *(vielsagend)* meaningful; meaning ⟨*look*⟩; 2. *adv.* meaningfully

bedienen 1. *tr. V.* a) serve; **werden Sie schon bedient?** are you being served?; b) *(handhaben)* operate ⟨*machine*⟩; 2. *itr. V.* serve; 3. *refl. V.* help oneself; **sich selbst ~** *(im Geschäft, Restaurant usw.)* serve oneself; **Bedienung** die; ~, ~en a) *o. Pl. (das Bedienen)* service; **~ inbegriffen** service included; b) *o. Pl. (das Handhaben)* operation; c) *(Servierer[in])* waiter/waitress; **Bedienungs·anleitung** die operating instructions *pl.*

bedingen *tr. V.* cause; **Bedingung** die; ~, ~en condition; **unter der ~, daß ...:** on condition that ...; **bedingungs·los** *Adj.* unconditional

bedrängen *tr. V.* a) besiege ⟨*town, fortress, person*⟩; put ⟨*opposing player*⟩ under pressure; b) *(belästigen)* pester; **bedrohen** *tr. V.* threaten; **bedrohlich** 1. *Adj. (unheilverkündend)* ominous; *(gefährlich)* dangerous; 2. *adv. (unheilverkündend)* ominously; *(gefährlich)* dangerously; **Bedrohung** die threat (*Gen.* to)

bedrucken *tr. V.* print

bedrücken *tr. V.* depress

Beduine der; ~n, ~n Bed[o]uin

bedürfen *unr. itr. V.* **jmds./einer Sache ~** *(geh.)* require *or* need sb./sth.; **Bedürfnis** das; ~ses, ~se need (**nach** for); **das ~ haben, etw. zu tun** feel a need to do sth.; **bedürftig** *Adj.* needy

Beef·steak ['bi:f-] das [beef]steak; deutsches ~: ≈ beefburger
beehren tr. V. (geh.) honour
beeiden tr. V. ~, daß ...: swear [on oath] that ...; eine Aussage ~: swear to the truth of a statement
beeilen refl. V. hurry [up (coll.)]
beeindrucken tr. V. impress; **beeindruckend** Adj. impressive
beeinflussen tr. V. influence; **Beeinflussung** die; ~, ~en influencing
beeinträchtigen tr. V. restrict ⟨sights, freedom⟩; detract from ⟨pleasure, enjoyment, value⟩; spoil ⟨appetite, good humour⟩; impair ⟨quality, reactions, efficiency, vision, hearing⟩; damage, harm ⟨sales, reputation⟩
beenden tr. V. end; finish ⟨piece of work etc.⟩; complete ⟨studies⟩
beengen tr. V. restrict
beerben tr. V. jmdn. ~: inherit sb's estate
beerdigen tr. V. bury; **Beerdigung die; ~, ~en** burial; (Trauerfeier) funeral; **Beerdigungs·institut das** [firm sing. of] undertakers pl.
Beere die; ~, ~n berry
Beet das; ~[e]s, ~e ⟨Blumen~⟩ bed; ⟨Gemüse~⟩ plot
befahrbar Adj. passable; **befahren** unr. tr. V. a) drive on ⟨road⟩; drive across ⟨bridge⟩; use ⟨railway line⟩; die Straße ist stark/wenig ~: the road is heavily/little used; eine stark ~e Straße a busy road; b) sail ⟨sea⟩; navigate, sail up/down ⟨river, canal⟩
befallen unr. tr. V. a) overcome; ⟨misfortune⟩ befall; von Panik/Angst ~ werden be seized with panic/fear; b) ⟨pests⟩ attack
befangen 1. Adj. a) self-conscious ⟨person⟩; b) (voreingenommen) biased; 2. adv. self-consciously; **Befangenheit die; ~ a)** self-consciousness; b) (Voreingenommenheit) bias
befassen refl. V. sich mit etw. ~: occupy oneself with sth.; ⟨article, book⟩ deal with sth.; (etw. studieren) study sth.
Befehl der; ~[e]s, ~e a) order; b) den ~ über jmdn./etw. haben be in command of sb./sth.; **befehlen 1.** unr. tr., itr. V. order; (Milit.) order; man befahl ihm zu warten he was told to wait; 2. unr. itr. V. über jmdn./etw. ~: have command of or be in command of sb./sth.; **Befehls·haber der; ~s, ~** (Milit.) commander

befestigen tr. V. a) fix; etw. an der Wand ~: fix sth. to the wall; b) (haltbar machen) stabilize ⟨bank, embankment⟩; make up ⟨road, path, etc.⟩; c) (sichern) fortify ⟨town etc.⟩; strengthen ⟨border⟩; **Befestigung die; ~, ~en a)** fixing; b) (Milit.) fortification
befeuchten tr. V. moisten; damp ⟨hair, cloth⟩
befiehlst, befiehlt 2., 3. Pers. Sg. Präsens v. befehlen
befinden unr. refl. V. be; **Befinden das; ~s** health; (eines Patienten) condition
beflecken tr. V. stain
befohlen 2. Part. v. befehlen
befolgen tr. V. follow, obey ⟨instruction, grammatical rule⟩; obey, comply with ⟨law, regulation⟩; follow ⟨advice, suggestion⟩
befördern tr. V. a) carry; transport; b) (aufrücken lassen) promote; **Beförderung die a)** o. Pl. carriage; transport; (Personen~) transport; b) (das Aufrückenlassen) promotion
befragen tr. V. a) question (über + Akk. about); b) (konsultieren) ask; **Befragung die; ~, ~en a)** questioning; b) (Konsultation) consultation; c) (Umfrage) opinion poll
befreien 1. tr. V. a) free; liberate ⟨country, people⟩ (von from); b) (freistellen) exempt (von from); c) jmdn. von Schmerzen ~: free sb. of pain; 2. refl. V. free oneself (von from); **Befreier** der liberator; **Befreiung die; ~ a)** s. befreien 1a: freeing; liberation; b) (Freistellung) exemption; c) die ~ von Schmerzen release from pain
befremden tr. V. jmdn. ~: put sb. off
befreunden refl. V. s. anfreunden; [gut od. eng] befreundet sein be [good or close] friends (mit with)
befriedigen tr. V. a) satisfy; gratify ⟨lust⟩; b) (ausfüllen) ⟨job, occupation, etc.⟩ fulfil; c) (sexuell) satisfy; sich [selbst] ~: masturbate; **befriedigend** 1. Adj. satisfactory; 2. adv. satisfactorily; **Befriedigung die; ~ a)** s. befriedigen a: satisfaction; gratification; b) (Genugtuung) satisfaction
befristet Adj. temporary ⟨visa⟩; fixed-term ⟨ban, contract⟩
befruchten tr. V. fertilize ⟨egg⟩; pollinate ⟨flower⟩; impregnate ⟨female⟩; **Befruchtung die; ~, ~en s.** befruchten: fertilization; pollination; impregnation

Befugnis die; ~, ~se authority
befühlen tr. V. feel
Befund der (bes. Med.) result[s pl.]
befürchten tr. V. fear; **ich befürchte, daß ...**: I am afraid that ...
befürworten tr. V. support
begabt Adj. talented; **Begabung** die; ~, ~en talent
begann 1. u. 3. Pers. Sg. Prät. v. beginnen
begatten tr. V. mate with; ⟨man⟩ copulate with; **sich** ~: mate; ⟨persons⟩ copulate; **Begattung** die mating; (bei Menschen) copulation
begeben unr. refl. V. (geh.) proceed; make one's way; go; **sich zu Bett** ~: retire to bed; **sich an die Arbeit** ~: commence work
begegnen itr. V.; mit sein jmdm. ~: meet sb.; **sich** (Dat.) ~: meet [each other]; **Begegnung** die; ~, ~en a) meeting; b) (Sport) match
begehen unr. tr. V. a) commit ⟨crime, adultery, indiscretion, sin, suicide, faux-pas, etc.⟩; make ⟨mistake⟩; **eine [furchtbare] Dummheit** ~: do something [really] stupid; b) (geh.: feiern) celebrate
begehren tr. V. desire; **begehrens·wert** Adj. desirable; **begehrlich** 1. Adj. greedy; 2. adv. greedily; **begehrt** Adj. much sought-after
begeistern 1. tr. V. jmdn. [für etw.] ~: fire sb. with enthusiasm [for sth.]; 2. refl. V. get enthusiastic (für about); **begeistert** 1. Adj. enthusiastic (von about); 2. adv. enthusiastically; **Begeisterung** die; ~: enthusiasm
Begierde die; ~, ~n desire (nach for); **begierig** 1. Adj. eager; 2. adv. eagerly
begießen unr. tr. V. water ⟨plants⟩
Beginn der; ~[e]s beginning; [gleich] zu ~: [right] at the beginning; **beginnen** 1. unr. itr. V. start; begin; **mit dem Bau** ~: start or begin building; **dort beginnt der Wald** the forest starts there; 2. unr. tr. V. start; begin; start ⟨argument⟩; ~, **etw. zu tun** start to do sth.
beglaubigen tr. V. certify; **Beglaubigung** die; ~, ~en certification
begleichen unr. tr. V. settle ⟨bill, debt⟩; pay ⟨sum⟩
begleiten tr. V. accompany; **jmdn. nach Hause** ~: see sb. home; **Begleiter** der; ~s, ~, **Begleiterin** die; ~, ~nen companion; (zum Schutz) escort; (Führer(in)) guide; **Begleitung**

die; ~, ~en a) o. Pl. **er bot uns seine** ~ **an** he offered to accompany us; **in** ~ **eines Erwachsenen** accompanied by an adult; b) (Musik) accompaniment
beglückwünschen tr. V. congratulate (zu on)
begnadet Adj. (geh.) divinely gifted;
begnadigen tr. V. pardon; reprieve;
Begnadigung die; ~, ~en reprieving; (Straferlaß) pardon; reprieve
begnügen refl. V. content oneself
Begonie [be'go:niə] die; ~, ~n begonia
begonnen 2. Part. v. beginnen
begraben unr. tr. V. bury; **Begräbnis** das; ~ses, ~se burial; (~feier) funeral
begreifen 1. unr. tr. V. understand; **er konnte nicht** ~, **was geschehen war** he could not grasp what had happened; 2. itr. V. understand; **schnell** od. **leicht/langsam** od. **schwer** ~: be quick/ slow on the uptake; **begreiflich** Adj. understandable
begrenzen tr. V. limit, restrict (**auf** + Akk. to)
Begriff der a) concept; (Terminus) term; (Auffassung) idea; **sich** (Dat.) **keinen** ~ **von etw. machen können** not be able to imagine sth.; **ein/kein** ~ **sein** be/not be well known; c) **im** ~ **sein** od. **stehen, etw. zu tun** be about to do sth.; **begriffs·stutzig** Adj. (abwertend) obtuse
begründen tr. V. a) give reasons for; b) (gründen) found; establish ⟨fame, reputation⟩; **Begründer** der founder; **begründet** Adj. well-founded; reasonable ⟨demand, objection, complaint⟩; **Begründung** die; ~, ~en reason[s]; **mit der** ~, **daß ...**: on the grounds that ...
begrüßen tr. V. a) greet; ⟨hostess, host⟩ welcome; b) (fig.) welcome; **Begrüßung** die; ~, ~en greeting; (von Gästen) welcoming; (Zeremonie) welcome (Gen. for)
begünstigen tr. V. favour
begutachten tr. V. a) examine and report on; b) (ugs.) have a look at
begütert Adj. wealthy
begütigen tr. V. placate
behäbig 1. Adj. slow and ponderous; 2. adv. slowly and ponderously
behagen itr. V. **etw. behagt jmdm.** sb. likes sth.; **Behagen** das; ~s pleasure; **behaglich** 1. Adj. comfortable; 2. adv. comfortably; **Behaglichkeit** die; ~: comfortableness

behalten *unr. tr. V.* **a)** keep; **etw. für sich ~**: keep sth. to oneself; **b)** *(zurück~)* be left with *(scar, defect, etc.)*; **c)** *(sich merken)* remember

Behälter der; **~s**, **~** container; *(für Abfälle)* receptacle

behandeln *tr. V. (auch Med.)* treat; handle *(matter, machine, device)*; deal with *(subject, question etc.)*

Behandlung die; **~**, **~en** treatment

behängen *tr. V.* hang

beharren *itr. V.* **auf etw.** *(Dat.)* **~** *(etw. nicht aufgeben)* persist in sth.; *(auf etw. bestehen)* insist on sth.; **beharrlich 1.** *Adj.* dogged; **2.** *adv.* doggedly; **Beharrlichkeit** die; **~**: doggedness

behauen *unr. tr. V.* hew

behaupten 1. *tr. V.* **a)** maintain; assert; **~**, **jmd. zu sein/etw. zu wissen** claim to be sb./know sth.; **b)** *(verteidigen)* maintain *(position)*; retain *(record)*; **2.** *refl. V.* **a)** assert oneself; *(nicht untergehen)* hold one's ground; *(dableiben)* survive; **b)** *(Sport)* win through; **Behauptung** die; **~**, **~en** assertion

Behausung die; **~**, **~en** dwelling

beheben *unr. tr. V.* remove *(danger, difficulty)*; repair *(damage)*; remedy *(abuse, defect)*; **Behebung** die; **~**, **~en** *s.* beheben: removal; repair; remedying

beheimatet *Adj.* **an einem Ort/in einem Land** *usw.* **~ sein** be native to a place/to a country *etc.*

beheizen *tr. V.* heat

behelfen *unr. refl. V.* make do

behelfs·mäßig 1. *Adj.* makeshift; **2.** *adv.* in a makeshift way

behelligen *tr. V.* bother; *(zudringlich werden gegen)* pester

behend, behende 1. *Adj. (geschickt)* deft; *(flink)* nimble; **2.** *adv.; s. Adj.:* deftly; nimbly

beherbergen *tr. V.* accommodate

beherrschen 1. *tr. V.* **a)** control; rule *(country, people)*; **b)** *(meistern)* control *(vehicle, animal)*; be in control of *(situation)*; **c)** *(bestimmen, dominieren)* dominate *(townscape, landscape, discussions)*; **d)** *(zügeln)* control *(feelings)*; control, curb *(impatience)*; **e)** *(gut können)* have mastered *(instrument, trade)*; have a good command of *(language)*; **2.** *refl. V.* control oneself; **beherrscht 1.** *Adj.* self-controlled; **2.** with self-control; **Beherrschung** die; **~ a)** control; *(eines*

Volks, La:des usw.) rule; **b)** *(das Meistern)* control; **c)** *(Beherrschtheit)* self-control; **d)** *(das Können)* mastery

beherzigen *tr. V.* take *(sth.)* to heart

beherzt 1. *Adj.* spirited; **2.** *adv.* spiritedly

behilflich *Adj.* **[jmdm.]** **~ sein** help [sb.] **(bei with)**

behindern *tr. V.* **a)** hinder; impede *(movement)*; hold up *(traffic)*; **b)** *(Sport, Verkehrsw.)* obstruct; **behindert** *Adj.* handicapped; **Behinderte** der/die; *adj. Dekl.* handicapped person; **die ~n** the handicapped; **WC für ~**: toilet for disabled persons; **Behinderung** die; **~**, **~en a)** hindrance; **b)** *(Sport, Verkehrsw.)* obstruction; **c)** *(Gebrechen)* handicap

Behörde die; **~**, **~n** authority; *(Amt, Abteilung)* department; **behördlich 1.** *Adj.* official; **2.** *adv.* officially

behüten *tr. V.* protect **(vor + Dat. from)**; *(bewachen)* guard

behutsam 1. *Adj.* careful; **2.** *adv.* carefully

bei *Präp. mit Dat.* **a)** *(nahe)* near; *(dicht an, neben)* by; **wer steht da ~ ihm?** who is standing there with him?; **etw. ~ sich haben** have sth. with *or* on one; **sich ~ jmdm. entschuldigen** apologize to sb.; **b)** *(unter)* among; **war heute ein Brief für mich ~ der Post?** was there a letter for me in the post today?; **c)** *(an)* by; **jmdn. ~ der Hand nehmen** take sb. by the hand; **d)** *(im Wohn-/Lebens-/Arbeitsbereich von)*; **~ uns tut man das nicht** we don't do that; **~ mir [zu Hause]** at my house; **~ uns um die Ecke/gegenüber** round the corner from us/opposite us; **~ seinen Eltern leben** live with one's parents; **wir sind ~ ihr eingeladen** we have been invited to her house; **wir treffen uns ~ uns/Peter** we'll meet at our/Peter's place; **~ uns in der Firma** in our company; **~ Schmidt** *(auf Briefen)* c/o Schmidt; **~ einer Firma sein** be with a company; **~ jmdm./einem Verlag arbeiten** work for sb./a publishing house; **e)** *(im Bereich eines Vorgangs)* at; **~ einer Hochzeit/einem Empfang** *usw.* be at a wedding/reception *etc.;* **~ einem Unfall** in an accident; **f)** *(im Werk von)* **~ Goethe** in Goethe; **g)** *(im Falle von)* in the case of; **wie ~ den Römern** as with the Romans; **~ der Hauskatze** in the domestic cat; **h)** *(modal)* **~ Tag/Nacht** by day/night; **~ Tageslicht** by daylight; **~ Nebel** in fog; **i)** *(im Falle des Auftretens von)* „~ Nässe Schleudergefahr" 'slippery when wet'; **j)** *(angesichts)* with; ~ dieser Hitze in this heat; ~ deinen guten Augen/ihrem Talent with your good eyesight/her talent; **k)** *(trotz)* ~ all seinem Engagement/seinen Bemühungen in spite of *or* despite *or* for all his commitment/efforts

bei|behalten *unr. tr. V.* keep; retain; keep up ⟨custom, habit⟩; keep to ⟨course, method⟩; preserve, maintain ⟨way of life; attitude⟩

bei|bringen *unr. tr. V.* **a)** jmdm. etw. ~: teach sb. sth.; **b)** *(ugs.: mitteilen)* jmdm. ~, daß ...: break it to sb. that ...; **c)** *(zufügen)* jmdm./sich etw. ~: inflict sth. on sb./oneself

Beichte die; ~, ~n confession *no def. art.;* **beichten 1.** *itr. V.* confess; **2.** *tr. V. (auch fig.)* confess

Beicht-: ~**stuhl der** confessional; ~**vater der** father confessor

beid... ** *Indefinitpron. u. Zahlw.* **1. *Pl.* ~e both; *(der/die/das eine oder der/die/das andere)* either *sing.;* die ~en the two; die/seine ~en Brüder the/his two brothers; die ~en ersten Strophen the first two verses; kennst du die ~en? do you know those two?; alle ~e both of us/you/them; ihr/euch ~e you two; ihr/euch ~e nicht neither of you; wir/uns ~e the two of us/both of us; er hat ~e Eltern verloren he has lost both [his] parents; mit ~en Händen with both hands; ich habe ~e gekannt I knew both of them; einer/eins von ~en one of the two; keiner/keins von ~en neither [of them]; **2.** *Neutr. Sg.;* ~es both *pl.; (das eine oder das andere)* either; ~es ist möglich either is possible; ich glaube ~es/~es nicht I believe both things/neither thing; das ist ~es nicht richtig neither of those is correct; **beiderlei** *Gattungsz., indekl.* ~ Geschlechts of both sexes; **beider·seits 1.** *Präp. mit Gen.* on both sides of; **2.** *Adv.* on both sides

bei·einander *Adv.* together; ~ Trost suchen seek comfort from each other

Bei·fahrer der, Bei·fahrerin die a) passenger; **b)** *(berufsmäßig)* co-driver; *(im LKW)* driver's mate; **Beifahrer·sitz der** passenger seat; *(eines Motorrads)* pillion

Bei·fall der; *o. Pl.* **a)** applause; **b)** *(Zustimmung)* approval; **bei·fällig 1.** *Adj.* approving; **2.** *adv.* approvingly

beige [be:ʃ] *Adj.* beige; **Beige das;** ~, ~ *od. (ugs.)* ~s beige

Bei·geschmack der: einen bitteren *usw.* ~ haben have a slightly bitter *etc.* taste [to it]

Bei·hilfe die a) aid; *(Zuschuß)* allowance; **b)** *o. Pl. (Rechtsw.: Mithilfe)* aiding and abetting

Beil das; ~[e]s, ~e axe; *(kleiner)* hatchet

Bei·lage die a) *(Zeitungs~)* supplement; **b)** *(zu Speisen)* side-dish; *(Gemüse~)* vegetables *pl.*

bei·läufig 1. *Adj.* casual; **2.** *adv.* casually

bei|legen *tr. V.* **a)** enclose; **b)** *(schlichten)* settle ⟨dispute etc.⟩

Bei·leid das sympathy; |mein| herzliches *od.* aufrichtiges ~! please accept my sincere condolences

bei|liegen *unr. itr. V.* einem Brief ~: be enclosed with a letter; **bei·liegend** *Adj.* enclosed; ~ senden wir ...: please find enclosed ...

beim *Präp. + Art.* **a)** = bei dem; **b)** ~ Film sein be in films; **c)** er will ~ Arbeiten nicht gestört werden he doesn't want to be disturbed when working; ~ Duschen sein be taking a shower

bei|messen *unr. tr. V.* attach

Bein das; ~[e]s, ~e leg; jmdm. ein ~ stellen trip sb.; *(fig.)* put *or* throw a spanner *or (Amer.)* a monkey-wrench in sb.'s works; wieder auf den ~en sein be back on one's feet again

bei·nah[e] *Adv.* almost

Bei·name der the epithet

Bein·bruch der: das ist |doch| kein ~ *(ugs.)* it's not the end of the world

beinhalten *tr. V. (Papierdt.)* involve

-beinig *adj.* -legged

bei|pflichten *itr. V.* agree (*Dat.* with)

beirren *tr. V.* sich durch nichts/von niemandem ~ lassen not be deterred by anything/anybody

beisammen *Adv.* together; **beisammen|haben** *unr. tr. V.* **a)** have got together; **b)** er hat |sie| nicht alle beisammen *(ugs.)* he's not all there *(coll.);* **Beisammen·sein das** get-together

Bei·schlaf der sexual intercourse

Bei·sein das: in jmds. ~: in the presence of sb. *or* in sb.'s presence

bei·seite *Adv.* aside

Beis[e]l das; ~s, ~ *od.* ~n *(österr.)* pub *(Brit. coll.);* bar *(Amer.)*

bei|setzen *tr. V.* lay to rest; inter ⟨ashes⟩; **Bei·setzung die;** ~, ~en funeral; burial

Bei·spiel das example (für of); **zum ~**: for example; **mit gutem ~ vorangehen** set a good example; **beispielhaft** Adj. exemplary; **beispiel·los** Adj. unparalleled; **beispiels·weise** Adv. for example

beißen 1. unr. tr., itr. V. (auch fig.) bite; 2. unr. refl. V. (ugs.) ⟨colours, clothes⟩ clash; **beißend** Adj. biting ⟨cold⟩; acrid ⟨smoke, fumes⟩; sharp ⟨frost⟩; **Beiß·zange** die s. Kneifzange

Bei·stand der o. Pl. (geh.: Hilfe) aid; **bei|stehen** unr. itr. V. jmdm. **~**: aid sb.

bei|steuern tr. V. contribute

Beitrag der; ~[e]s, Beiträge contribution; (Versicherungs~) premium; (Mitglieds~) subscription; **bei|tragen** unr. tr., itr. V. contribute (zu to)

bei|treten unr. itr. V.; mit sein join ⟨union, club, etc.⟩; **einem Abkommen/Pakt** accede to ⟨pact, agreement⟩; **Bei·tritt** der joining

Bei·wagen der side-car

Bei·werk das; o. Pl. accessories pl.

bei|wohnen itr. V. **einer Sache** (Dat.) **~** (geh.) be present at sth.

Beize die; ~, ~n (Holzbearb.) [wood]-stain

beizeiten Adv. in good time

beizen tr. V. (Holzbearb.) stain

bejahen [bə'ja:ən] tr. V. **a)** auch itr. answer ⟨sth.⟩ in the affirmative; **b)** (gutheißen) approve of; **das Leben ~**: have a positive or an affirmative attitude to life; **Bejahung** die; ~, ~en **a)** affirmative reply; **b)** (das Gutheißen) approval

bejammern tr. V. lament

bejubeln tr. V. cheer; acclaim

bekämpfen tr. V. **a)** fight against; **b)** combat ⟨disease, epidemic, pest, unemployment, crime, etc.⟩; **Bekämpfung** die; ~ **a)** fight (Gen. against); **b)** s. bekämpfen **b**: combating

bekannt Adj. well-known; **b)** jmd./etw. **ist jmdm. ~**: sb. knows sb./sth.; **Darf ich ~ machen? Meine Eltern** may I introduce my parents?; **Bekannte** der/die; adj. Dekl. acquaintance

Bekannt·gabe die; **~**: announcement; **bekannt|geben** unr. tr. V. announce

bekanntlich Adv. as is well known; **etw. ist ~ der Fall** sth. is known to be the case

bekannt|machen tr. V. announce; (der Öffentlichkeit) make public; **Be-**

kannt·machung die; ~, ~en announcement

Bekanntschaft die; ~, ~en acquaintance

bekannt|werden unr. itr. V.; mit sein (nur im Inf. und 2. Part. zusammengeschr.) become known

bekehren 1. tr. V. convert; 2. refl. V. become converted; **Bekehrung** die; ~, ~en (auch fig.) conversion (zu to)

bekennen 1. unr. tr. V. **a)** confess; **~, daß ...** admit that ...; **b)** (Rel.) profess; 2. refl. V. **sich zum Islam ~**: profess Islam; **sich zu Buddha ~**: profess one's faith in Buddha; **sich zu seiner Schuld ~**: confess one's guilt; **sich schuldig/nicht schuldig ~**: confess/not confess one's guilt; (vor Gericht) plead guilty/not guilty; **Bekenntnis** das; ~ses, ~se **a)** confession; **b)** (Eintreten) **ein ~ zum Frieden** a declaration for peace; **c)** (Konfession) denomination

beklagen 1. tr. V. (geh.) **a)** (betrauern) mourn; **b)** (bedauern) lament; 2. refl. V. complain

bekleckern tr. V. (ugs.) etw./sich |mit Soße usw.| **~**: drop or spill sauce etc. down sth./oneself

bekleiden tr. V. **a)** clothe; **mit etw. bekleidet sein** be wearing sth.; **b)** (geh.: innehaben) occupy ⟨office, position⟩; **Bekleidung** die clothing; clothes pl.

beklemmend Adj. oppressive; **Beklemmung** die; ~, ~en oppressive feeling; **beklommen** Adj. uneasy; (stärker) apprehensive

bekloppt Adj. (salopp) barmy (Brit. sl.); loony (sl.)

beknien tr. V. (ugs.) beg

bekommen 1. unr. tr. V. **a)** get; get, receive ⟨money, letter, reply, news, orders⟩; (erreichen) catch ⟨train, bus, flight⟩; **was ~ Sie?** (im Geschäft) can I help you?; (im Lokal, Restaurant) what would you like?; **was ~ Sie |dafür|?** how much is that?; **Hunger/Durst ~**: get hungry/thirsty; **Angst/Mut ~**: become frightened/take heart; **er bekommt einen Bart** he's growing a beard; **sie bekommt eine Brust** her breasts are developing; **Zähne ~**: ⟨baby⟩ teethe; **sie bekommt ein Kind** she's expecting a baby; **b)** etw. **durch die Tür/ins Auto ~**: get sth. through the door/into the car; 2. unr. V.; in der Funktion eines Hilfsverbs zur Umschreibung des Passivs get; **etw. geschenkt ~**: get [given] sth. or be given sth. as a present; **3.** *unr. itr. V.; mit sein* jmdm. **gut** ~: do sb. good; **jmdm. |gut|** ~: ⟨*food, medicine*⟩ agree with sb.; **wohl bekomm's!** your [very good] health!

bekömmlich *Adj.* easily digestible

beköstigen *tr. V.* cater for

bekräftigen *tr. V.* reinforce ⟨*statement*⟩; reaffirm ⟨*promise*⟩

bekreuzigen *refl. V. (kath. Kirche)* cross oneself

bekriegen *tr. V.* wage war on; *(fig.)* fight; **sich** ~: be at war; *(fig.)* fight

bekümmern *tr. V.* jmdn. ~: cause sb. worry; **bekümmert** *Adj.* worried; *(stärker)* distressed

bekunden *tr. V.* express

belächeln *tr. V.* smile [pityingly/ tolerantly *etc.*] at

beladen *unr. tr. V.* load ⟨*ship*⟩; load [up] ⟨*car, wagon*⟩; load up ⟨*horse, donkey*⟩

Belag der; ~|e|s, **Beläge a)** coating; **b)** *(Fußboden~)* covering; *(Straßen~)* surface; *(Brems~)* lining; **c)** *(von Kuchen, Scheibe Brot usw.)* topping; *(von Sandwiches)* filling

belagern *tr. V. (auch fig.)* besiege; **Belagerung** die; ~, ~en siege; *(fig.)* besieging

Belang der; ~|e|s, ~e a) von/ohne ~ sein be of importance/of no importance; **b)** *Pl. (Interessen)* interests

belangen *tr. V. (Rechtsw.)* sue; *(strafrechtlich)* prosecute

belang·los *Adj. (trivial)* trivial; *(unerheblich)* of no importance (für for); **Belanglosigkeit** die; ~, ~en unimportance; *(Trivialität)* triviality

belassen *unr. tr. V.* leave

belasten *tr. V.* **a)** etw. ~: put sth. under strain; *(durch Gewicht)* put weight on sth.; **b)** *(beeinträchtigen)* pollute ⟨*atmosphere*⟩; put pressure on ⟨*environment*⟩; **c)** *(in Anspruch nehmen)* burden (mit with); **d)** jmdn. ~ ⟨*responsibility, guilt*⟩ weigh upon sb.; ⟨*thought*⟩ weigh upon sb.'s mind; **e)** *(Rechtsw.)* incriminate

belästigen *tr. V.* bother; *(sehr aufdringlich)* pester; *(sexuell)* molest

Belastung die; ~, ~en a) strain; *(das Belasten)* straining; *(durch Gewicht)* loading; *(Last)* load; **b)** **die** ~ **der Atmosphäre/Umwelt durch Schadstoffe** the pollution of the atmosphere by harmful substances/the pressure on the environment caused by harmful substances; **c)** *(Bürde, Sorge)* burden

belaufen *unr. refl. V.* **sich auf ...**⟨*Akk.*⟩ ~: come to ...

belauschen *tr. V.* eavesdrop on

beleben 1. *tr. V.* enliven; stimulate ⟨*economy*⟩; **2.** *refl. V.* ⟨*market, economic activity*⟩ revive, pick up; **belebend 1.** *Adj.* invigorating; **2.** *adv.* ~ **wirken** have an invigorating effect; **belebt** *Adj.* busy ⟨*street, crossing, town, etc.*⟩

Beleg der; ~|e|s, ~e *(Beweisstück)* piece of [supporting] documentary evidence; *(Quittung)* receipt

belegen *tr. V.* **a)** *(Milit.: beschießen)* bombard; *(mit Bomben)* attack; **b)** *(mit Belag versehen)* cover ⟨*floor*⟩ (mit with); fill ⟨*flan base, sandwich*⟩; top ⟨*open sandwich*⟩; **eine Scheibe Brot mit Käse** ~: put some cheese on a slice of bread; **c)** *(in Besitz nehmen)* occupy ⟨*seat, room, etc.*⟩; **d)** *(Hochschulw.)* enrol for ⟨*seminar, lecture-course*⟩; **e)** **den ersten/letzten Platz** ~ *(Sport)* take first place/come last; **f)** *(nachweisen)* prove; give a reference for ⟨*quotation*⟩

Belegschaft die; ~, ~en staff

belegt *Adj.* **a)** **ein** ~**es Brot** an open *or (Amer.)* openface sandwich; *(zugeklappt)* a sandwich; **ein** ~**es Brötchen** a roll with topping; an open-face roll *(Amer.)*; *(zugeklappt)* a filled roll; a sandwich roll *(Amer.)*; **b)** *(mit Belag bedeckt)* furred ⟨*tongue, tonsils*⟩; **c)** *(heiser)* husky ⟨*voice*⟩; **d)** *(nicht mehr frei)* ⟨*room, flat*⟩ occupied

belehren *tr. V.* teach; instruct; *(aufklären)* enlighten; *(informieren)* inform; **ich lasse mich nicht gern** ~: I'm quite willing to believe otherwise; **Belehrung** die; ~, ~en instruction; *(Zurechtweisung)* lecture

beleibt *Adj. (geh.)* portly

beleidigen *tr. V.* insult; **beleidigt** *Adj.* insulted; *(gekränkt)* offended; **Beleidigung** die; ~, ~en a) insult; **b)** *(Rechtsw.) (schriftlich)* libel; *(mündlich)* slander

belesen *Adj.* well-read

beleuchten *tr. V.* light up; light ⟨*stairs, room, street, etc.*⟩; **Beleuchtung** die; ~, ~en a) lighting; *(Anstrahlung)* illumination

beleumdet *Adj.* übel/gut ~ sein have a bad/good reputation

Belgien ['bɛlgiən] **(das)**; ~s Belgium; **Belgier** der; ~s, ~ Belgian; **belgisch** *Adj.* Belgian

belichten *tr. V. (Fot.)* expose; *itr. richtig/falsch/kurz* ~: use the right/

wrong exposure/a short exposure time; **Belichtung** die *(Fot.)* exposure
Belieben das; ~s: **nach** ~: just as you/they *etc.* like
beliebig 1. *Adj.* any; **2.** *adv.* as you like/he likes *etc.;* ~ **lange/viele** as long/many as you like/he likes *etc.*
beliebt *Adj.* popular; favourite *attrib.;* **Beliebtheit** die; ~: popularity
beliefern *tr. V.* supply
bellen *itr. V.* bark
belohnen *tr. V.* reward ⟨*person, thing*⟩; **Belohnung** die; ~, ~en reward
belügen *unr. tr. V.* lie to
belustigen *tr. V.* amuse; **Belustigung** die; ~, ~en amusement
bemächtigen *refl. V.* sich jmds./einer Sache ~ *(geh.)* seize sb./sth.
bemalen *tr. V.* paint; *(verzieren)* decorate
bemängeln *tr. V.* find fault with
bemerkbar *Adj.* sich ~ **machen** attract attention [to oneself]; *(erkennbar werden)* become apparent; *(spürbar werden)* make itself felt; **bemerken** *tr. V.* **a)** *(wahrnehmen)* notice; **ich wurde nicht bemerkt** I was unobserved; **b)** *(äußern)* remark; **bemerkenswert 1.** *Adj.* remarkable; **2.** *adv.* remarkably; **Bemerkung** die; ~, ~en **a)** *(Äußerung)* remark; comment; **b)** *(Notiz)* note; *(Anmerkung)* comment
bemitleiden *tr. V.* pity; feel sorry for; **bemitleidens·wert** *Adj.* pitiable
bemogeln *tr. V. (ugs.)* cheat; diddle *(Brit. sl.)*
bemühen *refl. V.* make an effort; **sich ~, etw. zu tun** endeavour to do sth.; **sich um etw. ~:** try to obtain sth.; **sich um eine Stelle ~:** try to get a job; **sich um jmdn. ~** *(kümmern)* seek to help sb.; **Bemühung** die; ~, ~en effort
benachbart *Adj.* neighbouring *attrib.*
benachrichtigen *tr. V.* notify (von of); **Benachrichtigung** die; ~, ~en notification
benachteiligen *tr. V.* put at a disadvantage; *(diskriminieren)* discriminate against
benehmen *unr. refl. V.* behave; **Benehmen** das; ~s behaviour; **kein ~ haben** have no manners *pl.*
beneiden *tr. V.* envy; **jmdn. um etw. ~:** envy sb. sth.; **beneidens·wert** *Adj.* enviable
Benelux·länder *Pl.* Benelux countries
benennen *unr. tr. V.* name

Bengel der; ~s, ~ od. *(nordd.)* ~s **a)** *(abwertend: junger Bursche)* young rascal; **b)** *(fam.: kleiner Junge)* little lad
benommen *Adj.* dazed; *(durch Fieber, Alkohol)* muzzy
benoten *tr. V.* mark *(Brit.)*; grade *(Amer.)*; **einen Test mit „gut" ~:** mark a test 'good' *(Brit.)*; assign a grade of 'good' to a test *(Amer.)*
benötigen *tr. V.* need; require
benutzen *tr. V.* use; **Benutzer** der; ~s, ~: user; **Benutzung** die; ~: use
Benzin das; ~s petrol *(Brit.)*; gasoline *(Amer.)*; gas *(Amer. coll.)*; *(Wasch~)* benzine
Benzol das; ~s, ~e *(Chemie)* benzene
beobachten *tr. V.* observe; watch; **Beobachter** der; ~s, ~: observer; **Beobachtung** die; ~, ~en observation
bepacken *tr. V.* load
bepflanzen *tr. V.* plant
bequem 1. *Adj.* **a)** comfortable; **b)** *(abwertend: träge)* idle; **2.** *adv.* **a)** comfortably; **b)** *(leicht)* easily; **bequemen** *refl. V.* sich dazu ~, etw. zu tun *(geh.)* condescend to do sth.; **Bequemlichkeit** die; ~ **a)** comfort; **b)** *(Trägheit)* idleness
berappen *tr., itr. V. (ugs.) s.* blechen
beraten 1. *unr. tr. V.* **a)** advise; **jmdn. gut/schlecht ~:** give sb. good/bad advice; **b)** *(besprechen)* discuss ⟨*plan, matter*⟩; **2.** *unr. itr. V.* **über etw.** *(Akk.)* **~:** discuss sth. **3.** *unr. refl. V.* **sich mit jmdn. ~, ob ...:** discuss with sb. whether ...; **Berater** der; ~s, ~: adviser; **beratschlagen 1.** *tr. V.* discuss; **2.** *itr. V.* **über etw.** *(Akk.)* **~:** discuss sth.; **Beratung** die; ~, ~en **a)** advice *no indef. art.;* *(durch Arzt, Rechtsanwalt)* consultation; **b)** *(Besprechung)* discussion
berauben *tr. V. (auch fig.)* rob (Gen. of)
berauschen *(geh.)* **1.** *tr. V. (auch fig.)* intoxicate; **2.** *refl. V.* become intoxicated (an + Dat. with)
berechnen *tr. V.* **a)** *(auch fig.)* calculate; predict ⟨*behaviour, consequences*⟩; **b)** *(anrechnen)* charge; **jmdm. 10 Mark für etw.** *od.* **jmdm. etw. mit 10 Mark ~:** charge sb. 10 marks for sth.; **jmdm. zuviel ~:** overcharge sb.; **Berechnung** die **a)** calculation; **b)** *o. Pl. (Eigennutz)* [calculating] self-interest
berechtigen *tr. V.* entitle; *itr. die*

Karte berechtigt zum Eintritt the ticket entitles the bearer to admission; **berechtigt** *Adj.* **a)** *(gerechtfertigt)* justified; **b)** *(befugt)* authorized; **Berechtigung die;** ~, **~en a)** *(Befugnis)* entitlement; *(Recht)* right; **b)** *(Rechtmäßigkeit)* legitimacy

bereden *tr. V.* **a)** *(besprechen)* discuss; **b) jmdn.** ~, **etw. zu tun** talk sb. into doing sth.

Bereich der; ~|e|s, ~e area; **im privaten/staatlichen** ~: in the private/public sector

bereichern *refl. V.* get rich; **Bereicherung die;** ~, **~en a)** money-making; **b)** *(Nutzen)* valuable acquisition

bereifen *tr. V.* put tyres on ⟨*car*⟩; put a tyre on ⟨*wheel*⟩; **Bereifung die;** ~, **~en** [set *sing.* of] tyres *pl.*

bereinigen *tr. V.* clear up ⟨*misunderstanding*⟩; settle, resolve ⟨*dispute*⟩

bereit *Adj.* ready; ~ **sein, etw. zu tun** be ready *or* willing to do sth.

bereiten *tr. V.* **a)** prepare; make ⟨*tea, coffee*⟩; **b)** *(verursachen)* cause ⟨*trouble, sorrow, difficulty, etc.*⟩

bereit-: ~|**halten** *unr. tr. V.* have ready; ~|**legen** *tr. V.* lay out ready; ~|**liegen** *unr. itr. V.* be ready

bereits *Adv.* already

Bereitschaft die; ~: readiness; willingness; **Bereitschafts·dienst der:** ~**dienst haben** ⟨*doctor, nurse*⟩ be on call; ⟨*policeman, fireman*⟩ be on stand-by duty; ⟨*chemist's*⟩ be on rota duty *(for dispensing outside normal hours)*

bereit-: ~|**stehen** *unr. itr. V.* be ready; ~|**stellen** *tr. V.* place ready; get ready ⟨*food, drinks*⟩; ready, make ⟨*money, funds*⟩ available; ~**willig 1.** *Adj.* willing. **2.** *adv.* readily

bereuen 1. *tr. V.* regret; **2.** *itr. V.* be sorry; *(Rel.)* repent

Berg der; ~|e|s, ~e **a)** hill; *(im Hochgebirge)* mountain; **b)** *(Haufen)* huge pile; *(von Akten, Abfall auch)* mountain

berg-, Berg-: ~**ab** [-'-] *Adv.* downhill; ~**auf** [-'-] *Adv.* uphill; ~**bahn die** mountain railway; *(Seilbahn)* mountain cableway; ~**bau der;** *o. Pl.* mining

bergen *unr. tr. V.* **a)** rescue, save ⟨*person*⟩; salvage ⟨*ship, cargo, belongings*⟩; **b)** *(geh.: enthalten)* hold

Berg-: ~**führer der** mountain guide; ~**hütte die** mountain hut

bergig *Adj.* hilly; *(mit hohen Bergen)* mountainous

Berg-: ~**kristall der** rock crystal; ~**land das** hilly country *no indef. art.; (mit hohen Bergen)* mountainous country *no indef. art.;* ~**mann der;** *Pl.* ~**leute** miner; ~**station die** top station; ~**steigen das;** ~s mountaineering *no art.;* ~**steiger der** mountaineer

Bergung die; ~, **~en a)** rescue; **b)** *(von Schiffen, Gut)* salvaging

Berg-: ~**wacht die** mountain rescue service; ~**werk das** mine

Bericht der; ~|e|s, ~e report; **berichten** *tr., itr. V.* report

Bericht-: ~**erstatter der;** ~s, ~: reporter; ~**erstattung die** reporting *no indef. art.*

berichtigen *tr. V.* correct; **Berichtigung die;** ~, **~en** correction

berieseln *tr. V.* **a)** *(bewässern)* irrigate; **b) sich ständig mit Musik ~ lassen** *(ugs. abwertend)* constantly have music on in the background

Berlin *(das);* ~s Berlin; **Berliner 1.** *Adj.; nicht präd.* Berlin; **2. der;** ~s, ~: **a)** Berliner; **b)** (~ *Pfannkuchen)* [jam *(Brit.)* or *(Amer.)* jelly] doughnut; **berlinisch** *Adj.* Berlin *attrib.*

Bern *(das);* ~s Bern[e]

Bernhardiner der; ~s, ~: St. Bernard [dog]

Bern·stein der; *o. Pl.* amber

bersten *unr. itr. V.; mit sein (geh.)* ⟨*ice*⟩ break up; ⟨*glass*⟩ shatter [into pieces]; ⟨*wall*⟩ crack up

berüchtigt *Adj.* notorious **(wegen** for); *(verrufen)* disreputable

berücksichtigen *tr. V.* take into account; consider ⟨*applicant, application, suggestion*⟩; **Berücksichtigung die;** ~: **bei ~ aller Umstände** taking all the circumstances into account

Beruf der; ~|e|s, ~e occupation; *(akademischer)* profession; *(handwerklicher)* trade; **was sind Sie von ~?** what do you do for a living?

¹berufen 1. *unr. tr. V.* **a)** *(einsetzen)* appoint; **b) berufe es nicht!** *(ugs.)* don't speak too soon!; **2.** *unr. refl. V.* **sich auf etw. (Akk.)** ~: refer to sth.; **sich auf jmdn.** ~: quote *or* mention sb.'s name

²berufen *Adj.* **a)** competent; **aus ~em Munde** from somebody qualified to speak; **b) sich dazu ~ fühlen, etw. zu tun** feel called to do sth.

beruflich 1. *Adj.; nicht präd.* vocational ⟨*training etc.*⟩; *(bei akademischen Berufen)* professional ⟨*training etc.*⟩; **2.** *adv.* ~ erfolgreich sein be successful in one's career; **sich ~ weiterbilden** undertake further job training
berufs-, Berufs-: ~ausbildung die vocational training; **~beratung die** vocational guidance; **~erfahrung die;** *o. Pl.* [professional] experience; **~geheimnis das** professional secret; *(Schweigepflicht)* professional secrecy; **~krankheit die** occupational disease; **~leben das** working life; **~schule die** vocational school; **~soldat der** regular soldier; **~sportler der** professional sportsman; **~tätig** *Adj.* working *attrib.;* **~tätige der/die;** *adj. Dekl.* working person; **~tätige** *Pl.* working people; **~verkehr der** rush-hour traffic
Berufung die; ~, ~en a) *(für ein Amt)* offer of an appointment (**auf, in, an +** *Akk.* to); **b)** *(innerer Auftrag)* vocation; **c)** *(das Sichberufen)* **unter ~** *(Dat.)* **auf jmdn./etw.** referring *or* with reference to sb./sth.; **d)** *(Rechtsw.: Einspruch)* appeal; **~ einlegen** lodge an appeal
beruhen *itr. V.* **auf etw.** *(Dat.)* ~: be based on sth.; **etw. auf sich ~ lassen** let sth. rest
beruhigen [bə'ru:ıgn] **1.** *tr. V.* calm [down]; pacify ⟨*child, baby*⟩; salve ⟨*conscience*⟩; (trösten) soothe; *(von einer Sorge befreien)* reassure; **2.** *refl. V.* ⟨*person*⟩ calm down; ⟨*sea*⟩ become calm; **Beruhigung die;** ~ *s.* beruhigen 1: calming [down]; pacifying; salving; soothing; reassurance; **Beruhigungs·mittel das** tranquillizer
berühmt *Adj.* famous; **berühmt-berüchtigt** *Adj.* notorious; **Berühmtheit die;** ~, ~en **a)** *o. Pl. (Ruhm)* fame; **b)** *(Mensch)* celebrity
berühren *tr. V.* **a)** touch on ⟨*topic, issue. etc.*⟩; **sich ~:** touch; **b)** *(beeindrucken)* affect; **das berührt mich nicht** it's a matter of indifference to me; **Berührung die;** ~, ~en; **mit jmdm./etw. in ~** *(Akk.)* **kommen** *(auch fig.)* come into contact with sb./sth.
besagen *tr. V.* say; *(bedeuten)* mean
besänftigen *tr. V.* calm [down]; pacify; calm, soothe ⟨*temper*⟩
Besatz der *(Borte)* trimming *no indef. art.*
Besatzung die a) *(Mannschaft)* crew;

b) *(Milit.: Verteidigungstruppe)* garrison; **c)** *(Milit.: Okkupationstruppen)* occupying forces *pl.*
besaufen *unr. refl. V. (salopp)* get canned *(Brit. sl.) or* bombed *(Amer. sl.);* **Besäufnis das;** ~ses, ~se *(salopp)* booze-up *(Brit. sl.):* blast *(Amer. sl.)*
beschädigen *tr. V.* damage; **Beschädigung die a)** *o. Pl.* damaging; **b)** *(Schaden)* damage
¹beschaffen *tr. V.* obtain, get *(Dat.* for)
²beschaffen *Adj.* **so ~ sein, daß ...:** be such that ...; **Beschaffenheit die;** ~: properties *pl.*
Beschaffung die *s.* beschaffen: obtaining; getting
beschäftigen 1. *refl. V.* occupy oneself; **sich viel mit Musik/den Kindern ~:** devote a great deal of one's time to music/the children; **sehr beschäftigt sein** be very busy; **2.** *tr. V.* **a)** *(geistig in Anspruch nehmen)* jmdn. ~: preoccupy sb.; **b)** *(angestellt haben)* employ ⟨*workers, staff*⟩; **c)** *(zu tun geben)* occupy; **jmdn. mit etw. ~:** give sb. sth. to occupy him/her; **Beschäftigte der/die;** *adj. Dekl.* employee; **Beschäftigung die;** ~, ~en **a)** *(Tätigkeit)* activity; **b)** *(Anstellung, Stelle)* job; **c)** *(mit einer Frage, einem Problem)* consideration (**mit** of); *(Studium)* study (**mit** of); **d)** *o. Pl. (von Arbeitskräften)* employment
beschämen *tr. V.* shame; **beschämend 1.** *Adj.* **a)** *(schändlich)* shameful; **b)** *(demütigend)* humiliating; **2.** *adv.* shamefully; **beschämt** *Adj.* ashamed; **Beschämung die;** ~: shame
beschatten *tr. V.* **a)** *(geh.)* shade; **b)** *(überwachen)* shadow
beschaulich 1. *Adj.* peaceful ⟨*life, manner, etc.*⟩; **2.** *adv.* peacefully
Bescheid der; ~[e]s, ~e **a)** *(Auskunft)* information; *(Antwort)* answer; reply; **jmdm. ~ geben** *od.* **sagen[, ob ...]** let sb. know or tell sb. [whether ...]; **sage bitte im Hotel ~, daß ...:** please let the hotel know that ...; **[über etw.** *(Akk.)***] ~ wissen** know [about sth.]; **b)** *(Entscheidung)* decision
¹bescheiden 1. *unr. tr. V.* jmdn./etw. abschlägig ~: turn sb./sth. down; **2.** *unr. refl. V. (geh.)* be content
²bescheiden 1. *Adj.* modest; **2.** *adv.* modestly; **Bescheidenheit die;** ~: modesty

bescheinigen *tr. V.* confirm ⟨*sth.*⟩ in writing; **Bescheinigung** *die;* ~, ~en written confirmation *no indef. art.; (Schein, Attest)* certificate

beschenken *tr. V.* give ⟨*sb.*⟩ a present/presents

bescheren *tr. V.* jmdn. [mit etw.] ~: give sb. [sth. as] a Christmas present/ Christmas presents

Bescherung *die;* ~, ~en a) *(zu Weihnachten)* giving out of the Christmas presents; b) **das ist ja eine schöne** ~ *(ugs.)* this is a pretty kettle of fish

beschießen *unr. tr. V.* fire at; *(mit Artillerie)* bombard

beschimpfen *tr. V.* abuse; swear at; **Beschimpfung** *die;* ~, ~en insult; ~en abuse *sing.;* insults

Beschlag *der* a) fitting; b) **jmdn./etw. mit ~ belegen** *od.* **in ~ nehmen** monopolize sb./sth.; **¹beschlagen** 1. *unr. tr. V.* shoe ⟨*horse*⟩; 2. *unr. itr. V.; mit sein* ⟨*window*⟩ mist up *(Brit.),* fog up *(Amer.); (durch Dampf)* steam up

²beschlagen *Adj.* knowledgeable

Beschlag·nahme *die;* ~, ~n confiscation; **beschlag·nahmen** *tr. V.* confiscate

beschleunigen 1. *tr. V.* accelerate; speed up ⟨*work, delivery*⟩; quicken ⟨*pace, step[s], pulse*⟩; 2. *refl. V.* ⟨*heartrate*⟩ increase; ⟨*pulse*⟩ quicken; 3. *itr. V.* ⟨*driver, car, etc.*⟩ accelerate; **Beschleunigung** *die;* ~, ~en *s.* beschleunigen 1: acceleration; speeding up; quickening

beschließen *unr. tr. V.* a) decide; pass ⟨*law*⟩; ~, **etw. zu tun** decide *or* resolve to do sth.; b) *(beenden)* end

Beschluß *der* decision; *(gemeinsam gefaßt)* resolution; **einen ~ fassen** come to a decision/pass a resolution; **beschluß·fähig** *Adj.* quorate; **Beschluß·fähigkeit** *die;* o. *Pl.* presence of a quorum

beschmieren *tr. V.* etw./sich ~: get sth./oneself in a mess

beschmutzen *tr. V.* make ⟨*sth.*⟩ dirty

beschneiden *unr. tr. V.* a) cut ⟨*hedge*⟩; prune ⟨*bush*⟩; cut back ⟨*tree*⟩; **einem Vogel die Flügel ~:** clip a bird's wings; b) *(Med., Rel.)* circumcise; **Beschneidung** *die;* ~, ~en a) *s.* beschneiden a: cutting; pruning; cutting back; b) *(Med., Rel.)* circumcision

beschönigen *tr. V.* gloss over

beschränken 1. *tr. V.* restrict (**auf** + *Akk.* to); 2. *refl. V.* sich auf etw. *(Akk.)*

~: restrict oneself to sth.; **beschränkt** 1. *Adj.* a) *(dumm)* dullwitted; b) *(engstirnig)* narrowminded; 2. *adv.* narrow-mindedly; **Beschränktheit** *die;* ~: a) *(Dummheit)* lack of intelligence; b) *(Engstirnigkeit)* narrow-mindedness; **Beschränkung** *die;* ~, ~en restriction

beschreiben *unr. tr. V.* a) write on; *(vollschreiben)* write ⟨*page, side, etc.*⟩; b) *(darstellen)* describe; **Beschreibung** *die;* ~, ~en description

beschriften *tr. V.* label; inscribe ⟨*stone*⟩; letter ⟨*sign, label, etc.*⟩; *(mit Adresse)* address

beschuldigen *tr. V.* accuse (*Gen.* of); **Beschuldigte** *der/die; adj. Dekl.* accused; **Beschuldigung** *die;* ~, ~en accusation

beschummeln *tr. V.* *(ugs.)* cheat; diddle *(Brit. coll.)*

Beschuß *der* fire

beschützen *tr. V.* protect (**vor** + *Dat.* from); **Beschützer** *der;* ~s, ~, **Beschützerin** *die;* ~, ~nen protector

Beschwerde *die;* ~, ~n a) complaint (**gegen, über** + *Akk.* about); b) *Pl. (Schmerz)* pain *sing.; (Leiden)* trouble *sing.*

beschweren 1. *refl. V.* complain (**über** + *Akk.,* **wegen** about); 2. *tr. V.* weight down; **beschwerlich** *Adj.* arduous; *(ermüdend)* exhausting

beschwichtigen *tr. V.* pacify; mollify ⟨*anger etc.*⟩; **Beschwichtigung** *die;* ~, ~en pacification; *(des Zorns usw.)* mollification

beschwingt *Adj.* lively

beschwipst *Adj. (ugs.)* tipsy

beschwören *unr. tr. V.* a) swear to; ~, **daß** ...; swear that ...; **eine Aussage** ~: swear a statement on oath; b) charm ⟨*snake*⟩; c) *(erscheinen lassen)* invoke ⟨*spirit*⟩; d) *(bitten)* implore; **Beschwörung** *die;* ~, ~en a) *(Zauberspruch)* spell; incantation; b) *s.* beschwören c: invoking; c) *(Bitte)* entreaty

beseitigen *tr. V.* remove; eliminate ⟨*error, difficulty*⟩; dispose of ⟨*rubbish*⟩; **Beseitigung** *die;* ~: *s.* beseitigen: removal; elimination; disposal

Besen *der;* ~s, ~ broom; **ich fress' einen ~, wenn das stimmt** *(salopp)* I'll eat my hat if that's right *(coll.);* **neue ~ kehren gut** *(Spr.)* a new broom sweeps clean *(prov.)*

besessen *Adj.* a) possessed; b) *(fig.)* obsessive ⟨*gambler*⟩; **von einer Idee** ~

sein be obsessed with an idea; **Besessenheit die**; ~ a) possession; b) obsessiveness

besetzen *tr. V.* **a)** *(mit Pelz, Spitzen)* edge; trim; **mit Perlen besetzt** set with pearls; **b)** *(belegen; auch Milit.: erobern)* occupy; **c)** *(vergeben)* fill ⟨*post, position, role, etc.*⟩; **besetzt** *Adj.* occupied; ⟨*table, seat*⟩ taken *pred.; (gefüllt)* full; *(Fernspr.)* engaged; busy *(Amer.);* **Besetzung die**; ~, ~en a) *(einer Stellung)* filling; b) *(Film, Theater usw.)* cast; c) *(Eroberung)* occupation

besichtigen *tr. V.* see ⟨*sights*⟩; see the sights of ⟨*town*⟩; view ⟨*house etc. for sale*⟩; **Besichtigung die**; ~, ~en **zur ~ der Stadt/des Schlosses/der Wohnung** to see the sights of the town/to see the castle/to view the flat

besiedeln *tr. V.* settle

besiegen *tr. V.* defeat

besinnen *unr. refl. V.* **a)** think it over; **b) sich [auf jmdn./etw.] ~:** remember [sb./sth.]; **Besinnung die**; ~: consciousness; **die ~ verlieren** faint; **[wieder] zur ~ kommen** regain consciousness; **besinnungs·los 1.** *Adj.* unconscious; **2.** *adv.* mindlessly

Besitz der a) property; **b)** *(das Besitzen)* possession; **im ~ einer Sache** *(Gen.)* **sein** be in possession of sth.; **besitzen** *unr. tr. V.* own; have ⟨*quality, talent, etc.*⟩; *(nachdrücklicher)* possess; **Besitzer der**; ~s, ~, **Besitzerin die**; ~, ~nen owner

besoffen *Adj. (salopp)* canned *(Brit. sl.);* bombed *(Amer. sl.);* **Besoffene der/die;** *adj. Dekl. (salopp)* drunk

besohlen *tr. V.* sole; **neu ~:** resole

besonder... *Adj.; nicht präd.* special; **ein ~es Ereignis** an unusual or a special event; **keine ~e Leistung** no great achievement; **Besonderheit die**; ~, ~en special feature; *(Eigenart)* peculiarity; **besonders 1.** *Adv.* particularly; **2.** *Adj.; nicht attr.; nur verneint (ugs.)* **nicht ~ sein** be nothing special

besonnen 1. *Adj.* prudent; **2.** *adv.* prudently; **Besonnenheit die**; ~: prudence

besorgen *tr. V.* **a)** get; *(kaufen)* buy; **b)** *(erledigen)* take care of; **Besorgnis die**; ~, ~se concern; **besorgt 1.** *Adj.* concerned (um about); **2.** *adv.* with concern; **Besorgung die**; ~, ~en purchase

bespitzeln *tr. V.* spy on

besprechen *unr. tr. V.* discuss; *(re-*zensieren)* review; **Besprechung die**; ~, ~en discussion; *(Konferenz)* meeting; *(Rezension)* review

bespritzen *tr. V.* **a)** splash; *(mit einem Wasserstrahl)* spray; **b)** *(beschmutzen)* bespatter

besprühen *tr. V.* spray

besser 1. *Adj.* **a)** better; **um so ~:** so much the better; **b)** *(sozial höher gestellt)* superior; **2.** *adv.* **[immer] alles ~ wissen** always know better; **es ~ haben** be better off; **~ gesagt** to be [more] precise; **3.** *Adv. (lieber)* **das läßt du ~ sein** *od. (ugs.)* **bleiben** you'd better not do that

besser|gehen *unr. itr. V.; mit sein* **jmdm. geht es besser** sb. feels better

bessern 1. *refl. V.* improve; ⟨*person*⟩ mend one's ways; **2.** *tr. V.* improve; reform ⟨*criminal*⟩; **Besserung die**; ~: recovery; **gute ~!** get well soon

best... 1. *Adj.* **a)** best; **bei ~er Gesundheit/Laune sein** be in the best of health/spirits *pl.;* **im ~en Falle** at best; **in den ~en Jahren, im ~en Alter** in one's prime; **~e Grüße an ...** *(Akk.)* best wishes to ...; **mit den ~en Grüßen** *od.* **Wünschen** with best wishes; *(als Briefschluß)* ≈ yours sincerely; **b) es ist** *od.* **wäre das ~e, wenn ...:** it would be best if ...; **der/die/das nächste ~e ...:** the first ... one comes across; **einen Witz zum ~en geben** entertain [those present] with a joke; **das Beste vom Besten** the very best; **sein Bestes tun** do one's best; **zu deinem Besten** for your benefit; **2.** *adv.* **am ~en** best; **3.** *Adv.* **am ~en fährst du mit dem Zug** it would be best for you to go by train

Bestand der a) *o. Pl.* existence, *(Fort~)* continued existence; **b)** *(Vorrat)* stock **(an + Dat. of)**

bestanden *Adj.* **von** *od.* **mit etw. ~ sein** have sth. growing on it; **mit Tannen ~e Hügel** fir-covered hills

beständig 1. *Adj.; nicht präd.* constant; **b)** *(gleichbleibend)* constant; steadfast ⟨*person*⟩; settled ⟨*weather*⟩; **c)** *(widerstandsfähig)* resistant (gegen to); **2.** *adv.* constantly; **Beständigkeit die**; ~ a) constancy; b) steadfastness; **b)** *(Widerstandsfähigkeit)* resistance (gegen to)

Bestand·teil der component

bestärken *tr. V.* confirm

bestätigen 1. *tr. V.* confirm; endorse ⟨*document*⟩; acknowledge ⟨*receipt*⟩; **2.** *refl. V.* be confirmed; ⟨*rumour*⟩ prove to be true; **Bestätigung die**; ~, ~en

confirmation; *(des Empfangs)* acknowledgement; *(schriftlich)* letter of
confirmation

bestatten *tr. V. (geh.)* inter *(formal);*
bury; **Bestattung** die; ~, ~en *(geh.)*
interment *(formal);* burial; *(Feierlichkeit)* funeral

bestäuben *tr. V.* a) dust; b) *(Biol.)*
pollinate

bestaunen *tr. V.* marvel at

bestechen *unr. tr. V.* bribe; **bestechlich** *Adj.* corruptible; open to
bribery *postpos.;* **Bestechung** die;
~, ~en bribery *no indef. art.;* **Beste-
chungs·geld das** bribe

Besteck das; ~[e]s, ~e cutlery setting;
(ugs.: Gesamtheit der Bestecke) cutlery

bestehen 1. *unr. itr. V.* a) exist; **es besteht [die] Aussicht/Gefahr, daß ...:**
there is a prospect/danger that ...;
noch besteht die Hoffnung, daß ...:
there is still hope that ...; b) *(fortdau-
ern)* survive; last; c) **aus etw. ~:** consist of sth.; *(hergestellt sein)* be made
of sth.; d) **auf etw.** *(Dat.)* ~: insist on
sth.; 2. *unr. tr. V.* pass *(test, examination);* **Bestehen das; ~s** existence;
die Firma feiert ihr 10jähriges ~: the
firm is celebrating its tenth anniversary

bestehen|bleiben *unr. itr. V.; mit
sein* remain; *(regulation)* remain in
force

bestehend *Adj.* existing; current
(conditions)

bestehlen *unr. tr. V.* rob

besteigen *unr. tr. V.* a) climb; mount
(horse, bicycle); ascend *(throne);* b)
board *(ship, aircraft);* get on *(bus,
train);* **Besteigung** die ascent

bestellen *tr. V.* a) *auch itr.* order (bei
from); **würden Sie mir bitte ein Taxi
~?** would you order me a taxi?; b) *(reservieren lassen)* reserve *(tickets,
table);* c) jmdn. [für 10 Uhr] zu sich ~:
ask sb. to go/come to see one [at 10
o'clock]; d) *(ausrichten)* jmdm. etw. ~:
tell sb. sth.; **bestell deinem Mann schöne Grüße von mir** give your husband
my regards; **Bestellung** die a) order;
b) *(Reservierung)* reservation

besten·falls *Adv.* at best; **bestens**
Adv. extremely well

besteuern *tr. V.* tax

bestialisch 1. *Adj.* a) bestial; b) *nicht
präd. (ugs.: schrecklich)* ghastly *(coll.);*
2. *adv.* a) in a bestial manner; b) *(ugs.:
schrecklich)* awfully *(coll.);* **Bestiali-
tät** die; ~: bestiality

besticken *tr. V.* embroider

Bestie ['bɛstiə] die; ~, ~n beast

bestimmen 1. *tr. V.* a) *(festsetzen)* decide on; fix *(price, time, etc.);* b) *(vorsehen)* intend; **das ist für dich bestimmt** that is meant for you; c) *(identifizieren)* identify; determine *(age,
position);* define *(meaning);* d) *(prägen)* determine the character of; 2. *itr.
V.* a) make the decisions; b) **über
jmdn. ~:** tell sb. what to do; **[frei] über
etw.** *(Akk.)* ~: do as one wishes with
sth.; **bestimmend** 1. *Adj.* decisive;
2. *adv.* decisively; **bestimmt** 1. *Adj.*
a) *(speziell)* particular; *(gewiß)* certain; *(genau)* definite; *(festgelegt)*
fixed; given *(quantity);* c) *(Sprachw.)*
definite *(article etc.);* d) *(entschieden)*
firm; 2. *adv.* a) *(deutlich)* clearly; *(genau)* precisely; b) *(entschieden)* firmly
3. *Adv.* for certain; **du weißt es doch
[ganz] ~ noch** I'm sure you must remember it; **ich habe das ~ liegengelas-
sen** I must have left it behind; **Be-
stimmtheit** die; ~: firmness; *(im
Auftreten)* decisiveness; **Bestimmung** die a) *o. Pl. (das Festsetzen)* fixing; b) *(Vorschrift)* regulation; c) *o. Pl.
(Zweck)* purpose; d) *s.* bestimmen 1c:
identification; determination; definition; e) *(Sprachw.)* modifier; **adver-
biale ~:** adverbial qualification

best·möglich *Adj.* best possible

bestrafen *tr. V.* punish **(für, wegen**
for); **es wird mit Gefängnis bestraft** it
is punishable by imprisonment; **Bestrafung** die; ~, ~en punishment

bestrahlen *tr. V.* a) illuminate; floodlight *(building);* b) *(Med.)* treat *(tumour, part of body)* using radiother-
apy; **Bestrahlung** die; ~, ~en
(Med.) radiation [treatment] *no indef.
art.*

Bestreben das endeavour[s *pl.*]; **be-
strebt** *Adj.:* ~ sein, etw. zu tun endeavour to do sth.; **Bestrebung** die;
~, ~en effort; *(Versuch)* attempt

bestreichen *unr. tr. V.* A mit B ~:
spread B on A

bestreiten *unr. tr. V.* a) dispute;
(leugnen) deny; b) *(finanzieren)* finance *(studies);* pay for *(studies, sb.'s
keep);* meet *(costs, expenses);* c) *(gestalten)* carry *(programme, conversation, etc.)*

bestreuen *tr. V.* sprinkle

Bestseller der; ~s, ~: best seller

bestürzt 1. *Adj.* dismayed; 2. *adv.*
with dismay

Besuch der; ~|e|s, ~e **a)** visit *(Gen.,* **bei** to); **ein ~ bei jmdm.** a visit to sb.; *(kurz)* a call on sb.; **b)** *(Teilnahme)* attendance *(Gen.* at); **c)** *(Gast)* visitor; *(Gäste)* visitors *pl.;* ~ **haben** have visitors/a visitor; **besuchen** *tr. V.* **a)** visit; *(weniger formell)* go to see *(person)*; go to *(exhibition, theatre, museum, etc.)*; *(zur Besichtigung)* go to see *(church, exhibition, etc.)*; **b)** **die Schule/Universität** ~: go to school/ university; **Besucher** der; ~s, ~, **Besucherin** die; ~, ~nen visitor; **besucht** *Adj.* **gut/schlecht** ~: well/ poorly attended *(lecture, performance, etc.)*; much/little frequented *(restaurant etc.)*

betagt *Adj. (geh.)* elderly

betasten *tr. V.* feel [with one's fingers]

betätigen 1. *refl. V.* occupy oneself; **sich politisch/körperlich** ~: engage in political/physical activity; **2.** *tr. V.* operate *(lever, switch, flush, etc.)*; apply *(brake)*; **Betätigung** die; ~, ~en: **a)** activity; **b)** *o. Pl. s.* **betätigen 2:** operation; application

betäuben *tr. V.* **a)** *(Med.)* anaesthetize; deaden *(nerve)*; **jmdn. örtlich** ~: give sb. a local anaesthetic; **b)** *(unterdrücken)* deaden *(pain)*; still *(unease, fear)*; **c)** *(benommen machen)* daze; *(mit einem Schlag)* stun; **Betäubung** die; ~, ~en: **a)** *(Med.)* anaesthetization; *(Narkose)* anaesthesia; **b)** *(Benommenheit)* daze; **Betäubungsmittel das** narcotic; *(Med.)* anaesthetic

beteiligen 1. *refl. V.* take part **(an** + *Dat.* in); **2.** *tr. V.* **jmdn. |mit 10 %| an etw.** *(Dat.)* ~: give sb. a [10 %] share of sth.; **beteiligt** *Adj.* **a)** involved **(an** + *Dat.* in); **b)** *(finanziell)* **an einem Unternehmen/am Gewinn** ~ **sein** have a share in a business/in the profit; **Beteiligte** der/die; *adj. Dekl.* person involved; **Beteiligung** die; ~, ~en **a)** participation **(an** + *Dat.* in); **b)** *(Anteil)* share **(an** + *Dat.* in)

beten 1. *itr. V.* pray **(für, um** for); **2.** *tr. V.* say *(prayer)*

beteuern *tr. V.* affirm; protest *(one's innocence)*; **Beteuerung** die; ~, ~en *s.* **beteuern:** affirmation; protestation

Beton [be'tɔŋ, *bes. österrr.:* be'to:n] der; ~s, ~s [-ɔŋs] *od. (bes. österr.:)* ~e [-o:nə] concrete

betonen *tr. V.* **a)** stress *(word, syllable)*; **b)** *(hervorheben)* emphasize

betonieren *tr. V.* concrete; surface *(road etc.)* with concrete

betont 1. *Adj.* **a)** stressed; **b)** *(bewußt)* studied; **2.** *adv.* studiedly; **Betonung die;** ~, ~en **a)** stressing; **b)** *(Akzent)* stress; *(Intonation)* intonation; **c)** *(Hervorhebung)* emphasis

betören *tr. V. (geh.)* captivate

betr. *Abk.* betreffs, betrifft re; **Betr.** *Abk.* Betreff re

Betracht: jmdn./etw. in ~ **ziehen** consider sb./sth.; **jmdn./etw. außer** ~ **lassen** disregard sb./sth.; **betrachten** *tr. V.* **a)** look at; **b)** **jmdn./etw. als etw.** ~: regard sb./sth. as sth.; **c)** *(beurteilen)* consider; **Betrachter** der; ~s, ~: observer

beträchtlich 1. *Adj.* considerable; **2.** *adv.* considerably

Betrachtung die; ~, ~en **a)** contemplation; *(Untersuchung)* examination; **b)** *(Überlegung)* reflection

Betrag der; ~|e|s, **Beträge** amount; „**~ dankend erhalten**" 'received with thanks'; **betragen 1.** *unr. itr. V.* be; *(bei Geldsummen)* come to; **2.** *unr. refl. V.* behave; **Betragen das;** ~s behaviour

Betreff der; ~|e|s, ~e *(im Brief)* heading; **betreffen** *unr. tr. V.* concern; *(new rule, change, etc.)* affect; **betreffend** *Adj.* concerning; **der ~e Sachbearbeiter** the person dealing with this matter; **in dem ~en Fall** in the case in question; **betreffs** *Präp. mit Gen. (Amtsspr., Kaufmannsspr.)* concerning

betreiben *unr. tr. V.* **a)** proceed with, *(energisch)* press ahead with *(task, case, etc.)*; pursue *(policy, studies)*; carry on *(trade)*; go in for *(sport)*; **b)** run *(business, shop)*; **c)** *(in Betrieb halten)* operate

¹betreten *unr. tr. V. (hineintreten in)* enter; *(treten auf)* step on to; *(begehen)* walk on *(carpet, grass, etc.)*; „**Betreten verboten**" 'Keep off'; *(kein Eintritt)* 'Keep out'

²betreten 1. *Adj.* embarrassed; **2.** *adv.* with embarrassment

betreuen *tr. V.* look after; care for *(invalid)*; supervise *(youth group)*; see to the needs of *(tourists, sportsmen)*; **Betreuung** die; ~ care *no indef. art.*

Betrieb der; ~|e|s, ~e **a)** business; *(Firma)* firm; **b)** *o. Pl. (das In-Funktion-Sein)* operation; **außer** ~ **sein** not operate; *(wegen Störung)* be out of order; **in/außer** ~ **setzen** start up/stop

⟨machine etc.⟩; **c)** o. Pl. (ugs.: Treiben) bustle; (Verkehr) traffic; **es herrscht großer ~, es ist viel ~**: it's very busy; **betrieblich** Adj. firm's; company

Betriebs-: ~angehörige der/die employee; **~anleitung die, ~anweisung die** operating instructions pl.; **~ausflug der** staff outing; **~ferien** Pl. firm's annual close-down sing.; **„Wegen ~ferien geschlossen"** 'closed for annual holidays'; **~klima das** working atmosphere; **~rat der a)** works committee; **b)** (Person) member of a/the works committee; **~wirt der** graduate in business management; **~wirtschaft die**; o. Pl. business management

betrinken unr. refl. V. get drunk

betroffen 1. Adj. upset; (bestürzt) dismayed; **2.** adv. in dismay; **Betroffenheit die; ~**: dismay

betrüblich Adj. gloomy; **betrübt 1.** Adj. sad; gloomy ⟨face etc.⟩; **2.** sadly; (schwermütig) gloomily

Betrug der; ~[e]s deception; (Delikt) fraud; **betrügen 1.** unr. tr. V. deceive; be unfaithful to ⟨husband, wife⟩; (Rechtsw.) defraud; (beim Spielen) cheat; **jmdn. um 100 DM ~**: cheat or (coll.) do sb. out of 100 marks; (arglistig) swindle sb. out of 100 marks; **2.** unr. itr. V. cheat; (bei Geschäften) swindle people; **Betrüger der; ~s, ~**: swindler; (Hochstapler) con man (coll.); (beim Spielen) cheat; **Betrügerei die; ~, ~en** deception; (beim Spielen usw.) cheating; (bei Geschäften) swindling; **Betrügerin die; ~, ~nen** swindler; (beim Spielen) cheat

betrunken Adj. drunken attrib.; drunk pred.; **Betrunkene der/die**; adj. Dekl. drunk

Bett das; ~[e]s, ~en a) bed; **ins** od. **zu ~ gehen** go to bed; **die Kinder ins ~ bringen** put the children to bed; **b)** (Feder~) duvet

Bett-: ~bezug der duvet cover; **~decke die** blanket; (gesteppt) quilt

Bettelei die; ~, ~en begging no art.; **betteln** itr. V. beg (**um** for)

bett·lägerig Adj. bedridden

Bett·laken das sheet

Bettler der; ~s, ~, Bettlerin die; ~, ~nen beggar

Bett-: ~ruhe die bed rest; **~wäsche die** bed-linen; **~zeug das**; o. Pl. (ugs.) bedclothes pl.

betucht Adj. (ugs.) well-heeled (coll.); well-off

betupfen tr. V. dab

Beuge die; ~, ~n (Turnen) bend; **beugen 1.** tr. V. **a)** bend; bow ⟨head⟩; **b)** (Sprachw.: flektieren) inflect ⟨word⟩; **2.** refl. V. **a)** bend over; **sich nach vorn/hinten ~**: bend forwards/bend over backwards; **sich aus dem Fenster ~**: lean out of the window; **b)** (sich fügen) give way; **Beugung die; ~, ~en** (Sprachw.) inflexion

Beule die; ~, ~n bump; (Vertiefung) dent; **beulen** itr. V. bulge

beunruhigen tr., refl. V. worry

beurlauben tr. V. **a)** jmdn. [für zwei Tage] ~: give sb. [two days'] leave of absence; **b)** (suspendieren) suspend

beurteilen tr. V. judge; assess ⟨situation etc.⟩; **Beurteilung die; ~, ~en a)** judgement; (einer Lage usw.) assessment; **b)** (Gutachten) assessment

Beute die; ~, ~n (Gestohlenes) haul; loot no indef. art.; **b)** (von Raubtieren) prey; (eines Jägers) bag

Beutel der; ~s, ~ bag; (kleiner, für Tabak usw.) pouch

bevölkern tr. V. populate; **Bevölkerung die; ~, ~en** population; (Volk) people

bevollmächtigen tr. V. authorize; **Bevollmächtigte der/die**; adj. Dekl. authorized representative

bevor Konj. before; **~ du nicht unterschrieben hast** until you have signed; **bevor|stehen** unr. itr. V. be near; **unmittelbar ~**: be imminent; **jmdm. steht etw. bevor** sth. is in store for sb.; **bevorstehend** Adj. forthcoming; **unmittelbar ~**: imminent

bevorzugen tr. V. **a)** (vorziehen) prefer (**vor** + Dat. to); **b)** (begünstigen) favour; give preference or preferential treatment to (**vor** + Dat. over); **bevorzugt 1.** Adj. favoured; (privilegiert) privileged; preferential ⟨treatment⟩; **2.** adv. jmdn. **~ behandeln** give sb. preferential treatment; **Bevorzugung die; ~, ~en** (Begünstigung) preferential treatment

bewachen tr. V. guard; **bewachter Parkplatz** car park with an attendant; **Bewacher der; ~s, ~**: guard; **Bewachung die; ~, ~en** guarding

bewaffnen 1. tr. V. arm; **2.** refl. V. (auch fig.) arm oneself (**mit** with); **Bewaffnung die; ~, ~en a)** arming; **b)** (Waffen) weapons pl.

bewahren tr. V. **a)** protect (**vor** + Dat. from); **b)** (erhalten) seine Fassung ~: retain one's composure; **Stillschweigen** ~: remain silent

bewähren *refl. V.* prove oneself/itself; **bewährt** *Adj.* proven *(method, design, etc.)*; well-tried *(recipe, cure)*; reliable *(worker)*; **Bewährung die;** ~, ~**en** *(Rechtsw.)* probation

bewaldet *Adj.* wooded

bewältigen *tr. V.* cope with; overcome *(difficulty, problem)*; cover *(distance)*; **Bewältigung die;** ~, ~**en** *s.* bewältigen: coping with; overcoming; covering

bewandert *Adj.* well-versed

Bewandtnis die; ~, ~**se: mit etw. hat es [s]eine eigene/besondere** ~: there's a [special] story behind sth.

bewässern *tr. V.* irrigate; **Bewässerung die;** ~, ~**en** irrigation

¹bewegen 1. *tr. V.* a) move; b) *(ergreifen)* move; c) *(innerlich beschäftigen)* preoccupy; 2. *refl. V.* move

²bewegen *unr. tr. V.* **jmdn. dazu** ~, **etw. zu tun** *(thing)* induce sb. to do sth.; *(person)* prevail upon sb. to do sth.; **Beweg·grund der** motive

beweglich *Adj.* a) movable; moving *(target)*; b) *(rege)* agile *(mind)*; **bewegt** *Adj.* eventful; *(unruhig)* turbulent; **Bewegung die;** ~, ~ a) movement; *(bes. Technik, Physik)* motion; b) *(körperliche* ~*)* exercise; c) *(Ergriffenheit)* emotion; d) *(Bestreben, Gruppe)* movement; **Bewegungs·freiheit die;** *o. Pl.* freedom of movement; **bewegungs·los** *Adj.* motionless

Beweis der; ~**es,** ~**e** proof *(Gen., für* of); **belastende** ~**e** incriminating evidence; **beweisbar** *Adj.* provable; **beweisen** *unr. tr. V.* prove; **Beweis·material das** evidence

bewenden *unr. V.* **es bei** *od.* **mit etw.** ~ **lassen** content oneself with sth.

bewerben *unr. refl. V.* apply (**bei** to, **um** for); **Bewerber der** applicant; **Bewerbung die** application

bewerfen *unr. tr. V.* **jmdn./etw. mit etw.** ~: throw sth. at sb./sth.

bewerten *tr. V.* assess; rate; *(dem Geldwert nach)* value (**mit** at); **Bewertung die** assessment; *(dem Geldwert nach)* valuation

bewilligen *tr. V.* grant; **Bewilligung die;** ~, ~**en** granting

bewirken *tr. V.* bring about; cause

bewirten *tr. V.* feed; **jmdn. mit etw.** ~: serve sb. sth.

bewirtschaften *tr. V.* a) manage

(estate, farm, restaurant, business, etc.); b) farm *(fields, land)*

Bewirtung die; ~, ~**en** provision of food and drink

bewog *1. u. 3. Pers. Sg. Prät. v.* ²**bewegen**

bewohnbar *Adj.* habitable; **bewohnen** *tr. V.* inhabit, live in *(house, area)*; live in *(room, flat)*; **Bewohner der;** ~**s,** ~, **Bewohnerin die;** ~, ~**nen** *(eines Hauses, einer Wohnung)* occupant; *(einer Stadt, eines Gebietes)* inhabitant; **bewohnt** *Adj.* occupied *(house etc.)*; inhabited *(area)*

bewölken *refl. V.* cloud over; become overcast; **bewölkt** *Adj.* cloudy; overcast; **Bewölkung die;** ~, ~**en** cloud [cover]

Bewunderer der; ~**s,** ~, **Bewunderin die;** ~, ~**nen** admirer; **bewundern** *tr. V.* admire (**wegen, für** for); **bewunderns·wert** *Adj.* a) admirable; 2. *adv.* admirably; **Bewunderung die;** ~: admiration

bewußt 1. *Adj.* conscious *(reaction, behaviour, etc.)*; *(absichtlich)* deliberate *(lie, deception, attack, etc.)*; **etw. ist/wird jmdm.** ~: sb. is/becomes aware of sth.; sb. realizes sth.; **sich** *(Dat.)* **einer Sache** *(Gen.)* ~ **sein/werden** be/become aware of something; 2. *adv.* consciously; *(absichtlich)* deliberately; **bewußt·los** *Adj.* unconscious; **Bewußtlosigkeit die;** ~: unconsciousness; **Bewußt·sein das** a) consciousness; **das** ~ **verlieren/wiedererlangen** lose/regain consciousness; **bei vollem** ~ **sein** be fully conscious; b) *(deutliches Wissen)* awareness

bezahlbar *Adj.* affordable; **bezahlen** 1. *tr. V.* pay *(person, bill, taxes, rent, amount)*; pay for *(goods etc.)*; **das macht sich bezahlt** it pays off; 2. *itr. V.* pay; **Herr Ober, ich möchte** ~ *od.* **bitte** ~: waiter, the bill *or (Amer.)* check please; **Bezahlung die** payment; *(Lohn, Gehalt)* pay

bezaubernd 1. *Adj.* enchanting; 2. *adv.* enchantingly

bezeichnen *tr. V.* a) **jmdn./sich/etw. als etw.** ~: call sb./oneself/sth. sth.; b) *(Name, Wort sein für)* denote; **bezeichnend** *Adj.* characteristic *(für* of); **Bezeichnung die** a) marking; *(Angabe durch Zeichen)* indication; b) *(Name)* name

bezeugen *tr. V.* testify to

bezichtigen *tr. V.* accuse

beziehen 1. *unr. tr. V.* **a)** cover ⟨*seat, cushion, etc.*⟩; **die Betten frisch ~:** put clean sheets on the beds; **b)** *(einziehen in)* move into ⟨*house, office*⟩; **c)** *(Milit.)* take up ⟨*position, post*⟩; **d)** *(erhalten)* obtain ⟨*goods*⟩; take ⟨*newspaper*⟩; draw ⟨*pension, salary*⟩; **e)** *(in Beziehung setzen)* apply (**auf** + *Akk.* to); **2.** *unr. refl. V.* **a)** es/der Himmel bezieht sich it/the sky is clouding over *or* becoming overcast; **b) sich auf jmdn./ etw. ~** *(sich berufen auf)* ⟨*person, letter, etc.*⟩ refer to sb./sth.; *(betreffen)* ⟨*question, statement, etc.*⟩ relate to sb./sth.; **wir ~ uns auf Ihr Schreiben vom 28. 8.** with reference to your letter of 28 August; **Beziehung die a)** relation; *(Zusammenhang)* connection (**zu** with); **zwischen A und B besteht keine/eine ~:** there is no/a connection between A and B; **b)** *(Freundschaft, Liebes~)* relationship; **c)** *(Hinsicht* respect; **in mancher ~:** in many respects; **beziehungs·weise** *Konj.* and ... respectively; *(oder)* or

Bezirk der; ~[e]s, ~e district

bezug: in ~ auf jmdn./etw. regarding sb./sth.

Bezug der **a)** *(für Kissen usw.)* cover; *(für Polstermöbel)* loose cover; slipcover *(Amer.)*; *(für Betten)* duvet cover; *(für Kopfkissen)* pillowcase; **b)** *o. Pl. (Erwerb)* obtaining; *(Kauf)* purchase; **~ einer Zeitung** taking a newspaper; **c)** *Pl.* salary *sing.*; **d)** *(Papierdt.)* **mit** *od.* **unter ~ auf etw.** *(Akk.)* with reference to sth.; **~ nehmend auf unser Telex** with reference to our telex; **bezüglich** *Präp. mit Gen.* regarding

bezwecken *tr. V.* aim to achieve

bezweifeln *tr. V.* doubt

bezwingen *unr. tr. V.* conquer ⟨*enemy, mountain, pain, etc.*⟩; defeat ⟨*opponent*⟩; capture ⟨*fortress*⟩

BH [be:'ha:] der; ~[s], ~[s] *Abk.*: Büstenhalter bra

Bibel die; ~, ~n *(auch fig.)* Bible

Biber der; ~s, ~: beaver

Bibliothek die; ~, ~en library

biblisch *Adj.* biblical

Bidet [bi'de:] das; ~s, ~s bidet

bieder *Adj.* unsophisticated; *(langweilig)* stolid; *(treuherzig)* trusting

biegen 1. *unr. tr. V.* bend; **2.** *unr. refl. V.* bend; *(nachgeben)* give; **3.** *unr. itr. V.; mit sein* turn; **biegsam** *Adj.* flexible; pliable ⟨*material*⟩; **Biegung** die; ~, ~en bend

Biene die; ~, ~n bee

Bienen-: **~honig** der bees' honey; **~königin** die queen bee; **~korb** der straw hive; **~stock** der beehive

Bier das; ~[e]s, ~e beer

Bier-: **~deckel** der beer-mat; **~dose** die beer can; **~faß** das beer-barrel; **~flasche** die beer-bottle; **~garten** der beer garden; **~glas** das beer-glass; **~kasten** der beer-crate; **~zelt** das beer tent

Biest das; ~[e]s, ~er *(ugs. abwertend)* **a)** *(Tier, Gegenstand)* wretched thing; **b)** *(Mensch)* wretch

bieten 1. *unr. tr. V.* **a)** offer; put on ⟨*programme etc.*⟩; provide ⟨*shelter, guarantee, etc.*⟩; **b) ein schreckliches Bild ~:** present a terrible picture; **einen prächtigen Anblick ~:** be a splendid sight; **2.** *unr. refl. V.* **sich jmdm. ~:** present itself to sb.; **3.** *unr. itr. V.* bid

Bigamie die; ~: bigamy *no def. art.*

Bikini der; ~s, ~s bikini

Bilanz die; ~, ~en **a)** balance sheet; **b)** *(Ergebnis)* outcome; **~ ziehen** take stock

Bild das; ~[e]s, ~er **a)** picture; **b)** *(Anblick)* sight; **c)** *(Metapher)* image

bilden 1. *tr. V.* **a)** form (**aus** from); *(modellieren)* mould (**aus** from); **eine Gasse ~:** make a path; **sich** *(Dat.)* **ein Urteil ~:** form an opinion; **b)** *(ansammeln)* build up ⟨*fund, capital*⟩; **c)** *(darstellen)* be ⟨*exception etc.*⟩; **d)** *(erziehen)* educate; **2.** *refl. V.* **a)** form; **b)** *(lernen)* educate oneself

Bilder·buch das picture-book *(for children)*

Bild·hauer der sculptor

bild·hübsch *Adj.* really lovely; stunningly beautiful ⟨*girl*⟩

bildlich 1. *Adj.* pictorial; *(übertragen)* figurative; **2.** *adv.* pictorially; *(übertragen)* figuratively; **Bildnis** ['bɪltnɪs] das; ~ses, ~se portrait

Bild·schirm der *(Ferns., Informationst.)* screen; **Bildschirm·gerät** das VDU; visual display unit

bild·schön *Adj.* really lovely; stunningly beautiful ⟨*girl, woman*⟩

Bildung die; ~, ~en **a)** *(Erziehung)* education; *(Kultur)* culture; **b)** *(das Formen)* formation; **Bildungslücke** die gap in one's education

Billard ['bɪljart, *österr.:* bɪ'jaːɐ̯] das; ~s, ~e billiards

Billard-: **~kugel** die billiard-ball; **~stock** der billiard-cue; **~tisch** der billiard-table

Billett [bɪl'jɛt] **das;** ~|e|s, ~e *od.* ~s *(schweiz., veralt.)* ticket

Billiarde die; ~, ~n thousand million million; quadrillion *(Amer.)*

billig 1. *Adj.* **a)** cheap; **b)** *(abwertend: primitiv)* cheap ⟨*trick*⟩; feeble ⟨*excuse*⟩; **2.** *adv.* cheaply

billigen *tr. V.* approve; **Billigung die;** ~: approval

Billion die; ~, ~en million million; trillion *(Amer.)*

bimmeln *itr. V. (ugs.)* ring

bin *1. Pers. Sg. Präsens v.* ¹**sein**

Binde die; ~, ~n **a)** *(Verband)* bandage; *(Augen~)* blindfold; **b)** *(Arm~)* armband

Binde-: ~**gewebe das** *(Anat.)* connective tissue; ~**haut die** *(Anat.)* conjunctiva

binden 1. *unr. tr. V.* **a)** *(auch fig.)* tie; knot ⟨*tie*⟩; make up ⟨*wreath, bouquet*⟩; **jmdn. an sich** *(Akk.)* ~ *(fig.)* make sb. dependent on one; **b)** *(fesseln, festhalten, zusammenhalten, fig.: verpflichten, Buchw.)* bind; **c)** *(Kochk.: legieren)* thicken ⟨*sauce*⟩; **2.** *unr. refl. V.* tie oneself down; **Binder der;** ~s, ~ tie; **Bindestrich der** hyphen; **Bind·faden der** string

Bindung die; ~, ~en **a)** *(Beziehung)* relationship (**an** + *Akk.* to); **b)** *(Verbundenheit)* attachment (**an** + *Akk.* to); **c)** *(Ski~)* binding

binnen *Präp. mit Dat. od. (geh.) Gen.* within

Binsen·weisheit die truism

bio-, Bio: ~**chemie die** biochemistry; ~**graph der;** ~en, ~en biographer; ~**graphie die;** ~, ~n biography; ~**graphisch** *Adj.* biographical; ~**loge der;** ~n, ~n biologist; ~**logie die;** ~: biology *no art.;* ~**logisch** *Adj.* **a)** biological; **b)** *(natürlich)* natural ⟨*medicine, cosmetic, etc.*⟩; ~**top der** *od.* **das;** ~s, ~e *(Biol.)* biotope

Birke die; ~, ~n birch[-tree]; *(Holz)* birch[wood]

Birma (das); ~s Burma

Birn·baum der pear-tree; **Birne die;** ~, ~n **a)** pear; **b)** *(Glüh~)* [light-]bulb; **c)** *(salopp: Kopf)* nut *(sl.)*

bis 1. *Präp. mit Akk.* **a)** *(zeitlich)* until; till; *(die ganze Zeit über und bis zu einem bestimmten Zeitpunkt)* up until; up till; *(nicht später als)* by; **b)** *(räumlich)* to; **dieser Zug fährt nur** ~ **Offenburg** this train only goes as far as Offenburg; ~ **5 000 Mark** up to 5,000

marks; **c)** ~ **auf** *(einschließlich)* down to; *(mit Ausnahme von)* except for; **2.** *Adv.* ~ **zu 6 Personen** up to six people. **3.** *Konj.* **a)** *(nebenordnend)* to; **b)** *(unterordnend)* until; till; *(österr.: sobald)* when

Bisam·ratte die musk-rat

Bischof der; ~s, **Bischöfe** bishop; **bischöflich** *Adj.* episcopal

bis·her *Adv.* up to now; *(aber jetzt nicht mehr)* until now; till now; **bisherig** *Adj. (vorherig)* previous; *(momentan)* present

Biskaya [bɪs'kaːja] **die;** ~: the Bay of Biscay

Biskuit [bɪs'kviːt] **das** *od.* **der;** ~|e|s, ~s *od.* ~e **a)** sponge biscuit; **b)** *(~teig)* sponge

bis·lang *Adv.: s.* bisher

Bison der; ~s, ~s bison

Biß der; Bisses, Bisse bite

bißchen *indekl. Indefinitpron.* **a)** *adj.* **ein** ~ **Geld/Wasser** a bit of *or* a little money/a drop of *or* a little water; **ein/kein** ~ **Angst haben** be a bit/not a bit frightened; **b)** *adv.* **ein/kein** ~: a bit *or* a little/not a *or* one bit; **c)** *subst.* **ein** ~: a bit; a little; *(bei Flüssigkeiten)* a drop; a little; **das/kein** ~: the little [bit]/not a *or* one bit

Bissen der; ~s, ~: mouthful

bissig 1. *Adj.* **a)** ~ **sein** ⟨*dog*⟩ bite; **ein** ~**er Hund** a dog that bites; „**Vorsicht, ~er Hund**" 'beware of the dog'; **b)** cutting ⟨*remark, tone, etc.*⟩; **2.** *adv.* ⟨*say*⟩ cuttingly

Biß·wunde die bite

bist *2. Pers. Sg. Präsens v.* ¹**sein**

Bistum ['bɪstuːm] **das;** ~s, **Bistümer** bishopric; diocese

bis·weilen *Adv. (geh.)* from time to time

bitte 1. *Adv.* please; **2.** *Interj.* **a)** *(Bitte, Aufforderung)* please; **zwei Tassen Tee,** ~: two cups of tea, please; ~|, **nehmen Sie doch Platz|!** do take a seat; **Noch eine Tasse Tee?** – |**Ja**| ~! Another cup of tea? – Yes, please; **b)** *(Aufforderung, etw. entgegenzunehmen)* ~ |**schön** *od.* **sehr**|! there you are!; **c)** *(Ausdruck des Einverständnisses)* ~ |**gern**|! certainly; of course; **Entschuldigung!** – **Bitte!** [I'm] sorry! – That's all right!; **d)** ~ |**schön** *od.* **sehr**|! *(im Laden, Lokal)* yes, please?; **e)** |**wie**| ~? *(Nachfrage)* sorry; **f) Vielen Dank!** – **Bitte** |**schön** *od.* **sehr**| Many thanks! – Not at all *or* you're welcome

Bitte die; ~, ~n request; *(inständig)* plea; **bitten** *unr. tr. V.* a) *auch itr.* ask (um for); **darf ich Sie um Feuer/ein Glas Wasser ~?** could I ask you for a light/a glass of water, please?; b) *(einladen)* ask

bitter 1. *Adj.* a) bitter; plain *(chocolate)*; b) *(fig.) (verbittert)* bitter; c) *(schmerzlich)* bitter, painful, hard *(loss)*; hard *(time, fate, etc.)*; dire *(need)*; desperate *(poverty)*; grievous *(injustice, harm)*; **2.** *adv. (sehr stark)* desperately; *(regret)* bitterly

bitter-: ~böse 1. *Adj.* furious; **2.** *adv.* furiously; **~kalt** *Adj.; präd. getrennt geschr.* bitterly cold

bitterlich 1. *Adj.* slightly bitter *(taste)*. **2.** *adv. (heftig)* *(cry, complain, etc.)* bitterly; **bitter·süß** *Adj. (auch fig.)* bitter-sweet

Bitt·steller der; ~s, ~petitioner

Biwak das; ~s, ~s *(bes. Milit., Bergsteigen)* bivouac

bizarr 1. *Adj.* bizarre; **2.** *adv.* bizarrely

Bizeps der; ~|es|, ~e biceps

Blähung die; ~, ~en flatulence *no art., no pl.*

Blamage [bla'ma:ʒə] die; ~, ~n disgrace; **blamieren 1.** *tr. V.* disgrace; **2.** *refl. V.* disgrace oneself; *(sich lächerlich machen)* make a fool of oneself

blank *Adj.* shiny

Blanko-: ~scheck der *(auch fig.)* blank cheque; **~vollmacht** die *(auch fig.)* carte blanche

Bläschen ['blɛ:sçən] das; ~s, ~ a) [small] bubble; b) *(in der Haut)* [small] blister; **Blase** die; ~, ~n a) bubble; b) *(in der Haut)* blister; c) *(Harn~)* bladder; **Blase·balg** der bellows *pl.*; **blasen 1.** *unr. itr. V.* blow; **2.** *unr. tr. V.* a) blow; b) *(spielen)* play *(musical instrument, tune, melody, etc.)*; **Bläser** der; ~s, ~ *(Musik)* wind player

blasiert *(abwertend)* **1.** *Adj.* blasé; **2.** *adv.* in a blasé way

Blas-: ~instrument das wind instrument; **~kapelle** die brass band; **~musik** die brass-band music

Blasphemie [blasfe'mi:] die; ~, ~n blasphemy

Blas·rohr das blowpipe

blaß 1. *Adj.* pale; **2.** *adv.* palely; **Blässe** die; ~: paleness

Blatt das; ~|es|, Blätter a) *(von Pflanzen)* leaf; b) *(Papier)* sheet; c) *(Buchseite usw.)* page; etw. vom ~ spielen sight-read sth.; d) *(Zeitung)* paper; e)

(Spielkarten) hand; **f)** *(am Werkzeug, Ruder)* blade; **Blättchen** das; ~s, ~ a) *(von Pflanzen)* [small] leaf; b) *(Papier)* [small] sheet; **blättern** *itr. V.* **in einem Buch ~:** leaf through a book; **Blätter·teig** der puff pastry

Blatt-: ~gold das; *o. Pl.* gold leaf; **~laus** die aphid

blau *Adj.* blue; **ein ~er Fleck** a bruise; **~ sein** *(fig. ugs.)* be tight *(coll.)*; **das Blaue vom Himmel herunterlügen** *(ugs.)* lie like anything; **Blau** das; ~s, ~ *od. (ugs.:)* ~s blue

blau-, Blau-: ~äugig *Adj.* a) blue-eyed; b) *(naiv)* naive; **~beere** die bilberry; **~grau** *Adj.* blue-grey; **~grün** *Adj.* blue-green

bläulich *Adj.* bluish

blau-, Blau-: ~licht das flashing blue light; **~|machen** *itr. V. (ugs.)* skip work; **~mann** der; *Pl.* **~männer** *(ugs.)* boiler suit; **~säure** die; *o. Pl. (Chemie)* prussic acid; **~stichig** *Adj. (Fot.)* with a blue cast *postpos., not pred.;* **~stichig sein** have a blue cast

Blazer ['ble:zɐ] der; ~s, ~: blazer

Blech das; ~|e|s, ~e a) sheet metal; *(Stück Blech)* metal sheet; b) *(Back~)* [baking] tray

Blech-: ~büchse die, **~dose** die tin **blechen** *tr., itr. V. (ugs.)* cough up *(sl.)* **blechern** *Adj.* **1.** *(metallisch klingend)* tinny *(sound, voice)*; **2.** *adv.* tinnily

Blech-: ~musik die *(abwertend)* brass-band music; **~napf** der metal bowl

Blechner der; ~s, ~ *(südd.) s.* **Klempner**

Blech-: ~schaden der *(Kfz-W.)* damage *no indef. art.* to the bodywork; **~trommel** die tin drum

blecken *tr. V.* **die Zähne ~:** bare one's/its teeth

Blei das; ~|e|s, ~e lead

Bleibe die; ~, ~n place to stay; **bleiben** *unr. itr. V.; mit sein* a) stay; remain; **~ Sie bitte am Apparat** hold the line please; **wo bleibt er so lange?** where has he got to?; **auf dem Weg ~:** keep to the path; **sitzen ~:** stay *or* remain sitting down *or* seated; **bei etw. ~:** *(fig.: an etw. festhalten)* keep to sth.; b) *(übrigbleiben)* be left; remain; **bleibend** *Adj.* lasting; permanent *(damage)*; **bleiben|lassen** *unr. tr. V.* **etw. ~:** give sth. a miss

bleich *Adj.* pale; **¹bleichen** *tr. V.* bleach; **²bleichen** *regelm., veralt. auch unr. itr. V.* become bleached

blei-, Blei-: ~**frei** *Adj.* unleaded ⟨*fuel*⟩; ~**kristall das** lead crystal; ~**kugel die** lead ball; *(Geschoß)* lead bullet; ~**schwer** *Adj.* heavy as lead *postpos.*; ~**stift der** pencil; **mit** ~: in pencil; ~**stift·spitzer der** pencil-sharpener

Blende die; ~, ~**n a)** *(Lichtschutz)* shade; *(am Fenster)* blind; **b)** *(Optik, Film, Fot.)* diaphragm; *(Blendenzahl)* aperture setting; **blenden 1.** *tr. V.* **a)** *(auch fig.)* dazzle; **b)** *(blind machen)* blind; **2.** *itr. V.* ⟨*light*⟩ be dazzling; **blendend 1.** *Adj.* **es geht mir** ~: I feel wonderfully well; **2.** *adv.* **wir haben uns** ~ **amüsiert** we had a marvellous time

blich *1. u. 3. Pers. Sg. Prät. v.* ²**bleichen**

Blick der; ~|e|s, ~**e a)** look; *(flüchtig)* glance; **b)** *o. Pl. (Ausdruck)* look in one's eyes; **mit mißtrauischem** ~: with a suspicious look in one's eye; **c)** *(Aussicht)* view; **ein Zimmer mit** ~ **aufs Meer** a room with a sea view; **d)** *o. Pl. (Urteil[skraft])* eye; **blicken 1.** *itr. V.* look; *(flüchtig)* glance; **2.** *tr. V.* **sich** ~ **lassen** put in an appearance

Blick-: ~**feld das** field of vision; ~**punkt der** view; ~**winkel der a)** angle of vision; **b)** *(fig.)* point of view; viewpoint

blieb *1. u. 3. Pers. Sg. Prät. v.* **bleiben**

blies *1. u. 3. Pers. Sg. Prät. v.* **blasen**

blind 1. *Adj.* **a)** *(auch fig.)* blind; ~ **werden** go blind; **b)** *(trübe)* clouded ⟨*glass*⟩; **c) ein** ~**er Passagier** a stowaway; **d)** ~**er Alarm** a false alarm; **2.** *adv.* **a)** *(ohne hinzusehen)* without looking; *(wahllos)* blindly; **b)** *(unkritisch)* ⟨*trust*⟩ implicitly; ⟨*obey*⟩ blindly; **Blind·darm der a)** caecum; **b)** *(volkst.: Wurmfortsatz)* appendix; **Blinde der/die;** *adj. Dekl.* blind person; blind man/woman; **die** ~**n** the blind; **Blinde·kuh** *o. Art.* blind man's buff

Blinden-: ~**hund der** guide-dog; ~**schrift die** Braille

Blindheit die; ~ *(auch fig.)* blindness; **blindlings** *Adv.* blindly; ⟨*trust*⟩ implicitly; **Blind·schleiche die;** ~, ~**n** slowworm; **blind·wütig 1.** *Adj.* raging ⟨*anger, hatred, fury, etc.*⟩; wild ⟨*rage*⟩; **2.** *adv.* in a blind rage

blinken *itr. V.* **a)** ⟨*light, glass, crystal*⟩ flash; ⟨*star*⟩ twinkle; ⟨*metal, fish*⟩ gleam; **b)** *(Verkehrsw.)* indicate; **2.** *tr. V.* flash; **Blinker der;** ~**s,** ~ indicator [light]

Blink-: ~**licht das a)** flashing light; **b)** *s.* **Blinker;** ~**zeichen das** flashlight signal

blinzeln *itr. V.* blink; *(mit einem Auge, um ein Zeichen zu geben)* wink

Blitz der; ~**es,** ~**e a)** lightning *no indef. art.*; **ein** ~: a flash of lightning; |**schnell**| **wie der** ~: like lightning; **b)** *(~licht)* flash

blitz-, Blitz-: ~**ableiter der** lightning-conductor; ~**artig 1.** *Adj.* lightning; **2.** *adv.* like lightning; ⟨*disappear*⟩ in a flash; ~**blank** *Adj. (ugs.)* ~**blank** |**geputzt**| sparkling clean; brightly polished ⟨*shoes*⟩

blitzeblank *s.* **blitzblank; blitzen** *itr. V.* **a)** *unpers.* **es blitzte** *(einmal)* there was a flash of lightning; *(mehrmals)* there was lightning; **b)** *(glänzen)* ⟨*light, glass, crystal*⟩ flash; ⟨*metal*⟩ gleam

blitz-, Blitz-: ~**gerät das** flash [unit]; ~**licht das;** *Pl.* ~**lichter** flash[light]; ~**schnell 1.** *Adj.* lightning *attrib.*; ~**schnell sein** be like lightning; **2.** *adv.* like lightning; ⟨*disappear*⟩ in a flash; ~**start der** lightning start

Block der; ~|e|s, **Blöcke** *od.* ~**s a)** *Pl. nur* **Blöcke** *(Brocken)* block; **b)** *(Wohn~)* block; **c)** *Pl. nur* **Blöcke** *(Gruppierung von politischen Kräften, Staaten)* bloc; **d)** *(Schreib~)* pad

Blockade die; ~, ~**n** blockade

Block-: ~**flöte die** recorder; ~**haus das,** ~**hütte die** log cabin

blockieren *tr. V.* block; jam ⟨*telephone line*⟩; halt ⟨*traffic*⟩; lock ⟨*wheel, machine, etc.*⟩

Block·schrift die block capitals *pl.*

blöd[e] *(ugs.)* **1.** *Adj.* **a)** *(dumm)* stupid; idiotic *(coll.)*; **b)** *(unangenehm)* stupid; **2.** *adv.* stupidly; idiotically *(coll.)*; **Blödelei die;** ~, ~**en** silly joke; **blödeln** *itr. V.* make silly jokes; **Blödheit die;** ~, ~**en** stupidity

blöd-, Blöd-: ~**mann der;** *Pl.* ~**männer** *(salopp)* stupid idiot *(coll.)*; ~**sinn der;** *o. Pl. (ugs.)* nonsense; **mach doch keinen** ~**sinn!** don't be stupid; ~**sinnig** *(ugs.)* **1.** *Adj.* idiotic *(coll.)*; **2.** *adv.* idiotically *(coll.)*

blöken *itr. V.* ⟨*sheep*⟩ bleat; ⟨*cattle*⟩ low

blond *Adj.* fair-haired, blond ⟨*man, race*⟩; blonde ⟨*woman*⟩; blond/ blonde, fair ⟨*hair*⟩; **Blondine die;** ~, ~**n** blonde

bloß 1. *Adj.* **a)** *(nackt)* naked; **b)** *(nichts als)* mere ⟨*words, promises, tri-*⟩

viality, suspicion, etc.); der ~e Gedanke daran the mere thought of it; **2.** *Adv. (ugs.: nur)* only; **3.** *Partikel* was hast du dir ~ dabei gedacht? what on earth were you thinking of?; **Blöße** die; ~: sich *(Dat.)* eine/keine ~ geben show a/not show any weakness; **bloß|stellen** *tr. V.* show up; expose ⟨*swindler, criminal, etc.*⟩

Blouson [blu'zõ:] das *od.* der; ~[s], ~s blouson

blubbern *itr. V. (ugs.)* bubble

Bluejeans, Blue jeans ['blu:dʒi:ns] *Pl. od.* die; ~, ~: [blue] jeans *pl.*

Blues [blu:s] der; ~, ~: blues *pl.*

Bluff der; ~s, ~s bluff; **bluffen** *tr., itr. V.* bluff

blühen *itr. V.* **a)** ⟨*plant*⟩ flower, be in flower *or* bloom; ⟨*flower*⟩ be in bloom, be out; ⟨*tree*⟩ be in blossom; ~de Gärten gardens full of flowers; **b)** *(florieren)* thrive; **c)** *(ugs.: bevorstehen)* jmdm. ~: be in store for sb.; das kann dir auch noch ~: the same could happen to you; **blühend** *Adj.* **a)** *(frisch, gesund)* glowing ⟨*colour, complexion, etc.*⟩; radiant ⟨*health*⟩; **b)** *(übertrieben)* vivid ⟨*imagination*⟩

Blümchen das; ~s, ~: [little] flower; **Blume** die; ~, ~n **a)** flower; **b)** *(des Weines)* bouquet; **c)** *(des Biers)* head

blumen-, Blumen-: ~**beet** das flower-bed; ~**erde** die potting compost; ~**geschäft** das florist's; ~**geschmückt** *Adj.* flower-bedecked; adorned with flowers *postpos.*; ~**kasten** der flower-box; *(vor einem Fenster)* window box; ~**kohl** der cauliflower; ~**strauß** der bunch of flowers; *(Bukett)* bouquet of flowers; ~**topf** der flowerpot; ~**vase** die [flower] vase; ~**zwiebel** die bulb

Bluse die; ~, ~n blouse

Blut das; ~[e]s blood

blut-, Blut-: ~**arm** *Adj. (Med.)* anaemic; ~**armut** die *(Med.)* anaemia; ~**bad** das blood-bath; ~**bahn** die bloodstream; ~**befleckt** *Adj.* blood-stained; ~**beschmiert** *Adj.* smeared with blood *postpos.*; ~**buche** die copper beech; ~**druck** der blood pressure

Blüte die; ~, ~n **a)** flower; bloom; *(eines Baums)* blossom; ~n treiben flower; ⟨*tree*⟩ blossom; **b)** *(das Blühen)* flowering; *(Baum~)* blossoming

Blut·egel der leech; **bluten** *itr. V.* bleed (aus from)

blüten-, Blüten-: ~**blatt** das petal;

~**honig** der blossom honey; ~**staub** der pollen; ~**weiß** *Adj.* sparkling white

Bluter der; ~s, ~ *(Med.)* haemophiliac

Blut-: ~**erguß** der haematoma; *(blauer Fleck)* bruise; ~**fleck[en]** der blood-stain; ~**gefäß** das *(Anat.)* blood-vessel; ~**gerinnsel** das blood-clot; ~**gruppe** die blood group; ~**hochdruck** der high blood pressure; ~**hund** der bloodhound

blutig a) bloody; jmdn. ~ schlagen beat sb. to a pulp; **b)** *(fig. ugs.: völlig)* complete ⟨*beginner, layman, etc.*⟩

blut-, Blut-: ~**jung** *Adj.* very young; ~**konserve** die container of stored blood; ~**konserven** stored blood; ~**körperchen** das blood corpuscle; rote/weiße ~**körperchen** red/white corpuscles; ~**krebs** der leukaemia; ~**kreislauf** der blood circulation; ~**lache** die pool of blood; ~**leer** *Adj.* bloodless; ~**leere** die restricted blood supply; ~**orange** die blood orange; ~**probe** die **a)** *(~entnahme, ~untersuchung)* blood test; **b)** *(kleine ~menge)* blood sample; ~**rache** die blood revenge; ~**rot** *Adj.* blood-red; ~**rünstig 1.** *Adj.* bloodthirsty; **2.** *adv.* bloodthirstily; ~**schande** die incest; ~**spende** die *(das Spenden)* giving no *indef. art.* of blood; *(~menge)* blood-donation; ~**spender** der blood-donor; ~**spur** die trail of blood; ~**stillend** *Adj.* styptic

bluts-, Bluts-: ~**tropfen** der drop of blood; ~**verwandt** *Adj.* related by blood *postpos.*; ~**verwandtschaft** die blood relationship

Blut-: ~**tat** die *(geh.)* bloody deed; ~**transfusion** die blood-transfusion

Blutung die; ~, ~en **a)** bleeding no *indef. art., no pl.*; **b)** *(Regel~)* period

blut-, Blut-: ~**unterlaufen** *Adj.* suffused with blood *postpos.*; bloodshot ⟨*eyes*⟩; ~**vergießen** das; ~s bloodshed; ~**vergiftung** die bloodpoisoning no *indef. art., no pl.*; ~**wurst** die black pudding

Bö die; ~, ~en gust [of wind]

Bob der; ~, ~s bob[-sleigh]

Bob-: ~**bahn** die bob[-sleigh] run; ~**fahrer** der bobber

¹**Bock** der; ~[e]s, Böcke **a)** *(Reh~, Kaninchen~)* buck; *(Ziegen~)* billy-goat; he-goat; *(Schafs~)* ram; **b)** *(Gestell)* trestle; **c)** *(Turngerät)* buck

²**Bock** das; ~s *(Bier)* bock [beer]; **Bock·bier** das bock [beer]

bocken itr. V. refuse to go on; (vor einer Hürde) refuse; (sich aufbäumen) buck; **bockig** 1. Adj. stubborn and awkward; (coll.). 2. adv. stubbornly [and awkwardly]; **Bocks·horn das: sich ins ~horn jagen lassen** (ugs.) let oneself be browbeaten

Bock-: ~**springen** das (Turnen) vaulting [over the buck]; ~**wurst die** bockwurst

Boden der; ~s, Böden a) (Erd~) ground; (Fuß~) floor; **am ~ zerstört [sein]** (ugs.) [be] shattered (coll.); **bleiben wir doch auf dem ~ der Tatsachen** (fig.) let's stick to the facts; b) (unterste Fläche) bottom; (Torten~) base; c) (Dach~, Heu~) loft

boden-, Boden-: ~**belag** der floor-covering; ~**frost** der ground frost; ~**kammer die** attic; ~**los** Adj. a) bottomless; b) (ugs.: unerhört) incredible (foolishness, meanness, etc.); ~**nebel** der ground fog/mist; ~**satz** der sediment; ~**schätze** Pl. mineral resources

Boden·see der; o. Pl. Lake Constance

boden-, Boden-: ~**ständig** Adj. indigenous (culture, population, etc.); ~**turnen** das floor exercises pl.; ~**welle** die bump

Bodybuilding [bɔdibɪldɪŋ] das; ~s body-building no art.

Böe die; ~, ~n s. Bö

bog 1. u. 3. Pers. Sg. Prät. v. biegen

Bogen der; ~s, ~, (südd., österr.:) Bögen a) curve; (Math.) arc; b) (Archit.) arch; c) (Waffe, Musik: Geigen~ usw.) bow; d) (Papier~) sheet

bogen-, Bogen-: ~**fenster** das arched window; ~**förmig** Adj. arched; ~**schießen** das archery no art.

Boheme [bo'e:m] die; ~: bohemian society; **Bohemien** [boe'mjɛ̃:] der; ~s, ~s bohemian

Bohle die; ~, ~n [thick] plank

Böhnchen das; ~s, ~: [small] bean; **Bohne** die; ~, ~n bean; **nicht die ~** (ugs.) not one little bit

Bohnen-: ~**eintopf** der bean stew; ~**kaffee** der real coffee; ~**kraut** das savory; ~**stange die** (auch ugs.: Mensch) beanpole; ~**stroh** das: **dumm wie ~stroh** (ugs.) as thick as two short planks (coll.); ~**suppe die** bean soup

bohnern tr., itr. V. polish; **Bohnerwachs das** floor-polish

bohren 1. tr. V. a) bore; (mit Bohrer, Bohrmaschine) drill, bore (hole); sink (well, shaft, pole, post etc.) (in + Akk. into); b) (bearbeiten) drill (wood, concrete, etc.); c) (drücken in) poke (in + Akk. in[to]); 2. itr. V. a) drill; **in der Nase ~:** pick one's nose; **nach Öl/Wasser usw. ~:** drill for oil/water etc.; b) (ugs.: drängen, fragen) keep on; 3. refl. V. bore its way; **bohrend** Adj. a) gnawing (pain, hunger, remorse); b) (hartnäckig) piercing (look etc.); probing (question); **Bohrer** der; ~s, ~ drill; **Bohr·turm** der derrick; **Bohrung die;** ~, ~en drill-hole

böig Adj. gusty

Boiler ['bɔylɐ] der; ~s, ~: water-heater

Boje die; ~, ~n buoy

Bolivien [bo'li:viən] (das); ~s Bolivia

Böller·schuß der gun salute

Boll·werk das bulwark; (fig.) bulwark; bastion; stronghold

Bolschewik der; ~en, ~i, (abwertend:) ~en Bolshevik; **Bolschewismus** der; ~: Bolshevism no art.; **Bolschewist** der; ~en, ~en Bolshevist; **bolschewistisch** Adj. Bolshevik

bolzen (ugs.) itr. V. kick the ball about

Bolzen der; ~s, ~ bolt

bombardieren tr. V. a) bomb; b) (fig. ugs.) bombard; **Bombardierung die;** ~, ~en a) (Milit.) bombing; b) (fig. ugs.) bombardment

bombastisch 1. Adj. bombastic; 2. adv. bombastically

Bombe die; ~, ~n bomb

Bomben-: ~**angriff** der bomb attack; ~**anschlag** der bomb attack; ~**drohung die** bomb threat; ~**erfolg** der (ugs.) smash hit (sl.); ~**form die** (ugs.) top form

Bomber der; ~s, ~: bomber

Bon [bɔŋ] der; ~s, ~s a) voucher; coupon; b) (Kassenzettel) receipt

Bonbon [bɔŋ'bɔŋ] der od. (österr. nur) das; ~s, ~s sweet (Brit.); candy (Amer.); (fig.) treat

bongen tr. V. (ugs.) ring up

Bongo das; ~[s], ~s od. die; ~, ~s bongo [drum]

Bonmot [bõ'mo:] das; ~s, ~s bon mot

Bonze der; ~n, ~n bigwig (coll.)

Boom [bu:m] der; ~s, ~s boom

Boot das; ~[e]s, ~e boat

Boots-: ~**fahrt die** boat trip; ~**haus das** boathouse; ~**steg der** landing-stage; ~**verleih der** boat-hire

¹**Bord** das; ~[e]s, ~e shelf

²**Bord** der; ~[e]s, ~e (eines Schiffes)

side; **an** ~: on board; **über** ~: overboard

Bordẹll das; ~s, ~e brothel

Bọrd·stein der kerb

Bordüre die; ~, ~n edging

bọrgen tr. V.: s. leihen

Bọrke die; ~, ~n bark

bornịert 1. Adj. bigoted; 2. adv. in a bigoted way

Börse die; ~, ~n stock market; (Gebäude) stock exchange

Börsen-: ~**krach** der stock-market crash; ~**makler** der stockbroker

Bọrste die; ~, ~n bristle; **bọrstig** Adj. bristly

Bọrte die; ~, ~n braiding no indef. art.; edging no indef. art.

bös s. böse; **bös·artig** 1. Adj. a) malicious ⟨person, remark, etc.⟩; vicious ⟨animal⟩; b) (Med.) malignant; 2. adv. maliciously; **Bös·artigkeit** die a) maliciousness; (von Tieren) viciousness; b) (Med.) malignancy

Böschung die; ~, ~en embankment

böse 1. Adj. a) wicked; evil; b) (übel) bad ⟨times, illness, dream, etc.⟩; nasty ⟨experience, affair, situation, trick, surprise, etc.⟩; c) (ugs.) (wütend) mad (coll.); (verärgert) cross (coll.); d) (fam.: ungezogen) naughty; f) (ugs.: arg) terrible (coll.) ⟨pain, fall, shock, disappointment, storm, etc.⟩; 2. adv. a) (übel) ⟨end⟩ badly; **es war doch nicht ~ gemeint** I didn't mean it nastily; b) (ugs.) (wütend) angrily; (verärgert) crossly (coll.); c) (ugs.: sehr) terribly (coll.); **bọshaft** 1. Adj. malicious; 2. adv. maliciously; **Bọshaftigkeit** die; ~, ~en a) o. Pl. maliciousness; b) (Bemerkung) malicious remark; **Bọsheit** die; ~, ~en a) o. Pl. malice; b) (Bemerkung) malicious remark

Bọß der; Bọsses, Bọsse (ugs.) boss (coll.)

bös·willig 1. Adj. malicious; wilful ⟨desertion⟩; 2. adv. maliciously; wilfully ⟨desert⟩; **Bös·willigkeit** die; ~: malice; maliciousness

bot 1. u. 3. Pers. Sg. Prät. v. bieten

Botạnik die; ~ botany no art.; **botanisch** 1. Adj. botanical; 2. adv. botanically

Bötchen das; ~s, ~ little boat

Bote der; ~n, ~n a) messenger; b) (Laufbursche) errand-boy; **Botschaft** die; ~, ~en a) message; b) (diplomatische Vertretung) embassy; **Botschafter** der; ~s, ~: ambassador

Böttcher der; ~s, ~: cooper

Bọttich der; ~s, ~e tub

Bouillon [bul'jɔŋ] die; ~, ~s bouillon

Boulevard [bulə'vaːɐ̯] der; ~s, ~s boulevard

Bourgeoisie [burʒoa'ziː] die; ~, ~n bourgeoisie

Boutique [bu'tiːk] die; ~, ~s od. ~n boutique

Bowle ['boːlə] die; ~, ~n punch (made of wine, champagne, sugar, and fruit or spices)

bowlen ['boːlən] itr. V. bowl; **Bowling** ['boːlɪŋ] das; ~s, ~s [ten-pin] bowling; **Bowling·bahn** die bowling-alley

Box die; ~, ~en a) box; b) (Lautsprecher) speaker; c) (Pferde-) [loose] box; d) (Motorsport) pit

boxen 1. itr. V. box; **gegen jmdn.** ~: fight sb.; box [against] sb.; 2. tr. V. punch; **Boxer** der; ~s, ~ (Sportler, Hund) boxer

Box-: ~**handschuh** der boxing glove; ~**kampf** der boxing match; (im Streit) fist-fight; ~**ring** der boxing ring; ~**sport** der; o. Pl. boxing no art.

Boy [bɔy] der; ~s, ~s servant; (im Hotel) page-boy

Boykott [bɔy'kɔt] der; ~[e]s, ~s boycott; **boykottịeren** tr. V. boycott

¹**brạch** 1. u. 3. Pers. Sg. Prät. v. brechen

²**brạch** Adj. fallow; (auf Dauer) uncultivated

Brachiạl·gewalt die; o. Pl. brute force

Brạch·land das fallow [land]; (auf Dauer) uncultivated land; **brạch|liegen** unr. itr. V. (auch fig.) lie fallow; (auf Dauer) lie waste

brạchte 1. u. 3. Pers. Sg. Prät. v. bringen

Branche ['brãːʃə] die; ~, ~n [branch of] industry

Brạnd der; ~[e]s, Brände fire; **beim ~ der Scheune** when the barn caught fire; **etw. in ~ stecken** set fire to sth.

brạnden itr. V. (geh.) break

Brạnden·burg (das); ~s Brandenburg

brand-, Brạnd-: ~**marken** tr. V. brand ⟨person⟩; denounce ⟨thing⟩; ~**neu** Adj. (ugs.) brand-new; ~**salbe** die ointment for burns; ~**schaden** der fire damage no pl., no indef. art.; ~**stelle** die burn; ~**stifter** der arsonist; ~**stiftung** die arson

Brạndung die; ~, ~en surf

Brand·wunde die burn

brannte *1. u. 3. Pers. Sg. Prät. v.* **bren-nen**

Brannt·wein der spirits *pl.; (Sorte)* spirit

Brasilianer der; ~s, ~, Brazilian; **brasilianisch** *Adj.* Brazilian; **Brasilien** [bra'zi:liən] **(das);** ~s Brazil

brät *3. Pers. Sg. Präsens v.* braten; **Brat·apfel** der baked apple; **braten** *unr. tr., itr. V.* fry; *(im Backofen)* roast; **Braten** der; ~s, ~ a) joint; b) *o. Pl.* roast [meat] *no indef. art.*

Braten-: ~saft der meat juice[s *pl.*]; ~soße die gravy

Brat-: ~fett das [cooking] fat; ~fisch der fried fish; ~hähnchen das, *(südd., österr.)* ~hendl das roast chicken; *(gegrillt)* broiled chicken; ~hering der fried herring; ~kartoffeln *Pl.* fried potatoes; home fries *(Amer.);* ~pfanne die frying-pan; ~spieß der spit; ~wurst die [fried/grilled] sausage

Brauch der; ~[e]s, Bräuche custom

brauchbar *Adj.* useful; *(benutzbar)* usable; wearable 〈*clothes*〉; **brauchen** 1. *tr. V.* a) *(benötigen)* need; b) *(aufwenden müssen)* mit dem Auto braucht er zehn Minuten it takes him ten minutes by car; wie lange brauchst du dafür? how long will it take you?; *(im allgemeinen)* how long does it take you?; c) *(benutzen, gebrauchen)* use; ich könnte es gut ~: I could do with it; 2. *mod. V.; 2. Part~:* need; du brauchst nicht zu helfen there is no need [for you] to help; du brauchst doch nicht gleich zu weinen there's no need to start crying

Brauchtum ['brauxtu:m] das; ~s, Brauchtümer custom

Braue die; ~, ~n [eye]brow

brauen *tr. V.* brew; **Brauerei** die; ~, ~en brewery

braun *Adj.* brown; ~ werden *(sonnengebräunt)* get a tan; **Braun** das; ~s, ~, *(ugs.)* ~s brown; **Braun·bär** der brown bear; **Bräune** die; ~: [sun]tan; **bräunen** *tr. V.* a) tan; sich ~: get a tan; b) *(Kochk.)* brown; **braun·gebrannt** *Adj.* [sun-]tanned; **Braunkohle** die brown coal; lignite; **bräunlich** *Adj.* brownish; **Bräunung** die; ~, ~en browning

Braus *s.* Saus

Brause die; ~, ~n a) fizzy drink; *(~pulver)* sherbet; b) *(veralt.: Dusche)* shower; **brausen** 1. *itr. V.* a) 〈wind,

water, etc.〉 roar; b) *(sich schnell bewegen)* race; c) *auch refl.: s.* duschen 1; 2. *tr. V. s.* duschen 2

Brause-: ~pulver das sherbet; ~tablette die effervescent tablet

Braut die; ~, Bräute bride

Bräutigam der; ~s, ~e [bride]groom

Braut-: ~jungfer die bridesmaid; ~kleid das wedding dress; ~paar das bride and groom

brav 1. *Adj.* a) *(artig)* good; b) *(redlich)* honest; 2. *adv.* nun iß schön ~ deine Suppe be a good boy/girl and eat up your soup

bravo ['bra:vo] *Interj.* bravo; **Bravo** das; ~s, ~s cheer; **Bravo·ruf** der cheer

BRD [be:|ɛr'de:] die; ~ *Abk.* **Bundesrepublik Deutschland** FRG

Brech-: ~bohne die green bean; ~eisen das crowbar

brechen 1. *unr. tr. V.* a) break; sich *(Dat.)* den Arm/das Genick ~: break one's arm/neck; b) *(ablenken)* break 〈waves〉; refract 〈light〉; c) *(bezwingen)* overcome 〈resistance〉; break 〈will, silence, record, blockade, etc.〉; d) *(nicht einhalten)* break 〈agreement, contract, promise, the law, etc.〉; e) *(ugs.: erbrechen)* bring up. 2. *unr. itr. V.* a) mit sein break; brechend voll sein be full to bursting; b) mit jmdm. ~: break with sb.; c) mit sein durch etw. ~: break through sth.; d) *(ugs.: sich erbrechen)* throw up. 3. *unr. refl. V.* 〈waves etc.〉 break; 〈rays etc.〉 be refracted; **Brecher** der; ~s, ~: breaker

Brech-: ~mittel das emetic; ~reiz der nausea; ~stange die crowbar

Brei der; ~[e]s, ~e *(Hafer~)* porridge *(Brit.),* oatmeal *(Amer.) no indef. art.; (Reis~)* rice pudding; *(Grieß~)* semolina *no indef. art.;* **breiig** *Adj.* mushy

breit 1. *Adj.* a) wide; broad, wide 〈hips, face, shoulders, forehead, etc.〉; etw. ~er machen widen sth.; die Beine ~ machen open one's legs; ein 5 cm ~er Saum a hem 5 cm wide; b) *(groß)* die ~e Masse the general public; 2. *adv.* ~ gebaut sturdily built; **breit·beinig** 1. *Adj.* rolling 〈gait〉; 2. *adv.* with one's legs apart; **Breite** die; ~, ~n a) *s.* breit 1a: width; breadth; b) *(Geogr.)* latitude; **breiten** *(geh.) tr., refl. V.* spread

Breiten-: ~grad der degree of latitude; parallel *(~kreis);* ~kreis der parallel

breit-, Breit-: ~|**machen** *refl. V. (ugs.)* **a)** take up room; **b)** *(sich ausbreiten)* be spreading; ~**schult[e]-rig** *Adj.* broad-shouldered; ~**seite** die long side; *(eines Schiffes)* side; ~|**treten** *unr. tr. V. (ugs. abwertend)* go on about; ~**wand** die *(Kino)* big screen

Brems-: ~**backe** die brake-shoe; ~**belag** der brake lining

¹**Bremse** die; ~, ~n brake

²**Bremse** die; ~, ~n *(Insekt)* horse-fly

bremsen *tr. V.* **a)** *auch itr.* brake; **b)** *(fig.)* slow down ⟨*rate, development, production, etc.*⟩; restrict ⟨*imports etc.*⟩

Brems-: ~**klotz** der brake pad; ~**licht** das; *Pl.* ~**lichter** brake-light; ~**pedal** das brake-pedal; ~**spur** die skid-mark; ~**weg** der braking distance

brenn·bar *Adj.* combustible; **brennen** 1. *unr. itr. V.* **a)** burn; ⟨*house etc.*⟩ be on fire; **schnell/leicht** ~: catch fire quickly/easily; **es brennt!** fire!; **b)** *(glühen)* be alight; **c)** *(leuchten)* be on; **das Licht** ~ **lassen** leave the light on; **d) die Sonne brannte** the sun was burning down; **e)** *(schmerzen)* ⟨*wound etc.*⟩ sting; ⟨*feet etc.*⟩ be sore; **f) darauf** ~, **etw. zu tun** be dying to do sth.; 2. *unr. tr. V.* **a)** burn ⟨*hole, pattern, etc.*⟩; **einem Tier ein Zeichen ins Fell** ~: brand an animal; **b)** *(mit Hitze behandeln)* fire ⟨*porcelain etc.*⟩; distil ⟨*spirits*⟩; **c)** *(rösten)* roast ⟨*coffee-beans, almonds, etc.*⟩; **brennend** 1. *Adj. (auch fig.)* burning; lighted ⟨*cigarette*⟩; urgent ⟨*topic*⟩; 2. *adv.* **es interessiert mich** ~, **ob** ...: I'm dying to know whether ...; **Brennessel** die; ~, ~n stinging nettle

Brenn-: ~**glas** das burning-glass; ~**holz** das; *o. Pl.* firewood; ~**material** das fuel; ~**nessel** die *s.* Brennessel; ~**punkt** der focus; ~**spiritus** der methylated spirits; ~**stoff** der fuel

brenzlig *Adj.* **a)** ⟨*smell, taste, etc.*⟩ of burning *not pred.*; **b)** *(ugs.: gefährlich)* dicey *(sl.)*

Bresche die; ~, ~n gap; breach; |für jmdn.| in die ~ **springen** stand in [for sb.]

Brett das; ~|e|s, ~er **a)** board; *(lang und dick)* plank; *(Diele)* floorboard; **Schwarzes** ~: notice-board; **ein** ~ **vor dem Kopf haben** *(fig. ugs.)* be thick; **b)** *Pl. (Ski)* skis

Bretter-: ~**wand** die wooden partition; ~**zaun** der wooden fence

Brett·spiel das board game

Brezel die; ~, ~n pretzel

Bridge [britʃ] das; ~: bridge

Brief der; ~|e|s, ~e letter

Brief-: ~**beschwerer** der; ~s, ~: paperweight; ~**block** der; *Pl.* ~**blocks** writing-pad; ~**bogen** der sheet of writing-paper; ~**freund** der pen-friend; pen-pal *(coll.)*; ~**geheimnis** das privacy of the post; ~**karte** die correspondence card; ~**kasten** der **a)** post-box; **b)** *(privat)* letter-box; ~**kopf** der **a)** letter-heading; **b)** *(aufgedruckt)* letter-head; ~**kuvert** das *(veralt.) s.* ~umschlag

brieflich 1. *Adj.* written; 2. *adv.* by letter; **Brief·marke** die [postage] stamp

Briefmarken-: ~**album** das stamp-album; ~**sammler** der stamp-collector; ~**sammlung** die stamp-collection

Brief-: ~**öffner** der letter-opener; ~**papier** das writing-paper; ~**partner** der, ~**partnerin** die pen-friend; ~**schreiber** der [letter-]writer; ~**tasche** die wallet; ~**taube** die carrier pigeon; ~**träger** der postman; letter-carrier *(Amer.)*; ~**trägerin** die postwoman; [female] letter-carrier *(Amer.)*; ~**um·schlag** der envelope; ~**waage** die letter-scales *pl.*; ~**wahl** die postal vote; ~**wechsel** der correspondence

Bries das; ~es, ~e *(Kochk.)* sweetbreads *pl.*

briet *1. u. 3. Pers. Sg. Prät. v.* braten

Brigade die; ~, ~n *(Milit.)* brigade

Brikett das; ~s, ~s briquette

brillant [bril'jant] 1. *Adj.* brilliant. 2. *adv.* brilliantly; **Brillant** der; ~en, ~en brilliant

Brillant-: ~**ring** der *(brilliant-cut)* diamond ring; ~**schmuck** der; *o. Pl. (brilliant-cut)* diamond jewellery

Brillanz [bril'jants] die; ~ brilliance

Brille die; ~, ~n **a)** glasses *pl.*; spectacles *pl.*; **eine** ~: a pair of glasses *or* spectacles; **eine** ~ **tragen** wear glasses *or* spectacles; **b)** *(ugs.: Klosett~)* [lavatory] seat

Brillen-: ~**etui** das, ~**futteral** das glasses-case; spectacle-case; ~**glas** das [spectacle-]lens; ~**schlange** die spectacled cobra; ~**träger** der person who wears glasses; ~ **sein** wear glasses

Brimborium das; ~s *(ugs. abwertend)* hoo-ha *(coll.)*

bringen *unr. tr. V.* **a)** *(her~)* bring; *(hin~)* take; jmdm. Glück/Unglück ~: bring sb. [good] luck/bad luck; jmdm. eine Nachricht ~: bring sb. news; **b)** *(begleiten)* take; jmdn. nach Hause/zum Bahnhof ~: take sb. home/to the station; **c)** es zu etwas/nichts ~: get somewhere/get nowhere; **d)** jmdn. ins Gefängnis ~ ⟨*crime, misdeed*⟩ land sb. in gaol; jmdn. wieder auf den rechten Weg ~ *(fig.)* get sb. back on the straight and narrow; jmdn. zum Lachen/zur Verzweiflung ~: make sb. laugh/drive sb. to despair; jmdn. dazu ~, etw. zu tun get sb. to do sth.; etw. hinter sich ~ *(ugs.)* get sth. over and done with; **e)** jmdn. um seinen Besitz ~: do sb. out of his property; **f)** *(präsentieren)* present; *(veröffentlichen)* publish; *(senden)* broadcast; **g)** ein Opfer ~: make a sacrifice; **h)** einen großen Gewinn/hohe Zinsen ~: make a large profit/earn high interest; **i)** das bringt es mit sich, daß ...: that means that ...; **j)** *(verursachen)* cause

brisant *Adj.* explosive; **Brisanz** die; ~: explosiveness

Brise die; ~, ~n breeze

Britannien (das); ~s Britain; *(hist.)* Britannia; **Brite** der; ~n, ~n Briton; die ~n the British; er ist [kein] ~: he is [not] British; **Britin** die; ~, ~nen Briton; British girl/woman; **britisch** *Adj.* British; **die Britischen Inseln** the British Isles

bröckelig *Adj.* crumbly; **bröckeln** **1.** *itr. V.* **a)** crumble; **b)** *mit sein von der Wand* ~: crumble away from the wall; **2.** *tr. V.* crumble; **Brocken** der; ~s, ~ *(von Brot)* hunk; *(von Fleisch)* chunk; *(von Lehm, Kohle, Erde)* lump; ein paar ~ Englisch *(fig.)* a smattering of English

brodeln *itr. V.* bubble

Broiler ['brɔylɐ] der; ~s, ~ *(regional)* s. **Brathähnchen**

Brokat der; ~[e]s, ~e brocade

Brokkoli *Pl.* broccoli *sing.*

Brom·beere die blackberry

Bronchie ['brɔnçiə] die; ~, ~n bronchial tube; **Bronchitis** die; ~, bronchitis

Bronze ['brõːsə] die; ~: bronze; **Bronze·medaille** die bronze medal

Brosche die; ~, ~n brooch

Broschüre die; ~, ~n booklet

Brösel der; ~s, ~: breadcrumb; **bröselig** *Adj.* crumbly; **bröseln** *itr. V.* crumble

Brot das; ~[e]s, ~e bread *no pl., no indef. art.*; *(Laib ~)* loaf [of bread]; *(Scheibe ~)* slice [of bread]

Brot-: ~aufstrich der spread; ~belag der topping: *(im zusammengeklappten Brot)* filling

Brötchen das; ~s, ~: roll

Brot-: ~erwerb der way to earn a living; ~korb der bread-basket; ~laib der loaf [of bread]; ~messer das bread-knife; ~rinde die [bread] crust; ~zeit die *(südd.)* **a)** *(Pause)* [tea-/coffee-/lunch-]break; **b)** *o. Pl. (Vesper)* snack; *(Vesperbrot)* sandwiches *pl.*

Bruch der; ~[e]s, Brüche **a)** break; in die Brüche gehen *(zerbrechen)* get broken; *(fig.)* break up; **b)** *(Med.: Knochen~)* fracture; break; **c)** *(Med.: Eingeweide~)* hernia; **d)** *(fig.) (eines Versprechens)* breaking; *(eines Abkommens, Gesetzes)* violation; **e)** *(Math.)* fraction; **brüchig** *Adj.* **a)** brittle ⟨*rock, brickwork*⟩; **b)** *(fig.)* crumbling ⟨*relationship, marriage, etc.*⟩

Bruch-: ~landung die crash-landing; ~rechnen das fractions *pl.*; ~strich der fraction line; ~stück das fragment; ~teil der fraction; im ~teil einer Sekunde in a split second

Brücke die; ~, ~n **a)** *(auch: Kommando~, Zahnmed., Bodenturnen, Ringen)* bridge; **b)** *(Landungs~)* gangway; **c)** *(Teppich)* rug

Brücken-: ~bogen der arch [of a/the bridge]; ~geländer das parapet

Bruder der; ~s, Brüder brother; **Brüderchen** das; ~s, ~: little brother; **brüderlich** **1.** *Adj.* brotherly; **2.** *adv.* in a brotherly way; **Brüderlichkeit** die; ~: brotherliness; **Brüderschaft** die; ~: [mit jmdm.] ~ trinken drink to close friendship [with sb.] *(agreeing to use the familiar 'du' form)*

Brühe die; ~, ~n **a)** stock; *(als Suppe)* clear soup; **b)** *(ugs. abwertend) (Getränk)* muck; *(verschmutztes Wasser)* filthy water; **brühen** *tr. V.* **a)** blanch; **b)** *(auf~)* brew, make ⟨*tea*⟩; make ⟨*coffee*⟩

brüh-, Brüh-: ~warm *Adj.* etw. ~warm weitererzählen *(ugs.)* pass sth. on straight away; ~würfel der stock cube

brüllen **1.** *itr. V.* **a)** ⟨*bull, cow, etc.*⟩ bellow; ⟨*lion, tiger, etc.*⟩ roar; **b)** *(ugs.) (schreien)* roar; *(weinen)* howl; **2.** *tr. V.* yell

brummen *tr., itr. V.* **a)** ⟨*insect*⟩ buzz; ⟨*bear*⟩ growl; ⟨*engine etc.*⟩ drone; **b)**

(unmelodisch singen) drone; **c)** *(mürrisch sprechen)* mumble; **Brummer** der; ~s, ~ *(ugs.)* **a)** *(Fliege)* bluebottle; **b)** *(Lkw)* heavy lorry *(Brit.) or* truck; **brummig** *Adj. (ugs.)* grumpy

Brumm-: ~**kreisel** der humming top; ~**schädel** der *(ugs.)* thick head

brünett *Adj.* dark-haired *(person)*; dark *(hair)*; **Brünette** die; ~, ~n brunette

Brunnen der; ~s, ~ **a)** well; **b)** *(Spring~)* fountain; **Brunnen·kresse** die watercress

Brunst die; ~, **Brünste** *(von männlichen Tieren)* rut; *(von weiblichen Tieren)* heat; **Brunst·zeit** die *(bei männlichen Tieren)* rutting season; *(bei weiblichen Tieren)* [season of] heat

brüsk **1.** *Adj.* brusque; **2.** *adv.* brusquely; **brüskieren** *tr. V.* offend; *(stärker)* insult; *(schneiden)* snub

Brüssel (das); ~s Brussels

Brust die; ~, **Brüste** **a)** chest; **b)** *(der Frau)* breast; **c)** *(Hähnchen~)* breast; *(Rinder~)* brisket; **d)** *o. Pl. (~schwimmen)* breast-stroke

brüsten *refl. V.* sich mit etw. ~: boast about sth.

Brust-: ~**kasten** *(ugs.)* chest; ~**korb** der *(Anat.)* thorax *(Anat.)*; ~**krebs** der breast cancer; ~**schwimmen** *unr. itr. V.; nur im Inf.* do [the] breast-stroke; ~**schwimmen** das breast-stroke; ~**tasche** die breast pocket

Brüstung die; ~, ~en parapet; *(Balkon~)* balustrade

Brust·warze die nipple

Brut die; ~, ~en **a)** brooding; **b)** *(Jungtiere, auch fig. scherzh.: Kinder)* brood

brutal **1.** *Adj.* brutal; violent *(attack, programme, etc.)*; brute *(force, strength)*; **2.** *adv.* brutally; **Brutalität** die; ~, ~en **a)** *o. Pl.* brutality; **b)** *(Handlung)* act of brutality

brüten *itr. V.* **a)** brood; **b)** *(grübeln)* ponder (über + *Dat.* over); **brütend·heiß** *Adj. (ugs.)* boiling hot; **Brüter** der; ~s, ~ *(Kernphysik)* breeder

Brut-: ~**kasten** der incubator; ~**stätte** die *(auch fig.)* breeding-ground

brutto *Adv.* gross

Brutto-: ~**einkommen** das gross income; ~**gehalt** das gross salary; ~**sozialprodukt** das *(Wirtsch.)* gross national product

brutzeln **1.** *itr. V.* sizzle. **2.** *tr. V. (ugs.)* fry [up]

Bub der; ~en, ~en *(südd., österr., schweiz.)* boy; lad; **Bube** der; ~n, ~n *(Kartenspiele)* jack; knave; **Bubi** der; ~s, ~s **a)** [little] boy *or* lad; **b)** *(salopp: Schnösel)* young lad

Buch das; ~[e]s, **Bücher** book; *(Dreh~)* script; über etw. *(Akk.)* ~ führen keep a record of sth.

Buch-: ~**binder** der bookbinder; ~**druck** der; *o. Pl.* letterpress printing

Buche die; ~, ~n **a)** beech[-tree]; **b)** *o. Pl. (Holz)* beech[wood]

Buch·ecker die beech-nut

buchen *tr. V.* **a)** enter; **b)** *(vorbestellen)* book

Bücher·brett das bookshelf

Bücherei die; ~, ~en library

Bücher-: ~**regal** das bookshelves *pl.*; ~**schrank** der bookcase; ~**wurm** der *(scherzh.)* bookworm

Buch-: ~**fink** der chaffinch; ~**führung** die bookkeeping; ~**halter** der bookkeeper; ~**haltung** die **a)** accountancy; **b)** *(Abteilung)* accounts department; ~**händler** der bookseller; ~**handlung** die bookshop; ~**klub** der book club; ~**laden** der bookshop; ~**messe** die book fair; ~**rücken** der spine

Buchs·baum ['buks-] der box[-tree]

Buchse ['buksə] die; ~, ~n **a)** *(Elektrot.)* socket; **b)** *(Technik)* bush

Büchse ['byksə] die; ~, ~n **a)** tin; **b)** *(ugs.: Sammel~)* [collecting-]box; **c)** *(Gewehr)* rifle; *(Schrot~)* shotgun; **Büchsen-** *s.* Dosen-

Buchstabe der; ~ns, ~n letter; *(Druckw.)* character; ein großer/kleiner ~: a capital [letter]/small letter; **buchstabieren** *tr. V.* spell; **buchstäblich** *Adv.* literally

Bucht die; ~, ~en bay

Buchung die; ~, ~en **a)** entry; **b)** *(Vorbestellung)* booking

Buckel der; ~s, ~ **a)** hump; einen ~ machen *(cat)* arch its back; *(person)* hunch one's shoulders; **b)** *(ugs.: Rücken)* back; rutsch mir den ~ runter! *(salopp)* get lost! *(sl.)*; **buckeln** *itr. V. (ugs.)* bow and scrape; vor jmdm. ~: kowtow to sb.

bücken *refl. V.* bend down

bucklig *Adj.* hunchbacked; **Bucklige** der/die; *adj. Dekl.* hunchback

¹Bückling der; ~s, ~e *(ugs. scherzh.: Verbeugung)* bow

²Bückling der; ~s, ~e *(Hering)* bloater

buddeln *itr., tr. V. (ugs.)* dig
Buddha ['bʊda] *der;* ~s, ~s Buddha;
Buddhismus *der;* ~: Buddhism *no
art.;* **Buddhist** *der;* ~en, ~en Buddhist; **buddhistisch** *Adj.* Buddhist
attrib.
Bude *die;* ~, ~n a) kiosk; *(Markt~)*
stall; *(Jahrmarkts~)* booth; b) *(Bau~)*
hut; c) *(ugs.) (Haus)* dump *(coll.);*
(Zimmer) room; digs *pl. (Brit. coll.)*
Budget [by'dʒe:] *das;* ~s, ~s budget
Büfett *das;* ~[e]s, ~s *od.* ~e a) side-
board; b) *(Schanktisch)* bar; c) *(Ver-
kaufstisch)* counter; d) **kaltes** ~: cold
buffet
Büffel *der;* ~s, ~: buffalo
büffeln *(ugs.)* **1.** *itr. V.* swot *(Brit. sl.);*
cram; **2.** *tr. V.* swot up *(Brit. sl.);* cram
Buffet [by'fe:] *das;* ~s, ~s s. Büfett
Bug *der;* ~[e]s, ~e u. Büge bow
Bügel *der;* ~s, ~ a) *(Kleider~)* hanger;
b) *(Brillen~)* ear-piece; c) *(an einer
Tasche, Geldbörse)* frame
bügel-, Bügel-: ~**brett** *das* ironing-
board; ~**eisen** *das* iron; ~**falte** *die*
[trouser] crease; ~**frei** *Adj.* non-iron
bügeln *tr., itr. V.* iron
bugsieren [bʊ'ksi:rən] *tr. V. (ugs.)*
shift; manœuvre; steer *(person)*
buh *Interj.* boo; **Buh** *das;* ~s, ~s
(ugs.) boo; **buhen** *itr. V. (ugs.)* boo
buhlen *itr. V. (geh. abwertend)* **um**
jmds. Gunst ~: court sb.'s favour
Buh·mann *der; Pl.* **Buhmänner** *(ugs.)*
whipping-boy
Bühne *die;* ~, ~n a) stage; b) *(Theater)*
theatre
bühnen-, Bühnen-: ~**arbeiter** *der*
stage-hand; ~**bildner** *der;* ~s, ~:
stage designer; ~**reif** *Adj. (play etc.)*
ready for the stage; *(imitation etc.)*
worthy of the stage; dramatic *(entrance etc.)*
Buh·ruf *der* boo
buk *1. u. 3. Pers. Sg. Prät. v.* backen
Bukett *das;* ~s, ~s *od.* ~e *(geh.)* bouquet
Bulette *die;* ~, ~n *(bes. berl.)* rissole
Bulgare *der;* ~n, ~n Bulgarian; **Bulgarien** [bʊl'ga:rjən] *(das);* ~s Bulgaria; **bulgarisch** *Adj.* Bulgarian
Bull-: ~**auge** *das* circular porthole;
~**dogge** *die* bulldog; ~**dozer**
[-do:zɐ] *der;* ~s, ~: bulldozer
Bulle *der;* ~n, ~n a) bull; b) *(salopp:
Polizist)* cop *(sl.);* **Bullen·hitze** *die*
(ugs.) sweltering *or* boiling heat
Bulletin [byl'tɛ̃:] *das;* ~s, ~s bulletin
bullig 1. *Adj.* **a)** beefy *(person, appear-*

ance, etc.); chunky *(car);* b) *(drük-
kend)* sweltering *(heat);* **2.** *adv.* ~ **heiß**
boiling hot
Bull·terrier *der* bull-terrier
bum *Interj.* bang
Bumerang *der;* ~s, ~e *od.* ~s
boomerang
Bummel *der;* ~s, ~ a) stroll *(durch*
around); b) *(durch Lokale)* pub-crawl
(coll.); **Bummelei** *die;* ~, ~en *(ugs.)*
a) dawdling; b) *(Faulenzerei)* loafing
about; **bummelig** *(ugs.)* **1.** *Adj.* **a)**
slow; b) *(nachlässig)* slipshod; **2.** *adv.*
a) slowly; b) *(nachlässig)* in a slipshod
way; **bummeln** *itr. V.* **a)** *mit sein*
stroll *(durch* around); **durch die Knei-
pen** ~: go on a pub-crawl *(Brit. coll.);*
b) *(trödeln)* dawdle; c) *(faulenzen)*
laze about
bums *Interj.* bang; **Bums** *der;* ~es,
~e *(ugs.)* bang; *(dumpfer)* thud;
bumsen *itr. V. (ugs.)* a) bang; *(dumpfer)* thump; *unpers.* **es bumste ganz
furchtbar** there was a terrible bang/
thud; b) *mit sein (stoßen)* bang
¹**Bund** *der;* ~[e]s, Bünde a) *(Vereinigung)* association; *(Bündnis, Pakt)* alliance; b) *(föderativer Staat)* federation; c) *(an Röcken, Hosen)* waistband
²**Bund** *das;* ~[e]s, ~e bunch; **Bünd-
chen** *das;* ~s, ~ band; **Bündel** *das;*
~s, ~ bundle; **bündeln** *tr. V.* bundle
up *(newspapers, old clothes, rags, etc.);*
tie *(banknotes etc.)* into bundles/a
bundle; tie *(flowers, radishes, carrots,
etc.)* into bunches/a bunch; sheave
(straw, hay, etc.)
Bundes- federal; *(in Namen, Titeln)*
Federal
bundes-, Bundes-: ~**bürger** *der*
(veralt.) West German citizen;
~**deutsch** *Adj. (veralt.)* West German; ~**land** *das* [federal] state;
(österr.) province; ~**liga** *die* national
division; ~**rat** *der* Bundesrat; ~**republik** *die* federal republic; **die** ~**republik Deutschland** The Federal Republic of Germany; ~**straße** *die*
federal highway; ≈ A road *(Brit.);*
~**tag** *der* Bundestag
Bundestags-: ~**abgeordnete** *der/
die* member of parliament; member of
the Bundestag; ~**wahl** *die* parliamentary *or* general election
bundes-, Bundes-: ~**trainer** *der* national team manager; ~**wehr** *die*
[Federal] Armed Forces *pl.;* ~**weit**
Adj., adv. nation-wide

Bund-: ~**falten** *Pl.* pleats; ~**hose die** knee-breeches

bündig 1. *Adj.* **a)** succinct; **b)** *(schlüssig)* conclusive; **2.** *adv.* **a)** succinctly; **b)** *(schlüssig)* conclusively

Bündnis das; ~**ses,** ~**se** alliance

Bungalow ['bʊŋgalo] **der;** ~**s,** ~**s** bungalow

Bunker der; ~**s,** ~ **a)** bunker; **b)** *(Luftschutz~)* air-raid shelter

bunt 1. *Adj.* **a)** colourful; *(farbig)* coloured; ~**e Farben/Kleidung** bright colours/brightly coloured clothes; **b)** *(fig.)* varied *(programme etc.)*; **2.** *adv.* **a)** colourfully; **b)** *(fig.)* ein ~ **gemischtes Programm** a varied programme

bunt-, Bunt-: ~**bemalt** *Adj.* brightly painted; ~**papier das** coloured paper; ~**specht der** spotted woodpecker; ~**stift der** coloured pencil/crayon

Bürde die; ~, ~**n** *(geh.)* weight; load

Burg die; ~, ~**en a)** castle; **b)** *(Strand~)* wall of sand

Bürge der; ~**n,** ~**n** guarantor; **bürgen** *itr. V.* **a) für jmdn./etw.** ~: vouch for sb./sth.; **b)** *(fig.)* guarantee

Bürger der; ~**s,** ~, **Bürgerin die;** ~, ~**nen** citizen

Bürger-: ~**initiative die** citizens' action group; ~**krieg der** civil war

bürgerlich *Adj.* **a) nicht präd.** *(staats~)* civil *(rights, marriage, etc.)*; civic *(duties)*; **b)** *(dem Bürgertum zugehörig)* middle-class; **die** ~**e Küche** good plain cooking; **c)** *(Polit.)* nonsocialist; *(nicht marxistisch)* non-Marxist

bürger-, Bürger-: ~**meister der** mayor; ~**nah** *Adj.* which/who reflects the general public's interests *postpos., not pred.;* ~**pflicht die** duty as a citizen; ~**steig der** pavement *(Brit.);* sidewalk *(Amer.)*

Bürgertum ['--tu:m] **das;** ~**s a)** middle class; **b)** *(Groß~)* bourgeoisie

Bürgin die; ~, ~**nen** *s.* Bürge; **Bürgschaft die;** ~, ~**en a)** guarantee; **b)** *(Betrag)* penalty

Büro das; ~**s,** ~**s** office

Büro-: ~**angestellte der/die** officeworker; ~**artikel der** item of office equipment; ~**haus das** office-block; ~**klammer die** paper-clip

Bürokrat der; ~**en,** ~**en** bureaucrat; **Bürokratie die;** ~, ~**n** bureaucracy; **bürokratisch** 1. *Adj.* bureaucratic; **2.** *adv.* bureaucratically

Bürschchen ['bʏrʃçən] **das;** ~, ~:

little fellow; **Bursche der;** ~**n,** ~**n a)** boy; lad; **b)** *(abwertend: Kerl)* guy *(sl.)*

burschikos 1. *Adj.* **a)** sporty *(look, clothes)*; [tom]boyish *(behaviour, girl, haircut)*; **b)** *(ungezwungen)* casual *(comment, behaviour, etc.)*; **2.** *adv.* **a)** [tom]boyishly; **b)** *(ungezwungen)* in a colloquial way

Bürste die; ~, ~**n** brush; **bürsten** *tr. V.* brush

Bus der; ~**ses,** ~**se** bus; **Bus·bahnhof der** bus station

Busch der; ~**[e]s, Büsche** bush; **auf den** ~ **klopfen** *(fig. ugs.)* sound things out

Büschel das; ~**s,** ~: tuft; *(von Heu, Stroh)* handful

Busen der; ~**s,** ~ bust

Bus-: ~**fahrer der** bus-driver; ~**haltestelle die** bus-stop; ~**linie die** bus-route

Bussard der; ~**s,** ~**e** buzzard

Buße die; ~, ~**n** *(Rel.)* penance *no art.;* **büßen** 1. *tr. V.* **a)** atone for; **b)** *(fig.)* pay for; **2.** *itr. V.* **a) für etw.** ~: atone for sth.; **b)** *(fig.)* pay; **Buß·geld das** *(Rechtsw.)* fine

Büsten·halter der bra; brassière *(formal)*

Butan·gas das butane gas

Butt der; ~**[e]s,** ~**e** flounder; butt

Bütten·papier das handmade paper *(with deckle-edge)*

Butter die; ~: butter; **es ist alles in** ~ *(ugs.)* everything's fine

butter-, Butter-: ~**blume die** *(Sumpfdotterblume)* marsh marigold; *(Hahnenfuß)* buttercup; ~**brot das** slice of bread and butter; *(zugeklappt)* sandwich; ~**creme die** butter-cream; ~**milch die** buttermilk; ~**weich** *Adj.* beautifully soft

b.w. *Abk.* bitte wenden p.t.o.

bzw. *Abk.* beziehungsweise

C

c, C [tse:] **das;** ~, ~: **a)** *(Buchstabe)* c/C; **b)** *(Musik)* [key of] C

ca. *Abk.* cirka c.

Café das; ~**s,** ~**s** café

Cafeteria die; ~, ~s cafeteria
cal Abk. |Gramm|kalorie cal.
Callgirl ['kɔ:lgə:l] das; ~s, ~s call-girl
Camp [kɛmp] das; ~s, ~s camp; **campen** itr. V. camp; **Camping** das; ~s camping
Camping-: ~**bus** der motor caravan; camper; ~**platz** der campsite; campground (Amer.)
Canasta das; ~s canasta
Caravan ['ka(:)ravan] der; ~s, ~s (Wohnwagen) caravan; trailer (Amer.)
Cayenne·pfeffer [ka'jɛn-] der cayenne [pepper]
CD [tse:'de:] die; ~, ~s CD
CDU [tse:de:'u:] die; ~ Abk. Christlich-Demokratische Union |Deutschlands| [German] Christian Democratic Party
C-Dur ['tse:-] das; ~: C major
Cello ['tʃɛlo] das; ~s, ~s od. **Celli** cello
Celsius o. Art. 20 **Grad** ~: 20 degrees Celsius or centigrade
Cembalo ['tʃɛmbalo] das; ~s, ~s od. **Cembali** harpsichord
Ceylon ['tsailɔn] (das) (das); ~s (hist.) Ceylon (Hist.)
Champagner [ʃam'panjɐ] der; ~s, ~ champagne (from Champagne)
Champignon ['ʃampɪnjɔn] der; ~s, ~s mushroom
Chance ['ʃã:sə] die; ~, ~n a) chance; b) Pl. (Aussichten) prospects; |bei jmdm| ~n **haben** stand a chance [with sb.]
Chaos das; ~: chaos no art.
Charakter der; ~s, ~e [...'te:rə] character; **charakterisieren** tr. V. characterize; **charakteristisch** Adj. characteristic (für of); **charakterlich** 1. Adj. character attrib.; 2. adv. in [respect of] character; **charakter·los** Adj. unprincipled; (niederträchtig) despicable; (labil) spineless
charmant [ʃar'mant] 1. Adj. charming; 2. adv. charmingly; **Charme** [ʃarm] der; ~s charm
Charter- ['tʃartɐ-]: ~**flug** der charter flight; ~**maschine** die chartered aircraft
Chassis [ʃa'si:] das; ~ [ʃa'si:(s)], ~ [ʃa'si:s] chassis
Chauffeur [ʃɔ'fø:ɐ] der; ~s, ~e driver; (privat angestellt) chauffeur
Chef [ʃɛf] der; ~s, ~s, **Chefin** die; ~, ~nen (Leiter[in]) head; (der Polizei, des Generalstabs) chief; (einer Partei, Bande) leader; (Vorgesetzte[r]) superior; boss (coll.)

Chef-: ~**koch** der chef; head cook; ~**sekretärin** die director's secretary
Chemie die; ~ a) chemistry no art.; b) (ugs.: Chemikalien) chemicals pl.; **Chemiker** der; ~s, ~, **Chemikerin** die; ~, ~nen (graduate) chemist; **chemisch** 1. Adj. chemical; 2. adv. chemically
Chicorée ['ʃikore] der; ~s od. die; ~: chicory
Chiffon ['ʃɪfõ] der; ~s, ~s chiffon
Chiffre ['ʃɪfrə] die; ~, ~n a) (Zeichen) symbol; b) (Geheimzeichen) cipher; c) (in Annoncen) box number
Chile ['tʃi:le, 'çi:lə] (das); ~s Chile; **Chilene** [tʃi'le:nə, çi'le:nə] der; ~n, ~n, **Chilenin** die; ~, ~nen Chilean; **chilenisch** Adj. Chilean
Chili ['tʃi:li] der; ~s, ~es a) Pl. (Schoten) chillies; b) o. Pl. (Gewürz) chilli [powder]
China (das); ~s China; **Chinese** der; ~n, ~n, **Chinesin** die; ~, ~nen Chinese; **chinesisch** Adj. Chinese
Chip [tʃip] der; ~s, ~s a) (Spielmarke) chip; b) (Kartoffel~) [potato] crisp (Brit.) or (Amer.) chip; c) (Elektronik) [micro]chip
Chirurg der; ~en, ~en surgeon; **Chirurgie** die; ~, ~n a) o. Pl. surgery no art.; b) (Abteilung) surgical department; (Station) surgical ward; **chirurgisch** 1. Adj. surgical; 2. adv. surgically; by surgery
Chlor das; ~s chlorine; **Chloroform** das; ~s chloroform; **Chlorophyll** das; ~s chlorophyll
Cholera die; ~: cholera
cholerisch Adj. irascible; choleric (temperament)
Cholesterin das; ~s cholesterol
Chor der; ~[e]s, **Chöre** ['kø:rə] (auch Archit.) choir; (in Oper, Sinfonie, Theater; Komposition) chorus; im ~ **rufen** shout in chorus; **Choral** der; ~s, **Choräle** (Kirchenlied) chorale
Choreographie die; ~, ~n choreography
Chose ['ʃo:zə] die; ~, ~n (ugs.) stuff; die ganze ~: the whole lot (coll.) or (sl.) shoot
Chow-Chow [tʃau 'tʃau] der; ~s, ~s chow
Christ der; ~en, ~en Christian
Christ-: ~**baum** der (bes. südd.) Christmas tree; ~**demokrat** der (Politik) Christian Democrat
Christenheit die; ~ Christendom no art.; **Christentum** das; ~s Chris-

tianity *no art.; (Glaube)* Christian faith

Christin die; ~, ~**nen** Christian; **Christ·kind** das; *o. Pl.* Christ-child *(as bringer of Christmas gifts);* **christlich 1.** *Adj.* Christian. **2.** *adv.* in a [truly] Christian spirit

Christ-: ~**messe** die *(kath. Rel.)* Christmas Mass; ~**mette** die *(kath. Rel.)* Christmas Mass; *(ev. Rel.)* midnight service [on Christmas Eve]; ~**rose** die Christmas rose; ~**stollen** der [German] Christmas loaf *(with candied fruit, almonds, etc.)*

Christus (der); ~ *od.* Christi Christ

Chrom das; ~s chromium

Chromosom das; ~s, ~en *(Biol.)* chromosome

Chronik die; ~, ~en chronicle; **chronisch** *Adj.* chronic

Chrysantheme die; ~, ~n chrysanthemum

City ['sɪti] die; ~, ~s city centre

clever ['klɛvɐ] **1.** *Adj. (raffiniert)* shrewd; *(intelligent, geschickt)* clever; **2.** *adv.: s. Adj.:* shrewdly; cleverly

Clique ['klɪkə] die; ~, ~n **a)** *(abwertend)* clique; **b)** *(Freundeskreis)* set; *(größere Gruppe)* crowd *(coll.)*

Clown [klaun] der; ~s, ~s clown

Club *s.* Klub

cm *Abk.:* Zentimeter cm.

Co. *Abk.:* Compagnie Co.

Cockpit das; ~s, ~s cockpit

Cocktail ['kɔkteɪl] der; ~s, ~s cocktail

Cognac ⓦ der; ~s, ~s Cognac

Color- *(Fot.)* colour ⟨film, slide, etc.⟩

Colt ⓦ der; ~s, ~s Colt (P) [revolver]

Comic·heft das comic

Computer [kɔm'pjuːtɐ] der; ~s, ~: computer

Container [kɔn'teːnɐ] der; ~s, ~: container; *(für Müll)* [refuse] skip

cool [kuːl] *(ugs.)* **1.** *Adj.* cool; ~ bleiben keep one's cool *(sl.);* **2.** *adv.* coolly *(coll.)*

Cord der; ~[e]s, ~e *od.* ~s cord; *(~samt)* corduroy

Corned beef ['kɔːnd 'biːf] das; ~ ~: corned beef

Couch [kautʃ] die, *(schweiz. auch:)* der; ~, ~es sofa

Coup [kuː] der; ~s, ~s coup

Coupon [ku'põː] der; ~s, ~s coupon; voucher

Courage [ku'raːʒə] die; ~ *(ugs.)* courage

Cousin [ku'zɛ̃ː] der; ~s, ~s, **Cousine** die; ~, ~n cousin

Cowboy ['kaubɔy] der; ~s, ~s cowboy

Credo *s.* Kredo

Creme [kreːm] die; ~, ~s, *(schweiz.:)* ~n cream

ČSFR [tʃeːlɛslɛf'lɛr] die; ~: die ~: Czechoslovakia

CSU [tseːlɛs'luː] die; ~ *Abk.:* Christlich-Soziale Union CSU

Curry ['kœri] das; ~s, ~s curry-powder

D

d, D [deː] das; ~, ~ **a)** *(Buchstabe)* d/D; **b)** *(Musik)* [key of] D

D *Abk.* Damen

da 1. *Adv.* **a)** *(dort)* there; da draußen/drinnen/drüben/unten out/in/over/down there; da, wo where; **b)** *(hier)* here; **c)** *(zeitlich)* then; *(in dem Augenblick)* at that moment; **d)** *(deshalb)* der Zug war schon weg, da habe ich den Bus genommen the train had already gone, so I took the bus; **e)** *(ugs.: in diesem Fall)* da kann man nichts machen there's nothing one can do about it; **2.** *Konj. (weil)* as; since

da·bei *Adv.* **a)** with it/him/her/them; nahe ~: close by; **b)** *(währenddessen)* at the same time; *(bei diesem Anlaß)* then; on that occasion; die ~ entstehenden Kosten the expense involved; **c)** *(außerdem)* ~ [auch] what is more; **d)** *(hinsichtlich dessen)* about it/them; was hast du dir denn ~ gedacht? what 'were you thinking of?

dabei-: ~|**bleiben** *unr. itr. V.; mit sein* stay there; be there; ~|**haben** *unr. tr. V.* have with one; ~|**sein** *unr. itr. V.; mit sein (Zusschr. nur im Inf. u. 2. Part.)* **a)** *(anwesend sein)* be there; be present (bei at); *(teilnehmen)* take part (bei in); **b)** |gerade| ~sein, etw. zu tun be just doing sth.; ~|**stehen** *unr. itr. V.* stand there

da|bleiben *unr. itr. V.; mit sein* stay there; *(hier bleiben)* stay here

Dach das; ~[e]s, Dächer roof

Dach-: ~**decker** [~dɛkɐ] der; ~s, ~: roofer; ~**garten** der roof-garden; ~**kammer** die attic [room]; ~**luke**

die skylight; **~pappe** die roofing-felt; **~rinne** die gutter

Dachs [daks] der; ~es, ~e badger

dachte *1. u. 3. Pers. Sg. Prät. v. denken*

Dach-: **~terrasse** die roof-terrace; **~ziegel** der roof-tile

Dackel der; ~s, ~: dachshund

da·durch *Adv.* **a)** through it/them; **b)** *(durch diesen Umstand)* as a result; *(durch dieses Mittel)* by this [means]

da·für *Adv.* **a)** for it/them; ~, **daß** ... *(wenn man berücksichtigt, daß)* considering that ...; *(damit)* so that ...; ~ **sorgen |, daß ...|** see to it [that ...]; **b)** ~ **sein** be in favour [of it]; **ein Beispiel** ~ **ist ...**: an example of this is ...; **c)** *(als Gegenleistung)* in return [for it]; *(beim Tausch)* in exchange; *(statt dessen)* instead

dafür|können *unr. tr. V.* etwas/nichts ~: be/not be responsible

dagegen *Adv.* **a)** against it/them; etwas ~ **haben** have sth. against it; **ich habe nichts** ~: I've no objection; ~ **sein** be against it; **b)** *(im Vergleich dazu)* by *or* in comparison

da·heim *Adv.* *(bes. südd., österr., schweiz.)* **a)** *(zu Hause)* at home; *(nach Präp.)* home; **b)** *(in der Heimat)* [back] home

da·her *Adv.* **a)** from there; **b)** *(durch diesen Umstand)* hence; **c)** *(deshalb)* therefore; so

daher|kommen *unr. itr. V.* come along

da·hin a) there; **b)** *(fig.)* ~ mußte es kommen it had to come to that; **c)** bis ~: to there; *(zeitlich)* until then; **d)** ~ **sein** be *or* have gone; **e)** *(in diesem Sinne)* ~ |gehend|, **daß ...**: to the effect that ...

da·hinten *Adv.* over there

da·hinter *Adv.* behind it/them; *(folgend)* after it/them

Dahlie ['da:li̯ə] die; ~, ~n dahlia

da-: **~|lassen** *unr. tr. V. (ugs.)* leave [there]; *(hier lassen)* leave here; **~|liegen** *unr. itr. V.* lie there

dalli *Adv. (ugs.)* [~] ~! get a move on!

damalig *Adj.; nicht präd.* at that *or* the time *postpos.*; **damals** *Adv.* at that time

Damast der; ~|e|s, ~e damask

Dame die; ~, ~n **a)** *(Frau)* lady; **b)** *(Schach, Kartenspiele)* queen; **c)** *o. Pl. (Spiel)* draughts *(Brit.)*; checkers *(Amer.)*

Damen-: **~binde** die sanitary towel

(Brit.) or (Amer.) napkin; **~friseur** der ladies' hairdresser; **~rad** das lady's bicycle; **~toilette** die ladies' toilet

da·mit 1. *Adv.* **a)** with it/them; **b)** *(gleichzeitig)* with that; **c)** *(daher)* thus; **2.** *Konj.* so that

dämlich *(ugs. abwertend)* **1.** *Adj.* stupid; **2.** *adv.* stupidly

Damm der; ~|e|s, **Dämme** embankment; levee *(Amer.)*; *(Deich)* dike; *(Stau~)* dam

dämmern *itr. V.* es dämmert *(morgens)* it is getting light; *(abends)* it is getting dark; **Dämmerung** die; ~, ~en **a)** *(Abend~)* twilight; dusk; **b)** *(Morgen~)* dawn

Dämon der; ~s, ~en [dɛ'mo:nən] demon; **dämonisch** *Adj.* daemonic

Dampf der; ~|e|s, **Dämpfe** steam *no pl., no indef. art*; **dampfen** *itr. V.* steam (vor + *Dat.* with)

dämpfen *tr. V.* **a)** *(garen)* steam ⟨fish, vegetables, potatoes⟩; **b)** *(mildern)* muffle ⟨sound⟩; cushion, absorb ⟨blow, impact, shock⟩

Dampfer der; ~s, ~: steamer

Dampf-: **~maschine** die steam engine; **~nudel** die *(südd., Kochk.)* steamed yeast dumpling; **~walze** die steamroller

da·nach *Adv.* **a)** *(zeitlich)* after it/that; then; **b)** *(räumlich)* after it/them; **c)** *(entsprechend)* in accordance with it/them

Däne der; ~n, ~n Dane

da·neben *Adv.* **a)** beside him/her/it/them *etc.*; **b)** *(im Vergleich dazu)* in comparison

daneben-: **~|benehmen** *unr. refl. V. (ugs.)* blot one's copybook *(coll.)*; **~|gehen** *unr. itr. V.; mit sein* miss [the target]; **~|schießen** *unr. itr. V.* miss [the target]

Dänemark (das) ; ~s Denmark; **Dänin** die; ~, ~nen Dane; Danish woman/girl; **dänisch** *Adj.* Danish; *s. auch* deutsch, Deutsch

dank *Präp. mit Dat. u. Gen.* thanks to; **Dank** der; ~|e|s thanks *pl.; mit |vielem od. bestem|* ~ zurück thanks for the loan; *(bes. geschrieben)* returned with thanks!; **vielen/besten/herzlichen** ~! thank you very much; **dankbar 1.** *Adj.* grateful; *(anerkennend)* appreciative ⟨child, audience, etc.⟩; |jmdm.| **für etw.** ~ **sein** be grateful [to sb.] for sth.; **2.** *adv.* gratefully; **Dankbarkeit** die; ~: gratitude; **danke** *Höflich-*

keitsformel thank you; *(ablehnend)* no, thank you; ~ **schön/sehr/vielmals** thank you very much; **danken 1.** *itr. V. (Dank aussprechen)* thank; **ich danke Ihnen vielmals** thank you very much; **na, ich danke!** *(ugs.)* no, 'thank you!; **2.** *tr. V.* |**aber bitte,**| **nichts zu** ~: don't mention it; **Danke·schön das;** ~s thank-you

dann *Adv.* **a)** then; **was** ~? what happens then?; **noch drei Tage,** ~ **ist Ostern** another three days and it will be Easter; **bis** ~: see you then; ~ **und wann** now and then; **b)** *(in diesem Falle)* then; in that case; ~ **will ich nicht weiter stören** in that case I won't disturb you any further; |**na,**| ~ **eben nicht!** in that case, forget it!; **nur** ~, **wenn ...:** only if ...

daran [da'ran] *Adv.* **a)** *(an dieser/diese Stelle, an diesem/diesen Gegenstand)* on it/them; **dicht** ~: close to it/them; **nahe** ~ **sein, etw. zu tun** be on the point of doing sth.; **b)** *(hinsichtlich dieser Sache)* about it/them; ~ **ist nichts zu machen** there's nothing one can do about it; **kein Wort** ~ **ist wahr** not a word of it is true; **mir liegt viel** ~: it means a lot to me; **c) ich wäre beinahe** ~ **erstickt** I almost choked on it; **er ist** ~ **gestorben** he died of it

daran|setzen *tr. V.* devote ⟨*energy etc.*⟩ to it; summon up ⟨*ambition*⟩ for it; *(aufs Spiel setzen)* risk ⟨*one's life, one's honour*⟩ for it

darauf *Adv.* **a)** on it/them; *(oben* ~) on top of it/them; **b) er hat** ~ **geschossen** he shot at it/them; **c)** *(danach)* after that; **ein Jahr** ~ / **kurz** ~ **starb er** he died a year later/shortly afterwards

darauf-: ~**folgend** *Adj.* following; ~**hin** [--'-] *Adv.* **a)** thereupon; **b)** *(unter diesem Gesichtspunkt)* with a view to this/that

daraus *Adv.* **a)** from it/them; out of it/them; **b) mach dir nichts** ~ don't worry about it; **was ist** ~ **geworden?** what has become of it?

darf *1. u. 3. Pers. Sg. Präsens v.* **dürfen;** **darfst** *2. Pers. Sg. Präsens v.* **dürfen**

darin *Adv.* **a)** in it/them; **b)** *(in dieser Hinsicht)* in that respect

dar|legen *tr. V.* explain; set forth ⟨*reasons, facts*⟩

Darm *der;* ~|e|s, **Därme** intestines *pl.;* bowels *pl.*

dar|stellen *tr. V.* **a)** depict; portray; **etw. graphisch** ~: present sth. graphically; **b)** *(verkörpern)* play; act; **c)**

(schildern) describe ⟨*person, incident, etc.*⟩; present ⟨*matter, argument*⟩; **d)** *(sein, bedeuten)* represent

Darsteller *der;* ~s, ~ actor; **Darstellerin die;** ~, ~**nen** actress; **Darstellung die a)** representation; *(Schilderung)* portrayal ; *(Bild)* picture; **graphische/schematische** ~: diagram; *(Graph)* graph; **b)** *(Beschreibung, Bericht)* description; account

darüber *Adv.* **a)** over it/them; **b)** ~ **hinaus** in additon [to that]; *(noch obendrein)* what is more; **c)** *(über dieser/ diese Angelegenheit)* about it/them; **d)** *(über diese Grenze, dieses Maß hinaus)* over [that]

darüber|stehen *unr. itr. V. (fig.)* be above such things

darum *Adv.* **a)** [a]round it/them; **b)** *(diesbezüglich)* **ich sorge mich** ~: I worry about it; **c)** ['--] *(deswegen)* for that reason

darunter *Adv.* **a)** *(unter dem Genannten/das Genannte)* under it/them; **b)** *(unter dieser Grenze, diesem Maß)* less; **Bewerber im Alter von 40 Jahren und** ~: applicants aged 40 and under

das 1. *best. Art. Nom. u. Akk.* the; **2.** *Demonstrativpron.* **a)** *attr.* **das Kind war es** it was 'that child; **b) alleinstehend das** |**da**| that one; **das** |**hier**| this one [here]; **3.** *Relativpron. (Mensch)* who; that; *(Sache, Tier)* which; that

da|sein *unr. itr. V.; mit sein; Zusschr. nur im Inf. u. Part.* **a)** be there; *(hier sein)* be here; **noch** ~ *(übrig sein)* be left; **ist Herr X da?** is Mr X about or available?; **ich bin gleich wieder da** I'll be right back; **b)** *(fig.)* ⟨*case*⟩ occur; ⟨*moment*⟩ have arrived; ⟨*situation*⟩ have arisen

Da·sein *das* existence

da|sitzen *unr. itr. V.* sit there

dasjenige *s.* **derjenige**

daß *Konj.* **a)** that; **entschuldigen Sie bitte,** ~ **ich mich verspätet habe** please forgive me for being late; **ich verstehe nicht,** ~ **sie ihn geheiratet hat** I don't understand why she married him; **b)** *(nach Pronominaladverbien o.ä.)* [the fact] that; **das liegt daran,** ~ **du nicht aufgepaßt hast** that comes from your not paying attention; **c)** *(im Konsekutivsatz)* that; |**so**| ~: so that; **d)** *(im Finalsatz)* so that; **e)** *(im Ausruf)* ~ **mir das passieren mußte!** why did it have to [go and] happen to me!

dasselbe *s.* **derselbe**

da|stehen *unr. itr. V.* **a)** stand there;

b) *(fig.)* **gut** ~: be in a good position; |**ganz**| **allein** ~: be [all] alone in the world

Daten 1. *s.* **Datum; 2.** *Pl.* data

Daten-: ~**schutz der** data protection; ~**verarbeitung die** data processing *no def. art.*

datieren *tr. V.* date

Dativ der; ~**s,** ~**e** *(Sprachw.)* dative [case]; **Dativ·objekt das** *(Sprachw.)* indirect object

Dattel die; ~, ~**n** date; **Dattel·palme die** date-palm

Datum das; ~**s, Daten** date

Dauer die; ~ **a)** length; **für die** ~ **eines Jahres** *od.* **von einem Jahr** for a period of one year; **b)** *(Fortbestehen)* **von** ~ **sein** last [long]; **auf die** ~: in the long run; **auf** ~: permanently

dauer-, Dauer-: ~**auftrag der** *(Bankw.)* standing order; ~**haft 1.** *Adj.* **a)** [long-]lasting *(peace, friendship, etc.)*; **b)** *(haltbar)* durable; **2.** *adv.* lastingly; ~**karte die** season ticket; ~**lauf der** jogging *no art.*; **ein** ~**lauf** a jog

¹**dauern** *itr. V.* last; *(job etc.)* take; **einen Moment, es dauert nicht lange** just a minute, it won't take long

dauernd 1. *Adj.*constant *(noise, interruptions, etc.)*; permanent *(institution)*; **2.** *adv.* constantly; **er kommt** ~ **zu spät** he keeps on arriving late

Dauer-: ~**stellung die** permanent position; ~**welle die** perm; ~**wurst die** smoked sausage *(with good keeping properties, esp. salami)*

Daumen der; ~**s,** ~: thumb

Daune die; ~, ~**n** down [feather]; ~**n down** *sing.*

davon *Adv.* **a)** *(von dieser Stelle entfernt, weg)* from it/them; *(von dort)* from there; *(mit Entfernungsangabe)* away [from it/them]; **b)** *(hinsichtlich dieser Sache)* about it/them; **c)** *(durch diese Angelegenheit verursacht)* by it/them; **das kommt** ~! *(ugs.)* [there you are,] that's what happens; **d) ich hätte gern ein halbes Pfund** ~: I would like half a pound of that/those; **e)** ~ **kann man nicht leben** you can't live on that

davon-: ~|**fahren** *unr. itr. V.; mit sein* leave; *(mit dem Auto)* drive off; *(mit dem Fahrrad, Motorrad)* ride off; ~|**kommen** *unr. itr. V.; mit sein* get away; ~|**laufen** *unr. itr. V.; mit sein* run away; ~|**tragen** *unr. tr. V.* **a)** carry away; take away *(rubbish)*; **b)** *(geh.: erringen)* gain *(a victory, fame)*;

c) *(geh.: sich zuziehen)* receive *(injuries)*

da·vor *Adv.* **a)** in front of it/them; **b)** *(zeitlich)* before [it/them]

davor-: ~|**liegen** *unr. itr. V.* lie in front of it/them; ~|**schieben 1.** *unr. tr. V.* push in front of it/them; **2.** *unr. refl. V.* move in front of it/them; ~|**stehen** *unr. itr. V.* stand in front of it/them; ~|**stellen 1.** *tr. V.* put in front of it/them; **2.** *refl. V.* plant oneself in front of it/them

da·zu *Adv.* **a)** *(zusätzlich zu dieser Sache)* with it/them; *(gleichzeitig)* at the same time; *(außerdem)* what is more; **b)** *(diesbezüglich)* about it/them; **c)** *(zu diesem Zweck)* for it; **d)** *(zu diesem Ergebnis)* to it; ~ **reicht unser Geld nicht** we haven't enough money for that

dazu-: ~|**geben** *unr. tr. V.* add; ~|**gehören** *tr. V.* belong to it/them; ~|**kommen** *unr. itr. V.; mit sein* **a)** *(hinkommen)* arrive; **b)** *(hinzukommen)* **kommt noch etwas dazu?** is there anything else [you would like]?; ~ **kommt daß ...** *(fig.)* what's more, ...; on top of that ...; ~|**rechnen** *tr. V.* add on; ~|**tun** *unr. tr. V. (ugs.)* add

da·zwischen *Adv.* in between; between them; *(darunter)* among them

dazwischen-: ~|**kommen** *unr. itr. V.; mit sein* **a)** **mit dem Finger** ~**kommen** get one's finger caught [in it]; **b)** *(es verhindern)* prevent it; **es ist mir etwas** ~**gekommen** I had problems; ~|**reden** *itr. V.* interrupt

DDR [de:de:'|ɛr] **die;** ~ *Abk. (1949–1990)* Deutsche Demokratische Republik GDR; East Germany *(in popular use)*

Debatte die; ~, ~**n** debate *(über + Akk.* on); **zur** ~ **stehen** be under discussion

Debüt [de'by:] **das;** ~**s,** ~**s** debut

Deck das; ~|**e|s,** ~**s** deck

Deck·bett das *s.* **Oberbett; Decke die;** ~, ~**n a)** *(Tisch~)* tablecloth; **b)** *(Woll~, Pferde~, fig.)* blanket; *(Reise~)* rug; **c)** *(Zimmer~)* ceiling

Deckel der; ~**s,** ~ **a)** lid; *(auf Flaschen, Gläsern usw.)* top; *(Schacht~, Uhr~, Buch~ usw.)* cover; **b)** *(Bier~)* beer-mat

decken 1. *tr. V.* **a)** etw. **über** etw. *(Akk.)* ~: spread sth. over sth. **b)** roof *(house)*; cover *(roof)*; **c) den Tisch** ~: lay the table; **d)** *(schützen; Finanzw., Versicherungsw.)* cover; **e)** *(befriedigen)* meet *(need, demand)*; **2.** *itr. V.*

(den Tisch ~) lay the table; **Deckmantel** der; *o. Pl.* cover; **Deckung** die; ~, ~en a) *(Schutz; auch fig.)* cover *(esp. Mil.);* *(Boxen)* guard; *(bes. Fußball)* defence; in ~ gehen take cover; b) *(Befriedigung)* meeting; c) *(Finanzw., Versicherungsw.)* cover[ing]; **deckungs·gleich** *Adj.* *(Geom.)* congruent

defekt *Adj.* defective; faulty; ~ sein have a defect; be faulty; *(nicht funktionieren)* not be working; **Defekt** der; ~[e]s, ~e defect, fault (an + *Dat.* in)

defensiv 1. *Adj.* defensive; 2. *adv.* defensively; **Defensive** die; ~, ~n defensive; in der ~: on the defensive; die ~ *(Sport)* defensive play

definieren *tr. V.* define; **Definition** die; ~, ~en definition

definitiv 1. *Adj.* definitive; 2. *adv.* finally

Defizit das; ~s, ~e a) deficit; b) *(Mangel)* deficiency

deformieren *tr. V.* a) *(verformen)* distort; b) *(entstellen)* deform *(also fig.)*

deftig *Adj.* *(ugs.)* a) [good] solid *attrib.* ⟨*meal etc.*⟩; [nice] big ⟨*sausage etc.*⟩; b) *(derb)* crude, coarse ⟨*joke, speech, etc.*⟩

Degen der; ~s, ~ a) *(Waffe)* [light] sword; b) *(Fechtsport)* épée

degradieren *tr. V.* demote

dehnbar *Adj.* a) *(elastisch)* ⟨*material etc.*⟩ that stretches *not pred.*; elastic ⟨*waistband etc.*⟩; **Dehnbarkeit** die; ~: elasticity; **dehnen** *tr., refl. V.* stretch

Deich der; ~[e]s, ~e dike

Deichsel ['daiksl] die; ~, ~n shaft; **deichseln** *tr. V.* *(ugs.)* fix

dein *Possessivpron.* your; viele Grüße von Deinem Emil with best wishes, yours Emil; das Buch dort, ist das ~[e]s? that book over there, is it yours?; du und die Deinen *(geh.)* you and yours; **deiner** *Gen. des Personalpronomens du (geh.)* of you; **deinerseits** *Adv.* *(von deiner Seite)* on your part; *(auf deiner Seite)* for your part; **deinet·wegen** *Adv.* because of you; *(für dich)* on your behalf; *(dir zuliebe)* for your sake

dekadent *Adj.* decadent; **Dekadenz** die; ~: decadence

deklamieren *tr., itr. V.* recite

Deklination die; ~, ~en *(Sprachw.)* declension; **deklinieren** *tr. V.* *(Sprachw.)* decline

Dekolleté [dekɔl'te:] das; ~s, ~s low[-cut] neckline; décolletage

Dekor das; ~s, ~s *od.* ~e decoration; *(Muster)* pattern; **Dekorateur** [dekora'tø:ɐ̯] der; ~s, ~e, **Dekorateurin** die; ~, ~nen *(Schaufenster~)* window-dresser; *(von Innenräumen)* interior designer; **Dekoration** die; ~, ~en decorations *pl.*; *(Schaufenster~)* window display; **dekorativ** 1. *Adj.* decorative; 2. *adv.* decoratively; **dekorieren** *tr. V.* decorate ⟨*room etc.*⟩; dress ⟨*shop-window*⟩

Deko·stoff der furnishing fabric

Dekret das; ~[e]s, ~e decree

Delegation die; ~, ~en delegation; **delegieren** *tr. V.* a) send as a delegate/as delegates; b) delegate ⟨*task etc.*⟩ (an + *Akk.* to); **Delegierte** der/die; *adj. Dekl.* delegate

delikat *Adj.* a) delicious; *(fein)* delicate ⟨*bouquet, aroma*⟩; b) *(heikel)* delicate; **Delikatesse** die; ~, ~n delicacy

Delikt das; ~[e]s, ~e offence

Delinquent der; ~en, ~en offender

Delirium das; ~s, Delirien delirium

Delle die; ~, ~n *(ugs.)* dent

Delphin der; ~s, ~e dolphin

dem 1. *best. Art., Dat. Sg. v.* ¹der 1 *u.* das 1 to the; *(nach Präp.)* the; 2. *Demonstrativpron., Dat. Sg. v.* ¹der 2 *u.* das 2: a) *attr.* that; gib es dem Mann give it to 'that man; b) *alleinstehend* gib es nicht dem, sondern dem da! don't give it to him, give it to that man/child *etc.*; 3. *Relativpron., Dat. Sg. v.* ¹der 3 *u.* das 3 *(Person)* that/whom; *(Sache)* that/which; der Mann/das Kind, dem ich das Geld gab the man/the child I gave the money to

Demagoge der; ~n, ~n demagogue

demagogisch *Adj.* demagogic

Dementi das; ~s, ~s denial; **dementieren** 1. *tr. V.* deny; 2. *itr. V.* deny it

dem-: ~**entsprechend** 1. *Adj.* appropriate; 2. *adv.* accordingly; *(vor Adjektiven)* correspondingly; ~**gemäß** *Adv.* a) *(infolgedessen)* consequently; b) *(entsprechend)* accordingly; ~**jenigen** s. derjenige; ~**nach** *Adv.* therefore; ~**nächst** *Adv.* shortly

Demokrat der; ~en, ~en democrat; *(Parteimitglied)* Democrat; **Demokratie** die; ~, ~n democracy; **demokratisch** 1. *Adj.* democratic; 2. *adv.* democratically; **demokratisieren** *tr. V.* democratize

demolieren *tr. V.* wreck; smash up ⟨*furniture*⟩

Demonstrant der; ~en, ~en demonstrator; **Demonstrantin die**; ~, ~nen demonstrator; **Demonstration die**; ~, ~en demonstration (für in support of, **gegen** against); **demonstrativ 1.** *Adj.* **a)** pointed; **b)** *(Sprachw.)* demonstrative; **2.** *adv.* pointedly; **Demonstrativ · pronomen das** *(Sprachw.)* demonstrative pronoun; **demonstrieren 1.** *itr. V.* demonstrate (für in support of, **gegen** against); **2.** *tr. V.* demonstrate

dem · selben *s.* derselbe

Demut die; ~: humility; **demütig 1.** *Adj.* humble; **2.** *adv.* humbly; **demütigen 1.** *tr. V.* humiliate; **2.** *refl. V.* humble oneself; **Demütigung die**; ~, ~en humiliation

dem · zufolge *Adv.* consequently

¹**den 1.** *best. Art., Akk. Sg. v.* ¹der **1:** the; **2.** *Demonstrativpron., Akk. Sg. v.* ¹der **2: a)** *attr.* that; **ich meine den Mann** I mean 'that man; **b)** *alleinstehend* **wie meine den [da]** I mean 'that one; **3.** *Relativpron., Akk. Sg. v.* ¹der 3 *(Person)* that/whom; *(Sache)* that/which; **der Mann, den ich gesehen habe** the man that I saw

²**den 1.** *best. Art., Dat. Pl. v.* ¹der **1,** ¹**die 1, das 1** the; **2.** *Demonstrativpron. Dat. Pl. v.* ¹der 2a, ¹die 2a, das 2a those.

denen 1. *Demonstrativpron., Dat. Pl. v.* ¹der 2b, ¹die 2b, das 2b them; **gib es ~, nicht den anderen** give it to 'them, not to the others; **2.** *Relativpron., Dat. Pl. v.* ¹der 3, ¹die 3, das 3 *(Personen)* that/whom; *(Sachen)* that/which; **die Menschen,** ~ **wir Geld gegeben haben** the people to whom we gave money; **die Tiere,** ~ **er geholfen hat** the animals that he helped

denjenigen *s.* derjenige

denkbar 1. *Adj.* conceivable; **2.** *adv.* *(sehr, äußerst)* extremely; **denken 1.** *unr. itr. V.* think (**an** + *Akk.* of, **über** + *Akk.* about); **wie denkst du darüber?** what do you think about it?; what's your opinion of it?; **schlecht von jmdm.** ~: think badly of sb.; **denk daran, daß .../zu ...:** don't forget that .../to ...; **ich denke nicht daran!** no way!; not on your life!; **ich denke nicht daran, das zu tun** I've no intention of doing that; **2.** *unr. tr. V.* think; **wer hätte das gedacht?** who would have thought it?; **eine gedachte Linie** an imaginary line; **3.** *unr. refl. V.* **a)**

(sich vorstellen) imagine; **b) sich** *(Dat.)* **bei etw. etwas** ~: mean something by sth.; **ich habe mir nichts [Böses] dabei gedacht** I didn't mean any harm [by it]; **Denken das**; ~s thinking; *(Denkweise)* thought; **Denker der**; ~s, ~: thinker

denk-, Denk- ~**faul** *Adj.* mentally lazy; ~**mal das**; ~mals, ~mäler od. ~male monument; ~**vermögen** das ability to think [creatively]; ~**würdig** *Adj.* memorable; ~**zettel** der lesson

denn 1. *Konj. (kausal)* for; because; **b)** *(geh.: als)* than; **2.** *Adv.* **es sei** ~, ...: unless ...; **3.** *Partikel (in Fragesätzen)* **wie geht es dir** ~? tell me, how are you?; **wie heißt du** ~? tell me your name; **warum** ~ **nicht?** why ever not?

dennoch *Adv.* nevertheless

denselben *s.* derselbe

Denunziant der; ~en, ~en informer; grass *(sl.)*; **denunzieren** *tr. V.* denounce; *(bei der Polizei)* inform against; grass on *(sl.)* (**bei** to)

Deo das; ~s, ~s, **Deodorant das**; ~s, ~s *(auch:)* ~e deodorant

Deponie die; ~, ~n tip *(Brit.)*; dump; **deponieren** *tr. V.* put; *(im Safe o. ä.)* deposit

Deportation die; ~, ~en transportation; *(ins Ausland)* deportation; **deportieren** *tr. V.* transport; *(ins Ausland)* deport; **Deportierte der/die**; *adj. Dekl.* transportee; *(ins Ausland)* deportee

Depot [de'po:] **das**; ~s, ~s **a)** depot; *(Lagerhaus)* warehouse; *(für Möbel usw.)* depository; *(im Freien, für Munition o. ä.)* dump; *(in einer Bank)* strong-room; safe deposit; **b)** *(hinterlegte Wertgegenstände)* deposits *pl.*

Depp der; ~en *(auch:)* ~s, ~en *(auch:)* ~e *(bes. südd., österr., schweiz. abwertend)* *s.* **Dummkopf**

Depression die; ~, ~en depression; **depressiv 1.** *Adj.* depressive; **2.** *adv.* ~ **veranlagt sein** have a tendency towards depression; **deprimieren** *tr. V.* depress; **deprimierend** *Adj.* depressing; **deprimiert 1.** *Adj.* depressed; **2.** *adv.* dejectedly

¹**der 1.** *best. Art. Nom.* the; **der Tod** death; **der „Faust"** 'Faust'; **der Bodensee/Mount Everest** Lake Constance/Mount Everest; **der Iran/Sudan** Iran/the Sudan; **der Mensch/Mann ist ...:** man is .../men are ...; **2.** *Demonstrativpron.* **a)** *attr.* that; **der Mann war es** it was 'that man; **b)** *al-*

leinstehend he; d**er war es** it was 'him; der |da| *(Person)* that man/boy; *(Sache)* that one; d**er** |hier| *(Person)* this man/boy; *(Sache)* this one; 3. *Relativpron. (Person)* who/that; *(Sache)* which/that; **der Mann, der da drüben entlanggeht** the man walking along over there; 4. *Relativ- u. Demonstrativpron.* the one who

²**der** 1. *best. Art.* a) *Gen. Sg. v.* ¹**die** 1: **der Hut der Frau** the woman's hat; **der Henkel der Tasse** the handle of the cup; b) *Dat. Sg. v.* ¹**die** 1 to the; *(nach Präp.)* the; c) *Gen. Pl. v.* ¹**der** 1, ¹**die** 1, **das** 1: **das Haus der Freunde** our/their *etc.* friends' house; **das Bellen der Hunde** the barking of the dogs; 2. *Demonstrativpron.* a) *Gen. Sg. v.* ¹**die** 2: of the; of that; b) *Dat. Sg. v.* ¹**die** 2 *attr.* d**er Frau** |da/hier| **gehört es** it belongs to that woman there/this woman here; c) *Gen. Pl. v.* ¹**der** 2 a, ¹**die** 2 a, **das** 2 a of those; 3. *Relativpron.; Dat. Sg. v.* ¹**die** 3: **die Frau, der ich es gegeben habe** the woman I gave it to; **die Katze, der er einen Tritt gab** the cat [that] he kicked

d**er·art** *Adv.* so; **es hat lange nicht mehr ~ geregnet** it hasn't rained as hard as that for a long time; **sie hat ~ geschrien, daß ...:** she screamed so much that ...; d**er·artig** 1. *Adj.* such; 2. *adv. s.* **derart**

d**erb** 1. *Adj.* a) tough ⟨*material*⟩; stout, ⟨*shoes*⟩; b) *(kraftvoll, deftig)* earthy ⟨*scenes, humour*⟩; 2. *adv.* a) strongly ⟨*made, woven, etc.*⟩; b) *(kraftvoll, deftig)* earthily

d**eren** 1. *Relativpron.* a) *Gen. Sg. v.* ¹**die** 3 *(Personen)* whose; *(Sachen)* of which; b) *Gen. Pl. v.* ¹**der** 3, ¹**die** 3, **das** 3 *(Personen)* whose; *(Sachen)* **Maßnahmen, ~ Folgen wir noch nicht absehen können** measures, the consequences of which we cannot yet foresee; 2. *Demonstrativpron.* a) *Gen. Sg. v.* ¹**die** 2: **meine Tante, ihre Freundin und ~ Hund** my aunt, her friend and 'her dog; b) *Gen. Pl. v.* ¹**der** 2, ¹**die** 2, **das** 2: **meine Verwandten und ~ Kinder** my relatives and their children

d**erent-:** **~wegen** *Adv.* 1. *relativ (Personen)* because of whom; *(Sachen)* because of which; 2. *demonstrativ* because of them; **~willen** *Adv.* **um ~willen** *(Personen)* for whose sake; *(Sachen)* for the sake of which

d**erer** *Demonstrativpron.; Gen. Pl. v.* ¹**der** 2, ¹**die** 2, **das** 2 of those

der·gleichen *indekl. Demonstrativpron.* a) *attr.* such; like that *postpos., not pred.;* b) *alleinstehend* that sort of thing

der·jenige, die·jenige, das·jenige *Demonstrativpron.* a) *attr.* that; *Pl.* those; b) *alleinstehend* that one; *Pl.* those

derlei *indekl. Demonstrativpron.: s.* **dergleichen**

der·maßen *Adv.* **~ schön** *usw.*, **daß ...:** so beautiful *etc.* that ...

derselbe, dieselbe, dasselbe *Demonstrativpron.* a) *attr.* the same; b) *alleinstehend* the same one; *Pl.* the same people; **er sagt immer dasselbe** he always says the same thing; **noch einmal dasselbe, bitte** *(ugs.)* [the] same again please

der·zeit *Adv.* at present; **der·zeitig** *Adj.* present; current

d**es** 1. *best. Art.; Gen. Sg. v.* ¹**der** 1, **das** 1: **die Mütze des Jungen** the boy's cap; **das Klingeln des Telefons** the ringing of the telephone; 2. *Demonstrativpron.; Gen. Sg. v.* ¹**der** 2, **das** 2: **er ist der Sohn des Mannes, der ...:** he's the son of the man who ...

Deserteur [dezɛr'tø:ɐ̯] der; **~s, ~e** deserter; **desertieren** *itr. V.; mit sein* desert

d**es·gleichen** *Adv.* likewise; **er ist Arzt, ~ sein Sohn** he is a doctor, as is his son

d**es·halb** *Adv.* for that reason; **~ bin ich zu dir gekommen** that is why I came to you

Des·infektion die disinfection; **Desinfektions·mittel** das disinfectant; **des·infizieren** *tr. V.* disinfect

Des·interesse das lack of interest

Despot [dɛs'po:t] der; **~en, ~en** despot; *(fig. abwertend)* tyrant; **despotisch** 1. *Adj.* despotic; 2. *adv.* despotically

d**es·selben** *s.* **derselbe**

d**essen** 1. *Relativpron.; Gen. Sg. v.* ¹**der** 3, **das** 3 *attr. (Person)* whose; *(Sache)* of which; 2. *Demonstrativpron.; Gen. Sg. v.* ¹**der** 2, **das** 2: **mein Onkel, sein Sohn und ~ Hund** my uncle, his son, and 'his dog

Dessert [dɛ'se:ɐ̯] das; **~s, ~s** dessert

destillieren *tr. V. (Chemie)* distil

d**esto** *Konj., vor Komp.* **je eher, ~ besser** the sooner the better

d**es·wegen** *Adv. s.* **deshalb**

Detail [de'tai̯] das; **~s, ~s** detail; **de-**

tailliert 1. *Adj.* detailed; 2. *adv.* in detail; **sehr ~**: in great detail

Detektiv der; ~s, ~e [private] detective

Detonation die; ~, ~en detonation; explosion

Deut *in* **keinen ~**: not one bit

deuten 1. *itr. V.* point; |mit dem Finger| **auf jmdn./etw.** ~: point [one's finger] at sb./sth.; 2. *tr. V.* interpret

deutlich 1. *Adj.* clear; 2. *adv.* clearly; **Deutlichkeit** die; ~ **a)** clarity; **b)** *(Eindeutigkeit)* clearness

deutsch 1. *Adj.* German; **Deutsche Mark** Deutschmark; German mark; **auf** *od.* **in ~**: in German; **auf |gut| ~** *(ugs.)* in plain English; 2. *adv.* ~ **sprechen/schreiben** speak/write German; **Deutsch** das; ~|s| German; **gutes/fließend ~ sprechen** speak good/fluent German; ¹**Deutsche** der/die; *adj. Dekl.* German; ~|r| **sein** be German; ²**Deutsche** das; *adj. Dekl.* **das ~**: German; **aus dem ~n/ins ~ übersetzen** translate from/into German; **Deutschland (das)**; ~s Germany

deutsch-, **Deutsch-**: ~**lehrer** der German teacher; ~**sprachig** *Adj.* **a)** German-speaking; **b)** German-language *attrib.*; ~**unterricht** der German teaching; *(Unterrichtsstunde)* German lesson

Deutung die; ~, ~en interpretation

Devise die; ~, ~n motto; **Devisen** *Pl.* foreign currency *sing.*

Dezember der; ~s, ~: December

dezent 1. *Adj.* quiet *(colour, pattern, suit)*; subdued *(lighting, music)*; 2. *adv.* discreetly; *(dress)* unostentatiously

dezimal *Adj.* decimal

Dezimal-: ~**system** das decimal system; ~**zahl** die decimal [number]

dezimieren *tr. V.* decimate

dgl. *Abk.* dergleichen, desgleichen

d. h. *Abk.* das heißt i. e.

Di. *Abk.* Dienstag Tue[s].

Dia das; ~s, ~s slide

Diabetiker der; ~s, ~, **Diabetikerin**, die; ~, ~nen diabetic

Diagnose [dia'gno:zə] die; ~, ~n diagnosis

diagonal 1. *Adj.* diagonal; 2. *adv.* diagonally; **Diagonale** die; ~, ~n diagonal

Dialekt der; ~|e|s, ~e dialect

Dialog der; ~|e|s, ~e dialogue

Diamant der; ~en, ~en diamond

diät *adv.* ~ **kochen** cook according to

a/one's diet; ~ **essen** be on a diet; **Diät die**; ~, ~**en** diet; **eine ~ einhalten** keep to a diet; **Diäten** *Pl.* [parliamentary] allowance *sing.*

dich 1. *Akk. von* **du** you; 2. *Akk. des Reflexivpron. der 2. Pers. Sg.* yourself

dicht 1. *Adj.* **a)** thick; dense *(forest, hedge, crowd)*; heavy, dense *(traffic)*; **b)** *(undurchlässig)* *(für Luft)* airtight; *(für Wasser)* watertight; 2. *adv.* **a)** densely *(populated, wooded)*; **b)** *(undurchlässig)* tightly; **c)** *mit Präp.* *(nahe)* ~ **neben** right next to

dicht·besiedelt *Adj.* *(präd. getrennt geschrieben)* densely populated

Dichte die; ~ *(Physik, fig.)* density

dichten 1. *itr. V.* write poetry; 2. *tr. V.* *(verfassen)* write; compose; **Dichter** der; ~s, ~: poet; *(Schriftsteller)* writer; author; **Dichterin** die; ~, ~**nen** poet[ess]; *(Schriftstellerin)* writer; author[ess]; **dichterisch** *Adj.* poetic; *(schriftstellerisch)* literary

dicht|machen *tr.*, *itr. V.* *(ugs.)* shut; *(endgültig)* shut down

¹**Dichtung** die; ~, ~**en** seal; *(am Hahn usw.)* washer; *(am Vergaser, Zylinder usw.)* gasket

²**Dichtung** die; ~, ~**en a)** work of literature; *(in Versform)* poetic work; poem; **b)** *o. Pl.* *(Dichtkunst)* literature; *(in Versform)* poetry

dick 1. *Adj.* **a)** thick; stout *(tree)*; fat *(person, legs, etc.)*; swollen *(cheek, ankle, tonsils, etc.)*; ~ **werden** get fat; **5 cm ~ sein** be 5 cm thick; **b)** *(ugs.: groß)* big *(mistake)*; hefty, *(coll.)* fat *(salary)*; 2. *adv.* thickly; **etw.** ~ **unterstreichen** underline sth. heavily; **sich** ~ **anziehen** wrap up warm[ly]; **etw. 5 cm** ~ **schneiden** cut sth. 5 cm. thick; ~ **geschwollen** *(ugs.)* badly swollen; ¹**Dicke** die; ~: thickness; *(von Menschen, Körperteilen)* fatness; ²**Dicke** der/die; *adj. Dekl.* *(ugs.)* fatty *(coll.)*; **dick·fellig** *(ugs.)* *Adj.* thick-skinned; **Dickicht** ['dɪkɪçt] das; ~|e|s, ~e thicket

dick-, **Dick-**: ~**kopf** der *(ugs.)* mule *(coll.)*; **ein** ~**kopf sein** be stubborn as a mule; **einen** ~**kopf haben** be pigheaded; ~**köpfig** *Adj.* *(ugs.)* pigheaded; ~**milch** die sour milk

¹**die** 1. *best. Art. Nom.* the; **die Helga** *(ugs.)* Helga; **die Frau/Menschheit** women *pl.*/mankind; 2. *Demonstrativpron.* **a)** *attr.* **die Frau war es** it was 'that woman; **b)** *alleinstehend* she; **die war es** it was 'her; **die |da|** *(Person)* that

woman/girl; *(Sache)* that one; **3.** *Relativpron. Nom. (Person)* who; that; *(Sache, Tier)* which; that; **4.** *Relativ- u. Demonstrativpron.* the one who
²**die 1.** *best. Art.* **a)** *Akk. Sg. v.* ¹**die 1** the; **ich sah die Frau** I saw the women; **b)** *Nom. u. Akk. Pl. v.* ¹**der 1,** ¹**die 1, das 1** the; **2.** *Demonstrativpron. Nom. u. Akk. Pl. v.* ¹**der 1,** ¹**die 1, das 1:** *attr.* **ich meine die Männer, die ...** I mean those men who ...; *alleinstehend* **ich meine die |da|** I mean 'them; **3.** *Relativpron.* **a)** *Akk. Sg. v.* ¹**die 3** *(Person)* who; *(Sache)* that; **b)** *Nom. u. Akk. Pl. v.* ¹**der 3,** ¹**die 3, das 3** *(Personen)* whom; *(Sachen)* which; **die Männer, die ich gesehen habe** the men I saw
Dieb der; ~|e|s, ~e thief; **Diebin** die; ~, ~nen [woman] thief; **diebisch 1.** *Adj.* **a)** thieving; **b)** *(verstohlen)* mischievous; **2.** *adv.* mischievously; **Diebstahl** der; ~|e|s, **Diebstähle** theft
die·jenige *s.* derjenige
Diele die; ~, ~n hall[way]
dienen *itr. V.* serve; **womit kann ich ~?** what can I do for you?; **Diener** der; ~s, ~ servant; **einen ~ machen** *(ugs.)* bow; make a bow; **Dienerin** die; ~, ~nen maid; servant
dienlich *Adj.* helpful; **Dienst** der; ~|e|s, ~e **a)** *o. Pl. (Tätigkeit)* work; *(von Soldaten, Polizeibeamten, Krankenhauspersonal usw.)* duty; **seinen ~ antreten** start work/go on duty; **~ haben** be at work/on duty; *(doctor)* be on call; *(chemist)* be open; **b)** *(Arbeitsverhältnis)* post; **Major außer ~:** retired major; **c)** *o. Pl. (Tätigkeitsbereich)* service; *s. auch* **öffentlich; d)** *(Hilfe)* service
Diens·tag der Tuesday; **am ~:** on Tuesday; **~, der 1. Juni** Tuesday, 1 June; **er kommt ~:** he is coming on Tuesday; **ab nächsten ~:** from next Tuesday [onwards]; **~ in einer Woche** a week on Tuesday; **~ vor einer Woche** a week last Tuesday; **diens·tags** *Adv.* on Tuesday[s]
dienst-, Dienst-: ~**bereit** *Adj.* *(chemist)* open *pred.*; *(doctor)* on call; *(dentist)* on duty; ~**bote** der servant; ~**eifrig** *Adj.* zealous; ~**frei** *Adj.* free *(time)*; ~**geheimnis** das **a)** professional secret; *(im Staatsdienst)* official secret; **b)** *o. Pl.* professional secrecy; *(im Staatsdienst)* official secrecy; ~**grad** der *(Milit.)* rank; ~**leistung** die *(auch Wirtsch.)* service

dienstlich 1. *Adj.* business *(call)*; *(im Staatsdienst)* official *(letter, call, etc.)*; **2.** *adv.* on business; *(im Staatsdienst)* on official business
dienst-, Dienst-: ~**reise** die business trip; ~**stelle** die office; ~**wagen** der official car; *(Geschäftswagen)* company car; ~**weg** der official channels *pl.*; ~**zeit** die **a)** period of service; **b)** *(tägliche Arbeitszeit)* working hours *pl.*
dies *s.* dieser
dies·bezüglich *adv.* regarding this
diese *s.* dieser
Diesel der; ~|s|, ~: diesel
die·selbe *s.* derselbe
Diesel·motor der diesel engine
dieser, diese, dieses, dies *Demonstrativpron.* **a)** *attr.* this; *Pl.* these; **b)** *alleinstehend* this one; *Pl.* these; **dies alles** all this; **dies und das,** *(geh.)* **dieses und jenes** this and that
diesig *Adj.* hazy
dies-: ~**mal** *Adv.* this time; ~**seits 1.** *Präp. mit Gen.* on this side of; **2.** *Adv.* ~**seits von** on this side of
Dietrich der; ~s, ~e picklock
diffamieren *tr. V.* defame; **Diffamierung** die; ~, ~en defamation
Differenz die; ~, ~en difference; *(Meinungsverschiedenheit)* difference [of opinion]; **differenziert** *Adj.* complex; subtly differentiated *(methods, colours)*; sophisticated *(taste)*
diffus *Adj.* *(Physik, Chemie)* diffuse; **b)** *(geh.)* vague; vague and confused *(idea, statement, etc.)*; **2.** *adv.* in a vague and confused way
Digital- digital *(clock, display, etc.)*
digitalisieren *tr. V. (DV)* digitalize
Diktat das; ~|e|s, ~e dictation
Diktator der; ~s, ~en dictator; **diktatorisch 1.** *Adj.* dictatorial; **2.** *adv.* dictatorially; **Diktatur** die; ~, ~en dictatorship
diktieren *tr. V.* dictate
Diktier·gerät das dictating machine
Dilemma das; ~s, ~s dilemma
Dilettant die|ε'tant| der; ~en, ~en, **Dilettantin** die; ~, ~nen dilettante; **dilettantisch 1.** *Adj.* dilettante; amateurish; **2.** *adv.* amateurishly
Dill der; ~|e|s, ~e dill
Dimension die; ~, ~en *(Physik, fig.)* dimension
DIN [di:n] *Abk.* **Deutsche Industrie-Norm|en|** *German Industrial Standard[s]*; DIN; **DIN-A4-Format** A4

¹**Ding** das; ~|e|s, ~e a) thing; b) *meist Pl.* **nach Lage der** ~e the way things are; **persönliche/private** ~e personal/private matters; **ein** ~ **der Unmöglichkeit sein** be quite impossible; **vor allen** ~en above all; c) **guter** ~e **sein** *(geh.)* be in good spirits; ²**Ding** das; ~|e|s, ~er *(ugs.)* thing; **das ist ja ein** ~! that's really something

Diözese die; ~, ~n diocese

Dipl.-Ing. *Abk.* Diplomingenieur *academically qualified engineer*

Diplom das; ~s, ~e ≈ [first] degree *(in a scientific or technical subject); (für einen Handwerksberuf)* diploma; **Diplom-:** qualified

Diplomat der; ~en, ~en, **Diplomatin** die; ~, ~nen diplomat; **diplomatisch** 1. *Adj.* diplomatic; 2. *adv.* diplomatically

dir 1. *Dat. von* **du** to you; *(nach Präp.)* you; **Freunde von** ~: friends of yours; 2. *Dat. des Reflexivpron. der 2. Pers. Sg.* yourself

direkt 1. *Adj.* direct; 2. *adv.* straight; directly; **etw.** ~ **übertragen** broadcast sth. live; **Direkt·flug** der direct flight

Direktion die; ~, ~en management; *(Büroräume)* managers' offices *pl.*; **Direktor** der; ~s, ~en, **Direktorin** die; ~, ~nen director; *(einer Schule)* headmaster/headmistress; *(einer Strafanstalt)* governor; *(einer Abteilung)* manager

Direkt·übertragung die live broadcast

Dirigent der; ~en, ~en conductor; **dirigieren** *tr. V.* a) *auch itr.* conduct; b) *(führen)* steer

Disco ['dɪsko:] die; ~, ~s disco

Diskette die; ~, ~n *(DV)* floppy disc

Diskont·satz der *(Finanzw.)* discount rate

Diskothek die; ~, ~en discothèque

Diskrepanz die; ~, ~en discrepancy

diskret 1. *Adj. (vertraulich)* confidential; *(taktvoll)* discreet; tactful; 2. *adv. (vertraulich)* confidentially; *(taktvoll)* discreetly; tactfully; **Diskretion** die; ~ a) *(Verschwiegenheit, Takt)* discretion; b) *(Unaufdringlichkeit)* discreetness

diskriminieren *tr. V.* discriminate against; **Diskriminierung** die; ~, ~en discrimination

Diskussion die; ~, ~en discussion; **zur** ~ **stehen** be under discussion

Diskussions-: ~beitrag der contribution to a/the discussion; ~leiter der chairman [of the discussion]

diskutieren 1. *itr. V.* über etw. *(Akk.)* ~: discuss sth.; 2. *tr. V.* discuss

disqualifizieren *tr. V.* disqualify

Distanz die; ~, ~en *(auch fig.)* distance; **distanzieren** *refl. V.* sich von jmdm./etw. ~ *(fig.)* dissociate oneself from sb./sth.; **distanziert** *Adj.* reserved

Distel die; ~, ~n thistle

Distel·fink der goldfinch

Disziplin die; ~, ~en discipline; *(Selbstbeherrschung)* [self-]discipline; **disziplinieren** 1. *tr. V.* discipline; 2. *refl. V.* discipline oneself; **diszipliniert** 1. *Adj.* well-disciplined; *(beherrscht)* disciplined; 2. *adv.* in a well-disciplined way; *(beherrscht)* in a disciplined way

divers... [di'vɛrs...] *Adj.; nicht präd.* various; *(mehrer...)* several

Dividende [divi'dɛndə] die; ~, ~n *(Wirtsch.)* dividend

dividieren *tr. V.* divide; **Division** die; ~, ~en *(auch Milit.)* division

DM *Abk.* Deutsche Mark DM

Do. *Abk.* Donnerstag Thur[s].

doch 1. *Konj.* but; 2. *Adv.* a) *(jedoch)* but; b) *(dennoch)* all the same; still; c) *(geh.: nämlich)* wußte er ~, daß ...: because he knew that ...; d) *(entgegen allen gegenteiligen Behauptungen, Annahmen)* er war also ~ der Mörder! so he 'was the murderer!; e) *(ohnehin)* in any case; 3. *Interj.* Das stimmt nicht. – Doch! That's not right. – [Oh] yes it is!; Hast du keinen Hunger? – Doch! Aren't you hungry? – Yes [I am]!; 4. *Partikel* a) *(Ungeduld ausdrückend)* paß ~ auf! [oh] do be careful!; das ist ~ nicht zu glauben that's just incredible; b) *(Zweifel ausdrückend)* du hast ~ meinen Brief erhalten? you did get my letter, didn't you?; c) *(Überraschung ausdrückend)* das ist ~ Karl! there's Karl! d) *(verstärkt Bejahung/Verneinung ausdrückend)* gewiß/sicher ~: [why] certainly; of course; ja ~: [yes,] all right; nicht ~! *(abwehrend)* [no,] don't!; e) *(Wunsch verstärkend)* wäre es ~ ...: if only it were ...

Docht der; ~|e|s, ~ wick

Dock das; ~s, ~s dock

Dogge die; ~, ~n: |deutsche| ~: Great Dane

Dogma das; ~s, **Dogmen** *(auch fig.)* dogma; **dogmatisch** *(Theol., auch fig.) Adj.* dogmatic

Dohle die; ~, ~n jackdaw
Doktor der; ~s, ~en *(auch ugs. Arzt)* doctor; *(Titel)* Doctor; **Doktor·arbeit** die doctoral thesis
Doktrin die; ~, ~en doctrine
Dokument das; ~[e]s, ~e document
Dokumentar-: ~**bericht** der documentary report; ~**film** der documentary [film]
Dokumentation die; ~, ~en *o. Pl.* documentation; b) *(Bericht)* documentary report
dokumentieren *tr. V.* a) document; *(fig.)* demonstrate; b) *(festhalten)* record
Dolch der; ~[e]s, ~e dagger
Dolde die; ~, ~n *(Bot.)* umbel
doll *(bes. nordd., salopp)* 1. *Adj.* a) *(ungewöhnlich)* incredible; b) *(großartig)* great *(coll.)*; 2. *adv.* a) *(großartig)* fantastically [well] *(coll.)*; b) *(sehr)* ⟨hurt⟩ dreadfully *(coll.)*, like mad
Dollar der; ~[s], ~s dollar; zwei ~ : two dollars
dolmetschen *itr. V.* act as interpreter; **Dolmetscher** der; ~s, ~, **Dolmetscherin** die; ~, ~nen interpreter
Dom der; ~[e]s, ~e cathedral
dominieren *itr. V.* dominate
dominikanisch *Adj.* Dominican; die **Dominikanische Republik** the Dominican Republic
Domizil das; ~s, ~e *(geh.)* domicile; residence
Dom·pfaff der; ~en *od.* ~s, ~en *(Zool.)* bullfinch
Dompteur [dɔmpˈtøːɐ] der; ~s, ~e, **Dompteuse** [dɔmpˈtøːzə] die; ~, ~n tamer
Donau die; ~ : Danube
Donner der; ~s, ~ : thunder; **donnern** *itr. V.* a) *(unpers.)* thunder; b) *(fig.)* thunder; ⟨engine⟩ roar
Donners·tag der Thursday; *s. auch* Dienstag; **donnerstags** *Adv.* on Thursday[s]; *s. auch* dienstags
Donner·wetter das *(ugs.)* a) *(Krach)* row; b) ['--'--] zum ~ [noch einmal]! damn it!; ~! my word
doof *(ugs.)* 1. *Adj.* stupid; dumb *(coll.)*; 2. *adv.* stupidly
Doppel das, ~s, ~ a) *(Kopie)* duplicate; copy; b) *(Sport)* doubles *sing. or pl.*
doppel-, Doppel-: ~**bett** das double bed; ~**bock** das extra-strong bock beer; ~**decker** der; ~s, ~ : biplane; ~**deutig** [~dɔytɪç] 1. *Adj.* a)

ambiguous; b) *(anzüglich)* suggestive; 2. *adv.* a) ambiguously; b) *(anzüglich)* suggestively; ~**fenster** das double-glazed window; ~**gänger** der; ~s, ~, ~**gängerin** die; ~, ~nen double; ~**kinn** das double chin; ~**punkt** der colon
doppelt 1. *Adj.* double; die ~e Menge twice the quantity; mit ~er Kraft arbeiten work with twice as much energy; 2. *adv.* ~ so groß/alt wie ...: twice as large/old as ...; sich ~ anstrengen try twice as hard; **Doppelte** das; *adj. Dekl.* das ~ bezahlen pay twice as much; pay double
Doppel-: ~**tür** die double door; ~**zentner** der 100 kilograms; ~**zimmer** das double room
Dorf das; ~[e]s, Dörfer village; auf dem ~ : in the country; **Dorf·bewohner** der villager
Dorn der; ~[e]s, ~en thorn; jmdm. ein ~ im Auge sein annoy sb. intensely; **dornig** *Adj.* thorny; **Dorn·röschen** ⟨das⟩ the Sleeping Beauty
dörren *tr. V.* dry
Dörr-: ~**fleisch** das *(südd.)* lean bacon; ~**obst** das dried fruit
Dorsch der; ~[e]s, ~e cod
dort *Adv.* there; *s. auch* da 1 a
dort-: ~**|bleiben** *unr. itr. V.; mit sein* stay there; ~**her** *Adv.* [von] ~her from there; ~**hin** *Adv.* there
dortig *Adj.; nicht präd.* there
Dose die; ~, ~n a) *(Blech~)* tin; *(Pillen~)* box; *(Zucker~)* bowl; b) *(Konserven~)* can; tin *(Brit.)*; *(Bier~)* can
dösen *itr. V. (ugs.)* doze
Dosen-: ~**bier** das canned beer; ~**milch** die canned *or (Brit.)* tinned milk; ~**öffner** der can opener; tin-opener *(Brit.)*
dosieren *tr. V.* etw. ~ : measure out the required dose of sth.; **Dosis** die; ~, Dosen dose
Dotter der *od.* das; ~s, ~ : yolk
Dotter·blume die marsh marigold
Dozent der; ~en, ~en, **Dozentin** die; ~, ~nen lecturer (für in)
Dr. *Abk.:* Doktor Dr
Drache der; ~n, ~n *(Myth.)* dragon; **Drachen** der; ~s, ~ a) kite; b) *(Fluggerät)* hang-glider
Dragée, Dragee [draˈʒeː] das; ~s, ~s dragée
Draht der; ~[e]s, Drähte a) wire; b) *(Leitung)* wire; *(Telefonleitung)* line; wire; c) *(Telefonverbindung)* line
draht-, Draht-: ~**los** *(Nachrichtenw.)*

1. *Adj.* wireless; **2.** *adv.* etw. ~los telegrafieren/übermitteln radio sth.; ~**seil das** [steel] cable; ~**seil·bahn die** cable railway; ~**zieher der** *(fig.)* wire puller

drall *Adj.* strapping ⟨*girl*⟩; full, rounded ⟨*cheeks, face, bottom*⟩

Drama das; ~s, **Dramen** drama; *(fig., ugs.)* disaster; **dramatisch 1.** *Adj.* dramatic. **2.** *adv.* dramatically; **dramatisieren** *tr. V.* dramatize

dran *Adv. (ugs.)* **a)** häng das Schild ~! put the sign up!; **b)** arm ~ sein be in a bad way; gut/schlecht ~ sein be well off/badly off; früh/spät ~ sein be early/late; ich bin ~: it's my turn; **dran|bleiben** *unr. itr. V.; mit sein (ugs.)* (am Telefon) hang on *(coll.)*

drang *1. u. 3. Pers. Sg. Prät. v.* dringen

Drang der; ~|e|s, **Dränge** urge

dränge *1. u. 3. Pers. Sg. Konjunktiv II v.* dringen

drängeln *(ugs.)* **1.** *itr. V.* **a)** push [and shove]; **b)** *(auf jmdn. einreden)* go on *(coll.)*; **2.** *tr. V.* **a)** push; shove; **b)** *(einreden auf)* go on at *(coll.)*; **3.** *refl. V.* sich nach vorn ~: push one's way to the front

drängen 1. *itr. V.* **a)** push; **b)** die Zeit drängt time is pressing; **2.** *tr. V.* **a)** push; **b)** *(antreiben)* press; urge; **3.** *refl. V.* crowd

drangsalieren *tr. V. (quälen)* torment; *(plagen)* plague

dran-: ~|**halten** *unr. refl. V. (ugs.)* get a move on *(coll.)*; ~|**kommen** *unr. itr. V.; mit sein (ugs.)* have one's turn; ~|**nehmen** *unr. tr. V. (ugs.) (beim Friseur usw.)* see to; *(beim Arzt)* see

drastisch 1. *Adj.* drastic ⟨*measure, means*⟩; **2.** *adv.* drastically; ⟨*punish*⟩ severely

drauf *Adv. (ugs.)* on it

drauf-, Drauf-: ~**gänger der** daredevil; ~**gängerisch** *Adj.* daring; ~|**gehen** *unr. itr. V.; mit sein (ugs.)* **a)** *(umkommen)* kick the bucket *(sl.)*; **b)** *(verbraucht werden)* go *(für* on); ~|**zahlen** *(ugs.)* **1.** *tr. V.* noch etwas/ 1 250 DM ~zahlen fork out *(sl.)* or pay a bit more/an extra 1,250 marks; **2.** *itr. V. (Unkosten haben)* ich zahle dabei noch ~: it's costing me money

draußen *Adv.* outside; hier/da ~: out here/there; von/nach ~: from outside/ outside

Dreck der; ~|e|s **a)** *(ugs.)* dirt; *(sehr viel)* filth; *(Schlamm)* mud; **b)** *(salopp abwertend: Angelegenheit)* mach deinen ~ allein do it yourself; das geht dich einen |feuchten| ~ an *(salopp)* none of your damned business *(sl.)*; **c)** *(salopp abwertend: Zeug)* junk *no indef. art.*; **Dreck·arbeit die** *(auch fig.)* dirty work *no indef. art., no pl.*/dirty job; **dreckig 1.** *Adj.* **a)** *(ugs., auch fig.)* dirty; *(sehr schmutzig)* filthy; **b)** *(salopp: unverschämt)* cheeky; **2.** *adv.* **a)** es geht ihm ~ *(ugs.)* he's in a bad way; **b)** *(salopp: unverschämt)* cheekily

Dreck-: ~**sau die,** ~**schwein das** *(derb)* filthy swine

Dreh der; ~s, ~s *(ugs.)* **a)** den ~ heraushaben have [got] the knack; **b)** |so| um den ~: about that

Dreh-: ~**arbeiten** *Pl. (Film)* shooting *sing.* (zu of); ~**bank die** lathe; ~**buch das** screenplay; [film] script

drehen 1. *tr. V.* **a)** turn; **b)** *(formen)* twist ⟨*rope, thread*⟩; roll ⟨*cigarette*⟩; **c)** *(Film)* shoot ⟨*scene*⟩; film ⟨*report*⟩; make; ⟨*film*⟩; **2.** *itr. V.* **a)** ⟨*car*⟩ turn; ⟨*wind*⟩ change; **b)** an etw. *(Dat.)* ~: turn sth.; **c)** *(Film)* film; **3.** *refl. V.* **a)** turn; **b)** *(ugs.: zum Gegenstand haben)* sich um etw. ~: be about sth.

Dreh-: ~**orgel die** barrel-organ; ~**restaurant das** revolving restaurant; ~**stuhl der** swivel chair; ~**tür die** revolving door

Drehung die; ~, ~**en** turn; *(um einen Mittelpunkt)* revolution

drei *Kardinalz.* three; **Drei die;** ~, ~**en** three; eine ~ schreiben *(Schulw.)* get a C

drei-, Drei-: ~**eck das;** ~s, ~e *(Geom.)* triangle; ~**eckig** *Adj.* triangular; ~**ein·halb** *Bruchz.* three and a half

Dreier der; ~s, ~ *(ugs.)* three; **dreierlei** *Gattungsz.; indekl.* **a)** *attr.* three kinds *or* sorts of; three different; **b)** *subst.* three [different] things

drei-, Drei-: ~**fach** *Vervielfältigungsz.* triple; die ~**fache Menge** three times the amount; ~**fache das;** *adj. Dekl.* das ~fache kosten cost three times as much; das ~fache von 3 ist 9 three times three is nine; ~**hundert** *Kardinalz.* three hundred; ~**jährig** *Adj.* (3 Jahre alt) three-year-old *attrib.*; (3 Jahre dauernd) three-year *attrib.*; ~**kampf der** *(Sport)* triathlon; ~**klang der** triad; ~**köpfig** *Adj.* ⟨*family, crew*⟩ of three; ~**mal** *Adv.* three times; ~**malig** *Adj.* eine ~**malige Wiederholung** three repeats dr**ei**n *(ugs.) s.* **darein**

dr**ei**n-: ~|**blicken,** ~|**schauen** *itr. V.* look

dr**ei**-, **Drei**-: ~**rad** das tricycle; ~**satz** der; rule of three; ~**seitig** *Adj.* three-sided *⟨figure⟩;* three-page *⟨letter, leaflet, etc.⟩*

dr**ei**ßig *Kardinalz.* thirty; *s. auch* **achtzig;** dr**ei**ßig**jährig** *Adj. (30 Jahre alt)* thirty-year-old *attrib.; (30 Jahre dauernd)* thirty-year *attrib.;* **dreißigst...** *Ordinalz.* thirtieth; **Dreißigstel** das; ~s, ~: thirtieth

dr**ei**st 1. *Adj.* brazen; barefaced *⟨lie⟩;* 2. *adv.* brazenly

dr**ei**·**stellig** *Adj.* three-figure *attrib.*

Dr**ei**stigkeit die; ~, ~en a) *o. Pl.* brazenness; b) *(Handlung)* brazen act

dr**ei**-, **Drei**-: ~**tausend** *Kardinalz.* three thousand; ~**teilig** *Adj.* three-part *attrib.;* three-piece *attrib. ⟨suit⟩;* ~**viertel** *Bruchz.* three-quarters; ~**viertel·stunde** [---'--] die three-quarters of an hour; ~**viertel·takt** [-'---] der three-four time; ~**zehn** *Kardinalz.* thirteen; *s. auch* **achtzehn**

Dr**e**sche die; ~ *(salopp)* walloping *(sl.);* thrashing; **dr**e**schen** 1. *unr. tr. V.* a) thresh; b) *(salopp: schlagen)* wallop *(sl.);* thrash; 2. *unr. itr. V.* thresh

dress**ie**ren *tr. V.* train *⟨animal⟩;* **Dress**u**r** die; ~, ~en training

Dr**i**ll der; ~|e|s drilling; *(Milit.)* drill; **dr**i**llen** *tr. V. (auch Milit.)* drill

Dr**i**lling der; ~s, ~e triplet

dr**i**n *Adv. (ugs.)* a) in it; b) *s.* drinnen

dr**i**ngen *unr. itr. V.* a) *mit* sein *durch/in etw.* ~: penetrate sth.; b) *mit* sein *in* jmdn. ~ *(geh.)* press sb.; c) *auf etw. (Akk.)* ~: insist upon sth.; **dr**i**ngend** 1. *Adj.* urgent; strong *⟨suspicion, advice⟩;* 2. *adv.* urgently; *⟨advise, suspect⟩* strongly; ~ **erforderlich** essential; **dr**i**nglich** 1. *Adj.* urgent; 2. *adv.* urgently; **Dr**i**nglichkeit** die; ~: urgency

dr**i**nnen *Adv.* inside; *(im Haus)* indoors; inside

dr**i**tt *in* **wir waren zu** ~: there were three of us

dr**i**tt... *Ordinalz.* third; **Dr**i**ttel** das, *(schweiz. meist* der); ~s, ~: third; **dr**i**tteln** *tr. V.* split *or* divide three ways; **dr**i**ttens** *Adv.* thirdly

DRK [de:|ɛr'ka:] das; ~ *Abk.* **Deutsches Rotes Kreuz** German Red Cross

Dr. med. *Abk.* doctor medicinae MD

dr**o**ben *Adv. (südd., österr., sonst geh.)* up there

Dr**o**ge die; ~, ~n drug

dr**o**gen-: ~**abhängig,** ~**süchtig** *Adj.* addicted to drugs *postpos.*

Droger**ie** die; ~, ~n chemist's [shop] *(Brit.);* drugstore *(Amer.);* **Drog**i**st** der; ~en, ~en, **Drog**i**stin** die; ~, ~nen chemist *(Brit.);* druggist *(Amer.)*

dr**o**hen *itr., mod. V.* threaten; *(bevorstehen)* be threatening; **jmdm. droht etw.** sb. is threatened with sth.; **dr**o**hend** *Adj.* threatening; *(bevorstehend)* impending

Dr**o**hne die; ~, ~n drone

dr**ö**hnen *itr. V.* boom; *⟨machine⟩* roar

Dr**o**hung die; ~, ~en threat

dr**o**llig 1. *Adj.* funny, comical; *(niedlich)* sweet; cute *(Amer.);* 2. *adv.: s. Adj.:* comically; sweetly; cutely *(Amer.)*

Dromed**a**r das; ~s, ~e dromedary

Dr**o**ps der *od.* das; ~, ~: fruit *or (Brit.)* acid drop

dr**o**sch 1. *u.* 3. *Pers. Sg. Prät. v.* **dreschen**

Dr**o**ssel die; ~, ~n thrush

dr**o**sseln *tr. V.* a) turn down *⟨heating, air-conditioning⟩;* throttle back *⟨engine⟩;* b) *(herabsetzen)* reduce

Dr. phil. *Abk.* doctor philosophiae Dr

dr**ü**ben *Adv.* dort *od.* da ~: over there; ~ **auf der anderen Seite** over on the other side

¹Dr**u**ck der; ~|e|s, Drücke a) *(auch fig.)* pressure; b) *o. Pl.* **ein** ~ **auf den Knopf** a touch of the button; ²Dr**u**ck der; ~|e|s, ~e a) *o. Pl.* printing; **in** ~ **gehen** go to press; b) *(Produkt)* print; **Dr**u**ck·buchstabe** der printed letter; **dr**u**cken** *tr., itr. V.* print

dr**ü**cken 1. *tr. V.* a) press; press, push *⟨button⟩;* squeeze *⟨juice, pus⟩* **(aus** out of); **jmdm. die Hand** ~: squeeze sb.'s hand; b) *(liebkosen)* jmdn. ~: hug [and squeeze] sb.; c) *⟨shoe etc.⟩* pinch; d) *(herabsetzen)* push down *⟨price, rate⟩;* depress *⟨sales⟩;* bring down *⟨standard⟩;* 2. *itr. V.* a) press; **auf den Knopf** ~: press *or* push the button; „**bitte** ~": 'push'; b) *(Druck verursachen) ⟨shoe etc.⟩* pinch; 3. *refl. V. (ugs.: sich entziehen)* shirk; **sich vor etw.** *(Dat.)* ~: get out of sth.; **dr**u**ckend** *Adj.* heavy *⟨debt, taxes⟩;* serious *⟨worries⟩;* grinding *⟨poverty⟩;* b) *(schwül)* oppressive

Dr**u**cker der; ~s, ~: printer; **Drucker**ei die; ~, ~en printing-works; *(Firma)* printing-house; printer's

druck-, Druck-: ~**fehler** der misprint; printer's error; ~**knopf** der press-stud *(Brit.);* snap-fastener; ~**luft die** compressed air; ~**mittel das** means of bringing pressure to bear (gegenüber on); ~**reif** 1. *Adj.* ready for publication; *(~fertig)* ready for press; 2. *adv.* ⟨*speak*⟩ in a polished manner; ~**sache die** *(Postw.)* printed matter; ~**schrift die a)** printed writing; **b)** *(Schriftart)* type[-face]; **c)** *(Schriftwerk)* pamphlet

drum *Adv. (ugs.)* **a)** *s.* **darum; b)** [a]round; **alles** *od.* **das |ganze|** Drum **und Dran** *(bei einer Mahlzeit)* all the trimmings; *(bei einer Feierlichkeit)* all the palaver that goes with it ⟨*coll.*⟩; **Drum·herum das;** ~**s** everything that goes/went with it

drunter *Adv. (ugs.)* underneath; **es** *od.* **alles geht** ~ **und drüber** everything is topsy-turvy

Drüse die; ~, ~**n** gland

Dschungel ['dʒʊŋl̩] **der;** ~**s,** ~ *(auch fig.)* jungle

dt. *Abk.* **deutsch** G.

Dtzd. *Abk.* **Dutzend** doz.

du *Personalpron.;* 2. *Pers. Sg. Nom.* you; *(in Briefen)* **Du** you; **du zueinander sagen** use the familiar form in addressing one another; *s. auch (Gen.)* **deiner,** *(Dat.)* **dir,** *(Akk.)* **dich**

Dübel der; ~**s,** ~: plug

ducken 1. *refl. V.* duck; 2. *itr. V. (fig. abwertend)* humble oneself (**vor** + *Dat.* before)

Dudel·sack der bagpipes *pl.*

Duell das; ~**s,** ~**e** duel; **duellieren** *refl. V.* fight a duel

Duett das; ~|e|s, ~**e** *(Musik)* duet; **im** ~ **singen** sing a duet

Duft der; ~|e|s, **Düfte** scent; *(von Parfüm, Blumen)* scent; fragrance; *(von Kaffee usw.)* aroma; **duften** *itr. V.* smell (**nach** of)

dulden *tr. V.* tolerate; put up with; **duldsam** 1. *Adj.* tolerant (**gegen** towards); 2. *adv.* tolerantly

dumm, dümmer, dümmst... 1. *Adj.* **a)** stupid; **b)** *(unvernünftig)* foolish; **c)** *(ugs.: töricht, albern)* idiotic; silly; **d)** *(ugs.: unangenehm)* nasty ⟨*feeling*⟩; annoying ⟨*habit*⟩; **das wird mir jetzt zu** ~ *(ugs.)* I've had enough of it; 2. *adv. (ugs.)* idiotically; **Dumme der/die;** *adj. Dekl.* fool; **der** ~ **sein** *(ugs.)* be the loser; **dummer·weise** *Adv.* **a)** unfortunately; *(ärgerlicherweise)* annoyingly; **b)** *(törichterweise)* foolishly;

Dummheit die; ~, ~**en** *o. Pl.* stupidity; **b)** *(unkluge Handlung)* stupid thing; **Dumm·kopf der** *(ugs.)* nitwit *(coll.)*

dumpf 1. *Adj.* **a)** dull ⟨*thud, rumble of thunder*⟩; muffled ⟨*sound, thump*⟩; **b)** *(muffig)* musty; **c)** *(stumpfsinnig)* dull; 2. *adv.* **a)** ⟨*echo*⟩ hollowly; **b)** *(stumpfsinnig)* apathetically

Düne die; ~, ~**n** dune

düngen 1. *tr. V.* fertilize ⟨*soil, lawn*⟩; spread fertilizer on ⟨*field*⟩; scatter fertilizer around ⟨*plants*⟩; 2. *itr. V.* **gut** ~ ⟨*substance*⟩ be a good fertilizer; **Dünger der;** ~**s,** ~: fertilizer

dunkel 1. *Adj. (auch fig.)* dark; *(tief)* deep ⟨*voice, note*⟩; *(undeutlich)* vague; 2. *adv. (tief)* ⟨*speak*⟩ in a deep voice; **b)** *(undeutlich)* vaguely

Dünkel der; ~**s** *(geh.)* arrogance; *(Einbildung)* conceit[edness]

dunkel-: ~**blond** *Adj.* light brown ⟨*hair*⟩; ⟨*person*⟩ with light brown hair; ~**häutig** *Adj.* dark-skinned

Dunkelheit die; ~: darkness; **Dunkel·kammer die** dark-room; **dunkeln** *itr. V. (unpers.)* **es dunkelt** *(geh.)* it is growing dark; **Dunkel·ziffer die** number of unrecorded cases

dünn 1. *Adj.* thin; slim ⟨*book*⟩; fine ⟨*stocking*⟩; watery ⟨*coffee, tea, beer*⟩; 2. *adv.* thinly ⟨*sliced, populated*⟩; lightly ⟨*dressed*⟩

Dunst der; ~|e|s, **Dünste a)** *o. Pl.* haze; *(Nebel)* mist; **b)** *(Geruch)* smell; **dünsten** *tr. V.* steam ⟨*fish, vegetables*⟩; braise ⟨*meat*⟩; stew ⟨*fruit*⟩; **dunstig** *Adj.* hazy

Duo das; ~**s,** ~**s** *(Musik)* duet; *(fig. scherzh.)* duo; pair

Duplikat das; ~|e|s, ~**e** duplicate

Dur das; ~ *(Musik)* major [key]

durch 1. *Präp. mit Akk.* **a)** *(räumlich)* through; **b)** *(modal)* by; ~ **Boten** by courier; **zehn |geteilt|** ~ **zwei** ten divided by two; 2. *Adv.* **a)** *(hin~)* **das ganze Jahr** ~: throughout the whole year; **b)** *(ugs.: vorbei)* **es war 3 Uhr** ~: it was gone 3 o'clock; **c)** ~ **und** ~ **naß/überzeugt** wet through [and through]/completely totally convinced

durch|arbeiten 1. *tr. V.* work through; 2. *itr. V.* work through; **die Nacht** ~: work through the night

durch·aus *Adv.* absolutely; perfectly, quite ⟨*correct, possible, understandable*⟩; **das ist** ~ **richtig** that is entirely right; ~ **nicht** by no means

durch|beißen *unr. tr. V.* bite through

durch|blättern *tr. V.* leaf through

Durch·blick der *(ugs.)* **den [absoluten] ~ haben** know [exactly] what's going on; **durch|blicken** *itr. V.* **a)** look through; **durch etw. ~:** look through sth.; **b) ~ lassen, daß .../wie ...:** hint that .../at how ...

Durch·blutung die; *o. Pl.* flow of blood (+ *Gen.* to); [blood-]circulation

¹**durch|bohren** *tr. V.* drill through ⟨*wall, plank*⟩; drill ⟨*hole*⟩; ²**durch-bohren** *tr. V.* pierce

¹**durch|brechen 1.** *unr. tr. V.* **etw. ~:** break sth. in two; **2.** *unr. itr. V.; mit sein* **a)** break in two; **b)** *(hervorkommen)* ⟨*sun*⟩ break through; **c)** *(einbrechen)* fall through ⟨*ice, floor, etc.*⟩; ²**durch·brechen** *unr. tr. V.* break through

durch|brennen *unr. itr. V.; mit sein* **a)** ⟨*heating coil, light bulb*⟩ burn out; ⟨*fuse*⟩ blow; **b)** *(ugs.: weglaufen) (von zu Hause)* run away; *(mit der Kasse, mit dem Geliebten/der Geliebten)* run off

durch|bringen *unr. tr. V.* get through; *(bei Wahlen)* **jmdn. ~:** get sb. elected; **seine Familie/sich ~:** support one's family/ oneself

Durch·bruch der *(fig.)* breakthrough

durch|drehen 1. *tr. V.* put ⟨*meat*⟩ through the mincer *or (Amer.)* grinder; **2.** *itr. V. auch mit sein (ugs.)* crack up *(coll.)*

¹**durch|dringen** *unr. tr. V.; mit sein* ⟨*rain, sun*⟩ come through; ²**durch-dringen** *unr. tr. V.* penetrate; **jmdn. ~** ⟨*idea*⟩ take hold of sb. [completely]

durch·einander *Adv.* **~ sein** ⟨*papers, desk, etc.*⟩ be in a muddle; *(verwirrt sein)* be confused; *(aufgeregt sein)* be flustered; **Durcheinander** das; **~s a)** muddle; mess; **b)** *(Wirrwarr)* confusion

durcheinander|bringen *unr. tr. V.* **a)** get ⟨*room, flat*⟩ into a mess; get ⟨*papers, file*⟩ into a muddle; muddle up ⟨*papers, file*⟩; **b)** *(verwirren)* confuse; **c)** *(verwechseln)* confuse ⟨*names etc.*⟩; get ⟨*names etc.*⟩ mixed up

durch|fahren *unr. itr. V.; mit sein* **a)** [durch etw.] **~:** drive through [sth.]; **b)** *(nicht anhalten)* go straight through; *(mit dem Auto)* drive straight through; **der Zug fährt [in H.] durch** the train doesn't stop [at H.]; **Durch·fahrt** die **a)** „**~ verboten**" 'no entry except for

access'; **auf der ~ sein** be passing through; **b)** *(Weg)* thoroughfare; „**bitte [die] ~ freihalten**" 'please do not obstruct'

Durch·fall der diarrhoea *no art.;* **durch|fallen** *unr. itr. V.; mit sein* **a)** fall through; **b)** *(ugs.: nicht bestehen)* fail

durch|finden *unr. refl. V.* find one's way through

durchführbar *Adj.* practicable; **durch|führen 1.** *tr. V.* carry out; put into effect ⟨*decision, programme*⟩; perform ⟨*operation*⟩; hold ⟨*meeting, election, examination*⟩; **2.** *itr. V.* **durch etw./unter etw.** *(Dat.)* **~** ⟨*track, road*⟩ go through/under sth.; **Durch·führung** die carrying out; *(einer Operation)* performing; *(einer Versammlung, Wahl, Prüfung)* holding; *(eines Wettbewerbs)* staging

Durch·gang der **a)** passage[way]; „**kein ~**", „ **~ verboten**" 'no thoroughfare'; **b)** *(Phase)* stage; *(einer Versuchsreihe)* run; *(Sport, Wahlen)* round

Durchgangs-: **~straße** die through road; **~verkehr** der through traffic

durch|geben *unr. tr. V.* announce ⟨*news*⟩; give ⟨*results, weather report*⟩; **eine Meldung im Radio/Fernsehen ~:** make an announcement on the radio/ on television

durch·gefroren *Adj.* frozen stiff; chilled to the bone

durch|gehen 1. *unr. itr. V.; mit sein* **a)** [durch etw.] **~:** go *or* walk through [sth.]; **b)** *(hindurchdringen)* [durch etw.] **~:** ⟨*rain, water*⟩ come through [sth.]; **c)** *(direkt zum Ziel führen)* ⟨*train etc.*⟩ go [right] through (bis to); ⟨*flight*⟩ go direct; **d)** *(andauern)* go on (bis zu until); **e)** *(hingenommen werden)* ⟨*discrepancy*⟩ be tolerated; ⟨*mistake, discourtesy*⟩ be allowed to pass; **jmdm. etw. ~ lassen** let sb. get away with sth. **f)** ⟨*horse*⟩ bolt; **2.** *unr. tr. V.; mit sein* go through ⟨*newspaper, text*⟩

durch·gehend 1. *Adj.* **a)** continuous ⟨*line, pattern, etc.*⟩; constantly recurring ⟨*motif*⟩; **b)** *(direkt)* through *attrib.* ⟨*train, carriage*⟩; direct ⟨*flight, connection*⟩; **2.** *adv.* **~ geöffnet haben/bleiben** be/stay open all day

durch|greifen *unr. itr. V.* [hart] **~:** take drastic measures *or* steps

durch|halten 1. *unr. itr. V.* hold out; *(bei einer schwierigen Aufgabe)* see it through; **2.** *unr. tr. V.* stand

durch|hängen *unr. itr. V.* sag
durch|kämmen *tr. V.* **a)** comb ⟨hair⟩
through; **b)** *(durchsuchen)* comb ⟨area
etc.⟩

durch|kommen *unr. itr. V.; mit sein*
a) come through; *(mit Mühe)* get
through; **b)** *(ugs.: beim Telefonieren)*
get through; **c)** *(durchgehen, -fahren
usw.)* **durch etw. ~:** come through
sth.; **d)** *(ugs.: überleben)* pull through
¹**durch|kreuzen** *tr. V.* cross out;
²**durch·kreuzen** *tr. V. (vereiteln)*
frustrate

durch|lassen *unr. tr. V.* **a)** **jmdn.
[durch etw.] ~:** let sb. through [sth.]; **b)**
(durchlässig sein) let ⟨light, water, etc.⟩
through; **durchlässig** *Adj.* permeable; *(porös)* porous; *(undicht)*
leaky; ⟨raincoat, shoe⟩ that lets in
water

Durch·lauf der *(Sport, DV)* run;
¹**durch|laufen** **1.** *unr. itr. V.; mit sein*
a) **[durch etw.] ~:** run through [sth.];
(durchrinnen) trickle through [sth.]; **b)**
(passieren) ⟨runners⟩ run *or* pass
through; **c)** *(ohne Pause laufen)* run
without stopping; **2.** *unr. tr. V.* go
through ⟨soles⟩; ²**durch·laufen** *unr.
tr. V.* go through ⟨phase, stage⟩;
durchlaufend **1.** *Adj.* continuous; **2.**
adv. ⟨numbered, marked⟩ in sequence
durch|lesen *unr. tr. V.* **etw. [ganz] ~:**
read sth. [all the way] through
durch·leuchten *tr. V.* x-ray; *(fig.)*
investigate ⟨case, matter, problem,
etc.⟩ thoroughly
durch·löchern *tr. V.* make holes in
durch|machen *(ugs.)* **1.** *tr. V.* **a)**
undergo ⟨change⟩; complete ⟨training
course⟩; go through ⟨stage, phase⟩;
serve ⟨apprenticeship⟩; **b)** *(erleiden)* go
through; **c)** *(durcharbeiten)* work
through ⟨lunch-break etc.⟩; **2.** *itr. V.*
(durcharbeiten) work [right] through;
(durchfeiern) celebrate all night/day
etc.; keep going all night/day etc.

Durchmesser der; **~s, ~:** diameter
durch|nehmen *unr. tr. V. (Schulw.:
behandeln)* do
durch|probieren *tr. V.* taste ⟨wines,
cakes, etc.⟩ one after another
durch·queren *tr. V.* cross; travel
across ⟨country⟩; ⟨train⟩ go through
⟨country⟩
durch|rechnen *tr. V.* calculate ⟨costs
etc.⟩ [down to the last penny]; check
⟨bill⟩ thoroughly
Durch·reise die journey through;
durch|reisen *itr. V.; mit sein* travel

through; **Durchreise·visum** das
transit visa
durch|reißen **1.** *unr. tr. V.* **etw. ~:**
tear sth. in two *or* in half; **2.** *unr. itr.
V.; mit sein* ⟨fabric, garment⟩ rip, tear;
⟨thread, rope⟩ snap [in two]
durch|rosten *itr. V.; mit sein* rust
through
durchs *Präp. + Art.* **= durch das**
Durch·sage die announcement; *(an
eine bestimmte Person)* message
durchschaubar *Adj.* transparent;
leicht ~ easy to see through; **durch-
schauen** *tr. V.* **a)** see through ⟨per-
son, plan, etc.⟩; see ⟨situation⟩ clearly
durch|schlafen *unr. itr. V.* sleep
[right] through
Durch·schlag der **a)** *(Kopie)* carbon
[copy]; **b)** *(Küchengerät)* strainer;
durch|schlagen *unr. tr. V.* **etw. ~:**
chop sth. in two; **durchschlagend**
Adj. resounding ⟨success⟩; decisive
⟨effect, measures⟩; conclusive ⟨evid-
ence⟩
durch|schneiden *unr. tr. V.* cut
through ⟨thread, cable⟩; cut ⟨ribbon,
sheet of paper⟩ in two; cut ⟨throat, umbilical cord⟩; **etw. in der Mitte ~:** cut
sth. in half; **Durch·schnitt** der average; **im ~:** on average; **über/unter
dem ~ liegen** be above/below average;
durchschnittlich **1.** *Adj.* **a)** *nicht
präd.* average ⟨growth, performance,
output⟩; **b)** *(ugs.: nicht außergewöhnlich)* ordinary ⟨life, person, etc.⟩; **c)**
(mittelmäßig) modest; ordinary ⟨ap-
pearance⟩; **2.** *adv.* ⟨earn etc.⟩ on [an]
average; **~ groß** of average height
Durchschnitts-: ~alter das average
age; **~geschwindigkeit** die average speed
Durch·schrift die carbon [copy]
durch|sehen **1.** *unr. itr. V.* **[durch etw.]
~:** look through [sth.]; **2.** *unr. tr. V.*
look through
durch|sein *unr. itr. V., mit sein; nur im
Inf. u. Part. zusammengeschrieben*
(ugs.) **a)** **[durch etw.] ~:** be through
[sth.]; **b)** *(abgefahren sein)* ⟨train, bus,
etc.⟩ have gone; **c)** *(fertig sein)* have
finished; **durch etw. ~:** have got
through sth.; **d)** ⟨cheese⟩ be ripe;
⟨meat⟩ be well done
durch|setzen **1.** *tr. V.* carry through;
achieve ⟨objective⟩; enforce ⟨demand,
claim⟩; **2.** *refl. V.* assert oneself; ⟨idea
etc.⟩ find *or* gain acceptance
Durch·sicht die: **nach ~ der Unterlagen** after looking *or* checking through

the documents; **durchsichtig** *Adj.* *(auch fig.)* transparent

durch|sprechen *unr. tr. V.* talk ⟨*matter etc.*⟩ over; discuss ⟨*matter etc.*⟩ thoroughly

durch|stehen *unr. tr. V.* stand ⟨*pace, boring job*⟩; come through ⟨*difficult situation*⟩; get over ⟨*illness*⟩

durch|stellen *tr. V.* put ⟨*call*⟩ through (**in** + *Akk.*, **auf** + *Akk.* to)

durch|streichen *unr. tr. V.* cross out; *(in Formularen)* delete

durch · suchen *tr. V.* search (**nach** for); search, scour ⟨*area*⟩ (**nach** for); **Durchsuchung** die; ~, ~en search

durch|treten *unr. tr. V.* press ⟨*clutch-, brake-pedal*⟩ right down

durchtrieben *(abwertend)* 1. *Adj.* crafty; sly; 2. *adv.* craftily; slyly

durch · wachsen *Adj.* ~er Speck streaky bacon

Durchwahl die; a) direct dialling; **mein Apparat hat keine** ~: I don't have an outside line; b) *s.* **Durchwahlnummer; durch|wählen** *itr. V.* a) dial direct; b) *(bei Nebenstellenanlagen)* dial straight through; **Durchwahl · nummer** die number of the/one's direct line

durch|zählen *tr. V.* count; count up

durch|ziehen 1. *unr. tr. V.* jmdn./etw. |durch etw.| ~: pull sb./sth. through [sth.]; **ein Gummiband |durch etw.|** ~: draw an elastic through [sth.]; 2. *unr. itr. V.; mit sein* pass through; ⟨*soldiers*⟩ march through

Durch · zug der *o. Pl.* draught

dürfen 1. *unr. Modalverb; 2. Part.* ~ a) etw. tun ~: be allowed to do sth.; **darf ich rauchen?** may I smoke?; **was darf es sein?** can I help you?; b) *Konjunktiv II + Inf.* **das dürfte der Grund sein** that is probably the reason; 2. *unr. tr., itr. V.* **er hat nicht gedurft** he was not allowed to; **durfte,** *1. u. 3. Pers. Sg. Prät. v.* **dürfen; dürfte** *1. u. 3. Pers. Sg. Konjunktiv II v.* **dürfen**

dürr *Adj.* a) withered; arid, barren ⟨*ground, earth*⟩; b) *(mager)* scrawny; **Dürre** die; ~, ~n drought

Durst der; ~|e|s thirst; ~ haben be thirsty; **ich habe** ~ **auf ein Bier** I could just drink a beer; **durstig** *Adj.* thirsty; **durst · stillend** *Adj.* thirst-quenching

Dusche die; ~, ~n shower; **duschen** *itr., refl. V.* have a shower

Düse die; ~, ~n *(Technik)* nozzle; *(eines Vergasers)* jet

Düsen-: ~**flugzeug** das jet aircraft; ~**motor** der jet engine

düster 1. *Adj.* a) dark; gloomy; dim ⟨*light*⟩; b) *(fig.)* gloomy; sombre ⟨*colour, music*⟩; 2. *adv. (fig.)* gloomily

Dutzend das; ~s, ~e dozen; **zwei** ~: two dozen; **dutzend · weise** *Adv.* in [their] dozens *(coll.)*

duzen *tr. V.* call ⟨*sb.*⟩ 'du' *(the familiar form of address)*

dynamisch 1. *Adj.* a) *(auch fig.)* dynamic; 2. *adv.* dynamically

Dynamit das; ~s dynamite

Dynamo der; ~s, ~s dynamo

Dynastie die; ~, ~n dynasty

D-Zug ['de:-] der express train

E

e, E [e:] das; ~, ~ a) *(Buchstabe)* e/E; b) *(Musik)* [key of] E

Ebbe die; ~, ~n ebb tide; *(Zustand)* low tide; **es ist** ~: the tide is out

eben 1. *Adj.* a) flat; b) *(glatt)* level; 2. *adv.* a) *(gerade jetzt)* just; b) *(kurz)* [for] a moment; **Ebene** die, ~, ~n a) plain; **in der** ~: on the plain; b) *(Geom., Physik)* plane; c) *(fig.)* level

eben · falls *Adv.* likewise; as well; **danke,** ~: thank you, [and] [the] same to you

Eben · holz das ebony

eben · so *Adv.* a) *mit Adjektiven* just as; b) *mit Verben* in exactly the same way

ebenso-: ~**gern** *Adv.* ~**gern mag ich Erdbeeren |wie ...|** I like strawberries just as much [as ...]; ~**gern würde ich an den Strand gehen** I would just as soon go to the beach; ~**gut** *Adv.* just as well; ~**sehr** *Adv.* a) *mit Adjektiven* just as; b) *mit Verben* just as much

Eber der; ~s, ~: boar

Eber · esche die rowan; mountain ash

ebnen *tr. V.* level ⟨*ground*⟩

Echo das; ~s, ~s echo

echt 1. *Adj.* a) genuine; real ⟨*love, friendship*⟩; b) *nicht präd. (typisch)* real, typical; 2. *adv.* a) *(ugs. verstärkend)* really; b) *(typisch)* typically

Eck-: ~**ball** der *(Sport)* corner[-kick/ -hit/-throw]; **einen** ~**ball treten** take a corner; ~**bank** die corner seat

Ecke die; ~, ~**n** corner; **an der** ~: on *or* at the corner; **um die** ~: round the corner; **eckig** *Adj.* square; angular

Eck·zahn der canine tooth

edel *Adj.* a) *nicht präd.* thoroughbred ⟨*horse*⟩; species ⟨*rose*⟩; b) *(großmütig)* noble[-minded], high-minded ⟨*person*⟩; noble ⟨*thought, gesture, feelings, deed*⟩; honourable ⟨*motive*⟩

Edel-: ~**metall** das precious metal; ~**stahl** der stainless steel; ~**stein** der precious stone; gem[stone]

Edition die; ~, ~**en** edition

EDV *Abk.* **elektronische Datenverarbeitung** EDP

Efeu der; ~**s** ivy

Effekt der; ~[e]s, ~**e** effect; **effektvoll** *Adj.* effective; dramatic ⟨*pause, gesture, entrance*⟩

EG ['eː'geː] die; ~ *Abk.* a) **Europäische Gemeinschaft** EC; b) **Erdgeschoß**

egal *Adj. nicht attr. (ugs.: einerlei)* **es ist jmdm.** ~: it's all the same to sb.; [**ganz**] ~, **wie/wer** *usw.* ...: no matter how/who *etc.* ...

Egge die; ~, ~**n** harrow

ehe *Konj.* before

Ehe die; ~, ~**n** marriage

Ehe-: ~**bett** das marriage-bed; *(Doppelbett)* double bed; ~**frau** die wife; *(verheiratete Frau)* married woman; ~**krach** der *(ugs.)* row; ~**leute** *Pl.* married couple

ehelich *Adj.* marital; matrimonial; conjugal ⟨*rights, duties*⟩; legitimate ⟨*child*⟩

ehemalig *Adj.* former

Ehe-: ~**mann** der; *Pl.* ~**männer** husband; *(verheirateter Mann)* married man; ~**paar** das married couple

eher *Adv.* a) *(früher)* earlier; sooner; b) *(lieber)* rather; sooner

Ehe-: ~**ring** der wedding-ring; ~**scheidung** die divorce

Ehre die; ~, ~**n** honour; **jmdm.** ~ **antun** pay tribute to sb.; **ehren** *tr. V.* a) honour; **Sehr geehrter Herr Müller!/Sehr geehrte Frau Müller!** Dear Herr Müller/Dear Frau Müller; b) *(Ehre machen)* **deine Hilfsbereitschaft ehrt dich** your willingness to help does you credit; **ehrenhaft** *Adj.* honourable

ehren-, Ehren-: ~**rührig** *Adj.* defamatory ⟨*allegations*⟩; ~**sache** die: **das ist** ~**sache** that is a point of

honour; ~**sache!** you can count on me!; ~**voll** *Adj.* honourable; ~**wert** *Adj. (geh.)* worthy; ~**wort** das; *pl.* ~**worte:** ~**wort** [!/?] word of honour [!/?]

ehrerbietig *Adj. (geh.)* respectful

Ehr·furcht die reverence (**vor** + *Dat.* for); **ehrfürchtig** *Adj.* reverent

ehr-, Ehr-: ~**gefühl** das; *o. Pl.* sense of honour; ~**geiz** der ambition; ~**geizig** *Adj.* ambitious

ehrlich *Adj.* honest; genuine ⟨*concern, desire, admiration*⟩; upright ⟨*character*⟩; **Ehrlichkeit** die; ~ *s.* **ehrlich:** honesty; genuineness; uprightness

ehr·los *Adj.* dishonourable

Ehrung die; ~, ~**en:** **die** ~ **der Preisträger** the prize-giving *(Brit.)* or *(Amer.)* awards ceremony; **bei der** ~ **der Sieger** when the winners were awarded their medals/trophies

ehr·würdig *Adj.* venerable

Ei das; ~[e]s, ~**er** egg

Eiche die; ~, ~**n** oak[-tree]; *(Holz)* oak[-wood]

Eichel die; ~, ~**n** acorn

eichen *tr. V.* calibrate ⟨*measuring instrument, thermometer*⟩; standardize ⟨*weights, measures, containers, products*⟩; adjust ⟨*weighing-scales*⟩

Eich·hörnchen das squirrel

Eid der; ~[e]s, ~**e** oath

Eidechse ['aidɛksə] die; ~, ~**n** lizard

eide·stattlich *Adj. (Rechtsw.)* **eine** ~**e Erklärung** a statutory declaration

Ei·dotter der *od.* das egg yolk

Eier-: ~**becher** der egg-cup; ~**kuchen** der pancake; *(Omelett)* omelette; ~**likör** der egg-liqueur; ~**stock** der *(Physiol., Zool.)* ovary; ~**uhr** die egg-timer

Eifer der; ~**s** eagerness

Eifer·sucht die jealousy (**auf** + *Akk.* of); **eifer·süchtig** *Adj.* jealous (**auf** + *Akk.* of)

eifrig *Adj.* eager

Ei·gelb das; ~[e]s, ~**e** egg yolk

eigen *Adj.* own; *(selbständig)* separate

eigen-, Eigen-: ~**art** die *(Wesensart)* particular nature; *(Zug)* peculiarity; **eine** ~**art dieser Stadt** one of the characteristic features of this city; ~**artig** *Adj.* peculiar; strange; odd; ~**händig** 1. *Adj.* personal ⟨*signature*⟩; holographic ⟨*will, document*⟩; 2. *adv.* ⟨*present, sign*⟩ personally; ~**heim** das house of one's own

Eigenheit die; ~, ~**en** peculiarity

eigen-, Eigen-: ~**lob** das self-praise;

~**mächtig** *Adj.* unauthorized; ~**name** der proper name; ~**nützig** *Adj.* self-seeking; selfish ⟨*motive*⟩

eigens *Adv.* specially

Eigenschaft die; ~, ~**en** quality; characteristic; *(von Sachen, Stoffen)* property

Eigenschafts·wort das adjective

eigen-, Eigen-: ~**sinn** der; *o. Pl.* obstinacy; ~**sinnig** *Adj.* obstinate; ~**ständig** *Adj.* independent

eigentlich 1. *Adj.* *(wirklich)* actual; real; *(wahr)* true; *(ursprünglich)* original; 2. *Adv.* actually; 3. *Partikel* wie spät ist es ~? tell me, what time is it?; was willst du ~? what exactly do you want?

Eigen·tor das *(Ballspiele, fig.)* own goal

Eigentum das; ~**s** property; *(einschließlich Geld usw.)* assets *pl.* **Eigentümer** der; ~**s**, ~: owner; *(Hotel~, Geschäfts~)* proprietor; **Eigentümerin** die; ~, ~**nen** owner; *(Hotel~, Geschäfts~)* proprietress; proprietor; **Eigentums·wohnung** die owner-occupied flat *(Brit.)*; condominium apartment *(Amer.)*

eigen·willig *Adj.* self-willed

eignen *refl. V.* be suitable; **Eignung** die; ~: suitability; seine ~ zum Fliegen his aptitude for flying

Eignungs-: ~**prüfung** die, ~**test** der aptitude test

Eil-: ~**bote** der special messenger; „**durch** *od.* per ~**boten**" *(veralt.)* 'express'; ~**brief** der express letter

Eile die; ~: hurry; in ~ sein be in a hurry; **eilen** *itr. V.* a) *mit sein* hurry; *(besonders schnell)* rush; b) *(dringend sein)* be urgent; „**eilt!**" 'urgent'; **eilig** 1. *Adj.* a) hurried; es ~ haben be in a hurry; b) *(dringend)* urgent; 2. *adv.* hurriedly; **Eil·zug** der semi-fast train

Eimer der; ~**s**, ~: bucket; *(Milch~)* pail; *(Abfall~)* bin; ein ~ [voll] Wasser a bucket of water; im ~ sein *(salopp)* be up the spout *(sl.)*

ein 1. *Kardinalz.* one; 2. *unbest. Art.* a/an; 3. *Indefinitpron; s.* irgendein a; *s. auch* einer

ein *(elliptisch)* ~ – **aus** *(an Schaltern)* on – off

Einakter der; ~**s**, ~: one-act play

einander *reziprokes Pron.; Dat. u. Akk. (geh.)* each other; one another

ein|arbeiten *tr. V.* train ⟨*employee*⟩

ein·armig *Adj.* one-armed

ein|äschern *tr. V.* cremate

ein|atmen *tr., itr. V.* breathe in

ein·äugig *Adj.* one-eyed

Ein·bahn·straße die one-way street

Ein·band der; *Pl.* -**bände** binding

Ein·bau der; ~**s**, ~**ten** fitting; *(eines Motors)* installation; **ein|bauen** *tr. V.* build in, fit; install ⟨*engine, motor*⟩; **Einbau·küche** die fitted kitchen

ein·beinig *Adj.* one-legged

ein|berufen *unr. tr. V.* summon; call; **Ein·berufung** die a) *(das Einberufen)* calling; b) *(zur Wehrpflicht)* call-up; conscription; draft *(Amer.)*

Einbett·zimmer das single room

ein|beziehen *unr. tr. V.* include

ein|biegen *unr. itr. V.; mit sein* turn

ein|bilden *refl. V.* a) sich *(Dat.)* ~: imagine sth.; b) *(ugs.)* sich *(Dat.)* etwas ~: be conceited (**auf** + *Akk.* about); **Ein·bildung** die; ~, ~**en** a) imagination; b) *(falsche Vorstellung)* fantasy; c) *(Hochmut)* conceitedness

ein|binden *unr. tr. V.* bind ⟨*book*⟩; etw. neu ~: rebind sth.

ein|blenden *tr. V. (Rundf., Ferns., Film)* insert

Ein·blick der a) view; ~ in etw. *(Akk.)* haben be able to see into sth.; b) *(Durchsicht)* jmdm. ~ in etw. *(Akk.)* gewähren allow sb. to look at *or* examine sth.; c) *(Kenntnis)* insight

ein|brechen *unr. itr. V.* a) *mit haben od. sein* break in; in eine Bank ~: break into a bank; bei jmdm. ~: burgle sb.; b) *mit sein (einstürzen)* ⟨*roof, ceiling*⟩ cave in; c) *mit sein (durchbrechen)* fall through; **Einbrecher** der; ~**s**, ~: burglar

ein|bringen *unr. tr. V.* a) bring in ⟨*harvest*⟩; b) *(verschaffen)* Gewinn/Zinsen ~: yield a profit/bring in interest; jmdm. Ruhm ~: bring sb. fame; c) *(Parl.: vorlegen)* introduce ⟨*bill*⟩; d) invest ⟨*capital, money*⟩

Ein·bruch der a) burglary; ein ~ in eine Bank a break-in at a bank; b) *(das Einstürzen)* collapse

einbürgern 1. *tr. V.* naturalize; 2. *refl. V.* ⟨*custom, practice*⟩ become established; ⟨*person, plant, animal*⟩ become naturalized; **Einbürgerung** die; ~, ~**en** naturalization

Einbuße die loss; **ein|büßen** *tr. V.* lose; *(durch eigene Schuld)* forfeit

ein|checken *tr., itr. V. (Flugw.)* check in

ein|cremen *tr. V.* put cream on ⟨*hands etc.*⟩; sich ~: put cream on

ein|dämmen *tr. V. (fig.)* check; stem

ein|decken 1. *refl. V.* stock up; 2. *tr. V. (ugs.: überhäufen)* jmdn. mit Arbeit ~: swamp sb. with work

eindeutig *Adj.* clear; **Eindeutigkeit** die; ~, ~en clarity

ein|dringen *unr. itr. V.; mit sein* in etw. *(Akk.)* ~: penetrate into sth.; *(bullet)* pierce sth.; *(allmählich)* ⟨water, sand, etc.⟩ seep into sth.; **ein·dringlich** *Adj.* urgent; impressive ⟨voice⟩; forceful, powerful ⟨words⟩; **Eindringling** der; ~s, ~e intruder

Ein·druck der; ~[e]s, Eindrücke impression; **ein|drücken** *tr. V.* smash in ⟨mudguard, bumper⟩; stave in ⟨side of ship⟩; smash ⟨pier, column, support⟩; break ⟨window⟩; crush ⟨ribs⟩; flatten ⟨nose⟩; **eindrucks·voll** 1. *Adj.* impressive; 2. *adv.* impressively

eine *s.* ¹ein

ein|ebnen *tr. V.* level

eineiig ['ain|aiiç] *Adj.* identical ⟨twins⟩

ein·ein·halb *Bruchz.* one and a half; ~ Stunden an hour and a half

ein|engen *tr. V.* a) jmdn. ~: restrict sb.'s movement[s]; b) *(fig.)* restrict

einer, eine, eines, eins *Indefinitpron. (man)* one; *(jemand)* someone; somebody; *(fragend, verneint)* anyone; anybody; **kaum einer** hardly anybody; **ein[e]s ist sicher** one thing is for sure

Einer der; ~s, ~ a) *(Math.)* unit; b) *(Sport)* single sculler; **im ~:** in the single sculls

einerlei *Adj.* ~, ob/wo/wer *usw.* no matter whether/where/who *etc.;* **es ist ~:** it makes no difference

Einerlei das; ~s monotony

einerseits *Adv.* on the one hand

ein·fach 1. *Adj.* a) simple; b) *(nicht mehrfach)* single ⟨knot, ticket, journey⟩; 2. *Partikel* simply; just; **Einfachheit** die; ~: simplicity

ein|fädeln 1. *tr. V.* thread (in + *Akk.* into); 2. *refl. V. (Verkehrsw.)* filter in

ein|fahren 1. *unr. itr. V.; mit sein* come in; ⟨train⟩ pull in; **in den Bahnhof ~:** pull into the station; 2. *unr. tr. V.* a) bring in ⟨harvest⟩; b) *(beschädigen)* knock down ⟨wall⟩; smash in ⟨mudguard⟩; **Ein·fahrt** die a) *(das Hineinfahren)* entry; **Vorsicht bei der ~ des Zuges!** stand clear [of the edge of the platform], the train is approaching; b) *(Zufahrt)* entrance; *(Autobahn~)* slip road; **„keine ~"** 'no entry'

Ein·fall der a) *(Idee)* idea; b) *o. Pl.*

(Licht~) incidence *(Optics);* **ein|fallen** *unr. itr. V.; mit sein* a) jmdm. ~: occur to sb.; **was fällt dir denn ein!** what do you think you're doing?; b) *(in Erinnerung kommen)* **ihr Name fällt mir nicht ein** I cannot think of her name; **plötzlich fiel ihr ein, daß ...:** suddenly she remembered that ...; c) *(von Licht)* come in

einfalls-: ~los *Adj.* unimaginative; lacking in ideas; ~reich *Adj.* imaginative; full of ideas

Einfalt die; ~: simpleness; simple-mindedness; **einfältig** *Adj.* simple; naïve; naïve ⟨remarks⟩

ein|fangen *unr. tr. V.* catch

ein|fassen *tr. V.* border; edge; frame ⟨picture⟩; set ⟨gem⟩; edge ⟨grave, lawn, etc.⟩; **Ein·fassung** die *s.* einfassen: border; edging; frame; setting

ein|finden *unr. refl. V.* arrive; *(sich treffen)* meet; ⟨crowd⟩ gather

ein|fliegen *unr. tr. V.* fly in

ein|flößen *tr. V.* a) jmdm. Tee ~: pour tea into sb.'s mouth; b) *(fig.)* jmdm. Angst ~: put fear into sb.

Ein·fluß der influence; **Einfluß·bereich** der sphere of influence; **einfluß·reich** *Adj.* influential

ein·förmig *Adj.* monotonous

ein|frieren 1. *unr. itr. V.; mit sein* freeze; ⟨pipes⟩ freeze up; 2. *unr. tr. V.* a) deep-freeze ⟨food⟩; b) *(fig.)* freeze

ein|fügen *tr. V.* fit in; **etw. in etw.** *(Akk.)* ~: fit sth. into sth.

ein|fühlen *refl. V.* sich in jmdn. ~: empathize with sb.; **einfühlsam** *Adj.* understanding; **Ein·fühlung** die; ~: empathy (in + *Akk.* with)

Ein·fuhr die; ~, ~en *s.* Import; **ein|führen** *tr. V.* a) *(als Neuerung)* introduce ⟨fashion, method, technology⟩; b) *(importieren)* import; **Ein·führung** die introduction

Ein·gabe die petition; *(Beschwerde)* complaint

Ein·gang der entrance; **„kein ~"** 'no entry'; **ein·gängig** *Adj.* catchy; **eingangs** *Adv.* at the beginning; pedes

Eingangs-: ~halle die entrance hall; *(eines Hotels, Theaters)* foyer; ~tür die *(von Kaufhaus, Hotel usw.)* [entrance] door; *(von Wohnung, Haus usw.)* front door

ein·gebildet *Adj.* a) imaginary ⟨illness⟩; b) *(arrogant)* conceited

Eingeborene der/die; *adj. Dekl. (veralt.)* native

ein|gehen 1. *unr. itr. V.; mit sein* **a)** arrive; **b)** *(fig.)* **in die Geschichte ~:** go down in history; **c)** *(schrumpfen)* shrink; **d) auf eine Frage ~/nicht ~:** go into *or* deal with/ignore a question; **auf jmdn. ~:** be responsive to sb.; **auf jmdn. nicht ~:** ignore sb.'s wishes; **2.** *unr. tr. V.* enter into ⟨contract, matrimony⟩; take ⟨risk⟩; accept ⟨obligation⟩

eingehend *Adj.* detailed

Ein·gemachte *das;* **~n** preserved fruit/vegetables

ein|gemeinden *tr. V.* incorporate ⟨village⟩ **(in + Akk.** nach into)

ein·geschnappt *Adj. (ugs.)* huffy

Ein·geständnis *das* confession; admission; **ein|gestehen** *unr. tr. V.* admit

Eingeweide *das;* **~s, ~;** *meist Pl.* entrails *pl.;* innards *pl*

ein|gewöhnen *refl. V.* get used to one's new surroundings

ein|gießen *unr. tr., itr. V.* pour in

ein|gliedern *tr. V.* integrate **(in + Akk.** into); incorporate ⟨village, company⟩ **(in + Akk.** into); *(einordnen)* include **(in + Akk.** in)

ein|graben *unr. tr. V.* bury **(in + Akk.** in); sink ⟨pile, pipe⟩ **(in + Akk.** into)

ein|gravieren *tr. V.* engrave **(in + Akk.** on)

ein|greifen *unr. itr. V.* intervene **(in + Akk.** in); **Ein·griff der a)** intervention **(in + Akk.** in); **b)** *(Med.)* operation

ein|haken 1. *tr. V.* **a)** *(mit Haken befestigen)* fasten; **b) sich ~:** link arms; **2.** *refl. V.* **sich bei jmdm. ~:** link arms with sb.

Ein·halt der: jmdm./einer Sache ~ gebieten *od.* **tun** *(geh.)* halt sb./sth.; **ein|halten 1.** *unr. tr. V.* keep ⟨appointment⟩; meet ⟨deadline, commitments⟩; keep to ⟨diet, speed-limit, agreement⟩; observe ⟨regulation⟩; **2.** *unr. itr. V. (geh.)* stop

ein·heimisch *Adj.* native; home *attrib.* ⟨team⟩; **Einheimische der/die;** *adj. Dekl.* local

Einheit die; **~, ~en** unity; **einheitlich 1.** *Adj.* unified; *(unterschiedslos)* uniform ⟨dress⟩; standard ⟨procedure, practice⟩; **2.** *adv.* **~ gekleidet sein** be dressed the same

einhellig 1. *Adj.* unanimous; **2.** *adv.* unanimously

ein|holen 1. *tr. V.* **a)** catch up with ⟨person, vehicle⟩; **b)** make up ⟨arrears, time⟩; **2.** *itr. V. (ugs.)* s. **einkaufen 1**

ein·hundert *Kardinalz. s.* **hundert**

einig *Adj.* **sich (Dat.) ~ sein** be agreed; **sich (Dat.) ~ werden** reach agreement

einig... *Indefinitpron. u. unbest. Zahlwort* some; **~e wenige** a few; **~e hundert** several hundred

einigen 1. *tr. V.* unite; **2.** *refl. V.* reach an agreement

einigermaßen *Adv.* somewhat

Einigkeit die; **~ a)** unity; **b)** *(Übereinstimmung)* agreement

ein·jährig *Adj.* one-year-old *attrib.;* one year old *pred.;* *(ein Jahr dauernd)* one-year *attrib.*

Ein·kauf der a) Einkäufe machen do some shopping; **b)** *(eingekaufte Ware)* purchase; **c)** *o. Pl. (Abteilung)* purchasing department; **ein|kaufen 1.** *itr. V.* shop; **~ gehen** go shopping; **2.** *tr. V.* buy; purchase; **Ein·käufer der** buyer; purchaser

Einkaufs-: **~bummel der** [leisurely] shopping expedition; **~zentrum das** shopping centre

ein|kehren *itr. V.; mit sein* stop; **in einem Wirtshaus ~:** stop at an inn

ein|klammern *tr. V.* **etw. ~:** put sth. in brackets; bracket sth.

Ein·klang der harmony; **in** *od.* **im ~ stehen** accord

ein|kleben *tr. V.* stick in

ein|kleiden *tr. V.* clothe

ein|klemmen *tr. V.* **a)** *(quetschen)* catch; **b)** *(fest einfügen)* clamp

ein|kochen *tr. V.* preserve ⟨fruit etc.⟩

Einkommen das; **~s, ~:** income

ein|kreisen *tr. V.* **a) etw. ~:** put a circle round sth.; **b)** *(umzingeln)* surround

Einkünfte *Pl.* income *sing.;* **feste ~:** a regular income

¹ein|laden *unr. tr. V.* load ⟨goods⟩

²ein|laden *unr. tr. V.* invite ⟨person⟩ (zu for); **einladend** *Adj.* inviting; **Ein·ladung die** invitation

Ein·lage die a) *(in Brief)* enclosure; **b)** *(Kochk.)* vegetables, dumplings, etc. added to a clear soup; **c)** *(Schuh~)* arch-support; **d)** *(Programm~)* interlude

ein|lagern *tr. V.* store; lay in ⟨stores⟩

Einlaß der; **Einlasses, Einlässe** admission; **ein|lassen** *unr. tr. V.* **a)** *(hereinlassen)* admit; let in; **b)** *(einfüllen)* run ⟨water⟩

Ein·lauf der *(Med.)* enema; **ein|laufen 1.** *unr. itr. V.; mit sein* **a)** ⟨ship⟩ come in; **b)** *(kleiner werden)* shrink; **2.** *unr. tr. V.* wear in ⟨shoes⟩

ein|leben *refl. V.* settle down

ein|legen *tr. V.* a) load ⟨*film*⟩; engage ⟨*gear*⟩; b) *(Kochk.)* pickle

ein|leiten *tr. V.* a) introduce; b) induce ⟨*birth*⟩; c) lead in; **etw. in etw.** *(Akk.)* ~: lead sth. into sth.; **Ein·leitung die** a) introduction; b) *(einer Geburt)* induction

ein|leuchten *itr. V.* jmdm. ~: be clear to sb.; **ein·leuchtend** *Adj.* plausible

ein|liefern *tr. V.* take ⟨*letter, person*⟩ (**bei, in** + Abk. to)

ein|lösen *tr. V.* cash ⟨*cheque*⟩

ein|machen *tr. V.* preserve ⟨*fruit etc.*⟩; *(in Gläser)* bottle

einmal 1. *Adv.* a) once; **noch ~ so groß** |wie| twice as big [as]; **etw. noch ~ tun** do sth. again; b) ['-'-] *(später)* one day; *(früher)* once; **es war ~ ...:** once upon a time there was ...; 2. *Partikel* **nicht ~:** not even; **wieder ~:** yet again; **Einmal·eins das;** ~: [multiplication] tables *pl.;* **einmalig** 1. *Adj.* a) unique; one-off ⟨*payment, purchase*⟩; b) *(ugs.)* fantastic *(coll.);* 2. *adv. (ugs.)* really fantastically *(coll.)*

Ein·marsch der a) entry; b) *(Besetzung)* invasion (**in** + Akk. of); **ein|marschieren** *itr. V.; mit sein* march in

ein|mischen *refl. V.* interfere (**in** + Akk. in)

einmütig 1. *Adj.* unanimous; 2. *adv.* unanimously

ein|nähen *tr. V.* sew in

Einnahme die; ~, ~en a) income; *(Staats~)* revenue; *(Kassen~)* takings *pl.;* b) *(von Arzneimitteln)* taking; c) *(einer Stadt, Burg)* taking; **ein|nehmen** *unr. tr. V.* a) take; *(verdienen)* earn; b) *(ausfüllen)* take up ⟨*amount of room*⟩; c) *(beeinflussen)* **jmdn. für sich** ~: win sb. over

Ein·öde die barren waste

ein|ordnen 1. *tr. V.* arrange; put in order; 2. *refl. V.* a) *(Verkehrsw.)* get into the correct lane; „~" 'get in lane'; b) *(sich einfügen)* fit in

ein|packen 1. *tr. V.* pack (**in** + Akk. in); *(einwickeln)* wrap [up]; 2. *itr. V.* *(ugs.)* **er kann ~:** he's had it *(coll.)*

ein|parken *tr., itr. V.* park

ein|pflanzen *tr. V.* a) plant; b) *(Med., fig.)* implant

ein|prägen *tr. V.* a) stamp (**in** + Akk. into, on); b) *(fig.)* **sich** *(Dat.)* **etw.** ~: memorize sth.; **jmdm. etw.** ~: impress sth. on sb.; **einprägsam** *Adj.* easily remembered

ein|rahmen *tr. V.* frame

ein|räumen *tr. V.* a) put away; b) *(füllen)* **seinen Schrank** ~: put one's things away in one's cupboard; **ein Zimmer** ~: put the furniture into a room; c) *(zugestehen)* admit

ein|reden 1. *tr. V.* jmdm. etw. ~: talk sb. into believing sth.; **sich** *(Dat.)* ~, **daß** ...: persuade oneself that ...; 2. *itr. V.* **auf jmdn.** ~: talk insistently to sb.

ein|regnen *refl. V.; unpers.* **es hat sich eingeregnet** it's begun to rain steadily.

ein|reiben *tr. V.* rub ⟨*substance*⟩ in; **etw. mit Öl** ~: rub oil into sth.

ein|reichen *tr. V.* submit; lodge ⟨*complaint*⟩; tender ⟨*resignation*⟩

ein|reihen 1. *refl. V.* **sich in etw.** *(Akk.)* ~: join sth.; 2. *tr. V.* **jmdn. in eine Kategorie** ~: place sb. in a category

Einreiher der; ~s, ~: single-breasted suit/jacket

Ein·reise die entry; **Einreise·erlaubnis die** entry permit; **ein|reisen** *itr. V.; mit sein* enter; **nach Schweden** ~: enter Sweden

ein|reißen 1. *unr. tr. V.* a) pull down ⟨*building*⟩; b) *(einen Riß machen in)* tear; rip; 2. *unr. itr. V.; mit sein* tear; rip

ein|renken *tr. V.* a) *(Med.)* set; b) *(ugs.: bereinigen)* sort out

ein|richten 1. *refl. V.* **sich schön** ~: furnish one's home beautifully; **sich häuslich** ~: make oneself at home; 2. *tr. V.* furnish ⟨*flat, house*⟩; fit out ⟨*shop*⟩; equip ⟨*laboratory*⟩; **Ein·richtung die** a) *o. Pl.* furnishing; b) *(Mobiliar)* furnishings *pl*

ein|rollen 1. *tr. V.* roll up ⟨*carpet etc.*⟩; put ⟨*hair*⟩ in curlers; 2. *itr. V.; mit sein* roll in

ein|rosten *itr. V.; mit sein* go rusty

ein|rücken 1. *itr. V.; mit sein (einmarschieren)* move in; 2. *tr. V.* indent ⟨*line, heading, etc.*⟩

eins 1. *Kardinalz.* one; **es ist ~:** it is one o'clock; **~ zu null** one-nil; **~ zu ~:** one all; „~, **zwei, drei!**" 'ready, steady, go'; 2. *Adj.* **mir ist alles ~:** it's all the same to me; 3. *Indefinitpron. s.* **irgendein** a; **Eins die;** ~, ~en a) one; b) *(Schulnote)* one; A

einsam *Adj.* a) lonely ⟨*person, decision*⟩; b) *(einzeln)* solitary ⟨*tree, wanderer*⟩; c) *(abgelegen)* isolated; d) *(menschenleer)* deserted; **Einsamkeit die;** ~ a) loneliness; b) *(Alleinsein)* solitude; c) *(Abgeschiedenheit)* isolation

ein|sammeln *tr. V.* **a)** *(auflesen)* pick up; gather up; **b)** *(sich aushändigen lassen)* collect in; collect ⟨*tickets*⟩

Ein·satz der **a)** *(aus Stoff)* inset; *(in Kochtopf, Nähkasten usw.)* compartment; **b)** *(Betrag)* stake; **c)** *(Gebrauch)* use; *(von Truppen)* deployment

Einsatz-: **~befehl** der order to go into action; **den ~befehl haben** have operational command; **~leiter** der head of operations; **~wagen** der *(der Polizei)* police car; *(der Feuerwehr)* fire-engine; *(Notarztwagen)* ambulance

ein|saugen *unr. (auch regelm.) tr. V.* suck in; breathe [in] ⟨*fresh air*⟩

ein|schalten **1.** *tr. V.* **a)** switch on ⟨*radio, TV, electricity, etc.*⟩; **b)** *(fig.)* call in ⟨*press, police, expert, etc.*⟩; **2.** *refl. V.* **a)** switch [itself] on; **b)** *(eingreifen)* intervene (**in** + *Akk.* in)

ein|schärfen *tr. V.* **jmdm. etw. ~:** impress sth. [up]on sb.

ein|schätzen *tr. V.* judge ⟨*person*⟩; assess ⟨*situation, income, damages*⟩; *(schätzen)* estimate; **Ein·schätzung** die *s.* **einschätzen:** judging; assessment; estimation

ein|schenken *tr., itr. V.* **a)** *(eingießen)* pour [out]; **jmdm. etw. ~:** pour out sth. for sb.; **b)** *(füllen)* fill [up] ⟨*glass, cup*⟩

ein|scheren *itr. V.; mit sein* **auf eine Fahrspur ~:** get *or* move into a lane

ein|schicken *tr. V.* send in

ein|schieben *unr. tr. V.* **a)** push in; **b)** *(einfügen)* insert; put on ⟨*trains, buses*⟩

ein|schiffen *tr., refl. V.* embark

ein|schlafen *unr. itr. V.; mit sein* **a)** fall asleep; **b)** *(verhüll.: sterben)* pass away; **c)** *(gefühllos werden)* go to sleep; **ein|schläfern** *tr. V.* **a)** **jmdn. ~:** send sb. to sleep, *(betäuben)* put sb. to sleep; **b)** *(schmerzlos töten)* **ein Tier ~:** put an animal to sleep; **einschläfernd** **1.** *Adj.* soporific; **2.** *adv.* **~ wirken** have a soporific effect

ein|schlagen **1.** *unr. tr. V.* **a)** knock in; **b)** *(zertrümmern)* smash [in]; **c)** *(einwickeln)* wrap up ⟨*present*⟩; cover ⟨*book*⟩; **2.** *unr. itr. V.* **a)** ⟨*bomb*⟩ land; ⟨*lightning*⟩ strike; **b)** **auf jmdn./etw. ~:** rain blows on sb./sth.

einschlägig **1.** *Adj.* specialist ⟨*journal, shop*⟩; relevant ⟨*literature, passage*⟩; **2.** *adv.* **er ist ~ vorbestraft** he has previous convictions for a similar offence/similar offences

ein|schleichen *unr. refl. V.* steal in

ein|schließen *unr. tr. V.* **a)** **etw. in etw.** *(Dat.)* **~:** lock sth. up [in sth.]; **jmdn./sich ~:** lock sb./oneself in; **b)** *(umgeben)* surround; **einschließlich** **1.** *Präp. mit Gen.* including; **~ der Unkosten** including expenses; **2.** *adv.* **bis ~ 30. Juni** up to and including 30 June

ein|schmeicheln *refl. V.* **sich bei jmdm. ~:** ingratiate oneself with sb.

ein|schmuggeln *tr. V.* smuggle in

ein|schneiden *unr. tr. V.* **a)** make a cut in; **b)** *(einritzen)* carve; **einschneidend** *Adj.* drastic

ein|schneien *itr. V.; mit sein* get snowed in

Ein·schnitt der cut

ein|schränken **1.** *tr. V.* **a)** reduce, curb ⟨*expenditure, consumption*⟩; **b)** *(einengen)* limit; restrict; **jmdn. in seinen Rechten ~:** limit *or* restrict sb.'s rights; **2.** *refl. V.* economize; **Einschränkung** die **~, ~en a)** restriction; limitation; **b)** *(Vorbehalt)* reservation

ein|schreiben *unr. tr. V.* **a)** *(Postw.)* register ⟨*letter*⟩; **b)** *(eintragen)* **sich/jmdn. ~:** enter one's/sb's name; **Ein·schreiben** das *(Postw.)* registered letter; **per ~:** by registered mail

ein|schreiten *unr. itr. V.* intervene

ein|schüchtern *tr. V.* intimidate

ein|schulen *tr. V.* **eingeschult werden** start school

ein|sehen *unr. tr. V.* **a)** *(überblicken)* see into; **b)** *(prüfend lesen)* look at; **c)** *(erkennen)* realize; **d)** *(begreifen)* see

ein|seifen *tr. V.* lather

ein·seitig **1.** *Adj.* **a)** on one side *postpos.;* **b)** *(tendenziös)* one-sided; **2.** *adv.* **a)** on one side; **b)** *(tendenziös)* one-sidedly

ein|senden *unr. tr. V.* send [in]

ein|setzen **1.** *tr. V.* **a)** *(hineinsetzen)* put in; **b)** put on ⟨*special train etc.*⟩; **c)** *(ernennen)* appoint; **d)** *(in Aktion treten lassen)* use; **e)** *(aufs Spiel setzen)* stake ⟨*money*⟩; **f)** *(riskieren)* risk; **2.** *itr. V.* begin; ⟨*storm*⟩ break. **3.** *refl. V. (sich engagieren)* **ich werde mich dafür ~, daß ...:** I shall do what I can to see that ...; **sich nicht genug ~:** ⟨*pupil*⟩ lacking application; ⟨*minister*⟩ be lacking in commitment

Ein·sicht die **a)** view (**in** + *Akk.* into); **b)** *(Einblick)* **~ in die Akten nehmen** take *or* have a look at the files; **c)** *(Erkenntnis)* insight; **einsichtig** *Adj.* **a)**

(verständnisvoll) understanding; **b)** *(verständlich)* comprehensible

Ein·siedler der hermit

ein·silbig *Adj.* **a)** monosyllabic ⟨*word*⟩; **b)** *(fig.)* taciturn ⟨*person*⟩

ein|sinken *unr. itr. V.* sink in

Einsitzer der; ~s, ~: single-seater; **einsitzig** *Adj.* single-seater *attrib.*

ein|spannen *tr. V.* harness ⟨*horse*⟩; put in ⟨*paper*⟩: fix ⟨*fabric*⟩; clamp ⟨*work*⟩

ein|sparen *tr. V.* save

ein|sperren *tr. V.* lock up

einsprachig *Adj.* monolingual

ein|springen *unr. itr. V.; mit sein* stand in; *(aushelfen)* step in and help out

ein|spritzen *tr. V.* inject; **jmdm. etw. ~:** inject sb. with sth.; **Einspritz·motor** der fuel-injection engine

Ein·spruch der objection **(gegen** to)

einspurig 1. *Adj.* single-track ⟨*road*⟩; **2. adv. die Autobahn ist nur ~ befahr-bar** only one lane of the motorway is open

einst *Adv. (geh.)* once

ein|stampfen *tr. V.* pulp ⟨*books*⟩

Ein·stand der: **seinen ~ geben** celebrate starting a new job

ein|stecken *tr. V.* **a)** put in; **b)** *(mitnehmen)* put ⟨*sth.*⟩ in one's pocket/bag *etc.*

ein|stehen *unr. itr. V.* **für jmdn. ~:** vouch for sb.; **für etw. ~:** take responsibility for sth.

ein|steigen *unr. itr. V.; mit sein* **a)** *(in ein Fahrzeug)* get in; **in ein Auto ~:** get into a car; **in den Bus ~:** get on the bus; **b)** *(eindringen)* climb in

einstellbar *Adj.* adjustable; **ein|stellen 1.** *tr. V.* **a)** *(einordnen)* put away ⟨*books etc.*⟩; **b)** *(unterstellen)* put in ⟨*car, bicycle*⟩; **c)** *(beschäftigen)* take on ⟨*workers*⟩; **d)** *(regulieren)* adjust; **e)** *(beenden)* stop; call off ⟨*search, strike*⟩; **f)** *(Sport)* equal ⟨*record*⟩; **2. refl. V.** **a)** arrive; **b)** ⟨*pain, worry*⟩ begin; ⟨*success*⟩ come; ⟨*symptoms, consequences*⟩ appear; **c) sich auf etw. (Akk.) ~:** prepare oneself for sth.; **sich schnell auf neue Situationen ~:** adjust quickly to new situations

ein·stellig *Adj.* single-figure *attrib.*

Ein·stellung die **a)** *(von Arbeitskräften)* employment; **b)** *(Regulierung)* adjustment; **c)** *(Beendigung)* stopping; **d)** *(Sport)* **die ~ eines Rekordes** the equalling of a record; **e)** *(Ansicht)*

attitude; **ihre politische/religiöse ~:** her political/religious views *pl.;* **f)** *(Film)* take

Ein·stich der **a)** insertion; **b)** *(~stelle)* puncture; prick

Ein·stieg der; ~[e]s, ~e *(Eingang)* entrance; *(Tür)* door/doors

ein|stimmen 1. *itr. V.* join in. **2.** *tr. V.* **jmdn. auf etw. (Akk.) ~:** get sb. in the [right] mood for sth.

einstimmig 1. *Adj.* **a)** *(Musik)* for one voice; **b)** *(einmütig)* unanimous ⟨*decision, vote*⟩; **2. adv. a)** *(Musik)* in unison; **b)** *(einmütig)* unanimously

ein·stöckig *Adj.* single-storey *attrib.*

ein|studieren *tr. V.* rehearse

ein|stufen *tr. V.* classify; categorize

ein·stündig *Adj.* one-hour *attrib.*

ein|stürmen *itr. V.* **mit Fragen auf jmdn. ~:** besiege sb. with questions

Ein·sturz der collapse; **ein|stürzen** *itr. V.; mit sein* collapse

einst·weilen *Adv.* for the time being

eintägig *Adj.* one-day *attrib.;* **Eintags·fliege** die *(Zool.)* mayfly; *(fig. ugs.)* seven-day wonder

ein|tauchen 1. *tr. V.* dip; *(untertauchen)* immerse; **2.** *itr. V.; mit sein* dive in; ⟨*submarine*⟩ dive

ein|tauschen *tr. V.* exchange **(gegen** for)

ein·tausend *Kardinalz.: s.* **tausend**

ein|teilen *tr. V.* **a)** divide up; classify ⟨*plants, species*⟩; **b)** *(disponieren, verplanen)* organize

einteilig *Adj.* one-piece

eintönig 1. *Adj.* monotonous; **2. adv.** monotonously; **Eintönigkeit die;** ~: monotony

Ein·topf der stew

Ein·tracht die; *o. Pl.* harmony; **ein·trächtig** *Adj.* harmonious

Eintrag der; ~[e]s, **Einträge** entry; **ein|tragen** *unr. tr. V.* **a)** enter; **b)** *(Amtsspr.)* register

einträglich *Adj.* lucrative

ein|treffen *unr. itr. V.; mit sein* **a)** arrive; **b)** *(verwirklicht werden)* come true

ein|treiben *unr. tr. V.* collect ⟨*taxes, debts*⟩; *(durch Gerichtsverfahren)* recover ⟨*debts, money*⟩

ein|treten 1. *unr. itr. V.; mit sein* **a)** enter; **bitte, treten Sie ein!** please come in; **b)** *(Mitglied werden)* **in einen Verein/einen Orden ~:** join a club/enter a religious order; **c)** *(Raumfahrt)* enter; **2.** *unr. tr. V.* kick in ⟨*door, window, etc.*⟩

ein|trichtern *tr. V. (salopp)* jmdm. etw. ~: drum sth. into sb.

Ein·tritt der **a)** entry; entrance; **vor dem ~ in die Verhandlungen** *(fig.)* before entering into negotiations; **b)** *(Beitritt)* der ~ **in einen Verein/einen Orden** joining a club/entering a religious order; **c)** *(von Raketen)* entry; **d)** *(Zugang, Eintrittsgeld)* admission; **e)** *(Beginn)* onset; ~ **der Dunkelheit** nightfall

Eintritts-: ~**geld** das admission fee; ~**karte** die admission ticket; ~**preis** der admission charge

ein|trocknen *itr. V.; mit sein* dry; ⟨water, toothpaste⟩ dry up; ⟨leather⟩ dry out; ⟨berry, fruit⟩ shrivel

ein|üben *tr. V.* practise

einverstanden *Adj.* ~ **sein** agree; **mit jmdm./etw.** ~ **sein** approve of sb./sth.

Ein·verständnis das consent

ein|wachsen *unr. itr. V.; mit sein* grow into the flesh; **eingewachsen** ingrown ⟨toe-nail⟩

Einwand der; ~[e]s, **Einwände** objection **(gegen** to)

Ein·wanderer der immigrant; **ein|wandern** *itr. V.; mit sein* immigrate **(in** + *Akk.* into); **Ein·wanderung** die immigration

einwand·frei 1. *Adj.* flawless; impeccable ⟨behaviour⟩; indisputable ⟨proof⟩; **2.** *adv.* flawlessly; ⟨behave⟩ impeccably; ⟨prove⟩ beyond question

ein|wechseln *tr. V.* **a)** change ⟨money⟩; **b)** *(Sport)* substitute ⟨player⟩

ein|wecken *tr. V.* preserve; bottle

Ein·weg·flasche die non-returnable bottle

ein|weichen *tr. V.* soak

ein|weihen *tr. V.* open [officially] ⟨bridge, road⟩; dedicate ⟨monument⟩; **Einweihung** die; ~, ~**en** *s.* **einweihen:** [official] opening; dedication

ein|weisen *unr. tr. V.* **a)** *(in eine Tätigkeit)* introduce; **b)** *(in ein Amt)* install

ein|wenden *unr. (auch regelm.) tr. V.* **dagegen läßt sich vieles ~:** there is a lot to be said against that

ein|werfen *unr. tr. V.* **a)** mail ⟨letter⟩; insert ⟨coin⟩; **b)** smash ⟨window⟩; **c)** throw in ⟨ball⟩; **d)** *(bemerken, sagen)* throw in ⟨remark⟩

ein|wickeln *tr. V.* wrap [up]

ein|willigen *itr. V.* agree **(in** + *Akk.* to); **Einwilligung** die; ~, ~**en** agreement

ein|winken *tr. V. (Verkehrsw.)* guide in ⟨aircraft, car⟩

ein|wirken a) *(beeinflussen)* **auf jmdn.** ~: influence sb.; **b)** *(eine Wirkung ausüben)* have an effect **(auf** + *Akk.* on); **Ein·wirkung** die *(Einfluß)* influence; *(Wirkung)* effect

Einwohner der; ~**s**, ~, **Einwohnerin** die; ~, ~**nen** inhabitant

Ein·wurf der **a)** insertion; *(von Briefen)* mailing; **b)** *(Ballspiele)* throw-in; **c)** *(Bemerkung)* interjection

Ein·zahl die; *o. Pl.* singular

ein|zahlen *tr. V.* pay in; **Ein·zahlung** die payment

ein|zäunen *tr. V.* fence in; enclose; **Einzäunung** die; ~, ~**en** fencing-in

ein|zeichnen *tr. V.* draw *or* mark in

einzeilig *Adj.* one-line *attrib.*

Einzel das; ~**s**, ~, *(Sport)* singles *pl.*

Einzel-: ~**bett** das single bed; ~**fall** der **a)** particular case; **b)** *(Ausnahme)* isolated case; ~**gänger** [-gɛŋɐ] der; ~**s**, ~: loner; ~**haft** die solitary confinement; ~**handel** der retail trade; ~**händler** der retailer

Einzelheit die; ~, ~**en a)** detail; **b)** *(einzelner Umstand)* particular

Einzel·kind das only child

einzeln *Adj.* **a)** *(für sich allein)* individual; **b)** *(alleinstehend)* solitary ⟨building, tree⟩; single ⟨lady, gentleman⟩; **c)** ~**e** *(wenige)* a few; *(einige)* some; **d)** *substantivisch* **der/jeder** ~**e** the/each individual; ~**es** some things *pl.;* **das Einzelne** the particular

Einzel-: ~**teil** das individual part; ~**zelle** die single cell; ~**zimmer** das single room

ein|ziehen 1. *unr. tr. V.* **a)** put in; thread in ⟨tape, elastic⟩; **b)** *(einholen)* haul in ⟨net⟩; **c)** *(einatmen)* breathe in ⟨scent, fresh air⟩; inhale ⟨smoke⟩; **d)** *(einberufen)* call up ⟨recruits⟩; **e)** *(beitreiben)* collect; **2.** *unr. itr. V.; mit sein* **a)** *(liquid)* soak in; **b)** *(einkehren)* enter; **c)** *(in eine Wohnung)* move in

einzig 1. *Adj.* only; **kein** ~**es Wort** not a single word; **2.** *adv.* **a)** *intensivierend bei Adj.* extraordinarily; **b)** *(ausschließlich)* only; **das** ~ **Wahre** the only thing; **einzig·artig 1.** *Adj.* unique; **2.** *adv.* uniquely

Ein·zug der **a)** entry **(in** + *Akk.* into); **b)** *(in eine Wohnung)* move; **Einzugs·bereich** der catchment area

Eis das; ~**es a)** ice; **b)** *(Speise~)* ice-cream; **ein** ~ **am Stiel** an ice-lolly *(Brit.)* or *(Amer.)* ice pop

Eis-: ~**bahn** die ice-rink; ~**bär** der polar bear; ~**becher** der ice-cream sundae; ~**bein** das *(Kochk.)* knuckle of pork; ~**berg** der iceberg; ~**beutel** der ice-bag; ~**café** das ice-cream parlour

Ei·schnee der stiffly beaten egg-white

Eisen das; ~s, ~: iron

Eisen·bahn die a) railway; railroad *(Amer.)*; mit der ~ **fahren** go by train; b) *(Bahnstrecke)* railway line; railroad track *(Amer.)*; **Eisenbahner** der; ~s, ~: railwayman; railroader *(Amer.)*; **Eisenbahn·unglück** das train crash

Eisen-: ~**erz** das iron ore; ~**kette** die iron chain; ~**ring** der iron ring; ~**stange** die iron bar; ~**waren** *Pl.* ironmongery *sing.;* ~**zeit** die Iron Age

eisern 1. *Adj. (auch fig.)* iron; 2. *adv.* resolutely; *⟨save, train⟩* with iron determination; ~ **durchgreifen** take drastic measures

eis-, Eis-: ~**fach** das freezing compartment; ~**frei** *Adj.* ice-free; ~**gekühlt** *Adj.* iced; ~**glatt** *Adj.* icy; ~**glätte** die black ice; ~**hockey** das ice hockey

eisig 1. *Adj.* a) icy *⟨wind, cold⟩*; icy [cold] *⟨water⟩*; b) *(fig.)* frosty; 2. *adv.* a) ~ **kalt** sein be icy cold; b) *(fig.)* *⟨smile⟩* frostily; **eisig·kalt** *Adj. s.* eiskalt 1 a

eis-, Eis-: ~**kaffee** der iced coffee; ~**kalt** 1. *Adj.* a) ice-cold *⟨drink⟩*; freezing cold *⟨weather⟩*; b) *(gefühllos)* icy; ice-cold *⟨look⟩*; 2. *adv.* es lief mir ~**kalt** über den Rücken a cold shiver went down my spine; ~**kunstlauf** der figure skating; ~**kunst·läufer** der figure skater; ~**lauf** der ice-skating; ~**|laufen** *unr. itr. V.; mit sein* ice-skate; ~**laufen** das ice-skating; ~**läufer** der ice-skater; ~**schrank** der refrigerator; ~**sport** der ice sports *pl.;* ~**tanz** der *(Sport)* ice-dancing; ~**waffel** die [ice-cream] wafer; ~**wein** der *wine made from grapes frozen on the vine;* ~**würfel** der ice cube; ~**zapfen** der icicle; ~**zeit** die ice age

eitel *Adj.* vain; **Eitelkeit** die; ~, ~en vanity

Eiter der; ~s pus; **eitern** *itr. V.* suppurate; **eitrig** *Adj.* suppurating

Ei·weiß das a) egg-white; b) *(Protein)* protein

¹**Ekel** der; ~s revulsion; |einen| ~ vor etw. *(Dat.)* **haben** have a revulsion for

sth.; ²**Ekel** das; ~s, ~ *(ugs. abwertend)* horror; **er ist ein |altes|** ~: he is quite obnoxious; **ekelhaft** *Adj.* revolting *⟨sight⟩*; horrible *⟨weather, person⟩* **ekeln** 1. *refl. V.* be disgusted; sich vor etw. *(Dat.)* ~: find sth. repulsive; 2. *tr., itr. V. (unpers.)* es ekelt mich *od.* mir ekelt davor I find it revolting; **eklig** *Adj.* a) *s.* ekelhaft; b) *(ugs.: gemein)* nasty

Ekstase [εkˈstaːzə] die; ~, ~n ecstasy

Ekzem das; ~s, ~e *(Med.)* eczema

Elan der; ~s zest; vigour

elastisch *Adj.* elasticated *⟨material⟩*; springy *⟨surface⟩*; supple *⟨person, body⟩*; **Elastizität** die; ~: elasticity; *(Federkraft)* springiness; *(Geschmeidigkeit)* suppleness

Elch der; ~|e|s, ~e elk; *(in Nordamerika)* moose

Elefant der; ~en, ~en elephant

elegant 1. *Adj.* elegant; 2. *adv.* elegantly; **Eleganz** die; ~: elegance

elektrifizieren *tr. V.* electrify; **Elektrifizierung** die; ~, ~en electrification

Elektriker der; ~s, ~: electrician; **elektrisch** 1. *Adj.* electric; electrical *⟨resistance, wiring, system⟩*; 2. *adv.* ~ **kochen** cook with electricity; ~ **geladen** sein be electrically charged; **elektrisieren** 1. *tr. V. (Med.)* treat using electricity; 2. *refl. V.* get an electric shock; **Elektrizität** die; ~ electricity

Elektrizitäts·werk das power station

Elektro-: ~**artikel** der electrical appliance; ~**auto** das electric car; ~**gerät** das electrical appliance; ~**geschäft** das electrical shop *or (Amer.)* store; ~**herd** der electric cooker; ~**mobil** das electric car; ~**motor** der electric motor

Elektron das; ~s, ~en [-ˈtroːnən] electron

Elektronen-: ~**[ge]hirn** das *(ugs.)* electronic brain *(coll.);* ~**rechner** der electronic computer

Elektronik die; ~ a) electronics *sing., no art.;* b) *(Teile)* electronics *pl.;* **elektronisch** 1. *Adj.* electronic; 2. *adv.* electronically

Elektro-: ~**rasierer** der electric shaver; ~**technik** die electrical engineering *no art.;* ~**techniker** der a) electronics engineer; b) *(Elektriker)* electrician

Element das; ~|e|s, ~e element; ele-

mentar *Adj.* **a)** *(grundlegend)* fundamental; **b)** *(einfach)* elementary ⟨*knowledge*⟩; **c)** *(naturhaft)* elemental ⟨*force*⟩; **Elementar·teilchen das** *(Physik)* elementary particle

elend *Adj.* wretched; miserable; **Elend das**; ~s misery

Elends-: ~**quartier das** slum [dwelling]; ~**viertel das** slum area

elf *Kardinalz.* eleven; **Elf die**; ~, ~en **a)** eleven; **b)** *(Sport)* team; side

Elfe die; ~, ~n fairy

Elfen·bein das ivory

Elfenbein-: ~**schnitzerei die** *o. Pl.* ivory-carving; ~**turm der** *(fig.)* ivory tower

Elf·meter der *(Fußball)* penalty; **einen** ~ **schießen** take a penalty; **Elfmeter·schießen das** *(Fußball)* **durch** ~**schießen** by *or* on penalties

eliminieren *tr. V.* eliminate

Elite die; ~, ~n élite

Ell·bogen der; ~s, ~: elbow

Elle die; ~, ~n **a)** *(Anat.)* ulna; **b)** *(frühere Längeneinheit)* cubit; **c)** *(veralt.: Maßstock)* ≈ yardstick; **Ellen·bogen** *s.* Ellbogen

Ellipse die; ~, ~n ellipse

Elsaß das; ~ *od.* **Elsasses** Alsace

Elster die; ~, ~n magpie

elterlich *Adj.* parental; **Eltern** *Pl.* parents *pl.*

eltern-, **Eltern-:** ~**abend der** *(Schulw.)* parents' evening; ~**haus das** home; ~**los** *Adj.* orphaned; ~**teil der** parent

Email [e'maj] **das**; ~s, ~s, **Emaille** [e'maljə] **die**; ~, ~n enamel

Emanzipation die; ~, ~en emancipation; **emanzipieren** *refl. V.* emancipate; **emanzipiert** *Adj.* emancipated

Embargo das; ~s, ~s embargo

Emblem das; ~s, ~e emblem

Embryo der; ~s, ~nen [-'yo:nən] *od.* ~s embryo

Emigrant der; ~en, ~en emigrant; *(Flüchtling)* emigré; **Emigration die**; ~, ~en *(das Emigrieren)* emigration; **emigrieren** *itr. V.; mit sein* emigrate

Emotion die; ~, ~en emotion; **emotional 1.** *Adj.* emotional; emotive ⟨*topic, question*⟩; **2.** *adv.* emotionally

Empfang der; ~[e]s, **Empfänge** reception; *(Entgegennahme)* receipt; **empfangen** *unr. tr. V.* receive; **Empfänger der**; ~s, ~ **a)** recipient; *(eines Briefs)* addressee; **b)** *(Empfangsgerät)* receiver

empfänglich *Adj.* **a)** receptive (für to); **b)** *(beeinflußbar)* susceptible; **Empfängnis die**; ~: conception; **Empfängnis·verhütung die** contraception

empfangs-, **Empfangs-:** ~**berechtigt** *Adj.* authorized to receive payment/goods *postpos.;* ~**chef der** head receptionist; ~**dame die** receptionist; ~**halle die** reception lobby

empfehlen 1. *unr. tr. V.* recommend; **2.** *unr. refl. V.* **a)** take one's leave; **b)** *unpers.* es empfiehlt sich, ... zu ...: it's advisable to ...; **empfehlens·wert** *Adj.* **a)** to be recommended *postpos.;* recommendable; **b)** *(ratsam)* advisable; **Empfehlung die**; ~, ~en **a)** recommendation; **b)** *(Empfehlungsschreiben)* letter of recommendation; **empfiehl** *Imperativ Sg. v.* empfehlen; **empfiehlst** *2. Pers. Sg. Präsens v.* empfehlen; **empfiehlt** *3. Pers. Sg. Präsens v.* empfehlen

empfinden *unr. tr. V.* **a)** *(wahrnehmen)* feel; **b)** *(auffassen)* etw. als Beleidigung ~: feel sth. to be an insult; **Empfinden das**; ~s feeling; **für mein** *od.* **nach meinem** ~: to my mind; **empfindlich 1.** *Adj.* **a)** sensitive; fast ⟨*film*⟩; **b)** *(leicht beleidigt)* sensitive; **c)** *(anfällig)* zart und ~: delicate; **d)** *(spürbar)* severe ⟨*punishment, shortage*⟩; **2.** *adv.* ~ auf etw. *(Akk.)* reagieren *(sensibel)* be susceptible to sth.; *(beleidigt)* react oversensitively to sth.; **Empfindlichkeit die**; ~, ~en *s.* empfindlich: sensitivity; severity; *(eines Films)* speed

empfindsam *Adj.* sensitive ⟨*nature*⟩; **Empfindung die**; ~, ~en *(Gefühl)* feeling

empfing *1. u. 3. Pers. Sg. Prät. v.* empfangen

empfohlen 1. *2. Part. v.* empfehlen; **2.** *Adj.* recommended

empirisch 1. *Adj.* empirical; **2.** *adv.* empirically

empor *Adv. (geh.)* upwards

Empore die; ~, ~n gallery

empören 1. *tr. V.* fill with indignation; outrage; **2.** *refl. V.* become indignant *or* outraged; **empörend** *Adj.* outrageous; **empört** *Adj.* outraged

emsig 1. *Adj.* industrious ⟨*person*⟩; bustling ⟨*activity*⟩; **2.** *adv.* industriously

Emu der; ~s, ~s *(Zool.)* emu

Ende das; ~s, ~n end; **am** ~ **der Stra-**

ße/Stadt at the end of the road/town; am/bis/gegen ~ des Monats at/by/towards the end of the month; ~ April at the end of April; zu ~ sein ⟨patience, war⟩ be at an end; ⟨school⟩ be over; ⟨film, game⟩ have finished; ~ gut, alles gut all's well that ends well (prov.); **enden** itr. V. a) end; ⟨programme⟩ finish; b) in der Gosse ~: end up in the gutter; (dort sterben) die in the gutter

end·gültig 1. Adj. final ⟨consent, decision⟩; conclusive ⟨evidence⟩; **2.** adv. das ist ~ vorbei that's all over and done with; sich ~ trennen separate for good

End-: ~kampf der (Sport) final; (Milit.) final battle; ~lauf der (Sport) final

endlich 1. Adv. a) (nach langer Zeit) at last; b) (schließlich) in the end; **2.** Adj. finite

end-, End-: ~los **1.** Adj. a) (ohne Ende) infinite; (ringförmig) continuous; b) (nicht enden wollend) endless; interminable ⟨speech⟩; **2.** adv. ~los lange dauern be interminably long; ~runde die (Sport) final; ~spiel das (Sport) final; ~spurt der (bes. Leichtathletik) final spurt; ~stadium das final stage; (Med.) terminal stage; ~station die terminus

Endung die; ~, ~en (Sprachw.) ending
Energie die; ~, ~n energy

Energie-: ~politik die energy policy; ~quelle die energy source; ~versorgung die energy supply

energisch 1. Adj. a) energetic ⟨person⟩; firm ⟨action⟩; b) forceful ⟨voice, words⟩; **2.** adv. a) energetically; ~ durchgreifen take drastic action; b) ⟨reject, say⟩ forcefully; ⟨stress⟩ emphatically; ⟨deny⟩ strenuously

eng [ɛŋ] **1.** Adj. a) (schmal) narrow b) (dicht) close ⟨writing⟩; c) (fest anliegend) close-fitting; d) (beschränkt) narrow; e) (nahe) close ⟨friend⟩; **2.** adv. a) (dicht) ~ ⌊zusammen⌋ sitzen/ stehen sit/stand close together; b) (fest anliegend) ~ anliegen/sitzen fit closely; c) (beschränkt) etw. zu ~ auslegen interpret sth. too narrowly; d) (nahe) closely; **Enge** die; ~, ~n confinement

Engel der; ~s, ~: angel
eng·herzig Adj. petty

England (das); ~s England; **Engländer** der; ~s, ~: Englishman/English boy; er ist ~: he is English; die ~: the

English; **Engländerin** die; ~, ~nen Englishwoman/English girl; **englisch 1.** Adj. English; **die** ~e Sprache/ Literatur the English language/English literature; **2.** adv. ~ sprechen speak English; **Englisch das;** ~⌊s⌋ English

englisch-, Englisch-: ~lehrer der 'English teacher; ~sprachig Adj. a) English-language ⟨book, magazine⟩; b) (~ sprechend) English-speaking ⟨population, country⟩; ~unterricht der English teaching; (Unterrichtsstunde) English lesson

Eng·paß der a) defile; b) (fig.) bottleneck; **eng·stirnig** Adj. narrowminded

Enkel der; ~s, ~: grandson; **Enkelin** die; ~, ~nen granddaughter; **Enkel·kind** das grandchild

enorm 1. Adj. enormous ⟨sum, costs⟩; tremendous (coll.) ⟨effort⟩; immense ⟨strain⟩; **2.** adv. tremendously (coll.)

Ensemble [ã'sã:b|] das ensemble; (Theater~) company

entarten itr. V.; mit sein degenerate

entbehren tr. V. (verzichten auf) do without; **entbehrlich** Adj. dispensable; **Entbehrung** die; ~, ~en privation

entbinden 1. unr. tr. V. a) jmdn. von einem Versprechen ~: release sb. from a promise; seines Amtes od. von seinem Amt entbunden werden be relieved of [one's] office; b) jmdn. ~ (Med.) deliver sb.'s baby; **2.** unr. itr. V. give birth; **Entbindung** die (Med.) delivery

entblößen 1. refl. V. take one's clothes off; ⟨exhibitionist⟩ expose oneself; **2.** tr. V. uncover ⟨one's arm etc.⟩

entdecken tr. V. a) discover; b) (ausfindig machen) jmdn. ~: find sb.; etw. ~: find or discover sth.; **Entdecker** der; ~s, ~: discoverer; **Entdeckung** die; ~, ~en discovery

Ente die; ~, ~n duck

entehren tr. V. dishonour; ~d degrading

enteignen tr. V. expropriate; **Enteignung** die expropriation

enterben tr. V. disinherit

entern tr., itr. V. board ⟨ship⟩

entfachen tr. V. (geh.) a) kindle, light ⟨fire⟩; b) (fig.) provoke ⟨quarrel, argument⟩; arouse ⟨passion, enthusiasm⟩

entfallen unr. itr. V.; mit sein a) (aus dem Gedächtnis) es ist mir ~: it escapes me; **b)** *(zugeteilt werden)* **auf jmdn./etw.** ~ : be allotted to sb./sth.; **c)** *(wegfallen)* lapse

entfalten 1. *tr. V.* **a)** open [up]; unfold ⟨*map etc.*⟩; **b)** *(fig.)* display ⟨*ability, talent*⟩; **2.** *refl. V.* **a)** open [up]; **b)** *(fig.)* ⟨*personality, talent, etc.*⟩ develop; **Entfaltung die;** ~, ~**en** *(fig.)* **a)** *(Entwicklung)* development; **b)** *s.* **entfalten 1 b:** display

entfernen 1. *tr. V.* remove; take out ⟨*tonsils etc.*⟩; **2.** *refl. V.* go away; **entfernt 1.** *Adj.* **a)** *(fern)* remote; **das ist od. liegt weit** ~ **von der Stadt** it is a long way from the town; **10 km/zwei Stunden** ~ **:** 10 km/two hours away; **b)** slight ⟨*acquaintance*⟩; distant ⟨*relation*⟩; slight ⟨*resemblance*⟩; **2.** *adv.* **a)** *(fern)* remotely; **b)** slightly ⟨*acquainted*⟩; distantly ⟨*related*⟩; **Entfernung die;** ~, ~**en a)** *(Abstand)* distance; **b)** *(das Beseitigen)* removal

entfesseln *tr. V.* unleash

entflammen 1. *tr. V.* arouse ⟨*enthusiasm etc*⟩; **2.** *itr. V.;* **mit sein** flare up

entfliehen *unr. itr. V.;* **mit sein** escape; **jmdm.** ~ **:** escape from sb.

entfremden 1. *tr. V.* **a)** **etw. seinem Zweck** ~ **:** use sth. for a different purpose; **b)** *(Philos., Soziol.)* **entfremdet** alienated; **2.** *refl. V.* **sich jmdm./einer Sache** ~ **:** become estranged from sb./ unfamiliar with sth.; **Entfremdung die;** ~, ~**en** alienation; estrangement

entführen *tr. V.* kidnap ⟨*child etc.*⟩; hijack ⟨*plane, lorry, etc.*⟩; **Entführer der** *s.* **entführen:** kidnapper; hijacker; **Entführung die** *s.* **entführen:** kidnapping; hijacking

entgegen 1. *Adv.* towards; **2.** *Präp. mit Dat.* ~ **meinem Wunsch** against my wishes; ~ **dem Befehl** contrary to orders

entgegen-, Entgegen-: ~|**bringen** *unr. tr. V. (fig.)* show ⟨*love, understanding*⟩; ~|**fahren** *unr. itr. V.;* **mit sein jmdm.** ~**fahren** come/go to meet sb.; ~|**gehen** *unr. itr. V.;* **mit sein a) jmdm.** ~**gehen** go to meet sb.; **b)** *(fig.)* **be heading for** ⟨*catastrophe, hard times*⟩; ~**gesetzt 1.** *Adj.* **a)** *(umgekehrt)* opposite ⟨*end, direction*⟩; **b)** *(gegensätzlich)* opposing; **2.** *adv.* **genau** ~**gesetzt handeln/denken** do/think exactly the opposite; ~|**kommen** *unr. itr. V.;* **mit sein jmdm.** ~**kommen** come to meet sb.; *(Zugeständnisse machen)* be accommodating towards sb.; ~**kommen das** co-operation;

(Zugeständnis) concession; ~**kommend** *Adj.* obliging; ~|**nehmen** *unr. itr. V.* receive; ~|**treten** *unr. itr. V.;* **mit sein** go/come up to; *(fig.)* stand up to ⟨*difficulties*⟩

entgegnen *tr. V.* retort; reply

entgehen *unr. itr. V.;* **mit sein a)** *(entkommen)* escape; **b) jmdm. entgeht etw.** sb. misses sth.

entgeistert *Adj.* dumbfounded

Entgelt das; ~**[e]s,** ~**e** payment; fee

entgiften *tr. V.* decontaminate ⟨*substance etc.*⟩; detoxicate ⟨*body etc.*⟩

entgleisen *itr. V.;* **mit sein a)** be derailed; **b)** *(fig.)* make a/some faux pas

entgräten *tr. V.* fillet

enthaaren *tr. V.* remove hair from; **Enthaarungs·mittel das** hair remover

¹**enthalten 1.** *unr. tr. V.* contain; **2.** *unr. refl. V.* **sich einer Sache** *(Gen.)* ~ **:** abstain from sth.; **sich der Stimme** ~ **:** abstain; ²**enthalten** *Adj.* **in etw.** *(Dat.)* ~ **sein** be contained in sth.; **das ist im Preis** ~ **:** that is included in the price; **enthaltsam 1.** *Adj.* abstemious; *(sexuell)* abstinent; **2.** *adv.* ~ **leben** live in abstinence; **Enthaltsamkeit die;** ~ **:** abstinence; **Enthaltung die** abstention

enthaupten *tr. V. (geh.)* behead

enthäuten *tr. V.* skin

entheben *unr. tr. V. (geh.)* relieve

enthemmt *Adj.* uninhibited

enthüllen *tr. V.* unveil ⟨*monument etc.*⟩; reveal ⟨*face, truth, secret*⟩; **Enthüllung die;** ~, ~**en** *s.* **enthüllen:** unveiling; revelation

Enthusiasmus [ɛntu'ziasmʊs] **der;** ~ **:** enthusiasm; **enthusiastisch 1.** *Adj.* enthusiastic; **2.** *adv.* enthusiastically

entkalken *tr. V.* decalcify

entkleiden *tr. V. (geh.)* **a)** undress; **b)** *(berauben)* strip

entkommen *unr. itr. V.;* **mit sein** escape

entkorken *tr. V.* uncork ⟨*bottle*⟩

entkräften *tr. V.* **a)** weaken; **völlig** ~ **:** exhaust; **b)** *(fig.)* refute ⟨*argument etc.*⟩; **Entkräftung die;** ~, ~**en** debility; **völlige** ~ **:** exhaustion; **b)** *(fig.)* refutation

entladen 1. *unr. tr. V.* unload; **2.** *unr. refl. V.* **a)** ⟨*storm*⟩ break; **b)** *(fig.)* ⟨*anger etc.*⟩ erupt; ⟨*aggression etc.*⟩ be released

entlang 1. *Präp. mit Akk. u. Dat.* along; **2.** *Adv.* along; **hier/dort** ~, **bitte! this/that way please!**

entlang-: ~|fahren *unr. itr. V.; mit sein* a) drive along; b) *(streichen)* go along; ~|gehen *unr. itr. V.; mit sein* ⟨*person*⟩ go *or* walk along; ~|laufen *unr. itr. V.; mit sein* a) walk/run along; b) *(verlaufen)* go *or* run along

entlarven *tr. V.* expose

entlassen *unr. tr. V.* a) *(aus dem Gefängnis)* release; *(aus dem Krankenhaus, der Armee)* discharge; b) *(aus einem Arbeitsverhältnis)* dismiss; *(wegen Arbeitsmangels)* make redundant *(Brit.)*; lay off; Entlassung die; ~, ~en *s.* entlassen: release; discharge; dismissal; redundancy *(Brit.);* laying off

entlasten *tr. V.* a) relieve; b) *(Rechtsw.)* exonerate ⟨*defendant*⟩; Entlastung die; ~, ~en a) relief; b) *(Rechtsw.)* exoneration; defence

entlaufen *unr. itr. V.; mit sein* run away; ein ~er Sträfling/Sklave an escaped convict/a runaway slave

entlausen *tr. V.* delouse

entledigen *refl. V.* sich jmds./einer Sache *(Gen.)* ~ *(geh.)* rid oneself of sb./sth.

entleeren *tr. V.* empty; evacuate ⟨*bowels, bladder*⟩

entlegen *Adj.* remote

entleihen *unr. tr. V.* borrow

entlocken *tr. V. (geh.)* jmdm. etw. ~: elicit sth. from sb.

entlohnen *tr.V.* pay; Entlohnung die; ~, ~en payment; *(Lohn)* pay

entlüften *tr. V.* ventilate; Entlüfter der; ~s, ~: ventilator

entmachten *tr. V.* deprive of power

entmilitarisieren *tr. V.* demilitarize

entmündigen *tr. V.* incapacitate; Entmündigung die; ~, ~en incapacitation

entmutigen *tr. V.* discourage

Entnahme die; ~, ~n *(von Wasser)* drawing; *(von Blut)* extraction

entnehmen *unr. tr. V.* a) etw. [einer Sache *(Dat.)*] ~: take sth. [from sth.]; b) *(ersehen aus)* gather *(Dat.* from)

entnervend *Adj.* nerve-racking

entpuppen *refl. V.* sich als etw./jmd. ~: turn out to be sth./sb.

entrahmen *tr. V.* skim ⟨*milk*⟩

entreißen *unr. tr. V.* jmdm. etw. ~: snatch sth. from sb.

entrichten *tr. V. (Amtsspr.)* pay ⟨*fee*⟩

entrümpeln *tr. V.* clear out; Entrümpelung die; ~, ~en clear-out

entrüsten 1. *refl. V.* sich [über etw. *(Akk.)*] ~: be indignant [at *or* about

sth.]; 2. *tr. V. (empören)* jmdn. ~: make sb. indignant; Entrüstung die indignation *(über + Akk.* at, about)

Entsafter der; ~s, ~: juice-extractor

entsagen *itr. V.* einer Sache *(Dat.)* ~ *(geh.)* renounce sth.; Entsagung die; ~, ~en *(geh.)* renunciation

entschädigen *tr. V.* compensate (für for); jmdn. für etw. ~ *(fig.)* make up for sth.; Entschädigung die compensation

entschärfen *tr. V.* defuse; tone down ⟨*discussion, criticism*⟩

entscheiden 1. *unr. refl. V.* a) decide; b) *(unpers.)* morgen entscheidet es sich, ob ...: I/we/you will know tomorrow whether ...; 2. *unr. itr. V.* über etw. *(Akk.)* ~: settle sth; 3. *unr. tr. V.* decide on ⟨*dispute*⟩; decide ⟨*outcome, result*⟩; entscheidend 1. *Adj.* crucial; decisive ⟨*action*⟩; 2. *adv.* jmdn./etw. ~ beeinflussen have a decisive influence on sb./sth.; Entscheidung die decision

entschieden 1. *Adj.* a) *(entschlossen)* determined; resolute; b) *(eindeutig)* definite; 2. *adv.* resolutely; das geht ~ zu weit that is going much too far

entschlafen *unr. itr. V.; mit sein* pass away

entschließen *unr. refl. V.* decide; Entschließung die resolution; entschlossen *Adj.* determined; Entschlossenheit die; ~: determination; Entschluß der decision

entschlüsseln *tr. V.* decipher

entschuldigen 1. *refl. V.* apologize; 2. *tr. (auch itr.) V.* excuse ⟨*person*⟩; sich ~ lassen ask to be excused; ~ Sie [bitte]! *(bei Fragen, Bitten)* excuse me; *(bedauernd)* I'm sorry; Entschuldigung die; ~, ~en a) apology; b) *(Grund)* excuse; c) *(Höflichkeitsformel)* ~! *(bei Fragen, Bitten)* excuse me; *(bedauernd)* [I'm] sorry

entschwinden *unr. itr. V.; mit sein (geh.)* disappear; vanish

entsetzen 1. *refl. V.* be horrified; 2. *tr. V.* horrify; Entsetzen das; ~s horror; entsetzlich 1. *Adj.* a) horrible ⟨*accident, crime, etc.*⟩; b) *nicht präd. (ugs.: stark)* terrible ⟨*thirst, hunger*⟩; 2. *adv.* terribly *(coll.)*

entsinnen *unr. refl. V.* sich jmds./einer Sache ~: remember sb./sth.

entspannen 1. *tr. V.* relax; 2. *refl. V.* a) ⟨*person*⟩ relax; b) *(fig.)* ⟨*situation, tension*⟩ ease; Entspannung die; o. Pl. a) relaxation; b) *(politisch)* easing

of tension; détente; **Entspannungs·politik die** policy of détente
entsprechen *unr. itr. V.* **a)** *(übereinstimmen mit)* **einer Sache** *(Dat.)* ~: correspond to sth.; **b)** *(nachkommen)* **einem Wunsch** ~: comply with a request; **den Anforderungen** ~: meet the requirements; **entsprechend 1.** *Adj.* **a)** corresponding; *(angemessen)* appropriate; **b)** *nicht attr. (dem~)* in accordance *postpos.;* **2.** *adv.* **a)** *(angemessen)* appropriately; **b)** *(dem~)* accordingly; **3.** *Präp. mit Dativ:* ~ **einer Sache** in accordance with sth.
entspringen *unr. itr. V.; mit sein* **a)** ⟨*river*⟩ rise; **b)** *(entstehen aus)* **einer Sache** *(Dat.)* ~: spring from sth.
entstehen *unr. itr. V.; mit sein* **a)** originate; ⟨*quarrel, friendship, etc.*⟩ arise; **b)** *(gebildet werden)* be formed **(aus** from, **durch** by); **c)** *(sich ergeben)* occur; *(als Folge)* result; **Entstehung die;** ~: origin
entsteinen *tr. V.* stone
entstellen *tr. V.* **a)** disfigure; **b)** *(verfälschen)* distort ⟨*text, facts*⟩; **Entstellung die** **a)** disfigurement; **b)** *(Verfälschung)* distortion
entstören *tr. V. (Elektrot.)* suppress ⟨*engine, electrical appliance*⟩
enttarnen *tr. V.* uncover
enttäuschen *tr. V.* disappoint; **enttäuscht** *Adj.* disappointed; dashed ⟨*hopes*⟩; **Enttäuschung die** disappointment
entwachsen *unr. itr. V.; mit sein* **einer Sache** *(Dat.)* ~: grow out of sth.
entwaffnen *tr. V. (auch fig.)* disarm; **entwaffnend** *Adj.* disarming
entwarnen *itr. V.* sound the all-clear; **Entwarnung die** all-clear
entwässern *tr. V.* drain; **Entwässerung die;** ~, ~en drainage
entweder *Konj.:* ~ ... **oder** either ... or
entweichen *unr. itr. V.; mit sein* escape
entwenden *tr. V. (geh.)* purloin
entwerfen *unr. tr. V.* design ⟨*furniture, dress*⟩; draft ⟨*novel etc.*⟩; draw up ⟨*plans etc.*⟩
entwerten *tr. V.* **a)** cancel ⟨*ticket, postage stamp*⟩; **b)** devalue ⟨*currency*⟩
entwickeln 1. *refl. V.* develop; **2.** *tr. V.* produce ⟨*vapour, smell*⟩; display ⟨*ability, characteristic*⟩; develop ⟨*equipment, photograph, film*⟩; elaborate ⟨*theory, ideas*⟩; **Entwicklung die;** ~, ~en **a)** development; *(von Dämpfen usw.)* production; **in der** ~

sein ⟨*young person*⟩ be adolescent; **b)** *(Darlegung)* elaboration; **c)** *(Fot.)* developing
Entwicklungs-: ~**helfer der** development aid worker; ~**hilfe die** [development] aid; ~**land das;** *Pl.* ~**länder** developing country; ~**politik die** development aid policy
entwirren *tr. V.* disentangle
entwischen *itr. V.; mit sein (ugs.)* get away
entwöhnen *tr. V.* wean
entwürdigend *Adj.* degrading
Entwurf der **a)** design; **b)** *(Konzept)* draft
entwurzeln *tr. V.* uproot
entziehen 1. *unr. tr. V.* **a)** take away; **b)** *(nicht zugestehen)* withdraw; **2.** *unr. refl. V.* **sich seinen Pflichten** *(Dat.)* ~: evade one's duty; **das entzieht sich meiner Kontrolle** that is beyond my control; **Entziehung die** **a)** withdrawal; **b)** *(Entziehungskur)* withdrawal treatment *no indef. art.*
entziffern *tr. V.* decipher
entzückend *Adj.* delightful; **entzückt** *Adj.* delighted
Entzug der; ~[e]s withdrawal
entzündbar *Adj.* [in]flammable; **entzünden 1.** *tr. V.* light ⟨*fire*⟩; strike ⟨*match*⟩; **2.** *refl. V.* **a)** ignite; **b)** *(anschwellen)* become inflamed; **entzündlich** *Adj.* **a)** [in]flammable ⟨*substance*⟩; **b)** *(Med.)* inflammatory; **Entzündung die;** ~, ~en inflammation
entzwei *Adj. (geh.)* in pieces; **entzweien** *refl. V.* fall out; **entzwei**|**gehen** *unr. itr. V.; mit sein (geh.)* break
Enzian ['ɛntsi̯aːn] **der;** ~s, ~e gentian
Enzyklika die; ~, **Enzykliken** encyclical
Enzyklopädie die; ~, ~n encyclopaedia; **enzyklopädisch** *Adj.* encyclopaedic
Epen *s.* **Epos**
Epidemie die; ~, ~n epidemic
Epilepsie die; ~, ~n *(Med.)* epilepsy *no art.;* **Epileptiker der;** ~s, ~: epileptic; **epileptisch** *Adj.* epileptic
episch *Adj.* epic
Episode die; ~, ~n episode
Epoche die; ~, ~n epoch
Epos ['eːpɔs] **das;** ~, **Epen** epic [poem]; epos
er *Personalpron. 3. Pers. Sg. Nom. Mask.* he; *(betont)* him; *(bei Dingen/Tieren)* it; *s. auch* **ihm; ihn; seiner**

erachten *tr. V. (geh.)* consider; **etw. als** *od.* **für seine Pflicht** ~: consider sth. [to be] one's duty

erarbeiten *tr. V.* work for

Erb·anlage die hereditary disposition

erbarmen *refl. V. (geh.)* take pity (*Gen.* on); **Erbarmen das;** ~s pity; **erbärmlich 1.** *Adj.* **a)** *(elend)* wretched; **b)** *(unzulänglich)* pathetic; **c)** *(abwertend: gemein)* mean; wretched; **d)** *(sehr groß)* terrible 〈*hunger, fear, etc.*〉; **2.** *adv.* terribly

erbauen 1. *tr. V.* **a)** build; **b)** *(geh.: erheben)* uplift; **2.** *refl. V.* **sich an etw.** *(Dat.)* ~: be uplifted by sth.; **Erbauer der;** ~s, ~: architect

¹Erbe das; ~s **a)** inheritance; **b)** *(Vermächtnis)* legacy; **²Erbe der;** ~n ~n heir; **erben** *tr. (auch itr.) V.* inherit

erbetteln *tr. V.* get by begging

erbeuten *tr. V.* carry off, get away with 〈*valuables, prey, etc.*〉; capture 〈*enemy plane, tank, etc.*〉

Erb-: ~**folge die** succession; ~**gut das** *(Biol.)* genetic make-up

Erbin die; ~, ~**nen** heiress

erbitten *unr. tr. V. (geh.)* request

erbittern *tr. V.* enrage; **erbittert 1.** *Adj.* bitter; **2.** *adv.* ~ **kämpfen** wage a bitter struggle

erblassen *itr. V.; mit sein (geh.)* turn pale; blanch *(literary)*

erbleichen *itr. V.; mit sein (geh.) s.* **erblassen**

erblich *Adj.* hereditary 〈*title, disease*〉

erblicken *tr. V. (geh.)* catch sight of; *(fig.)* see

erblinden *itr. V.; mit sein* lose one's sight

erblühen *itr. V.; mit sein (geh.)* bloom; blossom

Erb·masse die *(Biol.)* genetic make-up

erbost *Adj.* furious

erbrechen 1. *unr. tr. V.* bring up 〈*food*〉; **2.** *unr. itr., refl. V.* vomit; **Erbrechen das;** ~s vomiting

erbringen *unr. tr. V.* produce

Erbschaft die; ~, ~**en** inheritance; **Erbschaft[s]·steuer die** estate *or* death duties *pl.*

Erbse die; ~, ~**n** pea

Erb-: ~**stück das** heirloom; ~**sünde die** original sin; ~**teil das** share of an/the inheritance

Erd-: ~**achse die** earth's axis; ~**apfel der** *(bes. österr.) s.* Kartoffel; ~**beben das** earthquake; ~**beere die** strawberry; ~**boden der** ground; earth;

etw. dem ~**boden gleichmachen** raze sth. to the ground

Erde die; ~, ~**n a)** *(Erdreich)* soil; earth; **b)** *o. Pl. (fester Boden)* ground; **c)** *o. Pl. (Welt)* earth; world; **d)** *o. Pl. (Planet)* Earth

erdenklich *Adj.* conceivable

Erd-: ~**gas das** natural gas; ~**geschoß das** ground floor; first floor *(Amer.);* ~**kunde die** geography; ~**nuß die** peanut; ~**oberfläche die** earth's surface; ~**öl das** oil

erdöl-, Erdöl-: ~**exportierend** *Adj.* oil-exporting 〈*country*〉; ~**gewinnung die** oil production; ~**leitung die** oil pipeline

erdrosseln *tr. V.* strangle

erdrücken *tr. V.* **a)** crush; **b)** *(fig.: belasten)* overwhelm; **erdrückend** *Adj.* overwhelming; oppressive 〈*heat, silence*〉

Erd-: ~**rutsch der** landslide; ~**teil der** continent

erdulden *tr. V.* endure 〈*sorrow, misfortune*〉; tolerate 〈*insults*〉; *(über sich ergehen lassen)* undergo

ereifern *refl. V.* get excited

ereignen *refl. V.* happen; 〈*accident, mishap*〉 occur; **Ereignis das;** ~**ses,** ~**se** event; occurrence

Eremit der; ~**en,** ~**en** hermit

ererbt *Adj.* inherited

¹erfahren *unr. tr. V.* **a)** find out; learn; *(hören)* hear; **b)** *(geh.: erleben)* experience; *(erleiden)* suffer; **²erfahren** *Adj.* experienced; **Erfahrung die;** ~, ~**en** experience; ~**en sammeln** gain experience *sing.;* **etw. in** ~ **bringen** discover sth.; **erfahrungs·gemäß** *Adv.* in our/my experience

erfassen *tr. V.* **a)** *(mitreißen)* catch; **b)** *(begreifen)* grasp 〈*situation, etc.*〉; **c)** *(registrieren)* record; **Erfassung die** registration

erfinden *unr. tr. V.* invent; **das ist alles erfunden** it is pure fabrication; **Erfinder der;** ~**s,** ~ **a)** inventor; **b)** *(Urheber)* creator; **erfinderisch** *Adj.* inventive; *(schlau)* resourceful; **Erfindung die;** ~, ~**en** invention

erflehen *tr. V. (geh.)* beg

Erfolg der; ~[e]s, ~**e** success; **keinen** ~ **haben** be unsuccessful

erfolgen *itr. V.; mit sein* take place; occur; **es erfolgte keine Reaktion** there was no reaction

erfolg-: ~**los 1.** *Adj.* unsuccessful; **2.** *adv.* unsuccessfully; ~**reich 1.** *Adj.* successful; **2.** *adv.* successfully

Erfolgs·erlebnis das feeling of achievement

erfolg·versprechend *Adj.* promising

erforderlich *Adj.* required; necessary; **erfordern** *tr. V.* require; demand

erforschen *tr. V.* discover ⟨*facts, causes, etc.*⟩; explore ⟨*country*⟩; **Erforschung die** research (+ *Gen.* into); *(eines Landes usw.)* exploration

erfreuen 1. *tr. V.* please; **2.** *refl. V.* **sich an etw.** *(Dat.)* ~: take pleasure in sth.; **erfreulich** *Adj.* pleasant

erfrieren 1. *unr. itr. V.; mit sein* freeze to death; ⟨*plant, harvest, etc.*⟩ be damaged by frost; **2.** *unr. refl. V.* **sich** *(Dat.)* **die Finger** ~: get frostbite in one's fingers

erfrischen 1. *tr. (auch itr.) V.* refresh; **2.** *refl. V.* freshen oneself up; **erfrischend** *(auch fig.) Adj.* refreshing; **Erfrischung die;** ~, ~en *(auch fig.)* refreshment; **Erfrischungs·raum der** refreshment room

erfüllen 1. *tr. V.* grant ⟨*wish, request*⟩; fulfil ⟨*contract*⟩; carry out ⟨*duty*⟩; meet ⟨*condition*⟩; **2.** *refl. V.* ⟨*wish*⟩ come true; **Erfüllung die: in** ~ **gehen** come true

erfunden *Adj.* fictional ⟨*story*⟩

ergänzen *tr. V.* **a)** *(vervollständigen)* complete; *(erweitern)* add to; **b)** *(hinzufügen)* add ⟨*remark*⟩; **Ergänzung die;** ~, ~en **a)** *(Vervollständigung)* completion; *(Erweiterung)* enlargement; **b)** *(Zusatz)* addition; *(zu einem Gesetz)* amendment

ergattern *tr. V. (ugs.)* manage to grab

ergaunern *tr. V.* get by underhand means

¹ergeben 1. *unr. refl. V.* **a) sich in etw.** *(Akk.)* ~: submit to sth.; **b)** *(kapitulieren)* surrender *(Dat.* to); **c)** *(folgen, entstehen)* arise **(aus** from); **2.** *unr. tr. V.* result in; **²ergeben** *Adj.* **a)** *(zugeneigt)* devoted; **b)** *(resignierend)* **mit** ~**er Miene** with an expression of resignation; **Ergebnis das;** ~ses, ~se result; **ergebnis·los** *Adj.* fruitless

ergehen *unr. refl. V.* **sich in etw.** *(Dat.)* ~: indulge in sth.

ergiebig *Adj.* rich ⟨*deposits, resources*⟩; fertile ⟨*topic*⟩

ergötzen *(geh.)* **1.** *tr. V.* enthral; **2.** *refl. V.* **sich an etw.** *(Dat.)* ~: be delighted by sth.

ergrauen *itr. V.; mit sein* go grey

ergreifen *unr. tr. V.* **a)** *(greifen)* grab;

b) *(festnehmen)* catch ⟨*thief etc.*⟩; **c)** *(fig.: erfassen)* seize; **d)** *(fig.: aufnehmen)* take up ⟨*career*⟩; take ⟨*initiative, opportunity*⟩; **e)** *(fig.: bewegen)* move; **ergreifend** *Adj.* moving; **ergriffen** *Adj.* moved

ergründen *tr. V.* ascertain; discover ⟨*cause*⟩

erhaben *Adj.* solemn ⟨*moment*⟩; awe-inspiring ⟨*sight*⟩; sublime ⟨*beauty*⟩; **über etw.** *(Akk.)* ~ **sein** be above sth.

Erhalt der; ~[e]s *(Amtsdt.)* receipt; **erhalten** *unr. tr. V.* **a)** receive ⟨*letter, news, gift*⟩; be given ⟨*order*⟩; get ⟨*good mark, impression*⟩; **b)** *(bewahren)* preserve ⟨*town, building*⟩ **erhältlich** *Adj.* obtainable; **Erhaltung die;** ~: preservation; *(des Friedens)* maintenance

erhängen *tr. V.* hang

erhärten *tr. V.* strengthen ⟨*suspicion, assumption*⟩; substantiate ⟨*claim*⟩

erheben 1. *unr. tr. V.* **a)** raise **b)** *(verlangen)* levy ⟨*tax*⟩; charge ⟨*fee*⟩; **2.** *unr. refl. V.* **a)** rise; **b)** *(rebellieren)* rise up **(gegen** against); **erhebend** *Adj.* uplifting; **erheblich 1.** *Adj.* considerable; **2.** *adv.* considerably

erheitern *tr. V.* **jmdn.** ~: cheer sb. up

erhellen *tr. V.* light up

erhitzen 1. *tr. V.* heat ⟨*liquid*⟩; **jmdn.** ~: make sb. hot; **2.** *refl. V.* heat up; ⟨*person*⟩ become hot

erhoffen *tr. V.* **sich** *(Dat.)* **viel/wenig von etw.** ~: expect a lot/little from sth.

erhöhen 1. *tr. V.* increase ⟨*prices, productivity, etc.*⟩; **2.** *refl. V.* ⟨*rent, prices*⟩ rise; **Erhöhung die;** ~, ~en increase *(Gen.* in)

erholen *refl. V. (auch fig.)* recover **(von** from); *(sich ausruhen)* have a rest; **erholsam** *Adj.* restful; **Erholung die;** ~: *s.* **erholen:** recovery; rest; ~ **brauchen** need a rest; **erholungs·bedürftig** *Adj.* in need of a rest *postpos.*

erhören *tr. V. (geh.)* hear

Erika die; ~, ~s *od.* **Eriken** *(Bot.)* erica

erinnern 1. *refl. V.* **sich an jmdn./etw.** ~: remember sb./sth.; **sich [daran]** ~, **daß** ...: remember *or* recall that ...; **2.** *tr. V.* **jmdn. an etw./jmdn.** ~: remind sb. of sth./sb.; **Erinnerung die;** ~, ~en memory **(an** + *Akk.* of); etw. **[noch gut] in** ~ **haben** [still] remember sth. [well]; **zur** ~ **an jmdn./etw.** in memory of sb./sth.

erjagen *tr. V.* **a)** catch; **b)** *(gewinnen)* win ⟨*fame*⟩; make ⟨*money, fortune*⟩

erkalten *tr. V.; mit sein* cool; **erkäl-**

ten *refl. V.* catch cold; **Erkältung die;** ~, ~en cold

erkämpfen *tr. V.* win; **den Sieg** ~: gain a victory

erkaufen *tr. V.* **a)** *(durch Opfer)* win; **b)** *(durch Geld)* buy

erkennbar *Adj.* recognizable; *(sichtbar)* visible; **erkennen** *unr. tr. V.* **a)** recognize; **b)** *(deutlich sehen)* make out; **erkenntlich** *Adj.* **a) sich |für etw.|** ~ **zeigen** show one's appreciation for sth.; **b)** *s.* **erkennbar; Erkenntnis die;** ~, ~se discovery; **zu der** ~ **kommen, daß ...**: come to the realization that ...

Erker *der;* ~s, ~: bay window; **Erker·fenster das** bay window

erklärbar *Adj.* explicable; **erklären 1.** *tr. V.* **a)** explain; **b)** *(mitteilen)* state; declare; **c)** **jmdn. für tot** ~: pronounce someone dead; **jmdn. zu etw.** ~: name sb. as sth; **2.** *refl. V.* **sich einverstanden/bereit** ~: declare oneself [to be] in agreement/willing; **erklärlich** *Adj.* understandable; **erklärt** *Adj.* declared; **Erklärung die;** ~, ~en **a)** *(Darlegung)* explanation; **b)** *(Mitteilung)* statement

erklimmen *unr. tr. V.* *(geh.)* climb

erklingen *unr. itr. V.; mit sein* ring out

erkranken *itr. V.; mit sein* become ill **(an** + *Dat.* with); **schwer erkrankt sein** be seriously ill; **Erkrankung die;** ~, ~en illness; *(eines Körperteils)* disease

erkunden *tr. V.* reconnoitre *(terrain)*; **erkundigen** *refl. V.* **sich nach jmdm./ etw.** ~: ask after sb./enquire about sth.; **Erkundigung die;** ~, ~en enquiry

erlahmen *itr. V.; mit sein* tire; *(strength)* flag

erlangen *tr. V.* gain; obtain *(credit, visa)*; reach *(age)*

Erlaß *der; Erlasses, Erlasse* decree; **erlassen** *unr. tr. V.* **a)** enact *(law)*; declare *(amnesty)*; issue *(warrant)*; **b)** *(verzichten auf)* remit *(sentence)*

erlauben 1. *tr. V.* **a)** allow; **b)** *(ermöglichen)* permit; **2.** *refl. V.* **sich** *(Dat.)* **etw.** ~: permit oneself sth.; **Erlaubnis die;** ~, ~se permission; *(Schriftstück)* permit

erläutern *tr. V.* explain; comment on *(picture etc.)*; annotate *(text)*; **Erläuterung die** explanation

Erle die; ~, ~n alder

erleben *tr. V.* experience; **etwas Schreckliches** ~: have a terrible experience; **er wird das nächste Jahr**

nicht mehr ~: he won't see next year; **du kannst was** ~! *(ugs.)* you won't know what's hit you!; **Erlebnis das;** ~ses, ~se experience

erledigen 1. *tr. V.* deal with *(task)*; settle *(matter)*; **ich muß noch einige Dinge erledigen** I must see to a few things; **sie hat alles pünktlich erledigt** she got everything done on time; **2.** *refl. V.* *(matter, problem)* resolve itself; **vieles erledigt sich von selbst** a lot of things sort themselves out; **erledigt** *Adj.* closed *(case)*; *(ugs.)* worn out *(person)*

erlegen *tr. V.* shoot *(animal)*

erleichtern *tr. V.* **a)** make easier; **b)** *(befreien)* relieve; **Erleichterung die;** ~, ~en **a) zur** ~ **der Arbeit** to make the work easier; **b)** *(Befreiung)* relief; **c)** *(Verbesserung, Milderung)* alleviation

erleiden *unr. tr. V.* suffer

erlernbar *Adj.* learnable; **erlernen** *tr. V.* learn

erlesen *Adj.* superior *(wine)*; choice *(dish)*

erleuchten *tr. V.* **a)** light; **b)** *(geh.: mit Klarheit erfüllen)* inspire; **Erleuchtung die;** ~, ~en inspiration

erliegen *unr. itr. V.; mit sein* succumb *(Dat.* to); **einem Irrtum** ~: be misled; **einer Krankheit** *(Dat.)* ~: die from an illness

erlogen *Adj.* made up

Erlös *der;* ~es, ~e proceeds *pl.*

erlöschen *unr. itr. V.; mit sein* *(fire)* go out; **ein erloschener Vulkan** an extinct volcano

erlösen *tr. V.* save, rescue **(von** from); **Erlöser** *der;* ~s, ~ **a)** saviour; **b)** *(christl. Rel.)* redeemer; **Erlösung die** release **(von** from)

ermächtigen *tr. V.* authorize; **Ermächtigung die;** ~, ~en authorization

ermahnen *tr. V.* admonish; tell *(coll.)*; *(warnen)* warn; **Ermahnung die** admonition; *(Warnung)* warning

Ermang[e]lung die; ~: **in** ~ (+ *Gen.*) *(geh.)* in the absence of

ermäßigen *tr. V.* reduce; **Ermäßigung die** reduction

ermatten *(geh.)* **1.** *itr. V.; mit sein* become exhausted; **2.** *tr. V.* exhaust, tire

ermessen *unr. tr. V.* estimate, gauge; **Ermessen das;** ~s estimation

ermitteln 1. *tr. V.* ascertain *(facts)*; discover *(culprit, address)*; establish

⟨*identity, origin*⟩; decide ⟨*winner*⟩; calculate ⟨*quota, rates, data*⟩; **2.** *itr. V. (Rechtsw.)* investigate; **Ermittlung die**; ~, ~**en a)** *(das Ermitteln) s.* **ermitteln a:** ascertainment; discovery; establishment; **b)** *(Untersuchung)* investigation

ermöglichen *tr. V.* enable

ermorden *tr. V.* murder; **Ermordung die**; ~, ~**en** murder

ermüden 1. *itr. V.; mit sein* tire; **2.** *tr. V.* tire; make tired; **ermüdend** *Adj.* tiring; **Ermüdung die**; ~, ~**en** tiredness

ermuntern *tr. V.* encourage; **ermunternd** *Adj.* encouraging

ermutigen *tr. V.* encourage; **Ermutigung die**; ~, ~**en** encouragement

ernähren 1. *tr. V.* **a)** feed ⟨*young, child*⟩; **b)** *(unterhalten)* keep ⟨*family, wife*⟩; **2.** *refl. V.* feed oneself; **Ernährer der**; ~s, ~, **Ernährerin die**; ~, ~**nen** breadwinner; **Ernährung die**; ~: feeding; *(Nahrung)* diet

ernennen *unr. tr. V.* appoint; **Ernennung die** appointment (**zu** as)

erneuern *tr. V.* **a)** replace; **b)** *(wiederherstellen)* renovate ⟨*roof, building*⟩; *(fig.)* thoroughly reform ⟨*system*⟩; **Erneuerung die a)** replacement; **b)** *(Wiederherstellung)* renovation; **erneut 1.** *Adj.* renewed; **2.** *adv.* once again

erniedrigen *tr. V.* humiliate; **Erniedrigung die**; ~, ~**en** humiliation

ernst 1. *Adj.* **a)** serious; **b)** *(aufrichtig)* genuine ⟨*intention, offer*⟩; **c)** *(gefahrvoll)* serious ⟨*injury*⟩; grave ⟨*situation*⟩; **2.** *adv.* seriously

Ernst der; ~|e|s **a)** seriousness; **das ist mein |voller|** ~: I mean that [quite] seriously; **b)** *(Wirklichkeit)* **daraus wurde |blutiger/bitterer|** ~: it became [deadly] serious; **der** ~ **des Lebens** the serious side of life

ernst-, Ernst-: ~**fall der: im** ~: when the real thing happens; ~**gemeint** *Adj. (präd. getrennt geschrieben)* serious; sincere ⟨*wish*⟩; ~**haft 1.** *Adj.* serious; **2.** *adv.* seriously; ~**haftigkeit die**; ~: seriousness

ernstlich 1. *Adj.* **a)** serious; **b)** *(aufrichtig)* genuine ⟨*wish*⟩; **2.** *adv.* **a)** seriously; **b)** *(aufrichtig)* genuinely ⟨*sorry, repentant*⟩

Ernte die; ~, ~**n a)** harvest; **b)** *(Ertrag)* crop; **die** ~ **einbringen** bring in the harvest; **Ernte·dank·fest das** harvest festival; **ernten** *tr. V.* harvest

ernüchtern *tr. V.* sober up; *(fig.)* bring down to earth; ~**d** sobering; **Ernüchterung die**; ~, ~**en** *(fig.)* disillusionment

Eroberer der; ~s, ~, **Eroberin die**; ~, ~**nen** conqueror; **erobern** *tr. V.* **a)** conquer; take ⟨*town, fortress*⟩; seize ⟨*power*⟩; **Eroberung die**; ~, ~**en** conquest; *(einer Stadt, Festung)* taking

eröffnen *tr. V.* **a)** open; start ⟨*business, practice*⟩; **b)** *(mitteilen)* **jmdm. etw.** ~: reveal sth. to sb.; **c) ein Testament** ~: read a will; **Eröffnung die a)** opening; *(einer Sitzung)* start; **b)** *(Mitteilung)* revelation; **c)** *(Testaments~)* reading

erörtern *tr. V.* discuss; **Erörterung die**; ~, ~**en** discussion

Erotik die; ~: eroticism; **erotisch** *Adj.* erotic

Erpel der; ~s, ~: drake

erpicht *Adj.* **in auf etw.** *(Akk.)* ~ **sein** be keen on sth.

erpressen *tr. V.* **a)** *(nötigen)* blackmail; **b)** *(erlangen)* extort ⟨*money etc.*⟩; **Erpresser der**; ~s, ~: blackmailer; **Erpressung die** blackmail *no indef. art.*; *(von Geld, Geständnis)* extortion

erproben *tr. V.* test ⟨*medicine*⟩ (**an** + *Akk.* on)

erraten *unr. tr. V.* guess

errechnen *tr. V.* calculate

erregen 1. *tr. V.* **a)** annoy; **b)** *(sexuell)* arouse; **c)** *(verursachen)* arouse; **2.** *refl. V.* get excited; **erregend** *Adj.* exciting; *(sexuell)* arousing; **Erreger der**; ~s, ~ *(Med.)* pathogen; **erregt** *Adj.* excited; *(sexuell)* aroused; **Erregung die** excitement

erreichbar *Adj.* **a)** within reach *postpos.*; **b) der Ort ist mit dem Zug** ~: the place can be reached by train; **erreichen** *tr. V.* **a)** reach; **den Zug** ~: catch the train; **er ist telefonisch zu** ~: he can be contacted by telephone; **b)** *(durchsetzen)* achieve ⟨*goal, aim*⟩

errichten *tr. V.* **a)** build ⟨*house, bridge, etc.*⟩; **b)** *(aufstellen)* erect

erringen *unr. tr. V.* gain ⟨*victory*⟩; reach ⟨*first etc. place*⟩

erröten *itr. V.; mit sein* blush

Errungenschaft die; ~, ~**en** achievement

Ersatz der; ~es **a)** replacement; **b)** *(Entschädigung)* compensation

Ersatz-: ~**kasse die** private health insurance company; ~**mann der**; *Pl.* ~**männer, ~leute** replacement; *(Sport)*

substitute; ~**rad** das spare wheel;
~**reifen** der spare tyre; ~**teil** das
(bes. Technik) spare part; spare *(Brit.)*
ersa**ufen** *unr. itr. V.; mit sein (salopp)*
drown; ers**äufen** *tr. V.* drown
ersch**affen** *unr. tr. V.* create; **Er-**
schaffung die creation
ersch**einen** *unr. itr. V.; mit sein*
(book) be published; **Erscheinung**
die; ~, ~en a) *(Vorgang)* phenomenon; b) *(äußere Gestalt)* appearance;
c) *(Vision)* apparition; **eine ~ haben**
see a vision
Erscheinungs-: ~**bild** das appearance; ~**form** die manifestation
ersch**ießen** *unr. tr. V.* shoot dead;
Erschießung die; ~, ~en shooting
ersch**laffen** *itr. V.; mit sein (muscle,*
limb) become limp; *(skin)* grow slack
¹ersch**lagen** *unr. tr. V.* strike dead;
kill; ²ersch**lagen** *Adj. (ugs.)* a) *(er-*
schöpft) worn out; b) *(verblüfft)* **wie ~**
sein be flabbergasted *(coll.)* or
thunderstruck
ersch**ließen** *unr. tr. V.* develop *(area,*
building land); tap *(resources)*
ersch**öpfen** *tr. V.* exhaust; **er-**
schöpfend *Adj.* exhaustive; **er-**
schöpft *Adj.* exhausted; **Erschöp-**
fung die exhaustion
¹ersch**recken** *unr. itr. V.; mit sein* be
startled; **vor etw.** *(Dat.) od.* **über etw.**
(Akk.) ~: be startled by sth.; ²**er-**
schrecken *tr. V.* frighten; scare;
³ersch**recken** *unr. od. regelm. refl.*
V. get a fright; **erschreckend** *Adj.*
alarming; **erschrocken 1.** *2. Part. v.*
¹**erschrecken; 2.** *Adj.* frightened
ersch**üttern** *tr. V. (auch fig.)* shake;
erschütternd *Adj.* deeply distressing; deeply shocking *(conditions)*; **Er-**
schütterung die; ~, ~en a) vibration; *(der Erde)* tremor; b) *(Ergriffen-*
heit) shock; *(Trauer)* distress
ersch**weren** *tr. V.* **etw.** ~: make sth.
more difficult; **erschwerend 1.** *Adj.*
complicating *(factor)*; **2.** *adv.* **es**
kommt ~ hinzu, daß er ...: to make
matters worse he ...
erschw**inglich** *Adj.* reasonable
ers**ehen** *unr. tr. V.* see; **aus etw. zu ~**
sein be evident from sth.
ers**etzen** *tr. V.* a) replace (**durch** by);
b) *(erstatten)* reimburse *(expenses)*;
jmdm. einen Schaden ~: compensate
sb. for damages
ers**ichtlich** *Adj.* apparent
ersp**ähen** *tr. V. (geh.)* espy *(literary)*;
catch sight of

ersp**aren** *tr. V.* save; **Ersparnis die;**
~, ~**se** saving
erspr**ießlich** *Adj. (geh.)* fruitful *(con-*
tacts, collaboration)
erst 1. *Adv.* a) *(zu~)* first; **~ einmal**
first [of all]; b) *(nicht eher als)* **eben ~:**
only just; **~ nächste Woche** not until
next week; **er war ~ zufrieden, als ...:**
he was not satisfied until ...; c) *(nicht*
mehr als) only; **2.** *Partikel* **so was lese**
ich gar nicht ~: I dont even start reading that sort of stuff
erst... *Ordinalz.* a) first; **etw. das ~e**
Mal tun do sth. for the first time; **am**
Ersten [des Monats] on the first [of the
month]; **als ~e/~er etw. tun** be the
first to do sth.; b) *(best...)* **das ~e Ho-**
tel the best hotel; **der/die Erste [der**
Klasse] the top boy/girl [of the class]
erst**arren** *itr. V.; mit sein (jelly, plas-*
ter) set; *(limbs, fingers)* grow stiff
erst**atten** *tr. V.* a) reimburse *(ex-*
penses); b) **Anzeige gegen jmdn. ~:** report sb. [to the police]; **Erstattung**
die; ~, ~en *(von Kosten)* reimbursement
Erst·aufführung die première
erst**aunen** *tr. V.* astonish; **Er-**
staunen das; ~s astonishment; **er-**
staunlich 1. *Adj.* astonishing; **2.**
adv. astonishingly
Erst·ausgabe die first edition
erst**echen** *unr. tr. V.* stab [to death]
erst**ehen** *(geh.)* **1.** *unr. tr. V. (kaufen)*
purchase; **2.** *unr. itr. V.; mit sein (diffi-*
culties, problems) arise
erst**eigen** *unr. tr. V.* climb
erst**eigern** *tr. V.* buy [at an auction]
erst**ellen** *tr. V. (Papierdt.)* a) *(bauen)*
build; b) *(anfertigen)* make *(assess-*
ment); draw up *(plan, report, list)*
erste·mal *Adv.* **das ~:** for the first
time; **ersten·mal** *Adv.* **zum ~:** for
the first time; **beim ~:** the first time
erstens *Adv.* firstly; in the first place;
erster... *Adj.* the former
erst·geboren *Adj.* first-born
erst**icken 1.** *itr. V.; mit sein* suffocate;
(sich verschlucken) choke; **2.** *tr. V.* a)
(töten) suffocate; b) smother *(flames)*
erst**klassig 1.** *Adj.* first-class; **2.** *adv.*
superbly; **erstmals** *Adv.* for the first
time; **erstrangig** *Adj.* a) first-class;
b) *(vordringlich)* of top priority *postp.*
erstr**eben** *tr. V.* strive for; **erstre-**
bens·wert *Adj. (ideals etc.)* worth
striving for; desirable *(situation)*
erstr**ecken** *refl. V.* a) *(sich ausdeh-*

nen) stretch; **b)** *(dauern)* **sich über 10 Jahre ~:** carry on for 10 years

erstürmen *tr. V.* take by storm

ersuchen *tr. V. (geh.)* ask; **jmdn. ~, etw. zu tun** request sb. to do sth.

ertappen *tr. V.* catch ⟨*thief, burglar*⟩

erteilen *tr. V.* give ⟨*advice, information*⟩; give, grant ⟨*permission*⟩; **Erteilung die** giving; *(einer Genehmigung)* granting

ertönen *itr. V.; mit sein* sound

Ertrag der; ~|e|s, Erträge **a)** yield; **b)** *(Gewinn)* return

ertragen *unr. tr. V.* bear; **erträglich** *Adj.* tolerable; bearable ⟨*pain*⟩

ertrag·reich *Adj.* lucrative ⟨*business*⟩; productive ⟨*land, soil*⟩

ertränken *tr. V.* drown; **ertrinken** *unr. itr. V.; mit sein* be drowned; drown

erübrigen 1. *tr. V.* spare ⟨*money, time*⟩; **2.** *refl. V.* be unnecessary

erwachen *itr. V.; mit sein (geh.)* awake

Erwachen das; ~s *(auch fig.)* awakening

¹**erwachsen** *unr. itr. V.; mit sein* **a)** grow **(aus** out of); ⟨*rumour*⟩ spread; **b)** *(sich ergeben)* ⟨*difficulties, tasks*⟩ arise; ²**erwachsen** *Adj.* grown-up *attrib.;* **~ sein** be grown up; **Erwachsene der/die;** *adj. Dekl.* adult; grown-up

erwägen *unr. tr. V.* consider; **Erwägung die; ~, ~en** consideration; **etw. in ~ ziehen** take sth. into consideration

erwählen *tr. V. (geh.)* choose

erwähnen *tr. V.* mention; **erwähnens·wert** *Adj.* worth mentioning *postpos.;* **Erwähnung die; ~, ~en** mention

erwärmen 1. *tr. V.* heat; **2.** *refl. V. (warm werden)* ⟨*air, water*⟩ warm up

erwarten *tr. V.* expect; **jmdn. am Bahnhof ~:** wait for sb. at the station; **Erwartung die; ~, ~en** expectation; **erwartungs-: ~gemäß** *Adv.* as expected; **~voll** *Adj.* expectant

erwecken *tr. V.* **a)** *(auf~)* wake; **b)** *(erregen)* arouse ⟨*longing, pity*⟩

erweichen *tr. V.* soften

erweisen 1. *unr. tr. V.* **a)** prove; **b)** *(bezeigen)* **jmdm. Achtung ~:** show respect to sb.; **2.** *unr. refl. V.* **sich als etw. ~:** prove to be sth.

erweitern 1. *tr. V.* widen ⟨*river, road*⟩; expand ⟨*library, business*⟩; enlarge ⟨*collection*⟩; dilate ⟨*pupil, blood vessel*⟩; **2.** *refl. V.* ⟨*road, river*⟩ widen;

⟨*pupil, blood vessel*⟩ dilate; **Erweiterung die; ~, ~en** *s.* **erweitern:** widening; expansion; enlargement; dilation

Erwerb der; ~|e|s **a)** *(Aneignung)* acquisition; **b)** *(Kauf)* purchase; **erwerben** *unr. tr. V.* **a)** *(verdienen)* earn; **b)** *(sich aneignen)* gain; **c)** *(kaufen)* acquire

erwerbs-: ~los *Adj.: s.* **arbeitslos; ~tätig** *Adj.* gainfully employed; **~unfähig** *Adj.* incapable of gainful employment *postpos.;* unable to work *postpos.*

Erwerbung die acquisition; *(Gekauftes)* purchase

erwidern *tr. V.* **a)** reply; **b)** *(reagieren auf)* return ⟨*greeting, visit*⟩; reciprocate ⟨*sb.'s feelings*⟩; **Erwiderung die; ~, ~en** **a)** reply **(auf +** *Akk.* to); **b)** *s.* **erwidern b:** return; reciprocation

erwiesen *Adj.* proved; proven ⟨*fact*⟩; **erwiesener·maßen** *Adv.* as has been proved

erwirken *tr. V.* obtain

erwirtschaften *tr. V.* **etw. ~:** obtain sth. by careful management

erwischen *tr. V. (ugs.)* **a)** catch ⟨*culprit, train, bus*⟩; **b)** *(greifen)* grab; **c)** *(bekommen)* manage to get; **d)** *(unpers.)* **es hat ihn erwischt** *(ugs.) (er ist tot)* he's bought it *(sl.); (er ist krank)* he's got it; *(er ist verletzt)* he's been hurt; *(scherzh.: er ist verliebt)* he's got it bad *(coll.)*

erwünscht *Adj.* wanted

erwürgen *tr. V.* strangle

Erz [ɛrts *od.* e:rts] **das; ~es, ~e** ore

erzählen *tr. (auch itr.) V.* tell ⟨*joke, story*⟩; **jmdm. etw. ~:** tell sb. sth.; **Erzähler der** story-teller; *(Autor)* writer [of stories]; narrative writer; **Erzählung die; ~, ~en** narration; *(Bericht)* account; *(Literaturw.)* story

Erz-: ~bischof der archbishop; **~bistum das, ~diözese die** archbishopric; archdiocese; **~engel der** archangel

erzeugen *tr. V.* produce; generate ⟨*electricity*⟩; **Erzeuger der; ~s, ~** *(Vater)* father; **Erzeugnis das** product; **Erzeugung die** *(von Lebensmitteln usw.)* production; *(von Industriewaren)* manufacture; *(Strom~)* generation

Erz·feind der arch enemy

erziehen *unr. tr. V.* bring up; *(in der Schule)* educate; **ein Kind zu Sauberkeit und Ordnung ~:** bring a child up

to be clean and tidy; **Erzi̯eher der;**
~s, ~, **Erzi̯eherin die;** ~, ~nen edu-
cator; *(Pädagoge)* educationalist;
(Lehrer) teacher; **Erzi̯ehung die;** *o.*
Pl. upbringing; *(Schul~)* education;
Erzi̯ehungs·berechtigte der/die;
adj. Dekl. parent or [legal] guardian
erzi̯elen *tr. V.* reach ⟨*agreement, com-*
promise, speed⟩; achieve ⟨*result, ef-*
fect⟩; make ⟨*profit*⟩; obtain ⟨*price*⟩
erzürnen *(geh.) tr. V.* anger; *(stärker)*
incense

erzwi̯ngen *unr. tr. V.* force
es *Personalpron.; 3. Pers. Sg. Nom. u.*
Akk. Neutr. **a)** *(s. auch Gen.* seiner;
Dat. ihm) *(Sache)* it; *(weibliche Per-*
son) she/her; *(männliche Person)* he/
him; **b)** *ohne Bezug auf ein bestimmtes*
Subst., mit unpers. konstruierten Ver-
ben, als formales Satzglied it; **ich bin**
es it's me; **wir sind traurig, ihr seid es**
auch we are sad, and so are you; **es sei**
denn, [daß] ...: unless ...; **es ist genug!**
that's enough!; **es hat geklopft** there
was a knock; **es klingelt** someone is
ringing; **es wird schöner** the weather is
improving; **es geht ihm gut/schlecht**
he is well/unwell; **es wird gelacht**
there is laughter; **es läßt sich aushalten**
it is bearable; **er hat es gut** he has it
good; **er meinte es gut** he meant well
Esche die; ~, ~n *(Bot.)* ash
Esel der; ~s, ~ **a)** donkey; ass; **b)**
(ugs.: Dummkopf) ass *(coll.)*
Esels-: ~**brücke die** *(ugs.)* mne-
monic; ~**ohr das** *(ugs.: umgeknickte*
Stelle) dog-ear
Eskalation die; ~, ~en escalation
Eskimo der; ~[s], ~[s] Eskimo
Eskorte die; ~, ~n escort; **eskor-**
ti̯eren *tr. V.* escort
Espe die; ~, ~n aspen
eßbar *Adj.* edible; **nicht** ~: inedible;
essen *unr. tr., itr. V.* eat; **etw. gern** ~:
like sth.; **sich satt** ~: eat one's fill; **gut**
~: have a good meal; *(immer)* eat
well; ~ **gehen** go out for a meal; **Es-**
sen das; ~s, ~ *(Mahlzeit)* meal; *(Spei-*
se) food; **[das]** ~ **machen/kochen** get/
cook the meal
Essen[s]-: ~**marke die** meal-ticket;
~**zeit die** mealtime
Essenz die; ~, ~en essence
Esser der; ~s, ~: **er ist ein schlechter**
~: he has a poor appetite
Essig der; ~s, ~e vinegar; **Essiggur-**
ke die pickled gherkin
Eß-: ~**kastanie die** sweet chestnut;
~**löffel der** *(Suppenlöffel)* soup-

spoon; *(für Nach-, Vorspeise)* dessert-
spoon; ~**stäbchen das** chopstick;
~**teller der** dinner plate; ~**tisch der**
dining-table; ~**waren** *Pl.* food *sing.;*
~**zimmer das** dining-room
Establishment [ɪs'tɛblɪʃmənt] **das;**
~s, ~s Establishment
Este der; ~n, ~n Estonian; **Est·land**
(das); ~s Estonia
Estragon ['ɛstragɔn] **der;** ~s tarragon
Estrich ['ɛstrɪç] **der;** ~s, ~e composition floor
etabli̯eren *tr. V.* establish; set up;
etabli̯ert *Adj.* established
Etage [e'ta:ʒə] **die;** ~, ~n floor; storey
Etappe die; ~, ~n stage
Etat [e'ta:] **der;** ~s, ~s budget
etepetete [e:təpe'te:tə] *Adj. (ugs.)*
fussy; finicky
Ethik die; ~, ~en **a)** ethics *sing.;* **b)** *o.*
Pl. (sittliche Normen) ethics *pl.;*
ethisch *Adj.* ethical
Etikett das; ~[e]s, ~en *od.* ~e *od.* ~s
label; **Etiketto die;** ~, ~n etiquette;
etiketti̯eren *tr. V.* label
etlich... *Indefinitpron. u. unbest. Zahl-*
wort: Sg. quite a lot of; *Pl.* quite a few
Etüde die; ~, ~n *(Musik)* étude
Etui [e'tviː] **das;** ~s, ~s case
etwa 1. *Adv.* **a)** *(ungefähr)* about; ~ **so**
groß wie ...: about as large as ...; ~ **so**
roughly like this; **b)** *(beispielsweise)*
for example; **2.** *Partikel* **störe ich** ~?
am I disturbing you at all?; **et-**
waig... ['ɛtva(ː)ɪg...] *Adj.* possible
etwas *Indefinitpron.* **a)** something;
(fragend, verneinend) anything; **irgend**
~: something; **b)** *(Bedeutsames)* **aus**
ihm wird ~: he'll make something of
himself; **c)** *(ein Teil)* some; *(fragend,*
verneinend) any; ~ **von dem Geld** some
of the money; **d)** *(ein wenig)* a little; ~
lauter/besser a little louder/better
Etymologi̯e die; ~, ~n etymology
euch 1. *Dat. u. Akk. Pl. des Personal-*
pron. **ihr** you; **2.** *Dat. u. Akk. Pl. des*
Reflexivpron. der 2. Pers. Pl. yourselves
¹euer *Possessivpron.* your; **Grüße von**
Eu[e]rer Helga/Eu[e]rem Hans Best
wishes, Yours, Helga/Hans; **²euer**
Gen. des Personalpron. **ihr** *(geh.)* **wir**
werden ~ **gedenken** we will remember
you
Eule die; ~, ~n owl; ~**n nach Athen**
tragen carry coals to Newcastle
Eunuch der; ~en, ~en eunuch
Euphori̯e die; ~, ~n *(bes. Med.,*
Psych.) euphoria

eure s. ¹**euer**; **eurer·seits** s. deinerseits; **euret·wegen** Adv. s. deinetwegen

Eurocheque ['ɔyroʃɛk] der; ~s, ~s
Eurocheque

Europa (das); ~s Europe

Europäer der; ~s, ~, **Europäerin**
die; ~, ~nen European; **europäisch**
Adj. European; **die Europäische Gemeinschaft** the European Community

Europa-: ~**meister** der *(Sport)* European champion; ~**meisterschaft**
die *(Sport)* a) *(Wettbewerb)* European
Championship; b) *(Sieg)* European
title; ~**parlament** das; o. Pl. European Parliament; ~**pokal** der *(Sport)*
European cup; ~**rat** der; o. Pl. Council of Europe; ~**straße** die European
long-distance road

Euro·scheck der s. Eurocheque

Euter das od. der; ~s, ~: udder

e. V., E. V. Abk. eingetragener Verein

ev. Abk. evangelisch ev.

evakuieren [evaku'iːrən] tr. V. evacuate; **Evakuierung** die; ~, ~en evacuation

evangelisch [evaŋ'geːlɪʃ] Adj. Protestant; **Evangelium** das; ~s, Evan-
gelien a) *(auch fig.)* gospel; b) *(christl.
Rel.)* Gospel

eventuell [evɛn'tuɛl] 1. Adj. possible;
2. adv. possibly; perhaps

Evolution [evolu'tsjoːn] die; ~, ~en
evolution

evtl. Abk. eventuell

EWG [eːveː'geː] die; ~: EEC

ewig 1. Adj. eternal; *(abwertend)*
never-ending; 2. adv. eternally; for
ever; **Ewigkeit** die; ~, ~en a) eternity; b) *(ugs.)* **es dauert eine** ~: it takes
ages *(coll.)*

ex Adv. *(ugs.)* **etw. ex trinken** drink sth.
down in one *(coll.)*; **Ex-** *(vor Personen-
bez.: vormalig)* ex-

exakt Adj. exact; precise

Examen das; ~s, ~ od. **Examina**
examination

Exekution die; ~, ~en execution;
Exekutive die; ~, ~n *(Rechtsw., Politik)* executive

Exempel das; ~s, ~: example; **Exemplar** das; ~s, ~e specimen; *(Buch,
Zeitung usw.)* copy

exerzieren tr., itr. V. drill

Exil das; ~s, ~e exile

Existenz die; ~, ~en a) existence; b)
(Lebensgrundlage) livelihood; c)
(Mensch) character

Existenz-: ~**grundlage** die basis of

one's livelihood; ~**minimum** das
subsistence level

existieren itr. V. exist

Exitus der; ~ *(Med.)* death

exkl. Abk. exklusiv[e] excl.

exklusiv 1. Adj. exclusive; 2. adv. exclusively; **exklusive** Präp. + Gen.
exclusive of

Ex·kommunikation die excom-
munication

Exkursion die; ~, ~en study trip

exotisch 1. Adj. exotic; 2. adv. ex-
otically

expandieren tr., itr. V. expand; **Expansion** die; ~, ~en expansion

Expedition die; ~, ~en expedition

Experiment das; ~[e]s, ~e experiment; **experimentell** 1. Adj. experimental; 2. adv. experimentally;
experimentieren itr. V. experiment

Experte der; ~n, ~n, **Expertin** die;
~, ~nen expert (für in)

explodieren itr. V.; mit sein *(auch
fig.)* explode; ⟨costs⟩ rocket; **Explosion** die; ~, ~en explosion; **explosiv** 1. Adj. *(auch fig.)* explosive; 2.
adv. explosively

Exponent der; ~en, ~en *(Math.)* exponent; **exponiert** Adj. exposed

Export der; ~[e]s, ~e export

Export-: ~**artikel** der export; ~**bier**
das export beer

Exporteur [ɛkspor'tøːɐ̯] der; ~s, ~e
(Wirtsch.) exporter

Export-: ~**firma** die exporter; ~**handel** der export trade

exportieren tr., itr. V. export

Expreß·gut das express freight

Expressionismus der; ~: expres-
sionism no art.; **expressionistisch**
Adj. expressionist

extra Adv. a) *(gesondert)* ⟨pay⟩ separately; b) *(zusätzlich, besonders)* extra;
c) *(eigens)* especially; **Extra** das; ~s,
~s extra; **Extra·blatt** das special
edition

Extrakt der; ~[e]s, ~e extract

extravagant [-va'gant] Adj. flamboyant; flamboyantly furnished ⟨flat⟩

extrem Adj. extreme; **Extrem** das;
~s, ~e extreme; **Extrem·fall** der extreme case; **Extremismus** der; ~:
extremism; **Extremist** der; ~en, ~en
extremist; **extremistisch** Adj. extremist

Exzellenz die; ~, ~en Excellency

exzentrisch 1. Adj. eccentric; 2. adv.
eccentrically

Exzeß der; Exzesses, Exzesse excess

F

---

f, F [ɛf] das; ~, ~ a) *(Buchstabe)* f/F; b) *(Musik)* [key of] F

f. *Abk.* folgend f.

Fa. *Abk.* Firma

Fabel die; ~, ~n fable; *(Kern einer Handlung)* plot

fabelhaft 1. *Adj. (ugs.: großartig)* fantastic *(coll.)*; **2.** *adv. (ugs.)* fantastically *(coll.)*

Fabrik die; ~, ~en factory

Fabrikant der; ~en, ~en manufacturer; **Fabrikat das;** ~[e]s, ~e product; *(Marke)* make; **Fabrikation die;** ~: production

Fabrik-: ~besitzer der factory-owner; ~direktor der works manager

fabrizieren *tr. V. (ugs. abwertend)* knock together *(coll.)*

Fach das; ~[e]s, Fächer a) compartment; *(für Post)* pigeon-hole; b) *(Studien~, Unterrichts~)* subject; *(Wissensgebiet)* field; *(Berufszweig)* trade; **ein Mann vom** ~: an expert

Fach-: ~arbeiter der skilled worker; ~arzt der specialist (für in); ~geschäft das specialist shop

fachlich *Adj.* specialist *(knowledge, work)*; technical *(problem, explanation, experience)*

Fach-: ~mann der expert; ~werk das *o. Pl. (Bauweise)* half-timbered construction; ~werk·haus das half-timbered house

Fackel die; ~, ~n torch

fade *Adj.* insipid

Faden der; ~s, Fäden thread; **ein** ~: a piece of thread

faden·scheinig *Adj.* threadbare; flimsy *(excuse)*

Fagott das; ~[e]s, ~e bassoon

fähig *Adj.* a) *(begabt)* able; capable; b) **zu etw.** ~ **sein** be capable of sth.; **Fähigkeit die;** ~, ~en a) *meist Pl.* ability; capability; **geistige** ~en intellectual faculties; b) *o. Pl. (Imstandesein)* ability **(zu** to)

fahl *Adj.* pale; pallid; wan *(light)*

fahnden *itr. V.* search **(nach** for)

Fahne die; ~, ~n flag

Fahr·bahn die carriageway

Fähre die; ~, ~n ferry

fahren 1. *unr. itr. V.; mit sein* a) *(als Fahrzeuglenker)* drive; *(mit dem Fahrrad, Motorrad usw.)* ride; b) *(als Mitfahrer; mit öffentlichem Verkehrsmittel)* go **(mit** by); *(mit dem Aufzug/der Rolltreppe/der Seilbahn)* take the lift *(Brit.)* or *(Amer.)* elevator/escalator/cable-car; *(per Anhalter)* hitch-hike; c) *(reisen)* go; **in Urlaub** ~: go on holiday; d) *(los~)* go; leave; e) *(motor vehicle, train, lift, cable-car)* go; *(ship)* sail; **mein Auto fährt nicht** my car won't go; f) *(verkehren) (train etc.)* run; **2.** *unr. tr. V.* a) *(fortbewegen)* drive *(car, lorry, train, etc.)*; ride *(bicycle, motor cycle)*; b) **50/80 km/h** ~: do 50/80 k.p.h.; **hier muß man 50 km/h** ~: you've got to keep to 50 k.p.h. here; sail *(boat)*; **Auto** ~: drive [a car]; **Kahn** *od.* **Boot/Kanu** ~: go boating/canoeing; **Ski** ~: ski; **U-Bahn** ~: ride on the underground *(Brit.)* or *(Amer.)* subway; c) *(befördern)* take

Fahrenheit *o. Art.* **70 Grad** ~: 70 degrees Fahrenheit

fahren|lassen *unr. tr. V.* let go; **Fahrer der;** ~s, ~: driver; **Fahrerflucht die:** ~ **begehen** fail to stop after [being involved ·in] an accident; **Fahrerin die;** ~, ~nen driver

Fahr-: ~gast der passenger; ~geld das fare

fahrig *Adj.* nervous

fahr-, Fahr-: ~karte die ticket; ~karten·automat der ticket machine; ~karten·schalter der ticket window; ~lässig **1.** *Adj.* negligent *(behaviour)*; ~e Tötung/Körperverletzung *(Rechtsw.)* causing death/injury through [culpable] negligence; **2.** *adv.* negligently; ~lehrer der driving instructor

Fähr·mann der ferryman

Fahr-: ~plan der timetable; schedule *(Amer.)*; ~preis der fare; ~prüfung die driving test; ~rad das bicycle; cycle; **mit dem** ~ **fahren** cycle; ride a bicycle; ~rad·ständer der bicycle rack; ~schein der ticket; ~schein·automat der ticket machine; ~schein·entwerter der ticket cancelling machine; ~schule die driving school; ~spur die traffic-lane

fährst *2. Pers. Sg. Präsens v.* **fahren**

Fahr-: ~**stuhl** der lift *(Brit.)*; elevator *(Amer.)*; *(für Lasten)* hoist; ~**stunde** die driving lesson

Fahrt die; ~, ~**en a)** journey; **freie** ~ **haben** have a clear run; *(Schiffsreise)* voyage; *(kurze Reise, Ausflug)* trip; **b)** o. Pl. *(Geschwindigkeit)* **in voller** ~: at full speed; **fährt** 3. Pers. Sg. Präsens v. **fahren**

Fährte die trail; jmds. ~ **verfolgen** track sb.

Fahrt·kosten Pl. *(für öffentliche Verkehrsmittel)* fare/fares; *(für Autoreisen)* travel costs; **Fahr·treppe** die escalator; **Fahrt·richtung** die direction; **in** ~ **parken** park in the direction of the traffic; **die** ~ **ändern** change direction; **fahr·tüchtig** Adj. *(driver)* fit to drive; *(vehicle)* roadworthy

Fahrt-: ~**wind** der airflow; ~**ziel** das destination

Fahr-: ~**werk** das *(Flugw.)* undercarriage; ~**zeit** die travelling time; ~**zeug** das vehicle; *(Luft~)* aircraft; *(Wasser~)* vessel

fair [fɛ:ɐ̯] **1.** Adj. fair **(gegen** to); **2.** adv. fairly

Fakten s. **Faktum**; **faktisch 1.** Adj. real; actual; **2.** adv. **das bedeutet** ~ ...: it means in effect ...

Faktor der; ~**s**, ~**en** *(auch Math.)* factor

Faktum das; ~**s**, **Fakten** fact

Fakultät die; ~, ~**en** *(Hochschulw.)* faculty

Falke der; ~**n**, ~**n** *(auch Politik fig.)* hawk

Fall der; ~**|e|s**, **Fälle a)** *(Sturz)* fall; **zu** ~ **kommen** have a fall; **jmdn. zu** ~ **bringen** *(fig.)* bring about sb.'s downfall; **b)** *(das Fallen)* descent; **der freie** ~: free fall; **c)** *(Ereignis; Rechtsw., Med., Grammatik)* case; *(zu erwartender Umstand)* eventuality; **es ist |nicht|** der ~: it is [not] the case; **gesetzt den** ~: assuming; **auf jeden** ~, **in jedem** ~, **auf alle Fälle** in any case; **auf keinen** ~: on no account; **Falle die;** ~, ~**n** *(auch fig.)* trap; **fallen** unr. itr. V.; mit sein **a)** fall; etw. ~ **lassen** drop sth.; **b)** *(hin-, stürzen)* fall [over], **über einen Stein** ~: trip over a stone; *(prices, light, glance, choise)* fall; *(temperature, water level)* fall, drop; *(fever)* subside; *(shot)* be fired; **d)** *(im Kampf sterben)* die; fall *(literary)*; **fällen** tr. V. **a)** fell *(tree, timber)*; **b)** **ein Urteil** ~ *(judge)* pass sentence; *(jury)* return a verdict; **fällig** Adj. due; **falls** *(Konj.)*

a) *(wenn)* if; **b)** *(für den Fall, daß)* in case; **Fall·schirm** der parachute; **mit dem** ~ **abspringen** *(im Notfall)* parachute out; *(als Sport)* make a [parachute] jump

falsch 1. Adj. **a)** *(unecht, imitiert)* false *(teeth, plait)*; imitation *(jewellery)*; **b)** *(gefälscht)* forged; assumed *(name)*; **c)** *(irrig, fehlerhaft)* wrong; **2.** adv. wrongly; **die Uhr geht** ~: the clock is wrong; **fälschen** tr. V. forge; **Fälscher** der; ~**s**, ~**forger**; **Falschgeld** das counterfeit money; **fälschlich 1.** Adj. false; **2.** adv. falsely; **Falschmeldung** die the false report; **Fälschung** die; ~, ~**en** fake

Falt·blatt das leaflet; *(in Zeitungen, Zeitschriften, Büchern)* insert; **Falte** die; ~, ~**n a)** crease; **b)** *(im Stoff)* fold; *(mit scharfer Kante)* pleat; **c)** *(Haut~)* wrinkle; **falten 1.** tr. V. fold; **die Hände** ~: fold one's hands; **2.** refl. V. *(auch Geol.)* fold; *(skin)* become wrinkled; **Falten·rock** der pleated skirt; **Falter** der; ~**s**, ~ *(Nacht~)* moth; *(Tag~)* butterfly; **faltig a)** Adj. *(clothes)* gathered [in folds]; wrinkled *(skin, hands)*; **b)** *(zerknittert)* creased **-fältig** Adj., adv. -fold

familiär Adj. **a)** family *(problems, worries)*; **b)** *(zwanglos)* familiar; informal; **Familie** [faˈmiːli̯ə] die; ~, ~**n** family; ~ **Meyer** the Meyer family

Familien-: ~**angehörige** der/die; adj. Dekl. member of the family; ~**feier** die family party; ~**leben** das; o. Pl. family life; ~**name** der surname; ~**planung** die; o. Pl. family planning no art.; ~**stand** der marital status; ~**vater** der: ~**vater sein** be the father of a family; **ein guter** ~**vater** a good husband and father

Fan [fɛn] der; ~**s**, ~**s** fan

Fanatiker der; ~**s**, ~: fanatic; *(religiös)* fanatic; zealot; **fanatisch 1.** Adj. fanatical; **2.** adv. fanatically

fand 1. u. 3. Pers. Sg. Prät. v. **finden**

Fanfare die; ~, ~**n** *(Signal)* fanfare

Fang der; ~**|e|s**, **Fänge a)** *(Tier~)* trapping; *(von Fischen)* catching; **b)** *(Beute)* bag; *(von Fischen)* catch; **fangen 1.** unr. tr. V. catch; capture *(fugitive etc.)*; **2.** unr. refl. V. **a)** *(in eine Falle geraten)* be caught; **b)** *(wieder in die normale Lage kommen)* **sich |gerade| noch** ~: [just] manage to steady oneself; **Fang·frage** die catch question

Farb-: ~**bild** das *(Foto)* colour photo; ~**dia** das colour slide

Farbe die; ~, ~n a) colour; b) *(für Textilien)* dye; *(zum Malen, Anstreichen)* paint; ~n mischen/auftragen mix/apply paint; **farb·echt** *Adj.* colourfast; **färben 1.** *tr. V.* dye; **2.** *refl. V.* change colour; **sich schwarz/rot** *usw.* ~: turn black/red *etc*; **3.** *itr. V. (ugs.: ab~)* ⟨*material, blouse etc.*⟩ run; **-farben** *Adj.* coloured

farben-: ~blind *Adj.* colour-blind; ~froh *Adj.* colourful; ~prächtig *Adj.* vibrant with colour *postpos.*

Farb-: ~fernsehen das colour television; ~fernseher der *(ugs.)* colour telly *(coll.)* or television; ~film der colour film; ~foto das colour photo

farbig 1. *Adj.* a) coloured; b) *(bunt, auch fig.)* colourful; **2.** *adv.* colourfully; **-farbig** *Adj.* -coloured; **Farbige** der/die; *adj. Dekl.* coloured man/woman; *Pl.* coloured people

farblich 1. *Adj.* in colour *postpos.*; as regards colour *postpos.*; **2.** *adv.* etw. ~ abstimmen match sth. in colour

farb-, Farb-: ~los *Adj. (auch fig.)* colourless; clear ⟨*varnish*⟩; neutral ⟨*shoe polish*⟩; ~stift der coloured pencil; ~stoff der a) *(Med., Biol.)* pigment; b) *(für Textilien)* dye; c) *(für Lebensmittel)* colouring; ~ton der shade

Färbung die; ~, ~en colouring

Farn der; ~[e]s, ~e, **Farn·kraut** das fern

Fasan der; ~[e]s, ~e[n] pheasant

Fasching der; ~s, ~e *od.* ~s [preLent] carnival

Faschismus der; ~: fascism *no art.* **Faschist** der; ~en, ~en fascist; **faschistisch** *Adj.* fascist

faseln *itr. V. (ugs. abwertend)* drivel

Faser die; ~, ~n fibre; **fasern** *itr. V.* fray

Faß das; Fasses, Fässer barrel; *(Öl~)* drum; *(kleines Bier~)* keg; *(kleines Sherry~ usw.)* cask; **Bier vom ~:** draught beer; **ein ~ ohne Boden** an endless drain on sb.'s resources

Fassade die; ~, ~n façade

faßbar *Adj.* a) tangible ⟨*results*⟩; b) *(verständlich)* comprehensible

Faß·bier das draught beer; beer on draught

fassen 1. *tr. V.* a) *(greifen)* grasp; take hold of; b) *(festnehmen)* catch ⟨*thief, culprit*⟩; c) *(aufnehmen können)* ⟨*hall, tank*⟩ hold; d) *(begreifen)* **ich kann es nicht ~:** I cannot take it in; e) **einen Entschluß ~:** make *or* take a decision; **2.** *itr. V.* a) *(greifen)* **nach etw. ~:** reach

for sth.; **in etw.** *(Akk.)* ~: put one's hand in sth.; **faßlich** *Adj.* comprehensible

Fasson [fa'sõ:] die; ~, ~s style; shape

Fassung die; ~, ~en a) *(Form)* version; b) *o. Pl. (Selbstbeherrschung)* composure; **die ~ bewahren** keep one's composure; **die ~ verlieren** lose one's self-control; **jmdn. aus der ~ bringen** upset sb.; c) *(für Glühlampen)* holder; **fassungs·los** *Adj.* stunned

fast *Adv.* almost; nearly; ~ **nie** hardly ever

fasten *itr. V.* fast; **Fast·nacht** die carnival

faszinieren *tr. V.* fascinate

fatal *Adj.* a) *(peinlich, mißlich)* awkward; b) *(verhängnisvoll)* fatal

fauchen *itr. V.* a) ⟨*cat*⟩ hiss; ⟨*tiger, person*⟩ snarl

faul *Adj.* a) *(verdorben)* rotten; bad ⟨*food, tooth*⟩; foul ⟨*water, air*⟩; b) *(träge)* lazy; **Fäule** die; ~: foulness; **faulen** *itr. V.; meist mit sein* rot; ⟨*water*⟩ go foul; ⟨*meat, fish*⟩ go off

faulenzen *itr. V.* laze about; loaf about *(derog.)*; **Faulenzer** der; ~s, ~: idler; lazy-bones *sing. (coll.)*

Faulheit die; ~: laziness; **faulig** *Adj.* stagnating ⟨*water*⟩; ~ **schmecken/riechen** taste/smell off; **Fäulnis** die; ~: rottenness

Faul-: ~pelz der *(fam.)* lazy-bones *sing. (coll.)*; ~tier das a) *(Zool.)* sloth; b) *(ugs.: Faulenzer)* s. ~pelz

Faust die; ~, Fäuste fist; **eine ~ machen** clench one's fist; **das paßt wie die ~ aufs Auge** *(ugs.) (paßt nicht)* that clashes horribly; *(paßt)* that matches perfectly; **auf eigene ~:** on one's own initiative; **Fäustchen** das; ~s, ~: **sich** *(Dat.)* **ins ~ lachen** laugh up one's sleeve; **faust·dick** *Adj.* as thick as a man's fist *postpos.*; *(fig.)* bare-faced ⟨*lie*⟩; **Fäustling** der; ~s, ~e mitten; **Faust·regel** die rule of thumb

Favorit [favo'ri:t] der; ~en, ~en favourite

Fax das; ~, ~[e] fax; **faxen** *tr. V.* fax

Faxen *Pl. (ugs.)* fooling around

Fazit ['fa:tsɪt] das; ~s, ~s *od.* ~e result

Februar der; ~[s], ~e February

fechten *unr. itr., tr. V.* fence; **Fechter** der; ~s, ~: fencer

Feder die; ~, ~n a) *(Vogel~)* feather; b) *(zum Schreiben)* nib; c) *(Technik)* spring

feder-, Feder-: ~ball der a) *(Spiel)* badminton; b) *(Ball)* shuttlecock;

~**bett** das duvet *(Brit.);* stuffed quilt *(Amer.);* ~**führend** *Adj.* in charge *postpos.;* ~**halter** der fountain-pen; ~**leicht** *Adj.* *(person)* as light as a feather; featherweight *(object);* ~**lesen** das: nicht viel ~**lesen[s]** mit jmdm./etw. machen give sb./sth. short shrift

federn 1. *itr. V.* *(springboard, floor, etc.)* be springy; 2. *tr. V.* *(mit einer Federung versehen)* spring; **das Bett ist gut gefedert** the bed is well-sprung; **Federung** die; ~, ~**en** *(Kfz-W.)* suspension

Fee die; ~, ~**n** fairy

Fege·feuer das purgatory; **fegen** 1. *tr. V.* **a)** *(bes. nordd.: säubern)* sweep; **b)** *(schnell entfernen)* brush; 2. *itr. V.* sweep up

fehl *Adv.* ~ **am Platz[e] sein** be out of place; **Fehl·anzeige** die: ~! *(ugs.)* no chance! *(coll.);* **fehlen** *itr. V.* **a)** *(nicht vorhanden sein)* **ihm fehlt das Geld** he has no money; **b)** *(ausbleiben)* be absent; **c)** *(verschwunden sein)* be missing; **in der Kasse fehlt Geld** money is missing from the till; **d)** *(vermißt werden)* **er/das wird mir** ~ : I shall miss him/that; *(erforderlich sein)* be needed; **ihm** ~ **noch zwei Punkte zum Sieg** he needs only two points to win; **es fehlte nicht viel, und ich wäre eingeschlafen** I all but fell asleep; **f)** *unpers. (mangeln)* **es fehlt an Lehrern** there is a lack of teachers; **g)** *(krank sein)* **was fehlt Ihnen?** what seems to be the matter?; **fehlt dir etwas?** is there something wrong?; **Fehler** der; ~**s**, ~ **a)** *(Irrtum)* mistake; error; *(Sport)* fault; **b)** *(schlechte Eigenschaft)* fault; **fehler·frei** *Adj.* faultless; **fehlerhaft** *Adj.* faulty; defective; imperfect *(pronunciation);* **Fehler·quelle** die source of error

fehl-, Fehl-: ~**geburt** die miscarriage; ~**schlag** der failure; ~**schlagen** *unr. itr. V.; mit sein* fail; ~**start** der *(Leichtathletik)* false start; ~**tritt** der *(fig. geh.)* slip; ~**zündung** die *(Technik)* misfire

Feier die; ~, ~**n a)** *(Veranstaltung)* party; *(aus festlichem Anlaß)* celebration; **b)** *(Zeremonie)* ceremony; **Feier·abend** der *(Arbeitsschluß)* finishing time; **nach** ~ : after work; ~ **machen** finish work; **feierlich** 1. *Adj.* ceremonial *(act etc.);* solemn *(silence);* 2. *adv.* solemnly; ceremoniously; **Feierlichkeit** die; ~, ~**en a)** *o.*

Pl. solemnity; **b)** *meist Pl. (Veranstaltung)* celebration; **feiern** 1. *tr. V.* **a)** celebrate *(birthday, wedding, etc.);* **b)** acclaim *(artist, sportsman, etc.);* 2. *itr. V.* celebrate

Feier·tag der holiday; **ein gesetzlicher/kirchlicher** ~ a public holiday/religious festival

feig[e] 1. *Adj.* cowardly; 2. *adv.* in a cowardly way

Feige die; ~, ~**n** fig

Feigheit die; ~: cowardice; **Feigling** der; ~**s**, ~**e** coward

Feile die; ~, ~**n** file; **feilen** *tr., itr. V.* file

feilschen *itr. V.* haggle **(um** over)

fein 1. *Adj.* **a)** fine; finely-ground *(flour);* finely-granulated *(sugar);* **b)** *(hochwertig)* high-quality *(fruit, soap, etc.);* fine *(silver, gold, etc.);* fancy *(cakes, pastries, etc.);* **c)** *(ugs.: erfreulich)* great *(coll.);* 2. *adv.* ~ **[he]raussein** *(ugs.)* be sitting pretty *(coll.)*

Feind der; ~**[e]s**, ~**e** enemy; **feindlich** 1. *Adj.* **a)** hostile; **b)** *(Milit.)* enemy *(attack, activity);* 2. *adv.* in a hostile manner; **Feindschaft** die; ~, ~**en** enmity; **feind·selig** *Adj.* hostile

Feinheit die; ~, ~**en a)** fineness; delicacy; **b)** *(Nuance)* subtlety

fein-, Fein-: ~**kost·geschäft** das delicatessen; ~**[machen** *refl. V.* *(ugs.)* dress up; ~**schmecker** der; ~**s**, ~: gourmet; ~**sinnig** *Adj.* sensitive and subtle; ~**waschmittel** das mild detergent

feist *Adj. (meist abwertend)* fat

Feld das; ~**[e]s**, ~**er a)** field; **b)** *(Sport: Spiel~)* pitch; field; **c)** *(auf Formularen)* box; space; *(auf Brettspielen)* space; *(auf dem Schachbrett)* square; **d)** *o. Pl. (Tätigkeitsbereich)* field; sphere

Feld-: ~**herr** der *(veralt.)* commander; ~**marschall** der Field Marshal; ~**salat** der corn salad; ~**stecher** der binoculars *pl.;* ~**webel** der; ~**s**, ~ *(Milit.)* sergeant; ~**weg** der path; track; ~**zug** der *(Milit., fig.)* campaign

Felge die; ~, ~**n** [wheel] rim

Fell das; ~**[e]s**, ~**e a)** *(Haarkleid)* fur; *(Pferde~, Hunde~, Katzen~)* coat; *(Schaf~)* fleece; **b)** *(Material)* fur; **c)** *(abgezogen)* hide; **ein dickes** ~ **haben** *(ugs.)* be thick-skinned

Fels der; ~**en**, ~**en** rock; **Felsen** der; ~**s**, ~: rock; *(an der Steilküste)* cliff; **felsen·fest** *Adj.* firm; unshakeable

⟨*opinion, belief*⟩; **fęlsig** *Adj.* rocky; **Fęls·wand** die rock face

feminįn *Adj.* feminine; **Feminįsmus** der; ~ feminism *no art.;* **Feminįstin** die; ~, ~nen feminist

Fęnchel der; ~s fennel

Fęnster das; ~s, ~: window

Fęnster-: ~**bank** die window-sill; ~**laden** der [window] shutter; ~**leder** das wash-leather; ~**platz** der window-seat; ~**putzer** der window-cleaner; ~**rahmen** der window-frame; ~**scheibe** die window-pane

Ferien ['fe:riǝn] *Pl.* **a)** holiday[s *pl.*] *(Brit.);* vacation *(Amer.);* **in die** ~ **fahren** go on holiday/vacation; ~ **haben** have a *or* be on holiday/vacation; **Ferien·haus** das holiday/vacation house

Fęrkel das; ~s, ~: piglet

fęrn 1. *Adj.* distant; **2.** *adv.* ~ **von der Heimat** far from home; **3.** *Präp. mit Dat. (geh.)* far [away] from; **fęrn|bleiben** *unr. itr. V.; mit sein (geh.)* stay away; **Fęrne** die; ~, ~n distance; **ferner** *Adv.* furthermore

fęrn-, Fęrn-: ~**fahrer** der long-distance lorry-driver *(Brit.)* or *(Amer.)* trucker; ~**gespräch** das long-distance call; ~**glas** das binoculars *pl.;* ~|**halten** *unr. tr., refl. V.* keep away; ~**heizung** die district heating system; ~**licht** das *(Kfz-W.)* full beam; ~**melde·amt** das telephone exchange; ~**ọst** *o. Art.* Far East; ~**rohr** das telescope; ~**ruf** der telephone number; ~**schreiben** das telex [message]; ~**schreiber** der telex [machine]

Fęrnseh-: ~**antenne** die television aerial *(Brit.)* or *(Amer.)* antenna; ~**apparat** der television [set]

fęrn|sehen *unr. itr. V.* watch television; **Fęrn·sehen** das; ~s television; **im** ~: on television; **Fęrn·seher** der; ~s, ~ *(ugs.)* telly *(Brit. coll.);* TV

Fęrnseh-: ~**gebühren** *Pl.* television licence fee; ~**gerät** das television [set]; ~**programm** das **a)** *(Sendungen)* television programmes *pl.;* **b)** *(Kanal)* television channel; **c)** *(Blatt, Programmheft)* television [programme] guide; ~**sendung** die television programme; ~**spiel** das television play; ~**zuschauer** der television viewer

Fęrn·sprecher der telephone

Fęrnsprech-: ~**gebühren** *Pl.* tele-

phone charges; ~**teilnehmer** der telephone subscriber; telephone customer *(Amer.)*

Fęrn-: ~**steuerung** die *(Technik)* remote control; ~**straße** die major road; ~**verkehr** der long-distance traffic; ~**zug** der long-distance train

Fęrse die; ~, ~n heel

fęrtig *Adj.* **a)** finished ⟨*manuscript, picture, etc.*⟩; **das Essen ist** ~: lunch/dinner *etc.* is ready; |**mit etw.**| ~ **sein/werden** have finished/finish [sth.]; **b)** *(bereit, verfügbar)* ready (**zu, für** for); **c)** *(ugs.: erschöpft)* shattered *(coll.)*

fęrtig-, Fęrtig-: ~**bau** der; *Pl.* ~ten prefabricated building; ~**bauweise** die prefabricated construction; prefabrication ~|**bringen** *unr. tr. V.* manage

fęrtigen *tr. V.* make

Fęrtig-: ~**gericht** das ready-to-serve meal; ~**haus** das prefabricated house; prefab *(coll.)*

Fęrtigkeit die; ~, ~en skill

fęrtig-, Fęrtig-: ~|**machen** *tr. V. (ugs.)* finish ⟨*task, job, etc.*⟩; get ⟨*meals, beds*⟩ ready; **jmdn.** ~**machen** *(erschöpfen)* wear sb. out; *(durch Schikanen)* wear sb. down; *(deprimieren)* get sb. down; ~|**stellen** *tr. V.* complete; ~**stellung** die completion

Fęssel die; ~, ~n fetter; shackle; *(Kette)* chain; **fęsseln** *tr. V.* **a)** tie up; **b)** *(faszinieren)* ⟨*book*⟩ grip; ⟨*work, person*⟩ fascinate

fęst 1. *Adj.* **a)** *(nicht flüssig od. gasförmig)* solid; **b)** firm ⟨*bandage*⟩; sound ⟨*sleep*⟩; sturdy ⟨*shoes*⟩; strong ⟨*fabric*⟩; solid ⟨*house, shell*⟩; steady ⟨*voice*⟩; **der** ~**en Überzeugung sein, daß ...**: be of the firm opinion that ...; **c)** *(dauernd)* permanent ⟨*address*⟩; fixed ⟨*income*⟩; **2.** *adv.* **a)** ⟨*tie, grip*⟩ tight[ly]; **b)** *(ugs. auch* ~**e)** ⟨*work*⟩ with a will; ⟨*eat*⟩ heartily; ⟨*sleep*⟩ soundly; **c)** ⟨*believe, be convinced*⟩ firmly; **sich auf jmdn./etw.** ~ **verlassen** rely one hundred per cent on sb./sth.; **d)** *(endgültig)* firmly; **etw.** ~ **vereinbaren** come to a firm arrangement about sth.; **e)** *(auf Dauer)* permanently; ~ **befreundet sein** be close friends; *(als Paar)* be going steady

Fęst das; ~[e]s, ~e **a)** celebration; *(Party)* party; **b)** *(Feiertag)* festival; **frohes** ~**!** happy Christmas/Easter!

fęst-: ~|**binden** *unr. tr. V.* tie [up]; ~|**bleiben** *unr. itr. V.; mit sein* stand firm; ~|**fahren** *unr. itr., refl. V. (itr. V.*

mit sein) get stuck; *(fig.)* get bogged down; ~|**halten 1.** *unr. tr. V.* **a)** *(halten, packen)* hold on to; **b)** *(nicht weiterleiten)* withhold *(letter, parcel, etc.)*; **c)** *(verhaftet haben)* hold, detain *(suspect)*; **2.** *unr. refl. V.* **sich an jmdm./ etw. ~halten** hold on to sb./sth.

festigen 1. *tr. V.* strengthen; consolidate *(position)*; **2.** *refl. V.* *(friendship, ties)* become stronger

Festival ['fɛstɪvəl] *das*; ~s, ~s festival

fest-, Fest-: ~|**kleben** *tr., itr. V.; mit sein* stick (**an** + *Dat.* to); ~|**land** *das*; *o. Pl.* (Kontinent) continent; *(im Gegensatz zu den Inseln)* mainland; ~|**legen** *tr. V.* **a)** fix *(time, deadline, price)*; arrange *(programme)*; **b)** *(verpflichten)* **sich [auf etw. *(Akk.)*] ~legen** [lassen] commit oneself [to sth.]; **jmdn. [auf etw. *(Akk.)*] ~legen** tie sb. down [to sth.]

festlich 1. *Adj.* festive *(atmosphere)*; formal *(dress)*; **2.** *adv.* festively; formally

fest-: ~|**machen** *tr. V.* **a)** *(befestigen)* fix; **b)** *(fest vereinbaren)* arrange *(meeting etc.)*; ~|**nageln** *tr. V.* **a)** *(befestigen)* nail (**an** + *Dat.* to); **b)** *(ugs.: festlegen)* **jmdn. [auf etw. *(Akk.)*]** ~**nageln** tie sb. down [to sth.]; ~|**nehmen** *unr. tr. V.* arrest

Fest·rede *die* speech

fest-, Fest-: |**schnallen** *tr. V.* tie (**an** + *Dat.* to); ~|**sitzen** *unr. itr. V.* be stuck; ~|**stehen** *unr. itr. V.* *(order, appointment, etc.)* have been fixed; *(decision)* be definite; *(fact)* be certain; ~|**stellen** *tr. V.* **a)** establish *(identity, age, facts)*; **b)** *(wahrnehmen)* detect; diagnose *(illness)*; ~**stellung** *die* **a)** establishment; **b)** *(Wahrnehmung)* realization; **die ~stellung machen, daß ...:** realize that ...

Fest·tag *der* holiday; *(Ehrentag)* special day

Festung *die*; ~, ~en fortress

Fest·zelt *das* marquee

fest|ziehen *unr. tr. V.* pull tight

Fete *die*; ~, ~n *(ugs.)* party

fett 1. *Adj.* **a)** fatty *(food)*; ~**er Speck** fat bacon; **b)** *(sehr dick)* fat; **2.** *adv.* ~ **essen** eat fatty foods; **Fett** *das*; ~[e]s, ~e fat; ~ **ansetzen** *(animal)* fatten up; *(person)* put on weight

fett-, Fett-: ~**arm** *Adj.* low-fat *(food)*; low in fat *pred.*; ~**auge** *das* speck of fat; ~**fleck[en]** *der* grease mark; ~**gedruckt** *Adj.* *(präd. getrennt geschrieben)* bold

fettig *Adj.* greasy

fett-, Fett-: ~**leibig** *Adj.* obese; ~**leibigkeit** *die*; ~: obesity; ~**näpfchen** *das*: **ins ~näpfchen treten** *(scherzh.)* put one's foot in it; ~**reich** *Adj.* high-fat

Fetzen *der*; ~s, ~: scrap

feucht *Adj.* damp; humid *(climate)*; **feucht·fröhlich** *Adj.* *(ugs. scherzh.)* merry *(company)*; boozy *(coll.)* *(evening)*; **Feuchtigkeit** *die* moisture

feucht-: ~**kalt** *Adj.* cold and damp; ~**warm** *Adj.* muggy

feudal *Adj.* **a)** feudal *(system)*; **b)** aristocratic *(regiment etc.)*; **c)** *(ugs.: vornehm)* plush *(hotel etc.)*

Feuer *das*; ~s, ~ **a)** fire; **jmdm. ~ geben** give sb. a light; **b)** *(Brand)* fire; blaze; ~! fire!; **c)** *o. Pl. (Milit.)* **das ~ einstellen** cease fire

feuer-, Feuer-: ~**eifer** *der* enthusiasm; zest; ~**fest** *Adj.* heat-resistant *(dish, plate)*; fire-proof *(material)*; ~**gefährlich** *Adj.* [in]flammable; ~**holz** *das*; *o. Pl.* firewood; ~**leiter** *die* *(bei Häusern)* fire escape; *(beim ~wehrauto)* [fireman's] ladder; ~**löscher** *der*; ~s, ~: fire extinguisher; ~**melder** *der* fire alarm

feuern 1. *tr. V.* **a)** *(ugs.: entlassen)* fire *(coll.)*; sack *(coll.)*; **b)** *(ugs.: schleudern, werfen)* fling; **2.** *itr. V. (Milit.)* fire (**auf** + *Akk.* at)

feuer-, Feuer-: ~**rot** *Adj.* fiery red; ~**schlucker** *der* fire-eater; ~**sirene** *die* fire siren; ~**stein** *der* flint; ~**versicherung** *die* fire insurance; ~**waffe** *die* firearm; ~**wehr** *die*; ~, ~**en** fire service; ~**wehr·auto** *das* fire engine; ~**wehr·mann** *der*; *Pl.* ~**männer** *od.* ~**leute** fireman; ~**werk** *das* firework display; *(~werkskörper)* fireworks *pl.*; ~**werks·körper** *der* firework; ~**zeug** *das* lighter

Feuilleton [fœjə'tõ:] *das*; ~s, ~s arts section

feurig *Adj.* fiery

ff. *Abk.* folgende [Seiten] ff.

Ffm. *Abk.* Frankfurt am Main

Fiaker ['fi̯akɐ] *der*; ~s, ~ *(österr.)* cab

Fiasko *das*; ~s, ~s fiasco

Fibel *die*; ~, ~n reader; primer

ficht [fɪçt] *Imperativ Sg. u. 3. Pers. Sg. Präsens v.* fechten

Fichte *die*; ~, ~n spruce

ficken *tr., itr. V. (vulg.)* fuck *(coarse)*

fidel *Adj. (ugs.)* jolly

Fieber *das*; ~s [high] temperature; *(über 38 °C)* fever; ~ **haben** have a

[high] temperature/a fever; **bei jmdm.**
~ messen take sb's temperature;
fieber·frei Adj. ⟨person⟩ free from
fever; **fieberhaft** Adj. feverish;
fieberig Adj. feverish; **fiebern** itr.
V. have a temperature; **Fieber·ther-**
mometer das [clinical] thermometer;
fiebrig Adj. feverish
Fiedel die; ~, ~n (veralt., scherzh.)
fiddle
fiel 1. u. 3. Pers. Sg. Prät. v. **fallen**
fiepen itr. V. ⟨dog⟩ whimper; ⟨bird⟩
cheep
fies 1. Adj. (ugs.) nasty ⟨person, charac-
ter⟩; 2. adv. in a nasty way
Figur die; ~, ~en a) (einer Frau)
figure; (eines Mannes) physique; b)
(Bildwerk) figure; c) (geometrisches
Gebilde) shape
fiktiv Adj. fictitious
Filet [fi'le:] das; ~s, ~s fillet
Filiale die; ~, ~n branch
Filigran das; ~s, ~e filigree
Film der; ~|e|s, ~e a) (Fot.) film; b)
(Kino~) film; movie (Amer. coll.); **fil-**
men tr., itr. V. film
Film-: ~**kamera** die film camera;
(Schmalfilmkamera)   cine-camera;
~**produzent** der film producer;
~**regisseur** der film director;
~**schau·spieler** der film actor
Filter der, ~s, ~: filter; **filtern** tr. V.
filter
Filter-: ~**papier** das filter paper; ~**zi-**
garette die [filter-]tipped cigarette
Filz der; ~es, ~e felt
Fimmel der; ~s, ~: **einen ~ für etw.**
haben (ugs. abwertend) have a thing
about sth. (coll.)
Finale das; ~s, ~|s| a) (Sport) final; b)
finale
Finanz die; ~: finance no art.
Finanz-: ~**amt das a)** (Behörde) ≈ Inland Revenue; b) (Gebäude) tax office; ~**beamte** der tax officer
Finanzen Pl. finances; **finanziell**
[finan'tsiɛl] Adj. financial; **finan-**
zieren tr. V. finance; **Finanzierung**
die; ~, ~en financing
Finanz-: ~**minister** der minister of
finance; ~**politik** die (des Staates, eines Unternehmens) financial policy;
(allgemeine) politics of finance
Findel·kind das foundling
finden unr. tr. V. find; **Freunde** ~:
make friends; **Finder der,** ~s, ~: finder; s. auch **ehrlich** 1 a; **Finder·lohn**
der reward [for finding sth.]; **findig**
Adj. resourceful; **Findling der;** ~s,

~e a) (Findelkind) foundling; b) (Geol.) erratic block
fing 1. u. 3. Pers. Sg. Prät. v. **fangen**
Finger der; ~s, ~: finger; **lange ~ ma-**
chen (ugs.) get itchy fingers
Finger-: ~**abdruck** der fingerprint;
~**fertigkeit** die; o. Pl. dexterity;
~**hut** der thimble
fingern itr. V. fiddle; **an etw.** (Dat.) ~:
fiddle with sth.; **nach etw. ~:** fumble
[around] for sth.
Finger-: ~**nagel** der fingernail;
~**spitze** die fingertip; ~**spit-**
zen·gefühl das; o. Pl. feeling
fingieren tr. V. fake; **ein fingierter**
Name a false name
Fink der; ~en, ~en finch
Finne der; ~n, ~n, **Finnin die;** ~,
~nen Finn; **finnisch** Adj. Finnish;
Finnland (das); ~s Finland
finster 1. Adj. dark; dimly-lit ⟨pub,
district⟩; 2. adv. **jmdn. ~ ansehen** give
sb. a black look; **Finsternis die;** ~,
~se darkness; (auch bibl., fig.) dark
Finte die; ~, ~n trick; **jmdn. durch eine**
~ täuschen deceive sb. by trickery
firm Adj. **in etw.** (Dat.) ~ **sein** be well
up in sth.
Firma die; ~, **Firmen** firm; company
Firmen-: ~**inhaber** der owner of
the/a company; ~**schild** das company's name plate; ~**zeichen** das
trademark
Firmung die; ~, ~en confirmation
First der; ~|e|s, ~e ridge
Fisch der; ~|e|s, ~e a) fish; |**fünf**| ~e
fangen catch [five] fish; **kleine** ~e (fig.)
small fry; b) (Astrol.) **die** ~e Pisces; **er**
ist |ein| ~: he is a Piscean; **fischen** 1.
tr. V. a) fish for; b) (ugs.) **etw. aus etw.**
~: fish sth. out of sth.; 2. itr. V. fish;
nach etw. ~: fish for sth.; **Fischer**
der; ~s, ~ fisherman
Fischer·boot das fishing boat
Fischerei die; ~: fishing
Fisch-: ~**fang** der; o. Pl. **vom ~ leben**
make a/one's living by fishing; **auf ~**
gehen go fishing; ~**geschäft** das
fishmonger's [shop] (Brit.); fish store
(Amer.);   ~**grät|en|·muster**   das
(Textilw.)   herringbone   pattern;
~**konserve** die canned fish; ~**kut-**
ter der fishing trawler; ~**stäbchen**
das (Kochk.) fish finger
Fiskus der; ~, **Fisken** od. ~se Government (as managing the State finances)
Fittich der; ~|e|s, ~e (dichter.) wing
fix 1. Adj. (ugs.) quick; **ein ~er Bursche**
a bright lad; ~ **und fertig** quite fin-

ished; *(völlig erschöpft)* completely shattered *(coll.)*; **2.** *adv. (ugs.)* quickly; **mach ~!** hurry up!

fixen *itr. V. (Drogenjargon)* fix *(sl.)*; **Fixer der;** ~s, ~ *(Drogenjargon)* fixer

fixieren *tr. V.* **a)** fix one's gaze on; **jmdn. scharf ~:** gaze sharply at sb.; **b)** *(geh.: schriftlich niederlegen)* take down

Fix·stern der *(Astron.)* fixed star

Fjord [fjɔrt] **der;** ~|e|s, ~e fiord

FKK [εf ka: 'ka:] *Abk.* Freikörperkultur nudism *no art.;* naturism *no art.;* **FKK-Strand der** nudist beach

flach *Adj.* **a)** flat; **b)** *(niedrig)* low; **c)** *(nicht tief)* shallow ⟨water, dish⟩; **Flä·che die;** ~, ~n **a)** area; **b)** *(Ober~)* surface; **c)** *(Geom.)* area; *(einer dreidimensionalen Figur)* side

Flächen-: ~inhalt der area; ~maß das unit of square measure

flach|fallen *itr. V.; mit sein (ugs.)* ⟨trip⟩ fall through; ⟨event⟩ be cancelled; **Flach·land das;** *o. Pl.* lowland

Flachs der; ~es flax

flachsen *itr. V. mit jmdm. ~ (ugs.)* joke with sb.

flackern *itr. V.* flicker

Fladen der; ~s, ~ *flat, round unleavened cake made with oat or barley flour*

Flagge die; ~, ~n flag; **flaggen** *itr. V.* put out the flags

flambieren *tr. V. (Kochk.)* flambé

Flamme die; ~, ~n **a)** flame; **b)** *(Brennstelle)* burner

Flanell der; ~s, ~e flannel

flanieren *itr. V.; mit Richtungsangabe mit sein* stroll

Flanke die; ~, ~n **a)** *(Weiche)* flank; **b)** *(Ballspiele: Vorlage)* centre; **c)** *(Teil des Spielfeldes)* wing

Flasche die; ~, ~n bottle; **eine ~ Wein** a bottle of wine; **dem Kind die ~ geben** feed the baby

Flaschen-: ~bier das bottled beer; ~öffner der bottle-opener; ~zug der block and tackle

flatterhaft *Adj.* fickle; **flattern** *itr. V. mit Richtungsangabe mit sein* flutter

flau *Adj.* **a)** slack ⟨breeze⟩; **b)** *(leicht übel)* queasy ⟨feeling⟩

Flaum der; ~|e|s fuzz

Flausch der; ~|e|s, ~e brushed wool; **flauschig** *Adj.* fluffy

Flause die; ~, ~n; *meist Pl. (ugs.)* **er hat nur ~n im Kopf** he can never think of anything sensible

Flaute die; ~, ~n **a)** *(Seemannsspr.)* calm; **b)** *(Kaufmannsspr.)* fall[-off] in trade

Flechte die; ~, ~n **a)** *(Bot.)* lichen; **b)** *(Med.)* eczema

flechten *unr. tr. V.* plait ⟨hair⟩; weave ⟨basket, mat⟩

Fleck der; ~|e|s, ~e **a)** stain; *(andersfarbige Stelle)* patch; **flecken** *itr. V.* stain; **flecken·los 1.** *Adj.* spotless; **2.** *adv.* spotlessly

Fleck·entferner der stain *or* spot remover

fleckig *Adj.* stained; blotchy ⟨face, skin⟩

Fleder·maus die bat

Flegel der; ~s, ~ *(abwertend)* lout; **flegelhaft** *Adj. (abwertend)* loutish

flehen ['fle:ən] *itr. V.* plead **(um** for)

Fleisch das; ~|e|s **a)** flesh; **b)** *(Nahrungsmittel)* meat; **Fleisch·brühe die** bouillon; consommé; **Fleischer der;** ~s, ~: butcher; **Fleischerei die;** ~, ~en butcher's shop

fleischig *Adj.* plump ⟨hands, face⟩; fleshy ⟨leaf, fruit⟩

Fleisch-: ~käse der meat loaf; ~klößchen das small meat ball; ~pastete die *(Kochk.)* pâté; ~salat der *(Kochk.)* meat salad; ~vergiftung die food poisoning [from meat]; ~waren *Pl.* meat products; ~wolf der mincer; ~wunde die fleshwound; ~wurst die pork sausage

Fleiß der; ~es hard work; *(Eigenschaft)* diligence; **fleißig 1.** *Adj.* hard-working; **2.** *adv.* hard; **~ lernen** learn as much as one can

flennen *itr. V. (ugs.)* blubber

fletschen *tr., itr. V.* **die Zähne** *od.* **mit den Zähnen ~:** bare one's teeth

Fleurop ⟨Wz⟩ ['flɔyrɔp] die Interflora **(P)**

flexibel 1. *Adj.* flexible; **2.** *adv.* flexibly

flicht *Imperativ Sg. u. 3. Pers. Sg. Präsens v.* **flechten**

flicken *tr. V.* mend; repair ⟨engine, cable⟩; **Flicken der;** ~s, ~: patch

Flick-: ~werk das; *o. Pl. (abwertend)* botched-up job; ~zeug das repair kit

Flieder der; ~s, ~: lilac

Fliege die; ~, ~n **a)** fly; **b)** *(Schleife)* bow-tie; **fliegen 1.** *unr. itr. V.; mit sein* **a)** fly; **b)** *(ugs.: fallen)* **vom Pferd/ Fahrrad ~:** fall off a/the horse/bicycle; **c)** *(ugs.: entlassen werden)* get the sack *(coll.)*; **von der Schule ~:** be chucked out [of the school] *(coll.)*; **2.** *unr. tr. V.* fly

Fliegen-: **~fenster das** wire-mesh window; **~gewicht das** *(Schwerathletik)* flyweight; **~pilz der** fly agaric; **Flieger der; ~s, ~** pilot; **Fliegeralarm der** air-raid warning; **fliegerisch** *Adj.* aeronautical

fliehen ['fli:ən] *unr. itr. V.; mit sein* flee (**vor** + *Dat.* from); *(aus dem Gefängnis usw.)* escape (**aus** from); **ins Ausland/über die Grenze ~:** flee the country/escape over the border

Flieh · kraft die *(Physik)* centrifugal force

Fliese die; ~, ~n tile

Fließ · band das conveyor belt; **am ~band arbeiten** *od. (ugs.)* **stehen** work on the assembly line

fließen *unr. itr. V.; mit sein* flow; **~des Wasser** running water; **eine Sprache ~d sprechen** speak a language fluently

flimmern *itr. V.; mit Richtungsangabe mit sein* shimmer

flink 1. *Adj.* nimble *(fingers)*; sharp *(eyes)*; quick *(hands)*; 2. *adv.* quickly

Flinte die; ~, ~n shotgun; **die ~ ins Korn werfen** *(fig.)* throw in the towel

Flirt der; ~s, ~s flirtation; **flirten** *itr. V.* flirt

Flitter der; ~s frippery; trumpery

Flitter · wochen *Pl.* honeymoon *sing.*

flitzen *itr. V.; mit sein (ugs.)* shoot; dart; **Flitzer der; ~s, ~** *(ugs.)* sporty job *(coll.)*

floaten ['floʊtn̩] *tr., itr. V. (Wirtsch.)* float

flocht *1. u. 3. Pers. Sg. Prät. v.* **flechten**

Flocke die; ~, ~n a) flake; **b)** *(Staub~)* piece of fluff; **flockig** *Adj.* fluffy

flog *1. u. 3. Pers. Sg. Prät. v.* **fliegen**

floh *1. u. 3. Pers. Sg. Prät. v.* **fliehen**

Floh der; ~|e|s, Flöhe flea

Floh-: **~markt der** flea market; **~zirkus der** flea-circus

Flora die; ~, Floren flora

Florett das; ~|e|s, ~e foil

florieren *itr. V.* flourish

Florist der; ~en, ~en, Floristin die; ~, ~nen [qualified] flower-arranger

Floskel die; ~, ~n cliché

floß *1. u. 3. Pers. Sg. Prät. v.* **fließen**

Floß das; ~es, Flöße raft

Flosse die; ~, ~n a) *(Zool., Flugw.)* fin; **b)** *(zum Tauchen)* flipper

flößen *tr., itr. V.* float

Flößer der; ~s, ~: raftsman

Flöte die; ~, ~n flute; *(Block~)* recorder; **flöten** 1. *itr. V.* *(bird)* flute; 2. *tr. V.* whistle; **flöten|gehen** *unr. itr.*

V.; mit sein (ugs.) *(money)* go down the drain; *(time)* be wasted

flott 1. *Adj.* **a)** *(schwungvoll)* lively; **b)** *(schick)* smart; 2. *adv.* *(work)* quickly; *(dance, write)* in a lively manner; *(be dressed)* smartly

Flotte die; ~, ~n fleet

flott|machen *tr. V.* refloat *(ship)*; get *(car)* back on the road

Flöz das; ~es, ~e *(Bergbau)* seam

Fluch der; ~|e|s, Flüche curse; oath; **fluchen** *itr. V.* curse; swear

Flucht die; ~: flight; **flucht · artig** 1. *Adj.* hurried; hasty; 2. *adv.* hurriedly; hastily; **flüchten** 1. *itr. V.; mit sein* **vor jmdm./etw. ~:** flee from sb./sth.; **vor der Polizei ~:** run away from the police; 2. *refl. V.* take refuge; **flüchtig** 1. *Adj.* **a)** fugitive; **b)** cursory; superficial *(insight)*; 2. *adv.* **a)** *(oberflächlich)* cursorily; **b)** *(eilig)* hurriedly; **Flüchtigkeit die; ~, ~en** cursoriness;

Flüchtigkeits · fehler der slip; **Flüchtling der; ~s, ~e** refugee; **Flucht · weg der** escape route

Flug der; ~|e|s, Flüge flight

Flug-: **~bahn die** trajectory; **~blatt das** pamphlet; leaflet

Flügel der; ~s, ~ a) wing; **b)** *(Klavier)* grand piano

Flug · gast der [air] passenger

flügge *Adj.* fully-fledged

Flug-: **~gesellschaft die** airline; **~hafen der** airport; **~linie die a)** *(Strecke)* air route; **b)** *(Gesellschaft)* airline; **~lotse der** air traffic controller; **~platz der** airfield; **~schein der** air ticket; **~verkehr der** air traffic

Flug · zeug das; ~|e|s, ~e aeroplane *(Brit.)*; airplane *(Amer.)*; aircraft

Flugzeug-: **~absturz der** plane crash; **~entführer der** [aircraft] hijacker; **~entführung die** [aircraft] hijack[ing]; **~träger der** aircraft carrier

Flunder die; ~, ~n flounder

flunkern *itr. V.* tell stories

Fluor das; ~s *(Chemie)* fluorine

¹Flur der; ~|e|s, ~e *(Korridor)* corridor; *(Diele)* [entrance] hall; **im/auf dem ~:** in the corridor/hall

²Flur die; ~, ~en farmland *no indef. art.*

Fluß der; Flusses, Flüsse river; *(fließende Bewegung)* flow

fluß-, Fluß-: **~ab[wärts]** *Adv.* downstream; **~auf[wärts]** *Adv.* upstream; **~bett das** river bed

Flüßchen das; ~s, ~: small river
flüssig 1. *Adj.* **a)** liquid; **b)** *(fließend, geläufig)* fluent; 2. *adv.* ⟨*write, speak*⟩ fluently; **Flüssigkeit** die; ~, ~en **a)** liquid; *(auch Gas)* fluid; **b)** *(Geläufigkeit)* fluency; **flüssig|machen** *tr. V.* make available ⟨*money, funds*⟩
Fluß·pferd das hippopotamus
flüstern *itr., tr. V.* whisper
Flut die; ~, ~en **a)** *o. Pl.* tide; **b)** *meist Pl. (geh.: Wassermasse)* flood; **fluten** *itr. V.; mit sein (geh.)* flood; **Flut·licht** das; *o. Pl.* floodlight
focht *1. u. 3. Pers. Sg. Prät. v.* **fechten**
Föderalismus der; ~: federalism *no art.;* **föderalistisch** *Adj.* federalist
Fohlen das; ~s, ~: foal
Föhn der; ~|e|s, ~e föhn
Folge die; ~, ~n **a)** *(Auswirkung)* consequence; *(Ergebnis)* consequence; result; **b)** *(Aufeinander~)* succession; *(zusammengehörend)* sequence; **c)** *(Fortsetzung) (einer Sendung)* episode; *(eines Romans)* instalment; **Folgeerscheinung** die consequence; **folgen** *itr. V.; mit sein* follow; **jmdm. im Amt/in der Regierung ~:** succeed sb. in office/in government; **auf etw.** *(Akk.)* ~: follow sth.; **aus etw.** ~: follow from sth.; **folgend** *Adj.* **der/die/das** ~e the next in order; **im** ~en *od.* **in** ~em in [the course of] the following discussion/passage *etc.;* **folgendermaßen** *Adv.* as follows; *(so)* in the following way; **folge·richtig** 1. *Adj.* logical; consistent ⟨*behaviour, action*⟩; 2. *adv.* logically; ⟨*act, behave*⟩ consistently; **folgern** 1. *tr. V.* **etw. aus etw.** ~: infer sth. from sth.; 2. *itr. V.* **richtig** ~: draw a/the correct conclusion; **Folgerung** die; ~, ~en conclusion
folglich *Adv.* consequently; **folgsam** 1. *Adj.* obedient; 2. *adv.* obediently
Folie ['foːliə] die; ~, ~n *(Metall~)* foil; *(Plastik~)* film
Folklore die; ~ **a)** folklore; **b)** *(Musik)* folk-music
Folter die; ~, ~n torture; **foltern** *tr. V.* torture; *(fig.)* torment; **Folterung** die; ~, ~en torture
Fön ⓦ der; ~|e|s, ~e hair-drier
Fond [fõː] der; ~s, ~s *(geh.)* back
Fonds [fõː] der; ~ [fõː(s)], ~ [fõːs] fund
Fondue [fõˈdyː] die; ~, ~s *od.* das; ~s, ~s *(Kochk.)* fondue
fönen *tr. V.* blow-dry
Fontäne die; ~, ~n jet; *(Springbrunnen)* fountain

forcieren [fɔrˈsiːrən] *tr. V.* step up ⟨*production*⟩; intensify ⟨*efforts*⟩; push forward ⟨*developments*⟩
Förderer der; ~s, ~: patron
fordern *tr. V.* **a)** demand; **b)** *(in Anspruch nehmen)* make demands on
fördern *tr. V.* **a)** promote; patronize, support ⟨*artist, art*⟩; further ⟨*investigation*⟩; foster ⟨*talent, tendency*⟩; improve ⟨*appetite*⟩; aid ⟨*digestion, sleep*⟩; **b)** *(Bergbau, Technik)* mine ⟨*coal, ore*⟩; extract ⟨*oil*⟩
Forderung die; ~, ~en **a)** demand; **b)** *(Kaufmannsspr.)* claim (**an** + *Akk.* against)
Förderung die; ~, ~en *o. Pl. s.* **fördern a:** promotion; patronage; support; furthering; fostering; improvement; aiding; **b)** *(Bergbau, Technik)* output; *(das Fördern)* mining; *(von Erdöl)* extraction
Forelle die; ~, ~n trout
Form die; ~, ~en **a)** shape; **in ~ von Tabletten** in the form of tablets; **b)** *(bes. Sport: Verfassung)* form; **in ~ sein** be on form; **c)** *(vorgeformtes Modell)* mould; *(Back~)* baking tin; **d)** *(Darstellungs~, Umgangs~)* form
formal 1. *Adj.* formal; 2. *adv.* formally; **formalisieren** *tr. V.* formalize
Formalität die; ~, ~en formality
Format das; ~|e|s, ~e **a)** size; *(Buch~, Papier~, Bild~)* format; **b)** *o. Pl. (Persönlichkeit)* stature
formbar *Adj.* malleable
Formel die; ~, ~n formula
formell *Adj.* formal
formen *tr. V.* **a)** *(gestalten)* form; shape; **b)** *(bilden, prägen)* mould, form ⟨*character, personality*⟩;
Form·fehler der irregularity
formieren *tr., refl. V.* form
förmlich 1. *Adj.* **a)** formal; **b)** *(regelrecht)* positive; 2. *adv.* **a)** formally; **b)** *(geradezu)* **sich ~ fürchten** be really afraid
form·los *Adj.* **a)** informal; **b)** *(gestaltlos)* shapeles
Form·sache die formality
Formular das; ~s, ~e form; **formulieren** *tr. V.* formulate; **Formulierung** die; ~, ~en **a)** *o. Pl. (das Formulieren)* formulation; *(eines Entwurfes, Gesetzes)* drafting; **b)** *(formulierter Text)* formulation
form·vollendet 1. *Adj.* perfectly executed ⟨*pirouette, bow, etc.*⟩; ⟨*poem*⟩ perfect in form; 2. *adv.* faultlessly
forsch *Adj.* forceful

forschen *itr. V.* **a) nach jmdm./etw. ~:** search *or* look for sb./sth.; **b)** *(als Wissenschaftler)* research; **Forscher** der; ~s, **Forscherin** die; ~, ~nen researcher; **Forschung** die; ~, ~en research; **Forschungs·reisende** der/die explorer

Forst der; ~|e|s, ~e|n| forest; **Förster** der; ~s, ~: forest warden

Forst·wirtschaft die forestry

Forsythie [fɔr'zy:tsiə] die; ~, ~n forsythia

fort *Adv.* **a)** *s.* weg; **b)** *(weiter)* und so ~: and so on

fort-, Fort-: ~**an** [-'-] *Adv.* from now/then on; ~**bestand** der; *o. Pl.* continuation; *(eines Staates)* continued existence; ~|**bewegen** 1. *tr. V.* move; shift; 2. *refl. V.* move [along]; ~|**bleiben** *unr. itr. V.; mit sein* fail to come; ~|**bringen** *unr. tr. V.: s.* wegbringen; ~|**dauern** *itr. V.* continue; ~|**fahren** 1. *unr. itr. V.* **a)** *mit sein* leave; **b)** *auch mit sein (weitermachen)* continue; go on; 2. *unr. tr. V.* drive away; ~|**führen** *tr. V.* **a)** lead away; **b)** *(fortsetzen)* continue; ~**gang** der; *o. Pl.* **a)** departure (aus from); **b)** *(Weiterentwicklung)* progress; ~|**gehen** *unr. itr. V.; mit sein* leave; geh ~! go away!; ~**geschritten** *Adj.* advanced; ~**geschrittene** der/die; *adj. Dekl.* advanced student/player; ~|**kommen** *unr. itr. V.; mit sein s.* wegkommen a, b; ~|**laufen** *unr. itr. V.; mit sein* **a)** *s.* weglaufen; **b)** *(sich ~setzen)* continue; ~**laufend** 1. *Adj.* continuous; 2. *adv.* continuously; ~|**pflanzen** *refl. V.* **a)** reproduce [oneself/itself]; **b)** *(sich verbreiten)* ⟨idea, mood⟩ spread; ⟨sound, light⟩ travel; ~**pflanzung** die reproduction; ~|**schaffen** *tr. V.* take away; ~|**schreiten** *unr. itr. V.; mit sein (process)* continue; ⟨time⟩ move on; ~**schritt** der progress; ~**schritte** progress *sing.;* ein ~schritt a step forward; ~**schrittlich** 1. *Adj.* progressive; 2. *adv.* progressively; ~**schrittlichkeit** die; ~: progressiveness; ~|**setzen** 1. *tr. V.* continue; 2. *refl. V.* continue; ~**setzung** die; ~, ~en a) *(das ~setzen)* continuation; **b)** *(anschließender Teil)* instalment; ~|**während** 1. *Adj.; nicht präd.* continual; 2. *adv.* continually; ~|**werfen** *unr. tr. V.: s.* wegwerfen

Foto das; ~s, ~s photo; ~s **machen** take photos

Foto-: ~**album** das photo album; ~**apparat** der camera

fotogen *Adj.* photogenic **Foto·graf** der; ~en, ~en photographer; **Fotografie** die; ~, ~n **a)** *o. Pl.* photography *no art.;* **b)** *(Lichtbild)* photograph; **fotografieren** *tr. V.* photograph; take a photograph/photographs of; **Fotografin** die; ~, ~nen photographer

foto-, Foto-: ~**kopie** die photocopy; ~**kopieren** *tr., itr. V.* photocopy; ~**kopierer** der photocopier; ~**labor** das photographic laboratory; ~**modell** das photographic model

Foul [faul] das; ~s, ~s *(Sport)* foul (an + *Dat.* on)

Foyer [foa'je:] das; ~s, ~s foyer

FPÖ *Abk.* Freiheitliche Partei Österreichs

Fr. *Abk.* **a)** Franken SFr.; **b)** Frau; **c)** Freitag Fri.

Fracht die; ~, ~en *(Schiffs~, Luft~)* cargo; freight; *(Bahn~, LKW~)* goods *pl.;* freight; **Fracht·brief** der consignment note; waybill; **Frachter** der; ~s, ~: freighter

Fracht-: ~**gut** das slow freight; slow goods *pl.;* ~**schiff** das cargo ship

Frack der; ~|e|s, Fräcke tails *pl.;* evening dress

Frage die; ~, ~n question; *(Angelegenheit)* issue; **in ~ kommen** be possible; **das kommt nicht in ~** *(ugs.)* that is out of the question; **Fragebogen** der questionnaire; *(Formular)* form; **fragen** 1. *tr., itr. V.* **a)** ask; **b)** *(sich erkundigen)* nach etw. ~: ask *or* inquire about sth.; **c)** *(nachfragen)* ask for; 2. *refl. V.* sich ~, ob ...: wonder whether ...; **Frage·zeichen** das question mark; **fraglich** *Adj.* **a)** doubtful; **b)** *nicht präd. (betreffend)* in question *postpos.;* relevant

Fragment das; ~|e|s, ~e fragment

frag·würdig *Adj.* **a)** questionable; **b)** *(zwielichtig)* dubious

Fraktion die; ~, ~en parliamentary party; *(mit zwei Parteien)* parliamentary coalition

Fraktions- *(Parl.):* ~**führer** der leader of the parliamentary party/coalition; ~**zwang** der obligation to vote in accordance with party policy

frank *Adv.* ~ **und frei** frankly and openly; openly and honestly

Franken der; ~s ~: [Swiss] franc

Frankfurter die; ~, ~ *(Wurst)* frankfurter

frankieren *tr. V.* frank

Frank·reich (das); ~s France

Franse die; ~, ~n strand [of a/the fringe]

Franzose der; ~n, ~n Frenchman; **er ist** ~: he is French; **die** ~**n** the French; **Französin die;** ~, ~**nen** Frenchwoman; **französisch** *Adj.* French; **Französisch das;** ~[s] French

Fräse die; ~, ~n (für Holz) moulding machine; (für Metall) milling machine

fraß *1. u. 3. Pers. Sg. Prät. v.* **fressen; Fraß der;** ~es (derb) muck

Fratze die; ~, ~n a) hideous face; b) (ugs.: Grimasse) grimace

Frau die; ~, ~en a) woman; b) (Ehe~) wife; c) (Titel, Anrede) ~ **Schulze** Mrs Schulze; (in Briefen) **Sehr geehrte** ~ **Schulze** Dear Madam; (bei persönlicher Bekanntschaft) Dear Mrs/Miss/Ms Schulze

Frauen-: ~**arzt der,** ~**ärztin die** gynaecologist; ~**bewegung die** *o. Pl.* women's movement; ~**rechtlerin die;** ~, ~**nen** feminist; Women's Libber (coll.)

Fräulein das; ~s, ~ (ugs. ~s) a) (junges ~) young lady; (ältliches ~) spinster; b) (Titel, Anrede) ~ **Mayer/ Schulte** Miss Mayer/Schulte

fraulich *1. Adj.* feminine; *2. adv.* in a feminine way

frech *1. Adj.* **a)** impertinent; cheeky; bare-faced (lie); b) (keck, keß) saucy; *2. adv.* impertinently; cheekily; **Frechheit die;** ~, ~en a) *o. Pl.* impertinence; cheek; b) (Äußerung) impertinent *or* cheeky remark

frei *1. Adj.* a) (unabhängig) free; b) (nicht angestellt) free-lance; c) (ungezwungen) free and easy; d) (nicht mehr in Haft) free; e) (offen) open; f) (unbesetzt) vacant; free; g) (kostenlos) free (food, admission); h) (verfügbar) spare; free (time); *2. adv.* freely

frei-, Frei-: ~**bad das** open-air swimming-pool; ~|**bekommen** 1. *unr. itr. V.* (ugs.) get time off; 2. *unr. tr. V.* jmdm./etw. ~bekommen get sb./ sth. released; ~**beruflich** 1. *Adj.* self-employed; free-lance (doctor, lawyer) in private practice; *2. adv.* ~**beruflich tätig sein/arbeiten** work free-lance/practise privately; ~**betrag der** (Steuerw.) [tax] allowance

Freier der; ~s, ~ (veralt.) suitor

frei-, Frei-: ~**exemplar das** (Buch) free copy; (Zeitung) free issue; ~|**geben** *unr. tr. V.* release; ~**gebig** *Adj.*

generous; open-handed; ~**gehege das** outdoor enclosure; ~**gepäck das** baggage allowance; ~**hafen der** free port; ~|**halten** *unr. tr. V.* a) treat; b) (offenhalten) keep (entrance, roadway) clear; **Einfahrt** ~**halten!** no parking in front of entrance; ~**handels·zone die** free-trade zone; ~**händig** *adv.* (cycle) without holding on

Freiheit die; ~, ~en a) freedom; ~, **Gleichheit, Brüderlichkeit** Liberty, Equality, Fraternity; b) (Vorrecht) freedom; privilege; **freiheitlich** 1. *Adj.* liberal (philosophy, conscience); ~ **und demokratisch** free and democratic; *2. adv.* liberally

Freiheits-: ~**beraubung die** (jur.) wrongful detention; ~**strafe die** term of imprisonment

frei-, Frei-: ~**herr der** baron; ~**karte die** complimentary ticket; ~|**kaufen** *tr. V.* ransom (hostage); buy the freedom of (slave); ~|**kommen** *unr. itr. V.* **aus dem Gefängnis** ~**kommen** be released from prison; ~**körper·kultur die;** *o. Pl.* nudism *no art.*; naturism *no art.*; ~|**lassen** *unr. tr. V.* set free; release; ~|**legen** *tr. V.* uncover

freilich *Adv.* of course

Frei·licht-: ~**bühne die,** ~**theater das** open-air theatre

frei-, Frei-: ~|**machen** 1. *refl. V.* (ugs.: frei nehmen) take time off; 2. *tr. V.* (Postw.) frank; **etw. mit 0,50 DM** ~**machen** put a 50-pfennig stamp on sth.; ~**marke die** postage stamp; ~**mütig** 1. *Adj.* frank; *2. adv.* frankly; ~**schaffend** *Adj.* free-lance; ~|**schwimmen** *unr. refl. V.* **sich** ~**schwimmen** pass the 15-minute swimming test; ~|**sprechen** *unr. tr. V.* a) (Rechtsw.) acquit; b) (für unschuldig erklären) exonerate (von from); ~**spruch der** (Rechtsw.) acquittal; ~|**stellen** *tr. V.* a) jmdm. etw. ~**stellen** leave sth. up to sb.; b) (befreien) release (person); jmdn. vom Wehrdienst ~**stellen** exempt sb. from military service; ~**stoß der** (Fußball) free kick

Frei·tag der Friday; *s. auch* **Dienstag, Dienstag-; freitags** *Adv.* on Friday[s]; *s. auch* **dienstags**

frei-, Frei-: ~**tod der** (verhüll.) suicide *no art.*; ~**treppe die** [flight of] steps; ~**übung die;** *meist Pl.* (Sport) keep-fit exercise; ~**wild das** fair game; ~**willig** 1. *Adj.* voluntary

⟨decision⟩; optional ⟨subject⟩; **2.** adv. voluntarily; sich **~willig melden** volunteer; **~zeichen das** ringing tone; **~zeit die**; o. Pl. spare time; **~zügig** Adj. **a)** generous; **b)** (gewagt, unmoralisch) risqué ⟨remark, film, dress⟩; **~zügigkeit die** generosity

fremd Adj. **a)** foreign; **b)** (nicht eigen) other people's; of others postpos.; **c)** (unbekannt) strange

fremd·artig Adj. strange

¹Fremde der/die; adj. Dekl. **a)** stranger; **b)** (Ausländer) foreigner; **²Fremde die;** ~ (geh.) die ~: foreign parts pl.

Fremden-: **~führer der** tourist guide; **~verkehr der** tourism no art.; **~zimmer das** room

fremd-, Fremd-: ~|**gehen** unr. itr. V.; mit sein (ugs.) be unfaithful; **~herrschaft die** foreign domination; **~ländisch** Adj. foreign; (exotisch) exotic

Fremdling der; ~s, ~e (veralt.) stranger

fremd-, Fremd-: **~sprache die** foreign language; **~sprachig** Adj. bilingual/multilingual ⟨staff, secretary⟩; foreign ⟨literature⟩; foreign-language ⟨edition, teaching⟩; **~sprachlich** Adj. foreign-language ⟨teaching⟩; foreign ⟨word⟩; **~wort das;** Pl. ~wörter foreign word

frenetisch 1. Adj. frenetic; **2.** adv. frenetically

Frequenz die; ~, ~en (Physik) frequency; (Med.: Puls~) rate

Fresse die; ~, ~n (derb) **a)** (Mund) gob (sl.); **b)** (Gesicht) mug (sl.); **fressen 1.** unr. tr. V. **a)** ⟨animal⟩ eat; (sich ernähren von) feed on; **b)** (ugs.: verschlingen) swallow up ⟨money, time, distance⟩; drink ⟨petrol⟩; **c)** (zerstören) eat away; **d)** (derb: von Menschen) guzzle; **2.** unr. itr. V. (von Tieren) feed; (derb: von Menschen) stuff one's face (sl.); **Fressen das;** ~s **a)** (für Hunde, Katzen usw.) food; (für Vieh) feed; **b)** (derb: Essen) grub (sl.); **Fresserei die;** ~, ~en (derb) guzzling

Freude die; ~, ~n joy; (Vergnügen) pleasure; ~ **an etw.** (Dat.) **haben** take pleasure in sth.

Freuden-: **~haus das** house of pleasure; **~tag der** happy day; **freudestrahlend** Adj. beaming with joy; **freudig** Adj. joyful; joyous ⟨heart⟩; delightful ⟨surprise⟩; **freud·los** Adj. joyless; **freuen 1.** refl. V. be glad

(über + Akk. about); (froh sein) be happy; **sich auf etw.** (Akk.) ~: look forward to sth.; **2.** tr. V. please

Freund der; ~es, ~e **a)** friend; **b)** (Verehrer, Geliebter) boy-friend; **Freundin die;** ~, ~nen **a)** friend; **b)** (Geliebte) girl-friend; (älter) lady-friend; **freundlich 1.** Adj. **a)** kind ⟨face⟩; friendly ⟨reception⟩; **b)** (angenehm) pleasant; **c)** (freundschaftlich) friendly; **2.** adv. jmdm. ~ **danken** thank sb. kindly; **Freundlichkeit die;** ~: kindness; **Freundschaft die;** ~, ~en friendship; **mit jmdm.** ~ **schließen** make friends with sb.; **freundschaftlich 1.** Adj. friendly; **2.** adv. in a friendly way

Frevel ['fre:fḷ] der; ~s, ~ (geh., veralt.) crime; outrage

Friede der; ~ns, ~n (älter, geh.) s. **Frieden; Frieden der;** ~s, ~: peace

Friedens-: **~forschung die** peace studies pl., no art.; **~konferenz die** peace conference; **~nobelpreis der** Nobel Peace Prize; **~pfeife die** pipe of peace; **~richter der** lay magistrate dealing with minor offences; ≈ Justice of the Peace; **~taube die** dove of peace; **~verhandlungen** Pl. peace negotiations; **~vertrag der** peace treaty

fried·fertig Adj. peaceable ⟨person, character⟩; **Fried·hof der** cemetery; (Kirchhof) graveyard; **friedlich 1.** Adj. peaceful; **2.** adv. peacefully; **fried·liebend** Adj. peace-loving

frieren unr. itr. V. **a)** be or feel cold; **b)** mit sein (ge~) freeze

Frikadelle die; ~, ~n rissole

frisch 1. Adj. fresh; new-laid ⟨egg⟩; clean ⟨linen, underwear⟩; wet ⟨paint⟩; **2.** adv. freshly; **Frische die;** ~ freshness; **geistige** ~: mental alertness; **körperliche** ~: physical fitness; **Frisch·halte·beutel der** airtight bag

Friseur [fri'zø:ɐ̯] der; ~s, ~e, **Friseuse** [fri'zø:zə] die; ~, ~n hairdresser; **frisieren** tr. V. jmdn./sich ~: do sb.'s/one's hair; **sich** ~ **lassen** have one's hair done

friß Imperativ Sg. v. **fressen**

frißt 2. u. 3. Pers. Sg. Präsens v. **fressen**

Frist die; ~, ~en **a)** time; period; **die** ~ **verlängern** extend the deadline; **b)** (begrenzter Aufschub) extension

frist-: **~gemäß, ~gerecht** Adj., adv. within the specified time postpos.; (bei Anmeldung usw.) before the closing date *postpos.;* ~**los** 1. *Adj.* instant; 2. *adv.* without notice

Frisur die; ~, ~en hairstyle

fritieren *tr. V.* deep-fry

frivol [fri'vo:l] *Adj.* **a)** *(schamlos)* suggestive ⟨*remark, picture, etc.*⟩; risqué ⟨*joke*⟩; earthy ⟨*man*⟩; flighty ⟨*woman*⟩; **b)** *(leichtfertig)* frivolous

froh *Adj.* **a)** happy; cheerful ⟨*person, mood*⟩; good ⟨*news*⟩; **b)** *(ugs.: erleichtert)* pleased, glad (**über** + *Akk.* about)

fröhlich *Adj.* cheerful; happy

Fröhlichkeit die; ~: cheerfulness; *(eines Festes, einer Feier)* gaiety

Froh·sinn der; *o. Pl.* cheerfulness; gaiety

fromm; ~**er** *od.* **frömmer**, ~**st**... *od.* **frömmst**... 1. *Adj.* pious, devout ⟨*person*⟩; devout ⟨*Christian*⟩; 2. *adv.* piously; **Frömmigkeit** die; ~: piety; devoutness

Fron·leichnam [fro:n-] *o. Art.* [the feast of] Corpus Christi

Front die; ~, ~en **a)** *(Gebäude~)* front; façade; **b)** *(Kampfgebiet)* front [line]; **frontal** 1. *Adj.* head-on ⟨*collision*⟩; frontal ⟨*attack*⟩; 2. *adv.* ⟨*collide*⟩ head-on; ⟨*attack*⟩ from the front; **Front·an·trieb** der *(Kfz-W.)* frontwheel drive

fror *1. u. 3. Pers. Sg. Prät. v.* **frieren**

Frosch der; ~[e]s, **Frösche** frog

Frosch-: ~**mann** der; *Pl.* ~**männer** frogman; ~**perspektive** die worm's-eye view; ~**schenkel** der frog's leg

Frost der; ~[e]s, **Fröste** frost; **Frost·beule** die chilblain; **frösteln** *itr. V.* feel chilly; **frostig** 1. *Adj.* *(auch fig.)* frosty; 2. *adv.* frostily; **Frost·schutz·mittel** das **a)** frost protection agent; **b)** *(Kfz-W.)* antifreeze

Frottee das u. der; ~s, ~s terry towelling; **Frottee·handtuch** das terry towel; **frottieren** *tr. V.* rub; towel

frozeln 1. *tr. V.* tease; 2. *itr. V.* über jmdn./etw. ~: make fun of sb./sth.

Frucht die; ~, **Früchte** fruit; **frucht·bar** *Adj.* fertile; fruitful ⟨*work, idea, etc.*⟩; **Fruchtbarkeit** die; ~: fertility; fruitfulness **Frucht·becher** der fruit sundae; **fruchten** *tr. V.* **nichts** ~: be no use; **fruchtig** *Adj.* fruity; **frucht·los** *Adj.* fruitless, vain ⟨*efforts*⟩; **Frucht·saft** der fruit juice

früh 1. *Adj.* **a)** early; **b)** *(vorzeitig)* premature; 2. *adv.* early; **heute** ~: this

morning; **früh·auf**: von ~**auf** from early childhood on[wards]; **Frühaufsteher** der; ~s, ~: early riser; **Frühe** die; ~: in aller ~: at the crack of dawn; **früher** ['fry:ɐ] 1. *Adj., nicht präd.* **a)** *(vergangen)* earlier; former; **b)** *(ehemalig)* former ⟨*owner, occupant, friend*⟩; 2. *adv.* formerly; ~ **war er ganz anders** he used to be quite different; **Früh·erkennung** die *(Med.)* early recognition

frühestens *Adv.* at the earliest; **Früh·geburt** die **a)** premature birth; **b)** *(Kind)* premature baby

Früh·jahr das spring; **Frühjahrsmüdigkeit** die springtime tiredness

Frühling der; ~s, ~e spring; **Frühlings·anfang** der first day of spring

früh-, Früh-: ~**reif** *Adj.* precocious ⟨*child*⟩; ~**schoppen** der morning drink; *(um Mittag)* lunchtime drink; ~**sport** der early-morning exercise

Früh·stück das; ~s, ~e breakfast; **frühstücken** *itr. V.* have breakfast; **Frühstücks·pause** die morning break; coffee break

früh·zeitig 1. *Adj.* early; *(vorzeitig)* premature; 2. *adv.* early; *(vorzeitig)* prematurely

Frustration die; ~, ~en *(Psych.)* frustration; **frustrieren** *tr. V.* frustrate

Fuchs der; ~es, **Füchse** fox; **fuchsen** *tr. V.* annoy; vex; **fuchs·teufels·wild** *Adj.* *(ugs.)* livid *(coll.)*

Fuchtel die; ~: **unter jmds.** ~ *(ugs.)* under sb.'s thumb; **fuchteln** *itr. V.* *(ugs.)* **mit etw.** ~: wave sth. about

Fuder das; ~s, ~: cart-load

¹**Fuge** die; ~, ~**n** joint; *(Zwischenraum)* gap; ²**Fuge** die; ~, ~**n** *(Musik)* fugue

fügen 1. *tr. V.* place; set; **etw zu etw.** ~ *(fig.)* add sth. to sth.; 2. *refl. V.* **a)** *(sich ein~)* **sich in etw.** *(Akk.)* ~: fit into sth.; **b)** *(gehorchen)* **sich** ~: fall into line; **fügsam** *Adj.* obedient

fühlbar *Adj.* noticeable; **fühlen** 1. *tr., itr. V.* feel; 2. *refl. V.* **sich krank** ~: feel sick; **Fühler** der; ~s, ~: feeler; antenna; **Fühlungnahme** die; ~: initial contact

fuhr *1. u. 3. Pers. Sg. Prät. v.* **fahren**

Fuhre die; ~, ~**n** load

führen 1. *tr. V.* **a)** lead; **b)** *(verkaufen)* stock, sell ⟨*goods*⟩; **c)** *(durch~)* **Gespräche/Verhandlungen** ~: hold conversations/negotiations; **eine glückliche Ehe** ~: be happily married; **d)** *(leiten)* manage, run ⟨*company, business,* *pub, etc.*⟩*;* lead ⟨*party, country*⟩*;* command ⟨*regiment*⟩*;* **e)** *(Amtsspr.)* drive ⟨*train, motor, vehicle*⟩*;* **f)** *(als Kennzeichnung, Bezeichnung haben)* bear; **einen Titel/Künstlernamen ~:** have a title/use a stage name; **g)** *(angelegt haben)* keep ⟨*diary, list, file*⟩*;* **h)** *(registrieren)* **jmdn. in einer Liste/Kartei ~:** have sb. on a list/on file; **i)** *(tragen)* **etw. bei** *od.* **mit sich ~:** have sth. on one; **eine Waffe/einen Ausweis bei sich ~:** carry a weapon/a pass; **2.** *itr. V.* **a)** lead; **b)** *(an der Spitze liegen)* lead; be ahead; **führend** *Adj.* leading; high-ranking ⟨*official*⟩*;* prominent ⟨*position*⟩

Führer der; ~s, ~ **a)** *(Leiter)* leader; **b)** *(Fremden~)* guide; **Führerin** die; ~, ~nen *s.* Führer; **führer·los 1.** *Adj.* leaderless; *(ohne Lenker)* driverless ⟨*car*⟩*;* **2.** *adv.: s.* **1;** without a leader; without a driver; **Führer·schein** der driving licence *(Brit.);* driver's license *(Amer.);* **Führung** die; ~, ~en **a)** *o. Pl. s.* **führen 1 d:** management; running; leadership; command; **b)** *(Fremden~)* guided tour; **c)** *o. Pl. (führende Position)* lead

Führungs-: **~kraft** die manager; **~spitze** die *(Politik)* top leadership; *(im Betrieb)* top management; **~zeugnis** das *document issued by police certifying that holder has no criminal record*

Fuhr-: **~unternehmer** der haulage contractor; **~werk** das cart

Fülle die; ~ **a)** wealth; abundance; **b)** *(Körper~)* corpulence; **füllen 1.** *tr. V.* fill; *(Kochk.)* stuff; **b)** *(fig.)* fill in ⟨*gap, time*⟩*;* **2.** *refl. V. (voll werden)* fill [up]; **Füller** der; ~s, ~ *(ugs.)* [fountain-]pen; **Füll·federhalter** der fountain-pen; **füllig** *Adj.* corpulent, portly ⟨*person*⟩*;* ample ⟨*figure, bosom*⟩*;* **Füllung** die; ~, ~en stuffing; *(Kochk.; Zahnmed.)* filling; *(in Schokolade)* centre

fummeln *itr. V. (ugs.)* **a)** *(fingern)* fiddle; **b)** *(erotisch)* pet

Fund der; ~|e|s, ~e *(auch Archäol.)* find

Fundament das; ~|e|s, ~e **a)** *(Bauw.)* foundations *pl.;* **b)** *(Basis)* base; basis; **fundamental** *Adj.* fundamental

Fund-: **~büro** das lost property office *(Brit.);* lost and found office *(Amer.);* **~grube** die treasure-house

fundieren *tr. V.* underpin

fündig *Adj.* **~ sein** yield something; **~ werden** make a find; *(bei Bohrungen)* make a strike

Fund·ort der place *or* site where sth. is/was found

fünf *Kardinalz.* five; **Fünf** die; ~, ~en five; *(Schulnote)* E

fünf-, Fünf-: **~eck** das; ~s, ~e pentagon; **~fach** *Vervielfältigungsz.* fivefold; **~fache** das; *adj. Dekl.* five times as much; **~hundert** *Kardinalz.* five hundred; **~kampf** der *(Sport)* pentathlon

Fünfling der; ~s, ~e quintuplet; quin *(coll.)*

fünf-: **~mal** *Adv.* five times; **~stellig** *Adj.* five-figure

fünft... *Ordinalz.* fifth; **Fünf·tagewoche** die five-day [working] week; **fünf·tausend** *Kardinalz.* five thousand

fünftel *Bruchz.* fifth; **Fünftel** das *(schweiz. meist* der*);* ~s, ~: fifth; **fünftens** *Adv.* fifthly; **fünf·zehn** *Kardinalz.* fifteen; **fünfzig** *Kardinalz.* fifty; **Fünfzig** die; ~: fifty; **fünfziger** *indekl. Adj.; nicht präd.* **die ~ Jahre** the fifties; **Fünfziger** der; ~s, ~ **a)** *(ugs.)* fifty-pfennig piece; **b)** *(50jähriger)* fifty-year-old; **fünfzigst...** *Ordinalz.* fiftieth

fungieren *itr. V.* **als etw. ~** ⟨*person*⟩ act as sth.; ⟨*word etc.*⟩ function as sth.

Funk der; ~s radio; **Funk·ausstellung** die radio and television exhibition

Funke der; ~ns, ~n *(auch fig.)* spark

funkeln *itr. V.* ⟨*light, star*⟩ twinkle; ⟨*gold, diamonds*⟩ glitter; ⟨*eyes*⟩ blaze

funken *tr. V.* radio; ⟨*transmitter*⟩ broadcast; **Funker** der; ~s, ~: radio operator

Funk-: **~gerät** das radio set; *(tragbar)* walkie-talkie; **~haus** das broadcasting centre; **~kolleg** das radio-based [adult education] course; **~sprech·gerät** das radiophone; *(tragbar)* walkie-talkie; **~spruch** der radio signal; *(Nachricht)* radio message; **~stille** die radio silence; **~streife** die [police] radio patrol; **~taxi** das radio taxi

Funktion die; ~, ~en function; **Funktionär** der; ~s, ~e official; functionary; **funktionieren** *itr. V.* work; function; **funktions·tüchtig** *Adj.* working; sound ⟨*organ*⟩

Funk-: **~turm** der radio tower; **~verbindung** die radio contact

Funzel die; ~, ~n *(ugs.)* useless light
für 1. *Präp. mit Akk.* for; etw. ~ ungültig erklären declare sth. invalid *s. auch* was 1
Furche die; ~, ~n a) furrow; b) *(Wagenspur)* rut
Furcht die; ~ : fear; ~ vor jmdm./etw. haben fear sb./sth.; **furchtbar** 1. *Adj.* a) dreadful; b) *(ugs.: unangenehm)* terrible *(coll.)*; 2. *adv. (ugs.)* terribly *(coll.)*; **fürchten** 1. *refl. V.* sich [vor jmdm./etw.] ~ : be afraid *or* frightened [of sb./sth.]; 2. *tr. V.* be afraid of; ich fürchte, [daß] ...: I'm afraid [that] ...; **fürchterlich** *Adj. s.* furchtbar; **furcht·los** 1. *Adj.* fearless; 2. *adv.* fearlessly; **furchtsam** 1. *Adj.* timid; 2. *adv.* timidly
für·einander *Adv.* for one another; for each other
Furie ['fu:riə] die; ~, ~n Fury
Furnier das; ~s, ~e veneer
Für·sorge die; ~ a) care; b) *(veralt.: Sozialhilfe)* welfare; c) *(veralt.: Sozialamt)* social services *pl.;* **für·sorglich** 1. *Adj.* considerate; 2. *adv.* considerately
Für·sprache die support; **Für·sprecher** der advocate
Fürst der; ~en, ~en prince; **Fürstentum** das; ~s, Fürstentümer principality; **fürstlich** 1. *Adj.* a) royal; b) *(fig.: üppig)* lavish; 2. *adv.* lavishly
Furt die; ~, ~en ford
Furunkel der *od.* das; ~s, ~ : boil; furuncle
Für·wort das; *Pl.* -wörter pronoun
Fusion die; ~, ~en amalgamation; *(von Konzernen)* merger; **fusionieren** *itr. V.* merge
Fuß der; ~es, Füße foot; *(einer Lampe, Säule)* base; *(von Möbeln)* leg; zu ~ gehen go on foot; walk; bei ~ ! heel!; *(fig.)* auf freiem ~ sein be at large; auf großem ~ leben live in great style
Fuß·ball der a) *o. Pl. (Ballspiel)* [Association] football; b) *(Ball)* football; **Fußballer** der; ~s, ~ : footballer
Fußball-: ~platz der football ground; *(Spielfeld)* football pitch; ~spiel das a) football match; b) *o. Pl. (Sportart)* football *no art.;* ~spieler der football player
Fuß·boden der floor; **fußen** *itr. V.* auf etw. *(Dat.)* ~ : be based on sth.
Fuß·ende das foot; **Fußgänger** der; ~s, ~, **Fußgängerin** die; ~, ~nen pedestrian

Fußgänger-: ~brücke die footbridge; ~übergang der, ~überweg der pedestrian crossing; ~unterführung die pedestrian subway; ~zone die pedestrian precinct
Fuß-: ~nagel der toe-nail; ~note die footnote; ~stapfen der; ~s, ~ : footprint; ~tritt der kick; ~volk das a) *(hist.)* footmen *pl.;* b) *(abwertend: Untergeordnete)* lower ranks *pl.;* ~weg der footpath
futsch *Adj.(salopp)* ~ sein have gone for a burton *(Brit. sl.)*
¹**Futter** das; ~s *(Tiernahrung)* feed; *(für Pferde, Kühe)* fodder
²**Futter** das; ~s *(von Kleidungsstücken)* lining
Futteral das; ~s, ~e case
¹**füttern** *tr. V.* feed
²**füttern** *tr. V. (mit* ²*Futter ausstatten)* line
Fütterung die; ~, ~en feeding
Futur das; ~s, ~e *(Sprachw.)* future [tense]

G

g, G [ge:] das; ~, ~ a) *(Buchstabe)* g/G; b) *(Musik)* [key of] G
g *Abk.* a) Gramm g; b) Groschen
gab 1. u. 3. *Pers. Sg. Prät. v.* geben
Gabe die; ~, ~n a) *(geh.: Geschenk, Talent)* gift; *(Almosen, Spende)* alms *pl.*
Gabel die; ~, ~n fork; *(Telefon~)* cradle; **gabeln** *refl. V.* fork; **Gabel·stapler** der; ~s, ~ : fork-lift truck; **Gabelung** die; ~, ~en fork
Gaben·tisch der gift table
gackern *itr. V.* a) cluck; b) *(ugs.: lachen)* cackle
gaffen *itr. V. (abwertend)* gape; gawp *(coll.)*
Gag [gɛk] der; ~s, ~s a) *(Theater, Film)* gag; b) *(Besonderheit)* gimmick
Gage ['ga:ʒə] die; ~, ~n salary; *(für einzelnen Auftritt)* fee
gähnen *itr. V. (auch fig.)* yawn
Gala ['ga:la, *auch* 'gala] die; ~ : formal dress

galant 1. *Adj.* gallant; *(amourös)* amorous; 2. *adv.* gallantly

Gala·vorstellung die gala performance

Galeere die; ~, ~n galley

Galerie die; ~, ~n gallery

Galgen der gallows *sing.*

Galgen-: ~frist die reprieve; ~humor der gallows humour

Galle die; ~, ~n a) *(Gallenblase)* gall[-bladder]; b) *(Sekret) (bei Tieren)* gall; *(bei Menschen)* bile

Galopp der; ~s, ~s *od.* ~e gallop; galoppieren *itr. V.; meist mit sein* gallop

galt *1. u. 3. Pers. Sg. Prät. v.* gelten

galvanisch [gal'va:nɪʃ] *Adj.* galvanic

Gamasche die; ~, ~n gaiter; *(bis zum Knöchel reichend)* spat

Gambe die; ~, ~n *(Musik)* viola da gamba

Gamma·strahlen *Pl. (Physik, Med.)* gamma rays

gammelig *Adj. (ugs.)* a) bad; rotten; b) *(unordentlich)* scruffy; gammeln *itr. V.* a) *(ugs.)* go off; b) *(nichts tun)* loaf around; bum around *(Amer. coll.)*; Gammler der; ~s, ~ *(ugs.)* drop-out *(coll.)*

gang: ~ und gäbe sein be quite usual

Gang der; ~[e]s, Gänge a) walk; gait; b) *(Besorgung)* errand; c) *o. Pl. (Verlauf)* course; d) *(Technik)* gear; e) *(Flur) (in Zügen, Gebäuden usw.)* corridor; *(Verbindungs~)* passage[-way]; *(im Theater, Kino, Flugzeug)* aisle; f) *(Kochk.)* course; gangbar *Adj.* passable; *(fig.)* practicable

Gängel·band das *in* jmdn. am ~ führen keep sb. in leading-reins; gängeln *tr. V. (ugs.)* jmdn. ~: boss sb. around

gängig *Adj.* a) *(üblich)* common; *(aktuell)* current; b) *(leicht verkäuflich)* popular

Gang·schaltung die *(Technik)* gear system; *(Art)* gear-change

Gangway ['gæŋweɪ] die; ~, ~s gangway

Ganove [ga'no:və] der; ~n, ~n *(ugs. abwertend)* crook *(coll.)*

Gans die; ~, Gänse goose

Gänse-: ~blümchen das daisy; ~braten der roast goose; ~füßchen das; *meist Pl. (ugs.) s.* Anführungszeichen; ~haut die *(fig.)* goose-flesh; goose pimples *pl.;* ~marsch *in* im ~marsch in single *or* Indian file

Gänserich der; ~s, ~e gander

ganz 1. *Adj.* a) *(gesamt)* whole; entire; den ~en Tag/das ~e Jahr all day/year; b) *(ugs.: alle)* die ~en Kinder/Leute/Gläser *usw.* all the children/people/glasses *etc.;* c) *(vollständig)* whole; d) *(ugs.: ziemlich [viel])* eine ~e Menge/ein ~er Haufen quite a lot/quite a pile; e) *(ugs.: unversehrt)* intact; etw. wieder ~ machen mend sth.; 2. *adv.* quite; Ganze das; *adj. Dekl.* a) whole; b) *(alles)* das ~: the whole thing; gänzlich *Adv.* entirely

ganz-: ~tägig 1. *Adj.* all-day; eine ~tägige Arbeit a full-time job; 2. *adv.* all day; ~tags *Adv.* ~ arbeiten work full-time

¹gar *Adj.* cooked; done *pred.*

²gar *Partikel* a) *(überhaupt)* ~ nicht [wahr] not [true] at all; ~ nichts nothing at all; ~ niemand *od.* keiner nobody at all; ~ keines not a single one; ~ kein Geld no money at all; b) *(südd., österr., schweiz.: verstärkend)* ~ zu only too; c) *(geh.: sogar)* even

Garage [ga'ra:ʒə] die; ~, ~n garage

Garant der; ~en, ~en guarantor; Garantie die; ~, ~n guarantee; garantieren 1. *tr. V.* guarantee; 2. *itr. V.* für etw. ~: guarantee sth.; garantiert *Adv. (ugs.)* wir kommen ~ zu spät we're dead certain to arrive late *(coll.);* Garantie·schein der guarantee [certificate]

Garaus ['ga:ʀlaus] jmdm. den ~ machen do sb. in *(coll.)*

Garbe die; ~, ~n a) sheaf; b) *(Geschoß~)* burst of fire

Garde die; ~, ~n guard

Garderobe die; ~, ~n a) *o. Pl.* wardrobe; clothes *pl.;* b) *(Flur~)* coatrack; c) *(im Theater o. ä.)* cloakroom; checkroom *(Amer.);* Garderobenfrau die cloakroom *or (Amer.)* checkroom attendant

Gardine die; ~, ~n a) net curtain; b) *(landsch., veralt.)* curtain

Gardinen-: ~predigt die *(ugs.)* telling-off *(coll.); (einer Ehefrau zu ihrem Mann)* curtain lecture; ~stange die curtain rail

garen *tr., itr. V.* cook

gären *regelm. (auch unr.) itr. V.* ferment; *(fig.)* seethe

Garn das; ~[e]s, ~e a) thread; *(Näh~)* cotton; b) *(Seew.)* yarn

Garnele die; ~, ~n shrimp

garnieren *tr. V.* a) decorate; b) *(Gastr.)* garnish

Garnison die; ~, ~en garrison

Garnitur die; ~, ~en a) set; *(Wäsche)* set of [matching] underwear; *(Möbel)* suite; b) *(ugs.)* **die erste/zweite ~**: the first/second-rate people *pl.*

garstig *Adj.* a) nasty; bad ⟨*behaviour*⟩ **Gärtchen** das; ~s, ~: little garden; **Garten** der; ~s, Gärten garden

Garten-: **~arbeit** die gardening; **~bau** der; *o. Pl.* horticulture; **~fest** das garden party; **~haus** das summer-house; **~laube** die summer-house; garden house; **~lokal** das beer garden; *(Restaurant)* open-air café; **~schau** die horticultural show; **~zwerg** der a) garden gnome; b) *(salopp abwertend)* little runt

Gärtner der; ~s, ~: gardener; **Gärtnerei** die; ~, ~en nursery; **Gärtnerin** die; ~, ~nen gardener

Gärung die; ~, ~en fermentation

Gas das; ~es, ~e a) gas; b) *(Treibstoff)* petrol *(Brit.)*; gasoline *(Amer.)*; gas *(Amer. coll.)*: **~ wegnehmen** take one's foot off the accelerator; **~ geben** accelerate; put one's foot down *(coll.)*

gas-, Gas-: **~flasche** die gas-cylinder; *(für einen Herd, Ofen)* gas bottle; **~förmig** *Adj.* gaseous; **~hahn** der gas tap; **~herd** der gas cooker; **~leitung** die gas pipe; *(Hauptrohr)* gas main; **~maske** die gas mask; **~pedal** das accelerator [pedal]; gas pedal *(Amer.)*; **~pistole** die pistol that fires gas cartridges

Gasse die; ~, ~n lane; *(österr.)* street; **Gassen·junge** der *(abwertend)* street urchin

Gast der; ~[e]s, Gäste a) guest; b) *(Besucher eines Lokals)* patron; c) *(Besucher)* visitor; **Gast·arbeiter** der immigrant *or* guest worker

Gäste-: **~buch** das guest book; **~zimmer** das *(privat)* guest room; spare room; *(im Hotel)* room

gast-, Gast-: **~freundlich** *Adj.* hospitable; **~freundschaft** die hospitality; **~geber** der host; **~geberin, die** hostess; **~haus** das, **~hof** der inn

gastieren *itr. V.* give a guest performance

gastlich *Adj.* hospitable; **Gastlichkeit** die; ~: hospitality

Gastronom der; ~en, ~en restaurateur; **Gastronomie** die; ~: catering *no art.*; *(Gaststättengewerbe)* restaurant trade

Gast-: **~spiel** das guest performance; **~stätte** die public house; *(Speiselokal)* restaurant; **~wirt** der publican;

landlord; *(eines Restaurants)* [restaurant] proprietor; *(Pächter)* restaurant manager; **~wirtschaft die** *s.* **~stätte**

Gas-: **~vergiftung** die gas-poisoning *no indef. art.*; **~versorgung** die gas supply; **~werk** das gasworks *sing.*; **~zähler** der gas meter

Gatte der; ~n, ~n husband

Gatter das; ~s, ~ a) *(Zaun)* fence; *(Lattenzaun)* fence; paling; b) *(Tor)* gate;

Gattin die; ~, ~nen *(geh.)* wife

Gattung die; ~, ~en a) kind; sort; *(Kunst~)* genre; form; b) *(Biol.)* genus

Gaudi das; ~s *(bayr., österr.)* die; ~ *(ugs.)* bit of fun

Gaukler der; ~s, ~ a) *(veralt.: Taschenspieler)* itinerant entertainer; b) *(geh.: Betrüger)* charlatan

Gaul der; ~[e]s, Gäule nag *(derog.)*

Gaumen der; ~s, ~: palate

Gauner der; ~s, ~ *(abwertend)* crook *(coll.)*; rogue; **Gaunerei** die; ~, ~en swindle; **Gauner·sprache die** thieves' cant *or* Latin

Gaze ['gaːzə] die; ~, ~n gauze

geachtet *Adj.* respected

Geäst das; ~[e]s branches *pl.*

geb. *Abk.* a) geboren; b) geborene

Gebäck das; ~[e]s, ~e cakes and pastries *pl.*; *(Kekse)* biscuits *pl.*; *(Törtchen)* tarts *pl.*

gebacken 2. *Part. v.* backen

Gebälk das; ~[e]s, ~e beams *pl.*; *(Dach~)* rafters *pl.*

gebar 1. u. 3. *Pers. Sg. Prät. v.* gebären

Gebärde die; ~, ~n gesture; **gebärden** *refl. V.* behave

gebären *unr. tr. V.* bear; give birth to; *s. auch* geboren

Gebäude das; ~s, ~ a) building; b) *(Gefüge)* structure

gebaut *Adj.* **gut ~ sein** have a good figure

Gebein das; ~[e]s, ~e *Pl. (geh.)* bones *pl.*; *(sterbliche Reste)* [mortal] remains

Gebell das; ~[e]s barking; *(der Jagdhunde)* baying

geben 1. *unr. tr. V.* give; **jmdm. die Hand ~**: shake sb.'s hand; **~ Sie mir bitte Herrn N.** please put me through to Mr N.; **Unterricht ~**: teach; **eins plus eins gibt zwei** one and one is *or* makes two; **etw. von sich ~**: utter sth.; 2. *unr. tr. V. (unpers.)* **es gibt** there is/ are; *(coll.)* **heute gibt's Fisch** we're having fish today; **morgen gibt es Schnee** it'll snow tomorrow; 3. *unr. itr.*

V. **a)** *(Karten austeilen)* deal; **b)** *(Sport: aufschlagen)* serve; **4.** *unr. refl. V.* **a)** sich |natürlich/steif| ~: act *or* behave [naturally/stiffly]; **b) das gibt sich noch** it will get better
Gebet das; ~|e|s, ~e prayer
gebeten *2. Part. v.* **bitten**
Gebets-: ~**mühle** die prayer wheel; ~**teppich** der *(islam. Rel.)* prayer mat
gebiert *3. Pers. Sg. Präsens v.* **gebären**
Gebiet das; ~|e|s, ~e region; area; *(Staats~)* territory; *(Bereich, Fach)* field
gebieten *(geh.)* **a)** command; order; **b)** *(erfordern)* demand; **Gebieter** der; ~s, ~ *(veralt.)* master; **gebieterisch** *(geh.) Adj.* imperious; *(herrisch)* domineering; peremptory ‹*tone*›
Gebilde das; ~s, ~: object; *(Bauwerk)* structure
gebildet *Adj.* educated
Gebirge das; ~s, ~: mountain range; **im ~:** in the mountains; **gebirgig** *Adj.* mountainous
Gebiß das; **Gebisses, Gebisse a)** set of teeth; teeth *pl.;* **b)** *(Zahnersatz)* denture; plate *(coll.); (für beide Kiefer)* dentures *pl.;* **gebissen** *2. Part. v.* **beißen**
geblasen *2. Part. v.* **blasen**
geblichen *2. Part. v.* **bleichen**
geblümt *Adj.* flowered
Geblüt das; ~|e|s *(geh.)* blood
gebogen *2. Part. v.* **biegen**
geboren 1. *2. Part. v.* **gebären; 2.** *Adj.* **blind/taub ~ sein** be born blind/deaf; **Frau Anna Schmitz ~e Meyer** Mrs Anna Schmitz née Meyer
geborgen 1. *2. Part. v.* **bergen; 2.** *Adj.* safe; secure; **Geborgenheit die; ~:** security
geborsten *2. Part. v.* **bersten**
gebot *1. u. 3. Pers. Sg. Prät. v.* **gebieten; Gebot** das; ~|e|s, ~e **a)** *(Grundsatz)* precept; **die Zehn ~e** *(Rel.)* the Ten Commandments; **b)** *(Vorschrift)* regulation; **geboten 1.** *2. Part. v.* **bieten, gebieten; 2.** *Adj. (ratsam)* advisable; *(notwendig)* necessary
Gebr. *Abk.* **Gebrüder** Bros.
gebracht *2. Part. v.* **bringen**
gebrannt *2. Part. v.* **brennen**
gebraten *2. Part. v.* **braten**
Gebrauch der **a)** *o. Pl.* use; **b)** *meist Pl. (Brauch)* custom; **gebrauchen** *tr. V.* use; **gebräuchlich** *Adj.* **a)** normal; customary; **b)** *(häufig)* common
gebrauchs-, Gebrauchs-: ~**an-**

weisung die instructions *pl.* [for use]; ~**fertig** *Adj.* ready for use *pred.;* ~**gegenstand** der item of practical use
gebraucht *Adj.* second-hand; used ‹*car*› **Gebraucht·wagen** der used car
Gebrechen das; ~s, ~ *(geh.)* affliction; **gebrechlich** *Adj.* infirm; **Gebrechlichkeit die; ~:** infirmity
gebrochen 1. *2. Part. v.* **brechen; 2.** *Adj.* ~**es Englisch/Deutsch** broken English/German; **3.** *adv.* ~ **Deutsch sprechen** speak broken German
Gebrüder *Pl.:* **die ~ Meyer** Meyer Brothers
Gebrüll das; ~|e|s roaring
Gebrumm das; ~|e|s *(von Bären)* growling; *(von Flugzeugen, Bienen)* droning; *(von Insekten)* buzz[ing]
gebückt *Adj.* **in ~er Haltung** bending forward
Gebühr die; ~, ~en charge; *(Maut)* toll; *(Anwalts~)* fee
gebühren *(geh.) itr. V.* **jmdm. gebührt Achtung** *usw.* sb. deserves respect *etc.;* **gebührend 1.** *Adj.* fitting; **2.** *adv.* fittingly
gebühren-, Gebühren-: ~**ermäßigung die** reduction of charges/fees; ~**frei 1.** *Adj.* free of charge *pred.;* **2.** *adv.* free of charge; ~**pflichtig** *Adj.* **eine ~pflichtige Verwarnung** a fine and a caution
gebunden 1. *2. Part. v.* **binden; 2.** *Adj. (verpflichtet)* bound
Geburt die; ~, ~en birth; **Geburten·kontrolle** die; *o. Pl.* birth control; **gebürtig** *Adj.* **ein ~er Schwabe** a Swabian by birth
Geburts-: ~**anzeige** die birth announcement; ~**datum das** date of birth; ~**helfer** der *(Arzt)* obstetrician; ~**ort** der place of birth; ~**tag** der birthday; **jmdm. zum ~ gratulieren** wish sb. many happy returns of the day; ~**ur·kunde** die birth certificate
Gebüsch das; ~|e|s, ~e bushes *pl.*
gedacht *2. Part. v.* **denken, gedenken**
Gedächtnis das; ~ses, ~se **a)** memory; **b)** *(Andenken)* memory
Gedächtnis-: ~**lücke** die gap in one's memory; ~**schwund** der loss of memory
gedämpft *Adj.* subdued ‹*mood*›; subdued, soft ‹*light*›; muffled ‹*sound*›
Gedanke der; ~ns, ~n **a)** thought; **der ~ an etw.** *(Akk.)* the thought of sth.; **b)** *Pl. (Meinung)* ideas; **c)** *(Einfall)* idea

gedanken-, Gedanken-: ~**gang** der train of thought; ~**los** 1. *Adj.* unconsidered; *(zerstreut)* absentminded; 2. *adv.* without thinking; *(zerstreut)* absent-mindedly; ~**losigkeit** die *(Zerstreutheit)* absent-mindedness; *(Unüberlegtheit)* lack of thought; ~**strich** der dash; ~**verloren** *Adv.* lost in thought; ~**voll** 1. *Adj.* pensive; 2. *adv.* pensively

gedanklich 1. *Adj.* intellectual; 2. *adv.* intellectually

Gedärm das; ~|e|s, ~e intestines *pl.*; bowels *pl.*, *(eines Tieres)* entrails *pl.*

Gedeck das; ~|e|s, ~e a) place setting; cover; b) *(Menü)* set meal; c) *(Getränk)* drink [with a cover charge]

gedeihen *unr. itr. V.; mit sein* a) thrive; b) *(fortschreiten)* progress

gedenken *unr. itr. V.* a) jmds./einer Sache ~ *(geh.)* remember sb./sth.; *(in einer Feier)* commemorate sb./sth.; b) etw. zu tun ~: intend to do *or* doing sth.

Gedenk·stätte die memorial

Gedicht das; ~|e|s, ~e poem

gediegen 1. *Adj.* solid *(furniture)*; sound *(piece of work)*; 2. *adv.* ~ gebaut/verarbeitet solidly built/made

gedieh *1. u. 3. Pers. Sg. Prät. v.* gedeihen; **gediehen** 2. *Part. v.* gedeihen

Gedränge das; ~s pushing and shoving; *(Menge)* crush; crowd

gedroschen 2. *Part. v.* dreschen

gedrungen 1. 2. *Part. v.* dringen; 2. *Adj.* stocky; thick-set

Geduld die; ~: patience; **gedulden** *refl. V.* be patient; **geduldig** 1. *Adj.* patient; 2. *adv.* patiently; **Geduldsspiel** das puzzle

gedurft 2. *Part. v.* dürfen

geeignet *Adj.* suitable; *(richtig)* right

Gefahr die; ~, ~en a) danger; *(Bedrohung)* danger; threat (für to); bei ~: in case of emergency; b) *(Risiko)* risk; auf eigene ~: at one's own risk; **gefährden** *tr. V.* endanger; jeopardize *(enterprise, success, position, etc.)*

gefahren 2. *Part. v.* fahren

gefährlich 1. *Adj.* dangerous; *(gewagt)* risky; 2. *adv.* dangerously

gefahr·los 1. *Adj.* safe; 2. *adv.* safely

Gefährt das; ~|e|s, ~e *(geh.)* vehicle

Gefährte der; ~n, ~n, **Gefährtin** die; ~, ~nen *(geh.)* companion; *(Ehemann/Ehefrau)* partner in life

Gefälle das; ~s, ~: slope; incline; *(einer Straße)* gradient

¹**gefallen** *unr. itr. V.* a) das gefällt mir

|gut| I like it [a lot]; b) sich *(Dat.)* etw. ~ lassen put up with sth.

²**gefallen** 2. *Part. v.* fallen, gefallen

¹**Gefallen** der; ~s, ~: favour

²**Gefallen** das; ~s pleasure

Gefallene der; *adj. Dekl.* soldier killed in action; die ~n the fallen

gefällig 1. *Adj.* a) obliging; helpful; b) *(anziehend)* pleasing; agreeable *(programme, behaviour)*; 2. *adv.* pleasingly; agreeably; **Gefälligkeit** die; ~, ~en favour; **gefälligst** *Adv.* *(ugs.)* kindly

gefangen 2. *Part. v.* fangen; **Gefangene** der/die; *adj. Dekl.* prisoner

gefangen-: ~|halten *unr. tr. V.* jmdn./ein Tier ~halten hold sb. prisoner/keep an animal in captivity; ~|nehmen *unr. tr. V.* jmdn. ~nehmen take sb. prisoner

Gefangenschaft die; ~, ~en captivity

Gefängnis das; ~ses, ~se a) prison; gaol; b) *(Strafe)* imprisonment

Gefängnis-: ~**strafe** die prison sentence; ~**wärter** der [prison] warder

Gefasel das; ~s *(ugs. abwertend)* twaddle *(coll.)*; drivel *(derog.)*

Gefäß das; ~es, ~e a) vessel; container; b) *(Anat.)* vessel

gefaßt *Adj.* a) calm; composed; b) in auf etw. *(Akk.)* |nicht| ~ sein |not| be prepared for sth.

Gefecht das; ~|e|s, ~e battle

Gefieder das; ~s, ~: plumage; feathers *pl.*; **gefiedert** *Adj.* feathered

geflissentlich 1. *Adj.* deliberate; 2. *adv.* deliberately

geflochten 2. *Part. v.* flechten

geflogen 2. *Part. v.* fliegen

geflohen 2. *Part. v.* fliehen

geflossen 2. *Part. v.* fließen

Geflügel das; ~s poultry

gefochten 2. *Part. v.* fechten

Gefolge das; ~s, ~: entourage

gefragt *Adj.* in great demand *postpos.*; sought-after

gefräßig *Adj.* *(abwertend)* greedy

Gefreite der; *adj. Dekl. (Milit.)* lancecorporal *(Brit.)*; private first class *(Amer.)*; *(Marine)* able seaman; *(Luftw.)* aircraftman first class *(Brit.)*; airman third class *(Amer.)*

gefressen 2. *Part. v.* fressen

gefrieren *unr. itr. V.; mit sein* freeze

gefrier-, Gefrier-: ~**fach** das freezing compartment; ~**punkt** der freezing-point; ~**schrank** der freezer;

~|**trocknen** *tr. V.; meist im Inf. u. 2. Part.* freeze-dry

gefroren *2. Part. v.* **frieren, gefrieren**

Gefüge das; ~s, ~: structure; **gefügig** *Adj.* compliant; docile ⟨animal⟩

Gefühl das; ~s, ~e a) sensation; feeling; b) *(Gemütsverfassung)* feeling; **gefühl·los** *Adj.* a) numb; b) *(herzlos, kalt)* unfeeling

gefühls-, Gefühls-: ~**betont** *Adj.* emotional; ~**duselei** die; ~ *(ugs. abwertend)* mawkishness; ~**mäßig** *Adj.* emotional ⟨reaction⟩; ⟨action⟩ based on emotion

gefühl·voll 1. *Adj.* sensitive; *(ausdrucksvoll)* expressive; 2. *adv.* sensitively; expressively

gefüllt *2. Part. v.* **füllen**

gefunden *2. Part. v.* **finden**; *s. auch* **Fressen b**

gegangen *2. Part. v.* **gehen**

gegeben *2. Part. v.* **geben**

gegen *Präp. mit Akk.* a) against; ~ etw. stoßen knock into sth.; ein Mittel ~ Krebs a cure for cancer; ~ die Abmachung contrary to the agreement; b) ~ Abend/Morgen towards evening/dawn; ~ vier Uhr around 4 o'clock; c) *(im Vergleich zu)* compared with; d) *(im Ausgleich für)* for; ~ Quittung against a receipt

Gegen-: ~**angriff** der counterattack; ~**argument** das counterargument; ~**besuch** der return visit

Gegend die; ~, ~en a) area; b) *(Körperregion)* region

Gegen-: ~**darstellung** die: eine ~darstellung |der Sache| an account [of the matter] from an opposing point of view; ~**druck** der counterpressure

gegen·einander *Adv.* against each other *or* one another

Gegen-: ~**gewicht** das counterweight; ein ~gewicht zu *od.* gegen etw. bilden *(fig.)* counterbalance sth.; ~**leistung** die service in return; ~**mittel** das *(gegen Gift)* antidote; *(gegen Krankheit)* remedy; ~**probe** die cross-check; ~**satz** der a) *(Gegenteil)* opposite; b) *(Widerspruch)* conflict; ~**sätzlich** *Adj.* conflicting; ~**seitig** 1. *Adj. (wechselseitig)* mutual; 2. *adv.* sich ~seitig helfen/überbieten help/outdo each other *or* one another; ~**seitigkeit** die reciprocity; auf ~seitigkeit *(Dat.)* beruhen be mutual; ~**spieler** der opponent; *(Sport)* opposite number

Gegen·stand der object; *(Thema)* subject; topic; **gegenständlich** *Adj.* *(Kunst)* representational; *(Philos.)* objective; **gegenstands·los** *Adj.* a) *(hinfällig)* invalid; b) *(grundlos, unbegründet)* unfounded ⟨accusation, complaint, jealousy⟩; baseless ⟨fear⟩

gegen-, Gegen-: ~**stimme** die vote against; ohne ~stimme unanimously; ~**stück** das companion piece; *(fig.)* counterpart; ~**teil** das opposite; im ~teil on the contrary; ~**teilig** *Adj.* opposite; contrary

gegen·über *Präp. mit Dat.* a) opposite; b) *(in bezug auf)* ~ jmdm. *od.* jmdm. ~ freundlich sein be kind to sb.; c) *(im Vergleich zu)* compared with

gegenüber-, Gegenüber-: ~|**stehen** *unr. itr. V.* a) jmdm./einer Sache ~stehen stand facing sb./sth.; *(fig.)* face sb./sth.; b) jmdm./einer Sache feindlich/wohlwollend ~stehen be ill/well disposed towards sb./sth.; ~|**stellen** *tr. V.* confront; ~**stellung** die confrontation; b) *(Vergleich)* comparison; ~|**treten** *unr. itr. V.; mit sein* jmdm./einer Sache treten *(auch fig.)* face sb./sth.

Gegen·verkehr der oncoming traffic

Gegenwart die; ~ a) present; b) *(Anwesenheit)* presence; c) *(Grammatik)* present [tense]; **gegenwärtig** 1. *Adj.* present; 2. *adv.* at present; at the moment

Gegen-: ~**wehr** die; *o. Pl.* resistance; ~**wind** der head wind; ~**zug** der *(Brettspiele, fig.)* countermove

gegessen *2. Part. v.* **essen**

geglichen *2. Part. v.* **gleichen**

geglitten *2. Part. v.* **gleiten**

Gegner der; ~s, ~ a) adversary; opponent; b) *(Sport)* opponent; **gegnerisch** *Adj.* opposing; opponents' ⟨goal⟩

gegolten *2. Part. v.* **gelten**

gegoren *2. Part. v.* **gären**

gegossen *2. Part. v.* **gießen**

gegriffen *2. Part. v.* **greifen**

gehabt *2. Part. v.* **haben**

¹**Gehalt** der; ~|e|s, ~e a) meaning; b) *(Anteil)* content

²**Gehalt** das, *österr. auch:* der; ~|e|s, Gehälter salary

gehalten *2. Part. v.* **halten**

Gehalts-: ~**empfänger** der salary earner; ~**erhöhung** die salary increase

gehalt·voll *Adj.* nutritious ⟨food⟩; ⟨novel, speech⟩ rich in substance

gehässig *Adj. (abwertend)* spiteful; **Gehässigkeit** die; ~, ~en a) *(Wesen)* spitefulness; b) *(Äußerung)* spiteful remark

gehauen *2. Part. v.* hauen

gehäuft *Adj.* ein ~er Teelöffel/Eßlöffel a heaped teaspoon/tablespoon

Gehäuse das; ~s, ~ *(einer Maschine)* casing; housing; *(einer Kamera, Uhr)* case

geh·behindert *Adj.* able to walk only with difficulty *postpos.;* disabled

Gehege das; ~s, ~ a) *(Jägerspr.)* preserve; b) *(im Zoo)* enclosure

geheim 1. *Adj.* a) secret; b) *(mysteriös)* mysterious; 2. *adv.* ~ abstimmen vote by secret ballot

geheim-, Geheim-: ~agent der secret agent; ~dienst der secret service; ~|halten *unr. tr. V.* keep secret

Geheimnis das; ~ses, ~se secret; **Geheimnis·tuerei** die; ~ *(ugs.)* secretiveness; **geheimnis·voll** *Adj.* mysterious

Geheiß das: auf jmds. ~ *(geh.)* at sb.'s behest

gehen *unr. itr. V.; mit sein* a) walk; go; **über die Straße** ~: cross the street; b) *(sich irgendwohin begeben)* go; c) *(regelmäßig besuchen)* attend; d) *(weg~)* go; leave; e) *(in Funktion sein)* work; **meine Uhr geht falsch** my watch is wrong; f) *(möglich sein)* **ja, das geht** yes, I/we can manage that; **das geht nicht** that can't be done; g) *(ugs.: gerade noch angehen)* **Hast du gut geschlafen? – Es geht** Did you sleep well? – Not too bad; h) *(sich entwickeln)* **der Laden/das Geschäft geht gut/gar nicht** the shop/business is doing well/not doing well at all; i) *(unpers.)* **wie geht es dir?** How are you?; **jmdm. geht es gut/schlecht** *(gesundheitlich)* sb. is well/not well; *(geschäftlich)* sb. is doing well/badly; j) *(unpers.) (sich um etw. handeln)* ; **worum geht es hier?** what is this all about?; 2. *unr. tr. V. (zurücklegen)* **10 km** ~: walk 10 km.

gehen|lassen *unr. refl. V. (sich nicht beherrschen)* lose control of oneself; *(sich vernachlässigen)* let oneself go

geheuer *Adj.* a) **in diesem Gebäude ist es nicht** ~: this building is eerie; b) **ihr war doch nicht |ganz|** ~: she felt [a little] uneasy; c) **die Sache ist |mir| nicht ganz** ~: [I feel] there's something odd about this business

Gehilfe der; ~n, ~n assistant

Gehirn das; ~|e|s, ~e brain

Gehirn-: ~erschütterung die concussion; ~schlag der stroke; ~wäsche die brainwashing *no indef. art.*

gehoben 1. *2. Part. v.* heben; 2. *Adj.* a) higher; senior *(position)*; b) *(gewählt)* elevated, refined

geholfen *2. Part. v.* helfen

Gehör das; ~|e|s [sense of] hearing

gehorchen *itr. V.* jmdm. ~: obey sb.

gehören 1. *itr. V.* a) jmdm. ~: belong to sb.; b) *(Teil eines Ganzen sein)* zu jmds. Freunden/Aufgaben ~: be one of sb.'s friends/part of sb.'s duties; c) *(passend sein)* **dein Roller gehört nicht in die Küche!** your scooter does not belong in the kitchen!; d) *(nötig sein)* **es hat viel Fleiß dazu gehört** it took a lot of hard work; **dazu gehört sehr viel** that takes a lot; 2. *refl. V. (sich schikken)* be fitting; **es gehört sich [nicht], ... zu ...:** it is [not] good manners to ...;

gehörig 1. *Adj.* a) proper; b) *(ugs.: beträchtlich)* ein ~er Schrecken/eine ~e Portion Mut a good fright/a good deal of courage; 2. *adv. (ugs.: beträchtlich)* ~ essen/trinken eat/drink heartily

gehorsam *Adj.* obedient; **Gehorsam** der; ~s obedience

Geh·steig der pavement *(Brit.);* sidewalk *(Amer.)*

Geier der; ~s, ~: vulture

Geige die; ~, ~n violin

Geiger·zähler der *(Physik)* Geiger counter

geil *Adj. (oft abwertend: sexuell erregt)* randy; horny *(sl.); (lüstern)* lecherous

Geisel die; ~, ~n hostage

Geißel die; ~, ~n *(hist., auch fig.)* scourge

Geist der; ~|e|s, ~er a) *o. Pl. (Verstand)* mind; b) *o. Pl. (Scharfsinn)* wit; c) *o. Pl. (innere Einstellung)* spirit; d) *(denkender Mensch)* mind; intellect; **ein großer/kleiner** ~: a great mind/a person of limited intellect; e) *(überirdisches Wesen)* spirit; **der Heilige** ~ *(christl. Rel.)* the Holy Ghost *or* Spirit; f) *(Gespenst)* ghost; **Geisterfahrer** der person driving on the wrong side of the road or the wrong carriageway; **geisterhaft** *Adj.* ghostly; eerie *(atmosphere)*

geistes-, Geistes-: ~abwesend 1. *Adj.* absent-minded; 2. *adv.* absentmindedly; ~blitz der *(ugs.)* brainwave; ~gegenwart die presence of mind; ~gegenwärtig 1. *Adj.* quickwitted; 2. *adv.* with great presence of mind; ~krank *Adj.* mentally ill; ~wissenschaften *Pl.* arts; humanities; ~zustand der; *o. Pl.* mental state

geistig 1. *Adj.* a) intellectual; *(Psych.)* mental; b) alcoholic ⟨*drinks*⟩; 2. *adv.* intellectually; *(Psych.)* mentally; geistlich *Adj.* sacred ⟨*song, music*⟩; religious ⟨*order, book, writings*⟩; Geistliche der; *adj. Dekl.* clergyman geist-: ~los *Adj.* dim-witted; *(trivial)* trivial; ~reich 1. *Adj.* witty; *(klug)* clever; 2. *adv.:* wittily; cleverly

Geiz der; ~es meanness; *(Knauserigkeit)* miserliness; geizen *itr. V.* be mean; Geiz·hals der *(abwertend)* skinflint; geizig *Adj.* mean; *(knauserig)* miserly

gekannt 2. *Part. v.* kennen
Gekicher das; ~s giggling
geklungen 2. *Part. v.* klingen
geknickt *Adj. (ugs.)* dejected
gekniffen 2. *Part. v.* kneifen
gekommen 2. *Part. v.* kommen
gekonnt 1. 2. *Part. v.* können; 2. *Adj.* accomplished; *(hervorragend ausgeführt)* masterly

gekrochen 2. *Part. v.* kriechen
gekünstelt 1. *Adj.* artificial; 2. *adv.* er lächelte ~: he gave a forced smile
Gelächter das; ~s, ~: laughter
geladen 2. *Part. v.* laden
Gelände das; ~s, ~ a) *(Landschaft)* ground; terrain; b) *(Grundstück)* site; *(von Schule, Krankenhaus usw.)* grounds *pl.*

Geländer das; ~s, ~: banisters *pl.;* handrail; *(am Balkon, an einer Brücke)* railing[s *pl.*]; *(aus Stein)* parapet

gelang 3. *Pers. Sg. Prät. v.* gelingen
gelangen *itr. V.;* mit sein an etw. *(Akk.)/zu etw.* ~: reach sth.; *(fig.)* zu Ansehen ~: gain esteem

gelassen 1. 2. *Part. v.* lassen; 2. *Adj.* calm; *(gefaßt)* composed; Gelassenheit die; ~: calmness; *(Gefaßtheit)* composure

Gelatine [ʒela'tiːnə] die; ~: gelatine
gelaufen 2. *Part. v.* laufen
geläufig *Adj. (vertraut)* common ⟨*expression, concept*⟩

gelaunt gut/schlecht ~ sein be in a good/bad mood

gelb *Adj.* yellow; Gelb das; ~s, ~ *od. (ugs.)* ~s yellow; gelblich *Adj.* yellowish; yellowed ⟨*paper*⟩; sallow ⟨*skin*⟩; Gelb·sucht die; *o. Pl. (Med.)* jaundice

Geld das; ~es, ~er money; großes ~: large denominations *pl.;* kleines/bares ~: change/cash

geld-, Geld-: ~automat der cash dispenser ~beutel der *(bes. südd.)* purse; ~börse die purse; ~gier die avarice; ~gierig *Adj.* avaricious; ~mittel *Pl.* financial resources; ~schein der banknote; bill *(Amer.);* ~schrank der the safe; ~strafe die fine; ~stück das coin; ~wechsel der exchanging of money; „~wechsel" 'bureau de change'

Gelee [ʒe'leː] der *od.* das; ~s, ~s jelly
gelegen 1. 2. *Part. v.* liegen; 2. *Adj.* a) *(passend)* convenient; Gelegenheit die; ~, ~en opportunity; *(Anlaß)* occasion

Gelegenheits-: ~arbeit die casual work; ~kauf der bargain

gelegentlich 1. *Adj.* occasional; 2. *adv.* occasionally

gelehrig *Adj.* ⟨*child*⟩ who is quick to learn; ⟨*animal*⟩ that is quick to learn; gelehrt *Adj.* learned; Gelehrte der/die; *adj. Dekl.* scholar

Geleit das; ~[e]s, ~e *(geh.)* sie bot uns ihr ~ an she offered to accompany us; geleiten *tr. V. (geh.)* escort; Geleit·schutz der *(Milit.)* escort

Gelenk das; ~[e]s, ~e joint; gelenkig 1. *Adj.* agile ⟨*person*⟩; supple ⟨*limb*⟩; 2. *adv.* agilely; Gelenkigkeit die; ~: agility; *(von Gliedmaßen)* suppleness

gelernt *Adj.* qualified
gelesen 2. *Part. v.* lesen
Geliebte der/die; *adj. Dekl.* lover/mistress

geliefert *Adj.:* ~ sein *(salopp)* have had it *(coll.)*

geliehen 2. *Part. v.* leihen
gelind[e] 1. *Adj.* mild; 2. *adv.* mildly; ~e gesagt to put it mildly

gelingen *unr. itr. V.;* mit sein succeed; Gelingen das; ~s success

gelitten 2. *Part. v.* leiden
gellen *itr. V.* a) *(hell schallen)* ring out; b) *(nachhallen)* ring

geloben *tr. V. (geh.)* vow; das Gelobte Land the Promised Land

gelogen 2. *Part. v.* lügen
gelöst *Adj.* relaxed
gelten 1. *unr. itr. V.* a) *(gültig sein)* be valid; ⟨*banknote, coin*⟩ be legal tender; ⟨*law etc.*⟩ be in force; b) *(angesehen werden)* als etw. ~: be regarded as sth.; c) (+ *Dat.*) *(bestimmt sein für)* be directed at; 2. *unr. tr. V.* a) *(wert sein)* sein Wort gilt viel/wenig his word car-

ries a lot of/little weight; b) *unpers.* es
gilt, etw. zu tun it is essential to do
sth.; **geltend: etw. ~ machen** assert
sth.; **Geltung die; ~ a)** validity; für
jmdn. ~ **haben** apply to sb.; b) *(Wirkung)* recognition; **zur ~ kommen**
show to [its best] advantage; **Gel-
tungs·bedürfnis das** need for recognition
gelungen 1. *2. Part. v.* **gelingen; 2.**
Adj. **a)** *(ugs.: spaßig)* priceless; b) *(ansprechend)* inspired
gemächlich [gə'mɛ(:)çlɪç] **1.** *Adj.* leisurely; **2.** *adv.* in a leisurely manner
gemacht *in ein ~er* **Mann sein** *(ugs.)*
be a made man
Gemahl der; ~s, ~e *(geh.)* consort;
husband; **Gemahlin die; ~, ~nen**
(geh.) consort; wife
Gemälde das; ~s, ~: painting
gemäß *Präp. + Dat.* in accordance
with
gemäßigt *Adj.* moderate; qualified
⟨*optimism*⟩; temperate ⟨*climate*⟩
gemein 1. *Adj.* **a)** vulgar ⟨*joke, expression*⟩; nasty ⟨*person*⟩; b) *(niederträchtig)* mean; dirty ⟨*lie*⟩; mean ⟨*trick*⟩; **2.**
adv. in a mean *or* nasty way
Gemeinde die; ~, ~n a) municipal-
ity; *(Bewohner)* community; **b)**
(Pfarr~) parish; **c)** *(versammelte Got-
tesdienstteilnahme)* congregation
Gemeinde-: ~rat der a) *(Gremium)*
local council; b) *(Mitglied)* local
councillor; **~schwester die** district
nurse; **~verwaltung die** local administration
gemein·gefährlich *Adj.* dangerous
to the public; **Gemein·gut das;** *o.
Pl. (geh.)* common property
Gemeinheit die; ~, ~en a) *o. Pl.*
meanness; b) *(Handlung)* mean trick
gemein·nützig *Adj.* serving the public good *postpos., not pred.; (wohltätig)*
charitable
gemeinsam 1. *Adj.* **a)** common ⟨*interests, characteristics*⟩; mutual ⟨*acquaintance, friend*⟩; joint ⟨*property,
account*⟩; shared ⟨*experience*⟩; b) *(miteinander unternommen)* joint; **2.** *adv.*
together; **Gemeinsamkeit die; ~,
~en** common feature
Gemeinschaft die; ~, ~en a) community; **b)** *o. Pl. (Verbundenheit)*
coexistence; **gemeinschaftlich** *s.*
gemeinsam
gemein·verständlich *Adj.* generally comprehensible; **Gemein·wohl
das** public good

gemessen 1. *2. Part. v.* **messen; 2.**
Adj. *(würdevoll)* measured ⟨*steps,
tones, language*⟩; deliberate ⟨*words,
manner of speaking*⟩
Gemetzel das; ~s, ~: massacre
gemieden *2. Part. v.* **meiden**
Gemisch das; ~[e]s, ~e mixture (**aus,
von** of)
gemocht *2. Part. v.* **mögen**
gemolken *2. Part. v.* **melken**
Gemse die; ~, ~n chamois
Gemurmel das; ~s murmuring
Gemüse das; ~s, ~: vegetables *pl.*
gemußt *2. Part. v.* **müssen**
Gemüt das; ~[e]s, ~er a) nature; **b)**
(Empfindungsvermögen) heart; **c)**
(Mensch) soul
gemütlich 1. *Adj.* snug; cosy; *(bequem)* comfortable; *(ungezwungen)*
informal; **2.** *adv.* cosily; *(bequem)*
comfortably; ~ **beisammensitzen** sit
pleasantly together; **Gemütlichkeit
die; ~:** snugness; *(Zwanglosigkeit)* in-
formality
gemüts·krank *Adj. (Med., Psych.)*
emotionally disturbed; **Ge-
müts·mensch der** *(ugs.)* even-tem-
pered person; **gemüt·voll** *Adj.*
warm-hearted; *(empfindsam)* sentimental
Gen das; ~s, ~e *(Biol.)* gene
genannt *2. Part. v.* **nennen**
genas *1. u. 3. Pers. Sg. Prät. v.* **genesen**
genau 1. *Adj.* **a)** *(exakt)* exact; precise; b) *(sorgfältig, gründlich)* meticulous, ⟨*person*⟩; careful ⟨*study*⟩; **2.** *adv.*
a) exactly; precisely; ~ **um 8⁰⁰** at 8
o'clock precisely; **b)** *(gerade, eben)*
just; **c)** *(als Verstärkung)* just; **d)** *(als
Zustimmung)* exactly; precisely; **e)**
(sorgfältig) ~ **arbeiten/etw. ~ durchdenken** work/think sth. out meticulously
genau·genommen *Adv.* strictly
speaking
Genauigkeit die; ~ a) *(Exaktheit)*
exactness; precision; *(einer Waage)*
accuracy; **b)** *(Sorgfalt)* meticulousness; **genau·so** *Adv.* **a)** mit *Adjektiven* just as; **b)** mit *Verben* in exactly
the same way; *(in demselben Maße)*
just as much
genehm *Adj.* in jmdm. ~ **sein** *(geh.)*
(jmdm. passen) be convenient to sb.;
(jmdm. angenehm sein) be acceptable
to sb.
genehmigen *tr. V.* approve ⟨*plan, alterations, application*⟩; authorize
⟨*stay*⟩; grant ⟨*request*⟩; give per-

mission for ⟨*demonstration*⟩; **sich** *(Dat.)* **etw. ~** *(ugs.)* treat oneself to sth.; **Genehmigung die; ~, ~en a)** s. **genehmigen**; approval; authorization; granting; permission *(Gen.* for); **b)** *(Schriftstück)* permit; *(Lizenz)* licence

geneigt *Adj. in* **~ sein, etw. zu tun** be inclined to do sth.

General der; ~s, ~e *od.* **Generäle** general

General-: ~direktor der chairman; president *(Amer.);* **~probe die** *(auch fig.)* dress rehearsal; **~streik der** general strike; **~vertreter der** general representative

Generation die; ~, ~en generation; **Generations·konflikt** der generation gap

Generator der; ~s, ~en generator

generell 1. *Adj.* general; **2.** *adv.* generally

genesen *unr. itr. V.; mit sein (geh.)* recover; **Genesung die; ~, ~en** *(geh.)* recovery

genetisch *(Biol.) Adj.* genetic

Genf (das); ~s Geneva; **Genfer 1.** der; ~s, ~: Genevese; **2.** *Adj.* Genevese; **der ~** See Lake Geneva

genial *Adj.* brillant; **Genialität die; ~:** genius

Genick das; ~[e]s, ~e back *or* nape of the neck

Genie [ʒe'niː] das; ~s, ~s genius

genieren [ʒe'niːrən] *refl. V.* be embarrassed

genießbar *Adj. (eßbar)* edible; *(trinkbar)* drinkable; **genießen** *unr. tr. V.* enjoy; **Genießer der; ~s, ~:** **er ist ein richtiger ~:** he is a regular 'bon viveur'

Genitale das; ~s, **Genitalien** [geni'taːliən], **Genital·organ** das genital organ

Genitiv der; ~s, ~e *(Sprachw.)* genitive [case]

genommen 2. *Part. v.* nehmen

genoß 1. u. 3. *Pers. Sg. Prät. v.* genießen

Genosse der; ~n, ~n comrade

genossen 2. *Part. v.* genießen

Genossenschaft die; ~, ~en cooperative; **Genossin die; ~, ~nen** comrade

genug *Adv.* enough

genügen *itr. V.* **a)** be enough; **b) einer Sache** *(Dat.)* **~:** satisfy sth.; **genügend 1.** *Adj.* **a)** enough; **b)** *(befriedigend)* satisfactory; **2.** *adv.* enough; **genügsam** *Adj.* modest

Genugtuung [-tuːʊŋ] **die; ~, ~en** satisfaction

Genus das; ~, **Genera** *(Sprachw.)* gender

Genuß der; **Genusses, Genüsse a)** o. *Pl.* consumption; **b)** *(Wohlbehagen)* **etw. mit ~ essen/lesen** eat sth. with relish/enjoy reading sth.

genüßlich *Adv.* ⟨*eat, drink*⟩ with relish

Geograph der; ~en, ~en geographer; **Geographie die; ~:** geography *no art.;* **geographisch** *Adj.* geographic[al]

Geologe der; ~n, ~n geologist; **Geologie die; ~:** geology *no art.;* **geologisch** *Adj.* geological

Geometrie die; ~: geometry *no art.;* **geometrisch** *Adj.* geometric[al]

Gepäck das; ~[e]s luggage *(Brit.);* baggage *(Amer.);* *(am Flughafen)* baggage

Gepäck-: ~annahme die a) checking in the luggage/baggage; **b)** *(Schalter)* [in-counter of the] luggage office *(Brit.)* or baggage office *(Amer.);* *(zur Aufbewahrung)* [in-counter of the] left-luggage office *(Brit.)* or checkroom *(Amer.);* *(am Flughafen)* baggage check-in; **~aufbewahrung die** left-luggage office *(Brit.);* checkroom *(Amer.);* *(Schließfächer)* luggage lockers *(Brit.);* baggage lockers *(Amer.);* **~ausgabe die** [out-counter of the] luggage office *(Brit.)* or *(Amer.)* baggage office; *(zur Aufbewahrung)* [out-counter of the] left-luggage office *(Brit.)* or *(Amer.)* checkroom; *(am Flughafen)* baggage reclaim; **~kontrolle die** baggage check; **~netz das** luggage rack *(Brit.);* baggage rack *(Amer.)* **~schalter der** s. **~annahme b;** **~schein der** luggage ticket *(Brit.);* baggage check *(Amer.);* **~träger der a)** porter; **b)** *(am Fahrrad)* carrier; rack

gepfeffert *Adj. (ugs.)* steep *(coll.)* ⟨*price, rent, etc.*⟩

gepfiffen 2. *Part. v.* pfeifen

gepflegt *Adj.* **a)** well-groomed spruce ⟨*appearance*⟩; neat ⟨*clothing*⟩; **b)** *(hochwertig)* choice ⟨*food, drink*⟩

Gepflogenheit die; ~, ~en *(geh.)* custom; *(Gewohnheit)* habit

gepriesen 2. *Part. v.* preisen

gequält *Adj.* forced ⟨*smile, gaiety*⟩; pained ⟨*expression*⟩

gequollen 2. *Part. v.* quellen

gerade, *(ugs.)* **grade 1.** *Adj.* **a)**

straight; **b)** *(nicht schief)* upright; **c)**
(aufrichtig) forthright; direct; **d)**
(Math.) even ⟨*number*⟩; **2.** *Adv.* just;
(direkt) right; **Gerade die;** ~**n,** ~**n**
(Geom.) straight line
gerade-: ~**aus** *Adv.* straight ahead;
~|**biegen** *unr. tr. V.* **a)** bend straight;
straighten [out]; **b)** *(ugs.: bereinigen)*
straighten out; ~**heraus** [----'-] *(ugs.)*
Adv. etw. ~**heraus sagen** say sth.
straight out; ~**so** *Adv.* ~**so groß/lang**
wie ...: just as big/long as ...; ~|**ste-**
hen *unr. itr. V.* **a)** stand up straight;
b) *(fig.: einstehen)* für etw. ~**stehen** accept responsibility for sth.; ~**zu** *Adv.*
really; *(beinahe)* almost
Geranie [ge'ra:niə] die; ~, ~**n** geranium
gerann *3. Pers. Sg. Prät. v.* **gerinnen**
gerannt *2. Part. v.* **rennen**
gerät *3. Pers. Sg. Präsens v.* **geraten**
Gerät das; ~|e|s, ~e **a)** piece of equip-
ment; *(Fernseher, Radio)* set; *(Gar-*
ten~) tool; **b)** *(Turnen)* piece of apparatus
¹**geraten** *unr. itr. V.; mit sein* **a)** *(gelan-*
gen) get; **b)** *(werden)* turn out; *(gut ~)*
turn out well
²**geraten** **1.** *2. Part. v.* **raten,** ¹**geraten;**
2. *Adj.* advisable
Geratewohl: **aufs** ~ *(ugs.)* ⟨*select*⟩ at
random; **wir fuhren aufs** ~ **los** *(ugs.)*
we went for a drive just to see where
we ended up
gerät *3. Pers. Sg. Präsens v.* ¹**geraten**
geraum *Adj. (geh.)* considerable
geräumig *Adj.* spacious ⟨*room*⟩;
roomy ⟨*cupboard etc.*⟩
Geräusch das; ~|e|s, ~e sound; *(uner-*
wünscht) noise
geräusch-: ~**arm** **1.** *Adj.* quiet; **2.**
adv. quietly ~**los** **1.** *Adj.* silent; **2.**
adv. **a)** silently; **b)** *(fig. ugs.)* without
[any] fuss; ~**voll** *Adj.* noisy
gerben *tr. V.* tan ⟨*hides, skins*⟩
gerecht **1.** *Adj.* just *(unparteiisch)*
fair; **2.** *adv.* justly
gerechtfertigt *Adj.* justified
Gerechtigkeit die; ~: justice; **Ge-**
rechtigkeits·sinn der the sense of justice
Gerede das; ~s *(abwertend)* **a)** *(ugs.)*
talk; **b)** *(Klatsch)* gossip
geregelt *Adj.* regular, steady ⟨*job*⟩
gereizt *Adj.* irritable
¹**Gericht** das; ~|e|s, ~e court; *(Richter)*
bench; *(Gebäude)* court[-house]; **das**
Jüngste ~ *(Rel.)* the Last Judgement
²**Gericht** das; ~|e|s, ~e dish

gerichtlich **1.** *Adj.* judicial; legal
⟨*proceedings*⟩; **2.** *adv.* jmdn. ~ **verfol-**
gen take sb. to court
Gerichts-: ~**hof der** Court of Justice;
~**kosten** *Pl.* legal costs; ~**saal der**
courtroom; ~**verfahren das** legal
proceedings *pl.;* ~**vollzieher der;**
~s, ~: bailiff
gerieben *2. Part. v.* **reiben**
gering *Adj.* **a)** low; little ⟨*value*⟩;
small ⟨*quantity, amount*⟩; short ⟨*dis-*
tance, time⟩; **b)** *(unbedeutend)* slight;
minor ⟨*role*⟩
geringfügig **1.** *Adj.* slight; minor ⟨*al-*
teration, injury⟩; trivial ⟨*amount, de-*
tail⟩; **2.** *adv.* slightly; **Geringfügig-**
keit die; ~, ~**en** triviality; **gering|-**
schätzen *tr. V.* think very little of
⟨*person, achievement*⟩; set little store
by ⟨*success, riches*⟩; **geringschätzig**
Adj. disdainful; disparaging ⟨*remark*⟩
gerinnen *unr. itr. V.; mit sein* ⟨*blood*⟩
clot; ⟨*milk*⟩ curdle
Gerippe das; ~s, ~: skeleton
gerippt *Adj.* ribbed; fluted ⟨*glass, col-*
umn⟩
gerissen **1.** *2. Part. v.* **reißen; 2.** *Adj.*
(ugs.) crafty
geritten *2. Part. v.* **reiten**
Germane der; ~**n,** ~**n** *(hist.)* ancient
German; Teuton; **germanisch** *Adj.*
(auch fig.) Germanic; Teutonic; **Ger-**
manistik die; ~: German studies *pl.,*
no art.
gern[e]; **lieber, am liebsten** *Adv.* **a)**
etw. ~ **tun** like *or* enjoy doing sth.; **er**
spielt lieber Tennis als Golf he prefers
playing tennis to golf; **etw.** ~/**am lieb-**
sten essen like sth./like sth. best; **ja,**
~/**aber** ~, of course; certainly!;
b) *(durchaus)* **das glaube ich** ~: I can
well believe that
gerochen *2. Part. v.* **riechen**
Geröll das; ~s, ~e debris; *(größer)*
boulders *pl.*
geronnen *2. Part. v.* **rinnen, gerinnen**
Gerste die; ~: barley; **Gersten-**
korn das *(Med.)* sty
Gerte die; ~, ~**n** switch
Geruch der; ~|e|s, **Gerüche** smell; *(von*
Blumen) scent
Gerücht das; ~|e|s, ~e rumour
gerufen *2. Part. v.* **rufen**
geruhsam **1.** *Adj.* peaceful; leisurely
⟨*stroll*⟩; **2.** *adv.* leisurely; quietly
Gerümpel das; ~s junk
gerungen *2. Part. v.* **ringen**
Gerüst das; ~|e|s, ~e scaffolding *no*
pl., no indef. art.

gesamt *Adj*.whole; entire; **gesamt-deutsch** *Adj.* all-German; **Gesamt·eindruck** der general impression; **Gesamtheit die: die ~ der** Bevölkerung the entire population

Gesamt-: ~schule die comprehensive [school]; **~werk** das œuvre; *(Bücher)* complete works *pl.*

gesandt 2. *Part. v.* senden

Gesandte der/die; *adj. Dekl.* envoy; **Gesandtschaft** die; ~, ~en legation

Gesang der; ~|e|s, Gesänge a) singing; b) *(Lied)* song

Gesang-: ~buch das hymn-book; **~verein** der choral society

Gesäß das; ~es, ~e backside; buttocks *pl.*

geschaffen 2. *Part. v.* schaffen 1

Geschäft das; ~|e|s, ~e a) business; *(Transaktion)* [business] deal; **ein gutes ~ machen** make a good profit; b) *(Laden)* shop; store *(Amer.)*

Geschäfte·macher der *(abwertend)* profit-seeker

geschäftig *Adj.* bustling

geschäftlich 1. *Adj.* business *attrib.*; 2. *adv.* on business

geschäfts-, Geschäfts-: ~freund der business associate; **~führer** der manager; *(Vereinswesen)* secretary; **~führung** die; *o. Pl.* management; **~inhaber** der owner of the/a business; **~jahr** das financial year; **~kosten** *Pl.* auf ~kosten on expenses; **~lage** die [business] position; **~leitung** die *s.* ~führung; ~leute *s.* ~mann; ~mann der; *Pl.* ~leute businessman; **~ordnung** die standing orders *pl.*; *(im Parlament)* [rules *pl.* of] procedure; **~partner** der business partner; **~reise** die business trip; **~schluß** der closing-time; **~stelle** die branch; *(einer Partei, eines Vereins)* office; **~straße** die shopping-street; **~tüchtig** *Adj.* able, ⟨*businessman, landlord, etc.*⟩; **~viertel** das business quarter; *(Einkaufszentrum)* shopping district; **~wagen** der company car; **~zeit** die business hours *pl.*; *(im Büro)* office hours *pl.*

geschah 3. *Pers. Sg. Prät. v.* geschehen

geschehen *unr. itr. V.*; *mit sein* happen; occur; *(ausgeführt werden)* be done; **jmdm. geschieht etw.** sth. happens to sb.

gescheit *Adj.* a) *(intelligent)* clever; b) *(ugs.: vernünftig)* sensible

Geschenk das; ~|e|s, ~e present; gift

Geschenk-: ~artikel der gift; **~packung** die gift pack

Geschichte die; ~, ~n a) history; b) *(Erzählung)* story; **geschichtlich** *Adj.* a) historical; b) *(bedeutungsvoll)* historic

¹**Geschick** das; ~|e|s, ~e *(geh.)* fate

²**Geschick** das; ~|e|s skill; **Geschicklichkeit** die; ~: skilfulness; skill; **geschickt** 1. *Adj.* a) skilful; b) *(klug)* clever; adroit; 2. *adv.* a) *(gewandt)* skilfully; b) *(klug)* cleverly; adroitly

geschieden 2. *Part. v.* scheiden

geschienen 2. *Part. v.* scheinen

Geschirr das; ~|e|s, ~e a) crockery; *(benutzt)* dishes *pl.*; b) *(für Zugtier)* harness

Geschirr-: ~spül·maschine die dishwasher; **~tuch** das; *Pl.* -tücher tea-towel; dish towel *(Amer.)*

geschissen 2. *Part. v.* scheißen

geschlafen 2. *Part. v.* schlafen

geschlagen 2. *Part. v.* schlagen

Geschlecht das; ~|e|s, ~er a) sex; b) *(Generation)* generation; c) *(Sippe)* family; d) *(Sprachw.)* gender; **geschlechtlich** *Adj.* sexual

geschlechts-, Geschlechts-: ~krank *Adj.* ⟨*person*⟩ suffering from VD; **~krankheit** die venereal disease; **~teil** das genitals *pl.*; **~verkehr** der sexual intercourse; **~wort** das *s.* Artikel a

geschlichen 2. *Part. v.* schleichen

geschliffen 1. 2. *Part. v.* schleifen; 2. *Adj.* polished

geschlossen 1. 2. *Part. v.* schließen; 2. *Adj.* united ⟨*action, front*⟩; unified ⟨*procedure*⟩; **eine ~e Ortschaft** a built-up area

geschlungen 2. *Part. v.* schlingen

Geschmack der; ~|e|s, Geschmäcke taste; **geschmacklos** 1. *Adj.* tasteless; 2. *adv.* tastelessly; **Geschmacklosigkeit** die; ~, ~en lack of [good] taste; bad taste; *(Äußerung)* tasteless remark; **Geschmack[s]·sache** die **in das ist ~**: that is a question *or* matter of taste

geschmack·voll 1. *Adj.* tasteful. 2. *adv.* tastefully

Geschmeide das; ~s, ~ *(geh.)* jewellery *no pl.*

geschmeidig 1. *Adj.* a) sleek ⟨*hair, fur*⟩; soft ⟨*leather, boots, skin*⟩; b) *(gelenkig)* supple ⟨*fingers*⟩; lithe ⟨*body, movement, person*⟩; 2. *adv.* *(gelenkig)* agilely

geschmissen *2. Part. v.* **schmeißen**
geschmolzen *2. Part. v.* **schmelzen**
Geschnetzelte das; *adj. Dekl.: small, thin slices of meat [cooked in sauce]*
geschnitten *2. Part. v.* **schneiden**
geschoben *2. Part. v.* **schieben**
geschollen *2. Part. v.* **schallen**
gescholten *2. Part. v.* **schelten**
Geschöpf das; ~[e]s, ~e creature
geschoren *2. Part. v.* **scheren**
¹Geschoß das; Geschosses, Geschosse projectile; *(Kugel)* bullet; *(Rakete)* missile
²Geschoß das; Geschosses, Geschosse floor; storey
geschossen *2. Part. v.* **schießen**
geschraubt *Adj. (ugs.)* stilted
Geschrei das; ~s a) shouting; *(von Verletzten, Tieren)* screaming; screams *pl.; (ugs. fig)* fuss
geschrieben *2. Part. v.* **schreiben**
geschrie[e]n *2. Part. v.* **schreien**
geschritten *2. Part. v.* **schreiten**
geschunden *2. Part. v.* **schinden**
Geschütz das; ~es, ~e [big] gun; **Geschütz·feuer** das artillery-fire; shell-fire
geschützt *Adj.* sheltered
Geschwader das; ~s, ~ *(Marine)* squadron; *(Luftwaffe)* wing *(Brit.);* group *(Amer.)*
Geschwätz das; ~es *(ugs. abwertend)* prattling; *(Klatsch)* gossip; **geschwätzig** *Adj. (abwertend)* talkative
geschwiegen *2. Part. v.* **schweigen**
geschwind *(bes. südd.)* 1. *Adj.* swift; quick; 2. *adv.* swiftly; quickly
Geschwindigkeit die; ~, ~en speed
Geschwindigkeits-: ~begrenzung die, ~beschränkung die speed limit
Geschwister *Pl.* brothers and sisters
geschwollen 1. *2. Part. v.* **schwellen**; 2. *Adj.* a) swollen; b) *(fig. abwertend)* pompous; 3. *adv.* pompously
geschwommen *2. Part. v.* **schwimmen**
geschworen *2. Part. v.* **schwören**; **Geschworene** der/die; *adj. Dekl.* juror
Geschwulst die; ~, Geschwülste tumour
geschwunden *2. Part. v.* **schwinden**
geschwungen 1. *2. Part. v.* **schwingen**; 2. *Adj.* curved
Geschwür das; ~s, ~e ulcer; *(Furunkel)* boil
gesehen *2. Part v.* **sehen**

Geselle der; ~n, ~n journeyman; *(Kerl)* fellow; **gesellen** *refl. V.* sich zu jmdm. ~: join sb.; **gesellig** *Adj.* sociable; **ein** ~**er Abend/**~**es Beisammensein** a convivial evening/a friendly get-together; **Geselligkeit** die; ~: die ~ lieben enjoy [good] company
Gesellschaft die; ~, ~en a) society; b) *(Veranstaltung)* party; c) *(Kreis von Menschen)* group of people; d) *(Wirtschaft)* company; **Gesellschafter** der; ~s, ~ a) ein guter ~ sein be good company; b) *(Wirtsch.)* partner; *(Teilhaber)* shareholder; **Gesellschafterin** die; ~, ~nen a) [lady] companion; b) *(Wirtsch.)* partner; *(Teilhaber)* shareholder; **gesellschaftlich** *Adj.* social
gesellschafts-, Gesellschafts-: ~fähig *Adj. (auch fig.)* socially acceptable; ~ordnung die social order; ~reise die group tour; ~schicht die stratum of society; ~spiel das party game
gesessen *2. Part. v.* **sitzen**
Gesetz das; ~es, ~e a) law; *(geschrieben)* statute; b) *(Regel)* rule
Gesetz-: ~buch das statute-book; ~geber der legislator; *(Organ)* legislature; ~gebung die; ~: legislation
gesetzlich 1. *Adj.* legal; statutory ⟨holiday⟩; lawful ⟨heir, claim⟩; 2. *adv.* legally; **gesetz·mäßig** 1. *Adj.* a) law-governed; ~ sein be governed by or obey a [natural] law/[natural] laws; b) *(gesetzlich)* legal; *(rechtmäßig)* lawful; 2. *adv.* in accordance with a [natural] law/[natural] laws; **Gesetz·mäßigkeit** die a) conformity to a [natural] law/[natural] laws; b) *(Gesetzlichkeit)* legality; *(Rechtmäßigkeit)* lawfulness
gesetzt *Adj.* staid
gesetz·widrig *Adj.* illegal; unlawful
Gesicht das; ~[e]s, ~er face; *(fig.)* das ~ einer Stadt the appearance of a town
Gesichts-: ~ausdruck der expression; look; ~creme die face-cream; ~punkt der point of view; ~wasser das face-lotion; ~züge *Pl.* features
Gesindel das; ~s *(abwertend)* rabble
gesinnt *Adj.* christlich/sozial ~ [sein] [be] Christian-minded/public-spirited; jmdm. freundlich ~ sein be well-disposed towards sb.; **Gesinnung** die; ~, ~en [basic] convictions *pl.;* [fundamental] beliefs *pl.;* **gesin-**

nungs·los *(abwertend) Adj.* unprincipled; **Gesinnungs·wandel** der change of attitude

gesittet 1. *Adj.* well-behaved; well-mannered

gesogen *2. Part. v.* saugen

gesondert 1. *Adj.* separate; **2.** *adv.* separately

gesonnen *Adj.* ~ sein, etw. zu tun feel disposed to do sth.

gesotten *2. Part. v.* sieden

Gespann das; ~|e|s, ~e a) *(Zugtiere)* team; b) *(Wagen)* horse and carriage; c) *(Menschen)* couple; pair

gespannt *Adj.* a) eager; rapt ⟨attention⟩; ~ zuhören listen with rapt attention; b) tense ⟨situation, atmosphere⟩; strained ⟨relationships⟩

Gespenst das; ~|e|s, ~er a) ghost; b) *(geh.: Gefahr)* spectre

gespenstig, gespenstisch *Adj.* ghostly; eerie ⟨building, atmosphere⟩

gespie[e]n *2. Part. v.* speien

gesponnen *2. Part. v.* spinnen

Gespött das; ~|e|s mockery; ridicule

Gespräch das; ~|e|s, ~e conversation; *(Diskussion)* discussion; *(Telefon~)* call (mit to); **gesprächig** *Adj.* talkative

Gesprächs-: ~**partner** der: mein heutiger ~partner wird X sein today I shall be talking to X; ~**stoff** der topics *pl.* of conversation; ~**thema** das topic of conversation

gesprochen *2. Part. v.* sprechen

gesprossen *2. Part. v.* sprießen

gesprungen *2. Part. v.* springen

Gespür das; ~s feel

gest. *Abk.* gestorben d.

Gestalt die; ~, ~en a) build; b) *(Mensch, Persönlichkeit)* figure; c) *(in der Dichtung)* character; d) *(Form)* form; **gestalten** *tr. V.* fashion; lay out ⟨public gardens⟩; shape ⟨character, personality⟩; arrange ⟨party, conference, etc.⟩; **Gestaltung** die; ~, ~en *s.* gestalten: fashioning; laying out; arranging

gestand *1. u. 3. Pers. Sg. Prät. v.* gestehen

gestanden *2. Part. v.* stehen, gestehen

geständig *Adj.:* ~ sein have confessed; **Geständnis** das; ~ses, ~se confession

Gestank der; ~|e|s *(abwertend)* stench; stink

gestatten 1. *tr., itr. V.* permit; allow; ~ Sie, daß ich ...?: may I ...?; **2.** *refl. V.* sich *(Dat.)* etw. ~: allow oneself sth.

Geste ['gɛstə, 'ge:stə] die; ~, ~n *(auch fig.)* gesture

Gesteck das; ~|e|s, ~e flower arrangement

gestehen *tr., itr. V.* confess

Gestein das; ~|e|s, ~e rock

Gestell das; ~|e|s, ~e a) *(für Weinflaschen)* rack; *(zum Wäschetrocknen)* horse; b) *(Unterbau)* frame

gestern *Adv.* yesterday

gestiegen *2. Part. v.* steigen

gestikulieren *itr. V.* gesticulate

Gestirn das; ~|e|s, ~e star

gestochen 1. *2. Part. v.* stechen; **2.** *Adj.* extremely neat ⟨handwriting⟩

gestohlen *2. Part. v.* stehlen

gestorben *2. Part. v.* sterben

gestoßen *2. Part. v.* stoßen

Gesträuch das; ~|e|s, ~e shrubbery; bushes *pl.*

gestreift *Adj.* striped

gestrichen 1. *2. Part. v.* streichen; **2.** *Adj.* level ⟨measure⟩

gestrig *Adj.* yesterday's

gestritten *2. Part. v.* streiten

Gestrüpp das; ~|e|s, ~e undergrowth

gestunken *2. Part. v.* stinken

Gestüt das; ~|e|s, ~e stud[-farm]

Gesuch das; ~|e|s, ~e request (um for); *(Antrag)* application (um for); **gesucht** *Adj.* a) [much] sought-after; b) *(gekünstelt)* laboured

gesund; gesünder, seltener: ~er, gesündest..., seltener: ~est... *Adj.* healthy; **wieder ~ werden** get better; **bleib ~!** look after yourself!; **Gesundheit** die; ~: health; ~! *(ugs.)* bless you!; **gesundheitlich 1.** *Adj.; nicht präd.* ~e Betreuung health care; sein ~er Zustand [the state of] his health; **2.** *adv.* wie geht es Ihnen ~? how are you?

gesundheits-, Gesundheits-: ~**amt** das [local] public health department; ~**schädlich** *Adj.* detrimental to [one's] health *postpos.*; ~**zeugnis** das certificate of health; ~**zustand** der state of health

gesungen *2. Part. v.* singen

gesunken *2. Part. v.* sinken

getan *2. Part. v.* tun

Getier das; ~|e|s *(geh.)* animals *pl.*

Getöse das; ~s [thunderous] roar; *(von vielen Menschen)* din

getragen *2. Part. v.* tragen

Getränk das; ~|e|s, ~e drink; beverage *(formal)*

getrauen *refl. V.* dare

Getreide das; ~s grain

Getreide-: ~**anbau** der growing of cereals; ~**handel** der corn-trade

getrennt 1. *Adj.* separate; 2. *adv.* ⟨pay⟩ separately; ⟨sleep⟩ in separate rooms

getreten 2. *Part. v.* **treten**

getreu 1. *Adj. (geh.)* exact; faithful ⟨image⟩; 2. *adv. (geh.)* ⟨report, describe⟩ faithfully

Getriebe das; ~s, ~ gears *pl.; (in einer Maschine)* gear system; **getrieben** 2. *Part. v.* **treiben**

getroffen 2. *Part. v.* **treffen, triefen**

getrogen 2. *Part. v.* **trügen**

getrost 1. *Adj.* confident; 2. *adv.* confidently; **du kannst es mir ~ glauben** you can take my word for it

getrunken 2. *Part. v.* **trinken**

Getto das; ~s, ~s ghetto

Getue das; ~s *(ugs. abwertend)* fuss (um about)

Getümmel das; ~s tumult

geübt *Adj.* accomplished; practised ⟨eye, ear⟩

Gewächs das; ~es, ~e plant; **gewachsen** 1. 2. *Part. v.* **wachsen**; 2. *in* jmdm./einer Sache ~ **sein** be a match for sb./be equal to sth.

gewagt *Adj.* daring; *(gefährlich)* risky; *(fast anstößig)* risqué ⟨joke etc.⟩

gewählt *Adj.* refined; 2. *adv.* in a refined manner

Gewähr die; ~: guarantee; **keine ~ übernehmen** be unable to guarantee sth.; **gewähren** *tr. V.* **a)** grant; give ⟨pleasure, joy⟩; **gewähr·leisten** *tr. V.* guarantee

Gewahrsam der; ~s **a)** *(Obhut)* safe-keeping; **b)** *(Haft)* custody

Gewährs·mann der; *Pl.* ~**männer** *od.* ~**leute** informant; source

Gewalt die; ~, ~en **a)** power; **b)** *o. Pl. (Willkür)* force; **c)** *o. Pl. (körperliche Kraft)* force; violence; **Gewalt·anwendung** die use of force *or* violence; **Gewalten·teilung** die separation of powers; **gewaltig** 1. *Adj.* **a)** *(immens)* huge; **b)** *(imponierend)* mighty, huge, massive ⟨building etc⟩; monumental ⟨literary work etc.⟩; 2. *adv. (ugs.)* very much; **gewalt·los** 1. *Adj.* non-violent; 2. *adv.* without violence; **Gewalt·losigkeit** die; ~: non-violence; **gewaltsam** 1. *Adj.* forcible ⟨expulsion⟩; enforced ⟨separation⟩; violent ⟨death⟩; 2. *adv.* forcibly; **gewalt·tätig** *Adj.* violent

Gewand das; ~|e|s, **Gewänder** *(geh.)* robe; gown

gewandt 1. 2. *Part. v.* **wenden**; 2. *Adj.* skilful; *(körperlich)* agile; 3. *adv.* skilfully; *(körperlich)* agilely; **Gewandtheit** die; ~: *s.* gewandt 2: skill; skilfulness; agility

gewann *1. u. 3. Pers. Sg. Prät. v.* **gewinnen**

gewaschen 2. *Part. v.* **waschen**

Gewässer das; ~s, ~: stretch of water

Gewebe das; ~s, ~ **a)** *(Stoff)* fabric; **b)** *(Med., Biol.)* tissue

Gewehr das; ~|e|s, ~e rifle; *(Schrot~)* shotgun

Geweih das; ~|e|s, ~e antlers *pl.*

Gewerbe das; ~s, ~: business; *(Handel, Handwerk)* trade

Gewerbe-: ~**freiheit** die right to carry on a business *or* trade; ~**ordnung** die laws *pl.* governing trade and industry; ~**schein** der licence to carry on a business *or* trade; ~**treibende** der/die; \ *adj. Dekl.* tradesman/tradeswoman; ~**zweig** der branch of trade

gewerblich 1. *Adj.* commercial; business *attrib.; (industriell)* industrial; 2. *adv.* ~ **tätig sein** work; **gewerbs·mäßig** *Adj.* professional

Gewerkschaft die; ~, ~en trade union; **Gewerkschaft|l|er** der; ~s, ~: trade unionist; **gewerkschaftlich** 1. *Adj.* [trade] union *attrib.;* 2. *adv.:* ~ **organisiert sein** belong to a [trade] union; **Gewerkschaftsfunktionär** der [trade] union official

gewesen 2. *Part. v.* ¹**sein**

gewichen 2. *Part. v.* **weichen**

Gewicht das; ~|e|s, ~e *(auch fig.)* weight; [nicht] **ins ~ fallen be** of [no] consequence; **Gewicht·heben** das; ~s weight-lifting; **gewichtig** *Adj.* weighty; **Gewichts·klasse** die *(Sport)* weight [division *or* class]

gewieft *Adj. (ugs.)* cunning

gewiesen 2. *Part. v.* **weisen**

gewillt *Adj.* **in ~ [nicht] ~ sein, etw. zu tun be** [un]willing to do sth.

Gewimmel das; ~s throng; *(von Insekten)* teeming mass

Gewinde das; ~s, ~ *(Technik)* thread

Gewinn der; ~|e|s, ~e **a)** profit; **b)** *(Preis einer Lotterie)* prize; *(beim Spiel)* winnings *pl.;* **c)** *(Sieg)* win; **Gewinn·beteiligung** die *(Wirtsch.)* profit-sharing; *(Betrag)* profit-sharing bonus; **gewinn·bringend** *Adj.* lucrative

gewinnen 1. *unr. tr. V.* win; gain

⟨*time, influence, validity, etc.*⟩; **2.** *unr. itr. V.* win (**bei** at); **gewinnend** *Adj.* winning; **Gewinner** der; ~s, ~: winner

Gewinn-: ~**spanne** die profit margin; ~**sucht** die greed for profit; ~**zahl** die winning number

Gewirr das; ~[e]s a) tangle; b) *(Durcheinander)* ein ~ von Ästen a maze of branches

gewiß 1. *Adj.* certain; **2.** *adv.* certainly

Gewissen das; ~s, ~: conscience; **gewissenhaft 1.** *Adj.* conscientious; **2.** *adv.* conscientiously; **gewissen·los** *Adj.* unscrupulous; **Gewissens·bisse** *Pl.* pangs of conscience

gewissermaßen *Adv. (sozusagen)* as it were; *(in gewissem Sinne)* to a certain extent; **Gewißheit** die; ~, ~en certainty

Gewitter das; ~s, ~: thunderstorm; **Gewitter·wolke** die thundercloud; **gewittrig** *Adj.* thundery

gewitzt *Adj.* shrewd

gewoben *2. Part. v.* weben

gewogen 1. *2. Part. v.* wiegen; **2.** *Adj. (geh.)* well disposed (+ *Dat.* towards)

gewöhnen 1. *tr. V.* jmdn. an jmdn./ etw. ~: get sb. used to sb./sth.; accustom sb. to sb./sth.; **2.** *refl. V.* sich an jmdn./etw. ~: get used or get or become accustomed to sb./sth.; accustom oneself to sb./sth.; **Gewohnheit** die; ~, ~en habit; **gewohnheits·mäßig 1.** *Adj.* habitual ⟨*drinker etc.*⟩; automatic ⟨*reaction etc.*⟩; **2.** *adv. (regelmäßig)* habitually; **gewöhnlich 1.** *Adj.* a) normal; ordinary; b) *(gewohnt, üblich)* usual; c) *(abwertend: ordinär)* common; **2.** *adv.* a) |für| ~: usually; wie ~: as usual; b) *(abwertend: ordinär)* in a common way

gewohnt *Adj.* a) usual; b) etw. *(Akk.)* ~ sein be used to sth.

Gewölbe das; ~s, ~: vault

gewonnen *2. Part. v.* gewinnen

geworben *2. Part. v.* werben

geworfen *2. Part. v.* werfen

gewrungen *2. Part. v.* wringen

Gewühl das; ~[e]s milling crowd

gewunden *2. Part. v.* winden

Gewürz das; ~es, ~e spice; *(würzende Zutat)* seasoning

Gewürz-: ~**gurke** die pickled gherkin; ~**nelke** die clove

gewußt *2. Part. v.* wissen

gez. *Abk.* gezeichnet sgd.

Gezeiten *Pl.* tides

gezielt 1. *Adj.* specific ⟨*questions, measures, etc.*⟩; deliberate ⟨*insult, indiscretion*⟩; well-directed ⟨*advertising campaign*⟩; **2.** *adv.* ⟨*proceed, act*⟩ purposefully

geziemen *(geh. veralt.)* **1.** *itr. V.* jmdm. |nicht| ~: [ill] befit sb; **2.** *refl. V.* be proper; sich für jmdn. ~: befit sb.

geziert 1. *Adj. (abwertend)* affected; **2.** *adv. (abwertend)* affectedly

gezogen *2. Part. v.* ziehen

Gezwitscher das; ~s twittering

gezwungen 1. *2. Part. v.* zwingen; **2.** *Adj.* forced; **gezwungenermaßen** *Adv.* of necessity

gib *Imperativ Sg. Präsens v.* geben; **gibst** *2. Pers. Sg. Präsens v.* geben; **gibt** *3. Pers. Sg. Präsens v.* geben

Gicht die; ~: gout

Giebel der; ~s, ~: gable

Gier die; ~: greed (**nach** for); **gierig 1.** *Adj.* greedy; **2.** *adv.* greedily

gießen 1. *unr. tr. V.* a) pour (in + *Akk.* into, über + *Akk.* over); b) *(verschütten)* spill (über + *Akk.* over); c) *(begießen)* water; **2.** *(unpers., ugs.)* pour [with rain]

Gießer der; ~s, ~: caster; **Gießerei** die; ~, ~en foundry

Gift das; ~[e]s, ~e a) poison; *(Schlangen~)* venom; **gift·grün** *Adj.* garish green; **giftig** *Adj.* poisonous; venomous ⟨*snake*⟩; toxic, poisonous ⟨*substance, gas, chemical*⟩; *(fig.)* venomous

Gift-: ~**müll** der toxic waste; ~**schlange** die venomous snake; ~**zahn** der poison fang

Gigant der; ~en, ~en giant; **gigantisch** *Adj.* gigantic

Gilde die; ~, ~n *(hist.)* guild

gilt *3. Pers. Sg. Präsens v.* gelten

Gimpel der; ~s, ~: bullfinch

Gin [dʒɪn] der; ~s gin

ging *1. u. 3. Pers. Sg. Prät. v.* gehen

Ginster der; ~s, ~: broom

Gipfel der; ~s, ~: peak; *(höchster Punkt des Berges)* summit; *(fig.)* height; **Gipfel·konferenz** die summit conference; **gipfeln** *itr. V.* in etw. *(Dat.)* ~: culminate in sth.

Gips der; ~es, ~e plaster; gypsum *(Chem.)*; **Gips·abdruck** der plaster cast; **gipsen** *tr. V.* plaster; put ⟨*leg, arm, etc.*⟩ in plaster; **Gips·verband** der plaster cast

Giraffe die; ~, ~n giraffe

Girlande die; ~, ~n festoon

Giro ['ʒiːro] das; ~s, ~s, österr. auch **Giri** (Finanzw.) giro; **Giro·konto** das (Finanzw.) current account

gis, Gis das; ~, ~ (Musik) G sharp

Gischt der; ~[e]s, ~e od. die; ~, ~en spray

Gitarre die; ~, ~n guitar

Gitter das; ~s, ~: bars pl.; (vor Fenster-, Türöffnungen) grille; (in der Straßendecke, im Fußboden) grating; (Geländer) railing[s pl.]; **Gitter·fenster** das barred window

Glacé·hand·schuh [gla'seː...] der kid glove

Gladiole die; ~, ~n gladiolus

Glanz der; ~es a) (von Licht, Sternen, Augen) brightness; (von Haar, Metall, Perlen, Leder usw.) lustre; sheen; b) (der Jugend, Schönheit) radiance; (des Adels usw.) splendour; **glänzen** itr. V. a) (Glanz ausstrahlen) shine; (hair, metal, etc.) gleam; (elbows, trousers, etc.) be shiny; b) (Bewunderung erregen) shine (bei at); **glänzend** (ugs.) 1. Adj. a) shining, gleaming (hair, metal, etc.); shiny (elbows, trousers, etc.); b) (bewundernswert) brilliant; splendid (references, marks, results, etc.); 2. adv. ~ mit jmdm. auskommen get on very well with sb.; **es geht mir/uns ~**: I am/we are very well

glanz-, Glanz-: ~leistung die (auch iron.) brilliant performance; ~los Adj. dull; lacklustre; ~nummer die star turn; ~voll 1. Adj. brilliant; sparkling (variety number); 2. adv. brilliantly

Glas das; ~es, Gläser a) o. Pl. glass; b) (Trinkgefäß) glass; **zwei ~ od. Gläser Wein** two glasses of wine; c) (Behälter) jar; **Glas·bläser** der glassblower; **Gläschen** ['glɛːsçən] das; ~s, ~ a) [little] glass; b) (kleines Gefäß) [little] [glass] jar; **Glaser** der; ~s, ~: glazier; **gläsern** Adj. glass; **Glas·faser** die; meist Pl. glass fibre; **glasieren** tr. V. a) glaze; b) (Kochk.) ice; glaze (meat); **glasig** Adj. a) glassy; b) (Kochk.) transparent; **Glas·malerei** die stained glass; **Glasur** die; ~, ~en a) glaze; b) (Kochk.) icing; (auf Fleisch) glaze

glatt 1. Adj. a) smooth; (rutschig) slippery; b) (ugs.: offensichtlich) downright (lie); outright (deception, fraud); flat (refusal); 2. adv. a) smoothly; b) (ugs.: rückhaltlos) **jmdm. etw. ~ ins Gesicht sagen** tell sb. sth. straight to

his/her face; (reject, deny) flatly; **Glätte** die; ~: smoothness; (Rutschigkeit) slipperiness; **Glatt·eis** das glaze; ice; (auf der Straße) black ice; **glätten** tr. V. smooth out (piece of paper, etc.); smooth [down] (feathers, fur, etc.); plane (wood etc.)

glatt-: ~[gehen unr. itr. V.; mit sein (ugs.) go smoothly; ~weg Adv. (ugs.) etw. ~weg ablehnen/ignorieren turn sth. down flat/simply ignore sth.; **das ist ~weg erlogen/erfunden** that's a downright lie/that's pure invention

Glatze die; ~, ~n bald head

Glaube der; ~ns faith (an + Akk. in); (Überzeugung, Meinung) belief (an + Akk. in); **glauben** 1. tr. V. (meinen) think; 2. itr. V. believe (an + Akk. in)

Glaubens-: ~bekenntnis das creed; ~freiheit die; o. Pl. religious freedom

glaubhaft 1. Adj. credible; 2. adv. convincingly; **gläubig** 1. Adj. devout; (vertrauensvoll) trusting; 2. adv. devoutly; (vertrauensvoll) trustingly; **Gläubige** der/die; adj. Dekl. believer; **Gläubiger** der; ~s, ~ creditor

glaub·würdig 1. Adj. credible; 2. adv. convincingly

gleich 1. Adj. a) (identisch, von derselben Art) same; (~berechtigt, ~wertig, Math.) equal; b) (ugs.: gleichgültig) **es ist mir völlig od. ganz ~**: I couldn't care less (coll.); **ganz ~, wer anruft, ...:** no matter who calls, ...; 2. adv. a) (übereinstimmend) ~ **groß/alt usw. sein** be the same height/age etc.; ~ **gut/schlecht usw.** equally good/bad etc.; b) (in derselben Weise) ~ **aufgebe/gekleidet** having the same structure/wearing identical clothes; c) (sofort) at once; straight away; (bald) in a moment; d) (räumlich) right; just; ~ **rechts/links** immediately on the right/left

gleich-, Gleich-: ~alt[e]rig [~alt[ə]rɪç] Adj. of the same age (mit as); ~artig 1. Adj. of the same kind postpos. (+ Dat. as); (sehr ähnlich) very similar (+ Dat. to); 2. adv. in the same way; ~berechtigt Adj. having equal rights postpos.; ~berechtigte Partner equal partners; ~berechtigung die equal rights pl.; ~[bleiben unr. itr. V.; mit sein remain the same; (speed, temperature, etc.) remain constant; ~bleibend Adj. constant, steady (temperature, speed, etc.); **gleichen** unr. itr. V. jmdm./einer Sa-

18*

che ~: be like *or* resemble sb./sth.; **gleichermaßen** *Adv.* equally

gleich-, Gleich-: ~falls *Adv. (auch)* also; *(ebenfalls)* likewise; **danke ~falls!** thank you, [and] the same to you; **~förmig 1.** *Adj.* **a)** *(einheitlich)* uniform; **b)** *(monoton)* monotonous; **2.** *adv.* **a)** *(einheitlich)* uniformly; **b)** *(monoton)* monotonously; **~geschlechtlich** *Adj.* homosexual; **~gewicht das;** *o. Pl.* balance; **~gewichts·störung die** disturbance of one's sense of balance; **~gültig 1.** *Adj.* indifferent (**gegenüber** towards); *(belanglos)* trivial; **das ist mir [vollkommen] ~:** it's a matter of [complete] indifference to me; **2.** *adv.* indifferently; **~gültigkeit die** indifference (**gegenüber** towards)

Gleichheit die; ~, ~en a) identity; *(Ähnlichkeit)* similarity; **b)** *o. Pl. (gleiche Rechte)* equality; **Gleichheitszeichen das** equals sign

gleich-, Gleich-: ~|kommen *unr. itr. V.; mit sein* **a)** *(entsprechen)* be tantamount to; **b)** *(die gleiche Leistung erreichen)* jmdm./einer Sache [an etw. *(Dat.)*] ~kommen equal sb./sth. [in sth.]; **~|machen** *tr. V.* make equal; **~mäßig 1.** *Adj.* regular ⟨*interval, rhythm*⟩; uniform ⟨*acceleration, distribution*⟩; even ⟨*heat*⟩; **2.** *adv.* ⟨*breathe*⟩ regularly; **etw. ~mäßig verteilen/auftragen** distribute sth. equally/apply sth. evenly; **~mut der** equanimity

Gleichnis das; ~ses, ~se *(Allegorie)* allegory; *(Parabel)* parable; **gleichsam** *Adv. (geh.)* as it were

gleich-, Gleich-: ~|schalten *tr. V.* force into line; **~schenk[e]lig** *Adj. (Math.)* isosceles; **~schritt der;** *o. Pl.* marching in step; **~seitig** *Adj. (Math.)* equilateral; **~|setzen** *tr. V.* equate; **~|stellen** *tr. V.* equate; **~strom der** *(Elektrot.)* direct current **Gleichung die; ~, ~en** equation

gleich-: ~wertig *Adj.* of the same value *postpos.;* **~wohl** [-'- *od.* '--] *Adv.* nevertheless; **~zeitig 1.** *Adj.* simultaneous; **2.** *adv.* at the same time

Gleis das; ~es, ~e track; *(Bahnsteig)* platform; *(einzelne Schiene)* rail **gleiten** *unr. itr. V.; mit sein* glide; ⟨*hand*⟩ slide; **Gleit·flug der** glide **Gletscher der; ~s, ~:** glacier; **Gletscher·spalte die** crevasse

glich *1. u. 3. Pers. Sg. Prät. v.* gleichen **Glied das; ~[e]s, ~er a)** limb; *(Finger~, Zehen~)* joint; **b)** *(Ketten~, auch fig.)*

link; **c)** *(Teil eines Ganzen)* section; *(Mitglied)* member; **gliedern 1.** *tr. V.* structure; organize ⟨*thoughts*⟩; **2.** *refl. V.* sich in Gruppen/Abschnitte *usw.* ~: be divided into groups/sections *etc.;* **Gliederung die; ~, ~en** structure **Glied-: ~maße** [-ma:sə] **die; ~, ~n** limb; **~satz der** *(Sprachw.)* subordinate clause

glimmen *unr. od. regelm. itr. V.* glow; **Glimm·stengel der** *(ugs. scherzh.)* fag *(sl.);* ciggy *(coll.)* **glimpflich 1.** *Adj.* **a)** der Unfall nahm ein ~es Ende the accident turned out not to be too serious; **b)** *(mild)* lenient ⟨*sentence, punishment*⟩; **2.** *adv.* **a)** *(ohne Schaden)* ~ davonkommen get off lightly; **b)** *(mild)* leniently **glitschig** *Adj. (ugs.)* slippery **glitt** *1. u. 3. Pers. Sg. Prät. v.* gleiten **glitzern** *itr. V.* ⟨*star*⟩ twinkle; ⟨*diamond, decorations*⟩ sparkle; ⟨*snow, eyes, tears*⟩ glisten

global 1. *Adj.* **a)** global; world-wide; **b)** *(umfassend)* all-round ⟨*education*⟩; overall ⟨*control, planning, etc.*⟩; **c)** *(allgemein)* general; **2.** *adv.* **a)** world-wide; **b)** *(umfassend)* in overall terms; **c)** *(allgemein)* in general terms; **Globen** *s.* Globus **Globetrotter der; ~s, ~:** globetrotter **Globus der; ~ od. ~ses, Globen** globe **Glöckchen das; ~s, ~:** [little] bell; **Glocke die; ~, ~n** bell **Glocken-: ~blume die** *(Bot.)* campanula; **~rock der** widely flared skirt; **~spiel das a)** carillon; *(mit einer Uhr gekoppelt auch)* chimes *pl.;* **b)** *(Instrument)* glockenspiel **glomm** *1. u. 3. Pers. Sg. Prät. v.* glimmen

Glorien·schein der glory; *(um den Kopf, fig.)* halo; **glorifizieren** *tr. V.* glorify; **Glorifizierung die; ~, ~en** glorification; **glor·reich 1.** *Adj.* glorious; **2.** *adv.* gloriously

Glossar das; ~s, ~e glossary **Glosse die; ~, ~n** commentary; *(spöttische Bemerkung)* sneering comment **glotzen** *itr. V. (abwertend)* goggle; gawp *(coll.)*

Glück das; ~[e]s a) luck; [es ist!] ein ~, daß ...: it's lucky that ...; [kein] ~ haben be [un]lucky; viel ~! [the] best of luck!; **b)** happiness **Glucke die; ~, ~n** brood-hen **glücken** *itr. V.; mit sein* succeed; etw. glückt jmdm. sb. is successful with sth. **gluckern** *itr. V.* gurgle; glug

glücklich 1. *Adj.* **a)** happy (über + *Akk.* about); **b)** *(erfolgreich)* lucky ⟨*winner*⟩; successful ⟨*outcome*⟩; safe ⟨*journey*⟩; **c)** *(vorteilhaft)* fortunate; 2. *adv.* **a)** *(erfolgreich)* successfully; **b)** *(vorteilhaft, zufrieden)* happily ⟨*chosen, married*⟩; **glücklicherweise** *Adv.* fortunately; luckily; **glück·selig** 1. *Adj.* blissfully happy; 2. *adv.* blissfully; **Glück·seligkeit** die; ~: bliss

glucksen *itr. V.* **a)** *s.* gluckern; **b)** *(lachen)* chuckle

Glücks-: **~klee** der four-leaf clover; **~pfennig** der lucky penny; **~pilz** der *(ugs.)* lucky devil *(coll.)*

Glück[s]·sache die: das ist ~: it's a matter of luck; **Glücks·spiel** das game of chance; **glück·strahlend** *Adj.* radiantly happy; **Glücks·zahl** die lucky number; **Glück·wunsch** der congratulations *pl.;* **herzlichen ~ zum Geburtstag!** happy birthday!

Glüh·birne die light-bulb; **glühen** *itr. V.* glow; **glühend** 1. *Adj.* red-hot ⟨*metal etc.*⟩; blazing ⟨*heat*⟩; ardent ⟨*admirer etc.*⟩; passionate ⟨*words, letter, etc.*⟩; 2. *adv.* ⟨*love*⟩ passionately; ⟨*admire*⟩ ardently; **~ heiß** blazing hot; **Glüh·wein** der mulled wine

Glut die; ~, ~en **a)** embers *pl.;* **b)** *(geh.: Leidenschaft)* passion; **glut·rot** *Adj.* fiery red

Glyzerin das; ~s glycerine

GmbH *Abk.* Gesellschaft mit beschränkter Haftung ≈ p.l.c.

Gnade die; ~, ~n *(Gunst)* favour; *(Rel.)* grace; *(Milde)* mercy

gnaden-, Gnaden-: **~brot** das: jmdm./einem Tier das ~brot geben keep sb./an animal in his/ her/its old age; **~frist** die reprieve; **~gesuch** das plea for clemency; **~los** *(auch fig.)* 1. *Adj.* merciless; 2. *adv.* mercilessly; **~schuß** der coup de grâce *(by shooting)*

gnädig *Adj.* gracious; *(glimpflich)* lenient ⟨*sentence etc.*⟩

Gnom der; ~en, ~en gnome

Gockel der; ~s, ~ *(bes. südd., sonst ugs. scherzh.)* cock

Gold das; ~[e]s gold; **Gold·barren** der gold bar; **golden** 1. *Adj.* *(aus Gold)* gold; *(herrlich)* golden ⟨*days, memories, etc.*⟩; 2. *adv.* like gold

Gold-: **~grube** die *(auch fig.)* goldmine; **~hamster** der golden hamster

goldig *Adj.* sweet

gold-, Gold-: **~richtig** *(ugs.) Adj.*

absolutely right; **~schmied** der goldsmith; **~schnitt** der gilt; **~währung,** die *(Wirtsch.)* currency tied to the gold standard

¹**Golf** der; ~[e]s, ~e gulf

²**Golf** das; ~s *(Sport)* golf

Golf-: **~platz** der golf-course; **~schläger** der golf club; **~spieler** der, **~spielerin** die golfer; **~strom** der Gulf Stream

Gondel die; ~, ~n gondola; **gondeln** *itr. V.;* mit sein *(ugs.)* **a)** *(mit einem Boot)* cruise; **b)** *(reisen)* travel around; **c)** *(herumfahren)* cruise around

Gong der; ~s, ~s gong; **gongen** *itr. V.* es hat gegongt the gong has sounded

gönnen *tr. V.* jmdm. etw. ~: not begrudge sb. sth.; sich/jmdm. etw. ~: allow oneself/sb. sth.; **Gönner** der; ~s, ~: patron; **gönnerhaft** *(abwertend) Adj.* patronizing

gor 3. *Pers. Sg. Prät. v.* gären

Göre die; ~, ~n *(nordd., oft abwertend)* kid *(coll.)*

Gorilla der; ~s, ~s gorilla

goß 1. u. 3. *Pers. Sg. Prät. v.* gießen

Gosse die; ~, ~n gutter

Gotik die; ~ *(Stil)* Gothic [style]; *(Epoche)* Gothic period; **gotisch** *Adj.* Gothic

Gott der; ~es, Götter **a)** *o. Pl.; o. Art.* God; grüß |dich| ~! *(landsch.)* hello!; um ~es Willen *(bei Erschrecken)* for God's sake; *(bei einer Bitte)* for heaven's sake; **b)** *(übermenschliches Wesen)* god

Gottes-: **~dienst** der service; **~haus** das *(geh.)* house of God; **~lästerung** die blasphemy

Gottheit die; ~, ~en deity; **Göttin** die; ~, ~nen goddess; **göttlich** 1. *Adj.* *(auch fig.)* divine; 2. *adv.* divinely

gott-, Gott-: **~lob** *adv.* thank goodness; **~los** 1. *Adj.* **a)** ungodly ⟨*life etc.*⟩; impious ⟨*words, speech, etc.*⟩; **b)** *(Gott leugnend)* godless ⟨*theory etc.*⟩; 2. *adv. (verwerflich)* irreverently; **~vater** der God the Father; **~vertrauen** das trust in God

Götze der; ~n, ~n *(auch fig.)* idol; **Götzen-:** **~bild** das idol; **~diener** der idolater

Gouverneur [guvɛr'nøːɐ̯] der; ~s, ~e governor

Grab das; ~[e]s, Gräber grave; das Heilige ~: the Holy Sepulchre; das ~

des Unbekannten Soldaten the tomb of
the Unknown Warrior; **graben** *unr.
tr., itr. V.* dig; **Graben** der; ~s, Grä-
ben ditch; *(Schützen~)* trench; *(Fe-
stungs~)* moat

Grab-: ~**kammer** die burial chamber; ~**mal das;** *Pl.* ~mäler, *geh.* ~ma-
le monument; ~**stein** der gravestone
gräbst *2. Pers. Sg. Präsens v.* graben;
gräbt *3. Pers. Sg. Präsens v.* graben
Gracht die; ~, ~en canal
Grad der; ~|e|s, ~e degree; *(Milit.)*
rank; **Grad·messer** der gauge,
yardstick **(für of)**
graduell 1. *Adj.* gradual; slight ⟨differ-
ence etc.⟩; **2.** *adv.* gradually; ⟨differ-
ent⟩ in degree; **graduiert** *Adj.* graduate; ein ~er Ingenieur an engineering
graduate
Graf der; ~en, ~en count; *(britischer
~)* earl
Grafik *usw. s.* Graphik *usw.*
Gräfin die; ~, ~nen countess; **Grafschaft** die; ~, ~en a) count's land;
(in Großbritannien) earldom; b) *(Verwaltungsbezirk)* county
Gram der; ~|e|s *(geh.)* grief; sorrow;
grämen 1. *tr. V.* grieve; **2.** *refl. V.*
grieve (über + *Akk.,* um over)
Gramm das; ~s, ~e gram
Grammatik die; ~, ~en grammar;
grammatisch 1. *Adj.* grammatical;
2. *adv.* grammatically
Grammophon Ⓦ **das;** ~s, ~e gramophone; phonograph *(Amer.)*
Granat der; ~|e|s, ~e *(Schmuckstein)*
garnet; **Granat·apfel** der pomegranate
Granate die; ~, ~n shell; *(Hand~)*
grenade
grandios 1. *Adj.* magnificent; **2.** *adv.*
magnificently
Granit der; ~s, ~e granite
grantig *(südd., österr. ugs.)* **1.** *Adj.*
bad-tempered; **2.** *adv.* bad-
temperedly
Graphik die; ~, ~en graphic art[s *pl.*];
(Kunstwerk) graphic; *(Druck)* print;
Graphiker der; ~s, ~, **Graphikerin**
die; ~, ~nen [graphic] designer;
(Künstler|in|) graphic artist; **gra-
phisch 1.** *Adj.* graphic; **2.** *adv.*
graphically
Gras das; ~es, Gräser grass; **über etw.
(Akk.)** ~ wachsen lassen *(ugs.)* let the
dust settle on sth.; **grasen** *itr. V.*
graze; **Gras·halm** der blade of grass
gräßlich 1. *Adj.* a) *(abscheulich)* horrible; terrible ⟨accident⟩; b) *(ugs.: un-*

angenehm) dreadful *(coll.);* c) *(ugs.:
sehr stark)* terrible *(coll.);* **2.** *adv.* a)
(abscheulich) horribly; terribly; b)
(ugs.: unangenehm) terribly *(coll.);* c)
(ugs.: sehr) terribly *(coll.)*
Grat der; ~|e|s, ~e ridge
Gräte die; ~, ~n [fish-]bone
Gratifikation die; ~, ~en bonus
gratis *Adv.* free [of charge]; gratis
Grätsche die; ~, ~n *(Turnen)*
straddle; *(Sprung)* straddle-vault
Gratulant der; ~en, ~en, **Gratulan-
tin** die; ~, ~nen well-wisher; **Gratu-
lation** die; ~, ~en congratulations
pl.; **gratulieren** *itr. V.* jmdm. ~: congratulate sb.; jmdm. zum Geburtstag
~: wish sb. many happy returns [of
the day]
grau *Adj.* grey; *(trostlos)* dreary; drab
¹**grauen** *itr. V. (geh.)* der Morgen/der
Tag graut morning/day is breaking
²**grauen** *itr. V. (unpers.)* ihm graut |es|
davor/vor ihr he dreads [the thought
of] it/he's terrified of her; **Grauen**
das; ~s, ~: horror (vor + *Dat.* of);
grauen·haft 1. *Adj.* horrifying;
(ugs.: sehr unangenehm) terrible
(coll.); **2.** *adv.* horrifyingly; *(ugs.: sehr
unangenehm)* terribly *(coll.)*
grau-: ~**haarig** *Adj.* grey-haired;
~**meliert** *Adj. (präd. getrennt geschrieben)* greying ⟨hair⟩
Graupe die; ~, ~n a) grain of pearl
barley; b) *Pl. (Gericht)* pearl barley
sing.
graupeln *itr. V. (unpers.)* es graupelt
there's soft hail falling
grausam 1. *Adj.* a) cruel; b) *(furchtbar)* terrible; dreadful; **2.** *adv.* a)
cruelly; b) *(furchtbar)* terribly, dreadfully; **Grausamkeit** die; ~, ~en a) *o.
Pl.* cruelty; b) *(Handlung)* act of
cruelty
grausen 1. *tr., itr. V. (unpers.)* es grau-
ste ihm *od.* ihn davor/vor ihr he
dreaded it/he was terrified of her; **2.**
refl. V. sich vor etw./jmdm. ~: dread
sth./be terrified of sb.; **Grausen** das;
~s horror; **grausig** *s.* grauenhaft
gravieren *tr. V.* engrave; **gra-
vierend** *Adj.* serious, grave; **Gra-
vierung** die; ~, ~en engraving
Gravitation die; ~ *(Physik, Astron.)*
gravitation
Gravur [gra'vuːɐ̯] die; ~, ~en engraving
Grazie ['graːtsiə] die; ~, ~n a) *o. Pl.
(Anmut)* gracefulness; b) *Pl. (Myth.)*
Graces

greifen 1. *unr. tr. V.* a) *(er~)* take hold of; grasp; *(rasch ~)* seize; b) *(fangen)* catch; 2. *unr. itr. V.* a) **in/unter/hinter etw./sich** *(Akk.)* ~: reach into/under/ behind sth./one; **nach etw.** ~: reach for sth.; *(hastig)* make a grab for sth.; b) *(Technik)* grip

Greis der; ~es, ~e old man; **Greisin** die; ~, ~nen old woman

grell 1. *Adj.* a) *(hell)* glaring, *(light, sun, etc.)*; b) *(auffallend)* garish *(colour etc.)*; loud *(dress, pattern, etc.)*; c) *(schrill)* shrill, *(cry, voice, etc.)*; 2. *adv.* a) *(hell)* with glaring brightness; b) *(auffallend)* **gegen** *od.* **von etw.** ~ **abstechen** contrast sharply with sth.; c) *(schrill)* shrilly

Gremium das; ~s, Gremien committee

Grenze die; ~, ~n a) boundary; *(Staats~)* border; *(gedachte Trennungslinie)* borderline; b) *(fig.)* limit; **grenzen** *itr. V.* **an etw.** *(Akk.)* ~: border [on] sth.; **grenzen·los** 1. *Adj.* boundless; *(fig.)* boundless, unbounded *(joy, wonder, jealousy, grief, etc.)*; unlimited *(wealth, power)*; limitless *(patience, ambition)*; extreme *(tiredness, anger, foolishness)*; 2. *adv.* endlessly; *(fig.)* beyond all measure; **Grenzen·losigkeit** die; ~: boundlessness

Grenz-: ~**übergang** der border crossing-point; ~**verkehr** der [cross-]border traffic

Greuel der; ~s, ~ a) **etw./jmd. ist jmdm. ein** ~: sb. loathes *or* detests sth./sb.; b) *meist Pl. (geh.) (~tat)* atrocity; **Greuel·tat** die atrocity; **greulich** 1. *Adj.* a) horrifying; b) *(unangenehm)* awful; 2. *adv.* a) horrifyingly; b) *(unangenehm)* terribly

Grieche der; ~n, ~n Greek; **Griechen·land** (das); ~s Greece; **griechisch** 1. *Adj.* Greek; 2. *adv.* *(speak, write)* in Greek; **Griechisch** das; ~[s] Greek *no art.*

griesgrämig 1. *Adj.* grumpy; 2. *adv.* in a grumpy manner

Grieß der; ~es, ~e semolina; **Grieß·brei** der semolina

griff 1. *u.* 3. *Pers. Sg. Prät. v.* greifen; **Griff** der; ~[e]s, ~e a) grip; grasp; b) *(Knauf, Henkel)* handle; **griff·bereit** *Adj.* ready to hand *postpos.*

Griffel der; ~s, ~: slate-pencil

griffig *Adj.* a) *(handlich)* handy; b) *(gut greifend)* that grips well *postpos., not pred.*; non-slip *(surface, floor)*

Grill der; ~s, ~s grill; *(Rost)* barbecue

Grille die; ~, ~n a) cricket; b) *(sonderbarer Einfall)* whim

grillen 1. *tr. V.* grill; 2. *itr. V.* **im Garten** ~: have a barbecue in the garden

Grimasse die; ~, ~n grimace

grimmig 1. *Adj.* furious *(person)*; grim *(expression)*; 2. *adv.* grimly

grinsen *itr. V.* grin; *(höhnisch)* smirk

Grippe die; ~, ~n a) influenza; flu *(coll.)*; b) *(volkst.: Erkältung)* cold

Grips der; ~es brains *pl.*

grob 1. *Adj.* a) coarse; thick *(wire)*; rough *(work)*; b) *(ungefähr)* rough; c) *(schwerwiegend)* gross; flagrant *(lie)*; d) *(barsch)* rude; 2. *adv.* a) coarsely; b) *(ungefähr)* roughly; c) *(schwerwiegend)* grossly; d) *(barsch)* rudely; **Grobheit** die; ~, ~en a) *o. Pl.* rudeness; b) *(Äußerung)* rude remark

Grobian der; ~[e]s, ~e lout

Grog der; ~s, ~s grog

grölen 1. *tr. V. (ugs. abwertend)* bawl [out]; roar, howl *(approval)*; 2. *itr. V.* bawl

Groll der; ~[e]s *(geh.)* rancour; **grollen** *itr. V. (geh.)* a) |mit| jmdm. ~: bear a grudge against sb.; b) *(thunder)* rumble

Grönland (das); ~s Greenland

Gros [gro:] das; ~ [gro:s], ~ [gro:s] bulk

Groschen der; ~s, ~ a) *(österreichische Münze)* groschen; b) *(ugs.: Zehnpfennigstück)* ten-pfennig piece; *(fig.)* penny; cent *(Amer.)*

groß; größer, größt... 1. *Adj.* a) big, large; great *(length, width, height)*; tall *(person)*; wide *(selection)*; **1 m² ~**: 1 m² in area; **im ~en und ganzen** by and large; b) *(älter)* big *(brother, sister)*; *(erwachsen)* grown-up; c) *(lange dauernd)* long, lengthy; d) intense *(heat, cold)*; high *(speed)*; great, major *(event, artist, work)*; 2. *adv.* **ein Wort ~ schreiben** write a word with a capital; *(ugs.: besonders)* greatly; **groß·artig** 1. *Adj.* magnificent; 2. *adv.* magnificently

Großbritannien (das); ~s the United Kingdom; [Great] Britain

Groß·buchstabe der capital [letter]

Größe die; ~, ~n size; *(Höhe, Körper~)* height; *(fig.)* greatness; **die ~ der Katastrophe** the [full] extent of the catastrophe

Groß·eltern *Pl.* grandparents; **Größen·wahn** der delusions *pl.* of grandeur; **größer** *s.* groß

Groß-: ~**fahndung** die large-scale

search; **~handel** der wholesale trade; **~händler** der wholesaler; **~industrielle der/die;** *adj. Dekl.* big industrialist

Grossist der; **~en, ~en** *(Kaufmannsspr.)* wholesaler

groß-, Groß-: **~macht** die great power; **~maul** das *(ugs. abwertend)* big-mouth *(coll.);* **~mut** die; **~:** generosity; **~mütig** *Adj.* generous; **~mutter** die grandmother; **~reinemachen** das *(ugs.)* thorough cleaning; **~|schreiben** *unr. tr. V. (ugs.) in* **~geschrieben werden** be stressed; *s. auch* groß 2; **~spurig** *(abwertend)* 1. *Adj.* boastful; *(hochtrabend)* pretentious; 2. *adv.* boastfully; *(hochtrabend)* pretentiously; **~stadt** die city; large town; **~städter** der city-dweller

größt... *s.* groß; **Groß·teil** der a) *(Hauptteil)* major part; b) *(nicht unerheblicher Teil)* large part; **größtenteils** *Adv.* for the most part; **größt·möglich** *Adj.* greatest possible

groß-, Groß-: **~|tun** *unr. itr. V.* boast; **~vater** der grandfather; **~|ziehen** *unr. tr. V.* bring up; raise; rear *⟨animal⟩;* **~zügig** 1. *Adj.* generous; grand and spacious *⟨building, garden, etc.⟩;* 2. *adv.* a) generously; **~zügigkeit** die generosity

grotesk 1. *Adj.* grotesque; 2. *adv.* grotesquely

Grotte die; **~, ~n** grotto

grub *1. u. 3. Pers. Sg. Prät. v.* graben; **Grübchen** das; **~s, ~:** dimple; **Grube** die; **~, ~n** pit; *(Bergbau)* mine

grübeln *itr. V.* ponder (über + *Dat.* on, over)

Gruben·arbeiter der miner; mineworker

grüezi *Adv. (schweiz.)* hallo

Gruft die; **~, Grüfte** vault; *(in einer Kirche)* crypt

grün *Adj.* green; **Grün** das; **~s, ~** *od.* *(ugs.)* **~s** a) green; b) *o. Pl. (Pflanzen)* greenery; **Grün·anlage** die green space; *(Park)* park

Grund der; **~[e]s, Gründe** a) ground; *(eines Gewässers)* bottom; b) *(Ursache, Veranlassung)* reason

Grund-: **~besitz** der a) *(Eigentum an Land)* ownership of land; b) *(Land)* land; **~buch** das land register

gründen 1. *tr. V.* a) found; set up, establish *⟨business⟩;* start [up] *⟨club⟩;* b) *(aufbauen)* base *⟨plan, theory, etc.⟩*

(auf + *Akk.* on); 2. *itr. V.* **auf** *od.* **in** etw. *(Dat.)* **~:** be based on sth. 3. *refl. V.* **sich auf etw.** *(Akk.)* **~:** be based on sth.; **Gründer** der; **~s, ~, Gründerin** die; **~, ~nen:** founder

grundieren *tr. V.* prime

grund-, Grund-: **~gesetz** das Basic Law; **~kenntnis** die; *meist Pl.* basic knowledge *no pl.* (in + *Dat.* of); **~lage** die basis; foundation; **~legend** 1. *Adj.* fundamental, basic (für to); seminal *⟨idea, work⟩;* 2. *adv.* fundamentally

gründlich 1. *Adj.* thorough; 2. *adv.* thoroughly; **Gründlichkeit** die; **~:** thoroughness

grund·los 1. *Adj.* groundless; 2. *adv.* sich **~los aufregen/ängstigen** be needlessly agitated/alarmed; **Grundnahrungs·mittel** das basic food[stuff]

Grün·donnerstag der Maundy Thursday

Grund-: **~prinzip** das fundamental principle; **~recht** das basic *or* constitutional right; **~riß** der a) *(Bauw.)* [ground-] plan; b) *(Leitfaden)* outline; **~satz** der principle

grund·sätzlich 1. *Adj.* a) fundamental *⟨difference, question, etc.⟩;* b) *(aus Prinzip)* *⟨opponent etc.⟩* on principle; c) *(allgemein)* *⟨agreement etc.⟩* in principle; 2. *adv.* a) fundamentally; b) *(aus Prinzip)* on principle; c) *(allgemein)* in principle

Grund-: **~schule** die primary school; **~stein** der foundation-stone; **~stück** das plot [of land]

Gründung die; **~, ~en** *s.* gründen 1 a: foundation; setting up; establishing; starting [up]

Grund-: **~wasser** das *(Geol.)* ground water; **~zug** der essential feature

Grüne das; **~n/ins ~:** [out] into the country

Grün-: **~fläche** die green space; *(im Park)* lawn; **~span** der verdigris; **~streifen** der central reservation *(grassed and often with trees and bushes)*

grunzen *tr., itr. V.* grunt

Gruppe die; **~, ~n** a) group; b) *(Klassifizierung)* class; category

Gruppen-: **~reise** die *(Touristik)* group travel *no pl., no art.;* **~sieg** der *(Sport)* top place in the group

gruppieren 1. *tr. V.* arrange; 2. *refl. V.* form a group/groups; **Gruppierung** die; **~, ~en** grouping

gruselig Adj. eerie; creepy; **gruseln** 1. tr., itr. V. (unpers.) es gruselt jmdm. od. jmdm. sb.'s flesh creeps; 2. refl. V. be frightened

Gruß der; ~es, Grüße a) greeting; (Milit.) salute; b) (im Brief) mit herzlichen Grüßen [with] best wishes; mit bestem ~/freundlichen Grüßen yours sincerely; **grüßen** 1. tr. V. a) greet; (Milit.) salute; b) (Grüße senden) grüße deine Eltern [ganz herzlich] von mir please give your parents my [kindest] regards; 2. itr. V. say hello; (Milit.) salute

Grütze die; ~, ~n groats pl.; rote ~: red fruit pudding (made with fruit juice, fruit and cornflour, etc.)

gucken itr. V. (ugs.) a) look; (heimlich) peep; b) (hervorsehen) stick out; c) (dreinschauen) look; **Guck·loch** das spy-hole

Guerilla [geˈrɪlja] die; ~, ~s guerrilla war; (Einheit) guerrilla unit

Gulasch [ˈgʊlaʃ, ˈguːlaʃ] das od. der; ~[e]s, ~e od. ~s goulash

Gulden der; ~s, ~: guilder

gültig Adj. valid; current (note, coin); **Gültigkeit** die; ~: validity; ~ haben/erlangen be/become valid

Gummi der od. das; ~s, ~[s] rubber

Gummi-: ~**band** das; Pl. ~bänder rubber or elastic band; (in Kleidung) elastic no indef. art.; ~**bärchen** das jelly baby; ~**baum** der rubber plant

gummieren tr. V. gum

Gummi-: ~**handschuh** der rubber glove; ~**knüppel** der [rubber] truncheon; ~**sohle** die rubber sole; ~**stiefel** der rubber boot; (für Regenwetter) wellington [boot] (Brit.)

Gunst die; ~: favour; goodwill; **günstig** 1. Adj. favourable; propitious (sign); auspicious (moment); beneficial (influence); good; 2. adv. favourably; etw. ~ beeinflussen have or exert a beneficial influence on sth.

Gurgel die; ~, ~n throat; jmdm. die ~ zudrücken throttle sb.; **gurgeln** itr. V. gargle

Gurke die; ~, ~n cucumber; (eingelegt) gherkin

gurren itr. V. (auch fig.) coo

Gurt der; ~[e]s, ~e strap; (im Auto, Flugzeug) [seat-]belt; **Gürtel** der; ~s, ~: belt

Gürtel-: ~**linie** die waist[line]; ~**reifen** der radial[-ply] tyre

GUS [geːʔuːˈʔɛs] Abk. **Gemeinschaft Unabhängiger Staaten** CIS

Guß der; Gusses, Güsse a) (das Gießen) casting; b) (ugs.: Regenschauer) downpour

Guß·eisen das cast iron; **guß·eisern** Adj. cast-iron

gut; besser, best... 1. Adj. good; fine (wine); ein ~es neues Jahr a happy new year; mir ist nicht ~: I'm not feeling well; ~en Appetit! enjoy your lunch/dinner etc.!; eine ~e Stunde [von hier] a good hour [from here]; 2. adv. a) well; b) (mühelos) easily; s. auch besser, best...

Gut das; ~[e]s, Güter a) property; (Besitztum, auch fig.) possession; b) (landwirtschaftlicher Grundbesitz) estate; c) (Fracht~, Ware) item; Güter goods; (Fracht~) freight sing.; goods (Brit.);

gut-, Gut-: ~**achten** das; ~s, ~: [expert's] report; ~**artig** Adj. a) good-natured; b) (nicht gefährlich) benign; ~**aussehend** Adj. good-looking; ~**bürgerlich** Adj. good middle-class; ~**bürgerliche Küche** good plain cooking; ~**dünken** das; ~s discretion

Güte die; ~: goodness; kindness; (Qualität) quality

Güter-: ~**abfertigung** die a) (Abfertigung von Waren) dispatch of freight or (Brit.) goods; b) (Annahmestelle) freight or (Brit.) goods office; ~**bahnhof** der freight depot; goods station (Brit.); ~**wagen** der goods wagon (Brit.); freight car (Amer.); ~**zug** der goods train (Brit.); freight train (Amer.)

gut-, Gut-: ~**[gehen** unr. itr. V.; mit sein a) (unpers.) es geht jmdm. gut sb. is well; b) (~ ausgehen) turn out well; ~**gelaunt** Adj. (präd. getrennt geschrieben) cheerful; ~**gemeint** Adj. (präd. getrennt geschrieben) well-meant; ~**gläubig** Adj. innocently trusting; ~**haben** das; ~s, ~: credit balance; ~**[heißen** unr. tr. V. approve of; ~**herzig** Adj. kind-hearted

gütig 1. Adj. kindly; 2. adv. ~ lächeln give a kindly smile; **gütlich** Adj. amicable

gut-, Gut-: ~**machen** tr. V. make good (damage); put right (omission, mistake, etc.); ~**mütig** Adj. good-natured; ~**mütigkeit** die; ~: good nature

Guts·besitzer der owner of a/the estate; landowner

gut-, Gut-: ~**schein** der voucher,

coupon (für, **auf** + *Akk.* for); ~|-
schreiben *unr. tr. V.* credit;
~**schrift** die credit
Guts·hof der estate; manor
gut-: ~|**tun** *unr. itr. V.* do good;
~**willig** 1. *Adj.* willing; *(entgegenkommend)* obliging; 2. *adv.* etw. ~**willig herausgeben/versprechen** hand sth.
over voluntarily/promise sth. willingly
Gymnasium das; ~s, **Gymnasien** ≈
grammar school
Gymnastik die; ~: physical exercises
pl.; (Turnen) gymnastics *sing.*
Gynäkologe der; ~n, ~n gynaecologist

H

h, H [ha:] das; ~, ~ **a)** *(Buchstabe)*
h/H; **b)** *(Musik)* [key of] B
h *Abk.* **a)** Uhr hrs; **b)** Stunde hr[s]
H *Abk.* **a)** Herren; **b)** Haltestelle
¹ha [ha(:)] *Interj.* **a)** *(Überraschung)* ah;
b) *(Triumph)* aha
²ha *Abk.* Hektar ha
Haar das; ~|e|s, ~e hair; blonde ~e *od.*
blondes ~ **haben** have fair hair; *(fig.)*
~**e auf den Zähnen haben** *(ugs.
scherzh.)* be a tough customer; **um ein
~** *(ugs.)* very nearly
Haar-: ~**ausfall** der hair loss; ~**bürste** die hairbrush; ~**büschel** das tuft
of hair
haaren *itr. V.* moult; **Haares·breite**
die *in* **um ~:** by a hair's breadth;
Haar·festiger der setting lotion;
haar·genau *(ugs.)* 1. *Adj.* exact; 2.
adv. exactly; **haarig** *Adj.* hairy
haar-, Haar-: ~**klemme** die hair-
grip; ~**nadel** die hairpin; ~**na-
del·kurve** die hairpin bend;
~**schnitt** der haircut; *(modisch)* hairstyle; ~**spange** die hair-slide;
~**sträubend** *Adj.* **a)** *(grauenhaft)*
hair-raising; **b)** *(empörend)* outrageous; shocking; ~**teil** das hair-
piece; ~**wasch·mittel** das shampoo; ~**wasser** das; *Pl.* ~wässer hair
lotion

Habe die; ~ *(geh.)* possessions *pl.;*
haben 1. *unr. tr. V.* have; have got;
heute ~ wir schönes Wetter the
weather is fine today; **es gut/schlecht/
schwer ~:** have it good *(coll.)*/have a
bad time [of it]/have a difficult time;
du hast zu gehorchen you must obey;
das Jahr hat 12 Monate there are 12
months in a year; 2. *refl. V. (ugs.: sich
aufregen)* make a fuss; 3. *Hilfsverb*
have; **ich habe/hatte ihn eben gesehen**
I've/I'd just seen him; **er hat es gewußt**
he knew it; 4. *mod. V.* **du hast zu gehorchen** you must obey; **er hat sich
nicht einzumischen** he's not to interfere; **Haben** das; ~s, ~ *(Kaufmannsspr.)* credit; **Habe·nichts**
der; ~, ~e pauper; **Hab·gier** die *(abwertend)* greed; **hab·gierig** 1. *Adj.
(abwertend)* greedy; 2. *adv.* greedily
Habicht der; ~s, ~e hawk
Hab-: ~**seligkeiten** *Pl.* [meagre] belongings; ~**sucht** die; ~ *(abwertend)*
greed; avarice
Hachse die; ~, ~n *(südd.)* knuckle
Hack das; ~s *(ugs., bes. nordd.)*
mince; **Hack·braten** der meat loaf
¹Hacke die; ~, ~n hoe; *(Pickel)*
pick[axe]
²Hacke die; ~, ~n *(bes. nordd. u. md.)*
heel
hacken 1. *itr. V.* **a)** hoe; **b)** *(picken)*
peck; 2. *tr. V.* **a)** hoe ⟨garden, flower-
bed, etc.⟩; **b)** *(zerkleinern)* chop; chop
[up] ⟨meat, vegetables, etc.⟩
Hack·fleisch das minced meat;
mince
Häcksel der *od.* das; ~s *(Landw.)*
chaff
hadern *itr. V. (geh.)* **mit etw. ~:** be at
odds with sth.
Hafen der; ~s, **Häfen** harbour; port
Hafen-: ~**arbeiter** der dock-worker;
docker; ~**kneipe** die dockland pub
(Brit. coll.) or *(Amer.)* bar; ~**rundfahrt** die trip round the harbour;
~**stadt** die port; ~**viertel** das dock
area
Hafer der; ~s oats *pl.*
Hafer-: ~**brei** der porridge;
~**flocken** *Pl.* porridge oats
Haff das; ~|e|s, ~s *od.* ~e lagoon
Haft die; ~ **a)** *(Gewahrsam)* custody;
(aus politischen Gründen) detention;
b) *(Freiheitsstrafe)* imprisonment
-haft *Adj., adv.* -like
haftbar *Adj. (bes. Rechtsspr.)* **für etw.
~ sein** be liable for sth.; **Haft·befehl**
der *(Rechtsw.)* warrant [of arrest]

¹**haften** *itr. V.* stick; *(sich festsetzen)* ⟨*smell, dirt, etc.*⟩ cling (**an** + *Dat.* to)

²**haften** *itr. V.* für jmdn./etw. ~: be responsible for sb./liable for sth.; *(Rechtsw., Wirtsch.)* be liable

haften|bleiben *unr. itr. V.; mit sein* stick (**an/auf** + *Dat.* to); ⟨*smell, smoke*⟩ cling (**an/auf** + *Dat.* to); *(ugs.: im Gedächtnis bleiben)* stick

Häftling *der;* ~s, ~e prisoner

Haft·pflicht die liability (für for); **Haftpflicht·versicherung** die personal liability insurance; *(für Autofahrer)* third party insurance

Haft·schale die contact lens

Haftung die; ~, ~en liability; Gesellschaft mit [un]beschränkter ~: [un]limited [liability] company

Hagebutte die; ~, ~n a) *(Frucht)* rose-hip; b) *(ugs.: Heckenrose)* dog-rose

Hagel der; ~s, ~ *(auch fig.)* hail; **hageln** *itr., tr. V. (unpers.)* hail

hager *Adj.* gaunt

haha [ha'ha(:)] *Interj.* ha ha

Häher der; ~s, ~: jay

¹**Hahn** der; ~[e]s, Hähne cock; *(Wetter~)* weathercock

²**Hahn** der; ~[e]s, Hähne, *fachspr.:* ~en a) tap; faucet *(Amer.)*; b) *(bei Waffen)* hammer

Hähnchen das; ~s, ~: chicken; **Hahnen·fuß** der buttercup

Hai der; ~s, ~e shark

Häkchen das; ~s, ~ a) [small] hook; b) *(Zeichen)* mark; *(beim Abhaken)* tick; **häkeln** *tr., itr. V.* crochet; **Häkel·nadel** die crochet-hook

haken 1. *tr. V.* hook (**an** + *Akk.* on to); 2. *itr. V. (klemmen)* be stuck; **Haken** der; ~s, ~ a) hook; b) *(Zeichen)* tick; c) *(ugs.: Schwierigkeit)* catch; d) *(Boxen)* hook; **Haken·kreuz** das swastika

halb 1. *Adj. u. Bruchz.* half; eine ~e Stunde/ein ~er Meter half an hour/a metre; zum ~en Preis [at] half price; ~ Europa/die ~e Welt half of Europe/half the world; es ist ~ eins it's half past twelve; die ~e Wahrheit half [of] the truth; [noch] ein ~es Kind sein be hardly more than a child; 2. *adv.* ~ voll/leer half-full/-empty; ~ angezogen half dressed; **Halb·dunkel** das semi-darkness; **Halbe** der od. die od. das; *adj. Dekl. (ugs.)* half litre *(of beer etc.)*; **Halb·edelstein** der *(veralt.)* semi-precious stone

halber *Präp. mit Gen.; nachgestellt*

(wegen) on account of; *(um ... willen)* for the sake of

halb-, Halb-: ~**finale** das *(Sport)* semi-final; ~**gar** *Adj.* half-cooked; ~**gefror[e]ne** das; *adj. Dekl.* soft ice cream

halbieren *tr. V.* cut/tear ⟨*object*⟩ in half; halve ⟨*amount, number*⟩

halb-, Halb-: ~**insel** die peninsula; ~**jahr** das six months *pl.;* half year; ~**jährlich** 1. *Adj.* six-monthly; 2. *adv.* every six months; ~**kreis** der semicircle; ~**kugel** die hemisphere; ~**lang** *Adj.* mid-length ⟨*hair*⟩; mid-calf length ⟨*coat, dress, etc.*⟩; ~**links** [-'-] *Adv. (Fußball)* ⟨*play*⟩ [at] inside left; ~**mast** *Adv.* at half-mast; ~**mond** der a) *(Mond)* half-moon; b) *(Figur)* crescent; ~**offen** *Adj. (präd. getrennt geschrieben)* half-open; ~**pension** die half-board; ~**rechts** [-'-] *Adv. (Fußball)* ⟨*play*⟩ [at] inside right; ~**schuh** der shoe; ~**starke** der; *adj. Dekl. (ugs. abwertend)* [young] hooligan; ~**tags** *Adv.* ⟨*work*⟩ part-time; *(morgens/nachmittags)* ⟨*work*⟩ [in the] mornings/afternoons; ~**voll** *Adj. (präd. getrennt geschrieben)* half-full; ~**wegs** *Adv.* to some extent; ~**wüchsig** [~vy:ksɪç] *Adj.* adolescent; ~**wüchsige** der/die; *adj. Dekl.* adolescent; ~**zeit** die *(bes. Fußball)* a) half; b) *(Pause)* half-time

Halde die; ~, ~n *(Bergbau)* slag-heap

half *1. u. 3. Pers. Sg. Prät. v.* helfen

Hälfte die; ~, ~n a) half; b) *(ugs.: Teil)* part

¹**Halfter** der od. das; ~s, ~: halter

²**Halfter** die; ~, ~n; *auch das;* ~s, ~: holster

Hall der; ~[e]s, ~e a) *(geh.)* reverberation; b) *(Echo)* echo

Halle die; ~, ~n hall; *(Fabrik~)* shed; *(Hotel~, Theater~)* foyer

hallen *itr. V.* a) reverberate; ⟨*shot, bell, cry*⟩ ring out; b) *(widerhallen)* echo

Hallen- indoor ⟨*swimming-pool, handball*⟩

Hallig die; ~, ~en small low island *(particularly one of those off Schleswig-Holstein)*

hallo *Interj.* hello; **Hallo** das; ~s, ~s cheering

Halluzination die; ~, ~en hallucination

Halm der; ~[e]s, ~e stalk; stem

Hals der; ~es, Hälse *(Kehle)* throat; ~ über Kopf *(ugs.)* in a rush

Hals-: ~**ab·schneider** der *(ugs. ab-*

wertend) shark; ~**band** das; *Pl.* ~**bänder** *(für Tiere)* collar; ~**bruch** der *s.* ~- und Beinbruch; ~**entzündung** die inflammation of the throat; ~-**Nasen-Ohren-Arzt** der ear, nose, and throat specialist; ~**schlagader** die carotid [artery]; ~**schmerzen** *Pl.* sore throat *sing.;* ~**starrig** [~ʃtariç] *Adj. (abwertend)* stubborn; obstinate; ~**tuch** das cravat; ~- und Beinbruch *Interj. (scherzh.)* good luck; ~**weh** das *(ugs.) s.* ~schmerzen

halt *Interj.* stop; **Halt** der; ~[e]s, ~e hold

haltbar *Adj.* a) ~ sein ⟨food⟩ keep [well]; ~ bis 5. 3. use by 5 March; b) *(nicht verschleißend)* hard-wearing ⟨material, clothes⟩; c) *(aufrechtzuerhalten)* tenable ⟨hypothesis etc.⟩; **Haltbarkeit** die; ~ *(Strapazierfähigkeit)* durability

halten 1. *unr. tr. V.* a) *(auch Milit.)* hold; **die Hand vor den Mund** ~: put one's hand in front of one's mouth; b) *(Ballspiele)* save ⟨shot, penalty, etc.⟩; c) *(bewahren)* keep; *(beibehalten, aufrechterhalten)* keep up ⟨speed etc.⟩; maintain ⟨temperature, equilibrium⟩; d) *(erfüllen)* keep; **sein Wort/ein Versprechen** ~: keep one's word/a promise; e) *(besitzen, beschäftigen, beziehen)* keep ⟨chickens etc.⟩; take ⟨newspaper, magazine, etc.⟩; f) *(einschätzen)* **jmdn. für reich/ehrlich** ~: think sb. is rich/honest; **viel von jmdm.** ~: think a lot of sb.; g) *(ab~, veranstalten)* give, ⟨speech, lecture⟩; 2. *unr. itr. V.* a) *(stehenbleiben)* stop; b) *(unverändert, an seinem Platz bleiben)* last; c) *(Sport)* save; d) *(beistehen)* **zu jmdm.** ~: stand by sb.; 3. *unr. refl. V.* a) *(sich durchsetzen, behaupten)* **wir werden uns/die Stadt wird sich nicht länger** ~ **können** we/the town won't be able to hold out much longer; b) *(sich bewähren)* **sich gut** ~: do well; c) *(unverändert bleiben)* ⟨weather, flowers, etc.⟩ last; ⟨milk, meat, etc.⟩ keep; d) *(Körperhaltung haben)* **sich schlecht/gerade** ~: hold oneself badly/straight; e) *(bleiben)* **sich auf den Beinen/im Sattel** ~: stay on one's feet/in the saddle; **sich links/rechts** ~: keep [to the] left/right; **sich an etw.** *(Akk.)* ~: keep to sth.

Halter der; ~s, ~ a) *(Fahrzeug~)* keeper; b) *(Tier~)* owner; c) *(Vorrichtung)* holder; **Halterung** die; ~, ~en support

Halte-: ~**stelle** die stop; ~**verbot** das a) „~verbot‟ 'no stopping'; **hier ist ~verbot** this is a no-stopping zone; b) *(Stelle)* no-stopping zone; ~**verbots·schild** das no-stopping sign

-**haltig,** *(österr.)* -**hältig:** vit-amin~/silber~ *usw.* containing vit-amins/silver *etc. postpos., not pred.;* **vitamin~ sein** contain vitamins

halt-: ~**los** *Adj.* a) *(labil)* ~los sein be a weak character; **ein ~loser Mensch** a weak character; b) *(unbegründet)* unfounded; ~**machen** *itr. V.* stop

Haltung die; ~, ~en a) *(Körper~)* posture; b) *(Pose)* manner; c) *(Einstellung)* attitude; d) *(Fassung)* composure

Halunke der; ~n, ~n scoundrel; villain

Hamburger der; ~s, ~ *od.* ~s *(Frikadelle)* hamburger

hämisch 1. *Adj.* malicious; 2. *adv.* maliciously

Hammel der; ~s, ~ a) wether; b) *(Fleisch)* mutton; **Hammel·fleisch** das mutton

Hammer der; ~s, **Hämmer** a) hammer; *(Holz~)* mallet; ~ **und Sichel** hammer and sickle; b) *(Technik)* ram; **hämmern** *itr., tr. V.* hammer

Hämorrhoiden [hɛmoro'iːdn̩] *Pl.* *(Med.)* haemorrhoids; piles

Hampel·mann der; ~[e]s, **Hampelmänner** a) jumping jack; b) *(ugs. abwertend)* puppet

hampeln *itr. V. (ugs.)* jump about

Hamster der; ~s, ~: hamster

hamstern *tr., itr. V.* a) *(horten)* hoard; b) *(Lebensmittel tauschen)* barter goods for [food]

Hand die; ~, **Hände** hand; **jmdm. die ~ geben** shake sb.'s hand; ~ **und Fuß/weder** ~ **noch Fuß haben** *(ugs.)* make sense/no sense; **alle** *od.* **beide Hände damit voll haben, etw. zu tun** *(ugs.)* have one's hands full doing sth.; **die Hände in den Schoß legen** sit back and do nothing; **etw. aus der** ~ **geben** let sth. out of one's hands; ~ **in** ~ **arbeiten** work hand in hand; **etw. zur** ~ **haben** have sth. handy; **zu Händen [von] Herrn Müller** attention Herr Müller

Hand-: ~**arbeit** die a) handicraft; **etw. in ~arbeit herstellen** make sth. by hand; b) *(Gegenstand)* handmade article; c) *(Nadelarbeit)* [piece of] needle-work; ~**ball** der handball; ~**besen** der brush; ~**betrieb** der; *o. Pl.* manual operation; ~**bewegung** die a)

movement of the hand; b) *(Geste)* gesture; **~bremse die** handbrake; **~buch das** handbook; *(technisches ~buch)* manual

Händchen das; ~s, ~: [little] hand; **Hände** s. **Hand**

Hände-: **~druck** der; *Pl.* ~drücke handshake; **~klatschen das**; ~s clapping

Handel der; ~s trade; **handeln** 1. *itr. V.* a) trade; deal; b) *(feilschen)* haggle; c) *(agieren)* act; d) *(sich verhalten)* behave; e) **von etw.** *od.* **über etw.** *(Akk.)* ~ *⟨book, film, etc.⟩* be about *or* deal with sth.; 2. *refl. V. (unpers.)* **es handelt sich um ...**: it is a matter of ...; *(es dreht sich um)* it's about ...

handels-, Handels-: **~abkommen** das trade agreement; **~bank** die merchant bank; **~bilanz die** a) *(eines Betriebes)* balance-sheet; b) *(eines Staates)* balance of trade; **~einig, ~eins in** mit jmdm. **~einig** *od.* **~eins werden/ sein** agree/have agreed terms with sb.; **~flotte** die merchant fleet; **~gesellschaft** die company; **~klasse die** grade; **~marine** die merchant navy; **~partner** der trading partner; **~register das** register of companies; **~schiff das** merchant ship; **~schule** die commercial college; **~straße die** *(hist.)* trade route; **~üblich** *Adj.* **~übliche Praktiken/Größen** standard business practices/standard [commercial] sizes; **~unternehmen das** trading concern; **~vertreter** der [sales] representative; travelling salesman/ saleswoman; **~vertretung die** trade mission; **~zentrum das** trading centre

hände·ringend *Adv. (ugs.: dringend)* *⟨need⟩* urgently; *⟨search for sb./sth.⟩* desperately

hand-, Hand-: **~feger** der brush; **~fest** *Adj.* robust; sturdy; substantial *⟨meal etc.⟩*; c) solid *⟨proof⟩*; concrete *⟨suggestion⟩*; complete *⟨lie⟩*; well-founded *⟨argument⟩*; **~fläche** die palm [of one's/the hand]; flat of one's/the hand; **~gas das** *(Kfz-W.)* hand throttle; **~gearbeitet** *Adj.* hand-made; **~gelenk das** wrist; **~gemenge das** fight; **~gepäck das** hand-baggage; **~geschrieben** *Adj.* handwritten; **~granate** die hand-grenade; **~greiflich** *Adj.* a) *(tätlich)* **~greiflich werden** start using one's fists; b) tangible *⟨success, advantage, proof, etc.⟩*; palpable *⟨contradiction,*

error⟩; obvious *⟨fact⟩*; **~griff der a)** **mit einem ~griff/wenigen ~griffen** in one movement/without much trouble; *(schnell)* in no time at all/next to no time; b) *(am Koffer, an einem Werkzeug)* handle; **~habe die**; ~, ~n: eine [rechtliche] ~habe [gegen jmdn.] a legal handle [against sb.]; **~haben** *tr. V.* a) handle; operate *⟨device, machine⟩*; b) *(praktizieren)* implement *⟨law etc.⟩*; **~habung die**; ~, ~en a) handling; *(eines Gerätes, einer Maschine)* operation; b) *(Durchführung)* implementation

Handikap ['hɛndikɛp] das; ~s, ~s *(auch Sport)* handicap; **handikapen** ['hɛndikɛpn] *tr. V.* handicap

Hand-: **~käse** der *(landsch.)* small, hand-formed curd cheese; **~koffer** der [small] suitcase; **~kuß** der kiss on sb.'s hand; **~langer der**; ~s, ~ *(ungelernter Arbeiter)* labourer; *(abwertend)* lackey; **~lauf** der handrail

Händler der; ~s, ~: trader

handlich *Adj.* handy; easily carried *⟨parcel, suitcase⟩*; easily portable *⟨television, camera⟩*

Handlung die; ~, ~en a) *(Vorgehen)* action; *(Tat)* act; b) *(Fabel)* plot

handlungs-, Handlungs-: **~fähig** *Adj.* able to act *pred.*; working *attrib.* *⟨majority⟩*; **~freiheit die**; *o. Pl.* freedom of action; **~reisende der/die** s. **Handelsvertreter**; **~weise die** conduct

hand-, Hand-: **puppe** die glove *or* hand puppet; **~schelle** die handcuff; **~schlag** der handshake; **~schrift** die handwriting; **~schriftlich** 1. *Adj.* hand-written; 2. *adv.* by hand; **~schuh** der glove; **~schuhfach das** glove compartment; **~signiert** *Adj.* signed; **~spiegel** der hand-mirror; **~stand** der *(Turnen)* handstand; **~tasche die** handbag; **~tuch das**; *Pl.* -tücher towel; **~umdrehen: im ~umdrehen** in no time at all; **~voll die**; ~ *(auch fig.)* handful

Hand·werk das craft; *(als Beruf)* trade; sein ~ **kennen** *od.* **verstehen/beherrschen** know one's job; **Handwerker** der; ~s, ~: tradesman; **handwerklich** *Adj.* ein ~er Beruf a [skilled] trade; **Handwerks·zeug das** tools *pl.*

Hand·zeichen das sign [with one's hand]; *(eines Autofahrers)* hand signal; *(Abstimmung)* show of hands

Hanf der; ~[e]s hemp

Hang der; ~|e|s, **Hänge** slope; *(Neigung)* tendency

Hänge-: ~**brücke** die suspension bridge; ~**lampe** die pendant-light; ~**matte** die hammock

¹**hängen** *unr. itr. V.; südd., österr., schweiz. mit sein* hang (an + *Dat.* from); *(an einem Fahrzeug)* be hitched (an + *Dat.* to); ²**hängen** 1. *tr. V.* a) hang (**in/über** + *(Akk.)* in/over; **an/ auf** + *Akk.* on); *(befestigen)* hitch up (an + *Akk.* to); couple on ⟨*railway carriage, etc.*⟩ (an + *Akk.* to); 2. *refl. V.* a) **sich an etw.** *(Akk.)* ~: hang on to sth.; b) *(sich festsetzen)* cling (an + *Akk.* to); **hängen|bleiben** *unr. itr. V.; mit sein (ugs.)* a) *(festgehalten werden)* [mit dem Ärmel *usw.*] **an/in etw.** *(Dat.)* ~: get one's sleeve *etc.* caught on/in sth.; b) *(verweilen)* get stuck *(coll.)*; c) *(haften)* **an/auf etw.** *(Dat.)* ~: stick to sth.; **hängend** *Adj.* hanging; **Hänge·schrank** der wall-cupboard

hänseln *tr. V.* tease

Hanse·stadt die Hanseatic city

Hantel die; ~, ~n *(Sport)* *(kurz)* dumbbell; *(lang)* barbell

hantieren *itr. V.* be busy

Häppchen das; ~s, ~ a) [small] morsel; b) *(Appetithappen)* canapé

Happen der; ~s, ~: morsel

happig *Adj. (ugs.)* ~e **Preise** fancy prices *(coll.)*

Happy-End ['hɛpi'|ɛnt] **das;** ~|s|, ~s happy ending

Harfe die; ~, ~n harp

Harke die; ~, ~n rake; **harken** *tr. V.* rake

harm·los 1. *Adj.* a) *(ungefährlich)* harmless; slight ⟨*injury, cold, etc.*⟩; mild ⟨*illness, etc.*⟩; safe ⟨*medicine, bend, road, etc.*⟩; b) *(arglos)* innocent; harmless ⟨*fun, pastime, etc.*⟩; 2. *adv.* a) *(ungefährlich)* harmlessly; b) *(arglos)* innocently; **Harmlosigkeit** die; ~ a) *(Ungefährlichkeit)* harmlessness; *(einer Krankheit)* mildness; *(eines Medikamentes)* safety; b) *(Arglosigkeit, harmloses Verhalten)* innocence

Harmonie die; ~, ~n *(auch fig.)* harmony; **harmonieren** *itr. V.* a) harmonize; b) *(miteinander auskommen)* get on well

Harmonika die; ~, ~s *od.* **Harmoniken** harmonica

harmonisch 1. *Adj.* harmonious; *(Musik)* harmonic; 2. *adv.* harmoniously; *(Musik)* harmonically

Harmonium das; ~s, **Harmonien** harmonium

Harn der; ~|e|s, ~e *(Med.)* urine; **Harn·blase** die bladder

Harnisch der; ~s, ~e armour

Harpune die; ~, ~n harpoon

harren *itr. V. (geh.)* **jmds./einer Sache** *od.* **auf jmdn./etw.** ~: await sb./sth.

harsch 1. *Adj.* a) *(vereist)* crusted; b) *(barsch)* harsh; 2. *adv.* harshly; **Harsch** der; ~|e|s crusted snow

hart; härter, härtest... 1. *Adj.* a) hard; b) tough ⟨*situation, job*⟩; harsh ⟨*reality, truth*⟩; c) *(streng)* harsh ⟨*penalty, punishment, judgement*⟩; tough ⟨*measure, law, course*⟩; d) *(rauh)* rough ⟨*game, opponent*⟩; 2. *adv.* hard chair; a) *(mühevoll)* ⟨*work*⟩ hard; b) *(streng)* harshly; c) *(nahe)* close (an + *Dat.* to); **Härte** die; ~, ~n a) *(auch Physik)* hardness; b) *(Widerstandsfähigkeit)* toughness; c) *(schwere Belastung)* hardship; d) *(Strenge)* harshness; e) *(Heftigkeit)* *(eines Aufpralls usw.)* force; *(eines Streits)* violence; f) *(Rauheit)* roughness; **Härte·fall** der a) case of hardship; b) *(ugs.: Person)* hardship case; **härten** *tr., itr. V.* harten; **härter** *s.* **hart; härtest...** *s.* hart

hart-, Hart-: ~**gekocht** *Adj.* hard-boiled ⟨*egg*⟩; ~**geld** das coins *pl.*; ~**herzig** 1. *Adj.* hard-hearted; 2. *adv.* hard-heartedly; ~**herzigkeit die;** ~: hard-heartedness; ~**näckig** 1. *Adj.* a) *(eigensinnig)* obstinate; stubborn; b) *(ausdauernd)* dogged; 2. *adv.* a) *(eigensinnig)* obstinately; stubbornly; b) *(ausdauernd)* doggedly; ~**näckig-keit die;** ~ a) *(Eigensinn)* obstinacy; stubbornness; b) *(Ausdauer)* doggedness; ~**wurst** die dry sausage

Harz das; ~es, ~e resin

Harzer Käse der; ~ ~s, ~ ~: Harz [Mountain] cheese

Haschee *(Kochk.)* das; ~s, ~s hash

¹**haschen** *tr. V. (veralt.)* catch

²**haschen** *itr. V. (ugs.)* smoke [hash] *(coll.)*

Häschen ['hɛːsçən] das; ~, ~: bunny

Haschisch das *od.* der; ~|s| hashish

Hase der; ~n, ~n a) hare; b) *(landsch.)* *s.* **Kaninchen**

Hasel·nuß die hazel-nut

Hasen-: ~**fuß** der *(spöttisch abwertend)* coward; chicken *(sl.)*; ~**scharte** die *(Med.)* harelip

Haspel die; ~, ~n *(Technik)* *(für Garn)* reel; *(für ein Seil, Kabel)* drum

Haß der; H**a**sses hatred (**auf** + *Akk.*, gegen of, for); **h**a**ssen** *tr., itr. V.* hate; **haß·erfüllt** *Adj.* filled with hatred *postpos.*

häßlich 1. *Adj.* **a)** ugly; **b)** *(gemein)* nasty; **c)** *(unangenehm)* awful ⟨*weather, cold, situation, etc.*⟩; **2.** *adv.* **a)** ⟨*dress*⟩ unattractively; **b)** *(gemein)* nastily; **Häßlichkeit** die; ~, ~en **a)** *o. Pl. (Aussehen)* ugliness; **b)** *o. Pl. (Gesinnung)* nastiness

ha**st** *2. Pers. Sg. Präsens v.* **haben**

Ha**st** die; ~: haste; **h**a**sten** *itr. V.; mit sein* hurry; **h**a**stig 1.** *Adj.* hasty; hurried; **2.** *adv.* hastily; hurriedly

hat *3. Pers. Sg. Präsens v.* **haben**

hä**tscheln** *tr. V.* caress

hatschi *Interj.* atishoo

ha**tte** *1. u. 3. Pers. Sg. Prät. v.* **haben**; **hätte** *1. u. 3. Pers. Sg. Konjunktiv II v.* **haben**

Ha**ube** die; ~, ~n **a)** bonnet; *(einer Krankenschwester)* cap; **b)** *(Kfz-W.)* bonnet *(Brit.);* hood *(Amer.)*

Ha**uch** der; ~[e]s, ~e *(geh.)* **a)** *(Atem, auch fig.)* breath; **b)** *(Luftzug)* breath of wind; **c)** *(leichter Duft)* delicate smell; **d)** *(dünne Schicht)* [gossamer-]thin layer; **h**a**uch·dünn** *Adj.* gossamer-thin ⟨*material, dress*⟩; wafer-thin ⟨*layer, slice, majority*⟩; **h**a**uchen** *itr. V.* breathe (**gegen, auf** + *Akk.* on)

Ha**ue** die; ~, ~n **a)** *(südd., österr.: Hakke)* hoe; **b)** *(ugs.: Prügel)* a hiding *(coll.);* **h**a**uen 1.** *unr. tr. V.* **a)** *(ugs.: schlagen)* belt; clobber *(coll.);* **b)** *(ugs.: auf einen Körperteil)* belt *(coll.);* hit; **c)** *(herstellen)* carve ⟨*figure, statue, etc.*⟩ **(in** + *Akk.* in); **2.** *unr. itr. V.* **a)** *(ugs.: prügeln)* **er haut immer gleich** he's quick to hit out; **b)** *(auf einen Körperteil)* belt *(coll.);* hit; **c)** *(ugs.: auf/gegen etw. schlagen)* thump; **3.** *unr. refl. V. (ugs.: sich prügeln)* have a punch-up *(coll.)*

Ha**ufen** der; ~s, ~: heap; pile; *(Gruppe)* bunch *(coll.);* **h**ä**ufen 1.** *tr. V.* heap, pile (**auf** + *Akk.* on to); *(aufheben)* hoard ⟨*money, supplies*⟩; **2.** *refl. V. (sich mehren)* pile up

hä**ufig 1.** *Adj.* frequent; **2.** *adv.* frequently; often; **Häufigkeit** die; ~, ~en frequency; **Häufung** die; ~, ~en increasing frequency

Ha**upt** das; ~[e]s, H**äupter** *(geh., auch fig.)* head

haupt-, Ha**upt-:** ~**bahnhof** der main station; ~**beruflich 1.** *Adj.* sei-
ne ~**berufliche Tätigkeit** his main occupation; **2.** *adv.* **er ist** ~**beruflich als Elektriker tätig** his main occupation is that of electrician; ~**darsteller** der *(Theater, Film)* male lead; ~**darstellerin** die *(Theater, Film)* female lead; ~**eingang** der main entrance; ~**fach** das the major; ~**figur** die main character; ~**film** der main feature; ~**gebäude** das main building; ~**gericht** das main course; ~**gewinn** der first prize

Häuptling der; ~s, ~e chief[tain]

haupt-, Ha**upt-:** ~**mahlzeit** die main meal; ~**mann** der; *Pl.* ~**leute** *(Milit.)* captain; ~**person** die central figure; ~**postamt** das main post office; ~**quartier** das *(Milit., auch fig.)* headquarters *sing. or pl.;* ~**rolle** die main role; lead; ~**sache** die main thing; ~**sächlich 1.** *Adv.* mainly; principally; **2.** *Adj.; nicht präd.* main; principal; ~**saison** die high season; ~**satz** der main clause; *(alleinstehend)* sentence; ~**schlagader** die aorta; ~**schul·abschluß** der ≈ secondary school leaving certificate; ~**schule** die ≈ secondary modern school; ~**stadt** die capital [city]; ~**städtisch** *Adj.* metropolitan; ~**straße** die main street; ~**verkehr** der bulk of the traffic

Hauptverkehrs-: ~**straße** die main road; ~**zeit** die rush hour

Ha**upt-:** ~**wache** die main police station; ~**wort** das *(Sprachw.)* noun

ha**u r**u**ck** *Interj.* heave[-ho]

Ha**us** das; ~es, H**äuser** house; *(Amts-, Firmengebäude usw.)* building; *(Heim)* home; **nach** ~**e** home; **zu** ~**e** at home; **das erste** ~ **am Platze** the best hotel in the town

haus-, Ha**us-:** ~**angestellte** der/die domestic servant; ~**apotheke** die medicine cabinet; ~**arbeit** die housework; *(Schulw.)* homework; ~**arrest** der house arrest; ~**arzt** der family doctor; ~**aufgabe** die homework; ~**backen 1.** *Adj.* plain; unadventurous ⟨*clothes*⟩; **2.** *adv.* ⟨*dress*⟩ unadventurously; ~**besetzer** der squatter; ~**besitzer** der house-owner; *(Vermieter)* landlord; ~**besitzerin** die house-owner; *(Vermieterin)* landlady; ~**besuch** der house-call; ~**boot** das houseboat

Häuschen ['hɔysçən] das; ~s, ~: small house; **aus dem** ~ **sein** *(ugs.)* be over the moon *(coll.);* **h**a**usen** *itr. V.*

(ugs. abwertend) live; **b)** *(Verwüstungen anrichten)* |furchtbar| ~: wreak havoc; **Häuser·block** der block [of houses]

haus-, Haus-: ~**flur** der hall[way]; *(im Obergeschoß)* landing; ~**frau** die housewife; **freund** der **a)** friend of the family; *(verhüll.: Liebhaber)* manfriend *(euphem.)*; ~**friedensbruch** der *(Rechtsw.)* trespass; ~**gebrauch** der domestic use; **das reicht für den** ~**gebrauch** *(ugs.)* it's good enough to get by *(coll.)*; ~**gehilfin** die [home] help; ~**gemacht** *Adj.* home-made

Haus·halt der **a)** household; **b)** *(Arbeit im ~)* housekeeping; **jmdm. den** ~ **führen** keep house for sb.; **c)** *(Politik)* budget; **haus|halten** *unr. itr. V.* be economical **(mit** with); **Haushälterin** die; ~, ~**nen** housekeeper

Haushalts-: ~**debatte,** die *(Politik)* budget debatte; ~**geld** das; *o. Pl.* housekeeping money; ~**jahr** das financial year; ~**kasse** die housekeeping money; ~**plan** der budget; ~**waren** *Pl.* household goods

Haus·herr der **a)** *(Familienoberhaupt)* head of the household; **b)** *(als Gastgeber)* host; **c)** *(Rechtsspr. (Eigentümer)* owner; *(Mieter)* occupier; **haushoch 1.** *Adj.* as high as a house; *(fig.)* overwhelming; **2.** *adv. (fig.)* ~ **gewinnen** win hands down

hausieren *itr. V.* |mit etw.| ~: hawk [sth.]; peddle [sth.]; „**Hausieren verboten**" 'no hawkers'; **Hausierer** der; ~s, ~: pedlar; hawker

häuslich *Adj.* **a)** domestic; **b)** *(das Zuhause liebend)* home-loving

Hausmacher·art die: **nach** ~: home-made-style *attrib.*

Hausmanns·kost die plain cooking

Haus-: ~**marke** die **a)** house wine; **b)** *(ugs.: bevorzugtes Getränk)* favourite tipple *(coll.)*; ~**meister** der, ~**meisterin** die caretaker; ~**mittel** das household remedy; ~**musik** die music at home; ~**nummer** die house number; ~**ordnung** die house rules *pl.*; ~**putz** der spring-clean; *(regelmäßig)* clean-out; ~**rat** der household goods *pl.*; ~**schlüssel** der front-door key; house-key; ~**schuh** der slipper

Haussuchung die; ~, ~**en** house search; **Haussuchungs·befehl** der search warrant

Haus-: ~**tier** das **a)** pet; **b)** *(Nutztier)* domestic animal; ~**tür** die front door; ~**verbot** das ban on entering the

house/pub/restaurant *etc.*; ~**verwalter** der manager [of the block]; ~**wirt** der landlord; ~**wirtin** die landlady; ~**wirtschaft** die; *o. Pl.* domestic science and home economics

Haut die; ~, **Häute** skin; **aus der ~ fahren** *(ugs.)* go up the wall *(coll.)*; **Haut·arzt** der skin specialist; **häuten 1.** *tr. V.* skin; flay; **2.** *refl. V.* shed its skin/their skins

haut-, Haut-: ~**eng** *Adj.* skin-tight; ~**farbe** die [skin] colour; ~**krankheit** die skin disease

Haxe die; ~, ~**n** *s.* **Hachse**

he *Interj. (ugs.)* hey

Heb·amme die midwife

Hebel der; ~**s**, ~: lever

heben *unr. tr. V.* **a)** lift; raise ⟨baton, camera, glass⟩; **b)** *(verbessern)* raise ⟨standard, level⟩; increase ⟨turnover, self-confidence⟩; improve ⟨mood⟩; enhance ⟨standing⟩; boost ⟨morale⟩

¹**hecheln** *itr. V. (ugs. abwertend)* gossip

²**hecheln** *itr. V.* pant [for breath]

Hecht der; ~|e|s, ~e pike; **Hechtsprung** der **a)** *(Turnen)* Hecht vault; **b)** *(Schwimmen)* racing dive; *(vom Sprungturm)* pike-dive

Heck das; ~|e|s, ~e *od.* ~s stern; *(Flugzeug~)* tail; *(Auto~)* rear

Hecke die; ~, ~**n a)** hedge; **b)** *(wildwachsend)* thicket

Hecken-: ~**rose** die dogrose; ~**schütze** der sniper

Heck·scheibe die rear window

Heer das; ~|e|s, ~e armed forces *pl.*; *(für den Landkrieg, fig.)* army

Hefe die; ~, ~**n** yeast

¹**Heft** das; ~|e|s, ~e *(geh.)* haft; handle

²**Heft** das; ~|e|s, ~e **a)** *(bes. Schule)* exercise-book; **b)** *(Nummer einer Zeitschrift)* issue; **Heftchen** das; ~s, ~: book [of tickets/stamps *etc.*]; **heften 1.** *tr. V.* **a)** *(mit einer Nadel)* pin; *(mit einer Klammer)* clip; *(mit Klebstoff)* stick; **b)** *(Schneiderei)* tack; **c)** *(Buchbinderei)* stitch; *(mit Klammern)* staple; **2.** *refl. V.* **sich an jmds. Fersen** *(Akk.)* ~: stick hard on sb.'s heels

heftig 1. *Adj.* violent; heavy ⟨rain, shower, blow⟩; severe ⟨pain⟩; ⟨person⟩ with a violent temper; **2.** *adv.* ⟨rain, snow, breathe⟩ heavily; ⟨hit⟩ hard; ⟨quarrel⟩ violently

Heft-: ~**klammer** staple; ~**pflaster** das sticking plaster; ~**zwecke** die *s.* **Reißzwecke**

hegen *tr. V.* **a)** *(bes. Forstw., Jagdw.)*

look after, tend; b) *(geh.: umsorgen)* look after; c) *(fig.)* feel ⟨contempt, hatred, mistrust⟩; cherish ⟨hope, wish, desire⟩; harbour ⟨grudge, suspicion⟩

Hehl der *od.* das: kein|en| ~ aus etw. machen make no secret of sth.; **Hehler** der; ~s, ~: receiver [of stolen goods]; **Hehlerei** die; ~, ~en *(Rechtsw.)* receiving [stolen goods] *no art.*

¹**Heide** der; ~n, ~n heathen

²**Heide** die; ~, ~n a) heath; *(~landschaft)* heathland; **Heide·kraut** das; *o. Pl.* heather

Heidel·beere die bilberry

heidnisch *Adj.* heathen

heikel *Adj.* a) *(schwierig)* delicate, ticklish ⟨matter, subject⟩; ticklish tricky ⟨problem, question, situation⟩; b) *(wählerisch)* fussy (**in bezug auf** + *Akk.* about)

heil *Adj. (nicht entzwei)* in one piece; **wieder** ~ **sein** ⟨injured part⟩ have healed [up]; **Heil** das; ~s a) *(Wohlergehen)* benefit; b) *(Rel.)* salvation; **Heiland** der; ~|e|s, ~e Saviour

Heil·anstalt die *(Anstalt für Kranke od. Süchtige)* sanatorium; *(psychiatrische Klinik)* mental hospital; **heilbar** *Adj.* curable

Heil·butt der halibut

heilen 1. *tr. V.* cure; heal ⟨wound⟩; 2. *itr. V.; mit sein* ⟨wound⟩ heal [up]; ⟨fracture⟩ mend

heil·froh *Adj.* very glad

heilig *Adj.* a) holy; **die Heilige Schrift** the Holy Scriptures *pl.;* **der Heilige Abend** Christmas Eve b) *(geh.: unantastbar)* sacred ⟨right, tradition, cause, etc.⟩; **Heilig·abend** der Christmas Eve; **Heilige** der/die; *adj. Dekl.* saint; **heiligen** *tr. V.* keep ⟨tradition, Sabbath, etc.⟩; **der Zweck heiligt die Mittel** the end justifies the means; **Heiligen·schein** der gloriole; *(um den Kopf)* halo; **Heiligkeit** die; ~: holiness; **Heiligtum** das; ~s, Heiligtümer shrine

Heil-: ~**kraut** das medicinal herb; ~**mittel** das *(auch fig.)* remedy (**gegen** for); *(Medikament)* medicament; ~**praktiker** der non-medical practitioner

heilsam *Adj.* salutary; **Heils·armee** die Salvation Army; **Heilung** die; ~, ~en *(einer Wunde)* healing; *(von Krankheit, Kranken)* curing

Heim das; ~|e|s, ~e a) *(Zuhause)* home; b) *(Anstalt, Alters~)* home; *(für*

Obdachlose) hostel; **Heim·arbeit** die outwork

Heimat die; ~, ~en a) *(~ort)* home; home town/village; *(~land)* home; homeland; b) *(Ursprungsland)* natural habitat

Heimat-: ~**kunde** die local history, geography, and natural history; ~**land** das native land; *(fig.)* home

heimatlich *Adj.* native ⟨dialect⟩; nostalgic ⟨emotions⟩

heimat-, Heimat-: ~**los** *Adj.* homeless; ~**museum** das museum of local history; ~**ort** der home town/village; ~**vertriebene** der/die; *adj. Dekl.* expellee [from his/her homeland]

heim-, Heim-: ~|**bringen** *unr. tr. V.* a) jmdn. ~: take *or.* see sb. home; b) bring home; ~|**fahren** 1. *unr. itr. V.; mit sein* drive home; 2. *unr. tr. V.* drive home; ~**fahrt** die journey home; *(mit dem Auto)* drive home; ~|**gehen** *unr. itr. V.; mit sein* go home

heimisch *Adj. (einheimisch)* indigenous, native ⟨plants, animals, etc.⟩ (**in** + *Dat.* to); domestic ⟨industry⟩; **sich** ~ **fühlen** feel at home; ~ **werden [in** (+ *Dat.*)] settle in[to]

heim-, Heim-: ~**kehr** die; ~: return home; homecoming: ~|**kehren** *itr. V.; mit sein* return home (**aus** from); ~|**kommen** *unr. itr. V.; mit sein* come home

heimlich 1. *Adj.* secret; 2. *adv.* secretly; **Heimlichkeit** die; ~, ~en; *meist Pl.* secret

heim-, Heim-: ~**reise** die journey home; ~|**suchen** *tr. V.* ⟨storm, earthquake, epidemic⟩ strike; ⟨disease⟩ afflict; ⟨nightmares, doubts⟩ plague; ~**tückisch** 1. *Adj. (bösartig)* malicious; *(fig.)* insidious ⟨disease⟩; 2. *adv.* maliciously; ~**wärts** *Adv. (nach Hause zu)* home; *(in Richtung Heimat)* homeward[s]; ~**weg** der way home–; ~**weh** das homesickness; ~**weh haben** be homesick (**nach** for); ~|**zahlen** *tr. V.* jmdm. etw. ~**zahlen** pay sb. back for sth.

Heinzel·männchen das brownie

Heirat die; ~, ~en marriage; **heiraten** 1. *itr. V.* get married; 2. *tr. V.* marry

Heirats-: ~**antrag** der: jmdm. einen ~**antrag machen** propose to sb.; ~**anzeige** die announcement of a/the forthcoming marriage; ~**schwindler** der *person who makes a spurious offer of marriage for purposes of fraud*

heiser 1. *Adj.* hoarse; 2. *adv.* in a hoarse voice; **Heiserkeit die**; ~ *s.* heiser: hoarseness

heiß 1. *Adj.* hot; jmdm. **ist** ~: sb. feels hot; etw. ~ **machen** heat sth. up; heated ⟨*debate, argument*⟩; fierce ⟨*fight, battle*⟩; ardent ⟨*wish, love*⟩; **ein** ~**es Thema** a controversial subject; 2. *adv.* ⟨*fight*⟩ fiercely; ⟨*love*⟩ dearly; ⟨*long*⟩ fervently

heißen *unr. itr. V.* *(den Namen tragen)* be called; *(bedeuten)* mean; *(lauten)* ⟨*saying*⟩ go; *(unpers.)* **es heißt, daß ...:** they say that ...; **in dem Artikel heißt es ...:** in the article it says that ...

heiter *Adj.* cheerful; fine ⟨*weather, day*⟩; **Heiterkeit die**; ~ a) *(Frohsinn)* cheerfulness; b) *(Belustigung)* merriment

heizbar *Adj.* heated; **Heiz·decke die** electric blanket; **heizen** 1. *itr. V.* have the heating on; 2. *tr. V.* heat ⟨*room etc.*⟩; **Heizer der**; ~s, ~ *(einer Lokomotive)* fireman; *(eines Schiffes)* stoker

Heiz-: ~**kissen das** heating pad; ~**körper der** radiator; ~**ofen der** stove; heater; ~**platte die** hotplate

Heizung die; ~, ~en a) *(zentral)* heating *no pl., no indef. art.;* b) *(ugs.: Heizkörper)* radiator

Hektar das *od.* der; ~s, ~e hectare

Hektik die; ~: hectic rush; *(des Lebens)* hectic pace; **hektisch** *Adj.* hectic

Held der; ~en, ~en hero; **heldenhaft** 1. *Adj.* heroic; 2. *adv.* heroically; **Heldentum das**; ~s heroism; **Heldin die**, ~, ~nen heroine

helfen *unr. itr. V.* help; jmdm. |bei etw.| ~: help sb. [with sth.]; *(unpers.)* **es hilft nichts** it's no use *or* good; **Helfer der**; ~s, ~: helper; *(Mitarbeiter)* assistant; *(eines Verbrechens)* accomplice

Helikopter der; ~s, ~: helicopter

hell 1. *Adj.* a) *(von Licht erfüllt)* light; well-lit ⟨*stairs*⟩; b) *(klar)* bright ⟨*day, sky, etc.*⟩; c) *(viel Licht spendend)* bright ⟨*light, lamp, star, etc.*⟩; d) *(blaß)* light ⟨*colour*⟩; fair ⟨*skin, hair*⟩; light-coloured ⟨*clothes*⟩; e) *(akustisch)* high, clear ⟨*sound, voice*⟩; ringing ⟨*laugh*⟩; f) *(klug)* bright; g) *(ugs.: absolut)* sheer, utter ⟨*madness, foolishness, despair*⟩; 2. *adv.* brightly

hell-: ~**blau** *Adj.* light blue; ~**blond** *Adj.* very fair; light blonde

Helle das; *adj. Dekl.* ≈ lager

Heller der; ~s, ~: heller; **bis auf den letzten** ~/**bis auf** ~ **und Pfennig** *(ugs.)* down to the last penny *or (Amer.)* cent

hell-: ~**grün** *Adj.* light green; ~**häutig** *Adj.* fair-skinned

Helligkeit die; ~, ~en *(auch Physik)* brightness

hell-, Hell-: ~**rot** *Adj.* light red; ~**sehen** *unr. itr. V.; nur im Inf.* ~**sehen können** have second sight; ~**seher der** clairvoyant; ~**wach** *Adj.* wide awake

Helm der; ~|e|s, ~e helmet

Hemd das; ~|e|s, ~en shirt; *(Unterhemd)* [under]vest; undershirt; **Hemds·ärmel der** shirt-sleeve

hemmen *tr. V.* a) *(verlangsamen)* slow [down]; b) *(aufhalten)* check; stem ⟨*flow*⟩; c) *(beeinträchtigen)* hinder; **Hemmung die**; ~, ~en a) *(Gehemmtheit)* inhibition; b) *(Bedenken)* scruple; **hemmungs·los** 1. *Adj.* unrestrained; 2. *adv.* unrestrainedly

Hendl das; ~s, ~|n| *(bayr., österr.)* chicken; *(Brathähnchen)* [roast] chicken

Hengst der; ~|e|s, ~e *(Pferd)* stallion

Henkel der; ~s, ~: handle

Henker der; ~s, ~: hangman; *(Scharfrichter, auch fig.)* executioner

Henne die; ~, ~n hen

her [he:ɐ̯] *Adv.* ~ **damit** give it to me; give it here *(coll.)*; **vom Fenster** ~: from the window; **von ihrer Kindheit** ~: since childhood; **von der Konzeption** ~: as far as the basic design is concerned

herab *Adv.* down; **von oben** ~ *(fig.)* condescendingly

herab-: ~|**hängen** *unr. itr. V.* hang [down] *(von from)*; ~**hängende Schultern** drooping shoulders; ~|**lassen** 1. *unr. tr. V.* let down; lower; 2. *unr. refl. V. (iron.: bereit sein)* **sich** ~**lassen, etw. zu tun** condescend to do sth.; ~|**lassend** 1. *Adj.* condescending; patronizing (**zu** towards); 2. *adv.* condescendingly; patronizingly; ~|**sehen** *unr. itr. V.* **auf** jmdn. ~**sehen** look down on sb.; ~|**setzen** *tr. V.* a) reduce; b) *(abwerten)* belittle

heran *Adv.* **an** etw. *(Akk.)* ~: right up to sth.

heran-, Heran-: ~|**bilden** *tr. V.* train [up]; *(auf der Schule, Universität)* educate; ~|**bringen** *unr. tr. V.* a) bring [up] (**an** + *Akk.,* **zu** to); b) *(vertraut machen)* jmdn. **an** etw. *(Akk.)* ~**bringen** introduce sb. to sth.; ~|**fahren**

unr. itr. V.; mit sein drive up (**an** + **Akk.** to); ~|**kommen** *unr. itr. V.; mit sein* **an etw.** *(Akk.)* ~**kommen** come near to sth.; *(erreichen)* reach sth.; *(erwerben)* obtain sth.; ~|**reifen** *itr. V.; mit sein* ⟨*fruit, crops*⟩ ripen; **zur Frau** ~**reifen** mature into a woman; ~|**treten** *unr. itr. V.; mit sein (sich wenden)* **an jmdn.** ~**treten** approach sb.; ~|**wachsen** *unr. itr. V.; mit sein* grow up; ~**wachsende der/die;** *adj. Dekl.* young person; ~|**ziehen** *unr. tr. V.* pull over; pull up ⟨*chair*⟩; **etw. zu sich** ~**ziehen** pull sth. towards one

herauf *Adv.* up

herauf-: ~|**beschwören** *tr. V.* a) *(verursachen)* cause ⟨*disaster, war, crisis*⟩; b) *(erinnern)* evoke ⟨*memories etc.*⟩; ~|**kommen** *unr. itr. V.; mit sein (nach oben kommen)* come up; ~|**setzen** *tr. V.* increase, put up ⟨*prices, rents, interest rates, etc.*⟩

heraus *Adv.* ~ **aus den Federn!/dem Bett!** rise and shine!/out of bed!

heraus-: ~|**bekommen** *unr. tr. V.* a) *(entfernen)* get out (**aus** of); b) *(ugs.: lösen)* work out ⟨*problem, answer, etc.*⟩; solve ⟨*puzzle*⟩; c) *(ermitteln)* find out; d) *(als Wechselgeld bekommen)* **5 DM** ~**bekommen** get back 5 marks change; **ich bekomme noch 5 DM** ~: I still have 5 marks [change] to come; ~|**bringen** *unr. tr. V.* a) *(nach außen bringen)* bring out (**aus** of); b) *(nach draußen begleiten)* show out; c) *(veröffentlichen)* bring out; *(aufführen)* put on, stage ⟨*play*⟩; screen ⟨*film*⟩; d) *(auf den Markt bringen)* bring out; e) *(populär machen)* make widely known; ~|**fahren** 1. *unr. itr. V.; mit sein* a) *(nach außen fahren)* **aus etw.** ~**fahren** drive/ride out of sth.; b) *(fahrend* ~**kommen**) come out; 2. *unr. tr. V.* **den Wagen [aus dem Hof]** ~**fahren** drive the car out [of the yard]; **jmdn.** ~**fahren** drive sb. out (**zu** to); ~|**finden** 1. *unr. tr. V.* find out; trace ⟨*fault*⟩; 2. *unr. itr. V.* find one's way out (**aus** of); ~|**fordern** 1. *tr. V.* a) *(auch Sport)* challenge; b) *(heraufbeschwören)* provoke ⟨*person, resistance, etc.*⟩; invite ⟨*criticism*⟩; court ⟨*danger*⟩; 2. *itr. V.* **zu etw.** ~**fordern** provoke sth.; ~|**forderung die** *(auch Sport)* challenge; *(Provokation)* provocation; ~|**geben** 1. *unr. tr. V.* a) *(aushändigen)* hand over ⟨*property, person, hostage, etc.*⟩; *(zurückgeben)* give back; b) *(als Wechselgeld zurückgeben)* **5 DM/zuviel** ~**ge-**

ben give 5 marks/too much change; c) *(veröffentlichen)* publish; d) issue ⟨*stamp, coin, etc.*⟩; 2. *itr. V.* give change; ~**geber der** publisher; *(Redakteur)* editor; ~|**gehen** *unr. itr. V.; mit sein* a) go out (**aus** of); b) *(sich entfernen lassen)* ⟨*stain etc.*⟩ come out; ~|**halten** *unr. refl. V.* keep out; ~|**hängen** *tr. V.* hang out (**aus** of); ~|**helfen** *unr. itr. V.* **jmdm.** ~**helfen** *(auch fig.)* help sb. out (**aus** of); ~|**holen** *tr. V.* a) *(nach außen holen)* bring out; b) *(ugs.: erwirken)* win ⟨*wage increase, advantage, etc.*⟩; ~|**kommen** *unr. itr. V.; mit sein* a) come out (**aus** of); b) *(erscheinen; ugs.: auf dem Markt kommen, bekannt werden)* come out; ~|**nehmen** *unr. tr. V.* a) take out (**aus** of); b) *(ugs.: entfernen)* take out ⟨*appendix, tonsils, tooth, etc.*⟩; ~|**reden** *refl. V. (ugs.)* talk one's way out (**aus** of); ~|**reißen** *unr. tr. V.* a) tear out (**aus** of); pull up ⟨*plant*⟩; ~|**rutschen** *itr. V.; mit sein (ugs.)* ⟨*remark etc.*⟩ slip out; ~|**stellen** *refl. V.* **es stellte sich** ~, **daß ...:** it turned out that ...; ~|**suchen** *tr. V.* pick out; look out ⟨*file*⟩

herb *Adj.* [slightly] dry ⟨*wine*⟩; [slightly] sharp ⟨*taste*⟩; dry ⟨*smell, perfume*⟩; bitter ⟨*disappointment*⟩; severe ⟨*face, features*⟩; austere ⟨*beauty*⟩; harsh ⟨*words, criticism*⟩

herbei-: ~|**eilen** *itr. V.; mit sein* hurry over; ~|**laufen** *unr. itr. V.; mit sein* come running up

Herberge die; ~, ~**n** *(veralt.: Gasthaus)* inn

her|bringen *unr. tr. V.* **etw.** ~: bring sth. [here]

Herbst der; ~[e]**s,** ~**e** autumn; fall *(Amer.);* s. auch **Frühling; Herbstanfang der** beginning of autumn; **herbstlich** *Adj.* autumn *attrib.;* autumnal

Herd der; ~[e]**s,** ~**e** cooker; *(fig.)* centre *(of disturbance/rebellion)*

Herde die; ~, ~**n** herd

herein-: ~|**bitten** *unr. tr. V.* **jmdn.** ~**bitten** ask *or* invite sb. in; ~|**brechen** *unr. tr. V.; mit sein (geh.)* ⟨*night, evening, dusk*⟩ fall; ⟨*winter*⟩ set in; ⟨*storm*⟩ strike, break; ~|**bringen** *unr. tr. V.* bring in; ~|**fallen** *unr. itr. V.; mit sein (ugs.)* be taken for a ride *(coll.);* be done *(coll.);* ~|**kommen** *unr. itr. V.; mit sein* come in; ~|**lassen** *unr. tr. V.* let in; ~|**legen** *tr. V. (ugs.)* **jmdn.** ~**legen** take sb. for a ride *(coll.)* (**mit, bei** with); ~|**platzen** *itr.*

V.; mit sein (ugs.) burst in; ~|**schnei-en** *unr. itr. V.; mit sein (ugs.)* turn up out of the blue *(coll.)*

her-, Her-: ~**fahrt** die journey here; ~|**fallen** *unr. itr. V.; mit sein* über jmdn. ~**fallen** attack sb.; *(gierig zu essen beginnen)* über etw. *(Akk.)* ~**fallen** fall upon sth.; ~**gang** der: der ~**gang** der Ereignisse the sequence of events; ~|**geben** *unr. tr. V.* hand over; *(weggeben)* give away; ~|**gehen** *unr. itr. V.; mit sein* neben/vor/hinter jmdm. ~**gehen** walk along beside/in front of/behind sb.; ~|**haben** *unr. tr. V. (ugs.)* wo hat er/sie das ~? where did he/she get that from?; ~|**halten** *unr. itr. V.* ~**halten müssen [für jmdn./etw.]** be the one to suffer [for sb./sth.]; ~|**hören** *itr. V.* listen

Hering der; ~s, ~e a) herring; b) *(Zeltpflock)* peg

her-: ~|**kommen** *unr. itr. V.; mit sein* come here; ~**kömmlich** *Adj.* conventional; traditional *(custom)*

Herkunft die; ~, **Herkünfte** origin

her-: ~|**laufen** *unr. itr. V.; mit sein* vor/hinter/neben jmdm. ~**laufen** run [along] in front of/behind/alongside sb.; *(nachlaufen)* hinter jmdm. ~**laufen** run after sb.; *(fig.)* chase sb. up; ~|**leiten** *tr., refl. V.* derive (aus, von from); ~|**machen** *(ugs.)* refl. V. sich über etw. *(Akk.)* ~**machen** get stuck into sth. *(coll.)*

Hermelin der; ~s, ~e *(Pelz)* ermine

hermetisch 1. *Adj.* hermetic; 2. *adv.* hermetically

Heroin das; ~s heroin

Herr der; ~n, ~en a) *(Mann)* gentleman; b) *(Titel, Anrede)* ~ Schulze Mr Schulze; Sehr geehrter ~ Schulze! Dear Sir; *(bei persönlicher Bekanntschaft)* Dear Mr Schulze; meine ~en gentlemen; c) *(Gebieter)* master

herren-, Herren-: ~**ausstatter** der [gentle]men's outfitter; ~**los** *Adj.* abandoned *(car, luggage)*; stray *(dog, cat)*; ~**salon** der men's hairdressing salon; ~**schuh** der man's shoe; ~**schuhe** men's shoes; ~**toilette** die [gentle]men's toilet

Herr·gott der; ~s: der [liebe]/unser ~: the Lord [God]; God; **Herrgottsfrühe** die *in* in aller ~: at the crack of dawn

her|richten *tr. V. (bereitmachen)* get *(room, refreshments, etc.)* ready; arrange *(table)*; *(in Ordnung bringen)* renovate

Herrin die; ~, ~en mistress; **herrisch** 1. *Adj.* overbearing; imperious; 2. *adv.* imperiously; **herrlich** 1. *Adj.* marvellous; magnificent *(view, clothes)*; 2. *adv.* marvellously; **Herrlichkeit** die; ~, ~en a) *o. Pl. (Schönheit)* magnificence; splendour; b) *meist Pl. (herrliche Sache)* marvellous thing; **Herrschaft** die; ~, ~en a) *o. Pl.* rule; *(Macht)* power; b) *Pl. (Damen u. Herren)* ladies and gentlemen; **herrschen** *itr. V.* rule; *(monarch)* reign, rule; **draußen ~ 30° Kälte** it's 30° below outside; **Herrscher** der; ~s, ~: ruler; **herrsch·süchtig** *Adj.* domineering

her-: ~|**rühren** *itr. V.* von jmdm./etw. ~**rühren** come from sb./stem from sth.; ~|**sein** *unr. itr. V.; mit sein* einen Monat/lange ~**sein** be a month/a long time ago; es ist lange ~, daß wir ...: it is a long time since we ...; von Köln ~**sein** be from Cologne; hinter jmdm. *(ugs.)/*etw. ~**sein** be after sb./sth.; ~|**stellen** *tr. V.* produce

Her·steller der; ~s, ~: producer; **Her·stellung** die production

herüber *Adv.* over

herum *Adv.* **um ... ~** *(Richtung)* round; *(Anordnung)* around; **um Weihnachten ~:** around Christmas

herum-: ~|**ärgern** refl. V. *(ugs.)* sich mit jmdm./etw. ~**ärgern** keep getting annoyed with sb./sth.; ~|**drehen** 1. *tr. V. (ugs.)* turn *(key)*; turn over *(coin, mattress, hand, etc.)*; 2. *refl. V.* turn [a]round; ~|**fahren** *(ugs.)* 1. *unr. itr. V.; mit sein (sich plötzlich herumdrehen)* spin round; 2. *unr. tr. V.* jmdn. [in der Stadt] ~**fahren** drive sb. around the town; ~|**führen** 1. *tr. V.* jmdn. [in der Stadt] ~**führen** show sb. around the town; 2. *itr. V.* um etw. ~**führen** *(road etc.)* go round sth.; ~|**gehen** *unr. itr. V.; mit sein (vergehen)* pass; um etw. ~**gehen** go round sth.; etw. ~**gehen lassen** circulate sth.; pass; ~|**kommen** *unr. itr. V.; mit sein (ugs.)* a) *(vermeiden können)* um etw. [nicht] ~**kommen** [not] be able to get out of sth.; b) *(viel reisen)* get around or about; in der Welt ~**kommen** see a lot of the world; ~|**laufen** *unr. itr. V.; mit sein* a) walk/*(schneller)* run around or about; um etw. ~**laufen** go round sth.; b) *(gekleidet sein)* wie ein Hippie ~**laufen** go about looking like a hippie; ~|**lungern** *itr. V. (salopp)* loaf around; ~|**schlagen** *unr. refl. V.*

(ugs.) sich mit Problemen/Einwänden ~schlagen grapple with problems/ battle against objections; ~|sein *unr. itr. V.; mit sein; Zusammenschreibung nur im Inf. und Part. (ugs.) (vergangen sein)* have passed; ~|sitzen *unr. itr. V. (ugs.)* sit around *or* about; ~|sprechen *unr. refl. V.* get around *or* about; ~|stöbern *itr. V. (ugs.)* keep rummaging around *or* about (in + *Dat.* in); ~|treiben *unr. refl. V. (ugs. abwertend)* sich auf den Straßen/in Discos ~treiben hang around the streets/in discos; sich in der Welt ~treiben roam about the world

her**unter** *Adv.* a) *(nach unten)* down; b) *(fort)* off; ~ vom Sofa! [get] off the sofa!

her**unter**-: ~|bringen *unr. tr. V.* bring down; ~|fallen *unr. itr. V.; mit sein* fall down; vom Tisch/Stuhl ~fallen fall off the table/chair; ~|gehen *unr. itr. V.; mit sein* a) come down; b) *(niedriger werden)* ⟨*temperature*⟩ drop; ⟨*prices*⟩ come down, fall; ~gekommen 1. 2. *Part. v.* ~kommen; 2. *Adj.* poor ⟨*health*⟩; dilapidated ⟨*building*⟩; run-down ⟨*area*⟩; down and out ⟨*person*⟩; ~|handeln *tr. V. (ugs.)* einen Preis ~handeln beat down a price; ~|hängen *unr. itr. V.* hang down; ~|hauen *unr. tr. V. (ugs.)* jmdm. eine ~hauen give sb. a clout round the ear *(coll.)*; ~|kommen *unr. itr. V.; mit sein* a) come down; b) *(ugs.: verfallen)* go to the dogs *(coll.)*; ~|lassen *unr. tr. V.* lower; ~|schlucken *tr. V.* swallow; ~|sein *unr. itr. V.; mit sein; Zusammenschreibung nur im Inf. und Part. (ugs.)* be down; |körperlich] ~sein be in poor health; ~|spielen *tr. V. (ugs.)* play down *(coll.)*

her**vor** *Adv.* aus ... ~: out of

her**vor**-: ~|heben *unr. tr. V.* stress; ~ragend 1. *Adj.* outstanding[ly good]; 2. *adv.* ~ragend geschult outstandingly well trained; ~ragend spielen/arbeiten play/work outstandingly well; ~|tun *unr. refl. V.* distinguish oneself; *(wichtig tun)* show off

Herz das; ~ens, ~en a) heart; *(Kartenspiel)* hearts *pl.;* von ~en kommen come from the heart; ein ~ für die Armen haben feel for the poor; ein ~ für Kinder haben have a love of children; schweren ~ens with a heavy heart; etw. auf dem ~en haben have sth. on one's mind; es nicht übers ~ bringen, etw. zu tun not have the heart to do sth.; sich *(Dat.)* etw. zu ~en nehmen take sth. to heart; **Herz·anfall** der heart attack; **herzens·gut** ['--'-] *Adj.* kindhearted; **Herzens·lust** die: nach ~: to one's heart's content; **herzhaft** 1. *Adj.* hearty; *(nahrhaft)* hearty ⟨*meal*⟩; *(von kräftigem Geschmack)* tasty; 2. *adv.* heartily; *(nahrhaft)* er ißt gern ~: he likes to have a hearty meal

her|ziehen *unr. itr. V.; mit sein od. haben (ugs.)* über jmdn./etw. ~: run sb./ sth. down

herzig 1. *Adj.* sweet; delightful; 2. *adv.* sweetly; delightfully

herz-, Herz-: ~infarkt der heart attack; ~klopfen das; ~s: jmd. hat ~klopfen sb.'s heart is pounding; ~krank *Adj.* ⟨*person*⟩ with a heart condition

herzlich 1. *Adj.* warm ⟨*smile, reception*⟩; kind ⟨*words, regards*⟩; *(ehrlich gemeint)* sincere; ~en Dank many thanks; 2. *adv.* warmly; *(ehrlich gemeint)* sincerely; ⟨*congratulate*⟩ heartily; ~ wenig very *or (coll.)* precious little; **Herzlichkeit** die warmth; kindness; *(Aufrichtigkeit)* sincerity; **herz·los** 1. *Adj.* heartless; 2. *adv.* heartlessly

Herzog der; ~s, Herzöge duke; **Herzogin** die; ~, ~nen duchess

herz-, Herz-: ~schlag der heartbeat; *(Herzversagen)* heart failure; ~schmerz der; *meist Pl.* pain in the region of the heart; ~transplantation die *(Med.)* heart transplantation; ~zerreißend 1. *Adj.* heart-rending; 2. *adv.* heart-rendingly

Hessen (das); ~s Hesse

Hetze die; ~ a) [mad] rush; b) *o. Pl. (abwertend)* smear campaign; **hetzen** 1. *tr. V.* a) hunt; b) *(antreiben)* rush; 2. *itr. V.* a) *(in großer Eile sein)* rush; b) *mit sein (hasten)* rush; *(rennen)* dash; race; **Hetz·rede** die *(abwertend)* inflammatory speech

Heu das; ~[e]s hay

Heuchelei die; ~: hypocrisy; **heucheln** 1. *itr. V.* be a hypocrite; 2. *tr. V.* feign; **Heuchler** der; ~s, ~: hypocrite; **heuchlerisch** 1. *Adj.* hypocritical; 2. *adv.* hypocritically

heuer *Adv. (südd., österr., schweiz.)* this year

Heuer die; ~, ~n *(Seemannsspr.)* pay; wages *pl.*

Heu·ernte die a) hay harvest; b) *(Ertrag)* hay crop

heulen *itr. V.* **a)** howl; ⟨siren etc.⟩ wail; **b)** *(ugs.: weinen)* howl; bawl

Heurige der; *adj. Dekl. (bes. österr.)* **a)** *(Wein)* new wine; **b)** *(Weinlokal)* inn with new wine on tap

Heu-: ~**schnupfen** der hay fever; ~**schrecke** die grasshopper

heute *Adv.* today; ~ **früh** early this morning; ~ **morgen/abend** this morning/evening; ~ **mittag** [at] midday today; ~ **nacht** tonight; *(letzte Nacht)* last night; ~ **in einer Woche** a week [from] today; today week; ~ **vor einer Woche** a week ago today; **heutig** *Adj.* **a)** *(von diesem Tag)* today's; der ~**e Tag** today; **b)** *(gegenwärtig)* today's; of today *postpos.;* **in der** ~**en Zeit** nowadays; **heut·zu·tage** *Adv.* nowadays

Hexe die; ~, ~**n** witch; **hexen** *itr. V.* work magic

Hexen·schuß der; *o. Pl.* lumbago *no indef. art.;* **Hexerei** die; ~, ~**en** witchcraft; *(von Kunststücken usw.)* magic

hieb *1. u. 3. Pers. Sg. Prät. v.* **hauen**; **Hieb** der; ~[e]s, ~e **a)** *(Schlag)* blow; *(mit der Peitsche)* lash; **b)** *Pl. (ugs.: Prügel)* hiding *sing.;* **hieb·fest** *Adj.:* hieb- und stichfest watertight; cast-iron

hielt *1. u. 3. Pers. Sg. Prät. v.* **halten**

hier *Adv.* **a)** here; |von| ~ **oben/unten** [from] up/down here; **b)** *(jetzt)* now; **von** ~ **an** from now on

hieran *Adv.* here; **sich** ~ **festhalten** hold on to this; *(fig.)* **im Anschluß** ~: immediately after this

Hierarchie [hierar'çi:] die; ~, ~**n** hierarchy

hierauf *Adv.* **a)** on here; *(darauf)* on this; **wir werden** ~ **zurückkommen** we'll come back to this; **b)** *(danach)* after that; then; **c)** *(infolgedessen)* whereupon; **hieraus** *Adv.* out of here; *(aus dieser Tatsache, Quelle)* from this

hier-: ~|**behalten** *unr. tr. V.* jmdn./ etw. ~: keep sb./sth. here; ~**bei** *Adv.* **a)** *(bei dieser Gelegenheit)* Diese Übung ist sehr schwierig. Man kann sich ~ leicht verletzen. This exercise is very difficult. You can easily injure yourself doing it; **b)** *(bei der erwähnten Sache)* here; ~|**bleiben** *unr. itr. V.;* mit sein stay here; ~**durch** *Adv.* through here; *(auf Grund dieser Sache)* because of this; ~**für** *Adv.* for this

hier·her *Adv.* here; **ich gehe bis** ~ **und nicht weiter** I'm going this far and no further

hierher-: ~|**gehören** *itr. V.* belong here; *(hierfür wichtig sein)* be relevant [here]; ~|**kommen** *unr. itr. V.; mit sein* come here

hier·hin *Adv.* here; **bis** ~: up to here

hier-: ~**in** *Adv.* **a)** *(räumlich)* in here; **b)** in this; ~|**lassen** *unr. tr. V.* etw. ~: leave sth. here; ~**mit** *Adv.* with this/ these; ~**mit ist der Fall erledigt** that puts an end to the matter; ~**nach** *Adv. (anschließend)* after that

Hieroglyphe die; ~, ~**n** hieroglyph

hier-: ~**sein** *unr. itr. V.; mit sein; Zusammenschreibung nur im Inf. und Part.* be here; ~**über** *Adv.* **a)** *(über dem Erwähnten)* above here; *(über das Erwähnte)* over here; **b)** *(das Erwähnte betreffend)* about this/these; ~**von** *Adv.* of this/these; ~**zu** *Adv.* with this; *(hinsichtlich dieser Sache)* about this; ~**zu gehört/gehören ...:** this includes/these include; ~**zu reicht mein Geld nicht** I haven't got enough money for that; ~**zu·lande** *Adv.* [here] in this country

hiesig *Adj.; nicht präd.* local

hieß *1. u. 3. Pers. Sg. Prät. v.* **heißen**

Hi-Fi-Anlage ['haifi] die hi-fi system

Hilfe die; ~, ~n **a)** help; *(für Notleidende)* aid; relief; **zu** ~! help!; **b)** *(Hilfskraft)* help; *(im Geschäft)* assistant

Hilfe-: ~**leistung** die help; ~**ruf** der cry for help; ~**stellung** die *(Turnen)* jmdm. ~**stellung geben** act as spotter for sb.

hilflos 1. *Adj.* helpless; **2.** *adv.* helplessly; **Hilflosigkeit** die; ~ helplessness

hilfs-, Hilfs-: ~**bedürftig** *Adj.* **a)** *(schwach)* in need of help *postpos.;* **b)** *(notleidend)* in need; needy; ~**bereit** *Adj.* helpful; ~**bereitschaft** die helpfulness; ~**kraft** die assistant; ~**mittel** das aid; ~**zeit·wort** das *(Sprachw.)* auxiliary [verb]

Himalaja der; ~[s]: der/im ~: the/in the Himalayas *pl.*

Him·beere die raspberry

Himmel der; ~s, ~ sky; *(Rel.)* heaven; ~ **noch |ein|mal!** for Heaven's sake!

Himmel·bett das four-poster bed; **himmel·blau** *Adj.* sky-blue; clear blue ⟨eyes⟩

Himmels-: ~**richtung** die point of the compass; ~**schlüsselchen** das cowslip

himmel·weit *Adj.* enormous, vast
⟨*difference*⟩; **himmlisch** *Adj.* *(auch
fig.)* heavenly

hin *Adv.* **a)** *(räumlich)* zur Straße ~ liegen face the road; **b)** *(zeitlich)* gegen
Mittag ~: towards midday; **c)** *(in Verbindungen)* nach außen ~: outwardly;
auf meinen Rat ~: on my advice; **auf
seine Bitte** ~: at his request; **d)** *(in
Wortpaaren)* ~ und zurück there and
back; **einmal Köln** ~ und zurück a return [ticket] to Cologne; ~ und her to
and fro; back and forth; ~ und wieder
[every] now and then

hinab *Adv. s.* **hinunter**

hinab|- *s.* **hinunter/-**

hinauf *Adv.* up; **bis** ~ **zu** up to

hinauf-: ~|**fahren** *unr. itr. V.; mit sein*
go up; *(im Auto)* drive up; *(mit einem
Motorrad)* ride up; ~|**gehen** *unr. itr.
V.; mit sein* **a)** *(nach oben gehen)* go
up; **b)** *(nach oben führen)* lead up; **c)**
(ugs.: steigen) ⟨*prices, taxes, etc.*⟩ go
up; rise; ~|**klettern** *itr. V.; mit sein*
climb up; ~|**steigen** *unr. itr. V.; mit
sein* climb up; ~|**ziehen** 1. *unr. tr. V.*
pull up; 2. *unr. itr. V.; mit sein* move
up; 3. *unr. refl. V. (sich erstrecken)*
stretch up

hinaus *Adv.* **a)** *(räumlich)* out; **b)** *(zeitlich)* **auf Jahre** ~: for years to come; **c)**
(etw. überschreitend) **über etw.** *(Akk.)*
~: in addition to

hinaus-: ~|**bringen** *unr. tr. V.* jmdn./
etw. ~**bringen** see sb. out/take sth. out
(aus of); ~|**fahren** 1. *unr. itr. V.; mit
sein* **aus etw.** ~**fahren** *(mit dem Auto)*
drive out of sth.; *(mit dem Zweirad)*
ride out of sth.; ⟨*car, bus*⟩ go out of
sth.; ⟨*train*⟩ pull out of sth.; **zum Flugplatz** ~**fahren** drive out to the airport;
2. *unr. tr. V.* jmdn./etw. ~**fahren** drive
sb./take sb. out; ~|**fallen** *unr. itr. V.;
mit sein* fall out (aus of); ~|**finden**
unr. itr. V. find one's way out (aus of);
~|**gehen** *unr. itr. V.; mit sein* **a)** go
out (aus of); **b)** *(gerichtet sein)* **das
Zimmer geht zum Garten/nach Westen**
~: the room looks out on to the gar-
den/faces west; ~|**kommen** *unr. itr.
V.; mit sein* come out (aus of); ~|**lau-
fen** *unr. itr. V.; mit sein* **a)** run out
(aus of); **b)** *(als Ergebnis haben)* **auf
etw.** *(Akk.)* ~**laufen** lead to sth.; ~|**se-
hen** *unr. itr. V.* look out; **zum Fenster**
~**sehen** look out of the window;
~|**sein** *unr. itr. V.; mit sein* **über etw.**
(Akk.) ~**sein** be past sth.; ~|**tragen**
unr. tr. V. jmdn./etw. ~**tragen** carry

sb./sth. out; ~|**werfen** *unr. tr. V.*
(auch ugs. fig.) throw out (aus of);
~|**ziehen** 1. *unr. tr. V.* **a)** *(nach drau-
ßen ziehen)* jmdn./etw. ~**ziehen** pull
sb./sth. out (aus of); tow ⟨*ship*⟩ out; **b)**
(verzögern) put off; delay; 2. *unr. refl.
V.* be delayed; ~|**zögern** 1. *tr. V.*
delay; 2. *refl. V.* be delayed

hin-, Hin-: ~|**blick** der *in* **im** *od.* **in
~blick auf etw.** *(Akk.)* *(wegen)* in view
of; *(hinsichtlich)* with regard to;
~|**bringen** *unr. tr. V.* jmdn./etw.
~**bringen** take sb./sth. [there]; ~|**den-
ken** *unr. itr. V.* **wo denkst du hin?**
(ugs.) whatever are you thinking of?

hinderlich *Adj.* ~ sein get in the way;
hindern *tr. V.* **a)** *(abhalten)* jmdn. ~:
stop sb. (**an** + *Dat.* from); **b)** *(behindern)* hinder; **Hindernis** das; ~**ses**,
~**se** obstacle

hin|deuten *itr. V.* **a)** **auf jmdn./etw.**
od. **zu jmdm./etw.** ~: point to sb./sth.;
b) **auf etw.** *(Akk.)* ~ *(fig.)* point to sth.

hin-durch *Adv.* **a)** *(räumlich)* **durch
den Wald** ~: through the wood; **b)**
(zeitlich) **das ganze Jahr** ~:
throughout the year

hinein *Adv.* **a)** *(räumlich)* in; **in etw.**
(Akk.) ~: into sth.; **b)** *(zeitlich)* **bis in
den Morgen/tief in die Nacht** ~: till
morning/far into the night

hinein-: ~|**bringen** *unr. tr. V.* take in;
~|**fahren** *(mit dem Auto)* drive in;
(mit dem Zweirad) ride in; **in etw.**
(Akk.) ~**fahren** drive/ride into sth.;
~|**fallen** *unr. itr. V.; mit sein* fall in; **in
etw.** *(Akk.)* ~**fallen** fall into sth.;
~|**gehen** *unr. itr. V.; mit sein* go in; **in
etw.** *(Akk.)* ~**gehen** go into sth.;
~|**gucken** *itr. V.* *(ugs.)* look in; **in
etw.** *(Akk.)* ~**gucken** look in[to] sth.;
~|**kommen** *unr. itr. V.; mit sein* **a)**
come in; **in etw.** *(Akk.)* ~**kommen**
come into sth.; **b)** *(gelangen, auch fig.)*
get in; **in etw.** *(Akk.)* ~**kommen** get
into sth.; ~|**reden** *itr. V.* jmdm. in sei-
ne Angelegenheiten/Entscheidungen
usw. ~**reden** interfere in sb.'s affairs/
decisions *etc.*; ~|**sehen** *unr. itr. V.*
look in; **in etw.** *(Akk.)* ~**sehen** look
into sth.; ~|**versetzen** *refl. V.* sich in
jmdn. *od.* jmds. Lage ~**versetzen** put
oneself in sb.'s position

hin-, Hin-: ~|**fahren** 1. *unr. itr. V.;
mit sein* go there; 2. *unr. tr. V.* jmdn.
~**fahren** drive sb. there; ~|**fahrt** die
journey there; *(Seereise)* voyage out;
~|**fallen** *unr. itr. V.; mit sein* **a)** fall
over; **b)** jmdm. fällt etw. ~: sb. drops

sth.; etw. ~**fallen lassen** drop sth.; ~**fällig** *Adj.* **a)** infirm; frail; **b)** *(ungültig)* invalid; ~|**fliegen** *unr. itr. V.; mit sein* fly there; ~**flug** der outward flight

hing *1. u. 3. Pers. Sg. Prät. v.* **hängen**
Hin·gabe die; ~: devotion; *(Eifer)* dedication; **Hingebung** die; ~: devotion; **hingebungs·voll 1.** *Adj.* devoted; **2.** *adv.* devotedly; with devotion; 〈*listen*〉 with rapt attention; 〈*dance, play*〉 with abandon

hin·gegen *Konj., Adv. (jedoch)* however; *(andererseits)* on the other hand

hin-: ~|**gehen** *unr. itr. V.; mit sein* **a)** go [there]; **zu jmdm./etw.** ~**gehen** go to sb./sth.; **b)** *(verstreichen)* 〈*time*〉 go by; ~|**halten** *unr. tr. V.* **a)** hold out; **b)** *(warten lassen)* jmdn. ~**halten** keep sb. waiting; ~|**hören** *itr. V.* listen

hinken ['hɪŋkn̩] *itr. V.* **a)** walk with a limp; **b)** *mit sein (hinkend gehen)* limp

hin-: ~|**kommen** *unr. itr. V.; mit sein* **a)** get there; **b)** *(an einen Ort gehören)* go; belong; **c)** *(ugs.: stimmen)* be right; ~**länglich 1.** *Adj.* sufficient; *(angemessen)* adequate; **2.** *adv.* sufficiently; *(angemessen)* adequately; ~|**legen 1.** *tr. V.* put; *(weglegen)* put down; **2.** *refl. V.* lie down; ~**reichend 1.** *Adj.* sufficient; *(angemessen)* adequate; **2.** *adv.* sufficiently; *(angemessen)* adequately; ~**reise die** journey there; *(mit dem Schiff)* voyage out; ~**reißend** *Adj.* enchanting 〈*person, picture, view*〉; captivating 〈*speaker, play*〉; ~|**richten** *tr. V.* execute; ~**richtung die** execution

Hinrichtungs·kommando das firing-squad

hin-, Hin-: ~|**sehen** *unr. itr. V.* look; ~|**sein** *unr. itr. V.; mit sein (nur im Inf. u. Part. zusammengeschrieben) (ugs.)* **a)** *(verloren sein)* be gone; **b)** *(nicht mehr brauchbar sein)* have had it *(coll.)*; 〈*car*〉 be a write-off; **c)** *(salopp: tot sein)* have snuffed it *(sl.)*; **d)** *(ugs.: hingerissen sein)* **von jmdm./etw. ganz** ~**sein** be mad about sb./bowled over by sth.; ~|**setzen 1.** *tr. V.* put; **2.** *refl. V.* sit down; ~**sicht die;** *o. Pl.* **in gewisser** ~**sicht** in a way/in some respect *or* ways; **in jeder** ~**sicht** in every respect; **in finanzieller** ~**sicht** financially; ~**sichtlich** *Präp. mit Gen. (Amtsspr.)* with regard to; *(in Anbetracht)* in view of; ~|**stellen 1.** *tr. V.* put; put up 〈*building*〉; *(absetzen)* put down; **2.** *refl. V.* stand

hinten *Adv.* at the back; **sich** ~ **anstellen** join the back of the queue *(Brit.) or (Amer.)* line; **weiter** ~: further back; *(in einem Buch)* further on; **die Adresse steht** ~ **auf dem Brief** the address is on the back of the envelope; **nach** ~ **hinaus liegen/gehen** be at the back; **die anderen sind ganz weit** ~: the others are a long way back

hinter 1. *Präp. mit Dat.* behind; *(nach)* after; **3 km** ~ **der Grenze** 3 km beyond the frontier; **eine Prüfung** ~ **sich haben** *(fig.)* have got an examination over [and done] with; **viele Enttäuschungen/eine Krankheit** ~ **sich haben** have experienced many disappointments/ have got over an illness; **2.** *Präp. mit Akk.* behind

hinter... *Adj.; nicht präd.* back

hinter-, Hinter-: ~**einander** *Adv.* **a)** *(räumlich)* one behind the other; **b)** *(zeitlich)* one after another *or* the other; ~**gedanke** der ulterior motive; ~**gehen** [-'--] *unr. tr. V.* deceive; ~**grund** der background; ~**gründig 1.** *Adj.* enigmatic; **2.** *adv.* enigmatically; ~**halt** der ambush; ~**hältig 1.** *Adj.* underhand; **2.** *adv.* in an underhand manner; ~**her** *Adv. (räumlich)* behind; *(nachher)* afterwards; ~**hof** der courtyard; ~**land** das hinterland; *(Milit.)* back area; ~**lassen** [-'--] *unr. tr. V.* leave; ~**legen** [-'--] *tr. V.* deposit (bei with); ~**list** die guile; deceit; ~**listig** *Adj.* deceitful; ~**mann** der; *Pl.* ~**männer** **a)** person behind; **b)** *(Gewährsmann)* [secret] informant

Hintern der; ~s, ~ *(ugs.)* backside; bottom

hinter-, Hinter-: ~**rad** das rear wheel; ~**teil** das backside; behind; ~**treffen das** *(ugs.)* **in ins** ~**treffen geraten** *od.* **kommen** fall behind; ~**treiben** [-'--] *unr. tr. V.* foil 〈*plan*〉; prevent 〈*marriage, promotion*〉; block 〈*law, investigation, reform*〉; ~**treppe** die back stairs *pl.*; ~**tür** die back door; ~**wäldler** [~vɛltlɐ] der; ~s, ~ *(spött.)* backwoodsman

hinüber *Adv.* over; across

Hin- und Rück·fahrt die journey there and back; round trip *(Amer.)*

hinunter *Adv.* down

hinunter-: ~|**fahren 1.** *unr. itr. V.; mit sein* go down; *(mit dem Auto)* drive down; *(mit dem Fahrrad)* ride down; **2.** *unr. tr. V.* **jmdn./ein Auto/eine Ladung** ~**fahren** drive sb. down/

drive a car down/take a load down;
~|**gehen** *unr. itr. V.; mit sein* go
down; ⟨*aircraft*⟩ descend; ~|**klettern**
itr. V.; mit sein climb down; ~|**rei-
chen** 1. *tr. V.* hand down; 2. *itr. V.*
(sich bis hinunter erstrecken) reach
down (**bis auf** + *Akk.* to)

Hin·weg der way there

hin·weg *Adv.* **a)** *(geh.)* ~ **mit dir!**
away with you!; **b)** **über etw.** ~: over
sth.

hinweg-: ~|**gehen** *unr. itr. V.; mit
sein* **über etw.** *(Akk.)* ~**gehen** pass over
sth.; ~|**kommen** *unr. itr. V.; mit sein*
über etw. *(Akk.)* ~**kommen** get over
sth.; ~|**setzen** *refl. V.* **sich über etw.**
(Akk.) ~**setzen** ignore sth.

Hinweis ['hɪnvaɪs] der; ~**es**, ~**e** hint;
unter ~ **auf** (+ *Akk.*) with reference
to

hin-: ~|**weisen** 1. *unr. itr. V.* **auf**
jmdn./etw. ~**weisen** point to sb./sth.;
2. *unr. tr. V.* **jmdn. auf etw.** *(Akk.)*
~**weisen** point sth. out to sb.; ~|**wei-
send** *Adj. (Grammatik)* demonstrat-
ive; ~|**werfen** *unr. tr. V.* throw
down; ~|**ziehen** 1. *unr. tr. V.* pull,
draw (**zu** to, **towards**); 2. *unr. itr. V.;*
mit sein **a)** *(umziehen)* move there; **wo**
ist sie ~**gezogen?** where did she move
to?; 3. *unr. refl. V.* **a)** *(sich erstrecken)*
drag on (**über** + *Akk.* for); **b)** *(sich*
verzögern) be delayed

hinzu-: ~|**fügen** *tr. V.* add; ~|**kom-
men** *unr. itr. V.; mit sein* **a)** come
along; **b)** *(hinzugefügt werden)* **zu etw.**
~**kommen** be added to sth.; **es kommt**
noch ~, **daß** ... *(fig.)* there is also the
fact that ...; ~|**tun** *unr. tr. V. (ugs.)* add

Hirn das; ~|**e|s**, ~**e** **a)** brain; **b)** *(Speise;*
ugs.: Verstand) brains *pl.*

Hirsch der; ~|**e|s**, ~**e** deer; *(Rothirsch)*
red deer; *(männlicher Rothirsch)* stag;
(Speise) venison

Hirse die; ~, ~**n** millet

Hirt der; ~**en**, ~**en**, **Hirte** der; ~**n**, ~**n**
herdsman; *(Schaf~)* shepherd

historisch *Adj.* **a)** historical; **b)** *(geschichtlich bedeutungsvoll)* historic

Hit der; ~|**s|**, ~**s** *(ugs.)* hit

Hitze die; ~: heat

hitze-, Hitze-: ~**beständig** *Adj.*
heat-resistant; ~**frei** *Adj.* ~**frei haben**
have the rest of the day off [school/
work] because of excessively hot
weather; ~**welle** die heat wave

hitzig *Adj.* **a)** hot-tempered; **b)** *(erregt)*
heated ⟨*discussion etc.*⟩

hitz-, Hitz-: ~**kopf** der hothead;
~**köpfig** *Adj.* hot-headed; ~**schlag**
der heat-stroke

hl *Abk.* Hektoliter hl

hob *1. u. 3. Pers. Sg. Prät. v.* heben

Hobby das; ~**s**, ~**s** hobby

Hobel der; ~**s**, ~ **a)** plane; **b)** *(Küchengerät)* [vegetable] slicer; **Ho-
bel·bank** die woodworker's bench;
hobeln *tr., itr. V.* **a)** plane; **b)** *(schneiden)* slice

hoch; höher, höchst... 1. *Adj.* high; tall
⟨*tree, mast*⟩; long ⟨*grass*⟩; deep ⟨*snow,*
water⟩; heavy ⟨*fine*⟩; large ⟨*sum,*
amount⟩; severe, extensive ⟨*damage*⟩;
senior ⟨*official, officer, post*⟩; high-
level ⟨*diplomacy, politics*⟩; **höchste Gefahr** extreme danger; **es ist höchste**
Zeit, daß ...: it is high time that ...; **das**
hohe C top C; **vier** ~ **zwei** *(Math.)* four
to the power [of] two; four squared; 2.
adv. (in großer Höhe) high; *(nach*
oben) up; *(zahlenmäßig viel, sehr)*
highly; ~ **verschuldet/versichert** heavily in debt/insured for a large sum [of
money]; **etw.** ~ **und heilig versprechen**
promise sth. faithfully; **Hoch** das;
~**s**, ~**s** **a)** *(Hochruf)* **ein |dreifaches|** ~
auf jmdn. ausbringen give three cheers
for sb.; **b)** *(Met.)* high

Hoch·achtung die great respect;
hochachtungs·voll *Adv. (Briefschluß)* yours faithfully

hoch-, Hoch-: ~**aktuell** *Adj.* highly
topical; ~**amt** das *(kath. Rel.)* high
mass; ~**arbeiten** *refl. V.* work one's
way up; ~**begabt** *Adj. (präd. getrennt*
geschrieben) highly gifted; ~**betagt**
Adj. aged; ~**betrieb** der; *o. Pl. (ugs.)*
es herrschte ~ **betrieb im Geschäft** the
shop was at its busiest; ~**blüte** die
golden age; ~**burg** die stronghold;
~**deutsch** *Adj.* High German;
~**deutsch** das, ~**deutsche** das
High German; ~**druck** der *(Physik,*
Met.) high pressure; ~**empfindlich**
Adj. highly sensitive ⟨*instrument, device, material, etc.*⟩; fast ⟨*film*⟩; ex-
tremely delicate ⟨*fabric*⟩; ~|**fahren**
unr. itr. V.; mit sein **a)** *(ugs.)* go up;
(mit dem Auto) drive up; *(mit dem*
Fahrrad, Motorrad) ride up; **b)** *(auffahren)* start up; **aus dem Sessel** ~**fahren** start [up] from one's chair; **c)** *(aufbrausen)* flare up; ~**finanz** die high
finance; ~**fliegend** *Adj.* ambitious;
~**form** die top form; ~**gebirge** das
[high] mountains *pl.*; ~**gefühl** das
[feeling of] elation; ~|**gehen** *unr. itr.*

V.; *mit sein (ugs.)* go up; *(zornig werden)* blow one's top *(coll.)*; explode; *(explodieren)* ⟨*bomb, mine*⟩ go off; **~genuß** der *in* ein ~genuß sein be a real delight; **~geschlossen** *Adj.* high-necked ⟨*dress*⟩; **~gestellt** *Adj.*; *nicht präd.* ⟨*person*⟩ in a high position; important ⟨*person*⟩; **~glanz der**: etw. auf ~glanz bringen give sth. a high polish; *(fig.)* make sth. spick and span; **~gradig** 1. *Adj.* extreme; 2. *adv.* extremely; **~|halten** *unr. tr. V.* hold up; **~haus das** high-rise-building; **~|heben** *unr. tr. V.* lift up; raise ⟨*arm, leg, hand*⟩; **~interessant** *Adj.* extremely interesting; **~kant** *Adv. (ugs.)* in jmdn. **~kant hinauswerfen** chuck sb. out *(sl.)*; throw sb. out on his/her ear *(coll.)*; **~|kommen** *unr. itr. V.; mit sein (ugs.)* come up; *(vorwärtskommen)* get on; **~|krempeln** *tr. V.* roll up; **~|leben** *itr. V. in* jmdn./etw. **~leben lassen** cheer sb./sth.; **er lebe ~!** three cheers for him; **~leistungs·sport der** top-level sport; **~modern** *Adj.* ultra-modern; **~mut der** arrogance; **~mütig** *Adj.* arrogant; **~näsig** *Adj. (abwertend)* stuck-up; **~|nehmen** *unr. tr. V. (ugs.: verspotten)* jmdn. **~nehmen** pull sb.'s leg; **~ofen der** blast furnace; **~prozentig** *Adj.* high-proof ⟨*spirits*⟩; **~rechnung die** *(Statistik)* projection; **~ruf der** cheer; **~saison die** high season; **~|schlagen** 1. *unr. tr. V.* turn up ⟨*collar, brim*⟩; 2. *unr. itr. V.; mit sein* ⟨*water, waves*⟩ surge up; *(flames)* leap up; **~schule die** college; *(Universität)* university

Hochsee·fischerei die deep-sea fishing *no art.*

hoch-, Hoch-: **~sitz der** *(Jagdw.)* raised hide; **~sommer der** high summer; **~spannung die** *(Elektrot.)* high voltage; **~|spielen** *tr. V.* blow up

höchst [hø:çst] *Adv.* extremely; most; **höchst...** *s.* hoch

Hoch·stapler [~ʃta:plɐ] der; ~s, ~ confidence trickster; con-man *(coll.)*; *(Aufschneider)* fraud

höchstens *Adv.* at most; *(bestenfalls)* at best

Höchst-: **~fall der** *in* im ~fall at [the] most; **~form die** *(bes. Sport)* peak form; **~geschwindigkeit die** top speed; *(Geschwindigkeitsbegrenzung)* speed limit

Hoch·stimmung die high spirits *pl.*

höchst-, Höchst-: **~leistung die** supreme performance; *(Ergebnis)* supreme achievement; **~maß das**: ein ~maß an etw. *(Dat.)* a very high degree of sth.; **~wahrscheinlich** *Adv.* very probably

hoch-, Hoch-: **~tour die**: auf ~touren laufen run at full speed; *(intensiv betrieben werden)* be in full swing; **~trabend** *(abwertend)* 1. *Adj.* high-flown; 2. *adv.* in a high-flown manner; **~|treiben** *unr. tr. V.* force up ⟨*prices etc.*⟩; **~verrat der** high treason; **~wasser das** *(Flut)* high tide; *(Überschwemmung)* flood; **~wertig** *Adj.* high-quality ⟨*goods*⟩; highly nutritious ⟨*food*⟩; **~würden** *o. Art.*; **~|s|** *(veralt.)* Reverend Father

Hochzeit die; ~, ~en wedding

Hochzeits-: **~feier die** wedding; **~nacht die** wedding night; **~reise die** honeymoon [trip]

Hocke die; ~, ~n **a)** *(Körperhaltung)* squat; crouch; **b)** *(Turnen)* squat vault; **hocken** 1. *itr. V.* **a)** *mit haben od. (südd.) sein* squat; crouch; **b)** *mit haben od. (südd.) sein (ugs.: sich aufhalten)* sit around; 2. *refl. V.* crouch down; **Hocker der**; ~s, ~: stool

Höcker der; ~s, ~: hump; *(auf der Nase)* bump; *(auf dem Schnabel)* knob

Hockey ['hɔki] **das**; ~s hockey

Hoden der; ~s, ~: testicle

Hof der; ~[e]s, Höfe **a)** courtyard; *(Schul~)* playground; *(Gefängnis~)* [prison] yard; **b)** *(Bauern~)* farm; **c)** *(Herrscher, Hofstaat)* court

Hof·dame die lady of the court; *(Begleiterin der Königin)* lady-in-waiting; **hof·fähig** *Adj.* presentable at court *pred.*

hoffen 1. *tr. V.* hope; 2. *itr. V.* hope; auf etw. *(Akk.)* ~: hope for sth.; *(Vertrauen setzen auf)* auf jmdn./etw. ~: put one's faith in sb./sth.; **hoffentlich** *Adv.* hopefully; ~! let's hope so; **Hoffnung die**; ~, ~en hope

hoffnungs-, Hoffnungs-: **~los** 1. *Adj.* hopeless; despairing ⟨*person*⟩; 2. *adv.* hopelessly; **~losigkeit die**; ~: despair; *(der Lage)* hopelessness; **~voll** 1. *Adj.* **a)** hopeful; full of hope *pred.*; **b)** *(erfolgversprechend)* promising; 2. *adv.* **a)** full of hope; **b)** *(erfolgversprechend)* promisingly

höflich 1. *Adj.* polite; 2. *adv.* politely; **Höflichkeit die**; ~: politeness

hohe ['ho:ə] *s.* hoch; **Höhe** ['ho:ə] **die**; ~, ~n height; etw. in die ~ heben lift sth. up; **das ist ja die ~!** *(fig. ugs.)* that's the limit

Hoheit die; ~, ~en sovereignty (**über** + *Akk.* over); **Seine/Ihre ~:** His/ Your Highness

Hoheits-: ~gebiet das [sovereign] territory; **~gewässer das;** *meist Pl.* territorial waters

Höhen-: ~flug der *(fig.)* flight; **~lage die** altitude; **~luft die;** *o. Pl.* mountain air; **~messer der** altimeter; **~sonne die** *(Med.)* sun lamp; **~unterschied der** difference in altitude

Höhepunkt der high point; *(einer Veranstaltung)* high spot; *(einer Laufbahn, des Ruhms)* pinnacle; *(Orgasmus)* climax

höher ['hø:ɐ] *s.* **hoch**

hohl *Adj.* hollow; **Höhle die; ~, ~n a)** cave; *(größer)* cavern; **b)** *(Tierbau)* lair

Hohl-: ~maß das measure of capacity; **~raum der** cavity; [hollow] space; **~spiegel der** concave mirror

Hohn der; ~[e]s scorn; derision; **höhnen** *(geh.) itr. V.* jeer; **höhnisch 1.** *Adj.* scornful; **2.** *adv.* scornfully

Hokuspokus der; ~: hocus-pocus; *(abwertend: Drum und Dran)* fuss

hold *Adj. (dichter. veralt.)* fair; lovely; lovely ⟨sight, smile⟩

holen 1. *tr. V.* **a)** fetch; get; **b)** *(ab~)* fetch; **c)** *(ugs.: erlangen)* get ⟨prize etc.⟩; carry off ⟨medal, trophy, etc.⟩; **2.** *refl. V. (ugs.: sich zuziehen)* catch; **sich** *(Dat.)* **[beim Baden] einen Schnupfen ~:** catch a cold [swimming]

Holland (das); ~s Holland; **Holländer der; ~s, ~:** Dutchman; **holländisch** *Adj.* Dutch

Hölle die; ~, ~n hell *no art.;* **Höllen·lärm der** *(ugs.)* diabolical noise or row *(coll.);* **höllisch 1.** *Adj.* **a)** infernal; ⟨spirits, torments⟩ of hell; **b)** *(ugs.: sehr groß)* tremendous *(coll.);* **2.** *adv. (ugs.: sehr)* hellishly *(coll.)*

Holm der; ~[e]s, ~e a) *(Turnen)* bar

holpern *itr. V. mit sein (fahren)* jolt; bump; **holprig** *Adj.* **a)** bumpy; rough; **b)** *(stockend)* halting ⟨speech⟩; clumsy ⟨verses, style, language, etc.⟩

Holunder der; ~s, ~: elder

Holz das; ~es, Hölzer wood; *(Bau~, Tischler~)* timber; wood; **Holz·bein das** wooden leg; **hölzern** *Adj. (auch fig.)* wooden; **Holz·fäller der** woodcutter; lumberjack *(Amer.);* **holzfrei** *Adj.* wood-free ⟨paper⟩; **holzig** *Adj.* woody

Holz-: ~klotz der block of wood; *(als Spielzeug)* wooden block; **~kohle die** charcoal; **~kopf der** *(salopp abwertend)* blockhead; **~pantoffel der** clog; **~scheit das** piece of wood; *(Brenn~)* piece of firewood; **~schnitt der a)** *o. Pl.* woodcutting *no art.;* **b)** *(Blatt)* woodcut; **~schuh der** clog; **~stoß der** pile of wood; **~weg der: auf dem ~weg sein** be on the wrong track *(fig.);* **~wolle die;** *o. Pl.* wood-wool; **~wurm der** woodworm

homogen *Adj.* homogeneous

homöopathisch *Adj.* homoeopathic

Homo·sexualität die; ~: homosexuality; **homo·sexuell 1.** *Adj.* homosexual; **2.** *adv.* **~ veranlagt sein** have homosexual tendencies

Honig der; ~s, ~e honey; **Honig·kuchen der** honey cake; **Honig·wabe die** honeycomb

Honorar das; ~s, ~e fee; *(Autoren~)* royalty; **Honoratioren** [honora-'tsio:rən] *Pl.* notabilities; **honorieren** *tr. V.* **a)** jmdn. **~:** pay sb. [a/his/her fee]; **b)** *(würdigen)* appreciate; *(belohnen)* reward

Hopfen der; ~s, ~: hop

hopp *Interj.* quick; look sharp; **hoppeln** *itr. V.; mit sein* hop; (**über** + *Akk.* across, over); **hoppla** *Interj.* oops; whoops; **hopsen** *itr. V.; mit sein (ugs.) (springen)* jump; *(hüpfen)* ⟨animal⟩ hop; ⟨child⟩ skip; ⟨ball⟩ bounce; **Hopser der; ~s, ~** *(ugs.)* [little] jump

Hör·apparat der hearing-aid; **hörbar 1.** *Adj.* audible; **2.** *adv.* audibly; *(geräuschvoll)* noisily; **horchen** *itr. V.* listen (**auf** + *Akk.* to); *(heimlich zuhören)* eavesdrop

Horde die; ~, ~n horde; *(von Halbstarken)* mob

hören 1. *tr. V.* hear; *(anhören)* listen to; **2.** *itr. V.* hear; *(zuhören)* listen; **auf jmdn./jmds. Rat ~:** listen to sb./sb.'s advice; **Hören·sagen das: vom ~:** from hearsay; **Hörer der; ~s, ~ a)** listener; **b)** *(Telefon~)* receiver

Hör-: ~fehler der a) das war ein ~fehler he/she *etc.* misheard; **b)** *(Schwerhörigkeit)* hearing defect; **~funk der** radio; **im ~funk** on the radio; **~gerät das** hearing-aid

hörig *Adj.:* jmdm. **~ sein** be submissively dependent on sb.; *(sexuell)* be sexually enslaved to sb.

Horizont der; ~[e]s, ~e *(auch Geol., fig.)* horizon; **horizontal 1.** *Adj.*

horizontal; **2.** *adv.* horizontally; **Horizontale die; ~, ~n a)** *(Linie)* horizontal line; **b)** *o. Pl. (Lage)* **die ~:** the horizontal
Hormon das; ~s, ~e hormone
Horn das; ~|e|s, Hörner horn; **Hörnchen das; ~s, ~** *(Gebäck)* croissant; **Horn·haut die a)** callus; hard skin *no indef. art.;* **b)** *(am Auge)* cornea
Hornisse die; ~, ~n hornet
Horoskop das; ~s, ~e horoscope
Hör·rohr das stethoscope
Horror der; ~s horror
Hör-: ~saal der lecture theatre *or* hall; **~spiel das** radio play
Horst der; ~|e|s, ~e eyrie
Hort der; ~|e|s, ~e *s.* **Kinderhort; horten** *tr. V.* hoard; stockpile ⟨*raw materials*⟩
Hortensie die; ~, ~n hydrangea
Hör·weite die: in/außer ~weite in/out of earshot
Höschen ['hø:sçən] **das; ~s, ~:** trousers *pl.;* pair of trousers; *(kurzes ~)* shorts *pl.;* pair of shorts; **Hose die; ~, ~n a)** trousers *pl.;* pants *pl. (Amer.); (Unter~)* pants *pl.; (Freizeit~)* slacks *pl.; (Bund~)* breeches *pl.; (Reit~)* riding breeches *pl.;* **eine ~:** a pair of trousers/pants/slacks *etc.*
Hosen-: ~an·zug der trouser suit *(Brit.);* pant suit; **~matz der** *(ugs. scherzh.)* toddler; **~rock der** culottes *pl.;* **~tasche die** trouser-pocket; pants pocket *(Amer.);* **~träger** *Pl.* braces; suspenders *(Amer.);* pair of braces/suspenders
Hospital das; ~s, ~e *od.* **Hospitäler** hospital
Hostie ['hɔstiə] **die; ~, ~n** *(christl. Rel.)* host
Hotel das; ~s, ~s hotel; **Hotel·bar die** hotel bar; **Hotel garni** [~ gar'ni:] **das; ~ ~, ~s ~s** bed-and-breakfast hotel; **Hotelier** [hotɛ'lie:] **der; ~s, ~s** hotelier
hüben *Adv.* over here
hübsch 1. *Adj.* pretty; nice ⟨*area, flat, voice, tune, etc.*⟩; nice-looking ⟨*boy, person*⟩; **ein ~es Sümmchen** *(ugs.)* a tidy sum *(coll.);* a nice little sum; **das ist eine ~e Geschichte** *(ugs. iron.)* this is a fine *or* pretty kettle of fish *(coll.);* **2.** *adv.* prettily; *(ugs.: sehr)* **~ kalt** perishing cold
Hub·schrauber der; ~s, ~: helicopter
huckepack *Adv.* **jmdn. ~ tragen** *(ugs.)* give sb. a piggyback

hudeln *itr. V. (bes. südd., österr.)* be sloppy **(bei in)**
Huf der; ~|e|s, ~e hoof
Huf-: ~eisen das horseshoe; **~schmied der** farrier
Hüfte die; ~, ~n hip
Hüft-: ~gelenk das *(Anat.)* hip-joint; **~gürtel der** girdle
Hügel der; ~s, ~ hill; **hügelig** *Adj.* hilly
Huhn das; ~|e|s, Hühner a) chicken; *(Henne)* chicken; hen; **Hühnchen das; ~s, ~:** small chicken; **mit jmdm. |noch| ein ~ zu rupfen haben** *(ugs.)* [still] have a bone to pick with sb.
Hühner-: ~auge das *(am Fuß)* corn; **~brühe die** chicken broth
hui [hui] *Interj.* whoosh
huldigen *itr. V.* **jmdm. ~:** pay tribute to sb.; **Huldigung die; ~, ~en** tribute
Hülle die; ~, ~n cover; **hüllen** *tr. V. (geh.)* wrap
Hülse die; ~, ~n a) case; **b)** *(Bot.)* pod
human *Adj.* humane; **Humanismus der; ~:** humanism; *(Epoche)* Humanism *no art.;* **humanitär** *Adj.* humanitarian
Humbug ['humbʊk] **der; ~s** *(ugs.)* humbug
Hummel die; ~, ~n bumble-bee
Hummer der; ~s, ~: lobster
Humor der; ~s humour; *(Sinn für ~)* sense of humour; **den ~ nicht verlieren** remain good-humoured; **Humorist der; ~en, ~en a)** *(Autor)* humorist; **b)** *(Vortragskünstler)* comedian; **humoristisch** *Adj.* humorous
humor-: ~los *Adj.* humourless; **~voll** *Adj.* humorous
humpeln *itr. V.* **a)** *auch mit* sein walk with a limp; **b)** *mit* sein *(sich ~d fortbewegen)* limp
Hund der; ~es, ~e a) dog; **auf den ~ kommen** *(ugs.)* go to the dogs *(coll.);* **vor die ~e gehen** *(ugs.)* go to the dogs *(coll.); (sterben)* kick the bucket *(sl.);* **b)** *(abwertend)* bastard *(coll.)*
hunde-, Hunde-: ~elend *Adj.; nicht attr. (ugs.)* [really] wretched *or* awful; **~hütte die** [dog-]kennel; **~kuchen der** dog-biscuit; **~müde** *Adj.; nicht attr. (ugs.)* dog-tired; **~rasse die** breed of dog
hundert *Kardinalz.* **a)** a *or* one hundred; **b)** *(ugs.: viele)* hundreds of; **[1]Hundert das; ~s, ~e** hundred; **[2]Hundert die; ~, ~en** hundred; **Hunderter der; ~s, ~** *(ugs.)* hundred-mark/-dollar *etc.* note; **hun-**

dert·mal *Adv.* a hundred times; auch wenn du dich ~ beschwerst *(ugs.)* however much you complain

Hundert-: ~mark·schein der hundred-mark note; ~meter·lauf der *(Leichtathletik)* hundred metres *sing.*

hundert·prozentig 1. *Adj.* a) *(von 100%)* [one-]hundred per cent *attrib.;* b) *(ugs.: völlig)* a hundred per cent; c) *(ugs.: ganz sicher)* absolutely reliable; 2. *adv. (ugs.)* ich bin nicht ~ sicher I'm not a hundred per cent sure; hundertst... ['hʊndɐtst...] *Ordinalz.* hundredth; hundertstel ['hʊndɐtstl] *Bruchz.* hundredth; Hundertstel das *(schweiz. meist* der*);* ~s, ~: hundredth; hundert·tausend *Kardinalz.* a or one hundred thousand

Hüne der; ~n, ~n giant; Hünen-grab das megalithic tomb; *(Hügelgrab)* barrow

Hunger der; ~s a) ~ bekommen/haben get/be hungry; b) *(geh.: Verlangen)* hunger; *(nach Ruhm, Macht)* craving; Hunger·kur die starvation diet; hungern *itr. V.* go hungry; starve; nach etw. ~: *(fig.)* hunger for sth.; Hungers·not die famine; Hunger·streik der hunger-strike; hungrig *Adj. (auch geh. fig.)* hungry (nach for)

Hupe die; ~, ~n horn; hupen *itr. V.* sound one's horn; dreimal ~: hoot three times

hüpfen *itr. V.; mit sein* hop; *⟨ball⟩* bounce

Hürde die; ~, ~n hurdle; Hürden-lauf der *(Leichtathletik)* hurdling; *(Wettbewerb)* hurdles *pl.*

Hure die; ~, ~n *(abwertend)* whore; huren *itr. V. (abwertend)* whore

hurra *Interj.* hurray; hurrah; ~ schreien cheer; Hurra das; ~s, ~s cheer

hurtig 1. *Adj.* rapid; 2. *adv.* quickly

huschen *itr. V.; mit sein (lautlos u. leichtfüßig) ⟨person⟩* steal; *(lautlos u. schnell) ⟨mouse, lizard, etc.⟩* dart; *⟨smile⟩* flit; *⟨light⟩* flash; *⟨shadow⟩* slide quickly

hüsteln *itr. V.* give a slight cough; husten 1. *itr. V.* cough; *(Husten haben)* have a cough; 2. *tr. V.* cough up *⟨blood, phlegm⟩;* Husten der; ~s, ~: cough

Husten-: ~bonbon das cough-drop; ~tropfen *Pl.* cough-drops

¹Hut der; ~es, Hüte hat; *(fig.)* da geht einem/mir der ~ hoch *(ugs.)* it makes you/me mad *(coll.);* das kann er sich *(Dat.)* an den ~ stecken *(ugs. abwertend)* he can keep it *(coll.)*

²Hut die; ~ *(geh.)* keeping;.care; auf der ~ sein be on one's guard; hüten 1. *tr. V.* look after; tend *⟨sheep, cattle, etc.⟩;* 2. *refl. V.* be on one's guard

Hut·schnur die *in das geht mir über die ~ (ugs.)* that's going too far

Hütte die; ~, ~n a) hut; *(ärmliches Haus)* shack; hut; b) *(Eisen~)* iron [and steel] works *sing. or pl.;* c) *(Jagd~)* [hunting-]lodge

Hütten-: ~käse der cottage cheese; ~schuh der slipper-sock

Hyäne die; ~, ~n hyena

Hyazinthe die; ~, ~n hyacinth

Hydrant der; ~en, ~en hydrant; Hydraulik die; ~ *(Technik)* a) *(Theorie)* hydraulics *sing., no art.;* b) *(Vorrichtungen)* hydraulics *pl.;* hydraulisch *(Technik)* 1. *Adj.* hydraulic; 2. *adv.* hydraulically; Hydro·kultur die; ~, ~en *(Gartenbau)* hydroponics *sing.*

Hygiene die; ~ a) *(Gesundheitspflege)* health care; b) *(Sauberkeit)* hygiene; hygienisch 1. *Adj.* hygienic; 2. *adv.* hygienically

Hymne ['hymnə] die; ~, ~n hymn; *(National~)* national anthem

Hypnose die; ~, ~n hypnosis; hypnotisieren *tr. V.* hypnotize

Hypochonder [hypo'xɔndɐ] der; ~s, ~: hypochondriac

Hypotenuse die; ~, ~n *(Math.)* hypotenuse

Hypothek die; ~, ~en *(Bankw.)* mortgage; *(fig.)* burden

Hypothese die; ~, ~n hypothesis; hypothetisch 1. *Adj.* hypothetical; 2. *adv.* hypothetically

Hysterie die; ~, ~n hysteria; hysterisch 1. *Adj.* hysterical; 2. *adv.* hysterically

I

i, I das; ~, ~: i/I; das Tüpfelchen auf dem i *(fig.)* the final touch

i *Interj.* ugh; i bewahre, i wo *(ugs.)* [good] heavens, no!

i. A.

574

i. A. *Abk.* im Auftrag[e] p.p.
IC *Abk.* Intercity IC
ich *Personalpron.; 1. Pers. Sg. Nom.* I;
immer ~ *(ugs.)* [it's] always me; ~
nicht not me; Menschen wie du und ~:
people like you and me; *s. auch (Gen.)*
meiner, *(Dat.)* mir, *(Akk.)* mich
Ich das; ~[s], ~[s] a) self; b) *(Psych.)*
ego
Ich-Form die; *o. Pl.* first person
ideal 1. *Adj.* ideal; 2. *adv.* ideally;
Ideal das; ~s, ~e ideal
Ideal-: ~bild das ideal; ~fall der
ideal case; ~gewicht das ideal
weight
idealisieren *tr. V.* idealize; **Idealismus** der; ~ *(auch Philos.)* idealism;
Idealist der; ~en, ~en idealist;
idealistisch *(auch Philos.)* 1. *Adj.*
idealistic; 2. *adv.* idealistically; **Idee**
die; ~, ~n a) idea; b) *(ein bißchen)* eine ~: a shade; eine ~ [Salz/Pfeffer] a
touch [of salt/pepper]; **ideell** *Adj.*
non-material; *(geistig-seelisch)* spiritual; **ideen·los** *Adj.* [completely]
lacking in ideas *postpos.*
Identifikation [identifika'tsio:n] die;
~, ~en *(auch Psych.)* identification;
identifizieren 1. *tr. V.* identify; 2.
refl. V. (auch Psych.) sich mit jmdm./
etw. ~: identify with sb./sth.; **identisch** *Adj.* identical; **Identität** die;
~: identity
Ideologe der; ~n, ~n ideologue;
Ideologie die; ~, ~n [-i:ən] ideology;
ideologisch 1. *Adj.* ideological; 2.
adv. ideologically
Idiot der; ~en, ~en *(auch ugs. abwertend)* idiot; **Idioten·hügel** der *(ugs.
scherzh.)* nursery slope; **Idiotie** die;
~, ~n [-i:ən] a) idiocy; b) *(ugs. abwertend: Dummheit)* madness; **Idiotin**
die; ~, ~nen *(auch ugs. abwertend)*
idiotisch 1. *Adj.* a) *(Psych.)* severely
subnormal; b) *(ugs. abwertend)* idiotic; 2. *adv. (auch ugs. abwertend)*
idiotically
Idol das; ~s, ~e *(auch bild. Kunst)* idol
Idyll das; ~s, ~e idyll; **Idylle** die; ~,
~n idyll; **idyllisch** *Adj.* idyllic
Igel der; ~s, ~: hedgehog
Iglu der *od.* das; ~s, ~s igloo
Ignoranz [igno'rants] die; ~: ignorance; **ignorieren** *tr. V.* ignore
ihm *Dat. von* er, es: *(bei männlichen
Personen)* him; *(bei weiblichen Personen)* her; *(bei Dingen, Tieren)* it; gib es
~: give it to him; give him it; Freunde
von ~: friends of his

ihn *Akk. von* er *(bei männlichen Personen)* him; *(bei Dingen, Tieren)* it
ihnen *Dat. von* sie, *Pl.* them; gib es ~:
give it to them; give them it; **Freunde
von ~:** friends of theirs
Ihnen *Dat. von* Sie you; ich habe es ~
gegeben I gave it to you; **Freunde von
~:** friends of yours
¹ihr [i:ɐ] *Dat. von* sie, *Sg. (bei Personen)*
her; *(bei Dingen, Tieren)* it
²ihr, *(in Briefen)* **Ihr** *Personalpron.; 2.
Pers. Pl. Nom.* you
³ihr *Possessivpron.* a) *Sg. (einer Person)*
her; *(eines Tieres, einer Sache)* its; b)
Pl. their
Ihr *Possessivpron. (Anrede)* your; ~
Hans Meier *(Briefschluß)* yours, Hans
Meier; welcher Mantel ist ~er? which
coat is yours?
ihrer a) *Gen. von* sie, *Sg. (geh.)* wir ge-
dachten ~: we remembered her; b)
Gen. von sie, *Pl. (geh.)* wir werden ~
gedenken we will remember them; es
waren ~ zwölf there were twelve of
them
Ihrer *Gen. von* Sie *(geh.)* wir werden ~
gedenken we will remember you
ihrerseits *Adv.* for her/their part;
(von ihr/ihnen) on her/their part
Ihrerseits *Adv. s.* deinerseits
ihres·gleichen *indekl. Pron.* people
pl. like her/them; *(abwertend)* the
likes of her/them
Ihresgleichen *indekl. Pron.* people
pl. like you; *(abwertend)* the likes of
you
ihret·wegen *Adv.: s.* meinetwegen:
because of her/them; for her/their
sake; about her/them; as far as she is/
they are concerned
Ihretwegen *Adv.: s.* deinetwegen
Ikone die; ~, ~n icon
illegal 1. *Adj.* illegal; 2. *adv.* illegally
Illegalität die; ~, ~en illegality; **ille-
gitim** ['ɪlegiti:m] *Adj. (geh.)* illegitimate
illuminieren *tr. V.* illuminate
Illusion die; ~, ~en illusion; **illusorisch** *Adj.* illusory; *(zwecklos)* pointless
Illustration [ɪlustra'tsio:n] die; ~, ~en
illustration; **illustrieren** *tr. V.* illustrate; **Illustrierte die;** *adj. Dekl.* magazine
Iltis der; ~ses, ~se polecat; *(Pelz)* fitch
im *Präp. + Art.* a) = in dem; b) *(räumlich)* in the; im Theater at the theatre;
im Fernsehen on television; im Bett in
bed; c) *(zeitlich)* im Mai in May; im

letzten **Jahr** last year; **im Alter von ...** at the age of ...; d) *(Verlauf)* etw. **im Sitzen tun** do sth. [while] sitting down; **im Gehen sein** be going

Image ['ɪmɪtʃ] **das; ~[s], ~s** ['ɪmɪtʃs] image; **imaginär** *Adj. (geh., Math.)* imaginary

Ịmbiß der; Ịmbisses, Ịmbisse a) *(kleine Mahlzeit)* snack; **b)** *s.* **Imbißstube; Ịmbiß·stube die** café

Imitation die; ~, ~en imitation; **imitieren** *tr. V.* imitate

Ịmker der; ~s, ~: bee-keeper

Immatrikulation [ɪmatrikulaˈtsi̯oːn] **die; ~, ~en** *(Hochschulw.)* registration; **immatrikulieren** *tr., refl. V. (Hochschulw.)* register

immer *Adv.* **a)** always; **schon ~:** always; **~ wieder** time and time again; **~, wenn** every time that; **b) ~ +** *Komp.* **~ dunkler** darker and darker; **~ mehr** more and more; **c)** *(ugs.: jeweils)* **~ drei Stufen auf einmal** three steps at a time; **d)** *(auch)* **wo/wer/wann/wie [auch] ~:** wherever/whoever/whenever/however; **e)** *(verstärkend)* **~ noch, noch ~:** still; **f)** *(ugs.: bei Aufforderung)* **~ geradeaus!** keep [going] straight on

immer-, Ịmmer-: ~fọrt *Adv.* all the time; **~grün** *Adj.* evergreen; **~grün das** periwinkle; **~hịn** *Adv.* **a)** *(wenigstens)* at any rate; **b)** *(trotz allem)* all the same; **c)** *(schließlich)* after all; **~zu** *Adv. (ugs.)* the whole time

Immigrạnt der; ~en, ~en immigrant; **Immigration** [ɪmigraˈtsi̯oːn] **die; ~, ~en** immigration; **immigrieren** *itr. V.; mit sein* immigrate

Immobịlien *Pl.* property *sing.;* real estate *sing.*

immụn a) *(Med., fig.)* immune (**gegen** to); **b)** *(Rechtsspr.)* **~ sein** have immunity; **Immunität die; ~, ~en a)** *(Med.)* immunity (**gegen** to); **b)** *(Rechtsspr.)* immunity (**gegen** from)

Ịmperativ der; ~s, ~e a) *(Sprachw.)* imperative; **b)** *(Philos.)* **[kategorischer] ~:** [categorical] imperative

Ịmperfekt das; ~s, ~e *(Sprachw.)* imperfect [tense]

Imperialịsmus der; ~: imperialism *no art.;* **imperialịstisch** *Adj.* imperialistic

Imperium das; ~s, Imperien *(hist., fig.)* empire

impfen *tr. V.* vaccinate; inoculate

Impf-: ~paß der vaccination certificate; **~stoff der** vaccine

Impfung die; ~, ~en vaccination

implantieren *tr. V. (Med.)* implant

imponieren *itr. V.* impress; **imponierend 1.** *Adj.* impressive; **2.** *adv.* impressively

Import der; ~[e]s, ~e import; **Importeur** [ɪmpɔrˈtøːɐ̯] **der; ~s, ~e** importer; **importieren** *tr., itr. V.* import

imposạnt 1. *Adj.* imposing; impressive ⟨*achievement*⟩; **2.** *adv.* imposingly

impotent *Adj.* impotent; **Ịmpotenz die; ~:** impotence

imprägnieren *tr. V.* impregnate; *(wasserdicht machen)* waterproof

Improvisation die; ~, ~en improvisation; **improvisieren** *tr., itr. V.* improvise

Impụls der; ~es, ~e stimulus; *(innere Regung)* impulse; **impulsịv 1.** *Adj.* impulsive; **2.** *adv.* impulsively

imstạnde *Adv.* **~ sein, etw. zu tun** be able to do sth.

ịn 1. *Präp. mit Dat.* **a)** *(auf die Frage: wo?/wann?/wie?)* in; **er hat ~ Tübingen studiert** he studied at Tübingen; *s. auch* **im; 2.** *Präp. mit Akk. (auf die Frage: wohin?)* into; *s. auch* **im**

Ịn·anspruchnahme die; ~, ~n *(starke Belastung)* demands *pl.*

Ịn·begriff der quintessence; **ịnbegriffen** *Adj.* included

Ịn·betriebnahme die; ~, ~n, Ịn·betriebsetzung die; ~, ~en *(Amtsspr.)* opening; *(von Maschinen)* bringing into service

Ịn·brunst die; ~ *(geh.)* fervour; *(der Liebe)* ardour; **ịn·brünstig** *(geh.)* **1.** *Adj.* fervent; ardent ⟨*love*⟩; **2.** *adv.* fervently; ⟨*love*⟩ ardently

in·dẹm *Konj.* **a)** *(während)* while; *(gerade als)* as; **b)** *(dadurch, daß)* **~ man etw. tut** by doing sth.

Inder ['ɪndɐ] **der; ~s, ~:** Indian

in·dẹssen 1. *Konj. (geh.)* **a)** *(während)* while; **b)** *(wohingegen)* whereas; **2.** *Adv.* **a)** *(inzwischen)* meanwhile; in the mean time; **b)** *(jedoch)* however

Index der; ~ od. ~es, ~e od. Indizes a) *Pl.* **~e od. Indizes** *(Register)* index; **b)** *Pl.* **~e** *(kath. Kirche)* Index

Indianer der; ~s, ~: [American] Indian; **Indianer·häuptling der** Indian chief

Indien ['ɪndi̯ən] **(das)** India

ịn·different *Adj.* indifferent

Ịndikativ der; ~s, ~e [-iːvə] *(Sprachw.)* indicative [mood] in·direkt 1. *Adj.; nicht präd.* indirect; 2. *adv.* indirectly

indisch *Adj.* Indian

in·diskret *Adj.* indiscreet; In·diskretion die; ~, ~en indiscretion

Individualist der; ~en, ~en *(geh.)* individualist; Individualität die; ~, ~en *(geh.)* a) *o. Pl.* individuality; b) *(Persönlichkeit)* personality; individuell 1. *Adj.* individual; private ⟨*property, vehicle, etc.*⟩; 2. *adv.* individually; Individuum das; ~s, Individuen *(auch Chemie, Biol.)* individual

Indiz das; ~es, ~ien a) *(Rechtsw.)* piece of circumstantial evidence; ~ien circumstantial evidence *sing.;* b) *(Anzeichen)* sign *(für of)*

indoktrinieren *tr. V.* indoctrinate

Indonesien [ɪndoˈneːziən] *(das)* Indonesia; Indonesier der; ~s, ~ Indonesian; indonesisch *Adj.* Indonesian

industrialisieren *tr. V.* industrialize; Industrialisierung die; ~: industrialization; Industrie die; ~, ~n industry

Industrie-: ~betrieb der industrial firm; ~gebiet das industrial area; ~kaufmann der *person with three years' business training employed on the business side of an industrial company*

industriell 1. *Adj.* industrial; 2. *adv.* industrially; Industrielle der/die; *adj. Dekl.* industrialist

Industrie-: ~staat der industrial nation; ~stadt die industrial town; ~zweig der branch of industry

in·einander *Adv.* ~ verliebt sein be in love with each other *or* one another; ~ verschlungene Ornamente intertwined decorations; ineinander|greifen *unr. itr. V.* mesh together *(lit. or fig.)*

infam 1. *Adj.* disgraceful; 2. *adv.* disgracefully

Infanterie die; ~, ~n *(Milit.)* infantry

Infarkt der; ~[e]s, ~e *(Med.)* infarction

Infekt der; ~[e]s, ~e *(Med.)* infection

Infektion [ɪnfɛkˈtsi̯oːn] die; ~, ~en *(Med.)* a) *(Ansteckung)* infection; b) *(ugs.: Entzündung)* inflammation

Infektions-: ~gefahr die *(Med.)* risk of infection; ~herd der *(Med.)* seat of the/an infection; ~krankheit die *(Med.)* infectious disease

Inferno das; ~s *(geh.)* inferno

Infinitiv der; ~s, ~e *(Sprachw.)* infinitive

infizieren 1. *tr. V.* infect; 2. *refl. V.* become infected; sich bei jmdm. ~: be infected by sb.

in flagranti *Adv. (geh.)* in flagrante [delicto]

Inflation die; ~, ~en *(Wirtsch.)* inflation; *(Zeit der ~)* period of inflation

in·folge 1. *Präp.* + *Gen.* as a result of; 2. *Adv.* ~ von etw. *(Dat.)* as a result of sth.

infolge·dessen *Adv.* consequently

Informatik die; ~: computer science *no art.;* Information die; ~, ~en a) information *no pl., no indef. art.* (über + *Akk.* about, on); eine ~: [a piece of] information; b) *(Büro)* information bureau; *(Stand)* information desk

Informations-: ~material das informational literature; ~quelle die source of information

informativ *Adj.* informative; informieren 1. *tr. V.* inform (über + *Akk.* about); 2. *refl. V.* inform oneself, find out (über + *Akk.* about)

Infra·rot das; ~s *(Physik)* infra-red radiation; Infra·struktur die infrastructure

Infusion die; ~, ~en *(Med.)* infusion

Ing. *Abk.* Ingenieur; Ingenieur [ɪnʒeˈni̯øːɐ̯] der; ~s, ~e [qualified] engineer

Ingwer der; ~s, ~ ginger

Inhaber der; ~s, ~ a) holder; b) *(Besitzer)* owner

inhaftieren *tr. V.* take into custody; detain; Inhaftierung die; ~, ~en detention

inhalieren *tr. V.* inhale

In·halt der; ~[e]s, ~e a) contents *pl.;* b) *(einer Geschichte usw.)* content; c) *(bes. Math.)* *(Flächen~)* area; *(Raum~)* volume

Inhalts-: ~angabe die summary [of contents]; synopsis; *(eines Films, Dramas)* synopsis; ~verzeichnis das table of contents; *(auf einem Paket)* list of contents

in·human *Adj.* a) *(unmenschlich)* inhuman; b) *(rücksichtslos)* inhumane

Initiale die; ~, ~n initial [letter]

Initiative die; ~, ~n initiative; Initiator [iniˈtsi̯aːtor] der; ~s, ~en initiator; *(einer Organisation)* founder

Injektion die; ~, ~en *(Med.)* injection; injizieren *tr. V. (Med.)* inject

inkl. *Abk.* inklusive incl.

inklusive [ɪnkluˈziːvə] 1. *Präp.* + *Gen.* *(bes. Kaufmannsspr.)* including; 2. *Adv.* inclusive

inkognito *Adv. (geh.)* incognito

in·kompetent *Adj.* incompetent; In·kompetenz die incompetence

in·konsequent 1. *Adj.* inconsistent; 2. *adv.* inconsistently; In·konsequenz die inconsistency

in·korrekt 1. *Adj.* incorrect; 2. *adv.* incorrectly

In·kraft·treten das; ~s: mit [dem] ~ des Gesetzes when the law comes/came into force

In·land das; ~[e]s a) im ~: at home; b) *(Binnenland)* interior; inland; im/ins ~: inland; inländisch *Adj.* domestic; home-produced *(goods)*

Inlands-: ~markt der domestic market; ~porto das inland postage

in·mitten 1. *Präp.* + *Gen.* *(geh.)* in the midst of; 2. *Adv.* ~ von in the midst of

inne|haben *unr. tr. V.* hold, occupy *(position)*; hold *(office)*

innen *Adv.* inside; *(auf/an der Innenseite)* on the inside

innen-, Innen-: ~architekt der interior designer; ~aufnahme die *(Fot.)* indoor photo[graph]; *(Film)* interior shot; ~einrichtung die furnishings *pl.*; ~hof der inner courtyard; ~leben das; *o. Pl.* a) [inner] thoughts and feelings *pl.*; b) *(oft scherzh.: Ausstattung)* inside; ~minister der Minister of the Interior; ≈ Home Secretary *(Brit.)*; ≈ Secretary of the Interior *(Amer.)*; ~politik die *(eines Staates)* home affairs *pl.*; *(einer Regierung)* domestic policy/policies *pl.*; ~politisch *s.* ~politik: 1. *Adj.* ~politische Fragen matters of domestic policy; 2. *adv.* as regards home affairs/domestic policy; ~stadt die town centre; downtown *(Amer.)*; *(einer Großstadt)* city centre

inner... *Adj.* inner; *(inländisch; Med.)* internal; inside *(pocket, lane)*; Innere das; *adj. Dekl.*; *o. Pl.* inside; *(eines Gebäudes, Wagens, Schiffes)* interior; inside; *(eines Landes)* interior; Innereien *Pl.* entrails; *(Kochk.)* offal *sing.*; inner·halb 1. *Präp.* + *Gen.* a) within; ~ der Familie/Partei *(fig.)* within the family/party; b) *(binnen)* within; ~ einer Woche within a week; 2. *Adv.* a) ~ von within; b) *(im Verlauf)* ~ von zwei Jahren within two years; innerlich 1. *Adj.* inner; 2. *adv.* inwardly; innerst... *Adj.* innermost; Innerste das; *adj. Dekl.*; *o. Pl.* innermost being

inne|wohnen *itr. V.* *(geh.)* etw. wohnt

jmdm./einer Sache ~: sb./sth. possesses sth.

innig 1. *Adj.* deep *(affection, sympathy)*; fervent *(wish)*; intimate *(friendship)*; mein ~ster Dank my sincerest thanks; 2. *adv.* *(love)* with all one's heart; Innigkeit die; ~: depth; *(einer Beziehung)* intimacy

Innung ['ɪnʊŋ] die; ~, ~en [trade] guild

in·offiziell 1. *Adj.* unofficial; 2. *adv.* unofficially

in puncto as regards

ins *Präp.* + *Art.* = in das

Insasse der; ~n, ~n a) *(Fahrgast)* passenger; b) *(Bewohner)* inmate

ins·besond[e]re *Adv.* particularly; in particular

In·schrift die inscription

Insekt [ɪn'zɛkt] das; ~s, ~en insect

Insel die; ~, ~n island

Inserat das; ~[e]s, ~e advertisement *(in a newspaper)*; Inserent der; ~en, ~en advertiser; inserieren *itr.V.* advertise

ins·geheim *Adv.* secretly

ins·gesamt *Adv.* in all; altogether; *(alles in allem)* all in all

insofern 1. *Adv.* [ɪn'zo:fɛrn] *(in dieser Hinsicht)* to this extent; 2. *Konj.* [ɪnzo-'fɛrn] *(falls)* provided [that]

insoweit [ɪn'zo:vait/ɪnzo'vait] *Adv./ Konj. s.* insofern

in spe [ɪn 'spe:] future *attrib.*; mein Schwiegersohn ~ ~: my future son-in-law

Inspektion [ɪnspɛk'tsɪo:n] die; ~, ~en inspection; *(Kfz-W.)* service

Inspiration [ɪnspira'tsɪo:n] die; ~, ~en inspiration; inspirieren *tr. V.* inspire

inspizieren *tr. V.* inspect

Installateur [ɪnstala'tø:ɐ̯] der; ~s, ~e plumber; *(Gas~)* [gas-]fitter; *(Heizungs~)* heating engineer; *(Elektro~)* electrician; Installation [ɪnstala-'tsɪo:n] die; ~, ~en installation; *(Rohre)* plumbing *no pl.*; installieren *tr. V.* install

in·stand *Adv.* etw. ist gut/schlecht ~: sth. is in good/poor condition; etw. ~ halten keep sth. in good condition; etw. ~ setzen/bringen repair sth.; In·stand·haltung die maintenance

in·ständig 1. *Adj.* urgent; 2. *adv.* urgently

Instanz [ɪn'stants] die; ~, ~en a) authority; b) *(Rechtsw.)* [die] erste/zweite/dritte ~: the court of original jurisdiction/the appeal court/the court of final appeal; **durch alle ~en gehen** go through all the courts

Instinkt [ɪn'stɪŋkt] *der*; ~[e]s, ~e instinct; **instinktiv 1.** *Adj.* instinctive; **2.** *adv.* instinctively

Institut *das*; ~[e]s, ~e a) institute; **Institution** [ɪnstitu'tsi̯oːn] *die*; ~, ~en *(auch fig.)* institution

Instruktion [ɪnstrʊk'tsi̯oːn] *die*; ~, ~en instruction

Instrument [ɪnstru'mɛnt] *das*; ~[e]s, ~e instrument; **instrumental** *(Musik)* **1.** *Adj.* instrumental; **2.** *adv.* instrumentally

Insulin *das*; ~s insulin

inszenieren *tr. V.* stage; put on; *(Regie führen bei)* direct; *(fig.) (einfädeln)* engineer; *(organisieren)* stage; **Inszenierung** *die*; ~, ~en staging; *(Regie)* direction; *(Aufführung)* production

intakt *Adj.* **a)** *(unbeschädigt)* intact; **b)** *(funktionsfähig)* in [proper] working order *postpos.*; healthy ⟨*economy*⟩

integer *Adj.* **eine integre Persönlichkeit** a person of integrity; **~ sein** be a person of integrity

Integral *das*; ~s, ~s *(Math.)* integral

integrieren *tr. V.* integrate

Intellekt *der*; ~[e]s intellect; **intellektuell** *Adj.* intellectual; **Intellektuelle** *der/die*; *adj. Dekl.* intellectual; **intelligent 1.** *Adj.* intelligent; **2.** *adv.* intelligently; **Intelligenz** *die*; ~a) intelligence; **b)** *(Gesamtheit der Intellektuellen)* intelligentsia; **Intelligenz·quotient** *der* intelligence quotient

Intendant *der*; ~en, ~en *(Theater)* manager and artistic director; *(Fernseh~, Rundfunk~)* director-general

Intensität *die*; ~: intensity

intensiv 1. *Adj.* **a)** *(gründlich)* intensive *(kräftig)* intense; **2.** *adv.* intensively; **intensivieren** *tr. V.* intensify; increase ⟨*exports*⟩; strengthen ⟨*connections*⟩; **Intensiv·station** *die* intensive-care unit

Intercity-Zug *der* inter-city train

interessant 1. *Adj.* interesting; **2.** *adv.* **~ schreiben** write in an interesting way; **interessanterweise** *Adv.* interestingly enough; **Interesse** *das*; ~s, ~n interest; **~ an jmdm./etw. haben** be interested in sb./sth.; **interesse·halber** *Adv.* out of interest; **Interessen·gebiet** *das* field of interest; **Interessent** *der*; ~en, ~en interested person; *(möglicher Käufer)* potential buyer; **Interessen·ver-**

band *der* [organized] interest group; **Interessen·vertretung** *die* **a)** representation; **b)** *(Vertreter von Interessen)* representative body; **interessieren 1.** *refl. V.* **sich für jmdn./etw. ~**: be interested in sb./sth. **2.** *tr. V.* interest; **das interessiert mich nicht** I'm not interested [in it]; **interessiert** *Adj.* interested (**an** + *Dat.* in)

Interjektion [ɪntɛjɛk'tsi̯oːn] *die*; ~, ~en *(Sprachw.)* interjection

Interkontinental·rakete *die* *(Milit.)* intercontinental ballistic missile

intern 1. *Adj.* internal; **2.** *adv.* internally

Internat *das*; ~[e]s, ~e boarding-school

inter·national 1. *Adj.* international; **2.** *adv.* internationally; **Inter·nationale** *die*; ~, ~n a) International; Internationale; **b)** *(Lied)* Internationale

Internats-: **~schüler** *der*, **~schülerin** *die* boarding-school pupil; boarder

internieren *tr. V.* *(Milit.)* intern; **Internierung** *die*; ~, ~en internment

Internist *der*; ~en, ~en *(Med.)* internist

Interpol *die*; ~: Interpol *no art.*

Interpret *der*; ~en, ~en interpreter *(of music, text, events, etc.)*; **Interpretation** [ɪntɛpreta'tsi̯oːn] *die*; ~, ~en interpretation *(of music, text, events, etc.)*; **interpretieren** *tr. V.* interpret ⟨*music, texts, events, etc.*⟩; **Interpretin** *die*; ~, ~nen *s.* Interpret

Interpunktion [ɪntɛpʊnk'tsi̯oːn] *die*; ~ *(Sprachw.)* punctuation

Intervall [ɪntɛ'val] *das*; ~s, ~e *(Musik, Math.)* interval

intervenieren *itr. V.* *(geh., Politik)* intervene; **Intervention** [ɪntɛvɛn'tsi̯oːn] *die*; ~, ~en *(geh., Politik)* intervention; *(Protest)* representations *pl.*

Interview [ɪntɛ'vju:] *das*; ~s, ~s interview; **interviewen** [ɪntɛ'vju:ən] *tr. V.* interview

intim 1. *Adj.* intimate; **2.** *adv.* **~ befreundet sein** be intimate friends; **Intimität** [ɪntimi'tɛːt] *die*; ~, ~en intimacy; **Intim·sphäre** *die* private life

in·tolerant *Adj.* intolerant

intransitiv 1. *Adj.* *(Sprachw.)* intransitive; **2.** *adv.* intransitively

Intrige [ɪn'tri:gə] *die*; ~, ~n intrigue

Intuition [ɪntu̯i'tsi̯oːn] *die*; ~, ~en intuition; **intuitiv 1.** *Adj.* intuitive; **2.** *adv.* intuitively

intus ['ɪntʊs] *in* etw. ~ **haben** *(ugs.) (begriffen haben)* have got sth. into one's head; *(gegessen od. getrunken haben)* have put sth. away *(coll.)*
Invalide der; *adj. Dekl.* invalid
Invasion die; ~, ~en invasion
Inventar das; ~s, ~e *(einer Firma)* fittings and equipment *pl.; (eines Hauses, Büros)* furnishings and fittings *pl.;* **Inventur** die; ~, ~en stocktaking
investieren *tr., itr. V. (auch fig.)* invest (**in** + *Akk.* in); **Investition** [ɪnvesti'tsio:n] die; ~, ~en investment; **Investitions·güter** *Pl. (Wirtsch.)* capital goods; **Investor** [ɪn'vestɔr] der; ~s, ~en [-'to:rən] *(Wirtsch.)* investor
in·wie·fern *Adv.* in what way; *(bis zu welchem Grade)* to what extent; **in·wie·weit** *Adv.* to what extent
Inzest der; ~|e|s, ~e incest; **In·zucht** die; ~: inbreeding
in·zwischen *Adv.* a) *(seither)* in the meantime; since [then]; b) *(bis zu einem Zeitpunkt) (in der Gegenwart)* by now; *(in der Vergangenheit/Zukunft)* by then; c) *(währenddessen)* meanwhile
IOK [i:o:'ka:] das; ~|s| Internationales Olympisches Komitee IOC
Ion das; ~s, ~en *(Physik, Chemie)* ion
Irak (das); ~s *od.* der; ~|s| Iraq; **Iraker** der; ~s, ~Iraqi; **irakisch** Iraqi
Iran (das); ~s *od.* der; ~|s| Iran; **Iraner** der; ~s, ~; **iranisch** *Adj.* Iranian
irden *Adj.* earthen[ware]; **irdisch** *Adj.* a) *earthly;* worldly *(goods, pleasures, possessions)*; b) *(zur Erde gehörig)* terrestrial; **das** ~e **Leben** life on earth
Ire der; ~n, ~n Irishman
irgend *Adv.* a) ~ **jemand** someone; somebody; *(fragend, verneinend)* anyone; anybody; ~ **etwas** something; *(fragend, verneinend)* anything; ~ **so etwas** something like that; b) *(irgendwie)* **wenn** ~ **möglich** if at all possible
irgend-: ~**ein** *Indefinitpron.* a) *(attr.)* some; *(fragend, verneinend)* any; b) *(subst.)* ~**einer**/~**eine** someone; somebody; *(fragend, verneinend)* anyone; anybody; ~**eines** od. *(ugs.)* ~**eins** any one; ~**einmal** *Adv.* sometime; ~**wann** *Adv.* [at] some time [or other]; *(zu jeder beliebigen Zeit)* [at] any time; ~**was** *Indefinitpron. (ugs.)*

something [or other]; *(fragend, verneinend)* anything; ~**welch** *Indefinitpron.* some; *(fragend, verneinend)* any; ~**wer** *Indefinitpron. (ugs.)* somebody or other *(coll.); (fragend, verneinend)* anyone; anybody; ~**wie** *Adv.* somehow; ~**wo** *Adv.* somewhere; *(fragend, verneinend)* anywhere; ~**woher** *Adv.* from somewhere; *(fragend, verneinend)* from anywhere; ~**wohin** *Adv.* somewhere; *(fragend, verneinend)* anywhere
Irin die; ~, ~nen Irishwoman
Iris die; ~, ~ *(Bot.; Anat.)* iris
irisch *Adj.* Irish; **Irland (das);** ~s Ireland
Ironie die; ~, ~n irony; **ironisch** 1. *Adj.* ironic; ironical; 2. *adv.* ironically
irre 1. *Adj.* insane; 2. *adv. (salopp)* terribly *(coll.);* **Irre** der/die; *adj. Dekl.* madman/madwoman; lunatic; *(fig.)* lunatic
irre|führen *tr. V.* mislead; *(täuschen)* deceive; **Irreführung** die; **eine bewußte** ~**führung** a deliberate attempt to mislead; ~**führung der Öffentlichkeit** misleading the public
irrelevant ['ɪrelevant] *Adj.* irrelevant (**für** to)
irre|machen *tr. V.* disconcert; put off; **irren** 1. *refl. V.* be mistaken; **Sie haben sich in der Nummer geirrt** you've got the wrong number; 2. *itr. V.* a) **da** ~ **Sie** you are wrong there; b) *mit sein (ziellos umherstreifen)* wander
Irren-: ~**anstalt** die *(veralt. abwertend)* mental home; ~**haus** das *(abwertend)* [lunatic] asylum
Irr·fahrt die wandering; **irriger·weise** *Adv.* mistakenly
irritieren *tr., itr. V.* a) *(verwirren)* put off; b) *(stören)* disturb
irr-, Irr-: ~**licht** das will o' the wisp; ~**sinn** der; o. Pl. a) insanity; madness; b) *(ugs. abwertend)* lunacy; ~**sinnig** 1. *Adj.* a) *(geistig gestört)* insane; mad; *(absurd)* idiotic; b) *(ugs.: extrem)* terrible *(coll.);* terrific *(coll.)* ⟨speed, heat, cold⟩; 2. *adv. (ugs.)* terribly *(coll.)*
Irrtum der; ~s, Irrtümer mistake; ~! wrong!; **im** ~ **sein** be wrong *or* mistaken; **irrtümlich** 1. *Adj.* incorrect; 2. *adv.* by mistake
Ischias ['ɪʃias] der *od.* das *od. Med.* die; ~: sciatica
Islam [ɪs'la:m *od.* 'ɪslam] der; ~|s|: der ~: Islam; **islamisch** *Adj.* Islamic

Island (das); ~s Iceland; **Isländer der**; ~s, ~ Icelander; **isländisch** *Adj.* Icelandic

Isolation die; ~, ~en *s.* **Isolierung**; **Isolator der**; ~s, ~en insulator; **Isolier·band das**; *Pl.* ~bänder insulating tape; **isolieren** *tr. V.* **a)** isolate; **b)** *(Technik)* insulate ⟨*wiring, wall, etc.*⟩; lag ⟨*boilers, pipes, etc.*⟩; **Isolier·station die** *(Med.)* isolation ward; **Isolierung die**; ~, ~en **a)** isolation; **b)** *(Technik) s.* **isolieren b:** insulation; lagging

Isotop das; ~s, ~e isotope

Israel ['ɪsraeːl] **(das)**; ~s Israel; **Israeli der**; ~[s], ~[s]/die; ~, ~[s] Israeli; **israelisch** *Adj.* Israeli; **Israelit der**; ~en, ~en Israelite; **israelitisch** *Adj.* Israelite

iß *Imperativ Sg. v.* **essen**

ißt *2. u. 3. Pers. Sg. Präsens v.* **essen**

ist *3. Pers. Sg. Präsens v.* **sein**

Italien [i'taːliən] **(das)**; ~s Italy; **Italiener** [ita'liːɛnɐ] **der**; ~s, ~: Italian; **italienisch** *Adj.* Italian

I-Tüpfel[chen] das; ~s, ~: final touch; **bis aufs [letzte]** ~: down to the last detail

i. V. [iː'faʊ] *Abk.* in Vertretung

J

j, J [jɔt, *österr.:* jeː] **das**; ~, ~: j/J

ja 1. *Interj.* yes; *(nachgestellt: nicht wahr?)* won't you/doesn't it *etc.*?; 2. *Partikel* Sie wissen ja, daß ...: you know, of course, that ...; da seid ihr ja! there you are!; **Ja das**; ~[s], ~[s] yes; **mit** ~ **stimmen** vote yes

Jacht die; ~, ~en yacht

Jacke die; ~, ~n jacket; *(gestrickt)* cardigan; **Jacken·kleid das** dress and jacket combination; **Jacketkrone** ['dʒɛkɪt-] **die** *(Zahnmed.)* jacket crown; **Jackett** [ʒa'kɛt] **das**; ~s, ~s jacket

Jade die; ~, ~: jade

Jagd die; ~, ~en **a)** *o. Pl.* **die** ~: shooting; hunting; **auf die** ~ **gehen** go hunting/shooting; **b)** *(Veranstaltung)* shoot; *(Hetzjagd)* hunt; **c)** *(Verfolgung)* hunt; *(Verfolgungsjagd)* chase; **auf jmdn./etw.** ~ **machen** hunt for sb./sth.

Jagd-: ~**beute die** bag; kill; ~**bomber der** *(Luftwaffe)* fighter-bomber; ~**flieger der** *(Luftwaffe)* fighter pilot; ~**flugzeug das** *(Luftwaffe)* fighter aircraft; ~**gewehr das** sporting gun; ~**horn das** hunting-horn; ~**hund der** gun-dog; ~**hütte die** shooting box; ~**revier das** preserve; shoot; ~**schein der** game licence; ~**wurst die** chasseur sausage; ~**zeit die** open season

jagen 1. *tr. V.* **a)** hunt ⟨*game, fugitive, criminal, etc.*⟩; shoot ⟨*game, game birds*⟩; *(hetzen)* chase ⟨*fugitive, criminal, etc.*⟩; **b)** *(treiben)* drive; **jmdn. aus dem Haus** ~: throw sb. out of the house; 2. *itr. V. (die Jagd ausüben)* go shooting *or* hunting; **Jäger der**; ~s, ~: **a)** hunter; **b)** *(Milit.)* rifleman; **c)** *(Soldatenspr.: Jagdflugzeug)* fighter; **Jäger·hut der** huntsman's hat

Jäger-: ~**latein das** *(scherzh.)* [hunter's] tall story/stories; **das ist das reinste** ~**latein** that's all wild exaggeration; ~**rock der** hunting jacket; ~**schnitzel das** *(Kochk.)* escalope chasseur

Jaguar der; ~s, ~e jaguar

jäh [jɛː] 1. *Adj. (geh.)* **a)** sudden; abrupt ⟨*change, movement, stop*⟩; sudden, sharp ⟨*pain*⟩; **b)** *(steil)* steep; precipitous; 2. *adv.* **a)** ⟨*change*⟩ abruptly; **b)** *(steil)* ⟨*fall, drop*⟩ steeply; **jählings** *Adv. (geh.)* **a)** *(plötzlich)* ⟨*change, end, stop*⟩ suddenly, abruptly; ⟨*die*⟩ suddenly; **b)** *(steil)* steeply

Jahr das; ~[e]s, ~e year; **ein halbes** ~: six months; **im** ~[e] **1908** in [the year] 1908; **er ist zwanzig** ~e [alt] he is twenty years old; **Kinder bis zu zwölf** ~en children up to the age of twelve; **zwischen den** ~en between Christmas and the New Year; **jahr·aus** *Adv.* ~, **jahrein** year in, year out; **jahre·lang** 1. *Adj.; nicht präd.* [many] years of; long-standing ⟨*feud, friendship*⟩; 2. *adv.* for [many] years

jähren *refl. V.* **heute jährt sich zum zehntenmal, daß** ...: it is ten years ago today that ...

Jahres-: ~**bilanz die** *(Wirtsch., Kaufmannsspr.)* annual balance [of accounts]; *(Dokument)* annual balance sheet; ~**einkommen das** annual income; ~**ende das** end of the year;

~**frist**; *o. Art.; o. Pl.* in *od.* **innerhalb** *od.* **binnen** ~**frist** within [a period of] a *or* one year; ~**hälfte die: die erste/ zweite** ~**hälfte** the first/secound half *or* six months of the year; ~**karte die** yearly season ticket; ~**tag** der anniversary; ~**urlaub** der annual holiday *or (formal)* leave *or (Amer.)* vacation; ~**wechsel** der turn of the year; **zum** ~**wechsel die besten Wünsche** best wishes for the New Year; ~**zahl die** date; ~**zeit die** season

Jahr·gang der a) *(Altersklasse)* year; **der** ~ **1900** those born in 1900; b) *(eines Weines)* vintage; c) *(einer Zeitschrift)* set [of issues] for a/the year; **Jahr·hundert** das century; **Jahr· hundert·wende** die turn of the century; -**jährig** a) *(... Jahre alt)* **ein elf· jähriges Kind** an eleven-year-old child; b) *(... Jahre dauernd)* ...year's/years'; **nach vierjähriger Vor· bereitung** after four years' preparation; **mit dreijähriger Verspätung** three years late; **jährlich** 1. *Adj.; nicht präd.* annual; yearly; 2. *adv.* annually; yearly; **zweimal** ~: twice a year

Jahr-: ~**markt** der fair; fun-fair; ~**tausend** das thousand years; millennium; ~**zehnt** das decade

jahrzehnte·lang 1. *Adj.; nicht präd.* decades of *⟨practice, experience, etc.⟩;* 2. *adv.* for decades

Jäh·zorn der violent anger; **jäh·zor· nig** 1. *Adj.* violent-tempered; 2. *adv.* in a blind rage

ja·ja *Part. (ugs.)* a) *(seufzend)* ~[, so ist das Leben] o well[, that's life]; b) *(ungeduldig)* ~[, ich komme schon]! all right, all right[, I'm coming]!

Jalousie [ʒalu'ziː] die; ~, ~n Venetian blind

Jamaika *(das);* -s Jamaica; **Jamai· kaner** der; ~s, ~ Jamaican

Jammer der; ~s [mournful] wailing; *(Elend)* misery; **jämmerlich** 1. *Adj.* pitiful; b) wretched *⟨appearance, existence, etc.⟩;* paltry, meagre *⟨quantity⟩;* 2. *adv.* pitifully; **jammern** *itr. V.* wail; *(sich beklagen)* moan; **jam· mer·schade** *Adj.; nicht attr. (ugs.)* **es ist** ~**schade, daß ...:** it's a crying shame that ...; **es ist** ~**schade um ihn** it's a great pity about him

Janker der; ~s, ~ *(südd., österr.)* Alpine jacket

Januar der; ~[s], ~e January

Japan *(das);* ~s Japan; **Japaner** der;

~s, ~**Japanese; japanisch** *Adj.* Japanese

japsen *itr. V. (ugs.)* pant

Jargon [jar'gõː] der; ~s, ~s jargon

Jasmin der; ~s, ~e jasmine

Ja·stimme die yes-vote

jäten *tr., itr. V.* weed; **Unkraut** ~: weed

Jauche die; ~, ~n liquid manure; **Jauche·grube** die liquid-manure reservoir

jauchzen *itr. V.* cheer; **vor Freude** ~: shout for joy; **Jauchzer** der; ~s, ~: cry of delight

jaulen *itr. V.* howl

Jause die; ~, ~n *(österr.)* a) snack; **ei· ne** ~ **machen** have a snack; b) *(Nach· mittagskaffee)* [afternoon] tea

ja·wohl *Part.* certainly; **Ja·wort** das consent; **jmdm. das** ~ **geben** consent to marry sb.

Jazz [dʒæz *od.* dʒɛs *od.* jats] der; ~: jazz; **Jazz·keller** der jazz cellar

¹**je** 1. *Adv.* a) *(jemals)* ever; **mehr/bes· ser denn je** more/better than ever; b) *(jeweils)* **je zehn Personen** ten people at a time; **sie kosten je 30 DM** they cost 30 DM each; c) *(entsprechend)* **je nach Gewicht** according to weight; 2. *Präp. mit Akk.* per; for each; 3. *Konj.* **je länger, je lieber** the longer the better; **je nachdem** it all depends

²**je** *Interj.* **ach je, wie schade!** oh dear, what a shame!

Jeans [dʒiːnz] *Pl. od.* die; ~, ~: jeans *pl.;* denims *pl.*

jede *s.* jeder; **jeden·falls** *Adv.* a) in any case; b) *(zumindest)* at any rate; **jeder, jede, jedes** *Indefinitpron. u. unbest. Zahlwort* 1. *attr.* a) *(alle)* every; b) *(alle einzeln)* each; c) *(jeglicher)* all; 2. *alleinstehend* a) *(alle)* everyone; everybody; b) *(alle einzeln)* **jedes der Kinder** each of the children

jeder-: ~**mann** *Indefinitpron. u. un· best. Zahlwort;* nur *alleinstehend* everyone; everybody; ~**zeit** *Adv.* [at] any time

jedes *s.* jeder; **jedes·mal** *Adv.* every time

je·doch *Konj., Adv.* however

je·her [*od.* '-'-] *Adv.* **seit** *od.* **von** ~: always; since time immemorial

jemals *Adv.* ever

jemand *Indefinitpron.* someone; somebody; *(fragend, verneinend)* anyone; anybody

Jemen *(das);* ~s *od.* der; ~[s] Yemen

jener, jene, jenes *Demonstrativpron.*

(geh.) **1.** *attr.* that; *(im Pl.)* those; **2.** *alleinstehend* that one; *(im Pl.)* those
jenseits 1. *Präp. mit Gen.* on the other side of; *(in größerer Entfernung)* beyond; **2.** *Adv.* on the other side; ~ **vom Rhein** on the other side of the Rhine; **Jenseits** das; ~: hereafter; beyond
¹**Jersey** [ˈdʒøːɐzi] der; ~|s|, ~s *(Textilind.)* jersey
²**Jersey** das; ~s, ~s *(Sport: Trikot)* jersey
Jesus (der); Jesu Jesus
Jet [dʒɛt] der; ~|s|, ~s jet; **mit einem ~ fliegen/reisen** fly/travel by jet
jetzig *Adj.; nicht präd.* current
jetzt *Adv.* **a)** just now; **bis ~:** up to now; **bis ~ noch nicht** not yet; **von ~ an** *od.* **ab** from now on[wards]; **erst ~** *od.* **~ erst** only just; **schon ~:** already; **b)** *(heutzutage)* now; nowadays
jeweilig *Adj.; nicht präd.* **a)** *(in einem bestimmten Fall)* particular; **b)** *(zu einer bestimmten Zeit)* current; of the time *postpos., not pred.;* **c)** *(zugehörig, zugewiesen)* respective; **jeweils** *Adv.* **a)** *(jedesmal)* ~ **am ersten/letzten Mittwoch des Monats** on the first/last Wednesday of each month; **b)** *(zur Zeit)* at the time
Jg. *Abk.* Jahrgang
Jh. *Abk.* Jahrhundert c.
jiddisch [ˈjɪdɪʃ] *Adj.* Yiddish
Job [dʒɔp] der; ~s, ~s *(ugs.; auch DV)* job; **jobben** [ˈdʒɔbn̩] *itr. V. (ugs.)* do a job/jobs
Joch das; ~|e|s, ~e yoke
Jockei, Jockey [ˈdʒɔke *od.* ˈdʒɔki] der; ~s, ~s jockey
Jod [joːt] das; ~|e|s iodine
jodeln *itr., tr. V.* yodel
jod·haltig *Adj.* iodiferous
Joga der *od.* das; ~|s| yoga
joggen [ˈdʒɔgn̩] *itr. V.; mit Richtungsangabe mit sein* jog
Joghurt [ˈjoːgʊrt] der *od.* das; ~|s|, ~|s| yoghurt; **Joghurt·becher** der yoghurt pot *(Brit.) or (Amer.)* container
Johannis·beere die currant; **rote/weiße/schwarze ~n** redcurrants/white currants/blackcurrants
johlen *itr. V.* yell; *(vor Wut)* howl
Joint [dʒɔɪnt] der; ~s, ~s *(ugs.)* joint *(sl.)*
Jolle die; ~, ~n keel-centre-board yawl
Jongleur [ʒɔŋˈløːɐ] der; ~s, ~e juggler; **jonglieren** *tr., itr. V.* juggle
Joppe die; ~, ~n heavy jacket
Jordanien (das); ~s Jordan; **Jorda-**

nier der; ~s, ~: Jordanian; **jordanisch** *Adj.* Jordanian
Jot das; ~, ~: j, J
Journalismus der; ~: journalism *no art.;* **Journalist** der; ~en, ~en journalist; **journalistisch 1.** *Adj.; nicht präd.* journalistic; **eine ~e Ausbildung** a training in journalism; **2.** *adv.* journalistically; ~ **tätig sein** be a journalist
jr. *Abk.* junior Jr.
Jubel der; ~s rejoicing; jubilation; *(laut)* cheering; **jubeln** *itr. V.* cheer; **über etw.** *(Akk.)* ~: rejoice over sth.; **Jubilar** der; ~s, ~e man celebrating his anniversary/birthday; **Jubiläum** das; ~s, Jubiläen anniversary; *(eines Monarchen)* jubilee; **jubilieren** *itr. V. (geh.)* jubilate *(literary);* rejoice
juchzen [ˈjʊxtsn̩] *itr. V. (ugs.)* shout with glee
jucken 1. *tr., itr. V.* **a) mir juckt die Haut** I itch; **es juckt mich hier** I've got an itch here; **b)** *(Juckreiz verursachen)* irritate; **2.** *tr. V. (reizen, verlocken)* **es juckt mich, das zu tun** I am itching to do it; **3.** *refl. V. (ugs.: sich kratzen)* scratch; **Juck·reiz** der itch
Jude der; ~n, ~n Jew; **Juden·stern** der *(ns.)* Star of David; **Judentum** das; ~s **a)** *(Volk)* Jewry; Jews *pl.;* **b)** *(Kultur u. Religion)* Judaism; **Jüdin** die; ~, ~nen Jewess; **jüdisch** *Adj.* Jewish
Judo [ˈjuːdo] das; ~|s| judo *no art.*
Jugend die; ~ **a)** youth; **b)** *(Jugendliche)* young people
jugend-, Jugend-: ~**amt** das youth office *(agency responsible for education and welfare of young people);* ~**arrest** der detention in a community home; ~**bewegung** die *(hist.)* [German] youth Movement; ~**buch** das book for young people; ~**frei** *Adj.* ⟨*film, book, etc.*⟩ suitable for persons under 18; **nicht ~frei** ⟨*film*⟩ not U-certificate *pred.;* ~**gefährdend** *Adj.* liable to have an undesirable influence on the moral development of young people *postpos.;* ~**heim** das youth centre; ~**herberge** die youth hostel; ~**kriminalität** die juvenile delinquency
jugendlich 1. *Adj.* **a)** *nicht präd.* young ⟨*offender, customer, etc.*⟩; **b)** *(für Jugendliche charakteristisch)* youthful; **Jugendliche** der/die; *adj. Dekl.* young person; **die ~n** the young people

Jugend-: ~liebe die sweetheart of one's youth; ~schutz der protection of young people; ~schutz·gesetz das laws pl. protecting young people; ~sprache die young people's language no art.; ~stil der art nouveau; (in Deutschland) Jugendstil; ~strafe die youth custody sentence; ~sünde die youthful folly; ~zeit die youth; ~zentrum das youth centre
Jugo·slawe der Yugoslav; Jugoslawien (das); ~s Yugoslavia; jugo·slawisch Adj. Yugoslav[ian]
Julei der; ~[s], ~s s. Juli
Juli der; ~[s], ~s July; s. auch April
jung Adj.; jünger, jüngst... a) young; new ⟨project, undertaking, sport, marriage, etc.⟩; b) (letzt...) recent; in jüngster Zeit recently
¹Junge der; ~n, ~n od. (ugs.) Jung[en]s boy; ²Junge das; adj. Dekl. ein ~s one of the young; ~ kriegen give birth to young; jungen itr. V. give birth; ⟨cat⟩ have kittens; ⟨dog⟩ have pups; jungenhaft Adj. boyish; jünger Adj. youngish; sie ist noch ~: she is still quite young; s. auch jung; Jünger der; ~s, ~: follower; Jungfer die; ~, ~n (abwertend: ältere ledige Frau) spinster; Jungfern·fahrt die maiden voyage; Jungfern·häutchen das hymen; Jung·frau die a) virgin; b) (Astrol.) Virgo; jung·fräulich Adj. (geh., auch fig.) virgin; Jung·geselle der bachelor; Junggesellin die; ~, ~nen bachelor girl
Jüngling der; ~s, ~e (geh., spött.) youth; boy; jüngst Adv. (geh.) recently; jüngst... s. jung; Jüngste der/die; adj. Dekl. youngest [one]
Jung-: ~verheiratete der/die; adj. Dekl., young married man/woman; die ~verheirateten the newly-weds; ~wähler der first-time voter
Juni der; ~[s], ~s June; s. auch April
junior indekl. Adj.; nach Personennamen junior; Junior der; ~s, ~en a) (oft scherzh.) junior (joc.); b) (Kaufmannsspr.) junior partner; Junior-chef der owner's or (coll.) boss's son
Juno der; ~[s], ~s s. Juni
Junta ['xʊnta] die; ~, Junten junta
Jura o. Art., o. Pl. law; ~ studieren read Law; Jurist der; ~en, ~en, Juristin die; ~, ~nen lawyer; jurist; juristisch Adj. legal
Jury [ʒy'ri:] die; ~, ~s a) (Preisrichter) panel [of judges]; jury; b) (Sachverständige) panel [of experts]

Justiz die; ~: justice; (Behörden) judiciary
Justiz-: ~irrtum der miscarriage of justice; ~minister der Minister of Justice; ~vollzugs·anstalt die (Amtsspr.) penal institution (formal); prison
Jute ['ju:tə] die; ~: jute
Juwel das od. der; ~s, ~en piece of jewellery; (Edelstein) jewel; Juwelier [juva'li:ɐ̯] der; ~s jeweller; Juwelier·geschäft das jeweller's shop
Jux der; ~es, ~e (ugs.) joke

K

k, K [ka:] das; ~, ~: k/K
Kabarett das; ~s, ~s od. ~e a) satirical revue; b) (Ensemble) cabaret act; Kabarettist der; ~en, ~en revue performer
kabbeln refl. V. (ugs.) bicker (mit with)
Kabel das; ~s, ~: cable; (für kleineres Gerät) flex
Kabeljau der; ~s, ~e od. ~s cod
kabeln itr., itr. V. (veralt.) cable
Kabine die; ~, ~n a) cabin; b) (Umkleideraum, abgeteilter Raum) cubicle; c) (einer Seilbahn) [cable-]car; Kabinett das; ~s, ~e Cabinet
Kabrio das; ~s, ~s, Kabriolett das; ~s, ~s convertible
Kachel die; ~, ~n [glazed] tile; kacheln tr. V. tile
Kadaver der; ~s, ~: carcass
Kader der od. (schweiz.) das; ~s, ~ cadre; b) (Sport) squad
Käfer der; ~s, ~: beetle
Kaff das; ~s, ~s od. Käffer (ugs. abwertend) dump (coll.)
Kaffee ['kafe od. (österr.) ka'fe:] der; ~s, ~s a) coffee; b) (Nachmittags~) afternoon coffee; ~ trinken have afternoon coffee
Kaffee-: ~kanne die coffee-pot; ~kränzchen das (veralt.) a) (Zusammentreffen) coffee afternoon; b) (Gruppe) coffee circle; ~maschine

die coffee-maker; ~**mühle** die coffee-grinder; ~**satz** der coffee-grounds *pl.;* ~**tante** die *(ugs. scherzh.)* coffee addict

Käfig der; ~s, ~e cage

kahl *Adj.* **a)** *(ohne Haare)* bald; **b)** *(ohne Grün, schmucklos)* bare

kahl-, Kahl-: ~|**fressen** *unr. tr. V.* etw. ~**fressen** strip sth. bare; ~**köpfig** *Adj.* bald[-headed]; ~|**scheren** *unr. tr. V.* jmdn. ~**scheren** shave sb.'s head; ~**schlag** der **a)** clear-felling *no indef. art.;* **b)** *(Waldfläche)* clear-felled area

Kahn der; ~|e|s, **Kähne a)** *(Ruder~)* rowing-boat; *(Stech~)* punt; **b)** *(Lastschiff)* barge

Kai der; ~s, ~s quay

Kaiser der; ~s, ~: emperor; **Kaiserin** die; ~, ~**nen** empress

Kaiser-: ~**krone** die imperial crown; ~**reich** das empire; ~**schnitt** der Caesarean section

Kajüte die; ~, ~**n** *(Seemannsspr.)* cabin

Kakao [ka'kau] der; ~s, ~s cocoa

Kakerlak der; ~s *od.* ~**en**, ~**en** cockroach

Kaktus der; ~, **Kakteen** cactus

Kalauer der; ~s, ~: corny joke *(coll.);* *(Wortspiel)* atrocious *or (coll.)* corny pun

Kalb das; ~|e|s, **Kälber a)** calf; **b)** *(ugs.:* ~*fleisch)* veal; **kalben** *itr. V.* calve; **Kalb·fleisch** das veal

Kalbs-: ~**braten** der *(Kochk.)* roast veal *no indef. art.;* *(Gericht)* roast of veal; ~**leder** das calfskin; ~**schnitzel** das veal cutlet

Kalender der; ~s, ~: calendar; *(Taschen~)* diary; **Kalender·jahr** das calendar year

Kalesche die; ~, ~**n** *(hist.)* barouche

Kali das; ~s, ~s potash

Kaliber das; ~s, ~: **a)** *(Technik, Waffenkunde)* calibre; **b)** *(ugs., oft abwertend)* sort; kind

Kalifornien [kali'fɔrniən] **(das);** ~s California

Kalium *(Chemie)* das; ~s potassium

Kalk der; ~|e|s, ~e calcium carbonate; *(Baustoff)* lime; quicklime; **kalken** *tr. V.* whitewash

Kalk-: ~**mangel** der; *o. Pl.* calcium deficiency; ~**stein** der limestone

Kalkül das *od.* der; ~s, ~e *(geh.)* calculation; **Kalkulation** die; ~, ~**en** *(auch Wirtsch.)* calculation; **kalkulieren** *tr. V.* calculate ‹cost, price›; cost ‹product, article›

Kalorie die; ~, ~**n** calorie; **kalorien·arm 1.** *Adj.* low-calorie *attrib.;* ~**arm** sein be low in calories; **2.** *adv.* ~**arm** kochen cook low-calorie meals

kalt; kälter, kältest... 1. *Adj.* cold; frosty ‹atmosphere, smile›; **2.** *adv.* **a)** ~ **duschen** have a cold shower Getränke/Sekt ~ **stellen** cool drinks/chill champagne; **b)** *(nüchtern)* coldly; **c)** *(abweisend, unfreundlich)* frostily

kalt-, Kalt-: ~|**bleiben** *unr. itr. V.;* *mit sein* remain unmoved; ~**blütig 1.** *Adj.* **a)** cool-headed; **b)** *(abwertend: skrupellos)* cold-blooded; **2.** *adv.* **a)** coolly; **b)** *(abwertend: skrupellos)* cold-bloodedly; ~**blütigkeit** die; ~: *s.* ~**blütig a, b:** cool-headedness; cold-bloodedness

Kälte die; ~ cold; *(fig.)* coldness

Kälte-: ~**einbruch** der *(Met.)* sudden onset of cold weather; ~**grad** der degree of frost

kälter *s.* kalt; **kältest...** *s.* kalt; **Kälte·welle** die cold spell

kalt-, Kalt-: ~**herzig** *Adj.* coldhearted; ~**lächelnd** *Adv. (ugs. abwertend)* etw. ~**lächelnd** tun take callous pleasure in doing sth.; ~|**lassen** *unr. tr. V. (ugs.)* jmdn. ~**lassen** leave sb. unmoved; *(nicht interessieren)* leave sb. cold *(coll.);* ~|**machen** *tr. V. (salopp)* jmdn. ~**machen** do sb. in *(sl.);* ~**miete** die rent exclusive of heating; ~**schale** die *cold sweet soup made with fruit, beer, wine, or milk;* ~**schnäuzig** [~ʃnɔytsɪç] *(ugs.)* **1.** *Adj.* cold and insensitive; *(frech)* insolent; **2.** *adv.* coldly and insensitively; *(frech)* insolently; ~|**stellen** *tr. V. (ugs.)* jmdn. ~**stellen** put sb. out of the way *(coll. joc.)*

kam *i. u. 3. Pers. Prät. v.* **kommen**

Kambodscha [kam'bɔdʒa] **(das);** ~s Cambodia

käme *1. u. 3. Pers. Konjunktiv II v.* **kommen**

Kamel das; ~s, ~e camel

Kamera die; ~, ~s camera

Kamerad der; ~**en**, ~**en** companion; *(Freund)* friend; *(Mitschüler)* mate; *(Soldat)* comrade; *(Sport)* team-mate; **Kameradschaft** die; ~: comradeship; **kameradschaftlich 1.** *Adj.* comradely; **2.** *adv.* in a comradely way

Kamera·mann der *Pl.* ~**männer** *od.* ~**leute** cameraman

Kamerun ['kaməru:n] **(das);** ~s Cameroon; the Cameroons *pl.*

Kamille die; ~, ~n camomile
Kamin der, *schweiz.: das*; ~s, ~e fireplace; **Kamin·feger** der *(bes. südd.)*
s. Schornsteinfeger
Kamm der; ~[e]s, **Kämme a)** comb; **b)**
(bei Hühnern usw.) comb; **c)** *(Gebirgs~)* ridge; **kämmen** *tr. V.* comb
Kammer die; ~, ~n **a)** store-room; **b)**
(Biol., Med., Technik, Waffenkunde)
chamber; **c)** *(Parl.)* chamber
Kammer-: ~**diener der** *(veralt.)*
valet; ~**jäger** der pest controller;
~**musik** die; *o. Pl.* chamber music;
~**sänger** der *title awarded to singer of
outstanding merit;* ~**zofe** die *(veralt.)*
lady's maid
Kamm·garn das worsted
Kampagne [kam'panjə] die; ~, ~n
campaign
Kampf der; ~[e]s, **Kämpfe a)** *(militärisch)* battle **(um** for); **b)** *(zwischen persönlichen Gegnern)* fight; *(fig.)*
struggle; **c)** *(Wett~)* contest; *(Boxen)*
contest; bout; **d)** *(Einsatz aller Mittel)*
fight **(um, für** for; **gegen** against);
kampf·bereit *Adj.* ready to fight
postpos.; ⟨*army, troops*⟩ ready for
battle; **kämpfen** *itr. V.* **a)** fight; **b)**
(Sport: sich messen) ⟨*team*⟩ play;
⟨*wrestler, boxer*⟩ fight
Kampfer der; ~s camphor
Kämpfer der; ~s, ~, **Kämpferin** die;
~, ~nen fighter
kampf-, Kampf-: ~**fähig** *Adj.*
⟨*troops*⟩ fit for action; ⟨*boxer etc.*⟩ fit
to fight; ~**handlungen** *Pl.* fighting
sing.; ~**richter** der *(Sport)* judge;
~**unfähig** *Adj.* ⟨*troops*⟩ unfit for action; ⟨*boxer etc.*⟩ unfit to fight
kampieren *itr. V.* camp
Kanada (das); ~s Canada; **Kanadier**
[ka'na:diɐ] der; ~s, ~ Canadian; **kanadisch** *Adj.* Canadian
Kanal der; ~s, **Kanäle a)** canal; **b)**
(Geogr.) der ~: the [English] Channel;
c) *(für Abwässer)* sewer; **d)** *(zur Entwässerung, Bewässerung)* channel;
(Graben) ditch; **e)** *(Rundf., Ferns.,
Weg der Information)* channel; **Kanalisation** die; ~, ~en sewerage system; sewers *pl.;* **kanalisieren** *tr. V.*
a) *(lenken)* channel ⟨*energies, goods,
etc.*⟩; **b)** *(schiffbar machen)* canalize
Kanaren *Pl.* Canaries; **Kanarienvogel** [ka'na:riən-] der canary; **Kanarische Inseln** *Pl.* Canary Islands
Kandare die; ~, ~n curb bit; **jmdn. an
die ~ nehmen** *(fig.)* take sb. in hand

Kandidat der; ~en, ~en **a)** candidate;
b) *(beim Quiz usw.)* contestant; **Kandidatur** die; ~, ~en candidature **(auf**
+ *Akk.* for); **kandidieren** *itr. V.*
stand [as a candidate] **(für** for)
kandieren *tr. V.* candy; **kandiert**
crystallized ⟨*orange, petal*⟩; glacé
⟨*cherry, pear*⟩; candied ⟨*peel*⟩; **Kandis** der; ~, **Kandis·zucker** der rock
candy
Känguruh ['kɛŋguru] das; ~s, ~s kangaroo
Kaninchen das; ~s, ~: rabbit
Kanister der; ~s, ~: can; [metal/plastic] container
kann *1. u. 3. Pers. Sg. Präsens v.* **können**
Kännchen das; ~s, ~: [small] pot;
(für Milch) [small] jug; **Kanne** die; ~,
~n **a)** pot; *(für Milch, Wein, Wasser)*
jug; **b)** *(Henkel~)* can; *(für Milch)*
pail; *(beim Melken)* churn
kannst *2. Pers. Sg. Präsens v.* **können**
kannte *1. u. 3. Pers. Sg. Prät. v.* **kennen**
Kanon der; ~s, ~s canon
Kanone die; ~, ~n cannon; *(fig. ugs.:
Könner)* ace
Kantate die; ~, ~n *(Musik)* cantata
Kante die; ~, ~n edge; **kantig** *Adj.*
square-cut ⟨*timber, stone*⟩; roughedged ⟨*rock*⟩; angular ⟨*face*⟩; square
⟨*chin*⟩
Kantine die; ~, ~n canteen
Kanton der; ~s, ~e canton
Kantor der; ~s, ~en choirmaster and
organist
Kanu das; ~s, ~s canoe
Kanüle die; ~, ~n *(Med.)* cannula
Kanzel die; ~, ~n **a)** pulpit; **b)**
(Flugw.) cockpit
Kanzlei die; ~, ~en **a)** *(veralt.: Büro)*
office; **b)** *(Anwalts~)* chambers *pl. (of
barrister);* office *(of lawyer)*
Kanzler der; ~s, ~ chancellor
Kap das; ~s, ~s cape
Kapazität die; ~, ~en **a)** capacity; **b)**
(Experte) expert
Kapelle die; ~, ~n **a)** *(Archit.)* chapel;
b) *(Musik~)* band; [light] orchestra
Kapell·meister der bandmaster; *(im
Orchester)* conductor; *(im Theater
usw.)* musical director
Kaper die; ~, ~n caper *usu. in pl.*
kapern *tr. V.* **a)** *(hist.)* capture; **b)**
(ugs.) **jmdn. [für etw.]** ~: rope sb. in[to
sth.]
kapieren *(ugs.)* **1.** *tr. V. (ugs.)* get
(coll.); **2.** *itr. V.* **kapiert?** got it? *(coll.)*

Kapital das; ~s, ~e *od.* ~ien **a)** capital; **b)** *(fig.)* asset; **Kapitalismus** der; ~: capitalism *no art.;* **Kapitalist** der; ~en, ~en capitalist; **kapitalistisch** *Adj.* capitalistic

Kapital·verbrechen das serious offence; *(mit Todesstrafe bedroht)* capital offence

Kapitän der; ~s, ~e *(Seew.)* captain

Kapitel das; ~s, ~: chapter

Kapitulation die; ~, ~en surrender; capitulation; **seine ~ erklären** admit defeat; **kapitulieren** *itr. V.* **a)** surrender; capitulate; **b)** *(fig.: aufgeben)* give up; **vor etw.** *(Dat.)* ~: give up in the face of sth.

Kaplan der; ~s, **Kapläne** *(kath. Kirche)* chaplain; *(Hilfsgeistlicher)* curate

Kappe die; ~, ~n cap

kappen *tr. V.* **a)** *(Seemannsspr.)* cut; **b)** *(beschneiden)* cut back *(hedge etc.);* *(abschneiden)* cut off *(branches etc.)*

Käppi das; ~s, ~s garrison cap

Kapsel die; ~, ~n capsule

Kapstadt (das) Cape Town

kaputt *Adj.* **a)** broken; **das Telefon ist ~:** the phone is not working; **b)** *(ugs.: erschöpft)* shattered *(coll.)*

kaputt-: ~|**gehen** *unr. itr. V.; mit sein (ugs.) (entzweigehen)* break; *(machine)* break down, *(sl.)* pack up; *(light-bulb)* go; *(zerbrechen)* be smashed; ~|**lachen** *refl. V. (ugs.)* kill oneself [laughing] *(coll.);* ~|**machen** (ugs.) 1. *tr. V.* break; spoil *(sth. made with effort);* ruin *(clothes, furniture, etc.);* finish *(person)* off; 2. *refl. V.* wear oneself out

Kapuze die; ~, ~n hood; *(bei Mönchen)* cowl; hood; **Kapuziner** der; ~s, ~: Capuchin [friar]

Karabiner der; ~s, ~: carbine

Karaffe die; ~, ~n carafe; *(mit Glasstöpsel)* decanter

Karambolage [karambo'la:ʒə] die; ~, ~n *(ugs.)* crash; collision

Karamel der *(schweiz.:* das); ~s caramel; **Karamel·bonbon** der *od.* das caramel [toffee]

Karat das; ~|e|s, ~e carat

Karate das; ~|s| karate

Karawane die; ~, ~n caravan

Kardinal der; ~s, **Kardinäle** *(kath. Kirche)* cardinal

Kardinal-: ~**tugend** die; *meist Pl.* cardinal virtue; ~**zahl** die cardinal [number]

Karenz die; ~, ~en, **Karenz·zeit** die waiting period

Kar·freitag der Good Friday

karg 1. *Adj.* meagre *(wages etc.);* frugal *(meal etc.);* poor *(light, accommodation);* *(wenig fruchtbar)* barren; 2. *adv.* ~ **bemessen sein** *(helping)* be mingy *(Brit. coll.);* *(supply)* be scanty; ~ **leben** live frugally; **kärglich** 1. *Adj.* meagre, poor *(wages etc.);* poor *(light);* frugal *(meal);* scanty *(supply);* 2. *adv.* poorly *(lit, paid, rewarded)*

karibisch *Adj.* Caribbean

kariert *Adj.* check, checked *(material, pattern);* check *(jacket etc.);* squared *(paper)*

Karies ['ka:riɛs] die; ~: caries

Karikatur die; ~, ~en cartoon; *(Porträt)* caricature; **Karikaturist** der; ~en, ~en cartoonist; *(Porträtist)* caricaturist; **karikieren** *tr. V.* caricature

karitativ *Adj.* charitable

Karl [karl] (der) Charles; ~ **der Große** Charlemagne

Karneval ['karnəval] der; ~s, ~e *od.* ~s carnival; ~ **feiern** join in the carnival festivities

Karnickel das; ~s, ~ *(landsch.)* rabbit

Kärnten (das); ~s Carinthia

Karo das; ~s, ~s **a)** square; *(auf der Spitze stehend)* diamond; **b)** *o. Pl. (~muster)* check; **c)** *o. Art.; o. Pl. (Kartenspiel: Farbe)* diamonds *pl.;* **d)** *(Kartenspiel: Karte)* diamond; **Karo·as** das ace of diamonds

Karosse die; ~, ~n [state-]coach; **Karosserie** die; ~, ~n bodywork

Karotte die; ~, ~n small carrot

Karpaten *Pl.* Carpathians; Carpathian Mountains

Karpfen der; ~s, ~: carp

Karre die; ~, ~n *(bes. nordd.)* **a)** *s.* **Karren; b)** *(abwertend: Fahrzeug)* [old] heap *(coll.)*

Karree das; ~s, ~s: **ums ~ gehen/fahren** walk/drive round the block

karren *tr. V.* **a)** cart; **b)** *(salopp: mit einem Auto)* run *(coll.);* **Karren** der; ~s, ~ *(bes. südd., österr.)* cart; *(zweirädrig)* barrow

Karriere [ka'riɛ:rə] die; ~, ~n career; ~ **machen** make a [successful] career for oneself

Kar·samstag der Easter Saturday

Karte die; ~, ~n card; *(Speise~)* menu; *(Fahr~, Flug~, Eintritts~)* ticket; *(Land~)* map; **alles auf eine ~ setzen** stake everything on one chance; **Kartei** die; ~, ~en card file

Kartei-: ~**karte** die file card; ~**kasten** der file-card box

Kartęll das; ~s, ~e *(Wirtsch., Politik)* cartel

Kartęll-: ~**amt das** *government body concerned with the control and supervision of cartels;* ≈ Monopolies and Mergers Commission *(Brit.);* ~**gesetz das** law relating to cartels; ≈ monopolies law *(Brit.)*

Karten-: ~**haus das** house of cards; ~**spiel das a)** *(Spiel mit Karten)* card-game; **b)** *(Satz Spielkarten)* pack or *(Amer.)* deck [of cards]; ~**vor·verkauf der;** *o. Pl.* advance booking

Kartoffel die; ~, ~n potato

Kartoffel-: ~**brei der** mashed potatoes *pl.;* mash *(coll.);* ~**chips** *Pl.* [potato] crisps *(Brit.)* or *(Amer.)* chips; ~**käfer der** Colorado beetle; ~**kloß der** potato dumpling; ~**puffer der** potato pancake *(made from grated raw potatoes);* ~**püree das;** *s.* ~**brei**

Karton [kar'tɔŋ] **der;** ~s, ~s **a)** *(Pappe)* card[board]; **b)** *(Schachtel)* cardboard box

Karussęll das; ~s, ~s *od.* ~e merry-go-round; carousel *(Amer.);* *(kleineres)* roundabout

Kar·woche die Holy Week

Karzinom das; ~s, ~e *(Med.)* carcinoma

kaschieren *tr. V.* conceal; hide; disguise ⟨*fault*⟩

¹**Kaschmir (das);** ~s Kashmir; ²**Kaschmir der;** ~s, ~e *(Textilw.)* cashmere

Käse der; ~s, ~: cheese; *(ugs. abwertend: Unsinn)* rubbish

Käse-: ~**blatt das** *(salopp abwertend)* rag; ~**glocke die** cheese dome

Kaserne die; ~, ~n barracks *sing.* or *pl.*

käse·weiß *Adj. (ugs.)* [as] white as a sheet; **käsig** *Adj. (ugs.)* pasty; pale

Kasino das; ~s, ~s **a)** *(Spiel~)* casino; **b)** *(Offiziers~)* [officers'] mess; **c)** *(Speiseraum)* canteen

Kasko·versicherung die *(Voll~)* comprehensive insurance; *(Teil~)* insurance against theft, fire, or act of God

Kasper der; ~s, ~: ≈ Punch; *(fig. ugs.)* clown; **Kasperl das;** ~s, ~[n] *(österr.),* **Kasperle das** *od.* **der;** ~s, ~: *s.* Kasper

Kasper-: ~**puppe die** ≈ Punch and Judy puppet; ~**theater das** ≈ Punch and Judy show; *(Puppenbühne)* ≈ Punch and Judy theatre

Kasse die; ~, ~n **a)** cash-box; *(Regi-*

strier~) till; **b)** *(Ort zum Bezahlen)* cash desk; *(im Supermarkt)* checkout; *(in einer Bank)* counter; **c)** *(Kassenraum)* cashier's office; **d)** *(Theater~, Kino~)* box-office

Kasseler das; ~s smoked loin of pork

Kassen-: ~**arzt der** *doctor who treats members of health insurance schemes;* ~**bon der** sales slip; receipt; ~**patient der** *patient who is a member of a health insurance scheme;* ~**zettel der** *s.* ~**bon**

Kassette die; ~, ~n **a)** box; case; **b)** *(mit Büchern, Schallplatten)* boxed set; *(Tonband~, Film~)* cassette; **Kassętten·recorder der** cassette recorder

kassieren 1. *tr. V.* **a)** collect; **b)** *(ugs.: wegnehmen)* confiscate; take away ⟨*driving licence*⟩; **2.** *itr. V.* **a) bei jmdm.** ~: give sb. his/her bill or *(Amer.)* check; *(ohne Rechnung)* settle up with sb.; **darf ich bei Ihnen** ~? would you like your bill?/can I settle up with you?; **Kassierer der;** ~s, ~, **Kassiererin die;** ~, ~nen cashier; *(bei einem Verein)* treasurer

Kastanie [kas'taːniə] **die;** ~, ~n chestnut; **kastanien·braun** *Adj.* chestnut

Kästchen das; ~s, ~ **a)** small box; **b)** *(vorgedrucktes Quadrat)* square; *(auf Fragebögen)* box

Kaste die; ~, ~n caste

kasteien *refl. V.* **a)** *(als Bußübung)* chastise oneself; **b)** *(sich Entbehrungen auferlegen)* deny oneself; **Kasteiung die;** ~, ~en **a)** *(als Bußübung)* self-chastisement; **b)** *(Auferlegung von Entbehrungen)* self-denial

Kastęll das; ~s, ~e **a)** *(hist.: röm. Lager)* fort; **b)** *(Burg)* castle

Kasten der; ~s, Kästen **a)** box; *(für Flaschen)* crate; **b)** *(ugs.: Briefkasten)* post-box; **c)** *(ugs. abwertend) (Gebäude)* barracks *sing.* or *pl.;* *(Auto)* heap *(coll.);* *(fig. ugs.)* **etw. auf dem** ~ **haben** have got it up top *(coll.);* **Kastenbrot das** tin[-loaf]

Kastration [kastra'tsioːn] **die;** ~, ~en castration; **kastrieren** *tr. V.* castrate

Katalog der; ~[e]s, ~e *(auch fig.)* catalogue; **katalogisieren** *tr. V.* catalogue

Katalysator der; ~s, ~en [-za'toːrən] *(Chemie, fig.)* catalyst; *(Kfz-W.)* catalytic converter

katapultieren *tr. V. (auch fig.)* catapult; eject ⟨*pilot*⟩

Katarrh [ka'tar] der; ~s, ~e *(Med.)* catarrh

katastrophal [katastro'fa:l] **1.** *Adj.* disastrous; *(stärker)* catastrophic; **2.** *adv.* disastrously; *(stärker)* catastrophically; **Katastrophe** [katas-'tro:fə] die; ~, ~n *(Unglück)* disaster; *(stärker, auch Literaturw.)* catastrophe

Katastrophen-: ~**alarm** der disaster alert; ~**gebiet** das disaster area; ~**schutz** der *(Organisation)* emergency services *pl.; (Maßnahmen)* disaster procedures *pl.*

Kategorie die; ~, ~n category; **kategorisch 1.** *Adj.* categorical; **2.** *adv.* categorically

Kater der; ~s, ~ **a)** tom-cat; **b)** *(ugs.)* hangover

Kathedrale die; ~, ~n cathedral

Katholik der; ~en, ~en, **Katholikin** die; ~, ~nen [Roman] Catholic; **katholisch** *Adj.* [Roman] Catholic; **Katholizismus** der; ~: [Roman] Catholicism *no art.*

Kätzchen das; ~s, ~ **a)** little cat; pussy; *(junge Katze)* kitten; **b)** *meist Pl.* catkin; **Katze** die; ~, ~n cat

katzen-, Katzen-: ~**auge** das reflector; Cat's-eye (P); ~**jammer** der **a)** *(Kater)* hangover; **b)** *(fig.)* mood of depression; ~**musik** die *(ugs. abwertend)* terrible row *(coll.);* ~**sprung** der stone's throw; ~**wäsche** die *(ugs.)* ~wäsche machen have a lick and a promise *(coll.)*

Kauderwelsch das; ~[s] gibberish *no indef. art.*

kauen *tr., itr. V.* chew; [die] Nägel ~: bite one's nails

kauern 1. *itr., refl. V.* crouch [down]; *(ängstlich)* cower

Kauf der; ~[e]s, Käufe **a)** *(das Kaufen)* buying; purchasing *(formal);* **b)** *(das Gekaufte)* purchase; **kaufen 1.** *tr. V.* buy; purchase; **2.** *itr. V. (einkaufen)* shop; **Käufer** der; ~s, ~: buyer; purchaser *(formal)*

Kauf-: ~**haus** das department store; ~**kraft** die *(Wirtsch.)* **a)** *(Wert des Geldes)* purchasing power; **b)** *(Zahlungsfähigkeit)* spending power

käuflich 1. *Adj.* **a)** for sale *postpos.;* **b)** *(bestechlich)* venal; ~ sein be easily bought; **2.** *adv.* etw. ~ erwerben/erstehen purchase sth.; **Kauf·mann** der; *Pl.* Kaufleute **a)** *(Geschäftsmann)* businessman; *(Händler)* trader; **b)** *(Besitzer)* shopkeeper; *(eines Lebens-*

mittelladens) grocer; **kaufmännisch** *Adj.* commercial; business *attrib.;* **Kauf·preis** der purchase price

Kau·gummi der *od.* das; ~s, ~s chewing gum

Kaukasus der; ~: the Caucasus

Kaulquappe die; ~, ~n tadpole

kaum *Adv.* hardly; scarcely; ~ hatte er Platz genommen, als ...: no sooner had he sat down than ...

kausal *Adj. (geh., Sprachw.)* causal

Kau·tabak der chewing tobacco

Kaution [kau'tsio:n] die; ~, ~en **a)** *(bei Freilassung eines Gefangenen)* bail; **b)** *(beim Mieten einer Wohnung)* deposit

Kautschuk der; ~s, ~e rubber

Kauz der; ~es, Käuze **a)** *(Wald~)* tawny owl; *(Stein~)* little owl; **b)** *(Sonderling)* strange fellow; oddball *(coll.)*

Kavalier [kava'li:ɐ] der; ~s, ~e gentleman; **Kavaliers·delikt** das trifling offence

Kavallerie die; ~, ~n *(Milit. hist.)* cavalry; **Kavallerist** der; ~en, ~en cavalryman

Kaviar ['ka:viar] der; ~s, ~e caviare

kcal *Abk.* Kilo[gramm]kalorie kcal

keck 1. *Adj.* **a)** cheeky; saucy *(Brit.);* **b)** *(veralt.: verwegen)* bold; **c)** *(flott)* jaunty, pert ⟨hat etc.⟩; **2.** *adv.* **a)** cheekily; saucily *(Brit.);* **b)** *(veralt.: verwegen)* boldly; **c)** *(flott)* jauntily; **Keckheit** die; ~, ~en **a)** cheek; sauce *(Brit.);* **b)** *(veralt.: Kühnheit)* boldness

Kegel der; ~s, ~ **a)** cone; **b)** *(Spielfigur)* skittle; *(beim Bowling)* pin; **Kegel·bahn** die skittle alley; **kegelförmig** *Adj.* conical; **kegeln 1.** *itr. V.* play skittles *or* ninepins; **2.** *tr. V.* eine Partie ~: play a game of skittles *or* ninepins; **eine Neun ~:** score a nine

Kehle die; ~, ~n throat; **Kehl·kopf** der *(Anat.)* larynx

Kehre die; ~, ~n sharp bend; ¹**kehren 1.** *tr. V.* turn; **2.** *refl. V.* turn

²**kehren 1.** *itr. V. (bes. südd.)* sweep; do the sweeping; **2.** *tr. V.* sweep; *(mit einem Handfeger)* brush; **Kehricht** der *od.* das; ~s *(schweiz.: Müll)* refuse; garbage *(Amer.)*

Kehr·seite die **a)** back; *(einer Münze, Medaille)* reverse; *(scherzh.) (Gesäß)* backside; **b)** *(nachteiliger Aspekt)* drawback; disadvantage; **kehrt|machen** *itr. V. (ugs.)* turn [round and go] back

keifen *itr. V. (abwertend)* nag

Keil der; ~|e|s, ~e a) *(zum Spalten)* wedge; b) *(zum Festklemmen)* chock; *(unter einer Tür)* wedge; **keilen** *refl. V. (ugs.: sich prügeln)* fight; scrap; **Keiler** der; ~s, ~ *(Jägerspr.)* wild boar; **Keilerei** die; ~, ~en *(ugs.)* punch-up *(coll.)*; fight

Keil-: ~**riemen** der *(Technik)* V-belt; ~**schrift** die cuneiform script

Keim der; ~|e|s, ~e *(Bot.)* shoot; *(Biol.)* embryo; **keimen** *itr. V.* germinate; *(fig.)* ⟨hope⟩ stir; **keim·frei** *Adj.* germ-free; sterile; **Keim·zelle** die nucleus

kein *Indefinitpron.* a) no; b) *(ugs.: nicht ganz, nicht einmal)* less than; **kein...** *Indefinitpron.* ~**er/~e** nobody; no one; ~**s** von beiden neither [of them]; **keinerlei** *indekl. unbest. Gattungsz.* no ... what[so]ever

keines-: ~**falls** *Adv.* on no account; ~**wegs** *Adv.* by no means

kein·mal *Adv.* not [even] once

Keks der; ~ *od.* ~es, ~ *od.* ~e biscuit *(Brit.)*; cookie *(Amer.)*

Kelch der; ~|e|s, ~e goblet; *(Rel.)* chalice

Kelle die; ~, ~n a) ladle; b) *(Signalstab)* signalling disc; c) *(Maurer~)* trowel

Keller der; ~s, ~: cellar; *(~geschoß)* basement; **Keller·assel** die woodlouse; **Kellerei** die; ~, ~en winery; *(Kellerräume)* [wine] cellars *pl.*; **Keller·geschoß** das basement

Kellner der; ~s, ~: waiter; **kellnern** *itr. V. (ugs.)* work as a waiter/waitress

Kelte der; ~n, ~n Celt

Kelter die; ~, ~n winepress; **keltern** *tr. V.* press ⟨grapes etc.⟩

keltisch *Adj.* Celtic

Kenia (das); ~s Kenya; **Kenianer** der; ~s, ~: Kenyan

kennen *unr. tr. V.* a) know; b) *(bekannt sein mit)* know; **kennen|lernen** *tr. V.* get to know; *(erstmals begegnen)* meet; *(in Berührung gebracht werden mit)* know; **Kenner** der; ~s, ~: expert (+ *Gen.* on); *(von Wein, Speisen)* connoisseur; **Kennerblick** der expert eye; **mit ~:** with an expert eye; **Kenn·marke** die [police] identification badge; ≈ [police] warrant card *or (Amer.)* ID card; **kenntlich** *Adj.*: ~ **sein** be recognizable (**an** by); **etw./jmdn.** ~ **machen** mark sth./make sb. [easily] identifiable; **Kenntnis** die; ~, ~se knowledge

kenn-, **Kenn**-: ~**wort** das; *Pl.* ~wörter code-word; *(Parole)* password; code-word; ~**zahl** die index; ~**zeichen** das a) sign; b) *(Erkennungszeichen)* badge; *(auf einem Behälter, einer Ware usw.)* label; *(am Fahrzeug)* registration number; ~**zeichnen** *tr. V.* a) mark; label; mark ⟨way⟩; b) *(charakterisieren)* characterize; ~**zeichnend** *Adj.* typical, characteristic (**für** of)

kentern *itr. V. mit sein* capsize

Keramik die; ~, ~en *o. Pl.* ceramics *pl.*; pottery; *(~gegenstand)* piece of pottery

Kerbe die; ~, ~n notch

Kerbel der; ~s chervil

Kerb·holz das: etwas auf dem ~holz haben *(ugs.)* have done a job *(sl.)*

Kerker der; ~s, ~ *(hist.)* dungeons *pl.*; *(einzelne Zelle)* dungeon

Kerl der; ~s, ~e *(nordd., md. auch:* ~s) *(ugs.)* fellow *(coll.)*; bloke *(Brit. sl.)*

Kern der; ~|e|s, ~e pip; *(von Steinobst)* stone; *(von Nüssen usw.)* kernel; *(Atom~)* nucleus; *(fig.)* der ~ einer Sache the heart of a matter; der harte ~: the hard core

kern-, **Kern**-: ~**energie** die nuclear energy *no art.*; ~**gehäuse** das core; ~**gesund** *Adj.* fit as a fiddle *pred.*

kernig *Adj.* earthy ⟨language⟩; forceful ⟨speech⟩; pithy ⟨saying⟩

kern-, **Kern**-: ~**kraft** die nuclear power; ~**kraftwerk** das nuclear power station *or* plant; ~**los** *Adj.* seedless; ~**obst** das pomaceous fruit; ~**physik** die nuclear physics *sing., no art.*; ~**reaktor** der nuclear reactor; ~**seife** die washing soap; ~**spaltung** die *(Physik)* nuclear fission *no art.*; ~**waffe** die; *meist Pl.* nuclear weapon

Kerze die; ~, ~n candle

kerzen-, **Kerzen**-: ~**gerade**, *(ugs.)* ~**grade** 1. *Adj.* dead straight; 2. *adv.* bolt upright; ~**halter** der candle-holder; ~**leuchter** der candlestick

keß 1. *Adj.* a) pert; jaunty ⟨hat, dress, etc.⟩; b) *(frech)* cheeky; 2. *adv.* a) *(flott)* jauntily; b) *(frech)* cheekily

Kessel der; ~s, ~ a) kettle; *(zum Kochen)* pot; *(Wasch~)* copper; b) *(Berg~)* basin-shaped valley; c) *(Milit.)* encircled area

Kessel-: ~**stein** der; *o. Pl.* scale; ~**treiben** das *(Hetzkampagne)* witch-hunt

Kette die; ~, ~n chain; *(Hals~)* necklace; *(von Ereignissen)* string; **ketten** *tr. V.* chain (**an** + *Akk.* to)

Ketten-: ~**hund** der guard-dog *(kept on a chain);* ~**raucher** der chainsmoker

Ketzer der; ~s, ~ *(auch fig.)* heretic; **Ketzerei** die; ~, ~en *(auch fig.)* heresy

keuchen *itr. V.* gasp for breath; **Keuch · husten** der whooping cough *no art.*

Keule die; ~, ~n a) club; b) *(Kochk.)* leg

keusch 1. *Adj.* chaste; 2. *adv.* ~ **leben** lead a chaste life; **Keuschheit** die; ~ chastity

Kfz [ka:ǀɛfˈtsɛt] *Abk.* Kraftfahrzeug

kg *Abk.* Kilogramm kg

KG *Abk.* Kommanditgesellschaft

kichern *itr. V.* giggle

kicken *(ugs.)* 1. *itr. V.* play football; 2. *tr. V.* kick

kidnappen [ˈkɪtnɛpn̩] *tr. V.* kidnap; **Kidnapper** der; ~s, ~: kidnapper

Kiebitz der; ~es, ~e lapwing; peewit

¹**Kiefer** der; ~s, ~: jaw; *(~knochen)* jaw-bone

²**Kiefer** die; ~, ~n pine[tree]

Kiefer · höhle die *(Anat.)* maxillary sinus

Kiefern · holz das pine[-wood]

Kiel der; ~ǀeǀs, ~e keel; **kiel · holen** *tr. V. (Seemannsspr.)* keel-haul *(person);* **Kiel · wasser** das wake

Kieme die; ~, ~n; *meist Pl.* gill

Kien der; ~ǀeǀs resinous wood

Kies der; ~es, ~e gravel; *(auf dem Strand)* shingle; **Kiesel** der; ~s, ~: pebble; **Kiesel · stein** der pebble

Kies-: ~**grube** die gravel pit; ~**weg** der gravel path

kiffen *itr. V. (ugs.)* smoke pot *(sl.)* or grass *(sl.);* **Kiffer** der; ~s, ~ *(ugs.)* pot-head *(sl.)*

kikeriki [kikəriˈkiː] *Interj. (Kinderspr.)* cock-a-doodle-doo

Killer der; ~s, ~ *(salopp)* killer; *(gegen Bezahlung)* hit man *(sl.)*

Kilo das; ~s, ~ǀsǀ kilo; **Kilo · gramm** das kilogram; **Kilometer** der; ~s, ~: kilometre; **kilometer · lang** 1. *Adj.* miles long *pred.;* 2. *adv.* for miles [and miles]; **Kilometer · stand** der mileage reading

Kimme die; ~, ~n sighting notch

Kimono der; ~s, ~s kimono

Kind das; ~ǀeǀs, ~er a) child; **ein** ~ **erwarten** be expecting; [~er,] ~er! my goodness!

Kinder-: ~**arzt** der paediatrician; ~**bett** das cot; *(für größeres Kind)* child's bed; ~**dorf** das children's village

Kinderei die; ~, ~en childishness *no indef. art., no pl.*

kinder-, Kinder-: ~**feindlich** *Adj.* hostile to children *pred.;* ~**freundlich** *Adj.* fond of children *pred.;* ⟨*town, resort*⟩ which caters for children; ⟨*planning, policy*⟩ which caters for the needs of children; ~**garten** der nursery school; ~**gärtnerin** die nusery-school teacher; ~**heilkunde** die paediatrics *sing., no art.;* ~**hort** der day-home for schoolchildren; ~**lähmung** die poliomyelitis; ~**leicht** *(ugs.) Adj.* childishly simple; dead easy; **das ist** ~**leicht** it's kid's stuff *(coll.);* it's child's play; ~**lieb** *Adj.* fond of children *pred.;* ~**los** *Adj.* childless; ~**reich** *Adj.* with many children *postpos., not pred.;* ~**sterblichkeit** die child mortality; ~**stube** die; *o. Pl.* **eine gute/schlechte** ~**stube gehabt haben** have been well/badly brought up; ~**teller** der *(auf der Speisekarte)* children's menu; ~**wagen** der pram *(Brit.);* baby carriage *(Amer.); (Sportwagen)* push-chair *(Brit.);* stroller *(Amer.)*

Kindes-: ~**alter** das; *o. Pl.* childhood; ~**mißhandlung** die *(Rechtsw.)* child abuse

Kindheit die; ~: childhood; **kindisch** 1. *Adj.* childish, infantile; naïve ⟨*ideas*⟩; 2. *adv.* childishly; **kindlich** 1. *Adj.* childlike; 2. *adv.* ⟨*behave*⟩ in a childlike way

Kinkerlitzchen *Pl. (ugs.)* trifles

Kinn das; ~ǀeǀs, ~e chin

Kinn-: ~**haken** der hook to the chin; ~**lade** die jaw

Kino das; ~s, ~s cinema *(Brit.);* movie theater *(Amer.);* **Kino · karte** die cinema ticket *(Brit.);* movie ticket *(Amer.)*

Kiosk der; ~ǀeǀs, ~e kiosk

¹**Kippe** die; ~, ~n *(ugs.)* cigarette end; dog-end *(sl.)*

²**Kippe** die; ~, ~n a) *(Bergmannsspr.)* slag-heap; **etw. steht auf der** ~ *(fig.)* it's touch and go with sth.; *(etw. ist noch nicht entschieden)* sth. hangs in the balance; **kippen** 1. *tr. V.* a) tip [up]; b) *(ausschütten)* tip [out]; 2. *itr. V.; mit sein* tip over; ⟨*top-heavy object*⟩ topple over; ⟨*person*⟩ topple; ⟨*boat*⟩ overturn; ⟨*car*⟩ roll over

Kipp-: ~**fenster das** horizontally pivoted window; ~**schalter der** tumbler switch

Kirche die; ~, ~**n** church; **in die** ~ **gehen** go to church

Kirchen-: ~**fest das** church festival; ~**lied das** hymn; ~**musik die** church music; ~**steuer die** church tax

Kirch·hof der *(veralt.)* churchyard; **kirchlich 1.** *Adj.* ecclesiastical; church *attrib.* ⟨*wedding, funeral*⟩; **2.** *adv.* ~ **getraut/begraben werden** have a church wedding/funeral

Kirch-: ~**turm der** [church] steeple; *(ohne Turmspitze)* church tower; ~**weih die;** ~, ~**en** fair *(held on the anniversary of the consecration of a church)*

Kirmes die; ~, **Kirmessen** *(bes. md., niederd.) s.* Kirchweih

Kirsch·baum der cherry[-tree]; **Kirsche die;** ~, ~**n** cherry

Kirsch-: ~**torte die** cherry gateau; *(mit Tortenboden)* cherry flan; ~**wasser das** kirsch

Kissen das; ~**s,** ~: cushion; *(Kopf~)* pillow

Kiste die; ~, ~**n** box; *(Truhe)* chest; *(Latten~)* crate

Kitsch der; ~|e|s kitsch; **kitschig** *Adj.* kitschy

Kitt der; ~|e|s, ~e putty; *(für Porzellan, Kacheln usw.)* cement

Kittchen das; ~**s,** ~ *(ugs.)* clink *(sl.)*

Kittel der; ~**s,** ~ **a)** overall; *(eines Arztes usw.)* white coat; **b)** *(hemdartige Bluse)* smock

kitten *tr. V.* cement [together]

Kitz das; ~**es,** ~**e** *(Reh~)* fawn; *(Ziegen~, Gemsen~)* kid

kitzeln *tr., itr. V.* tickle; **kitzlig** *Adj.* *(auch fig.)* ticklish

KKW [ka:ka:'|ve:] *Abk.* Kernkraftwerk

Klacks der; ~**es,** ~**e** *(ugs.)* dollop *(coll.);* (~ *Senf)* dab

Kladde die; ~, ~**n** rough book

Kladderadatsch der; ~|e|s, ~e *(ugs.)* unholy mess *(coll.)*

klaffen *itr. V.* yawn; ⟨*hole, wound*⟩ gape; **kläffen** *itr. V. (abwertend)* yap

Klafter ['klaftɐ] **der** *od.* **das;** ~**s,** ~ *(Raummaß für Holz)* cord

Klage die; ~, ~**n a)** *(Äußerung der Trauer)* lament; **b)** *(Beschwerde)* complaint; **c)** *(Rechtsw.)* action; *(im Strafrecht)* charge; **klagen 1.** *itr. V.* **a)** *(geh.: jammern)* wail; *(stöhnend)* moan; **b)** *(sich beschweren)* complain (**über** + *Akk.* about); **c)** *(bei Gericht)*

take legal action; **2.** *tr. V.* **jmdm. sein Leid/seine Not** ~: pour out one's sorrows *pl.*/troubles *pl.;* **Kläger der;** ~**s,** ~, **Klägerin die;** ~, ~**nen** *(im Zivilrecht)* plaintiff; *(im Strafrecht)* prosecuting party; *(bei einer Scheidung)* petitioner; **kläglich** *Adj.* **a)** *(mitleiderregend)* pitiful; **b)** *(minderwertig)* pathetic; **c)** *(erbärmlich)* despicable ⟨*behaviour, role, compromise*⟩; pathetic ⟨*result, defeat*⟩

Klamauk der; ~**s** *(ugs. abwertend)* fuss; *(Lärm, Krach)* row *(coll.)*

klamm *Adj.* **a)** *(feucht)* cold and damp; **b)** *(steif)* numb; **Klammer die;** ~, ~**n** *(Wäsche~)* peg; *(Haar~)* [hair-]grip; *(Zahn~)* brace; *(Büro~)* paper-clip; *(Heft~)* staple; *(Schriftzeichen)* bracket; **klammern 1.** *refl. V.* **sich an jmdn./etw.** ~ *(auch fig.)* cling to sb./sth.; **2.** *tr. V.* **a) eine Wunde** ~: close a wound with a clip/clips; **b)** *(mit einer Büroklammer)* clip; *(mit einer Heftmaschine)* staple; *(mit Wäscheklammern)* peg

Klamotten *Pl.* *(salopp)* *(Kleidung)* gear *sing. (sl.); (Kram)* stuff *sing.*

Klampfe die; ~, ~**n** *(volkst.: Gitarre)* guitar

klang *1. u. 3. Pers. Sg. Prät. v.* klingen

Klang der; ~|e|s, **Klänge a)** *(Ton)* sound; **b)** *(~farbe)* tone

Klapp·bett das folding bed; **Klappe die;** ~, ~**n a)** [hinged] lid; *(am LKW)* tail-gate; *(seitlich)* side-gate; *(am Kombiwagen)* back; *(am Ofen)* [drop-]door; **b)** *(an Musikinstrumenten)* key; *(an einer Trompete)* valve; **c)** *(Filmjargon)* clapper-board; **d)** *(salopp: Mund)* trap *(sl.);* **klappen 1.** *tr. V.* **nach oben/unten** ~: turn up/down ⟨*collar, hat-brim*⟩; lift up/put down ⟨*lid*⟩; **nach vorne/hinten** ~: tilt forward/back ⟨*seat*⟩; **2.** *itr. V.* **a)** ⟨*door, shutter*⟩ bang; **b)** *(stoßen)* bang; **c)** *(ugs.: gelingen)* work out all right; **klapperig** *Adj.* rickety; **klappern** *itr. V.* **a)** rattle; **b)** *(ein Klappern erzeugen)* make a clatter; **Klapperschlange die** rattlesnake; **klapprig** *Adj.* rickety

Klapp-: ~**sitz der** tip-up seat; ~**stuhl der** folding chair

Klaps der; ~**es,** ~**e** *(ugs.)* smack; slap; **Klaps·mühle die** *(salopp)* loony-bin *(sl.)*

klar 1. *Adj.* **a)** clear; straight ⟨*question, answer*⟩; **sich** *(Dat.)* **über etw.** *(Akk.)* **im** ~**en sein** realize sth.; **b)** *nicht attr.*

(fertig) ready; **2.** *adv.* clearly; **Klär-anlage die** sewage treatment plant; **Klare der;** ~n, ~n schnapps; **klären 1.** *tr. V.* **a)** settle ⟨*question, issue, matter*⟩*; clarify ⟨*situation*⟩; clear up ⟨*case, affair, misunderstanding*⟩; **b)** *(reinigen)* purify; treat ⟨*effluent, sewage*⟩; **2.** *refl. V.* **a)** ⟨*situation*⟩ become clear; ⟨*question, issue, matter*⟩ be settled; **b)** *(rein werden)* ⟨*liquid, sky*⟩ clear; ⟨*weather*⟩ clear [up]; **klar|gehen** *unr. itr. V.; mit sein (ugs.)* go OK *(coll.);* **Klarheit die;** ~: clarity; **sich** *(Dat.)* **über etw.** *(Akk.)* ~ **verschaffen** clarify sth.

Klarinette die; ~, ~n clarinet

klar-: ~**|machen** *tr. V.* **a)** *(ugs.)* make clear; **b)** *(Seemannsspr.)* get ready; ~**|sehen** *unr. itr. V.* understand the matter

Klarsicht·folie die transparent film; **klar|stellen** *tr. V.* clear up; clarify; **Klar·text der** *(auch DV)* clear text; **im** ~**text** *(fig.)* in plain language; **Klärung die;** ~, ~**en a)** clarification; **b)** *(Reinigung)* purification; *(von Abwässern)* treatment; **klar|werden** *unr. V.; mit sein; nur im Inf. und Part. zusammengeschrieben* **1.** *refl. V.* **sich** *(Dat.)* **über etw.** *(Akk.)* ~: realize sth.; **2.** *itr. V.* **jmdm. wird etw. klar** sth. becomes clear to sb.

klasse *(ugs.)* **1.** *indekl. Adj.* great *(coll.);* **2.** *adv.* marvellously; **Klasse die;** ~, ~n **a)** *(Schul~)* class; *(Raum)* class-room; *(Stufe)* year; grade *(Amer.);* **b)** *(Sport)* league; *(Boxen)* division; **c)** *(Fahrzeug~, Boots~, Qualitätsstufe)* class

klassen-, Klassen-: ~**arbeit die** *(Schulw.)* [written] class test; ~**gesellschaft die** *(Soziol.)* class society; ~**kampf der** *(marx.)* class struggle; ~**lehrer der,** ~**lehrerin die** *(Schulw.)* class teacher; ~**los** *Adj.* *(Soziol.)* classless; ~**sprecher der,** ~**sprecherin die** *(Schulw.)* class spokesman; ~**treffen das** *(Schulw.)* class reunion; ~**ziel das** *(Schulw.)* required standard *(for pupils in a particular class);* ~**zimmer das** *(Schulw.)* classroom

klassifizieren *tr. V.* classify (**als** as); **Klassifizierung die;** ~, ~en classification; **Klassik die;** ~ **a)** *(Antike)* classical antiquity *no art.;* **b)** *(Zeit kultureller Höchstleistung)* classical period

Klassiker der; ~s, ~: classical writer/

composer; **klassisch** *Adj.* classical; *(vollendet, zeitlos; auch iron.)* classic; **Klassizismus der;** ~: classicism

Klatsch der; ~[e]s, ~e **a)** *o. Pl. (ugs. abwertend)* gossip; **b)** *(Geräusch)* smack; **Klatsch·base die** *(ugs. abwertend)* gossip; **klatschen** *itr. V.* **a)** *auch mit sein* ⟨*waves, wet sails*⟩ slap; **b)** *(mit den Händen; applaudieren)* clap; **c)** *(schlagen)* slap; **d)** *(ugs. abwertend: reden)* gossip (**über** + *Akk.* about); **klatschhaft** *Adj.* gossipy; fond of gossip *pred.;* **Klatsch·mohn der** corn-poppy; **klatsch·naß** *Adj. (ugs.)* sopping wet; dripping wet ⟨*hair*⟩

Klaue die; ~, ~n **a)** claw; *(von Raubvögeln)* talon; *(salopp: Hand)* mitt *(sl.);* **b)** *o. Pl. (salopp abwertend: Schrift)* scrawl; **klauen** *(ugs.)* **1.** *tr. V.* pinch *(sl.);* **jmdm. etw.** ~: pinch sth. from sb.; **2.** *itr. V.* pinch *(sl.)* things

Klause die; ~, ~n hermitage; *(Klosterzelle)* cell; **Klausel die;** ~, ~n clause; *(Bedingung)* condition; *(Vorbehalt)* proviso

Klavier [kla'vi:ɐ] **das;** ~s, ~e piano

Klebe·folie die adhesive film; **kleben 1.** *itr. V.* **a)** stick (**an** + *Dat.* to); **b)** *(ugs.: klebrig sein)* be sticky (**von, vor** + *Dat.* with); **2.** *tr. V.* **a)** *(befestigen)* stick; *(mit Klebstoff)* glue; **jmdm. eine** ~ *(salopp)* belt sb. one *(coll.);* **b)** *(reparieren)* stick *or* glue ⟨*vase etc.*⟩ back together; **Kleber der;** ~s, ~: adhesive; glue; **klebrig** *Adj.* sticky

Kleb-: ~**stoff der** adhesive; glue; ~**streifen der** adhesive *or* sticky tape

kleckern *(ugs.)* *itr. V.* make a mess; **Klecks der;** ~es, ~e **a)** stain; *(nicht aufgesogen)* blob; *(Tintenfleck)* [ink-] blot; **b)** *(ugs.: kleine Menge)* spot; *(von Senf, Mayonnaise)* dab; **klecksen** *itr. V.* **a)** make a stain/stains; *(mit Tinte)* make a blot/blots; ⟨*pen*⟩ blot; **b)** *(ugs. abwertend: schlecht malen)* daub

Klee der; ~s clover; **Klee·blatt das** clover-leaf

Kleid das; ~es, ~er **a)** dress; **b)** *Pl. (Kleidung)* clothes; **kleiden 1.** *refl. V.* dress; **2.** *tr. V.* **a)** dress; **b)** *(jmdm. stehen)* suit

Kleider-: ~**bügel der** clothes-hanger; coat-hanger; ~**bürste die** clothes-brush; ~**haken der** coat-hook; ~**schrank der** wardrobe; ~**ständer der** coat-stand

kleidsam *Adj.* becoming; **Kleidung die;** ~: clothes *pl.*

Kleidungs·stück das garment
klein 1. *Adj.* **a)** little; small; **er ist ~er als ich** he is shorter than me; **b)** *(jung)* little; **von ~ auf** from an early age; **c)** *(von kurzer Dauer)* little, short ⟨*while*⟩; short ⟨*walk, break, holiday*⟩; brief ⟨*moment*⟩; **d)** *(von geringer Menge)* small; low ⟨*price*⟩; **~es Geld haben** have some [small] change; **e)** *(von geringem Ausmaß)* small ⟨*party, gift*⟩; scant ⟨*attention*⟩; slight ⟨*cold, indisposition, mistake, irregularity*⟩; minor ⟨*event, error*⟩; **f)** *(unbedeutend)* lowly ⟨*employee*⟩; minor ⟨*official*⟩; **~ anfangen** *(ugs.)* start off in a small way; **2.** *adv.* **die Heizung ~/~er einstellen** turn the heating down low/lower; **ein Wort ~ schreiben** write a word with a small initial letter
klein-, Klein-: **~anzeige** die *(Zeitungsw.)* small *or* classified advertisement; **~asien** (das) Asia Minor; **~bürgerlich** *Adj.* **a)** *(das Kleinbürgertum betreffend)* lower middle-class; **b)** *(abwertend: spießbürgerlich)* petit bourgeois
¹**Kleine** der; *adj. Dekl.* **a)** *(kleiner Junge)* little boy; **b)** *(ugs. Anrede)* little man; ²**Kleine** die; *adj. Dekl.* **a)** *(kleines Mädchen)* little girl; **b)** *(ugs. Anrede)* love; *(abwertend)* little madam
klein-, Klein-: **~familie** die *(Soziol.)* nuclear family; **~geld das;** *o. Pl.* [small] change; **~gläubig** sceptical
Kleinigkeit die; ~, ~en small thing; *(Einzelheit)* [small] detail; **ich habe noch eine ~ zu erledigen** I still have a small matter to attend to; **eine ~ essen** have a [small] bite to eat; **eine ~ für jmdn. sein** be no trouble for sb.
klein-, Klein-: **~kind** das small child; **~kram** der *(ugs.)* odds and ends *pl.*; *(unbedeutende Dinge)* trivial matters *pl.*; **~|kriegen** *tr. V. (ugs.)* **a)** *(zerkleinern)* crush [to pieces]; **b)** *(zerstören)* smash; break; **c)** *(aufbrauchen)* get through; **d)** jmdn. **~kriegen** get sb. down *(coll.)*; *(durch Drohungen)* intimidate sb.; *(gefügig machen)* bring sb. into line; **~laut** 1. *Adj.* subdued; *(verlegen)* sheepish; 2. *adv.* in a subdued fashion; *(verlegen)* sheepishly
kleinlich *(abwertend)* 1. *Adj.* pernickety; *(ohne Großzügigkeit)* mean; *(engstirnig)* small-minded; petty; 2. *adv.* meticulously
Kleinod das; ~[e]s, ~e *od.* ~ien *(geh.)* **a)** *(Schmuckstück)* piece of jewellery;

(Edelstein) jewel; **b)** *(Kostbarkeit)* gem
klein-, Klein-: ~|**schneiden** *unr. tr. V.* cut into small pieces; chop up ⟨*onion*⟩ [small]; **~stadt** die small town; **~städter** der small-town dweller
Kleinste der/die/das; *adj. Dekl.* youngest boy/girl/child
klein|stellen *tr. V.* turn down [low]
Kleister der; ~s, ~: paste
Klementine die; ~, ~n clementine
Klemme die; ~, ~n clip; **in der ~ sein** *od.* sitzen *(ugs.)* be in a fix *(coll.)*;
klemmen 1. *tr. V.* **a)** *(befestigen)* tuck; stick *(coll.)*; **b)** *(quetschen)* **sich** *(Dat.)* **den Fuß/die Hand ~:** get one's foot/hand caught *or* trapped; **2.** *refl. V.* **sich hinter etw.** *(Akk.)* **~** *(fig. ugs.)* put some hard work into sth.; **3.** *itr. V.* ⟨*door, drawer, etc.*⟩ stick
Klempner der; ~s, ~: tinsmith; *(~ und Installateur)* plumber
Kleptomanie die; ~ *(Psych.)* kleptomania *no art.*
klerikal *Adj. (auch abwertend)* clerical; church ⟨*property*⟩; **Klerus** der; ~: clergy
Klette die; ~, ~n bur; *(Pflanze)* burdock
klettern *itr. V.; mit sein (auch fig.)* climb; **auf einen Baum ~:** climb a tree; **Kletter·pflanze** die creeper; *(Bot.)* climbing plant; climber
klicken *itr. V.* click
Klient der; ~en, ~en, **Klientin** die; ~, ~nen client
Klima das; ~s, ~s *od.* **Klimate** climate; **Klima·an·lage** die air-conditioning *no indef. art.*; **klimatisch** *Adj.* climatic; **klimatisieren** *tr. V.* air-condition; **Klima·wechsel** der change of climate
Klimm·zug der *(Turnen)* pull-up
klimpern 1. *itr. V.* jingle; 2. *tr. V. (ugs. abwertend)* plunk out ⟨*tune etc.*⟩
Klinge die; ~, ~n blade
Klingel die; ~, ~n bell
Klingel-: **~beutel** der offertory-bag; collection-bag; **~knopf** der bell-push
klingeln *itr. V.* ring; ⟨*alarm clock*⟩ go off; **es klingelt** *(an der Tür)* there is a ring at the door; *(Telefon)* the telephone is ringing
klingen *unr. itr. V.* sound; **die Glokken klangen** the bells were ringing
Klinik die; ~, ~en hospital; *(spezialisiert)* clinic
Klinke die; ~, ~n door-handle
Klinker der; ~s, ~: [Dutch] clinker

klipp *Adv.*: ~ **und klar** *(ugs.)* quite plainly

Klippe die; ~, ~n rock

klirren *itr. V.* clink; ⟨*weapons in fight*⟩ clash; ⟨*window-pane*⟩ rattle; ⟨*chains, spurs*⟩ rattle; ⟨*harness*⟩ jingle

Klischee das; ~s, ~s cliché

klitsch·naß *Adj. (ugs.)* sopping wet; *(tropfnaß)* dripping wet

klitze·klein *Adj. (ugs.)* teeny[-weeny] *(coll.)*

Klo das; ~s, ~s *(ugs.)* loo *(Brit. coll.)*; john *(Amer. coll.)*

Kloake die; ~, ~n cesspit; *(Kanal)* sewer

klobig *Adj.* heavy and clumsy [-looking] ⟨*shoes, furniture*⟩; bulky ⟨*figure*⟩; *(plump)* clumsy

Klo·papier das *(ugs.)* loo-paper *(Brit. coll.)*; toilet-paper

klopfen 1. *itr. V.* a) *(schlagen)* knock; b) *(pulsieren)* ⟨*heart*⟩ beat; ⟨*pulse*⟩ throb; 2. *tr. V.* beat ⟨*carpet*⟩

Klöppel der; ~s, ~ *(Glocken~)* clapper; **klöppeln** *tr., itr. V.* |*etw.*| ~: make [sth. in] pillow-lace

Klops der; ~es, ~e *(nordostd.)* meat ball

Klosett das; ~s, ~s *od.* ~e lavatory

Kloß der; ~es, **Klöße** dumpling; *(Fleisch~)* meat ball

Kloster das; ~s, **Klöster** *(Mönchs~)* monastery; *(Nonnen~)* convent

Klotz der; ~es, **Klötze** block [of wood]; *(Stück eines Baumstamms)* log

Klub der; ~s, ~s club; **Klub·sessel** der club chair

¹**Kluft** die; ~, ~en *(ugs.)* gear *(coll.)*; *(Uniform)* garb

²**Kluft** die; ~, **Klüfte** *(veralt.) (Spalte)* cleft; *(im Gletscher)* crevasse; *(Abgrund)* chasm; *(fig.)* gulf

klug; **klüger,** **klügst...** *Adj.* clever; bright ⟨*child, pupil*⟩; intelligent ⟨*eyes*⟩; *(vernünftig)* wise; sound ⟨*advice*⟩; *(geschickt)* shrewd ⟨*politician, negotiator, question*⟩; astute ⟨*businessman*⟩; **klüger** *s.* klug; **Klugheit** die; ~ *s.* klug: cleverness; brightness; intelligence; wisdom; soundness; shrewdness; astuteness; **klügst...** *s.* klug

klumpen *itr. V.* go lumpy; **Klumpen** der; ~s, ~: lump; **ein ~ Gold** a gold nugget

km *Abk.* Kilometer km.

knabbern 1. *tr. V.* nibble; 2. *itr. V.* **an etw.** *(Dat.)* ~: nibble [at] sth.

Knabe der; ~n, ~n *(geh. veralt./ südd., österr., schweiz.)* boy; *(ugs.: Bursche)*

chap *(coll.)*; **knabenhaft** 1. *Adj.* boyish; 2. *adv.* boyishly

Knäcke·brot das crispbread; *(Scheibe)* slice of crispbread; **knacken** 1. *itr. V.* a) ⟨*bed, floor, etc.*⟩ creak; b) *mit sein (ugs.: zerbrechen)* snap; ⟨*window*⟩ crack; 2. *tr. V.* a) crack ⟨*nut, shell*⟩; *(salopp: aufbrechen)* crack ⟨*safe*⟩ [open]; break into ⟨*car, bank, etc.*⟩; **knackig** *Adj.* a) crisp; b) *(ugs.: attraktiv)* delectable; **Knacks** der; ~es, ~e *(ugs.)* crack; *(fig.: Defekt)* **einen ~ bekommen** ⟨*person*⟩ have a breakdown; ⟨*health*⟩ suffer

Knall der; ~|e|s, ~e bang; **knallen** 1. *itr. V.* a) ⟨*shot*⟩ ring out; ⟨*firework*⟩ go bang; ⟨*cork*⟩ pop; ⟨*door*⟩ slam; ⟨*whip, rifle*⟩ crack; **mit der Tür** ~: slam the door; b) *(ugs.: schießen)* shoot, fire **(auf + Akk.** at); c) *(Ballspiele ugs.)* **aufs Tor** ~: belt the ball/puck at the goal *(coll.)*; 2. *tr. V.* a) *(ugs.)* slam down; *(werfen)* sling *(coll.)*; b) *(ugs.: schlagen)* **jmdm. eine** ~ *(salopp)* belt sb. one *(coll.)*

knall-: ~**hart** *(ugs.)* 1. *Adj.* very tough ⟨*demands, measures, etc.*⟩; ⟨*person*⟩ as hard as nails; 2. *adv.* brutally; **gegen etw.** ~**hart vorgehen** take very tough action against sth.; ~**rot** *Adj.* bright *or* vivid red; **sie wurde** ~**rot** she turned as red as a beetroot

knapp 1. *Adj.* a) meagre; narrow ⟨*victory, lead*⟩; narrow, bare ⟨*majority*⟩; **die Vorräte wurden** ~: supplies ran short; **vor einer** ~**en Stunde** just under an hour ago; b) *(eng)* tight-fitting ⟨*garment*⟩; *(zu eng)* tight ⟨*garment*⟩; c) *(kurz)* terse ⟨*reply, greeting*⟩; succinct ⟨*description, account, report*⟩; 2. *adv.* a) ~ **bemessen sein** be meagre, ⟨*time*⟩ be limited; ~ **gewinnen/verlieren** win/ lose narrowly; **er ist** ~ **fünfzig** he is just this side of fifty; b) *(eng)* ~ **sitzen** fit tightly; *(zu eng)* be a tight fit; c) *(kurz)* ⟨*reply*⟩ tersely; ⟨*describe, summarize*⟩ succinctly; **Knappheit** die; ~ a) *(Mangel)* shortage **(an + Dat.** of); b) *(Kürze) (einer Antwort, eines Grußes)* terseness; *(einer Beschreibung, eines Berichts)* succinctness

knarren *itr. V.* creak

Knast der; ~|e|s, **Knäste** *od.* ~e *(ugs.)* a) *o. Pl. (Strafe)* bird *(sl.)*; time; b) *(Gefängnis)* clink *(sl.)*; prison

knattern *itr. V.* clatter; ⟨*sail*⟩ flap; ⟨*radio*⟩ crackle

Knäuel der *od.* das; ~s, ~ ball; *(wirres* ~*)* tangle

Knauf der; ~|e|s, **Knäufe** knob; *(eines Schwertes, Dolches)* pommel

knauserig *Adj.* *(ugs. abwertend)* stingy; tight-fisted; **knausern** *itr. V.* *(ugs. abwertend)* be stingy; skimp

knautschen *(ugs.)* 1. *tr. V.* crumple; crease ⟨*dress*⟩; 2. *itr. V.* ⟨*dress, material*⟩ crease

Knebel der; ~s, ~ a) gag; b) *(Griff)* toggle; **knebeln** *tr. V.* gag

Knecht der; ~|e|s, ~e farm-labourer; **knechten** *tr. V.* *(geh.)* reduce to slavery; enslave; *(unterdrücken)* oppress; **Knechtschaft** die; ~, ~en *(geh.)* bondage; slavery

kneifen 1. *unr. tr., itr. V.* pinch; 2. *unr. itr. V.* a) ⟨*clothes*⟩ be too tight; b) *(ugs.: sich drücken)* chicken *(sl.)* out (vor + *Dat.* of); **Kneif·zange** die pincers *pl.*

Kneipe die; ~, ~n *(ugs.)* pub *(Brit. coll.)*; bar *(Amer.)*

kneippen *itr. V.* *(ugs.)* take a Kneipp cure; **Kneipp·kur** die Kneipp cure

kneten *tr. V.* a) *(bearbeiten)* knead ⟨*dough, muscles*⟩; work ⟨*clay*⟩; b) *(formen)* model ⟨*figure*⟩; **Knet·masse** die Plasticine (P)

Knick der; ~|e|s, ~e sharp bend; *(Falz)* crease; **knicken** 1. *tr. V.* a) *(brechen)* snap; b) *(falten)* crease ⟨*page, paper, etc.*⟩; 2. *itr. V.; mit sein* snap

Knicks der; ~es, ~e curtsy; **knicksen** *itr. V.* curtsy (vor + *Dat.* to)

Knie das; ~s, ~ ['kni:(ə)] a) knee; b) *(Biegung)* sharp bend

knie-, Knie-: ~**beuge** die kneebend; ~**fall** der: einen ~fall tun *od.* machen *(auch fig.)* go down on one's knees (vor + *Dat.* before); ~**kehle** die the hollow of the knee

knien ['kni:(ə)n] 1. *itr. V.* kneel; 2. *refl. V.* kneel [down]

Knie-: ~**scheibe** die kneecap; ~**strumpf** der knee-length sock

Kniff der; ~|e|s, ~e a) pinch; b) *(Falte)* crease; c) *(Kunstgriff)* trick

knipsen *tr. V.* a) *(entwerten)* clip; punch; b) *(fotografieren)* take a snap[shot] of

Knirps der; ~es, ~e a) (Ⓦ *Taschenschirm)* telescopic umbrella; b) *(ugs.: Junge)* nipper *(coll.)*

knirschen *itr. V.* crunch; mit den Zähnen ~: grind one's teeth

knistern *itr. V.* rustle; ⟨*wood, fire*⟩ crackle

knittern *tr., itr. V.* crease; crumple

knobeln *itr. V.* *(mit Würfeln)* play dice

Knob·lauch der garlic; **Knob·lauch·zehe** die clove of garlic

Knöchel der; ~s, ~: ankle; *(am Finger)* knuckle; **Knochen** der; ~s, ~: bone

knochen-, Knochen-: ~**bau** der; o. *Pl.* bone structure; ~**bruch** der fracture; ~**hart** *Adj.* *(ugs.)* rock-hard; ~**mark** das bone marrow

knochig *Adj.* bony

Knödel der; ~s, ~ *(bes. südd., österr.)* dumpling

Knolle die; ~, ~n tuber

Knopf der; ~|e|s, **Knöpfe** button; *(Knauf)* knob; **knöpfen** *tr. V.* button [up]; **Knopf·loch** das buttonhole

Knorpel der; ~s, ~ *(Anat.)* cartilage; *(im Steak o. ä.)* gristle

knorrig *Adj.* gnarled

Knospe die; ~, ~n bud; **knospen** *itr. V.* bud

knoten *tr. V.* knot; **Knoten** der; ~s, ~: knot; *(Haartracht)* bun; knot; ⟨*Med.*⟩ lump; **Knoten·punkt** der junction; intersection

knuffen *tr. V.* poke

Knüller der; ~s, ~ *(ugs.)* sensation; *(Angebot, Verkaufsartikel)* sensational offer

knüpfen *tr. V.* a) tie (an + *Akk.* to); Bedingungen an etw. *(Akk.)* ~: attach conditions to sth.; b) *(durch Knoten herstellen)* knot; make ⟨*net*⟩

Knüppel der; ~s, ~ cudgel; *(Polizei~)* truncheon; **knüppel·dick** *Adv.* *(ugs.)* es kam ~dick it was one disaster after the other; **Knüppel·schaltung** die *(Kfz-W.)* floor[-type] gearchange

knurren 1. *itr. V.* a) ⟨*animal*⟩ growl; *(wütend)* snarl; *(fig.)* ⟨*stomach*⟩ rumble; b) *(murren)* grumble (über + *Akk.* about)

knusprig *Adj.* crisp; crusty ⟨*bread, roll*⟩

knutschen *(ugs.)* 1. *tr. V.* smooch with *(coll.)*; *(sexuell berühren)* pet; **sich** ~: smooch *(coll.)*/pet; 2. *itr. V.* smooch *(coll.)*; *(sich sexuell berühren)* pet

k. o. [ka:'|o:] *Adj.; nicht attr.* a) *(Boxen)* jmdn. k. o. schlagen knock sb. out; b) *(ugs.: übermüdet)* all in *(coll.)*

koalieren *itr. V.* *(Politik)* form a coalition (mit with); **Koalition** die; ~, ~en coalition

Kobalt das; ~s *(Chemie)* cobalt

Kobold der; ~|e|s, ~e goblin

Kobra die; ~, ~s cobra
Koch der; ~|e|s, **Köche** cook; *(Küchenchef)* chef; **Koch·buch** das cookery book *(Brit.)*; cookbook *(Amer.)*; **kochen 1.** *tr. V.* **a)** boil; *(zubereiten)* cook ‹*meal*›; make ‹*purée, jam*›; Tee ~: make some tea; **b)** *(waschen)* boil; **2.** *itr. V.* **a)** *(Speisen zubereiten)* cook; **b)** *(sieden)* ‹*water, milk, etc.*› boil; **Kocher** der; ~s, ~ [small] stove; *(Kochplatte)* hotplate
Köcher der; ~s, ~ *(für Pfeile)* quiver
Köchin die; ~, ~nen cook
Koch-: ~**löffel** der wooden spoon; ~**nische** die kitchenette; ~**topf** der [cooking] pot
Köder der; ~s, ~: bait; **ködern** *tr. V.* lure
Koffein das; ~s caffeine; **koffeinfrei** *Adj.* decaffeinated
Koffer der; ~s, ~: [suit]case
Koffer-: ~**kuli** der luggage trolley; ~**radio** das portable radio; ~**raum** der boot *(Brit.)*; trunk *(Amer.)*
Kognak ['kɔnjak] der; ~s, ~s brandy; *s. auch* **Cognac**
Kohl der; ~|e|s **a)** cabbage; **b)** *(ugs. abwertend: Unsinn)* rubbish; rot *(sl.)*; **Kohl·dampf** der; *o. Pl.* *(salopp)* |ei-nen| ~ haben be ravenously hungry
Kohle die; ~, ~n coal; **¹kohlen** *itr. V.* smoulder; ‹*wick*› smoke
²kohlen *itr. V. (fam.) (lügen)* tell fibs; *(übertreiben)* exaggerate
Kohlen-: ~**grube** die coal-mine; ~**händler** der coal merchant ~**monoxyd** [--'---] das *(Chemie)* carbon monoxide; ~**säure** die carbonic acid; ~**stoff** der; *o. Pl.* carbon
Kohle·papier das carbon paper
Köhler der; ~s, ~: charcoal burner
Kohl-: ~**kopf** der [head of] cabbage; ~**rübe** die swede
Koitus der; ~, **Koitus** *(geh.)* sexual intercourse; coitus *(formal)*
Koje die; ~, ~n **a)** *(Seemannsspr.)* bunk; berth; **b)** *(Ausstellungsstand)* stand
Kokain das; ~s cocaine
kokett 1. *Adj.* coquettish; **2.** *adv.* coquettishly
Kokos·nuß die coconut
Koks der; ~es coke
Kolben der; ~s, ~ **a)** *(Technik)* piston; **b)** *(Chemie: Glas~)* flask; **c)** *(Teil des Gewehrs)* butt
Kolchose [kɔl'çoːzə] die; ~, ~n kolkhoz; Soviet collective farm
Kolibri der; ~s, ~s humming-bird

Kolik die; ~, ~en colic
Kollaborateur [kɔlabora'tøːɐ̯] der; ~s, ~e collaborator
Kollaps der; ~es, ~e collapse
Kolleg das; ~s, ~s lecture
Kollege der; ~n, ~n colleague; **kollegial 1.** *Adj.* helpful and considerate; **2.** *adv.* ‹*act etc.*› like a good colleague/good colleagues; **Kollegium** das; ~s, **Kollegien a)** *(Gruppe)* group; *(unmittelbar zusammenarbeitend)* team; **b)** *(Lehrkörper)* [teaching] staff
Kollekte die; ~, ~n collection; **Kollektion** [kɔlɛk'tsi̯oːn] die; ~, ~en collection; *(Sortiment)* range; **kollektiv 1.** *Adj.* collective; **2.** *adv.* collectively
kollidieren *itr. V.* **a)** *mit sein* collide; **b)** *(fig.)* conflict
Kollier [kɔ'li̯eː] das; ~s, ~s necklace
Kollision die; ~, ~en collision
Köln (das); ~s Cologne; **Kölner 1.** *indekl. Adj.* Cologne *attrib.*; *(in Köln)* in Cologne *postpos., not pred;* ‹*suburb, archbishop, mayor, speciality*› of Cologne; **2.** der; ~s, ~: inhabitant of Cologne; *(von Geburt)* native of Cologne; **Kölnerin** die; ~, ~nen *s.* **Kölner 2**
Kolonialismus der; ~: colonialism *no art.;* **Kolonie** die; ~, ~n colony; **kolonisieren** *tr. V.* colonize
Kolonne die; ~, ~n column
Koloß der; **Kolosses, Kolosse** *(auch fig. ugs.)* giant; **kolossal 1.** *Adj.* **a)** colossal; gigantic; **b)** *(ugs.: sehr groß)* tremendous *(coll.);* incredible *(coll.)* ‹*rubbish, nonsense*›; **2.** *adv. (ugs.)* tremendously *(coll.)*
Kolumbianer der; ~s, ~: Colombian; **Kolumbien** [ko'lʊmbi̯ən] (das); ~s Colombia
Kombination die; ~, ~en **a)** combination; **b)** *(gedankliche Verknüpfung)* deduction; piece of reasoning; **c)** *(Kleidungsstücke)* ensemble; suit; *(Herren~)* suit; **kombinieren 1.** *tr. V.* combine; **2.** *itr. V.* deduce; reason
Kombi-: ~**wagen** der estate [car]; station wagon *(Amer.);* ~**zange** die combination pliers *pl.*
Komet der; ~en, ~en comet
Komfort der; ~s comfort; **komfortabel 1.** *Adj.* comfortable; **2.** *adv.* comfortably
Komik die; ~: comic effect; *(komisches Element)* comic element; **Komiker** der; ~s, ~ **a)** *(Vortragskünstler)* comedian; **b)** *(Darsteller)* comic actor; **komisch** *Adj.* **a)** comical; funny; **b)** *(seltsam)* funny

Komitee das; ~s, ~s committee
Komma das; ~s, ~s od. ~ta comma;
(Math.) decimal point; **zwei ~ acht**
two point eight
Kommandant der; ~en, ~en *(Milit.)*
commanding officer; **Kommandeur**
[kɔman'dø:ɐ̯] der; ~s, ~e *(Milit.)* s.
Kommandant; kommandieren 1. *tr.*
V. **a)** command; be in command of;
order *(retreat, advance)*; **b)** *(ugs.)*
jmdn. ~: boss sb. about *(coll.)*; 2. *itr.*
V. (ugs.) boss people about *(coll.)*
Kommandit·gesellschaft   die
(Wirtsch.) limited partnership
Kommando das; ~s, ~s command
kommen *unr. itr. V.; mit sein* **a)** come;
angelaufen ~: come running along;
(auf jmdn. zu) come running up; **b)**
(gelangen, geraten) get; **unter ein Auto**
~: be knocked down by a car; **wie**
kommst du darauf? what gives you
that idea?; **c)** ~ **lassen** *(bestellen)*
order *(taxi)*; **den Arzt/die Polizei** ~
lassen send for a doctor/the police; **d)**
(aufgenommen werden) **zur Schule/**
aufs Gymnasium ~: start school/
grammar school; **e)** *(auftauchen)*
(seeds, plants) come up; *(buds,*
flowers) come out; *(teeth)* come
through; **f)** *(seinen festen Platz haben)*
go; belong; **in die Schublade** ~: go or
belong in the drawer; *(seinen Platz er-*
halten) **in die Mannschaft** ~: get into
the team; **auf den ersten Platz** ~: go
into first place; **g)** *(Gelegenheit haben)*
dazu ~, **etw. zu tun** get round to doing
sth.; **h)** *(sich ereignen)* come about;
wie kommt es, daß ...: how is it that ...;
how come that ... *(coll.)*; **i)** *(etw. erlan-*
gen) **zu Geld** ~: become wealthy; **zu**
Erfolg/Ruhm *usw.* ~: gain success/
fame *etc.*
kommend *Adj.* **a)** *(folgend)* next; **b)**
(mit großer Zukunft) **der** ~**e Mann/**
Meister the coming man/future champion
Kommentar der; ~s, ~e commentt-
ary; *(Stellungnahme)* comment; **kein**
~**!** no comment!; **Kommentator**
der; ~s, ~en commentator; **kom-**
mentieren *tr. V.* **a)** *(erläutern)* furnish with a commentary *(text, work)*;
b) *(Stellung nehmen zu)* comment on
kommerziell 1. *Adj.* commercial; 2.
adv. commercially
Kommiß der; **Kommisses** *(Solda-*
tenspr.) army; **Kommissar** der; ~s,
~e **a)** *(Beamter der Polizei)* detective
superintendent; **b)** *(staatlicher Beauf-*

tragter) commissioner; **Kommissi-**
on die; ~, ~en **a)** *(Gremium)* committee; *(Prüfungs~)* commission; **b) etw.**
in ~ **nehmen/haben/geben** *(Wirtsch.)*
take/have sth. on commission/give
sth. to a dealer for sale on commission
Kommode die; ~, ~n chest of drawers
kommunal *Adj.* local; *(bei einer städ-*
tischen Gemeinde) municipal; local;
Kommunal·wahl die local [govern-
ment] elections *pl.;* **Kommunikati-**
on ['kɔmunika'tsio:n] die; ~, ~en
(Sprachw., Soziol.) communication;
Kommunion die; ~, ~en *(kath. Kir-*
che) [Holy] Communion; **Kommuni-**
qué [kɔmyni'ke:] das; ~s, ~s communiqué; **Kommunismus** der; ~:
communism; **Kommunist** der; ~en,
~en communist; **kommunistisch** 1.
Adj. communist; 2. *adv.* Communist-
(influenced, led, ruled, etc.); **kommu-**
nizieren *itr. V.* **a)** *(geh.)* communicate; **b)** *(kath. Kirche)* receive [Holy]
Communion
Komödiant der; ~en, ~en *(veralt.)*
actor; player; *(abwertend: Heuchler)*
play-actor; **Komödie** [ko'mø:djə]
die; ~, ~n comedy; *(Theater)* comedy
theatre
Kompagnon [kɔmpan'jõ:] der; ~s, ~s
(Wirtsch.) partner; associate
kompakt *Adj.* solid
Kompanie die; ~, ~n company
Komparativ der; ~s, ~e *(Sprachw.)*
comparative
Kompaß der; **Kompasses, Kompasse**
compass
kompensieren *tr. V.* etw. mit etw. *od.*
durch etw. ~: compensate for sth. by
sth.
kompetent *Adj.* competent; **Kom-**
petenz die; ~, ~en competence; *(bes.*
Rechtsw.) authority
komplett 1. *Adj.* complete; 2. *adv.*
fully *(furnished, equipped)*; *(ugs.: ganz*
und gar) completely
Komplex der; ~es, ~e *(auch Psych.)*
complex
Komplikation [kɔmplika'tsio:n] die;
~, ~en *(auch Med.)* complication
Kompliment das; ~[e]s, ~e compliment
Komplize der; ~n, ~n *(abwertend)* accomplice
komplizieren *tr. V.* complicate;
kompliziert 1. *Adj.* complicated; 2.
adv. ~ **aufgebaut sein** have a complicated *or* complex structure

Komplott das; ~|e|s, ~e plot; conspiracy

komponieren tr., itr. V. compose; **Komponist** der; ~en, ~en composer; **Komposition** [kɔmpozi-'tsio:n] die; ~, ~en composition; **Kompost** der; ~|e|s, ~e compost; **Kompott** das; ~|e|s, ~e stewed fruit; compote

Kompresse die; ~, ~n (Med.) a) (Umschlag) [wet] compress; b) (Mull) [gauze] pad; **Kompressor** der; ~s, ~en (Technik) compressor

Kompromiß der; Kompromisses, Kompromisse compromise

kompromiß-, Kompromiß-: ~bereit Adj. willing to compromise pred.; **~los** 1. Adj. uncompromising; 2. adv. uncompromisingly; **~vorschlag** der compromise proposal

kompromittieren tr. V. compromise

Kondensation [kɔndɛnza'tsio:n] die; ~, ~en (Physik, Chemie) condensation; **Kondensator** der; ~s, ~en (Elektrot.) capacitor; **kondensieren** tr., itr. V. (itr. auch mit sein) (Physik, Chemie) condense

Kondens-: ~milch die condensed milk; **~streifen** der condensation trail; **~wasser** das condensation

Kondition [kɔndi'tsio:n] die; ~, ~en condition; **Konditional·satz** der (Sprachw.) conditional clause; **Konditions·training** das fitness training

Konditor der; ~s, ~en pastry-cook; **Konditorei** die; ~, ~en cake-shop; (Lokal) café

kondolieren itr. V. offer one's condolences; jmdm. |zu jmds. Tod| ~: offer one's condolences to sb. [on sb.'s death]

Kondom das od. der; ~s, ~e condom

Konfekt das; ~|e|s a) confectionery; sweets pl. (Brit.); candies pl. (Amer.); b) (bes. südd., österr., schweiz.: Teegebäck) [small] fancy biscuits pl. (Brit.) or (Amer.) cookies pl.

Konfektion die; ~, ~en ready-made garments pl.

Konferenz die; ~, ~en conference; (Besprechung) meeting; **konferieren** 1. itr. V. confer (über + Akk. on, about)

Konfession die; ~, ~en denomination; **konfessionell** 1. Adj.; nicht präd. denominational; 2. adv. as regards denomination; ~ |un|gebunden sein have [no] denominational ties

Konfetti das; ~|s| confetti

Konfirmand der; ~en, ~en (ev. Rel.) confirmand; **Konfirmation** [kɔn-firma'tsio:n] die; ~, ~en (ev. Rel.) confirmation; **konfirmieren** tr. V. (ev. Rel.) confirm

konfiszieren tr. V. (bes. Rechtsw.) confiscate

Konfitüre die; ~, ~n jam

Konflikt der; ~|e|s, ~e conflict

Konföderation die; ~, ~en confederation

konform Adj. concurring attrib.; ~ gehen be in agreement; **Konformist** der; ~en, ~en conformist

Konfrontation die; ~, ~en confrontation; **konfrontieren** tr. V. confront

konfus 1. Adj. confused; 2. adv. in a confused fashion

¹Kongo der; ~|s| (Fluß) Congo; **²Kongo (das)**; ~s od. der; ~|s| (Staat) the Congo

Kongreß der; Kongresses, Kongresse congress; conference; **der ~ (USA):** Congress; **Kongreß·halle** die conference hall

König der; ~s, ~e king; **Königin** die; ~, ~nen queen; **königlich** 1. Adj. a) royal; b) (vornehm) regal; c) (reichlich) princely ⟨gift, salary, wage⟩; 2. adv. ⟨pay⟩ handsomely; (ugs.: außerordentlich) ⟨enjoy oneself⟩ immensely (coll.); **König·reich** das kingdom; **Königs·haus** das royal house; **Königtum** das; ~s, Königtümer a) o. Pl. (Monarchie) monarchy; b) (veralt.: Reich) kingdom

Konjugation [kɔnjuga'tsio:n] die; ~, ~en (Sprachw.) conjugation; **konjugieren** tr. V. (Sprachw.) conjugate; **Konjunktion** [kɔnjuŋk'tsio:n] die; ~, ~en (Sprachw.) conjunction; **Konjunktiv** der; ~s, ~e (Sprachw.) subjunctive; **Konjunktur** die; ~, ~en (Wirtsch.) a) (wirtschaftliche Lage) [level of] economic activity; economy; (Tendenz) economic trend; b) (Hoch~) boom; (Aufschwung) upturn [in the economy]; **konjunkturell** Adj. economic; **Konjunktur·politik** die (Wirtsch.) measures pl. aimed at avoiding violent fluctuations in the economy

konkav (Optik) 1. Adj. concave; 2. adv. concavely

konkret 1. Adj. concrete; 2. adv. in concrete terms

Konkurrent der; ~en, ~en, **Konkurrentin** die; ~, ~nen (Sport, Wirtsch.)

competitor; **Konkurrenz die;** ~, ~en *(Sport, Wirtsch.)* competition; **konkurrenz·fähig** *Adj.* competitive; **Konkurrenz·kampf** der competition; *(zwischen zwei Menschen)* rivalry; **konkurrieren** *itr. V.* compete; **Konkurs der;** ~es, ~e a) *(Bankrott)* bankruptcy; b) *(gerichtliches Verfahren)* bankruptcy proceedings *pl.*
können 1. *unr. Modalverb;* 2. *Part.* ~ a) be able to; **er kann gut reden/tanzen** he is a good talker/dancer; **ich kann nicht schlafen** I cannot *or (coll.)* can't sleep; **kann das explodieren?** could it explode?; **man kann nie wissen** you never know; **es kann sein, daß ...:** it could be that ...; **kann ich Ihnen helfen?** can I help you?; b) *(Grund haben)* **du kannst ganz ruhig sein** you don't have to worry; **das kann man wohl sagen!** you could well say that; c) *(dürfen)* **kann ich gehen?** can I go?; ~ **wir mit[kommen]?** can we come too?; 2. *unr. tr. V. (beherrschen) ⟨ein Sprache⟩;* be able to play *⟨game⟩;* **sie kann das [gut]** she can do that [well]; **etw./nichts für etw.~:** be/not be responsible for sth.; 3. *unr. itr. V.* **a)** *(fähig sein)* **er kann nicht anders** there's nothing else he can do; *(es ist seine Art)* he can't help it *(coll.);* b) *(Zeit haben)* **ich kann heute nicht** I can't today *(coll.);* c) *(ugs.: Kraft haben)* **kannst du noch [weiter]?** can you go on?; d) *(ugs.: umgehen ~)* **[gut] mit jmdm. ~:** get on [well] with sb.; **Können das;** ~s ability; **Könner** der; ~s, ~ expert; **konnte** *1. u. 3. Pers. Sg. Prät. v.* können; **könnte** *1. u. 3. Pers. Sg. Konjunktiv II v.* können
konsequent 1. *Adj.* consistent; *(folgerichtig)* logical; 2. *adv.* consistently; *(folgerichtig)* logically; **Konsequenz die;** ~, ~en a) *(Folge)* consequence; b) *o. Pl. (Unbeirrbarkeit)* determination
konservativ 1. *Adj.* conservative; 2. *adv.* conservatively; **Konservative** der/die; *adj. Dekl.* conservative; **Konservatorium** das; ~s, Konservatorien conservatoire; conservatory *(Amer.);* **Konserve** die; ~, ~n a) *(Büchse)* can; tin *(Brit.);* b) *(konservierte Lebensmittel)* preserved food; *(in Dosen)* canned *or (Brit.)* tinned food
Konserven-: ~**büchse** die, ~**dose** die can; tin *(Brit.)*
konservieren *tr. V.* preserve; conserve *⟨work of art⟩;* **Konservierung**

die; ~, ~en preservation; **Konservierungs·mittel** das preservative
konsolidieren *tr. V.* consolidate
Konsonant der; ~en, ~en consonant
Konsortium [kɔn'zɔrtsiʊm] das; ~s, Konsortien *(Wirtsch.)* consortium
konspirativ [kɔnspira'ti:f] *Adj.* conspiratorial
konstant [kɔn'stant] 1. *Adj.* a) constant; b) *(beharrlich)* persistent; 2. *adv.* a) constantly; b) *(beharrlich)* persistently
Konstellation [kɔnstɛla'tsio:n] die; ~, ~en a) *(von Parteien usw.)* grouping; *(von Umständen)* combination; b) *(Astron., Astrol.)* constellation
konstituieren [kɔnstitu'i:rən] 1. *tr. V. (gründen)* constitute; set up; 2. *refl. V.* be constituted; **Konstitution** [kɔnstitu'tsio:n] die; ~, ~en constitution
konstruieren [kɔnstru'i:rən] *tr. V.* a) *(entwerfen)* design; b) *(aufbauen, Geom, Sprachw.)* construct; c) *(abwertend)* fabricate; **Konstrukteur** [kɔnstrʊk'tø:ɐ] der; ~s, ~e designer; design engineer; **Konstruktion** [kɔnstrʊk'tsio:n] die; ~, ~en a) *(Aufbau, Geom., Sprachw.)* construction; *(das Entwerfen)* designing; b) *(Entwurf)* design; *(Bau)* construction; **konstruktiv** 1. *Adj.* constructive; 2. *adv.* constructively
Konsul der; ~s, ~n *(Dipl., hist.)* consul; **Konsulat** das; ~[e]s, ~e *(Dipl., hist.)* consulate; **konsultieren** *tr. V. (auch fig.)* consult
Konsum der; ~s consumption; **Konsument** der; ~en, ~en consumer; **Konsum·gesellschaft** die consumer society; **konsumieren** *tr. V.* consume
Kontakt der; ~[e]s, ~e contact; **mit od. zu jmdm. ~ haben/halten** be/remain in contact with sb.
Kontakt-: ~**linse** die contact lens; ~**mann** der; *Pl.:* ~**männer** od. ~**leute** *(Agent)* contact
Konten *s.* Konto
kontern *tr., itr. V. (Boxen, auch fig.)* counter; *(Ballspiele)* counter-attack; **Konter·revolution** die counter-revolution
Kontinent der; ~[e]s, ~e continent; **kontinental** *Adj.* continental
Kontingent das; ~[e]s, ~e quota
kontinuierlich 1. *Adj.* steady; 2. *adv.* steadily; **Kontinuität** die; ~: continuity
Konto das; ~s, Konten od. Konti ac-

count; **ein laufendes ~**: a current account

Konto-: **~aus·zug** der *(Bankw.)* [bank] statement; **~nummer** die account number

Kontor das; **~s**, **~e** branch; *(einer Reederei)* office

Konto·stand der *(Bankw.)* balance; state of an/one's account

kontra 1. *Präp. mit Akk. (Rechtsspr., auch fig.)* versus; 2. *Adv.* against

Kontra das; **~s**, **~s** *(Kartenspiele)* double; **jmdm. ~ geben** *(fig. ugs.)* flatly contradict sb.

Kontrahent der; **~en**, **~en** adversary; opponent

konträr *Adj.* contrary; opposite; **Kontrast** der; **~[e]s**, **~e** contrast

Kontroll·abschnitt der stub; **Kontrolle** die; **~**, **~n** a) *(Überwachung)* surveillance; b) *(Überprüfung)* check; *(bei Waren, bei Lebensmitteln)* inspection; c) *(Herrschaft)* control; **die ~ über etw.** *(Akk.)* **verlieren** lose control of sth.; **Kontrolleur** [kɔntrɔ'løːɐ̯] der; **~s**, **~e** inspector; **Kontrollgang** der tour of inspection; *(eines Nachtwächters)* round; *(eines Polizisten)* patrol; **kontrollieren** *tr. V.* **a)** *(überwachen)* check; monitor; b) *(überprüfen)* check; inspect ⟨goods, food⟩; c) *(beherrschen)* control; **Kontrollturm** der control tower

Kontroverse [kɔntro'vɛr] die; **~**, **~n** controversy (**um, über** + *Akk.* about)

Kontur die; **~**, **~en**; *meist Pl.* contour; outline

Konvention [kɔnvɛn'tsi̯oːn] die; **~**, **~en** convention; **konventionell** 1. *Adj.* a) conventional; b) *(förmlich)* formal; 2. *adv.* a) conventionally; b) *(förmlich)* formally

Konversation [kɔnvɛrza'tsi̯oːn] die; **~**, **~en** conversation; **Konversations·lexikon** das encyclopaedia

konvertieren [kɔnvɛr'tiːrən] *itr. V.; auch mit sein (Rel.)* be converted

konvex [kɔn'vɛks] *(Optik)* 1. *Adj.* convex; 2. *adv.* convexly

Konvoi [kɔn'vɔy̯] der; **~s**, **~s** *(bes. Milit.)* convoy

Konzentration [kɔntsɛntra'tsi̯oːn] die; **~**, **~en** concentration

Konzentrations-: **~fähigkeit** die; *o. Pl.* ability to concentrate; **~lager** das *(bes. ns.)* concentration camp

konzentrieren 1. *refl., tr. V.* concentrate; **sich auf etw.** *(Akk.)* **~**: concentrate on sth.; **konzentriert** 1. *Adj.*

concentrated; 2. *adv.* with concentration

Konzept das; **~[e]s**, **~e** a) [rough] draft; b) *(Programm)* programme; *(Plan)* plan

Konzern der; **~[e]s**, **~e** *(Wirtsch.)* group [of companies]

Konzert das; **~[e]s**, **~e** a) *(Komposition)* concerto; b) *(Veranstaltung)* concert; **Konzert·saal** der concert-hall

Konzession die; **~**, **~en** a) *(Amtsspr.)* licence; b) *(Zugeständnis)* concession

Konzil das; **~s**, **~e** *od.* **~ien** *(kath. Kirche)* council

konzipieren *tr. V.* draft; design ⟨device, car, etc.⟩

kooperativ 1. *Adj.* co-operative; 2. *adv.* co-operatively; **kooperieren** *tr. V.* co-operate

Koordinate die; **~**, **~n** coordinate; **Koordinaten·system** das *(Math.)* system of coordinates; **koordinieren** *tr. V.* coordinate

Kopenhagen (das); **~s** Copenhagen

Kopf der; **~[e]s**, **Köpfe** a) head; **ein ~ Salat** a lettuce; **~ an ~**: shoulder to shoulder; *(im Wettlauf)* neck and neck; *(fig.)* **nicht wissen, wo einem der ~ steht** not know whether one is coming or going; **~ hoch!** chin up!; **den ~ hängen lassen** become disheartened; b) *(Person)* person; **ein kluger/fähiger ~ sein** be a clever/able man/woman; **pro ~**: per head; **die führenden Köpfe der Wirtschaft** the leading minds in the field of economics; c) *(Wille)* **seinen ~ durchsetzen** make sb. do what one wants; d) *(Verstand)* mind; head; **sich** *(Dat.)* **den ~ zerbrechen** *(ugs.)* rack one's brains (**über** + *Akk.* over)

Kopf-: **~bahn·hof** der terminal station; **~bedeckung** die headgear; **ohne ~bedeckung** without anything on one's head

Köpfchen das; **~s**, **~**: brains *pl.;* **~ muß man haben** you've got to have it up here *(coll.);* **köpfen** *tr. V.* a) decapitate; *(hinrichten)* behead; b) *(Fußball)* head

Kopf-: **~ende** das head end; **~haut** die [skin of the] scalp; **~hörer** der headphones *pl.;* **~kissen** das pillow; **~lastig** *Adj.* down by the head *pred.;* **~los** 1. *Adj.* rash; *(in Panik)* panic-stricken; 2. *adv.* rashly; **~los davonrennen** flee in panic; **~rechnen** *itr. V.; nur im Inf. gebr.* do mental arithmetic; **~rechnen** das mental arithmetic; **~salat** der head lettuce;

~**schmerz** der; *meist Pl.* headache; ~**schmerzen haben** have a headache *sing.;* ~**sprung** der header; ~**stand** der headstand; ~|**stehen** *unr. itr. V. (ugs.: überrascht sein)* be bowled over; ~**stein·pflaster** das cobblestones *pl.;* ~**tuch** das headscarf; ~**weh** das; *o. Pl. (ugs.)* headache; ~**weh haben** have a headache; ~**zerbrechen** das; ~s: etw. bereitet *od.* macht jmdm. ~zerbrechen sb. has to rack his/her brains about sth.; *(etw. macht jmdm. Sorgen)* sth. is a worry to sb.

Kopie die; ~, ~n copy; *(Durchschrift)* carbon copy; *(Fotokopie)* photocopy; *(Fot., Film)* print; **kopieren** *tr. V.* copy; *(fotokopieren)* photocopy; *(Fot., Film)* print; **Kopier·gerät** das photocopier

¹**Koppel** das; ~s, ~, *österr.:* die; ~, ~n *(Gürtel)* [leather] belt *(as part of a uniform);* ²**Koppel** die; ~, ~n paddock

koppeln *tr. V.* couple (**an** + *Akk.* to); dock ⟨*spacecraft*⟩

Koppelung *s.* Kopplung; **Kopplung** die; ~, ~en coupling; *(Raumf.)* docking

kopulieren *itr. V.* copulate
Koralle die; ~, ~n coral
Koran der; ~s, ~e Koran
Korb der; ~es, Körbe a) basket; b) jmdm. einen ~ geben turn sb. down; **Korb·ball** der; *o. Pl.* netball
Kord der; ~[e]s a) corduroy; cord; b) *s.* Kordsamt
Kordel die; ~, ~n cord
Kord·samt der cord velvet
Korea (das); ~s Korea; **Koreaner** der; ~s, ~: Korean; **koreanisch** *Adj.* Korean
Korinthe die; ~, ~n currant
Kork der; ~s, ~e cork; **Korken** der; ~s, ~: cork; **Korken·zieher** der corkscrew
¹**Korn** das; ~[e]s, Körner a) *(Frucht)* grain; *(Getreide~)* grain [of corn]; *(Pfeffer~)* corn; b) *o. Pl. (Getreide)* corn; grain; c) *(Salz~, Sand~)* grain; *(Hagel~)* stone; ²**Korn** der; ~[e]s, ~ *(ugs.)* corn schnapps; corn liquor *(Amer.);* **Korn·blume** die cornflower; **Körnchen** das; ~s, ~: tiny grain; *(von Sand usw.)* [tiny] grain; granule; **Körner** *s.* Korn; **Korn·feld** das cornfield; **körnig** *Adj.* granular
Korona die; ~, Koronen crowd *(coll.)*
Körper der; ~s, ~: body
körper-, Körper-: ~**bau** der; *o. Pl.* physique; ~**behindert** *Adj.* physic-

ally handicapped; ~**behinderte** der/die physically handicapped person; ~**behinderte** *Pl.* physically handicapped people; ~**geruch** der body odour; BO *(coll.);* ~**größe** die height
körperlich 1. *Adj.* physical; 2. *adv.* physically
Körper-: ~**pflege** die body care *no art.;* ~**spray** der *od.* das deodorant spray; ~**teil** der part of the/one's body; ~**verletzung** die *(Rechtsw.)* bodily harm *no indef. art.*
Korps [koːɐ] das; ~ [koːɐ(s)], ~ [koːɐs] a) *(Milit.)* corps; b) *(Studentenverbindung)* student duelling society
korpulent *Adj.* corpulent
korrekt 1. *Adj.* correct; 2. *adv.* correctly; **korrekter·weise** *Adv.* to be [strictly] correct; **Korrektheit** die; ~: correctness; **Korrektor** der; ~s, ~en [-'toːrən] proof-reader; **Korrektur** die; ~, ~en correction
Korrespondent der; ~en, ~en correspondent; **Korrespondenz** die; ~, ~en correspondence; **korrespondieren** *itr. V.* correspond (mit with)
Korridor der; ~s, ~e corridor
korrigieren *tr. V.* correct; revise ⟨*opinion, view*⟩
korrupt *Adj.* corrupt; **Korruption** [kɔrʊp'tsi̯oːn] die; ~, ~en corruption
Korsett das; ~s, ~s *od.* ~e corset
Korsika (das); ~s Corsica
koscher *Adj.* kosher
Kose-: ~**form** die familiar form; ~**name** der pet name
Kosinus der; ~, ~ *od.* ~se *(Math.)* cosine
Kosmetik die; ~ a) beauty culture *no art.;* b) *(fig.)* cosmetic procedures *pl.;* **Kosmetikerin** die; ~, ~en cosmetician; beautician; **kosmetisch** 1. *Adj. (auch fig.)* cosmetic; 2. *adv.* jmdn. ~ beraten give sb. advice on beauty care; sich ~ behandeln lassen have beauty treatment
kosmisch *Adj.* cosmic ⟨*ray, dust, etc.*⟩; space ⟨*age, station, research, etc.*⟩; meteoric ⟨*iron*⟩; **Kosmos** der; ~: cosmos
Kost die; ~: food; ~ und Logis board and lodging
kostbar 1. *Adj.* valuable; precious ⟨*time*⟩; 2. *adv.* expensively ⟨*dressed*⟩; luxuriously ⟨*decorated*⟩; **Kostbarkeit** die; ~, ~en a) *(Sache)* treasure; b) *o. Pl. (Eigenschaft)* value
¹**kosten** 1. *tr. V.* a) taste; try; 2. *itr. V. (probieren)* have a taste

²**kosten** *tr. V.* **a)** cost; **b)** *(erfordern)* take; cost ⟨*lives*⟩; **Kosten** *Pl.* cost *sing.*; costs; *(Auslagen)* expenses; *(Rechtsw.)* costs

kosten-, Kosten-: ~**deckend** *Adj.* that covers/cover [one's] costs *postpos., not pred.;* ~**erstattung die** reimbursement of costs; ~**los** 1. *Adj.* free; 2. *adv.* free of charge; ~**pflichtig** *(Rechtsw.)* 1. *Adj.* eine ~pflichtige Verwarnung a fine and a caution; 2. *adv.* eine Klage ~pflichtig abweisen dismiss a case with costs; ein Auto ~pflichtig abschleppen tow a car away at the owner's expense; ~**voran·schlag der** estimate

Kost·geld das payment for [one's] board

köstlich 1. *Adj.* delicious; *(unterhaltsam)* delightful; 2. *adv.* ⟨*taste*⟩ delicious; **sich ~ amüsieren/unterhalten** enjoy oneself enormously *(coll.);* **Köstlichkeit die;** ~, ~**en** *(Sache)* delicacy

Kost·probe die; ~, ~**n** taste

kost·spielig *Adj.* costly

Kostüm das; ~**s,** ~**e a)** suit; **b)** *(Theater~, Verkleidung)* costume; **kostümieren** *tr. V.* dress up

Kot der; ~|e|s, ~**e** excrement

Kotangens der; ~, ~ *(Math.)* cotangent

Kotelett [kɔt'lɛt] **das;** ~**s,** ~**s** chop; *(vom Nacken)* cutlet; **Koteletten** *Pl.* side-whiskers

Köter der; ~**s,** ~ *(abwertend)* cur

Kot·flügel der *(Kfz-W.)* wing

kotzen *itr. V. (derb)* puke *(coarse)*

KP [ka:'pe:] *Abk.* Kommunistische Partei CP

Krabbe die; ~, ~**n a)** *(Zool.)* crab; **b)** *(ugs.: Garnele)* shrimp; *(größer)* prawn; **krabbeln** 1. *itr. V.;* mit sein crawl; 2. *tr. V. (ugs.: kraulen)* tickle

Krach der; ~|e|s, Kräche **a)** *o. Pl.* *(Lärm)* noise; row; **b)** *(lautes Geräusch)* crash; **c)** *(ugs.: Streit)* row; **krachen** 1. *itr. V.* **a)** *(Krach auslösen)* ⟨*thunder*⟩ crash; ⟨*shot*⟩ ring out; **b)** *mit sein (ugs.: bersten)* ⟨*ice*⟩ crack; ⟨*bed*⟩ collapse; **c)** *mit sein (ugs.: mit Krach auftreffen)* crash; 2. *refl. V. (ugs.)* row *(coll.);* **krächzen** *itr. V.* ⟨*raven, crow*⟩ caw; ⟨*parrot*⟩ squawk; ⟨*person*⟩ croak

kraft *Präp.* + *Gen. (Amtsspr.)* ~ |meines| Amtes by virtue of my office; ~ Gesetzes by law; **Kraft die;** ~, Kräfte strength; *(Wirksamkeit)* power; *(Physik)* force; *(Arbeits~)* employee; **mit**

letzter ~: with one's last ounce of strength; aus eigener ~: by one's own efforts; mit vereinten Kräften werden wir ...: if we join forces *or* combine our efforts we will ...; außer ~ setzen repeal ⟨*law*⟩; countermand ⟨*order*⟩; außer ~ sein/treten no longer be/cease to be in force; in ~ treten/sein/bleiben come into/be in/remain in force

Kraft-: ~**aufwand der** effort; ~**brühe die** strong meat broth; ~**fahrer der** driver; motorist; ~**fahrzeug das** motor vehicle

Kraftfahrzeug-: ~**brief der** vehicle registration document; log-book *(Brit.);* ~**schein der** vehicle registration document; ~**steuer die** vehicle tax

kräftig 1. *Adj.* strong; vigorous ⟨*plant, shoot*⟩; powerful, hefty ⟨*blow, kick, etc.*⟩; nourishing ⟨*soup, bread, meal, etc.*⟩; 2. *adv.* powerfully ⟨*built*⟩; ⟨*rain, snow*⟩ heavily; ⟨*eat*⟩ heartily; **kräftigen** *tr. V.* ⟨*holiday, air, etc.*⟩ invigorate; ⟨*food etc.*⟩ fortify

kraft-, Kraft-: ~**meier der;** ~**s,** ~ *(ugs.: abwertend)* muscleman; ~**probe die** trial of strength; ~**rad das** *(Amtsspr.)* motorcycle; ~**stoff der** *(Kfz-W.)* fuel; ~**stoff·verbrauch der** fuel consumption; ~**voll** 1. *Adj.* powerful; 2. *adv.* powerfully; ~**wagen der** motor vehicle; ~**werk das** power station

Kragen der; ~**s,** ~, *südd., österr. u. schweiz. auch:* Krägen collar; **Kragen·weite die** collar size

Krähe ['krɛ:ə] **die;** ~, ~**n** crow; **krähen** *itr. V. (auch fig.)* crow; **Krähen·füße** *Pl. (ugs.)* crow's feet

krakeelen *itr. V. (ugs.)* kick up a row *(coll.)*

krakeln *tr., itr. V. (ugs.)* scrawl; **kraklig** *Adj. (ugs. abwertend)* scrawly

Kralle die; ~, ~**n** claw; **krallen** 1. *refl. V.* **sich an etw.** *(Akk.)* ~ ⟨*cat*⟩ dig its claws into sth.; ⟨*person*⟩ clutch sth. [tightly]; 2. *tr. V. (fest greifen)* **die Finger in/um etw.** *(Akk.)* ~: dig one's fingers into sth./clutch sth. [tightly] with one's fingers

Kram der; ~|e|s *(ugs.)* **a)** stuff; *(Gerümpel)* junk; **b)** *(Angelegenheit)* affair; **kramen** 1. *itr. V.* in etw. *(Dat.)* ~: rummage about in sth.; 2. *tr. V. (ugs.)* etw. aus etw. ~: fish *(coll.)* sth. out of sth.; **Krämer der;** ~**s,** ~: grocer; **Kram·laden der** *(ugs. abwertend)* junk shop

Krampf der; ~[e]s, Krämpfe a) cramp; *(Zuckung)* spasm; b) *o. Pl.* painful strain; *(sinnloses Tun)* senseless waste of effort; **Krampf·ader** die varicose vein; **krampfhaft 1.** *Adj.* convulsive; *(verbissen)* desperate; **2.** *adv.* convulsively; *(verbissen)* desperately

Kran der; ~[e]s, Kräne a) crane; b) *(südwestd.: Wasserhahn)* tap; faucet *(Amer.)*

Kranich der; ~s, ~e crane

krank; kränker, kränkst... *Adj.* ill *usu. pred.;* sick; bad ⟨leg, tooth⟩; diseased ⟨plant, organ⟩; *(fig.)* ailing ⟨economy, business⟩; ~ werden be taken ill; jmdn. ~ schreiben give sb. a medical certificate; **Kranke** der/die; *adj. Dekl.* sick man/woman; *(Patient)* patient; **kränkeln** *itr. V.* be in poor health; **kränken** *tr. V.* jmdn. ~: hurt sb. *or* sb.'s feelings

Kranken-: ~geld das sickness benefit; ~haus das hospital; ~kasse die health insurance scheme; *(Körperschaft)* health insurance institution; *(privat)* health insurance company; ~pfleger der male nurse; ~schein der health insurance certificate; ~schwester die nurse; ~versicherung die a) *(Versicherung)* health insurance; b) *(Unternehmen)* health insurance company; ~wagen der ambulance

krank|feiern *itr. V. (ugs.)* skive off work *(sl.)* [pretending to be ill]; **kränker** *s.* krank; **krankhaft 1.** *Adj.* pathological; morbid ⟨growth, state, swelling, etc.⟩; **2.** *adv.* pathologically; morbidly ⟨swollen, sensitive⟩; **Krankheit** die; ~, ~en a) illness; *(bestimmte Art, von Pflanzen, Organen)* disease; b) *o. Pl. (Zeit des Krankseins)* illness; **Krankheits·erreger** der pathogen; **kränklich** *Adj.* ailing; **kränkst...** *s.* krank; **Kränkung** die; ~, ~en: eine ~: an injury to one's/sb.'s feelings

Kranz der; ~es, Kränze wreath; garland; *(auf einem Grab usw.)* wreath; **Kränzchen** das; ~s, ~: coffee circle; coffee klatch *(Amer.)*

Krapfen der; ~s, ~: doughnut

kraß 1. *Adj.* blatant ⟨case⟩; flagrant ⟨injustice⟩; stark ⟨contrast⟩; complete ⟨contradiction⟩; sharp ⟨difference⟩; out-and-out ⟨egoist⟩; **2.** *adv.* sich ~ ausdrücken put sth. bluntly; **sich von etw. ~ unterscheiden** be in stark contrast to sth.

Krater der; ~s, ~: crater

Kratz·bürste die *(ugs. scherzh.)* prickly so-and-so; **kratzen 1.** *tr. V.* scratch; *(entfernen)* scrape; **2.** *itr. V.* a) scratch; b) *(jucken)* itch; **Kratzer** der; ~s, ~ *(ugs.)* scratch; **kratzig** *Adj.* itchy ⟨material⟩

Kraul das; ~s *(Sport)* crawl; **¹kraulen 1.** *itr. V.* do the crawl; **2.** *tr. V.; auch mit sein* eine Strecke ~: cover a distance using the crawl

²kraulen *tr. V.* jmdm. das Kinn ~: tickle sb. under the chin; jmdn. in den Haaren ~: run one's fingers through sb.'s hair

kraus *Adj.* creased ⟨skirt etc.⟩; frizzy ⟨hair⟩; **Krause** die; ~, ~n *(Kragen)* ruff; *(am Ärmel)* ruffle

kräuseln 1. *tr. V.* ruffle ⟨water, surface⟩; gather ⟨material etc.⟩; frizz ⟨hair⟩; **2.** *refl. V.* ⟨hair⟩ go frizzy; ⟨water⟩ ripple; ⟨smoke⟩ curl up

Kraut das; ~[e]s, Kräuter a) herb; b) *o. Pl. (bes. südd., österr.: Kohl)* cabbage

Krawall der; ~s, ~e a) riot; b) *o. Pl. (ugs.: Lärm)* row *(coll.)*

Krawatte die; ~, ~n tie

kreativ 1. *Adj.* creative; **2.** *adv.* ~ veranlagt sein have a creative bent

Kreatur die; ~, ~en creature

Krebs der; ~es, ~e a) crustacean; *(Fluß~)* crayfish; *(Krabbe)* crab; b) *(Krankheit)* cancer

krebs-, Krebs-: ~erregend, ~erzeugend *Adj.* carcinogenic; ~krank *Adj.* ~krank sein have cancer; ~rot *Adj.* as red as a lobster *postpos.*

Kredit der; ~[e]s, ~e credit; *(Darlehen)* loan; **Kredit·karte** die credit card; **mit ~karte bezahlen** pay by credit card; **kredit·würdig** *Adj. (Finanzw.)* credit-worthy

Kreide die; ~, ~n chalk; **kreidebleich** *Adj.* as white as a sheet *postpos.;* **Kreide·felsen** der chalk cliff

kreieren [kre'iːrən] *tr. V.* create

Kreis der; ~es, ~e circle; *(Verwaltungsbezirk)* district; *(Wahl~)* ward; **Kreis·bahn** die orbit

kreischen *itr. V.* screech; ⟨door⟩ creak

Kreisel der; ~s, ~ *(Kinderspielzeug)* top; *(ugs.: Kreisverkehr)* roundabout; **kreisen** *itr. V.; auch mit sein* ⟨planet⟩ revolve (um around); ⟨satellite etc.⟩ orbit; ⟨aircraft, bird⟩ circle

kreis-, Kreis-: ~förmig *Adj.* circular; ~lauf der *(Physiol.)* circulation; *(der Natur, des Lebens usw.)* cycle;

~lauf·störungen *Pl. (Med.)* circulatory trouble *sing.;* ~rund *Adj.* [perfectly] round; ~säge die circular saw

Kreiß·saal der *(Med.)* delivery room

Kreis-: ~stadt die chief town of a/the district; ~verkehr der roundabout

Krem die; ~, ~s *s.* Creme

Krematorium das; ~s, Krematorien crematorium

kremig *s.* cremig

Krempe die; ~, ~n brim

Krempel der; ~s *(ugs. abwertend)* stuff; *(Gerümpel)* junk

krepieren *itr. V.; mit sein (salopp)* ⟨*person*⟩ snuff it *(sl.)*

Krepp der; ~s, ~s *od.* ~e crêpe

Kresse die; ~, ~n *(Bot.)* cress

Kreta (das); ~s Crete

Kreuz das; ~es, ~e a) cross; *(Kreuzzeichen)* sign of the cross; b) *(Teil des Rückens)* small of the back; jmdn. aufs ~ legen *(salopp)* take sb. for a ride *(sl.);* c) *o. Art., o. Pl. (Kartenspiel) (Farbe)* clubs *pl.; (Karte)* club; d) *(Autobahn)* interchange; e) *(Musik)* sharp;

kreuzen 1. *tr. V. (auch Biol.)* cross; 2. *refl. V.* a) *(überschneiden)* cross; b) *(zuwiderlaufen)* clash ⟨mit with⟩; 3. *itr. V.; mit haben od. sein (fahren)* cruise

Kreuz-: ~fahrer der *(hist.)* crusader; ~fahrt die cruise; ~feuer das *(Milit., auch fig.)* cross-fire; ~gang der cloister

kreuzigen *tr. V.* crucify; Kreuzigung die; ~, ~en crucifixion

Kreuz-: ~otter die adder; [common] viper; ~ritter der *(hist.)* crusader; ~schmerzen *Pl.* pain *sing.* in the small of the back

Kreuzung die; ~, ~en a) crossroads *sing.;* b) *(Biol.)* crossing; cross-breeding; *(Ergebnis)* cross

kreuz-, Kreuz-: ~verhör das cross-examination; ~weise crosswise; ~wort·rätsel das crossword [puzzle]; ~zug der *(hist., fig.)* crusade

kribbelig *Adj. (ugs.) (vor Ungeduld)* fidgety; *(nervös)* edgy; kribbeln *itr. V. (jucken)* tickle; *(prickeln)* tingle

kriechen *unr. itr. V.* a) *mit sein* ⟨*insect, baby*⟩ crawl; ⟨*plant*⟩ creep; ⟨*person, animal*⟩ creep, crawl; b) *auch mit sein (fig. abwertend)* crawl ⟨vor + *Dat.* to⟩; Kriecher der; ~s, ~ *(abwertend)* crawler; Kriech·spur die *(Verkehrsw.)* crawler lane

Krieg der; ~[e]s, ~e war

kriegen *tr. V. (ugs.)* get; *(erreichen)* catch ⟨*train, bus, etc.*⟩

Krieger der; ~s, ~: warrior; kriegerisch *Adj.* a) *(kampflustig)* warlike; b) *(militärisch)* military; eine ~e Auseinandersetzung an armed conflict

kriegs-, Kriegs-: ~beil das tomahawk; das ~beil begraben *(scherzh.)* bury the hatchet; ~bemalung die *(Völkerk.)* war-paint; ~beschädigt *Adj.* war-disabled; ~beschädigte der/die war invalid; ~dienst der a) *(im Krieg)* active service; b) *(Wehrdienst)* military service; den ~dienst verweigern be a conscientious objector; ~dienst·verweigerer der conscientious objector; ~erklärung die declaration of war; ~gefangene der prisoner of war; POW; ~gefangenschaft die captivity; ~schiff das warship; ~verbrechen das *(Rechtsw.)* war crime

Krimi der; ~[s], ~[s] *(ugs.)* crime thriller; Kriminal·beamte der [plain-clothes] detective; Kriminalität die; ~: crime *no art.*

Kriminal-: ~polizei die criminal investigation department; ~roman der crime novel; *(mit Detektiv als Held)* detective novel

kriminell 1. *Adj.* criminal; 2. *adv.* ~ veranlagt sein have criminal tendencies; ~ handeln act illegally; Kriminelle der/die; *adj. Dekl.* criminal

Krimskrams der; ~[es] *(ugs.)* stuff

Kringel der; ~s, ~ *(Kreis)* [small] ring; *(Kritzelei)* round squiggle; *(Gebäck)* [ring-shaped] biscuit; kringeln *refl. V.* curl [up]; ⟨*hair*⟩ go curly; sich ~ [vor Lachen] *(ugs.)* kill oneself [laughing] *(coll.)*

Kripo die; ~ *(ugs.)* die ~: ≈ the CID

Krippe die; ~, ~n a) *(Futtertrog)* manger; crib; b) *(Weihnachts~)* model of a nativity scene; c) *(Kinder~)* crèche

Krise die; ~, ~n *(auch Med.)* crisis; kriseln *itr. V. (unpers.)* es kriselt in ihrer Ehe/in der Partei their marriage is in trouble/the party is in a state of crisis; Krisen·herd der trouble spot

¹Kristall der; ~s, ~e crystal; ²Kristall das; ~s crystal *no indef. art.*

Kriterium das; ~s, Kriterien criterion

Kritik die; ~, ~en a) criticism *no indef. art.* (an + *Dat.* of); an jmdm./etw. ~ üben criticize sb./sth.; b) *(Besprechung)* review; Kritiker der; ~s, ~: critic; kritik·los 1. *Adj.* uncritical; 2. *adv.* uncritically; kritisch 1. *Adj.* critical; 2. *adv.* critically; kritisieren *tr. V.* criticize; review ⟨*book, play, etc.*⟩

kritzeln 1. *itr. V. (schreiben)* scribble; *(zeichnen)* doodle; **2.** *tr. V.* scribble
Kroatien [kro'a:tsi̯ən] **(das);** ~s Croatia; **kroatisch** *Adj.* Croatian
kroch *1. u. 3. Pers. Sg. Prät. v.* **kriechen**
Krokant der; ~s praline
Krokette die; ~, ~n *(Kochk.)* croquette
Krokodil das; ~s, ~e crocodile; **Krokodils·tränen** *Pl. (ugs.)* crocodile tears
Krokus der; ~, ~ *od.* ~se crocus
Krone die; ~, ~n crown; (eines Baumes) top; crown; *(einer Welle)* crest; **die** ~ **der Schöpfung** the pride of creation; **krönen** *tr. V. (auch fig.)* crown; **Kronen·korken** der crown cork
Kron-: ~**juwel** das *od.* der; *meist Pl.* die ~**juwelen** the crown jewels; ~**leuchter** der chandelier; ~**prinz** der crown prince
Krönung die; ~, ~en coronation; *(fig.)* culmination; **Kron·zeuge** der *(Rechtsw.)* person who turns Queen's/King's evidence; **als** ~ **auftreten** turn Queen's/King's evidence
Kropf der; ~[e]s, Kröpfe *(Med.)* goitre
Kröte die; ~, ~n a) toad; b) *Pl. (salopp: Geld)* **ein paar/eine ganze Menge** ~**n verdienen** earn a few bob *(Brit. sl.)*/a fair old whack *(sl.)*
Krücke die; ~, ~n crutch; **Krückstock** der walking-stick
Krug der; ~[e]s, Krüge jug; *(größer)* pitcher; *(Bier~)* mug
Krume die; ~, ~n crumb; **Krümel** der; ~s, ~: crumb; **krümeln** *itr. V.* a) crumble; b) *(Krümel machen)* make crumbs
krumm 1. *Adj.* a) bent ⟨*nail, back*⟩; crooked ⟨*stick, branch, etc.*⟩; bandy ⟨*legs*⟩; b) *nicht präd. (ugs.: unrechtmäßig)* crooked; **2.** *adv.* crookedly; **krümmen 1.** *tr. V.* bend; **2.** *refl. V.* a) *(sich winden)* writhe; b) *(krumm verlaufen)* ⟨*road, path, river*⟩ bend
krumm-: ~**lachen** *refl. V. (ugs.)* **sich über etw.** *(Akk.)* ~**lachen** fall about laughing over sth.; ~**nehmen** *unr. tr. V. (ugs.)* **etw.** ~**nehmen** take sth. the wrong way
Krümmung die; ~, ~en bend
Krüppel der; ~s, ~: cripple
Kruste die; ~, ~n crust; *(vom Braten)* crisp

Kruzifix das; ~es, ~e crucifix
Krypta die; ~, Krypten *(Archit.)* crypt
Kuba (das); ~s Cuba; **Kubaner** der; ~s, ~: Cuban
Kübel der; ~s, ~: pail
Kubik- cubic ⟨*metre, foot, etc.*⟩
Küche die; ~, ~n kitchen; *(Einrichtung)* kitchen furniture *no indef. art.; (Kochk.)* cooking; cuisine; **kalte/warme** ~: cold/hot food
Kuchen der; ~s, ~: cake; *(Obst~)* flan; *(Torte)* gateau
Kuchen-: ~**form** die cake-tin; ~**gabel** die pastry-fork
Küchen·gerät das kitchen utensil; *(als Kollektivum)* kitchen utensils *pl.*
Kuckuck der; ~s, ~e a) cuckoo; **zum** ~ **[noch mal]!** *(salopp)* for crying out loud! *(coll.)*; b) *(scherzh.: Pfandsiegel)* bailiff's seal *(placed on distrained goods)*; **Kuckucks·uhr** die cuckoo clock ·
Kufe die; ~, ~n runner; *(von Flugzeugen, Hubschraubern)* skid
Kugel die; ~, ~n a) ball; *(Geom.)* sphere; *(Kegeln)* bowl; *(beim Kugelstoßen)* shot; b) *(ugs.: Geschoß)* bullet; **Kugel·lager** das *(Technik)* ball-bearing; **kugeln** *tr. V.* roll; **2.** *refl. V.* **sich [vor Lachen]** ~ *(ugs.)* double *or* roll up [laughing]
kugel-, Kugel-: ~**rund** [--'-] *Adj.* round as a ball *postpos.; (scherzh.: dick)* rotund; tubby; ~**schreiber** der ball-pen; Biro (P); ~**sicher** *Adj.* bullet-proof; ~**stoßen** das; ~s shot[-put]; *(Disziplin)* putting the shot *no art.*
Kuh die; ~, Kühe cow
Kuh-: ~**fladen** der cow-pat; ~**haut** die: **das geht auf keine** ~**haut** *(fig. salopp)* it's absolutely staggering
kühl 1. *Adj.* cool; **etw.** ~ **lagern** keep sth. in a cool place; **2.** *adv.* coolly
Kuhle die; ~, ~n *(ugs.)* hollow
Kühle die; ~: coolness
kühlen *tr. V.* cool; chill ⟨*wine*⟩; refrigerate ⟨*food*⟩; **2.** *itr. V.* ⟨*cold compress, ointment, breeze, etc.*⟩ have a cooling effect; **Kühler** der; ~s, ~ a) *(am Auto)* radiator; *(~haube)* bonnet *(Brit.)*; hood *(Amer.)*; b) *(Sekt~)* ice-bucket; **Kühler·haube** die bonnet *(Brit.)*; hood *(Amer.)*
Kühl-: ~**schrank** der refrigerator; fridge *(Brit. coll.)*; icebox *(Amer.)*; ~**truhe** die [chest] freezer; *(im Lebensmittelgeschäft)* freezer [cabinet]
Kühlung die; ~, ~en cooling; *(Vor-*

richtung) cooling system; *(für Lebensmittel)* refrigeration system; **Kühl·wasser das** cooling water

kühn 1. *Adj.* bold; *(dreist)* audacious; **2.** *adv.* boldly; *(gewagt)* daringly; *(dreist)* audaciously; **Kühnheit die;** ~: boldness; *(Gewagtheit)* daringness; *(Dreistigkeit)* audacity

Kuh·stall der cowshed

Küken das; ~s, ~: chick

kulant *Adj.* obliging; fair ⟨*terms*⟩; **Kulanz die;** ~: willingness to oblige

Kuli der; ~s, ~s **a)** coolie; **b)** *(ugs.)* ball-point; Biro (P)

kulinarisch *Adj.* culinary

Kulisse die; ~, ~n piece of scenery; flat; *(Hintergrund)* backdrop; **die** ~n the scenery *sing.*

kullern *(ugs.) itr. V. mit sein* roll

Kult der; ~|e|s, ~e *(auch fig.)* cult; **kultivieren** *tr. V. (auch fig.)* cultivate; **kultiviert 1.** *Adj.* cultured; *(vornehm)* refined; **2.** *adv.* in a cultured manner; *(vornehm)* in a refined manner; **Kultur die;** ~, ~en **a)** *o. Pl.* culture; *(kultivierte Lebensart)* refinement; **ein Mensch von** ~: a cultured person; **b)** *(Zivilisation, Lebensform)* civilization

Kultur-: ~**abkommen das** cultural agreement; ~**austausch der** cultural exchange; ~**beutel der** sponge-bag *(Brit.);* toilet-bag

kulturell 1. *Adj.* cultural; **2.** *adv.* culturally

Kultur-: ~**film der** documentary film; ~**geschichte die a)** *o. Pl.* history of civilization; *(einer bestimmten Kultur)* cultural history; ~**politik die** cultural and educational policy

Kultus·minister der minister for education and cultural affairs

Kümmel der; ~s, ~: caraway [seed]; *(Branntwein)* kümmel

Kummer der; ~s sorrow; grief; *(Ärger, Sorgen)* trouble; ~ **um** *od.* **über jmdn.** grief for sb.; **jmdm.** ~ **machen** give sb. trouble; **kümmerlich** *Adj.* **a)** *(schwächlich)* puny; stunted ⟨*vegetation, plants*⟩; **b)** *(ärmlich)* wretched; miserable; **c)** *(gering)* miserable; meagre ⟨*knowledge, left-overs*⟩; **kümmern 1.** *refl. V.* **a)** **sich um jmdn./etw.** ~: take care of sb./sth.; **b)** *(sich befassen mit)* **sich nicht um Politik** ~: not be interested in politics; **2.** *tr. V.* concern

Kumpan der; ~s, ~e *(ugs.)* **a)** pal *(coll.);* buddy *(coll.);* **b)** *(abwertend:*

Mittäter) accomplice; **Kumpel der;** ~s, ~ **a)** *(Bergmannsspr.)* miner; **b)** *(salopp: Kamerad)* pal *(coll.);* buddy *(coll.)*

kündbar *Adj.* terminable ⟨*contract*⟩; redeemable ⟨*loan, mortgage*⟩; **¹Kunde der;** ~n, ~n customer; *(eines Architekten-, Anwaltbüros, einer Versicherung usw.)* client

²Kunde die; ~ *(geh.)* tidings *pl. (literary);* **Kunden·dienst der** *o. Pl.* service to customers; *(Wartung)* after-sales service; **Kundgebung die;** ~, ~en rally; **kundig** *Adj. (kenntnisreich)* knowledgeable; *(sachverständig)* expert; **kündigen 1.** *tr. V.* cancel ⟨*subscription, membership*⟩; terminate ⟨*contract, agreement*⟩; **seine Stellung** ~: hand in one's notice (**bei** to); **2.** *unr. itr. V.* **a)** *(ein Mietverhältnis beenden)* ⟨*tenant*⟩ give notice; **jmdm.** ~ ⟨*landlord*⟩ give sb. notice to quit; **zum 1. Juli** ~: give notice for 1 July; **b)** *(ein Arbeitsverhältnis beenden)* ⟨*employee*⟩ hand in one's notice (**bei** to); **jmdm.** ~ ⟨*employer*⟩ give sb. his/her notice; **Kündigung die;** ~, ~en **a)** *(der Mitgliedschaft, eines Abonnements)* cancellation; *(eines Vertrags)* termination; **b)** *(eines Arbeitsverhältnisses)* **jmdm. die** ~ **aussprechen** give sb. his/her notice; **Kundin die;** ~, ~nen customer/client; **Kundschaft die;** ~, ~en *o. Pl.; s.* ¹**Kunde a:** customers *pl.;* clientele; **Kundschafter der;** ~s, ~: scout; **kund|tun** *(geh.) unr. tr. V.* announce

künftig 1. *Adj.* future; **2.** *adv.* in future

Kunst die; ~, **Künste a)** art; **b)** *(das Können)* skill; **die ärztliche** ~: medical skill; **das ist keine** ~! *(ugs.)* there's nothing 'to it

kunst-, Kunst-: ~**aus·stellung die** art exhibition; ~**buch das** art book; ~**erzieher der,** ~**erzieherin die** art teacher; ~**faser die** synthetic fibre; ~**führer der** guide to cultural and artistic monuments [of an/the area]; ~**genuß der** enjoyment of art; *(Ereignis)* artistic treat; ~**gerecht 1.** *Adj.* expert; **2.** *adv.* expertly; ~**geschichte die** *o. Pl.* art history; ~**geschichtlich 1.** *Adj.* art historical ⟨*studies, evidence, expertise*⟩; ⟨*work*⟩ on art history; **2.** *adv.* ~**geschichtlich interessiert/versiert** interested/well versed in art history; ~**gewerbe das** arts and crafts *pl.;* ~**griff der** trick; dodge;

~**halle** die art gallery; ~**händler** der [fine-]art dealer; ~**handwerk** das craftwork; ~**kritiker** der art critic; ~**leder** imitation leather

Künstler der; ~s, ~, **Künstlerin** die; ~, ~nen a) artist; *(Zirkus~, Varieté~)* artiste; b) *(Könner)* genius (**in** + *Dat.* at); **künstlerisch** 1. *Adj.* artistic; 2. *adv.* artistically; **Künstler·name** der stage-name; **künstlich** 1. *Adj.* a) artificial; b) *(gezwungen)* forced ⟨*laugh, cheerfulness, etc.*⟩; 2. *adv.* artificially

kunst-, Kunst-: ~**los** *Adj.* plain; ~**post·karte** die art postcard; ~**sammler** der art collector; ~**sammlung** die art collection; ~**stoff** der synthetic material; plastic; ~**stück** das trick; **das ist kein** ~**stück** *(ugs.)* it's no great feat; ~**turnen** das gymnastics *sing.;* ~**voll** 1. *Adj.* ornate and artistic; *(kompliziert)* elaborate; 2. *adv.* a) ornately *or* elaborately and artistically; b) *(geschickt)* skilfully; ~**werk** das work of art

kunter·bunt 1. *Adj.* multi-coloured; *(abwechslungsreich)* varied; *(ungeordnet)* jumbled ⟨*confusion, muddle, etc.*⟩; 2. *adv.* ⟨*painted, printed*⟩ in many colours; ~ **durcheinander sein** be higgledy-piggledy

Kupfer das; ~s a) copper; b) *(~geschirr)* copperware; *(~geld)* coppers *pl.*

Kupfer-: ~**geld** das coppers *pl.;* ~**stich** der a) *o. Pl.* copperplate engraving *no art.;* b) *(Blatt)* copperplate print *or* engraving

Kuppe die; ~, ~n a) [rounded] hilltop; b) *(Finger~)* tip; end

Kuppel die; ~, ~n dome; *(kleiner)* cupola

Kuppelei die; ~: procuring; **kuppeln** *itr. V.* operate the clutch; **Kuppelung** *s.* Kupplung; **Kuppler** der; ~s, ~: procurer; **Kupplerin** die; ~, ~nen procuress; **Kupplung** die; ~, ~en a) *(Kfz-W.)* clutch; b) *(Technik: Vorrichtung zum Verbinden)* coupling

Kur die; ~, ~en [health] cure; *(ohne Aufenthalt im Badeort)* course of treatment

Kür die; ~, ~en *(Eiskunstlauf)* free programme; *(Turnen)* optional exercises *pl.*

Kurbel die; ~, ~n crank [handle]; *(an Spieldosen, Grammophonen)* winder; *(an einem Brunnen)* [winding-]handle; **kurbeln** *tr. V.* etw. nach oben/unten

~: wind sth. up/down; **Kurbel·welle** die *(Technik)* crankshaft

Kürbis der; ~ses, ~se pumpkin

Kurde der; ~n, ~n Kurd

Kur-: ~**fürst** der *(hist.)* Elector; ~**gast** der visitor to a/the spa; *(Patient)* patient at a/the spa

Kurier der; ~s, ~e courier

kurieren *tr. V. (auch fig.)* cure (**von** of)

kurios 1. *Adj.* curious; 2. *adv.* curiously; strangely; oddly; **Kuriosität** die; ~, ~en a) *o. Pl.* strangeness; b) *(Gegenstand)* curiosity; curio

Kur-: ~**konzert** das concert [at a spa]; ~**ort** der spa; ~**pfuscher** der *(ugs. abwertend)* quack

Kurs der; ~es, ~e a) *(Richtung)* course; **ein harter/weicher** ~ *(fig.)* a hard/soft line; b) *(von Wertpapieren)* price; *(von Devisen)* exchange rate; **der** ~ **des Dollars** the dollar rate; c) *(Lehrgang)* course; *(Teilnehmer)* class

Kürschner der; ~s, ~: furrier

kursieren *itr. V.; auch mit sein* circulate; **Kurs·teilnehmer** der the course participant; **Kursus** der; ~, **Kurse** *s.* Kurs; **Kurs·wagen** der *(Eisenb.)* through carriage

Kur·taxe die visitors' tax *(at a spa)*

Kurve die; ~, ~n a) *(einer Straße)* bend; b) *(Geom.)* curve; c) *(in der Statistik, Temperatur~ usw.)* graph; **kurven** *itr. V.; mit sein* a) ⟨*aircraft*⟩ circle; ⟨*tanks etc.*⟩ circle [round]; b) *(ugs.: fahren)* drive around; **kurven·reich** *Adj.* winding; twisting

kurz; kürzer, kürzest... 1. *Adj.* short; *(zeitlich; knapp)* short, brief; quick ⟨*look*⟩; 2. *adv.* a) *(zeitlich)* briefly; *(knapp)* ~ **gesagt** in a word; b) *(wenig)* just; ~ **vor/hinter der Kreuzung** just before/past the crossroads; ~ **vor/nach Pfingsten** just before/after Whitsun; **Kurz·arbeit** die short-time working; **kurz·ärm[e]lig** *Adj.* short-sleeved; **Kürze** die; ~ a) shortness; brevity; b) *(geringe Dauer)* shortness; brevity; **in** ~: shortly; c) *(Knappheit)* brevity; **Kürzel** das; ~s, ~: shorthand symbol; **kürzen** *tr. V.* shorten; abridge ⟨*article, book*⟩; cut ⟨*pension, budget*⟩; **kürzer** *s.* kurz; **kurzerhand** *Adv.* without more ado; **kürzest...** *s.* kurz

kurz-, Kurz-: ~**fristig** 1. *Adj.* a) ⟨*refusal, resignation, etc.*⟩ at short notice; b) *(für kurze Zeit)* short-term; 2. *adv.* a) at short notice; b) *(für kurze Zeit)* for a short time; *(auf kurze Sicht)* in the short term; *(in kurzer Zeit)* without delay; **~geschichte die** short story; **~lebig** *Adj. (auch fig.)* short-lived

kürzlich *Adv.* recently; not long ago

kurz-, Kurz-: **~parker der** short-stay *(Brit.) or* short-term parker; **~schluß der** *(Elektrot.)* short-circuit; **~sichtig** *(auch fig.)* **1.** *Adj.* short-sighted; **2.** *adv.* short-sightedly

Kürzung die; ~, **~en** cut

kurz-, Kurz-: **~waren** *Pl.* haberdashery *sing. (Brit.);* notions *(Amer.);* **~welle die** *(Physik, Rundf.)* short wave; **~zeitig 1.** *Adj.* brief; **2.** *adv.* briefly

kuscheln *refl. V.* sich an jmdn. ~: snuggle up to sb.

kuschen *itr. V.* knuckle under (vor + *Dat.* to)

Kusine die; ~, **~n** *s.* Cousine

Kuß der; Kusses, Küsse kiss; **kußecht** *Adj.* kissproof; **küssen** *tr., itr. V.* kiss; **Kuß · hand die: jmdm. eine ~ zuwerfen** blow sb. a kiss; **mit ~** *(ugs.)* gladly

Küste die; ~, **~n** coast; **Küsten · wache die** coastguard [service]

Küster der; ~s, ~: sexton

Kutsche die; ~, **~n** coach; **Kutscher der;** ~s, ~: coach-driver; **kutschieren** *itr. V.; mit sein* drive, ride [in a coach]; **2.** *tr. V.* **jmdn.** ~: drive sb. [in a coach]

Kutte die; ~, **~n** [monk's/nun's] habit

Kutter der; ~s, ~: cutter

Kuvert [ku've:ɐ̯] **das;** ~s, **~s** envelope; *(geh.: Gedeck)* cover

Kuwait [ku'vait] **(das);** ~s Kuwait

Kybernetik die; ~: cybernetics *sing.*

L

l, L [ɛl] **das;** ~, ~: l/L

l *Abk.* Liter l.

laben *(geh.)* **1.** *tr. V.* jmdn. ~: give sb. refreshment; **2.** *refl. V.* refresh oneself (**an** + *Dat.,* **mit** with)

labil *Adj.* **a)** *(Med.)* delicate ⟨*constitution, health*⟩; poor ⟨*circulation*⟩; **b)**

(auch Psych.) unstable ⟨*person, character, situation, etc.*⟩

Labor das; ~s, ~s, *auch:* ~e laboratory; **Laboratorium das;** ~s, **Laboratorien** laboratory

Labyrinth das; ~[e]s, ~e maze; labyrinth

¹**Lache die;** ~, ~n *(ugs.)* laugh

²**Lache** ['la(:)xə] **die;** ~, ~n puddle; *(von Blut, Öl)* pool

lächeln *itr. V.* smile (**über** + *Akk.* at); **Lächeln das;** ~s smile; **lachen 1.** *itr. V.* laugh (**über** + *Akk.* at); **2.** *tr. V.* **was gibt es denn zu ~?** what's so funny?; **Lachen das;** ~s laughter; **ein lautes ~:** a loud laugh; **lächerlich 1.** *Adj.* ridiculous; ludicrous ⟨*argument, statement*⟩; **2.** *adv.* ridiculously; **Lächerlichkeit die;** ~: ridiculousness; *(von Argumenten, Behauptungen usw.)* ludicrousness; **lachhaft** *Adj.* ridiculous

Lachs der; ~es, ~e salmon

Lack der; ~[e]s, ~e **a)** varnish; *(für Metall, Lackarbeiten)* lacquer; **lackieren** *tr. V.* varnish; spray ⟨*car*⟩; **Lack · leder das** patent leather

Lade die; ~, ~n *(landsch.)* drawer; **Lade · hemmung die** jam; ¹**laden 1.** *unr. tr. V.* load; *(Physik)* charge; **2.** *unr. itr. V.* load [up]

²**laden** *unr. tr. V.* **a)** *(Rechtsspr.)* summon; **b)** *(geh.: ein~)* invite

Laden der; ~s, **Läden a)** shop; store *(Amer.);* **der ~ läuft** *(ugs.)* business is good; **b)** *(Fenster~)* shutter

Laden-: **~diebstahl der** shop-lifting; **~schluß der** shop *or (Amer.)* store closing-time; **~tisch der** [shop-]counter

Lade-: **~rampe die** loading ramp; **~raum der** *(beim Auto)* luggage-space; *(beim Flugzeug, Schiff)* hold; *(bei LKWs)* payload space

lädieren *tr. V.* damage

lädst *2. Pers. Sg. Präsens v.* laden; **lädt** *3. Pers. Sg. Präsens v.* laden

Ladung die; ~, **~en a)** *(Schiffs~, Flugzeug~)* cargo; *(LKW~)* load; **b)** *(beim Sprengen, Schießen; Physik)* charge; **c)** *(Rechtsspr.: Vor~)* summons *sing.*

lag *1. u. 3. Pers. Sg. Prät. v.* liegen; **Lage die;** ~, **~n a)** situation; **eine gute ~ haben** be well situated; **b)** *(Art des Liegens)* position; **c)** *(Situation)* situation; **Lage · plan der** map of the area; **Lager das;** ~s, ~ **a)** camp; **b)** store-room; *(in Geschäften, Betrieben)* stock-room; **c)** *(Warenbestand)* stock

Lager-: ~**feuer** das camp-fire; ~**halle** die warehouse

lagern 1. *tr. V.* **a)** store; **b)** *(hinlegen)* lay down; 2. *itr. V.* **a)** camp; **b)** *(liegen)* lie; *(foodstuffs, medicines, etc.)* be kept

Lager-: ~**platz** der campsite; ~**raum** store-room; *(im Geschäft, Betrieb)* stock-room

Lagerung die; ~, ~en storage

Lagune die; ~, ~n lagoon

lahm *Adj.* **a)** *(gelähmt)* lame; *(ugs.: unbeweglich)* stiff; **b)** *(ugs.: unzureichend)* lame *(excuse, explanation, etc.)*; **c)** *(ugs. abwertend: matt)* dreary; **lahmen** *itr. V.* be lame; **lähmen** *tr. V.* paralyse; *(fig.)* paralyse *(economy, industry)*; bring *(traffic)* to a standstill; **Lähmung** die; ~, ~en paralysis; *(fig.) (der Wirtschaft, Industrie)* paralysis; **zu einer ~ des Verkehrs führen** bring traffic to a standstill

Laib der; ~|e|s, ~e loaf; **ein |halber| ~ Brot** [half] a loaf of bread

Laich der; ~|e|s, ~e spawn; **laichen** *itr. V.* spawn

Laie der; ~n, ~n *(Mann)* layman; *(Frau)* laywoman

Lakai der; ~en, ~en lackey; liveried footman

Lake die; ~, ~n brine

Laken das; ~s, ~ *(bes. nordd.)* sheet

Lakritze die; ~, ~n liquorice

lallen *tr., itr. V.* *(baby)* babble; *(drunk/ drowsy person)* mumble

Lamelle die; ~, ~n *(einer Jalousie)* slat; *(eines Heizkörpers)* rib

lamentieren *itr. V. (ugs.)* moan (über + *Akk.* about)

Lametta das; ~s lametta

Lamm das; ~|e|s, Lämmer lamb

lamm-, Lamm-: ~**fell** das lambskin; ~**fleisch** das lamb; ~**fromm** 1. *Adj.* *(person)* as meek as a [little] lamb; 2. *adv.* *(answer)* like a lamb

Lampe die; ~, ~n light; *(Tisch~, Öl~, Signal~)* lamp

Lampen~: ~**fieber** das stage fright; ~**schirm** der [lamp]shade

Lampion [lam'pjɔŋ] der; ~s, ~s Chinese lantern

Land das; ~es, Länder *od. (veralt.)* ~e **a)** *o. Pl.* land *no indef. art.; (dörfliche Gegend)* country *no indef. art.;* **an** ~: ashore; **auf dem ~ wohnen** live in the country; **b)** *(Staat)* country; **c)** *(Bundesland)* Land; state; *(österr.)* province; **Land·bevölkerung** die rural population

Lande-: ~**an·flug** der *(Flugw.)* [landing] approach; ~**bahn** die *(Flugw.)* [landing] runway

landen 1. *itr. V.; mit sein* **a)** land; *(ankommen)* arrive; **b)** *(ugs.: gelangen)* land up; 2. *tr. V.* **a)** land *(aircraft, troops, passengers, fish, etc.)*; **b)** *(ugs.: zustande bringen)* pull off *(victory, coup)*; have *(smash hit)*

Ländereien *Pl.* estates

Länder-: ~**kampf** der *(Sport)* international match; ~**spiel** das *(Sport)* international [match]

Landes-: ~**innere** das interior [of the country]; ~**kunde** die; *o. Pl.* regional studies *pl., no art.;* ~**regierung** die government of a/the Land/province; ~**sprache** die language of the country; ~**tracht** die national costume *or* dress; ~**verrat** der *(Rechtsw.)* treason; ~**währung** die currency of a/the country

land-, Land-: ~**flucht** die migration from the countryside [to the towns]; ~**gewinnung** die reclamation of land; ~**haus** das country house; ~**karte** die map; ~**kreis** der district; ~**läufig** *Adj.* widely accepted

ländlich *Adj.* rural; country *attrib.* *(life)*

Land-: ~**plage** die *(fig.)* pest; nuisance; ~**ratte** die *(ugs.)* landlubber

Landschaft die; ~, ~en landscape; *(ländliche Gegend)* countryside; **landschaftlich** 1. *Adj.* regional; 2. *adv.* ~ **herrlich gelegen sein** be in a glorious natural setting

Lands·mann der; *Pl.* ~**leute** fellow-countryman; compatriot

Land-: ~**straße** die country road; *(im Gegensatz zur Autobahn)* ordinary road; ~**streicher** der tramp; ~**strich** der area; ~**tag** der Landtag; state parliament; *(österr.)* provincial parliament; **Landung** die; ~, ~en landing; **Landungs·brücke** die [floating] landing-stage

land-, Land-: ~**weg** der overland route; **auf dem ~weg** overland; ~**wirt** der farmer; ~**wirtschaft** die *o. Pl.* agriculture *no art.;* farming *no art.;* ~**wirtschaftlich** 1. *Adj.* agricultural; 2. *adv.* ~ **genutzt werden** be used for agricultural purposes; ~**zunge** die *(Geogr.)* tongue of land

lang; länger, längst... 1. *Adj.* long; *(ugs.: groß)* tall; 2. *adv.* [for] a long time; **eine Sekunde/mehrere Stunden ~:** for a second/several hours

lang-: **~ärm[e]lig** *Adj.* long-sleeved; **~atmig** 1. *Adj.* long-winded; 2. *adv.* long-windedly; ⟨*relate*⟩ at great length

lange; länger, am längsten *Adv.* a) a long time; **bist du schon ~ hier?** have you been here long?; b) *(bei weitem)* **ich bin noch ~ nicht fertig** I'm nowhere near finished; **hier is es ~ nicht so schön** it isn't nearly as nice here; **Länge die; ~, ~n** length; *(Geogr.)* longitude

langen *(ugs.)* 1. *itr. V.* a) be enough; b) *(greifen)* reach **(in + Akk.** into; **auf + Akk.** on to; **nach** for); 2. *tr. V.* **jmdm. eine ~** *(ugs.)* give sb. a clout [around the ear] *(coll.)*

Längen·grad der *(Geogr.)* degree of longitude

länger 1. *s.* **lang, lange;** 2. *Adj.* **seit ~er Zeit** for quite some time

Lange·weile die; ~ *od.* **Langenweile** boredom; **~ haben** be bored

lang-, Lang-: **~fristig** 1. *Adj.* long-term; long-dated ⟨*loan*⟩; 2. *adv.* on a long-term basis; **~jährig** *Adj.* ⟨*customer, friend*⟩ of many years' standing; long-standing ⟨*friendship*⟩; **~jährige Erfahrung** many years of experience; **~lauf der** *(Skisport)* cross-country

länglich *Adj.* oblong; **längs** 1. *Präp.* + *Gen. od. (selten) Dat.* along; 2. *Adv.* lengthways; **Längs·achse die** longitudinal axis

langsam 1. *Adj.* slow; 2. *adv.* a) slowly; **~, aber sicher** *(ugs.)* slowly but surely; b) *(allmählich)* gradually

Lang-: **~schläfer der** late riser; **~spiel·platte die** long-playing record; LP

Längs·schnitt der longitudinal section

längst *Adv.* a) *(schon lange)* a long time ago; b) *(bei weitem)* **hier ist es ~ nicht so schön** it isn't nearly as nice here; **längst...** *s.* **lang; längstens** *Adv. (ugs.) (höchstens)* at [the] most; *(spätestens)* at the latest

Languste die; ~, ~n spiny lobster

lang-, Lang-: **~weilen** 1. *tr. V.* bore; 2. *refl. V.* be bored; **~weilig** 1. *Adj.* boring; dull ⟨*place*⟩; 2. *adv.* boringly; **~welle die** *(Physik, Rundf.)* long wave; **~wierig** *Adj.* lengthy; prolonged ⟨*search*⟩

Lanze die; ~, ~n lance; *(zum Werfen)* spear

Laos ['laːɔs] **(das); Laos'** Laos; **Laote** [la'oːtə] **der; ~n, ~n** Laotian

lapidar 1. *Adj. (kurz, aber wirkungsvoll)* succinct; *(knapp)* terse; 2. *adv.* succinctly/tersely

Lappalie die; ~, ~n trifle

Lappe der; ~n, ~n Lapp

Lappen der; ~s, ~: cloth; *(Fetzen)* rag; *(Wasch~)* flannel

läppisch *Adj.* silly

Lapp·land (das) Lapland

Lärche die; ~, ~n larch

Lärm der; ~[e]s noise; *(Krach)* din; row *(coll.)*; **Lärm·belästigung die** disturbance caused by noise; **lärmen** *itr. V.* make a noise *or (coll.)* row

Larve die; ~, ~n grub; larva

las *1. u. 3. Pers. Sg. Prät. v.* **lesen**

lasch 1. *Adj.* limp ⟨*handshake*⟩; feeble ⟨*action, measure*⟩; lax ⟨*upbringing*⟩; 2. *adv. s. Adj.:* limply; feebly; laxly

Lasche die; ~, ~n *(Gürtel~)* loop; *(eines Briefumschlags)* flap; *(Schuh~)* tongue

Laser ['leɪzɐ] **der; ~s, ~** *(Physik)* laser

laß *Imperativ Sg. v.* **lassen; lassen** 1. *unr. tr. V.* a) *mit Inf. + Akk. (2. Part. ~) (veranlassen)* **etw. tun/machen/bauen/waschen ~:** have *or* get sth. done/made/built/washed; **jmdn. warten ~:** keep sb. waiting; **jmdn. grüßen ~:** send one's regards to sb.; **jmdn. kommen/rufen ~:** send for sb.; b) *mit Inf. + Akk. (2. Part. ~) (erlauben)* **jmdn. etw. tun ~:** let sb. do sth.; allow sb. to do sth.; c) *(belassen)* **jmdn. in Frieden ~:** leave sb. in peace; d) *(hinein~/her-aus~)* let *or* allow **(in + Akk.** into, **aus** out of); e) *(unterlassen)* stop; f) *(zurück~; bleiben ~)* leave; g) *(über-lassen)* **jmdm. etw. ~:** let sb. have sth.; h) *(als Aufforderung)* **laß/laßt uns gehen/fahren!** let's go!; i) *(verlieren)* lose; *(ausgeben)* spend; 2. *unr. refl. V.* **die Tür läßt sich leicht öffnen** the door opens easily; **das läßt sich nicht beweisen** it can't be proved; 3. *unr. itr. V.* a) *(ugs.)* **Laß mal. Ich mache das schon** Leave it. I'll do it; b) *(veranlassen)* **ich lasse bitten** would you ask him/her/them to come in

lässig 1. *Adj.* casual; 2. *adv.* casually

läßt *3. Pers. Sg. Präsens v.* **lassen**

Last die; ~, ~en load; *(Gewicht)* weight; *(Bürde)* burden; **lasten** *itr. V.* be a burden; **auf jmdm./etw. ~:** weigh heavily [up]on sb./sth.; **¹Laster der; ~s, ~** *(ugs.: Lkw)* truck; lorry *(Brit.)*

²Laster das; ~s, ~: vice; **lasterhaft** *Adj. (abwertend)* depraved; **lästern**

1. *itr. V. (abwertend)* über jmdn./etw. ~: make malicious remarks about sb./ sth.; **2.** *tr. V. (veralt.)* blaspheme against

lästig *Adj.* tiresome; troublesome ⟨*illness, cough, etc.*⟩

Last-: **~schrift** die debit; **~wagen** der truck; lorry *(Brit.)*

Lasur die; ~, ~en varnish; *(farbig)* glaze

Latein das; ~s Latin; **Latein·amerika (das)** Latin America; **lateinisch** *Adj.* Latin

latent *Adj.* latent

Laterne die; ~, ~n a) *(Leuchte)* lamp; lantern *(Naut.);* b) *(Straßen~)* street light; **Laternen·pfahl** der lamppost

Latrine die; ~, ~n latrine

latschen *itr. V.; mit sein (salopp)* trudge; *(schlurfend)* slouch; **Latschen** der; ~s, ~ *(ugs.)* old worn-out shoe/slipper

Latte die; ~, ~n a) lath; *(Zaun~)* pale; b) *(Sport: Quer~ des Tores)* [cross]bar; c) *(Leichtathletik)* bar; **Latten·zaun** der paling fence

Latz der; ~es, Lätze bib; **Lätzchen** das; ~s, ~: bib

lau *Adj.* tepid, lukewarm ⟨*water etc.*⟩; mild ⟨*wind, air, evening, etc.*⟩

Laub das; ~[e]s leaves *pl.; dichtes* ~: thick foliage; **Laub·baum** der broad-leaved tree

Laube die; ~, ~n summer-house; *(überdeckter Sitzplatz)* bower; arbour

Laub-: **~frosch** der tree frog; **~säge** die fretsaw; **~wald** der deciduous wood/forest

Lauch der; ~[e]s *(Porree)* leek

Lauer die; ~: auf der ~ liegen *od.* sein *(ugs.) (jmdm. auflauern)* lie in wait; **lauern** *itr. V. (auch fig.)* lurk

Lauf der; ~[e]s, Läufe a) *o. Pl.* running; b) *(Sport: Wettrennen)* heat; c) *o. Pl. (Ver~)* course; **im** ~[e] der Zeit in the course of time; **im** ~[e] der Jahre/ des Tages over the years/during the day; d) *(von Schußwaffen)* barrel; **Lauf·bahn** die a) *(Werdegang)* career; b) *(Leichtathletik)* running-track; **laufen 1.** *unr. itr. V.; mit sein* a) run; *(beim Eislauf)* skate; *(beim Ski~)* ski; *(gehen)* go; *(zu Fuß gehen)* walk; **in** *(Akk.)***/gegen etw.** ~: walk into sth.; **dauernd zum Arzt** ~ *(ugs.)* keep running to the doctor; b) *(im Gang sein)* ⟨*machine*⟩ be running; ⟨*radio, television, etc.*⟩ be on; *(funktionie-*

ren) ⟨*machine*⟩ run; ⟨*radio, television, etc.*⟩ work; c) *(gelten)* ⟨*contract, agreement, engagement, etc.*⟩ run; d) *(gespielt werden)* ⟨*programme, play, etc.*⟩ be on; **2.** *unr. tr. u. itr. V.* a) *mit sein (zurücklegen) (zu Fuß)* walk; *(rennen)* run; b) *mit sein (erzielen)* **einen Rekord** ~: set up a record; c) *mit haben od. sein* **Ski/Schlittschuh/Rollschuh** ~: ski/skate/roller-skate; **laufend 1.** *Adj.* a) *(ständig)* regular ⟨*interest, income*⟩; recurring ⟨*costs*⟩; b) *(gegenwärtig)* current ⟨*issue, year, month, etc.*⟩; **2.** *adv.* constantly; ⟨*increase*⟩ steadily; **Läufer** der; ~s, ~ a) *(Sport)* runner; *(Handball; Fußball veralt.)* half-back; b) *(Teppich) (long narrow)* carpet; **Lauf·feuer** das brush fire; **wie ein** ~: like wildfire

Lauf-: **~masche** die ladder; **~paß** der: **er hat seiner Freundin den ~paß gegeben** *(ugs.)* he finished with his girl-friend *(coll.);* **~schritt** der: **im ~schritt, marsch, marsch!** at the double, quick march!

läufst *2. Pers. Sg. Präsens v.* laufen; **Lauf·stall** der playpen; **läuft** *3. Pers. Sg. Präsens v.* laufen

Lauge die; ~, ~n a) soapy water; b) *(Chemie)* alkaline solution; **Laugen·brezel** die *(südd.)* pretzel

Laune die; ~, ~n mood; **launenhaft** *Adj.* temperamental; *(unberechenbar)* capricious; **launig** witty; **launisch** *Adj.: s.* launenhaft

Laus die; ~, Läuse louse

Laus·bub der little rascal

lauschen *itr. V.* a) *(horchen)* listen; b) *(zuhören)* listen [attentively]; **Lauscher** der; ~s, ~: eavesdropper; **lauschig** *Adj.* cosy, snug ⟨*corner*⟩

lausig 1. *Adj. (ugs.)* a) *(abwertend: unangenehm, schäbig)* lousy *(sl.);* rotten *(coll.);* b) *(sehr groß)* perishing *(Brit. sl.),* freezing ⟨*cold*⟩; terrible *(coll.)* ⟨*heat*⟩; **2.** *adv.* terribly *(coll.)*

¹laut 1. *Adj.* loud; *(geräuschvoll)* noisy; **2.** *adv.* loudly; *(geräuschvoll)* noisily

²laut *Präp. + Gen. od. Dat. (Amtsspr.)* according to

Laut der; ~[e]s, ~e sound

Laute die; ~, ~n lute

lauten *itr. V.* ⟨*answer, instruction, slogan*⟩ be, run; ⟨*letter, passage, etc.*⟩ read, go; ⟨*law*⟩ state; **läuten 1.** *tr., itr. V.* ring; ⟨*alarm clock*⟩ go off; **2.** *itr. V. (bes. südd.: klingeln)* ring; **es läutete** the bell rang *or* went (**zu for**)

¹lauter *Adj. (geh.)* honourable ⟨*person, intentions, etc.*⟩; honest ⟨*truth*⟩

²lauter *indekl. Adj.* nothing but; sheer ⟨*nonsense, joy, etc.*⟩

läutern *tr. V. (geh.)* reform ⟨*character*⟩; purify ⟨*soul*⟩; **Läuterung** die; ~, ~en *(geh.)* reformation; *(der Seele)* purification

laut·hals *Adv.* at the top of one's voice; ~ **lachen** roar with laughter

lautlich 1. *Adj.* phonetic; **2.** *adv.* phonetically

laut-, Laut-: ~**los 1.** *Adj.* silent; soundless; *(wortlos)* silent; **2.** *adv.* silently; soundlessly; ~**schrift** die *(Phon.)* phonetic alphabet; *(Umschrift)* phonetic transcription; ~**sprecher** der loudspeaker; *(einer Stereoanlage usw.)* speaker; ~**stark 1.** *Adj.* loud; vociferous, loud ⟨*protest*⟩; **2.** *adv.* loudly; ⟨*protest*⟩ vociferously; ~**stärke** die volume

lau·warm *Adj.* lukewarm

Lava die; ~, **Laven** *(Geol.)* lava

Lavendel der; ~s, ~: lavender

Lawine die; ~, ~n *(auch fig.)* avalanche; **eine** ~ **von Protesten** *(fig.)* a storm of protest; **Lawinen·gefahr** die danger of avalanches

lax 1. lax; **2.** *adv.* laxly

Lazarett das; ~[e]s, ~e military hospital

leben *itr. V.* live; *(lebendig sein)* be alive; **leb[e] wohl!** farewell!; **von seiner Rente/seinem Gehalt** ~: live on one's pension/salary; **Leben** das; ~s, ~ **a)** life; **das** ~: life; **sich** *(Dat.)* **das** ~ **nehmen** take one's [own] life; **am** ~ **sein/ bleiben** be/stay alive; **ums** ~ **kommen** lose one's life; **b)** *(Betriebsamkeit)* **auf dem Markt herrschte ein reges** ~: the market was bustling with activity; **das** ~ **auf der Straße** the comings and goings in the street; **lebend** *Adj.* living; live ⟨*animal*⟩; **lebendig 1.** *Adj.* living; *(lebhaft)* lively; **2.** *adv. (lebhaft)* in a lively way

lebens-, Lebens-: ~**abend** der *(geh.)* evening of one's life *(literary)*; ~**art** die **a)** way of life; **b)** *o. Pl. (Umgangsformen)* manners *pl.*; ~**aufgabe** die life's work; ~**bejahend** *Adj.* ⟨*person*⟩ with a positive attitude to life; ~**bereich** der area of life; ~**dauer** die life-span; ~**ende** das end [of one's life]; ~**erinnerungen** *Pl.* memories of one's life; *(aufgezeichnet)* memoirs; ~**erwartung** die life expectancy; ~**fähig** *Adj. (auch fig.)*

viable; ~**freude** die; *o. Pl.* zest for life; ~**froh** *Adj.* full of zest for life *postpos.*; ~**gefahr** die mortal danger; „**Achtung,** ~**gefahr!**" 'danger'; ~**gefährlich 1.** *Adj.* highly dangerous; critical ⟨*injury*⟩; **2.** *adv.* critically ⟨*injured, ill*⟩; ~**geister** *Pl.* jmds. ~**geister |wieder| wecken** put new life into sb.; ~**groß** *Adj.* life-size; ~**größe** die: **eine Statue in** ~**größe** a life-size statue

Lebenshaltungs·kosten *Pl.* cost of living *sing.*

lebens-, Lebens-: ~**jahr** das year of [one's] life; ~**kraft** die vitality; ~**künstler** der: **ein |echter/wahrer| ** ~**künstler** a person who always knows how to make the best of things; ~**lage** die situation [in life]; ~**länglich 1.** *Adj.* ~**länglicher Freiheitsentzug** life imprisonment; **2.** *adv.* jmdn. ~**länglich gefangenhalten** keep sb. imprisoned for life; ~**lauf** der curriculum vitae; c.v.; ~**lustig** *Adj.* ⟨*person*⟩ full of the joys of life

Lebens·mittel das; *meist Pl.* food[stuff]; ~ *Pl.* food *sing.*; **Lebensmittel·geschäft** das food shop

lebens-: ~**müde** *Adj.* weary of life *pred.*; ~**notwendig** *Adj.* essential; ~**raum** der **a)** *(Umkreis)* lebensraum; **b)** *(Biol.) s.* Biotop; ~**retter** der rescuer; ~**standard** der standard of living; ~**unterhalt** der: **seinen** ~**unterhalt verdienen/bestreiten** earn one's living/support oneself; ~**versicherung** die life insurance; ~**wandel** der way of life; ~**weg** der [journey through] life; ~**weise** die way of life; ~**zeichen** das sign of life; ~**zeit** die life[-span]; **auf** ~**zeit** for life

Leber die; ~, ~n liver

Leber-: ~**fleck** der liver spot; ~**käse** der; *o. Pl.* meat loaf made with mincemeat, [minced liver,] eggs, and spices; ~**tran** der fish-liver oil; *(des Kabeljaus)* cod-liver oil; ~**wurst** die liver sausage

Lebe-: ~**wesen** das living being; ~**wohl** [--'-] das; ~[e]s, ~ od. ~e *(geh.)* farewell

lebhaft 1. *Adj.* **a)** lively; busy ⟨*traffic*⟩; brisk ⟨*business*⟩; **b)** *(deutlich)* vivid ⟨*idea, picture, etc.*⟩; **c)** *(kräftig)* bright ⟨*colour*⟩; vigorous ⟨*applause, opposition*⟩. **2.** *adv.* **a)** in a lively way; **b)** *(deutlich)* vividly; **c)** *(kräftig)* brightly ⟨*coloured*⟩

leb-, Leb-: **~kuchen** der ≈ ginger-bread; **~los** *Adj.* lifeless; **~zeiten** *Pl.* **bei** *od.* **zu** jmds. **~zeiten** during sb.'s lifetime

lechzen *itr. V. (geh.)* **nach einem Trunk** ~: long for a drink; **nach Rache** *usw.* ~: thirst for revenge *etc.*

leck *Adj.* leaky; ~ **sein** leak; **Leck** das; **~[e]s,** ~s leak

¹lecken 1. *tr. V.* lick; 2. *itr. V.* **an etw.** *(Dat.)* ~: lick sth.

²lecken *itr. V. (leck sein)* leak

lecker *Adj.* tasty *⟨meal⟩;* delicious *⟨cake etc.⟩;* good *⟨smell, taste⟩;* **Lekker·bissen** der delicacy; **Leckerei** die; ~, **~en** *(ugs.)* dainty; *(Süßigkeit)* sweet [meat]

led. *Abk.* ledig

Leder das; **~s,** ~: leather; **Leder-waren** *Pl.* leather goods

ledig *Adj.* single; **eine ~e Mutter** an unmarried mother; **Ledige** der/die; *adj. Dekl.* single person; **lediglich** *Adj.* merely

leer *Adj.* empty; clean *⟨sheet of paper⟩;* **Leere** die; ~ *(auch fig.)* emptiness; **leeren** *tr., refl. V.* empty

leer-, Leer-: **~gefegt** *Adj.* deserted *⟨street, town⟩;* **wie ~gefegt** deserted; **~lauf** der; *o. Pl.* **im ~lauf den Berg hinunterfahren** *⟨driver⟩* coast down the hill in neutral; *⟨cyclist⟩* freewheel down the hill; **~stehend** *Adj.* empty, unoccupied; **~taste** die space-bar

Leerung die; ~, **~en** emptying; *(von Briefkästen)* collection

Lefze die; ~, **~n** lip

legal 1. *Adj.* legal; 2. *adv.* legally; **legalisieren** *tr. V.* legalize; **Legalität** die; ~: legality

legen 1. *tr. V.* **a)** lay [down]; **b)** *(verlegen)* lay *⟨pipe, cable, carpet, tiles, etc.⟩;* 2. *tr., itr. V. ⟨hen⟩* lay; 3. *refl. V.* **a)** lie down; **b)** *(nachlassen)* die down; abate; *⟨enthusiasm⟩* wear off, subside

legendär *Adj.* legendary

Legende die; ~, **~n** legend

leger [le'ʒeːɐ̯] 1. *Adj.* casual; 2. *adv.* casually

legieren *tr. V.* alloy; **Legierung** die; ~, **~en** alloy

Legislative die; ~, **~n** *(Politik)* legis-lature; **Legislatur·periode** die legislative period; **legitim** *Adj.* legit-imate; **Legitimation** [legitima-'tsioːn] die; ~, **~en a)** legitimation; **b)** *(Ausweis)* proof of identity; **legiti-mieren** 1. *tr. V.* **a)** *(rechtfertigen)* jus-tify; **b)** *(bevollmächtigen)* authorize; **c)**

(für legitim erklären) legitimize *⟨child, relationship⟩;* 2. *refl. V.* show proof of one's identity

Lehm der; **~s** loam; *(Ton)* clay

Lehne die; ~, **~n** *(Rücken~)* back; *(Arm~)* arm; **lehnen** 1. *tr., refl. V.* lean **(an** + *Akk.,* **gegen** against); 2. *itr. V.* be leaning **(an** + *Dat.* against); **Lehn·stuhl** der armchair

Lehr·buch das textbook

Lehre die; ~, **~n a)** apprenticeship; **b)** *(Weltanschauung)* doctrine; **c)** *(Theo-rie, Wissenschaft)* theory; **d)** *(Erfah-rung)* lesson; **lehren** *tr., itr. V.* teach; **Lehrer** der; **~s,** ~ *(auch fig.)* teacher; *(Ausbilder)* instructor; **Lehrerin** die; ~, **~nen** teacher

Lehr-: **~gang** der course (**für, in** + *Dat.* in); **~jahr** das year as an ap-prentice; **~körper** der *(Amtsspr.)* teaching staff; faculty *(Amer.)*

Lehrling der; **~s,** **~e** apprentice; *(in kaufmännischen Berufen)* trainee

lehr-, Lehr-: **~reich** *Adj.* informat-ive; **~stelle** die apprenticeship; *(in kaufmännischen Berufen)* trainee post; **~stoff** der *(Schulw.)* syllabus

Leib der; **~[e]s,** **~er** *(geh.)* body; **Leib·gericht** das favourite dish; **leibhaftig** *Adj.* in person *postpos.;* *(echt)* real; **leiblich** *Adj.* physical *⟨well-being⟩; (blutsverwandt)* real

Leib-: **~schmerzen** *Pl.* abdominal pain *sing.;* **~wächter** der bodyguard

Leiche die; ~, **~n** [dead] body; corpse; **leichen·blaß** *Adj.* deathly pale; **Leichnam** der; **~s,** **~e** *(geh.)* body

leicht 1. *Adj.* light; lightweight *⟨suit, material⟩;* easy *⟨task, question, job, etc.⟩;* slight *⟨accent, illness, wound, doubt, etc.⟩;* mild *⟨cigar, cigarette⟩;* 2. *adv.* lightly *⟨built⟩; (einfach, schnell, spielend)* easily; *(geringfügig)* slightly

leicht-, Leicht-: **~athletik** die [track and field] athletics *sing.;* **~fal-len** *unr. itr. V.; mit sein* be easy; **das fällt mir ~:** its easy for me; **~fertig** 1. *Adj.* careless *⟨behaviour, person⟩;* rash *⟨promise⟩;* ill-considered, slapdash *⟨plan⟩;* 2. *adv.* carelessly; **~gläubig** *Adj.* gullible

Leichtigkeit die; ~ *(geringes Ge-wicht)* lightness; *(Mühelosigkeit)* ease

leicht-, Leicht-: **~|machen** *tr. V.* jmdm./sich etw. **~machen** make sth. easy for sb./oneself; **~nehmen** *unr. tr. V.* etw. **~nehmen** make light of sth.; **~sinn** der; *o. Pl.* carelessness *no in-*

def. art.; (mit Gefahr verbunden) recklessness *no indef. art.;* ~**sinnig** 1. *Adj.* careless; *(sich, andere gefährdend)* reckless; *(fahrlässig)* negligent; 2. *adv.* carelessly; *(gefährlich)* recklessly; *(promise)* rashly; ~**verletzt** *Adj.; präd. getrennt geschrieben* slightly injured

leid *Adj.; nicht attr.* **a)** es tut mir ~ [, **daß ...**] I'm sorry [that ...]; **er tut mir ~:** I feel sorry for him; **b)** *(überdrüssig)* etw./jmdn. ~ sein/werden *(ugs.)* be/get fed up with sth./sb. *(coll.);* **Leid** das; ~|e|s **a)** *(Schmerz)* suffering; *(Kummer)* grief; sorrow; **b)** *(Unrecht)* wrong; *(Böses)* harm; **leiden** 1. *unr. itr. V.* suffer **(an, unter** + *Dat.* from); 2. *unr. tr. V.* **a)** jmdn. |gut| ~ können *od.* mögen like sb.; **b)** *(geh.: ertragen müssen)* suffer *(hunger, thirst, etc.);* **Leiden** das; ~s, ~ **a)** *(Krankheit)* illness; *(Gebrechen)* complaint; **b)** *(Qual)* suffering; **leidend** *Adj.* **a)** *(krank)* ailing; **b)** *(schmerzvoll)* strained *(voice);* martyred *(expression)*

Leidenschaft die; ~, ~en passion **(zu, für** for); **leidenschaftlich** 1. *Adj.* passionate; vehement *(protest);* 2. *adv.* passionately; *(eifrig)* dedicatedly; **etw. ~ gern tun** adore doing sth.

leider *Adv.* unfortunately; **leidig** *Adj.* tiresome; **leidlich** *Adj.* reasonable

Leier die; ~, ~n lyre

leihen *unr. tr. V.* **a)** jmdm. etw. ~: lend sb. sth.; **b)** *(entleihen)* borrow

Leih-: ~**gebühr** die hire *or (Amer.)* rental charge; *(bei Büchern)* borrowing fee; ~**haus** das pawnbroker's; pawnshop; ~**mutter** die surrogate mother; ~**wagen** der hire *or (Amer.)* rental car

Leim der; ~|e|s glue; **leimen** *tr. V.* glue **(an** + *Akk.* to)

Leine die; ~, ~n rope; *(Wäsche~, Angel~)* line; *(Hunde~)* lead *(esp. Brit.);* leash; **Leinen** das; ~s **a)** *(Gewebe)* linen; **b)** *(Buchw.)* cloth; **Lein·wand** die **a)** *o. Pl.* linen; *(grob)* canvas; **b)** *(des Malers)* canvas; **c)** *(für Filme und Dias)* screen

leise 1. *Adj.* **a)** quiet; soft *(steps, music, etc.);* **b)** *(leicht)* faint; slight; slight, gentle *(touch);* 2. *adv.* **a)** quietly; **b)** *(leicht; kaum merklich)* slightly; *(touch, rain)* gently

Leiste die; ~, ~n strip; *(Holz~)* batten; *(profiliert)* moulding

leisten 1. *tr. V.* do *(work);* *(schaffen)* achieve *(a lot, nothing);* **jmdm. Hilfe ~:** help sb.; 2. *refl. V. (ugs.)* sich *(Dat.)* **etw. ~:** treat oneself to sth.; sich *(Dat.)* **etw. |nicht| ~ können** [not] be able to afford sth.; **Leistung** die; ~, ~en **a)** *o. Pl. (Qualität bzw. Quantität der Arbeit)* performance; **b)** *(Errungenschaft)* achievement; *(im Sport)* performance; **c)** *o. Pl. (Leistungsvermögen, Physik: Arbeits~)* power; **d)** *(Zahlung, Zuwendung)* payment; *(Versicherungsw.)* benefit; **e)** *(Dienst~)* service

leistungs-, Leistungs-: ~**fähig** *Adj.* capable *(person);* *(körperlich)* able-bodied; ~**gesellschaft** die competitive society; ~**prinzip** das; *o. Pl.* competitive principle; ~**sport** der competitive sport *no art.*

Leit·artikel der *(Zeitungsw.)* leading article; **leiten** *tr. V.* **a)** *(anführen)* lead; head; be head of *(school);* *(verantwortlich sein für)* be in charge of *(project, expedition, etc.);* manage *(factory, enterprise);* *(den Vorsitz führen bei)* chair; conduct *(orchestra, choir);* ~**der Angestellter** manager; **b)** *(begleiten, führen)* lead; **c)** *(lenken)* direct; route *(traffic);* *(um~)* divert; **¹Leiter** der; ~s, ~: leader; *(einer Abteilung)* head; *(eines Instituts)* director; *(einer Schule)* head teacher; headmaster *(Brit.);* principal *(esp. Amer.);* *(Vorsitzender)* chair[man]

²Leiter die; ~, ~n ladder

Leiterin die; ~, ~nen s. **¹Leiter;** *(einer Schule)* head teacher; headmistress *(Brit.);* principal *(esp. Amer.)*

Leit·planke die crash barrier; guardrail *(Amer.)*

Leitung die; ~, ~en **a)** *o. Pl. s.* leiten **a:** leading; heading; being in charge; management; chairing; **b)** *o. Pl. (einer Expedition usw.)* leadership; *(Verantwortung)* responsibility *(Gen.* for); *(eines Betriebes, Unternehmens)* management; *(einer Sitzung, Diskussion)* chairmanship; **c)** *(leitende Personen)* management; *(einer Schule)* head and senior staff; **d)** *(Rohr~)* pipe; *(Haupt~)* main; **e)** *(Draht, Kabel)* cable; *(für ein Gerät)* lead; **f)** *(Telefon~)* line

Leitungs·wasser das tap-water

Lektion [lɛkˈtsi̯oːn] die; ~, ~en lesson; **Lektüre** die; ~, ~n a) *o. Pl.* reading; **b)** *(Lesestoff)* reading [matter]

Lende die; ~, ~n loin

lenken *tr. V.* **a)** *auch itr.* steer; be at the controls of ⟨*aircraft*⟩; guide ⟨*missile*⟩; *(fahren)* drive ⟨*car etc.*⟩; **b)** direct ⟨*thoughts etc.*⟩ (**auf** + *Akk.* to); turn ⟨*attention*⟩ (**auf** + *Akk.* to); **c)** *(kontrollieren)* control ⟨*person, press, economy*⟩; govern ⟨*state*⟩; **Lenker der;** ~**s,** ~ **a)** handlebars *pl.*; **b)** *(Fahrer)* driver

Lenk-: ~**rad** das steering-wheel; ~**stange** die handlebars *pl.*

Lenz der; ~**es,** ~**e** *(dichter. veralt.)* spring

Leopard der; ~**en,** ~**en** leopard

Lepra die; ~: leprosy *no art.*

Lerche die; ~, ~**n** lark

lernen 1. *itr. V.* study; *(als Lehrling)* train; **2.** *tr. V.* learn (**aus** from)

lesbar *Adj.* legible; *(klar)* lucid ⟨*style*⟩; *(verständlich)* comprehensible

Lesbe die; ~, ~**n** *(ugs.)* Lesbian; **Lesbierin** ['lɛsbiərɪn] **die;** ~, ~**nen** Lesbian; **lesbisch** *Adj.* Lesbian

Lese·buch das reader; ¹**lesen** *unr. tr., itr. V.* read; ²**lesen** *unr. tr. V.* **a)** pick ⟨*grapes, berries, fruit*⟩; gather ⟨*firewood*⟩; **Ähren** ~: glean [ears of corn]; **b)** *(aussondern)* pick over; **Leser der;** ~**s,** ~, **Leserin die;** ~, ~**nen** reader; **leserlich 1.** *Adj.* legible; **2.** *adv.* legibly; **Lese·zeichen** das bookmark; **Lesung die;** ~, ~**en** reading

Lette der; ~**n,** ~**n, Lettin die;** ~, ~**nen** Latvian; **lettisch** *Adj.* Latvian; Lettish *⟨language⟩*; **Lett·land (das);** ~**s** Latvia

Letzt: zu guter ~: in the end; **letzt...** *Adj.* last; ~**en** Endes in the end; *(äußerst...)* ultimate; *(neuest...)* latest ⟨*news*⟩; **letzte·mal:** das ~: [the] last time; **letzten·mal:** beim ~: last time; **zum** ~: for the last time; **letzter...** *Adj.* latter; **letztlich** *Adv.* ultimately; in the end

Leuchte die; ~, ~**n** light; **leuchten** *itr. V.* **a)** ⟨*moon, sun, star, etc.*⟩ be shining; ⟨*fire, face*⟩ glow; **b)** shine a/the light; **jmdm.** ~: light the way for sb.; **leuchtend** *Adj.* **a)** shining ⟨*eyes*⟩; brilliant ⟨*colours*⟩; bright ⟨*blue, red, etc.*⟩; **b)** *(großartig)* shining ⟨*example*⟩; **Leuchter der;** ~**s,** ~: candelabrum; *(für eine Kerze)* candlestick

Leucht-: ~**reklame** die neon sign; ~**turm** der lighthouse; ~**zifferblatt** das luminous dial

leugnen 1. *tr. V.* deny; **2.** *itr. V.* deny it

Leukämie die; ~, ~**n** *(Med.)* leukaemia

Leumund der; ~**[e]s** *(geh.)* reputation

Leute *Pl.* people; **die reichen/alten** ~: the rich/the old

Leutnant der; ~**s,** ~**s** second lieutenant *(Milit.)*

leut·selig 1. *Adj.* affable; **2.** *adv.* affably

Lexikon das; ~**s, Lexika** *od.* **Lexiken** encyclopaedia *(Gen., für of)*

Libanese der; ~**n,** ~**n, Libanesin die;** ~, ~**nen** Lebanese; **Libanon (das)** *od.* **der;** ~**s** Lebanon

Libelle die; ~, ~**n** dragon-fly

liberal 1. *Adj.* liberal; **2.** *adv.* liberally; **Liberale der/die;** *adj. Dekl.* liberal; **liberalisieren** *tr. V.* liberalize; relax ⟨*import controls*⟩

Libero der; ~**s,** ~**s** *(Fußball)* sweeper

Libyen (das); ~**s** Libya; **libysch** *Adj.* Libyan

licht *Adj.* **a)** light; **b)** *(dünn bewachsen)* sparse; thin; **Licht das;** ~**[e]s,** ~**er a)** *o. Pl.* light; **b)** *(elektrisches* ~*)* light; **c)** *Pl. auch* ~**e** *(Kerze)* candle; **Licht-bild** das [small] photograph *(for passport etc.)*; **licht·empfindlich** *Adj.* sensitive to light; ¹**lichten 1.** *tr. V.* thin out ⟨*trees etc.*⟩; **2.** *refl. V.* ⟨*trees*⟩ thin out; ⟨*hair*⟩ grow thin; ⟨*fog, mist*⟩ lift

²**lichten** *tr. V. (Seemannsspr.)* **den/die Anker** ~: weigh anchor

lichterloh 1. *Adj.* blazing ⟨*fire*⟩; leaping ⟨*flames*⟩; **2.** *adv.* ~ **brennen** be blazing fiercely

Licht-: ~**hupe** die headlight flasher; ~**reklame** die neon sign; ~**schalter** der light-switch

Lichtung die; ~, ~**en** clearing

Lid das; ~**[e]s,** ~**er** eyelid

lieb 1. *Adj.* **a)** *(liebevoll)* kind ⟨*words, gesture*⟩; **b)** *(liebenswert)* likeable; nice; *(stärker)* lovable ⟨*child, girl, pet*⟩; ~ **aussehen** look sweet *or (Amer.)* cute; **c)** *(artig)* good ⟨*child, dog*⟩; **d)** *(geschätzt)* dear; **sein liebstes Spielzeug** his favourite toy; ~**er Hans/**~**e Else!** *(am Briefanfang)* dear Hans/ Else! **e)** *(angenehm)* welcome; **es wäre mir** ~**/**~**er, wenn ...:** I should be glad/ should prefer it if ...; **2.** *adv.* **a)** *(liebenswert)* kindly; **b)** *(artig)* nicely; **Liebe die;** ~, ~**n a)** *o. Pl.* love; ~ **zu jmdm./zu etw.** love for sb./of sth.; **aus** ~ **[zu jmdm.]** for love [of sb.]; **tu mir die** ~ **und ...:** do me a favour and ...; **mit** ~: lovingly; with loving care; **b)** *(ugs.:*

geliebter Mensch) love; **Liebelei die**; ~, ~en flirtation; **lieben 1.** *tr. V.* **a)** **jmdn.** ~: love sb.; *(sexuell)* make love to sb.; **sich** ~: be in love; *(sexuell)* make love; **b) etw.** ~: be fond of sth.; *(stärker)* love sth.; **2.** *itr. V.* be in love; **liebend** *Adv.* **etw.** ~ **gern tun** [simply] love doing sth.; **liebenswürdig** *Adj.* kind; charming ⟨*smile*⟩; **lieber** *Adv.* **a)** *s.* **gern; b)** better; **laß das** ~: better not do that

Liebes~: ~**brief der** love-letter; ~**paar das** courting couple; ~**roman der** romantic novel

liebe·voll 1. *Adj.* loving *attrib.*⟨*care*⟩; affectionate ⟨*embrace, gesture, person*⟩; **2.** *adv.* lovingly; affectionately; *(mit Sorgfalt)* lovingly; **lieb|haben** *unr. tr. V.* love; *(gern haben)* be fond of; **Liebhaber der;** ~**s,** ~ **a)** lover; **b)** *(Interessierter, Anhänger)* enthusiast *(Gen.* for); *(Sammler)* collector; **lieblich 1.** *Adj.* **a)** charming; *(angenehm)* sweet ⟨*scent, sound*⟩; **2.** *adv.* sweetly; *(angenehm)* pleasingly; **Liebling der;** ~**s,** ~**e** *(bes. als Anrede)* darling; *(bevorzugte Person)* favourite; **Lieblings-** favourite; **lieb·los 1.** *Adj.* loveless. **2.** *adv.* **a)** without affection; **b)** *(ohne Sorgfalt)* without proper care; **liebsten: am** ~: *s.* **gern**

Liechtenstein (das); ~**s** Liechtenstein

Lied das; ~|e|s, ~**er** song

liederlich *Adj.* slovenly; messy ⟨*hairstyle, person*⟩

Lieder-: ~**macher der;** ~**s,** ~, ~**macherin die;** ~, ~**nen** singer-songwriter

lief *1. u. 3. Pers. Sg. Prät. v.* **laufen**

Lieferant der; ~**en,** ~**en** supplier; **lieferbar** *Adj.* available; *(vorrätig)* in stock; **liefern** *tr. V.* **a)** *(bringen)* deliver **(an** + *Akk.* to); *(zur Verfügung stellen)* supply; **b)** *(hervorbringen)* produce; provide ⟨*eggs, honey, examples, raw material, etc.*⟩

Liefer-: ~**schein der** delivery note; ~**termin der** delivery date

Lieferung die; ~, ~**en** delivery

Liefer-: ~**wagen der** [delivery] van; ~**zeit die** delivery time

Liege die; ~, ~**n** day-bed; *(zum Ausklappen)* bed-settee; *(als Gartenmöbel)* sun-lounger; **liegen** *unr. itr. V.* lie; ⟨*person*⟩ be lying down; *(sich befinden)* be; ⟨*object*⟩ be [lying]; ⟨*town, house, etc.*⟩ be [situated]; **im Bett** ~:

lie in bed; **das liegt an ihm** *od.* **bei ihm** it is up to him; *(ist seine Schuld)* it is his fault; **es liegt mir nicht** it doesn't suit me; *(es spricht mich nicht an)* it doesn't appeal to me; *(ich mag es nicht)* I don't like it; **daran liegt ihm viel/wenig/nichts** he sets great/little/no store by that

liegen-: ~|**bleiben** *unr. itr. V.;* **mit sein a)** stay [lying]; |**im Bett|** ~: stay in bed; **b)** ⟨*things*⟩ stay, be left; *(vergessen werden)* be left behind; *(nicht erledigt werden)* be left undone; ~|**lassen** *unr. tr. V.* **a)** leave; *(vergessen)* leave [behind]; **b)** *(unerledigt lassen)* leave ⟨*work*⟩ undone; leave ⟨*letters*⟩ unposted/unopened

Liege-: ~**stuhl der** deck-chair; ~**wagen der** couchette car

lieh *1. u. 3. Pers. Sg. Prät. v.* **leihen**

lies *Imperativ Sg. v.* **lesen**

ließ *1. u. 3. Pers. Sg. Prät. v.* **lassen**

liest *3. Pers. Sg. Präsens v.* **lesen**

Lift der; ~|e|s, ~**e** *od.* ~**s a)** lift *(Brit.);* elevator *(Amer.);* **b)** *Pl.:* ~**e** *(Ski~, Sessel~)* lift

Liga die; ~, **Ligen** league; *(Sport)* division

Likör der; ~**s,** ~**e** liqueur

lila *indekl. Adj.* mauve; *(dunkel~)* purple; **Lila das;** ~**s** *od.* *(ugs.)* ~**s** mauve; *(Dunkel~)* purple

Lilie ['liːliə] **die;** ~, ~**n** lily

Liliputaner der; ~**s,** ~: dwarf

Limit das; ~**s,** ~**s** limit

Limo die, *auch:* **das;** ~, ~|**s|** *(ugs.)* fizzy drink; **Limonade die;** ~, ~**n** fizzy drink; *(Zitronen~)* lemonade

Linde die; ~, ~**n** lime[-tree]

lindern *tr. V.* relieve ⟨*suffering, pain*⟩; slake ⟨*thirst*⟩

Lineal das; ~**s,** ~**e** ruler

Linie ['liːniə] **die;** ~, ~**n** line; *(Verkehrsstrecke)* route; **die** ~ **12** *(Verkehrsw.)* the number 12; **auf die |schlanke|** ~ **achten** *(ugs. scherzh.)* watch one's figure; **auf der ganzen** ~ *(fig.)* all along the line

linien-, Linien-: ~**bus der** regular bus; ~**flug der** scheduled flight; ~**richter der** *(Fußball usw.)* linesman; *(Tennis)* line judge; *(Rugby)* touch judge; ~**treu 1.** *Adj.* loyal to the party line *postpos.;* **2.** *adv.* ⟨*act*⟩ in accordance with the party line

linieren, liniieren *tr. V.* rule

link... *Adj.* **a)** left; **b)** *(innen, nicht sichtbar)* wrong, reverse ⟨*side*⟩; **c)** *(in der Politik)* left-wing; **linkisch 1.**

Adj. awkward; **2.** *adv.* awkwardly; **links** *Adv.* on the left; *(Politik)* on the left wing

links-, Links-: ~**abbieger** der *(Verkehrsw.)* motorist/cyclist/car *etc.* turning left; ~**außen** der; ~, ~ *(Ballspiele)* left wing; outside left; ~**händer** der; ~s, ~: left-hander; ~**kurve** die left-hand bend; ~**verkehr** der driving *no art.* on the left

Linoleum das; ~s linoleum; lino

Linse die; ~, ~n **a)** *(Bot., Kochk.)* lentil; **b)** *(Med., Optik)* lens

Lippe die; ~, ~n lip; **Lippen·stift** der lipstick

liquid *Adj. (Wirtsch.)* liquid ⟨*funds, resources*⟩; solvent ⟨*business*⟩; **liquidieren** *(verhüll.: töten; Wirtsch.)* liquidate

lispeln *itr. V.* lisp

Lissabon (das); ~s Lisbon

List die; ~, ~en **a)** [cunning] trick; **b)** *(listige Art) o. Pl.* cunning

Liste die; ~, ~n list; **schwarze** ~: blacklist

listig 1. *Adj.* cunning; crafty; **2.** *adv.* cunningly; craftily

Litauen (das); ~s Lithuania; **Litauer** der; ~s, ~: Lithuanian; **litauisch** *Adj.* Lithuanian

Liter der, *auch:* das; ~s, ~: litre

literarisch *Adj.* literary; **Literatur** die; ~, ~en literature

Litfaß·säule die advertising column

Lithographie die; ~, ~n *(Druck)* lithograph

litt *1. u. 3. Pers. Sg. Prät. v.* leiden

Litze die; ~, ~n braid

Lizenz die; ~, ~en licence

Lkw, LKW [ɛlka:'ve:] der; ~|s|, ~|s| *Abk.* Lastkraftwagen truck; lorry *(Brit.)*

Lob das; ~|e|s, ~e praise *no indef. art.*

Lobby ['lɔbi] die; ~, ~s *od.* Lobbies lobby

loben *tr.V.* praise; **löblich** *Adj.* commendable; **Lob·lied** das song of praise

Loch das; ~|e|s, Löcher hole; **lochen** *tr. V.* punch holes/a hole in; punch ⟨*ticket*⟩; **Locher** der; ~s, ~: punch; **löcherig** *Adj.* full of holes *pred.*

Locke die; ~, ~n curl

locken *tr. V.* **a)** lure; **b)** *(reizen)* tempt

Locken·wickler der [hair] curler

locker 1. *Adj.* loose; *(entspannt)* relaxed ⟨*position, muscles*⟩; slack ⟨*rope, rein*⟩; **2.** *adv.* ~ **sitzen** ⟨*tooth, screw, nail*⟩ be loose; *(entspannt, ungezwun-*

gen) loosely; **locker|lassen** *unr. itr. V. (ugs.)* **nicht** ~: not give up; **lockern 1.** *tr. V.* loosen; slacken [off] ⟨*rope etc.*⟩; relax ⟨*muscles, limbs*⟩; **2.** *refl. V.* ⟨*brick, tooth, etc.*⟩ work itself loose; ⟨*person*⟩ loosen up

lockig *Adj.* curly

Lock·vogel der decoy

Loden·mantel der loden coat

Löffel der; ~s, ~: spoon; *(als Maßangabe)* spoonful; *(Jägerspr.)* ear; **löffeln** *tr. V.* spoon [up]

log *1. u. 3. Pers. Sg. Prät. v.* lügen

Logarithmus der; ~, Logarithmen *(Math.)* logarithm; log

Loge ['loːʒə] die; ~, ~n box; **logieren** *itr. V. (veralt.)* stay

Logik die; ~: logic; **logisch 1.** *Adj.* logical; **2.** *adv.* logically

Lohn der; ~|e|s, Löhne **a)** wage[s *pl.*]; pay *no indef. art., no pl.;* **b)** *o. Pl. (Belohnung)* reward; **Lohn·büro** das payroll office

lohnen 1. *refl., itr. V.* be worth it; **2.** *tr. V.* be worth; **lohnend** *Adj.* rewarding

Lohn·steuer die income tax; **Lohnsteuer·karte** die income-tax card; **Lohn·tüte** die pay-packet *(Brit.);* wage packet

Lokal das; ~s, ~e pub *(Brit. coll.);* bar *(Amer.); (Speise~)* restaurant; **Lokalität** die; ~, ~en locality

Lokal-: ~**blatt** das local paper; ~**termin** der *(Rechtsspr.)* visit to the scene [of the crime]

Lokomotive [lokomo'tiːvə] die; ~, ~n locomotive; **Lokomotiv·führer** der engine-driver *(Brit.);* engineer *(Amer.)*

Lokus der; ~ *od.* ~ses, ~ *od.* ~se *(salopp)* loo *(Brit. coll.);* john *(Amer. coll.)*

London (das); ~s London; **Londoner 1.** *indekl. Adj.* London; **2.** der; ~s, ~: Londoner

Lorbeer der; ~s, ~en **a)** laurel; **b)** *(Gewürz)* bay-leaf

Lore die; ~, ~n car; *(kleiner)* tub

los 1. *Adj.* **a)** *(gelöst, ab)* off; **b)** **es ist etwas** ~: there is something going on; **c)** **jmdn./etw.** ~ **sein** be rid of sb./sth.; **2.** *Adv. (als Aufforderung)* come on!

Los das; ~es, ~e **a)** lot; **b)** *(Lotterie~)* ticket

Lösch·blatt das piece of blotting-paper; **löschen** *tr. V.* **a)** put out; extinguish; **seinen Durst** ~ *(fig.)* quench one's thirst; **b)** *(tilgen)* delete ⟨*entry*⟩; erase ⟨*recording, memory, etc.*⟩;

Lösch·papier das blotting paper

lose 1. *Adj.* loose; **2.** *adv.* loosely
Löse·geld das ransom
losen *itr. V.* draw lots (**um** for)
lösen 1. *tr. V.* **a)** remove ⟨*stamp, wallpaper*⟩; **etw. von etw. ~**: remove sth. from sth.; **b)** *(lockern)* undo ⟨*screw, belt, tie*⟩; **c)** *(klären)* solve; resolve ⟨*contradiction, conflict*⟩; **d)** *(annullieren)* break off ⟨*engagement*⟩; cancel ⟨*contract*⟩; sever ⟨*relationship*⟩; **e)** *(kaufen)* buy, obtain ⟨*ticket*⟩; **2.** *refl. V.* **a)** *(lose werden)* come off; *(sich lockern)* ⟨*wallpaper, plaster*⟩ come off; ⟨*packing, screw*⟩ come loose; **b)** *(sich klären)* ⟨*puzzle, problem*⟩ be solved; **c)** *(sich auflösen)* dissolve
los-: ~|**fahren** *unr. itr. V.; mit sein* set off; *(wegfahren)* move off; ~|**gehen** *unr. itr. V.; mit sein* **a)** *(aufbrechen)* set off; **b)** *(ugs.: beginnen)* start; **c)** *(ugs.: abgehen)* ⟨*button, handle, etc.*⟩ come off; ~|**kommen** *unr. itr. V.; mit sein* *(ugs.)* **a)** get away; **b)** *(freikommen)* get free; ~|**lassen** *unr. tr. V.* **a)** *(nicht festhalten)* let go of; **b)** *(freilassen)* let ⟨*person, animal*⟩ go; ~|**legen** *itr. V.* *(ugs.)* get going
löslich *Adj.* soluble
los-: ~|**machen 1.** *tr. V.* *(ugs.)* let ⟨*animal*⟩ loose; untie ⟨*string, line, rope*⟩; unhitch ⟨*trailer*⟩; ~|**reißen** *unr. refl. V.* break free *or* loose; ~|**sagen** *refl. V.* **sich von jmdm./etw. ~sagen** break with sb./sth.; ~|**schlagen** *unr. itr. V.* *(bes. Milit.)* attack; launch one's attack
Losung die; ~, ~en slogan; *(Milit.: Kennwort)* password
Lösung die; ~, ~en **a)** solution *(Gen., für* to); **b)** *s.* lösen 1 d: breaking off; cancellation; severing
los|werden *unr. tr. V.; mit sein* get rid of
Lot das; ~[e]s, ~e plumb[-bob]; [**nicht**] **im ~ sein** be [out of] plumb
löten *tr. V.* solder
Lotion [lo'tsio:n] die; ~, ~en lotion
Lotse der; ~n, ~n *(Seew.)* pilot; **lotsen** *tr. V.* guide
Lotterie die; ~, ~n lottery; **Lotto** das; ~s, ~s national lottery
Lotto-: ~**schein** der national-lottery coupon; ~**zahlen** *Pl.* winning national-lottery numbers
Löwe der; ~n, ~n **a)** lion; **b)** *(Astrol.)* Leo; the Lion
Löwen-: ~**anteil** der lion's share; ~**mäulchen** das snapdragon; ~**zahn** der dandelion

Löwin die; ~, ~nen lioness
loyal [lŏa'ja:l] **1.** *Adj.* loyal; **2.** *adv.* loyally; **Loyalität** die; ~: loyalty
LP [ɛl'pe:] die; ~, ~[s] *Abk.* **Langspielplatte** LP
Luchs der; ~es, ~e lynx
Lücke die; ~, ~n gap; **Lücken·büßer** der; ~s, ~ *(ugs.)* stopgap; **lückenhaft** *Adj.* sketchy; **lückenlos** *Adj.* complete
lud *1. u. 3. Pers. Sg. Prät. v.* laden
Luft die; ~, **Lüfte** air; **an die frische ~ gehen** get out in[to] the fresh air; **die ~ anhalten** hold one's breath; **tief ~ holen** take a deep breath; **in die ~ gehen** *(fig. ugs.)* blow one's top *(coll.)*
luft-, Luft-: ~**ballon** der balloon; ~**brücke** die airlift; ~**dicht** *Adj.* airtight; ~**druck** der **a)** *(Physik)* air pressure; **b)** *(Druckwelle)* blast
lüften 1. *tr. V.* **a)** air ⟨*room, clothes, etc.*⟩; **b)** raise ⟨*hat*⟩; **c)** disclose ⟨*secret*⟩; **2.** *itr. V.* air the room/house *etc.*
luft-, Luft-: ~**fahrt** die; *o. Pl.* aviation *no art.*; ~**feuchtigkeit** die [atmospheric] humidity; ~**gekühlt** *Adj.* air-cooled; ~**gewehr** das air rifle; airgun
luftig *Adj.* airy ⟨*room, building, etc.*⟩; light ⟨*clothes*⟩
Luftkissen·boot das hovercraft
luft-, Luft-: ~**leer** *Adj.* **ein ~leerer Raum** a vacuum; ~**linie** die *o. Pl.* 1 000 km **~linie** 1,000 km. as the crow flies; ~**loch** das air-hole; ~**matratze** die air-bed; air mattress; Lilo (P); ~**pirat** der [aircraft] hijacker; ~**post** die airmail; **etw. per** *od.* **mit ~post schicken** send sth. [by] airmail; ~**pumpe** die air pump; *(für Fahrrad)* [bicycle-]pump; ~**röhre** die *(Anat.)* windpipe; ~**schiff** das airship; ~**schloß** das castle in the air; ~**schutz** der air-raid protection *no art.*; ~**schutz·keller** der air-raid shelter; ~**verschmutzung** die air pollution; ~**waffe** die air force
Lüge die; ~, ~n lie; **lügen 1.** *itr., tr. V.* lie; **das ist gelogen!** that's a lie!; **Lügner** der; ~s, ~: liar
Luke die; ~, ~n *(Dach~)* skylight; *(bei Schiffen)* hatch; *(Keller~)* trap-door
lukrativ 1. *Adj.* lucrative; **2.** *adv.* lucratively
Lümmel der; ~s, ~: lout; *(ugs., fam.: Bengel)* rascal
Lump der; ~en, ~en scoundrel; **lumpen** *(ugs.)* *tr. V.* **sich nicht ~ lassen** splash out *(coll.)*; **Lumpen** der; ~s,

~: rag; **Lumpen·sammler der** rag-and-bone man

Lunge die; ~, ~n lungs *pl.*

Lungen-: ~**entzündung die** pneumonia *no indef. art.;* ~**krebs der** lung cancer; ~**zug der** inhalation

Lunte die; ~, ~n fuse; match

Lupe die; ~, ~n magnifying glass

Lurch der; ~[e]s, ~e amphibian

Lust die; ~ a) ~ **haben, etw. zu tun** feel like doing sth.; b) *(Vergnügen)* pleasure; joy; **lustig 1.** *Adj.* a) merry; jolly; enjoyable ⟨*time*⟩; b) *(komisch)* funny; 2. *adv.* a) merrily; b) *(komisch)* funnily

lust-, Lust-: ~**los 1.** *Adj.* listless; 2. *adv.* listlessly; ~**spiel das** comedy

lutherisch *Adj.* Lutheran

lutschen 1. *tr. V.* suck; 2. *itr. V.* suck; **an etw.** *(Dat.)* ~: suck sth.

Luxemburg (das); ~s Luxembourg

luxuriös 1. *Adj.* luxurious; 2. *adv.* luxuriously; **Luxus der;** ~: luxury

Lymph·knoten der lymph node

lynchen *tr. V.* lynch

Lyrik die; ~: lyric poetry; **lyrisch** *Adj.* lyrical; lyric ⟨*poetry*⟩

Lyzeum das; ~s, Lyzeen girls' high school

M

m, M [ɛm] **das;** ~, ~: m/M

m *Abk.* Meter m

machen 1. *tr. V.* a) make; **aus Plastik/ Holz** *usw.* **gemacht** made of plastic/ wood *etc.;* **sich** *(Dat.)* **etw.** ~ **lassen** have sth. made; **etw. aus jmdm.** ~: make sb. into sth.; **jmdn. zum Präsidenten** *usw.* ~: make sb. president *etc.;* **jmdm./sich [einen] Kaffee** ~: make [some] coffee for sb./oneself; b) *(verursachen)* **jmdm. Arbeit** ~: make [extra] work for sb.; **das macht das Wetter** that's [because of] the weather; c) *(ausführen)* do ⟨*job, repair, etc.*⟩; **einen Spaziergang** ~: go for a walk; **eine Reise** ~: go on a journey; **einen Besuch [bei jmdm.]** ~: pay [sb.] a visit; d) *(tun)* do; **was machst du da?**

what are you doing?; **so etwas macht man nicht** that [just] isn't done; e) **was macht ...?** *(wie ist es um ... bestellt?)* how is ...?; **was macht die Gesundheit/ Arbeit?** how are you keeping/ how ist the job [getting on]?; f) *(ergeben) (beim Rechnen)* be; *(bei Geldbeträgen)* come to; **zwei mal zwei macht vier** two times two is four; **das macht 12 DM** that is 12 marks; *(Endsumme)* that comes to 12 marks; g) *(schaden)* **was macht das schon?** what does it matter?; **macht nichts!** *(ugs.)* it doesn't matter; h) *(teilnehmen an)* **einen Kursus** *od.* **Lehrgang** ~: take a course; i) **mach's gut!** *(ugs.)* look after yourself!; *(auf Wiedersehen)* so long!; 2. *refl. V.* a) **sich an etw.** *(Akk.)* ~: get down to sth.; b) *(ugs.: sich entwickeln)* do well; c) **mach dir nichts daraus!** *(ugs.)* don't let it bother you; 3. *itr. V.* a) **mach schon!** *(ugs.)* get a move on! *(coll.);* b) **das macht hungrig/durstig** it makes you hungry/thirsty; **das macht dick** it's fattening; **Machenschaften** *Pl. (abwertend)* wheeling and dealing *sing.*

Macht die; ~, **Mächte** power; **an die** ~ **kommen** come to power; **Macht·haber der;** ~s, ~: ruler; **mächtig 1.** *Adj.* a) powerful; b) *(beeindruckend groß)* mighty; 2. *adv. (ugs.)* terribly *(coll.)*

macht-, Macht-: ~**kampf der** power struggle; ~**los** *Adj.* powerless; **gegen etw.** ~**los sein** be powerless in the face of sth.; ~**probe die** trial of strength

Mädchen das; ~s, ~ a) girl; b) *(Haus~)* maid; **mädchenhaft** *Adj.* girlish; **Mädchen·name der** a) girl's name; b) *(Name vor der Ehe)* maiden name

Made die; ~, ~n maggot; **madig** *Adj.* maggoty; **jmdn./etw.** ~ **machen** *(ugs.)* run sb./sth. down

Madonna die; ~, Madonnen madonna

mag 1. u. 3. *Pers. Sg. Präsens v.* **mögen**

Magazin das; ~s, ~e a) *(Lager)* store; *(für Waren)* stock-room; b) *(für Patronen, Dias, Film usw.; Zeitschrift)* magazine

Magen der; ~s, Mägen *od.* ~: stomach

magen-, Magen-: ~**bitter der;** ~s, ~: bitters *pl.;* ~**schmerzen** *Pl.* stomach-ache *sing.*

mager *Adj.* a) thin; b) *(fettarm)* low-fat; low in fat *pred.;* lean ⟨*meat*⟩; c) *(fig.)* poor ⟨*soil, harvest*⟩; meagre ⟨*profit, increase, success, report, etc.*⟩; thin ⟨*programme*⟩

Mager-: **~milch** die skim[med] milk; **~quark** der low-fat curd cheese

Magie die; ~: magic; **Magier** ['ma:giɐ] der; ~s, ~ *(auch fig.)* magician; **magisch** Adj. magic ⟨*powers*⟩; *(geheimnisvoll)* magical

Magistrat der; ~|e|s, ~e City Council

Magnet der; ~en od. ~|e|s, ~e magnet; **magnetisch** 1. Adj. magnetic; 2. adv. magnetically; **Magnetismus** der; ~: magnetism; **Magnet·nadel** die [compass] needle

Mahagoni das; ~s mahogany

Mäh·drescher der combine harvester; **mähen** 1. tr. V. mow; cut ⟨*corn*⟩; 2. itr. V. mow; *(Getreide ~)* reap

Mahl das; ~|e|s, **Mähler** *(geh.)* meal; repast *(formal)*

mahlen unr. tr., itr. V. grind

Mahl·zeit meal

Mähne die; ~, ~n mane

mahnen tr. V. urge; remind ⟨*debtor*⟩

Mahn-: **~mal** das memorial *(erected as a warning to future generations)*; **~schreiben** das reminder

Mahnung die; ~, ~en a) exhortation; *(Warnung)* admonition; b) s. **Mahnschreiben**

Mai der; ~|e|s od. ~: May

Mai~: **~baum** der maypole; **~glöckchen** das lily of the valley; **~käfer** der May-bug

Mais der; ~es maize; corn *(esp. Amer.)*; *(als Gericht)* sweet corn; **Mais·kolben** der corn-cob; *(als Gericht)* corn on the cob

Majestät die; ~, ~en a) *(Titel)* Majesty; **Eure** ~: Your Majesty; b) *o. Pl. (geh.)* majesty; **majestätisch** 1. Adj. majestic; 2. adv. majestically

Major der; ~s, ~e *(Milit.)* major

Majoran der; ~s, ~e marjoram

makaber Adj. macabre

Makedonien [make'do:niən] **(das)**; ~s Macedonia

Makel der; ~s, ~ *(geh.)* a) *(Schmach)* stigma; b) *(Fehler)* blemish; **makel·los** 1. Adj. flawless; spotless ⟨*white, cleanness*⟩; 2. adv. immaculately; spotlessly ⟨*clean*⟩

Make-up [me:k'|ap] das; ~s, ~s make-up

Makkaroni Pl. macaroni sing.

Makler der; ~s, ~: estate agent *(Brit.)*; realtor *(Amer.)*

Makrele die; ~, ~n mackerel

Makrone die; ~, ~n macaroon

mal 1. Adv. times; *(bei Flächen)* by; 2. Partikel **komm — her!** come here!; **¹Mal** das; ~|e|s, ~e time; **mit einem ~|e|** all at once; **²Mal** das; ~|e|s, ~e od. **Mäler** mark; *(Muttermal)* birthmark; *(braun)* mole

Malaie der; ~n, ~n Malay; **Malaysia (das)**; ~s Malaysia

Mal·buch das colouring-book; **malen** tr., itr. V. paint; decorate ⟨*flat, room, walls*⟩; **Maler** der; ~s, ~: painter; **Malerei** die; ~, ~en painting; **malerisch** 1. Adj. picturesque; 2. adv. picturesquely

mal|nehmen unr. tr., itr. V. multiply *(mit by)*

Malz·bier das malt beer

Mama die; ~, ~s *(fam.)* mamma; **Mami** die; ~, ~s *(fam.)* mummy *(Brit. coll.)*; mommy *(Amer. coll.)*

Mammut das; ~s, ~e od. ~s mammoth

man Indefinitpron. im Nom. one; you *2nd person*; *(irgend jemand)* somebody; *(die Behörden; die Leute dort)* they pl.; *(die Menschen im allgemeinen)* people pl; ~ **hat mir gesagt ...**: I was told ...

Management ['mænidʒmənt] das; ~s, ~s management; **managen** ['mɛnidʒn] tr. V. a) *(ugs.)* fix; organize; b) *(betreuen)* manage ⟨*singer, artist, player*⟩; **Manager** ['mɛnidʒɐ] der; ~s, ~: manager; *(eines Fußballvereins)* club secretary

manch Indefinitpron. a) attr. many a; **in |so| ~er Beziehung** in many respects; b) alleinstehend ~er many a person/man; ~e Pl. some; *(viele)* many; |so| ~es a number of things; *(allerhand Verschiedenes)* all kinds of things; **mancherlei** unbest. Gattungsz. a) attr. various; a number of; b) alleinstehend various things; **manch·mal** Adv. sometimes

Mandant der; ~en, ~en client

Mandarine die; ~, ~n mandarin [orange]

Mandel die; ~, ~n a) almond; b) *(Anat.)* tonsil; **Mandel·entzündung** die tonsillitis *no indef. art.*

Manege [ma'ne:ʒə] die; ~, ~n *(im Zirkus)* ring; *(in der Reitschule)* arena

¹Mangel der; ~s, **Mängel** a) *o. Pl. (Fehlen)* lack **(an +** Dat. of); *(Knappheit)* shortage, lack **(an +** Dat. of); b) *(Fehler)* defect

²Mangel die; ~, ~n [large] mangle

mangelhaft 1. *Adj.* faulty ⟨*goods, German, English, etc.*⟩; *(unzulänglich)* inadequate ⟨*knowledge, lighting*⟩; *(Schulw.)* **die Note „~"** the mark 'unsatisfactory'; *(bei Prüfungen)* the fail mark; **2.** *adv.* faultily; *(unzulänglich)* inadequately; **¹mangeln** *itr. V.; unpers.* **es mangelt an etw.** *(Dat.) (etw. fehlt)* there is a lack of sth.; *(etw. ist unzureichend vorhanden)* there is a shortage of sth.; **jmdm./einer Sache mangelt es an etw.** *(Dat.)* sb./sth. lacks sth.

²mangeln *tr. V.* mangle

mangels *Präp. mit Gen.* in the absence of

Manie die; ~, ~n mania

Manier die; ~, ~en a) manner; b) *Pl. (Umgangsformen)* manners; **manierlich 1.** *Adj.* a) *(fam.)* well-mannered; well-behaved ⟨*child*⟩; b) *(ugs.: einigermaßen gut)* decent; **2.** *adv.* a) *(fam.)* nicely; b) *(ugs.: einigermaßen gut)* **ganz/recht** ~: quite/really nicely

Manifest das; ~[e]s, ~e manifesto

Maniküre die; ~: manicure; **maniküren** *tr. V.* manicure

manipulieren *tr. V.* manipulate; rig ⟨*election result etc.*⟩

Manko das; ~s, ~s shortcoming; deficiency

Mann der; ~[e]s, **Männer a)** man; b) *(Ehemann)* husband; **Männchen das;** ~s, ~ a) little man; b) *(Tier~)* male; ~ **machen** ⟨*animal*⟩ sit up and beg

Mannequin ['manəkɛ̃] **das;** ~s, ~s mannequin; [fashion] model

mannig·fach *Adj.* multifarious

männlich 1. *Adj.* a) male; b) *s.* **maskulin 1; 2.** *adv.* in a masculine way; **Mannschaft die;** ~, ~en *(Sport, auch fig.)* team; *(Schiffs-, Flugzeugbesatzung)* crew; *(Milit.)* unit

Manöver das; ~s, ~ a) *(Milit.)* exercise; ~ *Pl.* manœuvres; b) *(Bewegung; fig. abwertend: Trick)* manœuvre; **manövrieren** *itr., tr. V.* manœuvre

Mansarde die; ~, ~n attic; *(Zimmer)* attic room

Manschette die; ~, ~n cuff; **Manschetten·knopf der** cuff-link

Mantel der; ~s, **Mäntel** coat

Manuskript das; ~[e]s, ~e a) manuscript; *(Typoskript)* typescript; b) *(Notizen)* notes *pl.*

Mappe die; ~, ~n a) folder; b) *(Aktentasche)* briefcase; *(Schul~)* schoolbag

Marathon·lauf [...tɔn...] **der** marathon

Märchen das; ~s, ~: fairy story; fairy-tale; *(ugs.: Lüge)* [tall] story *(coll.)*; **Märchen·buch das** book of fairy stories; **märchenhaft 1.** *Adj.* magical; **2.** *adv.* magically; *(ugs.)* fantastically *(coll.)*

Margarine die; ~: margarine

Margerite die; ~, ~n ox-eye daisy

Maria (die); ~s *od. (Rel.)* **Mariä** Mary; **Marien·käfer der** ladybird

Marihuana das; ~s marijuana

Marinade die; ~, ~n *(Kochk.)* marinade; *(Salatsauce)* [marinade] dressing

Marine die; ~, ~n fleet; *(Kriegs~)* navy

Marionette die; ~, ~n puppet; marionette; **Marionetten·theater das** puppet theatre

¹Mark die; ~, ~: mark; **Deutsche ~:** Deutschmark

²Mark das; ~[e]s a) *(Knochen~)* marrow; b) *(Frucht~)* pulp

markant *Adj.* striking; prominent ⟨*figure, nose, chin*⟩; clear-cut ⟨*features, profile*⟩

Marke die; ~, ~n a) *(Waren~)* brand; *(Fabrikat)* make; b) *(Brief~, Rabatt~, Beitrags~)* stamp; c) *(Essen~)* meal-ticket; d) *(Erkennungs~)* [identification] disc; *(Dienst~)* [police] identification badge; ≈ warrant card *(Brit.)* or *(Amer.)* ID card

Marken-: ~**artikel der** proprietary or *(Brit.)* branded article; ~**zeichen das** trade mark

markieren *tr. V.* a) mark; b) *(ugs.: vortäuschen)* sham ⟨*illness, breakdown, etc.*⟩; **2.** *itr. V. (ugs.: simulieren)* put it on *(coll.)*; **Markierung die;** ~, ~en marking

Markt der; ~[e]s, **Märkte** market; *(~platz)* market-place or -square; **freitags ist** ~: Friday is market-day

Markt-: ~**forschung die** market research *no def. art.*; ~**frau die** market-woman; ~**halle die** covered market; ~**lücke die** gap in the market; ~**platz der** market-place; ~**stand der** market stall; ~**wirtschaft die** market economy

Marmelade die; ~, ~n jam; *(Orangen~)* marmalade

Marmor der; ~s marble

Marokkaner der; ~s, ~: Moroccan; **marokkanisch** *Adj.* Moroccan; **Marokko (das);** ~s Morocco

Marone die; ~, ~n [sweet] chestnut
Mars der; ~: Mars *no def. art.*
Marsch der; ~[e]s, Märsche march; *(Wanderung)* [long] walk; **marschieren** *itr. V.; mit sein* march; *(wandern)* walk
Mars·mensch der Martian
Marter die; ~, ~n *(geh.)* torture; *(seelisch)* torment; **martern** *tr. V. (geh.)* torture
Märtyrer der; ~s, ~: martyr; **Martyrium** das; ~s, Martyrien martyrdom
Marxismus der; ~: Marxism *no art.;* **Marxist** der; ~en, ~en Marxist; **marxistisch** *Adj.* Marxist
März der; ~[es] March
Marzipan das; ~s marzipan
Masche die; ~, ~n stitch; *(Lauf~)* run; ladder *(Brit.); (beim Netz)* mesh; **Maschen·draht** der wire netting
Maschine die; ~, ~n a) *(auch ugs.: Motorrad)* machine; b) *(ugs.: Automotor)* engine; c) *(Flugzeug)* [aero]plane; d) *(Schreib~)* typewriter; **maschine·geschrieben** *Adj.* typewritten; **maschinell** 1. *Adj.* machine *attrib.;* by machine *postpos.;* 2. *adv.* by machine; ~ hergestellt machine-made
Maschinen-: ~gewehr das machine-gun; ~pistole die sub-machine-gun
maschine|schreiben *unr. itr. V.; nur im Inf. u. Part.* type
Masern *Pl.* measles *sing. or pl.*
Maserung die; ~, ~en [wavy] grain
Maske die; ~, ~n mask; **Masken·ball** der masked ball; **Maskerade** die; ~, ~n [fancy-dress] costume; **maskieren** 1. *tr. V.* mask; 2. *refl. V.* put on a mask/masks
Maskottchen das; ~s, ~: [lucky] mascot
maskulin [*auch* '---] 1. *Adj. (auch Sprachw.)* masculine; 2. *adv.* in a masculine way
maß *1. u. 3. Pers. Sg. Prät. v.* messen; ¹**Maß** das; ~es, ~e a) measure (für of); *(fig.)* das ~ ist voll enough is enough; b) *(Größe)* measurement; c) *(Grad)* degree (an + Dat. of); in großem/gewissem ~e to a great/certain extent; ²**Maß** die; ~, ~[e] *(bayr., österr.)* litre [of beer]
Massage [ma'sa:ʒə] die; ~, ~n massage
Massaker das; ~s, ~: massacre
Maß-: ~anzug der made-to-measure suit; ~arbeit die a) custom-made item; *(Kleidungsstück)* made-to-

measure item; b) *(genaue Arbeit)* neat work
Masse die; ~, ~n a) mass; b) *(Gemisch)* mixture
Maß·einheit die unit of measurement
Massen·grab das mass grave; **massenhaft** 1. *Adj.; nicht präd.* in huge numbers *postpos.;* 2. *adv.* on a huge scale
massen-, Massen-: ~karambolage die multiple crash; ~medium das mass medium; ~mörder der mass murderer; ~produktion die mass production; ~weise *Adv.* in huge numbers
Masseur [ma'sø:ɐ̯] der; ~s, ~e masseur; **Masseurin** die; ~, ~nen, **Masseuse** [ma'sø:zə] die; ~, ~n masseuse
maß·gebend, maß·geblich 1. *Adj.* authoritative ⟨book, expert, opinion⟩; definitive ⟨text⟩; influential ⟨person, circles, etc.⟩; decisive ⟨factor, influence, etc.⟩; 2. *adv.* ⟨influence⟩ to a considerable extent; *(entscheidend)* decisively; **maß|halten** *unr. itr. V.* exercise moderation
massieren *tr. V.* massage
mäßig 1. *Adj.* moderate; *(mittel~)* mediocre; 2. *adv.* in moderation; moderately ⟨gifted, talented⟩; *(mittel~)* indifferently; **mäßigen** *refl. V. (geh.)* a) practise *or* exercise moderation; b) *(sich beherrschen)* control *or* restrain oneself; **Mäßigkeit** die; ~: moderation; **Mäßigung** die; ~: moderation
massiv 1. *Adj.* a) solid; b) *(heftig)* massive ⟨demand⟩; crude ⟨accusation, threat⟩; strong ⟨attack, criticism, pressure⟩; 2. *adv.* ⟨attack⟩ strongly; ⟨accuse, threaten⟩ crudely
maß-, Maß-; ~krug der *(südd., österr.)* litre beer-mug; *(aus Steingut)* stein; ~los 1. *Adj.* extreme; gross ⟨exaggeration, insult⟩; excessive ⟨demand, claim⟩; boundless ⟨ambition, greed, sorrow, joy⟩; 2. *adv.* extremely; ⟨exaggerate⟩ grossly; ~nahme die; ~, ~n measure; ~regel die regulation; *(Maßnahme)* measure; ~regeln *tr. V. (zurechtweisen)* reprimand; *(bestrafen)* discipline; ~stab der a) standard; b) *(einer Karte, eines Modells usw.)* scale; ~voll 1. *Adj.* moderate; 2. *adv.* in moderation
Mast der; ~[e]s, ~en, *auch:* ~e *(Schiffs~, Antennen~)* mast; *(Stange,*

Fahnen~) pole; *(Hochspannungs~)* pylon

mästen *tr. V.* fatten

masturbieren *itr., tr. V.* masturbate

Match [mɛtʃ] das *od.* der; ~[e]s, ~s *od.* ~e match

Material das; ~s, ~ien material; *(Bau~; Hilfsmittel)* materials *pl.;* **Materialismus** der; ~: materialism; **Materialist** der; ~en, ~en materialist; **materialistisch** 1. *Adj.* materialistic; 2. *adv.* materialistically

Materie die; ~, ~n a) matter; b) *(geh.: Thema, Gegenstand)* subject-matter; **materiell** 1. *Adj. (finanziell)* financial; 2. *adv.* materially; *(finanziell)* financially

Mathematik die; ~: mathematics *sing., no art.;* **mathematisch** 1. *Adj.* mathematical; 2. *adv.* mathematically

Matjes der; ~, ~: matie [herring]

Matratze die; ~, ~n mattress

Matrose der; ~n, ~n sailor; seaman

Matsch der; ~[e]s *(ugs.)* mud; *(breiiger Schmutz)* sludge; *(Schnee~)* slush; **matschig** *Adj. (ugs.)* a) muddy; slushy ⟨snow⟩; b) *(weich)* mushy; squashy ⟨fruit⟩

matt 1. *Adj.* a) weak; feeble ⟨applause, reaction⟩; b) *(glanzlos)* matt; dull ⟨metal, mirror, etc.⟩; c) *(undurchsichtig)* frosted ⟨glass⟩; pearl ⟨light-bulb⟩; d) subdued; *(Schach)* checkmated; ~! checkmate!; 2. *adv.* a) *(kraftlos)* weakly; b) *(mäßig)* ⟨protest, contradict⟩ feebly

Matte die; ~, ~n mat

Matt·scheibe die *(ugs.)* telly *(Brit. coll.);* box *(coll.)*

Mätzchen das; ~s, ~: ~ machen *(ugs.)* fod about *or* around

Mauer die; ~, ~n wall; **mauern** 1. *tr. V.* build; 2. *itr. V.* lay bricks; **Mauer·werk** das a) masonry; *(aus Ziegeln)* brickwork; b) *(Mauern)* walls *pl.*

Maul das; ~[e]s, Mäuler *(von Tieren)* mouth; *(derb: Mund)* gob *(sl.)*

Maul-: ~esel der mule; ~korb der *(auch fig.)* muzzle; ~tier das mule; ~wurf der mole

Maurer der; ~s, ~: bricklayer

Maus die; ~, Mäuse mouse

Mauschelei die; ~, ~en *(ugs. abwertend)* shady wheeling and dealing *no indef. art.;* **mauscheln** *itr. V. (ugs. abwertend)* engage in shady wheeling and dealing

Mäuschen das; ~s, ~: little mouse;

mäuschen·still *Adj.* ~ sein be as quiet as a mouse; **Mause·falle** die mousetrap

Maut die; ~, ~en toll

maximal 1. *Adj.* maximum; 2. *adv.* ~ zulässige Geschwindigkeit maximum permitted speed; **Maxime** die; ~, ~n maxim; **Maximum** das; ~s, Maxima maximum (**an** + *Dat.* of)

Mayonnaise [majɔ'nɛːzə] die; ~, ~n mayonnaise

Mäzen der; ~s, ~e *(geh.)* patron

MdB, M.d.B. *Abk.* Mitglied des Bundestages Member of the Bundestag

m.E. *Abk.* meines Erachtens in my opinion *or* view

Mechanik die; ~: mechanics *sing., no art.;* **Mechaniker** der; ~s, ~: mechanic; **mechanisch** 1. *Adj.* mechanical; power *attrib.* ⟨loom, press⟩; 2. *adv.* mechanically; **Mechanismus** der; ~, Mechanismen mechanism

meckern *itr. V.* a) *(auch fig.)* bleat; b) *(ugs.: nörgeln)* grumble; moan

Mecklenburg-Vorpommern (das); ~s Mecklenburg-Western Pomerania

Medaille [me'daljə] die; ~, ~n medal; **Medaillon** [medal'jõː] das; ~s, ~s a) locket; b) *(Kochk., bild. Kunst)* medallion

Medikament das; ~[e]s, ~e medicine; *(Droge)* drug

meditieren *itr. V.* meditate (**über** + *Akk.* [up]on)

Medium das; ~s, Medien medium

Medizin die; ~, ~en medicine; **Mediziner** der; ~s, ~: doctor; *(Student)* medical student; **medizinisch** 1. *Adj.* medical; medicinal ⟨bath etc.⟩; medicated ⟨toothpaste, soap, etc.⟩; 2. *adv.* medically

Meer das; ~[e]s, ~e *(auch fig.)* sea; **am** ~: by the sea; **Meer·enge** die straits *pl.;* strait

Meeres-: ~bucht die bay; ~früchte *Pl. (Kochk.)* seafood *sing.;* ~spiegel der sea-level

Meer-: ~jungfrau die mermaid; ~rettich der horse-radish; ~schweinchen das guinea-pig

Megaphon das; ~s, ~e megaphone; loud hailer

Mehl das; ~[e]s flour; **mehlig** *Adj.* a) floury; b) *(auch fig.)* mealy ⟨potato, apple, etc.⟩

mehr 1. *Indefinitpron.* more; 2. *Adv.* a) more; b) **nicht** ~: not ... any more; no longer; **es war niemand** ~ **da** there was no one left; **das wird nie** ~ **vorkommen** it will never happen again; **da ist**

nichts ~ zu machen there is nothing more to be done

mehr-: ~bändig *Adj.* in several volumes *postpos.; ~***deutig** 1. *Adj.* ambiguous; 2. *adv.* ambiguously

mehren *(geh.) refl. V.* increase; **mehrer...** *Indefinitpron. u. unbest. Zahlwort* **a)** *attr.* several; **b)** *alleinstehend* ~e several people; ~es several things *pl.;* **mehr·fach** 1. *Adj.* multiple; *(wiederholt)* repeated; 2. *adv.* several times; *(wiederholt)* repeatedly; **Mehrheit** die; ~, ~en majority

mehr-, Mehr-: ~jährig *Adj.* lasting several years *postpos.; ~***malig** *Adj.; nicht präd.* repeated; ~**mals** *Adv.* several times; *(wiederholt)* repeatedly; ~**sprachig** *Adj.* multilingual; ~**stimmig** *(Musik)* 1. *Adj.* for several voices *postpos.;* **ein ~stimmiges Lied** a part-song; 2. *adv.* ~**stimmig singen** sing in harmony; ~**teilig** *Adj.* in several parts *postpos.; ~***wert** der *(Wirtsch.)* surplus value; ~**wertsteuer** die *(Wirtsch.)* value added tax *(Brit.);* VAT *(Brit.);* sales tax *(Amer.); ~***zahl** die; *o. Pl.* **a)** *(Sprachw.)* plural; **b)** *(Mehrheit)* majority

meiden *unr. tr. V. (geh.)* avoid

Meile die; ~, ~n mile

mein *Possessivpron.* my; ~**e Damen und Herren** ladies and gentlemen; **das Buch dort, ist das ~[e]s?** that book over there, is it mine?

Mein·eid der perjury *no indef. art.;* **einen ~ schwören** commit perjury

meinen 1. *itr. V.* think; 2. *tr. V.* **a)** think; **b)** *(sagen wollen, im Sinn haben)* mean; **c)** *(beabsichtigen)* mean; intend; **es gut mit jmdm. ~:** mean well by sb.; **d)** *(sagen)* say

meiner *Gen. von ich (geh.)* **gedenkt ~:** remember me; **erbarme dich ~:** have mercy upon me; **meinerseits** *Adv.* for my part; **ganz ~:** the pleasure is [all] mine; **meinetwegen** *Adv.* **a)** because of me; *(mir zuliebe)* for my sake; *(um mich)* about me; **b)** *[auch --'--] (von mir aus)* as far as I'm concerned; **~!** if you like

Meinung die; ~, ~en opinion (**zu** on, **über** + *Akk.* about); **meiner ~ nach** in my opinion; **ganz meine ~:** I agree entirely; **einer ~ sein** be of the same opinion

Meinungs-: ~forschung die opinion research; ~**freiheit** die freedom to form and express one's own opinions; *(Redefreiheit)* freedom of speech; ~**umfrage** die [public] opinion poll; ~**verschiedenheit** die difference of opinion

Meise die; ~, ~n tit[mouse]

Meißel der; ~s, ~: chisel; **meißeln** *tr. V.* chisel; carve ⟨*statue, sculpture*⟩ with a chisel

meist *Adv.* mostly; **meist...** *Indefinitpron. u. unbest. Zahlw.* most; **die ~en Leute ...:** most people ...; **am ~en** most; **meistens** *Adv. s.* **meist**

Meister der; ~s, ~ **a)** master; **b)** *(Werk~, Polier)* foreman; **c)** *(Sport)* champion; **meisterhaft** 1. *Adj.* masterly; 2. *adv.* in a masterly manner; **meistern** *tr. V.* master; **Meisterschaft** die; ~, ~en **a)** *o. Pl.* mastery; **b)** *(Sport)* championship

Meister-: ~stück das masterpiece (**an** + *Dat.* of); ~**titel** der *(Sport)* championship [title]; ~**werk** das masterpiece (**an** + *Dat.* of)

Melancholie [melaŋko'li:] die; ~ *(Gemütszustand)* melancholy; *(Psych.)* melancholia; **melancholisch** 1. *Adj.* melancholy; melancholic, melancholic ⟨*person, temperament*⟩; 2. *adv.* melancholically

melden 1. *tr. V.* report; *(registrieren lassen)* register ⟨*birth, death, etc.*⟩ (*Dat.* with); 2. *refl. V.* **a)** report; **b)** *(am Telefon)* answer; **c)** *(ums Wort bitten)* put one's hand up; **d)** *(von sich hören lassen)* get in touch (**bei** with); **Meldung** die; ~, ~en **a)** report; *(Nachricht)* piece of news; **b)** *(Wort~)* request to speak

meliert *Adj.* mottled; **[grau] ~es Haar** hair streaked with grey

melken *regelm. (auch unr.) tr. V.* milk

Melodie die; ~, ~n melody; *(Weise)* tune; **melodisch** 1. *Adj.* melodic; 2. *adv.* melodically

Melone die; ~, ~n **a)** melon; **b)** *(ugs.: Hut)* bowler [hat]

Membran die; ~, ~en **a)** *(Technik)* diaphragm; **b)** *(Biol., Chemie)* membrane

Memoiren [me'moa:rən] *Pl.* memoirs

Menge die; ~, ~n **a)** quantity; amount; **b)** *(große ~)* lot *(coll.);* **eine ~** *(ugs.)* lots [of it/them] *(coll.);* **c)** *(Menschen~)* crowd; **d)** *(Math.)* set

Mengen-: ~lehre die; *o. Pl.* set theory *no art.; ~***rabatt** der bulk discount

Mensa die; ~, ~s *od.* **Mensen** refectory *(of university, college)*

Mensch der; ~en, ~en **a)** *(Gattung)*

der ~: man; **die** ~en man *sing.;* human beings; mankind *sing.;* b) *(Person)* person; man/woman; ~en people

menschen-, Menschen-: ~affe der anthropoid [ape]; ~auflauf der crowd [of people]; ~feind der misanthropist; ~fresser der *(ugs.)* cannibal; ~freund der philanthropist; ~handel der trade *or* traffic in human beings; ~kenner der judge of human nature; ~kenntnis die; *o. Pl.* ability to judge human nature; ~leben das life; ~leer *Adj.* deserted; ~menge die crowd [of people]; ~recht das human right; ~schlag der breed [of people]; ~seele die: keine ~seele not a [living] soul

Menschens·kind: ~! *(salopp) (erstaunt)* good heavens; good grief; *(vorwurfsvoll)* for heaven's sake

menschen·unwürdig 1. *Adj.* ⟨*accommodation*⟩ unfit for human habitation; ⟨*conditions*⟩ unfit for human beings; ⟨*behaviour*⟩ unworthy of a human being; **2.** *adv.* ⟨*treat*⟩ in a degrading and inhumane way; ⟨*live, be housed*⟩ in conditions unfit for human beings; **Menschen·verstand** der human intellect; **Menschheit die;** ~: mankind *no art.;* humanity *no art.;* human race; **menschlich 1.** *Adj.* a) human; b) *(annehmbar)* civilized; c) *(human)* humane ⟨*person, treatment, etc.*⟩; **2.** *adv.* a) er ist mir ~ sympathisch I like him as a person; b) *(human)* humanely; **Menschlichkeit die** humanity *no art.*

Mensen *s.* Mensa

Mentalität die; ~, ~en mentality

Menü das; ~s, ~s *(auch DV)* menu

merkbar 1. *Adj.* noticeable; **2.** *adv.* noticeably; **Merk·blatt** leaflet; **merken 1.** *tr. V.* notice; **2.** *refl. V.* sich *(Dat.)* etw. ~: remember sth.; **merklich** *s.* merkbar; **Merkmal das;** ~s, ~e feature

Merkur der; ~s Mercury

merkwürdig 1. *Adj.* strange; odd; **2.** *adv.* strangely; oddly

meßbar *Adj.* measurable

¹Messe die; ~, ~n *(Gottesdienst, Musik)* mass

²Messe die; ~, ~n *(Ausstellung)* [trade] fair

messen 1. *unr. tr. V.* **a)** *auch itr.* measure; **b)** *(beurteilen)* judge (**nach** by); **2.** *unr. refl. V. (geh.)* compete (**mit** with)

Messer das; ~s, ~: knife

messer-, Messer-: ~scharf **1.** *Adj.* razor-sharp; *(fig.)* incisive ⟨*logic*⟩; razor-sharp ⟨*wit, intellect*⟩; **2.** *adv. (fig. ugs.)* ⟨*argue*⟩ incisively; ~stich der knife-thrust; *(Wunde)* knife-wound

Messias der; ~, ~se Messiah

Messing das; ~s brass

Messung die; ~, ~en measurement

Metall das; ~s, ~e metal; **Metall·industrie** die metal-processing and metal-working industries *pl.;* **metallisch** *Adj.* metallic; metal *attrib.,* metallic ⟨*conductor*⟩

Metapher die; ~, ~n metaphor

Meta·physik die; ~: metaphysics *sing., no art.*

Meteor der; ~s, ~e meteor; **Meteorit der;** ~en *od.* ~s, ~e[n] meteorite

Meteorologe der; ~n, ~n meteorologist; **Meteorologie die;** ~: meteorology *no art.*

Meter der *od.* **das;** ~s, ~: metre

meter-, Meter-: ~dick *Adj. (sehr dick)* metres thick *postpos.;* ~hoch *Adj.* metres high *postpos.;* ⟨*snow*⟩ metres deep; ~maß das tape-measure; *(Stab)* [metre] rule

Methode die; ~, ~n method; **methodisch 1.** *Adj.* methodological; *(nach einer Methode vorgehend)* methodical; **2.** *adv.* methodologically; *(nach einer Methode)* methodically

Metier [me'tje:] **das;** ~s, ~s profession

Metrik die; ~, ~en metrics

Metropole die; ~, ~n metropolis

Mett·wurst die soft smoked sausage made of minced pork and beef

Metzger der; ~s, ~ *(bes. westmd., südd., schweiz.)* butcher; **Metzgerei die;** ~, ~en *(bes. westmd., südd., schweiz.)* butcher's [shop]

Meute die; ~, ~n **a)** *(Jägerspr.)* pack; **b)** *(ugs. abwertend)* mob; **Meuterei die;** ~, ~en mutiny; **meutern** *itr. V.* **a)** mutiny; ⟨*prisoners*⟩ riot; **b)** *(ugs.: Unwillen äußern)* moan

Mexikaner der; ~s, ~: Mexican; **mexikanisch** *Adj.* Mexican; **Mexiko (das);** ~s Mexico

MEZ *Abk.* mitteleuropäische Zeit CET

mg *Abk.* Milligramm mg

MG [εm'ge:] **das;** ~s, ~s *Abk.* Maschinengewehr

Mi. *Abk.* Mittwoch Wed.

miau *Interj.* miaow; **miauen** *itr. V.* miaow

mich 1. *Akk. von* ich me; **2.** *Akk. des Reflexivpron. der 1. Pers. Sg.* myself

mick[e]rig *Adj. (ugs.)* miserable; measly *(sl.);* puny ⟨*person*⟩
mied *1. u. 3. Pers. Sg. Prät. v.* meiden
Mieder·waren *Pl.* corsetry *sing.*
Miene die; ~, ~n expression
mies *(ugs.)* 1. *Adj.* lousy *(sl.);* 2. *adv.* lousily *(sl.)*
Mies·muschel die [common] mussel
Miete die; ~, ~n rent; *(für ein Auto, Boot)* hire charge; zur ~ wohnen live in rented accommodation; mieten *tr. V.* rent; *(für kürzere Zeit)* hire; Mieter der; ~s, ~: tenant; Miets·haus das block of rented flats *(Brit.)* or *(Amer.)* apartments
Miet-: ~vertrag der tenancy agreement; ~wagen der hire-car
Migräne die; ~, ~n migraine
mikro-, Mikro- micro-
Mikrobe die; ~, ~n microbe
mikro-, Mikro-: ~film der microfilm; ~phon [--'-] das; ~s, ~e microphone; ~skop [--'-] das; ~s, ~e microscope; ~skopisch --'--] 1. *Adj.* microscopic; 2. *adv.* microscopically
Milbe die; ~, ~n mite
Milch die; ~: milk; Milch·flasche die milk-bottle; milchig 1. *Adj.* milky; 2. *adv.* ~ weiß milky-white
Milch-: ~kaffee der coffee with plenty of milk; ~kännchen das milk-jug; ~reis der rice pudding; ~straße die Milky Way; Galaxy
mild, milde 1. *Adj.* mild; lenient ⟨*judge, judgement*⟩; soft ⟨*light*⟩; smooth ⟨*brandy*⟩; 2. *adv. (gütig)* leniently; *(gelinde)* mildly; Milde die; ~: mildness; *(Güte)* leniency; mildern *tr. V.* moderate; mitigate ⟨*punishment*⟩; Milderung die; ~: *s.* mildern: moderation; mitigation
Milieu [mi'liø:] das; ~s, ~s environment
militant *Adj.* militant; ¹Militär das; ~s armed forces *pl.;* military; *(Soldaten)* soldiers *pl.;* ²Militär der; ~s, ~s [high-ranking military] officer
Militär-: ~dienst der military service; ~diktatur die military dictatorship
militärisch *Adj.* military; militarisieren *tr. V.* militarize; Military ['mɪlɪtərɪ] die; ~, ~s *(Reiten)* three-day event; Miliz die; ~, ~en militia; *(Polizei)* police
Mill. *Abk.* Million m.
milli- Milli- milli-
Milliarde die; ~, ~n billion
Milli-: ~gramm das milligram;

~meter der *od.* das millimetre; ~meter·papier das [graph] paper ruled in millimetre squares
Million die; ~, ~en million; Millionär der; ~s, ~e millionaire
Millionen-: ~schaden der damage no *pl., no indef. art.* running into millions; ~stadt die town with over a million inhabitants
millionst... *Ordinalz.* millionth
Milz die; ~: spleen
Mimik die; ~: gestures and facial expressions *pl.*
Mimose die; ~, ~n a) mimosa; b) *(fig.)* over-sensitive person
minder *Adv. (geh.)* less; minder... *Adj.* inferior ⟨*goods, brand*⟩; minder·bemittelt *Adj.* without much money *postpos., not pred.;* ~bemittelt sein not have much money; geistig ~bemittelt *(fig. salopp abwertend)* not all that bright *(coll.);* Minderheit die; ~, ~en minority
minder·jährig *Adj.* ⟨*child etc.*⟩ who is/was a minor; Minder·jährige der/die *adj. Dekl.* minor; mindern *tr. V. (geh.)* reduce; Minderung die; ~, ~en reduction *(Gen.* in); minderwertig *Adj.* inferior; mindest... *Adj.* least; *(geringst...)* slightest; das ist das ~e, was du tun kannst it is the least you can do; mindestens *Adv.* at least
Mine die; ~, ~n a) *(Bergwerk, Sprengkörper)* mine; b) *(Bleistift~)* lead; *(Kugelschreiber~, Filzschreiber~)* refill
Mineral das; ~s, ~e *od.* Mineralien mineral; Mineralogie die; ~: mineralogy no *art.*
Mineral-: ~öl das mineral oil; ~wasser das mineral water
Mini das; ~s, ~s *(Mode)* mini *(coll.);* Mini- mini-; Miniatur die; ~, ~en miniature
minimal 1. *Adj.* minimal; marginal ⟨*advantage, lead*⟩; very slight ⟨*benefit, profit*⟩; 2. *adv.* minimally; Minimum das; ~s, Minima minimum (an + *Dat.* of)
Minister der; ~s, ~: minister (für for); *(eines britischen Hauptministeriums)* Secretary of State (für for); *(eines amerikanischen Hauptministeriums)* Secretary (für of); Ministerium das; ~s, ~s, Ministerien Ministry; Department *(Amer.);* Minister·präsident der a) *(eines deutschen Bundeslandes)* minister-president; b) *(Pre-*

mierminister) Prime Minister; **Ministrant** der; ~en, ~en *(kath. Kirche)* server

Minorität die; ~, ~en *s.* Minderheit

minus *Konj., Adv. (bes. Math.)* minus; **Minus** das; ~: deficit; **Minus·zeichen** das minus sign

Minute die; ~, ~n minute; **minuten·lang** 1. *Adj.* lasting [for] several minutes *postpos.;* 2. *adv.* for several minutes; **Minuten·zeiger** der minute-hand

Mio. *Abk.* Million[en] m.

mir 1. *Dat. von* ich to me; *(nach Präpositionen)* me; **Freunde von ~:** friends of mine; **gehen wir zu ~:** let's got to my place; **von ~ aus** as far as I'm concerned; 2. *Dat. des Reflexivpron. der 1. Pers. Sg.* myself

Mirabelle die; ~, ~n mirabelle

Misch-: ~**brot** das bread made from wheat and rye flour; ~**ehe** die mixed marriage

mischen 1. *tr. V.* mix; 2. *refl. V.* a) *(sich ver~)* mix (mit with); ⟨smell, scent⟩ blend (mit with); b) *(sich ein~)* sich in etw. *(Akk.)* ~: interfere in sth.; **Misch·farbe** die non-primary colour; **Mischling** der; ~s, ~e half-caste; **Mischmasch** der; ~[e]s, ~e *(ugs., meist abwertend)* hotchpotch; mishmash; **Mischung** die; ~, ~en mixture; *(Tee~, Kaffee~, Tabak~)* blend; *(Pralinen~)* assortment; **Misch·wald** der mixed [deciduous and coniferous] forest

miserabel *(ugs.)* 1. *Adj.* dreadful *(coll.);* 2. *adv.* dreadfully *(coll.);* ihm geht es gesundheitlich ~: he's in a bad way; **Misere** die; ~, ~n *(geh.)* wretched *or* dreadful state; *(Elend)* misery; *(Not)* distress

miß *Imperativ Sg. v.* messen

miß·achten *tr. V.* a) *(ignorieren)* disregard; ignore; b) *(geringschätzen)* be contemptuous of

miß·billigen *tr. V.* disapprove of; **Miß·billigung** die disapproval

Miß·brauch der *s.* mißbrauchen: abuse; misuse; **miß·brauchen** *tr. V.* abuse; misuse; abuse ⟨trust⟩

missen *tr. V. (geh.)* jmdn./etw. nicht ~ mögen not want to be without sb./sth.

Miß·erfolg der failure

Misse·tat die *(geh. veralt.)* misdeed

miß·fallen *unr. itr. V.* etw. mißfällt jmdn. sb. dislikes sth.; **Mißfallen** das; ~s displeasure; *(Mißbilligung)* disapproval

Miß·geschick das mishap

miß·glücken *itr. V.; mit sein* fail

miß·gönnen *tr. V.* jmdm. etw. ~: begrudge sb. sth.

Miß·griff der error of judgement

miß·handeln *tr. V.* maltreat; **Miß·handlung** die maltreatment

Mission die; ~, ~en mission; **Missionar** der; ~s, ~e, missionary

Miß·kredit der in jmdn./etw. in ~ bringen bring sb./sth. into discredit

mißlang *1. u. 3. Pers. Sg. Prät. v.* mißlingen

mißliebig *Adj.* unpopular

mißlingen *unr. itr. V.; mit sein* fail; **Mißlingen** das; ~s failure; **mißlungen** *2. Part. v.* mißlingen

Miß·mut der ill humour *no indef. art.;* **miß·mutig** 1. *Adj.* bad-tempered; sullen ⟨face⟩; 2. *adv.* bad-temperedly

Miß·stand der deplorable state of affairs *no pl.*

mißt *2. u. 3. Pers. Sg. Präsens v.* messen

miß·trauen *itr. V.* jmdm./einer Sache ~: mistrust *or* distrust sb./sth.; **Miß·trauen** das; ~s mistrust, distrust (gegen of); **mißtrauisch** 1. *Adj.* mistrustful; distrustful; 2. *adv.* mistrustfully; distrustfully

miß·verständlich 1. *Adj.* unclear; ⟨formulation, concept, etc.⟩ that could be misunderstood; 2. *adv.* ⟨express oneself, describe⟩ in a way that could be misunderstood; **Miß·verständnis** das misunderstanding; **miß·verstehen**[1] *unr. tr. V.* misunderstand

Mist der; ~[e]s a) dung; *(Dünger)* manure; *(mit Stroh usw. gemischt)* muck; b) *(~haufen)* dung/manure/ muck heap; c) *(ugs. abwertend) (Unsinn)* rubbish *no indef. art.; (Minderwertiges)* junk *no indef. art.*

Mistel die; ~, ~n mistletoe

Mist·haufen der dung/manure/ muck heap

mit 1. *Präp. mit Dat.* with; ein Zimmer ~ Frühstück a room with breakfast included; ~ 50 [km/h] fahren drive at 50 [k.p.h]; ~ der Bahn/dem Auto fahren go by train/car; ~ 20 [Jahren] at [the age of] twenty; 2. *Adv.* a) too; as well; b) seine Arbeit war ~ am besten *(ugs.)* his work was among the best

Mit·arbeit die; *o. Pl.* collaboration

---

[1] *ich mißverstehe, mißverstanden, mißzuverstehen*

(bei/an + Dat. on); *(Mithilfe)* assistance (bei, in + *Dat.* in); *(Beteiligung)* participation (in + *Dat.* in); **mịt|arbeiten** *itr. V.* collaborate (**bei/an +** *Dat.* on) *(sich beteiligen)* participate (in + *Dat.* in); **Mit·arbeiter der a)** collaborator; **freier** ~: free-lance worker; **b)** *(Angestellter)* employee

mịt|bekommen *unr. tr. V.* **a) etw.** ~: be given sth. to take with one; **b)** *(wahrnehmen)* be aware of; *(durch Hören, Sehen)* hear/see

mịt|bestimmen 1. *itr. V.* have a say; **2.** *tr. V.* have an influence on; **Mitbestimmung die**; *o. Pl.* participation (bei in); *(der Arbeitnehmer)* co-determination

mịt|bringen *unr. tr. V.* **a) etw.** ~: bring sth. with one; **jmdm./sich etw.** ~: bring sth. with one for sb./bring sth. back for oneself; **b)** *(haben)* have ⟨*ability, gift, etc.*⟩; **Mịtbringsel das**; ~s, ~: [small] present; *(Andenken)* [small] souvenir

mit·einạnder *Adv.* **a)** with each other *or* one another; ~ **sprechen** talk to each other *or* one another; **b)** *(gemeinsam)* together

mịt|erleben *tr. V.* **a)** witness ⟨*events etc.*⟩; **b)** *(mitmachen)* be alive during

mịt|fahren *unr. itr. V.; mit sein* **bei jmdm. |im Auto|** ~: go/travel with sb. [in his/her car]; *(mitgenommen werden)* get a lift with sb. [in his/her car]

mịt·fühlend 1. *Adj.* sympathetic; **2.** *adv.* sympathetically

mịt|führen *tr. V.* **a)** *(Amtsspr.: bei sich tragen)* **etw.** ~: carry sth. [with one]; **b)** *(transportieren)* ⟨*river, stream*⟩ carry along

mịt|geben *unr. tr. V.* **jmdm. etw.** ~: give sb. sth. to take with him/her; *(fig.)* provide sb. with sth.

Mịt·gefühl das; *o. Pl.* sympathy

mịt|gehen *unr. itr. V.; mit sein* **a)** go too; **mit jmdm.** ~: go with sb.; **b)** *(sich mitreißen lassen)* **begeistert** ~: respond enthusiastically

Mịt·gift die; ~, ~en *(veralt.)* dowry

Mịt·glied das member *(Gen.,* in + *Dat.* of)

mịt|halten *unr. itr. V.* keep up (bei in, mit with)

Mịt·hilfe die; *o. Pl.* help; assistance

mịt|hören 1. *tr. V.* listen to; *(zufällig)* overhear ⟨*conversation, argument, etc.*⟩; *(abhören)* listen in on; **2.** *itr. V.* listen; *(zufällig)* overhear

mịt|kommen *unr. itr. V.; mit sein* **a)**

come too; **kommst du mit?** are you coming [with me/us]?; **b)** *(Schritt halten)* keep up

Mịt·läufer der *(abwertend)* [mere] supporter

Mịt·laut der consonant

Mịt·leid das pity, compassion (mit for); *(Mitgefühl)* sympathy (mit for); **Mịt·leidenschaft die: jmdn./etw. in** ~ **ziehen** affect sb./sth.; **mịt·leidig 1.** *Adj.* compassionate; *(mitfühlend)* sympathetic; **2.** *adv.* compassionately; *(mitfühlend)* sympathetically

mịt|machen 1. *tr. V.* **a)** *(teilnehmen an)* go on ⟨*trip*⟩; join in ⟨*joke*⟩; follow ⟨*fashion*⟩; fight in ⟨*war*⟩; do ⟨*course, seminar*⟩; **das mache ich nicht mit** *(ugs.)* I can't go along with it; **b)** *(ugs.: erleiden)* **zwei Weltkriege/viele Bombenangriffe mitgemacht haben** have been through two world wars/many bomb attacks; **2.** *itr. V.* **a)** *(sich beteiligen)* join in; **b)** *(ugs.: funktionieren)* **mein Herz/Kreislauf macht nicht mit** my heart/circulation can't take it

Mịt·mensch der fellow human being

mịt|nehmen *unr. tr. V.* **a) jmdn.** ~: take sb. with one; **etw.** ~: take sth. with one; *(verhüll.: stehlen)* walk off with sth. *(coll.); (kaufen)* take sth.; **Essen/Getränke zum Mitnehmen** food/ drinks to take away *or (Amer.)* to go; **b)** *(in Mitleidenschaft ziehen)* **jmdn.** ~: take it out of sb.

mịt|reden *itr. V.* **a)** join in the conversation; **b)** *(mitbestimmen)* have a say

Mịt·reisende der/die fellow passenger

mịt|reißen *unr. tr. V.* **die Begeisterung/seine Rede hat alle Zuhörer mitgerissen** the audience was carried away with enthusiasm/by his speech

mit·sạmt *Präp. mit Dat.* together with

Mịt·schuld die share of the blame *or* responsibility (an + *Dat.* for)

Mịt·schüler der, Mịt·schülerin die schoolfellow

mịt|spielen *itr. V.* **a)** join in the game; **b) in einem Film** ~: be in a film; **in einem Orchester/in** *od.* **bei einem Fußballverein** ~: play in an orchestra/for a football club; **Mịt·spieler der, Mịt·spielerin die** player; *(in derselben Mannschaft)* team-mate

mịttag *Adv.* **heute/Montag** ~: at midday today/on Monday; **Mịttag der**; ~s, ~e **a)** midday *no art.;* **gegen** ~: around midday; **zu** ~ **essen** have

lunch; **b)** *o. Pl. (ugs.: Mittagspause)*
lunch-hour; **Mittag·essen das**
lunch; **mittags** *Adv.* at midday; **12
Uhr ~:** 12 noon

Mittags-: **~pause die** lunch-hour;
~ruhe die period of quiet after lunch;
~zeit die a) *o. Pl. (Zeit gegen 12 Uhr)*
lunch-time *no art.;* **b)** *(~pause)* lunch-
hour

Mitte die; ~, ~n middle; *(eines Krei-
ses, einer Kugel, Stadt)* centre; **~ des
Monats/Jahres** in the middle of the
month/year

mit|teilen *tr. V.* **jmdm. etw. ~:** tell sb.
sth.; *(informieren)* inform sb. of sth.;
mitteilsam *Adj.* communicative;
(gesprächig) talkative; **Mit·teilung
die** communication; *(Bekanntgabe)*
announcement

Mittel das; ~s, ~ a) means; *(Methode)*
way; method; *(Werbe~, Propagan-
da~ usw.)* device *(Gen.* for); **mit allen
~n versuchen, etw. zu tun** try by every
means to do sth.; **b)** *(Arznei)* **ein ~ gegen Husten** *usw.* a cure for coughs
etc.; **c)** *Pl. (Geld~)* funds; *(Privat~)*
means

Mittel·alter das; *o. Pl.* Middle Ages
pl.; **mittel·alterlich** *Adj.* medieval

mittelbar 1. *Adj.* indirect; **2.** *adv.* indirectly

mittel-, Mittel-: **~ding das;** *o. Pl.*
ein ~ding sein be something in between; **~europa (das)** Central
Europe; **~finger der** middle finger;
~gebirge das low mountains *pl.;*
~linie die centre line; *(Fußball)* half-
way line; **~los** *Adj.* without means
postpos.; **~mäßig** *Adj.* mediocre;
~meer das Mediterranean [Sea];
~punkt der a) *(Geom.)* centre; *(einer
Strecke)* midpoint; **b)** *(Mensch/Sache
im Zentrum)* centre of attention;
~scheitel der centre parting;
~schule die *s.* Realschule; **~stand
der;** *o. Pl.* middle class; **~weg der**
middle course; **~welle die** *(Physik,
Rundf.)* medium wave

mitten *Adv.* **~ an/auf etw.** *(Akk./Dat.)*
in the middle of sth.; **~ durch die
Stadt** right through the town

mitten-: **~drin** *Adv.* [right] in the
middle; **~durch** *Adv.* [right] through
the middle

Mitter·nacht die; *o. Pl.* midnight *no
art.;* **Mitternachts·sonne die** midnight sun

mittler... *Adj.* middle; moderate
(speed); medium-sized *(company,*

town); medium *(quality, size); (durchschnittlich)* average

mittler·weile *Adv.* since then; *(bis
jetzt)* by now; *(unterdessen)* in the
meantime

Mittwoch der; ~[e]s, ~e Wednesday;
mittwochs *Adv.* on Wednesday[s]

mit·unter *Adv.* from time to time

mit·wirken *itr. V.* **an etw.** *(Dat.)/bei
etw. ~:* collaborate on/be involved in
sth.; **in einem Orchester/Theaterstück
~:** play in an orchestra/act *or* appear
in a play; **Mitwirkende der/die** *adj.
Dekl. (an einer Sendung)* participant;
(in einer Show) performer; *(in einem
Theaterstück)* actor

Mit·wisser der; ~s ~: **~ einer Sache**
(Gen.) **sein** be an accessory to sth.

mixen *tr. V.* mix; **sich** *(Dat.)* **einen
Drink ~:** fix oneself a drink; **Mixer
der; ~s, ~ a)** *(Bar~)* barman; bar-
tender·*(Amer.);* **b)** *(Gerät)* blender and
liquidizer

mm *Abk.* Millimeter mm.

Mo. *Abk.* Montag Mon.

Mob der; ~s *(abwertend)* mob

Möbel das; ~s, ~ a) *Pl.* furniture
sing., no indef. art.; **b)** piece of furniture; **Möbel·wagen der** furniture
van; removal van; **mobil** *Adj.* **a)**
mobile; **~ machen** mobilize; **b)** *(ugs.)
(lebendig)* lively; **Mobiliar das; ~s**
furnishings *pl.;* **mobilisieren** *tr. V.*
a) *(Milit., fig.)* mobilize; **b)** *(aktivieren)* activate; **Mobilmachung die;
~, ~en** mobilization; **Mobil·telefon
das** cellular phone; **möblieren** *tr. V.*
furnish

mochte *1. u. 3. Pers. Sg. Prät. v.* mögen; **möchte** *1. u. 3. Pers. Sg. Konjunktiv II v.* mögen

Mode die; ~, ~n fashion; **Mode·far-
be die** fashionable colour

Modell das; ~s, ~e *(auch fig.)* model;
jmdm. ~ sitzen *od.* **stehen** sit for sb.;
modellieren *tr. V.* model, mould
(figures, objects); mould *(clay, wax);*
Modell·kleid das model dress

Moden·schau die fashion show

Moder der; ~s mould; *(~geruch)* mus-
tiness

Moderation die; ~, ~en *(Rundf.,
Ferns.)* presentation; **Moderator
der; ~s, ~en, Moderatorin die; ~,
~nen** *(Rundf., Ferns.)* presenter; **moderieren** *tr. V. (Rundf., Ferns.)* present *(programme)*

¹modern *itr. V.; auch mit sein* go
mouldy

²**modern** 1. *Adj.* modern; *(modisch)*
fashionable; 2. *adv.* in a modern manner; *(modisch)* fashionably; **modernisieren** *tr. V.* modernize
Mode-: ~**schöpfer der** couturier;
~**schöpferin die** couturière; ~**wort
das;** *Pl.* ~**wörter** vogue-word; ~**zeitschrift die** fashion magazine
modifizieren *tr. V. (geh.)* modify
modisch 1. *Adj.* fashionable; 2. *adv.*
fashionably
Mofa das; ~s, ~s [low-powered]
moped
Mogelei die; ~, ~en *(ugs.)* cheating
no pl.; **mogeln** *itr. V.* cheat
mögen 1. *unr. Modalverb;* 2. *Part.* ~:
a) *(wollen)* want to; **das hätte ich sehen
~:** I would have liked to see that; **b)**
(geh.: sollen) **das mag genügen** that
should be enough; **c)** *(Vermutung,
Möglichkeit)* **sie mag/mochte vierzig
sein** she must be/must have been
[about] forty; |**das| mag sein** maybe; **d)**
Konjunktiv II (den Wunsch haben) **ich/
sie möchte gern wissen ...:** I would/she
would like to know ...; 2. *unr. tr. V.*
like; **sie mag keine Rosen** she does not
like roses; **sie ~ sich** they're fond of
one another; **möchten Sie ein Glas
Wein?** would you like a glass of
wine?; **ich möchte lieber Tee** I would
prefer tea; 3. *unr. itr. V.* **a)** *(es wollen)*
like to; **b) ich möchte nach Hause** I
want to go home; **er möchte zu Herrn
A** he would like to see Mr A
möglich *Adj.* possible; **es war ihm
nicht ~ |zu kommen|** he was unable [to
come]; **alles ~e** *(ugs.)* all sorts of
things; |**das ist doch| nicht ~!** im-
possible!; **sein ~stes tun** do one's utmost; **möglicherweise** *Adv.* possibly; **Möglichkeit die;** ~, ~en **a)**
possibility; *(Methode)* way; **es besteht
die ~, daß ...:** there is a possibility
that ...; **b)** *(Gelegenheit)* opportunity;
chance; **möglichst** *Adv.* **a)** if [at all]
possible; **b)** ~ **schnell** as fast as possible
Mohammed (der) Muhammad; **Mo-
hammedaner der;** ~s, ~: Muslim;
Muhammadan; **mohammeda-
nisch** *Adj.* Muslim; Muhammadan
Mohn der; ~s poppy; *(Samen)* poppy
seed; *(auf Brot, Kuchen)* poppy seeds
pl.
Mohn-: ~**blume die** poppy; ~**brötchen das** poppy-seed roll; ~**kuchen
der** poppy-seed cake
Möhre die; ~, ~n carrot

Mohren·kopf der chocolate marshmallow
Mohr·rübe die carrot
mokieren *refl. V. (geh.)* **sich über etw.
(Akk.) ~:** scoff at sth.; **sich über jmdn.
~:** mock sb.
Mokka der; ~s strong black coffee
Molch der; ~|e|s, ~e newt
Mole die; ~, ~n [harbour] mole
Molekül das; ~s, ~e molecule
molk *1. u. 3. Pers. Sg. Prät. v.* **melken;
Molkerei die;** ~, ~en dairy
Moll das; ~ *(Musik)* minor [key]
mollig 1. *Adj.* **a)** *(rundlich)* plump; **b)**
(warm) snug; 2. *adv.* snugly; ~ **warm**
warm and snug
¹**Moment der;** ~|e|s, ~e moment; **je-
den ~** *(ugs.)* [at] any moment; **im ~:** at
the moment; ²**Moment das;** ~|e|s, ~e
factor, element **(für** in); **momentan**
1. *Adj.* **a)** present; **b)** *(vorübergehend)*
temporary; *(flüchtig)* momentary; 2.
adv. **a)** at present; **b)** *(vorübergehend)*
temporarily
Monaco (das); ~s Monaco
Monarch der; ~en, ~en monarch;
Monarchie die; ~, ~n monarchy;
Monarchin die; ~, ~nen monarch
Monat der; ~s, ~e month; **im ~ April**
in the month of April; **monatelang**
1. *Adj.* lasting for months *postpos., not
pred.;* 2. *adv.* for months [on end];
monatlich 1. *Adj.* monthly; 2. *adv.*
every month; *(pro Monat)* per month
Monats-: ~**erste der** first [day] of the
month; ~**hälfte die** half of the
month; ~**karte die** monthly season-
ticket; ~**letzte der** last day of the
month
Mönch der; ~|e|s, ~e monk
Mond der; ~|e|s, ~e moon; **auf** *od.*
hinter dem ~ leben *(fig. ugs.)* be a bit
behind the times *(coll.);* **nach dem ~
gehen** *(ugs.)* ⟨clock, watch⟩ be hopelessly wrong
Mond-: ~**finsternis die** eclipse of
the moon; ~**landung die** moon landing
Mongole der; ~n, ~n **a)** Mongol; **b)**
(Bewohner der Mongolei) Mongolian;
Mongolei die; ~: Mongolia
Monitor der; ~s, ~en monitor
Mono·gramm das; ~s, ~e mono-
gram; **Monographie die;** ~, ~n
monograph
Monolog der; ~s, ~e monologue
Monopol das; ~s, ~e monopoly **(auf**
+ *Akk.,* **für** in, of)
monoton 1. *Adj.* monotonous; 2. *adv.*

monotonously; **Monotonie** die; ~,
~n monotony

Monster das; ~s, ~: monster; *(häßlich)* [hideous] brute; **Monstren** *s.*
Monstrum; **monströs** *Adj.* monstrous; **Monstrum** das; ~s, **Monstren a)** monster; **b)** *(Sache)* hulking
great thing *(coll.)*

Mon·tag der Monday

Montage [mɔn'ta:ʒə] die; ~, ~n a)
(Zusammenbau) assembly; *(Einbau)*
installation; *(Aufstellen)* erection;
(Anbringen) fitting (**an** + *Akk. od.
Dat.* to); mounting (**auf** + *Akk. od.
Dat.* on); b) *(Film, bild. Kunst, Litera-
turw.)* montage

montags *Adv.* on Monday[s]

montieren *tr. V.* a) *(zusammenbauen)*
assemble (**aus** from); erect ⟨building⟩;
b) *(anbringen)* fit (**an** + *Akk. od. Dat.*
to; **auf** + *Akk. od. Dat.* on); *(einbau-
en)* install (**in** + *Akk.* in); *(befestigen)*
fix (**an** + *Akk. od. Dat.* to)

Monument das; ~|e|s, ~e monument;
monumental *Adj.* monumental

Moor das; ~|e|s, ~e bog; *(Bruch)*
marsh

Moos das; ~es, ~e moss

Moped ['mo:pɛt] das; ~s, ~s moped

Mops der; ~es, **Möpse** pug [dog]; *(salopp: dicke Person)* podge *(coll.)*

Moral die; ~ a) *(Norm)* morality; b)
(Sittlichkeit) morals *pl.;* c) *(Selbstvertrauen)* morale; d) *(Lehre)* moral;
moralisch 1. *Adj.* a) moral; b) *(tugendhaft)* virtuous; **2.** *adv.* a) morally;
b) *(tugendhaft)* virtuously; **morali-
sieren** *itr. V. (geh.)* moralize; **Moralist** der; ~en, ~en moralist

Morast der; ~|e|s, ~e od. **Moräste a)**
bog; swamp; b) *o. Pl. (Schlamm)* mud

Mord der; ~|e|s, ~e murder (**an** +
Dat. of); *(durch ein Attentat)* assassination; **einen ~ begehen** commit murder; **morden** *tr., itr. V.* murder;
Mörder der; ~s, ~: murderer *(esp.
Law);* killer; *(politischer ~)* assassin;
Mörderin die; ~, ~nen murderer;
murderess; *(politische ~)* assassin;
mörderisch 1. *Adj.* a) murderous; **2.**
adv. (ugs.) dreadfully *(coll.);* **Mord-
fall** der murder case; **mords-,
Mords-** *(ugs.)* terrific *(coll.)*

Mord-: ~**verdacht** der suspicion of
murder; ~**versuch** der attempted
murder; *(Attentat)* assassination attempt; ~**waffe** die murder weapon

morgen *Adv.* a) tomorrow; ~ **in einer
Woche** tomorrow week; a week to-

morrow; ~ **um diese Zeit** this time tomorrow; **bis ~!** until tomorrow!; **see
you tomorrow!;** b) *(am Morgen)* **heute
~**: this morning; |**am**| **Sonntag ~**: on
Sunday morning; **Morgen** der; ~s,
~: morning; **am ~**: in the morning;
am folgenden *od.* **nächsten ~**: next
morning; **früh am ~, am frühen ~**:
early in the morning; **morgendlich**
Adj. morning

Morgen-: ~**grauen** das daybreak;
~**mantel** der dressing-gown; ~**rot**
das *(geh.)* rosy dawn

morgens *Adv.* in the morning; *(jeden
Morgen)* every morning; **Dienstag** *od.*
dienstags ~: on Tuesday morning[s];
von ~ bis abends from morning to
evening; **morgig** *Adj.* tomorrow's

Morphium das; ~s morphine; **mor-
phium·süchtig** *Adj.* addicted to
morphine *pred.*

morsch *Adj. (auch fig.)* rotten

Mörser der; ~s, ~ *(Gefäß, Geschütz)*
mortar

Mörtel der; ~s mortar

Mosaik das; ~s, ~en *od.* ~e mosaic

Mosambik (das); ~s Mozambique

Moschee die; ~, ~n mosque

Moschus der; ~: musk

Mosel die; ~: Moselle; **Mosel·wein**
der Moselle [wine]

Moskau (das); ~s Moscow; **Mos-
kauer 1.** *indekl. Adj.* Moscow *attrib.;*
2. der; ~s, ~: Muscovite

Moskito der; ~s, ~s mosquito

Moslem der; ~s, ~s Muslim; **moslemisch** *Adj.* Muslim

Most der; ~|e|s, ~e a) [cloudy fer-
mented] fruit-juice; b) *(landsch.: neuer
Wein)* new wine; **Mostrich** der; ~s
(nordostd.) mustard

Motel das; ~s, ~s motel

Motiv das; ~s, ~e a) motive; b)
(fachspr.: Thema) motif; theme; *(bild.
Kunst)* subject

Motor der; ~s, ~en engine; *(Elek-
tro~)* motor; **Motor·haube** die
(Kfz-W.) bonnet *(Brit.);* hood *(Amer.);*
motorisieren *tr. V.* motorize; **Mo-
tor·rad** das motor cycle; **Motor-
rad·fahrer** der motor-cyclist

Motor-: ~**roller** der motor scooter;
~**schaden** der engine trouble *no indef. art.*

Motte die; ~, ~n moth; **Motten·ku-
gel** die moth-ball

Motto das; ~s, ~s motto; *(Schlagwort)* slogan

Möwe die; ~, ~n gull

Mrd. *Abk.* Milliarde bn.

Mücke die; ~, ~n midge; *(größer)* mosquito; **Mücken·stich** der midge/mosquito bite

Mucks der; ~es, ~e *(ugs.)* murmur [of protest]; **keinen** ~ **sagen** not utter a [single] word

müde 1. *Adj.* tired; *(ermattet)* weary; *(schläfrig)* sleepy; **jmdn./etw.** *od.* **jmds./einer Sache** ~ **sein** *(geh.)* be tired of sb./sth.; **2.** *adv.* wearily; *(schläfrig)* sleepily; **Müdigkeit die;** ~: tiredness

muffelig *(ugs.)* **1.** *Adj.* grumpy; **2.** *adv.* grumpily

muffig *Adj.* musty

Mühe die; ~, ~n trouble; **sich** *(Dat.)* **mit jmdm./etw.** ~ **geben** take [great] pains over sb./sth.; **mit Müh und Not** with great difficulty; **mühelos 1.** *Adj.* effortless; **2.** *adv.* effortlessly; **mühe·voll** *Adj.* laborious; painstaking ⟨work⟩

Mühle die; ~, ~n **a)** mill; *(Kaffee~)* [coffee-] grinder; **b)** *(Spiel)* o. Art., o. *Pl.* nine men's morris

Mühsal die; ~, ~e *(geh.)* tribulation; *(Strapaze)* hardship; **mühsam 1.** *Adj.* laborious; **2.** *adv.* laboriously; **müh·selig** *(geh.)* **1.** *Adj.* laborious; arduous ⟨journey, life⟩; **2.** *adv.* with [great] difficulty

Mulde die; ~, ~n hollow

Mull der; ~[e]s *(Stoff)* mull; *(Verband~)* gauze

Müll der; ~s refuse; rubbish; garbage *(Amer.)*; trash *(Amer.)*; *(Industrie~)* [industrial] waste

Mull·binde die gauze bandage

Müller der; ~s, ~: miller

Müll-: ~**halde die** refuse dump; ~**mann der;** *Pl.* ~**männer** *(ugs.)* dustman *(Brit.)*; garbage man *(Amer.)*; ~**sack der** refuse bag; ~**schlucker der** rubbish *or (Amer.)* garbage chute; ~**tonne die** dustbin *(Brit.)*; garbage *or* trash can *(Amer.)*; ~**wagen der** dust-cart *(Brit.)*; garbage truck *(Amer.)*

mulmig *Adj. (ugs.)* uneasy

Multiplikation die; ~, ~en *(Math.)* multiplication; **multiplizieren** *tr. V.* multiply (mit by)

Mumie ['mu:miə] **die;** ~, ~n mummy

Mumm der; ~s *(ugs.) (Mut)* guts *pl.* *(coll.)*; *(Tatkraft)* drive; zap *(sl.)*; *(Kraft)* muscle-power

Mumps der *od.* die; ~: mumps *sing.*

München (das); ~s Munich;

Münch[e]ner 1. *indekl. Adj.* Munich *attrib;* **2. der;** ~s, ~: inhabitant/ native of Munich

Mund der; ~[e]s, **Münder** mouth; **er küßte sie auf den** ~: he kissed her on the lips; **mit vollem** ~ **sprechen** speak with one's mouth full; **den** ~ **nicht aufmachen** *(fig. ugs.)* not say anything; **den** *od.* **seinen** ~ **halten** *(ugs.) (zu sprechen aufhören)* shut up *(coll.); (nichts sagen)* not say anything; *(nichts verraten)* keep quiet (**über** + *Akk.* about); **sie ist nicht auf den** ~ **gefallen** *(fig. ugs.)* she's never at a loss for words; **Mund·art die** dialect

münden *itr. V.; mit sein* **in etw.** *Akk.* ~: ⟨river⟩ flow into sth.; ⟨corridor, street⟩ lead into sth.

mund-, Mund-: ~**faul** *Adj. (ugs.)* uncommunicative; ~**gerecht** *Adj.* bitesized; ~**geruch der** bad breath *no indef. art.;* ~**harmonika die** mouth-organ

mündig *Adj.* of age *pred.;* ~ **werden** come of age

mündlich 1. *Adj.* oral; **2.** *adv.* orally; **Mund·stück das** mouthpiece; *(bei Zigaretten)* tip; **mund·tot** *Adj.* **jmdn.** ~ **machen** silence sb.; **Mündung die;** ~, ~en **a)** mouth; *(größere Trichter~)* estuary; **b)** *(bei Feuerwaffen)* muzzle

Mund-zu-Mund-Beatmung die mouth-to-mouth resuscitation

Munition die; ~: ammunition

munkeln *tr., itr. V. (ugs.)* **man munkelt, daß ...:** there is a rumour that ...

Münster das; ~s, ~: minster; *(Dom)* cathedral

munter 1. *Adj.* **a)** cheerful; *(lebhaft)* lively ⟨eyes, game⟩; **b)** *(wach)* awake; **2.** *adv.* cheerfully; **Munterkeit die;** ~: cheerfulness

Münz·automat der slot-machine; **Münze die;** ~, ~n coin

Münz-: ~**fernsprecher der** payphone; pay station *(Amer.);* ~**tankstelle die** coin-in-the-slot petrol *(Brit.)* or *(Amer.)* gas station; ~**wechsler der** change machine

mürbe *Adj.* crumbly ⟨biscuit, cake, etc.⟩; tender ⟨meat⟩; soft ⟨fruit⟩; **jmdn.** ~ **machen** *(fig.)* wear sb. down

Murmel die; ~, ~n marble

murmeln *tr., itr. V.* mumble; mutter; *(sehr leise)* murmur

Murmel·tier das marmot

murren *itr. V.* grumble; **mürrisch 1.** *Adj.* grumpy; **2.** *adv.* grumpily

Mus das *od.* der; ~es, ~e purée
Muschel die; ~, ~n a) mussel; *(Schale)* [mussel-]shell; b) *(am Telefon)* *(Hör~)* ear-piece; *(Sprech~)* mouth-piece
Muse die; ~, ~n muse
Museum das; ~s, Museen museum
Musik die; ~, ~en music; **musikalisch** 1. *Adj.* musical; 2. *adv.* musically; **Musikant** der; ~en, ~en musician; **Musik·box** die juke-box; **Musiker** der; ~s, ~, **Musikerin** die; ~, ~nen musician
Musik-: ~**hochschule** die college of music; ~**instrument** das musical instrument; ~**stunde** die music-lesson
musisch 1. *Adj.* artistic; *(education)* in the arts; 2. *adv.* artistically; **musizieren** *itr. V.* play music; *(bes. unter Laien)* make music
Muskat der; ~[e]s, ~e nutmeg; **Muskat·nuß** die nutmeg
Muskel der; ~s, ~n muscle
Muskel-: ~**kater** der stiff muscles *pl.;* ~**protz** der *(ugs.)* muscleman
Muskulatur die; ~, ~en musculature; muscular system; **muskulös** *Adj.* muscular
Müsli das; ~s, ~s muesli
muß *1. u. 3. Pers. Sg. Präsens v.* müssen; **Muß** das; ~: necessity; must *(coll.)*
Muße die; ~: leisure
müssen 1. *unr. Modalverb;* 2. *Part.* ~ a) have to; **er muß es tun** he must do it; he has to *or (coll.)* has got to do it; **das muß 1968 gewesen sein** it must have been in 1968; **er muß gleich hier sein** he will be here at any moment; b) *Konjunktiv II* **es müßte doch möglich sein** it ought to be possible; **reich müßte man sein!** how nice it would be to be rich!; 2. *unr. itr. V.* **ich muß nach Hause** I have to *or* must go home; **ich muß mal** *(fam.)* I need to spend a penny *(Brit. coll.) or (Amer. coll.)* go to the john
müßig 1. *Adj.* idle *(person);* *(hours, weeks, life)* of leisure; 2. *adv.* idly; **Müßig·gang** der *o. Pl.* leisure; *(Untätigkeit)* idleness
müßte *1. u. 3. Pers. Sg. Prät. v.* müssen
Muster das; ~s, ~ a) *(Vorlage)* pattern; b) *(Vorbild)* model **(an** + *Dat.* of); c) *(Verzierung)* pattern; d) *(Warenprobe)* sample; **muster·gültig** 1. *Adj.* exemplary; impeccable *(order);* 2. *adv.* in an exemplary fashion
mustern *tr. V.* a) eye; b) *(Milit.: ärzt-*

lich untersuchen) jmdn. ~: give sb. his medical; **Musterung** die; ~, ~en a) scrutiny; b) *(Milit.: von Wehrpflichtigen)* medical examination; medical
Mut der; ~[e]s courage; **mutig** 1. *Adj.* brave; 2. *adv.* bravely; **mut·los** *Adj.* dejected; *(entmutigt)* disheartened; **Mut·losigkeit** die; ~: dejection
mutmaßlich *Adj.* supposed; suspected *(murderer etc.)*
Mut·probe die test of courage
¹**Mutter** die; ~, Mütter mother;
²**Mutter** die; ~, ~n nut; **mütterlich** 1. *Adj.* a) maternal *(line, love, instincts, etc.);* b) *(fürsorglich)* motherly *(woman, care);* 2. *adv.* in a motherly way; **mütterlicher·seits** *Adv.* on the/ his/her *etc.* mother's side
Mutter-: ~**liebe** die motherly love *no art.;* ~**mal** das; *Pl.* ~**male** birthmark
Mutterschaft die; ~: motherhood
mutter-, Mutter-: ~**seelen·allein** *Adj.* all alone; ~**söhnchen** das mummy's *or (Amer.)* mama's boy; ~**sprache** die mother tongue; ~**tag** der; *o. Pl.* Mother's Day *no def. art.*
Mutti die; ~, ~s mummy *(Brit. coll.);* mum *(Brit. coll.);* mommy *(Amer. coll.);* mom *(Amer. coll.)*
mut·willig 1. *Adj.* wilful; wanton *(destruction);* 2. *adv.* wilfully
Mütze die; ~, ~n cap
MW *Abk. (Rundf.)* Mittelwelle MW
Mw.-St., MwSt. *Abk.* Mehrwertsteuer VAT
mysteriös 1. *Adj.* mysterious; 2. *adv.* mysteriously; **Mystik** die; ~: mysticism
Mythologie die; ~, ~n mythology; **Mythos** der; ~, Mythen myth

N

n, N [ɛn] das; ~, ~: n/N
N *Abk.* Nord[en] N
na *Interj. (ugs.)* well; **na so [et]was!** well I never!; **na und?** *(wennschon)* so what?; *(beschwichtigend)* **na, na, na!** now, now, come along; *(triumphierend)* **na also!** there you are!; *(unsi-*

cher) **na, ich weiß nicht** hmm, I'm not sure; *(ärgerlich)* **na, was soll das denn?** now what's all this about?; *(drohend)* **na warte!** just [you] wait!

Nabel der; ~s, ~: navel; **Nabelschnur** die umbilical cord

nach 1. *Präp. mit Dat.* a) *(räumlich)* to; **der Zug ~ München** the train for Munich *or* the Munich train; **~ Hause gehen** go home; **~ Osten |zu|** eastwards; [towards the] east; b) *(zeitlich)* after; **zehn |Minuten| ~ zwei** ten [minutes] past two; c) *(mit bestimmten Verben, bezeichnet das Ziel der Handlung)* for; d) *(bezeichnet [räumliche und zeitliche] Reihenfolge)* after; **~ Ihnen/dir!** after you; e) *(gemäß)* according to; **~ meiner Ansicht** *od.* **Meinung, meiner Ansicht** *od.* **Meinung ~:** in my view *or* opinion; **~ der neusten Mode gekleidet** dressed in [accordance with] the latest fashion; **dem Gesetz ~:** in accordance with the law; by law; **etw. schmecken/riechen** taste/smell of sth.; 2. *Adv.* a) *(räumlich)* **|alle| mir ~!** [everybody] follow me!; b) *(zeitlich)* **~ und ~:** little by little; gradually; **~ wie vor** still

nach|ahmen *tr. V.* imitate; **Nachahmung** die; ~, ~en imitation

Nachbar der; ~n , ~n neighbour; **Nachbarhaus** das house next door; **Nachbarin** die; ~, ~nen neighbour; **Nachbarschaft** die; ~ a) the whole neighbourhood; b) *(Beziehungen)* **gute ~:** good neighbourliness; c) *(Gegend)* neighbourhood; *(Nähe)* vicinity

nach|bestellen *tr. V.* **|noch| etw. ~:** order more of sth.; *(shop)* reorder sth.

Nachbildung die a) *o. Pl.* copying; b) *(Gegenstand)* copy

nach|blicken *tr. V. (geh.)* **jmdm./einer Sache ~:** gaze after sb./sth.

nach|datieren *tr. V.* backdate

nachdem *Konj.* a) after; b) *s.* ¹**je 3 b**

nach|denken *unr. itr. V.* think; **denk mal |gut** *od.* **scharf| nach** have a [good] think; **Nachdenken** das thought; **nachdenklich** 1. *Adj.* thoughtful; 2. *adv.* thoughtfully

Nachdruck der; *Pl.* ~e a) *o. Pl.* **mit ~:** emphatically; b) *(Druckw.)* reprint; **nachdrücklich** 1. *Adj.* emphatic; 2. *adv.* emphatically

nach|eifern *itr. V.* **jmdm. ~:** emulate sb.

nacheinander *Adv.* one after the other

nach|empfinden *unr. tr. V.* empathize with *(feeling)*; share *(delight, sorrow)*

Nach·erzählung die retelling [of a story]; *(Schulw.)* reproduction

Nachfahr der; ~en , ~en *(geh.)* descendant

Nach·folge die succession; **Nachfolger** der; ~s, ~, **Nachfolgerin** die; ~, ~nen successor

Nach·forschung die investigation

Nach·frage die demand **(nach** for)

nach|fühlen *tr. V.* empathize with

nach|füllen *tr. V.* top up; **Salz/Wein ~:** put [some] more salt/wine in

nach|geben *unr. itr. V.* give way

Nach·gebühr die excess postage

nach|gehen *unr. itr. V.; mit sein* a) **jmdm./einer Sache ~:** follow sb./sth.; **einer Sache ~** *(fig.)* look into a matter; **einem Beruf ~:** practise a profession; b) *(nicht aus dem Kopf gehen)* **jmdm. ~:** remain on sb.'s mind; c) *(clock, watch)* be slow; **|um| eine Stunde ~:** be an hour slow

Nach·geschmack der after-taste

nach·giebig *Adj.* indulgent; **Nachgiebigkeit** die; ~: indulgence

nach·haltig 1. *Adj.* lasting; 2. *adv.* *(auf längere Zeit)* for a long time

Nach·hause·weg der way home

nach|helfen *unr. itr. V.* help

nach·her [*auch:* '--] *Adv.* afterwards; *(später)* later [on]; **bis ~!** see you later!

Nachhilfe·unterricht der coaching

nach|holen *tr. V. (nachträglich erledigen)* catch up on *(work, sleep)*; make up for *(working hours missed)*

Nachkomme der; ~n, ~n descendant; **nach|kommen** *unr. itr. V.; mit sein* follow [later]; come [on] later; **Nachkommenschaft** die; ~: descendants *pl.*; **Nachkömmling** der; ~s, ~e much younger child *(than the rest)*

Nach·kriegs- post-war *(generation, period, etc.)*

Nach·laß der; **Nachlasses, Nachlasse** *od.* **Nachlässe** a) estate; b) *(Kaufmannsspr.: Rabatt)* discount; **nach|lassen** 1. *unr. itr. V.* let up; *(pain, stress, pressure)* ease; *(effect)* wear off; *(interest, enthusiasm, strength, courage)* wane; *(health, hearing, memory)* deteriorate; *(business)* drop off; 2. *unr. tr. V. (Kaufmannsspr.)* give a discount of; **nach·lässig** 1. *Adj.* careless; 2. *adv.* carelessly; **Nachlässigkeit** die; ~, ~en carelessness

nach|laufen *unr. itr. V.; mit sein* jmdm./einer Sache ~: run after sb./ sth.

nach|lesen *unr. tr. V.* look up

nach|lösen 1. *tr. V.* eine Fahrkarte ~: buy a ticket [on the train, bus, etc.]; **2.** *itr. V.* pay the excess [fare]

nach|machen *tr. V. (auch tun)* copy; *(imitieren)* imitate; *(genauso herstellen)* reproduce ⟨*period furniture etc.*⟩; forge ⟨*signature*⟩

nach·mittag *Adv.* heute ~: this afternoon; |am| Sonntag ~: on Sunday afternoon; **Nach·mittag** der afternoon; **am** ~: in the afternoon; **am späten** ~: late in the afternoon; **nachmittags** *Adv.* in the afternoon; dienstags *od.* Dienstag ~: on Tuesday afternoons; um vier Uhr ~: at four in the afternoon; at 4 p.m.

Nachnahme die; ~, ~n: per ~: cash on delivery; COD

Nach·name der surname

nachprüfbar *Adj.* verifiable; **nach|prüfen** *tr., itr. V.* check

nach|rechnen *tr. V.* check ⟨*figures*⟩

Nach·rede die: üble ~: malicious gossip; *(Rechtsw.)* defamation [of character]

Nachricht die; ~, ~en a) news *no pl.;* eine ~ hinterlassen leave a message; b) *Pl. (Ferns., Rundf.)* news *sing.;* ~en hören listen to the news

Nachrichten-: ~sprecher der, ~sprecherin die news-reader

nach|rücken *itr. V.; mit sein* move up

Nach·ruf der; ~[e]s, ~e obituary (auf + Akk. of); **nach|rufen** *unr. tr., itr. V.* jmdm. |etw.| ~: call [sth.] after sb.

nach|sagen *tr. V.* a) *(wiederholen)* repeat; b) man sagt ihm nach, er sei ...: he is said to be ...; jmdm. Schlechtes ~: speak ill of sb.

Nach·saison die late season

nach|schicken *tr. V.* a) *(durch die Post o. ä.)* forward; b) jmdm. jmdn. ~: send sb. after sb.

nach|schlagen 1. *unr. tr. V.* look up; **2.** *unr. itr. V.* im Lexikon/Wörterbuch ~: consult the encyclopaedia/dictionary; **Nachschlage·werk** das work of reference

Nach·schlüssel der duplicate key

Nach·schub der *(Milit.)* a) supply (an + Dat. of); b) *(~material)* supplies *pl.* (an + Dat. of)

nach|sehen 1. *unr. itr. V.* a) jmdm./einer Sache ~: gaze after sb./sth.; b) *(kontrollieren)* check; c) *(nachschla-*

gen) have a look; **2.** *unr. tr. V.* a) *(nachlesen)* look up; b) *(überprüfen)* check [over]

nach|senden *unr. od. regelm. tr. V.* forward

Nach·sicht die leniency; **nachsichtig 1.** *Adj.* lenient (gegen, mit towards); **2.** *adv.* leniently

nach|sitzen *unr. itr. V.* be in detention; |eine Stunde| ~ müssen have [an hour's] detention

Nach·speise die dessert; sweet

Nach·spiel das: die Sache wird noch ein ~ haben this affair will have repercussions; ein gerichtliches ~ haben result in court proceedings

nach|sprechen *unr. tr. V.* |jmdm.| etw. ~: repeat sth. [after sb.]

nächst... **1.** *Sup. zu* nahe; **2.** *Adj.* next; *(kürzest)* shortest ⟨*way*⟩; am ~en Tag the next day; beim ~en Mal, das ~e Mal the next time; der ~e bitte! next [one], please; wer kommt als ~er dran? whose turn is it next?; **Nächste** der; ~n, ~n *(geh.)* neighbour; **Nächsten·liebe** die charity [to one's neighbour]; **nächstens** *Adv.* a) shortly; b) *(ugs.: wenn es so weitergeht)* if it goes on like this

nächst-: ~liegend *Adj.; nicht präd.* first, immediate ⟨*problem*⟩; [most] obvious ⟨*explanation etc.*⟩; ~möglich *Adj.* earliest possible

nach|suchen *itr. V. (geh.)* um etw. ~: request sth.; *(bes. schriftlich)* apply for sth.

nacht *Adv.* gestern/morgen/Dienstag ~: last night/tomorrow night/on Tuesday night; heute ~: tonight; **Nacht** die; ~, Nächte night; bei ~, in der ~: at night[-time]; über ~ bleiben stay overnight

Nacht-: ~arbeit die; *o. Pl.* night work *no art.;* ~dienst der night duty; ~dienst haben be on night duty; ⟨*chemist's shop*⟩ be open late

Nach·teil der disadvantage; **nachteilig 1.** *Adj.* detrimental; harmful; **2.** *adv.* detrimentally; harmfully

Nacht-: ~essen das *(bes. südd., schweiz.)* s. Abendessen; ~hemd das night-shirt

Nachtigall die; ~, ~en nightingale

Nach·tisch der; *o. Pl.* dessert; sweet

nächtlich *Adj.* nocturnal; night ⟨*sky*⟩; ⟨*darkness, stillness*⟩ of the night; **Nacht·lokal** das night-spot *(coll.)*

nach|tragen *unr. tr. V. (schriftlich ergänzen)* insert; add; **nach·tragend**

Adj. unforgiving; *(rachsüchtig)* vindictive; **nachträglich 1.** *Adj.* later; subsequent ⟨*apology*⟩; *(verspätet)* belated ⟨*greetings, apology*⟩; **2.** *adv.* afterwards; subsequently; *(verspätet)* belatedly

nach|trauern *itr. V.* jmdm./einer Sache ~: bemoan the passing of sb./sth.

Nacht·ruhe die night's sleep; **nachts** *Adv.* at night; **Montag** *od.* **montags** ~: on Monday nights; **um 3 Uhr** ~: at ·3 o'clock in the morning

Nacht-: ~**schicht die** night-shift; ~**schwester die** night nurse; ~**tisch der** bedside table; ~**tischlampe die** bedside light; ~**topf der** chamber-pot; ~**wächter der** nightwatchman

Nach·untersuchung die follow-up examination; check-up

nach|vollziehen *unr. tr. V.* reconstruct; *(begreifen)* comprehend

nach|wachsen *unr. itr. V.; mit sein* |wieder| ~: grow again

Nach·wehen *Pl. (Med.)* afterpains; *(fig. geh.)* unpleasant after-effects

Nachweis der; ~**es,** ~**e** proof *no indef. art.* (*Gen.,* **über** + *Akk.* of); *(Zeugnis)* certificate (**über** + *Akk.* of); **nachweisbar 1.** *Adj.* demonstrable ⟨*fact, truth, error, defect, guilt*⟩; detectable ⟨*substance, chemical*⟩; **2.** *adv.* demonstrably; **nach|weisen** *unr. tr. V.* prove; **nachweislich** *Adv.* as can be proved

nach|winken *itr. V.* jmdm./einer Sache ~: wave after sb./sth.

Nach·wirkung die after-effect

Nach·wort das; *Pl.* ~**worte** afterword

Nach·wuchs der; *o. Pl* **a)** *(fam.: Kind[er])* offspring; **b)** *(junge Kräfte)* new blood; *(für eine Branche usw.)* new recruits *pl.; (in der Ausbildung)* trainees *pl.*

nach|zahlen *tr., itr. V.* **a)** pay later; **b)** *(zusätzlich zahlen)* **25 DM** ~: pay another 25 marks

nach|zählen *tr., itr. V.* [re]count

Nach·zahlung die additional payment

Nachzügler der straggler; *(spät Ankommender)* latecomer

Nackedei der; ~**s,** ~**s** *(fam. scherzh.)* |kleiner| ~: naked little thing

Nacken der; ~**s,** ~: back *or* nape of the neck; *(Hals)* neck

nackt *Adj.* naked; bare ⟨*feet, legs, arms, skin, fists*⟩; *(fig.)* plain ⟨*truth, fact*⟩; bare ⟨*existence*⟩; **Nackt·ba-**

de·strand der nudist beach; **Nackte der/die;** *adj. Dekl.* naked man/woman; **Nackt·foto das** nude photo

Nadel die; ~, ~**n** needle; *(Steck~, Hut~, Haar~)* pin

Nadel-: ~**baum der** conifer; coniferous tree; ~**wald der** coniferous forest

Nagel der; ~**s,** Nägel nail; **den** ~ **auf den Kopf treffen** *(fig. ugs.)* hit the nail on the head *(coll.)*

Nagel-: ~**bürste die** nailbrush; ~**feile die** nail-file; ~**lack der** nail varnish *(Brit.);* nail polish

nageln *tr. V.* nail **(an** + *Akk.* to, **auf** + *Akk.* on); *(Med.)* pin; **nagel·neu** *Adj. (ugs.)* brand-new; **Nagel·schere die** nail-scissors *pl.*

nagen 1. *itr. V.* gnaw; **an etw.** *(Dat.)* ~: gnaw [at] sth.; **2.** *tr. V.* gnaw off; **ein Loch ins Holz** ~: gnaw a hole in the wood

nah *s.* **nahe**

Näh·aufnahme die *(Fot.)* close-up [photograph]

nahe ['naːə]; **näher** ['nɛːɐ], **nächst...** **1.** *Adj.* **a)** *(räumlich)* near *pred.;* close *pred.;* nearby *attrib.;* **b)** *(zeitlich)* imminent; near *pred.;* **c)** *(eng)* close ⟨*relationship etc.*⟩; **2.** *adv.* **a)** *(räumlich)* ~ **an** (+ *Dat./Akk.*), ~ **bei** close to; ~ **gelegen** nearby; **von** ~**m** from close up; **b)** *(zeitlich)* ~ **an die achtzig** *(ugs.)* pushing eighty *(coll.);* **c)** *(eng)* closely; **3.** *Präp. mit Dat. (geh.)* near; close to; **Nähe die;** ~: closeness

nahe-: ~**bei** *Adv.* nearby; close by; ~|**gehen** *unr. itr. V.; mit sein* jmdm. ~**gehen** affect sb. deeply; ~|**kommen** *unr. itr. V.; mit sein* einer Sache *(Dat.)* ~**kommen** come close to sth.; ⟨*amount*⟩ approximate to sth.; jmdm./sich |menschlich| ~**kommen** get to know sb./one another well; ~|**legen** *tr. V.* suggest; give rise to ⟨*suspicion, supposition, thought*⟩; ~|**liegen** *unr. itr. V.* ⟨*thought*⟩ suggest itself; ⟨*suspicion, question*⟩ arise; ~**liegend** *Adj.* obvious ⟨*reason, solution*⟩

nähen 1. *itr. V.* sew; *(Kleider machen)* make clothes; **2.** *tr. V.* **a)** sew ⟨*seam, hem*⟩; make ⟨*dress etc.*⟩; **b)** *(Med.)* stitch

näher 1. *Komp. zu* **nahe; 2.** *Adj.* **a)** *(kürzer)* shorter ⟨*way, road*⟩; **b)** *(genauer)* more precise ⟨*information*⟩; closer ⟨*investigation, inspection*⟩; **3.** *adv.* **a)** **bitte treten Sie** ~! please come in/nearer/this way; **b)** *(genauer)* more closely; *(im einzelnen)* in [more] detail

näher|kommen *unr. itr. V.; mit sein*
jmdm. |menschlich| ~kommen get on
closer terms with sb.; **nähern** *refl. V.*
approach; sich jmdm./einer Sache ~:
approach sb./sth.

nahe-: ~|**stehen** *unr. itr. V.* jmdm.
~stehen be on intimate terms with sb.;
~**zu** *Adv.* almost; nearly; *(mit Zahlenangabe)* close on

Näh-: ~**garn** das [sewing] cotton;
~**kasten** der sewing-box

nahm *1. u. 3. Pers. Sg. Prät. v.* nehmen
Näh-: ~**maschine** die sewing-
machine; ~**nadel** die sewing-needle
nähren 1. *tr. V.* feed (mit on); 2. *refl.
V.* *(geh.)* sich von etw. ~: live on sth.;
⟨animal⟩ feed on sth.; **nahrhaft** *Adj.*
nourishing; **Nahrung** die; ~: food;
Nahrungs·mittel das food [item];
~**mittel** *Pl.* foodstuffs; **Nähr·wert**
der nutritional value

Näh·seide die sewing silk
Naht die; ~, **Nähte** seam
Nah-: ~**verkehr** der local traffic;
~**verkehrs·zug** der local train
Näh·zeug das sewing things *pl.*
naiv 1. *Adj.* naïve; 2. *adv.* naïvely;
Naivität die; ~: naïvety
Name der; ~ns, ~n name; **namens**
Adv. by the name of
Namens-: ~**schild** das a) *(an Türen
usw.)* name-plate; b) *(zum Anstecken)*
name-badge; ~**tag** der name-day
namentlich 1. *Adj.* by name *postpos.;*
2. *adv.* by name; 3. *Adv. (besonders)*
particularly; **namhaft** *Adj.* a) *(berühmt)* noted; b) *(ansehnlich)* note-
worthy ⟨sum, difference⟩; notable
⟨contribution, opportunity⟩; **nämlich**
Adv. a) er kann nicht kommen, er ist ~
krank he cannot come, as he is ill; b)
(und zwar) namely
nannte *1. u. 3. Pers. Sg. Prät. v.* nennen
nanu *Interj.* ~, was machst du denn
hier? hello, what are you doing here?;
~, Sie gehen schon? what, you're
going already?
Napf der; ~|e|s, **Näpfe** bowl *(esp. for
animal's food)*
Narbe die; ~, ~n scar; **narbig** *Adj.*
scarred
Narkose die; ~, ~n *(Med.)* narcosis
Narr der; ~en, ~en fool; **Narren-
freiheit** die freedom to do as one
pleases; **Närrin** die; ~, ~nen fool;
närrisch 1. *Adj.* crazy; carnival-
crazy ⟨season⟩; 2. *adv.* crazily
Narzisse die; ~, ~n narcissus

naschen 1. *itr. V. (Süßes essen)* eat
sweet things; *(heimlich essen)* have a
nibble; 2. *tr. V.* eat ⟨sweets, chocolate,
etc.⟩; er hat Milch genascht he has
been at the milk; **naschhaft** *Adj.*
sweet-toothed; ~ sein have a sweet
tooth
Nase die; ~, ~n nose; die ~ voll haben
(ugs.) have had enough
Nasen-: ~**bluten** das; ~s bleeding
from the nose; ~**loch** das nostril;
~**tropfen** *Pl.* nose-drops
nase·weis 1. *Adj.* precocious; pert
⟨remark, reply⟩; 2. *adv.* precociously;
Nas·horn das rhinoceros
naß; nasser *od.* nässer, nassest... *od.*
nässest...; *Adj.* wet; sich/das Bett ~
machen wet oneself/one's bed; **Nässe** die; ~: wetness; **naß·kalt** *Adj.*
cold and wet; **Naß·rasur** die wet
shaving *no art.*
Nation die; ~, ~en nation; **national**
1. *Adj.* a) national; 2. *adv.* nationally
National-: ~**elf** die *(Fußball)* national
side; ~**hymne** die national anthem
Nationalismus der; ~: nationalism
usu. no art.; **nationalistisch** 1. *Adj.*
nationalist; nationalistic; 2. *adv.* na-
tionalistically; **Nationalität** die; ~,
~en nationality
national-, National-: ~**mannschaft** die national team; ~**sozialismus** der National Socialism; ~**sozialist** der National Socialist; ~**sozialistisch** *Adj.* National Socialist
NATO, Nato die; ~: NATO, Nato *no
art.*
Natron das; ~s |doppeltkohlensaures|
~: sodium bicarbonate; |kohlensau-
res| ~: sodium carbonate
Natter die; ~, ~n colubrid
Natur die; ~, ~en nature; die freie ~:
[the] open countryside; **Naturalien**
[natu'ra:liən] *Pl.* natural produce *sing.*
(used as payment); in ~ *(Dat.)* bezahlen pay in kind; **Naturalismus** der;
~: naturalism; **naturalistisch** 1.
Adj. naturalistic; 2. *adv.* naturalist-
ically; **Naturell** das; ~s, ~e temperament
natur-, Natur-: ~**erscheinung** die
natural phenomenon; ~**farben** *Adj.*
natural-coloured; ~**freund** der
nature-lover; ~**gemäß** *Adv.* naturally; ~**geschichte** die; *o. Pl.* natural
history; ~**gesetz** das law of nature;
~**getreu** 1. *Adj.* lifelike ⟨portrait, im-
itation⟩; faithful ⟨reproduction⟩; 2.
adv. ⟨draw⟩ true to life; ⟨reproduce⟩

faithfully; ~**heilkunde die** naturo-
pathy *no art.;* ~**katastrophe die**
natural disaster
natürlich 1. *Adj.* natural; **2.** *adv.*
⟨*laugh, behave*⟩ naturally; **3.** *Adv.* **a)**
(selbstverständlich, wie erwartet)
naturally; of course; **b)** *(zwar)* of
course; **Natürlichkeit die;** ~: naturalness
Natur-: ~**park der** ≈ national park;
~**produkt das** natural product;
~**schutz der** [nature] conservation;
~**schutz·gebiet das** nature reserve;
~**verbunden** *Adj.* ⟨*person*⟩ in tune
with nature; ~**volk das** primitive
people; ~**wissenschaft die** natural
science *no art.;* ~**wissenschaftler
der** [natural] scientist; ~**wissenschaftlich 1.** *Adj.* scientific; **2.** *adv.*
scientifically; ~**wunder das** miracle
or wonder of nature
Navigation die; ~: navigation *no art.*
n. Chr. *Abk.* nach Christus AD
Neandertaler der; ~s, ~: Neander-
thal man
Nebel der; ~s, ~: fog; *(weniger dicht)*
mist; **nebelig** *s.* neblig
Nebel-: ~**scheinwerfer der** fog-
lamp; ~**wand die** wall of fog
neben 1. *Präp. mit Dat.* **a)** *(Lage)* next
to; beside; **b)** *(außer)* apart from;
aside from *(Amer.);* **c)** *(verglichen mit)*
beside; **2.** *Präp. mit Akk. (Richtung)*
next to; beside; **neben·an** *Adv.* next
door; **neben·bei** *Adv.* **a)** ⟨*work*⟩ on
the side; *(zusätzlich)* as well; **b)** *(beiläufig)* ⟨*remark, ask*⟩ by the way;
⟨*mention*⟩ in passing
neben-, Neben-: ~**beruf der** second
job; sideline; ~**beruflich 1.** *Adj.* **eine**
~**berufliche Tätigkeit** a second job; **2.**
adv. on the side; **er arbeitet** ~**beruflich als Übersetzer** he translates as a
sideline; ~**beschäftigung die** second job; sideline; ~**buhler der,**
~**buhlerin die** rival
neben·einander *Adv.* **a)** next to each
other; *(fig.: zusammen)* ⟨*live, exist*⟩
side by side; ~ **wohnen** live next door
to each other; **b)** *(gleichzeitig)*
together
nebeneinander-: ~|**legen** *tr. V.* lay
or place ⟨*objects*⟩ side by side; ~|**set-
zen** *tr. V.* put *or* place ⟨*persons, ob-
jects*⟩ next to each other; ~|**sitzen**
unr. itr. V. sit next to each other;
~|**stellen** *tr. V.* put *or* place ⟨*tables,
chairs, etc.*⟩ next to each other
Neben-: ~**erwerb der** secondary oc-

cupation; ~**fach das** subsidiary subject; minor *(Amer.);* ~**fluß der** tributary; ~**gebäude das a)** annexe; out-
building; **b)** *(Nachbargebäude)* neighbouring building; ~**geräusch das**
background noise; ~**haus das** house
next door
neben·her *Adv. s.* nebenbei
nebenher-: ~|**fahren** *unr. itr. V.; mit
sein* drive/ride alongside; ~|**gehen**
unr. itr. V.; mit sein walk alongside
neben-, Neben-: ~**kosten** *Pl.* **a)** additional costs; **b)** *(bei Mieten)* heating,
lighting, and services; ~**produkt das**
by-product; ~**rolle die** supporting
role; ~**sache die** minor matter; ~**sachen** inessentials; ~**sächlich** *Adj.* of
minor importance *postpos.;* unimportant; minor ⟨*detail*⟩; ~**satz der**
(Sprachw.) subordinate clause;
~**straße die** side street; ~**tätigkeit
die** second job; sideline; ~**tisch der**
next table; ~**verdienst der** additional income; ~**wirkung die** side-
effect; ~**zimmer das** next room
neblig *Adj.* foggy; *(weniger dicht)*
misty
Necessaire [nesɛ'sɛːɐ̯] **das;** ~s, ~s
sponge-bag *(Brit.);* toilet bag *(Amer.)*
necken *tr. V.* tease; **Neckerei die;**
~: teasing
nee *(ugs.)* no; nope *(Amer. coll.)*
Neffe der; ~n, ~n nephew
negativ 1. *Adj.* negative; **2.** *adv.* ⟨*answer*⟩ in the negative; **Negativ das;**
~s, ~e *(Fot.)* negative
Neger der; ~s, ~: Negro; **Negerin
die;** ~, ~**nen** Negress
nehmen *unr. tr. V.* take; **sich** *(Dat.)*
etw. ~: take sth.; *(sich bedienen)* help
oneself to sth.; **auf sich** *(Akk.)* ~: take
on ⟨*responsibility, burden*⟩; **jmdm./ei-
ner Sache etw.** ~: deprive sb./sth. of
sth.; **was nehmen Sie dafür?** how much
do you charge for it?
Neid der; ~|e|s envy; jealousy;
neiden *tr. V. (geh.)* **jmdm. etw.** ~:
envy sb. [for] sth.; **neidisch 1.** *Adj.*
envious; **2.** *adv.* enviously
neigen 1. *tr. V.* tip; tilt; incline ⟨*head,
upper part of body*⟩; **2.** *refl. V.:* ⟨*person*⟩ lean; ⟨*ship*⟩ heel over, list;
⟨*scales*⟩ tip; **3.** *itr. V.* **a) zu Erkältun-
gen/Krankheiten** ~: be prone to
colds/illnesses; **b)** *(tendieren)* tend;
Neigung die; ~, ~**en a)** *(Vorliebe)* inclination; **b)** *o. Pl. (Tendenz)* tendency
nein *Interj.* no; **Nein das;** ~|s|, ~|s|
no; **Nein·stimme die** no-vote

Nektar der; ~s, ~e *(Bot.)* nectar; **Nektarine** die; ~, ~n nectarine

Nelke die; ~, ~n a) pink; *(Dianthus caryophyllus)* carnation; b) *(Gewürz)* clove

nennen 1. *unr. tr. V.* a) call; b) *(angeben)* give ⟨*name, date of birth, address, reason, price, etc.*⟩; c) *(anführen)* give ⟨*example*⟩; *(erwähnen)* mention ⟨*person, name*⟩; 2. *unr. refl. V.* ⟨*person, thing*⟩ be called

neo-, Neo-: neo-

Neon das; ~s neon

Neon-: ~**licht** das neon light; ~**röhre** die neon tube

Nepal (das); ~s Nepal

Nepp der; ~s *(ugs. abwertend)* daylight robbery *no art.;* rip-off *(sl.);* **Nepp·lokal** das *(ugs. abwertend)* clip-joint *(sl.)*

Nerv der; ~s, ~en nerve; **die ~en verlieren** lose control [of oneself]; **jmdm. auf die ~en gehen** *od.* **fallen** get on sb.'s nerves

nerven-, Nerven-: ~**aufreibend** *Adj.* nerve-racking; ~**bündel** das *(ugs.)* bundle of nerves *(coll.);* ~**gift** das neurotoxin; ~**heilanstalt** die *(veralt.)* psychiatric hospital; ~**krank** *Adj.* ⟨*person*⟩ suffering from a nervous disease; ~**probe** die mental trial; ~**säge** die *(salopp)* pain in the neck *(coll.);* ~**zusammenbruch** der nervous breakdown

nervlich *Adj.* nervous ⟨*strain*⟩; **nervös** 1. *Adj. (auch Med.)* nervous; jittery ⟨*person*⟩; 2. *adv.* nervously; **Nervosität** die; ~ nervousness; **nervtötend** *Adj.* nerve-racking ⟨*wait*⟩; soul-destroying ⟨*activity, work*⟩

Nerz der; ~es, ~e mink; **Nerz·mantel** der mink coat

Nessel die; ~, ~n nettle

Nest das; ~[e]s, ~er a) nest; b) *(fam.: Bett)* bed; c) *(ugs. abwertend: kleiner Ort)* little place

nett 1. *Adj.* nice; *(freundlich)* kind; 2. *adv.* nicely; *(freundlich)* nicely; kindly; **netter·weise** *Adv.* kindly

netto *Adv.* ⟨*weigh, earn, etc.*⟩ net

Netto-: ~**einkommen** das net income; ~**gehalt** das net salary

Netz das; ~es, ~e a) net; *(Einkaufs~)* string bag; *(Gepäck~)* [luggage-]rack; b) *(Spinnen~)* web; c) *(Verteiler~, Verkehrs~ usw.)* network; *(für Strom, Wasser, Gas)* mains *pl.;* **Netz·haut** die *(Anat.)* retina

neu 1. *Adj.* new; **die ~este Mode** the latest fashion; **das ist mir ~:** that is news to me; **der/die Neue** the new man/woman/boy/girl; 2. *adv.* a) ~ **tapeziert/gestrichen** repapered/repainted; **sich ~ einrichten** refurnish one's home; b) *(gerade erst)* **diese Ware ist ~ eingetroffen** this item has just come in; **neu·artig** *Adj.* new; **Neu·bau** der; *Pl.* Neubauten new house/building; **Neubau·wohnung** die flat *(Brit.)* or *(Amer.)* apartment in a new block/house

neuerdings *Adv.* **er trägt ~ eine Brille** he has recently started wearing glasses; **neu·eröffnet** *Adj.* a) newly-opened; b) *(wiedereröffnet)* reopened; **Neu·eröffnung** die a) opening; b) *(Wiedereröffnung)* reopening; **Neuerung** die; ~, ~en innovation; **neu·geboren** *Adj.* newborn; **Neu·gier, Neugierde** die; ~: curiosity; *(Wißbegierde)* inquisitiveness; **neu·gierig** 1. *Adj.* curious; inquisitive; inquisitive ⟨*person*⟩; **ich bin ~, was er dazu sagt** I'm curious to know what he'll say about it; 2. *adv.* ⟨*ask*⟩ inquisitively; ⟨*peer*⟩ nosily *(coll. derog.);* **Neuheit** die; ~, ~en a) *o. Pl.* novelty; b) *(Neues)* new product/gadget/article *etc.;* **Neuigkeit** die; ~, ~en piece of news; ~en news *sing.;* **Neu·jahr** das New Year's Day; **Neu·land** das *(fig.)* new ground; **neulich** *Adv.* recently; ~ **morgens** the other morning; **Neuling** der; ~s, ~e newcomer; *(auf einem Gebiet)* novice; **Neu·mond** der new moon

neun *Kardinalz.* nine; **Neun** die; ~, ~en nine

neun-: ~**hundert** *Kardinalz.* nine hundred; ~**jährig** *Adj.* (9 Jahre alt) nine-year-old *attrib.;* (9 Jahre dauernd) nine-year *attrib.;* ~**mal** *Adv.* nine times

neunt... *Ordinalz.* ninth

neun·tausend *Kardinalz.* nine thousand; **Neuntel** das *(schweiz. meist* der*);* ~s, ~: ninth; **neuntens** *Adv.* ninthly; **neun·zehn** *Kardinalz.* nineteen; **neunzig** *Kardinalz.* ninety; **neunziger** *indekl. Adj.; nicht präd.* **die ~ Jahre** the nineties; **neunzigst...** *Ordinalz.* ninetieth

neu·reich *Adj.* nouveau riche

Neurose die; ~, ~n neurosis; **neurotisch** *Adj.* neurotic

Neu·see·land (das); ~s New Zealand; **Neuseeländer** der; ~s, ~: New Zealander

neutral 1. *Adj.* neutral; 2. *adv.* sich ~ verhalten remain neutral; **Neutralität die;** ~, ~en neutrality; **Neutron das;** ~s, ~en neutron; **Neutrum das;** ~s, Neutra *(österr. nur so) od.* Neutren *(Sprachw.)* neuter

neu-, Neu-: ~wert der value when new; ~wertig *Adj.* as new; ~zeit die; *o, Pl.* modern age; ~zeitlich *Adj.* modern

nicht *Adv.* **a)** not; ~! [no,] don't!; ~|wahr|? isn't it/he/she *etc.;* don't you/we/they *etc.;* du magst das, ~ |wahr|? you like that, don't you?; was du ~ sagst! you don't say!

nicht-, Nicht-: non-

Nicht·angriffs·pakt der non-aggression pact

Nichte die; ~, ~n niece

nichtig *Adj.* **a)** *(geh.)* vain *(things, pleasures, etc.);* trivial *(reason);* **b)** *(Rechtsspr.)* void; **Nicht·raucher** der non-smoker; „~raucher" 'no smoking'; **nicht·rostend** *Adj.* non-rusting *(blade);* stainless *(steel);* **nichts** *Indefinitpron.* nothing; ich möchte ~: I don't want anything

nichts-, Nichts-: ~nutz der; ~es, ~e *(veralt.)* good-for-nothing; ~nutzig *Adj. (veralt.)* good-for-nothing *attrib.;* worthless *(existence);* ~sagend 1. *Adj.* empty; *(fig.: ausdruckslos)* expressionless *(face);* 2. *adv.* meaninglessly *(formulated);* ~tun das idleness *no art.*

Nickel das; ~s nickel

nicken *itr. V.* nod

nie *Adv.* never

nieder 1. *Adj.; nicht präd.* lower *(class, intelligence);* minor *(official);* lowly *(family, origins, birth);* menial *(task);* 2. *Adv.* down

nieder-, Nieder-: ~gang der fall; decline; ~|gehen *unr. itr. V.; mit sein (plane etc., rain, avalanche)* come down; ~geschlagen *Adj.* dejected; ~geschlagenheit die; ~: dejection; ~lage die defeat

Nieder·lande *Pl.:* die ~: the Netherlands; **Niederländer der;** ~s, ~: Dutchman; **Niederländerin die;** ~, ~nen Dutchwoman; **niederländisch** *Adj.* Dutch; Netherlands *attrib. (government, embassy, etc.)*

nieder-, Nieder-: ~|lassen *unr. refl. V.* **a)** set up in business; *(doctor, lawyer)* set up in practice; **b)** *(seinen Wohnsitz nehmen)* settle; ~lassung die; ~, ~en *(Wirtsch.)* branch; ~|le-

gen *tr. V.* **a)** *(geh.: hinlegen)* lay *or* put down; lay *(wreath);* **b)** *(fig.)* resign [from] *(office);* relinquish *(command)*

Nieder·sachsen (das) Lower Saxony

nieder-, Nieder-: ~schlag der precipitation; ~|schlagen *unr. tr. V.* **a)** jmdn. ~schlagen knock sb. down; **b)** *(beenden)* suppress, put down *(revolt, uprising, etc.);* **c)** *(senken)* lower *(eyes, eyelids);* ~trächtig 1. *Adj.* malicious *(person, slander, lie, etc.);* *(verachtenswert)* despicable *(person);* base *(misrepresentation, slander, lie);* 2. *adv.* *(betray, lie, treat)* in a despicable way; ~trächtigkeit die; ~, ~en **a)** *o. Pl. s.* ~trächtig 1: maliciousness; despicableness; baseness; **b)** *(gemeine Handlung)* despicable act

Niederung die; ~, ~en low-lying area; *(an Flußläufen, Küsten)* flats *pl.;* *(Tal)* valley

niedlich 1. *Adj.* sweet; cute *(Amer. coll.);* 2. *adv.* sweetly

niedrig 1. *Adj.* low; lowly *(origins, birth);* base *(instinct, desire, emotion);* vile *(motive);* 2. *adv.* *(hang, fly)* low

niemals *Adv.* never; **niemand** *Indefinitpron.* nobody; no one

Niere die; ~, ~n kidney

Nieren-: ~entzündung die nephritis; ~stein der kidney stone

Niesel·regen der drizzle

niesen *itr. V.* sneeze

¹**Niete die;** ~, ~n **a)** *(Los)* blank; **b)** *(ugs.: Mensch)* dead loss *(coll.)* (in + *Dat.* at)

²**Niete die;** ~, ~n rivet; **nieten** *tr. V.* rivet

Nikolaus·tag der St Nicholas' Day

Nikotin das; ~s nicotine; **nikotinarm** *Adj.* low-nicotine *attrib.;* low in nicotine *pred.*

Nil der; ~|s| Nile; **Nil·pferd** das hippopotamus

nimm *Imperativ Sg. v.* nehmen

nippen *itr. V.* sip

nirgends, nirgend·wo *Adv.* nowhere

Nische die; ~, ~n niche; *(Erweiterung eines Raumes)* recess

nisten *itr. V.* nest

Nitrat das; ~|e|s, ~e nitrate

Niveau [ni'vo:] das; ~s, ~s level; *(Qualitäts-)* standard

Nixe die; ~, ~n nixie; *(mit Fischschwanz)* mermaid

nobel *Adj.* **a)** *(geh.)* noble;

noble[-minded] ⟨*person*⟩; **b)** *(oft spött.: luxuriös)* elegant; posh *(coll.)*

Nobel·preis der Nobel prize

noch 1. *Adv.* **a)** *([wie] bisher)* still; ~ **nicht** not yet; **sie sind immer ~ nicht da** they're still not here; **ich habe Großvater ~ gekannt** I'm old enough to have known grandfather; **er hat ~ Glück gehabt** he was lucky; **das geht ~:** that's [still] all right; **b)** *(als Rest einer Menge)* **ich habe [nur] ~ zehn Mark** I've [only] ten marks left; **es sind ~ 10 km bis zur Grenze** it's another 10 km. to the border; **c)** *(bevor etw. anderes geschieht)* just; **ich will ~ [schnell] duschen** I just want to have a [quick] shower; **d)** *(irgendwann einmal)* some time; one day; **er wird ~ anrufen/kommen** he will still call/come; **e)** *(womöglich)* if you're/he's *etc.* not careful; **du kommst ~ zu spät!** you'll be late if you're not careful; **f)** *(drückt eine geringe zeitliche Distanz aus)* only; ge**stern habe ich ihn ~ gesehen** I saw him only yesterday; **g)** *(nicht später als)* ~ **am selben Abend** the [very] same evening; **h)** *(außerdem, zusätzlich)* **wer war ~ da?** who else was there?; ~ **etwas Kaffee?** [would you like] some more coffee?; **Geld/Kleider** *usw.* ~ **und** ~ heaps and heaps of money/clothes *etc. (coll.)*; **i)** *(vor ist ~ größer [als Karl]* he is even taller [than Karl]; **er will ~ mehr haben** he wants even more; **jeder ~ so dumme Mensch versteht das** anyone, however stupid, can understand that; **j) wie heißt sie [doch] ~?** [now] what's her name again?; **2.** *Partikel* **das ist ~ Qualität!** that's what I call quality; **der wird sich ~ wundern** *(ugs.)* he's in for a surprise; **er kann ~ nicht einmal lesen** he can't even read; **3.** *Konj. (und auch nicht)* nor; **weder ... noch** neither ... nor; **noch·mals** *Adv.* again

Nominativ der; ~s, ~e *(Sprachw.)* nominative [case]

Nonne die; ~, ~n nun

Nord *o. Art.; o. Pl. (bes. Seemannsspr., Met.)* s. **Norden**

nord-, Nord-: ~**afrika (das)** North Africa; ~**amerika (das)** North America; ~**deutsch** *Adj.* North German

Norden der; ~s north; **der ~:** the North; **nach ~:** northwards; **Nordirland (das)** Northern Ireland; **nordisch** *Adj.* Nordic; **Nord·kap das** North Cape; **nördlich 1.** *Adj.* **a)** *(im*

Norden gelegen) northern; **b)** *(nach, aus dem Norden)* northerly; **c)** *(aus dem Norden kommend, für den Norden typisch)* Northern; **2.** *adv.* northwards; ~ **von ...:** [to the] north of ...; **3.** *Präp. mit Gen.* [to the] north of; **Nord·pol** ['--] der North Pole; **Nord·rhein-Westfalen (das);** ~s North Rhine-Westphalia; **Nord·see** die; *o. Pl.* North Sea; **nord·wärts** *Adv.* northwards; **Nord·wind** der northerly wind

Nörgelei die; ~ *(abwertend) o. Pl.* grumbling; **nörgeln** *itr. V. (abwertend)* moan, grumble **(an + Dat.** about)

Norm die; ~, ~en **a)** norm; **b)** *(geforderte Arbeitsleistung)* quota; **c)** *(Sport)* qualifying standard; **d)** *(technische, industrielle* ~) standard; **normal 1.** *Adj.* normal; **2.** *adv.* normally; **Normal·benzin** das ≈ two-star petrol *(Brit.)*; regular *(Amer.)*; **normalerweise** *Adv.* normally; **normalisieren 1.** *tr. V.* normalize; **2.** *refl. V.* return to normal

Normandie die; ~: Normandy

normen *tr. V.,* **normieren** *tr. V.* standardize

Norwegen (das); ~s Norway; **Norweger** der; ~s, ~, **Norwegerin** die; ~, ~nen Norwegian; **norwegisch** *Adj.* Norwegian

Nostalgie die; ~: nostalgia

Not die; ~, Nöte **a)** *(Gefahr)* **in ~ sein** be in desperate straits; **b)** *o. Pl. (Mangel, Armut)* need; poverty [and hardship]; ~ **leiden** suffer poverty [and hardship]; **in ~ geraten/sein** encounter hard times/be suffering want [and deprivation]; **c)** *o. Pl. (Verzweiflung)* distress; **d)** *(Sorge, Mühe)* trouble; **mit knapper ~:** by the skin of one's teeth; **f)** *o. Pl. (veralt.: Notwendigkeit)* necessity; **zur ~:** if need be

Notar der; ~s, ~e notary; **Notariat** das; ~[e]s, ~e **a)** *(Amt)* notaryship; **b)** *(Kanzlei)* notary's office

not-, Not-: ~**arzt** der doctor on [emergency] call; ~**ausgang** der emergency exit; ~**bremse** die emergency brake; ~**dienst** der s. Bereitschaftsdienst; ~**dürftig 1.** *Adj.* makeshift ⟨*shelter, repair*⟩; scanty ⟨*cover, clothing*⟩; **2.** *adv.* scantily ⟨*clothed*⟩

Note die; ~, ~n **a)** *(Zeichen)* note; **b)** *Pl. (Text)* music *sing.*; **c)** *(Schul~)* mark; **d)** *(Eislauf, Turnen)* score

not-, Not-: ~**fall** der **a)** emergency; **b)**

im ~**fall** *(nötigenfalls)* if need be;
~**falls** *Adv.* if need be; ~**gedrungen**
Adv. of necessity

notieren 1. *tr. V.* [sich *(Dat.)*] etw. ~:
make a note of sth.; 2. *itr. V. (Börsenw., Wirtsch.)* be quoted (mit at)

nötig 1. *Adj.* necessary; etw./jmdn. ~
haben need sth./sb.; 2. *adv.* er braucht
~ **Hilfe** he is in urgent need of help;
nötigen *tr. V.* compel; force;
(Rechtsspr.) coerce

Notiz die; ~, ~en note; *(Zeitungs~)*
brief report; von jmdm./etw. [keine] ~
nehmen take [no] notice of sb./sth.

Notiz-: ~**block** der; *Pl.* ~**blocks,**
schweiz.: ~**blöcke** notepad; ~**buch**
das notebook

not-, Not-: ~**lage** die serious difficulties *pl.;* ~**landen**[1] *itr. V.; mit sein*
do an emergency landing; ~**landung**
die emergency landing; ~**leidend**
Adj. needy; ~**lösung** die stopgap;
~**lüge** die evasive lie; *(aus Rücksichtnahme)* white lie

notorisch 1. *Adj.* notorious; 2. *adv.*
notoriously

Not-: ~**ruf der** a) *(Hilferuf)* emergency
call; *(eines Schiffes)* Mayday call; b)
(Nummer) emergency number;
~**ruf·nummer** die emergency number; ~**ruf·säule** die emergency telephone *(mounted in a pillar)*; ~**stand**
der crisis; *(Staatsrecht)* state of emergency; ~**unterkunft** die emergency
accommodation *no pl., no indef. art.;*
~**wehr** die self-defence

not·wendig *Adj.* necessary; **Notwendigkeit** die; ~, ~en necessity

Nougat ['nu:gat] der; *auch* das; ~s
nougat

Novelle die; ~, ~n *(Literaturw.)*
novella

November der; ~[s], ~: November

Nr. *Abk.* Nummer No

Nu der: im Nu in no time

Nuance ['nyã:sə] die; ~, ~n nuance;
(Grad) shade

nüchtern 1. *Adj. (nicht betrunken;
realistisch)* sober; *(ungeschminkt)*
bare, plain ⟨*fact*⟩; **der Patient muß** ~
sein the patient's stomach must be
empty; 2. *adv.* soberly

nuckeln *(ugs.)* itr. V. suck (an + *Dat.*
at)

Nudel die; ~, ~n piece of spaghetti/
vermicelli/tortellini *etc.; (als Suppen-*

einlage) noodle; ~n *(Teigwaren)* pasta
sing.; (als Suppeneinlage) noodles

nuklear 1. *Adj.* nuclear; 2. *adv.* ~ **angetrieben** nuclear-powered

null *Kardinalz.* nought; ~ **Komma**
sechs [nought] point six; **gegen** ~ **Uhr**
around twelve midnight; **Null die;** ~,
~**en** a) nought; zero; **in** ~ **Komma**
nichts *(ugs.)* in less than no time;
gleich ~ **sein** *(fig.)* be practically zero;
auf ~ **stehen** ⟨*indicator, needle, etc.*⟩ be
at zero; b) *(ugs.: Versager)* failure;
dead loss *(coll.);* **Null·punkt der**
zero

numerieren *tr. V.* number; **Numerierung** die; ~, ~en numbering;
Nummer die; ~, ~n a) number; **in**
Wagen mit [einer] Münchner ~: a car
with a Munich registration; **ich bin**
unter der ~ 24 26 79 **zu erreichen** I can
be reached on 24 26 79; b) *(Ausgabe)*
issue; c) *(Größe)* size; **Nummern·schild** das number-plate;
license plate *(Amer.)*

nun 1. *Adv.* now; 2. *Partikel* now; **das**
hast du ~ **davon!** it serves you right!;
kommst du ~ **mit oder nicht?** now are
you coming or not?; ~ **gut** [well,] all
right; ~, ~! now, come on; ~ **ja** ...:
well, yes ...

nur 1. *Adv.* a) *(nicht mehr als)* only;
just; b) *(ausschließlich)* only; **nicht**
~ ..., **sondern auch** ...: not only ..., but
also ...; ~ **so zum Spaß** just for fun; 2.
Konj. but; **ich kann dir das Buch leihen,** ~ **nicht heute** I can lend you the
book, only not today; 3. *Partikel* wenn
er ~ **hier wäre** if only he were here; ~
zu! go ahead; **laß dich** ~ **nicht erwischen** just don't let me/them *etc.* catch
you; **was sollen wir** ~ **tun?** what on
earth are we going to do?; **so schnell**
er ~ **konnte** just as fast as he could

Nürnberg (das); ~s Nuremberg

Nuß die; ~, Nüsse a) nut; **Nuß·baum**
der walnut-tree; **Nuß·knacker** der
nutcrackers *pl*

Nutte die; ~, ~n *(derb)* tart *(sl.);*
hooker *(Amer. sl.)*

nutz-: ~**bar** *Adj.* usable; exploitable,
utilizable ⟨*mineral resources, invention*⟩; cultivatable ⟨*land, soil*⟩; ~**bringend** 1. *Adj.* useful; *(gewinnbringend)* profitable; 2. *adv.* profitably

nutzen 1. *tr. V.* a) use; exploit, utilize
⟨*natural resources*⟩; cultivate ⟨*land,
soil*⟩; harness ⟨*energy source*⟩; exploit
⟨*advantage*⟩; b) *(be~, aus~)* use;
make use of; 2. *itr. V. s.* **nützen** 1;

---

[1] *ich notlande, notgelandet, notzulanden*

Nutzen der; ~s **a)** benefit; [jmdm.] von ~ **sein** be of use [to sb.]; **b)** *(Profit)* profit; **nutzen 1.** *itr. V.* be of use *(Dat.* to); **nichts ~:** be no use; **2.** *tr. V. s.* nutzen 1; **nützlich** *Adj.* useful; **nutzlos 1.** *Adj.* useless; *(vergeblich)* vain *attrib.;* in vain *pred.;* **2.** *adv.* uselessly; *(vergeblich)* in vain; **Nutz·losigkeit** die; ~: uselessness; *(Vergeblichkeit)* futility; **Nutznießer** der; ~s, ~, **Nutznießerin** die; ~, ~nen beneficiary; **Nutzung** die; ~, ~en use; *(des Landes, des Bodens)* cultivation; *(von Bodenschätzen)* exploitation; utilization; *(einer Energiequelle)* harnessing

Nylon Ⓦ ['nailɔn] das; ~s nylon
Nymphe die; ~, ~n *(Myth., Zool.)* nymph
Nymphomanin die; ~, ~nen *(Psych.)* nymphomaniac

O

o, O das; ~, ~: o/O
O *Abk.* Ost[en] E
ö, Ö das; ~, ~: o/O umlaut
o. ä. *Abk.* oder ähnlich[es] or similar
Oase die; ~, ~n *(auch fig.)* oasis
ob *Konj.* **a)** whether; **b)** und **ob!** of course!
OB *Abk.* Oberbürgermeister
Obacht die; ~ *(bes. südd.)* caution; ~ auf jmdn./etw. **geben** take care of sb./ sth.; *(aufmerksam sein)* pay attention to sb./sth.
Obdach das; ~[e]s *(geh.)* shelter; **obdach·los** *Adj.* homeless; **Obdachlose** der/die; *adj. Dekl.* homeless person/man/woman; **die** ~n the homeless
Obduktion die; ~, ~en *(Med., Rechtsw.)* post-mortem [examination]; autopsy
O-Beine *Pl.* bandy legs; bow-legs
oben *Adv.* **a)** hier/dort ~: up here/ there; **weiter** ~: further up; **nach** ~: upwards; **von** ~: from above; **von** ~ **herab** *(fig.)* condescendingly; **b)** *(im Gebäude)* upstairs; **nach** ~: upstairs;

c) *(am oberen Ende, zum oberen Ende hin)* at the top; **nach** ~ [hin] towards the top; **von** ~: from the top; **c)** *(an der Oberseite)* on top; **d)** *(in einer Hierarchie, Rangfolge)* at the top; **e)** *[weiter] vorn im Text)* above; **oben · genannt** *Adj.* above-mentioned
ober... *Adj.* upper *attrib.;* top *attrib.*
Ober der; ~s, ~: waiter; **Herr** ~! waiter!
Ober-: ~**arm** der upper arm; ~**bekleidung** die outer clothing; ~**bürgermeister** der mayor
Ober·fläche die surface; *(Flächeninhalt)* surface area; **oberflächlich 1.** *Adj.* superficial; **2.** *adv.* superficially
ober·halb 1. *Adv.* above; ~ **von** above; **2.** *Präp. mit Gen.* above
Ober-: ~**haupt** das head; *(einer Verschwörung)* leader; ~**hemd** das shirt; ~**kiefer** der upper jaw; ~**körper** der upper part of the body; ~**schenkel** der thigh; ~**schicht** die *(Soziol.)* upper class; ~**schule** die secondary school; ~**seite** die top
oberst... *s.* ober...
Ober·teil das *od.* der top [part]; *(eines Bikinis, Anzugs, Kleids usw.)* top [half]
ob·gleich *Konj. s.* obwohl
obig *Adj.* above
Objekt das; ~s, ~e object; *(Kaufmannsspr.: Immobilie)* property; **objektiv 1.** *Adj.* objective; **2.** *adv.* objectively; **Objektiv** das; ~s, ~e lens; **Objektivität** die; ~: objectivity
Obrigkeit die; ~, ~en authorities *pl.*
ob·schon *Konj. (geh.)* although
Obst das; ~[e]s fruit
Obst-: ~**baum** der fruit-tree; ~**garten** der orchard; ~**kuchen** der fruit flan; ~**saft** der fruit juice
obszön 1. *Adj.* obscene; **2.** *adv.* obscenely; **Obszönität** die; ~, ~en obscenity
ob·wohl *Konj.* although; though
Ochse ['ɔksə] der; ~n, ~n **a)** ox; bullock; **b)** *(salopp)* numskull *(coll.);* **Ochsen·schwanz·suppe** die oxtail soup
od. *Abk.* oder
öde *Adj.* **a)** deserted; desolate ⟨area, landscape⟩; **b)** *(unfruchtbar)* barren; **c)** *(langweilig)* tedious; dreary ⟨life, time, existence⟩; **Öde** die; ~: *s.* öde a–c: desertedness; desolateness; barrenness; tediousness; dreariness
oder *Konj.* or; *(in Fragen)* er ist doch hier, ~? he is here, isn't he? *(zweifelnd)* he is here – or isn't he?

OEZ *Abk.* osteuropäische Zeit EET
Ofen der; ~s, **Öfen** heater; *(Kohle~)* stove; *(Back~)* oven; *(Brenn~, Trokken~)* kiln; **Ofen·rohr** das [stove] flue

offen 1. *Adj.* **a)** open; **ein ~es Hemd** a shirt with the collar unfastened; **~ haben** *od.* **sein** be open; **~es Licht** a naked light; **b)** *(frei)* vacant ⟨*job, post*⟩; **c)** *(ungewiß, ungeklärt)* open ⟨*question*⟩; uncertain ⟨*result*⟩; **d)** *(noch nicht bezahlt)* outstanding ⟨*bill*⟩; **e)** *(freimütig, aufrichtig)* frank [and open] ⟨*person*⟩; frank, candid ⟨*look, opinion, reply*⟩; **2.** *adv.* openly; **~ gesagt** frankly; to be frank; **offen·bar 1.** *Adj.* obvious; **2.** *adv.* obviously; **Offenbarung** die; ~, ~en revelation; **offen|bleiben** *unr. itr. V.; mit sein* **a)** stay open; **b)** *(ungeklärt bleiben)* remain open; ⟨*decision*⟩ be left open; **Offen·heit** die; ~: *s.* offen e: frankness [and openness]; candour

offen-: **~kundig 1.** *Adj.* obvious; **2.** *adv.* obviously; **~lassen** *unr. tr. V.* etw. ~lassen leave sth. open; **~sichtlich 1.** *Adj.* obvious; **2.** *adv.* obviously

offensiv 1. *Adj.* **a)** offensive; **b)** *(Sport)* attacking; **2.** *adv.* **a)** offensively; **b)** *(Sport)* **~ spielen** play an attacking game; **Offensive** die; ~, ~n *(auch Sport)* offensive

offen|stehen *unr. itr. V.* be open

öffentlich 1. *Adj.* public; state *attrib.*, [state-] maintained ⟨*school*⟩; **der ~e Dienst** the civil service; **2.** *adv.* publicly; ⟨*perform, appear*⟩ in public; **Öffentlichkeit** die; ~: public

offiziell 1. *Adj.* official; **2.** *adv.* officially

Offizier der; ~s, ~e officer

öffnen 1. *tr. V.* open; turn on ⟨*tap*⟩; undo ⟨*coat, blouse, button, zip*⟩; **2.** *itr. V.* **a)** [jmdm.] ~: open the door [to sb.]; **b)** *(geöffnet werden)* ⟨*shop, bank, etc.*⟩ open; **3.** *refl. V.* open; **Öffner** der; ~s, ~: opener; **Öffnung** die; ~, ~en opening; **Öffnungs·zeiten** *Pl.* opening times

oft *Adv.* öfter, am öftesten often; **wie oft soll ich dir noch sagen, daß ...?** how many [more] times do I have to tell you that ...?; **öfter** *Adv.* now and then; **oftmals** *Adv.* often; frequently

OG *Abk.* Obergeschoß

ohne 1. *Präp. mit Akk.* without; **~ mich!** [you can] count me out!; **~ weiteres** *(leicht, einfach)* easily; *(ohne Ein-*

wand) readily; **2.** *Konj.* **~ zu zögern** without hesitation; **ohne·hin** *Adv.* anyway

Ohnmacht die; ~, ~en **a)** faint; **in ~ fallen** faint; **b)** *(Machtlosigkeit)* powerlessness; impotence; **ohnmächtig 1.** *Adj.* **a)** unconscious; **~ werden** faint; **~ sein** have fainted; **b)** *(machtlos)* powerless; impotent; **2.** *adv.* impotently; **~ zusehen** watch helplessly

Ohr das; ~[e]s, ~en ear; **gute/schlechte ~en haben** have good/poor hearing *sing.*; **jmdn. übers ~ hauen** *(fig. ugs.)* put one over on sb. *(coll.)*; **Öhr** das, ~[e]s, ~e eye; **Ohren·schmerz** der earache; **~schmerzen haben** have [an] earache *sing.*

ohr-, Ohr-: **~feige** die box on the ears; **~feigen** *tr. V.* jmdn. ~feigen box sb.'s ears; **ich könnte mich ~feigen!** *(ugs.)* I could kick myself!; **~läppchen** das ear-lobe; **~ring** der ear-ring

okay [o'ke] *(ugs.) Interj., Adj., adv.* OK *(coll.)*; okay *(coll.)*

öko-, Öko-: eco-
Ökologie die; ~: ecology; **ökologisch 1.** *Adj.* ecological; **2.** *adv.* ecologically

ökonomisch 1. *Adj.* **a)** economic; **b)** *(sparsam)* economical; **2.** *adv.* economically

Oktober der; ~[s], ~: October

Öl das; ~[e]s, ~e oil; **in Öl malen** paint in oils; **ölen** *tr. V.* oil; **Öl·farbe** die **a)** oil-based paint; **b)** *(zum Malen)* oil-paint; **ölig** *Adj.* oily

Olive die; ~, ~n olive

Öl-: **~ofen** der oil heater; **~sardine** die sardine in oil; **eine Dose ~n** a tin of sardines; **~wechsel** der *(bes. Kfz-W.)* oil-change

Olympiade die; ~, ~n Olympic Games *pl.*; Olympics *pl.*; **Olympiastadion** das Olympic stadium; **olympisch** *Adj.* Olympic; **die Olympischen Spiele** the Olympic Games; the Olympics

Oma die; ~, ~s *(fam.)* granny *(coll./child lang.)*

Omelett [ɔm[ə]'lɛt] das; ~[e]s, ~e *od.* ~s omelette

Omnibus der; ~ses, ~se omnibus *(formal)*; *(Privat- und Reisebus auch)* coach

Onkel der; ~s, ~ *od. (ugs.)* ~s uncle
OP [o:'pe:] der; ~[s], ~[s] *Abk.* Operationssaal

Opa der; ~s, ~s *(fam.)* grandad *(coll./ child lang.)*

Opal der; ~s, ~e opal

OPEC ['oːpɛk] die; ~ *Abk.* OPEC

Oper die; ~, ~n opera; *(Opernhaus)* Opera; opera-house

Operation die; ~, ~en operation; **Operations·saal** der operating-theatre *(Brit.)* or -room

Operette die; ~, ~n operetta

operieren 1. *tr. V.* operate on ⟨patient⟩; 2. *itr. V.* operate

Opern·glas das opera-glass[es *pl.*]

Opfer das; ~s, ~: a) sacrifice; b) *(Geschädigter)* victim; **opfern** *tr. V.* *(auch fig.)* sacrifice; offer up ⟨fruit, produce, etc.⟩

Opium das; ~s opium

opponieren *itr. V.* gegen jmdn./etw. ~: oppose sb./sth.; **Opposition** die; ~, ~en opposition; **oppositionell** *Adj.* opposition *attrib.* ⟨group, movement, etc.⟩; ⟨newspaper, writer, artist, etc.⟩ opposed to the government

Optik die; ~: optics *sing., no art.;* **Optiker** der; ~s, ~: optician

optimal 1. *Adj.* optimal; optimum *attrib.*; 2. *adv.* jmdn. ~ beraten give sb. the best possible advice; **Optimismus** der; ~: optimism; **Optimist** der; ~en, ~en, **Optimistin** die; ~, ~nen optimist; **optimistisch** 1. *Adj.* optimistic; 2. *adv.* optimistically

optisch 1. *Adj.* optical; visual ⟨impression⟩; eine ~e Täuschung an optical illusion; 2. *adv.* optically; visually ⟨impressive, effective⟩

orange [oˈrãːʒ(ə)] *indekl. Adj.* orange; **Orange** die; ~, ~n orange

Orangen-: ~**marmelade** die orange marmalade; ~**saft** der orange-juice

Orchester [ɔrˈkɛstɐ] das; ~s, ~: orchestra

Orden der; ~s, ~ a) order; b) *(Ehrenzeichen)* decoration

ordentlich 1. *Adj.* a) [neat and] tidy; neat ⟨handwriting, clothes⟩; b) *(anständig)* respectable; proper ⟨manners⟩; c) *(planmäßig)* ordinary ⟨meeting⟩; ~es Mitglied full member; d) *(ugs.: richtig)* proper; real; ein ~es Stück Kuchen a nice big piece of cake; e) *(ugs.: recht gut)* decent ⟨wine, flat, marks, etc.⟩; ganz ~: pretty good; 2. *adv.* a) tidily; neatly; ⟨write⟩ neatly; b) *(anständig)* properly; c) *(ugs.: gehörig)* ~ feiern have a real good celebration *(coll.);* d) *(ugs.: recht gut)* ⟨ski, speak, etc.⟩ really well

Ordinal·zahl die ordinal [number]; **ordinär** 1. *Adj.* vulgar; 2. *adv.* vulgarly; **Ordinate** die; ~, ~n *(Math.)* ordinate

ordnen *tr. V.* arrange; sein Leben/seine Finanzen ~: straighten out one's life/put one's finances in order; **Ordner** der; ~s, ~: file; **Ordnung** die; ~, ~en order; *(geregelter Ablauf)* routine; ~ halten keep things tidy; in ~ sein *(ugs.)* be OK *(coll.)* or all right; hier ist etw. nicht in ~: there's something wrong here; sie ist in ~ *(ugs.)* she's OK *(coll.);* in ~! *(ugs.)* OK! *(coll.);* all right!

ordnungs-, Ordnungs-: ~**gemäß** 1. *Adj.* ⟨conduct etc.⟩ in accordance with the regulations; 2. *adv.* in accordance with the regulations; ~**halber** *Adv.* as a matter of form; ~**widrig** *(Rechtsw.)* 1. *Adj.* ⟨actions, behaviour, etc.⟩ contravening the regulations; illegal ⟨parking⟩; 2. *adv.* ~widrig parken park illegally; ~**zahl** die ordinal [number]

Organ das; ~s, ~e organ; *(ugs.: Stimme)* voice; **Organisation** die; ~, ~en organization; **Organisator** der; ~s, ~en organizer; **organisatorisch** *Adj.* organizational; **organisch** 1. *Adj.* organic; 2. *adv.* organically; **organisieren** 1. *tr. V.* organize; 2. *itr. V.* gut ~ können be a good organizer; 3. *refl. V.* organize; **Organismus** der; **Organismen** organism

Organist der; ~en, ~en, **Organistin** die; ~, ~nen organist

Orgasmus der; ~, **Orgasmen** orgasm

Orgel die; ~, ~n organ

Orgie ['ɔrgiə] die; ~, ~n *(auch fig.)* orgy

Orient ['oːriɛnt] der; ~s Middle East and south-western Asia *(including Afghanistan and Nepal);* der Vordere ~: the Middle East; **orientalisch** *Adj.* oriental; **orientieren** 1. *refl. V.* a) get one's bearings; b) sich über etw. *(Akk.)* ~ *(fig.)* inform oneself about sth.; c) sich an etw. *(Dat.)* ~ *(fig.)* be oriented towards sth.; ⟨policy, advertising⟩ be geared towards sth; 2. *tr. V. (unterrichten)* inform (über + *Akk.* about); **Orientierung** die; ~ a) die ~ verlieren lose one's bearings; b) *(Unterrichtung)* zu Ihrer ~: for your information; **Orientierungs·sinn** der sense of direction

original 1. *Adj.* original; 2. *adv.* ~ italienischer Espresso genuine Italian es-

presso coffee; **etw. ~ übertragen**
broadcast sth. live; **Original das; ~s,
~e** original; **Original·fassung die**
original version; **Originalität die; ~:**
originality; **originell 1.** *Adj.* original;
2. *adv.* with originality
Orkan der; ~|e|s, ~e hurricane
Ornament das; ~|e|s, ~e ornament
¹**Ort der; ~|e|s, ~e** place; (Dorf) vil-
lage; *(Stadt)* town; **an ~ und Stelle**
there and then; ²**Ort:** vor ~ *(fig.)* on
the spot
Orthographie die; ~; ~n orthography; **orthographisch 1.** *Adj.* or-
thographic; **~e Fehler** spelling mistakes; **2.** *adv.* orthographically; **Or-
thopäde der; ~n, ~n** orthopaedic
specialist; **orthopädisch 1.** *Adj.* orthopaedic; **2.** *adv.* orthopaedically
örtlich 1. *Adj. (auch Med.)* local; **2.**
adv. (auch Med.) locally; **~ betäubt
werden** be given a local anaesthetic;
Ortschaft die; ~, ~en *(Dorf)* village;
(Stadt) town
Orts-: ~gespräch das *(Fernspr.)*
local call; **~name der** place-name;
~netz·kennzahl die *(Fernspr.)* dialling code; area code *(Amer.)*
Öse die; ~, ~n eye
Ossi der; ~s, ~s *(salopp)* East German
Ost *o. Art.; o. Pl. (bes. Seemannsspr.,
Met.) s.* Osten
ost-, Ost-: ~block der; *o. Pl.* East-
ern bloc; **~deutsch** *Adj.* Eastern
German; *(hist.: auf die DDR bezogen)*
East German; **~deutschland (das)**
Eastern Germany; *(hist.: DDR)* East
Germany
Osten der; ~s east; **der ~:** the East;
der Ferne ~: the Far East; **der Nahe ~:**
the Middle East
Oster-: ~ei das Easter egg; **~glocke
die** daffodil; **~hase der** Easter hare
*(said to bring children their Easter
Eggs)*
Ostern das; ~, ~: Easter; **Frohe** *od.*
Fröhliche ~: Happy Easter!; **zu ~:** at
Easter
Österreich (das); ~s Austria; **Öster-
reicher der; ~s, Österreicherin die;
~, ~nen** Austrian; **österreichisch**
Adj. Austrian
Ost·europa (das) Eastern Europe;
östlich 1. *Adj.* **a)** *(im Osten gelegen)*
eastern; **b)** *(nach, aus dem Osten)* easterly; **c)** *(aus dem Osten kommend, für
den Osten typisch; Politik)* Eastern;
(influence, policies) of the East; **2.** *adv.*
eastwards; **~ von ...:** [to the] east of ...;

3. *Präp. mit Gen.* [to the] east of;
Ost·see die; *o. Pl.* Baltic [Sea];
ost·wärts *Adv.* eastwards;
Ost·wind der easterly wind
¹**Otter der; ~s, ~** *(Fisch~)* otter
²**Otter die; ~, ~n** *(Viper)* adder; viper
Otto·motor der Otto engine
Ouvertüre [uvɛrˈtyːrə] **die; ~, ~n**
(auch fig.) overture *(Gen.* to)
oval *Adj.* oval
Ozean der; ~s, ~e ocean; **Oze-
an·dampfer der** ocean liner
Ozon der *od.* **das; ~s** ozone

P

p, P [pe] **das; ~, ~:** p/P
paar *indekl. Indefinitpron.* **ein ~ ...:** a
few ...; *(zwei od. drei)* a couple of ...;
Paar das; ~|e|s, ~e pair; *(Mann und
Frau)* couple; **ein ~ Würstchen** two
sausages; **paaren** *refl. V.* ⟨animals⟩
mate; ⟨people⟩ copulate; **Paar·lauf
der** pairs *pl.;* **paar·mal** *Adv.* **ein
~mal** a few times; *(zwei- oder dreimal)*
a couple of times; **paar·weise** *Adv.*
in pairs
Pacht die; ~, ~en a) lease; **etw. in ~
nehmen** lease sth.; **etw. in ~ haben**
have sth. on lease; **etw. in ~ geben**
lease sth.; **pachten** *tr. V.* lease;
**Pächter der; ~s, ~, Pächterin die;
~, ~nen** leaseholder; *(eines Hofes)*
tenant
¹**Pack der; ~|e|s, ~e** *od.* **Päcke a)** pack;
b) *s.* Packen; ²**Pack das; ~|e|s** *(ugs.
abwertend)* rabble; **Päckchen das;
~s, ~ a)** package; *(auch Postw.)* small
parcel; *(Bündel)* packet; **b)** *s.* **Pak-
kung a; packen 1.** *tr. V.* **a)** pack; **b)**
(fassen) grab [hold of]; *(fig.)* **Furcht
packte ihn/er wurde von Furcht ge-
packt** he was seized with fear; **2.** *itr. V.*
(Koffer usw. ~) pack; **Packen der;
~s, ~** pile; *(zusammengeschnürt)*
bundle; *(von Geldscheinen)* wad;
Pack·papier das [stout] wrapping-
paper; **Packung die; ~, ~en a)**
packet; pack *(esp. Amer.);* **b)** *(Med.,
Kosmetik)* pack

Pädagoge der; ~n, ~n *(Erzieher, Leh-rer)* teacher; *(Wissenschaftler)* educa-tionalist; **pädagogisch 1.** *Adj.* edu-cational; **seine ~en Fähigkeiten** his teaching ability *sing.*; **2.** *adv.* educa-tionally ⟨*sound, wrong*⟩

Paddel das; ~s, ~: paddle; **Pad-del·boot das** canoe; **paddeln** *itr. V.*; *mit sein* paddle; *(als Sport)* canoe

paffen 1. *tr. V.* puff at ⟨*pipe etc.*⟩; **2.** *itr. V.* puff away

Page ['pa:ʒə] der; ~n, ~n bellboy

Paket das; ~[e]s, ~e pile; *(zusammen-geschnürt)* bundle; *(Eingepacktes, Post~)* parcel; *(Packung)* packet; pack *(esp. Amer.)*

Paket-: ~**karte** die parcel dispatch form; ~**schalter** der parcels counter

Pakistan (das); ~s Pakistan; **Paki-staner** der; ~s, ~, **Pakistani** der; ~[s], ~[s] Pakistani; **pakistanisch** *Adj.* Pakistani

Pakt der; ~[e]s, ~e pact; **paktieren** *itr. V.* make *or* do a deal/deals

Palast der; ~[e]s, Paläste palace

Palästina (das); ~s Palestine; **Palä-stinenser** der; ~s, ~: Palestinian; **palästinensisch** *Adj.* Palestinian

Palette die; ~, ~n palette

Palme die; ~, ~n palm[-tree]

Pampelmuse die; ~, ~n grapefruit

Panama (das); ~s Panama; **Pana-ma·kanal der**; *o. Pl.* Panama Canal

panieren *tr. V.* bread; coat ⟨*sth.*⟩ with breadcrumbs; **Panier·mehl das** breadcrumbs *pl.*

Panik die; ~, ~en panic; **panisch** *Adj.* panic *attrib.* ⟨*fear, terror*⟩; panic-stricken ⟨*flight*⟩;

Panne die; ~, ~n **a)** breakdown; *(Rei-fen~)* puncture; flat [tyre]; **b)** *(Mißge-schick)* mishap; **Pannen·dienst der** breakdown service

Panorama das; ~s, **Panoramen** pan-orama

Panther der; ~s, ~: panther

Pantoffel der; ~s, ~n backless slipper

Pantomime die; ~, ~n mime

Panzer der; ~s, ~ **a)** *(Milit.)* tank; **b)** *(Zool.)* armour *no indef. art.*; *(von Schildkröten, Krebsen)* shell

Panzer-: ~**glas** das bullet-proof glass; ~**schrank** der safe

Papa der; ~s, ~s *(ugs.)* daddy *(coll.)*

Papagei der; ~en *od.* ~s, ~e[n] parrot

Papi der; ~s, ~s *(ugs.)* daddy *(coll.)*

Papier das; ~s, ~e **a)** paper; **b)** *Pl.* *(Ausweis[e])* [identity] papers; **c)** *(Fi-nanzw.: Wert~)* security

Papier-: ~**geld das** paper money; ~**korb der** waste-paper basket

Pappe die; ~, ~n cardboard

Pappel die; ~, ~n poplar

päppeln *tr. V.* feed up

Papp·karton der cardboard box

Paprika der; ~s, ~[s] **a)** pepper; **b)** *o. Pl. (Gewürz)* paprika

Papst der; ~[e]s, Päpste pope; **päpst-lich** *Adj.* papal

Parabel die; ~, ~n **a)** *(bes. Litera-turw.)* parable; **b)** *(Math.)* parabola

Parade die; ~, ~n parade

Paradies das; ~es, ~e paradise; **pa-radiesisch** *Adj.* paradisical; *(herr-lich)* heavenly

paradox *Adj.* paradoxical

Paragraph der; ~en, ~en section; *(in Vertrag)* clause

parallel 1. *Adj.* parallel; **2.** *adv.* ~ ver-laufen run parallel (mit, zu to); **Paral-lele die**; ~, ~n parallel; **Parallelo-gramm das**; ~s, ~e parallelogram; **Parallel·straße die** street running parallel *(Gen.* to)

Para·nuß die Brazil-nut

Parasit der; ~en, ~en *(auch fig.)* para-site

parat *Adj.* ready

Parfum [par'fœ:], **Parfüm das**; ~s, ~s perfume; **Parfümerie die**; ~, ~en perfumery; **parfümieren** *tr. V.* per-fume

Pariser 1. *indekl. Adj.* Parisian; Paris *attrib.*; **2.** der; ~s, ~ Parisian; **Pari-serin die**; ~, ~nen Parisian

Parität die; ~, ~en parity

Park der; ~s, ~s park; *(Schloß~ usw.)* grounds *pl.*

Parka der; ~s, ~s parka

parken *tr., itr. V.* park; „**Parken ver-boten!"** 'No Parking'

Parkett das; ~[e]s, ~e **a)** parquet floor; **b)** *(Theater)* [front] stalls *pl.*; parquet *(Amer.)*

Park-: ~**gebühr die** parking-fee; ~**haus** das multi-storey car-park; ~**lücke die** parking-space; ~**platz** der car-park; parking lot *(Amer.)*; *(für ein einzelnes Fahrzeug)* parking-space; ~**scheibe die** parking-disc; ~**schein** der car-park ticket; ~**uhr** die parking-meter; ~**verbot das** ban on parking; **im** ~**verbot stehen** be parked illegally; ~**verbots·schild** das no-parking sign

Parlament das; ~[e]s, ~e parliament; **Parlamentarier der**; ~s, ~, **Parla-mentarierin die**; ~, ~nen member of parliament; **parlamentarisch** *Adj.* parliamentary

Parodie die; ~, ~n parody (**auf** + *Akk.* of)

Parole die; ~, ~n a) *(Wahlspruch)* motto; *(Schlagwort)* slogan; b) *(bes. Milit.: Kennwort)* password

Partei die; ~, ~en a) *(Politik, Rechtsw.)* party; b) *(Gruppe, Mannschaft)* side; **für jmdn. ~ ergreifen** *od.* **nehmen** side with sb.; **parteiisch 1.** *Adj.* biased; 2. *adv.* in a biased manner; **partei·los** *Adj. (Politik)* independent ⟨*MP*⟩; **Partei·tag** der party conference *or (Amer.)* convention

Parterre das; ~s, ~s ground floor; first floor *(Amer.)*

Partie die; ~, ~n a) part; b) *(Spiel, Sport: Runde)* game; *(Golf)* round; c) **eine gute ~ |für jmdn.| sein** be a good match [for sb.]

Partisan der; ~s *od.* ~en, ~en, **Partisanin** die, ~, ~nen guerrilla; *(gegen Besatzungstruppen im Krieg)* partisan

Partitur die; ~, ~en *(Musik)* score

Partizip das; ~s, ~ien [-'tsi:piən] *(Sprachw.)* participle

Partner der; ~s, ~, **Partnerin** die; ~, ~nen partner; **Partnerschaft** die; ~, ~en partnership; **partnerschaftlich 1.** *Adj.* ⟨*co-operation etc.*⟩ on a partnership basis; 2. *adv.* in a spirit of partnership; **Partner·stadt** die twin town *(Brit.)*; sister city *or* town *(Amer.)*

Party ['pa:ɐ̯ti] die; ~, ~s *od.* **Parties** party

Parzelle die; ~, ~n [small] plot [of land]

Paß der; **Passes, Pässe** a) *(Reise~)* passport; b) *(Gebirgs~; Ballspiele)* pass

passabel 1. *Adj.* reasonable; presentable ⟨*appearance*⟩; 2. *adv.* reasonably well

Passage [pa'sa:ʒə] die; ~, ~n a) [shopping] arcade; b) *(Abschnitt)* passage; **Passagier** [pasa'ʒi:ɐ̯] der; ~s, ~e passenger; **blinder ~**: stowaway

Passagier-: ~**dampfer** der passenger steamer; ~**flugzeug** das passenger aircraft; ~**liste** die passenger list

Paß·amt das passport office

Passant der; ~en, ~en, **Passantin** die; ~, ~nen passer-by

Paß·bild das passport photograph

Pässe s. Paß

passen itr. V. a) *(die richtige Größe/*

Form haben) fit; b) *(geeignet sein)* be suitable (**auf** + *Akk.*, **zu** for); *(harmonieren)* ⟨*colour etc.*⟩ match; **zu etw./jmdm. ~**: go well with sth./be well suited to sb.; **zueinander ~** ⟨*things*⟩ go well together; ⟨*two people*⟩ be suited to each other; c) *(genehm sein)* **jmdm. ~** ⟨*time*⟩ suit sb.; d) *(Kartenspiel)* pass; **passend** *Adj.* a) *(geeignet)* suitable ⟨*dress, present, etc.*⟩; right ⟨*words, expression, moment*⟩; b) *(harmonierend)* matching ⟨*shoes etc.*⟩

Paß·foto das s. **Paßbild**

passierbar *Adj.* passable ⟨*road*⟩; navigable ⟨*river*⟩; negotiable ⟨*path*⟩; **passieren 1.** *tr. V.* pass; **die Grenze ~**: cross the border; 2. *itr. V.; mit sein* happen

Passion die; ~, ~en a) passion; b) *(christl. Rel.)* Passion

passioniert *Adj.* passionate ⟨*collector, card-player, huntsman*⟩

passiv 1. *Adj.* passive; 2. *adv.* passively; **Passiv** das; ~s, ~e *(Sprachw.)* passive; **Passivität** die; ~: passivity

Paß-: ~**kontrolle** die passport check; ~**zwang** der obligation to carry a passport

Paste die; ~, ~n paste

Pastell das; ~|e|s, ~e a) *(Farbton)* pastel shade; b) *o. Pl. (Maltechnik)* pastel *no art.*

Pastell-: ~**farbe** die pastel colour; ~**ton** der pastel shade

Pastete die; ~, ~n a) *(gefüllte ~)* vol-au-vent; b) *(in einer Schüssel o. ä. gegart)* pâté; *(in einer Hülle aus Teig gebacken)* pie

pasteurisieren [pastøri'zi:rən] *tr. V.* pasteurize

Pastille die; ~, ~n pastille

Pastor der; ~s, ~en, **Pastorin** die; ~, ~nen pastor

Pate der; ~n, ~n godfather; *(männlich od. weiblich)* godparent

Paten-: ~**kind** das godchild; ~**onkel** der godfather; ~**stadt** die s. **Partnerstadt**

patent *(ugs.)* 1. *Adj.* a) *(tüchtig)* capable; b) *(zweckmäßig)* ingenious; 2. *adv.* ingeniously; neatly ⟨*solved*⟩; **Patent** das; ~|e|s, ~e a) *(Schutz)* patent; **etw. zum** *od.* **als ~ anmelden** apply for a patent for sth.; b) *(Erfindung)* [patented] invention

Paten·tante die godmother

patentieren *tr. V.* patent; **Patentlösung** die patent remedy (**für, zu** for)

Pater der; ~s, ~ *od.* **Patres** *(kath. Kirche)* Father; **Paternoster** der; ~s, ~ *(Aufzug)* paternoster [lift]

pathetisch 1. *Adj.* emotional ⟨*speech, manner*⟩; melodramatic ⟨*gesture*⟩; pompous ⟨*voice*⟩; **2.** *adv.* emotionally; *(dramatisch)* [melo]dramatically; **Pathos** das; ~: emotionalism

Patient [pa'tsiɛnt] der; ~en, ~en, **Patientin** die; ~, ~nen patient

Patin die; ~, ~nen godmother

Patres *s.* Pater

Patriot der; ~en, ~en, **Patriotin** die; ~, ~nen patriot; **patriotisch 1.** *Adj.* patriotic; **2.** *adv.* patriotically; **Patriotismus** der; ~: patriotism

Patrone die; ~, ~n cartridge

Patrouille [pa'trʊljə] die; ~, ~n patrol; **patrouillieren** [patrʊl'jiːrən] *itr. V.; auch mit sein* be on patrol

Patsche die; ~, ~n *(ugs.) s.* **Klemme**; **patschen** *itr. V., mit sein (ugs.)* splash; **patsch·naß** *Adj. (ugs.)* sopping wet

patzig *(ugs.)* **1.** *Adj.* snotty *(coll.)*; *(frech)* cheeky; **2.** *adv.* snottily *(coll.)*; *(frech)* cheekily

Pauke die; ~, ~n kettledrum; **auf die ~ hauen** *(ugs.) (feiern)* paint the town red *(sl.)*; *(sich lautstark äußern)* come right out with it

pausbäckig *Adj.* chubby-faced; chubby ⟨*face*⟩

pauschal 1. *Adj.* **a)** all-inclusive ⟨*price, settlement*⟩; **b)** *(verallgemeinernd)* sweeping ⟨*judgement, criticism, statement*⟩; indiscriminate ⟨*prejudice*⟩; wholesale ⟨*discrimination*⟩; **2.** *adv.* **a)** ⟨*cost*⟩ all in all; ⟨*pay*⟩ in a lump sum; **b)** *(ohne zu differenzieren)* wholesale

Pauschale die; ~, ~n flat-rate payment

Pauschal-: ~**preis** der flat rate; *(Inklusivpreis)* all-in price; ~**reise die** package holiday; *(mit mehreren Reisezielen)* package tour

Pause die; ~, ~n break; *(Ruhe~)* rest; *(Theater)* interval *(Brit.)*; intermission *(Amer.)*

pausen *tr. V.* trace; *(eine Lichtpause machen)* Photostat *(Brit.* P)

pausen·los 1. *Adj.*; incessant ⟨*noise, moaning, questioning*⟩; continous ⟨*work, operation*⟩; **2.** *adv.* incessantly; ⟨*work*⟩ non-stop

Pavian ['paːvịaːn] der; ~s, ~e baboon

Pavillon ['pavịljɔn] der; ~s, ~s pavilion

Pazifik der; ~s Pacific; **pazifisch** *Adj.* Pacific ⟨*area*⟩; **der Pazifische Ozean** the Pacific Ocean

Pech das; ~[e]s, ~e **a)** pitch; **b)** *o. Pl. (Mißgeschick)* bad luck; **pech·schwarz** *Adj. (ugs.)* jet-black

Pedal das; ~s, ~e pedal

Pediküre die; ~, ~n pedicure; **pediküren** *tr. V.* pedicure

Pegel der; ~s, ~ **a)** water-level indicator; *(Tide~)* tide-gauge; **b)** *(Wasserstand)* water-level

peilen *tr. V.* take a bearing on ⟨*transmitter, fixed point*⟩

Pein die; ~ *(geh.)* torment; **peinigen** *tr. V. (geh.)* torment; *(foltern)* torture; **peinlich 1.** *Adj.* **a)** embarrassing; awkward ⟨*question, position, pause*⟩; **es ist mir sehr ~:** I feel very bad *(coll.)* or embarrassed about it; **b)** *(äußerst genau)* meticulous; **2.** *adv.* **a)** unpleasantly ⟨*surprised*⟩; **b)** *(überaus [genau])* meticulously; **Peinlichkeit** die; ~, ~en **a)** *o. Pl.* embarrassment; **die ~ der Situation** the awkwardness of the situation; **b)** *o. Pl. (Genauigkeit)* meticulousness; **c)** *(peinliche Situation)* embarrassing situation

Peitsche die; ~, ~n whip; **peitschen** *tr. V.* whip; *(fig.)* ⟨*storm, waves, rain*⟩ lash

Pelikan der; ~s, ~e pelican

Pelle die; ~, ~n *(bes. nordd.)* skin; *(abgeschält)* peel; **pellen** *(bes. nordd.) tr., refl. V.* peel; **Pell·kartoffel die** potato boiled in its skin

Pelz der; ~es, ~e **a)** fur; coat; *(des toten Tieres)* skin; pelt; **b)** *o. Pl. (Material)* fur; *(~mantel)* fur coat; **Pelz·mantel** der fur coat

Pendel das; ~s, ~: pendulum

pendeln *itr. V.* **a)** swing [to and fro]; *(mit weniger Bewegung)* dangle; **b)** *mit sein* ⟨*bus, ferry, etc.*⟩ operate a shuttle service; ⟨*person*⟩ commute

penetrant 1. *Adj.* **a)** penetrating ⟨*smell, taste*⟩; overpowering ⟨*stink, perfume*⟩; **b)** *(aufdringlich)* pushing, *(coll.)* pushy ⟨*person*⟩; overbearing ⟨*tone, manner*⟩; aggressive ⟨*question*⟩; **2.** *adv.* **a)** overpoweringly; **b)** *(aufdringlich)* overbearingly

penibel 1. *Adj.* over-meticulous ⟨*person*⟩; *(pedantisch)* pedantic; **2.** *adv.* painstakingly; over-meticulously ⟨*dressed*⟩

Penis der; ~, ~se penis

Penner der; ~s, ~ *(salopp)* tramp *(Brit.)*; hobo *(Amer.)*

Pensen s. Pensum
Pension [pãˈzi̯oːn] die; ~, ~en a) o. Pl.
(Ruhestand) in ~ gehen retire; in ~
sein be retired; b) (Ruhegehalt) [retire-
ment] pension; c) (Haus für
[Ferien]gäste) guest-house; d) o. Pl.
(Unterkunft u. Verpflegung) board;
Pensionär [pãzi̯oˈnɛːɐ̯] der; ~s, ~e,
Pensionärin die; ~, ~nen retired
civil servant; **pensionieren** tr. V.
pension off; retire; sich [vorzeitig] ~
lassen take [early] retirement; **Pen-
sionierung** die; ~, ~en retirement
Pensum das; ~s, **Pensen** work quota
per Präp. mit Akk. a) (mittels) by; ~
Adresse X care of X; c/o X; b) (Kaufmannsspr.: [bis] zum) by; (am) on; c)
(Kaufmannsspr.: pro) per
perfekt 1. Adj. a) perfect ⟨crime,
host⟩; faultless ⟨English, French, etc.⟩;
b) ~ sein (ugs.: abgeschlossen, fertig
sein) be finalized; 2. adv. perfectly;
Perfekt das; ~s (Sprachw.) perfect
Pergament·papier das grease-proof
paper
Periode die; ~, ~n period
Perle die; ~, ~n a) (auch fig.) pearl; b)
(aus Holz, Glas o. ä.) bead; **Perlmutt**
das; ~s mother-of-pearl
Perlon Ⓦ das; ~s ≈ nylon
Perser der; ~s, ~ a) Persian; b) s. Per-
serteppich; **Perserin** die; ~, ~nen
Persian; **Perser·teppich** der Persian carpet; **Persianer** der; ~s, ~
(~mantel) Persian lamb coat; **Per-
sien** (das); ~s Persia; **persisch** Adj.
Persian
Person die; ~, ~en person; (in der
Dichtung, im Film) character; **Personal** das; ~s (in einem Betrieb o. ä.)
staff; (im Haushalt) domestic staff pl.;
Personal·ausweis der identity
card; **Personalien** Pl. personal particulars; **Personal·pronomen** das
(Sprachw.) personal pronoun
Personen-: ~kraftwagen der (bes.
Amtsspr.) private car or (Amer.) automobile; ~name der personal name;
~wagen der (Auto) [private] car;
automobile (Amer.); (im Unterschied
zum Lastwagen) passenger car or
(Amer.) automobile; ~zug der stopping train
persönlich 1. Adj. personal; ~ werden get personal; 2. adv. personally;
(auf Briefen) 'private [and confiden-
tial]'; **Persönlichkeit** die; ~, ~en a)
personality; b) (Mensch) person of
character; eine ~ sein have a strong

personality; ~en des öffentlichen Le-
bens public figures
Perspektive die; ~, ~n perspective;
(Blickwinkel) angle; (Zukunftsaus-
sicht) prospect
Peru (das); ~s Peru; **Peruaner** der;
~s, ~: Peruvian; **peruanisch** Adj.
Peruvian
Perücke die; ~, ~n wig
pervers Adj. perverted
Pessimismus der; ~: pessimism;
Pessimist der; ~en, ~en, **Pessimi-
stin** die; ~, ~nen pessimist; **pessi-
mistisch** 1. Adj. pessimistic; 2. adv.
pessimistically
Pest die; ~: plague
Petersilie [petɐˈziːli̯ə] die; ~: parsley
Petroleum [peˈtroːleʊm] das; ~s par-
affin (Brit.); kerosene (Amer.)
Petrus (der); **Petri** (christl. Rel.: Apo-
stel) St Peter
Pf Abk. Pfennig
Pfad der; ~[e]s, ~e path
Pfad-: ~finder der Scout; ~finderin
die; ~, ~nen Guide (Brit.); girl scout
(Amer.)
Pfaffe der; ~n, ~n (abwertend) cleric;
Holy Joe (derog.)
Pfahl der; ~[e]s, Pfähle post; stake
Pfand das; ~[e]s, Pfänder a) security;
pledge (esp. fig.); b) (für Flaschen
usw.) deposit (auf + Dat. on); **pfän-
den** tr. V. seize [under distress] (Law)
⟨goods, chattels⟩; attach ⟨wages etc.⟩
(Law); **Pfänder** s. Pfand; **Pfän-
dung** die; ~, ~en seizure; distraint
(Law); (von Geldsummen, Vermögens-
rechten) attachment (Law)
Pfanne die; ~, ~n [frying-]pan;
Pfann·kuchen der a) pancake; b)
(Berliner ~) doughnut
Pfarrei die; ~, ~en a) (Bezirk) parish;
b) (Dienststelle) parish office; c) s.
Pfarrhaus; **Pfarrer** der; ~s, ~ pastor;
(anglikanisch) vicar; (von Freikirchen)
minister; **Pfarrerin** die; ~, ~nen
[woman] pastor; (in Freikirchen)
[woman] minister; **Pfarr·haus** das
vicarage; (katholisch) presbytery; (in
Schottland) manse
Pfau der; ~[e]s, ~en peacock;
Pfauen·auge das peacock butterfly
Pfd. Abk. Pfund lb.
Pfeffer der; ~s, ~: pepper; **Pfeffer·-
kuchen** der ≈ gingerbread; **Pfefferminz** o. Art., indekl. peppermint;
Pfeffer·minze die peppermint
[plant]; **Pfefferminz·tee** der peppermint tea

pfeffern *tr. V.* season with pepper
Pfeife die; ~, ~n pipe; *(Triller~)*
whistle; **pfeifen 1.** *unr. itr. V.*
whistle; ⟨*bird*⟩ sing; *(auf einer Triller-
pfeife o. ä.)* ⟨*policeman, referee, etc.*⟩
blow one's whistle; **auf jmdn./etw.** ~
(ugs.) not give a damn about sb./sth.;
2. *unr. tr. V.* whistle ⟨*tune etc.*⟩; ⟨*bird*⟩
sing ⟨*song*⟩; *(auf einer Pfeife)* pipe,
play ⟨*tune etc.*⟩
Pfeil der; ~|e|s, ~e arrow
Pfeiler der; ~s, ~: pillar; *(Brücken~)*
pier
Pfennig der; ~s, ~e pfennig; **es kostet
20** ~: it costs 20 pfennig[s]
pferchen *tr. V.* cram; pack
Pferd das; ~|e|s, ~e horse; *(Schachfi-
gur)* knight; **mit ihr kann man** ~**e stehlen** *(ugs.)* she's game for anything
Pferde-: ~**rennen das** horse-race;
(Sportart) horse-racing; ~**schwanz
der** *(Frisur)* pony-tail; ~**stall der**
stable
pfiff *1. u. 3. Pers. Sg. Prät. v.* pfeifen;
Pfiff der; ~|e|s, ~e **a)** whistle; **b)**
(ugs.: besonderer Reiz) style
Pfifferling der; ~s, ~e chanterelle;
keinen *od.* **nicht einen** ~ **wert sein**
(ugs.) be not worth a bean *(sl.)*
pfiffig 1. *Adj.* smart; bright ⟨*idea*⟩;
artful ⟨*smile, expression*⟩; **2.** *adv.* art-
fully
Pfingsten das; ~, ~: Whitsun
Pfingst-: ~**montag der** Whit Monday *no def. art.;* ~**sonntag der** Whit
Sunday *no def. art.*
Pfirsich der; ~s, ~e peach
Pflanze die; ~, ~n plant; **pflanzen**
tr. V. plant; **Pflanzen · öl das** vegetable oil; **pflanzlich** *Adj.* plant *attrib.*
⟨*life, motif*⟩; vegetable ⟨*dye, fat*⟩
Pflaster das; ~s, ~ **a)** *(Straßen~)*
road surface; *(auf dem Gehsteig)*
pavement; **ein teures/gefährliches** ~
(ugs.) an expensive/dangerous place
or spot to be; **b)** *(Wund~)* sticking-
plaster; **pflastern** *tr. (auch itr.) V.*
surface; *(mit Kopfsteinpflaster, Stein-
platten)* pave; **Pflaster · stein der**
paving-stone; *(Kopfstein)* cobble-
stone
Pflaume die; ~, ~n plum; **getrocknete**
~**n** [dried] prunes
Pflege die; ~: care; *(Maschinen~,
Fahrzeug~)* maintenance; *(fig.: von
Beziehungen, Kunst, Sprache)* cultivation; **jmdn./etw. in** ~ *(Akk.)* **nehmen**
look after sb./sth.
pflege-, Pflege-: ~**eltern** *Pl.* foster-

parents; ~**fall der: ein** ~ **sein** be in
[permanent] need of nursing; ~**kind
das** foster-child; ~**leicht** *Adj.* easy-
care *attrib.* ⟨*textiles, flooring*⟩
pflegen 1. *tr. V.* look after; care for;
take care of ⟨*skin, teeth, floor*⟩; look
after ⟨*bicycle, car, machine*⟩; look
after, tend ⟨*garden, plants*⟩; cultivate
⟨*relations, arts, interests*⟩; foster ⟨*con-
tacts, co-operation*⟩; pursue ⟨*hobby*⟩;
2. *mod. V.* **etw. zu tun** ~: usually do
sth.; **Pfleger der;** ~s, ~ **a)** *(Kran-
ken~)* [male] nurse; **b)** *(Tier~)*
keeper; **Pflegerin die;** ~, ~**nen a)**
(Kranken~) nurse; **b)** *(Tier~)* keeper
Pflicht die; ~, ~en duty
pflicht-, Pflicht-: ~**bewußt 1.** *Adj.*
conscientious; **2.** *adv.* with a sense of
duty; ~**bewußtsein das,** ~**gefühl
das;** *o. Pl.* sense of duty; ~**übung die**
(fig.) ritual exercise
Pflock der; ~|e|s, Pflöcke peg
pflücken *tr. V.* pick
Pflug der; ~|e|s, Pflüge plough; **pflü-
gen** *tr., itr. V.* plough
Pforte die; ~, ~n *(Tor)* gate; *(Tür)*
door; *(Eingang)* entrance; **Pförtner
der;** ~s, ~ porter; *(eines Wohnblocks,
Büros)* door-keeper; *(am Tor)* gate-
keeper
Pfosten der; ~s, ~ post
Pfote die; ~, ~n paw
Pfropf der; ~|e|s, ~e blockage;
pfropfen *tr. V. (ugs.)* cram; stuff; **ge-
pfropft voll** crammed [full]; packed;
Pfropfen der stopper; *(Korken)*
cork; *(für Fässer)* bung
pfui *Interj.* ugh; ~ **rufen** boo
Pfund das; ~|e|s, ~e pound
Pfütze die; ~, ~n puddle
Phänomen das; ~s, ~e phenomenon
Phantasie die; ~, ~n **a)** *o. Pl.* imagination; **b)** *meist Pl. (Produkt der* ~*)*
fantasy; **phantasie · los 1.** *Adj.* unimaginative; **2.** *adv.* unimaginatively;
phantasie · voll 1. *Adj.* imaginative;
2. *adv.* imaginatively; **phantastisch
1.** *Adj.* **a)** fantastic; ⟨*idea*⟩ divorced
from reality; **b)** *(ugs.: großartig)* fantastic *(coll.);* **2.** *adv. (ugs.)* fantastically
(coll.)
Phase die; ~, ~n phase
Philosoph der; ~**en,** ~**en** philosopher; **Philosophie die;** ~, ~**n**
philosophy; **philosophieren** *itr.
(auch tr.) V.* philosophize; **philosophisch 1.** *Adj.* philosophical; ⟨*dictionary, principles*⟩ of philosophy; **2.**
adv. philosophically

Photo 652

Photo das; ~s, ~s s. Foto

Phrase die; ~, ~n *(abwertend)* [empty] phrase; cliché

Physik die; ~: physics *sing., no art.;* **physikalisch** *Adj.* physics *attrib.* ⟨*experiment, formula, research, institute*⟩; physical ⟨*map, process*⟩; **Physiker** der; ~s, ~ physicist; **physisch** 1. *Adj.* physical; 2. *adv.* physically

Pianist der; ~en, ~en, **Pianistin** die; ~, ~nen pianist

Pickel der; ~s, ~: pimple

picken 1. *itr. V.* peck **(nach** at; **an +** *Akk.,* **gegen** on, against); 2. *tr. V.* ⟨*bird*⟩ peck; *(ugs.)* ⟨*person*⟩ pick

Picknick das; ~s, ~e *od.* ~s picnic

piep[s]en *itr. V. (ugs.)* squeak; ⟨*small bird*⟩ cheep

Pietät [pie'tɛ:t] die; ~: respect; *(Ehrfurcht)* reverence

Pik das; ~[s], ~[s] *(Kartenspiel)* **a)** *(Farbe)* spades *pl.;* **b)** *(Karte)* spade

pikant 1. *Adj.* **a)** piquant; **b)** *(fig.: witzig)* ironical; **c)** *(verhüll.: schlüpfrig)* racy ⟨*joke, story*⟩; 2. *adv.* piquantly ⟨*seasoned*⟩

pikiert 1. *Adj.* piqued; 2. *adv.* ⟨*reply, say*⟩ in an aggrieved tone

Pilger der; ~s, ~: pilgrim; **pilgern** *itr. V.* go on a pilgrimage

Pille die; ~, ~n pill

Pilot der; ~en, ~en pilot

Pils das; ~, ~: Pils

Pilz der; ~es, ~e fungus; *(Speise~, auch fig.)* mushroom

Pinguin der; ~s, ~e penguin

Pinie ['pi:niə] die; ~, ~n [stone- *or* umbrella] pine

pinkeln *itr. V. (salopp)* pee *(coll.)*

Pinsel der; ~s, ~: brush; *(Mal~)* paintbrush

Pinzette die; ~, ~n tweezers *pl.*

Pionier der; ~s, ~e *(Milit.)* sapper; *(fig.: Wegbereiter)* pioneer

Pirat der; ~en, ~en pirate

pissen *itr. V. (derb)* piss *(coarse)*

Pistazie [pɪs'ta:tsiə] die; ~, ~n pistachio

Piste die; ~, ~n *(Ski~)* piste; *(Renn~)* course; *(Flugw.)* runway

Pistole die; ~, ~n pistol

Pizza die; ~, ~s *od.* **Pizzen** pizza

Pkw, PKW ['pe:ka:ve:] der; ~[s], ~[s] [private] car; automobile *(Amer.)*

plädieren *itr. V. (Rechtsw.)* plead **(auf +** *Akk.* for); *(fig.)* argue; **Plädoyer** [plɛdɔa'je:] das; ~s, ~s *(Rechtsw.)* summing up *(for the defence/prosecution); (fig.)* plea

Plage die; ~, ~n **a)** nuisance; **b)** *(ugs.: Mühe)* bother; trouble; **plagen** 1. *tr. V.* **a)** torment; **b)** *(ugs.: bedrängen)* harass; *(mit Bitten, Fragen)* pester; 2. *refl. V.* **a)** *(sich abmühen)* slave away; **b)** *(leiden)* **sich mit etw.** ~: be bothered by sth.

Plakat das; ~[e]s, ~e poster; **Plakette** die; ~, ~n badge

Plan der; ~[e]s, **Pläne a)** plan; **b)** *(Karte)* map; plan

Plane die; ~, ~n tarpaulin

planen *tr., itr. V.* plan

Planet der; ~en, ~en planet

planieren *tr. V.* level; grade; **Planier·raupe** die bulldozer

Planke die; ~, ~n plank

plan-: ~**los** 1. *Adj.* aimless; *(ohne System)* unsystematic; 2. *adv. s.* **1:** aimlessly; unsystematically; ~**mäßig** 1. *Adj.* **a)** scheduled ⟨*service, steamer*⟩; ~**mäßige Ankunft/Abfahrt** scheduled time of arrival/departure; **b)** *(systematisch)* systematic; 2. *adv.* **a)** *(wie geplant)* according to plan; *(pünktlich)* on schedule; **b)** *(systematisch)* systematically

Plansch·becken das paddling-pool; **planschen** *itr. V.* splash [about]

Plantage [plan'ta:ʒə] die; ~, ~n plantation

Planung die; ~, ~en planning; **Plan·wirtschaft** die planned economy

¹**Plastik** die; ~, ~en sculpture; ²**Plastik** das; ~s *(ugs.)* plastic

Plastik-: ~**beutel** der, ~**tüte** die plastic bag

Platane die; ~, ~n plane-tree

Platin das; ~s platinum

plätschen *itr. V.* **a)** splash; **b)** *mit sein (~d auftreffen)* splash **(an +** *Akk.,* **gegen** against); **plätschern** *itr. V.* **a)** splash; ⟨*rain*⟩ patter; ⟨*stream*⟩ burble; **b)** *mit sein* ⟨*stream*⟩ burble along

platt *Adj.* flat; **ein Platter** *(ugs.)* a flat *(coll.)*

platt·deutsch *Adj.* Low German

Platte die; ~, ~n **a)** *(Stein~)* slab; *(Metall~)* plate; sheet; *(Span~, Hartfaser~ usw.)* board; *(Tisch~)* [table-]top; *(Grab~)* [memorial] slab; **b)** *(Koch~)* hotplate; **c)** *(Schall~)* [gramophone] record; **d)** *(Teller)* plate; *(zum Servieren, aus Metall)* dish; **kalte** ~: selection of cold meats [and cheese]; **Platten·spieler** der record-player; **Platt·fuß** der **a)** flat foot; **b)** *(ugs.: Reifenpanne)* flat *(coll.)*

Platz der; ~es, **Plätze a)** square; **b)** *(Sport~)* ground; *(Spielfeld)* field; *(Tennis~, Volleyball~ usw.)* court; *(Golf~)* course; **c)** *(Stelle, wo jmd., etw. hingehört)* place; **nicht** *od.* **fehl am ~\|e\| sein** *(fig.)* be out of place; **d)** *(Sitz~)* seat; *(am Tisch, Steh~ usw.)* place; ~ **nehmen** sit down; **e)** *(bes. Sport: Plazierung)* place; **f)** *(Ort)* place; **am ~** : in the town/village; **g)** *o. Pl. (Raum)* space; room; ~ **machen** make room *(Dat.* for); **Plätzchen das;** ~s, ~ **a)** little place; **b)** *(Keks)* biscuit *(Brit.);* cookie *(Amer.)*

platzen *itr. V.; mit sein* **a)** burst; *(explodieren)* explode; **b)** *(ugs.: scheitern)* fall through; **der Wechsel/das Treffen ist geplatzt** the bill has bounced *(sl.)*/the meeting is off; **c) in eine Versammlung ~** *(ugs.)* burst into a meeting

Platz-: ~**karte** die reserved-seat ticket; ~**konzert** das open-air concert *(by a military or brass band);* ~**mangel** der lack of space; ~**regen** der cloudburst; ~**wunde** die lacerated wound

plaudern *itr. V.* chat

plausibel *Adj.* plausible

pleite *(ugs.)* ~ **sein** ⟨person⟩ be broke *(coll.);* ⟨company⟩ have gone bust *(coll.);* ~ **gehen** go bust *(coll.);* **Pleite die;** ~, ~**n** *(ugs.)* **a)** *(Bankrott)* bankruptcy *no def. art.;* ~ **machen** go bust *(coll.);* **b)** *(Mißerfolg)* wash-out *(sl.)*

Plissee das; ~s, ~s accordion pleats *pl.*

Plombe die; ~, ~**n a)** *(Siegel)* [lead] seal; **b)** *(veralt.: Zahnfüllung)* filling; **plombieren** *tr. V.* **a)** *(versiegeln)* seal; **b)** *(veralt.)* fill ⟨tooth⟩

plötzlich 1. *Adj.* sudden; **2.** *adv.* suddenly

plump 1. *Adj.* **a)** *(dick)* plump; *(unförmig)* ungainly ⟨shape⟩; *(rundlich)* bulbous; **b)** *(schwerfällig)* clumsy ⟨movements, style⟩; **c)** *(fig.) (dreist)* crude ⟨lie, deception, trick⟩; *(leicht durchschaubar)* blatantly obvious; *(unbeholfen)* clumsy ⟨excuse, advances⟩; crude ⟨joke, forgery⟩; **2.** *adv.* **a)** *(schwerfällig)* clumsily; **b)** *(fig.)* in a blatantly obvious manner

plündern *itr., tr. V.* **a)** loot; plunder ⟨town⟩; **b)** *(scherzh.)* raid ⟨larder, fridge, account⟩

Plural der; ~s, ~e plural

plus *Konj., Adv.* plus; **Plus das;** ~ : surplus; *(Vorteil)* advantage

Plüsch der; ~\|e\|s, ~e plush

Plusquam·perfekt das pluperfect [tense]

PLZ *Abk.* Postleitzahl

Po der; ~s, ~s *(ugs.)* bottom

Pöbel der; ~s rabble

pochen *itr. V. (klopfen)* knock **(gegen/ an** + *Akk.* at, on); *(geh.: pulsieren)* ⟨heart⟩ pound

Pocken *Pl.* smallpox *sing.*

Podest das *od.* **der;** ~\|e\|s, ~e rostrum; **Podium das;** ~s, **Podien** *(Plattform)* platform; *(Bühne)* stage; *(trittartige Erhöhung)* rostrum

Poesie die; ~ : poetry; **Poet der;** ~en, ~en *(veralt.)* poet; bard *(literary);* **poetisch 1.** *Adj.* poetic\[al\]; **2.** *adv.* poetically

Pointe \['poɛ̃:tə\] die; ~, ~n *(eines Witzes)* punch line; *(einer Geschichte)* point; *(eines Sketches)* curtain line

Pokal der; ~s, ~e **a)** *(Trinkgefäß)* goblet; **b)** *(Siegestrophäe, ~wettbewerb)* cup

Pökel·fleisch das salt meat; **pökeln** *tr. V.* salt

Poker das *od.* **der;** ~s poker; **pokern** *itr. V.* play poker

Pol der; ~s, ~e pole

Pole der; ~n, ~n Pole

polemisch 1. *Adj.* polemic\[al\]; **2.** *adv.* polemically

Polen das; ~s Poland

Police \[po'li:sə\] die; ~, ~n *(Versicherungsw.)* policy

polieren *tr. V.* polish

Poli·klinik die out-patients' clinic

Polin die; ~n, ~nen Pole

Politik die; ~, ~en **a)** *o. Pl.* politics *sing., no art.;* **b)** *(eine spezielle ~)* policy; **Politiker der;** ~s, ~, **Politikerin die;** ~, ~nen politician; **politisch 1.** *Adj.* political; **2.** *adv.* politically; **politisieren 1.** *itr. V.* talk politics; **2.** *tr. V.* make politically active

Politur die; ~, ~en polish

Polizei die; ~, ~en police *pl.*

Polizei-: ~**auto** das police car; ~**beamte** der police officer; ~**kontrolle** die police check

polizeilich 1. *Adj.* police; ~e **Meldepflicht** obligation to register with the police; **2.** *adv.* by the police

Polizei-: ~**präsidium das** police headquarters *sing. or pl.;* ~**revier** das police station; ~**streife** die police patrol; ~**stunde** die closing time; ~**wache** die police station

Polizist der; ~en, ~en policeman

polnisch *Adj.* Polish
Polster das; ~s, ~: upholstery *no pl.,
no indef. art.;* Polster·möbel *Pl.*
upholstered furniture *sing.;* polstern *itr. V.* upholster ⟨*furniture*⟩
poltern *itr. V.* a) crash about; b) *mit
sein der Karren polterte über das Pflaster* the cart clattered over the cobblestones
Polyp der; ~en, ~en *(Zool., Med.)*
polyp
Pommern (das); ~s Pomerania
Pommes frites [pɔm'frit] *Pl.* chips
(Brit.); French fries *(Amer.)*
pompös 1. *Adj.* grandiose; 2. *adv.*
grandiosely
¹Pony ['pɔni] das; ~s, ~s pony; ²Pony
der; ~s, ~s *(Frisur)* fringe
Popeline·mantel der poplin coat
populär 1. *Adj.* popular (bei with); 2.
adv. popularly; Popularität die; ~:
popularity
Pore die; ~, ~n pore
Pornographie die; ~: pornography
Porree der; ~s leek
Portal das; ~s, ~e portal
Portemonnaie [pɔrtmɔ'neː] das; ~s,
~s purse
Porti *Pl. s.* Porto
Portier [pɔr'tie:] der; ~s, ~s, *österr.:*
[pɔr'tiːɐ] der; ~s, ~e porter
Portion [pɔr'tsioːn] die; ~, ~en a)
(beim Essen) portion; helping; b)
(ugs.: Anteil) amount
Porto das; ~s, ~s *od.* Porti postage
(für on, for)
Portugal (das); ~s Portugal; Portugiese der; ~n, ~n Portuguese; portugiesisch *Adj.* Portuguese
Portwein der port
Porzellan das; ~s porcelain; china
Posaune die; ~, ~n trombone
Position [pozi'tsioːn] die; ~, ~en position; positiv 1. *Adj.* positive; 2. *adv.*
positively; Positiv das; ~s, ~e *(Fot.)*
positive
Possessiv·pronomen das
(Sprachw.) possessive pronoun
Post die; ~, ~en a) post *(Brit.);* mail;
etw. mit der *od.* per ~ schicken send
sth. by post *or* mail; b) *(~amt)* post
office
Post-: ~amt das post office; ~anweisung die postal remittance form;
~auto das mail van; ~bote der
(ugs.) postman *(Brit.);* mailman
(Amer.)
Posten der; ~s, ~ a) post; b) *(bes. Milit.: Wachmann)* sentry

post-, Post-: ~fach das post-office
or PO box; *(im Büro, Hotel o. ä.)* pigeon-hole; ~karte die postcard;
~lagernd *Adj., adv.* poste restante;
general delivery *(Amer.);* ~leitzahl
die postcode; Zip code *(Amer.);*
~stempel der *(Abdruck)* postmark;
~wendend *Adv.* by return [of post]
potent *Adj.* potent; Potenz die; ~,
~en a) *o. Pl.* potency; b) *(Math.)*
power; potenzieren *tr. V. (Math.)*
mit 5 ~: raise to the power [of] 5
Pracht die; ~: splendour; prächtig,
pracht·voll 1. *Adj.* splendid; 2. *adv.*
splendidly
prädestiniert *Adj.* predestined
Prädikat das; ~[e]s, ~e a) *(Auszeichnung)* rating; b) *(Sprachw.)* predicate
Prag (das); ~s Prague
prägen *tr. V.* a) emboss; b) mint
⟨*coin*⟩; c) *(fig.: beeinflussen)* shape
prägnant 1. *Adj.* concise; succinct; 2.
adv. concisely; succinctly
Prägung die; ~, ~en embossing; *(von
Münzen)* minting
prahlen *itr. V.* boast, brag (mit about)
Praktik die; ~, ~en practice; Praktika *s.* Praktikum; Praktikant der;
~en, ~en, Praktikantin die; ~, ~nen
a) *(in einem Betrieb)* student trainee;
b) *(an der Hochschule)* physics/chemistry student *(doing a period of practical training);* Praktikum das; ~s,
Praktika period of practical training;
praktisch 1. *Adj.* practical; ~er Arzt
general practitioner; 2. *adv.* practically; *(auf die Praxis bezogen; wirklich)*
in practice; praktizieren *tr. V.* practise
Praline die; ~, ~n [filled] chocolate
prall *Adj.* a) hard ⟨*ball*⟩; bulging ⟨*sack,
wallet, bag*⟩; big strong *attrib.* ⟨*thighs,
muscles, calves*⟩; well-rounded
⟨*breasts*⟩; b) *(intensiv)* blazing ⟨*sun*⟩;
prallen *itr. V.; mit sein* crash (gegen/
auf/an + *Akk.* into); collide (gegen/
auf/an + *Akk.* with)
Prämie ['prɛːmiə] die; ~, ~n a) *(Leistungs~; Wirtschaft)* bonus; *(Belohnung)* reward; *(Spar~, Versicherungs~)* premium; b) *(einer Lotterie)*
[extra] prize; prämieren *tr. V.* award
a prize to ⟨*person, film*⟩; give an award
for ⟨*best essay etc.*⟩
Pranger der; ~s, ~ *(hist.)* pillory
Pranke die; ~, ~n paw
Präparat das; ~[e]s, ~e preparation
Präposition die; ~, ~en *(Sprachw.)*
preposition

Prärie die; ~, ~n prairie

Präsens ['prɛːzɛns] das; ~ *(Sprachw.)* present [tense]; **präsentieren** *tr. V.* present

Präservativ das; ~s, ~e condom

Präsident der; ~en, ~en; **Präsidentin** die; ~, ~nen president; **Präsidium** das; ~s, Präsidien a) committee; b) *(Vorsitz)* chairmanship; c) *(Polizei~)* police headquarters *sing. or pl.*

prasseln *itr. V.* pelt down; ⟨*shots*⟩ clatter; ⟨*fire*⟩ crackle

prassen *itr. V.* live extravagantly; *(schlemmen)* feast

Präteritum das; ~s *(Sprachw.)* preterite [tense]

Praxis die; ~, **Praxen** a) o. Pl. *(im Unterschied zur Theorie)* practice *no art.*; *(Erfahrung)* [practical] experience; b) *(eines Arztes, Anwalts usw.)* practice; *(~räume)(eines Arztes)* surgery *(Brit.)*; office *(Amer.);(eines Anwalts usw.)* office

präzise 1. *Adj.* precise; 2. *adv.* precisely; **Präzision** die; ~: precision

predigen 1. *itr. V.* deliver a/the sermon; 2. *tr. V.* preach; **Prediger** der; ~s, ~: preacher; **Predigt** die; ~, ~en sermon

Preis der; ~es, ~e a) *(Kauf~)* price (für of); b) *(Belohnung)* prize; **Preis·aus·schreiben** das [prize] competition

Preisel·beere die cowberry; cranberry *(Gastr.)*

preisen *unr. tr. V. (geh.)* praise

preis-, Preis-: ~**günstig** 1. *Adj.* ⟨*goods*⟩ available at unusually low prices; **das ist [sehr]** ~**günstig** that is [very] good value; 2. *adv.* at a low price; ~**nachlaß** der price reduction; ~**schild** das price-tag; ~**steigerung** die increase in prices; ~**träger** der prizewinner; ~**verleihung** die presentation [of prizes/awards]; ~**wert** 1. *Adj.* good value *pred.;* 2. *adv.* ⟨*eat*⟩ at a reasonable price; **dort kann man** ~**wert einkaufen** you get good value for money there

Prellung die; ~, ~en bruise

Premiere [prə'mi̯eːrə] die; ~, ~n opening night

Presse die; ~, ~n a) press; *(Zitronen~)* squeezer; b) o. Pl. *(Zeitungen)* press

Presse-: ~**freiheit** die freedom of the press; ~**meldung** die press report

pressen *tr. V.* press

Preß·luft-: ~**bohrer** der pneumatic drill; ~**hammer** der pneumatic hammer

Prestige [prɛs'tiːʒə] das; ~s prestige

prickeln *itr. V.* tingle

pries 1. u. 3. Pers. Sg. Prät. v. **preisen**

Priester der; ~s, ~: priest; **Priesterin** die; ~, ~nen priestess

prima *(ugs.)* 1. *indekl. Adj.* great *(coll.);* 2. *adv.* ⟨*taste*⟩ great *(coll.);* ⟨*sleep*⟩ fantastically well *(coll.)*

primär 1. *Adj.* primary; 2. *adv.* primarily

Primel die; ~, ~n primula; *(Schlüsselblume)* cowslip

primitiv 1. *Adj.* primitive; *(einfach, schlicht)* simple; 2. *adv.* primitively; *(einfach, schlicht)* in a simple manner

Prinz der; ~en, ~en prince; **Prinzessin** die; ~, ~nen princess

Prinzip das; ~s, ~ien [-'tsiːpi̯ən] principle; **aus** ~: on principle; **prinzipiell** 1. *Adj.* in principle *postpos., not pred.;* ⟨*rejection*⟩ on principle; 2. *adv.* *(im Prinzip)* in principle; *(aus Prinzip)* on principle

Prise die; ~, ~n pinch

privat 1. *Adj.* private; *(persönlich)* personal; 2. *adv.* privately

Privat-: ~**adresse** die private *or* home address; ~**angelegenheit** die private matter; ~**besitz** der private property; ~**eigentum** das private property; ~**leben** das; o. Pl. private life; ~**lehrer** der private tutor; ~**patient** der private patient; ~**unterricht** der private tuition

pro *Präp. mit Akk.* per; ~ **Stück** each; a piece

pro-: pro-; ~**westlich/**~**kommunistisch** pro-western/pro-communist

Probe die; ~, ~n a) test; b) *(Muster, Teststück)* sample; c) *(Theater~, Orchester~)* rehearsal

Probe-: ~**fahrt** die trial run; *(vor dem Kauf, nach einer Reparatur)* test drive; ~**jahr** das probationary year

proben *tr., itr. V.* rehearse; **probeweise** *Adv.* ⟨*employ*⟩ on a trial basis; **Probe·zeit** die probationary period; **probieren** 1. *tr. V.* a) try; have a go at; b) *(kosten)* taste; try; c) *(aus~)* try out; *(an~)* try on ⟨*clothes, shoes*⟩; 2. *itr. V.* a) *(versuchen)* try; b) *(kosten)* have a taste

Problem das; ~s, ~e problem; **problematisch** *Adj.* problematic[al]; **problem·los** 1. *Adj.* problem-free; 2. *adv.* without any problems

Produkt das; ~[e]s, ~e *(auch Math.,*

fig.) product; **Produktion** die; ~, ~en production; **produktiv** 1. productive; prolific *(writer, artist, etc.);* 2. *adv. (work, co-operate)* productively; **Produktivität** die; ~: productivity; **Produzent** der; ~en, ~en producer; **produzieren** *tr. V.* produce

Prof. *Abk.* Professor Prof.; **professionell** 1. *Adj.* professional; 2. *adv.* professionally; **Professor** der; ~s, ~en; **Professorin** die; ~, ~nen professor; **Profi** der; ~s, ~s *(ugs.)* pro *(coll.)*

Profil das; ~s, ~e a) *(Seitenansicht)* profile; **im** ~: in profile; b) *(von Reifen, Schuhsohlen)* tread

Profit der; ~[e]s, ~e profit; **profitieren** *itr. V.* profit (von, bei by)

Prognose die; ~, ~n prognosis; *(Wetter~, Wirtschafts~)* forecast

Programm das; ~s, ~e a) programme; program *(Amer., Computing); (Ferns.: Sender)* channel

Programm-: ~heft das programme; ~hinweis der programme announcement

programmieren *tr. V.* a) *(DV)* program; b) *(auf etw. festlegen)* programme

Programm-: ~vorschau die *(im Fernsehen)* preview [of the week's/evening's *etc.* viewing]; *(im Kino)* trailers *pl.;* ~zeitschrift die radio and television magazine

progressiv 1. *Adj.* progressive; 2. *adv.* progressively

Projekt das; ~[e]s, ~e project; **Projektor** der; ~s, ~en projector; **projizieren** *tr. V. (Optik)* project

proklamieren *tr. V.* proclaim

Prolet der; ~en, ~en *(abwertend)* peasant; **Proletariat** das; ~[e]s proletariat; **Proletarier** [proleˈtaːri̯ɐ] der; ~s, ~: proletarian; **proletarisch** *Adj.* proletarian

Promenade die; ~, ~n promenade

Promille das; ~s, ~: [part] per thousand; **er fährt nur ohne** ~ *(ugs.)* he never drinks and drives; **er hatte 1,8** ~: he had a blood alcohol level of 1.8 per thousand; **Promille·grenze die** *(ugs.)* legal [alcohol] limit

prominent *Adj.* prominent; **Prominenz** die; ~: prominent figures *pl.*

prompt 1. *Adj.* prompt; 2. *adv.* a) promptly; b) *(ugs., meist iron.: wie erwartet)* [and] sure enough

Pronomen das; ~s, ~ *od.* **Pronomina** *(Sprachw.)* pronoun

Propaganda die; ~: propaganda; **propagieren** *tr. V.* propagate

Propan·gas das; *o. Pl.* propane

Propeller der; ~s, ~ propeller

Prophet der; ~en, ~en prophet; **prophezeien** *tr. V.* prophesy *(Dat.* for); predict *(result, weather)*

Proportion die; ~, ~en proportion

Prosa die; ~: prose

prosit *Interj.* your [very good] health; ~ Neujahr! happy New Year!

Prospekt der *od. (bes. österr.)* das ~[e]s, ~e *(Werbeschrift)* brochure; *(Werbezettel)* leaflet

prost *Interj. (ugs.)* cheers *(Brit. coll.)*

Prostituierte die/der; *adj. Dekl.* prostitute; **Prostitution** die; ~: prostitution *no art.*

Protest der; ~[e]s, ~e protest; **Protestant** der; ~en, ~en, **Protestantin** die; ~, ~nen Protestant; **protestantisch** *Adj.* Protestant; **protestieren** *itr. V.* protest, make a protest (gegen against, about); **Protest·kundgebung** die protest rally

Prothese die; ~, ~n artificial limb; prosthesis *(Med.); (Zahn~)* set of dentures; dentures *pl.*

Protokoll das; ~s, ~e a) *(wörtlich mitgeschrieben)* transcript; *(Ergebnis~)* minutes *pl.; (bei Gericht)* record; **etw. zu** ~ **geben** make a statement about sth.; b) *(diplomatisches Zeremoniell)* protocol; **protokollieren** 1. *tr. V.* take down; take the minutes of *(meeting);* minute *(remark);* 2. *itr. V.* take the minutes; *(bei Gericht)* keep the record

Proviant der; ~s, ~e provisions *pl.*

Provinz die; ~, ~en province; **provinziell** 1. *Adj.* provincial; 2. *adv.* provincially

Provision die; ~, ~en *(Kaufmannsspr.)* commission; **provisorisch** 1. *Adj.* provisional; temporary; 2. *adv.* temporarily

Provokation die; ~, ~en provocation; **provozieren** *tr. V.* provoke

Prozedur die; ~, ~en procedure

Prozent das; ~[e]s, ~e a) *nach Zahlenangaben Pl.* ungebeugt per cent *sing.;* **fünf** ~: five per cent; b) *Pl. (ugs.: Gewinnanteil)* share *sing.* of the profits; *(Rabatt)* discount *sing.;* **auf etw.** *(Akk.)* ~e **bekommen** get a discount on sth.; **-prozentig** *adj.* -per-cent

Prozent-: ~rechnung die percentage calculation; ~satz der percentage

prozentual 1. *Adj.* percentage; 2. *adv.* ~ **am Gewinn beteiligt sein** have a percentage share in the profits

Prozeß der; **Prozesses, Prozesse a)** trial; *(Fall)* [court] case; **einen ~ gewinnen/verlieren** win/lose a case; **b)** *(Vorgang)* process; **prozessieren** *itr. V.* go to court; **gegen jmdn. ~:** bring an action against sb.; **Prozeß·kosten** *Pl.* legal costs

prüde *(abwertend)* 1. *Adj.* prudish; 2. prudishly

prüfen *tr. V.* **a)** *auch itr.* examine ⟨*pupil, student, etc.*⟩; **mündlich/schriftlich geprüft werden** have an oral/a written examination; **b)** *(untersuchen)* examine **(auf** + *Akk.* for); check ⟨*device, machine, calculation*⟩ **(auf** + *Akk.* for); investigate ⟨*complaint*⟩; *(testen)* test **(auf** + *Akk.* for); **c)** *(kontrollieren)* check; examine ⟨*accounts, books*⟩; **d)** *(vor einer Entscheidung)* check ⟨*price*⟩; examine ⟨*offer*⟩; consider ⟨*application*⟩; **Prüfer** der; ~s, ~, **Prüferin** die; ~, ~nen **a)** inspector; *(Buch~)* auditor; **b)** *(im Examen)* examiner; **Prüfung** die; ~, ~en **a)** examination; exam *(coll.);* **eine ~ machen** *od.* **ablegen** take an examination; **b)** *s.* **prüfen b–d:** examination; check; investigation; test; consideration

Prügel *Pl. (Schläge)* beating *sing.; (als Strafe für Kinder)* hiding *(coll.);* **prügeln** 1. *tr. (auch itr.)* V. beat; 2. *refl. V.* **sich ~:** fight; **sich mit jmdm. ǀum etw.ǀ ~:** fight sb. [over *or* for sth.]

Prunk der; ~ǀeǀs splendour; magnificence

PS [peːˈʔɛs] das; ~, ~ : *Abk.* **Pferdestärke** h.p.

Psalm der; ~s, ~en psalm

Psychiater der; ~s, ~ : psychiatrist; **Psychiatrie** die; ~ psychiatry *no art.;* **psychisch** 1. *Adj.* psychological; mental ⟨*process, illness*⟩; 2. *adv.* psychologically; ~ **gesund/krank sein** be mentally fit/ill

psycho-, Psycho- [psyːço-]: ~**loge** der; ~n, ~n psychologist; ~**logie** die; ~: psychology; ~**login** die psychologist; ~**logisch** 1. *Adj.* psychological; 2. *adv.* psychologically

Pubertät die; ~ : puberty

Publikum das; ~s **a)** *(Zuschauer, Zuhörer)* audience; *(bei Sportveranstaltungen)* crowd; **b)** *(Kreis von Interessierten)* public; *(eines Schriftstellers)* readership; **c)** *(Besucher)* clientele; **publizieren** *tr. (auch itr.)* V. publish

Pudding der; ~s, ~e *od.* ~s thick, usually flavoured, milk-based dessert; ≈ blancmange

Pudel der; ~s, ~ poodle

Puder der; ~s, ~ : powder; **Puderdose** die powder compact; **pudern** *tr. V.* powder; **Puder·zucker** der icing sugar *(Brit.);* confectioners' sugar *(Amer.)*

¹Puff der; ~ǀeǀs, **Püffe** *(ugs.)* **a)** *(Stoß)* thump; *(leichter/kräftiger Stoß mit dem Ellenbogen)* nudge/dig; **b)** *(Knall)* bang; **²Puff** der *od.* das; ~s, ~s *(salopp: Bordell)* knocking-shop *(Brit. sl.);* brothel; **puffen** *(ugs.)* *tr. V.: s.* **¹Puff a:** thump; nudge; dig

Pulli der; ~s, ~s *(ugs.),* **Pullover** der; ~s, ~ : pullover; sweater; **Pullunder** der; ~s, ~ : slipover

Puls der; ~es, ~e pulse; **Puls·ader** die artery

Pult das; ~ǀeǀs, ~e desk; *(Lese~)* lectern

Pulver das; ~s, ~ powder

pumm[e]lig *Adj. (ugs.)* chubby

Pumpe die; ~, ~n pump; **pumpen** *tr., itr. V.* **a)** *(auch fig.)* pump; **b)** *(salopp) s.* **leihen a, b**

Punkt der; ~ǀeǀs, ~e **a)** *(Tupfen)* dot; *(größer)* spot; **b)** *(Satzzeichen)* full stop; **c)** *(I-Punkt)* dot; **d)** *(Stelle)* point; **ein schwacher/wunder ~** *(fig.)* a weak/sore point; **e)** *(Gegenstand, Thema, Abschnitt)* point; *(einer Tagesordnung)* item; **f)** *(Bewertungs~)* point; *(bei einer Prüfung)* mark

pünktlich 1. *Adj.* punctual; 2. *adv.* punctually; on time; **Pünktlichkeit** die; ~ : punctuality

Punsch der; ~ǀeǀs, ~e *od.* **Pünsche** punch

Pupille die; ~, ~n pupil

Puppe die; ~, ~n **a)** doll[y]; **b)** *(Marionette)* puppet; marionette

Puppen-: ~**stube** die doll's house; dollhouse *(Amer.);* ~**wagen** der doll's pram

pur *Adj.* **a)** *(rein)* pure; **b)** *(unvermischt)* neat ⟨*whisky etc.*⟩; straight

Püree der; ~s, ~s **a)** purée; **b)** *s.* **Kartoffelbrei**

Purpur der; ~s crimson

Puste die; ~ *(salopp)* puff; breath

Pustel die; ~, ~n pimple; pustule *(Med.)*

pusten *(ugs.)* *tr., itr. V.* blow

Pute die; ~, ~n turkey hen; *(als Braten)* turkey; **Puter** der; ~s, ~ : turkeycock; *(als Braten)* turkey

Putsch der; ~[e]s, ~e putsch; coup [d'état]; **putschen** itr. V. organize a putsch or coup

Putz der; ~es plaster; (für Außenmauern) rendering; **putzen** tr. V. **a)** (blank reiben) polish; **b)** (säubern) clean; groom ⟨horse⟩; [sich (Dat.)] die Zähne/die Nase ~: clean or brush one's teeth/blow one's nose; **c)** auch itr. (saubermachen) clean ⟨room, shop, etc.⟩; ~ gehen work as a cleaner; **d)** (vorbereiten) wash and prepare ⟨vegetables⟩; **Putz·frau** die cleaner

Puzzle ['paz|] das; ~s, ~s, **Puzzlespiel** das jigsaw [puzzle]

Pyjama [pʏ'dʒaːma] der (österr., schweiz. auch: das); ~s, ~s pyjamas pl.

Pyramide die; ~, ~n pyramid

Q

q, Q [kuː] das; ~, ~: q, Q

Quadrat das; ~[e]s, ~e square; **quadratisch** Adj. square; **Quadratmeter** der od. das square metre

quaken itr. V. ⟨duck⟩ quack; ⟨frog⟩ croak

Qual die; ~, ~en **a)** o. Pl. torment; **b)** meist Pl. (Schmerzen) agony; ~en pain sing.; agony sing.; (seelisch) torment sing.; **quälen** tr. V. **a)** torment ⟨person, animal⟩; be cruel to ⟨animal⟩; (foltern) torture; **b)** (plagen) ⟨cough etc.⟩ plague; (belästigen) pester; **Quälerei** die; ~, ~en **a)** torment; (Folter) torture; (Grausamkeit) cruelty; **b)** (das Belästigen) pestering

Qualifikation die; ~, ~en **a)** (Ausbildung) qualifikations pl.; **b)** (Sport) qualification; **qualifizieren** refl. V. **a)** gain qualifications; **b)** (Sport) qualify

Qualität die; ~, ~en quality; **qualitativ** 1. Adj. qualitative; ⟨difference, change⟩ in quality; 2. adv. with regard to quality; **Qualitäts·erzeugnis** das quality product

Qualle die; ~, ~n jellyfish

Qualm der; ~[e]s [thick] smoke; **qual-** men itr. V. **a)** give off clouds of [thick] smoke; **b)** (ugs.: rauchen) puff away

qual·voll 1. Adj. agonizing; 2. adv. agonizingly

Quantität die; ~, ~en quantity; **Quantum** das; ~s, **Quanten** quota (an + Dat. of); (Dosis) dose

Quarantäne [karan'tɛːnə] die; ~, ~n quarantine

Quark der; ~s quark

Quartal das; ~s, ~e quarter [of the year]

Quartett das; ~[e]s, ~e **a)** quartet; **b)** (Spiel) ≈ Happy Families; (Satz von vier Karten) set [of four]

Quartier das; ~s, ~e accommodation no indef. art.; accommodations pl. (Amer.); place to stay; (Mil.) quarters pl.

Quarz der; ~es, ~e quartz

quasi Adv. |so| ~: more or less; (so gut wie) as good as

Quaste die; ~, ~n tassel

Quatsch der; ~[e]s (ugs.) **a)** (Äußerung) rubbish; **b)** (Handlung) nonsense; (Unfug) messing about; laß den ~: stop that nonsense

Queck·silber das mercury

Quelle die; ~, ~n spring; (eines Flusses; fig.) source; **quellen** unr. itr. V.; mit sein **a)** ⟨liquid⟩ gush, stream; (aus der Erde) well up; ⟨smoke⟩ billow; **b)** (sich ausdehnen) swell [up]

quer Adv. sideways; (schräg) diagonally; (rechtwinklig) at right angles; ~ durch/über (+ Akk.) straight through/across

quer-, Quer-: ~achse die transverse axis; ~schnitt der (auch fig.) cross-section; ~schnitt[s]·gelähmt Adj. (Med.) paraplegic; ~straße die intersecting road

quetschen tr. V. crush; sich (Dat.) die Hand ~: get one's hand caught

quietschen itr. V. squeak; ⟨brakes, tyres⟩ squeal, screech; (ugs.) ⟨person⟩ squeal, shriek

Quirl der; ~[e]s, ~e long-handled blender with a star-shaped head

quitt Adj. (ugs.) quits

Quitte die; ~, ~n quince

quittieren tr. V. **a)** auch itr. acknowledge, confirm ⟨receipt, condition⟩; give a receipt for ⟨sum, invoice⟩; **b)** etw. mit etw. ~: react or respond to sth. with sth.; **Quittung** die; ~, ~en **a)** receipt; **b)** (fig.) come-uppance (coll.)

Quiz [kvıs] das; ~, ~: quiz
quoll *1. u. 3. Pers. Sg. Prät. v.* **quellen**
Quote die; ~, ~n proportion; **Quoten·regelung** die *requirement that women should be adequately represented*

R

---

r, R [ɛr] das; ~, ~: r, R
Rabatt der; ~|e|s, ~e discount
Rabatte die; ~, ~n border
Rabe der; ~n, ~n raven
rabiat 1. *Adj.* violent; brutal; ruthless ⟨*methods*⟩; 2. *adv. (gewalttätig)* violently; brutally
Rache die; ~: revenge; |an jmdm.| ~ nehmen take revenge [on sb.]
Rachen der; ~s, ~ a) *(Schlund)* pharynx *(Anat.);* b) *(Maul)* mouth; maw *(literary); (fig.)* jaws *pl.*
rächen 1. *tr. V.* avenge ⟨*person, crime*⟩; take revenge for ⟨*insult, crime*⟩; 2. *refl. V.* a) take one's revenge; b) ⟨*mistake etc.*⟩ take its/their toll
Rachitis die; ~ *(Med.)* rickets *sing.*
Rach·sucht die; *o. Pl. (geh.)* lust for revenge; **rach·süchtig** *(geh.)* 1. *Adj.* vengeful; 2. *adv.* vengefully
Rad das; ~es, Räder ['rɛːdɐ] a) wheel; das fünfte ~ am Wagen sein *(fig. ugs.)* be superfluous; b) *(Fahr~)* bicycle; bike *(coll.)*
Radar der *od.* das; ~s radar
Radar: ~falle die *(ugs.)* [radar] speed trap; ~kontrolle die [radar] speed check
rad-, Rad-: **dampfer** der paddle-steamer; ~|fahren *unr. itr. V.* ; mit sein cycle; ride a bicycle *or (coll.)* bike; ~fahrer der cyclist
Radien *s.* Radius
radieren *tr. V. (auch itr.) V.* erase; **Radier·gummi** der rubber [eraser]
Radieschen das; ~s, ~: radish
radikal 1. *Adj.* radical; drastic ⟨*measure, method, cure*⟩; 2. *adv.* radically; *(vollständig)* totally; **Radikalismus** der; ~: radicalism

Radio das *(südd., schweiz. auch:* der); ~s, ~s radio; ~ hören listen to the radio; **Radio·wecker** der radio alarm clock
Radius der; ~, Radien radius
Rad·kappe die hub-cap
Radler der; ~s, ~: cyclist
Rad-: ~rennbahn die cycle-racing track; ~rennen das cycle race; *(Sport)* cycle-racing; ~sport der cycling *no def. art.;* ~tour die cycling tour; ~weg der cycle-path *or* -track
raffen *tr. V.* a) snatch; rake in *(coll.)* ⟨*money*⟩; etw. |an sich| ~: seize sth.; *(eilig)* snatch sth.; b) gather ⟨*material, curtain*⟩
Raffinerie die; ~, ~n refinery; **Raffinesse** die; ~, ~n a) *o. Pl. (Schlauheit)* guile; ingenuity; b) *meist Pl. (Finesse)* refinement; **raffiniert** 1. *Adj.* a) ingenious ⟨*plan, design*⟩; *(verfeinert)* refined, subtle ⟨*colour, scheme, effect*⟩; sophisticated ⟨*dish, cut (of clothes)*⟩; b) *(gerissen)* cunning ⟨*person, trick*⟩; 2. *adv.* a) ingeniously; *(verfeinert)* with great refinement/sophistication; b) *(gerissen)* cunningly
Rage ['raːʒə] die; ~ *(ugs.)* fury
ragen *itr. V.* a) *(vertikal)* rise [up]; ⟨*mountains*⟩ tower up; b) *(horizontal)* project, stick out (in + *Akk.* into; über + *Akk.* over)
Ragout [ra'guː] das; ~s, ~s ragout
Rahm der; ~|e|s cream
rahmen *tr. V.* frame; **Rahmen** der; ~s, ~ a) frame; *(Fahrgestell)* chassis; b) *(fig.)* framework
Rakete die; ~, ~n rocket; *(Lenkflugkörper)* missile
rammen *tr. V.* ram
Rampe die; ~, ~n a) *(Lade~)* [loading] platform; b) *(schiefe Fläche)* ramp; **Rampen·licht** das: im ~ |der Öffentlichkeit| stehen be in the limelight
Ramsch der; ~|e|s, ~e *(ugs.)* a) *(Ware)* trashy goods *pl.;* b) *(Kram)* junk
ran *Adv. (ugs.)* a) *s.* heran; b) *(fang[t] an)* off you go; *(fangen wir an)* let's go; c) *(greif[t] an)* go at him/them!
Rand der; ~|e|s, Ränder a) edge; *(Einfassung)* border; *(Hut~)* brim; *(Brillen~, Gefäß~, Krater~)* rim; *(eines Abgrunds)* brink; *(auf einem Schriftstück)* margin; *(Weg~)* verge; *(Stadt~)* outskirts *pl.;* b) *(Schmutz~)* mark; *(rund)* ring
randalieren *itr. V.* riot

Rand·bemerkung die marginal note *or* comment

rang *1. u. 3. Pers. Sg. Prät. v.* **ringen**

Rang der; ~|e|s, **Ränge a)** rank; *(in der Gesellschaft)* status; **b)** *(im Theater)* circle; **erster** ~: dress circle; **zweiter** ~: upper circle; **dritter** ~: gallery

rangieren [raŋ'ʒiːrən] *tr. V.* shunt *(trucks etc.)*; switch *(cars) (Amer.)*

Rang·ordnung die order of precedence; *(Verhaltensf.)* pecking order

Ranke die; ~, ~n *(Bot.)* tendril; **ranken** *refl. V.* climb, grow **(an** + *Dat.* up, **über** + *Akk.* over)

rann *1. u. 3. Pers. Sg. Prät. v.* **rinnen**

rannte *1. u. 3. Pers. Sg. Prät. v.* **rennen**

Ranzen der; ~s, ~: satchel

ranzig *Adj.* rancid

Rappe der; ~n, ~n black horse

Rappen der; ~s, ~: [Swiss] centime

Raps der; ~es *(Bot.)* rape

rar *Adj.* scarce; *(selten)* rare; **Rarität die;** ~, ~en rarity

rasant *(ugs.)* **1.** *Adj.* tremendously fast *(coll.) (car, horse, etc.)*; **2.** *adv.* at terrific speed *(coll.)*

rasch 1. *Adj.* quick; speedy, swift *(end, action, decision, progress)*; **2.** *adv.* quickly; *(decide, end, proceed)* swiftly, rapidly

rascheln *itr. V.* rustle; *(mouse etc.)* make a rustling noise

rasen *itr. V.* **a)** *mit sein (ugs.: eilen)* dash *or* rush [along]; *(fahren)* tear *or* race along; *(fig.) (pulse)* race; **b)** *(toben) (person)* rage

Rasen der; ~s, ~: grass *no indef. art.*; *(gepflegte ~fläche)* lawn

rasend 1. *Adj.* **a)** *(sehr schnell)* breakneck *attrib.* *(speed)*; **b)** *(tobend)* raging; **c)** *(heftig)* violent; **2.** *adv.* *(ugs.)* incredibly *(coll.)*

Rasen·mäher der; ~s, ~: lawnmower

Raserei die; ~, ~en *(ugs.)* tearing along *no art.*

Rasier·apparat der [safety] razor; *(elektrisch)* electric shaver; **rasieren** *tr. V.* shave; **sich** ~: shave; **sich naß/ trocken/elektrisch** ~: have a wet shave/ have a dry shave/use an electric shaver

Rasier-: ~**klinge die** razor-blade; ~**wasser das** aftershave; *(vor der Rasur)* pre-shave lotion

Rasse die; ~, ~n **a)** breed; **b)** *(Menschen~)* race

Rassel die; ~, ~n rattle; **rasseln** *itr. V.* rattle

Rassen-: ~**haß der** racial hatred *no art.*; ~**trennung die;** *o. Pl.* racial segregation *no art.*

Rassismus der; ~: racism; racialism; **Rassist der;** ~en, ~en racist; racialist

Rast die; ~, ~en rest; ~ **machen** stop for a break; **rasten** *itr. V.* rest; take a rest *or* break

Rast-: ~**haus das** roadside café; *(an der Autobahn)* motorway restaurant; ~**platz der a)** place to rest; **b)** *(an Autobahnen)* parking place *(with benches and WCs)*; picnic area; ~**stätte die** service area

Rasur die; ~, ~en shave

Rat der; ~|e|s, **Räte a)** *o. Pl.* advice; **ein** ~: a word of advice; **b)** *(Gremium)* council

rät *3. Pers. Sg. Präsens v.* **raten**

Rate die; ~, ~n **a)** *(Teilbetrag)* instalment; **etw. auf** ~**n kaufen** buy sth. by instalments *or (Brit.)* on hire purchase *or (Amer.)* on the installment plan; **b)** *(Statistik)* rate

raten 1. *unr. itr. V.* **a)** **jmdm.** ~: advise sb.; **b)** *(schätzen)* guess; **2.** *tr. V.* **a)** **jmdm.** ~, **etw. zu tun** advise sb. to do sth.; **b)** *(er~)* guess

Raten·zahlung die payment by instalments

Rat·haus das town hall

Ration die; ~, ~en ration; **rational** *Adj.* rational; **rationalisieren** *tr., itr. V.* rationalize

rationell 1. *Adj.* efficient; *(wirtschaftlich)* economic; **2.** *adv.* efficiently; *(wirtschaftlich)* economically; **rationieren** *tr. V.* ration

rat·los 1. *Adj.* baffled; helpless *(look)*; **2.** *adv.* helplessly; **Rat·losigkeit die;** ~: helplessness; **ratsam** *Adj.*; *nicht attr.* advisable; **Ratschlag der** [piece of] advice

Rätsel das; ~s, ~ **a)** riddle; *(Bilder~, Kreuzwort~ usw.)* puzzle; **b)** *(Geheimnis)* mystery; **rätselhaft 1.** *Adj.* mysterious; *(unergründlich)* enigmatic; **2.** *adv.* mysteriously; *(unergründlich)* enigmatically

Ratte die; ~, ~n *(auch fig.)* rat

Raub der; ~|e|s **a)** robbery; **b)** *(Beute)* stolen goods *pl.*; **rauben** *tr. V.* steal; kidnap *(person)*; **jmdm. etw.** ~: rob sb. of sth.; *(geh.: wegnehmen)* deprive sb. of sth.; **Räuber der;** ~s, ~: robber

Raub-: ~**fisch der** predatory fish; ~**mord der** *(Rechtsw.)* murder **(an** + *Dat.* of) in the course of a robbery *or* with robbery as motive; ~**tier das**

predator; ~**überfall** der robbery (**auf** + *Akk.* of); ~**vogel** der bird of prey
Rauch der; ~|e|s smoke; **rauchen 1.** *itr. V.* smoke; **2.** *tr. (auch itr.) V.* smoke ⟨*cigarette, pipe, etc.*⟩; „**Rauchen verboten**" 'No smoking'; **Raucher** der; ~s, ~: smoker; **Raucher·abteil** das smoking-compartment; smoker; **Raucherin** die; ~, ~**nen** smoker; **räuchern** *tr. V.* smoke ⟨*meat, fish*⟩; **rauchig** *Adj.* smoky; husky ⟨*voice*⟩; **Rauch·verbot** das ban on smoking
räudig *Adj.* mangy
rauf *Adv. (ugs.)* up; ~ **mit euch!** up you go!; *s. auch* **herauf; hinauf**
raufen 1. *itr., refl. V.* fight; **2.** *tr. V.* sich *(Dat.)* die **Haare/den Bart** ~: tear one's hair/at one's beard
rauh 1. *Adj.* **a)** *(nicht glatt)* rough; **b)** *(nicht mild)* harsh ⟨*climate, winter*⟩; raw ⟨*wind*⟩; **c)** *(kratzig)* husky, hoarse ⟨*voice*⟩; **d)** *(entzündet)* sore ⟨*throat*⟩; **e)** *(grob, nicht feinfühlig)* rough; harsh ⟨*words, tone*⟩; **2.** *adv.* **a)** *(kratzig)* ⟨*speak etc.*⟩ huskily, hoarsely; **b)** *(grob, nicht feinfühlig)* roughly
Rauh-: ~**faser·tapete** die woodchip wallpaper; ~**reif** der hoar-frost
Raum der; ~|e|s, **Räume a)** *(Wohn~, Nutz~)* room; **b)** *(Gebiet)* area; region; **c)** *o. Pl. (Platz)* room; space; **räumen** *tr. V.* **a)** clear [away]; clear ⟨*snow*⟩; **b)** *(an einen Ort)* clear; move; **c)** *(frei machen)* clear ⟨*street, building, warehouse, stocks, etc.*⟩; **d)** *(verlassen)* vacate; **Raum·fahrt** die; ~: space travel; **räumlich 1.** *Adj.* **a)** spatial; **aus** ~**en Gründen** for reasons of space; **b)** *(dreidimensional)* three-dimensional; stereoscopic ⟨*vision*⟩; **2.** *adv.* **a)** spatially; **b)** *(dreidimensional)* three-dimensionally; **Raum·schiff** das spaceship; **Räumung** die; ~, ~**en a)** clearing; **b)** *(das Verlassen)* vacation; vacating; **c)** *(wegen Gefahr)* evacuation; **d)** *(eines Lagers)* clearance
raunen *tr., itr. V. (geh.)* whisper
Raupe die; ~, ~**n** caterpillar
raus *Adv. (ugs.)* out; ~ **mit euch!** out you go!; *s. auch* **heraus; hinaus**
Rausch der; ~|e|s, **Räusche a)** state of drunkenness; **b)** *(starkes Gefühl)* transport; **der** ~ **der Geschwindigkeit** the exhilaration *or* thrill of speed; **rauschen** *itr. V.* ⟨*water, wind, torrent*⟩ rush; ⟨*trees, leaves*⟩ rustle; ⟨*skirt, curtains, silk*⟩ swish; ⟨*waterfall, strong wind*⟩ roar; ⟨*rain*⟩ pour down;

Rausch·gift das drug; narcotic; ~ **nehmen** take drugs; be on drugs
räuspern *refl. V.* clear one's throat
raus|schmeißen *unr. tr. V. (ugs.)* chuck *(coll.)* ⟨*objects*⟩ out *or* away; give ⟨*employee*⟩ the push *(coll.) or* sack *(coll.)*; chuck *(coll.) or* throw ⟨*customer, drunk, tenant*⟩ out (aus of)
Raute die; ~, ~**n** *(Geom.)* rhombus
Razzia die; ~, **Razzien** raid
reagieren *itr. V.* react (**auf** + *Akk.* to); **Reaktion** die; ~, ~**en** reaction (**auf** + *Akk.* to); **reaktionär** *Adj.* reactionary; **Reaktionär** der; ~s, ~**e** reactionary; **Reaktor** der; ~s, ~**en** [-'to:rən] reactor
real 1. *Adj.* real; **2.** *adv.* actually; **realisieren** *tr. V. (geh.)* realize; **Realismus** der; ~: realism; **Realist** der; ~**en**, ~**en** realist; **realistisch 1.** *Adj.* realistic; **2.** *adv.* realistically; **Realität** die; ~, ~**en** reality
Rebe die; ~, ~**n a)** vine shoot; **b)** *(Weinstock)* [grape] vine
Rebell der; ~**en**, ~**en** rebel; **rebellieren** *itr. V.* rebel (**gegen** against); **Rebellion** die; ~, ~**en** rebellion; **rebellisch** *Adj.* rebellious
Reb-: ~**huhn** das partridge; ~**stock** der vine
rechen *tr. V. (bes. südd.)* rake; **Rechen** der; ~s, ~ *(bes. südd.)* rake
Rechen-: ~**fehler** der arithmetical error; ~**maschine** die calculator
Rechenschaft die; ~: account; jmdn. **für etw. zur** ~ **ziehen** call *or* bring sb. to account for sth.
rechnen 1. *tr. V.* **a) eine Aufgabe** ~: work out a problem; **b)** *(veranschlagen)* reckon; estimate; **gut/rund gerechnet** at a generous/rough estimate; **c)** *(berücksichtigen)* take into account; **d)** *(einbeziehen)* count; **2.** *itr. V.* **a)** do *or* make a calculation/calculations; **gut/schlecht** ~ **können** be good/bad at figures; **b)** *(zählen)* reckon; **c)** *(ugs.: berechnen)* calculate; estimate; **d)** *(wirtschaften)* budget carefully; **e) auf** jmdn./**etw.** *od.* **mit** jmdn./**etw.** ~: count on sb./sth.; **f) mit etw.** ~ *(etw. einkalkulieren)* reckon with sth.; *(etw. erwarten)* expect sth.; **Rechnen** das; ~**s** arithmetic; **Rechner** der; ~**s**, ~: calculator; *(Computer)* computer; **rechnerisch** *Adj.* arithmetical; **Rechnung** die; ~, ~**en a)** calculation; **b)** *(schriftliche Kosten*~*)* bill; invoice *(Commerc.)*; |jmdm.| **etw. in** ~ **stellen** charge [sb.] for sth.

recht 1. *Adj.* a) *(geeignet, richtig)* right; b) *(gesetzmäßig, anständig)* right; proper; ~ **und billig** right and proper; c) *(wunschgemäß)* **jmdm.** ~ **sein** be all right with sb.; d) *(wirklich, echt)* real; 2. *adv.* a) *(geeignet)* **du kommst gerade** ~: you are just in time; b) *(richtig)* correctly; c) *(gesetzmäßig, anständig)* properly; d) *(wunschgemäß)* **es jmdm.** ~ **machen** please sb.; e) *(wirklich, echt)* really; f) *(ziemlich)* quite; rather; *s. auch* **Recht** d; **recht...** *Adj.* a) right; right[-hand] ⟨*edge*⟩; b) *(außen, sichtbar)* right ⟨*side*⟩; c) *(in der Politik)* right-wing; **Recht** das; ~[e]s, ~e a) *(Rechtsordnung)* law; b) *(Rechtsanspruch)* right; **sein** ~ **fordern** *od.* **verlangen** demand one's rights; c) *o. Pl. (Berechtigung)* right (**auf** + *Akk.* to); **gleiches** ~ **für alle!** equal rights for all!; **im** ~ **sein** be in the right; **zu** ~: rightly; d) **recht haben** be right; **jmdm. recht geben** admit that sb. is right

recht·fertigen *tr. V.* justify (**vor** + *Dat.* to); **Recht·fertigung** die justification

rechtlich 1. *Adj.* legal; 2. *adv.* legally; **recht·los** *Adj.* without rights *postpos.*; **rechtmäßig** 1. *Adj.* lawful; rightful; legitimate ⟨*claim*⟩; 2. *adv.* lawfully; rightfully; **Rechtmäßigkeit** die; ~: legality; *(eines Anspruchs)* legitimacy

rechts *Adv.* a) on the right; **von** ~: from the right; b) *(Politik)* on the right wing

Rechts-: ~**abbieger** der *(Verkehrsw.)* motorist/cyclist/car *etc.* turning right; ~**anwalt** der, ~**anwältin** die lawyer; solicitor *(Brit.)*; attorney *(Amer.)*; *(vor Gericht)* barrister *(Brit.)*; attorney[-at-law] *(Amer.)*; advocate *(Scot.)*; ~**außen** [-'--] der; ~, ~ *(Ballspiele)* right wing; outside right

recht-, Recht-: ~**schaffen** 1. *Adj.* honest; 2. *adv.* honestly; ~**schreibfehler** der spelling mistake; ~**schreibung** die orthography

rechts-, Rechts-: ~**händer** der; ~s, ~: right-hander; ~**kräftig** *(Rechtsw.)* 1. *Adj.* final [and absolute] ⟨*decision, verdict, etc.*⟩; 2. *adv.* **jmdn.** ~**kräftig verurteilen** pass a final sentence on sb.; ~**kurve** die right-hand bend

Recht·sprechung die; ~, ~en administration of justice; *(eines Gerichts)* jurisdiction

rechts-, Rechts-: ~**staat** der [constitutional] state founded on the rule of law; ~**staatlich** *Adj.* founded on the rule of law *postpos.*; ~**verkehr** der driving *no art.* on the right; ~**widrig** 1. *Adj.* unlawful; 2. *adv.* unlawfully

recht-: ~**wink[e]lig** *Adj.* right-angled; ~**zeitig** 1. *Adj.* timely; *(pünktlich)* punctual; 2. *adv.* in time; *(pünktlich)* on time

Reck das; ~[e]s, ~e *od.* ~s horizontal bar

recken 1. *tr. V.* stretch; 2. *refl. V.* stretch oneself

Redakteur [redak'tø:ɐ] der; ~s, ~e, **Redakteurin** die; ~, ~nen editor; **Redaktion** die; ~, ~en a) *(Redakteure)* editorial staff; b) *(Büro)* editorial department *or* office/offices *pl.*

Rede die; ~, ~n a) *(Ansprache)* address; speech; **eine** ~ **halten** give *or* make a speech; b) *o. Pl. (Vortrag)* rhetoric; c) *(Äußerung, Ansicht)* **nicht der** ~ **wert sein** be not worth mentioning; **jmdn. zur** ~ **stellen** make someone explain himself/herself; **reden** 1. *tr. V.* talk; **Unsinn** ~: talk nonsense; **kein Wort** ~: not say *or* speak a word; 2. *itr. V.* a) *(sprechen)* talk; speak; **viel/wenig** ~: talk a lot *(coll.)*/not talk much; b) *(sich äußern, eine Rede halten)* speak; **gut** ~ **können** be a good speaker; c) *(sich unterhalten)* talk; **mit jmdm./über jmdn.** ~: talk to/about sb.; **Redens·art** die a) expression; *(Sprichwort)* saying; b) *Pl. (Phrase)* empty *or* meaningless words

Rede·wendung die *(Sprachw.)* idiom

redlich 1. *Adj.* honest; 2. *adv.* honestly; **Redlichkeit** die; ~: honesty

Redner der; ~s, ~, **Rednerin** die ~, ~nen a) speaker; b) *(Rhetoriker)* orator; **red·selig** *Adj.* talkative

reduzieren 1. *tr. V.* reduce (**auf** + *Akk.* to); 2. *refl. V.* decrease; diminish

Reeder der; ~s, ~: shipowner; **Reederei** die; ~, ~en shipping firm

reell 1. *Adj.* honest, straight ⟨*person, deal, etc.*⟩; sound, solid ⟨*business, firm, etc.*⟩; straight ⟨*offer*⟩; 2. *adv.* honestly

Reet das; ~s *(nordd.)* reeds *pl.*

Referat das; ~[e]s, ~e a) paper; b) *(kurzer schriftlicher Bericht)* report; **referieren** *itr. V.* **über etw.** *(Akk.)* ~: present a paper on sth.; *(zusammenfassend)* give a report on sth.

reflektieren *tr. V.* reflect

Reflex der; ~es, ~e reflex; **Reflexiv·pronomen das** *(Sprachw.)* reflexive pronoun

Reform die; ~, ~en reform; **Reform·haus das** health-food shop; **reformieren** *tr. V.* reform

Refrain [rə'frɛ̃:] der; ~s, ~s chorus

Regal das; ~s, ~e [set *sing.* of] shelves *pl.*

rege 1. *Adj.* **a)** *(betriebsam)* busy ⟨traffic⟩; brisk ⟨demand, trade, business, etc.⟩; **b)** *(lebhaft)* lively; keen ⟨interest⟩; **2.** *adv.* **a)** *(betriebsam)* actively; **b)** *(lebhaft)* actively

Regel die; ~, ~n **a)** rule; **nach allen ~n der Kunst** *(fig.)* well and truly; **b)** rule; custom; **die ~ sein** be the rule; **in der od. aller ~:** as a rule; **c)** *(Menstruation)* period; **regel·mäßig 1.** *Adj.* regular; **2.** *adv.* regularly; **Regel·mäßigkeit die** regularity; **regeln 1.** *tr. V.* **a)** settle ⟨matter, question, etc.⟩; put ⟨finances, affairs, etc.⟩ in order; **b)** *(einstellen, regulieren)* regulate; *(steuern)* control; **2.** *refl. V.* take care of itself; **Regelung die;** ~, ~en **a)** o. *Pl. s.* **regeln 1 a,** b: settlement; putting in order; regulation; control; **b)** *(Vorschrift)* regulation

regen 1. *tr. V. (geh.)* move; **2.** *refl. V.* **a)** *(sich bewegen)* move; **b)** *(geh.)* ⟨hope, doubt, desire, conscience⟩ stir

Regen der; ~s, ~ **a)** rain; **vom od. aus dem ~ in die Traufe kommen** *(fig.)* jump out of the frying-pan into the fire; **b)** *(fig.)* shower

Regen-: **~bogen der** rainbow; **~mantel der** raincoat; mackintosh; **~schirm der** umbrella; **~tag der** rainy day; **~wetter das;** o. *Pl.* wet weather; **~wolke die** rain cloud; **~wurm der** earthworm

Regie [re'ʒi:] die; ~ **a)** *(Theater, Film, Ferns., Rundf.)* direction; **b)** *(Leitung, Verwaltung)* management

regieren 1. *itr. V.* rule (**über** + *Akk.* over); ⟨party, administration⟩ govern; **2.** *tr. V.* rule; govern; ⟨monarch⟩ reign over; **Regierung die;** ~, ~en **a)** o. *Pl. (Herrschaft)* rule; *(eines Monarchen)* reign; **b)** *(Kabinett)* government; **Regierungs·sitz der** seat of government

Regiment das; ~[e]s, ~e od. ~er **a)** *Pl.* ~e *(Herrschaft)* rule; **b)** *Pl.* ~er *(Milit.)* regiment

Region die; ~, ~en region; **regional 1.** *Adj.* regional; **2.** *adv.* regionally

Regisseur [reʒɪ'sø:ɐ̯] der; ~s, ~e, **Regisseurin die;** ~, ~nen director

Register das; ~s, ~ **a)** index; **b)** *(amtliche Liste)* register; **c)** *(Musik) (bei Instrumenten)* register; *(Orgel~)* stop; **registrieren** *tr. V.* **a)** register; **b)** *(bewußt wahrnehmen)* note; register

Regler der; ~s, ~ *(Technik)* regulator; *(Kybernetik)* control

reg·los *Adj.* motionless

regnen 1. *itr., tr. V. (unpers.)* rain; **es regnet** it is raining; **2.** *itr. V.; mit sein (fig.)* rain down; **regnerisch** *Adj.* rainy

regulär *Adj.* **a)** proper; normal ⟨working hours⟩; **b)** *(normal, üblich)* normal; **regulieren** *tr. V.* regulate; **Regulierung die;** ~, ~en regulation

Regung die; ~, ~en *(geh.: Gefühl)* stirring; **regungs·los** *Adj.* motionless

Reh das; ~[e]s, ~e roe-deer

Reh-: **~bock der** roebuck; **~kitz das** fawn [of a/the roe-deer]

Reibe die; ~, ~n, **Reib·eisen das** grater; **reiben 1.** *unr. tr. V.* **a)** rub; **b)** *(zerkleinern)* grate; **2.** *unr. itr. V.* rub (**an** + *Dat.* on); **Reibung die;** ~, ~en *(Physik, fig.)* friction; **reibungs·los 1.** *Adj.* smooth; **2.** *adv.* smoothly

reich 1. *Adj.* **a)** *(vermögend)* rich; **b)** *(prächtig)* costly ⟨goods, gifts⟩; rich ⟨décor, finery⟩; **c)** *(üppig)* rich; abundant ⟨harvest⟩; abundant ⟨mineral resources⟩; **~ an etw.** *(Dat.)* sein be rich in sth.; **d)** *(vielfältig)* rich ⟨collection, possibilities⟩; wide, large ⟨selection, choice⟩; wide ⟨knowledge, experience⟩; **2.** *adv.* richly

Reich das; ~[e]s, ~e **a)** empire; *(König~)* kingdom; realm; **das |Deutsche| ~** *(hist.)* the German Reich *or* Empire; **das Dritte ~** *(hist.)* the Third Reich; **b)** *(fig.)* realm

reichen 1. *itr. V.* **a)** *(aus~)* be enough; **das Geld reicht nicht** I/we *etc.* haven't got enough money; **jetzt reicht's mir aber!** now I've had enough!; **danke, es reicht** that's enough, thank you; **b)** *(sich erstrecken)* reach; ⟨forest, fields, etc.⟩ extend; **2.** *tr. V.* **a)** pass; hand; **jmdm. die Hand ~:** hold out one's hand to sb.; **sich** *(Dat.)* **die Hand ~:** shake hands; **b)** *(servieren)* serve ⟨food, drink⟩

reich·haltig *Adj.* extensive; varied ⟨programme⟩; substantial ⟨meal⟩; **reichlich 1.** *Adj.* large; ample ⟨space, time⟩; good ⟨hour, year⟩; **2.** *adv.* **a)** amply; **b)** *(mehr als)* over; more than;

c) *(ugs.: ziemlich, sehr)* a bit too ⟨*cheeky, dear, late*⟩; **Reichtum** der; ~s, **Reichtümer** a) *o. Pl.* wealth (**an** + *Dat.* of); b) *Pl. (Vermögenswerte)* riches

Reich·weite die reach; *(eines Geschützes, Senders, Flugzeugs)* range

reif *Adj.* a) ripe ⟨*fruit, grain, cheese*⟩; mature ⟨*brandy, cheese*⟩; ~ **für etw. sein** *(ugs.)* be ready for sth.; b) *(erwachsen, ausgewogen)* mature

¹**Reif** der; ~|e|s hoar-frost

²**Reif** der; ~|e|s, ~e *(geh.)* ring; *(Arm~)* bracelet; *(Diadem)* circlet

Reife die; ~ a) ripeness; *(von Menschen, Gedanken, Produkten)* maturity; b) *(Reifung)* ripening; **reifen** 1. *itr. V.; mit sein* a) ⟨*fruit, cereal, cheese*⟩ ripen; b) *(geh.: älter, reifer werden)* mature (**zu** into); c) ⟨*idea, plan, decision*⟩ mature; 2. *tr. V.* ripen ⟨*fruit, cereal*⟩;

Reifen der; ~s, ~ a) hoop; b) *(Gummi~)* tyre; c) *s.* ²**Reif**

Reifen-: ~**panne** die puncture; ~**wechsel** der tyre change

reiflich 1. *Adj.* [very] careful; 2. *adv.* [very] carefully

Reigen der; ~s, ~ a) round dance; b) *(fig.)* **den** ~ **eröffnen** start off

Reihe die; ~, ~n a) row; **in Reih und Glied** *(Milit.)* in rank and file; **aus der** ~ **tanzen** *(fig. ugs.)* be different; b) *o. Pl. (Reihenfolge)* series; **er/sie** *usw.* **ist an der** ~: it's his/her *etc.* turn; **der** ~ **nach, nach der** ~: in turn; c) *(größere Anzahl)* number; **reihen** *(geh.) tr. V.* string; thread

Reihen-: ~**folge** die order; ~**haus** das terraced house

Reiher der; ~s, ~: heron

Reim der; ~|e|s, ~e rhyme; **reimen** 1. *itr. V.* make up rhymes; 2. *tr., refl. V.* rhyme (**auf** + *Akk.* with)

¹**rein** *Adv. (ugs.)* ~ **mit dir!** in you go/come!

²**rein** 1. *Adj.* a) *(unvermischt)* pure; b) *(nichts anderes als)* pure; sheer; plain, unvarnished ⟨*truth*⟩; c) *(frisch, sauber)* clean; fresh ⟨*clothes, sheet of paper, etc.*⟩; pure, clean ⟨*water, air*⟩; clear ⟨*complexion*⟩; **etw. ins** ~**e schreiben** make a fair copy of sth.; **etw. ins** ~**e bringen** clear sth. up; 2. *Adv.* purely; ~ **gar nichts** *(ugs.)* absolutely nothing

Rein·fall der *(ugs.)* let-down

Rein·gewinn der net profit

Reinheit die; ~ a) purity; b) *(Sauberkeit)* cleanness; *(des Wassers, der*

Luft) purity; *(der Haut)* clearness; **reinigen** *tr. V.* clean; purify ⟨*effluents, air, water, etc.*⟩; **Kleider** |**chemisch**| ~ **lassen** have clothes [dry-] cleaned; **Reinigung** die; ~, ~en a) *s.* **reinigen:** cleaning; purification; drycleaning; b) *(Betrieb)* [dry-]cleaner's; **reinlich** *Adj.* cleanly; **Reinlichkeit** die; ~: cleanliness

rein·rassig *Adj.* thoroughbred ⟨*animal*⟩; **Rein·schrift** die fair copy

Reis der; ~es rice

Reise die; ~, ~n journey; *(kürzere Fahrt, Geschäfts~)* trip; *(Ausflug)* outing; trip; *(Schiffs~)* voyage; **eine** ~ **machen** go on a trip/an outing; **auf** ~**n sein** travel; *(nicht zu Hause sein)* be away; **glückliche** *od.* **gute** ~! have a good journey

Reise-: ~**an·denken** das souvenir; ~**büro** das travel agent's; travel agency; ~**bus** der coach; ~**führer** der a) *(~leiter)* courier; b) *(Buch)* guidebook; ~**führerin** die courier; ~**gepäck** das luggage *(Brit.)*; baggage *(Amer.); (am Flughafen)* baggage; ~**gesellschaft** die a) *(~gruppe)* party of tourists; b) *(ugs.: ~veranstalter)* tour operator; ~**kosten** *Pl.* travel expenses; ~**leiter** der, ~**leiterin** die courier

reisen *itr. V.; mit sein* a) travel; b) *(ab~)* leave; set off; **Reisende** der/die; *adj. Dekl.* traveller; *(Fahrgast)* passenger

Reise-: ~**paß** der passport; ~**scheck** der traveller's cheque; ~**tasche** die hold-all; ~**verkehr** der holiday traffic; ~**ziel** das destination

Reisig das; ~s brushwood

Reiß·brett das drawing-board

reißen 1. *unr. tr. V.* a) tear; *(in Stücke)* tear up; b) *(ziehen an)* pull; *(heftig)* yank *(coll.)*; c) *(werfen, ziehen)* jmdn. **zu Boden/in die Tiefe** ~: knock sb. to the ground/drag sb. down into the depths; d) *(töten)* ⟨*wolf, lion, etc.*⟩ kill ⟨*prey*⟩; e) etw. **an sich** ~ *(fig.)* seize sth.; 2. *unr. itr. V.* a) *mit sein* ⟨*paper, fabric*⟩ tear, rip; ⟨*rope, thread*⟩ break, snap; ⟨*film*⟩ break; ⟨*muscle*⟩ tear; b) *(ziehen)* **an etw.** *(Dat.)* ~: pull at sth.; 3. *unr. refl. V. (ugs.: sich bemühen um)* **sie** ~ **sich um die Eintrittskarten** they are fighting each other to get tickets; **reißend** *Adj.* rapacious ⟨*animal*⟩; raging ⟨*torrent*⟩; ~**en Absatz finden** sell like hot cakes

Reiß-: ~**leine** die *(Flugw.)* rip-cord;

~**nagel** der s. ~zwecke; ~**ver-schluß** der zip [fastener]; ~**zwecke** die drawing-pin *(Brit.);* thumbtack *(Amer.)*

reiten 1. *unr. itr. V.; meist mit sein* ride; 2. *unr. tr. V.; auch mit sein* ride; **Schritt/Trab/Galopp** ~: ride at a walk/trot/gallop; **Reiten** das; ~s riding *no art.;* **Reiter** der; ~s, ~, **Reite-rin** die; ~, ~**nen** rider

Reit-: ~**hose** die riding breeches *pl.;* ~**pferd** das saddle-horse; ~**stiefel** der riding boot

Reiz der; ~es, ~e a) *(Physiol.)* stimulus; b) *(Anziehungskraft)* attraction; appeal *no pl.; (des Verbotenen, der Ferne usw.)* lure; c) *(Zauber)* charm; **reizbar** *Adj.* irritable; **Reizbarkeit** die; ~: irritability; **reizen** 1. *tr. V.* a) annoy; tease *⟨animal⟩; (herausfor-dern, provozieren)* provoke; s. *auch* gereizt; b) *(Physiol.)* irritate; c) *(Inter-esse erregen bei)* jmdn. ~: attract sb.; appeal to sb.; d) *(Kartenspiele)* bid; 2. *itr. V. (Kartenspiele)* bid; **reizend** 1. *Adj.* charming; delightful, lovely *⟨child⟩;* 2. *adv.* charmingly; **reizlos** *Adj.* unattractive; *⟨*landscape, scenery*⟩* lacking in charm; **reizvoll** *Adj.* a) *(hübsch)* charming; b) *(interes-sant)* attractive

rekeln *refl. V. (ugs.)* stretch

Reklamation [reklama'tsio:n] die; ~, ~en complaint *(wegen about);* **Rekla-me** die; ~, ~**n** a) o. *Pl.* advertising *no indef. art.;* ~ für jmdn./etw. machen promote sb./advertise *or* promote sth.; b) *(ugs.: Werbemittel)* advert *(Brit. coll.);* ad *(coll.); (im Fernsehen, Radio auch)* commercial; **rekla-mieren** 1. *itr. V.* complain; 2. *tr. V.* a) complain about *(bei* to, *wegen* on account of); b) *(beanspruchen)* claim

rekonstruieren *tr. V.* reconstruct

Rekord der; ~[e]s, ~e record

Rekrut der; ~en, ~en *(Milit.)* recruit

Rektor der; ~s, ~en a) *(einer Schule)* head[master]; b) *(Universitäts~)* Rector; ≈ Vice-Chancellor *(Brit.); (einer Fachhochschule)* principal; **Rektorin** die; ~, ~**nen** a) *(einer Schu-le)* head[mistress]; b) s. **Rektor** b

Relation die; ~, ~en relation; **relativ** 1. *Adj.* relative; 2. *adv.* relatively

Relativ-: ~**pronomen** das *(Sprachw.)* relative pronoun; ~**satz** der *(Sprachw.)* relative clause

Relief das; ~s, ~s od. ~e relief

Religion die; ~, ~en religion; **religi-**

ös 1. *Adj.* religious; 2. *adv.* in a reli-gious manner

Relikt das; ~[e], ~e relic

Reling die; ~, ~s od. ~e [deck-]rail

Reliquie die; ~, ~**n** relic

Remis das; ~ [rə'mi:(s)], ~ [rə'mi:s] *(bes. Schach)* draw

Ren das; ~s, ~s od. ~e reindeer

Rendezvous [rãde'vu:] das; ~ [...'vu:(s)], ~ ['rãde'vu:s] rendezvous

Renn·bahn die *(Sport)* race-track; *(für Pferde)* racecourse; **rennen** *unr. itr. V.; mit sein* run; an/gegen jmdn./ etw. ~: run *or* bang into sb./sth.; **Rennen** das; ~s, ~: running; *(Pfer-de~, Auto~)* racing; *(Wettbewerb)* race

Renn-: ~**fahrer** der racing driver; ~**pferd** das racehorse; ~**rad** das racing cycle; ~**wagen** der racing car

renommiert *Adj.* renowned

renovieren *tr. V.* renovate; redec-orate *⟨room, flat⟩;* **Renovierung** die; ~, ~en renovation; *(eines Zim-mers, einer Wohnung)* redecoration

rentabel 1. *Adj.* profitable; 2. *adv.* profitably

Rente die; ~, ~**n** a) pension; b) *(Kapi-talertrag)* annuity

Ren·tier das reindeer

rentieren *refl. V.* be profitable; *⟨equipment, machinery⟩* pay its way

Rentner der; ~s, ~, **Rentnerin** die; ~, ~**nen** pensioner

Reparatur die; ~, ~en repair (an + Dat. to)

Reparatur·werkstatt die repair [work]shop; *(für Autos)* garage

reparieren *tr. V.* repair; mend

Repertoire [repɛ'toa:ɐ] das; ~s, ~s repertoire

Report der; ~[e]s, ~e, **Reportage** [repɔr'ta:ʒə] die; ~, ~n report; **Re-porter** der; ~s, ~, **Reporterin** die; ~, ~**nen** reporter

Repräsentant der; ~en, ~en, **Re-präsentantin** die; ~, ~**nen** repres-entative; **repräsentativ** *Adj.* rep-resentative; **repräsentieren** *tr. V.* represent

Repressalie die; ~, ~**n** repressive measure

Reproduktion die reproduction; **re-produzieren** *tr. V.* reproduce

Reptil das; ~s, ~ien reptile

Republik die; ~, ~en republic; **repu-blikanisch** *Adj.* republican

Reservat das; ~[e]s, ~e a) reserva-tion; b) *(Naturschutzgebiet)* reserve; **Reserve** die; ~, ~**n** reserve

Reserve-: ~**rad** das spare wheel; ~**reifen** der spare tyre

reservieren tr. V. reserve; **Reservoir** [rezɛr'voa:ɐ̯] das; ~s, ~e (auch fig.) reservoir (**an** + Dat. of)

Residenz die; ~, ~en a) residence; b) (Hauptstadt) [royal] capital

Resignation die; ~, ~en resignation; **resignieren** itr. V. give up

resolut 1. Adj. resolute; 2. adv. resolutely; **Resolution** die; ~, ~en resolution

Resonanz die; ~, ~en resonance

Respekt der; ~[e]s a) (Achtung) respect (**vor** + Dat. for); b) (Furcht) jmdm. ~ einflößen intimidate sb.; **respektieren** tr. V. respect; **respekt·los** 1. Adj. disrespectful; 2. adv. disrespectfully; **Respektlosigkeit** die; ~: disrespectfulness; **respekt·voll** 1. Adj. respectful; 2. adv. respectfully

Ressort [rɛ'so:ɐ̯] das; ~s, ~s area of responsibility; (Abteilung) department

Rest der; ~[e]s, ~e a) rest; **ein** ~ **von** a little bit of; b) (Endstück) remnant; c) (Math.) remainder

Restaurant [rɛsto'rã:] das; ~s, ~s restaurant; **restaurieren** tr. V. restore

restlich Adj. remaining; **rest·los** 1. Adj. complete; 2. adv. completely

Resultat das; ~[e]s, ~e result

Retorte die; ~, ~n retort

retten 1. tr. V. save; (vor Gefahr) save; rescue; (befreien) rescue; **jmdm. das Leben** ~: save sb.'s life; 2. refl. V. (fliehen) escape (**aus** from); **Retter** der; ~s, ~, **Retterin** die; ~, ~nen rescuer

Rettich der; ~s, ~e radish

Rettung die rescue; (vor Zerstörung) saving

rettungs-, Rettungs-: ~**boot** das lifeboat; ~**hubschrauber** der rescue helicopter; ~**los** 1. Adj. hopeless; inevitable ⟨disaster⟩; 2. adv. hopelessly; ~**ring** der lifebelt

Reue die; ~: remorse (**über** + Akk. for); (Rel.) repentance; **reuen** tr. V. etw. reut jmdn. sb. regrets sth.; **reu·mütig** Adj. remorseful; repentant ⟨sinner⟩

Reuse die; ~, ~n fish-trap

Revanche [re'vã:ʃ(ə)] die; ~, ~n revenge; (Sport) return match/fight/ game; **revanchieren** refl. V. **a)** get one's revenge, (coll.) get one's own back (**bei** on); b) **sich bei jmdm. für eine Einladung** ~ (ugs.) return sb.'s invitation

Revers [rə've:ɐ̯] das od. (österr.) der; ~ [rə'vɛ:ɐ̯(s)], ~ [rə'vɛ:ɐ̯s] lapel

Revier das; ~s, ~e a) (Aufgabenbereich) province; b) (Zool.) territory; c) (Polizei~) (Dienststelle) [police] station; (Bereich) district; (des einzelnen Polizisten) beat

Revision die; ~, ~en a) revision; (Änderung) amendment; b) (Rechtsw.) appeal [on a point/points of law]; ~ **einlegen, in die** ~ **gehen** lodge an appeal [on a point/points of law]

Revolte die; ~, ~n revolt; **Revolution** die; ~, ~en (auch fig.) revolution; **revolutionär** 1. Adj. revolutionary; 2. adv. in a revolutionary way; **Revolutionär** der; ~s, ~e, **Revolutionärin** die; ~, ~nen revolutionary

Revolver der; ~s, ~: revolver

Rezept das; ~[e]s, ~e a) (Med.) prescription; b) (Anleitung) recipe; **Rezeption** die; ~, ~en reception no art.; **rezept·pflichtig** Adj. ⟨drug etc.⟩ obtainable only on prescription

R-Gespräch ['ɛr-] das reverse-charge call (Brit.); collect call (Amer.)

Rhabarber der; ~s rhubarb

Rhein der; ~[e]s Rhine; **rheinisch** Adj. Rhenish; ⟨speciality etc.⟩ of the Rhine region; **Rhein·land** das; ~[e]s Rhineland; **Rheinland-Pfalz** (die); ~': the Rhineland-Palatinate

Rhetorik die; ~, ~en rhetoric

Rheuma das; ~s (ugs.) rheumatism; **rheumatisch** (Med.) 1. Adj. rheumatic; 2. adv. rheumatically; **Rheumatismus** der; ~, **Rheumatismen** (Med.) rheumatism

Rhinozeros das; ~[ses], ~se rhinoceros; rhino (coll.)

Rhododendron der od. das; ~s, **Rhododendren** rhododendron

rhythmisch 1. Adj. rhythmical; rhythmic; 2. adv. rhythmically; **Rhythmus** der; ~, **Rhythmen** (auch fig.) rhythm

richten 1. tr. V. **a)** direct ⟨gaze⟩ (**auf** + Akk. at, towards); turn ⟨eyes, gaze⟩ (**auf** + Akk. towards); point ⟨torch, telescope, gun⟩ (**auf** + Akk. at); aim ⟨gun, missile, telescope, searchlight⟩ (**auf** + Akk. on); (fig.) direct ⟨activity, attention⟩ (**auf** + Akk. towards); address ⟨letter, remarks, words⟩ (**an** + Akk. to); level ⟨criticism⟩ (**an** + Akk. at); b) (gerade~) straighten; c) (aburteilen) judge; (verurteilen) condemn; s. auch **zugrunde a;** 2. refl. V. **a)** (sich hinwenden) **sich auf jmdn./etw.** ~

(auch fig.) be directed towards sb./ sth.; **b) sich an jmdn./etw. ~** ⟨person⟩ turn on sb./sth.; ⟨appeal, explanation⟩ be directed at sb./sth.; **sich gegen jmdn./etw. ~** ⟨person⟩ criticize sb./ sth.; ⟨criticism, accusations, etc.⟩ be aimed *or* levelled at sb./sth.; **c)** *(sich orientieren)* **sich nach jmdn./jmds. Wünschen ~:** fit in with sb./sb.'s wishes; **d)** *(abhängen)* **sich nach jmdn./etw. ~:** depend on sb./sth.; **3.** *itr. V. (urteilen)* judge; **Richter** der; **~s, ~, Richterin** die; **~, ~nen** judge
Richt·geschwindigkeit die recommended maximum speed
richtig 1. *Adj.* **a)** right; *(zutreffend)* right; correct; accurate ⟨prophecy, premonition⟩; **b)** *(ordentlich)* proper; **c)** *(wirklich, echt)* real; **2.** *adv.* **a)** right; correctly; **b)** *(ordentlich)* properly; **c)** *(richtiggehend)* really
richtig|stellen *tr. V.* correct
Richt-: **~linie** die guideline; **~schnur** die; *Pl.* **~schnuren** *(fig.)* guiding principle
Richtung die; **~, ~en a)** direction; **b)** *(fig.: Tendenz)* movement; trend
rieb *1. u. 3. Pers. Prät. v.* **reiben**
riechen 1. *unr. tr. V.* **a)** smell; **b)** *(wittern)* ⟨dog etc.⟩ pick up the scent of; **2.** *unr. itr. V.* **a)** smell; **an jmdm./etw. ~:** smell sb./sth.; **b)** *(einen Geruch haben)* smell *(nach of)*
rief *1. u. 3. Pers. Sg. Prät. v.* **rufen**
Riegel der; **~s, ~ a)** bolt; **b) ein ~ Schokolade** a bar of chocolate
Riemen der; **~s, ~ a)** strap; *(Treib~, Gürtel)* belt; **sich am ~ reißen** *(ugs.)* pull oneself together; get a grip on oneself; **b)** *(Ruder)* [long] oar
Riese der; **~n, ~n** giant
rieseln *itr. V.; mit Richtungsangabe mit sein* trickle [down]; ⟨snow⟩ fall gently
Riesen- giant; enormous ⟨selection, profit, portion⟩; tremendous *(coll.)* ⟨effort, rejoicing, success⟩; terrific *(coll.)*, terrible *(coll.)* ⟨stupidity, scandal, fuss⟩
riesen·groß *Adj.* enormous; huge; terrific *(coll.)* ⟨surprise⟩; **Riesenschritt** der giant stride; **riesig 1.** *Adj.* enormous; huge; vast ⟨country⟩; tremendous ⟨effort, progress⟩; **2.** *adv.* *(ugs.)* tremendously *(coll.)*; terribly *(coll.)*
Riesling der; **~s, ~e** Riesling
riet *1. u. 3. Pers. Sg. Prät. v.* **raten**
Riff das; **~[e]s, ~e** reef
Rille die; **~, ~n** groove

Rind das; **~[e]s, ~er a)** cow; *(Stier)* bull; **~er** cattle *pl.*; **b)** *(~fleisch)* beef
Rinde die; **~, ~n a)** *(Baum~)* bark; **b)** *(Brot~)* crust; *(Käse~)* rind
Rinder·braten der roast beef *no indef. art.;* *(roh)* roasting beef *no indef. art.*
Rind-: **~fleisch** das beef; **~vieh** das cattle *pl.*
Ring der; **~[e]s, ~e** ring; **Ringel·natter** die ring-snake
ringen 1. *unr. tr. V. (Sport, fig.)* wrestle; *(fig.: kämpfen)* struggle, fight **(um** for; **gegen, mit** with); **nach Luft ~:** struggle for breath; **2.** *unr. tr. V.* **die Hände ~:** wring one's hands; **Ringen** das; **~s** *(Sport)* wrestling *no art.*
Ring-: **~finger** der ring-finger; **~kampf** der **a)** [stand-up] fight; **b)** *(Sport)* wrestling bout
rings *Adv.* all around; **rings·herum** *Adv.* all around [it/them etc.]
Ring·straße die ring road
rings-: **~um, ~umher** *Adv.* all around
Rinne die; **~, ~n** channel; *(Dach~, Rinnstein)* gutter; *(Abfluß)* drainpipe; **rinnen** *unr. itr. V.; mit sein* run; **Rinn·stein** der gutter
Rippchen das; **~s, ~** *(Kochk. südd.)* rib [of pork]; **Rippe** die; **~, ~n** rib
Risiko das; **~s, Risiken** risk; **riskant 1.** *Adj.* risky; **2.** *adv.* riskily; **riskieren** *tr. V.* risk
riß *1. u. 3. Pers. Sg. Prät. v.* **reißen**
Riß der; **Risses, Risse** tear; *(Spalt, Sprung)* crack; **rissig** *Adj.* cracked; chapped ⟨lips⟩
ritt *1. u. 3. Pers. Sg. Prät. v.* **reiten**; **Ritt** der; **~[e]s, ~e** ride; **Ritter** der; **~s, ~:** knight; **Ritter·sporn** der delphinium; **rittlings** *Adv.* astride
Ritze die; **~, ~n** crack; [narrow] gap; **ritzen** *tr. V.* scratch
Rivale der; **~n, ~n, Rivalin** die; **~, ~nen** rival; **Rivalität** die; **~, ~en** rivalry *no indef. art.*
Robbe die; **~, ~n** seal
Robe die; **~, ~n** robe; *(schwarz)* gown
Roboter der; **~s, ~:** robot
robust *Adj.* robust
roch *1. u. 3. Pers. Sg. Prät. v.* **riechen**
Rochade die; **~, ~n** *(Schach)* castling
röcheln *itr. V.* ⟨dying person⟩ give the death-rattle
Rock der; **~[e]s, Röcke** skirt
Rodel·bahn die toboggan-run; *(Sport)* luge-run; **rodeln** *itr. V.; mit sein* sledge; toboggan

roden *tr. V.* clear ⟨*wood, land*⟩; *(ausgraben)* grub up ⟨*tree*⟩

Rogen der; ~s, ~: roe

Roggen der; ~s rye

Roggen-: ~**brot** das rye bread; ein ~**brot** a loaf of rye bread; ~**brötchen** das rye-bread roll

roh 1. *Adj.* **a)** raw ⟨*food*⟩; unboiled ⟨*milk*⟩; unfinished ⟨*wood*⟩; **b)** *(ungenau)* rough; **c)** *(brutal)* brutish; brute *attrib.* ⟨*force*⟩; 2. *adv.* **a)** *(ungenau)* roughly; **b)** *(brutal)* brutishly; *(grausam)* callously; *(grob)* coarsely

Roh-: ~**bau** der shell [of a/the building]; ~**kost** die raw fruit and vegetables *pl.*; ~**material** das raw material; ~**öl** das crude oil

Rohr das; ~|e|s, ~e **a)** *(Leitungs~)* pipe; *(als Bauteil)* tube; **b)** *o. Pl. (Röhricht)* reeds *pl.*; **c)** *o. Pl. (Werkstoff)* reed; **Röhre** die; ~, ~n tube; *(Elektronen~)* valve *(Brit.)*; tube *(Amer.)*

Roh·stoff der raw material

Rokoko das; ~|s| rococo

Rolladen der; ~s, **Rolläden** [roller] shutter; **Roll·bahn** die *(Flugw.)* taxiway; **Rolle** die; ~, ~n **a)** *(Spule)* reel; **b)** *(zylindrischer [Hohl]körper; Zusammengerolltes)* roll; **c)** *(Walze)* roller; **d)** *(Rad)* [small] wheel; *(an Möbeln usw.)* castor; *(für Gardine, Schiebetür usw.)* runner; **e)** *(Turnen, Kunstflug)* roll; **f)** *(Theater, Film usw., fig.)* role; part; *(Soziol.)* role; **es spielt keine ~:** it is of no importance; *(es macht nichts aus)* it doesn't matter; **rollen** 1. *tr. V.* roll; 2. *itr. V.* **a)** mit sein ⟨*ball, wheel, etc.*⟩ roll; ⟨*vehicle*⟩ move; ⟨*aircraft*⟩ taxi; **Roller** der; ~s, ~: scooter

Roll-: ~**feld** das runway[s] and taxiway[s]; ~**kragen** der polo-neck; ~**laden** der *s.* Rolladen; ~**mops** der rollmops; ~**schuh** der roller-skate; ~**schuh laufen** roller-skate; ~**splitt** der loose chippings *pl.*; ~**stuhl** der wheelchair; ~**treppe** die escalator

Rom *(das)*; ~s Rome

Roman der; ~s, ~e novel

Romantik die; ~: romanticism; die ~: Romanticism; **romantisch** 1. *Adj.* romantic; 2. *adv.* romantically

Romanze die; ~, ~n romance

Römer der; ~s, ~: Roman

römisch-katholisch *Adj.* Roman Catholic

röntgen *tr. V.* X-ray

Röntgen-: ~**aufnahme** die, ~**bild** das X-ray [image/photograph *or* picture]; ~**strahlen** *Pl.* X-rays

rosa 1. *indekl. Adj.* pink; 2. *adv.* pink; **Rosa** das; ~s, ~ *od.* ~s pink; **Rose** die; ~, ~n rose

rosé 1. *indekl. Adj.* pale pink; **Rosé** der; ~s, ~s rosé [wine]

Rosen-: ~**kohl** der; *o. Pl.* [Brussels] sprouts *pl.*; ~**kranz** der *(kath. Kirche)* rosary; einen ~**kranz beten** say a rosary; ~**montag** der the day before Shrove Tuesday

rosig *Adj.* **a)** rosy; pink ⟨*piglet etc.*⟩; **b)** *(fig.)* rosy; optimistic ⟨*mood*⟩

Rosine die; ~, ~n raisin

Rosmarin der; ~s rosemary

Roß das; Rosses, Rosse *od.* Rösser horse; steed *(poet./joc.)*; **hoch zu ~:** on horseback; **auf dem** *od.* **seinem hohen ~ sitzen** *(fig.)* be on one's high horse

¹Rost der; ~|e|s, ~e **a)** *(Gitter)* grating; *(eines Ofens, einer Feuerstelle)* grate; *(Brat~)* grill; **b)** *(Bett~)* base

²Rost der; ~|e|s rust

Rost-: ~**braten** der grilled steak; ~**bratwurst** die grilled sausage

rosten *itr. V.; auch mit sein* rust

rösten ['rœstn̩, 'rø:stn̩] *tr. V.* roast; toast ⟨*bread*⟩

rost·frei *Adj.* stainless ⟨*steel*⟩

Rösti die; ~ *(schweiz. Kochk.)* thinly sliced fried potatoes *pl.*

rostig *Adj.* rusty

rot 1. *Adj.* red; ~ **werden** turn red; ⟨*person*⟩ blush; ⟨*traffic-light*⟩ change to red; 2. *adv.* red; **Rot** das; ~s, ~ *od.* ~s red; **Rot·barsch** der rose-fish; **Röte** die; ~: red[ness]; **röten** 1. *tr. V.* redden; 2. *refl. V.* go *or* turn red; **rot·haarig** *Adj.* red-haired; **Rot·hirsch** der red deer

rotieren *itr. V.* **a)** rotate; **b)** *(ugs.: hektisch sein)* get into a flap *(coll.)*

Rot-: ~**käppchen** (das) Little Red Riding Hood; ~**kehlchen** das; ~s, ~: robin [redbreast]; ~**kohl** der, *(bes. südd., österr.)* ~**kraut** das red cabbage

rötlich *Adj.* reddish; **Rot·stift** der red pencil; **Rötung** die; ~, ~en reddening; **Rot·wein** der red wine

Rotz der; ~es *(salopp)* snot *(sl.)*

Rouge [ru:ʒ] das; ~s, ~s rouge

Roulade [ru:la:də] die; ~, ~n *(Kochk.)* [beef/veal/pork] olive

Route ['ru:tə] die; ~, ~n route; **Routine** [ru'ti:nə] die; ~ **a)** *(Erfahrung)* experience; *(Übung)* practice; **b)** *(Gewohnheit)* routine *no def. art.*

Rübe die; ~, ~n turnip; **rote ~:** beetroot; **gelbe ~** *(südd.)* carrot

rüber *Adv. (ugs.)* over

Rubin der; ~s, ~e ruby

Rubrik die; ~, ~en column; *(fig.: Kategorie)* category

Ruck der; ~|e|s, ~e jerk

Rück·blick der look back (**auf** + *Akk.* at); retrospective view (**auf** + *Akk.* of)

rücken *itr., tr. V.* move

Rücken der; ~s, ~: back; *(Buch~)* spine
Rücken-: ~**deckung** die a) *(bes. Milit.)* rear cover; b) *(fig.)* backing; ~**lehne** die [chair/seat] back; ~**mark** das *(Anat.)* spinal cord; ~**schmerzen** *Pl.* backache *sing.;* ~**schwimmen** das backstroke; ~**wind** der tail wind

rück-, Rück-: ~|**erstatten** *tr. V.; nur im Inf. u. 2. Part.* repay; ~**erstattung** die repayment; ~**fahrkarte** die, ~**fahrschein** der return [ticket]; ~**fahrt** die return journey; ~**fall** der *(Med., auch fig.)* relapse; ~**fällig** *Adj. (Med., auch fig.)* relapsed *⟨patient, alcoholic, etc.⟩;* ~**fällig werden** have a relapse; *⟨alcoholic etc.⟩* go back to one's old ways; *⟨criminal⟩* commit a second offence; ~**flug** der return flight; ~**frage** die query; ~**gabe** die return; ~**gang** der drop, fall *(Gen.* in); ~**gängig** *Adj.* ~**gängig machen** cancel *⟨agreement, decision, etc.⟩;* ~**grat** das spine; *(bes. fig.)* backbone; ~**halt** der support; backing; ~**halt·los** 1. *Adj.* unreserved, unqualified *⟨support⟩;* 2. *adv.* unreservedly; ~**kehr** die; ~: return; ~**lage** die savings *pl.;* ~**läufig** *Adj.* decreasing *⟨number⟩;* declining *⟨economic growth etc.⟩;* falling *⟨rate, production, etc.⟩;* ~**licht** das rear- *or* taillight

rücklings *Adv.* on one's back

Rück-: ~**nahme** die taking back; ~**reise** die return journey; ~**ruf** der *(Fernspr.)* return call

Ruck·sack der rucksack; *(Touren~)* back-pack

rück-, Rück-: ~**schlag** der set-back; ~**schritt** der retrograde step; ~**seite** die back; *(einer Münze usw.)* reverse; far side; ~**sicht** die consideration; ~**sicht auf jmdn. nehmen** show consideration for *or* towards sb.; ~**sicht·nahme** die; ~: consideration; ~**sichts·los** 1. *Adj.* inconsiderate; thoughtless; *(verantwortungslos)* reckless *⟨driver⟩; (schonungslos)* ruthless; 2. *adv. s. Adj:* inconsiderately;

recklessly; ruthlessly; ~**sichts·losigkeit** die; ~, ~en *s.* rücksichtslos : lack of consideration; recklessness; ruthlessness; ~**sichts·voll** 1. *Adj.* considerate; 2. *adv.* considerately; ~**sitz** der back seat; ~**spiegel** der rear-view mirror; ~**sprache** die consultation; ~**stand** der a) *(Rest)* residue; b) *(ausstehende Zahlung)* arrears *pl.;* c) *(Zurückbleiben hinter dem gesetzten Ziel)* backlog; *(bes. Sport: hinter dem Gegner)* deficit; [**mit etw.**] **im ~stand sein/in ~stand** *(Akk.)* geraten be/get behind [with sth.]; ~**ständig** *Adj.* a) backward; *(schon länger fällig)* outstanding *⟨payment, amount⟩; ⟨wages⟩* still owing; ~**strahler** der reflector; ~**tritt** der resignation (**von** from); *(von einer Kandidatur, einem Vertrag usw.)* withdrawal (**von** from)

rückwärts *Adv.* backwards; **Rückwärts·gang** der *(Kfz-W.)* reverse [gear]

rück-, Rück-: ~**weg** der return journey; ~**wirkend** 1. *Adj.* retrospective; backdated *⟨pay increase⟩;* 2. retrospectively; ~**zahlung** die repayment; ~**zug** der retreat

Rüde der; ~n, ~n [male] dog

Rudel das; ~s, ~: herd; *(von Wölfen, Hunden)* pack

Ruder das; ~s, ~ a) *(Riemen)* oar; b) *(Steuer~)* rudder; **Ruder·boot** das row-boat; rowing-boat *(Brit.);* **rudern** 1. *itr. V.; mit sein* row; 2. *tr. V.* row

Ruf der; ~|e|s, ~e a) call; *(Schrei)* shout; cry; *(Tierlaut)* call; b) *o. Pl. (fig.: Forderung)* call (**nach** for); c) *o. Pl. (Telefonnummer)* telephone [number]; d) *(Leumund)* reputation; **rufen** 1. *unr. itr. V.* call (**nach** for); *(schreien)* shout (**nach** for); *⟨animal⟩* call; 2. *unr. tr. V.* a) *(aus~)* call; *(schreien)* shout; b) *(herbei~, an~)* **jmdn.** ~: call sb.; **jmdn. zu Hilfe** ~: call to sb. to help

Ruf-: ~**name** der first name *(by which one is generally known);* ~**nummer** die telephone number

Rüge die; ~, ~n reprimand; **rügen** *tr. V.* reprimand *⟨person⟩* (**wegen** for); censure *⟨carelessness etc.⟩*

Ruhe die; ~ a) *(Stille)* silence; ~ [**bitte**]! quiet *or* silence [please]!; b) *(Ungestörtheit)* peace; **jmdn. mit etw. in ~ lassen** stop bothering sb. with sth.; c) *(Unbewegtheit)* rest; d) *(Erholung)* rest *no def. art.;* e) *(Gelassenheit)*

calm[ness]; composure; |die| ~ bewah-ren/die ~ verlieren keep calm/lose one's composure; in |aller| ~: [really] calmly; ruhe·los 1. *Adj.* restless; 2. *adv.* restlessly; ruhen *itr. V.* a) *(aus~)* rest; b) *(geh.: schlafen)* sleep; c) *(stillstehen)* ⟨*work, business*⟩ have stopped; ⟨*production, firm*⟩ be at a standstill

Ruhe-: ~pause die break; ~stand der; *o. ·Pl.* retirement; ~störung die disturbance; *(Rechtsw.)* disturbance of the peace; ~tag der closing day; „Dienstag ~tag" 'closed on Tuesdays'

ruhig 1. *Adj.* a) *(still, leise)* quiet; b) *(friedlich, ungestört)* peaceful ⟨*times, life, valley, etc.*⟩; quiet ⟨*talk, reflection, life*⟩; c) *(unbewegt)* calm ⟨*sea, weather*⟩; still ⟨*air*⟩; *(fig.)* peaceful ⟨*melody*⟩; *(gleichmäßig)* steady ⟨*breathing, hand, steps*⟩; smooth ⟨*flight, crossing*⟩; d) *(gelassen)* calm ⟨*voice etc.*⟩; quiet, calm ⟨*person*⟩; 2. *adv.* a) *(still, leise)* quietly; sich ~ ver-halten keep quiet; b) *(friedlich, ohne Störungen)* peacefully; *(ohne Zwischenfälle)* uneventfully; ⟨*work, think*⟩ in peace; c) *(unbewegt)* ⟨*sit, lie, stand*⟩ still; *(gleichmäßig)* ⟨*burn, breathe*⟩ steadily; ⟨*run, fly*⟩ smoothly; d) *(gelassen)* ⟨*speak, watch, sit*⟩ calmly; 3. *Adv.* by all means

Ruhm der; ~|e|s fame; rühmen 1. *tr. V.* praise; 2. *refl. V.* boast (+ *Gen.* about; ruhm·reich *Adj.* glorious ⟨*victory, history*⟩; celebrated ⟨*general, army, victory*⟩

Ruhr die; ~, ~en dysentery *no art.*

Rühr·ei das scrambled egg[s *pl.*]; rüh-ren 1. *tr. V.* a) *(um~)* stir; (ein~) stir ⟨*egg, powder, etc.*⟩ ⟨an, in + *Akk.* into⟩; b) *(bewegen)* move ⟨*limb, fingers, etc.*⟩; c) *(fig.)* move; touch; 2. *itr. V.* a) *(um~)* stir; b) *(geh.: her~)* das rührt daher, daß ...: that stems from the fact that ...; 3. *refl. V.* a) *(sich bewegen)* move; b) *(Milit.)* rührt euch! at ease!; rührend 1. *Adj.* touching; 2. *adv.* touchingly; rühr·selig 1. *Adj.* a) emotional ⟨*person*⟩; b) *(allzu gefühlvoll)* over-sentimental; 2. *adv.* in an over-sentimental manner; Rührung die; ~: emotion

Ruine die; ~, ~n ruin; ruinieren *tr. V.* ruin

rülpsen *itr. V. (ugs.)* burp

rum *Adv. (ugs.) s.* herum

Rum der; ~s, ~s rum

Rumäne der; ~n, ~n Romanian; Ru-

mänien (das); ~s Romania; rumä-nisch *Adj.* Romanian

Rummel der; ~s *(ugs.)* a) commotion; *(Aufhebens)* fuss (um about); b) *(Jahrmarkt)* fair

Rumpf der; ~|e|s, Rümpfe a) trunk [of the body]; b) *(beim Schiff)* hull; c) *(beim Flugzeug)* fuselage

rümpfen *tr. V.* die Nase |bei etw.| ~: wrinkle one's nose [at sth.]; über jmdn./etw. die Nase rümpfen *(fig.)* look down one's nose at sb./turn up one's nose at sth.

rund 1. *Adj.* a) round; b) *(dicklich)* plump ⟨*arms etc.*⟩; chubby ⟨*cheeks*⟩; fat ⟨*stomach*⟩; c) *(ugs.: ganz)* round ⟨*dozen, number, etc.*⟩; 2. *Adv.* a) *(ugs.: etwa)* about; b) ~ um jmdn./etw. [all] around sb./sth.; Rund·blick der panorama; view in all directions; Runde die; ~, ~n a) *(Sport: Strecke)* lap; b) *(Sport: Durchgang usw.)* round; über die ~n kommen *(fig. ugs.)* get by; manage; c) *(Personenkreis)* circle; *(Gesellschaft)* company; d) *(Rundgang)* round; e) *(Lage)* round

rund-, Rund-: ~erneuern *tr. V.* *(Kfz-W.)* remould; ~fahrt die tour (durch of); ~funk der a) radio; b) *(Einrichtung, Gebäude)* radio station

Rundfunk-: ~anstalt die broadcasting corporation; ~gerät das radio set; ~sendung die radio programme; ~sprecher der radio announcer

rund-, Rund-: ~gang der round (durch of); ~herum *Adv.* a) *(ringsum)* all around; b) *(völlig)* completely

rundlich *Adj.* a) roundish; b) *(mollig)* plump

Rund-: ~reise die [circular] tour (durch of); ~weg der circular path *or* walk

runter *Adv. (ugs.)* ~ |da|! get off [there]; *s. auch* herunter; hinunter

Runzel die; ~, ~n wrinkle; runz[e]lig *Adj.* wrinkled; runzeln *tr. V.* die Stirn/die Brauen ~: wrinkle one's brow/knit one's brows; *(ärgerlich)* frown

rupfen *tr. V.* a) pluck ⟨*goose, hen, etc.*⟩; b) *(abreißen)* pull up ⟨*weeds, grass*⟩; pull off ⟨*leaves etc.*⟩

Rüsche die; ~, ~n ruche; frill

Ruß der; ~es soot

Russe der; ~n, ~n Russian

Rüssel der; ~s, ~ *(des Elefanten)* trunk; *(des Schweins)* snout; *(bei Insekten u. ä.)* proboscis

rußen *itr. V.* give off sooty smoke

Russin die; ~, ~nen Russian; **russisch** 1. *Adj.* Russian; 2. *adv. (auf ~)* in Russian; **Russisch das**; ~|s| Russian; **Ruß·land (das)**; ~s Russia

rüsten *itr. V.* arm

rüstig 1. *Adj.* sprightly; active

rustikal 1. *Adj.* country-style ⟨*food, inn, clothes, etc.*⟩; rustic ⟨*furniture*⟩; 2. *adv.* in [a] country style

Rüstung die; ~, ~en **a)** armament *no art.*; *(Waffen)* arms *pl.*; weapons *pl.*; **b)** *(hist.)* suit of armour

Rüstungs-: ~**industrie** die armaments *or* arms industry; ~**kontrolle** die arms control; ~**stopp** der arms freeze

Rute die; ~, ~n switch; *(Birken~, Angel~, Wünschel~)* rod

Rutsch·bahn die slide; **rutschen** *itr. V.; mit sein* slide; ⟨*clutch, carpet*⟩ slip; **rutschig** *Adj.* slippery

rütteln *tr., itr. V.* shake

S

---

s, S [ɛs] das; ~, ~: s, S

s *Abk.* Sekunde sec.; s.

S *Abk.* **a)** Süden S.; **b)** *(österr.)* Schilling Sch.

s. *Abk.* siehe

S. *Abk.* Seite p.

Sa. *Abk.* Samstag Sat.

Saal der; ~|e|s, Säle **a)** hall; *(Ball~)* ballroom; **b)** *(Publikum)* audience

Saar·land das; ~|e|s Saarland; Saar *(esp. Hist.)*

Saat die; ~, ~en **a)** *(das Gesäte)* [young] crops *pl.*; **b)** *o. Pl. (das Säen)* sowing; **c)** *(Samenkörner)* seed[s *pl.*]

Säbel der; ~s, ~: sabre

Sabotage [zabo'ta:ʒə] die; ~, ~n sabotage *no art.*; **sabotieren** *tr. V.* sabotage

sach·dienlich *Adj.* useful; **Sache** die; ~, ~n **a)** *Pl.* things; **b)** *(Angelegenheit)* matter; business *(esp. derog.)*; **zur ~ kommen** come to the point; **c)** *(Rechts~)* case; **d)** *o. Pl. (Anliegen)* cause

sach-, Sach-: ~**gemäß**, ~**gerecht** 1. *Adj.* proper; correct; 2. *adv.* properly; correctly; ~**kenntnis** die expertise; ~**kundig** 1. *Adj.* with a knowledge of the subject *postpos., not pred.*; 2. *adv.* expertly

sachlich 1. *Adj.* **a)** *(objektiv)* objective; *(nüchtern)* functional ⟨*building style, etc.*⟩; matter-of-fact ⟨*letter etc.*⟩; **b)** *nicht präd. (sachbezogen)* factual ⟨*error*⟩; 2. *adv. (objektiv)* objectively; ⟨*state*⟩ as a matter of fact; *(nüchtern) (furnished)* in a functional style; ⟨*written*⟩ in a matter-of-fact way; **b)** *(sachbezogen)* factually ⟨*wrong*⟩; **sächlich** *Adj. (Sprachw.)* neuter; **Sach·schaden** der damage [to property] *no indef. art.*

Sachse der; ~n, ~n Saxon; **Sachsen-Anhalt (das)**; ~s Saxony-Anhalt

sacht, sachte 1. *Adj.* **a)** *(behutsam)* gentle; **b)** *(leise)* quiet; 2. *adv.* **a)** gently; **b)** *(leise)* quietly

Sach-: ~**verhalt** der; ~|e|s, ~e facts *pl.* [of the matter]; ~**verstand** der expertise; grasp of the subject

Sack der; ~|e|s, Säcke sack; *(aus Papier, Kunststoff)* bag

Sack-: ~**gasse** die cul-de-sac; ~**hüpfen das**; ~s sack race

Sadismus der; ~: sadism *no art.*; **Sadist** der; ~en, ~en, **Sadistin** die; ~, ~nen sadist; **sadistisch** 1. *Adj.* sadistic; 2. *adv.* sadistically

säen *tr. (auch itr.) V.* sow

Saft der; ~|e|s, Säfte **a)** juice; **b)** *(in Pflanzen)* sap; **saftig** *Adj.* **a)** juicy; sappy ⟨*stem*⟩; lush ⟨*meadow, green*⟩; **b)** *(ugs.)* hefty ⟨*slap, blow*⟩; steep *(coll.)* ⟨*prices, bill*⟩; crude ⟨*joke, song, etc.*⟩; strongly-worded ⟨*letter etc.*⟩

Sage die; ~, ~n legend; *(bes. nordische)* saga

Säge die; ~, ~n saw

sagen 1. *tr. V.* **a)** say; **was ich noch ~ wollte** [oh] by the way; **unter uns gesagt** between you and me; **b)** *(mitteilen)* jmdm. etw. ~: say sth. to sb.; *(zur Information)* tell sb. sth.; **c)** *(nennen)* **zu jmdm./etw. X ~:** call sb./sth. X; **d)** *(anordnen, befehlen)* tell; 2. *refl. V.* **sich** *(Dat.)* **etw. ~:** say sth. to oneself

sägen *tr., itr. V.* saw

sah *1. u. 3. Pers. Sg. Prät. v.* sehen

Sahne die; ~: cream

Saison [zɛ'zõ:] die; ~, ~s season

Saite die; ~, ~n string; **Saiten·instrument das** stringed instrument

Sakko der od. das; ~s, ~s jacket
Sakrament das; ~|e|s, ~e sacrament; **Sakristei** die; ~, ~en sacristy
Salami die; ~, ~|s| salami
Salat der; ~|e|s, ~e a) salad; b) o. Pl. |grüner| ~: lettuce; ein Kopf ~: a [head of] lettuce
Salat-: ~besteck das salad-servers pl.; ~soße die salad-dressing
Salbe die; ~, ~n ointment
Salbei der od. die; ~: sage
Saldo der; ~s, ~s od. Saldi (Buchf., Finanzw.) balance
Säle s. **Saal**
Salmiak der od. das; ~: sal ammoniac
Salon [za'lō:] der; ~s, ~s a) (Raum) drawing-room; b) (Geschäft) [hair-etc.] salon
salopp 1. Adj. casual ⟨clothes⟩; informal ⟨behaviour⟩; 2. adv. ⟨dress⟩ casually
Salto der; ~s, ~s od. Salti somersault; (beim Turnen auch) salto
salutieren itr. V. (bes. Milit.) salute
Salve die; ~, ~n (Milit.) salvo; (aus Gewehren) volley
Salz das; ~es, ~e salt; **salzen** tr. V. salt; **salzig** Adj. salty
Salz-: ~kartoffel die; meist Pl. boiled potato; ~säure die; o. Pl. (Chemie) hydrochloric acid; ~stange die salt stick; ~streuer der; ~s, ~: salt-sprinkler; salt-shaker (Amer.); ~wasser das; Pl. ~wässer a) o. Pl. (zum Kochen) salted water; b) (Meerwasser) salt water
Sambia (das); ~s Zambia
Samen der; ~s, ~ a) (~korn) seed; b) o. Pl. (~körner) seed[s pl.]; c) o. Pl. (Sperma) sperm; semen
sammeln 1. tr. (auch itr.) V. a) collect; gather ⟨honey, firewood, fig.: experiences, impressions, etc.⟩; gather ⟨berries etc.⟩; b) (zusammenkommen lassen) gather ⟨people⟩ [together]; assemble ⟨people⟩; cause ⟨light rays⟩ to converge; 2. refl. V. gather [together]; **Sammler** der; ~s, ~: collector; **Sammlung** die; ~, ~en collection; b) |innere| ~: composure
Samstag der; ~|e|s, ~e Saturday; s. auch Dienstag; Dienstag-; **samstags** Adv. on Saturdays
samt 1. Präp. mit Dat. together with; 2. Adv. ~ und sonders one and all
Samt der; ~|e|s, ~e velvet
sämtlich Indefinitpron. u. unbest. Zahlwort all the
Sand der; ~|e|s sand

Sandale die; ~, ~n sandal
sandig Adj. sandy
Sand-: ~kasten der [child's] sand-pit; sand-box (Amer.); ~kuchen der Madeira cake; ~mann der, ~männchen das; o. Pl. sandman; ~stein der sandstone; ~strand der sandy beach
sandte 1. u. 3. Pers. Sg. Prät. v. senden
sanft 1. Adj. gentle; (leise) soft; (friedlich) peaceful; 2. adv. gently; (leise) softly; (friedlich) peacefully
sang 1. u. 3. Pers. Sg. Prät. v. singen; **Sänger** der; ~s, ~, **Sängerin** die; ~, ~nen singer
sanieren 1. tr. V. a) redevelop ⟨area⟩; rehabilitate ⟨building⟩; (renovieren) renovate [and improve] ⟨flat etc.⟩; b) (Wirtsch.) restore ⟨firm⟩ to profitability; 2. refl. V. ⟨company etc.⟩ restore itself to profitability; ⟨person⟩ get oneself out of the red; **Sanierung** die; ~, ~en a) s. sanieren a: redevelopment; rehabilitation; renovation; b) restoration to profitability; **sanitär** Adj. sanitary; **Sanitäter** der; ~s, ~: first-aid man; (im Krankenwagen) ambulance man
sank 1. u. 3. Pers. Sg. Prät. v. sinken
sann 1. u. 3. Pers. Sg. Prät. v. sinnen
Saphir der; ~s, ~e sapphire
Sardelle die; ~, ~n anchovy
Sardine die; ~, ~n sardine
Sarg der; ~|e|s, Särge coffin
saß 1. u. 3. Pers. Sg. Prät. v. sitzen
Satan der (bibl.) Satan no def. art.
Satellit der; ~en, ~en satellite
Satire die; ~, ~n satire
satt Adj. a) full [up] pred.; well-fed; sich ~ essen/trinken eat/drink as much as one wants; eat/drink one's fill; b) jmdn./etw. ~ haben (ugs.) be fed up with sb./sth. (coll.)
Sattel der; ~s, Sättel a) saddle; **satteln** 1. tr. V. saddle; 2. itr. V. saddle the/one's horse
sättigen itr. V. be filling
Sattler der; ~s, ~: saddler; (allgemein) leather-worker
Satz der; ~es, Sätze a) (sprachliche Einheit) sentence; b) (Musik) movement; c) (Tennis, Volleyball) set; (Tischtennis, Badminton) game; d) (Sprung) leap; jump; e) (Amtsspr.: Tarif) rate; f) (Set) set; g) (Boden~) sediment; (von Kaffee) grounds pl.
Satzung die; ~, ~en articles of association pl.; statutes pl.
Satz·zeichen das punctuation mark

Sau die; ~, Säue a) *(weibliches Schwein)* sow; b) *(bes. südd.: Schwein)* pig

sauber 1. *Adj.* a) clean; b) *(sorgfältig)* neat; 2. *adv.* a) *(sorgfältig)* neatly; b) *(fehlerlos)* |sehr| ~: [quite] perfectly; **Sauberkeit** die; ~: cleanness; **sauber|machen** 1. *tr. V.* clean; 2. *itr. V.* clean; do the cleaning; **säubern** *tr. V.* clean; **Säuberung** die; ~, ~en cleaning

Sauce *s.* Soße

Saudi [zaudi] der; ~s, ~s Saudi; **Saudi-Arabien (das)** Saudi Arabia

sauer 1. *Adj.* a) sour; pickled ⟨herring, gherkin, etc.⟩; acid[ic] ⟨wine, vinegar⟩; **saurer Regen** acid rain; b) *(ugs.: verärgert)* cross, annoyed **(auf + Akk.** with); 2. *adv.* in vinegar; **Sauer·braten** der *braised beef marinated in vinegar and herbs;* sauerbraten *(Amer.)*

Sauerei die; ~, ~en *(salopp abwertend)* a) *(Unflätigkeit)* obscenity b) *(Gemeinheit)* bloody scandal *(sl.)*

Sauer-: ~**kirsche** die sour cherry; ~**kraut** das *o. Pl.* sauerkraut

säuerlich *Adj.* |leicht| ~: slightly sour; slightly sharp ⟨sauce⟩

Sauer-: ~**stoff** der; *o. Pl.* oxygen; ~**stoffgerät** das oxygen apparatus; ~**stoffmangel** der; *o. Pl.* lack of oxygen; ~**teig** der leaven

saufen 1. *unr. itr. V. (salopp: trinken)* drink; swig *(coll.); (Alkohol trinken)* drink; booze *(coll.);* 2. *unr. tr. V. (salopp: trinken)* drink; **Säufer** der; ~s, ~ *(salopp)* boozer *(coll.);* **säuft** 3. *Pers. Sg. Präsens v.* saufen

saugen 1. *tr. V.* a) *auch unr.* suck; b) *auch itr. (staub~)* vacuum; hoover *(coll.);* 2. *regelm. (auch unr.) itr. V.* **an** etw. *(Dat.)* ~ suck [at] sth.; 3. *unr. (auch regelm.) refl. V.* **sich voll** etw. ~: become soaked with sth.; **säugen** *tr. V.* suckle; **Säuge·tier** das *(Zool.)* mammal; **Säugling** der; ~s, ~e baby; **Säuglings·pflege** die baby care

Säule die; ~, ~n column; *(nur als Stütze, auch fig.)* pillar

Saum der; ~|e|s, Säume hem; **säumen** *tr. V.* hem; *(fig. geh.)* line

säumig *(geh.) Adj.* tardy

Sauna die; ~, ~s *od.* Saunen sauna

Säure die; ~, ~n a) *o. Pl. (von Früchten)* sourness; *(von Wein, Essig)* acidity; *(von Soßen)* sharpness; b) *(Chemie)* acid; **Saure·gurken·zeit** die *(ugs.)* silly season *(Brit.)*

Saus: in ~ **und Braus leben** live the high life

säuseln 1. *itr. V.* ⟨leaves, branches, etc.⟩ rustle; ⟨wind⟩ murmur; 2. *tr. V. (iron.: sagen)* whisper; **sausen** *itr. V.* a) ⟨wind⟩ whistle; ⟨storm⟩ roar; ⟨head, ears⟩ buzz; b) *mit sein* ⟨person⟩ rush; ⟨vehicle⟩ roar; c) *mit sein* ⟨whip, bullet, etc.⟩ whistle

Savanne [za'vanə] die; ~, ~n savannah

Saxophon das; ~s, ~e saxophone

S-Bahn ['ɛs-] die city and suburban railway

SB- [ɛs'be:-] self-service *(attrib.)*

Schabe die; ~, ~n cockroach

schaben *tr., itr. V.* scrape; **Schaber** der; ~s, ~: scraper

schäbig 1. *Adj.* a) *(abgenutzt)* shabby; b) *(jämmerlich, gering)* pathetic; c) *(gemein)* shabby; 2. *adv.* a) *(abgenutzt)* shabbily; b) *(jämmerlich)* miserably; c) *(gemein)* meanly

Schach das; ~s, ~s a) *o. Pl. (Spiel)* chess; b) *(Stellung)* check; **jmdn./etw. in** ~ **halten** *(ugs. fig.)* keep sb./sth. in check

Schach-: ~**brett** das chessboard; ~**figur** die chess piece; ~**spiel** das a) *o. Pl. (Spiel)* chess; *(das Spielen)* chess-playing; b) *(Brett und Figuren)* chess set

Schacht der; ~|e|s, Schächte shaft

Schachtel die; ~, ~n a) box; **eine** ~ **Zigaretten** a packet *or (Amer.)* pack of cigarettes; b) **alte** ~ *(salopp abwertend)* old bag *(sl.)*

schade *Adj.* |ach, wie| ~! [what a] pity *or* shame; |es ist| ~ **um jmdn./etw.** it's a pity *or* shame about sb./sth.; **für jmdn./für** *od.* **zu etw. zu** ~ **sein** be too good for sb./sth.

Schädel der; ~s, ~: skull; *(Kopf)* head; **Schädel·bruch** der *(Med.)* skull fracture

schaden *itr. V.* **jmdm./einer Sache** ~: damage *or* harm sb./sth.; **Schaden** der; ~s, Schäden a) damage *no pl., no indef. art.;* **ein kleiner/großer** ~: little/major damage; b) *(Nachteil)* disadvantage

schaden-, Schaden-: ~**ersatz** der *(Rechtsw.)* damages *pl.;* ~**freude** die *o. Pl.* malicious pleasure; ~**froh** 1. *Adj.* gloating; ~**froh sein** gloat; 2. *adv.* with malicious pleasure

schadhaft *Adj.* defective; **schädigen** *tr. V.* damage ⟨health, reputation, interests⟩; harm, hurt ⟨person⟩; cause losses to ⟨firm, industry, etc.⟩; **Schädigung die;** ~, ~en damage no pl., no indef. art. (Gen. to); **schädlich** Adj. harmful; **Schädling der;** ~s, ~e pest

Schaf das; ~[e]s, ~e a) sheep; b) (ugs.: Dummkopf) twit (Brit. sl.); **Schaf·bock** der ram; **Schäfchen** das; ~s, ~: [little] sheep; (Lamm) lamb; **Schäfer** der; ~s, ~: shepherd; **Schäfer·hund** der sheep-dog; [deutscher] ~: Alsatian; **Schaf·fell** das sheepskin

schaffen 1. unr. tr. V. a) create; b) auch regelm. (herstellen) create ⟨conditions, jobs, situation, etc.⟩; make ⟨room, space, fortune⟩; 2. tr. V. a) (bewältigen) manage; es ~, etw. zu tun manage to do sth.; b) (ugs.: erschöpfen) wear out; c) etw. aus etw./in etw. (Akk.) ~: get sth. out of/into sth.; 3. itr. V. a) (südd.: arbeiten) work; b) sich (Dat.) zu ~ machen busy oneself; jmdm. zu ~ machen cause sb. trouble

Schaffner der; ~s ~ (im Bus) conductor; (im Zug) guard (Brit.); conductor (Amer.); **Schaffnerin** die; ~, ~nen (im Bus) conductress (Brit.); (im Zug) guard (Brit.); conductress (Amer.)

Schaffung die; ~: creation

Schafott das; ~[e]s, ~e scaffold

Schafs·käse der sheep's milk cheese; **Schaf·wolle** die sheep's wool

Schakal der; ~s, ~e jackal

schal Adj. stale ⟨drink, taste, smell, joke⟩; empty ⟨words, feeling⟩

Schal der; ~s, ~s od. ~e scarf

Schale die; ~, ~n a) (Obst~) skin; (abgeschälte ~) peel no pl.; b) (Nuß~, Eier~) shell; c) (Schüssel) bowl; (flacher) dish; d) sich in ~ werfen od. schmeißen (ugs.) get dressed [up] to the nines; **schälen 1.** tr. V. peel ⟨fruit, vegetable⟩; shell ⟨egg, nut, pea⟩; 2. refl. V. peel

Schall der; ~[e]s, ~e od. Schälle sound; **Schall·dämpfer** der a) silencer; b) (Musik) mute; **schalldicht** Adj. sound-proof; **schallen** regelm. (auch unr.) itr. V. ring out; ~des Gelächter ringing laughter

Schall-: ~geschwindigkeit die speed or velocity of sound; ~platte die record

Schalotte die; ~, ~n shallot

schalt 1. u. 3. Pers. Sg. Prät. v. schelten

schalten 1. tr. V. switch; 2. itr. V. a) (Schalter betätigen) switch, turn (auf + Akk. to); b) ⟨machine⟩ switch (auf + Akk. to); c) (im Auto) change [gear]; d) ~ und walten manage one's affairs; e) (ugs.: begreifen) twig (coll.); catch on (coll.); **Schalter** der; ~s, ~ a) switch; b) (Post~, Bank~ usw.) counter

Schalter-: ~beamte der counter clerk; (im Bahnhof) ticket clerk; ~halle die hall; (im Bahnhof) booking-hall (Brit.); ticket office

Schaltjahr das leap year; **Schaltung** die; ~, ~en (Elektrot.) circuit; wiring system

Scham die; ~: shame; **schämen** refl. V. be ashamed (Gen., für, wegen of); **Scham·gefühl** das; o. Pl. sense of shame; **schamhaft 1.** Adj. bashful; 2. adv. bashfully; **scham·los 1.** Adj. a) (skrupellos, dreist) shameless; b) (unanständig) indecent; shameless ⟨person⟩; 2. adv. a) (skrupellos, dreist) shamelessly; b) (unanständig) indecently

Schande die; ~: disgrace; **schändlich 1.** Adj. disgraceful; 2. adv. disgracefully

Schar die; ~, ~en crowd; horde; **scharen·weise** Adv. in swarms or hordes

scharf; schärfer, schärfst... 1. Adj. a) sharp; b) (stark gewürzt, brennend, stechend) hot; strong ⟨drink, vinegar, etc.⟩; caustic ⟨chemical⟩; pungent ⟨smell⟩; c) (durchdringend) shrill; (hell) harsh; (kalt) biting ⟨cold, wind, etc.⟩; sharp ⟨frost⟩; d) (deutlich wahrnehmend) keen; e) (schnell) fast; keen ⟨ride, gallop, etc.⟩; f) (explosiv) live; (Ballspiele) powerful ⟨shot⟩; g) das ~e S (bes. österr.) the German letter 'ß'; h) ~ auf jmdn./etw. sein (ugs.) really fancy sb. (coll.)/be really keen on sth; 2. adv. a) ~ würzen/abschmecken season/flavour highly; ~ riechen smell pungent; b) (durchdringend) shrilly; (hell) harshly; (kalt) bitingly; c) (deutlich wahrnehmend) ⟨listen, watch, etc.⟩ closely, intently; ⟨think, consider, etc.⟩ hard; d) (deutlich hervortretend) sharply; e) (schonungslos) ⟨attack, criticize, etc.⟩ sharply, strongly; ⟨watch, observe, etc.⟩ closely; f) (schnell) fast; ~ bremsen brake hard or sharply; **Schärfe die;** ~ a) sharpness; b) (von Geschmack) hotness; (von Chemikalien) causticity; (von Geruch) pungency; c) (Intensität) shrillness; (des

Frostes) sharpness; **schärfen 1.** *tr. V. (auch fig.)* sharpen; **2.** *refl. V.* become sharper *or* keener

scharf-: ~kantig *Adj.* sharp-edged; **~sichtig** *Adj.* sharp-sighted; perspicacious; **~sinnig 1.** *Adj.* astute; **2.** *adv.* astutely

Scharlach der; ~s *(Med.)* scarlet fever

Scharnier das; ~s, ~e hinge

scharren *itr. V.* a) scrape; b) *(wühlen)* scratch; **2.** *tr. V.* scrape, scratch out ⟨hole, hollow, etc.⟩

Schaschlik der *od.* das; ~s, ~s *(Kochk.)* shashlik

Schatten der; ~s, ~ a) shadow; b) *o. Pl. (schattige Stelle)* shade; **schattig** *Adj.* shady

Schatz der; ~es, Schätze treasure *no indef. art.;* **schätzen 1.** *tr. V.* a) estimate; **sich glücklich ~:** deem oneself lucky; b) *(ugs.: annehmen)* reckon; c) *(würdigen, hochachten)* jmdn. ~: hold sb. in high esteem; **2.** *itr. V.* guess; **Schätzung** die; ~, ~en estimate

Schau die; ~, ~en a) *(Ausstellung)* exhibition; b) *(Vorführung)* show; c) **zur** ~ **stellen** *(ausstellen)* exhibit; display; *(offen zeigen)* display

Schauder der; ~s, ~: shiver; **schauderhaft 1.** *Adj.* terrible; **2.** *adv.* terribly; **schaudern** *itr. V.* a) *(vor Kälte)* shiver; b) *(vor Angst)* shudder

schauen *(bes. südd., österr., schweiz.)* **1.** *itr. V.* a) look; b) *(sich kümmern um)* **nach jmdm./etw. ~:** take *or* have a look at sb./sth.; c) *(achten)* **auf etw.** *(Akk.)* ~: set store by sth.; d) *(ugs.: sich bemühen)* **schau, daß du ...:** see *or* mind that you ...; e) *(nachsehen)* have a look; **2.** *tr. V.* **Fernsehen ~:** watch television

Schauer der; ~s, ~: shower

Schauer·geschichte die horror story; **schauerlich 1.** *Adj.* a) horrifying; b) *(ugs.: fürchterlich)* terrible *(coll.);* **2.** *(ugs.: fürchterlich)* terribly *(coll.)*

Schaufel die; ~, ~n shovel; *(Kehr~)* dustpan; **schaufeln** *tr. V.* shovel; *(graben)* dig

Schau·fenster das shop-window; **Schaufenster·bummel** der: **einen** ~ **machen** go window-shopping

Schaukel die; ~, ~n a) swing; b) *(Wippe)* see-saw; **schaukeln 1.** *itr. V.* a) swing; *(im Schaukelstuhl)* rock; b)

(sich hin und her bewegen) sway [to and fro]; *(sich auf und ab bewegen)* ⟨ship, boat⟩ pitch and toss; ⟨vehicle⟩ bump [up and down]; **2.** *tr. V.* rock

Schaukel-: ~pferd das rocking-horse; **~stuhl** der rocking-chair

Schau·lustige der/die; adj. Dekl. curious onlooker

Schaum der; ~s, Schäume a) foam; *(von Seife usw.)* lather; *(von Getränken, Suppen usw.)* froth; b) *(Geifer)* foam; froth; **schäumen** *itr. V.* foam; froth; *(soap etc.)* lather; ⟨beer, fizzy drink, etc.⟩ froth [up]

Schaum-: ~gummi der foam rubber; **~wein** der sparkling wine

Schau-: ~spiel das a) *o. Pl. (Drama)* drama *no art.;* b) *(ernstes Stück)* play; c) *(geh.: Anblick)* spectacle; **~spieler** der actor; **~spielerin** die actress; **~steller** der; ~s, ~: showman

Scheck der; ~s, ~s cheque; **Scheckheft** das cheque-book; **Scheck·karte** die cheque card

scheel *(ugs.)* **1.** *Adj.* disapproving; *(neidisch)* envious; jealous; **2.** *adv.* disapprovingly; *(neidisch)* enviously; jealously

Scheibe die; ~, ~n a) disc; b) *(abgeschnittene ~)* slice; c) *(Glas~)* pane [of glass]; *(Fenster~)* [window-]pane; **Scheiben·wischer** der windscreen-wiper

Scheide die; ~, ~n a) sheath; b) *(Anat.)* vagina

scheiden *unr. tr. V.* dissolve ⟨marriage⟩; divorce ⟨married couple⟩; **sich** ~ **lassen** get divorced *or* get a divorce; **Scheidung** die; ~, ~en divorce

Schein der; ~[e]s, ~e a) *o. Pl. (Licht~)* light; b) *o. Pl. (An~)* appearances *pl., no art.;* *(Täuschung)* pretence; **etw. nur zum** ~ **tun** [only] pretend to do sth.; make a show of doing sth.; c) *(Geld~)* note; **scheinbar 1.** *Adj.* apparent; seeming; **2.** *adv.* seemingly; **scheinen** *unr. itr. V.* a) shine; b) *(den Eindruck erwecken)* seem; appear; **mir scheint, [daß] ...:** it seems *or* appears to me that ...

schein-, Schein-: ~heilig 1. *Adj.* hypocritical; **2.** *adv.* hypocritically; **~werfer** der floodlight; *(am Auto)* headlight

Scheiße die; ~ *(derb)* shit *(coarse);* crap *(coarse);* **scheißen** *unr. itr. V. (derb)* [have *or (Amer.)* take a] shit *(coarse);* crap *(coarse);* have a crap *(coarse)*

Scheitel der; ~s, ~: parting; **scheiteln** tr. V. part ⟨hair⟩
scheitern itr. V.; mit sein fail; ⟨talks, marriage⟩ break down; ⟨plan, project⟩ fail, fall through
Schelle die; ~, ~n bell; **schellen** itr. V. (westd.) s. klingeln
Schell·fisch der haddock
Schelm der; ~[e]s, ~e rascal; rogue; **schelmisch** 1. Adj. roguish; 2. adv. roguishly
Schelte die; ~, ~n (geh.) scolding; **schelten** (südd., geh.) 1. unr. itr. V. auf od. über jmdn./etw. ~: moan about sb./sth.; 2. unr. tr. V. scold
Schema das; ~s, ~s od. ~ta od. Schemen pattern; **schematisch** 1. Adj. a) diagrammatic; b) (mechanisch) mechanical; 2. adv. a) in diagram form; b) (mechanisch) mechanically
Schemel der; ~s, ~ a) stool; b) (südd.: Fußbank) footstool
Schenkel der; ~s, ~: thigh
schenken tr. V. a) give; jmdm. etw. [zum Geburtstag] ~: give sb. sth. or sth. to sb. [as a birthday present or for his/her birthday]; b) (ugs.: erlassen) jmdm./sich etw. ~: spare sb./oneself sth.
Scherbe die; ~, ~n fragment
Schere die; ~, ~n a) scissors pl.; eine ~: a pair of scissors; b) (Zool.) claw; ¹**scheren** unr. tr. V. crop; (von Haar befreien) shear, clip ⟨sheep⟩
²**scheren** tr., refl. V. sich um jmdn./ etw. nicht ~: not care about sb./sth.; **Schererein** Pl. (ugs.) trouble no pl.
Scherz der; ~es, ~e joke; **scherzen** itr. V. joke; **scherzhaft** 1. Adj. jocular; 2. adv. jocularly
scheu 1. Adj. shy; timid ⟨animal⟩; (ehrfürchtig) awed; 2. adv. a) shyly; b) (von Tieren) timidly; **Scheu** die; ~ a) shyness; (Ehrfurcht) awe; b) (von Tieren) timidity
scheuchen tr. V. shoo; drive
scheuen 1. tr. V. shrink from; shun ⟨people, light, company, etc.⟩; 2. refl. V. sich vor etw. (Dat.) ~: be afraid of or shrink from sth. 3. itr. V. ⟨horse⟩ shy (vor + Dat. at)
scheuern 1. tr., itr. V. a) (reinigen) scour; scrub; b) (reiben) rub; chafe; 2. tr. V. (reiben an) rub
Scheuer-: ~pulver das scouring powder; ~tuch das; Pl. ~tücher scouring cloth
Scheune die; ~, ~n barn
Scheusal das; ~s, ~e monster;

scheußlich 1. Adj. a) dreadful; b) (ugs.: äußerst unangenehm) dreadful (coll.); ghastly (coll.) ⟨weather, taste, smell⟩; 2. adv. a) dreadfully; b) (ugs.: sehr) dreadfully (coll.)
Schi usw.: s. **Ski** usw.
Schicht die; ~, ~en a) (Lage) layer; (Geol.) stratum; (von Farbe) coat; (sehr dünn) film; b) (Gesellschafts~) stratum; c) (Arbeits~) shift; ~ arbeiten work shifts; be on shift work; **schichten** tr. V. stack
schick 1. Adj. a) stylish; chic ⟨clothes, fashions⟩; smart ⟨woman, girl, man⟩; b) (ugs.: großartig, toll) great (coll.); fantastic (coll.); 2. adv. a) stylishly; smartly ⟨furnished, decorated⟩
schicken 1. tr. V. send; jmdm. etw. ~, etw. an jmdn. ~: send sth. to sb.; send sb. sth.; 2. itr. V. nach jmdm. ~: send for sb; 3. refl. V. (veralt.: sich ziemen) be proper or fitting
Schicksal das; ~s, ~e: [das] ~: fate; destiny; (schweres Los) fate; **Schicksals·schlag** der stroke of fate
Schiebe·dach das sunroof; **schieben** 1. unr. tr. V. a) push; (stecken) put; c) etw. auf jmdn./etw. ~: blame sb./sth. for sth.; 2. unr. refl. V. sich durch die Menge ~: push one's way through the crowd; 3. unr. itr. V. push; (heftig) shove; **Schiebe·tür** die sliding door; **Schiebung** die; ~, ~en (ugs.) a) (ugs.) shady deal; b) (o. Pl.: Begünstigung) pulling strings
schied 1. u. 3. Pers. Sg. Prät. v. **scheiden**
Schieds·richter der referee; (Tennis, Hockey, Kricket) umpire
schief 1. Adj. a) (schräg) leaning ⟨wall, fence, post⟩; (nicht parallel) crooked; sloping ⟨surface⟩; worn[-down] ⟨heels⟩; b) (fig.: verzerrt) distorted ⟨picture, presentation, view, impression⟩; false ⟨comparison⟩; 2. adv. a) (schräg) das Bild hängt/der Teppich liegt ~: the picture/carpet is crooked; der Tisch steht ~: the table isn't level; b) (fig.: verzerrt) etw. ~ darstellen give a distorted account of sth.
Schiefer der; ~s (Gestein) slate
schief-: ~|gehen, ~|laufen unr. itr. V.; mit sein (ugs.) go wrong
schielen itr. V. a) squint; auf dem rechten Auge ~: have a squint in one's right eye; b) (ugs.: blicken) look out of the corner of one's eye
schien 1. u. 3. Pers. Sg. Prät. v. **scheinen**

Schien·bein das shinbone;
Schiene die; ~, ~n a) rail; b)
(Gleit~) runner; c) *(Med.: Stütze)*
splint; **schienen** tr. V. jmds. Arm/
Bein ~: put sb.'s arm/leg in a splint/
splints

schießen 1. *unr. itr. V.* a) shoot; **auf**
jmdn./etw. ~: shoot/fire at sb./sth.; b)
mit sein (fließen, heraus~) gush;
(spritzen) spurt; c) *mit sein (schnell
wachsen)* shoot up; 2. *unr. tr. V.* a)
shoot; fire *(bullet, missile, rocket)*; b)
(Fußball) score *(goal)*; c) *(ugs.: fotografieren)* **einige Aufnahmen** ~: take a
few snaps; **Schießerei** die; ~, ~en
a) shooting *no indef. art., no pl.;* b)
(Schußwechsel) gun-battle

Schiff das; ~[e]s, ~e a) ship; **mit dem**
~: by ship *or* sea; b) *(Archit.: Kir-
chen~) (Mittel~)* nave; *(Quer~)* tran-
sept; *(Seiten~)* aisle; **Schiffahrt** die;
o. Pl. shipping *no indef. art.; (Schiff-
fahrtskunde)* navigation

Schiff-: ~**bruch** der *(veralt.)* ship-
wreck; ~**brüchige** der/die; *adj.
Dekl.* shipwrecked man/woman

Schiffer der; ~s, ~: boatman; *(eines
Lastkahns)* bargee; *(Kapitän)* skipper

Schiffs-: ~**arzt** der ship's doctor;
~**brücke** die pontoon bridge; ~**junge** der ship's boy; ~**reise** die voyage;
(Vergnügungsreise) cruise; ~**verkehr**
der shipping traffic

Schikane die; ~, ~n a) harassment *no
indef. art.;* b) **mit allen** ~n *(ugs.)* *(kitchen, house)* with all mod cons *(Brit.
coll.);* *(car, bicycle, stereo)* with all the
extras; **schikanieren** tr. V. jmdn. ~:
harass sb.

¹**Schild** der; ~[e]s, ~e shield; ²**Schild**
das; ~[e]s, ~er sign; *(Nummern~)*
number-plate; *(Namens~)* name-
plate; *(auf Denkmälern, Gebäuden
usw.)* plaque; *(Etikett)* label; **schildern** tr. V. describe; **Schild·kröte**
die tortoise; *(Seeschildkröte)* turtle

Schilf das; ~[e]s a) reed; b) *o. Pl. (Röh-
richt)* reeds *pl.*

schillern itr. V. shimmer

Schilling der; ~, ~e schilling

schilt 3. *Pers. Sg. Präsens v.* **schelten**

Schimmel der; ~s, ~ a) *o. Pl.* mould;
(auf Leder, Papier) mildew; b) *(Pferd)*
white horse; **schimmelig** *Adj.*
mouldy; mildewy *(paper, leather)*;
schimmeln itr. V.; *auch mit sein* go
mouldy; *(leather, paper)* get covered
with mildew

Schimmer der; ~s *(Schein)* gleam;

(von Seide) shimmer; sheen; **keinen** ~
[von etw.] haben *(ugs.)* not have the
faintest idea [about sth.] *(coll.);*
schimmern itr. V. gleam; *(water,
sea)* glisten, shimmer; *(metal)* glint,
gleam; *(silk etc.)* shimmer

schimmlig s. **schimmelig**

Schimpanse der; ~n, ~n chimpanzee

schimpfen 1. *itr. V.* a) carry on *(coll.)*
(auf, über + *Akk.* about); *(meckern)*
grumble, moan **(auf, über** + *Akk.* at);
b) **mit jmdm.** ~: tell sb. off; scold sb.;
2. *tr. V.* jmdn. ~: tell sb. off;
Schimpf·wort das *(Beleidigung)* insult; *(derbes Wort)* swear-word

schinden *unr. tr. V.* maltreat; ill-
treat; **Zeit** ~ *(ugs.)* play for time

Schinken der; ~s, ~ ham; **Schin-
ken·speck** der bacon

Schippe die; ~, ~n *(Schaufel)* shovel

Schirm der; ~[e]s, ~e umbrella; brolly
(Brit. coll.); (Sonnen~) sunshade

Schirm-: ~**herr** der patron; ~**herrin**
die patroness; ~**herrschaft** die patronage; ~**ständer** der umbrella
stand

schiß *1. u. 3. Pers. Sg. Prät. v.* **scheißen**

Schlacht die; ~, ~en battle;
schlachten tr. *(auch itr.)* V.
slaughter; kill *(rabbit, chicken, etc.);*
Schlachter der; ~s, ~ *(nordd.)*
butcher; **Schlachterei** die; ~, ~en
(nordd.) butcher's [shop]

Schlacht-: ~**hof** der abattoir; ~**vieh**
das animals *pl.* kept for meat; *(kurz
vor der Schlachtung)* animals *pl.* for
slaughter

Schlacke die; ~, ~n cinders *pl.;
(Hochofen~)* slag

Schlaf der; ~[e]s sleep; **einen leichten/
festen/gesunden** ~ **haben** be a light/
heavy/good sleeper; **Schlaf·an·zug**
der pyjamas *pl.;* **Schläfchen** das;
~s, ~: nap; snooze *(coll.)*

Schläfe die; ~, ~n temple

schlafen *unr. itr. V.* a) *(auch fig.)*
sleep; **tief** *od.* **fest** ~: be sound asleep;
lange ~: sleep for a long time; *(am
Morgen)* sleep in; ~ **gehen** go to bed;
b) *(ugs.: nicht aufpassen)* be asleep;
Schläfer der; ~s, ~: sleeper

schlaff 1. *Adj.* a) slack; flabby
(stomach, muscles); b) *(schlapp, matt)*
limp *(body, hand, handshake);* shaky
(knees); 2. *adv.* a) slackly; b) *(schlapp,
matt)* limply

Schlaf-: ~**gelegenheit** die place to
sleep; ~**mittel** das sleep-inducing
drug

schläfrig 1. *Adj.* sleepy; **2.** *adv.* sleepily
Schlaf-: ~**saal** der dormitory; ~**sack** sleeping-bag
schläft *3. Pers. Sg. Präsens v.* **schlafen**
Schlaf-: ~**tablette** die sleeping-pill; ~**wagen** der sleeping-car; sleeper; ~**zimmer** das bedroom
Schlag der; ~[e]s, Schläge **a)** blow; *(Faust~)* punch; *(Klaps)* slap; *(Tennis~, Golf~)* stroke; shot; ~ **auf** ~ *(fig.)* in quick succession; **b)** *(Auf~, Aufprall)* bang; *(dumpf)* thud; *(Klopfen)* knock; **c)** *o. Pl. (des Herzens, Pulses)* beating; *(eines Pendels)* swinging; **d)** *(einzelne rhythmische Bewegung) (Herz~, Puls~, Takt~)* beat; *(eines Pendels)* swing; **e)** *o. Pl. (Töne) (einer Uhr)* striking; *(einer Glocke)* ringing; **f)** *(einzelner Ton) (Stunden~)* stroke; *(Glocken~)* ring; ~ **acht Uhr** on the stroke of eight
schlag-, Schlag-: ~**ader** die artery; ~**anfall** der stroke; ~**artig 1.** *Adj.* very sudden; **2.** *adv.* quite suddenly; ~**baum** der barrier
schlagen 1. *unr. tr. V.* **a)** hit; beat; strike; *(mit der Faust)* punch; hit; *(mit der flachen Hand)* slap; **b)** *(mit Richtungsangabe)* hit ⟨*ball*⟩; **einen Nagel in etw.** *(Akk.)* ~: knock a nail into sth.; **c)** *(rühren)* beat ⟨*mixture*⟩; whip ⟨*cream*⟩; *(mit einem Schneebesen)* whisk; **d)** *(läuten)* ⟨*clock*⟩ strike; ⟨*bell*⟩ ring; **e)** *(legen)* throw; **f)** *(einwickeln)* wrap (in + *Akk.* in); **g)** *(besiegen, übertreffen)* beat; **2.** *unr. itr. V.* **a)** er schlug mit der Faust auf den Tisch he beat the table with his fist; **b)** mit den Flügeln ~ ⟨*bird*⟩ beat or flap its wings; **c)** *mit sein (prallen)* bang; **mit dem Kopf auf etw.** *(Akk.)/***gegen etw.** ~: bang one's head on/against sth.; **d)** *mit sein* **jmdm. auf den Magen** ~: affect sb.'s stomach; **e)** *(pulsieren)* ⟨*heart, pulse*⟩ beat; *(heftig)* ⟨*heart*⟩ pound; ⟨*pulse*⟩ throb; **f)** *(läuten)* ⟨*clock*⟩ strike; ⟨*bell*⟩ ring; **3.** *unr. refl. V.* fight; **sich mit jmdm.** ~: fight with sb.; **Schlager** der; ~s, ~ **a)** pop song; **b)** *(Erfolg) (Buch)* best seller; *(Ware)* best-selling line; *(Film, Stück, Lied)* hit
Schläger der; ~s, ~ **a)** *(Raufbold)* tough; thug; **b)** *(Tennis~, Federball~, Squash~)* racket; *(Tischtennis~, Krikket~)* bat; *([Eis]hockey~, Polo~)* stick; *(Golf~)* club; **Schlägerei** die; ~, ~en brawl; fight
Schlager·sänger der pop singer

schlag-, Schlag-: ~**fertig** *Adj.* quick-witted ⟨*reply*⟩; ⟨*person*⟩ who is quick at repartee; ~**fertigkeit** die; *o. Pl.* quickness at repartee; ~**loch** das pothole; ~**obers** das; ~ *(österr.)*, ~**rahm** der *(bes. südd., österr., schweiz.)*, ~**sahne** die whipping cream; *(geschlagen)* whipped cream; ~**zeile** die headline; ~**zeug** das drums *pl.*
schlaksig *(ugs.) Adj.* gangling; lanky
Schlamassel der *od.* das; ~s *(ugs.)* mess
Schlamm der; ~[e]s, ~e *od.* Schlämme **a)** mud; **b)** *(Schlick)* sludge; **schlammig** *Adj.* **a)** muddy; **b)** *(schlickig)* sludgy; muddy
Schlamperei die; ~, ~en *(ugs. abwertend)* sloppiness; **schlampig** *(ugs. abwertend)* **1.** *Adj.* **a)** *(liederlich)* slovenly; **b)** *(nachlässig)* sloppy; slipshod ⟨*work*⟩; **2.** *adv.* **a)** *(liederlich)* in a slovenly way; **b)** *(nachlässig)* sloppily
schlang *1. u. 3. Pers. Sg. Prät. v.* **schlingen; Schlange** die; ~, ~n **a)** snake; **b)** *(Menschen~)* queue; line *(Amer.)*; ~ **stehen** queue; stand in line *(Amer.)*; **c)** *(Auto~)* tailback *(Brit.)*; backup *(Amer.)*; **schlängeln** *refl. V.* ⟨*snake*⟩ wind [its way]; ⟨*road*⟩ wind, snake [its way]; **Schlangen·linie** die wavy line
schlank *Adj.* slim ⟨*person*⟩; slim, slender ⟨*build, figure*⟩; **Schlankheits-kur** die slimming diet
schlapp *Adj.* **a)** worn out; tired out; *(wegen Schwüle)* listless; *(wegen Krankheit)* run-down; **b)** *(ugs.: ohne Schwung)* wet *(sl.)*; feeble; **c)** slack ⟨*rope, cable*⟩; loose ⟨*skin*⟩; flabby ⟨*stomach, muscles*⟩; **Schlappe** die; ~, ~n setback; **schlapp|machen** *itr. V. (ugs.)* flag; *(zusammenbrechen)* flake out *(coll.)*; *(aufgeben)* give up
schlau 1. *Adj.* **a)** shrewd; astute; *(gerissen)* wily; crafty; cunning; **b)** *(ugs.: gescheit)* clever; bright; smart; **aus jmdm. nicht** ~ **werden** *(ugs.)* not be able to make sb. out; **2.** *adv.* shrewdly; astutely; *(gerissen)* craftily; cunningly
Schlauch der; ~[e]s, Schläuche **a)** hose; **b)** *(Fahrrad~, Auto~)* tube; **Schlauch·boot** das rubber dinghy; inflatable [dinghy]; **schlauchen** *(ugs.) tr., auch itr. V.* **jmdn.** ~: take it out of sb.; **schlauch·los** *Adj.* tubeless ⟨*tyre*⟩
Schläue die; ~: shrewdness; astute-

ness; *(Gerissenheit)* wiliness; craftiness; cunning

Schlaufe die; ~, ~n loop

schlecht 1. *Adj.* **a)** bad; poor, bad *(food, quality, style, harvest, health, circulation)*; poor *(salary, eater, appetite)*; poor-quality *(goods)*; bad, weak *(eyes)*; **um** jmdn./mit etw. steht es ~: sb./sth. is in a bad way; **b)** *(böse)* bad; wicked; **c)** *nicht attr. (ungenießbar)* off; **das Fleisch ist ~ geworden** meat has gone off; **2.** *adv.* **a)** badly; **er sieht/hört ~:** his sight is poor/he has poor hearing; **über** jmdn. *od.* **von** jmdm. ~ **sprechen** speak ill of sb.; **b)** *(schwer)* **heute geht es ~:** today is difficult; **c)** ~ **und recht, mehr ~ als recht** after a fashion

schlecht-: ~**bezahlt** *Adj. (präd. getrennt geschrieben)* badly or poorly paid; ~|**gehen** *unr. itr. V.; unpers.; mit sein* **es geht ihr/mir ~:** she is/I am doing badly; *(gesundheitlich)* she is/I am ill *or* unwell *or* poorly; ~**gelaunt** *Adj. (präd. getrennt geschrieben)* bad-tempered; ~|**machen** *tr. V.* jmdn. ~**machen** run sb. down; disparage sb.

schlecken *(bes. südd., österr.) tr. V.* lap up

schleichen 1. *unr. itr. V.; mit sein* creep; *(heimlich)* creep; sneak; *(cat)* slink, creep; *(langsam fahren)* crawl along; **2.** *unr. refl. V.* creep; sneak; *(cat)* slink, creep; **schleichend** *Adj.* insidious *(disease)*; slow[-acting] *(poison)*; creeping *(inflation)*; gradual *(crisis)*

Schleier der; ~s, ~: veil; **schleierhaft** *Adj.* jmdm. ~**haft sein/bleiben** be/remain a mystery to sb.

Schleife die; ~, ~n **a)** bow; *(Fliege)* bow-tie; **b)** *(starke Biegung)* loop

¹**schleifen** *unr. tr. V.* grind; cut *(diamond, glass)*; *(mit Schleifpapier usw.)* sand; *(schärfen)* sharpen

²**schleifen 1.** *tr. V.* **a)** *(auch fig.)* drag; **b)** *(niederreißen)* raze *(sth.)* [to the ground]; **2.** *itr. V.; auch mit sein* drag; **die Kupplung ~ lassen** *(Kfz-W.)* slip the clutch

Schleim der; ~|e|s, ~e mucus; *(im Hals)* phlegm; *(von Schnecken)* slime; **schleimig** *Adj. (auch fig.)* slimy; *(Physiol., Zool.)* mucous

schlemmen *itr. V.* have a feast; **Schlemmer** der; ~s, ~: gourmet

schlendern *itr. V.; mit sein* stroll

Schlenker der; ~s, ~ *(ugs.)* swerve; **einen ~ machen** swerve

schlenkern *tr., itr. V.* swing; **mit den Armen ~:** swing one's arms

Schleppe die; ~, ~n train; **schleppen 1.** *tr. V.* **a)** *(ziehen)* tow *(vehicle, ship)*; **b)** *(tragen)* carry; lug; **c)** *(ugs.: mitnehmen)* drag; **2.** *refl. V.* drag *or* haul oneself; **Schlepper** der; ~s, ~ **a)** *(Schiff)* tug; **b)** *(Traktor)* tractor

Schlepp-: ~**lift** der T-bar [lift]; ~**tau** das tow-line

Schleuder die; ~, ~n sling; *(mit Gummiband)* catapult *(Brit.)*; slingshot *(Amer.)*; **schleudern 1.** *tr. V.* hurl; **2.** *itr. V.; mit sein* *(vehicle)* skid

schleunigst *Adv.* **a)** *(auf der Stelle)* at once; immediately; straight away; **b)** *(eilends)* hastily; with all haste

Schleuse die; ~, ~n **a)** sluice[-gate]; **b)** *(Schiffs~)* lock

schlich *1. u. 3. Pers. Sg. Prät. v.* **schleichen**

schlicht 1. *Adj.* **a)** simple; plain *(pattern, furniture)*; **b)** *(unkompliziert)* simple, unsophisticated *(person, view, etc.)*; **2.** *adv.* simply; simply, plainly *(dressed, furnished)*

schlichten 1. *tr. V.* settle *(argument etc.)*; settle *(industrial dispute etc.)* by mediation; **2.** *itr. V.* mediate

Schlick der; ~|e|s, ~e silt

schlief *1. u. 3. Pers. Sg. Prät. v.* **schlafen**

Schließe die; ~, ~n clasp; *(Schnalle)* buckle; **schließen 1.** *unr. tr. V.* **a)** close; shut; turn off *(tap)*; fasten *(belt, bracelet)*; do up *(button, zip)*; close *(street, route, border, electrical circuit)*; fill, close *(gap)*; **b)** *(außer Betrieb setzen)* close [down] *(shop, school)*; **c)** *(ein~)* etw./jmdn./sich in etw. *(Akk.)* ~: lock sth./sb./oneself in sth.; **d)** *(beenden)* close *(meeting, proceedings, debate)*; end, conclude *(letter, speech, lecture)*; **e)** *(eingehen, vereinbaren)* conclude *(treaty, pact, cease-fire, agreement)*; reach *(settlement, compromise)*; enter into *(contract)*; **f)** *(folgern)* infer *(aus* from*)*; **2.** *unr. itr. V.* **a)** close, shut; **b)** *(enden)* end; conclude; **c)** |aus etw.| auf etw. *(Akk.)* ~: infer sth. [from sth.]; **3.** *unr. refl. V.* *(door, window)* close, shut; *(wound, circle)* close; **Schließ·fach** das locker; *(bei der Post)* PO box; *(bei der Bank)* safe-deposit box; **schließlich** *Adv.* **a)** finally; in the end; **b)** *(immerhin, doch)* after all

schliff *1. u. 3. Pers. Sg. Prät. v.* **schleifen**

Schliff der; ~[e]s, ~e a) *o. Pl.* cutting; *(von Messern, Sensen usw.)* sharpening; b) *(Art, wie etw. geschliffen wird)* cut; *(von Messern, Scheren usw.)* edge; c) *o. Pl.* **einem Brief/Text** *usw.* **den letzten ~ geben** put the finishing touches *pl.* to a letter/text *etc.*

schlimm 1. *Adj.* a) grave, serious ⟨*error, mistake, accusation, offence*⟩; bad, serious ⟨*error, mistake*⟩; b) *(übel)* bad; nasty, bad ⟨*experience*⟩; [**das ist alles] halb so ~!** it's not as bad as all that; **ist nicht ~!** [it] doesn't matter; 2. *adv.* **~ d[a]ran sein** be in a bad way; *(in einer ~en Situation)* be in dire straits; **schlimmsten·falls** *Adv.* if the worst comes to the worst

Schlinge die; ~, ~n a) loop; *(für den Arm)* sling; *(zum Aufhängen)* noose; b) *(Fanggerät)* snare

schlingen 1. *unr. tr. V.* **etw. um etw. ~:** loop sth. round sth.; 2. *unr. refl. V.* **sich um etw. ~:** wind itself round sth; 3. *unr. itr. V.* bolt one's food

schlingern *itr. V.; mit sein* ⟨*ship, boat*⟩ roll; ⟨*train, vehicle*⟩ lurch from side to side

Schlips der; ~es, ~e tie

Schlitten der; ~s, ~: sledge; sled; *(Pferde~)* sleigh; *(Rodel~)* toboggan; **~ fahren** go tobogganing

schlittern *itr. V.* slide

Schlitt-: **~schuh** der [ice-]skate; **~schuh laufen** *od.* **fahren** [ice-]skate; **~schuh·laufen das** [ice-]skating *no art.;* **~schuh·läufer der** [ice-]skater

Schlitz der; ·es, ~e a) slit; *(Briefkasten~, Automaten~)* slot; b) *(Hosen~)* flies *pl.;* fly

schloß *1. u. 3. Pers. Sg. Prät. v.* schließen

Schloß das; Schlosses, Schlösser a) lock; *(Vorhänge~)* padlock; **hinter ~ und Riegel** *(ugs.)* behind bars; b) *(Verschluß)* clasp; c) *(Wohngebäude)* castle; *(Palast)* palace; *(Herrschaftshaus)* mansion

Schlosser der; ~s, ~: metalworker; *(Maschinen~)* fitter; *(für Schlösser)* locksmith; *(Auto~)* mechanic

Schlot der; ~[e]s, ~e *od.* Schlöte chimney[-stack]; *(eines Schiffes)* funnel

schlottern *itr. V.* a) shake; b) ⟨*clothes*⟩ hang loose

Schlucht die; ~, ~en ravine

schluchzen *itr. V.* sob

Schluck der; ~[e]s, ~e *od.* Schlücke swallow; mouthful; *(großer ~)* gulp; *(kleiner ~)* sip; **Schluck·auf der; ~s**

hiccups *pl.;* **Schlückchen das; ~s, ~:** sip; **schlucken** 1. *tr. V.* swallow; **etw. hastig ~:** gulp sth. down; 2. *itr. V.* swallow; **Schlucker der; ~s, ~:** **armer ~** *(ugs.)* poor devil *or (Brit. coll.)* blighter

schluderig *s.* schludrig; **schludern** *itr. V. (ugs.)* work sloppily; **schludrig** *(ugs.)* 1. *Adj.* a) slipshod ⟨*work, examination*⟩; botched ⟨*job*⟩; slapdash ⟨*person, work*⟩; b) *(schlampig [aussehend])* scruffy; 2. *adv.* a) in a slipshod *or* slapdash way; b) *(schlampig)* scruffily

schlug *1. u. 3. Pers. Sg. Prät. v.* schlagen

Schlummer der; ~s *(geh.)* slumber *(poet./rhet.);* **schlummern** *itr. V. (geh.)* slumber *(poet./rhet.)*

Schlund der; ~[e]s, Schlünde [back of the] throat; pharynx *(Anat.)*

schlüpfen *itr. V.; mit sein* slip; [**aus dem Ei] ~:** ⟨*chick*⟩ hatch out

Schlüpfer der; ~s, ~ *(für Damen)* knickers *pl. (Brit.);* panties *pl.; (für Herren)* [under]pants *pl. or* trunks *pl.*

schlüpfrig *Adj.* a) slippery; b) *(anstößig)* lewd

schlurfen *itr. V.; mit sein* shuffle

schlürfen 1. *tr. V.* slurp [up] *(coll.);* 2. *itr. V.* slurp *(coll.)*

Schluß der; Schlusses, Schlüsse a) end; *(eines Vortrags o. ä.)* conclusion; *(eines Buchs, Schauspiels usw.)* ending; **am** *od.* **zum ~:** at the end; *(schließlich)* in the end; b) *(Folgerung)* conclusion

Schlüssel der; ~s, ~: key

Schlüssel-: **~bein das** collar-bone; clavicle *(Anat.);* **~blume die** cowslip; *(Primel)* primula; **~bund der** *od.* **das** bunch of keys; **~loch das** keyhole

schlüssig 1. *Adj.* a) conclusive ⟨*proof, evidence*⟩; convincing, logical ⟨*argument, conclusion*⟩; b) **sich** *(Dat.)* **~ werden** make up one's mind; 2. *adv.* conclusively

Schluß-: **~licht das;** *Pl.* **~lichter** tail- *or* rear-light; **~strich der** [bottom] line; **~verkauf der** [end-of-season] sale[s *pl.*]

schmächtig *Adj.* slight

schmackhaft *Adj.* tasty

schmal; **~er** *od.* **schmäler, ~st...** *od.* **schmälst...** *Adj.* narrow; slim, slender ⟨*hips, hands, figure, etc.*⟩; thin ⟨*lips, face, nose, etc.*⟩; **schmälern** *tr. V.* diminish; restrict ⟨*rights*⟩

¹**Schmalz** das; ~es dripping; *(Schwei-*

ne~) lard; **²Schmalz** der; ~es *(abwertend)* schmaltz *(coll.);* **Schmalz·brot das** slice of bread and dripping; **schmalzig** *(abwertend)* 1. *Adj.* schmaltzy *(coll.);* 2. *adv.* with slushy sentimentality

schmarotzen *itr. V. (fig.)* sponge; free-load *(sl.)*

Schmarren der; ~s, ~ *(österr., auch südd.)* pancake broken up with a fork after frying

schmatzen *itr. V.* smack one's lips; *(geräuschvoll essen)* eat noisily

Schmaus der; ~es, Schmäuse *(veralt., scherzh.)* [good] spread *(coll.)*

schmecken 1. *itr. V.* taste (**nach** of); |gut| ~: taste good; **schmeckt es |dir|?** are you enjoying it *or* your meal?; 2. *tr. V.* taste; *(kosten)* sample

schmeicheln *itr. V.* jmdm. ~: flatter sb.; **Schmeichler** der; ~s, ~: flatterer

schmeißen *(ugs.)* 1. *unr. tr. V.* chuck *(coll.);* sling *(coll.); (schleudern)* fling; hurl; 2. *unr. refl. V.* throw oneself; *(mit Wucht)* hurl oneself; 3. *unr. itr. V.* **mit etw. |nach jmdm.|** ~: chuck sth. [at sb.] *(coll.)*

Schmeiß·fliege die blowfly; *(blaue ~)* bluebottle

schmelzen 1. *unr. itr. V.; mit sein* melt; *(fig.)⟨doubts, apprehension, etc.⟩* dissolve, fade away; 2. *unr. tr. V.* melt; smelt ⟨*ore*⟩; render ⟨*fat*⟩; **Schmelz·käse** der processed cheese

Schmerz der; ~es, ~en a) *(physisch)* pain; *(dumpf u. anhaltend)* ache; **wo haben Sie ~en?** where does it hurt?; **~en haben** be in pain; b) *(psychisch)* pain; *(Kummer)* grief; **schmerz-empfindlich** *Adj.* sensitive to pain *pred.;* **schmerzen** 1. *tr. V.* jmdn. ~: hurt sb.; *(jmdm. Kummer bereiten)* grieve sb.; cause sb. sorrow; 2. *itr. V.* hurt; **schmerzhaft** *Adj.* painful; **schmerzlich** 1. *Adj.* painful; distressing; 2. *adv.* painfully

schmerz-, Schmerz-: ~**los** 1. *Adj.* painless; 2. *adv.* painlessly; ~**stillend** *Adj.* pain-killing; ~**tablette die** pain-killing tablet

Schmetterling der; ~s, ~e butterfly **schmettern** 1. *tr. V.* a) hurl (**an** + Akk. at, **gegen** against); b) *(laut spielen, singen usw.)* blare out ⟨*march, music*⟩; ⟨*person*⟩ sing lustily ⟨*song*⟩; c) *(Tennis usw.)* smash ⟨*ball*⟩; 2. *itr. V.* ⟨*trumpet, music, etc.*⟩ blare out

Schmied der; ~|e|s, ~e blacksmith; **Schmiede die** ~, ~en smithy; forge; **schmieden** *tr. V. (auch fig.)* forge

schmiegen 1. *refl. V.* snuggle, nestle (**in** + Akk. in); **sich an jmdn.** ~: snuggle [close] up to sb.; 2. *tr. V.* press (**an** + Akk. against)

schmieren 1. *tr. V.* a) lubricate; b) *(streichen)* spread ⟨*butter, jam, etc.*⟩ (**auf** + Akk. on); **Brote** ~: spread slices of bread; 2. *itr. V.* a) ⟨*oil, grease*⟩ lubricate; b) *(ugs.: unsauber schreiben)* ⟨*person*⟩ scrawl, scribble; ⟨*pen, ink*⟩ smudge, make smudges; **schmierig** *Adj.* greasy; **Schmier-seife die** soft soap

schmilzt 2. *u.* 3. *Pers. Sg. Präsens v.* schmelzen

Schminke die ~, ~n make-up; **schminken** 1. *tr. V.* make up ⟨*face, eyes*⟩; 2. *refl. V.* make oneself up

Schmirgel·papier das emery-paper; *(Sandpapier)* sandpaper

schmiß 1. *u.* 3. *Pers. Sg. Prät. v.* schmeißen

Schmöker der; ~s, ~ *(ugs.)* lightweight adventure story/romance; **schmökern** *(ugs.)* 1. *itr. V.* bury oneself in a book; 2. *tr. V.* bury oneself in ⟨*book*⟩

schmollen *itr. V.* sulk; **Schmoll-mund** der pouting mouth

schmolz 1. *u.* 3. *Pers. Sg. Prät. v.* schmelzen

Schmor·braten der braised beef; **schmoren** 1. *tr. V.* braise; 2. *itr. V.* a) braise; b) *(ugs.: schwitzen)* swelter

schmuck *Adj.* attractive

Schmuck der; ~|e|s a) *(~stücke)* jewelry; jewellery *(esp. Brit.);* b) *s.* ~stück; c) *(Zierde)* decoration

schmücken *tr. V.* decorate; embellish ⟨*writings, speech*⟩

schmuck-, Schmuck-: ~**kästchen das, ~kasten** der jewelry *or (esp. Brit.)* jewellery box; ~**los** *Adj.* plain; bare ⟨*room*⟩; ~**stück das** piece of jewelry *or (esp. Brit.)* jewellery

schmuddelig *Adj. (ugs.)* grubby; mucky *(coll.); (schmutzig u. unordentlich)* messy; grotty *(Brit. sl.)*

Schmuggel der; ~s smuggling *no art.;* **schmuggeln** *tr., itr. V.* smuggle (**in** + Akk. into; **aus** out of); **Schmuggler** der; ~s, ~: smuggler

schmunzeln *itr. V.* smile to oneself

schmusen *itr. V. (ugs.)* cuddle; ⟨*couple*⟩ kiss and cuddle

Schmutz der; ~es dirt; *(Schlamm)* mud; **schmutzen** *itr. V.* get dirty; **schmutzig** *Adj.* dirty

Schnabel der; ~s, Schnäbel a) beak; b) *(ugs.: Mund)* gob *(sl.)*

Schnake die; ~, ~n a) daddy-long-legs; b) *(bes. südd.: Stechmücke)* mosquito

Schnalle die; ~, ~n *(Gürtel~)* buckle; **schnallen** *tr. V.* a) *(mit einer Schnalle festziehen)* buckle *(shoe, belt)*; fasten *(strap)*; b) *(mit Riemen/Gurten befestigen)* strap (**auf** + *Akk.* on to)

schnalzen *itr. V.* |mit der Zunge/den Fingern| ~: click one's tongue/snap one's fingers

schnappen 1. *itr. V.* nach jmdm./etw. ~ *(animal)* snap at sb./sth.; **nach Luft** ~: gasp for breath; 2. *tr. V.* *(dog, bird, etc.)* snatch; [sich *(Dat.)*] jmdn./etw. ~ *(ugs.)* *(person)* grab sb./sth.; *(mit raschem Zugriff)* snatch sb./sth.; **Schnapp·schuß** der snapshot

Schnaps der; ~es, Schnäpse a) spirit; *(Klarer)* schnapps; b) *o. Pl. (Spirituosen)* spirits *pl.*

schnarchen *itr. V.* snore

schnattern *itr. V.* a) *(goose etc.)* cackle, gaggle; b) *(ugs.: eifrig schwatzen)* jabber [away]; chatter

schnauben *itr. V.* snort (vor with)

schnaufen *itr. V.* puff (vor with)

Schnauze die; ~, ~n a) *(von Tieren)* muzzle; *(der Maus usw.)* snout; *(Maul)* mouth; b) *(derb: Mund)* gob *(sl.)*; |halt die| ~! shut your trap! *(sl.)*; **schnauzen** *tr. V., itr. V. (ugs.)* bark; *(ärgerlich)* snap; snarl

Schnecke die snail; *(Nackt~)* slug; **Schnecken·haus** das snail-shell

Schnee der; ~s snow

Schnee-: ~**ball** der snowball; ~**besen** der whisk; ~**flocke** die snowflake; ~**gestöber** das snow flurry; ~**glöckchen** das snowdrop; ~**kette** die snow-chain; ~**matsch** der slush; ~**pflug** der snow-plough; ~**sturm** der snowstorm

Schneewittchen (das) Snow White

Schneide die; ~, ~n [cutting] edge; **schneiden** 1. *unr. itr. V.* cut (**in** + *Akk.* into); 2. *unr. tr. V.* a) cut; (in Scheiben) slice *(bread, sausage, etc.)*; *(klein ~)* cut up, chop *(wood, vegetables)*; *(stutzen)* prune *(tree, bush)*; trim *(beard)*; cut, mow *(grass)*; **sich** *(Dat.)* **die Haare ~ lassen** have one's hair cut; b) **eine Kurve ~:** cut a corner

Schneider der; ~s, ~: tailor; *(Damen~)* dressmaker; **Schneiderei** die; ~, ~en tailor's shop; *(Damen~)* dressmaker's shop; **Schneiderin** die; ~, ~nen *s.* Schneider; **schneidern** *tr. V.* make; make, tailor *(suit)*

Schneide·zahn der incisor

schneien 1. *itr. v., tr. V. (unpers.)* snow; **es schneit** it is snowing; 2. *itr. V.; mit sein (fig.)* rain down; fall like snow

Schneise die; ~, ~n *(Wald~)* aisle; *(als Feuerschutz)* firebreak

schnell 1. *Adj.* quick *(journey, decision, service, etc.)*; fast *(car, skis, road, track, etc.)*; quick, swift *(progress, movement, blow, action)*; 2. *adv.* quickly; *(drive, move, etc.)* fast, quickly; *(spread)* quickly, rapidly; *(bald)* soon *(sold, past, etc.)*; **mach ~!** *(ugs.)* move it! *(coll.)*; **schnellen** *itr. V.; mit sein* shoot (**aus** + *Dat.* out of; **in** + *Akk.* into); **Schnelligkeit** die; ~, ~en speed; **Schnell·imbiß** der snack-bar; **schnellstens** *Adv.* as quickly as possible

Schnell-: ~**straße** die expressway; ~**zug** der express [train]

Schnepfe die; ~, ~n snipe

schneuzen 1. *tr. V.* sich *(Dat.)*/einem Kind die Nase ~: blow one's/a child's nose; 2. *refl. V.* blow one's nose

schnippeln *(ugs.)* 1. *itr. V.* snip [away] (**an** + *Dat.* at); 2. *tr. V.* shred *(vegetables)*; chop *(beans etc.)* [finely]

schnippen 1. *itr. V.* snap one's fingers (**nach** at); 2. *tr. V.* flick (**von** off, from)

schnippisch 1. *Adj.* pert *(reply, tone, etc.)*; 2. *adv.* pertly

Schnipsel der *od.* das; ~s, ~: scrap; *(Papier~, Stoff~)* snippet; shred

schnipseln *s.* schnippeln

schnitt *1. u. 3. Pers. Sg. Prät. v.* schneiden

Schnitt der; ~|e|s, ~e a) cut; b) *(das Mähen) (von Gras)* mowing; *(von Getreide)* harvest

Schnitt-: ~**blume** die cut flower; ~**bohne** die French bean

Schnitte die; ~, ~n slice; **eine ~ |Brot|** a slice of bread; **schnittig** 1. *Adj.* stylish, smart *(suit, appearance, etc.)*; *(sportlich)* racy *(car, yacht, etc.)*; 2. *adv.* stylishly; *(sportlich)* racily

Schnitt-: ~**lauch** der chives *pl.*; ~**wunde** die the cut; *(lang u. tief)* gash

Schnitzel das; ~s, ~ a) *(Fleisch)* [veal/pork] escalope; b) *(von Papier)* scrap; *(von Holz)* shaving; **schnitzeln** *tr. V.*

chop up ⟨*vegetables*⟩ [into small pieces]; shred ⟨*cabbage*⟩; **schnitzen** *tr., itr. V.* carve

schnodderig *(ugs.)* **1.** *Adj.* brash; **2.** *adv.* brashly

schnöde *(geh.)* **1.** *Adj.* **a)** *(verachtenswert)* contemptible; **b)** *(gemein)* contemptuous, scornful ⟨*glance, reply, etc.*⟩. **2.** *adv. (gemein)* contemptuously; ⟨*exploit, misuse*⟩ flagrantly

Schnörkel der; ~s, ~: scroll; *(der Handschrift, in der Rede)* flourish

schnorren *tr., itr. V. (ugs.)* scrounge *(coll.)* **(bei, von** + *Dat.* off); **Schnorrer** der; ~s, ~ *(ugs.)* scrounger *(coll.)*

schnüffeln *itr. V.* **a)** sniff; **b)** *(ugs.: spionieren)* snoop [about] *(coll.);* **Schnüffler** der; ~s, ~ *(ugs.)* Nosey Parker; *(Spion)* snooper *(coll.)*

schnupfen 1. *tr. V.* sniff; **Tabak ~:** take snuff; **2.** *itr. V.* take snuff; **Schnupfen** der; ~s, ~: [head] cold; **Schnupf·tabak** der snuff

schnuppe: das/er ist mir ~/mir völlig ~ *(ugs.)* I don't care/I couldn't care less about it/him *(coll.)*

schnuppern *itr. V.* sniff; **an etw.** *(Dat.)* ~: sniff sth.

Schnur die; ~, Schnüre **a)** *(Bindfaden)* piece of string; *(Kordel)* piece of cord; **b)** *(ugs.: Kabel)* flex *(Brit.);* lead; cord *(Amer.);* **schnüren** *tr. V.* tie ⟨*bundle, string, etc.*⟩; tie, lace up ⟨*shoe, corset, etc.*⟩

Schnurr·bart der moustache; **schnurren** *itr. V.* ⟨*cat*⟩ purr; ⟨*machine*⟩ hum

Schnür-: ~schuh der lace-up shoe; ~senkel der; ~s, ~ *(bes. nordd.)* [shoe-]lace; *(für Stiefel)* bootlace

schob *1. u. 3. Pers. Prät. v.* schieben

Schock der; ~[e]s, ~s shock; **schockieren** *tr. V.* shock; **über etw.** *(Akk.)* **schockiert sein** be shocked at sth.

Schöffe der; ~n, ~n lay judge *(acting together with another lay judge and a professional judge);* **Schöffen·gericht** das *court presided over by a professional judge and two lay judges*

Schokolade die; ~, ~n **a)** chocolate; **b)** *(Getränk)* [drinking] chocolate

Schokolade[n]-: ~eis das chocolate ice-cream; ~guß der chocolate icing; ~pudding der chocolate blancmange; ~torte die chocolate cake *or* gateau

scholl *1. u. 3. Pers. Sg. Prät. v.* schallen

Scholle die; ~, ~n **a)** *(Erd~)* clod [of

earth]; **b)** *(Eis~)* [ice-]floe; **c)** *(Fisch)* plaice

schon 1. *Adv.* **a)** *(bereits) (oft nicht übersetzt)* already; *(in Fragen)* yet; **wie lange bist du ~ hier?** how long have you been here?; **b)** *(fast gleichzeitig)* there and then; **c)** *(jetzt)* ~ |mal| now; *(inzwischen)* meanwhile; **d)** *(selbst, sogar)* even; *(nur)* only; **e)** *(ohne Ergänzung, ohne weiteren Zusatz)* on its own; |allein| ~ **der Gedanke daran** the mere thought of it; ~ **deshalb** for this reason alone; **f)** *(wohl)* really; **Lust hätte ich ~, aber ...:** I'd certainly like to, but ...; **2.** *Partikel* **a)** *(ugs. ungeduldig: endlich)* **nun komm ~!** come on!; hurry up!; **b)** *(beruhigend: bestimmt)* all right; **c)** *(durchaus)* **das ist ~ möglich** that is quite possible

schön 1. *Adj.* **a)** beautiful; handsome ⟨*youth, man*⟩; **b)** *(angenehm)* pleasant, nice ⟨*day, holiday, dream, relaxation, etc.*⟩; fine ⟨*weather*⟩; *(nett)* nice; **das war eine ~e Zeit** those were wonderful days; **c)** *(gut)* good; **d)** *(in Höflichkeitsformeln)* ~e **Grüße** best wishes; **recht ~en Dank für ...:** thank you very much for ...; **e)** ~! *(ugs.: einverstanden)* OK *(coll.);* all right; **f)** *(iron.: leer)* ~e **Worte** fine[-sounding] words; *(schmeichlerisch)* honeyed words; **g)** *(ugs.: beträchtlich)* handsome, *(coll.)* tidy ⟨*sum, fortune, profit*⟩; considerable ⟨*quantity, distance*⟩; pretty good ⟨*pension*⟩; **h)** *(iron.: unerfreulich)* nice *(coll. iron.);* **das sind ja ~e Aussichten!** this is a fine look-out *sing. (iron.);* **2.** *adv.* **a)** beautifully; **b)** *(angenehm, erfreulich)* nicely; ~ **warm/weich/langsam** nice and warm/soft/slow; **c)** *(gut)* well; **d)** *(in Höflichkeitsformeln)* **bitte ~, können Sie mir sagen, ...:** excuse me, could you tell me ...; **e)** *(iron.)* **wie es so ~ heißt, wie man so ~ sagt** as they say; **f)** *(ugs.: beträchtlich)* really; *(vor einem Adjektiv)* pretty; **ganz ~ arbeiten müssen** have to work jolly hard *(Brit. coll.);* **3.** *Partikel (ugs.)* **bleib ~ liegen!** lie there and be good

schonen 1. *tr. V.* treat ⟨*clothes, books, furniture, etc.*⟩ with care; *(schützen)* protect ⟨*hands, furniture*⟩; *(nicht strapazieren)* spare ⟨*voice, eyes, etc.*⟩; conserve ⟨*strength*⟩; **2.** *refl. V.* take things easy

Schönheit die; ~, ~en beauty

Schönheits-: ~chirurgie die cosmetic surgery *no art.;* ~pflege die beauty care *no art.*

Schon·kost die light food

schön|machen *(ugs.)* **1.** *tr. V.* smarten ⟨*person, thing*⟩ up; make ⟨*person, thing*⟩ look nice; **2.** *refl. V.* smarten oneself up

schonungs|los 1. *Adj.* unsparing, ruthless ⟨*criticism etc.*⟩; blunt ⟨*frankness*⟩; **2.** *adv.* unsparingly; ⟨*say*⟩ without mincing one's words

Schopf der; ~|e|s, Schöpfe shock of hair

schöpfen *tr. V.* **a)** scoop [up] ⟨*water, liquid*⟩; *(mit einer Kelle)* ladle ⟨*soup*⟩; **b)** *(geh.: einatmen)* draw, take ⟨*breath*⟩

Schöpfer der; ~s, ~: creator; *(Gott)* Creator; **schöpferisch 1.** *Adj.* creative; **2.** *adv.* creatively

Schöpf-: ~kelle die, ~löffel der ladle

Schöpfung die; ~, ~en *(geh.)* creation; **die ~** *(die Welt)* Creation

Schoppen der; ~s, ~: [quarter-litre/half-litre] glass of wine/beer

schor *1. u. 3. Pers. Sg. Prät. v.* scheren

Schorf der; ~|e|s, ~e scab

Schorn·stein der chimney; *(Lokomotiv~, Schiffs~)* funnel; **Schornstein·feger der; ~s, ~:** chimney-sweep

schoß *1. u. 3. Pers. Sg. Prät. v.* schießen

Schoß der; ~es, Schöße lap

Schote die; ~, ~n pod

Schotte der; ~n, ~n Scot; Scotsman; **die ~n** the Scots; the Scottish; **Schotten·rock der** tartan skirt; *(Kilt)* kilt; **Schottin die; ~, ~nen** Scot; Scotswoman; **schottisch** *Adj.* Scottish; **~er Whisky** Scotch whisky; **Schottland (das); ~s** Scotland

schräg 1. *Adj.* diagonal ⟨*line, beam, cut, etc.*⟩; sloping ⟨*surface, roof, wall, side, etc.*⟩; slanting, slanted ⟨*writing, eyes, etc.*⟩; tilted ⟨*position of the head etc., axis*⟩; **2.** *adv.* at an angle; *(diagonal)* diagonally; **Schräge die; ~, ~n a)** *(schräge Fläche)* sloping surface; **b)** *(Neigung)* slope

schrak *1. u. 3. Pers. Sg. Prät. v.* schrecken

Schramme die; ~, ~n scratch; **schrammen** *tr. V.* scratch

Schrank der; ~|e|s, Schränke cupboard; closet *(Amer.)*; *(Glas~; kleiner Wand~)* cabinet; *(Kleider~)* wardrobe; *(Bücher~)* bookcase; **Schränkchen das; ~s, ~:** cabinet

Schranke die; ~, ~n a) *(auch fig.)* barrier; **b)** *(fig.: Grenze)* limit

Schraube die; ~, ~n bolt; *(Holz~, Blech~)* screw; **schrauben** *tr. V.* **a)** *s.* **Schraube**: bolt/screw **(an, auf +** *Akk.* on to); **b)** *(drehen)* screw ⟨*nut, hook, light-bulb, etc.*⟩ **(auf +** *Akk.* on to; **in +** *Akk.* into)

Schrauben-: ~schlüssel der spanner; **~zieher der; ~s, ~:** screwdriver

Schraub·verschluß der screw-top

Schreber·garten der ≈ allotment *(cultivated primarily as a garden)*

Schreck der; ~|e|s, ~e fright; scare; *(Schock)* shock; **jmdm. einen ~ einjagen** give sb. a fright; **schrecken** *regelm. (auch unr.) itr. V.* start [up]; **aus dem Schlaf ~:** awake with a start; start from one's sleep; **Schrecken der; ~s, ~:** fright; scare; *(Entsetzen)* horror; *(große Angst)* terror; **jmdm. einen ~ einjagen** give sb. a fright; **schreckhaft** *Adj.* easily scared; **schrecklich 1.** *Adj.* terrible; **2.** *adv.* terribly

Schrei der; ~|e|s, ~e cry; *(lauter Ruf)* shout; *(durchdringend)* yell; *(gellend)* scream; *(kreischend)* shriek

Schreib·block der; *Pl.* **~blocks** *od.* **~blöcke** writing-pad

schreiben 1. *unr. itr. V.* write; *(mit der Schreibmaschine)* type; **an einem Roman** *usw.* **~:** be writing a novel *etc.*; **jmdm.** *od.* **an jmdn. ~:** write to sb.; **2.** *unr. tr. V.* **a)** write; *(mit der Schreibmaschine)* type; **wie schreibt man dieses Wort?** how is this word spelt?; **3.** *unr. refl. V.* be spelt; **Schreiben das; ~s, ~ a)** *o. Pl.* writing *no def. art.*; **b)** *(Brief)* letter; **Schreiber der; ~s, ~:** writer; *(Verfasser)* author; **Schreiberin die; ~, ~en** writer; *(Verfasserin)* authoress

Schreib-: ~maschine die typewriter; **~maschinen·papier das** typing paper; **~papier das** writing-paper; **~tisch der** desk

Schreibung die; ~, ~en spelling

Schreib-: ~waren *Pl.* stationery *sing.;* **~waren·geschäft das** stationer's

schreien *unr. itr. V.* ⟨*person*⟩ cry [out]; *(laut rufen/sprechen)* shout; *(durchdringend)* yell; *(gellend)* scream; ⟨*baby*⟩ yell, bawl; **zum Schreien sein** *(ugs.)* be a scream *(sl.)*

Schreiner der; ~s, ~ *(bes. südd.) s.* **Tischler**

schreiten *unr. itr. V.; mit sein (geh.)* walk; *(mit großen Schritten)* stride

schrickst *2. Pers. Sg. Präsens v.*

schrecken; **schrickt** 3. Pers. Sg. Präsens v. schrecken
schrie 1. u. 3. Pers. Sg. Prät. v. schrei-
en
schrieb 1. u. 3. Pers. Sg. Prät. v.
schreiben; **Schrieb** der; ~|e|s, ~e
(ugs.) missive (coll.)
Schrift die; ~, ~en a) (System) script;
(Alphabet) alphabet; b) (Hand~)
[hand]writing; c) (Werk) work;
schriftlich 1. Adj. written; 2. adv. in
writing
Schrift-: ~steller der; ~s, ~: writer;
~stück das [official] document;
~wechsel der correspondence
schrill 1. Adj. shrill; 2. adv. shrilly;
schrillen itr. V. shrill; sound shrilly
schritt 1. u. 3. Pers. Sg. Prät. v. schrei-
ten; **Schritt** der; ~|e|s, ~e a) step; ei-
nen ~ machen od. tun take a step; b)
Pl. (Geräusch) footsteps; c) (Entfernung) pace; d) (Gleich~) aus dem ~
kommen get out of step; e) o. Pl.
(Gangart) walk; seinen ~ verlang-
samen/beschleunigen slow/quicken
one's pace; |mit jmdm./etw.| ~ halten
(auch fig.) keep up or keep pace [with
sb./sth.]; f) (~geschwindigkeit) walking pace; „~ fahren" 'dead slow'; g)
(fig.: Maßnahme) step; measure;
Schritt·geschwindigkeit die
walking pace
schroff 1. Adj. a) precipitous ⟨rock
etc.⟩; b) (plötzlich) sudden ⟨transition,
change⟩; (kraß) stark ⟨contrast⟩; c)
(barsch) curt ⟨refusal, manner⟩;
brusque ⟨manner, behaviour, tone⟩; 2.
adv. a) ⟨rise, drop⟩ sheer; ⟨fall away⟩
precipitously; b) (plötzlich, unvermittelt) suddenly; c) (barsch) curtly; ⟨in-
terrupt⟩ abruptly; ⟨treat⟩ brusquely
schröpfen tr. V. (ugs.) fleece
Schrot der od. das; ~|e|s, ~e a) coarse
meal; (aus Getreide) whole meal
(Brit.); whole grain; b) (Munition)
shot; **schroten** tr. V. grind ⟨grain
etc.⟩ [coarsely]; crush ⟨malt⟩ [coarsely]
Schrot-: ~flinte die shotgun; ~ku-
gel die pellet
Schrott der; ~|e|s, ~e a) scrap
[-metal]; ein Auto zu ~ fahren (ugs.)
write a car off; b) o. Pl. (salopp fig.)
rubbish; **schrott·reif** Adj. ready for
the scrap-heap postpos.
schrubben tr. (auch itr.) V. scrub;
Schrubber der; ~s, ~: [long-
handled] scrubbing-brush
Schrulle die; ~, ~n cranky idea; (Marotte) quirk

schrumpelig Adj. (ugs.) wrinkly;
schrumpeln itr. V.; mit sein (ugs.)
⟨skin⟩ go wrinkled; ⟨apple etc.⟩ shrivel
schrumpfen itr. V.; mit sein shrink;
⟨metal, rock⟩ contract; ⟨apple etc.⟩
shrivel; ⟨skin⟩ go wrinkled; (abnehmen) decrease; ⟨supplies, capital,
hopes⟩ dwindle
Schub der; ~|e|s, Schübe a) (Physik:
~kraft) thrust; b) (Med.: Phase)
phase; stage; c) (Gruppe, Anzahl)
batch
Schuber der; ~s, ~: slip-case
Schub-: ~karre die, ~karren der
wheelbarrow; ~lade die drawer
Schubs der; ~es, ~e (ugs.) shove;
schubsen tr. (auch itr.) V. (ugs.)
push; shove
schüchtern 1. Adj. a) shy ⟨person,
smile, etc.⟩; shy, timid ⟨voice, knock,
etc.⟩; b) (fig.: zaghaft) tentative, cautious ⟨attempt, beginnings, etc.⟩; 2.
adv. shyly; ⟨knock, ask, etc.⟩ timidly;
Schüchternheit die; ~: shyness
Schuft der; ~|e|s, ~e scoundrel
schuften (ugs.) itr. V. slave away
Schuh der; ~|e|s, ~e shoe; (hoher ~,
Stiefel) boot; jmdm. etw. in die ~e
schieben (fig. ugs.) pin the blame for
sth. on sb.
Schuh-: ~creme die shoe-polish;
~größe die shoe size; welche ~größe
hast du? what size shoe[s] do you
take?; ~macher der; ~s, ~: shoemaker; ~sohle die sole [of a/one's
shoe]
Schul-: ~abschluß der school-
leaving qualification; ~buch das
school-book; ~bus der school bus
schuld s. Schuld b; **Schuld** die; ~,
~en a) o. Pl. guilt; er ist sich (Dat.)
keiner ~ bewußt he is not conscious of
having done any wrong; b) o. Pl. (Ver-
antwortlichkeit) blame; es ist |nicht|
seine ~: it is [not] his fault; |an etw.
(Dat.)| schuld haben od. sein be to
blame [for sth.]; c) (Verpflichtung zur
Rückzahlung) debt; 5 000 Mark ~en
haben have debts of 5,000 marks, owe
5,000 marks; **schuld·bewußt** 1.
Adj. guilty ⟨look, face, etc.⟩; 2. adv.
guiltily; **schulden** tr. V. owe; was
schulde ich Ihnen? how much do I owe
you?; **Schuld·gefühl** das feeling of
guilt; **schuldig** Adj. a) guilty; der |an
dem Unfall| ~e Autofahrer the driver
to blame [for the accident]; b) jmdm.
etw. ~ sein/bleiben owe sb. sth.; c) (gebührend) due; proper; **Schuldige**

der/die; *adj. Dekl.* guilty person; *(im Strafprozeß)* guilty party; **schuld·los** *Adj.* innocent (**an** + *Dat.* of); **Schuld·spruch** der verdict of guilty

Schule die; ~, ~n a) school; **zur** *od.* **in die ~ gehen, die ~ besuchen** go to school; **auf** *od.* **in der ~:** at school; **schulen** *tr. V.* train; **Schüler** der; ~s, ~: pupil; *(Schuljunge)* schoolboy; **Schülerin** die; ~, ~nen pupil; *(Schulmädchen)* schoolgirl

schul-, Schul-: ~**ferien** *Pl.* school holidays *or (Amer.)* vacation *sing.*; ~**frei** *Adj.* ⟨day⟩ off school; ~**hof** der school yard; ~**jahr** das a) school year; b) *(Klasse)* year; ~**junge** der schoolboy; ~**kind** das schoolchild; ~**klasse** die [school] class; ~**mädchen** das schoolgirl; ~**ranzen** der [school] satchel; ~**tag** der school day; ~**tasche** die school-bag; *(Ranzen)* [school] satchel

Schulter die; ~, ~n shoulder; **jmdm. auf die ~ klopfen** pat sb. on the shoulder *or (fig.)* back; **Schulterblatt** das *(Anat.)* shoulder-blade; **schultern** *tr. V.* shoulder; **das Gewehr ~:** shoulder arms

Schul-: ~**weg** der way to school; ~**zeit** die school-days *pl.*

schummerig *Adj.* dim ⟨light etc.⟩; dimly lit ⟨room etc.⟩

Schund der; ~[e]s trash

Schuppe die; ~, ~n a) scale; b) *Pl.* *(auf dem Kopf)* dandruff *sing.*; *(auf der Haut)* flaking skin *sing.*; **schuppen** 1. *tr. V.* scale ⟨fish⟩; 2. *refl. V.* ⟨skin⟩ flake; ⟨person⟩ have flaking skin

Schuppen der; ~s, ~ a) shed; b) *(ugs.: Lokal)* joint *(sl.)*

schüren *tr. V.* a) poke ⟨fire⟩; b) *(fig.)* stir up ⟨hatred, envy, etc.⟩

schürfen 1. *itr. V.* scrape; 2. *tr. V.* a) **sich** *(Dat.)* **das Knie** *usw.* ~: graze one's knee *etc.*; b) *(Bergbau)* mine ⟨ore etc.⟩ open-cast *or (Amer.)* opencut; **Schürf·wunde** die graze; abrasion

Schurke der; ~n, ~n rogue

Schur·wolle die new wool

Schürze die; ~, ~n apron; *(Frauen~, Latz~)* pinafore

Schuß der; **Schusses, Schüsse** a) shot (**auf** + *Akk.* at); **weit** *od.* **weitab vom ~** *(fig. ugs.)* well away from the action; b) *(Menge Munition/Schießpulver)* round; **drei ~ Munition** three rounds of ammunition; c) *(~wunde)* gunshot

wound; d) *(kleine Menge)* dash; e) *(Drogenjargon)* shot; fix *(sl.)*; f) *(Skisport)* schuss; ~ **fahren** schuss; g) *(ugs.)* etw. in ~ **bringen/halten** get sth. into/keep sth. in [good] shape

Schüssel die; ~, ~n bowl; *(flacher)* dish

schusselig *(ugs.)* 1. *Adj.* scatterbrained; 2. *adv.* in a scatter-brained way

Schuß-: ~**linie** die line of fire; **in die/ jmds. ~linie geraten** *od.* **kommen** *(auch fig.)* come under fire/come under fire from sb.; ~**verletzung** die gunshot wound; ~**waffe** die weapon *(firing a projectile)*; *(Gewehr usw.)* firearm

Schuster der; ~s, ~ *(ugs.)* shoemaker; *(Flick~)* shoe-repairer

Schutt der; ~[e]s rubble; „~ **abladen verboten"** 'no tipping'; 'no dumping'

Schüttel·frost der [violent] shivering fit

schütteln 1. *tr. V.* a) shake; **den Kopf** [**über etw.** *(Akk.)*] ~: shake one's head [over sth.]; **jmdm. die Hand** ~: shake sb.'s hand; shake sb. by the hand; b) *(unpers.)* **es schüttelte ihn [vor Kälte]** he was shaking [with *or* from cold]; 2. *refl. V.* shake oneself/itself; 3. *itr. V.* **mit dem Kopf** ~: shake one's head

schütten 1. *tr. V.* pour ⟨liquid, flour, etc.⟩; *(unabsichtlich)* spill ⟨liquid, flour, etc.⟩; tip ⟨rubbish, coal, etc.⟩; 2. *itr. V. (unpers.) (ugs.: regnen)* pour [down]

schütter *Adj.* sparse; thin

Schutz der; ~es protection (**vor** + *Dat.*, **gegen** against); *(Zuflucht)* refuge

schutz-, Schutz-: ~**bedürftig** *Adj.* in need of protection *postpos.*; ~**blech** das mudguard; ~**brief** der *(Kfz-W.)* travel insurance; *(Dokument)* travel insurance certificate

Schütze der; ~n, ~n a) marksman; b) *(Fußball usw.: Tor~)* scorer; c) *(Milit.: einfacher Soldat)* private; d) *(Astrol.)* Sagittarius

schützen 1. *tr. V.* protect (**vor** + *Dat.* from, **gegen** against); safeguard ⟨interest, property, etc.⟩ (**vor** + *Dat.* from); **gesetzlich geschützt** registered [as a trade-mark]; 2. *itr. V.* provide *or* give protection (**vor** + *Dat.* from, **gegen** against); *(vor Wind, Regen)* give shelter (**vor** + *Dat.* from)

Schützen·fest das *shooting competition with fair*

Schutz·engel der guardian angel

Schützen-: ~**graben** der trench; ~**panzer** der armoured personnel carrier; ~**verein** der rifle club

Schutz-: ~**helm** der helmet; *(bei Motorradfahrern usw.)* crash-helmet; *(bei Bauarbeitern usw.)* safety helmet; ~**hütte** die a) *(Unterstand)* shelter; b) *(Berghütte)* mountain hut; ~**impfung** die vaccination

Schützling der; ~s, ~e protégé; *(Anvertrauter)* charge

schutz-, Schutz-: ~**los** *Adj.* defenceless; ~**mann** der; *Pl.* ~**männer** *od.* ~**leute** *(ugs. veralt.)* [police] constable; copper *(Brit. coll.);* ~**patron** der patron saint; ~**suchend** *Adj.* seeking protection *postpos.;* ~**umschlag** der dust-jacket

schwabbelig *Adj.* flabby ⟨*stomach, person, etc.*⟩; wobbly ⟨*jelly etc.*⟩; **schwabbeln** *itr. V. (ugs.)* wobble

Schwabe der; ~n, ~n Swabian; **Schwaben (das);** ~s Swabia; **Schwäbin** die; ~, ~**nen** Swabian; **schwäbisch** *Adj.* Swabian

schwach; schwächer, schwächst... 1. *Adj.* a) weak; weak, delicate ⟨*child, woman*⟩; frail ⟨*invalid, old person*⟩; low-powered ⟨*engine, bulb, amplifier, etc.*⟩; weak, poor ⟨*eyesight, memory, etc.*⟩; poor ⟨*hearing*⟩; delicate ⟨*health, constitution*⟩; ~ **werden** grow weak; *(fig.: schwanken)* weaken; *(fig.: nachgeben)* give in; b) *(nicht gut)* poor ⟨*pupil, player, performance, result, etc.*⟩; weak ⟨*argument, opponent, play, film, etc.*⟩; c) *(gering, niedrig)* poor, low ⟨*attendance etc.*⟩; slight ⟨*effect, resistance, gradient, etc.*⟩; light ⟨*wind, rain, current*⟩; faint ⟨*voice, pressure, hope, smile, smell*⟩; weak, faint ⟨*pulse*⟩; faint, dim ⟨*light*⟩; pale ⟨*colour*⟩; d) *(wenig konzentriert)* weak ⟨*solution, coffee, poison, etc.*⟩; e) *(Sprachw.)* weak; 2. *adv.* a) weakly; b) *(nicht gut)* poorly; c) *(in geringem Maße)* poorly ⟨*attended, developed*⟩; slightly ⟨*poisonous, sweetened, inclined*⟩; ⟨*rain*⟩ slightly; ⟨*remember, glow, smile*⟩ faintly; d) *(Sprachw.)* ~ **gebeugt** weak; **Schwäche** die; ~, ~n weakness; **eine** ~ **für jmdn./etw. haben** have a soft spot for sb./a weakness for sth.; **Schwäche·anfall** der sudden feeling of faintness; **schwächen** *tr. V.* weaken; **schwächlich** *Adj.* weakly ⟨*person*⟩; frail ⟨*old person, constitution*⟩; **Schwächling** der; ~s, ~e weakling

schwach-, Schwach-: ~**sinn** der; *o. Pl.* a) *(Med.)* mental deficiency; b) *(ugs.)* [idiotic *(coll.)*] rubbish; ~**sinnig** 1. *Adj.* a) *(Med.)* mentally deficient; b) *(ugs.)* idiotic *(coll.),* nonsensical ⟨*measure, policy, etc.*⟩; rubbishy ⟨*film etc.*⟩; 2. *adv. (ugs.)* idiotically *(coll.);* stupidly

schwafeln *(ugs.)* 1. *itr. V.* rabbit on *(Brit. sl.),* waffle (von about); 2. *tr. V.* blether ⟨*nonsense*⟩

Schwager der; ~s, Schwäger brother-in-law; **Schwägerin** die; ~, ~**nen** sister-in-law

Schwalbe die; ~, ~n swallow

Schwall der; ~[e]s, ~e torrent

schwamm *1. u. 3. Pers. Sg. Prät. v.* **schwimmen**

Schwamm der; ~[e]s, Schwämme a) sponge; ~ **drüber!** *(ugs.)* [let's] forget it; b) *(südd., österr.: Pilz)* mushroom; **Schwammerl** das; ~s, ~[n] *(bayr., österr.)* mushroom; **schwammig** 1. *Adj.* a) spongy; b) *(aufgedunsen)* flabby, bloated ⟨*face, body, etc.*⟩; 2. *adv. (unpräzise)* vaguely

Schwan der; ~[e]s, Schwäne swan

schwand *1. u. 3. Pers. Sg. Prät. v.* **schwinden**

schwang *1. u. 3. Pers. Sg. Prät. v.* **schwingen**

schwanger *Adj.* pregnant (von by); **Schwangere** die; *adj. Dekl.* expectant mother; pregnant woman; **schwängern** *tr. V.* make ⟨*woman*⟩ pregnant; **Schwangerschaft** die; ~, ~**en** pregnancy

Schwank der; ~[e]s, Schwänke comic tale; *(auf der Bühne)* farce

schwanken *itr. V.; mit Richtungsangabe mit sein* a) sway; ⟨*boat*⟩ rock; *(heftiger)* roll; ⟨*ground, floor*⟩ shake; b) *(fig.: unbeständig sein)* ⟨*prices, temperature, etc.*⟩ fluctuate; ⟨*number, usage, etc.*⟩ vary; c) *(fig.: unentschieden sein)* waver; *(zögern)* hesitate

Schwanz der; ~es, Schwänze a) tail; b) *(salopp: Penis)* prick *(coarse);* cock *(coarse)*

schwänzeln *itr. V.* wag its tail/their tails

schwänzen *tr., itr. V. (ugs.)* skip, cut ⟨*lesson etc.*⟩; |die Schule| ~: play truant *or (Amer.)* hookey

schwappen *itr. V.* slosh

Schwarm der; ~[e]s, Schwärme a) swarm; b) *(fam.: Angebetete[r])* idol; heart-throb; **schwärmen** *itr. V.* a) *mit Richtungsangabe mit sein* swarm;

b) *(begeistert sein)* für jmdn./etw. ~:
be mad about *or* really keen on sb./
sth.; von etw. ~: go into raptures
about sth.; **schwärmerisch 1.** *Adj.*
rapturous; **2.** *adv.* rapturously

Schwarte die; ~, ~n a) rind; b) *(ugs.:*
dickes Buch) tome

schwarz; schwärzer, schwärzest... **1.**
Adj. **a)** black; Black *(person);*
filthy[-black] *(hands, finger-nails,*
etc.); mir wurde ~ vor den Augen
everything went black; der ~e Erdteil
od. Kontinent the Dark Continent;
das Schwarze Meer the Black Sea; ins
Schwarze treffen *(fig.)* hit the nail on
the head; *(illegal)* illicit *(deal, ex-*
change, etc.); der ~e Markt the black
market; **2.** *adv. (illegal)* illegally;
Schwarz das; ~|es|, ~: black

Schwarz·brot das black bread

Schwarze der/die; *adj. Dekl.* Black;
schwärzen *tr. V.* blacken

schwarz-, Schwarz-: ~|fahren
unr. itr. V.; mit sein dodge paying the
fare; ~**fahrer** der fare-dodger;
~**haarig** *Adj.* black-haired; ~**han-**
del der black market (**mit** in); *(Tätig-*
keit) black marketeering (**mit** in);
~**markt** der black market; ~|**sehen**
unr. itr. V. **a)** *(pessimistisch sein)* look
on the black side; be pessimistic (**für**
about; **b)** *(schwarz fernsehen)* watch
television without a licence; ~**seher**
der **a)** *(ugs.)* pessimist; **b)** *(jmd, der*
schwarz fernsieht) [television] licence
dodger; ~**wald** der; ~|e|s Black
Forest; **schwarz·weiß** *Adj.* black
and white; **Schwarzweiß·foto** das
black and white photo; **Schwarz-**
wurzel die black salsify

schwatzen, *(bes. südd.)* **schwätzen**
1. *itr. V.* chat; *(über belanglose Dinge)*
chatter; natter *(coll.);* **2.** *tr. V.* say; talk
(nonsense, rubbish); **Schwätzer** der;
~s, ~: chatterbox; *(klatschhafter*
Mensch) gossip; **schwatzhaft** *Adj.*
talkative; *(klatschhaft)* gossipy

Schwebe die; in der ~ sein/bleiben
(fig.) be/remain in the balance

Schwebe-: ~**bahn** die cableway;
~**balken** der *(Turnen)* [balance] beam

schweben *itr. V.* **a)** *(bird, balloon,*
etc.) hover; *(cloud, balloon, mist)*
hang; **in Gefahr** ~ *(fig.)* be in danger;
b) mit sein *(durch die Luft)* float

Schwede der; ~n, ~n Swede;
Schweden (das); ~s Sweden;
Schwedin die; ~, ~nen Swede;
schwedisch *Adj.* Swedish

Schwefel der; ~s sulphur

Schweif der; ~|e|s, ~e tail;
schweifen *itr. V.; mit sein (geh.;*
auch fig.) wander

Schweige·geld das hush money

schweigen *unr. itr. V.* remain *or* stay
silent; say nothing; **ganz zu** ~ **von** ...:
not to mention ...; **Schweigen** das;
~s silence; **schweigsam** *Adj.* silent;
quiet

Schwein das; ~|e|s, ~e a) pig; b) *o. Pl.*
(Fleisch) pork; **c)** *(salopp: gemeiner*
Mensch) swine; *(Schmutzfink)* mucky
devil *(coll.);* mucky pig *(coll.);* **d)** *(sa-*
lopp: Mensch) ein armes ~: a poor
devil; kein ~ war da there wasn't a
bloody *(Brit. sl.) or (coll.)* damn soul
there; **e)** *(ugs.: Glück)* |großes| ~ haben
have a [big] stroke of luck; *(davon-*
kommen) get away with it *(coll.)*

Schweine-: ~**braten** der roast pork
no indef. art.; ~**fleisch** das pork;
~**kotelett** das *(Kochk.)* pork chop

Schweinerei die; ~, ~en *(ugs.)* **a)**
(Schmutz) mess; **b)** *(Gemeinheit)*
mean *or* dirty trick

Schweine-: ~**schnitzel** das escalope of pork; ~**stall** der *(auch fig.)*
pigsty; pigpen *(Amer.)*

schweinisch *(ugs.) Adj.* **a)** *(schmut-*
zig) filthy; **b)** *(unanständig)* dirty;
smutty

Schweins·leder das pigskin

Schweiß der; ~es sweat; mir brach
der ~ aus I broke out in a sweat

Schweiß·brenner der welding
torch; **schweißen** *tr., itr. V.* weld;
Schweißer der; ~s, ~: welder

Schweiß-: ~**fuß** sweaty foot; ~**per-**
le die bead of sweat

Schweiz die; ~: Switzerland *no art.;*
Schweizer der; ~s, ~: Swiss;
schweizer·deutsch *Adj.* Swiss
German; **Schweizerin** die; ~, ~nen
Swiss; **schweizerisch** *Adj.* Swiss

schwelen *(auch fig.)* smoulder

schwelgen *itr. V.* feast

Schwelle die; ~, ~n a) threshold; b)
(Eisenbahn~) sleeper *(Brit.);* [cross-]
tie *(Amer.)*

schwellen *unr. itr. V.; mit sein* swell;
(limb, face, cheek, etc.) swell [up];
Schwellung die; ~, ~en swelling

Schwemme die; ~, ~n glut (an +
Dat. of)

Schwengel der; ~s, ~ a) *(Glocken~)*
clapper; b) *(Pumpen~)* handle

schwenken 1. *tr. V.* **a)** swing; wave
(flag, handkerchief); **b)** *(spülen)* rinse;

2. *itr. V.; mit sein* ⟨*marching column*⟩ swing, wheel; ⟨*camera*⟩ pan; ⟨*path, road, car*⟩ swing

schwer 1. *Adj.* **a)** heavy; **2** Kilo ~ sein weigh two kilos; **b)** *(mühevoll)* heavy ⟨*work*⟩; hard, tough ⟨*job*⟩; hard ⟨*day*⟩; difficult ⟨*birth*⟩; es ~/nicht ~ **haben** have it hard/easy; **c)** *(schlimm)* severe ⟨*shock, disappointment, strain, storm*⟩; serious, grave ⟨*wrong, injustice, error, illness, blow, reservation*⟩; serious ⟨*accident, injury*⟩; heavy ⟨*punishment, strain, loss, blow*⟩; **2.** *adv.* **a)** heavily ⟨*built, laden, armed*⟩; ~ **tragen** be carrying sth. heavy [with difficulty]; **b)** ⟨*work*⟩ hard; ⟨*breathe*⟩ heavily; ~ **hören** be hard of hearing; **c)** *(schwierig)* with difficulty; **d)** *(sehr)* seriously ⟨*injured*⟩; greatly, deeply ⟨*disappointed*⟩; ⟨*punish*⟩ severely, heavily; ~ **verunglücken** have a serious accident

Schwer-: ~**arbeiter** der worker engaged in heavy physical work; ~**behinderte** der/die severely handicapped person; *(körperlich auch)* severely disabled person; die ~**behinderten** the severely handicapped/disabled; ~**beschädigte** der/die; *adj. Dekl.* severely disabled person

Schwere die; ~ **a)** weight; **b)** *(Schwerkraft)* gravity; **c)** *s.* **schwer 1c:** severity; seriousness; gravity; heaviness; **schwere·los** *Adj.* weightless; **Schwerelosigkeit** die; ~: weightlessness

schwer-, Schwer-: ~|**fallen** *unr. itr. V.; mit sein* jmdm. fällt etw. ~: sb. finds sth. difficult; ~**fällig 1.** *Adj. (auch fig.)* ponderous; cumbersome ⟨*bureaucracy, procedure*⟩; **2.** *adv.* ponderously; ~**gewicht** das **a)** *(Sport)* heavyweight; **b)** *o. Pl. (Schwerpunkt)* main focus; ~**hörig** *Adj.* hard of hearing *pred.;* ~**industrie** die heavy industry; ~**kraft** die; *o. Pl.* gravity; ~**krank** *Adj.; präd. getrennt geschrieben* seriously ill

schwerlich *Adv.* hardly

schwer-, Schwer-: ~|**machen** *tr. V.* jmdm./sich etw. ~**machen** make sth. difficult for sb./oneself; ~**metall** das heavy metal; ~**mütig 1.** *Adj.* melancholic; **2.** *adv.* melancholically; ~|**nehmen** *unr. tr. V.* take ⟨*sth.*⟩ seriously; ~**punkt** der centre of gravity; *(fig.)* main focus; *(Hauptgewicht)* main stress

Schwert das; ~[e]s, ~er sword; **Schwert·lilie** die iris

schwer-, Schwer-: ~|**tun** *unr. refl. V. (ugs.)* sich mit *od.* bei etw. ~**tun** *(ugs.)* have trouble with sth.; ~**verbrecher** der serious offender; ~**verdaulich** *Adj.; präd. getrennt geschrieben (auch fig.)* hard to digest *pred.;* ~**verletzt** *Adj.; präd. getrennt geschrieben* seriously injured; ~**wiegend** *Adj.* serious; momentous ⟨*decision*⟩

Schwester die; ~, ~n **a)** sister; **b)** *(Kranken~)* nurse; **schwesterlich 1.** *Adj.* sisterly; **2.** *adv.* ~ **handeln** act in a sisterly way

schwieg *1. u. 3. Pers. Prät. v.* **schweigen**

Schwieger-: ~**eltern** *Pl.* parents-in-law; ~**mutter** die mother-in-law; ~**sohn** der son-in-law; ~**tochter** die daughter-in-law; ~**vater** der father-in-law

Schwiele die; ~, ~n callus; ~n an den Händen horny hands

schwierig *Adj.* difficult; **Schwierigkeit** die; ~, ~en difficulty

Schwimm-: ~**bad** das swimming-baths *pl. (Brit.);* swimming-pool; ~**becken** das swimming-pool

schwimmen 1. *unr. itr. V.* **a)** *meist mit sein* swim; **b)** *meist mit sein (treiben, nicht untergehen)* float; **c)** *(ugs.: unsicher sein)* be all at sea; **ins Schwimmen geraten** start to flounder; **2.** *unr. tr. V.; auch mit sein* swim; **Schwimmen** das; ~: swimming *no art.;* **Schwimmer** der; ~s, ~ **a)** swimmer; **b)** *(Technik)* float

Schwimm-: ~**flosse** die flipper; ~**lehrer** der swimming instructor; ~**weste** die life-jacket

Schwindel der; ~s **a)** dizziness; giddiness; **b)** *(Betrug)* swindle; *(Lüge)* lie; **schwindel·frei** *Adj.* ~ sein have a head for heights; **schwindelig** *s.* **schwindlig; schwindeln** *itr. V.* **a)** *(unpers.)* mich *od.* mir schwindelt I feel dizzy *or* giddy; **b)** *(lügen)* tell fibs

schwinden *unr. itr. V.; mit sein* fade; ⟨*supplies, money*⟩ run out; ⟨*effect*⟩ wear off; ⟨*fear, mistrust*⟩ lessen; ⟨*powers, influence*⟩ wane

Schwindler der; ~s, ~ *(Lügner)* liar; *(Betrüger)* swindler; *(Hochstapler)* con man *(coll.)*

schwindlig *Adj.* dizzy; giddy; jmdm. wird es ~: sb. gets dizzy *or* giddy

schwingen 1. *unr. itr. V.* **a)** *mit sein* swing; **b)** *(vibrieren)* vibrate; **2.** *unr. tr. V.* swing; wave ⟨*flag, wand*⟩; brandish ⟨*sword, axe, etc.*⟩; **3.** *unr. refl. V.* **sich aufs Pferd/Fahrrad ~:** leap on to one's horse/bicycle; **Schwingung die; ~, ~en a)** swinging; *(Vibration)* vibration; **b)** *(Physik)* oscillation

Schwips der; ~es, ~e *(ugs.)* **einen ~ haben** be tipsy

schwirren *itr. V. mit sein* ⟨*arrow, bullet, etc.*⟩ whiz; ⟨*bird*⟩ whirr; ⟨*insect*⟩ buzz

schwitzen *itr. V. (auch fig.)* sweat

schwor *1. u. 3. Pers. Sg. Prät. v.* schwören; **schwören 1.** *unr. tr., itr. V.* swear ⟨*fidelity, friendship*⟩; swear, take ⟨*oath*⟩; **2.** *unr. itr. V.* swear an/the oath

schwul *Adj. (ugs.)* gay *(coll.)*

schwül *Adj.* sultry; close

Schwule der; *adj. Dekl. (ugs.)* gay *(coll.); (abwertend)* queer *(sl.)*

Schwüle die; ~: sultriness

schwülstig 1. *Adj.* bombastic; pompous; over-ornate ⟨*art, architecture*⟩; **2.** *adv.* bombastically; pompously

Schwund der; ~[e]s decrease, drop *(Gen.* in); *(an Interesse)* waning; falling off

Schwung der; ~[e]s, Schwünge **a)** *(Bewegung)* swing; **b)** *(Linie)* sweep; **c)** *o. Pl. (Geschwindigkeit)* momentum; ~ **holen** build *or* get up momentum; **d)** *o. Pl. (Antrieb)* drive; energy; **e)** *o. Pl. (mitreißende Wirkung)* sparkle; **schwung·haft** *Adj.* thriving; brisk, flourishing ⟨*trade, business*⟩; **schwung·voll a)** lively; **b)** *(kraftvoll)* vigorous; sweeping ⟨*movement, gesture*⟩; bold ⟨*handwriting, line, stroke*⟩; **2.** *adv.* spiritedly; *(kraftvoll)* with great vigour

Schwur der; ~[e]s, Schwüre **a)** *(Gelöbnis)* vow; **b)** *(Eid)* oath; **Schwur·gericht** das *court with a jury*

sechs *Kardinalz.* six; **Sechs** die; ~, ~en six

sechs-, Sechs-: ~**eck** das hexagon; ~**eckig** *Adj.* hexagonal; ~**fach** *Vervielfältigungsz.* sixfold; ~**hundert** *Kardinalz.* six hundred; ~**mal** *Adv.* six times

sechst... *Ordinalz.* sixth

sechs·tausend *Kardinalz.* six thousand

sechstel *Bruchz.* sixth; **Sechstel** das, *schweiz. meist* der; ~s, ~: sixth; **sechstens** *Adv.* sixthly; **sechzehn** *Kardinalz.* sixteen; **sechzig** *Kardinalz.* sixty; **sechzigst...** *Ordinalz.* sixtieth

SED [ɛs|e:'de:] die; ~: *Abk. (ehem. DDR)* Sozialistische Einheitspartei Deutschlands Socialist Unity Party of Germany

¹**See** der; ~s, ~n lake; ²**See** die; ~: die ~: the sea; **an die ~ fahren** go to the seaside; **auf hoher ~:** on the high seas

See-: ~**bad** das seaside health resort; ~**fahrt** die *o. Pl.* seafaring *no art.;* sea travel *no art.;* ~**gang** der; *o. Pl.* leichter/starker *od.* hoher *od.* schwerer ~**gang** light/heavy *or* rough sea; ~**hund** der [common] seal; *(Pelz)* seal[skin]; ~**igel** der sea-urchin; ~**krank** *Adj.* seasick; ~**krankheit** die; *o. Pl.* seasickness; ~**lachs** der pollack

Seele die; ~, ~n soul; *(Psyche)* mind; **seelen·ruhig 1.** *Adj.* calm; **2.** *adv.* calmly; **seelisch 1.** *Adj.* psychological ⟨*cause, damage, tension*⟩; mental ⟨*equilibrium, breakdown, illness, health*⟩; **2.** *adv.* ~ **bedingt sein** have psychological causes; ~ **krank** mentally ill; **Seel·sorge** die; *o. Pl.* pastoral care; **Seelsorger** der; ~s, ~: pastoral worker; *(Geistlicher)* pastor

See-: ~**macht** die sea power; ~**mann** der; *Pl.* ~leute seaman; sailor; ~**meile** die nautical mile; ~**not** die; *o. Pl.* distress [at sea]; **in ~ geraten** get into difficulties *pl.;* ~**pferd[chen]** das sea-horse; ~**räuber** der pirate; ~**reise** die voyage; *(Kreuzfahrt)* cruise; ~**rose** die water-lily; ~**stern** der starfish; ~**tüchtig** *Adj.* seaworthy; ~**zunge** die sole

Segel das; ~s, ~: sail

Segel-: ~**boot** das sailing-boat; ~**flieger** der glider pilot; ~**flugzeug** das glider

segeln *itr. V.; mit sein* sail

Segel-: ~**schiff** das sailing ship; ~**tuch** das sailcloth

Segen der; ~s, ~: blessing; *(Gebet in der Messe)* benediction

Segler der; ~s, ~: yachtsman

segnen *tr. V.* bless

sehen 1. *unr. itr. V.* **a)** see; **schlecht/gut ~:** have bad/good eyesight; **mal ~, wir wollen od. werden ~** *(ugs.)* we'll see; **siehst!** *(ugs.)* there, you see!; **b)** *(hin~)* look **(auf** + *Akk.* at); **sieh mal** *od.* **doch!** look!; **siehe da!** lo and behold!; **2.** *unr. tr. V.* **a)** *(auch fig.)* see; **jmdn./etw. [nicht] zu ~ bekommen** [not] get to see sb./sth.; **ich habe ihn kommen [ge]~:** I saw him coming; **b)** *(an~)* watch ⟨*television programme*⟩;

sehens·wert *Adj.* worth seeing *postpos.;* **Sehens·würdigkeit die;** ~, ~en sight; **Seher der;** ~s, ~: seer; prophet; **Seh·fehler der** sight defect

Sehne die; ~, ~n **a)** tendon; **b)** *(Bogen~)* string

sehnen *refl. V.* sich nach jmdm./etw. ~: long *or* yearn for sb./sth.

sehnig *Adj.* **a)** stringy ⟨meat⟩; **b)** *(kräftig)* sinewy ⟨figure, legs, etc.⟩

sehnlichst 1. *Adj.* **das ist mein ~es Verlangen/mein ~er Wunsch** that's what I long for most/that's my dearest wish; **2.** *adv.* **etw. ~ herbeiwünschen** look forward longingly to sth.; **Sehn·sucht die** longing; ~ **nach jmdm. haben** long to see sb.; **sehn·süchtig** *Adj.* longing *attrib.,* yearning *attrib.* ⟨desire, look, gaze, etc.⟩

sehr *Adv.* **a)** *mit Adj. u. Adv.* very; ~ **viel** a great deal; **jmdn. ~ gern haben** like sb. a lot *(coll.) or* a great deal; **b)** *mit Verben* very much; greatly; **danke ~!** thank you *or* thanks [very much]; **bitte ~, Ihr Steak!** here's your steak, sir/madam

Seh·test der eye test

sei *1. u. 3. Pers. Sg. Präsens Konjunktiv u. Imperativ Sg. v.* sein

seicht 1. *Adj. (auch fig.)* shallow; **2.** *adv. (fig.)* shallowly

seid *2. Pers. Pl. Präsens u. Imperativ Pl. v.* sein

Seide die; ~, ~n silk

Seidel das; ~s, ~: beer-mug

seiden *Adj.; nicht präd.* silk; **Seiden·papier das** tissue paper; **seidig 1.** *Adj.* silky; **2.** *adv.* silkily

Seife die; ~, ~n soap

Seifen-: ~**blase die** soap bubble; ~**schale die** soap-dish; ~**schaum der;** *o. Pl.* lather

Seil das; ~s, ~e rope; *(Draht~)* cable

Seil-: ~**bahn die** cableway; ~**tänzer der** tightrope-walker; ~**winde die** cable winch

¹sein 1. *unr. itr. V.* be; *(existieren)* be; exist; *(sich ereignen)* be; happen; **wie dem auch sei** be that as it may; **er ist Schwede/Lehrer** he is Swedish *or* a Swede/a teacher; **bist du es?** is that you?; **mir ist kalt/besser** I am *or* feel cold/better; **mir ist schlecht** I feel sick; **drei und vier ist** *od.* **(ugs.) sind sieben** three and four is *or* makes seven; **es ist drei Uhr/Mai/Winter** it is three o'clock/May/winter; **er ist aus Berlin** he is *or* comes from Berlin; **was**

darf es ~? *(im Geschäft)* what can I get you?; **es war einmal ein Prinz** once upon a time there was a prince; **2.** *mod. V. (in der Funktion von* können/müssen + *Passiv)* **es ist niemand zu sehen** there's no one to be seen; **das war zu erwarten** that was to be expected; **die Schmerzen sind kaum zu ertragen** the pain is hardly bearable; **die Richtlinien sind strengstens zu beachten** the guidelines are to be strictly followed; **3.** *Hilfsverb* **a)** *(zur Bildung des Perfekts usw. im Aktiv)* have; **er ist gestorben** he has died; **b)** *(zur Bildung des Perfekts usw. im Passiv und des Zustandspassivs)* be; **wir sind gerettet worden/wir waren gerettet** we were saved

²sein *Possessivpron. (einer männlichen Person)* his; *(einer weiblichen Person)* her; *(einer Sache, eines Tiers)* its; *(nach man)* one's; his *(Amer.)*

seiner *(geh.) Gen. von* er: **sich ~ erbarmen** have pity on him; ~ **gedenken** remember him

seiner-: ~**seits** *Adv.* for his part; *(von ihm)* on his part; ~**zeit** *Adv.* at that time

seines·gleichen *indekl. Pron.* his own kind

seinet·wegen *Adv. s.* meinetwegen: because of him; for his sake; about him; as far as he is concerned

Seismo·graph der; ~en, ~en seismograph

seit 1. *Präp. mit Dat. (Zeitpunkt)* since; *(Zeitspanne)* for; **ich bin ~ zwei Wochen hier** I've been here [for] two weeks; **2.** *Konj.* since; ~ **du hier wohnst** since you have been living here; **seit·dem 1.** *Adv.* since then; **2.** *Konj. s.* seit **2**

Seite die; ~, ~n **a)** side; **zur** *od.* **auf die ~ gehen** move aside *or* to one side; ~ **an ~:** side by side; **jmdm. zur ~ stehen** stand by sb.; **von allen ~n** *(auch fig.)* from all sides; **nach allen ~n** in all directions; *(fig.)* on all sides; **b)** *(Buch~, Zeitungs~)* page

Seiten-: ~**ansicht die** side view; ~**hieb der** *(fig.)* side-swipe **(auf +** *Akk.* at); ~**ruder das** *(Flugw.)* rudder

seitens *Präp. mit Gen. (Papierdt.)* on the part of

Seiten-: ~**sprung der** infidelity; ~**straße die** side-street; ~**wind der;** *o. Pl.* side wind; cross-wind; ~**zahl die** **a)** page number; **b)** *(Anzahl der Seiten)* number of pages

seit·her *Adv.* since then
seitlich 1. *Adj.* at the side *(postpos.);* 2. *adv. (an der Seite)* at the side; *(von der Seite)* from the side; *(nach der Seite)* to the side; **seit·wärts** *Adv.* sideways
Sekretär der; ~s, ~e a) secretary; b) *(Schreibschrank)* bureau *(Brit.);* **Sekretariat** das; ~|e|s, ~e [secretary's/secretaries'] office; **Sekretärin** die; ~, ~nen secretary
Sekt der; ~|e|s, ~e high-quality sparkling wine; ≈ champagne
Sekte die; ~, ~n sect
sekundär 1. *Adj.* secondary; 2. *adv.* secondarily; **Sekunde** die; ~, ~n a) *(auch Math., Musik)* second; b) *(ugs.: Augenblick)* second; moment; **Sekunden·zeiger** der second hand
selb... *Demonstrativpron.* same; **selber** *indekl. Demonstrativpron.* s. **selbst** 1; **selbst** 1. *indekl. Demonstrativpron.* myself / yourself / himself / herself / itself / ourselves / yourselves / themselves; **von** ~: automatically; 2. *Adv.* even
Selbst·achtung die self-respect
selb·ständig 1. *Adj.* independent; self-employed *(business man, tradesman, etc.);* **sich** ~ **machen** set up on one's own; 2. *adv.* independently; ~ **denken** think for oneself; **Selbständigkeit** die; ~: independence
selbst-, Selbst-: ~**auslöser** der *(Fot.)* delayed-action shutter release; ~**bedienung** die self-service *no art.;* ~**befriedigung** die masturbation *no art.;* ~**beherrschung** die self-control *no art.;* ~**bewußt** 1. *Adj.* self-confident; 2. *adv.* self-confidently; ~**bewußtsein** das self-confidence *no art.;* ~**erkenntnis** die; *o. Pl.* self-knowledge *no art.;* ~**gefällig** 1. *Adj.* self-satisfied; smug; 2. *adv.* smugly; ~**gefälligkeit** die; *o. Pl.* self-satisfaction; smugness; ~**gemacht** *Adj.* home-made; ~**gespräch** das conversation with oneself; ~**los** 1. *Adj.* selfless; 2. *adv.* selflessly; unselfishly; ~**mord** der suicide *no art.;* ~**mörder** der suicide; ~**sicher** 1. *Adj.* self-confident; 2. *adv.* in a self-confident manner; ~**süchtig** 1. *Adj.* selfish; 2. *adv.* selfishly; ~**tätig** 1. *Adj.* automatic; 2. *adv.* automatically; ~**verständlich** 1. *Adj.* natural; **etw. für** ~**verständlich halten** regard sth. as a matter of course; *(für gegeben hinnehmen)* take sth. for granted; 2. *adv.*

naturally; of course; ~**vertrauen** das self-confidence; ~**verwaltung** die self-government *no art.;* ~**zweck** der; *o. Pl.* end in itself
selig 1. *Adj.* a) *(Rel.)* blessed; b) *(tot)* late [lamented]; c) *(glücklich)* blissful *(idleness, slumber, etc.);* blissfully happy *(person);* 2. *adv.* blissfully; **Seligkeit** die; ~, ~en bliss *no pl.;* [blissful] happiness *no pl.*
Sellerie der; ~s, ~|s| *od.* die; ~, ~: celeriac; *(Stangen~)* celery
selten 1. *Adj.* rare; infrequent *(visit, visitor);* 2. *adv.* a) rarely; b) *(sehr)* exceptionally; uncommonly; **Seltenheit** die; ~, ~en rarity; **Seltenheits·wert** der; ~|es| rarity value
Selters·wasser das seltzer [water]
seltsam 1. *Adj.* strange; odd; 2. *adv.* strangely
Semester das; ~s, ~: semester
Semikolon das; ~s, ~s semicolon
Seminar das; ~s, ~e a) seminar (über + *Akk.* on); b) *(Institut)* department
Semmel die; ~, ~n *(bes. österr., bayr., ostmd.)* [bread] roll; **Semmel·knödel** der *(bayr., österr.)* bread dumpling
Senat der; ~|e|s, ~e senate; **Senator** der; ~s, ~en
[1]**senden** *unr. (auch regelm.) tr. V. (geh.)* send
[2]**senden** *regelm. (schweiz. unr.) tr., itr. V.* broadcast *(programme, play, etc.);* transmit *(signals, Morse, etc.);* **Sender** der; ~s, ~: [broadcasting] station; *(Anlage)* transmitter
Sende-: ~**reihe** die series [of programmes]; ~**schluß** der close-down
Sendung die; ~, ~en a) consignment; b) *(Rundf., Ferns.)* programme
Senf der; ~|e|s, ~e mustard
senior *indekl. Adj.; nach Personennamen* senior; **Senior** der; ~s, ~en a) *(Kaufmannsspr.)* senior partner; b) *(Sport)* senior [player]; c) *(Rentner)* senior citizen; **Senioren·heim** das home for the elderly
Senke die; ~, ~n hollow; **senken** 1. *tr. V.* lower; 2. *refl. V. (curtain, barrier, etc.)* fall, come down; *(ground, building, road)* subside, sink; *(water-level)* fall, sink
senk-, Senk-: ~**fuß** der flat foot; ~**recht** 1. *Adj.* vertical; ~ **zu etw.** perpendicular to sth.; 2. *adv.* vertically; ~**rechte** die vertical; *(Geom.: Gerade)* perpendicular
Sensation die; ~, ~en sensation;

sensationell 1. *Adj.* sensational; **2.** *adv.* sensationally

Sense die; ~, ~n scythe

sensibel 1. *Adj.* sensitive; **2.** *adv.* sensitively; **Sensibilität die**; ~: sensitivity

sentimental 1. *Adj.* sentimental; **2.** *adv.* sentimentally; **Sentimentalität die**; ~, ~en sentimentality

separat 1. *Adj.* separate; self-contained ⟨*flat etc.*⟩; **2.** *adv.* separately

September der; ~|s|, ~: September

Serbe der; ~n, ~n Serb; Serbian; **Serbien (das)**; ~s Serbia; **serbisch** *Adj.* Serbian

Serenade die; ~, ~n serenade

Serie die; ~, ~n series; **serien·mäßig 1.** *Adj.* standard ⟨*product, model, etc.*⟩; **2.** *adv.* **a)** ~ gefertigt *od.* gebaut produced in series; **b)** *(nicht als Sonderausstattung)* ⟨*fitted, supplied, etc.*⟩ as standard

seriös *Adj.* respectable ⟨*person, hotel, etc.*⟩; trustworthy ⟨*firm, partner, etc.*⟩; serious ⟨*offer, applicant, artist, etc.*⟩

Serpentine die; ~, ~n hairpin bend

Serum das; ~s, Seren serum

¹Service [zɛr'viːs] **das**; ~, ~: |dinner *etc.*| service; **²Service** ['zøːɐ̯vɪs] **der**; ~, ~s ['zøːɐ̯vɪsɪs] *(Bedienung, Kundendienst)* service; **servieren** *tr. V.* serve; **Serviererin die**; ~, ~nen waitress; **Serviette** [zɛr'vi̯ɛtɐ] **die**; ~, ~n napkin; serviette *(Brit.)*

Servo·lenkung die power [-assisted] steering *no indef. art.*

Sesam der; ~s sesame seeds *pl.*

Sessel der; ~s, ~ **a)** armchair; **b)** *(österr.: Stuhl)* chair; **Sessel·lift der** chair-lift

seßhaft *Adj.* settled; ~ werden settle down

Set das *od.* **der**; ~|s|, ~s **a)** set, combination (aus of); **b)** *(Deckchen)* table- *or* place-mat

setzen 1. *refl. V.* **a)** sit [down]; **setzen Sie sich** sit down; take a seat; **sich aufs Sofa** *usw.* ~: sit on the sofa *etc.*; **b)** ⟨*coffee, froth, etc.*⟩ settle; ⟨*sediment*⟩ sink to the bottom; **2.** *tr. V.* **a)** put; **b)** *(einpflanzen)* plant ⟨*tomatoes, potatoes, etc.*⟩; **c)** *(aufziehen)* hoist ⟨*flag etc.*⟩; set ⟨*sails, navigation lights*⟩; **d)** *(Druckw.)* set ⟨*manuscript etc.*⟩; **3.** *itr. V.* **a)** *meist mit sein (springen)* leap, jump; **b)** **über einen Fluß** ~ *(mit einer Fähre o. ä.)* cross a river; **c)** *(beim Wetten)* bet; **auf ein Pferd/auf Rot** ~:

back a horse/put one's money on red; **Setzer der**; ~s, ~, **Setzerin die**; ~, ~nen *(Druckw.)* |type|setter; **Setzling der**; ~s, ~e seedling

Seuche die; ~, ~n epidemic

seufzen *itr., tr. V.* sigh; **Seufzer der**; ~s, ~: sigh

Sex der; ~|es| sex *no art.*; **Sexualität die**; ~: sexuality *no art.*; **sexuell 1.** *Adj.* sexual; **2.** *adv.* sexually

sezieren *tr. V.* dissect ⟨*corpse*⟩

sfr., *(schweiz. nur:)* **sFr.** *Abk.* Schweizer Franken

Shampoo [ʃam'puː], **Shampoon** [ʃam'poːn] **das**; ~s, ~s shampoo

Sherry ['ʃɛrɪ] **der**; ~s, ~s sherry

Show [ʃoʊ] **die**; ~, ~s show

siamesisch *Adj.* Siamese; **Siam·katze die** Siamese cat

Sibirien (das); ~s Siberia

sich *Reflexivpron. der 3. Pers. Sg. und Pl. Akk. und Dat.* **a)** himself/herself/itself/themselves; *(auf man bezogen)* oneself; *(auf das Anredepronomen Sie bezogen)* yourself/yourselves; ~ **freuen/wundern/schämen/täuschen** be pleased/surprised/ashamed/mistaken; ~ **sorgen** worry; **b)** *(reziprok)* one another, each other

Sichel die; ~, ~n sickle

sicher 1. *Adj.* **a)** safe ⟨*road, procedure, etc.*⟩; secure ⟨*job, investment, etc.*⟩; **b)** reliable ⟨*evidence, source*⟩; certain ⟨*proof*⟩; reliable, sure ⟨*judgment, taste, etc.*⟩; **c)** *(selbstbewußt)* [self-]assured ⟨*person, manner*⟩; **d)** *(gewiß)* certain; sure; **2.** *adv.* **a)** safely; **b)** *(zuverlässig)* reliably; ~ **|Auto| fahren** be a safe driver; **c)** *(selbstbewußt)* [self-]confidently; **3.** *Adv.* certainly; **sicher|gehen** *unr. itr. V.; mit sein* play safe; **Sicherheit die**; ~, ~en **a)** *o. Pl.* safety; *(der Öffentlichkeit)* security; **jmdn./etw. in** ~ **|vor etw. *(Dat.)*|** bringen save *or* rescue sb./sth. [from sth.]; **b)** *o. Pl. (Gewißheit)* certainty; **c)** *(Wirtsch.: Bürgschaft)* security

sicherheits-, Sicherheits-: ~**abstand der** *(Verkehrsw.)* safe distance between vehicles; ~**gurt der** seat-belt; ~**halber** *Adv.* to be on the safe side; ~**nadel die** safety-pin; ~**schloß das** safety lock

sicherlich *Adv.* certainly; **sichern** *tr. V.* make ⟨*door etc.*⟩ secure; *(garantieren)* safeguard ⟨*rights, peace*⟩; *(schützen)* protect ⟨*rights etc.*⟩; **sich** *(Dat.)* **etw.** ~: secure sth.; **sicher|stellen** *tr. V.* **a)** impound ⟨*goods, vehicle*⟩; **b)**

guarantee ⟨*supply, freedom, etc.*⟩; **Si-cherung** die; ~, ~**en a)** *o. Pl.* safeguarding; *(das Schützen)* protection; **b)** *(Elektrot.)* fuse; **c)** *(techn. Vorrichtung)* safety-catch

Sicht die; ~: view (**auf** + *Akk.*, **in** + *Akk.* of); **gute** *od.* **klare/schlechte** ~: good/poor visibility; **sichtbar 1.** *Adj.* visible; *(fig.)* apparent ⟨*reason*⟩; **2.** *adv.* visibly; **sichten** *tr. V.* sight; **sichtlich 1.** *Adj.* obvious; evident; **2.** *adv.* obviously; evidently; visibly ⟨*impressed*⟩

Sicht-: ~**verhältnisse** *Pl.* visibility *sing.*; ~**vermerk** der visa; ~**weite** die visibility *no art.*; **außer/in** ~**weite sein** be out of/in sight

sickern *itr. V.; mit sein* seep; *(spärlich fließen)* trickle

sie 1. *Personalpron.; 3. Pers. Sg. Nom. Fem.* she; *(betont)* her; *(bei Dingen, Tieren)* it; s. auch ¹**ihr; ihrer a.; 2.** *Personalpron.; 3. Pers. Pl. Nom.* they; *(betont)* them; *s. auch* **ihnen; ihrer b; 3.** *Akk. von* sie 1 her; *(bei Dingen, Tieren)* it; **4.** *Akk. von* sie 2 a them

Sie *Personalpron.; 3. Pers. Pl. Nom. u. Akk; Anrede an eine od. mehrere Personen* you; *s. auch* **Ihnen; Ihrer**

Sieb das; ~(e)s, ~e sieve; *(für Tee)* strainer; ¹**sieben** *tr. V.* **a)** sieve ⟨*flour etc.*⟩; riddle ⟨*sand, gravel, etc.*⟩; **b)** *(auswählen)* screen ⟨*candidates*⟩

²**sieben** *Kardinalz.* seven; **Sieben** die; ~, ~en seven

sieben-, Sieben-: ~**fach** *Vervielfältigungsz.* sevenfold; ~**mal** *Adj.* seven times; ~**sachen** *Pl. (ugs.)* **meine/deine** usw. ~**sachen** my/your *etc.* belongings *or (coll.)* bits and pieces

siebt... *Ordinalz.* seventh; **siebtel** *Bruchz.* seventh; **Siebtel** das, *schweiz. meist* der; ~s, ~: seventh; **siebtens** *Adv.* seventhly; **siebzehn** *Kardinalz.* seventeen; **siebzig** *Kardinalz.* seventy; **siebzigst...** *Ordinalz.* seventieth

siedeln *itr. V.* settle

sieden *unr. od. regelm. itr. V.* boil; **Siede·punkt** der *(auch fig.)* boiling-point

Siedler der; ~s, ~: settler; **Siedlung** die; ~, ~en **a)** *(Wohngebiet)* [housing] estate; **b)** *(Niederlassung)* settlement

Sieg der; ~(e)s, ~e victory, *(bes. Sport)* win (**über** + *Akk.* over)

Siegel das; ~s, ~: seal; *(von Behörden)* stamp

siegen *itr. V.* win; **über jmdn.** ~: gain

or win a victory over sb.; *(bes. Sport)* win against sb.; beat sb.; **Sieger** der; ~s, ~: winner; *(Mannschaft)* winners *pl.; (einer Schlacht)* victor; **Sieger-ehrung** die presentation ceremony; awards ceremony; **sieges·sicher 1.** *Adj.* confident of victory *pred.;* **2.** *adv.* confident of victory; **sieg-reich** *Adj.* victorious; winning ⟨*team*⟩; successful ⟨*campaign*⟩

sieh, siehe *Imperativ Sg. v.* sehen; **siehst** *2. Pers. Sg. Präsens v.* sehen; **sieht** *3. Pers. Sg. Präsens v.* sehen

Signal das; ~s, ~e signal; **signalisieren** *tr. V.* indicate ⟨*danger, change, etc.*⟩

Signatur die; ~, ~en **a)** initials *pl.; (Kürzel)* abbreviated signature; *(des Künstlers)* autograph; **b)** *(Unterschrift)* signature; **c)** *(in einer Bibliothek)* shelf-mark; **signieren** *tr. V.* sign; autograph ⟨*one's own work*⟩

Silbe die; ~, ~n syllable

Silber das; ~s **a)** silver; **b)** *(silbernes Gerät)* silver[ware]; **Silber·medaille** die silver medal; **silbern 1.** *Adj.* silver; silvery ⟨*moonlight, shade, gleam, etc.*⟩; **2.** *adv.* ⟨*shine, shimmer, etc.*⟩ with a silvery lustre; **Silber·papier** das silver paper

Silhouette [zi'luɛtə] die; ~, ~n silhouette

Silo der *od.* das; ~s, ~s silo

Silvester der *od.* das; ~s, ~: New Year's Eve

Simbabwe (das); ~s Zimbabwe

simpel 1. *Adj.* **a)** simple ⟨*question, task*⟩; **b)** *(beschränkt)* simple-minded ⟨*person*⟩; simple ⟨*mind*⟩; **2.** *adv.* **a)** simply; **b)** *(beschränkt)* in a simple-minded manner; **Simpel** der; ~s, ~ *(bes. südd. ugs.)* simpleton; fool

Sims der *od.* das; ~es, ~e ledge; sill; *(Kamin~)* mantelpiece

Simulant der; ~en, ~en malingerer; **simulieren 1.** *tr. V.* feign, sham ⟨*illness, emotion, etc.*⟩; simulate ⟨*situation, condition, etc.*⟩; **2.** *itr. V.* feign illness

simultan 1. *Adj.* simultaneous; **2.** *adv.* simultaneously

sind *1. u. 3. Pers. Pl. Präsens v.* ¹**sein**

Sinfonie die; ~, ~n symphony; **Sinfonie·orchester** das symphony orchestra

singen *unr. tr., itr. V.* sing

Singular der; ~s singular

Sing·vogel der songbird

sinken *unr. itr. V.; mit sein* **a)** ⟨*ship,*

sun⟩ sink, go down; ⟨*plane, balloon*⟩ descend, go down; **b)** *(nieder~)* fall; **c)** *(niedriger werden)* ⟨*temperature, level*⟩ fall, drop; **d)** *(an Wert verlieren; nachlassen; abnehmen)* fall, go down

Sinn der; ~|e|s, ~e **a)** sense; **b)** *Pl. (geh.: Bewußtsein)* senses; mind *sing.;* **nicht bei** ~**en sein** be out of one's senses *or* mind; **c)** *o. Pl. (Gefühl, Verständnis)* feeling; **d)** *o. Pl. (~gehalt, Bedeutung)* meaning; **e)** *(Ziel u. Zweck)* point; **Sinn·bild das** symbol

Sinnes-: ~**organ das** sense-organ; sensory organ; ~**täuschung die** trick of the senses

sinn·gemäß 1. *Adj.* **eine** ~**e** Übersetzung a translation which conveys the general sense; **2.** *adv.* **etw.** ~ **übersetzen/wiedergeben** translate the general sense of sth./give the gist of sth.;

sinnlich *Adj.* sensory ⟨*impression, perception, stimulus*⟩; sensual ⟨*love, mouth*⟩; sensuous ⟨*pleasure, passion*⟩; **Sinnlichkeit die;** ~: sensuality;

sinn·los 1. *Adj.* **a)** senseless; **b)** *(zwecklos)* pointless; **2.** *adv.* **a)** senselessly; **b)** *(zwecklos)* pointlessly; **Sinnlosigkeit die;** ~ **a)** senselessness; **b)** *(Zwecklosigkeit)* pointlessness; **sinn·voll 1.** *Adj.* **a)** *(vernünftig)* sensible; **b)** *(einen Sinn ergebend)* meaningful; **2.** *adv.* **a)** *(vernünftig)* sensibly; **b)** *(einen Sinn ergebend)* meaningfully

Sint·flut die Flood; Deluge; **sintflut·artig 1.** *Adj.* torrential; **2.** *adv.* in torrents

Sippe die; ~, ~n **a)** *(Völkerk.)* sib; **b)** *(ugs.: Verwandtschaft)* clan; **Sippschaft die;** ~, ~en *(ugs.) s.* Sippe b

Sirene die; ~, ~n siren

Sirup der; ~s, ~e syrup

Sitte die; ~, ~n **a)** *(Brauch)* custom; tradition; **b)** *(moralische Norm)* common decency; **c)** *Pl. (Benehmen)* manners; **sittlich 1.** *Adj.* moral; **2.** *adv.* morally; **Sittlichkeit die;** *o. Pl.* morality

Sittlichkeits-: ~**verbrechen das** sexual crime; ~**verbrecher der** sex offender

Situation die; ~, ~en situation

Sitz der; ~es, ~e **a)** seat; **b)** *(Verwaltungs~)* headquarters *sing. or pl.;* **c)** *(von Kleidungsstücken)* fit

sitzen *unr. itr. V.; südd., österr., schweiz. mit sein* **a)** sit; **b)** *(sein)* be; **c)** *([gut] passen)* fit

sitzen-: ~|**bleiben** *unr. itr. V. (ugs.)*

a) *(nicht versetzt werden)* stay down [a year]; **b)** *(unverheiratet bleiben)* be left on the shelf; **c) auf etw.** *(Dat.)* ~**bleiben** *(für etw. keinen Käufer finden)* be left *or (coll.)* stuck with sth.; ~|**lassen** *unr. tr. V. (ugs.)* **a)** *(nicht heiraten)* jilt; **b)** *(im Stich lassen)* leave in the lurch; **c) etw. nicht auf sich** *(Dat.)* ~**lassen** not take sth.

Sitzplatz der seat; **Sitzung die;** ~, ~en meeting; *(Parlaments~)* sitting; session; **Sitzungs·saal der** conference hall

Skala die; ~, **Skalen** scale

Skalp der; ~s, ~e scalp

Skalpell das; ~s, ~e scalpel

skalpieren *tr. V.* scalp

Skandal der; ~s, ~e scandal; **skandalös** *Adj.* scandalous

Skandinavien (das); ~s Scandinavia; **Skandinavier der;** ~s, ~: Scandinavian; **skandinavisch** *Adj.* Scandinavian

Skat der; ~|e|s, ~e *od.* ~s skat

Skelett das; ~|e|s, ~e skeleton

Skepsis die; ~: scepticism; **skeptisch 1.** *Adj.* sceptical; **2.** *adv.* sceptically

Ski [ʃiː] **der;** ~s, ~er *od.* ~: ski; ~ **laufen** *od.* **fahren** ski

Ski-: ~**läufer der** skier; ~**lehrer der** ski-instructor; ~**lift der** ski-lift; ~**springen das;** ~s ski-jumping *no art.*

Skizze die; ~, ~n sketch; **Skizzen·block der** sketch-pad; **skizzieren** *tr. V.* sketch

Sklave der; ~n, ~n slave; **Sklaven·händler der** slave-trader; **Sklaverei die;** ~: slavery *no art.;* **Sklavin die;** ~, ~nen slave; **sklavisch 1.** *Adj.* slavish; **2.** *adv.* slavishly

Skonto der *od.* **das;** ~s, ~s *(Kaufmannsspr.)* [cash] discount

Skorbut der; ~|e|s scurvy *no art.*

Skorpion der; ~s, ~e scorpion

Skrupel der; ~s, ~: scruple; **skrupel·los 1.** *Adj.* unscrupulous; **2.** *adv.* unscrupulously; **Skrupellosigkeit die;** ~: unscrupulousness

Skulptur die; ~, ~en sculpture

Slalom der; ~s, ~s slalom

Slawe der; ~n, ~n Slav; **slawisch** *Adj.* Slav[ic]; Slavonic

Slip der; ~s, ~s briefs *pl.*

Slowake der; ~n, ~n Slovak; **Slowakei die;** ~: Slovakia *no art.*

Smaragd der; ~|e|s, ~e emerald

Smoking der; ~s, ~s dinner-jacket *or*
(Amer.) tuxedo and dark trousers

so 1. *Adv.* **a)** *(auf diese Weise; in, von
dieser Art)* like this/that; this/that
way; **weiter so!** carry on in the same
way!; **b)** *(dermaßen, überaus)* so; **c)**
(genauso) as; **so gut ich konnte** as best
I could; **d)** *(ugs.: solch)* such; **so ein
Idiot!** what an idiot!; **so einer/eine/
eins** one like that; **e)** *betont (eine Zäsur
ausdrückend)* right; OK *(coll.);* **g)**
(ugs.: schätzungsweise) about; **2.**
Konj. **so daß ...** *(damit)* so that ...; *(und
deshalb)* and so ...; **3.** *Partikel* **a)** just;
ach, das hab' ich nur so gesagt oh, I
didn't mean anything by that; **b)** *(in
Aufforderungssätzen verstärkend)* **so
komm doch** come on now

So. *Abk.* Sonntag Sun.

s. o. *Abk.* **siehe oben**

sobald *Konj.* as soon as

Socke die; ~, ~n sock

Sockel der; ~s, ~ **a)** *(einer Säule,
Statue)* plinth; **b)** *(unterer Teil eines
Hauses, Schrankes)* base

Soda·wasser das; *Pl.* **Sodawässer**
soda; soda-water

Sod·brennen das; ~s heartburn

so·eben *Adv.* just

Sofa das; ~s, ~s sofa; settee

so·fern *Konj.* provided [that]

soff *1. u. 3. Pers. Sg. Prät. v.* **saufen**

so·fort *Adv.* immediately; at once;
sofortig *Adj. (unmittelbar)* immediate

sog *1. u. 3. Pers. Sg. Prät. v.* **saugen**;
Sog der; ~[e]s, ~e suction; *(bei Schiffen)* wake; *(bei Fahr-, Flugzeugen)*
slip-stream; *(von Wasser, auch fig.)*
current

so·gar *Adv.* even

so·genannt *Adj.* so-called

so·gleich *Adv.* immediately; at once

Sohle die; ~, ~n **a)** *(Schuh~)* sole;
(Einlege~) insole; **b)** *(Fuß~)* sole [of
the foot]

Sohn der; ~es, **Söhne** son

Soja-: ~**bohne** die soy[a] bean; ~**soße** die soy[a] sauce

so·lang[e] *Konj.* so *or* as long as

Solarium das; ~s, **Solarien** solarium

solch *Demonstrativpron.* **a)** *attr.* such;
das macht ~en Spaß! it's so much
fun!; **b)** *alleinstehend* ~**e wie die**
people like that

Sold der; ~[e]s, ~e [military] pay

Soldat der; ~en, ~en soldier; **Solda-
ten·friedhof** der military *or* war
cemetery; **Soldatin** die; ~, ~nen [fe-

male *or* woman] soldier; **soldatisch**
1. *Adj.* military ⟨*discipline, expression,
etc.*⟩; soldierly ⟨*figure, virtue*⟩; **2.** *adv.*
in a military manner

Söldner der; ~s, ~: mercenary

solidarisch 1. *Adj.* ~**es Verhalten** zeigen show one's solidarity; **2.** *adv.* ~
handeln/sich ~ verhalten act in/show
solidarity; **solidarisieren** *refl. V.*
show [one's] solidarity; **Solidarität**
die; ~: solidarity

solide 1. *Adj.* **a)** solid; sturdy ⟨*shoes,
material*⟩; [good-]quality ⟨*goods*⟩; **b)**
(gut fundiert) sound ⟨*work, education,
knowledge*⟩; solid ⟨*firm*⟩; **c)** *(anständig)* respectable ⟨*person, life, profession*⟩; **2.** *adv.* **a)** solidly ⟨*built*⟩;
sturdily ⟨*made*⟩; **b)** *(gut fundiert)*
soundly ⟨*educated, constructed*⟩; **c)**
(anständig) ⟨*live*⟩ respectably, steadily

Solist der; ~en, ~en soloist

Soll das; ~[s], ~[s] **a)** *(Bankw.)* debit; **b)**
(Arbeits~) quota; **sein ~ erfüllen** *od.*
erreichen achieve one's target

sollen 1. *unr. Modalverb;* **2.** *Part.* ~ **a)**
*(bei Aufforderung, Anweisung, Auf-
trag)* **was soll ich als nächstes tun?**
what should I do next?; |**sagen Sie
ihm,|** **er soll hereinkommen** tell him to
come in; **b)** *(bei Wunsch, Absicht, Vorhaben)* **das sollte ein Witz sein** that
was meant to be a joke; **was soll denn
das heißen?** what is that supposed to
mean?; **c)** *(bei Ratlosigkeit)* **was soll
ich nur machen?** what am I to do?; **d)**
(Notwendigkeit ausdrückend) **man soll
so etwas nicht unterschätzen** it
shouldn't be taken so lightly; **e)** *häufig im Konjunktiv II (Erwartung, Wün-
schenswertes ausdrückend)* **du solltest
dich schämen** you ought to be
ashamed of yourself; **das hättest du
besser nicht tun ~:** it would have been
better if you hadn't done that; **f)**
(jmdm. beschieden sein) **er sollte seine
Heimat nicht wiedersehen** he was
never to see his homeland again; **g)** *im
Konjunktiv II (eine Möglichkeit ausdrückend)* **wenn du ihn sehen solltest,
sage ihm bitte ...:** if you should see
him, please tell him ...; **h)** *im Präsens
(sich für die Wahrheit nicht verbür-
gend)* **das Restaurant soll sehr teuer
sein** the restaurant is supposed *or* said
to be very expensive; **i)** *im Konjunktiv
II (Zweifel ausdrückend)* **sollte das
sein Ernst sein?** is he really being serious?; **j)** *(können)* **mir soll es gleich
sein** it's all the same to me; **2.** *tr., itr.*

V. was soll das? what's the idea?; was soll ich dort? what would I do there?

Solo das; ~s, ~s *od.* Soli solo

so·mit [auch: '--] *Adv.* consequently; therefore

Sommer der; ~s, ~: summer

Sommer·ferien *Pl.* summer holidays; sommerlich 1. *Adj.* summer; summery ⟨*warmth, weather*⟩; summer's *attrib.* ⟨*day, evening*⟩; 2. *adv.* es war ~ warm it was as warm as summer

sommer-, Sommer-: ~reifen der standard tyre; ~schluß·verkauf der summer sale/sales; ~sprosse die freckle; ~sprossig *Adj.* freckled; ~zeit die *(Uhrzeit)* summer time

Sonate die; ~, ~n *(Musik)* sonata

Sonde die; ~, ~n probe; *(zur Ernährung)* tube

Sonder·angebot das special offer; sonderbar 1. *Adj.* strange; odd; 2. *adv.* strangely; oddly; Sonder·fall der special case

sonder·gleichen *Adv., nachgestellt* eine Frechheit/Unverschämtheit ~: the height of cheek/impudence

sonderlich *Adv.* particularly; Sonderling der; ~s, ~e strange *or* odd person; Sonder·müll der hazardous waste

¹sondern *tr. V. (geh.)* separate **(von** from)

²sondern *Konj.* but; **nicht nur ...,** ~ [auch] ...: not only ... but also ...

Sonder-: ~schule die special school; ~zug der special train

sondieren *tr. V.* sound out

Sonett das; ~[e]s, ~e sonnet

Sonn·abend der *(bes. nordd.)* Saturday; sonn·abends *Adv.* on Saturday[s]

Sonne die; ~, ~n sun; *(Licht der ~)* sun[light]; sonnen *refl. V.* sun oneself

sonnen-, Sonnen-: ~aufgang der sunrise; ~baden *itr. V.* sunbathe; ~blume die sunflower; ~brand der sunburn *no indef. art.;* ~brille die sun-glasses *pl.;* ~energie die solar energy; ~finsternis die solar eclipse; ~hut der sun-hat; ~licht das sunlight; ~öl das sun-oil; ~schein der *o. Pl.* sunshine; ~schirm der sunshade; ~stich der sunstroke *no indef. art.;* ~strahl der ray of sun[shine]; ~uhr die sundial; ~untergang der sunset

sonnig *Adj.* sunny

Sonn·tag der Sunday; sonn·täg-lich 1. *Adj.* Sunday *attrib.;* 2. *adv.* ~ gekleidet dressed in one's Sunday best; sonntags *Adv.* on Sunday[s]

sonst *Adv.* **a) der ~ so freundliche Mann ...:** the man, who is/was usually so friendly, ...; **alles war wie ~:** everything was [the same] as usual; **~ noch was?** *(ugs., auch iron.)* anything else?; **wer/was/wie/wo [denn]** ~? who/what/how/where else?; **b)** *(andernfalls)* otherwise; or; **sonstig...** *Adj.; nicht präd.* other; further

sonst-: ~**was** *Indefinitpron. (ugs.)* anything else; ~**wer** *Indefinitpron. (ugs.)* somebody else; *(fragend, verneinend)* anybody else; ~**wo** *Adv. (ugs.)* somewhere else; *(fragend, verneinend)* anywhere else

so·oft *Konj.* whenever

Sopran der; ~s, ~e *(Musik)* soprano *(im Chor)* sopranos *pl.;* Sopranistin die; ~, ~nen soprano

Sorge die; ~, ~n worry; **keine ~!** don't [you] worry!; sorgen 1. *refl. V.* worry (um about); 2. *itr. V.* **für jmdn./etw. ~:** take care of sb./sth.

sorgen-, Sorgen-: ~frei 1. *Adj.* carefree; 2. *adv.* ~frei leben live in a carefree manner; ~kind das *(auch fig.)* problem child; ~voll 1. *Adj.* worried; 2. *adv.* worriedly

Sorg·falt die; ~: care; sorg·fältig 1. *Adj.* careful; 2. *adv.* carefully

sorg·los 1. *Adj.* **a)** *(ohne Sorgfalt)* careless; **b)** *(unbekümmert)* carefree; 2. *adv.* ~ mit etw. umgehen treat sth. carelessly; Sorglosigkeit die; ~ **a)** *(Mangel an Sorgfalt)* carelessness; **b)** *(Unbekümmertheit)* carefreeness; sorgsam 1. *Adj.* careful; 2. *adv.* carefully

Sorte die; ~, ~n **a)** sort; type; kind; **b)** *Pl. (Devisen)* foreign currency *sing.*

sortieren *tr. V.* sort [out] ⟨*pictures, letters, washing, etc.*⟩; grade ⟨*goods etc.*⟩

Sortiment das; ~[e]s, ~e range **(an +** *Dat.* of)

so·sehr *Konj.* however much

Soße die; ~, ~n sauce; *(Braten~)* gravy; sauce; *(Salat~)* dressing

sott *1. u. 3. Pers. Sg. Prät. v.* sieden

Souffleur [zu'flø:ɐ̯] der; ~s, ~e, Souffleuse [zu'flø:zə] die; ~, ~n prompter; soufflieren [zu'fli:rən] *tr. V.* prompt

Souvenir [suvə'ni:ɐ̯] das; ~s, ~s souvenir

souverän [zuvə'rɛ:n] *Adj.* sovereign; Souveränität die; ~: sovereignty

so·viel 1. *Konj.* as *or* so far as; **2.** *Indefinitpron.* ~ **wie** *od.* **als** as much as; **halb/doppelt** ~: half/twice as much
so·weit 1. *Konj.* **a)** as *or* so far as; **b)** *(in dem Maße, wie)* [in] so far as; **2.** *Adv.* by and large; *(bis jetzt)* up to now; ~ **sein** *(ugs.)* be ready
so·wenig *Indefinitpron.* ~ **wie** *od.* **als** **möglich** as little as possible
so·wie *Konj.* **a)** *(und)* as well as; **b)** *(sobald)* as soon as
so·wie·so *Adv.* anyway
sowjetisch *Adj.* Soviet
Sowjet·union die *(1922–1991)* Soviet Union
so·wohl *Konj.* ~ ... **als** *od.* **wie** |auch|...: both ... and ...; ... as well as ...
sozial 1. *Adj.* social; **2.** *adv.* socially
sozial-, Sozial-: ~**abgaben** *Pl.* social welfare contributions; ~**arbeiter der** social worker; ~**demokrat der** Social Democrat; ~**demokratisch** *Adj.* social democratic; ~**hilfe die** social welfare
Sozialismus der; ~: socialism *no art.;* **Sozialist der;** ~**en,** ~**en, Sozialistin die;** ~, ~**nen** socialist; **sozialistisch 1.** *Adj.* socialist; **2.** ~ **regierte** **Länder** countries with socialist governments
Sozial-: ~**politik die** social policy; ~**produkt das** *(Wirtsch.)* national product; ~**staat der** welfare state
Soziologe der; ~**n,** ~**n** sociologist; **Soziologie die;** ~: sociology; **soziologisch 1.** *Adj.* sociological; **2.** *adv.* sociologically
Sozius der; ~, ~**se a)** *Pl. auch:* **Sozii** *(Wirtsch.: Teilhaber)* partner; **b)** *(beim Motorrad)* pillion
so·zu·sagen *Adv.* as it were
Spachtel der; ~**s,** ~ *od.* **die;** ~, ~**n** putty-knife; *(zum Malen)* palette-knife; **spachteln** *tr. V.* **a)** stop, fill 〈*hole, crack, etc.*〉; smooth over 〈*wall, panel, surface, etc.*〉; **b)** *(ugs.: essen)* put away *(coll.)* 〈*food, meal*〉
Spagat der *od.* **das;** ~|e|s, ~e splits *pl.*
Spaghetti *Pl.* spaghetti *sing.*
spähen *itr. V.* peer; *(durch ein Loch, eine Ritze usw.)* peep; **Späher der;** ~**s,** ~ *(Milit.)* scout; *(Posten)* lookout; *(Spitzel)* informer
Spalier das; ~**s,** ~**e a)** trellis; **b)** *(Ehren-)* guard of honour; ~ **stehen** line the route; 〈*soldiers*〉 form a guard of honour
Spalt der; ~|e|s, ~e opening; *(im Fels)* fissure; crevice; *(zwischen Vorhängen)*

chink; gap; *(langer Riß)* crack; **Spalte die;** ~, ~**n a)** crack; *(Fels~)* crevice **b)** *(Druckw.)* column; **spalten** *unr.* *(auch regelm.) tr., refl. V.* split
Span der; ~|e|s, **Späne** *(Hobel~)* shaving
Span·ferkel das sucking pig
Spange die; ~, ~**n** clasp; *(Haar~)* hair-slide *(Brit.);* barrette *(Amer.);* *(Arm~)* bracelet; bangle
Spaniel ['ʃpa:niəl] **der;** ~**s,** ~**s** spaniel
Spanien ['ʃpa:niən] **(das);** ~**s** Spain; **Spanier der;** ~**s,** ~: Spaniard; **spanisch** *Adj.* Spanish
Span·korb der chip basket; chip
spann *1. u. 3. P. Sing. Prät. v.* **spinnen**
spannen 1. *tr. V.* **a)** tighten 〈*violin string, violin bow, etc.*〉; draw 〈*bow*〉; tension 〈*spring, tennis net, drumhead, saw-blade*〉; stretch 〈*fabric, shoe, etc.*〉; draw *or* pull 〈*line*〉 tight *or* taut; flex 〈*muscle*〉; cock 〈*gun, camera shutter*〉; **b)** *(befestigen)* put up 〈*washing-line*〉; stretch 〈*net, wire, tarpaulin, etc.*〉 (**über** + *Akk.* over); **c)** *(schirren)* harness (**vor, an** + *Akk.* to); **2.** *refl. V.* **a)** become *or* go taut 〈*muscles*〉 tense; **b)** *(geh.: sich wölben)* **sich über** etw. *(Akk.)* ~: span sth.; **3.** *itr. V.* 〈*clothing*〉 be [too] tight; 〈*skin*〉 be taut; **spannend 1.** *Adj.* exciting; *(stärker)* thrilling; **2.** *adv.* excitingly; *(stärker)* thrillingly; **Spannung die;** ~, ~**en a)** *o. Pl.* excitement; *(Neugier)* suspense; **b)** *o. Pl. (eines Romans, Films usw.)* suspense; **c)** *(Zwistigkeit, Nervosität)* tension; **d)** *(Elektrot.)* voltage;
Spann·weite die [wing-]span
Spar·buch das savings book
sparen 1. *tr. V.* save; **2.** *itr. V.* **a)** save; **für** *od.* **auf** etw. *(Akk.)* ~: save up for sth.; **b)** *(sparsam wirtschaften)* economize (**mit** on); **an** etw. *(Dat.)* ~: be sparing with sth.; *(beim Einkauf)* economize on sth.; **Sparer der;** ~**s,** ~: saver
Spargel der; ~**s,** ~, *schweiz. auch* **die;** ~, ~**n** asparagus *no pl., no indef. art.*
Spar-: ~**groschen der** *(ugs.)* nest-egg; savings *pl.;* ~**kasse die** savings bank; ~**konto das** savings *or* deposit account
spärlich 1. *Adj.* sparse 〈*vegetation, beard, growth*〉; thin 〈*hair, applause*〉; scanty 〈*left-overs, knowledge, news, evidence, clothing*〉; poor 〈*lighting*〉; **2.** *adv.* sparsely, thinly 〈*populated, covered*〉; poorly 〈*lit, attended*〉; scantily 〈*dressed*〉

sparsam 1. *Adj.* thrifty ⟨*person*⟩; *(wirtschaftlich)* economical; **mit etw. ~ sein** be economical with sth.; **2.** *adv.* ~ **mit der Butter/dem Papier umgehen** use butter/paper sparingly; economize on butter/paper; **Sparsamkeit die;** ~: thrift[iness]; *(Wirtschaftlichkeit)* economicalness

Sparte die; ~, ~**n a)** *(Teilbereich)* area; *(eines Geschäfts)* line [of business]; **b)** *(Rubrik)* section

Spaß der; ~**es, Späße a)** *o. Pl. (Vergnügen)* fun; ~ **an etw.** *(Dat.)* **haben** enjoy sth.; [jmdm.] ~ **machen** be fun [for sb.]; **viel ~!** have a good time!; **b)** *(Scherz)* joke; *(Streich)* prank; **er macht nur ~:** he's only joking; ~ **beiseite!** joking aside; ~ **muß sein!** there's no harm in a joke; **im** *od.* **zum** *od.* **aus** ~: as a joke; for fun; **spaßen** *itr. V.* **a)** *(Spaß machen)* joke; **b) er läßt nicht mit sich ~:** he won't stand for any nonsense; **mit ihm/damit ist nicht zu ~:** he/it is not to be trifled with; **spaßes·halber** *Adv.* for the fun of it; for fun; **spaßig** *Adj.* funny; comical; amusing

spät 1. *Adj.* late; **wie ~ ist es?** what time is it?; **2.** *adv.* late; ~ **am Abend** late in the evening

Spaten der; ~**s,** ~: spade

später 1. *Adj.* **a)** later ⟨*years, generations, etc.*⟩; **b)** *(zukünftig)* future ⟨*owner, wife, etc.*⟩; **2.** *Adv.* later; **spätestens** *Adv.* at the latest

Spatz der; ~**en,** ~**en a)** sparrow; **b)** *(fam.: Liebling)* pet

Spätzle *Pl.* spaetzle; *kind of noodles*

spazieren *itr. V.; mit sein* stroll

spazieren-: ~|**fahren 1.** *unr. itr. V.; mit sein* go for a ride; **2.** *tr. V.* **ein Kind [im Kinderwagen]** ~**fahren** take a baby for a walk [in a pram]; ~|**gehen** *unr. itr. V.; mit sein* go for a walk

Spazier-: ~**gang** der walk; ~**gänger** der; ~**s,** ~: person out for a walk

SPD [ɛspe:'de:] die; ~ *Abk.* Sozialdemokratische Partei Deutschlands SPD

Specht der; ~|e|**s,** ~**e** woodpecker

Speck der; ~|e|**s,** ~**e a)** bacon fat; *(Schinken~)* bacon; **b)** *(ugs. scherzh.: Fettpolster)* fat; flab *(sl.);* **speckig** *Adj.* greasy

Spediteur [ʃpedi'tø:ɐ] der; ~**s,** ~**e** carrier; haulage contractor; *(Möbel~)* furniture-remover

Speer der; ~|e|**s,** ~**e a)** spear; **b)** *(Sportgerät)* javelin

Speichel der; ~**s** saliva

Speicher der; ~**s,** ~ **a)** storehouse; *(Lagerhaus)* warehouse; **b)** *(südd.: Dachboden)* loft; **c)** *(Elektronik)* memory; **speichern** *tr. V.* store

speien *(geh.) unr. tr., itr. V.* spit

Speise die; ~, ~**n a)** *(Gericht)* dish; **b)** *o. Pl. (geh.: Nahrung)* food

Speise-: ~**gaststätte** die restaurant; ~**kammer** die larder; ~**karte** die menu; ~**lokal** das restaurant

speisen *(geh.)* **1.** *itr. V.* eat; *(dinieren)* dine; **2.** *tr. V.* eat; *(dinieren)* dine on

Speise-: ~**saal** der dining-hall; *(im Hotel, in einer Villa usw.)* dining-room; ~**wagen** der restaurant car *(Brit.);* ~**zettel** der menu

Spektakel der; ~**s,** ~ *(ugs.) (Lärm)* row *(coll.);* rumpus *(coll.);* **spektakulär 1.** *Adj.* spectacular; **2.** *adv.* spectacularly

Spekulation die; ~, ~**en** speculation; **spekulieren** *itr. V.* **a)** *(ugs.)* **darauf** ~**, etw. tun zu können** count on being able to do sth.; **b)** *(Wirtsch.)* speculate **(mit in)**

Spelunke die; ~, ~**n** *(ugs. abwertend)* dive *(coll.)*

Spelze die; ~, ~**n** *(des Getreidekorns)* husk

Spende die; ~, ~**n** donation; contribution; **spenden** *tr., itr. V.* **a)** donate; give; **b)** *(fig. geh.)* give ⟨*light*⟩; afford; give ⟨*shade*⟩; give off ⟨*heat*⟩; **Spender der;** ~**s,** ~, **Spenderin die;** ~, ~**nen** donor; donator; **spendieren** *tr. V. (ugs.)* get, buy ⟨*drink, meal, etc.*⟩; stand ⟨*round*⟩

Spengler der; ~**s,** ~ *(südd., österr., schweiz.) s.* **Klempner**

Sperling der; ~**s,** ~**e** sparrow

Sperma das; ~**s, Spermen** sperm; semen

sperr·angel·weit *Adv. (ugs.)* ~ **offen** *od.* **geöffnet** wide open

Sperre die; ~, ~**n a)** barrier; *(Straßen~)* road-block; *(Milit.)* obstacle; **b)** *(fig.) (Handels~)* embargo; *(Import~, Export~)* blockade; *(Nachrichten~)* [news] black-out; **sperren 1.** *tr. V.* **a)** close; close off ⟨*area*⟩; block ⟨*entrance, access, etc.*⟩; lock ⟨*mechanism etc.*⟩; **b)** cut off ⟨*water, gas, electricity, etc.*⟩; **c)** *(Bankw.)* stop ⟨*cheque, overdraft facility*⟩; freeze ⟨*bank account*⟩; **d)** *(ein~)* **ein Tier/jmdn. in etw.** *(Akk.)* ~: shut an animal/sb. in sth.; **e)** *(Sport: von der Teilnahme ausschließen)* ban; **f)** *(Druckw.: spationieren)* print ⟨*word,*

text⟩ with the letters spaced; **2.** *refl. V.*
sich [gegen etw.] ~: balk [at sth.];
Sperr·holz das plywood; **sperrig**
Adj. unwieldy

Sperr-: ~**müll der** bulky refuse *(for
which there is a separate collection service);* ~**sitz der** *(im Kino)* seat in the
back stalls; *(im Zirkus)* front seat; *(im
Theater)* seat in the front stalls;
~**stunde die** closing time

Spesen *Pl.* expenses; **auf** ~: on expenses

Spezi der; ~**s,** ~[s] *(südd., österr.,
schweiz. ugs.)* [bosom] pal *(coll.);*
chum *(coll.)*

spezialisieren *refl. V.* specialize (**auf**
+ *Akk.* in); **Spezialist der;** ~**en,** ~**en**
specialist; **Spezialität die;** ~, ~**en**
speciality; **speziell 1.** *Adj.* special;
specific ⟨*question, problem, etc.*⟩; **2.**
Adv. especially; *(eigens)* specially;
spezifisch 1. *Adj.* specific; characteristic ⟨*smell, style*⟩; **2.** *adv.* specifically

spicken *tr. V.* lard

spie *1. u. 3. Pers. Sg. Prät. v.* **speien**

Spiegel der; ~**s,** ~ **a)** mirror; **b)** *(Was-
ser~, fig.: Konzentration)* level

spiegel-, Spiegel-: ~**bild das** reflection; ~**blank** *Adj.* shining; ~**ei das**
fried egg; ~**glatt** *Adj.* like glass *postpos.;* as smooth as glass *postpos.*

spiegeln 1. *itr. V.* **a)** *(glänzen)* shine;
gleam; **b)** *(als Spiegel wirken)* reflect
the light; **2.** *tr. V.* reflect; mirror; **3.**
refl. V. be mirrored *or* reflected

Spiegel·reflex·kamera die reflex
camera

Spiel das; ~[e]s, ~e **a)** play; **b)**
(Glücks~; Gesellschafts~) game;
(Wett~) game; match; **auf dem** ~ **stehen be** at stake; **etw. aufs** ~ **setzen** put
sth. at stake; risk sth.; **Spiel·bank
die;** *Pl.* **...banken** casino; **spielen 1.**
itr. V. **a)** play; **auf der Gitarre** ~: play
the guitar; **um Geld** ~: play for
money; **b)** *(als Schauspieler)* act; perform; **c) der Roman/Film spielt im 17.
Jahrhundert/in Berlin** the novel/film
is set in the 17th century/in Berlin; **d)**
(fig.) **das Blau spielt ins Violette** the
blue is tinged with purple; **2.** *tr. V.* **a)**
play; **Cowboy** ~: play at being a cowboy; **Geige** *usw.* ~: play the violin
etc.; **b)** *(aufführen, vorführen)* put on
⟨*play*⟩; show ⟨*film*⟩; perform ⟨*piece of
music*⟩; play ⟨*record*⟩; **den Beleidig-
ten/Unschuldigen** ~ *(fig.)* act of-
fended/play the innocent; **spielend**

Adv. easily; **Spieler der;** ~**s,** ~:
player; *(Glücks~)* gambler; **Spiele-
rei die;** ~, ~**en a)** *o. Pl.* playing *no
art.; (im Glücksspiel)* gambling *no art.;*
b) eine ~ **mit Worten/Zahlen** playing
[around] with words/numbers;
Spielerin die; ~, ~**nen** *s.* **Spieler**

Spiel-: ~**feld das** field; pitch *(Brit.);
(Tennis, Squash, Volleyball usw.)*
court; ~**film der** feature film; ~**kamerad der** playmate; ~**karte die**
playing-card; ~**plan der** programme;
~**platz der** playground; ~**raum der**
room to move *(fig.);* scope; latitude;
~**sachen** *Pl.* toys; ~**verderber der;**
~**s,** ~: spoil-sport; ~**waren** *Pl.* toys;
~**zeug das a)** toy; *(fig.)* toy; play-
thing; **b)** *o. Pl.* *(~sachen, ~waren)* toys
pl.

Spieß der; ~**es,** ~**e a)** *(Waffe)* spear;
den ~ **umdrehen** *od.* **umkehren** *(ugs.)*
turn the tables; **b)** *(Brat~)* spit; **c)**
(Fleisch~) kebab; **d)** *(Soldatenspr.)*
[company] sergeant-major

Spießer der; ~**s,** ~ *(abwertend)* [petit]
bourgeois; **spießig** *(abwertend)* **1.**
Adj. [petit] bourgeois; **2.** *adv.* ⟨*think,
behave, etc.*⟩ in a [petit] bourgeois way

Spinat der; ~[e]s, ~e spinach

Spind der *od.* **das;** ~[e]s, ~e locker

Spindel die; ~, ~**n** spindle

Spinne die; ~, ~**n** spider; **spinnen 1.**
unr. tr. V. **a)** spin *(fig.);* plot ⟨*intrigue*⟩; think up ⟨*idea*⟩; hatch ⟨*plot*⟩; **2.**
unr. itr. V. **a)** spin; **b)** *(ugs.: verrückt
sein)* be crazy *or (sl.)* nuts; **Spin-
nen·netz das** spider's web; **Spin-
ner der;** ~**s,** ~ **a)** *(Beruf)* spinner; **b)**
(ugs. abwertend) nut-case *(sl.);* idiot;
Spinnerei die; ~, ~**en** spinning mill;
Spinnerin die; ~, ~**nen** *s.* **Spinner**

Spinn-: ~**rad das** spinning-wheel;
~**webe die;** ~, ~**n** cobweb

Spion der; ~**s,** ~**e a)** spy; **b)** *(Guckloch)* spyhole; **Spionage** [ʃpio'na:ʒə]
die; ~: spying; espionage; **spionieren** *itr. V.* spy; **Spionin die;** ~,
~**nen** spy

Spirale die; ~, ~**n** spiral

Spiral·feder die coil spring

Spirituose die; ~, ~**n** spirit *usu. in pl.*

Spiritus der; ~, ~**se** spirit; ethyl alcohol

Spiritus·kocher der spirit stove

Spital das; ~**s, Spitäler** *(bes. österr.,
schweiz.)* hospital

spitz 1. *Adj.* **a)** pointed; sharp ⟨*pencil,
needle, stone, etc.*⟩; fine ⟨*pen nib*⟩; *(Geom.)* acute ⟨*angle*⟩; **b)** *(schrill)* shrill

⟨*cry etc.*⟩; **c)** *(boshaft)* cutting ⟨*remark etc.*⟩; **2.** *adv.* **a)** ~ **zulaufen** taper to a point; ~ **zulaufend** pointed; **b)** *(boshaft)* cuttingly

Spitz der; ~es, ~e spitz

spitz-, Spitz-: ~**bart** der goatee; ~**bube** der *(scherzh.: Schlingel)* rascal; ~**bübisch** 1. *Adj.* mischievous; **2.** *adv.* mischievously

spitze *indekl. Adj. (ugs.) s.* **klasse; Spitze** die; ~, ~n **a)** point; *(Pfeil~, Horn~ usw.)* tip; **b)** *(Turm~, Baum~, Mast~ usw.)* top; *(eines Berges)* summit; **c)** *(Zigarren~, Haar~, Zweig~)* end; *(Schuh~)* toe; *(Finger~, Nasen~)* tip; **d)** *(vorderes Ende)* front; **an der ~ liegen** *(Sport)* be in the lead *or* in front; **e)** *(führende Position)* top; **f)** *(einer Firma, Organisation usw.)* head; *(einer Hierarchie)* top; *(leitende Gruppe)* management; **g)** *(Höchstwert)* maximum; peak; **h)** |absolute/einsame| ~ sein *(ugs.)* be [absolutely] great *(coll.)*; **i)** *(fig.: Angriff)* dig (**gegen** at); **j)** *(Textilwesen)* lace

Spitzel der; ~s, ~: informer

spitzen *tr. V.* sharpen ⟨*pencil*⟩; purse ⟨*lips, mouth*⟩; prick up ⟨*ears*⟩

Spitzen-: ~**erzeugnis das** top-quality product; ~**klasse die** top class; ~**qualität** die top quality; ~**sportler** top sportsman

spitz-, Spitz-: ~**findig** *Adj.* hair-splitting; ~**hacke** die pick; ~|**kriegen** *tr. V. (ugs.)* tumble to *(coll.)*; ~**name** der nickname

Spleen [ʃpliːn] der; ~s, ~e *od.* ~s strange habit; eccentricity

Splitt der; ~|e|s, ~e [stone] chippings *pl.; (zum Streuen)* grit

Splitter der; ~s, ~: splinter; *(Granat~, Bomben~)* splinter; **splittern** *itr. V.* **a)** *(Splitter bilden)* splinter; **b)** *mit sein (in Splitter zerbrechen)* ⟨*glass, windscreen, etc.*⟩ shatter; **splitter-nackt** *Adj. (ugs.)* stark naked; starkers *pred. (Brit. sl.);* **Splitter--partei** *die* splinter party

sponsern *tr. V.* sponsor; **Sponsor** der; ~s, ~en sponsor

spontan 1. *Adj.* spontaneous; **2.** *adv.* spontaneously

sporadisch 1. *Adj.* sporadic; **2.** *adv.* sporadically

Spore die; ~, ~n spore

Sporn der; ~|e|s, **Sporen** *(des Reiters)* spur; **einem Pferd die Sporen geben** spur a horse

Sport der; ~|e|s **a)** sport; *(als Unter-*

richtsfach) sport; PE; ~ **treiben** do sport; **b)** *(Hobby, Zeitvertreib)* hobby; pastime

Sport-: ~**fest das** sports festival; *(einer Schule)* sports day; ~**flugzeug** das sports plane; ~**geist** der; *o. Pl.* sportsmanship; ~**journalist** der sports journalist; ~**kleidung** die sportswear

Sportler der; ~s, ~: sportsman; **Sportlerin** die; ~, ~nen sportswoman; **sportlich** 1. *Adj.* **a)** sporting *attrib.;* **b)** *(fair)* sportsmanlike; sporting; **c)** *(fig.: flott, rasant)* sporty ⟨*car, driving, etc.*⟩; **d)** *(zu sportlicher Leistung fähig)* sporty, athletic ⟨*person*⟩; **e)** *(jugendlich wirkend)* sporty, smart but casual ⟨*clothes*⟩; smart but practical ⟨*hair-style*⟩; **2.** *adv.* **a)** as far as sport is concerned; **b)** *(fair)* sportingly; **c)** *(fig.: flott, rasant)* in a sporty manner

Sport-: ~**platz** der sports field; *(einer Schule)* playing field/fields *pl.;* ~**schuh** der sports shoe; ~**stadion** das [sports] stadium; ~**verein** der sports club; ~**wagen** der **a)** *(Auto)* sports car; **b)** *(Kinderwagen)* pushchair *(Brit.);* stroller *(Amer.)*

Spott der; ~|e|s mockery; *(höhnischer)* ridicule; derision; **spott·billig** *Adj., adv. (ugs.)* dirt cheap; **spötteln** *itr. V.* mock [gently]; poke *or* make [gentle] fun; **spotten** *itr. V.* **a)** mock; poke *or* make fun; *(höhnischer)* ridicule; be derisive; **b) einer Sache** *(Gen.)* ~: be contemptuous of *or* scorn sth.; **Spötter** der; ~s, ~: mocker; **spöttisch** 1. *Adj.* mocking; *(höhnischer)* derisive; **2.** *adv.* mockingly; **Spott·preis** der *(ugs.)* ridiculously low price

sprach *1. u. 3. Pers. Sg. Prät. v.* **sprechen; Sprache** die; ~, ~n **a)** language; **in englischer ~:** in English; **b)** *(Sprechweise)* way of speaking; speech; *(Stil)* style; **c) etw. zur ~ bringen** bring sth. up; raise sth.; **heraus mit der ~!** come on, out with it!; **Sprachen·schule** die language school

Sprach-: ~**fehler** der speech impediment *or* defect; ~**führer** der phrasebook; ~**kenntnisse** *Pl.* knowledge *sing.* of a language/languages; ~**kurs** der language course

sprachlich 1. *Adj.* linguistic; **2.** *adv.* linguistically

sprach-, Sprach-: ~**los** *Adj. (über-*

rascht) speechless; ~**rohr** das *(Repräsentant)* spokesman; *(Propagandist)* mouthpiece; ~**unterricht** der language teaching

sprang *1. u. 3. Pers. Sg. Prät. v.* springen

Spray [ʃpreː] das *od.* der; ~s, ~s spray; **Spray·dose** die aerosol [can]; **sprayen** *tr., itr. V.* spray

Sprech-: ~**anlage** die intercom *(coll.);* ~**chor** der chorus

sprechen *1. unr. itr. V.* speak (über + *Akk.* about; von about, of); *(sich unterhalten, sich besprechen auch)* talk (über + *Akk.,* von about); ⟨*parrot etc.*⟩ talk; **deutsch/flüsternd** ~: speak German/in a whisper; **für/gegen etw.** ~: speak in favour of/against sth.; **mit jmdm.** ~: speak *or* talk with *or* to sb.; **mit wem spreche ich?** who is speaking please?; *2. unr. tr. V.* **a)** speak ⟨*language, dialect*⟩; say ⟨*word, sentence*⟩; „**Hier spricht man Deutsch**" 'German spoken'; **b)** *(rezitieren)* say, recite ⟨*poem, text*⟩; say ⟨*prayer*⟩; **c)** **jmdn.** ~: speak to sb.; **d)** *(aus~)* pronounce ⟨*name, word, etc.*⟩; **Sprecher** der; ~s, ~ **a)** spokesman; **b)** *(Ansager)* announcer; *(Nachrichten~)* newscaster; news-reader; **c)** *(Kommentator, Erzähler)* narrator

sprech-, Sprech-: ~**funk·gerät** das radio-telephone; *(Walkie-talkie)* walkie-talkie; ~**stunde** die consultation hours *pl.; (eines Arztes)* surgery; ~**stunden·hilfe** die *(eines Arztes)* receptionist; *(eines Zahnarztes)* assistant; ~**zimmer** das consulting-room

spreizen *tr. V.* spread ⟨*fingers, toes, etc.*⟩; **die Beine** ~: spread one's legs apart; open one's legs

Spreiz·fuß der *(Med.)* spread foot

sprengen *tr. V.* **a)** blow up; blast ⟨*rock*⟩; **etw. in die Luft** ~: blow sth. up; **b)** *(gewaltsam öffnen, aufbrechen)* force [open] ⟨*door*⟩; force ⟨*lock*⟩; burst, break ⟨*bonds, chains*⟩; *(fig.)* break up ⟨*meeting, demonstration*⟩; **c)** *(be~)* water ⟨*flower-bed, lawn*⟩; sprinkle ⟨*street, washing*⟩ with water; *(verspritzen)* sprinkle; *(mit dem Schlauch)* spray

Sprenkel der; ~s, ~: spot; dot; speckle; **sprenkeln** *tr. V.* sprinkle spots of ⟨*colour*⟩; sprinkle ⟨*water*⟩

Spreu die; ~: chaff

sprich *Imperativ Sg. v.* sprechen; **sprichst** *2. Pers. Sg. Präsens v.* sprechen; **spricht** *3. Pers. Sg. Präsens v.*

sprechen; Sprich·wort das; *Pl.* Sprichwörter proverb

sprießen *unr. itr. V.; mit sein* ⟨*leaf, bud*⟩ shoot, sprout; ⟨*seedlings*⟩ come *or* spring up; ⟨*beard*⟩ sprout

Spring·brunnen der fountain; **springen** *1. unr. itr. V.* **a)** *mit sein (auch Sport)* jump; *(mit Schwung)* leap; spring; jump; ⟨*frog, flea*⟩ hop, jump; *(sich in Sprüngen fortbewegen)* bound; **b)** *mit sein (fig.)* ⟨*pointer, milometer, etc.*⟩ jump (**auf** + *Akk.* to); ⟨*traffic-lights*⟩ change (**auf** + *Akk.* to); ⟨*spark*⟩ leap; ⟨*ball*⟩ bounce; **c)** *mit sein* ⟨*string, glass, porcelain, etc.*⟩ break; *(Risse, Sprünge bekommen)* crack; *2. unr. tr. V.; auch mit sein (Sport)* perform ⟨*somersault, twist dive, etc.*⟩

Springer der; ~s, ~ **a)** *(Sport)* jumper; **b)** *(Schachfigur)* knight

spring·lebendig *Adj.* extremely lively; full of beans *pred. (coll.)*

Spring·reiten das show-jumping *no art.*

sprinten *itr. (auch tr.) V.; mit sein* sprint; **Sprinter** der; ~s, ~, **Sprinterin** die; ~, ~**nen** *(Sport)* sprinter

Sprit der; ~[e]s, ~e **a)** *(ugs.: Treibstoff)* gas *(Amer. coll.);* juice *(sl.);* petrol *(Brit.);* **b)** *(ugs.: Schnaps)* shorts *pl.*

Spritze die; ~, ~**n a)** syringe; **b)** *(Injektion)* injection; **c)** *(Feuer~)* hose; *(Löschfahrzeug)* fire engine

spritzen *1. tr. V.* **a)** *(versprühen)* spray; *(ver~)* splash; *(in Form eines Strahls)* spray, squirt ⟨*water, foam, etc.*⟩; pipe ⟨*cream etc.*⟩; **b)** *(be~, besprühen)* water ⟨*lawn, tennis-court*⟩; water, spray ⟨*street, yard*⟩; spray ⟨*plants, crops, etc.*⟩; *(mit Lack)* spray ⟨*car etc.*⟩; **jmdn. naß** ~: splash sb.; *(mit Wasserpistole, Schlauch)* spray sb.; **c)** *(injizieren)* inject ⟨*drug etc.*⟩; *(ugs.: einer Injektion unterziehen)* **jmdn./sich** ~: give sb. an injection/inject oneself; *2. itr. V.; mit Richtungsangabe mit sein* ⟨*hot fat*⟩ spit; ⟨*mud etc.*⟩ spatter, splash; ⟨*blood, water*⟩ spurt; **Spritzer** der; ~s, ~ *(kleiner Tropfen)* splash; *(von Farbe)* splash; spot; **spritzig** *1. Adj.* **a)** sparkling ⟨*wine*⟩; tangy ⟨*fragrance, perfume*⟩; **b)** lively ⟨*show, music, article*⟩; sparkling ⟨*performance*⟩; racy ⟨*style*⟩; nippy *(coll.);* zippy ⟨*car, engine*⟩; agile ⟨*person*⟩; *2. adv.* sparklingly ⟨*produced, performed, etc.*⟩; racily ⟨*written*⟩; **Spritz·tour** die *(ugs.)* spin

spröd, spröde *Adj.* **a)** brittle ⟨*glass, plastic, etc.*⟩; dry ⟨*hair, lips, etc.*⟩; *(rissig)* chapped ⟨*lips, skin*⟩; *(rauh)* rough ⟨*skin*⟩; **b)** *(fig.: abweisend)* aloof ⟨*person, manner, nature*⟩

sproß *1. u. 3. Pers. Sg. Prät. v.* **sprießen; Sproß** der; **Sprosses, Sprosse** *(Bot.)* shoot

Sprosse die; ~, ~n **a)** *(auch fig.)* rung; **b)** *(eines Fensters)* glazing bar

Sprößling der; ~s, ~e *(ugs. scherzh.)* offspring; **seine** ~ *e* his offspring *pl.*

Sprotte die; ~, ~n sprat

Spruch der; ~|e|s, **Sprüche** *(Wahl~)* motto; *(Sinn~)* maxim; *(Aus~)* saying; aphorism; *(Zitat)* quotation; **spruch·reif** *Adj.* **das ist noch nicht** ~: that's not definite, so people mustn't start talking about it yet

Sprudel der; ~s, ~ **a)** sparkling mineral water; **b)** *(österr.)* fizzy drink; **sprudeln** *itr. V.;* mit sein bubble; ⟨*lemonade, champagne, etc.*⟩ fizz, effervesce; **Sprudel·wasser das;** *Pl.* -wässer sparkling mineral water

Sprüh·dose die aerosol [can]; **sprühen 1.** *tr. V.* spray; **2.** *itr. V.;* mit Richtungsangabe mit sein ⟨*sparks, spray*⟩ fly; *(fig.)* ⟨*eyes*⟩ sparkle (vor + *Dat.* with); ⟨*intellect, wit*⟩ sparkle

Sprüh·regen der drizzle; fine rain

Sprung der; ~|e|s, **Sprünge a)** *(auch Sport)* jump; *(schwungvoll)* leap; *(Satz)* bound; *(fig.)* leap; **keine großen Sprünge machen können** *(fig. ugs.)* not be able to afford many luxuries; **auf dem ~|e| sein** *(fig. ugs.)* be in a rush; **b)** *(ugs.: kurze Entfernung)* stone's throw; **c)** *(Riß)* crack

Sprung·brett das *(auch fig.)* springboard; **sprunghaft 1.** *Adj.* **a)** erratic ⟨*person, character, manner*⟩; disjointed ⟨*conversation, thoughts*⟩; **b)** *(unvermittelt)* sudden; **c)** *(ruckartig)* rapid ⟨*change*⟩; sharp ⟨*increase*⟩; **2.** *adv.; s.* **1 b–c:** disjointedly; suddenly; rapidly; sharply

Spucke die; ~: spit; **spucken 1.** *itr. V.* spit; **in die Hände** ~ *(fig.: an die Arbeit gehen)* go to work with a will; **2.** *tr. V.* spit; cough up ⟨*blood, phlegm*⟩

Spuk der; ~|e|s, ~e [ghostly *or* supernatural] manifestation; **spuken** *itr. V.; unpers.* **hier/in dem Haus spukt es** this place/the house is haunted

Spule die; ~, ~n spool; *(für Tonband, Film)* spool; reel

Spüle die; ~, ~n sink unit; *(Becken)* sink

spulen *tr., itr. V.* spool; *(am Tonbandgerät)* wind

spülen 1. *tr. V.* **a)** rinse; bathe ⟨*wound*⟩; **b)** *(landsch.: abwaschen)* wash up ⟨*dishes, glasses, etc.*⟩; **Geschirr** ~: wash up; **2.** *itr. V.* **a)** *(beim WC)* flush [the toilet]; **b)** *(den Mund ausspülen)* rinse out [one's mouth]; **c)** *(landsch.) s.* **abwaschen 2**

Spül-: ~**maschine** die dishwasher; ~**mittel** das washing-up liquid

Spur die; ~, ~en **a)** *(Abdruck im Boden)* track; *(Folge von Abdrücken)* tracks *pl.;* **eine heiße** ~ *(fig.)* a hot trail; **jmdm./einer Sache auf der** ~ **sein** be on to the track *or* trail of sb./sth.; **b)** *(Anzeichen)* trace; *(eines Verbrechens)* clue (Gen. to); **c)** *(sehr kleine Menge; auch fig.)* trace; **d)** *(Verkehrsw.: Fahr~)* lane; **die** ~ **wechseln** change lanes

spürbar 1. *Adj.* noticeable; distinct, perceptible ⟨*improvement*⟩; evident ⟨*relief, embarrassment*⟩; **2.** *adv.* noticeably; perceptibly; *(sichtlich)* clearly ⟨*relieved, on edge*⟩; **spüren** *tr. V.* feel; *(instinktiv)* sense

spur·los 1. *Adj.* total, complete ⟨*disappearance*⟩; **2.** *adv.* ⟨*disappear*⟩ completely *or* without trace

Spür·sinn der; *o. Pl.* *(feiner Instinkt)* intuition

Spurt der; ~|e|s, ~s *od.* ~e spurt; **spurten** *itr. V.* **a)** *mit Richtungsangabe be mit sein* spurt; **b)** mit sein *(ugs.: schnell laufen)* sprint

sputen *refl. V. (veralt.)* make haste

St. *Abk.* **a)** Sankt St.; **b)** Stück

Staat der; ~|e|s, ~en state; **staatlich 1.** *Adj.* state *attrib.;* ⟨*power, unity, etc.*⟩ of the state; state-owned ⟨*factory etc.*⟩; **2.** *adv.* by the state; ~ **anerkannt/geprüft** state-approved/-certified

staats-, Staats-: ~**angehörige** der/die national; ~**angehörigkeit** die nationality; ~**anwalt** der public prosecutor; ~**bürger** der citizen; **er ist deutscher** ~**bürger** he is a German citizen *or* national; ~**bürgerlich** *Adj.* civil ⟨*rights*⟩; civic ⟨*duties, loyalty*⟩; ⟨*education, attitude*⟩ as a citizen; ~**bürgerschaft** die *s.* ~**angehörigkeit;** ~**grenze** die state frontier *or* border; ~**mann** der; *Pl.* -**männer** statesman; ~**oberhaupt** das head of state; ~**präsident** der [state] president

Stab der; ~|e|s, **Stäbe a)** rod; *(länger)*

pole; *(eines Käfigs, Gitters, Geländers)* bar; **b)** *(Milit.)* staff; **c)** *(Team)* team
stabil 1. *Adj.* sturdy ⟨*chair, cupboard*⟩; robust, sound ⟨*health*⟩; stable ⟨*prices, government, economy, etc.*⟩; **2.** *adv.* ~ **gebaut** solidly built; **stabilisieren 1.** *tr. V.* stabilize; **2.** *refl. V.* **a)** stabilize; **b)** ⟨*health, circulation, etc.*⟩ become stronger
Stab·lampe die torch *(Brit.);* flashlight *(Amer.)*
Stabs·arzt der *(Milit.)* medical officer, MO *(with the rank of captain)*
stach *1. u. 3. Pers. Sg. Prät. v.* stechen
Stachel der; ~s, ~n **a)** spine; *(Dorn)* thorn; **b)** *(Gift~)* sting; **c)** *(spitzes Metallstück)* spike; *(an ~draht)* barb
Stachel-: ~**beere** die gooseberry; ~**draht** der barbed wire
stachelig *Adj.* prickly
Stadion das; ~s, Stadien stadium
Stadium das; ~s, Stadien stage
Stadt die; ~, Städte **a)** town; *(Groß~)* city; **die** ~ **Basel** the city of Basel; **in die** ~ **gehen** go into town; go downtown *(Amer.);* **b)** *(Verwaltung)* town council; *(in der Großstadt)* city council; city hall *no art. (Amer.)*
Stadt-: ~**bahn** die urban railway; ~**bummel** der *(ugs.)* **einen** ~**bummel machen** take a stroll through the town/city centre
Städter der; ~s, ~, **Städterin** die; ~, ~**nen a)** town-dweller; *(Großstädter, -städterin)* city-dweller; **b)** *(Stadtmensch)* townie *(coll.)*
Stadt-: ~**führer** der town/city guidebook; ~**gespräch** das: ~**gespräch sein** be the talk of the town
städtisch 1. *Adj.* **a)** *(kommunal)* municipal; **b)** *(urban)* urban ⟨*life, way of life, etc.*⟩; **2.** *adv. (kommunal)* municipally
Stadt-: ~**mauer** die town/city wall; ~**mitte** die town centre; *(einer Großstadt)* city centre; downtown area *(Amer.);* ~**park** der municipal park; ~**plan** der [town/city] street plan *or* map; ~**rand** der outskirts *pl.* of the town/city; **am** ~: on the outskirts of the town/city; ~**rundfahrt** die sightseeing tour round a/the town/city; ~**teil** der district; part [of a/the town]; ~**tor** das town/city gate; ~**viertel** das district
Staffel die; ~, ~**n a)** *(Sport: Mannschaft)* relay team; **b)** *(Sport: ~lauf)* relay race; **c)** *(Luftwaffe: Einheit)* flight; **d)** *(Eskorte)* escort formation

Staffelei die; ~, ~en easel
stahl *1. u. 3. Pers. Sg. Prät. v.* stehlen
Stahl der; ~[e]s, Stähle *od.* ~e steel
Stahl-: ~**beton** der reinforced concrete; ~**blech** das sheet steel
stählern *Adj.; nicht präd.* steel
stak *1. u. 3. Pers. Sg. Prät. v.* stecken
Stall der; ~[e]s, Ställe *(Pferde~, Renn~)* stable; *(Kuh~)* cowshed; *(Hühner~)* [chicken-]coop; *(Schweine~)* [pig]sty; *(für Kaninchen, Kleintiere)* hutch; *(für Schafe)* pen; **Stallung** die; ~, ~en *(Pferdestall)* stable; *(Kuhstall)* cow-shed; *(Schweinestall)* [pig]sty
Stamm der; ~[e]s, Stämme **a)** *(Baum~)* trunk; **b)** *(Volks~)* tribe; **Stamm·baum** der family tree; *(eines Tieres)* pedigree
stammeln *tr., itr. V.* stammer
stammen *itr. V.* come (**aus, von** from); *(datieren)* date (**aus, von** from)
Stamm-: ~**gast** der *(im Lokal/Hotel)* regular customer/visitor; regular *(coll.);* ~**tisch** der **a)** *(Tisch)* regulars' table *(coll.);* **b)** *(~tischrunde)* group of regulars *(coll.);* **c)** *(Treffen)* get-together with the regulars *(coll.)*
stampfen 1. *itr. V.* **a)** *(laut auftreten)* stamp; **b)** *mit sein (sich fortbewegen)* tramp; *(mit schweren Schritten)* trudge; **2.** *tr. V.* **a) mit den Füßen den Rhythmus** ~: tap the rhythm with one's feet; **b)** *(fest~)* compress; **c)** *(zerkleinern)* mash ⟨*potatoes*⟩
stand *1. u. 3. Pers. Sg. Prät. v.* stehen; **Stand** der; ~[e]s, Stände **a)** *o. Pl. (das Stehen)* standing position; **[bei jmdm.** *od.* **gegen jmdn.] einen schweren** ~ **haben** *(fig.)* have a tough time [of it] [with sb.]; **b)** *(~ort)* position; **c)** *(Verkaufs~; Box für ein Pferd)* stall; *(Messe~, Informations~)* stand; *(Zeitungs~)* [newspaper] kiosk; **d)** *o. Pl. (erreichte Stufe; Zustand)* state; **etw. auf den neu[e]sten** ~ **bringen** bring sth. up to date; **e)** *(des Wassers, Flusses)* level; *(des Thermometers, Zählers, Barometers)* reading; *(der Kasse, Finanzen)* state; *(eines Himmelskörpers)* position; **f)** *o. Pl. (Familien~)* status; **g)** *(Gesellschaftsschicht)* class; *(Berufs~)* trade; *(Ärzte, Rechtsanwälte)* [professional] group
Standard der; ~s, ~s standard
Ständchen das; ~s, ~: serenade; **jmdm. ein** ~ **bringen** serenade sb.
Ständer der; ~s, ~: stand; *(Kleider~)* coat-stand; *(Wäsche~)* clothes-horse

standes-, Standes-: ~amt das registry office; ~amtlich 1. *Adj.; nicht präd.* registry office ⟨wedding, document⟩; 2. *adv.* ~amtlich heiraten get married in a registry office; ~beamte der registrar

stand-, Stand-: ~fest *Adj.* steady; stable; strong ⟨stalk, stem⟩; ~haft 1. *Adj.* steadfast; 2. *adv.* steadfastly; ~haftigkeit die; ~: steadfastness; ~|halten *unr. itr. V.* stand firm; einer Sache *(Dat.)* ~halten withstand sth.

ständig 1. *Adj.* constant ⟨noise, worry, pressure, etc.⟩; permanent ⟨residence, correspondent, staff, member, etc.⟩; standing ⟨committee⟩; regular ⟨income⟩; 2. *adv.* constantly

Stand-: ~licht das *(Kfz-W.)* sidelights *pl.;* ~ort der; *Pl.* ~orte a) position; *(eines Betriebes o. ä.)* location; site; b) *(Milit.: Garnison)* garrison; base; ~punkt der *(fig.)* point of view; viewpoint; auf dem ~punkt stehen, daß ...: take the view that ...; ~spur die *(Verkehrsw.)* hard shoulder; ~uhr die grandfather clock

Stange die; ~, ~n pole; *(aus Metall)* bar; *(dünner)* rod; *(Kleider~)* rail; *(Vogel~)* perch; ein Anzug von der ~ *(ugs.)* an off-the-peg-suit

Stangen-: ~brot das French bread; ~spargel der asparagus spears *pl.*

stank *1. u. 3. Pers. Sg. Prät. v.* stinken

Stapel der; ~s, ~: pile; ein ~ Holz a pile *or* stack of wood; **stapeln** 1. *tr. V.* pile up; stack; 2. *refl. V.* pile up

stapfen *itr. V.; mit sein* tramp

¹Star der; ~|e|s, ~e *od. (schweiz.)* ~en *(Vogel)* starling

²Star der; ~s, ~s *(berühmte Persönlichkeit)* star

³Star der; ~|e|s *(Med.)* grauer ~: cataract; grüner ~: glaucoma

starb *1. u. 3. Pers. Sg. Prät. v.* sterben

stark; stärker, stärkst... 1. *Adj.* a) strong; potent ⟨drink, medicine, etc.⟩; powerful ⟨engine, lens, voice, etc.⟩; *(ausgezeichnet)* excellent; *s. auch* Stück c; b) *(dick)* thick; stout ⟨rope, string⟩; *(verhüll.: korpulent)* well-built *(euphem.);* c) *(zahlenmäßig groß, umfangreich)* sizeable, large; big ⟨demand⟩; eine 100 Mann ~e Truppe a 100-strong unit; d) *(heftig, intensiv)* heavy; severe ⟨frost, pain⟩; strong ⟨impression, current, resistance, dislike⟩; grave ⟨doubt, reservations⟩; great ⟨exaggeration, interest⟩; loud ⟨applause⟩; e) *(Jugendspr.: großartig)* great *(coll.);*

fantastic *(coll.);* 2. *adv.* a) *(sehr, überaus, intensiv) (mit Adj.)* very; heavily ⟨indebted, stressed⟩; greatly ⟨increased, reduced, enlarged⟩; strongly ⟨emphasized, characterized⟩; badly ⟨damaged, worn, affected⟩; *(mit Verb)* heavily; ⟨exaggerate, impress⟩ greatly; ⟨enlarge, reduce, increase⟩ considerably; ⟨support, oppose, suspect⟩ strongly; ⟨remind⟩ very much; ~ erkältet sein have a heavy *or* bad cold; b) *(Jugendspr.: großartig)* fantastically *(coll.);* **Stark·bier** das strong beer; **Stärke** die; ~, ~n a) *o. Pl.* strength; *(eines Motors)* power; *(einer Glühbirne)* wattage; b) *(Dicke)* thickness; *(Technik)* gauge; c) *o. Pl. (zahlenmäßige Größe)* strength; d) *(besondere Fähigkeit, Vorteil)* strength; jmds. ~/nicht jmds. ~ sein be sb.'s forte/not be sb.'s strong point; e) *(Intensität)* strength; *(von Sturm, Schmerzen, Abneigung)* intensity; *(von Frost)* severity; *(von Lärm, Verkehr)* volume; f) *(organischer Stoff)* starch; **stärken** 1. *tr. V.* a) strengthen; boost ⟨power, prestige⟩; ⟨drink, food, etc.⟩ fortify ⟨person⟩; b) *(steif machen)* starch ⟨washing etc.⟩; 2. *refl. V.* refresh oneself; **Stärkung** die; ~, ~en a) *o. Pl.* strengthening; b) *(Erfrischung)* refreshment

starr 1. *Adj.* a) rigid; *(steif)* stiff (vor + *Dat.* with); fixed ⟨expression, smile, stare⟩; b) *(nicht abwandelbar)* inflexible, rigid ⟨law, rule, principle⟩; c) *(unnachgiebig)* inflexible ⟨person, attitude, etc.⟩; 2. *adv.* rigidly; *(steif)* stiffly

starren *itr. V.* a) stare (in + *Akk.* into, auf, an, gegen + *Akk.* at); jmdm. ins Gesicht ~: stare sb. in the face; b) vor/ von Schmutz ~: be filthy

Starr·sinn der; *o. Pl.* pig-headedness

starr·sinnig *Adj.* pig-headed

Start der; ~|e|s, ~s start; *(eines Flugzeugs)* take-off; *(einer Rakete)* launch; **Start·bahn** die [take-off] runway; **start·bereit** *Adj.* ready to start *postpos.;* ⟨aircraft⟩ ready for take-off; **starten** 1. *itr. V.; mit sein* a) start; ⟨aircraft⟩ take off; ⟨rocket⟩ blast off, be launched; b) *(den Motor anlassen)* start the engine; 2. *tr. V.* start; launch ⟨rocket, satellite, attack⟩; start [up] ⟨engine, machine, car⟩

Station die; ~, ~en a) station; b) *(Haltestelle)* stop; c) *(Zwischen~, Aufenthalt)* stopover; ~ machen stop over *or* off; **d)** *(Kranken~)* ward; **stationär 1.** *Adj.* *(Med.)* ⟨*treatment*⟩ in hospital, as an in-patient; **2.** *adv.* *(Med.)* in hospital; *jmdn.* ~ **behandeln** treat sb. as an in-patient; **stationieren** *tr. V.* station ⟨*troops*⟩; deploy ⟨*weapons, bombers, etc.*⟩

Stations-: **~arzt** der ward doctor; **~schwester die** ward sister; **~taste die** *(Rundf.)* preset [tuning] button; preset

Statistik die; ~: statistics *sing., no art.*

statt 1. *Präp. mit Gen.* instead of; ~ **dessen** instead [of this]; **2.** *Konj.: s.* **anstatt**

statt-: ~|**finden** *unr. itr. V.* take place; ⟨*process, development*⟩ occur; **~haft** *Adj.* permissible

stattlich 1. a) well-built; imposing ⟨*figure, stature, building, etc.*⟩; fine ⟨*farm, estate*⟩; impressive ⟨*trousseau, collection*⟩; **b)** *(beträchtlich)* considerable; **2.** *adv.* impressively

Statue die; ~, **~n** statue

Statur die; ~, **~en** build

Status der; ~, ~ ['ʃtaːtuːs] status

Statut das; ~|e|s, **~en** statute

Stau der; ~|e|s, **~s** *od.* **~e a)** build-up; **b)** *(von Fahrzeugen)* tailback *(Brit.)*; backup *(Amer.)*

Staub der; ~|e|s dust; ~ **wischen** dust; ~ **saugen** vacuum *or (Brit. coll.)* hoover; **sich aus dem ~|e| machen** *(fig. ugs.)* make oneself scarce *(coll.)*; **stauben** *itr. V.* cause dust; **staubig** *Adj.* dusty

staub-, Staub-: ~**saugen** *itr., tr. V.* vacuum, *(Brit. coll.)* hoover; **~sauger der** vacuum cleaner; Hoover *(Brit. P)*; **~tuch das;** *Pl.* **~tücher** duster

Staude die; ~, **~n** *(Bot.)* herbaceous perennial

stauen 1. *tr. V.* dam [up] ⟨*stream, river*⟩; staunch ⟨*blood*⟩; **2.** *refl. V.* ⟨*water, blood, etc.*⟩ accumulate, build up; ⟨*people*⟩ form a crowd; ⟨*traffic*⟩ form a tailback/tailbacks *(Brit.)* or *(Amer.)* backup/backups

staunen *itr. V.* be amazed *or* astonished **(über + Akk.** at); *(beeindruckt sein)* marvel **(über + Akk.** at); **~d** with *or* in amazement; **Staunen das;** ~**s** amazement **(über + Akk.** at); *(Bewunderung)* wonderment

Stauung die; ~, **~en a)** *(eines Bachs, Flusses)* damming; *(des Blutes, Wassers)* stemming the flow; *(das Sich-*

stauen) build-up; **b)** *(Verkehrsstau)* tailback *(Brit.)*; backup *(Amer.)*; jam

Std. *Abk.* **Stunde** hr.

stechen 1. *unr. itr. V.* **a)** prick; ⟨*wasp, bee*⟩ sting; ⟨*mosquito*⟩ bite; *(hinein~)* **mit etw. in etw.** *(Akk.)* ~: stick *or* jab sth. into sth.; **2.** *unr. tr. V.* *(mit dem Messer, Schwert)* stab; *(mit der Nadel, mit einem Dorn usw.)* prick; ⟨*bee, wasp*⟩ sting; ⟨*mosquito*⟩ bite; **sich in den Finger** ~: prick one's finger

Stech-: **~mücke die** mosquito; gnat; **~uhr die** time clock

Steck-: **~brief der** description [of a/the wanted person]; *(Plakat)* 'wanted' poster; **~dose die** socket; power point

stecken 1. *tr. V.* **a)** put; **b)** *(mit Nadeln)* pin ⟨*hem, lining, etc.*⟩; pin [on] ⟨*badge*⟩; pin up ⟨*hair*⟩; **2.** *itr. V.* be; **wo steckt meine Brille?** *(ugs.)* where have my glasses got to *or* gone?; **hinter etw.** *(Dat.)* ~ *(fig. ugs.)* be behind sth.

stecken-, Stecken-: ~|**bleiben** *unr. itr. V.; mit sein* get stuck; ~|**lassen** *unr. tr. V.* leave; **~pferd das a)** *(Spielzeug)* hobby-horse; **b)** *(Liebhaberei)* hobby

Stecker der; ~**s,** ~: plug; **Steck·nadel die** pin

Steg der; ~|e|s, **~e** *(Brücke)* [narrow] bridge; *(Laufbrett)* gangplank; *(Boots~)* landing-stage

Steg·reif der: aus dem ~: impromptu

stehen *unr. itr. V.; südd., österr., schweiz. mit sein* **a)** stand; **b)** *(sich befinden)* be; ⟨*upright object, building*⟩ stand; **c)** *(einen bestimmten Stand haben)* **auf etw.** *(Dat.)* ~ ⟨*needle, hand*⟩ point to sth.; **das Barometer steht tief/ auf Regen** the barometer is reading low/indicating rain; **das Spiel/es steht 1 : 1** *(Sport)* the score is one all; **die Sache steht gut/schlecht** things are going well/badly; **d)** *(einen bestimmten Kurs, Wert haben)* ⟨*currency*⟩ stand **(bei** at); **wie steht das Pfund?** what is the rate for the pound?; **e)** *(nicht in Bewegung sein)* be stationary; ⟨*machine etc.*⟩ be at a standstill; **meine Uhr steht** my watch has stopped; **f)** *(geschrieben, gedruckt sein)* be; **in der Zeitung steht, daß ...:** it says in the paper that ...; **g)** *(Sprachw.: gebraucht werden)* ⟨*subjunctive etc.*⟩ occur; be found; **h)** *jmdm.* |**gut**| ~ ⟨*dress etc.*⟩ suit sb. [well]

stehen: ~|**bleiben** *unr. itr. V.; mit sein* **a)** stop; ⟨*traffic*⟩ come to a standstill; **b)** *(stehengelassen werden)* stay; be left; *(zurückgelassen werden)* be left behind; *(der Zerstörung entgehen)* ⟨*building*⟩ be left standing; ~**lassen** *unr. tr. V.* **a)** leave; **b)** *(vergessen)* leave [behind]

Steh·lampe die standard lamp *(Brit.)*; floor lamp *(Amer.)*

stehlen *unr. tr., itr. V.* steal; *s. auch* gestohlen 2

Steh·platz der *(im Theater usw.)* standing place; *(im Bus)* space to stand

Steiermark die; ~: Styria *no art.*

steif 1. *Adj.* stiff; *(förmlich)* stiff; formal; **2.** *adv.* stiffly

steigen 1. *unr. itr. V.; mit sein* **a)** climb; ⟨*mist, smoke, sun*⟩ rise; ⟨*balloon*⟩ climb, rise; **auf die Leiter** ~: get on to the ladder; **in den/aus dem Bus/ Zug** ~: board *or* get on/get off *or* out of the bus/train; **b)** *(ansteigen, zunehmen)* rise; ⟨*price, cost, salary, output*⟩ increase, rise; ⟨*debts, tension*⟩ increase, mount; ⟨*chances*⟩ improve; **2.** *unr. tr. V.; mit sein* climb ⟨*stairs, steps*⟩; **Steiger** der; ~s, ~ *(Bergbau)* overman

steigern 1. *tr. V.* **a)** increase ⟨*speed, value, sales, consumption, etc.*⟩ (**auf** + *Akk.* to); step up ⟨*demands, production, etc.*⟩; raise ⟨*standards, requirements*⟩; *(verstärken)* intensify ⟨*fear, tension*⟩; heighten ⟨*effect*⟩; **b)** *(Sprachw.)* compare ⟨*adjective*⟩; **2.** *refl. V.* ⟨*confusion, speed, profit, etc.*⟩ increase; ⟨*pain, excitement, tension, etc.*⟩ become more intense; ⟨*costs*⟩ escalate; ⟨*effect*⟩ be heightened; **Steigerung** die; ~, ~en **a)** increase *(Gen.* in); *(Verstärkung)* intensification; *(einer Wirkung)* heightening; *(Verbesserung)* improvement *(Gen.* in); *(bes. Sport: Leistungs~)* improvement [in performance]; **b)** *(Sprachw.)* comparison

Steigung die; ~, ~en gradient

steil 1. *Adj.* steep; meteoric ⟨*career*⟩; rapid ⟨*rise*⟩; **2.** *adv.* steeply; **Steil·hang** der steep escarpment

Stein der; ~[e]s, ~e stone; *(Fels)* rock; *(Bau~)* [stone]block; **mir fällt ein ~ vom Herzen** that's a weight off my mind; **Stein·bock** der **a)** ibex; **b)** *(Astrol.)* Capricorn; the Goat; **steinern** *Adj.* stone; **Stein·gut** das earthenware; **stein·hart** *Adj.* rockhard; **steinig** *Adj.* stony

Stein-: ~**kohle** die [hard] coal;

~**metz** der; ~en, ~en stonemason; ~**obst** das stone-fruit; ~**pilz** der cep; ~**schlag** der rock fall; „Achtung ~schlag" 'beware falling rocks'; ~**zeit** die Stone Age; *(fig.)* stone age

Steiß·bein das *(Anat.)* coccyx

Stelle die; ~, ~n **a)** place; **an jmds. ~ treten** take sb.'s place; **ich an deiner ~ ...: ...** if I were you; **an achter ~ liegen** be in eighth place; **die erste ~ hinter** *od.* **nach dem Komma** *(Math.)* the first decimal place; **an ~** (+ *Gen.*) instead of; **auf der ~:** immediately; **b)** *(begrenzter Bereich)* patch; *(am Körper)* spot; **c)** *(Passage)* passage; *(Punkt im Ablauf einer Rede usw.)* point; **d)** *(Arbeits~)* job; post; **eine freie ~:** a vacancy; **e)** *(Dienst~)* office; *(Behörde)* authority; **stellen 1.** *tr. V.* **a)** put; *(mit Sorgfalt)* place; *(aufrecht hin~)* stand; *(ein~)* set ⟨*points, clock, scales*⟩; **den Wecker auf 6 Uhr** ~: set the alarm for 6 o'clock; **die Heizung höher/niedriger** ~: turn the heating up/down; **c)** *(bereit~)* provide; **d) jmdn. besser** ~: *(firm)* improve sb.'s pay; **gut/schlecht/besser gestellt** comfortably/badly/better off; **e)** *verblaßt* put ⟨*question*⟩; set ⟨*task, topic, condition*⟩; make ⟨*application, demand, request*⟩; **jmdm. eine Frage** ~: ask sb. a question; **2.** *refl. V.* **a)** place oneself; **sich auf die Zehenspitzen** ~: stand on tiptoe; **b) sich schlafend/taub/tot** *usw.* ~: feign sleep/ deafness/death *etc.;* pretend to be asleep/deaf/dead *etc.*

stellen-, Stellen-: ~**angebot** das offer of a job; *(Inserat)* job advertisement; „~angebote" 'situations vacant'; ~**gesuch** das 'situation wanted' advertisement; ~**weise** *Adv.* in places

Stellung die; ~, ~en position; **zu etw. ~ nehmen** express one's opinion on sth.; **Stellungnahme** die; ~, ~n opinion; *(kurze Äußerung)* statement; **Stell·vertreter** der deputy

Stelze die; ~, ~n stilt; **stelzen** *itr. V.; mit sein* strut; stalk

stemmen 1. *tr. V.* **a)** *(hoch~)* lift [above one's head]; **b)** *(drücken)* brace ⟨*feet, knees*⟩ (**gegen** against); **2.** *refl. V.* **sich gegen etw.** ~: brace oneself against sth.

Stempel der; ~s, ~: stamp; *(Post~)* postmark; **stempeln** *tr. V.* stamp ⟨*passport, form*⟩; postmark ⟨*letter*⟩; cancel ⟨*postage stamp*⟩

Stengel der; ~s, ~: stem; stalk
steno-, Steno-: ~gramm das shorthand text; ~graph der; ~en, ~en stenographer; ~graphie die; ~, ~n stenography *no art.;* shorthand *no art.;* ~graphieren *itr. V.* do shorthand; ~typistin die shorthand typist
Stepp·decke die quilt
Steppe die; ~, ~n steppe
steppen *tr. (auch itr.) V.* backstitch
sterben *unr. itr. V.; mit sein* die; **im Sterben liegen** lie dying; **sterbens·krank** *Adj.* mortally ill; **sterblich** *Adj.* mortal
stereo *Adv.* in stereo; **Stereo** das; ~s stereo; **Stereo·anlage** die stereo [system]
steril *Adj.* sterile
Sterling ['stɛːlɪŋ]: **Pfund** ~: pound/pounds sterling
Stern der; ~[e]s, ~e star; **Sternchen** das; ~s, ~ *(Druckw.)* asterisk; **Stern·schnuppe** die; ~, ~n shooting star
Stethoskop [ʃteto'skoːp] das; ~s, ~e *(Med.)* stethoscope
¹**Steuer** das; ~s, ~: [steering-]wheel; *(von Schiffen)* helm; ²**Steuer** die; ~, ~n tax
steuer-, Steuer-: ~berater der tax consultant *or* adviser; ~bord das *od.* österr. der; *o. Pl. (Seew., Flugw.)* starboard; ~erklärung die tax return; ~frei *Adj.* tax-free; ~mann der; *Pl.* ~leute *od.* ~männer *(Rudersport)* cox
steuern 1. *tr. V. (fahren)* steer; *(fliegen)* pilot, fly *(aircraft)*; fly *(course)*; 2. *itr. V.* a) be at the wheel; *(auf dem Schiff)* be at the helm; b) *mit sein (Kurs nehmen, ugs.: sich hinbewegen; auch fig.)* head; **Steuerung** die; ~, ~en a) *(System)* controls *pl.;* b) *o. Pl. s.* steuern 1: steering; piloting; flying
Steward ['stjuːɐt] der; ~s, ~s steward; **Stewardeß** ['stjuːɐdɛs] die; ~, Stewardessen stewardess
stich *Imper. Sg. v.* stechen
Stich der; ~[e]s, ~e a) *(mit einer Waffe)* stab; b) *(Dornen~, Nadel~)* prick; *(von Wespe, Biene usw.)* sting; *(Mükken~ usw.)* bite; c) *(~wunde)* stab wound; d) *(beim Nähen)* stitch; e) *(Schmerz)* stabbing *or* shooting pain; f) *(Kartenspiel)* trick; g) jmdn./etw. im ~ lassen leave sb. in the lurch/abandon sth.; **sticheln** *itr. V.* make snide remarks *(coll.)* (gegen about)
stich-, Stich-: ~flamme die tongue of flame; ~haltig *Adj.* sound (argu-

ment, reason); valid (assertion, reply); conclusive (evidence); ~probe die [random] sample; *(bei Kontrollen)* spot check
stichst 2. *Pers. Sg. Präsens v.* stechen; **sticht** 3. *Pers. Sg. Präsens v.* stechen
Stich-: ~tag der set date; deadline; ~wunde die stab wound
sticken 1. *itr. V.* do embroidery; 2. *tr. V.* embroider; **Stickerei** die; ~, ~en embroidery *no pl.;* *(gestickte Arbeit)* piece of embroidery; **Stick·garn** das embroidery thread
stickig *Adj.* stuffy; stale (air); **Stick·stoff** der nitrogen
Stief- step (brother, child, mother, etc.)
Stiefel der; ~s, ~ boot
Stief·mütterchen das *(Bot.)* pansy; **stief·mütterlich** 1. *Adj.* poor, shabby (treatment); 2. *adv.* ~ behandeln treat (person) poorly *or* shabbily; neglect (pet, flowers, doll, problem)
stieg 1. u. 3. *Pers. Sg. Prät. v.* steigen
Stieglitz der; ~es, ~e goldfinch
stiehl *Imp. Sg. v.* stehlen; **stiehlst** 2. *Pers. Sg. Präsens v.* stehlen; **stiehlt** 3. *Pers. Sg. Präsens v.* stehlen
Stiel der; ~[e]s, ~e *(Griff)* handle; *(Besen~)* [broom-]stick; *(für Süßigkeiten)* stick; *(bei Gläsern)* stem; *(bei Blumen)* stem; *(an Obst usw.)* stalk
Stier der; ~[e]s, ~e bull
stieren *itr. V.* stare [vacantly] (auf + Akk. at)
Stier·kampf der bullfight
stieß 1. u. 3. *Pers. Sg. Prät. v.* stoßen
Stift der; ~[e]s, ~e a) *(aus Metall)* pin; *(aus Holz)* peg; b) *(Blei~)* pencil; *(Mal~)* crayon; *(Schreib~)* pen
stiften *tr. V.* a) found, establish (monastery, hospital, etc.); endow (prize, scholarship); *(als Spende)* donate, give (für·to); b) *(herbeiführen)* cause, create (unrest, confusion, strife, etc.); bring about (peace, order, etc.); arrange (marriage); **Stifter** der; ~s, ~: founder; *(Spender)* donor
Stift·zahn der *(Zahnmed.)* post crown
Stil der; ~[e]s, ~e style; **stilistisch** 1. *Adj.* stylistic; 2. *adv.* stylistically
still 1. *Adj.* quiet; *(ohne Geräusche)* silent; still; *(reglos)* still; *(wortlos)* silent; *(heimlich)* secret; der Stille Ozean the Pacific [Ocean]; 2. *adv.* quietly; *(geräuschlos)* silently; *(wortlos)* in silence; **Stille** die; ~: quiet; *(Geräuschlosigkeit)* silence; stillness; **stillegen** *tr. V.* close *or* shut down;

close ⟨railway line⟩; **stillen 1.** tr. V. **a)**
ein Kind ~: breast-feed a baby; **b)** (befriedigen) satisfy; quench ⟨thirst⟩; **c)**
(eindämmen) stop ⟨bleeding, tears,
pain⟩; **2.** itr. V. breast-feed
still-, Still-: ~|**halten** unr. itr. V.
keep or stay still; ~|**legen** s. **stille-**
gen; ~**schweigen das** silence;
~**schweigen bewahren** maintain
silence; keep silent; ~**schweigend**
1. Adj. silent; (ohne Abmachung) tacit
⟨assumption, agreement⟩; **2.** adv. in
silence; (ohne Abmachung) tacitly;
~|**sitzen** unr. itr. V. sit still; ~**stand**
der; o. Pl. standstill; ~|**stehen** unr.
itr. V. **a)** ⟨factory, machine⟩ stand idle;
⟨traffic⟩ be at a standstill; ⟨heart etc.⟩
stop; **b)** (Milit.) stand to attention
Stimm·bruch der: er ist im ~: his
voice is breaking; **Stimme die;** ~, ~**n**
a) voice; **b)** (bei Wahlen) vote
stimmen 1. itr. V. **a)** be right or correct; **stimmt es, daß ...?** is it true
that ...?; **b)** (seine Stimme geben) vote;
mit Ja ~: vote yes or in favour; **2.** tr.
V. **a)** (in eine Stimmung versetzen)
make; **b)** (Musik) tune ⟨instrument⟩
Stimm-: ~**enthaltung** die abstention; ~**recht das** right to vote
Stimmung die; ~, ~**en a)** mood; **b)**
(Atmosphäre) atmosphere
Stink·bombe die stink-bomb; **stin-**
ken unr. itr. V. stink (**nach** of); **stin-**
kig Adj. (salopp abwertend) stinking;
smelly
stirb Imp. Sg. v. **sterben; stirbst**
2. Pers. Sg. Präsens v. **sterben; stirbt**
3. Pers. Sg. Präsens v. **sterben**
Stirn die; ~, ~**en** forehead; brow
stöbern itr. V. (ugs.) rummage
stochern itr. V. poke
¹**Stock der;** ~|e|s, Stöcke **a)** stick; (Zei-
ge~) pointer; stick; (Takt~) baton;
(Ski~) pole; stick; **b)** (Pflanze) (Ro-
sen~) [rose-]bush; (Reb~) vine;
²**Stock der;** ~|e|s, ~ (Etage) floor;
storey; **in welchem ~?** on which
floor?; **stock·dunkel** Adj. (ugs.)
pitch-dark; **stocken** itr. V. **a)** ⟨traf-
fic⟩ be held up; ⟨conversation, produc-
tion⟩ stop; ⟨business⟩ slacken; ⟨jour-
ney⟩ be interrupted; **b)** (innehalten)
falter; **stock·finster** Adj. (ugs.)
pitch-dark
-**stöckig** -storey attr.; -storeyed
Stockung die; ~, ~**en** hold-up (Gen.
in); **Stockwerk das** floor; storey
Stoff der; ~|e|s, ~**e a)** material; fabric;
b) (Materie) substance; **c)** o. Pl. (Phi-

los.) matter; **d)** (Thema) subject[-mat-
ter]; (Gesprächsthema) topic; **Stoff-**
wechsel der; o. Pl. metabolism
stöhnen itr. V. moan; (vor Schmerz)
groan
Stola die; ~, Stolen shawl; (Pelz~)
stole
Stollen der; ~s, ~ **a)** (Kuchen) Stollen; **b)** (Bergbau) gallery; **c)** (bei Sport-
schuhen) stud
stolpern itr. V.; mit sein stumble; trip
stolz 1. Adj. proud (**auf** + Akk. of);
eine ~e Summe (ugs.) a tidy sum; **2.**
adv. proudly; **Stolz der;** ~**es** pride
(**auf** + Akk. in); **stolzieren** itr. V.;
mit sein strut
stop Interj. stop; (Verkehrsw.) halt
stopfen tr. V. **a)** darn; **b)** (hineintun)
stuff; **c)** (füllen) stuff ⟨cushion, quilt,
etc.⟩; fill ⟨pipe⟩; plug, stop [up] ⟨hole,
leak⟩
Stopf-: ~**garn das** darning-cotton;
~**nadel die** darning-needle
Stopp der; ~s, ~s stop; (Einstellung)
freeze (Gen. on)
Stoppel die; ~, ~**n** stubble no pl.;
stoppelig Adj. stubbly
stoppen tr., itr. V. stop
Stopp-: ~**licht das;** Pl. ~**er** stop-
light; ~**schild das** stop sign; ~**uhr**
die stop-watch
Stöpsel der; ~s, ~ plug
Stör der; ~s, ~**e** sturgeon
Storch der; ~|e|s, Störche stork
stören 1. tr. V. **a)** disturb; disrupt
⟨court proceedings, lecture, church service, etc.⟩; interfere with ⟨transmitter,
reception⟩; **b)** (mißfallen) bother; **2.**
itr. V. **a)** disturb; **b)** (Unruhe stiften)
make or cause trouble; **3.** refl. V. **sich**
an jmdm./etw. ~: take exception to
sb./sth.; **Störenfried der;** ~|e|s, ~**e**
trouble-maker
störrisch 1. Adj. stubborn; **2.** adv.
stubbornly
Störung die; ~, ~**en a)** disturbance;
(einer Gerichtsverhandlung, Vorlesung,
eines Gottesdienstes usw.) disruption;
bitte entschuldigen Sie die ~, aber ...:
I'm sorry to bother you, but ...; **b) eine**
technische ~: a technical fault
Stoß der; ~**es,** Stöße **a)** (mit der Faust)
punch; (mit dem Fuß) kick; (mit dem
Kopf, den Hörnern) butt; (mit dem Ellbogen) dig; **b)** (mit einer Waffe) (Stich)
thrust; (Schlag) blow; **c)** (beim
Schwimmen, Rudern) stroke; **d)** (Stapel) pile; stack; **stoßen 1.** unr. tr. V.
a) auch itr. (mit der Faust) punch; (mit

dem Fuß) kick; *(mit dem Kopf, den Hörnern)* butt; *(mit dem Ellbogen)* dig; b) *(hineintreiben)* plunge, thrust ⟨*dagger, knife*⟩; push ⟨*stick, pole*⟩; c) *(schleudern)* push; **die Kugel ~**: put the shot; 2. *unr. itr. V.* a) *mit sein (auftreffen)* bump **(gegen** into); **mit dem Kopf gegen etw. ~**: bump one's head on sth.; b) *mit sein (fig.)* **auf etw.** *(Akk.)* ~ *(etw. entdecken)* come upon sth.; **auf Ablehnung** ~ *(abgelehnt werden)* meet with disapproval; c) *(grenzen)* **an etw.** *(Akk.)* ~ ⟨*room, property, etc.*⟩ be [right] next to sth.; 3. *unr. refl. V.* bump *or* knock oneself; **sich an etw.** *(Dat.)* ~ *(fig.)* object to sth.

Stoß-: **~seufzer** der heartfelt groan; **~stange** die bumper

stößt 3. *Pers. Sg. Präsens v.* stoßen; **stoß·weise** *Adv.* a) spasmodically; b) *(in Stapeln)* by the pile; in piles

Stotterer der; **~s, ~**: stutterer; **stottern** 1. *itr. V.* stutter; 2. *tr. V.* stutter [out]

Str. *Abk.* Straße St./Rd.

stracks *Adv.* a) *(direkt)* straight; b) *(sofort)* straight away

straf·bar *Adj.* punishable; **Strafe** die; **~, ~n** *(Rechtsspr.)* punishment; *(Freiheits~)* penalty; *(Geld~)* fine; **strafen** *tr. V.* punish

straff 1. *Adj.* a) tight, taut ⟨*rope, lines, etc.*⟩; firm ⟨*breasts, skin*⟩; b) *(energisch)* tight ⟨*organization, planning, etc.*⟩; strict ⟨*discipline, leadership, etc.*⟩; 2. *adv.* a) [zu] ~ **sitzen** ⟨*clothes*⟩ be [too] tight; b) *(energisch)* tightly, strictly

straf·fällig *Adj.* ~ werden commit a criminal offence

straffen *tr. V.* a) tighten; firm ⟨*skin*⟩; b) *(fig.)* tighten up ⟨*text, procedure, organization, etc.*⟩

straf-, Straf-: **~frei** *Adj.* **~frei** ausgehen go unpunished; **~gefangene** der/die prisoner; **~gesetz·buch das** penal code

sträflich 1. *Adj.* criminal; 2. *adv.* criminally; **Sträfling** der; **~s, ~e** prisoner

straf-, Straf-: **~los** *Adj.* unpunished; **~tat** die criminal offence; **~täter** der offender; **~zettel** der *(ugs.)* [parking-, speeding-, *etc.*] ticket

Strahl der; **~[e]s, ~en** *(auch Phys., Math., fig.)* ray; *(von Scheinwerfern, Taschenlampen)* beam; *(von Flüssigkeit)* jet; **strahlen** *itr. V.* a) shine; **bei ~dem Wetter/Sonnenschein** in glori-

ous sunny weather/in glorious sunshine; **~d weiß** sparkling white; b) *(glänzen)* sparkle; c) *(lächeln)* beam **(vor** + *Dat.* with); **Strahler** der; **~s, ~** a) radiator; b) *(Heiz~)* radiant heater; **Strahlung** die; **~, ~en** radiation

Strähne die; **~, ~n** strand; **eine graue ~**: a grey streak; **strähnig** 1. *Adj.* straggly ⟨*hair*⟩; 2. *adv.* in strands

stramm 1. *Adj.* a) *(straff)* tight, taut ⟨*rope, line, etc.*⟩; tight ⟨*clothes*⟩; b) *(kräftig)* strapping ⟨*girl, boy*⟩; sturdy ⟨*legs, body*⟩; c) *(gerade)* upright, erect ⟨*posture, etc.*⟩; 2. *adv.* a) *(straff)* tightly; b) *(kräftig)* sturdily ⟨*built*⟩

strampeln *itr. V.* ⟨*baby*⟩ kick [his/her feet]

Strand der; **~[e]s, Strände** beach; **am ~**: on the beach; **Strand·bad** das bathing beach *(on river, lake)*; **stranden** *itr. V.*; *mit sein* ⟨*ship*⟩ run aground; **Strand·korb** der basket chair

Strang der; **~[e]s, Stränge** rope

Strapaze die; **~, ~n** strain *no pl.*; **strapazieren** *tr. V.* be a strain on ⟨*person, nerves*⟩; **strapazier·fähig** *Adj.* hard-wearing ⟨*clothes, shoes*⟩; durable ⟨*material*⟩

Straße die; **~, ~n** *(in Ortschaften)* street; road; *(außerhalb)* road

Straßen-: **~bahn** die tram *(Brit.)*; streetcar *(Amer.)*; **~ecke** die street corner; **~feger** der *(bes. nordd.)* road-sweeper; **~graben** der ditch [at the side of the road]; **~karte** die road-map; **~sperre** die road-block

sträuben 1. *tr. V.* ruffle [up] ⟨*feathers*⟩; bristle ⟨*fur, hair*⟩; 2. *refl. V.* ⟨*hair, fur*⟩ bristle, stand on end; ⟨*feathers*⟩ become ruffled; b) *(sich widersetzen)* resist

Strauch der; **~[e]s, Sträucher** shrub

straucheln *itr. V.*; *mit sein (geh.)* stumble

¹Strauß der; **~es, Sträuße** bunch of flowers; bouquet [of flowers]

²Strauß der; **~es, ~e** *(Vogel)* ostrich

Sträußchen das; **~s, ~**: posy

streben *itr. V.* a) *mit sein* make one's way briskly; b) *(trachten)* strive **(nach** for); **Streber** der; **~s;** ~ *(abwertend)* pushy person *(coll.)*; *(in der Schule)* swot *(Brit. sl.)*; grind *(Amer. sl.)*; **strebsam** *Adj.* ambitious and industrious

Strecke die; **~, ~n** distance; *(Abschnitt, Route)* route; *(Eisenbahn~)*

line; **strẹcken 1.** *tr. V. (gerade machen)* stretch ⟨*arms, legs*⟩; *(dehnen)* stretch [out] ⟨*arms, legs, etc.*⟩; **den Kopf aus dem Fenster** ~: stick one's head out of the window *(coll.)*; **2.** *refl. V.* stretch out; **strẹcken·weise** *Adv.* in places; *(fig.: zeitweise)* at times

Streich der; ~|e|s, ~e trick; prank; jmdm. einen ~ **spielen** play a trick on sb.; **strẹicheln** *tr. V.* stroke; **strẹichen 1.** *unr. tr. V.* **a)** stroke; **b)** *(an~)* paint; „**frisch gestrichen**" 'wet paint'; **c)** *(auftragen)* spread ⟨*butter, jam, ointment, etc.*⟩; *(be~)* **ein Brötchen mit Butter/mit Honig** ~: butter a roll/spread honey on a roll; **d)** *(aus~, tilgen)* delete; cancel ⟨*train, flight*⟩; **2.** *unr. itr. V.* **a)** stroke; **jmdm. über den Kopf** ~: stroke sb.'s head; **b)** *(an~)* paint

Streich-: ~**holz** das match; ~**instrument** das string[ed] instrument; ~**käse** der cheese spread; ~**wurst** die [soft] sausage for spreading; ≈ meat spread

Streife die; ~, ~n **a)** *(Personen)* patrol; **b)** *(Streifengang)* patrol; **strẹifen 1.** *tr. V.* **a)** *(leicht berühren)* touch; ⟨*shot*⟩ graze; **b)** *(kurz behandeln)* touch [up]on ⟨*problem, subject, etc.*⟩; **c) den Ring vom Finger** ~: slip the ring off one's finger; **die Ärmel nach oben** ~: pull/push up one's sleeves; **2.** *itr. V. mit sein* roam; **Streifen** der; ~s, ~ **a)** stripe; **b)** *(Stück, Abschnitt)* strip; **Streifen·wagen** der patrol car; **strẹifig** *Adj.* streaky

Streik der; ~|e|s, ~s strike; **Streik·brecher** der strike-breaker; blackleg *(derog.)*; **strẹiken** *itr. V.* **a)** strike; be on strike; *(in den Streik treten)* come out *or* go on strike; strike; **b)** *(ugs.: nicht mitmachen)* go on strike; **c)** *(ugs.: nicht funktionieren)* pack up *(coll.)*; **Strẹikende** der/die; *adj. Dekl.* striker; **Strẹik·posten** der picket

Streit der; ~|e|s, ~e *(Zank)* quarrel; *(Auseinandersetzung)* dispute; argument; **strẹiten** *unr. itr., refl. V.* quarrel; argue; *(sich zanken)* quarrel; **Streiterei** die; ~, ~en arguing *no pl., no indef. art.*; *(Gezänk)* quarrelling *no pl.*; **Streitigkeit** die; ~, ~en *meist Pl.* **a)** quarrel; argument; **b)** *(Streitfall)* dispute

streng 1. *Adj.* **a)** strict; severe ⟨*punish-*

ment⟩; stringent, strict ⟨*rule, regulation, etc.*⟩; stringent ⟨*measure*⟩; rigorous ⟨*examination, check, test, etc.*⟩; stern ⟨*reprimand, look*⟩; absolute ⟨*discretion*⟩; complete ⟨*rest*⟩; **b)** *(schmucklos, herb)* austere, severe ⟨*cut, collar, style, etc.*⟩; severe ⟨*face, features, hairstyle, etc.*⟩; **c)** *(durchdringend)* pungent, sharp ⟨*taste, smell*⟩; **d)** *(rauh)* severe ⟨*winter*⟩; sharp, severe ⟨*frost*⟩; **2.** *adv.* ⟨*mark, judge, etc.*⟩ strictly, severely; ⟨*punish*⟩ severely; ⟨*look, reprimand*⟩ sternly; ⟨*smell*⟩ strongly; **Strẹnge** die; ~ **a)** *s.* **streng a:** strictness; severity; stringency; rigour; sternness; **b)** *(von [Gesichts]zügen)* severity; **c)** *(von Geruch, Geschmack)* pungency; sharpness; **d)** *s.* **streng d:** severity; sharpness; **strẹngstens** *Adv.* [most] strictly

Streß der; Strẹsses stress

Streu die; ~, ~en straw; **strẹuen** *tr. V.* **a)** spread ⟨*manure, sand, grit*⟩; sprinkle ⟨*salt, herbs, etc.*⟩; strew, scatter ⟨*flowers*⟩; **b)** *auch itr.* **die Straßen |mit Sand/Salz|** ~: grit/salt the roads

strẹunen *itr. V.; meist mit sein* wander *or* roam about *or* around; ~**de Katzen/Hunde** stray cats/dogs

Streusel·kuchen der streusel cake

strich *1. u. 3. Pers. Sg. Prät. v.* **streichen**

Strich der; ~|e|s, ~e *(Linie)* line; *(Gedanken~)* dash; *(Schräg~)* diagonal; *(Binde~, Trennungs~)* hyphen; **auf den** ~ **gehen** *(salopp)* walk the streets; **strịcheln** *tr. V.* **a)** sketch in [with short lines]; **b)** *(schraffieren)* hatch

Strich-: ~**junge** der *(salopp)* [young] male prostitute; ~**mädchen** das *(salopp)* street-walker; hooker *(Amer. sl.)*; ~**punkt** der semicolon

Strick der; ~|e|s, ~e cord; *(Seil)* rope; **strịcken** *tr., itr. V.* knit

Strick-: ~**jacke** die cardigan; ~**nadel** die knitting-needle; ~**zeug** das knitting

striegeln *tr. V.* groom ⟨*horse*⟩

strịkt 1. *Adj.* strict; **2.** *adv.* strictly

Strippe die; ~, ~n *(ugs.)* string; **an der** ~ **hängen** *(fig.)* be on the phone *(coll.)*; *(dauernd)* hog the phone *(coll.)*

Stripperin die; ~, ~nen *(ugs.)* stripper

stritt *1. u. 3. Pers. Sg. Prät. v.* **streiten**; **strịttig** *Adj.* contentious ⟨*point, problem*⟩; disputed ⟨*territory*⟩; ⟨*question*⟩ in dispute, at issue

Stroh das; ~|e|s straw

Stroh-: ~**blume** die a) *(Immortelle)* immortelle; b) *(Korbblütler)* straw- flower; ~**halm** der straw; ~**witwe** die *(ugs. scherzh.)* grass widow; ~**wit- wer** der *(ugs. scherzh.)* grass widower
Strolch der; ~|e|s, ~e *(fam. scherzh.: Junge)* rascal
Strom der; ~|e|s, Ströme river; *(fig.)* stream; *(Strömung; Elektrizität)* cur- rent; *(~versorgung)* electricity; **unter** ~ **stehen** be live
strom-: ~**abwärts** *Adv.* down- stream; ~**auf[wärts]** *Adv.* upstream
strömen *itr. V.; mit sein* stream; **Strömung** die; ~, ~en current; *(Met.)* airstream; *(fig.)* trend
Strophe die; ~, ~n verse; *(einer Ode)* strophe
strotzen *itr. V.* von *od.* vor etw. *(Dat.)* ~: be full of sth.; von *od.* vor Gesund- heit ~: be bursting with health
strubbelig *Adj.* tousled
Strudel der; ~s, ~ a) whirlpool; b) *(bes. südd., österr.: Gebäck)* strudel
Strumpf der; ~|e|s, Strümpfe stock- ing; *(Socke, Knie~)* sock
Strumpf-: ~**band** das garter; *(Straps)* suspender *(Brit.)*; garter *(Amer.)*; ~**hose** die tights *pl. (Brit.)*; pantyhose *(esp. Amer.)*
Strunk der; ~|e|s, Strünke stem; stalk; *(Baum~)* stump
struppig *Adj.* shaggy; tangled, tousled *(hair)*
Stube die; ~, ~n a) *(veralt.: Wohn- raum)* [living-]room; parlour *(dated)*; b) *Milit.)* [barrack-]room; **Stuben- fliege** die [common] house-fly
Stück das; ~|e|s, ~e a) piece; *(kleines)* bit; *(Teil, Abschnitt)* part; **ein** ~ **Ku- chen** a piece *or* slice of cake; **ein** ~ **Zucker/Seife** a lump of sugar/ a piece *or* bar of soap; **im** *od.* **am** ~: unsliced *(sausage, cheese, etc.)*; b) *(Einzel~)* item; *(Exemplar)* specimen; **ich nehme 5** ~: I'll take five [of them]; **30 Pfennig das** ~: thirty pfennigs each; ~ **für** ~: piece by piece; *(eins nach dem andern)* one by one; **das ist |ja| ein starkes** ~ *(ugs.)* that's a bit much; **ein faules/fre- ches** ~ *(salopp)* a lazy/cheeky thing *or* devil; c) *(Bühnen~)* play; *(Musik~)* piece; **Stückchen** das; ~s, ~: [little] piece; bit; **stückeln** *tr. V.* put together *(sleeve, curtain)* with patches
Student der; ~en, ~en, **Studentin** die; ~, ~nen a) student; b) *(österr.: Schüler)* [secondary-school] pupil; **Studie** ['ʃtuːdi̯ə] die; ~, ~n study

Studien-: ~**aufenthalt** der study visit (**in** + *Dat.* to); ~**freund** der university/college friend; ~**reise** die study trip
studieren *tr., itr. V.* study; **Stu- dierende** der/die; *adj. Dekl.* stu- dent; **Studio** das; ~s, ~s studio; **Studium** das; ~s, Studien study; *(Studiengang)* course of study
Stufe die; ~, ~n a) step; *(einer Treppe)* stair; „**Vorsicht,** ~!" 'mind the step'; b) *(Raketen~, Geol., fig.: Stadium)* stage; *(Niveau)* level; *(Steigerungs~, Grad)* degree; *(Rang)* grade
Stuhl der; ~|e|s, Stühle chair
Stuhl-: ~**gang** der; *o. Pl.* bowel movement[s]; *(Kot)* stool; ~**lehne** die *(Rückenlehne)* chair-back; *(Armlehne)* chair-arm
stülpen *tr. V.* etw. auf *od.* über etw. *(Akk.)* ~: pull/put sth. on to *or* over sth.
stumm *Adj.* dumb *(person)*; *(schweig- sam)* silent; *(wortlos)* wordless; mute *(glance, gesture)*; **Stumme** der/die; *adj. Dekl.* mute; **die** ~**n** the dumb
Stummel der; ~s, ~: stump; *(Blei- stift~)* stub; *(Zigaretten-/Zigarren~)* [cigarette-/cigar-]butt
Stümper der; ~s, ~: botcher; bun- gler; **stümperhaft 1.** *Adj.* incompet- ent; botched *(job)*; *(laienhaft)* ama- teurish *(attempt, drawing)*; **2.** *adv.* in- competently; *(laienhaft)* amateur- ishly; **stümpern** *itr. V.* work incom- petently; *(pfuschen)* bungle
stumpf *Adj.* a) blunt *(pin, needle, knife, etc.)*; b) *(glanzlos, matt)* dull *(paint, hair, metal, colour, etc.)*; **Stumpf** der; ~|e|s, Stümpfe stump
Stumpf·sinn der; *o. Pl.* a) apathy; b) *(Monotonie)* monotony; tedium; **stumpf·sinnig 1.** *Adj.* a) apathetic; vacant *(look)*; b) *(monoton)* tedious; souldestroying *(job, work)*; **2.** *adv.* a) apathetically; *(stare)* vacantly; b) *(monoton)* tediously
Stunde die; ~, ~n hour; *(Unter- richts~)* lesson; **eine** ~ **Aufenthalt/ Pause** an hour's stop/break; a stop/ break of an hour
stünde *1. u. 3. Pers. Sg. Konjunktiv II v.* **stehen**
stunden *tr. V.* jmdm. einen Betrag *usw.* ~: allow sb. to defer payment of a sum *etc.*
stunden-, Stunden-: ~**kilometer** der kilometre per hour; k.p.h.; ~**lang 1.** *Adj.* lasting hours *postpos.*; **2.** *adv.*

for hours; **~lohn** der hourly wage; **~plan** der timetable; **~zeiger der** hour-hand

-stündig adj. -hour; **-stündlich** adj. -hourly; **zwei~/halb~**: two-hourly/ half-hourly; adv. every two hours/ half an hour; **stündlich** Adj., adv. hourly

Stups der; ~es, ~e (ugs.) push; shove; (leicht) nudge; **stupsen** tr. V. (ugs.) push; shove; (leicht) nudge; **Stups·nase** die snub nose

stur (ugs.) 1. Adj. **a)** obstinate; dogged ⟨insistence⟩; (phlegmatisch) dour; **b)** (unbeirrbar) dogged; persistent; **c)** (stumpfsinnig) tedious; 2. adv. **a)** obstinately; **b)** (unbeirrbar) doggedly; **c)** (stumpfsinnig) tediously; ⟨learn, copy⟩ mechanically

stürbe 1. u. 3. Pers. Sg. Konjunktiv II v. sterben

Sturheit die; ~ (ugs.) **a)** obstinacy; (phlegmatisches Wesen) dourness; **b)** (Stumpfsinnigkeit) deadly monotony

Sturm der; ~[e]s, Stürme **a)** storm; (heftiger Wind) gale; **b)** (Milit.) assault (**auf** + Akk. on); ~ **klingeln** ring the [door]bell like mad; **stürmen** 1. itr. V. **a)** unpers. es stürmt [heftig] it's blowing a gale; **b)** mit sein (rennen) rush; (verärgert) storm; 2. tr. V. (Milit.) storm ⟨town, position, etc.⟩; (fig.) besiege ⟨booking-office, shop, etc.⟩; **Stürmer der:** ~s, ~ (Sport) striker; forward; **stürmisch** 1. Adj. **a)** stormy; (fig.) tempestuous, turbulent; **b)** (ungestüm) tumultuous ⟨applause, welcome, reception⟩; wild ⟨enthusiasm⟩; passionate ⟨lover, embrace, temperament⟩; vehement ⟨protest⟩; 2. adv. ⟨protest⟩ vehemently; ⟨embrace⟩ impetuously, passionately; ⟨demand⟩ clamorously; ⟨applaud⟩ wildly

Sturz der; -es, Stürze **a)** fall; (Unfall) accident; **b)** (fig.: von Preis, Temperatur usw.) [sharp] fall, drop (Gen. in); **c)** (Verlust des Amtes, der Macht) fall; (Absetzung) overthrow; (Amtsenthebung) removal from office; **stürzen** 1. itr. V.; mit sein **a)** fall; (fig.) ⟨temperature, exchange rate, etc.⟩ drop [sharply]; ⟨prices⟩ tumble; ⟨government⟩ fall, collapse; **b)** (laufen) rush; dash; **c)** (fließen) stream; pour; 2. refl. V. sich auf jmdn./etw. ~ (auch fig.) pounce on sb./sth.; sich in etw. (Akk.) ~: throw oneself into sth.; 3. tr. V. **a)** throw; (mit Wucht) hurl; **b)** (umdrehen) upturn ⟨mould⟩; turn out ⟨pud-

ding, cake, etc.⟩; **c)** (des Amtes entheben) oust ⟨person⟩ [from office]; (gewaltsam) overthrow ⟨leader, government⟩; **Sturz·helm der** crash-helmet

Stute die; ~, ~n mare

Stütze die; ~, ~n (auch fig.) support

¹stutzen itr. V. stop short

²stutzen tr. V. trim; dock ⟨tail⟩; clip ⟨ear, hedge, wing⟩; prune ⟨tree, bush⟩

stützen 1. tr. V. support; (mit Pfosten o. ä.) prop up; (aufstützen) rest ⟨head, hands, arms, etc.⟩; 2. refl. V. **sich auf** jmdn./etw. ~: lean or support oneself on sb./sth.

stutzig Adj. ~ **werden** begin to wonder; jmdn. ~ **machen** make sb. wonder

s.u. Abk. siehe unten see below

Subjekt das; ~[e]s, ~e **a)** subject; **b)** (abwertend: Mensch) creature; **subjektiv** 1. Adj. subjective; 2. adv. subjectively; **Subjektivität die;** ~: subjectivity

Substantiv das; ~s, ~e (Sprachw.) noun; **Substanz die;** ~, ~en **a)** (auch fig.) substance; 2. (Grundbestand) die ~: the reserves pl.

sub·tropisch Adj. subtropical

Suche die; ~, ~n search (**nach** for); **auf der** ~ [**nach** jmdn./etw.] sein be looking/(intensiver) searching [for sb./ sth.]; **suchen** 1. tr. V. **a)** look for; (intensiver) search for; „Leerzimmer gesucht" 'unfurnished room wanted'; **b)** (bedacht sein auf, sich wünschen) seek ⟨protection, advice, company, warmth, etc.⟩; look for ⟨adventure⟩; 2. itr. V. search; nach jmdn./etw. ~: look/ search for sb./sth.

Sucht die; ~, Süchte od. ~en **a)** addiction (**nach** to); [bei jmdm.] zur ~ **werden** (auch fig.) become addictive [in sb.'s case]; **b)** Pl. Süchte (übermäßiges Verlangen) craving (**nach** for); **süchtig** Adj. **a)** addicted; **b)** (fig.) **nach etw.** ~ **sein** be obsessed with sth.

Süd o. Art.; o. Pl. (bes. Seemannsspr., Met.) s. Süden

Süd-: ~afrika (das) South Africa; **~amerika (das)** South America

Sudan (das); ~s od. der; ~s Sudan

Süden der; ~s south; der ~: the South; **Süd·frucht** die tropical [or sub-tropical] fruit; **Südländer der;** ~s, ~: Southern European; **südländisch** Adj. Southern [European]; Latin ⟨temperament⟩; ~ **aussehen** have Latin looks; **südlich** 1. Adj. **a)** southern; **b)** (nach, von Süden) southerly; **c)** (aus dem Süden) Southern; 2. adv.

southwards; **3.** *Präp. mit Gen.* [to the] south of

süd-, Süd-: ~**pol** der South Pole; ~**see die;** ~: **die** ~: the South Seas *pl.;* ~**see·insel die** South Sea island; ~**tirol (das)** South Tirol ~**wärts** *Adv.* southwards; ~**wind der** south *or* southerly wind

Sues·kanal ['zu:ɛs-] **der;** ~s Suez Canal

Sühne die; ~, ~**n** *(geh.)* atonement; expiation; **sühnen** *tr., itr. V.* |für| etw. ~: atone for *or* pay the penalty for sth.

Sultanine die; ~, ~**n** sultana

Sülze die; ~, ~**n a)** diced meat/fish in aspic; *(vom Schweinskopf)* brawn; **b)** *(Aspik)* aspic

Summe die; ~, ~**n** sum

summen 1. *itr. V.* hum; *(lauter, heller)* buzz; **2.** *tr. V.* hum ⟨*tune, song, etc.*⟩

summieren *refl. V.* add up (**auf** + *Akk.* to)

Sumpf der; ~|e|s, **Sümpfe** marsh; *(bes. in den Tropen)* swamp; **sumpfig** *Adj.* marshy

Sund der; ~|e|s, ~**e** *(Geogr.)* sound

Sünde die; ~, ~**n** sin; *(fig.)* misdeed; transgression; **Sünden·bock der** *(ugs.)* scapegoat; **Sünder der;** ~s, ~, **Sünderin die;** ~, ~**nen** sinner; **sündigen** *itr. V.* sin

Super das; ~s, ~: four star *(Brit.)*; premium *(Amer.)*; **super-** ultra-⟨*long, high, fast, modern, masculine, etc.*⟩; **Super-** super-⟨*hero, figure, car, group, etc.*⟩; terrific *(coll.)*, tremendous *(coll.)* ⟨*success, offer, chance, idea, etc.*⟩; **Superlativ** ['zu:pɐlati:f] **der;** ~s, ~**e** *(Sprachw.)* superlative; **Super·markt der** supermarket

Suppe die; ~; ~**n** soup; **Suppen·löffel der** soup-spoon

Surf·brett ['sɔːf-] **das** surf-board; **surfen** ['sɔːfn̩] *itr. V.* surf; **Surfer** ['sɔːfɐ] **der;** ~s, ~: surfer

surren *itr. V.* **a)** *(summen)* hum; ⟨*camera, fan*⟩ whirr; **b)** *mit sein (schwirren)* whirr

suspekt 1. *Adj.* suspicious; jmdm. ~ **sein** arouse sb.'s suspicions; **2.** *adv.* suspiciously

süß 1. *Adj.* sweet; **2.** *adv.* sweetly; **süßen** *tr. V.* sweeten; **Süßigkeit die;** ~, ~**en** sweet *(Brit.)*; candy *(Amer.)*; ~**en** sweets *(Brit.)*; candy *sing. (Amer.)*; *(als Ware)* confectionery *sing.*; **süßlich 1.** *Adj.* **a)** [slightly] sweet; on the sweet side *pred.*; **b)** *(sen-*

timental) sickly mawkish; **2.** *adv.* ⟨*write, paint*⟩ mawkishly

süß-, Süß-: ~**most der** unfermented fruit juice; ~**sauer 1.** *Adj.* sweet-and-sour; *(fig.)* wry ⟨*smile, face*⟩; **2.** *adv.* **a)** etw. ~-**sauer zubereiten** give sth. a sweet-and-sour flavour; **b)** *(fig.)* ⟨*smile*⟩ wryly; ~**speise die** sweet; dessert; ~**stoff der** sweetener; ~**wasser das;** *Pl.* ~**wasser** fresh water

svw. *Abk.* soviel wie

Symbol das; ~s, ~**e** symbol; **symbolisch 1.** *Adj.* symbolic; **2.** *adv.* symbolically

Sympathie [zʏmpa'ti:] **die;** ~, ~**n** sympathy (**für** with); **sympathisch 1.** *Adj.* congenial, likeable ⟨*person, manner*⟩; appealing ⟨*voice, appearance, material*⟩; **2.** *adv.* in an appealing way; *(angenehm)* agreeably

Symphonie *usw. s.* **Sinfonie** *usw.*

Synagoge die; ~, ~**n** synagogue

Syrer der; ~s, ~, **Syrerin die;** ~, ~**nen** Syrian; **Syrien** ['zy:riən] **(das)** ~s Syria; **syrisch** *Adj.* Syrian

System das; ~, ~**e** system; **systematisch 1.** *Adj.* systematic; **2.** *adv.* systematically

Szene ['stse:nə] **die;** ~, ~**n** *(auch fig.)* scene

T

t, T [te:] **das;** ~, ~: t, T

t *Abk.* Tonne t

Tab. *Abk.* Tabelle

Tabak ['ta(:)bak] **der;** ~s, ~**e** tobacco; **Tabaks·pfeife die** [tobacco-]pipe

Tabelle die; ~, ~**n** table

Tabernakel das *od.* **der;** ~s, ~: tabernacle

Tablett das; ~|e|s, ~s *od.* ~**e** tray; **Tablette die;** ~, ~**n** tablet

tabu *Adj.* taboo; **Tabu das;** ~s, ~s taboo

Tacho der; ~s, ~s *(ugs.)* speedo *(coll.)*; **Tacho·meter der** *od.* **das** speedometer

Tadel der; ~s, ~ **a)** censure; **b)** *(im*

Klassenbuch) black mark; **tadel·los**
1. *Adj.* impeccable; immaculate ⟨*hair,*
clothing, suit, etc.⟩; perfect ⟨*condition,*
teeth, pronunciation, German, etc.⟩; **2.**
adv. ⟨*dress*⟩ impeccably; ⟨*fit, speak,*
etc.⟩ perfectly; ⟨*live, behave, etc.*⟩ irre-
proachably; **tadeln** *tr. V.* jmdn. |für
od. wegen etw.| ~ : rebuke sb. [for sth.]
Tafel die; ~, ~n a) *(Schiefer~)* slate;
(Wand~) blackboard; **b)** *(plattenför-*
miges Stück) slab; **eine ~ Schokolade**
a bar of chocolate; **c)** *(Gedenk~)*
plaque; **d)** *(geh.: festlicher Tisch)*
table; **Täfelchen das; ~s, ~** : *s.* **Tafel**
b: [small] slab; [small] bar; **tafeln** *itr.*
V. (geh.) feast; **täfeln** *tr. V.* panel
Tafel-: ~spitz der *(österr.)* boiled fillet of beef; **~wasser das;** *Pl.* ~wäs-
ser [bottled] mineral water; **~wein**
der table wine
Taft der; ~|e|s, ~e taffeta
Tag der; ~|e|s, ~e day; **am ~|e|** during
the day[time]; **guten ~!** hello; *(bei Vor-*
stellung) how do you do?; **an diesem**
~ : on this day; **dreimal am ~** : three
times a day; **am folgenden ~** : the next
day; **eines ~es** one day; some day;
tag·aus *Adv.* ~, **tagein** day in, day
out; day after day; **Tage·buch das**
diary; **tag·ein** *Adv. s.* **tagaus; ta-**
ge·lang 1. *Adj.* lasting for days *post-*
pos.; **nach ~em Regen** after days of
rain; **2.** *adv.* for days [on end]; **tagen**
itr. V. meet; **das Gericht/Parlament**
tagt the court/parliament is in session
Tages-: ~karte die a) *(Gastron.)*
menu of the day; **b)** *(Fahr-, Eintritts-*
karte) day ticket; **~kasse die a)** box-
office *(open during the day);* **b)** *(~ein-*
nahme) day's takings *pl.;* **~licht das;**
o. Pl. daylight; **~zeit die** time of day;
~zeitung die daily newspaper
-tägig a) *(... Tage alt)* **ein sechstägiges**
Küken a six-day-old chick; **b)** *(... Tage*
dauernd) **nach dreitägiger Vorberei-**
tung after three days' preparation;
täglich 1. *Adj.* daily; **2.** *adv.* every
day; **zweimal ~** : twice a day; **~ drei**
Tabletten einnehmen take three tablets
daily; **tags** *Adv.* **a)** by day; in the
daytime; **b) ~ zuvor/davor** the day before; **~ darauf** the next *or* following
day; the day after; **tags·über** *Adv.*
during the day; **tag·täglich 1.** *Adj.*
day-to-day; daily; **2.** *adv.* every single
day; **Tagung die; ~, ~en** conference
Taifun der; ~s, ~e typhoon
Taille ['taljə] **die; ~, ~n** waist
Taiwan (das); ~s Taiwan

Takt der; ~|e|s, ~e a) *(Musik)* time;
(Einheit) bar; measure *(Amer.);* **aus**
dem ~ kommen lose the beat; **b)**
o. Pl. (rhythmischer Bewegungsablauf)
rhythm; **c)** *o. Pl. (Feingefühl)* tact
Taktik die; ~, ~en: |eine| ~ : tactics *pl.;*
taktisch 1. *Adj.* tactical; **2.** *adv.* tac-
tically
takt-: ~los 1. *Adj.* tactless; **2.** *adv.*
tactlessly; **~voll 1.** *Adj.* tactful; **2.**
adv. tactfully
Tal das; ~|e|s, Täler valley
Talent das; ~|e|s, ~e talent (zu, für
for); *(Mensch)* talented person
Talg der; ~|e|s, ~e suet; *(zur Herstel-*
lung von Seife, Kerzen usw.) tallow
Talisman der; ~s, ~e talisman
Tampon der; ~s, ~s tampon
Tamtam das; ~s *(ugs. abwertend)*
|großes| ~ : [a big] fuss
Tang der; ~|e|s, ~e seaweed
Tangente die; ~, ~n *(Math.)* tangent
Tank der; ~s, ~s tank; **tanken** *tr.,* itr.
V. fill up; **Öl ~** : fill up with oil
Tank-: ~säule die petrol-pump
(Brit.); gasoline pump *(Amer.);*
~stelle die petrol station *(Brit.);* gas
station *(Amer.);* **~wart der; ~s, ~e**
petrol-pump attendant *(Brit.)*
Tanne die; ~, ~n fir[-tree]
Tannen-: ~baum der *(ugs.)* fir-tree;
(Weihnachtsbaum) Christmas tree;
~grün das; *o. Pl.* fir sprigs *pl.;*
~zweig der fir branch
Tansania [tan'za:ni̯a] **(das); ~s** Tanza-
nia
Tante die; ~, ~n a) aunt; **b)** *(Kin-*
derspr.: Frau) lady; **c)** *(ugs.: Frau)*
woman
Tanz der; ~es, Tänze dance
Tanz-: ~abend der evening dance;
~bar die night-spot *(coll.)* with dancing; **~café das** coffee-house with
dancing
tanzen *itr., tr. V.* dance; **Tänzer der;**
~s, ~, Tänzerin die; ~, ~nen dancer;
(Ballett~) ballet-dancer
Tanz-: ~fläche die dance-floor; **~lo-**
kal das café/restaurant with dancing;
~orchester das dance band;
~stunde die a) *(~kurs)* dancing-
class; **b)** *(einzelne Stunde)* dancing
lesson
Tapete die; ~, ~n wallpaper; **tape-**
zieren *tr. V.* [wall]paper
tapfer 1. *Adj.* brave; **2.** *adv.* bravely;
Tapferkeit die; ~ : courage; bravery
tappen *itr. V.* **a)** *mit sein)* patter; **b)** *(ta-*
stend greifen) grope **(nach** for); **Taps**

der; ~es, ~e *(ugs. abwertend)* clumsy oaf

Tarif der; ~s, ~e charge; *(Post~, Wasser~)* rate; *(Verkehrs~)* fares *pl.; (Zoll~)* tariff; *(Lohn~)* [wage] rate; *(Gehalts~)* [salary] scale

tarnen 1. *tr., itr. V.* camouflage; 2. *refl. V.* camouflage oneself

Tasche die; ~, ~n bag; *(in Kleidung, Rucksack usw.)* pocket; **jmdm. auf der ~ liegen** *(fig. ugs.)* live off sb.

Taschen-: ~**buch** das paperback; ~**lampe** die [pocket] torch *(Brit.)* or *(Amer.)* flashlight; ~**messer** das penknife; ~**rechner** der pocket calculator; ~**tuch** das; *Pl.* ~**tücher** handkerchief; ~**uhr** die pocket-watch

Tasse die; ~, ~n cup

Taste die; ~, ~n a) *(eines Musikinstruments, einer Schreibmaschine)* key; b) *(Fuß~)* pedal [key]; c) *(am Telefon, Radio, Fernsehgerät, Taschenrechner usw.)* button; **tasten** 1. *itr. V. (fühlend suchen)* grope, feel (**nach** for); 2. *refl. V. (sich tastend bewegen)* grope or feel one's way; **Tasten·telefon** das push-button telephone

tat *1. u. 3. Pers. Sg. Prät. v.* tun; **Tat** die; ~, ~en act; *(das Tun)* action; **eine gute ~**: a good deed; **in der ~** *(verstärkend)* actually; *(zustimmend)* indeed

Tatar das; ~[s] steak tartare

Täter der; ~s, ~, **Täterin** die; ~, ~nen culprit; **tätig** *Adj.* a) ~ **sein** work; b) *(rührig, aktiv)* active; **tätigen** *tr. V. (Kaufmannsspr., Papierdt.)* transact ⟨business, deal, etc.⟩; **Tätigkeit** die; ~, ~en activity; *(Arbeit)* job; **Tatkraft** die energy; drive; **tat·kräftig** 1. *Adj.* energetic ⟨person⟩; 2. *adv.* energetically

tätowieren *tr. V.* tattoo; **Tätowierung** die; ~, ~en tattoo

Tat·sache die fact; **tatsächlich** 1. *Adj.* actual; real; 2. *adv.* actually; really

tätscheln *tr. V.* pat

Tatze die; ~, ~n paw

¹**Tau** der; ~[e]s dew

²**Tau** das; ~[e]s, ~e *(Seil)* rope

taub *Adj.* a) deaf; b) *(wie abgestorben)* numb; c) *(leer, unbefruchtet usw.)* empty ⟨nut⟩; dead ⟨rock⟩

¹**Taube** die; ~, ~n pigeon; *(Turtel~; auch Politikfig.)* dove

²**Taube** der/die; *adj. Dekl.* deaf person; deaf man/woman; **die ~n** the deaf; **Taubheit** die; ~: deafness; **taub·stumm** *Adj.* deaf and dumb;

Taub·stumme der/die; *adj. Dekl.* deaf mute

tauchen 1. *itr. V.* a) *auch mit sein* dive (**nach** for); b) *mit sein (ein~)* dive; *(auf~)* rise; emerge; 2. *tr. V.* a) *(ein~)* dip; b) *(unter~)* duck; **Taucher** der; ~s, ~, **Taucherin** die; ~, ~nen diver; *(mit Flossen und Atemgerät)* skin-diver; **Tauch·sieder** der; ~s, ~: portable immersion heater

tauen 1. *itr. V.* a) *unpers.* **es taut** it's thawing; b) *mit sein (schmelzen)* melt; 2. *tr. V.* melt; thaw

Taufe die; ~, ~n *(christl. Rel.)* a) *o. Pl. (Sakrament)* baptism; b) *(Zeremonie)* christening; baptism; **taufen** *tr. V.* a) baptize; b) *(einen Namen geben)* christen

taugen *itr. V.* **nichts/nicht viel/etwas ~**: be no/not much/some good or use; **tauglich** *Adj.* [nicht] ~: [un]suitable; *(für Militärdienst)* fit [for service]

Taumel der; ~s a) [feeling of] dizziness; b) *(Rausch)* frenzy; fever; **taumelig** *Adj.* dizzy; giddy; **taumeln** *itr. V.* a) *auch mit sein (wanken)* reel, sway (**vor** + *Dat.* with); b) *mit sein (sich ~d bewegen)* stagger

Tausch der; ~[e]s, ~e exchange; **ein guter/schlechter ~**: a good/bad deal; **tauschen** 1. *tr. V.* exchange (**gegen** for); **sie tauschten die Plätze** they changed places; 2. *itr. V.* **mit jmdm. ~** *(fig.)* change places with sb.

täuschen 1. *tr. V.* deceive; **wenn mich nicht alles täuscht** unless I'm completely mistaken; 2. *itr. V.* be deceptive; 3. *refl. V.* be wrong or mistaken (**in** + *Dat.* about); **täuschend** 1. *Adj.* remarkable, striking ⟨similarity, imitation⟩; 2. *adv.* remarkably; **Täuschung** die; ~, ~en deception; *(Selbst~)* delusion

tausend *Kardinalz.* a) a or one thousand; b) *(ugs.: sehr viele)* thousands of; ~ **Dank/Küsse** a thousand thanks/kisses; **Tausend** das; ~s, ~e od. ~ a) *nicht in Verbindung mit Kardinalzahlen; Pl.:* ~ thousand; b) *Pl. (eine unbestimmte große Zahl)* thousands; **tausend·ein[s]** *Kardinalz.* a or one thousand and one; **Tausender** der; ~s, ~ *(ugs.) (Tausendmarkschein usw.)* thousand-mark/-dollar etc. note; *(Betrag)* thousand marks/dollars etc.; **tausenderlei** *Gattungsz.; indekl. (ugs.)* a thousand and one different ⟨answers, kinds, etc.⟩; **tausend·mal** *Adv.* a thousand times; **Tausend-**

mark·schein der thousand-mark note; **tausendst...** Ordinalz. thousandth; s. auch **acht...**; **tausendstel** Bruchz. thousandth; **Tausendstel** das (schweiz. meist der); ~s, ~: thousandth

Tau·wetter das thaw

Taxi das; ~s, ~s taxi; **Taxi·fahrer** der taxi-driver

Tb, Tbc [te:'be:, te:be:'tse:] die; ~ Abk. Tuberkulose TB

Technik die; ~, ~en a) o. Pl. technology; (Studienfach) engineering no art.; b) o. Pl. (technische Ausrüstung) equipment; c) (Arbeitsweise, Verfahren) technique; **Techniker** der; ~s, ~, **Technikerin** die; ~, ~nen technical expert; **technisch** ['tɛçnɪʃ] 1. Adj. technical; technological (progress, age); 2. adv. technically; technologically (advanced); **Technologie** die; ~, ~n technology

TEE [te:|e:'|e:] der; ~|s|, ~|s| Abk. Trans-Europ-Express TEE

Tee der; ~s, ~s tea

Tee-: ~**beutel** der tea-bag; ~**kanne** die teapot; ~**löffel** der teaspoon; ~**sieb** das tea-strainer; ~**tasse** die teacup

Teich der; ~|e|s, ~e pond

Teig der; ~|e|s, ~e dough; (Kuchen~, Biskuit~) pastry; (Pfannkuchen~, Waffel~) batter; **Teig·waren** Pl. pasta sing.

Teil a) der; ~|e|s, ~e part; fünfter ~: fifth; b) der od. das; ~|e|s, ~e (Anteil; Beitrag) share; c) der; ~|e|s, ~e (beteiligte Person|en|; Rechtsw.: Partei) party; d) das; ~|e|s, ~e (Einzel~) part; **teil·bar** Adj. divisible (durch by); **Teilchen** das; ~s, ~ a) (kleines Stück) [small] part; b) (Partikel) particle; **teilen** 1. tr. V. a) divide (durch by); b) (auf~; teilhaben [lassen] an) share (unter + Dat. among); 2. refl. V. sich (Dat.) etw. |mit jmdm.| ~: share sth. [with sb.]; **teil|haben** unr. itr. V. share (an + Dat. in); **Teil·kaskoversicherung** die insurance giving limited cover

Teilnahme die; ~, ~n a) participation (an + Dat. in); ~ an einem Kurs attendance at a course; b) (Interesse) interest (an + Dat. in); c) (geh.: Mitgefühl) sympathy; **teilnahms·los** Adj. indifferent; **Teilnahmslosigkeit** die indifference; **teilnahms·voll** 1. Adj. compassionate; 2. adv. compassionately; **teil|nehmen** unr. itr. V.

[an etw. (Dat.)] ~: take part [in sth.]; [an einem Lehrgang] ~: attend [a course]; **Teilnehmer** der; ~s, ~ a) participant (Gen., an + Dat. in); (bei Wettbewerb auch) competitor, contestant (an + Dat. in); b) (Fernspr.) subscriber

teils Adv. partly; **Teilung** die; ~, ~en division; **teil·weise** 1. Adv. partly; 2. adj. partial; **Teilzeit·arbeit** die part-time work no indef. art.

Teint [tɛ̃:] der; ~s, ~s complexion

Telefon ['te:lefo:n, auch tele'fo:n] das; ~s, ~e telephone; phone (coll.); ans ~ gehen answer the [tele]phone

Telefon-: ~**anruf** der [tele]phone call; ~**anschluß** der telephone; line; ~**apparat** der telephone

Telefonat das; ~|e|s, ~e telephone call

Telefon-: ~**buch** das [tele]phone book or directory; ~**gespräch** das telephone conversation

telefonieren itr. V. make a [tele]phone call; mit jmdm. ~: talk to sb. [on the telephone]; **telefonisch** 1. Adj. telephone; 2. adv. by telephone; **Telefonist** der; ~en, ~en, **Telefonistin** die; ~, ~nen telephonist; (in einer Firma) switchboard operator

Telefon-: ~**nummer** die [tele]phone number; ~**verzeichnis** das telephone list; ~**zelle** die [tele]phonebooth or (Brit.) -box; call-box (Brit.)

Telegraf der; ~en, ~en telegraph; **Telegrafie** die; ~: telegraphy no art.; **telegrafieren** itr., tr. V. telegraph; **telegrafisch** 1. Adj. telegraphic; 2. adv. by telegraph or telegram

Telegramm das telegram

Tele·objektiv das (Fot.) telephoto lens

Teller der; ~s, ~: plate

Temperament das; ~|e|s, ~e a) (Wesensart) temperament; b) o. Pl. (Schwung) eine Frau mit ~: a woman with spirit; das ~ geht oft mit mir durch I often lose my temper; **temperament·voll** Adj. spirited (person, speech, dance, etc.)

Temperatur die; ~, ~en temperature

Temperatur-: ~**anstieg** der rise in temperature; ~**rückgang** der drop or fall in temperature

Tempo das; ~s, ~s od. Tempi a) Pl. ~s speed; b) (Musik) tempo; time

Tempus das; ~, Tempora (Sprachw.) tense

Tendenz die; ~, ~en trend; **tendieren** itr. V. tend (**zu** towards)

Teneriffa (das); ~s Tenerife

Tennis das; ~: tennis no art.

Tennis-: ~**ball der** tennis-ball; ~**platz** der tennis-court; ~**schläger** der tennis-racket; ~**spieler** der tennis-player

Tenor der; ~s, Tenöre, (österr. auch:) ~**e** (Musik) tenor; (im Chor) tenors pl.; tenor voices pl.

Teppich der; ~s, ~e carpet; (kleiner) rug; **Teppich·boden** der fitted carpet

Termin der; ~s, ~e date; (Anmeldung) appointment; (Verabredung) engagement; (Rechtsw.) hearing; **Terminal** ['tø:ɐmɪnəl] das; ~s, ~s terminal; **Termin·kalender** der appointments book

Terpentin das, (österr. meist:) der; ~s **a)** (Harz) turpentine; **b)** (ugs.: Terpentinöl) turps sing. (coll.); **Terpentinöl** das oil of turpentine

Terrain [tɛ'rɛ̃:] das; ~s, ~s terrain

Terrasse die; ~, ~n terrace

Terrier ['tɛrɪɐ] der; ~s, ~: terrier

Terrine die; ~, ~n tureen

Territorium das; ~s, Territorien territory

Terror der; ~s terrorism no art.; **terrorisieren** tr. V. **a)** terrorize; **b)** (ugs.: belästigen) pester; **Terrorist** der; ~en, ~en terrorist

Terz die; ~, ~en (Musik) third

Test der; ~[e]s, ~s od. ~e test

Testament das; ~[e]s, ~e **a)** will; **b)** (christl. Rel.) Testament

testen tr. V. test (**auf** + Akk. for)

teuer 1. Adj. expensive; dear usu. pred.; **wie ~ war das?** how much did that cost?; 2. adv. expensively; dearly; **etw. ~ kaufen/verkaufen** pay a great deal for sth./sell sth. at a high price; **Teuerung** die; ~, ~en rise in prices

Teufel der; ~s, ~: devil; **teuflisch** 1. Adj. **a)** devilish, fiendish ⟨plan, trick, etc.⟩; diabolical ⟨laughter, pleasure, etc.⟩; **b)** (ugs.: groß, intensiv) terrible (coll.); dreadful (coll.); 2. adv. **a)** diabolically; **b)** (ugs.) terribly (coll.)

Text der; ~[e]s, ~e text; (Wortlaut) wording; (eines Theaterstücks) script; (einer Oper) libretto; (eines Liedes, Chansons usw.) words pl.; (eines Schlagers) words pl.; lyrics pl.; (zu einer Abbildung) caption; **texten** tr. V. write ⟨song, advertisement, etc.⟩

Textilien Pl. **a)** textiles; **b)** (Fertigwaren) textile goods

Thailand (das); ~s Thailand

Theater das; ~s, ~ **a)** theatre; **ins ~ gehen** go to the theatre; **im ~:** at the theatre; ~ **spielen** act; (fig.) play-act; pretend; **b)** o. Pl. (fig. ugs.) fuss

Theater-: ~**abonnement** das theatre subscription [ticket]; ~**stück** das [stage] play

Theke die; ~, ~n **a)** (Schanktisch) bar; **b)** (Ladentisch) counter

Thema das; ~s, Themen subject; topic; (einer Abhandlung) subject; theme; (Leitgedanke) theme

Themse die; ~: Thames

Theologe der; ~n, ~n theologian; **Theologie** die; ~, ~n theology no art.; **theologisch** 1. Adj. theological; 2. adv. theologically

Theorie die; ~, ~n theory

Therapeut der; ~en, ~en, **Therapeutin** die; ~, ~nen therapist; therapeutist; **therapeutisch** 1. Adj. therapeutic; 2. adv. therapeutically

Therapie die; ~, ~n therapy (**gegen** for)

Thermo·meter das (österr. u. schweiz. der od. das) thermometer; **Thermos·flasche** ⓦ die Thermos flask (P); vacuum flask; **Thermostat** der; ~[e]s od. ~en, ~e od. ~en thermostat

Thron der; ~[e]s, ~e throne

Thun·fisch der tuna

Thüringen (das); ~s Thuringia; **Thüringer Wald** der Thuringian Forest

Thymian der; ~s, ~e thyme

ticken itr. V. tick

tief 1. Adj. (auch fig.) deep; (niedrig) low; low ⟨neckline, bow⟩; deep; intense ⟨pain, suffering⟩; 2. adv. deep; (niedrig) low; (intensiv) deeply; ⟨stoop, bow⟩ low; ⟨breathe, inhale⟩ deeply; **Tief** das; ~s, ~s (Met.) low

tief-, Tief-: ~**bewegt** Adj. (präd. getrennt geschrieben) deeply moved; ~**blau** Adj. deep blue; ~**druck** der; o. Pl. (Met.) low pressure

Tiefe die; ~, ~n depth; **in die ~ stürzen** plunge into the depths

tief-, Tief-: ~**garage** die underground car park; ~**greifend; tiefer greifend, am tiefsten greifend** od. **tiefstgreifend** 1. Adj. profound, deep ⟨crisis⟩; far-reaching ⟨improvement⟩; 2. adv. profoundly; ~**gründig** Adj. profound; ~**kühlen** tr. V. [deep-]freeze

Tief·kühl-: ~**fach** das freezer [compartment]; ~**kost** die frozen food
tief-, Tief-: ~**punkt** der low [point]; ~**see** die *(Geogr.)* deep sea; ~**sinnig** 1. *Adj.* profound; 2. *adv.* profoundly
Tiegel der; ~**s, ~** *(zum Kochen)* pan; *(Schmelz~)* crucible; *(Behälter)* pot
Tier das; ~|e|s, ~e animal
Tier-: ~**arzt** der veterinary surgeon; vet; ~**garten** der zoo; zoological garden; ~**heim** das animal home
tierisch 1. *Adj.* a) animal *attrib.;* savage ⟨*cruelty, crime*⟩; b) *(ugs.: unerträglich groß)* terrible *(coll.);* ~**er** Ernst deadly seriousness; 2. *adv.* a) ⟨*roar*⟩ like an animal; savagely ⟨*cruel*⟩; b) *(ugs.: unerträglich)* terribly *(coll.)*
tier-, Tier-: ~**kreis** der; *o. Pl. (Astron., Astrol.)* zodiac; ~**kreis·zeichen** das *(Astron., Astrol.)* sign of the zodiac; ~**lieb** *Adj.* animal-loving *attrib.;* fond of animals *postpos.;* ~**park** der zoo; ~**pfleger** der animal-keeper; ~**quälerei** [---'-] die cruelty to animals; ~**reich** das; *o. Pl.* animal kingdom
Tiger der; ~**s, ~**: tiger
tilgen *tr. V.* a) *(geh.)* delete ⟨*word, letter, error*⟩; erase ⟨*record, endorsement*⟩; *(fig.)* wipe out ⟨*shame, guilt, traces*⟩; b) *(Wirtsch., Bankw.)* repay; pay off
Tilsiter der; ~**s, ~**: Tilsit [cheese]
Tinte die; ~, ~**n** ink; **in der ~ sitzen** *(ugs.)* be in the soup *(coll.);* **Tinten·fisch** der cuttlefish; *(Krake)* octopus
Tip der; ~**s, ~s** a) *(ugs.)* tip; b) *(bei Toto, Lotto usw.)* [row of] numbers; **tippen** 1. *itr. V.* a) **an/gegen etw.** *(Akk.)* ~: tap sth.; b) *(ugs.: maschineschreiben)* type; c) *(wetten)* do the pools/lottery *etc.;* **im Lotto ~**: do the lottery; 2. *tr. V.* a) tap; b) *(ugs.: mit der Maschine schreiben)* type; c) *(setzen auf)* choose; **sechs Richtige ~**: make six correct selections
tipp·topp *(ugs.)* 1. *Adj. (tadellos)* immaculate; *(erstklassig)* tip-top; 2. *adv.* immaculately
Tirol (das); ~**s** [the] Tyrol; **Tiroler** der; ~**s, ~, Tirolerin** die; ~, ~**nen** Tyrolese; Tyrolean
Tisch der; ~|e|s, ~e table; **reinen ~ machen** *(ugs.)* sort things out
Tisch-: ~**dame** die dinner partner; ~**decke** die table-cloth; ~**gebet** das grace; ~**herr** der dinner partner; ~**lampe** die table-lamp

Tischler der; ~**s, ~**: joiner; *(bes. Kunst~)* cabinet-maker; **Tischlerei** die; ~, ~**en** a) *(Werkstatt)* joiner's/cabinet-maker's [workshop]; b) *o. Pl. (Handwerk)* joinery/cabinet-making
Tisch-: ~**nachbar** der person next to one [at table]; ~**platte** die table-top; ~**tennis** das table tennis; ~**tuch** das; *Pl.* ~**tücher** table-cloth; ~**wäsche** die table-linen; ~**wein** der table wine; ~**zeit** die lunch-time
Titel der; ~**s, ~** a) title; b) *(ugs.: Musikstück, Song usw.)* number
Titel-: ~**bild** das cover picture; ~**blatt** das title-page; ~**rolle** die title-role; ~**seite** die a) *(einer Zeitung, Zeitschrift)* [front] cover; b) *(eines Buchs)* title-page
titulieren *tr. V.* call
tja [tja(:)] *Interj.* [yes] well; *(Resignation ausdrückend)* oh, well
Toast [to:st] der; ~|e|s, ~e *od.* ~s toast; **Toast·brot** das; *o. Pl.* [sliced white] bread for toasting; **toasten** *tr. V.* toast; **Toaster** der; ~**s, ~**: toaster
toben *itr. V.* a) go wild (vor + *Dat.* with); *(fig.)* ⟨*storm, sea, battle*⟩ rage; b) *(tollen)* romp *or* charge about; c) *mit sein (laufen)* charge
Tochter die; ~, **Töchter** daughter
Tod der, ~|e|s, ~e death; **eines natürlichen/gewaltsamen ~es sterben** die a natural/violent death; **jmdn. zum ~e verurteilen** sentence sb. to death; **tod·ernst** 1. *Adj.* deadly serious; 2. *adv.* deadly seriously
Todes-: ~**anzeige** die a) *(in einer Zeitung)* death notice; b) *(Karte)* card announcing a person's death; ~**fall** der death; *(in der Familie)* bereavement; ~**nachricht** die news of his/her/their *etc.* death; ~**opfer** das death; fatality; ~**strafe** die death penalty; ~**ursache** die cause of death; ~**urteil** das death sentence
Tod·feind der deadly enemy; **tod·krank** *Adj.* critically ill; **tödlich** 1. *Adj.* a) fatal ⟨*accident, illness, outcome, etc.*⟩; lethal, deadly ⟨*poison, bite, shot, trap, etc.*⟩; lethal ⟨*dose*⟩; b) *(sehr groß, ausgeprägt)* deadly ⟨*hatred, seriousness, certainty, boredom*⟩; 2. *adv.* a) fatally; b) *(sehr)* terribly *(coll.)*
tod-, Tod-: ~**müde** *Adj.* dead tired; ~**sicher** *(ugs.)* 1. *Adj.* sure-fire *(coll.);* 2. *adv.* for certain *or* sure; ~**sünde** die *(auch fig.)* deadly *or* mortal sin; ~**unglücklich** *Adj. (ugs.)* extremely *or* desperately unhappy

Toilette [tŏa'lɛtə] **die;** ~, ~**n** toilet
Toiletten·papier das toilet paper
toi, toi, toi ['tɔy 'tɔy 'tɔy] *Interj.* good
luck!; *(unberufen!)* touch wood!
Tokio (das); ~s Tokyo
tolerant 1. *Adj.* tolerant **(gegen** of); **2.**
adv. tolerantly; **Toleranz die;** ~:
tolerance; **tolerieren** *tr. V.* tolerate
toll 1. *Adj.* **a)** *(ugs.) (großartig)* great
(coll.); fantastic *(coll.); (erstaunlich)*
amazing; *(heftig, groß)* enormous ⟨*re-*
spect⟩; terrific *(coll.)*⟨*noise, storm*⟩; **b)**
(wild) wild; **2.** *adv.* **a)** *(ugs.: großartig)*
terrifically well *(coll.);* **b)** *(ugs.: heftig)*
⟨*rain, snow*⟩ like billy-o *(coll.); c)*
(wild) **bei dem Fest ging es** ~ **zu** it was a
wild party; **tollen** *itr. V.* **a)** romp
about; **b)** *mit sein* romp
toll-, Toll-: ~kühn 1. *Adj.* daredevil
attrib.; daring; **2.** *adv.* daringly;
~**wut die** rabies *sing.;* ~**wütig** *Adj.*
rabid
Tolpatsch der; ~|e|s, ~**e** *(ugs.)* clumsy
or awkward creature; **tolpatschig**
(ugs.) **1.** *Adj.* clumsy; awkward; **2.**
adv. clumsily; awkwardly
Tölpel der; ~s, ~: fool; **tölpelhaft 1.**
Adj. foolish; **2.** *adv.* foolishly
Tomate die; ~, ~**n** tomato; **Toma-
ten·mark das** tomato purée
Tombola die; ~, ~s raffle
¹Ton der; ~|e|s, ~**e** clay
²Ton der; ~|e|s, **Töne a)** *(auch Physik,*
Musik; beim Telefon) tone; *(Klang)*
note; **b)** *(Film, Ferns. usw.,* ~*wiederga-*
be) sound; **c)** *(ugs.: Äußerung)* word;
d) *(Farb*~*)* shade; **e)** *(Akzent)* stress
ton-, Ton-: ~angebend *Adj.* predominant; ~**art die a)** *(Musik)* key; **b)**
(fig.) tone; ~**band das;** *Pl.* ~**bänder**
tape
Ton·band·gerät das tape recorder
tönen 1. *itr. V. (geh.)* sound; ⟨*bell*⟩
sound, ring; *(schallen, widerhallen)* resound; **2.** *tr. V. (färben)* tint
Ton·fall der tone; *(Intonation)* intonation
Tonne die; ~, ~**n a)** *(Behälter)* drum;
(Müll~*)* bin; *(Regen*~*)* water-butt; **b)**
(Gewicht) tonne; **tonnen·weise**
Adv., adj. by the ton
Tönung die; ~, ~**en** tint; shade
Topf der; ~**es, Töpfe a)** pot; *(Braten*~,
Schmor~*)* casserole; *(Stielkasserolle)*
saucepan; **b)** *(zur Aufbewahrung)* pot;
c) *(Krug)* jug; **d)** *(Nacht*~*)* chamber
pot; *(für Kinder)* potty *(Brit. coll.);* **e)**
(Blumen~*)* [flower]pot; **Topf·blume**
die [flowering] pot plant

Töpfchen das; ~s, ~: potty *(Brit.*
coll.); **Töpfer der;** ~s, ~: potter;
Töpferei die; ~, ~**en a)** *o. Pl. (Hand-*
werk) pottery *no art.;* **b)** *(Werkstatt)*
pottery; potter's workshop; **c)** *(Er-*
zeugnis) piece of pottery; ~**en** pottery
sing.
Topf-: ~lappen der oven cloth;
~**pflanze die** pot plant
Tor das; ~|e|s, ~**e a)** gate; *(einer Gara-*
ge, Scheune) door; *(fig.)* gateway; **b)**
(Ballspiele) goal; **c)** *(Ski)* gate
Torf der; ~|e|s, ~**e** peat
Torheit die; ~, ~**en** *(geh.)* **a)** *o. Pl.*
foolishness; **b)** *(Handlung)* foolish act
Tor·hüter der *(Ballspiele)* goalkeeper
töricht *(geh.)* **1.** *Adj.* foolish; **2.** *adv.*
foolishly
torkeln *itr. V.; mit sein* stagger
Tor·mann der; *Pl.* ~**männer** *od.* ~**leute** *(Ballspiele)* goalkeeper
Tornister [tɔr'nɪstə] **der;** ~s, ~: knapsack; *(Schulranzen)* satchel
torpedieren *tr. V. (Milit., fig.)* torpedo; **Torpedo der;** ~s, ~s torpedo
Törtchen das; ~s, ~: tartlet; **Torte**
die; ~, ~**n** *(Creme*~, *Sahne*~*)* gateau;
(Obst~*)* [fruit] flan
Torten-: ~boden der flan case; *(ohne*
Rand) flan base; ~**guß der** glaze;
~**heber der** cake-slice
Tortur die; ~, ~**en a)** ordeal; **b)** *(ver-*
alt.: Folter) torture
Tor-: ~wart der; ~|e|s, ~**e** *(Ballspiele)*
goalkeeper; ~**weg der** gateway
tosen *itr. V.* roar; ⟨*storm*⟩ rage
tot *Adj.* dead; ~ **umfallen** drop dead
total 1. *Adj.* total; **2.** *adv.* totally; **totalitär** *(Politik)* **1.** *Adj.* totalitarian; **2.**
adv. in a totalitarian way; ⟨*organized,*
run⟩ along totalitarian lines; **To-
tal·schaden der** *(Versicherungsw.)*
an beiden Fahrzeugen entstand ~: both
vehicles were a write-off
tot|ärgern *refl. V. (ugs.)* get livid
(coll.); **Tote der/die;** *adj. Dekl.* dead
person; **die** ~**n** the dead; **töten** *tr., itr.*
V. kill; deaden ⟨*nerve etc.*⟩
toten-, Toten-: ~blaß, ~bleich
Adj. deathly pale; ~**gräber der**
grave-digger; ~**kopf der a)** skull; **b)**
(als Symbol) death's head; *(mit ge-*
kreuzten Knochen) skull and cross-
bones; ~**schädel der** skull; ~**still**
Adj. deathly quiet; ~**stille die**
deathly silence; ~**wache die** vigil by
the body
tot-, Tot-: ~|fahren *unr. tr. V.* [run
over and] kill; ~**geboren** *Adj. (präd.*

getrennt geschrieben) stillborn; **~ge-burt** die still birth; ~|**lachen** *refl. V. (ugs.)* kill oneself laughing; **zum Totlachen sein** be killing *(coll.)*
Toto das *od.* der; ~s, ~s a) *(Pferde~)* tote *(sl.);* im ~: on the tote; b) *(Fußball~)* [football] pools *pl.;* |im| ~ spielen do the pools; **Toto·schein** der pools coupon/*(sl.)* tote ticket
tot-, Tot-: ~|**schießen** *unr. tr. V. (ugs.)* jmdn. ~**schießen** shoot sb. dead; ~**schlag der** *(Rechtsw.)* manslaughter *no indef. art.;* ~|**schlagen** *unr. tr. V.* beat to death; ~|**stellen** *refl. V.* pretend to be dead; play dead; ~|**treten** *unr. tr. V.* trample ⟨person⟩ to death; step on and kill ⟨insect⟩
Tötung die; ~, ~en killing; **fahrlässige** ~ *(Rechtsspr.)* manslaughter by culpable negligence
Toupet [tu'pe:] *das;* ~s, ~s toupee; **toupieren** [tu'pi:rən] *tr. V.* backcomb
Tour [tu:ɐ̯] die; ~, ~en tour (durch of); *(kürzere Fahrt, Ausflug)* trip; *(mit dem Auto)* drive; *(mit dem Fahrrad)* ride; *(feste Strecke)* route; **in einer** ~ *(ugs.)* the whole time; **Tourismus** [tu'rɪsmʊs] der; ~: tourism *no art.;* **Tourist** der; ~en, ~en tourist; **Touristenklasse** die tourist class; **Touristin** die; ~, ~nen tourist
Tournee [tʊr'ne:] die; ~, ~s *od.* ~n [tʊr'ne:ən] tour; **auf** ~ **sein/gehen** be/go on tour
Trab der; ~|e|s trot; **im** ~: at a trot; **im** ~ **reiten** trot; **traben** *itr. V.; mit sein (auch ugs.: laufen)* trot
Tracht die; ~, ~en a) *(Volks~)* national costume; *(Berufs~)* uniform; b) **eine** ~ **Prügel** a thrashing; *(als Strafe)* a hiding
trachten *itr. V. (geh.)* strive (nach for, after)
Tradition die; ~, ~en tradition; **traditionell 1.** *Adj.* traditional; **2.** *adv.* traditionally
traf *1. u. 3. Pers. Sg. Prät. v.* treffen; **träfe** *1. u. 3. Pers. Sg. Konjunktiv II v.* treffen
Trafik die; ~, ~en *(österr.)* tobacconist's [shop]
Trag·bahre die stretcher; **tragbar** *Adj.* **a)** portable; **b)** wearable ⟨clothes⟩; **c)** *(finanziell)* supportable ⟨cost, debt, etc.⟩; **d)** *(erträglich)* bearable; tolerable
träge 1. *Adj.* **a)** sluggish; **2.** *adv.* sluggishly

tragen 1. *unr. tr. V.* **a)** carry; **b)** *(bringen)* take; **c)** *(ertragen)* bear ⟨fate, destiny⟩; bear, endure ⟨suffering⟩; **d)** *(halten)* hold; **einen/den linken Arm in der Schlinge** ~: have one's arm/one's left arm in a sling; **e)** *(von unten stützen)* support; **f)** *(belastbar sein durch)* be able to carry *or* take ⟨weight⟩; **g)** *(übernehmen, aufkommen für)* bear, carry ⟨costs etc.⟩; take ⟨blame, responsibility, consequences⟩; **h)** *(am Körper)* wear ⟨clothes, wig, glasses, jewellery, etc.⟩; have ⟨false teeth, beard, etc.⟩; **j)** *(hervorbringen)* ⟨tree⟩ bear ⟨fruit⟩; ⟨field⟩ produce ⟨crops⟩; **2.** *unr. itr. V.* **a)** carry; **b)** *(am Körper)* **man trägt |wieder| kurz/lang** short/long skirts are in fashion [again]; **c) der Baum trägt gut** the tree produces a good crop;
tragend *Adj. (Stabilität gebend)* load-bearing; supporting ⟨wall, column, function, etc.⟩
Träger der; ~s, ~ **a)** porter; **b)** *(Zeitungs~)* paper boy/girl; delivery boy/girl; **c)** *(Bauw.)* girder; [supporting] beam; **d)** *(an Kleidung)* strap; *(Hosen~)* braces *pl.;* **e)** *(Inhaber) (eines Amts)* holder; *(eines Namens, Titels)* bearer; *(eines Preises)* winner; **Trägerin** die; ~, ~nen *s.* Träger a, b, e
Trage·tasche die carrier-bag
Trag-: ~**fähigkeit** die load-bearing capacity; ~**fläche** die wing; ~**flügel·boot** das hydrofoil
Trägheit die; ~, ~en sluggishness
Tragik die; ~: tragedy; **tragi·komisch 1.** *Adj.* tragicomic; **2.** *adv.* tragicomically; **tragisch 1.** *Adj.* tragic; **das ist nicht |so|** ~ *(ugs.)* it's not the end of the world *(coll.);* **2.** *adv.* tragically; **Tragödie die;** ~, ~n tragedy
Trag·weite die; *o. Pl.* consequences *pl.*
Trainer ['trɛːnɐ] der; ~s, ~: coach; trainer; *(einer Fußballmannschaft)* manager; **trainieren 1.** *tr. V.* **a)** train; coach ⟨swimmer, tennis-player⟩; manage ⟨football team⟩; exercise ⟨muscles etc.⟩; **b)** *(üben, einüben)* practise ⟨exercise, jump, etc.⟩; **Fußball** ~: do football training; **2.** *itr. V.* train; **Training** ['trɛːnɪŋ] das; ~s, ~s training *no indef. art.*
Trainings-: ~**anzug** der track suit; ~**hose** die track-suit bottoms *pl.*
Trakt der; ~|e|s, ~e section; *(Flügel)* wing; **Traktor** der; ~s, ~en tractor

trällern *itr., tr. V.* warble
trampeln 1. *itr. V.* **a)** |mit den Füßen|
~: stamp one's feet; **b)** *mit sein (treten)* trample (**auf** + *Akk.* on); **2.** *tr. V.*
trample; **Trampel·pfad der** [beaten]
path
trampen ['trɛmpn̩] *itr. V. mit sein*
hitch-hike
Tramway ['tramve] **die**; ~, ~s *(österr.)*
tram *(Brit.);* streetcar *(Amer.)*
Tran der; ~|e|s train-oil
tranchieren [trãˈʃiːrən] *tr. V.* carve
Träne die; ~, ~n tear; ~n lachen laugh
till one cries; **tränen** *itr. V.* ⟨*eyes*⟩
water
tranig *Adj. (ugs. abwertend: langsam)*
sluggish; slow
trank *1. u. 3. Pers. Sg. Prät. v.* **trinken**;
Tränke die; ~, ~n watering-place;
tränken *tr. V.* **a)** water; **b)** *(sich voll-
saugen lassen)* soak
Transfer der; ~s, ~s *(bes. Wirtsch.,
Sport)* transfer
Trans·formator der; ~s, ~en trans-
former
Transistor der; ~s, ~en transistor
Transit [tran'ziːt, *auch:* 'tranzɪt] **das**;
~s, ~s transit visa; **transitiv**
(Sprachw.) **1.** *Adj.* transitive; **2.** *adv.*
transitively; **Transit·verkehr der**
transit traffic
transparent *Adj.* transparent; *(Licht
durchlassend)* translucent; **Transpa-
rent das**; ~|e|s, ~e *(Spruchband)* ban-
ner; *(Bild)* transparency; **Transpa-
renz die**; ~ transparency
Transport der; ~|e|s, ~e **a)** trans-
portation; **b)** *(beförderte Lebewesen
od. Sachen) (mit dem Zug)* train-load;
(mit mehreren Fahrzeugen) convoy;
(Fracht) consignment; **transporta-
bel** *Adj.* transportable; *(tragbar)*
portable; **Transporteur** [...'tøːɐ̯] **der**;
~s, ~e carrier; **transport·fähig**
Adj. moveable; **transportieren** *tr.
V.* transport ⟨*goods, people*⟩*;* move
⟨*patient*⟩*;* **Transport·kosten** *Pl.*
carriage *sing.;* transport costs
Transvestit der; ~en, ~en transvestite
Trapez das; ~es, ~e **a)** *(Geom.)* tra-
pezium *(Brit.);* trapezoid *(Amer.);* **b)**
(im Zirkus o. ä.) trapeze
trappeln *itr. V.; mit sein* patter
[along]; ⟨*feet*⟩ patter; ⟨*hoofs*⟩ go clip-
clop
Trara das; ~s *(ugs.)* razzmatazz *(coll.)*
trat *1. u. 3. Pers. Sg. Prät. v.* **treten**
Tratsch der; ~|e|s *(ugs.)* gossip; tittle-

tattle; **tratschen** *itr. V. (ugs.)* gossip;
(schwatzen) chatter
Traube die; ~, ~n **a)** *(Beeren)* bunch;
(von Johannisbeeren o. ä.) cluster; **b)**
(Wein~) grape; **c)** *(Menschenmenge)*
bunch; cluster
trauen 1. *itr. V.* jmdm./einer Sache ~:
trust sb./sth.; **2.** *refl. V.* dare; **3.** *tr. V.*
(verheiraten) ⟨*vicar, registrar, etc.*⟩
marry
Trauer die; ~ **a)** grief (**über** + *Akk.*
over); *(um einen Toten)* mourning (**um**
+ *Akk.* for); **b)** *(~zeit)* [period of]
mourning; **c)** ~ **tragen** be in mourning
Trauer-: ~**fall der** bereavement;
~**feier die** memorial ceremony; *(beim
Begräbnis)* funeral ceremony; ~**karte die** [pre-printed] card of condolence; ~**kleidung die** mourning
clothes *pl.*
trauern *itr. V.* mourn; **um jmdn.** ~:
mourn for sb.
Trauer-: ~**spiel das** tragedy; *(fig.
ugs.)* deplorable business; ~**weide
die** weeping willow
träufeln *tr. V.* [let] trickle (**in** + *Akk.*
into); drip ⟨*ear-drops etc.*⟩
Traum der; ~|e|s, Träume ['trɔymə]
dream; **träumen** 1. *itr. V.* dream (**von**
of, about); *(unaufmerksam sein)* [day-]
dream; **2.** *tr. V.* dream; **Träumer der**;
~s, ~, **Träumerin die**; ~, ~nen
dreamer; **träumerisch** **1.** *Adj.*
dreamy; **2.** *adv.* dreamily;
traumhaft *(ugs.)* **1.** *Adj.* marvellous;
fabulous *(coll.);* **2.** *adv.* fabulously
(coll.)
traurig **1.** *Adj.* **a)** sad; unhappy ⟨*childhood, youth*⟩*;* painful ⟨*duty*⟩*;* **b)** *(kümmerlich)* sorry ⟨*state etc.*⟩*;* miserable
⟨*result*⟩*;* **2.** *adv.* sadly; **Traurigkeit
die**; ~: sadness; sorrow
Trau-: ~**ring der** wedding-ring;
~**schein der** marriage certificate
Trauung die; ~, ~en wedding [cere-
mony]; **Trau·zeuge der** witness *(at
wedding ceremony)*
Trecker der; ~s, ~: tractor
Treff der; ~s, ~s *(ugs.)* rendezvous;
(Ort) meeting-place; **treffen 1.** *unr.
tr. V.* **a)** hit; *(punch, blow, object)*
strike; **ihn trifft keine Schuld** he is in
no way to blame; **b)** *(erschüttern)* affect [deeply]; *(verletzen)* hurt; **c)** *(begegnen)* meet; **d)** *(vorfinden)* come
upon, find ⟨*anomalies etc.*⟩*;* **es gut/
schlecht** ~: be *or* strike lucky/be unlucky; **e)** *(als Funktionsverb)* make ⟨*ar-
rangements, choice, preparations, de-*

cision, *etc.*)*; 2. unr. itr. V.* **a)** ⟨*person,
shot, etc.*⟩ hit the target; **nicht** ~: miss
[the target]; **b)** *mit sein* auf etw. *(Akk.)*
~: come upon sth.; **auf Widerstand/
Ablehnung/Schwierigkeiten** ~: meet
with resistance/rejection/difficulties;
3. *unr. refl. V.* **a)** **sich mit** jmdm. ~:
meet sb.; **b)** *unpers.* **es trifft sich gut/
schlecht** it is convenient/inconveni-
ent; **Treffen** das; ~s, ~: meeting;
treffend 1. *Adj.* apt; **2.** *adv.* aptly;
Treffer der; ~s, ~ **a)** *(Milit., Boxen,
Fechten usw.)* hit; *(Schlag)* blow;
(Ballspiele) goal; **b)** *(Gewinn)* win;
(Los) winner; **trefflich** *(geh.)* **1.** *Adj.*
excellent; splendid ⟨*person*⟩; **2.** *adv.*
excellently; splendidly
treff-, Treff-: ~**punkt** der meeting-
place; ~**sicher** 1. *Adj.* accurate
⟨*language, mode of expression*⟩; unerring ⟨*judgement*⟩; **2.** *adv.* accurately;
~**sicherheit** die; *o. Pl.* accuracy
Treib·eis das drift-ice
treiben 1. *unr. tr. V.* **a)** drive; **b)** *(sich
beschäftigen mit)* go in for *(farming,
cattle-breeding, etc.)*; study ⟨*French
etc.*⟩; carry on, pursue ⟨*studies, trade,
craft*⟩; **viel Sport** ~: do a lot of sport;
es wüst/übel/toll ~ *(ugs.)* lead a dis-
solute/bad life/live it up; **2.** *unr. itr. V.
meist, mit Richtungsangabe nur, mit
sein* drift; **Treiben** das; ~s **a)** *(Durcheinander)* bustle; **b)** *(Tun)* activities
pl.; doings *pl.*
Treib-: ~**haus** das hothouse; ~**haus-
effekt** der greenhouse effect; ~**stoff**
der fuel
Trenchcoat ['trɛntʃkoʊt] der; ~|s|, ~s
trench coat
Trend der; ~s, ~s trend **(zu** + *Dat.* towards); *(Mode)* vogue
trennen 1. *tr. V.* **a)** separate **(von**
from); sever ⟨*head, arm*⟩; **b)** *(auf~)*
unpick ⟨*dress, seam*⟩; **c)** *(teilen)* divide
⟨*word, parts of a room etc., fig.:
people*⟩; **2.** *refl. V.* **a)** *(voneinander
weggehen)* part [company]; **b)** *(eine
Partnerschaft auflösen)* ⟨*couple, partners*⟩ split up; **c) sich von etw.** ~: part
with sth.; **Trennung** die; ~, ~**en** *(von
Menschen)* separation **(von** from);
(von Gegenständen) parting; *(von
Wörtern)* division
trepp- ~**ab** *Adv.* down the stairs;
~**auf** *Adv.* up the stairs
Treppe die; ~, ~**n** staircase; [flight
sing. of] stairs *pl.; (im Freien, auf der
Bühne)* [flight *sing.* of] steps *pl.*
Treppen-: ~**absatz** der half-landing;

~**geländer** das banisters *pl.;* ~**haus**
das stair-well; ~**stufe** die stair; *(im
Freien)* step
Tresen der; ~s, ~ *(bes. nordd.)* bar;
(Ladentisch) counter
Tresor der; ~s, ~e safe
Tret·boot das pedalo; **treten** 1. *unr.
itr. V.* **a)** *mit sein* step **(in** + *Akk.* into,
auf + *Akk.* on to); **b)** *(seinen Fuß setzen)* **auf etw.** *(Akk.)* ~ tread on sth.; **c)**
(ausschlagen) kick; **2.** *unr. tr. V.* **a)**
(Tritt versetzen) kick ⟨*person, ball,
etc.*⟩; **b)** *(trampeln)* trample ⟨*path*⟩; **c)**
(mit dem Fuß niederdrücken) step on
⟨*brake, pedal*⟩; operate ⟨*bellows,
clutch*⟩
treu 1. *Adj.* faithful; loyal; faithful
⟨*husband, wife*⟩; loyal ⟨*ally, subject*⟩;
jmdm. ~ **sein** be true to sb.; **sich selbst**
(Dat.)/**seinem Glauben** ~ **bleiben** be
true to oneself/one's faith; **2.** *adv.*
faithfully; loyally; **Treue** die; ~ **a)**
loyalty; *(von [Ehe]partnern)* fidelity;
b) *(Genauigkeit)* accuracy
treu-, Treu-: ~**hand[anstalt]** die *o.
Pl. (Wirtschaft)* German privatization
agency; ~**herzig** 1. *Adj.* ingenuous;
(naiv) naïve; *(unschuldig)* innocent; **2.**
adv. ingenuously; *(naiv)* naïvely; *(unschuldig)* innocently; ~**los** 1. *Adj.* disloyal, faithless ⟨*friend, person*⟩; unfaithful ⟨*husband, wife, lover*⟩; **2.** *adv.*
faithlessly
Tribunal das; ~s, ~e tribunal; **Tribüne** die; ~, ~**n** [grand]stand
Trichter der; ~s, ~ funnel
Trick der; ~s, ~s trick; *(fig.: List)* ploy
trieb 1. *u.* 3. *Pers. Sg. Prät. v.* **treiben**;
Trieb der; ~|e|s, ~e **a)** *(innerer An-
trieb)* impulse; *(Drang)* urge; *(Verlangen)* [compulsive] desire; **b)** *(Sproß)*
shoot
trieb-, Trieb-: ~**feder** die main-
spring; *(fig.)* driving *or* motivating
force; ~**haft** 1. *Adj.* compulsive; carnal ⟨*sensuality*⟩; **2.** *adv.* compulsively;
~**wagen** der *(Eisenb.)* railcar
triefen *unr. od. regelm. itr. V.* **a)** *mit
sein (fließen) (in Tropfen)* drip; *(in kleinen Rinnsalen)* trickle; **b)** *(naß sein)*
be dripping wet; ⟨*nose*⟩ run
triff *Imperativ Sg. v.* **treffen**; **trifft**
3. *Pers. Sg. Präsens v.* **treffen**
triftig *Adj.* good ⟨*reason, excuse*⟩;
valid, convincing ⟨*motive, argument*⟩
¹Trikot [tri'ko] der *od.* das; ~s, ~s
(Stoff) cotton jersey; **²Trikot** das; ~s,
~ *(ärmellos)* singlet; *(eines Tänzers)*
leotard; *(eines Fußballspielers)* shirt

Triller der; ~s, ~: trill; **trillern** 1. *itr. V.* trill; 2. *tr. V.* warble ⟨*song*⟩; **Triller·pfeife** die police/referee's whistle

Trimm-dich-Pfad der keep-fit trail; **trimmen** *tr. V. (durch Sport) get ⟨person⟩ into shape*

trinken 1. *unr. itr. V.* drink; **auf jmdn./ etw.** ~: drink to sb./sth.; 2. *unr. tr. V.* drink; **einen Kaffee/ein Bier** ~: have a coffee/beer; **Trinker** der; ~s, ~: alcoholic; **Trinkerei** die; ~, ~en drinking *no art.*

Trink-: ~**geld** das tip; ~**wasser** das; *Pl.* ~wässer drinking-water; „kein ~wasser" 'not for drinking'

Trio das; ~s, ~s *(Musik, fig.)* trio

trippeln *itr. V.; mit sein* trip; ⟨*child*⟩ patter

trist *Adj.* dreary; dismal

tritt *Imperativ Sg. u. 3. Pers. Sg. Präsens v. treten;* **Tritt** der; ~|e|s, ~e *(Schritt; Trittbrett)* step; *(Fuß~)* kick; **Tritt·brett** das step

Triumph der; ~|e|s, ~e triumph; **triumphieren** *itr. V.* a) exult; b) *(siegen)* be triumphant; triumph *(lit. or fig.)* **(über** + *Akk.* over)

trivial 1. *Adj.* a) *(platt)* banal; trite; *(unbedeutend)* trivial; b) *(alltäglich)* humdrum ⟨*life, career*⟩; 2. *adv. (platt)* banally; ⟨*say etc.*⟩ tritely

trocken 1. *Adj. (auch fig.)* dry; 2. *adv.* drily; **Trocken·haube** die [hoodtype] hair-drier; **Trockenheit** die; ~, ~en a) *o. Pl.* dryness; b) *(Dürreperiode)* drought

trocken-, Trocken-: ~|legen *tr. V.* a) **ein Baby** ~legen change a baby's nappies *(Brit.)* or *(Amer.)* diapers; b) *(entwässern)* drain ⟨*marsh, pond, etc.*⟩; ~**milch** die dried milk; ~|reiben *unr. tr. V.* rub ⟨*hair, child, etc.*⟩ dry; wipe ⟨*crockery, window, etc.*⟩ dry

trocknen 1. *itr. V.; meist mit sein* dry; 2. *tr. V.* dry

Troddel die; ~, ~n tassel

Trödel der; ~s *(ugs.)* junk; *(für den Flohmarkt)* jumble; **trödeln** *itr. V.* a) *(ugs.)* dawdle **(mit** over); b) *mit sein (ugs.: schlendern)* saunter; **Trödler** der; ~s, ~ *(ugs.)* junk-dealer

troff *1. u. 3. Pers. Sg. Prät. v. triefen*

trog *1. u. 3. Pers. Sg. Prät. v. trügen*

Trog der; ~|e|s, **Tröge** trough

trollen *(ugs.) refl. V.* push off *(coll.)*

Trommel die; ~, ~n drum; **trommeln** 1. *itr. V.* a) beat the drum; *(als Beruf, Hobby usw.)* play the drums; b) *[auf etw.]* schlagen, auftreffen) drum **(auf** + *Akk.* on, **an** + *Akk.* against); **Trommel·wirbel** der drum-roll; **Trommler** der; ~s, ~drummer

Trompete die; ~, ~n trumpet; **trompeten** 1. *itr. V.* play the trumpet; *(fig.) ⟨elephant⟩* trumpet; 2. *tr. V.* play ⟨*piece*⟩ on the trumpet; **Trompeter** der; ~s, ~trumpeter

Tropen *Pl.* tropics; **Tropen-** tropical; **Tropen·helm** der sun-helmet

Tropf der; ~|e|s, ~e *(Med.)* drip; **Tröpfchen** das; ~s, ~: droplet; *(kleine Menge)* drop; **tröpfeln** 1. *itr. V.* a) *mit sein* drip **(auf** + *Akk.* on to, **aus, von** from); b) *unpers. (ugs.: leicht regnen)* es tröpfelt it's spitting [with rain]; 2. *tr. V.* let ⟨*sth.*⟩ drip **(in** + *Akk.* into, **auf** + *Akk.* on to); **tropfen** 1. *itr. V.; mit Richtungsangabe mit sein* drip; ⟨*tears*⟩ fall; *unpers.* es tropft [vom Dach usw.] water is dripping from the roof *etc.;* 2. *tr. V.* let ⟨*sth.*⟩ drip **(in** + *Akk.* into, **auf** + *Akk.* on to); **Tropfen** der; ~s, ~ drop; **ein guter/edler** ~: a good/fine vintage; **Tropfstein·höhle** die limestone cave with stalactites and/or stalagmites

Trophäe die; ~, ~n *(hist., Jagd, Sport)* trophy

tropisch *Adj.* tropical

Troß der; **Trosses, Trosse** a) *(Milit.)* baggage train; b) *(Gefolge)* retinue; *(fig.: Zug)* procession [of hangers-on]

Trost der; ~|e|s consolation; *(bes. geistlich)* comfort; **nicht [ganz od. recht] bei** ~ **sein** *(ugs.)* be out of one's mind; **trösten** 1. *tr. V.* comfort, console **(mit** with); 2. *refl. V.* console oneself; **tröstlich** *Adj.* comforting; **trost·los** *Adj.* a) hopeless; *(verzweifelt)* in despair *postpos.;* b) *(deprimierend, öde)* miserable; dreary; hopeless ⟨*situation*⟩; **Trostpreis** der consolation prize

Trott der; ~|e|s, ~e trot; *(fig.)* routine; **Trottel** der; ~s, ~ *(ugs.)* fool; **trottelig** *(ugs.)* 1. *Adj.* doddery; 2. *adv.* in a feeble-minded way

trotten *itr. V.; mit sein* trot [along]

trotz *Präp. mit Gen., seltener mit Dat.* in spite of; despite; **Trotz** der; ~es defiance; **trotz·dem** [*auch:* '-'-'] *Adv.* nevertheless; **trotzen** *itr. V.* a) *(geh.: widerstehen)* jmdm./einer Sache ~ *(auch fig.)* defy sb./sth.; b) *(trotzig sein)* be contrary; **trotzig** 1. *Adj.* defiant; *(widerspenstig)* contrary; difficult ⟨*child*⟩; 2. *adv.* defiantly

trüb[e] 1. *Adj.* **a)** *(nicht klar)* murky ⟨*stream, water*⟩; cloudy ⟨*liquid, wine, juice*⟩; *(schlammig)* muddy ⟨*puddle*⟩; *(schmutzig)* dirty ⟨*glass, windowpane*⟩; dull ⟨*eyes*⟩; **b)** *(nicht hell)* dim ⟨*light*⟩; dull, dismal ⟨*day, weather*⟩; grey, overcast ⟨*sky*⟩; 2. *adv.* ⟨*shine, light*⟩ dimly
Trubel der; ~s [hustle and] bustle
trüben 1. *tr. V.* **a)** make ⟨*liquid*⟩ cloudy; cloud ⟨*liquid*⟩; **b)** *(beeinträchtigen)* dampen ⟨*mood*⟩; mar ⟨*relationship*⟩; cloud ⟨*judgement*⟩; 2. *refl. V.* ⟨*liquid*⟩ become cloudy; ⟨*eyes*⟩ become dull; ⟨*sky*⟩ darken; **Trübsal** die; ~, ~e *(geh.)* **a)** *(Leiden)* affliction; **b)** *o. Pl. (Kummer)* grief; ~ **blasen** *(ugs.)* mope (**wegen** over, about)
trüb-, Trüb-: ~**selig** 1. *Adj.* **a)** *(öde)* dreary, depressing ⟨*place, area, colour*⟩; **b)** *(traurig)* gloomy; 2. *adv. (traurig)* gloomily; ~**sinn** der; *o. Pl.* melancholy; ~**sinnig** 1. *Adj.* melancholy; 2. *adv.* gloomily
Trübung die; ~, ~en **a)** clouding; *(des Auges)* dimming; **b)** *(Beeinträchtigung)* deterioration; *(der Stimmung)* dampening
trudeln *itr. V. mit sein* roll
Trüffel die; ~, ~n truffle
trug *1. u. 3. Pers. Prät. v.* **tragen; trüge** *1. u. 3. Pers. Sg. Konjunktiv II v.* **tragen**
trügen 1. *unr. tr. V.* deceive; 2. *unr. itr. V.* be deceptive; ⟨*feeling, deception*⟩ be a delusion; **trügerisch** 1. *Adj.* deceptive; false ⟨*hope, sign, etc.*⟩; treacherous ⟨*ice*⟩; 2. *adv.* deceptively
Truhe die; ~, ~n chest
Trümmer *Pl. (eines Gebäudes)* rubble *sing.; (Ruinen)* ruins; *(eines Flugzeugs usw.)* wreckage *sing.; (kleinere Teile)* debris *sing.;* **Trümmer·haufen** der pile *or* heap of rubble
Trumpf der; ~[e]s, **Trümpfe** *(auch fig.)* trump [card]; *(Farbe)* trumps *pl.;* ~ **sein** *(fig.: Mode sein)* be the in thing; **trumpfen** *itr. V.* play a trump
Trunk der; ~[e]s, **Trünke** *(geh.) (Getränk)* drink; beverage *(formal);* **Trunkenheit** die; ~: drunkenness; ~ **am Steuer** drunken driving; **Trunk·sucht** die; *o. Pl.* alcoholism *no art.*
Trupp der; ~s, ~s troop; *(von Arbeitern, Gefangenen)* gang; *(von Soldaten, Polizisten)* squad; **Truppe** die; ~, ~n **a)** *(Einheit der Streitkräfte)* unit; **b)** *Pl. (Soldaten)* troops; **c)** *o. Pl. (Streit-*

kräfte) [armed] forces *pl.; (Heer)* army; **d)** *(Gruppe von Schauspielern, Artisten)* troupe; *(von Sportlern)* squad
Trut·hahn der turkey [cock]
tschau *Interj. (ugs.)* ciao *(coll.)*
Tscheche der; ~n, ~n Czech; **tschechisch** *Adj.* Czech; **Tschechoslowakei** die; ~: Czechoslovakia *no art.;* **tschechoslowakisch** *Adj.* Czechoslovak[ian]
tschüs *Interj. (ugs.)* bye *(coll.)*
Tsd. *Abk.* Tausend
T-Shirt ['tiːʃɔːt] das; ~s, ~s T-shirt
Tube die; ~, ~n tube
Tuberkulose die; ~, ~n *(Med.)* tuberculosis *no art.*
Tuch das; ~[e]s, **Tücher** *od.* ~e **a)** *Pl.* Tücher cloth; *(Kopf~, Hals~)* scarf; **b)** *Pl.* ~e *(Gewebe)* cloth
tüchtig 1. *Adj.* **a)** efficient; *(fähig)* capable, competent (**in** + *Dat.* at); **b)** *(ugs.: beträchtlich)* sizeable ⟨*piece, portion*⟩; big ⟨*gulp*⟩; hearty ⟨*eater, appetite*⟩; 2. *adv.* **a)** efficiently; *(fähig)* competently; **b)** *(ugs.: sehr)* really ⟨*cold, warm*⟩; ⟨*snow, rain*⟩ good and proper *(coll.);* ⟨*eat*⟩ heartily; **Tüchtigkeit** die; ~: efficiency; *(Fähigkeit)* ability; competence; *(Fleiß)* industry
Tücke die; ~, ~n **a)** *o. Pl. (Hinterhältigkeit)* deceit[fulness]; *(List)* guile; **b)** *meist Pl. ([verborgene] Gefahr/Schwierigkeit)* [hidden] danger/difficulty
tuckern *itr. V.; mit Richtungsangabe mit sein* chug
tückisch 1. *Adj.* **a)** *(hinterhältig)* wily; *(betrügerisch)* deceitful; **b)** *(gefährlich)* treacherous ⟨*bend, slope, spot, etc.*⟩; 2. *adv.* craftily
tüfteln *itr. V. (ugs.)* fiddle (**an** + *Dat.* with); do finicky work (**an** + *Dat.* on); *(geistig)* rack one's brains (**an** + *Dat.* over)
Tugend die; ~, ~en virtue; **tugendhaft** 1. *Adj.* virtuous; 2. *adv.* virtuously
Tüll der; ~s, ~e tulle
Tülle die; ~, ~n *(bes. nordd.)* spout
Tulpe die; ~, ~n tulip
tummeln *refl. V.* romp [about]; **Tummel·platz** der *(auch fig.)* playground
Tumor der; ~s, ~en *(Med.)* tumour
Tümpel der; ~s, ~: pond
Tumult der; ~[e]s, ~e tumult; commotion; *(Protest)* uproar
tun 1. *unr. tr. V.* **a)** do; **so etwas tut man nicht** that is just not done; [**etwas**] **mit etw./jmdm. zu** ~ **haben** be concerned

with sth./have dealings with sb.; **b)** *als Funktionsverb* make ⟨*remark, catch, etc.*⟩; take ⟨*step, jump*⟩; do ⟨*deed*⟩; **c)** *(bewirken)* work, perform ⟨*miracle*⟩; **d)** *(an~)* jmdm. etw. ~: do sth. to sb.; **e)** es ~ *(ugs.: genügen)* be good enough; **f)** *(ugs.: irgendwohin bringen)* put; **2.** *unr. itr. V.* **a)** *(ugs.: funktionieren)* work; **b)** freundlich/geheimnisvoll ~: pretend to be or *(coll.)* act friendly/ act mysteriously; **3.** *unr. refl. V.; unpers.* es hat sich einiges getan quite a bit has happened

Tünche die; ~, ~n distemper; wash; |weiße| ~: whitewash; **tünchen** *tr. (auch itr.) V.* distemper; weiß ~: whitewash

Tunell das; ~s, ~s *(südd., österr., schweiz.) s.* **Tunnel**

Tunesien [tu'neːzi̯ən] (das); ~s Tunisia; **tunesisch** *Adj.* Tunisian

Tunke die; ~, ~n *(bes. ostmd.)* sauce; *(Bratensoße)* gravy; **tunken** *tr. V. (bes. ostmd.)* dip

Tunnel der; ~s, ~ *od.* ~s tunnel

tupfen *tr. V.* **a)** dab; **b)** *(mit Tupfen versehen)* dot; **Tupfen** der; ~s, ~: dot; *(größer)* spot; **Tupfer** der; ~s, ~ *(Med.)* swab

Tür die; ~, ~en door; *(Garten~)* gate; an die ~ gehen *(öffnen)* [go and] answer the door; vor die ~ gehen go outside

Turban der; ~s, ~e turban

Turbine die; ~, ~n turbine

turbulent **1.** *Adj. (auch fachspr.)* turbulent; **2.** *adv. (auch fachspr.)* turbulently

Tür·griff der door-handle

Türke der; ~n, ~n Turk; **Türkei** die; ~: Turkey *no art.*

türkis *indekl. Adj.* turquoise; **Türkis** der; ~es, ~e turquoise

türkisch *Adj.* Turkish

Tür·klinke die door-handle

Turm der; ~|e|s, Türme **a)** tower; *(spitzer Kirch~)* spire; steeple; **b)** *(Schach)* rook; **c)** *(Sprung~)* diving-platform; **Türmchen** das; ~s, ~: turret; ¹**türmen** **1.** *tr. V. (stapeln)* stack up; *(häufen)* pile up; **2.** *refl. V.* be piled up; ⟨*clouds*⟩ gather

²**türmen** *itr. V.; mit sein (salopp)* scarper *(Brit. sl.)*

Turm·falke der kestrel

turnen **1.** *itr. V.* do gymnastics; *(Schulw.)* do gym; **2.** *tr. V.* do, perform ⟨*exercise, routine*⟩; **Turnen** das; ~s gymnastics *sing., no art.; (Schulw.)*

gym *no art.; PE no art.;* **Turner** der; ~s, ~, **Turnerin** die; ~, ~nen gymnast

Turn-: ~**halle** die gymnasium; ~**hemd** das [gym] singlet; ~**hose** die gym shorts *pl.*

Turnier das; ~s, ~e *(auch hist.)* tournament; *(Reit~)* show; *(Tanz~)* competition

Turn·schuh der gym shoe

Turnus der; ~, ~se regular cycle

Turn·verein der gymnastics club

Tusch der; ~|e|s, ~e fanfare

Tusche die; ~, ~n Indian *(Brit.)* or *(Amer.)* India ink

tuscheln *itr., tr. V.* whisper

Tüte die; ~, ~n bag

tuten *itr. V.* hoot; ⟨*siren, [fog-]horn*⟩ sound

Typ der; ~s, ~en **a)** type; **b)** *Gen. auch* ~en *(ugs.: Mann)* bloke *(Brit. sl.);* **Type** die; ~, ~n *(Druck~, Schreibmaschinen~)* type

Typhus der; ~ typhoid [fever]

typisch **1.** *Adj.* typical *(für of);* **2.** *adv.* typically

Tyrann der; ~en, ~en *(auch fig.)* tyrant; **Tyrannei** die; ~, ~en *(auch fig.)* tyranny; **tyrannisch** **1.** *Adj.* tyrannical; **2.** *adv.* tyrannically; **tyrannisieren** *tr. V.* tyrannize

U

u, U [uː] das; ~, ~: u, U

ü, Ü [yː] das; ~, ~: u umlaut

u. *Abk.* und

u. a. *Abk.* unter anderem

U-Bahn die underground *(Brit.);* subway *(Amer.); (bes. in London)* tube; **U-Bahn-Station** die underground station *(Brit.);* subway station *(Amer.); (bes. in London)* tube station

übel *Adj.* **a)** foul, nasty ⟨*smell, weather*⟩; bad, nasty ⟨*headache, cold, taste*⟩; nasty ⟨*consequences, situation*⟩; sorry ⟨*state, affair*⟩; foul, *(coll.)* filthy ⟨*mood*⟩; nicht ~ *(ugs.)* not bad at all; **b)** *(unwohl)* jmdm. ist/wird ~: sb. feels sick; **c)** *(verwerflich)* bad;

wicked; nasty, dirty ⟨trick⟩. **Übel** das;
~s, ~ evil; **Übelkeit** die; ~, ~en
nausea
übel|nehmen unr. tr. V. jmdm. etw.
~: hold sth. against sb.; etw. ~: take
offence at sth.; **Übel·täter** der
wrongdoer
üben tr. V. **a)** (auch itr.) practise; rehearse ⟨scene, play⟩; practise on ⟨mu-
sical instrument⟩; **b)** (trainieren, schulen) exercise ⟨fingers⟩; train ⟨memory⟩
über 1. Präp. mit Dat. **a)** (Lage, Standort) over; above; (in einer Rangfolge)
above; ~ jmdm. **wohnen** live above
sb.; **zehn Grad ~ Null** ten degrees
above zero; **sie trug eine Jacke ~ dem
Kleid** she wore a jacket over her dress;
b) (während) during; ~ **dem Lesen/der
Arbeit einschlafen** fall asleep over
one's book/magazine etc./over one's
work; 2. Präp. mit Akk. **a)** (Richtung)
over; (quer hinüber) across; ~ **Ulm
nach Stuttgart** via Ulm to Stuttgart; **b)**
(während) over; (für die Dauer von)
for; **c)** (betreffend) about; ~ **etw. re-
den/schreiben** talk/write about sth.;
**ein Scheck/eine Rechnung ~ 1 000
Mark** a cheque/bill for 1,000 marks;
d) Kinder ~ 10 Jahre children over ten
[years of age]; 3. Adv. **a)** (mehr als)
over; **b)** ~ **und** ~: all over
über·all [od. --'-] Adv. **a)** everywhere;
b) (bei jeder Gelegenheit) always
über·anstrengen tr. V. overtax ⟨per-
son, energy⟩; strain ⟨eyes, nerves,
heart⟩; **sich** ~: over-exert oneself
über·arbeiten 1. tr. V. rework; revise
⟨text, edition⟩; 2. refl. V. overwork
über·aus Adv. (geh.) extremely
über·backen unr. tr. V. etw. mit Käse
usw. ~: top sth. with cheese etc. and
brown it lightly [under the grill/in a
hot oven]
überbelichten[1] tr. V. (Fot.) over-
expose
über·bieten unr. tr. V. **a)** outbid (um
by); **b)** (übertreffen) surpass; outdo
⟨rival⟩; break ⟨record⟩ (um by); exceed
⟨target⟩ (um by)
Über·blick der a) view; **einen guten ~
über etw.** (Akk.) **haben** have a good
view over sth.; **b)** (Abriß) survey; **c)** o.
Pl. (Einblick) overall view; **überblicken** tr. V. s. **übersehen a, b**
über·bringen unr. tr. V. deliver; convey ⟨greetings, congratulations⟩

---

[1] ich überbelichte, überbelichtet, über-
zubelichten

über·brücken tr. V. bridge ⟨gap,
gulf⟩; reconcile ⟨difference⟩; **Über-
brückung** die; ~, ~en (fig.)
bridging; (von Gegensätzen) recon-
ciliation
überdacht Adj. covered ⟨terrace,
station platform, etc.⟩
über·dauern tr. V. survive ⟨war, separation, hardship⟩
über·dies Adv. moreover
Über·druck der; Pl. ~drücke excess
pressure
Überdruß der; Überdrusses surfeit
(an + Dat. of); **überdrüssig** Adj.
jmds./einer Sache ~ sein/werden be/
grow tired of sb./sth.
über·eilen tr. V. rush; **übereilt** over-
hasty
über·einander Adv. **a)** one on top of
the other; **b)** ⟨talk etc.⟩ about each
other
übereinander-: ~|legen tr. V. Holz-
scheite usw. ~legen lay pieces of wood
etc. one on top of the other; ~|schla-
gen unr. tr. V. **die Arme/Beine
~schlagen** fold one's arms/cross one's
legs
überein|kommen unr. itr. V.; mit
sein agree; come to an agreement;
**Überein·kommen das; ~s, ~,
Übereinkunft die; ~, Übereinkünfte**
agreement
überein|stimmen itr. V. **a)** (einer
Meinung sein) agree (in + Dat. on);
b) (sich gleichen) ⟨colours, styles⟩
match; ⟨figures, statements, reports,
results⟩ tally, agree; ⟨views, opinions⟩
coincide; **Überein·stimmung die
a)** agreement (in + Dat. on; Gen. between)
über·empfindlich 1. Adj. over-
sensitive **(gegen** to); (Med.) hyper-
sensitive **(gegen** to); 2. adv. oversensitively; (Med.) hypersensitively
[1]über|fahren 1. unr. tr. V. jmdn. ~:
ferry or take sb. over; 2. unr. itr. V.;
mit sein cross over; **[2]über·fahren**
unr. tr. V. **a)** run over; **b)** (hinwegfah-
ren über) cross; go over ⟨crossroads⟩;
Über·fahrt die crossing (über +
Akk. of)
Über·fall der attack (auf + Akk. on);
(aus dem Hinterhalt) ambush (auf +
Akk. on); (mit vorgehaltener Waffe)
hold-up; (auf eine Bank o. ä.) raid
(auf + Akk. on); **über·fallen** unr. tr.
V. **a)** attack; raid ⟨bank, enemy position, village, etc.⟩; (hinterrücks) ambush; (mit vorgehaltener Waffe) hold

up; b) *(überkommen)* ⟨*tiredness, home-sickness, fear*⟩ come over; **über·fäl-lig** *Adj.* overdue
über·fliegen *unr. tr. V.* **a)** fly over; overfly *(formal);* b) *(flüchtig lesen)* skim [through]
über·flügeln *tr. V.* outshine; outstrip
Über·fluß der; *o. Pl.* abundance (**an** + *Dat.* of); *(Wohlstand)* affluence; **über·flüssig** *Adj.* superfluous; unnecessary ⟨*purchase, words, work*⟩
über·fluten *tr. V. (auch fig.)* flood
über·fordern *tr. V.* jmdn. |mit etw.| ~: overtax sb. [with sth.]; ask *or* demand too much of sb. [with sth.]
¹über|führen *tr. V.* transfer; **²über·führen** *tr. V.* **a)** s. **¹überführen;** b) jmdn. |eines Verbrechens| ~: find sb. guilty [of a crime]; convict sb. [of a crime]; **Über·führung** die **a)** transfer; b) *(eines Verdächtigen)* conviction; c) *(Brücke)* bridge; *(Hochstraße)* overpass; *(Fußgänger~)* [foot-]bridge
über·füllt *Adj.* crammed full (**von** with); *(mit Menschen)* overcrowded (**von** with); over-subscribed ⟨*course*⟩
Über·gabe die a) handing over (**an** + *Akk.* to); *(von Macht)* handing over; b) *(Auslieferung an den Gegner)* surrender (**an** + *Akk.* to)
Über·gang der a) crossing; b) *(Stelle zum Überqueren)* crossing; *(Bahn~)* level crossing *(Brit.);* grade crossing *(Amer.);* *(Grenz~)* crossing-point; c) *(Wechsel, Überleitung)* transition (**zu, auf** + *Akk.* to)
über·geben 1. *unr. tr. V.* **a)** hand over; pass ⟨*baton*⟩; b) *(übereignen)* transfer, make over *(Dat.* to); c) *(ausliefern)* surrender *(Dat.,* **an** + *Akk.* to); d) **eine Straße dem Verkehr** ~: open a road to traffic; **2.** *unr. refl. V. (sich erbrechen)* vomit
¹über|gehen *unr. itr. V.; mit sein* **a)** pass; b) **zu etw.** ~: go over to sth.; c) **in etw.** *(Akk.)* ~ *(zu etw. werden)* turn into sth.
²über·gehen *unr. tr. V.* **a)** *(nicht beachten)* ignore; b) *(auslassen, überspringen)* skip [over]; c) *(nicht berücksichtigen)* pass over
über·geordnet *Adj.* higher ⟨*court, authority, position*⟩; greater ⟨*significance*⟩; superordinate ⟨*concept*⟩
Über·gewicht das a) excess weight; *(von Person)* overweight; b) *(fig.)* predominance
über·glücklich *Adj.* blissfully happy; *(hoch erfreut)* overjoyed

über|greifen *unr. itr. V.* **auf etw.** *(Akk.)* ~: spread to sth.
Über·griff der *(unrechtmäßiger Eingriff)* encroachment (**auf** + *Akk.* on); infringement (**auf** + *Akk.* of); *(Angriff)* attack (**auf** + *Akk.* on)
Über·größe die outsize
überhand|nehmen *unr. itr. V.* get out of hand; ⟨*attacks, muggings, etc.*⟩ increase alarmingly; ⟨*weeds*⟩ run riot
über|hängen *tr. V.* **sich** *(Dat.)* **eine Jacke** ~: put a jacket round one's shoulders; **sich** *(Dat.)* **das Gewehr/die Tasche** ~: hang the rifle/bag over one's shoulder
über·häufen *tr. V.* jmdn. mit etw. ~: heap *or* shower sth. on sb.
überhaupt *Adv.* **a)** in general; **b)** ~ **nicht** not at all; ~ **keine Zeit haben** have no time at all; ~ **nichts** nothing at all
überheblich 1. *Adj.* arrogant; supercilious ⟨*grin*⟩; **2.** *adv.* arrogantly; ⟨*grin*⟩ superciliously
Überheblichkeit die; ~: arrogance
über·holen 1. *tr. V.* **a)** overtake *(esp. Brit.);* pass *(esp. Amer.);* b) *(übertreffen)* outstrip; c) *(wieder instand setzen)* overhaul; **2.** *itr. V.* overtake *(esp. Brit.);* pass *(esp. Amer.);* **Überhol-spur die** overtaking lane *(esp. Brit.);* pass lane *(esp. Amer.);* **überholt** *Adj. (veraltet)* outdated; **Überholung die;** ~, ~**en** overhaul
Überhol·verbot das prohibition of overtaking
über·hören *tr. V.* not hear
über·irdisch 1. *Adj.* celestial; heavenly; *(übernatürlich)* supernatural; **2.** *adv.* celestially; *(übernatürlich)* supernaturally
über|kochen *itr. V.; mit sein (auch fig. ugs.)* boil over
über·kommen *unr. tr. V.* **Mitleid/ Ekel/Furcht überkam mich** I was overcome by pity/revulsion/fear
über·laden *unr. tr. V. (auch fig.)* overload
über·lassen *unr. tr. V.* **a)** jmdm. etw. ~: let sb. have sth.; b) **sich** *(Dat.)* **selbst** ~ sein be left to one's own devices; c) etw. jmdm. ~ *(etw. jmdn. entscheiden/tun lassen)* leave sth. to sb.
über·lasten *tr. V.* overload; overtax ⟨*person*⟩; *(mit Arbeit)* overwork ⟨*person*⟩
Über·lauf der overflow; **¹über|laufen** *unr. itr. V.; mit sein* **a)** overflow; b) *(auf die gegnerische Seite überwech-*

seln) defect; ⟨partisan⟩ go over to the other side; ²über·laufen unr. tr. V. seize; ein Frösteln/Schauer überlief mich, es überlief mich ⌊eis⌋kalt my shiver ran down my spine; ³überlaufen Adj. overcrowded; Überläufer der (auch fig.) defector

über·leben tr. V. survive; Über·lebende der/die; adj. Dekl. survivor

¹über|legen tr. V. jmdm. etw. ~: put sth. over sb.; ²über·legen 1. tr. V. consider; think about; etw. ~: change one's mind; 2. itr. V. think; ³überlegen 1. Adj. a) superior; clear, convincing ⟨win, victory⟩; jmdm. ~ sein be superior to sb. (an + Dat. in); b) (herablassend) supercilious; 2. adv. a) in a superior manner; ⟨play⟩ much the better: ⟨win, argue⟩ convincingly; b) (herablassend) superciliously; Überlegenheit die; ~ superiority; überlegt 1. Adj. carefully considered; 2. adv. in a carefully considered way; Überlegung die; ~, ~en a) o. Pl. thought; b) (Gedanke) idea; ~en (Gedankengang) thoughts

über·liefern tr. V. hand down; Über·lieferung die tradition

überlisten tr. V. outwit

überm Präp. + Art. = über dem

Über·macht die; o. Pl. superior strength; (zahlenmäßig) superior numbers pl.

über·mannen tr. V. overcome

Über·maß das; o. Pl. excessive amount, excess (an + Dat. of); über·mäßig 1. Adj. excessive; 2. adv. excessively

über·menschlich Adj. superhuman

über·mitteln tr. V. send; (als Mittler weitergeben) pass on, convey ⟨greetings, regards, etc.⟩

über·morgen Adv. the day after tomorrow

Übermüdung die; ~: overtiredness

Über·mut der high spirits pl.; übermütig 1. Adj. high-spirited; 2. adv. high-spiritedly

über·nächst... Adj. im ~en Jahr, ~es Jahr the year after next; am ~en Tag two days later

über·nachten itr. V. stay overnight; übernächtigt Adj. ⟨person⟩ tired or worn out [through lack of sleep]; tired ⟨face, look, etc.⟩; Übernachtung die; ~, ~en overnight stay; ~ und Frühstück bed and breakfast

Übernahme die; ~ (von Waren, einer Sendung) taking delivery no art.; (ei

ner Idee usw.) adoption, taking over no indef. art.; (der Macht, einer Praxis usw.) take over

über·natürlich Adj. supernatural

über·nehmen 1. unr. tr. V. take delivery of ⟨goods, consignment⟩; take over ⟨power, practice, business, etc.⟩; take on ⟨job, position, etc.⟩; undertake to pay ⟨costs⟩; b) (sich zu eigen machen) adopt ⟨ideas, methods, subject, etc.⟩ (von from); borrow ⟨word, phrase⟩ (von from); 2. unr. refl. V. overdo things or it; sich mit etw. ~: take on too much with sth.

über·prüfen tr. V. check (auf + Akk. for); review ⟨issue, situation, results⟩; Über·prüfung die a) o. Pl. checking no indef. art. (auf + Akk. for); b) (Kontrolle) check; (einer Lage, Frage usw.) review

über·queren tr. V. cross

über·ragen tr. V. a) jmdn./etw. ~: tower above sb./sth.; b) (übertreffen) jmdn. an etw. (Dat.) ~: be head and shoulders above sb. in sth.; überragend 1. Adj. outstanding; 2. adv. outstandingly

überraschen tr. V. surprise; Überraschung die; ~, ~en surprise

über·reden tr. V. persuade

über·reichen tr. V. [jmdm.] etw. ~: present sth. [to sb.]

über·rumpeln tr. V. jmdn. ~: take sb. by surprise

über·runden tr. V. a) (Sport) lap; b) (übertreffen) outstrip

übers Präp. + Art. = über das

Überschall-: ~flugzeug das supersonic aircraft; ~geschwindigkeit die supersonic speed

über·schätzen tr. V. overestimate; overrate ⟨artist, talent, etc.⟩

überschaubar Adj. eine ~e Menge/ Zahl a manageable quantity/number

Über·schlag der a) rough calculation or estimate; b) (Turnen) handspring; c) s. Looping; ¹über|schlagen 1. unr. tr. V. die Beine ~: cross one's legs; 2. unr. itr. V.; mit sein ⟨wave⟩ break; ²über·schlagen 1. unr. tr. V. a) skip ⟨chapter, page, etc.⟩; b) (ungefähr berechnen) calculate or estimate roughly; 2. unr. refl. V. go head over heels; ⟨car⟩ turn over

über·schnappen itr. V.; mit sein (ugs.) go crazy

über·schneiden unr. refl. V. cross, intersect; (fig.) overlap

über·schreiben unr. tr. V. a) entitle;

head ⟨*chapter, section*⟩; **b) etw. jmdm.**
od. **auf jmdn.** ∼: transfer sth. to sb.
über·schreiten *unr. itr. V.* cross;
(fig.) exceed
Über·schrift die heading; *(in einer
Zeitung)* headline; *(Titel)* title
Über·schuß der surplus **(an** + *Dat.*
of); **überschüssig** *Adj.* surplus
über·schütten *tr. V.* cover
Überschwang der; ∼-[e]s exuberance
über·schwemmen *tr. V. (auch fig.)*
flood; **Überschwemmung die;** ∼,
∼-en flood; *(das Überschwemmen)*
flooding *no pl.*
über·schwenglich 1. *Adj.* effusive
⟨*words etc.*⟩; wild ⟨*joy, enthusiasm*⟩; **2.**
adv. effusively
Über·see *o. Art.* **aus** *od.* **von** ∼: from
overseas; **in/nach** ∼: overseas
über·sehen *unr. tr. V.* **a)** look out
over; **b)** *(abschätzen)* assess ⟨*damage,
situation, consequences, etc.*⟩; **c)** *(nicht
sehen)* overlook; miss; miss ⟨*turning,
signpost*⟩; **d)** *(ignorieren)* ignore
über·senden *unr. (auch regelm.) tr.
V.* send
¹**über|setzen 1.** *tr. V.* ferry over; **2.** *itr.
V.; auch mit sein* cross [over]; ²**übersetzen** *tr., itr. V. (auch fig.)* translate;
**Über·setzer der, Übersetzerin
die;** ∼, ∼-nen translator; **Übersetzung die;** ∼, ∼-en translation
Über·sicht die a) *o. Pl.* overall view,
overview **(über** + *Akk.* of); **b)** *(Darstellung)* survey; *(Tabelle)* summary;
über·sichtlich 1. *Adj.* clear; ⟨*crossroads*⟩ which allows a clear view; **2.**
adv. clearly
¹**über|siedeln,** ²**über·siedeln** *itr. V.;
mit sein* move **(nach** to)
über·spielen *tr. V.* **a)** *(hinweggehen
über)* cover up; smooth over ⟨*difficult
situation*⟩; **b)** *(aufnehmen)* [auf ein
Tonband] ∼: transfer ⟨*record*⟩ to tape;
put ⟨*record*⟩ on tape
über·spitzen *tr. V.* **etw.** ∼: push *or*
carry sth. too far
über·springen *unr. tr. V.* **a)** jump
⟨*obstacle*⟩; **b)** *(auslassen)* miss out
¹**über|stehen** *unr. itr. V.; südd.,
österr., schweiz. mit sein* jut out
²**über·stehen** *unr. tr. V.* come
through ⟨*danger, war, operation*⟩; get
over ⟨*illness*⟩
über·steigen *unr. tr. V.* **a)** climb
over; **b)** *(fig.)* exceed
über·stimmen *tr. V.* outvote
Über·stunde die: ∼-n **machen** do
overtime

über·stürzen 1. *tr. V.* rush; **2.** *refl. V.*
rush; *(rasch aufeinanderfolgen)*
⟨*events, news, etc.*⟩ come thick and
fast; **überstürzt 1.** *Adj.* hurried
⟨*escape, departure*⟩; over-hasty
⟨*decision*⟩; **2.** *adv.* ⟨*decide, act*⟩ over-
hastily; ⟨*depart*⟩ hurriedly
übertölpeln *tr. V.* dupe; con *(coll.)*
über·tönen *tr. V.* drown out
Übertrag der; ∼-[e]s, **Überträge** *(bes.
Buchf.)* carry-over; **über·tragbar**
Adj. transferable **(auf** + *Akk.* to);
(auf etw. anderes anwendbar) applicable **(auf** + *Akk.* to); *(übersetzbar)*
translatable; *(ansteckend)* infectious
⟨*disease*⟩; **über·tragen** *unr. tr. V.* **a)**
transfer **(auf** + *Akk.* to); transmit
⟨*power, torque, etc.*⟩ **(auf** + *Akk.* to);
communicate ⟨*disease, illness*⟩ **(auf** +
Akk. to); carry over ⟨*subtotal*⟩; *(auf
etw. anderes anwenden)* apply **(auf** +
Akk. to); *(übersetzen)* translate; **b)**
(senden) broadcast ⟨*concert, event,
match, etc.*⟩; *(im Fernsehen)* televise;
c) *(geben)* **jmdm. Aufgaben/Pflichten**
usw. ∼: hand over tasks/duties *etc.* to
sb.; *(anvertrauen)* entrust sb. with
tasks/duties *etc.;* **Übertragung die;**
∼, ∼-en **a)** *s.* **übertragen a:** trans-
ference; transmission; communication; carrying over; application;
translation; **b)** *(das Senden)* broadcasting; *(Sendung)* broadcast; *(im
Fernsehen)* televising/television
broadcast
über·treffen *unr. tr. V.* **a)** surpass,
outdo **(an** + *Dat.* in); break ⟨*record*⟩;
b) *(übersteigen)* exceed
über·treiben *unr. tr. V.* **a)** *auch itr.*
exaggerate; **b)** *(zu weit treiben)*
overdo; **Übertreibung die;** ∼, ∼-en
exaggeration
¹**über|treten** *unr. itr. V.; mit sein*
change sides; **zum Katholizismus/Is-
lam** ∼: convert to Catholicism/Islam;
²**über·treten** *unr. tr. V.* contravene
⟨*law*⟩; violate ⟨*regulation, prohibition*⟩; **Übertretung die;** ∼, ∼-en **a)** *s.*
²**übertreten:** contravention; violation;
b) *(Vergehen)* misdemeanour
übertrieben *Adj.* **1.** exaggerated;
(übermäßig) excessive ⟨*care, thrift,
etc.*⟩; **2.** *adv.* excessively
Über·tritt der change of allegiance,
switch **(zu** to); *(Rel.)* conversion **(zu**
to)
über·trumpfen *tr. V.* outdo
über·vor·teilen *tr. V.* cheat
über·wachen *tr. V.* keep under sur-

veillance ⟨*suspect, agent, area, etc.*⟩; supervise ⟨*factory, workers, process*⟩; control ⟨*traffic*⟩; monitor ⟨*progress, production process, experiment, patient*⟩; **Überwachung** die; ~, ~en *s.* **überwachen:** surveillance; supervision; controlling; monitoring

überwältigen *tr. V.* **a)** overpower; **b)** *(fig.)* ⟨*sleep, emotion, fear, etc.*⟩ overcome; ⟨*sight, impressions, beauty, etc.*⟩ overwhelm; **überwältigend 1.** *Adj.* overwhelming ⟨*sight, impression, victory, majority, etc.*⟩; overpowering ⟨*smell*⟩; stunning ⟨*beauty*⟩; **2.** *adv.* stunningly ⟨*beautiful*⟩

über·weisen *unr. tr. V.* **a)** transfer ⟨*money*⟩ (**an, auf** + *Akk.* to); **b)** refer ⟨*patient*⟩ (**an** + *Akk.* to); **Überweisung** die **a)** *o. Pl.* transfer (**an, auf** + *Akk.* to); **b)** *(Summe)* remittance; **c)** *(eines Patienten)* referral (**an** + *Akk.* to)

überwiegend 1. [*auch* --'--] *Adj.* overwhelming; **2.** *adv.* mainly

über·winden 1. *unr. tr. V.* overcome; get past ⟨*stage*⟩; **2.** *unr. refl. V.* overcome one's reluctance; **sich** |**dazu**| ~, **etw. zu tun** bring oneself to do sth.; **Über·windung** die **a)** *s.* überwinden **1:** overcoming; getting past; **b)** *(das Sichüberwinden)* **es war eine große** ~ **für ihn** it cost him a great effort

Über·zahl die; *o. Pl.* majority; **über·zählig** *Adj.* surplus

überzeugen 1. *tr. V.* convince; **2.** *itr. V.* be convincing; **überzeugend 1.** *Adj.* convincing; **2.** *adv.* convincingly; **überzeugt** *Adj.* convinced; **Über·zeugung** die *(feste Meinung)* conviction

¹über|ziehen *unr. tr. V.* pull on; **²über·ziehen** *unr. tr. V.* **a)** etw. mit etw. ~: cover sth. with sth.; **b)** overdraw ⟨*account*⟩ (um by); **Überzug** der **a)** *(Beschichtung)* coating; **b)** *(Bezug)* cover

üblich *Adj.* usual; *(normal)* normal; *(gebräuchlich)* customary

U-Boot das submarine; sub *(coll.)*

übrig *Adj.* remaining *attrib.*; *(ander...)* other; **alle** ~**en Gäste** ...: all the other guests ...; **im** ~**en** besides; **es ist etwas** ~: there is some left; **übrig|bleiben** *unr. itr. V.*; *mit sein* be left; ⟨*food, drink*⟩ be left over; **übrigens** *Adv.* by the way; **übrig|lassen** *unr. tr. V.* leave; leave ⟨*food, drink*⟩ over

Übung die; ~, ~en **a)** exercise; **b)** *o. Pl. (das Üben, Geübtsein)* practice

UdSSR [u:de:|ɛs|ɛs|'ɛr] *Abk.* **die**; ~ *(1922–1991)* **Union der Sozialistischen Sowjetrepubliken** USSR

Ufer das; ~s, ~: bank; *(des Meers)* shore

UG *Abk.* **Untergeschoß**

Uganda (das); ~s Uganda

Uhr die; ~, ~en **a)** clock; *(Armband~, Taschen~)* watch; *(Wasser~, Gas~)* meter; *(an Meßinstrumenten)* dial; gauge; **auf die** *od.* **nach der** ~ **sehen** look at the time; **rund um die** ~ *(ugs.)* round the clock; **b)** *o. Pl.* **acht** ~: eight o'clock; **wieviel** ~ **ist es?** what's the time?; what time is it?

Uhr-: ~**armband** das watch-strap; ~**kette** die watch-chain; ~**macher** der watchmaker/clockmaker; ~**werk** das clock/watch mechanism; ~**zeiger** der clock-/watch-hand; ~**zeiger·sinn** der: **im/entgegen dem** ~**zeigersinn** clockwise/anticlockwise; ~**zeit** die time; **jmdn. nach der** ~**zeit fragen** ask sb. the time

Uhu der; ~s, ~s eagle owl

Ukraine die; ~: Ukraine; **Ukrainer** der; ~s, ~, **Ukrainerin** die; ~, ~nen Ukrainian

UKW [u:ka:'ve:] *o. Art.; Abk.* **Ultrakurzwelle** VHF; **UKW-Sender** der VHF station; ≈ FM station

Ulk der; ~s, ~e lark *(coll.)*; *(Streich)* trick; *[practical]* joke; **ulkig** *(ugs.)* **1.** *Adj.* funny; **2.** *adv.* in a funny way

Ulme die; ~, ~n elm

Ultimatum das; ~s, **Ultimaten** ultimatum

Ultra·kurz·welle die ultra-short wave; *(Rundf.: Wellenbereich)* very high frequency; VHF

Ultra·schall der *(Physik, Med.)* ultrasound; **ultra·violett** *Adj.* ultraviolet

um 1. *Präp. mit Akk.* **a)** *(räumlich)* [a]round; **um die Ecke** round the corner; **b)** *(zeitlich)* *(genau)* at; *(etwa)* around [about]; **c)** **Tag um Tag/Stunde um Stunde** day after day/hour after hour; **d)** *(bei Maß- u. Mengenangaben)* by; **2.** *Adv.* around; about; **um |die| 10 Mark/50 Personen |herum|** around *or* about ten marks/50 people; **3.** *Konj.* **a)** *(final)* **um ... zu** [in order] to; **b)** *(konsekutiv)* **er ist groß genug/ist noch zu klein, um ... zu ...:** he is big enough/is still too young to ...; **c) je ... um so** the ..., the; **um so besser/schlimmer!** all the better/worse!

um|ändern *tr. V.* change; revise ⟨*text, novel*⟩; alter ⟨*garment*⟩

umạrmen *tr. V.* embrace; *(an sich drücken)* hug; **Umạrmung die; ~, ~en** embrace; hug

Ụm·bau der; ~[e]s, ~ten *s.* **umbauen:** rebuilding; alteration; conversion; *(fig.)* reorganization; **ụm|bauen** *tr., auch itr. V.* rebuild; *(leicht ändern)* alter; *(zu etw. anderem)* convert **(zu** into); *(fig.)* reorganize ⟨*system, administration, etc.*⟩

ụm|benennen *unr. tr. V.* change the name of; rename

ụm|biegen 1. *unr. tr. V.* bend; 2. *unr. itr. V.; mit sein* turn

ụm|binden *unr. tr. V.* put on

ụm|blättern 1. *tr. V.* turn [over]; 2. *itr. V.* turn the page/pages

ụm|blicken *refl. V.* **a)** look around; **b)** *(zurückblicken)* [turn to] look back **(nach** at)

ụm|bringen *unr. tr. V.* kill

Ụm·bruch der **a)** radical change; *(Umwälzung)* upheaval; **b)** *o. Pl.* *(Druckw.)* make-up; *(Ergebnis)* page proofs *pl.*

ụm|buchen 1. *tr. V.* change **(auf +** *Akk.* to); 2. *itr. V.* change one's booking **(auf +** *Akk.* to)

ụm|drehen 1. *tr. V.* turn round; turn over ⟨*coin, hand, etc.*⟩; turn ⟨*key*⟩; 2. *refl. V.* turn round; *(den Kopf wenden)* turn one's head; 3. *itr. V.; auch mit sein (ugs.: umkehren)* turn back; *(ugs.: wenden)* turn round; **Ụm·drehung die** turn; *(eines Motors usw.)* revolution; rev *(coll.)*

um·einạnder *Adv.* sich ~ kümmern/ sorgen take care of/worry about each other *or* one another

¹**ụm|fahren** *unr. tr. V.* knock down; ²**um·fạhren** *unr. tr. V.* go round; make a detour round ⟨*obstruction etc.*⟩; *(im Auto)* drive round; *(im Schiff)* sail round; *(auf einer Umgehungsstraße)* bypass ⟨*town, village, etc.*⟩

ụm|fallen *unr. itr. V.; mit sein* **a)** fall over; **b)** *(zusammenbrechen)* collapse; **tot ~:** fall down dead

Ụm·fang der **a)** circumference; *(eines Quadrats usw.)* perimeter; *(eines Baums, Menschen usw.)* girth; **b)** *(Größe)* size; **c)** *(Ausmaß)* extent; **ụm·fang·reich** *Adj.* extensive; substantial ⟨*book*⟩

um·fạssen *tr. V.* **a)** grasp; *(umarmen)* embrace; **b)** *(enthalten)* contain; *(einschließen)* include; span, cover ⟨*period*⟩; **umfạssend** 1. *Adj.* full ⟨*re-*

ply, information, survey, confession⟩; extensive, wide ⟨*knowledge, powers*⟩; 2. *adv.* ⟨*inform*⟩ fully

ụm|formen *tr. V.* reshape; revise ⟨*poem, novel*⟩; transform ⟨*person*⟩

Ụm·frage die survey; *(Politik)* opinion poll

ụm|füllen *tr. V.* **etw. in etw.** *(Akk.)* ~: transfer sth. into sth.

Ụm·gang der; o. Pl. **a)** *(gesellschaftlicher Verkehr)* contact; **b)** *(das Umgehen)* **den ~ mit Pferden lernen** learn how to handle horses; **ụmgänglich** *Adj.* affable; *(gesellig)* sociable

Ụmgangs-: **~form die** gute/schlechte/keine ~formen haben have good/ bad/no manners; **~sprache die** colloquial language

um·geben *unr. tr. V.* **a)** surround; ⟨*hedge, fence, wall, etc.*⟩ enclose; **b)** **etw. mit etw. ~:** surround sth. with sth.; *(einfrieden)* enclose sth. with sth.; **Umgebung die; ~, ~en** surroundings *pl.; (Nachbarschaft)* neighbourhood; *(eines Ortes)* surrounding area

¹**ụm|gehen** *unr. itr. V.; mit sein* **a)** *(im Umlauf sein)* ⟨*list, rumour, etc.*⟩ round, circulate; ⟨*illness, infection*⟩ go round; **b)** *(spuken)* **hier geht ein Gespenst um** this place is haunted; **c)** *(behandeln)* **mit jmdm. freundlich/liebevoll usw. ~:** treat sb. kindly/lovingly *etc.;* **er kann mit Geld nicht ~:** he can't handle money

²**um·gehen** *unr. tr. V.* **a)** go round; make a detour round; *(auf einer Umgehungsstraße)* bypass ⟨*town etc.*⟩; **b)** *(vermeiden)* avoid; evade ⟨*question, issue*⟩; **c)** *(nicht befolgen)* circumvent ⟨*law, restriction, etc.*⟩; evade ⟨*obligation, duty*⟩; **ụmgehend** 1. *Adj.* immediate; 2. *adv.* immediately; **Umgehung die; ~, ~en** **a)** **durch ~ der Innenstadt** by bypassing *or* avoiding the town centre; **b)** *s.* ²**umgehen c:** circumvention; evasion; **Umgehungs·straße die** bypass

ụmgekehrt 1. *Adj.* inverse ⟨*ratio, proportion*⟩; reverse ⟨*order*⟩; opposite ⟨*sign*⟩; 2. *adv.* inversely ⟨*proportional*⟩

ụm|graben *unr. tr. V.* dig over

Ụm·hang der cape; **ụm|hängen** *tr. V.* **a)** etw. ~: hang sth. somewhere else; **b)** jmdm./sich einen Mantel/eine Decke ~: drape a coat/blanket round sb.'s/one's shoulders

ụm|hauen *unr. tr. V.* fell; *(fig.)* knock down

um·her *Adv.* around
umher-: *s.* herum-
um|hören *refl. V.* keep one's ears
open; *(direkt fragen)* ask around
um·jubeln *tr. V.* cheer
um|kehren 1. *itr. V.; mit sein* turn
back; 2. *tr. V.* turn upside down; turn
over ⟨*sheet of paper*⟩; *(nach links drehen)* turn ⟨*garment etc.*⟩ inside out;
(nach rechts drehen) turn ⟨*garment
etc.*⟩ right side out
um|kippen 1. *itr. V.; mit sein* **a)** fall
over; ⟨*boat*⟩ capsize, turn over;
⟨*vehicle*⟩ overturn; **b)** *(ugs.: ohnmächtig werden)* keel over; 2. *tr. V.* tip
over; knock over ⟨*lamp, vase, glass,
cup*⟩; capsize ⟨*boat*⟩; turn ⟨*boat*⟩ over;
overturn ⟨*vehicle*⟩
um|klappen *tr. V.* fold down
Umkleide·kabine **die** changing-
cubicle
um|knicken *itr. V.; mit sein* **a)** |mit
dem Fuß| ~: go over on one's ankle; **b)**
bend; ⟨*branch*⟩ bend and snap
um|kommen *unr. itr. V.; mit sein* die;
(bei einem Unglück, durch Gewalt) get
killed; die; ⟨*food*⟩ go off
Um·kreis **der** *o. Pl.* surrounding area;
im ~ von 5 km within a radius of 5
km.; **um·kreisen** *tr. V.* circle;
⟨*spacecraft, satellite*⟩ orbit; ⟨*planet*⟩ revolve [a]round
Um·lauf **der** **a)** *(von Planeten)* revolution; **b)** *o. Pl. (Zirkulation)* circulation; **in** *od.* **im ~ sein** be circulating;
⟨*coin, banknote*⟩ be in circulation; **in ~
bringen** circulate; bring ⟨*coin, banknote*⟩ into circulation; **Umlauf-
bahn** **die** *(Astron., Raumf.)* orbit
Um·laut **der** *(Sprachw.)* umlaut
um|legen *tr. V.* **a)** *(um einen Körper-
teil)* put on; **b)** *(verlegen)* transfer ⟨*patient, telephone call*⟩; **c)** *(salopp: ermorden)* jmdn. ~: bump sb. off *(sl.)*
um|leiten divert; **Um·leitung** **die**
diversion
umliegend *Adj.* surrounding ⟨*area*⟩;
(nahe) nearby ⟨*building*⟩
um|räumen 1. *tr. V.* rearrange; 2. *itr.
V.* rearrange things
um|rechnen *tr. V.* convert (**in** + *Akk.*
into)
¹um|reißen *unr. tr. V.* pull ⟨*mast, tree*⟩
down; knock ⟨*person*⟩ down; ⟨*wind*⟩
tear ⟨*tent etc.*⟩ down
²um·reißen *unr. tr. V.* outline; summarize ⟨*subject, problem, situation*⟩
um|rennen *unr. tr. V.* [run into and]
knock down

um·ringen *tr. V.* surround
Um·riß **der** *(auch fig.)* outline
um|rühren *tr. (auch itr.) V.* stir
um|rüsten *tr. V. (Technik)* convert
(**auf** + *Akk.* to, **zu** into)
ums [ʊms] *Präp.* + *Art.* **a)** = **um das**;
b) **~ Leben kommen** lose one's life
um|satteln *itr. V. (ugs.)* change jobs;
⟨*student*⟩ change courses
Um·satz **der** turnover; *(Verkauf)* sales
pl. (**an** + *Dat.* of); **~ machen** *(ugs.)*
make money
um|säumen *tr. V.* hem
um|schalten 1. *tr. V. (auch. fig.)*
switch [over] (**auf** + *Akk.* to); move
⟨*lever*⟩; 2. *itr. V.* switch *or* change over
(**auf** + *Akk.* to)
Um·schlag **der** **a)** cover; **b)** *(Brief~)*
envelope; **c)** *(Schutz~)* jacket; *(einer
Broschüre, eines Heftes)* cover; **d)**
(Med.: Wickel) compress; *(warm)*
poultice; **um|schlagen** 1. *unr. tr. V.*
a) turn up ⟨*sleeve, collar, trousers*⟩;
turn over ⟨*page*⟩; **b)** *(umladen, verla-
den)* turn round, trans-ship ⟨*goods*⟩;
2. *unr. itr. V.; mit sein* change (**in** +
Akk. into); ⟨*wind*⟩ veer [round]
¹um|schreiben *unr. tr. V.* rewrite;
²um·schreiben *unr. tr. V.* **a)** *(in
Worte fassen)* describe; *(definieren)*
define ⟨*meaning, sb.'s task, etc.*⟩; *(pa-
raphrasieren)* paraphrase ⟨*word, expression*⟩; **b)** *(Sprachw.)* construct (**mit**
with); **Um·schreibung** **die** description; *(Definition)* definition; *(Verhül-
lung)* circumlocution (*Gen.* for);
Um·schrift **die** *(Sprachw.)* transcription
um|schulen 1. *tr. V. (beruflich)* re-
train; 2. *itr. V.* retrain (**auf** + *Akk.* as)
um|schütten *tr. V.* **a)** pour [into an-
other container]; decant ⟨*liquid*⟩; **b)**
(verschütten) spill
Um·schwung **der** complete change;
(in der Politik usw.) U-turn
um|sehen *unr. refl. V.* **a)** look; **sich im
Zimmer ~:** look [a]round the room; **b)**
(zurücksehen) look round *or* back
umseitig *Adj., adv.* overleaf
um|setzen *tr. V.* **a)** move; *(auf ande-
ren Posten usw.)* move, transfer (**in** +
Akk. to); *(umpflanzen)* transplant; *(in
anderen Topf)* repot; **b)** *(verwirklichen)*
implement ⟨*plan*⟩; translate ⟨*plan, intention, etc.*⟩ into action *or* reality;
realize ⟨*ideas*⟩; **c)** *(Wirtsch.)* turn over,
have a turnover of ⟨*x marks etc.*⟩; sell
⟨*shares, goods*⟩
Um·sicht **die**; *o. Pl.* circumspection;

um·sichtig 1. *Adj.* circumspect; 2. *adv.* circumspectly

um|siedeln 1. *tr. V.* resettle; 2. *itr. V.; mit sein* move (in + *Akk.,* nach to)

um·sonst *Adv.* a) *(unentgeltlich)* free; for nothing; b) *(vergebens)* in vain

Um·stand der a) *(Gegebenheit)* circumstance; *(Tatsache)* fact; unter Umständen possibly; b) *(Aufwand)* business; macht keine |großen| Umstände please don't go to any bother

umständlich 1. *Adj.* involved, elaborate *(procedure, method, description, explanation, etc.);* elaborate, laborious *(preparation, check, etc.);* awkward, difficult *(journey, job); (weitschweifig)* long-winded; *(Umstände machend)* awkward *(person);* 2. *adv.* in an involved *or* roundabout way; *(weitschweifig)* at great length

Umstands·kleid das maternity dress

umstehend *Adj.* standing round *postpos.*

um|steigen *unr. itr. V.* change (in + *Akk.* [on] to)

¹um|stellen 1. *tr. V.* a) rearrange, change round *(furniture, books, etc.);* reorder *(words etc.);* transpose *(two words);* b) *(anders einstellen)* reset *(lever, switch, points, clock);* c) *(ändern)* change *or* switch over (auf + *Akk.* to); 2. *refl. V.* adjust (auf + *Akk.* to);

²um·stellen *tr. V.* surround

um|stimmen *tr. V.* win *(person)* round

um|stoßen *unr. tr. V.* a) knock over; b) *(rückgängig machen)* change *(plan, decision); (zunichte machen)* upset, wreck *(plan, theory)*

umstritten *Adj.* disputed; controversial *(book, author, policy, etc.)*

Um·sturz der coup; um|stürzen 1. *tr. V.* overturn; *(fig.)* topple, overthrow *(political system, government);* 2. *itr. V.* overturn; *(wall, building, chimney)* fall down; umstürzlerisch *Adj.* subversive

Um·tausch der exchange; um|tauschen *tr. V.* exchange *(goods, article)* (gegen for); change *(dollars, pounds, etc.)* (in + *Akk.* into)

Um·trunk der communal drink

um|tun *unr. refl. V. (ugs.)* look [a]round; sich nach etw. ~: be on the look-out for sth.

um|wandeln *tr. V.* convert *(substance, building, etc.)* (in + *Akk.* into); *(ändern)* change; alter

Um·weg der detour

Um·welt die a) environment; b) *(Menschen)* people *pl.* around sb.

umwelt-, Umwelt-: ~bedingt *Adj.* caused by the *or* one's environment *postpos.;* ~freundlich 1. *Adj.* environment-friendly; 2. *adv.* in an ecologically desirable way; ~schutz der environmental protection *no art.;* ~schützer der environmentalist; conservationist; ~verschmutzung die pollution [of the environment]

um|wenden *regelm. (auch unr.) tr. V.* a) turn over *(page, joint, etc.);* b) turn round *(vehicle, horse)*

um|werfen *unr. tr. V.* a) knock over; knock *(person)* down *or* over; *(fig. ugs.: aus der Fassung bringen)* bowl *(person)* over; stun *(person);* b) *(fig. ugs.: umstoßen)* knock *(plan)* on the head *(coll.);* umwerfend *(ugs.)* 1. *Adj.* fantastic *(coll.);* stunning *(coll.);* 2. *adv.* fantastically [well] *(coll.);* brilliantly

um·wickeln *tr. V.* wrap; bind; *(mit einem Verband)* bandage

Umzäunung die ~, ~en fence, fencing (*Gen.* round)

um|ziehen 1. *unr. itr. V.; mit sein* move (an + *Akk.,* in + *Akk.,* nach to); 2. *unr. tr. V.* jmdn. ~: change sb. *or* get sb. changed; sich ~: change *or* get changed

um·zingeln *tr. V.* surround; encircle

Um·zug der a) move; *(von Möbeln)* removal; b) *(Festzug)* procession

UN [u:'ɛn] *Pl.* UN *sing.*

unabänderlich 1. *Adj.* unalterable; irrevocable *(decision);* 2. *adv.* irrevocably

unabhängig 1. *Adj.* independent (von of); *(unbeeinflußt)* unaffected (von by); 2. *adv.* independently (von of); ~davon, ob .../was .../wo ... usw. irrespective *or* regardless of whether .../ what .../where ... *etc.;* Unabhängigkeit die independence

unabkömmlich *Adj.* indispensable; sie ist im Moment ~: she is otherwise engaged

unablässig 1. *Adj.* incessant; 2. *adv.* incessantly

unabsichtlich 1. *Adj.* unintentional; 2. *adv.* unintentionally

unabwendbar *Adj.* inevitable

unachtsam 1. *Adj.* a) inattentive; b) *(nicht sorgfältig)* careless; 2. *adv. (ohne Sorgfalt)* carelessly; Unachtsamkeit die; ~ a) inattentiveness; b) *(mangelnde Sorgfalt)* carelessness

ụnangebracht *Adj.* inappropriate
ụnangefochten *Adj.* unchallenged;
(Rechtsw.) uncontested ⟨*verdict, will,
etc.*⟩
ụnangenehm 1. *Adj.* unpleasant
(*Dat.* for); *(peinlich)* embarrassing
⟨*question, situation*⟩; **2.** *adv.* unpleasantly
unannẹhmbar *Adj.* unacceptable;
Ụnannehmlichkeit die trouble
ụnansehnlich *Adj.* unprepossessing;
plain ⟨*girl*⟩
ụnanständig 1. *Adj.* improper; *(anstößig)* indecent; dirty ⟨*joke*⟩; rude
⟨*word, song*⟩; **2.** *adv.* improperly;
Ụnanständigkeit die impropriety;
indecency; *(Obszönität)* obscenity
ụnappetitlich 1. *Adj.* unappetizing;
(fig.) unsavoury ⟨*joke*⟩; disgusting
⟨*wash-basin, nails, etc.*⟩; **2.** *adv.* unap-
petizingly
Ụnart die bad habit; **ụnartig** *Adj.*
naughty
ụnästhetisch *Adj.* unpleasant ⟨*sight
etc.*⟩; ugly ⟨*building etc.*⟩
ụnauffällig 1. *Adj.* inconspicuous;
unobtrusive ⟨*scar, defect, skill, behaviour, surveillance, etc.*⟩; discreet ⟨*signal, elegance*⟩; **2.** *adv.* inconspicu-
ously; unobtrusively
ụnaufgefordert *Adv.* without being
asked
unaufhạltsam 1. *Adj.* inexorable; **2.**
adv. inexorably
ụnaufmerksam *Adj.* inattentive (gegenüber to); careless ⟨*driver*⟩
ụnaufrichtig *Adj.* insincere; **Ụnauf-
richtigkeit** die insincerity
unausblẹiblich *Adj.* inevitable
ụnbändig 1. *Adj.* **a)** boisterous; **b)**
(überaus groß/stark) unbridled; **2.**
adv. **a)** wildly; **b)** *(sehr, äußerst)* unrestrainedly; tremendously *(coll.)*
ụnbarmherzig *Adj.* merciless
ụnbeabsichtigt 1. *Adj.* unintentional; **2.** *adv.* unintentionally
ụnbeachtet *Adj.* unnoticed
ụnbedenklich *adv.* without second
thoughts
ụnbedeutend 1. *Adj.* insignificant;
minor ⟨*artist, poet*⟩; slight, minor ⟨*improvement, change, error*⟩; **2.** *adv.*
slightly
ụnbedingt 1. *Adj.* absolute; **2.** *adv.*
absolutely; **3.** *Adv. (auf jeden Fall)*
whatever happens
ụnbefangen *Adj.* **a)** *(ungehemmt)* uninhibited; **b)** *(unvoreingenommen)* impartial

ụnbefristet 1. *Adj.* for an indefinite
period *postpos.;* indefinite ⟨*strike*⟩;
unlimited ⟨*visa*⟩; **2.** *adv.* for an indefinite period
ụnbefugt 1. *Adj.* unauthorized; **2.**
adv. without authorization
unbegrẹiflich *Adj.* incomprehensible *(Dat.,* für to); incredible ⟨*love,
goodness, stupidity, carelessness, etc.*⟩
ụnbegrenzt 1. *Adj.* unlimited; **2.** *adv.*
⟨*stay, keep, etc.*⟩ indefinitely
Ụnbehagen das uneasiness, disquiet;
(Sorge) concern (**an** + *Dat.* about);
ụnbehaglich 1. *Adj.* uneasy ⟨*feeling,
atmosphere*⟩; uncomfortable ⟨*thought,
room*⟩; **2.** *adv.* uneasily
ụnbeholfen 1. *Adj.* clumsy; **2.** *adv.*
clumsily
ụnbekannt *Adj.* **a)** unknown; *(nicht
vertraut)* unfamiliar; unidentified
⟨*caller, donor*⟩; „**Empfänger** ~" 'not
known at this address'; **b)** *(nicht vielen
bekannt)* little known; obscure ⟨*poet,
painter, etc.*⟩; ¹**Ụnbekannte der/die;**
adj. Dekl. unknown *or* unidentified
man/woman; *(Fremde[r])* stranger;
²**Ụnbekannte die;** *adj. Dekl. (Math.;
auch fig.)* unknown
ụnbekleidet *Adj.* without any clothes
on *postpos.;* bare ⟨*torso etc.*⟩; naked
⟨*corpse*⟩
ụnbekümmert 1. *Adj.* carefree; *(ohne Bedenken, lässig)* casual; **2.** *adv.* **a)**
in a carefree way; **b)** *(ohne Bedenken)*
without caring *or* worrying
ụnbeleuchtet *Adj.* unlit ⟨*street, corridor, etc.*⟩; ⟨*vehicle*⟩ without [any]
lights
ụnbeliebt *Adj.* unpopular (**bei** with)
ụnbemannt *Adj.* unmanned
ụnbemerkt *Adj., adv.* unnoticed
ụnbenutzt *Adj.* unused
ụnbequem 1. *Adj.* **a)** uncomfortable;
b) *(lästig)* awkward, embarrassing
⟨*question, opinion*⟩; troublesome ⟨*politician etc.*⟩; unpleasant ⟨*criticism,
truth, etc.*⟩; **2.** *adv.* uncomfortably
unberẹchenbar 1. *Adj.* unpredict-
able; **2.** *adv.* unpredictably
ụnberechtigt *Adj.* **a)** *(ungerechtfertigt)* unjustified; **b)** *(unbefugt)* unauthorized
ụnberührt *Adj.* untouched; **sie ist
noch** ~: she is still a virgin
ụnbeschrankt *Adj.* ⟨*crossing*⟩ without gates, with no gates
unbeschrẹiblich 1. *Adj.* indescribable; unimaginable ⟨*fear, beauty*⟩;
⟨*fear, beauty*⟩ beyond description; **2.**

adv. indescribably ⟨*beautiful*⟩; unbelievably ⟨*busy*⟩

unbesorgt *Adj.* unconcerned; **seien Sie ~**: don't [you] worry

unbeständig *Adj.* changeable ⟨*weather*⟩; *fickle* ⟨*lover etc.*⟩

unbestimmt 1. *Adj.* **a)** indefinite; indeterminate ⟨*age, number*⟩; *(ungewiß)* uncertain; **b)** *(ungenau)* vague; **c)** *(Sprachw.)* indefinite ⟨*article, pronoun*⟩; **2.** *adv. (ungenau)* vaguely

unbewacht *Adj.* unsupervised; unattended ⟨*car-park*⟩

unbewaffnet *Adj.* unarmed

unbeweglich *Adj.* motionless; still ⟨*air, water*⟩; fixed ⟨*gaze, expression*⟩

unbewußt *Adj.* unconscious

unbrauchbar *Adj.* unusable; *(untauglich)* useless ⟨*method, person*⟩

und *Konj.* and; *(folglich)* [and] so; **ich ~ tanzen?** what, me dance?; **sei so gut ~ mach das Fenster zu** be so good as to shut the window

Undank der ingratitude; **undankbar** *Adj.* ungrateful ⟨*person, behaviour*⟩

undeutlich 1. *Adj.* unclear; indistinct; *(ungenau)* vague ⟨*idea, memory, etc.*⟩; **2.** *adv.* indistinctly; *(ungenau)* vaguely

undicht *Adj.* leaky; leaking; **~e Fenster** windows which do not fit tightly

undurchführbar *Adj.* impracticable

undurchlässig *Adj.* impermeable; *(wasserdicht)* watertight; waterproof; *(luftdicht)* airtight

unehelich *Adj.* illegitimate ⟨*child*⟩; unmarried ⟨*mother*⟩

unehrlich 1. *Adj.* dishonest; **2.** *adv.* dishonestly; by dishonest means

uneigennützig *Adj.* unselfish

uneinig *Adj.* ⟨*party*⟩ divided by disagreement; **[sich *(Dat.)*] ~ sein** disagree; **Uneinigkeit die** disagreement (**in** + *Dat.* on); **uneins** *Adj.; nicht attr.* **~ sein** be divided (**in** + *Dat.* on); ⟨*persons*⟩ be at variance *or* at cross purposes (**in** + *Dat.* over)

unempfindlich *Adj.* **a)** insensitive (**gegen** to); **b)** *(immun)* immune (**gegen** to, against); **c)** *(strapazierfähig)* hardwearing

unendlich 1. *Adj.* infinite; boundless; *(zeitlich)* endless; *(Math.)* infinite; **2.** *adv.* infinitely ⟨*lovable, sad*⟩; immeasurably ⟨*happy*⟩; ⟨*happy*⟩ beyond measure

unentbehrlich *Adj.* indispensable *(Dat.,* **für** to)

unentgeltlich [*od.* '----] **1.** *Adj.* free;

2. *adv.* free of charge; ⟨*work*⟩ for nothing, without pay

unentschieden 1. *Adj.* unsettled; undecided ⟨*question*⟩; *(Sport, Schach)* drawn; **2.** *adv.* **~ spielen** draw

unentwegt [*od.* --'-] **1.** *Adj.* **a)** *(beharrlich)* persistent ⟨*fighter, champion, efforts*⟩; **b)** *(unaufhörlich)* constant; incessant; **2.** *adv.* **a)** *(beharrlich)* persistently; **b)** *(unaufhörlich)* constantly; incessantly

unerbittlich 1. *Adj. (auch fig.)* inexorable; unsparing ⟨*critic*⟩; relentless ⟨*battle, struggle*⟩; implacable ⟨*hate, enemy*⟩; **2.** *adv. (auch fig.)* inexorably

unerfahren *Adj.* inexperienced

unerfreulich 1. *Adj.* unpleasant; bad ⟨*news*⟩; **2.** *adv.* unpleasantly

unerheblich *Adj.* insignificant

unerhört 1. *Adj. (empörend)* outrageous; **2.** *adv.* outrageously

unerlaubt 1. *Adj.* unauthorized; **2.** *adv.* without authorization

unermüdlich 1. *Adj.* tireless, untiring (**bei, in** + *Dat.* in); **2.** *adv.* tirelessly

unerreichbar *Adj.* inaccessible; *(fig.)* unattainable; **unerreicht** *Adj.* unequalled

unerschöpflich *Adj.* inexhaustible

unersetzlich *Adj.* irreplaceable

unerträglich [*od.* '----] *Adj.* unbearable; intolerable ⟨*situation, conditions, etc.*⟩

unerwartet 1. *Adj.* unexpected; **es kam für alle ~**: it came as a surprise to everybody; **2.** *adv.* unexpectedly

unerwünscht *Adj.* unwanted; unwelcome ⟨*interruption, visit, visitor*⟩; undesirable ⟨*side-effects*⟩

unfähig *Adj.* **a)** **~ sein, etw. zu tun** *(ständig)* be incapable of doing sth.; *(momentan)* be unable to do sth.; **b)** *(inkompetent)* incompetent

unfair 1. *Adj.* unfair (**gegen** to); **2.** *adv.* unfairly

Un·fall der accident

Unfall-: **~arzt** der casualty doctor; **~stelle** die scene of an/the accident; **~versicherung** die accident insurance

unförmig *Adj.* shapeless; huge ⟨*legs, hands, body*⟩; bulky, ungainly ⟨*shape, shoes, etc.*⟩

unfrei *Adj.* not free *pred.*; subject, dependent ⟨*people*⟩; ⟨*life*⟩ of bondage;

unfreiwillig 1. *Adj.* involuntary; *(erzwungen)* enforced ⟨*stay*⟩; *(nicht beabsichtigt)* unintended ⟨*publicity, joke, humour*⟩; **2.** *adv.* involuntarily; without wanting to; *(unbeabsichtigt)* unintentionally

unfreundlich 1. *Adj.* unfriendly (**zu, gegen** to); unkind ⟨*words, remark*⟩; **2.** *adv.* in an unfriendly way

unfrisiert *Adj.* ungroomed ⟨*hair*⟩

unfruchtbar *Adj.* infertile; *(fig.)* unproductive; **Unfruchtbarkeit die** infertility; *(fig.)* unproductiveness

Unfug der; ~|e|s a) [piece of] mischief; **grober ~:** public nuisance; **b)** *(Unsinn)* nonsense

Ungar der; ~n, ~n Hungarian; **ungarisch** *Adj.* Hungarian; **Ungarn (das); ~s** Hungary

ungeachtet *Präp. mit Gen. (geh.)* notwithstanding; despite

ungebildet *Adj.* uneducated

ungebräuchlich *Adj.* uncommon; rare; rarely used ⟨*method, process*⟩

ungedeckt *Adj.* uncovered ⟨*cheque*⟩

Ungeduld die impatience; **ungeduldig 1.** *Adj.* impatient; **2.** *adv.* impatiently

ungeeignet *Adj.* unsuitable; *(für eine Aufgabe)* unsuited (**für, zu** to, for)

ungefähr 1. *Adj.* approximate; rough ⟨*idea, outline*⟩; **2.** *adv.* approximately; roughly

ungefährlich *Adj.* safe; harmless ⟨*animal, person, illness, etc.*⟩

ungeheizt *Adj.* unheated

ungeheuer 1. *Adj.* enormous; tremendous ⟨*strength, energy, effort, enthusiasm, fear, success, pressure, etc.*⟩; vast, immense ⟨*fortune, knowledge*⟩; *(schrecklich)* terrible *(coll.)*, terrific *(coll.)* ⟨*pain, rage*⟩; **2.** *adv.* tremendously; terribly *(coll.)* ⟨*difficult, clever*⟩; **Ungeheuer das; ~s, ~** *(auch fig.)* monster

ungehindert *Adj.* unimpeded

ungehörig 1. *Adj.* improper; *(frech)* impertinent; **2.** *adv.* improperly; *(frech)* impertinently

ungehorsam *Adj.* disobedient (**gegenüber** to); **Ungehorsam der** disobedience (**gegenüber** to)

ungekürzt *Adj.* unabridged ⟨*edition, book*⟩; uncut ⟨*film, speech*⟩

ungelegen 1. *Adj.* **das kommt mir sehr ~/nicht ~:** that is very inconvenient *or* awkward/quite convenient for me; **2.** *adv.* inconveniently

ungelernt *Adj.* unskilled

ungemütlich 1. *Adj.* uninviting, cheerless ⟨*room, flat*⟩; uncomfortable, unfriendly ⟨*atmosphere*⟩; **2.** *adv.* uncomfortably ⟨*furnished*⟩

ungenau 1. *Adj.* inaccurate; imprecise, inexact ⟨*definition, formulation, etc.*⟩; *(undeutlich)* vague ⟨*memory, idea, impression*⟩; **2.** *adv.* inaccurately; ⟨*define*⟩ imprecisely, inexactly; ⟨*remember*⟩ vaguely

ungeniert ['ʊnʒeni:ɐ̯t] **1.** *Adj.* free and easy; uninhibited; **2.** *adv.* openly; ⟨*yawn*⟩ unconcernedly; ⟨*undress etc.*⟩ without any embarrassment

ungenießbar *Adj. (nicht eßbar)* inedible; *(nicht trinkbar)* undrinkable; *(fig. ugs.)* unbearable

ungenügend 1. *Adj.* inadequate; **die Note „~"/ein Ungenügend** *(Schulw.)* the/an 'unsatisfactory' [mark]; **2.** *adv.* inadequately

ungepflegt *Adj.* neglected ⟨*garden, park, car, etc.*⟩; unkempt ⟨*person, appearance, hair*⟩; uncared-for ⟨*hands*⟩

ungerade *Adj.* odd ⟨*number*⟩

ungerecht 1. *Adj.* unjust, unfair (**gegen, zu, gegenüber** to); **2.** *adv.* unjustly; unfairly; **Ungerechtigkeit die; ~, ~en** injustice

ungern *Adv.* reluctantly; **etw. ~ tun** not like *or* dislike doing sth.

ungeschält *Adj.* unpeeled ⟨*fruit*⟩

ungeschickt 1. *Adj.* clumsy; awkward; **2.** *adv.* clumsily; awkwardly

ungesetzlich 1. *Adj.* unlawful; illegal; **2.** *adv.* unlawfully; illegally

ungestempelt *Adj.* uncancelled ⟨*stamp*⟩

ungestört *Adj.* undisturbed; uninterrupted ⟨*development*⟩

ungesund *Adj. (auch fig.)* unhealthy

Ungetüm das; ~s, ~e monster

ungewiß *Adj.* uncertain; **über etw.** *(Akk.)* **im ungewissen sein** be uncertain *or* unsure about sth.; **Ungewißheit die** uncertainty

ungewöhnlich 1. *Adj.* **a)** unusual; **b)** *(sehr groß)* exceptional ⟨*strength, beauty, ability, etc.*⟩; outstanding ⟨*achievement, success*⟩; **2.** *adv.* **a)** ⟨*behave*⟩ abnormally, strangely; **b)** *(enorm)* exceptionally

ungewohnt 1. *Adj.* unaccustomed; *(nicht vertraut)* unfamiliar ⟨*method, work, surroundings, etc.*⟩; **2.** *adv.* unusually

Ungeziefer das; ~s vermin *pl.*

ungezogen 1. *Adj.* naughty; badly behaved; bad ⟨*behaviour*⟩; *(frech)* cheeky; **2.** *adv.* naughtily; ⟨*behave*⟩ badly

ungläubig 1. *Adj.* **a)** disbelieving; **b)** *(Rel.)* unbelieving; **2.** *adv.* in disbe-

lief; **unglaublich 1.** *Adj.* incredible;
2. *adv. (ugs.: äußerst)* incredibly
(coll.); **unglaubwürdig** *Adj.* implausible; untrustworthy, unreliable
⟨*witness etc.*⟩
ungleich 1. *Adj.* unequal; odd, un-
matching ⟨*socks, gloves, etc.*⟩; *(unähnlich)* dissimilar; **2.** *adv.* **a)** unequally;
b) *(ungleichmäßig)* unevenly
Unglück das; ~|e|s, ~e **a)** *(Unfall)* ac-
cident; *(Flugzeug~, Zug~)* crash; accident; **b)** *o. Pl. (Not)* misfortune;
(Leid) suffering; **c)** *(Pech)* bad luck; ~
haben be unlucky; **das bringt** ~ **:** that's
unlucky; **d)** *(Schicksalsschlag)* misfortune; **unglücklich 1.** *Adj.* **a)** unhappy; **b)** *(nicht vom Glück begünstigt)*
unfortunate ⟨*person*⟩; *(bedauernswert,
arm)* hapless ⟨*person, animal*⟩; **c)** *(ungünstig, ungeschickt)* unfortunate ⟨*moment, combination, meeting, etc.*⟩; unhappy ⟨*end, choice, solution*⟩; **2.** *adv.*
a) unhappily; **b)** *(ungünstig)* unfortunately; *(ungeschickt)* unhappily,
clumsily ⟨*translated, expressed*⟩; **un-
glücklicherweise** *Adv.* unfortunately; **Unglücks · fall der** accident
ungültig *Adj.* invalid; void *(esp.
Law);* spoilt ⟨*vote, ballot-paper*⟩; dis-
allowed ⟨*goal*⟩
ungünstig 1. *Adj.* **a)** unfavourable;
unfortunate, bad ⟨*shape, layout*⟩; **b)**
(unpassend) inconvenient ⟨*time*⟩; *(ungeeignet)* inappropriate, inconvenient
⟨*time, place*⟩; **2.** *adv.* **a)** unfavourably;
badly ⟨*designed, laid out*⟩; **b)** *(unpassend)* inconveniently
Unheil das disaster; **unheilbar 1.**
Adj. incurable; **2.** *adv.* incurably; **unheil · voll** *Adj.* disastrous; *(verhängnisvoll)* fateful
unheimlich 1. *Adj.* **a)** eerie; **b)** *(ugs.)*
(schrecklich) terrible *(coll.)* ⟨*hunger,
headache, etc.*⟩ terrific *(coll.) (fun
etc.)*; **2.** *adv.* **a)** eerily; **b)** *(ugs.: äußerst)* terribly *(coll.);* incredibly *(coll.)*
⟨*quick, long*⟩
unhöflich 1. *Adj.* impolite; **2.** *adv.*
impolitely; **Unhöflichkeit die** im-
politeness
unhygienisch 1. *Adj.* unhygienic; **2.**
adv. unhygienically
Uniform die; ~, ~en uniform
uninteressant *Adj.* uninteresting;
(nicht von Belang) of no interest *postpos.;* unimportant
Union [u'nịo:n] **die;** ~, ~en union
Universität die; ~, ~en university
Universum das; ~s universe

Unkenntnis die; *o. Pl.* ignorance
unklar *Adj.* unclear; **sich** *(Dat.)* **über**
etw. *(Akk.)* **im** ~en **sein** be unclear *or*
unsure about sth.
Unkosten *Pl.* **a)** [extra] expense *sing.;*
expenses; **b)** *(ugs.: Ausgaben)* costs;
expenditure *sing.*
Unkraut das weeds *pl.*
unleserlich 1. *Adj.* illegible; **2.** *adv.*
illegibly
unmäßig 1. *Adj.* immoderate; excessive; **2.** *adv.* excessively; ⟨*eat,
drink*⟩ to excess
Unmensch der brute; **unmenschlich 1.** *Adj.* **a)** inhuman; brutal; appalling ⟨*conditions*⟩; **b)** *(entsetzlich)*
appalling; **2.** *adv.* **a)** in an inhuman
way; **b)** *(entsetzlich)* appallingly *(coll.)*
unmißverständlich 1. *Adj.* **a)** *(eindeutig)* unambiguous; **b)** *(offen, direkt)* blunt ⟨*answer, refusal*⟩; unequi-
vocal ⟨*language*⟩; **2.** *adv.* **a)** *(eindeutig)* unambiguously; **b)** *(offen, direkt)*
bluntly; unequivocally
unmittelbar 1. *Adj.* immediate; direct ⟨*contact, connection, influence,
etc.*⟩; **2.** *adv.* immediately; directly
unmöbliert *Adj.* unfurnished
unmodern 1. *Adj.* old-fashioned;
(nicht modisch) unfashionable; **2.** *adv.*
in an old-fashioned way; *(nicht modisch)* unfashionably
unmöglich 1. *Adj.* impossible; *(ugs.:
seltsam)* incredible; **2.** *adv. (ugs.) (behave)* impossibly; ⟨*dress*⟩ ridicu-
lously; **3.** *Adv. (ugs.)* **ich/es** *usw.* **kann**
~ **...:** I/it *etc.* can't possibly ...
unmoralisch 1. *Adj.* immoral; **2.** *adv.*
immorally
unmündig *Adj.* under-age
unnatürlich 1. *Adj.* unnatural; forced
⟨*laugh*⟩; **2.** *adv.* unnaturally; ⟨*laugh*⟩
in a forced way; ⟨*speak*⟩ affectedly
unnötig 1. *Adj.* unnecessary; **2.** *adv.*
unnecessarily
UNO ['u:no] **die;** ~: UN
unordentlich 1. *Adj.* **a)** untidy; **b)**
(ungeregelt) disorderly ⟨*life*⟩; **2.** *adv.*
untidily; ⟨*tie, treat, etc.*⟩ carelessly;
Unordnung die disorder; mess
unparteiisch 1. *Adj.* impartial; **2.**
adv. impartially
unpassend 1. *Adj.* inappropriate; unsuitable ⟨*dress etc.*⟩; **2.** *adv.* inappro-
priately; unsuitably ⟨*dressed etc.*⟩
unpersönlich 1. *Adj.* impersonal;
distant, aloof ⟨*person*⟩; **2.** *adv.* imper-
sonally; ⟨*answer, write*⟩ in impersonal
terms

ụnpraktisch 1. *Adj.* unpractical; **2.** *adv.* in an unpractical way

ụnpünktlich 1. *Adj.* unpunctual ⟨*person*⟩; late, unpunctual ⟨*payment*⟩; **2.** *adv.* late

Ụnrecht das; *o. Pl.* wrong; **zu ~:** wrongly; **unrecht haben** be wrong; **jmdm. unrecht tun** do sb. an injustice; **ụnrechtmäßig 1.** *Adj.* unlawful; **2.** *adv.* unlawfully

ụnregelmäßig 1. *Adj.* irregular; **2.** *adv.* irregularly

ụnreif *Adj.* **a)** unripe; **b)** *(nicht erwachsen)* immature

Ụnruhe die *(auch fig.)* unrest; *(Lärm)* noise; *(Unrast)* restlessness; *(Besorgnis)* anxiety; **ụnruhig 1.** *Adj.* **a)** restless; *(besorgt)* anxious; unsettled, troubled ⟨*time*⟩; **b)** *(laut)* noisy; **c)** *(ungleichmäßig)* uneven ⟨*breathing, pulse, etc.*⟩; fitful ⟨*sleep*⟩; disturbed ⟨*night*⟩; **2.** *adv.* **a)** restlessly; *(besorgt)* anxiously; **b)** *(ungleichmäßig)* unevenly; ⟨*sleep*⟩ fitfully

ụns 1. a) *Akk. von* **wir** us; **b)** *Dat. von* **wir; gib es ~:** give it to us; **bei ~:** at our home *or (coll.)* place; **2.** *Reflexivpron. der 1. Pers.* **a)** *refl.* ourselves; **b)** *reziprok* one another

ụnsachlich 1. *Adj.* unobjective; **2.** *adv.* without objectivity

ụnsauber 1. *Adj.* **a)** dirty; **b)** *(nachlässig)* untidy; sloppy; **2.** *adv. (nachlässig)* untidily

ụnschädlich *Adj.* harmless

ụnscharf *Adj.* blurred ⟨*photo, picture*⟩

ụnscheinbar *Adj.* inconspicuous

Ụnschuld die; *o. Pl.* innocence; **ụnschuldig 1.** *Adj.* innocent; **2.** *adv.* innocently

ụnselbständig *Adj.* dependent [on other people]

[1]**ụnser** *Possessivpron. der 1. Pers. Pl.* our; **das ist ~s** that is ours; [2]**ụnser** *Gen. von* **wir** *(geh.)* of us; **in ~ aller/ beider Interesse** in the interest of all/ both of us; **ụnser·einer, ụnser·eins** *Indefinitpron. (ugs.)* the likes of us *pl.;* our sort *(coll.);* **ụnserer·seits** *Adv.* for our part; *(von uns)* on our part; **ụnser[e]s·gleichen** *indekl. Indefinitpron.* people *pl.* like us; **ụnsert·wegen** *Adv., s.* meinetwegen: because of us; for our sake; about us; as far as we are concerned

ụnsicher 1. *Adj.* uncertain; *(nicht selbstsicher)* insecure; **2.** *adv.* ⟨*walk, stand, etc.*⟩ unsteadily; *(nicht selbstsicher)* ⟨*smile, look*⟩ diffidently

ụnsichtbar *Adj.* invisible (für to)

Ụnsinn der nonsense; **~ machen** mess *or* fool about; **ụnsinnig** *Adj.* nonsensical ⟨*statement, talk, etc.*⟩; absurd, ridiculous ⟨*demand etc.*⟩

Ụnsitte die bad habit; **ụnsittlich 1.** *Adj.* indecent; **2.** *adv.* indecently

unsr- *s.* [1]**unser**

unstẹrblich *Adj.* immortal

ụnsympathisch *Adj.* uncongenial, disagreeable ⟨*person*⟩; unpleasant ⟨*characteristic, nature, voice*⟩

Ụntat die misdeed; evil deed

ụntauglich *Adj.* unsuitable; *(für Militärdienst)* unfit [for service] *postpos.*

ụnten *Adv.* **a)** down; **hier/da ~:** down here/there; **von ~:** from below; **b)** *(in Gebäuden)* downstairs; **nach ~:** downstairs; **c)** *(am unteren Ende, zum unteren Ende hin)* at the bottom; **~ [links] auf der Seite/im Schrank** at the bottom [left] of the page/cupboard; **d)** *(an der Unterseite)* underneath; **e)** *(im Text)* below; **ụnten·genannt** *Adj.* undermentioned *(Brit.);* mentioned below *postpos.*

ụnter 1. *Präp. mit Dat. (Lage, Standort)* under; *(zwischen)* among[st]; **Mengen ~ 100 Stück** quantities of less than 100; **~ Angst/Tränen** in *or* out of fear/in tears; **2.** *Präp. mit Akk.* under; *(zwischen)* among[st]; **~ Null sinken** drop below zero; **3.** *Adv.* less than; **~ 30 [Jahre alt] sein** be under 30 [years of age]

ụnter... *Adj.* lower; bottom; *(ganz unten)* bottom; *(in der Rangfolge o. ä.)* lower

Ụnter·arm der forearm; **ụnterbelichten**[1] *tr. V. (Fot.)* underexpose

unter·blẹiben *unr. itr. V.; mit sein* **etw. unterbleibt** sth. does not occur *or* happen; **unter·brẹchen** *unr. tr. V.* interrupt; break ⟨*journey, silence*⟩; **Unter·brẹchung die** *s.* unterbrecher: interruption; break ⟨*Gen.* in⟩

ụnter|bringen *unr. tr. V.* **a)** put; **b)** *(beherbergen)* put up; **Ụnterbringung die;** **~, ~en** accommodation *no indef. art.*

unter·der·hand *Adv.* on the quiet

unter·dẹssen *s.* inzwischen; **unterdrücken** *tr. V.* suppress; hold back ⟨*comment, question, answer, criticism, etc.*⟩; oppress ⟨*minority etc.*⟩; **Unterdrückung die;** **~, ~en a)** *(das Unter-*

---

[1] *ich unterbelichte, unterbelichtet, unterzubelichten*

drücken) suppression; **b)** *(das Unterdrücktwerden, -sein)* oppression

unter·einander *Adv.* **a)** *(räumlich)* one below the other; **b)** *(miteinander)* among[st] ourselves/themselves *etc.*

unter·ernährt *Adj.* undernourished;

Unter·ernährung die malnutrition

Unter·führung die underpass; *(für Fußgänger)* subway *(Brit.);* [pedestrian] underpass *(Amer.)*

unter-, Unter-: ~**gang** der **a)** *(Sonnen~, Mond~ usw.)* setting; **b)** *(von Schiffen)* sinking; **c)** *(das Zugrundegehen)* decline; ~|**gehen** *unr. itr. V.; mit sein* **a)** ⟨*sun, star, etc.*⟩ set; ⟨*ship*⟩ sink, go down; ⟨*person*⟩ drown, go under; **b)** *(zugrunde gehen)* come to an end; ~**geordnet** *Adj.* secondary ⟨*role, importance, etc.*⟩; subordinate ⟨*position, post, etc.*⟩; ~**gewicht** das; *o. Pl.* underweight; ~**grund** der *o. Pl. (bes. Politik)* underground

Untergrund·bahn die underground [railway] *(Brit.);* subway *(Amer.)*

unter-, Unter-: ~|**haken** *tr. V. (ugs.)* jmdn. ~**haken** take sb.'s arm; ~**halb** **1.** *Adv.* below; ~**halb von** below; **2.** *Präp. mit Gen.* below; ~**halt** der; *o. Pl.* **a)** living; **b)** *(~haltszahlung)* maintenance; **c)** *(Instandhaltung[skosten])* upkeep; ~**halten** **1.** *unr. tr. V.* **a)** support; **b)** *(instand halten)* maintain ⟨*building*⟩; **c)** *(betreiben)* run, keep ⟨*car, hotel*⟩; **d)** *(pflegen)* maintain, keep up ⟨*contact, correspondence*⟩; **e)** entertain ⟨*guest, audience*⟩; **2.** *unr. refl. V.* **a)** talk; converse; **b)** *(sich vergnügen)* enjoy oneself; ~**haltsam** *Adj.* entertaining; ~**haltung** die **a)** *o. Pl. (Versorgung)* support; **b)** *o. Pl. (Instandhaltung)* maintenance; **c)** *(Gespräch)* conversation; **e)** *(Zeitvertreib)* entertainment; ~**händler** der *(bes. Politik)* negotiator; ~**hemd** das vest *(Brit.);* undershirt *(Amer.);* ~**hose** die *(Herren~)* briefs *pl.;* [under]pants *pl.; (Damen~)* panties; knickers *(Brit.);* ~**irdisch** **1.** *Adj.* underground; **2.** *adv.* underground; ~**kiefer** der lower jaw; ~|**kommen** *unr. itr. V.; mit sein* find accommodation

unter·kühlt *Adj.* ~ sein be suffering from hypothermia *or* exposure

Unter-: ~**kunft** die; ~, ~**künfte** accommodation *no indef. art.; (Logis)* lodging *no indef. art.;* ~**kunft und Frühstück** bed and breakfast; ~**kunft und Verpflegung** board and lodging; ~**lage** die **a)** *(Schreib~)* pad; *(für eine*

Schreibmaschine usw.) mat; **b)** *Pl.* documents; papers

unter-: ~**lassen** *unr. tr. V.* refrain from [doing]; ~**laufen** *unr. itr. V.; mit sein* occur; jmdm. ist ein Fehler/Irrtum ~ : sb. made a mistake; ~**legen** *Adj.* inferior; jmdm. ~ sein be inferior to sb. **(an +** *Dat.* in)

Unter·leib der lower abdomen

unter·liegen *unr. itr. V.* **a)** *mit sein (besiegt werden)* lose; be beaten *or* defeated; **b)** *(unterworfen sein)* be subject to

unterm *Präp.* + *Art.* = **unter dem**

unter·mauern *tr. V. (mit Argumenten, Fakten absichern)* back up

Unter-: ~**miete** die subtenancy; sublease; ~**mieter** der subtenant; lodger

untern *(ugs.) Präp.* + *Art.* = **unter den**

unter-, Unter-: ~**nehmen** *unr. tr. V.* **a)** *(durchführen)* undertake; make; take ⟨*steps*⟩; **b)** etwas ~**nehmen** do something; ~**nehmen** das; ~**nehmens, ~nehmen a)** *(Vorhaben)* enterprise; **b)** *(Firma)* concern; ~**nehmer** der; ~**nehmers, ~nehmer** employer; ~**nehmungs·lustig** *Adj.* active; sie ist sehr ~nehmungslustig she is always out doing things

Unter·offizier der **a)** non-commissioned officer; **b)** *(Dienstgrad)* corporal

unter·ordnen **1.** *tr. V.* subordinate; **2.** *refl. V.* accept a subordinate role

Unterredung die ~, ~**en** discussion

Unterricht der; ~[e]s, ~e instruction; *(Schul~)* teaching; *(Schulstunden)* classes *pl.;* ~**en** **unterrichten 1.** *tr. V.* **a)** teach; **b)** *(informieren)* inform ⟨über + Akk. of, about⟩; **2.** *itr. V. (Unterricht geben)* teach; **3.** *refl. V. (sich informieren)* inform oneself (über + Akk. about); **Unterrichts·stunde** die lesson; period

Unter·rock der [half] slip

unter|rühren *tr. V.* stir in

unters *Präp.* + *Art.* = **unter das**

unter·sagen *tr. V.* forbid; prohibit

Unter·satz der *s.* Untersetzer

unter-, Unter-: ~**schätzen** *tr. V.* underestimate ⟨*amount, effect, etc.*⟩; underrate ⟨*talent, ability, etc.*⟩; ~**scheiden** **1.** *unr. tr. V.* distinguish; **2.** *unr. refl. V.* differ **(durch** in, **von** from); ~**scheidung** die *(Vorgang)* differentiation; *(Resultat)* distinction

Unter-: ~**schenkel** der shank; lower leg; ~**schicht** die *(Soziol.)* lower class

Unter·schied der; ~[e]s, ~e difference; **unterschiedlich 1.** *Adj.* different; *(uneinheitlich)* variable; varying; **2.** *adv.* |sehr/ganz| ~: in [very/quite] different ways; **unterschieds·los 1.** *Adj.* uniform; equal ⟨*treatment*⟩; **2.** *adv.* ⟨*treat*⟩ equally; *(ohne Benachteiligung)* without discrimination

unter·schlagen *unr. tr. V.* embezzle ⟨*money, funds, etc.*⟩; *(unterdrücken)* intercept ⟨*letter*⟩; withhold ⟨*fact, news, information, etc.*⟩

Unter·schlupf der; ~[e]s, ~e shelter; *(Versteck)* hiding-place; hide-out; **unter|schlüpfen** *itr. V.; mit sein (ugs.)* hide out

unter·schreiben *unr. itr., tr. V.* sign; **Unter·schrift** die signature; *(Bild~)* caption

Unter-: ~**see·boot** das submarine; ~**setzer** der mat; *(für Gläser)* coaster

untersetzt *Adj.* stocky

Unter·stand der *(Schutzbunker)* dug-out; *(Unterschlupf)* shelter

unter|stehen 1. *unr. itr. V.* jmdm. ~: be subordinate *or* answerable to sb.; **2.** *unr. refl. V.* dare

¹**unter|stellen 1.** *tr. V. (zur Aufbewahrung)* keep; store ⟨*furniture*⟩; **2.** *refl. V.* take shelter

²**unter·stellen** *tr. V.* **a)** jmdm. eine Abteilung ~: put sb. in charge of a department; **die Behörde ist dem Ministerium unterstellt** the office is under the ministry; **b)** *(unterschieben)* jmdm. böse Absichten *usw.* ~: insinuate that sb.'s intentions *etc.* are bad; **Unter·stellung** die *(falsche Behauptung)* insinuation

unter·streichen *unr. tr. V.* **a)** underline; **b)** *(hervorheben)* emphasize

unter·stützen *tr. V.* support; **Unter·stützung** die **a)** support; **b)** *(finanzielle Hilfe)* allowance; *(für Arbeitslose)* [unemployment] benefit *no art.*

unter·suchen *tr. V.* examine; *(überprüfen)* test **(auf** + *Akk.* for); *(aufzuklären suchen)* investigate; *(durchsuchen)* search **(auf** + *Akk.,* **nach** for); **Untersuchung** die; ~, - en **a)** *s.* untersuchen: examination; test; investigation; search; **b)** *(wissenschaftliche Arbeit)* study; **Untersuchungs·haft** die imprisonment *or* detention while awaiting trial

Unter·tasse die saucer

unter|tauchen 1. *itr. V.; mit sein* **a)**

(im Wasser) dive [under]; **b)** *(verschwinden)* disappear; **2.** *tr. V.* duck

Unter·teil das *od.* der bottom part; **unter·teilen** *tr. V.* divide; *(gliedern)* subdivide

unter·treiben *unr. itr. V.* play things down

Unter·wäsche die underwear

unterwegs *Adv.* on the way; *(nicht zu Hause)* out [and about]

unter·weisen *unr. tr. V. (geh.)* instruct

Unter·welt die; ~: underworld

unter·werfen 1. *unr. tr. V.* **a)** subjugate ⟨*people, country*⟩; **b)** *(unterziehen)* subject (*Dat.* to); **2.** *unr. refl. V.* sich |jmdm./einer Sache| ~: submit [to sb./sth.]; **unterwürfig 1.** *Adj.* obsequious; **2.** *adv.* obsequiously

unter·zeichnen *tr. V.* sign

unter·ziehen 1. *unr. tr. V.* etw. einer Untersuchung/Überprüfung *(Dat.)* ~: examine/check sth.; **2.** *unr. refl. V.* sich einer Operation *(Dat.)* ~: undergo *or* have an operation

untragbar *Adj.* unbearable

untreu *Adj.* disloyal; *(in der Ehe, Liebe)* unfaithful; **Untreue** die disloyalty; *(in der Ehe, Liebe)* unfaithfulness

untröstlich *Adj.* inconsolable

Untugend die bad habit

unüberlegt 1. *Adj.* rash; **2.** *adv.* rashly

unübersehbar 1. *Adj.* **a)** *(offenkundig)* conspicuous; **b)** *(sehr groß)* enormous; **2.** *adv. (sehr)* extremely

unübersichtlich 1. *Adj.* unclear; confusing ⟨*arrangement*⟩; blind ⟨*bend*⟩; broken ⟨*country etc.*⟩; **2.** *adv.* unclearly; confusingly ⟨*arranged*⟩

unübertrefflich 1. *Adj.* superb; **2.** *adv.* superbly; **unübertroffen** *Adj.* unsurpassed

unumgänglich *Adj.* [absolutely] necessary

unumwunden 1. *Adj.* frank; **2.** *adv.* frankly; openly

ununterbrochen 1. *Adj.* incessant; **2.** *adv.* incessantly

unveränderlich *Adj.* unchangeable

unverantwortlich 1. *Adj.* irresponsible; **2.** *adv.* irresponsibly

unverbesserlich *Adj.* incorrigible

unverbindlich 1. *Adj.* **a)** not binding *pred.;* without obligation *postpos;* **b)** *(reserviert)* non-committal ⟨*answer, words*⟩; impersonal ⟨*attitude, person*⟩; **2.** *adv.* ⟨*send, reserve*⟩ without obligation

unverblümt 1. *Adj.* blunt; **2.** *adv.* bluntly

unverbraucht *Adj.* untouched; unspent ⟨*energy*⟩; fresh ⟨*air*⟩

unverdaut *Adj.* undigested

unverdorben *Adj.* unspoilt

unverfroren *Adj.* insolent; impudent

unvergänglich *Adj.* immortal ⟨*fame*⟩; unchanging ⟨*beauty*⟩; abiding ⟨*recollection*⟩

unvergeßlich *Adj.* unforgettable

unvergleichlich 1. *Adj.* incomparable; **2.** *adv.* incomparably

unverheiratet *Adj.* unmarried

unverhofft 1. *Adj.* unexpected; **2.** *adv.* unexpectedly

unverkäuflich *Adj.* **diese Vase ist ~:** this vase is not for sale; *(nicht absetzbar)* unsaleable

unvermeidlich *Adj.* unavoidable; *(sich als Folge ergebend)* inevitable

Unvermögen das lack of ability

unvermutet 1. *Adj.* unexpected; **2.** *adv.* unexpectedly

unvernünftig *Adj.* stupid; foolish

unverrichtet *Adj.* **~er Dinge** without having achieved anything

unverschämt 1. *Adj.* **a)** impertinent ⟨*person, manner, words, etc.*⟩; barefaced ⟨*lie*⟩; **b)** *(ugs.: sehr groß)* outrageous ⟨*price, luck, etc.*⟩; **2.** *adv.* impertinently; ⟨*lie*⟩ barefacedly; blatantly; **Unverschämtheit die; ~, ~en** impertinence

unversehens *Adv.* suddenly

unversehrt *Adj.* unscathed; *(unbeschädigt)* undamaged

unverständlich *Adj.* incomprehensible; **Unverständnis das** lack of understanding

unverträglich *Adj.* **a)** quarrelsome; **b)** incompatible ⟨*blood groups, medicines, transplant tissue*⟩

unverwechselbar *Adj.* unmistakable; distinctive

unverwüstlich *Adj.* indestructible

unverzeihlich *Adj.* unforgivable

unverzüglich 1. *Adj.* prompt; **2.** *adv.* promptly

unvollkommen 1. *Adj.* **a)** imperfect; **b)** *(unvollständig)* incomplete; **2.** *adv.* **a)** imperfectly; **b)** *(unvollständig)* incompletely; **Unvollkommenheit die a)** imperfectness; **b)** *(Unvollständigkeit)* incompleteness

unvollständig *Adj.* incomplete; **Unvollständigkeit die** incompleteness

unvorhergesehen *Adj.* unforeseen; unexpected ⟨*visit*⟩

unvorsichtig 1. *Adj.* careless; *(unüberlegt)* rash; **2.** *adv.* carelessly; *(unüberlegt)* rashly

unvorstellbar 1. *Adj.* inconceivable; **2.** *adv.* unimaginably

unvorteilhaft *Adj.* **a)** unattractive ⟨*figure, appearance*⟩; **b)** *(ohne Vorteil)* unfavourable, poor ⟨*purchase, exchange*⟩; unprofitable ⟨*business*⟩

Unwahrheit die a) *o. Pl.* untruthfulness; **b)** *(Äußerung)* untruth; **unwahrscheinlich 1.** *Adj.* **a)** improbable; unlikely; **b)** *(ugs.: sehr viel)* incredible *(coll.);* **2.** *adv. (ugs.: sehr)* incredibly *(coll.)*

unweiblich *Adj.* unfeminine

unweigerlich 1. *Adj.* inevitable; **2.** *adv.* inevitably

Unwetter das [thunder]storm

unwichtig *Adj.* unimportant

unwiderruflich 1. *Adj.* irrevocable; **2.** *adv.* irrevocably

unwiderstehlich *Adj.* irresistible

Unwille[n] der; *o. Pl.* displeasure

unwillig 1. *Adj.* indignant; *(widerwillig)* unwilling; **2.** *adv.* indignantly; *(widerwillig)* unwillingly

unwillkürlich 1. *Adj.* **a)** spontaneous ⟨*cry, sigh*⟩; instinctive ⟨*reaction, movement, etc.*⟩; **b)** *(Physiol.)* involuntary ⟨*movement etc.*⟩; **2.** *adv.* **a)** ⟨*shout etc.*⟩ spontaneously; ⟨*react, move, etc.*⟩ instinctively; **b)** *(Physiol.)* ⟨*move etc.*⟩ involuntarily

unwirklich *(geh.) Adj.* unreal

unwirsch 1. *Adj.* surly; ill-natured; **2.** *adv.* ill-naturedly

unwirtschaftlich 1. *Adj.* uneconomic ⟨*procedure etc.*⟩; *(nicht sparsam)* uneconomical ⟨*driving etc.*⟩; **2.** *adv.* ⟨*work, drive, etc.*⟩ uneconomically

Unwissenheit die; ~: ignorance; **unwissentlich 1.** *Adj.* unconscious; **2.** *adv.* unknowingly; unwittingly

unwohl *Adv.* unwell; **mir ist ~:** I don't feel well; **Unwohlsein das; ~s** indisposition

unwürdig *Adj.* **a)** undignified ⟨*person, behaviour*⟩; degrading ⟨*treatment*⟩; **b)** *(unangemessen)* unworthy

unzählig *Adj.* innumerable; countless

Unze die; ~, ~n ounce

unzeitgemäß *Adj.* anachronistic

unzerbrechlich *Adj.* unbreakable

unzertrennlich *Adj.* inseparable

Unzucht die: ~ treiben fornicate; **gewerbsmäßige ~:** prostitution; **unzüchtig 1.** *Adj.* obscene ⟨*letter, ges-*

ture⟩; **2.** *adv.* ⟨*touch, approach, etc.*⟩ indecently; ⟨*speak*⟩ obscenely

unzufrieden *Adj.* dissatisfied; *(stärker)* unhappy; **Unzufriedenheit die** dissatisfaction; *(stärker)* unhappiness

unzugänglich *Adj.* inaccessible ⟨*area, building, etc.*⟩; unapproachable ⟨*character, person, etc.*⟩

unzulänglich *(geh.)* **1.** *Adj.* insufficient; **2.** *adv.* insufficiently

unzumutbar *Adj.* unreasonable

unzurechnungsfähig *Adj.* not responsible for one's actions *pred.*; *(geistesgestört)* of unsound mind *postpos.*

unzustellbar *Adj. (Postw.)* „~": 'not known [at this address]'

unzutreffend *Adj.* inappropriate; *(falsch)* incorrect

unzuverlässig *Adj.* unreliable; **Unzuverlässigkeit die** unreliability

unzweckmäßig 1. *Adj.* unsuitable; *(unpraktisch)* impractical; **2.** *adv.* unsuitably; *(unpraktisch)* impractically

üppig 1. *Adj.* lush ⟨*vegetation*⟩; thick ⟨*hair, beard*⟩; full ⟨*bosom, lips*⟩; voluptuous ⟨*figure, woman*⟩; *(fig.)* sumptuous, opulent ⟨*meal*⟩; **2.** *adv.* luxuriantly; *(fig.)* sumptuously

Ur·abstimmung die [*esp.* strike] ballot

Ural der; ~|s| Urals *pl.*; Ural Mountains *pl.*

ur·alt *Adj.* very old; ancient

Uran das; ~s uranium

urbar *Adj.* **ein Stück Land ~ machen** cultivate a piece of land

Ur·einwohner der native inhabitant; **Ur·enkel der** great-grandson; **Ur·groß·eltern** *Pl.* great-grandparents

Ur·heber der; ~s, ~ originator; initiator; *(bes. Rechtsspr.: Verfasser, Autor)* author

urig *Adj.* natural ⟨*person*⟩; real ⟨*beer*⟩; cosy ⟨*pub*⟩

Urin der; ~s, ~e *(Med.)* urine; **urinieren** *itr. V.* urinate

Ur·kunde die; ~, ~n document; *(Bescheinigung, Sieger~, Diplom~ usw.)* certificate

Urlaub der; ~|e|s, ~e holiday[s] *(Brit.)*; vacation; *(bes. Milit.)* leave

Urlaubs-: ~reise **die** holiday [trip]; ~zeit **die** holiday period *or* season

Urne die; ~, ~n urn; *(Wahl~)* [ballot-]box

Ur·sache die cause

Ur·sprung der origin; **ur·sprünglich 1.** *Adj.* **a)** original ⟨*plan, price, form, material, etc.*⟩; **b)** *(natürlich)*

natural; **2.** *adv.* **a)** originally; **b)** *(natürlich)* naturally

Urteil das; ~s, ~e judgement; *(Strafe)* sentence; *(Gerichts~)* verdict; **urteilen** *itr. V.* form an opinion; judge; **über etw./jmdn.** ~: judge sth./sb.; **Urteils·vermögen das;** *o. Pl.* competence to judge

Ur·wald der primeval forest; *(tropisch)* jungle

USA [u:|es'|a:] *Pl.* USA

usw. *Abk.* und so weiter etc.

Utensil das; ~s, ~ien [... |ən] piece of equipment; ~ien equipment *sing.*

Utopie die; ~, ~n utopian dream; **utopisch** *Adj.* utopian

UV *Abk.* Ultraviolett UV

V

v, V [vau] **das;** ~, ~: v, V

v. *Abk.* von

vage 1. *Adj.* vague; **2.** *adv.* vaguely

vakuum·verpackt *Adj.* vacuum-packed

Vanille [va'nɪljə] **die;** ~: vanilla; **Vanille·zucker der** vanilla sugar

variabel 1. *Adj.* variable; **2.** *adv.* variably

variieren *tr., itr. V.* vary

Vase ['va:zə] **die;** ~, ~n vase

Vater der; ~s, Väter father; **Gott ~:** God the Father; **Vater·land das;** *Pl.* ~länder fatherland; **väterlich 1.** *Adj.* **a)** paternal ⟨*line, love, instincts, etc.*⟩; **b)** *(fürsorglich)* fatherly; **2.** *adv.* in a fatherly way; **väterlicherseits** *Adv.* on the/his/her *etc.* father's side; **Vaterschaft die;** ~, ~en fatherhood; **Vaterunser das;** ~s, ~: Lord's Prayer; **Vati der;** ~s, ~s *(fam.)* dad[dy] *(coll.)*

Vatikan [vati'ka:n] **der;** ~s Vatican

v. Chr. *Abk.* vor Christus BC

Vegetarier [vege'ta:riɐ] **der;** ~s, ~: vegetarian; **vegetarisch 1.** *Adj.* vegetarian; **2.** *adv.* **er ißt** *od.* **lebt** ~: he is a vegetarian; **Vegetation die;** ~, ~en vegetation *no indef. art.*; **vegetieren** *itr. V.* vegetate

Veilchen das; ~s, ~: violet
Vene ['ve:nə] die; ~, ~n vein
Venedig [ve'ne:dɪç] (das); ~s Venice
Venezolaner [venet̮so'la:nɐ] der; ~s, ~: Venezuelan; **venezolanisch** Adj. Venezuelan; **Venezuela** (das); ~s Venezuela
Ventil [vɛn'ti:l] das; ~s, ~e valve; **Ventilator** [vɛnti'la:tɔr] der; ~s, ~en ventilator
Venus ['ve:nʊs] die; ~: Venus no def. art.
verabreden 1. tr. V. arrange; 2. refl. V. sich im Park/zum Tennis/für den folgenden Abend ~: arrange to meet in the park/for tennis/next evening; **Verabredung** die; ~, ~en a) arrangement; b) (verabredete Zusammenkunft) appointment; **eine ~ absagen** call off a meeting
verabscheuen tr. V. detest; loathe
verabschieden 1. tr. V. a) say goodbye to; b) (aus dem Dienst) retire ⟨general, civil servant, etc.⟩; 2. refl. V. sich [von jmdm.] ~: say goodbye [to sb.]; **Verabschiedung** die; ~, ~en a) leave-taking; b) (aus dem Dienst) retirement
verachten tr. V. despise; **verächtlich** 1. Adj. a) contemptuous; b) (verachtenswürdig) contemptible; 2. adv. contemptuously; **Verachtung** die; ~: contempt
verallgemeinern tr., itr. V. generalize; **Verallgemeinerung** die; ~, ~en generalization
veralten itr. V.; mit sein become obsolete
Veranda [vɛ'randa] die; ~, **Veranden** veranda; porch
veränderlich Adj. changeable; **verändern** tr., refl. V. change; **Veränderung** die change (Gen. in)
verängstigen tr. V. frighten; scare
verankern tr. V. fix ⟨tent, mast, pole, etc.⟩; (mit einem Anker) anchor
veranlagen tr.V. (Steuerw.) assess (mit at); **veranlagt** Adj. künstlerisch/praktisch ~ sein have an artistic bent/be practically minded; **Veranlagung** die; ~, ~en [pre]disposition
veranlassen tr. V. cause; induce; ~, daß ... see to it that ... **Veranlassung** die; ~, ~en reason
veranschaulichen tr. V. illustrate
veranschlagen tr. V. estimate (mit at)
veranstalten tr. V. organize; hold,

give ⟨party⟩; hold ⟨auction⟩; do ⟨survey⟩; **Veranstalter** der; ~s, ~ organizer; **Veranstaltung** die; ~, ~en a) (das Veranstalten) organizing; organization; b) (etw., was veranstaltet wird) event
verantworten 1. tr. V. etw. ~: take responsibility for sth.; 2. refl. V. sich für etw. ~: answer for sth.; sich vor jmdm. ~: answer to sb.; **verantwortlich** Adj. responsible; **Verantwortung** die; ~, ~en responsibility (für for)
verantwortungs-: ~**bewußt** Adj. responsible; ~**los** Adj. irresponsible; ~**voll** Adj. responsible
verarbeiten tr. V. use; **etw. zu etw. ~:** make sth. into sth.; (geistig bewältigen) assimilate ⟨film, experience, impressions⟩
verärgern tr. V. annoy
verarzten tr. V. (ugs.) patch up (coll.) ⟨person⟩; fix (coll.) ⟨wound etc.⟩
veräußern tr. V. dispose of ⟨property⟩
Verb [vɛrp] das; ~s, ~en verb
Verband der a) (Binde) bandage; dressing; b) (von Vereinen, Clubs o. ä.) association
Verband[s]-: ~**kasten** der first-aid-box; ~**material** das dressing materials pl.
Verband·zeug das first-aid things pl.
Verbannung die; ~, ~en banishment
verbergen unr. tr. V. hide; conceal
verbessern 1. tr. V. a) improve; reform ⟨schooling, world⟩; b) (korrigieren) correct; 2. refl. V. a) improve; b) ([beruflich] aufsteigen) better oneself; **Verbesserung** die a) improvement; b) (Korrektur) correction
verbeugen refl. V. bow (vor + Dat. to); **Verbeugung** die; ~, ~en bow
verbieten unr. tr. V. a) forbid; **jmdm. etw.~:** forbid sb. sth.; „Betreten des Rasens/Rauchen verboten" 'keep off the grass'/'no smoking'; b) (für unzulässig erklären) ban
verbinden 1. unr. tr. V. a) (bandagieren) bandage; dress; b) (zubinden) bind; **jmdm. die Augen ~:** blindfold sb.; c) (zusammenfügen) join; d) (in Beziehung bringen) connect (durch by); link ⟨towns, lakes, etc.⟩ (durch by); e) (verknüpfen) combine ⟨abilities, qualities, etc.⟩; f) auch itr. (telefonisch) **jmdn. [mit jmdm.] ~:** put sb. through [to sb.]; 2. unr. refl. V. a) (auch Chemie) combine (mit with); b) (sich zusammentun) join [together]; join

forces; **verbindlich 1.** *Adj.* **a)**
friendly; **b)** *(bindend)* obligatory;
compulsory; binding ⟨*agreement, de-
cision, etc.*⟩; **2.** *adv.* **a)** *(freundlich)* in a
friendly manner; **b)** ~ **zusagen** definitely agree; **jmdm. etw.** ~ **zusagen**
make sb. a firm offer of sth.; **Verbindung** die **a)** *(das Verknüpfen)* linking;
b) *(Zusammenhalt)* join; connection;
c) *(verknüpfende Strecke)* link; **d)**
(durch Telefon, Funk, Verkehrs~) connection **(nach** to); **e)** *(Kombination)*
combination; **in** ~ **mit etw.** in conjunction with sth.; **f)** *(Kontakt)* contact; **sich mit jmdm. in** ~ **setzen** get in
touch *or* contact with sb.; **g)** *(Zusammenhang)* connection
verbissen 1. *Adj.* dogged; doggedly
determined; **2.** *adv.* doggedly
verbitten *unr. refl. V.* **sich** *(Dat.)* **etw.**
~: refuse to tolerate sth.
verbittern *tr. V.* embitter
verblassen *itr. V.; mit sein (auch fig.
geh.)* fade
Verbleib der; ~|e|s *(geh.)* where-
abouts *pl.;* **verbleiben** *unr. itr. V.;
mit sein* remain; **wie seid ihr verblie-
ben?** what did you arrange?
Verblendung die; ~, ~en blindness
verblüffen *tr. (auch itr.) V.* amaze;
verblüffend 1. *Adj.* amazing; **2.** *adv.*
amazingly
verblühen *itr. V.; mit sein (auch fig.)*
fade
verbluten *itr. (auch refl.) V.; mit sein*
bleed to death
verbohrt *Adj.* pigheaded
verborgen *Adj. (abgelegen)* secluded;
(nicht sichtbar) hidden
Verbot das; ~|e|s, ~e ban *(Gen., von*
on); **Verbots·schild** das; *Pl.*
~schilder sign *(prohibiting sth.); (Verkehrsw.)* prohibitive sign
Verbrauch der; ~|e|s consumption;
(von, an + *Dat.* of); **verbrauchen** *tr.
V.* use; consume ⟨*food, drink*⟩; use up
⟨*provisions*⟩; spend ⟨*money*⟩; consume, use ⟨*fuel*⟩; *(fig.)* use up
⟨*strength, energy*⟩; **Verbraucher**
der; ~s, ~: consumer
Verbrechen das; ~s, ~: crime **(an** +
Dat., **gegen** against); **Verbrecher**
der; ~s, ~: criminal; **verbreche-
risch** *Adj.* criminal
verbreiten 1. *tr. V.* spread; radiate
⟨*optimism, calm, etc.*⟩; **2.** *refl. V.*
spread; **Verbreitung** die; ~, ~en **a)**
s. verbreiten 1: spreading; radiation;
b) *(Ausbreitung)* spread

verbrennen 1. *unr. itr. V.; mit sein*
burn; **2.** *tr. V.* burn; cremate ⟨*dead
person*⟩; **sich** *(Dat.)* **den Mund** ~ *(fig.)*
say too much; **Verbrennung** die; ~,
~en **a)** *s.* verbrennen 2: burning; cremation; **b)** *(Wunde)* burn
verbringen *unr. tr. V.* spend
verbummeln *tr. V. (ugs.)* **a)** waste
⟨*time*⟩; **b)** *(vergessen)* forget [all]
about; clean forget; *(verlieren)* lose
verbünden *refl. V.* form an alliance;
Verbündete der/die; *adj. Dekl.* ally
verbüßen *tr. V.* serve ⟨*sentence*⟩
Verdacht der; ~|e|s, ~e *od.* **Verdächte**
suspicion; **verdächtig 1.** *Adj.* suspicious; **2.** *adv.* suspiciously; **Verdächtige** der/die; *adj. Dekl.* suspect;
verdächtigen *tr. V.* suspect
verdammen *tr. V.* condemn; *(Rel.)*
damn ⟨*sinner*⟩
verdampfen 1. *itr. V.; mit sein* evaporate; **2.** *tr. V.* evaporate
verdanken *tr. V.* **jmdm./einer Sache
etw.** ~: owe sth. to sb./sth.
verdarb *1. u. 3. Pers. Sg. Prät. v.* **verderben**
verdattert *(ugs.) Adj.* flabbergasted;
(verwirrt) dazed; stunned
verdauen 1. *tr. V. (auch fig.)* digest; **2.**
itr. V. digest [one's food]; **ver-
daulich** *Adj.* digestible; **Verdauung** die; ~: digestion
Verdeck das; ~|e|s, ~e top; hood
(Brit.); (bei Kinderwagen) hood; **verdecken** *tr. V.* hide; cover
verderben 1. *unr. itr. V.; mit sein* go
bad *or* off, spoil; **2.** *unr. tr. V.* spoil;
(stärker) ruin; spoil ⟨*appetite, enjoyment, fun, etc.*⟩; **3.** *unr. refl. V.* **sich**
(Dat.) **den Magen/die Augen** ~: give
oneself an upset stomach/ruin one's
eyesight; **Verderben** das; ~s ruin;
verderblich *Adj.* perishable ⟨*food*⟩;
pernicious ⟨*influence, effect, etc.*⟩
verdeutlichen *tr. V.* etw. ~: make
sth. clear; *(erklären)* explain sth.
verdichten *refl. V.* ⟨*fog, smoke*⟩
thicken, become thicker; *(fig.)* ⟨*suspicion, rumour*⟩ grow; ⟨*feeling*⟩ intensify
verdienen 1. *tr. V.* **a)** earn; **b)** *(wert
sein)* deserve; **2.** *itr. V.* **beide Eheleute**
~: husband and wife are both earn-
ing; **Verdiener** der; ~s, ~: wage-
earner; ¹**Verdienst** der income;
earnings *pl.;* ²**Verdienst** das; ~|e|s,
~e merit
verdienst·voll 1. *Adj.* commendable; ⟨*person*⟩ of outstanding merit; **2.**
adv. commendably; **verdient** *Adj.*

⟨*person*⟩ of outstanding merit; **sich um etw. ~ machen** render outstanding services to sth.

verdoppeln 1. *tr. V.* double; *(fig.)* double, redouble ⟨*efforts etc.*⟩; **2.** *refl. V.* double

verdorben 2. *Part. v.* **verderben**

verdorren *itr. V.; mit sein* wither [and die]; ⟨*meadow*⟩ scorch

verdrängen *tr. V.* **a)** drive out ⟨*inhabitants*⟩; *(fig.: ersetzen)* displace; **b)** *(Psych.)* repress; *(bewußt)* suppress

verdrehen *tr. V.* **a)** twist ⟨*joint*⟩; roll ⟨*eyes*⟩; **b)** *(ugs. abwertend: entstellen)* twist ⟨*words, facts, etc.*⟩

verdrießen *unr. tr. V. (geh.)* irritate; annoy; **verdrießlich 1.** *Adj.* morose; **2.** *adv.* morosely; **verdroß** *1. u. 3. Pers. Sg. Prät. v.* **verdrießen; verdrossen 1.** *Adj. (mißmutig)* morose; *(mißmutig und lustlos)* sullen; **2.** *adv. (mißmutig)* morosely; *(mißmutig und lustlos)* sullenly; **Verdruß der; Verdrusses, Verdrusse** annoyance

verdunkeln *tr. V.* darken; *(vollständig)* black out ⟨*room, house, etc.*⟩

Verdunk[e]lung die; ~, ~en darkening; *(vollständig)* black-out

verdünnen *tr. V.* dilute

verdunsten *itr. V.; mit sein* evaporate; **Verdunstung die; ~:** evaporation

verdursten *itr. V.; mit sein* die of thirst

verdutzt *Adj.* taken aback *pred.;* nonplussed; *(verwirrt)* baffled

verehren *tr. V.* **a)** venerate; **b)** *(geh.: bewundern)* admire; *(ehrerbietig lieben)* worship; **Verehrer der; ~s, ~, Verehrerin die; ~, ~en** admirer; **Verehrung die; o. Pl.** **a)** veneration; **b)** *(Bewunderung)* admiration

vereidigen *tr. V.* swear in; **Vereidigung die; ~, ~en** swearing in

Verein der; ~s, ~e organization; *(der Kunstfreunde usw.)* association; society; *(Sport~)* club; **vereinbar** *Adj.; nicht attr.* compatible; **vereinbaren** *tr. V.* agree; arrange ⟨*meeting etc.*⟩; **Vereinbarung die; ~, ~en** **a)** agreeing; *(eines Termins usw.)* arranging; **b)** *(Abmachung)* agreement

vereinfachen *tr. V.* simplify

vereinheitlichen *tr. V.* standardize

vereinigen *tr., refl. V.* unite; *(in der Wirtschaft)* merge; **vereinigt** *Adj.* united; **Vereinigung die** **a)** organization; **b)** *(das Vereinigen)* uniting; *(von Unternehmen)* merging

vereinzelt 1. *Adj.; nicht präd.* occasional; **2.** *adv. (zeitlich)* occasionally; *(örtlich)* here and there

vereisen *itr. V.; mit sein* freeze *or* ice over; ⟨*wing*⟩ ice up; ⟨*lock*⟩ freeze up

vereiteln *tr. V.* thwart

vereitern *itr. V.; mit sein* go septic

verenden *itr. V.; mit sein* perish; die

verengen *refl. V.* narrow; ⟨*pupils*⟩ contract

vererben *tr. V.* leave, bequeath ⟨*property*⟩ (*Dat.*, **an** + *Akk.* to)

Vererbung die; ~, ~en heredity *no art.*

verfahren 1. *unr. refl. V.* lose one's way; **2.** *unr. itr. V.; mit sein* proceed; **Verfahren das; ~s, ~** **a)** procedure; *(Technik)* process; *(Methode)* method; **b)** *(Rechtsw.)* proceedings *pl.*

Verfall der; o. Pl. **a)** decay; *(fig.: der Preise, einer Währung)* collapse; **b)** *(Auflösung)* decline; **verfallen** *unr. itr. V.; mit sein* **a)** *(baufällig werden)* fall into disrepair; **b)** *(körperlich)* ⟨*strength*⟩ decline; **c)** *(untergehen)* ⟨*empire*⟩ decline; ⟨*morals, morale*⟩ deteriorate; **d)** *(ungültig werden)* expire

verfassen *tr. V.* write; draw up ⟨*resolution*⟩; **Verfasser der; ~s, ~, Verfasserin die; ~, ~nen** writer; *(eines Buchs, Artikels usw.)* author; writer; **Verfassung die** **a)** *(Politik)* constitution; **b)** *o. Pl. (Zustand)* state [of health/mind]; **in guter/schlechter ~ sein** be in good/poor shape

verfaulen *itr. V.; mit sein* rot

verfehlen *tr. V.* miss; **Verfehlung die; ~, ~en** misdemeanour; *(Rel.: Sünde)* transgression

verfeinden *refl. V.* **sich ~ mit** make an enemy of

verfeinern *tr. V.* improve; refine ⟨*method, procedure*⟩

verfertigen *tr. V.* produce

verfilmen *tr. V.* film; make a film of; **Verfilmung die; ~, ~en** **a)** *(das Verfilmen)* filming; **b)** *(Film)* film [version]

verflixt *(ugs.)* **1.** *Adj.* **a)** *(ärgerlich)* awkward, unpleasant ⟨*situation, business, etc.*⟩; **b)** *(verdammt)* blasted *(Brit.);* blessed; confounded; **~ [noch mal]!** [damn and] blast! *(Brit. coll.);* **c)** *nicht präd. (sehr groß)* **er hat ~es Glück gehabt** he was damned lucky *(coll.);* **2.** *adv. (sehr)* damned *(coll.)*

verflossen *Adj. (ugs.)* former

verfluchen *tr. V.* curse; **verflucht 1.**

Adj. (salopp) damned *(coll.);* bloody *(Brit. sl.);* ~ |noch mal|! damn [it]! *(coll.);* 2. *adv. (sehr)* damned *(coll.)*

verfolgen *tr. V.* pursue; hunt, track ⟨*animal*⟩; etw. |strafrechtlich| ~: prosecute sth.; **Verfolgung die;** ~, ~en pursuit; *(eines Ziels, Plans usw.)* pursuance

verfressen *Adj. (salopp)* greedy

verfügen 1. *tr. V. (anordnen)* order; *(dekretieren)* decree; 2. *itr. V.* über etw. *(Akk.)* |frei| ~ können be free to decide what to do with sth.; über etw. *(Akk.)* ~ *(etw. haben)* have sth. at one's disposal; **Verfügung die;** ~, ~en a) *(Anordnung)* order; *(Dekret)* decree; b) o. *Pl. (Disposition)* etw. zur ~ haben have sth. at one's disposal; jmdm. etw. zur ~ stellen put sth. at sb.'s disposal

verführen *tr. V.* a) *(verleiten)* tempt; b) *(sexuell)* seduce; **Verführer der** seducer; **verführerisch** 1. *Adj.* a) *(verlockend)* tempting; b) *(aufreizend)* seductive; 2. *adv.* a) *(verlockend)* temptingly; b) *(aufreizend)* seductively; **Verführung die** a) temptation; b) *(sexuell)* seduction

vergangen 1. *Adj.* a) *(vorüber, vorbei)* bygone, former ⟨*times, years, etc.*⟩; b) *(letzt...)* last ⟨*year, week, etc.*⟩; **Vergangenheit die;** ~ a) past; b) *(Grammatik: Präteritum)* past tense; **vergänglich** *Adj.* transient; transitory; ephemeral; **Vergänglichkeit die;** ~: transience

Vergaser der; ~s, ~: carburettor

vergaß *1. u. 3. Pers. Sg. Prät. v.* vergessen

vergeben *unr. tr. V.* a) *auch itr. (geh.: verzeihen)* forgive; jmdm. etw. ~: forgive sb. [for] sth.; b) throw away ⟨*chance, goal, etc.*⟩; c) *(geben)* place ⟨*order*⟩ (an + *Akk.* with); award ⟨*grant, prize*⟩ (an + *Akk.* to); **vergebens** 1. *Adv.* in vain; vainly; 2. *adj.* es war ~: it was of *or* to no avail; **vergeblich** 1. *Adj.* futile; vain, futile ⟨*attempt, efforts*⟩; 2. *adv.* in vain; **Vergebung die;** ~, ~en *(geh.)* forgiveness

vergehen *unr. itr. V.;* mit sein ⟨*time*⟩ pass [by], go by; ⟨*pain*⟩ wear off, pass; ⟨*pleasure*⟩ fade; **Vergehen das;** ~s, ~: crime; *(Rechtsspr.)* offence

vergelten *unr. tr. V.* repay

vergessen 1. *unr. tr. (auch itr.) V.* forget; **Vergessenheit die;** ~: oblivion; **vergeßlich** *Adj.* forgetful

vergeuden *tr. V.* waste; **Vergeudung die;** ~, ~en waste

vergewaltigen *tr. V.* rape; **Vergewaltigung die;** ~, ~en rape

vergewissern *refl. V.* make sure *(Gen.* of)

vergießen *unr. tr. V.* spill; Tränen ~: shed tears

vergiften *tr. V. (auch fig.)* poison; **Vergiftung die;** ~, ~en poisoning

vergiß *Imper. Sg. v.* vergessen; **Vergiß·mein·nicht das;** ~|e|s, ~|e| forget-me-not; **vergißt** *2. u. 3. Pers. Sg. Präs. v.* vergessen

Vergleich der; ~|e|s, ~e a) comparison; b) *(Rechtsw.)* settlement; **vergleichbar** *Adj.* comparable; **vergleichen** *tr. V.* compare; **Vergleichs·form die** *(Sprachw.)* comparative/superlative form

vergnügen *refl. V.* enjoy oneself; have a good time; **Vergnügen das;** ~s, ~: pleasure; *(Spaß)* fun; viel ~! *(auch iron.)* have fun!; **vergnügt** 1. *Adj.* cheerful; 2. *adv.* cheerfully; **Vergnügungs·viertel das** pleasure district

vergolden *tr. V.* gold-plate ⟨*jewellery etc.*⟩; *(mit Blattgold)* gild

vergraben *unr. tr. V.* bury

vergrämt *Adj.* care-worn

vergreifen *unr. refl. V.* sich an jmdm. ~: assault sb.; **vergriffen** *Adj.* out of print *pred.*

vergrößern 1. *tr. V.* a) *(erweitern)* extend ⟨*room, area, building, etc.*⟩; b) *(vermehren)* increase; c) *(größer reproduzieren)* enlarge ⟨*photograph etc.*⟩; 2. *refl. V.* a) *(größer werden)* ⟨*firm, business, etc.*⟩ expand; b) *(zunehmen)* increase; 3. *itr. V.* ⟨*lens etc.*⟩ magnify; **Vergrößerung die;** ~, ~en a) *s. v.* vergrößern 1, 2: extension; increase; enlargement; expansion; b) *(Foto)* enlargement; **Vergrößerungs·glas das** magnifying glass

Vergünstigung die; ~, ~en privilege

vergüten *tr. V.* a) *(erstatten)* jmdm. etw. ~: reimburse sb. for sth.; b) *(bes. Papierdt.: bezahlen)* remunerate, pay for ⟨*work, services*⟩; **Vergütung die;** ~, ~en a) *(Rückerstattung)* reimbursement; b) *(Geldsumme)* remuneration

verhaften *tr. V.* arrest; Sie sind verhaftet you are under arrest; **Verhaftung die;** ~, ~en arrest

verhalten *unr. refl. V.* a) behave; *(reagieren)* react; b) *(beschaffen sein)* be; **Verhalten das;** ~s behaviour

Verhaltens·weise die behaviour;
Verhältnis das; ~ses, ~se a) ein ~
von drei zu eins a ratio of three to one;
b) *(persönliche Beziehung)* relationship (zu with); mit jmdm. ein ~ haben
(ugs.) have an affair with sb.; c) *Pl.
(Umstände)* conditions; **verhält-
nis·mäßig** *Adv.* relatively; com-
paratively; **Verhältnis·wort** das;
Pl. ~wörter *(Sprachw.)* preposition
verhandeln 1. *itr. V.* a) negotiate
(über + Akk. about); b) *(strafrecht-
lich)* try a case; *(zivilrechtlich)* hear a
case; 2. *tr. V.* a) etw. ~: negotiate over
sth.; b) *(strafrechtlich)* try ⟨case⟩; *(zivilrechtlich)* hear ⟨case⟩; **Verhandlung**
die a) ~en negotiations; b) *(strafrecht-
lich)* trial; *(zivilrechtlich)* hearing; die
~ gegen X the trial of X
verhängen *tr. V.* impose ⟨fine, punish-
ment⟩ (über + Akk. on); declare
⟨state of emergency, state of siege⟩;
(Sport) award, give ⟨penalty etc.⟩;
Verhängnis das; ~ses, ~se undoing;
verhängnis·voll *Adj.* disastrous
verharmlosen *tr. V.* play down
verharren *itr. V. (geh.)* remain
verhärten 1. *tr. V.* harden; make ⟨per-
son⟩ hard; 2. *refl. V.* ⟨tissue⟩ become
hardened
verhaßt *Adj.* hated; detested
verhätscheln *tr. V. (ugs.)* pamper
verhauen *(ugs.) unr. tr. V.* beat up;
(als Strafe) beat
verheben *unr. refl. V.* do oneself an
injury [while lifting sth.]
verheeren *tr. V.* devastate; lay waste
[to]; **verheerend** *Adj.* a) devastating; b) *(ugs.: scheußlich)* ghastly *(coll.)*
verhehlen *tr. V. (geh.)* conceal *(Dat.
from)*
verheilen *itr. V.; mit sein* ⟨wound⟩
heal [up]
verheimlichen *tr. V.* [jmdm.] etw. ~:
keep sth. secret [from sb.]
verheiraten *refl. V.* get married; sich
mit jmdm. ~: marry sb.; get married to
sb.; **Verheiratete der/die**; *adj. Dekl.*
married person; married man/
woman; **Verheiratung die**; ~, ~en
marriage
verhelfen *unr. itr. V.* jmdm./einer Sache zu etw. ~: help sb./sth. to get/
achieve sth.
verherrlichen *tr. V.* glorify
verheult *Adj. (ugs.)* ⟨eyes⟩ red from
crying; ⟨face⟩ puffy *or* swollen from
crying
verhexen *tr. V. (auch fig.)* bewitch

verhindern *tr. V.* prevent; **Verhinderung die**; ~, ~en prevention
verhöhnen *tr. V.* mock
Verhör das; ~[e]s, ~e interrogation;
questioning; *(bei Gericht)* examination; **verhören** 1. *tr. V.* interrogate;
question; *(bei Gericht)* examine; 2.
refl. V. mishear
verhüllen *tr. V.* cover; *(fig.)* disguise
verhungern *itr. V.; mit sein* die of
starvation; starve [to death]
verhüten *tr. V.* prevent; **Verhütung
die**; ~, ~en prevention; *(Empfäng-
nis~)* contraception; **Verhütungsmittel** das contraceptive
verirren *refl. V.* a) get lost; lose one's
way; ⟨animal⟩ stray; b) *(irgendwohin
gelangen)* stray (in, an + Akk. into)
verjagen *tr. V.* chase away
verkalken *itr. V.; mit sein* a) ⟨tissue⟩
calcify; ⟨arteries⟩ become hardened;
b) *(ugs.: senil werden)* become senile
Verkauf der sale; **verkaufen** *tr. V.
(auch fig.)* sell *(Dat., an + Akk. to)*;
„zu ~" 'for sale'; **Verkäufer der,
Verkäuferin die** a) seller; vendor
(formal); b) *(Berufsbez.)* sales *or* shop
assistant; *(im Außendienst)* salesman/
saleswoman; **verkäuflich** *Adj. (zum
Verkauf geeignet)* saleable; *(zum Verkauf bestimmt)* for sale *postpos.*; **ver-
kaufs·offen** *Adj.* der ~e Samstag
Saturday on which the shops are open
all day; **Verkaufs·preis** der retail
price
Verkehr der; ~s a) traffic; b) *(Kontakt)* contact; communication; c) *(Ge-
schlechts~)* intercourse; **verkehren**
itr. V. a) *auch mit sein (fahren)* run;
⟨aircraft⟩ fly; b) *(in Kontakt stehen)*
mit jmdm. ~: associate with sb.; c) *(zu
Gast sein)* bei jmdm. ~: visit sb. regularly
Verkehrs-: ~ampel die traffic lights
pl.; ~aufkommen das volume of
traffic; ~hindernis das obstruction
to traffic; ~knotenpunkt der [traf-
fic] junction; ~kontrolle die traffic
check; ~meldung die traffic announcement; ~mittel das means of
transport; die öffentlichen ~mittel
public transport *sing.;* ~schild das;
Pl. ~schilder traffic sign; road sign;
~teilnehmer der road-user; ~un-
fall der road accident; ~zeichen das
traffic sign; road sign
verkehrt 1. *Adj.* wrong; 2. *adv.*
wrongly; alles ~ machen do everything wrong

verkẹnnen *unr. tr. V.* fail to recognize; misjudge ⟨*situation*⟩

verklạgen *tr. V.* sue; take to court; **eine Firma auf Schadenersatz ~:** sue a company for damages

verklẹben **1.** *itr. V.; mit sein* stick together; **2.** *tr. V.* *(zukleben)* seal up ⟨*hole*⟩; *(festkleben)* stick [down] ⟨*floor-covering etc.*⟩

verklẹiden *tr. V.* disguise; *(kostümieren)* dress up; **sich ~:** disguise oneself/dress [oneself] up; **Verklẹidung die a)** *o. Pl.* disguising; *(das Kostümieren)* dressing up; **b)** *(Kleidung)* disguise; *(bei einer Party)* fancy dress

verklẹinern **1.** *tr. V.* **a)** make smaller; **b)** *(verringern)* reduce ⟨*size, number, etc.*⟩; **c)** *(kleiner reproduzieren)* reduce ⟨*photograph etc.*⟩; **2.** *refl. V.* become smaller; ⟨*number*⟩ decrease; **Verklẹinerungs·form die** *(Sprachw.)* diminutive form

verknọten *tr. V.* tie; knot

verknüpfen *tr. V.* **a)** *(knoten)* tie; knot; **b)** *(in Beziehung setzen)* link

verkọchen *itr. V.; mit sein* boil away

verkọhlen *itr. V.* char

¹verkọmmen *unr. itr. V.; mit sein* go to the dogs; *(moralisch, sittlich)* go to the bad; **²verkọmmen** *Adj.* depraved

verkọ̈stigen *tr. V.* feed; provide with meals

verkrạften *tr. V.* cope with

verkrạmpfen *refl. V.* ⟨*muscle*⟩ become cramped; ⟨*person*⟩ tense up; **Verkrạmpfung die; ~, ~en** tenseness; tension

verkrẹichen *unr. refl. V.* ⟨*animal*⟩ creep [away]; ⟨*person*⟩ hide [oneself away]

verkrụ̈mmt *Adj.* bent ⟨*person*⟩; crooked ⟨*finger*⟩; curved ⟨*spine*⟩; **Verkrụ̈mmung die** crookedness

verkrụ̈ppeln *tr. V.* cripple

verkụ̈mmern *itr. V.; mit sein* ⟨*person, animal*⟩ go into a decline; ⟨*plant etc.*⟩ become stunted; ⟨*talent, emotional life, etc.*⟩ wither away

verkụ̈nden *tr. V.* announce; pronounce ⟨*judgement*⟩; promulgate ⟨*law, decree*⟩; **verkụ̈ndigen** *tr. V.* *(geh.)* announce; proclaim; **Verkụ̈ndigung die** announcement; proclamation

verkụ̈rzen *tr. V.* **a)** *(verringern)* reduce; *(abkürzen)* shorten; **b)** *(abbrechen)* cut short ⟨*stay, life*⟩; put an end to, end ⟨*suffering*⟩

verlạden *unr. tr. V.* load

Verlạg der; ~[e]s, ~e publishing house *or* firm; publisher's

verlạgern *tr. V.* shift; *(an einen anderen Ort)* move; *(fig.)* transfer; shift ⟨*emphasis*⟩

verlạngen *tr. V.* demand; *(nötig haben)* ⟨*task etc.*⟩ require, call for ⟨*patience, knowledge, experience, skill, etc.*⟩; *(berechnen)* charge; *(sehen/sprechen wollen)* ask for; **du wirst am Telefon verlangt** you're wanted on the phone *(coll.)*; **Verlạngen das; ~s, ~ a)** desire **(nach** for); **b) auf ~:** on request

verlạ̈ngern *tr. V.* extend; lengthen, make longer ⟨*skirt, sleeve, etc.*⟩; renew ⟨*passport, driving-licence, etc.*⟩

Verlạ̈ngerung die; ~, ~en *s.* verlängern: extension; lengthening; renewal

verlạngsamen *tr. V.* **das Tempo/seine Schritte ~:** reduce speed/slacken one's pace; slow down

¹verlạssen **1.** *unr. refl. V.* rely, depend **(auf +** *Akk.* on); **2.** *unr. tr. V.* leave; **²verlạssen** *Adj.* deserted ⟨*street etc.*⟩; empty ⟨*house*⟩; *(öd)* desolate ⟨*region etc.*⟩

verlạ̈ßlich **1.** *Adj.* reliable; **2.** *adv.* reliably

Verlạuf der; ~[e]s, Verlạ̈ufe course; **verlạufen** **1.** *unr. itr. V.; mit sein* **a)** *(sich erstrecken)* run; **b)** *(ablaufen)* ⟨*test, rehearsal, etc.*⟩ go; ⟨*party etc.*⟩ go off; **2.** *unr. refl. V.* get lost; lose one's way; **Verlạufs·form die** *(Sprachw.)* progressive *or* continuous form

verlạutbaren *tr. V.* announce [officially]; **verlạuten** *itr. V.; mit sein* be reported; **wie verlautet** according to reports

verlẹben *tr. V.* spend; **verlẹbt** *Adj.* dissipated

¹verlẹgen *tr. V.* **a)** mislay; **b)** *(verschieben)* postpone **(auf +** *Akk.* until); *(vor~)* bring forward **(auf +** *Akk.* to); **einen Termin ~:** alter an appointment; **c)** *(verlagern)* move; transfer ⟨*patient*⟩; **d)** *(legen)* lay ⟨*cable, pipe, carpet, etc.*⟩; **²verlẹgen** **1.** *Adj.* embarrassed; **2.** *adv.* in embarrassment; **Verlẹgenheit die; ~, ~en a)** *o. Pl.* *(Befangenheit)* embarrassment; **jmdn. in ~ bringen** embarrass sb.; **b)** *(Unannehmlichkeit)* embarrassing situation

Verlẹger der; ~s, ~, Verlẹgerin die; ~, ~nen publisher

Verlẹih der; ~[e]s, ~e a) *o. Pl.* hiring out; *(von Autos)* renting *or* hiring out;

b) *(Unternehmen)* hire firm; *(Film~)* distribution company; *(Video~)* video library; *(Auto~)* rental *or* hire firm; **verleihen** *unr. tr. V.* **a)** hire out; rent *or* hire out *⟨car⟩;* *(umsonst)* lend [out]; **b)** *(überreichen)* award; confer ⟨award, honour⟩
Verleihung *die; ~, ~en a) s.* **verleihen a:** hiring out; renting out; lending [out]; **b)** *s.* **verleihen b:** awarding; conferring; *(Zeremonie)* award; conferment
verleiten *tr. V.* **jmdn. dazu ~, etw. zu tun** lead *or* induce sb. to do sth.
verlernen *tr. V.* forget
verlesen **1.** *unr. tr. V.* read out; **2.** *unr. refl. V. (falsch lesen)* make a mistake/ mistakes in reading
verletzen *tr. V.* **a)** injure; *(durch Schuß, Stich)* wound; **b)** *(kränken)* hurt ⟨person, feelings⟩; **c)** *(verstoßen gegen)* violate; infringe *⟨regulation⟩;* break ⟨agreement, law⟩; **verletzlich** *Adj.* vulnerable; **Verletzte der/die;** *adj. Dekl.* casualty; *(durch Schuß, Stich)* wounded person; **Verletzung die; ~, ~en a)** *(Wunde)* injury; **b)** *(Kränkung)* hurting; **c)** *s.* **verletzen c:** violation; infringement; breaking
verleugnen *tr. V.* deny; disown ⟨friend, relation⟩
verleumden *tr. V.* slander; *(schriftlich)* libel; **Verleumdung die; ~, ~en** slander; *(in Schriftform)* libel
verlieben *refl. V.* fall in love (**in +** **Akk.** with); **Verliebte der/die;** *adj. Dekl.* lover
verlieren *unr. tr. V., itr. V.* lose; **Verlierer der; ~s, ~:** loser
verloben *refl. V.* get engaged; **verlobt sein** be engaged; **Verlobte der/die;** *adj. Dekl.* fiancé/fiancée
verlockend *Adj.* tempting; **Verlokkung die** temptation
verlogen *Adj.* lying, mendacious ⟨person⟩; false ⟨morality etc.⟩
verlor *1. u. 3. Pers. Sg. Prät. v.* **verlieren; verloren** *2. Part. v.* **verlieren; verloren|gehen** *unr. itr. V.; mit sein* get lost
verlosen *tr. V.* raffle; **Verlosung die; ~, ~en** raffle; draw
verlottern *itr. V.; mit sein* ⟨person⟩ go to seed
Verlust der; ~|e|s, ~e loss (**an +** *Dat.* of)
vermachen *tr. V.* **jmdm. etw. ~:** leave *or* bequeath sth. to sb.; *(fig.: schenken, überlassen)* give sth. to sb.

vermählen *refl. V. (geh.)* **sich |jmdm. od. mit jmdm.| ~:** marry *or* wed [sb.]; **Vermählung die; ~, ~en** *(geh.)* **a)** marriage; **b)** *(Fest)* wedding ceremony
vermehren 1. *tr. V.* increase (**um** by); **2.** *refl. V.* **a)** increase; **b)** *(sich fortpflanzen)* reproduce; **Vermehrung die; ~, ~en a)** increase (*Gen.* in); **b)** *(Fortpflanzung)* reproduction
vermeiden *unr. tr. V.* avoid
vermeintlich *Adj.* supposed
vermengen *tr. V.* mix (**miteinander** together)
Vermerk der; ~|e|s, ~e note; *(amtlich)* remark; **vermerken** *tr. V.* make a note of; note [down]; *(in Akten, Wachbuch usw.)* record
¹vermessen *unr. tr. V.* measure; survey ⟨land, site⟩; **²vermessen** *Adj. (geh.)* presumptuous
vermieten *tr. (auch itr.) V.* rent [out], let [out] (**an +** *Akk.* to); hire [out] ⟨boat, car, etc.⟩; „Zimmer zu ~" 'room to let'; **Vermieter der** landlord; **Vermieterin die** landlady
vermindern 1. *tr. V.* reduce; decrease; reduce, lessen ⟨danger, stress⟩; lower ⟨resistance⟩; reduce ⟨debt⟩; **2.** *refl. V.* decrease; ⟨resistance⟩ diminish
vermischen 1. *tr. V.* mix (**miteinander** together); blend ⟨teas, tobaccos, etc.⟩; **2.** *refl. V.* mix; *(fig.)* mingle; ⟨races, animals⟩ interbreed; **Vermischung die** *s.* **vermischen:** mixing; blending; *(fig.)* mingling
vermissen *tr. V.* **a)** miss; **b)** *(nicht haben)* **ich vermisse meinen Ausweis** my identity card is missing; **Vermißte der/die;** *adj. Dekl.* missing person
vermitteln 1. *itr. V.* mediate, act as [a] mediator (**in +** *Dat.* in); **2.** *tr. V.* **a)** *(herbeiführen)* arrange; negotiate ⟨transaction, cease-fire, compromise⟩; **b)** *(besorgen)* **jmdm. eine Stelle ~:** find sb. a job; **c)** *(weitergeben)* impart ⟨knowledge, insight, values, etc.⟩; communicate ⟨message, information, etc.⟩; convey ⟨feeling⟩; pass on ⟨experience⟩; **Vermittler der; ~s, ~ a)** *(Mittler)* mediator; **b)** *s.* **vermitteln 2 c:** imparter; communicator; conveyer; **c)** *(von Berufs wegen)* agent; **Vermittlung die; ~, ~en a)** *(Schlichtung)* mediation; **b)** *s.* **vermitteln 2 a:** arrangement; negotiation; **c)** *s.* **vermitteln 2 c:** imparting; communicating; conveying; **d)** *(Telefonzentrale)* exchange; *(in einer Firma)* switchboard

vermögen *(geh.) unr. tr. V.* etw. zu tun ~: be able to do sth.; be capable of doing sth.; **Vermögen das; ~s, ~ a)** *o. Pl. (geh.: Fähigkeit)* ability; **b)** *(Besitz)* fortune; **er hat ~:** he has money; **vermögend** *Adj.* wealthy; well-off; **Vermögen[s]·steuer die** wealth tax

vermummen *tr. V.* wrap up [warmly]; *(verbergen)* disguise

vermuten *tr. V.* suspect; **das ist zu ~:** that is what one would suppose *or* expect; we may assume that; **vermutlich 1.** *Adj.* probable; **2.** *Adv.* presumably; *(wahrscheinlich)* probably; **Vermutung die; ~, ~en** supposition

vernachlässigen *tr. V.* neglect; *(unberücksichtigt lassen)* ignore; disregard; **Vernachlässigung die; ~, ~en** neglect

vernarben *itr. V.; mit sein* [form a] scar; heal *(lit. or fig.)*

vernehmbar *Adj. (geh.)* audible; **vernehmen** *unr. tr. V.* **a)** *(geh.: hören, erfahren)* hear; **b)** *(verhören)* question; **vernehmlich 1.** *Adj.* [clearly] audible; **2.** *adv.* audibly; **Vernehmung die; ~, ~en** questioning

verneigen *refl. V. (geh.)* bow (vor + Dat. to, *(literary)* before)

verneinen *tr. (auch itr.) V.* **a)** say 'no' to *(question)*; answer *(question)* in the negative; **b)** *(Sprachw.)* negate; **Verneinung die; ~, ~en** *(Sprachw.)* negation

vernichten *tr. V.* destroy; exterminate *(pests, vermin)*; **Vernichtung die; ~, ~en** destruction; *(von Schädlingen)* extermination

Vernunft die; ~: reason; **vernünftig 1.** *Adj.* **a)** sensible; **b)** *(ugs.: ordentlich, richtig)* decent; **2.** *adv.* **a)** sensibly; **b)** *(ugs.: ordentlich, richtig)* *(talk, eat)* properly; *(dress)* sensibly

veröffentlichen *tr. V.* publish; **Veröffentlichung die; ~, ~en** publication

verordnen *tr. V.* [jmdm. etw.] ~: prescribe [sth. for sb.]; **Verordnung die** prescribing

verpachten *tr. V.* lease

verpacken *tr. V.* pack; wrap up *(present, parcel)*; **Verpackung die a)** *o. Pl.* packing; **b)** *(Umhüllung)* packaging *no pl.*; wrapping

verpassen *tr. V.* miss

verpflanzen *tr. V. (auch Med.)* transplant; graft *(skin)*

verpflegen *tr. V.* cater for; feed; **Verpflegung die; ~, ~en a)** *o. Pl.* catering *no indef. art.* *(Gen.* for); **b)** *(Nahrung)* food; **Unterkunft und ~:** board and lodging

verpflichten 1. *tr. V.* **a)** oblige; commit; *(festlegen, binden)* bind; **b)** *(einstellen, engagieren)* engage *(manager, actor, etc.)*; **2.** *refl. V.* undertake; promise; **sich vertraglich ~:** sign a contract; **Verpflichtung die; ~, ~en a)** obligation; commitment; **b)** *(Engagement)* engaging; engagement

verprügeln *tr. V.* beat up; *(zur Strafe)* thrash

Verputz der plaster; *(auf Außenwänden)* rendering; **verputzen** *tr. V.* plaster; render *(outside wall)*

verquollen *Adj.* swollen

Verrat der; ~[e]s betrayal (an + Dat. of); **verraten** *unr. tr. V.* **a)** betray (an + Akk. to); **b)** *(ugs.: mitteilen)* jmdm. den Grund usw. ~: tell sb. the reason etc.; **c)** *(erkennen lassen)* show, betray *(feelings, surprise, fear, etc.)*; show *(influence, talent)*; **Verräter der; ~s, ~:** traitor; **Verräterin die; ~, ~:** traitress; **verräterisch** *Adj.* treacherous *(plan, purpose, act, etc.)*

verrechnen 1. *tr. V.* include *(amount etc.)*; *(gutschreiben)* credit *(cheque etc.)* to another account; **2.** *refl. V.* miscalculate; **Verrechnungsscheck der** crossed cheque

verregnen *itr. V.; mit sein* be spoilt *or* ruined by rain

verreiben *unr. tr. V.* rub in

verreisen *itr. V.; mit sein* go away

verrenken *tr. V.* dislocate; **Verrenkung die; ~, ~en** dislocation

verrichten *tr. V.* perform

verriegeln *tr. V.* bolt

verringern 1. *tr. V.* reduce; **2.** *refl. V.* decrease; **Verringerung die; ~:** reduction; decrease *(Gen., von* in)

verrosten *itr. V.; mit sein* rust; **verrostet** rusty

verrückt *(ugs.)* **1.** *Adj.* **a)** mad; ~ werden go mad *or* insane; **b)** *(überspannt, ausgefallen)* crazy *(idea, fashion, prank, day, etc.)*; **2.** *adv.* crazily; *(behave)* crazily *or* like a madman; *(dress etc.)* in a mad *or* crazy way; **Verrückte der/die;** *adj. Dekl. (ugs.)* madman/madwoman; lunatic

verrühren *tr. V.* stir together; mix

verrutschen *itr. V.* slip

Vers der; ~es, ~e verse

versagen *itr. V.* fail; *(machine, en-*

gine⟩ stop [working]; **menschliches Versagen** human error; **Versager** der; ~s, ~: failure

versalzen *unr. tr. V.* put too much salt in/on; *(fig. ugs.)* spoil

versammeln *tr., refl. V.* assemble; **Versammlung die a)** meeting; **b)** *(Gremium)* assembly

versäumen *tr. V.* **a)** *(verpassen)* miss; lose ⟨*time, sleep*⟩; **b)** *(vernachlässigen, unterlassen)* neglect ⟨*duty, task*⟩

verschaffen *tr. V.* **jmdm. etw. ~:** provide sb. with sth.; get sb. sth.; **sich** *(Dat.)* **etw. ~:** get hold of sth.; obtain sth.

verschämt [fɛɐ̯ʃɛːmt] **1.** *Adj.* bashful; **2.** *adv.* bashfully

verschenken *tr. V.* give away

verscheuchen *tr. V.* chase away

verschicken *tr. V. s.* **versenden**

verschieben 1. *unr. tr. V.* **a)** shift; move; **b)** *(aufschieben)* put off, postpone (**auf** + *Akk.* till); **2.** *unr. refl. V.* be postponed (**um** for); ⟨*start*⟩ be put back *or* delayed (**um** by); **Verschiebung die** postponement

verschieden 1. *Adj.* **a)** different (**von** from); **b)** *(vielfältig)* various; **die ~sten ...:** all sorts of ...; **die ~en ...:** the various ...: **c)** ~es various things *pl.;* **2.** *adv.* differently; **verschieden·artig 1.** *Adj.* different in kind *pred.; (mehr als zwei)* diverse; **2.** *adv.* diversely; **Verschiedenheit die** ~, ~**en** difference; *(unter mehreren)* diversity; **verschiedentlich** *Adv.* on various occasions

verschimmeln *itr. V.; mit sein* go mouldy; **verschimmelt** mouldy

¹verschlafen 1. *unr. itr. (auch refl.)V.* oversleep; **2.** *unr. tr. V.* **a)** *(schlafend verbringen)* sleep through ⟨*morning, journey, etc.*⟩; **b)** *(versäumen)* not wake up in time for ⟨*appointment*⟩; not wake up in time to catch ⟨*train, bus*⟩; **c)** *(ugs.: vergessen)* forget about ⟨*appointment etc.*⟩; **²verschlafen** *Adj.* half-asleep; *(fig.)* sleepy ⟨*town*⟩

Verschlag der shed

¹verschlagen *unr. tr. V.* **die Seite ~:** lose one's place *or* page; **jmdm. die Sprache ~:** leave sb. speechless; **²verschlagen 1.** *Adj.* sly; shifty; **2.** *adv.* slyly; shiftily

verschlechtern 1. *tr. V.* make worse; **2.** *refl. V.* get worse; deteriorate; **Verschlechterung die;** ~, ~**en** worsening, deterioration (*Gen.* in)

Verschleiß der; ~**es,** ~**e a)** wear *no*

indef. art.; **b)** *(Verbrauch)* consumption (**an** + *Dat.* of); **verschleißen 1.** *unr. itr. V.; mit sein* wear out; **2.** *unr. tr. V.* wear out; *(fig.)* run dowr., ruin ⟨*one's nerves, one's health*⟩; use up ⟨*energy, ability, etc.*⟩

verschleppen *tr. V.* **a)** carry off; take away ⟨*person*⟩; **b)** *(weiterverbreiten)* carry, spread ⟨*disease, bacteria, mud, etc.*⟩; **c)** *(verzögern)* delay; *(in die Länge ziehen)* draw out; let ⟨*illness*⟩ drag on [and get worse]

verschleudern *tr. V.* **a)** sell dirt cheap; *(mit Verlust)* sell at a loss; **b)** *(verschwenden)* squander

verschließbar *Adj.* closable; lockable ⟨*suitcase, drawer, etc.*⟩; **|luftdicht|** ~: sealable ⟨*container etc.*⟩; **verschließen** *unr. tr. V.* **a)** close; stop, *(mit einem Korken)* cork ⟨*bottle*⟩; **b)** *(abschließen)* lock; lock up ⟨*house etc.*⟩; **c)** *(wegschließen)* lock away (**in** + *Dat. od. Akk.* in)

verschlimmern 1. *tr. V.* make worse; **2.** *refl. V.* get worse; ⟨*position, conditions*⟩ deteriorate, worsen

verschlingen *unr. tr. V.* **a)** [inter]twine ⟨*threads etc.*⟩ (**zu** into); **b)** *(essen, fressen)* devour ⟨*food*⟩; *(fig.)* devour ⟨*novel, money, etc.*⟩

verschlissen *2. Part. v.* **verschleißen** 2

verschlossen *Adj. (wortkarg)* taciturn; *(zurückhaltend)* reserved

verschlucken 1. *tr. V.* swallow; **2.** *refl. V.* choke

Verschluß der *(am BH, an Schmuck usw.)* fastener; fastening; *(an Taschen, Schmuck)* clasp; *(an Schuhen, Gürteln)* buckle; *(am Schrank, Fenster, Koffer usw.)* catch; *(an Flaschen)* top; *(Stöpsel)* stopper

verschmähen *tr. V. (geh.)* spurn

verschmerzen *tr. V.* get over

verschmieren *tr. V.* smear ⟨*window etc.*⟩; *(beim Schreiben)* mess up ⟨*paper*⟩; scrawl all over ⟨*page*⟩; smudge ⟨*ink*⟩

verschmitzt 1. *Adj.* mischievous; **2.** *adv.* mischievously

verschmutzen 1. *itr. V.; mit sein* get dirty; ⟨*river etc.*⟩ become polluted; **2.** *tr. V.* dirty; soil; pollute ⟨*air, water, etc.*⟩; **Verschmutzung die;** ~, ~**en** *(der Umwelt)* pollution; *(von Stoffen, Teppichen usw.)* soiling

verschnaufen *itr. V. (auch refl.) V.* have *or* take a breather

verschneit *Adj.* snow-covered *attrib.;* covered with snow *postpos.*

verschnörkelt *Adj.* ornate
verschnüren *tr. V.* tie up
verschollen *Adj.* missing
verschonen *tr. V.* spare; **jmdn. mit etw. ~**: spare sb. sth.
verschränken *tr. V.* fold ⟨*arms*⟩; cross ⟨*legs*⟩; clasp ⟨*hands*⟩
verschreiben 1. *unr. tr. V. (Med.: verordnen)* prescribe; 2. *unr. refl. V.* **a)** make a slip of the pen; **b) sich einer Sache** *(Dat.)* **~**: devote oneself to sth.; **verschreibungs·pflichtig** *Adj.* available only on prescription *postpos.*
verschrie[e]n *Adj.* notorious **(wegen** for)
verschulden 1. *tr. V.* be to blame for ⟨*accident, death, etc.*⟩; 2. *refl. V.* get into debt; **Verschulden das; ~s** guilt; **durch eigenes ~**: through one's own fault; **verschuldet** *Adj.* in debt *postpos.* **(bei** to); **hoch ~**: deeply in debt
verschütten *tr. V.* **a)** spill; **b)** *(begraben)* bury ⟨*person*⟩ [alive]
verschwägert *Adj.* related by marriage *postpos.*
verschweigen *unr. tr. V.* conceal *(Dat.* from)
verschwenden *tr. V.* waste **(an +** *Akk.* on); **Verschwender der; ~s, ~** *(von Geld)* spendthrift; *(von Dingen)* wasteful person; **verschwenderisch** 1. *Adj.* wasteful ⟨*person*⟩; ⟨*life*⟩ of extravagance; 2. *adv.* wastefully; **Verschwendung die; ~, ~en** wastefulness; extravagance
verschwiegen *Adj.* discreet; *(still, einsam)* secluded; **Verschwiegenheit die; ~**: secrecy; *(Diskretion)* discretion
verschwimmen *unr. itr. V.; mit sein* blur
verschwinden *unr. itr. V.; mit sein* disappear; vanish; **verschwinde |hier|!** off with you!; go away!; hop it! *(sl.);* **ich muß mal ~** *(ugs. verhüll.)* I have to pay a visit *(coll.)* or *(Brit. coll.)* spend a penny
verschwommen 1. *Adj.* blurred ⟨*photograph, vision*⟩; blurred, hazy ⟨*outline*⟩; vague, woolly ⟨*idea, concept, formulation, etc.*⟩; 2. *adv.* vaguely; ⟨*remember*⟩ hazily
versehen 1. *unr. tr. V.* **a)** *(ausstatten)* provide; equip ⟨*car, factory, machine, etc.*⟩; **b)** *(ausüben, besorgen)* perform ⟨*duty etc.*⟩; 2. *unr. refl. V.* make a slip; slip up; **Versehen das; ~s, ~**: over-

sight; slip; **aus ~**: by mistake; inadvertently; **versehentlich** 1. *Adv.* by mistake; inadvertently; 2. *adj.; nicht präd.* inadvertent
Versehrte der/die; *adj. Dekl.* disabled person; **die ~n** the disabled
versenden *unr. (auch regelm.) tr. V.* send ⟨*letter, parcel*⟩; send out ⟨*invitations*⟩; dispatch ⟨*goods*⟩
versetzen 1. *tr. V.* **a)** move; transfer; move ⟨*employee*⟩; *(in die nächsthöhere Klasse)* move ⟨*pupil*⟩ up, *(Amer.)* promote ⟨*pupil*⟩ **(in +** *Akk.* to); *(umpflanzen)* transplant, move ⟨*plant*⟩; *(fig.)* transport **(in +** *Akk.* to); **b)** *(nicht geradlinig anordnen)* stagger; **c)** *(verpfänden)* pawn; **d)** *(verkaufen)* sell; **e)** *(ugs.: vergeblich warten lassen)* stand ⟨*person*⟩ up *(coll.);* **f)** *(vermischen)* mix; **g)** *(erwidern)* retort; **h) etw. in Bewegung/Tätigkeit ~**: set sth. in motion/operation; **jmdn. in die Lage ~, etw. zu tun** put sb. in a position to do sth.; **jmdm. einen Stoß/Fußtritt/Schlag** *usw.* **~**: give sb. a push/kick/deal sb. a blow *etc.*; 2. *refl. V.* **sich in jmds. Lage** *(Akk.)* **~**: put oneself in sb.'s position *or* place; **Versetzung die; ~, ~en** *(eines Schülers)* moving up, *(Amer.)* promotion **(in +** *Akk.* to); *(eines Angestellten)* transfer
verseuchen *tr. V. (auch fig.)* contaminate; **radioaktiv ~**: contaminate with radioactivity
versichern *tr. V.* **a)** assert ⟨*sth.*⟩; **b)** *(vertraglich schützen)* insure **(bei** with); **Versicherte der/die;** *adj. Dekl.* insured [person]; **Versicherung die a)** *(Beteuerung)* assurance; **b)** *(Schutz durch Vertrag)* insurance; *(Vertrag)* insurance [policy] **(über +** *Akk.* for); *(Gesellschaft)* insurance [company]
Versicherungs-: ~beitrag der insurance premium; **~gesellschaft die** insurance company; **~police die** insurance policy
versickern *itr. V.; mit sein* ⟨*river etc.*⟩ drain *or* seep away
versiegeln *tr. V.* seal
versiegen *itr. V.; mit sein (geh.)* dry up; run dry
versinken *unr. itr. V.; mit sein* sink; **im Schlamm ~**: sink into the mud
versöhnen 1. *refl. V.* **sich |miteinander| ~**: become reconciled; **sich mit jmdm. ~**: make it up with sb.; 2. *tr. V.* reconcile; **Versöhnung die; ~, ~en** reconciliation

versonnen 1. *Adj.* dreamy; **2.** *adv.* dreamily

versorgen *tr. V.* **a)** supply; **b)** *(unterhalten, ernähren)* provide for ⟨*children, family*⟩; **c)** *(sorgen für)* look after; **jmdn. ärztlich ~:** give sb. medical care; *(kurzzeitig)* give sb. medical attention; **Versorger** der; **~s, ~, Versorgerin** die; **~, ~nen** breadwinner; **Versorgung** die; **~, ~en a)** *o. Pl.* supply[ing]; **b)** *(Unterhaltung, Ernährung)* support[ing]; **c)** *(Bedienung, Pflege)* care; **ärztliche ~:** medical care *or* treatment; *(kurzzeitig)* medical attention

Verspannung die *(Med.: der Muskulatur)* tension

verspäten *refl. V.* be late; **verspätet** *Adj.* late ⟨*arrival etc.*⟩; belated ⟨*greetings, thanks*⟩; **~ eintreffen** arrive late; **Verspätung** die; **~, ~en** lateness; *(verspätetes Eintreffen)* late arrival; |**fünf Minuten**| **~ haben** be [five minutes] late

versperren *tr. V.* block; obstruct ⟨*view*⟩

verspielen *tr. V.* gamble away; *(fig.)* squander, throw away ⟨*opportunity, chance*⟩; forfeit ⟨*right, credibility, etc*⟩; **verspielt 1.** *Adj. (auch fig.)* playful; fanciful, fantastic ⟨*form, design, etc.*⟩; **2.** *adv.* playfully *(lit. or fig.)*; ⟨*dress, designed*⟩ fancifully, fantastically

verspotten *tr. V.* mock; ridicule

versprechen 1. *unr. tr. V.* promise; **sich** *(Dat.)* **etw. von etw./jmdm. ~:** hope for sth. *or* to get sth. from sth./sb.; **2.** *unr. refl. V.* make a slip/slips of the tongue; **Versprechen** das; **~s, ~, Versprechung** die; **~, ~en** promise

versprühen *tr. V.* spray

verspüren *tr. V.* feel

Verstand der; **~[e]s** *(Fähigkeit zu denken)* reason *no art.*; *(Fähigkeit, Begriffe zu bilden)* mind; *(Vernunft)* [common] sense *no art.*; **hast du denn den ~ verloren?** *(ugs.)* have you taken leave of your senses?; **verständig 1.** *Adj.* sensible; **2.** *adv.* sensibly

verständigen 1. *tr. V.* notify, inform (**von, über** + *Akk.* of); **2.** *refl. V.* **a)** make oneself understood; **sich mit jmdm. ~:** communicate with sb.; **b)** *(sich einigen)* **sich** [**mit jmdm.**] **über/auf etw.** *(Akk.)* **~:** come to an understanding [with sb.] about *or.* on sth.; **Verständigkeit** die; **~:** understanding; intelligence; **Verständigung** die; **~,**

~en a) notification; **b)** *(das Sichverständlichmachen)* communication *no art.*; **c)** *(Einigung)* understanding; **Verständigungs·schwierigkeit** die difficulty of communication; **verständlich 1.** *Adj.* **a)** comprehensible; *(deutlich)* clear ⟨*pronunciation, presentation, etc.*⟩; **sich ~ machen** make oneself understood; **jmdm. etw. ~ machen** make sth. clear to sb.; **b)** *(begreiflich, verzeihlich)* understandable; **2.** *adv.* comprehensibly; *(deutlich)* ⟨*speak, express oneself, present*⟩ clearly; **verständlicher·weise** *Adv.* understandably; **Verständlichkeit** die; **~:** comprehensibility; clarity; **Verständnis** das; **~ses, ~se** understanding; **ich habe volles ~ dafür, daß ...:** I fully understand that ...; **für die Unannehmlichkeiten bitten wir um** |**Ihr**| **~:** we apologize for the inconvenience caused

verständnis-: ~los 1. *Adj.* uncomprehending; **2.** *adv.* uncomprehendingly; **~voll 1.** *Adj.* understanding; **2.** *adv.* understandingly

verstärken 1. *tr. V.* **a)** strengthen; **b)** *(zahlenmäßig)* reinforce ⟨*troops etc.*⟩ (**um** by); enlarge ⟨*orchestra, choir*⟩ (**um** by); **c)** *(intensiver machen)* intensify, increase ⟨*effort, contrast*⟩; strengthen, increase ⟨*impression, suspicion*⟩; *(größer machen)* increase ⟨*pressure, voltage, effect, etc.*⟩; *(lauter machen)* amplify ⟨*signal, sound, guitar, etc.*⟩; **2.** *refl. V.* increase; **Verstärker** der; **~s, ~:** amplifier; **Verstärkung** die; **~, ~en a)** strengthening; **b)** *(zahlenmäßig)* reinforcement *(esp. Mil.)*; **c)** *(Zunahme)* increase *(Gen.* in); *(der Lautstärke)* amplification; **d)** *(zusätzliche Person[en])* reinforcements *pl.*

verstauben *itr. V.; mit sein* get dusty; gather dust *(lit. or fig.)*

verstauchen *tr. V.* sprain; **sich** *(Dat.)* **den Fuß/die Hand ~:** sprain one's ankle/wrist; **Verstauchung** die; **~, ~en** sprain

verstauen *tr. V.* pack (**in** + *Dat. od. Akk.* in[to]); *(bes. im Boot/Auto)* stow (**in** + *Dat. od. Akk.* in)

Versteck das; **~[e]s, ~e** hiding-place: **~ spielen** play hide-and-seek; **verstecken 1.** *tr. V.* hide (**vor** + *Dat.* from); **2.** *refl. V.* **sich** |**vor jmdm./etw.**| **~:** hide [from sb./sth.]; **versteckt** *Adj.* hidden; *(heimlich)* secret ⟨*malice, activity, etc.*⟩; disguised ⟨*foul*⟩

verstehen 1. *unr. tr. V.* understand;

wie soll ich das ~? how am I to interpret that?; **jmdn./etw. falsch ~:** misunderstand sb./sth.; **2.** *unr. refl. V.* **sich mit jmdm. ~:** get on with sb.; **das versteht sich [von selbst]** that goes without saying

versteigern *tr. V.* auction; **etw. ~ lassen** put sth. up for auction; **Versteigerung die** auction

verstellbar *Adj.* adjustable; **verstellen 1.** *tr. V.* **a)** *(falsch plazieren)* misplace; **b)** *(anders einstellen)* adjust ⟨*seat etc.*⟩; alter [the adjustment of] ⟨*mirror etc.*⟩; reset ⟨*alarm clock, points, etc.*⟩; **c)** *(versperren)* block, obstruct; **d)** *(zur Täuschung verändern)* disguise ⟨*voice, handwriting*⟩; **2.** *refl. V.* pretend; **Verstellung die** pretence; *(der Stimme, Schrift)* disguising

verstimmen *tr. V.* put ⟨*person*⟩ in a bad mood; *(verärgern)* annoy; **verstimmt** *Adj.* **a)** *(Musik)* out of tune *pred.;* **b)** *(verärgert)* put out, peeved, disgruntled (**über** + *Akk.* by, about); **ein ~er Magen** an upset stomach; **Verstimmung die** bad mood

verstohlen 1. *Adj.* furtive; **2.** *adv.* furtively

verstopfen 1. *tr. V.* block; **verstopft sein** ⟨*pipe, drain, jet, nose, etc.*⟩ be blocked (**durch, von** with); **2.** *itr. V.;* **mit sein** become blocked; **Verstopfung die;** ~, ~en *(Med.)* constipation

verstorben *Adj.* late; **Verstorbene der/die;** *adj. Dekl. (geh.)* deceased

verstören *tr. V.* distress; **verstört** *Adj.* distraught

Verstoß der violation (**gegen** of); **verstoßen 1.** *unr. tr. V.* disown; **2.** *unr. itr. V.* **gegen etw. ~:** infringe sth.

verstreichen 1. *unr. tr. V.* apply, put on ⟨*paint*⟩; spread ⟨*butter etc.*⟩; **2.** *unr. itr. V.;* **mit sein** *(geh.)* ⟨*time*⟩ pass [by]

verstreuen *tr. V.* scatter; put down ⟨*bird food, salt*⟩; *(versehentlich)* spill

verstricken 1. *tr. V.* **jmdn. in etw.** *(Akk.)* **~:** involve sb. in sth.; draw sb. into sth.; **2.** *refl. V.* **sich in etw.** *(Akk.)* **~:** become entangled *or* caught up in sth.

verstümmeln *tr. V.* mutilate; *(fig.)* garble ⟨*report*⟩; chop, mutilate ⟨*text*⟩

verstummen *itr. V.;* **mit sein** *(geh.)* fall silent; ⟨*music, noise, conversation*⟩ cease

Versuch der; ~[e]s, ~e attempt; *(Experiment)* experiment (**an** + *Dat.* on); *(Probe)* test; **versuchen** *tr. V.* **a)** try; attempt; **b)** *(probieren)* try ⟨*cake etc.*⟩

versündigen *refl. V.* **sich an jmdm./etw. ~:** sin against sb./sth.

versüßen *tr. V.* **jmdm./sich etw. ~** *(fig.)* make sth. more pleasant for sb./oneself

vertauschen *tr. V.* exchange; switch; reverse ⟨*roles, poles*⟩; **etw. mit** *od.* **gegen etw. ~:** exchange sth. for sth.

verteidigen *tr. V.* defend; **Verteidiger der;** ~s, ~, **Verteidigerin, die;** ~, ~nen *(auch Sport)* defender; *(Rechtsw.)* defence counsel; **Verteidigung die,** ~, ~en defence; **Verteidigungs·minister der** minister of defence

verteilen *tr. V.* distribute, hand out ⟨*leaflets, prizes, etc.*⟩ (**an** + *Akk.* to, **unter** + *Akk.* among); share [out], distribute ⟨*money, food*⟩ (**an** + *Akk.* to, **unter** + *Akk.* among); allocate ⟨*work*⟩; distribute ⟨*weight etc.*⟩ (**auf** + *Akk.* over); spread ⟨*cost*⟩ (**auf** + *Akk.* among); distribute, spread ⟨*butter, seed, dirt, etc.*⟩; **Verteilung die** distribution; *(der Rollen, der Arbeit)* allocation

verteuern 1. *tr. V.* make ⟨*goods*⟩ more expensive; **2.** *refl. V.* become more expensive

verteufeln *tr. V.* condemn; denigrate

vertiefen 1. *tr. V. (auch fig.)* deepen (**um** by); **2.** *refl. V.* **sich ~ in** (+ *Akk.*) bury oneself in ⟨*book, work, etc.*⟩; **in etw.** *(Akk.)* **vertieft sein** be engrossed in sth.; **Vertiefung die;** ~, ~en *(Mulde)* depression; hollow

vertikal 1. *Adj.* vertical; **2.** *adv.* vertically; **Vertikale die;** ~; ~n *s.* Senkrechte

vertilgen *tr. V.* **a)** *(vernichten)* exterminate ⟨*vermin*⟩; kill off ⟨*weeds*⟩; **b)** *(ugs.: verzehren)* devour, *(joc.)* demolish ⟨*food*⟩

vertonen *tr. V.* set ⟨*text, poem*⟩ to music; **Vertonung die;** ~, ~en setting

Vertrag der; ~[e]s, Verträge contract; *(zwischen Staaten)* treaty; **vertragen 1.** *unr. tr. V.* endure; tolerate *(esp. Med.);* *(aushalten, leiden können)* stand; bear; **ich vertrage keinen Kaffee** coffee disagrees with me; **2.** *unr. refl. V.* **sich mit jmdm. ~:** get on *or* along with sb.; *(passen)* **sich mit etw. ~:** go with sth.; **verträglich 1.** *Adj.* contractual; **2.** *adv.* contractually; by contract; **verträglich** *Adj.* **a)** digestible ⟨*food*⟩; **b)** *(umgänglich)* good-natured; easy to get on with *pred.*

vertrauen *itr. V.* **jmdm./einer Sache ~:** trust sb./sth.; **auf etw.** *(Akk.)* ~: [put one's] trust in sth.; **Vertrauen das;** ~s trust; confidence; **jmdn. ins ~ ziehen** take sb. into one's confidence; **vertrauen·erweckend** *Adj.* inspiring

vertrauens-, Vertrauens-: ~**bruch der** breach of trust; ~**person die** person in a position of trust; ~**sache die** matter *or* question of trust; ~**selig** *Adj.* all too trusting; ~**voll** 1. *Adj.* trusting *(relationship)*; *(collaboration, co-operation)* based on trust; *(zuversichtlich)* confident; 2. *adv.* trustingly; *(zuversichtlich)* confidently; ~**würdig** *Adj.* trustworthy

vertraulich 1. *Adj.* **a)** confidential; **b)** *(freundschaftlich, intim)* familiar *(manner, tone, etc.)*; intimate *(conversation)*; 2. *adv.* **a)** confidentially; **b)** *(freundschaftlich, intim)* in a familiar way; **Vertraulichkeit die;** ~, ~**en a)** *o. Pl.* confidentiality; **b)** *(vertrauliche Information)* confidence; **c)** *o. Pl. (distanzloses Verhalten)* familiarity; *(Intimität)* intimacy; **vertraut** *Adj.* **a)** close *(friend etc.)*; intimate *(circle, conversation, etc.)*; **b)** *(bekannt)* familiar; **jmdn./sich mit etw. ~ machen** familiarize sb./oneself with sth.; **Vertraute der/die;** *adj. Dekl.* close friend

vertreiben *unr. tr. V.* **a)** drive out (**aus** of); drive away *(animal, smoke, clouds)* (**aus** from); fight off *(tiredness, troubles)*; **b)** *(verkaufen)* sell

vertreten 1. *unr. tr. V.* **a)** stand in or deputize for *(colleague etc.)*; *(teacher)* cover for *(colleague)*; **b)** *(eintreten für, repräsentieren)* represent *(person, firm, interests, constituency, country, etc.)*; *(Rechtsw.)* act for *(person, prosecution, etc.)*; ~ **sein** be represented; **c)** *(einstehen für, verfechten)* support *(point of view, principle)*; hold *(opinion)*; advocate *(thesis etc.)*; 2. *unr. refl. V.* **sich** *(Dat.)* **die Füße** *od.* **Beine** ~ *(ugs.)* stretch one's legs; **Vertreter der;** ~s, ~ **a)** *(Stell~)* deputy; stand-in; **b)** *(Interessen~, Repräsentant)* representative; *(Handels~)* sales representative; commercial traveller; **c)** *(Verfechter, Anhänger)* supporter; advocate; **Vertretung die;** ~, ~**en** deputy; *(Delegierte[r])* representative; *(Delegation)* delegation *(Handels~)* [sales] agency; **eine diplomatische ~:** a diplomatic mission

Vertriebene der/die; *adj. Dekl.* expellee [from his/her homeland]

vertrocknen *itr. V.; mit sein* dry up

vertrödeln *tr. V. (ugs. abwertend)* dawdle away, waste *(time)*

vertrösten *tr. V.* put *(person)* off (**auf** + *Akk.* until)

vertun 1. *unr. tr. V.* waste; 2. *unr. refl. V. (ugs.)* make a slip

vertuschen *tr. V.* hush up *(scandal etc.)*; keep *(truth etc.)* secret

verübeln *tr. V.* **jmdm. eine Äußerung** *usw.* ~: take sb.'s remark *etc.* amiss

verüben *tr. V.* commit *(crime etc.)*

verunglücken *itr. V.; mit sein* have an accident; *(car etc.)* be involved in an accident; **mit dem Auto/Flugzeug** ~: be in a car/an air accident *or* crash; **Verunglückte der/die;** *adj. Dekl.* accident victim; casualty

verunreinigen *tr. V.* pollute; contaminate *(water, milk, flour, oil)*

verunsichern *tr. V.* **jmdn.** ~: make sb. feel unsure *or* uncertain

verunstalten *tr. V.* disfigure

verursachen *tr. V.* cause

verurteilen *tr. V.* pass sentence on; sentence; *(fig.)* condemn *(behaviour, action)*; **jmdn. zum Tode** ~: sentence *or* condemn sb. to death; **Verurteilte der/die;** *adj. Dekl.* convicted man/woman; **Verurteilung die;** ~, ~**en** sentencing; *(fig.)* condemnation

vervollkommnen *tr. V.* perfect

vervollständigen *tr. V.* complete

verwachsen *Adj.* deformed

verwählen *refl. V.* misdial

verwahren 1. *tr. V.* keep [safe]; 2. *refl. V.* protest; **verwahrlosen** *itr. V.; mit sein* get in a bad state; *(house, building)* fall into disrepair; *(garden, hedge)* become overgrown; *(person)* let oneself go; **verwahrlost** neglected; overgrown *(hedge, garden)*; dilapidated *(house, building)*; unkempt *(person, appearance, etc.)*; *(in der Kleidung)* ragged *(person)*; **Verwahrlosung die;** ~: *(eines Gebäudes)* dilapidation; *(einer Person)* advancing decrepitude

verwaisen *itr. V.* be orphaned

verwalten *tr. V.* **a)** administer *(estate, property)*; run *(house)*; hold *(money)* in trust; **b)** *(leiten)* run, manage *(hostel, kindergarten, etc.)*; *(regieren)* administer *(area, colony, etc.)*; govern *(country)*; **Verwalter der;** ~s, ~, **Verwalterin die;** ~, ~**nen** administrator; *(eines Amts usw.)* manager; *(ei-*

nes Nachlasses) trustee; **Verwaltung die**; ~, ~en **a)** administration; *(eines Landes)* government; *(eines Amtes)* tenure; *(einer Aufgabe)* performance; **b)** *(Organ)* administration

verwandeln 1. *tr. V.* convert (**in** + *Akk., * **zu** into); *(völlig verändern)* transform (**in** + *Akk., * **zu** into); **2.** *refl. V.* sich in etw. *(Akk.)* od. **zu etw. ~:** turn *or* change into sth.; *(bei chemischen Vorgängen usw.)* be converted into sth.; **Verwandlung die;** ~, ~en conversion (**in** + *Akk., * **zu** into); *(völlige Veränderung, das Sichverwandeln)* transformation (**in** + *Akk., * **zu** into)

¹**verwandt 2.** *Part. v.* verwenden

²**verwandt** *Adj.* related (**mit** to); *(fig.)* similar ⟨*views, ideas, forms*⟩; **Verwandte der/die;** *adj. Dekl.* relative; relation; **Verwandtschaft die;** ~, ~en **a)** relationship (**mit** to); *(fig.)* affinity; **b)** *o. Pl. (Verwandte)* relatives *pl.;* relations *pl.;* **die ganze ~:** all one's relatives; **verwandtschaftlich** *Adj.* family ⟨*ties, relationships, etc.*⟩

verwarnen *tr. V.* warn, caution (**wegen** for); **Verwarnung die;** ~, ~en warning; caution

verwechseln *tr. V.* **a)** [miteinander] ~: confuse ⟨*two things/people*⟩; **etw. mit etw./jmdn. mit jmdm. ~:** mistake sth. for sth./sb. for sb.; confuse sth. with sth./sb. with sb.; **b)** *(vertauschen)* mix up; **Verwechslung die;** ~, ~en **a)** [case of] confusion; **b)** *(Vertauschung)* mixing up; **eine ~:** a mix-up

verwegen 1. *Adj.* daring; *(auch fig.)* audacious; **2.** *adv. (auch fig.)* audaciously; **Verwegenheit die;** ~: daring; *(auch fig.)* audacity

verwehren *tr. V.* **jmdm. etw. ~:** refuse *or* deny sb. sth.

Verwehung die; ~, ~en [snow]drift

verweigern *tr. V.* refuse; **Verweigerung die;** ~, ~en refusal

Verweis der; ~es, ~e **a)** reference (**auf** + *Akk.* to); *(Quer-)* cross-reference; **b)** *(Tadel)* reprimand; **verweisen** *unr. tr. V.* **a)** jmdn./einen Fall *usw.* an jmdn./etw. ~ *(auch Rechtsspr.)* refer sb./a case *etc.* to sb./sth.; **b)** *(wegschicken)* jmdn. von der Schule/aus dem Saal ~: expel sb. from the school/send sb. out of the room; **einen Spieler vom Platz ~:** send a player off [the field]; **c)** *auch itr. (hinweisen)* [jmdn.] auf etw. *(Akk.)* ~: refer [sb.] to sth.

verwelken *itr. V.; mit sein* wilt

verwendbar *Adj.* usable; **Verwend-**

barkeit die; ~: usability; **verwenden** *unr. od. regelm. tr. V.* **a)** use (**zu, für** for); **b)** *(aufwenden)* spend ⟨*time*⟩ (**auf** + *Akk.* on); **Verwendung die;** ~, ~en use

verwerfen *unr. tr. V.* reject; dismiss ⟨*thought*⟩; **verwerflich** *(geh.)* **1.** *Adj.* reprehensible; **2.** *adv.* reprehensibly

verwertbar *Adj.* utilizable; usable; **verwerten** *tr. V.* utilize, use (**zu** for); make use of ⟨*suggestion, experience, knowledge, etc.*⟩

verwesen *itr. V.; mit sein* decompose; **Verwesung die;** ~: decomposition

verwickeln 1. *refl. V.* get tangled up *or* entangled; **sich in etw.** *(Akk. od. Dat.)* ~: get caught [up] in sth.; **2.** *tr. V.* involve; **Verwicklung die;** ~, ~en complication

verwildern *itr. V.* ⟨*garden*⟩ become overgrown; ⟨*domestic animal*⟩ return to the wild

verwirklichen 1. *tr. V.* realize ⟨*dream*⟩; realize, put into practice ⟨*plan, proposal, idea, etc.*⟩; carry out ⟨*project, intention*⟩; **2.** *refl. V.* ⟨*hope, dream*⟩ be realized; **Verwirklichung die;** ~, ~en realization; *(eines Wunsches, einer Hoffnung)* fulfilment

verwirren *tr. (auch itr.) V.* confuse; **verwirrt** confused; **~d** bewildering; **Verwirrung die;** ~, ~en confusion

verwischen *tr. V.* smudge ⟨*signature, writing, etc.*⟩; smear ⟨*paint*⟩; *(fig.)* cover up ⟨*tracks*⟩

verwitwet *Adj.* widowed

verwöhnen *tr. V.* spoil; **verwöhnt** *Adj.* spoilt; *(anspruchsvoll)* discriminating; ⟨*taste, palate*⟩ of a gourmet

verworren *Adj.* confused, muddled ⟨*ideas, situation, etc.*⟩

verwunden *tr. V.* wound; injure; **Verwundete der/die;** *adj. Dekl.* casualty; **die ~n** the wounded; **Verwundung die;** ~, ~en wound

verwünschen *tr. V.* curse

verwüsten *tr. V.* devastate; **Verwüstung die;** ~, ~en devastation

verzählen *refl. V.* miscount

verzaubern *tr. V.* cast a spell on; bewitch; *(fig.)* enchant; **jmdn. in etw.** *(Akk.)* ~: transform sb. into sth.

Verzehr der; ~[e]s consumption; **verzehren** *tr. V.* consume

Verzeichnis das; ~ses, ~se list; *(Register)* index

verzeihen *unr. tr., itr. V.* forgive; *(entschuldigen)* excuse ⟨*behaviour, re-*

mark, *etc.*)*; ~ Sie |bitte|, können Sie mir sagen ...?* excuse me, could you tell me ...?; **Verzeihung** die; ~: forgiveness; ~! sorry!; jmdn. um ~ bitten apologize to sb.

verzęrren 1. *tr. V.* **a)** contort ⟨*face etc.*⟩ (zu into); **b)** *(akustisch, optisch)* distort ⟨*sound, image*⟩; **etw. verzerrt darstellen** *(fig.)* present a distorted account *or* picture of sth.

Verzicht der; ~|e|s, ~e **a)** renunciation (auf + *Akk.* of); **b)** *(auf Reichtum, ein Amt usw.)* relinquishment (auf + *Akk.* of); **verzichten** *itr. V.* do without; ~ auf (+ *Akk.*) do without; *(sich enthalten)* refrain from; *(aufgeben)* give up ⟨*share, smoking, job, etc.*⟩*;* renounce ⟨*inheritance*⟩*;* relinquish ⟨*right, privilege*⟩*; (opfern)* sacrifice ⟨*holiday, salary*⟩

¹**verziehen 2.** *Part. v.* verzeihen

²**verziehen 1.** *unr. tr. V.* **a)** screw up ⟨*face, mouth, etc.*⟩*;* **b)** *(schlecht erziehen)* spoil; **2.** *unr. refl. V.* **a)** *(aus der Form geraten)* go out of shape; ⟨*wood*⟩ warp; **b)** *(wegziehen)* ⟨*clouds, storm*⟩ move away, pass over; ⟨*fog, mist*⟩ disperse; **c)** *(ugs.: weggehen)* take oneself off; **3.** *unr. itr. V.; mit sein* move [away]; „**Empfänger |unbekannt| verzogen**" 'no longer at this address'

verzieren *tr. V.* decorate; **Verzierung** die; ~, ~en decoration

verzögern 1. *tr. V.* **a)** delay (um by); **b)** *(verlangsamen)* slow down; **2.** *refl. V.* be delayed (um by); **Verzögerung** die; ~, ~en delay *(Gen.* in); *(Verlangsamung)* slowing down

Verzug der; ~|e|s delay; **im ~ sein/in ~ kommen** be/fall behind

verzweifeln *itr. V.; mit sein* despair; **über etw./jmdn. ~:** despair at sth./of sb.; **verzweifelt 1.** *Adj.* despairing ⟨*person*⟩*;* desperate ⟨*situation, attempt, effort, struggle, etc*⟩*; ~ sein* be in despair; **2.** *adv.* desperately; **Verzweiflung** die; ~ despair

verzweigen *refl. V.* branch [out]

Veteran [vete'ra:n] der; ~en, ~en *(auch fig.)* veteran

Vetter der; ~s, ~n cousin

vgl. *Abk.* vergleiche cf.

v. H. *Abk.* vom Hundert per cent

via ['vi:a] *Präp.* via

Viadukt [via'dʊkt] das *od.* der; ~|e|s, ~e viaduct

vibrieren [vi'bri:rən] *itr. V.* vibrate

video-, Video- ['vi:deo-]: video; **Video** das; ~s, ~s *(ugs.)* video

Vieh das; ~|e|s **a)** *(Nutztiere)* livestock *sing. or pl.;* **b)** *(Rind~)* cattle *pl.;* **Vieh·zucht** die; *o. Pl.* [live]stock/cattle breeding *no art.*

viel 1. *Indefinitpron. u. unbest. Zahlw.* **a)** *Sg.* a great deal of; a lot of *(coll.);* **wie/nicht/zu ~:** how/not/too much; ~|es| *(vielerlei)* much; **der ~e Regen** all the rain; **um ~es jünger** a great deal younger; **b)** *Pl.* many; **gleich ~|e|** the same number of; **die ~en Menschen** all the people; **2.** *Adv.* **a)** *(oft, lange)* a great deal; a lot *(coll.);* **b)** *(wesentlich)* much; a great deal *(coll.);* **~ zu klein** much too small; **vielerlei** *indekl. unbest. Gattungsz.* **a)** *attr.* many different; all kinds *or* sorts of; **b)** *subst.* all kinds of things

viel-, Viel-: **~fach 1.** *Adj.* **a)** multiple; **die ~fache Menge** many times the amount; **b)** *(vielfältig)* many kinds of; **2.** *adv.* many times; **~falt** die; ~: diversity; **~fältig 1.** *Adj.* many and diverse; **2.** *adv.* in many different ways

vielleicht *Adv.* perhaps; maybe

viel-: **~mals** *Adv.* **ich bitte ~mals um Entschuldigung** I'm very sorry; **danke ~mals** thank you very much; **~mehr** [*od.* -'-] *Konj. u. Adv.* rather; **~sagend 1.** *Adj.* meaningful; **2.** *adv.* meaningfully; **~seitig** *Adj.* versatile ⟨*person*⟩*;* **~versprechend 1.** *Adj.* [very] promising; **2.** *adv.* [very] promisingly

vier *Kardinalz.* four; **Vier** die; ~, ~en four*;* **eine ~ schreiben/bekommen** *(Schulw.)* get a D

vier-, Vier-: *(s. auch* acht-, Acht-*);* **~beiner** der; ~s, ~ *(ugs.)* four-legged friend; **~beinig** *Adj.* four-legged; **~eck** das quadrilateral; *(Rechteck)* rectangle; *(Quadrat)* square; **~eckig** *Adj.* quadrilateral; *(rechteckig)* rectangular; **~fach** *Vervielfältigungsz.* fourfold; quadruple; **~fache das;** *adj. Dekl.* **um das ~fache:** fourfold; by four times the amount; **~hundert** *Kardinalz.* four hundred

Vierling der; ~s, ~e quadruplet

vier-, Vier-: **~mal** *Adv.* four times; **~spurig** *Adj.* four-lane ⟨*road, motorway*⟩*;* **~spurig sein** have four lanes; **~stellig** *Adj.* four-figure *attrib.;* **~sterne·hotel** [-'----] das four-star hotel

viert... *Ordinalz.* fourth; **viertausend** *Kardinalz.* four thousand; **viertel** ['fɪrtl̩] *Bruchz.* quarter; **ein ~**

Pfund a quarter of a pound; **Viertel** ['fɪrt]] **das** (*schweiz. meist* der); ~s, ~ **a)** quarter; ~ **vor/nach eins** [a] quarter to/ past one; **drei** ~: three-quarters; **b)** (*Stadtteil*) quarter; district

viertel-, Viertel-: ~**finale das** (*Sport*) quarter-final; ~**jahr das** three months *pl.*; ~**jährlich 1.** *Adj.* quarterly; **2.** *adv.* quarterly; ~**liter der** quarter of a litre; ~**note die** (*Musik*) crotchet (*Brit.*); quarter note (*Amer.*); ~**pfund das** quarter [of a] pound; ~**stunde die** quarter of an hour; ~**stündig** *Adj.* quarter-of-an-hour; ~**stündlich 1.** *Adj.* every quarter of an hour *postpos.*; **2.** *adv.* every quarter of an hour

viertens *Adv.* fourthly; **viertürig** *Adj.* four-door *attrib.*; ~ **sein** have four doors

Vierwaldstätter See, (*schweiz.:*) **Vierwaldstättersee der** Lake Lucerne

vier- ['fɪr-]: ~**zehn** *Kardinalz.* fourteen; ~**zehn·tägig** *Adj.* two-week; ~**zehn·täglich 1.** *Adj.* fortnightly; **2.** *adv.* fortnightly

vierzig ['fɪrtsɪç] *Kardinalz.* forty; *s. auch* **achtzig; vierzigst ...** *Ordinalz.* fortieth; *s. auch* **acht ...**

Vikar der; ~s, ~e **a)** (*kath. Kirche*) locum tenens; **b)** (*ev. Kirche*) ≈ [trainee] curate

Villa ['vɪla] **die;** ~, **Villen** villa; **Villen·viertel das** exclusive residential district

violett [vjo'lɛt] purple; violet; **Violett das;** ~s, ~e *od. ugs.* ~s purple; violet; (*im Spektrum*) violet

Violine [vjo'liːnə] **die;** ~, ~n (*Musik*) violin

Viper ['viːpɐ] **die;** ~, ~n viper; adder

Viren *s.* Virus

Virtuose [vɪr'tuoːzə] **der;** ~n, ~n virtuoso; **Virtuosität die;** ~: virtuosity

Virus ['viːrʊs] **das;** ~, **Viren** virus

Visa *s.* Visum; **Visen** *s.* Visum

Visier [vi'ziːɐ] **das;** ~s, ~e (*am Helm*) visor; (*an der Waffe*) backsight

Vision [vi'zjoːn] **die;** ~, ~en vision

Visite [vi'ziːtə] **die;** ~, ~n round; ~ **machen** do one's round; **Visiten·karte die** visiting-card

Visum ['viːzʊm] **das;** ~s, **Visa** *od.* **Visen** visa

Vitamin [vita'miːn] **das;** ~s, ~e vitamin

vitamin-, Vitamin-: ~**arm** *Adj.* low in vitamins *postpos.*; ~**mangel der;**

o. Pl. vitamin deficiency; ~**reich** *Adj.* rich in vitamins *postpos.*

Vitrine [vi'triːnə] **die;** ~, ~n display case; (*Möbel*) display cabinet

Vize- vice-

Vogel der; ~s, **Vögel** bird; **einen** ~ **haben** (*salopp*) be off one's rocker (*sl.*)

Vogel-: ~**käfig der** birdcage; ~**nest das** bird's nest; ~**perspektive die** bird's eye view; ~**scheuche die;** ~, ~n scarecrow

Vokabel [vo'kaːb]] **die;** ~, ~n word; ~n vocabulary *sing.*

Vokal [vo'kaːl] **der;** ~s, ~e (*Sprachw.*) vowel

Volk das; ~[e]s, **Völker** people

volks-, Volks-: ~**abstimmung die** plebiscite; ~**eigen** *Adj.* (*ehem. DDR*) publicly *or* nationally owned; ~**entscheid der** (*Politik*) referendum; ~**fest das** public festival; (*Jahrmarkt*) fair; ~**hochschule die** adult education centre; ~**kunde die** folklore; ~**lied das** folk-song; ~**musik die** folk-music; ~**polizei die;** *o. Pl.* (*ehem. DDR*) People's Police; ~**republik die** People's Republic; ~**stamm der** tribe; ~**tanz der** folkdance; ~**tracht die** traditional costume; (*eines Landes*) national costume

volkstümlich 1. *Adj.* popular; **2.** *adv.* ~ **schreiben** write in terms readily comprehensible to the layman; **Volks·wirtschaft die** national economy; (*Fach*) economics *sing., no art.*; **volks·wirtschaftlich 1.** *Adj.* economically; **2.** *adv.* economically

voll 1. *Adj.* full; ample ⟨*bosom*⟩; (*salopp: betrunken*) plastered (*sl.*); ~ **von** *od.* **mit etw. sein** be full of sth.; **jmdn. nicht für** ~ **nehmen** not take sb. seriously; **2.** *adv.* fully; ~ **und ganz** completely; **voll·auf** [*od.* '--] *Adv.* completely; **vollaufen** *unr. itr. V., trennbar* fill up; **etw.** ~ **lassen** fill sth. [up]

voll-, Voll-: ~**automatisch 1.** *Adj.* fully automatic; **2.** *adv.* fully automatically; ~**bad das** bath; ~**bart der** full beard; ~**bringen** [-'--] *unr. tr. V.* (*geh.*) accomplish; achieve

voll·enden *tr. V.* complete; **voll·endet 1.** *Adj.* accomplished ⟨*performance*⟩; perfect ⟨*gentleman, host, manners, reproduction*⟩; **2.** *adv.* ⟨*play*⟩ in an accomplished manner; **vollends** *Adv.* completely; **Voll·endung die** completion; **voller** *indekl. Adj.* full of; ~ **Flecken** covered with stains

Volley·ball ['vɔlibal] **der** volleyball

voll-, Voll-: ~**führen** [-'--] *tr. V.* perform; ~|**füllen** *tr. V.* fill up; ~**gas das;** *o. Pl.* ~**gas geben** put one's foot down; **mit** ~**gas** at full throttle; ~|**gießen** *unr. tr. V.* fill [up]

völlig 1. *Adj.* complete; total; 2. *adv.* completely; totally; **du hast** ~ **recht** you are absolutely right

voll-, Voll-: ~**jährig** *Adj.* of age *pred.;* ~**jährig werden** come of age; ~**jährigkeit die;** ~: majority *no art.;* ~**kasko·versicherung die** fully comprehensive insurance

voll·kommen 1. *Adj.* a) [-'-- *od.* '---] *(vollendet)* perfect; b) ['---] *(vollständig)* complete; total; 2. ['---] *adv.* completely; totally

voll-, Voll-: ~**korn·brot das** wholemeal *(Brit.) or (Amer.)* wholewheat bread; ~|**laufen** *s.* vollaufen; ~|**machen** *tr. V.* fill up; [sich *(Dat.)*] **die Hosen/Windeln** ~**machen** *(ugs.)* mess one's pants/nappy; ~**macht die;** ~, ~**en a)** authority; b) *(Urkunde)* power of attorney; ~**milch die** full-cream milk; ~**milch·schokolade die** full-cream milk chocolate; ~**mond der;** *o. Pl.* full moon; ~**pension die;** *meist o. Art.;* *o. Pl.* full board *no art.;* ~**ständig** 1. *Adj.* complete; full ⟨*text, address, etc.*⟩; 2. *adv.* completely; ⟨*list*⟩ in full; ~**ständigkeit die;** ~: completeness; ~**strecken** [-'--] *tr. V.* enforce ⟨*penalty, fine, law*⟩; carry out ⟨*sentence*⟩ **(an** + *Dat.* on); ~|**tanken** *tr. (auch itr.) V.* fill up; **bitte** ~**tanken** fill it up, please; ~**treffer der** direct hit; **ein** ~**treffer sein** *(fig.)* hit the bull's eye; ~**zählig** *Adj.* complete

voll·ziehen *unr. tr. V.* carry out **(an** + *Dat.* on); execute, carry out ⟨*order*⟩; perform ⟨*sacrifice, ceremony, sexual intercourse*⟩; **Voll·zug der** *s.* **vollziehen:** carrying out; execution; performance

Volt [vɔlt] **das;** ~ *od.* ~|**e|s,** ~: *(Physik, Elektrot.)* volt

Volumen [vo'lu:mən] **das;** ~**s,** ~: volume

vom *Präp.* + *Art.* **a) = von dem; b)** *(räumlich)* from the; **links/rechts** ~ **Eingang** to the left/right of the entrance; ~ **Stuhl aufspringen** jump up out of one's chair; **c)** *(zeitlich)* ~ **Morgen bis zum Abend** from morning till night; ~ **ersten Januar an** [as] from the first of January; **d)** *(zur Angabe der Ursache)* **das kommt** ~ **Rauchen/Alko-**

hol that comes from smoking/drinking alcohol; **jmdn.** ~ **Sehen kennen** know sb. by sight; **von** *Präp. mit Dat.* **a)** *(räumlich)* from; **nördlich/südlich** ~ **Mannheim** to the north/south of Mannheim; **rechts/links** ~ **mir** on my right/left; ~ **hier an** *od. (ugs.)* **ab** from here on[ward]; ~ **Mannheim aus** from Mannheim; **b)** *(zeitlich)* from; ~**jetzt an** *od. (ugs.)* **ab** from now on; ~ **heute/ morgen an** [as] from today/tomorrow; starting today/tomorrow; **in der Nacht** ~ **Freitag auf** *od.* **zu Samstag** during Friday night; **das Brot ist** ~ **gestern** it's yesterday's bread; **c)** *(anstelle eines Genitivs)* of; **acht** ~ **hundert/zehn** eight out of a hundred/ten; **d)** *(zur Angabe des Urhebers, der Ursache, beim Passiv)* by; **der Roman ist** ~ **Fontane** the novel is by Fontane; **müde** ~ **der Arbeit sein** be tired from work[ing]; **sie hat ein Kind** ~ **ihm** she has a child by him; **e)** *(zur Angabe von Eigenschaften)* of; **eine Fahrt** ~ **drei Stunden** a three-hour drive; **von·einander** *Adv.* from each other *or* one another; **vonstatten** *Adv.* ~ **gehen** proceed

vor 1. *Präp. mit Dat.* **a)** *(räumlich)* in front of; *(weiter vorn)* ahead of; in front of; *(nicht ganz so weit wie)* before; *(außerhalb)* outside; **kurz** ~ **der Abzweigung** just before the turn-off; ~ **der Stadt** outside the town; **etw.** ~ **sich haben** *(fig.)* have sth. before one; **das liegt noch** ~ **mir** *(fig.)* I still have that to come *or* have that ahead of me; **b)** *(zeitlich)* before; **es ist fünf [Minuten]** ~ **sieben** it is five [minutes] to seven; **c)** *(bei Reihenfolge, Rangordnung)* before; **knapp** ~ **jmdm. siegen** win just ahead *or* in front of sb.; **d)** *(auf Grund von)* with; ~ **Freude strahlen** beam with joy; ~ **Hunger/Durst umkommen** *(ugs.)* die of hunger/thirst; **e)** ~ **fünf Minuten/10 Jahren/Wochen** *usw.* five minutes/ten years/weeks ago; **heute** ~ **einer Woche** a week ago today; 2. *Präp. mit Akk.* in front of; ~ **sich hin** to oneself

Vor·abend der evening before; *(fig.)* eve

vor·an *Adv.* forward[s] ahead; first

voran-: ~|**gehen** *unr. itr. V.; mit sein* **a)** go first; **b)** *(Fortschritte machen)* make progress; ~|**kommen** *unr. itr. V.; mit sein* **a)** make headway; **b)** *(Fortschritte machen)* make progress

Vor·arbeiter der foreman

vor·aus 1. [-'-] *Präp. mit Dat., nachge-**

stellt in front; **jmdm./seiner Zeit ~ sein** *(fig.)* be ahead of sb./one's time; **2.** *Adv.* **im ~** ['--] in advance

voraus-, Voraus-: ~|**gehen** *unr. itr. V.; mit sein* **a)** go [on] ahead; **b)** *(zeitlich)* **einem Ereignis ~gehen** precede an event; ~**sage die** *s.* Vorhersage; ~|**sagen** *tr. V.* predict; ~|**sehen** *unr. tr. V.* foresee; ~|**setzen** *tr. V.* **a)** *(als gegeben ansehen)* assume; ~**gesetzt, |daß| ...:** provided [that] ...; **b)** *(erfordern)* require ⟨*skill, experience, etc.*⟩; presuppose ⟨*good organization, planning, etc.*⟩; ~**setzung die;** ~, ~**en a)** *(Annahme)* assumption; *(Prämisse)* premiss; **b)** *(Vorbedingung)* prerequisite; **unter** *or* **on the pre-condition that ...;** ~**sichtlich 1.** *Adj.* anticipated; **2.** *adv.* probably

Vor·bau der; *Pl.* ~**ten** porch

Vorbehalt der; ~|**e|s, ~e** reservation; **unter dem ~, daß ...:** with the reservation that ...; **vor|behalten** *unr. tr. V.* **sich** *(Dat.)* **etw. ~:** reserve oneself sth.; „**Änderungen ~**" 'subject to alterations'

vor·bei *Adv.* **a)** *(räumlich)* past; by; **an etw.** *(Dat.)* ~ past sth.; **b)** *(zeitlich)* past; over; *(beendet)* finished; over; **es ist acht Uhr ~** *(ugs.)* it is past *or* gone eight o'clock

vorbei-: ~|**fahren 1.** *unr. itr. V.; mit sein* **a)** drive/ride past; pass; **an jmdm. ~fahren** drive/ride past *or* pass sb.; **b)** *(ugs.: einen kurzen Besuch machen)* **|bei jmdm./der Post|** ~**fahren** drop in *(coll.)* [at sb.'s/at the post office]; ~|**gehen** *unr. itr. V.; mit sein* **a)** pass; go past; **an jmdm./etw. ~gehen** pass *or* go past sb./sth.; **der Schuß ist ~gegangen** the shot missed; **b)** *(ugs.: einen kurzen Besuch machen)* **|bei jmdm./der Post|** ~**gehen** drop in *(coll.)* [at sb.'s/at the post office]; **c)** *(vergehen)* pass; ~|**kommen** *unr. itr. V.; mit sein* pass; **an etw.** *(Dat.)* ~**kommen** pass sth.; ~|**reden** *itr. V.* **an etw.** *(Dat.)* ~**reden** talk round sth. without getting to the point; **aneinander ~reden** talk at cross purposes; ~|**schießen** *unr. itr. V.* miss

vor|bereiten *tr. V.* prepare; **jmdn./ sich auf** *od.* **für etw. ~:** prepare sb./ oneself for sth.; **Vor·bereitung die;** ~, ~**en** preparation; ~**en |für etw.| treffen** make preparations for sth.

vor|bestellen *tr. V.* order in advance; **Vor·bestellung die** advance order

vor·bestraft *Adj.* with a previous conviction/previous convictions *postpos., not pred.*

vor|beugen 1. *tr. V.* bend ⟨*head, upper body*⟩ forward; **sich ~:** lean forward; **2.** *itr. V.* **einer Sache** *(Dat.)* *od.* **gegen etw. ~:** prevent sth.; **Vor·beugung die** prevention **(gegen** of); **zur ~:** as a preventive

Vor·bild das model; **jmdm. ein gutes ~ sein** be a good example to sb.; **vor·bildlich 1.** *Adj.* exemplary; **2.** *adv.* in an exemplary way

vor|bringen *unr. tr. V.* say; **eine Forderung/ein Anliegen ~:** make a demand/express a desire; **Argumente ~:** present arguments

vor·christlich *Adj.* pre-Christian

vor|datieren *tr. V.* postdate

vorder... *Adj.* front; **der Vordere Orient** the Middle East

Vorder-: ~**grund der** foreground; **im ~grund stehen** *(fig.)* be prominent *or* to the fore; ~**mann der;** *Pl.* ~**männer** person in front; **jmdn. auf ~mann bringen** *(ugs.)* lick sb. into shape

vor|drängen *refl. V.* push [one's way] forward *or* to the front; *(fig.)* push oneself forward

vor|dringen *unr. itr. V.; mit sein* push forward; advance

vor·dringlich 1. *Adj.* **a)** priority *attrib.* ⟨*treatment*⟩; **b)** *(dringlich)* urgent; **2.** *adv.* **a)** as a matter of priority; **b)** *(dringlich)* as a matter of urgency

Vor·druck der; *Pl.* **Vordrucke** form

vor·eilig 1. *Adj.* rash; **2.** *adv.* rashly

vor·einander *Adv.* **a)** one in front of the other; **b)** *(einer dem anderen gegenüber)* opposite each other; face to face; **c) Angst ~ haben** be afraid of each other

vor·eingenommen *Adj.* prejudiced; biased; **für/gegen jmdn. ~ sein** be prejudiced in sb.'s favour/against sb.

vorenthalten[1] *unr. tr. V.* **jmdm. etw. ~:** withhold sth. from sb.

vor·erst [*od.* '-'] *Adv.* for the present

Vorfahr der; ~**en , ~en** forefather; **vor|fahren** *unr. itr. V.; mit sein* **a)** *(ankommen)* drive/ride up; **b)** *(weiter nach vorn fahren)* ⟨*person*⟩ drive *or* move forward; ⟨*car*⟩ move forward; **c)** *(vorausfahren)* drive *or* go on ahead; **Vor·fahrt die;** *o. Pl.* right of way; „**~ beachten/gewähren**" 'give way'

---

[1] *ich enthalte vor (od. seltener: vorenthalte), vorenthalten, vorzuenthalten*

Vorfahrt[s]-: ~**schild** das right-of-way sign; ~**straße** die main road
Vor·fall der incident; occurrence; **vor|fallen** unr. itr. V.; mit sein **a)** (sich ereignen) happen; occur; **b)** (nach vorn fallen) fall forward
vor|finden unr. tr. V. find
Vor·freude die anticipation
vor|führen tr. V. show ⟨film, slides, etc.⟩; present ⟨circus act, programme⟩; perform ⟨play, trick, routine⟩; (demonstrieren) demonstrate; **jmdn. dem Richter** ~: bring sb. before the judge; **Vor·führung** die show; (eines Theaterstücks) performance
Vor·gang der occurrence; (Amtsspr.) file; **Vorgänger** der; ~s, ~, **Vorgängerin** die; ~, ~**nen** predecessor
Vor·garten der front garden
vor|geben unr. tr. V. pretend
Vor·gebirge das promontory
vor·gefaßt Adj. preconceived
vor|gehen unr. itr. V.; mit sein **a)** (ugs.: nach vorn gehen) go forward; **b)** (vorausgehen) go on ahead; **jmdn.** ~ **lassen** let sb. go first; **c)** ⟨clock⟩ be fast; **d)** (einschreiten) **gegen jmdn./etw.** ~: take action against sb./sth.; **e)** (verfahren) proceed; **f)** (sich abspielen) happen; go on; **g)** (Vorrang haben) have priority; come first
Vor·geschmack der; o. Pl. foretaste
Vor·gesetzte der/die; adj. Dekl. superior
vor·gestern Adv. the day before yesterday
vor|greifen unr. itr. V. **jmdm.** ~: anticipate sb. or jump in ahead of sb.
vor|haben unr. tr. V. intend; (geplant haben) plan; **Vor·haben** das; ~s, ~: plan; (Projekt) project
Vor·halle die entrance hall; (eines Theaters, Hotels) foyer
vor|halten unr. tr. V. **a)** hold up; **mit vorgehaltener Schußwaffe** at gunpoint; **b)** (zum Vorwurf machen) **jmdm. etw.** ~: reproach sb. for sth.; **Vor·haltungen** Pl. **jmdm. [wegen etw.]** ~ **machen** reproach sb. [for sth.]
vorhanden Adj. existing; (verfügbar) available; ~ **sein** exist or be in existence/be available
Vor·hang der (auch Theater) curtain; **Vorhänge·schloß** das padlock
Vor·haut die foreskin
vor·her [od. -'-] beforehand; (davor) before; **vorher|gehen** unr. itr. V.; mit sein **in den** ~**den Wochen** in the preceding weeks

Vor·herrschaft die supremacy; **vorherrschen** itr. V. predominate
vorher-, Vorher-: ~**sage** die prediction; (des Wetters) forecast; ~|**sagen** tr. V. predict; forecast ⟨weather⟩; ~|**sehen** unr. tr. V. s. **voraussehen**
vor·hin [od. -'-] Adv. a short time or while ago
vorig... Adj. last
Vor·jahr das previous year; **vor·jährig** Adj. of the previous year
Vor·kämpfer der pioneer
Vorkehrungen Pl. precautions
Vor·kenntnis die background knowledge
vor|kommen unr. itr. V.; mit sein **a)** (sich ereignen) happen; **b)** (vorhanden sein) occur; **c)** (erscheinen) seem; **das Lied kommt mir bekannt vor** I seem to know the song; **Vorkommnis** das; ~**ses**, ~**se** incident; occurrence
vor|laden unr. itr. V. summon; **Vor·ladung** die summons
Vor·lage die **a)** o. Pl.; s. **vorlegen**: presentation; showing; production; submission; tabling; **b)** (Entwurf) draft; **c)** (Muster) pattern; (Modell) model
Vor·läufer der precursor; forerunner; **vor·läufig 1.** Adj. temporary; provisional; interim ⟨order, agreement⟩; **2.** adv. for the time being
vor·laut 1. Adj. forward; **2.** adv. forwardly
vor|legen tr. V. present; show; produce ⟨certificate, identity card, etc.⟩; show ⟨sample⟩; submit ⟨evidence⟩; table ⟨parliamentary bill⟩
vor|lesen unr. tr., itr. V. read aloud or out; read ⟨story, poem, etc.⟩ aloud; **jmdm. [etw.]** ~: read [sth.] to sb.; **Vor·lesung** die lecture; (~sreihe) series or course of lectures
vor·letzt... Adj. last but one; penultimate ⟨page, episode, etc.⟩
Vor·liebe die preference; **vorlieb|nehmen** unr. itr. V. **mit jmdm./etw.** ~: put up with sb./sth.; (sich begnügen) make do with sb./sth.
vor|liegen unr. itr. V. **jmdm.** ~: be with sb.; **die Ergebnisse liegen uns noch nicht vor** we do not have the results yet; **im** ~**den Fall** in the present case
vorm Präp. + Art. **a)** = vor dem; **b)** (räumlich) in front of the; **c)** (zeitlich, bei Reihenfolge) before the
vor|machen tr. V. (ugs.) **jmdm. etw.** ~: show sb. sth.; (vortäuschen) kid (coll.) or fool sb.

vormalig *Adj.* former; **vormals** *Adv.*
formerly
Vor·marsch der *(auch fig.)* advance
vor|merken *tr. V.* make a note of; **ich
habe Sie für den Kurs vorgemerkt** I've
put you down for the course
vor·mittag *Adv.* **heute/morgen/Frei-
tag** ~: this/tomorrow/Friday morning; **Vor·mittag** der morning; **vor-
mittags** *Adv.* in the morning
Vor·mund der; *Pl.* **Vormunde** *od.*
Vormünder guardian
vorn[e] *Adv.* at the front; **nach** ~: to
the front; **von** ~: from the front; **noch
einmal von** ~ **anfangen** start afresh;
von ~ **bis hinten** *(ugs.)* from beginning
to end
vornehm 1. *Adj. (nobel; adelig)* noble;
(kultiviert) distinguished; *(elegant)* exclusive *(district, hotel, restaurant, resort)*; elegant *(villa, clothes)*; 2. *adv.*
nobly; *(elegant)* elegantly
vor|nehmen *unr. refl. V.* **sich** *(Dat.)*
etw. ~: plan sth.; **sich** *(Dat.)* ~, **mit
dem Rauchen aufzuhören** resolve to
give up smoking
vorn-: ~**herein** *in* **von** ~**herein** from
the outset; ~**über** *Adv.* forwards
Vor·ort der suburb
Vor·rang der; *o. Pl.* **a)** priority (vor +
Dat. over); **b)** *(bes. österr.: Vorfahrt)*
right of way
Vor·rat der supply, stock (**an** + *Dat.*
of); **vorrätig** *Adj.* in stock *postpos.*
Vor·recht das privilege
Vor·richtung die device
vor|rücken 1. *tr. V.* move forward;
advance *(chess piece)*; 2. *itr. V.; mit
sein* move forward; **auf den 5. Platz**
~: move up to fifth place
Vor·ruhestand der early retirement
vors *Präp.* + *Art.* = vor das
vor|sagen *tr. V.* **a)** *auch itr.* **jmdm.|die
Antwort|** ~: tell sb. the answer; *(flü-
sternd)* whisper the answer to sb.; **b)**
(aufsagen) recite
Vor·saison die start of the season;
early [part of the] season
Vor·satz der of intention; **vorsätzlich**
1. *Adj.* intentional; wilful *(murder, arson, etc.)*; 2. *adv.* intentionally
Vor·schau die preview
Vor·schein der: **zum** ~ **kommen** appear; *(entdeckt werden)* come to light
vor|schieben *unr. tr. V.* **a)** push *(bolt)*
across; **b)** *(nach vorn schieben)* push
forward
vor|schießen *unr. tr. V.* **jmdm. Geld**
~: advance sb. money

Vorschlag der suggestion; proposal;
vor|schlagen *unr. tr. V.* **|jmdm.|** etw.
~: suggest *or* propose sth. [to sb.]
vor·schreiben *unr. tr. V.* stipulate,
set *(conditions)*; lay down *(rules)*;
prescribe *(dose)*; **Vor·schrift** die instruction; order; *(gesetzliche od. amtliche Bestimmung)* regulation; **vor-
schrifts·mäßig** 1. *Adj.* correct;
proper; 2. *adv.* correctly; properly
Vor·schuß der advance
vor|sehen 1. *unr. tr. V.* **a)** plan; etw.
für/als etw. ~: intend sth. for/as sth.;
b) *(law, plan, contract, etc.)* provide
for; 2. *unr. refl. V.* **sich |vor jmdm./
etw.|** ~: be careful [of sb./sth.]
vor|setzen *tr. V.* **jmdm. etw.** ~: serve
sb. sth.; *(fig.)* serve *or* dish sb. up sth.
Vor·sicht die; *o. Pl.* care; *(bei Risiko,
Gefahr)* caution; care; **zur** ~: as a precaution; ~! be careful!; „~, **Stufe!"**
'mind the step!'; **vorsichtig** 1. *Adj.*
careful; *(bei Risiko, Gefahr)* cautious;
sei ~! be careful!; take care!; 2. *adv.*
carefully; with care; **vorsichts·hal-
ber** *Adv.* as a precaution; to be on the
safe side; **Vorsichts·maßnahme**
die precautionary measure; precaution
Vor·silbe die [monosyllabic] prefix
vor|singen *unr. tr. V.* **|jmdm.|** etw. ~:
sing sth. [to sb.]
Vor·sitz der chairmanship; **Vorsitzende** der/die; *adj. Dekl.* chair[per-
son]; *(bes. Mann)* chairman; *(Frau
auch)* chairwoman
Vor·sorge die; *o. Pl.* precautions *pl.;*
(für den Todesfall, Krankheit, Alter)
provisions *pl.;* **vor|sorgen** *itr. V.* **für
etw.** ~: make provisions for sth.; provide for sth.; **Vorsorge·untersu-
chung** die *(Med.)* medical check-up;
vorsorglich *adv.* as a precaution
Vor·spann der *(Film, Ferns.)* opening
credits *pl.*
Vor·speise die starter; hors d'œuvre
Vor·spiel das *(Theater)* prologue;
(Musik) prelude; **vor|spielen** *tr. V.*
a) play *(piece of music)* (*Dat.* to, for);
act out, perform *(scene)* (*Dat.* for, in
front of) **b)** *(vorspiegeln)* **jmdm. etw.** ~:
feign sth. to sb.
vor|sprechen 1. *unr. tr. V.* **a)** *(zum
Nachsprechen)* **jmdm. etw.** ~: pronounce *or* say sth. first for sb.; **b)** *(zur
Prüfung)* recite; 2. *unr. itr. V.* audition
Vor·sprung der lead (vor + *Dat.*
over)
Vor·stadt die suburb

Vor·stand der *(einer Firma)* board [of directors]; *(eines Vereins, einer Gesellschaft)* executive committee; *(einer Partei)* executive

vor|stehen *unr. itr. V.* **a)** project, jut out; ⟨*teeth, chin*⟩ stick out; ~de Zähne buck-teeth; projecting teeth; **b)** *(geh.: leiten)* **einer Institution** ~: be the head of an institution

vor|stellen **1.** *tr. V.* jmdn./sich jmdm. ~: introduce sb./oneself to sb.; *(bei Bewerbung)* sich ~: come/go for [an] interview; **die Uhr [um eine Stunde]** ~: put the clock forward [one hour]; **2.** *refl. V.* sich *(Dat.)* etw. ~: imagine sth.; **Vor·stellung** die **a)** *(Begriff)* idea; **b)** *o. Pl. (Phantasie)* imagination; **c)** *(Aufführung)* performance; *(im Kino)* showing

Vor·stoß der advance; **vor|stoßen** *unr. itr. V.; mit sein* advance; push forward

Vor·strafe die previous conviction

vor|strecken *tr. V.* stretch ⟨*arm, hand*⟩ out; advance ⟨*money, sum*⟩

Vor·tag der day before

vor|täuschen *tr. V.* feign; simulate ⟨*reality etc.*⟩; fake ⟨*crime*⟩

Vor·teil [*od.* 'fɔrtai̯l] der advantage; **vorteilhaft** **1.** *Adj.* advantageous; **2.** *adv.* advantageously

Vortrag der; ~[e]s, Vorträge talk; *(wissenschaftlich)* lecture; **einen** ~ **halten** give a talk/lecture; **vor|tragen** *unr. tr. V.* **a)** sing ⟨*song*⟩; perform, play ⟨*piece of music*⟩; recite ⟨*poem*⟩; **b)** *(darlegen)* present ⟨*case, matter, request, demands*⟩; lodge, make ⟨*complaint*⟩; express ⟨*wish, desire*⟩

vor·trefflich **1.** *Adj.* excellent; **2.** *adv.* excellently

vorüber *Adv.* over; *(räumlich)* past; **vorüber|gehen** *unr. itr. V.; mit sein* **a)** go *or* walk past; pass by; **an jmdm./etw.** ~: go past sb./sth.; pass sb./sth.; *(achtlos)* pass sb./sth. by; **b)** *(vergehen)* pass; ⟨*pain*⟩ go; **vorübergehend** **1.** *Adj.* temporary; passing ⟨*interest, infatuation*⟩; brief ⟨*illness, stay*⟩; **2.** *adv.* temporarily; *(für kurze Zeit)* for a short time; briefly

Vor·urteil das bias; *(voreilige Schlußfolgerung)* prejudice

Vor·vergangenheit die *(Sprachw.)* pluperfect

Vor·verkauf der advance sale of tickets

vor|verlegen *tr. V. (zeitlich)* bring forward (**auf** + *Akk.* to; **um** by)

Vorwahl die, **Vorwähl·nummer** die *(Fernspr.)* dialling code

Vorwand der; ~[e]s, Vorwände pretext; *(Ausrede)* excuse

vor·wärts *Adv.* forwards; *(weiter)* onwards; **vorwärts|kommen** *unr. itr. V.; mit sein* make progress; *(im Beruf, Leben)* get on; get ahead

vor·weg *Adv.* beforehand; **vor|weg|nehmen** *unr. tr. V.* anticipate

vor|weisen *unr. tr. V.* produce

vor|werfen *unr. tr. V.* jmdm. etw. ~: reproach sb. with sth.; *(beschuldigen)* accuse sb. of sth.

vor·wiegend *Adv.* mainly

vor·witzig *Adj.* bumptious; pert ⟨*child*⟩

Vor·wort das; *Pl.* ~e foreword

Vor·wurf der reproach; *(Beschuldigung)* accusation; **vorwurfs·voll** **1.** *Adj.* reproachful; **2.** *adv.* reproachfully

Vor·zeichen das **a)** *(Omen)* omen; **b)** *(Math.)* [algebraic] sign

vor|zeigen *tr. V.* produce; show

Vor·zeit die prehistory; **vorzeitig** **1.** *Adj.* premature; early ⟨*retirement*⟩; **2.** *adv.* prematurely

vor|ziehen *unr. tr. V.* prefer

Vor·zimmer das outer office

Vor·zug der **a)** *o. Pl.* preference (gegenüber over); **b)** *(gute Eigenschaft)* good quality; merit; **vorzüglich** **1.** *Adj.* excellent; first-rate; **2.** *adv.* excellently

vulgär **1.** *Adj.* vulgar; **2.** *adv.* in a vulgar way

Vulkan [vʊl'kaːn] der; ~s, ~e volcano; **vulkanisch** *Adj.* volcanic; **vulkanisieren** *tr. V.* vulcanize

v. u. Z. *Abk.* vor unserer Zeit[rechnung] BC

W

w, W [veː] das; ~s, ~: w, W

W *Abk.* **a)** West, Westen W.; **b)** Watt W.

Waage die; ~, ~n [pair *sing.* of] scales *pl.;* **waage·recht** **1.** *Adj.* horizontal;

2. *adv.* horizontally; **Waage·rech-te die** horizontal; **Waag·schale die** scale pan
Wabe die; ~, ~n honeycomb
wach 1. *Adj.* awake; **2.** *adv.* alertly; attentively; **Wache die**; ~, ~n **a)** *(Milit.)* guard *or* sentry duty; *(Seew.)* watch [duty]; **b)** *(Wächter, Milit.)* guard; *(Seew.)* watch; **c)** *(Polizei~)* police station; **wachen** *itr. V. (geh.)* be awake; **bei jmdm.** ~: stay up at sb.'s bedside; sit up with sb.; **Wach-hund der** guard-dog
Wacholder der; ~s, ~: juniper
Wach·posten der *(Milit.)* guard
Wachs das; ~es, ~e wax
wachsam *Adj.* watchful; vigilant
¹**wachsen** *unr. itr. V.; mit sein* grow
²**wachsen** *tr. V.* wax
Wachs-: ~**figur die** waxwork; ~**fi-guren·kabinett das** waxworks *sing. or pl.;* waxworks museum
wächst 2. u. 3. Pers. Sg. Präsens v. wachsen; Wachstum das; ~s growth
Wachtel die; ~, ~n quail
Wächter der; ~s, ~: guard; *(Nacht~, Turm~)* watchman; *(Park~)* [park-]keeper; **Wach[t]·turm der** watchtower
wackelig *Adj.* **a)** wobbly ⟨*chair, table, etc.*⟩; loose ⟨*tooth*⟩; **b)** *(ugs.: kraftlos, schwach)* frail; **Wackel·kontakt der** *(Elektrot.)* loose connection; **wackeln** *itr. V.* wobble; ⟨*tooth etc.*⟩ be loose; ⟨*house, window, etc.*⟩ shake; **mit dem Kopf/den Ohren** ~: waggle one's head/ears
wacker *(veralt.)* **1.** *Adj.* upright; **2.** *adv.* valiantly; **sich** ~ **halten** put up a good show
Wade die; ~, ~n *(Anat.)* calf; **Wa-den·krampf der** cramp in one's calf
Waffe die; ~, ~n weapon
Waffel die; ~, ~n waffle; *(dünne ~, Eis~)* wafer; *(Eistüte)* cone
Waffen-: ~**gewalt die**; *o. Pl.* **mit** ~**gewalt** by force of arms; ~**handel der** arms trade; ~**händler der** arms dealer; ~**schein der** firearms licence; ~**stillstand der** armistice
Wage·mut der daring; **wage·mu-tig** *Adj.* daring; **wagen 1.** *tr. V.* risk; |es| ~, **etw. zu tun** dare to do sth.; **2.** *refl. V.* **sich irgendwohin/nicht irgend-wohin** ~: venture somewhere/not dare to go somewhere
Wagen der; ~s, ~: *(PKW)* car; *(Pfer-de~)* cart; *(Eisenbahn~) (Personen~)* coach; *(Güter~)* truck; *(Straßen-bahn~)* car; *(Kinder~, Puppen~)* pram *(Brit.);* baby carriage *(Amer.); (Sport~)* push-chair *(Brit.);* stroller *(Amer.);* **Wagen·heber der** jack;
Waggon [va'gɔŋ, *südd., österr.:* va-'go:n] **der**; ~s, ~s, *südd., österr.:* ~s, ~e wagon; truck *(Brit.);* car *(Amer.)*
waghalsig 1. *Adj.* daring; *(leichtsin-nig)* reckless; **2.** *adv.* daringly; ⟨*specu-late*⟩ riskily; *(leichtsinnig)* recklessly; **Wagnis das**; ~ses, ~se daring ex-ploit *or* feat; *(Risiko)* risk
Wahl die; ~, ~en **a)** *o. Pl.* choice; **eine/ seine** ~ **treffen** make a/one's choice; **b)** *(in ein Gremium, Amt usw.)* elec-tion; **geheime** ~: secret ballot; **wahl·berechtigt** *Adj.* eligible *or* entitled to vote *postpos.;* **Wahl·be-teiligung die** turn-out; **wählen 1.** *tr. V.* **a)** choose; *(aus~)* select; **b)** *(Fernspr.)* dial ⟨*number*⟩; **c)** *(durch Stimmabgabe)* elect; **d)** *(stimmen für)* vote for ⟨*party, candidate*⟩; **2.** *itr. V.* **a)** choose; **b)** *(Fernspr.)* dial; **c)** *(stim-men)* vote; **Wähler der**; ~s, ~: voter; **Wahl·ergebnis das** election result; **Wählerin die**; ~, ~nen voter; **wäh-lerisch** *Adj.* choosy; particular **(in +** *Dat.* about)
wahl-, Wahl-: ~**gang der** ballot; ~**geheimnis das** secrecy of the bal-lot; ~**kabine die** polling-booth; ~**kampf der** election campaign; ~**kreis der** constituency; ~**lokal das** polling-station; ~**los 1.** *Adj.* indis-criminate; **2.** *adv.* indiscriminately; ~**recht das** *o.Pl.* right to vote
Wähl·scheibe die *(Fernspr.)* dial
Wahl-: ~**sieg der** election victory; ~**spruch der** motto; ~**urne die** ballot-box
Wahn der; ~|e|s mania delusion; **Wahn·sinn der**; *o. Pl.* **a)** insanity; madness; **b)** *(ugs.: Unvernunft)* mad-ness; lunacy; **wahnsinnig 1.** *Adj.* **a)** *(geistesgestört)* insane; mad; **b)** *(ugs.: ganz unvernünftig)* mad; crazy; **c)** *(ugs.: groß, heftig, intensiv)* terrific *(coll.)* ⟨*effort, speed, etc.*⟩; terrible *(coll.)* ⟨*fright, job, pain*⟩; **2.** *adv.* *(ugs.)* incredibly *(coll.);* terribly *(coll.)*
wahr *Adj.* **a)** true; **nicht** ~? *translation depends on preceding verb-form;* **du hast Hunger, nicht** ~? you're hungry, aren't you?; **nicht** ~, **er weiß es doch?** he does know, doesn't he?; **b)** *(wirk-lich)* real ⟨*reason, motive, feelings, joy, etc.*⟩; actual ⟨*culprit*⟩; *(echt)* true, real ⟨*friend, friendship, love, art*⟩

wahren *tr. V. (geh.)* preserve ⟨*balance, equality, neutrality, etc.*⟩; maintain ⟨*authority, right*⟩; *(verteidigen)* defend
währen *itr. V. (geh.)* last; **während** 1. *Konj.* **a)** *(zeitlich)* while; **b)** *(adversativ)* whereas; 2. *Präp. mit Gen.* during; *(über einen Zeitraum von)* for
wahr|haben *unr. tr. V. in etw.* **nicht ~ wollen** not want to admit sth.; **wahrhaft** *(geh.)* 1. *Adj.* true; 2. *adv.* truly; **wahrhaftig** 1. *Adj. (geh.)* truthful ⟨*person*⟩; 2. *adv.* really; genuinely; **Wahrheit** die; ~, ~en truth; **wahrheits·getreu** 1. *Adj.* truthful; faithful ⟨*account*⟩; 2. *adv.* truthfully; ⟨*portray*⟩ faithfully
wahr|nehmen *unr. tr. V.* **a)** *(mit den Sinnen erfassen)* perceive; *(spüren)* feel; detect ⟨*sound, smell*⟩; *(bemerken)* notice; *(erkennen, ausmachen)* make out; **b)** *(nutzen)* take advantage of ⟨*opportunity*⟩; exploit ⟨*advantage*⟩; exercise ⟨*right*⟩; **c)** *(vertreten)* look after ⟨*sb.'s interests, affairs*⟩; **d)** *(erfüllen, ausführen)* carry out, perform *(function, task, duty)*; fulfil ⟨*responsibility*⟩; **Wahrnehmung** die; ~, ~en **a)** perception; *(eines Sachverhalts)* awareness; *(eines Geruchs, eines Tons)* detection; **b)** *(Nutzung) (eines Rechts)* exercise; *(einer Gelegenheit, eines Vorteils)* exploitation; *(Vertretung)* representation; **d)** *(einer Funktion, Aufgabe, Pflicht)* performance; execution; *(einer Verantwortung)* fulfilment
wahr·sagen 1. *itr. V.* tell fortunes; 2. *tr. V.* predict, foretell ⟨*future*⟩; **Wahrsager** der; ~s, ~, **Wahrsagerin** die; ~, ~nen fortune-teller
wahrscheinlich 1. *Adj.* probable; likely; 2. *adv.* probably; **Wahrscheinlichkeit** die; ~, ~en probability; likelihood
Währung die; ~, ~en currency; **Währungs·reform** die currency reform
Wahr·zeichen das symbol; *(einer Stadt, einer Landschaft)* [most famous] landmark
Waise die; ~, ~n orphan; **Waisen·haus** das orphanage
Wal der; ~[e]s, ~e whale
Wald der; ~[e]s, Wälder wood; *(größer)* forest; **Wald·brand** der forest fire; **Wäldchen** das copse; **Waldmeister** der; *o. pl. (Bot.)* woodruff
Waliser der; ~s, ~ Welshman; **Waliserin** die; ~, ~nen Welshwoman; **walisisch** *Adj.* Welsh

Wall der; ~[e]s, Wälle earthwork; embankment; rampart *(esp. Mil.)*
Wall-: ~**fahrer** der pilgrim; ~**fahrt** die pilgrimage
Wal·nuß die walnut
Wal·roß das; *Pl.* -rosse walrus
walten *itr. V. (geh.)* ⟨*good sense, good spirit*⟩ prevail; ⟨*peace, silence, harmony, etc.*⟩ reign
Walze die; ~, ~n roller; *(Straßen~)* [road-]roller; *(Schreib~)* platen; **walzen** *tr. V.* roll ⟨*field, road, steel, etc.*⟩; **wälzen** 1. *tr. V.* roll; heave ⟨*heavy object*⟩; *(fig.)* shove ⟨*blame, responsibility*⟩ **(auf** + *Akk.* on); **etw. in Mehl** *usw.* ~ *(Kochk.)* toss sth. in flour *etc.*; **Probleme** ~ *(fig. ugs.)* mull over problems; 2. *refl. V.* roll; *(auf der Stelle)* roll about *or* around; *(im Krampf, vor Schmerzen)* writhe around; **Walzer** der; ~s, ~: waltz
wand *1. u. 3. Pers. Sg. Prät. v.* **winden**
Wand die; ~, Wände wall; *(Trenn~)* partition; *(bewegliche Trenn~)* screen; *(eines Behälters, Schiffs)* side
Wandel der; ~s change; **wandeln** *refl., tr. V.* change **(in** + *Akk.* into)
Wanderer der; ~s, ~: rambler; hiker; **Wander·karte** die rambler's [path] map; **wandern** *itr. V.; mit sein* **a)** hike; ramble; **b)** *(ugs.: gehen; fig.)* wander *(lit. or fig.)*; **c)** *(ziehen, reisen)* travel; *(ziellos)* roam ⟨*exhibition, circus, theatre*⟩ tour, travel; ⟨*animal, people, tribe*⟩ migrate
Wanderung die; ~, ~en **a)** hike; walking tour; **eine ~ machen** go on a hike/tour/trek; **b)** *(Zool., Soziol.)* migration
Wander·weg der footpath *(constructed for ramblers)*
Wandlung die; ~, ~en change; *(grundlegend)* transformation
Wand-: ~**malerei** die *(Bild)* mural; ~**schrank** der wall cupboard *or (Amer.)* closet
wandte *1. u. 3. Pers. Prät. v.* **wenden**
Wange die; ~, ~n *(geh.)* cheek
wankelmütig *Adj. (geh.)* vacillating;
wanken *itr. V.* **a)** sway; ⟨*person*⟩ totter; *(unter einer Last)* stagger; **b)** *mit sein (unsicher gehen)* stagger; totter
wann *Adv.* when; **seit ~ wohnst du dort?** how long have you been living there?
Wanne die; ~, ~n bath[tub]
Wanze die; ~, ~n bug *(coll.)*
Wappen das; ~s, ~: coat of arms
war *1. u. 3. Pers. Sg. Prät. v.* **sein**

warb *1. u. 3. Pers. Sg. Prät. v.* **werben**
ward *(geh.) 1. u. 3. Pers. Sg. Prät. v.*
werden
Ware die; ~, ~n a) ~|n| goods *pl.*; b)
(einzelne ~) article; commodity
(Econ., fig.); (Erzeugnis) product
Waren-: ~**haus** das department
store; ~**lager** das *(einer Fabrik o.ä.)*
stores *pl.; (eines Geschäftes)* stock-
room; *(größer)* warehouse; ~**muster**
das, ~**probe** die sample; ~**zeichen**
das trade mark
warf *1. u. 3. Pers. Sg. Prät. v.* **werfen**
warm; **wärmer, wärmst** ... 1. *Adj. (auch*
fig.) warm; hot ⟨*meal, food, bath,*
spring⟩; das Essen ~ **machen** heat up
the food; „~" *(auf Wasserhahn)* 'hot';
keen, lively ⟨*interest*⟩; 2. *adv.* warmly;
~ **essen/duschen** have a hot meal/
shower; **Wärme** die; ~: warmth;
(Hitze; auch Physik) heat; **wärmen** 1.
tr. V. warm; *(aufwärmen)* warm up
⟨*food, drink*⟩; 2. *itr. V.* be warm;
(warm halten) keep one warm;
Wärm·flasche die hot-water bottle
Warm·wasser-: ~**bereiter** der; ~s,
~: water-heater; ~**heizung** die hot-
water heating
Warn-: ~**blinkanlage** die *(Kfz-W.)*
hazard warning lights *pl.;* ~**dreieck**
das *(Kfz-W.)* hazard warning triangle
warnen *tr. (auch itr.) V.* warn (vor +
Dat. of, about); jmdn. |davor| ~, etw.
zu tun warn sb. against doing sth.
Warn-: ~**schild** das warning sign;
~**schuß** der warning shot; ~**signal**
das warning signal; ~**streik** der
token strike
Warnung die; ~, ~en warning (vor +
Dat. of, about)
Warschau (das); ~s Warsaw
Warte-: ~**halle** die waiting room;
(Flugw.) departure lounge; ~**liste** die
waiting list
warten 1. *itr. V.* wait (**auf** + *Akk.*
for); 2. *tr. V.* service ⟨*car etc.*⟩
Wärter der; ~s, ~: attendant; *(Tier~,*
Zoo~, Leuchtturm~) keeper; *(Kran-*
ken~) orderly; *(Gefängnis~)* warder
Warte-: ~**saal** der waiting-room;
~**zimmer** das waiting-room
Wartung die; ~, ~en service; *(das*
Warten) servicing; *(Instandhaltung)*
maintenance
warum *Adv.* why
Warze die; ~, ~n wart; *(Brust~)*
nipple
was 1. *Interrogativpron. Nom. u. Akk.*
u. (nach Präp.) Dat. Neutr.; ~ kostet

das? what *or* how much does that
cost?; **ach** ~! *(ugs.)* oh, come on!; ~
für ein .../~ für ...: what sort *or* kind
of ...; 2. *Relativpron. Nom. u. Akk. u.*
(nach Präp.) Dat. Neutr.; |das,| ~:
what; **alles,** ~ ...: everything *or* all
that ...; **vieles/nichts/etwas,** ~ ...:
much/nothing/something that ...; ~
mich betrifft, |so| ...: as far as I'm con-
cerned, ...; 3. *Indefinitpron. Nom. u.*
Akk. u. (nach Präp.) Dat. Neutr. (ugs.)
s. **etwas**; 4. *Adv. (ugs.) (warum, wozu)*
why; what ... for
Wasch-: ~**anlage** die car-wash;
~**automat** der washing-machine;
~**becken** das wash-basin
Wäsche die; ~, ~n a) *o. Pl. (zu wa-*
schende Textilien) washing; *(für die*
Wäscherei) laundry; b) *o. Pl. (Unter~)*
underwear; c) *(das Waschen)* washing
no pl.; (einmalig) wash; **in der ~ sein**
be in the wash; **wasch·echt** *Adj.* a)
colour-fast ⟨*textile, clothes*⟩; fast ⟨*col-*
our⟩; b) *(fig.)* genuine
Wäsche-: ~**klammer** die clothes-
peg *(Brit.);* clothes-pin *(Amer.);*
~**korb** der laundry-basket; ~**leine**
die clothes-line
waschen 1. *unr. tr.V.* wash; **sich** ~:
wash [oneself]; have a wash; **Wäsche**
~: do the/some washing; 2. *unr. itr.*
V. do the washing; **Wäscherei** die;
~, ~en laundry
Wäsche-: ~**schleuder** die spin-
drier; ~**trockner** der a) *(Maschine)*
tumble-drier; b) *(Gestell)* clothes-airer
Wasch-: ~**gelegenheit** die washing
facilities *pl.;* ~**küche** die laundry-
room; ~**lappen** der [face] flannel;
washcloth *(Amer.);* ~**maschine** die
washing-machine; ~**mittel** das detergent; ~**pulver** das washing-powder;
~**straße** die [automatic] car-wash
wäscht *3. Pers. Sg. Präsens v.* **waschen**
Wasser das; ~s, ~/Wässer a) *o. Pl.*
water; b) *(Mineral~, Tafel~)* mineral
water; *(Heil~)* water; c) *o. Pl. (Gewäs-*
ser) **ein fließendes/stehendes** ~: a
moving/stagnant stretch of water; d)
o. Pl. (Urin) water; urine; ~ **lassen**
pass water
wasser-, Wasser-: ~**ball** der a)
beach-ball; b) *o. Pl. (Spiel)* water
polo; ~**dicht** *Adj.* waterproof ⟨*clo-*
thing, watch, etc.⟩; watertight ⟨*con-*
tainer, seal, etc.⟩; ~**fall** der waterfall;
~**farbe** die water-colour; ~**hahn** der
water-tap; faucet *(Amer.)*
wässerig *s.* **wäßrig**

Wasser-: ~**kessel** der kettle; ~**leitung** die water-pipe; *(Hauptleitung)* water-main

wassern *itr. V.; mit sein* land [on the water]; **wässern** *tr. V.* soak; *(Phot.)* wash ⟨*negative, print*⟩

wasser-, Wasser-: ~**pflanze** die aquatic plant; ~**rohr** das water-pipe; ~**schlauch** der [water-]hose; ~**schutz·polizei** die river/lake police; [1]~**ski** der water-ski; ~ **fahren** water-ski; [2]~**ski** das; ~s water-skiing *no art.;* ~**spiegel** der a) *(Oberfläche)* surface [of the water]; b) *(Niveau)* water-level; ~**sport** der water-sport *no art.;* ~**spülung** die flush

Wasser·stoff der; *o. Pl.* hydrogen; **Wasser·stoff·bombe** die hydrogen bomb

Wasser-: ~**strahl** der jet of water; ~**straße** die waterway; ~**temperatur** die water-temperature; ~**tiefe** die depth of the water; ~**tropfen** der drop of water; ~**turm** der watertower; ~**werfer** der water-cannon; ~**werk** das waterworks *sing.;* ~**zeichen** das watermark

wäßrig *Adj.* watery

waten *itr. V.; mit sein* wade

watscheln *itr. V.; mit sein* waddle

[1]**Watt** das; ~[e]s, ~en mud-flats *pl.*

[2]**Watt** das; ~s, ~ *(Technik, Physik)* watt

Watte die; ~, ~n cotton wool; **Watte·bausch** der wad of cotton wool

Watten·meer das tidal shallows *pl.*

wattiert *Adj.* quilted; padded ⟨*shoulder etc., envelope*⟩

WC [veːˈtseː] das; ~[s], ~[s] toilet; WC

weben *tr., itr. V.* weave; **Weber** der; ~s, ~: weaver; **Web·stuhl** der loom

Wechsel der; ~s, ~ a) *(das Auswechseln)* change; *(Geld~)* exchange; b) *(Aufeinanderfolge)* alternation; **im ~:** alternately; *(bei mehr als zwei)* in rotation; c) *(das Überwechseln)* move; *(Sport)* transfer; d) *(Bankw.)* bill of exchange *(über + Akk.* for)

wechsel-, Wechsel-: ~**geld** das; *o. Pl.* change; ~**haft** *Adj.* changeable; ~**jahre** *Pl.* change of life *sing.;* menopause *sing.;* ~**kurs** der exchange rate

wechseln 1. *tr. V.* a) change; **das Hemd ~:** change one's shirt; **die Wohnung ~:** move home; b) *([aus]tauschen)* exchange ⟨*letters, glances, etc.*⟩; c) *(um~)* change ⟨*money, note, etc.*⟩ *(in + Akk.* into); 2. *itr. V.* change

wechsel-, Wechsel-: ~**seitig** 1. *Adj.* mutual; 2. *adv.* mutually; ~**strom** der *(Elektrot.)* alternating current; ~**stube** die bureau de change; ~**wirkung** die interaction

wecken *tr. V.* jmdn. [aus dem Schlaf] ~: wake sb. [up]; *(fig.: hervorrufen)* arouse ⟨*interest, curiosity, anger*⟩; **Wecker** der; ~s, ~ alarm clock

wedeln *itr. V.* ⟨*tail*⟩ wag; [mit dem Schwanz] ~ ⟨*dog*⟩ wag its tail

weder *Konj.* ~ A noch B neither A nor B

weg *Adv.* away; *(verschwunden, ~gegangen)* gone; **er ist schon seit einer Stunde ~:** he left an hour ago; **weit ~:** far away; a long way away

Weg der; ~[e]s,~e a) *(Fuß~)* path; *(Feld~)* track; b) *(Zugang)* way; *(Passage, Durchgang)* passage; **sich** *(Dat.)* **einen ~ durch etw. bahnen** clear a path *or* way through sth.; c) *(Route, Verbindung)* way; route; d) *(Strecke, Entfernung)* distance; *(Gang)* walk; *(Reise)* journey; **auf dem kürzesten ~:** by the shortest route; **auf halbem ~[e]** *(auch fig.)* half-way; **sich auf den ~ machen** set off; **etw. in die ~e leiten** get sth. under way; e) *(ugs.: Besorgung)* errand; f) *(Methode)* way; *(Mittel)* means

weg-: ~|**bleiben** *unr. itr. V.; mit sein (nicht kommen)* stay away; *(nicht nach Hause kommen)* stay out; ~|**bringen** *unr. tr. V.* take away; *(zur Reparatur, Wartung usw.)* take in

wegen *Präp. mit Gen.* a) because of; ~ **Umbau[s] geschlossen** closed for alterations; b) *(um... willen)* for the sake of; ~ **der Kinder/***(ugs.)* **dir** for the children's/your sake; c) *(bezüglich)* about; regarding

weg-: ~|**fahren** 1. *unr. itr. V.; mit sein* a) leave; *(im Auto)* drive off; *(losfahren)* set off; b) *(irgendwohin fahren)* go away; 2. *unr. tr. V.* drive away; *(mit dem Handwagen usw.)* take away; ~|**fallen** *unr. itr. V.; mit sein* be discontinued; *(nicht mehr zutreffen)* no longer apply; ~|**fliegen** *unr. itr. V.; mit sein* fly away; *(~geblasen werden)* fly off; ~|**gehen** *unr. itr. V.* a) leave; *(ugs.: ausgehen)* go out; *(ugs.: ~ziehen)* move away; b) *(verschwinden)* ⟨*spot, fog, etc.*⟩ go away; c) *(sich entfernen lassen)* come out; ~|**jagen** *tr. V.* chase away; ~|**kommen** *unr. itr. V.; mit sein* a) get away; b) *(abhanden kommen)* go missing; c) **gut/**

schlecht *usw.* |bei etw.| ~kommen *(ugs.)* come off well/badly *etc.* [in sth.].; ~|kriegen *tr. V.* get rid of ⟨*cold, pain, etc.*⟩; get out, get rid of ⟨*stain*⟩; ~|lassen *unr. tr. V.* **a)** jmdn. ~lassen let sb. go; *(ausgehen lassen)* let sb. go out; **b)** *(auslassen)* leave out; omit; ~|laufen *unr. itr. V.; mit sein* run away **(von, vor** + *Dat.* from); ~|legen *tr. V.* put aside; *(an seinen Platz legen)* put away; ~|nehmen *unr. tr. V.* **a)** take away; move ⟨*head, arm*⟩; **b)** jmdm. etw. ~nehmen take sth. away from sb.; ~|schicken *tr. V.* **a)** send off ⟨*letter, parcel*⟩; **b)** send ⟨*person*⟩ away; ~|schmeißen *unr. tr. V. (ugs.)* chuck away *(coll.)*; ~|schütten *tr. V.* pour away; ~|sehen *unr. itr. V.* look away; ~|stellen *tr. V.* put away; *(beiseite stellen)* put aside; ~|stoßen *unr. tr. V.* push *or* shove away; ~|tragen *unr. tr. V.* carry away

Wegweiser der; ~s, ~ signpost

weg-: ~|werfen *unr. tr. V. (auch fig.)* throw away; ~werfend *Adj.* dismissive ⟨*gesture, remark*⟩; ~|wischen *tr. V.* wipe away; ~|ziehen **1.** *unr. tr. V.* pull away; draw back ⟨*curtain*⟩; pull off ⟨*blanket*⟩; **2.** *unr. itr. V.; mit sein* **a)** *(umziehen)* move away; **b)** *(wandern)* ⟨*animals, nomads, etc.*⟩ leave [on their migration]

weh *(ugs.)* **1.** *Adj.* sore; **2.** *adv.* ~ tun hurt; mir tut der Magen/Kopf ~: my stomach/head is aching *or* hurts; jmdm./sich ~ tun hurt sb./oneself; **Wehe** die; ~, ~n: ~n haben have contractions; in den ~n liegen be in labour

wehen *itr. V.* **a)** *(blasen)* blow; **b)** *(flattern)* flutter

weh-, Weh-: ~leidig *(abwertend)* **1.** *Adj. (überempfindlich)* soft; *(weinerlich)* whining *attrib.;* **2.** *adv.* self-pityingly; *(weinerlich)* whiningly; ~mut die; ~ *(geh.)* wistful nostalgia; ~mütig *Adj.* wistfully nostalgic

¹**Wehr** die; ~, ~en sich |gegen jmdn./ etw.| zur ~ setzen make a stand [against sb./sth.]; resist [sb./sth.];

²**Wehr** das; ~|e|s, ~e weir

Wehr·dienst der; o. Pl. military service *no art.;* seinen ~ ableisten do one's military service

Wehr·dienst-: ~verweigerer der; ~s, ~: conscientious objector; ~verweigerung die conscientious objection

wehren *refl. V.* defend oneself

wehr-, Wehr-: ~los *Adj.* defenceless; ~pflicht die; *o. Pl.* military service; **die allgemeine ~pflicht** compulsory military service; ~pflichtig *Adj.* liable for military service *postpos.*

Weib das; ~|e|s, ~er **a)** *(veralt., ugs.)* woman; female *(derog.);* **Weibchen** das; ~s, ~ female; **weiblich 1.** *Adj.* **a)** female; **b)** *(für die Frau typisch; Sprachw.)* feminine; **2.** *adv.* femininely

weich 1. *Adj. (auch fig.)* soft; **ein ~es Ei** a soft-boiled egg; **2.** *adv.* softly

¹**Weiche** die; ~, ~n *(Flanke)* flank

²**Weiche** die; ~, ~n points *pl. (Brit.);* switch *(Amer.)*

weichen *unr. itr. V.; mit sein* move; vor jmdm./einer Sache ~: give way to sb./sth.

weich·gekocht *Adj. (präd. getrennt geschrieben)* soft-boiled ⟨*egg*⟩;

weichlich 1. *Adj.* soft; *(ohne innere Festigkeit)* weak; **2.** *adv.* softly

¹**Weide** die; ~, ~n willow

²**Weide** die; ~, ~n pasture; **weiden** *itr., tr. V.* graze

Weiden·kätzchen das willow catkin

weigern *refl. V.* refuse; **Weigerung** die; ~, ~en refusal

Weih·bischof der *(kath. Kirche)* suffragan bishop; **Weihe** die; ~, ~n *(Rel.)* consecration; *(kath. Kirche: Priester~, Bischofs~)* ordination; **weihen** *tr. V.* **a)** *(Rel.)* consecrate; *(zueignen)* dedicate *(Dat.* to); **b)** *(kath. Kirche: ordinieren)* ordain

Weiher der; ~s, ~: [small] pond

Weihnachten das; ~, ~: Christmas; **frohe** *od.* **fröhliche** *od.* **gesegnete ~!** Merry *or* Happy Christmas!; **weihnachtlich** *Adj.* Christmassy

Weihnachts-: ~baum der Christmas tree; ~feiertag der: der erste/ zweite ~feiertag Christmas Day/Boxing Day; ~fest das Christmas; ~geschenk das Christmas present *or* gift; ~lied das Christmas carol; ~mann der; *Pl.* ~männer Father Christmas; Santa Claus; ~markt der Christmas fair; ~zeit die Christmas time

Weih-: ~rauch der incense; ~wasser das *(kath. Kirche)* holy water

weil *Konj.* because

Weile die; ~: while; **weilen** *itr. V. (geh.) (ver~)* stay; *(sein)* be

Wein der; ~|e|s, ~e wine

Wein-: ~berg der vineyard; ~brand der brandy

weinen *itr. V.* cry (über + *Akk.* over, about); *(aus Trauer, Kummer)* cry, weep (um for); **weinerlich 1.** *Adj.* tearful; weepy; **2.** *adv.* tearfully

wein-, Wein-: ~essig der wine vinegar; **~flasche** winebottle; **~glas** das wineglass; **~handlung** die wine-merchant's; **~karte** die wine-list; **~lokal** das wine bar; **~probe** die wine-tasting [session]; **~rot** *Adj.* wine-red; **~stube** die wine bar; **~traube** die grape

weise 1. *Adj.* wise; **2.** *adv.* wisely

Weise die; ~, ~n a) *(Art, Verfahren)* way; b) *(Melodie)* tune; melody

weisen 1. *unr. tr. V. (geh.: zeigen)* show; **jmdn. aus dem Zimmer ~:** send sb. out of the room; **2.** *unr. itr. V. (irgendwohin zeigen)* point

Weisheit die; ~, ~en a) *o. Pl.* wisdom; b) *(Erkenntnis)* wise insight; *(Spruch)* wise saying; **Weisheits·zahn** der wisdom tooth; **weis|machen** *tr. V. (ugs.)* **das kannst du mir nicht ~!** you can't expect me to swallow that!

¹weiß *1. u. 3. Pers. Sg. Präsens v.* **wissen**

²weiß *Adj.* white; **Weiß** das; ~|e|s, ~: white

weis·sagen *tr. V.* prophesy; **Weissagung** die; ~, ~en prophecy

Weiß-: ~bier das weiss beer; **~brot** das white bread; **~dorn** der hawthorn

Weiße der/die; *adj. Dekl.* white; white man/woman; **weißen** *tr. V.* paint white; *(tünchen)* whitewash

weiß-, Weiß-: ~gold das white gold; **~herbst** der ≈ rosé wine; **~kohl** der, *(bes. südd., österr.)* **~kraut** das white cabbage

weißlich *Adj.* whitish

weißt *2. Pers. Sg. Präsens v.* **wissen**

Weiß-: ~wein der white wine; **~wurst** die veal sausage

Weisung die; ~, ~en *(geh., sonst Amtsspr.)* instruction; *(Direktive)* directive

weit 1. *Adj.* wide; long ⟨*way*⟩; **jmdm. zu ~ sein** ⟨*clothes*⟩ be too loose on sb.; **2.** *adv.* a) *(räumlich ausgedehnt)* **~ geöffnet** wide open; **~ und breit war niemand zu sehen** there was no one to be seen anywhere; b) *(lang)* far; **~er** further; farther; **am ~esten** [the] furthest *or* farthest; **~ |entfernt *od.* weg| wohnen** live a long way away *or* off; live far away; **von ~em** from a distance; **das geht zu ~** *(fig.)* that is going too

far; c) *(zeitlich entfernt)* **~ nach Mitternacht** well past midnight; d) *(in der Entwicklung)* far; **Weit·blick** der; *o. Pl.* far-sightedness; **Weite** die; ~, ~n a) *(räumliche Ausdehnung)* expanse; b) *(bes. Sport: Entfernung)* distance; c) *(eines Kleidungsstückes)* width; **weiten 1.** *tr. V.* widen; **2.** *refl. V.* widen; ⟨*pupil*⟩ dilate; **weiter** *Adv.* a) s. **weit 2;** b) **und so ~:** and so on; c) *(~hin, anschließend)* then; d) *(außerdem, sonst)* ~ **nichts** nothing more *or* else; **weiter...** *Adj.* further; **bis auf ~es** for the time being; *s. auch* **ohne**

weiter-, Weiter-: ~|bringen *unr. tr. V.* **die Diskussion brachte uns nicht ~:** the discussion did not get us any further [forward]; **~|erzählen** *tr. V.* a) continue telling; *itr.* **erzähl weiter!** do carry *or* go on; b) *(~sagen)* pass on; **~|fahren** *unr. itr. V.; mit sein* continue [on one's way]; *(~ reisen)* travel on; **~|führen** *tr., itr. V.* continue; **~|geben** *unr. tr. V.* pass on; **~|gehen** *unr. itr. V.; mit sein* go on; **bitte ~gehen!** please move along *or* keep moving!; **~hin** *Adv.* a) *(immer noch)* still; b) *(künftig)* in future; c) *(außerdem)* in addition; **~|kommen** *unr. itr. V.; mit sein* a) get further; b) *(Fortschritte machen)* make progress; **im Beruf ~kommen** get on in one's career; **~|machen** *(ugs.) itr. V.* carry on; go on; **~|reichen** *tr. V.* pass on; **~|sagen** *tr. V.* pass on; **~|sehen** *unr. itr. V.* see; **~|verarbeiten** *tr. V.* process; **~verarbeitung** die processing

weit-, Weit-: ~gehend 1. *Adj.* extensive, wide, sweeping ⟨*powers*⟩; far-reaching ⟨*support, concessions, etc.*⟩; wide ⟨*support, agreement, etc.*⟩; general ⟨*renunciation*⟩; **2.** *adv.* to a large *or* great extent; **~gereist** *Adj.* widely travelled; **~hin** *Adv.* for miles around; **~läufig 1.** *Adj.* a) *(ausgedehnt)* extensive; *(geräumig)* spacious; b) *(entfernt)* distant; **2.** *adv.* a) *(ausgedehnt)* spaciously; b) *(entfernt)* distantly; **~räumig 1.** *Adj.* spacious ⟨*room, area, etc.*⟩; wide ⟨*gap, space*⟩; **2.** *adv.* spaciously; **~reichend 1.** *Adj. (fig.)* far-reaching ⟨*importance, consequences*⟩; sweeping ⟨*changes, powers*⟩; extensive ⟨*relations, influence*⟩; **2.** *adv.* extensively; **~sichtig** *Adj.* long-sighted; **~sichtigkeit** die; ~ long-sightedness; **~sprung** der *(Sport)* long jump *(Brit.)*; broad jump *(Amer.)*; **~verbreitet** *Adj.* widespread; common; common ⟨*plant, animal*⟩; **~winkel·objektiv das** wide-angle lens

Weizen der; ~s wheat

welch 1. *Interrogativpron. (bei Wahl aus einer unbegrenzten Menge)* what; *(bei Wahl aus einer begrenzten Menge) (adj.)* which; *(subst.)* which one; **2.** *Relativpron. (bei Menschen)* who; *(bei Sachen)* which; **3.** *Indefinitpron.* some; *(in Fragen)* any

welk *Adj.* withered ⟨*skin, hands, etc.*⟩; wilted ⟨*leaves, flower*⟩; limp ⟨*lettuce*⟩; **welken** *itr. V.; mit sein* ⟨*plant, flower*⟩ wilt

Well·blech das corrugated iron; **Welle die;** ~, ~n **a)** *(auch fig.)* wave; *(Rundf.:* ~*nlänge)* wavelength; **b)** *(Technik)* shaft

wellen-, Wellen-: **~bad das** artificial wave pool; **~brecher der** breakwater; **~gang der;** *o. Pl.* swell; **bei starkem ~gang** in heavy seas; **~länge die** wavelength; **~sittich der** budgerigar

Well·fleisch das boiled belly pork; **wellig** *Adj.* wavy ⟨*hair*⟩; undulating ⟨*scenery, hills, etc.*⟩; uneven ⟨*surface, track, etc.*⟩; **Well·pappe die** corrugated cardboard

Welt die; ~, ~en **a)** *o. Pl.* world; **auf der** ~: in the world; **die Alte/Neue** ~: the Old/New World; **die dritte/vierte** ~ the Third/Fourth World; **auf die od. zur** ~ **kommen** be born; **alle** ~ *(fig. ugs.)* the whole world; everybody; **b)** *(~all)* universe

welt-, Welt-: **~all das** universe; **~anschauung die** world-view; **~ausstellung die** world fair; **~berühmt** *Adj.* world-famous

Welten·bummler der; ~s, ~ globetrotter

welt-, Welt-: **~fremd 1.** *Adj.* unworldly; **2.** *adv.* unrealistically; **~frieden der** world peace; **~karte die** map of the world; **~krieg der** world war; **der erste/zweite ~krieg** the First/Second World War

weltlich *Adj.* **a)** worldly **b)** *(nicht geistlich)* secular

welt-, Welt-: **~literatur die** world literature *no art.;* **~macht die** world power; **~markt der** *(Wirtsch.)* world market; **~meister der** world champion; **~meisterschaft die** world championship; **~raum der** space *no art.;* **~reise die** world tour; **~rekord der** world record; **~stadt die** cosmo-

politan city; **~weit 1.** *Adj.* worldwide; **2.** *adv.* throughout the world; **~wirtschaft die** world economy

wem *Dat. von* wer **1.** *Interrogativpron.* to whom; who ... to; **mit/von/zu** ~: with/from/to whom; who ... with/from/to; **2.** *Relativpron.* the person to whom ...; the person who ... to; **3.** *Indefinitpron. (ugs.: jemandem)* to somebody *or* someone; *(fragend od. verneint)* to anybody *or* anyone

wen *Akk. von* wer **1.** *Interrogativpron.* whom; who *(coll.);* **an/für** ~: to/for whom ...; who ... to/for; **2.** *Relativpron.* the person whom; **3.** *Indefinitpron. (ugs.: jemanden)* somebody; someone; *(fragend od. verneint)* anybody; anyone

Wende die; ~, ~n change (**zu** for); **Wende·kreis der a)** *(Geogr.)* tropic; **b)** *(Kfz-W.)* turning circle; **Wendel·treppe die** spiral staircase; **¹wenden 1.** *tr., auch itr. V. (auf die andere Seite)* turn [over]; *(in die entgegengesetzte Richtung)* turn [round]; **bitte** ~! please turn over; **2.** *itr. V.* turn [round]; **3.** *refl. V.* **sich zum Besseren/Schlechteren** ~: take a turn for the better/worse; **²wenden 1.** *unr. (auch regelm.) tr. V.* turn; **2.** *unr. (auch regelm.) refl. V.* **a)** ⟨*person*⟩ turn; **b)** *(sich richten)* **sich an jmdn. [um Rat]** ~: turn to sb. [for advice]; **wendig 1.** *Adj.* **a)** agile; manœuvrable ⟨*vehicle, boat, etc.*⟩; **b)** *(gewandt)* astute; **2.** *adv.* **a)** *(beweglich)* agilely; **b)** *(gewandt)* astutely; **Wendung die;** ~, ~en **a)** *(Änderung der Richtung)* turn; **b)** *(Veränderung)* change

wenig 1. *Indefinitpron. u. unbest. Zahlw. a) Sing.* little; **das ist** ~: that isn't much; **zu** ~ **Zeit/Geld haben** not have enough time/money; **ein Exemplar/50 Mark zu** ~: one copy too few/50 marks too little; **b)** *Pl.* a few; **mit** ~**en Worten** in a few words; **2.** *Adv.* little; ~ **mehr** not much more; **weniger 1.** *Komp. von* wenig; *Indefinitpron. u. unbest. Zahlw.* (+ *Sg.*) less; (+ *Pl.*) fewer; **immer** ~: less and less; **2.** *Komp. von* wenig; *Adv.* less; **das ist** ~ **angenehm/erfreulich/schön** that is not very pleasant/pleasing/nice; *s. auch* mehr 1; **3.** *Konj.* less; **fünf** ~ **drei** five, take away three; **wenigst... 1.** *Sup. von* wenig; *Indefinitpron. u. unbest. Zahlw.* least; **am** ~**en** least; **2.** *Sup. von* wenig; **am** ~**en** the least; **wenigstens** *Adv.* at least

wenn *Konj.* **a)** *(konditional)* if; **außer
~:** unless; **~ es nicht anders geht** if
there's no other way; **b)** *(temporal)*
when; **jedesmal,** *od.* **immer, ~:** whenever; **c)** *(konzessiv)* **wenn ... auch** even
though; **d)** *(in Wunschsätzen)* if only

wer *Nom. Mask. u. Fem.; s. auch
(Gen.)* **wessen;** *(Dat.)* **wem;** *(Akk.)* **wen
1.** *Interrogativpron.* who; **~ von ...:**
which of; **2.** *Relativpron.* the person
who; *(jeder, der)* anyone *or* anybody
who; **3.** *Indefinitpron. (ugs.: jemand)*
someone; *(in Fragen, Konditionalsätzen)* anyone; anybody

Werbe-: **~agentur die** advertising
agency; **~fernsehen das** television
commercials *pl.;* **~funk der** radio
commercials *pl.*

werben 1. *unr. itr. V.* advertise; **für
etw. ~:** advertise sth.; **2.** *unr. tr. V.* attract ⟨*readers, customers, etc.*⟩; recruit
⟨*soldiers, members, etc.*⟩; **Werbung
die;** **~:** advertising; **für etw. ~ machen**
advertise sth.

Werde·gang der career; **werden 1.**
unr. itr. V.; mit sein become; get; **älter
~:** get *or* grow old[er]; **wahnsinnig** *od.*
verrückt ~: go mad; **das muß anders
~:** things have to change; **wach ~:**
wake up; **rot ~:** go *or* turn red; **Arzt/
Professor ~:** become a doctor/professor; **zu etw. ~:** become sth.; **es wird
|höchste| Zeit** it is [high] time; **es wird
10 Uhr** it is nearly 10 o'clock; **es wird
Herbst** autumn is coming; **sind die Fotos |etwas| geworden?** *(ugs.)* have the
photos turned out [well]?; **2.** *Hilfsverb;* **2.** *Part.* **worden a)** *(zur Bildung
des Futurs)* **wir ~ uns um ihn kümmern**
we will take care of him; **es wird gleich
regnen** it is going to rain any minute;
es wird um die 80 Mark kosten *(ich vermute, es kostet um die 80 Mark)* it will
cost around 80 marks; **b)** *(zur Bildung
des Passivs)* **du wirst gerufen** you are
being called; **er wurde gebeten** he was
asked

werfen 1. *unr. tr. V.* throw; drop
⟨*bombs*⟩; **2.** *unr. itr. V.* **a)** throw; **mit
etw. ~:** throw sth.; **b)** *(Junge kriegen)*
give birth; ⟨*dog, cat*⟩ litter; **3.** *unr. refl.
V.* throw oneself; **sich vor einen Zug
~:** throw oneself under a train

Werft die; **~,** **~en** shipyard

Werk das; **~|e|s,** **~e a)** work; **b)** *(Betrieb, Fabrik)* factory; works *sing. or
pl.;* **ab ~:** ex works

Werk·bank die; *Pl.* **~bänke** workbench

Werk[s]-: **~angehörige** der/die
factory *or* works employee; **~arzt der**
factory *or* works doctor

werk-, Werk-: **~statt die, ~statt,
~stätten** workshop; *(Kfz-W.)* garage;
~stoff der material; **~tag der** working day; workday; **~tags** *Adv.* on
weekdays; **~tätig** *Adj.* working;
~tätige der/die; *adj. Dekl.* worker;
~zeug das; *Pl.* **~zeuge** *(auch fig.)* tool

Werkzeug·kasten der tool-box

Wermut der; **~|e|s, ~s a)** *(Pflanze)*
wormwood; **b)** *(Wein)* vermouth

wert *Adj. (geh.)* esteemed; *(als Anrede)* my dear ...; **etw./nichts ~ sein** be
worth sth./be worthless; **Wert der;
~|e|s, ~e** value; **im ~|e| von ...:**
worth ...; **~ auf etw.** *(Akk.)* **legen** set
great store by *or* on sth.; **wert·beständig** *Adj.* of lasting value *postpos.;* **werten** *tr., itr. V.* judge; assess

wert-, Wert-: **~gegenstand der**
valuable object; **~gegenstände** valuables; **~los** *Adj.* worthless; valueless;
~papier das *(Wirtsch.)* security;
~sache die valuable item; **~sachen**
valuables; **~sendung die** *(Postw.)*
registered item

Wertung die; **~,** **~en** judgement;
wert·voll *Adj.* valuable; *(moralisch)*
estimable

Wesen das; **~s** nature; **wesentlich
1.** *Adj.* fundamental (**für** to); **im ~en**
essentially; **2.** *adv. (erheblich)* considerably; much

wes·halb *Adv. s.* **warum**

Wespe die; **~,** **~n** wasp

wessen *Interrogativpron.* **a)** *Gen. von*
wer whose; **b)** *Gen. von* **was: ~ wird er
beschuldigt?** what is he accused of?

Wessi der; **~s,** **~s** *(salopp)* West German

West *o. Art.; o. Pl. (bes. Seemannsspr.,
Met.)* *s.* **Westen;** **west·deutsch**
Adj. Western German; *(hist.: auf die
alte BRD bezogen)* West German;
West·deutschland (das) Western
Germany; *(hist.: alte BRD)* West Germany

Weste die; **~,** **~n** waistcoat *(Brit.);*
vest *(Amer.)*

Westen der; **~s** west; **der ~:** the
West; **Western der;** **~|s|,** **~:** western; **West·europa (das)** Western
Europe; **Westfalen (das);** **~s** Westphalia; **westfälisch** *Adj.* Westphalian; **West·indien (das)** the West
Indies *pl.;* **westlich 1.** *Adj.* **a)** western; **b)** *(nach Westen)* westerly; **c)** *(aus*

dem Westen) Western; **2.** *adv.* west-
wards; **3.** *Präp. mit Gen.* [to the] west
of; **wẹst·wärts** *Adv.* [to the] west;
Wẹst·wind der west[erly] wind
wes·wẹgen *Adv. s.* **warum**
Wẹtt·bewerb der; ~|e|s, ~e **a)** com-
petition; **b)** *o. Pl. (Wirtsch.)* competition *no indef. art.;* **Wẹtte** die; ~, ~n
bet; **eine ~** |mit jmdm.| abschließen
make a bet [with sb.]; **mit jmdm. um
die ~** laufen race sb.; **wẹtt·eifern**
itr. V. **mit jmdm.** |um etw.| ~: compete
with sb. [for sth.]; **wẹtten** *itr. V.* bet;
mit jmdm. ~: have a bet with sb.; **mit
jmdm. um etw.** ~: bet sb. sth.
Wẹtter das; ~s weather
Wẹtter-: ~**aussichten** *Pl.* weather
outlook *sing.;* ~**bericht** der weather
report; *(Vorhersage)* weather forecast;
~**karte** die weather-chart; weather-
map; ~**lage** die weather situation;
~**vorhersage** die weather forecast;
~**warte** die weather station
wẹtt-, Wẹtt-: ~**kampf** der competi-
tion; ~**lauf** der race; ~|**machen** *tr.
V.* make up for (**durch** with); ~**rennen** das race; ~**rüsten** das; ~s arms
race; ~**streit** der contest
wẹtzen *tr. V.* sharpen; whet
WEZ *Abk.* **Westeuropäische Zeit** GMT
Whiskey ['vɪski] der; ~s ~s whiskey;
Whisky ['vɪski] der; ~s, ~s whisky
wịch *1. u. 3. Pers. Sg. Prät. v.* **weichen**
wịchtig *Adj.* important; **Wịchtig-
keit** die; ~ importance
Wịcke die; ~, ~n vetch; *(im Garten)*
sweet pea
Wịckel der; ~s, ~: compress;
wịckeln *tr. V.* wind; *(ein~)* wrap (**in**
+ *Akk.* in); *(aus~)* unwrap (**aus** +
Dat. from); *(ab~)* unwind (**von** from);
ein Kind ~: change a baby's nappy
Wịdder der; ~s, ~: **a)** ram; **b)** *(Astrol.)*
Aries
wịder *Präp. mit Akk. (geh.)* against
wider-: ~**fạhren** *unr. itr. V.; mit sein
(geh.)* etw. ~**fährt** jmdm. sth. happens
to sb.; ~**legen** *tr. V.* etw. ~**legen** refute sth.; **jmdn.** ~**legen** prove sb.
wrong
wịderlich 1. *Adj.* revolting; repulsive
(person, behaviour, etc.); awful
(headache etc.); **2.** *adv.* revoltingly;
(behave) in a repugnant *or* repulsive
manner; awfully *(cold, sweet, etc.)*
wịder-, Wịder-: ~**rede** die: keine
~**rede!** don't argue!; ~**ruf** der retrac-
tion; |**bis**| **auf** ~**ruf** until revoked;
~**rufen** [--'--] *unr. tr., auch itr. V.* re-

tract *(statement, claim, confession,
etc.);* ~**setzen** [--'--] *refl. V.* **sich**
jmdm./einer Sache ~**setzen** oppose
sb./sth.; ~**spenstig 1.** *Adj.* unruly;
stubborn *(horse, mule, etc.);* **2.** *adv.*
wilfully; ~|**spiegeln,** ~**spiegeln**
[--'--] **1.** *tr. V.* mirror; *(fig.)* reflect; **2.**
refl. V. be mirrored; *(fig.)* be reflected; ~**sprechen** [--'--] *unr. itr. V.*
contradict; ~**spruch der a)** *o. Pl. (Widerrede, Protest)* opposition; protest;
b) *(etw. Unvereinbares)* contradiction;
~**sprüchlich** *Adj.* contradictory
(news, statements, etc.); inconsistent
(behaviour, attitude, etc.)
Wider·stand der a) resistance (**gegen**
to); **b)** *(Hindernis)* opposition
widerstands-: ~**fähig** *Adj.* robust;
resistant *(material etc.);* hardy *(animal, plant);* ~**los** *Adj., adv.* without
resistance postpos.
wider-: ~**stehen** [--'--] *unr. itr. V.* **a)**
(nicht nachgeben) |jmdm./einer Sache|
~**stehen** resist [sb./sth.]; **b)** *(standhalten)* jmdm./einer Sache ~**stehen** with-
stand sb./sth.; ~**streben** [--'--] *itr. V.*
etw. ~**strebt** jmdm. sb. dislikes *or* detests sth.; ~**wärtig 1.** *Adj.* revolting,
repugnant *(smell, taste, etc.);* offensive *(person, behaviour, etc.);* **2.** *adv.*
(behave etc.) in an offensive manner;
~**wille** der aversion (**gegen** to); ~**willig** *adv.* reluctantly; unwillingly
wịdmen 1. *tr. V.* **a)** dedicate; **b)** *(verwenden für/auf)* devote; **2.** *refl. V.* **sich**
jmdm./einer Sache ~: attend to sb./
sth.; *(ausschließlich)* devote oneself to
sb./sth.; **Wịdmung** die; ~, ~en dedication (**an** + *Akk.* to)
wịdrig *Adj.* unfavourable; adverse
wie 1. *Interrogativadv.* how; ~ |**bitte**|?
[I beg your] pardon?; ~ **spät ist es?**
what time is it?; **2.** *Relativadv.* ~ **er es
tut** the way *or* manner in which he
does it; **3.** *Konj.* **a)** *Vergleichspartikel*
as; |**so**| **... ~ ...:** as ... as ...; **ich fühlte
mich ~ ...:** I felt as if I were ...; „**N“** ~
„**Nordpol“** N for November; **b)** *(zum
Beispiel)* like; such as; **c)** *(und, sowie)*
as well as; both
wieder *Adv.* again; **alles ist ~ beim al-
ten** everything is back as it was before; **ich bin gleich ~ da** I'll be right
back *(coll.)*
wieder-, Wieder-: ~|**bekommen**
unr. tr. V. get back; ~|**beleben** *tr. V.*
revive, resuscitate *(person);* ~**bele-
bungs·versuch** der attempt at re-
suscitation; ~|**erkennen** *unr. tr. V.*

recognize; ~|**finden** *unr. tr. V.* find again; ~**gabe** *die (Bericht)* report; *(Übersetzung)* rendering; *(Reproduktion)* reproduction; ~|**geben** *unr. tr. V.* **a)** *(zurückgeben)* give back; **b)** *(berichten)* report; *(wiederholen)* repeat

wieder·gut|machen *tr. V.* make good; put right; **den Schaden ~** *(bezahlen)* pay for the damage

wieder|haben *untr. tr. V. (auch fig.)* have back

wieder-: ~**her|stellen** *tr. V.* **a)** re-establish ⟨*contact, peace*⟩; **b)** *(reparieren)* restore ⟨*building*⟩; ~**holen** 1. *tr. V.* repeat; *(repetieren)* revise ⟨*lesson, vocabulary, etc.*⟩; **2.** *refl. V.* **a)** *(wieder dasselbe sagen)* repeat oneself; **b)** *(erneut geschehen)* happen again; **c)** *(wiederkehren)* be repeated; recur

wieder|holen *tr. V.* fetch or get back

wiederholt 1. *Adj.* repeated; **2.** *adv.* repeatedly; **Wiederholung** *die;* ~, ~**en** repetition; *(eines Fußballspiels usw.)* replay; *(einer Sendung)* repeat; *(einer Aufführung)* repeat performance; *(von Lernstoff)* revision

Wieder·hören *das;* |**auf**| ~! goodbye! *(at end of telephone call)*

wieder-, Wieder-: ~**kehr** *die;* ~ *(geh.)* return; ~|**kehren** *itr. V.; mit sein (geh.)* return; ~|**kommen** *unr. itr. V.; mit sein* **a)** *(zurückkommen)* return; come back; **b)** *(noch einmal kommen)* come back or again; **c)** *(sich noch einmal ereignen)* ⟨*opportunity, past*⟩ come again; ~|**kriegen** *tr. V. (ugs.)* get back; ~|**schauen** *das:* |**auf**| ~**schauen!** *(südd., österr.)* goodbye!; ~|**sehen** *unr. tr. V.* see again; ~**sehen** *das;* ~**s,** ~: reunion; |**auf**| ~**sehen!** goodbye!; ~**um** *Adv.* **a)** *(erneut)* again; **b)** *(andererseits)* on the other hand; ~**wahl** *die* re-election; ~|**wählen** *tr. V.* re-elect

Wiege *die;* ~, ~**n** *(auch fig.)* cradle

¹**wiegen** *unr. itr., tr. V.* weigh;

²**wiegen** *tr. V.* rock; shake ⟨*head*⟩

Wiegen·lied *das* lullaby; cradle-song

wiehern *itr. V.* whinny; *(lauter)* neigh

Wien *(das);* ~**s** Vienna; ¹**Wiener** *der;* ~**s,** ~: Viennese; ²**Wiener** *Adj.* Viennese; *s. auch* Würstchen; **Wienerin** *die;* ~, ~**nen** Viennese; **wienerisch** *Adj.* Viennese

wies *1. u. 3. Pers. Sg. Prät. v.* weisen

Wiese *die;* ~, ~**n** meadow; *(Rasen)* lawn

wie·so *Interrogativadv.* why

wie·viel [*od.* '--] *Interrogativpron.*

(+ *Sg.*) how much; (+ *Pl.*) how many; ~ **Uhr ist es?** what time is it?

wie·viel·mal [*od.* ·'--] *Interrogativadv.* how many times

wievielt... [*od.* '--] *Interrogativadj.* **der ~e Band?** which number volume?; **der Wievielte ist heute?** what is the date today?

wie·weit *Interrogativadv.* to what extent; how far

wild 1. *Adj. (auch fig.)* wild; *(wütend)* furious ⟨*cursing, shouting, etc.*⟩; ~**es Parken** illegal parking; ~**er Streik** wildcat strike; ~ **auf etw./jmdn. sein** *(ugs.)* be mad or crazy about sth./sb. *(coll.)*; ~ **werden** get furious; **jmdn. ~ machen** infuriate sb.; **2.** *adv.* **a)** wildly; **wie** ~ *(ugs.)* like mad *(coll.)*; **b)** *(ordnungswidrig)* illegally; **Wild** *das;* ~|**e**|**s** *(Tiere, Fleisch)* game; **b)** *(einzelnes Tier)* [wild] animal; **Wild·bret** [~brɛt] *das;* ~**s** *(geh.)* game; **Wilde** *der/die; adj. Dekl.* savage; **Wilderer** *der;* ~**s,** ~: poacher; **wild·fremd** *Adj.* completely strange; ~**e Leute** complete strangers; **Wildheit** *die;* ~: wildness; **Wild·leder** *das* suede; **Wildnis** *die;* ~, ~**se** wilderness

Wild-: ~**schwein** *das* wild boar; ~**wechsel** *der o. Pl.* game crossing; ~**west·film** *der* western

will [vɪl] *1. u. 3. Pers. Sg. Präsens v.* wollen

Wille *der;* ~**ns** will; *(Wunsch)* wish

willen *Präp. mit Gen.* **um jmds./einer Sache** ~: for sb.'s/sth.'s sake; **Willen** *der;* ~**s** *s.* **Wille; willen·los** 1. *Adj.* will-less; **2.** *adv.* will-lessly; **willens** *Adj.* ~ **sein, etw. zu tun** *(geh.)* be willing to do sth.; **willens·stark** *Adj.* strong-willed; **willentlich** 1. *Adj.* deliberate; **2.** *adv.* deliberately; on purpose; **willig** 1. *Adj.* willing; **2.** *adv.* willingly

will·kommen *Adj.* welcome; **jmdn.** ~ **heißen** welcome sb.

Will·kür *die;* ~: arbitrary use of power; *(Handlung o. ä.)* arbitrariness; **willkürlich** 1. *Adj.* arbitrary; *(vom Willen gesteuert)* voluntary ⟨*muscle, movement, etc.*⟩; **2.** *adv.* arbitrarily; *(vom Willen gesteuert)* voluntarily

wimmeln *itr. V.* **von Fehlern** ~: be teeming with mistakes

wimmern *itr. V.* whimper

Wimpel *der;* ~**s,** ~: pennant

Wimper *die;* ~, ~**n** [eye]lash

Wind *der;* ~|**e**|**s,** ~**e** wind; **Windbeutel** *der* cream puff

Winde die; ~, ~n winch
Windel die; ~, ~n nappy *(Brit.)*;
diaper *(Amer.)*; **Windel·höschen**
das nappy pants *pl.*
winden 1. *unr. tr. V. (geh.)* make
⟨*wreath, garland*⟩; etw. um etw. ~:
wind sth. around sth.; 2. *unr. refl. V.*
⟨*plant, tendrils*⟩ wind (um around);
⟨*snake*⟩ coil [itself], wind itself (um
around); sich vor Schmerzen~: writhe
in pain
Windes·eile die: in ~: in next to no
time; **Wind·hund** der greyhound;
windig *Adj.* windy
Wind-: ~**mühle** die windmill;
~**pocken** *Pl.* chicken-pox *sing.*;
~**schutz·scheibe** die windscreen
(Brit.); windshield *(Amer.)*; ~**stärke**
die: ~**stärke** 7/9 *usw.* wind force 7/9
etc.; ~**still** *Adj.* windless; still;
~**stoß** der gust of wind; ~**surfing**
das windsurfing *no art.*
Windung die; ~, ~en a) bend; b) *(spiralförmiger Verlauf)* spiral; *(einer Spule o. ä.)* winding
Wink der; ~[e]s, ~e sign; *(Hinweis)*
hint; *(Ratschlag)* tip; hint
Winkel der; ~s, ~ a) *(Math.)* angle;
toter ~: blind spot; b) *(Ecke; auch
fig.)* corner; **winkelig** *Adj.* twisty
⟨*streets*⟩
winken 1. *itr. V.* a) wave; mit etw. ~:
wave sth.; b) *(auffordern heranzukommen)* jmdm. ~: beckon sb. over; einem
Taxi ~: hail a taxi; 2. *tr. V.* beckon;
jmdn. zu sich ~: beckon sb. over [to
one]
winklig *Adj. s.* winkelig
winseln *itr. V.* ⟨*dog*⟩ whimper
Winter der; ~s, ~: winter; **Winteranfang** der beginning of winter;
winterlich 1. *Adj.* wintry; winter *attrib.* ⟨*clothing, break*⟩; 2. *adv.* ~ kalt
cold and wintry
Winter-: ~**reifen** der winter tyre;
~**schlußverkauf** der winter sale[s
pl.]; ~**sport** der winter sports *pl.*;
~**zeit** die; *o. Pl.* winter-time
Winzer der; ~s, ~winegrower
winzig 1. *Adj.* tiny; 2. *adv.* ~ klein
tiny; minute
Wipfel der; ~s, ~: tree-top
Wippe die; ~, ~n see-saw; **wippen**
itr. V. bob up and down; *(hin und her)*
bob about; *(auf einer Wippe)* see-saw
wir *Personalpron.*; 1. *Pers. Pl. Nom.*
we; *s. auch (Gen.)* unser; *(Dat.)* uns;
(Akk.) uns
wirb *Imperativ Sg. v.* werben

Wirbel der; ~s, ~ a) *(kreisende Bewegung) (im Wasser)* whirlpool; *(in der
Luft)* whirlwind; *(kleiner)* eddy; *(von
Rauch, beim Tanz)* whirl; b) *(Trubel)*
hurly-burly; c) *(Aufsehen)* fuss; d)
(Anat.) vertebra; **wirbeln** 1. *itr. V.
mit sein* whirl; ⟨*water, snowflakes*⟩
swirl; 2. *tr. V.* swirl ⟨*leaves, dust*⟩;
whirl ⟨*dancer*⟩
Wirbel-: ~**säule** die spinal column;
~**sturm** der cyclone
wirbt 3. *Pers. Sg. Präsens v.* werben
wird 3. *Pers. Sg. Präsens v.* werden
wirf *Imperativ Sg. v.* werfen; **wirft** 3.
Pers. Sg. Präsens v. werfen
wirken *itr. V.* a) *(eine Wirkung haben)*
have an effect; gegen etw. ~: be effective against sth.; b) *(erscheinen)* seem;
appear
wirklich 1. *Adj.* real; 2. *Adv.* really;
Wirklichkeit die; ~, ~en reality
wirksam 1. *Adj.* effective; 2. *adv.* effectively; **Wirksamkeit** die; ~: effectiveness; **Wirk·stoff** der active
agent; **Wirkung** die; ~, ~en effect
(auf + *Akk.* on); mit ~ vom 1. Juli
(Amtsspr.) with effect from 1 July
wirkungs-: ~**los** 1. *Adj.* ineffective;
2. *adv.* ineffectively; ~**voll** 1. *Adj.* effective; 2. *adv.* effectively
wirr *Adj. (unordentlich)* tousled ⟨*hair,
beard*⟩; tangled ⟨*ropes, roots*⟩; *(unklar, verwirrt)* confused; **Wirren** *Pl.*
turmoil *sing.*; **Wirrwarr** der; ~s
chaos; *(von Stimmen)* clamour
Wirsing der; ~s, **Wirsingkohl** der
savoy [cabbage]
Wirt der; ~[e]s, ~e landlord; **Wirtin**
die; ~, ~nen landlady
Wirtschaft die; ~, ~en a) economy;
(Geschäftsleben) commerce and industry; b) *(Gast~)* public house; pub
(Brit. coll.); bar *(Amer.)*; c) *(Haushalt)*
household; d) *o. Pl. (ugs. abwertend:
Unordnung)* mess; shambles *sing.*;
wirtschaften *itr. V.* mit dem Geld
gut ~: manage one's money well; mit
Verlust/Gewinn ~: run at a loss/
profit; **wirtschaftlich** 1. *Adj.* a)
economic; b) *(finanziell)* financial; c)
(sparsam, rentabel) economical; 2.
adv.; s. Adj.: economically; financially; **Wirtschaftlichkeit** die; ~:
economic viability
Wirtschafts-: ~**hilfe** die economic
aid *no indef. art.*; ~**krise** die economic crisis; ~**minister** der minister
for economic affairs; ~**politik** die
economic policy

Wirts: ~**haus** das pub *(Brit. coll.);*
~**leute** *Pl.* landlord and landlady
Wisch der; ~|e|s, ~e *(salopp)* piece *or*
bit of paper; **wischen** *itr., tr. V.*
wipe; **Staub** ~: do the dusting; dust
wispern *itr., tr. V.* whisper
wiß-, Wiß-: ~**begier, ~begierde**
die; *o. Pl.* thirst for knowledge; ~**begierig** *Adj.* eager for knowledge;
⟨*child*⟩ eager to learn
wissen 1. *unr. tr. V.* know; **von jmdm./
etw.** nichts |mehr| ~ wollen want to
have nothing [more] to do with sb./
sth.; **2.** *unr. itr. V.* **von etw./um etw.** ~ :
know about sth.; **Wissen** das; ~s
knowledge; **meines/unseres** ~s to my/
our knowledge; **Wissenschaft** die;
~, ~en science; **Wissenschaftler**
der; ~s ~, **Wissenschaftlerin** die;
~, ~nen academic; *(Natur~)* scientist; **wissenschaftlich 1.** *Adj.* scholarly; *(natur~)* scientific; **2.** *adv.* in a
scholarly manner; *(natur~)* scientifically; **wissens·wert** *Adj.* ~ sein be
worth knowing; **wissentlich 1.** *Adj.*
deliberate; **2.** *adv.* knowingly; deliberately
wittern 1. *itr. V.* sniff the air; **2.** *tr. V.*
get wind of; *(fig.: ahnen)* sense; **Wit-
terung** die; ~, ~en a) *(Wetter)*
weather *no indef. art;* b) *(Jägerspr.)*
(Geruchssinn) sense of smell; *(Geruch)*
scent
Witwe die; ~, ~n widow; ~ **werden** be
widowed; **Witwer** der; ~s, ~ : widower
Witz der; ~es, ~e joke
Witz-: ~**blatt** das humorous magazine; ~**bold** der; ~es, ~e joker
witzig 1. *Adj.* funny; **2.** *adv.* amus-
ingly; **witz·los** *Adj.* a) dull; b) *(ugs.:
sinnlos)* pointless
wo 1. *Adv.* where; **2.** *Konj.* a) *(da, weil)*
seeing that; b) *(obwohl)* although;
when; **wo·anders** *Adv.* somewhere
else; **wo·bei** *Adv.* a) *(interrogativ)* ~
hast du sie ertappt? what did you
catch her doing?; b) *(relativisch)* er
gab sechs Schüsse ab, ~ einer der Täter
getötet wurde he fired six shots – one
of the criminals was killed
Woche die; ~, ~n week; **in dieser/der
nächsten/der letzten** ~ : this/next/last
week; **heute in/vor einer** ~ : a week
today/a week ago today
wochen-, Wochen-: ~**bett** das: **im**
~**bett liegen** be lying in; ~**ende** das
weekend; ~**lang 1.** *Adj.* lasting weeks
postpos; **2.** *adv.* for weeks [on end];

~**tag** der weekday *(including Saturday);* ~**tags** *Adv.* on weekdays [and
Saturdays]
wöchentlich *Adj., adv.* weekly; **Wo-
chen·zeitung** die weekly newspaper; -**wöchig** a) *(... Wochen alt)*
... -week-old; b) *(... Wochen dauernd)*
... week's/weeks'; ...-week; **Wöchne-
rin** die; ~, ~nen woman who has just
given birth
Wodka der; ~s, ~s vodka
wo·durch *Adv.* a) *(interrogativ)* how;
b) *(relativisch)* as a result of which;
wo·für *Adv.* a) *(interrogativ)* for
what; b) *(relativisch)* for which
wog *1. u. 3. Pers. Sg. Prät. v.* wiegen
Woge die; ~, ~n wave
wo·gegen 1. *Adv.* a) *(interrogativ)*
against what; what ... against; b) *(relativisch)* against which; which ...
against; **2.** *Konj.* whereas
wogen *itr. V. (geh.)* ⟨*sea*⟩ surge; *(fig.)*
⟨*corn*⟩ wave
wo·her *Adv.* a) *(interrogativ)* where ...
from; ~ weißt du das? how do you
know that?; b) *(relativisch)* where ...
from; **wo·hin** *Adv.* a) *(interrogativ)*
where [... to]; b) *(relativisch)* where;
wo·hingegen *Konj.* whereas
wohl 1. *Adv.* a) well; **jmdm. ist nicht** ~,
jmd. fühlt sich nicht ~ : sb. does not
feel well; b) *(behaglich)* at ease;
happy; **leb** ~!/**leben Sie** ~! farewell!;
c) *(durchaus)* well; d) *(ungefähr)*
about; **2.** *Partikel* probably; ~ **kaum**
hardly; **Wohl** das; ~|e|s welfare; **auf
jmds.** ~ **trinken** drink sb.'s health; **zum**
~! cheers!
wohl-, Wohl-: ~**auf** [-'-] *Adj.(geh.)*
~**auf sein** be well; ~**befinden** das
well-being; ~**behagen** das sense of
well-being; ~**behalten** *Adj.* safe and
well ⟨*person*⟩; undamaged ⟨*thing*⟩;
~**fahrts·staat** der welfare state;
~**gefallen** das pleasure; ~**gemerkt**
Adv. please note; ~**habend** *Adj.*
prosperous
wohlig 1. *Adj.*pleasant; agreeable; **2.**
adv. ⟨*sigh, purr, etc.*⟩ with pleasure
wohl, Wohl-: ~**klang** der *(geh.)* melodious sound; ~**schmeckend** *Adj.*
(geh.) delicious; ~**stand** der; *o. Pl.*
prosperity; ~**stands·gesellschaft**
die *o. Pl.* affluent society; ~**tat die a)**
(gute Tat) good deed; *(Gefallen)* favour; b) *o. Pl. (Genuß)* blissful relief;
~**tätig** *Adj.* charitable; ~**tuend** *Adj.*
agreeable; ~|**tun** *unr. itr. V.* etw. tut
jmdm. ~ : sth. does sb. good; ~**ver-**

dient *Adj.* well-earned; **~weislich**
Adv. deliberately; **~wollen das; ~s**
goodwill; **~wollend 1.** *Adj.* benevolent; favourable ⟨*judgement, opinion*⟩;
2. *adv.* benevolently; ⟨*judge, consider*⟩
favourably
Wohn·anhänger der caravan; trailer
(Amer.); **wohnen** *itr. V.* live; *(kurzfristig)* stay
wohn-, Wohn-: **~gemeinschaft**
die group sharing a flat *(Brit.)* or
(Amer.) apartment/house; **~haft** *Adj.*
resident (**in** + *Dat.* in); **~heim das**
(für Alte, Behinderte) home; *(für Obdachlose, Lehrlinge)* hostel; *(für Stu-
denten)* hall of residence
wohnlich *Adj.* homely
Wohn-: **~mobil das; ~s,** **~e** motor
home; **~ort der;** *Pl.* **~e** place of
residence; **~sitz der** place of
residence; **ohne festen ~sitz** of no
fixed abode
Wohnung die; ~, ~en a) flat *(Brit.);*
apartment *(Amer.);* **b)** *o. Pl. (Unterkunft)* lodging
Wohn-: **~verhältnisse** *Pl.* living
conditions; **~wagen der** caravan;
trailer *(Amer.);* **~zimmer das** living-
room
wölben 1. *tr. V.* curve; vault, arch
⟨*roof, ceiling*⟩; **2.** *refl. V.* curve;
⟨*bridge, ceiling*⟩ arch; **Wölbung die;**
~, ~en curve; *(einer Decke)* arch;
vault
Wolf der; **~[e]s, Wölfe** wolf
Wolke die; ~, ~n cloud
wolken-, Wolken-: **~bruch der;** *Pl.*
~brüche cloudburst; **~bruch·artig**
Adj. torrential; **~kratzer der** sky-
scraper; **~los** *Adj.* cloudless
wolkig *Adj.* cloudy
Wolle die; ~, ~n wool; **¹wollen** *Adj.*
woollen
²wollen 1. *unr. Modalverb;* **2. Part. ~
etw. tun ~** *(den Wunsch haben, etw. zu
tun)* want to do sth.; *(die Absicht haben, etw. zu tun)* be going to do sth.;
die Wunde will nicht heilen the wound
[just] won't heal; **2.** *unr. itr. V.* **du mußt
nur ~, dann** ... you only have to want
to enough, then ... **ganz wie du willst**
just as you like; *(ugs.)* **ich will nach
Hause** I want to go home; **zu wem ~
Sie?** whom do you want to see?; **3.**
unr. tr. V. want; **das habe ich nicht ge-
wollt** I never meant that to happen
wo·mit *Adv.* **a)** *(interrogativ)* **~
schreibst du?** what do you write
with?; **b)** *(relativisch)* **~ du schreibst**

which *or* that you write with; *(more
formal)* with which you write; **womöglich** *Adv.* possibly; **wo·nach**
Adv. **a)** *(interrogativ)* after what;
what ... after; **~ suchst du?** what are
you looking for?; **b)** *(relativisch)* after
which; which ... after
Wonne die; ~, ~n *(geh.)* bliss *no pl.;*
ecstasy; *(etw., was Freude macht)* joy;
wonnig *Adj.* sweet
woran *Adv.* **a)** *(interrogativ)* **~ denkst
du?** what are you thinking of?; **b)** *(relativisch)* **nichts, ~ man sich anlehnen
könnte** nothing one could lean
against; **worauf a)** *(interrogativ)* **~
wartest du?** what are you waiting for?;
b) *(relativisch)* **etwas, ~ man sich verlassen kann** something one can rely
on; **c)** *(relativisch: woraufhin)* whereupon
woraus *Adv.* **a)** *(interrogativ)* **~
schließt du das?** what do you infer that
from?; **b)** *(relativisch)* **es gab nichts, ~
wir den Wein hätten trinken können**
there was nothing for us to drink the
wine out of
worden *2. Part. v.* werden 2
worin *Adv.* **a)** *(interrogativ)* in what;
what ... in; **b)** *(relativisch)* in which;
which ... in
Wort das; ~[e]s, Wörter/~e a) *Pl.*
Wörter, *(auch:)* **~e** word; **~ für ~:**
word for word; **DM 1 000 (in ~en:
tausend)** DM 1,000 (in words: one
thousand); **b)** *Pl.* **~e** *(Äußerung)*
word; **mir fehlen die ~e** I'm lost for
words; **Dr. Meyer hat das ~:** it's Dr
Meyer's turn to speak; **c)** *Pl.* **~e**
(Spruch) saying; *(Zitat)* quotation; **d)**
Pl. **~e** *(geh.: Text)* words *pl.;* **in ~ und
Bild** in words and pictures; **e)** *Pl.* **~e**
(Versprechen) word; **[sein] ~ halten**
keep one's word; **wort·brüchig**
Adj. **~ werden** break one's word;
Wörter·buch das dictionary
wort-, Wort-: **~getreu** *Adj.* word-
for-word; **~karg 1.** *Adj.* taciturn ⟨*person*⟩; **2.** *adv.* taciturnly; **~laut der**
wording; **im [vollen] ~laut** verbatim
wörtlich 1. *Adj.* **a)** word-for-word; **b)**
*(der eigentlichen Bedeutung entspre-
chend)* literal; **2.** *adv.: s. Adj.:* word
for word; literally
wort-, Wort-: **~los 1.** *Adj.* silent;
wordless; **2.** *adv.* without saying a
word; **~spiel das** play on words;
pun; **~wechsel der** exchange of
words; **~wörtlich** *Adj.* word-for-
word

worüber *Adv.* **a)** *(interrogativ)* over what ...; what ... over; **b)** *(relativisch)* over which; which ... over; **worum** *Adv.* **a)** *(interrogativ)* around what; what ... around; **b)** *(relativisch)* around which; which ... around; **worunter** *Adv.* **a)** *(interrogativ)* under what; what ... under; **b)** *(relativisch)* under which; which ... under; **wo·von** *Adv.* **a)** *(interrogativ)* from where; where ... from; **b)** *(relativisch)* from which; which ... from; **wo·vor** *Adv.* **a)** *(interrogativ)* in front of what; what ... in front of; **b)** *(relativisch)* in front of which; which ... in front of; **wo·zu** *Adv.* **a)** *(interrogativ)* to what; what ... to; *(wofür)* what ... for; **b)** *(relativisch)* ~ du dich auch entschließt whatever you decide on

Wrack das; ~|e|s, ~s *od.* ~e wreck

wrang *1. und 3. Pers. Sg. Prät. v.* **wringen**; **wringen** *unr. tr. V. (bes. nordd.)* wring

Wucher der; ~s profiteering; *(beim Verleihen von Geld)* usury; **wuchern** *itr. V.* **a)** *auch mit sein* ⟨*plants, weeds, etc.*⟩ proliferate, run wild; **b)** *(Wucher treiben)* |mit etw.| ~: profiteer [on sth.]; *(beim Verleihen von Geld)* lend [sth.] at extortionate interest rates; **Wucherung** die; ~, ~en growth

wuchs *1. u. 3. Pers. Sg. Prät. v.* **wachsen**; **Wuchs** der; ~es *(Gestalt)* stature

Wucht die; ~ force; *(von Schlägen)* power; weight; **wuchtig 1.** *Adj.* **a)** *(voller Wucht)* powerful; mighty; **b)** *(schwer, massig)* massive; **2.** *adv.* powerfully

wühlen 1. *itr. V.* **a)** dig; *(mit der Schnauze, dem Schnabel)* root (**nach** for); ⟨*mole*⟩ tunnel, burrow; **b)** *(ugs.: suchen)* rummage [around] (**nach** for); **2.** *tr. V.* burrow; tunnel out ⟨*burrow*⟩

wulstig *Adj.* bulging

wund *Adj.* sore; **Wunde die;** ~, ~n wound

wunder *Adv. (ugs.)* **er denkt, er sei** ~ **wer** he thinks he's really something; **Wunder das;** ~s, ~ a) miracle; ~ **wirken** *(fig. ugs.)* work wonders; **ein/kein** ~ **sein** *(ugs.)* be a/no wonder; **b)** *(etw. Erstaunliches)* wonder; **wunderbar 1.** *Adj.* **a)** miraculous; **b)** *(sehr schön, herrlich)* wonderful; marvellous; **2.** *adv. (sehr schön, herrlich)* wonderfully; marvellously; **b)** *(ugs.: sehr)* wonderfully

Wunder-: ~**kerze** die sparkler; ~**kind** das child prodigy

wunderlich 1. *Adj.* strange; odd; **2.** *adv.* strangely; oddly; **wundern 1.** *tr. V.* surprise; **mich wundert** *od.* **es wundert mich, daß** ...: I'm surprised that ...; **2.** *refl. V.* **sich über jmdn./etw.** ~: be surprised at sb./sth.

wunder-: ~**schön 1.** *Adj.* simply beautiful; *(herrlich)* simply wonderful; **2.** *adv.* quite beautifully; ~**voll 1.** *Adj.* wonderful; **2.** *adv.* wonderfully

wund|liegen *unr. refl. V.* get bedsores (**an** + *Dat.* on); **Wund·starrkrampf** der *(Med.)* tetanus

Wunsch der; ~|e|s, Wünsche wish (**nach** to have); *(Sehnen)* desire (**nach** for); **haben Sie |sonst| noch einen** ~? will there be anything else?; **auf jmds.** ~: at sb.'s wish; **mit den besten/herzlichsten Wünschen** with best/warmest wishes; **wünschen** *tr. V.* **a) sich** *(Dat.)* **etw.** ~: want sth.; *(im stillen)* wish for sth.; **b)** *(in formelhaften Wünschen)* wish; **jmdm. alles Gute/frohe Ostern** ~: wish sb. all the best/a happy Easter; **c)** *auch itr. V. (begehren)* want; **was** ~ **Sie?, Sie** ~? *(im Lokal)* what would you like?; *(in einem Geschäft)* can I help you?

Wunsch-: ~**kind** das wanted child; ~**konzert** das request concert; *(im Rundfunk)* request programme; ~**zettel** der *(zum Geburtstag o. ä.)* list of presents one would like

wurde *1. u. 3. Pers. Prät. v.* **werden**; **würde** *1. u. 3. Pers. Sg. Konjunktiv II v.* **werden**

Würde die; ~ dignity; **würde·los 1.** *Adj.* undignified; *(schimpflich)* disgraceful; **2.** *adv.* in an undignified way; *(schimpflich)* disgracefully; **Würden·träger** der dignitary; **würde·voll 1.** *Adj.* dignified; **2.** *adv.* with dignity; **würdig 1.** *Adj.* **a)** dignified; **b)** *(wert)* worthy; **2.** *adv.* **a)** with dignity; **b)** *(angemessen)* worthily; **würdigen** *tr. V.* **a)** *(anerkennen, beachten)* recognize; *(schätzen)* appreciate; *(lobend hervorheben)* acknowledge; **b)** *(für wert halten)* **jmdn. keines Blickes/keiner Antwort** ~: not deign to look at/answer sb.

Wurf der; ~|e|s, Würfe **a)** throw; *(beim Kegeln)* bowl; **b)** *o. Pl. (das Werfen)* throwing/pitching/bowling; **c)** *(Zool.)* litter

Würfel der; ~s, ~ cube; *(Spiel~)* dice; die *(formal)*; **Würfel·becher** der dice-cup; **würfeln 1.** *itr. V.* throw the dice; **um etw.** ~: play dice for sth.; **2.**

tr. V. **a)** throw; **b)** *(in Würfel schneiden)* dice
Würfel-: ~**spiel das** dice; *(Brettspiel)* dice game; ~**zucker der;** *o. Pl.* cube sugar
würgen 1. *tr. V.* strangle; throttle; **2.** *itr. V. (Brechreiz haben)* retch
Wurm der; ~|e|s, Würmer worm; *(Made)* maggot; **wurmig** *Adj.,* **wurmstichig** *Adj.* worm-eaten; *(madig)* maggoty
Wurst die; ~, Würste sausage; **es geht um die** ~ *(fig. ugs.)* the crunch has come; **jmdm. ist jmd./etw.** ~ *(ugs.)* sb. doesn't care about sb./sth.; **Würstchen das;** ~s, ~ **a)** [small] sausage; **Frankfurter/Wiener** ~: frankfurter/ wienerwurst; **b)** *(fig. ugs.)* nobody; *(hilfloser Mensch)* poor soul; **Würstchen·bude die** sausage-stand
Würze die; ~, ~n spice; seasoning
Wurzel die; ~, ~n *(auch fig.)* root; **wurzeln** *itr. V.* take root
würzen *tr. V.* season; **würzig** *Adj.* tasty; full-flavoured ⟨*beer, wine*⟩; aromatic ⟨*fragrance*⟩; tangy ⟨*air*⟩
wusch *1. u. 3. Pers. Sg. Prät. v.* **waschen**
wußte *1. und 3. Pers. Sg. Prät. v.* **wissen; wüßte** *1. und 3. Pers. Sg. Konjunktiv II v.* **wissen**
wüst 1. *Adj.* **a)** *(öde)* desolate; **b)** *(unordentlich)* chaotic; **c)** *(ungezügelt)* wild; *(unanständig)* rude; **2.** *adv.* **a)** *(unordentlich)* chaotically; **b)** *(ungezügelt)* wildly
Wüste die; ~, ~n desert
Wut die; ~: rage; fury; **wüten** *itr. V. (auch fig.)* rage; *(zerstören)* wreak havoc; **wütend** *1. Adj.* furious; angry ⟨*voice, mob*⟩; **2.** *adv.* furiously; in a fury

X

¹**x, X** [ɪks] **das;** ~, ~: x, X
²**x** *unbest. Zahlwort (ugs.)* umpteen *(coll.)*
x-Achse die *(Math.)* x-axis
X-Beine *Pl.* knock-knees

x-beliebig *Adj. (ugs.)* irgendein ~er/ irgendeine ~e/irgendein ~es any old *(coll. attrib.);* jeder ~e Ort any old place *(coll.)*
x-fach 1. *Vervielfältigungsz.* die ~e Menge *(Math.)* x times the amount; *(ugs.)* umpteen times the amount *(coll.);* **2.** *adv. (ugs.)* ~ erprobt sein ⟨*tested etc.*⟩ umpteen times *(coll.);* **x-mal** *Adv. (ugs.)* umpteen times *(coll.)*
x-t... *Ordinalz. (ugs.)* umpteenth *(coll.)*

Y

y, Y ['ʏpsilɔn] **das;** ~, ~: y, Y
y-Achse die *(Math.)* y-axis
Yacht *s.* **Jacht**
Yoga *s.* **Joga**
Ypsilon das; ~|s|, ~s y, Y; *(im griechischen Alphabet)* upsilon

Z

z, Z [tsɛt] **das;** ~, ~: z, Z
Zacke die; ~, ~n point; peak; *(einer Säge, eines Kamms)* tooth; *(einer Gabel, Harke)* prong; **Zacken der;** ~s, ~ *s.* **Zacke**
zaghaft 1. *Adj.* timid; *(zögernd)* hesitant; **2.** *adv.* timidly; *(zögernd)* hesitantly; **Zaghaftigkeit die;** ~: timidity; *(Zögern)* hesitancy
zäh 1. *Adj.* **a)** tough; heavy ⟨*dough, soil*⟩; *(dickflüssig)* glutinous; viscous ⟨*oil*⟩; **b)** *(widerstandsfähig)* tough ⟨*person*⟩; **c)** *(beharrlich)* tenacious; tough ⟨*negotiations*⟩; dogged ⟨*resistance*⟩; **2.** *adv. (beharrlich)* tenaciously; ⟨*resist*⟩ doggedly; **Zähigkeit die;** ~ **a)** *(Widerstandsfähigkeit)* toughness; **b)** *(Be-*

harrlichkeit) tenacity; **mit** ~: tenaciously

Zahl die; ~, ~**en** number; *(Ziffer)* numeral; *(Zahlenangabe, Geldmenge)* figure; **in den roten/schwarzen** ~**en** in the red/black; **zahlbar** *Adj. (Kaufmannsspr.)* payable; **zahlen 1.** *tr. V.* pay **(an** + *Akk.* to); **2.** *itr. V.* pay; ~ **bitte!** *(im Lokal)* [can I/we have] the bill, please!; **zählen 1.** *itr. V.* **a)** count; **zu einer Gruppe** *usw.* ~: be one of *or* belong to a group *etc.;* **b) auf jmdn./etw.** ~: count on sb./sth; **2.** *tr. V.* count; **jmdn. zu seinen Freunden** ~: count sb. among one's friends

zahl-, Zahl-: ~**karte die** *(Postw.)* paying-in slip; ~**los** *Adj.* countless; ~**reich** *Adj.* numerous

Zahlung die; ~, ~**en** payment; **Zählung die;** ~, ~**en** counting; eine ~: a count; **Zahlungs·mittel das** means of payment; **Zahl·wort das;** *Pl.* ~**wörter** *(Sprachw.)* numeral

zahm 1. *Adj.* tame; **2.** *adv.* tamely; **zähmen** *tr. V. (auch fig.)* tame

Zahn der; ~**[e]s, Zähne** tooth; *(Raubtier~)* fang; *(an einer Briefmarke usw.)* serration

Zahn-: ~**arzt** der dentist; *(mit chirurgischer Ausbildung)* dental surgeon; ~**bürste** die toothbrush

zahnen *itr. V. ⟨baby⟩* be teething

zahn-, Zahn-: ~**fleisch das** gum; *(als Ganzes)* gums *pl.;* ~**los** *Adj.* toothless; ~**lücke die** gap in one's teeth; ~**pasta die;** ~, ~**pasten** toothpaste; ~**prothese die** dentures *pl.;* [set *sing.* of] false teeth *pl.;* ~**schmerzen** *Pl.* toothache *sing.;* ~**stocher der;** ~**s,** ~: toothpick; ~**weh das;** *o. Pl. (ugs.)* toothache

Zange die; ~, ~**n a)** *(Werkzeug)* pliers *pl.; (Eiswürfel~, Zucker~)* tongs *pl.; (Geburts~)* forceps *pl.; (Kneif~)* pincers *pl.;* **eine** ~: a pair of pliers/tongs/forceps/pincers; **b)** *(bei Tieren)* pincer

Zank der; ~**[e]s** squabble; row; **zanken** *refl. (auch itr.) V.* squabble, bicker **(um** *od.* **über** + *Akk.* over); **zänkisch** *Adj.* quarrelsome

Zäpfchen das; ~**s,** ~ suppository; **zapfen** *tr. V.* tap, draw *⟨beer, wine⟩;* **Zapfen der;** ~**s,** ~ **a)** *(Bot.)* cone; **b)** *(Stöpsel)* bung; **Zapf·säule die** petrol-pump *(Brit.);* gasoline pump *(Amer.)*

zappeln *itr. V.* wriggle; *⟨child⟩* fidget

Zar der; ~**en,** ~**en** *(hist.)* Tsar; **Zarin die;** ~, ~**nen** *(hist.)* Tsarina

zart 1. *Adj. (auch fig.)* delicate; soft *⟨skin⟩;* tender *⟨bud, shoot; meat, vegetables⟩;* fine *⟨biscuits⟩;* gentle *⟨kiss, touch⟩;* soft *⟨pastel colours⟩;* **2.** *adv. (empfindlich)* delicately; *⟨kiss, touch⟩* gently; **zärtlich 1.** *Adj.* tender; **2.** *adv.* tenderly; **Zärtlichkeit die;** ~, ~**en a)** *o. Pl. (Zuneigung)* tenderness; affection; **b)** *meist Pl. (Liebkosung)* caress

Zauber der; ~**s,** ~ **a)** *(auch fig.)* magic; *(Bann)* [magic] spell; **b)** *o. Pl. (ugs. abwertend: Aufheben)* fuss; **Zauberei die;** ~, ~**en a)** *o. Pl. (das Zaubern)* magic; **b)** *(Zaubertrick)* magic trick; **Zauberer der;** ~**s,** ~: magician; **zauber·haft 1.** *Adj.* enchanting; **2.** *adv.* enchantingly; **Zauberin die;** ~, ~**nen a)** sorceress; **b)** *(Zauberkünstlerin)* conjurer; **Zauber·künstler der** conjurer; magician; **zaubern 1.** *itr. V.* **a)** do magic; **b)** *(Zaubertricks ausführen)* do conjuring tricks; **2.** *tr. V. (auch fig.)* conjure

zaudern *itr. V. (geh.)* delay

Zaum der; ~**[e]s, Zäume** bridle; **zäumen** *tr. V.* bridle; **Zaum·zeug das** bridle

Zaun der; ~**[e]s, Zäune** fence; **Zaun·könig der** wren

z. B. *Abk.* zum Beispiel e.g.

ZDF [tsɛt|deː'|ɛf] *das;* ~ *Abk.* Zweites Deutsches Fernsehen Second German Television Channel

Zebra das; ~**s,** ~**s** zebra; **Zebra·streifen der** zebra crossing *(Brit.);* pedestrian crossing

Zeche die; ~, ~**n a)** *(Rechnung)* bill *(Brit.);* check *(Amer.);* **b)** *(Bergwerk)* pit; mine; **zechen** *itr. V. (veralt., scherzh.)* tipple

Zeh der; ~**s,** ~**en, Zehe die;** ~, ~**n a)** toe; **b)** *(Knoblauch~)* clove; **Zehen·spitze die: auf** ~**n** on tiptoe

zehn *Kardinalz.* ten; **Zehn die;** ~, ~**en** ten; **Zehner der;** ~**s,** ~ **a)** *(ugs.: Geldschein, Münze)* ten; **b)** *(ugs.: Autobus)* number ten; **c)** *(Math.)* ten; **zehn·fach** *Vervielfältigungsz.* tenfold; **Zehnfache das;** *adj. Dekl.* **das** ~: ten times as much

zehn-, Zehn-: ~**kampf der** *(Sport)* decathlon; ~**mal** *Adv.* ten times; ~**mark·schein der** ten-mark note; ~**pfennig·[brief]marke die** ten-pfennig stamp; ~**pfennig·stück das** ten-pfennig piece

zehnt... *Ordinalz.* tenth; **zehn·tausend** *Kardinalz.* ten thousand;

zehntel *Bruchz.* tenth; **Zehntel das**
(schweiz. meist der); ~s, ~: tenth;
zehntens *Adv.* tenthly
zehren *itr. V.* **von etw.** ~: live on *or* off
sth.
Zeichen das; ~s, ~ sign; *(Markierung)* mark; *(Chemie, Math., auf
Landkarten usw.)* symbol; **jmdm. ein
~ geben** signal to sb.
Zeichen-: ~**setzung** die punctuation; ~**sprache** die sign language
zeichnen 1. *tr. V.* draw; *(fig.)* portray
⟨*character*⟩; **2.** *itr. V.* draw; **Zeichner**
der; ~s, ~, **Zeichnerin** die; ~, ~nen
graphic artist; *(Technik)* draughts-
man/-woman; **Zeichnung** die; ~,
~en drawing
Zeige·finger der index finger; forefinger; **zeigen 1.** *itr. V.* point; **2.** *tr.
V.* show; **3.** *refl. V.* **a)** *(sich sehen lassen)* appear; **b)** *(sich erweisen)* prove
to be; **es wird sich** ~, ...: time will
tell ...; **Zeiger** der; ~s, ~: pointer;
(Uhr~) hand
Zeile die; ~, ~n line; *(Reihe)* row
zeit *Präp. mit Gen.* ~ **meines** *usw./*un-
seres *usw.* **Lebens** all my *etc.* life/our
etc. lives; **Zeit** die; ~, ~en **a)** *o. Pl.*
time *no art.;* **mit der** ~: with *or* in
time; *(allmählich)* gradually; **b)**
(~punkt) time; **zur** ~: at the moment;
c) *(~abschnitt, Lebensabschnitt)* time;
period; *(Geschichtsabschnitt)* age;
period; **d)** *(Sprachw.)* tense
zeit-, Zeit-: ~**alter** das age; era;
~**gemäß** *Adj. (modern)* up-to-date;
(aktuell) topical ⟨*theme*⟩; contemporary ⟨*views*⟩; ; ~**genosse** der, ~**genossin** die contemporary; ~**genös-
sisch** *Adj.* contemporary; ~**ge-
schehen das: das** |aktuelle| ~**gesche-
hen** current events *pl.*
zeitig *Adj., adv.* early
Zeit·lang die: **eine** ~: for a while;
zeit·lebens *Adv.* all one's life;
zeitlich 1. *Adj.* ⟨*length, interval*⟩ in
time; chronological ⟨*order, sequence*⟩;
2. *adv.* with regard to time
zeit-, Zeit-: ~**los 1.** *Adj.* timeless;
classic ⟨*fashion, shape*⟩; **2.** *adv.* time-
lessly; ~**lupe** die; *o. Pl.* slow motion;
~**punkt** der moment; ~**raubend**
Adj. time-consuming; ~**raum** der
period; ~**schrift** die magazine; *(bes.
wissenschaftlich)* journal; periodical;
~**spanne** die period
Zeitung die; ~, ~en [news]paper; *(fig.)*
Zeitungs·notiz die newspaper item
zeit-, ~**Zeit-:** ~**verschwendung**

die waste of time; ~**vertreib** der;
~|e|s, ~e pastime; **zum** ~**vertreib** to
pass the time; ~**weilig 1.** *Adj.* tem-
porary; **2.** *adv.* temporarily; ~**weise**
Adv. (gelegentlich) occasionally; *(von
Zeit zu Zeit)* from time to time;
~**wort das;** *Pl.* ~**wörter** *(Sprachw.)*
verb
Zelle die; ~, ~n cell
Zelluloid [tsɛlu'lɔyt] **das;** ~|e|s cellu-
loid
Zelt das; ~|e|s, ~e tent; *(Fest~)* mar-
quee; *(Zirkus~)* big top; **zelten** *itr.
V.* camp
Zelt-: ~**lager** das camp; ~**plane** die
tarpaulin
Zement der; ~|e|s, ~e cement
Zensur die; ~, ~en mark; grade
(Amer.)
Zenti- [tsɛnti-]: ~**meter** der, *auch:*
das centimetre; ~**meter·maß** das
[centimetre] measuring-tape
Zentner der; ~s, ~ **a)** metric hundredweight; **b)** *(österr., schweiz.) s.* **Dop-
pelzentner**
zentral 1. *Adj.* central; **2.** *adv.* cent-
rally; **Zentrale** die; ~, ~n **a)** *(zentrale
Stelle)* head *or* central office; *(der Polizei, einer Partei)* headquarters *sing.
or pl.; (Funk~)* control centre; **b)** *(Te-
lefon~)* [telephone] exchange; *(eines
Hotels, einer Firma o. ä.)* switchboard;
Zentral·heizung die central heating
Zentren *s.* **Zentrum**
Zentrifugal·kraft die *(Physik)* centrifugal force; **Zentrifuge** die; ~, ~n
centrifuge
Zentrum das; ~s, **Zentren** centre; **im**
~: at the centre; *(im Stadt~)* in the
town/city centre
Zeppelin der; ~s, ~e Zeppelin
Zepter das, *auch:* der; ~s, ~: sceptre
zerbeißen *unr. tr. V.* bite in two
zerbersten *unr. itr. V.; mit sein* burst
apart
zerbrechen 1. *unr. itr. V.; mit sein*
break [into pieces]; smash [to pieces];
⟨*glass*⟩ shatter; *(fig.)* ⟨*marriage, relationship*⟩ break up; **2.** *unr. tr. V.*
break; smash, shatter ⟨*dishes, glass*⟩;
zerbrechlich *Adj.* fragile; *(fig.)* frail
zerbröckeln 1. *itr. V.; mit sein*
crumble away; **2.** *tr. V.* break into
small pieces
zerdrücken *tr. V.* mash
Zeremonie die; ~, ~n ceremony;
(fig.) ritual; **Zeremoniell das;** ~s,
~e ceremonial
zerfallen *unr. itr. V.; mit sein* **a)** *(auch*

fig.) disintegrate (**in** + *Akk., zu* into); ⟨*building*⟩ fall into ruin, decay; ⟨*corpse*⟩ decompose, decay

zerfetzen *tr. V.* rip *or* tear to pieces; *(fig.)* tear apart ⟨*body, limb*⟩

zerfleischen *tr. V.* tear ⟨*person, animal*⟩ limb from limb

zerfressen *unr. tr. V.* **a)** eat away; ⟨*moth etc.*⟩ eat holes in; **b)** *(zersetzen)* corrode ⟨*metal*⟩; eat away ⟨*bone*⟩

zergehen *unr. itr. V.; mit sein* melt; *(in Wasser, im Mund)* ⟨*tablet etc.*⟩ dissolve

zerhacken *tr. V.* chop up (**zu** into)

zerhauen *unr. tr. V.* chop up

zerkleinern *tr. V.* chop up *(zermahlen)* crush ⟨*rock etc.*⟩

zerknautschen *tr. V. (ugs.)* crumple

zerknirscht **1.** *Adj.* remorseful; **2.** *adv.* remorsefully

zerknittern *tr. V.* crease; crumple

zerknüllen *tr. V.* crumple up [into a ball]

zerkratzen *tr. V.* scratch

zerkrümeln *tr. V.* crumble up

zerlegen *tr. V.* **a)** dismantle; take to pieces; **b)** *(zerschneiden)* cut up ⟨*animal, meat*⟩; carve ⟨*joint*⟩

zerplatzen *itr. V.; mit sein* burst

Zerr·bild *das* distorted image

zerreiben *unr. tr. V.* crush

zerreißen **1.** *unr. tr. V.* **a)** tear up; *(in kleine Stücke)* tear to pieces; break ⟨*thread*⟩; **b)** *(beschädigen)* tear ⟨*stocking, trousers, etc.*⟩ (**an** + *Dat.* on); **2.** *unr. itr. V.; mit sein* ⟨*thread, string, rope*⟩ break; ⟨*paper, cloth, etc.*⟩ tear

zerren **1.** *tr. V.* **a)** drag; **b) sich** *(Dat.)* **einen Muskel/eine Sehne** ~: pull a muscle/tendon; **2.** *itr. V.* **an etw.** *(Dat.)* ~: tug *or* pull at sth.; **Zerrung** die; ~, ~en pulled muscle/tendon

zerrütten *tr. V.* ruin; shatter ⟨*nerves*⟩

zerschellen *itr. V.; mit sein* be dashed *or* smashed to pieces

zerschlagen **1.** *unr. tr. V.* smash ⟨*plate, windscreen, etc.*⟩; smash up ⟨*furniture*⟩; *(fig.)* smash ⟨*spy ring etc.*⟩; **2.** *unr. refl. V.* ⟨*plan, deal*⟩ fall through

zerschmettern *tr. V.* smash; shatter ⟨*glass, leg, bone*⟩

zerschneiden *unr. tr. V.* cut; *(in Stükke)* cut up; *(in zwei Teile)* cut in two

zersetzen *tr. V.* corrode ⟨*metal*⟩; decompose ⟨*organism*⟩

zersplittern *itr. V.; mit sein* ⟨*wood, bone*⟩ splinter; ⟨*glass*⟩ shatter

zerspringen *unr. itr. V.; mit sein* shatter; *(Sprünge bekommen)* crack

zerstäuben *tr. V.* spray

zerstören *tr. V.* destroy; ⟨*hooligan*⟩ smash up, vandalize; *(fig.)* ruin ⟨*health, life*⟩; **Zerstörung** die *s.* zerstören: destruction; smashing up; vandalization; *(fig.)* ruin[ation]

zerstreuen **1.** *tr. V.* scatter; disperse ⟨*crowd*⟩; **jmdn./sich** ~ *(ablenken)* take sb.'s/one's mind off things; **2.** *refl. V.* disperse; *(schneller)* scatter; **zerstreut** **1.** *Adj.* distracted; *(vergeßlich)* absent-minded; **2.** *adv.* absentmindedly; **Zerstreuung** die; ~, ~en *(Ablenkung)* diversion

zerstückeln *tr. V.* break ⟨*sth.*⟩ up into small pieces; *(zerschneiden)* cut *or* chop ⟨*sth.*⟩ up into small pieces; dismember ⟨*corpse*⟩

zerteilen *tr. V.* divide into pieces; *(zerschneiden)* cut into pieces; cut up

Zertifikat *das;* ~[e]s, ~e certificate

zertrampeln *tr. V.* trample all over ⟨*flower-bed etc.*⟩; trample ⟨*child etc.*⟩ underfoot

zertreten *unr. tr. V.* stamp on; stamp out ⟨*cigarette, match*⟩

zertrümmern *tr. V.* smash; smash, shatter ⟨*glass*⟩; smash up ⟨*furniture*⟩; wreck ⟨*car, boat*⟩; reduce ⟨*building*⟩ to ruins

Zerwürfnis *das;* ~ses, ~se *(geh.)* quarrel; dispute; *(Bruch)* rift

zerzausen *tr. V.* ruffle; **zerzaust aussehen** look dishevelled

zetern *itr. V.* scold [shrilly]; *(sich beklagen)* moan (**über** + *Akk.* about)

Zettel *der;* ~s, ~: slip *or* piece of paper; *(mit einigen Zeilen)* note; *(Bekanntmachung)* notice; *(Formular)* form; *(Kassen~)* receipt; *(Hand~)* leaflet

Zeug *das;* ~[e]s, ~e **a)** *o. Pl. (ugs.)* stuff; **dummes** ~: nonsense; rubbish; **b)** *(Kleidung)* things *pl.;* **Zeuge** der; ~n, ~n witness

zeugen *tr. V.* procreate; ⟨*man*⟩ father ⟨*child*⟩

Zeugen·aussage die testimony; **Zeugin** die; ~, ~nen witness; **Zeugnis** *das;* ~ses, ~se **a)** *(Schulw.)* report; **b)** *(Arbeits~)* reference; testimonial; **c)** *(Gutachten)* certificate

Zeugung die; ~, ~en procreation; *(eines Kindes)* fathering; **zeugungsfähig** *Adj.* fertile

z. Hd. *Abk.* zu Händen attn.

Zickzack *der;* ~[e]s, ~e zigzag

Ziege die; ~, ~n goat; *(Schimpfwort: Frau)* cow *(sl. derog.)*

Ziegel der; ~s, ~ brick; *(Dach~)* tile;
Ziegel·stein der brick
Ziegen-: ~**bock** der he- *or* billy-goat;
~**käse** der goat's cheese
ziehen 1. *unr. tr. V.* **a)** pull; *(sanfter)*
draw; *(zerren)* tug; *(schleppen)* drag;
etw. nach sich ~ *(fig.)* result in sth.; entail sth.; **b)** *(heraus~)* extract ⟨tooth⟩;
take out, remove ⟨stitches⟩; draw
⟨cord, sword, pistol⟩; **den Hut** ~: raise
one's hat; **die |Quadrat|wurzel** ~
(Math.) extract the square root; **c)**
(dehnen) stretch ⟨elastic etc.⟩; stretch
out ⟨sheets etc.⟩; **d)** *(Gesichtspartien
bewegen)* make ⟨face, grimace⟩; **e)** *(bei
Brettspielen)* move ⟨chess-man etc.⟩; **f)**
(zeichnen) draw ⟨line etc.⟩; **g)** *(anlegen)* dig ⟨trench⟩; build ⟨wall⟩; erect
⟨fence⟩; put up ⟨washing-line⟩; run,
lay ⟨cable, wires⟩; draw ⟨frontier⟩; **h)**
(auf~) grow ⟨plants, flowers⟩; breed
⟨animals⟩; **2.** *unr. itr. V.* **a)** *(reißen)*
pull; **an etw.** *(Dat.)* ~: pull on sth.; **b)**
(funktionieren) ⟨stove, pipe, chimney⟩
draw; **c)** *mit sein (um~)* move (**nach,
in** + *Akk.* to); **d)** *mit sein (gehen)* go;
(marschieren) march; *(umherstreifen)*
roam; *(weggehen)* go away; leave;
⟨fog, clouds⟩ drift; **e)** *(saugen)* draw;
an einer Zigarette/Pfeife ~: draw on a
cigarette/pipe; **f)** *⟨tea, coffee⟩* draw; **g)**
(Kochk.) simmer; **h)** *unpers.* **es zieht**
there's a draught; **3.** *unr. refl. V.*
⟨road⟩ run, stretch; ⟨frontier⟩ run;
Zieh·harmonika die piano accordion; **Ziehung** die; ~, ~en draw
Ziel das; ~|e|s, ~e **a)** destination; **b)**
(Sport) finish; *(~linie)* finishing-line;
(Pferderennen) finishing-post; **c)**
(~scheibe; auch Milit.) target; **d)**
(Zweck) aim; goal; **sein** ~ **erreichen**
achieve one's objective *or* aim;
ziel·bewußt 1. *Adj.* determined; **2.**
adv. determinedly; **zielen** *itr. V.* aim
(auf + *Akk.,* at); *(fig.)* **auf jmdn./etw.**
~ ⟨reproach, efforts, etc.⟩ be aimed at
sb./sth.
ziel-, Ziel-: ~**los 1.** *Adj.* aimless; **2.**
adv. aimlessly; ~**scheibe** die *(auch
fig.)* target (*Gen.* for); ~**strebig 1.**
Adj. **a)** purposeful; **b)** *(energisch)*
single-minded ⟨person⟩; **2.** *adv.* **a)**
purposefully; **b)** *(energisch)* single-
mindedly
ziemlich 1. *Adj. (ugs.)* fair, sizeable
⟨quantity, number⟩; **2.** *adv.* **a)** quite;
fairly; **b)** *(ugs.: fast)* pretty well
Zierde die; ~, ~n *(auch fig.)* ornament; **zieren** *refl. V.* be coy; **zierlich**

1. *Adj.* dainty; petite, dainty ⟨woman,
figure⟩; **2.** *adv.* daintily
Ziffer die; ~, ~n numeral; *(in einer
mehrstelligen Zahl)* digit; figure; **Zif-
fer·blatt das** dial; face
-zig, zig *unbest. Zahlwort (ugs.)* umpteen *(coll.)*
Zigarette die; ~, ~n cigarette; **Ziga-
rillo** der *od.* das; ~, ~s cigarillo;
small cigar; **Zigarre** die; ~, ~n cigar
Zigeuner der; ~s, ~, **Zigeunerin**
die; ~, ~nen gypsy
zig·mal *Adv. (ugs.)* umpteen times
(coll.); **zig·tausend** *unbest. Zahlwort (ugs.)* umpteen thousand *(coll.)*
Zimmer das; ~s, ~: room; **Zimmermädchen das** chambermaid
zimmern *tr. V.* make ⟨shelves etc.⟩;
Zimmer·suche die room-hunt
zimperlich 1. *Adj.* timid; *(leicht angeekelt)* squeamish; *(prüde)* prissy; **2.**
adv.: s. *Adj.:* timidly; squeamishly;
prissily
Zimt der; ~|e|s, ~e cinnamon
Zink das; ~|e|s zinc
Zinke die; ~, ~n prong; *(eines Kam-
mes)* tooth
Zinn das; ~|e|s tin; *(Gegenstände)*
pewter[ware]
Zins der; ~es, ~en interest; **Zin-
ses·zins** der compound interest
zins·los 1. *Adj.* interest-free; **2.** *adv.*
free of interest; **Zins·satz** der interest rate
Zipfel der; ~s, ~ *(einer Decke, eines
Tisch-, Handtuchs usw.)* corner;
(Wurst~, eines Halstuchs) [tail-]end;
Zipfel·mütze die [long-]pointed cap
zirka *Adv.* about; approximately
Zirkulation die; ~, ~en circulation;
zirkulieren *itr. V.; auch mit sein* circulate
Zirkus der; ~, ~se **a)** circus; **b)** *(ugs.)
o. Pl. (Trubel)* hustle and bustle;
(Krach) to-do
zirpen *itr. V.* chirp
zischeln *tr. V.* whisper angrily
zischen *itr. V.* **a)** hiss; ⟨hot fat⟩ sizzle;
b) *mit sein* hiss
Zitat das; ~|e|s, ~e quotation (**aus**
from)
zitieren *tr., itr. V.* **a)** quote;
(Rechtsspr.) cite; **b)** *(rufen)* summon
Zitronat das; ~|e|s candied lemon-
peel; **Zitrone** die; ~, ~n lemon
Zitronen-: ~**limonade die** lemonade; ~**presse die** lemon-squeezer;
~**saft** der lemon-juice
Zitrus·frucht die citrus fruit

zittern itr. V. tremble (vor + Dat. with); (vor Kälte) shiver; (beben) ⟨walls, windows⟩ shake; vor jmdm./ etw. ~: be terrified of sb./sth.; **zittrig** Adj. shaky; doddery ⟨old man⟩
Zitze die; ~, ~n teat
zivil 1. Adj. **a)** civilian; non-military ⟨purposes⟩; civil ⟨aviation, marriage, law, defence⟩; **b)** (annehmbar) decent; **2.** adv. (annehmbar) decently; **Zivil das**; ~s civilian clothes pl.; **Zivil·bevölkerung die** civilian population; **Zivilisation** [t̠sivilizaʹtsi̠oːn] die; ~, ~en civilization; **zivilisieren** tr. V. civilize; **zivilisiert 1.** Adj. civilized; **2.** adv. in a civilized way; **Zivilist der**; ~en, ~en civilian; **Zivil·kleidung die** civilian clothes pl.
Zofe die; ~, ~n (hist.) lady's maid
zog 1. u. 3. Pers. Sg. Prät. v. ziehen
zögern itr. V. hesitate; **ohne zu ~:** without hesitation
Zoll der; ~[e]s, Zölle **a)** [customs] duty; **b)** o. Pl. (Behörde) customs pl.
zoll-, Zoll-: ~**amt das** customs house or office; ~**beamte der** customs officer; ~**frei 1.** Adj. duty-free; free of duty pred.; **2.** adv. free of duty; ~**kontrolle die** customs examination or check; ~**stock der** folding rule
Zone die; ~, ~n zone
Zoo der; ~s, ~s zoo; **Zoologe der**; ~n, ~n zoologist; **Zoologie die**; ~: zoology no art.; **zoologisch** Adj. zoological; ~**er Garten** zoological gardens pl.
Zoom das; ~s, ~s (Film, Fot.: Objektiv) zoom; **Zoom·objektiv das** (Film, Fot.) zoom lens
Zopf der; ~[e]s, Zöpfe plait; (am Hinterkopf) pigtail
Zorn der; ~[e]s anger; (stärker) wrath; fury; **zornig 1.** Adj. furious; **2.** adv. furiously
Zote die; ~, ~n dirty joke; **zotig 1.** Adj. smutty; dirty ⟨joke⟩; **2.** adv. smuttily
zottig Adj. shaggy
zu 1. Präp. mit Dat. **a)** (Richtung) to; **zu ... hin** towards ...; **b)** (zusammen mit) with; **zu dem Käse gab es Wein** there was wine with the cheese; **c)** (Lage) at; **zu beiden Seiten** on both sides; **d)** (zeitlich) at; **zu Weihnachten** at Christmas; **e)** (Art u. Weise) **zu meiner Zufriedenheit/Überraschung** to my satisfaction/surprise; (bei Mengenangaben o. ä.) **zu Dutzenden/zweien** by the dozen/in twos; **f)** (ein Zahlen-

verhältnis ausdrückend) **ein Verhältnis von 3 zu 1** a ratio of 3 to 1; **g)** (einen Preis zuordnend) at; for; **h)** (Zweck) for; **i)** (Ziel, Ergebnis) into; **zu etw. werden** turn into sth.; **j)** (über) about; on; **sich zu etw. äußern** comment on sth.; **k)** (gegenüber) **freundlich/häßlich zu jmdm. sein** be friendly/nasty to sb.; s. auch **zum; zur; 2.** Adv. **a)** (allzu) too; **zu sehr** too much; **b)** nachgestellt (Richtung) towards; **3.** Konj. **a)** (mit Infinitiv) to; **was gibt's da zu lachen?** what is there to laugh about?; **b)** (mit 1. Part.) **die zu erledigende Post** the letters pl. to be dealt with
Zubehör das; ~[e]s, ~e od. schweiz. ~den accessories pl.; (eines Staubsaugers, Mixers o. ä.) attachments pl.; (Ausstattung) equipment
zu|bereiten tr. V. prepare ⟨meal etc.⟩; make up ⟨medicine, ointment⟩; (kochen) cook ⟨fish, meat, etc.⟩
zu|billigen tr. V. **jmdm. etw. ~:** grant or allow sb. sth.
zu|binden unr. tr. V. tie [up]
zu|blinzeln itr. V. **jmdm. ~:** wink at sb.
zu|bringen unr. tr. V. spend; **Zubringer der**; ~s, ~ **a)** (Straße) access road; **b)** (Verkehrsmittel) shuttle
Zucht die; ~, ~en **a)** breeding; (von Pflanzen) cultivation; **ein Pferd aus deutscher ~:** a German-bred horse; **b)** o. Pl. (geh.: Disziplin) discipline; **züchten** tr. V. (auch fig.) breed; cultivate ⟨plants⟩; culture ⟨bacteria, pearls⟩; **Züchter der**; ~s, ~, **Züchterin die**; ~, ~nen breeder; (von Pflanzen) grower [of new varieties]; **Züchtung die**; ~, ~en **a)** breeding; (von Pflanzen) cultivation; **b)** (Zuchtergebnis) strain
zucken itr. V.; mit Richtungsangabe mit sein twitch; ⟨body, arm, leg, etc.⟩ jerk; (vor Schreck) start; ⟨flames⟩ flicker; **mit den Achseln/Schultern ~:** shrug one's shoulders; **zücken** tr. V. draw ⟨sword, dagger, knife⟩
Zucker der; ~s, ~ **a)** sugar; **b)** o. Pl. (ugs.: ~krankheit) diabetes; ~ **haben** be a diabetic
zucker-, Zucker-: ~**dose die** sugar bowl; ~**hut der** sugar loaf; ~**krank** Adj. diabetic
zuckern tr. V. sugar
Zuckung die; ~, ~en twitch
zu|decken tr. V. cover up; cover [over] ⟨well, ditch⟩; **jmdn./sich ~:** tuck sb./oneself up

zu|drehen *tr. V.* **a)** *(abdrehen)* turn off; **b)** *(zuwenden)* **jmdm. den Rücken** ~: turn one's back on sb.

zu·dringlich **1.** *Adj.* pushy *(coll.)*, pushing 〈*person, manner*〉; *(sexuell)* importunate 〈*person, manner*〉; prying 〈*glance*〉; **2.** *adv.* importunately; **Zudringlichkeit die;** ~, ~**en a)** *o. Pl.* pushiness *(coll.)*; *(in sexueller Hinsicht)* importunate manner; **b)** *(Handlung)* ~**en** insistent advances *or* attentions

zu|drücken *tr. V.* press shut; push 〈*door*〉 shut; **jmdm. die Kehle** ~: choke *or* throttle sb.

zu·einander *Adv.* to one another

zu·erst *Adv.* **a)** first; **b)** *(anfangs)* at first; to start with; **c)** *(erstmals)* first

Zu·fahrt die a) *o. Pl.* access [for vehicles]; **b)** *(Straße, Weg)* access road; *(zum Haus)* driveway; **Zufahrts·straße die** access road

Zu·fall der chance; *(zufälliges Zusammentreffen von Ereignissen)* coincidence; **durch** ~: by chance; **zu|fallen** *unr. itr. V.; mit sein* **a)** 〈*door etc.*〉 slam shut; 〈*eyes*〉 close; **b)** *(zukommen)* **jmdm.** ~ 〈*task*〉 fall to sb.; 〈*prize, inheritance*〉 go to sb.; **zu·fällig 1.** *Adj.* accidental; chance *attrib.* 〈*meeting, acquaintance*〉; random 〈*selection*〉; **2.** *adv.* by chance; **wissen Sie** ~**, wie spät es ist?** *(ugs.)* do you by any chance know the time?; **Zufallstreffer der** fluke

zu|fassen *itr. V.* make a snatch *or* grab

zu|fliegen *unr. itr. V.; mit sein (ugs.)* 〈*door, window, etc.*〉 slam shut

Zu·flucht die refuge 〈vor + *Dat.* from); *(vor Unwetter o. ä.)* shelter 〈vor + *Dat.* from); **Zufluchts·ort der** place of refuge; sanctuary

Zu·fluß der a) *o. Pl. (das Zufließen)* inflow; supply; **b)** *(Gewässer)* feeder stream/river

zu|flüstern *tr. V.* **jmdm. etw.** ~: whisper sth. to sb.

zu·folge *Präp. mit Dat.; nachgestellt* according to

zu·frieden 1. *Adj.* contented; *(befriedigt)* satisfied; **mit etw.** ~ **sein** be satisfied with sth.; **2.** *adv.* contentedly; **zufrieden|geben** *unr. refl. V.* be satisfied; **Zufriedenheit die;** ~: contentment; *(Befriedigung)* satisfaction; **zufrieden|stellen** *tr. V.* satisfy; **zufriedenstellend 1.** *Adj.* satisfactory; **2.** *adv.* satisfactorily

zu|frieren *unr. itr. V.; mit sein* freeze over

zu|fügen *tr. V.* **jmdm. etw.** ~: inflict sth. on sb.; **jmdm. Schaden/[ein] Unrecht** ~: do sb. harm/an injustice

Zufuhr die; ~: supply; *(Material)* supplies *pl.*; **zu|führen 1.** *itr. V.* **auf etw.** *(Akk.)* ~: lead towards sth.; **2.** *tr. V.* **a)** *(zuleiten)* **einer Sache** *(Dat.)* **etw.** ~: supply sth. to sth.; **b)** *(bringen)* **einer Partei Mitglieder** ~: bring new members to a party

Zug der; ~[e]s, **Züge a)** *(Bahn)* train; **b)** *(Kolonne)* column; *(Umzug)* procession; *(Demonstrations~)* march; **c)** *(das Ziehen)* pull; traction *(Phys.)*; **d)** *(Vorrichtung)* pull; **e)** *(Wanderung)* migration; **f)** *(beim Brettspiel)* move; **g)** *(Schluck)* swig *(coll.)*; mouthful; *(großer Schluck)* gulp; **das Glas auf einen** *od.* **in einem** ~ **leeren** empty the glass at one go; **h)** *(beim Rauchen)* pull; drag *(coll.)*; **i)** *(Atem~)* breath; **j)** *o. Pl. (Zugluft; beim Ofen)* draught; **k)** *(Gesichts~)* feature; *(Wesens~)* characteristic; trait

Zu·gabe die a) *(Geschenk)* [free] gift; **b)** *(im Konzert, Theater)* encore

Zu·gang der a) *(Weg, auch fig.)* access; *(Eingang)* entrance; **b)** *o. Pl. (das Hinzukommen) (von Personen)* intake; *(von Patienten)* admission; *(Zuwachs)* increase **(von** in); **zu·gänge:** ~ **sein** *(ugs.)* be busy *or* occupied; **zugänglich** *Adj.* **a)** accessible; *(geöffnet)* open; **b)** *(zur Verfügung stehend)* available *(Dat., für* to); *(verständlich)* accessible *(Dat., für* to); **c)** *(aufgeschlossen)* approachable 〈*person*〉

zu·geben *unr. tr. V.* admit; admit to 〈*deed, crime*〉

zu·gegen *Adj.* ~ **sein** be present

zu|gehen *unr. itr. V.; mit sein* **a) auf jmdn./etw.** ~: approach sb./sth.; **b) jmdm.** ~ *(zugeschickt werden)* be sent to sb.; **c)** *(ugs.: sich schließen)* close; shut; **die Tür geht nicht zu** the door will not shut

Zügel der; ~s, ~: rein; **zügel·los** *(fig.)* **1.** *Adj.* unrestrained; unbridled 〈*rage, passion*〉; **2.** *adv.* without restraint; **zügeln** *tr. V.* rein [in] 〈*horse*〉; *(fig.)* curb, restrain 〈*desire etc.*〉

zu|gesellen *refl. V.* **sich jmdm./einer Sache** ~: join sb./sth.

Zu·geständnis das concession; **zu·gestehen** *unr. tr. V.* admit; concede

zu·getan *Adj.* **jmdm. [herzlich]** ~ **sein** *(geh.)* be [very] attached to sb.

zugig *Adj.* draughty, *(im Freien)* windy *⟨corner etc.⟩*

zügig 1. *Adj.* speedy; rapid; 2. *adv.* speedily; rapidly

zu·gleich *Adv.* at the same time

Zug·luft die; *o. Pl.* draught

zu|greifen *unr. itr. V.* **a)** take hold; **b)** *(sich bedienen)* help oneself; **c)** *(fleißig arbeiten)* |hart *od.* kräftig| ~: [really] knuckle down to it; **Zu·griff** der *(Zugang)* access **(auf** + *Akk.* to)

zu·grunde *Adv.* **a)** ~ gehen *(sterben)* die **(an** + *Dat.* of); *(zerstört werden)* be destroyed **(an** + *Dat.* by); ~ richten destroy; *(finanziell)* ruin *⟨company, person⟩*; **b)** etw. einer Sache *(Dat.)* ~ legen base sth. on sth.; etw. liegt einer Sache ~: sth. is based on sth.

zu|gucken *itr. V.* *(ugs.)* s. zusehen

zu·gunsten 1. *Präp. mit Gen.* in favour of; 2. *Adv.* ~ von in favour of

zu·gute *Adv.* jmdm. seine Unerfahrenheit *usw.* ~ halten *(geh.)* make allowances for sb.'s inexperience *etc.; sich (Dat.) etwas/viel auf etw. (Akk.)* ~ tun *od.* halten *(geh.)* be proud/very proud of sth.; jmdm./einer Sache ~ kommen stand sb./sth. in good stead

zu|haben *unr. itr. V.* *(ugs.) ⟨shop, office⟩* be shut *or* closed

zu|halten *unr. tr. V.* hold closed; *(nicht öffnen)* keep closed

zu|hängen *tr. V.* cover *⟨window, cage⟩*

zu|hauen *(ugs.)* 1. *unr. itr. V.* bang *or* slam *⟨door, window⟩* shut; 2. *unr. itr. V.* hit *or* strike out

Zu·hause das; ~s home

zu|hören *itr. V.* jmdm./einer Sache ~: listen to sb./sth.; **Zu·hörer** der, **Zu·hörerin** die listener

zu|kleben *tr. V.* seal *⟨letter, envelope⟩*

zu|knallen *(ugs.)* 1. *tr. V.* slam; 2. *itr. V.; mit sein* slam

zu|knöpfen *tr. V.* button up

zu|kommen *itr. V.; mit sein* auf jmdn. ~: approach sb.

Zukunft die; ~: future

Zulage die extra pay *no indef. art.;* additional allowance *no indef. art.*

zu|lassen *unr. tr. V.* **a)** allow; permit; **b)** *(teilnehmen lassen)* admit; **c)** *(mit einer Lizenz usw. versehen)* jmdn. als Arzt ~: register sb. as a doctor; **d)** *(Kfz-W.)* register *⟨vehicle⟩*; **e)** *(geschlossen lassen)* leave closed *or* shut *⟨door, window, etc.⟩*; **zu·lässig** *Adj.* permissible; admissible *⟨appeal⟩*; **Zulassung** die; ~, ~en registration

Zu·lauf der *o. Pl.* ~ haben *⟨shop, restaurant, etc.⟩* enjoy a large clientele; *⟨doctor, lawyer⟩* have a large practice; **zu|laufen** *unr. itr. V.; mit sein* **a)** auf jmdn./etw. ~ *(auch fig.)* run towards sb./sth.; **b)** jmdm. ~ *⟨cat, dog, etc.⟩* adopt sb. as a new owner

zu|legen *refl. V.* sich *(Dat.)* etw. ~: get oneself sth.

zu·letzt *Adv.* **a)** last [of all]; **b)** *(als letzter/letzte/letztes)* last; **c)** *(fig.: am wenigsten)* least of all; **d)** *(schließlich, am Ende)* in the end; **bis** ~: [right up] to *or* until the end

zum *Präp.* + *Art.* **a)** = zu dem; **b)** *(räumlich: Richtung)* to the; **c)** *(räumlich: Lage)* etw. ~ Fenster hinauswerfen throw sth. out of the window; **d)** *(Hinzufügung)* Milch ~ Tee nehmen take milk with [one's] tea **e)** *(zeitlich)* at the; **spätestens** ~ 15. April by 15 April at the latest; **f)** *(Zweck)* ~ Spaß/Vergnügen for fun/pleasure; **g)** *(Folge)* ~ Ärger seines Vaters to the annoyance of his father

zu|machen *tr. V.* close; fasten, do up *⟨dress⟩*; seal *⟨envelope, letter⟩*; turn off *⟨tap⟩*; put the top on *⟨bottle⟩*; *(stillegen)* close *or* shut down *⟨factory, mine, etc.⟩*

zu·mal 1. *Adv.* especially; particularly; 2. *Konj.* especially *or* particularly since

zumindest *Adv.* at least

zu·mute *Adj.* jmdm. ist unbehaglich *usw.* ~: sb. feels uncomfortable *etc.;* mir war nicht danach ~: I didn't feel like it *or* in the mood

zu|muten *tr. V.* jmdm. etw. ~ *(abverlangen)* expect *or* ask sth. of sb.; *(antun)* expect sb. to put up with sth.; **Zumutung** die; ~, ~en unreasonable demand; eine ~ sein be unreasonable

zu·nächst *Adv.* **a)** *(als erstes)* first; *(anfangs)* at first; **b)** *(im Moment, vorläufig)* for the moment

Zunahme die; ~, ~n increase *(Gen.,* an + *Dat.* in)

Zu·name der surname; last name

zünden 1. *tr. V.* ignite *⟨gas, fuel, etc.⟩*; detonate *⟨bomb, explosive device, etc.⟩*; let off *⟨fireworks⟩*; fire *⟨rocket⟩*; 2. *itr. V. ⟨rocket, engine⟩* fire; *⟨lighter, match⟩* light; *⟨gas, fuel, explosive⟩* ignite

Zünd-: ~holz das *(bes. südd., österr.)* match; ~schlüssel der *(Kfz-W.)* ignition key

Zündung die; ~, ~en **a)** *s.* zünden 1:

ignition; detonation; letting off; firing; b) *(Kfz-W.: Anlage)* ignition
zu|**nehmen** *unr. itr. V.* **a)** increase **(an + Dat.** in); ⟨*moon*⟩ wax; **b)** *(schwerer werden)* put on *or* gain weight
Zu·neigung die; ~, ~, ~en affection
Zunge die; ~, ~n tongue; ⌊jmdm.⌋ **die ~ herausstrecken** put one's tongue out [at sb.]
zu·nichte *Adj.* etw. **~ machen** ruin sth.
zu·oberst *Adv.* [right] on [the] top
zupfen 1. *itr. V.* **an etw.** *(Dat.)* ~: pluck *or* pull at sth.; **2.** *tr. V.* **a) etw. aus/von** *usw.* **etw.** ~: pull sth. out of/ from *etc.* sth.; **b)** *(auszupfen)* pull out; pluck *(eyebrows)*; **c)** pluck ⟨*string, guitar, tune*⟩; **d)** jmdn. **am Ärmel** ~: pull *or* tug [at] sb.'s sleeve
zur Präp. + Art. a) = zu der; **b)** *(räumlich, fig.: Richtung)* to the; **~ Schule/ Arbeit gehen** go to school/work; **c)** *(räumlich: Lage)* **~ Tür hereinkommen** come [in] through the door; **d)** *(Zusammengehörigkeit, Hinzufügung)* with; **e)** *(zeitlich)* at the; **~ Zeit** at the moment; at present; **f)** *(Zweck)* **~ Entschuldigung** by way of [an] excuse; **g)** *(Folge)* **~ vollen Zufriedenheit** to the complete satisfaction
zurechnungs·fähig *Adj.* sound of mind *pred.*
zurecht-: ~|**finden** *unr. refl. V.* find one's way [around]; ~|**kommen** *unr. itr. V.; mit sein* get on (with with); ~|**legen** *tr. V.* lay out [ready]; jmdm. etw. ~**legen** lay sth. out ready for sb.; ~|**machen** *tr. V. (ugs.)* **a)** *(vorbereiten)* get ready; **b)** *(herrichten)* do up; **c)** jmdn./sich ~: get sb. ready/get [oneself] ready; *(schminken)* make sb. up/put on one's make-up; ~|**weisen** *unr. tr. V.* rebuke; reprimand ⟨*pupil, subordinate, etc.*⟩
zu|reden *itr. V.* jmdm. ~: persuade sb.; *(ermutigen)* encourage sb.
Zürich (das); ~s Zurich
zu·rück *Adv.* back; *(weiter hinten)* behind; **einen Schritt ~:** a step backwards; ~! get *or* go back!
zurück-, Zurück-: ~|**behalten** *unr. tr. V.* **a)** keep [back]; retain; **b)** be left with ⟨*scar, heart defect, etc.*⟩; ~|**bekommen** *unr. tr. V.* get back; **Sie bekommen 10 Mark ~:** you get 10 marks change; ~|**bleiben** *unr. itr. V.; mit sein* **a)** remain; **b)** *(nicht mithalten)* lag behind; *(fig.)* fall behind; **c)** *(bleiben)* remain; ~|**erstatten** *tr. V.* refund;

jmdm. etw. ~**erstatten** refund sth. to sb.; ~|**fahren** *unr. itr. V.; mit sein* **a)** go back; return; **b)** *(nach hinten fahren)* go back[wards]; ~|**fallen** *unr. itr. V.; mit sein* **a)** *(in Rückstand geraten)* fall behind; **b)** *(auf einen niedrigeren Rang)* drop **(auf + Akk.** to); **c) an** jmdn. ~**fallen** ⟨*property*⟩ revert to sb.; **d) auf** jmdn. ~**fallen** ⟨*actions, behaviour*⟩ reflect [up]on sb.; ~|**fliegen** *unr. itr. V.; mit sein* fly back; ~|**führen** *tr. V.* etw. **auf etw.** *(Akk.)* ~**führen** attribute sth. to sth.; ~|**geben** *unr. tr. V.* give back; return; take back ⟨*defective goods*⟩; ~|**gehen** *unr. itr. V.; mit sein* **a)** go back; return; **b)** *(nach hinten)* go back; **c)** *(verschwinden)*⟩ disappear; ⟨*swelling, inflammation*⟩ go down; ⟨*pain*⟩ subside; **d)** *(sich verringern)* decrease; ⟨*fever*⟩ abate; ⟨*flood*⟩ subside; ⟨*business*⟩ fall off; **e)** *(zurückgeschickt werden)* be returned *or* sent back; ~|**greifen** *unr. itr. V.* **auf** jmdn./etw. ~**greifen** fall back on sb./sth.; ~|**halten 1.** *unr. tr. V.* **a)** jmdn. ~**halten** hold sb. back; *(von etw. abhalten)* stop sb.; **b)** *(am Vordringen hindern)* keep back ⟨*crowd, mob, etc.*⟩; **c)** *(behalten)* withhold ⟨*news, letter, etc.*⟩; **d)** *(nicht austreten lassen)* hold back ⟨*tears etc.*⟩; **2.** *unr. refl. V.* restrain *or* control oneself; **sich in einer Diskussion** ~**halten** keep in the background in a discussion; ~**haltend 1.** *Adj.* **a)** reserved; **b)** *(kühl, reserviert)* cool, restrained ⟨*reception, response*⟩; **c)** *(Wirtsch.: schwach)* slack ⟨*demand*⟩; **2.** *adv.* ⟨*behave*⟩ with reserve *or* restraint; *(kühl, reserviert)* coolly; ~**haltung die;** *o. Pl.* reserve; *(Kühle, Reserviertheit)* coolness; *(Wirtsch.)* caution; ~|**kehren** *itr. V.; mit sein* return; come back; ~|**kommen** *unr. itr. V.; mit sein* come back; return; *(zurückgelangen)* get back; ~**kommen auf** **(+ Akk.)** come back to ⟨*subject, question, point, etc.*⟩; ~|**kriegen** *tr. V. s.* ~**bekommen**; ~|**lassen** *unr. tr. V.* leave; ~|**legen** *tr. V.* **a)** put back; **b)** *(reservieren)* put aside, keep *(Dat.,* für for); **c)** *(sparen)* put away; **d)** *(hinter sich bringen)* cover ⟨*distance*⟩; ~|**lehnen** *refl. V.* lean back; ~|**nehmen** *unr. tr. V. (auch fig. widerrufen)* take back; ~|**rufen** *unr. tr. V.* **a)** call back; recall ⟨*ambassador*⟩; **b)** *auch itr. (telefonisch)* call *or (Brit.)* ring back; ~|**schicken** *tr. V.* send back; ~|**schrecken** *regelm., veralt. unr. itr.*

V.; mit sein vor etw. (Dat.) ~**schrecken**
(fig.) shrink from sth.; **er schreckt vor
nichts** ~: he will stop at nothing;
~|**senden** *unr. od. regelm. tr. V. (geh.)*
s. ~**schicken;** ~|**treten** *unr. itr. V.;
mit sein* step back; *(von einem Amt)*
resign; step down; *(government)*
resign; *(von einem Vertrag usw.)* withdraw (**von** from); back out (**von** of);
(fig.: in den Hintergrund treten)
become less important; ~|**weisen**
unr. tr. V. reject *(proposal, question,
demand, application, etc.)*; turn down,
refuse *(offer, request, help, etc.)*; turn
away *(petitioner, unwelcome guest)*;
repudiate *(accusation, claim, etc.)*;
~|**werfen** *unr. tr. V.* throw back; re-
flect *(light, sound)*; repulse *(enemy)*;
(fig.: in einer Entwicklung) set back;
~|**zahlen** *tr. V.* pay back; ~|**ziehen**
1. *unr. tr. V.* **a)** pull back; draw back
(bolt, curtains, one's hand, etc.); **b)**
(abziehen, zurückbeordern) withdraw
(troops); recall *(ambassador)*; **c)**
(rückgängig machen) withdraw; can-
cel *(order, instruction)*; **2.** *unr. refl. V.*
withdraw

Zu·ruf der shout; **zu|rufen** *unr. tr. V.*
jmdm. etw. ~: shout sth. to sb.

Zu·sage die **a)** *(auf eine Einladung
hin)* acceptance; *(auf eine Stellenbe-
werbung hin)* offer; **b)** *(Versprechen)*
promise; undertaking; **zu|sagen 1.**
itr. V. **a)** accept; **b) jmdm.** ~ *(gefallen)*
appeal to sb.; **2.** *tr. V.* promise

zusammen *Adv.* together

zusammen-, Zusammen-: ~|**ar-
beiten** *itr. V.* co-operate; ~|**binden**
unr. tr. V. tie together; ~|**brechen**
unr. itr. V.; mit sein collapse; *(fig.)(order, communications, system, telephone network)* break down; *(traffic)*
come to a standstill; ~**bruch** der collapse; *(fig., auch psychisch, nervlich)*
breakdown; ~|**drücken** *tr. V.* press
together; ~|**fahren** *unr. itr. V.; mit
sein (~zucken)* start; jump; ~|**fallen**
unr. itr. V.; mit sein **a)** collapse; **b)**
|**zeitlich|** ~**fallen** coincide; ~|**fassen**
tr. V. summarize; ~**fassung** die sum-
mary; ~|**fegen** *tr. V. (bes. nordd.)*
sweep together; ~|**fließen** *unr. itr.
V.; mit sein (rivers, streams)* flow into
each other; ~**fluß** der confluence;
~|**fügen** *tr. V.* fit together; ~|**führen**
tr. V. bring together; ~|**gehören** *itr.
V.* belong together; ~**gehörig** *Adj.*
[closely] related *or* connected *(subjects, problems, etc.)*; matching *attrib.*

(pieces of tea service, cutlery, etc.);
~**gehörigkeit** die; ~: **ein starkes Gefühl der** ~**gehörigkeit** a strong sense of
belonging together; ~**hang** der connection; *(einer Geschichte, Rede)* co-
herence; *(Kontext)* context; ~|**hän-
gen** *unr. itr. V.* **a)** be joined
[together]; **b) mit etw.** ~**hängen** *(fig.)*
be related to sth.; *(durch etw. [mit] verursacht sein)* be the result of sth.;
~|**kehren** *tr. V. (bes. südd.) s.* ~**fegen;**
~**klappbar** *Adj.* folding; ~|**klappen**
tr. V. fold up; ~|**kommen** *unr. itr. V.;
mit sein* **a)** meet; **mit jmdm.** ~**kommen**
meet sb.; **b)** *(zueinanderkommen; auch
fig.)* get together; *(gleichzeitig auftreten)* occur *or* happen together;
~**kunft** die ~, ~**künfte** meeting;
~|**laufen** *unr. itr. V.; mit sein* **a)**
(people, crowd) gather, congregate; **b)**
(rivers, streams) flow into each other,
join up; ~|**leben** *itr. V.* live together;
~|**leben** das; *o. Pl.* living together *no
art.*; ~|**legen 1.** *tr. V.* **a)** put *or* gather
together; **b)** *(zusammenfalten)* fold
[up]; **c)** *(miteinander verbinden)* amalgamate, merge *(classes, departments,
etc.)*; combine *(events)*; **d)** put *(pa-
tients, guests, etc.)* together [in the
same room]; **2.** *itr. V.* club together;
~|**nehmen 1.** *unr. tr. V.* **a)** summon
up *(courage, strength, understanding)*;
2. *unr. refl. V.* get *or* take a grip on
oneself; **nimm dich** ~! pull yourself
together!; ~|**passen** *itr. V.* go
together; *(persons)* be suited to each
other; ~**prall** der; ~|**e|s,** ~**e** collision;
~|**prallen** *itr. V.; mit sein* collide (**mit**
with); ~|**sein** *unr. itr. V.; mit sein;
Zusschr. nur im Inf. u. Part.* **a)** be
together; **b)** *(zusammenleben)* be *or*
live together; ~|**setzen 1.** *tr. V.* put
together; **2.** *refl. V.* **a) sich aus etw.**
~**setzen** be made up *or* composed of
sth.; **b)** *(sich zueinander setzen)* sit
together; *(zu einem Gespräch)* get
together; ~|**stehen** *unr. itr. V.* stand
together; ~|**stellen** *tr. V.* put
together; draw up *(list)*; ~**stoß** der
collision; *(fig.)* clash (**mit** with);
~|**stoßen** *unr. itr. V.; mit sein* collide
(**mit** with); ~|**treffen** *unr. itr. V.; mit
sein* meet; **mit jmdm.** ~**treffen** meet
sb.; *(zeitlich)* coincide; ~|**zählen** *tr.
V.* add up; ~|**zucken** *itr. V.; mit sein*
start; jump

Zu·satz der addition; *(Zugesetztes,
Additiv)* additive; **zusätzlich 1.** *Adj.*
additional; **2.** *adv.* in addition

zu|schauen *itr. V. (südd., österr., schweiz.)* s. zusehen; **Zu·schauer** der, *Zu·schauerin* die; ~, ~nen spectator; *(im Theater, Kino)* member of the audience; *(an einer Unfallstelle)* onlooker; *(Fernseh~)* viewer; die ~: *(im Theater, Kino)* the audience *sing.*

zu|schicken *tr. V.* send

zu|schieben *unr. tr. V.* **a)** push ⟨*drawer, door*⟩ shut; **b)** *(fig.)* jmdm. die Schuld ~: lay the blame on sb.

Zu·schlag der **a)** additional *or* extra charge; *(für Nacht-, Feiertagsarbeit usw.)* additional *or* extra payment; **b)** *(Eisenb.)* supplement ticket; **zu|schlagen** **1.** *unr. tr. V.* bang *or* slam ⟨*door, window, etc.*⟩ shut; close ⟨*book*⟩; *(heftig)* slam ⟨*book*⟩ shut; **2.** *unr. itr. V.* **a)** *mit sein* ⟨*door, trap*⟩ slam *or* bang shut; **b)** *(einen Schlag führen)* throw a blow/blows; *(losschlagen)* hit *or* strike out; *(fig.)* ⟨*army, police, murderer*⟩ strike

zu|schließen **1.** *unr. tr. V.* lock; **2.** *unr. itr. V.* lock up

zu|schnüren *tr. V.* tie up

zu|schrauben *tr. V.* screw the lid *or* top on ⟨*jar, flask*⟩; screw ⟨*lid, top*⟩ on

Zu·schrift die letter; *(auf eine Anzeige)* reply

Zu·schuß der contribution (**zu** towards)

zu|sehen *unr. itr. V.* **a)** watch; jmdm. [beim Arbeiten *usw.*] ~: watch sb. [working *etc.*]; **b)** *(dafür sorgen)* make sure; see to it

zu|senden *unr. od. regelm. tr. V.: s.* zuschicken; **Zu·sendung** die sending

zu|spitzen *refl. V.* become aggravated

zu|sprechen **1.** *unr. tr. V.* **a)** er sprach ihr Trost/Mut zu his words gave her comfort/courage; **b)** jmdm. ein Erbe *usw.* ~: award sb. an inheritance *etc.*; **2.** *unr. itr. V.* jmdm. ermutigend/tröstend *usw.* ~: speak encouragingly/comfortingly to sb.

Zu·stand der **a)** condition; *(bes. abwertend)* state; **b)** *(Stand der Dinge)* state of affairs; **zu·stande** Adv. etw. ~ bringen [manage to] bring about sth.; ~ kommen come into being; *(geschehen)* take place; **zu·ständig** Adj. appropriate relevant ⟨*authority, office, etc.*⟩; |für etw.| ~ sein *(verantwortlich)* be responsible [for sth.]

zu|stehen *unr. itr. V.* etw. steht jmdm. zu sb. is entitled to sth.

zu|steigen *unr. itr. V.; mit sein* get on;

ist noch jemand zugestiegen? *(im Bus)* ≈ any more fares, please?; *(im Zug)* ≈ tickets, please!

zu|stellen *tr. V.* deliver ⟨*letter, parcel, etc.*⟩

zu|stimmen *itr. V.* agree; jmdm. |in einem Punkt| ~: agree with sb. [on a point]; einer Sache *(Dat.)* ~: agree to sth.; **Zu·stimmung** die *(Billigung)* approval (**zu** of); *(Einverständnis)* agreement (**zu** to, with)

zu|stoßen *unr. itr. V.; mit sein* jmdm. ~: happen to sb.

Zu·tat die ingredient

zu·teil Adv. jmdm./einer Sache ~ werden *(geh.)* be granted to sb./sth.; **zu|teilen** *tr. V.* jmdm. jmdn./etw. ~: allot *or* assign sb./sth. to sb.; jmdm. seine Portion ~: mete out his/her share to sb.

zu|tragen *unr. refl. V. (geh.)* occur; **zuträglich** Adj. healthy ⟨*climate*⟩; jmdm./einer Sache ~ sein be good for sb./sth.; be beneficial to sb./sth.

zu|trauen *tr. V.* jmdm. etw. ~: believe sb. [is] capable of [doing] sth.; sich *(Dat.)* etw. ~: think one can do *or* is capable of doing sth.; **Zutrauen** das; ~s confidence, trust (**zu** in); **zutraulich** **1.** Adj. trusting; **2.** adv. trustingly; **Zutraulichkeit** die; ~: trust[fulness]

zu|treffen *unr. itr. V.* **a)** be correct; **b)** auf *od.* für jmdn./etw. ~: apply to sb./sth.; **zutreffend** **1.** Adj. **a)** correct; **b)** *(geltend)* applicable; relevant; **2.** adv. correctly

zu|trinken *unr. itr. V.* jmdm. ~: raise one's glass and drink to sb.

Zu·tritt der entry; admittance; ,,kein ~", ,,~ verboten" 'no entry'; 'no admittance'; ~ |zu etw.| haben have access [to sth.]

zu·unterst Adv. right at the bottom

zuverlässig **1.** Adj. reliable; *(verläßlich)* dependable ⟨*person*⟩; **2.** adv. reliably; **Zuverlässigkeit** die; ~: reliability; *(Verläßlichkeit)* dependability

zuversichtlich **1.** Adj. confident; **2.** adv. confidently

zuviel **1.** indekl. Indefinitpron. too much; *(ugs.: zu viele)* too many; **2.** adv. too much

zu·vor Adv. before

zuvor|kommen *unr. itr. V.; mit sein* **a)** jmdm. ~: beat sb. to it; **b)** einer Sache *(Dat.)* ~: anticipate sth.; **zuvorkommend** **1.** Adj. obliging; *(höflich)* courteous; **2.** *adv.* obligingly; *(höflich)* courteously

zu·weilen *Adv. (geh.)* now and again

zu|weisen *unr. tr. V.* jmdm. etw. ~ : allocate *or* allot sb. sth.

zu|wenden *unr. od. regelm. refl. V.* sich jmdm./einer Sache ~ *(auch fig.)* turn to sb./sth.

zu·wenig 1. *indekl. Indefinitpron.* too little; *(ugs.: zu wenige)* too few; **2.** *adv.* too little

zuwider *Adj.* jmdm. ~ sein be repugnant to sb.

zu|winken *itr. V.* jmdm./einander ~ : wave to sb./one another

zu|zahlen *tr. V.* pay ⟨*five marks etc.*⟩ extra

zu|ziehen 1. *unr. tr. V.* pull ⟨*door*⟩ shut; draw ⟨*curtain*⟩; do up ⟨*zip*⟩; **2.** *unr. refl. V.* sich ⟨*Dat.*⟩ eine Krankheit ~ : catch an illness; **3.** *unr. itr. V.; mit sein* move into the area

zuzüglich *Präp. mit Gen.* plus

zwang *1. u. 3. Pers. Sg. Prät. v.* **zwingen**; **Zwang** *der;* ~[e]s, **Zwänge a)** compulsion; **b)** *(unwiderstehlicher Drang)* irresistible urge; **zwängen 1.** *tr. V.* squeeze; **2.** *refl. V.* squeeze [oneself]; **zwanglos 1.** *Adj.* **a)** informal; casual ⟨*behaviour*⟩; **b)** *(unregelmäßig)* haphazard ⟨*arrangement*⟩; **2.** *adv.* **a)** informally; **b)** *(unregelmäßig)* haphazardly ⟨*arranged*⟩; **Zwangs·lage** die predicament; **zwangs·läufig 1.** *Adj.* inevitable; **2.** *adv.* inevitably

zwanzig *Kardinalz.* twenty; *s. auch* **achtzig**; **zwanziger** *indekl. Adj.; nicht präd.* die ~ Jahre the twenties; **Zwanzig·mark·schein** der twenty-mark note; **zwanzigst ...** *Ordinalz.* twentieth

zwar *Adv.* **a)** admittedly; **b)** und ~ : to be precise

Zweck *der;* ~[e]s, ~e purpose; *(Sinn)* point; es hat keinen ~ : it's pointless; es hat keinen ~, das zu tun there is no point in doing that

zweck-, Zweck-: ~los *Adj.* pointless; ~mäßig **1.** *Adj.* appropriate; expedient ⟨*behaviour, action*⟩; functional ⟨*building, fittings, furniture*⟩; **2.** *adv.* appropriately ⟨*arranged, clothed*⟩; ⟨*act*⟩ expediently; ⟨*equip, furnish*⟩ functionally; ~mäßigkeit die appropriateness; *(einer Handlung)* expediency; *(eines Gebäudes)* functionalism

zwecks *Präp. mit Gen. (Papierdt.)* for the purpose of

zwei *Kardinalz.* two; *s. auch* ¹**acht**; **Zwei** die; ~, ~en **a)** *(Zahl)* two; **b)** *(Schulnote)* B

zwei-, Zwei-: ~bettzimmer das twin-bedded room; ~deutig **1.** *Adj.* ambiguous; *(fig.: schlüpfrig)* suggestive ⟨*remark, joke*⟩; **2.** *adv.* ambiguously; *(fig.)* suggestively; ~deutigkeit die; ~, ~en ambiguity; *(fig.)* suggestiveness; ~dimensional **1.** *Adj.* two-dimensional; **2.** *adv.* two-dimensionally; ~ein·halb *Bruchz.* two and a half

zweierlei *Gattungsz.; indekl.* **a)** *attr.* two sorts *or* kinds of; two different ⟨*sizes, kinds, etc.*⟩; odd ⟨*socks, gloves*⟩; **b)** *(alleinstehend)* two [different] things; **zwei·fach** *Vervielfältigungsz.* double; *(~mal)* twice; **Zwei·fache das;** *adj. Dekl.* das ~ : twice as much

Zweifel *der;* ~s, ~ : doubt (an + *Dat.* about); etw. in ~ ziehen question sth.; **zweifelhaft** *Adj.* **a)** doubtful; **b)** *(fragwürdig)* dubious; *(suspekt)* suspicious; **zweifel·los** *Adv.* undoubtedly; **zweifeln** *itr. V.* doubt; an jmdm./etw. ~ : doubt sb./sth.; have doubts about sb./sth.

Zweig *der;* ~[e]s, ~e [small] branch; *(meist ohne Blätter)* twig

zwei-, Zwei-: ~hundert *Kardinalz.* two hundred; ~mal *Adv.* twice; ~mark·stück das two-mark piece; ~pfennig·stück das two-pfennig piece; ~reiher der double-breasted suit/coat/jacket; ~schneidig *Adj.* double-edged; ~sprachig **1.** *Adj.* bilingual; ⟨*sign*⟩ in two languages; **2.** *adv.* bilingually; ⟨*written*⟩ in two languages; ⟨*published*⟩ in a bilingual edition; ~spurig *Adj.* **a)** two-lane ⟨*road*⟩; **b)** two-track ⟨*vehicle*⟩; **c)** two- *or* twin-track ⟨*recording*⟩; ~stellig *Adj.* two-figure *attrib.* ⟨*number, sum*⟩; ~stöckig *Adj.* two-storey *attrib.*; ~stöckig sein have two storeys

zweit ... *Ordinalz.* second; jeder ~e every other one; *s. auch* **erst...**

zwei·tägig *Adj. (2 Tage alt)* two-day-old *attrib.*; *(2 Tage dauernd)* two-day *attrib.*; **zweit·ältest ...** *Adj.* second oldest; **zwei·tausend** *Kardinalz.* two thousand; **zweit·best ...** *Adj.* second best

zweite·mal *Adv.* das ~ : for the second time; **zweiten·mal** *Adv.* zum ~ : for the second time; **beim** ~ : the second time [round]; **zweitens** *Adv.*

secondly; in the second place; **Zweite[r]-Klasse-Abteil das** second-class compartment; **zweit·rangig** *Adj.* of secondary importance *postpos.;* *(~klassig)* second-rate; **zweitürig** *Adj.* two-door ⟨*car*⟩

Zweit-: ~**wagen** der second car; ~**wohnung die** second home

Zwei·zimmerwohnung die two-room flat *(Brit.) or (Amer.)* apartment

Zwerg der; ~|e|s, ~e dwarf; *(Garten~)* gnome

Zwetsche die; ~, ~n damson plum

Zwieback der; ~|e|s, ~e *od.* Zwiebäkke rusk; *(unzählbar)* rusks *pl.*

Zwiebel die; ~, ~n onion; *(Blumen~)* bulb

zwie-, **Zwie-:** ~**gespräch das** *(geh.)* dialogue; ~**spalt der;** ~|e|s, ~e *od.* ~**spälte** [inner] conflict; ~**spältig** *Adj.* conflicting ⟨*mood, feelings*⟩; discordant ⟨*impression*⟩; *(widersprüchlich)* contradictory ⟨*nature, attitude, person, etc.*⟩

Zwilling der; ~s, ~e twin

Zwillings-: ~**bruder der** twin brother; ~**paar das** pair of twins; ~**schwester die** twin sister

zwingen 1. *unr. tr. V.* force; **jmdn. zu etw.** ~, **jmdn.** |dazu| ~, **etw. zu tun** force *or* compel sb. to do sth.; **zwingend** *Adj.* compelling ⟨*reason, logic*⟩; conclusive ⟨*proof, argument*⟩; imperative ⟨*necessity*⟩

zwinkern *itr. V.* |mit den Augen| ~: blink; *(als Zeichen)* wink

Zwirn der; ~|e|s, ~e [strong] thread *or* yarn

zwischen *Präp. mit Dat./Akk.* between; *(mitten unter)* among[st]

zwischen-, Zwischen-: ~**durch** [-'-] *Adv.* **a)** *(zeitlich)* between times;

(zwischen zwei Zeitpunkten) in between; *(von Zeit zu Zeit)* from time to time; ~**fall der** incident; ~|**landen** land in X on the way; ~**mahlzeit die** snack [between meals]; ~**menschlich 1.** *Adj.* interpersonal ⟨*relations*⟩; ⟨*contacts*⟩ between people; **2.** *adv.* on a personal level; ~**raum der** space; gap; *(Lücke)* gap; ~**zeit die** interim

Zwist der; ~|e|s, ~e *(geh.)* strife *no indef. art.; (Fehde)* feud; dispute; **Zwistigkeit die;** ~, ~en *(geh.)* dispute

zwitschern *itr. (auch tr.) V.* chirp

Zwitter der; ~s, ~ *(Biol.)* hermaphrodite

zwo *Kardinalz. (ugs.; bes. zur Verdeutlichung)* two

zwölf *Kardinalz.* twelve; ~ **Uhr mittags/nachts** [twelve o'clock] midday/midnight; *s. auch* ¹**acht; zwölft...** *Ordinalz.* twelfth; *s. auch* **acht...;** **zwölftel** *Bruchz.* twelfth; *s. auch* **achtel; Zwölftel das** *(schweiz. meist* der) ~s, ~: twelfth

zwot... *Ordinalz. (ugs.; bes. bei Datumsangaben)* second; **zwotens** *Adv. (ugs.)* secondly

Zylinder [tsi'lɪndɐ] der; ~s, ~ **a)** cylinder; **b)** *(Hut)* top hat; **zylindrisch 1.** *Adj.* cylindrical; **2.** *adv.* cylindrically

zynisch 1. *Adj.* cynical; **2.** *adv.* cynically

Zynismus der; ~: cynicism

Zypern (das); ~s Cyprus; **Zyprer der;** ~s, ~, **Zyprerin die;** ~, ~nen Cypriot

Zypresse die; ~, ~n cypress

Zypriot der; ~en, ~en, **Zypriotin die;** ~, ~nen Cypriot; **zypriotisch, zyprisch** *Adj.* Cypriot

Zyste die; ~, ~n *(Med.)* cyst

Englische unregelmäßige Verben

Ein Sternchen (*) weist darauf hin, daß die korrekte Form von der jeweiligen Bedeutung abhängt.

| Infinitive | Past Tense | Past Participle | Infinitive | Past Tense | Past Participle |
Infinitiv	*Präteritum*	*2. Partizip*	*Infinitiv*	*Präteritum*	*2. Partizip*
arise	arose	arisen	flee	fled	fled
awake	awoke	awoken	fling	flung	flung
be	was *sing.*,	been	floodlight	floodlit	floodlit
	were *pl.*		fly	flew	flown
bear	bore	borne	forbid	forbade,	forbidden
beat	beat	beaten		forbad	
become	became	become	forecast	forecast,	forecast,
begin	began	begun		forecasted	forecasted
bend	bent	bent	foretell	foretold	foretold
bet	bet, betted	bet, betted	forget	forgot	forgotten
bid	*bade, bid	*bidden, bid	forgive	forgave	forgiven
bind	bound	bound	forsake	forsook	forsaken
bite	bit	bitten	freeze	froze	frozen
bleed	bled	bled	get	got	got, *(Amer.)*
blow	blew	blown			gotten
break	broke	broken	give	gave	given
breed	bred	bred	go	went	gone
bring	brought	brought	grind	ground	ground
broadcast	broadcast	broadcast	grow	grew	grown
build	built	built	hang	*hung,	*hung,
burn	burnt, burned	burnt, burned		hanged	hanged
burst	burst	burst	have	had	had
bust	bust, busted	bust, busted	hear	heard	heard
buy	bought	bought	hew	hewed	hewn, hewed
cast	cast	cast	hide	hid	hidden
catch	caught	caught	hit	hit	hit
choose	chose	chosen	hold	held	held
cling	clung	clung	hurt	hurt	hurt
come	came	come	keep	kept	kept
cost	*cost, costed	*cost, costed	kneel	knelt,	knelt,
creep	crept	crept		*(esp. Amer.)*	*(esp. Amer.)*
cut	cut	cut		kneeled	kneeled
deal	dealt	dealt	know	knew	known
dig	dug	dug	lay	laid	laid
dive	dived,	dived	lead	led	led
	(Amer.) dove		lean	leaned,	leaned,
do	did	done		*(Brit.)* leant	*(Brit.)* leant
draw	drew	drawn	leap	leapt, leaped	leapt, leaped
dream	dreamt,	dreamt,	learn	learnt,	learnt,
	dreamed	dreamed		learned	learned
drink	drank	drunk	leave	left	left
drive	drove	driven	lend	lent	lent
dwell	dwelt	dwelt	let	let	let
eat	ate	eaten	²lie	lay	lain
fall	fell	fallen	light	lit, lighted	lit, lighted
feed	fed	fed	lose	lost	lost
feel	felt	felt	make	made	made
fight	fought	fought	mean	meant	meant
find	found	found	meet	met	met

Infinitive	Past Tense	Past Participle	Infinitive	Past Tense	Past Participle
Infinitiv	_Präteritum_	_2. Partizip_	_Infinitiv_	_Präteritum_	_2. Partizip_
mow	mowed	mown, mowed	spend	spent	spent
			spill	spilt, spilled	spilt, spilled
overhang	overhung	overhung	spin	spun	spun
pay	paid	paid	spit	spat, spit	spat, spit
prove	proved	proved, proven	split	split	split
			spoil	spoilt, spoiled	spoilt, spoiled
put	put	put			
quit	quitted, _(Amer.)_ quit	quitted, _(Amer.)_ quit	spread	spread	spread
			spring	sprang, _(Amer.)_ sprung	sprung
read [ri:d]	read [red]	read [red]			
rid	rid	rid	stand	stood	stood
ride	rode	ridden	steal	stole	stolen
²ring	rang	rung	stick	stuck	stuck
rise	rose	risen	sting	stung	stung
run	ran	run	stink	stank, stunk	stunk
saw	sawed	sawn, sawed	strew	strewed	strewed, strewn
say	said	said			
see	saw	seen	stride	strode	stridden
seek	sought	sought	strike	struck	struck
sell	sold	sold	string	strung	strung
send	sent	sent	strive	strove	striven
set	set	set	sublet	sublet	sublet
sew	sewed	sewn, sewed	swear	swore	sworn
shake	shook	shaken	sweep	swept	swept
shear	sheared	shorn, sheared	swell	swelled	swollen, swelled
shed	shed	shed			
shine	shone	shone	swim	swam	swum
shit	shitted, shit	shitted, shit	swing	swung	swung
shoe	shod	shod	take	took	taken
shoot	shot	shot	teach	taught	taught
show	showed	shown	tear	tore	torn
shrink	shrank	shrunk	tell	told	told
shut	shut	shut	think	thought	thought
sing	sang	sung	thrive	thrived, throve	thrived, thriven
sink	sank, sunk	sunk			
sit	sat	sat	throw	threw	thrown
slay	slew	slain	thrust	thrust	thrust
sleep	slept	slept	tread	trod	trodden, trod
slide	slid	slid	understand	understood	understood
sling	slung	slung	undo	undid	undone
slink	slunk	slunk	wake	woke	woken
slit	slit	slit	wear	wore	worn
smell	smelt, smelled	smelt, smelled	¹weave	wove	woven
			weep	wept	wept
sow	sowed	sown, sowed	wet	wet, wetted	wet, wetted
speak	spoke	spoken	win	won	won
speed	*sped, speeded	*sped, speeded	²wind [waɪnd]	wound [waʊnd]	wound [waʊnd]
			wring	wrung	wrung
spell	spelled, _(Brit.)_ spelt	spelled, _(Brit.)_ spelt	write	wrote	written

German irregular verbs

Irregular and partly irregular verbs are listed alphabetically by infinitive. 1st, 2nd, and 3rd person present and imperative forms are given after the infinitive, and preterite subjunctive forms after the preterite indicative, where they take an umlaut, change *e* to *i*, etc.

Verbs with a raised number in the German-English section of the Dictionary have the same number in this list.

Compound verbs (including verbs with prefixes) are only given if a) they do not take the same forms as the corresponding simple verb, e.g. *befehlen,* or b) there is no corresponding simple verb, e.g. *bewegen.*

An asterisk (*) indicates a verb which is also conjugated regularly.

Infinitive *Infinitiv*	Preterite *Präteritum*	Past Participle *2. Partizip*
abwägen	wog (wöge) ab	abgewogen
backen (du bäckst, er bäckt; *auch:* du backst, er backt)	backte, *älter:* buk (büke)	gebacken
befehlen (du befiehlst, er befiehlt; befiehl!)	befahl (beföhle, befähle)	befohlen
beginnen	begann (begänne, *seltener:* begönne)	begonnen
beißen	biß	gebissen
bergen (du birgst, er birgt; birg!)	barg (bärge)	geborgen
bersten (du birst, er birst; birst!)	barst (bärste)	geborsten
besinnen	besann (besänne)	besonnen
²bewegen	bewog (bewöge)	bewogen
biegen	bog (böge)	gebogen
bieten	bot (böte)	geboten
binden	band (bände)	gebunden
bitten	bat (bäte)	gebeten
blasen (du bläst, er bläst)	blies	geblasen
bleiben	blieb	geblieben
bleichen*	blich	geblichen
braten (du brätst, er brät)	briet	gebraten
brechen (du brichst, er bricht; brich!)	brach (bräche)	gebrochen
brennen	brannte (brennte)	gebrannt
bringen	brachte (brächte)	gebracht
denken	dachte (dächte)	gedacht
dreschen (du drischst, er drischt; drisch!)	drosch (drösche)	gedroschen
dringen	drang (dränge)	gedrungen
dürfen (ich darf, du darfst, er darf)	durfte (dürfte)	gedurft

empfehlen (du empfiehlst,
  er empfiehlt, empfiehl!)
	empfahl (empföhle,
	  seltener: empfähle)
		empfohlen

erklimmen	erklomm (erklömme)	erklommen
erlöschen (du erlischst, er erlischt; erlisch!)	erlosch (erlösche)	erloschen
erschallen*	erscholl (erschölle)	erschollen
¹˒³erschrecken (du erschrickst, er erschrickt; erschrick!)	erschrak (erschräke)	erschrocken
erwägen	erwog (erwöge)	erwogen

Infinitive *Infinitiv*	Preterite *Präteritum*	Past Participle *2. Partizip*
essen (du ißt, er ißt; iß!)	aß (äße)	gegessen
fahren (du fährst, er fährt)	fuhr (führe)	gefahren
fallen (du fällst, er fällt)	fiel	gefallen
fangen (du fängst, er fängt)	fing	gefangen
fechten (du fichtst, er ficht; ficht!)	focht (föchte)	gefochten
finden	fand (fände)	gefunden
flechten (du flichtst, er flicht; flicht!)	flocht (flöchte)	geflochten
fliegen	flog (flöge)	geflogen
fliehen	floh (flöhe)	geflohen
fließen	floß (flösse)	geflossen
fressen (du frißt, er frißt; friß!)	fraß (fräße)	gefressen
frieren	fror (fröre)	gefroren
gären*	gor (göre)	gegoren
gebären (*geh.:* du gebierst, sie gebiert; gebier!)	gebar (gebäre)	geboren
geben (du gibst, er gibt; gib!)	gab (gäbe)	gegeben
gedeihen	gedieh	gediehen
gehen	ging	gegangen
gelingen	gelang (gelänge)	gelungen
gelten (du giltst, er gilt; gilt!)	galt (gölte, gälte)	gegolten
genesen	genas (genäse)	genesen
genießen	genoß (genösse)	genossen
geschehen (es geschieht)	geschah (geschähe)	geschehen
gewinnen	gewann (gewönne, gewänne)	gewonnen
gießen	goß (gösse)	gegossen
gleichen	glich	geglichen
gleiten	glitt	geglitten
glimmen	glomm (glömme)	geglommen
graben (du gräbst, er gräbt)	grub (grübe)	gegraben
greifen	griff	gegriffen
haben (du hast, er hat)	hatte (hätte)	gehabt
halten (du hältst, er hält)	hielt	gehalten
[1]hängen	hing	gehangen
hauen	haute, *geh.:* hieb	gehauen
heben	hob (höbe)	gehoben
heißen	hieß	geheißen
helfen (du hilfst, er hilft; hilf!)	half (hülfe, *selten:* hälfe)	geholfen
kennen	kannte (kennte)	gekannt
klingen	klang (klänge)	geklungen
kneifen	kniff	gekniffen
kommen	kam (käme)	gekommen
können (ich kann, du kannst, er kann)	konnte (könnte)	gekonnt
kriechen	kroch (kröche)	gekrochen
[1,2]laden (du lädst, er lädt)	lud (lüde)	geladen
lassen (du läßt, er läßt)	ließ	gelassen
laufen (du läufst, er läuft)	lief	gelaufen
leiden	litt	gelitten
leihen	lieh	geliehen
[1,2]lesen (du liest, er liest; lies!)	las (läse)	gelesen
liegen	lag (läge)	gelegen
lügen	log (löge)	gelogen
mahlen	mahlte	gemahlen
meiden	mied	gemieden

797

Infinitive *Infinitiv*	Preterite *Präteritum*	Past Participle 2. *Partizip*
melken* (du milkst, er milkt; milk!; du melkst, er melkt; melke!)	molk (mölke)	gemolken
messen (du mißt, er mißt; miß!)	maß (mäße)	gemessen
mißlingen	mißlang (mißlänge)	mißlungen
mögen (ich mag, du magst, er mag)	mochte (möchte)	gemocht
müssen (ich muß, du mußt, er muß)	mußte (müßte)	gemußt
nehmen (du nimmst, er nimmt; nimm!)	nahm (nähme)	genommen
nennen	nannte (nennte)	genannt
pfeifen	pfiff	gepfiffen
preisen	pries	gepriesen
quellen (du quillst, er quillt; quill!)	quoll (quölle)	gequollen
raten (du rätst, er rät)	riet	geraten
reiben	rieb	gerieben
reißen	riß	gerissen
reiten	ritt	geritten
rennen	rannte (rennte)	gerannt
riechen	roch (röche)	gerochen
ringen	rang (ränge)	gerungen
rinnen	rann (ränne, *seltener:* rönne)	geronnen
rufen	rief	gerufen
salzen*	salzte	gesalzen
saufen (du säufst, er säuft)	soff (söffe)	gesoffen
saugen*	sog (söge)	gesogen
schaffen*	schuf (schüfe)	geschaffen
schallen*	scholl (schölle)	geschallt
scheiden	schied	geschieden
scheinen	schien	geschienen
scheißen	schiß	geschissen
schelten (du schiltst, er schilt; schilt!)	schalt (schölte)	gescholten
¹scheren	schor (schöre)	geschoren
schieben	schob (schöbe)	geschoben
schießen	schoß (schösse)	geschossen
schinden	schindete	geschunden
schlafen (du schläfst, er schläft)	schlief	geschlafen
schlagen (du schlägst, er schlägt)	schlug (schlüge)	geschlagen
schleichen	schlich	geschlichen
¹schleifen	schliff	geschliffen
schließen	schloß (schlösse)	geschlossen
schlingen	schlang (schlänge)	geschlungen
schmeißen	schmiß	geschmissen
schmelzen (du schmilzt, er schmilzt; schmilz!)	schmolz	geschmolzen
schneiden	schnitt	geschnitten
schrecken* (du schrickst, er schrickt; schrick!)	schrak (schräke)	geschreckt
schreiben	schrieb	geschrieben
schreien	schrie	geschrie[e]n
schreiten	schritt	geschritten
schweigen	schwieg	geschwiegen
schwellen (du schwillst, er schwillt; schwill!)	schwoll (schwölle)	geschwollen
schwimmen	schwamm (schwömme, *seltener:* schwämme)	geschwommen

Infinitive *Infinitiv*	Preterite *Präteritum*	Past Participle *2. Partizip*
schwinden	schwand (schwände)	geschwunden
schwingen	schwang (schwänge)	geschwungen
schwören	schwor (schwüre)	geschworen
sehen (du siehst, er sieht; sieh[e]!)	sah (sähe)	gesehen
sein (ich bin, du bist, er ist, wir sind, ihr seid, sie sind; sei!)	war (wäre)	gewesen
senden*	sandte (sendete)	gesandt
sieden*	sott (sötte)	gesotten
singen	sang (sänge)	gesungen
sinken	sank (sänke)	gesunken
sitzen	saß (säße)	gesessen
sollen (ich soll, du sollst, er soll)	sollte	gesollt
spalten*	spaltete	gespalten
speien	spie	gespie[e]n
spinnen	spann (spönne, spänne)	gesponnen
sprechen (du sprichst, er spricht; sprich!)	sprach (spräche)	gesprochen
sprießen	sproß (sprösse)	gesprossen
springen	sprang	gesprungen
stechen (du stichst, er sticht; stich!)	stach (stäche)	gestochen
stehen	stand (stünde, *auch:* stände)	gestanden
stehlen (du stiehlst, er stiehlt; stiehl!)	stahl (stähle, *seltener:* stöhle)	gestohlen
steigen	stieg	gestiegen
sterben (du stirbst, er stirbt; stirb!)	starb (stürbe)	gestorben
stinken	stank (stänke)	gestunken
stoßen (du stößt, er stößt)	stieß	gestoßen
streichen	strich	gestrichen
streiten	stritt	gestritten
tragen (du trägst, er trägt)	trug (trüge)	getragen
treffen (du triffst; er trifft; triff!)	traf (träfe)	getroffen
treiben	trieb	getrieben
treten (du trittst, er tritt; tritt!)	trat (träte)	getreten
triefen*	troff (tröffe)	getroffen
trinken	trank (tränke)	getrunken
trügen	trog (tröge)	getrogen
tun	tat (täte)	getan
verderben (du verdirbst, er verdirbt; verdirb!)	verdarb (verdürbe)	verdorben
verdrießen	verdroß (verdrösse)	verdrossen
vergessen (du vergißt, er vergißt, vergiß!)	vergaß (vergäße)	vergessen
verlieren	verlor (verlöre)	verloren
verschleißen*	verschliß	verschlissen
verzeihen	verzieh	verziehen
¹wachsen (du wächst, er wächst)	wuchs (wüchse)	gewachsen
waschen (du wäschst, er wäscht)	wusch (wüsche)	gewaschen
weichen	wich	gewichen
weisen	wies	gewiesen
²wenden*	wandte (wendete)	gewandt
werben (du wirbst, er wirbt; wirb!)	warb (würbe)	geworben
werden (du wirst, er wird; werde!)	wurde, *dichter.:* ward (würde)	geworden; *als Hilfsv.:* worden
werfen (du wirfst, er wirft; wirf!)	warf (würfe)	geworfen
¹wiegen	wog (wöge)	gewogen

Infinitive	Preterite	Past Participle
Infinitiv	*Präteritum*	*2. Partizip*
winden	wand (wände)	gewunden
wissen (ich weiß, du weißt, er weiß)	wußte (wüßte)	gewußt
wollen (ich will, du willst, er will)	wollte	gewollt
wringen	wrang (wränge)	gewrungen
ziehen	zog (zöge)	gezogen
zwingen	zwang (zwänge)	gezwungen

Weights and Measures / Maße und Gewichte

Weight / Gewichte

1,000 milligrams (mg)	= 1 gram (g)	
1 000 Milligramm (mg)	*= 1 Gramm (g)*	= 15.43 grains
1,000 grams	= 1 kilogram (kg)	
1 000 Gramm	*= 1 Kilogramm (kg)*	= 2.205 pounds
1,000 kilograms	= 1 tonne (t)	= 19.684 hun-
1 000 Kilogramm	*= 1 Tonne (t)*	dredweight
	1 grain (gr.)	= 0.065 g
437½ grains	= 1 ounce (oz.)	= 28.35 g
16 ounces	= 1 pound (lb.)	= 0.454 kg
14 pounds	= 1 stone (st.)	= 6.35 kg
112 pounds	= 1 hundredweight	= 50.8 kg
20 hundredweight	= 1 ton (t.)	= 1,016.05 kg

Length / Längenmaße

10 millimetres (mm)	= 1 centimetre (cm)	
10 Millimeter (mm)	*= 1 Zentimeter (cm)*	= 0.394 inch
100 centimetres	= 1 metre (m)	= 39.4 inches /
100 Zentimeter	*= 1 Meter (m)*	1.094 yards
1,000 metres	= 1 kilometre (km)	= 0.6214 mile ≈
1 000 Meter	*= 1 Kilometer (km)*	⅝ mile
	1 inch (in.)	= 25.4 mm
12 inches	= 1 foot (ft.)	= 30.48 cm
3 feet	= 1 yard (yd.)	= 0.914 m
220 yards	= 1 furlong	= 201.17 m
8 furlongs	= 1 mile (m.)	= 1.609 km
1,760 yards	= 1 mile	= 1.609 km

Square measure / Flächenmaße

| 100 square metres (sq. m) | = 1 are | |
| *100 Quadratmeter (m²)* | *= 1 Ar (a)* | = 0.025 acre |

| 100 ares | = 1 hectare (ha) | |
| *100 Ar* | *= 1 Hektar (ha)* | = 2.471 acres |

| 100 hectares | = 1 square kilometre (sq. km) | = 0.386 square |
| *100 Hektar* | *= 1 Quadratkilometer (km²)* | miles |

	1 square inch	= 6.452 cm²
144 square inches	= 1 square foot	= 929.03 cm²
9 square feet	= 1 square yard	= 0.836 m²
4,840 square yards	= 1 acre	= 0.405 ha
640 acres	= 1 square mile	= 2.59 k²/ 259 ha

Cubic measure / Raummaße

| 1 cubic centimetre (cc) | | = 0.06 |
| *1 Kubikzentimeter (cm³)* | | cubic inches |

| 1,000,000 cubic centimetres | = 1 cubic metre (cu. m) | = 35.714 cubic feet / |
| *1 000 000 Kubikzentimeter* | *= 1 Kubikmeter (m³)* | 1.307 cubic yards |

	1 cubic inch	= 16.4 cm³
1,728 cubic inches	= 1 cubic foot	= 0.028 m³
27 cubic feet	= 1 cubic yard	= 0.764 m³

Capacity / Hohlmaße

| 10 millilitres (ml) | = 1 centilitre (cl) | |
| *10 Milliliter (ml)* | *= 1 Zentiliter (cl)* | |

| 100 centilitres | = 1 litre (l) | = 1.76 pints |
| *100 Zentiliter* | *= 1 Liter (l)* | (2.1 US pints) / 0.22 gallons (0.264 US gallons) |

4 gills	= 1 pint (pt.) (1.201 US pints)	= 0.568 l
2 pints	= 1 quart (qt.) (1.201 US quarts)	= 1.136 l
4 quarts	= 1 gallon (gal.) (1.201 US gallons)	= 4.546 l

Revisions to German spelling / die neue Regelung der Rechtschreibung

In July 1996, after much debate, wide-ranging changes to the spelling of German were agreed and ratified by the governments of Germany, Austria, and Switzerland. The following list, whilst not all-encompassing, details those changes which may be of interest to the user of this dictionary. It is worth noting that although these reforms are valid immediately, they will not be expected to be reflected in all written texts until 2005. Until that date, both old and new spellings will be acceptable.

The following list contains words which are not included in the A–Z text of this dictionary. Nonetheless, it is the editors' view that the learner of German will gain a better overview of the systematic changes involved by studying a more comprehensive list of words affected by the reforms.

alt	neu
A	
[gestern, heute, morgen] abend	[gestern, heute, morgen] Abend
aberhundert	*auch:* Aberhundert
Aberhunderte	*auch:* aberhunderte
abertausend	*auch:* Abertausend
Abertausende	*auch:* abertausende
Abfluß	Abfluss
abgeblaßt	abgeblasst
Abguß	Abguss
Ablaß	Ablass
Abriß	Abriss
Abschluß	Abschluss
Abschuß	Abschuss
absein	ab sein
Abszeß	Abszess
abwärtsgehen	abwärts gehen
in acht nehmen	in Acht nehmen
außer acht lassen	außer Acht lassen
8achser	8-Achser
der/die achte, den/die ich sehe	der/die Achte, den/die ich sehe

alt	neu
jeder/jede achte kommt mit	jeder/jede Achte kommt mit
achtgeben	Acht geben
achthaben	Acht haben
8jährig	8-jährig
der/die 8jährige	der/die 8-Jährige
8mal	8-mal
achtmillionenmal	acht Millionen Mal
8tonner	8-Tonner
achtunggebietend	Achtung gebietend
über Achtzig	über achtzig
Mitte [der] Achtzig	Mitte [der] achtzig
in die Achtzig kommen	in die achtzig kommen
die achtziger Jahre	*auch:* die Achtzigerjahre*
die Achtzigerjahre	*auch:* die achtziger Jahre
ackerbautreibende Völker	Ackerbau treibende Völker
Action-painting	Actionpainting
	auch: Action-Painting
ade sagen	*auch:* Ade sagen*
Aderlaß	Aderlass
Adhäsionsverschluß	Adhäsionsverschluss
Adreßbuch	Adressbuch
afro-amerikanisch	afroamerikanisch
afro-asiatisch	afroasiatisch
Afro-Look	Afrolook
After-shave	Aftershave
ich habe ähnliches erlebt	ich habe Ähnliches erlebt
und/oder ähnliches (u. ä./o. ä.)	und/oder Ähnliches (u. Ä./o. Ä.)
Alkoholmißbrauch	Alkoholmissbrauch
alleinerziehend	allein erziehend
alleinseligmachend	allein selig machend
alleinstehend	allein stehend
es ist das allerbeste, daß ...	es ist das Allerbeste, dass ...
im allgemeinen	im Allgemeinen
allgemeingültig	allgemein gültig
allgemeinverständlich	allgemein verständlich
allzubald	allzu bald

alt	neu
allzufrüh	allzu früh
allzugern	allzu gern
allzulange	allzu lange
allzuoft	allzu oft
allzusehr	allzu sehr
allzuviel	allzu viel
allzuweit	allzu weit
Alma mater	Alma Mater
Alpdruck	*auch:* Albdruck
Alptraum	*auch:* Albtraum
als daß	als dass
aus alt mach neu	aus Alt mach Neu
für alt und jung	für Alt und Jung
er ist immer der alte geblieben	er ist immer der Alte geblieben
alles beim alten lassen	alles beim Alten lassen
Alter ego	Alter Ego
altwienerisch	alt-wienerisch
Amboß	Amboss
Anbiß	Anbiss
andersdenkend	anders denkend
andersgeartet	anders geartet
anderslautend	anders lautend
aneinanderfügen	aneinander fügen
aneinandergeraten	aneinander geraten
aneinandergrenzen	aneinander grenzen
aneinanderlegen	aneinander legen
aneinanderreihen	aneinander reihen
angepaßt	angepasst
Angepaßtheit	Angepasstheit
Anglo-Amerikaner	Angloamerikaner
jmdm. angst machen	jmdm. Angst machen
anheimfallen	anheim fallen
anheimstellen	anheim stellen
Anlaß	Anlass
anläßlich	anlässlich
Anriß	Anriss

alt	neu
Anschiß	Anschiss
Anschluß	Anschluss
ansein	an sein
der Archimedische Punkt	der archimedische Punkt
im argen liegen	im Argen liegen
bei arm und reich	bei Arm und Reich
Armee-Einheit	*auch:* Armeeeinheit
Aschantinuß	Aschantinuss
As	Ass
aufeinanderbeißen	aufeinander beißen
aufeinanderfolgen	aufeinander folgen
aufeinandertreffen	aufeinander treffen
aufgepaßt!	aufgepasst!
aufgerauht	aufgeraut
Aufguß	Aufguss
Auflösungsprozeß	Auflösungsprozess
aufrauhen	aufrauen
Aufriß	Aufriss
Aufschluß	Aufschluss
aufschlußreich	aufschlussreich
ein aufsehenerregendes Ereignis	ein Aufsehen erregendes Ereignis
aufsein	auf sein
auf seiten	aufseiten
	auch: auf Seiten
der aufsichtführende Lehrer	der Aufsicht führende Lehrer
aufwärtsgehen	aufwärts gehen
aufwendig	*auch:* aufwändig
auseinanderbiegen	auseinander biegen
auseinanderfallen	auseinander fallen
auseinandergehen	auseinander gehen
auseinanderhalten	auseinander halten
auseinanderleben	auseinander leben
auseinanderreißen	auseinander reißen
auseinandersetzen	auseinander setzen
Ausfluß	Ausfluss
Ausguß	Ausguss

alt	neu
Ausschluß	Ausschluss
Ausschuß	Ausschuss
aussein	aus sein
aufs äußerste gespannt	*auch:* aufs Äußerste gespannt
außerstande	*auch:* außer Stande

B

alt	neu
Bajonettverschluß	Bajonettverschluss
Ballettänzerin	Balletttänzerin
	auch: Ballett-Tänzerin
Ballokal	Balllokal
	auch: Ball-Lokal
Bänderriß	Bänderriss
jmdm. [angst und] bange machen	jmdm. [Angst und] Bange machen
bankrott gehen	Bankrott gehen
Baroneß	Baroness
baselstädtisch	basel-städtisch
baß erstaunt	bass erstaunt
Baß	Bass
Baßgeige	Bassgeige
Baßsänger	Basssänger
	auch: Bass-Sänger
Baukostenzuschuß	Baukostenzuschuss
beeinflußbar	beeinflussbar
Beeinflußbarkeit	Beeinflussbarkeit
beeinflußt	beeinflusst
befaßt	befasst
Begrüßungskuß	Begrüßungskuss
behende	behände
Behendigkeit	Behändigkeit
beieinanderhaben	beieinander haben
beieinandersein	beieinander sein
beieinandersitzen	beieinander sitzen
beieinanderstehen	beieinander stehen
beifallheischend	Beifall heischend
beisammensein	beisammen sein

alt	neu
Beischluß	Beischluss
belemmert	belämmert
jeder beliebige	jeder Beliebige
Beschiß	Beschiss
Beschluß	Beschluss
beschlußfähig	beschlussfähig
Beschlußfassung	Beschlussfassung
Beschuß	Beschuss
ich will im besonderen	ich will im Besonderen
erwähnen ...	erwähnen ...
bessergehen	besser gehen
es ist das beste, wenn ...	es ist das Beste, wenn ...
aufs beste geregelt sein	*auch:* aufs Beste geregelt sein
zum besten geben	zum Besten geben
zum besten haben/halten	zum Besten haben/halten
das erste beste	das erste Beste
bestehenbleiben	bestehen bleiben
Bestelliste	Bestellliste
	auch: Bestell-Liste
bestgehaßt	bestgehasst
bestußt	bestusst
Betelnuß	Betelnuss
um ein beträchtliches höher	um ein Beträchtliches höher
in betreff	in Betreff
betreßt	betresst
Bettuch *[zu: Bett]*	Betttuch
	auch: Bett-Tuch
bevorschußt	bevorschusst
bewußt	bewusst
bewußtlos	bewusstlos
Bewußtlosigkeit	Bewusstlosigkeit
Bewußtsein	Bewusstsein
in bezug auf	in Bezug auf
bezuschußt	bezuschusst
Bibliographie	*auch:* Bibliografie
Bierfaß	Bierfass

alt	neu
die Bismarckschen Sozialgesetze	die bismarckschen Sozialgesetze
	auch: die Bismarck'schen Sozialgesetze
Biß	Biss
bißchen	bisschen
du sollst bitte sagen	*auch:* du sollst Bitte sagen*
es ist bitter kalt	es ist bitterkalt
Bittag	Bitttag
	auch: Bitt-Tag
Blackout	*auch:* Black-out*
blankpoliert	blank poliert
blaß	blass
Bläßhuhn/Bleßhuhn	Blässhuhn/Blesshuhn
bläßlich	blässlich
blaßrosa	blassrosa
Blattschuß	Blattschuss
der blaue Planet *[die Erde]*	der Blaue Planet
blaugestreift	blau gestreift
bläulichgrün	bläulich grün
bleibenlassen	bleiben lassen
blendendweiß	blendend weiß
blondgefärbt	blond gefärbt
Bluterguß	Bluterguss
Bonbonniere	*auch:* Bonboniere
Börsentip	Börsentipp
im bösen wie im guten	im Bösen wie im Guten
Boß	Boss
Bouclé	*auch:* Buklee
braungebrannt	braun gebrannt
bräunlichgelb	bräunlich gelb
des langen und breiten	des Langen und Breiten
breitgefächert	breit gefächert
Brennessel	Brennnessel
	auch: Brenn-Nessel
Bruderkuß	Bruderkuss
Brummbaß	Brummbass

alt	neu
brütendheiß	brütend heiß
buntgefiedert	bunt gefiedert
buntschillernd	bunt schillernd
Büroschluß	Büroschluss
Butterfaß	Butterfass

C

alt	neu
Cashewnuß	Cashewnuss
Centre Court	Centrecourt
	auch: Centre-Court
Chansonnier	*auch:* Chansonier
Choreographie	*auch:* Choreografie
Cleverneß	Cleverness
Comeback	*auch:* Come-back*
Common sense	Commonsense
	auch: Common Sense
Corned beef	Cornedbeef
	auch: Corned Beef
Corpus delicti	Corpus Delicti
Countdown	*auch:* Count-down*

D

alt	neu
dabeisein	dabei sein
Dachgeschoß	Dachgeschoss *[in Österreich weiterhin mit ß]*
dahinterklemmen	dahinter klemmen
dahinterkommen	dahinter kommen
Dampfschiffahrt	Dampfschifffahrt
Danaidenfaß	Danaidenfass
darauffolgend	darauf folgend
Darmverschluß	Darmverschluss
darüberstehen	darüber stehen
dasein	da sein
daß	dass
daß-Satz	dass-Satz
	auch: Dasssatz

alt	neu
datenverarbeitend	Daten verarbeitend
Dein [in Briefen]	dein
mein und dein verwechseln	Mein und Dein verwechseln
die Deinen	auch: die deinen
die Deinigen	auch: die deinigen
Dekolleté	auch: Dekolletee
Delikateßgurke	Delikatessgurke
Delikateßsenf	Delikatesssenf
	auch: Delikatess-Senf
Delphin	auch: Delfin
Denkprozeß	Denkprozess
wir haben derartiges nicht bemerkt	wir haben Derartiges nicht bemerkt
dessenungeachtet	dessen ungeachtet
des weiteren	des Weiteren
auf deutsch	auf Deutsch
deutschsprechend	Deutsch sprechend
das d'Hondtsche System	das d'hondtsche System
	auch: das d'Hondt'sche System
diät leben	Diät leben
Dich [in Briefen]	dich
dichtbehaart	dicht behaart
dichtgedrängt	dicht gedrängt
Differential	auch: Differenzial*
Diktaphon	auch: Diktafon
Dir [in Briefen]	dir
Doppelpaß	Doppelpass
dortbleiben	dort bleiben
dortzulande	auch: dort zu Lande
draufsein	drauf sein
Dreß	Dress
etwas aufs dringendste fordern	auch: etwas aufs Dringendste fordern
drinsein	drin sein
jeder dritte, der mitwollte	jeder Dritte, der mitwollte
zum dritten	zum Dritten

alt	neu
die dritte Welt	die Dritte Welt
drückendheiß	drückend heiß
Du *[in Briefen]*	du
auf du und du stehen	auf Du und Du stehen
im dunkeln tappen	im Dunkeln tappen
im dunkeln bleiben	im Dunkeln bleiben
dünnbesiedelt	dünn besiedelt
Dünnschiß	Dünnschiss
durcheinanderbringen	durcheinander bringen
durcheinandergeraten	durcheinander geraten
durcheinanderlaufen	durcheinander laufen
Durchfluß	Durchfluss
Durchlaß	Durchlass
durchnumerieren	durchnummerieren
Durchschuß	Durchschuss
durchsein	durch sein
dußlig	dusslig
Dußligkeit	Dussligkeit
Dutzende Reklamationen	*auch:* dutzende Reklamationen
Dutzende von Reklamationen	*auch:* dutzende von Reklamationen

E

alt	neu
ebensogut	ebenso gut
ebensosehr	ebenso sehr
ebensoviel	ebenso viel
ebensowenig	ebenso wenig
an Eides Statt	an Eides statt
sein eigen nennen	sein Eigen nennen
sich zu eigen machen	sich zu Eigen machen
einbleuen	einbläuen
aufs eindringlichste warnen	*auch:* aufs Eindringlichste warnen
das einfachste ist, wenn ...	das Einfachste ist, wenn ...
Einfluß	Einfluss
einflußreich	einflussreich

810

alt	neu
aufs eingehendste untersuchen	*auch:* aufs Eingehendste untersuchen
einiggehen	einig gehen
Einlaß	Einlass
einläßlich	einlässlich
Einriß	Einriss
Einschluß	Einschluss
Einschuß	Einschuss
Einschußstelle	Einschussstelle
	auch: Einschuss-Stelle
Einsendeschluß	Einsendeschluss
einwärtsgebogen	einwärts gebogen
der/die/das einzelne kann ...	der/die/das Einzelne kann ...
jeder einzelne von uns	jeder Einzelne von uns
bis ins einzelne geregelt	bis ins Einzelne geregelt
ins einzelne gehend	ins Einzelne gehend
einzelnstehend	einzeln stehend
der/die/das einzige wäre ...	der/die/das Einzige wäre ...
kein einziger war gekommen	kein Einziger war gekommen
er als einziger/sie als einzige hatte ...	er als Einziger/sie als Einzige hatte ...
das einzigartige ist, daß ...	das Einzigartige ist, dass ...
Eisenguß	Eisenguss
die eisenverarbeitende Industrie	die Eisen verarbeitende Industrie
eisigkalt	eisig kalt
eislaufen	Eis laufen
Eisschnellauf	Eisschnelllauf
Eisschnelläufer	Eisschnellläufer
energiebewußt	energiebewusst
aufs engste verflochten	*auch:* aufs Engste verflochten
engbefreundet	eng befreundet
engbedruckt	eng bedruckt
Engpaß	Engpass
nicht im entferntesten beabsichtigen	*auch:* nicht im Entferntesten beabsichtigen

alt	neu
auf das entschiedenste zurückweisen	*auch:* auf das Entschiedenste zurückweisen
Entschluß	Entschluss
ein Entweder-Oder gibt es hier nicht	ein Entweder-oder gibt es hier nicht
Entwicklungsprozeß	Entwicklungsprozess
erblaßt	erblasst
Erdgeschoß	Erdgeschoss *[in Österreich weiterhin mit ß]*
Erdnuß	Erdnuss
die erdölexportierenden Länder	die Erdöl exportierenden Länder
erfaßbar	erfassbar
erfaßt	erfasst
Erguß	Erguss
erholungsuchende Großstädter	Erholung suchende Großstädter
Erlaß	Erlass
ermeßbar	ermessbar
ernstgemeint	ernst gemeint
ernstzunehmend	ernst zu nehmend
erpreßbar	erpressbar
nicht den erstbesten nehmen	nicht den Erstbesten nehmen
der erste, der gekommen ist	der Erste, der gekommen ist
das reicht fürs erste	das reicht fürs Erste
zum ersten, zum zweiten, zum dritten	zum Ersten, zum Zweiten, zum Dritten
die Erste Hilfe	die erste Hilfe
das erstemal	das erste Mal
zum erstenmal	zum ersten Mal
Erstkläßler	Erstklässler
die Erstplazierten	die Erstplatzierten
eßbar	essbar
Eßbesteck	Essbesteck
Eßecke	Essecke
essentiell	*auch:* essenziell*
Eßlöffel	Esslöffel
eßlöffelweise	esslöffelweise

alt	neu
Eßtisch	Esstisch
etlichemal	etliche Mal
Euch *[in Briefen]*	euch
Euer *[in Briefen]*	euer
die Euren	*auch:* die euren
die Eurigen	*auch:* die eurigen
Existentialismus	*auch:* Existenzialismus*
existentialistisch	*auch:* existenzialistisch*
existentiell	*auch:* existenziell*
Exportüberschuß	Exportüberschuss
Exposé	*auch:* Exposee
expreß	express
Expreßreinigung	Expressreinigung
Expreßzug	Expresszug
Exzeß	Exzess

F

alt	neu
Fabrikationsprozeß	Fabrikationsprozess
fahrenlassen	fahren lassen
Fairneß	Fairness
Fair play	Fairplay
	auch: Fair Play
fallenlassen	fallen lassen
Fallinie	Falllinie
	auch: Fall-Linie
Fallout	*auch:* Fall-out*
Familienanschluß	Familienanschluss
Fangschuß	Fangschuss
Faß	Fass
faßbar	fassbar
Faßbier	Fassbier
Fäßchen	Fässchen
faßlich	fasslich
du faßt	du fasst
Fast food	Fastfood
	auch: Fast Food

alt	neu
Faxanschluß	Faxanschluss
Feedback	*auch:* Feed-back*
Fehlpaß	Fehlpass
Fehlschuß	Fehlschuss
jmdm. feind sein	jmdm. Feind sein
feingemahlen	fein gemahlen
fernliegen	fern liegen
fertigbringen	fertig bringen
fertigstellen	fertig stellen
Fertigungsprozeß	Fertigungsprozess
festangestellt	fest angestellt
festumrissen	fest umrissen
festverwurzelt	fest verwurzelt
fettgedruckt	fett gedruckt
feuerspeiende Drachen	Feuer speiende Drachen
die fischverarbeitende Industrie	die Fisch verarbeitende Industrie
Fitneß	Fitness
Flachschuß	Flachschuss
fleischfressende Pflanzen	Fleisch fressende Pflanzen
Flohbiß	Flohbiss
das Bier floß in Strömen	das Bier floss in Strömen
flötengehen	flöten gehen
Fluß	Fluss
flußabwärts	flussabwärts
flußaufwärts	flussaufwärts
Flußbett	Flussbett
Flüßchen	Flüsschen
Flußdiagramm	Flussdiagramm
flüssigmachen	flüssig machen
Flußsand	Flusssand
	auch: Fluss-Sand
Flußschiffahrt	Flussschifffahrt
	auch: Fluss-Schifffahrt
Flußspat	Flussspat
	auch: Fluss-Spat
die Haare fönen	die Haare föhnen

alt	neu
folgendes ist zu beachten	Folgendes ist zu beachten
wie im folgenden erläutert	wie im Folgenden erläutert
Fraktionsausschuß	Fraktionsausschuss
Fraktionsbeschluß	Fraktionsbeschluss
Free climbing	Freeclimbing
	auch: Free Climbing
Free Jazz	*auch:* Freejazz
Freßgier	Fressgier
Freßpaket	Fresspaket
Freßsack	Fresssack
	auch: Fress-Sack
Friedensschluß	Friedensschluss
frischgebacken	frisch gebacken
fritieren	frittieren
frohgelaunt	froh gelaunt
frühverstorben	früh verstorben
Full-time-Job	Fulltimejob
	auch: Full-Time-Job
Fünfpaß	Fünfpass
funkensprühend	Funken sprühend
Funkmeßtechnik	Funkmesstechnik
fürbaß	fürbass
fürliebnehmen	fürlieb nehmen
Fußballänderspiel	Fußballländerspiel
	auch: Fußball-Länderspiel

G

alt	neu
Gangsterboß	Gangsterboss
im ganzen gesehen	im Ganzen gesehen
im großen und ganzen	im Großen und Ganzen
Gärungsprozeß	Gärungsprozess
Gäßchen	Gässchen
gefangenhalten	gefangen halten
gefangennehmen	gefangen nehmen
gefaßt	gefasst
gefirnißt	gefirnisst

alt	neu
es ist das gegebene, schnell zu handeln	es ist das Gegebene, schnell zu handeln
gegeneinanderprallen	gegeneinander prallen
gegeneinanderstoßen	gegeneinander stoßen
von allen gehaßt	von allen gehasst
geheimhalten	geheim halten
gehenlassen	gehen lassen
Gelaß	Gelass
gutgelaunt	gut gelaunt
gelblichgrün	gelblich grün
Gemse	Gämse
wir haben gemußt	wir haben gemusst
die Wunde hat genäßt	die Wunde hat genässt
aufs genaueste festgelegt	*auch:* aufs Genaueste festgelegt
genaugenommen	genau genommen
genausogut	genauso gut
genausowenig	genauso wenig
Generalbaß	Generalbass
sie genoß den Sonnenschein	sie genoss den Sonnenschein
Genuß	Genuss
genüßlich	genüsslich
Genußmittel	Genussmittel
genußsüchtig	genusssüchtig
Geographie	*auch:* Geografie
es hat gut gepaßt	es hat gut gepasst
wir haben gepraßt	wir haben geprasst
frisch gepreßter Saft	frisch gepresster Saft
geradehalten	gerade halten
geradesitzen	gerade sitzen
geradestellen	gerade stellen
Gerichtsbeschluß	Gerichtsbeschluss
um ein geringes weniger	um ein Geringes weniger
es geht ihn nicht das geringste an	es geht ihn nicht das Geringste an
nicht im geringsten stören	nicht im Geringsten stören
geringachten	gering achten
geringschätzen	gering schätzen

alt	neu
Geruchsverschluß	Geruchsverschluss
Geschäftsschluß	Geschäftsschluss
er wurde geschaßt	er wurde geschasst
Geschichtsbewußtsein	Geschichtsbewusstsein
Geschirreiniger	Geschirrreiniger
	auch: Geschirr-Reiniger
Geschoß	Geschoss *[in Österreich weiterhin mit ß]*
gestern abend/morgen/nacht	gestern Abend/Morgen/Nacht
alle waren gestreßt	alle waren gestresst
getrenntlebend	getrennt lebend
Gewinnummer	Gewinnnummer
	auch: Gewinn-Nummer
gewiß	gewiss
Gewissensbiß	Gewissensbiss
Gewißheit	Gewissheit
gewißlich	gewisslich
ich habe es gewußt	ich habe es gewusst
Ginkgo	*auch:* Ginko
Glacéhandschuh	*auch:* Glaceehandschuh
glänzendschwarz	glänzend schwarz
glattgehen	glatt gehen
glatthobeln	glatt hobeln
glattschleifen	glatt schleifen
glattstreichen	glatt streichen
das gleiche tun	das Gleiche tun
aufs gleiche hinauskommen	aufs Gleiche hinauskommen
gleich und gleich gesellt sich gern	Geich und Gleich gesellt sich gern
gleichlautend	gleich lautend
Gleisanschluß	Gleisanschluss
Glimmstengel	Glimmstängel
glühendheiß	glühend heiß
Gnadenerlaß	Gnadenerlass
die Goetheschen Dramen	die goetheschen Dramen
	auch: die Goethe'schen Dramen
Graphit	*auch:* Grafit

alt	neu
Graphologie	*auch:* Grafologie
gräßlich	grässlich
graugestreift	grau gestreift
grellbeleuchtet	grell beleuchtet
Grenzfluß	Grenzfluss
Greuel	Gräuel
greulich	gräulich
grifffest	grifffest
jmdn. aufs gröbste beleidigen	*auch:* jmdn. aufs Gröbste beleidigen
grobgemahlen	grob gemahlen
ein Programm für groß und klein	ein Programm für Groß und Klein
im großen und ganzen	im Großen und Ganzen
das größte wäre, wenn ...	das Größte wäre, wenn ...
Großschiffahrtsweg	Großschifffahrtsweg
groß schreiben *[mit großem Anfangsbuchstaben]*	großschreiben
grünlichgelb	grünlich gelb
Guß	Guss
Gußeisen	Gusseisen
gußeisern	gusseisern
guten Tag sagen	*auch:* Guten Tag sagen*
es im guten versuchen	es im Guten versuchen
gutaussehend	gut aussehend
gutbezahlt	gut bezahlt
gutgehen	gut gehen
gutgehend	gut gehend
gutgelaunt	gut gelaunt
gutgemeint	gut gemeint
guttun	gut tun
gutunterrichtet	gut unterrichtet

H

alt	neu
haftenbleiben	haften bleiben
haltmachen	Halt machen
Hämorrhoide	*auch:* Hämorride

alt	neu
händchenhaltend	Händchen haltend
handeltreibend	Handel treibend
Handkuß	Handkuss
Handout	*auch:* Hand-out*
hängenbleiben	hängen bleiben
hängenlassen	hängen lassen
Happy-End	Happyend
	auch: Happy End
Haraß	Harass
Hard cover	Hardcover
	auch: Hard Cover
Hard-cover-Einband	Hardcovereinband
	auch: Hard-Cover-Einband
hartgekocht	hart gekocht
Haselnuß	Haselnuss
Haselnußstrauch	Haselnussstrauch
	auch: Haselnuss-Strauch
Haß	Hass
haßerfüllt	hasserfüllt
häßlich	hässlich
Häßlichkeit	Hässlichkeit
Haßliebe	Hassliebe
du haßt	du hasst
Hauptschulabschluß	Hauptschulabschluss
nach Hause	*in Österreich und der Schweiz*
	auch: nachhause
zu Hause	*in Österreich und der Schweiz*
	auch: zuhause
haushalten	*auch:* Haus halten
Haushaltsausschuß	Haushaltsausschuss
Hawaii-Insel	*auch:* Hawaiiinsel
heiligsprechen	heilig sprechen
Heilungsprozeß	Heilungsprozess
heimlichtun	heimlich tun
heißbegehrt	heiß begehrt
heißgeliebt	heiß geliebt

819

alt	neu
heißumkämpft	heiß umkämpft
helleuchtend	hell leuchtend
hellicht	helllicht
hellila	helllila
hellodernd	hell lodernd
heransein	heran sein
heraussein	heraus sein
herbstlichgelb	herbstlich gelb
Heringsfaß	Heringsfass
hersein	her sein
herumsein	herum sein
heruntersein	herunter sein
Herzas	Herzass
jmdn. auf das herzlichste begrüßen	*auch:* jmdn. auf das Herzlichste begrüßen
heute abend/mittag/nacht	heute Abend/Mittag/Nacht
Hexenschuß	Hexenschuss
hierbleiben	hier bleiben
hierlassen	hier lassen
hiersein	hier sein
hierzulande	*auch:* hier zu Lande
High-Fidelity	Highfidelity
	auch: High Fidelity
High-Society	Highsociety
	auch: High Society
hilfesuchend	Hilfe suchend
hinaussein	hinaus sein
es wurde etwas hineingeheimnißt	es wurde etwas hineingeheimnisst
hinsein	hin sein
hintereinanderfahren	hintereinander fahren
hintereinandergehen	hintereinander gehen
hintereinanderschalten	hintereinander schalten
hinterhersein	hinterher sein
hinübersein	hinüber sein
er hißt die Flagge	er hisst die Flagge

alt	neu
Hochgenuß	Hochgenuss
Hochschulabschluß	Hochschulabschluss
aufs höchste erfreut sein	*auch:* aufs Höchste erfreut sein
hofhalten	Hof halten
die Hohe Schule	die hohe Schule
hohnlachen	*auch:* Hohn lachen
das holzverarbeitende Gewerbe	das Holz verarbeitende Gewerbe
Hosteß	Hostess
Hot dog	Hotdog
	auch: Hot Dog
ein paar hundert	*auch:* ein paar Hundert
viele Hunderte	*auch:* viele hunderte
Hunderte von Zuschauern	*auch:* hunderte von Zuschauern
Hungers sterben	hungers sterben
hurra schreien	*auch:* Hurra schreien*

I

auch Ihr seid herzlich eingeladen *[in Briefen]*	auch ihr seid herzlich eingeladen
im allgemeinen	im Allgemeinen
im besonderen	im Besonderen
Imbiß	Imbiss
Imbißstand	Imbissstand
	auch: Imbiss-Stand
im einzelnen	im Einzelnen
im nachhinein	im Nachhinein
Impfpaß	Impfpass
imstande	*auch:* im Stande
im übrigen	im Übrigen
im voraus	im Voraus
im vorhinein	im Vorhinein
in betreff	in Betreff
in bezug auf	in Bezug auf
Indizes	*auch:* Indices
Indizienprozeß	Indizienprozess

821

alt	neu
ineinanderfließen	ineinander fließen
ineinandergreifen	ineinander greifen
inessentiell	*auch:* inessenziell*
Informationsfluß	Informationsfluss
in Frage stellen	*auch:* infrage stellen
in Frage kommen	*auch:* infrage kommen
innesein	inne sein
insektenfressende Pflanzen	Insekten fressende Pflanzen
instand halten	*auch:* in Stand halten
instand setzen	*auch:* in Stand setzen
I-Punkt	i-Punkt
irgend etwas	irgendetwas
irgend jemand	irgendjemand
I-Tüpfelchen	i-Tüpfelchen

J

alt	neu
ja sagen	*auch:* Ja sagen*
Jagdschloß	Jagdschloss
Jäheit	Jähheit
Jahresabschluß	Jahresabschluss
2jährig, 3jährig, 4jährig ...	2-jährig, 3-jährig, 4-jährig ...
ein 2jähriger, 3jähriger, 4jähriger kann das noch nicht verstehen	ein 2-Jähriger, 3-Jähriger, 4-Jähriger kann das noch nicht verstehen
Jaß	Jass
du jaßt	du jasst
Jauchefaß	Jauchefass
jedesmal	jedes Mal
Job-sharing	Jobsharing
Joghurt	*auch:* Jogurt
Joint-venture	Jointventure
	auch: Joint Venture
Judaskuß	Judaskuss
Julierpaß	Julierpass
Jumbo-Jet	Jumbojet
für jung und alt	für Jung und Alt

K

alt	neu
Kabelanschluß	Kabelanschluss
Kabinettsbeschluß	Kabinettsbeschluss
Kaffee-Ernte	*auch:* Kaffeeernte
Kaffee-Ersatz	*auch:* Kaffeeersatz
Kalligraphie	*auch:* Kalligrafie
kalorienbewußt	kalorienbewusst
kaltlächelnd	kalt lächelnd
Kameraverschluß	Kameraverschluss
Kammacher	Kammmacher
	auch: Kamm-Macher
Kämmaschine	Kämmmaschine
	auch: Kämm-Maschine
Kammuschel	Kammmuschel
	auch: Kamm-Muschel
Känguruh	Känguru
Kanonenschuß	Kanonenschuss
Kapselriß	Kapselriss
Karamel	Karamell
karamelisieren	karamellisieren
2karäter, 3karäter, 4karäter ...	2-Karäter, 3-Karäter, 4-Karäter ...
2karätig, 3karätig, 4karätig ...	2-karätig, 3-karätig, 4-karätig ...
Karoas	Karoass
Kartographie	*auch:* Kartografie
Kaßler	Kassler
Katarrh	*auch:* Katarr
kegelschieben	Kegel schieben
Kellergeschoß	Kellergeschoss *[in Österreich weiterhin mit ß]*
kennenlernen	kennen lernen
Kennummer	Kennnummer
	auch: Kenn-Nummer
die Keplerschen Gesetze	die keplerschen Gesetze
	auch: die Kepler'schen Gesetze
keß	kess
Keßheit	Kessheit

823

alt	neu
Ketchup	*auch:* Ketschup*
Kickdown	*auch:* Kick-down*
Kick-off	*auch:* Kickoff
an Kindes Statt	an Kindes statt
Kindesmißhandlung	Kindesmisshandlung
Kißchen	Kisschen
sich über etwas im klaren sein	sich über etwas im Klaren sein
klardenkend	klar denkend
klarsehen	klar sehen
klarwerden	klar werden
Klassenbewußtsein	Klassenbewusstsein
Klassenhaß	Klassenhass
klatschnaß	klatschnass
Klausenpaß	Klausenpass
klebenbleiben	kleben bleiben
Klee-Einsaat	*auch:* Kleeeinsaat
Klee-Ernte	*auch:* Kleeernte
bis ins kleinste geregelt	bis ins Kleinste geregelt
ein Staat im kleinen	ein Staat im Kleinen
ein Programm für groß und klein	ein Programm für Groß und Klein
kleingedruckt	klein gedruckt
kleinschneiden	klein schneiden
klein schreiben *[mit kleinem Anfangsbuchstaben]*	kleinschreiben
Klemmmappe	Klemmmappe *auch:* Klemm-Mappe
Klettverschluß	Klettverschluss
klitschnaß	klitschnass
es wäre das klügste, wenn ...	es wäre das Klügste, wenn ...
knapphalten	knapp halten
Knockout	*auch:* Knock-out*
kochendheiß	kochend heiß
kohleführende Flöze	Kohle führende Flöze
Kolanuß	Kolanuss
Kollektivbewußtsein	Kollektivbewusstsein
Kolophonium	*auch:* Kolofonium

alt	neu
Koloß	Koloss
Kombinationsschloß	Kombinationsschloss
Kommiß	Kommiss
Kommißbrot	Kommissbrot
Kommißstiefel	Kommissstiefel
	auch: Kommiss-Stiefel
Kommuniqué	*auch:* Kommunikee
Kompaß	Kompass
kompreß	kompress
Kompromiß	Kompromiss
kompromißbereit	kompromissbereit
kompromißlos	kompromisslos
Kompromißlösung	Kompromisslösung
Komteß	Komtess
Konferenzbeschluß	Konferenzbeschluss
Kongreß	Kongress
Kongreßhalle	Kongresshalle
Kongreßsaal	Kongresssaal
	auch: Kongress-Saal
Kongreßstadt	Kongressstadt
	auch: Kongress-Stadt
Königsschloß	Königsschloss
Kontrabaß	Kontrabass
Kontrollampe	Kontrolllampe
	auch: Kontroll-Lampe
Kontrolliste	Kontrollliste
	auch: Kontroll-Liste
Kopfnuß	Kopfnuss
Kopfschuß	Kopfschuss
kopfstehen	Kopf stehen
Koppelschloß	Koppelschloss
krank schreiben	krankschreiben
kraß	krass
Kraßheit	Krassheit
krebserregende Substanzen	Krebs erregende Substanzen
Kreiselkompaß	Kreiselkompass

825

alt	neu
Kreppapier	Krepppapier
	auch: Krepp-Papier
Kreuzas	Kreuzass
die kriegführenden Parteien	die Krieg führenden Parteien
Kriminalprozeß	Kriminalprozess
Kristallüster	Kristallüster
	auch: Kristall-Lüster
kroß	kross
krummnehmen	krumm nehmen
KSZE-Schlußakte	KSZE-Schlussakte
Kunststoffolie	Kunststofffolie
	auch: Kunststoff-Folie
Küraß	Kürass
den kürzeren ziehen	den Kürzeren ziehen
kürzertreten	kürzer treten
kurzgebraten	kurz gebraten
kurzhalten	kurz halten
Kurzpaß	Kurzpass
Kurzschluß	Kurzschluss
kurztreten	kurz treten
Kuß	Kuss
Küßchen	Küsschen
kußecht	kussecht
Kußhand	Kusshand
du/er/sie küßt	du/er/sie küsst
Küstenschiffahrt	Küstenschifffahrt
Kwaß	Kwass
L	
Ladenschluß	Ladenschluss
die La-Fontaineschen Fabeln	die la-fontaineschen Fabeln
	auch: die la-Fontaine'schen Fabeln
Lamé	*auch:* Lamee
Lamellenverschluß	Lamellenverschluss

826

alt	neu
etwas des langen und breiten erklären	etwas des Langen und Breiten erklären
langgestreckt	lang gestreckt
länglichrund	länglich rund
langstengelig	langstängelig
langziehen	lang ziehen
Lapsus linguae	Lapsus Linguae
läßlich	lässlich
du läßt	du lässt
zu Lasten	*auch:* zulasten
Lattenschuß	Lattenschuss
laubtragende Bäume	Laub tragende Bäume
auf dem laufenden sein	auf dem Laufenden sein
laufenlassen	laufen lassen
Laufpaß	Laufpass
Layout	*auch:* Lay-out*
Lebensgenuß	Lebensgenuss
Leberabszeß	Leberabszess
die lederverarbeitende Industrie	die Leder verarbeitende Industrie
leerstehend	leer stehend
leichenblaß	leichenblass
es ist mir ein leichtes, das zu tun	es ist mir ein Leichtes, das zu tun
leichtentzündlich	leicht entzündlich
leichtfallen	leicht fallen
leichtmachen	leicht machen
leichtnehmen	leicht nehmen
leichtverderblich	leicht verderblich
leichtverständlich	leicht verständlich
jmdm. leid tun	jmdm. Leid tun
Lenkradschloß	Lenkradschloss
Lernprozeß	Lernprozess
der letzte, der gekommen ist	der Letzte, der gekommen ist
als letzter fertig sein	als Letzter fertig sein
das letzte, was sie tun würde	das Letzte, was sie tun würde
bis ins letzte geklärt	bis ins Letzte geklärt
letzteres trifft zu	Letzteres trifft zu

alt	neu
zum letztenmal	zum letzten Mal
leuchtendblau	leuchtend blau
Lichtmeß	Lichtmess
es wäre uns das liebste, wenn ...	es wäre uns das Liebste, wenn ...
liebenlernen	lieben lernen
liebgewinnen	lieb gewinnen
liebhaben	lieb haben
liegenbleiben	liegen bleiben
liegenlassen	liegen lassen
Live-Mitschnitt	*auch:* Livemitschnitt
Lizentiat	*auch:* Lizenziat*
Lorbaß	Lorbass
Löß	*auch:* Löss *[bei Aussprache mit kurzem ö]*
Lößboden	*auch:* Lössboden *[bei Aussprache mit kurzem ö]*
Lößschicht	*auch:* Lössschicht oder Löss-Schicht *[bei Aussprache mit kurzem ö]*
Lötschenpaß	Lötschenpass
Love-Story	*auch:* Lovestory
Luftschiffahrt	Luftschifffahrt
Luftschloß	Luftschloss

M

alt	neu
Magistratsbeschluß	Magistratsbeschluss
2mal, 3mal, 4mal ...	2-mal, 3-mal, 4-mal ...
Malaise	*auch:* Maläse
Marschkompaß	Marschkompass
maschineschreiben	Maschine schreiben
maßhalten	Maß halten
Matrizes	*auch:* Matrices
Maulkorberlaß	Maulkorberlass
Megaphon	*auch:* Megafon
Mehrheitsbeschluß	Mehrheitsbeschluss
Meldeschluß	Meldeschluss

alt	neu
Meniskusriß	Meniskusriss
wir haben das menschenmögliche getan	wir haben das Menschenmögliche getan
Mesner	*auch:* Messner
Meßband	Messband
meßbar	messbar
Meßbecher	Messbecher
Meßbuch	Messbuch
Meßdaten	Messdaten
Meßdiener	Messdiener
Meßfühler	Messfühler
Meßgewand	Messgewand
Meßinstrument	Messinstrument
Meßopfer	Messopfer
Meßstab	Messstab
	auch: Mess-Stab
Meßtischblatt	Messtischblatt
Metallguß	Metallguss
Metallegierung	Metalllegierung
	auch: Metall-Legierung
die metallverarbeitende Industrie	die Metall verarbeitende Industrie
Midlife-crisis	Midlifecrisis
	auch: Midlife-Crisis
Milchgebiß	Milchgebiss
millionenmal	Millionen Mal
Milzriß	Milzriss
nicht im mindesten	nicht im Mindesten
mißachten	missachten
Mißbildung	Missbildung
mißbilligen	missbilligen
Mißbrauch	Missbrauch
Mißerfolg	Misserfolg
Mißernte	Missernte
mißfallen	missfallen
Mißfallenskundgebung	Missfallenskundgebung
Mißgeburt	Missgeburt

alt	neu
Mißgeschick	Missgeschick
mißglücken	missglücken
Mißgunst	Missgunst
mißgünstig	missgünstig
Mißklang	Missklang
Mißkredit	Misskredit
mißlich	misslich
mißlingen	misslingen
mißmutig	missmutig
mißraten	missraten
Mißstand	Missstand
Mißtrauen	Misstrauen
mißtrauisch	misstrauisch
Mißverständnis	Missverständnis
Mißwirtschaft	Misswirtschaft
mit Hilfe	*auch:* mithilfe
[gestern, heute, morgen] mittag	[gestern, heute, morgen] Mittag
Mixed Pickles	*auch:* Mixedpickles*
modebewußt	modebewusst
wir sprachen über alles mögliche	wir sprachen über alles Mögliche
sein möglichstes tun	sein Möglichstes tun
3monatig, 4monatig, 5monatig ...	3-monatig, 4-monatig, 5-monatig ...
3monatlich, 4monatlich, 5monatlich ...	3-monatlich, 4-monatlich, 5-monatlich ...
Monographie	*auch:* Monografie
Mop	Mopp
Mordprozeß	Mordprozess
morgen abend, mittag, nacht	morgen Abend, Mittag, Nacht
[gestern, heute] morgen	[gestern, heute] Morgen
Moto-Cross	*auch:* Motocross
Mückenschiß	Mückenschiss
Mulläppchen	Mulläppchen
	auch: Mull-Läppchen
Multiple-choice-Verfahren	Multiplechoiceverfahren
	auch: Multiple-Choice-Verfahren

alt	neu
Muskatnuß	Muskatnuss
Muskelriß	Muskelriss
ich muß	ich muss
du mußt	du musst
ich müßte	ich müsste
du müßtest	du müsstest
Mußheirat	Mussheirat
müßiggehen	müßig gehen
Musterprozeß	Musterprozess
Myrrhe	*auch:* Myrre

N

alt	neu
nachfolgendes gilt auch ...	Nachfolgendes gilt auch ...
nach Hause	*in Österreich und der*
	Schweiz auch: nachhause
im nachhinein	im Nachhinein
Nachlaß	Nachlass
Nachlaßverwalter	Nachlassverwalter
[gestern, heute, morgen]	[gestern, heute, morgen]
nachmittag	Nachmittag
Nachschuß	Nachschuss
der nächste, bitte!	der Nächste, bitte!
als nächstes wollen wir ...	als Nächstes wollen wir ...
im nachstehenden heißt es ...	im Nachstehenden heißt es ...
[gestern, heute, morgen] nacht	[gestern, heute, morgen] Nacht
nahebringen	nahe bringen
nahelegen	nahe legen
naheliegen	nahe liegen
naheliegend	nahe liegend
etwas des näheren erläutern	etwas des Näheren erläutern
näherliegen	näher liegen
nahestehen	nahe stehen
nahestehend	nahe stehend
Narziß	Narziss
Narzißmus	Narzissmus
narzißtisch	narzisstisch

alt	neu
naß	nass
naßforsch	nassforsch
naßgeschwitzt	nass geschwitzt
naßkalt	nasskalt
Naßrasur	Nassrasur
Naßschnee	Nassschnee
	auch: Nass-Schnee
nationalbewußt	nationalbewusst
Nationaldreß	Nationaldress
Nebelschlußleuchte	Nebelschlussleuchte
Nebenanschluß	Nebenanschluss
nebeneinandersitzen	nebeneinander sitzen
nebeneinanderstehen	nebeneinander stehen
nebeneinanderstellen	nebeneinander stellen
Nebenfluß	Nebenfluss
im nebenstehenden wird gezeigt ...	im Nebenstehenden wird gezeigt ...
Necessaire	*auch:* Nessessär
Negligé	*auch:* Negligee
nein sagen	*auch:* Nein sagen*
Netzanschluß	Netzanschluss
es aufs neue versuchen	es aufs Neue versuchen
auf ein neues!	auf ein Neues!
neueröffnet	neu eröffnet
New Yorker	*auch:* New-Yorker
nichtrostend	*auch:* nicht rostend
Nichtseßhafte	Nichtsesshafte
nichtssagend	nichts sagend
No-future-Generation	No-Future-Generation
die notleidende Bevölkerung	die Not leidende Bevölkerung
in Null Komma nichts	in null Komma nichts
das Thermometer steht auf Null	das Thermometer steht auf null
Nullage	Nulllage
	auch: Null-Lage
Nulleiter	Nullleiter
	auch: Null-Leiter

alt	neu
Nullösung	Nulllösung
	auch: Null-Lösung
numerieren	nummerieren
Numerierung	Nummerierung
Nuß	Nuss
Nüßchen	Nüsschen
Nußknacker	Nussknacker
Nußschale	Nussschale
	auch: Nuss-Schale
Nußschinken	Nussschinken
	auch: Nuss-Schinken
Nußschokolade	Nussschokolade
	auch: Nuss-Schokolade
Nußstrudel	Nussstrudel
	auch: Nuss-Strudel
Nußtorte	Nusstorte

O

O-beinig	*auch:* o-beinig
obenerwähnt	oben erwähnt
obenstehend	oben stehend
Obergeschoß	Obergeschoss *[in Österreich weiterhin mit ß]*
offenbleiben	offen bleiben
offenlassen	offen lassen
offenstehen	offen stehen
O-förmig	*auch:* o-förmig
des öfteren	des Öfteren
Ölmeßstab	Ölmessstab
Ordonnanz	*auch:* Ordonanz
Orthographie	*auch:* Orthografie

P

Panther	*auch:* Panter
die papierverarbeitende Industrie	die Papier verarbeitende Industrie
Pappmaché	*auch:* Pappmaschee

alt	neu
parallellaufend	parallel laufend
parallelschalten	parallel schalten
Paranuß	Paranuss
Parlamentsbeschluß	Parlamentsbeschluss
Parnaß	Parnass
Parteikongreß	Parteikongress
Parteitagsbeschluß	Parteitagsbeschluss
Paß	Pass
Paßbild	Passbild
passé	*auch:* passee
Paßform	Passform
Paßgang	Passgang
paßgerecht	passgerecht
Paßkontrolle	Passkontrolle
Paßstelle	Passstelle
	auch: Pass-Stelle
Paßstraße	Passstraße
	auch: Pass-Straße
Paßwort	Passwort
Patentverschluß	Patentverschluss
patschnaß	patschnass
Perkussionsschloß	Perkussionsschloss
Personenschiffahrt	Personenschifffahrt
Petitionsausschuß	Petitionsausschuss
Pfeffernuß	Pfeffernuss
Pferdegebiß	Pferdegebiss
pflichtbewußt	pflichtbewusst
Pflichtbewußtsein	Pflichtbewusstsein
Pfostenschuß	Pfostenschuss
Pikas	Pikass
Pimpernuß	Pimpernuss
er pißt	er pisst
Pistolenschuß	Pistolenschuss
pitschnaß	pitschnass
Platitüde	Plattitüde
	auch: Platitude

alt	neu
Playback	*auch:* Play-back*
plazieren	platzieren
pleite gehen	Pleite gehen
polyphon	*auch:* polyfon
Pornographie	*auch:* Pornografie
Portemonnaie	*auch:* Portmonee
Potemkinsche Dörfer	potemkinsche Dörfer
	auch: Potemkin'sche Dörfer
potentiell	*auch:* potenziell*
potthäßlich	potthässlich
Poussierstengel	Poussierstängel
präferentiell	*auch:* präferenziell*
er praßt	er prasst
preisbewußt	preisbewusst
Preisnachlaß	Preisnachlass
Preßform	Pressform
Preßluftbohrer	Pressluftbohrer
Preßsack	Presssack
	auch: Press-Sack
Preßschlag	Pressschlag
	auch: Press-Schlag
Preßspan	Pressspan
	auch: Press-Span
du preßt	du presst
Preßwehe	Presswehe
Prinzeßbohne	Prinzessbohne
privatversichert	privat versichert
probefahren	Probe fahren
Problembewußtsein	Problembewusstsein
Produktionsprozeß	Produktionsprozess
Profeß	Profess
Programmusik	Programmmusik
	auch: Programm-Musik
Progreß	Progress
Prozeß	Prozess
Prozeßkosten	Prozesskosten

alt	neu
Prozeßbevollmächtigte	Prozessbevollmächtigte
prozeßführend	prozessführend
Prozeßkosten	Prozesskosten
Prozeßrechner	Prozessrechner
pudelnaß	pudelnass
Pulverfaß	Pulverfass
pußlig	pusslig

Q

Quadrophonie	*auch:* Quadrofonie
qualitätsbewußt	qualitätsbewusst
Quartalsabschluß	Quartalsabschluss
Quellfluß	Quellfluss
Quentchen	Quäntchen
Querpaß	Querpass
Quickstep	Quickstepp

R

radfahren	Rad fahren
Radikalenerlaß	Radikalenerlass
radschlagen	Rad schlagen
Rammaschine	Rammmaschine
	auch: Ramm-Maschine
zu Rande kommen	*auch:* zurande kommen
Rassenhaß	Rassenhass
ich raßle mit den Ketten	ich rassle mit den Ketten
zu Rate ziehen	*auch:* zurate ziehen
Räterußland	Räterussland
Ratsbeschluß	Ratsbeschluss
Ratschluß	Ratschluss
Rauchfaß	Rauchfass
rauh	rau
rauhbeinig	raubeinig
Rauhfasertapete	Raufasertapete
Rauhfrost	Raufrost
Rauhhaardackel	Rauhaardackel

836

alt	neu
Rauhnächte	Raunächte
Rauhputz	Rauputz
Rauhreif	Raureif
Rausschmiß	Rausschmiss
recht haben	Recht haben
recht behalten	Recht behalten
recht bekommen	Recht bekommen
jmdm. recht geben	jmdm. Recht geben
Rechtens sein	rechtens sein
Rechtsbewußtsein	Rechtsbewusstsein
Redaktionsschluß	Redaktionsschluss
Regenguß	Regenguss
regennaß	regennass
Regreß	Regress
Regreßanspruch	Regressanspruch
Regreßpflicht	Regresspflicht
regreßpflichtig	regresspflichtig
reichgeschmückt	reich geschmückt
reichverziert	reich verziert
Reifungsprozeß	Reifungsprozess
Reisepaß	Reisepass
Reißverschluß	Reißverschluss
Reißverschlußsystem	Reißverschlusssystem
	auch: Reißverschluss-System
Reschenpaß	Reschenpass
Rettungsschuß	Rettungsschuss
Rezeß	Rezess
Rhein-Main-Donau-	Rhein-Main-Donau-
Großschiffahrtsweg	Großschifffahrtsweg
das ist genau das richtige	das ist genau das Richtige
für mich	für mich
mit etwas richtigliegen	mit etwas richtig liegen
richtigstellen	richtig stellen
Riß	Riss
rißfest	rissfest
Roheit	Rohheit

alt	neu
Rolladen	Rollladen
	auch: Roll-Laden
Rommé	*auch:* Rommee
rosigweiß	rosig weiß
Roß	Ross
Roßbreiten	Rossbreiten
Roßhaarmatratze	Rosshaarmatratze
Roßkastanie	Rosskastanie
Roßkur	Rosskur
Rößl	Rössl
Roßtäuscherei	Rosstäuscherei
der rote Planet *[Mars]*	der Rote Planet
rotgestreift	rot gestreift
rotglühend	rot glühend
rötlichbraun	rötlich braun
die Rubensschen Gemälde	die rubensschen Gemälde
	auch: die Rubens'schen
	Gemälde
Rückfluß	Rückfluss
Rückpaß	Rückpass
Rückschluß	Rückschluss
rückwärtsgewandt	rückwärts gewandt
Ruhegenuß	Ruhegenuss
ruhenlassen	ruhen lassen
ruhigstellen	ruhig stellen
Runderlaß	Runderlass
Rußland	Russland

S

alt	neu
Säbelraßler	Säbelrassler
Saisonnier	*auch:* Saisonier
Saisonschluß	Saisonschluss
Salutschuß	Salutschuss
Salzfaß	Salzfass
Samenerguß	Samenerguss
Sammelanschluß	Sammelanschluss

alt	neu
Sankt Gallener	*auch:* Sankt-Gallener
sanktgallisch	sankt-gallisch
Sanmarinese	San-Marinese
sanmarinesisch	san-marinesisch
sauberhalten	sauber halten
saubermachen	sauber machen
sausenlassen	sausen lassen
Saxophon	*auch:* Saxofon
sein Schäfchen ins trockene bringen	sein Schäfchen ins Trockene bringen
Schalenguß	Schalenguss
Schallehre	Schalllehre
	auch: Schall-Lehre
Schalloch	Schallloch
	auch: Schall-Loch
Schalterschluß	Schalterschluss
etwas auf das schärfste verurteilen	*auch:* etwas auf das Schärfste verurteilen
er schaßte ihn	er schasste ihn
ein schattenspendender Baum	ein Schatten spendender Baum
schätzenlernen	schätzen lernen
Schauprozeß	Schauprozess
Scheidungsprozeß	Scheidungsprozess
schießenlassen	schießen lassen
Schiffahrt	Schifffahrt
	auch: Schiff-Fahrt
Schippenas	Schippenass
Schiß	Schiss
Schlachtroß	Schlachtross
Schlagfluß	Schlagfluss
Schlammasse	Schlammmasse
	auch: Schlamm-Masse
schlechtgehen	schlecht gehen
schlechtgelaunt	schlecht gelaunt
das schlimmste ist, daß ...	das Schlimmste ist, dass ...

alt	neu
sie haben ihn auf das schlimmste getäuscht	*auch:* sie haben ihn auf das Schlimmste getäuscht
er schliß Federn	er schliss Federn
Schlitzverschluß	Schlitzverschluss
Schloß	Schloss
Schlößchen	Schlösschen
Schloßherr	Schlossherr
Schloßpark	Schlosspark
Schluß	Schluss
Schlußbemerkung	Schlussbemerkung
schlußendlich	schlussendlich
schlußfolgern	schlussfolgern
Schlußfolgerung	Schlussfolgerung
Schlußlicht	Schlusslicht
Schlüßpfiff	Schlusspfiff
Schlußpunkt	Schlusspunkt
Schlußsatz	Schlusssatz
	auch: Schluss-Satz
Schlußspurt	Schlussspurt
	auch: Schluss-Spurt
Schlußstrich	Schlussstrich
	auch: Schluss-Strich
Schlußverkauf	Schlussverkauf
Schlußwort	Schlusswort
Schmerfluß	Schmerfluss
sie schmiß mit Steinen	sie schmiss mit Steinen
Schmiß	Schmiss
Schmuckblattelegramm	Schmuckblatttelegramm
	auch: Schmuckblatt-Telegramm
schmutziggrau	schmutzig grau
Schnappschloß	Schnappschloss
Schnappschuß	Schnappschuss
Schnee-Eifel	*auch:* Schneeeifel
Schnee-Eule	*auch:* Schneeeule
Schneewächte	Schneewechte
Schnellimbiß	Schnellimbiss

alt	neu
Schnelläufer	Schnellläufer
	auch: Schnell-Läufer
schnellebig	schnelllebig
Schnellebigkeit	Schnelllebigkeit
Schnellschuß	Schnellschuss
Schnepper	*auch:* Schnäpper
schneppern	*auch:* schnäppern
schneuzen	schnäuzen
Schokoladenguß	Schokoladenguss
aufs schönste übereinstimmen	*auch:* aufs Schönste übereinstimmen
er schoß	er schoss
Schoß *[einer Pflanze]*	Schoss
schräglaufend	schräg laufend
Schraubverschluß	Schraubverschluss
schreckensblaß	schreckensblass
Schreckschußpistole	Schreckschusspistole
Schrittempo	Schritttempo
	auch: Schritt-Tempo
Schrotschuß	Schrotschuss
Schulabschluß	Schulabschluss
an etwas schuld haben	an etwas Schuld haben
sich etwas zuschulden kommen lassen	*auch:* sich etwas zu Schulden kommen lassen
schuldbewußt	schuldbewusst
Schuldenerlaß	Schuldenerlass
Schulschluß	Schulschluss
Schulstreß	Schulstress
Schulterschluß	Schulterschluss
Schuß	Schuss
schußbereit	schussbereit
schußfest	schussfest
schußlig	schusslig
Schußlinie	Schusslinie
Schußschwäche	Schussschwäche
	auch: Schuss-Schwäche

841

alt	neu
Schußwaffe	Schusswaffe
Schußwechsel	Schusswechsel
schwachbetont	schwach betont
schwachbevölkert	schwach bevölkert
aus schwarz weiß machen	aus Schwarz Weiß machen
Schwarze Magie	schwarze Magie
schwarzgefärbt	schwarz gefärbt
schwarzrotgolden	*auch:* schwarz-rot-golden
schwerfallen	schwer fallen
schwernehmen	schwer nehmen
schwertun	schwer tun
schwerverständlich	schwer verständlich
Schwimmeister	Schwimmmeister
	auch: Schwimm-Meister
Science-fiction	Sciencefiction
	auch: Science-Fiction
Sechspaß	Sechspass
See-Elefant	*auch:* Seeelefant
jedem das Seine	*auch:* jedem das seine
das Seine beitragen	*auch:* das seine beitragen
die Seinen	*auch:* die seinen
die Seinigen	*auch:* die seinigen
seinlassen	sein lassen
Seismograph	*auch:* Seismograf
auf seiten	aufseiten
	auch: auf Seiten
von seiten	vonseiten
	auch: von Seiten
selbständig	*auch:* selbstständig
Selbständigkeit	*auch:* Selbstständigkeit
selbstbewußt	selbstbewusst
Selbstbewußtsein	Selbstbewusstsein
selbsternannt	selbst ernannt
selbstgebacken	selbst gebacken
selbstgemacht	selbst gemacht
selbstgestrickt	selbst gestrickt

alt	neu
Selbstschuß	Selbstschuss
selbstverdient	selbst verdient
seligpreisen	selig preisen
seligsprechen	selig sprechen
Senatsbeschluß	Senatsbeschluss
Sendeschluß	Sendeschluss
Sendungsbewußtsein	Sendungsbewusstsein
Sensationsprozeß	Sensationsprozess
Séparée	*auch:* Separee
sequentiell	*auch:* sequenziell*
seßhaft	sesshaft
Seßhaftigkeit	Sesshaftigkeit
S-förmig	*auch:* s-förmig
die Shakespeareschen Sonette	die shakespeareschen Sonette
	auch: die Shakespeare'schen Sonette
Short story	Shortstory
	auch: Short Story
Showbusineß	Showbusiness
Showdown	*auch:* Show-down*
Shrimp	*auch:* Schrimp
auf Nummer Sicher gehen	*auch:* auf Nummer sicher gehen
das sicherste ist, wenn ...	das Sicherste ist, wenn ...
Sicherheitsschloß	Sicherheitsschloss
Sicherheitsverschluß	Sicherheitsverschluss
siedendheiß	siedend heiß
siegesbewußt	siegesbewusst
siegesgewiß	siegesgewiss
Simplonpaß	Simplonpass
die Singende Säge	die singende Säge
Siphonverschluß	Siphonverschluss
sitzenbleiben	sitzen bleiben
sitzenlassen	sitzen lassen
Skipaß	Skipass
Small talk	Smalltalk
	auch: Small Talk

alt	neu
so daß	sodass
	auch: so dass
Sommerschlußverkauf	Sommerschlussverkauf
alles sonstige besprechen wir morgen	alles Sonstige besprechen wir morgen
Soufflé	*auch:* Soufflee
soviel du willst	so viel du willst
soviel wie	so viel wie
noch einmal soviel	noch einmal so viel
es ist soweit	es ist so weit
soweit wie möglich	so weit wie möglich
ich kann das sowenig wie du	ich kann das so wenig wie du
Sowjetrußland	Sowjetrussland
hier gilt kein Sowohl-Als-auch	hier gilt kein Sowohl-als-auch
Spaghetti	*auch:* Spagetti
Spantenriß	Spantenriss
spazierenfahren	spazieren fahren
spazierengehen	spazieren gehen
Speichelfluß	Speichelfluss
Sperrad	Sperrrad
	auch: Sperr-Rad
Sperriegel	Sperrriegel
	auch: Sperr-Riegel
Spliß	Spliss
du splißt	du splisst
eine sporenbildende Pflanze	eine Sporen bildende Pflanze
Sportdreß	Sportdress
Sprenggeschoß	Sprenggeschoss *[in Österreich weiterhin mit ß]*
Spritzguß	Spritzguss
es sproß neues Grün	es spross neues Grün
Sproß	Spross
Sproßachse	Sprossachse
Sprößchen	Sprösschen
Sprößling	Sprössling
staatenbildende Insekten	Staaten bildende Insekten

alt	neu
Stahlroß	Stahlross
Stallaterne	Stalllaterne
	auch: Stall-Laterne
Stammutter	Stammmutter
	auch: Stamm-Mutter
standesbewußt	standesbewusst
Standesbewußtsein	Standesbewusstsein
Startschuß	Startschuss
steckenbleiben	stecken bleiben
steckenlassen	stecken lassen
Steckschloß	Steckschloss
Steckschuß	Steckschuss
stehenbleiben	stehen bleiben
stehenlassen	stehen lassen
Stehimbiß	Stehimbiss
Steilpaß	Steilpass
Stemmeißel	Stemmmeißel
	auch: Stemm-Meißel
Stendelwurz	Ständelwurz
Stengel	Stängel
Step	Stepp
Steptanz	Stepptanz
Stereophonie	*auch:* Stereofonie
Steuererlaß	Steuererlass
Steuermeßbetrag	Steuermessbetrag
Stewardeß	Stewardess
stiftengehen	stiften gehen
etwas im stillen vorbereiten	etwas im Stillen vorbereiten
Stilleben	Stillleben
	auch: Still-Leben
stillegen	stilllegen
Stillegung	Stilllegung
Stoffarbe	Stofffarbe
	auch: Stoff-Farbe
Stoffetzen	Stofffetzen
	auch: Stoff-Fetzen

845

alt	neu
Stoffülle	Stofffülle
	auch: Stoff-Fülle
Stop	Stopp
Straferlaß	Straferlass
Strafprozeß	Strafprozess
Strafprozeßordnung	Strafprozessordnung
Straß	Strass
Streifschuß	Streifschuss
Streitroß	Streitross
strenggenommen	streng genommen
strengnehmen	streng nehmen
aufs strengste unterschieden	*auch:* aufs Strengste
	unterschieden
Streß	Stress
der Lärm streßt	der Lärm stresst
Streßsituation	Stresssituation
	auch: Stress-Situation
2stündig, 3stündig, 4stündig ...	2-stündig, 3-stündig, 4-stündig ...
2stündlich, 3stündlich,	2-stündlich, 3-stündlich,
4stündlich ...	4-stündlich ...
Stuß	Stuss
substantiell	*auch:* substanziell*
Sustenpaß	Sustenpass
T	
Tablettenmißbrauch	Tablettenmissbrauch
tabula rasa machen	Tabula rasa machen
zutage treten	*auch:* zu Tage treten
2tägig, 3tägig, 4tägig ...	2-tägig, 3-tägig, 4-tägig ...
Tankschloß	Tankschloss
Tarifabschluß	Tarifabschluss
Täßchen	Tässchen
ein paar tausend	*auch:* ein paar Tausend
Tausende von Zuschauern	*auch:* tausende von Zuschauern
T-bone-Steak	T-Bone-Steak
Tee-Ei	*auch:* Teeei

846

alt	neu
Tee-Ernte	*auch:* Teeernte
Teerfaß	Teerfass
Telephon	Telefon
Telephonanschluß	Telefonanschluss
Thunfisch	*auch:* Tunfisch
Tie-Break	*auch:* Tiebreak
aufs tiefste gekränkt	*auch:* aufs Tiefste gekränkt
tiefbewegt	tief bewegt
tiefempfunden	tief empfunden
tiefverschneit	tief verschneit
Tintenfaß	Tintenfass
Tip	Tipp
todblaß	todblass
Todesschuß	Todesschuss
Tolpatsch	Tollpatsch
tolpatschig	tollpatschig
Tomatenketchup	*auch:* Tomatenketschup
Topographie	*auch:* Topografie
Torschlußpanik	Torschlusspanik
Torschuß	Torschuss
totenblaß	totenblass
totgeboren	tot geboren
traditionsbewußt	traditionsbewusst
Tränenfluß	Tränenfluss
tränennaß	tränennass
Traß	Trass
Trekking	*auch:* Trecking
treuergeben	treu ergeben
triefnaß	triefnass
auf dem trockenen sitzen	auf dem Trockenen sitzen
sein Schäfchen ins trockene bringen	sein Schäfchen ins Trockene bringen
tropfnaß	tropfnass
Troß	Tross
im trüben fischen	im Trüben fischen
Truchseß	Truchsess

847

alt	neu
Trugschluß	Trugschluss
Trumpfas	Trumpfass
Tuffelsen	Tufffelsen
	auch: Tuff-Felsen
Türschloß	Türschloss

U

alt	neu
übelgelaunt	übel gelaunt
übelnehmen	übel nehmen
übelriechend	übel riechend
Überbiß	Überbiss
Überdruß	Überdruss
übereinanderlegen	übereinander legen
übereinanderliegen	übereinander liegen
übereinanderwerfen	übereinander werfen
Überfluß	Überfluss
Überflußgesellschaft	Überflussgesellschaft
Überguß	Überguss
überhandnehmen	überhand nehmen
übermorgen abend, nachmittag	übermorgen Abend, Nachmittag
Überschuß	Überschuss
überschwenglich	überschwänglich
überwächtet	überwechtet
ein übriges tun	ein Übriges tun
im übrigen wissen wir doch alle ...	im Übrigen wissen wir doch alle ...
alles übrige später	alles Übrige später
die übrigen kommen nach	die Übrigen kommen nach
übrigbehalten	übrig behalten
übrigbleiben	übrig bleiben
übriglassen	übrig lassen
U-förmig	*auch:* u-förmig
Ultima ratio	Ultima Ratio
Umdenkprozeß	Umdenkprozess
die Liste umfaßt alles Wichtige	die Liste umfasst alles Wichtige
Umriß	Umriss
Umrißzeichnung	Umrisszeichnung

alt	neu
Umschichtungsprozeß	Umschichtungsprozess
Umschluß	Umschluss
umsein	um sein
um so [mehr, größer, weniger ...]	umso [mehr, größer, weniger ...]
Umstellungsprozeß	Umstellungsprozess
Umwandlungsprozeß	Umwandlungsprozess
Umwelteinfluß	Umwelteinfluss
sich ins unabsehbare ausweiten	sich ins Unabsehbare ausweiten
unangepaßt	unangepasst
Unangepaßtheit	Unangepasstheit
unbeeinflußbar	unbeeinflussbar
unbeeinflußt	unbeeinflusst
Anzeige gegen Unbekannt	Anzeige gegen unbekannt
unbewußt	unbewusst
und ähnliches (u. ä.)	und Ähnliches (u. Ä.)
unendlichemal	unendliche Mal
unerläßlich	unerlässlich
unermeßlich	unermesslich
Unfairneß	Unfairness
unfaßbar	unfassbar
unfaßlich	unfasslich
ungewiß	ungewiss
Ungewißheit	Ungewissheit
unigefärbt	uni gefärbt
im unklaren bleiben	im Unklaren bleiben
im unklaren lassen	im Unklaren lassen
unmißverständlich	unmissverständlich
unpäßlich	unpässlich
Unpäßlichkeit	Unpässlichkeit
unplaziert	unplatziert
unrecht haben	Unrecht haben
unrecht behalten	Unrecht behalten
unrecht bekommen	Unrecht bekommen
Unrechtsbewußtsein	Unrechtsbewusstsein
unselbständig	*auch:* unselbstständig
Unselbständigkeit	*auch:* Unselbstständigkeit

alt	neu
die Unseren	*auch:* die unseren
die Unsrigen	*auch:* die unsrigen
untenerwähnt	unten erwähnt
untenstehend	unten stehend
unterbewußt	unterbewusst
Unterbewußtsein	Unterbewusstsein
unterderhand	unter der Hand
untereinanderstehen	untereinander stehen
Untergeschoß	Untergeschoss *[in Österreich weiterhin mit ß]*
ohne Unterlaß	ohne Unterlass
Untersuchungsausschuß	Untersuchungsausschuss
unvergeßlich	unvergesslich
unerläßlich	unerlässlich
unzähligemal	unzählige Mal

V

alt	neu
va banque spielen	*auch:* Vabanque spielen
Varieté	*auch:* Varietee
veranlaßt	veranlasst
verantwortungsbewußt	verantwortungsbewusst
Verantwortungsbewußtsein	Verantwortungsbewusstsein
Verbiß	Verbiss
verblaßt	verblasst
verbleuen	verbläuen
im verborgenen blühen	im Verborgenen blühen
das verdroß uns	das verdross uns
Verdruß	Verdruss
du verfaßt	du verfasst
vergeßlich	vergesslich
Vergeßlichkeit	Vergesslichkeit
Vergißmeinnicht	Vergissmeinnicht
du vergißt	du vergisst
verhaßt	verhasst
auf jmdn. ist Verlaß	auf jmdn. ist Verlass
verläßlich	verlässlich

alt	neu
Verläßlichkeit	Verlässlichkeit
verlorengehen	verloren gehen
vermißt	vermisst
Vermißtenanzeige	Vermisstenanzeige
er hat den Zug verpaßt	er hat den Zug verpasst
das Geld wurde verpraßt	das Geld wurde verprasst
Verriß	Verriss
verschiedenes war noch unklar	Verschiedenes war noch unklar
verschiedenemal	verschiedene Mal
Verschiß	Verschiss
Verschluß	Verschluss
Verschlußkappe	Verschlusskappe
Verschlußsache	Verschlusssache
	auch: Verschluss-Sache
verselbständigen	*auch:* verselbstständigen
Versorgungsengpaß	Versorgungsengpass
Vertragsabschluß	Vertragsabschluss
Vertragsschluß	Vertragsschluss
V-förmig	*auch:* v-förmig
Vibraphon	*auch:* Vibrafon
viel zuviel	viel zu viel
viel zuwenig	viel zu wenig
vielbefahren	viel befahren
vielgelesen	viel gelesen
Vierpaß	Vierpass
aus dem vollen schöpfen	aus dem Vollen schöpfen
voneinandergehen	voneinander gehen
von seiten	vonseiten
	auch: von Seiten
vorangehendes gilt auch ...	Vorangehendes gilt auch ...
im vorangehenden heißt es ...	im Vorangehenden heißt es ...
im voraus	im Voraus
vorgefaßt	vorgefasst
vorgestern abend, mittag, morgen	vorgestern Abend, Mittag, Morgen
Vorhängeschloß	Vorhängeschloss

alt	neu
vorhergehendes gilt auch ...	Vorhergehendes gilt auch ...
im vorhergehenden heißt es ...	im Vorhergehenden heißt es ...
im vorhinein	im Vorhinein
das vorige gilt auch ...	das Vorige gilt auch ...
im vorigen heißt es ...	im Vorigen heißt es ...
Vorlegeschloß	Vorlegeschloss
vorliebnehmen	vorlieb nehmen
[gestern, heute, morgen]	[gestern, heute, morgen]
vormittag	Vormittag
Vorschlußrunde	Vorschlussrunde
Vorschuß	Vorschuss
Vorschußlorbeeren	Vorschusslorbeeren
vorstehendes gilt auch ...	Vorstehendes gilt auch ...
im vorstehenden heißt es ...	im Vorstehenden heißt es ...
vorwärtsgehen	vorwärts gehen
vorwärtskommen	vorwärts kommen

W

alt	neu
ein wachestehender Soldat	ein Wache stehender Soldat
Wachsabguß	Wachsabguss
Wächte	Wechte
Waggon	*auch:* Wagon
Wahlausschuß	Wahlausschuss
Walkie-talkie	Walkie-Talkie
Walnuß	Walnuss
Walroß	Walross
Wandlungsprozeß	Wandlungsprozess
Warnschuß	Warnschuss
Wasserschloß	Wasserschloss
wäßrig	wässrig
Wehrpaß	Wehrpass
weichgekocht	weich gekocht
Weinfaß	Weinfass
aus schwarz weiß machen	aus Schwarz Weiß machen
weißgekleidet	weiß gekleidet
Weißrußland	Weißrussland

852

alt	neu
des weiteren wurde gesagt ...	des Weiteren wurde gesagt ...
weitgereist	weit gereist
weitreichend	weit reichend
weitverbreitet	weit verbreitet
Werkstattage	Werkstatttage
	auch: Werkstatt-Tage
Werkstoffforschung	Werkstoffforschung
	auch: Werkstoff-Forschung
es besteht im wesentlichen aus ...	es besteht im Wesentlichen aus ...
Wetteufel	Wettteufel
	auch: Wett-Teufel
Wetturnen	Wettturnen
	auch: Wett-Turnen
widereinanderstoßen	widereinander stoßen
wieviel	wie viel
Winterschlußverkauf	Winterschlussverkauf
Wißbegierde	Wissbegierde
wißbegierig	wissbegierig
ihr wißt	ihr wisst
du wußtest	du wusstest
wir wüßten gern ...	wir wüssten gern ...
Witterungseinfluß	Witterungseinfluss
Wollappen	Wolllappen
	auch: Woll-Lappen
Wollaus	Wolllaus
	auch: Woll-Laus
als ob er wunder was getan hätte	als ob er Wunder was getan hätte
sich wundliegen	sich wund liegen
Wurfgeschoß	Wurfgeschoss *[in Österreich weiterhin mit ß]*

X, Y

alt	neu
X-beinig	*auch:* x-beinig
X-förmig	*auch:* x-förmig
zum x-tenmal	zum x-ten Mal

alt	neu
Z	
Zäheit	Zähheit
Zahlenschloß	Zahlenschloss
Zäpfchen-R	*auch:* Zäpfchen-r
Zaubernuß	Zaubernuss
Zechenstillegung	Zechenstilllegung
Zeilengußmaschine	Zeilengussmaschine
2zeilig, 3zeilig, 4zeilig ...	2-zeilig, 3-zeilig, 4-zeilig ...
eine Zeitlang	eine Zeit lang
zur Zeit *[derzeit]*	zurzeit
Zellehre	Zelllehre
	auch: Zell-Lehre
Zellstoffabrik	Zellstofffabrik
	auch: Zellstoff-Fabrik
Zersetzungsprozeß	Zersetzungsprozess
zielbewußt	zielbewusst
Zierat	Zierrat
zigtausend	*auch:* Zigtausend
Zigtausende	*auch:* zigtausende
Zippverschluß	Zippverschluss
Zirkelschluß	Zirkelschluss
Zivilprozeß	Zivilprozess
Zivilprozeßordnung	Zivilprozessordnung
Zoo-Orchester	*auch:* Zooorchester
sich zu eigen machen	sich zu Eigen machen
zueinanderfinden	zueinander finden
Zufluß	Zufluss
sich zufriedengeben	sich zufrieden geben
zufriedenlassen	zufrieden lassen
zufriedenstellen	zufrieden stellen
zugrunde gehen	*auch:* zu Grunde gehen
zugrunde legen	*auch:* zu Grunde legen
zugrunde liegen	*auch:* zu Grunde liegen
zugrundeliegend	zugrunde liegend
	auch: zu Grunde liegend
zugrunde richten	*auch:* zu Grunde richten

alt	neu
zugunsten	*auch:* zu Gunsten
zu Hause	*in Österreich und der Schweiz*
	auch: zuhause
bei uns zulande	bei uns zu Lande
zulasten	*auch:* zu Lasten
jmdm. etwas zuleide tun	*auch:* jmdm. etwas zu Leide tun
zumute sein	*auch:* zu Mute sein
Zündschloß	Zündschloss
Zungenkuß	Zungenkuss
Zungen-R	*auch:* Zungen-r
sich etwas zunutze machen	*auch:* sich etwas zu Nutze machen
jmdm. zupaß kommen	jmdm. zupass kommen
zugepreßt	zugepresst
zu Rande kommen	*auch:* zurande kommen
jmdn. zu Rate ziehen	*auch:* jmdn. zurate ziehen
sie hat zurückgemußt	sie hat zurückgemusst
zur Zeit *[derzeit]*	zurzeit
Zusammenfluß	Zusammenfluss
zusammengefaßt	zusammengefasst
zusammengepaßt	zusammengepasst
zusammengepreßt	zusammengepresst
Zusammenschluß	Zusammenschluss
zusammensein	zusammen sein
zuschanden werden	*auch:* zu Schanden werden
sich etwas zuschulden kommen lassen	*auch:* sich etwas zu Schulden kommen lassen
Zuschuß	Zuschuss
Zuschußbetrieb	Zuschussbetrieb
zusein	zu sein
zustande bringen	*auch:* zu Stande bringen
zustande kommen	*auch:* zu Stande kommen
zutage fördern	*auch:* zu Tage fördern
zutage treten	*auch:* zu Tage treten
zuungunsten	*auch:* zu Ungunsten
zuviel	zu viel
zuwege bringen	*auch:* zu Wege bringen

alt	neu
zuwenig	zu wenig
die zwanziger Jahre	*auch:* die Zwanzigerjahre*
die Zwanzigerjahre	*auch:* die zwanziger Jahre
das Zweite Gesicht	das zweite Gesicht
er hat wie kein zweiter gearbeitet	er hat wie kein Zweiter gearbeitet
jeder zweite war krank	jeder Zweite war krank
Zweitkläßler	Zweitklässler
Zwischengeschoß	Zwischengeschoss *[in Österreich weiterhin mit ß]*